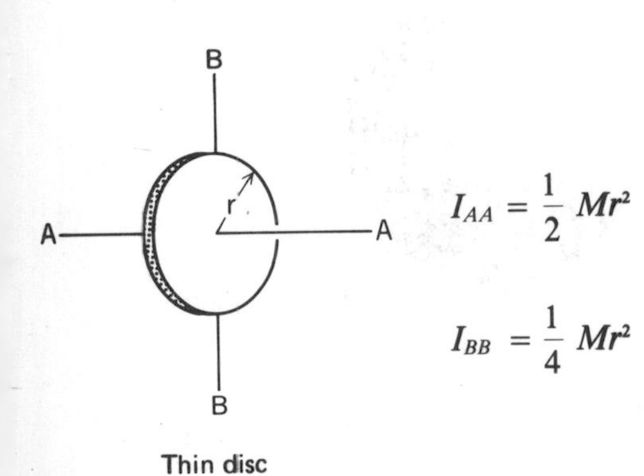

Thin disc

$$I_{AA} = \frac{1}{2} Mr^2$$

$$I_{BB} = \frac{1}{4} Mr^2$$

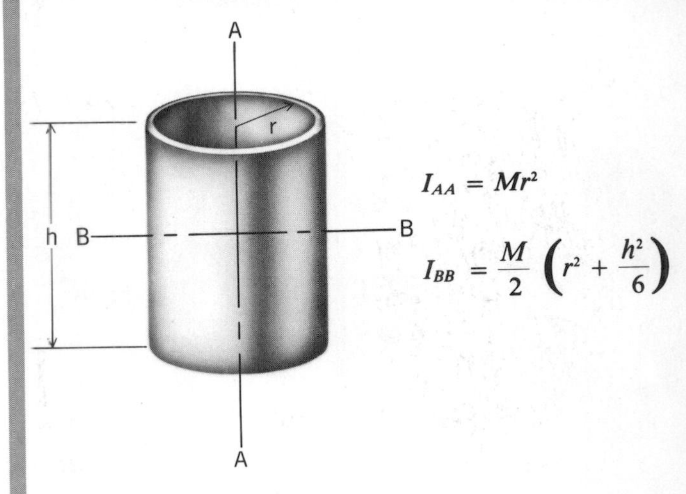

Thin-walled cylinder

$$I_{AA} = Mr^2$$

$$I_{BB} = \frac{M}{2}\left(r^2 + \frac{h^2}{6}\right)$$

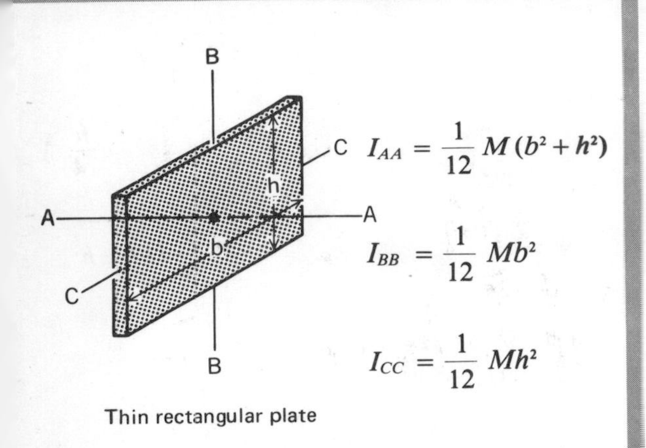

Thin rectangular plate

$$I_{AA} = \frac{1}{12} M (b^2 + h^2)$$

$$I_{BB} = \frac{1}{12} Mb^2$$

$$I_{CC} = \frac{1}{12} Mh^2$$

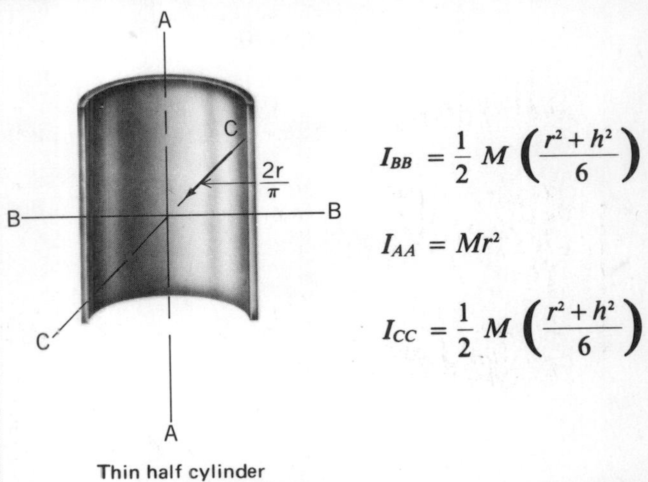

Thin half cylinder

$$I_{BB} = \frac{1}{2} M \left(\frac{r^2 + h^2}{6}\right)$$

$$I_{AA} = Mr^2$$

$$I_{CC} = \frac{1}{2} M \left(\frac{r^2 + h^2}{6}\right)$$

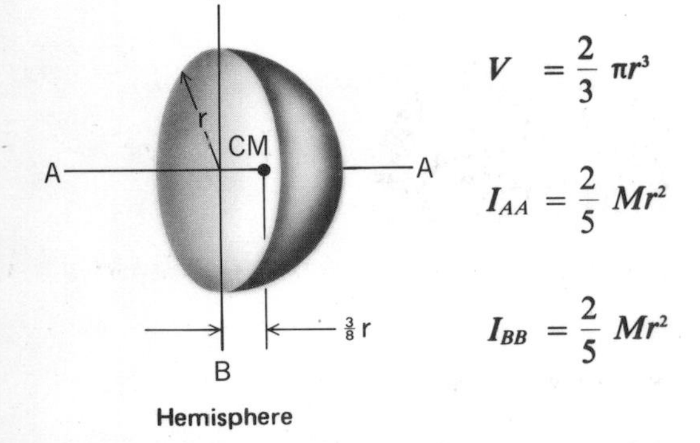

Hemisphere

$$V = \frac{2}{3} \pi r^3$$

$$I_{AA} = \frac{2}{5} Mr^2$$

$$I_{BB} = \frac{2}{5} Mr^2$$

SELECTED DIMENSIONAL EQUIVALENTS

LENGTH	1 m ≡ 3.281 ft ≡ 39.37 in. 1 mi ≡ 5280 ft ≡ 1.609 km 1 km ≡ .6214 mi
TIME	1 hr ≡ 60 min ≡ 3600 sec
MASS	1 kg ≡ 2.2046 lbm ≡ .068521 slug
FORCE	1 N ≡ .2248 lbf 1 dyne ≡ 1 μN
SPEED	1 mi/hr ≡ 1.609 km/hr ≡ 1.467 ft/sec 1 km/hr = .6214 mi/hr 1 knot = 1.152 mi/hr ≡ 1.853 km/hr ≡ 1.689 ft/sec
ENERGY	1 J ≡ 1 N-m 1 Btu ≡ 778.16 ft-lbf ≡ 1.055 kJ 1 watt-hour ≡ 2.778×10^{-4} J
VOLUME	1 gal ≡ .16054 ft^3 ≡ .0045461 m^3 1 liter ≡ .03531 ft^3 = .2642 gal
POWER	1 w ≡ 1 J/S 1 hp ≡ 550 ft-lb/sec ≡ .7068 Btu/sec ≡ 746 w

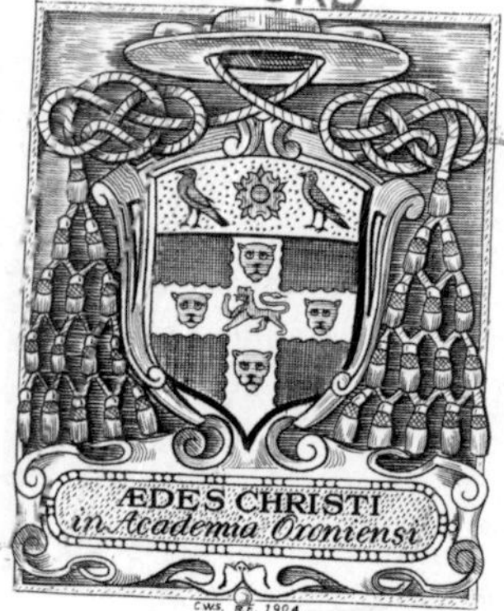
ÆDES CHRISTI
in Academia Oxoniensi

IRVING H. SHAMES
Faculty Professor
Engineering and Applied Science
State University of New York at Buffalo
third edition
ENGINEERING
MECHANICS
STATICS
AND
DYNAMICS
PRENTICE-HALL, INC., Englewood Cliffs, New Jersey 07632

Library of Congress Cataloging in Publication Data

SHAMES, IRVING HERMAN, (date)
Engineering mechanics: statics and dynamics.

Includes index.
1. Mechanics, Applied. I. Title.
TA350.S493 1980 620 80-11902
ISBN 0-13-279166-8

Editorial/production supervision and interior design by Leon J. Liguori and Suzanne Behnke with the assistance of Karen Mulé
Manufacturing buyer: Anthony Caruso

Printed in the United States of America

10 9 8 7 6 5 4 3 2 1

Prentice-Hall International, Inc. *London*
Prentice-Hall of Australia Pty. Limited, *Sydney*
Prentice-Hall of Canada, Ltd., *Toronto*
Prentice-Hall of India Private Limited, *New Delhi*
Prentice-Hall of Japan, Inc., *Tokyo*
Prentice-Hall of Southeast Asia Pte. Ltd., *Singapore*
Whitehall Books Limited, Wellington, *New Zealand*

I am happy and grateful to acknowledge the collaboration
on the Statics Volume of Dr. Robert M. Jones
from Southern Methodist University.
Dr. Jones went over with me the entire Statics Volume
with the aim of making the treatment
as clear and as simple as possible.
In addition, Dr. Jones made available to me 200 fine statics problems.
Finally, Bob gave me the benefit of a careful review
of the Dynamics Volume.

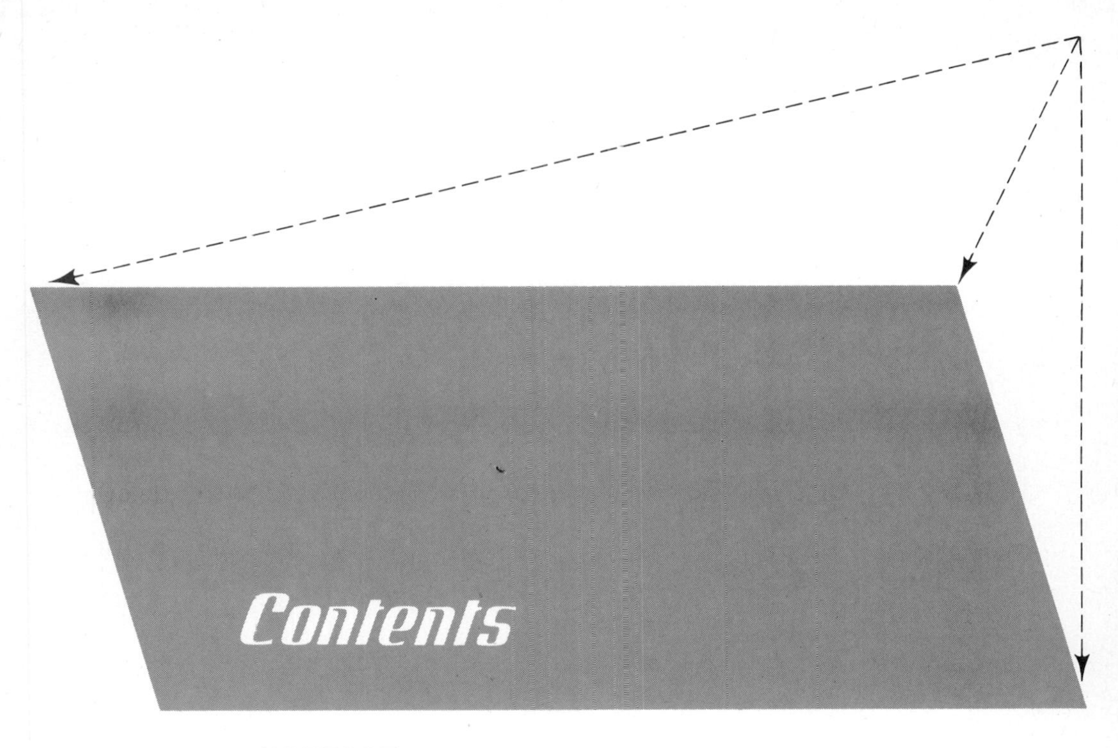

VOLUME I
STATICS

2

ELEMENTS OF VECTOR ALGEBRA 22

IMPORTANT VECTOR QUANTITIES 56

EQUIVALENT FORCE SYSTEMS 85

5

EQUATIONS OF EQUILIBRIUM 134

6

INTRODUCTION TO STRUCTURAL MECHANICS 195

7

FRICTION FORCES 250

PROPERTIES OF SURFACES 291

MOMENTS AND PRODUCTS OF INERTIA 335

PARTICLE DYNAMICS 454

ENERGY METHODS FOR PARTICLES 507

16

KINETICS OF PLANE MOTION OF RIGID BODIES 678

17

ENERGY AND IMPULSE-MOMEMTUM METHODS FOR RIGID BODIES 734

*DYNAMICS OF GENERAL RIGID-BODY MOTION 791

VIBRATIONS 841

APPENDICES

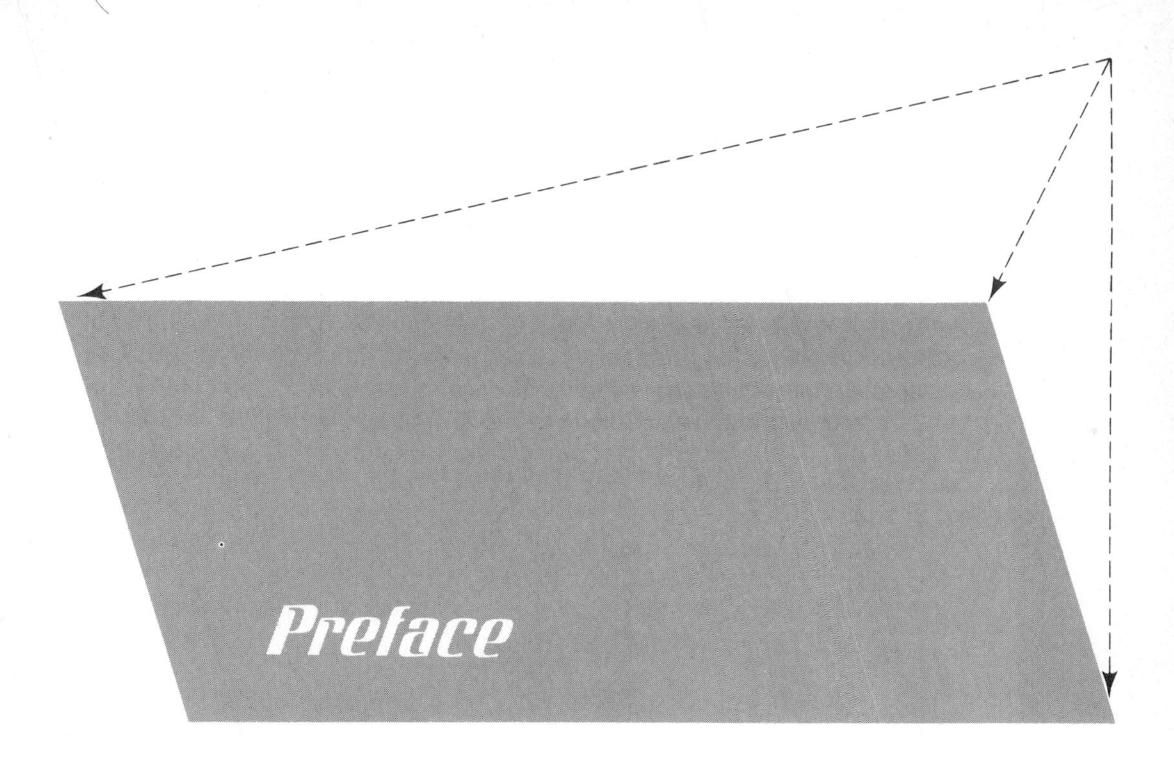

My main effort in writing the third edition has been to simplify and streamline the treatment and to make the text more practical and down to earth. In addition, I have sought to make the book more flexible to allow a range of coverage and degree of depth that will suit the wide range of present engineering programs.* These changes have been developed over a period of six years as a result of teaching mechanics to large classes from 100 to 300 students from all fields.

Specifically, the primary changes are listed as follows:

1. Over 40 percent of the problems are new. In this regard, I have endeavored to present interesting, practical, topical problems. The others have been overhauled from the best problems of the second edition. Furthermore, the problems have been moved in from the ends of the chapters to appropriate places within each chapter. At the end of each chapter, I have selected about 10 problems that represent the key concepts and techniques of that chapter. I have called this group of problems "Review Problems" and I urge prospective instructors not to use these problems as assignments but to leave them for students for review purposes in preparation for tests. The answers to all these review problems are given at the end of the book. For the other problems of the chapter, the even numbered problems have answers. I have used SI units at the rate of 50 percent of all the problems at the beginning of Statics and at the beginning of Dynamics. As one progresses in each area, there is a shift towards SI units to comprise 65 to 70 percent of the problems at the ends of Statics and Dynamics.

*In the instructor's manual, I have listed a number of coverages having varying degrees of pace and depth starting from minimum coverage of topics and maximum stress on problem solving.

2. I now have a separate Chapter (16) on *plane motion*. Unless he chooses, the instructor will not have to go through the somewhat difficult three-dimensional Euler equations first before reaching this very important class of problems, as was the case in earlier editions. However, I have not simply rehashed freshman physics. The treatment is general enough to include rotation of arbitrary bodies, and the development sets the stage for a smooth continuation to the three-dimensional Euler's equations (now a starred chapter) should time permit.
3. It has been brought to my attention from various users around the country that students are not well versed in multiple integration and that one cannot assume that they have fully mastered this technique by the early part of the sophomore fall semester. Accordingly, I have set forth procedures in Example problems for carrying out multiple integrations first with constant limits and then later with variable limits. The arguments are based primarily on physical reasoning; more rigorous discussions are left for math classes.
4. I have moved the chapter on the inertia tensor from Dynamics volume to the Statics volume to follow Chapter 8 on properties of surfaces including second moments and products of area. In this way, the inertia tensor can be more easily related to second moments and products of area. Additionally, by discussing the case of the thin plate, we can extend the concept of principal axes, first brought up in Chapter 8, to the inertia tensor of Chapter 9. Also, I have made clear how, for a body having two orthogonal planes of symmetry, the principal axes for points in the intersection of the planes of symmetry can be determined by inspection. This suffices for most of the problems in Dynamics and so the instructor, if he is short of time, can delete the subsequent discussion on the transformation equations for inertia components, as well as the discussion of elipsoid of inertia.
5. In Chapter 17, I have presented a much more detailed and stronger presentation of angular impulse and momentum for rigid bodies. Some of this material has not appeared in earlier editions. The text material of the book is not longer in length than the second edition, however, since I have deleted or put into homework problems some of the less vital or particularly advanced material that has appeared in earlier editions. Examples include electron ballistics, the analysis of the fast spinning top, the chapter on continuum mechanics, and the gyrocompass.
6. In the first two chapters I have made use of the dagger (†) for those sections which I believe have been covered at various times at various levels in courses such as chemistry, physics and mathematics. For such sections, I have included at the ends of these chapters a series of simple questions that the instructor may wish to assign to assure himself that these sections, after being assigned to be read by the students during the first week of the course, are well understood. By using this procedure of assigning readings at the early stages so as to bring together in a unified way the various and sundry approaches that the student has had before, it is possible (I and my colleagues do this) to begin Chapter 3 by the second week of the course and still have a rational, sound foundation to develop the subject. For those who wish to spend more time on vector algebra in Chapter 2, I have

presented an array of problems that will help the students quickly master these operations.

7. A great effort has been made to make the book more flexible as to possible usage by instructors having varied interests and concerns. For this reason I have starred much more material than the earlier edition indicating that such material can be bypassed with no loss in continuity. Any essential feature of material from starred sections needed later in an unstarred section has been deliberately repeated when needed. For instance, the trajectory discussion on space mechanics has been starred in Chapter 12. If an instructor decides to bypass this material, he can still deal with space mechanics problems in unstarred discussions in Chapter 14 using energy and angular momentum considerations. At that point, a short discussion is presented to give the student enough backgourd to study this material without having studied the entire trajectory theory of Chapter 12. As another example, the key material on conservative forces first presented in the starred Chapter 10 on virtual work has been presented again in the Chapter 13 for unstarred material on energy methods. Finally, I have made greater use of fine print for material which is less likely to be covered in class but which may be of interest and value to the reader.
8. In general, I have overhauled the entire second edition and have rewritten much of it to attain greater clarity and simplicity. I have had the good fortune of having my friend, Dr. Robert M. Jones, of Southern Methodist University, go over the entire text, including problems, line by line with the mission of testing each phrase and each expression for maximum clarity and continuity.

The following is a more detailed description of the contents of the text and will further illustrate some of the changes made in the third edition.

In Chapter 1, we begin by introducing certain fundamental ideas underlying mechanics, such as the concepts of dimensions and units, permitting us then to discuss certain common idealizations employed in mechanics. This sets the stage for a brief discussion of the basic laws of mechanics. A self-contained treatment of vector algebra is then presented in Chapter 2 for students who have as yet not been introduced to this subject matter. This chapter may also be used for purposes of review for those students who have had this material in their mathematics and physics courses. In Chapter 3 we then examine carefully the position vector, the moment of a force, and the couple. This permits us in Chapter 4 to present a thorough discussion of the equivalence of force systems for rigid-body mechanics. In particular, the simplest resultant force system is set forth for general, coplanar, parallel, and concurrent force systems. In Chapter 5, we then can establish the necessary equations of equilibrium for each of the aforementioned force systems by seeing what is necessary in each case to render the simplest resultant equal to zero. At the end of the chapter, we explain why the rigid body equation of equilibrium must still be satisfied when considering statically indeterminate problems where deformation must be taken into account.

Application of these equations is continued in Chapter 6 to simple trusses, beams, chains and cables. In Chapter 7 there are still further applications of these laws coupled now with the laws of Coulomb friction. In Chapter 8 we next present the concepts of the first moment of areas, masses, and volumes, and then go on to the second moment of areas, wherein we carefully present transformation properties with respect to a rotation of axes. The concept of principal axes is then presented. In Chapter 9, we consider first the definition of moments and products of inertia. Considering the thin plate, we are able to extend the principal axis concept for areas to that of masses. Those programs with sufficient time to spend can then study the transformation equations for the general inertia components as well as the ellipsoid of inertia. In Chapter 10, there is a rather careful development of virtual work. I feel this should be done carefully or not at all; "quickee" treatments do more harm than good. This conclusion stems from teaching follow on courses in variational mechanics.*

We begin in Chapter 11 by computing time derivatives of a vector in the presence of a single reference using Cartesian coordinates, cylindrical coordinates, and path coordinates. We are then able to present the kinematics of a particle in the presence of a single reference. The concept of relative motion is next presented but is restricted, until Chapter 15 to references that are translating relative to each other. We are able to define carefully, at this time, what is meant by the motion of a particle "relative to a point." In Chapter 12 we then examine the dynamics of a particle for rectilinear translation, for central force motion, and other curvilinear motions. The concept of vibration is set forth and made ready for more careful study later in Chapter 19. The chapter closes with an examination of a system of particles and, using the center of mass concept, we are able, at this time, to understand better the particle concept that we have been using so often. In Chapters 13 and 14, respectively, the powerful methods of energy and linear momentum are employed on particles and systems of particles. At the end of Chapter 14, in the interest of continuity, we now examine the application of the equation $\boldsymbol{M} = \dot{\boldsymbol{H}}$ for simple, rigid-body, plane-motion problems of the type studied in earlier physics courses. Here we employ kinematical formulations of rigid bodies as given in earlier course work. These simple problems then set the stage for the remainder of the text by motivating the need for a careful study of the kinematics of a rigid body in Chapter 15.

Accordingly in Chapter 15 we present Chasle's theorem for a rigid body and, with this, are now able to extend the relative motion relations, introduced in Chapter 11, to cover the general case of references moving arbitrarily relative to each other. We are thus able in this chapter to wind up (except for vibrations) our efforts in particle dynamics and simultaneously to form the basis for a careful treatment of rigid-body dynamics. In Chapter 16, we study the dynamics of plane motion starting with the simplest form and going to the most general form. We consider balancing of rotating bodies. In Chapter 17, we consider energy methods and impulse-momentum methods for rigid bodies as an extension of the work done in Chapter 13 and 14 for particles. As a result of our teaching experience at Buffalo, we find that here we could develop the general three-dimensional formulation first and then solve many two- and three-dimensional problems. I believe time is saved and greater overall understanding is developed by this approach.

---

*See the author's text with C. Dym, "Solid Mechanics—A Variational Approach". McGraw-Hill Book Co., 1973.

We are now ready, in Chapter 18, to examine carefully the equation $M = \dot{H}$ as applied to a rigid body having general motions. In particular, the very useful Euler equations are presented. A careful procedure is presented for properly and effectively using Euler's equations for three dimensional problems of a wide range. Euler angles are presented and, in a series of examples, various gyroscope applications are presented. Finally, as a starred section, torque-free motion is developed with space applications. Finally, in Chapter 19, we study vibrations of one degree of freedom with a short introduction to multidegree of freedom systems.

In the first edition of this book, I had a chapter on momentum methods for deformable media (mostly fluids). Such coverage has become common practice in many current mechanics test books. However, I have deleted this chapter in the third edition as I have in the second edition believing that such material is best covered in a fluids course where the control volume can be considered carefully not only for momentum considerations but also for continuity and the first law of thermodynamics.*

I am indebted to many people for valuable assistance in writing the third edition. First, I wish to thank Dr. Robert M. Jones of Southern Methodist University and his lovely wife, Donna. Bob and I went over the statics volume line by line together alternating at my home in Buffalo and his home in Dallas. The main effort was to achieve maximum simplicity and clarity. In addition, Bob made available to me 200 of his best statics problems. Finally, Bob carefully reviewed the dynamics manuscript again with the purpose of achieving greatest lucidity. I was fortunate also in having careful reviews from Profs. David McGill and Wilton King of Georgia Institute of Technology. As a result of their keen perception, I was able to make improvements in the text. My sincere thanks goes to these gentlemen. Professor William Lee of the Naval Academy gave me a detailed, line by line review of the entire text including problems. I found his suggestions to be extremely helpful and I wish to extend my sincere appreciation to him for such a useful and valuable input. Profs. I. McIvor of U. of Michigan, J.S. Chen of U. of Pittsburgh and W.E. Clausen of Ohio State University also were helpful viewers of the manuscript to whom I wish to extend my thanks for their efforts. At home in Buffalo, I wish to thank my colleagues, Profs. P. Culkowski, C. Fogel, R. Mates, S. Prawel, T. Rancv and H. Reismann. They have been a constant source of encouragement and of valuable assistance as we taught different classes of the mechanics course. I wish to thank my son Bruce for the photographs used in Chapter 6. Finally, I wish to thank Mrs. K. Ward and Mrs. G. Huck for their excellent typing efforts.

Irving H. Shames

*See the author's text, "Mechanics of Fluids". McGraw-Hill Book Co., 1962.

VOLUME I

STATICS

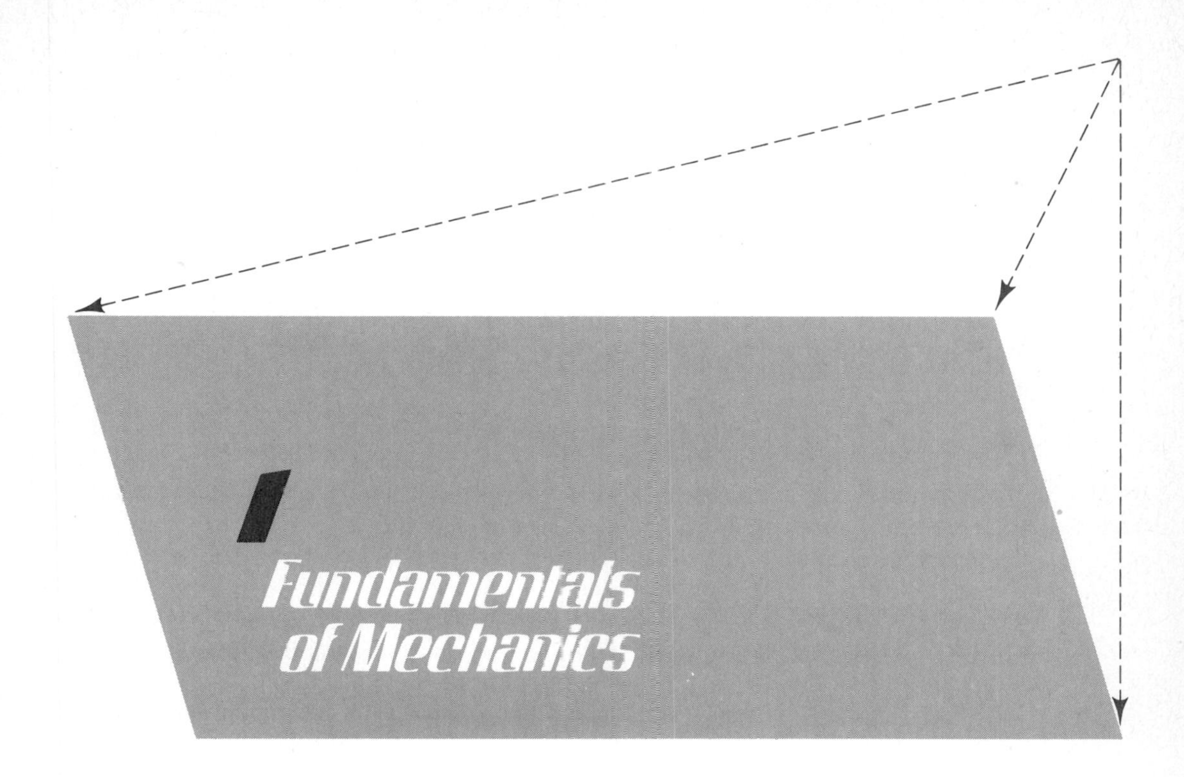

†1.1 Introduction

Mechanics is the physical science concerned with the dynamical behavior (as opposed to chemical and thermal behavior) of bodies that are acted on by mechanical disturbances. Since such behavior is involved in virtually all the situations that confront an engineer, mechanics lies at the core of much engineering analysis. In fact, no physical science plays a greater role in engineering then does mechanics, and it is the oldest of all the physical sciences. The writings of Archimedes covering buoyancy and the lever were recorded before 200 B.C. Our modern knowledge of gravity and motion was established by Isaac Newton (1642–1727), whose laws founded Newtonian mechanics, the subject matter of this text.

In 1905, Einstein placed limitations on Newton's formulations with his theory of relativity and thus set the stage for the development of relativistic mechanics. The newer theories, however, give results that depart from those of Newton's formulations only when the speed of a body approaches the speed of light (186,000 miles/sec). These speeds are encountered in the large-scale phenomena of dynamical astronomy and the small-scale phenomena involving subatomic particles. Despite these limitations, it remains nevertheless true that, in the great bulk of engineering problems, Newtonian mechanics still applies.

†1.2 Basic Dimensions and Units of Mechanics

To study mechanics, we must establish abstractions to describe those characteristics of a body that interest us. These abstractions are called *dimensions*. The dimensions that we pick, which are independent of all dimensions, are termed *primary* or *basic dimensions*, and the ones that are then developed in terms of the basic dimensions we call *secondary dimensions*. Of the many possible sets of basic dimensions that we could use, we will confine ourselves at present to the set that includes the dimensions of length, time, and mass. Another convenient set will be examined later.

Length—A Concept for Describing Size Quantitatively. In order to determine the size of an object, we must place a second object of known size next to it. Thus, in pictures of machinery, a man often appears standing disinterestedly beside the apparatus. Without him, it would be difficult to gage the size of the unfamiliar machine. Although the man has served as some sort of standard measure, we can, of course, only get an approximate idea of the machine's size. Men's heights vary, and, what is even worse, the shape of a man is too complicated to be of much help in acquiring a precise measurement of the machine's size. What we need, obviously, is an object that is constant in shape and, moreover, simple in concept. Thus, instead of a three-dimensional object, we choose a one-dimensional object.[1] Then, we can use the known mathematical concepts of geometry to extend the measure of size in one dimension to the three dimensions necessary to characterize a general body. A straight line scratched on a metal bar that is kept at uniform thermal and physical conditions (as, e.g., the meter bar kept at Sèvres, France) serves as this simple invariant standard in one dimension. We can now readily calculate and communicate the distance along a certain direction of an object by counting the number of standards and fractions thereof that can be marked off along this direction. We commonly refer to this distance as length, although the term "length" could also apply to the more general concept of size. Other aspects of size, such as volume and area, can then be formulated in terms of the standard by the methods of plane, spherical, and solid geometry.

A *unit* is the name we give an accepted measure of a dimension. Many systems of units are actually employed around the world, but we shall only use the two major systems, the American system and the SI system. The unit of length in the American system is the foot, whereas the unit of length in the SI system is the meter.

Time—A Concept for Ordering the Flow of Events. In observing the picture of the machine with the man standing close by, we can sometimes tell approximately when the picture was taken by the style of clothes the man is wearing. But how do we determine this? We may say to ourselves: "During the thirties, people wore the type of straw hat that the fellow in the picture is wearing." In other words, the "when" is tied to certain events that are experienced by, or otherwise known to, the observer. For a more accurate description of "when," we must find an action that appears to be completely repeatable. Then, we can order the events under study by counting the

---

[1]We are using the word "dimensional" here in its everyday sense and not as defined above.

number of these repeatable actions and fractions thereof that occur while the events transpire. The rotation of the earth gives rise to an event that serves as a good measure of time—the day. But we need smaller units in most of our work in engineering, and thus, generally, we tie events to the second, which is an action repeatable 86,400 times a day.

Mass—A Property of Matter. The student ordinarily has no trouble understanding the concepts of length and time because he is constantly aware of the size of things through his senses of sight and touch, and is always conscious of time by observing the flow of events in his daily life. The concept of mass, however, is not as easily grasped since it does not impinge as directly on our daily experience.

Mass is a property of matter that can be determined from *two* different actions of bodies. To study the first action, suppose that we consider two hard bodies of entirely different composition, size, shape, color, and so on. If we attach the bodies to identical springs, as shown in Fig. 1.1, each spring will extend some distance as a result of the attraction of gravity for the bodies. By grinding off some of the material on the body that causes the greater extension, we can make the deflections that are induced on both springs equal. Even if we raise the springs to a new height above the earth's surface, thus lessening the deformation of the springs, the extensions induced by the pull of gravity will be the same for both bodies. And since they are, we can conclude that the bodies have an equivalent innate property. This property of each body that manifests itself in the *amount of gravitational attraction* we call *mass*.

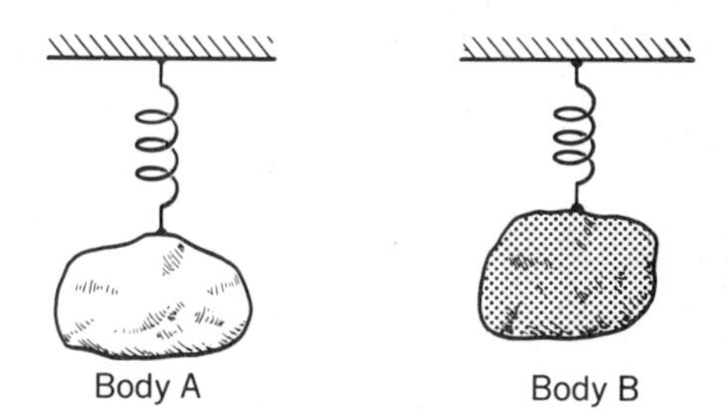

Figure 1.1. Bodies restrained by identical springs.

The equivalence of these bodies can be indicated in yet a second action. If we move both bodies an equal distance downward, by stretching each spring, and then release them at the same time, they will begin to move in an identical manner (except for small variations due to differences in wind friction and local deformations of the bodies). We have imposed, in effect, the same mechanical disturbance on each body and we have elicited the same dynamical response. Hence, despite many obvious differences, the two bodies again show an equivalence.

> *The property of mass, then, characterizes a body both in the action of gravitational attraction and in the response to a mechanical disturbance.*

To communicate this property quantitatively, we may choose some convenient body and compare other bodies to it in either of the two abovementioned actions. The two units commonly used in much American engineering practice to measure mass are the *pound mass*, which is defined in terms of the attraction of gravity for a standard body at a standard location, and the *slug*, which is defined in terms of the dynamical response of a standard body to a standard mechanical disturbance. A similar duality of mass units does not exist in the SI system. There only the *kilogram* is used as a measure of mass. The kilogram is measured in terms of response of a body to a mechanical disturbance. Both systems of units will be discussed further in a subsequent section.

We have now established three basic independent dimensions to describe certain physical phenomena. It is convenient to identify these dimensions in the following manner:

length $[L]$

time $[t]$

mass $[M]$

These formal expressions of identification for basic dimensions and the more complicated groupings to be presented in Section 1.3 for secondary dimensions are called "dimensional representations."

Often, there are occasions when we want to change units during computations. For instance, we may wish to change feet into inches or millimeters. In such a case, we must replace the unit in question by a *physically equivalent* number of new units. Thus, a foot is replaced by 12 inches or 305 millimeters. A listing of common systems of units is given in Table 1.1, and a table of equivalences between these and other units is given

TABLE 1.1 Common Systems of Units

cgs		SI	
Mass	Gram	Mass	Kilogram
Length	Centimeter	Length	Meter
Time	Second	Time	Second
Force	Dyne	Force	Newton
English		**American Practice**	
Mass	Pound mass	Mass	Slug or pound mass
Length	Foot	Length	Foot
Time	Second	Time	Second
Force	Poundal	Force	Pound force

on the inside covers. Such relations between units will be expressed in this way:

$$1 \text{ ft} \equiv 12 \text{ in.} \equiv 305 \text{ mm}$$

The three horizontal bars are not used to denote *algebraic* equivalence; instead, they are used to indicate physical equivalence. Here is another way of expressing the relations above:

$$\begin{aligned} \left(\frac{1 \text{ ft}}{12 \text{ in.}}\right) &\equiv 1, \qquad \left(\frac{1 \text{ ft}}{305 \text{ mm}}\right) \equiv 1 \\ \left(\frac{12 \text{ in.}}{1 \text{ ft}}\right) &\equiv 1, \qquad \left(\frac{305 \text{ mm}}{1 \text{ ft}}\right) \equiv 1 \end{aligned} \tag{1.1}$$

The unity on the right side of these relations indicates that the numerator and denominator on the left side are physically equivalent, and thus have a 1:1 relation. This notation will prove convenient when we consider the change of units for secondary dimensions in the next section.

†1.3 Secondary Dimensional Quantities

When physical characteristics are described in terms of basic dimensions by the use of suitable definitions (e.g., velocity is defined[2] as a distance divided by a time interval), such quantities are called *secondary dimensional quantities*. In Section 1.4, we will see that these quantities may also be established as a consequence of natural laws. The dimensional representation of secondary quantities is given in terms of the basic dimensions that enter into the formulation of the concept. For example, the dimensional representation of velocity is

$$[\text{velocity}] \equiv \frac{[L]}{[t]}$$

That is, the dimensional representation of velocity is the dimension length divided by the dimension time. The units for a secondary quantity are then given in terms of the units of the constituent basic dimensions. Thus,

$$\text{velocity units} \equiv \frac{[\text{ft}]}{[\text{sec}]}$$

A *change* of units from one system into another usually involves a change in the scale of measure of the secondary quantities involved in the problem. Thus, one scale unit of velocity in the American system is 1 foot per second, while in the SI system it is 1 meter per second. How may these scale units be correctly related for complicated secondary quantities? That is, for our simple case, how many meters per second are equivalent to 1 foot per second? The formal expressions of dimensional representation may be put to good use for such an evaluation. The procedure is as follows. Express the dependent quantity dimensionally; substitute existing units for the basic dimensions; and finally, change these units to the equivalent numbers of units in the new system. The result gives the number of scale units of the quantity in the new system of units that is equivalent to 1 scale unit of the quantity in the old system. Performing these operations for velocity, we would thus have

$$1\left(\frac{\text{ft}}{\text{sec}}\right) \equiv 1\left(\frac{.305\ \text{m}}{\text{sec}}\right) \equiv .305\left(\frac{\text{m}}{\text{sec}}\right)$$

which means that .305 scale unit of velocity in the SI system is equivalent to 1 scale unit in the American system.

Another way of changing units when secondary dimensions are present is to make use of the formalism illustrated in relations 1.1. To change a unit in an expression,

---

[2]A more precise definition will be given in the chapters on dynamics.

multiply this unit by a ratio physically equivalent to unity, as we discussed earlier, so that the old unit is canceled out, leaving the desired unit with the proper numerical coefficient. In the example of velocity used above, we may replace ft/sec by m/sec in the following manner:

$$1\left(\frac{\text{ft}}{\text{sec}}\right) \equiv \left(\frac{1\,\cancel{\text{ft}}}{\text{sec}}\right) \cdot \left(\frac{.305\text{ m}}{1\,\cancel{\text{ft}}}\right) \equiv .305\left(\frac{\text{m}}{\text{sec}}\right)$$

It should be clear that, when we multiply by such ratios to accomplish a change of units as shown above, we do not alter the magnitude of the *actual physical quantity* represented by the expression. Students are strongly urged to employ the above technique in their work, for the use of less formal methods is generally an invitation to error.

†1.4 Law of Dimensional Homogeneity

Now that we can describe certain aspects of nature in a quantitative manner through basic and secondary dimensions, we can by careful observation and experimentation learn to relate certain of the quantities in the form of equations. In this regard, there is an important law, the law of *dimensional homogeneity*, which imposes a restriction on the formulation of such equations. This law states that, because natural phenomena proceed with no regard for man-made units, *basic equations representing physical phenomena must be valid for all systems of units*. Thus, the equation for the period of a pendulum, $t = 2\pi\sqrt{L/g}$, must be valid for all systems of units, and is accordingly said to be *dimensionally homogeneous*. It then follows that the fundamental equations of physics are dimensionally homogeneous; and all equations derived analytically from these fundamental laws must also be dimensionally homogeneous.

What restriction does this condition place on an equation? To answer this, let us examine the following arbitrary equation:

$$x = ygd + k$$

For this equation to be dimensionally homogeneous, the numerical equality between both sides of the equation must be maintained for all systems of units. To accomplish this, the change in the scale of measure of each group of terms must be the same when there is a change of units. That is, if the numerical measure of one group such as ygd is doubled for a new system of units, so must that of the quantities x and k. *For this to occur under all systems of units, it is necessary that every grouping in the equation have the same dimensional representation.*

In this regard, consider the dimensional representation of the above equation expressed in the following manner:

$$[x] = [ygd] + [k]$$

From the previous conclusion for dimensional homogeneity, we require that

$$[x] \equiv [ygd] \equiv [k]$$

As a further illustration, consider the dimensional representation of an equation that is *not* dimensionally homogeneous:

$$[L] = [t]^2 + [t]$$

When we change units from the American to the SI system, the units of feet give way to units of meters, but there is no change in the unit of time, and it becomes clear that the numerical value of the left side of the equation changes while that of the right side does not. The equation, then, becomes invalid in the new system of units and hence is not derived from the basic laws of physics. Throughout this book, we shall invariably be concerned with dimensionally homogeneous equations. Therefore, we should dimensionally analyze our equations to help spot errors.

†1.5 Dimensional Relation Between Force and Mass

We shall now employ the law of dimensional homogeneity to establish a new secondary dimension—namely *force*. A superficial use of Newton's law will be employed for this purpose. In a later section, this law will be presented in greater detail, but it will suffice at this time to state that the acceleration of a particle[3] is inversely proportional to its mass for a given disturbance. Mathematically, this becomes

$$a \propto \frac{1}{m} \tag{1.2}$$

where $\propto$ is the proportionality symbol. Inserting the constant of proportionality, F, we have, on rearranging the equation,

$$F = ma \tag{1.3}$$

The mechanical disturbance, represented by F and called *force*, must have the following dimensional representation, according to the law of dimensional homogeneity:

$$[F] \equiv [M]\frac{[L]}{[t]^2} \tag{1.4}$$

The type of disturbance for which relation 1.2 is valid is usually the action of one body on another by direct contact. However, other actions, such as magnetic, electrostatic, and gravitational actions of one body on another, also create mechanical effects that are valid in Newton's equation.

We could have initiated the study of mechanics by considering *force* as a basic dimension, the manifestation of which can be measured by the elongation of a standard spring at a prescribed temperature. Experiment would then indicate that for a given body the acceleration is directly proportional to the applied force. Mathematically,

$$F \propto a; \quad \text{therefore, } F = ma$$

[3]We shall define particles in Section 1.7.

from which we see that the proportionality constant now represents the property of mass. Here, mass is now a secondary quantity whose dimensional representation is determined from Newton's law:

$$[M] \equiv [F]\frac{[t]^2}{[L]} \tag{1.5}$$

As was mentioned earlier, we now have a choice between two systems of basic dimensions—the *MLt* or the *FLt* system of basic dimensions. Physicists prefer the former, whereas engineers usually prefer the latter.

1.6 Units of Mass

As we have already seen, the concept of mass arose from two types of actions—those of motion and gravitational attraction. In American engineering practice, units of mass are based on both actions, and this sometimes leads to confusion. Let us consider the *FLt* system of basic dimensions for the following discussion. The unit of force may be taken to be the pound, which is defined as a force that extends a standard spring a certain distance. Using Newton's law, we then define the *slug* as the amount of mass that a 1-pound force will cause to accelerate at the rate of 1 foot per second per second.

On the other hand, another unit of mass can be stipulated if we use the gravitational effect as a criterion. Here, the *pound mass* (lbm) is defined as the amount of matter that is drawn by gravity toward the earth by a force of 1 pound (lbf) at a specified position on the earth's surface.

We have formulated two units of mass by two different actions, and to relate these units we must subject them to the *same* action. Thus, we can take 1 pound mass and see what fraction or multiple of it will be accelerated 1 ft/sec^2 under the action of 1 pound of force. This fraction or multiple will then represent the number of units of pound mass that are equivalent to 1 slug. It turns out that this coefficient is g_0, where g_0 has the value corresponding to the acceleration of gravity at a position on the earth's surface where the pound mass was standardized. To three significant figures, the value of g_0 is 32.2. We may then make the statement of equivalence that

$$1 \text{ slug} \equiv 32.2 \text{ pounds mass}$$

To use the pound-mass unit in Newton's law, it is necessary to divide by g_0 to form units of mass that have been derived from Newton's law. Thus,

$$F = \frac{m}{g_0} a \tag{1.6}$$

where m has the units of pound mass. Having properly introduced into Newton's law the pound-mass unit from the viewpoint of physical equivalence, let us now consider the dimensional homogeneity of the resulting equation. The right side of Eq. 1.6 must have the dimensional representation of F and, since the unit here for F is the pound force, the right side must then have this unit. Examination of the units on the right side of the equation then indicates that the units of g_0 must be

$$[g_0] \equiv \frac{[\text{lbm}][\text{ft}]}{[\text{lbf}][\text{sec}]^2} \tag{1.7}$$

How does *weight* fit into this picture? Weight is defined as *the force of gravity on a body*. Its value will depend on the position of the body relative to the earth's surface. At a location on the earth's surface where the pound mass is standardized, a mass of 1 pound (lbm) has the weight of 1 pound (lbf), but with increasing altitude the weight will become smaller than 1 pound (lbf). The mass, however, remains at all times a 1-pound mass (lbm). If the altitude is not exceedingly large, the measure of weight, in lbf, will practically equal the measure of mass, in lbm. Therefore, it is unfortunately the practice in engineering erroneously to think of weight at positions other than on the earth's surface as the measure of mass, and consequently to use the symbol W to represent either lbm or lbf. In this age of rockets and missiles, it behooves us to be careful about the proper usage of units of mass and weight throughout the entire text.

If we know the weight of a body at some point, we can determine its mass in slugs very easily, provided that we know the acceleration of gravity, g, at that point. Thus, according to Newton's law,

$$W\ (\text{lbf}) = m\ (\text{slugs}) \times g\ (\text{ft/sec}^2)$$

Therefore,

$$m\ (\text{slugs}) = \frac{W\ (\text{lbf})}{g\ (\text{ft/sec}^2)} \tag{1.8}$$

Up to this point, we have only considered the American system of units. In the SI system of units, a *kilogram* is the amount of mass that will accelerate 1 m/sec^2 under the action of a force of 1 newton. Here we do not have the problem of 2 units of mass; the kilogram is the basic unit of mass. However, we do have another kind of problem —that the kilogram is unfortunately also used as a measure of force, as is the newton. One kilogram of force is the weight of 1 kilogram of mass at the earth's surface, where the acceleration of gravity (i.e., the acceleration due to the force of gravity) is 9.81 m/sec^2. A newton, on the other hand, is the force that acclerates 1 kilogram of mass only 1 m/sec^2. Hence, 9.81 newtons are equivalent to 1 kilogram of force. That is,

$$9.81 \text{ newtons} \equiv 1 \text{ kilogram (force)} \equiv 2.205 \text{ lbf}$$

Note from the above that the newton is a comparatively small force, equaling approximately one-fifth of a pound. A kilonewton (1000 newtons), which will be used often, is about 200 lb. In this text, we shall not use the kilogram as a unit of force. However, you should be aware that many people do.

Note that at the earth's surface the weight W of a mass M is:

$$W\ (\text{newtons}) = [M\ (\text{kilograms})]\ (9.81) \tag{1.9}$$

Hence:

$$M\ (\text{kilograms}) = \frac{W\ (\text{newtons})}{9.81} \tag{1.10}$$

Away from the earth's surface, use the acceleration of gravity g rather than 9.81 in the above equations.

1.7 Idealizations of Mechanics

As we have pointed out, basic and secondary dimensions may sometimes be related in equations to represent a physical action that we are interested in. We want to represent an action using the known laws of physics, and also to be able to form equations simple enough to be susceptible to mathematical computational techniques. Invariably in our deliberations, we must replace the actual physical action and the participating bodies with hypothetical, highly simplified substitutes. We must be sure, of course, that the results of our substitutions have some reasonable correlation with reality. All analytical physical sciences must resort to this technique, and, consequently, their computations are not cut and dried but involve a considerable amount of imagination, ingenuity, and insight into physical behavior. We shall, at this time, set forth the most fundamental idealizations of mechanics and a bit of the philosophy involved in scientific analysis.

Continuum. Even the simplification of matter into molecules, atoms, electrons, and so on, is too complex a picture for many problems of engineering mechanics. In most problems, we are interested only in the average measurable manifestations of these elementary bodies. Pressure, density, and temperature are actually the gross effects of the actions of the many molecules and atoms, and they can be conveniently assumed to arise from a hypothetically continuous distribution of matter, which we shall call the *continuum*, instead of from a conglomeration of discrete, tiny bodies. Without such an artifice, we would have to consider the action of each of these elementary bodies—a virtual impossibility for most problems.

Rigid Body. In many cases involving the action on a body by a force, we simplify the continuum concept even further. The most elemental case is that of a rigid body, which is a continuum that undergoes theoretically no deformation whatever. Actually, every body must deform to a certain degree under the actions of forces, but in many cases the deformation is too small to affect the desired analysis. It is then preferable to consider the body as rigid, and proceed with the simplified computations. For example, assume that we are to determine the forces transmitted by a beam to the earth as the result of a load P (Fig. 1.2). If P is small enough, the beam will undergo little deflection, and we can carry out a straightforward simple analysis using the undeformed geometry as if the body were indeed rigid. If we were to attempt a more accurate analysis—even

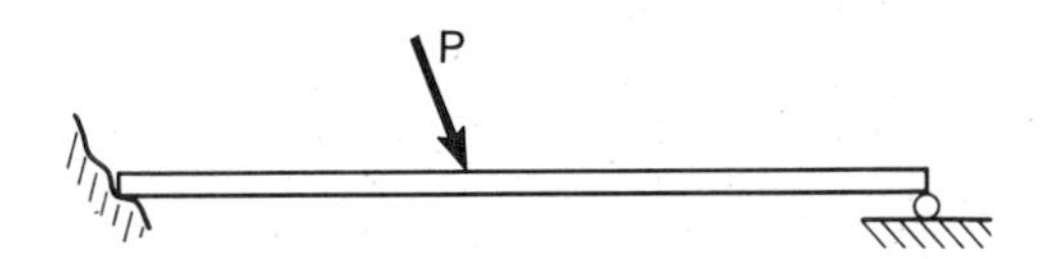

Figure 1.2. Rigid-body assumption—use original geometry.

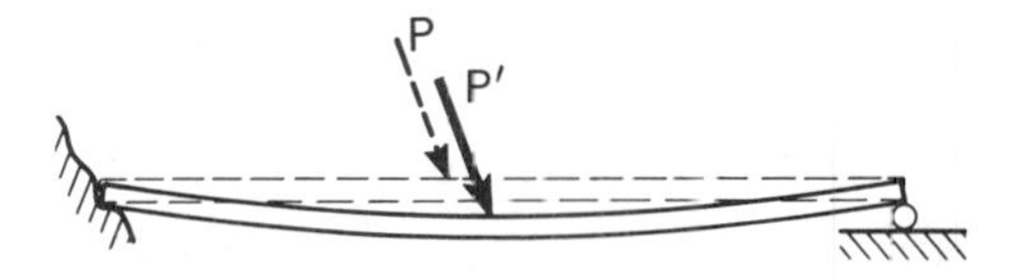

Figure 1.3. Deformable body.

though a slight increase in accuracy is not required—we would then need to know the exact position that the load assumes relative to the earth *after* the beam has ceased to deform, as shown in an exaggerated manner in Fig. 1.3. To do this accurately is a hopelessly difficult task, especially when we consider that the support must also "give" in a certain way. Although the alternative to a rigid-body analysis here leads us to a virtually impossible calculation, situations do arise in which more realistic models must be employed to yield the required accuracy. For example, when determining the internal force distribution in a body, we must often take the deformation into account, however small it might be. *The guiding principle is to make such simplifications as are consistent with the required accuracy of the results.*

Point Force. A finite force exerted on one body by another must cause a finite amount of local deformation, and always creates a finite area of contact between the bodies through which the force is transmitted. However, since we have formulated the concept of the rigid body, we should also be able to imagine a finite force to be transmitted through an infinitesimal area or point. This simplification of a force distribution is called a *point force.* In many cases where the actual area of contact in a problem is very small but is not known exactly, the use of the concept of the point force results in little sacrifice in accuracy. In Figs. 1.2 and 1.3, we actually employed the graphical representation of the point force.

Particle. The *particle* is defined as an object that has no size but that has a mass. Perhaps this does not sound like a very helpful definition for engineers to employ, but it is actually one of the most useful in mechanics. For the trajectory of a planet, for example, it is the mass of the planet and not its size that is significant. Hence, we can consider planets as particles for such computations. On the other hand, take a figure skater spinning on the ice. Her revolutions are controlled beautifully by the orientation of the body. In this motion, the size and distribution of the body are significant, and since a particle, by definition, can have no distribution, it is patently clear that a particle cannot represent the skater in this case. If, however, the skater should be billed as the "human cannonball on skates" and be shot out of a large gun, it would be possible to consider her as a single particle in ascertaining her trajectory, since arm and leg movements that were significant while she was spinning on the ice would have little effect on the arc traversed by the main portion of her body.

Many other simplifications pervade mechanics. The perfectly elastic body, the frictionless fluid, and so on, will become quite familiar as you study various phases of mechanics.

†1.8 Vector and Scalar Quantities

We have now proposed sets of basic dimensions and secondary dimensions to describe certain aspects of nature. However, more than just the dimensional identification and the number of units are often needed to convey adequately the desired information. For instance, to specify fully the motion of a car, which we may represent as a particle at this time, we must answer the following questions:

1. How fast?
2. Which way?

The concept of velocity entails the information desired in questions 1 and 2. The first question, "How fast?", is answered by the speedometer reading, which gives the value of the velocity in miles per hour or kilometers per hour. The second question, "Which way?", is more complicated, because two separate factors are involved. First, we must specify the angular orientation of the velocity relative to a reference frame. Second, we must specify the sense of the velocity, which tells us whether we are moving *toward or away from* a given point. The concepts of angular orientation of the velocity and sense of the velocity are often collectively denoted as the *direction* of the velocity. Graphically, we may use a *directed line segment* (an arrow) to describe the velocity of the car. The *length* of the directed line segment gives information as to "how fast" and is the *magnitude* of the velocity. The angular orientation of the directed line segment and the position of the arrowhead give information as to "which way"—that is, as to the *direction* of the velocity. The directed line segment itself is called the *velocity*, whereas the length of the directed line segment—that is, the magnitude—is called the *speed.*

There are many physical quantities that are represented by a directed line segment and thus are describable by specifying a magnitude and a direction. The most common example is force, where the magnitude is a measure of the intensity of the force and the direction is evident from how the force is applied. Another example is the *displacement vector* between two points on the path of a particle. The magnitude of the dis-

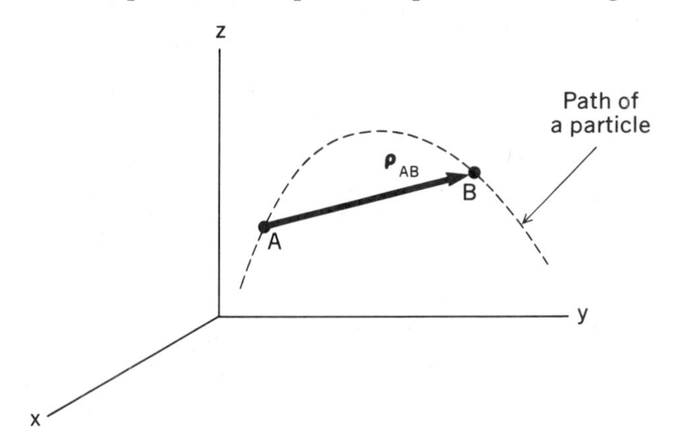

Figure 1.4. Displacement vector $\boldsymbol{\rho}_{AB}$.

placement vector corresponds to the distance moved along a *straight line* between two points, and the direction is defined by the orientation of this line relative to a reference, with the sense corresponding to which point is being approached. Thus, $\boldsymbol{\rho}_{AB}$ (see Fig. 1.4) is the displacement vector from A to B (while $\boldsymbol{\rho}_{BA}$ goes from B to A).

Certain quantities having magnitude and direction combine their effects in a special way. Thus, the combined effect of two forces acting on a particle, as shown in Fig. 1.5, corresponds to a single force that may be shown by experiment to be equal to the diagonal of a parallelogram formed by the graphical representation of the forces. That is, the quantities add according to the *parallelogram law.* All quantities that have magnitude and direction and that add according to the parallelogram law are called *vector quantities*. Other quantities that have only magnitude, such as temperature and work, are called *scalar quantities*. A vector quantity will be denoted with a boldface italic letter, which in the case of force becomes $\boldsymbol{F}$.[4]

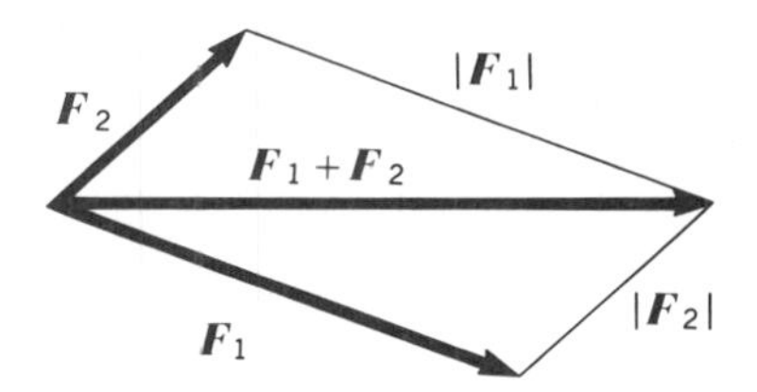

Figure 1.5. Parallelogram law.

The reader may ask: Don't all quantities having magnitude and direction combine according to the parallelogram law and, therefore, become vector quantities? No, not all of them do. One very important example will be pointed out after we reconsider Fig. 1.5. In the construction of the parallelogram it matters not which force is laid out first. In other words, "$\boldsymbol{F}_1$ combined with $\boldsymbol{F}_2$" gives the same result as "$\boldsymbol{F}_2$ combined with $\boldsymbol{F}_1$." In short, the combination is *commutative.* If a combination is not commutative, it cannot in general be represented by a parallelogram operation and is thus not a vector. With this in mind, consider the finite angle of rotation of a body about an axis. We can associate a magnitude (degrees or radians) and a direction (the axis and a stipulation of clockwise or counterclockwise) with this quantity. However, the finite angle of rotation cannot be considered a vector because in general two finite rotations about different axes cannot be replaced by a single finite rotation consistent with the parallelogram law. The easiest way to show this is to demonstrate that the combination of such rotations is not commutative. In Fig. 1.6(a) a book is to be given two rotations —a 90° counterclockwise rotation about the x axis and a 90° clockwise rotation about the z axis, both looking in toward the origin. This is carried out in Figs. 1.6(b) and (c). In Fig. 1.6(c), the sequence of combination is reversed from that in Fig. 1.6(b), and you can see how it alters the final orientation of the book. Finite angular rotation, therefore, is not a vector quantity, since the parallelogram law is not valid for such a combination.[5]

---

[4]Your instructor on the blackboard and you in your homework will not be able to use boldface notation for vectors. Accordingly, you may choose to use a superscript arrow or bar, e.g., $\vec{F}$ or $\bar{F}$ ($\underline{F}$ or $\underset{\sim}{F}$ are other possibilities).

[5]However, *vanishingly small* rotations can be considered as vectors since the commutative law applies for the combination of such rotations. This will be an important consideration when we discuss the angular velocity vector in Chapter 15.

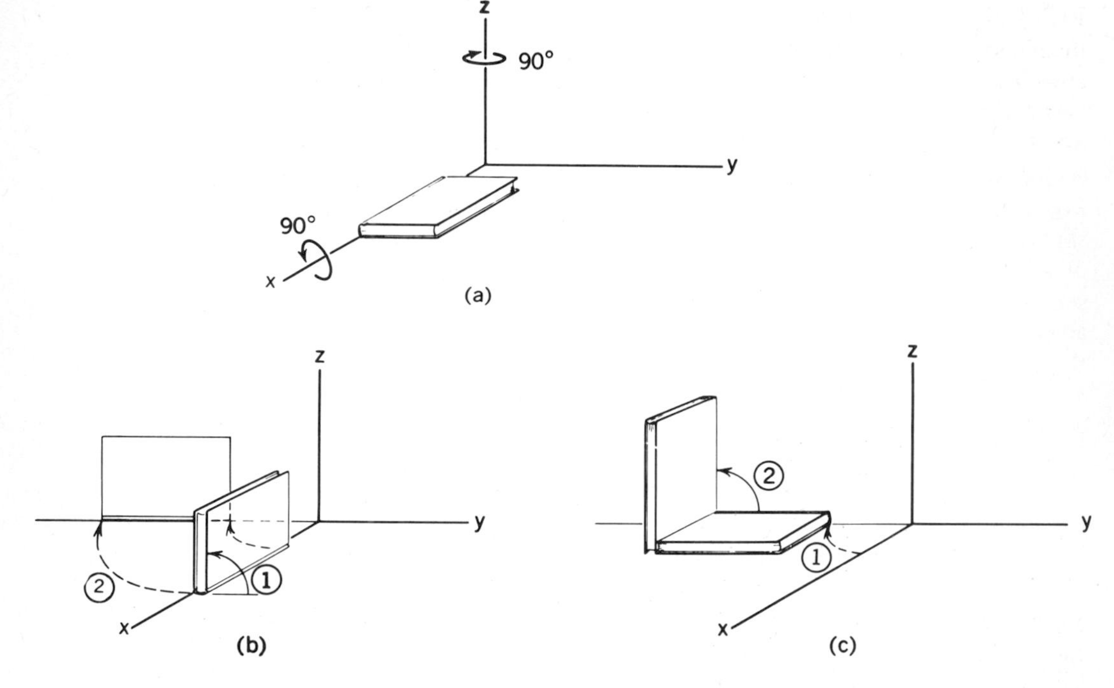

Figure 1.6. Successive rotations are not commutative.

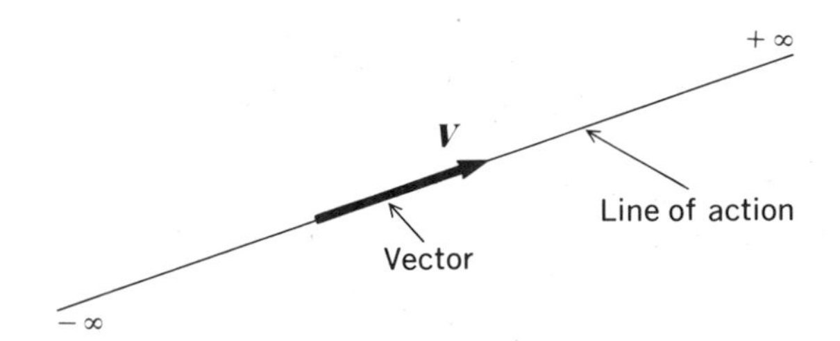

Figure 1.7. Line of action of a vector.

Before closing the section, we will set forth one more definition. The *line of action* of a vector is a hypothetical infinite straight line collinear with the vector (see Fig. 1.7). Thus, the velocities of two cars moving on different lanes of a straight highway have different lines of action. Keep in mind that the line of action involves no connotation as to sense. Thus, a vector $\boldsymbol{V}'$ collinear with $\boldsymbol{V}$ in Fig. 1.7 and with opposite sense would nevertheless have the same line of action.

1.9 Equality and Equivalence of Vectors

We shall avoid many pitfalls in the study of mechanics if we clearly make a distinction between the equality and the equivalence of vectors.

Two vectors are equal if they have the same dimensions, magnitude, and direction.

In Fig. 1.8, the velocity vectors of three particles have equal length, are identically inclined toward the reference *xyz*, and have the same sense. Although they have different lines of action, they are nevertheless equal according to the definition.

Two vectors are equivalent in a certain capacity if each produces the very same effect in this capacity. If the criterion in Fig. 1.8 is change of elevation of the particles or total distance traveled by the particles, all three vectors give the same result. They are, in addition to being equal, also equivalent for these capacities. If the absolute

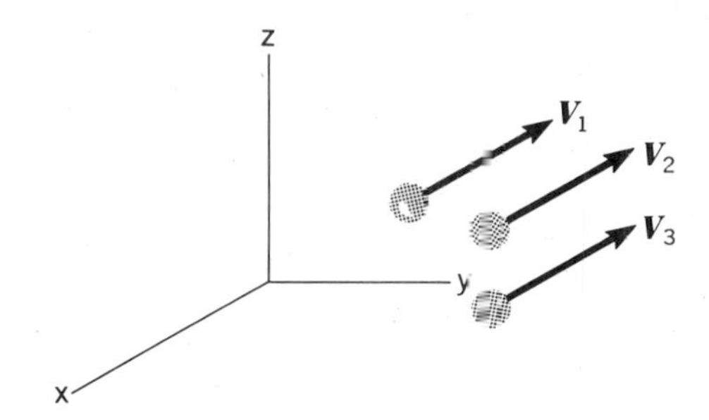

Figure 1.8. Equal-velocity vectors.

height of the particles above the *xy* plane is the question in point, these vectors will not be equivalent despite their equality. Thus, it must be emphasized that *equal vectors need not always be equivalent; it depends entirely on the situation at hand.* Furthermore, vectors that are not equal may still be equivalent in some capacity. Thus, in the beam in Fig. 1.9, forces F_1 and F_2 are unequal, since their magnitudes are 10 lb and 20 lb, respectively. However, it is clear from elementary physics that their moments about the base of the beam are equal, and so the forces have the same "turning" action at the fixed end of the beam. In that capacity, the forces are equivalent. If, however, we

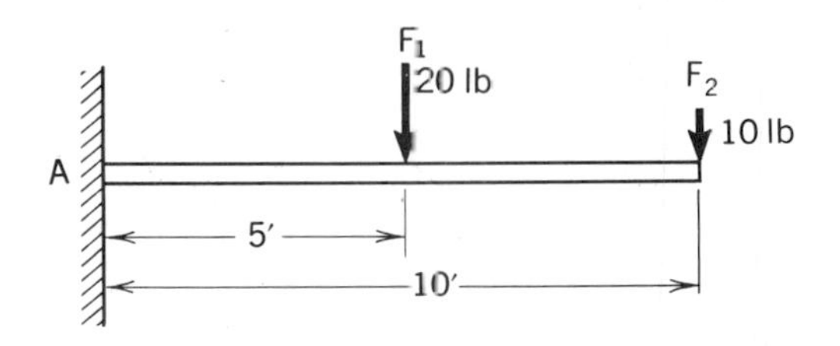

Figure 1.9. F_1 and F_2 equivalent for moment about *A*.

are interested in the deflection of the free end of the beam resulting from each force, there is no longer an equivalence between the forces, since each will give a different deflection.

To sum up, the *equality* of two vectors is determined by the vectors themselves, and the *equivalence* between two vectors is determined by the task involving the vectors.

In problems of mechanics, we can profitably delineate three classes of situations concerning equivalence of vectors:

1. *Situations in which vectors may be positioned anywhere in space without loss or change of meaning provided that magnitude and direction are kept intact.* Under such circumstances the vectors are called *free vectors*. For example, the velocity vectors in Fig. 1.8 are free vectors as far as total distance traveled is concerned.

2. *Situations in which vectors may be moved along their lines of action without change of meaning.* Under such circumstances the vectors are called *transmissible vectors*. For example, in towing the object in Fig. 1.10, we may apply the force anywhere along the rope AB or may push at point C. The resulting motion is the same in all cases, so the force is a transmissible vector for this purpose.

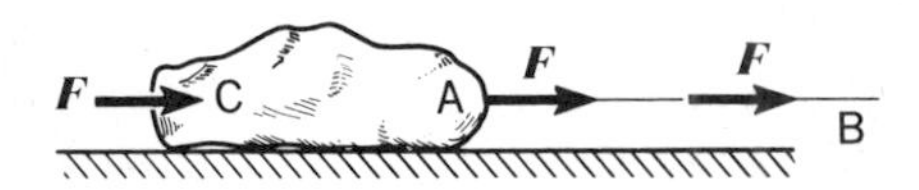

Figure 1.10. F transmissible for towing.

3. *Situations in which the vectors must be applied at definite points.* The point may be represented as the tail or head of the arrow in the graphical representation. For this case, no other position of application leads to equivalence. Under such circumstances, the vector is called a *bound vector*. For example, if we are interested in the deformation induced by forces in the body in Fig. 1.10, we must be more selective in our actions than we were when all we wanted to know was the motion of the body. Clearly, force $\mathbf{F}$ will cause a different deformation when applied at point C than it will when applied at point A. The force is thus a bound vector for this problem.

We shall be concerned throughout this text with considerations of equivalence.

†1.10 Laws of Mechanics

The entire structure of mechanics rests on relatively few basic laws. Nevertheless, for the student to comprehend these laws sufficiently to undertake novel and varied problems, much study will be required.

We shall now discuss briefly the following laws, which are considered to be the foundation of mechanics:

1. Newton's first and second laws of motion.
2. Newton's third law.
3. The gravitational law of attraction.
4. The parallelogram law.

Newton's First and Second Laws of Motion. These laws were first stated by Newton as

> *Every particle continues in a state of rest or uniform motion in a straight line unless it is compelled to change that state by forces imposed on it.*

> *The change of motion is proportional to the natural force impressed and is made in a direction of the straight line in which the force is impressed.*

Notice that the words "rest," "uniform motion," and "change of motion" appear in the statements above. For such information to be meaningful, we must have some frame of reference relative to which these states of motion can be described. We may then ask: relative to what reference in space does every particle remain at "rest" or "move uniformly along a straight line" in the absence of any forces? Or, in the case of a force acting on the particle, relative to what reference in space is the "change in motion proportional to the force"? Experiment indicates that the "fixed" stars act as a reference for which the first and second laws of Newton are highly accurate. Later, we will see that any other system that moves uniformly and without rotation relative to the fixed stars may be used as a reference with equal accuracy. All such references are called *inertial references*. The earth's surface is usually employed as a reference in engineering work. Because of the rotation of the earth and the variations in its motion around the sun, it is not, strictly speaking, an inertial reference. However, the departure is so small for most situations (exceptions are the motion of guided missiles and spacecraft) that the error incurred is very slight. We shall, therefore, usually consider the earth's surface as an inertial reference, but will keep in mind the somewhat approximate nature of this step.

As a result of the preceding discussion, we may define *equilibrium* as *that state of a body in which all its constituent particles are at rest or moving uniformly along a straight line relative to an inertial reference*. The converse of Newton's first law, then, stipulates for the equilibrium state that there must be no force (or equivalent action of no force) acting on the body. Many situations fall into this category. The study of bodies in equilibrium is called *statics*, and it will be an important consideration in this text.

In addition to the reference limitations explained above, a serious limitation was brought to light at the turn of this century. As pointed out earlier, the pioneering work of Einstein revealed that the laws of Newton become increasingly more approximate as the speed of a body increases. Near the speed of light, they are untenable. In the vast majority of engineering computations, the speed of a body is so small compared to the speed of light that these departures from Newtonian mechanics, called *relativistic effects*, may be entirely disregarded with little sacrifice in accuracy. In considering the motion of high-energy elementary particles occurring in nuclear phenomena, however, we cannot ignore relativistic effects. Finally, when we get down to very small distances, such as those between the protons and neutrons in the nucleus of an atom, we find that Newtonian mechanics cannot explain many observed phenomena. In this case, we must resort to quantum mechanics, and then Newton's laws give way to the Schrödinger equation as the key equation.

Newton's Third Law. Newton stated in his third law:

> *To every action there is always opposed an equal reaction, or the mutual actions of two bodies upon each other are always equal and directed to contrary points.*

This is illustrated graphically in Fig. 1.11, where the action and reaction between two bodies arise from direct contact. Other important actions in which Newton's third law holds are gravitational attractions (to be discussed next) and electrostatic forces between charged particles. It should be pointed out that there are actions that do not follow this law, notably the electromagnetic forces between charged moving bodies.[6]

Law of Gravitational Attraction. It has already been pointed out that there is an attraction between the earth and the bodies at its surface, such as A and B in Fig. 1.11. This attraction is mutual and Newton's third law applies. There is also an attraction between the two bodies A and B themselves, but this force because of the small size of

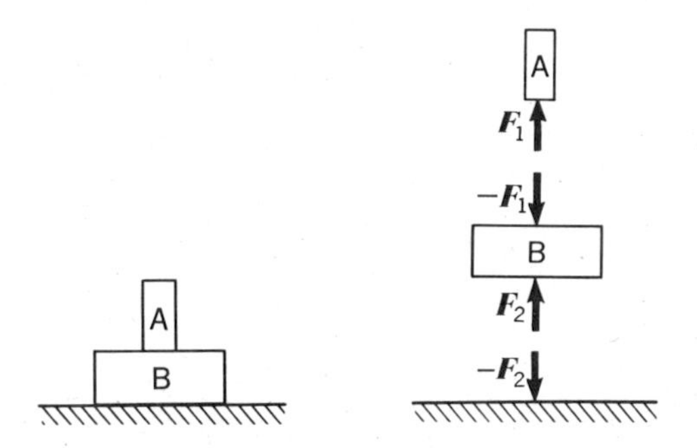

Figure 1.11. Newton's third law.

both bodies is extremely weak. However, the mechanism for the mutual attraction between the earth and each body is the same as that for the mutual attraction between the bodies. These forces of attraction may be given by the *law of gravitational attraction*:

> Two particles will be attracted toward each other along their connecting line with a force whose magnitude is directly proportional to the product of the masses and inversely proportional to the distance squared between the particles.

Avoiding vector notation for now, we may thus say that

$$F = G\frac{m_1 m_2}{r^2} \tag{1.11}$$

where G is called the *universal gravitational constant*. In the actions involving the earth and the bodies discussed above, we may consider each body as a particle, with its entire mass concentrated at its center of gravity.[7] Hence, if we know the various constants in formula 1.11, we can compute the weight of a given mass at different altitudes above the earth.

Parallelogram Law. Stevinius (1548–1620) was the first to demonstrate that forces could be combined by representing them by arrows to some suitable scale, and then

---

[6]Electromagnetic forces between charged moving particles are equal and opposite but are not collinear and hence are not "directed to contrary points."

[7]To be studied in detail in Chapter 4.

forming a parallelogram in which the diagonal represents the sum of the two forces. As we pointed out, all vectors must combine in this manner.

1.11 Closure

In this chapter, we have introduced the basic dimensions by which we can describe in a quantitative manner certain aspects of nature. These basic, and from them secondary, dimensions may be related by dimensionally homogeneous equations which, with suitable idealizations, can represent certain actions in nature. The basic laws of mechanics were thus introduced. Since the equations of these laws relate vector quantities, we shall introduce a useful and highly descriptive set of vector operations in Chapter 2 in order to learn to handle these laws effectively and to gain more insight into mechanics in general. These operations are generally called *vector algebra.*

Check-Out for Sections with†

1.1 What are two kinds of limitations on Newtonian mechanics?
1.2 What are the two phenomena wherein mass plays a key role?
1.3 If a pound force is defined by the extension of a standard spring, define the pound mass and the slug.
1.4 Express mass density dimensionally. How many scale units of mass density in the SI units are equivalent to 1 scale unit in the American system using (a) slugs, ft, sec and (b) lbm, ft, sec.
1.5 (a) What is a necessary condition for *dimensional homogeneity* in an equation?
(b) In the Newtonian viscosity law, the frictional resistance τ (force per unit area) in a fluid is proportional to the distance rate of change of velocity dV/dy. The proportionality constant μ is called the *coefficient of viscosity*. What is its dimensional representation?
1.6 Define a vector and a scalar.
1.7 What is meant by *line of action* of a vector?
1.8 What is a *displacement* vector?
1.9 What is an *inertial reference*?

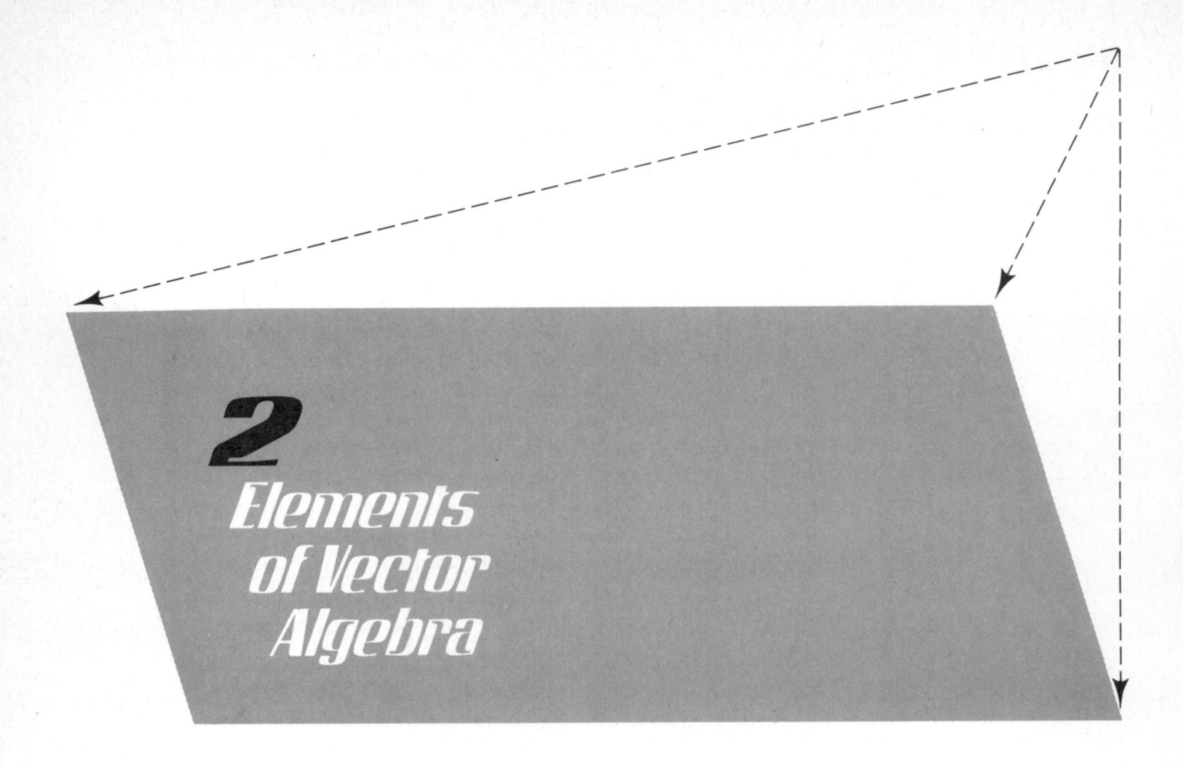

†2.1 Introduction

In Chapter 1, we saw that a scalar quantity is adequately given by a magnitude, while a vector quantity requires the additional specification of a direction. The basic algebraic operations for the handling of scalar quantities are those familiar ones studied in grade school, so familiar that you now wonder even that you had to be "introduced" to them. For vector quantities, these methods may be cumbersome since the directional aspects must be taken into account. Therefore, an algebra has evolved that clearly and concisely allows for certain very useful manipulations of vectors. It is not merely for elegance or sophistication that we employ vector algebra. Indeed, we can achieve greater insight into the subject matter—particularly into dynamics—by employing the more powerful and descriptive methods introduced in this chapter. These methods will at first appear rather arbitrary and artificial. However, you must remember that many years ago, perhaps when you were six or seven years of age, addition and subtraction were rather puzzling and later, when you were eight or nine, multiplication and division were by no means perfectly "natural." Now, of course, you perform those operations almost without thinking about them. Similarly, constant use of vector algebra in mechanics and in other disciplines will bring a comfortable familiarity in a surprisingly short time.

†2.2 Magnitude and Multiplication of a Vector by a Scalar

The magnitude of a quantity, in strict mathematical parlance, is always a *positive* number of units whose value corresponds to the numerical measure of the quantity. Thus, the magnitude of a quantity of measure −50 units is +50 units. Note that the magnitude of a quantity is its absolute value. The mathematical symbol for indicating the magnitude of a quantity is a set of vertical lines enclosing the quantity. That is,

$$|-50 \text{ units}| = \text{absolute value } (-50 \text{ units}) = +50 \text{ units}$$

Similarly, the magnitude of a vector quantity is a positive number of units corresponding to the length of the vector in those units. Using our vector symbols, we can say that

$$\text{magnitude of vector } \boldsymbol{A} = |\boldsymbol{A}|$$

Thus, $|\boldsymbol{A}|$ is a positive scalar quantity. We may now discuss the multiplication of a vector by a scalar.

The definition of the product of vector $\boldsymbol{A}$ by scalar m, written simply as $m\boldsymbol{A}$, is given in the following manner:

> $m\boldsymbol{A}$ is a vector having the same direction as $\boldsymbol{A}$ and a magnitude equal to the ordinary scalar product between the magnitudes of m and $\boldsymbol{A}$. If m is negative, it means simply that the vector $m\boldsymbol{A}$ has a direction directly opposite to that of $\boldsymbol{A}$.

The vector $-\boldsymbol{A}$ may be considered as the product of the scalar −1 and the vector $\boldsymbol{A}$. Thus, from the statement above we see that $-\boldsymbol{A}$ differs from $\boldsymbol{A}$ in that it has an opposite sense. Furthermore, these operations have nothing to do with the line of action of a vector, so $\boldsymbol{A}$ and $-\boldsymbol{A}$ may have different lines of action. This will be the case of the couple to be studied in Chapter 3.

†2.3 Addition and Subtraction of Vectors

In adding a number of vectors, we may repeatedly employ the parallelogram construction. We can do this graphically by scaling the lengths of the arrows according to the magnitudes of the vector quantities they represent. The magnitude of the final arrow can then be interpreted in terms of its length by employing the chosen scale factor. As an example, consider the coplanar[1] vectors $\boldsymbol{A}$, $\boldsymbol{B}$, and $\boldsymbol{C}$ shown in Fig. 2.1(a). The addition of the vectors $\boldsymbol{A}$, $\boldsymbol{B}$, and $\boldsymbol{C}$ has been accomplished in two ways. In Fig. 2.1(b) we first add $\boldsymbol{B}$ and $\boldsymbol{C}$ and then add the resulting vector (shown dashed) to $\boldsymbol{A}$. This combination can be represented by the notation $\boldsymbol{A} + (\boldsymbol{B} + \boldsymbol{C})$. In Fig.

[1]Coplanar, meaning "same plane," is a word used often in mechanics.

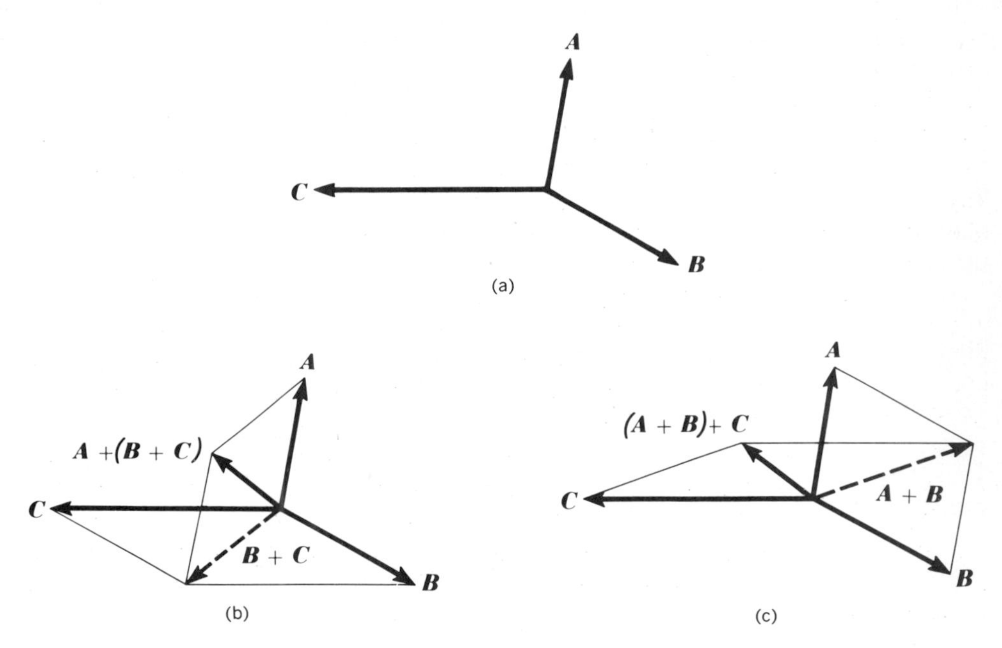

Figure 2.1. Addition by parallelogram law.

2.1(c), we add $\boldsymbol{A}$ and $\boldsymbol{B}$, and then add the resulting vector (shown dashed) to $\boldsymbol{C}$. The representation of this combination is given as $(\boldsymbol{A}+\boldsymbol{B})+\boldsymbol{C}$. Note that the final vector is identical for both procedures. Thus,

$$\boldsymbol{A}+(\boldsymbol{B}+\boldsymbol{C})=(\boldsymbol{A}+\boldsymbol{B})+\boldsymbol{C} \tag{2.1}$$

When the quantities involved in an algebraic operation can be grouped without restriction, the operation is said to be *associative*. Thus, the addition of vectors is both commutative, as explained earlier, and associative.

To determine a summation of, let us say, two vectors without recourse to graphics, we need only make a simple sketch of the vectors approximately to scale. By using familiar trigonometric relations, we can get a direct evaluation of the result. This is illustrated in the following examples.

EXAMPLE 2.1

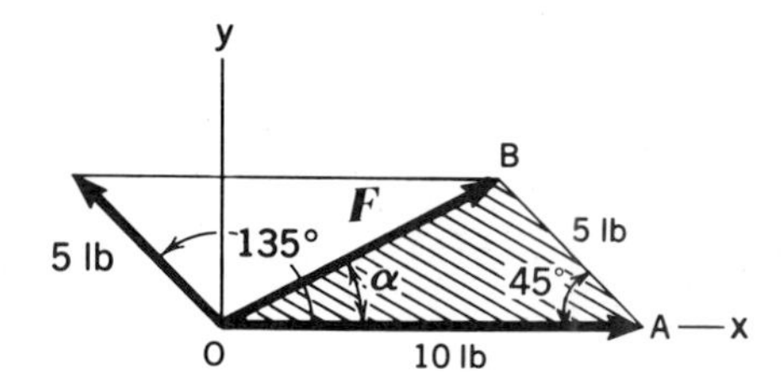

Figure 2.2. Find F and α using trigonometry.

Add the forces acting on a particle situated at the origin of a two-dimensional reference frame (Fig. 2.2). One force has a magnitude of 10 lb acting in the positive x direction, whereas the other has a magnitude of 5 lb acting at an angle of 135° with a sense directed away from the origin.

To get the sum (shown as $\boldsymbol{F}$), we may use the law of cosines[2] for one of the triangular portions of the sketched parallelogram. Thus, using triangle OBA,

$$|\boldsymbol{F}| = [10^2 + 5^2 - (2)(10)(5)\cos 45°]^{1/2}$$
$$= (100 + 25 - 70.7)^{1/2} = \sqrt{54.3} = 7.37 \text{ lb}$$

The direction of the vector may be described by giving the angle and the sense. This is done by employing the law of sines for triangle OBA.[3]

$$\frac{5}{\sin \alpha} = \frac{7.37}{\sin 45°}$$

$$\sin \alpha = \frac{(5)(0.707)}{7.37} = 0.480$$

Therefore,

$$\alpha = 28.6°$$

EXAMPLE 2.2

A light cable from a Jeep is tied to the peak of an A-frame and exerts a force of 450 N along the cable (see Fig. 2.3). A 1000-kg log is suspended from a second cable, which is fastened to the peak. What is the total force from the cables on the A-frame?

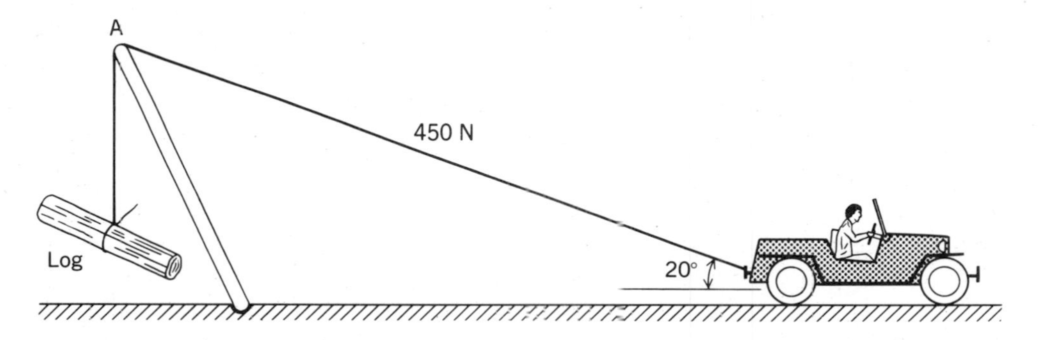

Figure 2.3. An A-frame supports a log.

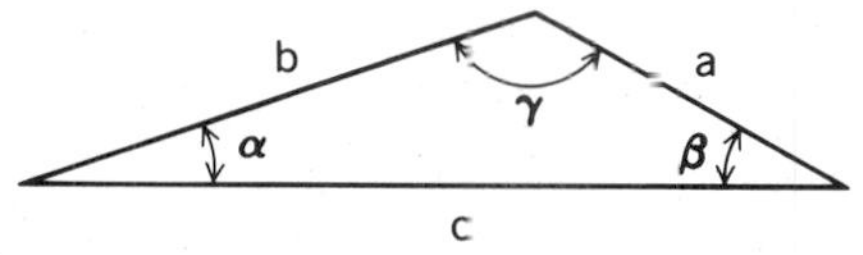

[2]You will recall from trigonometry that the *law of cosines* for side b of a triangle is given as

$$b^2 = a^2 + c^2 - 2ac\cos\beta$$

[3]The *law of sines* is given as follows for a triangle:

$$\frac{a}{\sin\alpha} = \frac{b}{\sin\beta} = \frac{c}{\sin\gamma}$$

We first note that the force F on the log from gravity (i.e., the weight) is

$$F = Mg = (1000)(9.81) = 9810 \text{ N} \qquad \text{(a)}$$

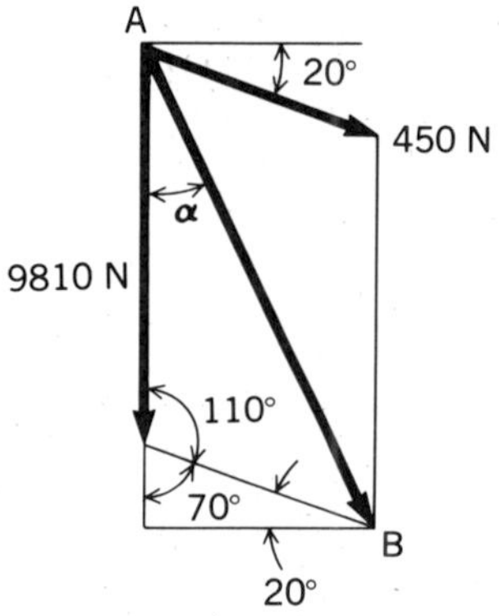

The force parallelogram is now drawn (see Fig. 2.4). From the law of cosines, we can say that

$$(AB)^2 = 450^2 + 9810^2 - (2)(450)(9810) \cos 110°$$

$$AB = 9973 \text{ N}$$

The angle α is next determined using the law of sines. Thus,

$$\frac{450}{\sin \alpha} = \frac{9973}{\sin 110°}$$

$$\alpha = 2.43°$$

Figure 2.4. Parallelogram law.

We may also add the vectors by moving them successively to parallel positions so that the head of one vector connects to the tail of the next vector, and so on. The sum of the vectors will then be a vector whose tail connects to the tail of the first vector and whose head connects to the head of the last vector. This last step will form a polygon from the vectors, and we say that the vector sum then "closes the polygon." Thus, adding the 10-lb vector to the 5-lb vector in Fig. 2.2, we would form the sides OA and AB of a triangle. The sum $\boldsymbol{F}$ then closes the triangle and is OB. Also, in Fig. 2.5(a), we have shown three coplanar vectors $\boldsymbol{F}_1$, $\boldsymbol{F}_2$, and $\boldsymbol{F}_3$. The vectors are connected in Fig. 2.5(b)

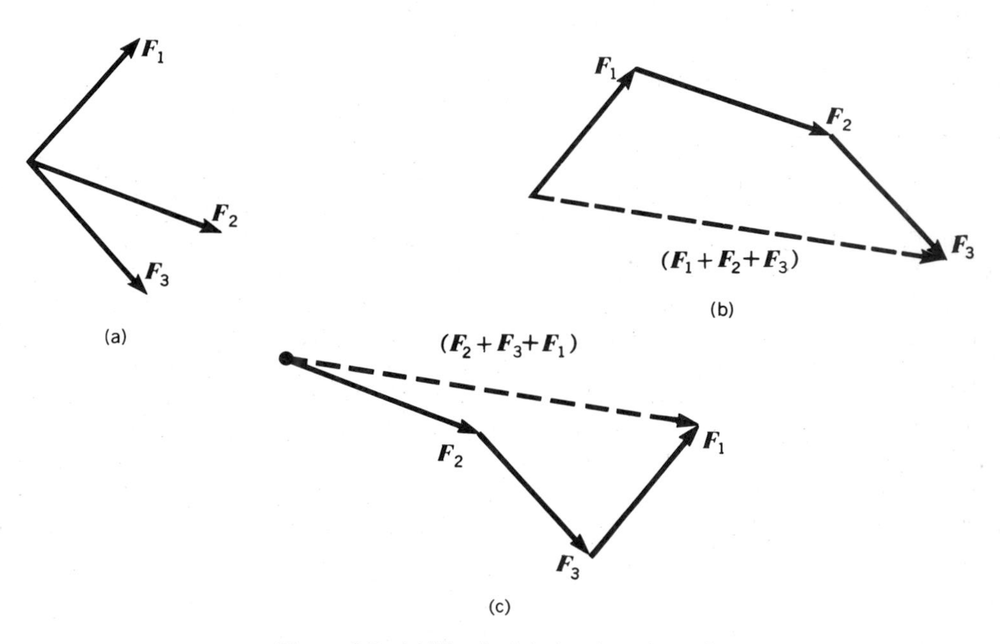

Figure 2.5. Addition by "closing the polygon."

as described. The sum of the vectors then is the dashed vector that closes the polygon. In Fig. 2.5(c), we have laid off the vectors $\boldsymbol{F}_1$, $\boldsymbol{F}_2$, and $\boldsymbol{F}_3$ in a different sequence. Nevertheless, it is seen that the sum is the same vector as in Fig. 2.5(b). Clearly, the *order* of laying off the vectors is not significant.

A simple physical interpretation of the above vector sum can be formed for vectors each of which represents a movement of a certain distance and direction (i.e., a *displacement* vector). Then, traveling along the system of given vectors you start from one point (the tail of the first vector) and end at another point (the head of the last vector). The vector *sum* that closes the polygon is equivalent to the system of given vectors, in that it takes you from the same initial to the same final point.

The polygon summation process, like the parallelogram of addition, can be used as a graphical process, or, still better, can be used to generate analytical computations with the aid of trigonometry. The extension of this procedure to any number of vectors is obvious.

The process of *subtraction* of vectors is defined in the following manner: to subtract vector $\boldsymbol{B}$ from vector $\boldsymbol{A}$, we reverse the direction of $\boldsymbol{B}$ (i.e., multiply by -1) and then add this new vector to $\boldsymbol{A}$ (Fig. 2.6).

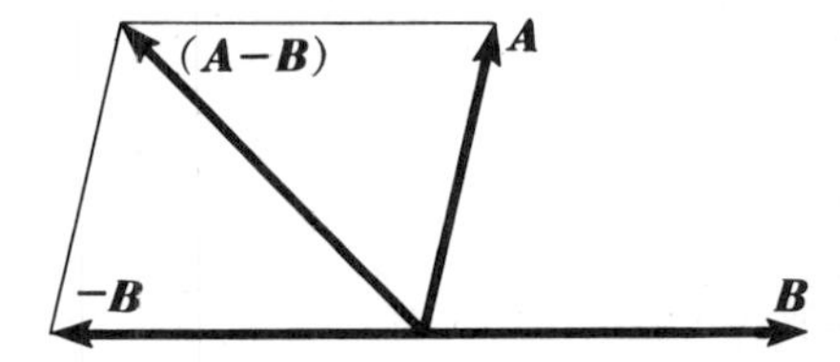

Figure 2.6. Subtraction of vectors.

This process may also be used in the polygon construction. Thus, consider coplanar vectors $\boldsymbol{A}$, $\boldsymbol{B}$, $\boldsymbol{C}$, and $\boldsymbol{D}$ in Fig. 2.7(a). To form $\boldsymbol{A}+\boldsymbol{B}-\boldsymbol{C}-\boldsymbol{D}$, we proceed as shown in Fig. 2.7(b). Again, the order of the process is not significant, as can be seen in Fig. 2.7(c).

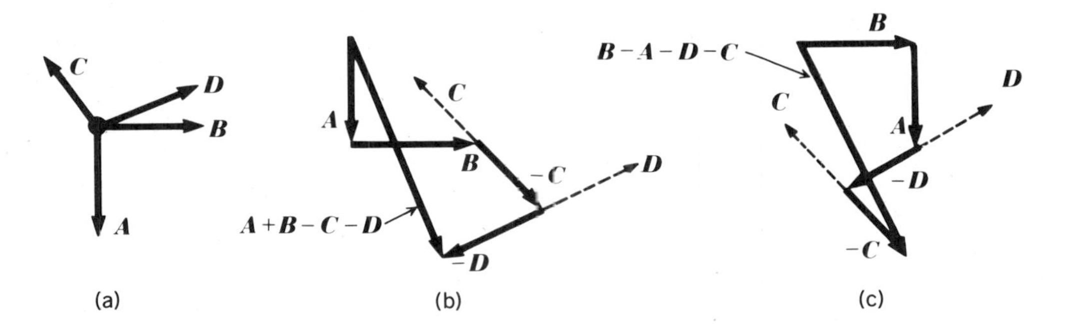

Figure 2.7. Addition and subtraction using polygon construction.

Problems

2.1. Add a 20-N force pointing in the positive x direction to a 50-N force at an angle 45° to the x axis in the first quadrant and directed away from the origin.

2.2. Subtract the 20-N force in Problem 2.1 from the 50-N force.

2.3. Add the vectors in the xy plane. Do this first graphically, using the force polygon and then do it analytically.

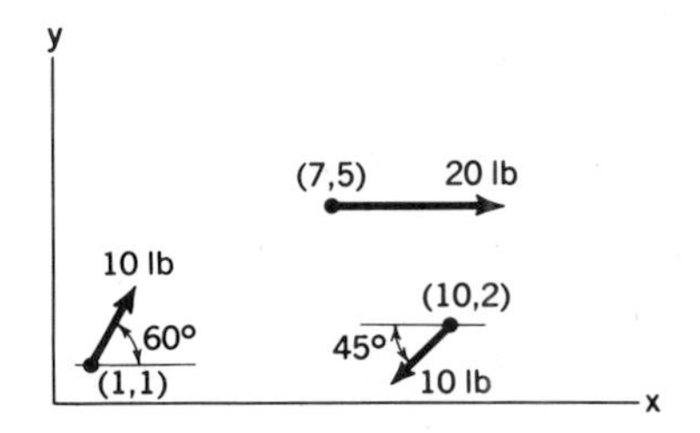

Figure P.2.3

2.4. A homing pigeon is released at point A and is observed. It flies 10 km due south, then goes due east for 15 km. Next it goes southeast for 10 km and finally goes due south 5 km to reach its destination B. Graphically determine the shortest distance between A and B. Neglect the earth's curvature.

2.5. Force $\boldsymbol{A}$ (given as a horizontal 10-N force) and $\boldsymbol{B}$ (vertical) add up to a force $\boldsymbol{C}$ that has a magnitude of 20 N. What is the magnitude of force $\boldsymbol{B}$ and the direction of force $\boldsymbol{C}$? (For the simplest results, use the force polygon, which for this case is a right triangle, and perform analytical computations.)

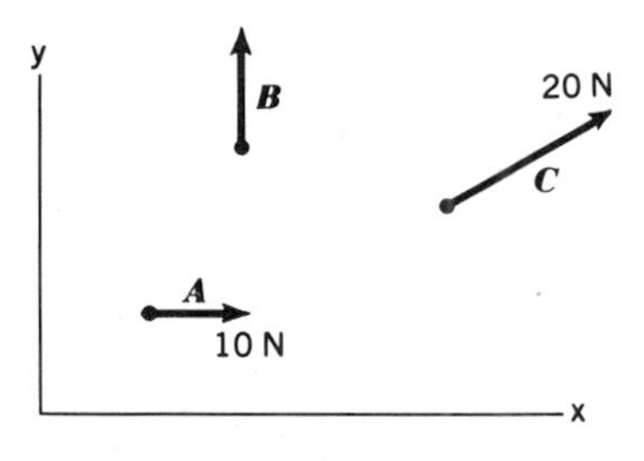

Figure P.2.5

2.6. If the difference between forces $\boldsymbol{B}$ and $\boldsymbol{A}$ in Fig. P.2.5 is a force $\boldsymbol{D}$ having a magnitude of 25 N, what is the magnitude of $\boldsymbol{B}$ and the direction of $\boldsymbol{D}$?

2.7. What is the sum of the forces transmitted by the structural members to the pin at A?

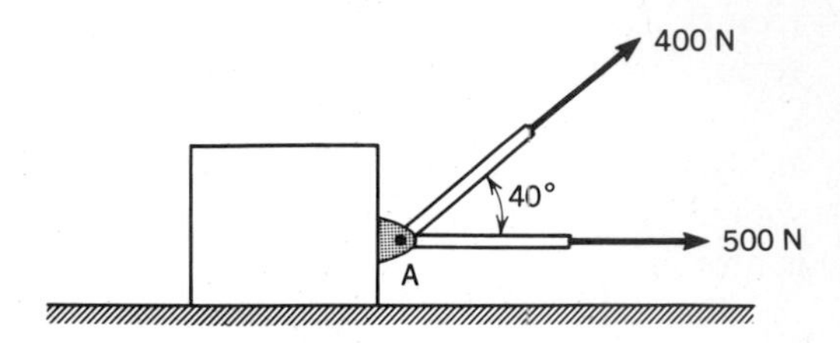

Figure P.2.7

2.8. Suppose in Problem 2.7 we require that the total force transmitted by the members to pin A be inclined 12° to the horizontal. If we do not change the force transmitted by the horizontal member, what must be the new force for the other member whose direction remains at 40°? What is the total force?

2.9. A man pulls with force W on a rope through a simple frictionless pulley to raise a weight W. What total force is exerted on the pulley?

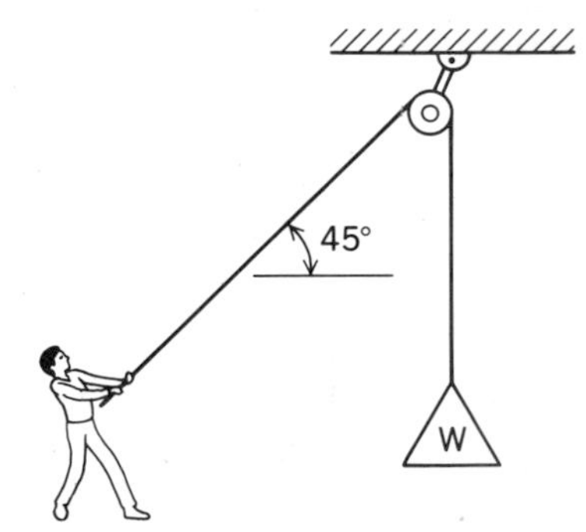

Figure P.2.9

2.10. A mass M is supported by cables (1) and (2). The tension in cable (1) is 200 N, whereas the tension in (2) is such as to maintain the configuration shown. What is the mass of

M in kilograms? (You will learn very shortly that the weight of M must be equal and opposite to the vector sum of the supporting forces for equilibrium.)

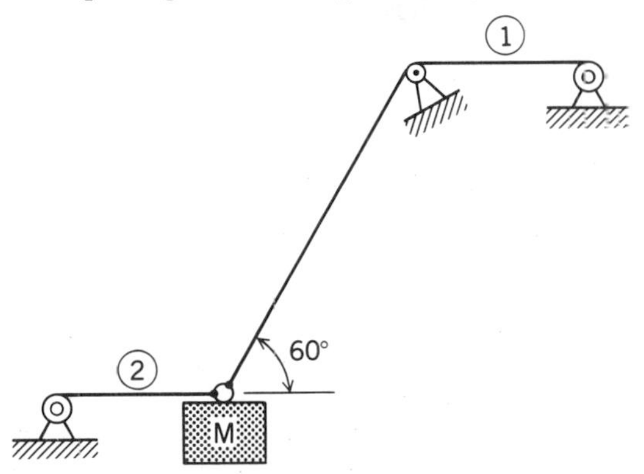

Figure P.2.10

2.11. Two football players are pushing a blocking dummy. Player A pushes with 100-lb force while player B pushes with 150-lb force toward bow C of the dummy. What is the total force exerted on the dummy by the players?

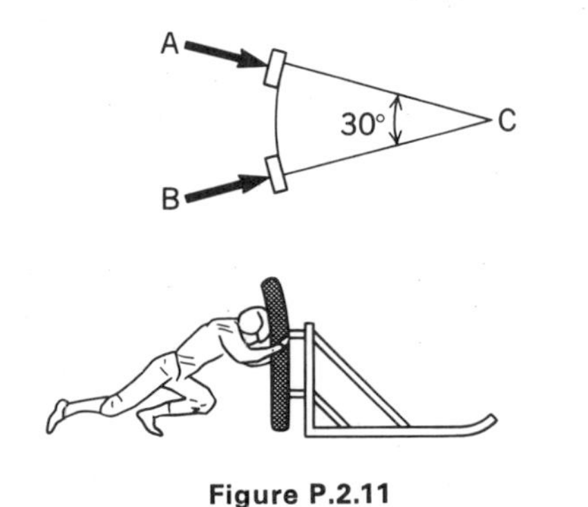

Figure P.2.11

2.12. A simple slingshot is about to be "fired." If the entire rubber band has a stretch of 1 in./3 lb, what force does the band exert on the right hand? The total unstretched length of the rubber band is 5 in. Note, if you use half the rubber band, you double the resistance to stretching.

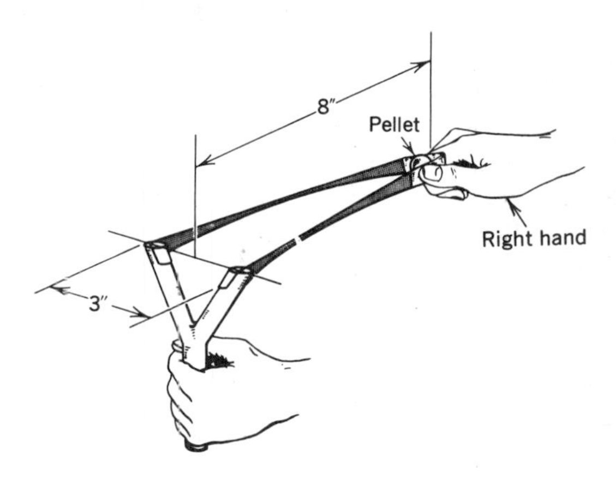

Figure P.2.12

2.13. Two soccer players approach a stationary ball 10 ft away from the goal. Simultaneously, a player on team O (offense) kicks the ball with force 100 lb for a split second while a player on team D (defense) kicks with force 70 lb during the same time interval. Does the offense score (assuming that the goalie is asleep)?

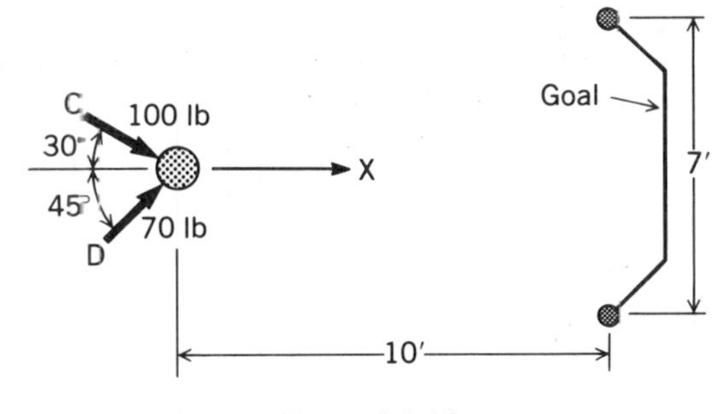

Figure P.2.13

2.4 Resolution of Vectors; Scalar Components

The opposite action of addition of vectors is called *resolution*. Thus, for a given vector C, we may find a pair of vectors in any two stipulated directions coplanar with C such that the two vectors, called *components*, sum to the original vector. This is a *two-dimensional* resolution involving two component vectors *coplanar* with the original vector. We shall discuss three-dimensional resolution involving three noncoplanar component vectors later in the section. The two-dimensional resolution can be accomplished by graphical construction, or by using simple helpful sketches and then employing trigonometric relations. An example of two-dimensional resolution is shown in Fig. 2.8. The two vectors $\boldsymbol{C}_1$ and $\boldsymbol{C}_2$ formed in this way are the *component* vectors. We often replace a vector by its components since the components are always equivalent in rigid-body mechanics to the original vector. When this is done, it is often helpful to indicate that the original vector is no longer operative by drawing a wavy line through the original vector as shown in Fig. 2.9.

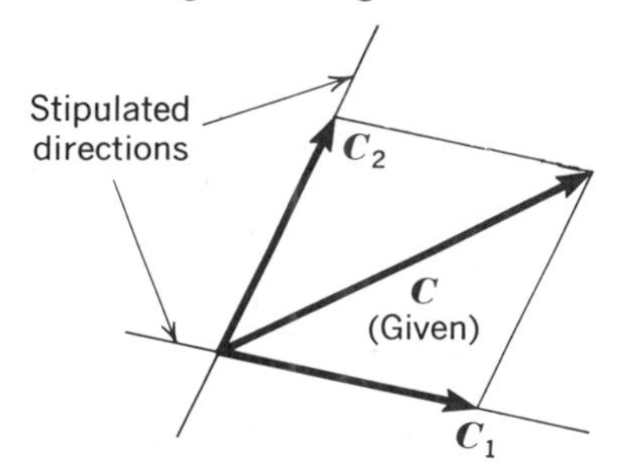

Figure 2.8. Two-dimensional resolution of vector $\boldsymbol{C}$.

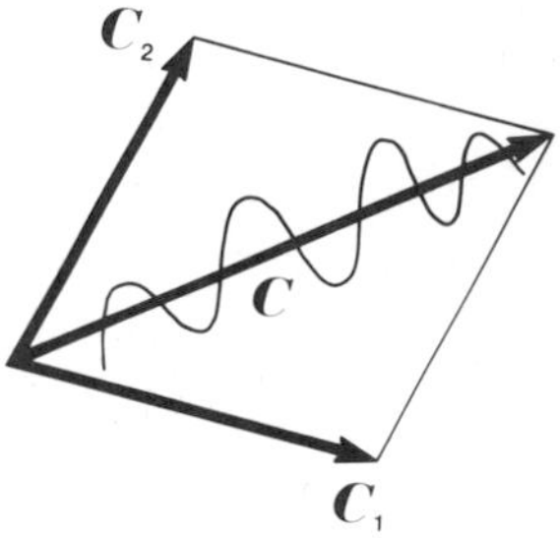

Figure 2.9. Vector $\boldsymbol{C}$ is replaced by its components and is no longer operative.

EXAMPLE 2.3

Flight 304 from Dallas is flying NE to Chicago 900 miles away (see Fig. 2.10). To avoid a massive storm front, the pilot decides instead to fly due north to Topeka, Kansas, and then ENE (see Fig. 2.11 for compass settings) to Chicago. What are the distances that he must travel from Dallas to Topeka and from Topeka to Chicago?

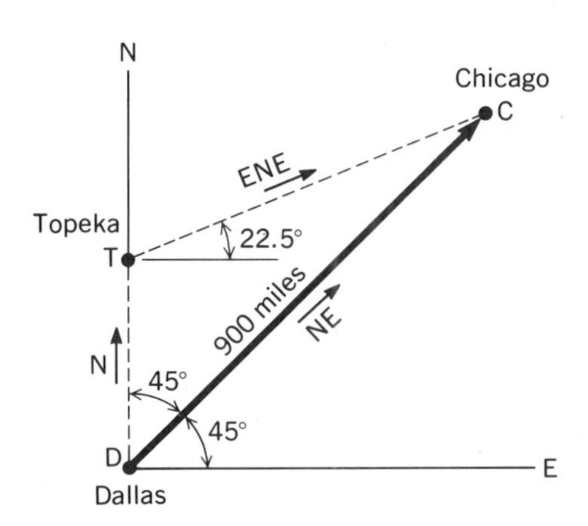

Figure 2.10. Compass headings.

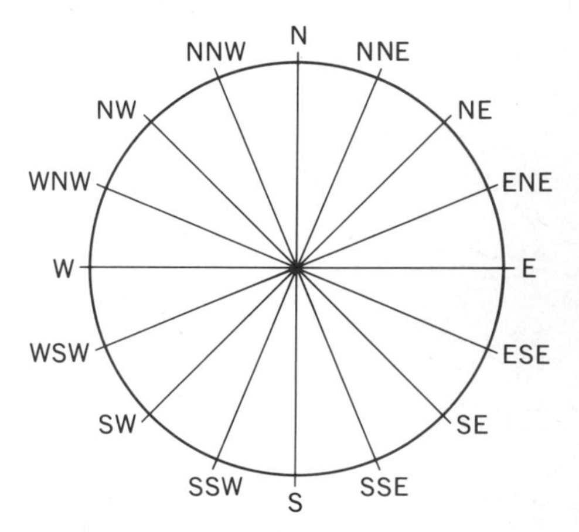

Figure 2.11. Compass settings.

We know the vector from Dallas to Chicago, and we want to resolve this vector into two components whose directions are known (see Fig. 2.10). Note from Fig. 2.11 that NE corresponds to a 45° angle to the north axis, and ENE corresponds to a 22.5° angle to the east axis. Now, form a parallelogram from DC and the two directions (see Fig. 2.12). We can consider triangle DTC of the paralleologram.[4] Note angle $\measuredangle DTC = 180° - 45° - 22.5° = 112.5°$, so that using the law of sines, we can say

$$\frac{900}{\sin 112.5°} = \frac{DT}{\sin 22.5°} = \frac{TC}{\sin 45°}$$

Therefore,

$$DT = 373 \text{ miles}$$
$$TC = 689 \text{ miles}$$

The alternate route is thus 162 miles longer.

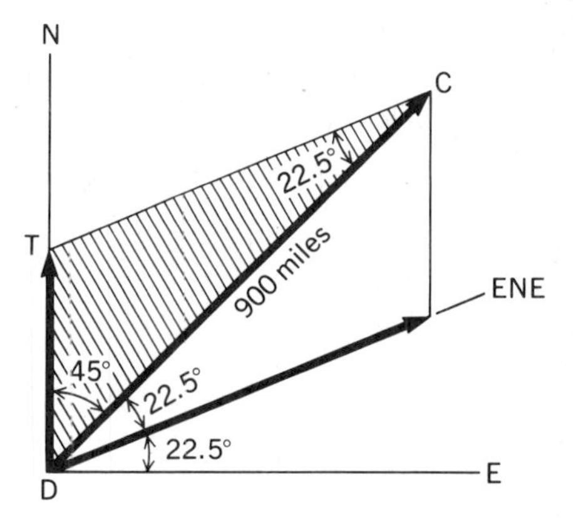

Figure 2.12. Form parallelogram.

It is also readily possible to find *three components not in the same plane as* **C**. This is the aforementioned three-dimensional resolution. Consider the specification of three *orthogonal* directions[5] for the resolution of **C**, as is shown in Fig. 2.13. The

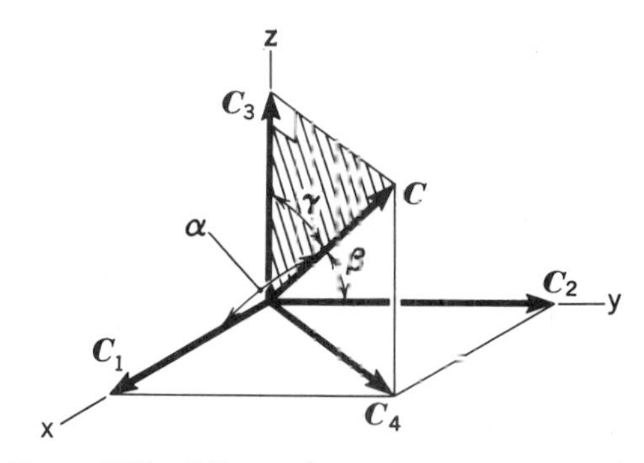

Figure 2.13. Orthogonal or rectangular components.

resolution may be accomplished in two steps. Resolve **C** along the z direction, and along the intersection of the xy plane and the plane formed by **C** and the z axis. This gives vectors $\mathbf{C}_3$ and $\mathbf{C}_4$. Since $\mathbf{C}_3$ is perpendicular to the xy plane, it must be perpendicular to $\mathbf{C}_4$. Thus, $\mathbf{C}_3$ and $\mathbf{C}_4$ form the sides of a *rectangle*. Now resolve $\mathbf{C}_4$ along the x and y directions, forming the other two component vectors $\mathbf{C}_1$ and $\mathbf{C}_2$. It is clear that the vectors $\mathbf{C}_1$, $\mathbf{C}_2$, and $\mathbf{C}_3$ add up to the vector **C** and are all mutually perpendicular to one another. Hence, $\mathbf{C}_1$, $\mathbf{C}_2$, and $\mathbf{C}_3$ are called the *orthogonal* (or *rectangular*) *component vectors* of **C**.

---

[4]This triangle could have been reached directly using the concept of the vector polygon of Section 2.3.

[5]Although the vector can be resolved along three *skew* directions (hence nonorthogonal), the orthogonal directions are used most often in engineering practice.

The direction of a vector $\boldsymbol{C}$ relative to an orthogonal reference is given by the cosines of the angles formed by the vector and the respective coordinate axes. These are called *direction cosines* and are denoted as

$$\begin{aligned}\cos(\boldsymbol{C}, x) &= \cos\alpha \equiv l \\ \cos(\boldsymbol{C}, y) &= \cos\beta \equiv m \\ \cos(\boldsymbol{C}, z) &= \cos\gamma \equiv n\end{aligned} \tag{2.2}$$

where α, β, and γ are associated with the x, y, and z axes, respectively. Now let us consider the right triangle, whose sides are $\boldsymbol{C}$ and the component vector, $\boldsymbol{C}_3$, shown shaded in Fig. 2.13. It then becomes clear, from trigonometric considerations of the right triangle, that

$$|\boldsymbol{C}_3| = |\boldsymbol{C}|\cos\gamma = |\boldsymbol{C}|n \tag{2.3}$$

If we had decided to resolve $\boldsymbol{C}$ first in the y direction instead of the z direction, we would have produced a geometry from which we could conclude that $|\boldsymbol{C}_2| = |\boldsymbol{C}|m$. Similarly, we can say that $|\boldsymbol{C}_1| = |\boldsymbol{C}|l$. We can then express $|\boldsymbol{C}|$ in terms of its orthogonal components in the following manner, using the Pythagorean theorem[6]:

$$|\boldsymbol{C}| = [(|\boldsymbol{C}|l)^2 + (|\boldsymbol{C}|m)^2 + (|\boldsymbol{C}|n)^2]^{1/2} \tag{2.4}$$

From this equation we can define the *orthogonal* or *rectangular scalar components* of the vector $\boldsymbol{C}$ as

$$C_x = |\boldsymbol{C}|l, \qquad C_y = |\boldsymbol{C}|m, \qquad C_z = |\boldsymbol{C}|n \tag{2.5}$$

Note that C_x, C_y, and C_z may be negative, depending on the sign of the direction cosines. Finally, it must be pointed out that *although C_x, C_y, and C_z are associated with certain axes and hence certain directions, they have been developed as scalars and must be handled as scalars.* Thus, an equation such as $10\boldsymbol{V} = V_x \cos\beta$ is not correct, because the left side is a vector and the right side is a scalar. This should spur you to observe care in your notation.

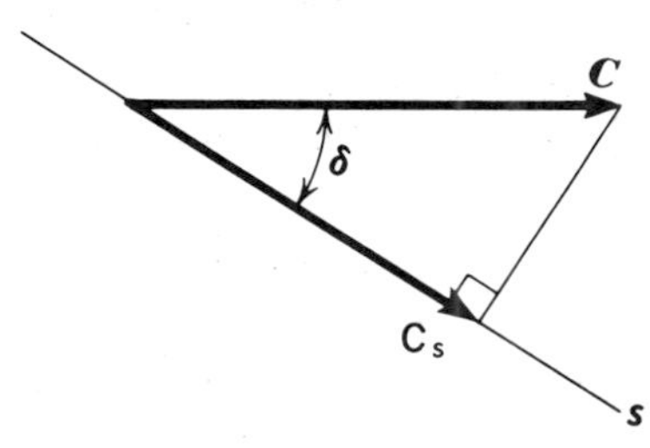

Figure 2.14. Rectangular component of ***C***.

Sometimes only *one* of the scalar orthogonal components of a vector (often called a rectangular component) is desired. Then, just one direction is prescribed, as shown in Fig. 2.14. Thus, the scalar rectangular component C_s is $|\boldsymbol{C}|\cos\delta$. It is always the case that the triangle formed by the vector and its scalar rectangular component is a right triangle. In establishing C_s we speak, therefore, of "dropping a perpendicular from $\boldsymbol{C}$ to s" or of "projecting $\boldsymbol{C}$ along s."

As a final consideration, let us examine vectors $\boldsymbol{A}$ and $\boldsymbol{B}$, which, along with direction s, form a plane as is shown in Fig. 2.15. The sum of the vectors $\boldsymbol{A}$ and $\boldsymbol{B}$ is found by the parallelogram

---

[6]From Eq. 2.4 one readily can conclude that $l^2 + m^2 + n^2 = 1$, which is a well-known geometric relation.

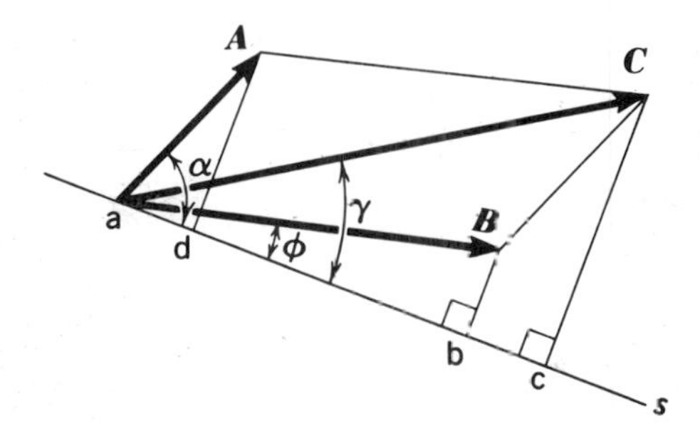

Figure 2.15. $C_s = A_s + B_s$.

law to be C. We shall now show that the projection of C along s is the same as the sum of the projections of its components, A and B, along s. That is,

$$C_s = A_s + B_s$$

On the diagram, then, the following relation must be verified:

$$ac = ad + ab \tag{a}$$

But

$$ac = ab + bc \tag{b}$$

Also, it is clear that

$$ad = bc \tag{c}$$

By substituting from Eqs. (b) and (c) into Eq. (a), we reduce Eq. (a) to an identity which shows that the projection of the sum of two vectors is the same as the sum of the projections of the two vectors.

†2.5 Unit Vectors

It is sometimes convenient to express a vector C as the product of its magnitude and a vector a of unit magnitude and having direction corresponding to the vector C. The vector a is called a *unit vector*. (You will write it as $\hat{a}$.) It has no dimensions. We formulate this vector as follows:

$$a \text{ (unit vector in direction } C) = \frac{C}{|C|} \tag{2.6}$$

Clearly, this development fulfills the requirements that have been set forth for this vector. We can then express the vector C in the form

$$C = |C|a \tag{2.7}$$

The unit vector, once established, does not have, per se, an inherent line of action. This will be determined entirely by its use. In the preceding equation, the unit vector a is collinear with the vector C. However, we can represent the vector D, shown in Fig. 2.16 parallel to C, by using the unit vector a as follows:

$$D = |D|a \tag{2.7a}$$

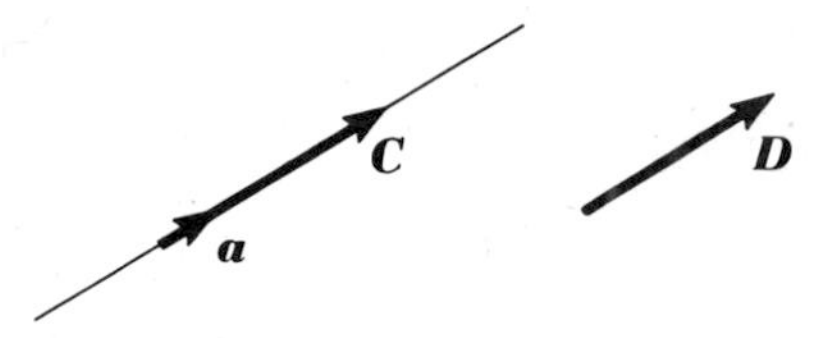

Figure 2.16. Unit vector $\boldsymbol{a}$.

In this operation, the unit vector $\boldsymbol{a}$ has been moved to have a line of action collinear with the vector $\boldsymbol{D}$. Occasionally, it is useful to label a unit vector that has the line of action of a certain vector with the lowercase letter of the capital letter associated with the vector. Thus, in Eqs. 2.7 and 2.7(a) we might have employed in the place of $\boldsymbol{a}$ the letters $\boldsymbol{c}$ and $\boldsymbol{d}$, (in your case $\hat{c}$ and $\hat{d}$) respectively. Next, if a given vector is represented using a lowercase letter, such as the vector $\boldsymbol{r}$, then we often make use of the circumflex mark to indicate the associated unit vector. Thus,

$$\boldsymbol{r} = |\boldsymbol{r}|\hat{\boldsymbol{r}} \tag{2.7b}$$

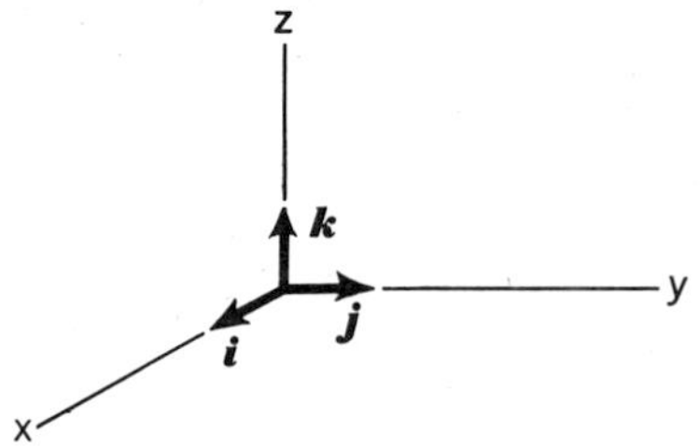

Figure 2.17. Unit vectors for *xyz* axes.

Unit vectors that are of particular use are those directed along the coordinate axes of a rectangular reference, where $\boldsymbol{i}$, $\boldsymbol{j}$, and $\boldsymbol{k}$ (your instructor will probably use the notation $\hat{\imath}$, $\hat{\jmath}$, and $\hat{k}$) correspond to the x, y, and z directions, as shown in Fig. 2.17.[7]

Since the sum of a set of concurrent vectors is equivalent in all situations to the original vector, we can always replace the vector $\boldsymbol{C}$ by its rectangular scalar components in the following manner:

$$\boldsymbol{C} = C_x\boldsymbol{i} + C_y\boldsymbol{j} + C_z\boldsymbol{k} \tag{2.8}$$

In Chapter 1, we saw that vectors that are equal have the same magnitude and direction. Hence, if $\boldsymbol{A} = \boldsymbol{B}$, we can say that

$$A_x\boldsymbol{i} + A_y\boldsymbol{j} + A_z\boldsymbol{k} = B_x\boldsymbol{i} + B_y\boldsymbol{j} + B_z\boldsymbol{k} \tag{2.9}$$

Then, since the unit vectors have mutually different directions, we conclude that

$$A_x\boldsymbol{i} = B_x\boldsymbol{i}$$
$$A_y\boldsymbol{j} = B_y\boldsymbol{j}$$
$$A_z\boldsymbol{k} = B_z\boldsymbol{k}$$

It then follows that

$$A_x = B_x$$
$$A_y = B_y$$
$$A_z = B_z$$

---

[7]Curvilinear coordinate systems have associated sets of unit vectors just as do the rectangular coordinate systems. As will be seen later, however, these unit vectors do not all have fixed directions in space for a given reference as do the vectors $\boldsymbol{i}, \boldsymbol{j}$, and $\boldsymbol{k}$.

Hence, the vector equation, $\boldsymbol{A} = \boldsymbol{B}$, has resulted in three scalar equations that in totality are equivalent in every way to the vector statement of equality. Thus, in Newton's law we would have

$$\boldsymbol{F} = m\boldsymbol{a} \tag{2.10a}$$

as the vector equation, and

$$F_x = ma_x, \qquad F_y = ma_y, \qquad F_z = ma_z \tag{2.10b}$$

as the corresponding scalar equations.

We now present a series of homework problems. In a few of these problems, we have inserted a rectangular parallelepiped along a vector (see Fig. P.2.24 for Prob. 2.24). The purpose of this rectangular parallelepiped clearly is to convey sufficient information concerning the *direction* of the vector. In this regard, note first in Fig. P.2.24 that the vector $\overrightarrow{DE}$ between points D and E can be given in terms of rectangular components as follows

$$\begin{aligned}\overrightarrow{DE} &= 4\boldsymbol{i} - 3\boldsymbol{k} + 5\boldsymbol{j}\\ &= 4\boldsymbol{i} + 5\boldsymbol{j} - 3\boldsymbol{k}\end{aligned}$$

by simply moving from D to E *along the coordinate directions*. Since from Eq. 2.5

$$4 = (DE)l$$

then

$$l = \frac{4}{DE} = \frac{4}{\sqrt{4^2 + 5^2 + 3^2}}$$

This value of l must then be the direction cosine for the 100-lb force collinear with DE. We get m and n the same way. The 100-N force in Fig. P.2.24 can now be given vectorially as

$$\boldsymbol{F} = 100(l\boldsymbol{i} + m\boldsymbol{j} + n\boldsymbol{k}) = 100\frac{(4\boldsymbol{i} + 5\boldsymbol{j} - 3\boldsymbol{k})}{\sqrt{4^2 + 5^2 + 3^2}} = 100\frac{\overrightarrow{DE}}{|\overrightarrow{DE}|}$$

Problems

2.14. Resolve the 100-lb force into a set of components along the slot shown and in the vertical direction.

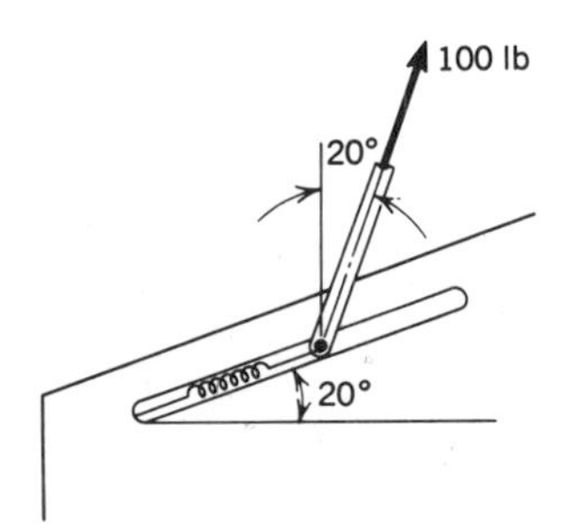

Figure P.2.14

2.15. A farmer needs to build a fence from the corner of his barn to the corner of his chicken house 30 m NE away. However, he wants to enclose as much of the barnyard as possible. Thus, he runs the fence east, from the corner of his barn to the property line and then NNE to the corner of his chicken house. How long is the fence?

2.16. Resolve the force F into a component perpendicular to AB and a component parallel to BC.

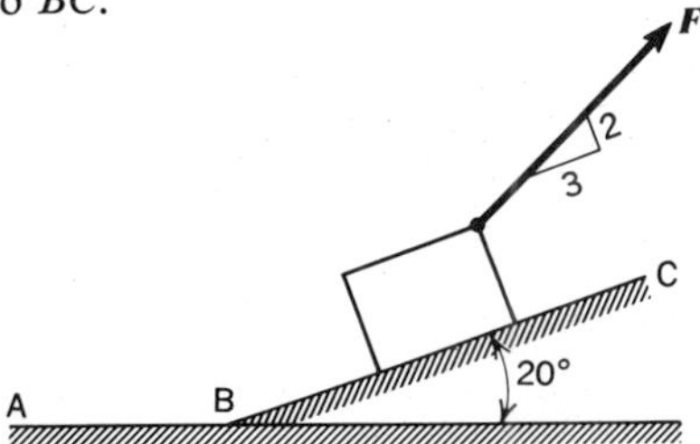

Figure P.2.16

2.17. A 1000-N force is resolved into components along AB and AC. If the component along AB is 700 N, determine the angle α and the value of the component along AC.

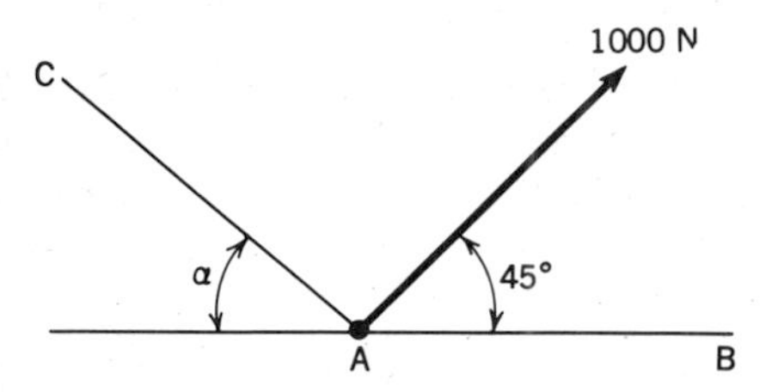

Figure P.2.17

2.18. Two men are trying to pull a crate which will not move until a 150-lb total force is applied in any one direction. Man A can pull only at 45° to the desired direction of crate motion, whereas man B can pull only at 60° to the desired motion. What force must each man exert to start the box moving as shown?

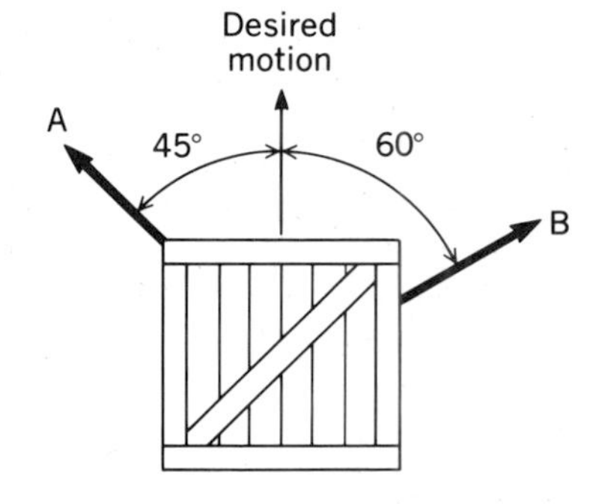

Figure P.2.18

2.19. The 500-N force is to be resolved into components along the AB and AC directions measured by the angles α and β. If the component along AC is to be 1000 N and the component along AB is to be 800 N, compute α and β.

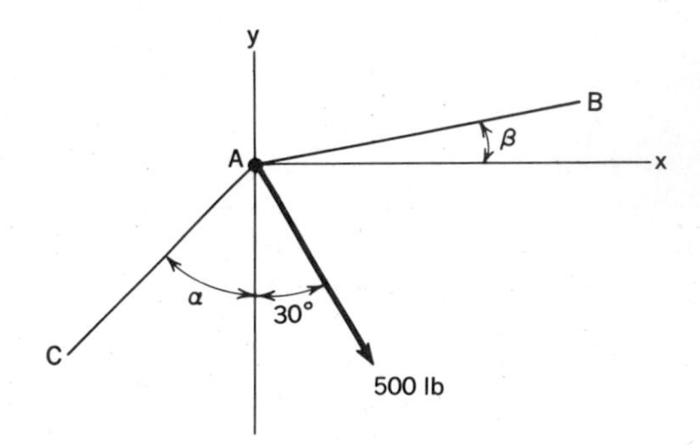

Figure P.2.19

2.20. The orthogonal components of a force are:
x component 10 lb in positive x direction
y component 20 lb in positive y direction
z component 30 lb in negative z direction

(a) What is the magnitude of the force itself?

(b) What are the direction cosines of the force?

2.21. What are the rectangular components of the 100-lb force? What are the direction cosines for this force?

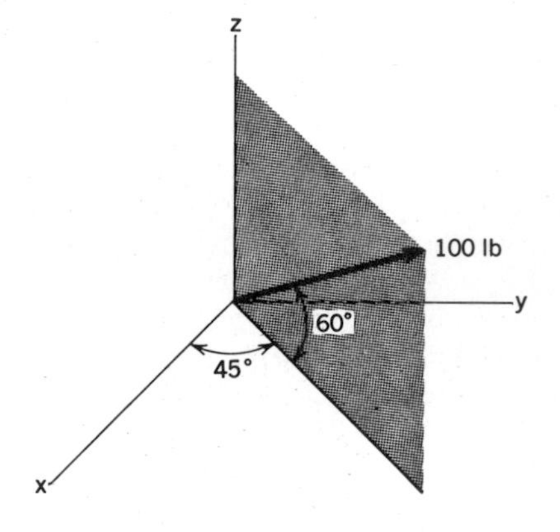

Figure P.2.21

2.22. A 50-m-long diagonal member OE in a space frame is inclined at $\alpha = 70°$ and $\beta = 30°$ to the x and y axes, respectively. What is γ? How long must members OA, AC, OB, BC, and CE be to support end E of OE?

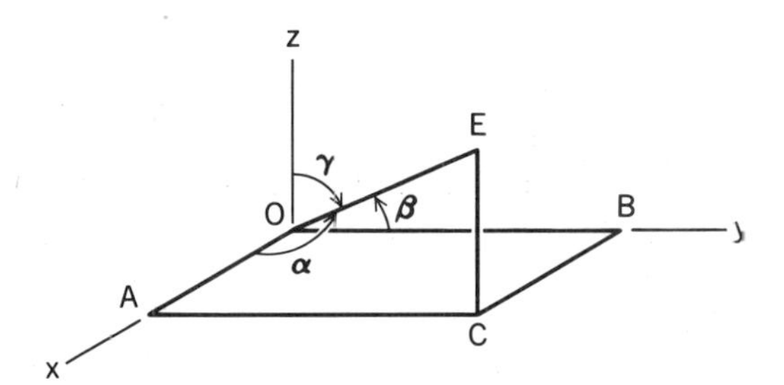

Figure P.2.22

2.23. What is the orthogonal total force component in the x direction of the force transmitted to pin A of a roof truss by the four members? What is the total component in the y direction?

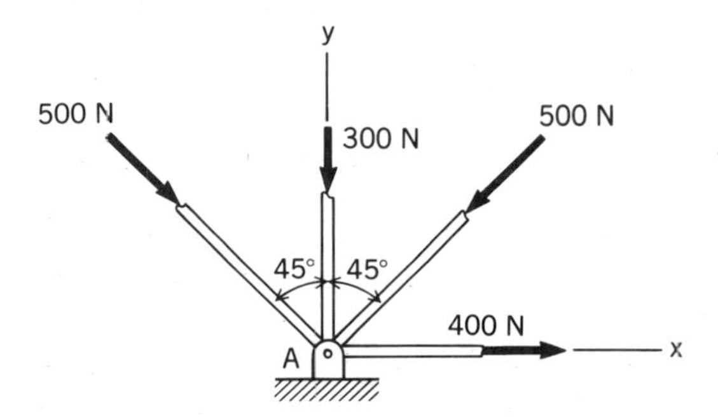

Figure P.2.23

2.24. A man pulls with a force of 100 N on a rope attached to a ring. The ring is supported by three linkages, A, B, and C, that can take forces only in the x, y, and z directions, respectively. What are the forces in the direction of the linkages stemming from the 100-N force?

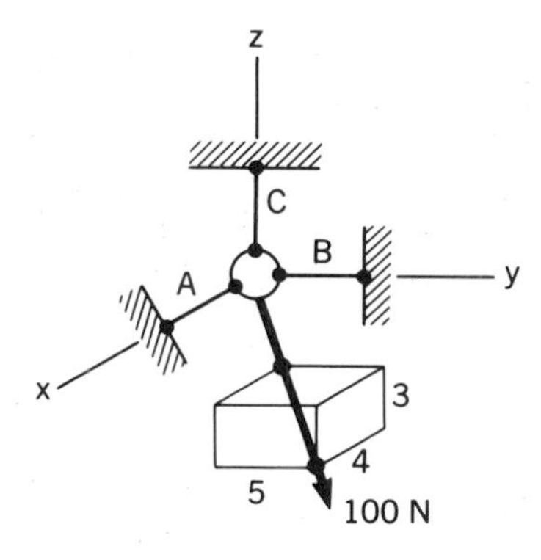

Figure P.2.24

2.25. Given the following force expressed as a function of position:

$$\boldsymbol{F} = (10x - 6)\boldsymbol{i} + x^2z\boldsymbol{j} + xy\boldsymbol{k}$$

What are the direction cosines of the force at position (1, 2, 2)? What is the position along the x coordinate where $F_x = 0$? Plot F_y versus the x coordinate for an elevation $z = 1$.

2.26. What is the sum of the following set of three vectors?
$\boldsymbol{A} = 6\boldsymbol{i} + 10\boldsymbol{j} + 16\boldsymbol{k}$ lb
$\boldsymbol{B} = 2\boldsymbol{i} - 3\boldsymbol{j}$ lb
$\boldsymbol{C}$ is a vector in the xy plane at an inclination of 45° to the positive x axis and directed away from the origin; it has a magnitude of 25 lb

2.27. What is the unit vector for the displacement vector from point (2, 1, 9) to point (7, 4, 2)? Express a 10-m displacement vector in the same direction in terms of $\boldsymbol{i}, \boldsymbol{j}$, and $\boldsymbol{k}$.

2.28. A vector $\boldsymbol{A}$ has a line of action that goes through the coordinates (0, 2, 3) and (−1, 2, 4). If the magnitude of this vector is 10 units, express the vector in terms of the unit vectors $\boldsymbol{i}, \boldsymbol{j}$, and $\boldsymbol{k}$.

2.29. Express the force $\boldsymbol{F}$ in terms of the unit vectors $\boldsymbol{i}, \boldsymbol{j}$, and $\boldsymbol{k}$.

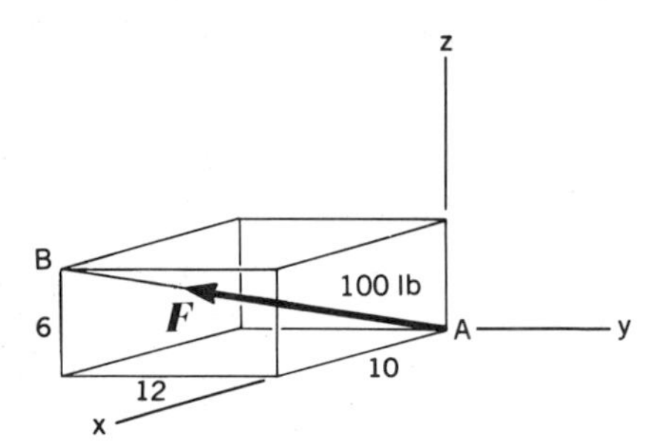

Figure P.2.29

2.30. Express the 100-N force in terms of the unit vectors $\boldsymbol{i}, \boldsymbol{j}$, and $\boldsymbol{k}$. What is the unit vector in the direction of the 100-N force? The force lies along diagonal AB.

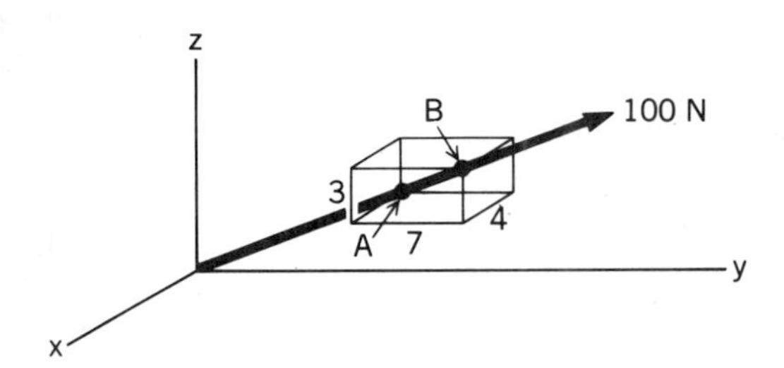

Figure P.2.30

2.31. Express the unit vectors $\boldsymbol{i}, \boldsymbol{j}$, and $\boldsymbol{k}$ in terms of unit vectors $\boldsymbol{\epsilon}_r$, $\boldsymbol{\epsilon}_\theta$, and $\boldsymbol{\epsilon}_z$. (These are unit vectors for *cylindrical coordinates.*) Express the 1000-lb force going through the origin and through point (2, 4, 4) in terms of the unit vectors $\boldsymbol{i}, \boldsymbol{j}, \boldsymbol{k}$ and $\boldsymbol{\epsilon}_r$, $\boldsymbol{\epsilon}_\theta$, $\boldsymbol{\epsilon}_z$ with $\theta = 60°$. (See the footnote on p. 34.)

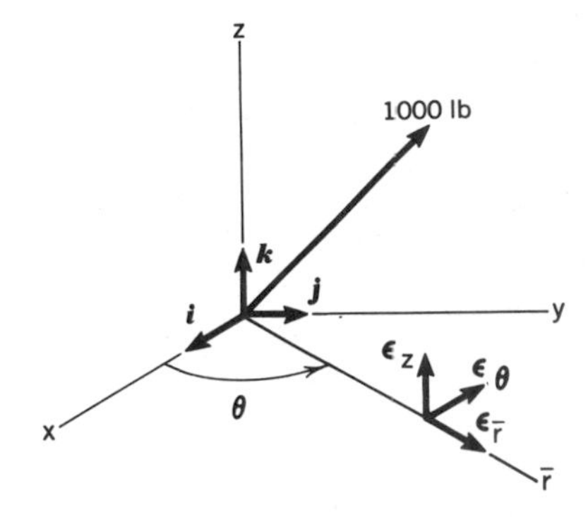

Figure P.2.31

2.6 Scalar or Dot Product of Two Vectors

In elementary physics, work was defined as the product of the force component, in the direction of a displacement, times the displacement. In effect, two vectors, force and displacement, are employed to give a scalar, work. In other physical problems, vectors are associated in this same manner so as to result in a scalar quantity. A vector operation that represents such operations concisely is the scalar product (or dot product), which, for the vector $\boldsymbol{A}$ and $\boldsymbol{B}$ in Fig. 2.18, is defined as[8]

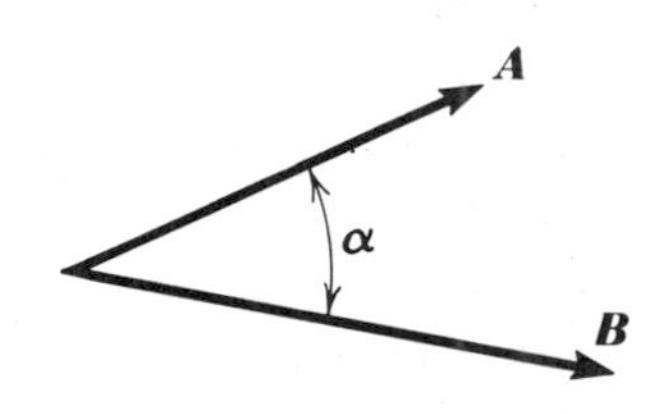

Figure 2.18. α is smallest angle between $\boldsymbol{A}$ and $\boldsymbol{B}$.

$$\boldsymbol{A} \cdot \boldsymbol{B} = |\boldsymbol{A}||\boldsymbol{B}| \cos \alpha \tag{2.11}$$

[8]To ensure that there is no confusion between the dot product of two vectors and the ordinary product of two scalars that you have used up to now, we urge you to read $\boldsymbol{A} \cdot \boldsymbol{B} = C$ as "$\boldsymbol{A}$ dotted into $\boldsymbol{B}$ yields C." Also, note that we can move the vectors so they intersect as in Fig. 2.18.

where α is the smaller angle between the two vectors. Note that the dot product may involve vectors of different dimensional representation, and may be positive or negative, depending on whether the smaller included angle α is less than or greater than 90°. Note also that $\boldsymbol{A} \cdot \boldsymbol{B}$ is equivalent to first projecting vector $\boldsymbol{A}$ onto the line of action of vector $\boldsymbol{B}$ (this gives us $|\boldsymbol{A}| \cos \alpha$), and than multiplying by the magnitude of vector $\boldsymbol{B}$ (or vice versa). The appropriate sign must, of course, be assigned positive if the projected component of $\boldsymbol{A}$ and the vector $\boldsymbol{B}$ point in the same direction; negative, if not.

The work concept for a force $\boldsymbol{F}$ acting on a particle moving along a path described by $\boldsymbol{s}$ can now be given as

$$W = \int \boldsymbol{F} \cdot d\boldsymbol{s} \tag{2.12}$$

where $d\boldsymbol{s}$ is a displacement on the path along which the particle is moved.

Let us next consider the scalar product of $m\boldsymbol{A}$ and $n\boldsymbol{B}$. If we carry it out according to our definitions:

$$\begin{aligned}(m\boldsymbol{A}) \cdot (n\boldsymbol{B}) &= |m\boldsymbol{A}||n\boldsymbol{B}| \cos (m\boldsymbol{A}, n\boldsymbol{B}) \\ &= (mn)|\boldsymbol{A}||\boldsymbol{B}| \cos (\boldsymbol{A}, \boldsymbol{B}) = (mn)(\boldsymbol{A} \cdot \boldsymbol{B})\end{aligned} \tag{2.13}$$

Hence, the scalar coefficients in the dot product of two vectors multiply in the ordinary way, while only the vectors themselves undergo the vectorial operation as we have defined it.

From the definition, clearly the dot product is *commutative*, since the number $|\boldsymbol{A}||\boldsymbol{B}| \cos (\boldsymbol{A}, \boldsymbol{B})$ is independent of the order of multiplication of its terms. Thus,

$$\boldsymbol{A} \cdot \boldsymbol{B} = \boldsymbol{B} \cdot \boldsymbol{A} \tag{2.14}$$

Let us now consider $\boldsymbol{A} \cdot (\boldsymbol{B} + \boldsymbol{C})$. By definition, we may project the vector $(\boldsymbol{B} + \boldsymbol{C})$ onto the line of action of $\boldsymbol{A}$ and then, assigning the appropriate sign, multiply the magnitude of $\boldsymbol{A}$ times the projection of $\boldsymbol{B} + \boldsymbol{C}$. However, we have shown that the projection of the sum of two vectors is the same as the sum of the projections of the vectors, which means that

$$\boldsymbol{A} \cdot (\boldsymbol{B} + \boldsymbol{C}) = \boldsymbol{A} \cdot \boldsymbol{B} + \boldsymbol{A} \cdot \boldsymbol{C} \tag{2.15}$$

An operation on a sum of quantities that is the same as the sum of the operations on the quantities is called a *distributive operation.* Thus, the dot product is distributive.

The scalar product between unit vectors will now be carried out. The product $\boldsymbol{i} \cdot \boldsymbol{j}$ is 0, since the angle α in Eq. 2.11 is 90°, which makes $\cos \alpha = 0$. On the other hand, $\boldsymbol{i} \cdot \boldsymbol{i} = 1$. We can thus conclude that the dot product of equal orthogonal unit vectors for a given reference is unity and that of unequal orthogonal unit vectors is zero.

If we express the vectors $\boldsymbol{A}$ and $\boldsymbol{B}$ in Cartesian components when taking the dot product, we get

$$\begin{aligned}\boldsymbol{A} \cdot \boldsymbol{B} &= (A_x\boldsymbol{i} + A_y\boldsymbol{j} + A_z\boldsymbol{k}) \cdot (B_x\boldsymbol{i} + B_y\boldsymbol{j} + B_z\boldsymbol{k}) \\ &= A_xB_x + A_yB_y + A_zB_z\end{aligned} \tag{2.16}$$

Thus, we see that a scalar product of two vectors is the sum of the ordinary products of the respective components.

If a vector is multiplied by itself as a dot product, the result is the square of the magnitude of the vector. That is,

$$\boldsymbol{A} \cdot \boldsymbol{A} = |\boldsymbol{A}||\boldsymbol{A}| = A^2 \tag{2.17}$$

Conversely, the square of a number may be considered to be the dot product of two equal vectors having a magnitude equal to the number. Note also that

$$\boldsymbol{A} \cdot \boldsymbol{A} = A_x^2 + A_y^2 + A_z^2 \tag{2.18}$$

We can conclude from Eqs. 2.17 and 2.18 that

$$A = \sqrt{A_x^2 + A_y^2 + A_z^2}$$

which checks with the Pythagorean theorem.

The dot product may be of immediate use in expressing the scalar rectangular component of a vector along a given direction as discussed in Section 2.4. If you refer back to Fig. 2.14, you will recall that the component of $\boldsymbol{C}$ along the direction $\boldsymbol{s}$ is given as

$$C_s = |\boldsymbol{C}| \cos \delta$$

Now let us consider a unit vector $\boldsymbol{s}$ along the direction of the line s. If we carry out the dot product of $\boldsymbol{C}$ and $\boldsymbol{s}$ according to our fundamental definition, the result is

$$\boldsymbol{C} \cdot \boldsymbol{s} = |\boldsymbol{C}||\boldsymbol{s}| \cos \delta$$

But since $|\boldsymbol{s}|$ is unity, when we compare the preceding two equations, it is apparent that

$$C_s = \boldsymbol{C} \cdot \boldsymbol{s}$$

Similarly, the following useful relations are valid:

$$C_x = \boldsymbol{C} \cdot \boldsymbol{i}, \qquad C_y = \boldsymbol{C} \cdot \boldsymbol{j}, \qquad C_z = \boldsymbol{C} \cdot \boldsymbol{k}$$

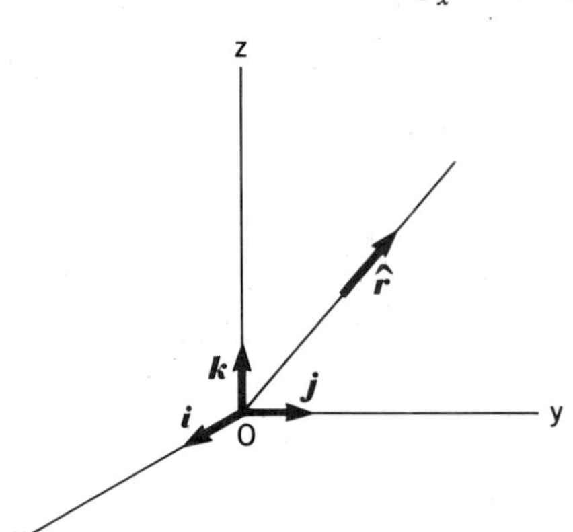

Figure 2.19. Unit vector $\hat{\boldsymbol{r}}$ directed from O.

Finally, express the unit vector $\hat{\boldsymbol{r}}$ directed out from the origin (see Fig. 2.19) in terms of the orthogonal scalar components:

$$\hat{\boldsymbol{r}} = (\hat{\boldsymbol{r}} \cdot \boldsymbol{i})\boldsymbol{i} + (\hat{\boldsymbol{r}} \cdot \boldsymbol{j})\boldsymbol{j} + (\hat{\boldsymbol{r}} \cdot \boldsymbol{k})\boldsymbol{k}$$

But

$$\hat{\boldsymbol{r}} \cdot \boldsymbol{i} = |\hat{\boldsymbol{r}}||\boldsymbol{i}| \cos(\hat{\boldsymbol{r}}, x) = l$$

Similarly, $\hat{\boldsymbol{r}} \cdot \boldsymbol{j} = m$ and $\hat{\boldsymbol{r}} \cdot \boldsymbol{k} = n$. Hence, we can say that

$$\hat{\boldsymbol{r}} = l\boldsymbol{i} + m\boldsymbol{j} + n\boldsymbol{k} \tag{2.19}$$

Thus, *the orthogonal scalar components of a unit vector are the direction cosines of the direction of the unit vector.*

Now, computing the square of the magnitude of $\hat{\boldsymbol{r}}$, we have

$$|\hat{\boldsymbol{r}}|^2 = 1 = l^2 + m^2 - n^2 \tag{2.20}$$

We thus arrive at the familiar geometrical relation that the sum of the squares of the direction cosines of a vector is unity.

EXAMPLE 2.4

Cables GA and GB (see Fig. 2.20) are part of a guy-wire system supporting two radio transmission towers. What are the lengths of GA and GB and the angle α between them?

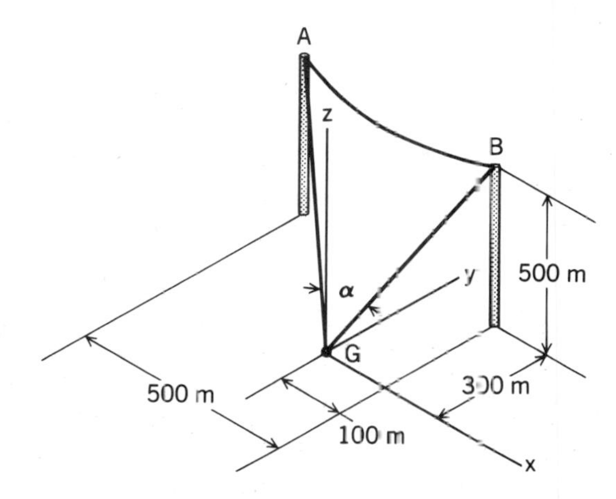

Figure 2.20. Radio transmission towers.

We may directly set up the vectors $\overrightarrow{GA}$ and $\overrightarrow{GB}$ by inspecting the diagram. Thus, it is easy to see that

$$\overrightarrow{GA} = 300\boldsymbol{j} - 400\boldsymbol{i} + 500\boldsymbol{k} \text{ m}$$
$$\overrightarrow{GB} = 300\boldsymbol{j} + 100\boldsymbol{i} + 500\boldsymbol{k} \text{ m}$$

Using the Pythagorean theorem, we can say for the lengths of $\overrightarrow{GA}$ and $\overrightarrow{GB}$:

$$GA = (300^2 + 400^2 + 500^2)^{1/2} = 707 \text{ m}$$
$$GB = (300^2 + 100^2 + 500^2)^{1/2} = 592 \text{ m}$$

Now we use the dot product definition to find the angle.

$$\overrightarrow{GA} \cdot \overrightarrow{GB} = (GA)(GB)\cos\alpha$$

Therefore,

$$\cos\alpha = \frac{\overrightarrow{GA} \cdot \overrightarrow{GB}}{(GA)(GB)} = \frac{90{,}000 - 40{,}000 + 250{,}000}{(707)(592)}$$
$$= .717$$

Hence,

$$\alpha = 44.18^\circ$$

Problems

2.32. Given the vectors

$$A = 10i + 20j + 3k$$
$$B = -10j + 12k$$

what is $A \cdot B$? What is $\cos(A, B)$? What is the projection of A along B?

2.33. Given the vectors

$$A = 16i + 3j, \quad B = 10k - 6i, \quad C = 4j$$

compute
(a) $C(A \cdot C) + B$
(b) $-C + [B \cdot (-A)]C$

2.34. Given the vectors

$$A = 6i + 3j + 10k$$
$$B = 2i - 5j + 5k$$
$$C = 5i - 2j + 7k$$

what vector D gives the following results?

$$D \cdot A = 20$$
$$D \cdot B = 5$$
$$D \cdot i = 10$$

2.35. Show that

$$\cos(A, B) = ll' + mm' + nn'$$

where l, m, n and l', m', n' are direction cosines of A and B, respectively, with respect to the given xyz reference.

2.36. Explain why the following operations are meaningless:
(a) $(A \cdot B) \cdot C$
(b) $(A \cdot B) + C$

2.37. A block A is constrained to move along a 20° incline in the yz plane. How far does the block have to move if the force F is to do 10 ft-lb of work?

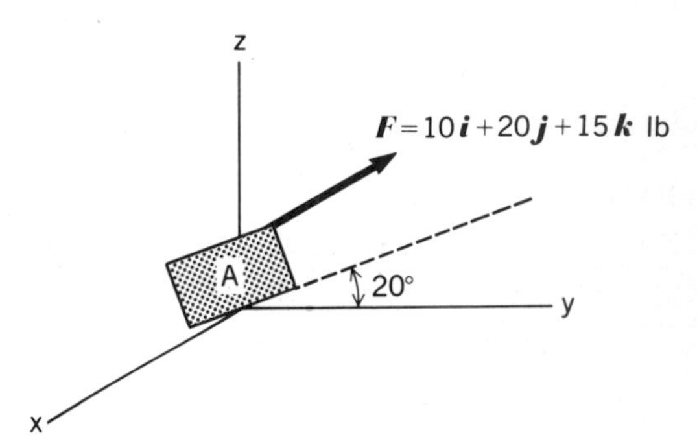

Figure P.2.37

2.38. An electrostatic field E exerts a force on a charged particle of qE, where q is the charge of the particle. If we have for E:

$$E = 6i + 3j + 2k \text{ dynes/coulomb}$$

what work is done by the field on a particle with a unit charge moving along a straight line from the origin to position $x = 20$ mm, $y = 40$ mm, $z = -40$ mm?

2.39. A force vector of magnitude 100 N has a line of action with direction cosines $l = .7$, $m = .2$, $n = .59$ relative to a reference xyz. The vector points away from the origin. What is the component of the force vector along a direction a having direction cosines $l = -.3$, $m = .1$, and $n = .95$ for the xyz reference? (*Hint:* Whenever simply a component is asked for, it is virtually always the *rectangular* component that is desired.)

2.40. Given a force $F = 10i + 5j + Ak$ N. If this force is to have a rectangular component of 8 N along a line having a unit vector $\hat{r} = .6i + .8k$, what should A be? What is the angle between F and $\hat{r}$?

2.41. Given a force $Ai + Bj + 20k$ N, what must A and B be to give a rectangular component of 10 N in the direction

$$\hat{r}_1 = .3i + .6j + .742k$$

as well as a component of 18 N in the direction

$$\hat{r}_2 = .4i + .9j + .1732k?$$

2.42. Find the dot product of the vectors represented by the diagonals from A to F and from D to G. What is the angle between them?

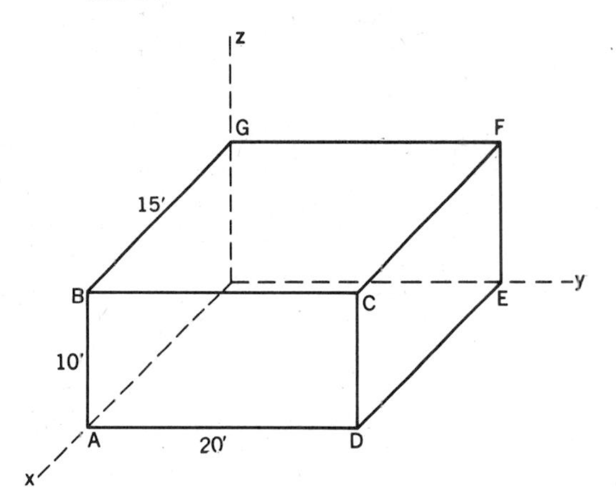

Figure P.2.42

2.43. What is the rectangular component of the 500-N force along the diagonal from B to A?

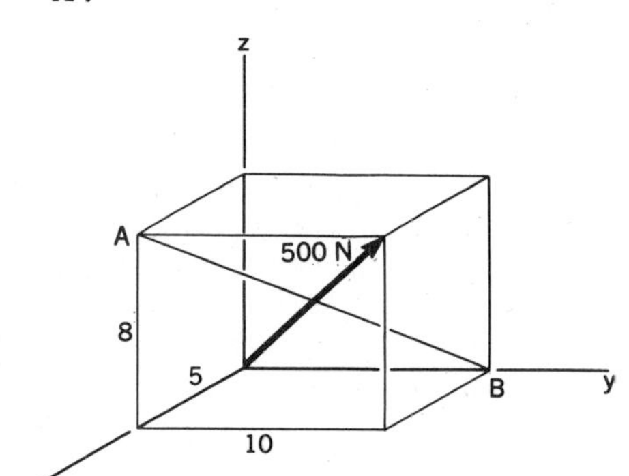

Figure P.2.43

2.44. A radio tower is held by guy wires. If AB were to be moved to intersect CD while remaining parallel to its original position, what is the angle between AB and CD?

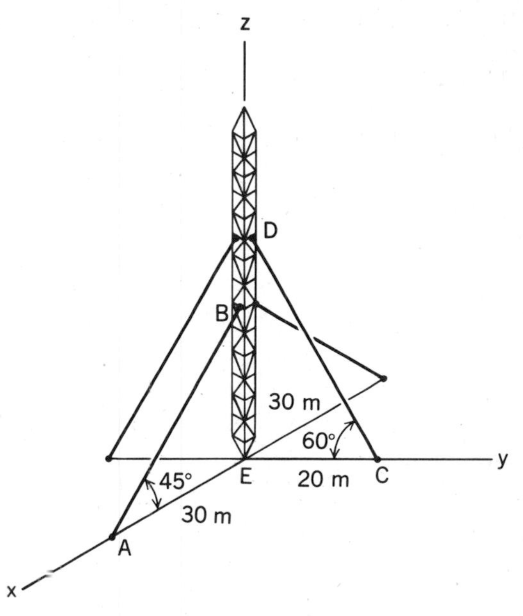

Figure P.2.44

2.7 Cross Product of Two Vectors

There are interactions between the vector quantities that result in vector quantities. One such interaction is the moment of a force, which involves a special product of the force and a position vector (to be studied in Chapter 3). To set up a convenient operation for these situations, the *vector cross product* has been established. For the

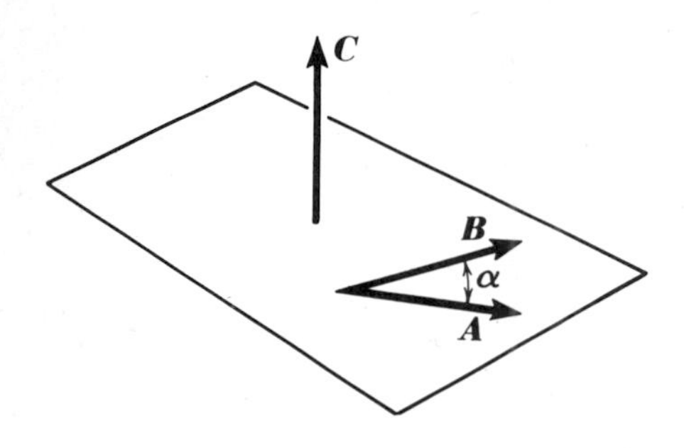

Figure 2.21. $A \times B = C$.

two vectors[9] (having possibly different dimensions) shown in Fig. 2.21 as $\boldsymbol{A}$ and $\boldsymbol{B}$, the operation[10] is defined as

$$\boldsymbol{A} \times \boldsymbol{B} = \boldsymbol{C} \tag{2.21}$$

where $\boldsymbol{C}$ has a magnitude that is given as

$$|\boldsymbol{C}| = |\boldsymbol{A}||\boldsymbol{B}| \sin \alpha \tag{2.22}$$

The angle α is the smaller of the two angles between the vectors, thus making $\sin \alpha$ always positive. The vector $\boldsymbol{C}$ has an orientation normal to the plane of the vectors $\boldsymbol{A}$ and $\boldsymbol{B}$. The sense, furthermore, corresponds to the advance of a right-hand screw rotated about $\boldsymbol{C}$ as an axis while turning from $\boldsymbol{A}$ to $\boldsymbol{B}$ through α—that is, from the first stated vector to the second stated vector through the smaller angle between them. In Fig. 2.21, the screw would advance upward in rotating from $\boldsymbol{A}$ to $\boldsymbol{B}$, whether the procedure is viewed from above or below the plane formed by $\boldsymbol{A}$ and $\boldsymbol{B}$. The reader can easily verify this for himself. The description of vector $\boldsymbol{C}$ is now complete, since the magnitude and direction are fully established. The line of action of $\boldsymbol{C}$ is not determined by the cross product; it depends on the use of the vector $\boldsymbol{C}$.

As in the previous case, the coefficients of the vectors will multiply as ordinary scalars. This may be deduced from the nature of the definition. However, the *commutative* law breaks down for this product. We can verify, by carefully considering the definition of the cross product, that

$$(\boldsymbol{A} \times \boldsymbol{B}) = -(\boldsymbol{B} \times \boldsymbol{A}) \tag{2.23}$$

We can readily show that the cross product, like the dot product, is a distributive operation. To do this, consider in Fig. 2.22 a prism *mnopqr* with edges coinciding with the vectors $\boldsymbol{A}$, $\boldsymbol{B}$, $\boldsymbol{C}$, and $(\boldsymbol{A} + \boldsymbol{B})$. We can represent the area of each face of the prism as a vector whose magnitude equals the area of the face and whose direction is normal to the face with a sense pointing out (by convention) from the body. It will be left to the student to justify the given formulation for each of the vectors in Fig. 2.23. Since the prism is a closed surface, the net projected area in any direction must be zero, and this, in turn, means that the total area vector must be zero. We then get

$$(\boldsymbol{A} + \boldsymbol{B}) \times \boldsymbol{C} + \tfrac{1}{2}\boldsymbol{A} \times \boldsymbol{B} + \tfrac{1}{2}\boldsymbol{B} \times \boldsymbol{A} + \boldsymbol{C} \times \boldsymbol{A} + \boldsymbol{C} \times \boldsymbol{B} = 0$$

Noting that the second and third expressions cancel each other, we get, on rearranging the terms,

$$\boldsymbol{C} \times (\boldsymbol{A} + \boldsymbol{B}) = \boldsymbol{C} \times \boldsymbol{A} + \boldsymbol{C} \times \boldsymbol{B} \tag{2.24}$$

We have thus demonstrated the *distributive* property of the cross product.

---

[9]In carrying out the cross product between any two vectors $\boldsymbol{A}$ and $\boldsymbol{B}$, we may move the vectors without changing their directions so that the vectors come together at a point, such as shown in Fig. 2.21.

[10]Again, we urge you to read $\boldsymbol{A} \times \boldsymbol{B} = \boldsymbol{C}$ as "$\boldsymbol{A}$ crossed into $\boldsymbol{B}$ yields $\boldsymbol{C}$."

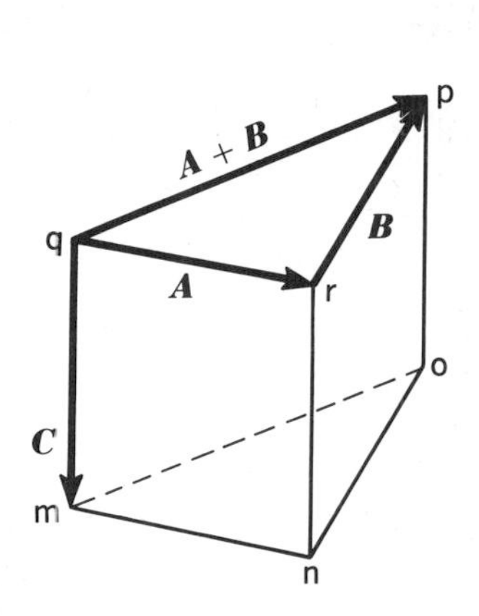

Figure 2.22. Prism using ***A***, ***B***, and ***C***.

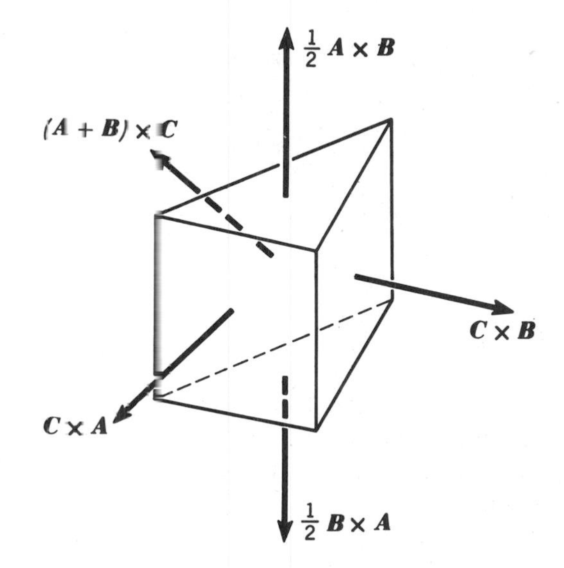

Figure 2.23. Area vectors for prism faces.

Next, consider the cross product of rectangular unit vectors. Here, the product of equal vectors is zero because α and, consequently, sin α are zero. The product $\boldsymbol{i} \times \boldsymbol{j}$ is unity in magnitude, and because of the right-hand-screw rule must be parallel to the z axis. If the z axis has been erected in a sense consistent with the right-hand-screw rule when rotating from the x to the y direction, the reference is called a *right-hand triad* [see Fig. 2.24(a)] and we can write

$$\boldsymbol{i} \times \boldsymbol{j} = \boldsymbol{k}$$

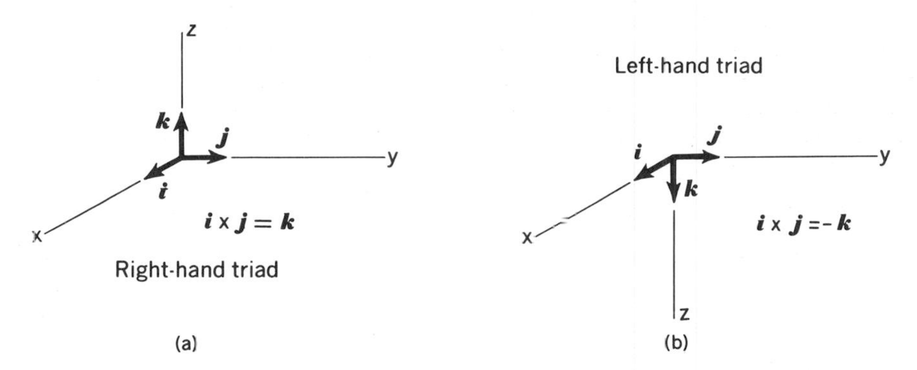

Figure 2.24. Different kinds of references.

If a left-hand triad is used, the result is a $-\boldsymbol{k}$ for the cross product above [see Fig. 2.24(b)]. In this text, we will use a right-hand triad as a reference. For ease in evaluation of unit cross products for such references, a simple permutation scheme is helpful.

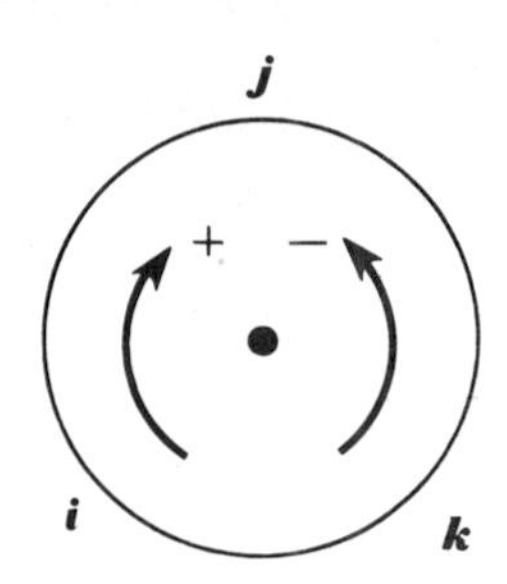

Figure 2.25. Permutation scheme.

In Fig. 2.25, the unit vectors $\boldsymbol{i}, \boldsymbol{j}$, and $\boldsymbol{k}$ are indicated on a circle in a clockwise sequence. Any cross product of a pair of unit vectors results in a positive third unit vector if going from the first vector to the second vector involves a clockwise motion on this circle. Otherwise, the vector is negative. Thus,

$$\boldsymbol{k} \times \boldsymbol{j} = -\boldsymbol{i}, \qquad \boldsymbol{k} \times \boldsymbol{i} = \boldsymbol{j}, \qquad \text{etc.}$$

Next, the cross product of two vectors in terms of their rectangular components is

$$\begin{aligned} \boldsymbol{A} \times \boldsymbol{B} &= (A_x\boldsymbol{i} + A_y\boldsymbol{j} + A_z\boldsymbol{k}) \times (B_x\boldsymbol{i} + B_y\boldsymbol{j} + B_z\boldsymbol{k}) \\ &= (A_yB_z - A_zB_y)\boldsymbol{i} + (A_zB_x - A_xB_z)\boldsymbol{j} + (A_xB_y - A_yB_x)\boldsymbol{k} \end{aligned} \tag{2.25}$$

Another method of carrying out this long computation is to evaluate the following determinant:

$$\begin{vmatrix} A_x & A_y & A_z \\ B_x & B_y & B_z \\ \boldsymbol{i} & \boldsymbol{j} & \boldsymbol{k} \end{vmatrix} \tag{2.26}$$

The determinant may easily be evaluated in the following manner. Repeat the first two rows below the determinant, and then form products along diagonals.

$$\begin{matrix} A_x & A_y & A_z \\ B_x & B_y & B_z \\ \boldsymbol{i} & \boldsymbol{j} & \boldsymbol{k} \\ A_x & A_y & A_z \\ B_x & B_y & B_z \end{matrix} \tag{2.27}$$

For the products along the dashed diagonals, we must remember in this method to multiply by -1. We then add all six products as follows:

$$\begin{aligned} &A_xB_y\boldsymbol{k} + B_xA_z\boldsymbol{j} + A_yB_z\boldsymbol{i} - A_zB_y\boldsymbol{i} - B_zA_x\boldsymbol{j} - A_yB_x\boldsymbol{k} \\ &\qquad = (A_yB_z - A_zB_y)\boldsymbol{i} + (A_zB_x - A_xB_z)\boldsymbol{j} + (A_xB_y - A_yB_x)\boldsymbol{k} \end{aligned}$$

Clearly, this is the same result as in Eq. 2.25. It must be cautioned that this method of evaluating a determinant is correct only for 3×3 determinants. If the cross product of two vectors involves less than six nonzero components, such as in the cross product

$$(6\boldsymbol{i} + 10\boldsymbol{j}) \times (5\boldsymbol{j} - 3\boldsymbol{k})$$

then it is advisable to multiply the components directly and collect terms, as in Eq. 2.25.

2.8 Scalar Triple Product

A very useful quantity is the *scalar triple product*, which for a set of vectors $\boldsymbol{A}$, $\boldsymbol{B}$, and $\boldsymbol{C}$ is defined as

$$(\boldsymbol{A} \times \boldsymbol{B}) \cdot \boldsymbol{C} \tag{2.28}$$

This clearly is a scalar quantity.

A simple geometric meaning can be associated with this operation. In Fig. 2.26, we have shown $\boldsymbol{A}$, $\boldsymbol{B}$, and $\boldsymbol{C}$ as an arbitrary set of concurrent vectors. We have set up

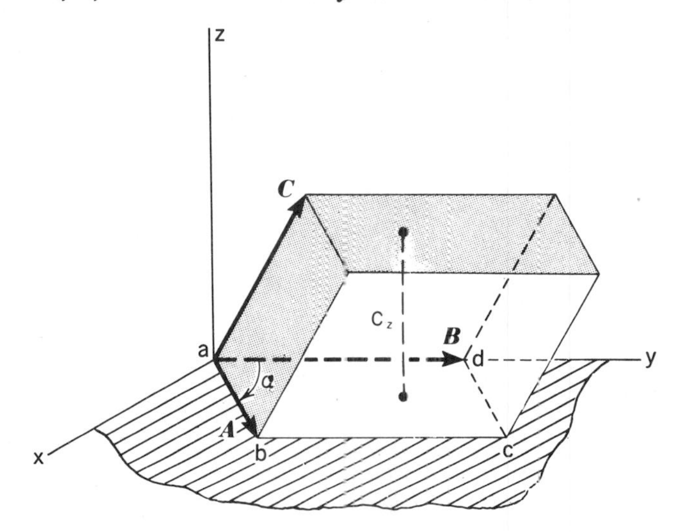

Figure 2.26. ***A*** and ***B*** in *xy* plane.

an *xyz* reference such that the $\boldsymbol{A}$ and $\boldsymbol{B}$ vectors are in the *xy* plane. We have formed a parallelogram *abcd* in the *xy* plane as shown in the diagram. We can say that

$$|\boldsymbol{A} \times \boldsymbol{B}| = |\boldsymbol{A}||\boldsymbol{B}| \sin \alpha = \text{area of } abcd$$

Furthermore, the direction of $\boldsymbol{A} \times \boldsymbol{B}$ is in the z direction. Clearly, when we carry out Eq. 2.28, we are thus multiplying the component of $\boldsymbol{C}$ in the z direction by the area of the aforementioned parallelogram. Thus, we have, for Eq. 2.28:

$$(\boldsymbol{A} \times \boldsymbol{B}) \cdot \boldsymbol{C} = (\text{area of } abcd)(C_z)$$

But C_z is the *slant height* of the parallelepiped formed by vectors $\boldsymbol{A}$, $\boldsymbol{B}$, and $\boldsymbol{C}$. We then conclude from solid geometry that *the scalar triple product is the volume of the parallelepiped formed by the concurrent vectors of the scalar triple product.*

Using this geometrical interpretation of the scalar triple product, the reader can easily conclude that

$$(\boldsymbol{A} \times \boldsymbol{B}) \cdot \boldsymbol{C} = -(\boldsymbol{A} \times \boldsymbol{C}) \cdot \boldsymbol{B} = -(\boldsymbol{C} \times \boldsymbol{B}) \cdot \boldsymbol{A} \tag{2.29}$$

The computation of the scalar triple product is a very straightforward process. It will be left as an exercise (Problem 2.54) for you to demonstrate that

$$(\boldsymbol{A} \times \boldsymbol{B}) \cdot \boldsymbol{C} = \begin{vmatrix} A_x & A_y & A_z \\ B_x & B_y & B_z \\ C_x & C_y & C_z \end{vmatrix} \tag{2.30}$$

In later chapters, we shall employ the scalar triple product, although we shall not always want to associate the preceding geometric interpretation of this product.

Another operation involving three vectors is the *vector triple product* defined for vectors $\boldsymbol{A}$, $\boldsymbol{B}$, and $\boldsymbol{C}$ as $\boldsymbol{A} \times (\boldsymbol{B} \times \boldsymbol{C})$. The vector triple product is a vector quantity and will appear quite often in studies of dynamics. It will be left for you to demonstrate that

$$\boldsymbol{A} \times (\boldsymbol{B} \times \boldsymbol{C}) = \boldsymbol{B}(\boldsymbol{A} \cdot \boldsymbol{C}) - \boldsymbol{C}(\boldsymbol{A} \cdot \boldsymbol{B}) \tag{2.31}$$

Notice here that the vector triple product can be carried out by using only dot products.

EXAMPLE 2.5

A pyramid is shown in Fig. 2.27. If the height of the pyramid is 300 ft, find the angle between planes ADB and BDC (i.e., find the angle between the normals to these planes).

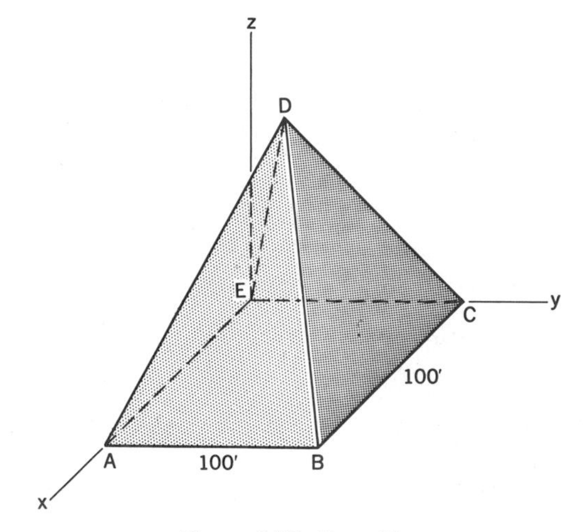

Figure 2.27. Pyramid.

We shall first find the unit normals to the aforestated planes. Then, using the dot product between these normals, we can easily find the desired angle.

To get the unit normal $\boldsymbol{n}_1$ to plane ABD, we first compute the area vector $\boldsymbol{A}_1$ for this plane. Thus, from simple trigonometry and the definition of the cross product,

$$\boldsymbol{A}_1 = \tfrac{1}{2}\overrightarrow{AB} \times \overrightarrow{AD}$$

Next, note that

$$\overrightarrow{AB} = 100\boldsymbol{j} \text{ ft}$$
$$\overrightarrow{AD} = \overrightarrow{ED} - \overrightarrow{EA} = (50\boldsymbol{i} + 50\boldsymbol{j} + 300\boldsymbol{k}) - 100\boldsymbol{i}$$
$$= -50\boldsymbol{i} + 50\boldsymbol{j} + 300\boldsymbol{k} \text{ ft}$$

Hence,

$$\boldsymbol{A}_1 = \tfrac{1}{2}(100\boldsymbol{j}) \times (-50\boldsymbol{i} + 50\boldsymbol{j} + 300\boldsymbol{k})$$
$$= 15{,}000\boldsymbol{i} + 2500\boldsymbol{k} \text{ ft}^2$$

Accordingly,

$$\boldsymbol{n}_1 = \frac{\boldsymbol{A}_1}{|\boldsymbol{A}_1|} = \frac{15{,}000\boldsymbol{i} + 2500\boldsymbol{k}}{\sqrt{15{,}000^2 + 2500^2}} \qquad \text{(a)}$$
$$= .986\boldsymbol{i} + .1644\boldsymbol{k}$$

As for unit normal $\boldsymbol{n}_2$ corresponding to plane BDC, whose area vector we denote as $\boldsymbol{A}_2$, we have

$$\boldsymbol{A}_2 = \tfrac{1}{2}\overrightarrow{BC} \times \overrightarrow{BD}$$

Note that

$$\overrightarrow{BC} = -100\boldsymbol{i} \text{ ft}$$
$$\overrightarrow{BD} = \overrightarrow{ED} - \overrightarrow{EB} = (50\boldsymbol{i} + 50\boldsymbol{j} + 300\boldsymbol{k}) - (100\boldsymbol{i} + 100\boldsymbol{j})$$
$$= -50\boldsymbol{i} - 50\boldsymbol{j} + 300\boldsymbol{k} \text{ ft}$$

Hence,

$$\boldsymbol{A}_2 = \tfrac{1}{2}(-100\boldsymbol{i}) \times (-50\boldsymbol{i} - 50\boldsymbol{j} + 300\boldsymbol{k})$$
$$= 15{,}000\boldsymbol{j} + 2500\boldsymbol{k} \text{ ft}^2$$

Accordingly,

$$\boldsymbol{n}_2 = \frac{\boldsymbol{A}_2}{|\boldsymbol{A}_2|} = \frac{15{,}000\boldsymbol{j} + 2500\boldsymbol{k}}{\sqrt{15{,}000^2 + 2500^2}} \qquad \text{(b)}$$
$$= .986\boldsymbol{j} + .1644\boldsymbol{k}$$

Now, we use the dot product of $\boldsymbol{n}_1$ and $\boldsymbol{n}_2$. Thus,

$$\boldsymbol{n}_1 \cdot \boldsymbol{n}_2 = \cos\beta \qquad \text{(c)}$$

where β is the angle between the normals to the planes. Substituting from Eqs. (a) and (b) into (c), we get

$$\cos\beta = .0270$$

Therefore,

$$\beta = 88.5°$$

EXAMPLE 2.6

In Example 2.5, what is the area projected by plane ADE onto an infinite plane that is inclined equally to the x, y, and z axes?

The normal $\boldsymbol{n}$ to the infinite plane must have three equal direction cosines. Hence, noting Eq. 2.20 for the sum of the squares of a set of direction cosines, we can say that

$$l^2 = m^2 = n^2 = \frac{1}{3}$$

Therefore,

$$l = m = n = \frac{1}{\sqrt{3}}$$

Hence,

$$\boldsymbol{n} = \frac{1}{\sqrt{3}}\boldsymbol{i} + \frac{1}{\sqrt{3}}\boldsymbol{j} + \frac{1}{\sqrt{3}}\boldsymbol{k}$$

The projected area then is given as

$$A_n = \left(\frac{1}{2}\overrightarrow{AD} \times \overrightarrow{AE}\right) \cdot \boldsymbol{n}$$

$$= \left[\frac{1}{2}(-50\boldsymbol{i} + 50\boldsymbol{j} + 300\boldsymbol{k}) \times (-100\boldsymbol{i})\right] \cdot \frac{1}{\sqrt{3}}(\boldsymbol{i} + \boldsymbol{j} + \boldsymbol{k})$$

The preceding result is a scalar triple product which can readily be solved as follows (disregarding the final sign):

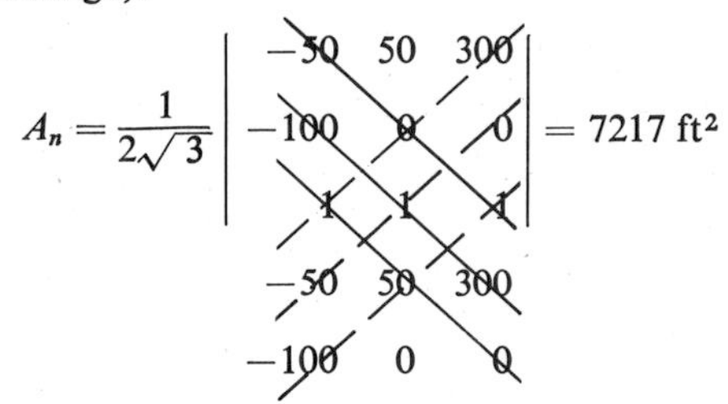

$$A_n = \frac{1}{2\sqrt{3}} \begin{vmatrix} -50 & 50 & 300 \\ -100 & 0 & 0 \\ 1 & 1 & 1 \end{vmatrix} = 7217 \text{ ft}^2$$

2.9 A Note on Vector Notation

When expressing *equations*, we must at all times clearly denote scalar and vector quantities and handle them accordingly. When we are simply identifying quantities in a *discussion* or in a *diagram*, however, instead of using the vector representation, $\boldsymbol{F}$, we can use just F. On the other hand, F will be understood to represent in an equation the magnitude of the vector $\boldsymbol{F}$. Thus, using $\boldsymbol{f}$ as the unit vector in the direction of $\boldsymbol{F}$, we can say:

$$\boldsymbol{F} = F\boldsymbol{f}$$

$$= F[\cos(\boldsymbol{F}, x)\boldsymbol{i} + \cos(\boldsymbol{F}, y)\boldsymbol{j} + \cos(\boldsymbol{F}, z)\boldsymbol{k}]$$

As another example, we might want to employ the force F, which is shown in the coplanar diagram of Fig. 2.28(a) at a known inclination and acting at a point a. A correct representation of this force in a vector equation would be $F(-\cos\alpha\boldsymbol{i} + \sin\alpha\boldsymbol{j})$.

As for scalar components of any vector $\boldsymbol{F}$, we shall adopt the following understanding. The notation F_x, F_y, or F_z labeling some vector component in a *diagram* will be understood to represent the *magnitude* of that particular component. Thus, in Fig. 2.28(b) the two components shown are equal in magnitude but opposite in sense. Nevertheless, they are both labeled F_x. However, in an equation involving these quantities, the sense must properly be accounted for by the appropriate use of signs.

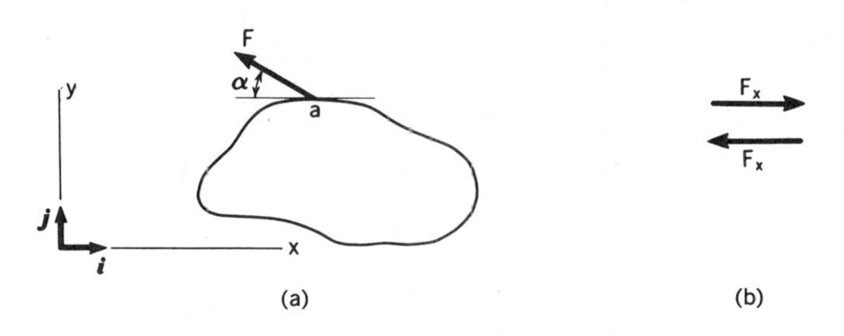

Figure 2.28. Notation in diagrams.

Problems

2.45. If $A = 10i + 6j - 3k$ and $B = 6i$, find $A \times B$ and $B \times A$. What is the magnitude of the resulting vector? What are its direction cosines relative to the xyz reference in which A and B are expressed?

2.46. What are the cross and dot products for the vectors A and B given as:

$$A = 6i + 3j + 4k$$
$$B = 8i - 3j + 2k?$$

2.47. If vectors A and B in the xy plane have a dot product of 50 units, and if the magnitudes of these vectors are 10 units and 8 units, respectively, what is $A \times B$?

2.48. (a) If $A \cdot B = A \cdot B'$, does B necessarily equal B'? Explain.

(b) If $A \times B = A \times B'$, does B necessarily equal B'? Explain.

2.49. What is the cross product of the displacement vector from A to B times the displacement vector from C to D?

2.50. Making use of the cross product, give the unit vector n normal to the inclined surface ABC.

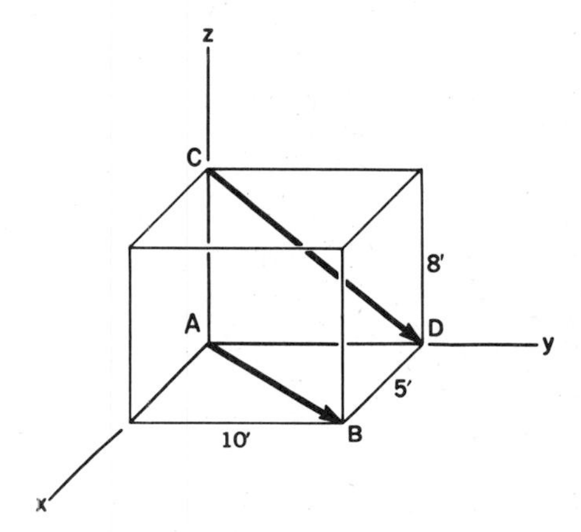

Figure P.2.49

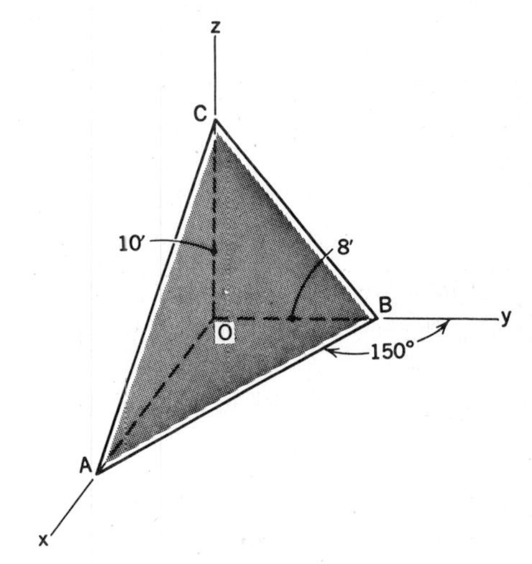

Figure P.2.50

2.51. If coordinates of vertex E of the inclined pyramid are (5, 50, 80) m, what is the angle between faces ADE and BCE?

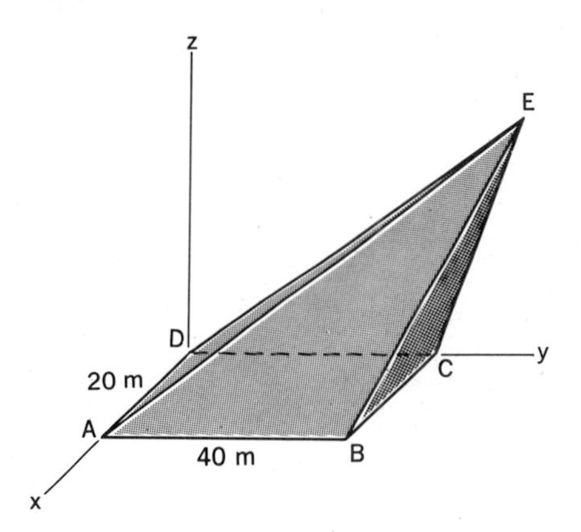

Figure P.2.51

2.52. In Problem 2.51, what is the area of face ADE of the pyramid? What is the projection of the area of face ADE onto a plane whose normal is along the direction $\boldsymbol{\epsilon}$ where:

$$\boldsymbol{\epsilon} = 0.6\boldsymbol{i} - 0.8\boldsymbol{j}$$

2.53. (a) Compute the product

$$(\boldsymbol{A} \times \boldsymbol{B}) \cdot \boldsymbol{C}$$

in terms of orthogonal components.

(b) Compute $(\boldsymbol{C} \times \boldsymbol{A}) \cdot \boldsymbol{B}$ and compare with the result in part (a).

2.54. Compute the determinant

$$\begin{vmatrix} A_x & A_y & A_z \\ B_x & B_y & B_z \\ C_x & C_y & C_z \end{vmatrix}$$

where each row represents the scalar components of $\boldsymbol{A}$, $\boldsymbol{B}$, and $\boldsymbol{C}$. Compare the result with the computation of $(\boldsymbol{A} \times \boldsymbol{B}) \cdot \boldsymbol{C}$ by using the dot-product and cross-product operations.

2.55. What is the component of the cross product $\boldsymbol{A} \times \boldsymbol{B}$ along the direction $\boldsymbol{n}$, where

$$\boldsymbol{A} = 10\boldsymbol{i} + 16\boldsymbol{j} + 3\boldsymbol{k}$$

$$\boldsymbol{B} = 5\boldsymbol{i} - 2\boldsymbol{j} + 2\boldsymbol{k}$$

$$\boldsymbol{n} = 0.8\boldsymbol{i} + 0.6\boldsymbol{k}$$

2.56. The surface *abcd* of the parallelepiped is in the xz plane. Compute the volume using vector analysis.

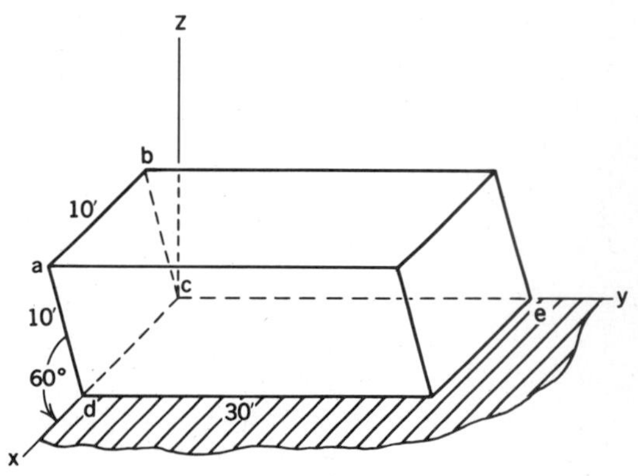

Figure P.2.56

2.57. Given the vectors

$$\boldsymbol{A} = 10\boldsymbol{i} + 6\boldsymbol{j}$$

$$\boldsymbol{B} = 3\boldsymbol{i} + 5\boldsymbol{j} + 10\boldsymbol{k}$$

$$\boldsymbol{C} = \boldsymbol{i} + \boldsymbol{j} - 3\boldsymbol{k}$$

find

(a) $(\boldsymbol{A} + \boldsymbol{B}) \times \boldsymbol{C}$

(b) $(\boldsymbol{A} \times \boldsymbol{B}) \cdot \boldsymbol{C}$

(c) $\boldsymbol{A} \cdot (\boldsymbol{B} \times \boldsymbol{C})$

***2.58.** A mirror system is used to relay a laser-beam signal from mountain M to hill H. The mountain is 5000 m high and 20,000 m NW from the mirror site S, while the hill is 200 m high and 1500 m ENE from the mirror site S. Set up, but do not necessarily solve, the equations to find the direction of the mirror to properly relay the signal. Recall that the angle of reflection for a mirror equals the angle of incidence and that the incident ray, reflecting ray, and normal to the mirror are coplanar.

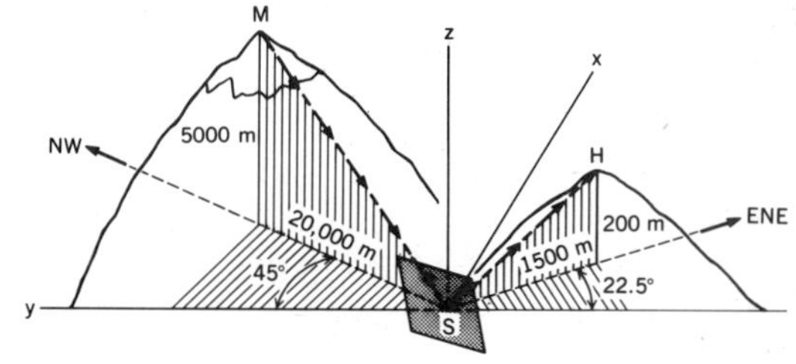

Figure P.2.58

2.10 Closure

In this chapter, we have presented symbols and notations that are associated with vectors. Also, various vector operations have been set forth that enable us to represent certain actions in nature mathematically. With this background, we shall now be able to study certain vector quantities that are of essential importance in mechanics. Some of these vectors will be formulated in terms of the operations contained in this chapter.

Check-Out for Sections with †

2.1 What is meant by the *magnitude* of a vector? What sign must it have?
2.2 Can you multiply a vector $\boldsymbol{C}$ by a scalar s? If so, describe the result.
2.3 What are the *law of cosines* and the *law of sines*?
2.4 What is meant by the *associative* law of addition?
2.5 Describe two ways to add any three vectors graphically.
2.6 How do you subtract vector $\boldsymbol{D}$ from vector $\boldsymbol{F}$?
2.7 Given a vector $\boldsymbol{D}$, how would you form a *unit* vector collinear with $\boldsymbol{D}$?
2.8 What are the scalar equations of the following vector equation?

$$D\boldsymbol{i} + E\boldsymbol{j} - 16\boldsymbol{k} = 20\boldsymbol{i} + (15 + G)\boldsymbol{k}$$

Review Problems

2.59. A bridge truss has bar forces as shown in the cutaway sketch. What is the total force on the supporting pin at point A from the members?

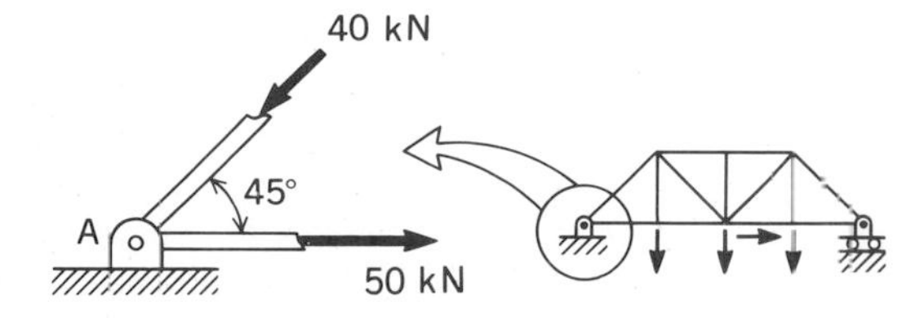

Figure P.2.59

2.60. Forces are transmitted by two members to pin A. If the sum of these forces is 700 lb directed vertically, what are the angles α and β?

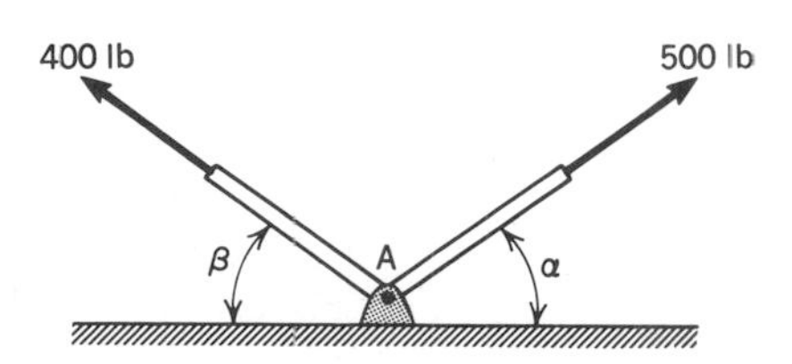

Figure P.2.60

2.61. Contractors encountered an impassable swamp while building a road from town T to city C 50 km SE. To avoid the swamp, they built the road SSW from T and then ENE to C. How long is the road? (*Hint:* See the compass-settings diagram, Fig. 2.11.)

2.62. Four members of a space frame are loaded as shown. What are the orthogonal scalar components of the forces on the ball joint at O? The 1000-N force goes through points D and E of the rectangular parallelepiped.

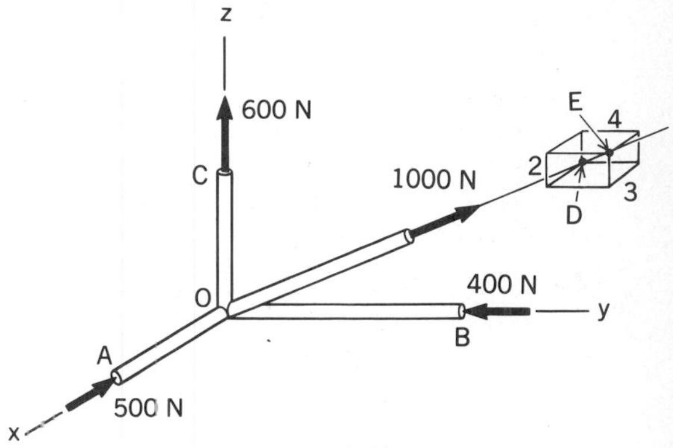

Figure P.2.62

2.63. The x and z components of the force F are known to be 100 lb and -30 lb, respectively. What is the force F and what are its direction cosines?

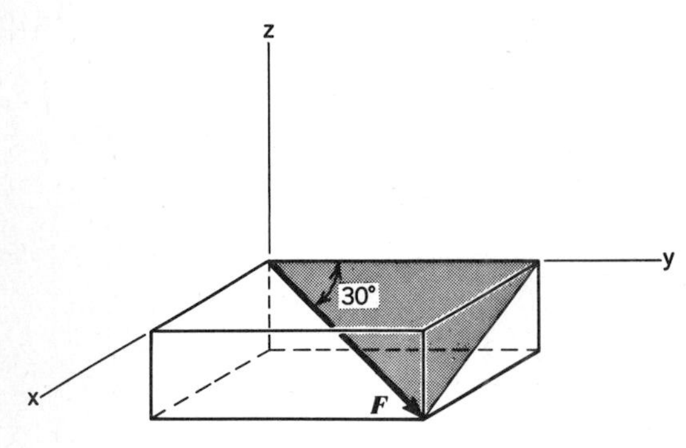

Figure P.2.63

2.64. A constant force given as $2\boldsymbol{i} + 3\boldsymbol{k}$ N moves a particle along a straight line from position $x = 10, y = 20, z = 0$ to position $x = 3, y = 0, z = -10$. If the coordinates of the xyz reference are given in meter units, how much work does the force do in ft-lb?

2.65. The force on a charge moving through a magnetic field $\boldsymbol{B}$ is given as

$$\boldsymbol{F} = q\boldsymbol{V} \times \boldsymbol{B}$$

where q = magnitude of the charge, coulombs
$\boldsymbol{F}$ = force on the body, newtons
$\boldsymbol{V}$ = velocity vector of the particle, meters per second
$\boldsymbol{B}$ = magnetic flux density, webers per meter^2

Suppose that an electron moves through a uniform magnetic field of 10^6 Wb/m^2 in a direction inclined 30° to the field, as shown, with a speed of 100 m/sec. What are the force components on the electron? The charge of the electron is 1.6018×10^{-19} coulomb.

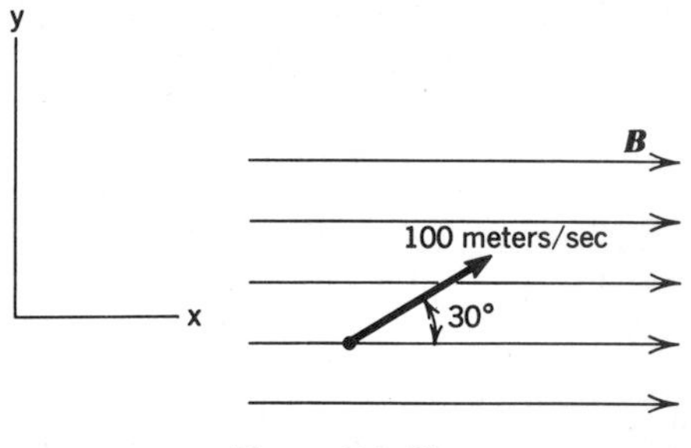

Figure P.2.65

2.66. The velocity of a particle of flow is given as

$$\boldsymbol{V} = 10\boldsymbol{i} + 16\boldsymbol{j} + 2\boldsymbol{k} \text{ m/sec}$$

What is the cross product $\boldsymbol{r} \times \boldsymbol{V}$, where $\boldsymbol{r}$ is given as

$$\boldsymbol{r} = 3\boldsymbol{i} + 2\boldsymbol{j} + 10\boldsymbol{k} \text{ m}$$

Give the proper units.

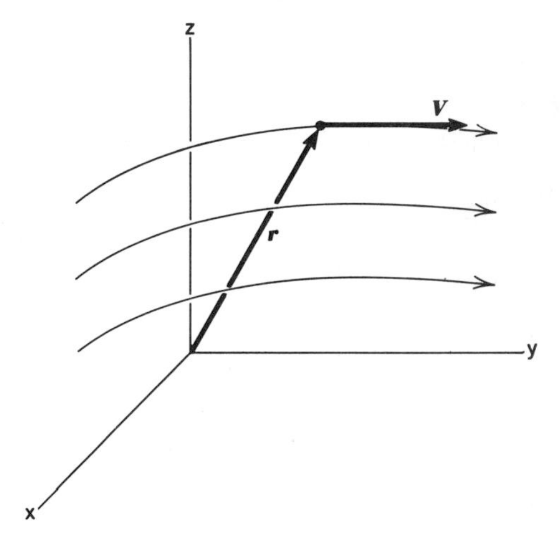

Figure P.2.66

2.67. Using the scalar triple product, find the area projected onto the plane *N* from the surface *ABC*. Plane *N* is infinite and is normal to the vector

$$\boldsymbol{r} = 50\boldsymbol{i} + 40\boldsymbol{j} + 30\boldsymbol{k} \text{ ft}$$

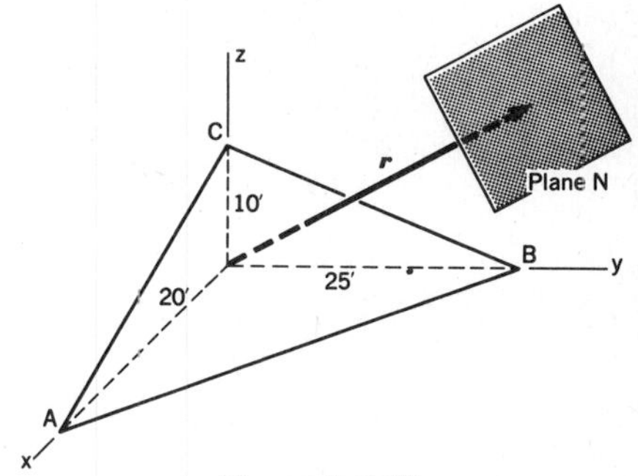

Figure P.2.67

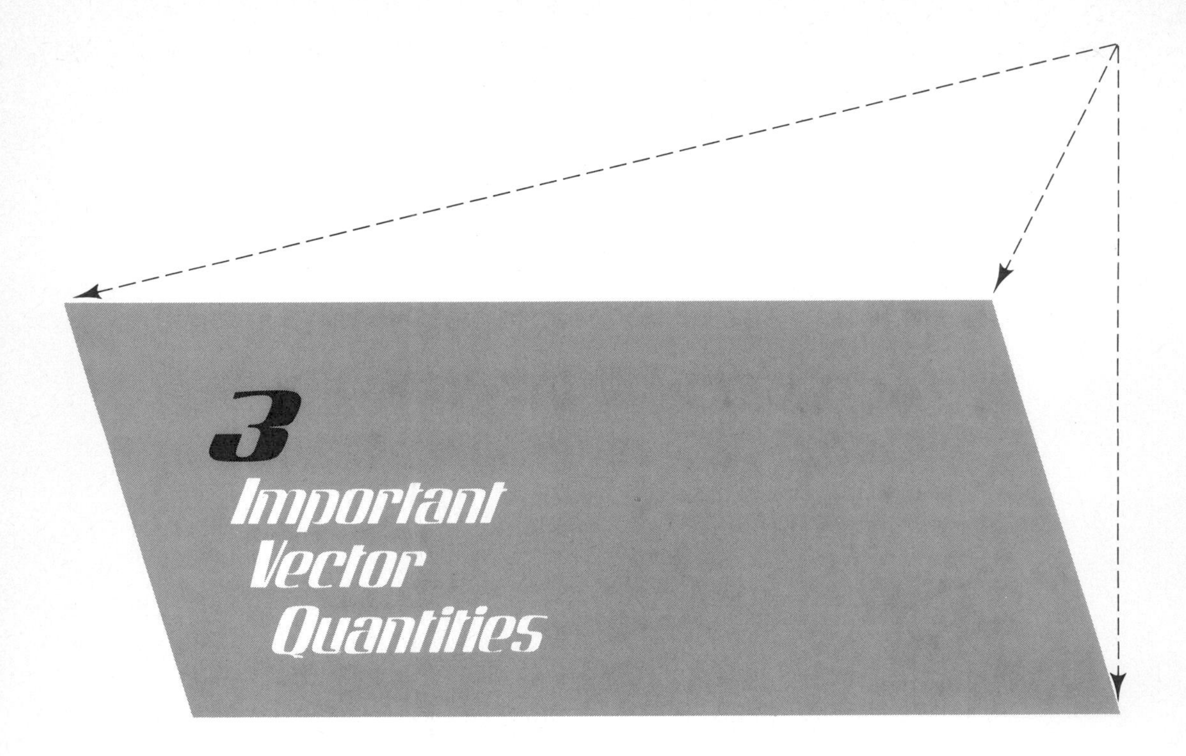

3 Important Vector Quantities

3.1 Position Vector

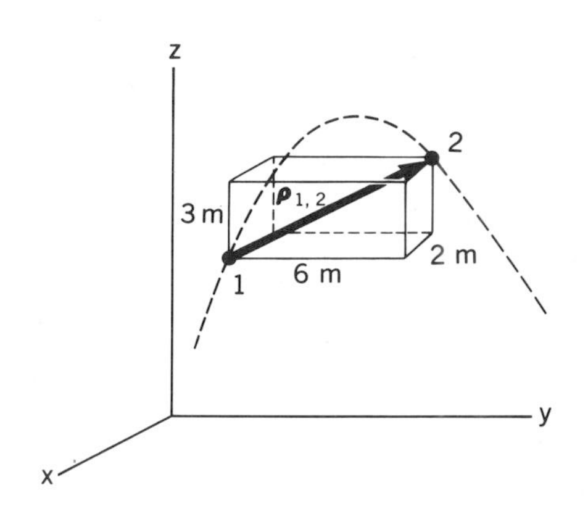

Figure 3.1. Displacement vector $\boldsymbol{\rho}$ between points 1 and 2.

In this chapter, we shall discuss a number of useful vector quantities. Consider first the path of motion of a particle shown dashed in Fig. 3.1. As indicated in Chapter 1, the *displacement vector* $\boldsymbol{\rho}$ is a directed line segment connecting any two points on the path of motion, such as points 1 and 2 in Fig. 3.1. The displacement vector thus represents the shortest movement of the particle to get from one position on the path of motion to another. The purpose of the rectangular parallelepiped shown in the diagram is to convey the magnitude and direction of $\boldsymbol{\rho}$. We can readily express $\boldsymbol{\rho}$ between points 1 and 2 in terms of rectangular components by noting the distance in the coordinate directions needed to go from 1 to 2. Thus, in Fig. 3.1, $\boldsymbol{\rho}_{1,2} = -2\boldsymbol{i} + 6\boldsymbol{j} + 3\boldsymbol{k}$ m.

The directed line segment $\boldsymbol{r}$ from the origin of a coordinate system to a point P in space (Fig. 3.2) is called the *position vector*. The notations $\boldsymbol{R}$ and $\boldsymbol{\rho}$ are also used for position vectors. You can conclude from Chapter 2 that the magnitude of the position vector is the distance between the origin O and point P. The scalar components of a position vector are simply the coordinates of the point P. To express $\boldsymbol{r}$ in Cartesian

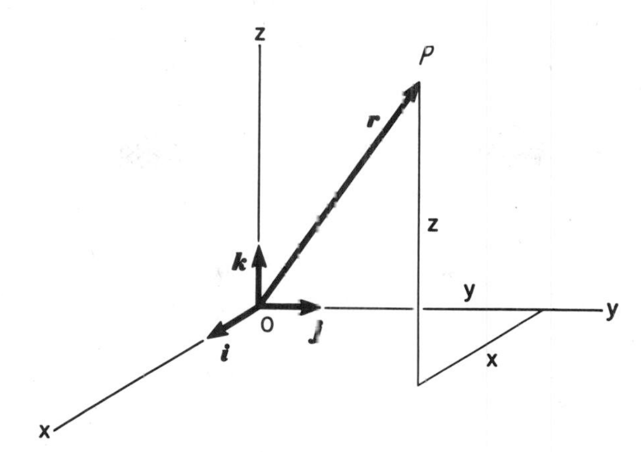

Figure 3.2. Position vector.

components, we then have

$$r = xi + yj + zk \tag{3.1}$$

We can obviously express a displacement vector $\boldsymbol{\rho}$ between points 1 and 2 (see Fig. 3.3) in terms of position vectors for points 1 and 2 (i.e., $\boldsymbol{r}_1$ and $\boldsymbol{r}_2$) as follows:

$$\boldsymbol{\rho} = \boldsymbol{r}_2 - \boldsymbol{r}_1 = (x_2 - x_1)\boldsymbol{i} + (y_2 - y_1)\boldsymbol{j} + (z_2 - z_1)\boldsymbol{k} \tag{3.2}$$

EXAMPLE 3.1

Two sets of references, xyz and XYZ, are shown in Fig. 3.4. The position vector of the origin O of xyz relative to XYZ is given as

$$\boldsymbol{R} = 10\boldsymbol{i} + 6\boldsymbol{j} + 5\boldsymbol{k} \text{ m} \tag{a}$$

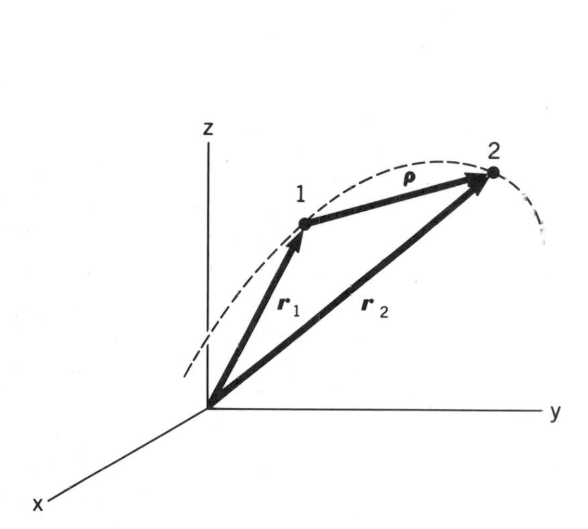

Figure 3.3. Relation between a displacement vector and position vectors.

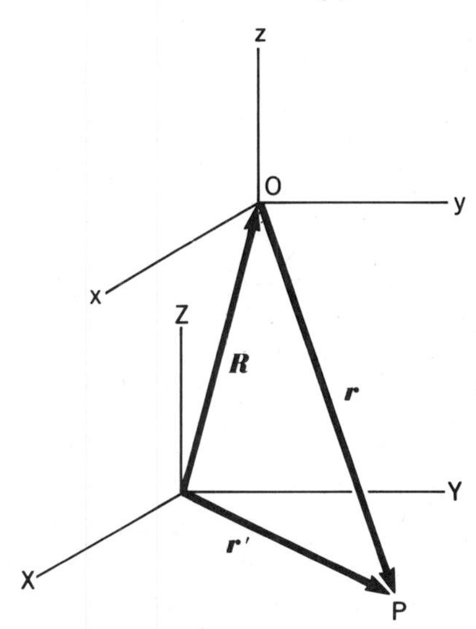

Figure 3.4. References xyz and XYZ separated by position vector $\boldsymbol{R}$.

The position vector, $\boldsymbol{r}'$, of a point P relative to XYZ is

$$\boldsymbol{r}' = 3\boldsymbol{i} + 2\boldsymbol{j} - 6\boldsymbol{k} \text{ m} \tag{b}$$

What is the position vector $\boldsymbol{r}$ of point P relative to xyz? What are the coordinates x, y, and z of P?

From Fig. 3.4, it is clear that

$$\boldsymbol{r}' = \boldsymbol{R} + \boldsymbol{r} \tag{c}$$

Therefore,

$$\begin{aligned} \boldsymbol{r} = \boldsymbol{r}' - \boldsymbol{R} &= (3\boldsymbol{i} + 2\boldsymbol{j} - 6\boldsymbol{k}) - (10\boldsymbol{i} + 6\boldsymbol{j} + 5\boldsymbol{k}) \\ &= -7\boldsymbol{i} - 4\boldsymbol{j} - 11\boldsymbol{k} \text{ m} \end{aligned} \tag{d}$$

We can then conclude that

$$\begin{aligned} x &= -7 \text{ m} \\ y &= -4 \text{ m} \\ z &= -11 \text{ m} \end{aligned} \tag{e}$$

3.2 Moment of a Force About a Point

The moment of a force about a point O (see Fig. 3.5), you will recall from physics, is a vector $\boldsymbol{M}$ whose magnitude equals the product of the force magnitude times the perpendicular distance d from O to the line of action of the force. And the direction of this vector is perpendicular to the plane of the point and the force, with a sense determined from the familiar right-hand-screw rule.[1] The line of action of $\boldsymbol{M}$ is

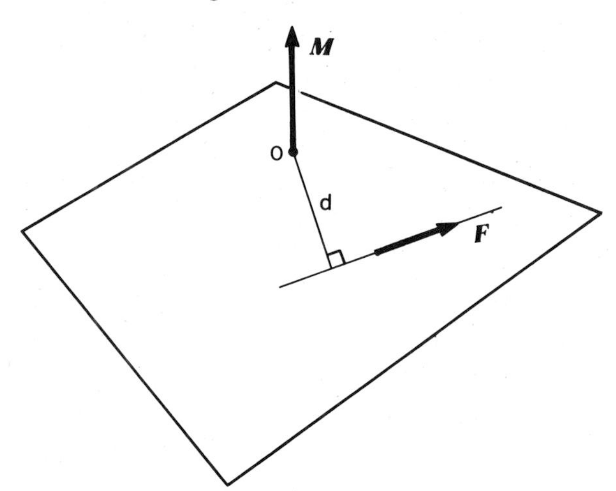

Figure 3.5. Moment of force $\boldsymbol{F}$ about O is Fd.

[1]The sense of $\boldsymbol{M}$ would be that of the direction of advance of a screw at O, oriented normal to the plane of O and $\boldsymbol{F}$, when this screw is turned with a sense of rotation corresponding to that of $\boldsymbol{F}$ around O.

determined by the problem at hand. In Fig. 3.5, the line of action of $\boldsymbol{M}$ is taken for simplicity through point O.

Another approach is to employ a position vector $\boldsymbol{r}$ from point O to *any point P* along the line of action of force $\boldsymbol{F}$ as shown in Fig. 3.6. The moment of $\boldsymbol{F}$ about point O is then defined as

$$\boxed{\boldsymbol{M} = \boldsymbol{r} \times \boldsymbol{F}} \tag{3.3}$$

For the purpose of forming the cross product, the vectors in Fig. 3.6 can be moved to the configuration shown in Fig. 3.7. Then the cross product between $\boldsymbol{r}$ and $\boldsymbol{F}$ obviously has the magnitude

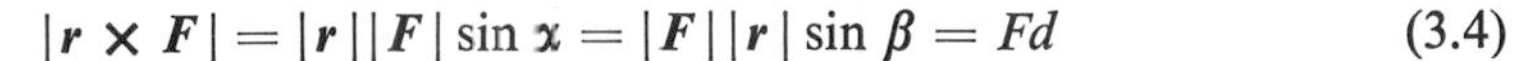

$$|\boldsymbol{r} \times \boldsymbol{F}| = |\boldsymbol{r}||\boldsymbol{F}| \sin \alpha = |\boldsymbol{F}||\boldsymbol{r}| \sin \beta = Fd \tag{3.4}$$

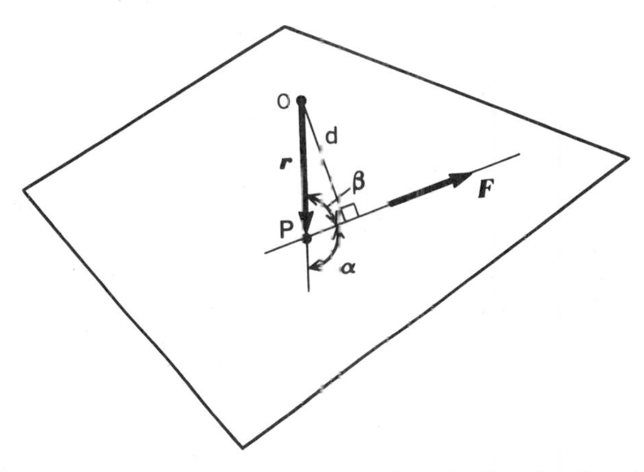

Figure 3.6. Put $\boldsymbol{r}$ from O to any point along the line of action of $\boldsymbol{F}$.

where $|\boldsymbol{r}| \sin \beta = d$, the perpendicular distance from O to the line of action of $\boldsymbol{F}$, as can readily be seen in Fig. 3.7. Thus, we get the same *magnitude* of $\boldsymbol{M}$ as with the elementary definition. Also, note that the vector definition of the *direction* of $\boldsymbol{M}$ is identical to that of the elementary definition. Thus, we have the same result for the vector definition as for the elementary definition in all pertinent respects. We shall use both definitions. The first one will be used generally for cases where the force and point are in a convenient plane, and where the perpendicular distance between the point and the line of action of the force is easily measured. As an example, we have shown in Fig. 3.8 a system of coplanar forces acting on a beam. The moment of the forces about point A is then[2]

$$\begin{aligned} \boldsymbol{M}_A &= -(5)(1000)\boldsymbol{k} - (4)(600)\boldsymbol{k} + (11)R_B\boldsymbol{k} \text{ ft-lb} \\ &= (11R_B - 7400)\boldsymbol{k} \text{ ft-lb} \end{aligned}$$

[2]Please note that we still use the right-hand-screw rule in determining the signs of the respective moments.

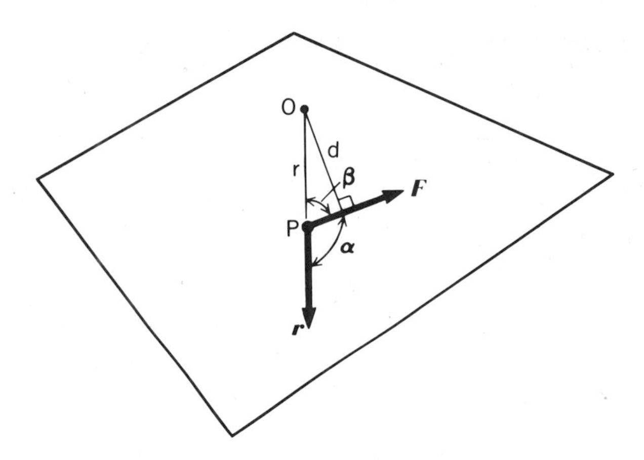

Figure 3.7. Move vector $\boldsymbol{r}$.

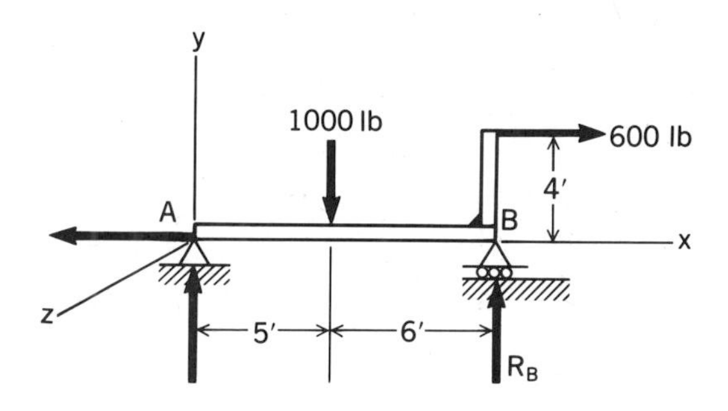

Figure 3.8. Coplanar forces on a beam.

For a coplanar force system such as this, we may simply give the scalar form of the equation above, as follows:

$$M_A = 11R_B - 7400 \text{ ft-lb}$$

The second definition of the moment about a point, namely $\boldsymbol{r} \times \boldsymbol{F}$, is used for complicated coplanar cases and for three-dimensional cases. We shall illustrate such a case in Example 3.2.

Consider next a system of n concurrent forces in Fig. 3.9 whose total moment about point O (where we have established reference xyz) is desired. We can say that

$$\boldsymbol{M} = \boldsymbol{M}_1 + \boldsymbol{M}_2 + \boldsymbol{M}_3 + \ldots + \boldsymbol{M}_n$$
$$= \boldsymbol{r} \times \boldsymbol{F}_1 + \boldsymbol{r} \times \boldsymbol{F}_2 + \boldsymbol{r} \times \boldsymbol{F}_3 + \ldots + \boldsymbol{r} \times \boldsymbol{F}_n \quad (3.5)$$

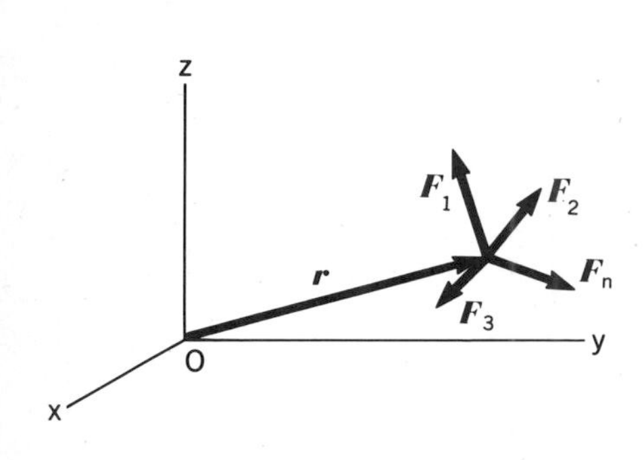

Figure 3.9. Concurrent forces.

Now, because of the distributive property of the cross product, Eq. 3.5 can be written

$$\boldsymbol{M} = \boldsymbol{r} \times (\boldsymbol{F}_1 + \boldsymbol{F}_2 + \boldsymbol{F}_3 + \ldots + \boldsymbol{F}_n) \quad (3.6)$$

We can conclude from the preceding equations that the sum of the moments about a point of a system of concurrent forces is the same as the moment about the point of the sum of the forces. This result is known as *Varignon's theorem*, which you may well recall from physics.

As a special case of Varignon's theorem, we may find it convenient to decompose a force $\boldsymbol{F}$ into its rectangular components (Fig. 3.10), and then to use these components for taking moments about a point. We can then say that

$$\boldsymbol{M} = \boldsymbol{r} \times \boldsymbol{F} = \boldsymbol{r} \times (F_x\boldsymbol{i} + F_y\boldsymbol{j} + F_z\boldsymbol{k}) \quad (3.7)$$

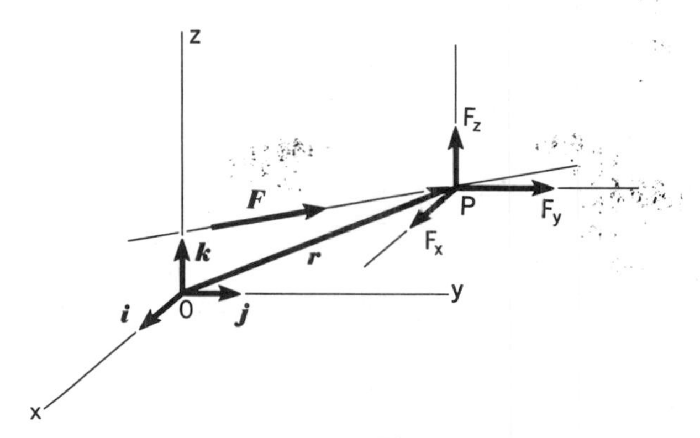

Figure 3.10. Decompose *F* into components.

Now replacing $\boldsymbol{r}$ by its components, we get

$$\begin{aligned} \boldsymbol{M} &= (x\boldsymbol{i} + y\boldsymbol{j} + z\boldsymbol{k}) \times (F_x\boldsymbol{i} + F_y\boldsymbol{j} + F_z\boldsymbol{k}) \\ &= (yF_z - zF_y)\boldsymbol{i} + (zF_x - xF_z)\boldsymbol{j} + (xF_y - yF_x)\boldsymbol{k} \end{aligned} \quad (3.8)$$

The scalar rectangular components of $\boldsymbol{M}$ are then

$$M_x = yF_z - zF_y \quad (3.9a)$$

$$M_y = zF_x - xF_z \quad (3.9b)$$

$$M_z = xF_y - yF_x \quad (3.9c)$$

As a final note, it should be apparent that, because we can choose $\boldsymbol{r}$ so as to terminate anywhere along the line of action of $\boldsymbol{F}$ in computing $\boldsymbol{M}$, we are, in effect, stipulating that $\boldsymbol{F}$ is a *transmissible* vector (defined in Chapter 1) in the computation of $\boldsymbol{M}$.

EXAMPLE 3.2

Determine the moment of the 100-lb force *F*, shown in Fig. 3.11, about points *A* and *B*, respectively.

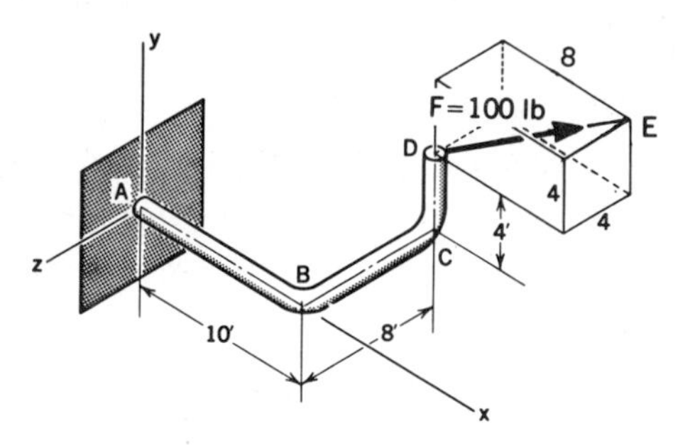

Figure 3.11. Find moments at A and B.

As a first step, let us express force F vectorially. Note that the force is collinear with the vector $\boldsymbol{\rho}_{DE}$ from D to E, where

$$\boldsymbol{\rho}_{DE} = 8\boldsymbol{i} + 4\boldsymbol{j} - 4\boldsymbol{k} \tag{a}$$

To get a unit vector $\hat{\boldsymbol{\rho}}$ in the direction of $\boldsymbol{\rho}$, we proceed as follows:

$$\begin{aligned} \hat{\boldsymbol{\rho}}_{DE} = \frac{\boldsymbol{\rho}_{DE}}{|\boldsymbol{\rho}_{DE}|} &= \frac{8\boldsymbol{i} + 4\boldsymbol{j} + 4\boldsymbol{k}}{\sqrt{8^2 + 4^2 - 4^2}} \\ &= .816\boldsymbol{i} + .408\boldsymbol{j} - .408\boldsymbol{k} \end{aligned} \tag{b}$$

We can then express the force $\boldsymbol{F}$ in the following manner:

$$\begin{aligned} \boldsymbol{F} = F\hat{\boldsymbol{\rho}}_{DE} &= (100)(.816\boldsymbol{i} + .408\boldsymbol{j} - .408\boldsymbol{k}) \\ &= 81.6\boldsymbol{i} + 40.8\boldsymbol{j} - 40.8\boldsymbol{k} \end{aligned} \tag{c}$$

To get the moment $\boldsymbol{M}_A$ about point A, we choose a position vector from point A to point D which is on the line of action of force F. Thus, we have, for $\boldsymbol{r}_{AD}$,

$$\boldsymbol{r}_{AD} = 10\boldsymbol{i} + 4\boldsymbol{j} - 8\boldsymbol{k} \text{ ft} \tag{d}$$

and for $\boldsymbol{M}_A$, we then get

$$\boldsymbol{M}_A = \boldsymbol{r}_{AD} \times \boldsymbol{F} = (10\boldsymbol{i} + 4\boldsymbol{j} - 8\boldsymbol{k}) \times (81.6\boldsymbol{i} + 40.8\boldsymbol{j} - 40.8\boldsymbol{k})$$

$$= \begin{vmatrix} 10 & 4 & -8 \\ 81.6 & 40.8 & -40.8 \\ \boldsymbol{i} & \boldsymbol{j} & \boldsymbol{k} \end{vmatrix}$$

$$\begin{matrix} 10 & 4 & -8 \\ 81.6 & 40.8 & -40.8 \end{matrix}$$

$$\begin{aligned} &= (10)(40.8)\boldsymbol{k} + (81.6)(-8)\boldsymbol{j} + (4)(-40.8)\boldsymbol{i} \\ &\quad -(-8)(40.8)\boldsymbol{i} - (-40.8)(10)\boldsymbol{j} - (4)(81.6)\boldsymbol{k} \end{aligned}$$

Therefore,

$$\boldsymbol{M}_A = 163.2\boldsymbol{i} - 245\boldsymbol{j} + 81.6\boldsymbol{k} \text{ ft-lb} \tag{e}$$

As for the moment about reference point B, we employ the position vector $\boldsymbol{r}_{BD}$ from B to position D, again on the line of action of force $\boldsymbol{F}$. Thus, we have

$$\boldsymbol{r}_{BD} = 4\boldsymbol{j} - 8\boldsymbol{k} \text{ ft}$$

Accordingly,

$$\begin{aligned} \boldsymbol{M}_B &= \boldsymbol{r}_{BD} \times \boldsymbol{F} = (4\boldsymbol{j} - 8\boldsymbol{k}) \times (81.6\boldsymbol{i} + 40.8\boldsymbol{j} - 40.8\boldsymbol{k}) \\ &= (4)(81.6)(-\boldsymbol{k}) + (4)(-40.8)(\boldsymbol{i}) + (-8)(81.6)(\boldsymbol{j}) + (-8)(40.8)(-\boldsymbol{i}) \qquad \text{(f)} \\ &= 163.2\boldsymbol{i} - 653\boldsymbol{j} - 326\boldsymbol{k} \text{ ft-lb} \end{aligned}$$

Problems

3.1. What is the position vector $\boldsymbol{r}$ from the origin (0, 0, 0) to the point (3, 4, 5) ft? What are its magnitude and direction cosines?

3.2. What is the displacement vector from position (6, 13, 7) ft to position (10, −3, 4) ft?

3.3. A surveyor determines that the top of a radio transmission tower is at position $\boldsymbol{r}_1 = (1000\boldsymbol{i} + 1000\boldsymbol{j} + 1000\boldsymbol{k})$ m relative to her position. Similarly, the top of a second tower is located by $\boldsymbol{r}_2 = (2000\boldsymbol{i} + 500\boldsymbol{j} + 700\boldsymbol{k})$ m. What is the distance between the two tower tops?

3.4. Reference xyz is rotated 30° about its x axis relative to reference XYZ. What is the position vector $\boldsymbol{r}$ for reference xyz of a point having a position vector $\boldsymbol{r}'$ for reference XYZ given as

$$\boldsymbol{r}' = 6\boldsymbol{i}' + 10\boldsymbol{j}' + 3\boldsymbol{k}' \text{ m?}$$

Use $\boldsymbol{i}, \boldsymbol{j}$, and $\boldsymbol{k}$ (no primes) for unit vectors associated with reference xyz.

3.5. A particle moves along a circular path in the xy plane. What is the position vector $\boldsymbol{r}$ of this particle as a function of the coordinate x?

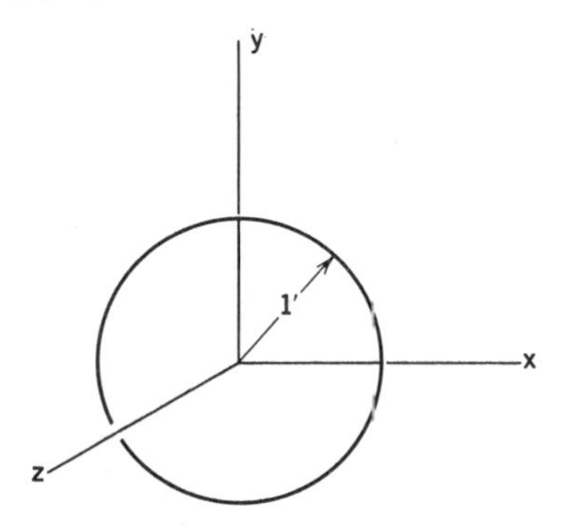

Figure P.3.5

3.6. A particle moves along a parabolic path in the yz plane. If the particle has at one point a position vector $\boldsymbol{r} = 4\boldsymbol{j} + 2\boldsymbol{k}$, give the position vector at any point on the path as a function of the z coordinate.

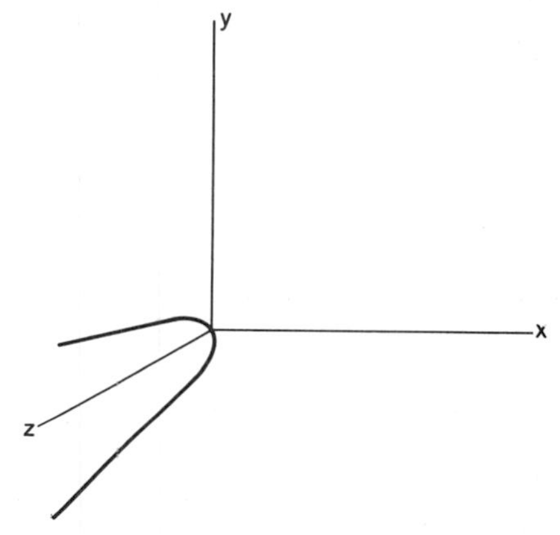

Figure P.3.6

3.7. An artillery spotter on Hill 350 (350 m high) estimates the position of an enemy tank as 3000 m NE of him at an elevation 200 m below his position. A 105-mm howitzer unit with a range of 11,000 m is 10,000 m due south of the spotter, and a 155-mm howitzer unit with a range of 15,000 m is 13,000 m SSE of the spotter. Both gun units are located at an elevation of 150 m. Can either or both gun units hit the tank, or must an air strike be called in?

3.8. The crew of a submarine patrol plane, with three-dimensional radar, sights a surfaced submarine 10,000 yards north and 5000 yards east while flying at an elevation of 3000 ft above sea level. Where should the pilot instruct a second patrol plane flying at an elevation of 4000 ft at a position 40,000 yards east of the first plane to look for confirmation of the sighting?

3.9. A power company lineman can comfortably trim branches 1 m from his waist at an angle of 45° above the horizontal. His waist coincides with the pivot of the work capsule. How high a branch can he trim if the maximum elevation angle of the arm is 75° and the maximum extended length is 12 m?

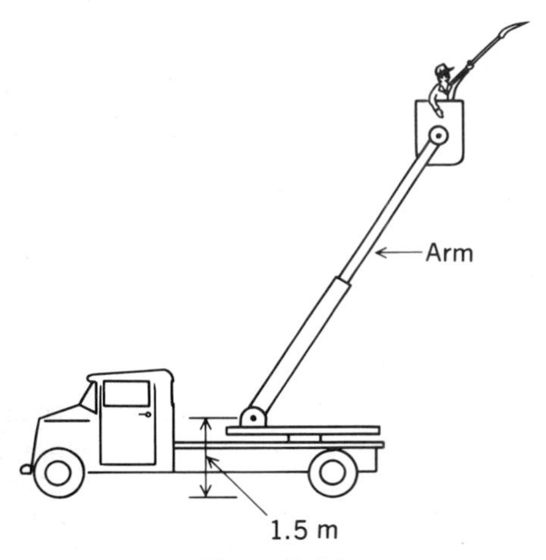

Figure P.3.9

3.10. The total equivalent forces from water and gravity are shown on the dam. (We will soon be able to compute such equivalents.) Compute the moment of these forces about the toe of the dam in the right-hand corner.

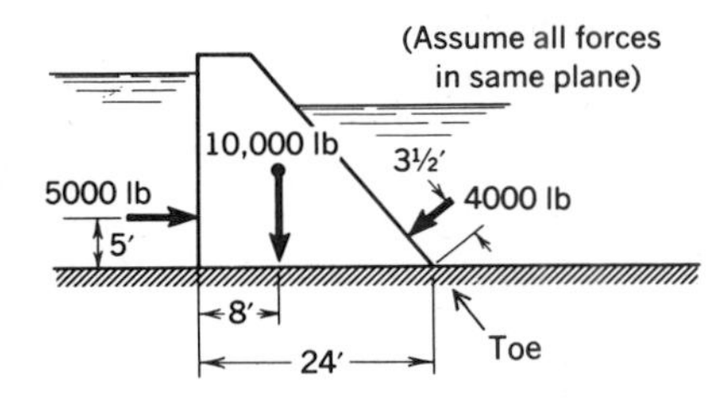

Figure P.3.10

3.11. Three transmission lines are placed unsymmetrically on a power-line pole. For each pole, the weight of a single line when covered with ice is 2000 N. What is the moment at the base of a pole?

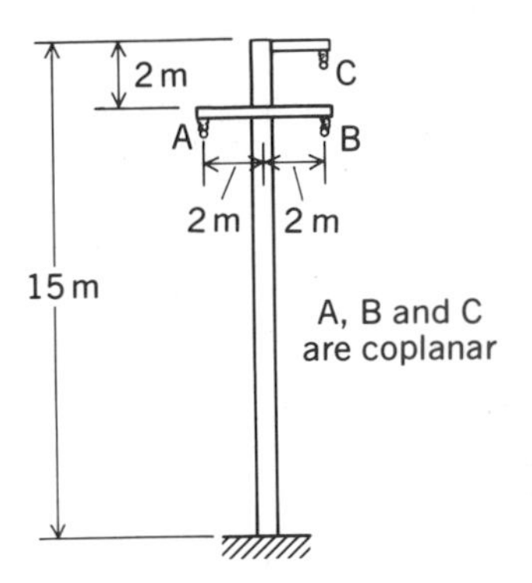

Figure P.3.11

3.12. Find the moment of the 50-lb force about the support at A and about support B of the simply supported beam.

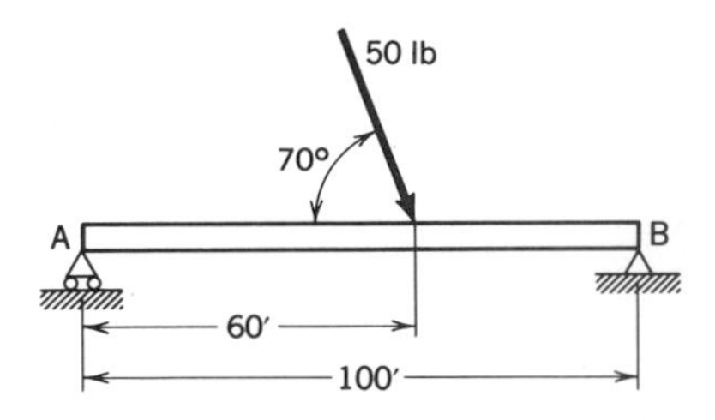

Figure P.3.12

3.13. Compute the moment of the 1000-lb force about points A, B, and C. Use the transmissibility property of force and rectangular components to make the computations simplest.

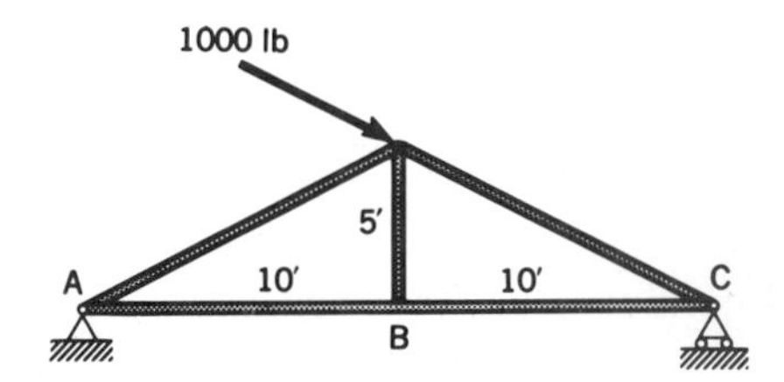

Figure P.3.13

3.14. A truck-mounted crane has a 20-m boom inclined at 60° to the horizontal. What is the moment about the boom pivot due to a lifted weight of 30 kN? Do by vector and by scalar methods.

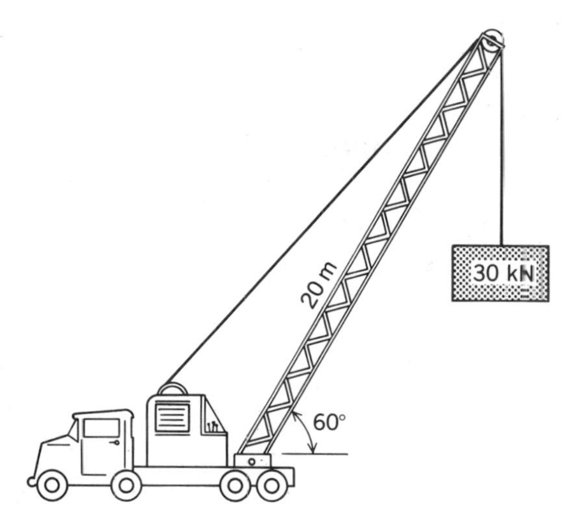

Figure P.3.14

3.15. A force $\boldsymbol{F} = 10\boldsymbol{i} + 6\boldsymbol{j} - 6\boldsymbol{k}$ N acts at position (10, 3, 4) m relative to a coordinate system. What is the moment of the force about the origin?

3.16. What is the moment of the force in Problem 3.15 about the point (6, −4, −3) m?

3.17. Two forces $\boldsymbol{F}_1$ and $\boldsymbol{F}_2$ have magnitudes of 10 lb and 20 lb, respectively. $\boldsymbol{F}_1$ has a set of direction cosines $l = 0.5$, $m = 0.707$, $n = -0.5$. $\boldsymbol{F}_2$ has a set of direction cosines $l = 0$, $m = 0.6$, $n = 0.8$. If $\boldsymbol{F}_1$ acts at point (3, 2, 2) and $\boldsymbol{F}_2$ acts at (1, 0, −3), what is the sum of these moments about the origin?

3.18. What is the moment of a 10-lb force $\boldsymbol{F}$ directed along the diagonal of a cube about the corners of the cube? The side of the cube is a ft.

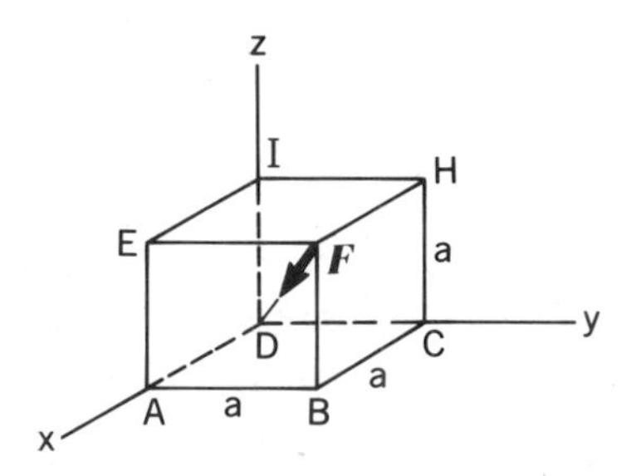

Figure P.3.18

3.19. Compute the moment of the 300-lb force about points P_1 and P_2.

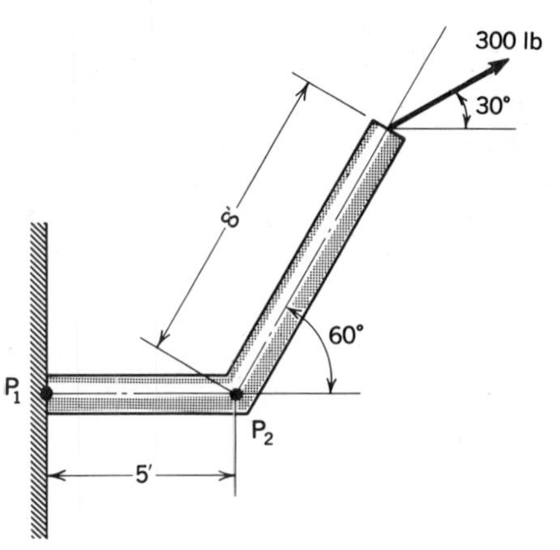

Figure P.3.19

3.20. Three guy wires are used in the support system for a television transmission tower that is 600 m tall. Wires A and B are tightened to a tension of 60 kN, whereas wire C has only 30 kN of tension. What is the moment of the wire forces about the base of the tower? The y axis is collinear with AO.

Figure P.3.20

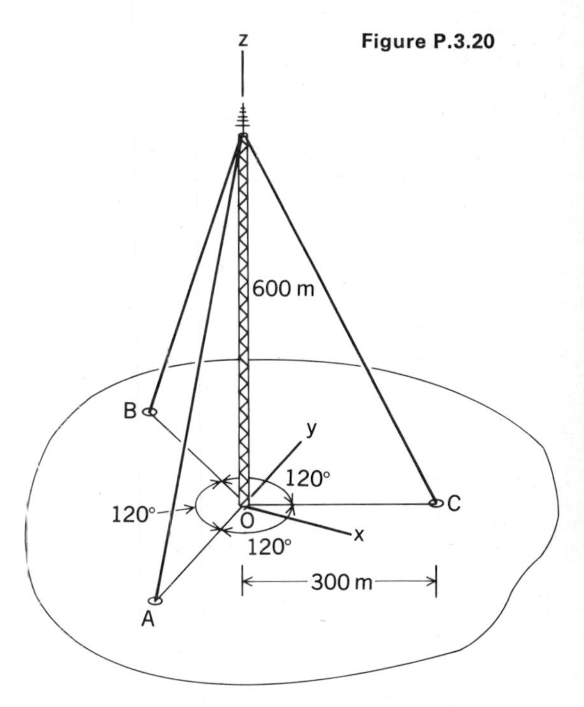

3.21. Cables CD and AB help support the member ED and the 1000-lb load at D. At E there is a ball-and-socket joint which also supports the member. Denoting the forces from the cables as F_{CD} and F_{AB}, respectively, compute moments of the three forces about the point E.

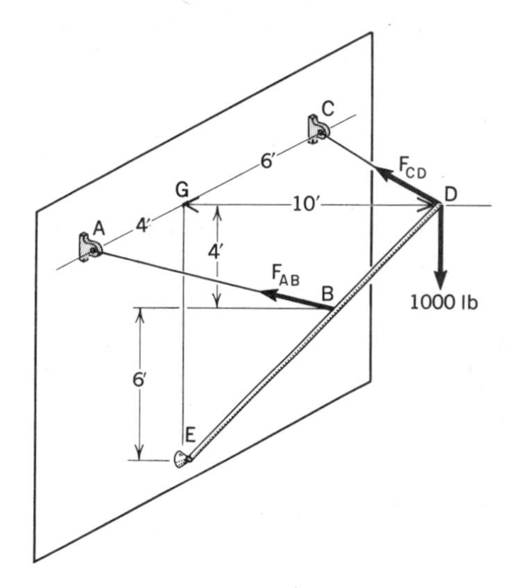

Figure P.3.21

3.3 Moment of a Force About an Axis

To compute the moment of (or torque) a force $\boldsymbol{F}$ about an axis B–B [Fig. 3.12(a)], we pass any plane A perpendicular to the axis. This plane cuts B–B at a and the line of action of force $\boldsymbol{F}$ at some point P. The force $\boldsymbol{F}$ is then projected to form a rectangular component F_B along a line at P normal to plane A and thus parallel to B–B, as shown in the diagram. The intersection of plane A with the plane of forces F_B and $\boldsymbol{F}$ (the latter plane is shown shaded and is a plane through $\boldsymbol{F}$ and perpendicular to plane A) gives a direction C–C along which the other rectangular component of $\boldsymbol{F}$, denoted as F_A, can be projected.[3] The moment of $\boldsymbol{F}$ about the line B–B is then defined as the moment of the component F_A about point a—a coplanar problem discussed at the beginning of the previous section. Thus, in accordance with the definition, the component F_B, which is parallel to the axis B–B, contributes no moment about the axis, and we may say:

$$\text{moment about the axis } B\text{–}B = (F_A)(d) = |\boldsymbol{F}|(\cos\alpha)(d)$$

The moment about an axis is a scalar, even though this moment is associated with a particular axis that has a distinct direction. The situation is the same as it is with the scalar components V_x, V_y, V_z, etc., which are associated with certain directions but which are, nevertheless, scalars. Before continuing, we wish to point out that F_A in Fig. 3.12(a) can be decomposed into pairs of components in plane A. From Varignon's theorem we can employ these components instead of F_A in computing the moment

[3]Notice that we are decomposing $\boldsymbol{F}$ into only *two* rectangular components, which, to replace $\boldsymbol{F}$, must be coplanar with $\boldsymbol{F}$.

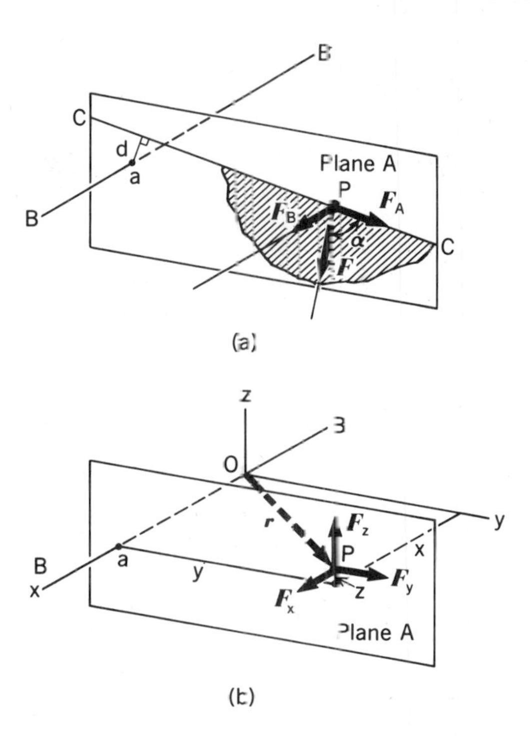

Figure 3.12. Moment about an axis.

about the *B–B* axis. For each force component, we multiply the force times the perpendicular distance from *a* to the line of action of the force component using the right-hand-screw rule to determine the sense and thus the sign.

By means of a simple situation, we can easily show why we set forth the definition above for moment of a force about an axis. Suppose that a disc is mounted on a shaft that is free to rotate in a set of bearings, as shown in Fig. 3.13. A force *F*, inclined to the plane *A* of the disc, acts on the disc. We decompose the force into two rectangular components, one normal to the plane *A* of the disc and the other tangent to the plane *A* of the disc, that is, into forces F_B and F_A, respectively. We know from experience

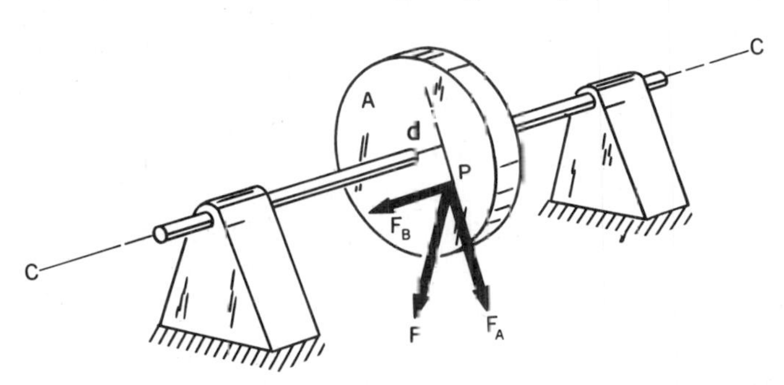

Figure 3.13. F_A turns disc.

that F_B does not cause the disc to rotate. And we know from physics that the rotational motion of the disc is determined by the product of F_A and the perpendicular distance d from the centerline of the shaft to the line of action F_A. But in accordance with our definition, this product is nothing more than the moment of force F about the axis of the shaft. Later, in more general dynamics problems, the moments of forces about a point as well as about an axis will enter into the dynamics relations.

What is the relation between a moment about a point and a moment about an axis? To answer this most simply, consider B–B of Fig. 3.12(a) to be an x axis, as shown in Fig. 3.12(b). Now choose *any* point O along this axis, and set up coordinate axes y and z, as shown in the diagram. The coordinate distances x, y, and z for point P are shown for this reference. The position vector $\boldsymbol{r}$ to P is also shown. The force component F_B of Fig. 3.12(a) now becomes force component F_x. And, instead of using F_A, we shall decompose it into components F_y and F_z in plane A as shown in Fig. 3.12(b). We now compute the moment about the x axis for force $\boldsymbol{F}$ using this new arrangement. Clearly, F_x contributes no moment, as before. The force components F_y and F_z are in plane A that is perpendicular to the axis of interest and so, as before in the case of F_A, we multiply each of these forces by the perpendicular distance of point a to the respective lines of action of these forces. For force F_z, this perpendicular distance is clearly y, as can readily be seen from the diagram, and, for force F_y, this perpendicular distance is z. Using the right-hand-screw rule for ascertaining the sense of each of the moments, we can say:

$$\text{moment about } x \text{ axis} = (yF_z - zF_y) \tag{3.10}$$

Were we to take moments of $\boldsymbol{F}$ about the origin O, we would get (see Eq. 3.8)

$$\begin{aligned} \boldsymbol{M} &= M_x\boldsymbol{i} + M_y\boldsymbol{j} + M_z\boldsymbol{k} = \boldsymbol{r} \times \boldsymbol{F} \\ &= (yF_z - zF_y)\boldsymbol{i} + (zF_x - xF_z)\boldsymbol{j} + (xF_y - yF_x)\boldsymbol{k} \end{aligned} \tag{3.11}$$

Comparing Eqs. 3.10 and 3.11, we can conclude that the moment about the x axis is simply M_x, the x component of $\boldsymbol{M}$ about O. We can thus conclude that the moment about the x axis of the force $\boldsymbol{F}$ is the component in the x direction of the moment of $\boldsymbol{F}$ about a point O positioned *anywhere* along the x axis. That is,

$$\text{moment about } x \text{ axis} = M_x = \boldsymbol{M}_o \cdot \boldsymbol{i} = (\boldsymbol{r} \times \boldsymbol{F}) \cdot \boldsymbol{i} \tag{3.12}$$

We may generalize the preceding discussion as follows. Consider an arbitrary axis n–n to which we have assigned a unit vector $\boldsymbol{n}$ (Fig. 3.14). An arbitrary force $\boldsymbol{F}$ is also shown. To get the moment M_n of force $\boldsymbol{F}$ about axis n–n, we choose any point O along n–n. Then draw a position vector $\boldsymbol{r}$ from point O to any point along the line of action of $\boldsymbol{F}$. This has been shown in the diagram. We can then say, from our previous discussion,

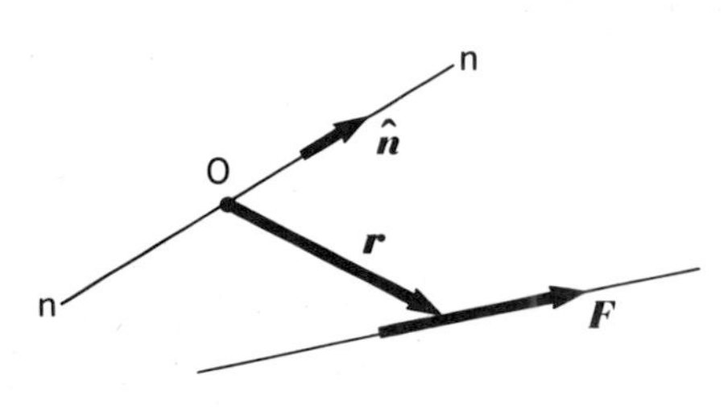

Figure 3.14. $M_n = (\boldsymbol{r} \times \boldsymbol{F}) \cdot \boldsymbol{n}$.

$$\boxed{M_n = (\boldsymbol{r} \times \boldsymbol{F}) \cdot \boldsymbol{n}} \tag{3.13}$$

(Notice from Eqs. 3.12 and 3.13 that the moment of a force about an axis involves a *scalar triple product.*) Equation 3.13 stipulates in words that:

> *The moment of a force about an axis equals the scalar component in the direction of the axis of the moment vector taken about any point along the axis.*[4]

Note that the unit vector $\boldsymbol{n}$ can have two opposite senses along the axis n, in contrast to the usual unit vectors $\boldsymbol{i}, \boldsymbol{j}$, and $\boldsymbol{k}$ associated with the coordinate axes. A moment M_n about the n axis determined from $\boldsymbol{M} \cdot \boldsymbol{n}$ has a sense consistent with the sense chosen for $\boldsymbol{n}$. That is, a positive moment M_n has a sense corresponding to that of $\boldsymbol{n}$, and a negative moment M_n has a sense opposite to that of $\boldsymbol{n}$. If the opposite sense had been chosen for $\boldsymbol{n}$, the sign of $\boldsymbol{M} \cdot \boldsymbol{n}$ would be opposite to that found in the first case. However, the same physical moment is obtained in both cases.

If we specify the moments of a force about *three* orthogonal concurrent axes, we then single out *one* possible point in space for O along the axes. Point O, of course, is the origin of the axes. These three moments then become the orthogonal scalar components of the moment of $\boldsymbol{F}$ about point O, and we can say:

$$\begin{aligned} \boldsymbol{M} = &\text{(moment about the } x \text{ axis)}\boldsymbol{i} \\ &+ \text{(moment about the } y \text{ axis)}\boldsymbol{j} \\ &+ \text{(moment about the } z \text{ axis)}\boldsymbol{k} = M_x\boldsymbol{i} + M_y\boldsymbol{j} + M_z\boldsymbol{k} \end{aligned} \tag{3.14}$$

From this relation, we can conclude that:

> *The three orthogonal components of the moment of a force about a point are the moments of this force about the three orthogonal axes that have the point as an origin.*

You may now ask what the physical differences are in applications of moments about an axis and moments about a point. The simplest example is in the dynamics of rigid bodies. If a body of revolution is constrained so it can only spin about its axis, as in Fig. 3.13, the rotary motion will depend on the moment of the forces about the axis of rotation, as related by a scalar equation. The less familiar concept of moment about a point is illustrated in the motion of bodies that have no constraints, such as missiles and rockets. In these cases, the motion of the body is related by a vector equation to the moment of forces acting on the body about a point called the *center of mass.* (The center of mass will be defined completely later.)

EXAMPLE 3.3

Compute the moment of a force $\boldsymbol{F} = 10\boldsymbol{i} + 6\boldsymbol{j}$ N, which goes through position $\boldsymbol{r}_a = 2\boldsymbol{i} + 6\boldsymbol{j}$ m (see Fig. 3.15), about a line going through points 1 and 2 having the

---

[4]If the force is in a plane *perpendicular* to the axis about which we are taking the moment like force F_A in Fig. 3.13, where plane A is perpendicular to axis C–C about which we desire the moment (or torque) of F_A, remember we can use the elementary definition of moment of a force about an axis that was presented at the outset of this section. Hence for F_A we then have $M_{CC} = F_A d$.

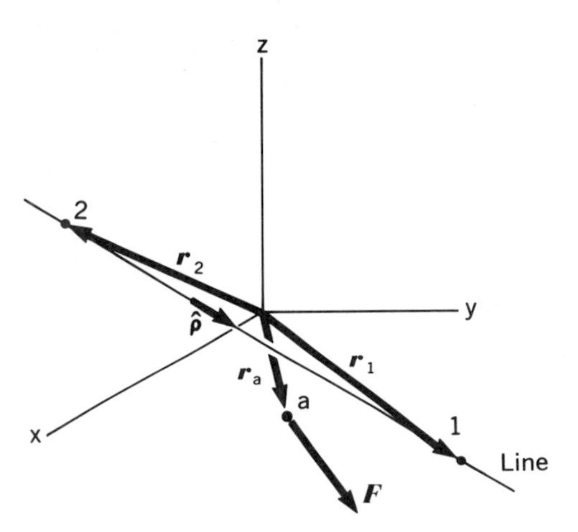

Figure 3.15. Find moment of F about line.

respective position vectors

$$r_1 = 6i + 10j - 3k \text{ m}$$
$$r_2 = -3i - 12j + 6k \text{ m}$$

To compute this moment, we can take the moment of F about either point 1 or point 2, and then find the component of this vector along the direction of the displacement vector between 1 and 2 or between 2 and 1. Mathematically, we have, using a position vector from point 1 to point a, namely $(r_a - r_1)$,

$$M_p = [(r_a - r_1) \times F] \cdot \hat{\rho} \tag{a}$$

where $\hat{\rho}$ is the unit vector along the line chosen to have a sense going from point 2 to point 1. The formulation above is the scalar triple product examined in Chapter 2 and we can use the determinant approach for the calculation once the components of the vectors $(r_a - r_1)$, F, and $\hat{\rho}$ have been determined. Thus, we have

$$r_a - r_1 = (2i + 6j) - (6i + 10j - 3k)$$
$$= -4i - 4j + 3k \text{ m}$$
$$F = 10i + 6j \text{ N}$$
$$\hat{\rho} = \frac{r_1 - r_2}{|r_1 - r_2|} = \frac{9i + 22j - 9k}{\sqrt{81 + 484 + 81}}$$
$$= .354i + .866j - .354k$$

We then have, for M_p:

$$M_p = \begin{vmatrix} -4 & -4 & 3 \\ 10 & 6 & 0 \\ .354 & .866 & -.354 \end{vmatrix} = 13.94 \text{ N-m} \tag{b}$$

Because M_p is positive, we have a clockwise moment about the line as we look from point 2 to point 1. If we had chosen $\hat{\rho}$ to have an opposite sense, then M_p would have been computed as -13.94 N-m. Then, we would conclude that M_p is a counterclockwise moment about the line as one looks from point 1 to point 2. Note that the same physical moment is determined in both cases.

Problems

3.22. Disc A has a radius of 600 mm. What is the moment of the forces about the center of the disc? What is the torque of these forces about the axis of the shaft?

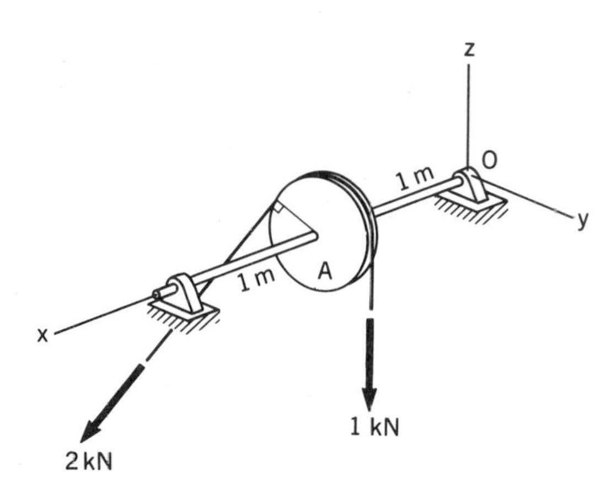

Figure P.3.22

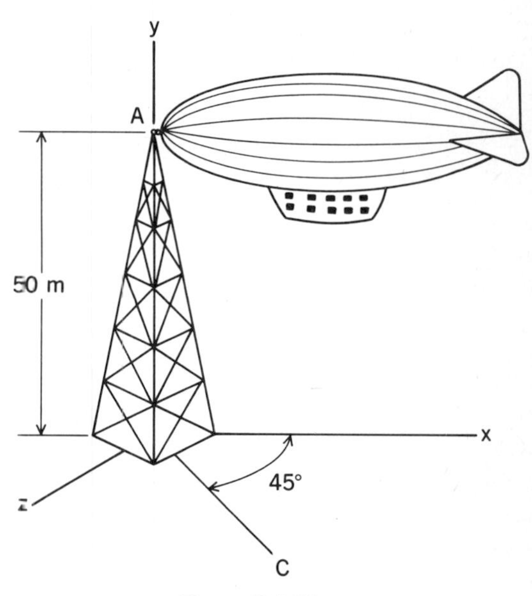

Figure P.3.25

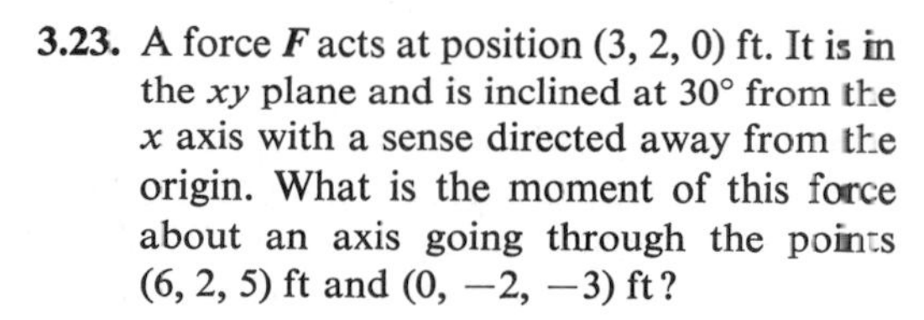

3.23. A force F acts at position (3, 2, 0) ft. It is in the xy plane and is inclined at 30° from the x axis with a sense directed away from the origin. What is the moment of this force about an axis going through the points (6, 2, 5) ft and (0, −2, −3) ft?

3.24. A force $\boldsymbol{F} = 10\boldsymbol{i} + 16\boldsymbol{j}$ N goes through the origin of the coordinate system. What is the moment of this force $\boldsymbol{F}$ about an axis going through points 1 and 2 with position vectors

$$\boldsymbol{r}_1 = 6\boldsymbol{i} + 3\boldsymbol{k} \text{ m}$$

$$\boldsymbol{r}_2 = 16\boldsymbol{j} - 4\boldsymbol{k} \text{ m}$$

3.25. A blimp is moored to a tower at A. A force on A from this blimp is

$$\boldsymbol{F} = 5\boldsymbol{i} + 3\boldsymbol{j} + 1.8\boldsymbol{k} \text{ kN}$$

What is the moment about axis C on the ground? Knowledge of this moment and other moments at the base is needed to properly design the foundation of the tower.

3.26. Compute the thrust of the applied forces shown along the axis of the shaft and the torque of the forces about the axis of the shaft.

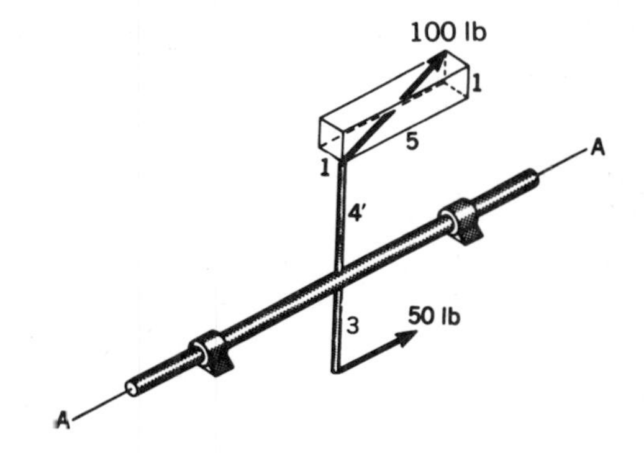

Figure P.3.26

3.27. A deep-submergence vessel is connected to its mother ship by a cable. The vessel becomes snagged on some rocks, and the mother ship steams away in an attempt to free the submerged vessel. The connecting cable is suspended from a crane sticking out over the water 20 m above the mother ship's center of mass and 15 m out from the longitudinal axis of the mother ship. The cable develops a force of 200 kN. It is inclined at 50° from the vertical in a vertical plane oriented 20° from the longitudinal axis of the ship. What is the moment tending to cause the mother ship to roll over?

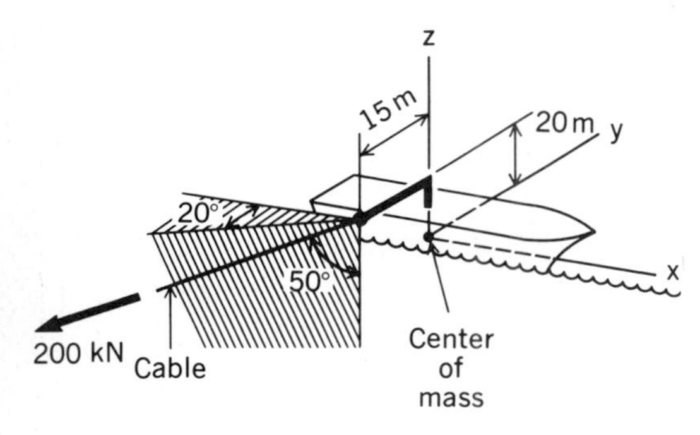

Figure P.3.27

3.28. Find the moment of the 1000-lb force about an axis going between points D and C.

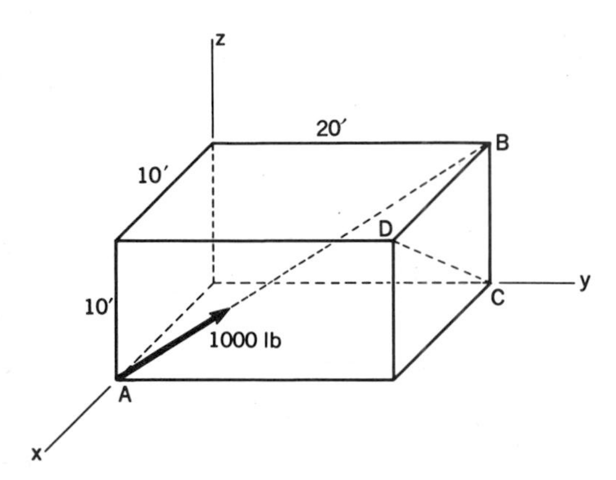

Figure P.3.28

3.29. In Problem 3.21, what is the moment of the three indicated forces about axis GD?

3.30. A tow truck is pointed at 45° to the edge A–A of a ravine with sides sloping at 45° to the vertical. The operator attaches a cable to

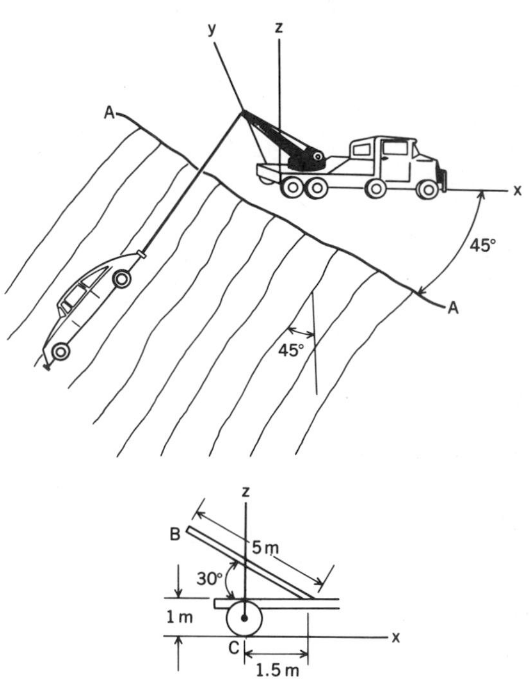

Figure P.3.30

a wrecked car in the ravine and starts the winch. The cable is oriented normal to A–A and develops a force of 15 kN. What are the moments tending to tip over the tow truck about rear wheels (rocking backwards) and also the moment for rolling sideways. (*Hint:* Use the position vector from C to B.)

3.31. The base of a fire truck extension ladder is rotated 75° from the front of the truck. The 25-m ladder is elevated 60° from the horizontal. The ladder weight is 20 kN and is regarded as concentrated at a point 10 m up from the base (the lower part of the ladder weighs much more than the upper part). A 900-N fireman and the 500-N young lady he is rescuing are at the top of the ladder. (a) What is the moment at the base of the ladder tending to tip over the fire truck? (b) What is the moment about the horizontal axis $\hat{\mathbf{p}}$ about which the ladder rotates?

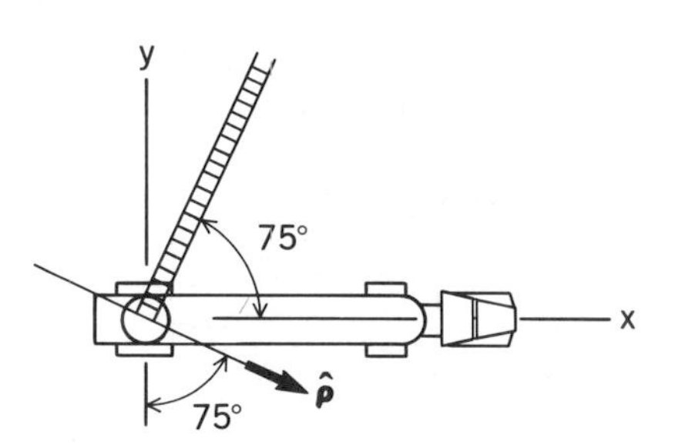

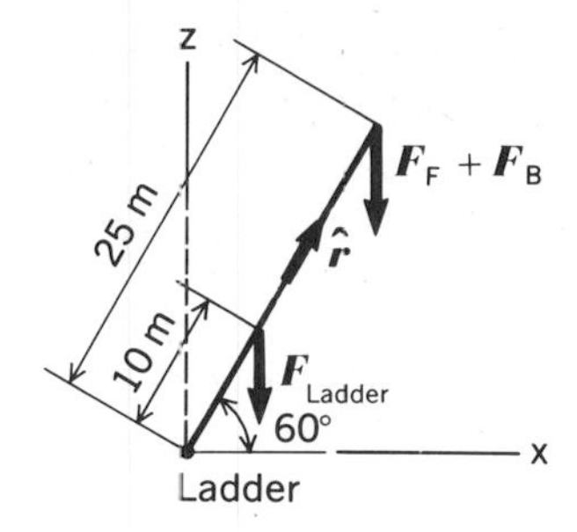

Figure P.3.31

3.4 The Couple and Couple Moment

A special arrangement of forces that is of great importance is the *couple. The couple is formed by any two equal parallel forces that have opposite senses* (Fig. 3.16). On a rigid body, a couple has only *one* effect, a "turning" action. Individual forces or combinations of forces that do not constitute couples may "push" or "pull" as well as "turn" a body. The turning action is given quantitatively by the moment of forces about a point or an axis. We shall, accordingly, be most concerned with the moment of a couple, or what we shall call the *couple moment.*

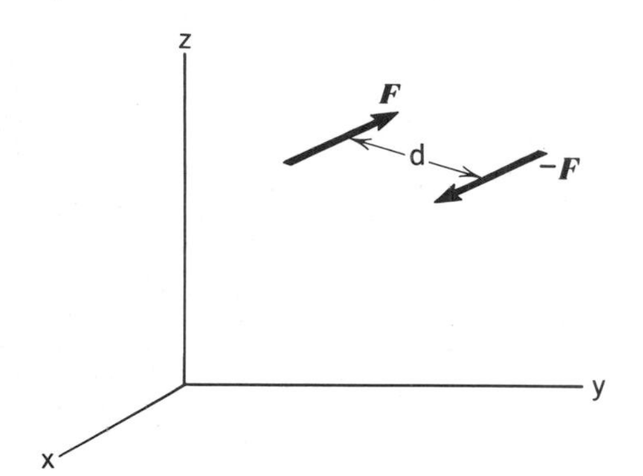

Figure 3.16. A couple.

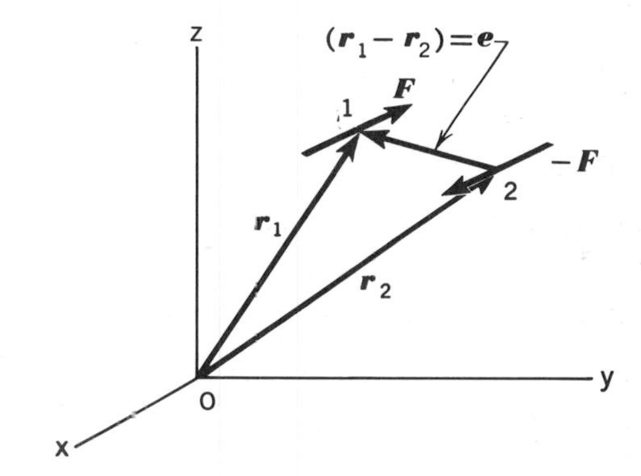

Figure 3.17. Compute moment of couple about O.

Let us now evaluate the moment of the couple about the origin. Position vectors have been drawn in Fig. 3.17 to points 1 and 2 anywhere along the respective line of action of each force. Adding the moment of each force about O, we have for the couple moment $\boldsymbol{M}$

$$\begin{aligned} \boldsymbol{M} &= \boldsymbol{r}_1 \times \boldsymbol{F} + \boldsymbol{r}_2 \times (-\boldsymbol{F}) \\ &= (\boldsymbol{r}_1 - \boldsymbol{r}_2) \times \boldsymbol{F} \end{aligned} \tag{3.15}$$

We can see that $(\boldsymbol{r}_1 - \boldsymbol{r}_2)$ is a position vector between points 2 and 1, and if we call this vector $\boldsymbol{e}$, the formulation above becomes

$$\boldsymbol{M} = \boldsymbol{e} \times \boldsymbol{F} \tag{3.16}$$

Since $\boldsymbol{e}$ is in the plane of the couple, it is clear from the definition of a cross product that $\boldsymbol{M}$ is in an orientation normal to the plane of the couple. The sense in this case may be seen in Fig. 3.18 to be directed downward, in accordance with the right-hand-screw rule. Note the use of the double arrow to represent the couple moment. Note also that the rotation of $\boldsymbol{e}$ to $\boldsymbol{F}$, as stipulated in the cross-product formulation, is in the same direction as the "turning" action of the two force vectors, and from now on we shall use the latter criterion for determining the sense of rotation to be used with the right-hand-screw rule.

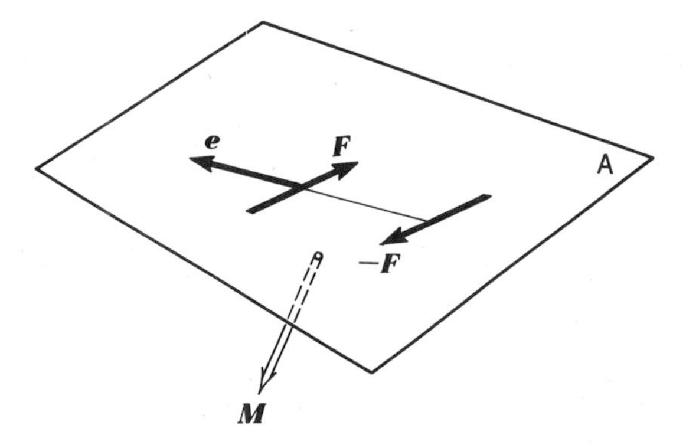

Figure 3.18. The couple moment $\boldsymbol{M}$.

Now that the direction of couple moment $\boldsymbol{M}$ has been established for the couple, we need only compute the magnitude for a complete description. Points 1 and 2 may be chosen anywhere along the lines of action of the forces without changing the resulting moment, since the forces are transmissible for this computation. Therefore, to compute the magnitude of the couple moment vector it will be simplest to choose positions 1 and 2 so that $\boldsymbol{e}$ is *perpendicular* to the lines of action of the forces ($\boldsymbol{e}$ is then denoted as $\boldsymbol{e}_\perp$). From the definition of the cross product, we can then say:

$$|\boldsymbol{M}| = |\boldsymbol{e}_\perp||\boldsymbol{F}| \sin 90° = |\boldsymbol{e}_\perp||\boldsymbol{F}| = |\boldsymbol{F}|d \tag{3.17}$$

where the more familiar notation, d, has been used in place of $|\boldsymbol{e}_\perp|$ as the perpendicular distance between the lines of action of the forces.

> To summarize the preceding discussions, we may say that: The moment of a couple is a vector whose orientation is normal to the plane of the couple and whose sense is determined in accordance with the right-hand-screw rule, using the "turning" action of the forces to give the proper rotation. The magnitude of the couple moment equals the product of either force magnitude comprising the couple times the perpendicular distance between the forces.

3.5 The Couple Moment as a Free Vector

Had we chosen any other position in space as the origin, and had we computed the moment of the couple about it, we would have formed the same moment vector. To understand this, note that although the position vectors to points 1 and 2 will change for a new origin, the *difference* between these vectors (which has been termed $\boldsymbol{e}$) does *not* change, as can readily be observed in Fig. 3.19. Since $\boldsymbol{M} = \boldsymbol{e} \times \boldsymbol{F}$, we

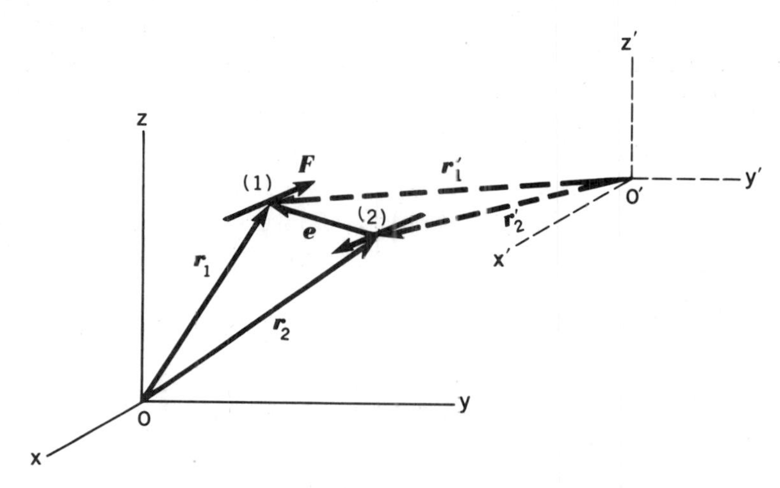

Figure 3.19. Vector $\boldsymbol{e}$ is the same for both references.

can conclude that *the couple has the same moment about every point in space.* The particular line of action of the vector representation of the couple moment that is illustrated in Fig. 3.18 is then of little significance and can be moved anywhere. In short, the *couple moment is a free vector.* That is, we may move this vector anywhere in space without changing its meaning, provided that we keep the direction and magnitude intact. Consequently, *for the purpose of taking moments,* we may move the couple itself anywhere in its own or a parallel plane, provided that the direction of turning is not altered—i.e., we cannot "flip" the couple over. In any of these possible planes, we can also change the magnitude of the forces of the couple to other equal values, provided that the distance d is simultaneously changed so that the product $|\boldsymbol{F}|d$ remains the same. Since none of these steps changes the direction or magnitude of the couple moment, all of them are permissible.

As we pointed out earlier, the only effect of a couple on a rigid body is its turning action, which is represented quantitatively by the moment of the couple—i.e., the couple moment. Since this is so often its sole effect, it is only natural to represent the couple by specifications of its moment; its mag-

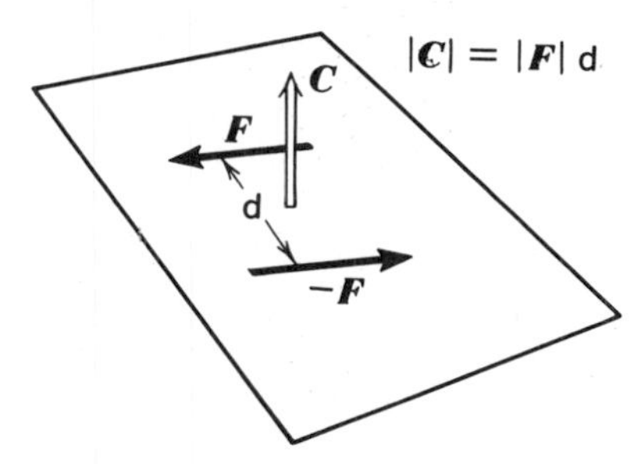

Figure 3.20. $\boldsymbol{C}$ represents couple.

nitude, then, becomes $|\boldsymbol{F}|d$ and its direction that of its moment. This is the same as identifying a person by her/his job (i.e., as a teacher, plumber, etc.). Thus, in Fig. 3.20, the couple moment $\boldsymbol{C}$ may be used to represent the indicated couple.

3.6 Addition and Subtraction of Couples

Since couples themselves have zero net forces, addition per se of couples always yields zero force. For this reason, the addition and subtraction of couples is interpreted to mean addition and subtraction of the *moments* of the couples. Since couple moments are free vectors, we can always arrange to have a concurrent system of vectors. We shall now take the opportunity to illustrate many of the earlier remarks about couples by adding the two couples shown on the face of the cube in Fig. 3.21. Notice that the couple moment vectors of the couples have been drawn. Since these vectors are free, they may be moved to a convenient position and then added. The total couple moment then becomes 103.2 lb-ft at an angle of 76° with the horizontal, as shown in Fig. 3.22. The couple that creates this turning action is in a plane at right angles to this orientation with a clockwise sense as observed from below.

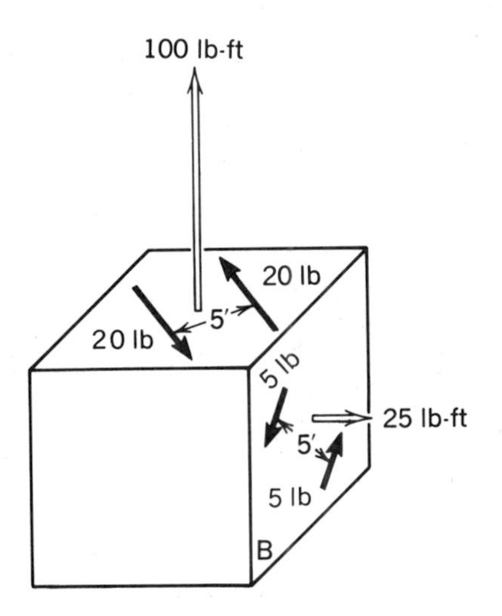

Figure 3.21. Add couples.

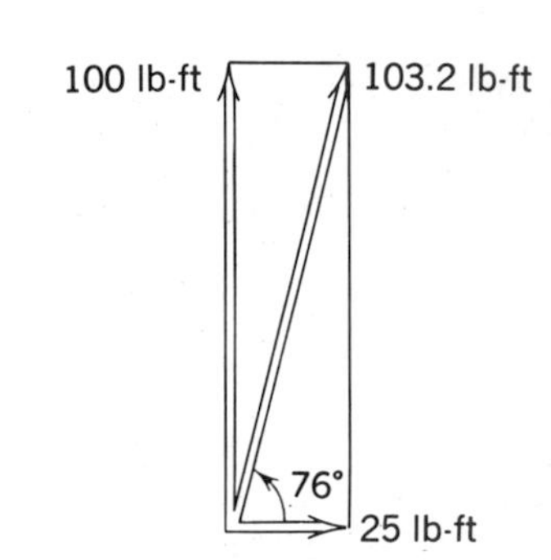

Figure 3.22. Add couple moments.

This addition may be shown to be valid by the following more elementary procedure. The couples of the cube are moved in their respective planes to the positions shown in Fig. 3.23, which does not alter the moment of the couples, as pointed out in Section 3.5. If the couple on plane B is adjusted to have a force magnitude of 20 lb and if the separating distance is decreased to $\frac{5}{4}$ ft, the couple moment is not changed (Fig. 3.24). We thus form a system of forces in which two of the forces are equal, opposite, and collinear and, since these two forces cannot contribute moment, they

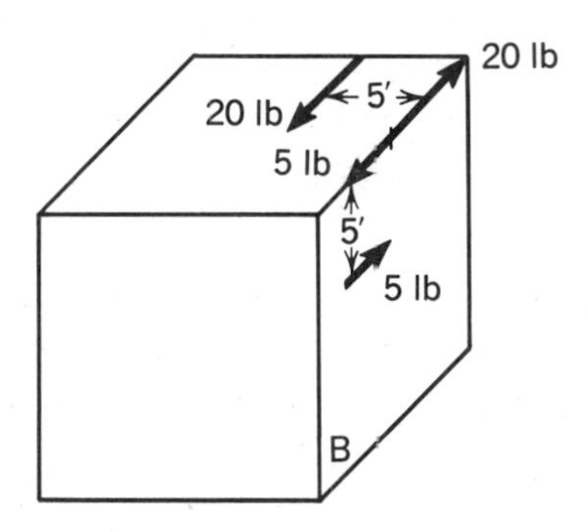

Figure 3.23. Move couples.

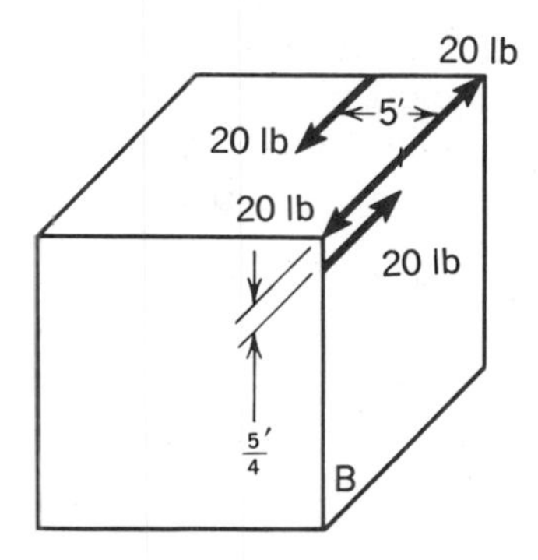

Figure 3.24. Change values of two forces.

may be deleted, leaving a single couple on a plane inclined to the original planes (Fig. 3.25). The distance between the remaining forces is

$$\sqrt{25 + \tfrac{25}{16}}\ \text{ft} = 5.16\ \text{ft}$$

and so the magnitude of the couple moment may then be computed to be 103.2 lb-ft. The orientation of the normal to the plane of the couple is readily evaluated as 76° with the horizontal, making the total couple moment identical to our preceding result.

A common notation for couples is shown in Fig. 3.26. The values given will be that of the couple moments.

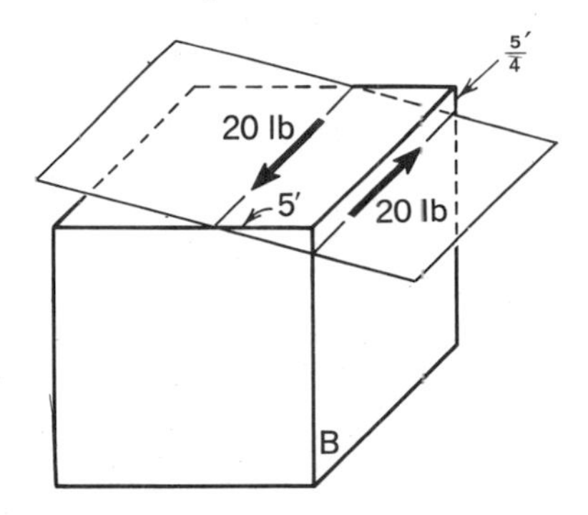

Figure 3.25. Eliminate collinear 20-lb forces.

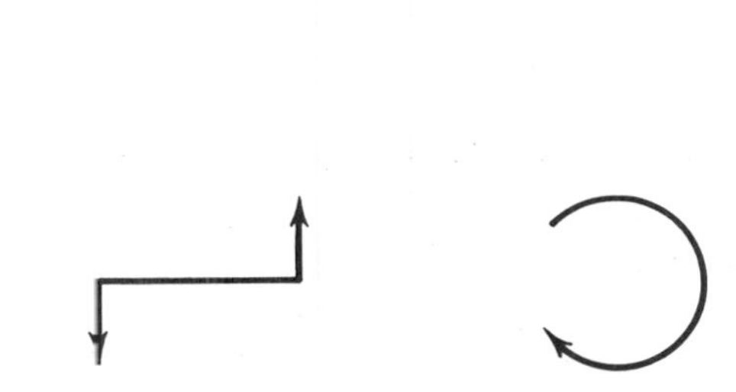

Figure 3.26. Representation of couples.

3.7 Moment of a Couple About a Line

In a previous section, we pointed out that the moment of a force $\boldsymbol{F}$ about a line A–A (see Fig. 3.27) is found by first taking the moment of $\boldsymbol{F}$ about *any* point P on A–A and then dotting this vector into $\boldsymbol{a}$, the unit vector along the line. That is,

$$M_{AA} = (\boldsymbol{r} \times \boldsymbol{F}) \cdot \boldsymbol{a} \tag{3.18}$$

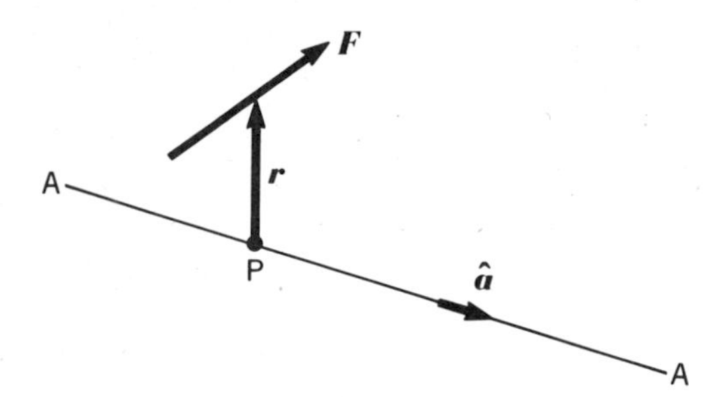

Figure 3.27. To find moment of ***F*** about *A–A*.

Consider now the moment of a couple about a line. For this purpose, we show a couple moment ***C*** and line *A–A* in Fig. 3.28. As before, we want first the moment of the couple about any point *P* along *A–A*. But the moment of ***C*** about *every* point in space is simply ***C*** itself. Therefore, to get the moment about the line *A–A* all we need do is dot ***C*** into ***a***. Thus,

$$M_{AA} = \mathbf{C} \cdot \mathbf{a} \tag{3.19}$$

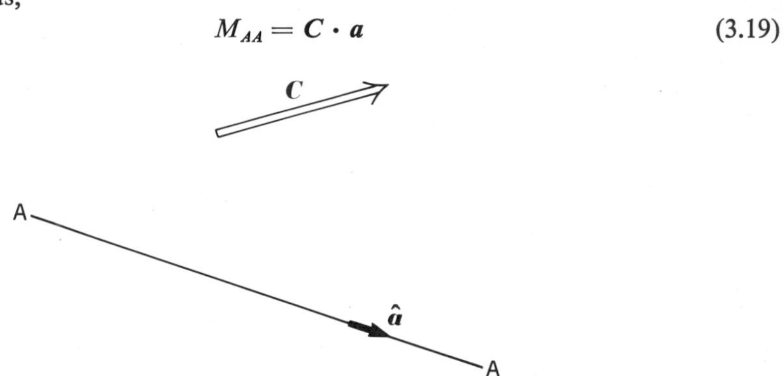

Figure 3.28. To find moment of couple about *A–A*.

Since ***C*** is a free vector, the moments of ***C*** about all lines parallel to *A–A* must have the same value.

Problems

3.32. A truck driver, while changing a tire, must tighten the nuts holding on the wheel using a torque of 80 lb-ft. If his tire tool has a length such that the forces from his hand are 22 in. apart, how much force must he exert with each hand? To remove the nuts, he exerted 70 lb with each hand. What torque did he apply?

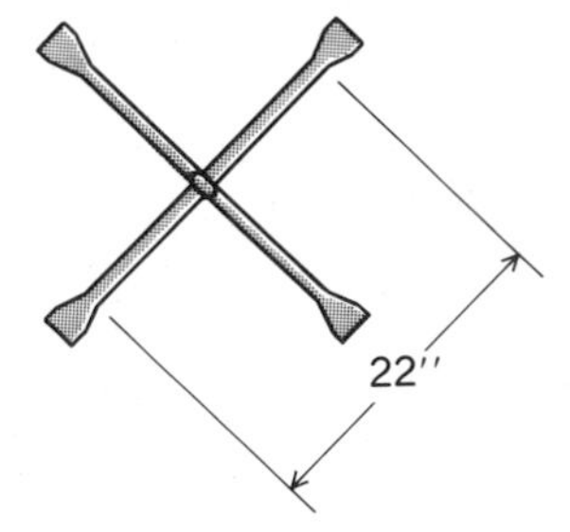

Figure P.3.32

3.33. Equal couples in the plane of a wheel are shown in Fig. P.3.33. Explain why they are equivalent for the purpose of turning the wheel. Are they equivalent from the viewpoint of the deformation of the wheel? Explain.

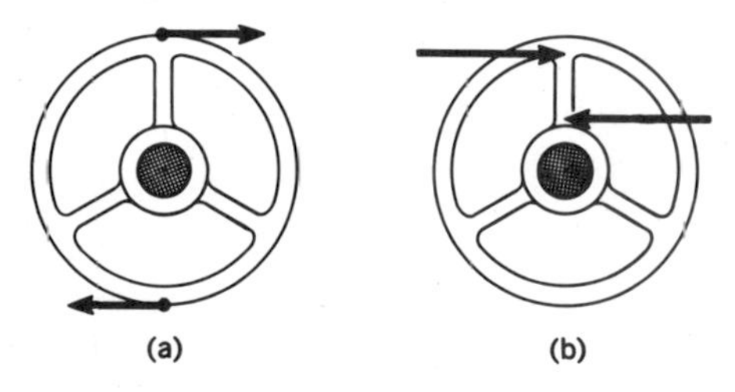

Figure P.3.33

3.34. Oil-field workers can exert between 50 lb and 125 lb with each hand on a valve wheel (one hand on each side). If a couple moment of 100 lb-ft is required to close the valve, what diameter d must the wheel have?

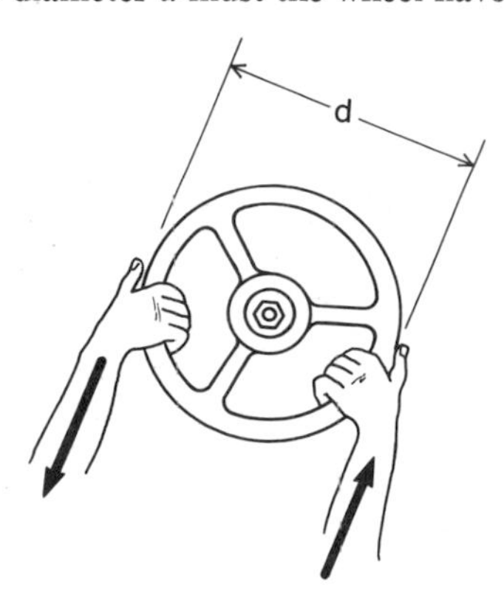

Figure P.3.34

3.35. Two children push with 30 lb of force each on the rail of a 10-ft-diameter merry-go-round. What couple moment do they produce? They fasten a 20-ft-long 2-in. by 4-in. board to the merry-go-round so that the middle of the board is at the middle of the merry-go-round. What is the resulting couple moment if they push on the ends of the board? What moment about the merry-go-round axis would they generate by fastening one end of the board to the middle of the merry-go-round and both pushing in the same direction on the other end?

3.36. A posthole digger has a 2-ft-long handle on a 5-ft-long shaft fastened to the digging (scraping) base. From tests, we know that a couple moment of 100 lb-ft is required to dig a posthole in clay, but only 65 lb-ft is needed in sandy soil. What force F must be applied in each case to dig a hole if the distance between the forces from a person's hands is 20 in.?

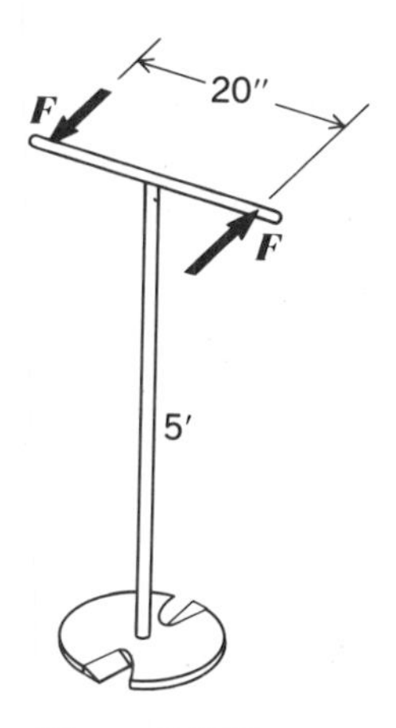

Figure P.3.36

3.37. While stopping, a truck develops a 350 N-m of torque at the rear axle due to the action of the brake drum on the axle. What forces are generated at the front and rear supports of the springs to which the axle is attached?

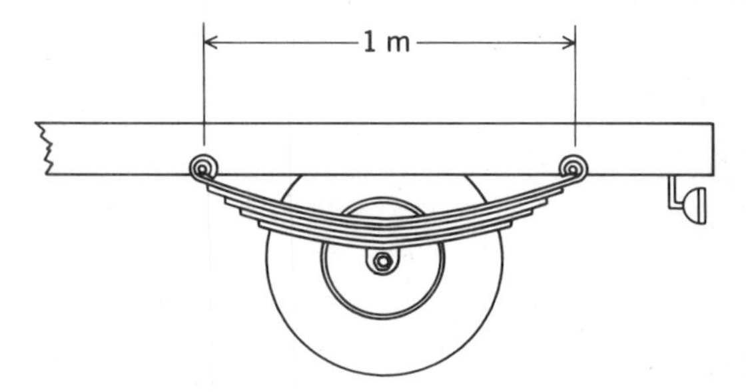

Figure P.3.37

3.38. A couple is shown in the yz plane. What is the moment of this couple about the origin?

About point (6, 3, 4) m? What is the moment of the couple about a line through the origin with direction cosines $l = 0$, $m = .8$, $n = -.6$? If this line is shifted to a parallel position so that it goes through point (6, 3, 4) m, what is the moment of the couple about this line?

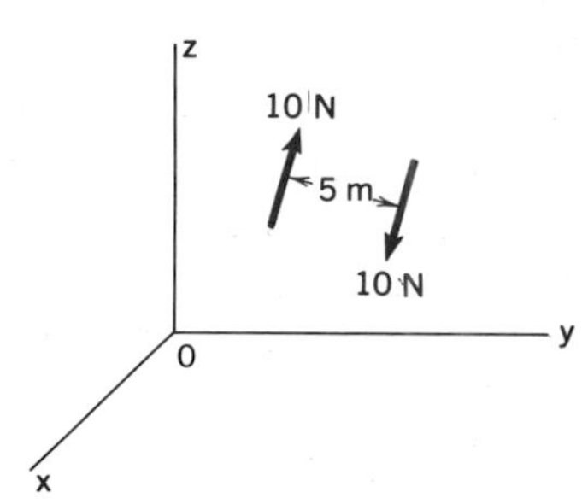

Figure P.3.38

3.39. Given the indicated forces, what is the moment of these forces about points *A* and *B*?

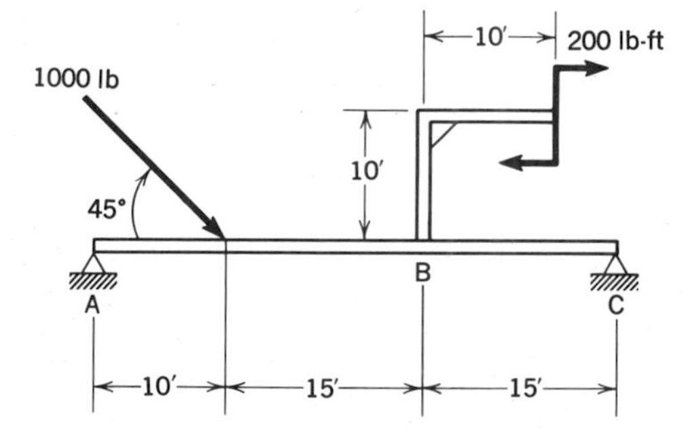

Figure P.3.39

3.40. Consider the steering mechanism for a go-cart. The linkages are all in a plane at 45° to the horizontal, which is perpendicular to the steering shaft. In a hard turn, the driver exerts 30 lb with each hand on the 12-in.-diameter steering wheel. What forces develop at the joints B and E of the steering mechanism? What couple moment is applied to each wheel normal to the ground? Assume half the transmitted torque goes to each wheel.

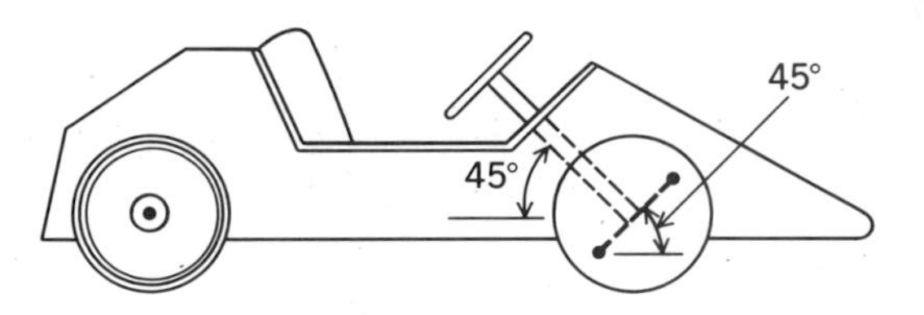

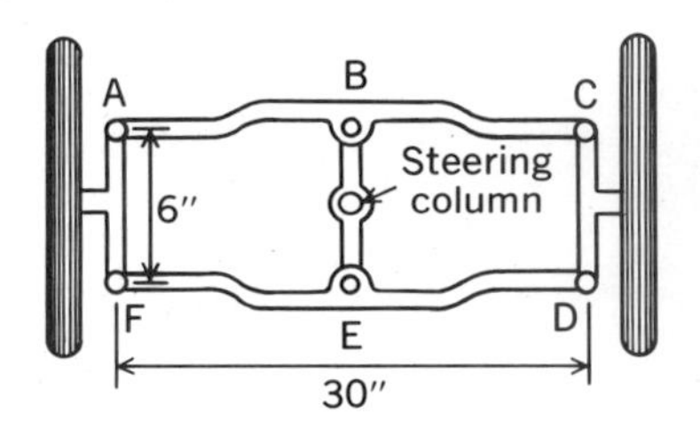

View looking down steering column

Figure P.3.40

3.41. An eight-bladed windmill used for power generation and pumping water stops turning because a bearing on the blade shaft has "frozen up." However, the wind still blows, so each blade is subjected to a 25-lb force perpendicular to the (flat) blade surface. The force effectively acts at 2 ft from the shaft to which the blades are attached. The blades are inclined at 30° to the axis of rotation. What is the total thrust of all the blade forces on the windmill shaft? What is the moment on the stalled shaft?

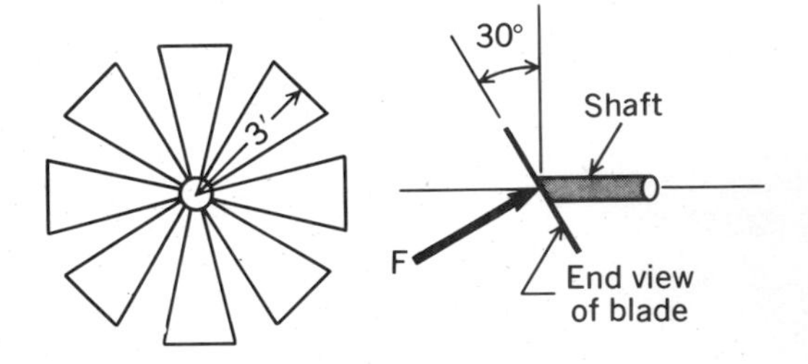

Figure P.3.41

3.42. What is the moment of the forces shown about point *A* and about a point *P* having a position vector

$$\boldsymbol{r}_p = 10\boldsymbol{i} + 7\boldsymbol{j} + 15\boldsymbol{k} \text{ m?}$$

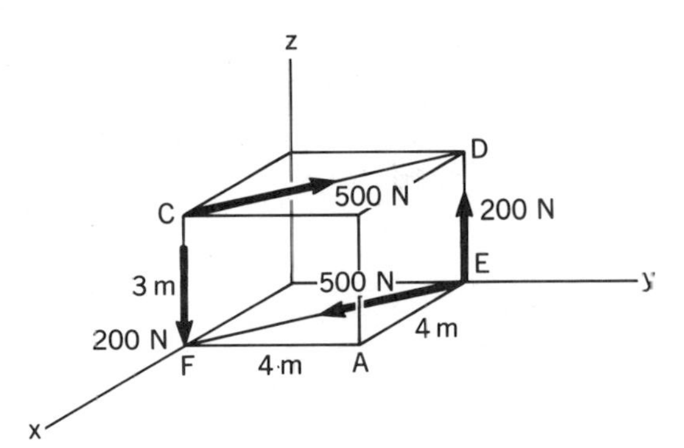

Figure P.3.42

3.43. A force $F_1 = 10i + 6j + 3k$ N acts at position (3, 0, 2) m. At point (0, 2, −3) m, an equal but opposite force $-F_1$ acts. What is the couple moment? What are the direction cosines of the normal to the plane of the couple?

3.44. Force $F_1 = 16i - 10j + 5k$ N acts at the origin while $F_2 = -F_1$ acts at the end of a rod of length 12 m protruding from the origin with direction cosines $l = .6$, $m = .8$. What is the moment about point P at

$$r_p = 3i + 10j + 15k \text{ m?}$$

What is the twist about an axis going through P having the unit vector

$$\epsilon = .2i + .8j + .566k?$$

3.45. Equal and opposite forces are directed along diagonals on the faces of a cube. What is the couple moment if $a = 3$ m and $F = 10$ N? What is the moment of this couple about a diagonal from A to D?

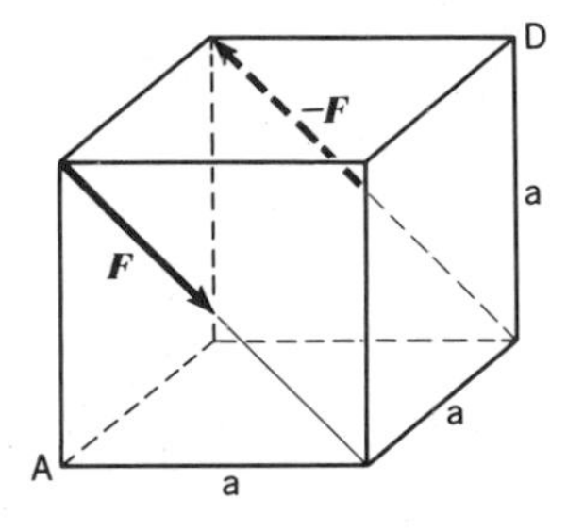

Figure P.3.45

3.46. What is the turning action of the forces shown about the diagonal A–D?

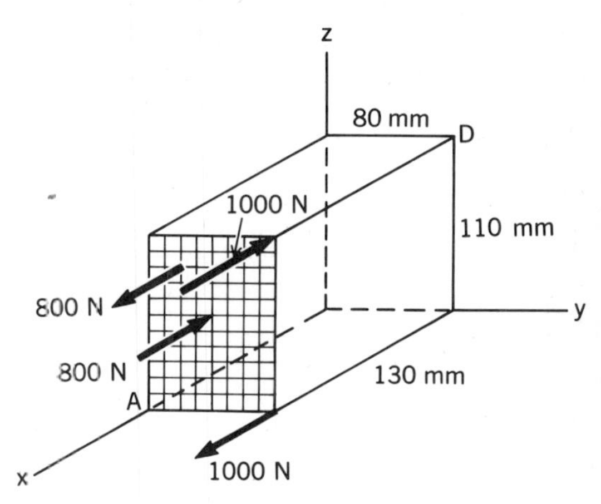

Figure P.3.46

3.47. What is the total couple moment of the three couples shown? What is the moment of this force system about point (3, 4, 2) ft? What is the moment of this force system about the position vector $r = 3i + 4j + 2k$ ft? What is the total force of this system?

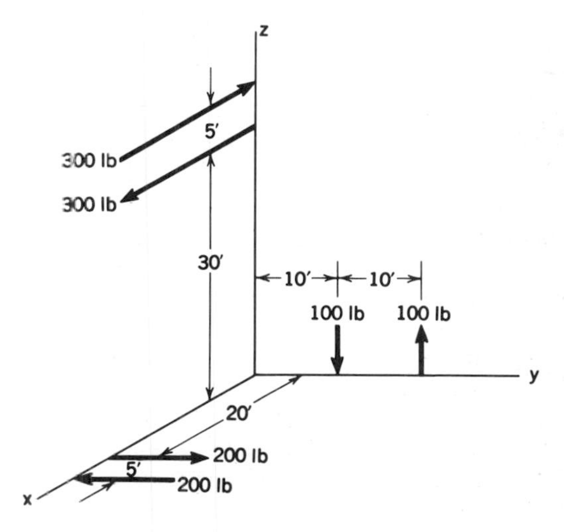

Figure P.3.47

3.48. An oil-field pump has two valves, one on top and one on the side, that must be closed simultaneously. The valve wheels are each 27 in. in diameter and are turned with both hands by workers who can exert between 50 lb and 125 lb with each hand. If a weak worker turns the side wheel and a strong worker turns the top wheel, what is the total twisting moment (couple moment) on the pump?

3.49. What is the total moment of the force system shown about the origin?

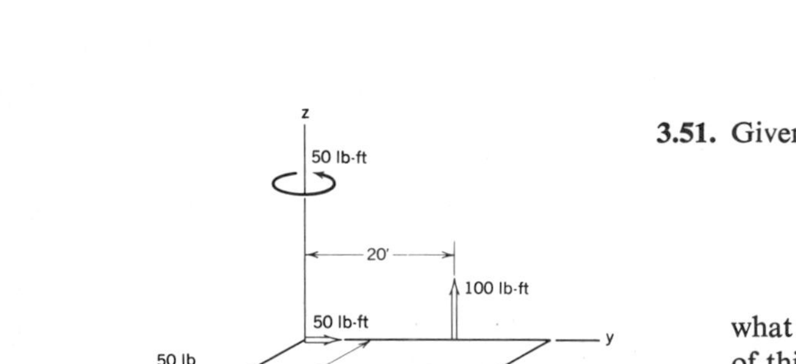

Figure P.3.49

3.50. Add the couples whose forces act along diagonals of the sides of the rectangular parallelepiped.

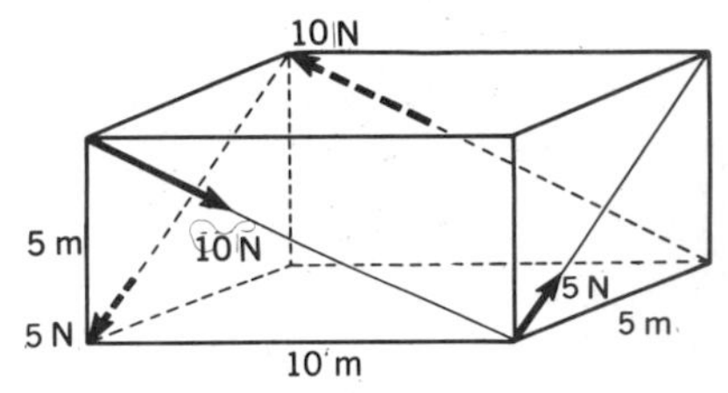

Figure P.3.50

3.51. Given the couple moments

$$C_1 = 100\boldsymbol{i} + 30\boldsymbol{j} + 82\boldsymbol{k} \text{ lb-ft}$$

$$C_2 = -16\boldsymbol{i} + 42\boldsymbol{j} \text{ lb-ft}$$

$$C_3 = 15\boldsymbol{k} \text{ lb-ft}$$

what couple will restrain the twisting action of this system about an axis going from

$$\boldsymbol{r}_1 = 6\boldsymbol{i} + 3\boldsymbol{j} + 2\boldsymbol{k} \text{ ft}$$

to

$$\boldsymbol{r}_2 = 10\boldsymbol{i} - 2\boldsymbol{j} + 3\boldsymbol{k} \text{ ft}$$

while giving a moment of 100 lb-ft about the x axis and 50 lb-ft about the y axis?

3.8 Closure

In this chapter, we have considered several important vector quantities and their properties. In particular, for rigid bodies we could take certain liberties with a couple without invalidating the results. We are now ready to pursue in greater detail the important subject of equivalence of force systems for rigid-body considerations.

Review Problems

3.52. A surveyor on a 100-m-high hill determines that the corner of a building at the base of the hill is 600 m east and 1500 m north of her position. What is the position of the building corner relative to another surveyor on top of a 5000-m-high mountain that is 10,000 m west and 3000 m south of the hill? What is the distance from the second surveyor to the building corner?

3.53. Compute the moment of the 1000-lb force about supporting points A and B.

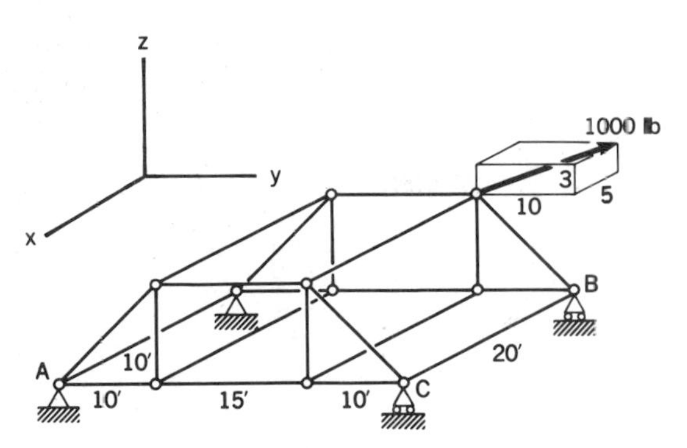

Figure P.3.53

3.54. An A-frame for hoisting and dragging equipment is held in the position shown by a cable C. To determine the cable force, the moment of the applied force about axis B–B must be known. What is that moment when a 1000-N force is applied as shown?

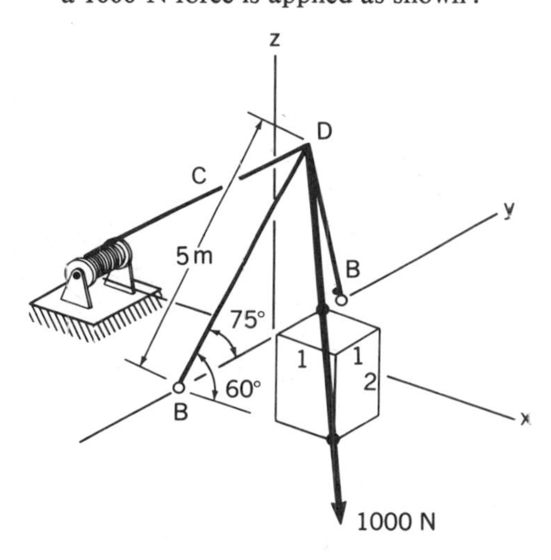

Figure P.3.54

3.55. A plumber places his hands 18 in. apart on a pipe threader and can push (and pull) with 80 lb of force. What couple moment does he exert? How much could he exert if he moved his hands to the ends so that his hands are 24 in. apart? What force must he apply at the ends to achieve the same couple moment as when he held his hands 18 in. apart?

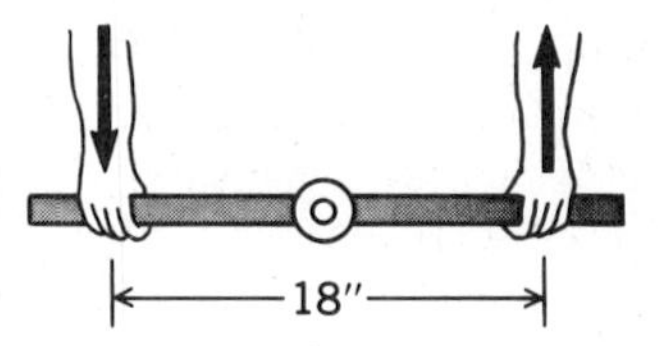

Figure P.3.55

3.56. What is the moment about A of the 500-N force and the 3000-N-m couple acting on the cantilever beam?

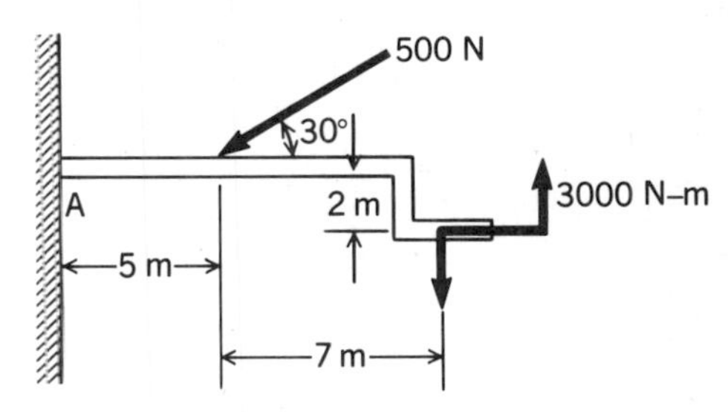

Figure P.3.56

3.57. A force $\boldsymbol{F} = 16\boldsymbol{i} + 10\boldsymbol{j} - 3\boldsymbol{k}$ lb goes through point a having a position vector $\boldsymbol{r}_a = 16\boldsymbol{i} - 3\boldsymbol{j} + 12\boldsymbol{k}$ ft. What is the moment about an axis going through points 1 and 2 having respective position vectors given as

$$\boldsymbol{r}_1 = 6\boldsymbol{i} + 3\boldsymbol{j} - 2\boldsymbol{k} \text{ ft}$$

$$\boldsymbol{r}_2 = 3\boldsymbol{i} - 4\boldsymbol{j} + 12\boldsymbol{k} \text{ ft?}$$

3.58. Replace the system of forces and couples by a single couple. Note that the 1000-N-m couple is in the diagonal plane $ABCD$.

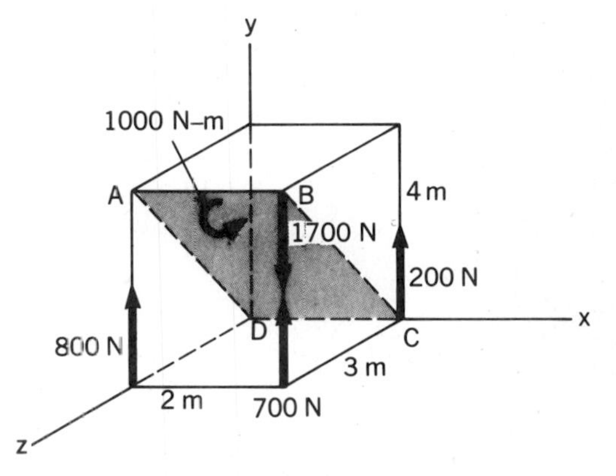

Figure P.3.58

3.59. Find the torque of the force system about axis AB.

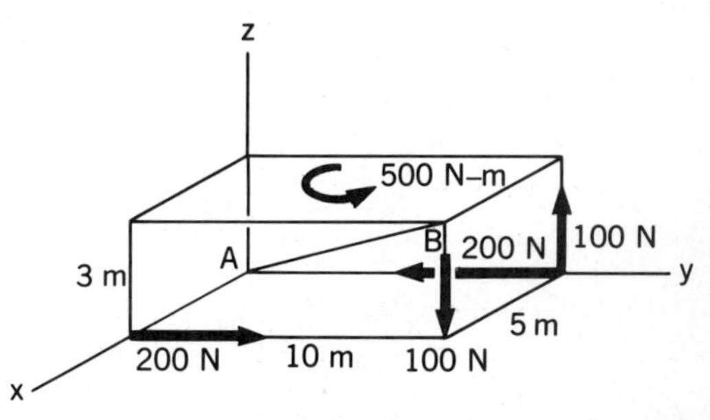

Figure P.3.59

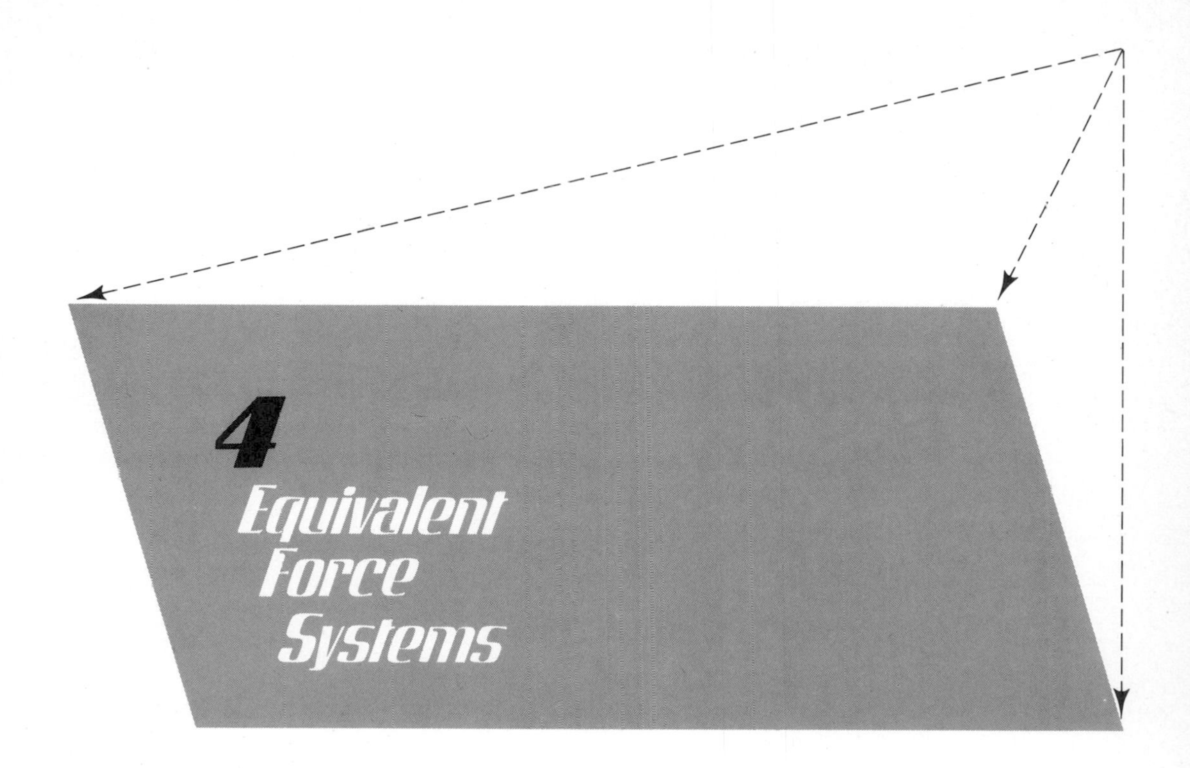

4.1 Introduction

In Chapter 1, we defined equivalent vectors as those that have the same capacity in some given situation. We shall now investigate an impórtant class of situations, those in which a rigid-body model can be employed. We will be concerned with the equivalence requirements for force systems acting on a rigid body.

The effect that forces have on a rigid body is only manifested in the motion (or lack of motion) of the body induced by the forces. Two force systems, then, are equivalent if they are capable of *initiating* the same motion of the rigid body. The conditions required to give two force systems this equal capacity are:

1. Each force system must exert an equal "push" or "pull" on the body in any direction. For two concurrent force systems, this requirement is satisfied if the add tion of the forces in each system results in equal force vectors.
2. Each force system must exert an equal "turning" action about any point in space. This means that the moment vectors of the force systems for any chosen point must be equal.

Although these conditions will most likely be intuitively acceptable to the reader, we shall later prove them to be necessary and, for certain situations, sufficient for equivalence when we study dynamics.

As a beginning here, we shall reiterate several basic force equivalences for rigid bodies that will serve as a foundation for more complex cases. You should subject them to the tests listed above.

1. The sum of a set of concurrent forces is a single force that is equivalent to the original system. Conversely, a single force is equivalent to any set of its components.
2. A force may be moved along its line of action (i.e., forces are transmissible vectors).
3. The only effect that a couple develops on a rigid body is embodied in the couple moment. Since the couple moment is always a free vector, for our purposes at present the couple may be altered in any way as long as the couple moment is not changed.

In succeeding sections, we shall present other equivalence relations for rigid bodies and then examine perfectly general force systems with a view to replacing them with more convenient and simpler equivalent force systems. These simpler replacements are often called *resultants* of the more general systems.

4.2 Translation of a Force to a Parallel Position

In Fig. 4.1, let us consider the possibility of moving a force $\boldsymbol{F}$ (solid arrow) acting on a rigid body to a parallel position at point a while maintaining rigid-body equivalence. If at position a we apply equal and opposite forces, one of which is $\boldsymbol{F}$ and the other $-\boldsymbol{F}$, a system of three forces is formed that is clearly equivalent to the single force $\boldsymbol{F}$. Note that the original force $\boldsymbol{F}$ and the new force in the opposite sense form a couple

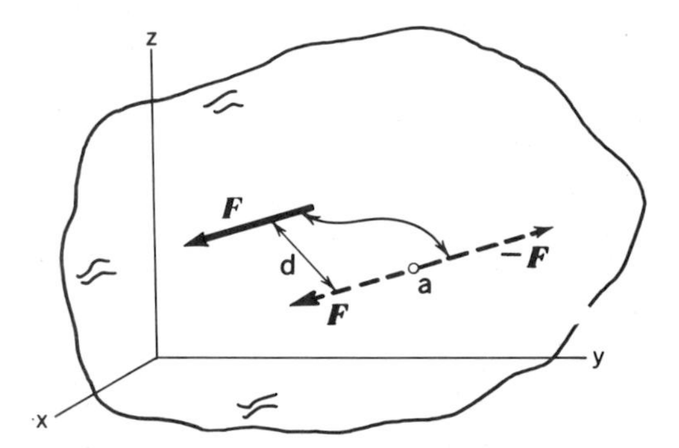

Figure 4.1. Insert equal and opposite forces at *a*.

(the pair is identified by a wavy connecting line). As usual, we represent the couple by its moment $\boldsymbol{C}$, as shown in Fig. 4.2, normal to the plane A of point a and the original force $\boldsymbol{F}$. The magnitude of the couple moment $\boldsymbol{C}$ is $|\boldsymbol{F}|d$, where d is the perpendicular distance between point a and the original line of action of the force. The couple moment may be moved to any parallel position, including the origin, as indicated in Fig. 4.2.

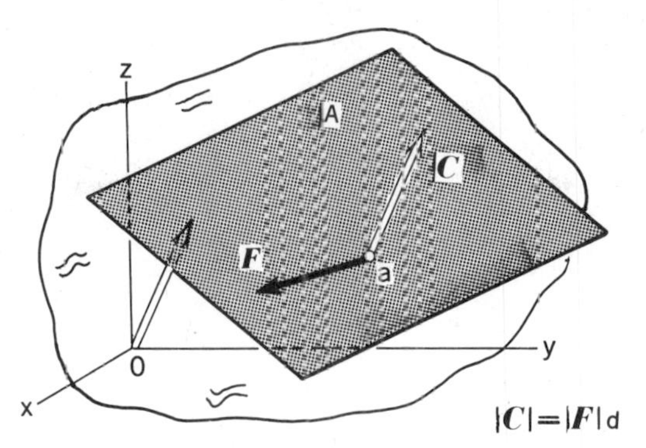

Figure 4.2. Equivalent system at *a*.

Thus, we see that *a force may be moved to any parallel position, provided that a couple moment of the correct orientation and size is simultaneously provided.* There are, then, an infinite number of arrangements possible to get the equivalent effects of a single force on a rigid body.

The reverse procedure may also be instituted in reducing a force and a couple *in the same plane* to a *single* equivalent force. This is illustrated in Fig. 4.3, where a couple composed of forces $\boldsymbol{B}$ and $-\boldsymbol{B}$ a distance d_1 apart and a force $\boldsymbol{A}$ are shown in plane N. The moment representation of the couple is shown with force $\boldsymbol{A}$ in Fig. 4.4.

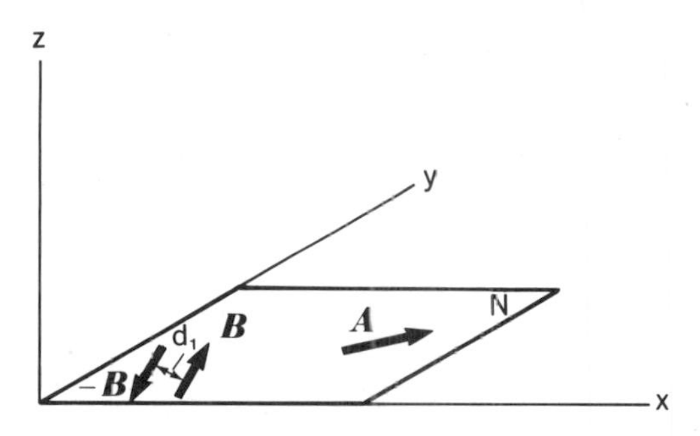

Figure 4.3. Force *A* and couple in same plane.

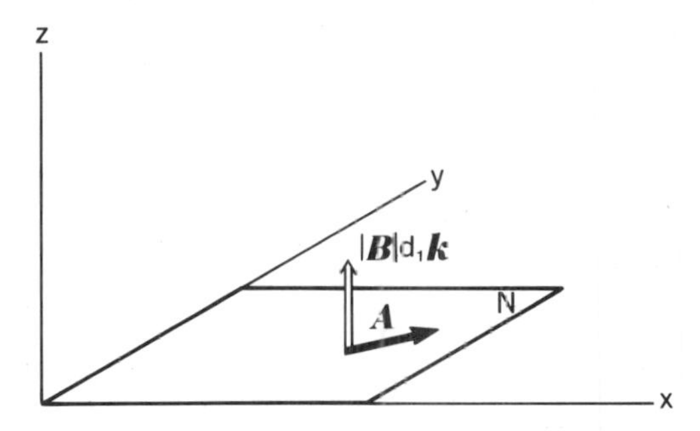

Figure 4.4. Force *A* and couple moment.

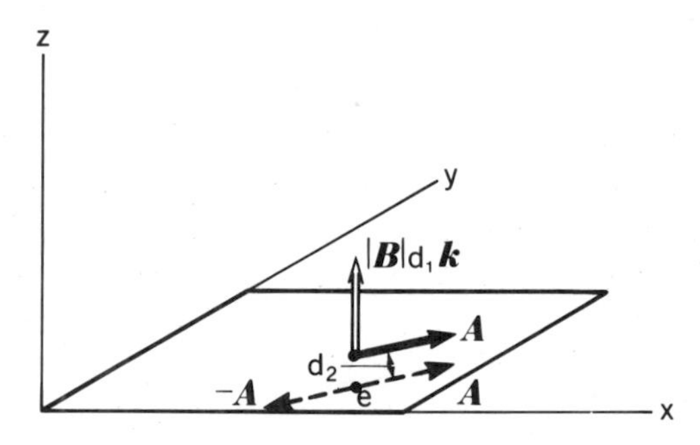

Figure 4.5. Equal and opposite forces placed at *e*.

Equal and opposite forces $\boldsymbol{A}$ and $-\boldsymbol{A}$ may next be added to the system at a position *e* (see Fig. 4.5). The purpose of this step is to form another couple moment with a magnitude $|\boldsymbol{A}|d_2$ equal to $|\boldsymbol{B}|d_1$ and with a direction of turning opposite to the original couple moment (see Fig. 4.6). The couple moments then cancel each other out, and we have, in effect, only the single force $\boldsymbol{A}$ going through point *e*.

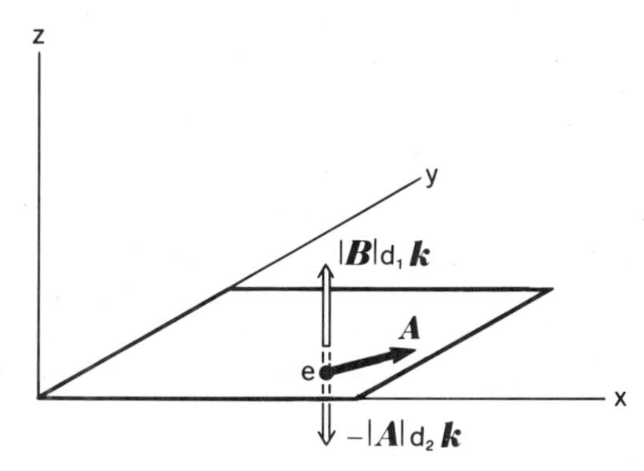

Figure 4.6. Adjust d_2 so that couple moments cancel.

We now present a simple method for computing the couple moment developed on moving a force to a parallel position. Return to Fig. 4.1 and compute the moment $\boldsymbol{M}$ of the force $\boldsymbol{F}$ about point *a*. We can express this as [see Fig. 4.7(a)]

$$\boldsymbol{M} = \boldsymbol{\rho} \times \boldsymbol{F} \tag{4.1}$$

where $\boldsymbol{\rho}$ is a position vector from *a* to any point along the line of action of $\boldsymbol{F}$. Now the equivalent force system, shown in Fig. 4.7(b), must have the *same moment*, $\boldsymbol{M}$, about point *a* as the original system. Clearly, the moment about point *a* is due only to the couple moment $\boldsymbol{C}$. That is,

$$\boldsymbol{M} = \boldsymbol{C} \tag{4.2}$$

Accordingly, we conclude, on comparing the previous two equations, that

$$\boldsymbol{C} = \boldsymbol{\rho} \times \boldsymbol{F}$$

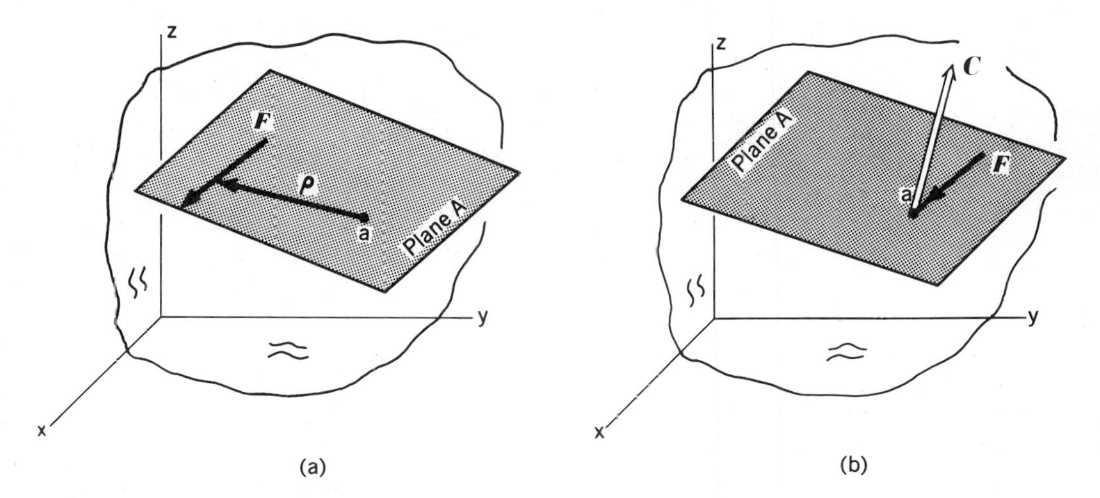

Figure 4.7. Couple moment on moving $\boldsymbol{F}$ is $\boldsymbol{\rho} \times \boldsymbol{F}$.

Thus, in shifting a force to pass through some new point, *we introduce a couple whose couple moment equals the moment of the force about this new point.*

We illustrate this in the following example.

EXAMPLE 4.1

A force $\boldsymbol{F} = 6\boldsymbol{i} + 3\boldsymbol{j} + 6\boldsymbol{k}$ lb goes through a point whose position vector is $\boldsymbol{r}_1 = 2\boldsymbol{i} + \boldsymbol{j} + 10\boldsymbol{k}$ ft (see Fig. 4.8). Replace this force by an equivalent force system, for purposes of rigid-body mechanics, going through position P, whose position vector is $\boldsymbol{r}_2 = 6\boldsymbol{i} + 10\boldsymbol{j} + 12\boldsymbol{k}$ ft.

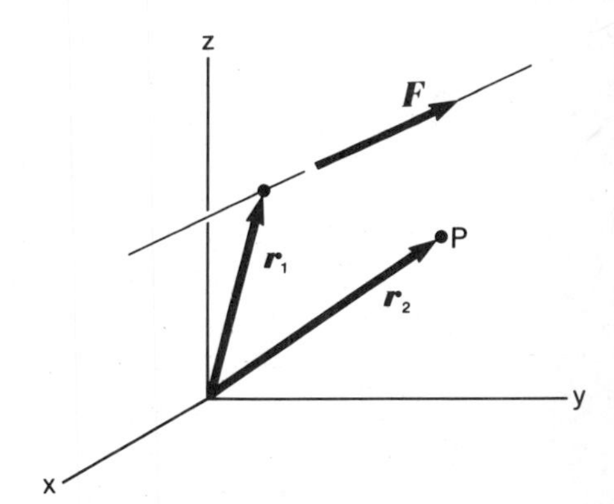

Figure 4.8. Move $\boldsymbol{F}$ to point P.

The new system will consist of this force $\boldsymbol{F}$ going through the position $\boldsymbol{r}_2$ and, in addition, there will be a couple moment $\boldsymbol{C}$ given as

$$\boldsymbol{C} = \boldsymbol{\rho} \times \boldsymbol{F} = (\boldsymbol{r}_1 - \boldsymbol{r}_2) \times \boldsymbol{F}$$

Inserting values, we have

$$\begin{aligned} \boldsymbol{C} &= [(2\boldsymbol{i} + \boldsymbol{j} + 10\boldsymbol{k}) - (6\boldsymbol{i} + 10\boldsymbol{j} + 12\boldsymbol{k})] \\ &\quad \times (6\boldsymbol{i} + 3\boldsymbol{j} + 6\boldsymbol{k}) \\ &= (-4\boldsymbol{i} - 9\boldsymbol{j} - 2\boldsymbol{k}) \times (6\boldsymbol{i} + 3\boldsymbol{j} + 6\boldsymbol{k}) \end{aligned}$$

$$= \begin{vmatrix} -4 & -9 & -2 \\ 6 & 3 & 6 \\ \boldsymbol{i} & \boldsymbol{j} & \boldsymbol{k} \end{vmatrix} \begin{matrix} -4 & -9 & -2 \\ 6 & 3 & 6 \end{matrix}$$

Therefore,

$$C = -12k - 12j - 54i + 6i + 24j + 54k$$
$$= -48i + 12j + 42k \text{ ft-lb}$$

EXAMPLE 4.2

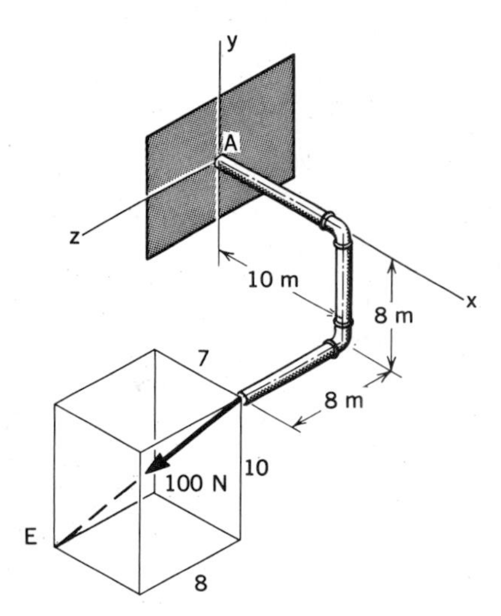

Figure 4.9. Find equivalent force system at A.

What is the equivalent force system at position A for the 100-N force shown in Fig. 4.9?

The 100-N force can be expressed vectorially as follows:

$$F = F\frac{\overrightarrow{BE}}{|\overrightarrow{BE}|} = 100\left(\frac{-7i - 10j + 8k}{\sqrt{7^2 + 10^2 + 8^2}}\right) \qquad \text{(a)}$$
$$= -48.0i - 68.5j + 54.8k \text{ N}$$

We then have the force given above at A. And in addition, we have a couple moment, C, found using a position vector, r, from A to any point along the line of action of the 100-N force. Thus, choosing point B for r, we have

$$C = (10i - 8j + 8k) \times (-48.0i - 68.5j + 54.8k)$$

$$= \begin{vmatrix} 10 & -8 & 8 \\ -48.0 & -68.5 & 54.8 \\ i & j & k \end{vmatrix}$$
$$\begin{matrix} 10 & -8 & 8 \\ -48.0 & -68.5 & 54.8 \end{matrix}$$

$$= (10)(-68.5)k + (-48)(8)j + (-8)(54.8)i$$
$$- (8)(-68.5)i - (54.8)(10)j - (-8)(-48.0)k$$

Therefore,

$$C = 109.6i - 932j - 1069k \text{ N-m} \qquad \text{(b)}$$

Problems

In several of the problems of this set we shall concentrate the weight of a body at its center of gravity. Most likely you are used to doing this from an earlier physics course. In Section 4.5 we shall justify this procedure.

4.1. Replace the 100-lb force by an equivalent system, from the rigid-body point of view, at A. Do the same for point B. Do this problem by the technique of adding equal and opposite collinear forces and also by using the cross product.

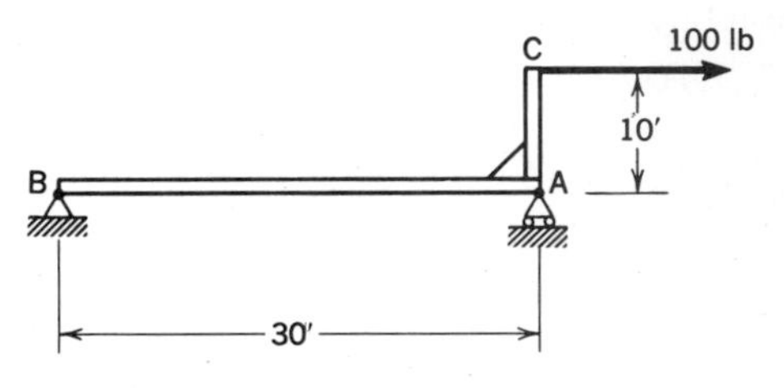

Figure P.4.1

4.2. To back an airplane away from the boarding gate, a tractor pushes with a force of 15 kN on the nose wheels. What is the equivalent force system on the landing-gear pivot point which is 2 m above the point where the tractor pushes?

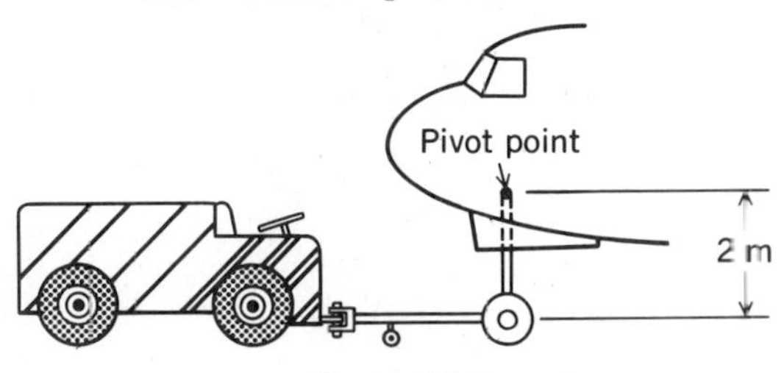

Figure P.4.2

4.3. Replace the 1000-lb force by equivalent systems at points A and B. Do so by using the addition of equal and opposite collinear force components and by using the cross product.

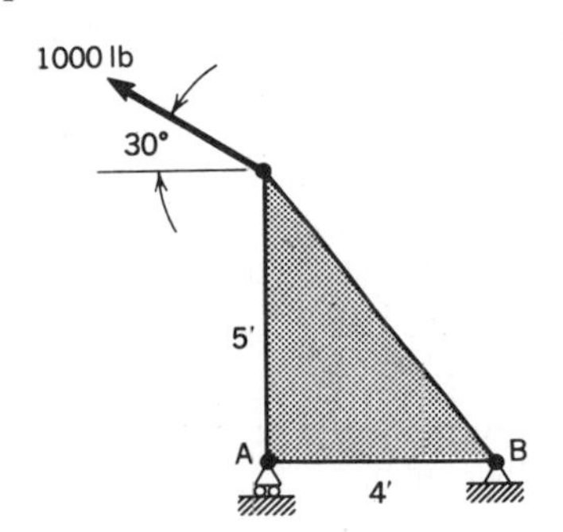

Figure P.4.3

4.4. A parking-lot gate arm weighs 150 N. Because of the taper, the weight is concentrated at a point $1\frac{1}{4}$ m from the pivot point. What is the equivalent force system at the gate-arm pivot point?

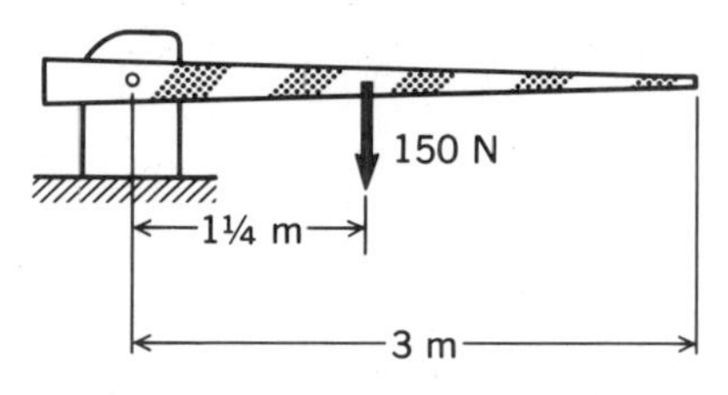

Figure P.4.4

4.5. A plumber exerts a vertical 60-lb force on a pipe wrench inclined at 30° to the horizontal. What force and couple moment on the pipe are equivalent to the plumber's action?

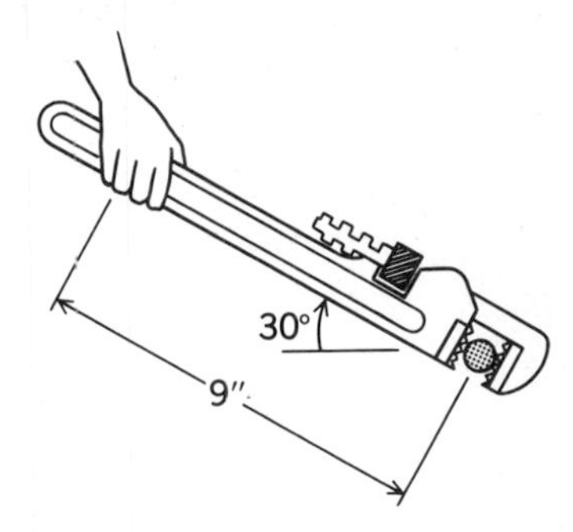

Figure P.4.5

4.6. A tractor operator is attempting to lift a 10-kN boulder. What are the equivalent force systems at A and at B from the boulder?

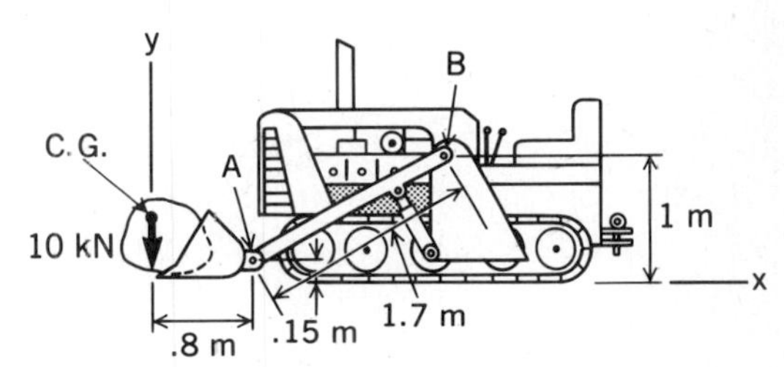

Figure P.4.6

4.7. A small hoist has a lifting capacity of 20 kN. What are the largest and smallest equivalent force systems at A for the rated maximum capacity?

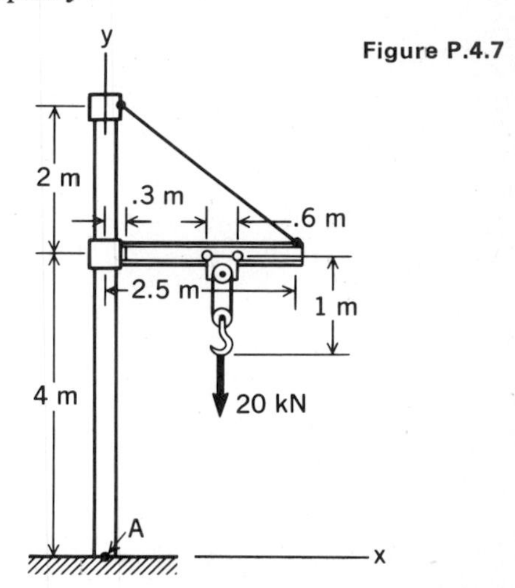

Figure P.4.7

4.8. Replace the forces by a single equivalent force.

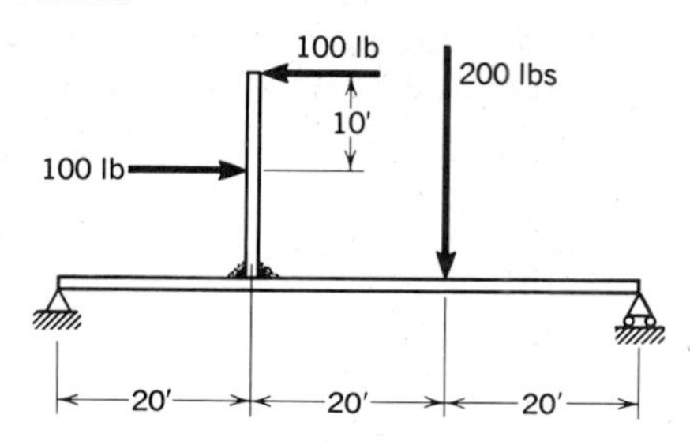

Figure P.4.8

4.9. Replace the forces and torques shown acting on the apparatus by a single force. Carefully give the line of action of this force.

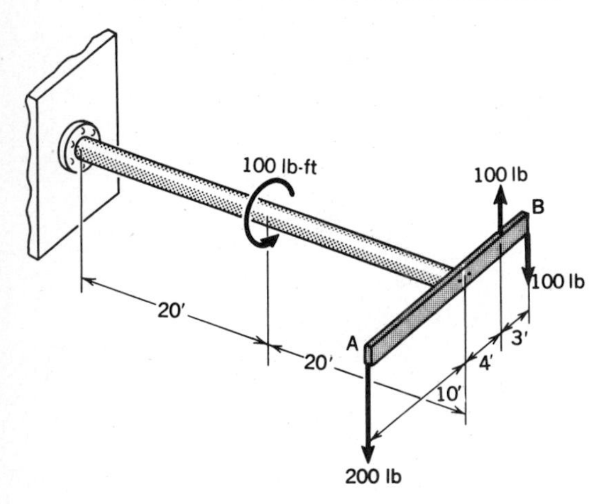

Figure P.4.9

4.10. A carpenter presses down on a brace-and-bit with a 150-N force while turning the brace with a 200-N force oriented for maximum twist. What is the equivalent force system on the end of the bit at A?

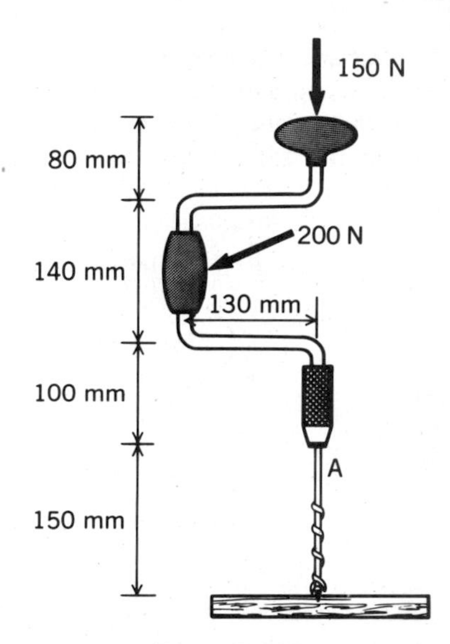

Figure P.4.10

4.11. A force $\boldsymbol{F} = 3\boldsymbol{i} - 6\boldsymbol{j} + 4\boldsymbol{k}$ lb goes through point (6, 3, 2) ft. Replace this force by an equivalent system where the force goes through point (2, −5, 10) ft.

4.12. A force $\boldsymbol{F} = 20\boldsymbol{i} - 60\boldsymbol{j} + 30\boldsymbol{k}$ N goes through a point (10, −5, 4) m. What is the equivalent system at point A having position vector $\boldsymbol{r}_A = 20\boldsymbol{i} + 3\boldsymbol{j} - 15\boldsymbol{k}$ m?

4.13. Find the equivalent force system at the base of the cantilever pipe system stemming from force $F = 1000$ lb.

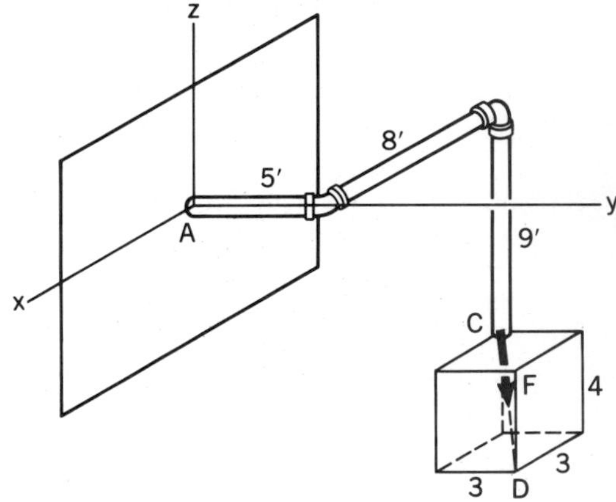

Figure P.4.13

4.14. In Problem 4.13, the pipe weighs 20 lb/ft. What is the equivalent force system at A from the weight of the pipe? [*Hint:* Concentrate the weights of the pipe sections at the respective centers of gravity (geometric centers in this case).]

4.15. The operator of a small boom-type crane is trying to drag a chunk of concrete. The boom is 10° above the horizontal and rotated 30° from the front of the crane. The cable is directed as shown in diagram and has 60 kN of tension. What is the equivalent force system at the boom pivot point?

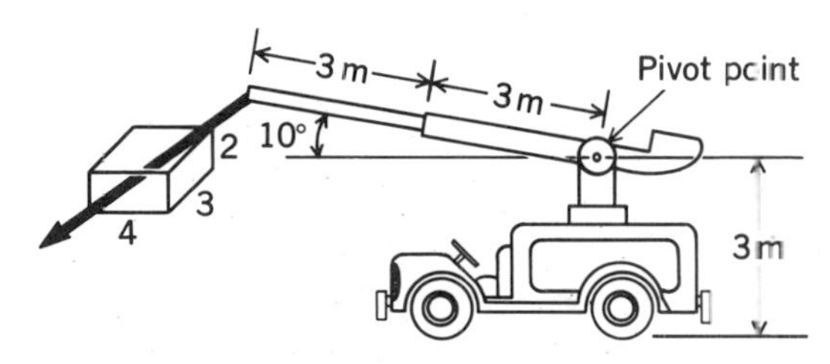

Figure P.4.15

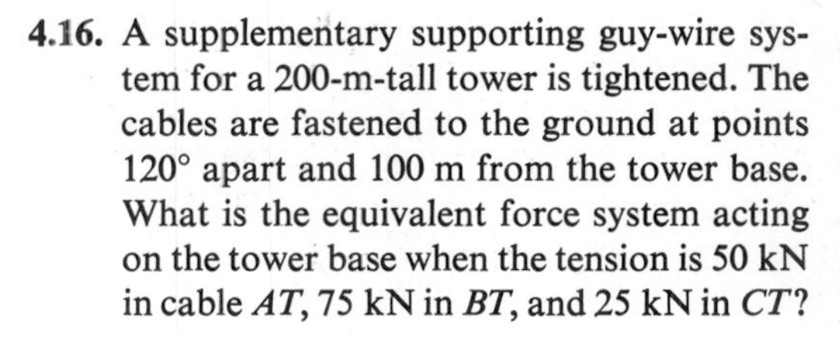

4.16. A supplementary supporting guy-wire system for a 200-m-tall tower is tightened. The cables are fastened to the ground at points 120° apart and 100 m from the tower base. What is the equivalent force system acting on the tower base when the tension is 50 kN in cable AT, 75 kN in BT, and 25 kN in CT?

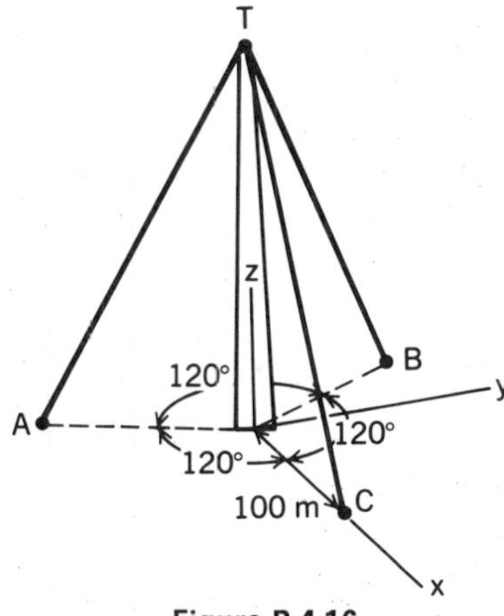

Figure P.4.16

4.3 Resultant of a Force System

As defined at the beginning of the chapter, a *resultant of a force system* is a simpler equivalent force system. In many computations it is desirable first to establish the resultant before entering into other computations.

For a general arrangement of forces, no matter how complex, we can always move all forces and couple moments, the latter including both those given and those formed from the movement of forces, to proceed through any single point. The result is then a system of concurrent forces at the point and a system of concurrent couple moments. These systems may then be combined into a single force and a single couple moment. Thus, in Fig. 4.10 we have shown some arbitrary system of forces

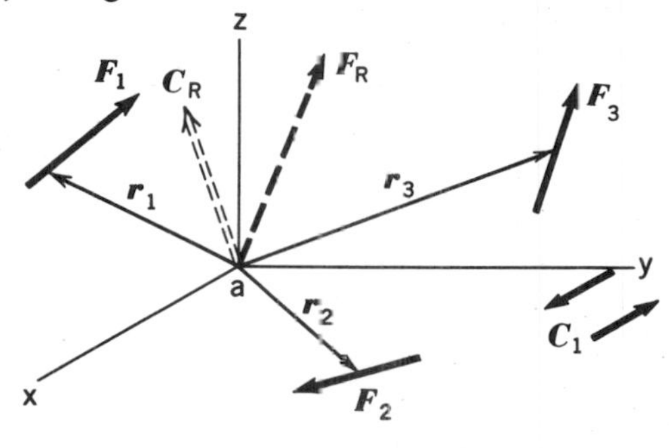

Figure 4.10. Resultant of general force system.

and couples using full lines. The resultant force and couple-moment combination at the origin of a rectangular reference is shown as dashed lines.

Thus, *any force system can be replaced at any point by equivalents no more complex than a single force and a single couple moment.* In special cases, which we shall examine shortly, we may have simpler equivalents such as a single force or a single couple moment. Finally, for *equilibrium of a body*, it is necessary that at any point the resultant system of forces and couple moments acting on the body be zero vectors—a fact that will be discussed in dynamics.

The methods of finding a resultant of forces involve nothing new. In moving to any new point, you will recall, there is no change in the force itself other than a shift of line of action; thus, any component of the *resultant* force, such as the x component, can simply be taken as the sum of the respective x components of all the forces in the system. We may then say for the resultant force

$$\boldsymbol{F}_R = [\sum_i (F_i)_x]\boldsymbol{i} + [\sum_i (F_i)_y]\boldsymbol{j} + [\sum_i (F_i)_z]\boldsymbol{k} \tag{4.3}$$

The couple moment accompanying $\boldsymbol{F}_R$ for a chosen point a may then be given as

$$\boldsymbol{C}_R = [\boldsymbol{r}_1 \times \boldsymbol{F}_1 + \boldsymbol{r}_2 \times \boldsymbol{F}_2 + \dots] + [\boldsymbol{C}_1 + \boldsymbol{C}_2 + \dots] \tag{4.4}$$

where the first bracketed quantities result from moving the noncouple forces to a, and the second are simply the sum of the given couple moments. The vectors $\boldsymbol{r}$ are from a to arbitrary points along the lines of action of the forces. In more compact form, the equation above becomes

$$\boldsymbol{C}_R = \sum_i \boldsymbol{r}_i \times \boldsymbol{F}_i + \sum_i \boldsymbol{C}_i \tag{4.5}$$

The following example is an illustration of the procedure.

EXAMPLE 4.3

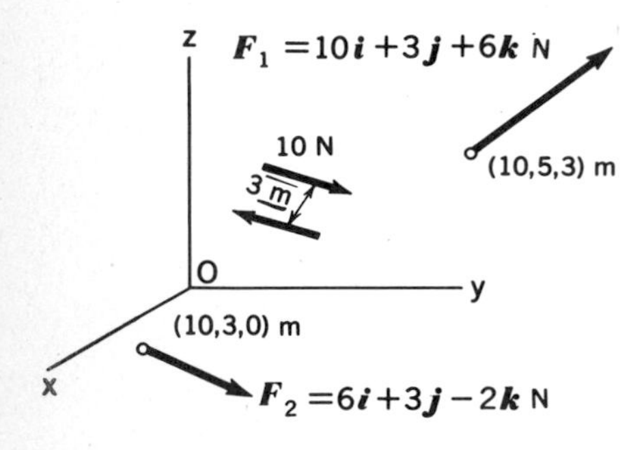

Figure 4.11. Find resultant at O.

Two forces and a couple are shown in Fig. 4.11, the couple being positioned in plane zy. We shall find the resultant of the system at the origin O.

At O we will have a set of two concurrent forces, which may be added to give $\boldsymbol{F}_R$:

$$\boldsymbol{F}_R = (10 + 6)\boldsymbol{i} + (3 + 3)\boldsymbol{j} + (6 - 2)\boldsymbol{k}$$
$$= 16\boldsymbol{i} + 6\boldsymbol{j} + 4\boldsymbol{k} \text{ N}$$

The resultant couple moment at point O is the vector sum of the couple-moment vectors developed by moving the two forces, plus the couple moment of the couple in the zy plane. Thus,

$$\boldsymbol{C}_R = \boldsymbol{r}_1 \times \boldsymbol{F}_1 + \boldsymbol{r}_2 \times \boldsymbol{F}_2 - 30\boldsymbol{i} \text{ N-m}$$

Now

$$\boldsymbol{r}_1 \times \boldsymbol{F}_1 = (10\boldsymbol{i} + 5\boldsymbol{j} + 3\boldsymbol{k}) \times (10\boldsymbol{i} + 3\boldsymbol{j} + 6\boldsymbol{k})$$
$$= 21\boldsymbol{i} - 30\boldsymbol{j} - 20\boldsymbol{k} \text{ N-m}$$
$$\boldsymbol{r}_2 \times \boldsymbol{F}_2 = (10\boldsymbol{i} + 3\boldsymbol{j}) \times (6\boldsymbol{i} + 3\boldsymbol{j} - 2\boldsymbol{k})$$
$$= -6\boldsymbol{i} + 20\boldsymbol{j} + 12\boldsymbol{k} \text{ N-m}$$

Hence,

$$\boldsymbol{C}_R = -15\boldsymbol{i} - 10\boldsymbol{j} - 8\boldsymbol{k} \text{ N-m}$$

The resultant is shown in Fig. 4.12.

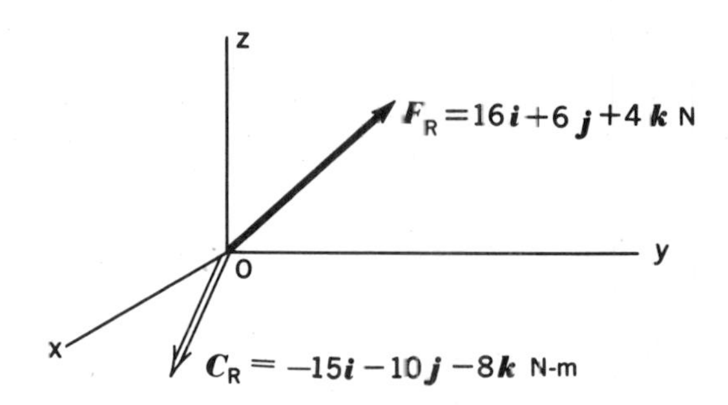

Figure 4.12. Resultant at *O*.

EXAMPLE 4.4

What is the resultant at A of the applied loads acting in Fig. 4.13? We first express the loads vectorially. Thus,

$$\boldsymbol{F}_1 = F_1\hat{\boldsymbol{d}}_1 = 150\left(\frac{-10\boldsymbol{k} - 3\boldsymbol{i} + 4\boldsymbol{j}}{\sqrt{10^2 + 3^2 + 4^2}}\right)$$
$$= -40.2\boldsymbol{i} + 53.7\boldsymbol{j} - 134.1\boldsymbol{k} \text{ lb}$$
$$\boldsymbol{F}_2 = F_2\hat{\boldsymbol{d}}_2 = 200\left(\frac{-13\boldsymbol{k} + 7\boldsymbol{i}}{\sqrt{13^2 + 7^2}}\right)$$
$$= 94.8\boldsymbol{i} - 176\boldsymbol{k} \text{ lb}$$
$$\boldsymbol{F}_3 = -100\boldsymbol{j} \text{ lb}$$
$$\boldsymbol{C} = -50\boldsymbol{k} \text{ ft-lb}$$

We can now readily find the resultant force system at A. Thus,

$$\boldsymbol{F}_R = (-40.2 + 94.8)\boldsymbol{i} + (53.7 - 100)\boldsymbol{j} + (-134.1 - 176.0)\boldsymbol{k}$$
$$= 54.6\boldsymbol{i} - 46.3\boldsymbol{j} - 310\boldsymbol{k} \text{ lb}$$
$$\boldsymbol{C}_R = (-11\boldsymbol{k}) \times \boldsymbol{F}_1 + (-8\boldsymbol{k}) \times (\boldsymbol{F}_2 + \boldsymbol{F}_3) + (-50\boldsymbol{k})$$
$$= -11\boldsymbol{k} \times (-40.2\boldsymbol{i} + 53.7\boldsymbol{j} - 134.1\boldsymbol{k}) + (-8\boldsymbol{k})$$
$$\times (94.8\boldsymbol{i} - 176.0\boldsymbol{k} - 100\boldsymbol{j}) - 50\boldsymbol{k}$$
$$= -209\boldsymbol{i} - 316\boldsymbol{j} - 50\boldsymbol{k} \text{ ft-lb}$$

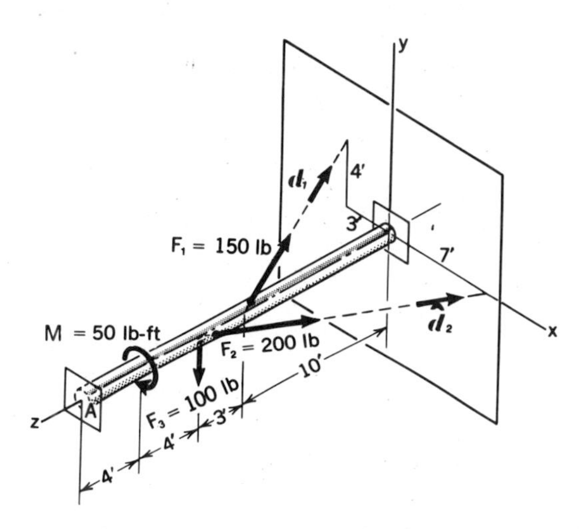

Figure 4.13. Find resultant at A; F_2 and F_3 are concurrent.

4.4 Simplest Resultants of Special Force Systems

We shall now consider special but important force systems in order to establish the *simplest* resultants possible. Examples will serve to illustrate the method of procedure.

Case A. Coplanar Force Systems. In Fig. 4.14 is shown a system of forces and couples in plane M. By moving the forces to a common point a in plane M, we will form only couples in the plane. The force portion of the equivalent system at such point will be given as

$$\boldsymbol{F}_R = [\sum_i (F_i)_x]\boldsymbol{i} + [\sum_i (F_i)_y]\boldsymbol{j} \tag{4.6}$$

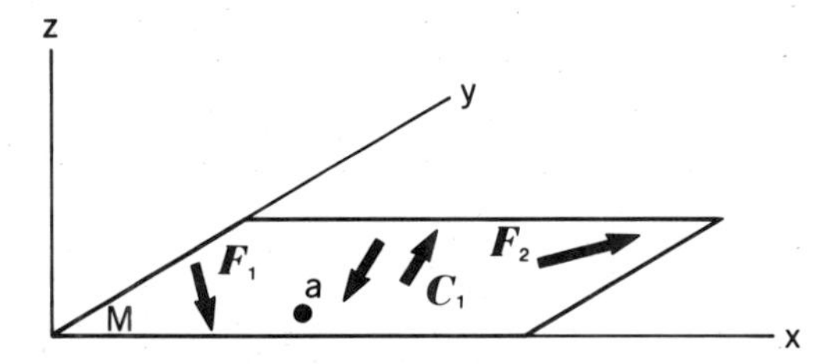

Figure 4.14. Coplanar force system.

The couple-moment portion of the equivalent system can be given as:

$$\boldsymbol{C}_R = (F_1 d_1 + F_2 d_2 + \dots)\boldsymbol{k} + (C_1 + C_2 + \dots)\boldsymbol{k} \tag{4.7}$$

where d_1, d_2, etc., are perpendicular distances from point a to the lines of action of the noncouple forces, and C_1, C_2, etc., are the values of the given couple moments. The resultant at a is shown in Fig. 4.15.

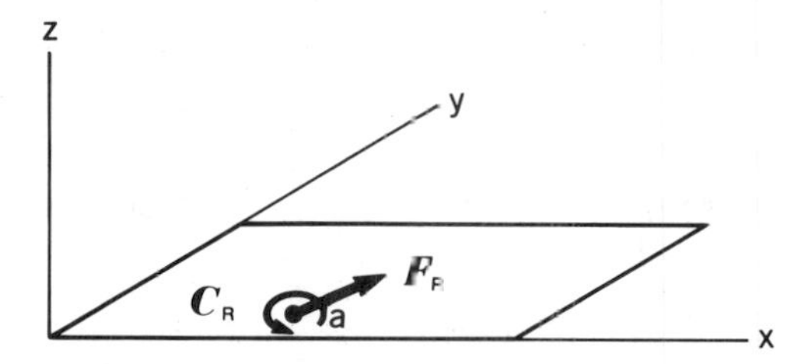

Figure 4.15. Resultant at point *a*.

If $\boldsymbol{F}_R \neq \boldsymbol{0}$—that is, if $\sum_i F_x \neq 0$ and/or $\sum_i F_y \neq 0$—we can move the force from *a* to yet a new position so as to introduce a second couple moment to cancel $\boldsymbol{C}_R$ of Fig. 4.15 in the manner described earlier in Section 4.2. Since the *x* and *y* directions used are arbitrary, except for the condition that they be in the plane of the forces, we can make the following conclusion. *If the force components in any direction in the plane add to other than zero, we may replace the entire coplanar system by a single force with a specific line of action.*

What happens if $\sum_i F_x = 0$ and $\sum_i F_y = 0$? Without a force at point *a*, we can no longer eliminate a couple in plane *M*. Thus, our second conclusion is that *if $\sum_i F_x$ and $\sum_i F_y$ are zero, the resultant must be a couple moment or be zero.*

In the coplanar case, therefore, the simplest equivalent force system must be a single force along a specific line of action, or a single couple moment, or the zero vector. The following example is used to illustrate the method of determining such a resultant directly without the intermediate steps followed in this discussion.

EXAMPLE 4.5

Consider a coplanar force system shown in Fig. 4.16. The *simplest* resultant is to be found. Since $\sum_i F_x$ and $\sum_i F_y$ are not zero, we know that we can replace the

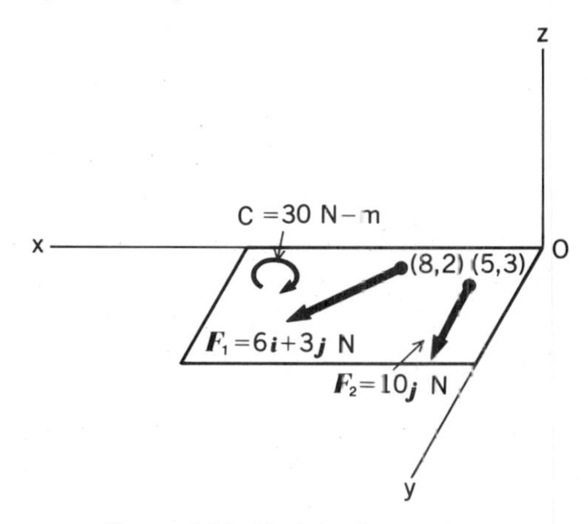

Figure 4.16. Find simplest resultant.

system by a single force, which is

$$\boldsymbol{F}_R = 6\boldsymbol{i} + 13\boldsymbol{j}\text{ N} \tag{a}$$

We now need to find the line of action in the plane that will make this single force equivalent to the given system. To be equivalent for rigid-body mechanics, this force without a couple moment must have the same turning action about any point or axis in space as that of the given system. Now the simplest resultant force must intercept the x axis at some point $\bar{x}$.[1] We can determine $\bar{x}$ by equating the moment of the resultant force without a couple moment about the origin with that of the original system of forces and couples. Using the vector $\bar{x}\boldsymbol{i}$ as a position vector from the origin to the line of action of $\boldsymbol{F}_R$ (see Fig. 4.17), we accordingly have

$$\bar{x}\boldsymbol{i} \times (6\boldsymbol{i} + 13\boldsymbol{j}) = (8\boldsymbol{i} + 2\boldsymbol{j}) \times (6\boldsymbol{i} + 3\boldsymbol{j}) + (5\boldsymbol{i} + 3\boldsymbol{j}) \times (10\boldsymbol{j}) - 30\boldsymbol{k} \tag{b}$$

Carrying out the cross products,

$$24\boldsymbol{k} - 12\boldsymbol{k} + 50\boldsymbol{k} - 30\boldsymbol{k} = 13\bar{x}\boldsymbol{k} \tag{c}$$

Hence,

$$\bar{x} = 2.46\text{ ft}$$

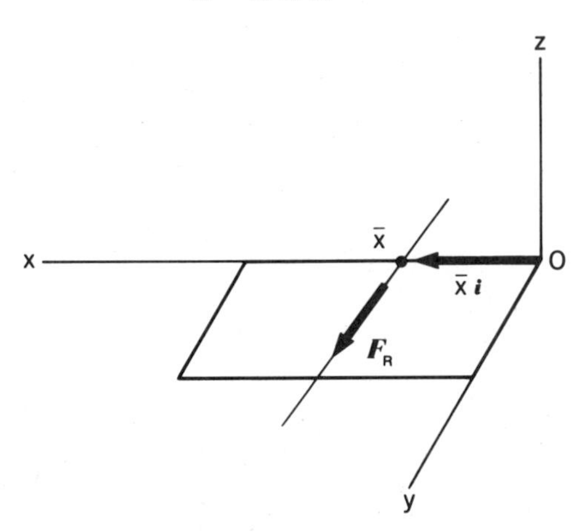

Figure 4.17. Simplest resultant.

By specifying the x intercept, $\bar{x}$, we fully determine the line of action of the simplest resultant force. We could have also used the intercept with the y axis, $\bar{y}$, for this purpose. In that case, the position vector from the origin out to the line of action is $\bar{y}\boldsymbol{j}$, and we have, on equating moments about O of the resultant without a couple moment with that of the original system:

$$\bar{y}\boldsymbol{j} \times (6\boldsymbol{i} + 13\boldsymbol{j}) = (8\boldsymbol{i} + 2\boldsymbol{j}) \times (6\boldsymbol{i} + 3\boldsymbol{j}) + (5\boldsymbol{i} + 3\boldsymbol{j}) \times (10\boldsymbol{j}) - 30\boldsymbol{k}$$

$$y = -5.35\text{ ft}$$

EXAMPLE 4.6

Compute the *simplest* resultant for the loads shown acting on the beam in Fig. 4.18(a). Give the intercept with the x axis.

---

[1]If the resultant force is parallel to the x axis, the intercept will be at infinity.

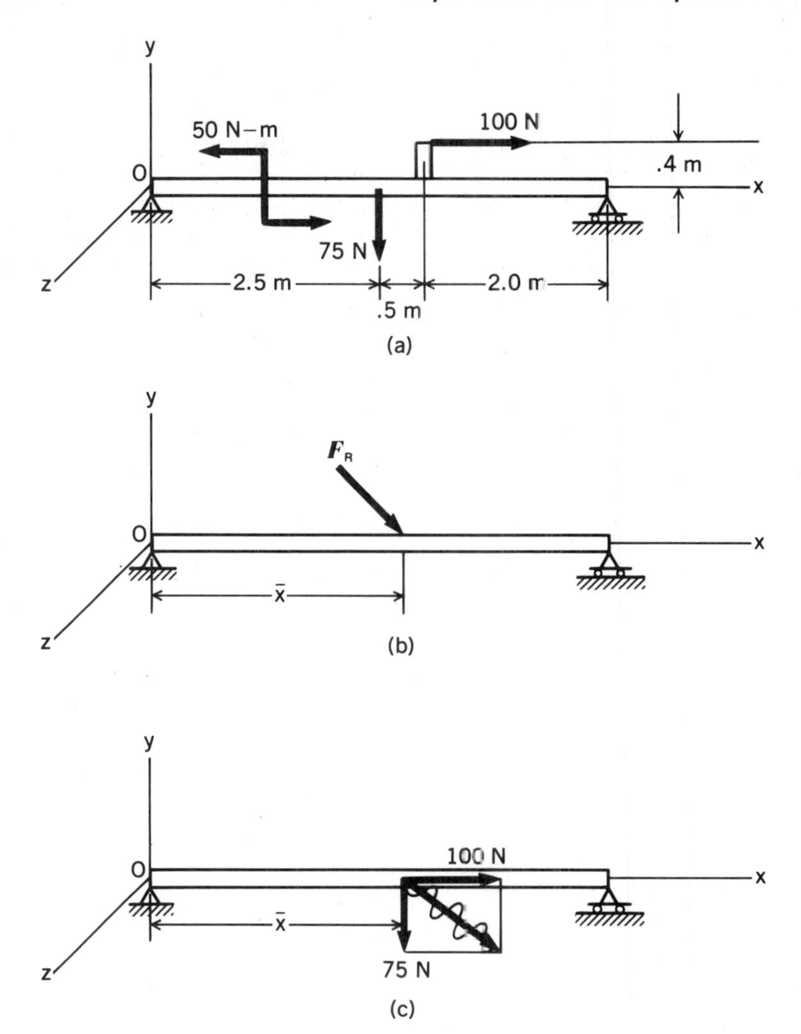

Figure 4.18. Find simplest resultant.

It is immediately apparent on inspection of the diagram that

$$\boldsymbol{F}_R = 100\boldsymbol{i} - 75\boldsymbol{j}\,\text{N} \tag{a}$$

Let $\bar{x}$ be the intercept with the x axis of the line of action of $\boldsymbol{F}_R$ when this line of action corresponds to zero couple moment $\boldsymbol{C}_R$ [see Fig. 4.18(b)]. In Fig. 4.18(c), we have decomposed $\boldsymbol{F}_R$ along this line of action into rectangular components so as to permit simple calculations of moments about the origin O (i.e., about the z axis). Accordingly, equating moments about the z axis of $\boldsymbol{F}_R$, without a couple moment, with that of the original system of loads, we get, using scalar components:

$$-(75)(\bar{x}) = 50 - (2.5)(75) - (.4)(100)$$

$$\bar{x} = 2.37\text{ m}$$

Thus, the simplest resultant is a force $100\boldsymbol{i} - 75\boldsymbol{j}$ N intercepting the beam axis at a position $\bar{x} = 2.37$ m.

As pointed out earlier, in the instance wherein $\boldsymbol{F}_R = \boldsymbol{0}$, we then possibly have as the simplest resultant a couple moment normal to the plane of the coplanar force system. There is the possibility that there is also zero couple moment, in which case the members of the coplanar force system *completely cancel* each other's effects on a rigid body. To find the couple moment for the case where $\boldsymbol{F}_R = \boldsymbol{0}$, we simply take moments of the coplanar force system about *any point* in space. This moment, if it be not zero, is clearly the couple-moment vector sought. We leave this straightforward kind of a problem to the exercises.

Case B. Parallel Force Systems in Space. Now, consider the system of n parallel forces in Fig. 4.19, where the z direction has been selected parallel to the forces. We also include m couples whose planes are parallel to the z direction because such couples

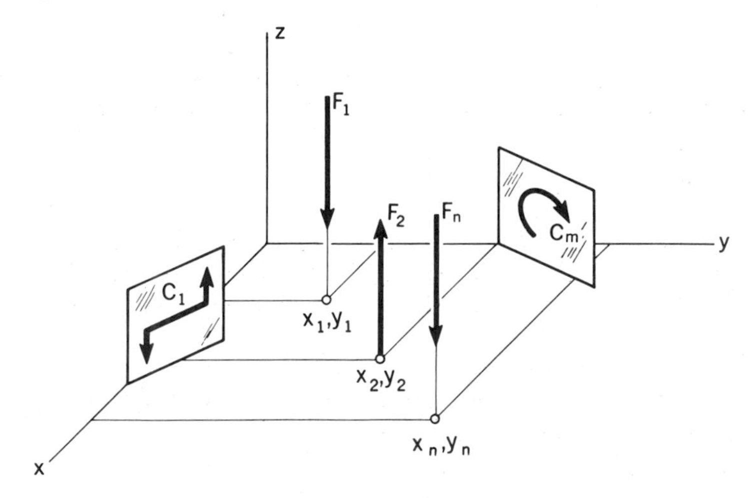

Figure 4.19. Parallel system of forces.

can be considered to be composed of equal and opposite forces parallel to the z direction. We can move the forces so that they all pass through the origin of the xyz axes; the force portion of the equivalent system is then

$$\boldsymbol{F}_R = \left(\sum_{i=1}^{n} F_i\right)\boldsymbol{k} \tag{4.8}$$

The couple-moment portion of the equivalent system is found by applying Eq. 4.5 to this case:

$$\boldsymbol{C}_R = \sum_{i=1}^{n} [(x_i\boldsymbol{i} + y_i\boldsymbol{j}) \times F_i\boldsymbol{k}] + \sum_{i=1}^{m} [(C_i)_x\boldsymbol{i} + (C_i)_y\boldsymbol{j}] \tag{4.9}$$

where F_i represents the noncouple force magnitudes. Carrying out the cross product, we get

$$\boldsymbol{C}_R = \sum_{i=1}^{n} [(F_i y_i)\boldsymbol{i} - (F_i x_i)\boldsymbol{j}] + \sum_{i=1}^{m} [(C_i)_x\boldsymbol{i} + (C_i)_y\boldsymbol{j}] \tag{4.10}$$

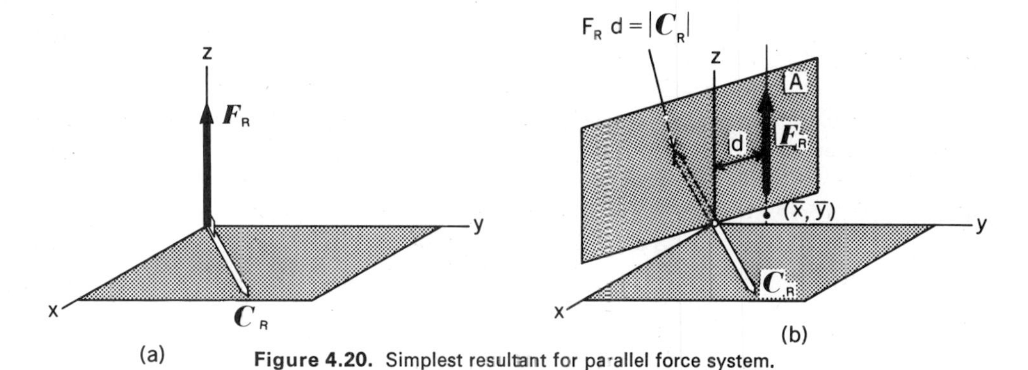

Figure 4.20. Simplest resultant for parallel force system.

From this, we see that the couple moment must always be parallel to the xy plane (i.e., perpendicular to the direction of the forces). We then have at the origin a single force and a single couple moment at right angles to each other [see Fig. 4.20(a)]. If $\boldsymbol{F}_R \neq \boldsymbol{0}$, we can move $\boldsymbol{F}_R$ again to another line of action in a plane A perpendicular to $\boldsymbol{C}_R$ and, choosing the proper value of d, such that $F_R d = |\boldsymbol{C}_R|$, we can eliminate the couple moment [see Fig. 4.20(b)]. We thus end up with a *single* force having a particular line of action specified by the intercept $\bar{x}\bar{y}$ of the line of action of the force with the xy plane. If the summation of forces should happen to be zero, the equivalent system must then be a couple moment or be zero.

Thus, *the simplest resultant system of a parallel force system is either a force or a couple moment.* The following example will illustrate how we can directly determine the simplest resultant.

EXAMPLE 4.7

Find the simplest resultant of the parallel force system in Fig. 4.21(a).

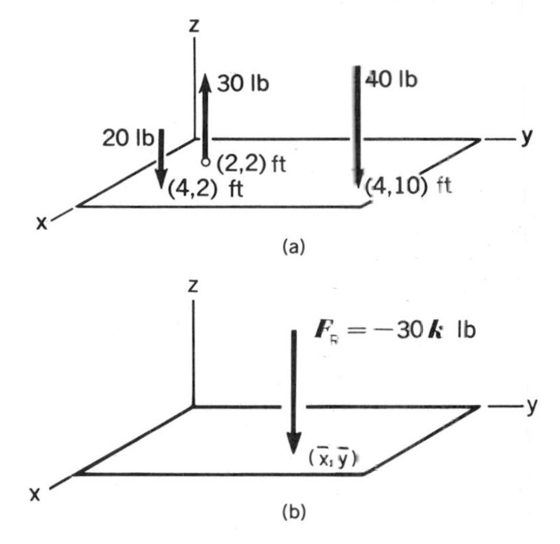

Figure 4.21. Find simplest resultant.

The sum of the forces is 30 lb in the negative z direction. Hence, a position can be found in which a single force is equivalent to the original system. Assume that this resultant force without a couple moment proceeds through the point $\bar{x}$, $\bar{y}$ [Fig. 4.21(b)]. We can equate the moment of this resultant force about the x and y axes with the corresponding moments of the original system and thus form the scalar equations that yield the proper value of $\bar{x}$ and $\bar{y}$. Equating moments about the x axis,[2] we get

$$(30)(2) - (20)(2) - (40)(10) = -30\bar{y}$$

Therefore,

$$\bar{y} = 12.7 \text{ ft}$$

Equating moments about the y axis, we have

$$-(30)(2) + (20)(4) + (40)(4) = 30\bar{x}$$

Therefore,

$$\bar{x} = 6 \text{ ft}$$

You can also show, as an exercise, that the same result can be reached for $\bar{x}$, $\bar{y}$ by equating moments of the resultant force without a couple moment about the origin with that of the original system about the origin.

EXAMPLE 4.8

Consider the parallel force system in Fig. 4.22. What is the simplest resultant?

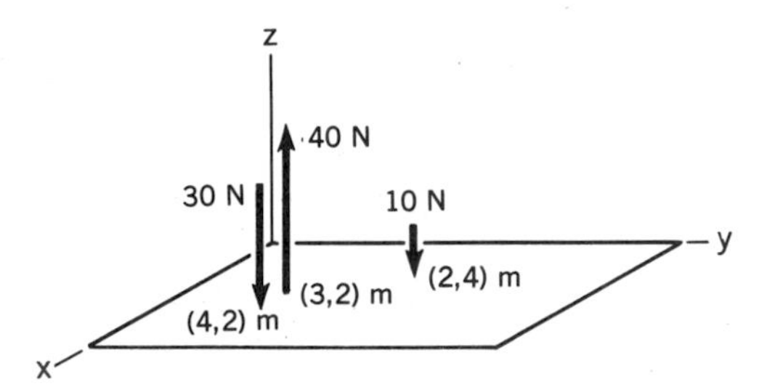

Figure 4.22. Parallel force system.

Here we have a case where the sum of the forces is zero and so $\boldsymbol{F}_R = \mathbf{0}$. Therefore, the simplest resultant must be a couple moment or be zero. To get this couple moment, $\boldsymbol{C}_R$, we can take moments of the forces about *any point* in space. This moment vector then equals the desired couple moment $\boldsymbol{C}_R$. The simplest procedure is to use the origin of the reference as the point about which to take moments. Then we

[2]To get the moment of a force, such as the 40-lb force, about the x axis we use the elementary definition of moment about an axis given at the outset of Section 3.3. Thus, consider a plane A perpendicular to the x axis and containing the 40-lb force. The perpendicular distance from the x axis to this force clearly is the y coordinate of the intercept (4, 10) of the line of action of the force and the xy plane. Hence, M_x for the 40-lb force must be $-(40)(10) = -400$ ft-lb.

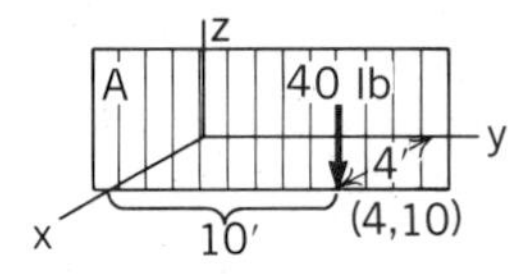

can say that

$$\begin{aligned} \boldsymbol{C}_R &= (4\boldsymbol{i} + 2\boldsymbol{j}) \times (-30\boldsymbol{k}) + (3\boldsymbol{i} + 2\boldsymbol{j}) \times (40\boldsymbol{k}) + (2\boldsymbol{i} + 4\boldsymbol{j}) \times (-10\boldsymbol{k}) \\ &= -20\boldsymbol{i} + 20\boldsymbol{j} \text{ N-m} \end{aligned} \tag{a}$$

The rectangular components of $\boldsymbol{C}_R$ along the x and y axes are the moments of the force system about these axes. Thus,

$$\begin{aligned} (C_R)_x &= -20 \text{ N-m} \\ (C_R)_y &= 20 \text{ N-m} \end{aligned} \tag{b}$$

We can get the moments of the forces about the x and y axes directly and thus generate the components of the desired couple moment $\boldsymbol{C}_R$. Accordingly, using the elementary definition of the moment of a force about a line as presented earlier, we have

$$\begin{aligned} (C_R)_x &= -(10)(4) + (40)(2) - (30)(2) = -20 \text{ N-m} \\ (C_R)_y &= (10)(2) - (40)(3) + (30)(4) = 20 \text{ N-m} \end{aligned}$$

Thus, the moment of the force system about the origin, and hence about any point, is then the desired couple moment (Fig. 4.23).

$$\boldsymbol{C}_R = -20\boldsymbol{i} + 20\boldsymbol{j} \text{ N-m}$$

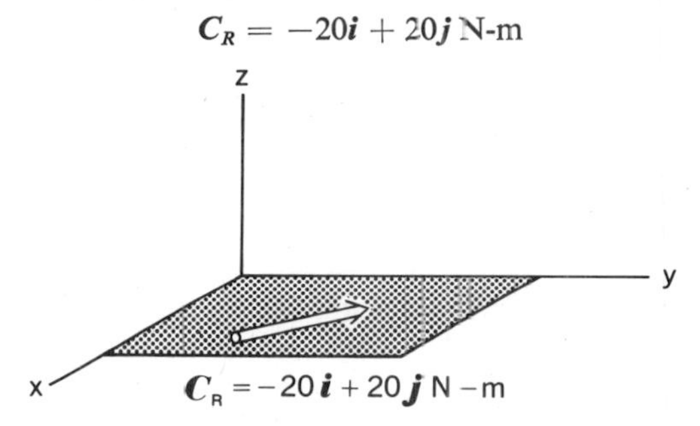

Figure 4.23. Simplest resultant is a couple moment.

Now that we have considered the concept of the simplest resultant for coplanar and parallel force systems, we wish to go back to the *general force* systems for a moment. We learned earlier that we can always replace such a system in rigid body mechanics by a single force $\boldsymbol{F}_R$ and a single couple moment $\boldsymbol{C}_R$ at any chosen point. Is this always the very simplest system for rigid body mechanics? No, it is not. To show this, decompose the couple moment $\boldsymbol{C}_R$ into two rectangular components $C_\perp$ and $C_{||}$, perpendicular to the force and collinear with the force respectively. We can now move the force to a chosen parallel position and can eliminate $C_\perp$, the component of couple moment normal to the force. However, there is nothing that we can do about the $C_{||}$ component of couple moment collinear (or parallel) to the force. The reason for this is that any movement of the force to a parallel position *always* introduces a couple moment *perpendicular* to the force. Thus the component $C_{||}$ cannot be affected. By eliminating $C_\perp$ we end up with the force $\boldsymbol{F}_R$ and $\boldsymbol{C}_{||}$ collinear with $\boldsymbol{F}_R$. This system is the simplest in the general case and it is called a *wrench*. However, we shall not use the wrench concept in this text and will work instead with the resultant force $\boldsymbol{F}_R$ and the couple moment $\boldsymbol{C}_R$ at any chosen point.

Problems

In several of the problems of this set we shall concentrate the weight of a body at its center of gravity. Most likely you are used to doing this from your previous physics course. In Section 4.5 we shall justify this procedure.

4.17. Compute the resultant force system of the applied loads at positions A and B.

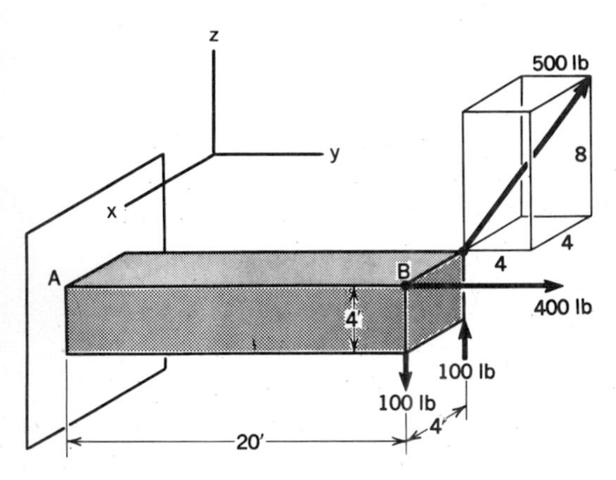

Figure P.4.17

4.18. Compute the resultant force system at A stemming from the indicated 50-lb force. What is the twist developed about the axis of the shaft at A?

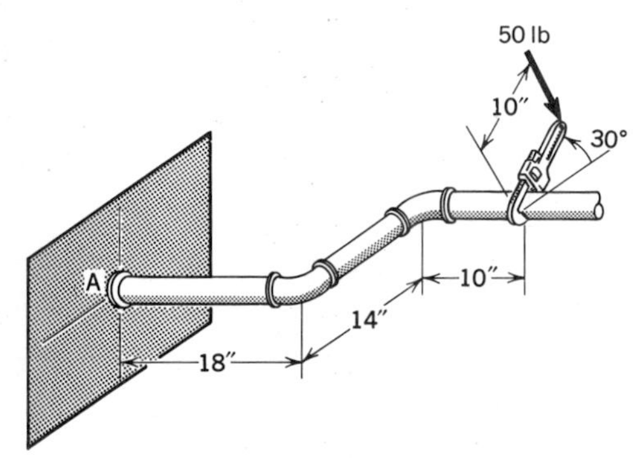

Figure P.4.18

4.19. Find the resultant of the force system at point A. The 300-N, 200-N, and 900-N loads are at the centers of the pipe sections.

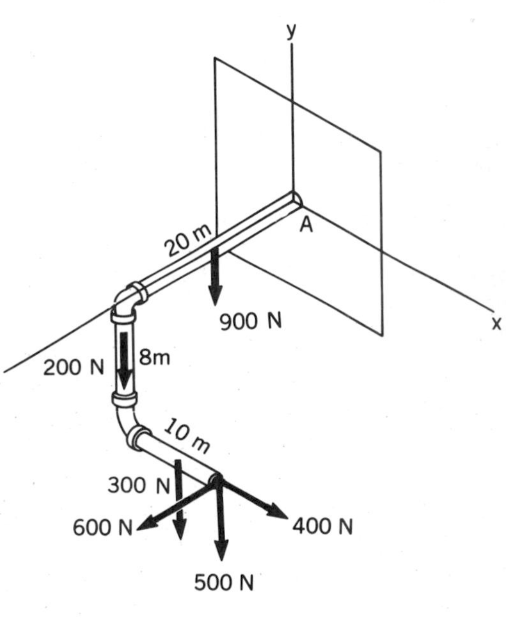

Figure P.4.19

4.20. A 20-kN car and an 80-kN truck are stopped on a bridge. What is the resultant force system of these vehicles at the center of the bridge? At the center of the left end of the bridge? The distances given to truck and car are to respective centers of gravity where we can concentrate the weights.

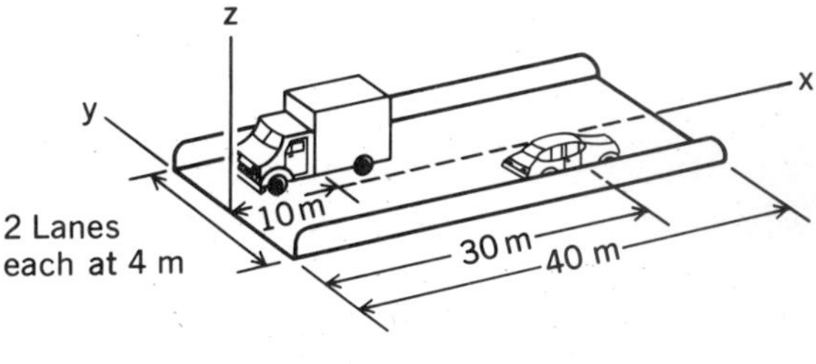

Figure P.4.20

4.21. Two heavy machinery crates (A weighs 20 kN and B weighs 30 kN) are placed on a truck. What is the resultant force system at the center of the rear axle? The centers of gravity of the crates, where we can concen-

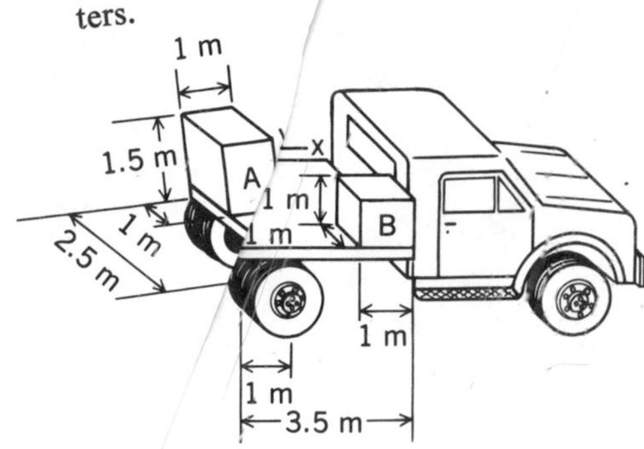

trate the weights, are at the geometric centers.

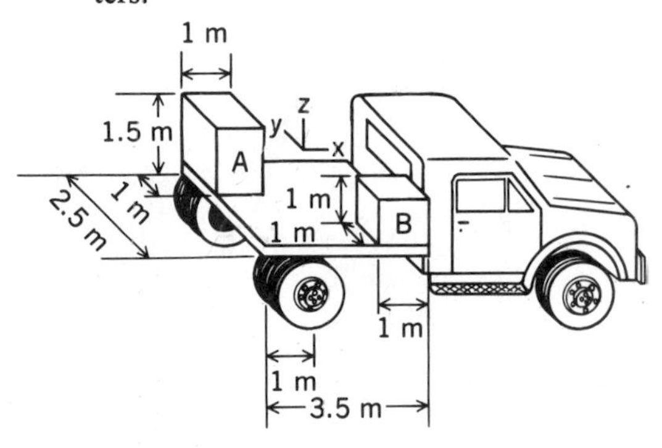

Figure P.4.21

4.22. Replace the system of forces by a resultant at A.

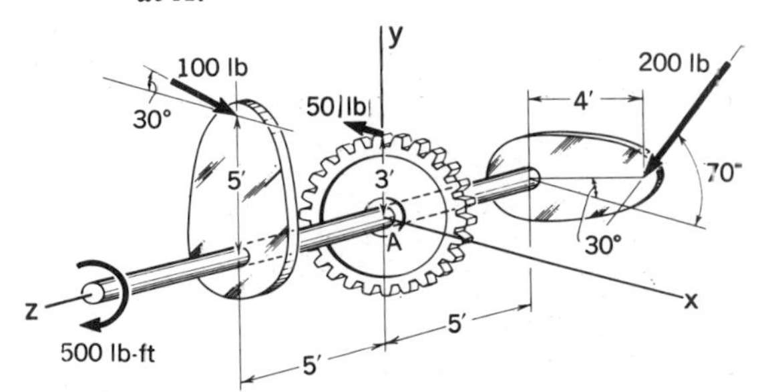

Figure P.4.22

4.23. Evaluate forces F_1, F_2, and F_3 so that the resultant of the forces and torque acting on the plate is zero in both force and couple moment. (*Hint:* If the resultant is zero for one point, will it not be zero for any point? Explain why.)

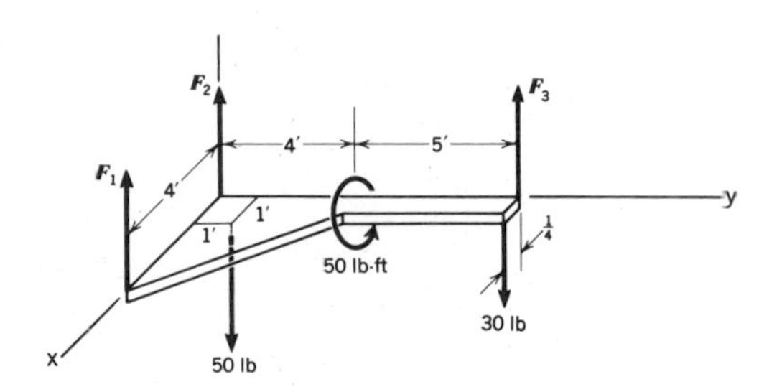

Figure P.4.23

4.24. Find the *simplest* resultant of the forces shown acting on the beam. Give the intercept with the axis of the beam.

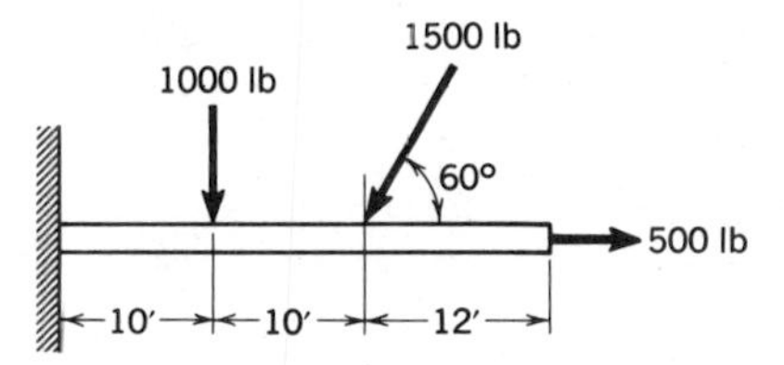

Figure P.4.24

4.25. Find the *simplest* resultant of the forces shown acting on the pulley. Give the intercept with the x axis.

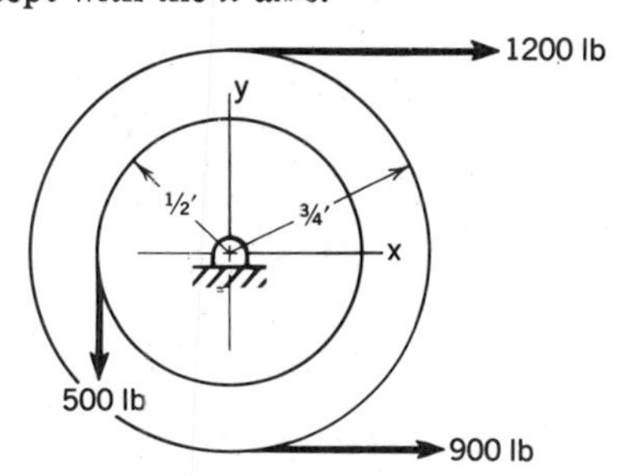

Figure P.4.25

4.26. A man raises a 50-lb bucket of water to the top of a bricklayer's scaffold. Also, a Jeep winch is used to raise a 200-lb load of bricks. What is the *simplest* resultant force system on the scaffold? Give the x intercept. Consider the pulleys to be frictionless so that the 50-lb force and the 200-lb force are transmitted respectively to the man and to the Jeep.

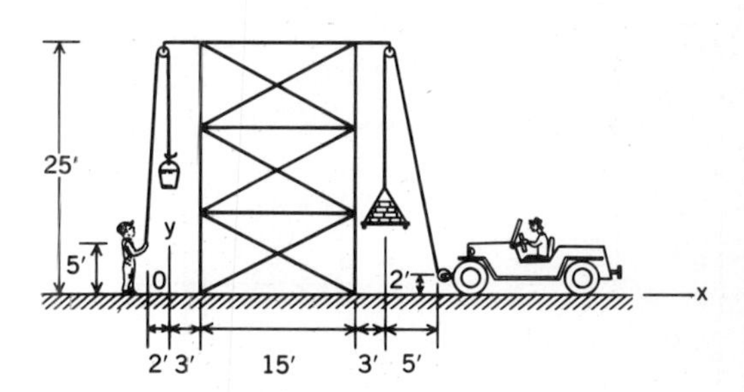

Figure P.4.26

4.27. Compute the *simplest* resultant for the loads acting on the beam. Give the intercept with the axis of the beam.

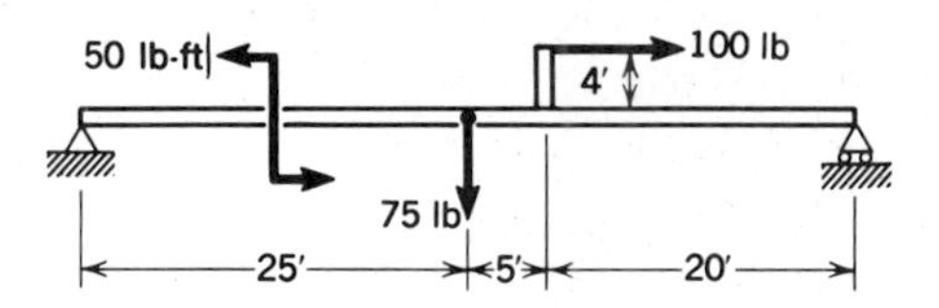

Figure P.4.27

4.28. Find the *simplest* resultant for the forces. Give the location of this resultant clearly.

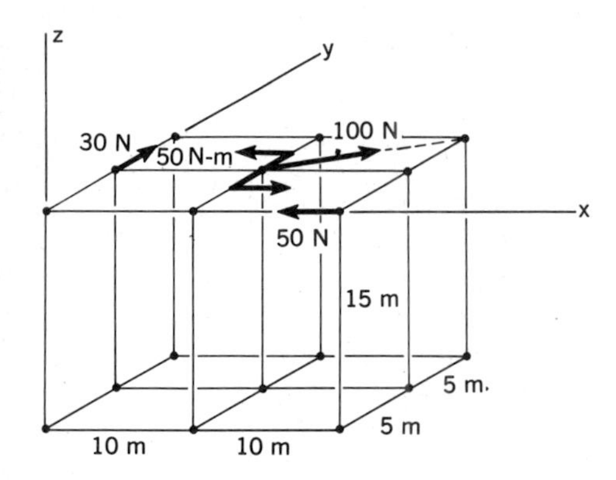

Figure P.4.28

4.29. Replace the system of forces acting on the rivets of the plate by the *simplest* resultant. Give the intercept of this resultant with the x axis.

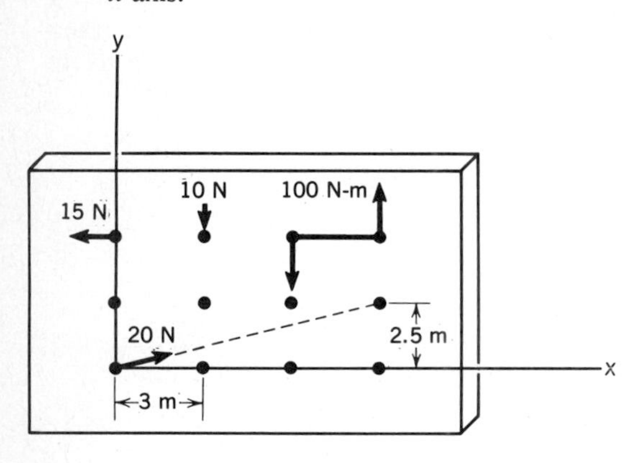

Figure P.4.29

4.30. A parallel system of forces is such that: a 20-N force acts at position $x = 10, y = 3$ m; a 30-N force acts at position $x = 5, y = -3$ m; a 50 N force acts at position $x = -2, y = 5$ m.

(a) If all forces point in the negative z direction, give the *simplest* resultant force and its line of action.

(b) If the 50-N force points in the plus z direction and the others in the negative z direction, what is the *simplest* resultant?

4.31. What is the *simplest* resultant of the three forces and couple shown acting on the shaft and disc? The disc radius is 5 ft.

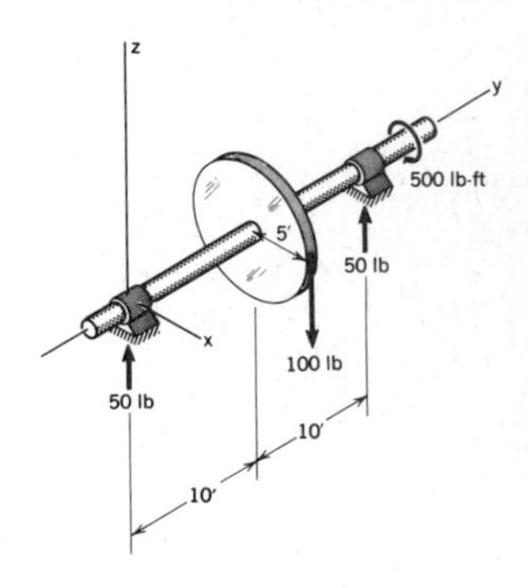

Figure P.4.31

4.32. What is the *simplest* resultant for the system of forces? Each square is 10 mm on edge.

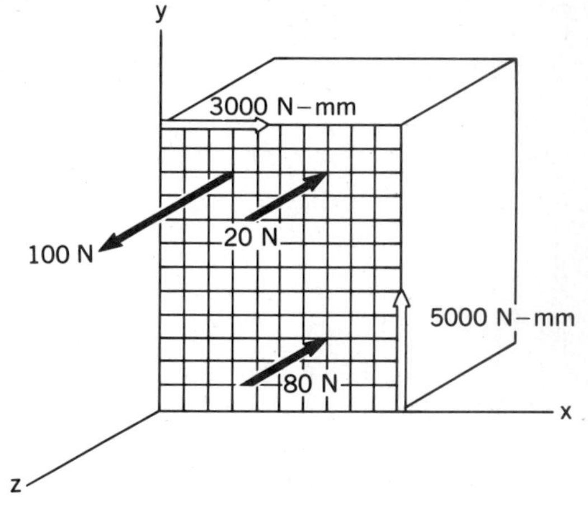

Figure P.4.32

4.33. Two hoists are operated on the same overhead track. Hoist A has a 3000-kN load, and hoist B has a 4000-kN load. What is the resultant force system at the left end O of the track? Where does the *simplest* resultant force act?

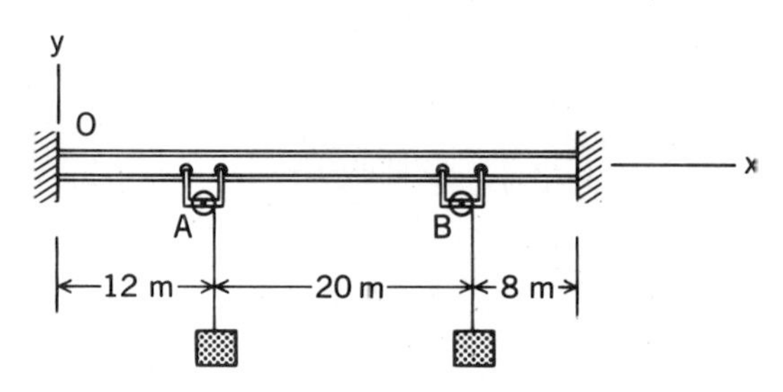

Figure P.4.33

4.34. A lo-boy trailer weighs 16,000 lb and is loaded with a 15,000-lb tractor and a 12,000-lb front-end loader. What is the simplest resultant force and where does it act? The weights of the machines and trailer act at their respective centers of gravity (C.G.).

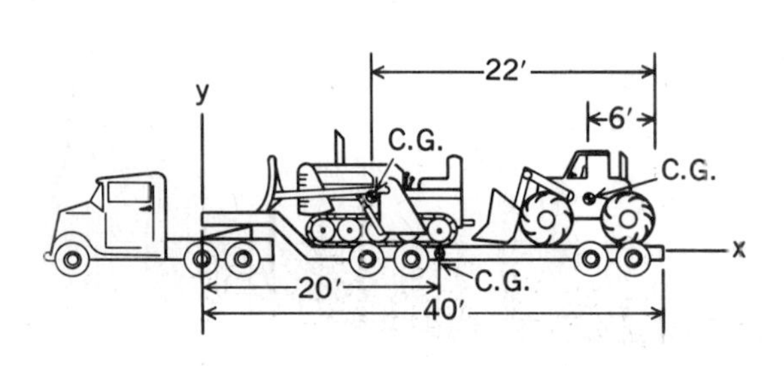

Figure P.4.34

4.35. Where should a 100-N force in a downward direction be placed for the *simplest* resultant of all shown forces to be at position (5, 5) m?

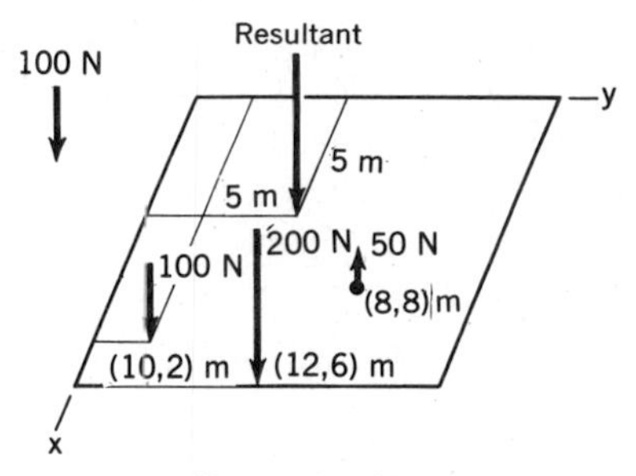

Figure P.4.35

4.36. A barge must be evenly loaded so it does not list in any direction. Where can the three large machinery crates be placed (without either hanging over the edge, stacking, or standing on end)? Each crate is as tall as it is wide. Is there only one solution to this problem? Centers of gravity of crates correspond to geometric centers.

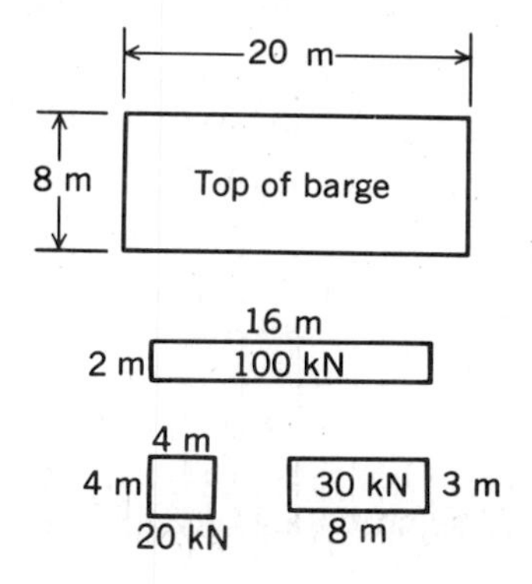

Figure P.4.36

4.5 Distributed Force Systems

Our discussions up to now have been restricted to discrete vectors—in particular, to point forces. Scalars and vectors may also be continuously distributed throughout a finite volume so that at each position in space in the volume there is a definite scalar or vector quantity. Such distributions are called *scalar* and *vector fields*, respectively. A simple example of a scalar field is the temperature distribution, expressed as $T(x, y, z, t)$, where the variable t indicates that the field may be changing with time. Thus, if a position x_0, y_0, z_0 and a time t_0 are specified, we can determine the temperature at this position and time provided that we know the temperature distribution function (i.e., how T depends on the independent variables $x, y, z,$ and t). A vector field is sometimes expressed in the form $\boldsymbol{F}(x, y, z, t)$. A common example of a vector field is the gravitational force field of the earth—a field that is known to vary with elevation above sea level, among other factors. Note however that the gravitational field is virtually constant with time.

In place of the vector field, it is more convenient at times to employ three scalar fields that represent the orthogonal scalar components of a vector field at all points. Thus, for a force field we can say:

$$\text{force component in } x \text{ direction} = g(x, y, z, t)$$

$$\text{force component in } y \text{ direction} = h(x, y, z, t)$$

$$\text{force component in } z \text{ direction} = k(x, y, z, t)$$

where g, h, and k represent functions of the coordinates and time. If we substitute coordinates of a special position and the time into these functions, we get the force components F_x, F_y, and F_z for that position and time. The force field and its component scalar fields are then related in this way:

$$\boldsymbol{F}(x, y, z, t) = g(x, y, z, t)\boldsymbol{i} + h(x, y, z, t)\boldsymbol{j} + k(x, y, z, t)\boldsymbol{k}$$

More often, the notation for the equation above is written

$$\boldsymbol{F}(x, y, z, t) = F_x(x, y, z, t)\boldsymbol{i} + F_y(x, y, z, t)\boldsymbol{j} + F_z(x, y, z, t)\boldsymbol{k} \tag{4.11}$$

Vector fields are not restricted to forces but include such other quantities as velocity fields and heat-flow fields.

Force distributions, such as gravitational force, that exert influence directly on the elements of mass distributed throughout the body are termed *body force distributions* and are usually given per unit of mass that they directly influence. Thus, if $\boldsymbol{B}(x, y, z, t)$ is such a body force distribution, the force on an element dm would be $\boldsymbol{B}(x, y, z, t)dm$.

Force distributions over a *surface* are called *surface force distributions*[3] and are given per unit area of the surface directly influenced. A simple example is the force distribution on the surface of a body submerged in a fluid. In the case of a static fluid or of a frictionless fluid, the force from the fluid on an area element is always normal to the area element and directed in toward the body. The force per unit area stemming from such fluid action is called *pressure* and is denoted as p. Pressure is a scalar quantity. The direction of the force resulting from a pressure on a surface is given by the orientation of the surface. [You will recall from Chapter 2 that an area element can be considered as a vector which is normal to the area element and directed outward from the enclosed body (Fig. 4.24).] The infinitesimal force on the area element is then given as

$$d\boldsymbol{f} = -p\, d\boldsymbol{A}$$

[3]Surface forces are often called *surface tractions* in solid mechanics.

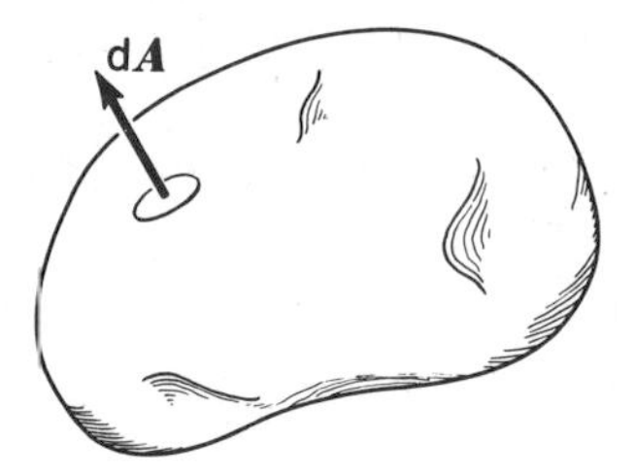

Figure 4.24. Area vector.

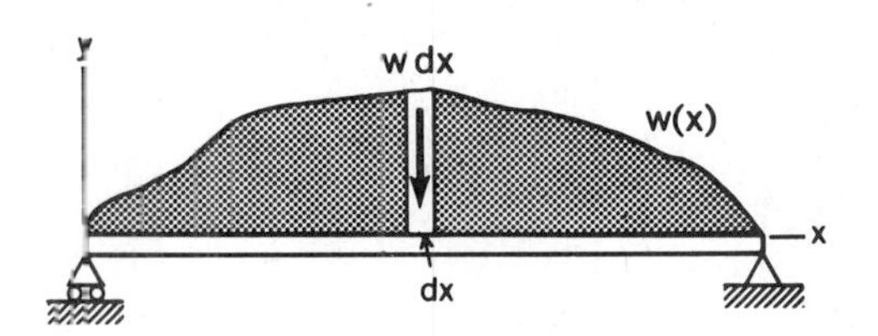

Figure 4.25. Loading on a beam.

A more specialized, but nevertheless common, force distribution is that of a continuous load on a beam. This is often a parallel loading distribution that is symmetrical about the center plane xy of a beam, as illustrated in Fig. 4.25. Various heights of bricks stacked on a beam would be an example of this kind of loading. We can replace such a loading by an equivalent coplanar distribution that acts at the center plane. The loading is given per unit length and is denoted as w, the *intensity of loading*. The force on an element dx of the beam, then, is $w\,dx$.

We have thus presented force systems distributed throughout volumes (body forces), over surfaces (surface forces), and over lines. The conclusions about resultants that were reached earlier for general, parallel, and coplanar point-force systems are also valid for these distributed-force systems. These conclusions are true because each distributed force system can be considered as an infinite number of infinitesimal point forces of the type used heretofore. We shall illustrate the handling of force distributions in the following examples.

Case A. Parallel Body Force System—Center of Gravity. Consider a rigid body (Fig. 4.26) whose density (mass/unit volume) is given as $\rho(x, y, z)$. It is acted on by gravity, which, for a small body, may be considered to result in a distributed parallel force field.

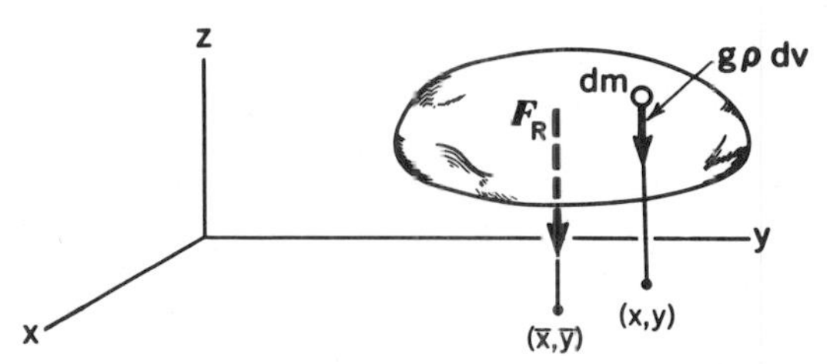

Figure 4.26. Gravity body force distribution.

Since we have here a parallel system of forces in space with the same sense, we know that a single force without a couple moment along a certain line of action will be equivalent to the distribution. The gravity body force $\boldsymbol{B}(x, y, z)$ given per unit mass is $-g\boldsymbol{k}$. The infinitesimal force on a differential mass element dm, then, is $-g(\rho\, dv)\boldsymbol{k}$, where dv is the volume of the element.[4] We find the resultant force on the system by

[4]Note that $g\rho$ is the weight per unit volume which is often given as γ, the so-called *specific weight*.

replacing the summation in Eq. 4.8 with an integration. Thus,

$$\boldsymbol{F}_R = -\int_V g(\rho\, dv)\boldsymbol{k} = -g\boldsymbol{k}\int_V \rho\, dv = -gM\boldsymbol{k}$$

where, with g as a constant, the second integral becomes simply the entire mass of the body M.

Next, we must find the line of action of this single equivalent force without a couple moment. Let us denote the intercept of this line of action with the xy plane as $\bar{x}, \bar{y}$ (see Fig. 4.26). The resultant at this position must have the same moments as the distribution about both the x and y axes:

$$-F_R\bar{x} = -g\int_V x\rho\, dv, \qquad F_R\bar{y} = g\int_V y\rho\, dv$$

Hence, we have

$$\bar{x} = \frac{\int x\rho\, dv}{M}, \qquad \bar{y} = \frac{\int y\rho\, dv}{M}$$

Thus, we have fully established the simplest resultant. Now, the body is reoriented in space, keeping with it the line of action of the resultant as shown in Fig. 4.27. A new computation of the line of action of the simplest resultant for the second orientation yields a line that intersects the original line at a point C. It can be shown that lines of action for simplest resultants for all other orientations of the body must intersect at the same point. We call this point the *center of gravity*. Effectively, we can say for rigid-body considerations that all the weight of the body can be assumed to be concentrated at the center of gravity.

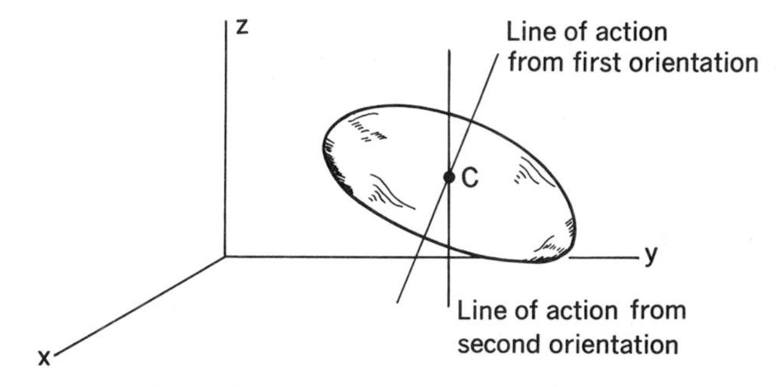

Figure 4.27. Location of center of gravity.

EXAMPLE 4.9

Find the center of gravity of the triangular block having a uniform density ρ shown in Fig. 4.28.

The total weight of the body is easily evaluated as

$$F_R = g\rho\frac{abc}{2} \tag{a}$$

To find $\bar{y}$, we will equate the moment of F_R about the x axis with that of the weight distribution of the block. To facilitate the latter, we shall choose within the block *infinitesimal* elements whose weights are easily computed. Also, the moment

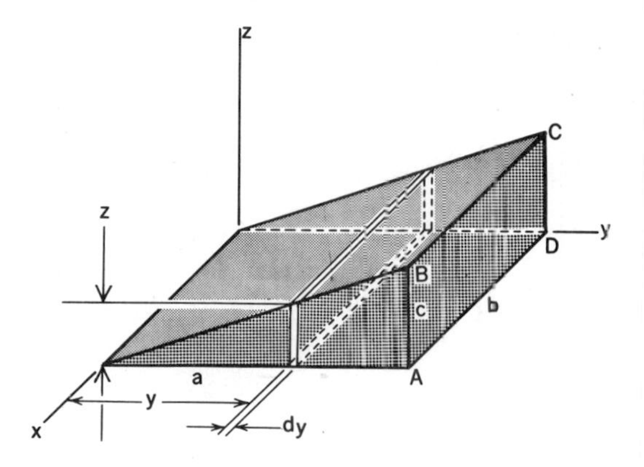

Figure 4.28. Find center of gravity.

of the weight of each element about the x axis is to be likewise easily computed. Infinitesimal slices of thickness dy parallel to the xz plane fulfill our requirements nicely. The weight of such a slice is simply $(zb\,dy)\rho g$, where z is the height of the slice (see Fig. 4.28). Because all points of the slice are a distance y from the x axis, clearly the moment of the weight of the slice is easily computed as $-y(zb\,dy)\rho g$. By letting y run from 0 to a during an integration, we can account for all the slices in the body. Thus, we have

$$-F_R\bar{y} = -\int_0^a y(zb\,dy)\rho g \qquad \text{(b)}$$

The term z can be expressed with the aid of similar triangles in terms of the integration variable y as follows:

$$\frac{z}{c} = \frac{y}{a}$$

$$z = \left(\frac{y}{a}\right)c \qquad \text{(c)}$$

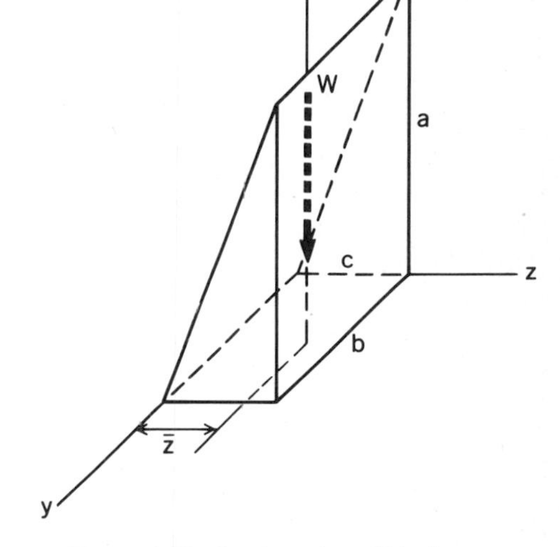

Figure 4.29. Reorientation of block.

We then have for Eq. (b), on replacing F_R using Eq. (a),

$$\bar{y} = \frac{1}{g\rho(abc/2)}g\int_0^a \rho y^2\frac{bc}{a}dy = \frac{2}{3}a \qquad \text{(d)}$$

To find the coordinate in the z direction to the center of gravity, we could reorient the body as shown in Fig. 4.29. A computation simliar to the preceding one would give the result that $\bar{z} = \frac{2}{3}c$. You are urged to verify this yourself.

Finally, it should be clear by inspection of Fig. 4.28 that $\bar{x} = \frac{1}{2}b$.

EXAMPLE 4.10

Find the center of gravity for the body of revolution shown in Fig. 4.30. The radial distance of the surface from the y axis is given as

$$r = \tfrac{1}{20}y^2 \text{ ft} \qquad \text{(a)}$$

The body has constant density ρ, is 10 ft long, and has a cylindrical hole at the right end of length 2 ft and diameter of 1 ft.

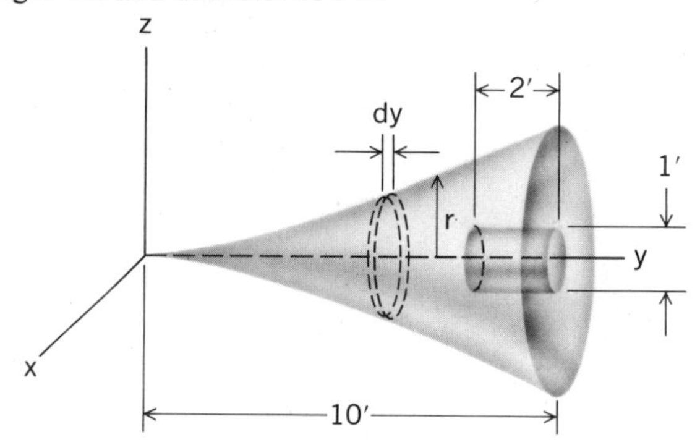

Figure 4.30. Body of revolution. Find center of gravity.

We need only compute $\bar{y}$, since it is clear that $\bar{z} = \bar{x} = 0$, owing to symmetry. We first compute the weight of the body. Using slices of thickness dy, as shown in the diagram, we sum the weight of all slices in the body assuming it is whole by letting y run from 0 to 10 in an integration. We then subtract the weight of a 2-ft cylinder of diameter 1 ft to take into account the cylindrical cavity inside the body. Thus, we have, noting that the area of a circle is πr^2 or $\pi D^2/4$

$$W = \int_0^{10} (\pi r^2)\, dy\, \rho g - \frac{\pi(1^2)}{4}(2)(\rho g) \tag{b}$$

Using Eq. (a) to replace r^2 in terms of y, we get

$$W = \rho g\left(\pi \int_0^{10} \frac{y^4}{400}\, dy - \frac{\pi}{2}\right) = g\rho\pi\left(50 - \frac{1}{2}\right) \tag{c}$$
$$= 49.5\, \pi\rho g \text{ lb}$$

To get $\bar{y}$, we equate the moment of W about the x axis with that of the weight distribution. For the latter, we sum the moments about the x axis of the weight of all slices, assuming first no inside cavity. Then, we subtract from this the moment about the x axis of the weight of a cylinder forming the cavity in the body. Because ρ is constant, the center of gravity of this latter cylinder is at its geometric center so that the moment arm from the x axis for the weight of the cylinder is clearly 9 ft. Thus, we have

$$-(49.5\pi\rho g)\bar{y} = -\rho g\left\{\int_0^{10} y(\pi r^2\, dy) - \left[\frac{\pi(1^2)}{4}(2)\right]9\right\}$$

$$\bar{y} = \frac{1}{49.5}\left(\int_0^{10} \frac{y^5}{400}\, dy - 4.5\right) = 8.33 \text{ ft}$$

Suppose as will be the case in Problem 4.40 that γ $(= \rho g)$, which is the *specific weight* (giving weight per unit volume), varies with y. That is, $\gamma = \gamma(y)$. Then for this problem, note that:

$$W = \int_0^{10} \pi r^2 \gamma \, dy - \int_8^{10} \pi \left(\frac{1}{2}\right)^2 \gamma \, dy$$

also

$$-W\bar{y} = -\left[\int_0^{10} y\pi r^2 \gamma \, dy - \int_8^{10} y\pi \left(\frac{1}{2}\right)^2 \gamma \, dy\right]$$

Note here we cannot take the short cuts used in the original problem where γ was a constant.

EXAMPLE 4.11

A plate is shown in Fig. 4.31 lying flat on the ground. The plate is 60 mm thick and has a uniform density. The curved edge is that of a parabola with zero slope at the origin. Find the coordinates of the center of gravity.

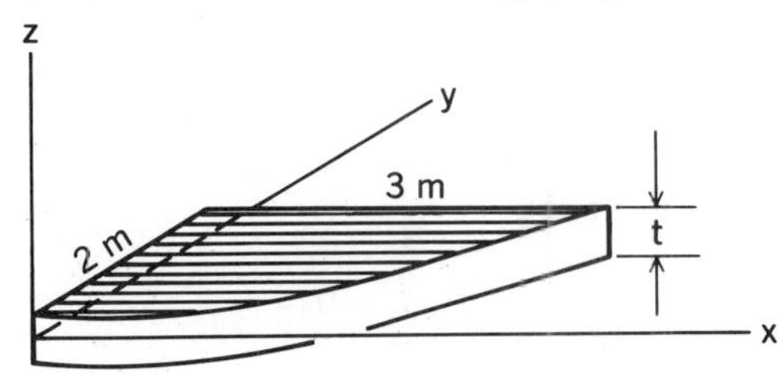

Figure 4.31. Find center of gravity of plate.

The equation of a parabola oriented like that of the curved edge of the plate is

$$y = Cx^2 \tag{a}$$

We can determine C by noting that $y = 2m$ when $x = 3m$. Hence,

$$2 = C \cdot 9 \tag{b}$$

Therefore,

$$C = \frac{2}{9}$$

The desired curve then is

$$y = \frac{2}{9}x^2 \tag{c}$$

Therefore,

$$x = \frac{3}{\sqrt{2}}y^{1/2}$$

We shall consider horizontal strips of the plate of width dy (see Fig. 4.32).

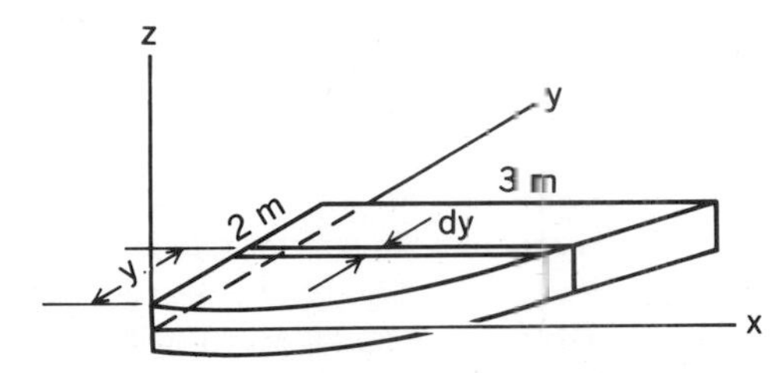

Figure 4.32. Use of horizontal strips.

Using the specific weight, γ, which has units of weight per volume and is equal to ρg, we have for the total weight W of the plate:

$$W = \int_0^2 (dy)(t)(x)\gamma$$

We replace x using Eq. (c) to get

$$W = t\gamma \int_0^2 \left(\frac{3}{\sqrt{2}} y^{1/2}\right) dy$$

where t is the thickness. Integrating, we get

$$W = t\gamma \frac{3}{\sqrt{2}} (y^{3/2})\left(\frac{2}{3}\right)\Big|_0^2 = t\gamma\sqrt{2}(2)^{3/2} = 4t\gamma \text{ N} \tag{d}$$

We next take moments about the x axis in order to get $\bar{y}$. Thus,

$$\begin{aligned} -W\bar{y} &= -\int_0^2 y(t\,dy\,x)\gamma \\ &= -\gamma t \int_0^2 (y)\left(\frac{3}{\sqrt{2}} y^{1/2}\right) dy \\ &= -\gamma t \frac{3}{\sqrt{2}} (y^{5/2})\left(\frac{2}{5}\right)\Big|_0^2 \\ &= -\gamma t\left(\frac{3}{\sqrt{2}}\right)\left(\frac{2}{5}\right)[(2^2)(2^{1/2})] \\ &= -\frac{24}{5}\gamma t \end{aligned} \tag{e}$$

Using $4t\gamma$ for W from Eq. (d), we get, for $\bar{y}$:

$$\bar{y} = \tfrac{6}{5} \text{ m} \tag{f}$$

To get $\bar{x}$, we take moments about the y axis, still utilizing the horizontal strips of Fig. 4.32. The center of gravity of a strip is at its center since γ is constant and so the moment arm about the y axis is $x/2$.

$$W\bar{x} = \int_0^2 \frac{x}{2}(t\gamma\,dy\,x) \tag{g}$$

Continuing with the calculations, we have

$$\begin{aligned} W\bar{x} &= \frac{t\gamma}{2}\int_0^2 x^2\,dy = \frac{t\gamma}{2}\int_0^2 \left(\frac{9}{2}y\right) dy \\ &= \frac{t\gamma}{2}\,\frac{9}{2}\,\frac{y^2}{2}\Big|_0^2 = \frac{9t\gamma}{2} \end{aligned}$$

On replacing W according to Eq. (d), we get, for $\bar{x}$:

$$\bar{x} = \tfrac{9}{8} \text{ m} \tag{h}$$

Finally, is clear that the $\bar{z}$ coordinate is zero for reference xy at the center plane of the plate.

As an exercise (Problem 4.39) you may be asked to solve this problem using vertical strips.

In the previous problems, we used slices of the body having a thickness dy. If the specific weight were a function of position, $\gamma(x, y, z)$, we could not readily use such slices, since we cannot easily express the weight of such slices in a simple manner. The reason for this is that in the x and z directions the dimensions of the element are finite, and so γ would vary in these directions throughout the element. If, however, we choose an element that is infinitesimal in *all directions*, such as an infinitesimal rectangular parallelepiped, $dx\, dy\, dz$, then γ can be assumed to be constant throughout the element. The weight of the element is then easily seen to be $\gamma(dx\, dy\, dz)$, where the coordinates of γ correspond to the position of the element. We now illustrate a simple case.

*EXAMPLE 4.12

Consider a block (see Fig. 4.33) wherein the specific weight γ at corner A is 200 lbf/ft³. The specific weight in the block does not change in the x direction. However, it decreases linearly by 50 lbf/ft³ in the y direction, and increases linearly by 50 lbf/ft³ in the z direction, as has been shown in the diagram. What are the coordinates $\bar{x}$, $\bar{y}$ of the center of gravity for this block?

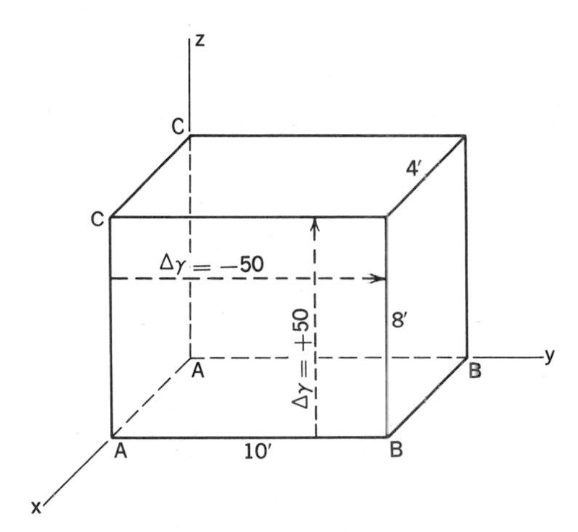

Figure 4.33. Block with varying γ

We must first express γ at any position $P(x, y, z)$. Using simple proportions, we can say

$$\begin{aligned}\gamma &= 200 - \frac{y}{10}(50) + \frac{z}{8}(50) \\ &= 200 - 5y + 6.25z \text{ lbf/ft}^3\end{aligned} \qquad \text{(a)}$$

We shall first compute the weight of the block (i.e., the resultant force of gravity). We do not use an infinitesimal slice or rectangular rod of the block, as we have done heretofore. With the specific weight varying with both y and z, it would not be an easy matter to compute the weight and moment of a slice or a rod. Instead, we shall use an infinitesimal rectangular parallelepiped, $dx\, dy\, dz$, at position xyz as has been shown in Fig. 4.34(a). Because of the vanishingly small size of this element, the

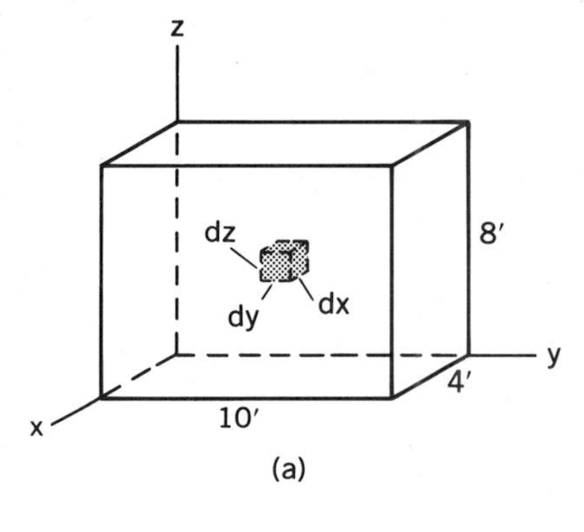

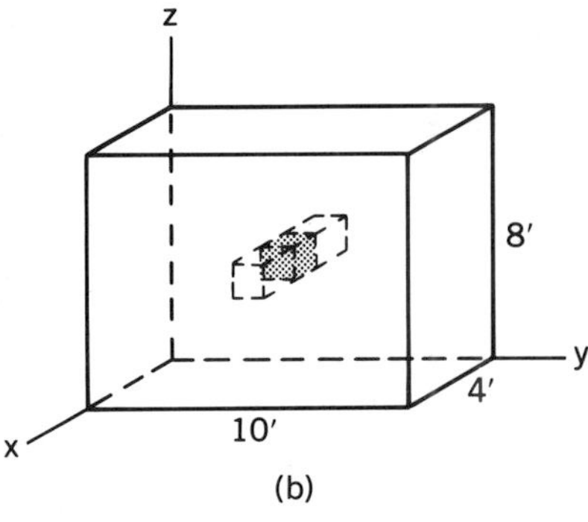

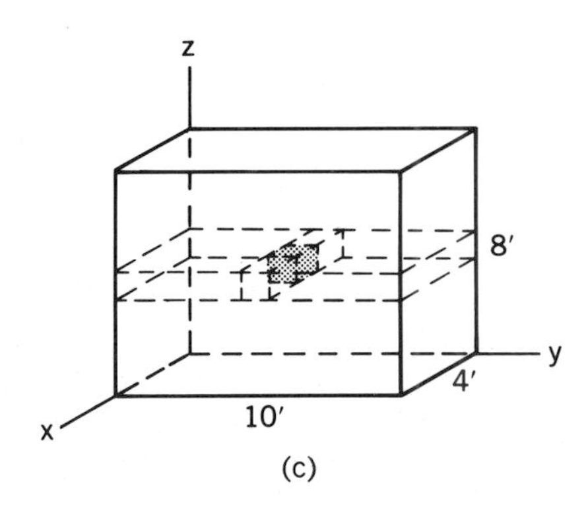

Figure 4.34. (a) Element $dx\, dy\, dz$ at $P\,(x, y, z)$; (b) x runs from 0 to 4, while z and y are fixed, to form rectangular rod; (c) y runs from 0 to 10, while holding z, to form slice.

specific weight γ can be considered constant inside the element, and so the weight dW of the element can be given as

$$dW = \gamma(dx\,dy\,dz) = (200 - 5y + 6.25z)\,dx\,dy\,dz$$

To include the weight of *all* such elements in the block, we first let x "run" from 0 to 4 ft while holding y and z fixed. The rectangular parallelepiped of Fig. 4.34(a) then becomes a rectangular rod as shown in Fig. 4.34(b). Having run its course, x is no longer a variable in this summation process. Next, let y "run" from 0 to 10 while holding z constant. The rectangular rod of Fig. 4.34(b) then becomes an infinitesimal slice, as shown in Fig. 4.34(c). The variable y has thus run its course and is no longer a variable. This leaves only the variable z, and now we let z "run" from 0 to 8. Clearly, we cover the entire block by this process.

We can do this mathematically by a process called *multiple integration.* We perform three integrations, paralleling the three steps outlined in the previous paragraph. Thus, we can formulate W as follows:

$$-W\mathbf{k} = \int_0^8 \int_0^{10} \int_0^4 (200 - 5y + 6.25z)\,dx\,dy\,dz\,(-\mathbf{k})$$

We first consider the integration.

$$\int_0^4 (200 - 5y + 6.25z)\,dx$$

As in the first step set forth in the previous paragraph, to go from a rectangular parallelepiped to a rectangular rod, we integrate with respect to x from $x = 0$ to $x = 4$ while holding y and z constant. Thus,

$$\int_0^4 (200 - 5y + 6.25z)\,dx = (200x - 5yx + 6.25zx)\Big|_0^4$$
$$= 800 - 20y + 25z$$

With x no longer a variable (since it has run its course), the equation for W becomes

$$W = \int_0^8 \int_0^{10} (800 - 20y + 25z)\,dy\,dz$$

Now, we hold z constant and integrate with respect to y from 0 to 10. (This takes us from a rectangular rod to a slice.) Thus,

$$\int_0^{10} (800 - 20y + 25z)\,dy = \left(800y - 20\frac{y^2}{2} + 25zy\right)\Big|_0^{10}$$
$$= 8000 - 1000 + 250z$$

Now y has run its course, and we have

$$W = \int_0^8 (7000 + 250z)\, dz$$

By integrating with respect to z, we sum up all the slices, and we have covered the entire block. Thus,

$$W = \left(7000z + 250\frac{z^2}{2}\right)\Big|_0^8 = 64{,}000 \text{ lb}$$

To get $\bar{y}$, we equate the moment about the x axis of the resultant force without a couple moment with the moment of the distribution. Thus, using multiple integration as described above:

$$-(64{,}000)\bar{y} = -\int_0^8 \int_0^{10} \int_0^4 y(200 - 5y + 6.25z)\, dx\, dy\, dz$$

Therefore,

$$64{,}000\bar{y} = \int_0^8 \int_0^{10} (200yx - 5y^2x + 6.25yzx)\Big|_0^4 dy\, dz$$

$$= \int_0^8 \int_0^{10} (800y - 20y^2 + 25yz)\, dy\, dz$$

$$= \int_0^8 \left(800\frac{y^2}{2} - \frac{20y^3}{3} + \frac{25y^2}{2}z\right)\Big|_0^{10} dz$$

$$= \int_0^8 (40{,}000 - 6667 + 1250z)\, dz$$

$$= \left(33{,}333z + 1250\frac{z^2}{2}\right)\Big|_0^8 = 307{,}000$$

and

$$\bar{y} = 4.79 \text{ ft}$$

Because γ does not depend on x, we can directly conclude by inspection that $\bar{x} = 2$.

Case B. Parallel Force Distribution over a Plane Surface—Center of Pressure. Let us now consider a normal pressure distribution over a *plane* surface A in the xy plane in Fig. 4.35. The vertical ordinate is taken as a pressure ordinate, so that over the area A we have a pressure distribution $p(x, y)$ represented by the pressure surface. Since in this case there is a parallel force system with one sense of direction, we know that the

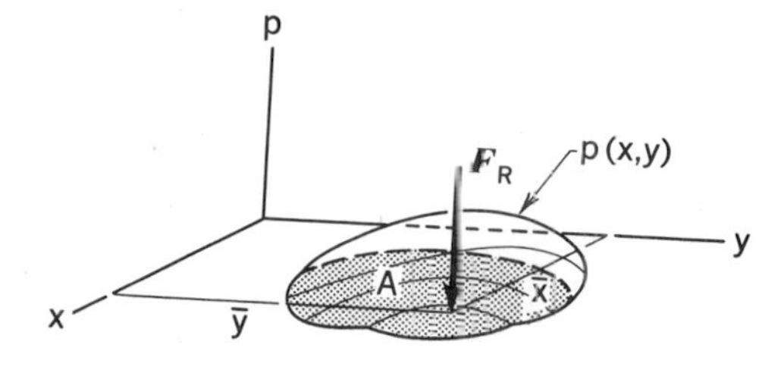

Figure 4.35. Pressure distribution.

simplest resultant is a single force, which is given as

$$\boldsymbol{F}_R = -\int p\, dA = -\left(\int p\, dA\right)\boldsymbol{k} \tag{4.12}$$

The position $\bar{x}, \bar{y}$ can be computed by equating the moments about the x and y axes of the resultant force without a couple moment with the corresponding moments of the distribution. Solving for $\bar{x}$ and $\bar{y}$,

$$\bar{x} = \frac{\int px\, dA}{\int p\, dA}$$

$$\bar{y} = \frac{\int py\, dA}{\int p\, dA}$$

Since we know that p is a function of x and y over the surface, we can carry out the preceding integrations either analytically or numerically. The point thus determined is called the *center of pressure*.

(In later chapters, we shall consider distributed frictional forces over plane and curved surfaces. In these cases, the simplest resultant is not necessarily a single force as it was in the special case above.)

EXAMPLE 4.13

A plate *ABCD* on which both distributed and point force systems act is shown in Fig. 4.36. The pressure distribution is given as

$$p = -4y^2 + 100 \text{ psf} \tag{a}$$

Find the simplest resultant for the system.

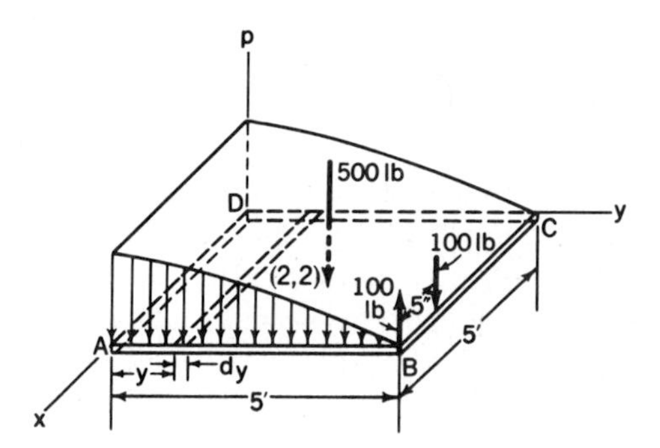

Figure 4.36. Find simplest resultant.

To get the resultant force, we consider a strip dy along the plate as shown in Fig. 4.36. The reason for using such a strip is that the pressure p is uniform along this strip, as can be seen from the diagram. Hence, the force from the pressure on the strip is simply $p\,dA = p(dy)(5)$. In summing forces, we simply integrate the forces $p(dy)(5)$ over all the strips of the plate. Thus, we can say that

$$\begin{aligned} F_R &= -\int_0^5 p(5)(dy) - 500 \\ &= -\int_0^5 (-4y^2 + 100)5\,dy - 500 \qquad \text{(b)} \\ &= \left(20\frac{y^3}{3} - 500y\right)\Big|_0^5 - 500 = -2167 \text{ lb} \end{aligned}$$

To get the position $\bar{x}, \bar{y}$ of the resultant force F_R without a couple moment, we equate moments of F_R about the x and y axes with that of the original system. Thus, starting with the x axis, we have using strip dy as before:

$$\begin{aligned} -2167\bar{y} &= -\int_0^5 yp(5dy) - (500)(2) \\ &= -\int_0^5 5y(-4y^2 + 100)\,dy - 1000 \\ &= \left(20\frac{y^4}{4} - 500\frac{y^2}{2}\right)\Big|_0^5 - 1000 = -4125 \end{aligned}$$

Therefore,

$$\bar{y} = 1.904 \text{ ft}$$

Now, considering the y axis, we still use the strips dy because p is uniform along such strips. However, the force $df = p\,dA = p(5)(dy)$ may be considered acting at the center of the strip, and accordingly has a moment arm about the y axis equal to $\frac{5}{2}$ for each strip. Hence, we can say that

$$\begin{aligned} 2167\bar{x} &= \int_0^5 \frac{5}{2}p(5dy) + (500)(2) - \frac{500}{12} \\ &= \frac{25}{2}\int_0^5 (-4y^2 + 100)\,dy + 1000 - 41.7 = 5125 \end{aligned}$$

Therefore,

$$\bar{x} = 2.36 \text{ ft}$$

***EXAMPLE 4.14**

What is the simplest resultant and the center of pressure for the pressure distribution shown in Fig. 4.37?

Notice that the pressure varies linearly in the x and y directions. The pressure at any point x, y in the distribution can be given as follows with the aid of similar triangles[5]:

$$\begin{aligned} p &= \left(\frac{y}{10}\right)(20) + \left(\frac{x}{5}\right)(30) \qquad \text{(a)} \\ &= 2y + 6x \text{ Pa} \end{aligned}$$

[5]The unit of pressure in SI units is the pascal, where

$$1 \text{ Pa} \equiv 1 \text{ N/m}^2$$

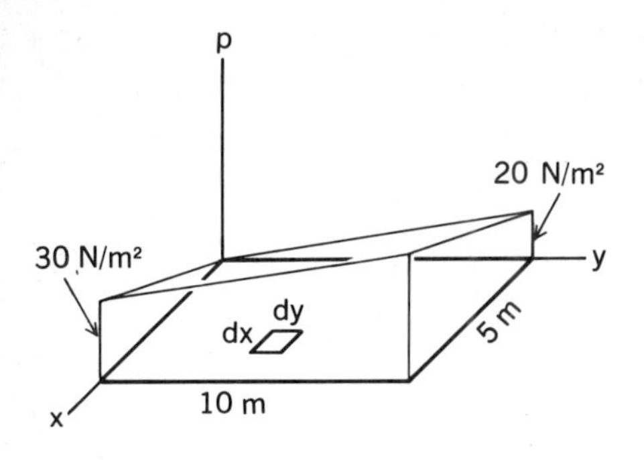

Figure 4.37. Nonuniform pressure distribution.

We cannot employ a convenient strip here along which the pressure is uniform, as in Example 4.13. For this reason we consider rectangular area element $dx\,dy$ to work with (see Fig. 4.37). For such small area, we can assume the pressure as constant so that $p\,dx\,dy$ is the force on the element. To find the resultant force, we must integrate over the 10×5 rectangle. This integration involves two variables and is again a case of *multiple integration*. Thus, we can say that

$$F_R = \int_0^{10} \int_0^5 p\,dx\,dy$$

wherein we first integrate with respect to x while holding y constant and then integrate with respect to y (in this way we cover the entire 10×5 rectangular area). Thus, we have

$$\begin{aligned}
F_R &= \int_0^{10} \int_0^5 (2y + 6x)\,dx\,dy \\
&= \int_0^{10} \left(2yx + \frac{6x^2}{2}\right)\Bigg|_0^5 dy \\
&= \int_0^{10} (10y + 75)\,dy \\
&= \frac{10y^2}{2} + 75y \Bigg|_0^{10} = 1250 \text{ N}
\end{aligned}$$

To find $\bar{y}$ for F_R without a couple moment, we equate moments of F_R about the x axis with that of the distribution. Thus,

$$-(\bar{y})(1250) = -\int_0^{10} \int_0^5 py\,dx\,dy$$

Therefore,

$$\begin{aligned}
\bar{y} &= \frac{1}{1250} \int_0^{10} \int_0^5 (2y + 6x)y\,dx\,dy \\
&= \frac{1}{1250} \int_0^{10} \left(2y^2x + 6y\frac{x^2}{2}\right)\Bigg|_0^5 dy \\
&= \frac{1}{1250} \int_0^{10} (10y^2 + 75y)\,dy \\
&= \frac{1}{1250}\left(\frac{10y^3}{3} + 75\frac{y^2}{2}\right)\Bigg|_0^{10} \\
&= 5.67 \text{ m}
\end{aligned}$$

As for $\bar{x}$, we proceed as follows:

$$(\bar{x})(1250) = \int_5^{10} \int_0^5 px\,dx\,dy$$

Therefore,

$$\begin{aligned}\bar{x} &= \frac{1}{1250}\int_0^{10}\int_0^5 (2y + 6x)x\,dx\,dy \\ &= \frac{1}{1250}\int_0^{10}\left(2y\frac{x^2}{2} - \frac{6x^3}{3}\right)\Bigg|_0^5 dy \\ &= \frac{1}{1250}\int_0^{10} (25y + 250)\,dy \\ &= \frac{1}{1250}\left(25\frac{y^2}{2} + 250y\right)\Bigg|_0^{10} \\ &= 3.00 \text{ m}\end{aligned}$$

The center of pressure is thus at (3.00, 5.67) m.

Case C. Coplanar Parallel Force Distribution. As we pointed out earlier, this type of loading may be considered for beams loaded symmetrically over the longitudinal midplane of the beam. The loading is represented by an intensity function $w(x)$ as shown in Fig. 4.25. This coplanar parallel force distribution can be replaced by a single force given as

$$\boldsymbol{F}_R = -\int w(x)\,dx\,\boldsymbol{j}$$

We find the position of $\boldsymbol{F}_R$ without a couple moment by equating moments of $\boldsymbol{F}_R$ and the distribution w about a convenient point of the beam, usually one of the ends. Solving for $\bar{x}$, we get

$$\bar{x} = \frac{\int xw(x)\,dx}{\int w(x)\,dx}$$

EXAMPLE 4.15

A simply supported beam is shown in Fig. 4.38 supporting a 1000-lb point force, a 500 lb-ft couple, and a coplanar, parabolic, distributed load w lb/ft. Find the simplest resultant of this force system.

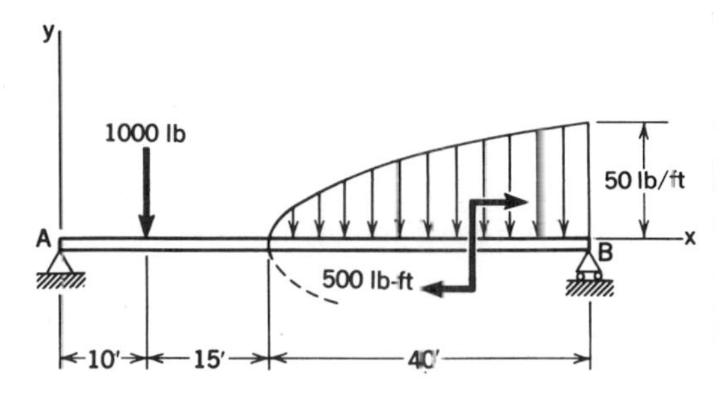

Figure 4.38. Find simplest resultant.

To express the intensity of loading for the coordinate system shown in the diagram, we begin with the general formulation

$$w^2 = ax + b \tag{a}$$

Note from the diagram that when $x = 25$ we have $w = 0$, and when $x = 65$, we have $w = 50$. Subjecting Eq. (a) to these conditions, we can determine a and b. Thus,

$$0 = a(25) + b \tag{b}$$

$$2500 = a(65) + b \tag{c}$$

Subtracting, we can get a as follows:

$$-2500 = -40a$$

Therefore,

$$a = 62.5$$

From Eq. (b), we get

$$b = -(25)(62.5) = -1562.5$$

Thus, we have

$$w^2 = 62.5x - 1562.5 \text{ lb/ft} \tag{d}$$

Summing forces, we get for F_R,

$$F_R = -1000 - \int_{25}^{65} \sqrt{62.5x - 1562.5}\, dx \tag{e}$$

To integrate this, we may change variables as follows:

$$\mu = 62.5x - 1562.5 \tag{f}$$

Therefore,

$$d\mu = 62.5\, dx$$

Substituting into the integral in Eq. (e), we have[6]

$$\begin{aligned} F_R &= -1000 - \int_0^{2500} \mu^{1/2} \frac{d\mu}{62.5} \\ &= -1000 - \frac{1}{62.5}\mu^{3/2}\left(\frac{2}{3}\right)\Bigg|_0^{2500} \\ &= -1000 - \frac{1}{62.5}(2500)^{3/2}\left(\frac{2}{3}\right) \\ &= -2333 \text{ lb} \end{aligned}$$

We now compute $\bar{x}$ for the resultant without a couple as follows:

$$-2333\bar{x} = -(10)(1000) - \int_{25}^{65} x\sqrt{62.5x - 1562.5}\, dx - 500 \tag{g}$$

We can evaluate the integral most readily by consulting the mathematical formulas in Appendix I. We find the following formula:

[6]Do not forget to change the limits for μ. Thus, from Eq. (f), the upper limit is $(62.5)(65) - 1562.5 = 2500$, whereas the lower limit is $(62.5)(25) - 1562.5 = 0$.

$$\int x\sqrt{a+bx}\,dx = -\frac{2(2a-3bx)\sqrt{(bx+a)^3}}{15b^2}$$

In our case $b = 62.5$ and $a = -1562.5$, so the indefinite integral for our case is

$$\int x\sqrt{62.5x - 1562.5}\,dx = -\frac{(2)(-3125 - 187.5x)\sqrt{(62.5x - 1562.5)^3}}{(15)(3906)}$$

Putting in limits, we have

$$\int_{25}^{65} x\sqrt{62.5x - 1562.5}\,dx = -\frac{(2)(-3125 - 187.5x)\sqrt{(62.5x - 1562.5)^3}}{(15)(3906)}\Bigg|_{25}^{65}$$

$$= 65{,}333 - 0 = 65{,}333$$

Going back to Eq. (g), we can now solve easily for $\bar{x}$. Thus,

$$\bar{x} = -\frac{1}{2330}[-(10)(1000) - 65{,}300 - 500]$$

$$= 32.5 \text{ ft}$$

Before closing, it will be pointed out that, for a loading function $w(x)$, the resultant, $\int_0^x w\,dx$, equals the *area* under the loading curve. This fact is particularly useful for the case of a triangular loading function such as is shown in Fig 4.39. Hence, we can say on inspection that the resultant force has the value

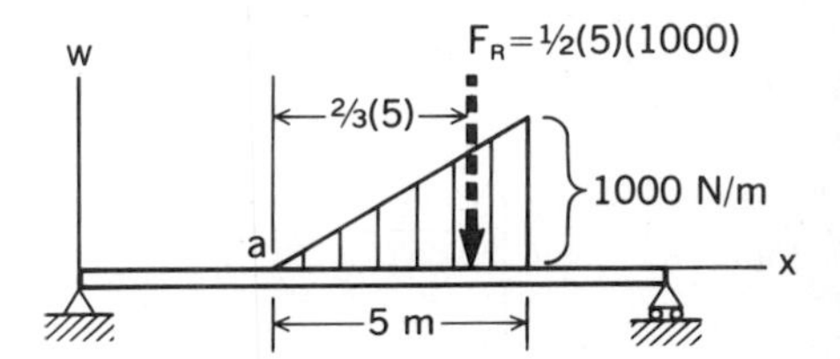

Figure 4.39. Triangular loading resultant.

$$F_R = \tfrac{1}{2}(5)(1000) = 2500 \text{ N}$$

Furthermore, you can readily show that the *simplest* resultant has a line of action that is $(\frac{2}{3}) \times$ (length of loading) from the toe of the loading.[7] Thus, F_R without a couple moment is at a position $(\frac{2}{3})(5)$ to the right of a (see Fig. 4.39). You are urged to use this information when needed.

Finally, in the case of a body made up of simple shapes (subbodies) such as cones and cubes, we can find the center of gravity by using the centers of gravity of the known shapes. Thus, we can say on taking moments about the y axis that

$$W_{\text{total}}(x_c) = \sum_i W_i(x_c)_i \tag{4.13}$$

where W_i is the weight of the ith subbody and where $(x_c)_i$ is the x coordinate to the center of gravity of the ith subbody. Such bodies are called *composite bodies* (see Fig. P.4.51 for an example).

[7] In Chapter 8, you will learn that the simplest resultant force for a distribution $w(x)$ goes through the *centroid* of the area under $w(x)$. The centroid will be carefully defined at that time.

Problems

4.37. A force field is given as

$$F(x, y, z, t) = (10x + 5)\boldsymbol{i} + (16x^2 + 2z)\boldsymbol{j} + 15\boldsymbol{k}\ \text{N}$$

What is the force at position (3, 6, 7) m? What is the difference between the force at this position and that at the origin?

4.38. A magnetic field is developed such that the body force on the rectangular parallelepiped of metal is given as

$$\boldsymbol{f} = (.01x + \tfrac{1}{8})\boldsymbol{k}\ \text{oz/lbm}$$

If the specific weight of the metal is 450 lb/ft³, what is the *simplest* resultant body force from such a field? Note that at the earth's surface the number of pounds mass equals the number of pounds force.

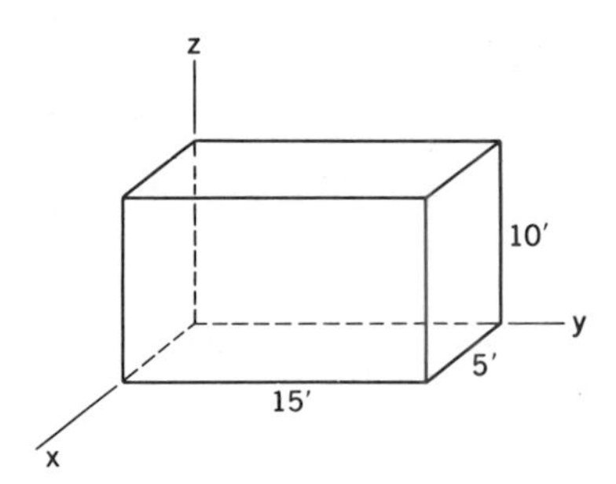

Figure P.4.38

4.39. Do the problem given in Example 4.11 using vertical strips.

4.40. A body of revolution has a variable specific weight such that $\gamma = (36 + .01x^2)$ kN/m³ with x in meters. A hole of diameter 3 m and length 6 m is cut from the body as shown. Where is the center of gravity?

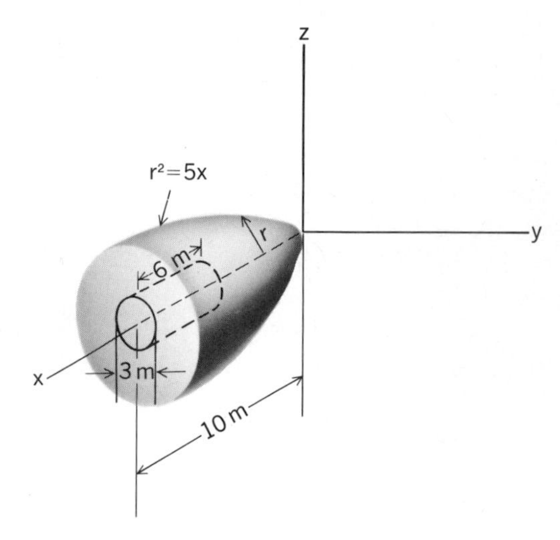

Figure P.4.40

4.41. The specific weight γ of the material in the solid cylinder varies linearly as one goes from face A to face B. If

$$\gamma_A = 400\ \text{lbf/ft}^3, \qquad \gamma_B = 500\ \text{lbf/ft}^3$$

what is the position of the center of gravity of the cylinder?

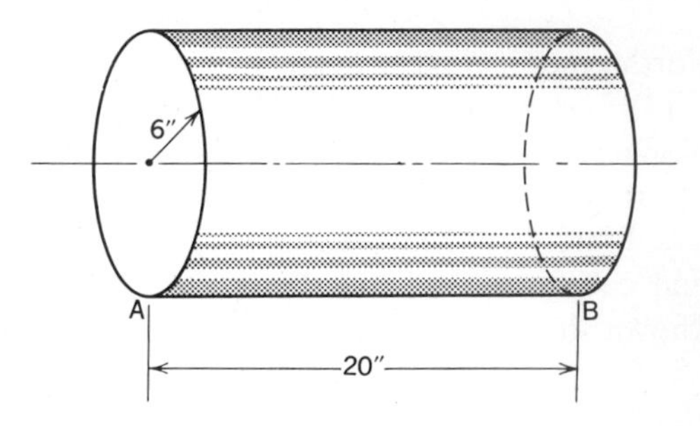

Figure P.4.41

4.42. The specific weight of the material in a right circular cone is constant. What is the center of gravity of the cone? *Hint*: Rotate cone 90° so that gravity is perpendicular to the z axis. Use concept of similar triangles to show that $r/R = (h - z)/h$ and solve for r needed for the integration.

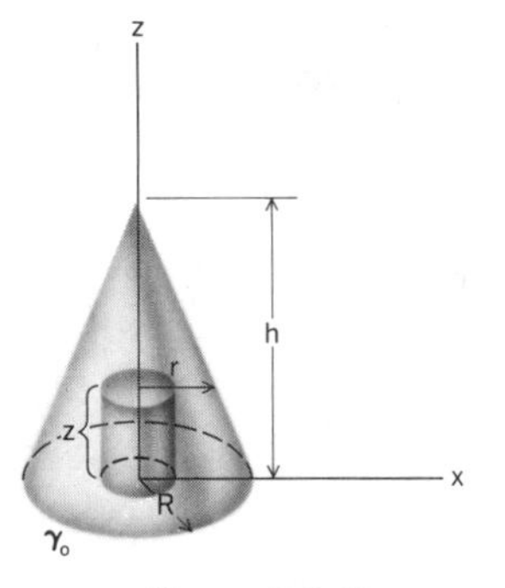

Figure P.4.42

4.43. Show that the center of gravity of the right triangular plate of thickness t is at $x = a/3$ and $y = b/3$.

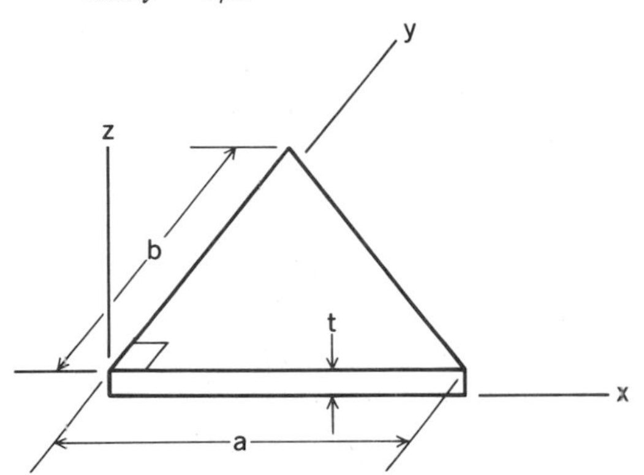

Figure P.4.43

4.44. Show that the volume and center of gravity of the conical frustum are, respectively,

$$\frac{\pi h}{3}(r_2^2 + r_1 r_2 + r_1^2)$$

and

$$\frac{h}{4}\,\frac{3r_2^2 + r_1^2 + 2r_1 r_2}{r_2^2 + r_1^2 + r_1 r_2}$$

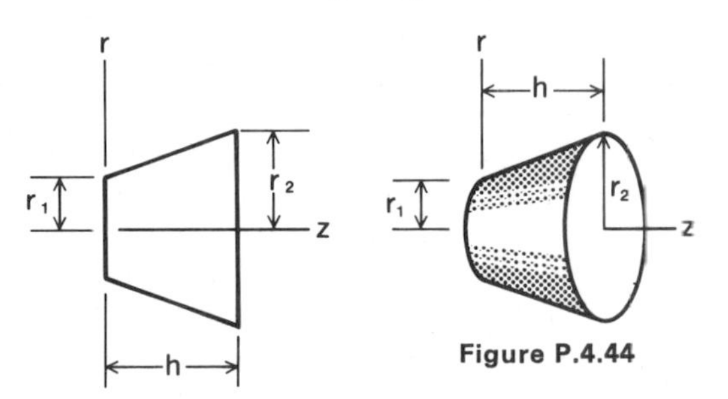

Figure P.4.44

4.45. Find the center of gravity of the plate bounded by a straight line and a parabola.

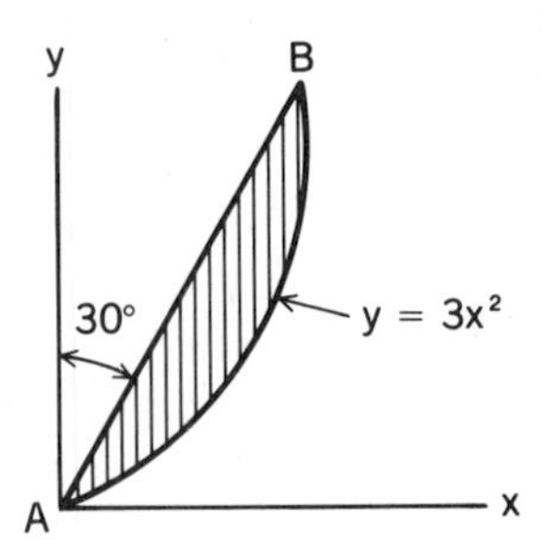

Figure P.4.45

4.46. A massive radio-wave antenna for detection of signals from outer space is a body of revolution with a parabolic face (see the diagram). These antennas may be carved from rock in a valley away from other disturbing signals. What would the antenna weigh if made from concrete (23. 6 kN/m³) for location in a remote desert area?

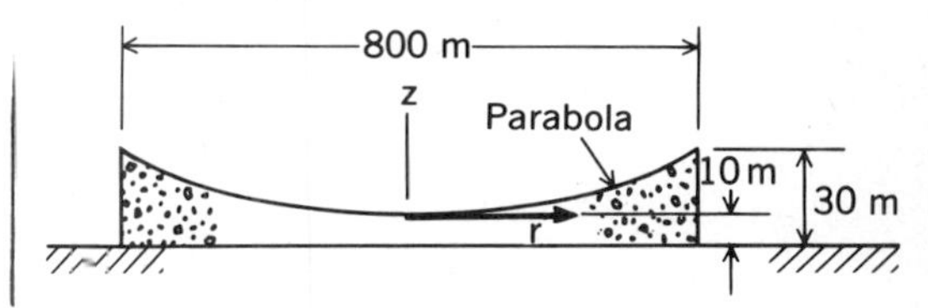

Figure P.4.46

4.47. In Problem 4.46, find the distance from the ground to the center of gravity if the total weight is 2.37×10^8 kN.

***4.48.** A plate of thickness 30 mm has a specific weight γ that varies linearly in the x direction from 26 kN/m³ at A to 36 kN/m³ at B, and varies as the square of y from 26 kN/m³ at A to 40 kN/m³ at C. Where is the center of gravity of the plate?

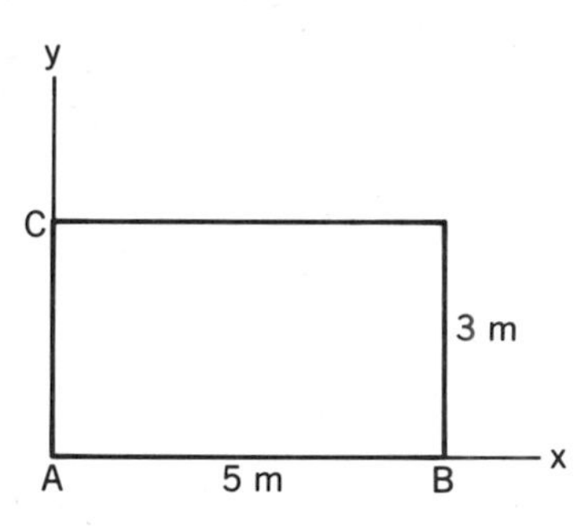

Figure P.4.48

***4.49.** Suppose in Problem 4.38 that

$$f = (.01x + .2y + .3z)\boldsymbol{k} \text{ oz/lbm}$$

Find the *simplest* resultant for $\gamma = 450$ lb/ft³. Find the proper line of action.

4.50. After a fast stop and swerve to the left, the load of sand (specific weight = 15 kN/m³) in a dump truck is in the position shown. What is the simplest resultant force on the truck from the sand and where does it act? If the truck was full (with a level top) before the stop, how much sand spilled? Use the results of Problem 4.43.

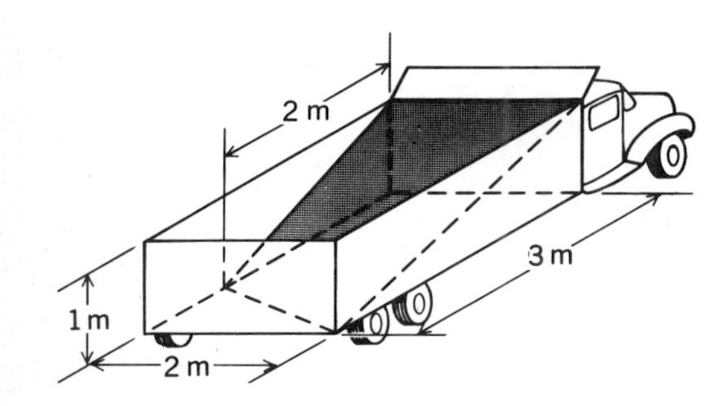

Figure P.4.50

In Problems 4.51 through 4.53, use the known positions of centers of gravity of simple shapes.

4.51. Find the weight and center of gravity of a large steam turbine for power generation needed for earthquake safety calculations. The specific weights of each turbine component are shown. The big cylinder having a radius r_2 of 5 m is 14 m long. Half of this cylinder is embedded in the large block.

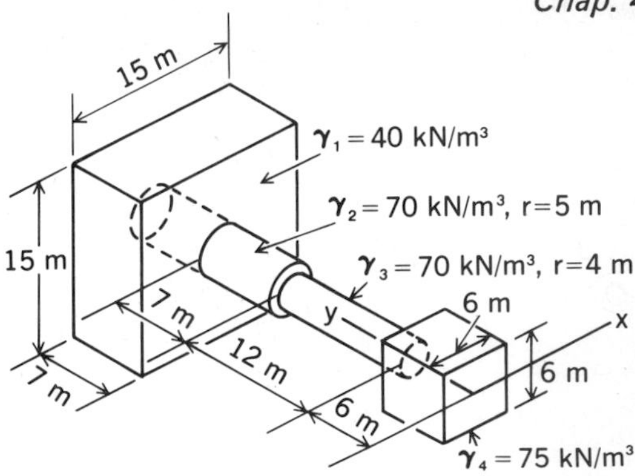

Figure P.4.51

4.52. An I-beam cantilevered out from a wall weighs 30 lb/ft and supports a 300-lb hoist. Steel (487 lb/ft³) cover plates 1 in. thick are welded on the beam near the wall to increase the carrying capacity of the beam. What is the moment at the wall due to the weight of the reinforced beam and the hoisted load of 4000 lb at the outermost position of the hoist? What is the simplest resultant force and its location?

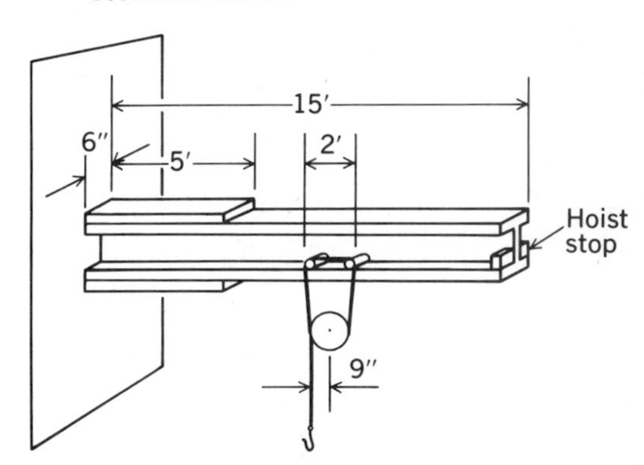

Figure P.4.52

4.53. The bulk materials trailer weighs 10,000 lb and is filled with cement ($\gamma = 94$ lb/ft³) in the front compartment (sections 1 and 2), and half-filled with water ($\gamma = 62.5$ lb/ft³) in the rear compartment (sections 3 and 4). What is the simplest resultant force, and where does it act? What is the resultant when the water is drained? Use the center of gravity and volume results from Problem 4.44 (conical frustum).

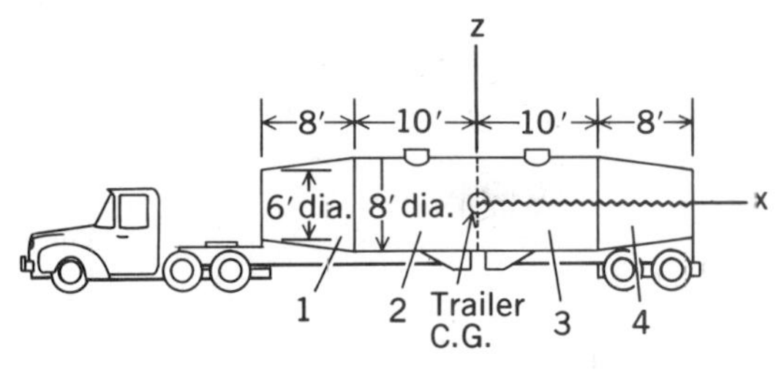

Figure P.4.53

4.54. Find the simplest resultant of a normal pressure distribution over the rectangular area with sides a and b. Give the coordinates of the center of pressure.

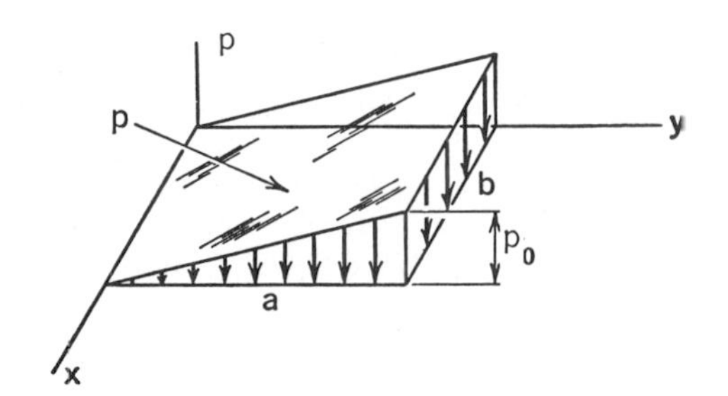

Figure P.4.54

4.55. Find the simplest resultant acting on wall $ABCD$. Give the coordinates of the center of pressure. The pressure varies such that $p = A/(y + 1) + B$ psi, with y in feet, from 10 psi to 50 psi, as indicated in the diagram.

4.56. One floor of a warehouse is divided into four areas. Area 1 is stacked high with TV sets such that the distributed load is $p = 120\ \text{lb/ft}^2$. Area 2 has refrigerators with $p = 65\ \text{lb/ft}^2$. Area 3 has stereos stacked so that $p = 80\ \text{lb/ft}^2$. Area 4 has washing machines with $p = 50\ \text{lb/ft}^2$. What is the simplest resultant force and where does it act?

4.57. Consider a pressure distribution p forming a hemispherical surface over a domain of radius 5 m. If the maximum pressure is 5 Pa, what is the *simplest* resultant from this pressure distribution?

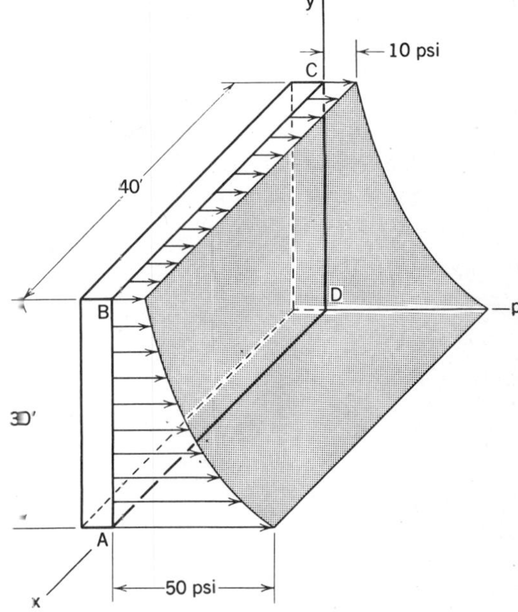

Figure P.4.55

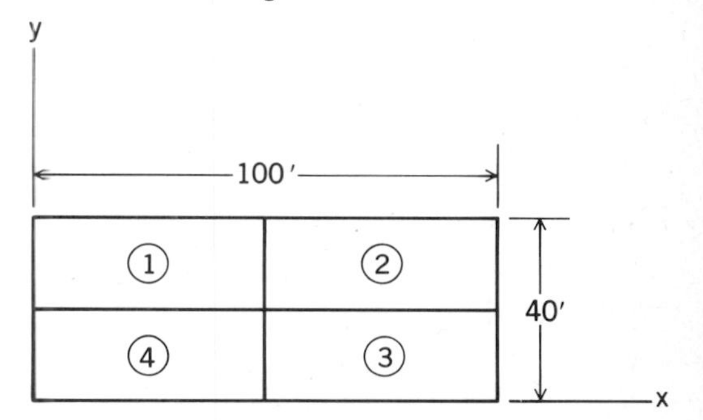

Figure P.4.56

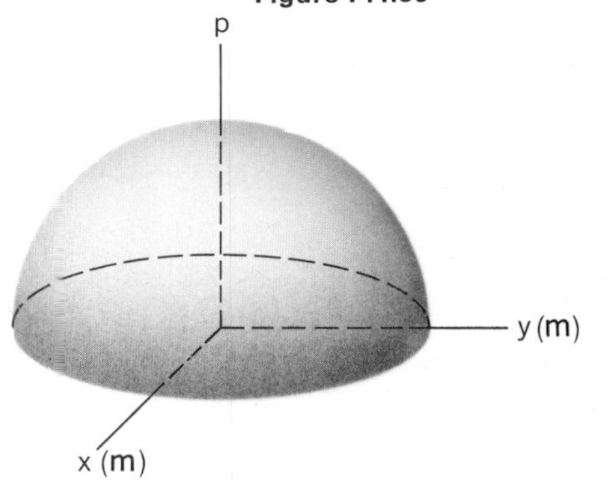

Figure P.4.57

***4.58.** The pressure p_0 at the corner O of the plate is 50 Pa and increases linearly in the y direction by 5 Pa/m. In the x direction, it increases parabolically starting with zero slope so that in 20 m the pressure has gone from 50 Pa to 500 Pa. What is the simplest resultant for this distribution? Give the coordinates of the center of pressure.

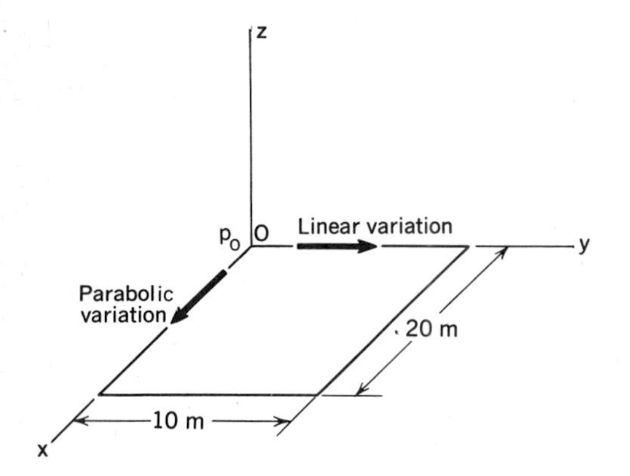

Figure P.4.58

In Problems 4.59 through 4.63, we are concerned with pressure from water on a submerged surface. These are hydrostatic problems. You learned in physics that the pressure in water is γd, where γ is the specific weight and d is the vertical depth below the free surface.

4.59. A sluice-gate door in a dam is 3 m wide and 3 m high. The water level in the dam is 4 m above the top of the door. The gate is opened until the water level falls 4 m. What is the simplest resultant force on the closed door at both water levels? Where do the forces act (i.e., where is the "center of pressure" in each case)? Water weighs 9818 N/m³.

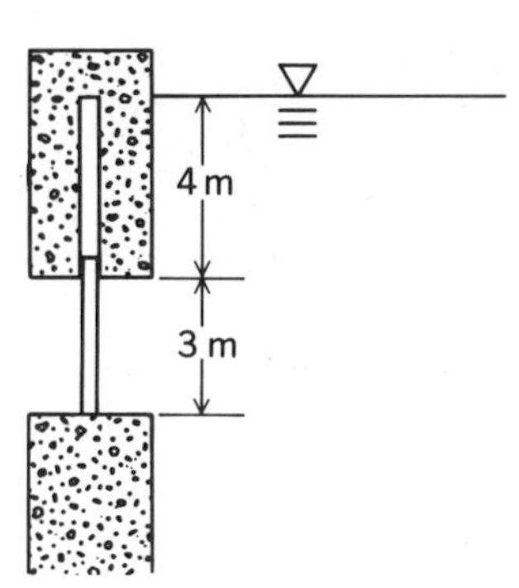

Figure P.4.59

4.60. A cylindrical tank of water is rotated at constant angular speed ω until the water ceases to change shape. The result is a free surface which, from fluid mechanics considerations, is that of a paraboloid. If the pressure varies directly as the depth below the free surface, what is the resultant force on a quadrant of the base of the cylinder? Take $\gamma = 62.4$ lb/ft³. [*Hint:* Use circular strip in quadrant having area $\frac{1}{4}(2\pi r\, dr)$.]

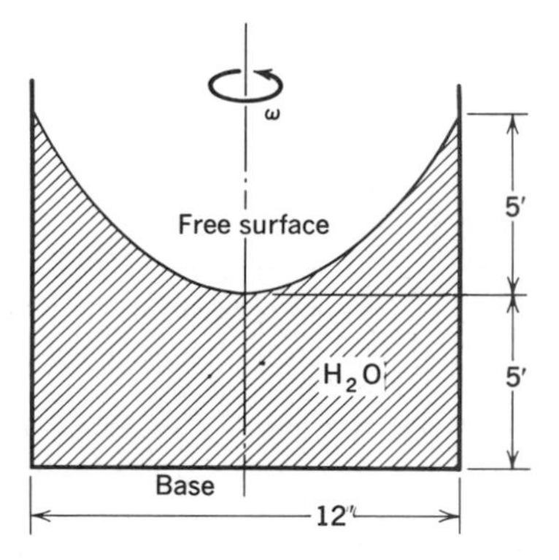

Figure P.4.60

4.61. What is the simplest resultant force from the water and where does it act on the 60-m-high 800-m-long straight earthfill dam? (Water weighs 9818 N/m³.)

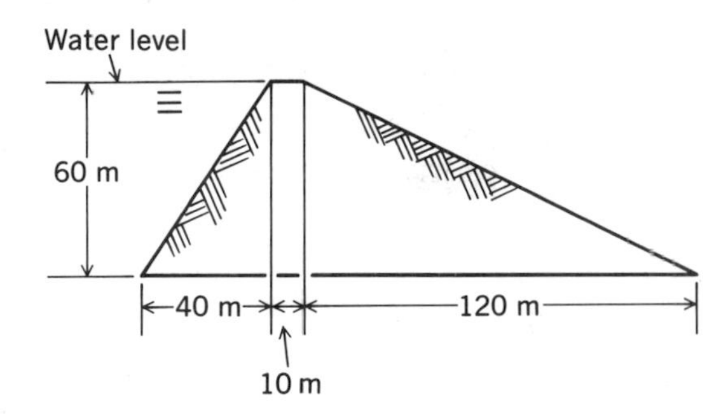

Figure P.4.61

4.62. A block 1 ft thick is submerged in water. Compute the simplest resultant force and the center of pressure on the bottom surface. Take $\gamma = 62.4$ lb/ft³.

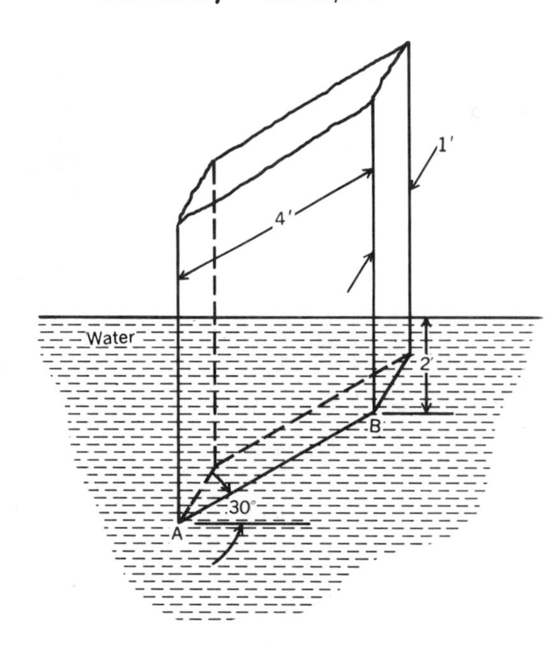

Figure P.4.62

***4.63.** What is the resultant force from water and where does it act on the 40-m-high circular concrete dam between two walls of a rocky gorge? (Water weighs 9818 N/m³.)

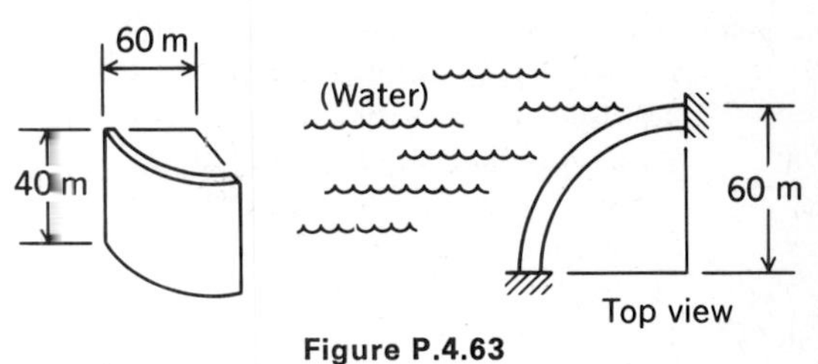

Figure P.4.63

4.64. The weight of the wire *ABCD* per unit length, *w*, increases linearly from 4 oz/ft at *A* to 20 oz/ft at *D*. Where is the center of gravity of the wire?

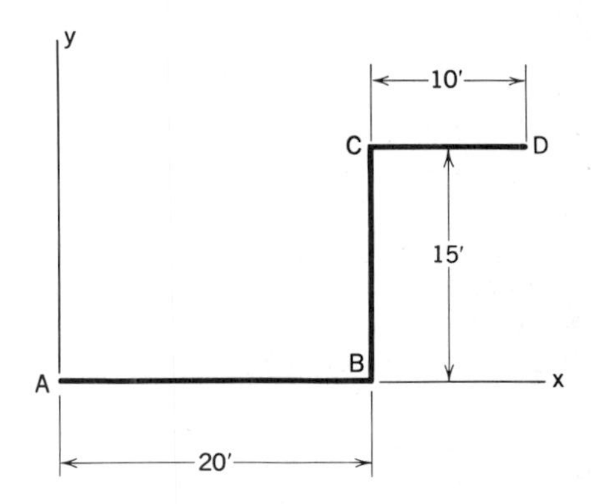

Figure P.4.64

4.65. Find the center of gravity of the wire. The weight per unit length increases as the square of the length of wire from a value of 3 oz/ft at *A* until it reaches the value of 8 oz/ft at *C*. It then decreases 1 oz/ft for every 10 ft of length.

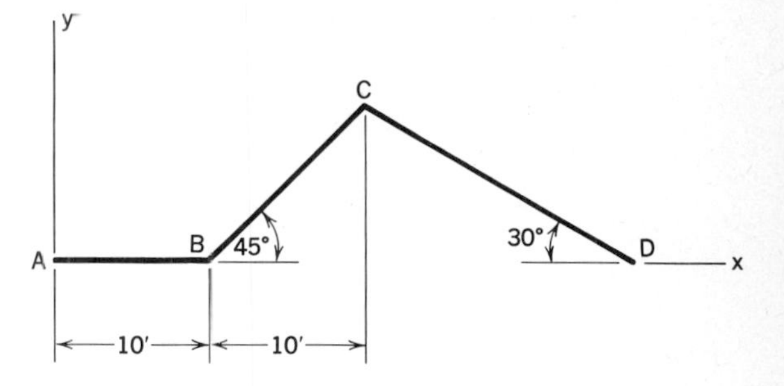

Figure P.4.65

4.66. Sandbags are piled on a beam. Each bag is 1 ft wide and weighs 100 lb. What is the simplest resultant force and where does it act? What linear mathematical function of the distributed load can be used to represent the sandbags over the left 3 ft of the beam?

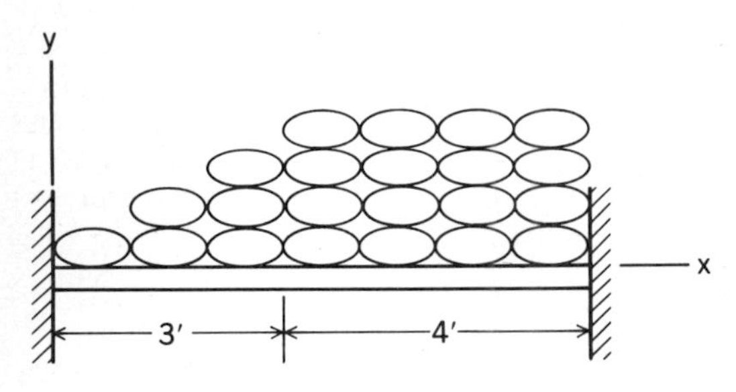

Figure P.4.66

4.67. A cantilever beam is subjected to a linearly varying load over part of its length. What is the *simplest* resultant force, and where does it act? What is the moment at the supported end?

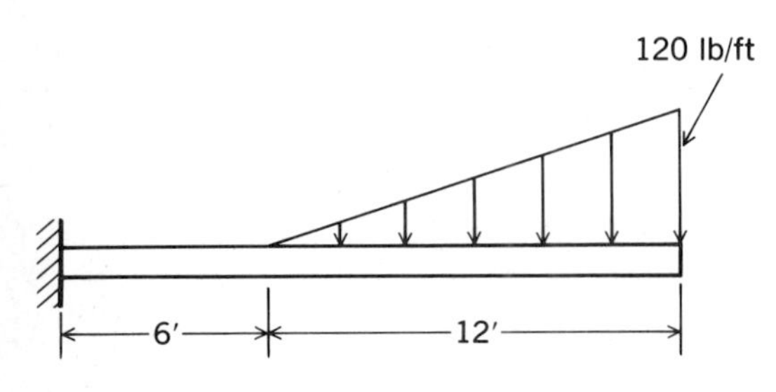

Figure P.4.67

4.68. Compute the *simplest* resultant force for the loads acting on the cantilever beam.

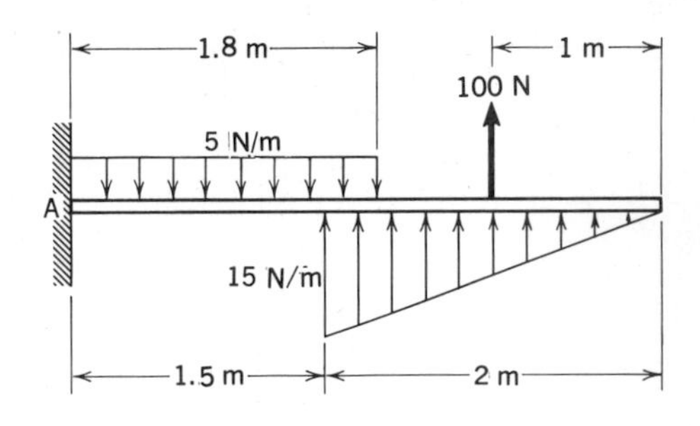

Figure P.4.68

4.69. Find the resultant force system at A for the forces on the bent cantilever beam.

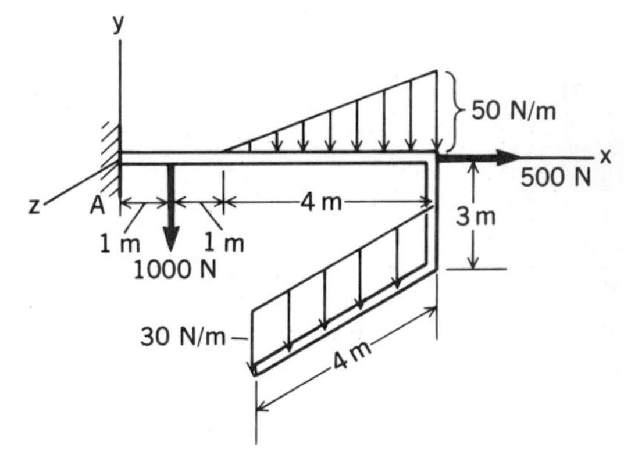

Figure P.4.69

4.6 Closure

We now have the tools that enable us to replace, for purposes of rigid-body mechanics, any system of forces by a resultant consisting of a force and a couple moment. These tools will prove very helpful in our computations. More important at this time, however, is the fact that in considering conditions of equilibrium for rigid bodies we need only concern ourselves with this resultant to reach conclusions valid for any force system, no matter how complex. From this viewpoint, we shall develop the fundamental equations of statics in Chapter 5 and then employ them to solve a large variety of problems.

Review Problems

4.70. Replace the force and couples acting on the plate by a single force. Give the intercept of the line of action of this force with the vertical edge BC of the plate.

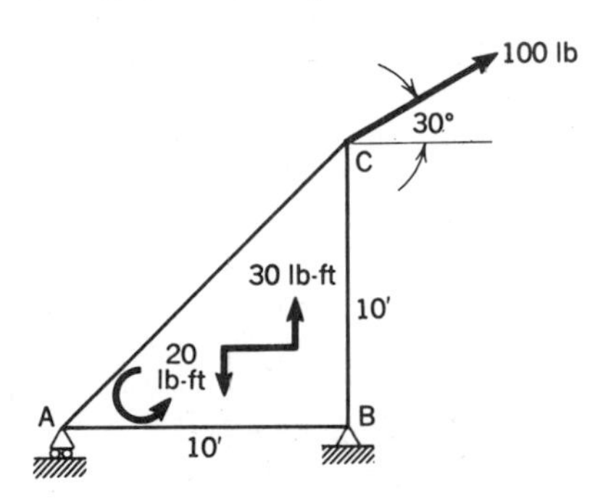

Figure P.4.70

4.71. A 100-kN bridge pier supports a 10-m segment of roadway weighing 150 kN and a 150-kN truck. The truck is located at the same position along the roadway as the pier. What is the equivalent force system acting on the base of the bridge pier when the truck is (a) in the center of the outside lane and (b) in the center of the inside lane?

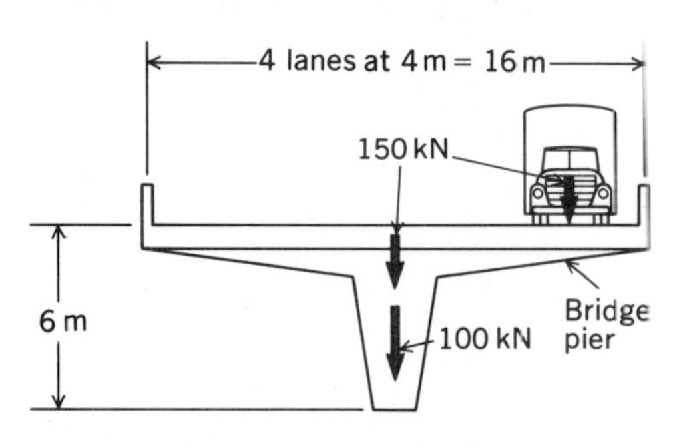

Figure P.4.71

4.72. A force $\boldsymbol{F} = 10\boldsymbol{i} + 3\boldsymbol{j} - 2\boldsymbol{k}$ lb goes through a point whose position vector is $\boldsymbol{r} = 6\boldsymbol{i} - 2\boldsymbol{j}$ ft. Find an equivalent system such that the force goes through position $\boldsymbol{r} = 2\boldsymbol{i} + 3\boldsymbol{k}$ ft.

4.73. A Jeep weighs 11 kN and has both a front winch and a rear power take-off. The tension in the winch cable is 5 kN. The power take-off develops 300 N-m of torque T about an axis parallel to the x axis. If the driver weighs 800 N, what is the resultant force system at the indicated center of gravity of the Jeep where we can consider the weight of the Jeep to be concentrated?

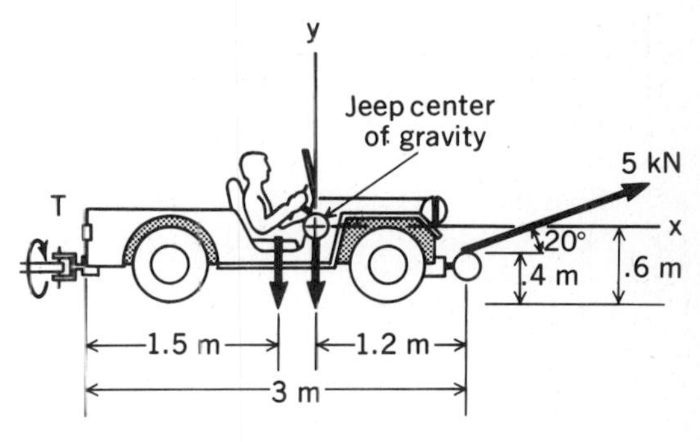

Figure P.4.73

4.74. What is the *simplest* resultant for the forces and couple acting on the beam?

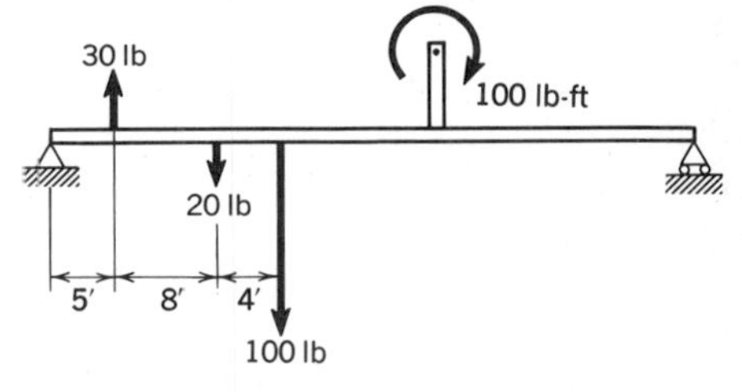

Figure P.4.74

4.75. A heavy duty off-the-road dump truck is loaded with iron ore that weighs 51 kN/m³. What is the *simplest* resultant force on the truck and where does it act?

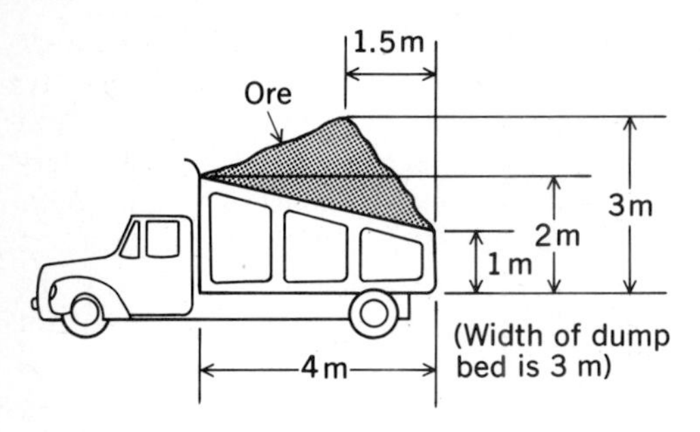

Figure P.4.75

4.76. The L-shaped concrete post supports an elevated railroad. The concrete weighs 150 lb/ft³. What is the simplest resultant force from the weight and the load and where does it act?

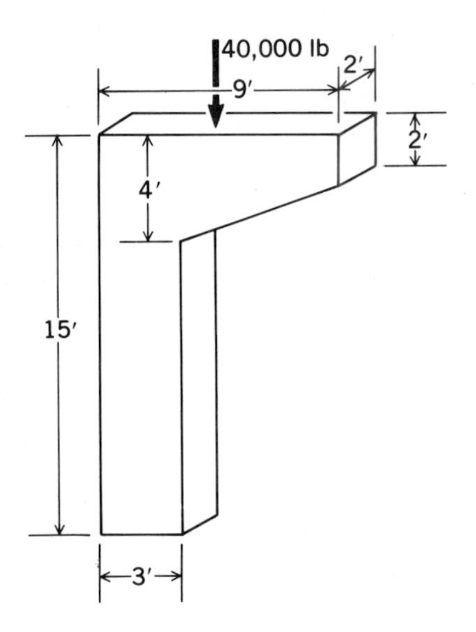

Figure P.4.76

4.77. Explain why the system shown can be considered a system of parallel forces. Find the *simplest* resultant for this system. The grid is composed of 1-m squares.

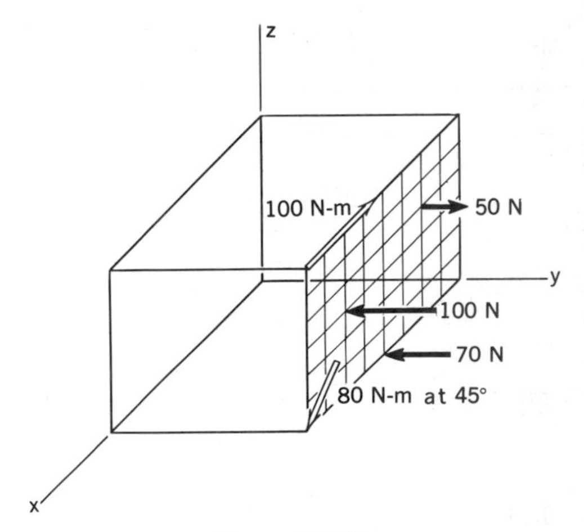

Figure P.4.77

4.78. A plate of thickness t has as the upper edge a parabolic curve with infinite slope at the origin. Find the x, y coordinates of the center of gravity for this plate.

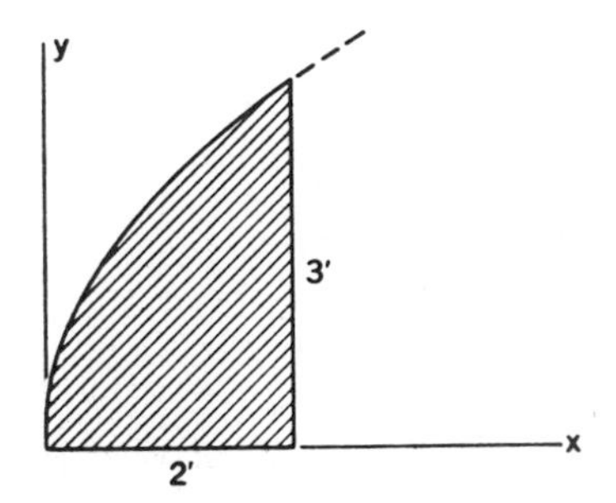

Figure P.4.78

4.79. A rectangular tank contains water. At the top of the water there is a pressure of .1380 N/mm² absolute. What is the simplest resultant force in the inside surface of the door AB? Where is the center of pressure relative to the bottom of the door? (*Hint:* It is known that the pressure in the water equals γd, where d is the distance below the surface of the water, plus the pressure on the surface. For water, $\gamma = 8190$ N/m³.)

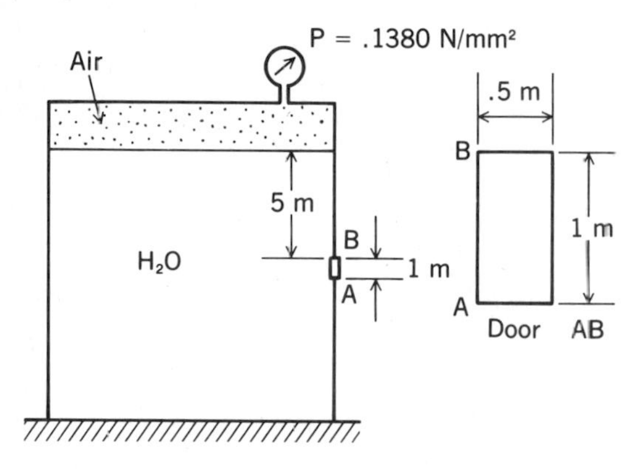

Figure P.4.79

4.80. The specific weight of the material in a right circular cone varies directly as the square of the distance y from the base. If $\gamma_0 = 50$ lb/ft³ is the specific weight at the base, and if $\gamma' = 70$ lb/ft³ is the specific weight at the tip, where is the center of gravity of the cone? (See hint in Problem 4.42.)

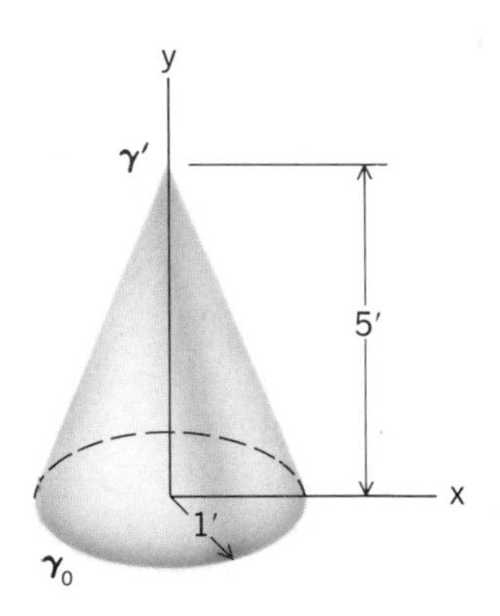

Figure P.4.80

***4.81.** A block has a rectangular portion removed (darkened region). If the specific weight is given as

$$\gamma = (2.0x + y + 3xyz)\ \text{kN/m}^3$$

find the $\bar{x}$ for the center of gravity.

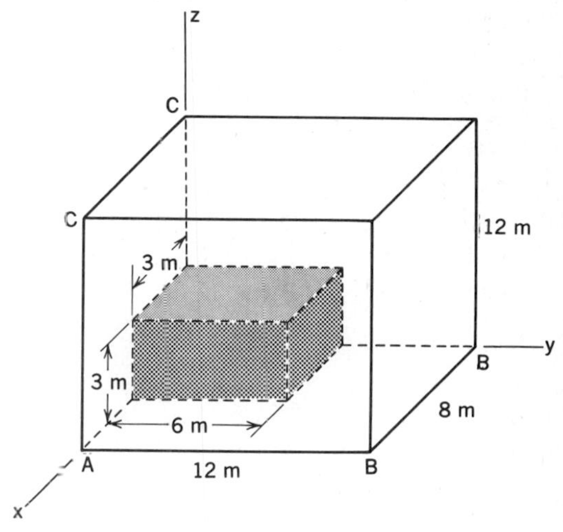

Figure P.4.81

4.82. Compute the *simplest* resultant for the loads shown acting on the simply supported beam. Give the line of action.

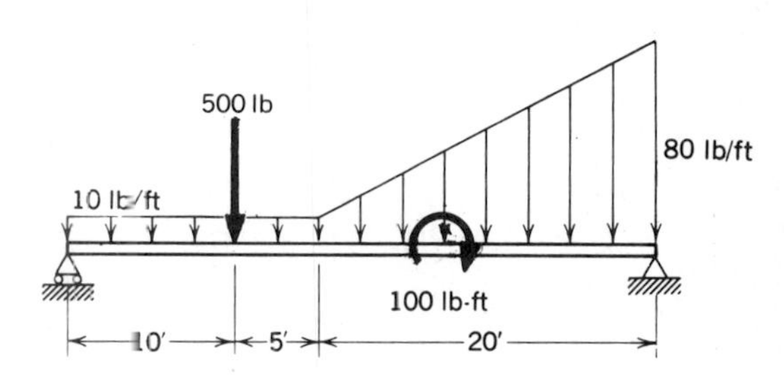

Figure P.4.82

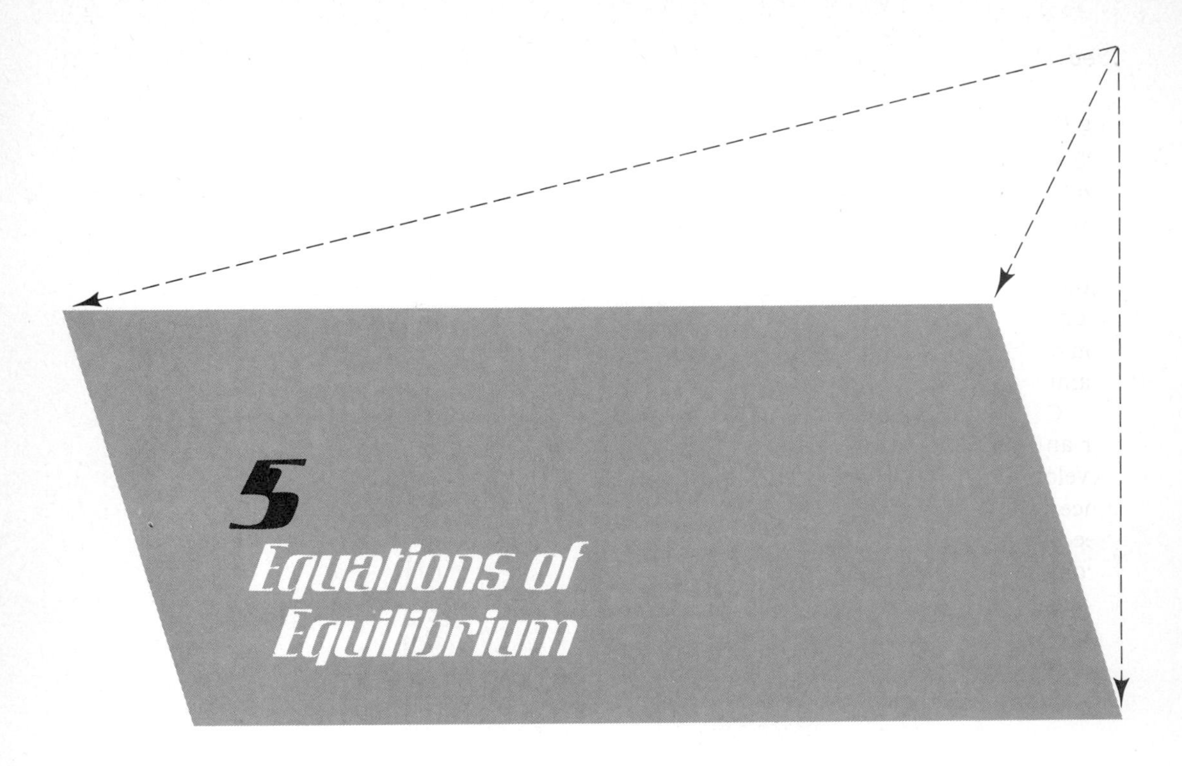

5.1 Introduction

You will recall from Section 1.10 that a *particle* in equilibrium is one that is stationary or that moves uniformly relative to an inertial reference. A *body* is in equilibrium if all the particles that may be considered to comprise the body are in equilibrium. It follows, then, that a rigid body in equilibrium cannot be rotating relative to an inertial reference. In this chapter, we shall consider bodies in equilibrium for which the rigid-body model is valid. For these bodies, there are certain simple equations that relate all the surface and body forces, or their equivalents, that act on the body. With these equations, we can sometimes ascertain the value of a certain number of unknown forces. For instance, in the beam shown in Fig. 5.1, we know the loads F_1 and F_2 and also the weight W of the beam, and we want to determine the forces transmitted to the earth so that we can design a foundation to support the structure properly. Know-

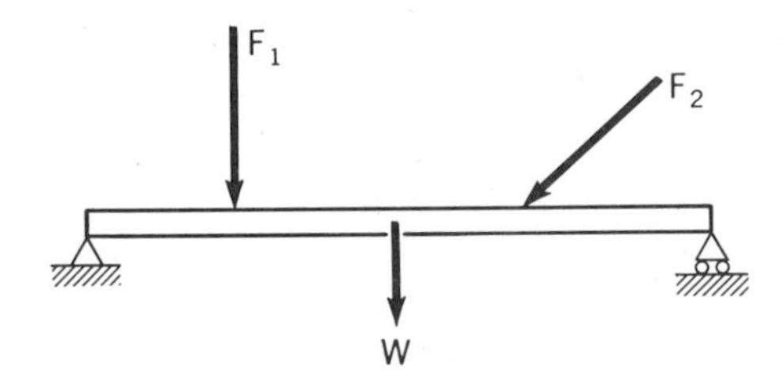

Figure 5.1. Loaded beam.

ing that the beam is in equilibrium and that the small deflection of the beam will not appreciably affect the forces transmitted to the earth, we can write rigid-body equations of equilibrium involving the unknown and known forces acting on the beam and thus arrive at the desired information.

Note in the beam problem above that a number of steps are implied. First, there is the singling out of the beam itself for discussion. Then, we express certain equations of equilibrium for the beam, which we take as a rigid body. Finally, there is the evaluation of the unknowns and interpretation of the results. In this chapter, we will carefully examine each of these steps.

Of critical importance is the need to be able to isolate a body or part of a body for analysis. Such a body is called a *free body*. We will first carefully investigate the development of free-body diagrams. We urge you to pay special heed to this topic, since *it is the most important step in the solving of mechanics problems.* An incorrect free-body diagram means that all ensuing work, no matter how brilliant, will lead to wrong results. More than just a means of attacking statics problems, the free-body concept is your first exposure to the overridingly important topic of *engineering analysis* in general.[1] We now examine this critical step.

5.2 The Free-Body Diagram

Since the equations of equilibrium for a particular body actually stem from the dynamic considerations of the body, we must be sure to include *all* the forces (or their equivalents) acting *on* this body, because they all affect the motion of the body and must be accounted for. To help identify all the forces and so ensure the correct use of the equations of equilibrium, we isolate the body in a simple diagram and show *all* the forces from the *surroundings* that act *on* the body. Such a diagram is called a *free-body diagram.* When we isolate the beam in our problem from its surroundings, we get Fig. 5.2. On the left end, there is an unknown force from the ground that has a magnitude denoted as R_1 and a direction denoted as θ, with a line of action going

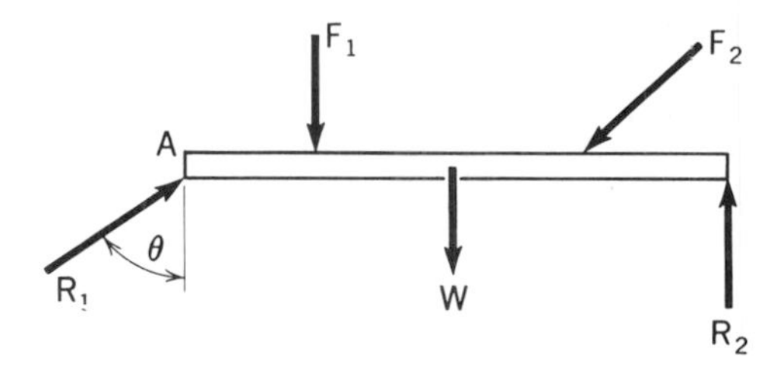

Figure 5.2. Free-body diagram of beam.

[1]The author has found through many years of experience that the absence of a free-body diagram in a student's work on a particular problem signifies that:

1. There will most likely be errors in the analysis of the problem, or
2. Even worse, the student does not have a good grasp of the problem.

through a known point A. [We may also use components $(R_1)_x$ and $(R_1)_y$ as unknowns rather than R_1 and θ.] The right side involves a force in the vertical direction with an unknown magnitude denoted as R_2. The direction is vertical because the beam is on rollers to allow for thermal expansion and to relieve stretching of the beam in the axial direction. As a result, the ground exerts a negligibly small horizontal force there. Once all the forces acting on the beam have been identified, including the three unknown quantities R_1, R_2, and θ, we can, by using three equations of equilibrium, solve for these unknowns.

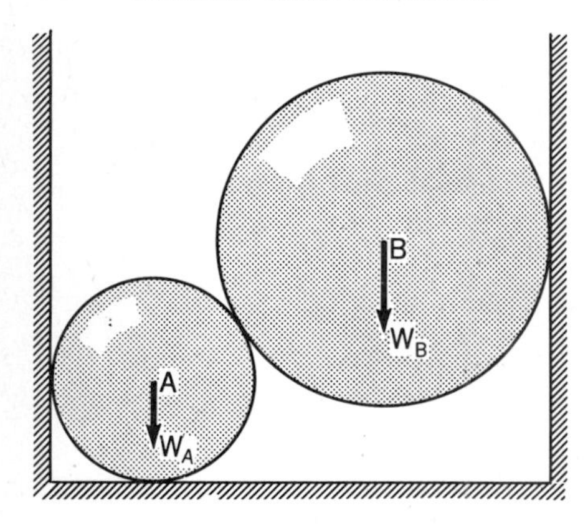

Figure 5.3. Smooth spheres in equilibrium.

Consider now the spheres shown in Fig. 5.3 in a condition of equilibrium with surfaces smooth and hard enough to permit us to neglect friction completely. The contact forces thus must be in a direction normal to the surface of contact. The free bodies of the spheres are shown in Fig. 5.4. Notice that F_3 is the magnitude of the force from sphere B on sphere A, while the reaction, also shown as F_3 according to Newton's third law, is the magnitude of the force from sphere A on sphere B.

You might be tempted to consider a portion of the container as a free body in the manner shown in Fig. 5.5. But even if this diagram did clearly depict a body (which it does not!), it would not qualify as a free body, since all the forces acting on the body have not been shown.

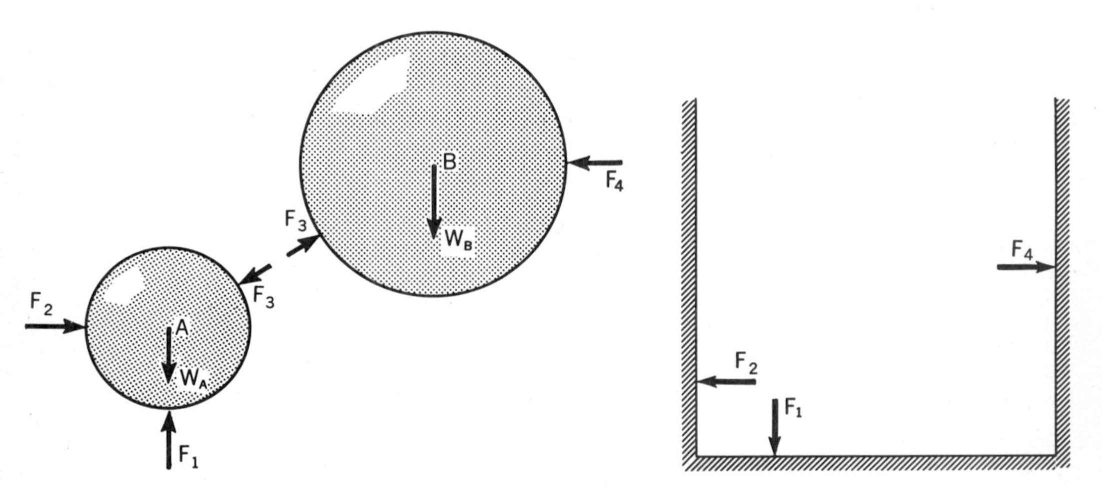

Figure 5.4. Free-body diagrams.

Figure 5.5. This is *not* a free-body diagram.

In engineering problems, bodies are often in contact in a number of standard ways. In Fig. 5.6, you will find the types of forces transmitted from body M to body

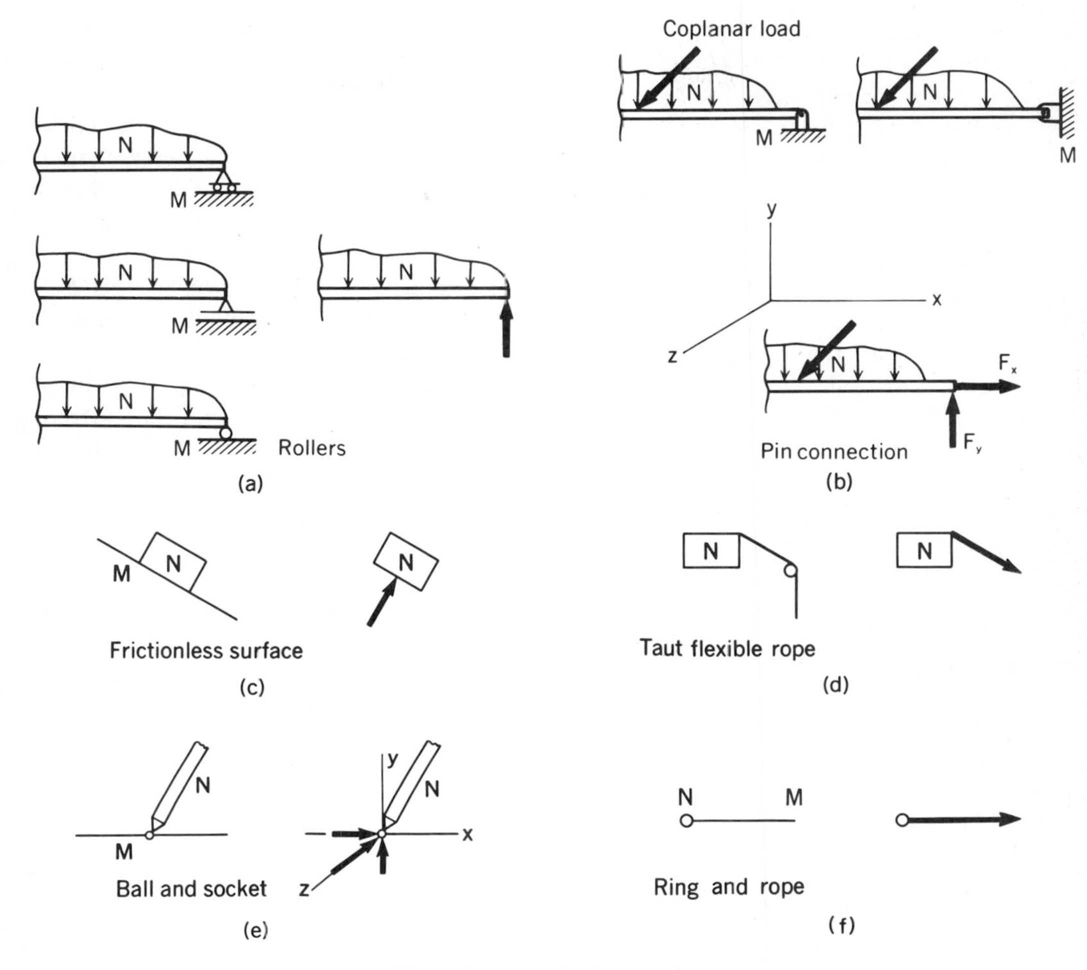

Figure 5.6. Standard connections.

N for body connections that are often found in practice. (These are not free-body diagrams, since all the forces on any body have not been shown.)

In general, to ascertain the nature of the force system that a body *M* is capable of transmitting to a second body *N* through some connector or support, we may proceed in the following manner. Mentally move the bodies relative to each other in each of three orthogonal directions. In those directions where relative motion is impeded or prevented by the connector or support, there can be a force component at this connector or support in a free-body diagram of either body *M* or *N*. Next, mentally rotate bodies *M* and *N* relative to each other about the orthogonal axes. In each direction about which relative rotation is impeded or prevented by the connector or support, there can be a couple-moment component at this connector or support in a free-body diagram of body *M* or *N*. Now as a result of equilibrium considerations of body *M* or *N*, certain force and couple-moment components that are capable of being generated at a support or connector will be zero for the particular loadings at hand. Indeed, one can often readily recognize this by inspection.

For instance, consider the pin-connected beam shown in Fig. 5.7. If we mentally move the beam relative to the ground in the x, y, and z directions, we get resistance from the pin for each direction, and so the ground at A can transmit force components A_x, A_y, and A_z. However, because the loading is coplanar in the xy plane, the force component A_z must be zero and can be deleted. Next, mentally rotate the beam relative to the ground at A about the three orthogonal axes. Because of the smooth pin connection, there is no resistance about the z axis and so $M_z = 0$. But there is resistance about the x and y axes. However, the coplanar loading in the xy plane cannot exert moments about the x and y axes, and so the couple moments M_x and M_y are zero. All told, then, we just have force components A_x and A_y at the pin connection, as has been shown earlier in Fig. 5.6(b), wherein we relied on physical reasoning for this result.

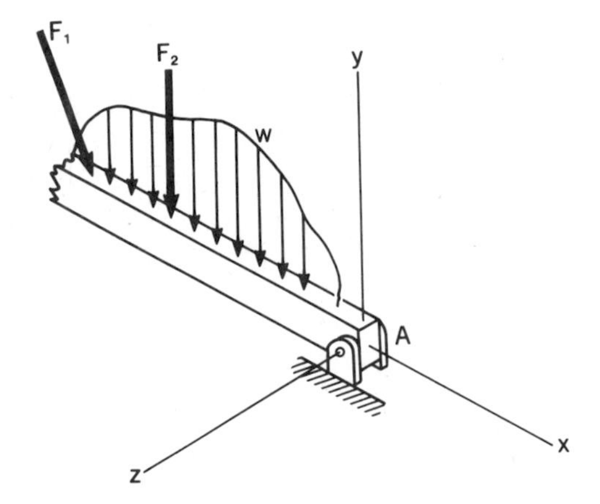

Figure 5.7. Pin connection.

5.3 Free Bodies Involving Interior Sections

Let us consider a rigid body in equilibrium as shown in Fig. 5.8. Clearly, every portion of this body must also be in equilibrium. If we consider the body as two parts A and B, we can present either part in a free-body diagram. To do this, we must include on the portion chosen to be the free body the forces *from the other part* that arise at the common section (Fig. 5.9). The surface between both sections may be any curved or plane surface, and over it there will be a continuous force distribution. In the general case, we know that such a distribution can be replaced by a single force and a single couple moment, and this has been done in the free-body diagram of parts A and B in Fig. 5.9. Notice that Newton's third law has been observed.

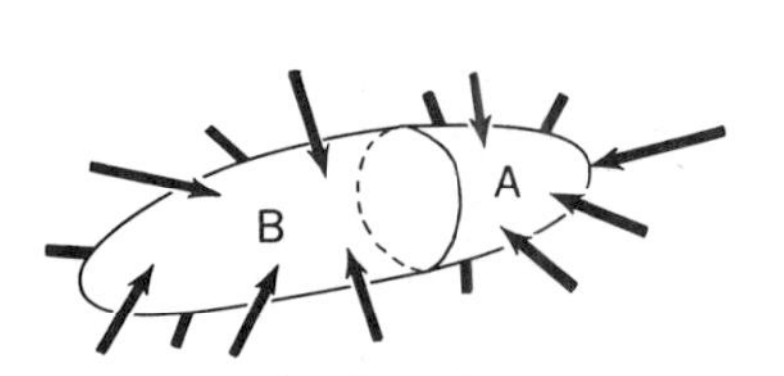

Figure 5.8. Rigid body in equilibrium.

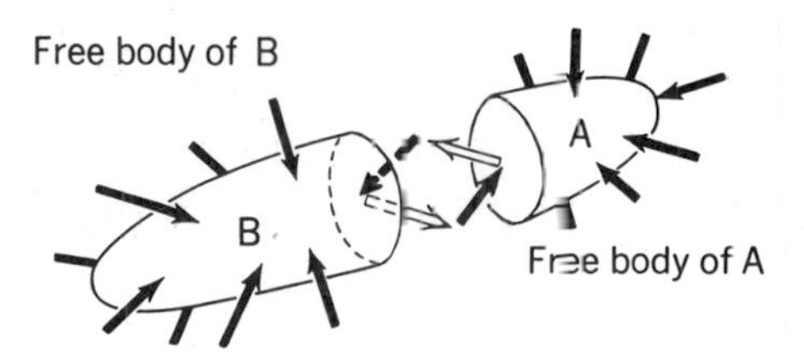

Figure 5.9. Free bodies of parts *A* and *B*.

As a special case, consider a beam with one end embedded in a massive wall (cantilever beam) and loaded along the xy plane (Fig. 5.10). A free body of the portion of the beam extending from the wall is shown in Fig. 5.11. Because of the geometric symmetry about the xy plane and the fact that the loads are in this plane, the exposed

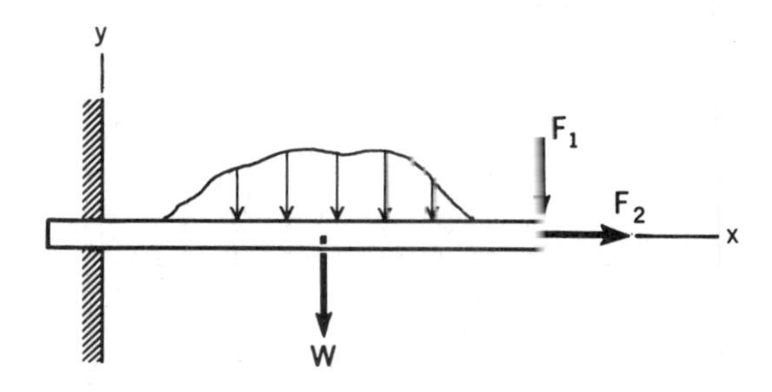

Figure 5.10. Cantilever beam.

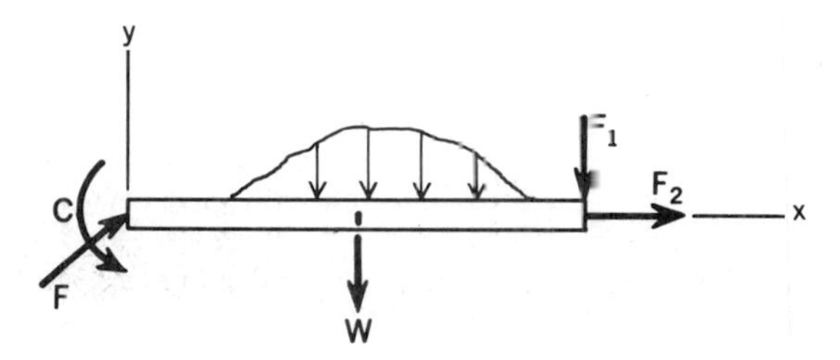

Figure 5.11. Free-body diagram of cantilever beam.

forces in the cut section can be considered coplanar. Hence, this distribution can be replaced by a force and a couple moment in the center plane, and it is the usual practice to decompose the force into components F_y and F_x. Although a line of action for the force can be found that would enable us to eliminate the couple moment, it is desirable in structural problems to work with an equivalent system that has the force passing through the center of the beam cross section, and thus to have a couple moment. In the next section, we will see how F and C can be ascertained.

EXAMPLE 5.1

As a further illustration of a free-body diagram, we shall now consider the frame[2] shown in Fig. 5.12, which consists of members connected by frictionless pins. The force systems acting on the assembly and its parts will be taken as coplanar. We shall now sketch free-body diagrams of the assembly and its parts.

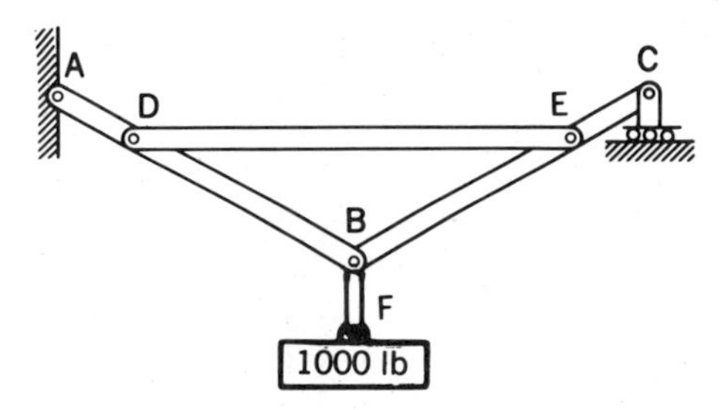

Figure 5.12. A frame.

Free-body diagram of the entire assembly. The magnitude and direction of the force at A from the wall onto the assembly is not known. However, we know that this force is in the plane of the system. Therefore, two components are shown at this point (Fig. 5.13). Since the direction of the force C is known, there are then three unknown scalar quantities, A_y, A_x, and C, for the free body.

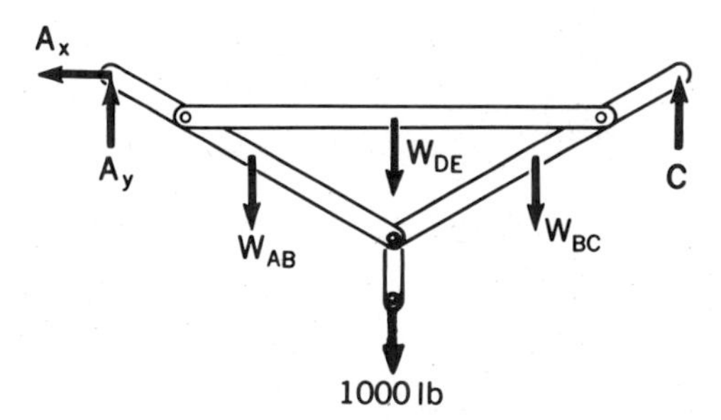

Figure 5.13. Free-body diagram of frame.

Free-body diagram of the component parts. When two members are pinned together, such as members DE and AB or DE and BC, we usually consider the pin to be part of one of the bodies. However, when more than two members are connected at a pin, such as members AB, BC, and BF at B, we often isolate the pin and consider that all members act on the pin rather than directly on each other, as illustrated in Fig. 5.14. Notice the forces that form pairs of reactions have been enclosed with dashed lines.

Do not be concerned about the proper sense of an unknown force component that you draw on the free-body diagram, for you may choose either a positive or negative sense for these components. When the values of these quantities are ascer-

[2] A *frame* is a system of connected straight or bent, long, slender members.

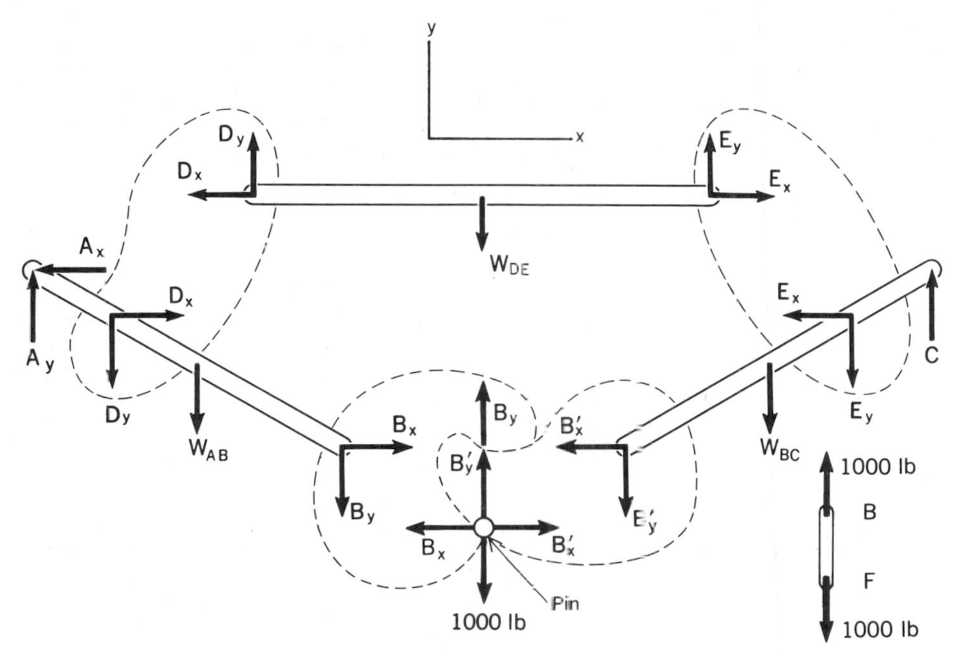

Figure 5.14. Free-body diagrams of parts.

tained by methods of statics, the proper sense for each component can then be established; but, having chosen a sense for a component, you must be sure that the *reaction* to this component has the *opposite* sense—else you will violate Newton's third law.

Free-body diagram of portion of the assembly to the right of *M—M*. In making a free body of the portion to the right of section *M—M* (see Fig. 5.15), we must remember

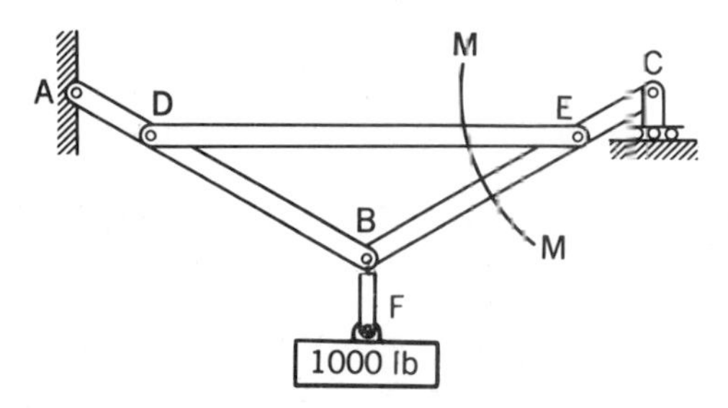

Figure 5.15. Cut along *M—M*.

to put in the weight of the portions of the members remaining *after* the cut has been made. At the two cuts made by *M—M* we must replace coplanar force distributions by resultants, as in the case of the previously considered cantilever beam. This is accomplished by inserting two force components and a couple moment as was done for the cantilever beam. Note in Fig. 5.16 that there are seven unknown scalar quantities

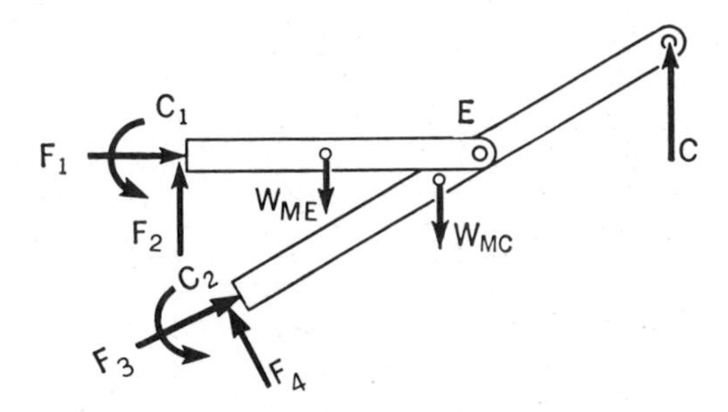

Figure 5.16. Free-body diagram.

for this free-body diagram. They are C_1, C_2, F_1, F_2, F_3, and F_4. Apparently, the number of unknowns varies widely for the various free bodies that may be drawn for the system. For this reason, you must choose the free-body diagram that is suitable for your needs with some discretion in order to effectively solve for the desired unknowns.

EXAMPLE 5.2

Draw a free-body diagram of the beam *AB* and the pulley in Fig. 5.17. The weight of the pulley is W_D, and the weight of the beam is W_{AB}.

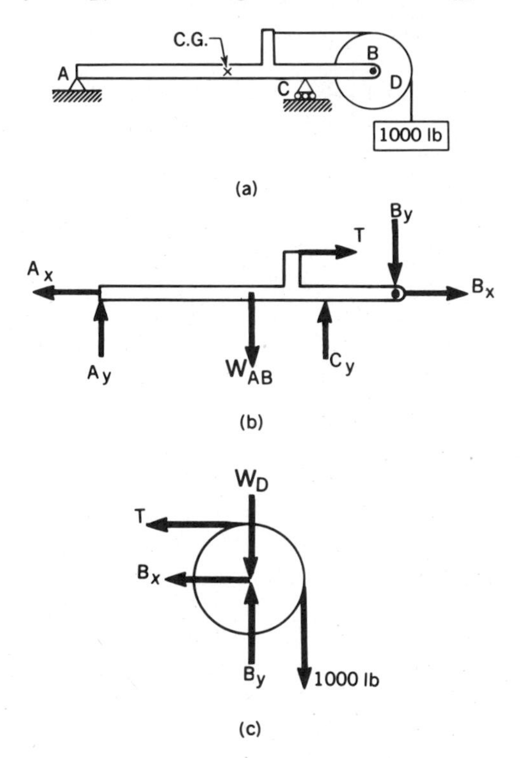

Figure 5.17. Free body diagrams.

The free-body diagram of beam AB is shown in Fig. 5.17(b). The weight of the beam has been shown at the center of gravity. Components B_x and B_y are forces from the pulley D acting on the beam through the pin at B. The free-body diagram of the pulley is shown in Fig. 5.17(c).

Some students may be tempted to put the weight of the pulley at B in the free-body diagram of beam AB. The argument given is that this weight "goes through B." To put the pulley weight at B on free body AB is strictly speaking an error! The fact is that the weight of the pulley is a body force acting throughout the *pulley* and *does not* act on the *beam BD*. It so happens that the simplest resultant of this body force distribution on D goes through a *position* corresponding to pin B. This does not alter the fact that this weight acts *on the pulley* and *not on the beam*. The beam can only feel forces B_x and B_y transmitted from the pulley to the beam through pin B. These forces are related to the pulley weight as well as the tension in the cord around the pulley through equations of equilibrium for the free body of the pulley itself.

Problems

5.1. Draw the free-body diagram when the gas-grill lid is lifted at the handle to a 45° open position.

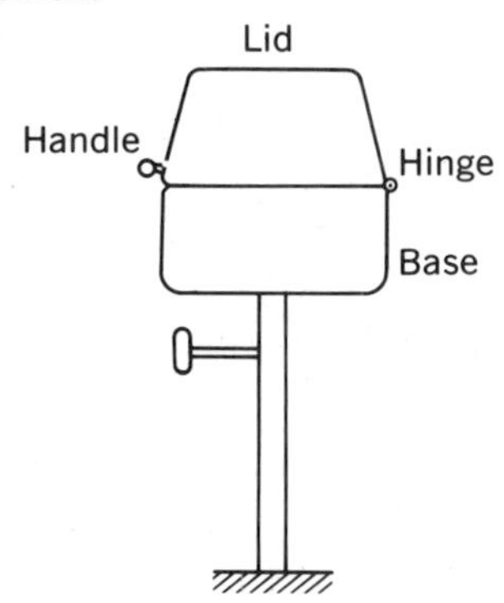

Figure P.5.1

5.2. A large antenna is supported by three guy wires and rests on a large spherical ball. Draw the free-body diagram of the antenna.

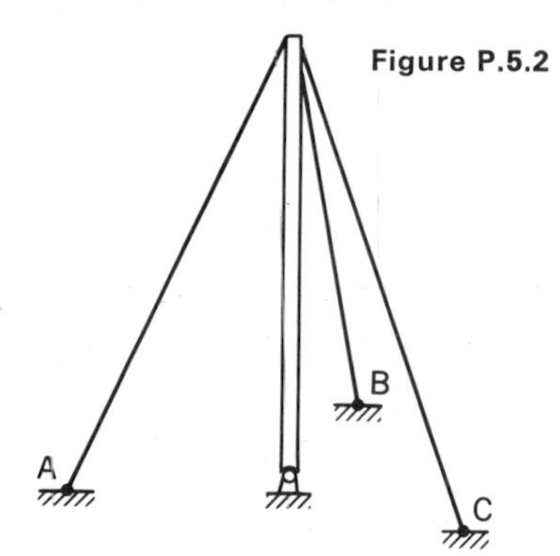

Figure P.5.2

5.3. Draw a free-body diagram of the A-frame.

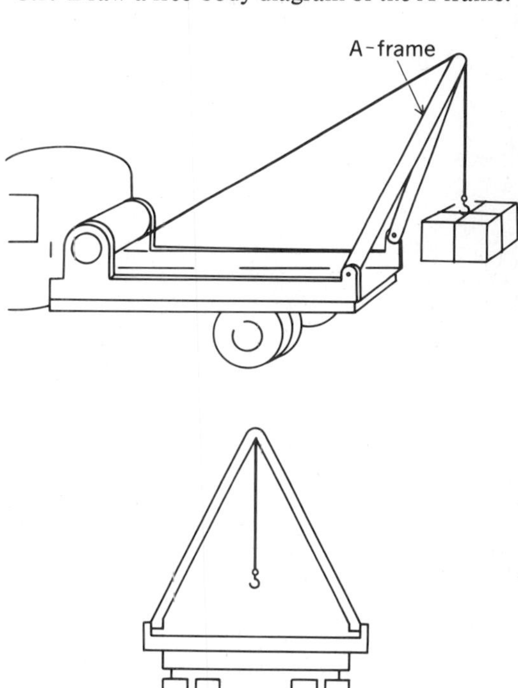

Figure P.5.3

5.4. Draw complete free-body diagrams for the member AB and for cylinder D. Neglect friction at the contact surfaces of the cylinder. The weights of the cylinder and the member are denoted as W_D and W_{AB}, respectively.

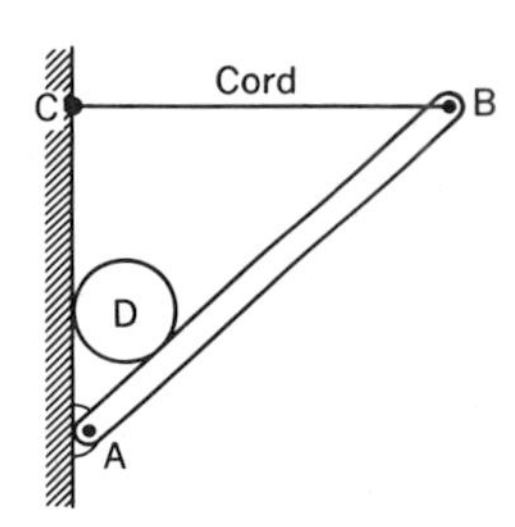

Figure P.5.4

5.5. Draw free-body diagrams of the plate $ABCD$ and the bar EG. Assume that there is no friction at the pulley H or at the contact surface C.

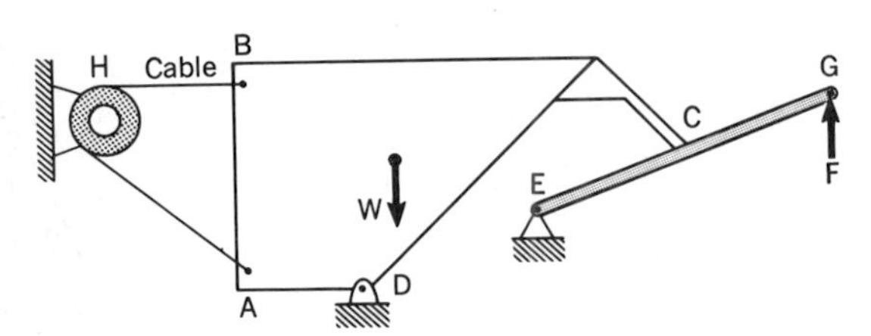

Figure P.5.5

5.6. Draw the free-body diagram of one part of the two-piece posthole digger.

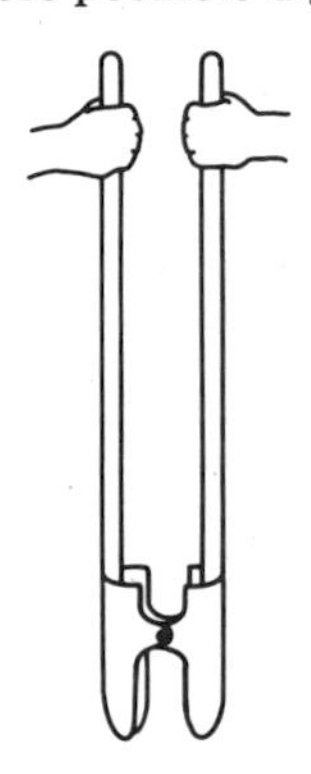

Figure P.5.6

5.7. Draw a free-body diagram for each member of the system. Neglect the weights of the members. Replace the distributed load by a resultant.

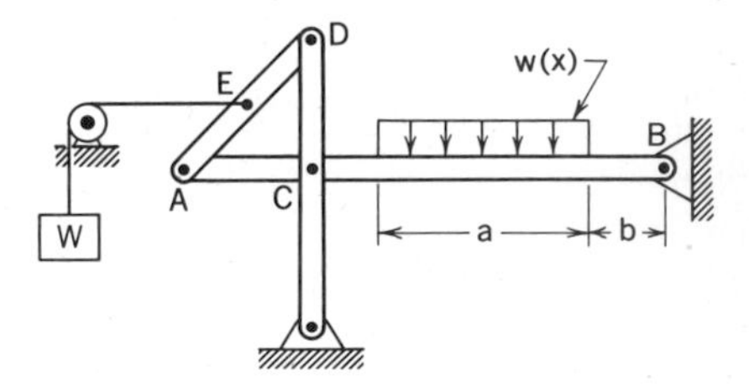

Figure P.5.7

5.8. Draw the free-body diagrams for the oars of a rowboat when the rower pushes with one hand and pulls with the other (i.e., turns the boat).

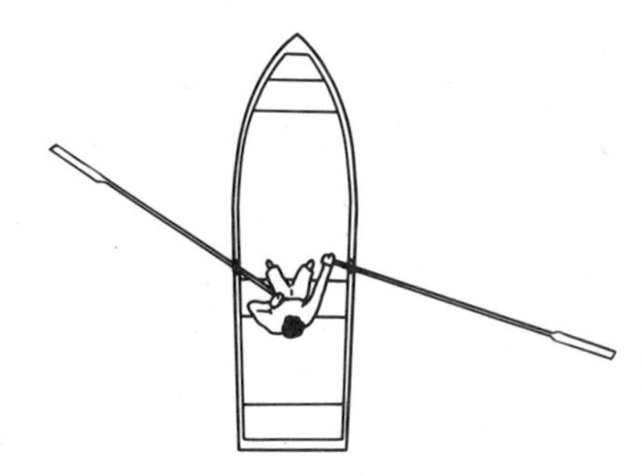

Figure P.5.8

5.9. Make a free-body diagram of the portion of the beam that is exposed from the wall. Replace all distributions by simplest equivalent force systems. Neglect the weight of the beam.

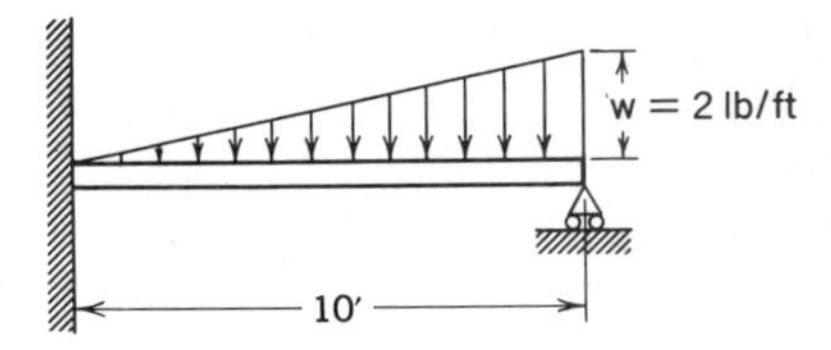

Figure P.5.9

5.10. Two cantilever beams are pinned together at A. Draw free-body diagrams of each cantilever beam.

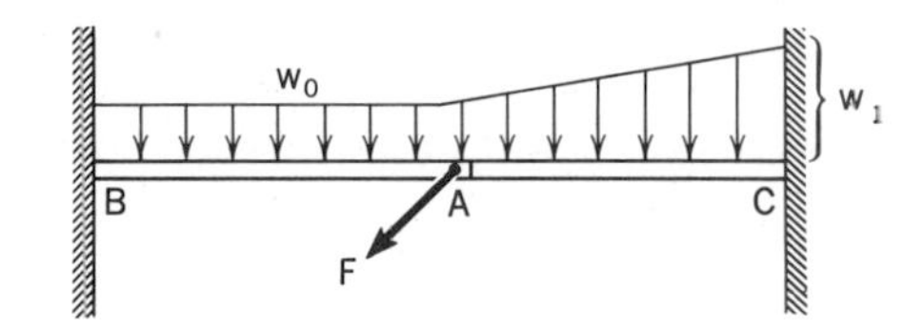

Figure P.5.10

5.11. Draw free-body diagrams of each part of the tree-branch trimmer.

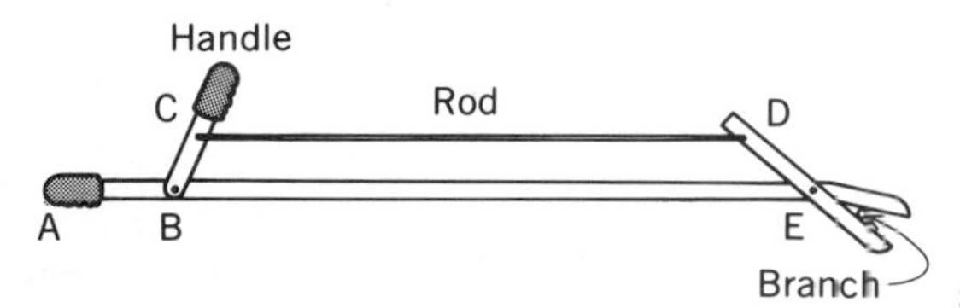

Figure P.5.11

5.12. Draw free-body diagrams for the two booms and the body E of the power shovel. Consider the weight of each part to act at a central location. (Regard the shovel and payload as concentrated forces, W_S and W_{PL}, respectively.)

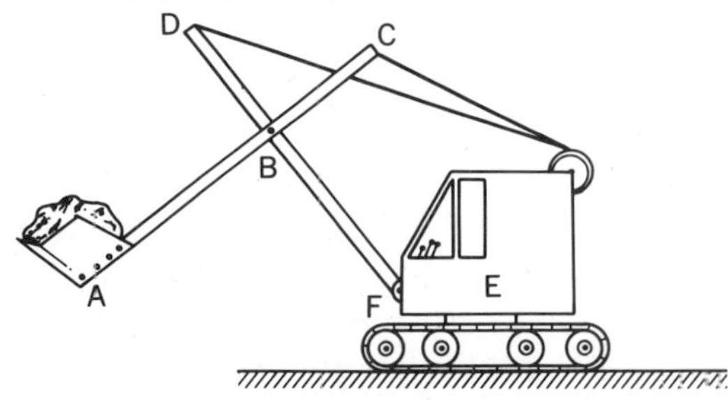

Figure P.5.12

5.13. Draw the free-body diagram for the bulldozer, B, hydraulic ram, R, and tractor, T. Consider the weight of each part B, R, and T.

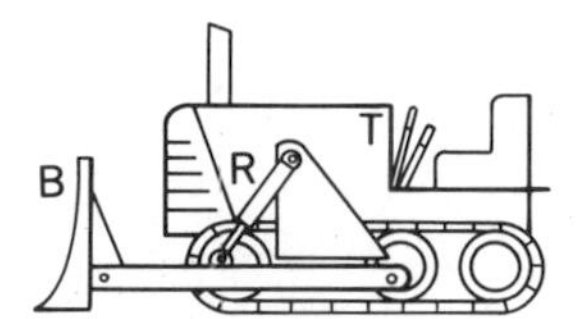

Figure P.5.13

5.14. Draw a free-body diagram first of the whole apparatus, then of each of its parts: AB, AC, BC, and D. Include the weights of all bodies. Label forces.

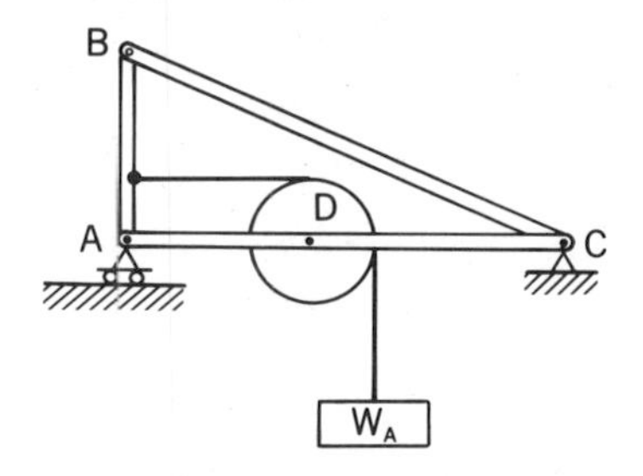

Figure P.5.14

5.15. Draw a free-body diagram of members CG, AG, the disc B, and the pin at G. Include as the only weight that of disc B. Label all forces. (*Hint:* Consider the pin at G as a separate free body.)

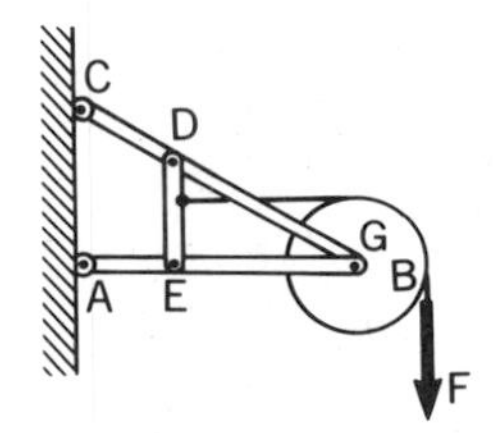

Figure P.5.15

5.16. Draw the free-body diagram of the horizontally bent cantilevered beam. Use only xyz components of all vectors drawn.

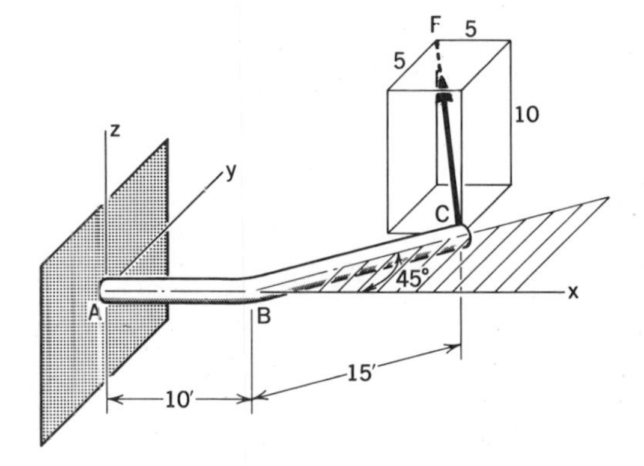

Figure P.5.16

5.4 Equations of Equilibrium

For every free-body diagram, we can replace the system of forces and couples acting on the body by a single force and a single couple moment at a point a. The force will have the same magnitude and direction, no matter where point a is chosen to move the entire system by methods discussed earlier. However, the couple-moment vector will depend on the point chosen. We will prove in dynamics that:

> *The necessary conditions for a rigid body to be in equilibrium are that the resultant force $\mathbf{F}_R$ and the resultant couple moment $\mathbf{C}_R$ for any point a be zero vectors.*

That is,

$$\mathbf{F}_R = \mathbf{0} \tag{5.1a}$$

$$\mathbf{C}_R = \mathbf{0} \tag{5.1b}$$

We shall prove in dynamics, furthermore, that the conditions above are *sufficient* to maintain an *initially stationary* body in a state of equilibrium. These are the fundamental equations of statics. You will remember from Section 4.3 that the resultant $\mathbf{F}_R$ is the sum of the forces moved to the common point, and that the couple moment $\mathbf{C}_R$ is equal to the sum of the moments of all the original forces and couples taken about this point. Hence, the equations above can be written

$$\sum_i \mathbf{F}_i = \mathbf{0} \tag{5.2a}$$

$$\sum_i \boldsymbol{\rho}_i \times \mathbf{F}_i + \sum_i \mathbf{C}_i = \mathbf{0} \tag{5.2b}$$

where the $\boldsymbol{\rho}_i$'s are displacement vectors from the common point a to any point on the lines of action of the respective forces. From this form of the equations of statics, we can conclude that for equilibrium to exist, *the vector sum of the forces must be zero and the moment of the system of forces and couples about any point in space must be zero.*

Now that we have summed forces and have taken moments about a point a, we will demonstrate that we cannot find another *independent* equation by taking mo-

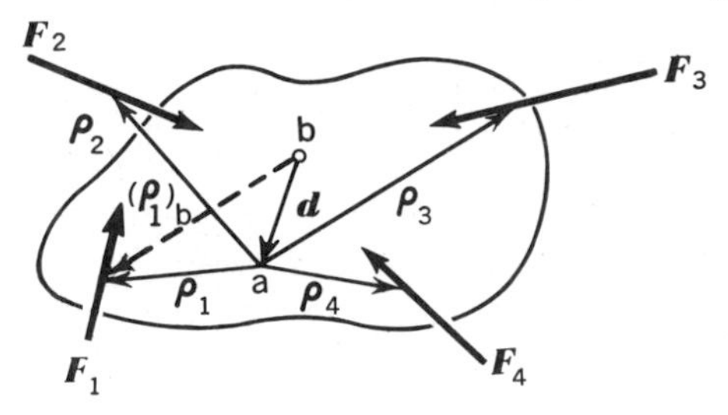

Figure 5.18. Consider moments about point *b*.

ments about a *different* point b. For the body in Fig. 5.18, we have initially the following equations of equilibrium using point a:

$$\boldsymbol{F}_1 + \boldsymbol{F}_2 + \boldsymbol{F}_3 + \boldsymbol{F}_4 = \boldsymbol{0} \quad (5.3)$$

$$\boldsymbol{\rho}_1 \times \boldsymbol{F}_1 + \boldsymbol{\rho}_2 \times \boldsymbol{F}_2 + \boldsymbol{\rho}_3 \times \boldsymbol{F}_3 + \boldsymbol{\rho}_4 \times \boldsymbol{F}_4 = \boldsymbol{0} \quad (5.4)$$

The new point b is separated from a by the position vector $\boldsymbol{d}$. The position vector (shown dashed) from b to the line of action of the force $\boldsymbol{F}_1$ can be given in terms of $\boldsymbol{d}$ and the displacement vector $\boldsymbol{\rho}_1$ as follows. Similarly for $(\boldsymbol{\rho}_2)_b$, which is not shown, and others.

$$(\boldsymbol{\rho}_1)_b = (\boldsymbol{d} + \boldsymbol{\rho}_1)$$

$$(\boldsymbol{\rho}_2)_b = (\boldsymbol{d} + \boldsymbol{\rho}_2), \quad \text{etc.}$$

The moment equation for point b can then be given as

$$(\boldsymbol{\rho}_1 + \boldsymbol{d}) \times \boldsymbol{F}_1 + (\boldsymbol{\rho}_2 + \boldsymbol{d}) \times \boldsymbol{F}_2 + (\boldsymbol{\rho}_3 + \boldsymbol{d}) \times \boldsymbol{F}_3 + (\boldsymbol{\rho}_4 + \boldsymbol{d}) \times \boldsymbol{F}_4 = \boldsymbol{0}$$

Using the distributive rule for cross products, we can restate this equation as

$$(\boldsymbol{\rho}_1 \times \boldsymbol{F}_1 + \boldsymbol{\rho}_2 \times \boldsymbol{F}_2 + \boldsymbol{\rho}_3 \times \boldsymbol{F}_3 + \boldsymbol{\rho}_4 \times \boldsymbol{F}_4) + \boldsymbol{d} \times (\boldsymbol{F}_1 + \boldsymbol{F}_2 + \boldsymbol{F}_3 + \boldsymbol{F}_4) = \boldsymbol{0} \quad (5.5)$$

Since the expression in the second set of parentheses is zero, in accordance with Eq. 5.3, the remaining portion degenerates to Eq. 5.4, and thus we have not introduced a new equation. Therefore, *there are only two independent vector equations of equilibrium for any single free body.*

We shall now show that instead of using Eqs. 5.3 and 5.4 as the equations of equilibrium, we can instead use Eqs. 5.4 and 5.5. That is, instead of summing forces and then taking moments about a point for equilibrium, we can instead take moments about *two* points. Thus, if Eq. 5.4 is satisfied for point a, then for point b we end up in Eq. 5.5 with

$$\boldsymbol{d} \times (\boldsymbol{F}_1 + \boldsymbol{F}_2 + \boldsymbol{F}_3 + \boldsymbol{F}_4) = \boldsymbol{0} \quad (5.6)$$

If point b can be any point in space making $\boldsymbol{d}$ arbitrary, then the above equation indicates that the vector sum of forces is zero. We thus have equilibrium since $\boldsymbol{F}_R = \boldsymbol{0}$ and $\boldsymbol{C}_R = \boldsymbol{0}$.

Using the vector Eqs. 5.2, we can now express the scalar equations of equilibrium. Since, as you will recall, the rectangular components of the moment of a force about a point are the moments of the force about the orthogonal axes at the point, we may state these equations in the following manner:

$$\begin{aligned}
&\sum_i (F_x)_i = 0 \quad \text{(a)} \qquad &&\sum_i (M_x)_i = 0 \quad \text{(d)} \\
&\sum_i (F_y)_i = 0 \quad \text{(b)} \qquad &&\sum_i (M_y)_i = 0 \quad \text{(e)} \\
&\sum_i (F_z)_i = 0 \quad \text{(c)} \qquad &&\sum_i (M_z)_i = 0 \quad \text{(f)}
\end{aligned} \quad (5.7)$$

From this set of equations, it is clear that *no more then six unknown scalar quantities in the general case can be solved by methods of statics for a single free body.*[3]

[3]Keep in mind that we can also take moments about two sets of axes just as we could take moments about two points for the vector equations of equilibrium.

We can easily express *any number* of scalar equations of equilibrium for a free body by selecting references that have different axis directions, along which we can sum forces and about which we can take moments. However, in choosing six *independent* equations, we will find that the remaining equations will be dependent on these six. That is, these equations will be sums, differences, etc. of the independent set and so will be of no use in solving for desired unknowns.

5.5 Special Cases of Equilibrium

The preceding conditions for equilibrium apply to the general case. We shall now consider a number of important *special* cases of equilibrium primarily to ascertain the number of scalar equations that are necessary for equilibrium for these special cases. With this information, we will then know the number of unknown scalar quantities for any free body that can be solved by methods of statics. If there are more such unknowns than available independent equations, no amount of algebraic perseverance will lead to the solution of the unknowns for the chosen free body.

The type of *simplest* resultant for each special system of forces is most useful in determining the number of scalar equations available in a given problem. The procedure is to classify the force system, note what simplest resultant force system is associated with the classification, and then consider the number of scalar equations necessary and sufficient to guarantee this resultant to be zero. The following cases exemplify this procedure.

Case A. Concurrent System of Forces. In this case, since the simplest resultant is a single force at the point of concurrency, the only requirement for equilibrium is that this force be zero. We can ensure this condition if the orthogonal components of this force are separately equal to zero. Thus, we have *three* equations of equilibrium of the form

$$\sum_i (F_x)_i = 0, \qquad \sum_i (F_y)_i = 0, \qquad \sum_i (F_z)_i = 0 \tag{5.8}$$

As was pointed out in the general vector discussion, there are other ways of ensuring a zero resultant. Suppose that the moments of the concurrent force system are zero about three nonparallel axes: α, β, and γ. That is,

$$\sum_i (M_\alpha)_i = 0, \qquad \sum_i (M_\beta)_i = 0, \qquad \sum_i (M_\gamma)_i = 0 \tag{5.9}$$

Any one of the following three conditions must then be true:

1. The resultant force F_R is zero.
2. F_R cuts all three axes (see Fig. 5.19).
3. F_R cuts two axes and is parallel to the third (see Fig. 5.20).

We can guarantee condition 1 and thus equilibrium if we select axes α, β, and γ so that no straight line can intersect all three axes or can cut two axes and be parallel to

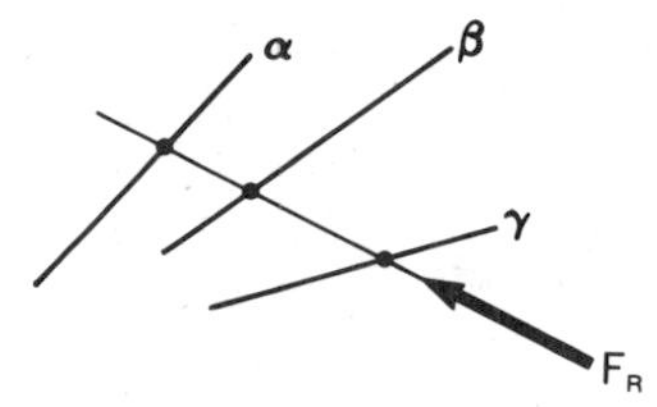

Figure 5.19. F_R cuts three axes.

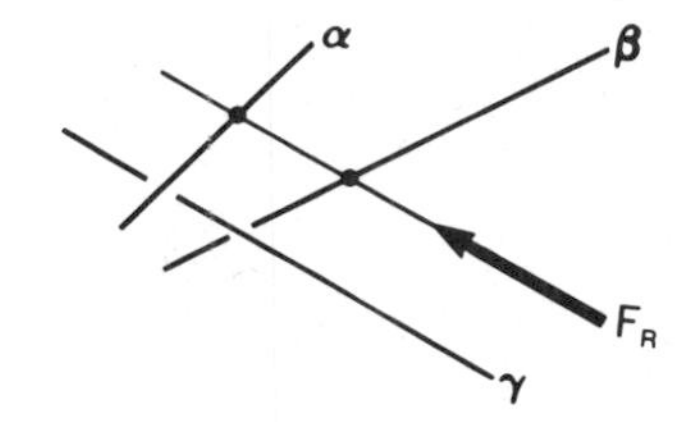

Figure 5.20. F_R cuts two axes and is parallel to third.

the third. Then we can use Eqs. 5.9 as the equations of equilibrium under the afore-stated conditions rather than using Eqs. 5.8. What happens if an axis used violates these conditions? The resulting equation will either be an *identity* $0 = 0$ or will be dependent on a previous independent equation of equilibrium for one of the axes. No harm is done. One should use other axes until three independent equations are found.

Similarly, one can sum forces in one direction and take moments about two axes. Setting these equal to zero can yield three independent equations of equilibrium. If not, use other axes.

The essential conclusion to be drawn is that *there are three independent equations of equilibrium for a concurrent force system.*

Case B. Coplanar Force System. We have shown that the simplest resultant for a coplanar force system is a single force or a single couple moment. Thus, to ensure that the resultant force is zero, we require for a coplanar system in the xy plane:

$$\sum_i (F_x)_i = 0, \qquad \sum_i (F_y)_i = 0 \tag{5.10}$$

To ensure that the resultant couple moment is zero, we require for moments about any axis parallel to the z axis:

$$\sum_i (M_z)_i = 0 \tag{5.11}$$

We conclude that there are *three* scalar equations of equilibrium for a coplanar force system. Other combinations, such as two moment equations for two axes parallel to the z axis and a single force summation, if properly chosen, may be employed to give the three independent scalar equations of equilibrium, as was discussed in case A.

Case C. Parallel Forces in Space. In the case of parallel forces in space, we already know that the simplest resultant can be either a single force or a couple moment. If the forces are in the z direction, then

$$\sum_i (F_z)_i = 0 \tag{5.12}$$

ensures that the resultant force is zero. Also,

$$\sum_i (M_x)_i = 0, \qquad \sum_i (M_y)_i = 0 \tag{5.13}$$

guarantees that the resultant couple moment is zero, where the x and y axes may be chosen in any plane perpendicular to the direction of the forces.[4] Thus, three independent scalar equations are available for equilibrium of parallel forces in space.

A summary of the special cases discussed in this section is given below. For even simpler systems such as the concurrent-coplanar and the parallel-coplanar systems, clearly, there is one less equation of equilibrium.

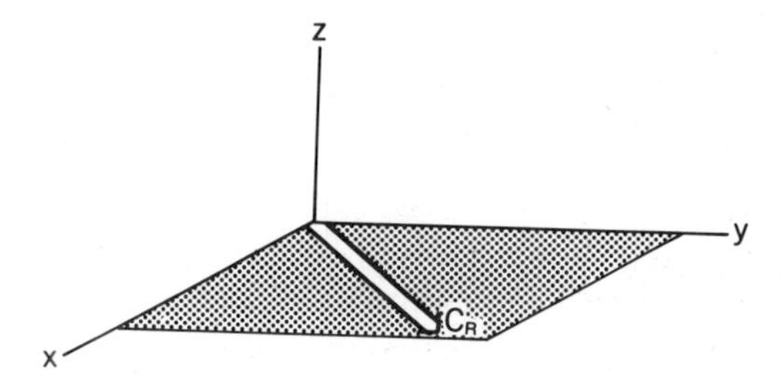

Figure 5.21. Resultant couple moment put in *xy* plane.

SUMMARY FOR SPECIAL CASES

System	*Simplest Resultant*	*Number of Equations for Equilibrium*
Concurrent (three-dimensional)	Single force	3
Coplanar	Single force or single couple moment	3
Parallel (three-dimensional)	Single force or single couple moment	3

5.6 Problems of Equilibrium

We shall now examine problems of equilibrium in which the rigid-body assumption is valid. To solve such problems, we must find the value of certain unknown forces and couple moments. We first draw a free-body diagram of the entire system or portions thereof to clearly *expose* pertinent unknowns for analysis. We then write the equilibrium equations in terms of the unknowns along with the known forces and geometry. As we have seen, for any free body there is a limited number of independent scalar equations of equilibrium. Thus, at times we must employ several free-body diagrams for portions of the system to produce enough independent equations to solve all the unknowns.

[4]For parallel forces in the z direction, a simplest resultant consisting of a couple moment only must have this couple moment parallel to the xy plane (see Fig. 5.21). Recall from Chapter 3 that the orthogonal xyz components of $\boldsymbol{C_R}$ equals the torques of the system about these axes. Hence, by setting $\sum_i (M_x)_i = \sum_i (M_y)_i = 0$, we are ensuring that $\boldsymbol{C_R} = \boldsymbol{0}$.

For any free body, we may proceed by expressing two basic vector equations of statics. After carrying out such vector operations as cross products and additions in the equations, we form scalar equations. These scalar equations are then solved simultaneously (together with scalar equations from other free-body diagrams that may be needed) to find the unknown forces and couple moments. We can also express the scalar equations immediately by using the alternative scalar equilibrium relations that we formulated in previous sections. In the first case, we start with more compact vector equations and arrive at the expanded scalar equations by the formal procedures of vector algebra. In the latter case, we evaluate the expanded scalar equations by carrying out arithmetic operations on the free-body diagram as we write the equations. Which procedure is more desirable? It all depends on the problem and the investigator's skill in vector manipulation. It is true that many statics problems submit easily to a direct scalar approach, but the more challenging problems of statics and dynamics definitely favor an initial vector approach. In this text, we shall employ the particular procedure that the occasion warrants.

In statics problems, we must assign a sense to each component of an unknown force or couple moment in order to write the equations. If, on solving the equations, *we obtain a negative sign for a component, then we have guessed the wrong sense for that component*. Nothing need be redone should this occur. Continue with the remainder of the problem, retaining the minus sign (or signs). At the end of the problem, report the correct sense of your force components and couple-moment components.

We shall now solve and discuss a number of problems of equilibrium. These problems are divided into four classes of force systems:

1. Concurrent.
2. Coplanar.
3. Parallel.
4. General.

Concurrent Force Systems. Recall that there are three independent equations of equilibrium for a concurrent force system acting on a body. If the concurrent forces are coplanar, there are but two independent equations. We shall illustrate both cases, starting for simplicity with the coplanar case.

EXAMPLE 5.3

A 500-lb weight is suspended by flexible cables as shown in Fig. 5.22. Determine the tension in the cables.

A suitable free body that exposes the desired unknown quantities is the ring C, which may be considered as a particle for this computation because of its comparatively small size (Fig. 5.23). Physical intuition indicates that the cables should be in tension and hence pulling away from C, and we have so indicated in the diagram. The force system acting on a particle must always be a concurrent system. Here we have the additional fact that it is coplanar as well, and therefore we may solve for the two unknowns. We shall proceed directly to the scalar equations of equilibrium.

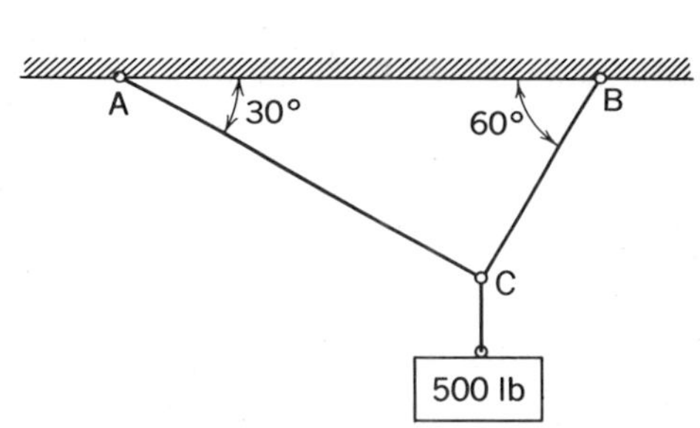

Figure 5.22. Find tensions in cables.

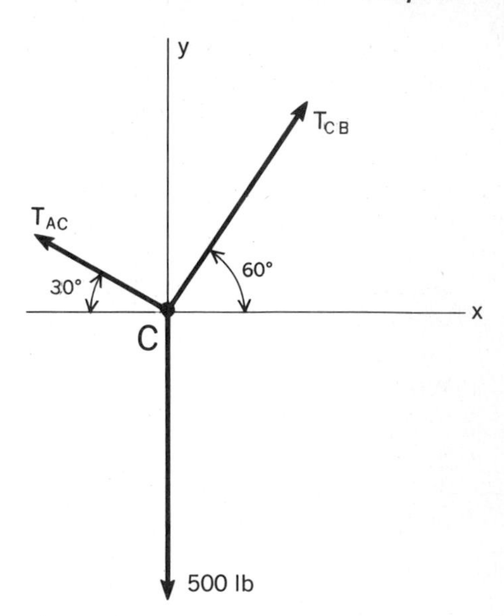

Figure 5.23. Free-body diagram of C.

Thus,

$\underline{\sum F_y = 0:}$

$$-500 + T_{CB} \sin 60° + T_{AC} \sin 30° = 0 \qquad \text{(a)}$$

$\underline{\sum F_x = 0:}$

$$-T_{AC} \cos 30° + T_{CB} \cos 60° = 0 \qquad \text{(b)}$$

By solving these equations simultaneously, we get the desired results:

$$T_{CB} = 433 \text{ lb}, \qquad T_{AC} = 250 \text{ lb}$$

Since the signs for T_{CB} and T_{AC} are positive, we have chosen the correct senses for the forces in the free-body diagram. Thus, the cables are in tension, confirming the physical intuition that guided our initial choices.

Another way of arriving at the solution is to consider the *force polygon* that was discussed in Section 2.3. Because the forces are in equilibrium, the polygon must close; that is, the head of the final force must coincide with the tail of the initial force. In this case, we have a right triangle, as shown in Fig. 5.24, drawn approximately to scale.

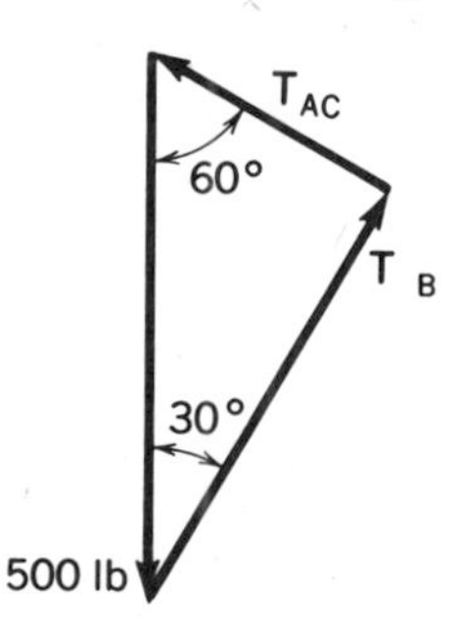

Figure 5.24. Force polygon.

From trigonometric considerations of this right triangle, we can state:

$$T_{CB} = 500 \cos 30° = 433 \text{ lb}$$

$$T_{AC} = 500 \sin 30° = 250 \text{ lb}$$

The force polygon may thus be used to good advantage when three concurrent coplanar forces are in equilibrium.

As a final alternative, let us now initiate the computations for the unknown tensions directly from the basic *vector* equations of statics. First, we must express all forces in vector notation:

$$\boldsymbol{T}_{CB} = T_{CB}(.500\boldsymbol{i} + .866\boldsymbol{j})$$
$$\boldsymbol{T}_{AC} = T_{AC}(-.866\boldsymbol{i} + .500\boldsymbol{j})$$

We get the following equation when the vector sum of the forces is set equal to zero:

$$T_{CB}(.500\boldsymbol{i} + .866\boldsymbol{j}) + T_{AC}(-.866\boldsymbol{i} + .500\boldsymbol{j}) - 500\boldsymbol{j} = \boldsymbol{0}$$

Choosing point C, the point of concurrency, we see clearly that the sum of moments of the forces about this point is zero, so the second basic equation of equilibrium is intrinsically satisfied. We now regroup the preceding equation in the following manner:

$$(.500T_{CB} - .866T_{AC})\boldsymbol{i} + (.866T_{CB} + .500T_{AC} - 500)\boldsymbol{j} = \boldsymbol{0}$$

To satisfy this equation, each of the quantities in parentheses must be zero. This gives the scalar equations (a) and (b) stated earlier, from which the scalar quantities T_{CB} and T_{AC} can be solved.

The three alternative methods of solution are apparently of equal usefulness in this simple problem. However, the force polygon is only practicably useful for three concurrent coplanar forces, where the trigonometric properties of the triangle can be directly used. The other two methods can be readily extended to more complex concurrent problems.

EXAMPLE 5.4

What are the forces in the cables shown in Fig. 5.25 supporting a 50-kg mass? Note that cable BD lies in the zy plane.

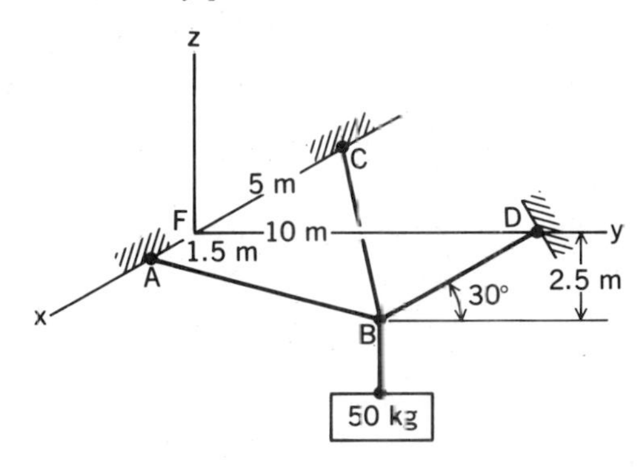

Figure 5.25. Find forces in cables.

If we take the connecting point B of the cables as a free body (see Fig. 5.26), we have three unknown concurrent forces which can be solved from the equations of equilibrium. The 50-kg mass has a weight of (9.81)(50) = 490.5 N, as shown in the diagram.

We shall first express the forces vectorially. For this purpose, note from Fig. 5.26 that

$$\frac{2.5}{ED} = \tan 30^\circ$$

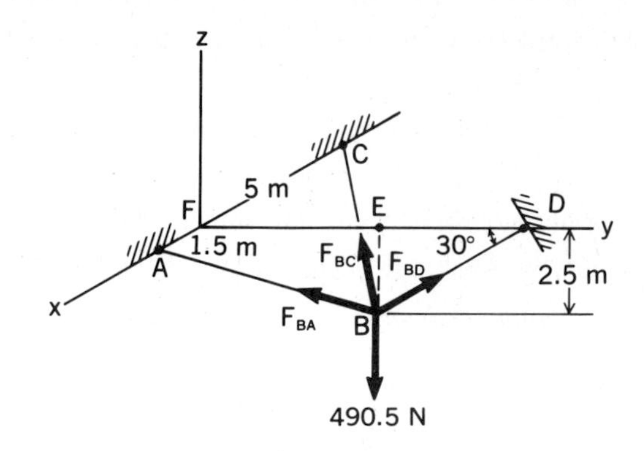

Figure 5.26. Free-body diagram of B.

Therefore,

$$ED = \frac{2.5}{\tan 30°} = 4.33 \text{ m} \tag{a}$$

Accordingly, we have

$$FE = 10 - 4.33 = 5.67 \text{ m} \tag{b}$$

We can now give the forces as follows using unit vectors for the first two:

$$\begin{aligned}\boldsymbol{F}_{BA} &= F_{BA}\left(\frac{1.5\boldsymbol{i} - 5.67\boldsymbol{j} + 2.5\boldsymbol{k}}{\sqrt{1.5^2 + 5.67^2 + 2.5^2}}\right)\\ &= F_{BA}(.235\,\boldsymbol{i} - .889\boldsymbol{j} + .392\boldsymbol{k}) \text{ N}\\ \boldsymbol{F}_{BC} &= F_{BC}\left(\frac{-5\boldsymbol{i} - 5.67\boldsymbol{j} + 2.5\boldsymbol{k}}{\sqrt{5^2 + 5.67^2 + 2.5^2}}\right)\\ &= F_{BC}(-.628\boldsymbol{i} - .712\boldsymbol{j} + .314\boldsymbol{k}) \text{ N}\\ \boldsymbol{F}_{BD} &= F_{BD}(\cos 30°\,\boldsymbol{j} + \sin 30°\boldsymbol{k})\\ &= F_{BD}(.866\boldsymbol{j} + .500\boldsymbol{k}) \text{ N}\end{aligned}$$

Now, summing the forces acting on B equal to zero, we get

$$\begin{aligned}F_{BA}(.235\boldsymbol{i} - .889\boldsymbol{j} + .392\boldsymbol{k}) + F_{BC}(-.628\boldsymbol{i} - .712\boldsymbol{j} + .314\boldsymbol{k})&\\ + F_{BD}(.866\boldsymbol{j} + .500\boldsymbol{k}) - 490.5\boldsymbol{k} = \boldsymbol{0}&\end{aligned} \tag{c}$$

The scalar equations are

$$.235F_{BA} - .628F_{BC} = 0 \tag{d}$$

$$-.889F_{BA} - .712F_{BC} + .866F_{BD} = 0 \tag{e}$$

$$.392F_{BA} + .314F_{BC} + .500F_{BD} - 490.5 = 0 \tag{f}$$

From Eq. (d), we may solve for F_{BC} in terms of F_{AB}. Thus,

$$F_{BC} = \frac{.235}{.628}F_{BA} = .374F_{BA} \tag{g}$$

Substituting Eq. (g) into Eqs. (e) and (f), we get

$$-1.155F_{BA} + .866F_{BD} = 0 \tag{h}$$

$$.509F_{BA} + .500F_{BD} = 490.5 \tag{i}$$

Now multiplying Eq. (i) by 1.155/.509 = 2.269, we have for the equations above,

$$-1.155F_{BA} + .866F_{BD} = 0 \qquad \text{(j)}$$

$$1.155F_{BA} + 1.135F_{BD} = 1113 \qquad \text{(k)}$$

Adding these equations, we may directly determine F_{BD}. Thus,

$$F_{BD} = 556 \text{ N}$$

From Eq. (h), we then may determine F_{BA} to be

$$F_{BA} = \frac{.866}{1.155}(556) = 417 \text{ N}$$

Finally, from Eq. (g) we get F_{BC}:

$$F_{BC} = (.374)(417) = 156.0 \text{ N}$$

Coplanar Force System. The simplest resultant for a coplanar force distribution is a single force or a single couple moment. Hence, there are three independent equations of equilibrium for a given free body. We shall first examine a problem for which only one free-body diagram is needed, and then we shall consider a problem involving several free-body diagrams.

EXAMPLE 5.5

A crane weighing 3000 lb supports a 10,000-lb load as shown in Fig. 5.27. Determine the supporting forces at A, which is a pinned connection, and at B, which is a roller.

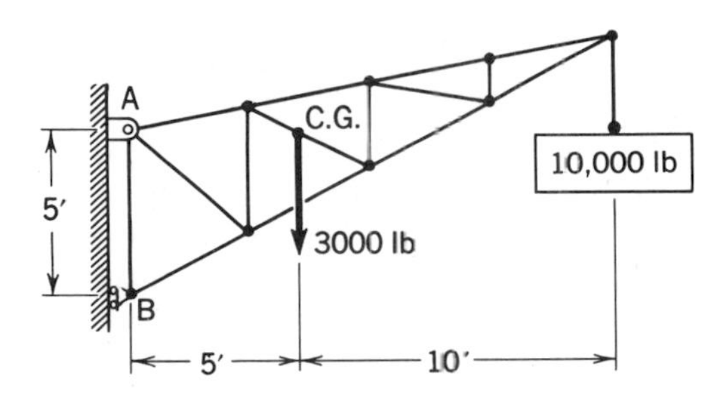

Figure 5.27. Loaded crane.

A free-body diagram of the main structure exposes the desired unknowns (Fig. 5.28). Note that since the system of forces may be taken as coplanar, we have three equations of equilibrium available and we may accordingly solve for the three unknown quantities from this single free-body diagram. Hence:

$\underline{\sum F_x = 0:}$

$$A_x + B = 0$$

Therefore,

$$A_x = -B$$

$\underline{\sum F_y = 0:}$

$$A_y - 3000 - 10{,}000 = 0$$

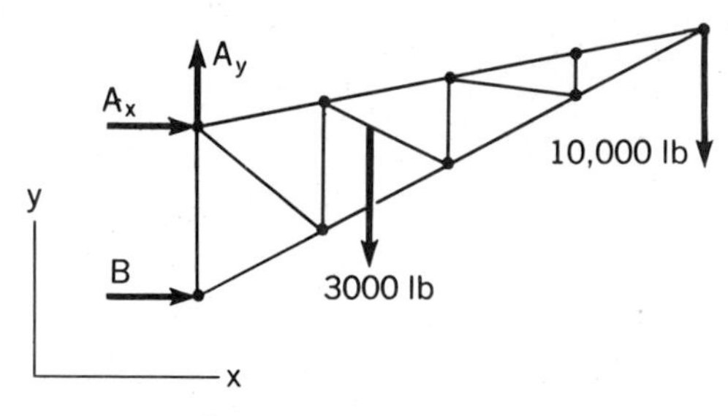

Figure 5.28. Free-body diagram of crane.

Therefore,

$$A_y = 13{,}000 \text{ lb}$$

$\underline{\sum M_B = 0:}$

$$-(5)(3000) - (15)(10{,}000) - 5A_x = 0$$

Therefore,

$$A_x = -33{,}000 \text{ lb}$$

The results are

$$A_x = -33{,}000 \text{ lb}, \qquad A_y = 13{,}000 \text{ lb}, \qquad B = 33{,}000 \text{ lb}$$

Note that A_x has a negative sign. As noted earlier, this is the result of having chosen the wrong sense for A_x at the outset of the computations. All we need do now is recognize that A_x has a sense opposite to what is shown in Fig. 5.28.

We have now solved the forces from the *wall onto the structure*. The forces from the *structure onto the wall* are the *reactions* to these forces and are equal and opposite.

EXAMPLE 5.6

In Fig. 5.29 is shown a frame where the pulley at D has a mass of 200 kg. Neglecting the weights of the bars, find the force transmitted from one bar to another at C.

To expose force components C_x and C_y, we form the free body of bar BD. This is shown in F.B.D. I in Fig. 5.30. It is clear that for this free body we have six unknowns and only three independent equations of equilibrium.[5] The free-body diagram of the bent bar AC is then drawn (F.B.D. II in Fig. 5.30). Here, we have three more equations but we bring in three more unknowns. Finally, the free-body diagram of the pulley (F.B.D. III in Fig. 5.30) gives three more equations with no additional unknowns. We now have nine equations available and nine unknowns and can proceed with confidence. Since only two of the unknowns are desired, we

[5]It should be noted that it is possible to have situations wherein there are more unknowns than independent equations of equilibrium for a given free body, but wherein some of the unknowns—perhaps the desired ones—can be still determined by the equations available. However, not all the unknowns of the free body can be solved. Accordingly, be alert for such situations, so as to minimize the work involved. In this case, we must consider other free-body diagrams.

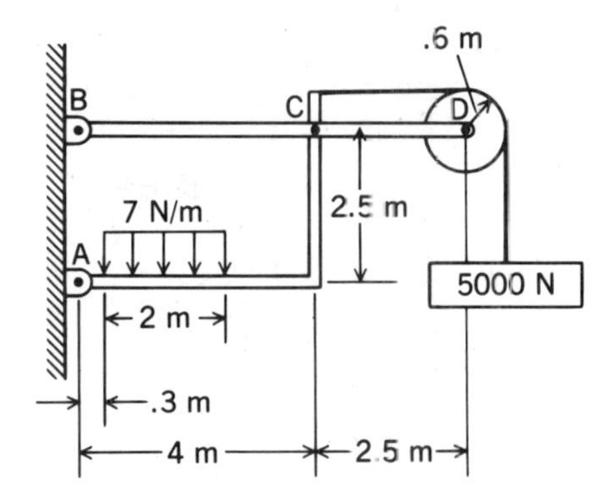

Figure 5.29. Loaded frame.

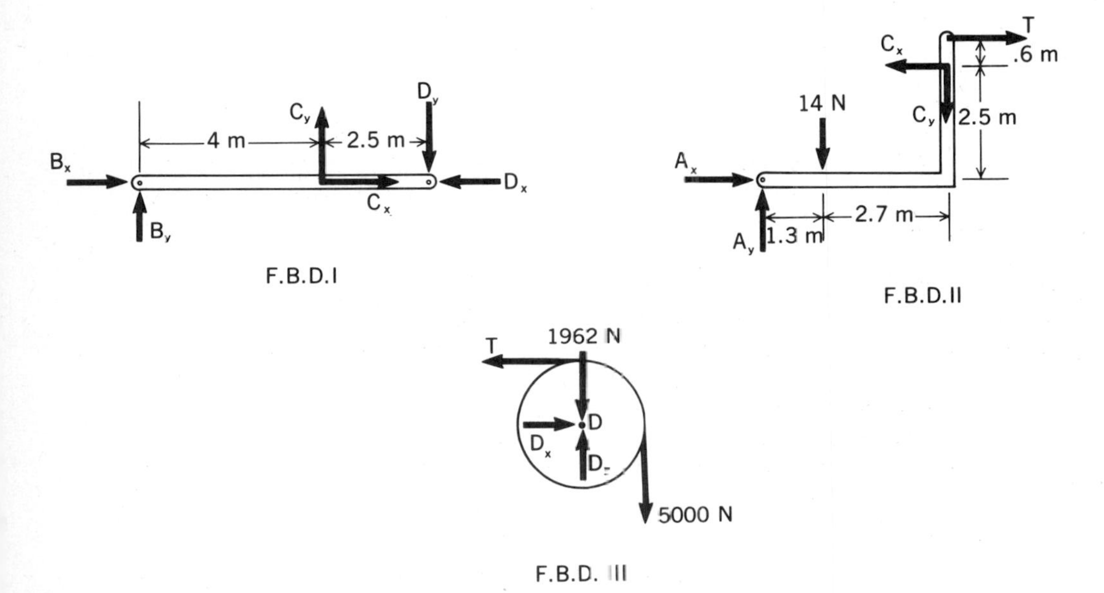

Figure 5.30. Free-body diagrams of frame parts.

shall take select scalar equations from each of the free-body diagrams to arrive at the components C_x and C_y most quickly.

From F.B.D. III:

$\sum M_D = 0$:

$$(T)(.6) - (5000)(.6) = 0$$

Therefore,

$$T = 5000 \text{ N}$$

$\sum F_x = 0$:

$$-T + D_x = 0$$

Therefore,

$$D_x = 5000 \text{ N}$$

$\underline{\sum F_y = 0:}$

$$-1962 - 5000 + D_y = 0$$

Therefore,

$$D_y = 6962 \text{ N}$$

From F.B.D. I:

$\underline{\sum M_B = 0:}$

$$(4)(C_y) - (6.5)(D_y) = 0$$

Therefore,

$$C_y = 11{,}313 \text{ N}$$

From F.B.D. II:

$\underline{\sum M_A = 0:}$

$$-(1.3)(14) - (T)(3.1) - C_y(4) + C_x(2.5) = 0$$

Therefore,

$$C_x = 24{,}300 \text{ N}$$

We can give the force at C as

$$\mathbf{C} = 24{,}300\boldsymbol{i} + 11.313\boldsymbol{j} \text{ N}$$

Problems

5.17. In a tug of war, when team B pulls with 400-lb force, how much force must team C exert for a draw? With what force does team A pull?

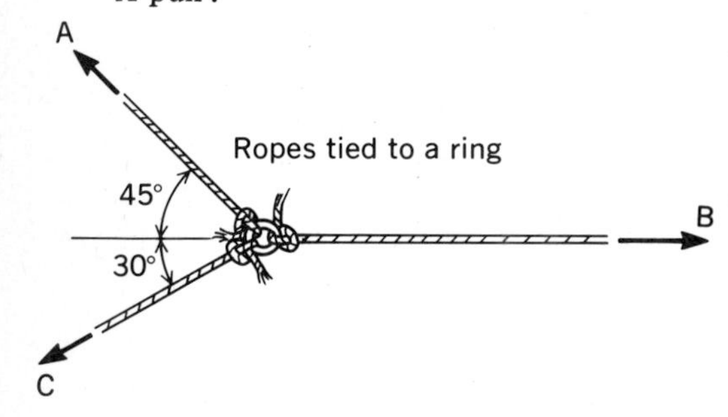

Figure P.5.17

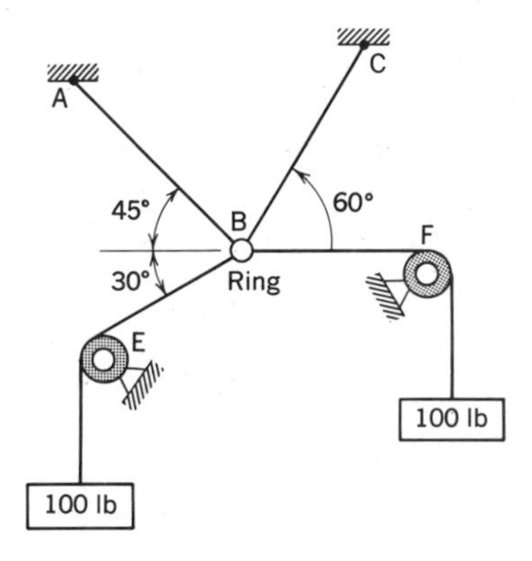

Figure P.5.18

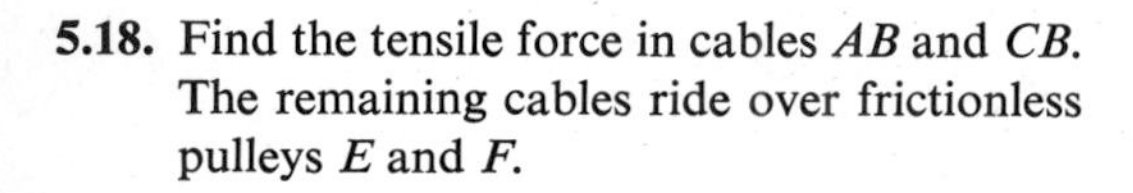

5.18. Find the tensile force in cables AB and CB. The remaining cables ride over frictionless pulleys E and F.

5.19. Find the force transmitted by wire BC. The pulley E can be assumed to be frictionless in this problem.

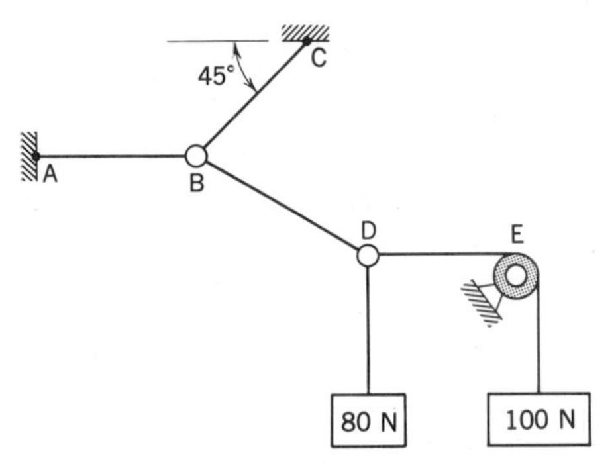

Figure P.5.19

5.20. A 700-N circus performer causes a .15-m sag in the middle of a 12-m tightrope with a 5000-N initial tension. What additional tension is induced in the cable? What is the cable tension when the performer is 3 m from the end and the sag is .12 m?

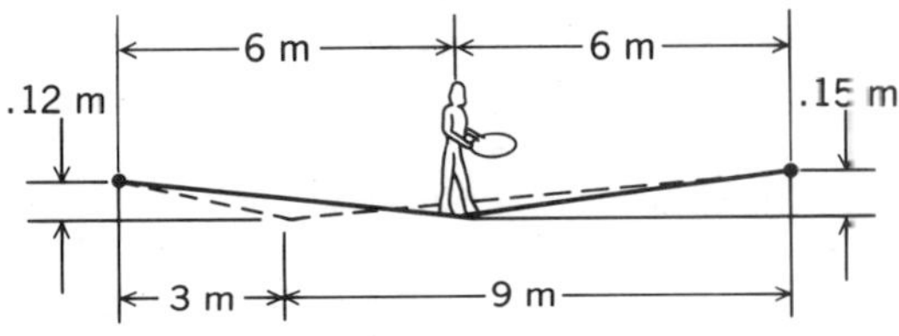

Figure P.5.20

5.21. A 27-lb mirror is held up by a wire fastened to two hooks on the mirror frame. (a) What is the force on the wall hook and the tension in the wire? (b) If the wire will break at a tension of 32 lb, must the wall hook be moved (i.e., the wire lengthened or shortened and the 4 in. rise distance changed)? If so, to what point?

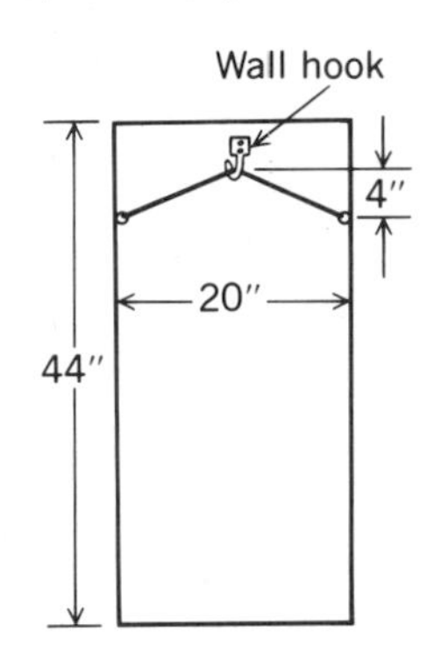

Figure P.5.21

5.22. Explain why equilibrium of a concurrent force system is guaranteed by having $\sum_i (F_y)_i = 0$, $\sum_i (M_d)_i = 0$, and $\sum_i (M_e)_i = 0$. Axes d and e are not parallel to the xz plane. Moreover, the axes are oriented so that the line of action of the resultant force cannot intersect both axes.

5.23. Cylinders A and B weigh 500 N each and cylinder C weighs 1000 N. Compute all contact forces.

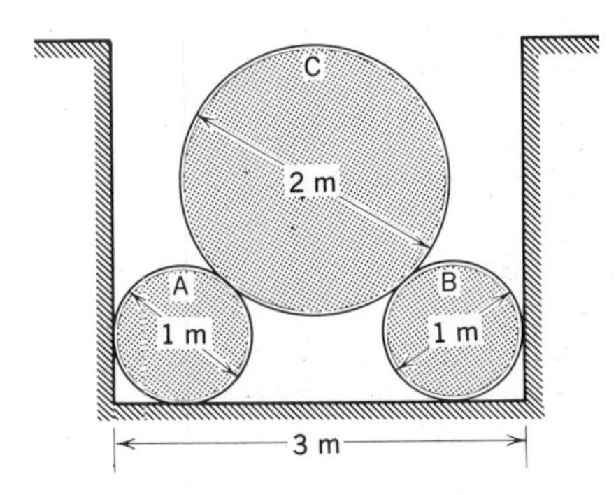

Figure P.5.23

5.24. A block having a mass of 500 kg is held by five cables. What are the tensions in these cables?

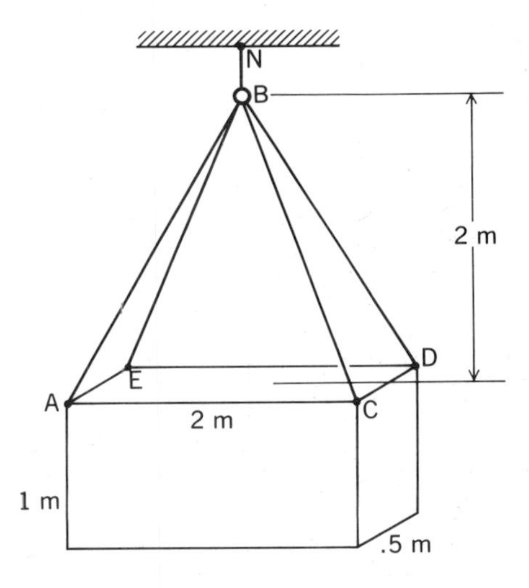

Figure P.5.24

5.25. What are the tensions in cables AB, BC, and BD? Points A and B are in the yz plane.

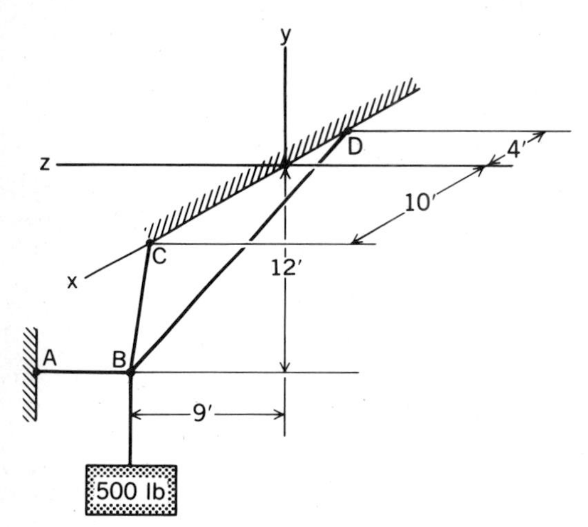

Figure P.5.25

5.26. An elastic cord AB is just taut before the 1000-N force is applied. If it takes 5.0 N/mm of elongation of the cord, what is the tension T in the cord after the 1000-N force is applied? Set up the equation for T but do not solve.

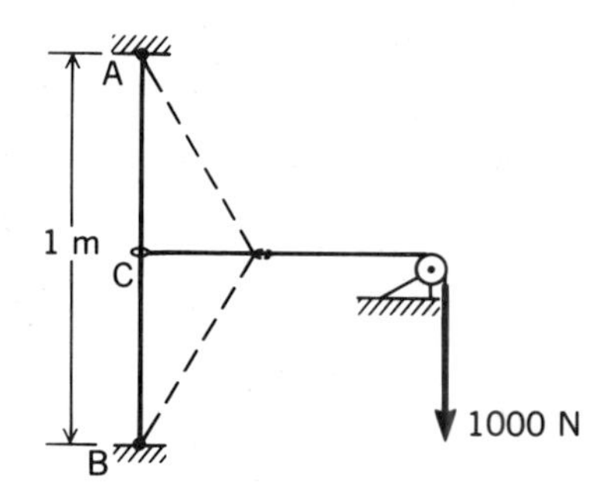

Figure P.5.26

5.27. A thin-walled cylinder of outside radius 1 m and weight 500 N rests on an incline. What friction force f at A is needed for this configuration? What is the tension in wire CB?

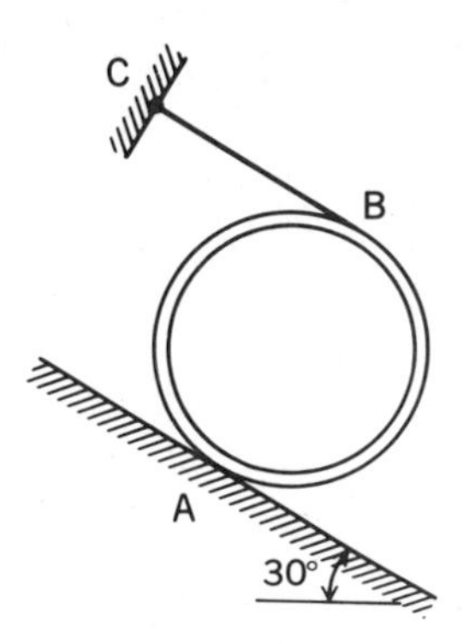

Figure P.5.27

5.28. A stepped cylinder is pulled down an incline by a force F which is increased from zero to 20 N very very slowly while always maintaining the 30° inclination shown. If the cylinder is in equilibrium when $F = 0$, how far does O move as a result of F after equilibrium has been established with $F = 20$ N. The stepped cylinder has a mass of 10 kg. There is no slipping at the base. The force from the spring is K times the extension of the spring. For this spring, $K = 5$ N/mm.

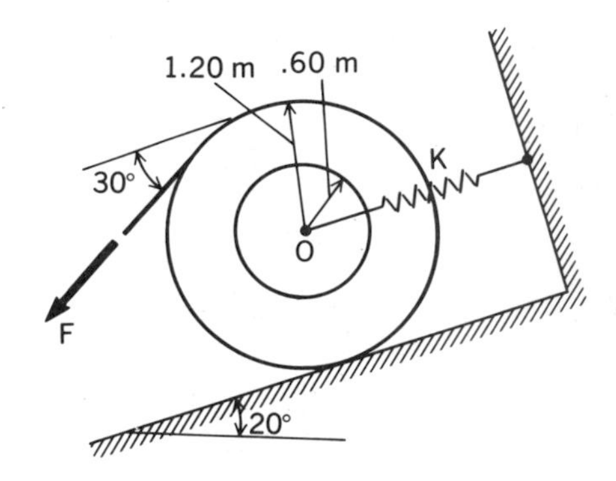

Figure P.5.28

5.29. A 10-kg ring is supported by a smooth surface E and a wire AB. A body D having a mass of 3 kg is fixed to the ring at the orientation shown. What is the tension in the wire AB? What is its orientation α?

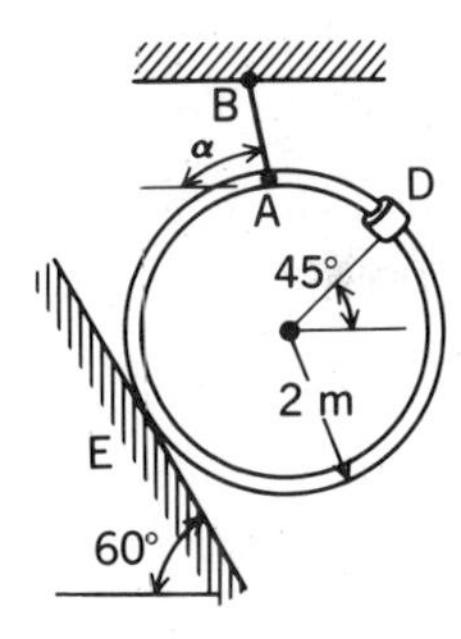

Figure P.5.29

5.30. What is the tension in the cables of a 10-ft-wide 12-ft-long 6000-lb castle drawbridge when the bridge is first raised? When the bridge is at 45°? What are the reactions at the hinge pin?

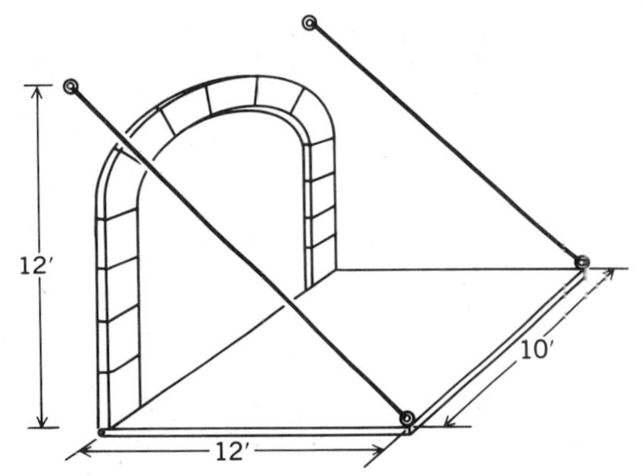

Figure P.5.30

5.31. A small hoist has a lifting capacity of 20 kN. What is the maximum cable tension and the corresponding reactions at C? Do not consider weight of beam.

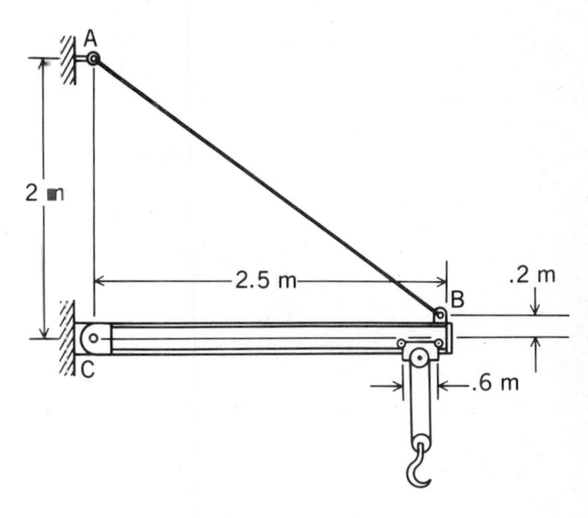

Figure P.5.31

5.32. A uniform block weighing 500 lb is constrained by three wires. What are the tensions in these wires?

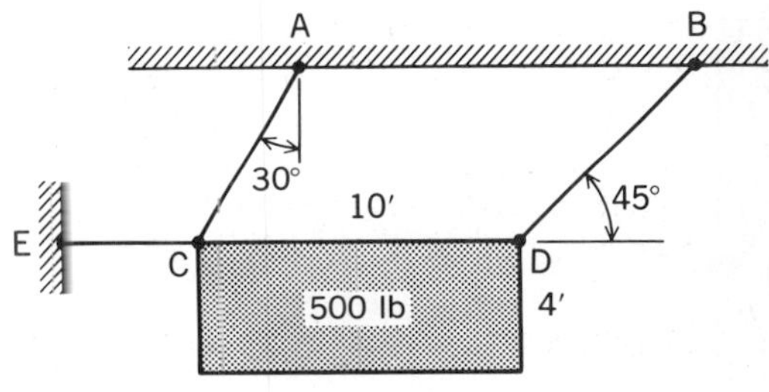

Figure P.5.32

5.33 Find the supporting forces for the frame shown.

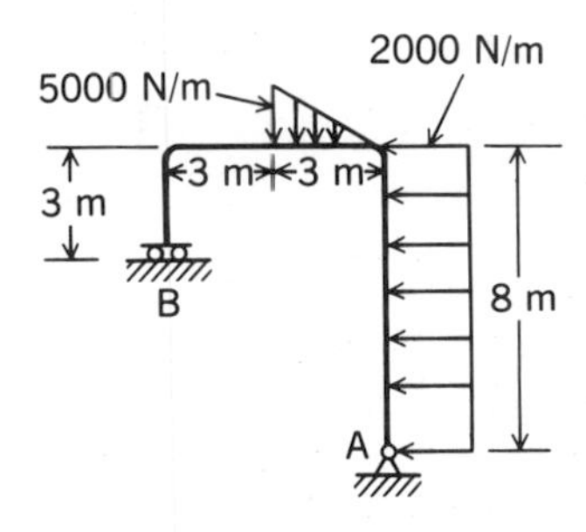

Figure P.5.33

5.34. A 300-kN tank is climbing up a 30° incline at constant speed. What is the torque developed on the rear drive wheels to accomplish this? Assume that all other wheels are free-turning.

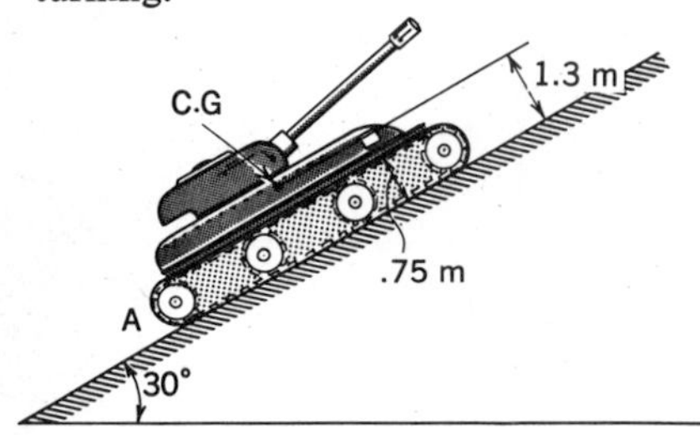

Figure P.5.34

5.35. Find the components of the forces acting on pins A, B, and C connecting and supporting the blocks shown. Block I weighs 10 kN, and block II weighs 30 kN.

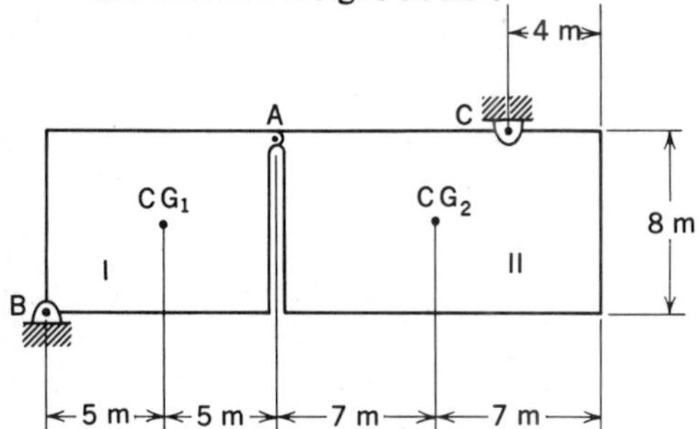

Figure P.5.35

5.36. What force F do the pliers develop on the pipe section D? Neglect friction.

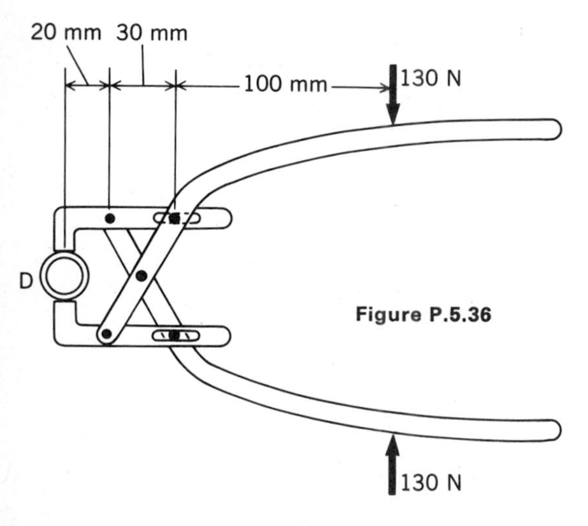

Figure P.5.36

5.37. What are the supporting forces for the frame? Neglect all weights except the 10-kN weight.

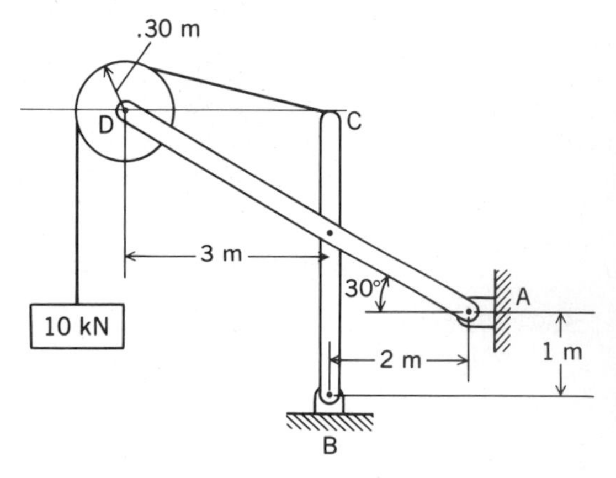

Figure P.5.37

5.38. A 20-m circular arch must withstand a wind load given for $0 < \theta < \pi/2$ as

$$f = 5000\left(1 - \frac{\theta}{\pi/2}\right) \text{ N/m}$$

where θ is measured in radians. Note that for $\theta > \pi/2$, there is no loading. What are the supporting forces? (*Hint:* What is the point for which taking moments is simplest?)

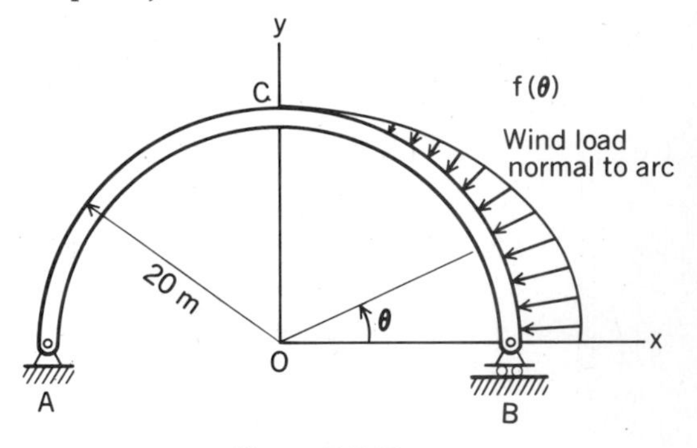

Figure P.5.38

5.39. An arch is formed by uniform plates A and B. Plate A weighs 5 kN and plate B weighs 2 kN. What are the supporting forces at C, D, and E?

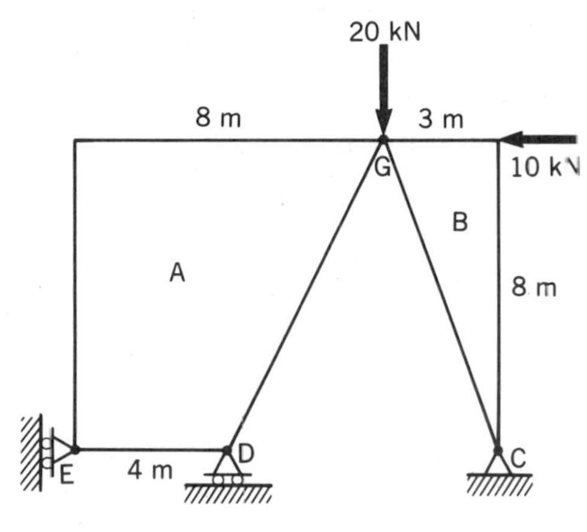

Figure P.5.39

5.40. Find the supporting forces on the beam *EF* and the supporting forces at *A*, *B*, *C*, and *D*.

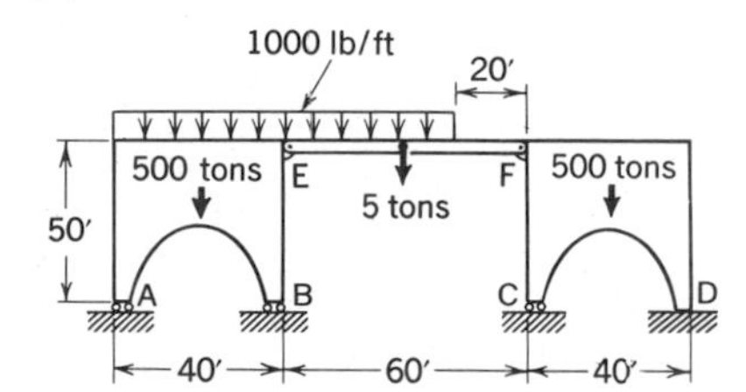

Figure P.5.40

5.41. What is the supporting force system at *A* for the cantilever beam? Neglect the weight of the beam.

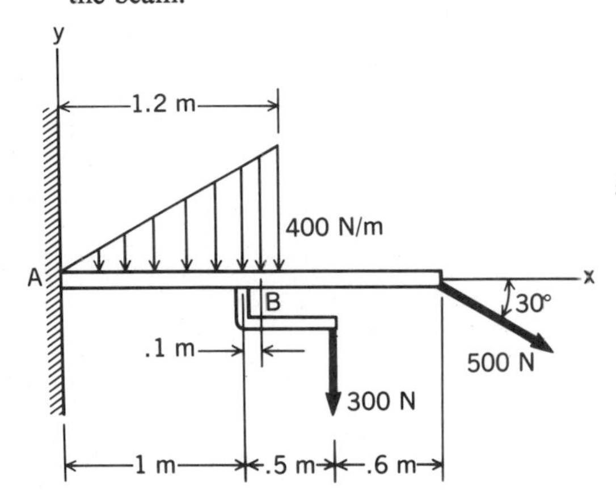

Figure P.5.41

5.42. In Problem 5.41, find the force system transmitted through the cross section at *B*.

5.43. A cantilever beam *AB* is pinned at *B* to a simply supported beam *BC*. For the loads given, find the supporting force system at *A*. Determine force components normal and axial to the beam *AB*. Neglect the weights of the beams.

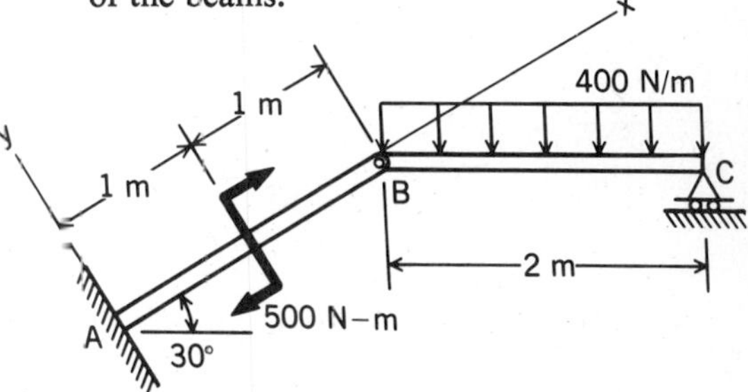

Figure P.5.43

5.44 Find the supporting force system at *A*.

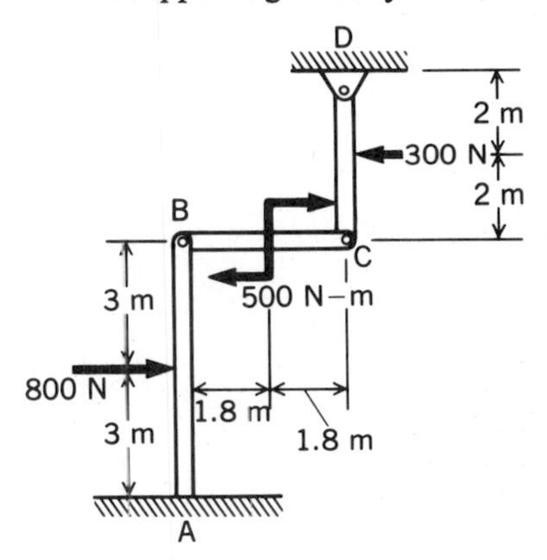

Figure P.5.44

5.45. A light bent rod *AD* is pinned to a straight light rod *CB* at *C*. The bent rod supports a uniform load. A spring is stretched to connect the two rods. The spring has a spring constant of 10^4 N/m, and its unstretched length is .8 m. Find the supporting forces at *A* and *B*. The force in the spring is 10^4 times the elongation in meters.

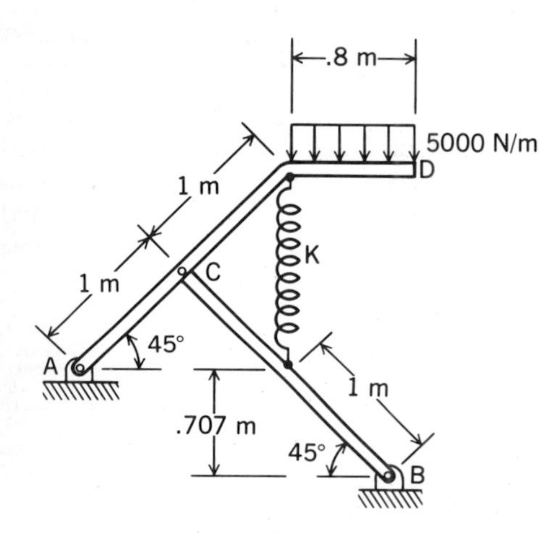

Figure P.5.45

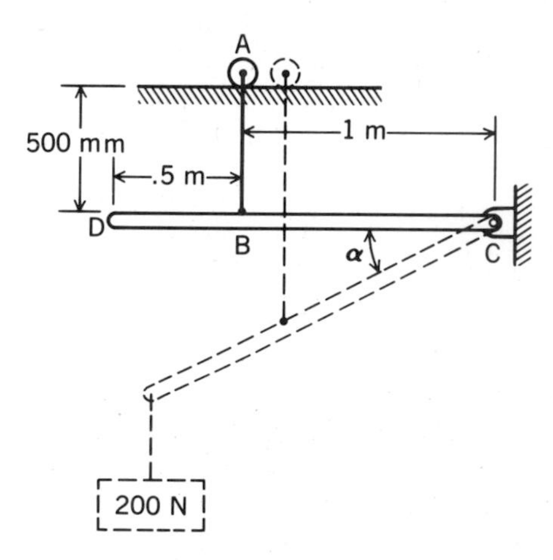

Figure P.5.47

5.46. Light rods AD and AC are pinned together at C and support a 300-N and a 100-N load. What are the supporting forces at A and B?

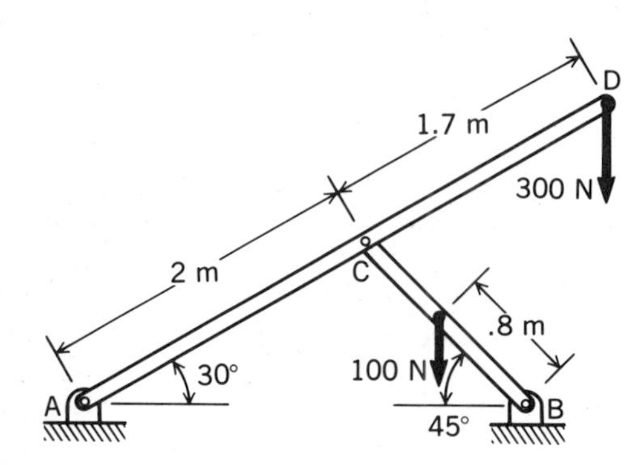

Figure P.5.46

5.47. A light rod CD is held in a horizontal position by a strong elastic band AB (shock cord) which acts like a spring in that it takes 10^3 N per meter of elongation of the band. The upper part of the band is connected to a small wheel free to roll on a horizontal surface. What is the angle α needed to support a 200-N load as shown?

5.48. Solve for the supporting forces at A and C. AB weighs 100 lb, and BC weighs 150 lb.

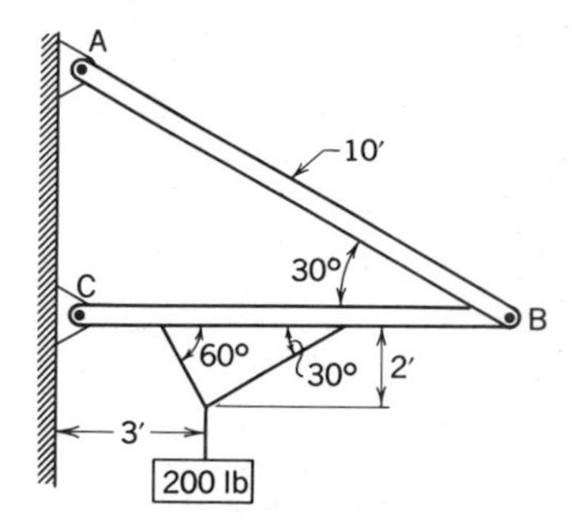

Figure P.5.48

5.49. What torque T is needed to maintain the configuration shown for the compressor if $p_1 = 5$ psig? The system lies horizontally.

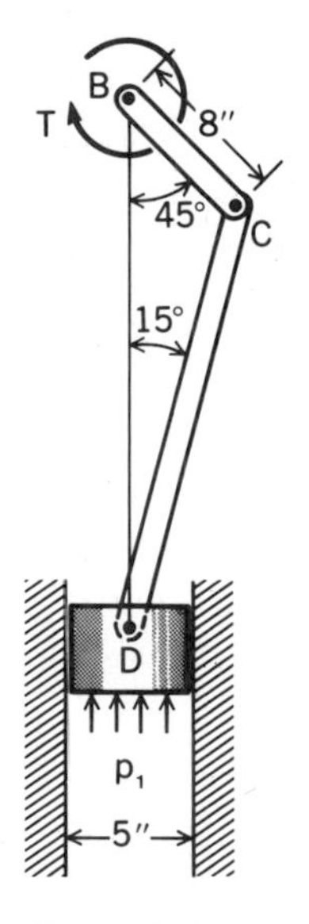

Figure P.5.49

5.50. Work Problem 5.49 for the system oriented vertically with BC weighing 3 lb and CD weighing 5 lb.

5.51. Neglecting friction, find the angle β of line AB for equilibrium in terms of α_1, α_2, W_1, and W_2.

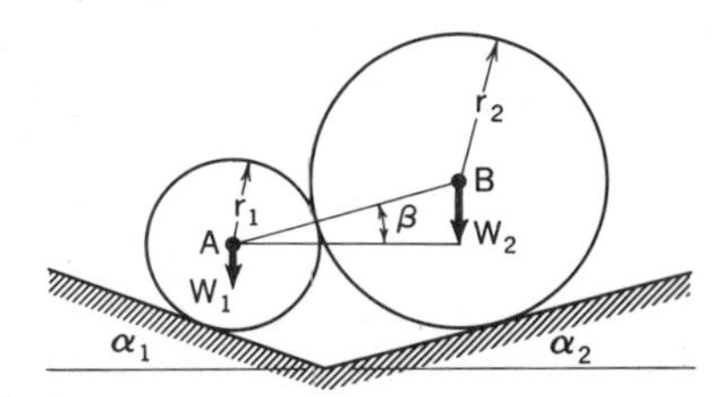

Figure P.5.51

5.52. If the rod CD weighs 20 lb, what torque T is needed to maintain equilibrium? The system is in a vertical plane. Cylinder A weighs 10 lb and cylinder B weighs 5 lb. Disregard friction. At D there is a slot.

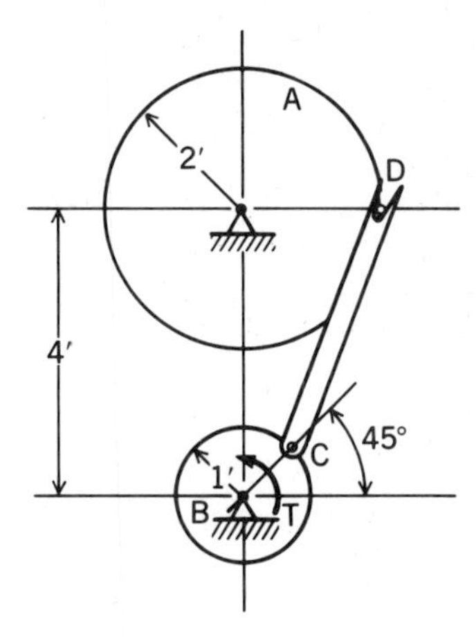

Figure P.5.52

5.53. Find the supporting forces at A and G. The weight of W is 500 N and the weight of C is 200 N. Neglect all other weights. The cord connecting C and D is vertical.

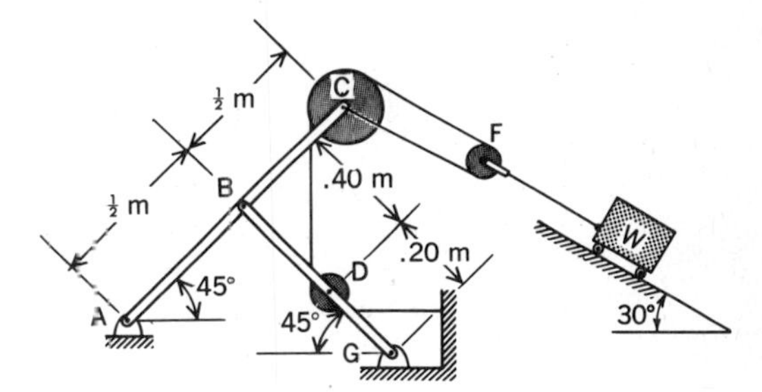

Figure P.5.53

5.54. What torque T is needed for equilibrium if cylinder B weighs 500 N and CD weighs 300 N?

Figure P.5.54

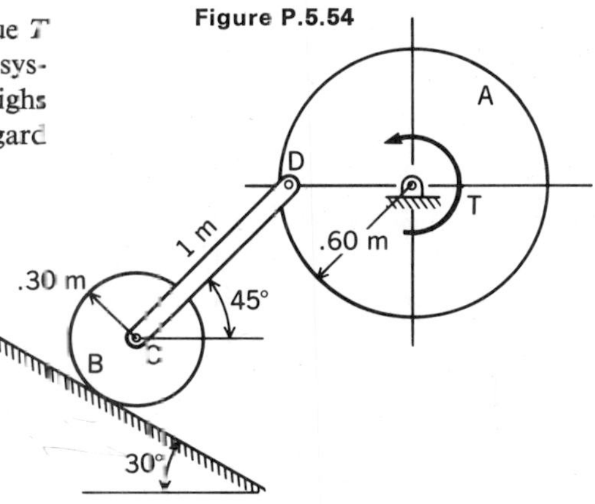

5.55. A bar AB is pinned to two identical planetary gears each of diameter .30 m. Gear E is pinned to bar AB and meshes with the two planetary gears, which in turn mesh with stationary gear D. If a torque T of 100 N-m is applied to the bar AB, what external torque is needed to be applied to the upper gear D to maintain equilibrium? The system is horizontal.

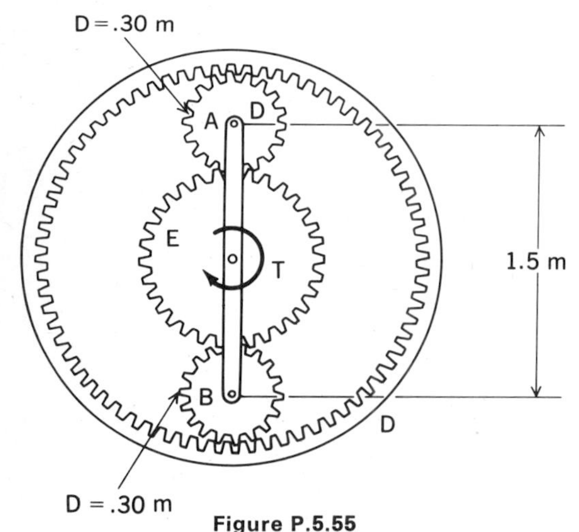

Figure P.5.55

5.56. In Problem 5.55, equilibrium is maintained by applying a torque on gear E rather than gear D. What is this torque?

5.57. A Bucyrus–Erie transit crane is holding a chimney having a weight of 20 kN. The chimney is held by a cable that goes over a pulley at A, then goes over a second pulley at D, and then to a winch at K. The position of boom AH (on top) is maintained by two separate cables, one from A to B, and the other from B to pulley C. Find the tensions in cables AB and BC. Note that BC is vertical for the setup shown. Consider only the weight of the load and neglect friction.

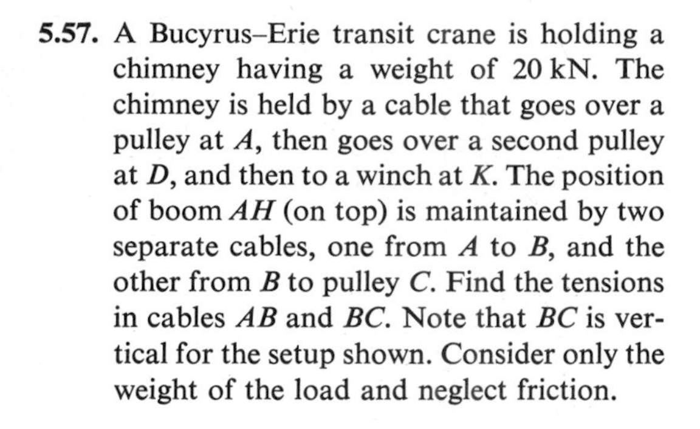

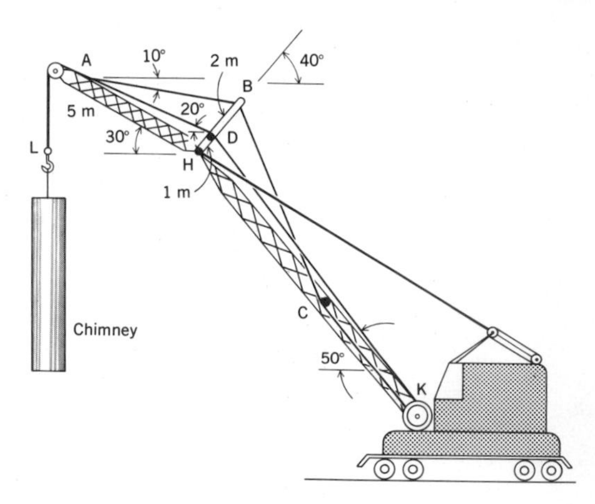

Figure P.5.57

5.58. What are the forces at the arm connections at B and the cable tensions when the power shovel is in the position shown? Arm AC weighs 13,000 N, arm DF weighs 11,000 N, and the shovel and payload together weigh 9000 N and act at the center of gravity as shown. B is at the same elevation as G.

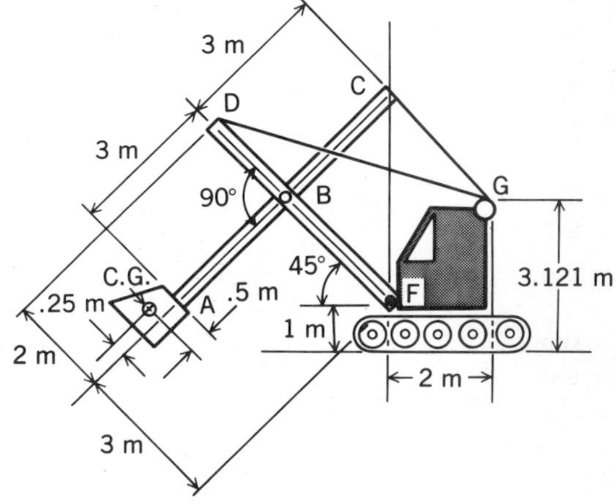

Figure P.5.58

Parallel Force Systems. The simplest resultant is a single force or a single couple moment. There are then three independent equations of equilibrium. If the forces are also coplanar, then we have only two independent equations of equilibrium. We shall examine the latter case in the following example.

EXAMPLE 5.7

Determine the forces required to support the uniform beam in Fig. 5.31 shown loaded with a couple, a point force, and a parabolic distribution of load. The weight of the beam is 100 lb.

Since a couple can be rotated without affecting the equilibrium of the body, we can orient the couple so that the forces are vertical. Accordingly, we have here a beam loaded by a system of parallel coplanar loads. Clearly, the supporting forces must be vertical, as shown in Fig. 5.32, where we have a free-body diagram of the beam. Since there are only two unknown quantities, we can handle the problem by statical consideration of this free body.

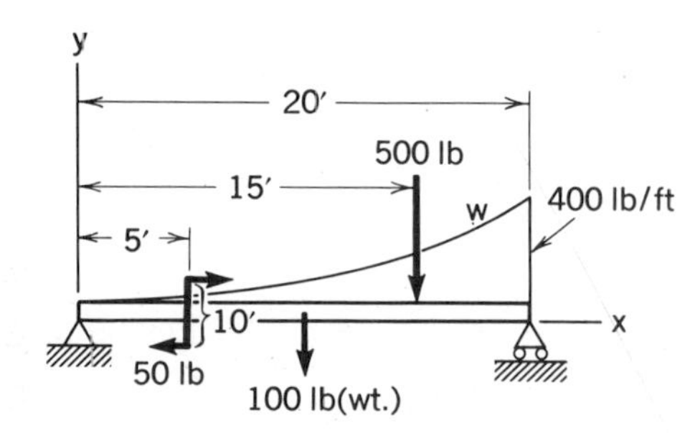

Figure 5.31. Find supporting forces.

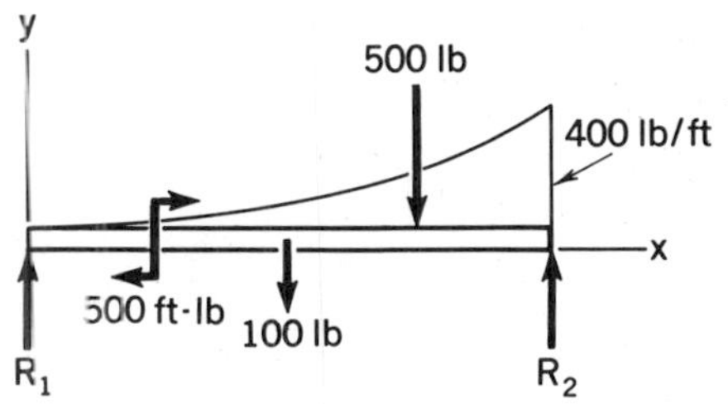

Figure 5.32. Free-body diagram.

The equation for the loading curve must be $w = ax^2 + b$, where a and b are to be determined from the loading data and the choice of reference. With an xy reference at the left end, as shown, we then have the conditions:

1. When $x = 0$, $w = 0$.
2. When $x = 20$, $w = 400$.

To satisfy these conditions, b must be zero and a must be unity; the loading function is thus given as

$$w = x^2 \text{ lb/ft}$$

In this problem, we shall again work directly with the scalar equations. By summing moments about the left and right ends of the beam, we can then solve the unknowns directly:

$\underline{\sum M_1 = 0:}$

$$-500 - (10)(100) - (15)(500) - \int_0^{20} x^3\, dx + 20R_2 = 0$$

Integrating and canceling terms, we get

$$-9000 - \frac{x^4}{4}\bigg|_0^{20} + 20R_2 = 0$$

By inserting limits and solving, we get one of the unknowns:

$$R_2 = 2450 \text{ lb}$$

Next,

$\underline{\sum M_2 = 0:}$

$$-20(R_1) - 500 + (10)(100) + (5)(500) + \int_0^{20} (20 - x)x^2\, dx = 0$$

Solving for R_1, we have

$$R_1 = 817 \text{ lb}$$

As a *check* on these computations, we can sum forces in the vertical direction. The result must be zero (or as close to zero as the accuracy of our calculations permits):

$\underline{\sum F_y = 0:}$

$$R_1 + R_2 - 100 - 500 - \int_0^{20} x^2\, dx = 0$$

$$3267 - 600 - \frac{x^3}{3}\bigg|_0^{20} = 0$$

Therefore,

$$2667 - 2667 = 0$$

Always take the opportunity to check a solution in this manner (i.e., by using a redundant equilibrium equation). In later problems, we shall rely heavily on calculated reactions; thus, we must make sure they are correct.

General Force Systems. The simplest resultant in the general case is a force and a couple moment. Six equations of equilibrium can be given for each free-body diagram. We now examine two examples for this case.

EXAMPLE 5.8

A derrick is shown in Fig. 5.33 supporting a 1000-lb load. The vertical beam has a ball-and-socket connection into the ground at *d* and is held by guy wires. Neglect the weight of the members and guy wires, and find the tensions in the guy wires *ac*, *bc*, and *ce*.

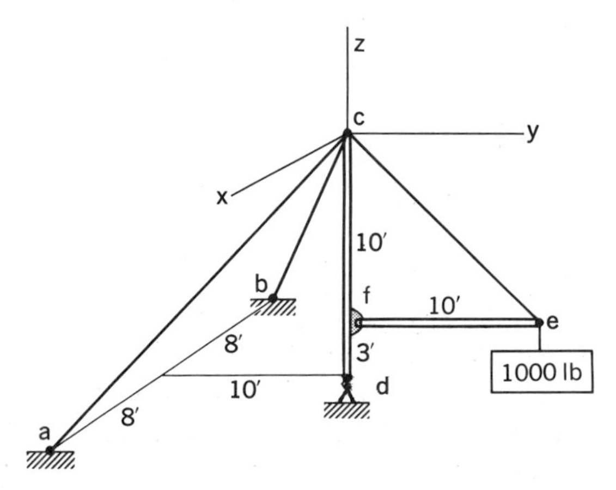

Figure 5.33. Loaded derrick.

If we select as a free body both members and the interconnecting guy wire *ce*, we shall expose two of the desired unknowns (Fig. 5.34). Note that this is a general three-dimensional force system with only five unknowns. Although all these unknowns can be solved by statical considerations of this free body, you will notice that, if we take moments about point *d*, we will involve in a vector equation only the desired unknowns T_{bc} and T_{ac}. Accordingly, all unknown forces need not be computed for this free-body diagram. You should always look for such short cuts in situations such as these.

To determine the unknown tension T_{ce}, we must employ another free-body diagram. Either the vertical or horizontal member will expose this unknown in a manner susceptible to solution. The latter has been selected and is shown in Fig.5.35. Note that we have here a coplanar force system with three unknowns. Again, you can see that, by taking moments about point *f*, we will involve only the desired unknown.

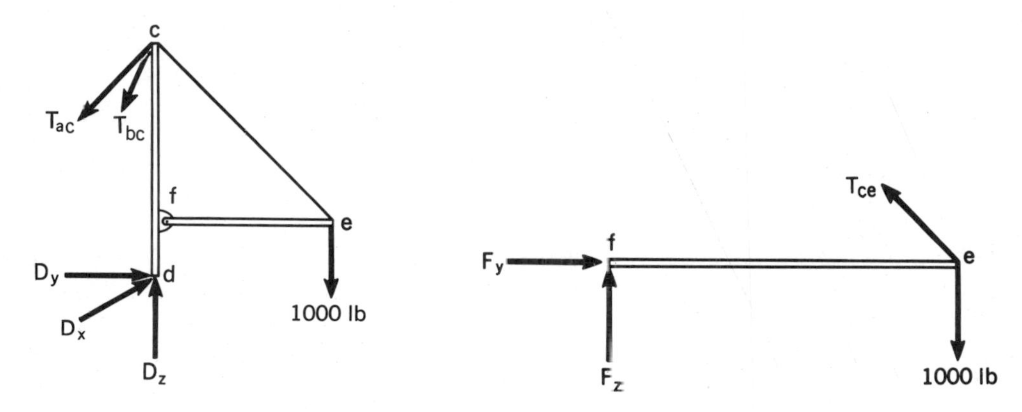

Figure 5.34. Free-body diagram 1.

Figure 5.35. Free-body diagram 2.

The vector $\boldsymbol{T}_{ac}$ may then be given as

$$\boldsymbol{T}_{ac} = T_{ac}\left[\frac{1}{\sqrt{333}}(8\boldsymbol{i} - 10\boldsymbol{j} - 13\boldsymbol{k})\right] \tag{a}$$

Similarly, we have for $\boldsymbol{T}_{bc}$,

$$\boldsymbol{T}_{bc} = T_{bc}\left[\frac{1}{\sqrt{333}}(-8\boldsymbol{i} - 10\boldsymbol{j} - 13\boldsymbol{k})\right] \tag{b}$$

Using the free-body diagram in Fig. 5.34, we now set the sum of moments about point *d* equal to zero. Thus, employing the relations above, we get

$$13\boldsymbol{k} \times \frac{T_{ac}}{\sqrt{333}}(8\boldsymbol{i} - 10\boldsymbol{j} - 13\boldsymbol{k}) + 13\boldsymbol{k} \times \frac{T_{bc}}{\sqrt{333}}(-8\boldsymbol{i} - 10\boldsymbol{j} - 13\boldsymbol{k}) + 10\boldsymbol{j} \times (-1000\boldsymbol{k}) = \boldsymbol{0} \tag{c}$$

When we make the substitution of variable $t_1 = T_{ac}/\sqrt{333}$ and $t_2 = T_{bc}/\sqrt{333}$, the preceding equation becomes

$$[130(t_1 + t_2) - 10{,}000]\boldsymbol{i} + [104(t_1 - t_2)]\boldsymbol{j} = \boldsymbol{0} \tag{d}$$

The scalar equations,

$$130(t_1 + t_2) - 10{,}000 = 0$$
$$104(t_1 - t_2) = 0$$

can now be readily solved to give $t_1 = t_2 = 38.5$. Hence, we get $T_{ac} = 38.5\sqrt{333} = 702$ lb and $T_{bc} = 38.5\sqrt{333} = 702$ lb.[6]

Turning finally to the free-body diagram 2 in Fig. 5.35, we see that, in summing moments about f, the horizontal component of the tension T_{ce} has a zero-moment arm. Thus,

$$(10)(0.707)T_{ce} - (10)(1000) = 0$$

Hence,

$$T_{ce} = 1414 \text{ lb}$$

EXAMPLE 5.9

A blimp is shown in Fig. 5.36 fixed at the mooring tower D by a ball-joint connection, and held by cables AB and AC. The blimp has a mass of 1500 kg. The simplest resultant force $\boldsymbol{F}$ from air pressure (including the effects of wind) is

$$\boldsymbol{F} = 17{,}500\boldsymbol{i} + 1000\boldsymbol{j} + 1500\boldsymbol{k} \text{ N}$$

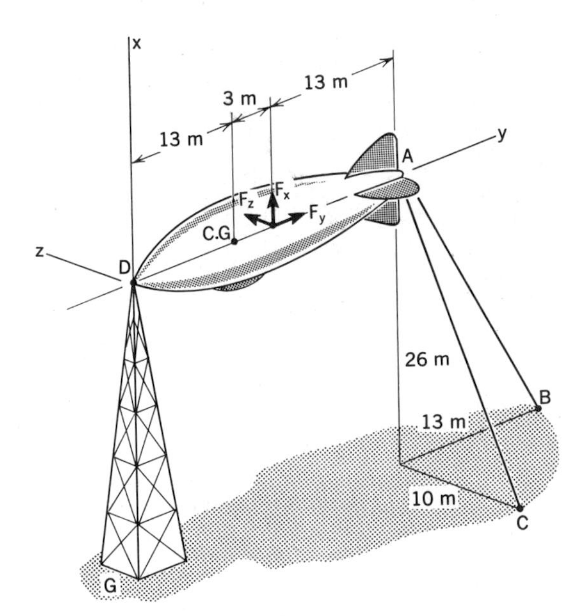

Figure 5.36. Tethered blimp.

at a position shown in the diagram. Compute the tension in the cables as well as the force transmitted to the ball joint at the top of the tower at D. Also, what force sys-

[6]By taking moments about the line connecting points a and d, we could get T_{bc} directly using the scalar triple product. We suggest that you try this.

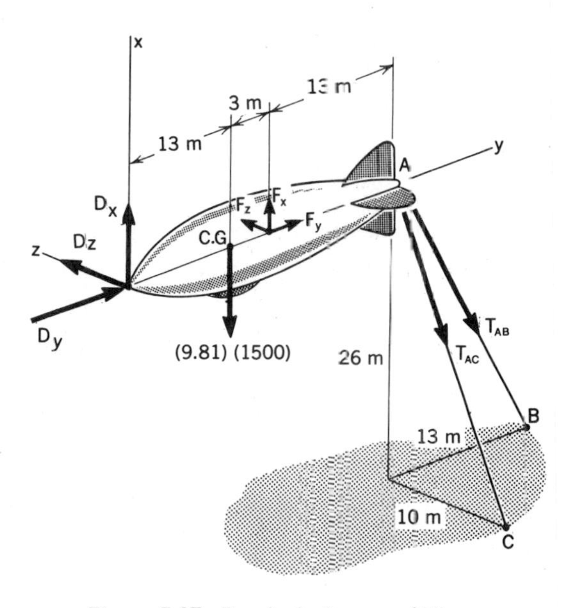

Figure 5.37. Free-body diagram of blimp.

tem is transmitted to the ground at G through the mooring tower? The tower weighs 5000 N.

We shall first consider a free-body diagram of the blimp, as shown in Fig. 5.37. We have five unknown forces here, and we can solve all of them by using equations of equilibrium for this free body. As a first step, we express the cable tensions vectorially. That is,

$$\boldsymbol{T}_{AC} = T_{AC}\left(\frac{-26\boldsymbol{i} - 10\boldsymbol{k}}{\sqrt{26^2 + 10^2}}\right) = T_{AC}(-.933\boldsymbol{i} - .359\boldsymbol{k})$$

$$\boldsymbol{T}_{AB} = T_{AB}\left(\frac{-26\boldsymbol{i} + 13\boldsymbol{j}}{\sqrt{26^2 + 13^2}}\right) = T_{AB}(-.894\boldsymbol{i} + .447\boldsymbol{j})$$

We now go back to the basic vector equations of equilibrium. Thus:

$\underline{\sum \boldsymbol{F}_i = \boldsymbol{0}}$:

$$D_x\boldsymbol{i} + D_y\boldsymbol{j} + D_z\boldsymbol{k} - (1500)(9.81)\boldsymbol{i} + 17{,}500\boldsymbol{i} + 1000\boldsymbol{j} + 1500\boldsymbol{k}$$
$$+ T_{AC}(-.933\boldsymbol{i} - .359\boldsymbol{k}) + T_{AB}(-.894\boldsymbol{i} + .447\boldsymbol{j}) = \boldsymbol{0}$$

The scalar equations are:

$$D_x + 2785 - .933T_{AC} - .894T_{AB} = 0 \qquad \text{(a)}$$

$$D_y + 1000 + .447T_{AB} = 0 \qquad \text{(b)}$$

$$D_z + 1500 - .359T_{AC} = 0 \qquad \text{(c)}$$

Next take moments about point D.

$\underline{\sum (\boldsymbol{M}_i)_D = \boldsymbol{0}}$:

$$13\boldsymbol{j} \times (9.81)(1500)(-\boldsymbol{i}) + 16\boldsymbol{j} \times (17{,}500\boldsymbol{i} - 1000\boldsymbol{j} + 1500\boldsymbol{k})$$
$$+ 29\boldsymbol{j} \times T_{AC}(-.933\boldsymbol{i} - .359\boldsymbol{k}) + 29\boldsymbol{j} \times T_{AB}(-.894\boldsymbol{i} + .447\boldsymbol{j}) = \boldsymbol{0}$$

Carrying out the various cross products, we end up only with $\boldsymbol{k}$ and $\boldsymbol{i}$ components, thus generating two scalar equations.[7] They are:

$$-10.41T_{AC} + 24{,}000 = 0 \tag{d}$$

$$25.9T_{AB} + 27.1T_{AC} - 88{,}700 = 0 \tag{e}$$

We now have five independent equations for five unknowns. We can thus solve these equations simultaneously. From Eq. (d), we have

$$T_{AC} = 2305 \text{ N}$$

From Eq. (e), we have

$$T_{AB} = 1012 \text{ N}$$

From Eq. (c), we have

$$D_z = -673 \text{ N}$$

From Eq. (b), we have

$$D_y = -1452 \text{ N}$$

From Eq. (a), we have

$$D_x = 270 \text{ N}$$

Next, consider the mooring tower as a free body (Fig. 5.38). Notice that in showing the forces at the ball joint D, we have taken into account both of the negative signs shown above for D_y and D_z as well as Newton's third law.

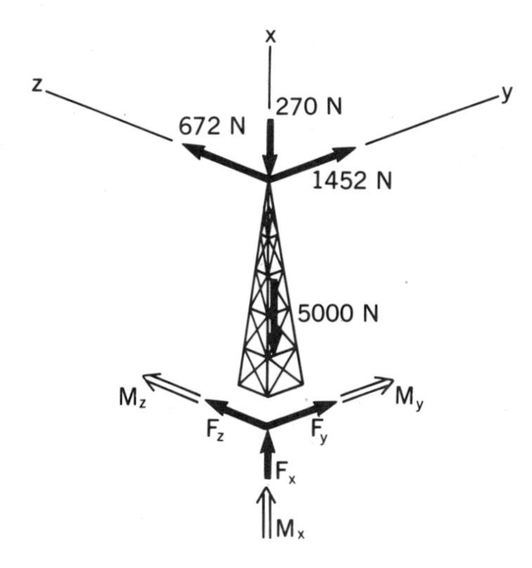

Figure 5.38. Free-body diagram of mooring tower.

Again, using the basic vector equations of statics, we have

$$-270\boldsymbol{i} + 1452\boldsymbol{j} + 672\boldsymbol{k} + F_x\boldsymbol{i} + F_y\boldsymbol{j} + F_z\boldsymbol{k} - 5000\boldsymbol{i} = \boldsymbol{0}$$

---

[7]The third equation is $0 = 0$. That is, there are no moments about the y axis, because all forces pass through the y axis.

Hence,

$$F_x = 5270 \text{ N}$$
$$F_y = -1452 \text{ N}$$
$$F_z = -672 \text{ N}$$

Now take moments about the base at F. We get

$$26\boldsymbol{i} \times (-270\boldsymbol{i} + 1452\boldsymbol{j} + 672\boldsymbol{k}) + M_x\boldsymbol{i} + M_y\boldsymbol{j} + M_z\boldsymbol{k} = \boldsymbol{0}$$

From this, we get

$$M_x = 0$$
$$M_y = 17{,}470 \text{ N-m}$$
$$M_z = -37{,}800 \text{ N-m}$$

We can conclude that, at the center of the base, the force system from the ground is

$$\boldsymbol{F} = 5270\boldsymbol{i} - 1452\boldsymbol{j} - 672\boldsymbol{k} \text{ N}$$
$$\boldsymbol{C} = 17{,}470\boldsymbol{j} - 37{,}800\boldsymbol{k} \text{ N-m}$$

The force system acting on the ground at the center of the base is the reaction to the system above. Thus,

$$\boldsymbol{F}_{\text{ground}} = -5270\boldsymbol{i} + 1452\boldsymbol{j} + 672\boldsymbol{k} \text{ N}$$
$$\boldsymbol{C}_{\text{ground}} = -17{,}470\boldsymbol{j} + 37{,}300\boldsymbol{k} \text{ N-m}$$

Problems

5.59. The triple pulley sheave and the double pulley sheave weigh 15 lb and 10 lb, respectively. What rope force is necessary to lift a 350-lb engine? What is the force on the ceiling hook?

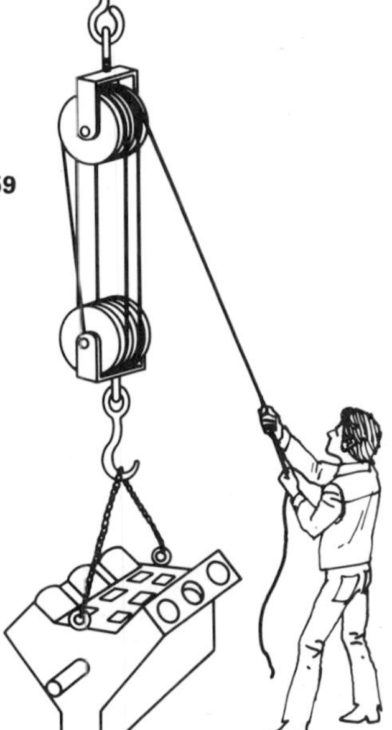

Figure P.5.59

5.60. A multipurpose pry bar can be used to pull nails in the three positions. If a force of 400 lb is required to remove a nail and a carpenter can exert 50 lb, which position(s) must he use?

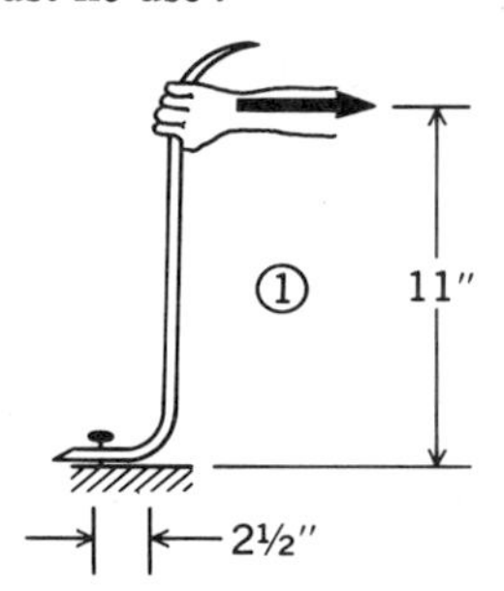

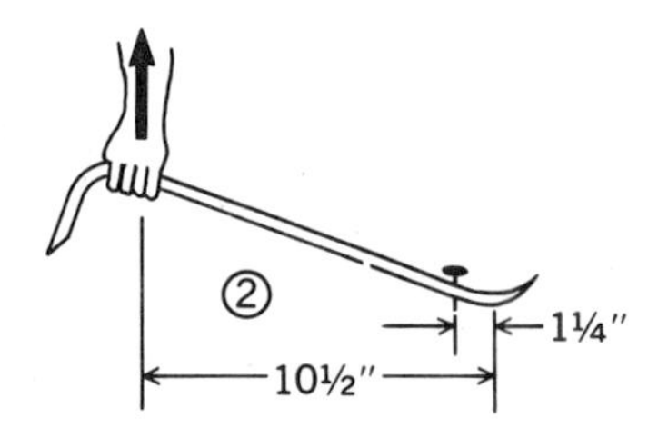

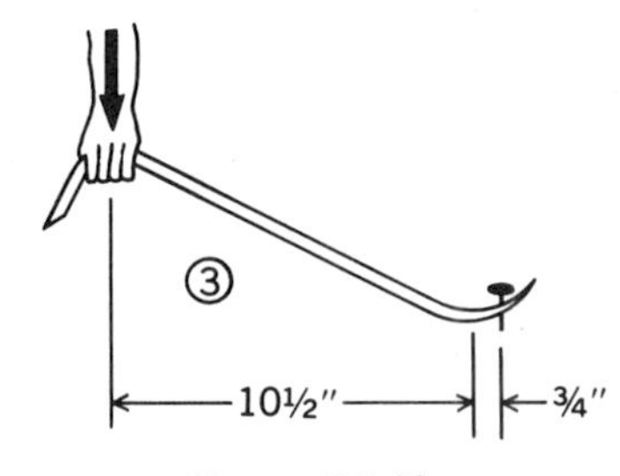

Figure P.5.60

5.61. At what position must the operator of the counterweight crane locate the 50-kN counterweight when he lifts a 10-kN load of steel?

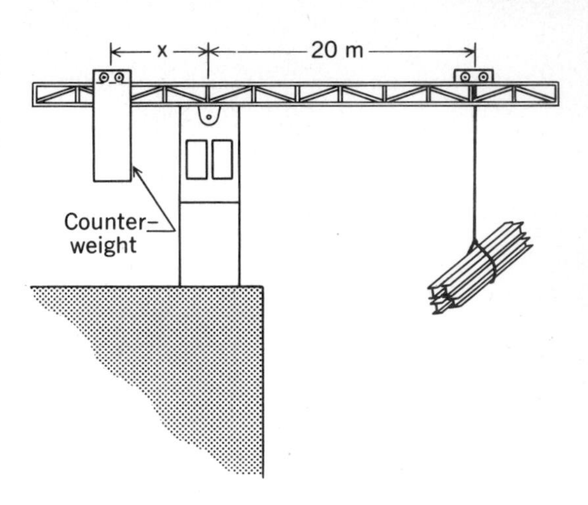

Figure P.5.61

5.62. A Jeep winch is used to raise itself by a force of 2 kN. What are the reactions at the Jeep tires with and without the winch load? The driver weighs 800 N, and the Jeep weighs 11 kN. The center of gravity of the Jeep is shown.

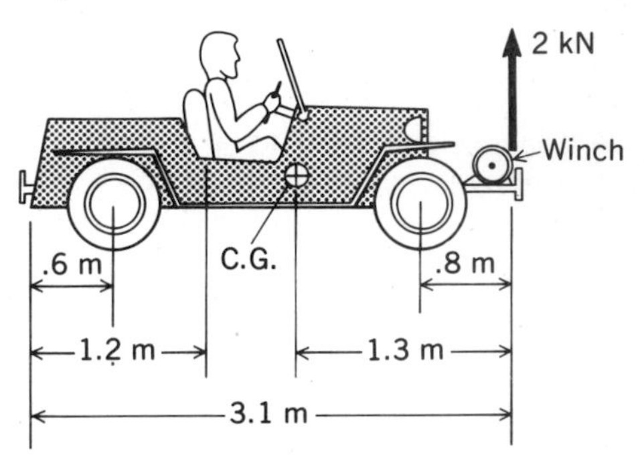

Figure P.5.62

5.63. A *differential pulley* is shown. Compute F in terms of W, r_1, and r_2.

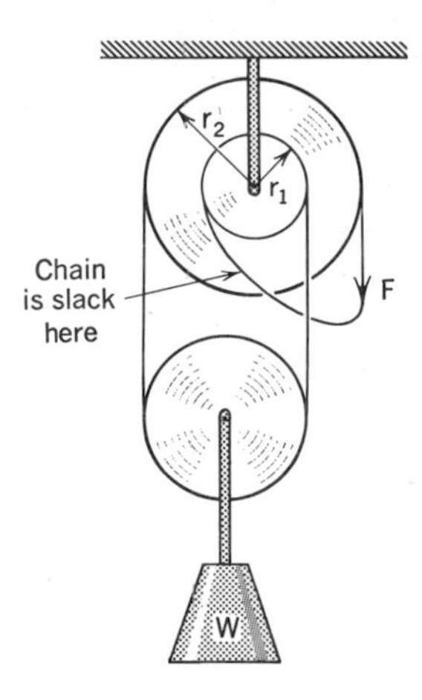

Figure P.5.63

5.64. What is the longest portion of pipe weighing 400 lb/ft that can be lifted without tipping the 12,000-lb tractor? Take the center of gravity of the tractor at the geometric center.

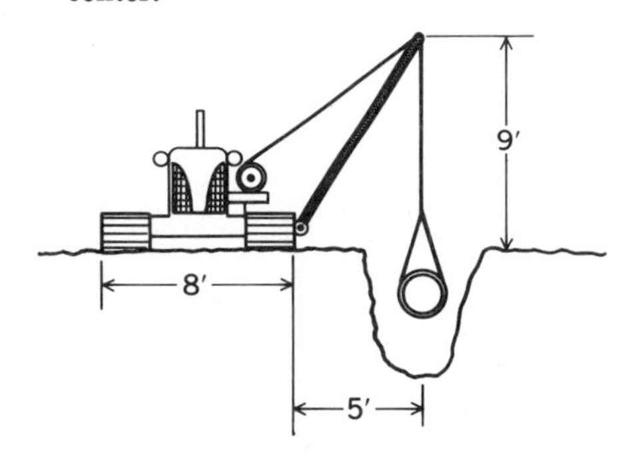

Figure P.5.64

5.65. The L-shaped concrete post supports an elevated railroad. The concrete weighs 150 lb/ft³. What are the reactions at the base of the post?

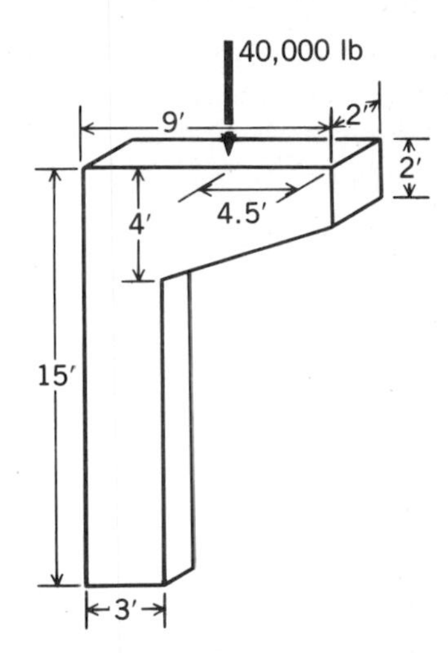

Figure P.5.65

5.66. Two hoists are operated on the same overhead track. Hoist A has a 3000-lb load, and hoist B has a 4000-lb load. What are the reactions at the ends of the track when the hoists are in the position shown?

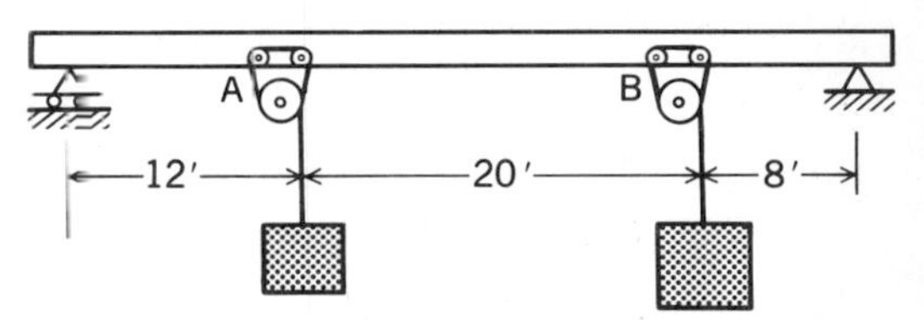

Figure P.5.66

5.67. An I-beam cantilevered out from a wall weighs 30 lb/ft and supports a 300-lb hoist. Steel (487 lb/ft³) cover plates 1 in. thick are welded on the beam near the wall to increase the moment-carrying capacity of the beam. What are the reactions at the wall when a 400-lb load is hoisted at the outermost position of the hoist?

Figure P.5.67

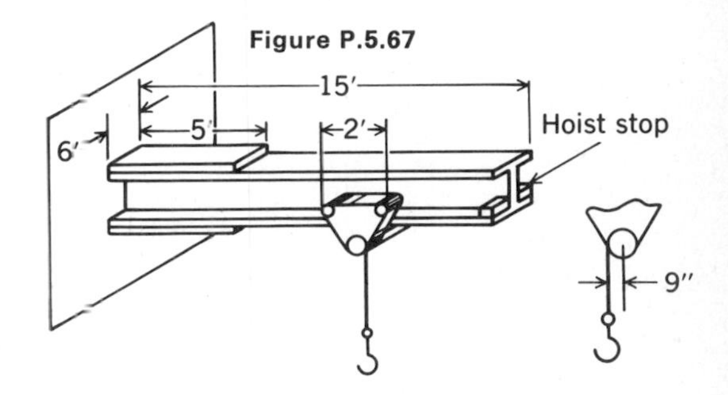

5.68. Find the supporting force system for the cantilever beam shown pinned at C.

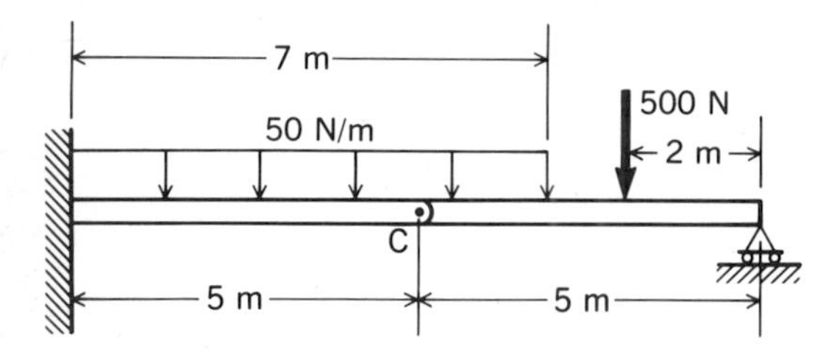

Figure P.5.68

5.69. Find the supporting force system for the cantilever beams connected to bar AB by pins.

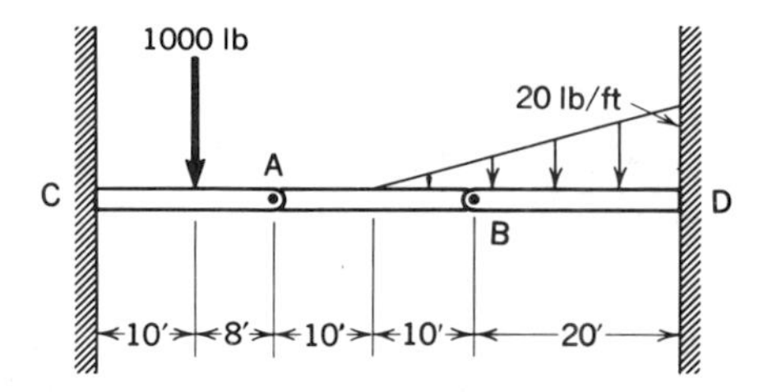

Figure P.5.69

5.70. The trailer weighs 50 kN and is loaded with crates weighing 90 kN and 40 kN. What are the reactions at the rear wheel and on the tractor at A?

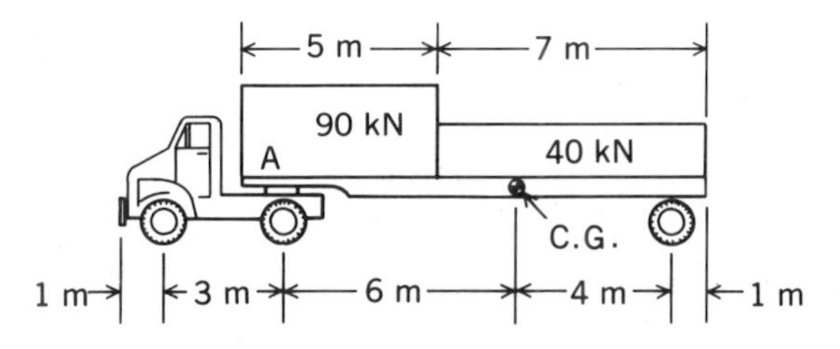

Figure P.5.70

5.71. What load W will a pull P of 100 lb lift in the pulley system? Sheaves A, B, and C weigh 20 lb, 15 lb, and 30 lb, respectively. Assume first that the three sheaves are frictionless and find W. Then, calculate W that can be raised at constant speed for the case where the resisting torque in each of sheaves A and B is .01 times the total force at the bearing of each of sheaves A and B.

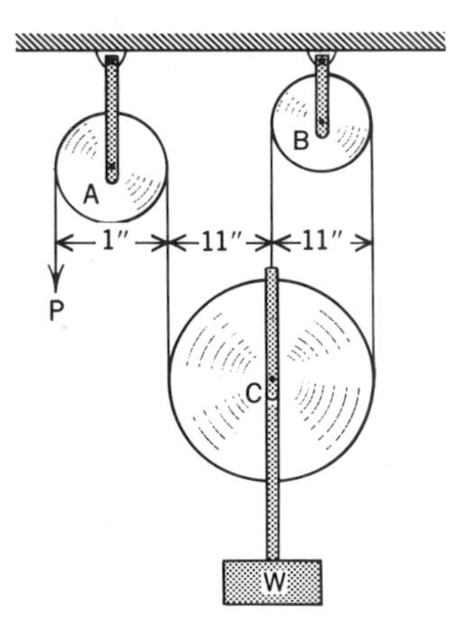

Figure P.5.71

5.72. A piece of pop art is being developed. The weight of the body enclosed by the full lines is 2 kN. What is the smallest distance d that the artist can use for cutting a .5-m diameter hole and still avoid tipping? The body is uniform in thickness.

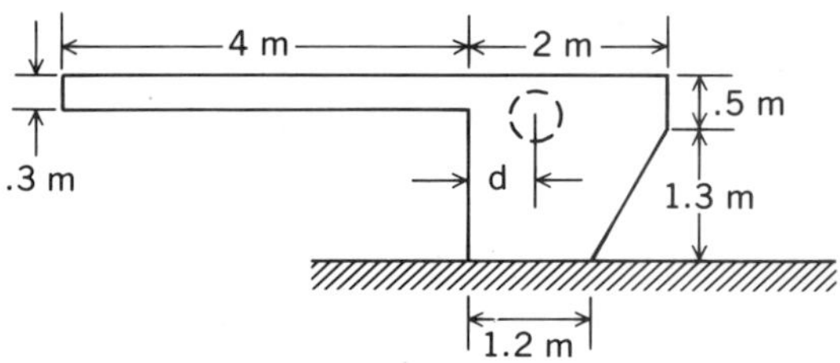

Figure P.5.72

5.73. What is the largest weight W that the crane can lift without tipping? What are the supporting forces when the crane lifts this load? What is the force and couple-moment system transmitted through section C of the beam? Compute the force and couple-moment system transmitted through section D. The crane weighs 10 tons, having a center of gravity as shown in the diagram.

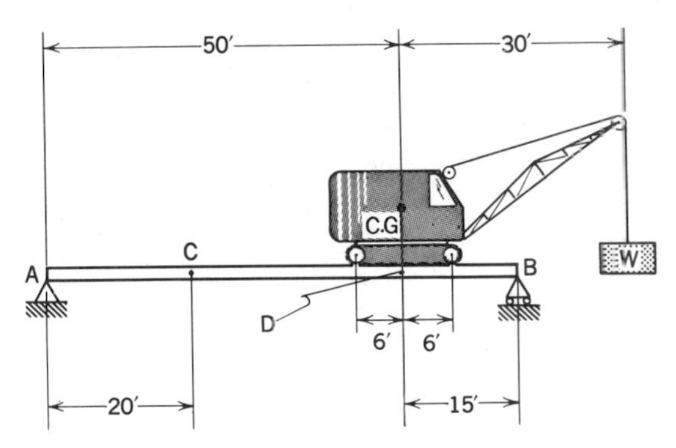

Figure P.5.73

5.74. A 20-kN block is being raised at constant speed. If there is no friction in the three pulleys, what are forces F_1, F_2, and F_3 needed for the job? The block is not rotating in any way. The line of action of the weight vector passes through point C as shown.

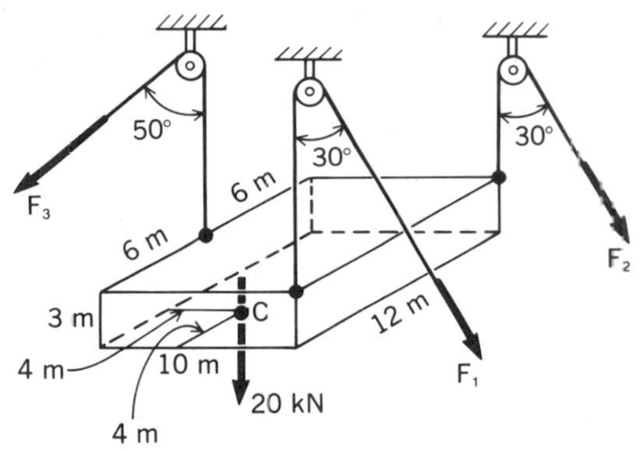

Figure P.5.74

5.75. A 10-ton sounding rocket (used for exploring outer space) has a center of gravity shown as C.G.$_1$. It is mounted on a launcher whose weight is 50 tons with a center of gravity at C.G.$_2$. The launcher has three identical legs separated 120° from each other. Leg AB is in the same plane as the rocket and supporting arms CDE. What are the supporting forces from the ground? What torque is transmitted from the horizontal arm CD to the ramp ED by the rack and pinion at hinge D to counteract the weight of the rocket?

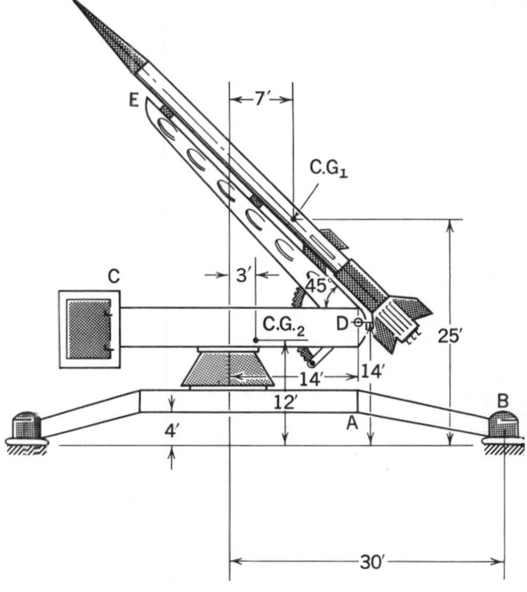

Figure P.5.75

5.76. A door is hinged at A and B and contains water whose specific weight γ is 62.5 lb/ft³. A force F normal to the door keeps the door closed. What are the forces on the hinges A and B and the force F to counteract the water? As noted in Chapter 4, the pressure in the water above atmosphere is given as γd, where d is the perpendicular distance from the free surface of the water.

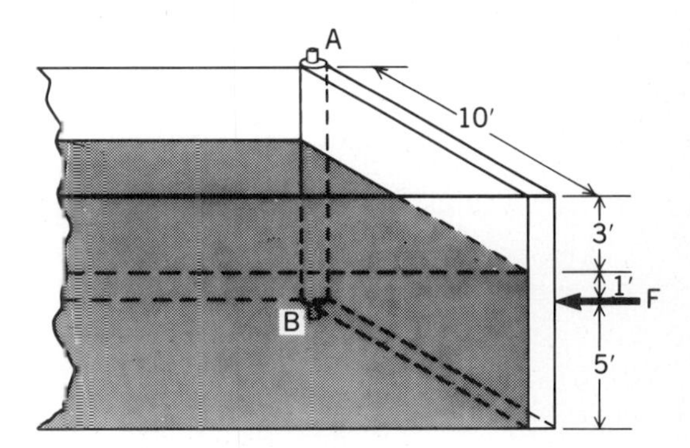

Figure P.5.76

5.77. A row of books of length 800 mm and weighing 200 N sits on a three-legged table as shown. The legs are equidistant from each other with one leg coinciding with the y axis. If the table weighs 400 N, will it tip? If not, what are the forces on the legs?

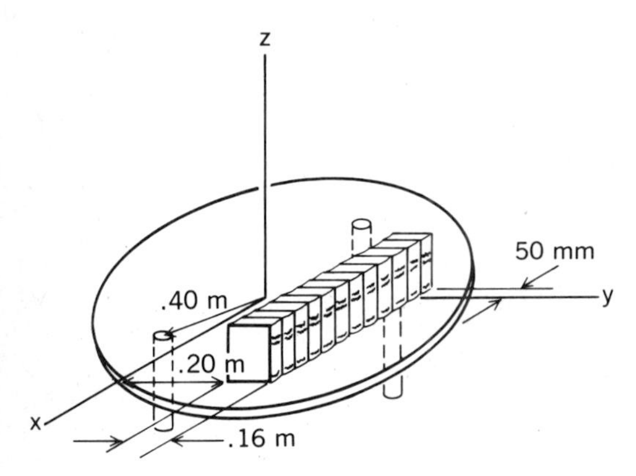

Figure P.5.77

5.78. A small helicopter is in a hovering maneuver. The helicopter rotor blades give a lifting force F_1 but there results from the air forces on the blades a torque C_1. The rear rotor prevents the helicopter from rotating about the z axis but develops a torque C_2. Compute the force F_1 and couple C_2 in terms of the weight W. How are F_3 and C_1 related?

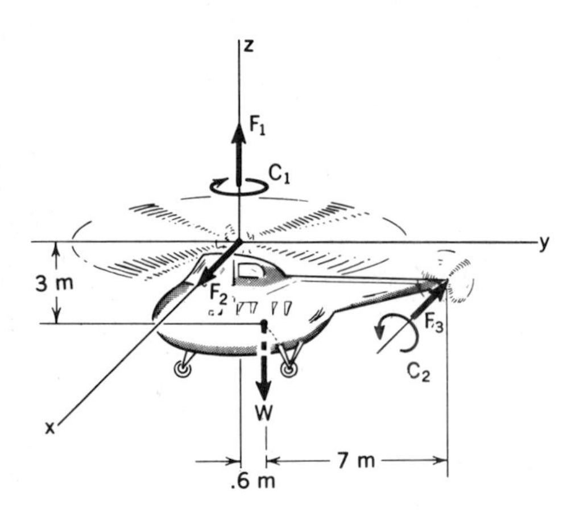

Figure P.5.78

5.79. Find the supporting force and couple-moment system for the cantilever beam. What is the force and couple-moment system transmitted through a cross section of the beam at B?

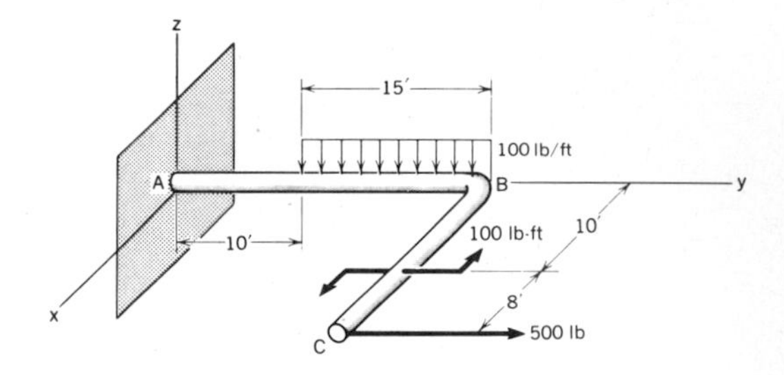

Figure P.5.79

5.80. A structure is supported by a ball-and-socket joint at A, a pin connection at B offering no resistance in the direction AB, and a simple roller support at C. What are the supporting forces for the loads shown?

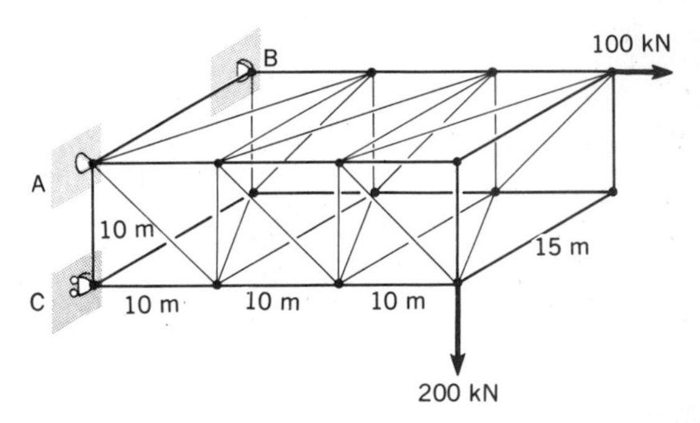

Figure P.5.80

5.81. Compute the value of F to maintain the 200-lb weight shown. Assume that the bearings are frictionless, and determine the forces from the bearings on the shaft at A and B.

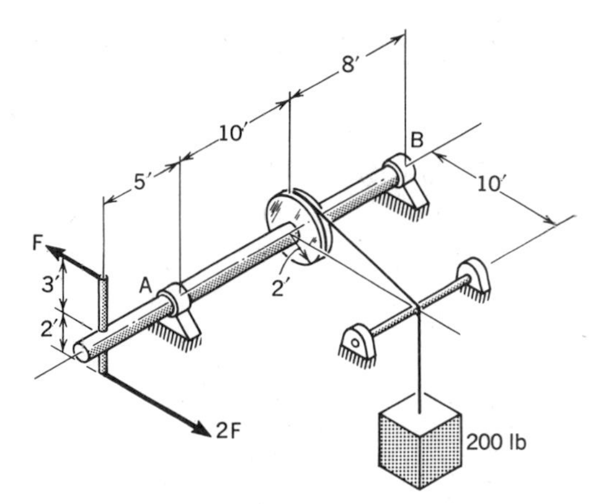

Figure P.5.81

5.82. A bar with two right-angle bends supports a force $\boldsymbol{F}$ given as

$$\boldsymbol{F} = 10\boldsymbol{i} + 3\boldsymbol{j} + 100\boldsymbol{k} \text{ N}$$

If the bar has a weight of 10 N/m, what is the supporting force system at A?

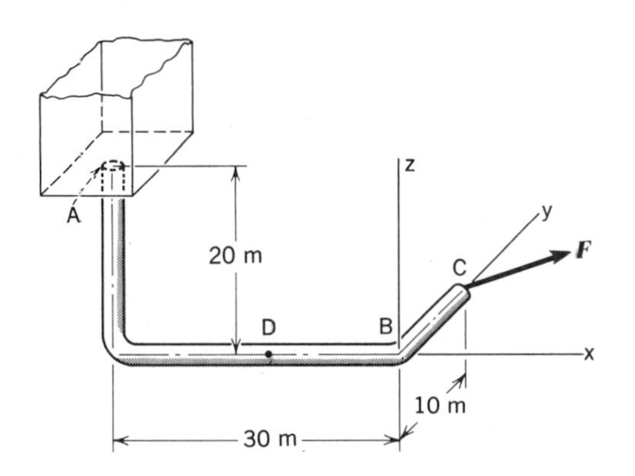

Figure P.5.82

5.83. What is the resultant of the force system transmitted across the section at A? The couple is parallel to plane M.

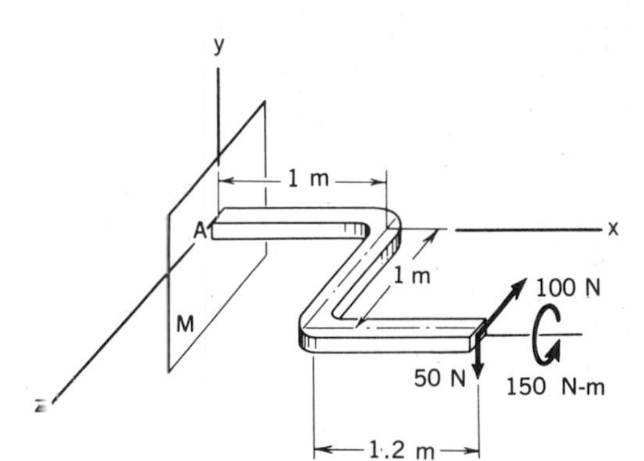

Figure P.5.83

5.84. Determine the vertical force F that must be applied to the windlass to maintain the 100-lb weight. Also, determine the supporting forces from the bearings onto the shaft. The handle DE on which the force is applied is in the indicated xz plane.

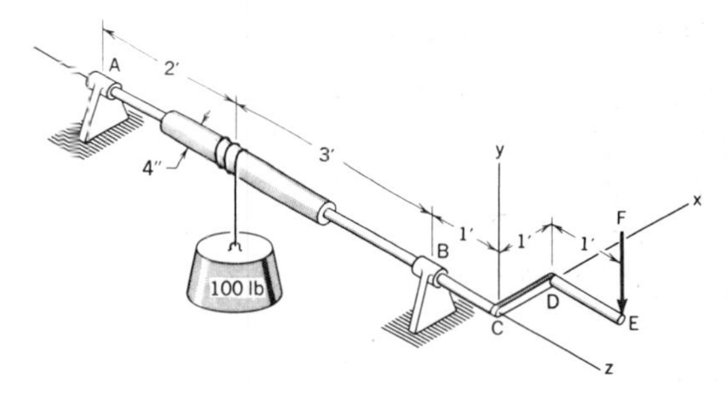

Figure P.5.84

5.85. A transport plane has a gross weight of 70,000 lb with a center of gravity as shown. Wheels A and B are locked by the braking system while an engine is being tested under load prior to take off. A thrust T of 3000 lb is developed by this engine. What are the supporting forces?

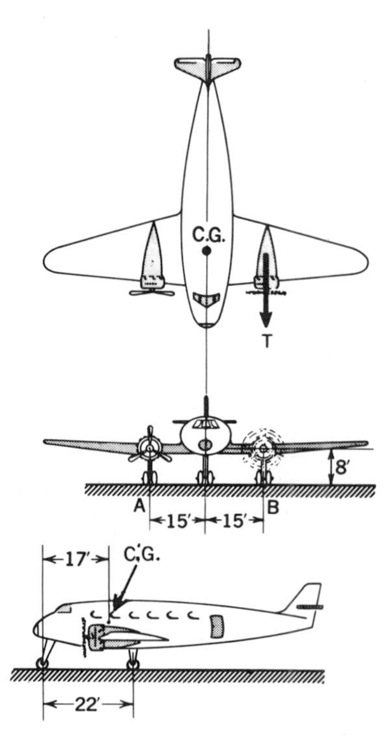

Figure P.5.85

5.86. Two cables GH and KN support a rod AB which connects to a ball-and-socket joint support at A and supports a 500-kg body C at B. What are the tensions in the cable and the supporting forces at A?

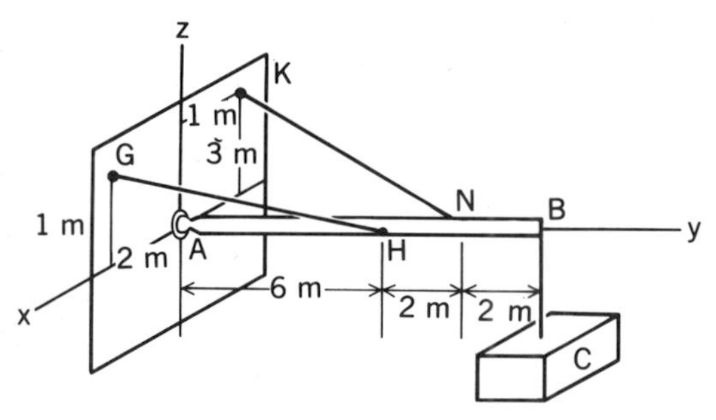

Figure P.5.86

5.87. What change in elevation for the 100-lb weight will a couple of 300 lb-ft support if we neglect friction in the bearings at A and B? Also, determine the supporting force components at the bearings for this configuration.

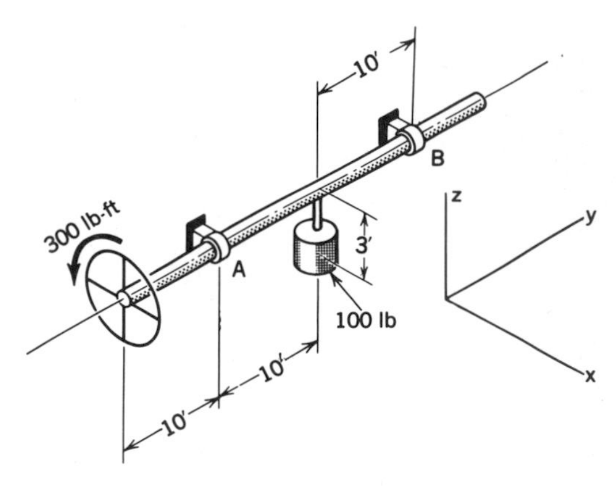

Figure P.5.87

5.88. Determine the force P required to keep the 150-N door of an airplane open 30° while in flight. The force P is exerted in a direction normal to the fuselage. There is a net pressure increase on the outside surface of .02 N/mm². Also, determine the supporting forces at the hinges. Consider that the top hinge supports any vertical force on the door.

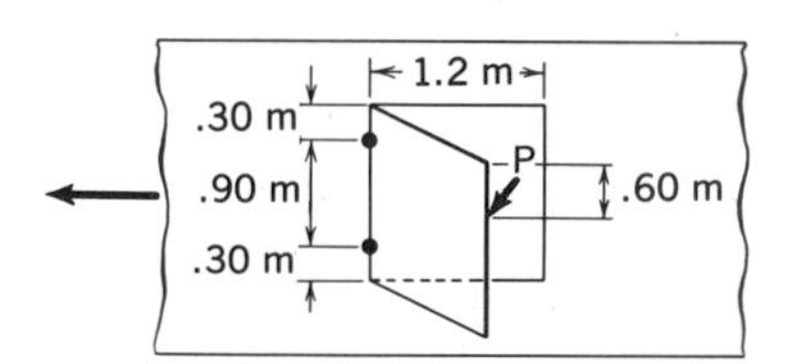

Figure P.5.88

5.89. What force P is needed to hold the door in a horizontal position? The door weighs 50 lb. Determine the supporting forces at A and B. At A there is a pin and at B there is a ball-and-socket joint.

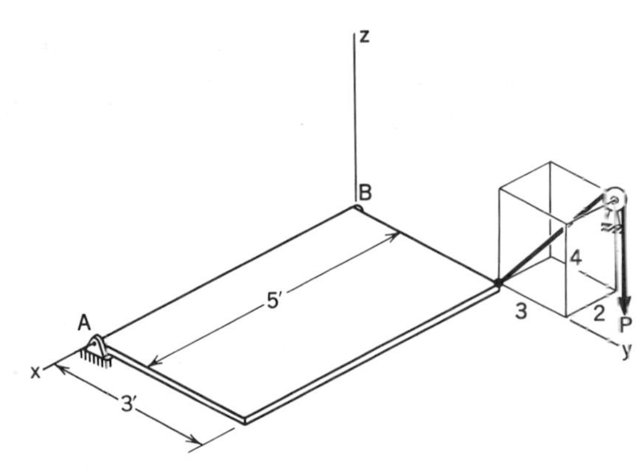

Figure P.5.89

5.90. A rod AB is held by a ball-and-socket joint at A and supports a 100-kg mass C at B. This rod is in the zy plane and is inclined to the y axis by an angle of 15°. The rod is 16 m long and F is at its midpoint. Find the forces in cables DF and EB.

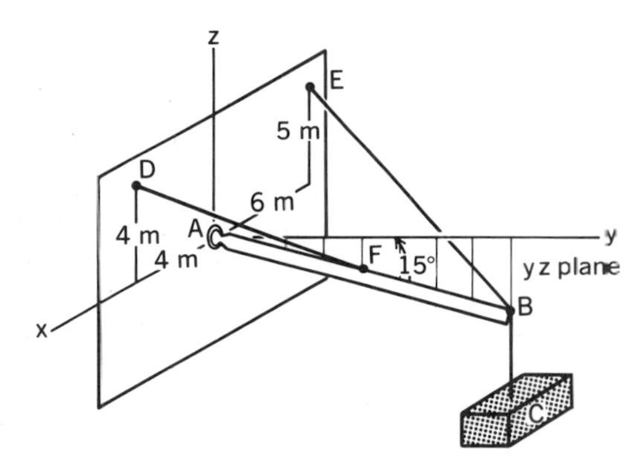

Figure P.5.90

5.91. A bent rod $ADGB$ supports two weights—one at the center of AD and one at the center of DG. There are ball-and-socket joint supports at A and B. With one scalar equation using the triple scalar product, determine the tension in cable DC.

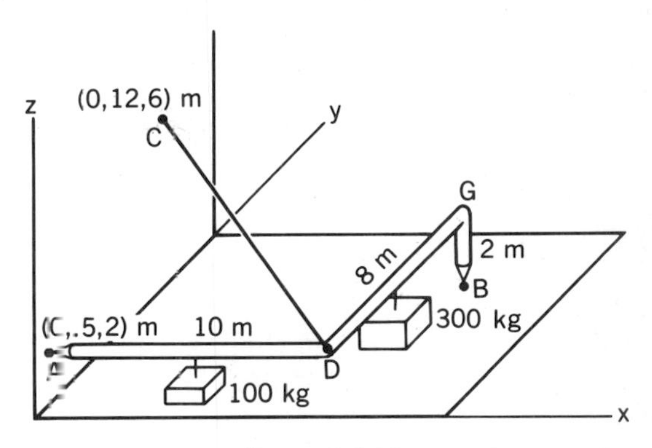

Figure P.5.91

5.92. Four cables support a block of weight 5000 N. The edges of the block are parallel to the coordinate axes. Point B is at (7, 7, −15). What are the forces in the cables and the direction cosines for cable

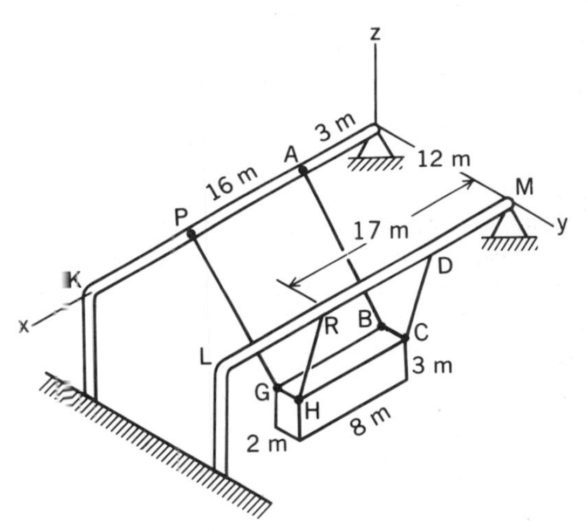

Figure P.5.92

5.93. A bar can rotate parallel to plane A about an axis of rotation normal to the plane at O. A weight W is held by a cord that is attached to the bar over a small pulley that can rotate freely as the bar rotates. Find the value of C for equilibrium if $h = 300$ mm, $W = 30$ N, $\phi = 30°$, $l = 700$ mm, and $d = 500$ mm.

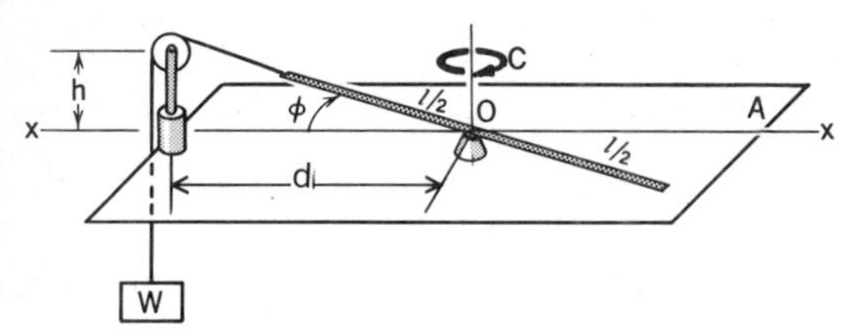

Figure P.5.93

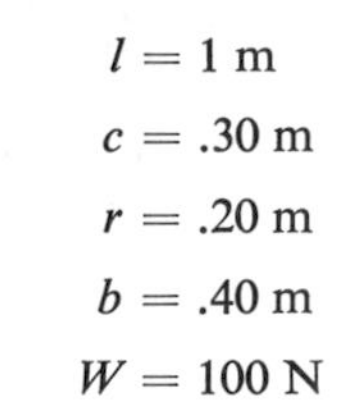

$l = 1$ m
$c = .30$ m
$r = .20$ m
$b = .40$ m
$W = 100$ N

5.94. A uniform bar of length l and weight W is connected to the ground by a ball-and-socket joint, and rests on a semicylinder from which it is not allowed to slip down by a wall at B. If we consider the wall and cylinder to be frictionless, determine the supporting forces at A for the following data:

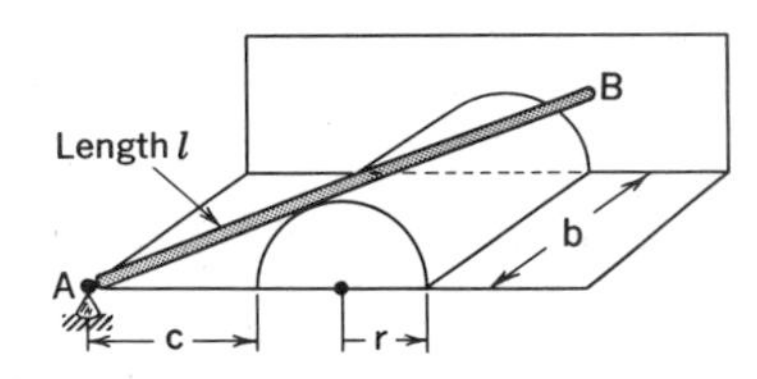

Figure P.5.94

5.7 Two-Force Members

We shall now consider a simple case of equilibrium that occurs quite often and from which simple, useful conclusions may readily be drawn.

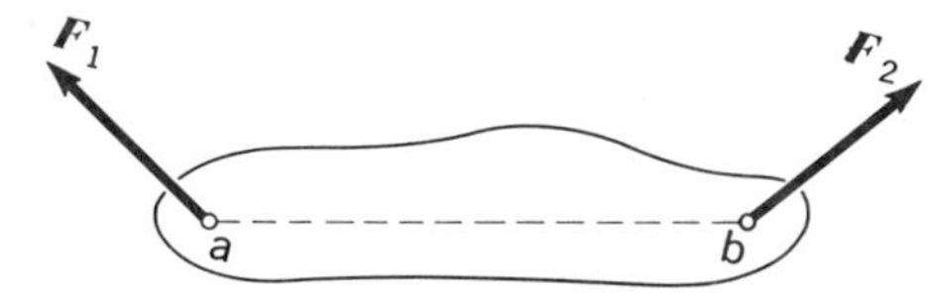

Figure 5.39. Two-force member.

Consider a rigid body on which *two forces* are, respectively, acting at points a and b (see Fig. 5.39). If the body is in equilibrium, the first basic equation of statics, 5.1(a), stipulates that $\boldsymbol{F}_1 = -\boldsymbol{F}_2$; that is, the forces must be *equal* and *opposite*. The second fundamental equation of statics, 5.1(b), requires that $\boldsymbol{M} = \boldsymbol{0}$, indicating that the forces be *collinear* so as not to form a nonzero couple. With points a and b given as points of application for the two forces in Fig. 5.39, clearly the *common line of action for the forces must coincide with the line segment ab*. Such bodies, where there are only *two points of loading*, are called "two-force" members.

We often have to deal with pin-connected structural members with loads applied at the pins. If we neglect friction at the pins and also the weight of the members, we can conclude that only two forces act on each member. These forces, then, must be equal and opposite and must have lines of action that are collinear, with the line joining the points of application of the forces. If the member is straight (see Fig. 5.40), the common line of action of the two forces coincides with the centerline of the member.[8] The top member in Fig. 5.40 is a *compression* member, the one below a *tensile* member.

Before considering an example, it should be pointed out that the forces F_1 and F_2 in Fig. 5.39 may be the resultants of systems of concurrent forces at a and b respec-

[8]Note that the bent member in Fig. 5.41, if weightless, is also a two-force member. The line of action of the forces must coincide with the line ab connecting the points of application of the two forces.

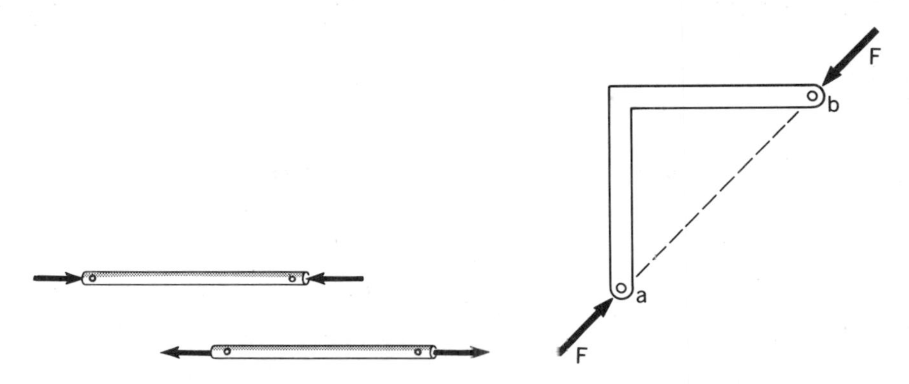

Figure 5.40. Compression and tension members.

Figure 5.41. Line of action of F collinear with ab.

tively. Since concurrent forces are always equivalent to their resultant at the point of concurrency, the member in Fig. 5.39 is still a two-force member with the resulting restrictions on the resultants F_1 and F_2.

EXAMPLE 5.10

A device for crushing rocks is shown in Fig. 5.42. A piston D having an 8-in. diameter is activated by a pressure p of 50 psig (above that of the atmosphere). Rods AB, BC, and BD can be considered weightless for this problem. What is the horizontal force transmitted at A to the trapped rock shown in the diagram?

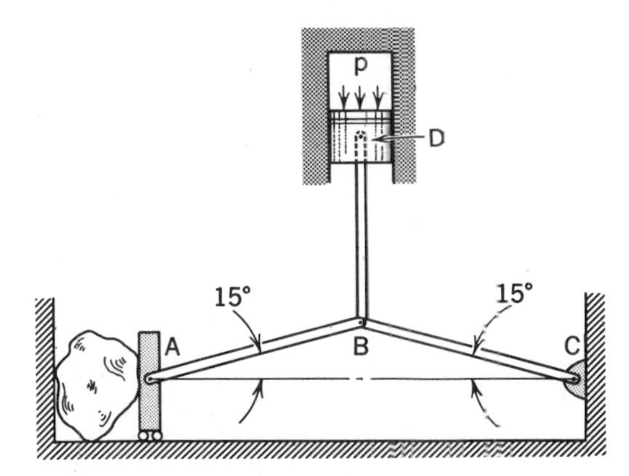

Figure 5.42. Rock crusher.

We have here three two-force members coming together at B. Accordingly, if we isolate pin B as a free body, we will have three forces acting on the pin. These forces must be collinear with the centerlines of the respective members, as explained earlier (Fig. 5.43).

The force F_D is easily computed by considering the action of the piston. Thus, we get

$$F_D = (50)\frac{\pi 8^2}{4} = 2510 \text{ lb}$$

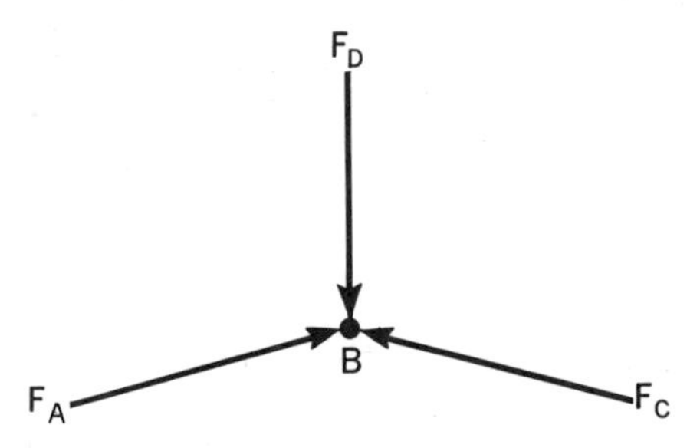

Figure 5.43. Free-body diagram of B.

Summing forces at pin B:

$\underline{\sum F_x = 0:}$

$$F_A \cos 15° - F_C \cos 15° = 0$$

$$F_A = F_C$$

$\underline{\sum F_y = 0:}$

$$2F_A \sin 15° - 2510 \text{ lb} = 0$$

$$F_A = \frac{2510}{(2)(0.259)} = 4850 \text{ lb}$$

The force transmitted to the rock in the horizontal direction is then 4850 cos 15° = 4690 lb.

Of less direct use is the *three-force* theorem. It states that a system of three forces in equilibrium must be *coplanar* and either be *concurrent* or be *parallel.*[9]

5.8 Static Indeterminacy

Examine the simple beam in Fig. 5.44, with known external loads and weight. If the deformation of the beam is small, and the final positions of the external loads after deformation differ only slightly from their initial positions, we can assume the beam to be rigid and, using the undeformed geometry, we can solve for the supporting forces A, B_x, and B_y. This is possible since we have three equations of equilibrium available. Suppose, now, that an additional support is made available to the beam, as indicated in Fig. 5.45. The beam can still be considered a rigid body, since the applied loads will shift even less because of deformation. Therefore, the equivalent force coming from the ground to counteract the applied loads and weight of the beam must be the *same* as before. In the first case, in which two supports were given, however, a unique set of values for the forces A, B_x, and B_y gave us the required resistance. In other words,

[9]To prove this, assume that two of the forces intersect at a point A. Show from the basic equations of equilibrium that the forces must be coplanar and concurrent. Now assume the forces do *not* intersect. Setting moments of the system equal to zero about two points along the line of action of one of the forces, show that the system must be coplanar. Now since the forces do not intersect, they must be parallel. The theorem will thus have been proved.

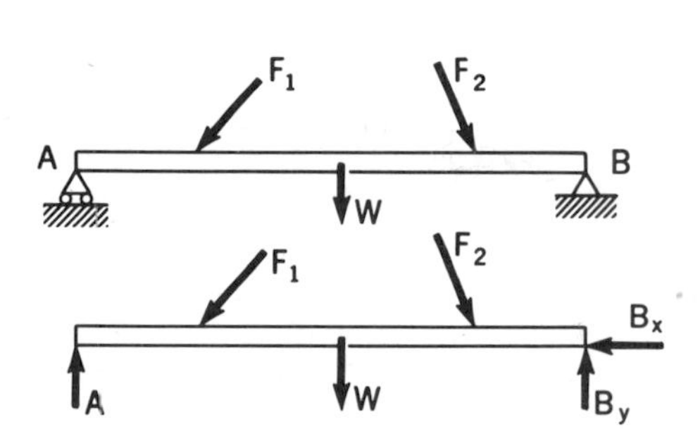

Figure 5.44. Statically determinate problem.

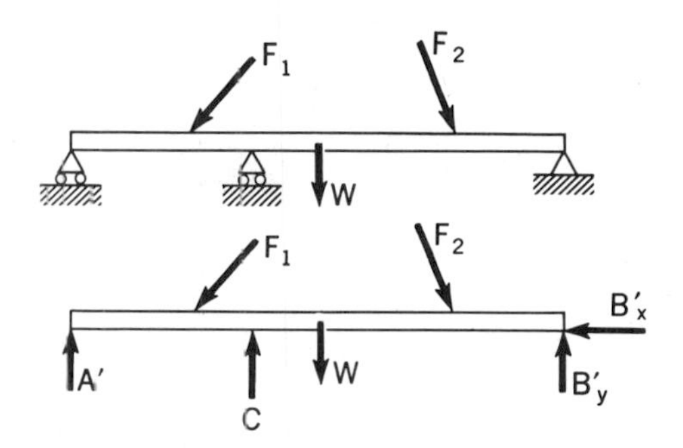

Figure 5.45. Statically indeterminate problem.

we were able to solve for these forces by statics alone, without further considerations. In the second case, rigid body statics will give the required *same* equivalent supporting force system, but now there are an infinite number of possible combinations of values of the supporting forces that will give us this equivalent system demanded by equilibrium of rigid bodies. To decide on the proper combination of supporting forces requires additional computation. Although the deformation properties of the beam were unimportant up to this point, they now become the all-important criterion in apportioning the supporting forces. These problems are termed *statically indeterminate*, in contrast to the statically determinate type, in which statics and the rigid-body assumption suffice. For a given system of loads and masses, two models—the rigid-body model and models taught in other courses involving elastic behavior—are accordingly both employed to achieve a desired end. In summary:

> *In statically indeterminate problems, we must satisfy both the equations of equilibrium for rigid bodies and the equations that stem from deformation considerations. In statically determinate problems, we need only satisfy the equations of equilibrium.*

In the discussion thus far, we used a beam as the rigid body and discussed the statical determinacy of the supporting system. Clearly, the same conclusions apply to any structure that, without the aid of the external constraints, can be taken as a rigid body. If, for such a structure as a free body, there are as many unknown supporting force and couple-moment components as there are equations of equilibrium, and if these equations can be solved for these unknowns, we say that the structure is *externally statically determinate*.

On the other hand, should we desire to know the forces transmitted *between* internal members of this kind of structure (i.e., one that does not depend on the external constraints for rigidity), we then examine free bodies of these members. When all the unknown force and couple-moment components can be found by the equations of equilibrium for these free bodies, we then say that the structure is *internally statically determinate*.

There are structures that depend on the external constraints for rigidity (see the

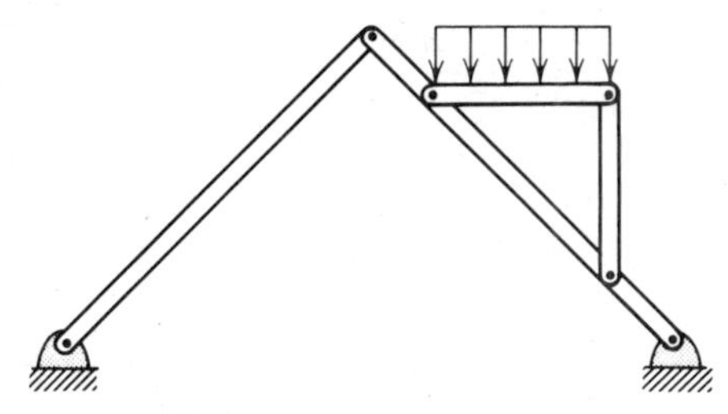

Figure 5.46. Nonrigid structure.

structure shown in Fig. 5.46). Mathematically speaking, we can say for such structures that the supporting force system always depends on both the internal forces and the external loads. (This is in contrast to the previous case, where the supporting forces could, for the externally statically determinate case, be related directly with the external loads without consideration of the internal forces.) In this case, we do not distinguish between internal and external statical determinacy, since the evaluation of supporting forces will involve free bodies of some or all of the internal members of the structure; hence, some or all of the internal forces and moments will be involved. For such cases, we simply state that the structure is statically determinate if, for all the unknown force and couple-moment components, we have enough equations of equilibrium that can be solved for these unknowns.

Problems

5.95. Draw free-body diagrams for the hoe, arms, and tractor of the backhoe. Consider the weight of each part to act at a central location. The backhoe is not digging at the instant shown. Neglect the weights of the hydraulic systems *CE*, *AB*, and *FH*.

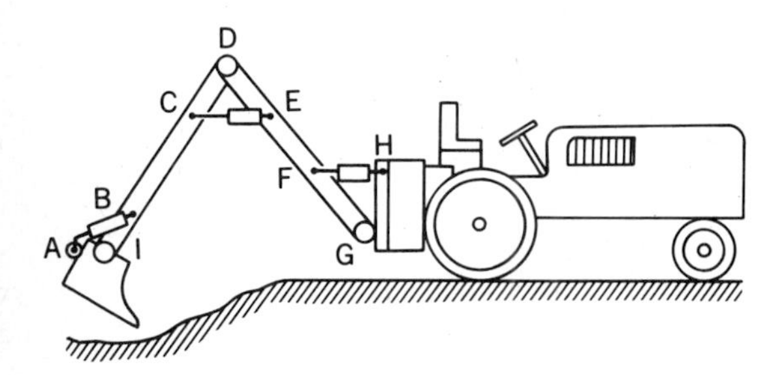

Figure P.5.95

5.96. A parking-lot gate arm weighs 150 N. Because of the taper, the weight can be regarded as concentrated at a point 1.25 m from the pivot point. What force must be exerted by the solenoid to lift the gate? What solenoid force is necessary if a 300-N counterweight is placed .25 m to the left of the pivot point?

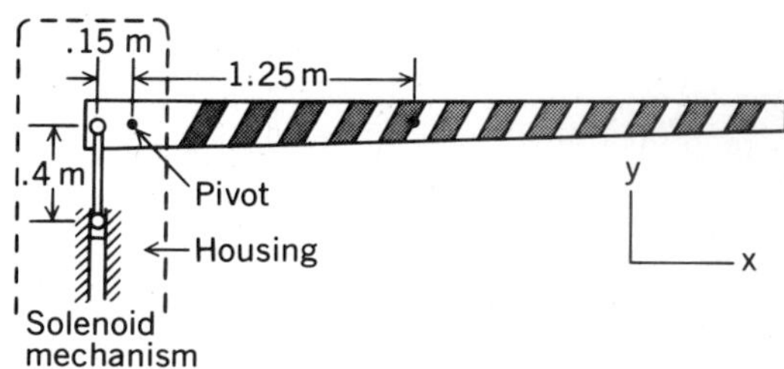

Figure P.5.96

5.97. Find the force delivered at *C* in a horizontal direction to crush the rock. Pressure $p_1 = 100$ psig and $p_2 = 60$ psig (pressures measured above atmospheric pressure). The diameters of the pistons are 6 in. each. Neglect the weight of the rods.

Figure P.5.97

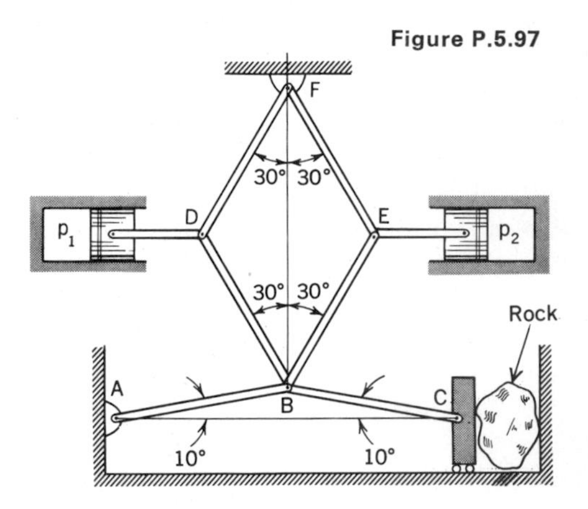

5.98. A Broyt X-20 digger carries a 20-kN load as shown. If hydraulic ram *CB* is normal to *BA*, where *A* is the axis of rotation for member *E*, find the force needed by ram *CB*. Do not consider the weights of the members.

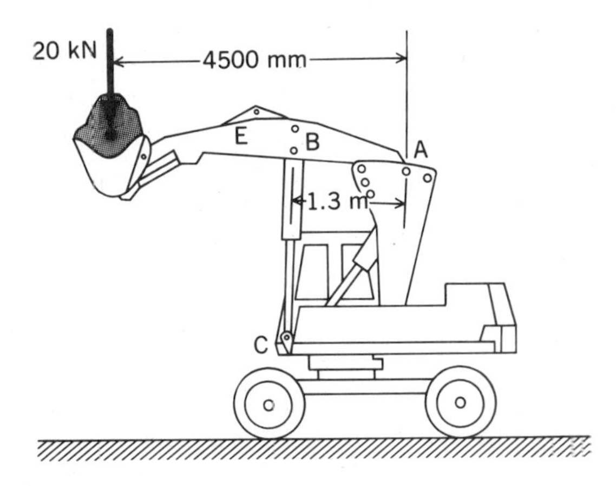

Figure P.5.98

5.99. What force F_1 will be developed by the 500-lb load? Neglect friction. The design is symmetrical.

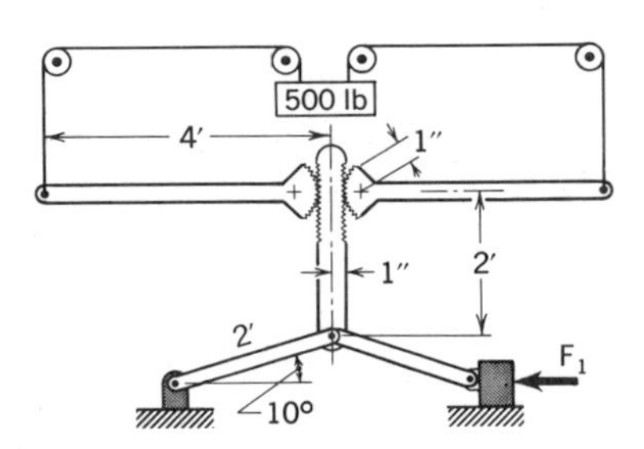

Figure P.5.99

5.100. Find the forces in the cables *DB* and *CB* as well as the compression member *AB*. The 500-N force is parallel to the *y* axis.

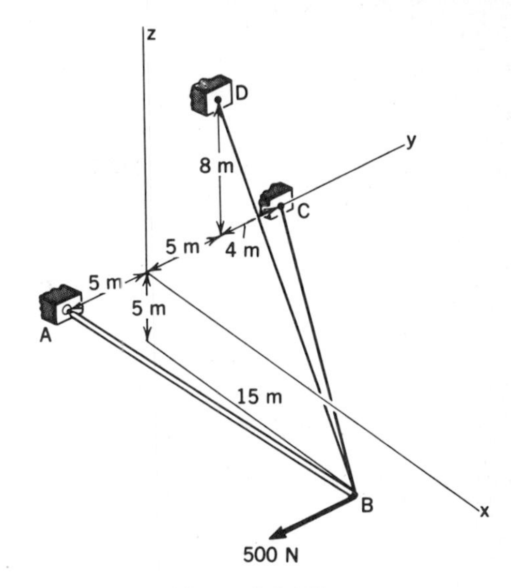

Figure P.5.100

5.101. Find the values of *F* and *C* so that members *AB* and *CD* fail simultaneously. The maximum load for *AB* is 15 kN and for *CD* is 22 kN. Neglect the weight of the members.

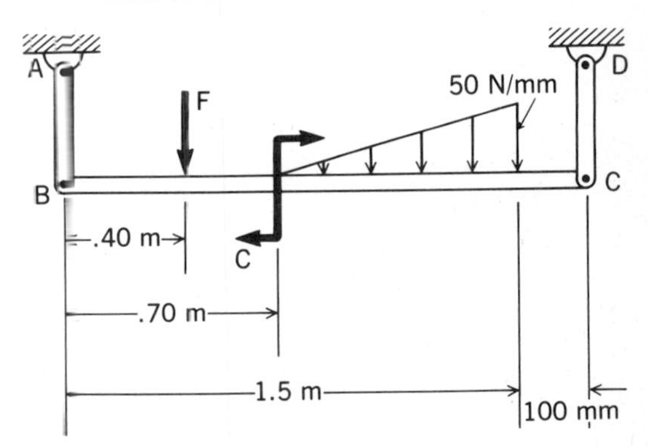

Figure P.5.101

5.102. The landing carriage of a transport plane supports a stationary vertical load of 50 kN per wheel. There are two wheels on each side of shock strut *AB*. Find the force in member *EC*, and the forces transmitted to the fuselage at *A*, if the brakes are locked and the engines are tested resulting in a thrust of 5 kN, 40% of which is resisted by this landing gear.

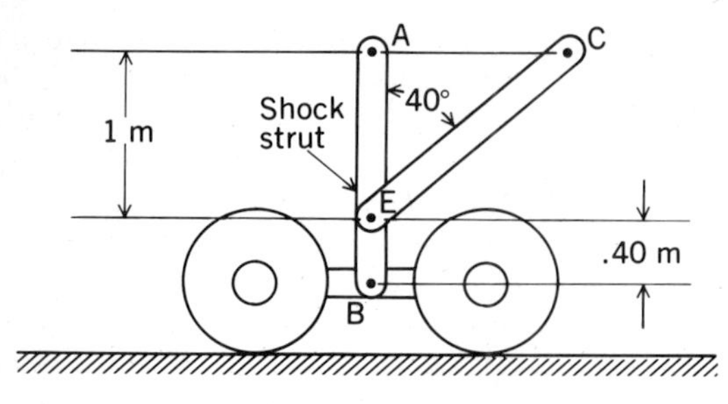

Figure P.5.102

5.103. Find the supporting forces at A and B in the frame. Neglect weights of members.

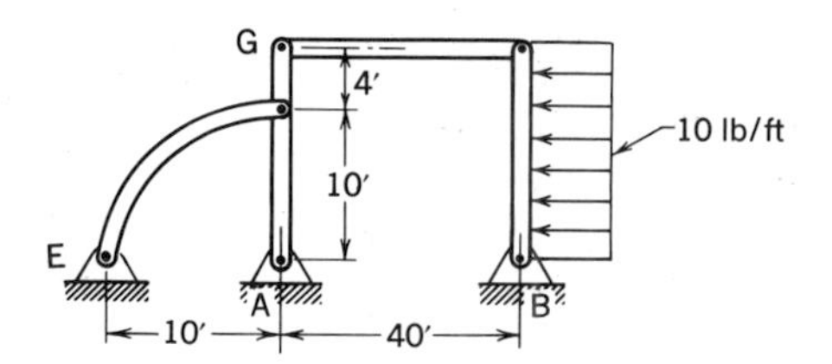

Figure P.5.103

5.104. A bolt cutter has a force of 130 N applied at each handle. What is the force on the bolt from the cutter edge?

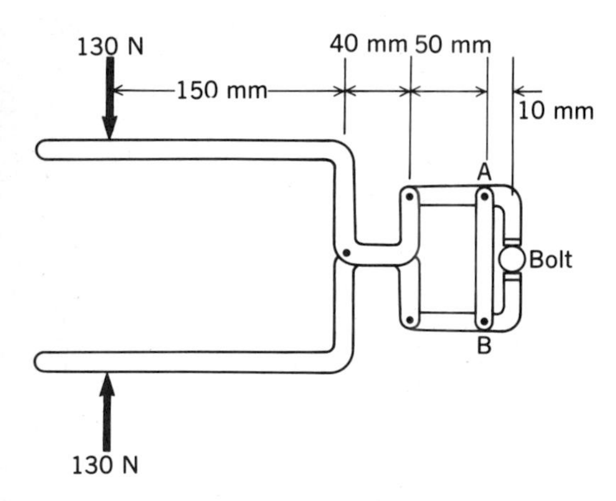

Figure P.5.104

5.105. A steam locomotive is developing a pressure of .20 N/mm² gage. If the train is stationary, what is the total traction force from the two wheels shown? Neglect the weight of the various connecting rods. Neglect friction in piston system and connecting rod pins.

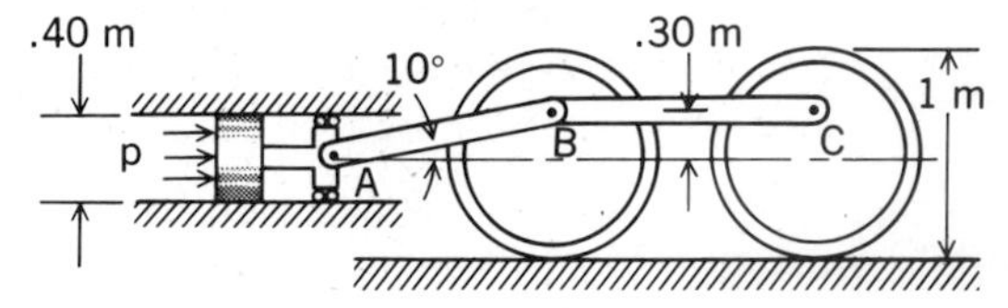

Figure P.5.105

5.106. Find the supporting forces at A, B, and C. Neglect the weight of the rod.

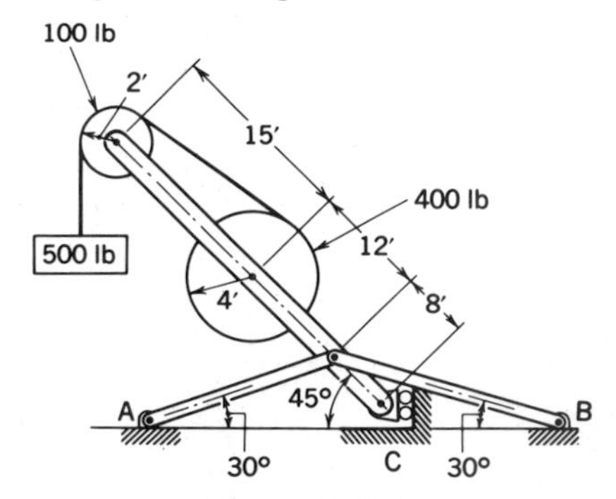

Figure P.5.106

5.107. The 5000-lb van of an airline food-catering truck rises straight up until its floor is level with the airplane floor. What forces exist at each joint of the scissors assembly? What force must be exerted by the hydraulic ram?

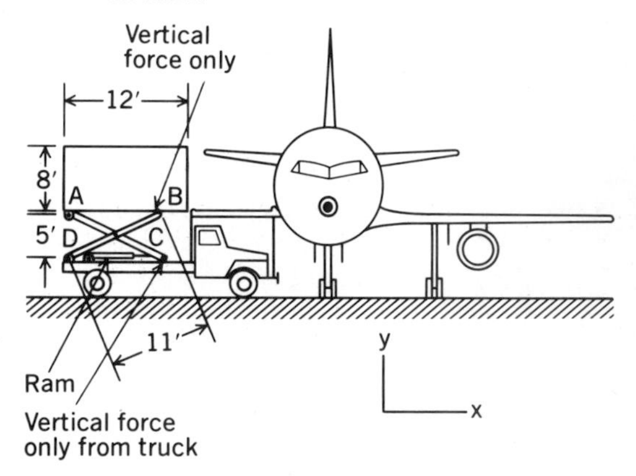

Figure P.5.107

5.108. The pavement exerts a force of 1000 lb on the tire. The tire, brakes, and so on, weigh 100 lb; the center of gravity is taken at the center plane of the tire. Determine the force from the spring and the compression force in CD.

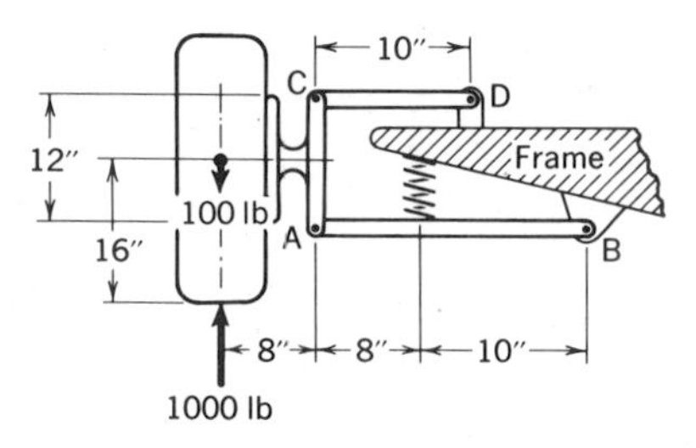

Figure P.5.108

5.109. A *flyball governor* is shown rotating at a constant speed ω of 500 rpm. The weights C and D are each of mass 500 g and are pin-connected to light rods. The centrifugal force on the weights, you will recall from physics, is given as $mr\omega^2$, where r is the radial distance to the particle from the axis of rotation and ω is in rad/sec. Using this centrifugal force, and, imagining that we are rotating with the system, we can consider that we have equilibrium. (This is the D'Alembert principle that you learned in physics.) What is the tension in the rods and the downward force F at B needed to maintain the configuration shown for the given ω?

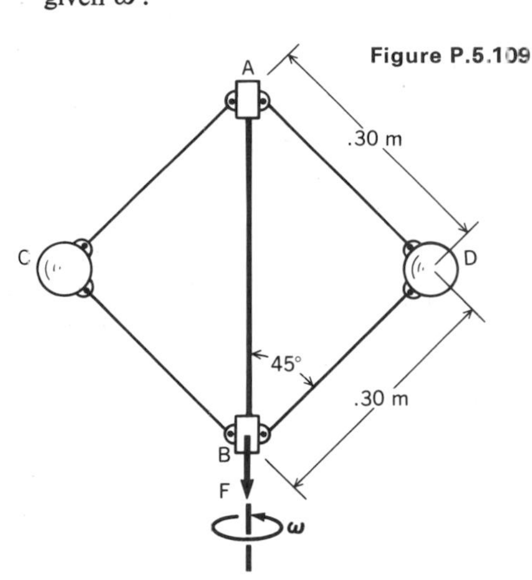

Figure P.5.109

5.110. Another kind of flyball governor is shown. If $\omega_1 = 3$ rad/sec, compute the tension in AG and AE. Neglect the weights of the members but assume that HC and HB are stiff. The weights C and B each have a mass of 200 g. What is the force F needed to maintain the configuration? (*Hint:* Read the discussion of centrifugal force and D'Alembert's principle in Problem 5.109.)

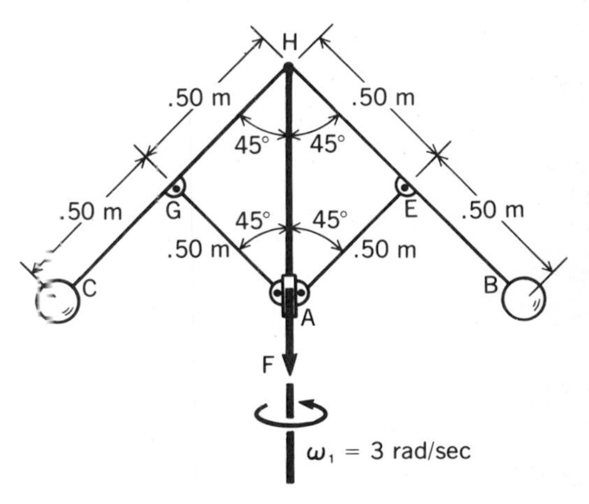

Figure P.5.110

5.111. Determine the supporting forces at A, C, D, G, F, and H for the structure.

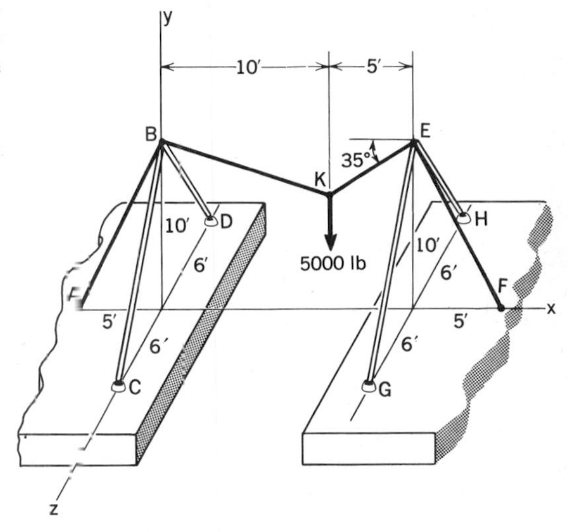

Figure P.5.111

5.112. Find the supporting forces at the ball-and-socket connections A, D, and C. Members AB and DB are pinned together through member EC at B.

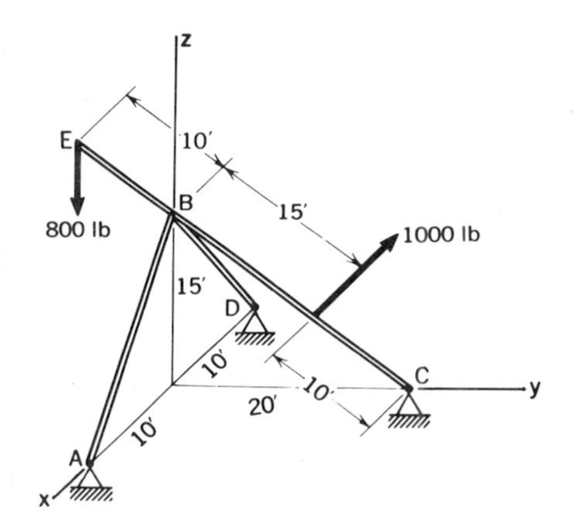

Figure P.5.112

5.113. A tractor with a bulldozer is used to push an earthmover picking up dirt. If the tractor force on the earthmover is 150 kN, what are the reactions of the bulldozer on the tractor at B and A?

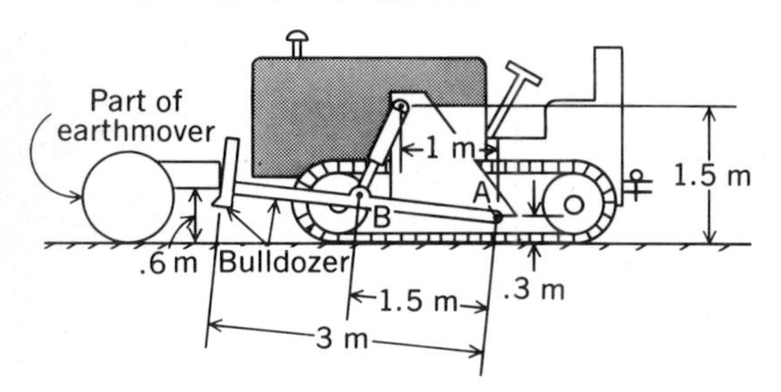

Figure P.5.113

5.114. A hydraulic-lift platform for loading t ucks supports a weight W of 5000 lb. Only one side of the system has been shown; the other side is identical. If the diameter of the pistons in the cylinders is 4 in., what pressure p is needed to support W when $\theta = 60°$. The following data apply.

$l = 24$ in., $\quad d = 60$ in., $\quad e = 10$ in.

Neglect friction everywhere. (*Hint:* Only two free-body diagrams need be drawn.)

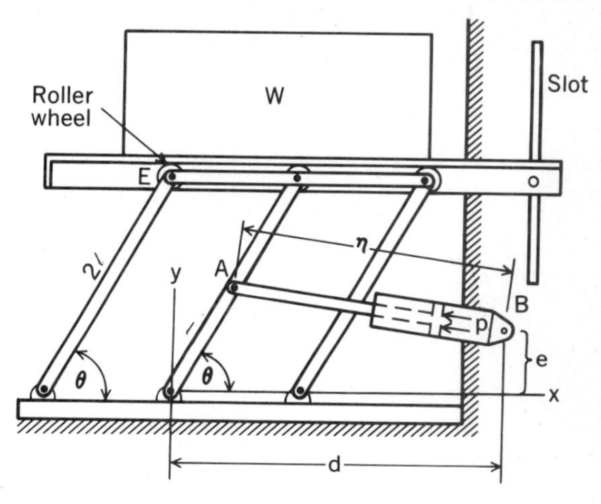

Figure P.5.114

5.115. A Bucyrus–Erie Dynahoe digger is partially shown. To develop the indicated forces in the bucket, what forces must hydraulic cylinders HB and CD develop? Consider only the 3-kN and 5-kN loads and not the weights of the members.

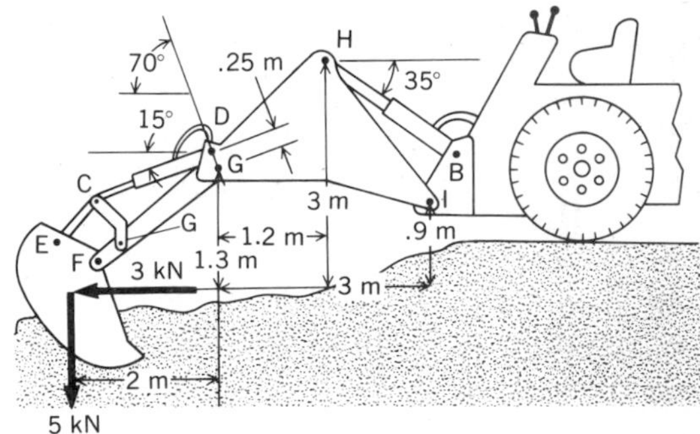

Figure P.5.115

5.116. A block of material weighing 200 lb is supported by members KC and HB, whose weight we neglect, a ball-and-socket-joint support at A, and a smooth, frictionless support at E. Members KC and HB have directions collinear with diagonals of the block as shown. What are the supporting forces for this block?

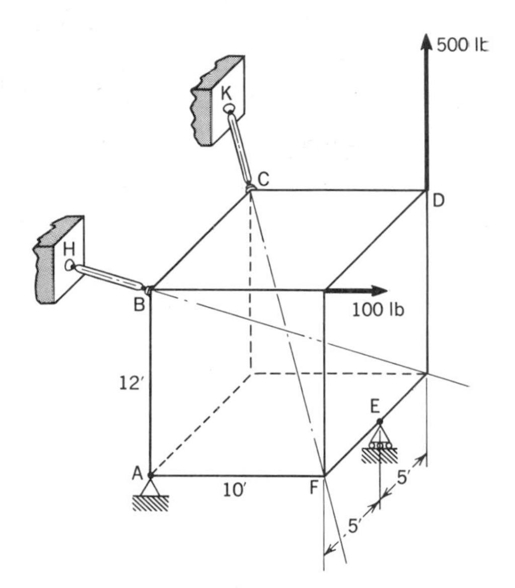

Figure P.5.116

5.117. A trap door is kept open by a rod *CD*, whose weight we shall neglect. The door has hinges at *A* and *B* and has a weight of 200 lb. A wind blowing against the outside surface of the door creates a pressure increase of 2 lb/ft². Find the force in the rod, assuming that it cannot slip from the position shown. Also determine the forces transmitted to the hinges. Only hinge *B* can resist motion along direction *AB*.

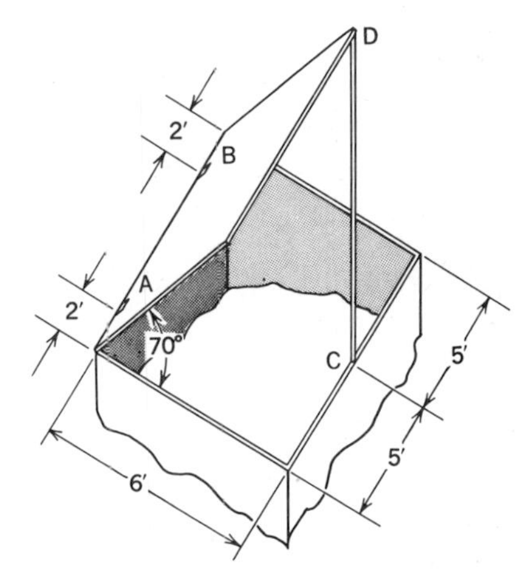

Figure P.5.117

5.9 Closure

We can now draw a free-body diagram that exposes a system of forces which, for equilibrium, must satisfy certain equations. By noting the type of simplest resultant for the system, we can readily deduce how many unknown quantities can be found for the free body. We thus have the direct means of solving statically determinate problems. And we also have available some of the conditions (the rigid-body equations of equilibrium) that must be satisfied in certain statically indeterminate problems. However, considerations beyond the scope of this text are necessary for the solution of these latter problems.[10]

In Chapter 6, we shall consider certain types of bodies that are of great engineering interest. The problems will be statically determinate and will involve nothing that is fundamentally new. We devote a separate chapter to these problems because they contain sign conventions and techniques that are important and complex enough to warrant such a study. We therefore proceed to an introduction of statically determinate structural mechanics problems.

---

[10]For such problems, see I. H. Shames, *Introduction to Solid Mechanics*, Prentice-Hall, Inc., Englewood Cliffs, N.J., 1975.

Review Problems

5.118. Determine the tensions in all the cables. Block A has a mass of 600 kg. Note that GH is in the yz plane.

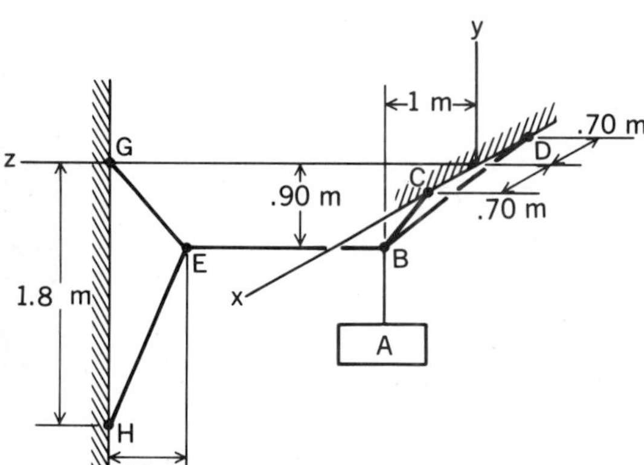

Figure P.5.118

5.119. Determine the force components at G. E weighs 300 lb.

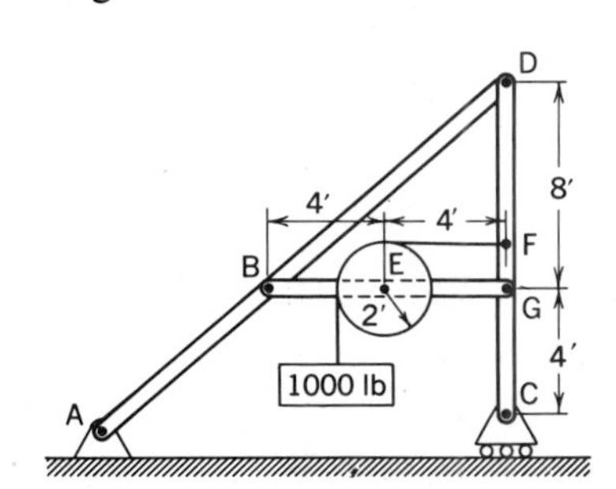

Figure P.5.119

5.120. A scenic excursion train with cog wheels for steep inclines weighs with load 30 tons. If the cog wheels have a mean radius to the contact points of the teeth of 2 ft, what torque must be applied to the driver wheels A if wheels B run free? What force do wheels B transmit to the ground?

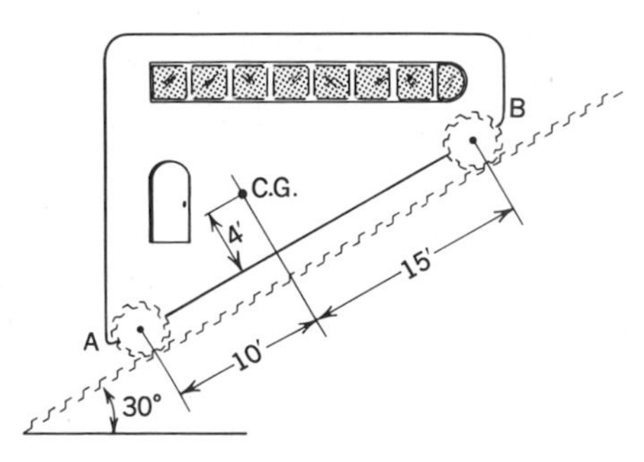

Figure P.5.120

5.121. Find the forces on the block of ice from the hooks at A and F.

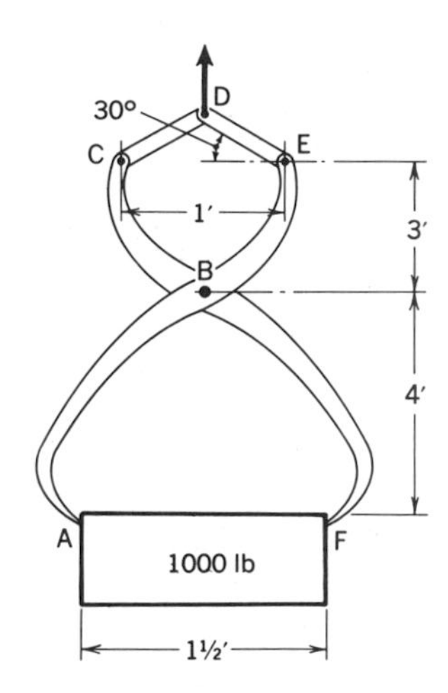

Figure P.5.121

5.122. Members AB and BC weighing, respectively, 50 N and 200 N are connected to each other by a pin. BC connects to a disc K on which a torque $T_K = 200$ N-m is applied. What torque T is needed on AB to keep the system in equilibrium at the configuration shown?

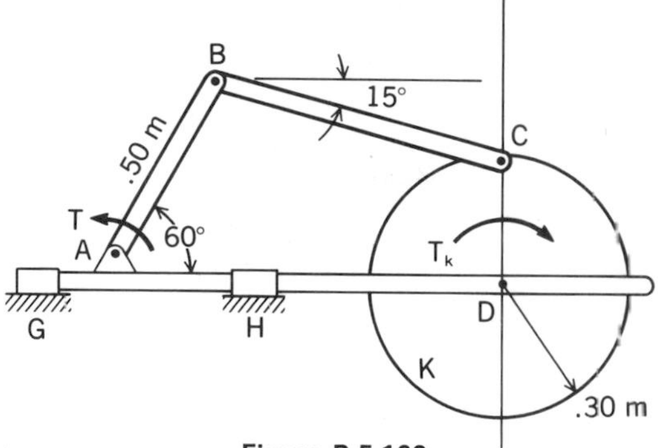

Figure P.5.122

5.123. A transport jet plane has a weight without fuel of 220 kN. If one wing is loaded with 50 kN of fuel, what are the forces in each of the three landing gear?

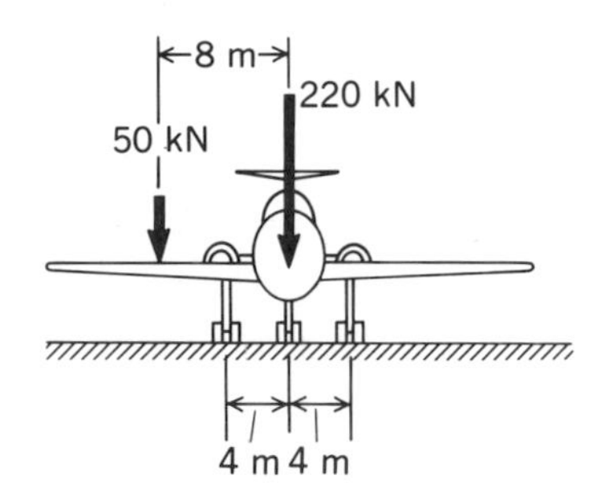

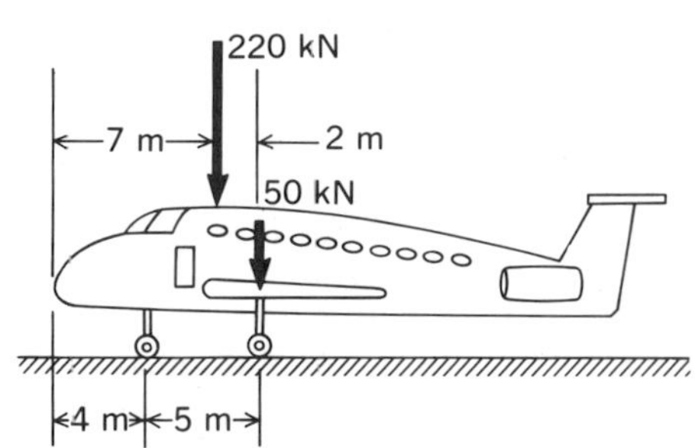

Figure P.5.123

5.124. A rod AB is connected by a ball-and-socket joint to a frictionless sleeve at A, and by a ball-and-socket joint to a fixed position at B. What are the supporting forces at B and at A if we neglect the weight of AB? The 100-lb load is connected to the center of AB.

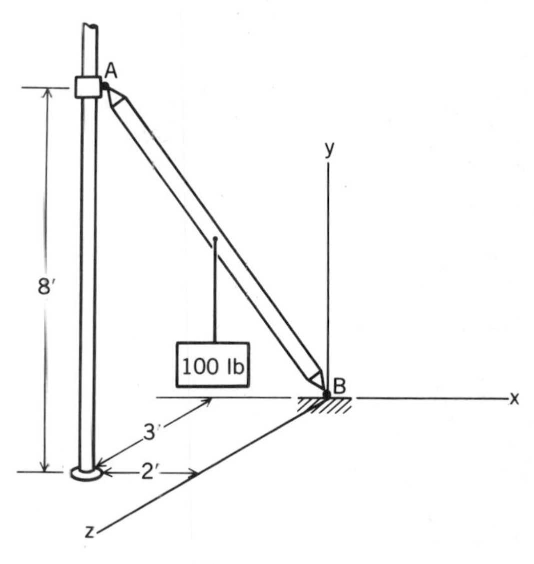

Figure P.5.124

5.125. A beam weighing 400 lb is held by a ball-and-socket joint at A and by two cables CD and EF. Find the tension in the cables. They are attached at opposite ends of the beam as shown.

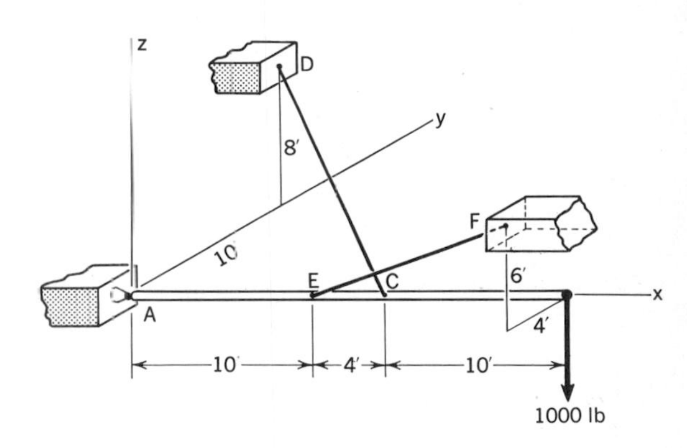

Figure P.5.125

5.126. What should the values of R and C be if the supporting rods AB and CD are to fail simultaneously? Rod AB can withstand a 5000-lb force, and rod DC can withstand an 8000-lb force. Neglect the weights of the members.

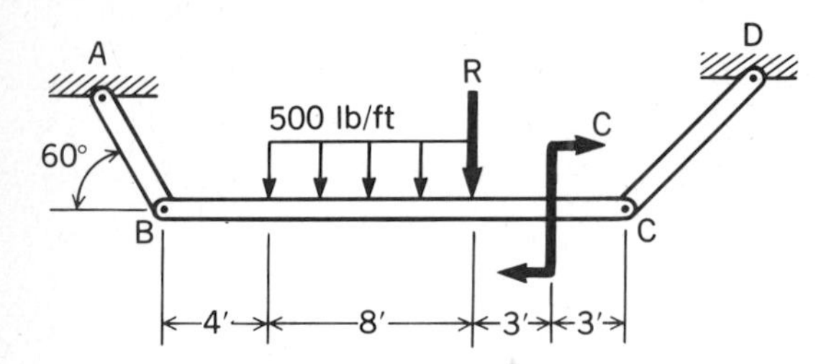

Figure P.5.126

5.127. A mechanism consists of two weights W each of weight 50 N, four light linkage rods each of length a equal to 200 mm, and a spring K whose spring constant is 8 N/mm. The spring is unextended when $\theta = 45°$. If held vertically, what is the angle θ for equilibrium? Neglect friction. The force from the spring equals K times the compression of the spring.

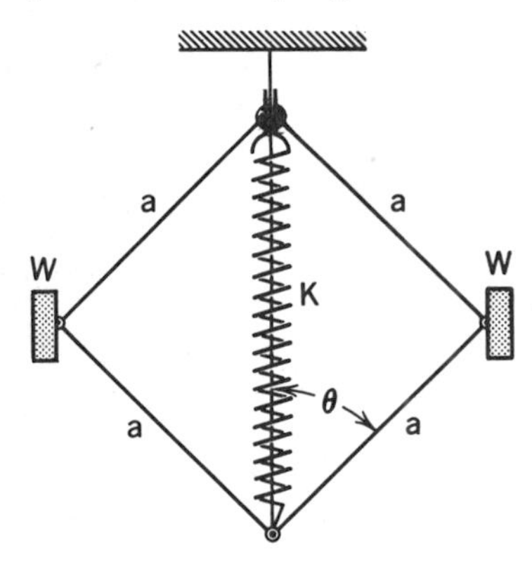

Figure P.5.127

5.128. Find the compressive force in pawl AB. What is the resultant supporting force system at E?

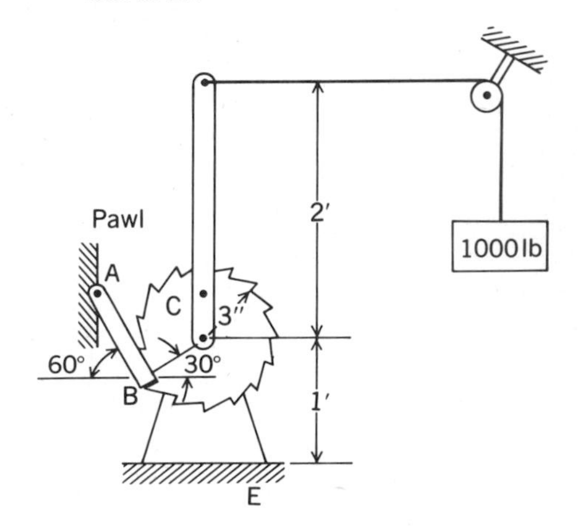

Figure P.5.128

5.129. A 10-kN load is lifted in the front loader bucket. What are the forces at the connections to the bucket and to arm AE? Hydraulic ram DF is perpendicular to arm AE, and BC is horizontal. Points A and F are at the same height above the ground.

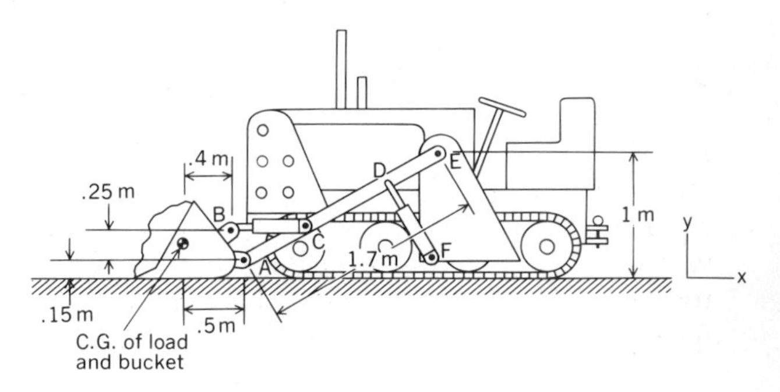

Figure P.5.129

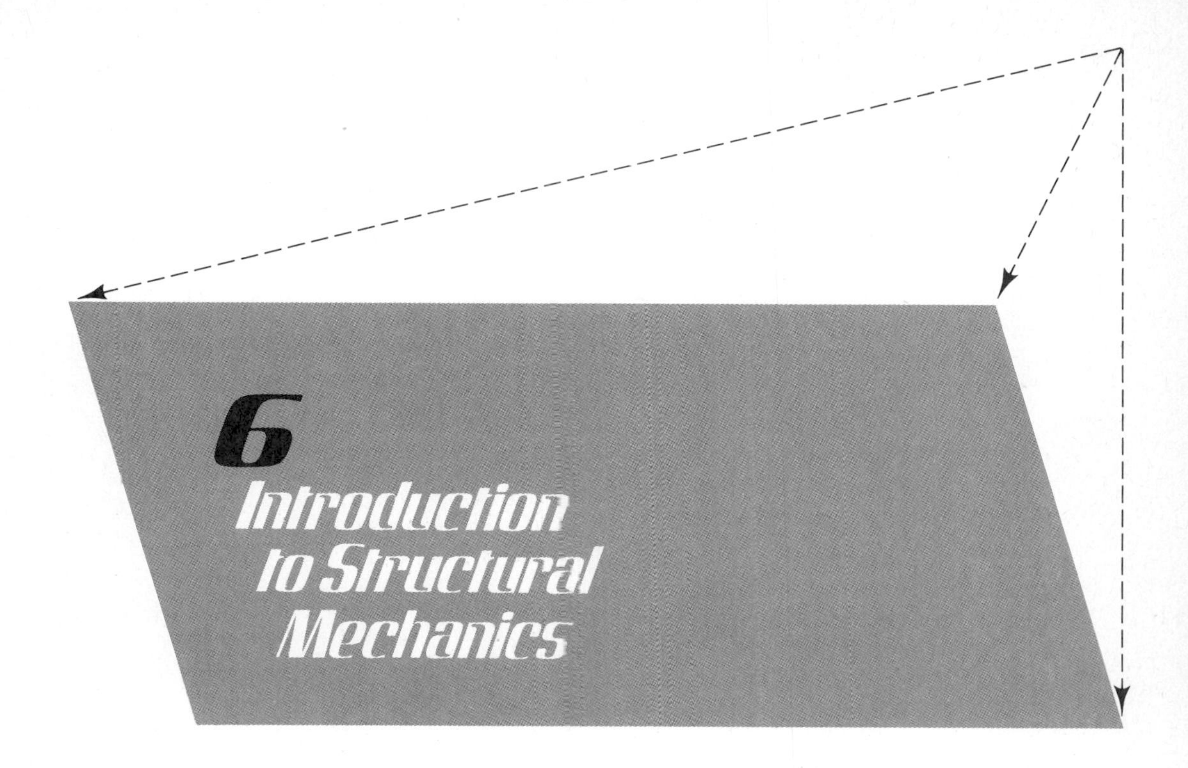

Part A Trusses

6.1 The Structural Model

A *truss* is a system of members that are fastened together to support stationary and moving loads.[1] Everyday examples of trusses are shown in Figs. 6.1 and 6.2. Each member of a truss is usually of uniform cross section along its length; however, the various members typically have different cross-sectional areas because they must transmit different forces. Our purpose in Part A of this chapter is to set forth methods for determining forces in members of an elementary class of trusses.

As a first step, we shall divide trusses into two main categories according to geometry. A truss consisting of a coplanar system of members is called a *plane truss*. Examples of plane trusses are the sides of a bridge (see Fig. 6.1) and a roof truss (see Fig. 6.2). A three-dimensional system of members, on the other hand, is called a *space truss*. A common example of a space truss is the tower from an electric power transmission system (see Fig. 6.3). Both plane trusses and space trusses consist of members having cross sections resembling the letters H, I, and L. Such members are commonly

[1] A truss is different than a frame (see the footnote on p. 140) in that the members of a truss are always connected together at the ends of the members, as will soon become evident.

Figure 6.1. Foot bridge near author's home. Sides of structure are plane trusses.

Figure 6.2. Roof trusses that are simple plane trusses.

Figure 6.3. Space trusses supporting transmission lines sending power into the northeast grid of the U.S.

used in many structural applications. These members are fastened together to form a truss by being welded, riveted, or bolted to intermediate structural elements called *gusset plates* such as has been shown in Fig. 6.4(a) for the case of a plane truss. The analysis of forces and moments in such connections is clearly quite complicated. Fortunately, there is a way of simplifying these connections such as to incur very little loss in accuracy in determining forces in the members. Specifically, if the centerlines of the members are *concurrent* at the connections, such as is shown in Fig. 6.4(a) for the coplanar case, then we can replace the complex connection at the points of concurrency by a simple pin connection in the coplanar truss and a simple ball-and-socket connection for the space truss. Such a replacement is called an *idealization* of the system. This is illustrated for a plane truss in Fig. 6.4, where the actual connector or joint is shown in (a) and the idealization as a pinned joint is shown in (b).

In order to maximize the load-carrying capacity of a truss, the external loads must be applied at the joints. The prime reason for this rule is the fact that the members

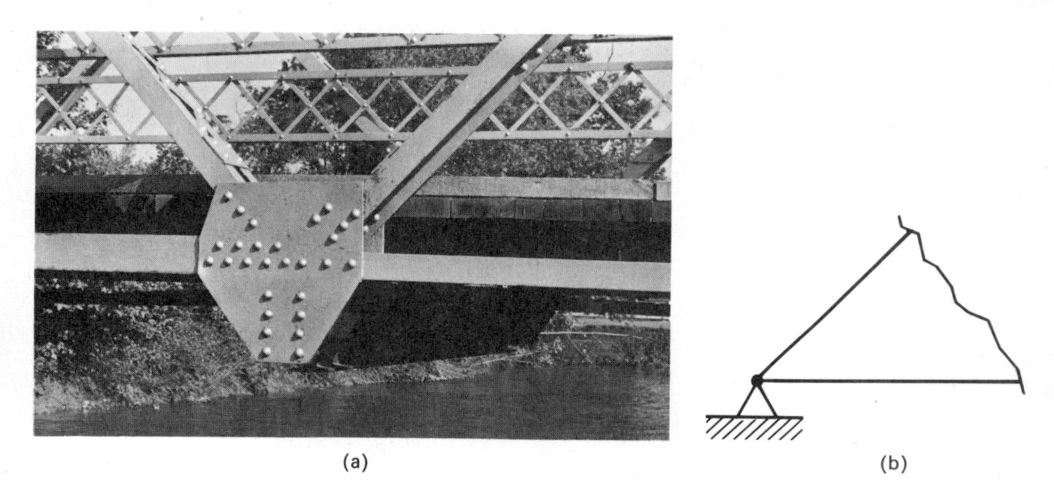

Figure 6.4. (a) Gusset plate; (b) idealization.

of a truss are long and slender, thus rendering them less able to carry loads transverse to their centerlines away from the ends.[2] If the weights of the members are neglected, as is sometimes the case, it should be apparent that each member is a *two-force member*, and accordingly is either a tensile member or a compression member. If the weight is not negligible, the common practice as an approximation is to apply half the weight of a member to each of its two joints. Thus, the idealization of a member as a two-force member is still valid.

6.2 The Simple Truss

An idealized truss as described in Section 6.1 is termed *just-rigid* if the removal of any of its members destroys its rigidity. If removing a member does not destroy rigidity, the structure is said to be *over-rigid*. We shall be concerned with just-rigid trusses in Part A of this chapter.[3]

The most elementary just-rigid truss is one with three members connected to form a triangle. Just-rigid space trusses may be built up from this triangle by adding for each new joint three new members, as is shown in Fig. 6.5.[4] Trusses constructed in this manner are called *simple space trusses*. The *simple plane truss* is built up from an elementary triangle by adding two new members for each new pin as shown in Fig. 6.6. Clearly, the simple plane truss is just-rigid.

---

[2]You will understand these limitations more clearly when you study buckling in your strength of materials course.

[3]Over-rigid structures are studied in courses of strength of materials and structural mechanics. They are internally statically indeterminate and deformation must be taken into account when computing forces in the members.

[4]To ensure that the space truss is just-rigid, no set of three new members can be coplanar. Why?

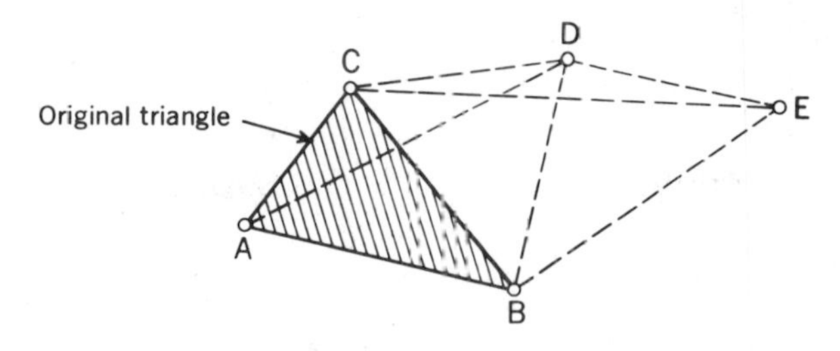

Figure 6.5. Simple space truss.

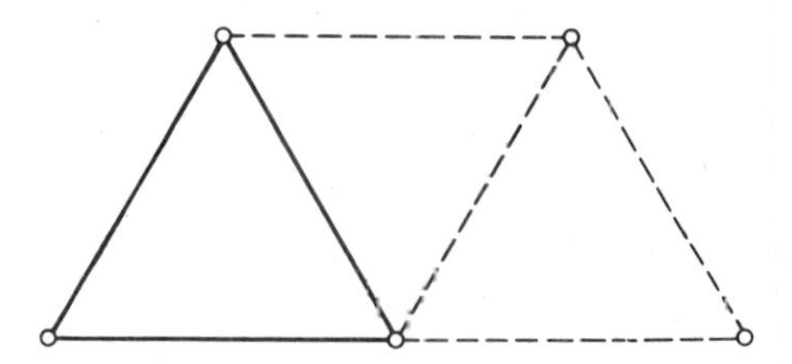

Figure 6.6. Simple plane truss.

A simple relationship exists between the number of joints j and the number of members m in a simple truss. You can directly verify by examining the simple space truss in Fig. 6.5 that m is related to j as follows:

$$m = 3j - 6 \tag{6.1}$$

Similarly, for the simple plane truss in Fig. 6.6 you can verify that

$$m = 2j - 3 \tag{6.2}$$

You will learn in more advanced structures courses that Eqs. 6.1 and 6.2 hold generally for just-rigid space trusses and for just-rigid plane trusses, respectively.

We now show that if the supporting force system is statically determinate, we can compute the forces in all the members of simple trusses. Specifically, in examining the ball joints of simple space trusses, we can see that in the general three-dimensional case, a ball joint with only three unknown forces acting on it from the members can always be found. (One such joint is the last joint formed.) Each unknown force from a member onto this joint must have a direction collinear with that member, and hence has a known direction. There are, then, only three unknown scalars, and since we have a concurrent force system they can be determined by statics alone. We then find another joint with only three unknowns and so carry on the computations until the forces in the entire structure have been evaluated. For the simple plane truss, a similar procedure can be followed. The free body of at least one joint has only two unknown forces. We have a concurrent, coplanar force system, and we accordingly can solve the corresponding two equilibrium equations in two unknowns at that joint. We then proceed to the other joints, thereby evaluating all member forces by the use of statics alone.

6.3 Solution of Simple Trusses

Generally, the first step in a truss analysis is to compute the supporting forces in the overall truss. This calculation of the external forces or reactions that must exist to keep the truss in equilibrium is independent of whether the truss, internally, is statically determinate or statically indeterminate. Simply regard the truss as a rigid body to which forces are applied, some known (given applied forces) and some unknown (reactions),[5] and solve for the reactions as we did in Chapter 5. We have shown a simple plane truss in Fig. 6.7(a) and have shown the features of the truss in Fig. 6.7(b) that are essential for the calculations of the reactions. Note that members *CB*, *DB*, and *DE* need not be shown in the free body since they provide *internal* forces for the body.

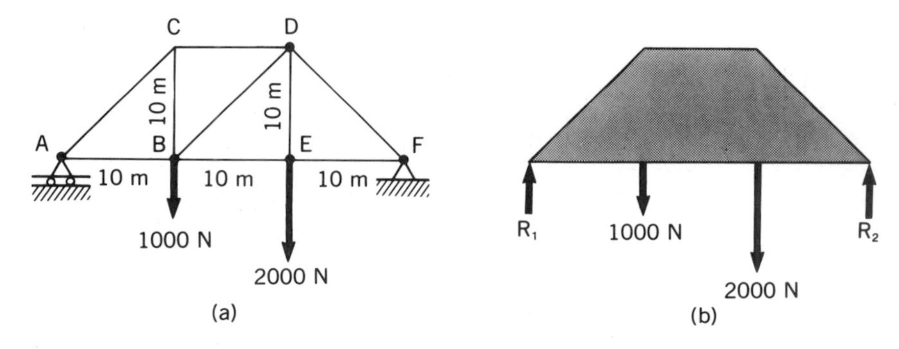

Figure 6.7. Free-body analysis of truss.

Once the free-body diagram has been carefully drawn, use three equations of equilibrium to determine the reactions of a plane truss (six equations for a space truss). It is highly advisable to then check your results by using another (dependent) equation of equilibrium. You will be using the computed reactions for many subsequent calculations involving forces in internal members. Accordingly, with much work at stake, it is important to start off with a correct set of reactions.

We shall present two methods for determining the forces in the members of the truss. One is called the *method of joints* and the other is called the *method of sections.* As will be seen in the following sections, the prime difference between these methods lies in the choice of the free bodies to be used.

6.4 Method of Joints

In the method of joints, the free-body diagrams to be used, once the reactions are determined, are the pins or ball joints and the forces applied to them by the attached members and external loads. Note that we have already alluded to this method in

[5]Supporting forces are often called *reactions* in structural mechanics.

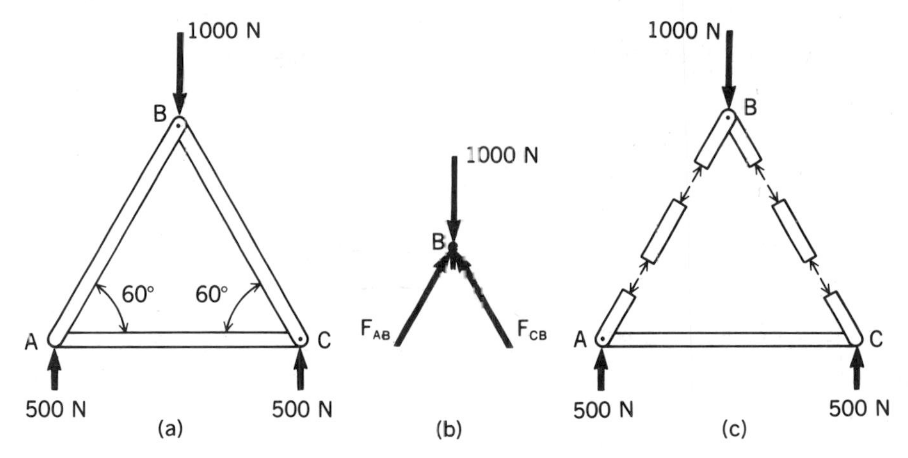

Figure 6.8. Method of joints—joint *B*.

Section 6.2. Consider first the triangular plane truss shown in Fig. 6.8(a). Notice we have already determined the reactions.

Next, consider the free body of pin *B* [Fig. 6.8(b)]. The unknown forces from the members are shown collinear with these members since they are two-force members. We can solve for these forces by setting the sum of forces equal to zero in the horizontal and vertical directions, to get

$$F_{BA} = F_{CB} = 577 \text{ N}$$

Because both forces are *pushing* against pin *B*, the corresponding members are *compressive* rather than tension members. We can most readily see this fact by considering Fig. 6.8(c), where members *AB* and *CB* have been cut at various places. Notice that *AB* is also pushing against pin *A* as does *BC* against pin *C*. Thus, once having decided that the members are compressive members as a result of considerations at a pin at one end of the member, we can conclude that the member is pushing with equal force against the pin at the other end. To make for speed and accuracy as we go from one joint to another, we recommend that, once the nature of the loading in a member has been established by considerations at a pin, we mark down this value using a T for tension or a C for compression after it on the truss diagram, as shown in Fig. 6.9(a). Note also that appropriate arrows are drawn in the members. These arrows represent forces developed by the members on the pins. Hence, for *compression* the arrows point *toward* the pins, and for *tension* they point *away* from the pins. Accordingly, if we now consider the free body of pin *A* as shown in Fig. 6.9(b), we know the direction and value of the force on *A* from member *BA*.

If a negative value is found for a force at a pin, the sense of the force has been taken incorrectly at the outset. With this in mind, we decide whether the member associated with the force is a tension or compression member. And we label the member accordingly, as shown in Fig. 6.9(a) for use later in examining the pin at the other end of the member as a free body.

We now consider the solution of a plane truss problem by the method of joints in greater detail.

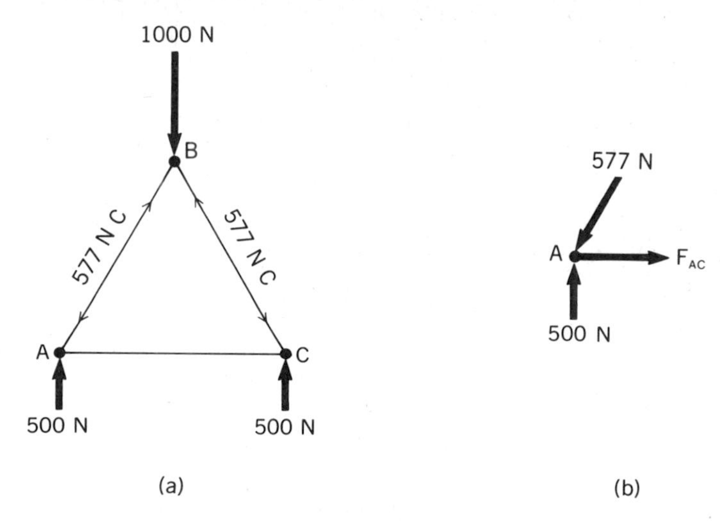

Figure 6.9. Procedure for method of joints. (a) Notation for members *AB* and *BC*; (b) free body diagram of *A*.

EXAMPLE 6.1

A simple plane truss is shown in Fig. 6.10. Two 1000-lb loads are shown acting on pins *C* and *E*. We are to determine the force transmitted by each member. Neglect the weight of the members.

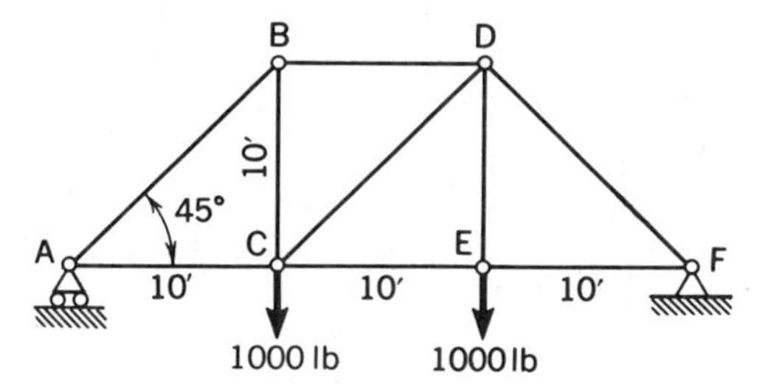

Figure 6.10. Plane truss.

In this simple loading, we see by inspection that there are 1000-lb vertical forces at each support. We shall begin, then, by studying pin *A*, for which there are only two unknowns.

Pin *A*. The forces on pin *A* are the known 1000-lb supporting force and two unknown forces from the members *AB* and *AC*. The orientation of these forces is known from the geometry of the truss, but the magnitude and sense must be determined. To help in interpreting the results, put the forces in the same position as the corresponding bars in the space diagram (Fig. 6.11). That is, avoid the force diagram in Fig. 6.12, which is equivalent to the one in Fig. 6.11 but which may lead to errors in interpretation. There are two unknowns for the concurrent coplanar force sys-

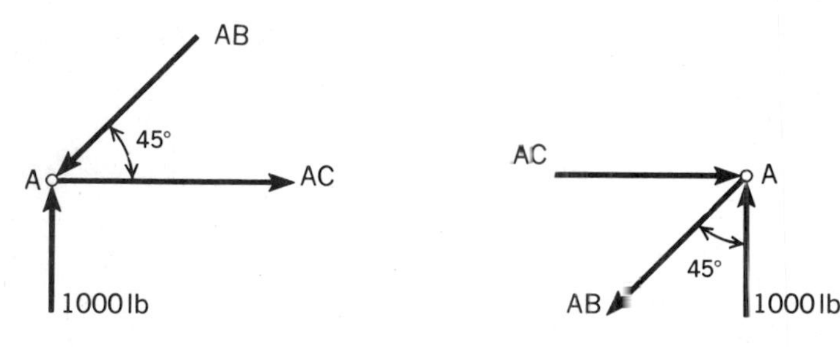

Figure 6.11. Pin *A*.

Figure 6.12. Pin *A*—avoid this diagram.

tem in Fig. 6.11 and thus, if we use the scalar equations of equilibrium, we may evaluate AB and AC:

$\sum F_x = 0$:

$$AC - 0.707AB = 0$$

$\sum F_y = 0$:

$$-0.707AB + 1000 = 0$$

Therefore,

$$AB = 1414 \text{ lb}; \quad AC = 1000 \text{ lb}$$

Since both results are positive, we have chosen the proper senses for the forces. We can then conclude on examining Fig. 6.11 that AB is a compression member, whereas AC is a tension member.[6] In Fig. 6.13, we have labeled the members accordingly.

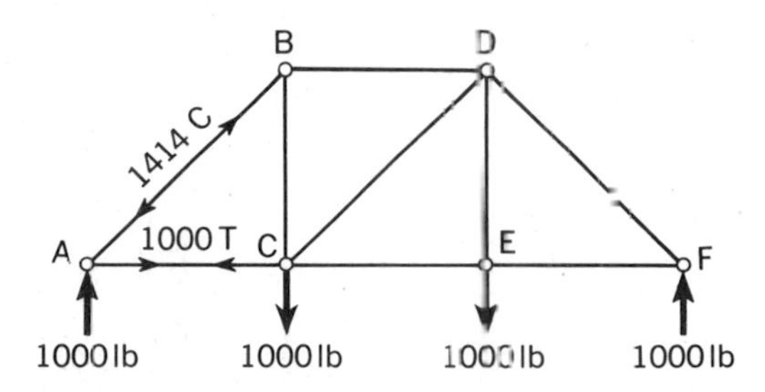

Figure 6.13. Notation for members *AB* and *AC*.

If we next examine pin C, clearly since there are three unknowns involved for this pin, we cannot solve the forces by equilibrium equations at this time. However, pin B can be handled, and once BC is known, the forces on pin C can be solved.

Pin *B*. Since AB is a compression member (see Fig. 6.13) we know that it exerts a force of 1414 lb directed against pin B as has been shown in Fig. 6.14. As for members BC and BD, we assign senses as shown.

[6]Had we used Fig. 6.12 as a free body, the state of loading in the members (i.e., tension or compression) would not be clear. Therefore, we strongly recommend putting forces representing members in positions coinciding with the members.

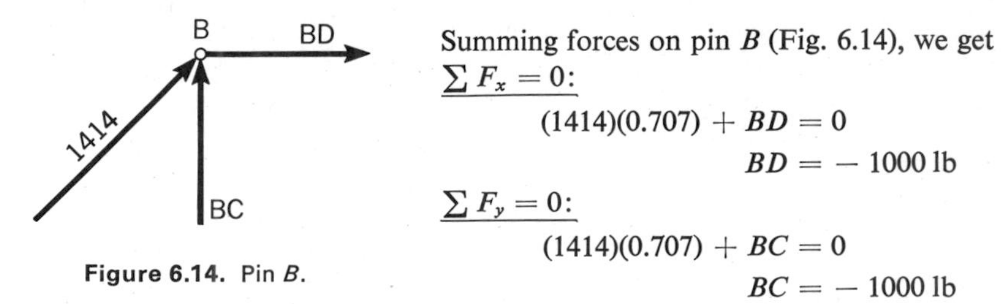

Figure 6.14. Pin B.

Summing forces on pin B (Fig. 6.14), we get

$\underline{\sum F_x = 0:}$

$$(1414)(0.707) + BD = 0$$
$$BD = -1000 \text{ lb}$$

$\underline{\sum F_y = 0:}$

$$(1414)(0.707) + BC = 0$$
$$BC = -1000 \text{ lb}$$

Here we have obtained two negative quantities, indicating that we have made incorrect choices of sense. Keeping this in mind, we can conclude that member BD is a compression member, whereas member BC is a tension member.

We can proceed in this manner from joint to joint. At the last joint all the forces will have been computed without using it as a free body. Thus, it is available to be

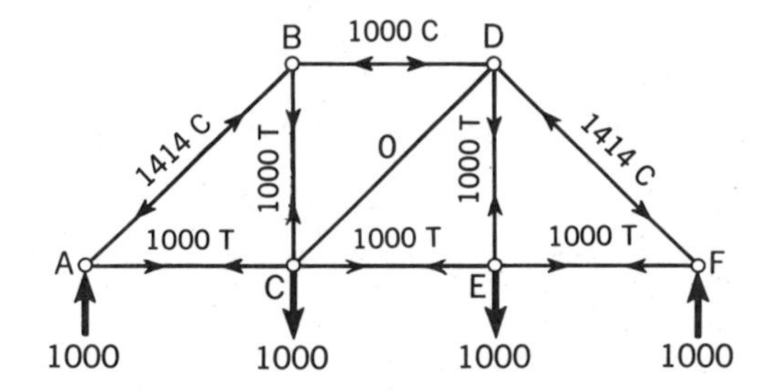

Figure 6.15. Solution for truss.

used as a check on the solution. That is, the sum of the known forces for the last joint in the x and y directions should be zero or close to zero, depending on the accuracy of your calculations. We urge you to take advantage of this check. The final solution is shown in Fig. 6.15. Notice that member CD has zero load. This does not mean that we can get rid of this member. Other loadings expected for the truss will result in nonzero force for CD. Furthermore, without CD the truss will not be rigid.

*EXAMPLE 6.2

Ascertain the forces transmitted by each member of the three-dimensional truss [Fig. 6.16(a)].

We can readily find the supporting forces for this simple structure by considering the whole structure as a free body and by making use of the symmetry of the loading and geometry. The results are shown in Fig. 6.16(b).

Joint *F*. It is clear, on an inspection of the forces in the x direction acting on joint F, that the force FE must be zero, since all other forces are in a plane at right angles to it. These other forces are shown in Fig. 6.17. Summing forces in the y and z directions, we get

$\underline{\sum F_y = 0:}$

$$-FD \frac{20}{\sqrt{20^2 + 10^2}} + 2000 = 0$$

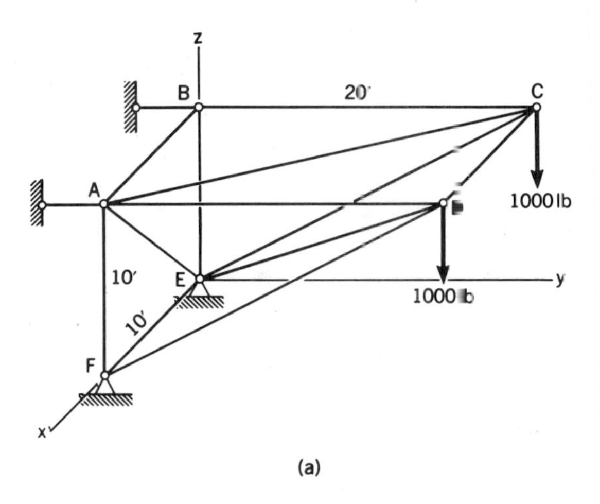

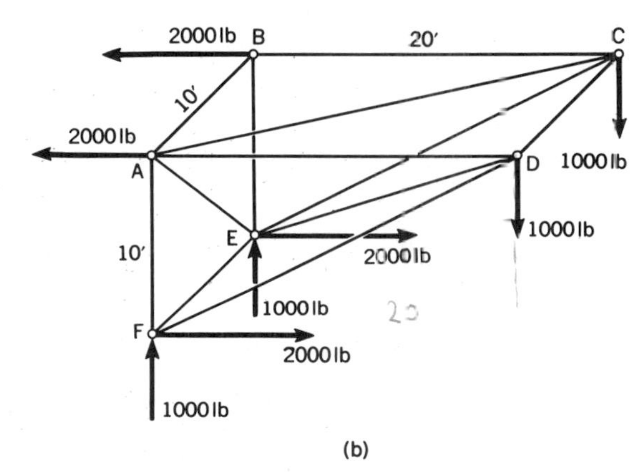

Figure 6.16. (a) Space truss and (b) free-body diagram.

Therefore,

$$FD = 2240 \text{ lb of compression}$$

$\underline{\sum F_z = 0:}$

$$-AF + 1000 - 2240 \frac{10}{\sqrt{500}} = 0$$

Therefore,

$$AF = 1000 - 1000 = 0$$

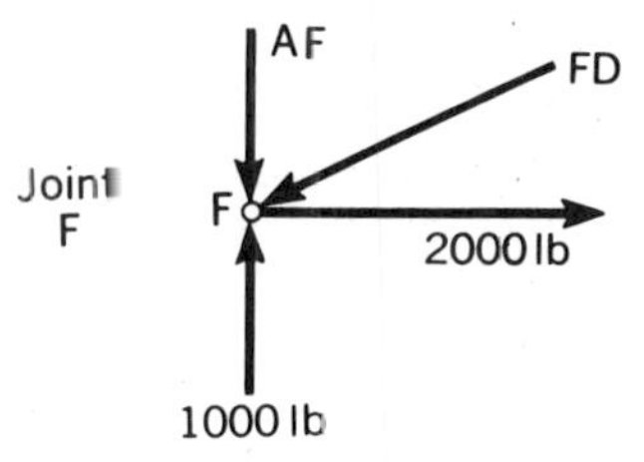

Figure 6.17. Pin *F*.

Joint *B*. Going to joint *B*, we see [Fig. 6.16(b)] that $AB = 0$ and $BE = 0$, since there are no other force components on pin *B* in the directions of these members. Finally, $BC = 2000$ lb of tension.

Joint *A*. Let us next consider joint *A* (Fig. 6.18). We can express force $\overrightarrow{AC}$ and $\overrightarrow{AE}$ vectorially. Thus,

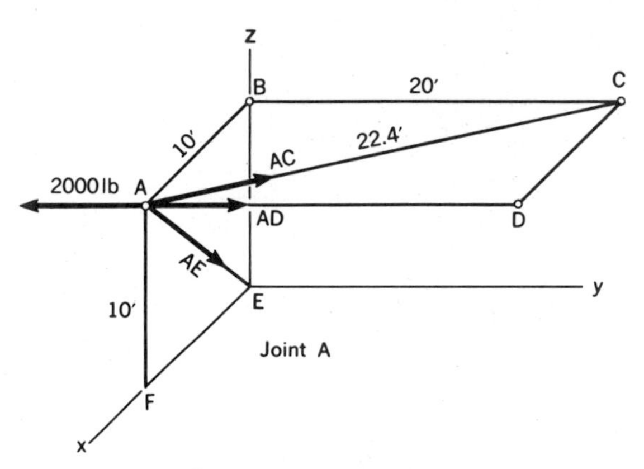

Figure 6.18. Joint *A*.

$$\overrightarrow{AC} = AC\frac{-10\boldsymbol{i} + 20\boldsymbol{j}}{\sqrt{10^2 + 20^2}} = AC\,(-.447\boldsymbol{i} + .894\boldsymbol{j})\text{ lb}$$

$$\overrightarrow{AE} = AE\frac{-10\boldsymbol{i} - 10\boldsymbol{k}}{\sqrt{10^2 + 10^2}} = AE\,(-.707\boldsymbol{i} - .707\boldsymbol{k})\text{ lb}$$

Summing forces, we have

$$-2000\boldsymbol{j} + AD\boldsymbol{j} + AC(-.447\boldsymbol{i} + .894\boldsymbol{j}) + AE(-.707\boldsymbol{i} - .707\boldsymbol{k}) = \boldsymbol{0}$$

Hence,

$$.894AC + AD = 2000 \qquad \text{(a)}$$

$$-.447AC - .707AE = 0 \qquad \text{(b)}$$

$$-.707AE = 0 \qquad \text{(c)}$$

We see that $AE = AC = 0$ and $AD = 2000$ lb of tension.

Joint *D*. We now consider joint *D* (Fig. 6.19). Forces $\overrightarrow{FD}$ and $\overrightarrow{ED}$ are expressed as follows:

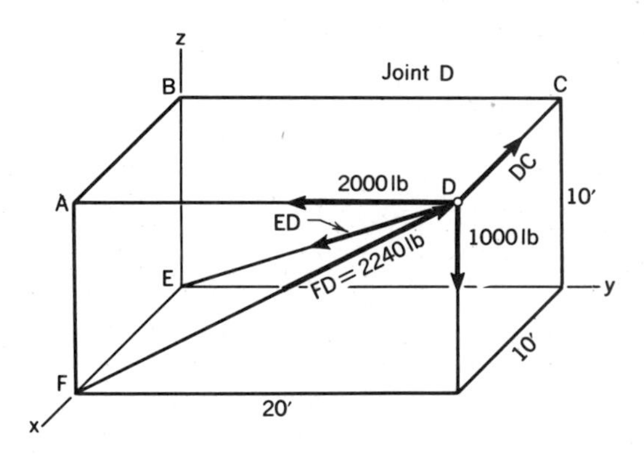

Figure 6.19. Joint *D*.

$$\overrightarrow{ED} = ED\frac{-10\boldsymbol{i} - 20\boldsymbol{j} - 10\boldsymbol{k}}{\sqrt{10^2 + 20^2 + 10^2}} = ED(-.408\boldsymbol{i} - .816\boldsymbol{j} - .408\boldsymbol{k}) \text{ lb}$$

$$\overrightarrow{FD} = FD\frac{20\boldsymbol{j} + 10\boldsymbol{k}}{\sqrt{20^2 + 10^2}} = 2240(.894\boldsymbol{j} - .447\boldsymbol{k}) \text{ lb}$$

Hence, summing forces, we get

$$\begin{aligned} -2000\boldsymbol{j} - 1000\boldsymbol{k} - DC\boldsymbol{i} &+ 2240(.894\boldsymbol{j} + .447\boldsymbol{k}) \\ &+ ED(-.408\boldsymbol{i} - .816\boldsymbol{j} - .408\boldsymbol{k}) = \boldsymbol{0} \end{aligned} \tag{d}$$

Thus,

$$-2000 + 2000 - .816ED = 0 \tag{e}$$

$$DC + .408ED = 0 \tag{f}$$

$$-1000 + 1000 - .408ED = 0 \tag{g}$$

We see here that $ED = 0$ and $DC = 0$.

Joint *E*. The only nonzero forces on joint E are the supporting forces and CD, as shown in Fig. 6.20(a). We may solve for CE directly and get 2240 lb of compression.

Joint *C*. As a check on our problem, we can examine joint C. The only nonzero forces are shown on the joint [Fig. 6.20(b)]. The reader may readily verify that the solution checks.

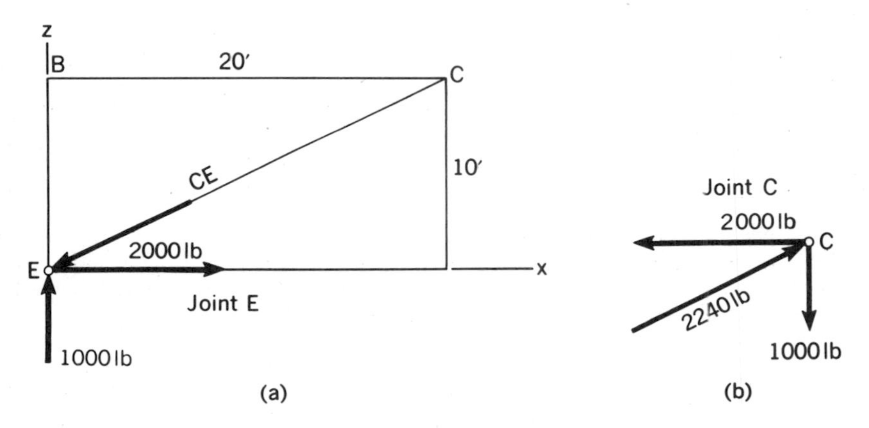

Figure 6.20. (a) Joint E; (b) check at joint C.

Before proceeding with the problems, it will be well to comment on the loading of plane roof trusses. Usually there will be a series of separated parallel trusses supporting the loading from the roof such as is shown in Fig. 6.21, where a wind pressure is shown on a roof as p. Now the inside truss can be considered to support the loading over a region extending halfway to each neighboring truss (shown as distance d). Furthermore, pins A and B support the force exerted on area $lhmk$ while pins B and C support the forces exerted on area $lrvh$. When dealing with the entire inside truss as a free body, you can use the resultant force from pressure over $krvm$. However, when dealing with the pins as a free body you must use the forces coming on to each pin as

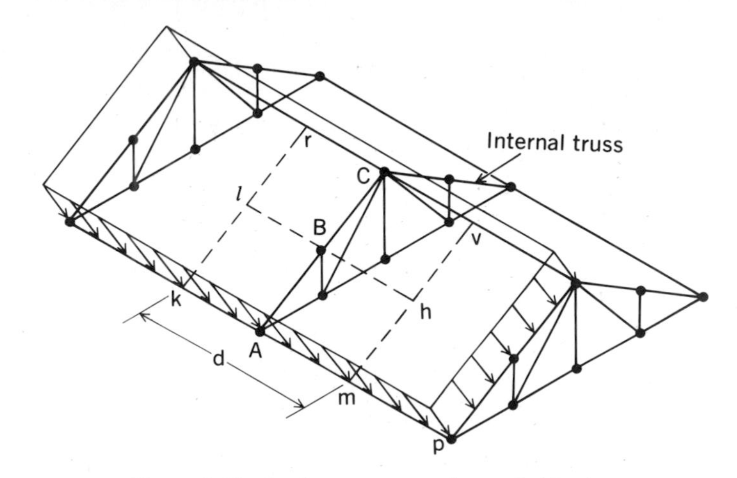

Figure 6.21. Roof trusses supporting a wind load.

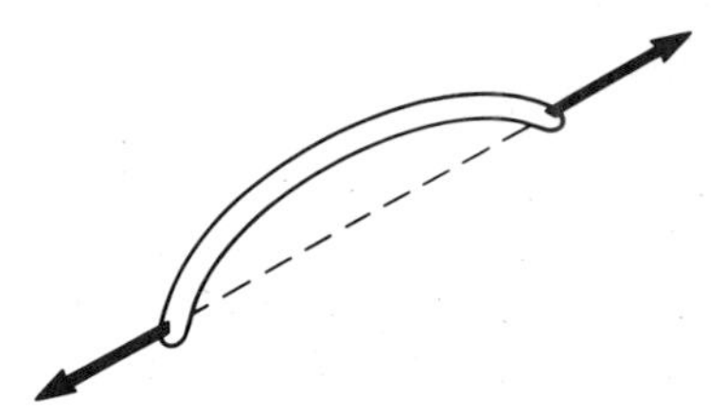

Figure 6.22. Curved two-force member.

described above and *not* the total resultant which was used for the free body of the entire internal truss.

Finally, we wish to remind you that a curved member in a truss, such as appears in Problems 6.5 and 6.9, is a two-force member with forces coming only from the pins. Recall that, for such members, the force transmitted to the pins must be collinear with the line connecting the points of application of the pins, such as is shown in Fig. 6.22.

Problems

6.1. State which of the trusses shown are simple trusses and which are not.

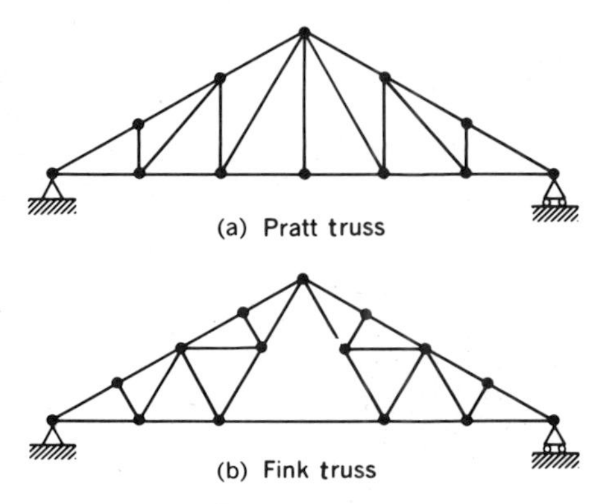

Figure P.6.1

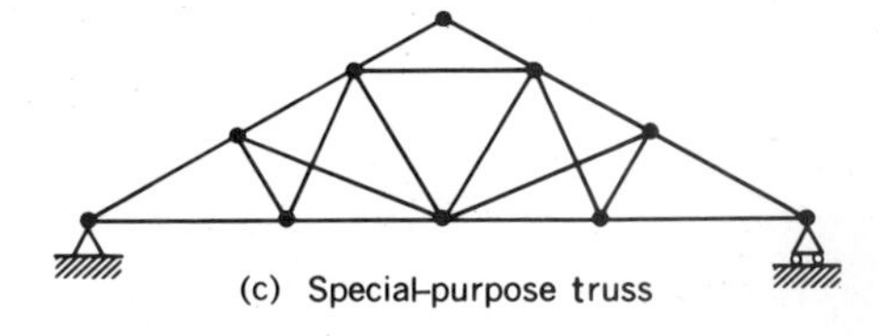

Figure P.6.1. (cont.)

6.2. Find the forces transmitted by each member of the truss.

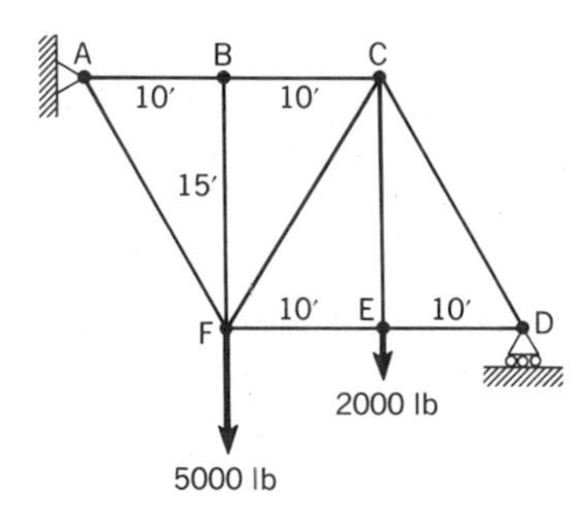

Figure P.6.2

6.3. The simple country-road bridge has floor beams to carry vehicle loads to the truss joints. Find the forces in all members for a truck-loaded weight of 160 kN. Floor beams 1 are supported by pins A and B, while floor beams 2 are supported by pins B and C.

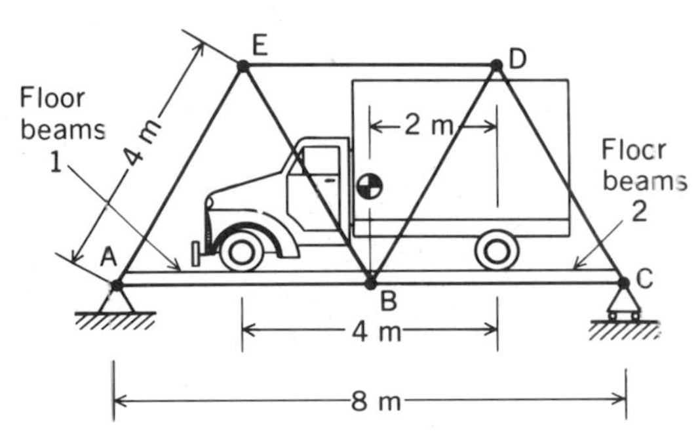

Figure P.6.3

6.4. A rooftop pond is filled with cooling water from an air conditioner and is supported by a series of parallel plane trusses. What are the forces in each member of an inside truss? The roof trusses are spaced at 10 ft apart. Water weighs 62.4 lb/ft³.

Figure P.6.4

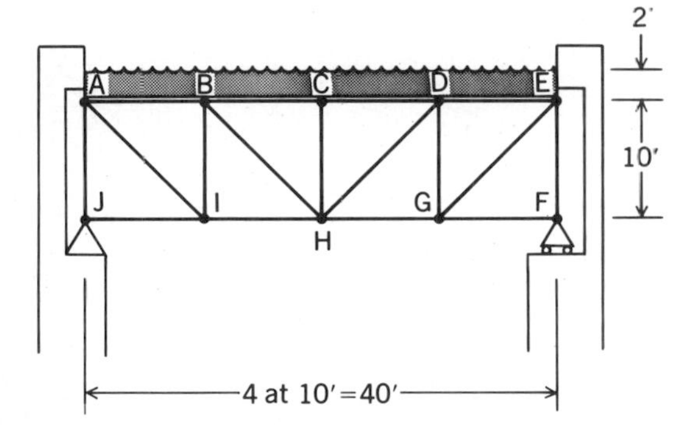

6.5. Find the forces transmitted by the straight members of the truss. DC is circular.

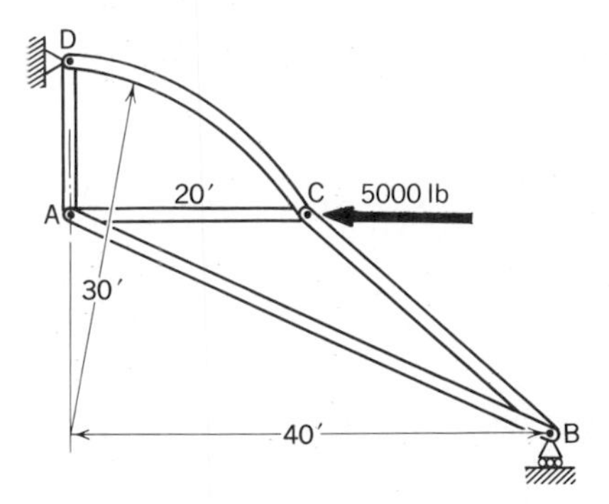

Figure P.6.5

6.6. Roof trusses such as the one shown are spaced 6 m apart in a long, rectangular building. During the winter, snow loads of up to 1 kN/m² (or 1 kPa) accumulate on the central portion of the roof. Find the force in each member for a truss not at the ends of the building.

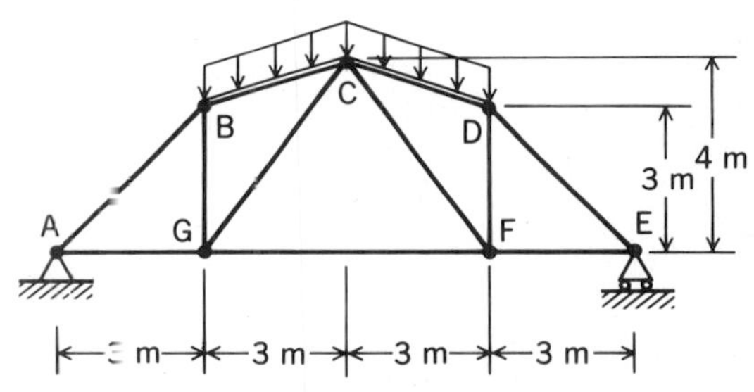

Figure P.6.6

6.7. The bridge supports a roadway load of 1000 lb/ft for each of the two trusses. Each member weighs 30 lb/ft. Compute the forces in the members, accounting approximately for the weight of the members.

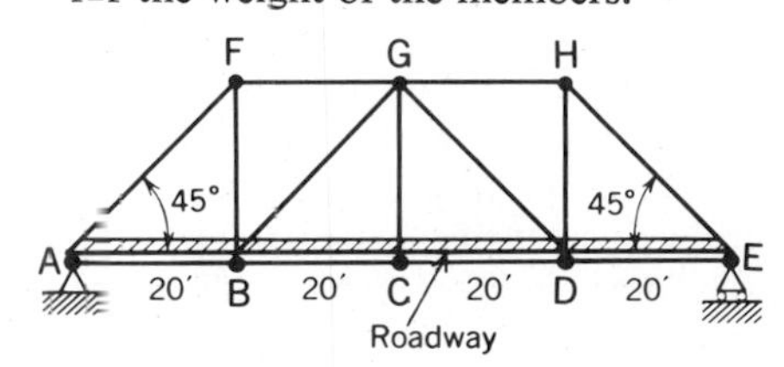

Figure P.6.7

6.8. Roadway and vehicle loads are transmitted to the highway bridge truss as the idealized forces shown. What are the forces in members?

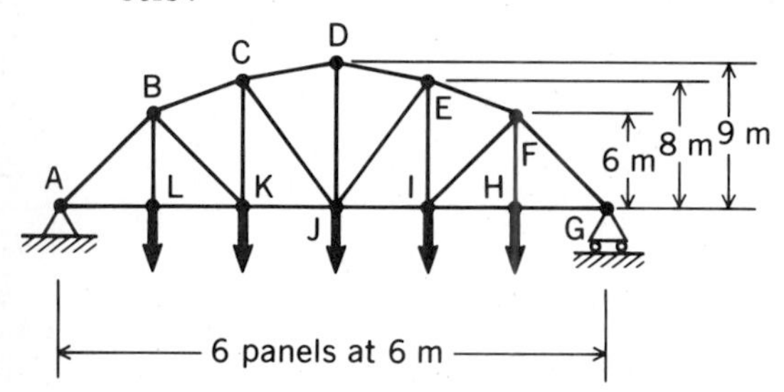

Figure P.6.8

6.9. A 5-kN hoist to lift railroad cars for truck repair has a 150-kN capacity and hangs from a truss with an L-shaped member to clear boxcars. What are the forces in the straight members?

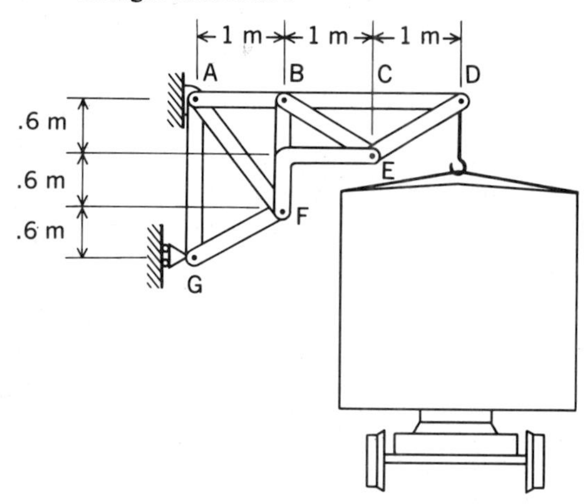

Figure P.6.9

6.10. A 5-kN traveling hoist has a 50-kN capacity and is suspended from a beam weighing 1 kN/m, which, in turn, is fastened to the roof truss at *I* and *G* as shown. In addition, wind pressures of up to 2 kN/m² (or 2 kPa) act on the side of the roof. The resulting force is transmitted to pins *A* and *J*. If the trusses are spaced 10 m apart, what are the forces in each member for an internal truss when the hoist is in the middle of the span?

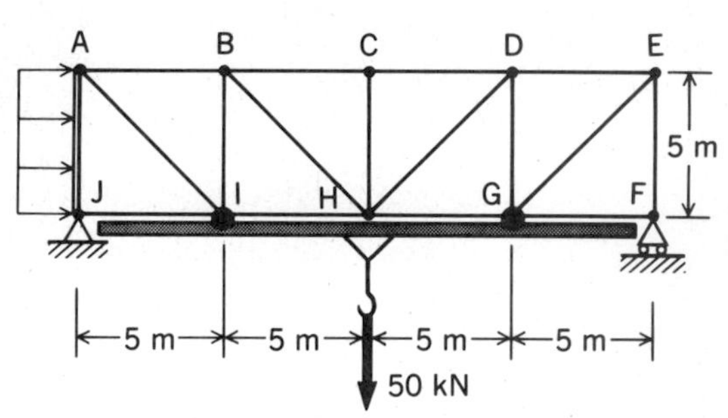

Figure P.6.10

6.11. Find the forces in the members of the truss. The 1000-lb force is parallel to *y*, and the 500-lb force is parallel to *z*.

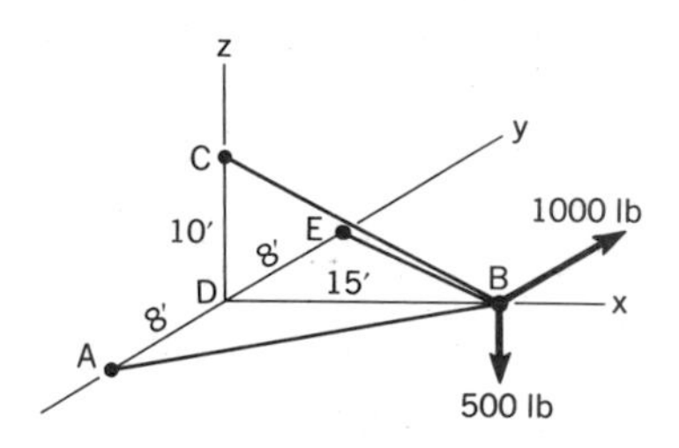

Figure P.6.11

6.12. Find the forces in the members and the supporting forces for the space truss *ABCD*. Note that *BDC* is in the *xz* plane.

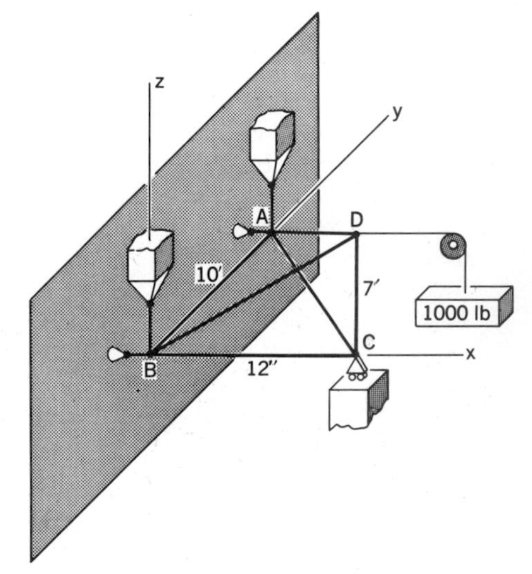

Figure P.6.12

6.13. A space truss *ABCDE* supports a 50-kN vertical load as well as a 10-kN horizontal load and rests on smooth, mutually perpendicular surfaces. Assume that the contact between the space truss and the smooth surface is at the ball joints. What are the forces in the members?

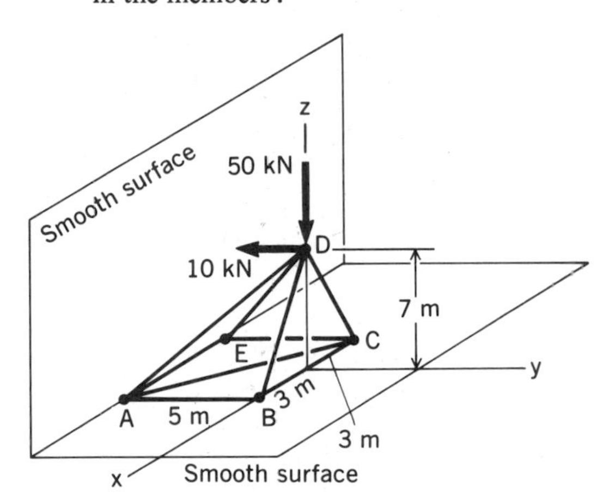

Figure P.6.13

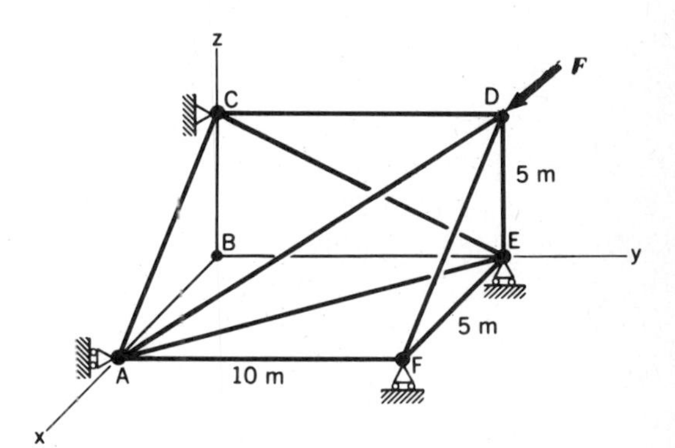

Figure P.6.14

6.14. Find the forces in the members of the space truss under the action of a force $\boldsymbol{F}$ given as

$$\boldsymbol{F} = 10\boldsymbol{i} - 6\boldsymbol{j} - 12\boldsymbol{k} \text{ kN}$$

Note that C is a ball-and-socket joint while A, F, and E are on rollers.

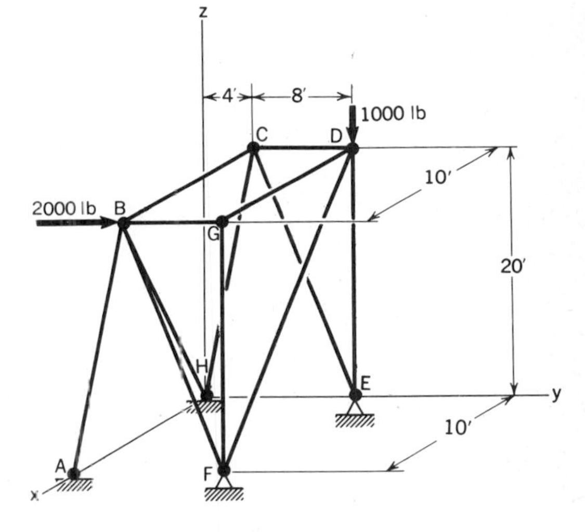

Figure P.6.15

6.15. The plane of ball-and-socket joints *CDHE* of the space truss is in the zy plane, while the plane of *FGDE* is parallel to the xz plane. Note that this is *not* a simple space truss. Nevertheless, the forces in the members can be ascertained by choosing a desirable starting joint and proceeding by statics from joint to joint. Determine the forces in all the members and then determine the supporting forces.

6.5 Method of Sections

In the method of sections that we shall use for plane trusses, we employ free-body diagrams that are generally different than that of the method of joints, as was pointed out earlier. *A free body in this method is formed by cutting away a portion of a truss and including at the cut sections the forces that are transmitted across these sections.* We then use the equations of equilibrium for these free bodies. In this way, we can expose for calculation individual members well inside a truss and avoid the laborious process of proceeding joint by joint until reaching a joint on which the desired unknown force acts.

Generally, a free body is created by passing a section (or cut) through the truss such as section *A–A* or section *B–B* in Fig. 6.23(a). Note that the section can be straight or curved. The corresponding free-body diagrams [see Fig. 6.23(b) for cut *A–A* and Fig. 6.23(c) for cut *B–B*] involve coplanar force systems. We have, accordingly, three equations of equilibrium available for each free body. Note that in contrast to the method

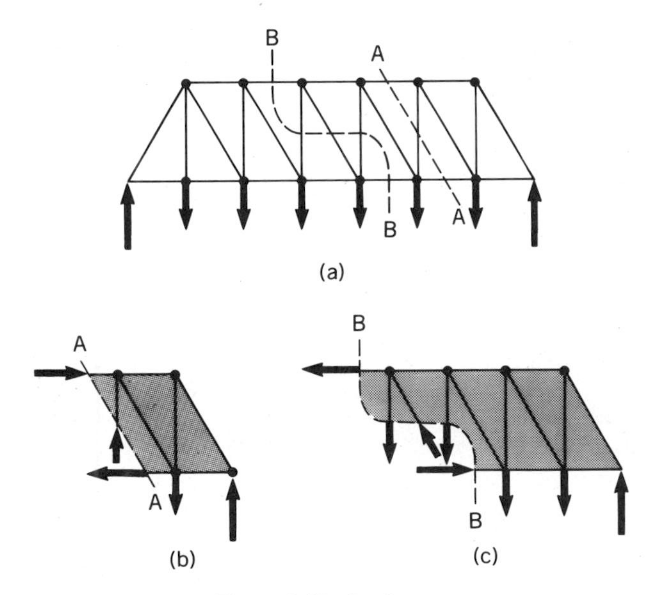

Figure 6.23. Section cuts.

of joints, one or more equilibrium equations can most profitably be moment equations. The choice of the section (or sections) to find the desired unknowns inside a truss involves ingenuity on the part of the engineer. He will want the fewest and simplest sections to find desired forces for one or more members inside the truss. The method of sections is used for finding limited information. The method of joints for such problems is by contrast one of "brute force."

We now illustrate the method of sections in the following examples.

EXAMPLE 6.3

In Example 6.1 suppose that we wish to know the force in member *CE* only. To avoid the laborious joint-by-joint procedure, we employ a portion of the truss to the left of cut *K–K*, as shown in Fig. 6.24. Notice that the forces from the other part of the truss acting on this part through the cut members have been included, and in this way the desired force has been exposed. The sense of these exposed forces is not known, but we do know the orientations, as explained in our earlier discussions. Using the equations of equilibrium and taking advantage of the fact that the lines of action of some of the exposed unknown forces are concurrent at certain joints, we may readily solve for the unknowns if they number three or less. To determine CE, we take moments about a point corresponding to joint D through which the lines of action of forces BD and CD pass:

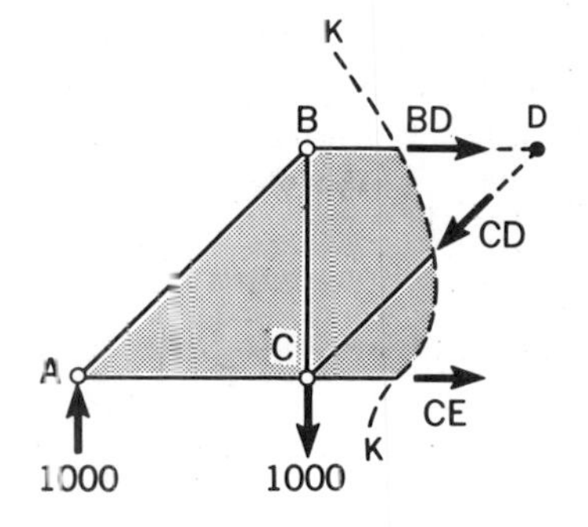

Figure 6.24. Cut K–K.

$\underline{\sum M_D = 0:}$

$$-(1000)(20) + (1000)(10) + 10CE = 0$$

Therefore,

$$CE = 1000 \text{ lb}$$

Our ingenuity here has led us to one equation with only one unknown, the desired force CE. By observing the free-body diagram in Fig. 6.24, we can clearly see that CE is a tension member.

If we desire BD also, we can take moments about point C through which the lines of action of CE and CD pass. However, BD now comes out negative, indicating that we have made an incorrect choice of sense. With this in mind, we can conclude that BD is in compression.

Perhaps a suitable single section with sufficient unknowns for a solution cannot be found. We may then have to take several sections before we can expose the desired force in a free body with enough simultaneous equations to effect a solution. These problems are no different from the ones we studied in Chapter 5, where several free-body diagrams were needed to generate a complete set of equations containing the unknown quantity. We now consider such a problem.

EXAMPLE 6.4

A plane truss is shown in Fig. 6.25 for which only the force in member AB is desired. The supporting forces have been determined and are shown in the diagram.

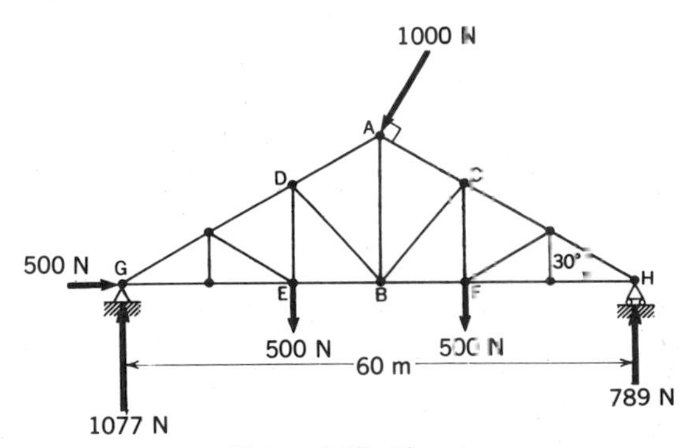

Figure 6.25. Plane truss.

In Fig. 6.26, we have shown a cut J–J of the truss exposing force AB. (This is the same force diagram as that which results from the free-body diagram of pin A.) We have here three unknown forces for which only two equations of equilibrium are available. We must use an additional free body.

Thus, in Fig. 6.27 we have shown a second cut K–K. Note that by taking moments about joint B, we can solve for AC directly. With this information, we can

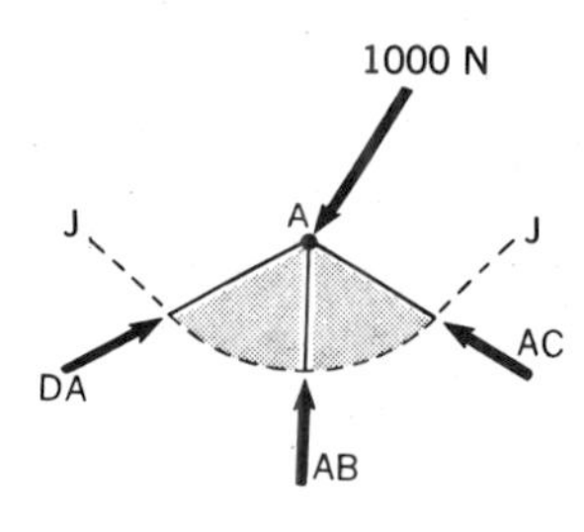

Figure 6.26. Free body I from cut J–J.

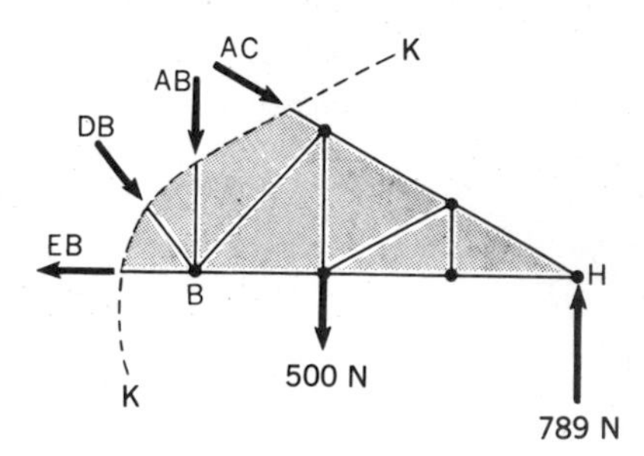

Figure 6.27. Free body II from cut K–K.

then return to the first cut to get the desired unknown AB. Accordingly, we have, for free body II:

$\underline{\sum M_B = 0:}$

$$-(10)(500) + (30)(789) - (AC)(\sin 30°)(30) = 0$$

(Note we have transmitted AC to joint H in evaluating its moment contribution.) Solving for AC, we get

$$AC = 1245 \text{ N}$$

Summing forces for free body I, we have

$\underline{\sum F_x = 0:}$

$$DA \cos 30° - AC \cos 30° - 1000 \sin 30° = 0$$

Therefore,

$$DA = 1822 \text{ N}$$

$\underline{\sum F_y = 0:}$

$$DA \sin 30° + AC \sin 30° + AB - 1000 \cos 30° = 0$$

Therefore,

$$AB = -667 \text{ N}$$

We see that member AB is a tension member rather than a compression member as was our initial guess in drawing the free-body diagrams.

In retrospect, you will note that, in the method of joints, errors made early will of necessity propagate through the calculations. There is, on the other hand, much less likelihood of this occurring in the method of sections, since many of the equations will be independent. However, for simple trusses with many members, we may profitably use the method of joints in conjunction with a computer for which the brute-force approach of the method of joints is ideally suited.

Problems

6.16. Find the forces in members CB and BE of the plane truss.

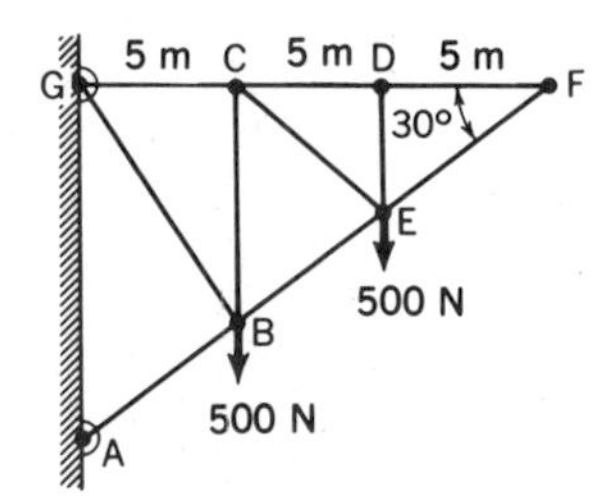

Figure P.6.16

6.17. For the roof truss: (a) Find the forces transmitted by member DC. (b) What is the force transmitted by DE?

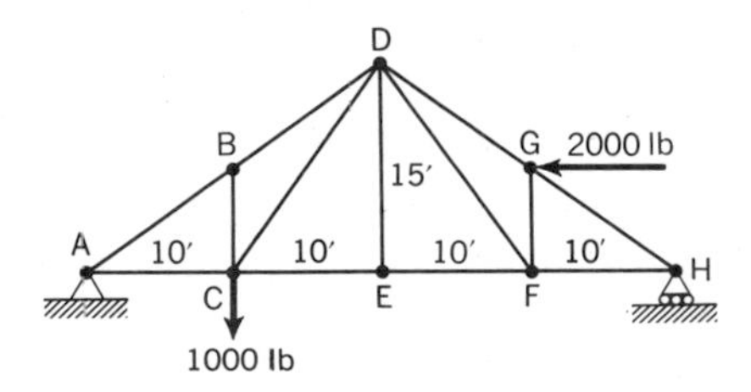

Figure P.6.17

6.18. Determine the force transmitted by member KU in the plane truss.

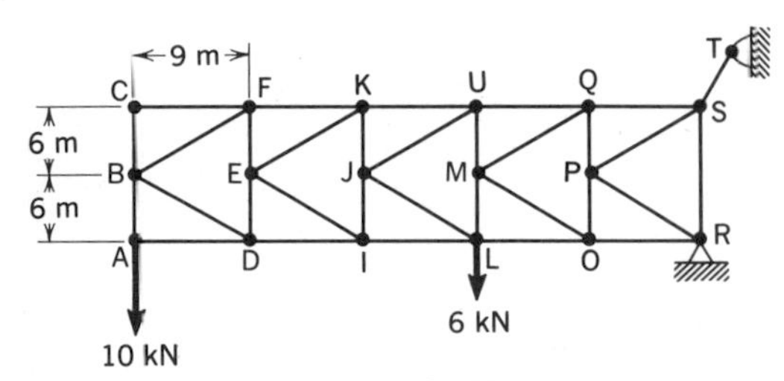

Figure P.6.18

6.19. In the roof truss of Problem 6.6, find the force in member GF. Remember loads are applied to the pins.

6.20. Find the forces in members CD, DG, and HG in the plane truss.

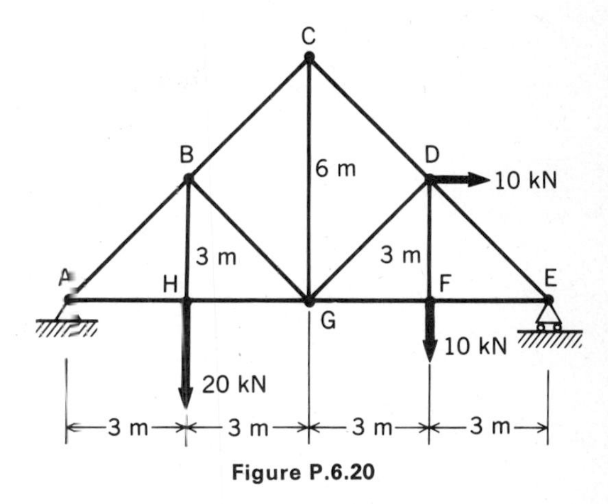

Figure P.6.20

6.21. In Problem 6.9, find the force in members BF and AB.

6.22. The roof is subjected to a wind loading of 20 lb/ft². Find the forces in members LK and KJ if the trusses are spaced 10 ft apart.

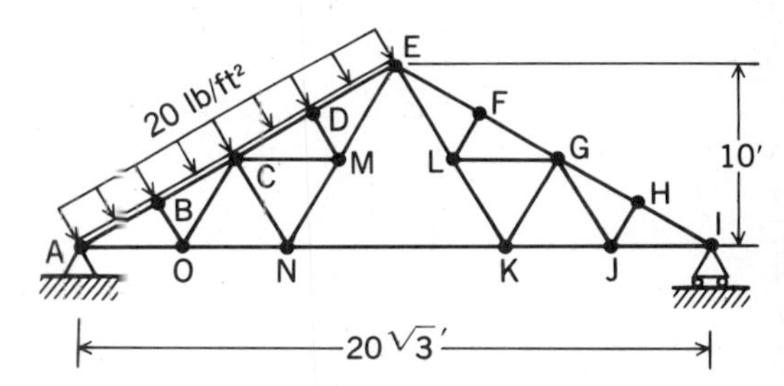

Figure P.6.22

6.23. The guideways for a large overhanging crane are suspended from certain joints of the truss (M, K, J, and G). Find the forces in members BC, BK, DE, DI, and EF. Neglect the truss and guideway weights. Guideways only transmit supported loads to pins and are not considered part of the truss structurally.

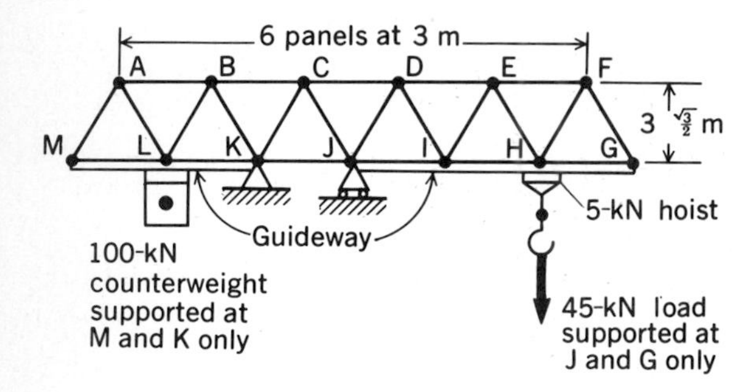

Figure P.6.23

6.24. Find the force in members *HE*, *FH*, *FE*, and *FC* of the truss.

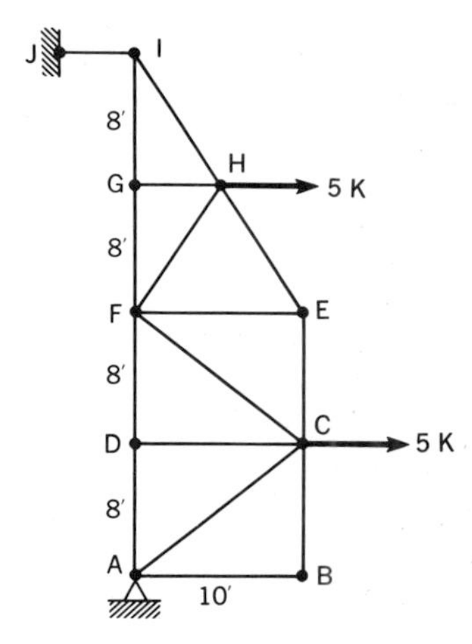

Figure P.6.24

6.25. Find the forces in member *JF* in the truss.

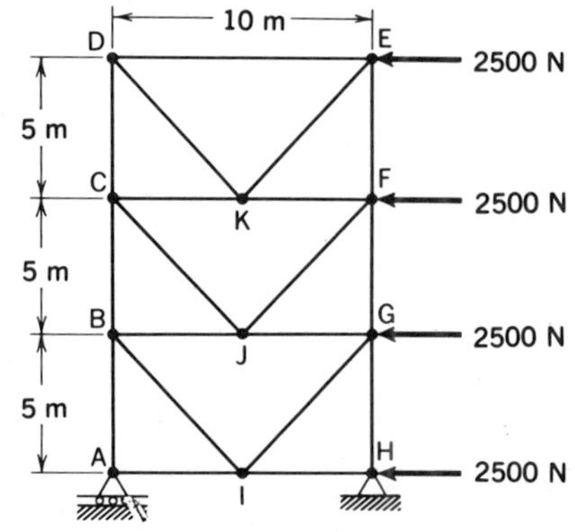

Figure P.6.25

6.26. Find the force in members *FI*, *EF*, and *DH* in the truss. Neglect the weight of the pulleys.

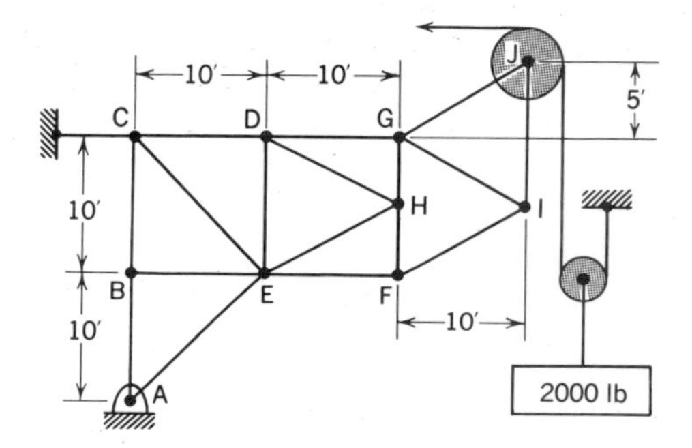

Figure P.6.26

6.27. A railroad engine is starting to cross the deck-type truss bridge shown. If the weight of the engine is idealized by the four 50-kip loads,[7] find the forces in members *AB*, *BL*, *CK*, *CL*, *LK*, *DK*, *KJ*, and *DJ*.

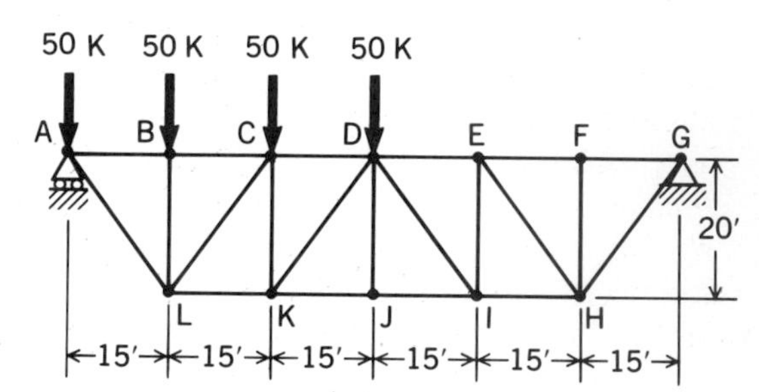

Figure P.6.27

[7] A kip is a kilopound, or 1000 lb.

6.28. A truss supports a roadway load of 800 lb/ft per truss. Concentrated loads have been shown representing approximations of vehicle loading for each truss at some instant of time. The bridge has six 20-ft panels. Determine the forces in members *EG*, *FH*, and *IJ*.

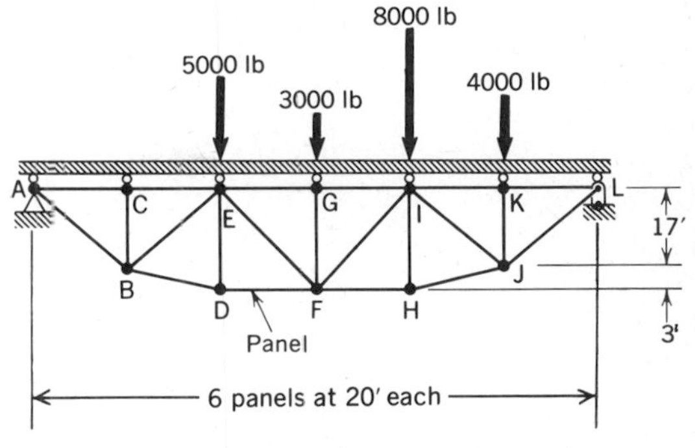

Figure P.6.28

Part B
Section Forces in Beams

6.6 Introduction

In Part A, we considered a number of problems involving members loaded axially along the axes of the members. The resultant force at any section was easily established as a single axial force. We shall now consider thin prismatic members that are loaded *transversely* as well as axially. Generally, when such members are loaded transversely, we call them *beams*. Of considerable use will be certain components of the resultant force system acting on *cross sections* of the beams. We shall set forth methods in this section for computing these quantities. We consider beams with a vertical plane of symmetry along the axis of the beam.

6.7 Shear Force, Axial Force, and Bending Moment

Consider first a beam with an arbitrary intensity of loading $w(x)$ in the plane of symmetry and a load P along the direction of the beam applied at the end A as shown in Fig. 6.28(a). It will be assumed that the supporting forces have been determined. To find the force transmitted across the cross-sectional interface at position x, we take a portion of the beam as a free body so as to "expose" the section of the beam at x as shown in Fig. 6.28(b). Since we have a coplanar-loading distribution, we know from rigid-body mechanics that, depending on the problem, we can replace the distribution at section x most simply by a single force or a single couple moment in the plane of the external loads. If the resultant is most simply a single force, we know that it must have a particular line of action. This line of action does not usually go through the center of the cross section. Since the actual position of the intersection of this force with the

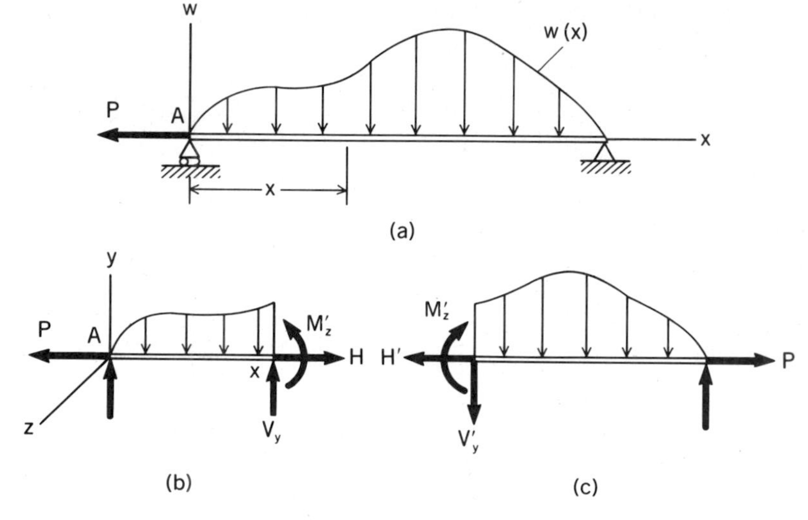

Figure 6.28. Resultant at a section.

cross section is of little interest in beam theory, we deliberately take the position of the resultant force to be at the center of the cross section at all times and include the proper couple moment M_z to accompany the force. Furthermore, we decompose the force into orthogonal components—in this case, a vertical force V_y and a horizontal force H. These quantities are shown in Fig. 6.28(b). Since these quantities are used to such a great extent in structural work, we have associated names with them. They are $V_y \equiv$ *shear-force* component, $H \equiv$ *axial-force* component, $M_z \equiv$ *bending-moment* component.[8] If we had a three-dimensional load, there would have been one additional shear

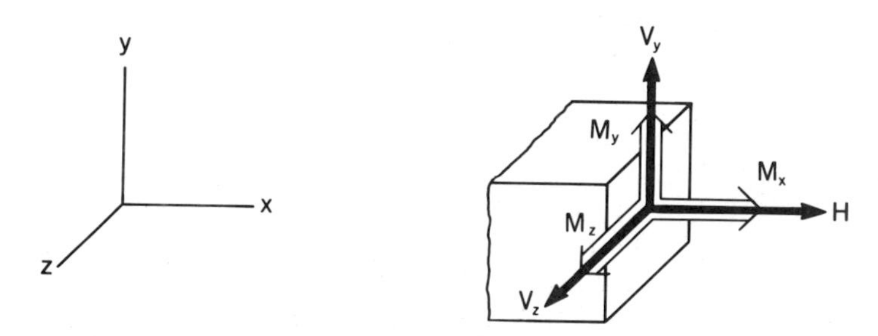

Figure 6.29. Section resultant for three-dimensional loading.

component V_z (see Fig. 6.29), one additional bending-moment component M_y, and a couple moment along the axis of the beam M_x, which we shall call the *twisting moment*.

Notice in Fig. 6.28(c) that a second free-body diagram has been drawn which

---

[8]For curved beams, shear forces V are always tangent to the cross section, whereas axial force H is always normal to the cross section.

exposes the "other side" of the cross section at position x. The shear force, axial force, and bending moment for this section have been primed in the diagram. We know from Newton's third law that they should be equal and opposite to the corresponding unprimed quantities in part (b) of the diagram. We can thus choose for our computations either a left-hand or a right-hand free-body diagram. But this poses somewhat of a problem for us when we come to reporting the signs of the transmitted forces and couple moments at a section. We cannot use the direction of a force or couple moment at the section. Clearly, this would be inadequate since the sense of the force or couple moment at a section would depend on whether a left-hand or a right-hand free-body diagram was used. To associate an unambiguous sign for shear force, axial force, and bending moment at a section, we adopt the following convention:

> *A force component for a section is positive if the area vector of the cross section and the force component both have senses either in the positive or in the negative directions of any one or two of the reference axes.*[9]

The same is true for the bending moment.

Thus, consider Fig. 6.28. For the left-hand free-body diagram, the area vector for section x points in the positive x direction. Note also that H, V_y, and the vectorial representation of M_z also point in positive directions of the xyz axes. Hence, according to our convention we have drawn a positive shear force, a positive axial force, and a positive bending moment at the section at x. For the right-hand free-body diagram, the cross-sectional area vector points in the negative x direction. And, since H', V'_y, and M'_z point in negative directions of the x, y, and z axes, these components are again positive for the section at x according to our convention. Clearly, by employing this convention, we can easily and effectively specify the force system at a section without the danger of ambiguity.

As pointed out earlier, we can solve for V_y, H, and M_z at section x using rigid-body mechanics for either a left-hand or a right-hand free-body diagram provided that we know all the external forces. The quantities V_y, H, and M_z will depend on x, and for this reason, it is the practice to sketch shear-force and bending-moment diagrams to convey this information for the entire beam.

We now illustrate the computation of V and M.

EXAMPLE 6.5

We shall express the shear-force and bending-moment equations for the simply supported beam shown in Fig. 6.30(a), whose weight we shall neglect. The support forces obtained from equilibrium are 500 N each.

To get the shear force at a section x, we isolate either the left or right side of the beam at x and employ the equations of equilibrium on the resulting free body. If x

---

[9]Some authors employ the reverse convention for shear force from the one that we have proposed. Our convention is consistent with the usual convention used in the theory of elasticity for the sign of stress at a point, and it is for this reason that we have employed this convention rather than the other one.

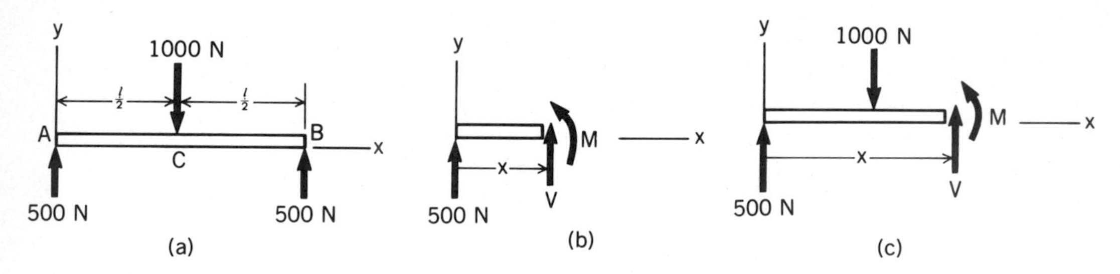

Figure 6.30. Simply supported beam.

lies between A and C of the beam, the only noninternal force present for a left-hand free body is the left supporting force [see Fig. 6.30(b)]. Notice that we have used directions for V and M (there is no need for subscripts in the simple problem) corresponding to the *positive* states from the point of view of our convention. Clearly, the *algebraic* sign we get for these quantities from equilibrium calculations will then correspond to the *convention* sign. If x is between C and B for such a free body, two external forces appear [see Fig. 6.30(c)]. Therefore, if the shear force is to be expressed as a function of x, clearly separate equations covering the two ranges, $0 < x < l/2$ and $l/2 < x < l$, are necessary.[10] Summing forces we then get

$\underline{0 < x < l/2:}$

$$500 + V = 0; \qquad \text{therefore, } V = -500 \text{ N} \tag{a}$$

$\underline{l/2 < x < l:}$

$$500 - 1000 + V = 0; \qquad \text{therefore, } V = 500 \text{ N} \tag{b}$$

Now let us turn to the bending-moment equations. Again, we must consider two discrete regions. Taking moments about position x, we get

$\underline{0 \leq x \leq l/2:}$

$$-500x + M = 0; \qquad \text{therefore, } M = 500x \text{ N-m} \tag{c}$$

$\underline{l/2 \leq x \leq l:}$

$$-500x + 1000\left(x - \frac{l}{2}\right) + M = 0; \qquad \text{therefore, } M = 500(l - x) \text{ N-m} \tag{d}$$

EXAMPLE 6.6

Determine the shear-force and bending-moment equations for the simply supported beam shown in Fig. 6.31. Neglect the weight of the beam.

We must first find the supporting forces for the beam. Hence, we have:

$\underline{\sum M_B = 0:}$

$$-R_1(22) + (50)(8)(14) + (1000)(14) - 500 = 0$$

Therefore,

$$R_1 = 868 \text{ lb}$$

$\underline{\sum M_A = 0:}$

$$R_2(22) - 500 - (50)(8)(8) - (1000)(8) = 0$$

---

[10] We exclude points A, C, and B because, as you will soon see in the shear-force diagrams, V is *indeterminate* at locations of point forces. Bending moment, however, will be continuous except at a point couple moment.

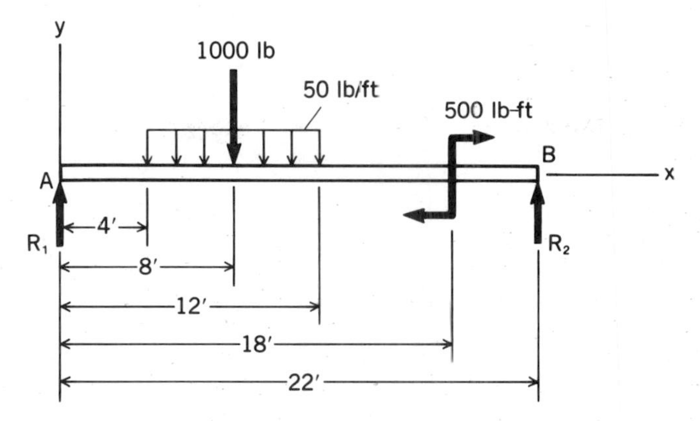

Figure 6.31. Simply supported beam.

Therefore,

$$R_2 = 532 \text{ lb}$$

In Fig. 6.32(a) we have shown a free-body diagram exposing sections between the left support and the uniform load. Summing forces and taking moments about a

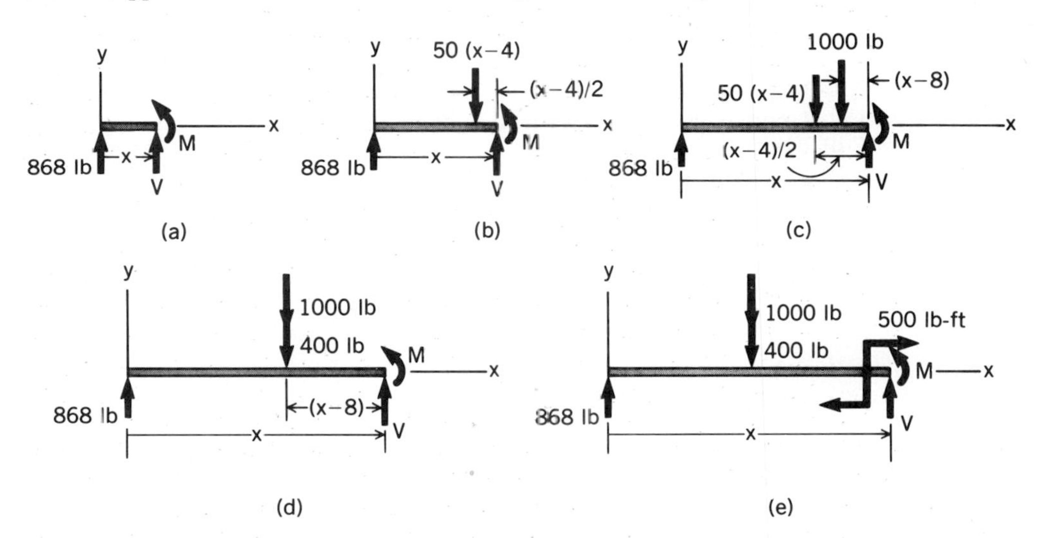

Figure 6.32. Free-body diagrams for various ranges.

point in the section, where we have drawn V and M as positive according to our convention, we get

$0 < x \leq 4$:

$$868 + V = 0; \qquad \text{therefore, } V = -868 \text{ lb}$$

$$-868x + M = 0; \qquad \text{therefore, } M = 868x \text{ ft-lb}$$

The next interval is between the beginning of the uniform load and the point force. Thus, observing Fig. 6.32(b):

$\underline{4 < x \leq 8:}$

$$868 - 50(x - 4) + V = 0$$

Therefore,

$$V = 50x - 1068 \text{ lb}$$

$$-868x + \frac{50(x - 4)^2}{2} + M = 0$$

Therefore,

$$M = -25x^2 + 1068x - 400 \text{ ft-lb}$$

We now consider the interval between the point force and the end of the uniform load. Thus, observing Fig. 6.32(c):

$\underline{8 \leq x < 12:}$

$$868 - 50(x - 4) - 1000 + V = 0$$

Therefore,

$$V = 50x - 68 \text{ lb}$$

$$-868x + \frac{50(x - 4)^2}{2} + 1000(x - 8) + M = 0$$

Therefore,

$$M = -25x^2 + 68x + 7600 \text{ ft-lb}$$

The next interval is between the end of uniform loading and the point couple. We can now replace the uniform loading by its resultant of 400 lb, as shown in Fig. 6.32(d). Thus,

$\underline{12 < x \leq 18:}$

$$868 - 400 - 1000 + V = 0$$

Therefore,

$$V = 532 \text{ lb}$$

$$-868x + 1400(x - 8) + M = 0$$

Therefore,

$$M = -532x + 11{,}200 \text{ ft-lb}$$

The last interval goes from the point couple to the right support. It is to be pointed out that the point couple *does not* contribute directly to the shear force and we could have used the above formulation for V for interval $18 < x \leq 22$. However, the couple *does* contribute directly to the bending moment, thus requiring the additional interval. Accordingly, using Fig. 6.32(e), we get

$$V = 532 \text{ lb} \quad \text{(as in previous interval)}$$

Whereas for M we have

$$-868x + 1400(x - 8) - 500 + M = 0$$

Therefore,

$$M = -532x + 11{,}700 \text{ ft-lb}$$

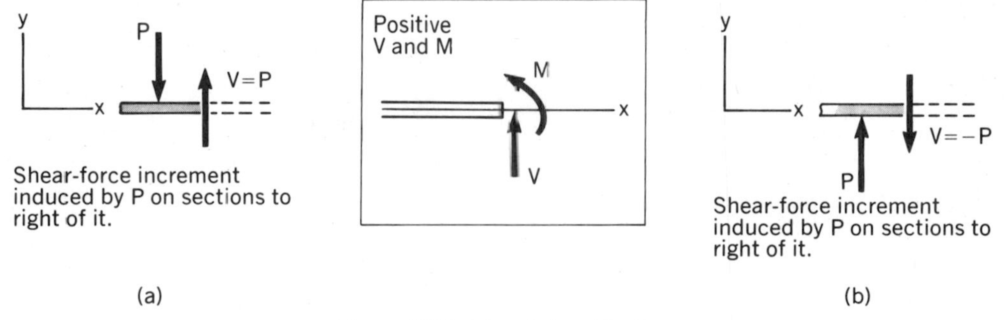

Figure 6.33. Shear induced by P.

We wish to point out now that we can determine shear-force and bending-moment equations in a less formal manner than what has been shown thus far. In this connection, it will be useful to note that a downward force P as shown in Fig. 6.33(a), induces on sections to the right a positive shear force (see insert) of value $+P$, whereas an upward force of P induces on sections to the right of it a negative shear force $-P$ [see Fig. 6.33(b)]. Also, an upward force P induces on sections at a distance ξ to the right of it a positive bending moment $P\xi$ [see Fig. 6.34(a)], whereas a downward force

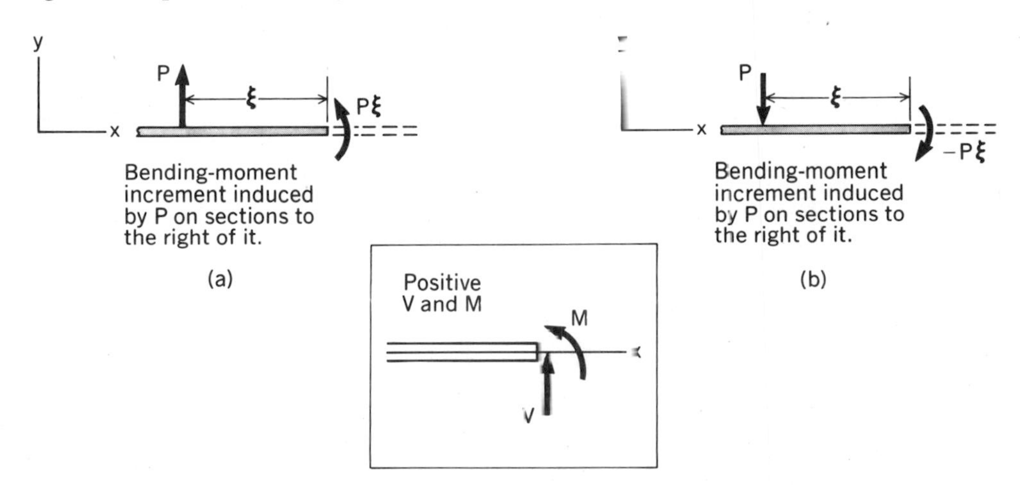

Figure 6.34. Bending moment induced by P.

P induces on sections ξ to the right of it a negative bending moment $-P\xi$ [Fig. 6.34(b)]. Finally, as can be seen in Fig. 6.35(a), a clockwise couple moment C induces a positive bending moment $+C$ on the sections to the right of it (it does not induce a shear force), whereas a counterclockwise couple moment C [Fig. 6.35(b)] induces a negative bending moment $-C$ on sections to the right of it. In the following example, we shall show how by this reasoning we may more directly formulate the shear-force and bending-moment equations.

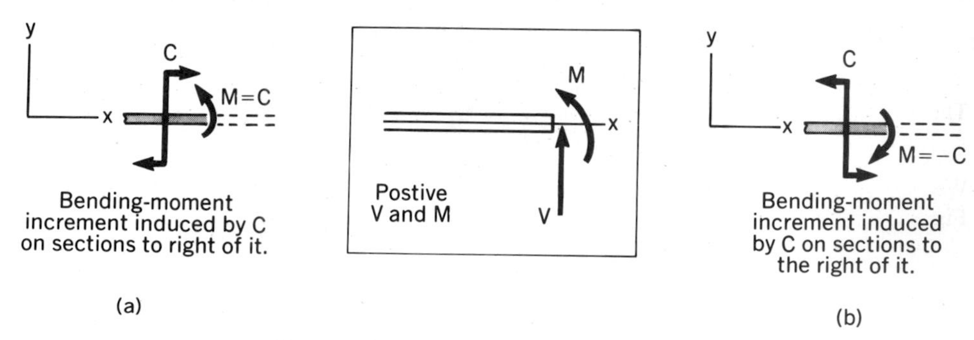

Figure 6.35. Bending moment induced by C.

EXAMPLE 6.7

Evaluate the shear-force and bending-moment equations for the beam shown in Fig. 6.36.

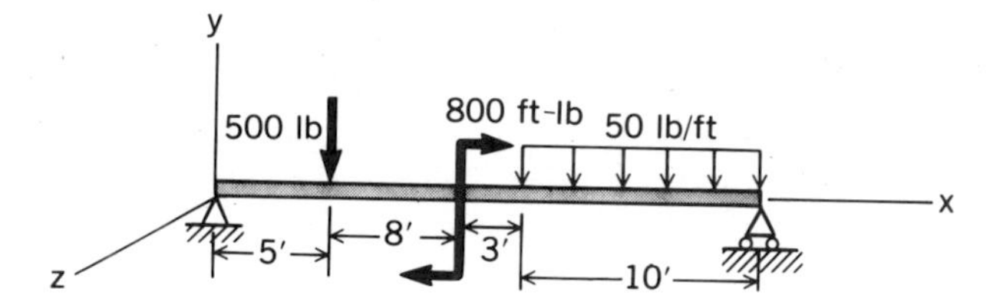

Figure 6.36. Simply supported beam.

A free-body diagram of the beam is shown in Fig. 6.37. We can immediately compute the supporting forces as follows:

$\underline{\sum M_2 = 0:}$

$$-R_1(26) + (500)(21) - 800 + (500)(5) = 0$$

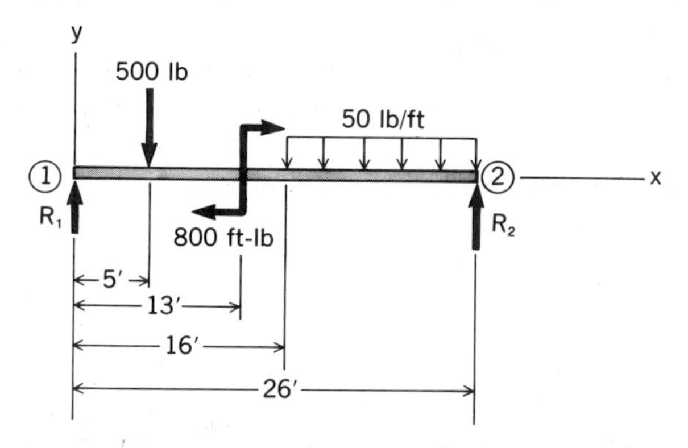

Figure 6.37. Free-body diagram of beam.

Therefore,

$$R_1 = 469 \text{ lb}$$

$\underline{\sum M_1 = 0:}$

$$R_2(26) - (500)(21) - 800 - (500)(5) = 0$$

Therefore,

$$R_2 = 531 \text{ lb}$$

We shall now directly give the shear force V and bending moment M while viewing Fig. 6.37. Thus,

$\underline{0 < x < 5:}$

$$V = -469 \text{ lb}$$

$$M = 469x \text{ ft-lb}$$

$\underline{5 < x < 13:}$

$$V = -469 + 500 = 31 \text{ lb}$$

$$M = 469x - 500(x - 5) = -31x + 2500 \text{ ft-lb}$$

$\underline{13 < x \leq 16:}$

$$V = 31 \text{ lb} \quad \text{(same as previous interval)}$$

$$M = 469x - 500(x - 5) + 800 = -31x + 3300 \text{ ft-lb}$$

$\underline{16 \leq x < 26:}$

$$V = -469 + 500 + 50(x - 16) = -769 + 50x \text{ lb}$$

$$M = 469x - 500(x - 5) + 800 - \frac{50(x - 16)^2}{2} = -25x^2 + 769x - 3100 \text{ ft-lb}$$

We shall present effective methods of sketching the shear-force and bending-moment diagrams in Section 6.7.

Before we proceed further, it must be carefully pointed out that the replacement of a distributed load by a single resultant force is only meaningful for the particular free body on which the force distribution acts. Thus, to compute the reactions for the entire beam taken as a free body (Fig. 6.38), we can replace the weight distribution w_0 by the total weight at position $L/2$ (Fig. 6.39). For the bending moment at x, the resultant of the loading for the free body shown in Fig. 6.40 becomes wx and is midway at position $x/2$. In other words, *in making shear-force and bending-moment equations and diagrams, we cannot replace loading distributions over the entire beam by a resultant and then proceed*; there is inherent in these equations an infinite number of free bodies, each smaller than the beam itself, which makes the abovementioned replacements invalid for shear-force and bending-moment considerations.

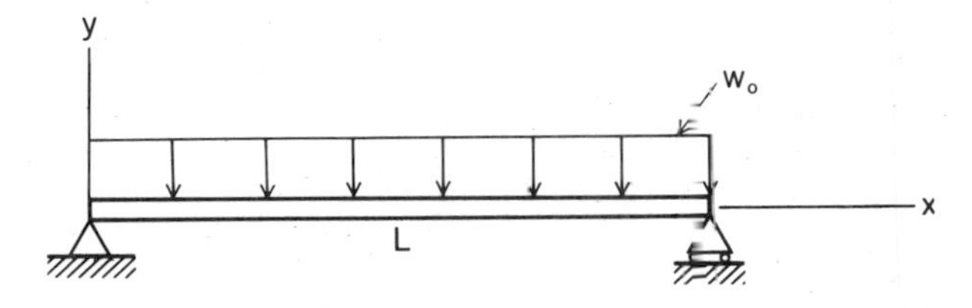

Figure 6.38. Uniform loading.

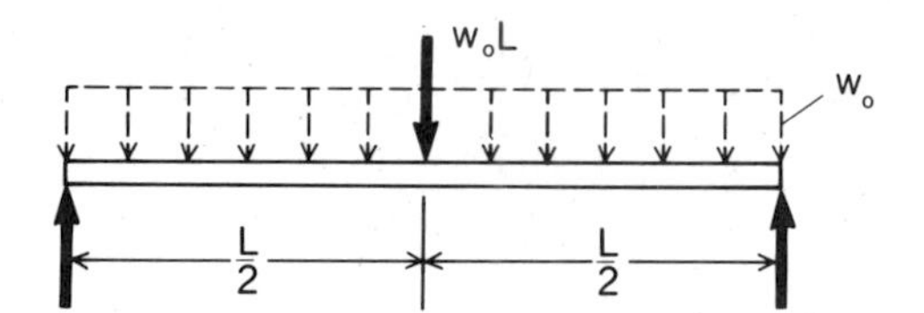

Figure 6.39. Resultant for w_0 for entire beam.

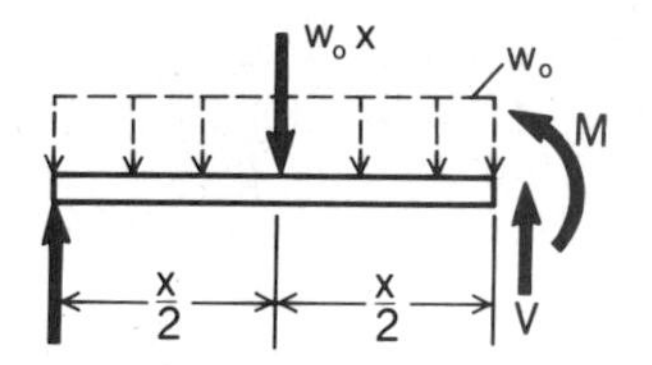

Figure 6.40. Resultant for w_0 for portion x of beam.

Problems

In Problems 6.29 through 6.40 make use of free-body diagrams.

6.29. Formulate the shear-force and bending-moment equations for the simply supported beam. Do not include the weight of the beam.

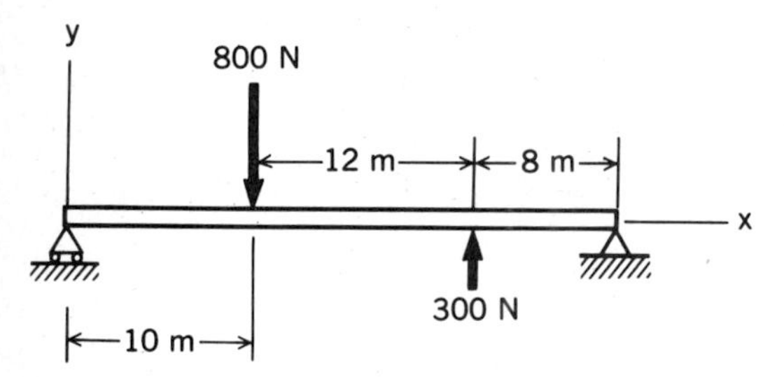

Figure P.6.29

6.30. Formulate the shear-force and bending-moment equations for the cantilever beam. Do not include the weight of the beam.

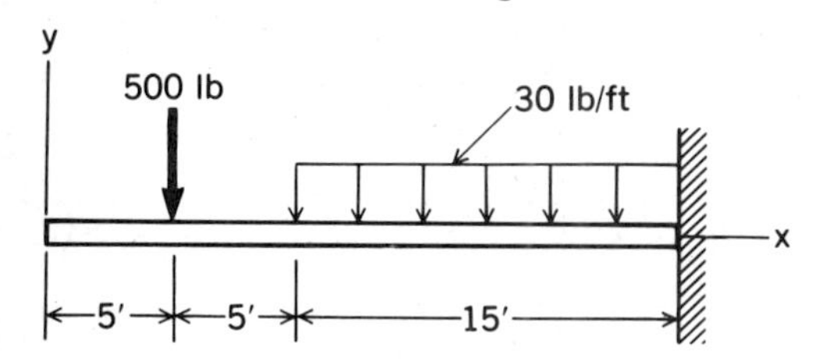

Figure P.6.30

6.31. Determine the shear-force and bending-moment equations for the simply supported beam.

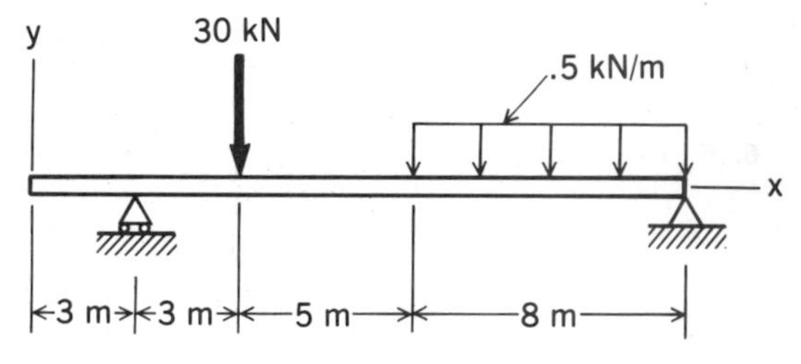

Figure P.6.31

6.32. For the beam shown, what is the shear force and bending-moment at the following positions?

(a) 5 ft from the left end
(b) 12 ft from the left end
(c) 5 ft from the right end

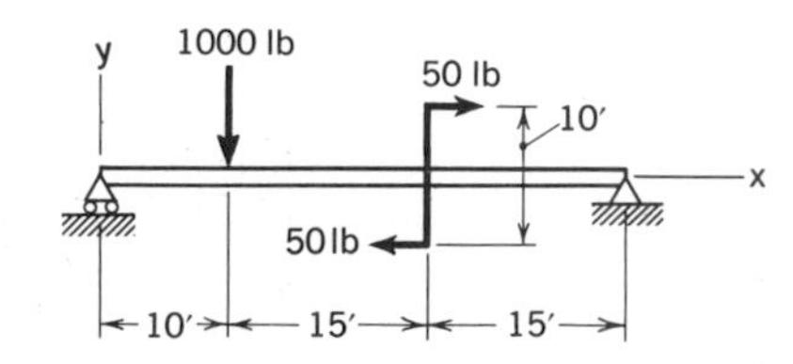

Figure P.6.32

6.33. Formulate the shear-force and b nding-moment equations for the simply supported beam.

Figure P.6.33

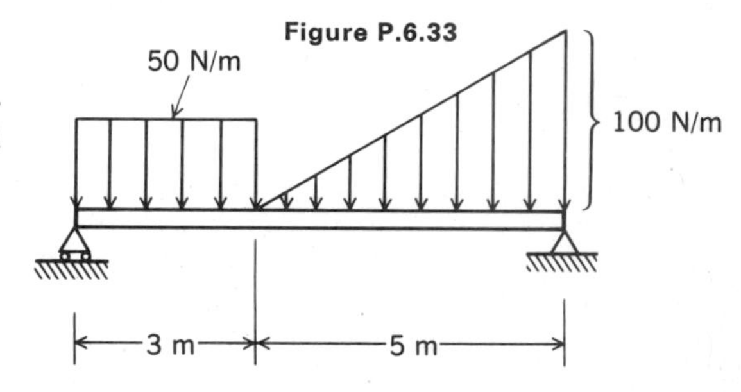

6.34. Compute shear force and bending moments for the bent beam as functions of s along the centerline of the beam.

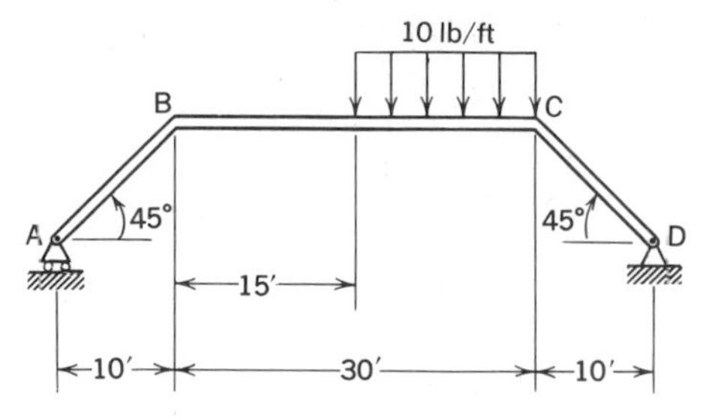

Figure P.6.34

6.35. A simply supported beam is loaded in two planes. This means there will be shear-force components V_y and V_z and bending-moment components M_z and M_y. Compute these as functions of x. The beam is 40 ft in length.

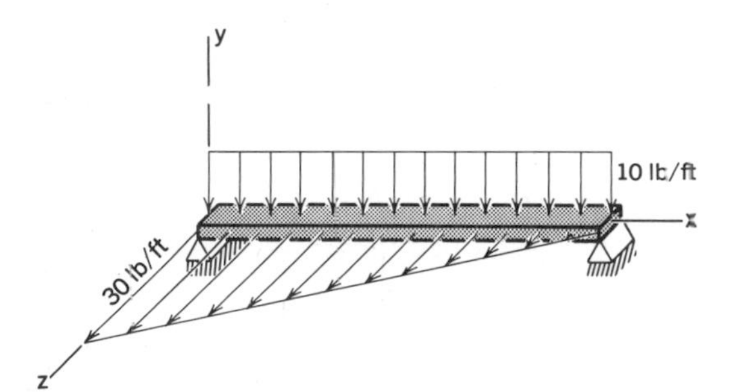

Figure P.6.35

6.36. What are the shear force, bending moment, and axial force for the three-dimensional cantilever beam? Give your results separately for the three portions AB, BC, and CD. Neglect the weight of the member. Use s as the distance along the centerline from D.

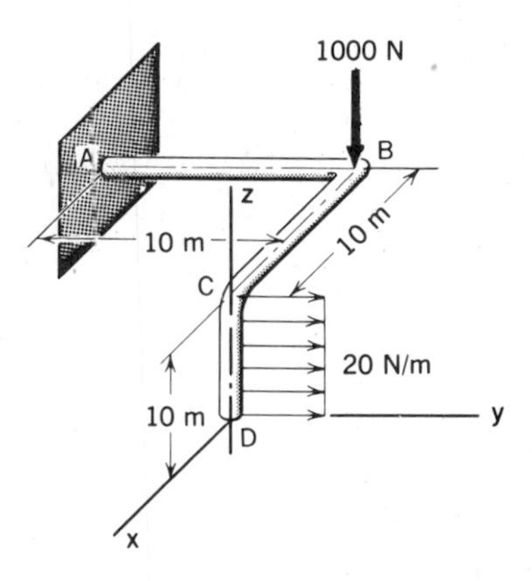

Figure P.6.36

6.37. Oil flows from a tank through a pipe AB. The oil weighs 40 lb/ft³ and, in flowing, develops a drag on the pipe of 1 lb/ft. The pipe has an inside diameter of 3 in. and a length of 20 ft. Flow conditions are assumed to be the same along the entire length of the pipe. What are the shear force, bending moment, and axial force along the pipe? The pipe weighs 10 lb/ft.

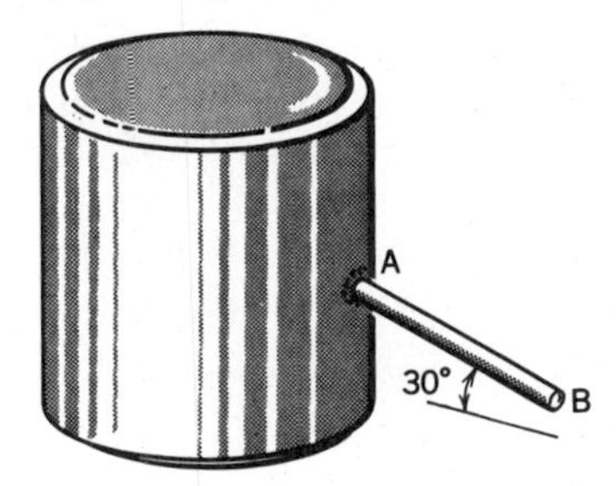

Figure P.6.37

6.38. Determine the shear force, bending moment, and axial force as functions of θ for the circular beam.

Figure P.6.38

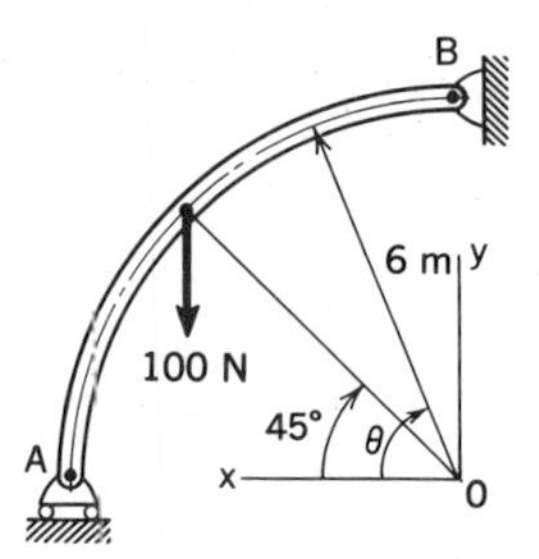

6.39. A hoist can move along a beam while supporting a 10,000-lb load. If the hoist starts at the left and moves from $\bar{x} = 3$ to $\bar{x} = 12$, determine the shear force and bending moment at A in terms of $\bar{x}$. At what position $\bar{x}$ do we get the maximum shear force at A and the maximum bending moment at A? What are their values?

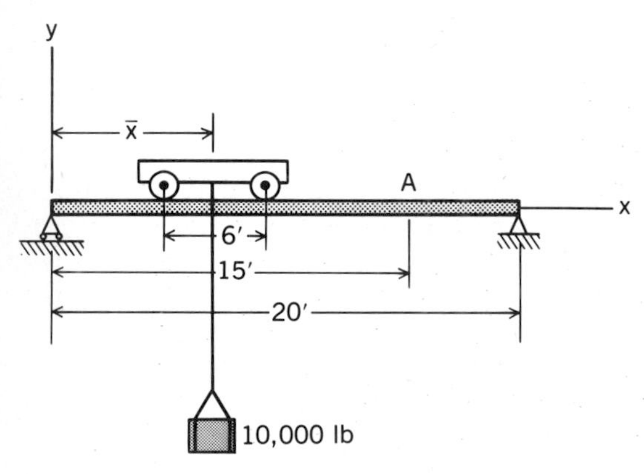

Figure P.6.39

6.40. A pipe weighs 10 lb/ft and has an inside diameter of 2 in. If it is full of water and the pressure of the water is that of the atmosphere at the entrance A, compute the shear force, axial force, and bending moment of the pipe from A to D. Use coordinate s measured from A along the centerline of the pipe.

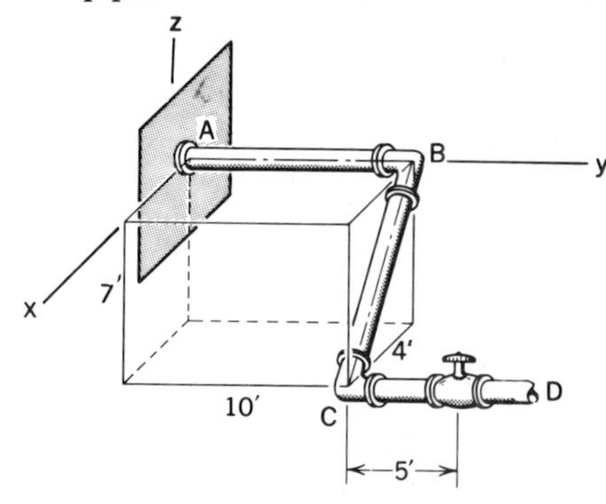

Figure P.6.40

6.41. After finding the supporting forces, determine for Problem 6.29 the shear-force and bending-moment equations without the further aid of free-body diagrams.

6.42. Determine the shear-force and bending-moment equations for Problem 6.30 without the aid of free-body diagrams.

6.43. In Problem 6.31, after determining the supporting forces, determine the shear-force and bending-moment equations without the aid of free-body diagrams.

6.44. In Problem 6.32, after finding the supporting forces, write the shear force and bending moment as a function of x for the beam without the aid of free-body diagrams.

6.45. Give the shear-force and bending-moment equations for the cantilever beam. Except for determining the supporting forces, do not use free-body diagrams.

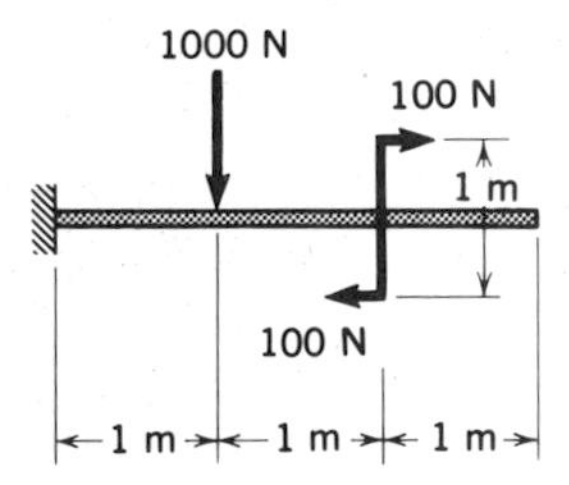

Figure P.6.45

6.46. Formulate the shear-force and bending-moment equations for the simply supported beam. (*Suggestion:* For the domain $5 < x < 10$, it is simplest to replace the indicated downward triangular load, going from 400 N/m to zero, by a uniform 400-N/m uniform downward load from $x = 5$ to $x = 10$ plus a triangular upward load going from zero to 400 N/m in the interval.)

Figure P.6.46

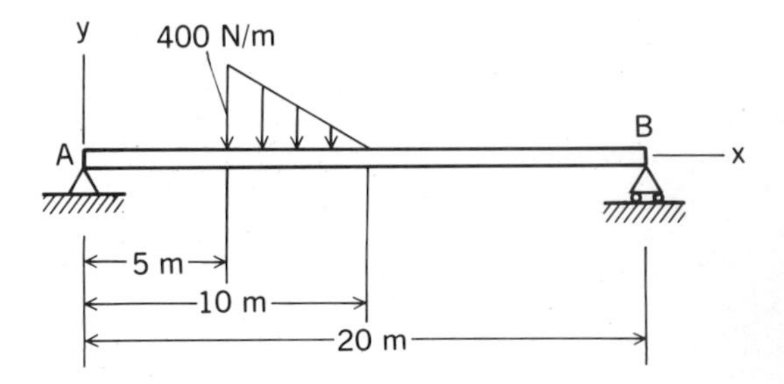

6.47. After finding the supporting forces for the simply supported beam AB, express the shear-force and bending-moment equations without the aid of free-body diagrams. The 10-kN load is applied to a bracket welded to the beam AB.

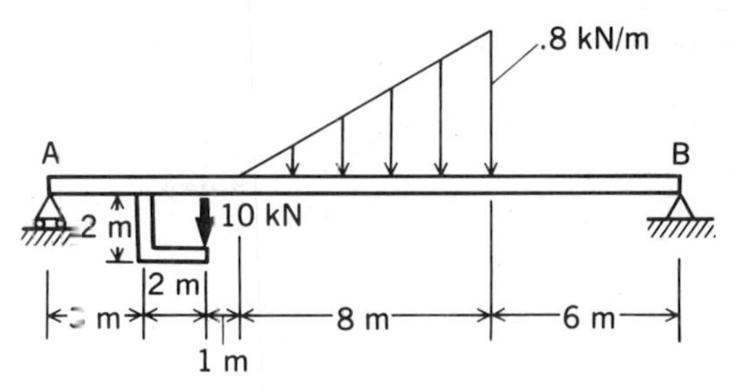

Figure P.6.47

6.8 Differential Relations for Equilibrium

In Section 6.7, we considered free bodies of *finite size* comprising variable portions of a beam in order to ascertain the resultant force system at sections along the beam. We shall now proceed in a different manner by examining an *infinitesimal slice* of the beam. Equations of equilibrium for this slice will then yield *differential equations* rather than algebraic equations for the variables V and M.

Consider a slice Δx of the beam shown in Fig. 6.41. We adopt the convention that intensity of loading w in the positive coordinate direction is positive. We shall assume here that the weight of the beam has been included in the intensity of loading so that all forces acting on the element have been shown on the free-body diagram of the element in Fig. 6.42. Note we have employed positive shear-force and bending-moment

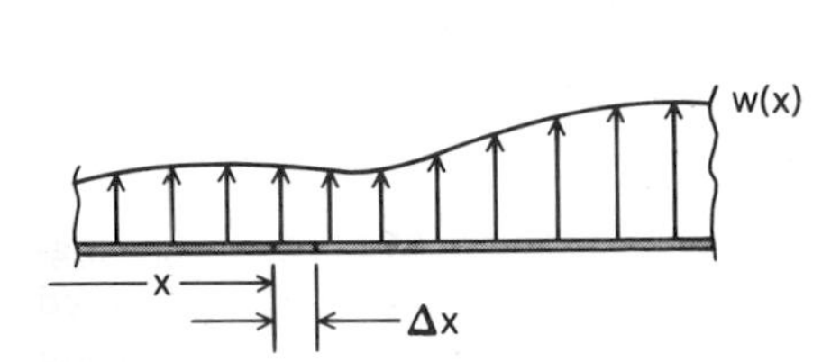

Figure 6.41. Element Δx of beam.

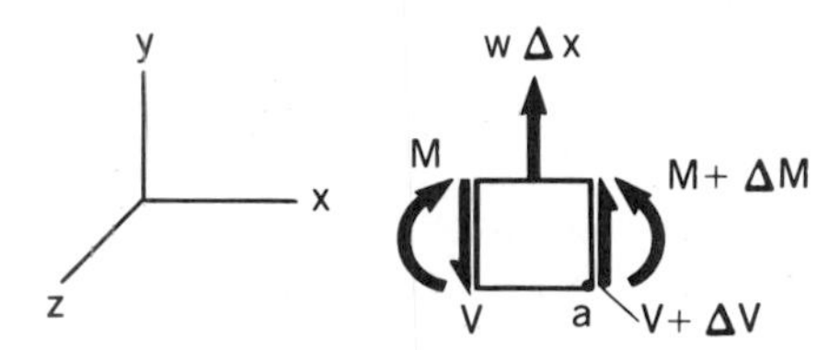

Figure 6.42. Free-body diagram of element

conventionwise in the free-body diagram for reasons explained earlier. We now apply the equations of equilibrium. Thus, summing forces:

$\sum F_y = 0$:

$$-V + (V + \Delta V) + w\,\Delta x = 0$$

Taking moments about edge a of the element, we get

$\sum M_a = 0$:

$$-M + V\,\Delta x - (w\,\Delta x)(\beta\,\Delta x) + (M + \Delta M) = 0$$

where β is some fraction which, when multiplied by Δx, gives the proper moment arm of the force $w\,\Delta x$ about edge a. These equations can be written in the following manner after we cancel terms and divide through by Δx

$$\frac{\Delta V}{\Delta x} = -w$$

$$\frac{\Delta M}{\Delta x} = -V + w\beta\,\Delta x$$

In the limit as $\Delta x \longrightarrow 0$, we get the following differential equations:

$$\boxed{\begin{aligned} \frac{dV}{dx} &= -w \qquad \text{(a)} \\ \frac{dM}{dx} &= -V \qquad \text{(b)} \end{aligned}} \tag{6.3}$$

We may next integrate Eqs. 6.3(a) and 6.3(b) from position 1 along the beam to position 2. Thus, we have

$$(V)_2 - (V)_1 = -\int_1^2 w\,dx$$

Therefore,

$$\boxed{(V)_2 = (V)_1 - \int_1^2 w\,dx} \tag{6.4}$$

$$(M)_2 - (M)_1 = -\int_1^2 V\,dx$$

Therefore,

$$\boxed{(M)_2 = (M)_1 - \int_1^2 V\,dx} \tag{6.5}$$

Equation (6.4) means that the change in the shear force between two points on a beam equals minus the area under the loading curve between these points provided that there is no point force present in the interval.[11] Note that, if $w(x)$ is positive in an interval, the area under this curve is positive in this interval; if $w(x)$ is negative in an interval, the area under this curve is negative in this interval. Similarly, Eq. 6.5 indicates that the change in bending moment between two points on a beam equals minus the area of the shear-force diagram between these points provided that there are no point couple moments applied in the interval. If $V(x)$ is positive in an interval, the area under this curve is positive in this interval; if $V(x)$ is negative in an interval, the area under the curve is negative for this interval. In sketching the diagram, we shall make use of Eqs. 6.4 and 6.5 as well as the differential equations 6.3.

EXAMPLE 6.8

Sketch the shear-force and bending-moment distributions for the simply supported beam shown in Fig. 6.43 and label the key points.

---

[11]The differential equation 6.3(a) is only meaningful with a continuous loading present, while Eq. 6.3(b) is only valid in the absence of point couple moments.

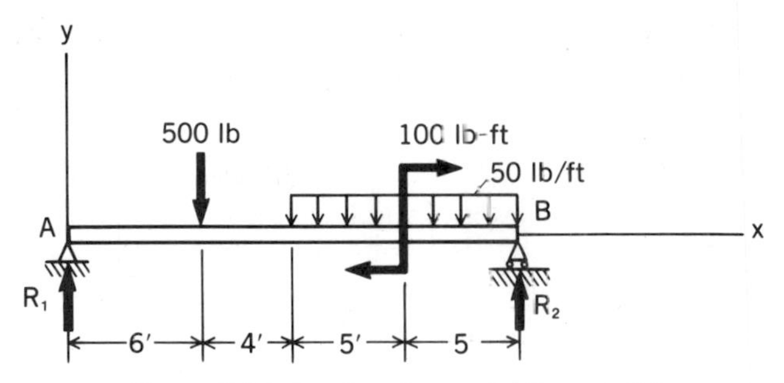

Figure 6.43. Loading diagram for Example 6.8.

The supporting forces R_1 and R_2 are found by rules of statics. Thus,

$\underline{\sum M_B = 0:}$

$$-R_1(20) + (500)(14) + (50)(10)(10/2) - 100 = 0$$

Therefore,

$$R_1 = 470 \text{ lb}$$

$\underline{\sum M_A = 0:}$

$$R_2(20) - (500)(6) - (50)(10)(15) - 100 = 0$$

Therefore,

$$R_2 = 530 \text{ lb}$$

In sketching the diagrams, we shall employ Eqs. 6.3, 6.4, and 6.5—i.e., the differential equations of equilibrium and their integrals. Accordingly, we first draw the loading diagram in Fig. 6.44(a), and we shall then sketch the shear-force and

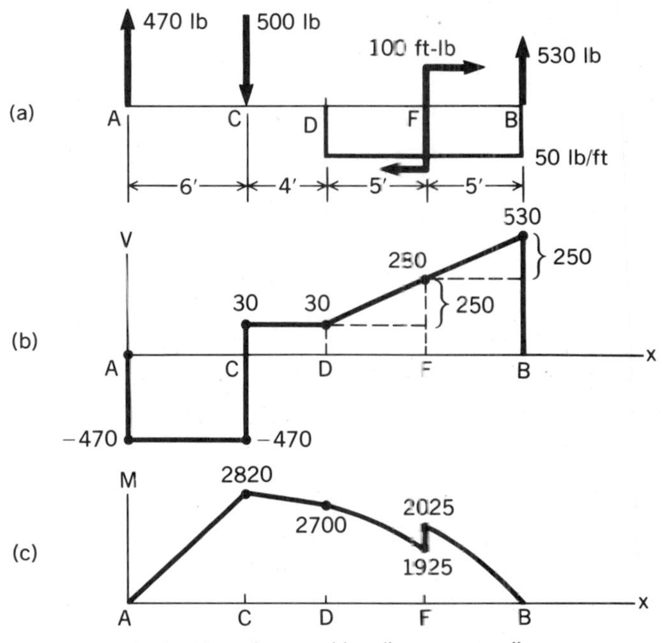

Figure 6.44. Shear-force and bending-moment diagrams.

bending-moment diagrams directly below without the aid of the shear-force and bending-moment equations, evaluating key points as we go.

Note as we start on the shear diagram that the 470-lb supporting force induces a negative shear of -470 lb just to the right of the support. Now from A to C, the area under the loading curve is zero and so, in accordance with Eq. 6.4, there is no change in the value of shear between A and C. Hence, $V_C = -470$ lb, as shown in Fig. 6.44(b). Also, since $w = 0$ between A and C, the slope of the shear curve should be zero, in accordance with Eq. 6.3(a). And so we have a horizontal line for V between A and C. Now as we cross C, the 500-lb downward force will induce a positive increment of shear of value 500 on sections to the right of it. Accordingly, V *jumps* from -470 lb to $+30$ lb as we cross C. Between C and D there is no loading w, so $V_D = V_C$ and we have a 30-lb shear force at point D. Again, since $w = 0$ in this interval, the slope of the shear curve is zero and we have a horizontal line for the shear curve between C and D. Since there is no concentrated load at D, there is no sudden change in shear as we cross this point. Next, the change in shear between D and B is minus the area of the loading curve[12] in this interval in accordance with Eq. 6.4. But this area is $(-50)(10) = -500$. Hence, from Eq. 6.4 the value of V_B (just to the left of the support) is $V_D - (-500) = 530$ lb. Also, since w is negative and constant between D and B, the slope of the shear curve should be positive and constant, in accordance with Eq. 6.3(a). Hence, we can draw a straight line between $V_D = 30$ lb and $V_B = 530$ lb. As we now cross the right support force, we see that it induces a negative shear of 530 lb on sections to the right of the support, and so at B the shear curve comes back to zero.

We now proceed with the bending-moment curve. With no point couple moment present at A, the value of M_A must be zero. The change in moment between A and C is then minus the area underneath the shear curve in this interval. We can then say from Eq. 6.5 that $M_C = M_A - (-470)(6) = 0 + 2820 = 2820$ ft-lb, and we denote this in the moment diagram. Furthermore the value of V is a negative constant in the interval and, accordingly [see Eq. 6.3(b)], the slope of the moment curve is positive and constant. We can then draw a straight line between M_A and M_C. Between C and D the area for the shear diagram is 120 lb-ft, and so we can say that $M_D = M_C - (120) = 2820 - 120 = 2700$ ft-lb. Again, with V constant and positive in the interval, the slope of the moment curve must be negative and constant in the interval and has been so drawn. Between D and F the area under the shear curve is readily seen to be $(30)(5) + \frac{1}{2}(5)(250) = 775$ ft-lb. Hence, the bending moment goes from 2700 ft-lb at D to 1925 ft-lb at F. Now the shear curve is positive and *increasing* in value as we go from D to F. This means that the slope of the bending-moment curve is negative and becoming *steeper* as we go from D to F. As we go by F we encounter the 100-ft-lb point couple moment and we can say that this point couple moment induces a positive 100-ft-lb moment on sections to the right of point F. Accordingly, there is a sudden increase in bending moment of 100 ft-lb at F, as has been shown in the diagram. The area of the shear diagram between F and B is readily seen from Fig. 6.44(b) to be $(280)(5) + \frac{1}{2}(5)(250) = 2025$ ft-lb. We see then that the

[12]Note that the point couple moment has a zero net force and so need not be of concern in the interval from D to B as far as shear is concerned. However, it will be a point where sudden change occurs in the bending-moment diagram.

bending moment goes to zero at B. Since the shear force is positive and *increasing* between F and B, we conclude that the slope of the bending-moment curve is negative and becoming *steeper* as we approach B. We have thus drawn the shear-force and bending-moment diagram and have labeled all key points.

Note that to be correct both the shear-force and bending-moment curves must go to zero at the end of the beam to the right of the right support. This serves as a check on the correctness of the calculations.

In Example 6.8, we can get equations and diagrams of shear force and bending moment independently of each other. With simple loadings such as point forces, point couples, and uniform distributions, this can readily be done. Indeed, this covers many problems that occur in practice. Usually, all that is needed are the labeled diagrams of the kind that we set forth in the previous problem. In problems with more complex loadings, we usually set forth the equations in the customary manner and then sketch the curves using the *equations* to give key values of V and M (the areas for the various curves are no longer the simple familiar ones, thus precluding advantageous use of Eqs. 6.4 and 6.5); the key points are then connected by curves sketched by making use of the slope relations as in Example 6.8.

It will be helpful to remember that if a curve has *increasing* negative[13] or positive values, the subsequent curve must have a *steepening* slope over the corresponding range. On the other hand, if a curve has *decreasing* negative or positive values, the subsequent curve must have a *flattening* slope over the corresponding range.

You will note in the preceding examples that the key points of the shear-force and bending-moment diagrams were evaluated and marked. The maximum value of both the shear force and bending moment were easily depicted from these diagrams. We wish to note, in this regard, that at points on shear-force and bending-moment curves where there is zero value of slope, there may be possible maximum values of shear

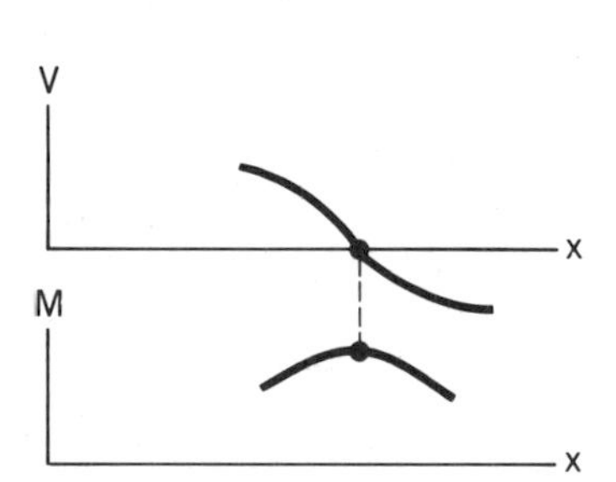

Figure 6.45. At $w = 0$, possible maximum for V.

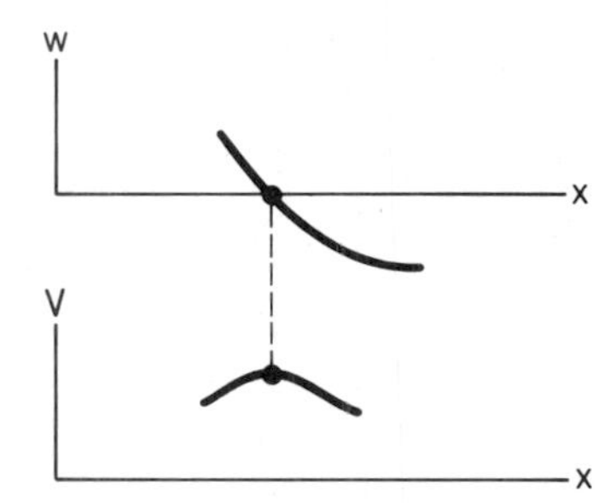

Figure 6.46. At $V = 0$, possible maximum for M.

force and bending moment, respectively, for the beam. This is illustrated for shear force in Fig. 6.45 and for bending moment in Fig. 6.46. Note that where the loading

[13]By an increasing negative value, we mean here an increasing absolute value. That is, -200 is considered larger negatively than -100.

curve w crosses the x axis, we accordingly have the position of a possible maximum shear force; similarly, where the shear curve V crosses the x axis, we accordingly have the position of a possible maximum bending moment. These respective positions and corresponding values of shear force and bending moment should be evaluated and marked in the diagram.

Problems

6.48. After finding the supporting forces of the cantilever beam, sketch the shear-force and bending-moment diagrams labeling key points.

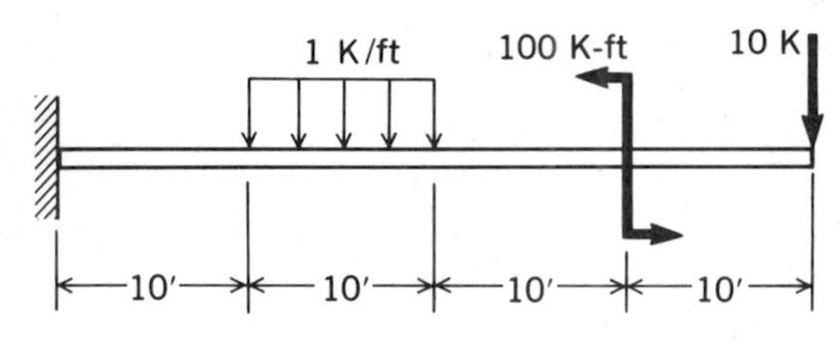

Figure P.6.48

6.49. What is the maximum negative bending moment in the region between the supports for the simply supported beam?

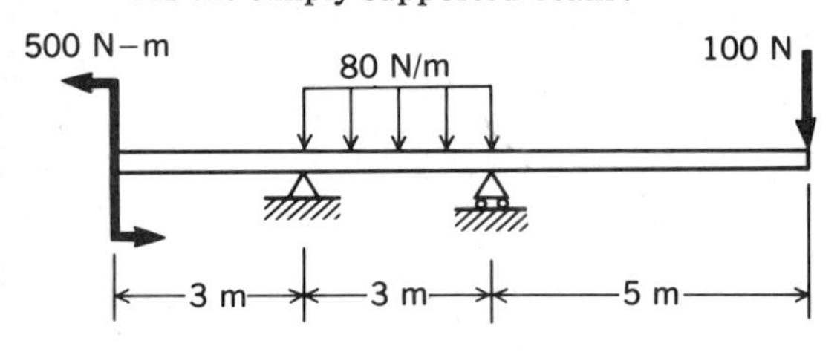

Figure P.6.49

6.50. Find the supporting forces for the simply supported beam. Then sketch the shear-force and bending-moment diagrams, labeling key points.

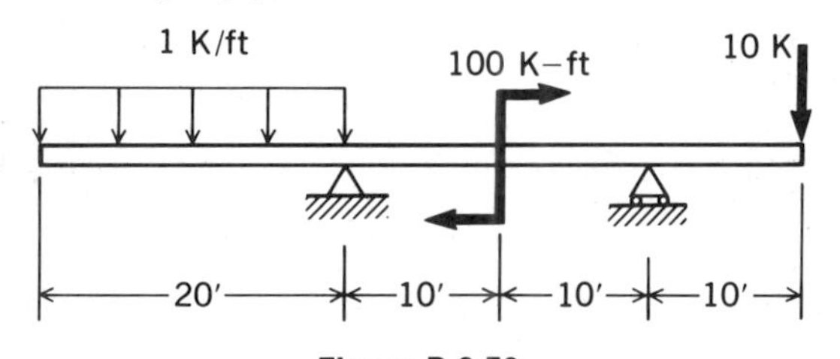

Figure P.6.50

6.51. Sketch the shear-force and bending-moment diagrams and compute key points for the overhanging beam.

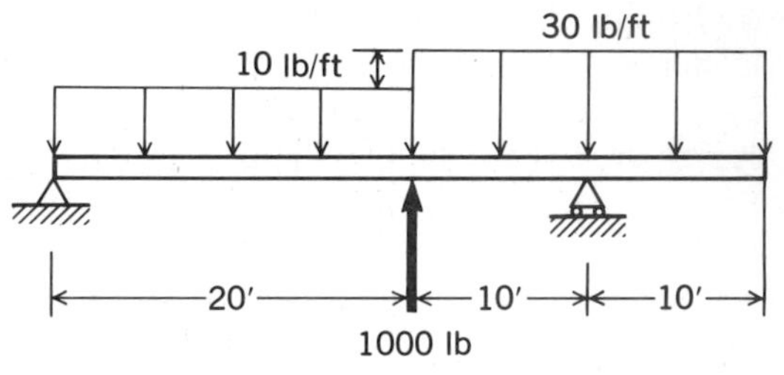

Figure P.6.51

6.52. A simply supported beam AB is shown. A bar CD is welded to the beam. After determining the supporting forces, sketch the shear-force and bending-moment diagram and determine the maximum bending moment. (*Hint:* Find the position for $V = 0$ using similar triangles.)

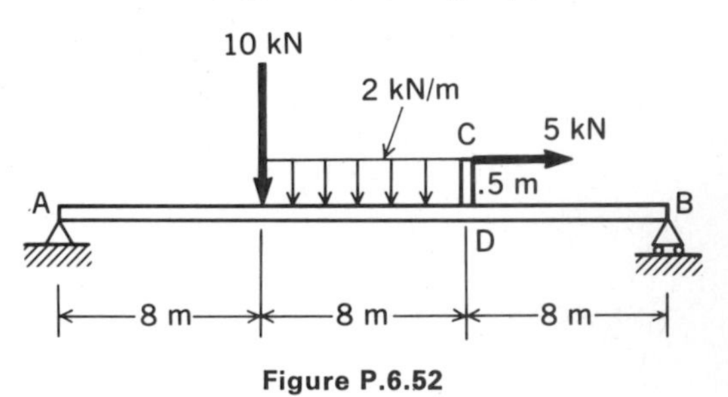

Figure P.6.52

6.53. Show the shear-force and bending-moment diagrams and evaluate key points only for the cantilever beam.

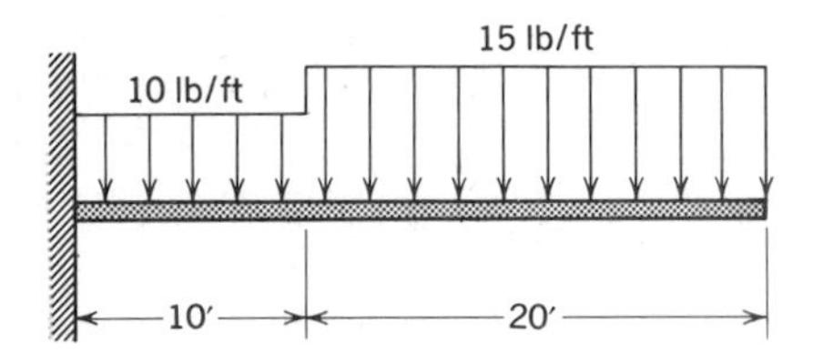

Figure P.6.53

6.54. Sketch the shear-force and bending-moment diagrams for the sinusoidally loaded beam. What is the maximum bending moment?

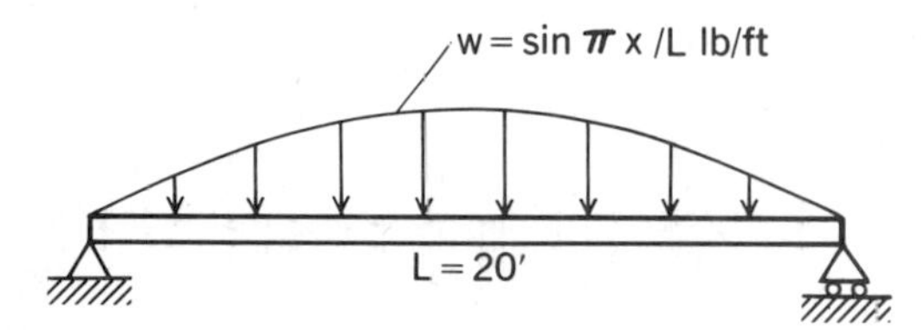

Figure P.6.54

6.55. Formulate the shear-force and bending-moment equations for the beam. Sketch the shear and moment diagrams.

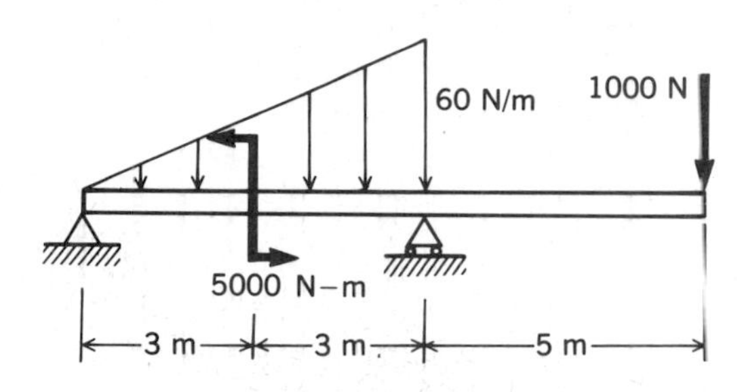

Figure P.6.55

6.56. A simply supported I-beam is shown. A hole must be cut through the web to allow passage of a pipe that runs horizontally at right angles to the beam.

(a) Where, within the marked 24-ft section, would the hole least affect the moment-carrying capacity of the beam?

(b) In the same marked section, where should the hole go to least affect the shear-carrying capacity of the beam?

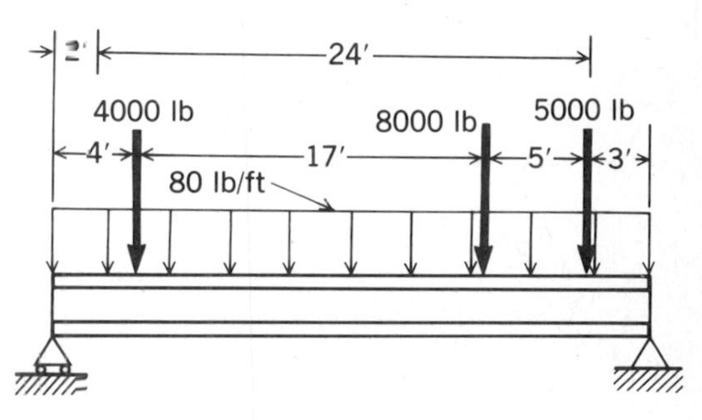

Figure P.6.56

***6.57.** A cantilever beam supports a parabolic and a triangular load. What are the shear-force and bending-moment equations? Sketch the shear-force and bending-moment diagrams. See the suggestion in Problem 6.46 regarding the triangular load.

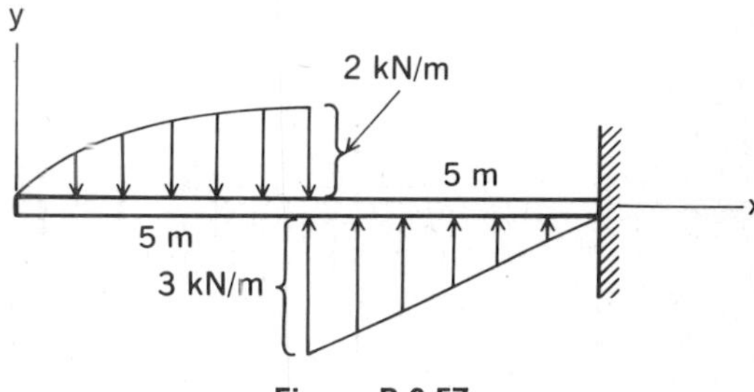

Figure P.6.57

6.58. Determine the shear-force and bending-moment equations for the beam. Then sketch the diagrams using the aforementioned equations if necessary to ascertain key points in the diagrams, such as the position between the supports where $V = 0$. What is the bending moment there?

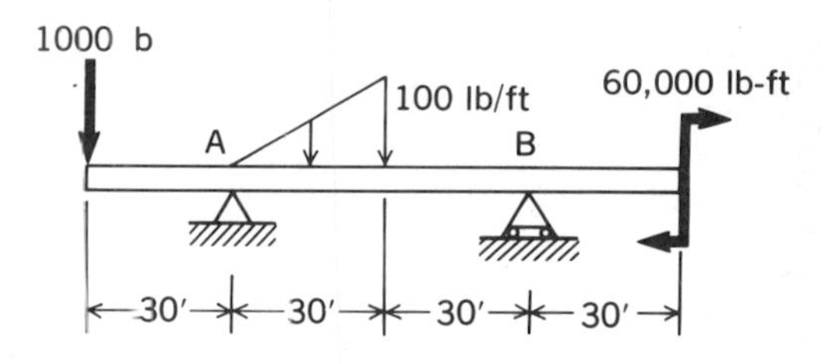

Figure P.6.58

Part C
Chains and Cables

6.9 Introduction

We often encounter relatively flexible cables or chains that are used to support loads. In suspension bridges, for example, we find a coplanar arrangement in which a cable supports a large load. The weight of the cable itself in such cases may often be considered negligible. In transmission lines, on the other hand, the principal force is the weight of the cable itself. In this section, we shall evaluate the shape of and the tension in the cables for both these cases.

To facilitate computations, the model of the structural system will be assumed to be perfectly flexible and inextensible. The flexibility assumption means that at the center of any cross section of the cable only a tensile force is transmitted and there can be no bending moment there. The force transmitted through the cable must, under these conditions, be tangent to the cable at all positions along the cable. The inextensibility assumption means that the length of the cable is constant.

6.10 Coplanar Cables

We shall now consider the case of a cable suspended between two rigid supports A and B under the action of a loading function $w(x)$ given per unit length as measured in the *horizontal* direction. This loading will be considered to be coplanar with the cable and directed vertically, as shown in Fig. 6.47. Consider an element of the cable of length

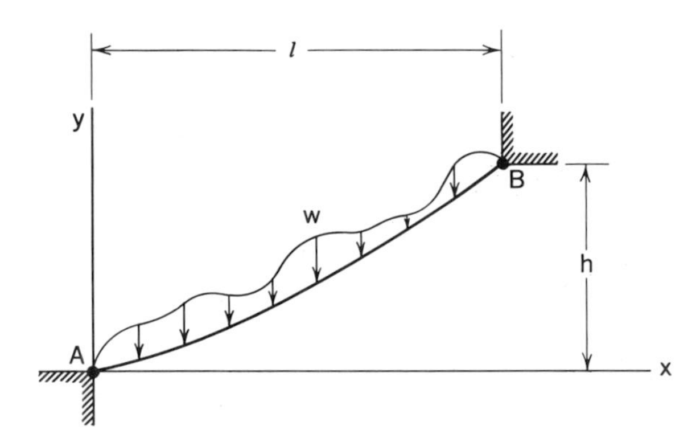

Figure 6.47. Coplanar cable; $w = w(x)$.

Δs as a free body (Fig. 6.48). Summing forces in the x and y directions, respectively, we get

$$-T\cos\theta + (T + \Delta T)\cos(\theta + \Delta\theta) = 0 \quad (6.6a)$$

$$-T\sin\theta + (T + \Delta T)\sin(\theta + \Delta\theta) - w_{av}\,\Delta x = 0 \quad (6.6b)$$

where w_{av} is the average loading. Dividing by Δx and taking the limit as $\Delta x \longrightarrow 0$, we have

$$\lim_{\Delta x \to 0}\left[\frac{(T + \Delta T)\cos(\theta + \Delta\theta) - T\cos\theta}{\Delta x}\right] = 0$$

$$\lim_{\Delta x \to 0}\left[\frac{(T + \Delta T)\sin(\theta + \Delta\theta) - T\sin\theta}{\Delta x}\right] = w$$

The term w is now the loading at position x. The left sides of the equations above are derivatives in accordance with elementary calculus, and so we can say for these equations:

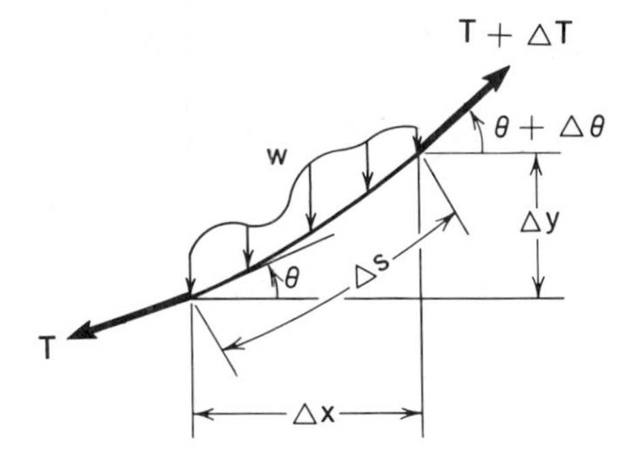

Figure 6.48. Element of cable.

$$\frac{d(T\cos\theta)}{dx} = 0 \quad (6.7a)$$

$$\frac{d(T\sin\theta)}{dx} = w \quad (6.7b)$$

From Eq. 6.7(a), we conclude that

$$T\cos\theta = \text{constant} = H \quad (6.8)$$

where clearly the constant H represents the *horizontal* component of the tensile force anywhere along the cable. Integrating Eq. 6.7(b), we get

$$T\sin\theta = \int w(x)\,dx + C_1' \quad (6.9)$$

where C_1' is a constant of integration. Solving for T in Eq. 6.8 and substituting into Eq. 6.9, we get

$$\frac{\sin\theta}{\cos\theta} = \frac{1}{H}\int w(x)\,dx + C_1$$

Noting that $\sin\theta/\cos\theta = \tan\theta = dy/dx$, we have, on carrying out a second integration:

$$\boxed{y = \frac{1}{H}\int\left[\int w(x)\,dx\right]dx + C_1 x + C_2} \quad (6.10)$$

Equation 6.10 is the deflection curve for the cable in terms of H, $w(x)$, and the constants of integration. The constants of integration must be determined by the boundary conditions at the supports A and B.

EXAMPLE 6.9

A cable is shown in Fig. 6.49 terminating at points at the same elevation. The loading distribution is uniform, given by constant w. Other known data are the span,

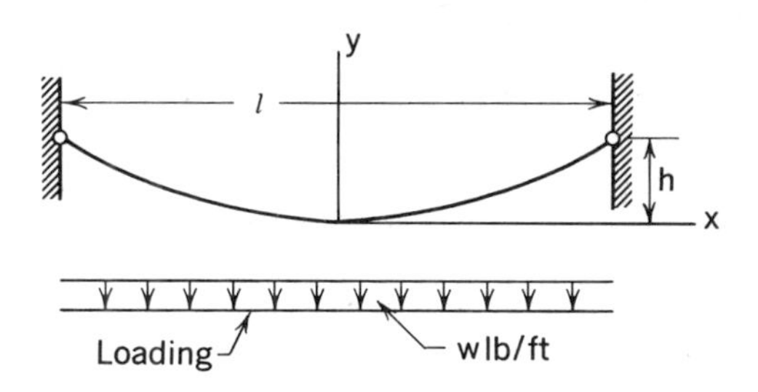

Figure 6.49. Cable with sag h.

l, and the sag, h. The maximum force in the cable, the shape of the cable, and the length of the cable are desired. Neglect the weight of the cable itself.

We have placed the reference at the center of the cable for simplicity as shown in the diagram. Noting that $w(x) = w =$ constant for this problem, we can proceed directly with the integrations in Eq. 6.10. Thus, we have

$$y = \frac{1}{H}\int\left[\int w\,dx\right]dx + C_1x + C_2 = \frac{1}{H}\int wx\,dx + C_1x + C_2$$

Therefore,

$$y = \frac{1}{H}\frac{wx^2}{2} + C_1x + C_2 \tag{a}$$

The deflection curve is thus a *parabola*. We now require that $y = dy/dx = 0$, when $x = 0$. Thus, the constants $C_1 = C_2 = 0$. The deflection curve then is simply

$$y = \frac{w}{2H}x^2 \tag{b}$$

To get the constant H, we set $y = h$ for $x = l/2$. Thus,

$$h = \frac{w}{2H}\frac{l^2}{4}$$

Therefore,

$$H = \frac{wl^2}{8h} \tag{c}$$

The deflection curve is now fully established in terms of the data of the problem in the form

$$y = \frac{w}{2(wl^2/8h)}x^2 = 4\frac{hx^2}{l^2} \tag{d}$$

We next compute the *maximum tension* in the cable. Equation 6.8 can be used for this purpose. Solving for T, we get

$$T = \frac{H}{\cos\theta} \tag{e}$$

from which is apparent that the maximum value of T occurs where θ is greatest. Examining the slope of the deflection curve,

$$\frac{dy}{dx} = \frac{w}{H}x \tag{f}$$

it is apparent that the largest θ occurs at $x = l/2$ (i.e., at the supports). Hence, from above we have, for θ_{max}:

$$\theta_{max} = \tan^{-1}\left(\frac{dy}{dx}\right)_{x=l/2} = \tan^{-1}\left(\frac{w}{H}\frac{l}{2}\right) \tag{g}$$

Consequently, we get for T_{max}:

$$T_{max} = \frac{H}{\cos\,[\tan^{-1}(wl/2H)]} \tag{h}$$

From trigonometric consideration of the denominator,

$$T_{max} = \frac{H(4H^2 + w^2l^2)^{1/2}}{2H} = H\left[1 + \left(\frac{wl}{2H}\right)^2\right]^{1/2} \tag{i}$$

Substituting for H using Eq. (c), we then get, on rearranging the terms,

$$T_{max} = \frac{wl}{2}\sqrt{1 + \left(\frac{l}{4h}\right)^2} \tag{j}$$

Finally, to determine the *length of the cable* for the given conditions, we must perform the following integration:

$$L = 2\int_0^{s_{max}} ds = 2\int_0^{s_{max}} \sqrt{dx^2 + dy^2} = 2\int_0^{l/2} \sqrt{1 + \left(\frac{dy}{dx}\right)^2}\, dx \tag{k}$$

Now the slope, dy/dx, equals wx/H [see Eq. (f)], which on substituting for H [see Eq.(c)] becomes $8hx/l^2$. Therefore,

$$L = 2\int_0^{l/2} \sqrt{1 + \left(\frac{8hx}{l^2}\right)^2}\, dx$$

This may be integrated using a formula to be found in Appendix I to give

$$L = \left[x\sqrt{1 + \left(\frac{8hx}{l^2}\right)^2} + \frac{l^2}{8h}\sinh^{-1}\frac{8hx}{l^2}\right]_0^{l/2}$$

Substituting limits, we have

$$L = \left[\frac{l}{2}\sqrt{1 + \left(\frac{4h}{l}\right)^2} + \frac{l^2}{8h}\sinh^{-1}\frac{4h}{l}\right]$$

Rearranging so that the result is given as a function of the sag ratio h/l and the span l, we get finally

$$L = \frac{l}{2}\left[\sqrt{1 + 16\left(\frac{h}{l}\right)^2} + \frac{1}{4h/l}\sinh^{-1}\frac{4h}{l}\right] \tag{l}$$

Another possible approach to determining the length of the cable is to expand the integrand in Eq. (k) as a power series using the *binomial theorem*. Thus, we have

$$L = 2\int_0^{l/2} \left[1 + \frac{1}{2}\left(\frac{dy}{dx}\right)^2 - \frac{1}{8}\left(\frac{dy}{dx}\right)^4 + \cdots\right] dx \tag{m}$$

provided that $|dy/dx| < 1$ at all positions along the interval.[14] Now, employ Eq.

[14]Otherwise, the series diverges. Hence, this approach is limited to cases where the slope of the cable is less than 45°.

(f) to replace dy/dx in Eq. (m) to get

$$L = 2\int_0^{l/2}\left(1 + \frac{1}{2}\frac{w^2}{H^2}x^2 - \frac{1}{8}\frac{w^4}{H^4}x^4 + \ldots\right)dx \tag{n}$$

We can integrate a power series term by term and so we have, for L:

$$L = l\left[1 + \frac{1}{24}\left(\frac{w^2}{H^2}\right)l^2 - \frac{1}{640}\left(\frac{w^4}{H^4}\right)l^4 + \ldots\right] \tag{o}$$

For cables having small slopes, the series converges rapidly and only the first few terms need generally be employed.

In Example 6.9, the supports are at the same level and consequently the position of zero slope is known (i.e., it is at the midpoint). We found it simplest to set our reference xy at this point. In problems at the end of this section, the supports may not be at the same level. For such cases the reference is best taken at one of the supports. Also, the slope of the cable is often known at some point, and the problem may then be solved in much the same way as Example 6.9.

In the previous development, the loading was given as a function of x. Let us now consider the case of a cable *loaded only by its own weight*. The loading function is now most easily expressed as a function of s, the position along the cable. Equations 6.6 apply to this case provided that we replace Δx by Δs. Dividing through by Δs and taking the limit as $\Delta s \longrightarrow$ zero, we get equations analogous to Eqs. 6.7.

$$\frac{d(T\cos\theta)}{ds} = 0$$

$$\frac{d(T\sin\theta)}{ds} = w(s)$$

Integrating, we have

$$T\cos\theta = H \tag{6.11a}$$

$$T\sin\theta = \int w(s)\,ds + C_1' \tag{6.11b}$$

Eliminating T from Eqs. 6.11, we get, as in the previous development:

$$\frac{dy}{dx} = \frac{1}{H}\int w(s)\,ds + C_1 \tag{6.12}$$

The right side of the equation is a function of s. Thus, we cannot directly integrate as a next step. Accordingly, note that

$$dy = (ds^2 - dx^2)^{1/2}$$

Hence, from this equation,

$$\frac{dy}{dx} = \left[\left(\frac{ds}{dx}\right)^2 - 1\right]^{1/2} \tag{6.13}$$

Substituting for dy/dx in Eq. 6.12 using the preceding result, we get

$$\left[\left(\frac{ds}{dx}\right)^2 - 1\right]^{1/2} = \frac{1}{H}\int w(s)\,ds + C_1$$

Solving for ds/dx, we have

$$\frac{ds}{dx} = \left\{1 + \left[\frac{1}{H}\int w(s)\,ds + C_1\right]^2\right\}^{1/2}$$

Separating variables and integrating, we get

$$x = \int \frac{ds}{\left\{1 + [(1/H)\int w(s)\,ds + C_1]^2\right\}^{1/2}} + C_2 \tag{6.14}$$

As a first step, determine if possible the constant C_1 by applying a slope-boundary condition to Eq. 6.12. With this C_1 in Eq. 6.14, solve for s as a function of x. Next, substitute for s in Eq. 6.12 using this relation. Finally, integrate Eq. 6.12 with respect to x to get y as a function of x. Boundary conditions must then be used to determine H as well as the remaining constant of the integration. The following examples will illustrate how these steps are carried out.

EXAMPLE 6.10

Consider a uniform cable having a span l and a sag h as shown in Fig. 6.50. The weight per unit length w of the cable is a constant.

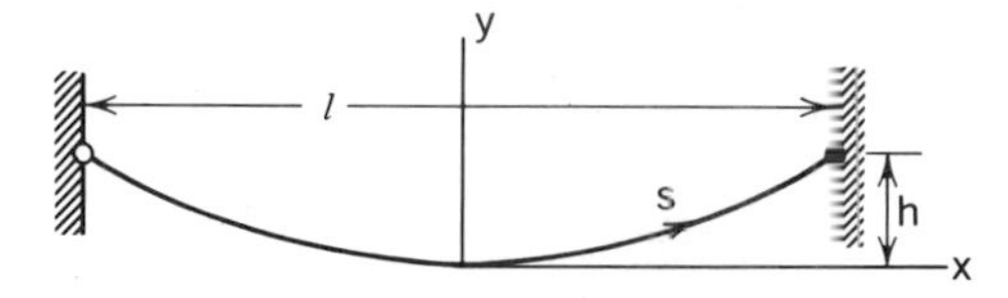

Figure 6.50. Uniform cable loaded by its own weight.

For simplicity, we have placed a reference at the center of the span where the slope of the cable is zero. Accordingly, consider Eq. 6.12 for this case:

$$\frac{dy}{dx} = \frac{1}{H}\int w\,ds + C_1 = \frac{w}{H}s + C_1 \tag{a}$$

When $s = 0$ we require that $dy/dx = 0$, whereupon C_1 is zero. Now consider Eq. 6.14:

$$\begin{aligned} x &= \int \frac{ds}{\left\{1 + [(1/H)\int w\,ds]^2\right\}^{1/2}} + C_2 \\ &= \int \frac{ds}{\{1 + [(w/H)s]^2\}^{1/2}} + C_2 \end{aligned} \tag{b}$$

Integrating the right side of the equation using an integration formula from Appendix I, we get

$$x = \frac{H}{w} \sinh^{-1} \frac{sw}{H} + C_2 \tag{c}$$

The constant C_2 must also be zero, since $x = 0$ at $s = 0$. Solving for s from Eq. (c), we get

$$s = \frac{H}{w} \sinh \frac{xw}{H} \tag{d}$$

Substituting for s in Eq. (a) using the preceding result, we have

$$\frac{dy}{dx} = \sinh \frac{w}{H} x \tag{e}$$

Integrating, we get

$$y = \frac{H}{w} \cosh \frac{w}{H} x + C_3$$

Since $y = 0$ at $x = 0$, the constant C_3 becomes $-H/w$. We then have the deflection curve:

$$y = \frac{H}{w} \left(\cosh \frac{w}{H} x - 1 \right) \tag{f}$$

This curve is called a *catenary curve*.[15]

To determine H, we set $y = h$ when $x = l/2$. Thus,

$$h = \frac{H}{w} \left(\cosh \frac{wl}{2H} - 1 \right) \tag{g}$$

This equation can be solved by trial and error. We may then proceed to determine the maximum force in the cable as well as the length of the cable in the manner followed in Example 6.9.

EXAMPLE 6.11

A water skier is shown in Fig. 6.51 dangling from a kite that is towed by a powerboat at a speed of 30 mph. The boat develops a thrust of 200 lb. The drag on

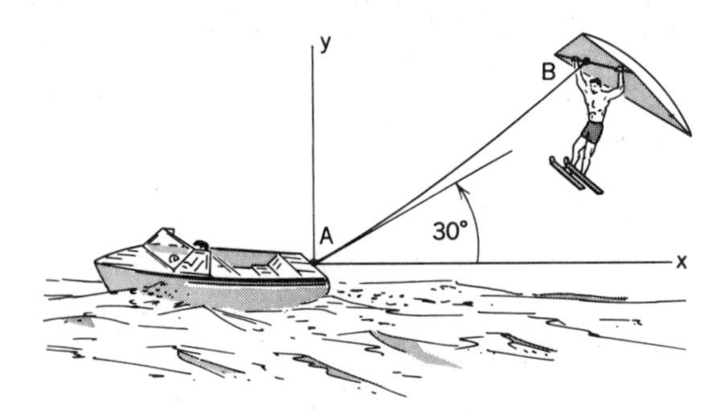

Figure 6.51. Analyze tow rope *AB*.

the boat from the water is estimated as 100 lb. At the support A, the rope has a tangent of 30°. If the man weighs 150 lb, find the height and the lift of the kite as

[15]The Latin for chain is *catena*.

well as the maximum tension in the rope. The kite weighs 25 lb. The uniform rope is 50 ft long and weighs .5 lb/ft. Neglect aerodynamic effects on the rope.

We start with Eq. 6.12, which becomes for this case:

$$\frac{dy}{dx} = \frac{w}{H}s + C_1 \tag{a}$$

Using a reference at A as shown in the diagram, we know that $dy/dx = \tan 30° = .577$ when $s = 0$. Thus, we get for C_1,

$$C_1 = .577$$

Equation 6.14 is considered next. We have

$$x = \int \frac{ds}{\{1 + [(w/H)s + .577]^2\}^{1/2}} + C_2$$

Integrating by making a change in variable and using the proper integration formula in Appendix I, we get

$$x = \frac{H}{w}\sinh^{-1}\left(\frac{w}{H}s + .577\right) + C_2 \tag{b}$$

Solving for s, we have

$$s = \frac{H}{w}\left\{\sinh\left[(x - C_2)\frac{w}{H}\right] - .577\right\} \tag{c}$$

Substituting for s in Eq. (a) using Eq. (c), we get

$$\frac{dy}{dx} = \sinh\left[(x - C_2)\frac{w}{H}\right] \tag{d}$$

Integrating again, we have

$$y = \frac{H}{w}\cosh\left[(x - C_2)\frac{w}{H}\right] + C_3 \tag{e}$$

We must now evaluate the unknown constants C_2, C_3, and H using the boundary conditions and data of the problem. First, since H is the horizontal component of force transmitted by the rope, we know that H is the thrust of the boat minus the drag of the water. Thus,

$$H = 100 \text{ lb}$$

Also, $x = 0$ when $s = 0$, so that from Eq. (c), we get C_2 as follows:

$$\sinh\left(-\frac{.5}{100}C_2\right) = .577$$

Therefore,

$$-\frac{.5}{100}C_2 = \sinh^{-1} .577 = .549$$

Hence,

$$C_2 = -109.8$$

Finally, note that $x = 0$ when $y = 0$. From Eq. (e), we can then get constant C_3 in the following manner:

$$C_3 = -\frac{100}{.5}\cosh\left[\frac{-.5}{100}(-109.8)\right]$$
$$= -200 \cosh .548 = -231$$

We may now evaluate the position x', y' of point B of the kite. To get x', we insert for s in Eq. (b) the value of 50 ft. Thus,

$$x' = \frac{100}{.5}\sinh^{-1}\left(\frac{.5}{100}50 + .577\right) - 109.8 = 40.9 \text{ ft}$$

Now from Eq. (e) we can get y' and consequently the desired height.

$$y' = \frac{100}{.5}\cosh\left[(40.9 + 109.8)\frac{.5}{100}\right] - 231$$
$$= 28.6 \text{ ft} \tag{f}$$

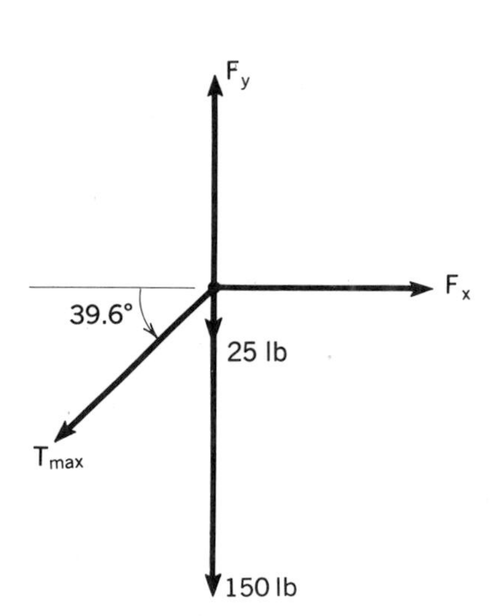

Figure 6.52. Free-body diagram of kite support.

The maximum tension in the rope occurs at point B, where θ is greatest. To get θ_{max}, we go back to Eq. (a). Thus,

$$\left(\frac{dy}{dx}\right)_{max} = \tan\theta_{max} = \frac{.5}{100}(50) + .577 = .827$$

Therefore,

$$\theta_{max} = 39.6° \tag{g}$$

Hence, from Eq. 6.11(a) we have for T_{max}:

$$T_{max} = \frac{100}{\cos 39.6°} = 130 \text{ lb} \tag{h}$$

To get the lifting force of the kite, we draw a free-body diagram of point B of the kite as shown in Fig. 6.52. Note that F_y and F_x are, respectively, the aerodynamic lift and drag forces on the kite. The lift force F_y of the kite then becomes

$$F_y = 175 + T_{max}\sin 39.6° = 258 \text{ lb} \tag{i}$$

Problems

6.59. Find the length of a cable stretched between two supports at the same elevation with span $l = 200$ ft and sag $h = 50$ ft, if it is subjected to a vertical load of 4 lb/ft uniformly distributed in the horizontal direction. (Assume that the weight of the cable is either negligible or included in the 4-lb/ft distribution.) Find the maximum tension.

6.60. A cable supports a 8000-km uniform bar. What is the equation of the cable and what is the maximum tension in the cable?

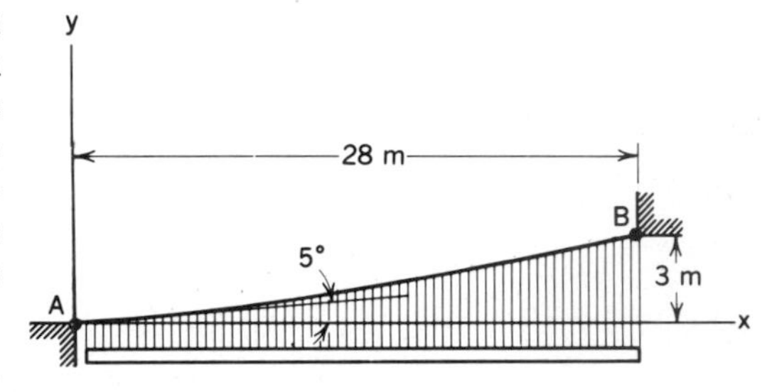

Figure P.6.60

6.61. A cable supports a uniform loading of 100 lb/ft. If the lowest point of the cable occurs 20 ft from point A as shown, what is the maximum tension in the cable and its length? Use A as the origin of reference.

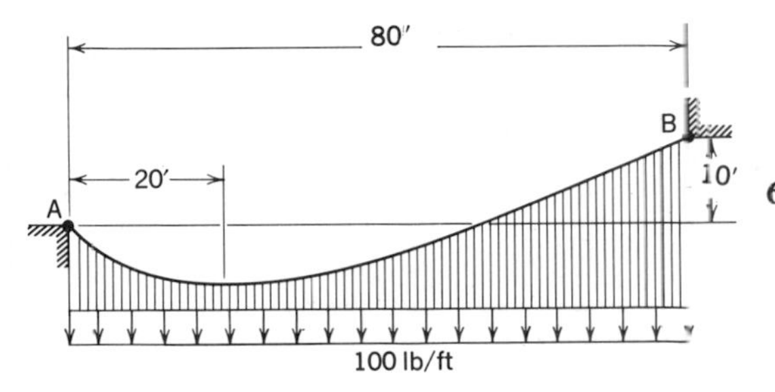

Figure P.6.61

6.62. A uniform cable is shown whose weight we shall neglect. If a loading given as $5x$ N/m is imposed on the cable, what is the deflection curve of the cable if there is a zero slope of the curve at point A? What is the maximum tension?

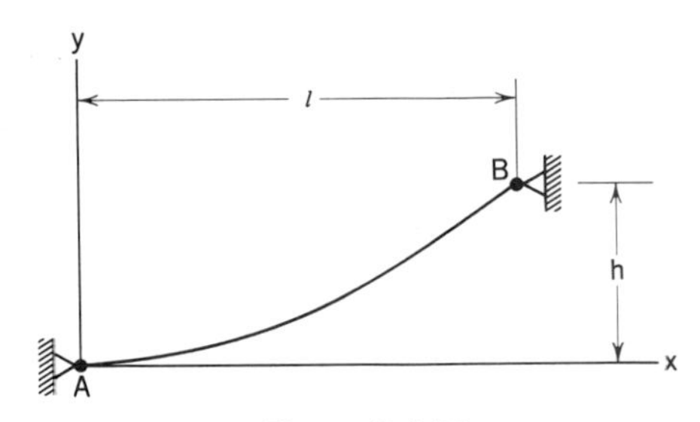

Figure P.6.62

6.63. The left side of a cable is mounted at an elevation 7 m below the right side. The sag, measured from the left support, is 7 m. Find the maximum tension if the cable has a *uniform* loading in the vertical direction of 1500 N/m. (*Suggestion:* Place reference at position of zero slope and determine the location of this point from the boundary conditions.)

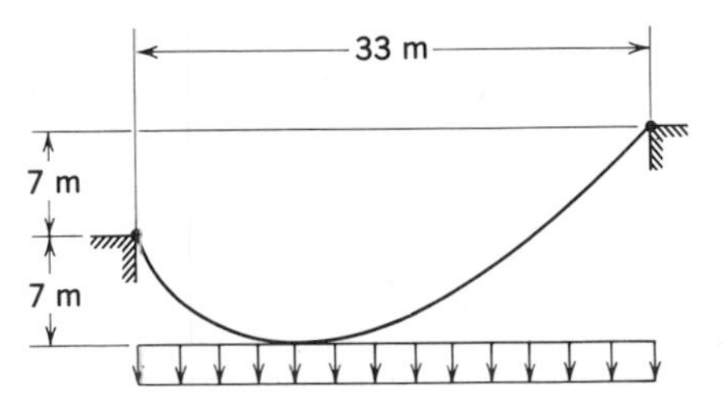

Figure P.6.63

6.64. A blimp is dragging a chain of length 500 ft and weight 10 lb/ft. A thrust of 300 lb is developed by the blimp as it moves against an air resistance of 200 lb. How much chain is on the ground and how high is the blimp? The vertical lift of the blimp on the cable is taken as 1000 lb.

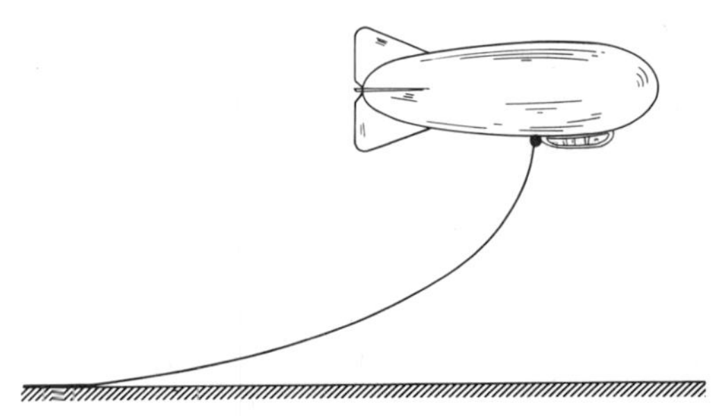

Figure P.6.64

6.65. A large balloon has a buoyant force of 100 lb. It is held by a 150-ft cable whose weight is .5 lb/ft. What is the height h of the balloon above the ground when a steady wind causes it to assume the position shown? What is the maximum tension on the cable?

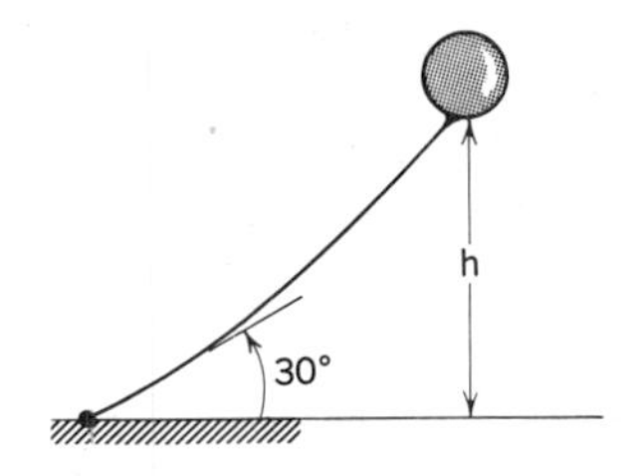

Figure P.6.65

6.66. What is the deflection curve for the uniform cable shown weighing 30 N/m? Find the maximum tension. Compute the height h of the support B.

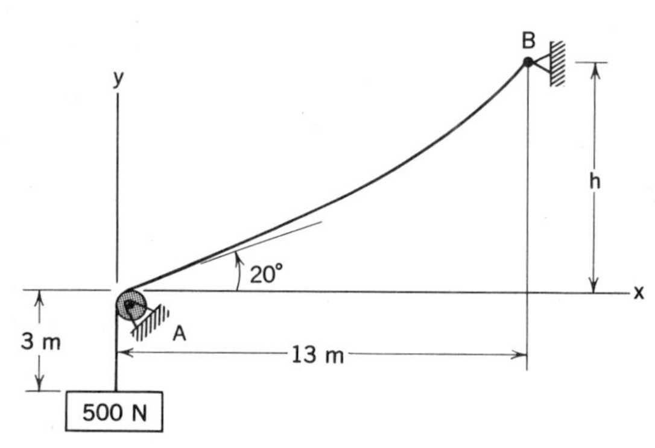

Figure P.6.66

6.67. A search boat is dragging the lake floor for stolen merchandise using a 100-m chain weighing 100 N/m. The tension of the chain at support point B is 5000 N and the chain makes an angle of 50° there. What is the height of point B above the lake bed? Also, what length of chain is dragging along on the bottom? Do not consider buoyant effects.

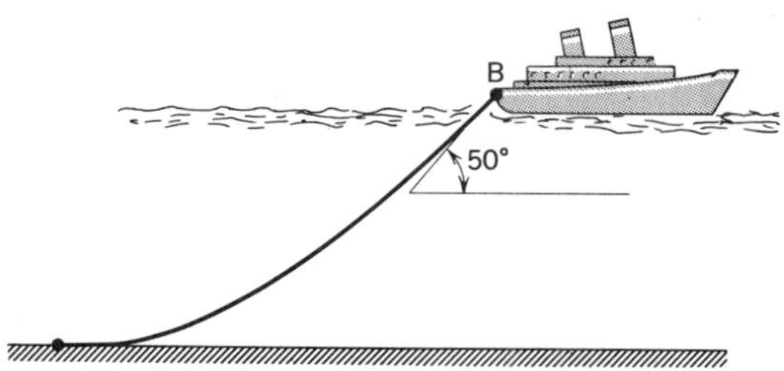

Figure P.6.67

6.68. A cable weighing 3 lb/ft is stretched between two points on the same level. If the length of the cable is 450 ft and the tension at the points of support is 1500 lb, find the sag and the distance between the points of support. Put reference at left support.

6.69. A flexible, inextensible cable is loaded by concentrated forces. If we neglect the weight of the cable, what are the supporting forces at A and B? What are the tensions in the chord AC and the angle α? (*Hint:* Proceed by using finite free bodies and working from first principles.)

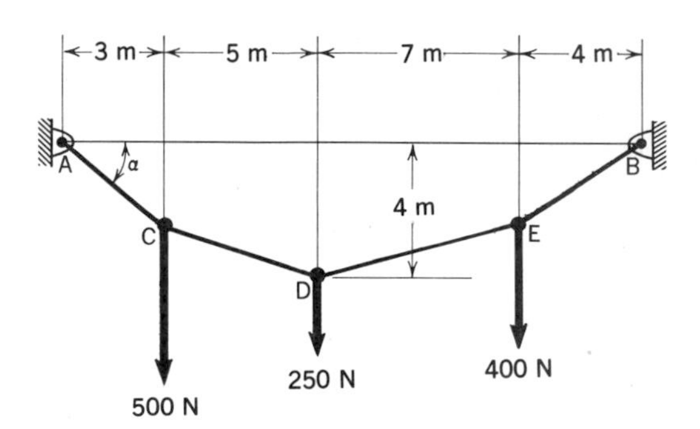

Figure P.6.69

6.70. A system of two inextensible, flexible cables is shown supporting a 2000-lb platform in a horizontal position. What are the inclinations of the cable segments AB, BC, and DE to accomplish this and what lengths should they be? Neglect the weight of the segments and note the hint in Problem 6.69.

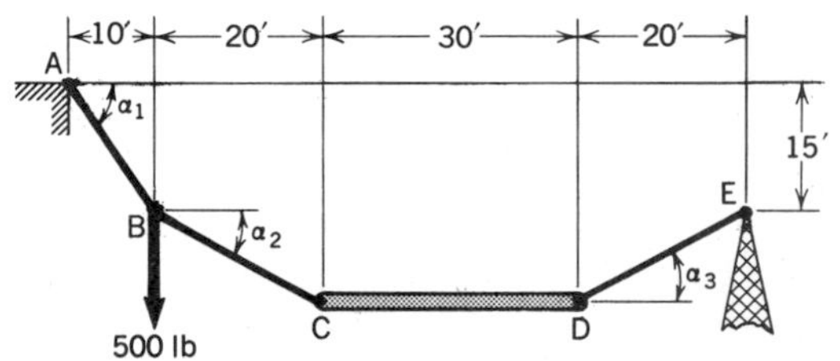

Figure P.6.70

6.11 Closure

Essentially what we have done in this chapter is to apply previously developed material to situations of singular importance in engineering. Further information on structures can be found in books on strength of materials and structural mechanics.[16] We turn again to new material in Chapter 7, where we will discuss the Coulomb laws of friction.

Review Problems

6.71. A 3-kN traveling hoist has a 27-kN capacity and is suspended from a beam weighing .5 kN/m. The beam is fastened to several trusses spaced 8 m apart. What are the forces in each truss member when the fully loaded hoist is located at point C directly under the truss shown? Assume that the hoist acts on pin C and that pin C also supports half of the I-beam G between each of the adjacent trusses.

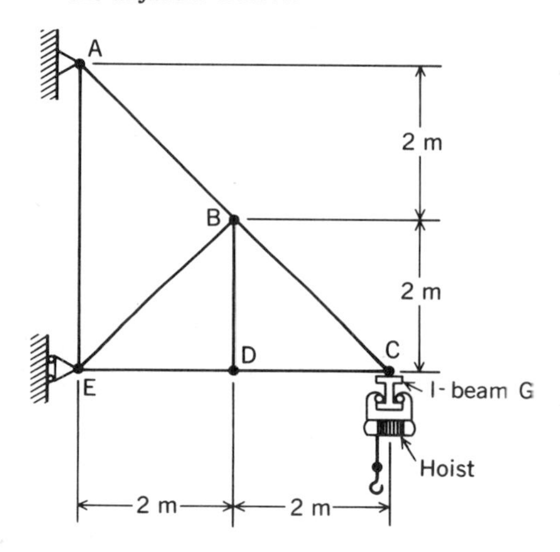

Figure P.6.71

6.72. The truss is used to support the roof of a low-clearance train-car repair shed (hence the curved members). The roof is subjected to a snow load of 1 kN/m². What are the forces in the straight members if the trusses are 10 m apart?

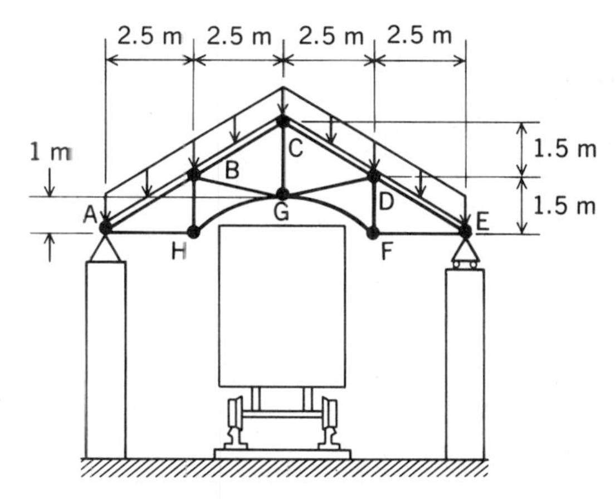

Figure P.6.72

6.73. Find forces in all the members of the space truss. Note that ACE is in the xz plane.

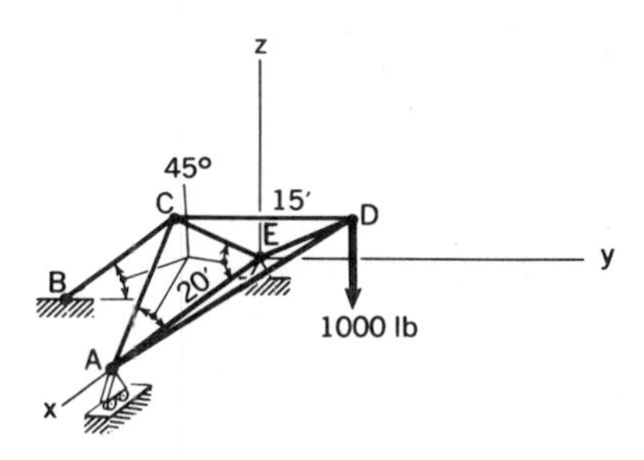

Figure P.6.73

6.74. Determine the forces in members BG, BF, and CE for the plane truss.

[16]See I. H. Shames, *Introduction to Solid Mechanics*, Prentice-Hall, Inc., Englewood Cliffs, N.J., 1975.

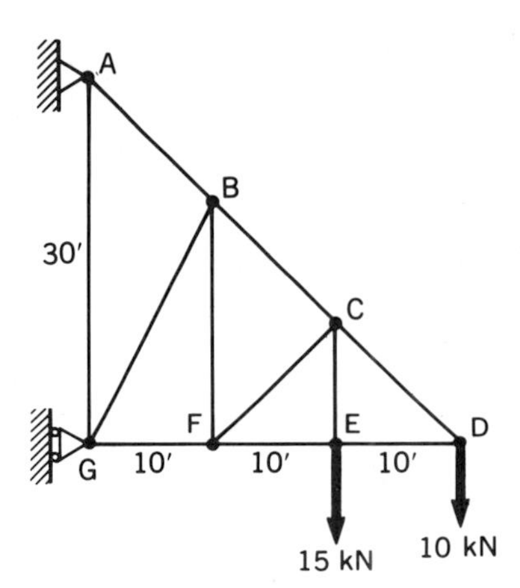

Figure P.6.74

6.75. Express the shear-force and bending-moment equations with the aid of free-body diagrams. Then express V and M without the diagrams.

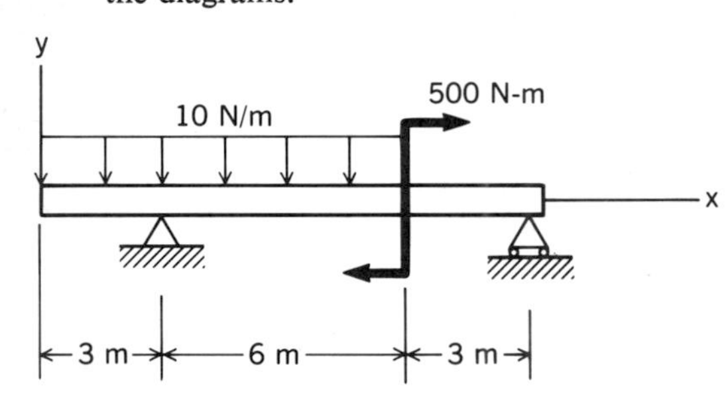

Figure P.6.75

6.76. Express the shear-force and bending-moment equations without the aid of free-body diagrams.

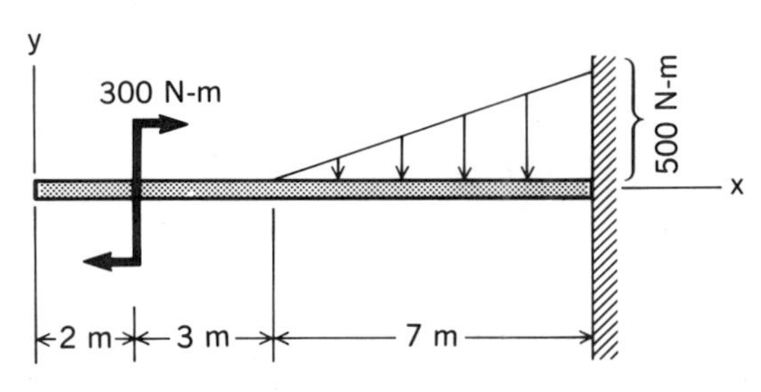

Figure P.6.76

6.77. Give the shear-force and bending-moment equations for the beam, and sketch shear-force and bending-moment diagrams. At what position between supports is the bending moment equal to zero?

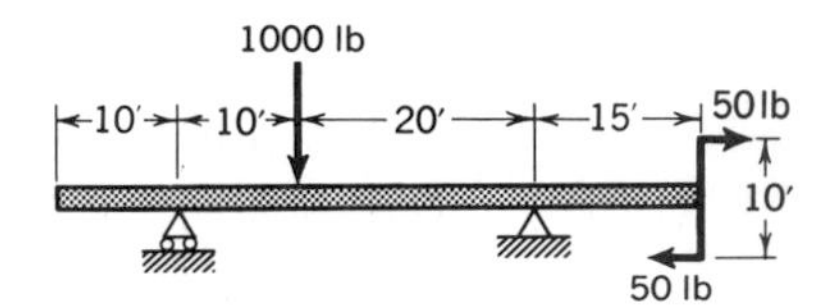

Figure P.6.77

6.78. Sketch the shear-force and bending-moment diagrams labeling key points.

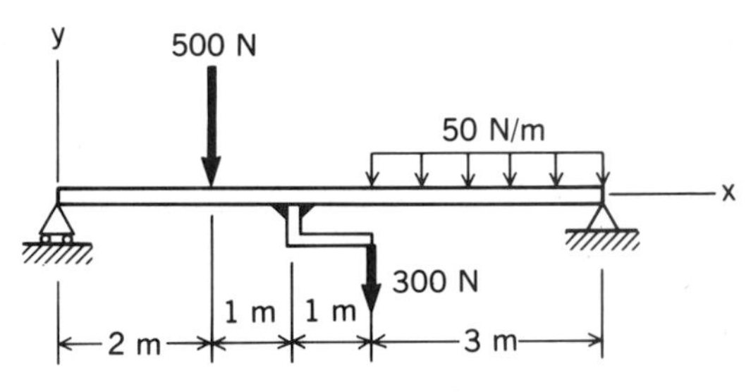

Figure P.6.78

6.79. Find the shape of a cable stretched between two points on the same level, l units apart with sag h, and subjected to a vertical loading of

$$w(x) = 5 \cos \frac{\pi x}{l} \text{ N/m}$$

distributed in the horizontal direction. The coordinate x is measured from the zero slope position of the cable.

6.80. A uniform cable weighs 1 lb/ft. It is connected to a uniform rod at *B*. This rod is free to swing about hinge *C*. If a force component F_x of -200 lb is exerted at *A* as shown, what is the resulting angle of inclination α? The cable is 50 ft long and the rod is 20 ft long. What is the weight of the rod?

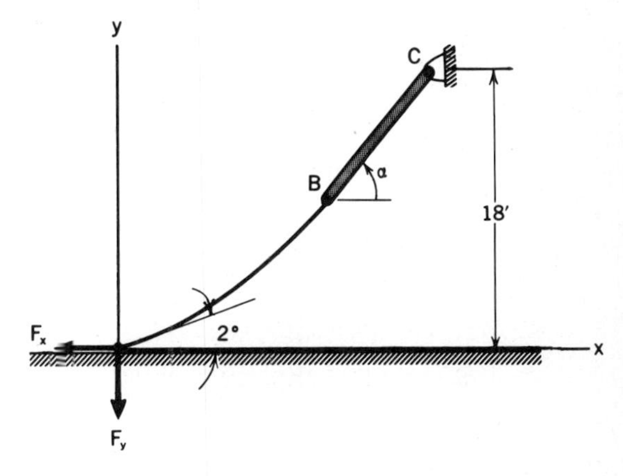

Figure P.6.80

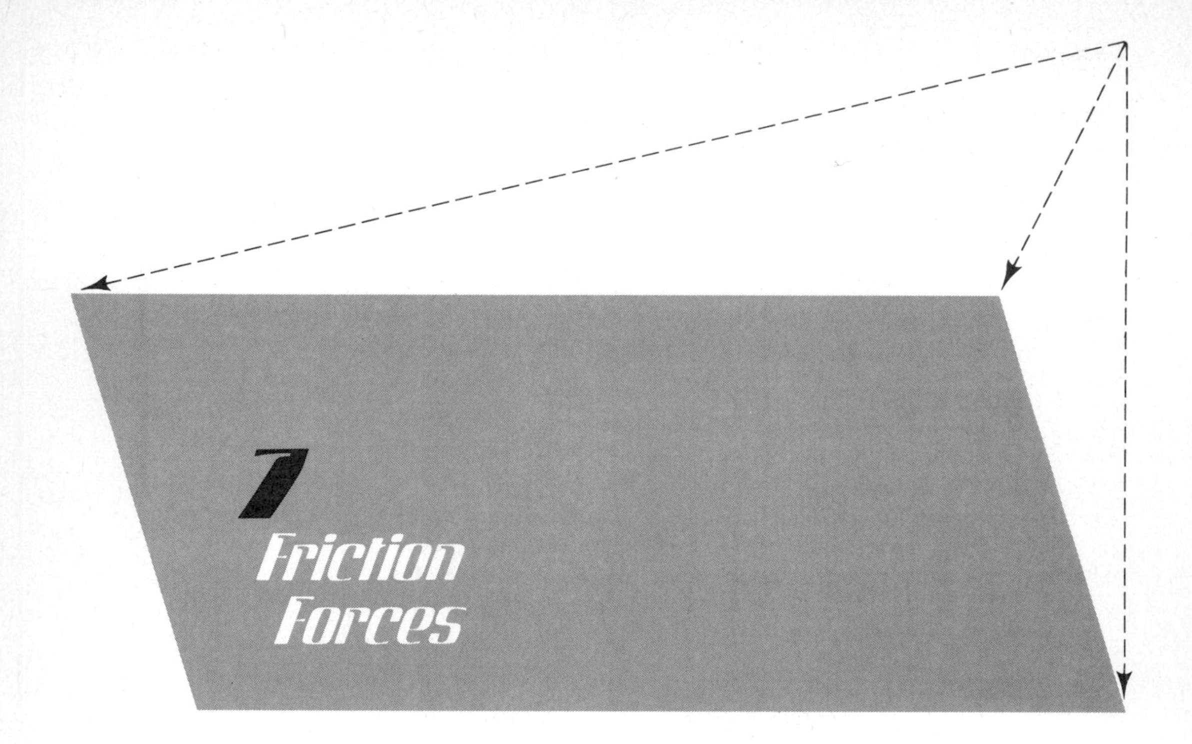

7.1 Introduction

Friction is the force distribution at the surface of contact between two bodies that prevents or impedes sliding motion of one body relative to the other. This force distribution is tangent to the contact surface and has, for the body under consideration, a direction at every point in the contact surface that is in opposition to the possible or existing slipping motion of the body at that point.

Frictional effects are associated with energy dissipation and are therefore sometimes considered undesirable. At other times, however, this means of changing mechanical energy to heat is a beneficial one, as for example in brakes, where the kinetic energy of a vehicle is dissipated into heat. In statics applications, frictional forces are often necessary to maintain equilibrium.

Coulomb friction is that friction which occurs between bodies having dry contact surfaces, and is not to be confused with the action of one body on another separated by a film of fluid such as oil. These latter problems are termed *lubrication problems* and are studied in the fluid mechanics courses. Coulomb, or *dry*, friction is a complicated phenomenon, and actually not much is known about its true nature.[1] The major cause of dry friction is believed to be the microscopic roughness of the surfaces of contact. Interlocking microscopic protuberances oppose the relative motion between the surfaces. When sliding is present between the surfaces, some of these protuberances

[1]For a more complete discussion of friction, see F. P. Bowden and D. Tabor, *The Friction and Lubrication of Solids*, Oxford University Press, New York, 1950.

either are sheared off or are melted by high local temperatures. This is the reason for the high rate of "wear" for dry-body contact and indicates why it is desirable to separate the surfaces by a film of fluid.

We have previously employed the terms "smooth" and "rough" surfaces of contact. A "smooth" surface can only support a normal force. On the other hand, a "rough" surface in addition can support a force tangent to the contact surface (i.e., a friction force). In this chapter, we shall consider situations whereby the friction force can be directly related to the normal force at a surface of contact. Other than including this new relationship, we use only the usual static equilibrium equations.

7.2 Laws of Coulomb Friction

Everybody has gone through the experience of sliding furniture along a floor. We exert a continuously increasing force which is completely resisted by friction until the object begins to move—usually with a lurch. The lurch occurs because once the object begins to move, there is a decrease in frictional force from the maximum force attained under static conditions. An idealized plot of this force as a function of time is shown in Fig. 7.1. There the force P applied to the furniture, idealized as a block in Fig. 7.2, is shown to drop from the highest or limiting value to a lower value which is constant with time. This latter constant value is independent of the velocity of the object. The condition corresponding to the maximum value is termed the condition of *impending motion* or *impending slippage*.

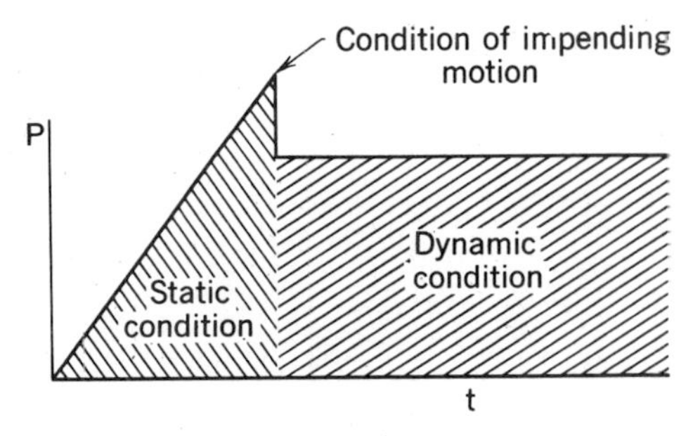

Figure 7.1. Idealized plot of friction force P.

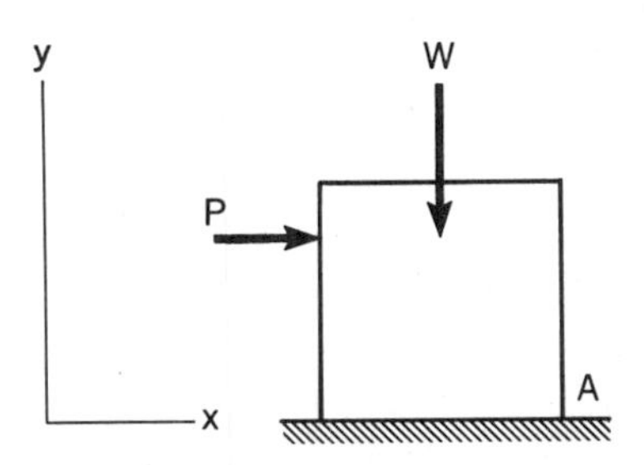

Figure 7.2. Idealization of furniture.

By carrying out experiments on blocks tending to move without rotation or actually moving without rotation on flat surfaces, Coulomb in 1781 presented certain conclusions which are applicable at the condition of *impending slippage* or once *slippage has begun*. These have since become known as Coulomb's laws of friction. For block problems, he reported that:

1. The total force of friction that can be developed is independent of the magnitude of the area of contact.

2. For low relative velocities between sliding objects, the frictional force is practically independent of velocity. However, the sliding frictional force is less than the frictional force corresponding to impending slippage.
3. The total frictional force that can be developed is proportional to the normal force transmitted across the surface of contact.

Conclusions 1 and 2 may come as a surprise to most of you and be contrary to your "intuition." Nevertheless, they are accurate enough statements for many engineering applications. More precise studies of friction, as was pointed out earlier, are complicated and involved. We can express conclusion 3 mathematically as:

$$f \propto N$$

Therefore,

$$f = \mu N \tag{7.1}$$

where μ is called the *coefficient of friction*.

Equation 7.1 is valid *only at conditions of impending slippage or while the body is slipping*. Since the limiting static friction force exceeds the dynamic force friction, we differentiate between coefficients of friction for those conditions. Thus, we have coefficients of *static* friction and coefficients of *dynamic* friction, μ_s and μ_d, respectively. The accompanying table is a small list of static coefficients. The corresponding coefficients of friction for dynamic conditions are about 25% less.

STATIC COEFFICIENTS OF FRICTION[2]

Steel on cast iron	.40
Copper on steel	.36
Hard steel on hard steel	.42
Mild steel on mild steel	.57
Rope on wood	.70
Wood on wood	.20–.75

Let us consider carefully the simple block problem used to develop the laws of Coulomb. Note that we have:

1. A plane surface of contact.
2. An impending or actual motion which is in the same direction for all area elements of the contact surface. Thus, there is no impending or actual rotation between the bodies in contact.
3. The further implication that the properties of the respective bodies are uniform at the contact surface. Thus, the coefficient of friction μ is constant for all area elements of the contact surface.

---

[2]F. P. Bowden and D. Tabor, *The Friction and Lubrication of Solids*, Oxford University Press, New York, 1950.

What do we do if any of these conditions is violated? We can always choose an *infinitesimal* part of area of contact between the bodies. Such an infinitesimal area can be considered plane even though the general surface of contact of which it is an infinitesimal part is not. Furthermore, the relative motion at this infinitesimal contact surface may be considered as along a straight line even though the finite surface of which it is a part may not have such a simple straight motion. Finally, for the infinitesimal area of contact, we may consider the materials to be uniform even though the properties of the material vary over the finite area of contact. In short, when conditions 1 through 3 do not prevail, we can still use Coulomb's law *in the small* (i.e., at infinitesimal contact areas) and then integrate the results. We shall call such problems *complex* surface contact problems and we shall examine a series of such problems in Section 7.4.

We now examine simple contact problems where Coulomb's laws apply to the contact surface *as a whole* without requiring integration procedures. We shall thus consider uniform blocklike bodies akin to those used by Coulomb. Also, we shall consider bodies which however complex have very *small* contact surfaces, such as in Fig. 7.3(a). Clearly, the whole contact surface can be considered an infinitesimal plane area and for reasons set forth earlier, we shall directly use Coulomb's laws when appropriate as has been shown in Fig. 7.3(b).

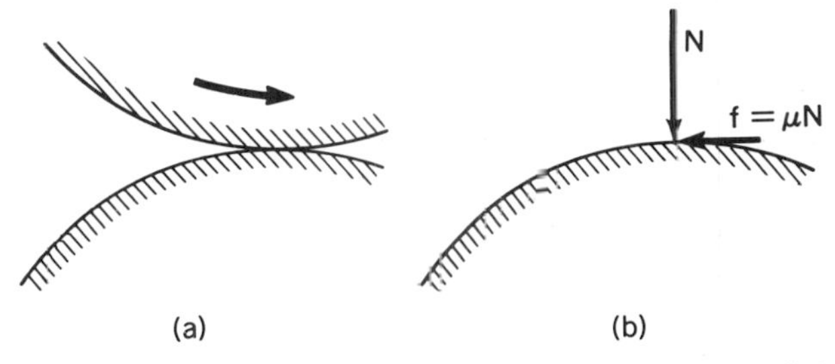

Figure 7.3. (a) Small contact surface; (b) Coulomb's laws applied.

Before proceeding to the problems, we have one additional point to make. For a finite simple surface of contact, such as the block shown in Fig. 7.2, we must note that we do not generally know the line of action for the simplest resultant supporting force N, since we do not generally know the normal force distribution between the two bodies. Hence, we cannot take moments for such free-body diagrams without introducing additional unknown distances in the equation. Consequently, for such problems we limit ourselves to summing forces only. This is not true, however, when we have a *point* contact surface such as in Fig. 7.3(a). The line of action of the supporting force must be at the point of contact, and we can thus take moments without introducing additional unknown distances.

7.3 Simple Contact Friction Problems

Two common classes of statics problems involve dry friction. In one class, we know that motion is impending, or has been established and is uniform, and we desire information about certain forces that are present. We can then express friction forces at surfaces of contact where there is impending or actual slippage as μN according to Coulomb's law and, using f_i for other friction forces, proceed by methods of statics. However, the proper direction must be given to *all* friction forces. That is, *they must oppose possible, impending, or actual relative motion at the contact surfaces.* In the second class of problems, external loads on a body are given, and we desire to determine whether the friction forces present are sufficient to maintain equilibrium. One way to attack this latter type of problem is to assume that impending motion exists in the various possible directions, and to solve for the external forces required for such conditions. By comparing the actual external forces present with those required for the various impending motions, we can then deduce whether the body can be restrained by frictional forces from sliding.

The following examples are used to illustrate the two classes of problems.

EXAMPLE 7.1

In Fig. 7.4(a) is shown an automobile on a roadway inclined at an angle θ with the horizontal. If the coefficients of static and dynamic friction between the tires and the road are taken as 0.6 and 0.5, respectively, what is the maximum inclination θ_{max} that the car can climb at uniform speed? It has rear-wheel drive and has a total loaded weight of 3600 lb. The center of gravity for this loaded condition has been shown in the diagram.

Let us assume that the drive wheels do not "spin"; that is, there is zero relative velocity between the tire surface and the road surface at the point of contact. Then, clearly, the maximum friction force possible is μ_s times the normal force at this contact surface, as has been indicated in Fig. 7.4(b).

We can consider this to be a coplanar problem with three unknowns, N_1, N_2, and θ_{max}. Accordingly, since the friction force is restricted to a point, three equations of equilibrium are available. Using the reference xy shown in the diagram, we have:

$\underline{\sum F_x = 0:}$

$$.6N_1 - 3600 \sin \theta_{max} = 0 \tag{a}$$

$\underline{\sum F_y = 0:}$

$$N_1 + N_2 - 3600 \cos \theta_{max} = 0 \tag{b}$$

$\underline{\sum M_A = 0:}$

$$10N_2 - (3600 \cos \theta_{max})(5) + (3600 \sin \theta_{max})(1) = 0 \tag{c}$$

To solve for θ_{max}, we eliminate N_1 from Eqs. (a) and (b), getting as a result the equation

$$N_2 = 3600 \cos \theta_{max} - 6000 \sin \theta_{max} \tag{d}$$

Now, eliminating N_2 from Eq. (c) using Eq. (d), we get

$$18{,}000 \cos \theta_{max} - 56{,}400 \sin \theta_{max} = 0$$

Therefore,

$$\tan \theta_{max} = .320 \tag{e}$$

Hence,

$$\theta_{max} = 17.7° \tag{f}$$

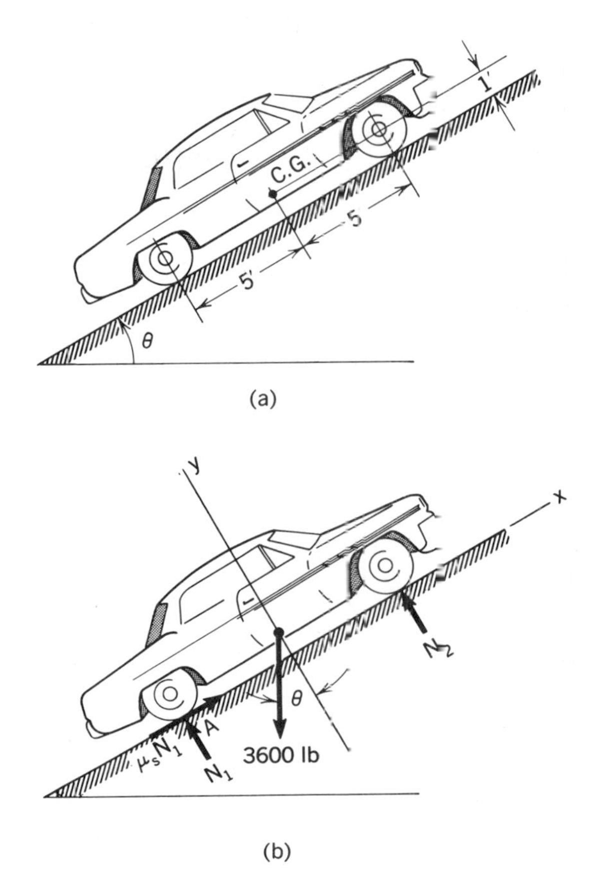

Figure 7.4. (a) Find maximum θ; (b) free-body diagram using Coulomb's law.

If the drive wheels were caused to spin, we would have to use μ_d in place of μ_s for this problem. We would then arrive at a smaller θ_{max}, which for this problem would be 14.7°.

EXAMPLE 7.2

Using the data of Example 7.1, compute the torque needed by the drive wheels to move the car at a uniform speed up an incline where $\theta = 15°$. Also, assume that the brakes have "locked" while the car is in a parked position on the incline. What force is then needed to tow the car either up the incline or down the incline with the brakes in this condition? The diameter of the tire is 25 in.

A free-body diagram for the first part of the problem is shown in Fig. 7.5(a). Note that the friction force f will now be determined by Newton's law and not by Coulomb's law, since we do not have impending slippage between the wheel and the road for this case. Accordingly, we have, for f:

$\underline{\sum F_x = 0:}$

$$f - 3600 \sin 15° = 0$$

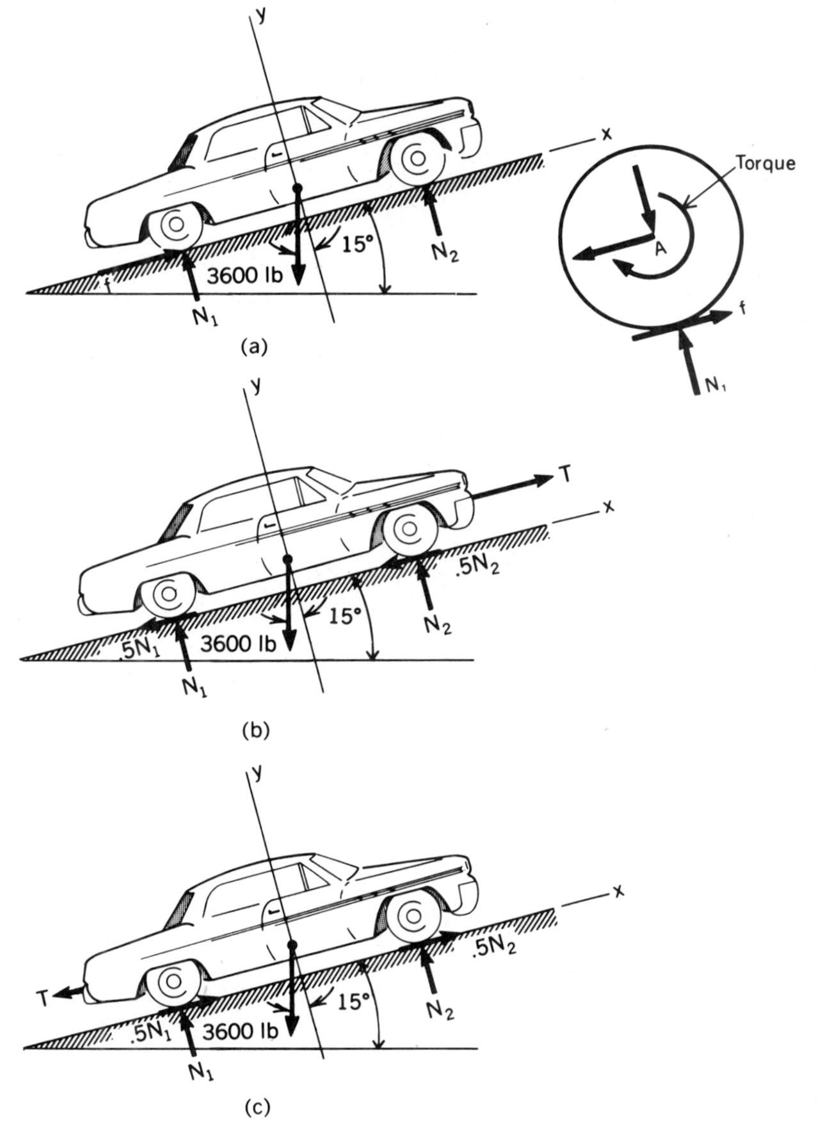

Figure 7.5. Free-body diagrams: (a) climbing at uniform speed; (b) under tow upward; (c) under tow downward.

Therefore,

$$f = 932 \text{ lb}$$

The torque needed is then computed using the rear wheels as a free body [see Fig. 7.5(a)]. Taking moments about A, we have

$$\text{torque} = (f)(r) = (932)\left(\frac{25/2}{12}\right) = 971 \text{ ft-lb}$$

For the second part of the problem, we have shown the required free body in Fig. 7.5(b). Note that we have used Coulomb's law for the friction forces with the *dynamic* friction coefficient μ_d. We now write the equations of equilibrium for this free body.

$\sum F_x = 0$:

$$T - .5(N_1 + N_2) - 3600 \sin 15° = 0 \tag{a}$$

$\sum F_y = 0$:

$$(N_1 + N_2) - 3600 \cos 15° = 0 \tag{b}$$

Solving for $N_1 + N_2$ from Eq. (b), and substituting into Eq. (a), we can now solve for T. Hence,

$$T = (.5)(3600)(.966) + 932 = 2670 \text{ lb} \tag{c}$$

For towing the car down the incline we must reverse the direction of the friction force as shown in Fig. 7.5(c). Solving for T as in the previous calculation, we get

$$T = (.5)(3600)(.966) - 932 = 807 \text{ lb} \tag{d}$$

EXAMPLE 7.3

In Fig. 7.6, a strongbox of mass 75 kg rests on a floor. The static coefficient for the contact surface is .20. What is the highest position h for a horizontal load $\boldsymbol{P}$ at which the box will move without tipping?

The free-body diagram for the strongbox is shown in Fig. 7.7. The condition of impending motion has been recognized by the use of Coulomb's law. Furthermore, by concentrating the supporting and friction forces at the left corner, we are stipulating *impending tipping* for the problem. This latter condition imposes the desired *largest* possible value of h for equilibrium without tipping.

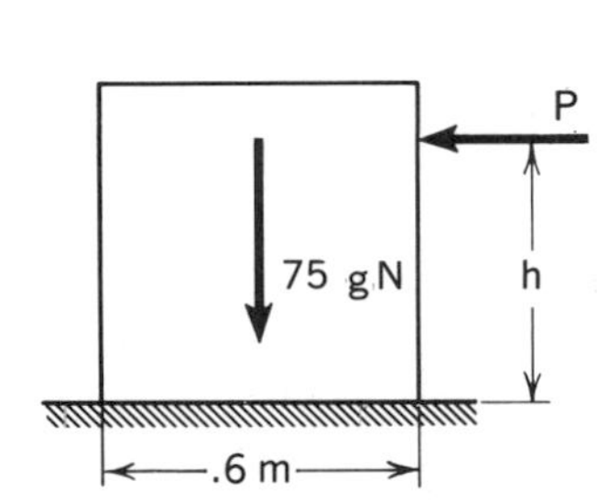

Figure 7.6. Strongbox being pushed.

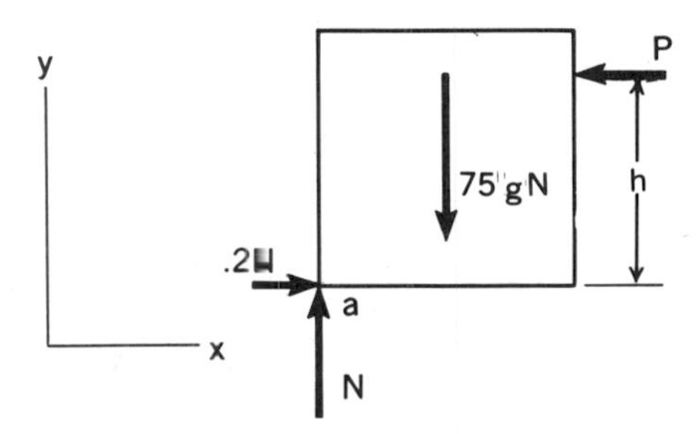

Figure 7.7 Impending tipping and slipping.

The pertinent forces constitute a coplanar system of forces at the midplane of the strongbox. We proceed with the scalar equations of equilibrium:

$\sum F_y = 0$:

$$N = 75\,g = 736 \text{ N}$$

$\sum F_x = 0$:

$$P = .2\,N = 147.15 \text{ N}$$

$\sum M_a = 0$:

$$-(75\,g)(.3) + (147.15)h = 0$$

Therefore,

$$h = 1.500 \text{ m}$$

Thus, the height of the applied load must be less than 1.50 m in order to avoid tipping.

The three examples presented illustrated the *first* type of friction problem wherein we know the nature of the motion or impending motion present in the system and we determine certain forces or positions of certain forces. In the last example of this series, we illustrate the *second* type of friction problem set forth earlier—namely the problem of deciding whether bodies will move or not move under prescribed external forces.

EXAMPLE 7.4

The coefficient of static friction for all contact surfaces in Fig. 7.8 is .2. Does the 50-lb force move the block A up, hold it in equilibrium, or is it too small to prevent A from coming down and B from moving out? The 50-lb force is exerted at the midplane of the blocks so that we can consider this a coplanar problem.

We can compute a force $\mathbf{P}$ in place of the 50-lb force to cause impending motion of block B to the left, and a force $\mathbf{P}$ for impending motion of block B to the right. In this way, we can judge by comparison the action that the 50-lb force will cause.

The free-body diagrams for impending motion of block B to the left have been shown in Fig. 7.9, which contains the unknown force $\mathbf{P}$ mentioned above. We need not be concerned about the correct location of the centers of gravity of the blocks, since we shall only add forces in the analysis. (We do not know the line of action of

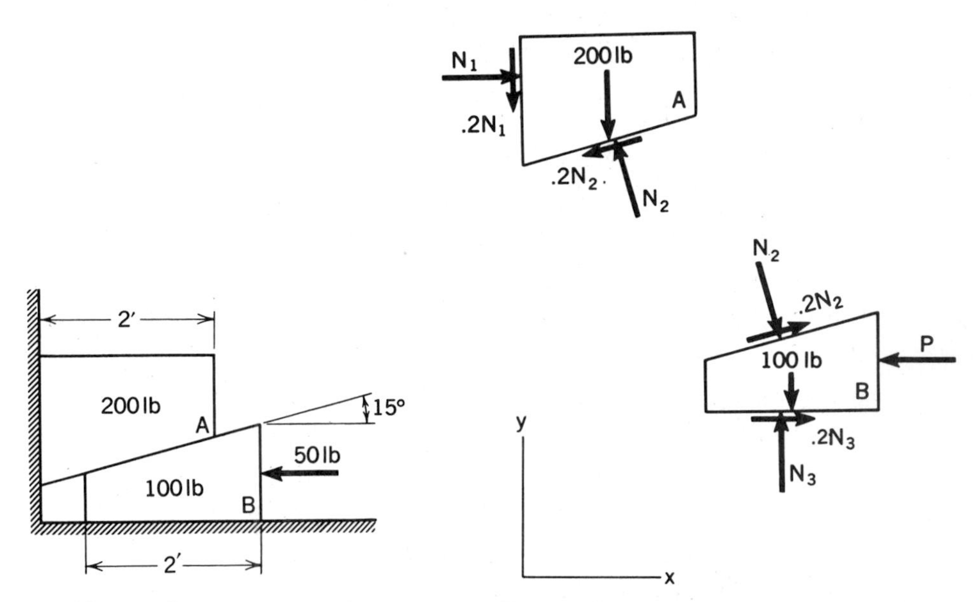

Figure 7.8. Do blocks move?

Figure 7.9. Impending motion of B to the left.

the normal forces at the contact surfaces and therefore cannot take moments.) Summing forces on block A, we get

$$N_2 \cos 15°\boldsymbol{j} - N_2 \sin 15°\boldsymbol{i} - .2N_1\boldsymbol{j} - 200\boldsymbol{j}$$
$$+ N_1\boldsymbol{i} - .2N_2 \cos 15°\boldsymbol{i} - .2N_2 \sin 15°\boldsymbol{j} = \boldsymbol{0}$$

The scalar equations are:

$$N_1 - .259N_2 - .1932N_2 = 0$$
$$.966N_2 - .2N_1 - 200 - .0518N_2 = 0$$

Solving simultaneously, we get

$$N_2 = 243 \text{ lb}, \qquad N_1 = 109.8 \text{ lb}$$

For the free-body diagram of B, we have, on summing forces,

$$-N_2 \cos 15°\boldsymbol{j} + N_2 \sin 15°\boldsymbol{i} - P\boldsymbol{i} + .2N_3\boldsymbol{i} + N_3\boldsymbol{j}$$
$$- 100\boldsymbol{j} + .2N_2 \cos 15°\boldsymbol{i} + .2N_2 \sin 15°\boldsymbol{j} = \boldsymbol{0}$$

This yields the following scalar equations:

$$-P + 62.9 + .2N_3 - 46.9 = 0$$
$$-235 + N_3 - 100 - 12.6 = 0$$

Solving simultaneously, we have

$$P = 174 \text{ lb}$$

Clearly, the stipulated force of 50 lb is insufficient to induce a motion on block B to the left, so further computation is necessary.

Next, we reverse the direction of force P and compute what its value must be to move the block B to the right. The frictional forces in Fig. 7.9 are all reversed, and the vector equation of equilibrium for block A becomes

$$N_2 \cos 15°\boldsymbol{j} - N_2 \sin 15°\boldsymbol{i} + .2N_1\boldsymbol{j} - 200\boldsymbol{j}$$
$$+ N_1\boldsymbol{i} + .2N_2 \cos 15°\boldsymbol{i} + .2N_2 \sin 15°\boldsymbol{j} = \boldsymbol{0}$$

The scalar equations are:

$$N_1 - .259N_2 + .1932N_2 = 0$$
$$.966N_2 + .2N_1 - 200 + .0518N_2 = 0$$

Solving simultaneously, we get

$$N_2 = 194.1 \text{ lb}, \qquad N_1 = 12.80 \text{ lb}$$

For free body B we have, on summing forces,

$$-N_2 \cos 15°\boldsymbol{j} + N_2 \sin 15°\boldsymbol{i} + P\boldsymbol{i} - .2N_3\boldsymbol{i} + N_3\boldsymbol{j}$$
$$-100\boldsymbol{j} - .2N_2 \cos 15°\boldsymbol{i} - .2N_2 \sin 15°\boldsymbol{j} = \boldsymbol{0}$$

The following are the scalar equations:

$$P - .2N_3 + 50.3 - 37.5 = 0$$
$$-100 + N_3 - 187.5 - 10.05 = 0$$

Solving, we get $P = 46.7$ lb. Thus, we would have to *pull to the right* to get block B to move in this direction. We can now conclude from this study that the blocks are in equilibrium.

Problems

7.1. A block has a force F applied to it. If this force has a time variation as shown in the diagram, draw a simple sketch showing the friction force variation with time. Take $\mu_s = .3$ and $\mu_d = .2$ for the problem.

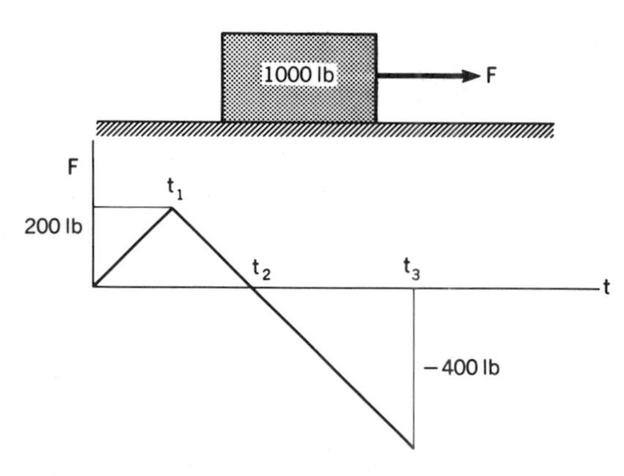

Figure P.7.1

7.2. Show that by increasing the inclination ϕ on an inclined surface until there is impending slippage of supported bodies, we reach the *angle of repose* ϕ_s, so that $\tan \phi_s = \mu_s$.

7.3. To what angle must the driver elevate the dump bed of the truck to cause the wooden crate of weight W to slide out? For wood on steel, $\mu_s = .6$ and $\mu_d = .4$.

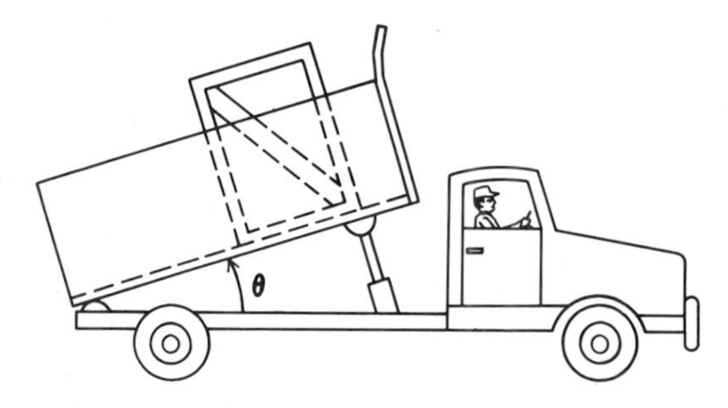

Figure P.7.3

7.4. A platform is suspended by two ropes which are attached to blocks that can slide horizontally. At what value of W does the platform begin to descend? Will W start tipping?

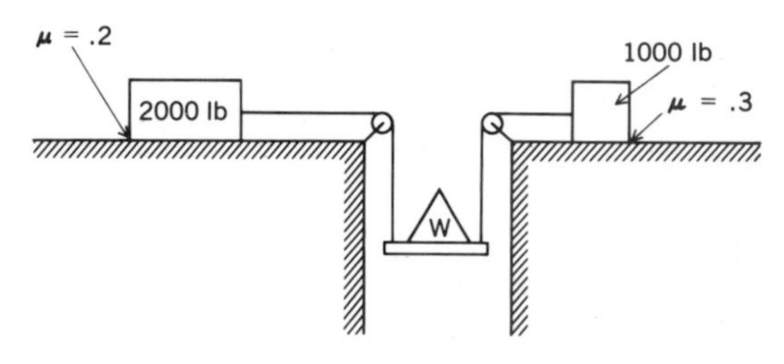

Figure P.7.4

7.5. Explain how a violin bow, when drawn over a string, maintains the vibration of the string. Do this in terms of friction forces and the difference in static and dynamic coefficients of friction.

7.6. What is the value of the force F, inclined at 30° to the horizontal, needed to get the block just started up the incline? What is the force F needed to keep it just moving up at a constant speed? The coefficients of static and dynamic friction are .3 and .275, respectively.

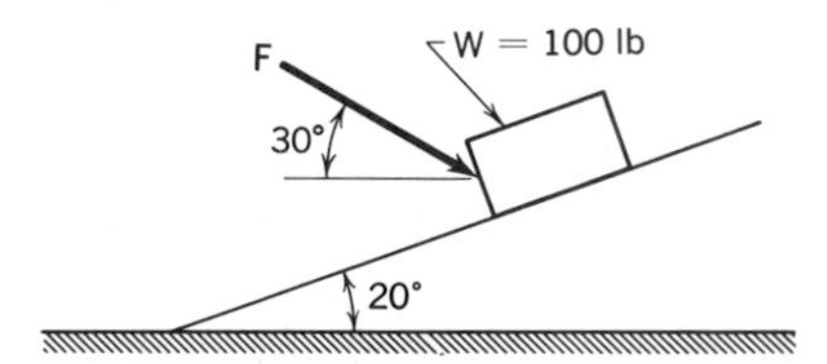

Figure P.7.6

7.7. Bodies A and B weigh 500 N and 300 N, respectively. The platform on which they are placed is raised from the horizontal position to an angle θ. What is the *maximum* angle that can be reached before the bodies slip down the incline? Take μ_s for body B and the plane as .2 and μ_s for body A and the plane as .3.

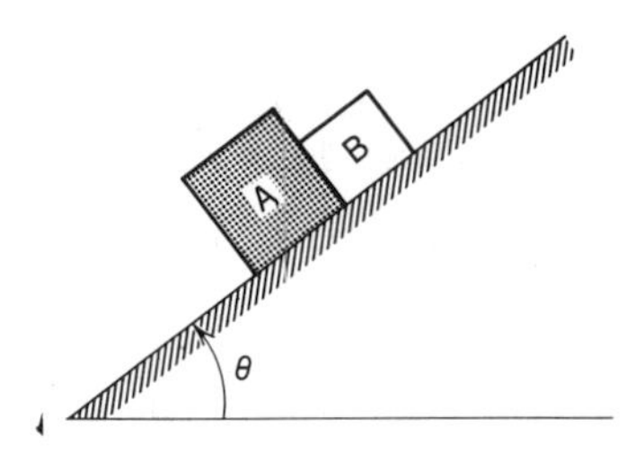

Figure P.7.7

7.8. A 30-ton tank is moving up a 30° incline. If $\mu_s = .6$ for the contact surface between tread and ground, what *maximum* torque can be developed at the rear drive sprocket with no slipping? What maximum towing force F can the tank develop? Take the mean diameter of the rear sprocket as 2 ft.

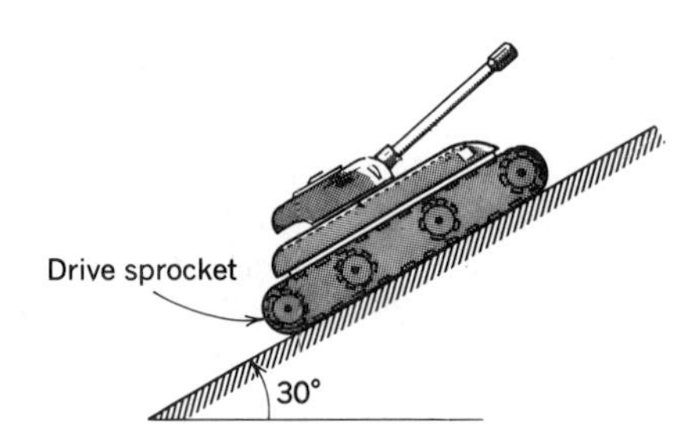

Figure P.7.8

7.9. A 500-lb crate A rests on a 1000-lb crate B. The centers of gravity of the crates are at the geometric centers. The coefficients of static friction between contact surfaces are shown in the diagram. The force T is increased from zero. What is the first action to occur?

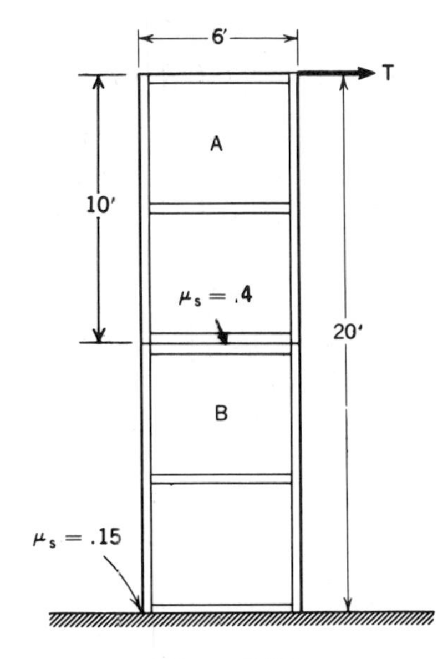

Figure P.7.9

7.10. What force F is needed to get the 300-kg block moving to the right? The coefficient of static friction for all surfaces is .3.

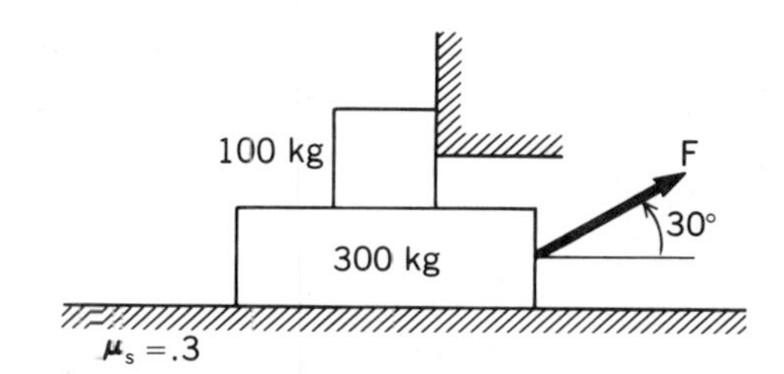

Figure P.7.10

7.11. A chute is shown having sides that are at right angles to each other. The chute is 30 ft in length with end A 10 ft higher than end B. Cylinders weighing 200 lb are to slide down the chute. What is the *maximum* allowable coefficient of friction so there cannot be sticking of the cylinders along the chute?

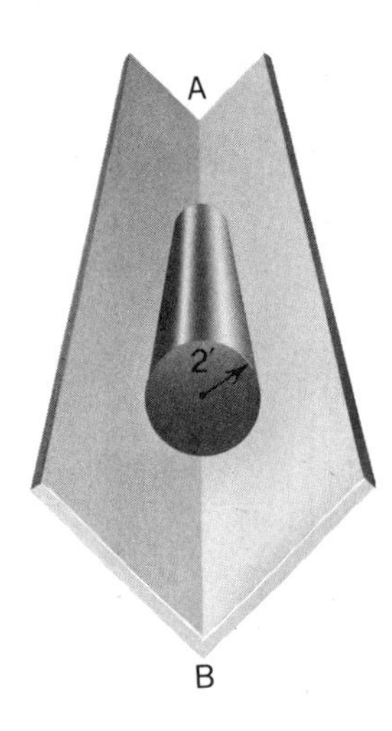

Figure P.7.11

7.12. A block rests on a surface for which there is a coefficient of friction $\mu_s = .2$. Over what range of angle β will there be no movement of the block for the 150-N force? (You will have to solve an equation by trial and error.)

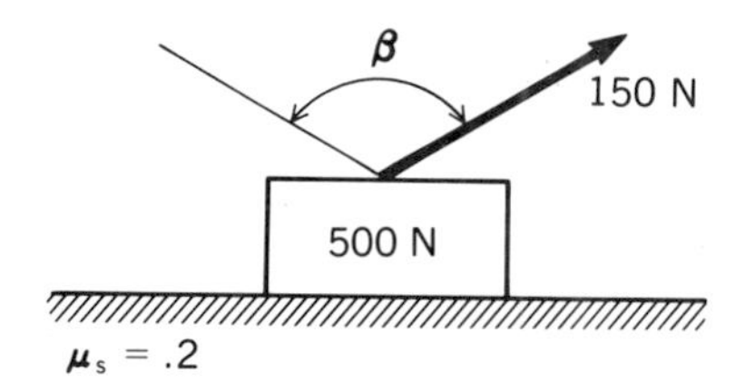

Figure P.7.12

7.13. What is the *largest* load that can be suspended without moving blocks A and B? The coefficient of friction for all plane surfaces of contact is .3. Block A weighs 500 N and block B weighs 700 N. Neglect friction in the pulley system.

Figure P.7.13

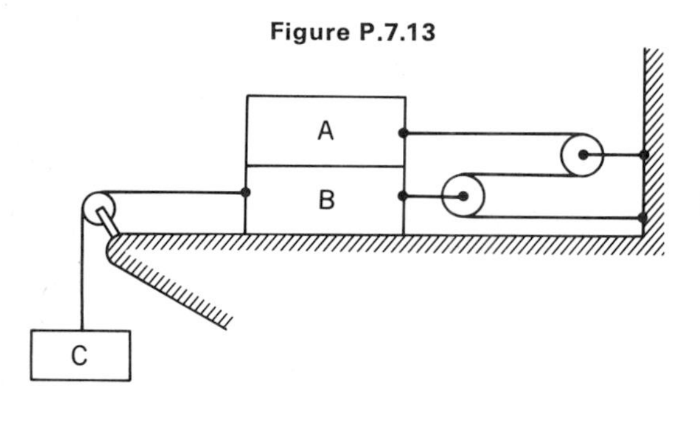

7.14. What is the *minimum* force F to hold the cylinders, each weighing 100 lb? Take $\mu_s = .2$ for all surfaces of contact.

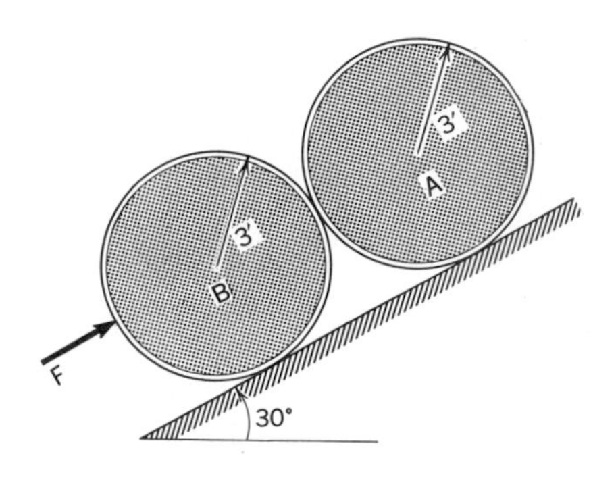

Figure P.7.14

7.15. A compressor is shown. If the pressure in the cylinder is 1.40 N/mm² above atmosphere (gage), what *minimum* torque T is needed to start the system to move? Neglect the weight of the crank and connecting rod as far as their contribution toward moving the system. Consider friction only on the piston where with the cylinder walls there is a coefficient of friction $\mu = .15$.

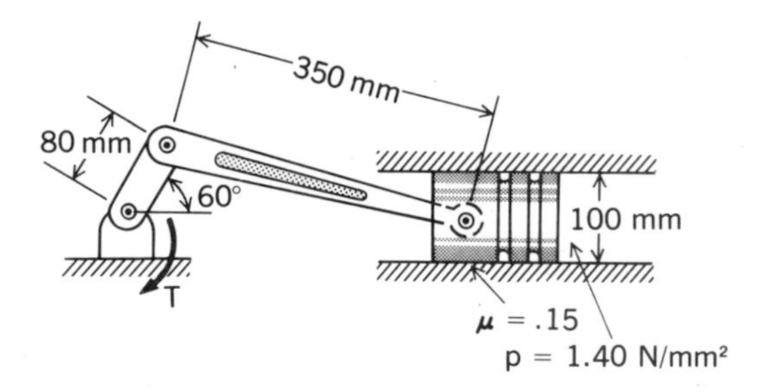

Figure P.7.15

7.16. Find the cord tension, the friction forces, and whether the blocks move if $\mu = .2$.

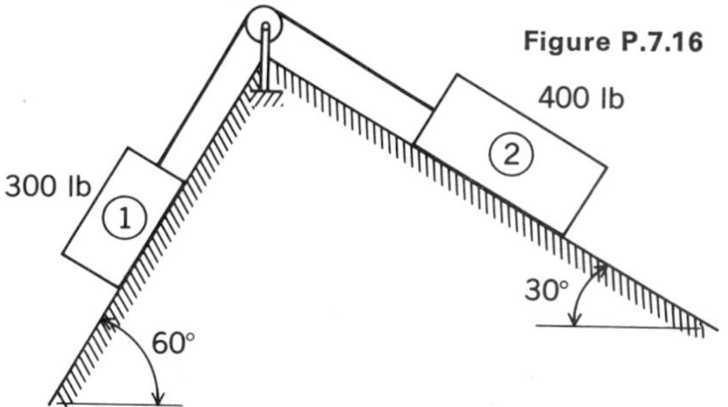

Figure P.7.16

7.17. Given that $\mu_s = .2$ for all surfaces, find the force P needed to start the block A to the right.

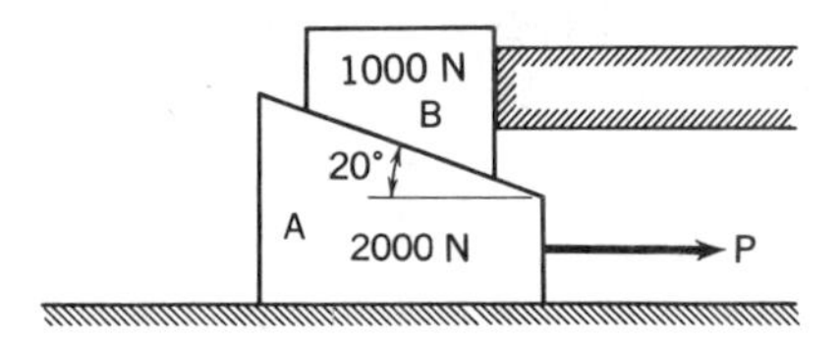

Figure P.7.17

7.18. The cylinder shown weighs 200 N and is at rest. What is the friction force at A? If there is impending slippage, what is the friction coefficient? The supporting plane is inclined at 60° to the horizontal.

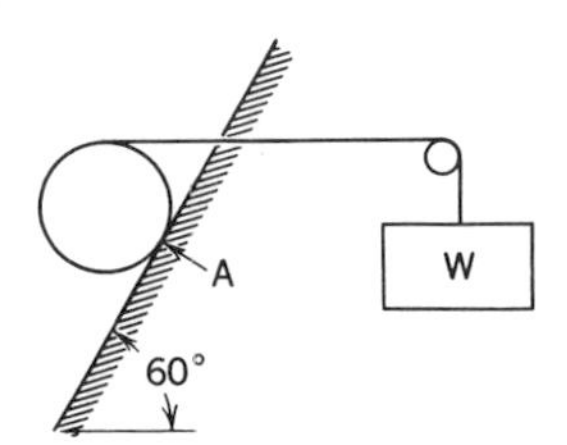

Figure P.7.18

7.19. Armature B is stationary while rotor A rotates with angular speed ω. In armature B there is a braking system. If $\mu_d = .4$, what is the braking torque on A for a force F of 300 N? Note that the rod on which F is applied is pinned at C to the armature B. Neglect friction between B and the brake pads G and H.

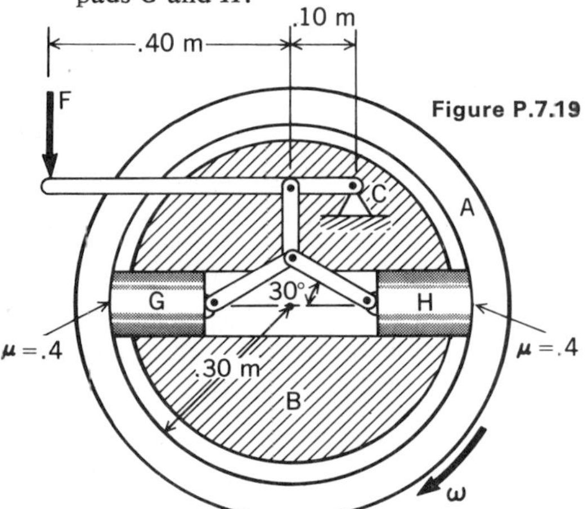

Figure P.7.19

7.20. A 200-lb load is placed on the luggage rack of the 4500-lb station wagon. Will the station wagon climb the hill easier or harder with the luggage than without? Explain. The coefficient of friction is .55.

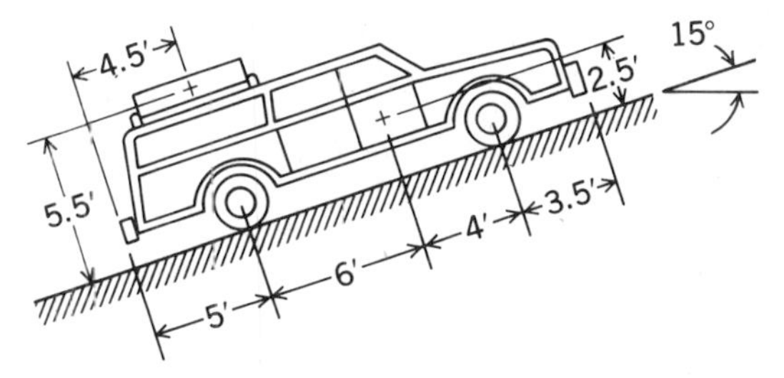

Figure P.7.20

7.21. An insect tries to climb out of a hemispherical bowl of radius 600 mm. If the coefficient of friction between insect and bowl is .4, how high up does the insect go? If the bowl is spun about a vertical axis, the bug gets pushed out in a radial direction by the force $mr\omega^2$, as you learned in physics. At what speed ω will the bug just be able to get out of the bowl?

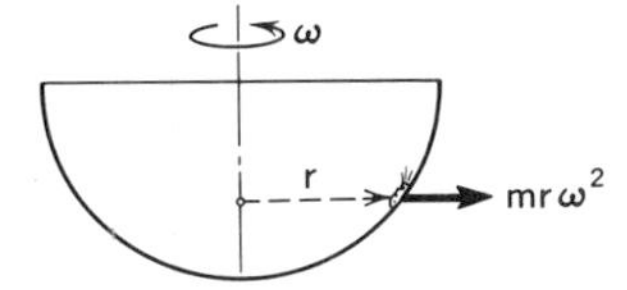

Figure P.7.21

7.22. A block A of mass 500 kg rests on a stationary support B where the static coefficient of friction $\mu_s = .4$. On the right side, support C is on rollers. The dynamic coefficient of friction μ_d of the support C with body A is .2. If C is moved at constant speed to the left, how far does it move before body A begins to move?

Figure P.7.22

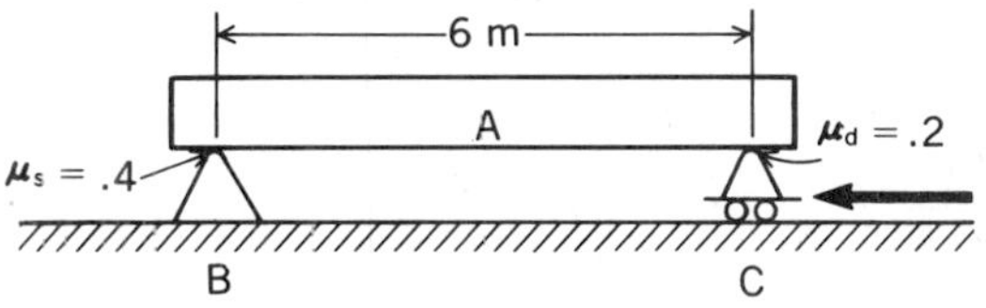

7.23. In a preliminary grinding operation for a 1500-N car engine block, the grinding wheel is pushed against the block with a 500-N force. What force must be exerted by the hydraulic ram to move the block to the right if (a) the wheel rotates clockwise and (b) the wheel rotates counterclockwise? The coefficient of friction between the grinding wheel and the block is .7 and between the table and the block is .2.

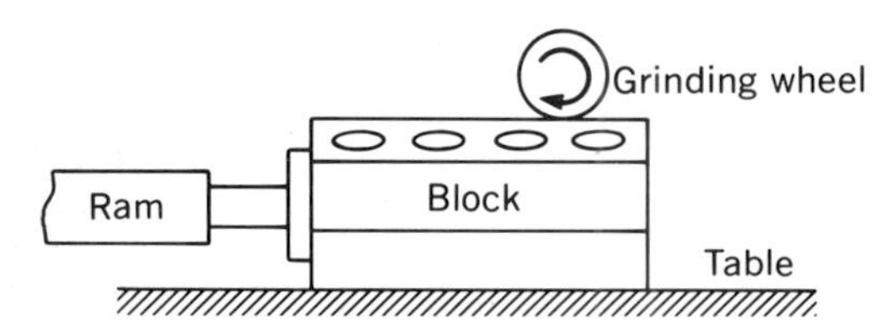

Figure P.7.23

7.24. An 8000-lb tow truck with four-wheel drive develops a torque of 750 lb-ft at each axle. What is the *heaviest* car that can be towed up a 10° slope if $\mu_s = .3$?

truck tire radius = 18 in.

7.25. A 7-m ladder weighing 250 N is being pushed by force F. What is the *minimum* force needed to get the ladder to move? The coefficient of friction for all contact surfaces is .4.

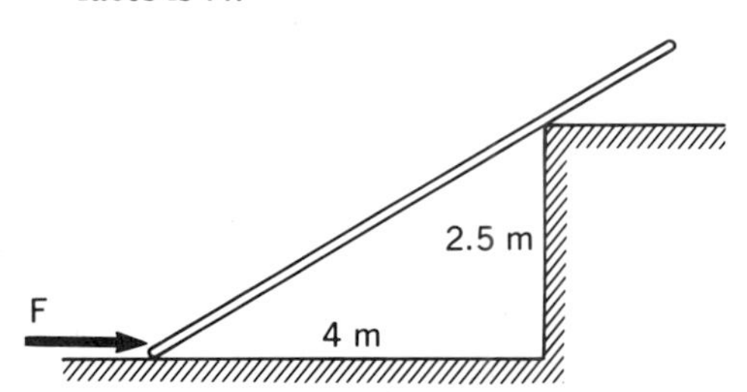

Figure P.7.25

7.26. In Problem 7.25 if F is released will the ladder begin to slide down?

7.27. Can a force P roll the 50-lb cylinder over the step? The coefficient of static friction is .4. What is the value of P if this can be done?

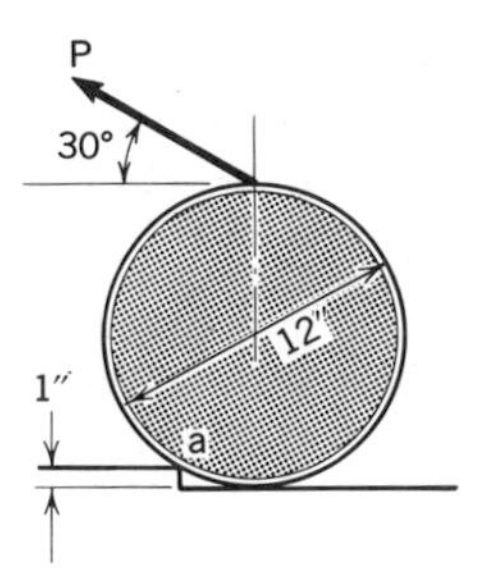

Figure P.7.27

7.28. The block of weight W is to be moved up an inclined plane. A rod of length c with negligible weight is attached to the block and the force F is applied to the top of this rod. If the coefficient of starting friction is μ_s, determine in terms of a, d, and μ_s the *maximum* length c for which the block will begin to slide rather than tip.

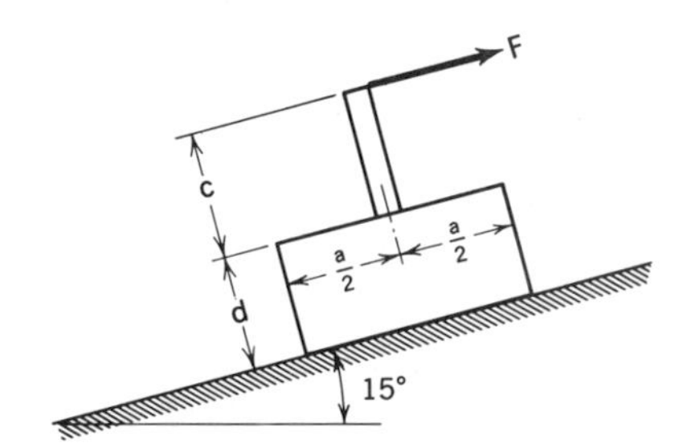

Figure P.7.28

7.29. Determine the range of values of W_1 for which the block will either slide up the plane or slide down the plane. At what value of W_1 is the friction force zero? $W_2 = 100$ lb.

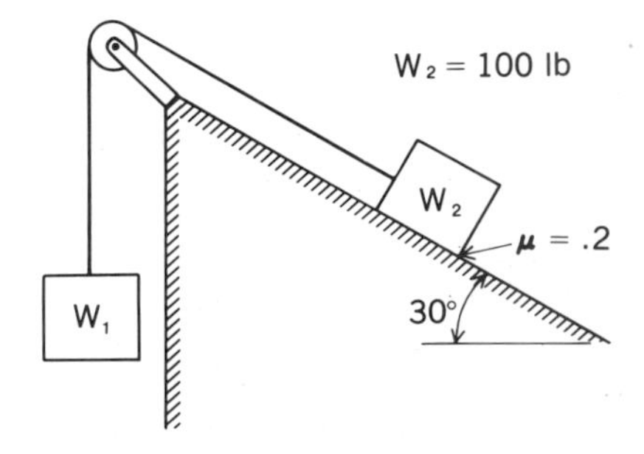

Figure P.7.29

7.30. A 200-kN tractor is to push a 60-kN concrete beam up a 15° incline at a construction site. If $\mu_d = .5$ between beam and dirt, and if μ_s is .6 between tractor tread and dirt, can the tractor do the job? If so, what torque must be developed on the tractor drive sprocket which is .8 m in diameter? What force P is then developed to push the beam?

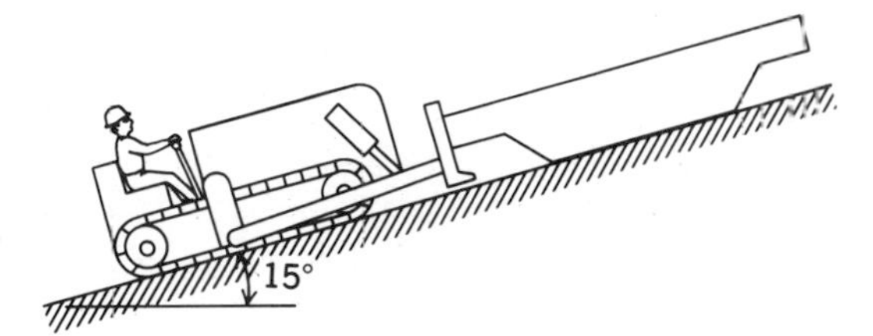

Figure P.7.30

7.31. What is the *minimum* coefficient of friction required just to maintain the bracket and its 500-lb load in a static position? (Assume point contacts at the horizontal centerlines of the arms.) The center of gravity is 7 in. from the shaft centerline.

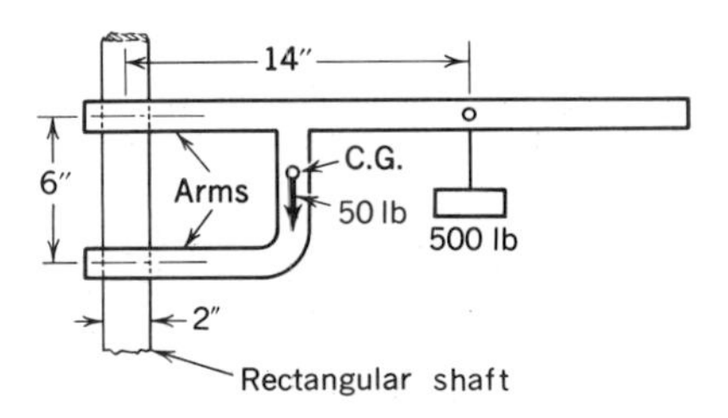

Figure P.7.31

7.32. If the coefficient of friction in Problem 7.31 is .2, at what *minimum* distance from the centerline of the vertical shaft can we support the 500-lb load without slipping?

7.33. A rod is held by a cord at one end. If the force $F = 200$ N, and if the rod weighs 450 N, what is the *maximum* angle α that the rod can be placed for μ_s between the rod and the floor equal to .4? The rod is 1 m in length.

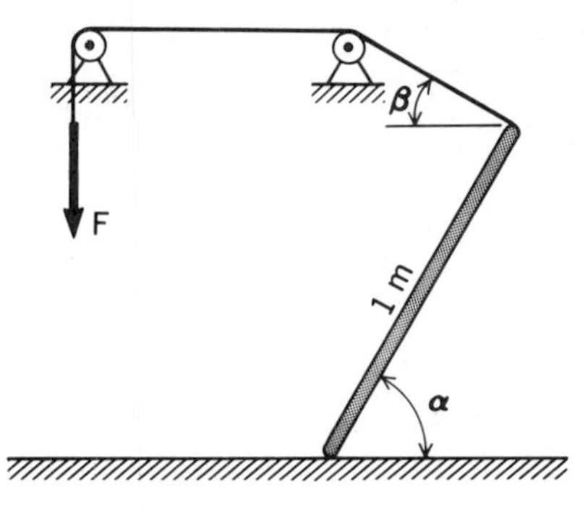

Figure P.7.33

7.34. Suppose that the ice lifter is used to support a hard block of material by friction only. What is the *minimum* coefficient of static friction, μ_s, to accomplish this for any weight W and for the geometry shown in the diagram?

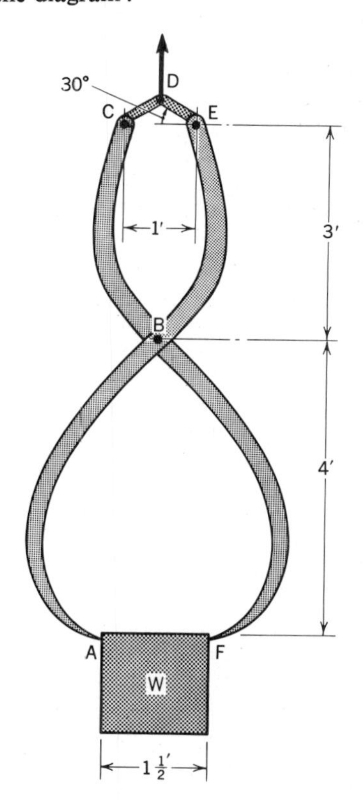

Figure P.7.34

7.35. A rectangular case is loaded with uniform vertical thin rods such that when it is full, as shown in (a), the case has a total weight of 1000 lb. The case weighs 100 lb when empty and has a coefficient of static friction of .3 with the floor as shown in the diagram. A force T of 200 lb is maintained on the case. If the rods are unloaded as shown in (b), what is the limiting value of x for equilibrium to be maintained?

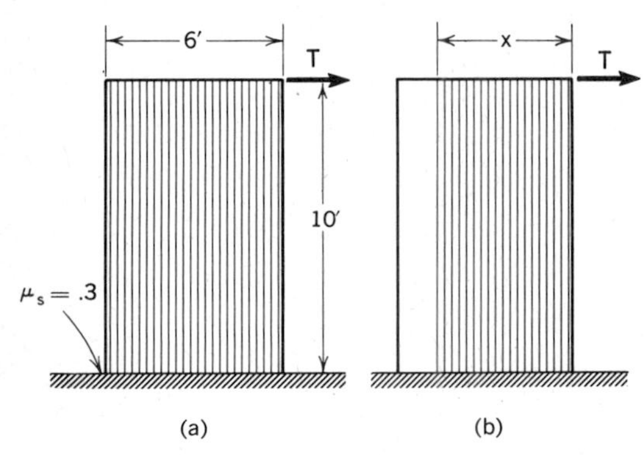

Figure P.7.35

7.36. A beam supports load C weighing 500 N. At supports A and B, the coefficient of frictions is .2. At the contact surface between load C and the beam, the coefficient of friction is .75. If force F moves C steadily to the left, how far does it move before the beam begins to move? The beam weighs 200 N. Neglect the height t of the beam in your calculations.

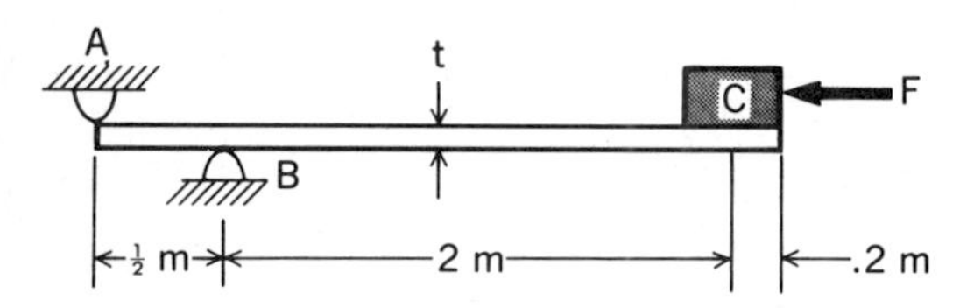

Figure P.7.36

7.37. Do Problem 7.36 for the case where the height t is taken into account. Take $t =$ 120 mm.

7.38. A rod is supported by two wheels spinning in opposite directions. If the wheels were horizontal, the rod would be placed centrally over the wheels for equilibrium. However, the wheels have an inclination of 20° as shown, and the rod must be placed at a position off center for equilibrium. If the coefficient of friction is $\mu_d = .8$, how many feet off center must the rod be placed?

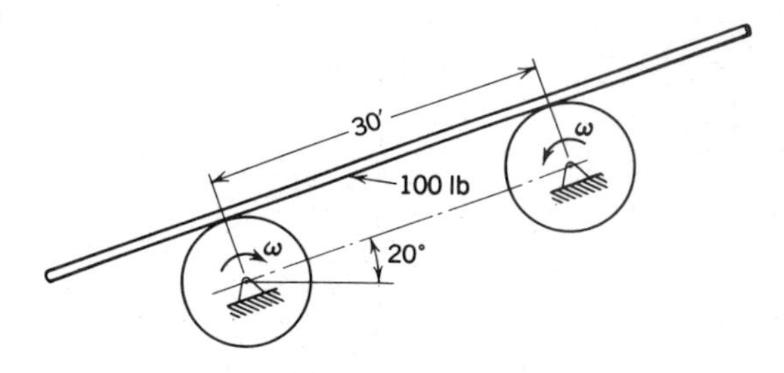

Figure P.7.38

7.39. How much force F must be applied to the wedge to begin to raise the crate? Neglect changes in geometry. What force must the stopper block provide to prevent the crate from moving to the left? The coefficient of friction between all surfaces is .3.

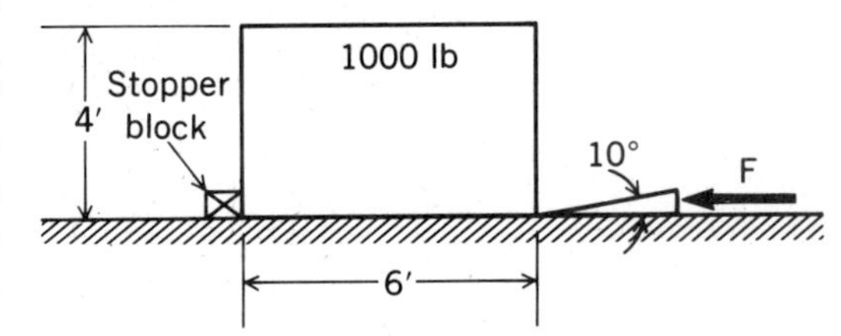

Figure P.7.39

7.40. What is the *maximum* height x of a step so that the force P will roll the 50-lb cylinder over the step with no slipping at a? Take $\mu_s = .3$.

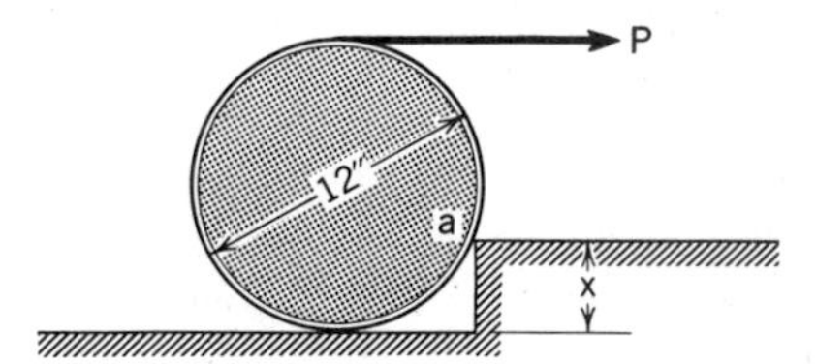

Figure P.7.40

7.41. Two identical light rods are pinned together at B. End C of rod BC is pinned while end A of rod AB rests on a rough floor having a coefficient of friction with the rod of $\mu_d = .5$. The spring requires a force of 5 N/mm of stretch. A load $F = 300$ N is applied slowly at B and then maintained constant. What is the angle θ when the system ceases to move? The spring is unstretched when $\theta = 45°$. (*Hint:* You will have to solve an equation by trial and error.)

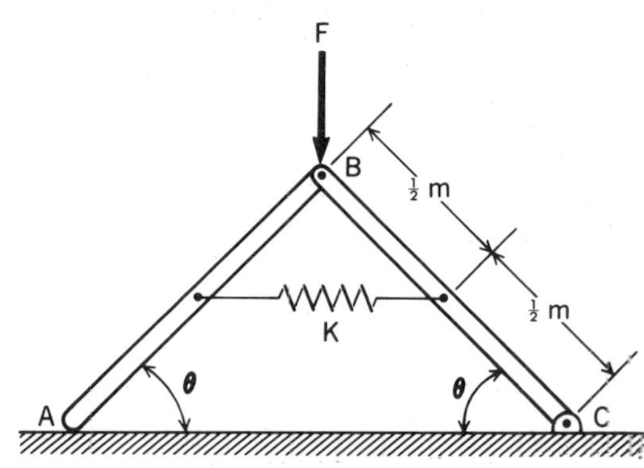

Figure P.7.41

7.42. A device for "throwing" baseballs is shown. This device is to be found in amusement parks for batting practice. The two wheels are inflated automobile tires that rotate as shown with a constant speed ω. A baseball is fed to a point where it is touching the wheels. What is the *minimum* separation d of the wheels if the ball is to be drawn into the slot and then ejected on the other side as a pitched ball? The coefficient of friction at the contact surfaces is .4.

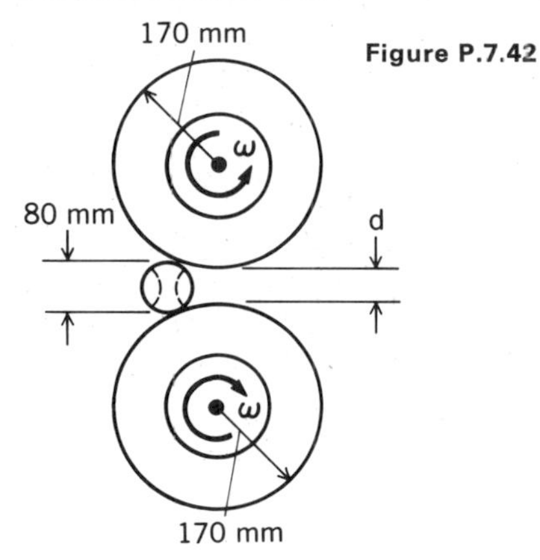

Figure P.7.42

7.43. What is the *maximum* angle α for which there will be equilibrium if A has a mass of 50 kg. The coefficient of friction at the supports is equal to .3. What is the force in each of the supporting members?

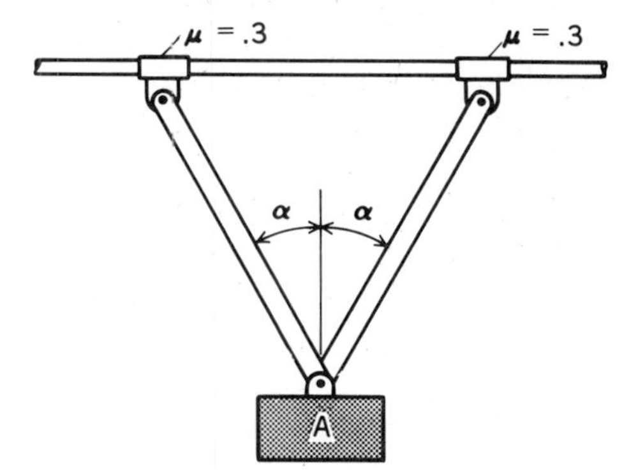

Figure P.7.43

7.44. What is the *maximum* angle α for which there will be equilibrium if A weighs 1000 N and if μ at the supporting surface is .3? The rods are each 1.3 m long. You will have to solve a transcendental equation by trial and error.

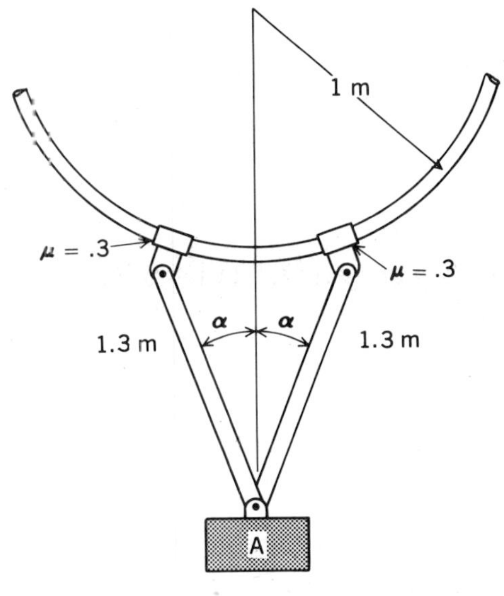

Figure P.7.44

7.45. The rod AD is pulled at A and it moves to the left. If the coefficient of dynamic friction for the road at A and B is .4, what must the *minimum* value of W_2 be to prevent the block from tipping when $\alpha = 20°$? With this value of W_2, determine the minimum coefficient of static friction between the block and the supporting plane needed to just prevent the block from sliding. W_1 is 100 N.

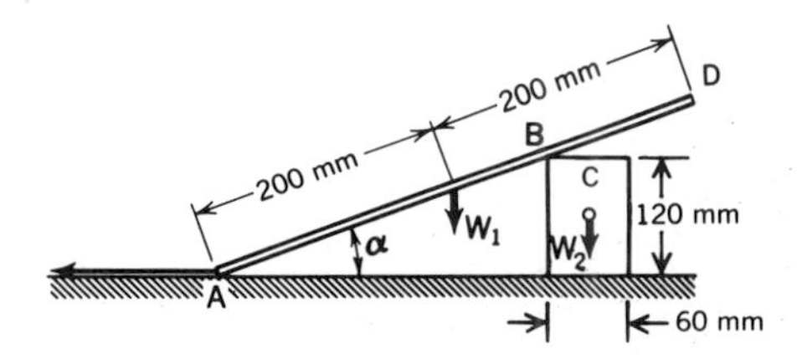

Figure P.7.45

7.46. If we neglect friction at the rollers, and if the coefficient of static friction is .2 for all surfaces, ascertain whether the 5000-lb weight will go up, go down, or stay stationary.

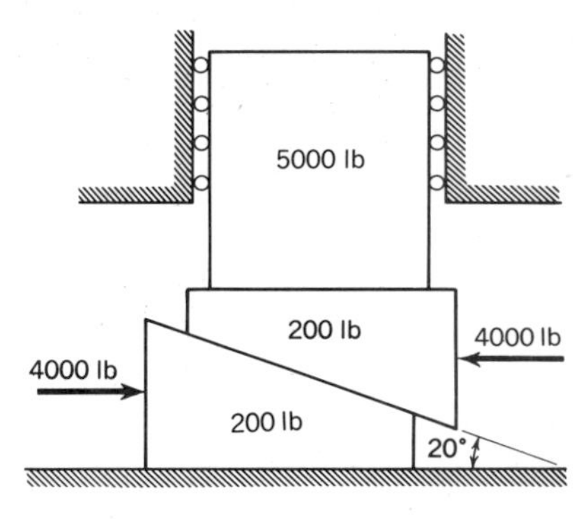

Figure P.7.46

7.4 Complex Surface Contact Friction Problems

In the examples undertaken heretofore, the nature of the relative impending or actual motion between the plane surfaces of contact was quite simple—that of motion without rotation. We shall now examine more general types of contacts between bodies. In Example 7.5, we have a plane contact surface but with varying direction of impending or slipping motion for the area elements as a result of rotation. In such problems we shall have to apply Coulomb's laws locally to infinitesimal areas of contact and to integrate the results, for reasons explained in Section 7.2. To do this, we must ascertain the distribution of the normal force at the contact surface, an undertaking that is usually difficult and well beyond the capabilities of rigid-body statics, as explained in Chapter 5. However, we can at times *approximately* compute frictional effects by *estimating* the manner of distribution of the normal force at the surface of contact. We now illustrate this.

EXAMPLE 7.5

Compute the frictional resistance to rotation of a rotating solid cylinder with an attached pad A pressing against a flat dry surface with a force P (see Fig. 7.10). The pad A and the stationary flat dry surface consitute a dry *thrust bearing*.

The direction of the frictional forces distributed over the contact surface is no longer simple. We therefore take an infinitesimal area for examination. This area is shown in Fig. 7.10, where the element has been formed from polar-coordinate differentials so as to be related simply to the boundaries. The area dA is equal to

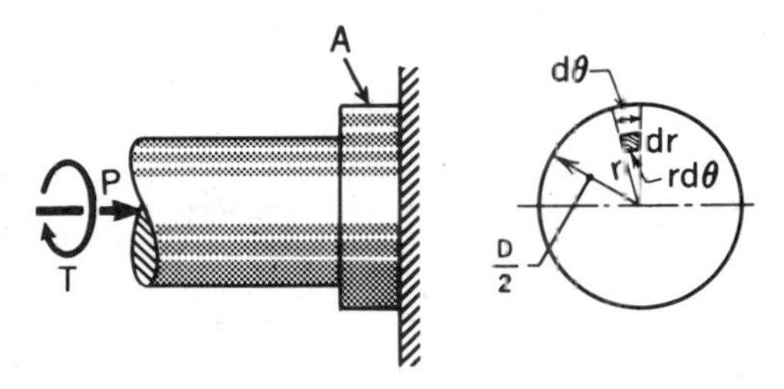

Figure 7.10. Dry thrust bearing.

$r\,d\theta\,dr$. We shall *assume* that the normal force P is uniformly distributed over the entire area of contact. The normal force on the area element is then

$$dN = \frac{P}{\pi D^2/4} r\,d\theta\,dr \tag{a}$$

The friction force associated with this force during motion is

$$df = \mu_d \frac{P}{\pi D^2/4} r\,d\theta\,dr \tag{b}$$

The direction of df must oppose the relative motion between the surfaces. The relative motion is rotation of concentric circles about the centerline, so the direction of a force df_1 (Fig. 7.11) must lie tangent to a circle of radius r. At 180° from the position of the area element for df_1, we may carry out a similar calculation for a force df_2, which for the same r must be equal and opposite to df_1, thus forming a couple. Since the entire area may be decomposed in this way, we can conclude that there are only couples in the plane of contact. If we take moments of all infinitesimal forces about the center, we get the magnitude of the total frictional couple moment. The direction of the couple moment is along the shaft axis. First, consider area elements on the ring of radius r:

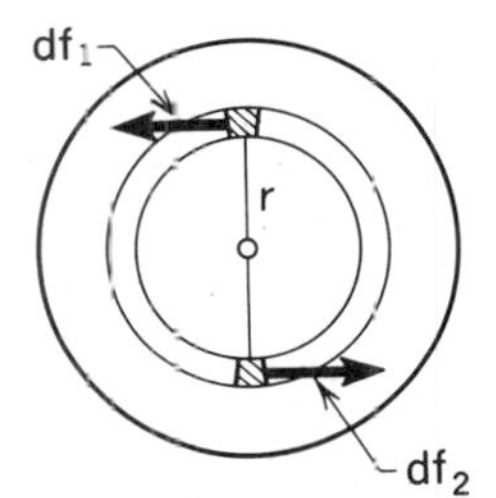

Figure 7.11. Friction forces form couples.

$$dM = \int_0^{2\pi} r\,\mu_d \frac{P}{\pi D^2/4} r\,d\theta\,dr \tag{c}$$

Taking μ_d as constant and holding r constant, we have on integration with respect to θ:

$$dM = \mu_d \frac{P}{\pi D^2/4} 2\pi r^2\,dr$$

We thus account for all area elements on the ring of radius r. To account for all the rings of the contact surface, we next integrate with respect to r from zero to $D/2$. Clearly, this gives us the total resisting torque M. Thus,

$$M = \mu_d \frac{8P}{D^2} \int_0^{D/2} r^2\,dr = \frac{PD\mu_d}{3} \tag{d}$$

What we have performed in the last three steps is *multiple integration*, which we introduced in Chapter 4 when dealing with rectangular coordinates.

7.5 Belt Friction

A flexible belt is shown in Fig. 7.12 wrapped around a portion of a drum, with the amount of wrap indicated by angle β. The angle β is called the *angle of wrap*. Assume that the drum is stationary and tensions T_1 and T_2 are such that motion is impending between the belt and the drum. We shall take the impending motion of the belt to be clockwise relative to the drum, and therefore the tension T_1 exceeds tension T_2.

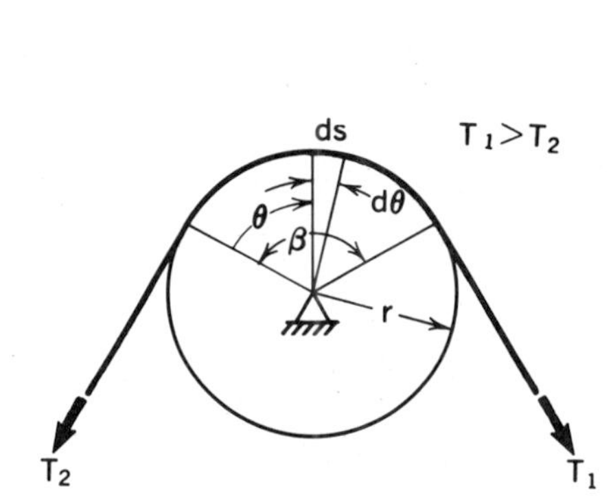

Figure 7.12. Belt wrapped around drum.

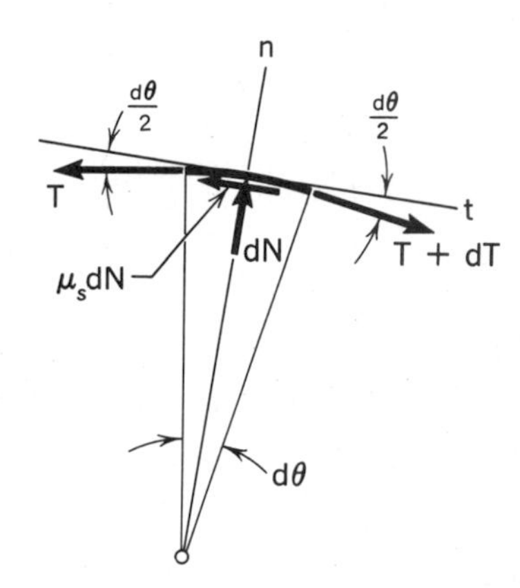

Figure 7.13. Free-body diagram of segment of belt; impending slippage.

Consider an infinitesimal segment of the belt as a free body. This segment subtends an angle $d\theta$ at the drum center as shown in Fig. 7.13. Summing force components in the radial and transverse directions and equating them to zero, we get the following scalar equations:

$\underline{\sum F_t = 0:}$

$$-T\cos\frac{d\theta}{2} + (T + dT)\cos\frac{d\theta}{2} - \mu_s\,dN = 0$$

Therefore,

$$dT\cos\frac{d\theta}{2} = \mu_s\,dN$$

$\underline{\sum F_n = 0:}$

$$-T\sin\frac{d\theta}{2} - (T + dT)\sin\frac{d\theta}{2} + dN = 0$$

Therefore,

$$-2T\sin\frac{d\theta}{2} - dT\sin\frac{d\theta}{2} + dN = 0$$

The sine of a very small angle approximately equals the angle itself in radians. Furthermore, to the same degree of accuracy, the cosine of a small angle approaches unity. (That these relations are true may be seen by expanding the sine and cosine in a power series and then retaining only the first terms.) The preceding equilibrium equations then become

$$dT = \mu_s\, dN \tag{7.2a}$$

$$-T\, d\theta - dT\frac{d\theta}{2} + dN = 0 \tag{7.2b}$$

In the last equation, we have an expression involving the product of two infinitesimals. This quantity may be considered negligible compared to the other terms of the equation involving only one differential. Thus, we have for this equation:

$$T\, d\theta = dN \tag{7.3}$$

From Eqs. 7.2a and 7.3, we may form an equation involving T and θ. Thus, by eliminating dN from the equations, we have

$$dT = \mu_s T\, d\theta$$

Hence,

$$\frac{dT}{T} = \mu_s\, d\theta$$

Integrating both sides around the portion of the belt in contact with the drum,

$$\int_{T_2}^{T_1} \frac{dT}{T} = \int_0^{\beta} \mu_s\, d\theta$$

we get

$$\ln \frac{T_1}{T_2} = \mu_s \beta$$

or

$$\frac{T_1}{T_2} = e^{\mu_s \beta} \tag{7.4}$$

We therefore have established a relation between the tensions on each part of the belt at a condition of impending motion between the belt and the drum. The same relation can be reached for a *rotating* drum with impending slippage between the belt and the drum *if we neglect centrifugal effects on the belt*. Furthermore, by using the *dynamic* coefficient of friction in the formula above, we have the case of the belt slipping over either a rotating or stationary drum (again neglecting centrifugal effects on the belt). Thus, for all such cases, we have

$$\boxed{\frac{T_1}{T_2} = e^{\mu \beta}} \tag{7.5}$$

where the proper coefficient of friction must be used to suit the problem, and the angle β must be expressed in *radians*. Note that *the ratio of tensions depends only on the angle*

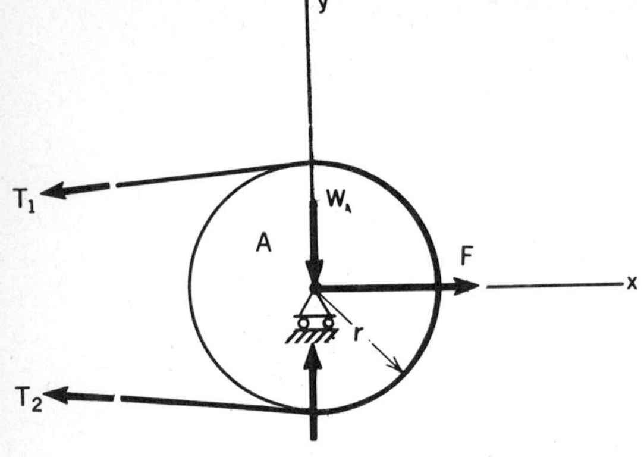

Figure 7.14. Force F affects T_1 and T_2 but not T_1/T_2.

of wrap β and the coefficient of friction μ. Thus, if the drum A is forced to the right, as shown in Fig. 7.14, the tensions will increase, but if β is not affected by the action, the ratio of T_1/T_2 for impending or actual slippage is not affected by this action. However, the *torque* developed by belt on drum as a result of friction *is* affected by the force F. The torque is easily determined by using the drum and the portion of the belt in contact with the drum as a free body, as is shown in Fig. 7.14. Thus,

$$\text{torque} = T_1 r - T_2 r = (T_1 - T_2)r \tag{7.6}$$

If we pull the drum to the right without disturbing the angle of wrap, we can see from Eq. 7.5 that the tensions T_1 and T_2 must increase by the same factor. And if we call this factor H, the new tensions become HT_1 and HT_2, respectively. Substituting into Eq. 7.6, we see that the frictional torque is also increased by the same factor:

$$\text{torque} = H(T_1 - T_2)r = H(\text{torque})_{\text{original}}$$

Sometimes we know the force F (Fig. 7.14) which acts on the drum support to maintain the belt tension. Summing forces on the free-body diagram in Fig. 7.14, we get in the x direction:

$$(T_1)_x + (T_2)_x = F_{\text{known}} \tag{7.7}$$

At impending slippage or at slippage, Eq. 7.5 is valid, and with Eq. 7.7, we can solve for T_1 and T_2. With Eq. 7.6, the torque that the belt is capable of developing on the drum now becomes a simple computation.

EXAMPLE 7.6

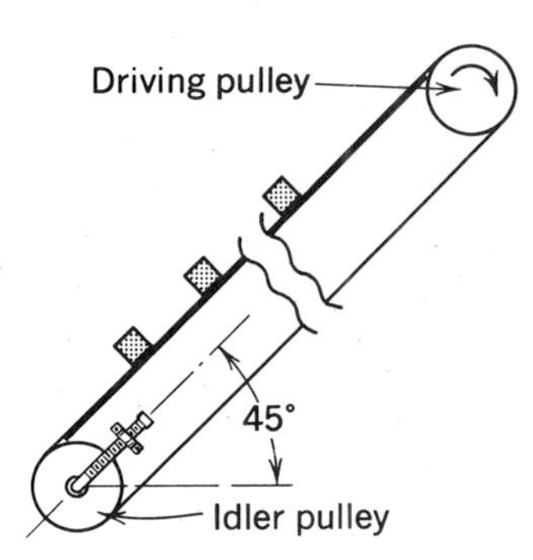

Figure 7.15. Conveyor.

A conveyor is moving ten 50-lb boxes at a 45° setting (Fig. 7.15). The coefficient of friction between the belt and the bed of the coveyor is .05. Furthermore, the coefficient of friction between the driving pulley and the belt is .4. The idler pulley is moved along the direction of the conveyor by a crank mechanism so that the idler pulley is subject to a force F of 500 lb. Compute the maximum tension found in the belt and ascertain if there will be slipping on the driving pulley. Neglect the weight of the belt.

In Fig. 7.16, we have shown free-body diagrams of various parts of the conveyor.[3] For the portion of the belt on the conveyor frame, we can sum forces normal and tangent to the belt:

$\underline{\sum F_n = 0:}$

$$N - (10)(50)(.707) = 0$$

[3]The weights of the pulleys have been counteracted by supporting forces at the axles and have not been shown.

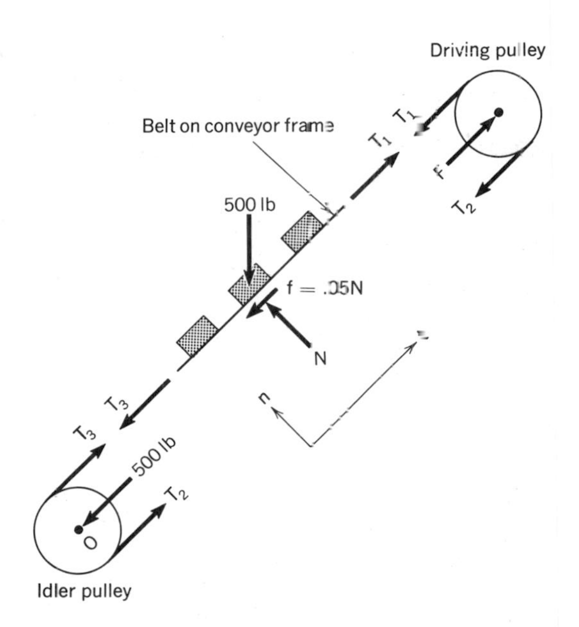

Figure 7.16. Various free-body diagrams of parts of conveyor.

Therefore,

$$N = 354 \text{ lb} \tag{a}$$

$\underline{\sum F_t = 0:}$

$$T_1 - T_3 - (10)(50)(.707) - (.05)(354) = 0$$

Therefore,

$$T_1 - T_3 = 371 \text{ lb} \tag{b}$$

For the idler pulley, we have

$\underline{\sum M_0 = 0:}$

$$T_3 = T_2 \tag{c}$$

$\underline{\sum F_t = 0:}$

$$T_3 + T_2 = 500 \tag{d}$$

From Eqs. (c) and (d), we conclude that

$$T_2 = T_3 = 250 \text{ lb} \tag{e}$$

From Eq. (b) we now get for the maximum tension T_1:

$$T_1 = T_3 + 371 = 621 \text{ lb} \tag{f}$$

We must next check the driving pulley to ensure that there is no slippage occurring. For the condition of impending slippage, we have using as T_2 the value of 250 lb and solving for T_1

$$T_1 = T_2 e^{.4\pi} = (250)(3.51) = 878 \text{ lb}$$

Clearly, since the T_1 needed is only 621 lb, we do not have slippage at the driving pulley, and we conclude that the maximum tension is indeed 621 lb.

EXAMPLE 7.7

An electric motor (not shown) in Fig. 7.17 drives at constant speed the pulley B, which connects to pulley A by a belt. Pulley A is connected to a compressor (not shown) which requires 700 N-m torque to drive it at constant speed ω_A. If μ_s for the belt and either pulley is .4, what minimum value of the indicated force F is required to have no slipping anywhere?

As a first step, we determine the angles of wrap β for the respective pulleys. For this purpose, we first compute α (Fig. 7.18). Note that the radii O_AD and O_BE, being perpendicular to the same line DE, are therefore parallel to each other. Drawing EC parallel to O_AO_B, we then form α in the cross-hatched triangle. Hence, we can say:

$$\alpha = \sin^{-1}\frac{CD}{CE} = \sin^{-1}\frac{r_A - r_B}{O_A O_B} = \sin^{-1}\frac{.50 - .30}{2} = 5.74°$$

Hence,

$$\beta_A = 180° + 2(5.74) = 191.5°$$

$$\beta_B = 180° - 2(5.74) = 168.5°$$

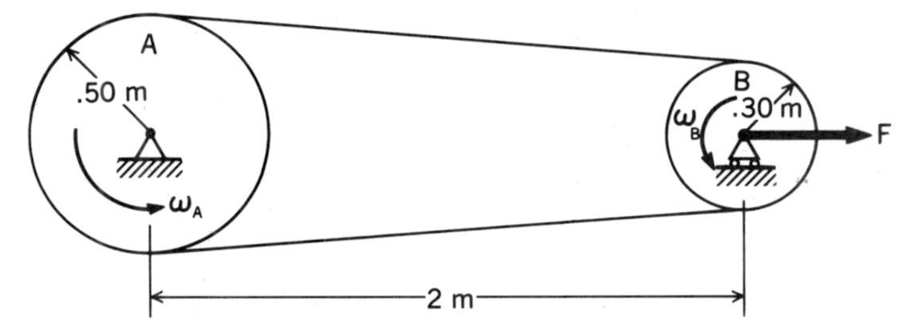

Figure 7.17. Belt-driven compressor.

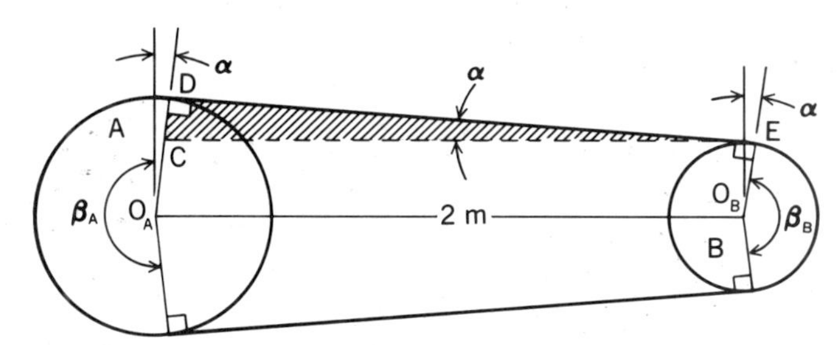

Figure 7.18. Find angles of wrap.

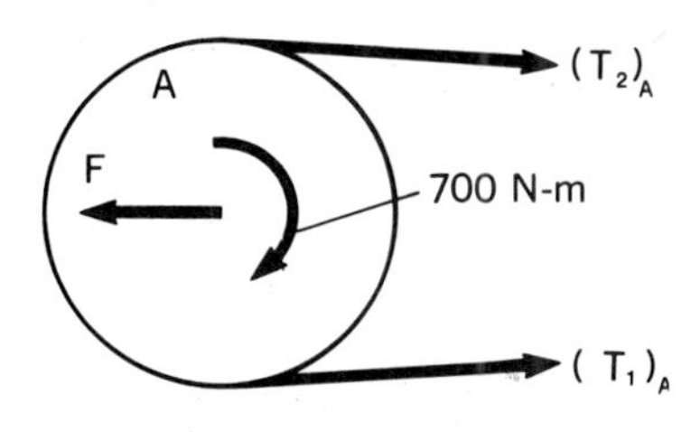

Figure 7.19. Free-body diagram of A.

Now consider pulley A as a free body in Fig. 7.19. Note that the minimum force F corresponds to the condition of impending slippage. Accordingly, for this condition at A, we have

$$\frac{(T_1)_A}{(T_2)_A} = e^{\mu_s \beta_A} = e^{(.4)[(191.5/360)2\pi]} = 3.81 \qquad \text{(a)}$$

Also, summing moments about the center of the pulley, we have

$$[(T_1)_A - (T_2)_A](.50) - 700 = 0 \qquad \text{(b)}$$

Therefore,

$$(T_1)_A - (T_2)_A = 1400$$

Solving Eqs. (a) and (b) simultaneously, we get

$$(T_1)_A = 1898 \text{ N}; \qquad (T_2)_A = 498 \text{ N}$$

From equilibrium, we can compute force F as follows:

$$(1898 + 498)\cos 5.74° - F = 0 \qquad \text{(c)}$$

Therefore,

$$F = 2384 \text{ N}$$

Now go to pulley B to see what minimum force F is needed so that the belt does not slip on it during operations. Consider in Fig. 7.20 the free-body diagram of pulley B. For impending slipping on pulley B, we have

$$\frac{(T_1)_B}{(T_2)_B} = e^{\mu_s \beta_B} = e^{(.4)[(168.5/360)2\pi]} = 3.24 \qquad \text{(d)}$$

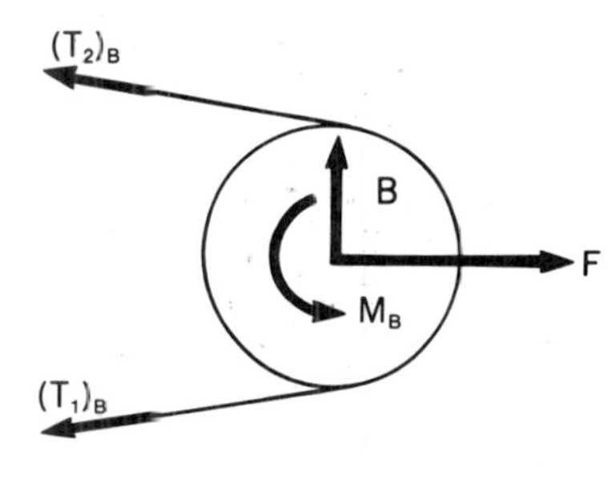

Figure 7.20. Free-body diagram of *B*.

The torque for pulley B needed to develop 700 N-m on pulley A is next computed. Thus,[4]

$$M_B = \frac{r_B}{r_A} M_A = \frac{.30}{.50} M_A = \frac{.30}{.50}(700)$$

Therefore,

$$M_B = 420 \text{ N-m}$$

Summing moments in Fig. 7.20 about the center of B, we then have on using the above result

$$-[(T_1)_B - (T_2)_B](.30) - 420 = 0$$

Therefore,

$$(T_1)_B - (T_2)_B = 1400 \qquad \text{(e)}$$

Solving Eqs. (d) and (e) simultaneously, we get

$$(T_1)_B = 2025 \text{ N}; \qquad (T_2)_B = 625 \text{ N}$$

Hence, the minimum F needed for pulley B is

$$F = (2025 + 625)\cos 5.74° = 2637 \text{ N}$$

Thus, for no slipping on *either* pulley we require $F = 2637$ N as a minimum value.

[4]Note that the ratio of transmitted torques M_2/M_1 between directly connected pulleys and gears will equal r_2/r_1 or D_2/D_1 of the pulleys or gears. Can you verify this yourself?

Problems

7.47. Compute the frictional resisting torque for the concentric dry thrust bearing. The coefficient of friction is taken as μ_d.

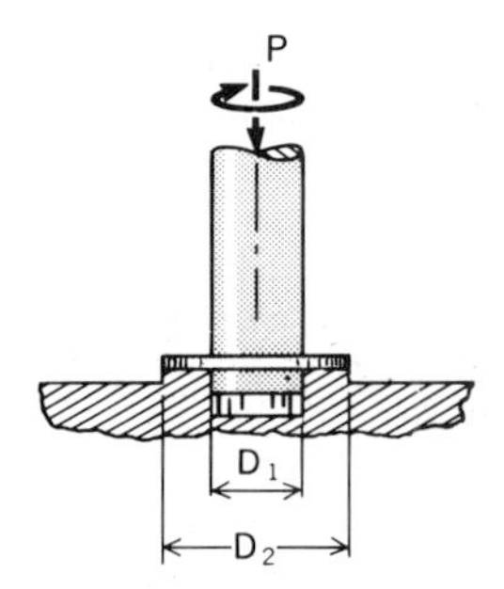

Figure P.7.47

7.48. The support end of a dry thrust bearing is shown. Four pads form the contact surface. If a shaft creates a 100-N thrust uniformly distributed over the pads, what is the resisting torque for a coefficient of friction of .1?

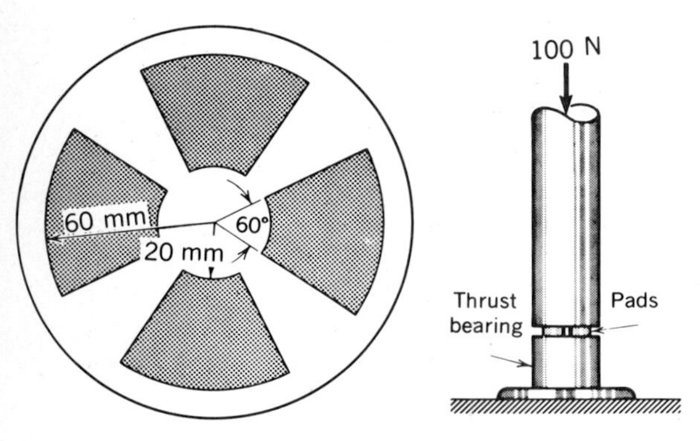

Figure P.7.48

7.49. In Example 7.5, the normal force distribution at the contact surface is not uniform but, as a result of wear, is inversely proportional to the radius r. What, then, is the resisting torque M?

7.50. Compute the frictional torque needed to rotate the truncated cone relative to the fixed member. The cone has a 20-mm-diameter base and a 60° cone angle and is cut off 3 mm from the cone tip. The coefficient of dynamic friction is .2.

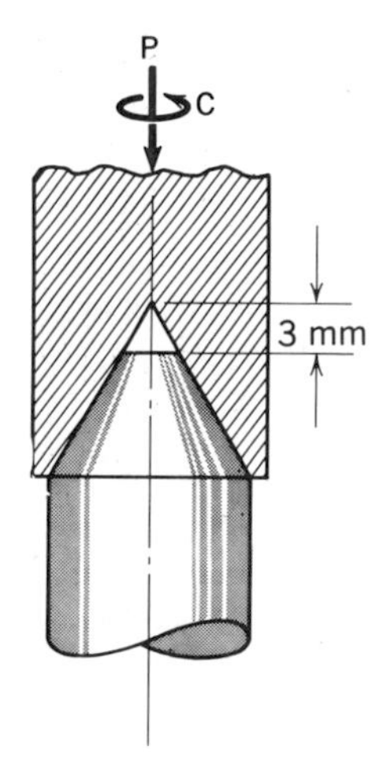

Figure P.7.50

7.51. A 1000-N block is being lowered down an inclined surface. The block is pinned to the incline at C, and at B a cord is played out so as to cause the body to rotate at uniform speed about C. Taking μ_d to be .3 and assuming the contact pressure is uniform along the base of the block, compute T for the configuration shown in the diagram.

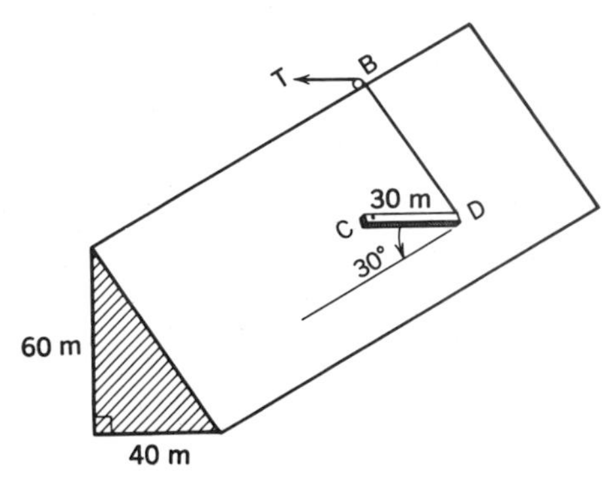

Figure P.7.51

7.52. A pulley requires 200 N-m torque to get it rotating. The angle of wrap is π radians, and μ_s is known to be 5.2. What is the *minimum* horizontal force F required to create enough tension in the belt so that it can rotate the pulley?

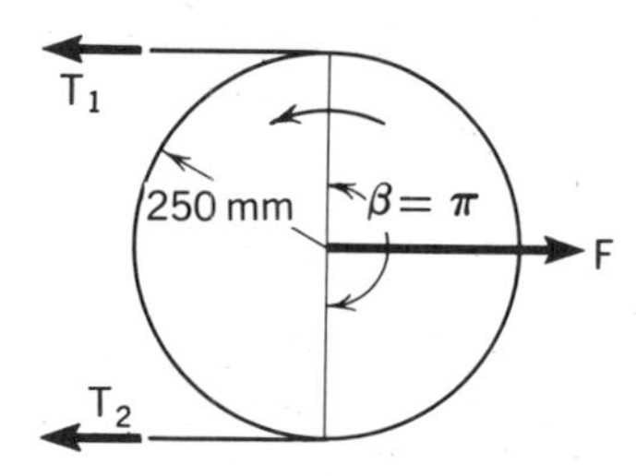

Figure P.7.52

7.53. If in Problem 7.52 the belt is wrapped $2\frac{1}{2}$ times around the pulley, what is the *minimum* horizontal force F needed to rotate the pulley?

7.54. The seaman pulls with 100-N force and wants to stop the motorboat from moving away from the dock under power. How few wraps n of the rope must he make around the post if the motorboat develops 3500 N of thrust and the coefficient of friction between the rope and the post is .2?

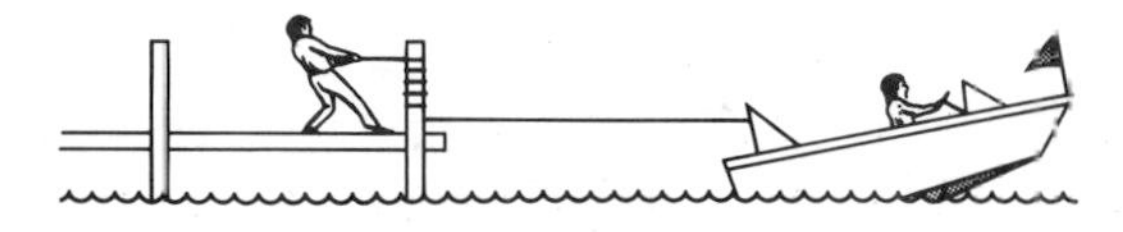

Figure P.7.54

7.55. A length of belt rests on a flat surface and runs over a quarter of the drum. A load W rests on the horizontal portion of the belt, which in turn is supported by a table. If the coefficient of friction for all surfaces is .3, compute the *maximum* weight W that can be moved by rotating the drum.

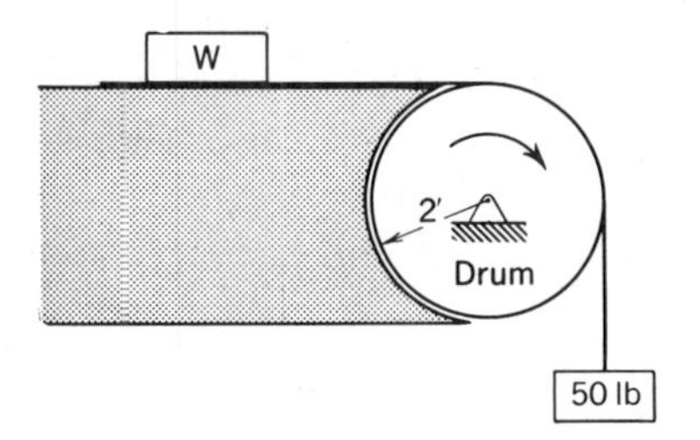

Figure P.7.55

7.56. The rope holding the 50-lb weight E passes over the drum and is attached at A. The weight of C is 60 lb. What is the *minimum* coefficient of friction between the rope and the drum to maintain equilibrium?

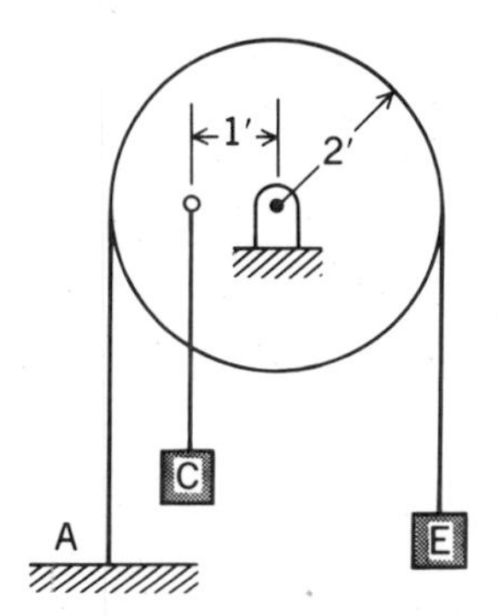

Figure P.7.56

7.57. What is the *maximum* weight that can be supported by the system in the position shown? Pulley B *cannot* turn. Bar AC is fixed to cylinder A, which weighs 500 N. The coefficient of static friction for all contact surfaces is .3.

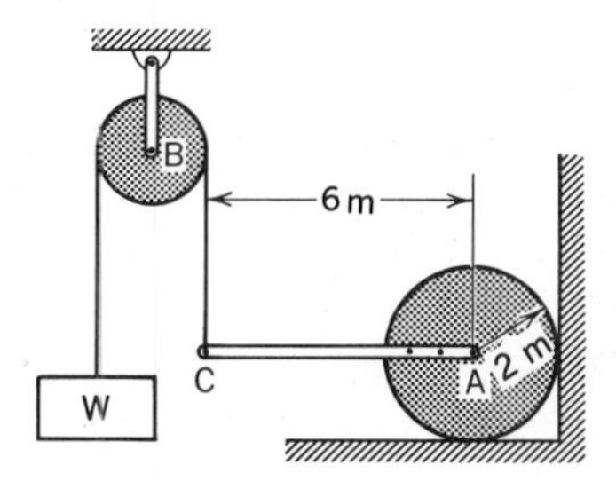

Figure P.7.57

7.58. A mountain climber of weight W hangs freely suspended by one rope that is fastened at one end to his waist, wrapped one-half turn about a rock with $\mu_s = .2$, and held at the other end in his hand. What *minimum* force in terms of W must he pull with to maintain his position? What minimum force must he pull with to gain altitude?

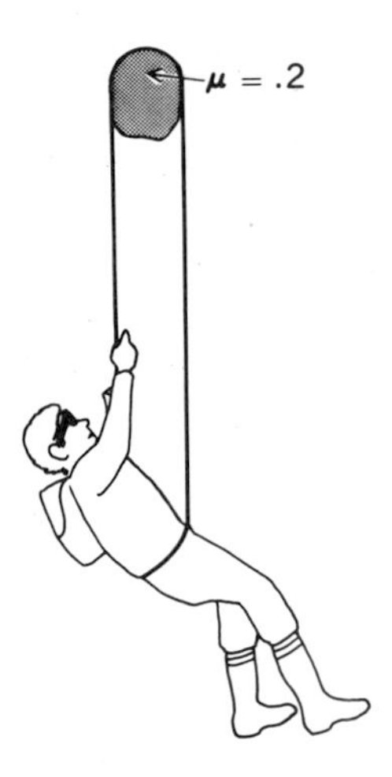

Figure P.7.58

7.59. Pulley B is turned by a diesel engine and drives pulley A connected to a generator. If the torque that A must transmit to the generator is 500 N-m, what is the *minimum* coefficient of friction between the belt and pulleys for the case where the force F is 2000 N?

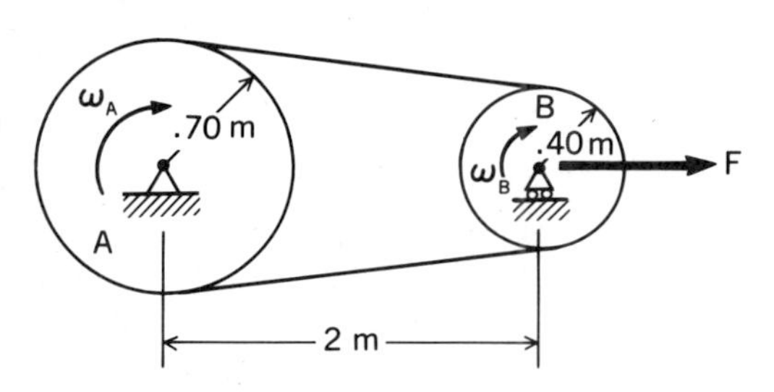

Figure P.7.59

7.60. A hand brake is shown. If $\mu_d = .4$, what is the resisting torque when the shaft is rotating?

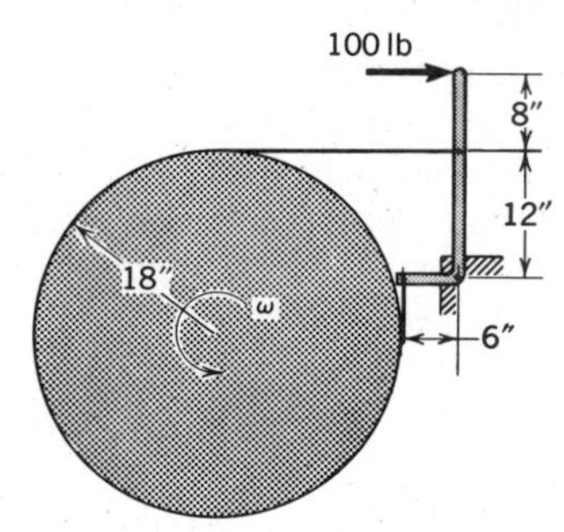

Figure P.7.60

7.61. A conveyor is shown with two driving pulleys A and B. Driver A has an angle of wrap of 330°, whereas B has a wrap of 180°. If the coefficient of friction between the belt and the bed of the conveyor is .1, and the weight to be transported is 10,000 N, what is the *smallest* coefficient of friction between the belt and the driving pulleys? One-fifth of the load can be assumed to be between the active pulleys at all times, and the tension in the slack side (underneath) is 2000 N. There is a free-wheeling pulley at the left end of the conveyor. You will have to solve an equation by trial and error.

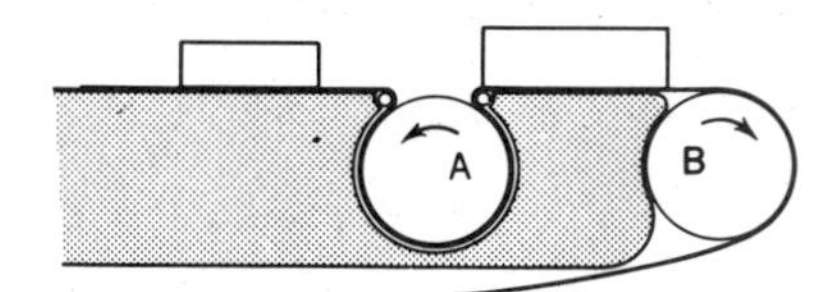

Figure P.7.61

7.62. A freely turning idler pulley is used to increase the angle of wrap for the pulleys shown. If the tension in the slack side below is 200 lb, find the *maximum* torque that can be transmitted by the pulleys for a coefficient of friction of .3.

Figure P.7.62

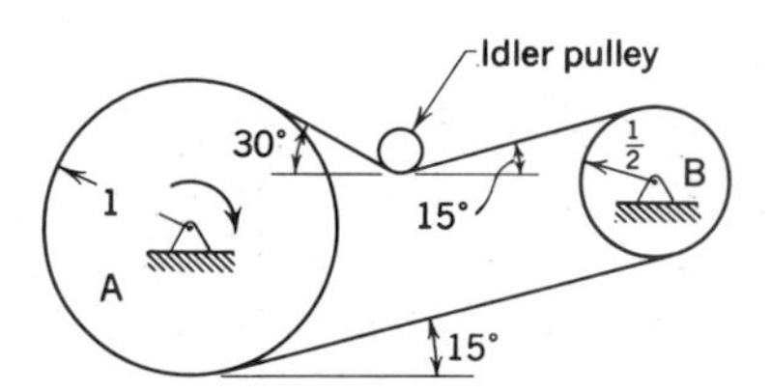

7.63. Rod AB weighing 200 N is supported by a cable wrapped around a semicylinder having a coefficient of friction μ_s equal to .2. A weight A having a mass of 10 kg can slide on rod AB. What is the maximum range x from the centerline that the center of A can be placed without causing slippage?

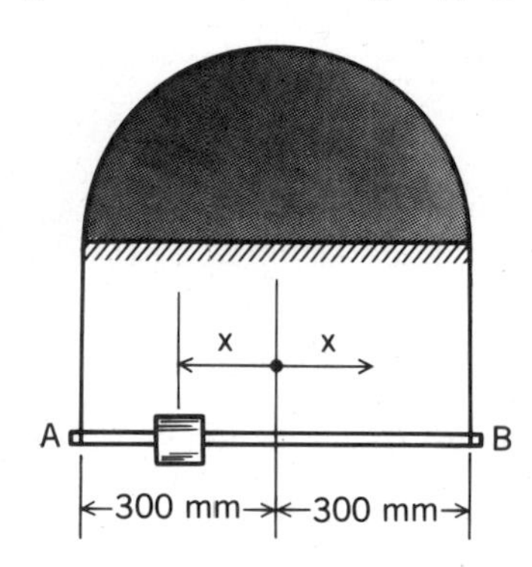

Figure P.7.63

7.64. The cable mechanism shown is similar to that used to move the station indicator on a radio. If the indicator jams, what force is developed at the indicator base to free the jam when the torque applied to the handle is 10 lb-in.? Also, what are the forces in the various regions of the cable? The coefficient of friction is .15.

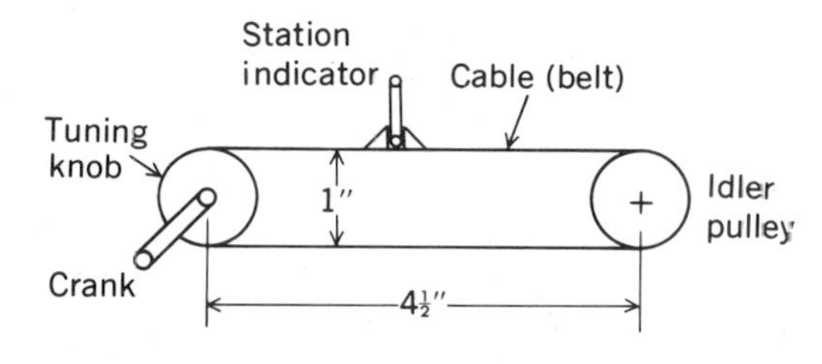

Figure P.7.64

7.65. What are the *minimum* possible supporting force components needed for pulley B as a result of the action of the belt? The coefficient of friction for the belt and pulley B is .3 and for the belt and pulley A is .4. The torque that the belt delivers to A is 200 N-m.

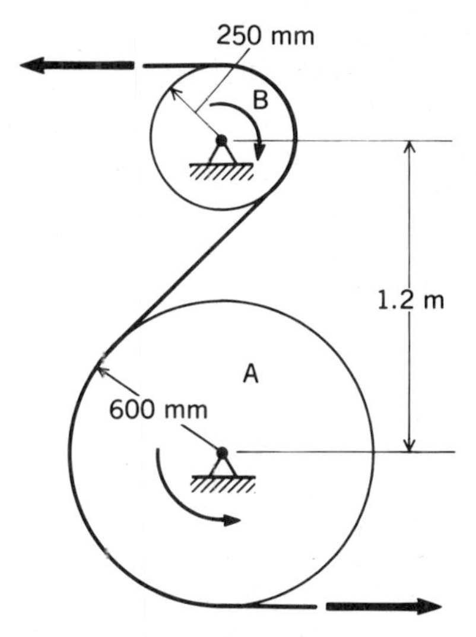

Figure P.7.65

7.66. From first principles, show that the normal force per unit length, w, acting on a drum from a belt is given as

$$w = \frac{T_2}{r} e^{\mu\theta}$$

Use the indicated diagram as an aid. [*Hint:* Start with Eq. 7.2(a) and use Eq. 7.4 for any point *a*.]

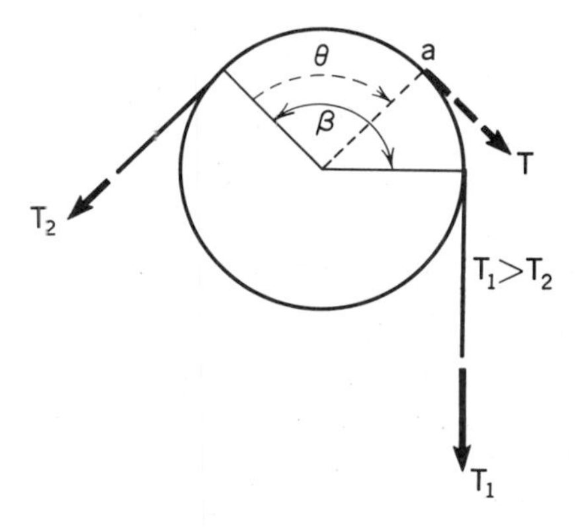

Figure P.7.66

7.67. What *minimum* force F is needed so that drum A can transmit a clockwise torque of 500 N-m without slipping? The coefficient of friction, μ_s, for A and the belt is .4. What *minimum* coefficient of friction is needed for B for no slipping?

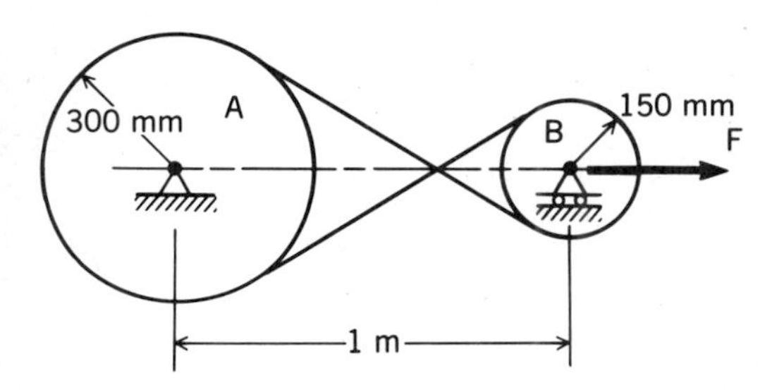

Figure P.7.67

7.68. What is the *minimum* weight B that will prevent rotation induced by body C weighing 500 N? The weight of A is 100 N. The coefficient of friction between the belts and A is .4, and between A and the walls is .1. Neglect friction at pulley G.

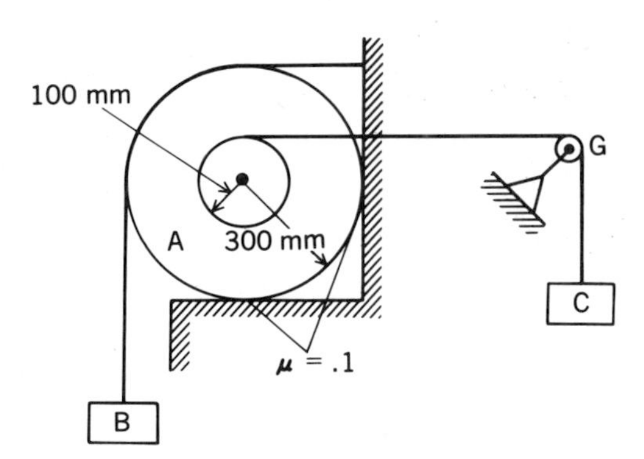

Figure P.7.68

7.69. A V-belt is shown. Show that

$$\frac{T_1}{T_2} = e^{\mu_s \beta / \sin(\alpha/2)}$$

for impending slippage. Use a development analogous to that of the flat belt in Section 7.5.

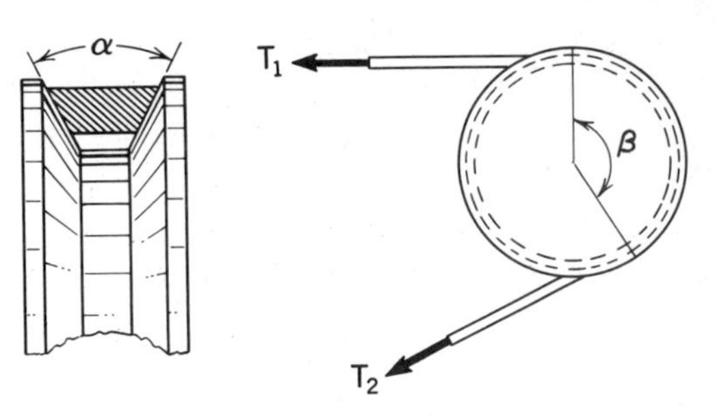

Figure P.7.69

7.70. An electric motor drives a pulley B which drives three V-belts having the cross section shown. These V-belts then drive a compressor through pulley A. If the torque needed to drive the compressor is 1000 N-m, what *minimum* force F is needed to do the job? The coefficient of friction between belts and pulleys is .5. See Problem 7.69 before doing this problem.

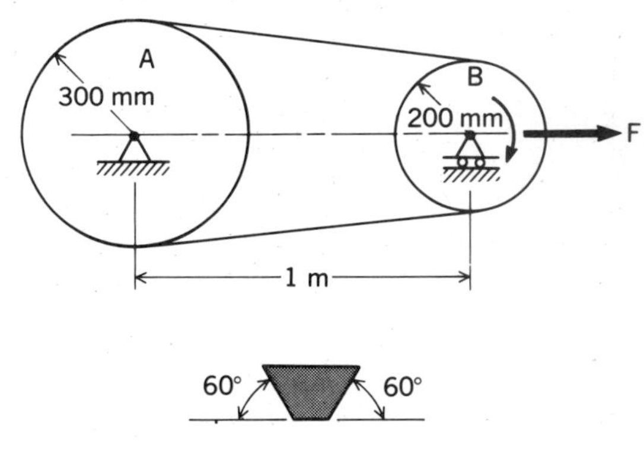

Figure P.7.70

7.71. A drum of radius r is rotated by a belt with a constant speed ω rad/sec. What is the relation between T_1 and T_2 for the case of impending slipping between drum and belt if *centrifugal* effects are counted? The belt has a mass per unit length of m kg/m. Recall from physics that the centrifugal force of a particle of mass M is $Mr\omega^2$, where r is the distance from the axis of rotation. Assume the belt is thin compared to the radius of the drum. The desired result is

$$\frac{T_1 - r^2\omega^2 m}{T_2 - r^2\omega^2 m} = e^{\mu\beta}$$

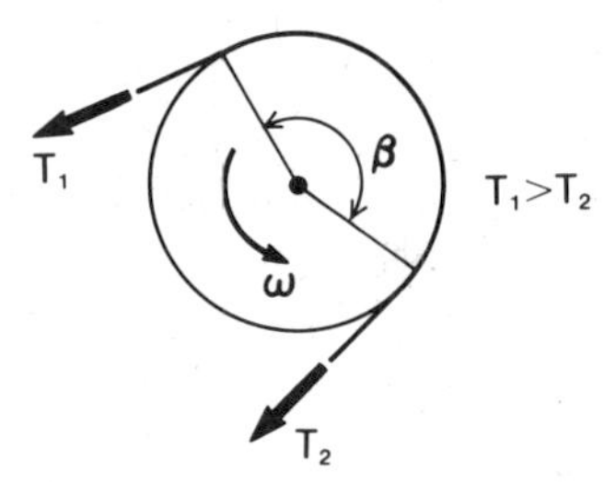

Figure P.7.71

7.72. A pulley A is driven by an outside agent at a speed ω of 100 rpm. A belt weighing 30 N/m is driven by the pulley. If $T_2 = 200$ N, what is the *maximum* possible tension T_1 computed without considering centrifugal effects? Compute T_1 counting centrifugal effects, and give the percentage error incurred by not including centrifugal effects. The coefficient of friction between the belt and the pulley is .3. See Problem 7.71 before doing this problem.

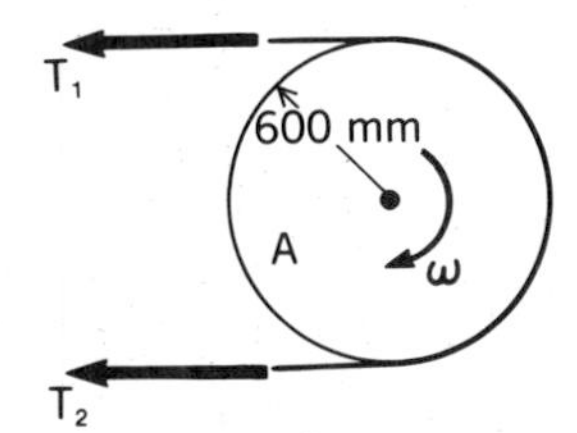

Figure P.7.72

7.6 The Square Screw Thread

We shall now consider the action of a nut on a screw that has square threads (Fig. 7.21). Let us take r as the mean radius from the centerline of the screw to the thread. The *pitch*, p, is the distance along the screw between adjacent threads, and the *lead*, L, is the distance that a nut will advance in the direction of the axis of the screw in one revolution. For screw threads that are single-threaded, L equals p. For an n-threaded screw, the lead L is np.

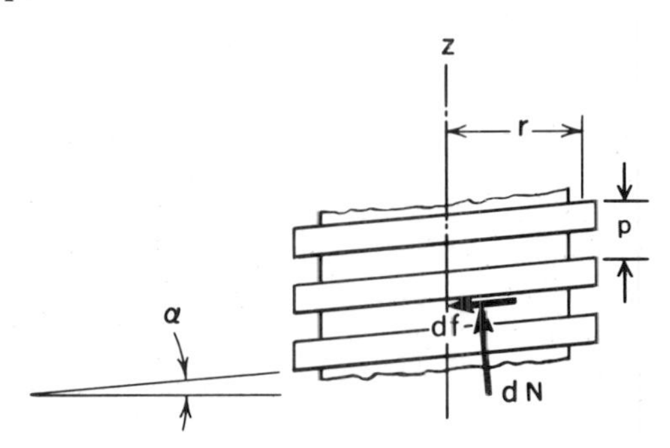

Figure 7.21. Square screw thread.

Forces are transmitted from screw to nut over several revolutions of thread, and hence we have a distribution of normal and friction forces. However, because of the narrow width of the thread, we may consider the distribution to be confined at a distance r from the centerline, thus forming a "loading" strip winding around the centerline of the screw. Figure 7.21 illustrates infinitesimal normal and frictional forces on an infinitesimal part of the strip. The local slope $\tan\alpha$ as one looks in radially is determined by considering the definition of L, the lead. Thus,

$$\text{slope} = \tan\alpha = \frac{L}{2\pi r} = \frac{np}{2\pi r}$$

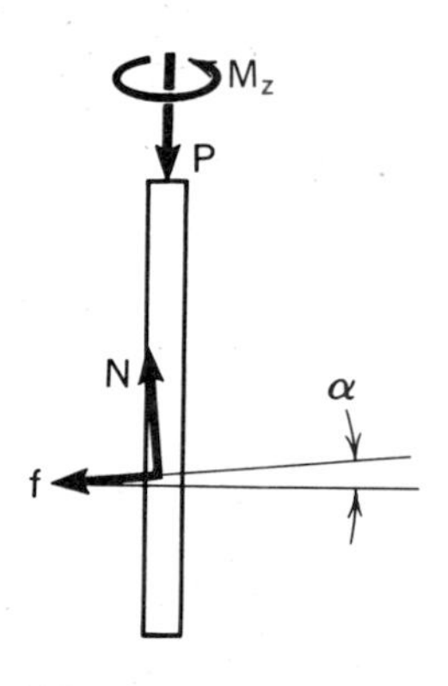

Figure 7.22. Free-body diagram.

All elements of the proposed distribution have the same inclination (direction cosine) relative to the z direction. In the summation of forces in this direction, therefore, we can consider the distribution to be replaced by a single normal force N and a single friction force f at the inclinations shown in Fig. 7.22 at a position anywhere along the thread. And, since the elements of the distribution have the same moment arm about the centerline in addition to the common inclination, we may use the concentrated forces mentioned above in taking moments about the centerline. There is thus a "limited equivalence" between N and f and the force distribution from the nut onto the screw. The other forces on the screw will be considered as an axial load P and a torque M_z collinear with P (Fig. 7.22). For equilibrium at a condition of impending motion to raise the screw, we then have the following scalar equations:[5]

$\underline{\sum F_z = 0}$:

$$-P + N\cos\alpha - \mu N \sin\alpha = 0 \tag{a}$$

$\underline{\sum M_z = 0}$:

$$-\mu N \cos\alpha\, r - N \sin\alpha\, r + M_z = 0 \tag{b}$$

These equations may be used to eliminate the force N and so get a relation between P and M_z that will be of practical significance. This may readily be done by solving for N in (a) and substituting into (b). The result is

$$\boxed{M_z = \frac{Pr\,(\mu\cos\alpha + \sin\alpha)}{\cos\alpha - \mu\sin\alpha}} \tag{7.8}$$

An important question arises when we employ the screw and nut in the form of a jack as shown in Fig. 7.23. Once having raised a load P by applying the torque M_z to the jackscrew, does the device maintain the load at the raised position when the applied torque is released, or does the screw unwind under the action of the load and thus lower the load? In other words, is this a *self-locking* device? To examine this, we go back to the equations of equilibrium. Setting $M_z = 0$ and changing the direction of the friction forces, we have the condition for impending "unwinding" of the screw. Eliminating N from the equations, we get

[5]The equations also apply to *steady rotation* of the nut on the screw, in which case one uses the dynamic coefficient of friction μ_d in the equations.

$$\frac{Pr\,(-\mu_s \cos \alpha + \sin \alpha)}{\cos \alpha + \mu_s \sin \alpha} = 0$$

This requires that

$$-\mu_s \cos \alpha + \sin \alpha = 0$$

Therefore,

$$\mu_s = \tan \alpha \qquad (7.9)$$

We can conclude that, if the coefficient of friction μ_s equals or exceeds $\tan \alpha$, we will have a self-locking condition. If μ_s is less than $\tan \alpha$, the screw will unwind and will not support a load P without the proper external torque.

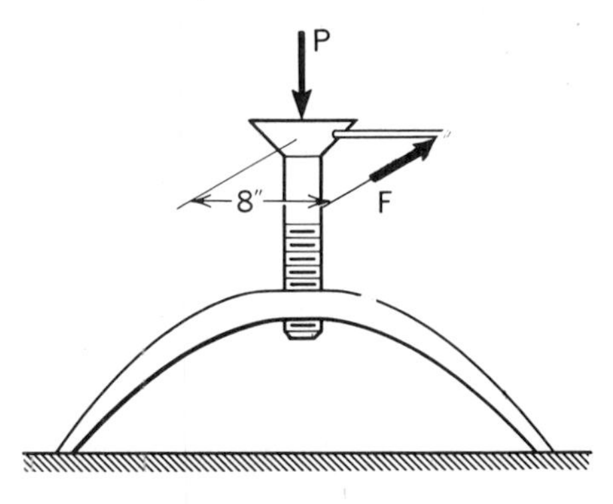

Figure 7.23. Jackscrew.

EXAMPLE 7.8

A jackscrew with a double thread of mean diameter 2 in. is shown in Fig. 7.23. The pitch is .2 in. If a force F of 40 lb is applied to the device, what load W can be raised? With this load on the device, what will happen if the applied force F is released? Take $\mu_s = .3$ for the surfaces of contact.

The applied torque M_z is clearly:

$$M_z = \tfrac{8}{12}(40) = 26.7 \text{ lb-ft} \qquad \text{(a)}$$

The angle α for this screw is given as

$$\tan \alpha = \frac{(2)(.2)}{(2\pi)(1)} = .0636 \qquad \text{(b)}$$

Therefore,

$$\alpha = 3.64°$$

Using Eq. 7.8 we can solve for P. Thus,

$$\begin{aligned} P &= \frac{M_z(\cos \alpha - \mu_s \sin \alpha)}{r(\mu_s \cos \alpha + \sin \alpha)} \\ &= \frac{(26.7)[.998 - (.3)(.0635)]}{\frac{1}{12}[(.3)(.998) + .0635]} = 864 \text{ lb} \end{aligned} \qquad \text{(c)}$$

The load W is 864 lb. The device is self-locking since μ_s exceeds $\tan \alpha = .0636$. To lower the load requires a reverse torque. We may readily compute this torque by using Eq. 7.8 with the friction forces reversed. Thus,

$$\begin{aligned} (M_z)_{\text{down}} &= \frac{864(\frac{1}{12})[-(.3)(.998) + .0635]}{.998 + (.3)(.0635)} \\ &= -16.71 \text{ lb-ft} \end{aligned} \qquad \text{(d)}$$

*7.7 Rolling Resistance

Let us now consider the situation where a hard roller moves without slipping along a horizontal surface while supporting a load W at the center. Since we know from experience that a horizontal force P is required to maintain uniform motion, some sort of

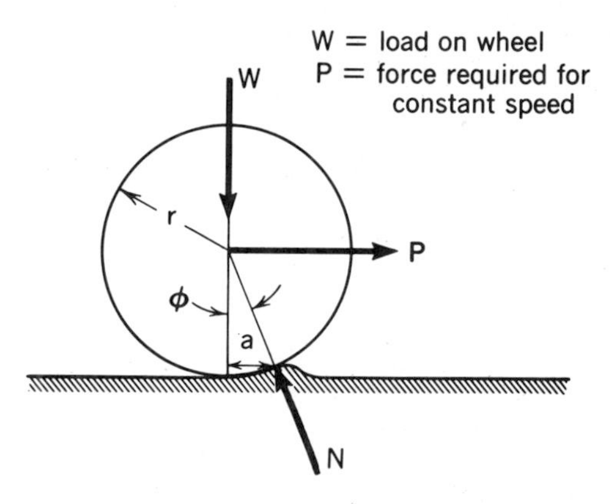

Figure 7.24. Rolling resistance model.

resistance must be present. We can understand this resistance if we examine the deformation shown in an exaggerated manner in Fig. 7.24. If force P is along the centerline as shown, the equivalent force system coming onto the roller from the region of contact must be that of a force N whose line of action also goes through the center of the roller since, you will recall from Chapter 5, three nonparallel forces must be concurrent for equilibrium. In order to develop a resistance to motion, clearly N must be oriented at an angle ϕ with the vertical direction, as is shown in Fig. 7.24. The scalar equations of equilibrium become

$$W = N\cos\phi; \qquad P = N\sin\phi$$

Therefore,

$$\frac{P}{W} = \tan\phi \tag{7.10}$$

Since the area of contact is small, we note that ϕ is a small angle and that $\tan\phi \approx \sin\phi$. The $\sin\phi$ is seen to be a/r from Fig. 7.24. Therefore, we may say that

$$\frac{P}{W} = \frac{a}{r} \tag{7.11a}$$

Solving for P, we get

$$P = \frac{Wa}{r} \tag{7.11b}$$

The distance a in these equations is called the *coefficient of rolling resistance.*

Coulomb suggested that for *variable* loads W the ratio P/W is constant for given materials and a given geometry ($r =$ constant). Looking at Eq. 7.11a, we see that a must then be a constant for given geometry and materials. Coulomb added that, for given materials and *variable* radius, the ratio P/W varies inversely as r; that is, as the radius of the cylinder is increased, the resistance to uniform motion for a given load W decreases. Thus, considering Eq. 7.11a again, we may conclude that, for given materials, a is also constant for all sizes of rollers and loads. However, other investigators have contested both statements, particularly the latter one, and there is a need for further investigation in this area. Lacking better data, we present the following list of rolling coefficients for your use, but we must caution that you should not expect great accuracy from this general procedure.

COEFFICIENTS OF ROLLING RESISTANCE

	a (in.)
Steel on steel	.007 –.015
Steel on wood	.06 –.10
Pneumatic tires on smooth road	.02 –.03
Pneumatic tires on mud road	.04 –.06
Hardened steel on hardened steel	.0002–.0005

EXAMPLE 7.9

What is the rolling resistance of a railroad freight car weighing 100 tons? The wheels have a diameter of 30 in. The coefficient of rolling resistance between wheel and track is .001 in. Compare the resistance to that of a truck and trailer having the same total weight and with tires having a diameter of 4 ft. The coefficient of rolling resistance a for the truck tires and road is .025 in.

We can use Eq. 7.11b directly for the desired results. Thus, for the railroad freight car, we have[6]

$$P_1 = \frac{(100)(2000)(.001)}{15} = 13.33 \text{ lb} \quad \text{(a)}$$

For the truck, we get

$$P_2 = \frac{(100)(2000)(.025)}{24} = 208 \text{ lb} \quad \text{(b)}$$

We see a decided difference between the two vehicles, with clear advantage toward the railroad freight car.

Problems

7.73. A simple C-clamp is used to hold two pieces of metal together. The clamp has a single square thread with a pitch of .12 in. and a mean diameter of .75 in. The coefficient of friction is .30. Find the torque required if a 1000-lb compressive load is required on the blocks. If the thread is a double thread, what is the required torque?

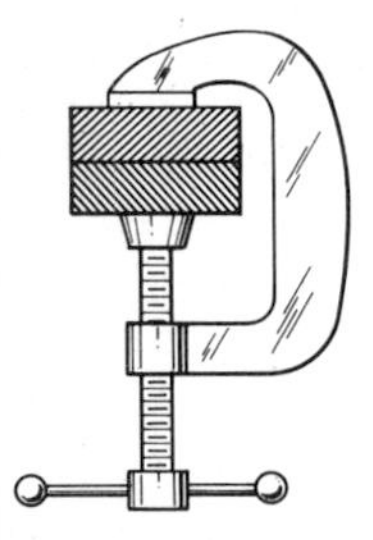

Figure P.7.73

[6]The number of wheels n plays no role here since we divide the load by n to get the load per wheel and later multiply by n to get the total resistance.

7.74. The mast of a sailboat is held by wires called shrouds, as shown in the diagram. Racing sailors are careful to get the proper tension in the shrouds by adjusting the turnbuckle at the bottom of the shrouds. When we do this we say we are "tuning" the boat. If a tension of 150 N exists in the shroud, what torque is needed to start tightening further by turning the turnbuckle? The pitch of the single threaded screw is 1.5 mm and the mean diameter is 8.0 mm. The coefficient of friction is .2.

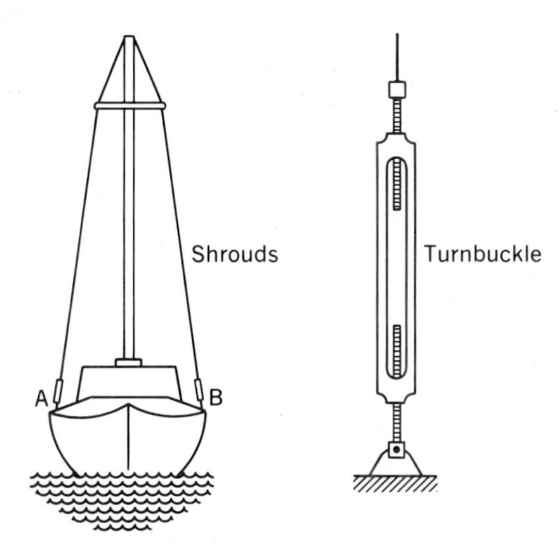

Figure P.7.74

7.75. Forces F of 50 lb are applied to the jack-screw shown. The thread diameter is 2 in. and the pitch is $\frac{1}{2}$ in. The coefficient of friction for the thread is .05. The weight W is not permitted to rotate and so the collar must rotate on the shaft of the screw. If the coefficient of friction for the collar and shaft is .1, determine the weight W that can be lifted by this system.

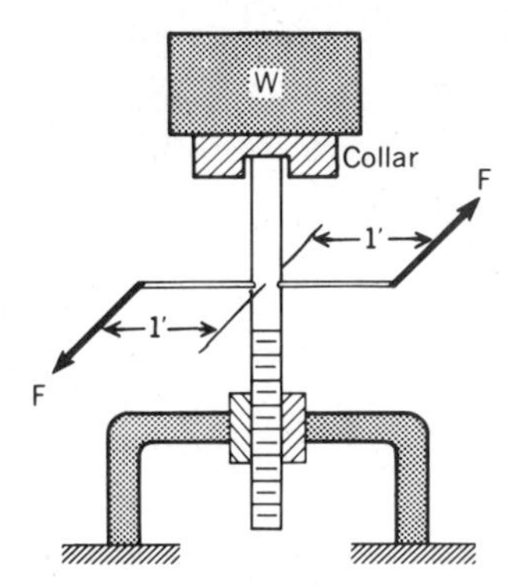

Figure P.7.75

7.76. A brake is shown. Force is developed at the brake shoes by turning A, which has a single right-handed square screw thread at B and a single left-handed screw thread at C. The diameter of the screw thread is $1\frac{1}{2}$ in. and the pitch is .3 in. If the coefficient of friction is .1 for the thread and .4 for the brake shoes, what resisting torque is developed on the wheel by a 100 in.-lb torque at A?

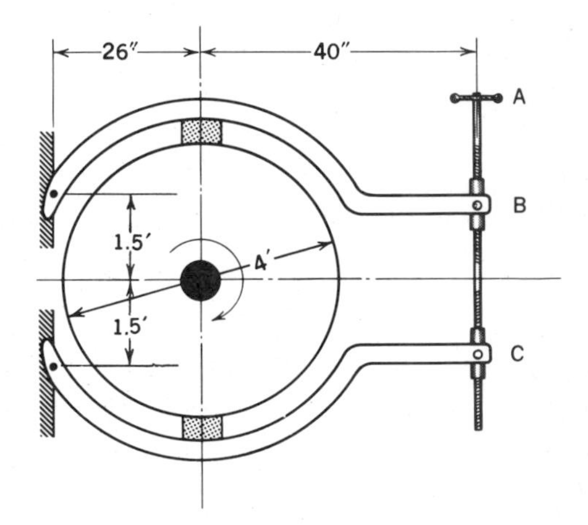

Figure P.7.76

7.77. A triangular-threaded screw is shown. In a manner paralleling the development of the square-thread formulation in Section 7.6, show that

$$M_z = \frac{rP(\mu \cos \alpha + \cos \theta \tan \alpha)}{\cos \theta - \mu \sin \alpha}$$

where

$$\cos\theta = \frac{1}{\sqrt{\tan^2\alpha + \tan^2\gamma + 1}}$$

and

$$\gamma = \beta - \alpha$$

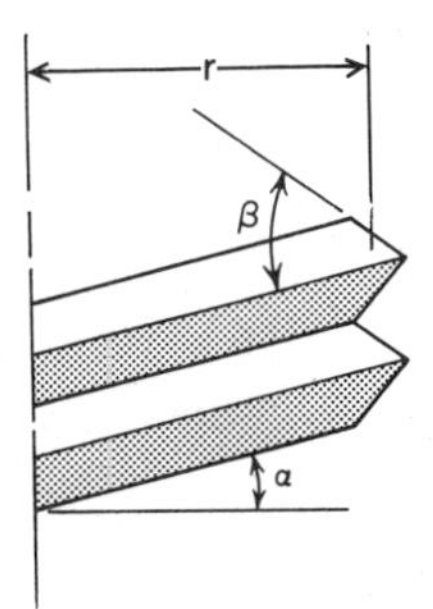

Figure P.7.77

7.78. Consider a single-threaded screw where the pitch $p = 4.5$ mm and the mean radius is 20 mm. For a coefficient of friction $\mu = .3$, what torque is needed on the nut for it to turn under a load of 1000 N? Compute this for a square thread and then do it for a triangular thread where the angle β is 30°. See Problem 7.77 before doing this problem.

7.79. If the coefficient of rolling friction of a cylinder on a flat surface is .05 in., at what inclination of the surface will the cylinder of radius $r = 1$ ft roll with uniform velocity?

7.80. A 65-kN vehicle designed for polar expeditions is on a very slippery ice surface for which the coefficient of friction between tires and ice is .005. Also, the coefficient of rolling friction is known to be .8 mm. Will the vehicle be able to move? The vehicle has four-wheel drive.

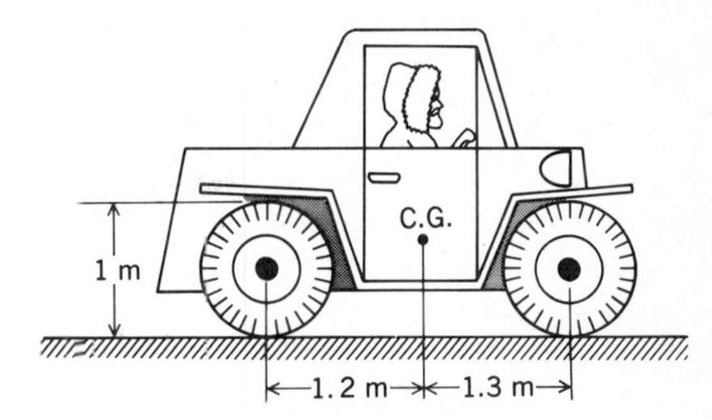

Figure P.7.80

7.81. In Problem 7.80, suppose there is only rear-wheel drive available. What is the minimum coefficient of friction needed between tires and ground for the vehicle to move?

7.82. A roller thrust bearing is shown supporting a force P of 2.5 kN. What torque T is needed to turn the shaft A at constant speed if the only resistance is that from the ball bearings? The coefficient of rolling for the balls and the bearing surfaces is .01270 mm. The mean radius from the centerline of the shaft and the balls is 30 mm.

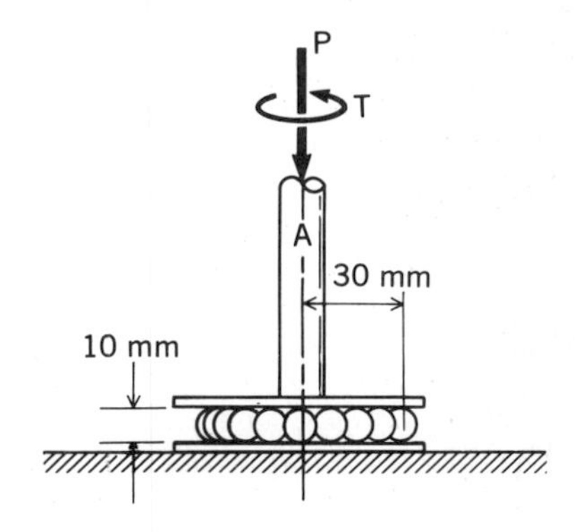

Figure P.7.82

7.8 Closure

In this chapter, we have examined the results of two independent experiments: that of impending or actual sliding of one body over another and that of a cylinder or sphere rolling at constant speed over a flat surface. Without any theoretical basis, the results of such experiments must be used in situations that closely parallel the experiments themselves.

In the case of a rolling cylinder, both rolling resistance and sliding resistance are present. However, for a cylinder accelerating with any appreciable magnitude, only sliding friction need be accounted for. With no acceleration on a horizontal surface, only rolling resistance need be considered. Most situations fall into these categories. For very small accelerations, both effects are present and must be taken into account. We can then expect only a crude result for such computations.

Before going further, we must carefully define certain properties of plane surfaces in order to facilitate later computations in mechanics where such properties are most useful. This will be done in Chapter 8.

Review Problems

7.83. Find the force F needed to start the 200-N weight moving to the right, if the coefficient of friction is $\mu_s = .35$.

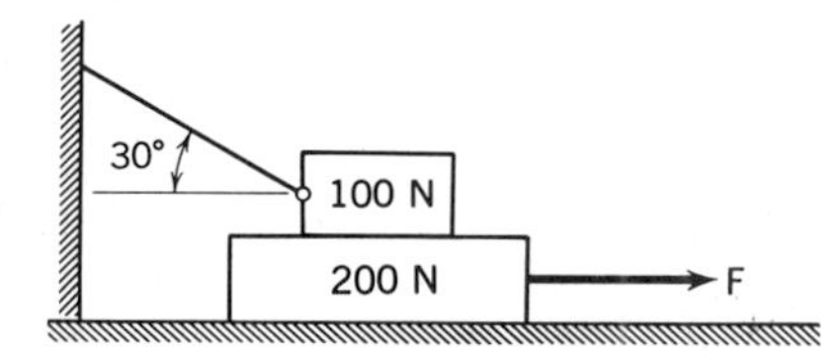

Figure P.7.83

7.84. A loaded crate is shown. The crate weighs 500 lb with a center of gravity at its geometric center. The contact surface between crate and floor has a static coefficient of friction of .2. If $\theta = 90°$, show that the crate will slide before one can increase T large enough for tipping to occur. If a stop is to be inserted in the floor at A to prevent slipping so that the crate could be tipped, what *minimum* horizontal force will be exerted on the stop?

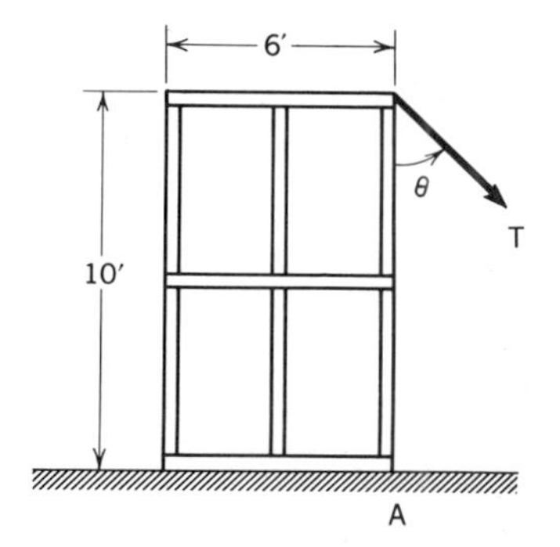

Figure P.7.84

7.85. In Problem 7.84, compute a value of θ and T where slipping and tipping will occur simultaneously. If the actual angle θ is smaller than this value of θ, is there any further need of the stop at A to prevent slipping?

7.86. A friction drive is shown with A the driver disc and B the driven disc. If force F pressing B onto A is 150 N, what is the *maximum* torque M_2 that can be developed? For this torque, what is the torque M_1 needed for the drive disc A? The coefficient of friction between A and B is .7. What vertical force must rod G withstand for the action above?

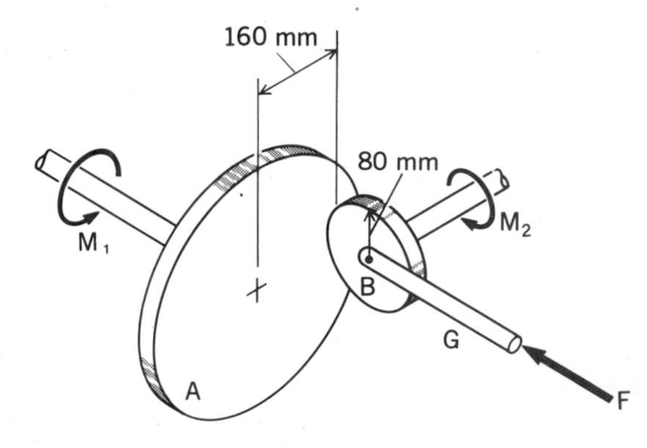

Figure P.7.86

7.87. A tug is pushing a barge into a berth. After the barge turns and touches the sides of the pilings, what thrust must the tug develop to move it at uniform speed of 2 knots farther into the berth? The coefficient of friction between the barge and the sides of the berth is .4. The drag from the water is 3000 N along the centerline of the barge.

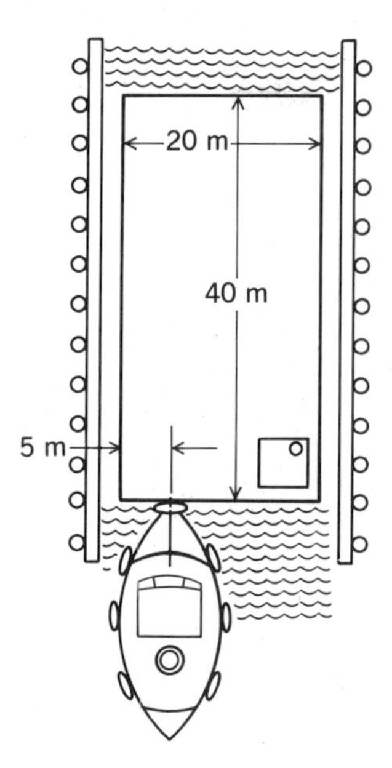

Figure P.7.87

7.88. The static and dynamic coefficients of friction for the upper surface of contact A of the cylinder are $\mu_s = .4$, $\mu_d = .3$, and for the lower surface of contact B are $\mu_s = .1$ and $\mu_d = .08$. What is the *minimum* force P needed to just get the cylinder moving?

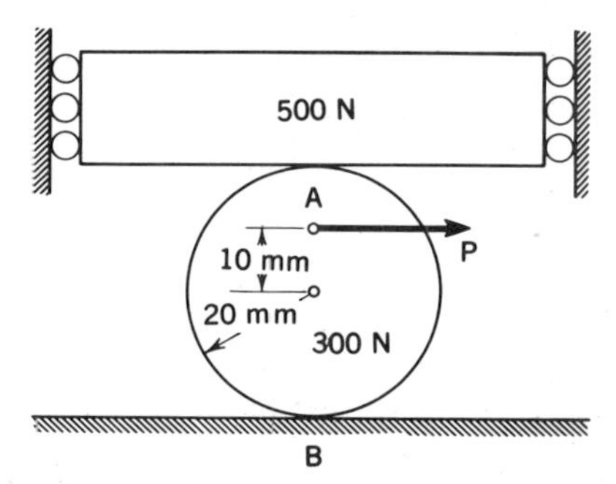

Figure P.7.88

7.89. A hot rectangular metal ingot is to be flattened by passing through cylindrical rollers. If the ingot is to be drawn into the rollers by friction once it touches the rollers, what is the *minimum* thickness t of the ingot that can be achieved by this process on one pass? The coefficient of friction for the contact between ingot and cylinder is .3. The cylinders rotate as shown with angular speed ω.

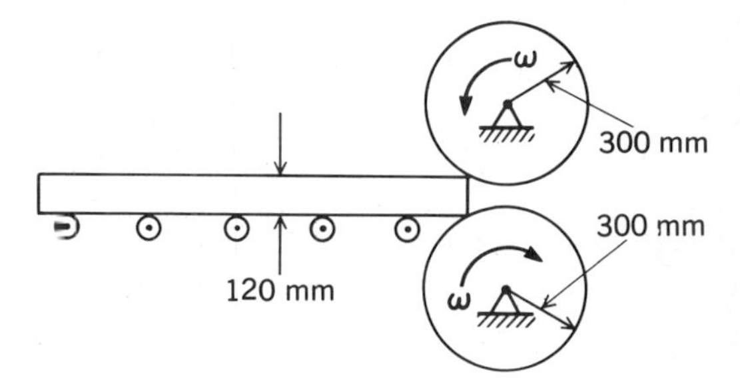

Figure P.7.89

7.90. A cone clutch is shown. Assuming that uniform pressures exist between the contact surfaces, compute the *maximum* torque that can be transmitted. The coefficient of friction is .30 and the activating force F is 100 lb. (*Hint:* Assume that the moving cone transmits its 100-lb axial force to the stationary cone by pressure primarily. That is, we will neglect the friction-force component on the cone surface in the axial direction.)

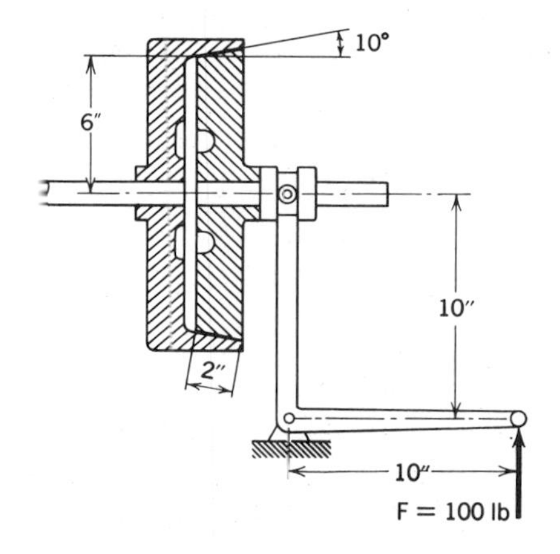

Figure P.7.90

7.91. The drum is driven by a motor with a maximum torque capability of 500 lb-ft. The coefficient of friction between the drum and the braking strap (belt) is .4. How much force P must an operator exert to stop the drum if it rotates (1) clockwise and (2) counterclockwise? What are the belt forces in each case?

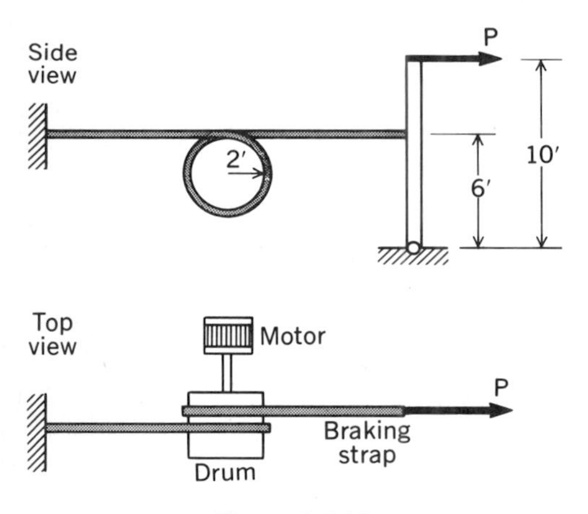

Figure P.7.91

7.92. The four drive pulleys shown are used to transmit a torque from pulley A to pulley D on an electric typewriter. If the coefficient of friction between the belts and the pulleys is .3, what is the torque available at pulley D if 10 lb-in. of torque is input to the shaft of pulley A? What are the belt forces?

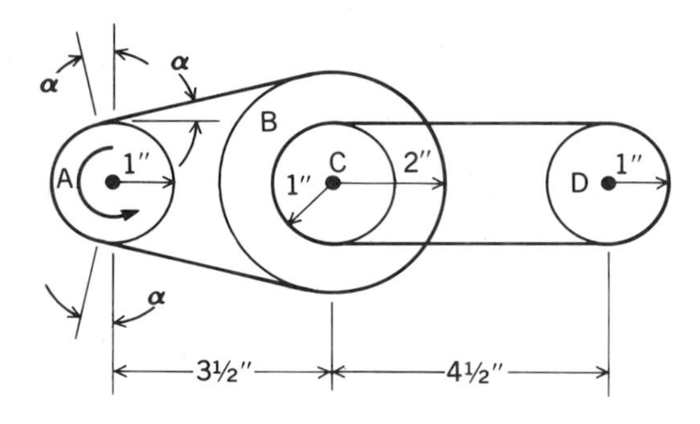

Figure P.7.92

7.93. A scissors jack is shown lifting the end of a car so that $R = 6.67$ kN. What torque T is needed for this operation? Note that A is merely a bearing and at B we have a nut. The screw is single-threaded with a pitch of 3 mm and a mean diameter of 20 mm. The coefficient of friction is .3. Neglect the weight of the members and evaluate T for $\theta = 45°$ and for $\theta = 60°$.

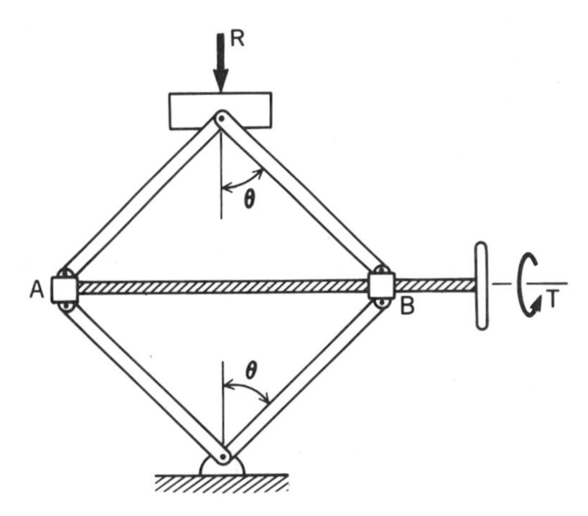

Figure P.7.93

7.94. A block C weighing 10 kN is being moved on rollers A and B each weighing 1 kN. What force P is needed to maintain steady motion? Take the coefficient of rolling resistance between the rollers and the ground to be .6 mm and between block C and the rollers to be .4 mm.

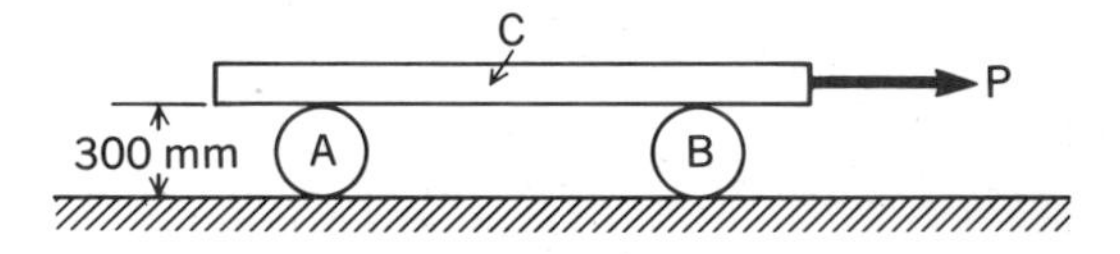

Figure P.7.94

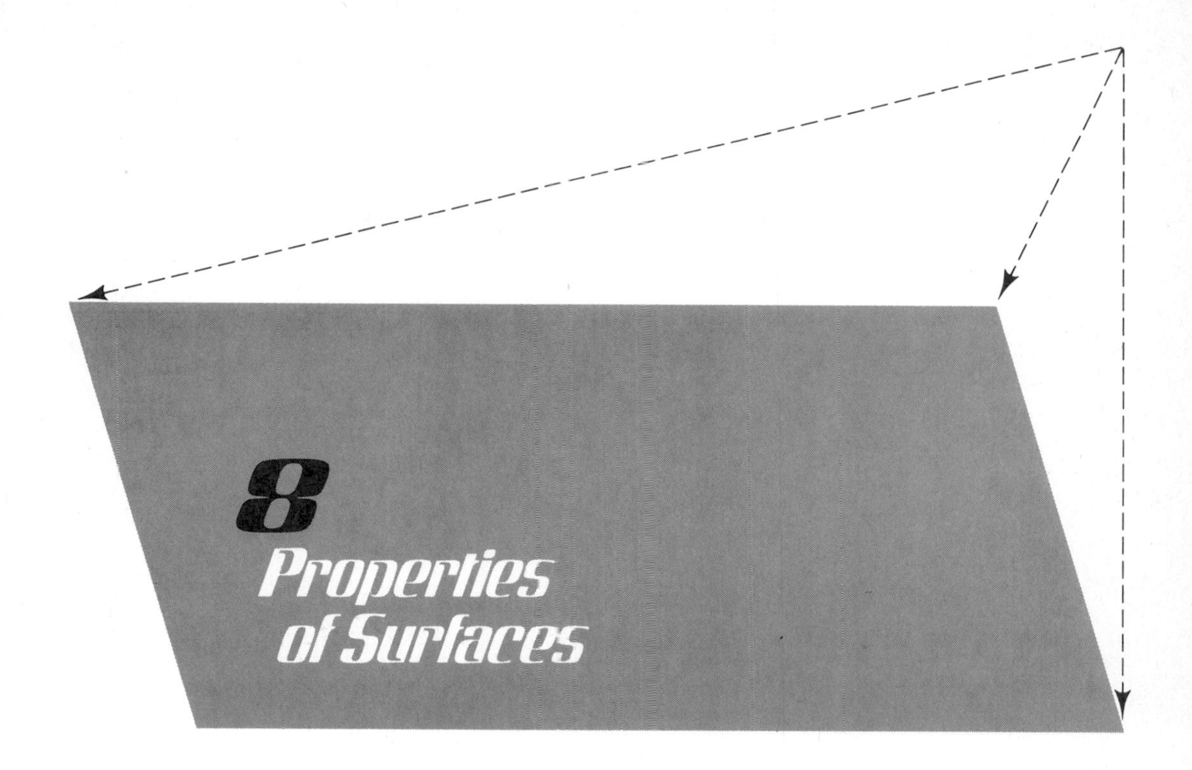

8.1 Introduction

If we are buying a tract of land, we certainly want to consider the size and, with equal interest, the shape and orientation of the earth's surface, and possibly its agricultural, geological, or aesthetic potentials. The size of a surface (i.e., the area) is a familiar concept and has been used in the previous sections. Certain aspects of the shape and orientation of a surface will be examined in this chapter. There are a number of formulations that convey meaning about the shape and disposition of a surface relative to some reference. To be sure, these formulations are not used by real estate people, but in engineering work, where a variety of quantitative descriptions are necessary, they will prove most useful. In general, we shall restrict our attention to coplanar surfaces.

8.2 First Moment of an Area and the Centroid

A coplanar surface of area A and a reference xy in the plane of the surface are shown in Fig. 8.1. We define the *first moment* of area A about the x axis as

$$M_x = \int_A y\,dA \tag{8.1}$$

and the first moment about the y axis as

$$M_y = \int_A x\,dA \tag{8.2}$$

These two quantities convey a certain knowledge of the shape, size, and orientation of the area which we can use in many analyses of mechanics.

You will no doubt notice the similarity of the preceding integrals to those which would occur for computing moments about the x and y axes from a parallel force distribution oriented normal to the area A in Fig. 8.1. The moment of such a force distribution has been shown for the purposes of rigid-body calculations to be equivalent to that of a single resultant force located at a particular point $\bar{x}, \bar{y}$. Similarly, we can concentrate the entire area A at a position x_c, y_c, called the *centroid*,[1] where, for

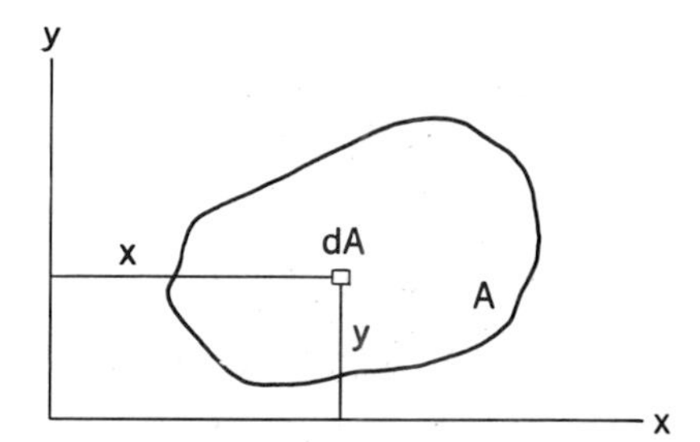

Figure 8.1. Plane area.

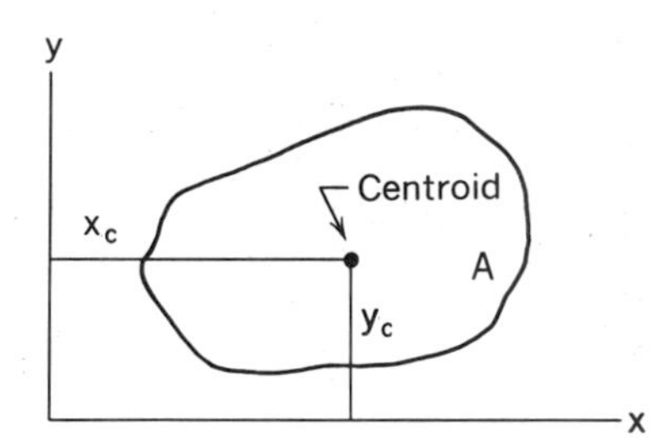

Figure 8.2. Centroidal coordinates.

computations of first moments, this new arrangement is equivalent to the original distribution (Fig. 8.2). The coordinates x_c and y_c are usually called the *centroidal coordinates*. To compute these coordinates, we simply equate moments of the distributed area with that of the concentrated area about both axes:

$$Ay_c = \int_A y\,dA; \qquad \text{therefore, } y_c = \frac{\int_A y\,dA}{A} \tag{8.3a}$$

$$Ax_c = \int_A x\,dA; \qquad \text{therefore, } x_c = \frac{\int_A x\,dA}{A} \tag{8.3b}$$

The location of the centroid of an area can readily be shown to be independent of the reference axes employed. That is, the centroid is a property only of the area itself. We have asked the reader to prove this in Problem 8.1.

If the axes xy have their origin at the centroid, then these axes are called *centroidal axes* and clearly the first moments about these axes must be zero.

Finally, we point out that all axes going through the centroid of an area are called *centroidal axes* for that area. Clearly, the *first moments of an area about any of its centroidal axes must be zero.*

EXAMPLE 8.1

A plane surface is shown in Fig. 8.3 bounded by the x axis, the curve $y^2 = 25x$, and a line parallel to the y axis.

[1]The concept of the centroid can be used for any geometric quantity. In the next section, we shall consider centroids of volumes and arcs.

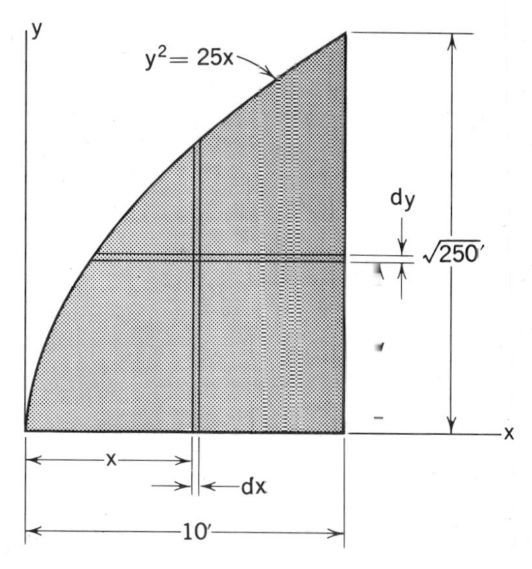

Figure 8.3. Find centroid.

We shall first compute M_x and M_y for this area. Using vertical infinitesimal area elements of width dx and height y, we have

$$M_y = \int_0^{10} x(y\,dx) = \int_0^{10} x(5\sqrt{x})\,dx$$

$$= \frac{5x^{5/2}}{\frac{5}{2}}\Big|_0^{10} = 632 \text{ ft}^3$$

To compute M_x, we use horizontal area elements of width dy as shown in the diagram. Thus,

$$M_x = \int_0^{\sqrt{250}} y[(10 - x)dy]$$

$$= \int_0^{\sqrt{250}} \left(10y - \frac{y^3}{25}\right)dy$$

$$= \left(5y^2 - \frac{y^4}{100}\right)\Big|_0^{\sqrt{250}} = 625 \text{ ft}^3$$

We could also have used vertical strips for computing M_x as follows:

$$M_x = \int_0^{10} \frac{y}{2}(y\,dx) = \int_0^{10} \frac{25x}{2}\,dx$$

$$= (12.5)\left(\frac{x^2}{2}\right)\Big|_0^{10} = 625 \text{ ft}^3$$

To compute the position of the centroid (x_c, y_c), we will need the area A of the surface. Thus, using vertical strips:

$$A = \int_0^{10} y\,dx = \int_0^{10} 5\sqrt{x}\,dx = \frac{5x^{3/2}}{\frac{3}{2}}\Big|_0^{10}$$

$$= 105.4 \text{ ft}^2$$

The centroidal coordinates are, accordingly,

$$x_c = \frac{M_y}{A} = \frac{632}{105.4} = 6.00 \text{ ft}$$

$$y_c = \frac{M_x}{A} = \frac{625}{105.4} = 5.93 \text{ ft}$$

To get the moment of the area about an axis y', which is 15 ft to the left of the y axis, we simply proceed as follows:

$$M_{y'} = (A)(x_c + 15) = 105.4(6.00 + 15) = 2213 \text{ ft}^3$$

Consider now a plane area with an *axis of symmetry* such as is shown in Fig. 8.4, where the y axis is collinear with the axis of symmetry. In computing x_c for this area, we have

$$x_c = \frac{1}{A}\int_A x\, dA$$

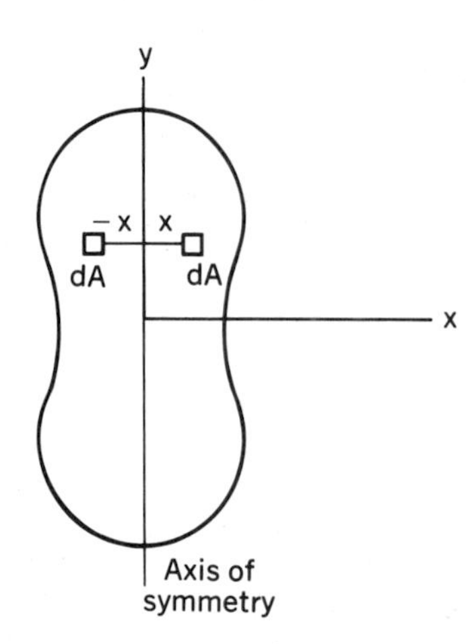

Figure 8.4. Area with one axis of symmetry.

In evaluating the integral above, we can consider area elements in symmetric pairs such as shown in Fig. 8.4, where we have shown a pair of area elements which are mirror images of each other about the axis of symmetry. Clearly, the first moment of such a pair about the axis of symmetry is zero. And, since the entire area can be considered as composed of such pairs, we can conclude that $x_c = 0$. Thus, the centroid of an area with one axis of symmetry must therefore lie somewhere along this axis of symmetry. With two orthogonal axes of symmetry, the centroid must lie at the intersection of these axes. Thus, for such areas as circles and rectangles, the centroid is easily determined by inspection.

In many problems, the area of interest can be considered formed by the addition or subtraction of simple familiar areas whose centroids are known by inspection as well as by other familiar areas, such as triangles and sectors of circles whose centroids and areas are given in handbooks. We call areas made up of such simple areas *composite* areas. (A listing of familiar areas is given for your convenience on the inside covers of this text.) For such problems, we can say that

$$x_c = \frac{\sum_i A_i \bar{x}_i}{A}$$

$$y_c = \frac{\sum_i A_i \bar{y}_i}{A}$$

where $\bar{x}_i$ and $\bar{y}_i$ (with proper signs) are the centroidal coordinates to simple area A_i and where A is the total area.

We now illustrate how we can use the composite-area approach for finding the centroid of an area composed of familiar parts as described above.

EXAMPLE 8.2

Find the centroid of the shaded section shown in Fig. 8.5.

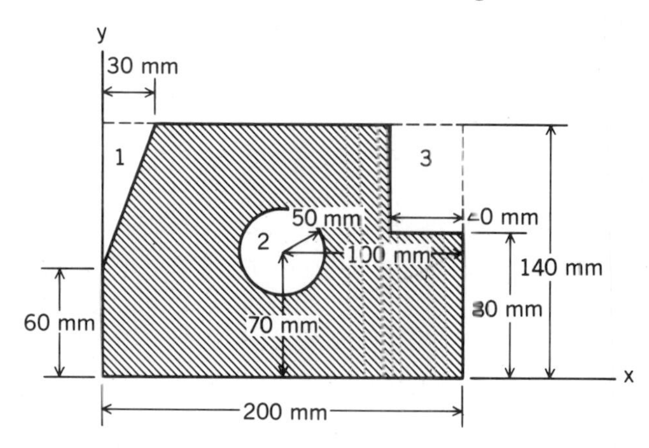

Figure 8.5. Composite area.

We may consider four separate areas. These are the triangle (1), the circle (2), and the rectangle (3) all cut from an original rectangular 200 × 140 mm² area which we denote as area (4). In composite-area problems, we urge you to set up a format of the kind we shall now illustrate. Using the positions of the centroid of a right triangle as given in the inside covers of this text, we have:

A_i	$\bar{x}_i$	$A_i\bar{x}_i$	$\bar{y}_i$	$A_i\bar{y}_i$
$A_1 = -\frac{1}{2}(30)(80) = -1{,}200$	10	$-12{,}000$	113.3	$-136{,}000$
$A_2 = -\pi 50^2 = -7{,}850$	100	$-785{,}000$	70	$-549{,}780$
$A_3 = -(40)(60) = -2{,}400$	180	$-432{,}000$	110	$-264{,}000$
$A_4 = (200)(140) = 28{,}000$	100	$2{,}800{,}000$	70	$1{,}960{,}000$
$A = 16{,}550 \text{ mm}^2$		$\sum_i A_i\bar{x}_i = 1.571\times 10^6 \text{ mm}^3$		$\sum_i A_i\bar{y}_i = 1.011\times 10^6 \text{ mm}^3$

Therefore,

$$x_c = \frac{\sum A_i\bar{x}_i}{A} = \frac{1.571 \times 10^6}{16{,}550} = 94.9 \text{ mm}$$

$$y_c = \frac{\sum A_i\bar{y}_i}{A} = \frac{1.011 \times 10^6}{16{,}550} = 61.1 \text{ mm}$$

In closing, we would like to point out that the centroid concept can be of use in finding the simplest resultant of a distributed loading. Thus, consider the distributed loading $w(x)$ shown in Fig. 8.6. The resultant force F_R of this loading, also shown in the diagram, is given as

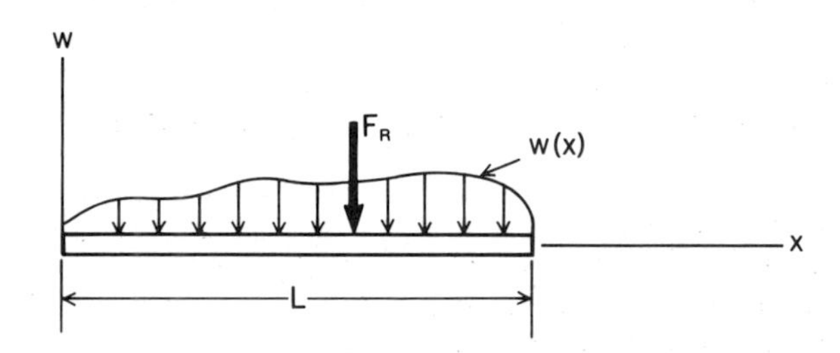

Figure 8.6. Loading curve $w(x)$ and its resultant F_R.

$$F_R = \int_0^L w(x)\,dx \tag{8.4}$$

From the equation above, we can readily see that the *resultant force equals the area under the loading curve.* To get the position of the *simplest* resultant for the loading, we then say that

$$F_R\bar{x} = \int_0^L xw(x)\,dx$$

Therefore,

$$\bar{x} = \frac{\int_0^L xw(x)\,dx}{F_R} \tag{8.5}$$

The preceding result shows that $\bar{x}$ is actually the centroidal coordinate of the loading curve area from reference xy. Thus, the *simplest resultant force of a distributed load acts at the centroid of the area under the loading curve.* Accordingly, for a triangular load such as is shown in Fig. 8.7, we can replace the loading for free bodies on which the entire loading acts by a force F equal to $(w_0)(b - a)(\frac{1}{2})$ at a position $\frac{2}{3}(b - a)$ from the left end of the loading. You will recall that we pointed this out in Chapter 4.

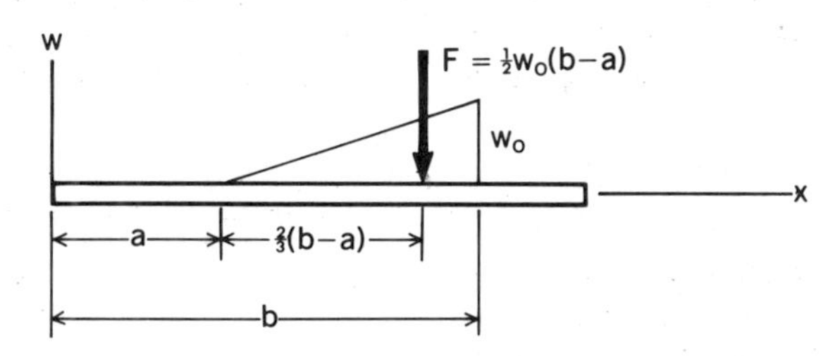

Figure 8.7. Triangular loading with simplest resultant.

Problems

8.1. Show that the centroid of area A is the same point for axes xy and $x'y'$. Thus, the position of the centroid of an area is a property only of the area.

8.2. Show that the centroid of the right triangle is $x_c = 2a/3$, $y_c = b/3$.

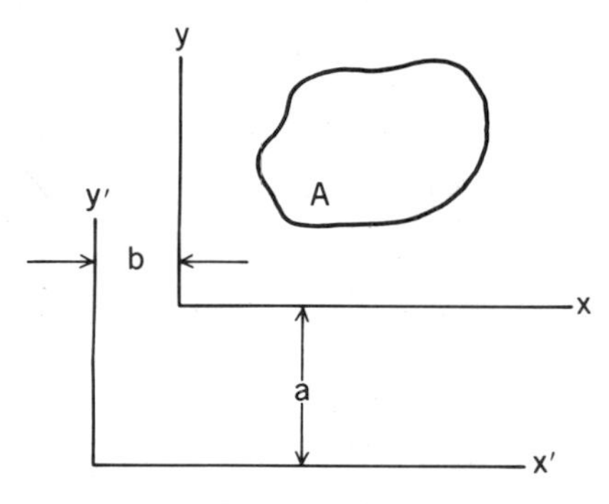

Figure P.8.1

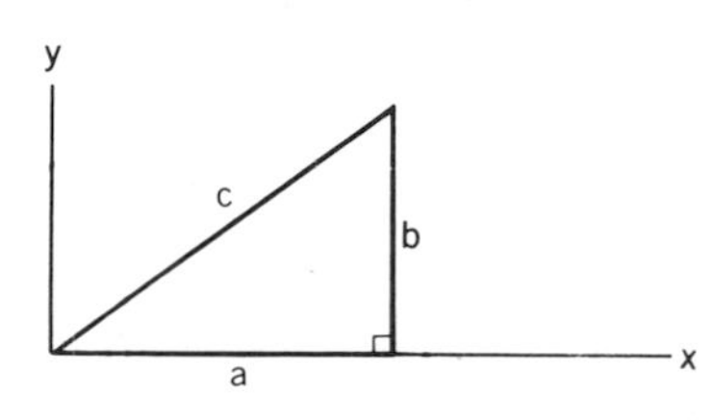

Figure P.8.2

8.3. Find the centroid of the area under the half-sine wave. What is the first moment of this area about axis A–A?

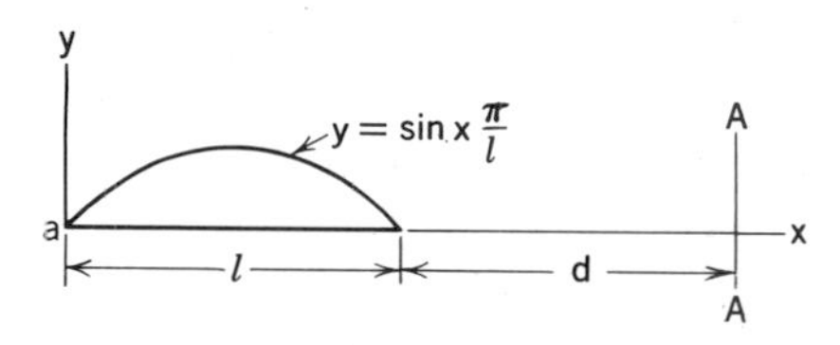

Figure P.8.3

8.4. What are the first moments of the area about the x and y axes? The curved boundary is that of a parabola. (*Hint:* The general equation for parabolas of the shape shown is $y^2 = ax + b$.)

Figure P.8.4

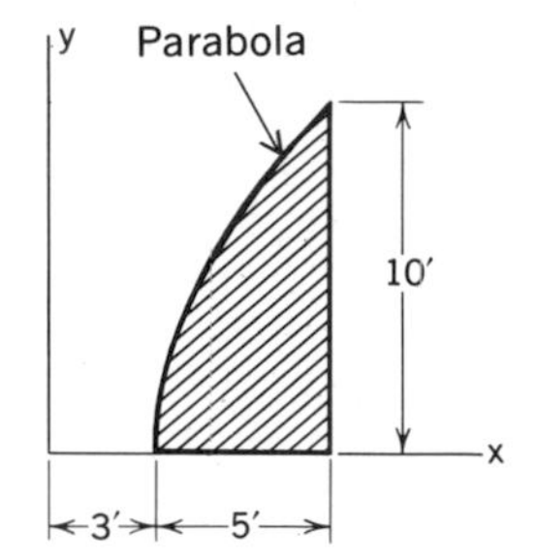

8.5. What are the centroidal coordinates for the shaded area? The curved boundary is that of a parabola. (*Hint:* The general equation for parabolas of the shape shown is $y = ax^2 + b$.)

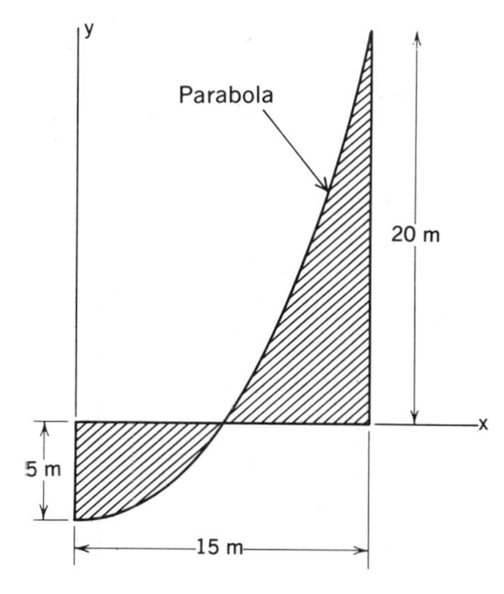

Figure P.8.5

8.6. Show that the centroid of the area under a semicircle is as shown in the diagram.

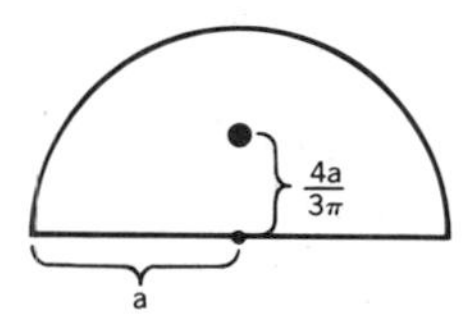

Figure P.8.6

8.7. What is the first moment of the area under the parabola about an axis through the origin and going through point $\boldsymbol{r} = 6\boldsymbol{i} + 7\boldsymbol{j}$ m. Take $l = 10$ m.

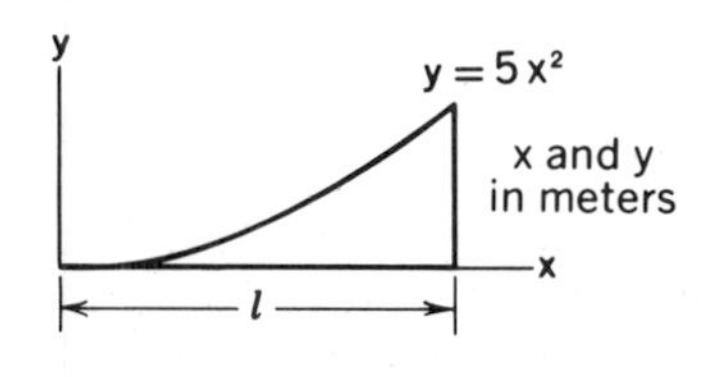

Figure P.8.7

8.8. Show that the centroid of the triangle is at $x_c = (a + b)/3$, $y_c = h/3$. (*Hint:* Break the triangle into two right triangles for which the centroids are known from Problem 8.2.)

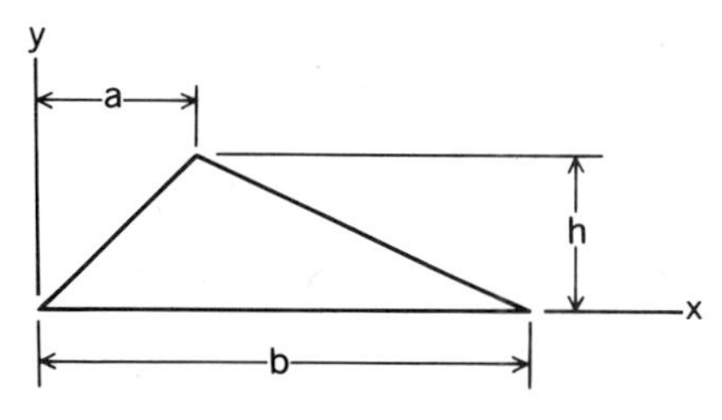

Figure P.8.8

8.9. What are the centroidal coordinates for the shaded area? The outer boundary is that of a circle having a radius of 1 m.

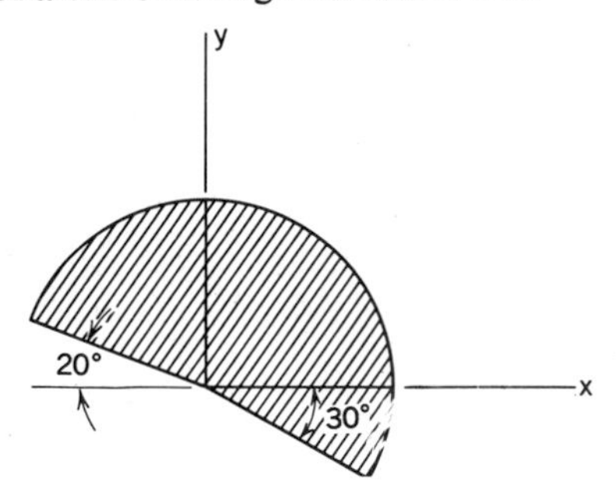

Figure P.8.9

8.10. What are the coordinates of the centroid of the shaded area? The parabola is given as $y^2 = 2x$ with y and x in millimeters.

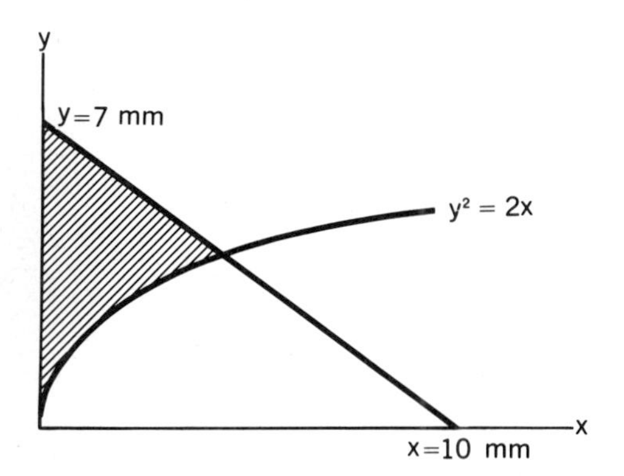

Figure P.8.10

8.11. Find the centroid of the shaded area. The equation of the curve is $y = 5x^2$ with x and y in millimeters. What is the first moment of the area about line AB?

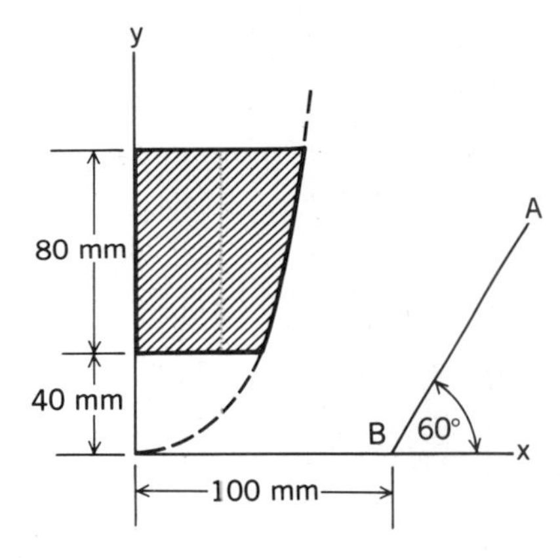

Figure P.8.11

8.12. Find the centroid of the shaded area. What is the first moment of this area about line A–A. The upper boundary is a parabola $y^2 = 3x$ with x and y in millimeters.

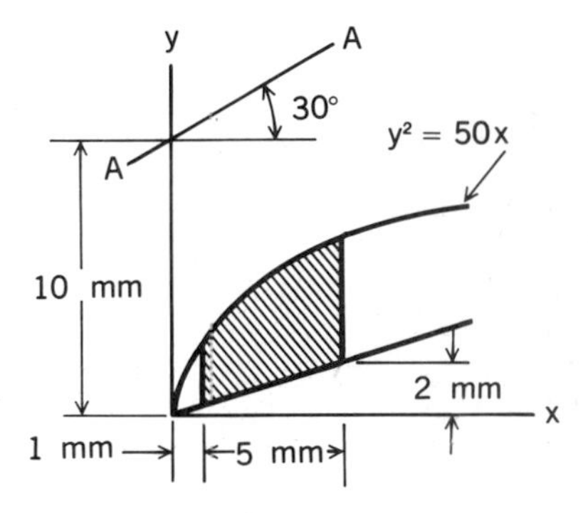

Figure P.8.12

In the remaining problems of this section, use centroidal positions of simple areas as found in the inside covers.

8.13. Find the centroid of the end shield of a bulldozer blade.

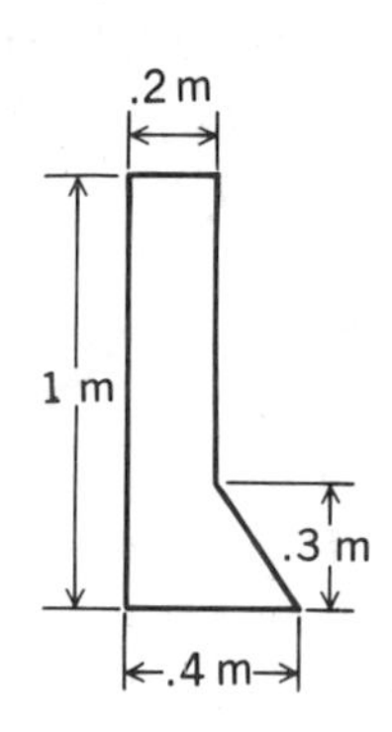

Figure P.8.13

8.14. Find the centroid of the truss gusset plate.

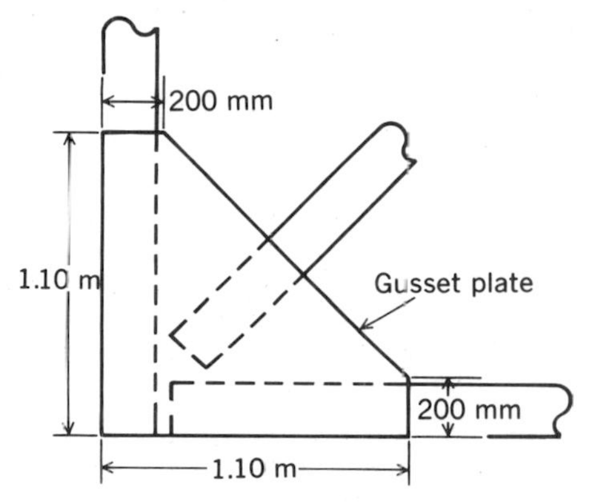

Figure P.8.14

8.15. Find the centroid of the indicated area.

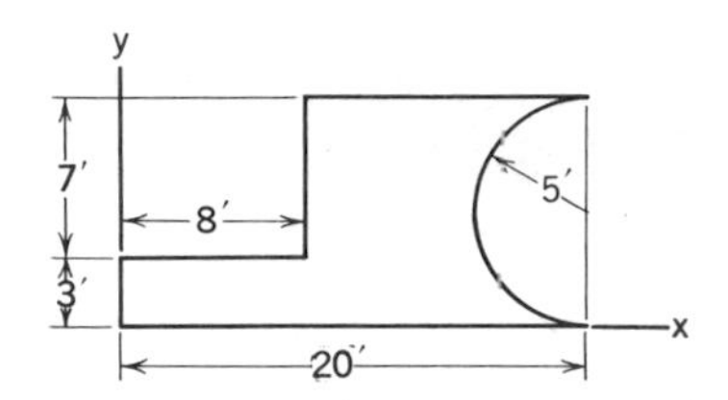

Figure P.8.15

8.16. Find the centroidal coordinates for the shaded area shown. Give the results in meters. (*Hint:* See Fig. P.8.6.)

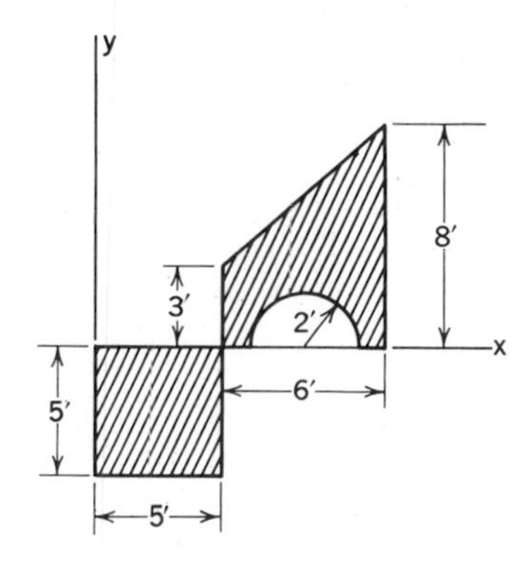

Figure P.8.16

8.17. Find the centroid of the end of the bucket of a small front-end loader.

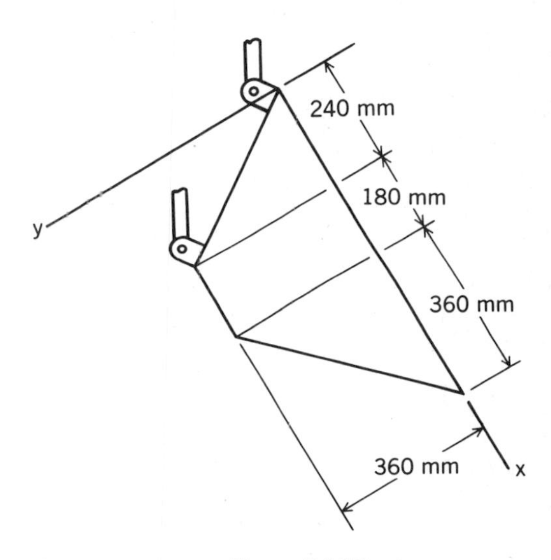

Figure P.8.17

8.18. Where is the centroid of the airplane vertical stabilizer (whole area)?

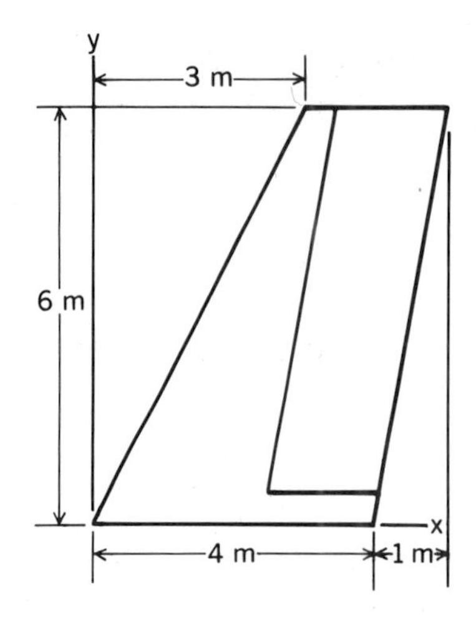

Figure P.8.18

8.19. What is the first moment of the shaded area about the diagonal *A–A*? (*Hint:* Consider symmetry.)

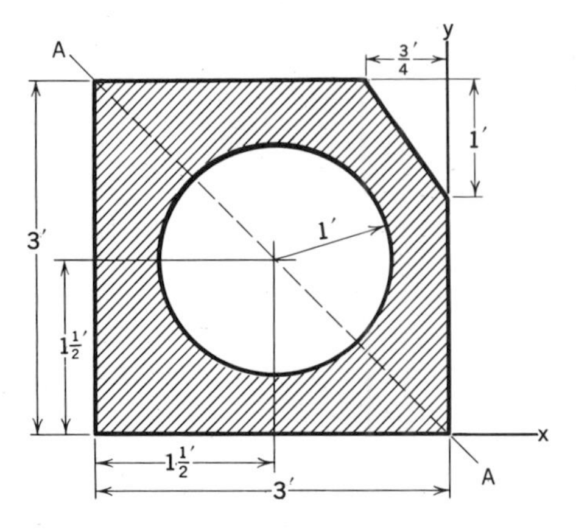

Figure P.8.19

8.20. A built-up beam is shown with four 120-mm by 120-mm by 20-mm angles. Find the vertical distance above the base for the centroid of the cross section.

8.21. A wide-flange I-beam (identified as 14 WF 202 I-beam) is shown with two reinforcing plates on top. At what height above the bottom is the centroid located?

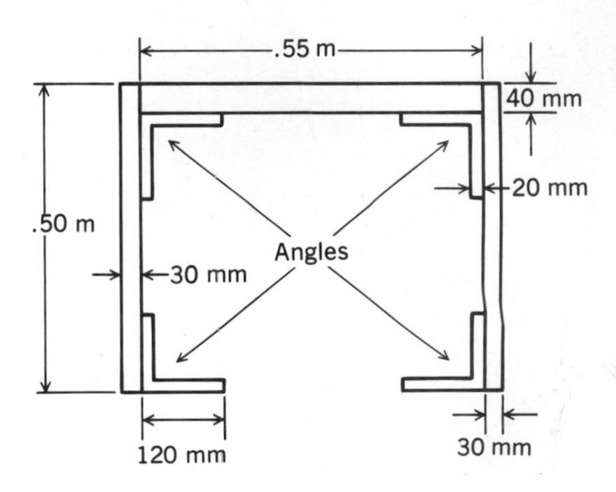

Figure P.8.20

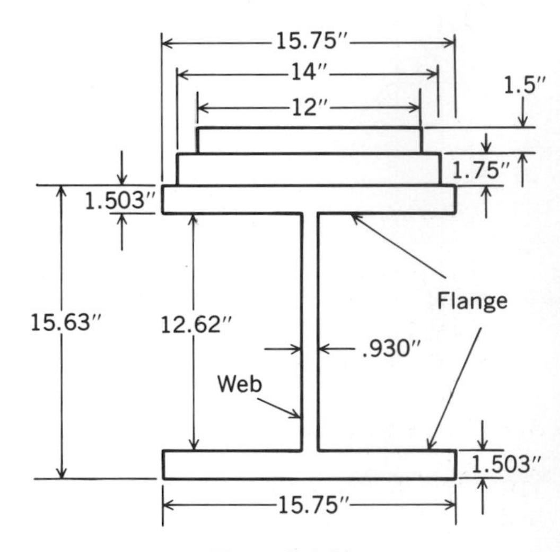

Figure P.8.21

8.22. Compute the position of the centroid of the shaded area. (*Hint:* See Fig. P.8.6.)

Figure P.8.22

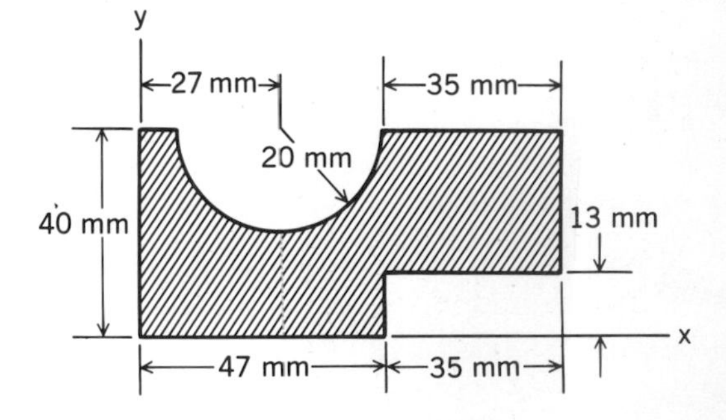

8.23. Find the centroid of the sheet metal cover of a centrifugal blower (shown shaded).

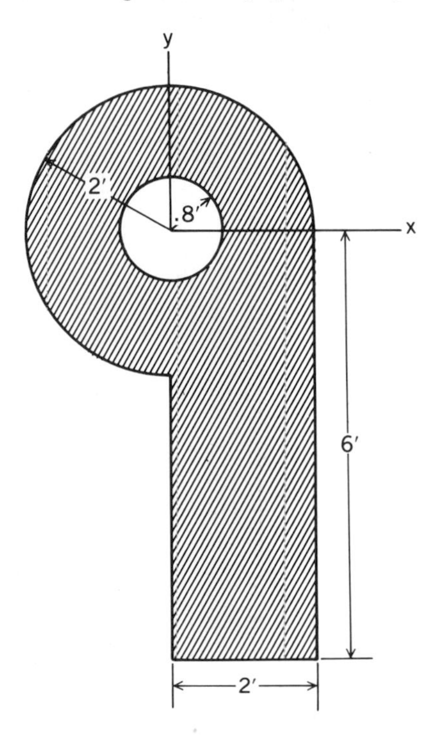

Figure P.8.23

8.24 What is the position from the left end of the simplest resultant force of the distribution shown?

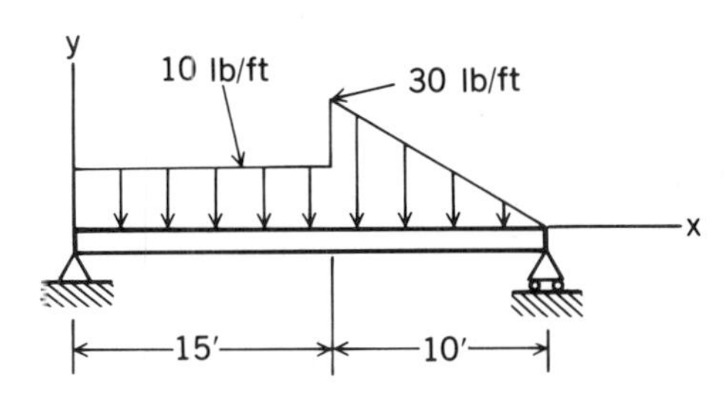

Figure P.8.24

8.3 Other Centers

We employ the concepts of moments and centroids in mechanics for three-dimensional bodies as well as for plane areas. Thus, we introduce now the first moment of a volume, V, of a body (see Fig. 8.8) about a point O where we have shown a reference xyz. We say that the first *moment of volume* V about O is

$$\text{moment vector of volume} \equiv \iiint_V \mathbf{r}\, dv \tag{8.6}$$

The *center of volume*, $\mathbf{r}_c$, is then defined as follows:

$$V\mathbf{r}_c = \iiint_V \mathbf{r}\, dv$$

Therefore,

$$\mathbf{r}_c = \frac{1}{V}\iiint_V \mathbf{r}\, dv \tag{8.7}$$

We see that the center of volume is the point where we could hypothetically concentrate the entire volume of a body for purposes of computing the first moment of the

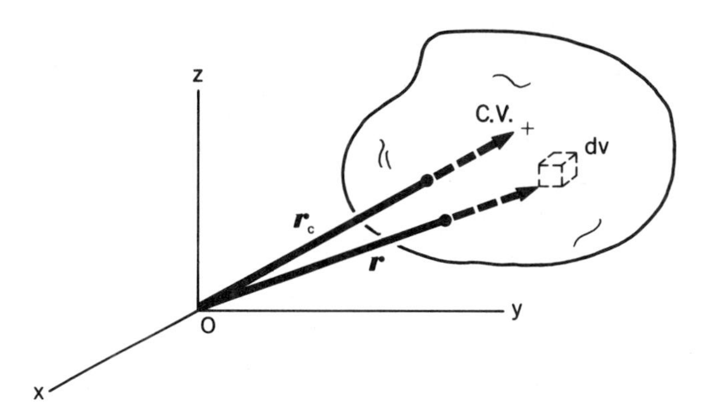

Figure 8.8. Center of volume, C.V., of a body.

volume of the body about some point O. The components of Eq. 8.7 give the *centroid distances* of volume x_c, y_c, and z_c. Thus, we have

$$x_c = \frac{\iiint x\, dv}{\iiint dv}, \qquad y_c = \frac{\iiint y\, dv}{\iiint dv}, \qquad z_c = \frac{\iiint z\, dv}{\iiint dv} \tag{8.8}$$

The integral $\iiint x\, dv$, it should be noted, gives the first moment of volume about the yz plane, etc.

If we replace dv by $dm = \rho\, dv$ in Eq. 8.6, where ρ is the mass *density*, we get the *first moment of mass* about O. That is,

$$\text{moment vector of mass} \equiv \iiint_V \mathbf{r}\, \rho\, dv \tag{8.9}$$

The *center of mass* $\mathbf{r}_c$ is then given as

$$\mathbf{r}_c = \frac{1}{M} \iiint_V \mathbf{r} \rho\, dv \tag{8.10}$$

where M is the total mass of the body. The center of mass is the point in space where hypothetically we could concentrate the entire mass for purposes of computing the first moment of mass about a point O. Using components of Eq. 8.10, we can say that

$$x_c = \frac{\iiint x\rho\, dv}{\iiint \rho\, dv}, \qquad y_c = \frac{\iiint y\rho\, dv}{\iiint \rho\, dv}, \qquad z_c = \frac{\iiint z\rho\, dv}{\iiint \rho\, dv}$$

In our work in dynamics, we shall consider the center of mass of a system of n particles (see Fig. 8.9). We will then say:

$$\left(\sum_{i=1}^{n} m_i\right)\mathbf{r}_c = \sum_{i=1}^{n} m_i \mathbf{r}_i$$

Therefore,

$$\boldsymbol{r}_c = \frac{\sum_{i=1}^{n} m_i \boldsymbol{r}_i}{M} \tag{8.11}$$

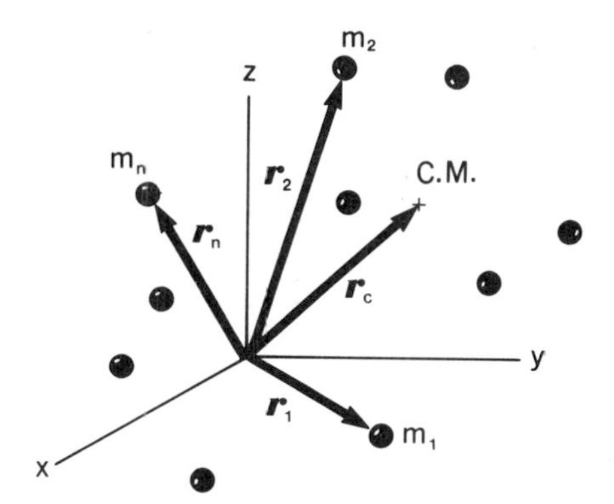

Figure 8.9. System of n particles showing center of mass, C.M.

where M is the total mass of the system. Clearly, if the particles are of infinitesimal mass and constitute a continuous body, we get back Eq. 8.10.

Finally, if we replace dv by $\gamma\ dv$, where γ ($= \rho g$) is the *specific weight*, we arrive at the concept of *center of gravity* discussed in Chapter 4. We have used the center of gravity of a body in many calculations thus far as a point to concentrate the entire weight of a body.

You should have no trouble in concluding from Eq. 8.10 that if ρ is constant throughout a body, the center of mass coincides with the center of volume. Furthermore, if γ ($= \rho g$) is constant throughout a body, the center of gravity of the body corresponds to the center of volume of the body. If, finally, ρ and g are each constant for a body, all three points coincide for the body.

We now illustrate the computation of the center of volume. Computation for the center of mass follows similar lines, and we have already computed centers of gravity in Chapter 4.

EXAMPLE 8.3

Consider a volume of revolution formed by revolving the area shown in Fig. 8.3 about the x axis. This volume has been shown in Fig. 8.10. Clearly, the centroid of this volume must lie somewhere along the x axis. We therefore need only com-

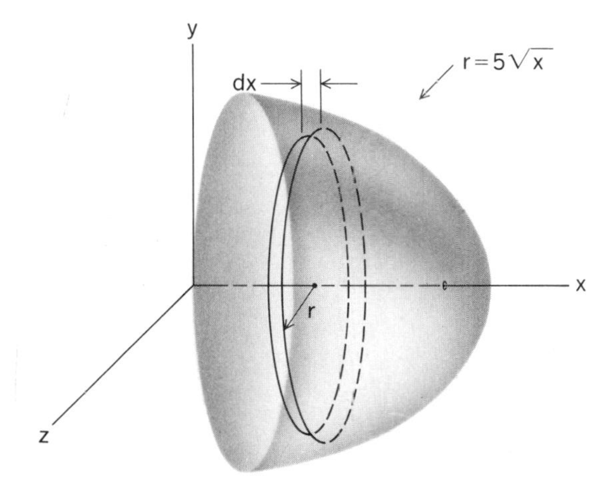

Figure 8.10. Body of revolution.

ponent x_c. Using r, θ, and x as coordinates (cylindrical coordinates), we then have, using slices of thickness dx as volume elements:

$$V = \int_0^{10} (\pi r^2)\, dx = \int_0^{10} (\pi)(25x)\, dx$$

where we have replaced r^2 by $25x$ according to the equation for the boundary of the generating area. Integrating, we get

$$V = 25\pi \frac{x^2}{2}\bigg|_0^{10} = 3927 \text{ ft}^3$$

Now we compute x_c by using infinitesimal slices of the body of the kind employed for the computation of V. The centroid of each such slice is at the intercept of the slice with the x axis. Thus, we have

$$x_c = \frac{1}{V}\int_0^{10} x(\pi r^2 dx) = \frac{1}{3927}\int_0^{10} x(\pi)(25x)\, dx$$

$$= \frac{25\pi}{3927}\frac{x^3}{3}\bigg|_0^{10} = 6.67 \text{ ft}$$

Many volumes are composed of a number of simple familiar shapes whose centers of volume are either known by inspection or can be found in handbooks (also see the inside cover pages). Such volumes may be called *composite volumes*. To find the centroid of such a volume, we use the known centroids of the composite parts. Thus, for x_c of the composite body whose total volume is V, we have

$$x_c = \frac{\sum_i \bar{x}_i V_i}{V}$$

where $\bar{x}_i$ is the x coordinate to the centroid of the ith composite body of volume V_i. Similarly,

$$y_c = \frac{\sum_i \bar{y}_i V_i}{V}$$

$$z_c = \frac{\sum_i \bar{z}_i V_i}{V}$$

We now illustrate the use of these formulas.

EXAMPLE 8.4

What is the coordinate x_c for the center of volume of the body of revolution shown in Fig. 8.11? Note that a cone has been cut away from the left end while, at the right end, we have a hemispherical region.

We have a composite body consisting of three simple domains—a cone (body 1), a cylinder (body 2), and a hemisphere (body 3). Using formulas from the inside covers, we have:

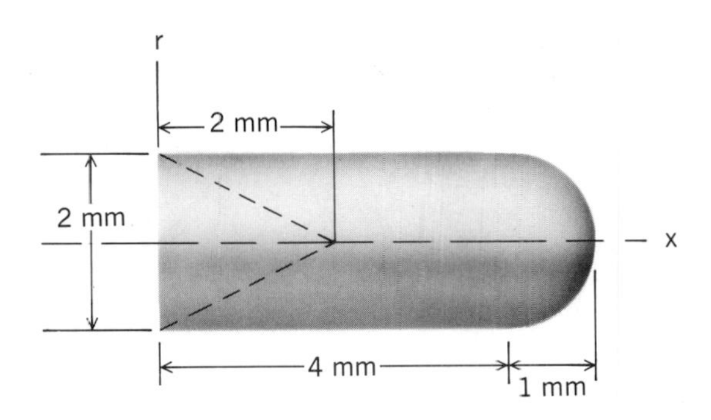

Figure 8.11. Composite volume.

	V_i (mm³)	$\bar{x}_i$ (mm)	$V_i\bar{x}_i$ (mm⁴)
1.	$-(\frac{1}{3})(\pi)(1^2)(2) = -2.09$	$\frac{1}{2}$	-1.047
2.	$(\pi)(1^2)(4) = 12.57$	2	25.14
3.	$\frac{2}{3}(\pi)(1^3) = 2.09$	$4 + \frac{3}{8}(1) = 4.38$	9.15
	$V = 12.57$		$\sum_i V_i\bar{x}_i = 33.24$

Therefore,

$$x_c = \frac{\sum_i V_i\bar{x}_i}{V} = \frac{33.24}{12.57} = 2.64 \text{ mm}$$

In closing, we wish to point out further that curved surfaces and lines have centroids. Since we shall have occasion in the next section to consider the centroid of a line, we simply point out now (see Fig. 8.12) that

$$x_c = \frac{\int x\, dl}{L} \tag{8.12a}$$

$$y_c = \frac{\int y\, dl}{L} \tag{8.12b}$$

where L is the length of the line. Note that the centroid C will not generally lie along the line.

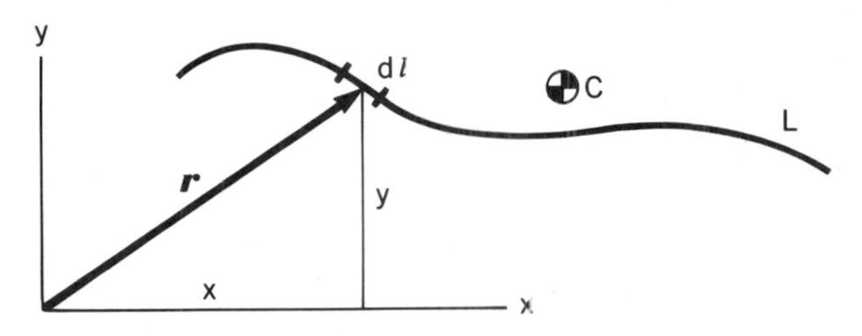

Figure 8.12. Centroid for curved line.

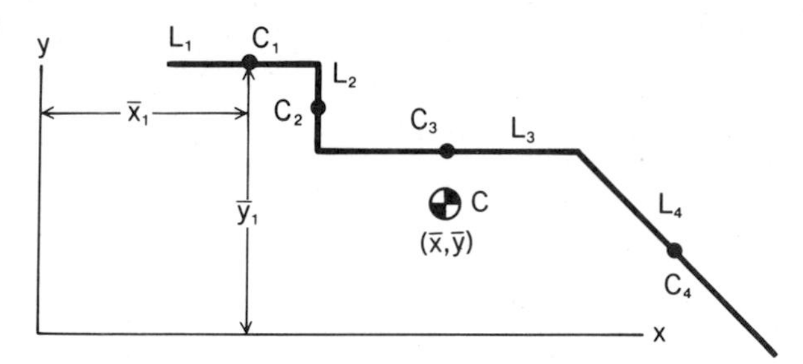

Figure 8.13. Centroid for composite line.

Consider next a curve made up of simple curves each of whose centroids is known. Such is the case shown in Fig. 8.13, made up of straight lines. The line segment L_1 has centroid C_1 with coordinates $\bar{x}_1\bar{y}_1$, as has been shown in the diagram. We can then say for the entire curve that

$$x_c = \frac{\sum_i \bar{x}_i L_i}{L}$$
$$y_c = \frac{\sum_i \bar{y}_i L_i}{L} \tag{8.13}$$

*8.4 Theorems of Pappus–Guldinus

The theorems of Pappus–Guldinus were first set forth by Pappus about 300 A.D. and then restated by the Swiss mathematician Paul Guldin about 1640. These theorems are concerned with the relation of a surface of revolution to its generating curve, and the relation of a volume of revolution to its generating area.

The first of the theorems may be stated as follows:

> *Consider a coplanar generating curve and an axis of revolution in the plane of this curve (see Fig. 8.14). The generating curve can touch but must not cross the axis of revolution. The surface of revolution developed by revolving the generating curve about the axis of revolution has an* area *equal to the product of the* length *of the* generating curve *times the* circumference *of the* circle *formed by the* centroid *of the* generating curve *in the process of generating a surface of revolution.*

To prove this theorem, consider first an element dl of the generating curve shown in Fig. 8.14. For a single revolution of the generating curve about the x axis, the line segment dl traces an area

$$dA = 2\pi y\, dl$$

For the entire curve this area becomes the surface of revolution given as

$$A = 2\pi \int y\, dl = 2\pi y_c L \tag{8.14}$$

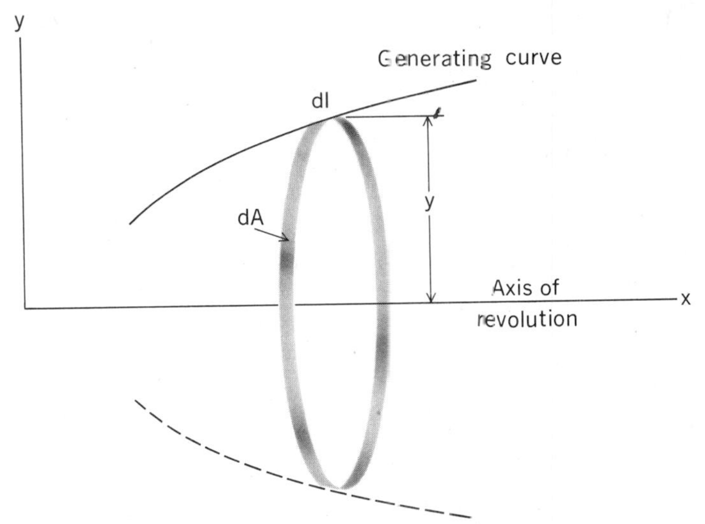

Figure 8.14. Coplanar generating curve.

where L is the length of the curve and y_c is the centroidal coordinate of the curve. But $2\pi y_c$ is the circumferential length of the circle formed by having the centroid of the curve rotate about the x axis. The first theorem is thus proved.

Another way of interpreting Eq. 8.14 is to note that the area of the body of revolution is equal to 2π times the *first moment* of the generating curve about the axis of revolution. If the generating curve is composed of simple curves, L_i, whose centroids are known, such as the case shown in Fig. 8.13, then we can express A as follows:

$$A = 2\pi(\sum_i L_i \bar{y}_i) \tag{8.15}$$

where $\bar{y}_i$ is the centroidal coordinate to the ith line segment L_i.

The second theorem may be stated as follows:

Consider a plane surface and an axis of revolution coplanar with the surface but oriented such that the axis can intersect the surface only as a tangent at the boundary or have no intersection at all. The volume of the body of revolution developed by rotating the plane surface about the axis of revolution equals the product of the area *of the surface times the* circumference *of the* circle *formed by the* centroid *of the surface in the process of generating the body of revolution.*

To prove the second theorem, consider a plane surface A as shown in Fig. 8.15. The volume generated by revolving dA of this surface about the x axis is

$$dV = 2\pi y \, dA$$

The volume of the body of revolution formed from A is then

$$V = 2\pi \int_A y \, dA = 2\pi y_c A \tag{8.16}$$

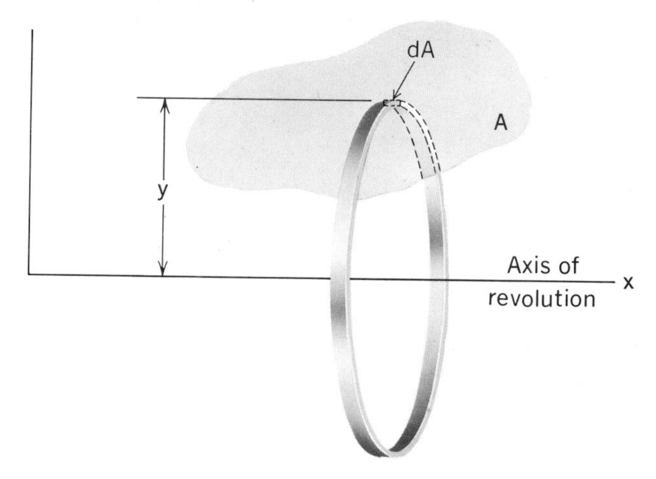

Figure 8.15. Plane surface A coplanar with x.

Thus, the volume V equals the area of the generating surface A times the circumferential length of the circle of radius y_c. The second theorem is thus also proved.[2]

Another way to interpret Eq. 8.16 is to note that V equals 2π times the *first moment* of the generating area A about the axis of revolution. If this area A is made up of simple areas A_i, we can say that

$$V = 2\pi(\sum_i A_i \bar{y}_i) \tag{8.17}$$

where $\bar{y}_i$ is the centroidal coordinate to the ith area A_i.

We now illustrate the use of theorems of Pappus and Guldinus. As we proceed, it will be helpful to remember the theorems by noting that you multiply a length (or area) of the generator by the distance moved by the centroid of the generator.

EXAMPLE 8.5

Determine the surface area and volume of the bulk materials trailer shown in Fig. 8.16.

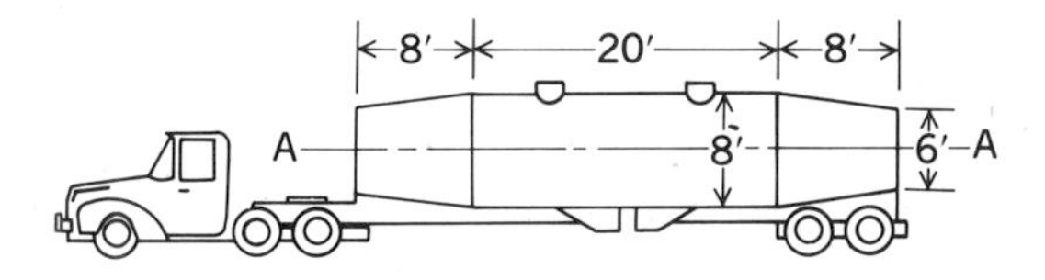

Figure 8.16. Bulk materials trailer.

We shall first determine the surface area by considering the first moment about the centerline A–A (see Fig. 8.17) of the generating curve of the surface of revolution. This curve is a set of 5 straight lines each of whose centroids is easily known by

---

[2]It is to be pointed out that the centroid of a volume of revolution will not be coincident with the centroid of a longitudinal cross section taken along the axis of the volume. Example: a cone and its triangular, longitudinal cross section.

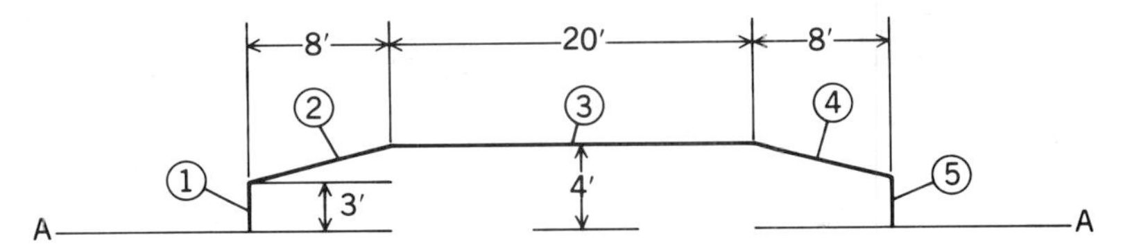

Figure 8.17. Generating curve for surface of revolution.

inspection. Accordingly we may use Eq. 8.15. For clarity, we use a column format for the data as follows:

	L_i (ft)	$\bar{y}_i$ (ft)	$L_i\bar{y}_i$ (ft^2)
1.	3	1.5	4.5
2.	$\sqrt{8^2+1^2} = 8.06$	3.5	28.21
3.	20	4	80
4.	8.06	3.5	28.21
5.	3	1.5	4.5
			$\sum_i L_i\bar{y}_i = 145.43$

Therefore,

$$A = (2\pi)(145.43) = 914 \text{ ft}^2$$

To get the volume, we next show in Fig. 8.18 the generating area for the body of revolution. Notice it has been decomposed into simple composite areas. We shall

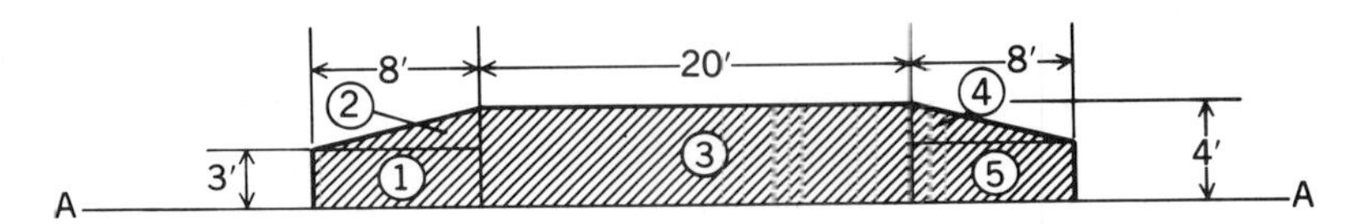

Figure 8.18. Generating area for body of revolution.

employ Eq. 8.17 and hence we shall need the first moment about the axis A–A of the composite areas. Again, we shall employ a column format for the data.

	A_i (ft^2)	$\bar{y}_i$ (ft)	$A_i\bar{y}_i$ (ft^3)
1.	24	1.5	26
2.	$(\frac{1}{2})(8)(1) = 4$	$3 + \frac{1}{3} = 3.33$	13.33
3.	80	2	160
4.	4	3.333	13.33
5.	24	1.5	36
			$\sum_i A_i\bar{y}_i = 258.7$

Therefore,

$$V = 2\pi \sum_i A_i\bar{y}_i = (2\pi)(258.7) = 1625 \text{ ft}^3$$

The theorems of Pappus and Guldinus have enabled us to compute the surface area and the volume of bulk materials trailer quickly and easily.

Problems

8.25. If $r^2 = ax$ in the body of revolution shown, compute the centroidal distance x_c of the body.

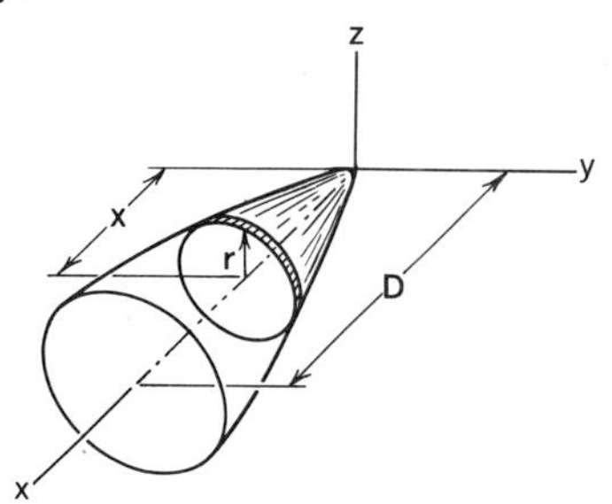

Figure P.8.25

8.26. Using vertical elements of volume as shown, compute the centroidal coordinates x_c, y_c of the body. Then, using horizontal elements, compute z_c.

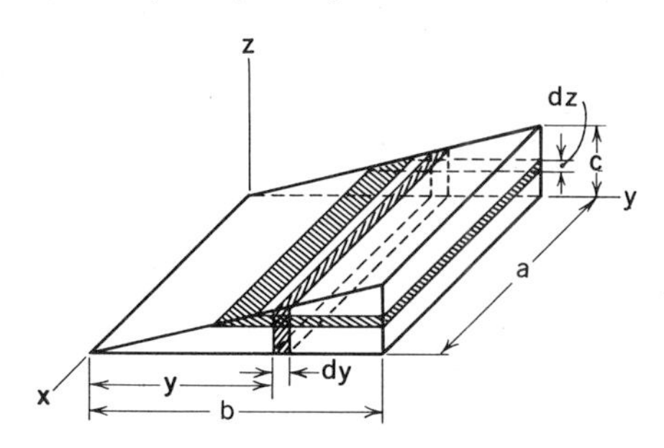

Figure P.8.26

8.27. Compute the center of volume of a right circular cylinder of height h and radius at the base r.

Figure P.8.27

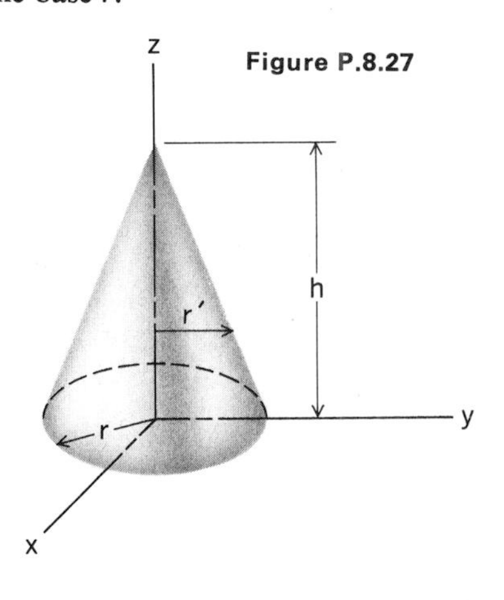

8.28. Determine the position of the center of mass of the solid hemisphere having a uniform mass density ρ and with a radius a.

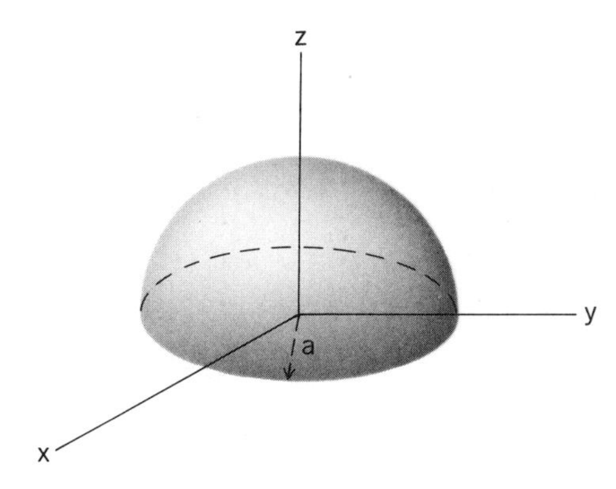

Figure P.8.28

8.29. Find the center of mass for the paraboloid of revolution having a uniform density ρ.

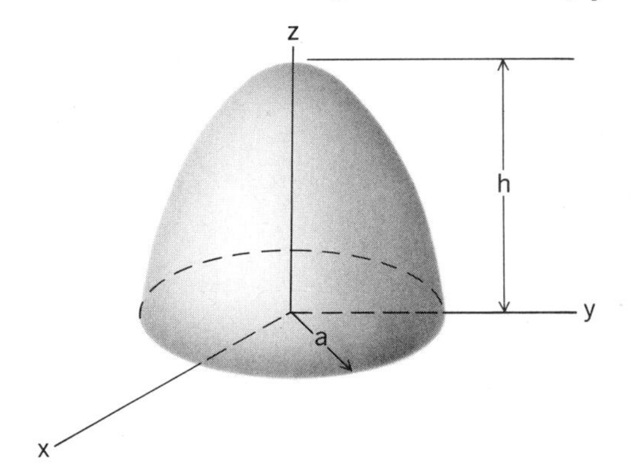

Figure P.8.29

8.30. A small bomb has exploded at position O. Four pieces of the bomb move off at high speed. At $t = 3$ sec, the following data apply:

	m (kg)	$\boldsymbol{r}$ (m)
1.	.2	$2\boldsymbol{i} + 3\boldsymbol{j} + 4\boldsymbol{k}$
2.	.1	$4\boldsymbol{i} + 4\boldsymbol{j} - 6\boldsymbol{k}$
3.	.15	$-3\boldsymbol{i} + 2\boldsymbol{j} - 3\boldsymbol{k}$
4.	.22	$2\boldsymbol{i} - 3\boldsymbol{j} + 2\boldsymbol{k}$

What is the position of the center of mass?

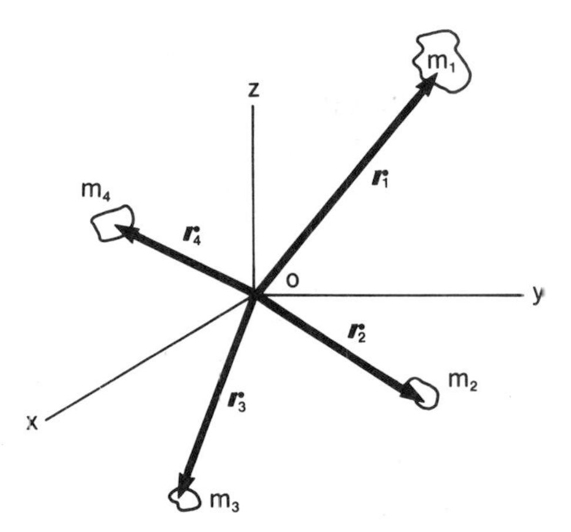

Figure P.8.30

8.31. A plate of uniform thickness and density has for its curved edge a rectangular hyperbola (xy = constant). Find the centroid of the upper surface. Find the centers of mass, volume, and gravity for the plate.

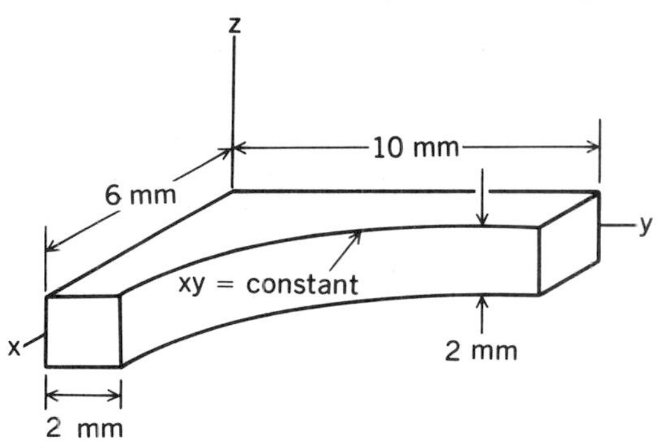

Figure P.8.31

In Problems 8.32 through 8.38, use the formulas on the inside covers for simple shapes.

8.32. Where must a lifting hook be cast in a tapered concrete beam so that the beam always stays horizontal when lifted?

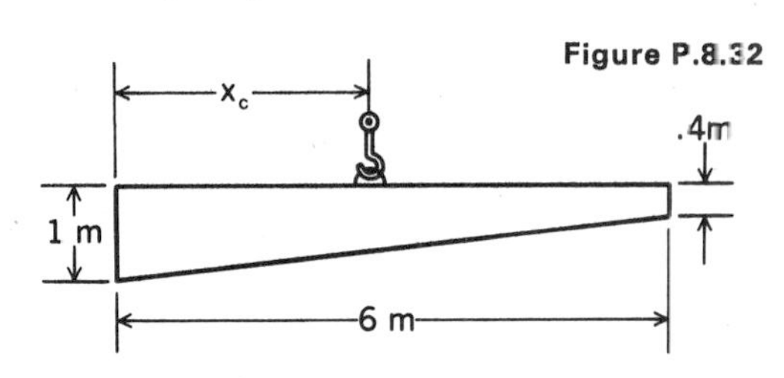

Figure P.8.32

8.33. Two solid semicylinders are glued together. Body A has a uniform mass density of 6.54 kN/m³, while body B has a uniform mass density of 10 kN/m³. Determine:
(a) Center of volume
(b) Center of mass
(c) Center of gravity

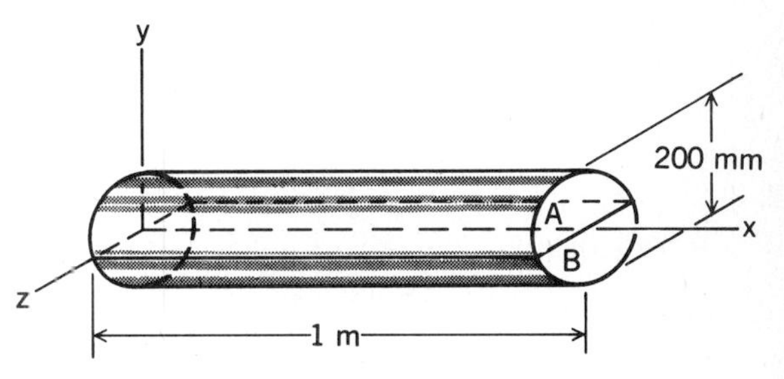

Figure P.8.33

8.34. What is the centroid of the body shown? It consists of a cylinder A of length 2 m and diameter 6 m, a shaft B of diameter 2 m and length 8 m, and a block C of length 4 m and height and width of 7 m. The x axis is a centerline for the arrangement. Origin O is at the geometric center of cylinder A.

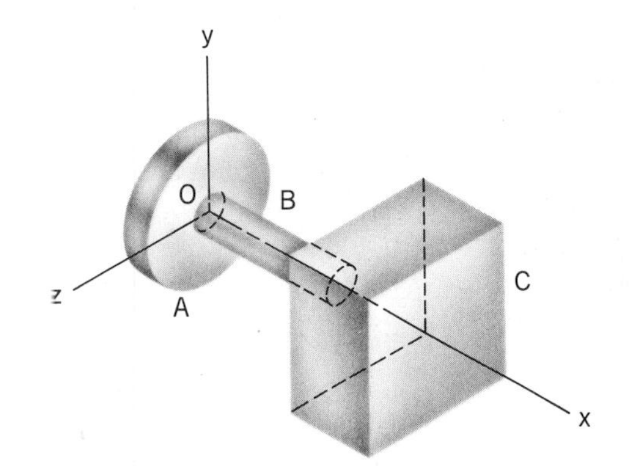

Figure P.8.34

8.35. Find the center of volume for the cone–cylinder shown. Note that there is a cylindrical hole of length 16 ft and diameter 4 ft cut into the body.

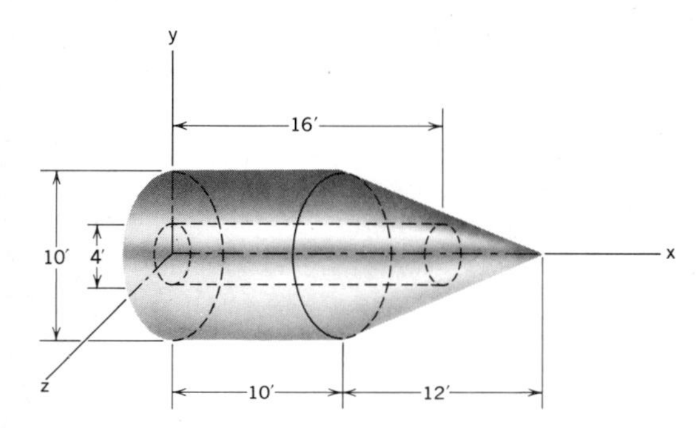

Figure P.8.35

8.36. A bent aluminum rod weighing 30 N/m is fitted into a plastic cylinder weighing 200 N, as shown. What are the centers of volume, mass, and gravity?

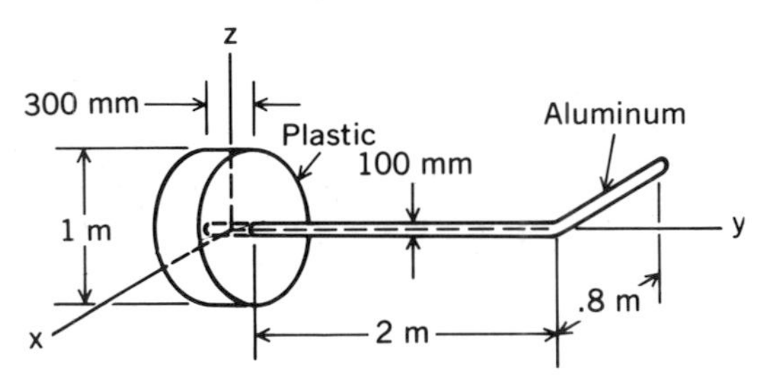

Figure P.8.36

8.37. A brass cylinder fits snugly into an aluminum block. The brass weighs 43.2 kN/m³ and the aluminum weighs 30 kN/m³. Find the center of volume, the center of mass, and the center of gravity.

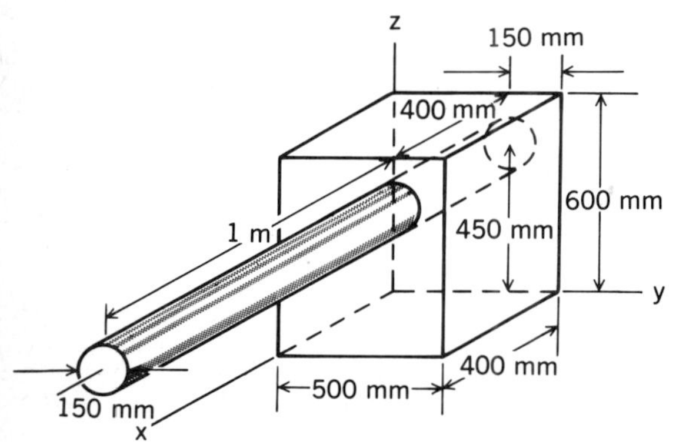

Figure P.8.37

8.38. Two thin plates are welded together. One has circle of radius 200 mm cut out as shown. If each plate weighs 450 N/m², what is the position of the center of mass?

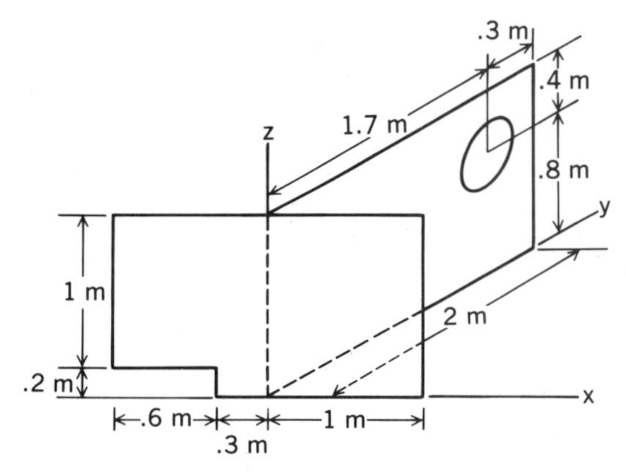

Figure P.8.38

8.39. What is the center of mass of the bent wire if it weighs 10 N/m?

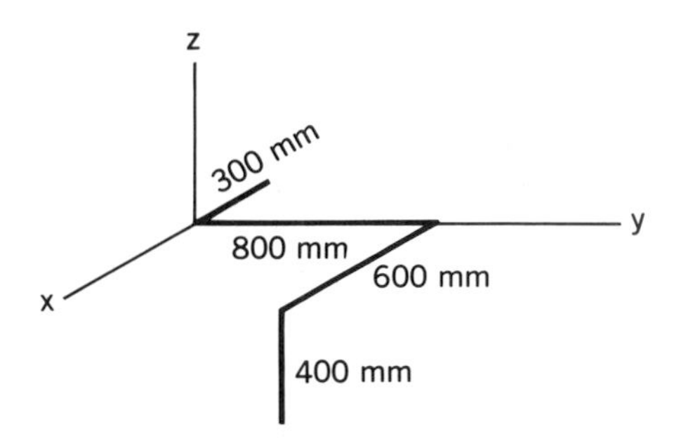

Figure P.8.39

8.40. Find the center of mass of the bent wire shown in the *zy* plane. The wire weighs 15 N/m.

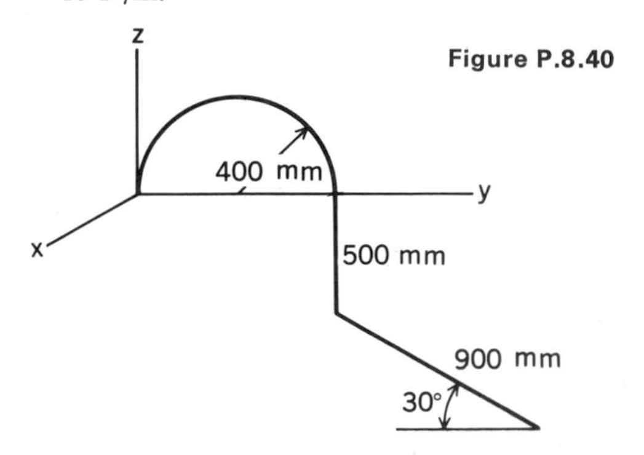

Figure P.8.40

8.41. In Problem 8.35, involving a wooden cone–cylinder with a cylindrical hole, find the center of mass for the case where the cylinder has a density of 46.0 lbm/ft³ and the cone has a density of 30.0 lbm/ft³.

8.42. The volume of an ellipsoidal body of revolution is known from the calculus to be $\frac{1}{6}\pi ab^2$. If the area of an ellipse is $\pi ab/4$, find the centroid of the area for a semiellipse.

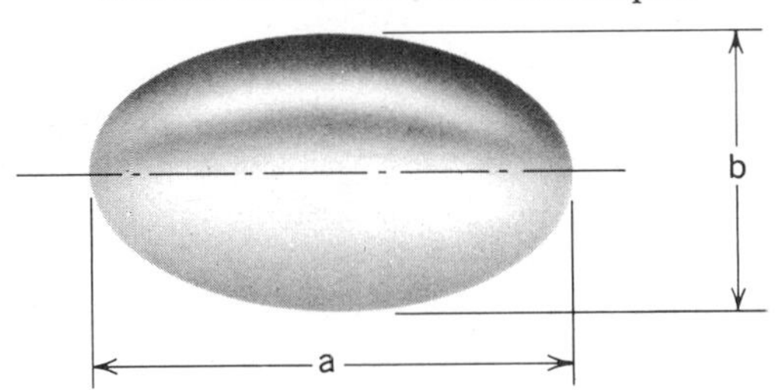

Figure P.8.42

8.43. Find the centroidal coordinate y_c of the shaded area shown, using the theorems of Pappus and Guldinus.

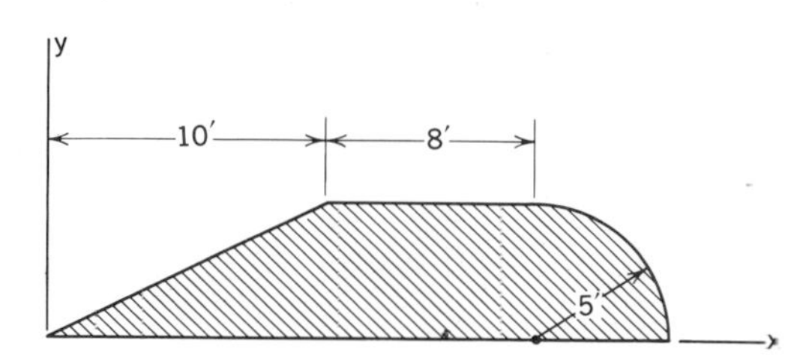

Figure P.8.43

8.44. A cutting tool of the lathe is programmed to cut along the dashed line as shown. What are the volume and the area of the body of revolution formed on the lathe?

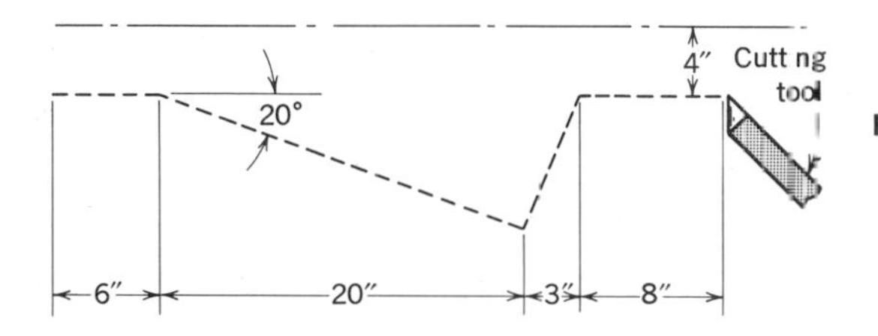

Figure P.8.44

8.45. Find the surface area and volume of the right conical frustum.

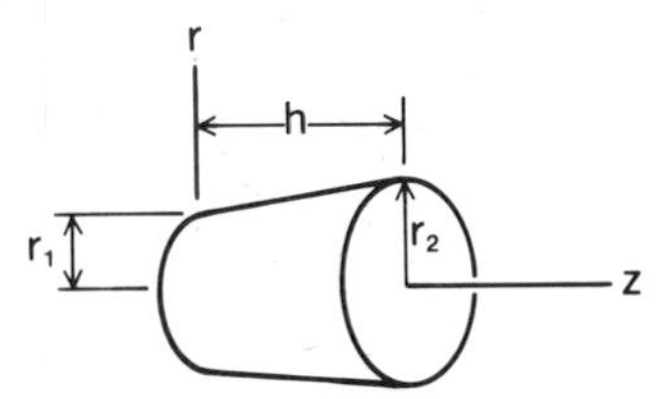

Figure P.8.45

8.46. Find the surface area and volume of the earth entry capsule for an unmanned Mars sampling mission. Approximate the rounded nose with a pointed nose as shown with the dashed lines.

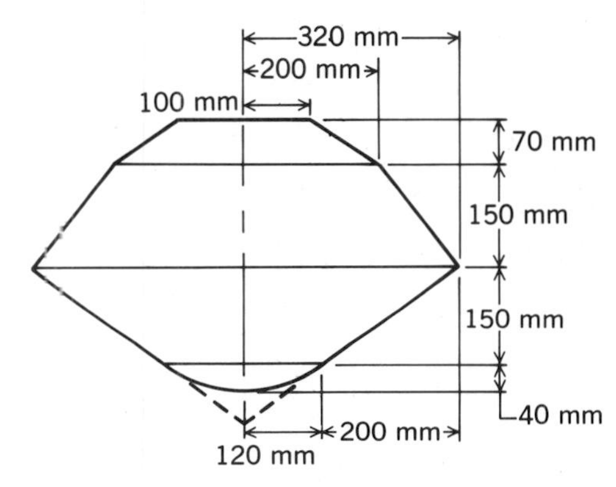

Figure P.8.46

8.47. Find the volume and surface area of the Apollo spaceship used for lunar exploration.

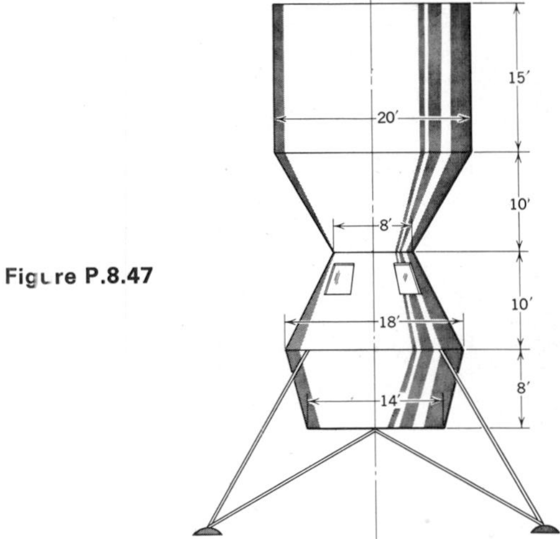

Figure P.8.47

8.5 Second Moments and the Product of Area[3] of a Plane Area

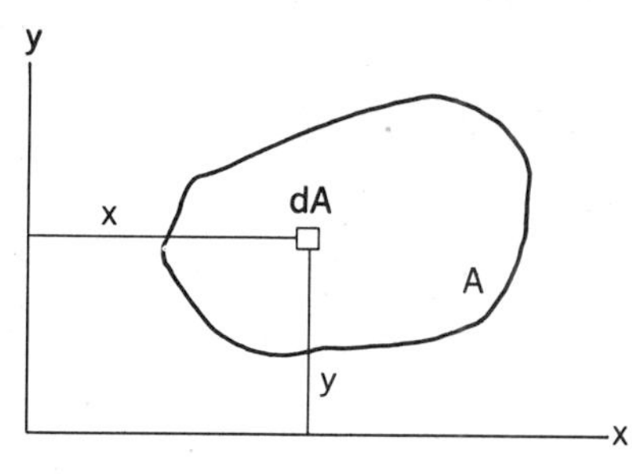

Figure 8.19. Plane surface.

We shall now consider other properties of a plane area relative to a given reference. The *second moments* of the area A about the x and y axes (Fig. 8.19), denoted as I_{xx} and I_{yy}, respectively, are defined as:

$$I_{xx} = \int_A y^2 \, dA \tag{8.18a}$$

$$I_{yy} = \int_A x^2 \, dA \tag{8.18b}$$

The second moment of area cannot be negative, in contrast to the first moment. Furthermore, because the square of the distance from the axis is used, elements of area that are farthest from the axis contribute most to the second moment of area.

In an analogy to the centroid, the entire area may be concentrated at a single point (k_x, k_y) to give the same second moment of area for a given reference. Thus,

$$\begin{aligned} Ak_x^2 &= I_{xx} = \int_A y^2 \, dA; \qquad \text{therefore, } k_x^2 = \frac{\int_A y^2 \, dA}{A} \\ Ak_y^2 &= I_{yy} = \int_A x^2 \, dA; \qquad \text{therefore, } k_y^2 = \frac{\int_A x^2 \, dA}{A} \end{aligned} \tag{8.19}$$

The distances k_x and k_y are called the *radii of gyration. This point will have a position that depends not only on the shape of the area but also on the position of the reference.* This situation is unlike the centroid, whose location is independent of the reference position.

The *product of area* relates an area directly to a set of axes and is defined as

$$I_{xy} = \int_A xy \, dA \tag{8.20}$$

This quantity may be negative. We shall soon show that second moments and products of area are related for a given reference.

If the area under consideration has an axis of symmetry, the product of area for this axis and any axis orthogonal to this axis must be zero. You can readily reach this conclusion by considering the area in Fig. 8.20 which is symmetrical about the axis A–A. Notice that the centroid is somewhere along this axis. (Why?) The axis of symmetry has been indicated as the y axis, and an arbitrary x axis coplanar with the area has been shown. Also indicated are two elemental areas that are positioned as mirror images about the y axis. The contribution to the product of area of each element is

[3]We often use the expressions *moment* and *product of inertia* for second moment and product of area, respectively. However, we shall also use the former expressions in Chapter 9 in connection with mass distributions.

$xy\,dA$, but with opposite signs, and so the net result is zero. Since the entire area can be considered to be composed of such pairs, it becomes evident that the product of area for such cases is zero. This *should not* be taken to mean that a nonsymmetric area cannot have a zero product of area about a set of axes. We shall discuss this last condition in more detail later.

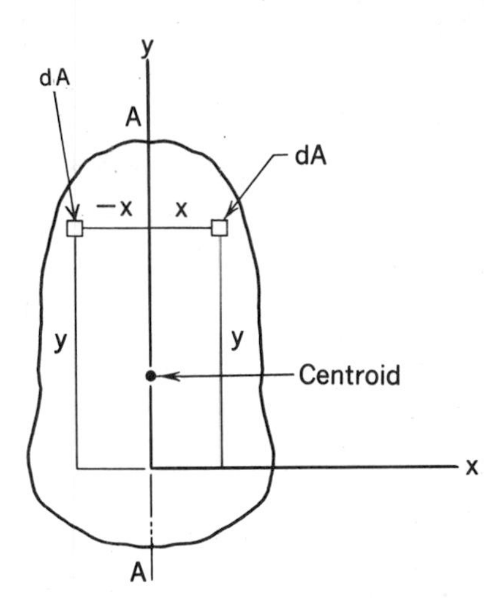

Figure 8.20. Area symmetric about y axis.

8.6 Transfer Theorems

We shall now set forth a theorem that will be of great use in computing second moments and products of area for areas that can be decomposed into simple parts (composite areas). With this theorem, we can find second moments or products of area about any axis in terms of second moments or products of area about a parallel set of axes going through the centroid of the area in question.

An x axis is shown in Fig. 8.21 parallel to and at a distance d from an axis x' going through the centroid of the area. The latter axis you will recall is a *centroidal axis*. The second moment of area about the x axis is

$$I_{xx} = \int_A y^2\,dA = \int_A (y' + d)^2\,dA$$

where the distance y has been replaced by $(y' + d)$. Integration leads to the result

$$I_{xx} = \int_A y'^2\,dA + 2d\int_A y'\,dA + Ad^2$$

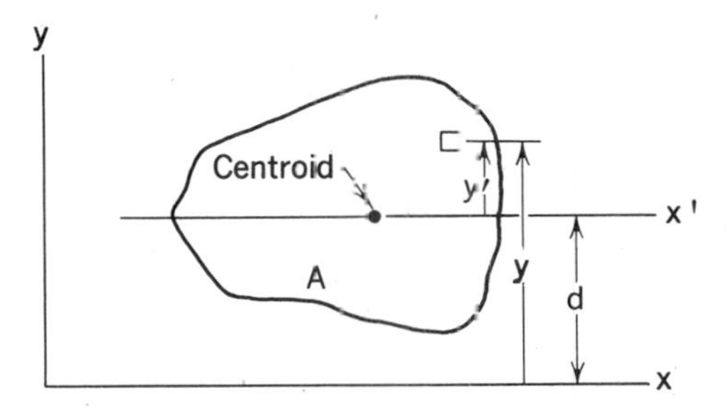

Figure 8.21. x and x' are parallel axes.

The first term on the right-hand side is clearly $I_{x'x'}$. The second term involves the first moment of area about the x' axis. But the x' axis here is a centroidal axis, and so the second term is zero. We can now state the transfer theorem (frequently called the parallel-axis theorem):

$$\boxed{I_{\text{about any axis}} = I_{\substack{\text{about a parallel}\\ \text{axis at centroid}}} + Ad^2} \tag{8.21}$$

where d is the perpendicular distance between the axis for which I is being computed and the parallel centroidal axis.

In strength of materials, a course generally following statics, second moments of area about noncentroidal axes are commonly used. The areas involved are complicated and not subject to simple integration. Accordingly, in structural handbooks, the areas and second moments about various centroidal axes are listed for many of the practical configurations with the understanding that designers will use the parallel-axis theorem for axes not at the centroid.

Let us now examine the product of area in order to establish a parallel-axis theorem for this quantity. Accordingly, two references are shown in Fig. 8.22, one (x', y')

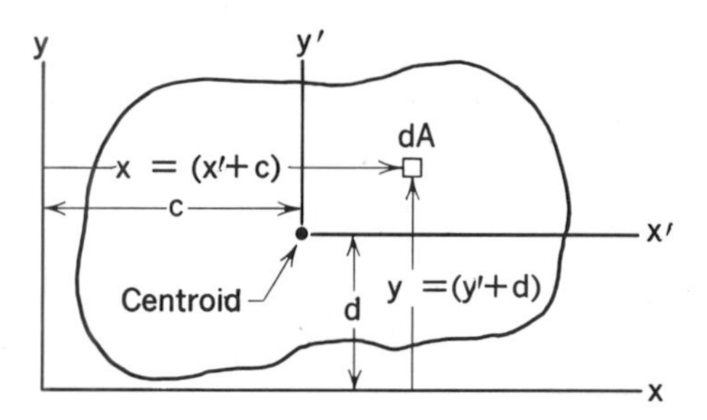

Figure 8.22. *c* and *d* measured from *xy*.

at the centroid and the other (x, y) positioned arbitrarily but *parallel* relative to xy. Note that c and d are the x and y *coordinates* of the centroid of A as measured from reference xy. These coordinates accordingly must have the proper signs, dependent on what quadrant the centroid of A is in relative to xy. The product of area about the noncentroidal axes xy can then be given as

$$I_{xy} = \int_A xy\, dA = \int_A (x' + c)(y' + d) dA$$

Carrying out the multiplication, we get

$$I_{xy} = \int_A x'y'\, dA + c \int_A y'\, dA + d \int_A x'\, dA + Adc$$

Clearly, the first term on the right side is $I_{x'y'}$, whereas the next two terms are zero, since x' and y' are centroidal axes. Thus, we arrive at a parallel-axis theorem for products of area of the form:

$$\boxed{I_{xy \text{ for any set of axes}} = I_{x'y' \substack{\text{ for a parallel set of}\\ \text{axes at centroid}}} + Adc} \tag{8.22}$$

It is important to remember that c and d are measured *from the xy axes to the centroid* and must have the appropriate sign. This will be carefully pointed out again in the examples of Section 8.7.

8.7 Computations Involving Second Moments and Products of Area

We shall now examine examples for the computation of second moments and products of area.

EXAMPLE 8.6

A rectangle is shown in Fig. 8.23. Compute the second moment and product of area about the centroidal $x'y'$ axes as well as about the xy axes.

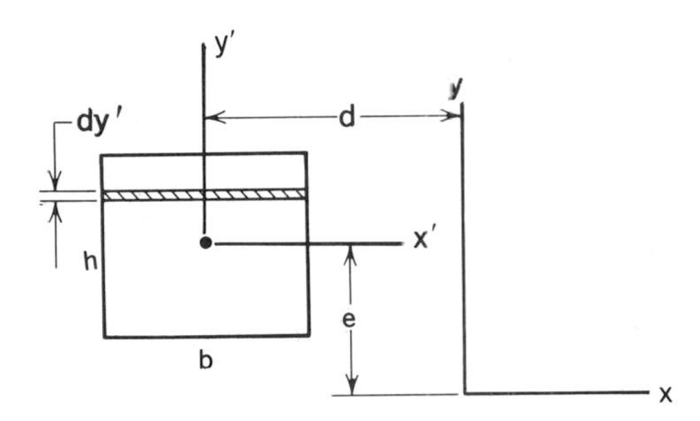

Figure 8.23. Rectangle: base *b*, height *h*.

$I_{x'x'}, I_{y'y'}, I_{x'y'}$. For computing $I_{x'x'}$, we can use a strip of width dy' at a distance y' from the x' axis. The area dA then becomes $b\,dy'$. Hence, we have

$$I_{x'x'} = \int_{-h/2}^{+h/2} y'^2 b\,dy = b\frac{y'^3}{3}\Bigg|_{-h/2}^{+h/2} = \frac{b}{3}\left(\frac{h^3}{8} + \frac{h^3}{8}\right) = \frac{1}{12}bh^3 \tag{a}$$

This is a common result and should well be remembered since it occurs so often. Verbally, for such an axis, the second moment is equal to $\frac{1}{12}$ the base b times the height h cubed. The second moment of area for the y' axis can immediately be written as

$$I_{y'y'} = \tfrac{1}{12}hb^3 \tag{b}$$

where the base and height have simply been interchanged.

As a result of the previous statements on symmetry, we immediately note that

$$I_{x'y'} = 0 \tag{c}$$

I_{xx}, I_{yy}, I_{xy}. Employing the transfer theorems, we get

$$I_{xx} = \tfrac{1}{12}bh^3 + bhe^2$$
$$I_{yy} = \tfrac{1}{12}hb^3 + bhd^2$$

In computing the product of area, we must be careful to employ the proper signs for the transfer distances. In checking the derivation of the transfer theorem, we see that these distances are measured from the noncentroidal axes to the centroid C. Therefore, in this problem the transfer distances are $(+e)$ and $(-d)$. Hence, the computation of I_{xy} becomes

$$I_{xy} = 0 + (bh)(+e)(-d) = -bhed$$

and is thus a negative quantity.

EXAMPLE 8.7

What are I_{xx}, I_{yy}, and I_{xy} for the area under the parabolic curve shown in Fig. 8.24?

To find I_{xx}, we may use horizontal strips of width dy as shown in Fig. 8.25. We can then say for I_{xx}:

$$I_{xx} = \int_0^{10} y^2[dy(10 - x)]$$

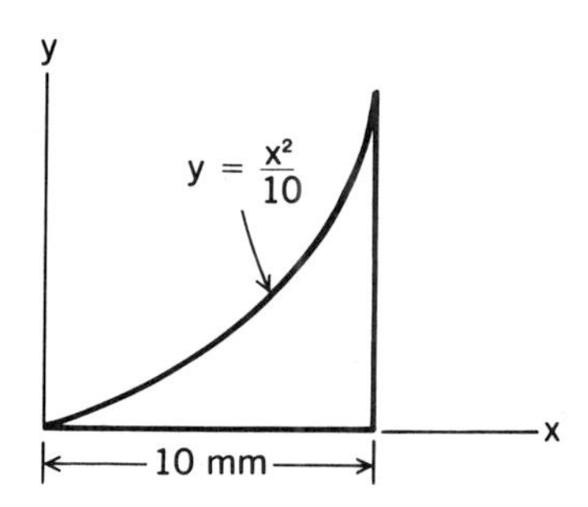

Figure 8.24. Plane area.

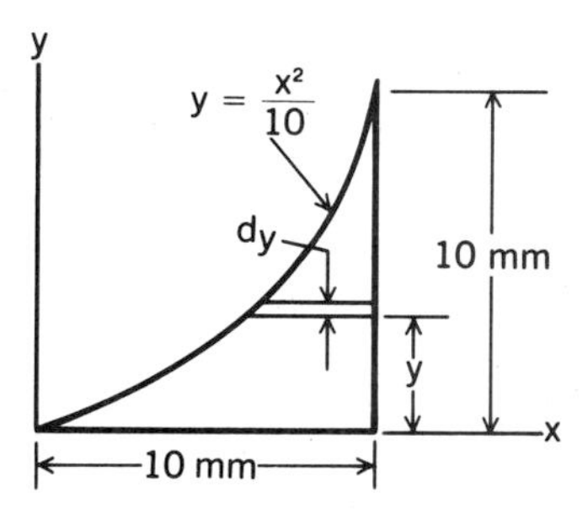

Figure 8.25. Horizontal strip.

But

$$x = \sqrt{10}\, y^{1/2}$$

Therefore,

$$I_{xx} = \int_0^{10} y^2(10 - \sqrt{10}\, y^{1/2})\, dy$$

$$= \left[10\frac{y^3}{3} - \sqrt{10}\, y^{7/2}\left(\frac{2}{7}\right)\right]\Bigg|_0^{10}$$

$$= \frac{10(10^3)}{3} - \sqrt{10}(10^{7/2})\left(\frac{2}{7}\right) = 476.2 \text{ mm}^4$$

As for I_{yy}, we use vertical infinitesimal strips as shown in Fig. 8.26. We can, accordingly, say:

$$I_{yy} = \int_0^{10} x^2(y\, dx) = \int_0^{10} \frac{x^4}{10}\, dx$$

$$= \frac{x^5}{50}\Bigg|_0^{10} = 2000 \text{ mm}^4$$

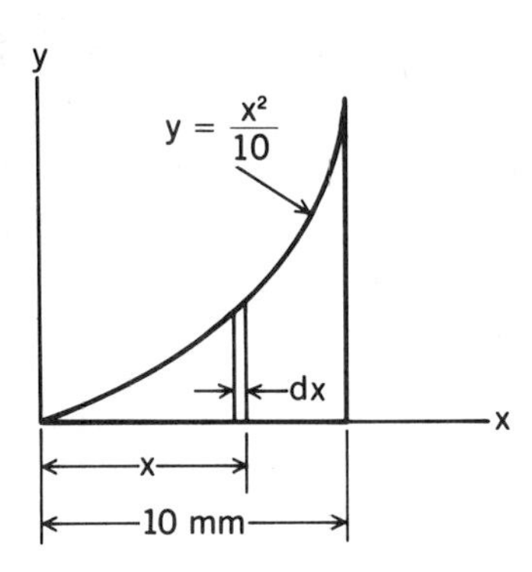

Figure 8.26. Vertical strip.

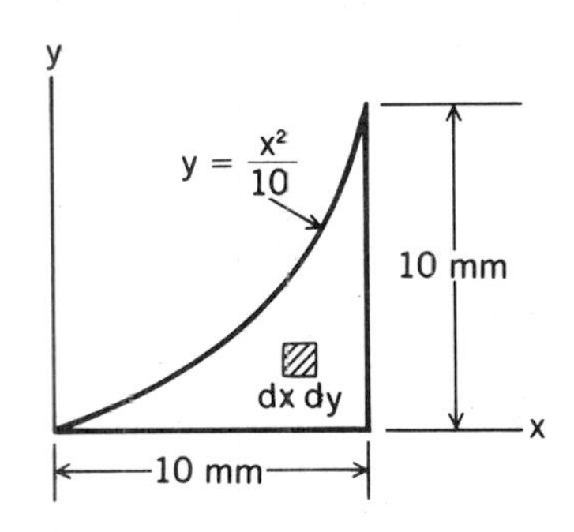

Figure 8.27. Element for multiple integration.

Finally, for I_{xy} we use an infinitesimal area element $dx\, dy$ shown in Fig. 8.27. We must now perform multiple integration.[4] Thus, we have

$$I_{xy} = \int_0^{10} \int_{y=0}^{y=x^2/10} xy\, dy\, dx$$

Notice by holding x constant and letting y first run from $y = 0$ to the curve $y = x^2/10$ we cover the vertical strip of thickness dx at position x such as is shown in Fig. 8.26. Then by letting x run from zero to 10, we cover the entire area. Accordingly, we first integrate with respect to y holding x constant. Thus,

$$I_{xy} = \int_0^{10} x\left(\frac{y^2}{2}\right)\Bigg|_0^{x^2/10} dx = \int_0^{10} \frac{x^5}{200}\, dx$$

Next, integrating with respect to x, we have

$$I_{xy} = \frac{x^6}{1200}\Bigg|_0^{10} = 833 \text{ mm}^4$$

EXAMPLE 8.8

Compute the second moment of area of a circular area about a diameter (Fig. 8.28).

Using polar coordinates, we have[5] for I_{xx}:

$$I_{xx} = \int_0^{D/2} \int_0^{2\pi} (r \sin \theta)^2 r\, d\theta\, dr = \int_0^{D/2} \pi r^3\, dr$$

---

[4]This multiple integration involves boundaries requiring some variable limits, in contrast to previous multiple integrations.

[5]The integral $\int_0^{2\pi} \sin^2 \theta\, d\theta$ may be evaluated by methods of substitution or may readily be seen in the following manner. $\int_0^{2\pi} \sin^2 \theta\, d\theta$ equals the area under the curve shown, which is half the area of the rectangle. Hence, this integral equals π.

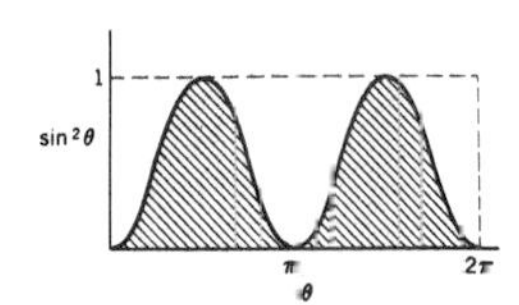

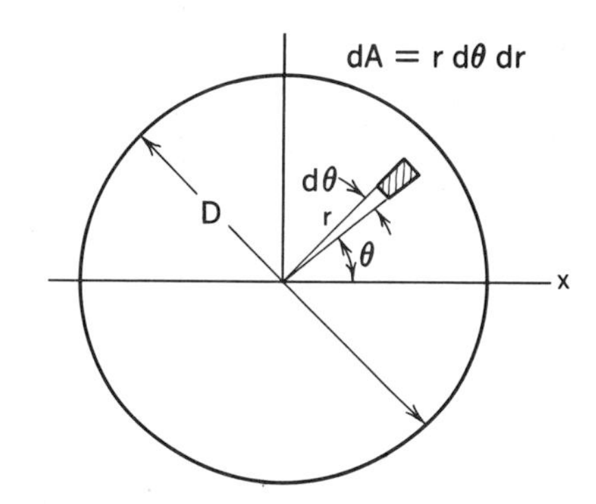

Figure 8.28. Circular area with polar coordinates.

Completing the integration, we have

$$I_{xx} = \frac{r^4}{4}\pi \Big|_0^{D/2} = \pi \frac{D^4}{64}$$

The product of area I_{xy} must be zero, owing to symmetry of the area about the xy axes.

In the previous examples, we computed second moments and products of area using the calculus. Many problems of interest involve an area that may be subdivided into simpler component areas. Such an area has been referred to in earlier discussions as a *composite* area. The second moments and products of area for certain centroidal axes of many simple areas may be found in engineering handbooks (also see the inside covers). Using these formulas plus the parallel-axis theorems, we can easily compute desired second moments and products of area for composite areas as we have done earlier for first moments of area. The following example illustrates this procedure.

EXAMPLE 8.9

Find the centroid of the area of the unequal-leg Z section shown in Fig. 8.29. Next, determine the second moment of area about the centroidal axes parallel to the sides of the Z section. Finally, determine the product of area for the aforementioned centroidal axes.

We shall subdivide the Z section into three rectangular areas, as shown in Fig. 8.30. Also, we shall insert a convenient reference xy, as shown in the diagram. To find the centroid, we proceed in the following manner:

	A_i (in.^{2})	$\bar{x}_i$ (in.)	$\bar{y}_i$ (in.)	$A_i\bar{x}_i$ (in.^{3})	$A_i\bar{y}_i$ (in.^{3})
1.	(2)(1) = 2	1	7.50	2	15
2.	(8)(1) = 8	2.50	4	20	32
3.	(4)(1) = 4	5	.50	20	2
	$\sum_i A_i = 14$			$\sum_i A_i\bar{x}_i = 42$	$\sum_i A_i\bar{y}_i = 49$

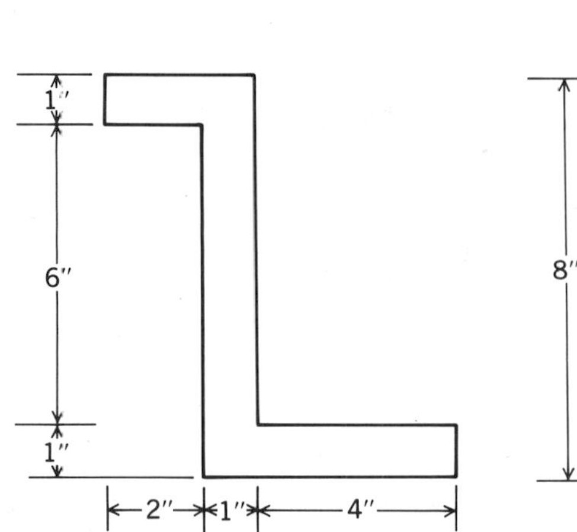

Figure 8.29. Unequal-leg Z section.

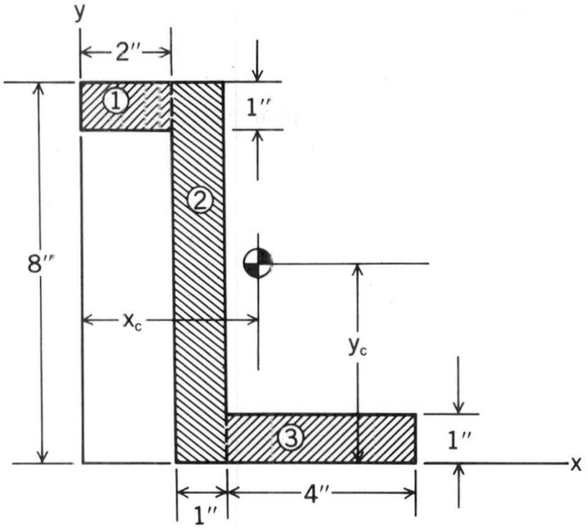

Figure 8.30. Composite area.

Therefore,

$$x_c = \frac{\sum_i A_i \bar{x}_i}{\sum_i A_i} = \frac{42}{14} = 3 \text{ in.}$$

$$y_c = \frac{\sum_i A_i \bar{y}_i}{\sum_i A_i} = \frac{49}{14} = 3.5 \text{ in.}$$

We have shown the centroidal axes $x_c y_c$ in Fig. 8.31. We now find $I_{x_c x_c}$ and $I_{y_c y_c}$ using the parallel-axis theorem and the formula $\frac{1}{12} bh^3$ for the second moment of area about a centroidal axis of symmetry of a rectangle.

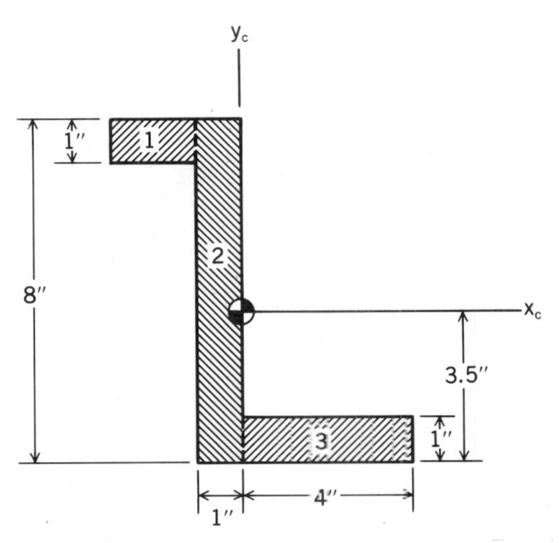

Figure 8.31. Centroidal axes $x_c y_c$.

$$I_{x_cx_c} = \underbrace{[(\tfrac{1}{12})(2)(1^3) + (2)(4^2)]}_{①} + \underbrace{[(\tfrac{1}{12})(1)(8^3) + (8)(\tfrac{1}{2})^2]}_{②}$$

$$+ \underbrace{[(\tfrac{1}{12})(4)(1^3) + (4)(3^2)]}_{③} = 113.2 \text{ in.}^4$$

$$I_{y_cy_c} = \underbrace{[(\tfrac{1}{12})(1)(2^3) + (2)(2^2)]}_{①} + \underbrace{[(\tfrac{1}{12})(8)(1^3) + (8)(\tfrac{1}{2})^2]}_{②}$$

$$+ \underbrace{[(\tfrac{1}{12})(1)(4^3) + (4)(2^2)]}_{③} = 32.67 \text{ in.}^4$$

Finally, we consider the product of area $I_{x_cy_c}$. Here we must be cautious in using the parallel-axis theorem. Remember that x_cy_c are centroidal axes for the *entire* area of the Z section. In using the parallel-axis theorem for a *subarea*, we must note that x_cy_c are *not* centroidal axes for the subarea. The centroidal axes to be used in this problem for subareas are the axes of symmetry of each subarea. In short, x_cy_c are simply axes about which we are computing the product of area of each subarea. Therefore, in the parallel axis theorem, the transfer distances c and d are measured *from the x_cy_c axes to the centroid* in each subarea, as noted in the development of the parallel-axis theorem. The proper sign must be assigned each time to the transfer distances with this in mind. We have for $I_{x_cy_c}$:

$$I_{x_cy_c} = \underbrace{[0 + (2)(-2)(4)]}_{①} + \underbrace{[0 + (8)(-\tfrac{1}{2})(\tfrac{1}{2})]}_{②}$$

$$+ \underbrace{[0 + (4)(2)(-3)]}_{③} = -42 \text{ in.}^4$$

Problems

8.48. Find I_{xx}, I_{yy}, and I_{xy} for the triangle shown. Give the results in feet.

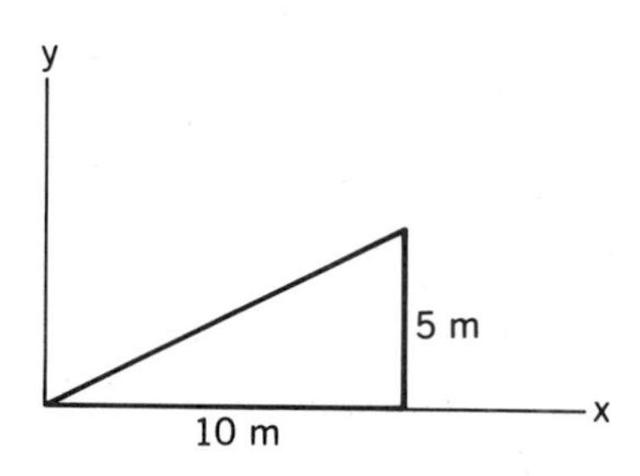

Figure P.8.48

8.49. What are the second moments and products of area of the ellipse for reference xy? (*Hint:* Can you work with one quadrant and then multiply by 4 for the second moments?)

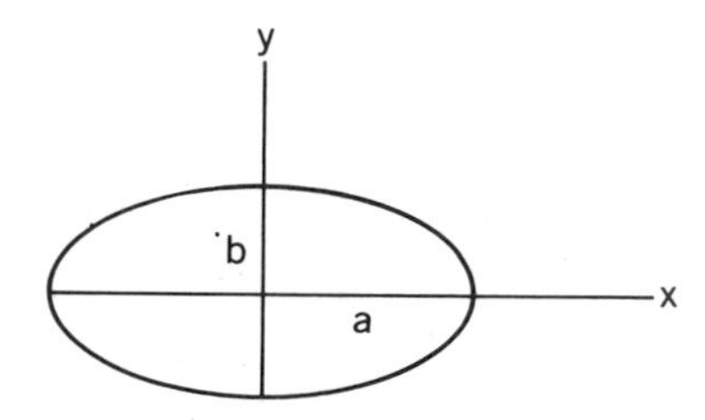

Figure P.8.49

8.50. Find I_{xx} and I_{yy} for the quarter circle of radius 5 m.

Figure P.8.50

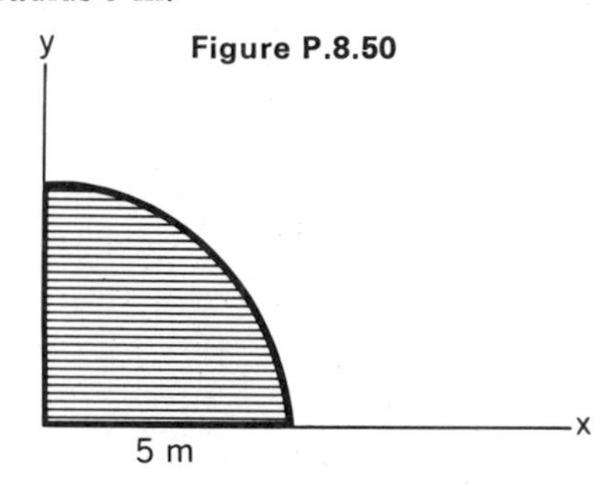

8.51. Find I_{xx}, I_{yy}, and I_{xy} for the shaded area.

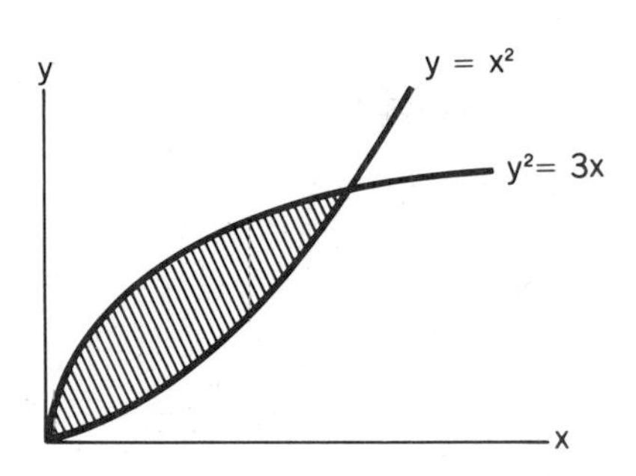

Figure P.8.51

8.52. Find I_{yy} for the shaded area.

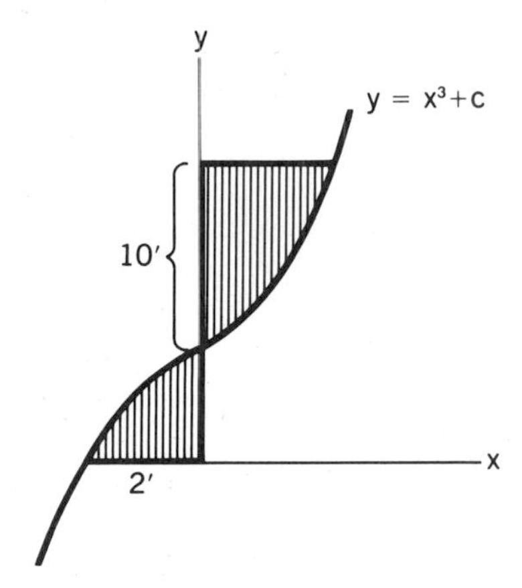

Figure P.8.52

8.53. Find I_{yy} for the area between the curves

$$y = 2 \sin x \text{ ft}$$

$$y = \sin 2x \text{ ft}$$

from $x = 0$ to $x = \pi$ ft.

8.54. Find I_{yy} for the areas enclosed between curves $y = \cos x$ and $y = \sin x$ and the lines $x = 0$ and $x = \pi/2$.

8.55. Show that $I_{xx} = bh^3/12$, $I_{yy} = b^3h/12$, and $I_{xy} = b^2h^2/24$ for the right triangle.

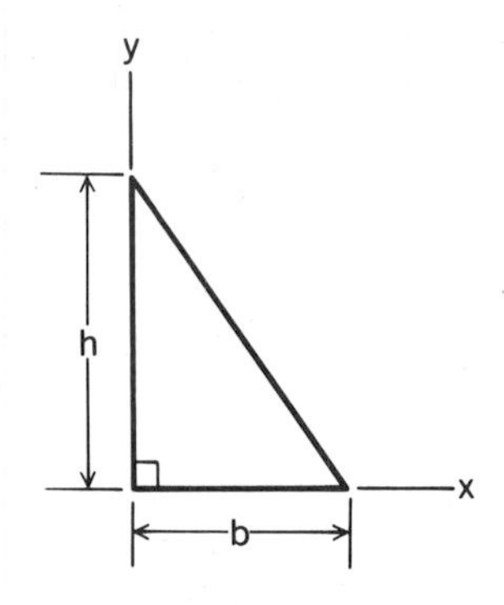

Figure P.8.55

8.56. Find I_{xx}, I_{yy}, and I_{xy} for the shaded area.

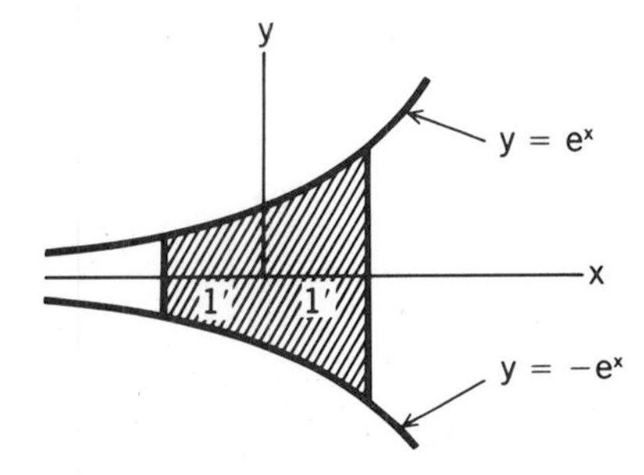

Figure P.8.56

8.57. Find I_{xx}, I_{yy}, and I_{xy} for the area of Problem 8.4. The equation of the curve is $y^2 = 20x - 60$.

8.58. Find I_{xx}, I_{yy}, and I_{xy} for the cross section shown.

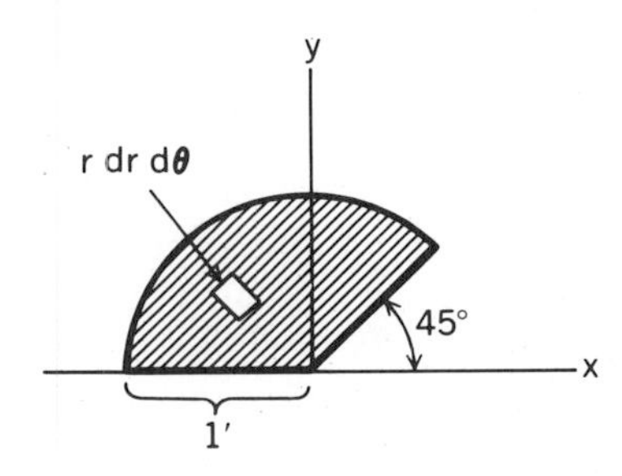

Figure P.8.58

8.59. Find I_{xx} and I_{yy} for the area of Problem 8.5. The equation of the parabola is $y = (x^2/9) - 5$. (*Hint:* The area of a vertical element in the region below the x axis is $(0 - y)\,dx$.)

8.60. In Problem 8.59, determine I_{xy} using multiple integration.

8.61. Find the second moments of area about axes xy for the shaded area shown.

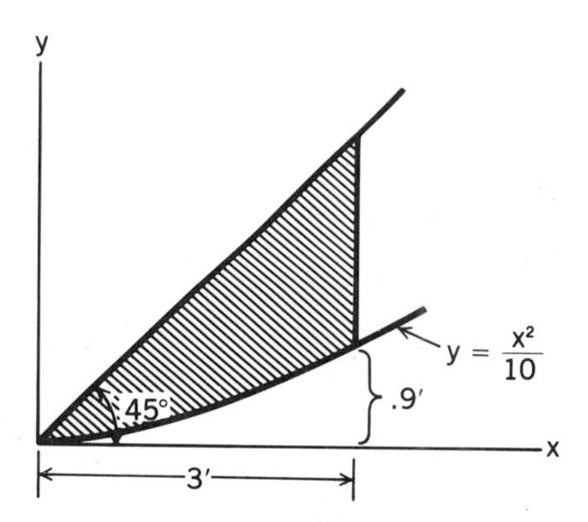

Figure P.8.61

8.62. If the second moment of area about axis A–A is known to be 600 ft^4, what is the second moment of area about a parallel axis B–B a distance 3 ft from A–A, for an area of 10 ft^2? The centroid of this area is 4 ft from B–B.

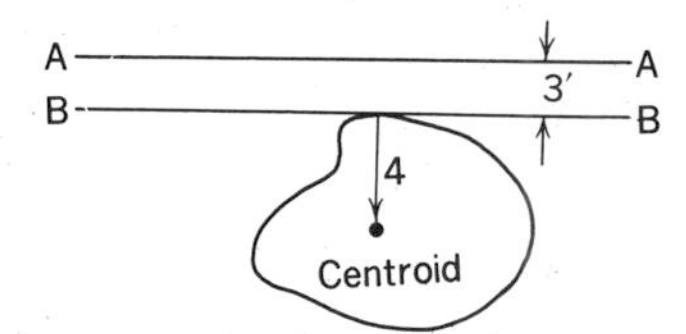

Figure P.8.62

8.63. Using the results of Problem 8.55, show that $I_{x_cx_c} = bh^3/36$, $I_{y_cy_c} = hb^3/36$, and $I_{x_cy_c} = -b^2h^2/72$ for the right triangle shown.

Figure P.8.63

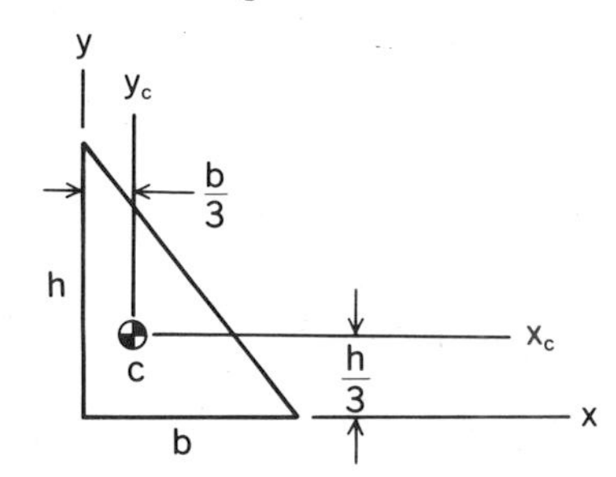

8.64. Show that $I_{xx} = bh^3/12$, $I_{yy} = (hb/12)(b^2 + ab + a^2)$, and $I_{xy} = (h^2b/24)(2a + b)$ for the triangle. (*Hint:* Break the triangle into two right triangles for which the various moments are known. (See Problem 8.63).)

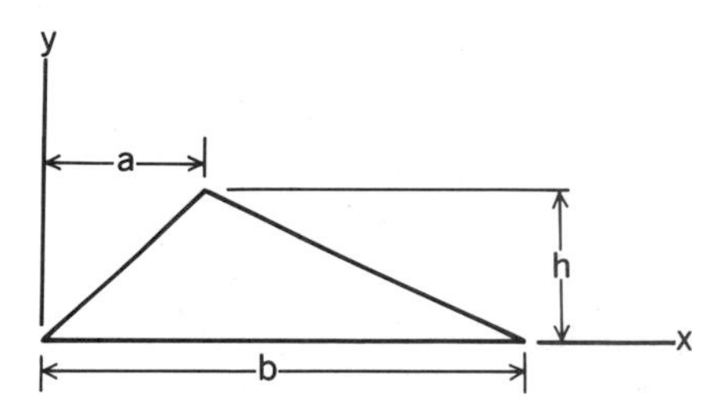

Figure P.8.64

8.65. In Problem 8.64, show that $I_{x_cx_c} = bh^3/36$, $I_{y_cy_c} = (bh/36)(b^2 - ab + a^2)$, and $I_{x_cy_c} = (h^2b/72)(2a - b)$ for the triangle. (*Hint:* Use the results of Problems 8.8 and 8.64 and the parallel-axis theorem.)

8.66. Find I_{xx}, I_{yy}, and I_{xy} of the extruded section. Disregard all rounded edges.

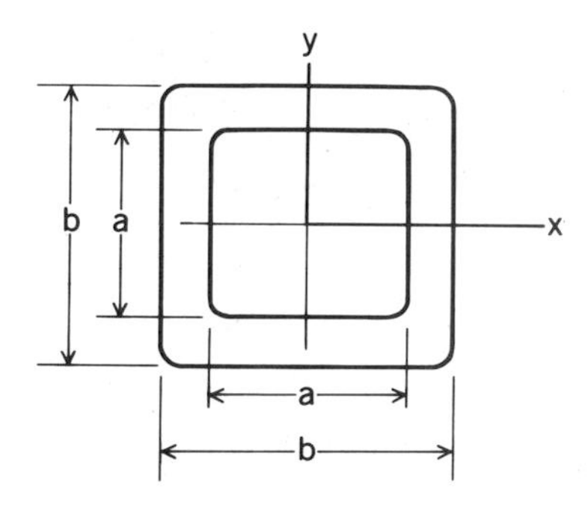

Figure P.8.66

8.67. Find the second moment of area of the rectangle (with a hole) about the base of the rectangle. Also, determine the product of area about the base and left side.

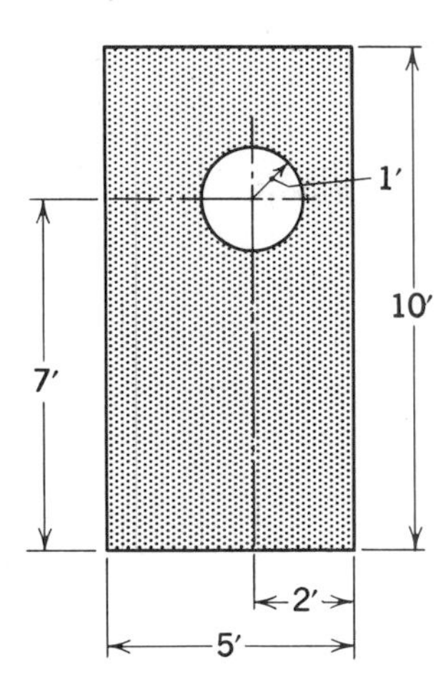

Figure P.8.67

8.68. Find I_{xx}, I_{yy}, $I_{x_cx_c}$, and $I_{y_cy_c}$ for the structural "hat" section. Disregard all rounded edges.

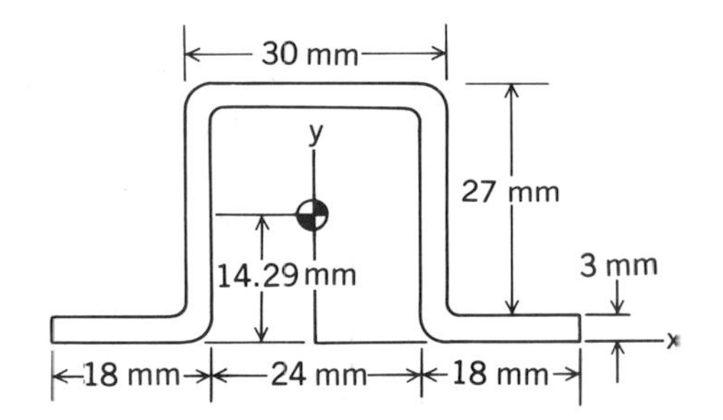

Figure P.8.68

8.69. Find I_{xx}, I_{yy}, and I_{xy} of the hexagon.

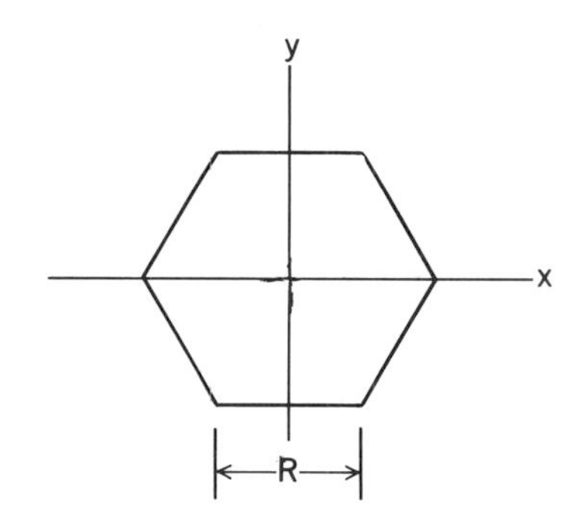

Figure P.8.69

8.70. A beam cross section is made up of an I-shaped section with an additional thick plate welded on. Find the second moments of area for the centroidal axes x_cy_c of the beam cross section. What is $I_{x_cy_c}$? Give the results in millimeters.

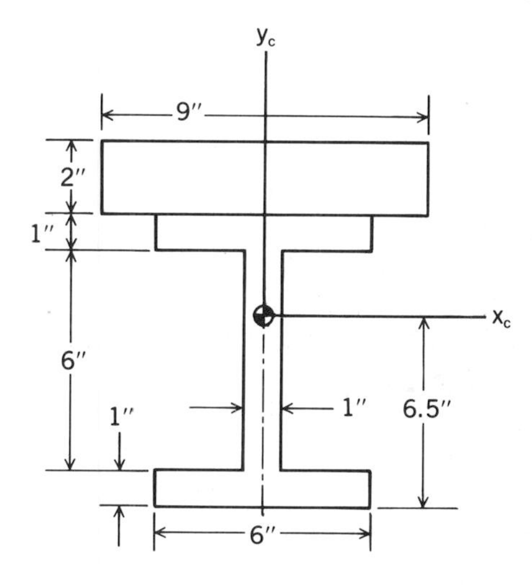

Figure P.8.70

8.71. Find the second moments of the area shown about centroidal axes parallel to the x and y axes. That is, find $I_{x_cx_c}$ and $I_{y_cy_c}$. Give the results in millimeters.

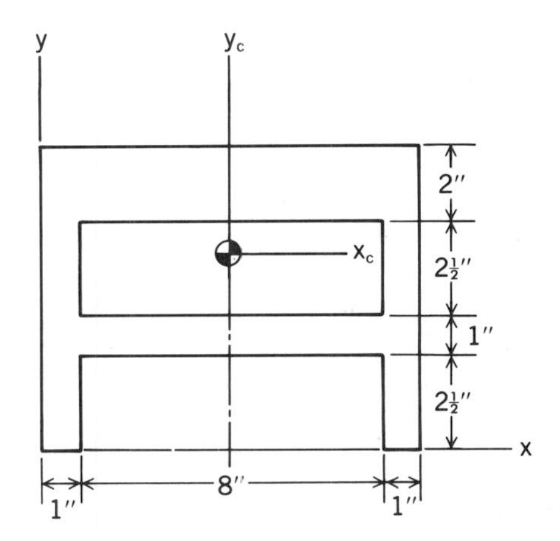

Figure P.8.71

8.72. Find I_{xx}, I_{yy}, I_{xy}, $I_{x_cx_c}$, $I_{y_cy_c}$, and $I_{x_cy_c}$ of the unequal-leg rolled channel section. Disregard all rounded edges.

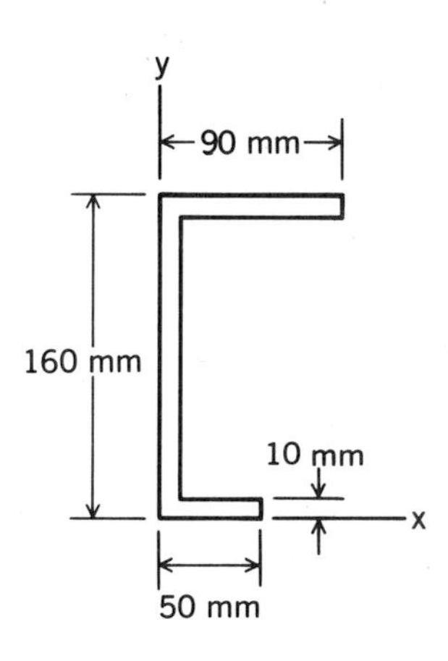

Figure P.8.72

8.8 Relation Between Second Moments and Products of Area

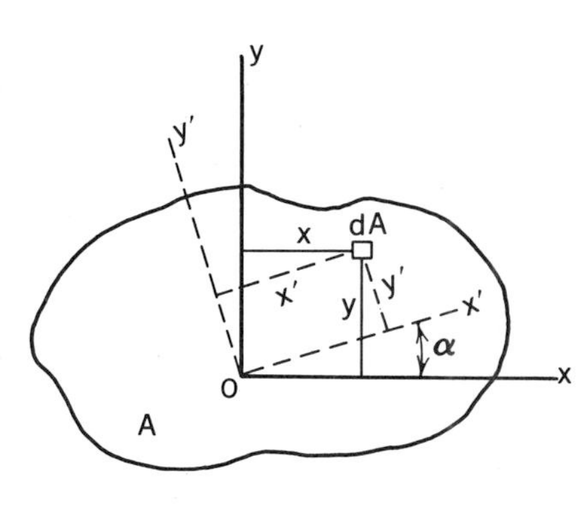

Figure 8.32. Rotation of axes.

We shall now show that we can ascertain second moments and product of area relative to a rotated reference $x'y'$ if we know these quantities for reference xy that has the *same origin*. Such a reference $x'y'$ rotated an angle α from xy is shown in Fig. 8.32. We shall assume that the second moments and product of area for the unprimed reference are known.

Before proceeding, we must know the relation between the coordinates of the area elements dA for the two references. From Fig. 8.32, it is clear that

$$x' = x \cos \alpha + y \sin \alpha \tag{8.23a}$$

$$y' = -x \sin \alpha + y \cos \alpha \tag{8.23b}$$

With relation 8.23b, we can express $I_{x'x'}$ in the following manner:

$$I_{x'x'} = \int_A (y')^2 \, dA = \int_A (-x \sin \alpha + y \cos \alpha)^2 \, dA \tag{8.24}$$

Carrying out the square, we have

$$I_{x'x'} = \sin^2 \alpha \int_A x^2 \, dA - 2 \sin \alpha \cos \alpha \int_A xy \, dA + \cos^2 \alpha \int_A y^2 \, dA$$

Therefore,

$$I_{x'x'} = I_{yy} \sin^2 \alpha + I_{xx} \cos^2 \alpha - 2I_{xy} \sin \alpha \cos \alpha \tag{8.25}$$

A more common form of the desired relation can be formed by using the following trigonometric identities:

$$\cos^2 \alpha = \tfrac{1}{2}(1 + \cos 2\alpha) \tag{a}$$

$$\sin^2 \alpha = \tfrac{1}{2}(1 - \cos 2\alpha) \tag{b}$$

$$2 \sin \alpha \cos \alpha = \sin 2\alpha \tag{c}$$

We then have

$$\boxed{I_{x'x'} = \frac{I_{xx} + I_{yy}}{2} + \frac{I_{xx} - I_{yy}}{2} \cos 2\alpha - I_{xy} \sin 2\alpha} \tag{8.26}$$

To determine $I_{y'y'}$, we need only replace the α in the preceding result by $(\alpha + \pi/2)$. Thus,

$$I_{y'y'} = \frac{I_{xx} + I_{yy}}{2} + \frac{I_{xx} - I_{yy}}{2} \cos (2\alpha + \pi) - I_{xy} \sin (2\alpha + \pi)$$

Note that $\cos (2\alpha + \pi) = -\cos 2\alpha$ and $\sin (2\alpha + \pi) = -\sin 2\alpha$. Hence, the equation above becomes

$$\boxed{I_{y'y'} = \frac{I_{xx} + I_{yy}}{2} - \frac{I_{xx} - I_{yy}}{2} \cos 2\alpha + I_{xy} \sin 2\alpha} \tag{8.27}$$

Next, the product of area $I_{x'y'}$ can be computed in a similar manner:

$$I_{x'y'} = \int_A x'y'\, dA = \int_A (x \cos \alpha + y \sin \alpha)(-x \sin \alpha + y \cos \alpha)\, dA$$

This becomes

$$I_{x'y'} = \sin \alpha \cos \alpha\, (I_{xx} - I_{yy}) + (\cos^2 \alpha - \sin^2 \alpha) I_{xy}$$

Utilizing the previously defined trigonometric identities, we get

$$\boxed{I_{x'y'} = \frac{I_{xx} - I_{yy}}{2} \sin 2\alpha + I_{xy} \cos 2\alpha} \tag{8.28}$$

Thus, we see that, if we know the quantities I_{xx}, I_{yy}, and I_{xy} for some reference xy at point O, the second moments and products of area for *every* set of axes at point O can be computed. And if, in addition, we employ the transfer theorems, we can compute second moments and products of area for *any* reference in the plane of the area.

8.9 Polar Moment of Area

In the previous section, we saw that the second moments and product of area for an orthogonal reference determined all such quantities for *any* orthogonal reference

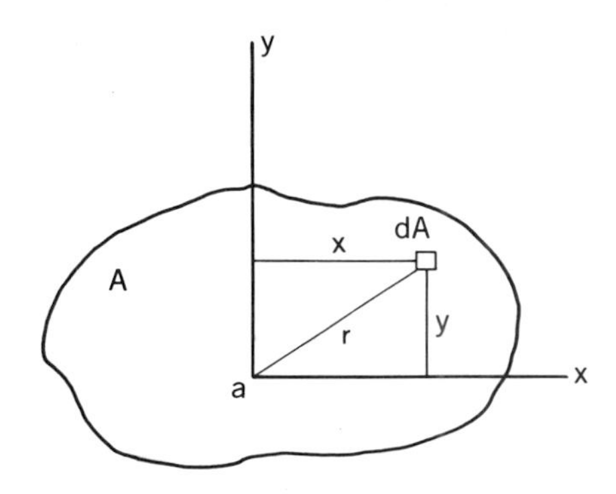

Figure 8.33. $J = I_{xx} + I_{yy}$.

having the same origin. We shall now show that the sum of the pairs of second moments of area is a constant for all such references at a point. Thus, in Fig. 8.33 we have a reference xy associated with point a. Summing I_{xx} and I_{yy}, we have

$$I_{xx} + I_{yy} = \int_A y^2\, dA + \int_A x^2\, dA$$

$$= \int_A (x^2 + y^2)\, dA = \int_A r^2\, dA$$

Since r^2 is independent of the inclination of the coordinate system, the sum $I_{xx} + I_{yy}$ is independent of the inclination of the reference. Therefore, the sum of second moments of area about orthogonal axes is a function only of the position of the origin a for the axes. This sum is termed the *polar moment of area, J.*[6] We can then consider J to be a scalar field. Mathematically, this statement is expressed as

$$J = J(x', y') \tag{8.29}$$

where x' and y' are the coordinates of the origin a in the plane as measured from some convenient reference $x'y'$.

That $(I_{xx} + I_{yy})$ does not change on rotation of axes can also be deduced from the transformation equations 8.26 and 8.27. This group of terms is accordingly termed an *invariant*. We can similarly show that $(I_{xx}I_{yy} - I_{xy}^2)$ is also invariant under a rotation of axes.

8.10 Principal Axes

Still other conclusions may be drawn about second moments and products of area associated with a point in an area. In Fig. 8.34 is an area with a reference xy having its origin at point a. We shall assume that I_{xx}, I_{yy}, and I_{xy} are known for this reference, and shall ask at what angle α we shall find an axis having the *maximum* second moment of area. Since the sum of the second moments of area is constant for any reference with origin at a, the *minimum* second moment of area must then correspond to an axis at *right angles* to the axis having the maximum second moment. Since second moments of area have been expressed in Eqs. 8.26 and 8.27 as functions of the variable α at a point, these extremes may readily be determined by setting the partial derivative of $I_{x'x'}$ with respect to α equal to zero. Thus,

$$\frac{\partial I_{x'x'}}{\partial \alpha} = (I_{xx} - I_{yy})(-\sin 2\alpha) - 2I_{xy}\cos 2\alpha = 0$$

[6]Quite often I_p is used for the polar moment of area.

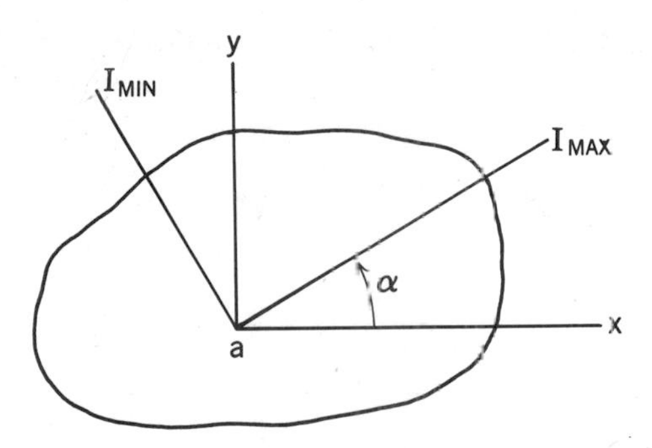

Figure 8.34. Principal axes.

If we denote the value of α that satisfies the equation above as $\tilde{\alpha}$, we have

$$(I_{yy} - I_{xx}) \sin 2\tilde{\alpha} - 2I_{xy} \cos 2\tilde{\alpha} = 0$$

Hence,

$$\boxed{\tan 2\tilde{\alpha} = \frac{2I_{xy}}{I_{yy} - I_{xx}}} \tag{8.30}$$

This formulation gives us the angle $\tilde{\alpha}$, which corresponds to an extreme value of $I_{x'x'}$ (i.e., to a maximum or minimum value). Actually, there are two possible values of $2\tilde{\alpha}$ which are π radians apart that will satisfy the equation above. Thus,

$$2\tilde{\alpha} = \beta \qquad \text{where } \beta = \tan^{-1} \frac{2I_{xy}}{I_{yy} - I_{xx}}$$

or

$$2\tilde{\alpha} = \beta + \pi$$

This means that we have two values of $\tilde{\alpha}$, given as

$$\tilde{\alpha}_1 = \frac{\beta}{2}, \qquad \tilde{\alpha}_2 = \frac{\beta}{2} + \frac{\pi}{2}$$

Thus, there are two axes orthogonal to each other having extreme values for the second moment of area at a. On one of these axes is the maximum second moment of area and, as pointed out earlier, the minimum second moment of area must appear on the other axis. These axes are called the *principal axes*.

Let us now substitute the angle $\tilde{\alpha}$ into Eq. 8.28 for $I_{x'y'}$:

$$I_{x'y'} = \frac{I_{xx} - I_{yy}}{2} \sin\left(\tan^{-1} \frac{2I_{xy}}{I_{yy} - I_{xx}}\right) + I_{xy} \cos\left(\tan^{-1} \frac{2I_{xy}}{I_{yy} - I_{xx}}\right)$$

This becomes

$$I_{x'y'} = -(I_{yy} - I_{xx}) \frac{I_{xy}}{[(I_{yy} - I_{xx})^2 + 4I_{xy}^2]^{1/2}} + I_{xy} \frac{I_{yy} - I_{xx}}{[(I_{yy} - I_{xx})^2 + 4I_{xy}^2]^{1/2}}$$

Hence,

$$I_{x'y'} = 0 \tag{8.31}$$

Thus, we see that the *product of area corresponding to the principal axes is zero.* If we set $I_{x'y'}$ equal to zero in Eq. 8.28, you can demonstrate the converse of the preceding statement by solving for α and comparing the result with Eq. 8.30. That is, if the product of area is zero for a set of axes at a point, these axes must be principal axes. Consequently, if one axis of a set of axes is symmetrical for the area, the axes are principal axes.

The concept of principal axes will appear again in the following chapter in connection with the inertia tensor. Thus, the concept is not an isolated occurrence but is characteristic of a whole family of quantities. We shall, then, have further occasion to examine some of the topics introduced in this chapter from a more general viewpoint.

EXAMPLE 8.10

Find the principal second moments of area at the centroid of the Z section of Example 8.9.

We have from this example the following results that will be of use to us:

$$I_{x_cx_c} = 113.2 \text{ in.}^4$$
$$I_{y_cy_c} = 32.67 \text{ in.}^4$$
$$I_{x_cy_c} = -42.0 \text{ in.}^4$$

Hence, we have

$$\tan 2\tilde{\alpha} = \frac{2I_{x_cy_c}}{I_{y_cy_c} - I_{x_cx_c}} = \frac{(2)(-42.0)}{32.67 - 113.2} = 1.043$$
$$2\tilde{\alpha} = 46.21°; \quad 226.2°$$

For $2\tilde{\alpha} = 46.21°$:

$$I_1 = \frac{113.2 + 32.67}{2} + \frac{113.2 - 32.67}{2} \cos(46.21°) - (-42) \sin 46.21°$$
$$= 72.9 + 27.9 + 30.3 = 131.1 \text{ in.}^4$$

For $2\tilde{\alpha} = 226.2°$:

$$I_2 = 72.9 - 27.9 - 30.3 = 14.75 \text{ in.}^4$$

As a check on our work, we note that the sum of the second moments of area are invariant at a point for a rotation of axes. This means that

$$I_{x_cx_c} + I_{y_cy_c} = I_1 + I_2$$
$$113.2 + 32.7 = 131.1 + 14.75$$

Therefore,

$$145.9 = 145.9$$

We thus have a check on our work.

Before closing, we wish to point out that there is a graphical construction called *Mohr's circle* relating second moments and products of area for all possible axes at a point. However, in this text we shall use the analytical relations thus far presented rather than Mohr's circle. You will see Mohr circle construction in your strength of materials course where its use in conjunction with plane stress and plane strain is very helpful.[7]

Problems

8.73. It is known that area A is 10 ft² and has the following moments and products of area for the centroidal axes shown:
$I_{xx} = 40 \text{ ft}^4$, $I_{yy} = 20 \text{ ft}^4$, $I_{xy} = -4 \text{ ft}^4$
Find the moments and products of area for the $x'y'$ reference at point ⓐ.

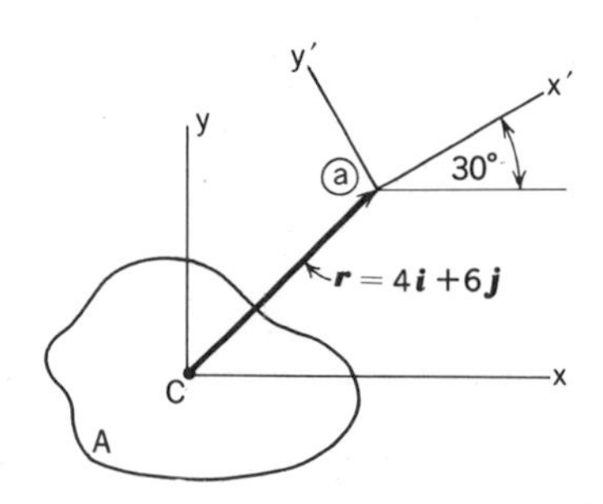

Figure P.8.73

8.74. The cross section of a beam is shown. Compute $I_{x'x'}$, $I_{y'y'}$, and $I_{x'y'}$ in the simplest way without using formulas for second moments and products of area for a triangle.

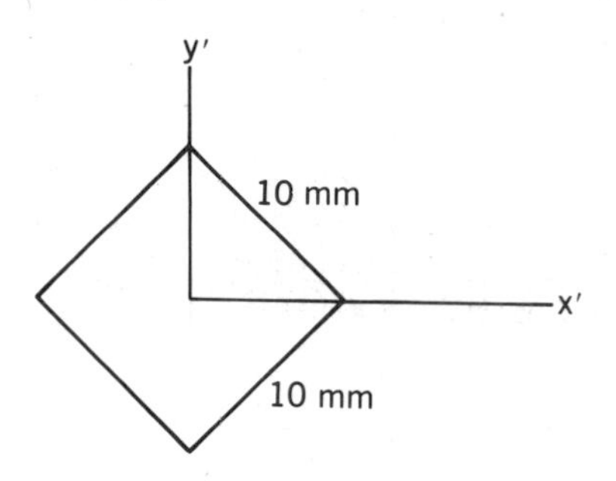

Figure P.8.74

8.75. Find $I_{x'x'}$, $I_{y'y'}$, and $I_{x'y'}$ for the cross section of the beam shown. The origin of $x'y'$ is at the centroid of the cross section.

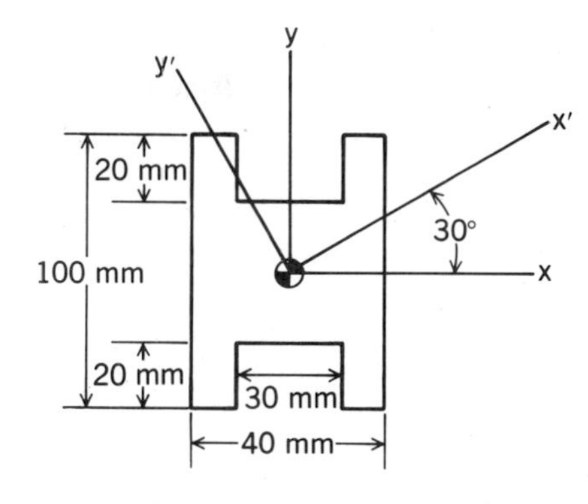

Figure P.8.75

8.76. Find I_{xx}, I_{yy}, and I_{xy} for the rectangle. Also, compute the polar moment of area at points ⓐ and ⓑ.

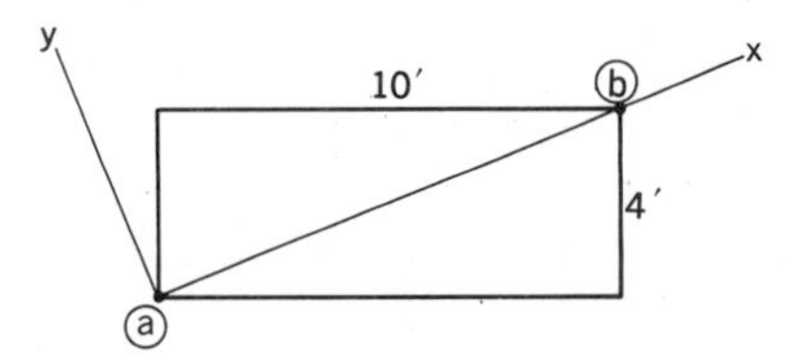

Figure P.8.76

8.77. Express the polar moment of area of the square as a function of x, y, the coordinates of points about which the polar moment is taken.

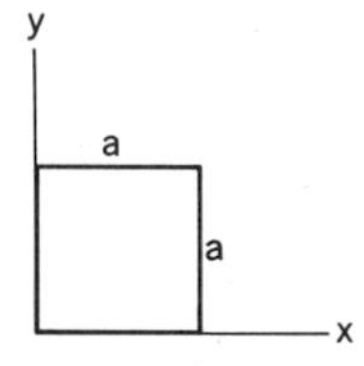

Figure P.8.77

[7] See I. H. Shames, *Introduction to Solid Mechanics*, Prentice-Hall, Inc., Englewood Cliffs, N.J., 1975.

8.78. Use the calculus to show that the polar moment of area of a circular area of radius r is $\pi r^4/2$ at the center.

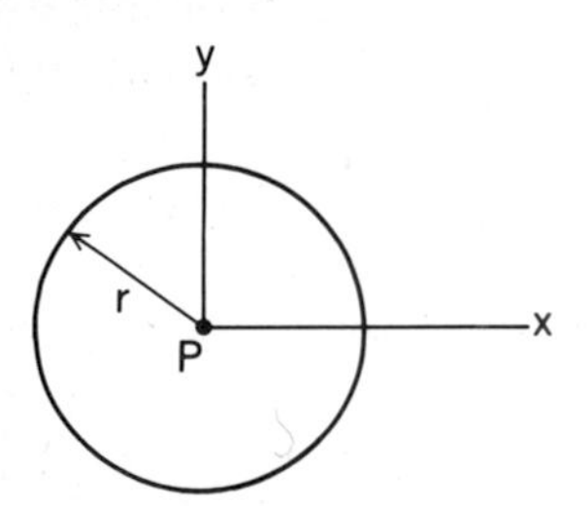

Figure P.8.78

8.79. Find the direction of the principal axes for the angle section about point A.

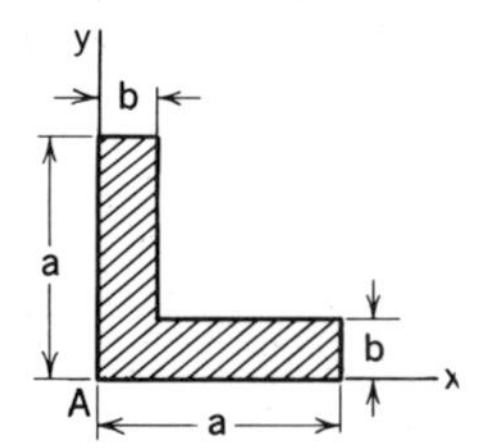

Figure P.8.79

8.80. What are the principal second moments of area for the area of Example 8.7?

8.81. Find the principal second moments of area at the centroid for the area shown.

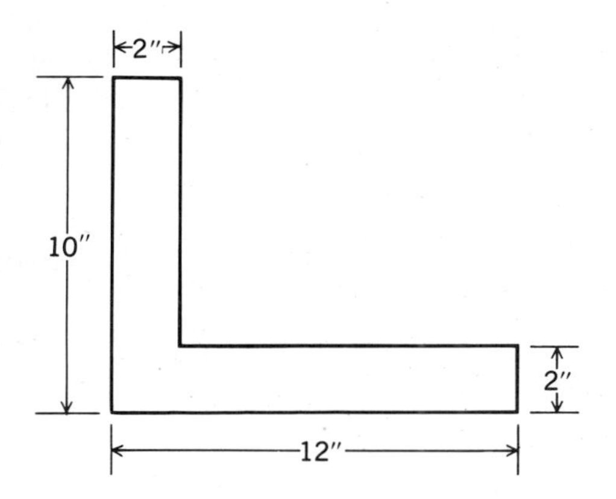

Figure P.8.81

8.82. Determine the principal second moments of area at point A.

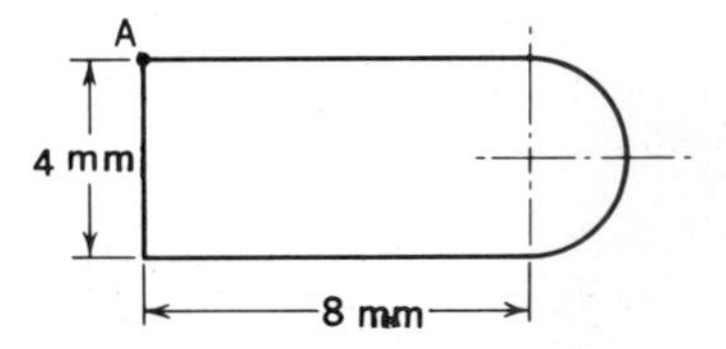

Figure P.8.82

8.83. A rectangular area has two holes cut out. What is the maximum second moment of area at A? What is it at B at the center of the area?

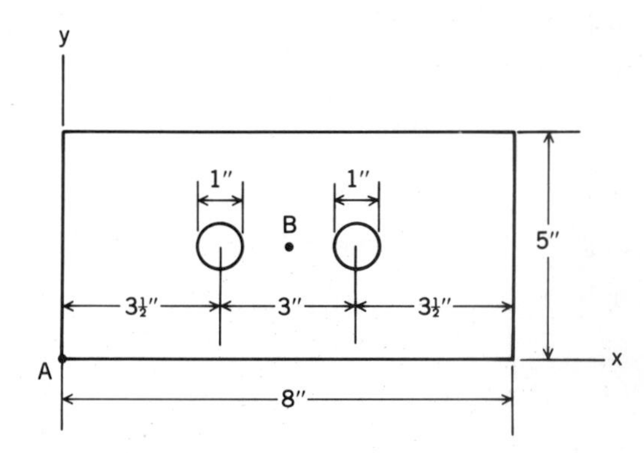

Figure P.8.83

8.84. Show that the axes for which the product of area is a maximum are rotated from xy by an angle α so that

$$\tan 2\alpha = \frac{I_{xx} - I_{yy}}{2I_{xy}}$$

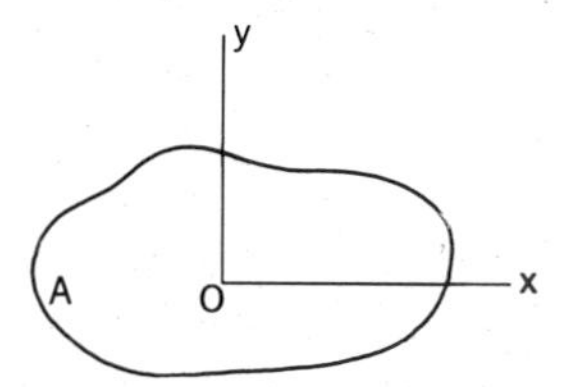

Figure P.8.84

8.11 Closure

In this chapter, we discussed primarily the first and second moments of plane areas as well as the product of plane areas. These formulations give certain kinds of evaluations of the distribution of area relative to a plane reference xy. You will most certainly make much use of these quantities in your later courses in strength of materials.

We also briefly discussed in this chapter the first moment of mass. Another set of quantities that will prove to be indispensible in rigid-body dynamics are the second moments of *mass* (or inertia) as well as products of *mass* (or inertia). They, like the first moment of mass, represent certain measures of mass distribution relative to a reference xyz. We shall consider such quantities in Chapter 9 and we shall see that the second moments and products of area represent a special case of second moments and products of mass (or inertia).

Review Problems

8.85. Find the position of the centroid of the shaded area under the curve $y = \sin^2 x$ m. Find $M_{x'}$ and $M_{y'}$ of this area.

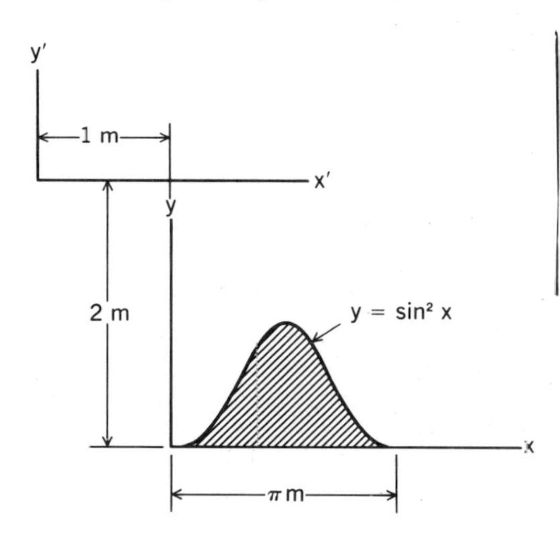

Figure P.8.85

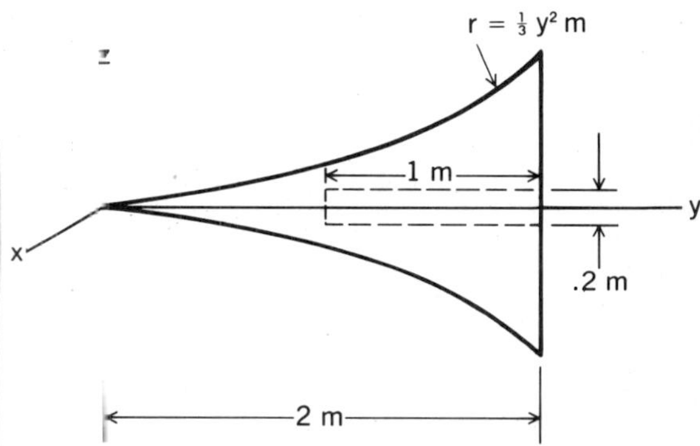

Figure P.8.86

8.86. Find the center of volume of the body of revolution with a cylindrical cavity.

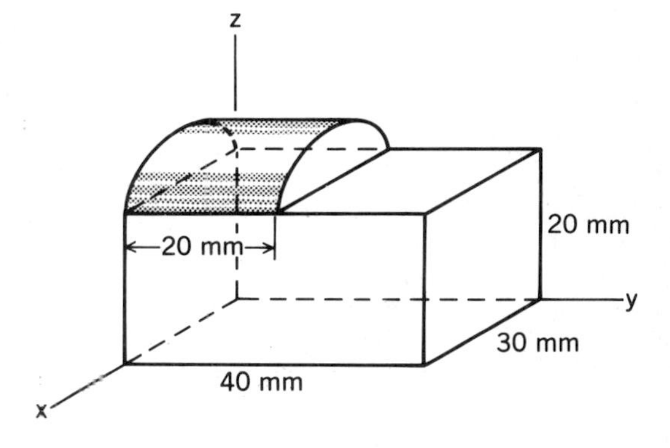

Figure P.8.87

8.87. Locate the center of volume, center of mass, and center of gravity of the wooden rectangular block and the plastic semicylinder. The wood weighs .0003 N/mm³ and the plastic weighs .0005 N/mm³.

8.88. Using the theorems of Pappus and Guldinus, find the centroid of the area of a quarter-circle.

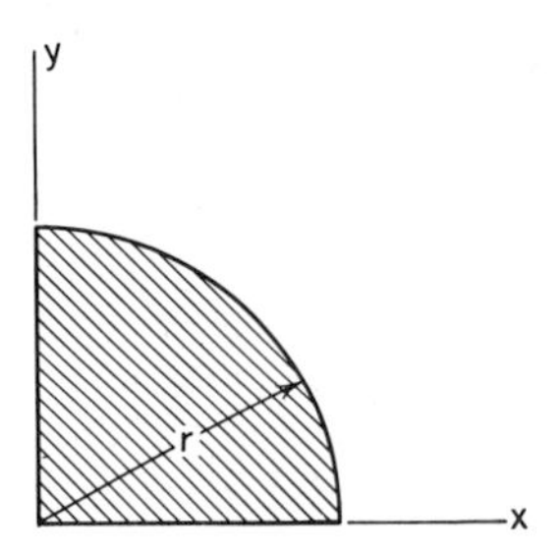

Figure P.8.88

8.89. A tank has a semispherical dome at the left end. Using the theorems of Pappus and Guldinus, compute the surface and volume of the tank. Give the results in meters.

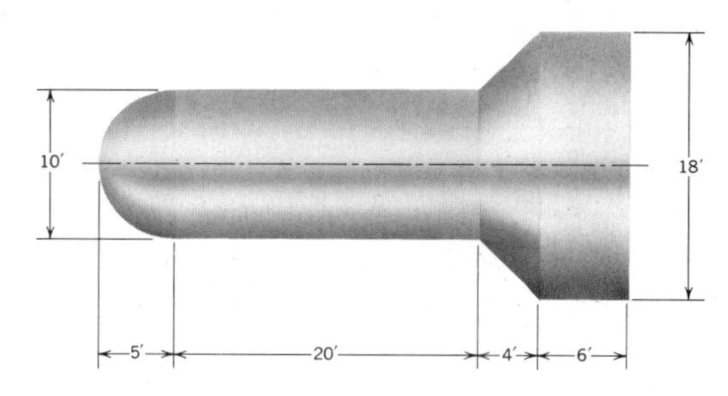

Figure P.8.89

8.90. Find $I_{x'x'}$, $I_{y'y'}$, and $I_{x'y'}$ at point A for the rectangular area.

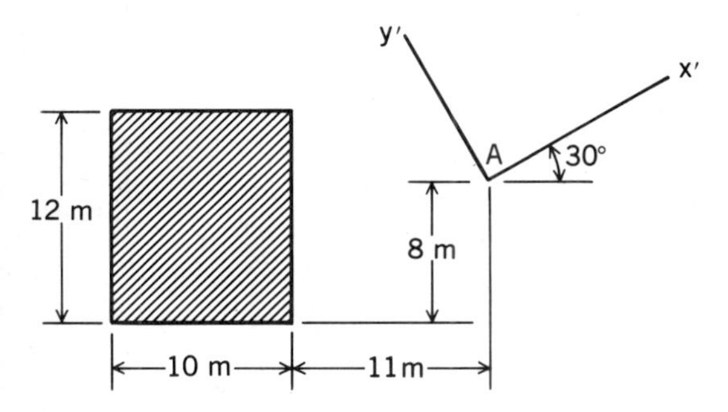

Figure P.8.90

8.91. Find the centroid of the area, and then find the second moments of the area about centroidal axes parallel to the sides of the area.

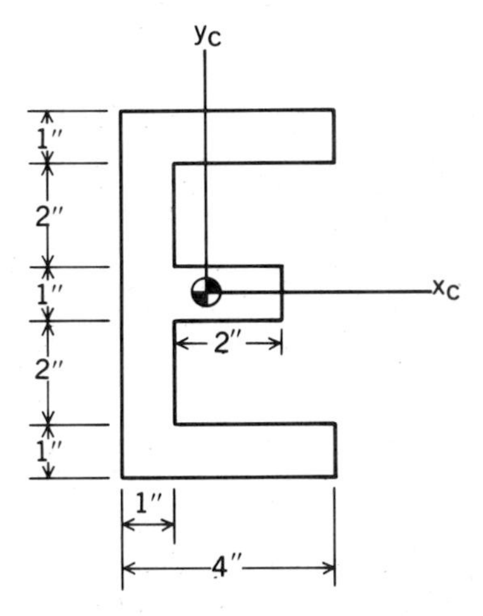

Figure P.8.91

8.92. Find the principal second moments of area at a point where $I_{xy} = 321$ in.^{4}, $I_{xx} = 118.4$ in.^{4} and $I_{yy} = 1028$ in.^{4}.

8.93. Find the polar moment of area at O for the shaded area.

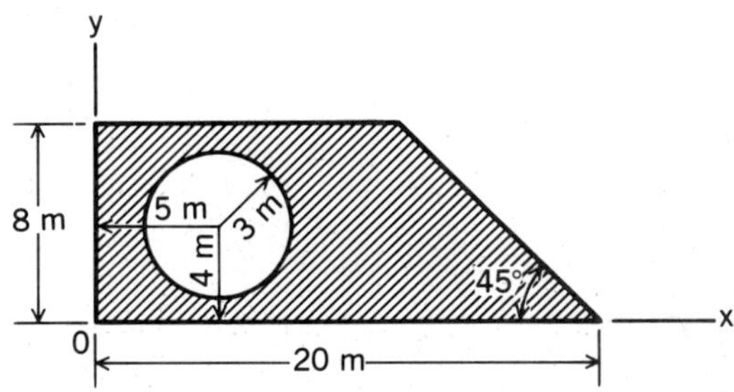

Figure P.8.93

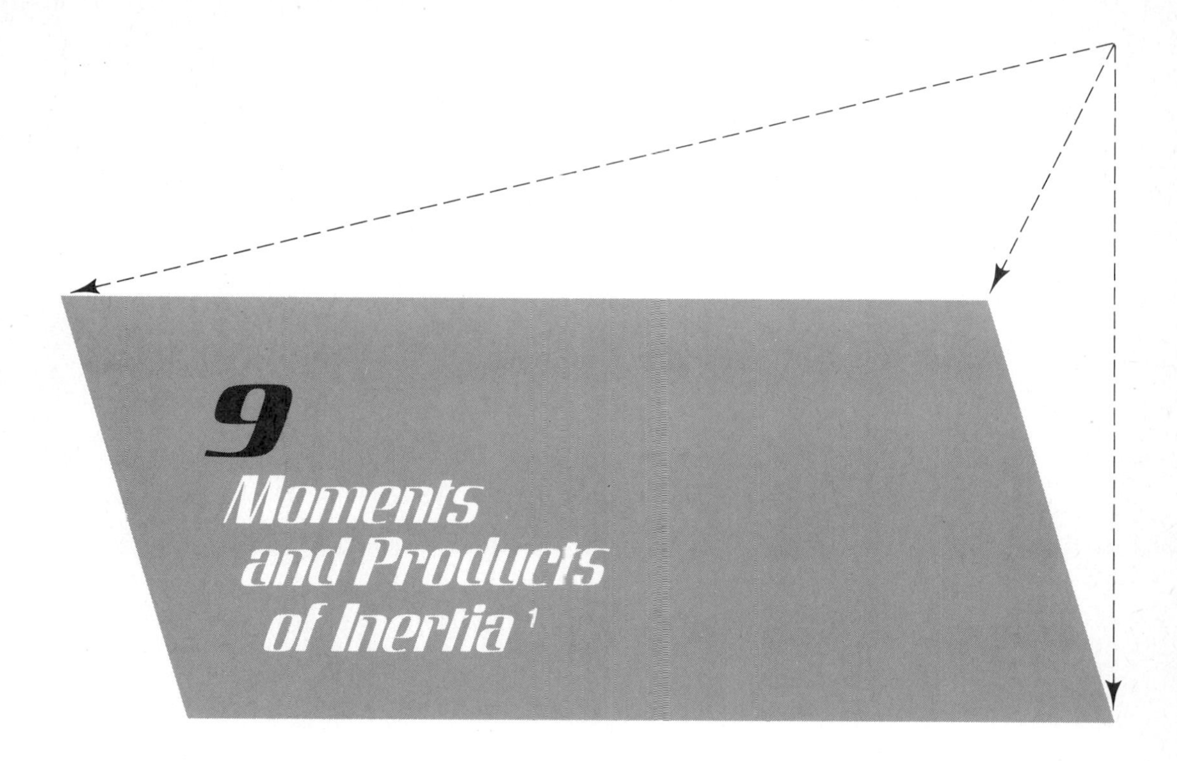

9.1 Introduction

In this chapter, we shall consider certain measures of mass distribution relative to a reference. These quantities are vital for the study of the dynamics of rigid bodies. Because these quantities are so closely related to second moments and products of area, we shall consider them at this early stage rather than wait for dynamics. We shall also discuss the fact that these measures of mass distribution—the second moments of inertia of mass and the products of inertia of mass—are components of what we call a second-order tensor. Recognizing this fact early will make more simple and understandable your future studies of stress and strain, since these quantities also happen to be second-order tensors.

9.2 Formal Definition of Inertia Quantities

We shall now formally define a set of quantities that give information about the distribution of mass of a body relative to a Cartesian reference. For this purpose, a body of mass M and a reference xyz are presented in Fig. 9.1. This reference and the body may have any motion whatever relative to each other. The ensuing discussion then holds for the instantaneous orientation shown at time t. We shall consider that the body is composed of a continuum of particles each of which has a mass given by $\rho\, dv$, where

[1] This chapter may be covered at a later stage when studying dynamics. In that case, it should be covered directly after Chapter 15.

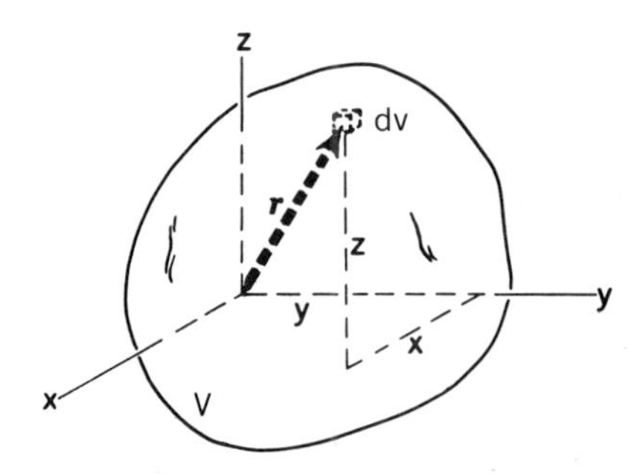

Figure 9.1. Body and reference at time t.

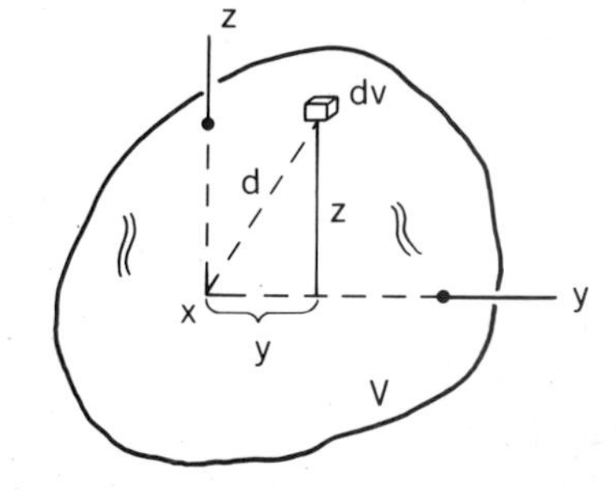

Figure 9.2. View of body along x axis.

ρ is the mass density and dv is a volume element. We define the second moments and products of inertia of the body M for the reference xyz at time t in the following manner[2]:

$$I_{xx} = \iiint_V (y^2 + z^2)\rho \, dv \tag{9.1a}$$

$$I_{yy} = \iiint_V (x^2 + z^2)\rho \, dv \tag{9.1b}$$

$$I_{zz} = \iiint_V (x^2 + y^2)\rho \, dv \tag{9.1c}$$

$$I_{xy} = \iiint_V xy\rho \, dv \tag{9.1d}$$

$$I_{xz} = \iiint_V xz\rho \, dv \tag{9.1e}$$

$$I_{yz} = \iiint_V yz\rho \, dv \tag{9.1f}$$

The terms I_{xx}, I_{yy}, and I_{zz} in the set above are called the *mass moments of inertia* of the body about the x, y, and z axes, respectively. Note that in each such case we are integrating the mass elements, $\rho \, dv$, times the *perpendicular distance squared* from the mass elements to the coordinate axis about which we are computing the moment of inertia. Thus, if we look along the x axis toward the origin in Fig. 9.1, we would have the view shown in Fig. 9.2. The quantity $y^2 + z^2$ used in Eq. 9.1a for I_{xx} is clearly d^2, the perpendicular distance squared from dv to the x axis (now seen as a dot). Each of the terms with mixed indices is called the *mass product of inertia* about the pair of axes given by the indices. Clearly, from the definition of the product of inertia, we

[2]We use the same notation as was used for second moments and products of area, which are also sometimes called moments and products of inertia. This is standard practice in mechanics. There need be no confusion in using these quantities if we keep the context of discussions clearly in mind.

could reverse indices and thereby form three additional products of inertia for a reference. The additional three quantities formed in this way, however, are equal to the corresponding quantities of the original set. That is,

$$I_{xy} = I_{yx}, \qquad I_{xz} = I_{zx}, \qquad I_{yz} = I_{zy}$$

We now have nine inertia terms at a point for a given reference at this point. The values of the set of six independent quantities will, for a given body, depend on the *position* and *inclination* of the reference relative to the body. You should also understand that the reference may be established anywhere in space and *need not* be situated in the rigid body of interest. Thus there will be nine inertia terms for reference xyz at point O outside the body (Fig. 9.3) computed using Eqs. 9.1, where the domain of integration is the volume V of the body. As will be explained later, the nine moments and products of inertia are components of the inertia tensor.

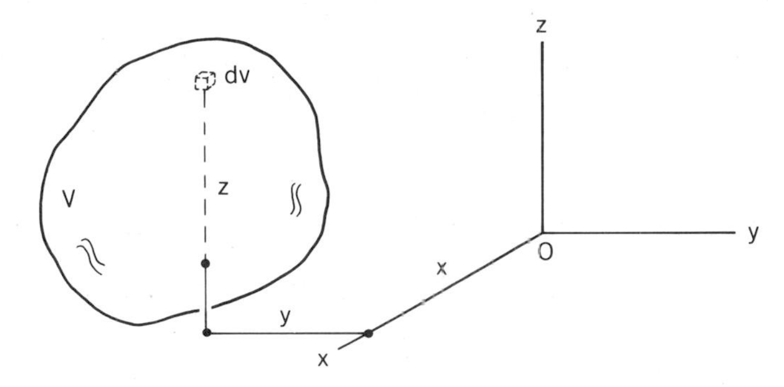

Figure 9.3. Origin of *xyz* outside body.

It will be convenient, when referring to the nine moments and products of inertia for reference xyz at a point, to list them in a matrix array, as follows:

$$I_{ij} = \begin{pmatrix} I_{xx} & I_{xy} & I_{xz} \\ I_{yx} & I_{yy} & I_{yz} \\ I_{zx} & I_{zy} & I_{zz} \end{pmatrix}$$

Notice that the first subscript gives the row and the second subscript gives the column in the array. Furthermore, the left-to-right downward diagonal in the array is composed of mass moment of inertia terms while the products of inertia, oriented at mirror-image positions about this diagonal, are equal. For this reason we say that the array is *symmetric*.

We shall now show that the sum of the mass moments of inertia for a set of orthogonal axes is independent of the orientation of the axes and depends only on the position of the origin. Examine the sum of such a set of terms:

$$I_{xx} + I_{yy} + I_{zz} = \iiint_V (y^2 + z^2)\rho\, dv + \iiint_V (x^2 + z^2)\rho\, dv + \iiint_V (x^2 + y^2)\rho\, dv$$

Combining the integrals and rearranging, we get

$$I_{xx} + I_{yy} + I_{zz} = \iiint_V 2(x^2 + y^2 + z^2)\rho\, dv = \iiint_V 2|\mathbf{r}|^2 \rho\, dv \tag{9.2}$$

But the magnitude of the position vector from the origin to a particle is *independent* of the inclination of the reference at the origin. Thus, *the sum of the moments of inertia at a point in space for a given body clearly is an invariant with respect to rotation of axes.*

Clearly, on inspection of the definitions 9.1, the moments of inertia must always exceed zero, while the products of inertia may have any value. Of interest is the case where one of the coordinate planes is a *plane of symmetry* for the mass distribution of the body. Such a plane is the zy plane shown in Fig. 9.4 cutting a body into two parts, which, by definition of symmetry, are mirror images of each other. For the computation of I_{xz}, each half will give a contribution of the same magnitude but of opposite sign. We can most readily see that this is so by looking along the y axis toward the origin. The plane of symmetry then appears as a line coinciding with the z axis (see Fig. 9.5). We can consider the body to be composed of pairs of mass elements dm which are mirror images of each other with respect to position and shape about the plane of symmetry. The product of inertia I_{xz} for such a pair is then

$$xz\, dm - xz\, dm = 0$$

Thus, we can conclude that

$$I_{xz} = \underbrace{\int xz\, dm}_{\substack{\text{right}\\ \text{domain}}} - \underbrace{\int xz\, dm}_{\substack{\text{left}\\ \text{domain}}} = 0$$

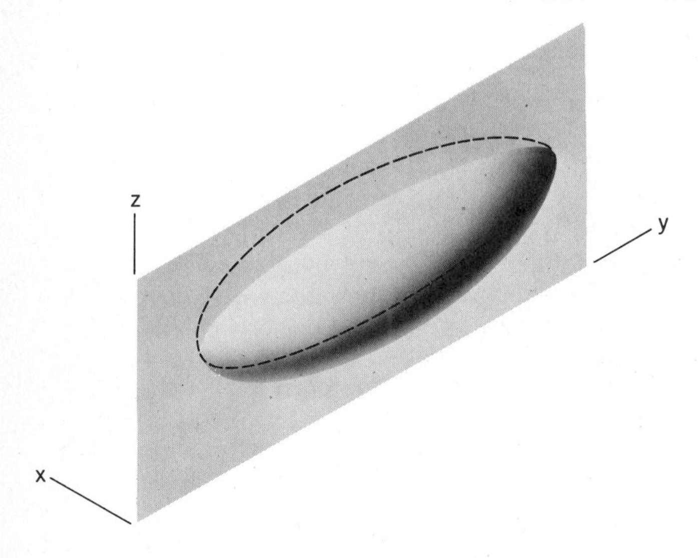

Figure 9.4. zy is plane of symmetry.

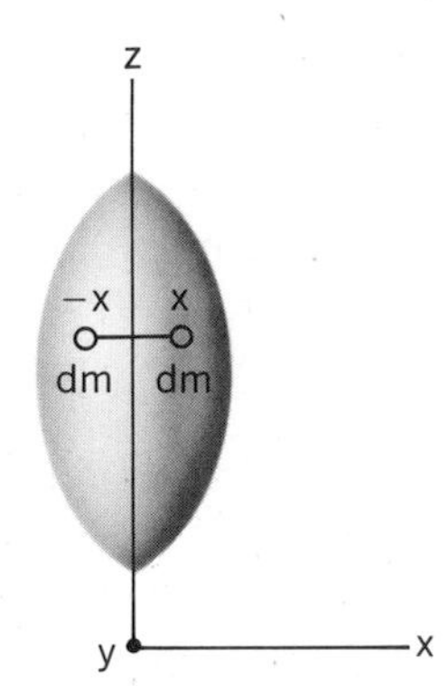

Figure 9.5. View along y axis.

This conclusion is also true for I_{xy}. We can say that $I_{xy} = I_{xz} = 0$. But on consulting Fig. 9.4, you should be able to readily decide that the term I_{zy} will have a positive value. Note that those products of inertia having x as an index are zero and that the x coordinate axis is normal to the plane of symmetry. Thus, we can conclude that *if two axes form a plane of symmetry for the mass distribution of a body, the products of inertia having as an index the coordinate that is normal to the plane of symmetry will be zero.*

Consider next a body of *revolution*. Take the z axis to coincide with the axis of symmetry. It is easy to conclude for the origin O of xyz anywhere along the axis of symmetry that

$$I_{xz} = I_{yz} = 0$$

$$I_{xx} = I_{yy} = \text{constant}$$

for all possible xy axes formed by rotating about the z axis at O. Can you justify these conclusions?

Finally, we define *radii of gyration* in a manner analogous to that used for second moments of area in Chapter 8. Thus:

$$I_{xx} = k_x^2 M$$

$$I_{yy} = k_y^2 M$$

$$I_{zz} = k_z^2 M$$

where k_x, k_y, and k_z are the radii of gyration and M is the total mass.

EXAMPLE 9.1

Find the nine components of the inertia tensor of a rectangular body of uniform density ρ about point O for a reference xyz coincident with the edges of the block as shown in Fig. 9.6.

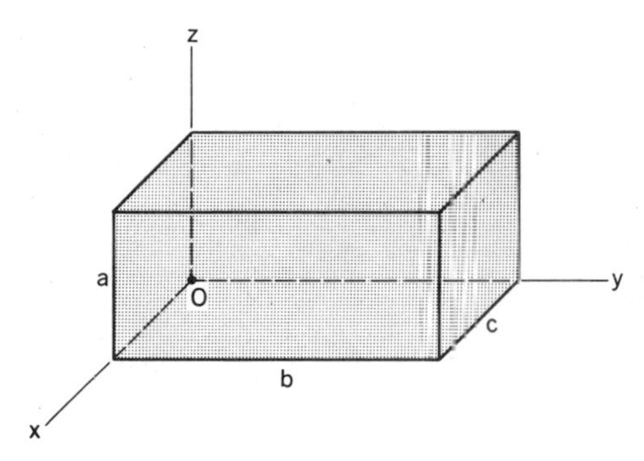

Figure 9.6. Find I_{ij} at O.

We first compute I_{xx}. Using volume elements $dv = dx\,dy\,dz$, we get on using simple multiple integration:

$$I_{xx} = \int_0^a \int_0^b \int_0^c (y^2 + z^2)\rho \, dx \, dy \, dz$$

$$= \int_0^a \int_0^b (y^2 + z^2)c\rho \, dy \, dz = \int_0^a \left(\frac{b^3}{3} + z^2 b\right) c\rho \, dz \tag{a}$$

$$= \left(\frac{ab^3c}{3} + \frac{a^3bc}{3}\right)\rho = \frac{\rho V}{3}(b^2 + a^2)$$

where V is the volume of the body. Permuting the terms, we can get I_{yy} and I_{zz} by inspection as follows:

$$I_{yy} = \frac{\rho V}{3}(c^2 + a^2) \tag{b}$$

$$I_{zz} = \frac{\rho V}{3}(b^2 + c^2) \tag{c}$$

We next compute I_{xy}.

$$I_{xy} = \int_0^a \int_0^b \int_0^c xy\rho \, dx \, dy \, dz = \int_0^a \int_0^b \frac{c^2}{2} y\rho \, dy \, dz$$

$$= \int_0^a \frac{c^2b^2}{4}\rho \, dz = \frac{ac^2b^2}{4}\rho = \frac{\rho V}{4} cb \tag{d}$$

Permuting the terms, we get

$$I_{xz} = \frac{\rho V}{4} ac \tag{e}$$

$$I_{yz} = \frac{\rho V}{4} ab \tag{f}$$

We accordingly have, for the inertia tensor:

$$I_{ij} = \begin{pmatrix} \frac{\rho V}{3}(b^2 + a^2) & \frac{\rho V}{4} cb & \frac{\rho V}{4} ac \\ \frac{\rho V}{4} cb & \frac{\rho V}{3}(c^2 + a^2) & \frac{\rho V}{4} ab \\ \frac{\rho V}{4} ac & \frac{\rho V}{4} ab & \frac{\rho V}{3}(b^2 + c^2) \end{pmatrix} \tag{g}$$

EXAMPLE 9.2

Compute the components of the inertia tensor at the center of a solid sphere of uniform density ρ as shown in Fig. 9.7.

We shall first compute I_{yy}. Using spherical coordinates, we have[3]

$$I_{yy} = \iiint_V (x^2 + z^2)\rho \, dv$$

[3]For those unfamiliar with spherical coordinates, we have shown in Fig.9.8 a more detailed study of the volume element used. The volume dv is simply the product of the three edges of the element shown in the diagram.

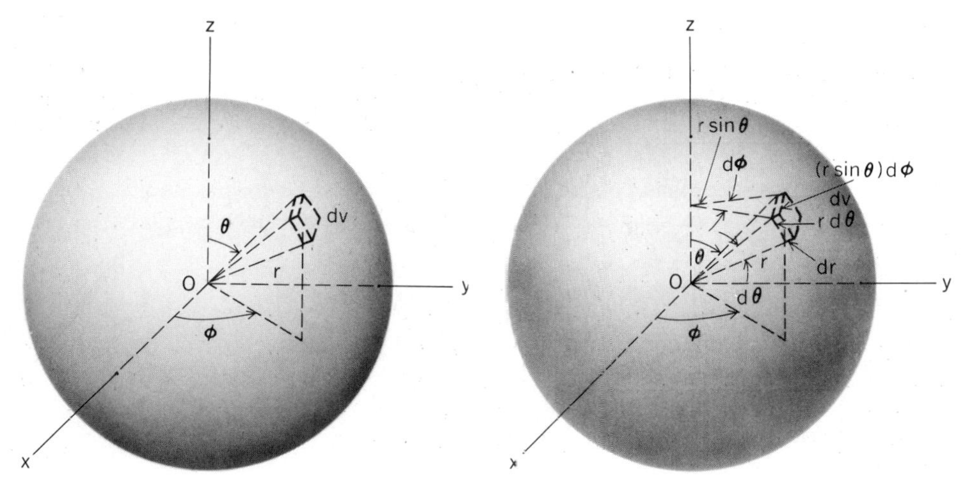

Figure 9.7. Find I_{ij} at O.

Figure 9.8. $dv = (r\sin\theta\, d\phi)(dr)(r\, d\theta) = r^2 \sin\theta\, d\theta\, d\phi\, dr$.

$$= \int_0^R \int_0^{2\pi} \int_0^{\pi} [(r\sin\theta\cos\phi)^2 + (r\cos\theta)^2]\rho(r^2\sin\theta\, d\theta\, d\phi\, dr)$$

$$= \int_0^R \int_0^{2\pi} \int_0^{\pi} (r^4 \sin^3\theta\cos^2\phi)\rho\, d\theta\, d\phi\, dr + \int_0^R \int_0^{2\pi} \int_0^{\pi} (r^4\cos^2\theta\sin\theta)\rho\, d\theta\, d\phi\, dr$$

$$= \rho \int_0^R \int_0^{2\pi} (r^4\cos^2\phi)\left(\int_0^{\pi} \sin^3\theta\, d\theta\right) d\phi\, dr$$

$$+ \rho \int_0^R \int_0^{2\pi} r^4 \left(\int_0^{\pi} \cos^2\theta\sin\theta\, d\theta\right) d\phi\, dr$$

With the aid of integration formulas from Appendix I, we have

$$I_{yy} = \rho \int_0^R \int_0^{2\pi} r^4\cos^2\phi \left[-\tfrac{1}{3}\cos\theta\,(\sin^2\theta + 2)\right]\Big|_0^{\pi} d\phi\, dr$$

$$+ \rho \int_0^R \int_0^{2\pi} r^4 \left(-\frac{\cos^3\theta}{3}\right)\Big|_0^{\pi} d\phi\, dr$$

$$= \rho \int_0^R \int_0^{2\pi} r^4\cos^2\phi\, \tfrac{4}{3}\, d\phi\, dr + \rho \int_0^R \int_0^{2\pi} (r^4)(\tfrac{2}{3})\, d\phi\, dr$$

Integrating next with respect to ϕ, we get

$$I_{yy} = \rho \int_0^R (r^4)(\tfrac{4}{3})(\pi)\, dr + \rho \int_0^R r^4(\tfrac{2}{3})(2\pi)\, dr$$

Finally, we get

$$I_{yy} = \rho\frac{R^5}{5}\frac{4}{3}\pi + \rho\frac{R^5}{5}\frac{4}{3}\pi$$

$$= \frac{8}{15}\rho\pi R^5$$

But

$$M = \rho\,\tfrac{4}{3}\pi R^3$$

Hence,

$$I_{yy} = \tfrac{2}{5} MR^2$$

Because of the point symmetry about point O, we can also say that

$$I_{xx} = I_{zz} = \tfrac{2}{5} MR^2$$

Because the coordinate planes are all planes of symmetry for the mass distribution, the products of inertia are zero. Thus, the inertia tensor can be given as

$$I_{ij} = \begin{pmatrix} \tfrac{2}{5}MR^2 & 0 & 0 \\ 0 & \tfrac{2}{5}MR^2 & 0 \\ 0 & 0 & \tfrac{2}{5}MR^2 \end{pmatrix}$$

9.3 Relation Between Mass-Inertia Terms and Area-Inertia Terms

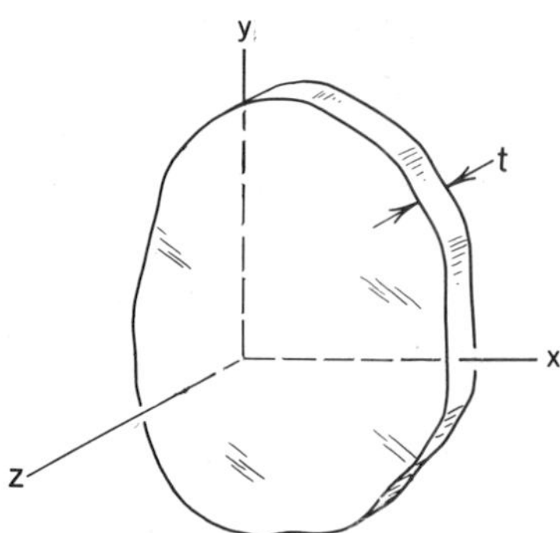

Figure 9.9. Plate of thickness t.

We now relate the second moment and product of area studied in Chapter 8 with the inertia tensor. To do this, consider a plate of constant thickness t and uniform density ρ (Fig. 9.9). A reference is picked so that the xy plane is in the midplane of this plate. The components of the inertia tensor are rewritten for convenience as

$$\begin{aligned} I_{xx} &= \rho \iiint_V (y^2 + z^2)\,dv, & I_{xy} &= \rho \iiint_V xy\,dv \\ I_{yy} &= \rho \iiint_V (x^2 + z^2)\,dv, & I_{xz} &= \rho \iiint_V xz\,dv \\ I_{zz} &= \rho \iiint_V (x^2 + y^2)\,dv, & I_{yz} &= \rho \iiint_V yz\,dv \end{aligned} \tag{9.3}$$

Now consider that the thickness t is *small* compared to the lateral dimensions of the plate. This means that z is restricted to a range of values having a small magnitude. As a result, we can make two simplifications in the equations above. First, we shall set z equal to zero whenever it appears on the right side of the equations above. Second, we shall express dv as

$$dv = t\,dA$$

where dA is an area element on the *surface* of the plate, as shown in Fig. 9.10. Equations 9.3 then become

$$I_{xx} = \rho t \iint_A y^2\, dA, \qquad I_{xy} = \rho t \iint_A xy\, dA$$

$$I_{yy} = \rho t \iint_A x^2\, dA, \qquad I_{xz} = 0$$

$$I_{zz} = \rho t \iint_A (x^2 + y^2)\, dA, \qquad I_{yz} = 0$$

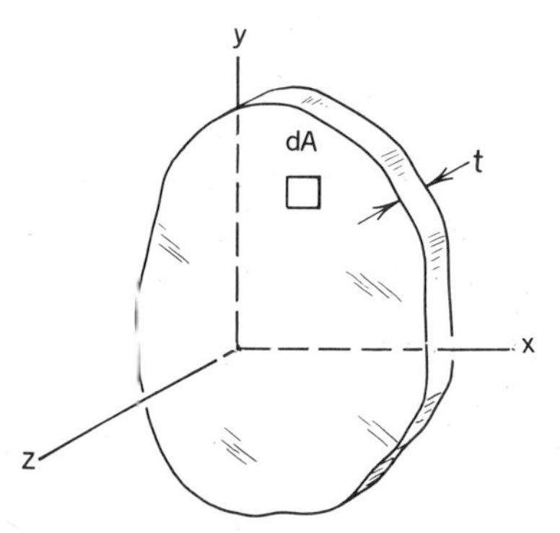

Figure 9.10. Use volume elements $t\,dA$.

Notice, now, that the integrals on the right sides of the equations above are moments and products of *area* presented in Chapter 8. Denoting mass-inertia terms with a subscript M and area-inertia terms with a subscript A, we can then say for the nonzero expressions:

$$(I_{xx})_M = \rho t (I_{xx})_A$$
$$(I_{yy})_M = \rho t (I_{yy})_A$$
$$(I_{zz})_M = \rho t (J)_A$$
$$(I_{xy})_M = \rho t (I_{xy})_A$$

Thus, for a thin plate with a constant product ρt throughout, we can compute the inertia tensor components for reference xyz (see Fig. 9.9) by using the moments and product of area of the surface of the plate relative to axes xy. It should be pointed out, in this regard, that ρt is the *mass per unit area* of the plate.[4] We now illustrate this procedure in the following example.

EXAMPLE 9.3

Determine the inertia tensor components for the thin plate (Fig. 9.11) relative to the indicated axes xyz. The weight of the plate is .002 N/mm². For the top edge, $y = 2\sqrt{x}$ with x and y in millimeters.

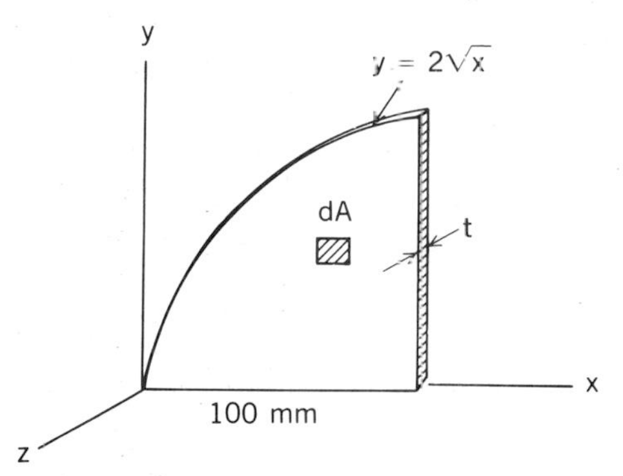

Figure 9.11. Plate of thickness t

[4]If we let $t \longrightarrow 0$ and $\rho \longrightarrow \infty$ such that $\rho t \longrightarrow 1$, we see that the nonzero mass-inertia terms degenerate to the moments and product of area.

It is clear that for pt we have

$$pt = \frac{.002}{9.81} = .000204 \text{ kg/mm}^2 \tag{a}$$

We now examine the moments and products of area for the surface of the plate about axes xy. Thus,[5]

$$(I_{xx})_A = \int_0^{100}\int_{y=0}^{y=2\sqrt{x}} y^2\,dy\,dx$$

$$= \int_0^{100} \frac{y^3}{3}\bigg|_0^{2\sqrt{x}} dx = \int_0^{100} \frac{8}{3}x^{3/2}\,dx$$

$$= \frac{8}{3}\frac{x^{5/2}}{\frac{5}{2}}\bigg|_0^{100} = (\tfrac{8}{3})(\tfrac{2}{5})(100^{5/2})$$

$$= 1.067 \times 10^5 \text{ mm}^4$$

$$(I_{yy})_A = \int_0^{100}\int_{y=0}^{y=2\sqrt{x}} x^2dy\,dx$$

$$= \int_0^{100} x^2y\bigg|_0^{2\sqrt{x}} dx = \int_0^{100} x^2(2\sqrt{x})\,dx$$

$$= 2\frac{x^{7/2}}{\frac{7}{2}}\bigg|_0^{100} = 2(\tfrac{2}{7})(100)^{7/2}$$

$$= 5.71 \times 10^6 \text{ mm}^4$$

$$(I_{xy})_A = \int_0^{100}\int_{y=0}^{y=2\sqrt{x}} xy\,dy\,dx$$

$$= \int_0^{100} x\frac{y^2}{2}\bigg|_0^{2\sqrt{x}} dx = \int_0^{100} 2x^2\,dx$$

$$= 2\left(\frac{100^3}{3}\right) = 6.67 \times 10^5 \text{ mm}^4$$

Using Eq. (a), we can then say for the nonzero inertia tensor components:

$$(I_{xx})_M = (.0204)(1.067 \times 10^5) = 2176 \text{ kg-mm}^2$$
$$(I_{yy})_M = (.0204)(5.71 \times 10^6) = 116{,}500 \text{ kg-mm}^2$$
$$(I_{xy})_M = (.0204)(6.67 \times 10^5) = 13{,}610 \text{ kg-mm}^2$$

Note that the nonzero inertia tensor components for a reference xyz on a plate (see Fig. 9.9) are *proportional* through pt to the corresponding area-inertia terms for the plate surface. This means that all the formulations of Chapter 8 apply to the aforementioned nonzero inertia tensor components. Thus, on rotating the axes about the z axis we may use the transformation equations of Chapter 8. Consequently, the concept of *principal axes* in the midplane of the plate at a point applies. For such axes,

[5]Note we have multiple integration where one of the boundaries is variable. The procedure to follow should be evident from the example.

the product of inertia is zero. One such axis then gives the maximum moment of inertia for all axes in the midplane at the point, the other the minimum moment of inertia. We have presented such problems at the end of the section.

What about principal axes for the inertia tensor at a point in a general three-dimensional body? Those students who have time to study Section 9.6 will learn that there are *three principal axes* at a point in the general case. These axes are *mutually orthogonal* and the *products of inertia are all zero* for such a set of axes at a point.[6] Furthermore, one of the axes will have a maximum moment of inertia, another axis will have a minimum moment of inertia, while the third axis will have an intermediate value. The sum of these three inertia terms must have the value that is common for all sets of axes at the point.

If, perchance, a set of axes xyz at a point is such that xy and xz form *two planes of symmetry* for the mass distribution of the body, then, as we learned earlier, since the z axis and the y axis are normal to planes of symmetry, $I_{xy} = I_{xz} = I_{yz} = 0$. Thus, all products of inertia are zero. This would also be true for *any* two sets of axes of xyz forming two planes of symmetry. Clearly, axes forming two planes of symmetry must be *principal axes.* This information will suffice most instances when we have to identify principal axes. On the other hand, consider the case where there is only *one plane of symmetry* for the mass distribution of a body at some point A. Let the xy plane at A form this plane of symmetry. Then, clearly, the products of inertia between the z axis that is normal to the plane of symmetry xy and *any axis* in the xy plane at A must be zero, as pointed out earlier. Obviously, the z axis must be a principal axis. The other two principal axes must be in the plane of symmetry, but generally cannot be located by inspection.

Problems

9.1. A uniform homogeneous slender rod of mass M is shown. Compute I_{xx} and $I_{x'x'}$.

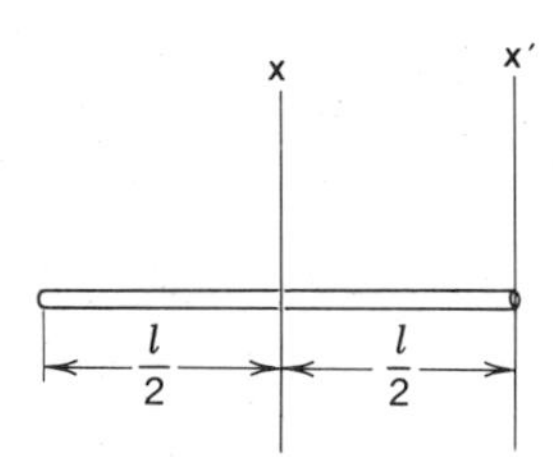

Figure P.9.1

9.2. Find I_{xx} and $I_{x'x'}$ for the thin rod of Problem 9.1 for the case where the mass per unit length at the left end is 5 lbm/ft and increases linearly so that at the right end it is 8 lbm/ft. The rod is 20 ft in length.

9.3. Compute I_{xy} for the thin homogeneous hoop of mass M.

---

[6]The third principal axis for the plate at a point in the midplane is the z axis normal to the plate. Note that $(I_{zz})_M$ must always equal $(I_{xx})_M + (I_{yy})_M$. Why?

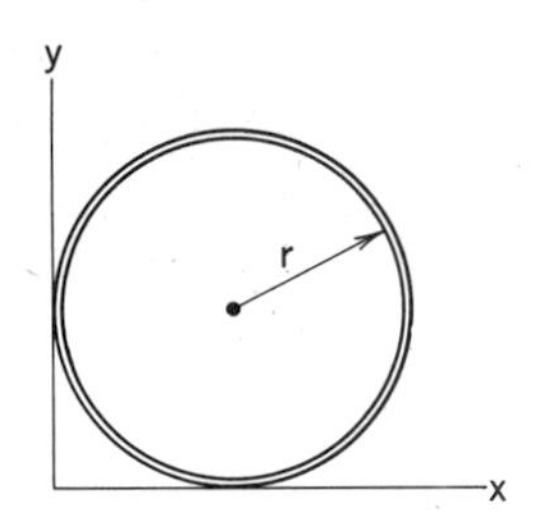

Figure P.9.3

9.4. Compute I_{xx}, I_{yy}, I_{zz}, and I_{xy} for the homogeneous rectangular parallelepiped.

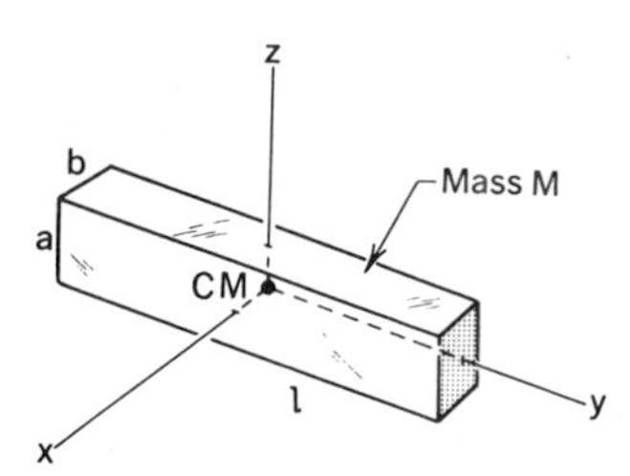

Figure P.9.4

9.5. A wire having the shape of a parabola is shown. The curve is in the yz plane. If the mass of the wire is .3 N/m, what are I_{yy} and I_{xz}? [*Hint:* Replace ds along the wire by $\sqrt{(dy/dz)^2 + 1}\, dz$.]

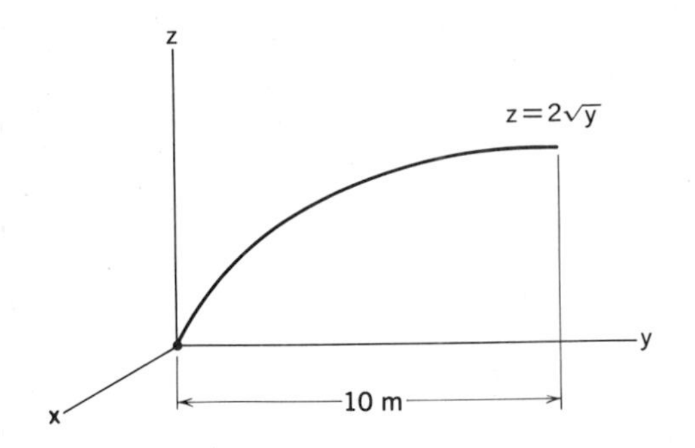

Figure P.9.5

9.6. Compute the moment of inertia, I_{BB}, for the half-cylinder shown. The body is homogeneous and has a mass M.

9.7. Find I_{zz} and I_{xx} for the homogeneous right circular cylinder of mass M.

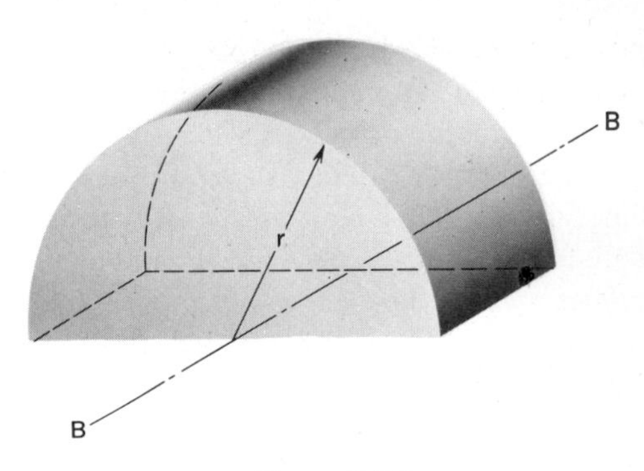

Figure P.9.6

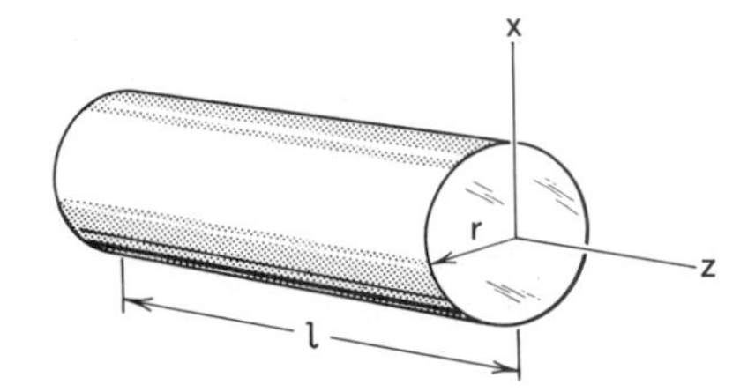

Figure P.9.7

9.8. For the cylinder in Problem 9.7, the density increases linearly in the z direction from a value of .100 grams/mm³ at the left end to a value of .180 grams/mm³ at the right end. Take $r = 30$ mm and $l = 150$ mm. Find I_{xx} and I_{zz}.

9.9. Show that I_{zz} for the homogeneous right circular cone is $\frac{3}{10} MR^2$.

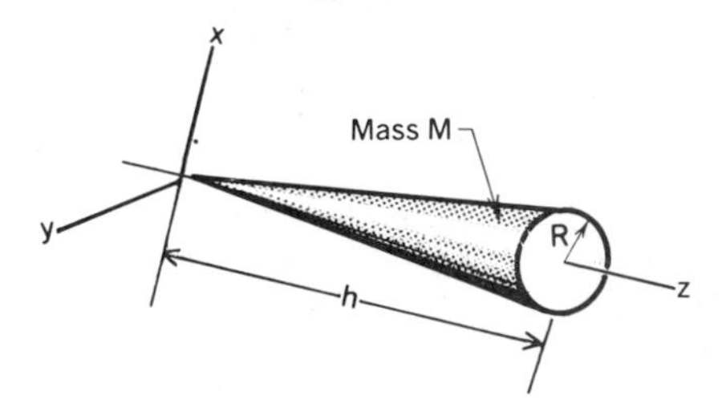

Figure P.9.9

9.10. In Problem 9.9, the density increases as the square of z in the z direction from a value of .200 grams/mm³ at the left end to a value of .400 grams/mm³ at the right end. If $r = 20$ mm and the cone is 100 mm in length, find I_{zz}.

9.11. A body of revolution is shown. The radial distance r of the boundary from the x axis is given as $r = .2x^2$ m. What is I_{xx} for a uniform density of 1600 kg/m³?

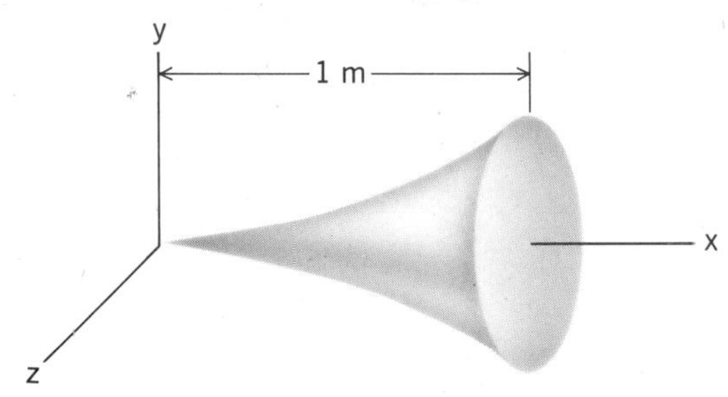

Figure P.9.11

9.12. A thick hemispherical shell is shown with an inside radius of 40 mm and an outside radius of 60 mm. If the density ρ is 7000 kg/m³, what is I_{yy}?

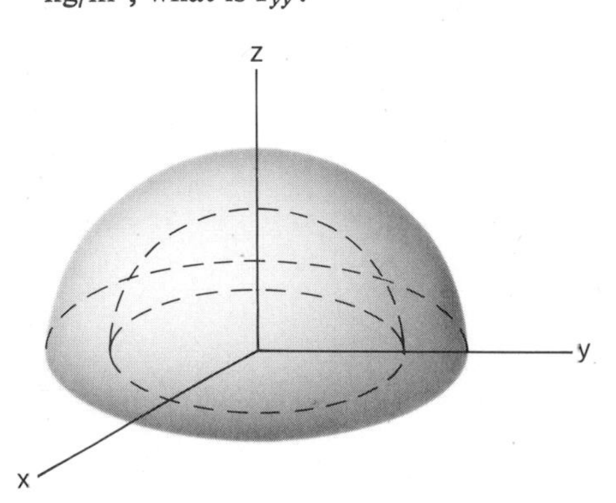

Figure P.9.12

9.13. Find the mass moment of inertia I_{xx} for a very thin plate forming a quarter-sector of a circle. The plate weighs .4 N. What is the second moment of area about the x axis? What is the product of inertia?

Figure P.9.13

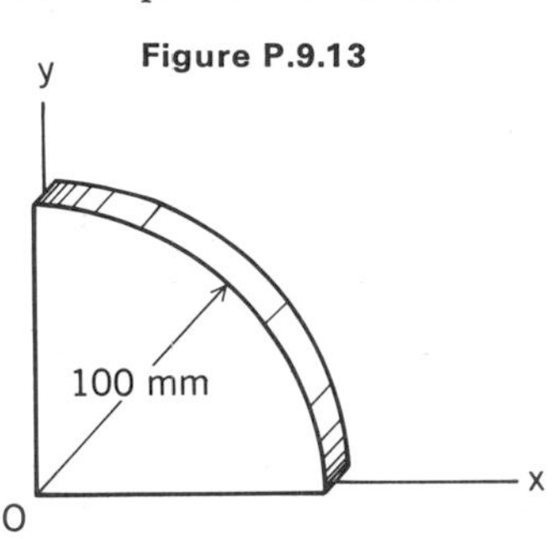

9.14. Find the second moment of area about the x axis for the top surface of a very thin plate. If the weight of the plate is .02N/mm², find the second moments of mass about the xy axes. What is the mass product of inertia I_{xy}?

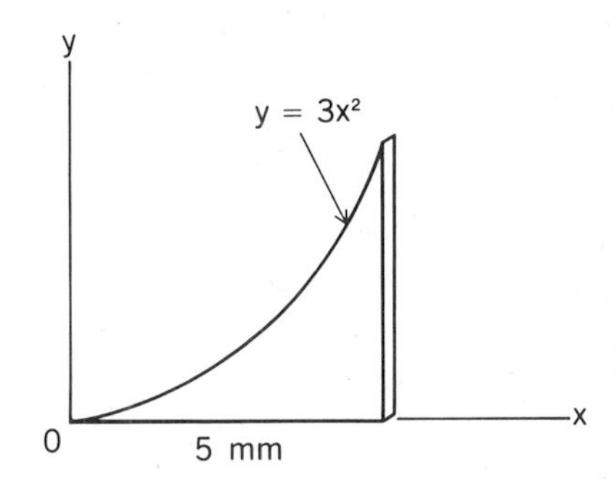

Figure P.9.14

***9.15.** A uniform tetrahedron is shown having sides of length a, b, and c, respectively, and a mass M. Show that $I_{yz} = \frac{1}{20} Mac$. (*Suggestion:* Let z run from zero to surface ABC. Let x run from zero to AB. Finally, let y run from zero to B. Note that the equation of a plane surface is $z = \alpha x + \beta y + \gamma$, where α, β, and γ are constants. The mass of the tetrahedron is $\rho abc/6$. It will be simplest in expanding $(1 - x/b - y/c)^2$ to proceed in the form $[(1 - y/c) - (x/b)]^2$, keeping $(1 - y/c)$ intact. In the last integration replace y by $[-c(1 - y/c) + c]$, etc.)

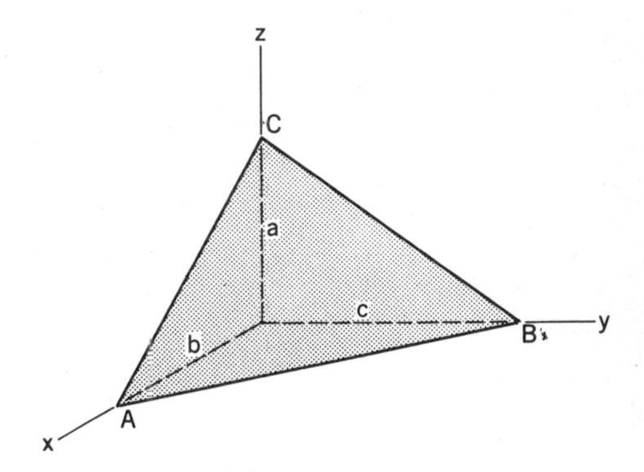

Figure P.9.15

9.16. In Problem 9.13, find the three principal mass moments of inertia at O. Use the results from this problem, that

$$(I_{xx})_M = 101.9 \text{ kg-mm}^2$$
$$(I_{xy})_M = 64.9 \text{ kg-mm}^2$$

9.17. In Problem 9.14, compute the values of the three principal mass moments of inertia at O. From Problem 9.14 we have the results

$$(I_{xx})_M = 20{,}500 \text{ kg-mm}^2$$
$$(I_{yy})_M = 382 \text{ kg-mm}^2$$
$$(I_{xy})_M = 2390 \text{ kg-mm}^2$$

9.18. Can you identify by inspection any of the principal axes of inertia at A? At B? Explain. The density of the material is uniform.

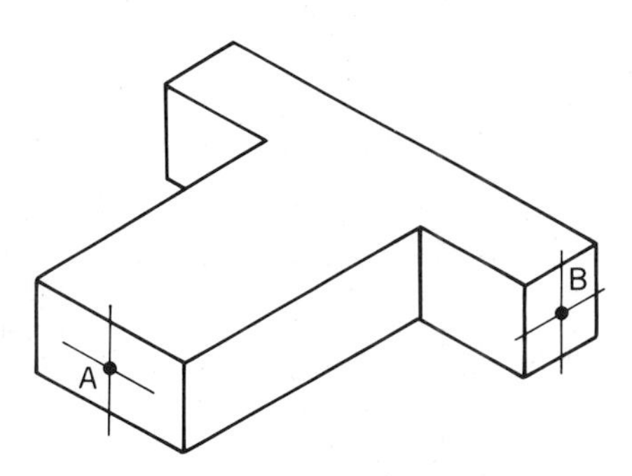

Figure P.9.18

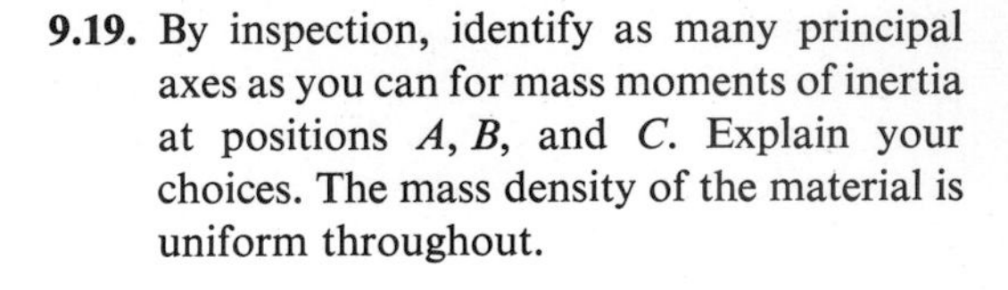

9.19. By inspection, identify as many principal axes as you can for mass moments of inertia at positions A, B, and C. Explain your choices. The mass density of the material is uniform throughout.

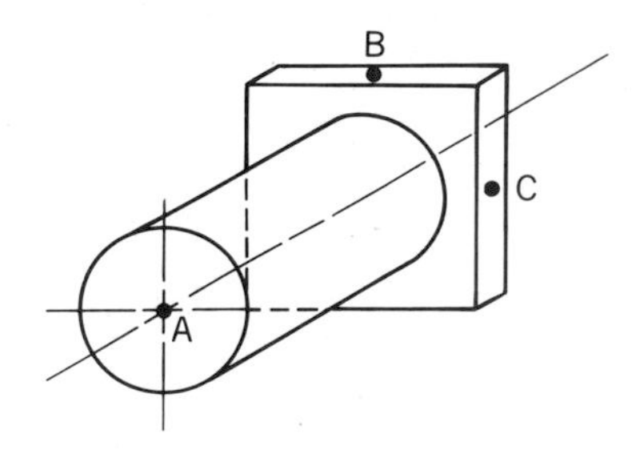

Figure P.9.19

9.4 Translation of Coordinate Axes

In this section, we will compute mass moment and product of inertia quantities for a reference xyz that is displaced under a translation (no rotation) from a reference $x'y'z'$ at the center of mass (Fig. 9.12) for which the inertia terms are presumed known. Let us first compute the moment of inertia I_{zz}. Observing Fig. 9.12, we see that

$$\boldsymbol{r} = \boldsymbol{r}_c + \boldsymbol{r}'$$

Hence,

$$x = x_c + x'$$
$$y = y_c + y'$$
$$z = z_c + z'$$

We can now formulate I_{zz} in the following way:

$$I_{zz} = \iiint_V (x^2 + y^2)\rho \, dv = \iiint_V [(x_c + x')^2 + (y_c + y')^2]\rho \, dv \tag{9.4}$$

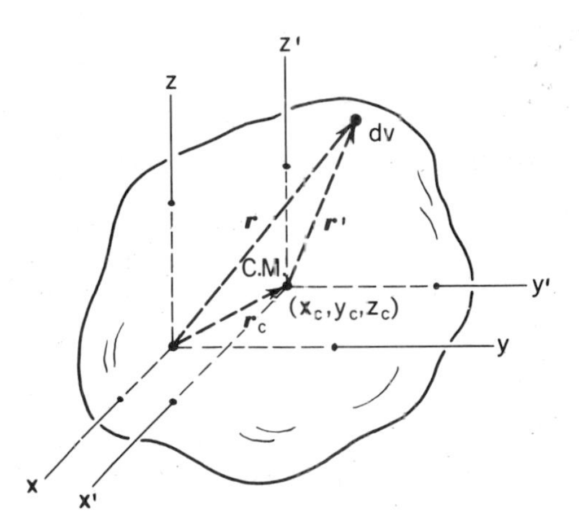

Figure 9.12. *xyz* translated from *x'y'z'* at C.M.

Carrying out the squares and rearranging, we have

$$I_{zz} = \iiint_V (x_c^2 + y_c^2)\rho\, dv + 2\iiint_V x_c x' \rho\, dv + 2\iiint_V y_c y' \rho\, dv + \iiint_V (x'^2 + y'^2)\rho\, dv \tag{9.5}$$

Note that the quantities bearing the subscript c are constant for the integration and can be extracted from under the integral sign. Thus,

$$I_{zz} = M(x_c^2 + y_c^2) + 2x_c \iiint_V x'\, dm + 2y_c \iiint_V y'\, dm + \iiint_V (x'^2 + y'^2)\rho\, dv \tag{9.6}$$

where $\rho\, dv$ has been replaced in some terms by dm, and the integration $\iiint_V \rho\, dv$ in the first integration has been evaluated as M, the total mass of the body. The origin of the primed reference being at the center of mass requires of the first moments of mass that $\iiint x'\, dm = \iiint y'\, dm = \iiint z'\, dm = 0$. The middle two terms accordingly drop out of the expression above, and we recognize the last expression to be $I_{z'z'}$. Thus, the desired relation is

$$I_{zz} = I_{z'z'} + M(x_c^2 + y_c^2) \tag{9.7}$$

By observing the body in Fig. 9.12 along the z and z' axes (i.e., from directly above), we get a view as is shown in Fig. 9.13. From this diagram, we can see that $y_c^2 + x_c^2 = d^2$, where d is the perpendicular distance between the z' axis through the centroid and the z axis about which we are taking moments of inertia. We may then give the result above as

$$\boxed{I_{zz} = I_{z'z'} + Md^2} \tag{9.8}$$

Let us generalize from the above statement.

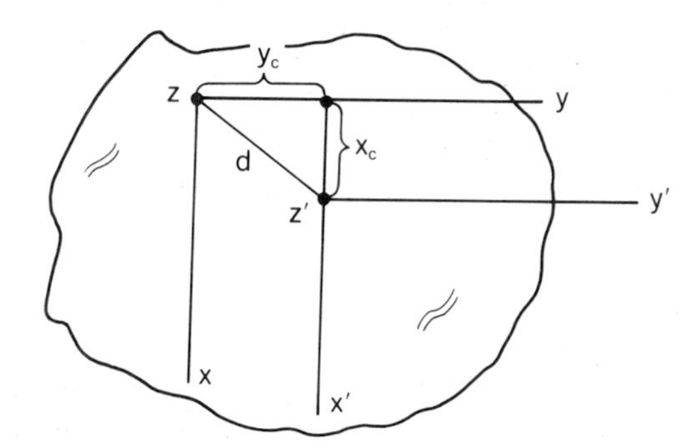

Figure 9.13. View along z direction (from above).

The moment of inertia of a body about any axis equals the moment of inertia of the body about a parallel axis that goes through the center of mass, plus the total mass times the perpendicular distance between the axes squared.

We leave it to you to show that for products of inertia a similar relation can be reached. For I_{xy}, for example, we have

$$\boxed{I_{xy} = I_{x'y'} + Mx_c y_c} \tag{9.9}$$

Here, we must take care to put in the proper signs of x_c and y_c as measured *from* the xyz reference. Equations 9.8 and 9.9 comprise the well-known *parallel-axis theorems* analogous to those formed in Chapter 8 for areas. You can use them to advantage for bodies composed of simple familiar shapes, as we now illustrate.

EXAMPLE 9.4

Find I_{xx} and I_{xy} for the body shown in Fig. 9.14. Take ρ as constant for the body. Use the formulations for moments and products of inertia at the center of mass as given in the inside cover pages.

We shall consider first a solid rectangular prism having the outer dimensions given in Fig. 9.14, and we shall then subtract the contribution of the cylinder and the rectangular block that have been cut away. Thus, we have, for the overall rectangular block which we consider as body 1,

$$\begin{aligned}(I_{xx})_1 &= (I_{xx})_c + Md^2 = \tfrac{1}{12}M(a^2 + b^2) + Md^2 \\ &= \tfrac{1}{12}[(\rho)(20)(8)(15)](8^2 + 15^2) + [(\rho)(20)(8)(15)](4^2 + 7.5^2) \qquad \text{(a)} \\ &= 231{,}200\rho\end{aligned}$$

From this, we shall take away the contribution of the cylinder, which we denote as body 2. Use formulas from inside cover pages.

$$\begin{aligned}(I_{xx})_2 &= \tfrac{1}{12}M(3r^2 + h^2) + Md^2 = \tfrac{1}{12}[\rho\pi(1)^2(15)][3(1^2) + 15^2] \\ &\qquad + [\rho\pi(1)^2(15)](6^2 + 7.5^2) \qquad \text{(b)} \\ &= 5243\rho\end{aligned}$$

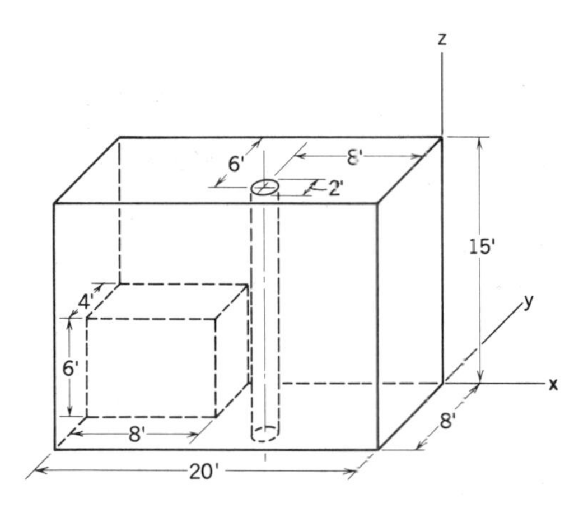

Figure 9.14. Find I_{xx} and I_{xy}.

Also, we shall take away the contribution of the rectangular cutout (body 3):

$$(I_{xx})_3 = \frac{1}{12}M(a^2 + b^2) + Md^2 = \frac{1}{12}[(\rho)(8)(6)(4)](4^2 + 6^2) + [(\rho)(8)(6)(4)](2^2 + 3^2) \quad \text{(c)}$$

$$= 3328\rho$$

We get, accordingly,

$$I_{xx} = (231{,}200 - 5243 - 3328)\rho = 223{,}000\rho \quad \text{(d)}$$

We do the same for I_{xy}. Thus, for the block as a whole, we have

$$(I_{xy})_1 = (I_{xy})_c + Mx_c y_c$$

At the center of mass of the block, both the $(x')_1$ and $(y')_1$ axes are normal to planes of symmetry. Accordingly, $(I_{xy})_c = 0$. Hence,

$$(I_{xy})_1 = 0 + [\rho(20)(8)(15)](-4)(-10) = 96{,}000\rho \quad \text{(e)}$$

For the cylinder, we note that both the $(x')_2$ and $(y')_2$ axes at the center of mass are normal to planes of symmetry. Hence, we can say that

$$(I_{xy})_2 = 0 + [\rho(\pi)(1^2)(15)](-8)(-6) = 2262\rho \quad \text{(f)}$$

Finally, for the small cutout rectangular parallelepiped, we note that the $(x')_3$ and $(y')_3$ axes at the center of mass are perpendicular to planes of symmetry. Hence, we have

$$(I_{xy})_3 = 0 + [(\rho)(8)(6)(4)](-2)(-16) = 6144\rho \quad \text{(g)}$$

The quantity I_{xy} for the body with the rectangular and cylindrical cavities is then

$$I_{xy} = (96{,}000 - 2262 - 6144)\rho = 87{,}600\rho \quad \text{(h)}$$

If ρ is given in units of lbm/ft^3, the inertia terms have units lbm-ft^2.

*9.5 Transformation Properties of the Inertia Terms

Let us assume that the six independent inertia terms are known for a given reference. What is the mass moment of inertia for an axis going through the origin of the reference and having the direction cosines l, m, and n relative to the axes of this reference? The axis is designated as kk in Fig. 9.15. From previous conclusions, we can say that

$$I_{kk} = \iiint_V [|\boldsymbol{r}|(\sin\phi)]^2 \rho\, dv \tag{9.10}$$

where ϕ is the angle between kk and $\boldsymbol{r}$. We shall now put $\sin^2\phi$ into a more useful form by considering the right triangle formed by the position vector $\boldsymbol{r}$ and the axis kk. This triangle is shown enlarged in Fig. 9.16. The side a of the triangle has a magnitude that can be given by the dot product of $\boldsymbol{r}$ and the unit vector $\boldsymbol{\epsilon}_k$ along kk. Thus,

$$a = \boldsymbol{r} \cdot \boldsymbol{\epsilon}_k = (x\boldsymbol{i} + y\boldsymbol{j} + z\boldsymbol{k}) \cdot (l\boldsymbol{i} + m\boldsymbol{j} + n\boldsymbol{k}) \tag{9.11}$$

Hence,

$$a = lx + my + nz$$

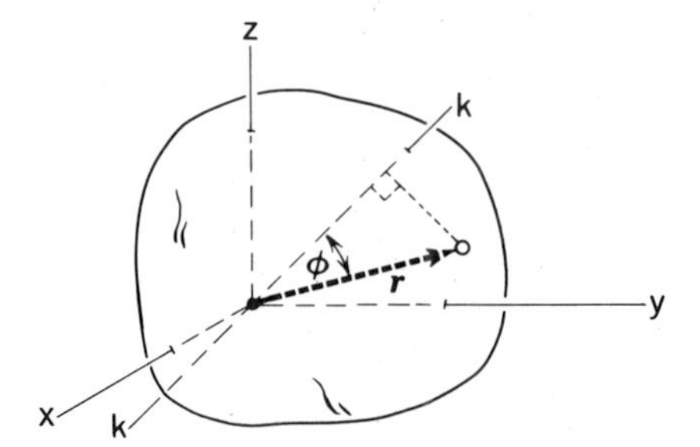

Figure 9.15. Find I_{kk}.

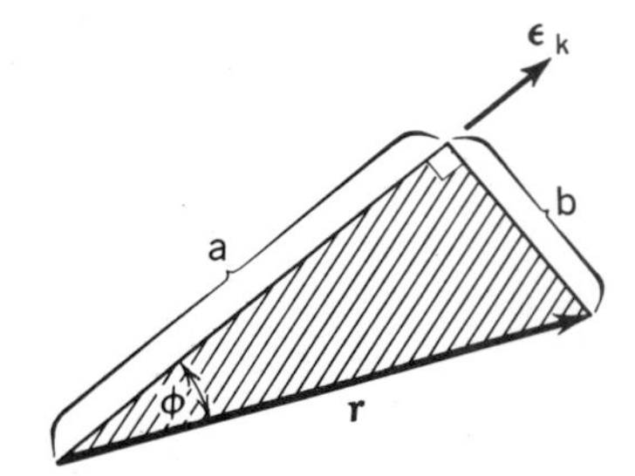

Figure 9.16. Right triangle formed by $\boldsymbol{r}$ and kk.

Using the Pythagorean theorem, we can now give side b as

$$b^2 = |\boldsymbol{r}|^2 - a^2 = (x^2 + y^2 + z^2) - (l^2x^2 + m^2y^2 + n^2z^2 + 2lmxy + 2lnxz + 2mnyz)$$

The term $\sin^2\phi$ may next be given as

$$\sin^2\phi = \frac{b^2}{r^2} = \frac{(x^2 + y^2 + z^2) - (l^2x^2 + m^2y^2 + n^2z^2 + 2lmxy + 2lnxz + 2mnyz)}{x^2 + y^2 + z^2} \tag{9.12}$$

Substituting back into Eq. 9.10, we get, on canceling terms,

$$I_{kk} = \iiint_V [(x^2 + y^2 + z^2) - (l^2x^2 + m^2y^2 + n^2z^2 + 2lmxy + 2lnxz + 2mnyz)]\rho\, dv$$

Since $l^2 + m^2 + n^2 = 1$, we can multiply the first bracketed expression in the integral by this sum:

$$I_{kk} = \iiint_V [(x^2 + y^2 + z^2)(l^2 + m^2 + n^2) - (l^2x^2 + m^2y^2 - n^2z^2 + 2lmxy + 2lnxz + 2mnyz)]\rho\, dv$$

Carrying out the multiplication and collecting terms, we get the relation

$$I_{kk} = l^2 \iiint_V (y^2 + z^2)\rho\, dv + m^2 \iiint_V (x^2 + z^2)\rho\, dv + n^2 \iiint_V (x^2 + y^2)\rho\, dv - 2lm \iiint_V (xy)\rho\, dv - 2ln \iiint_V (xz)\rho\, dv - 2mn \iiint_V (yz)\rho\, dv$$

Referring back to the definitions presented by relations 9.1, we reach the desired transformation equation:

$$I_{kk} = l^2 I_{xx} + m^2 I_{yy} + n^2 I_{zz} - 2lmI_{xy} - 2lnI_{xz} - 2mnI_{yz} \tag{9.13}$$

We next put this in a more useful form of the kind you will see in later courses in mechanics. Note first that l is the direction cosine between the k axis and the x axis. It is common practice to identify this cosine as a_{kx} instead of l. Note that the subscripts identify the axes involved. Similarly, $m = a_{ky}$ and $n = a_{kz}$. We can now express Eq. 9.13 in a form similar to a matrix array as follows on noting that $I_{xy} = I_{yx}$, etc.

$$\boxed{\begin{aligned} I_{kk} = \quad & I_{xx}a_{kx}^2 && - I_{xy}a_{kx}a_{ky} && - I_{xz}a_{kx}a_{kz} \\ & - I_{yx}a_{ky}a_{kx} && + I_{yy}a_{ky}^2 && - I_{yz}a_{ky}a_{kz} \\ & - I_{zx}a_{kz}a_{kx} && - I_{zy}a_{kz}a_{ky} && + I_{zz}a_{kz}^2 \end{aligned}} \tag{9.14}$$

This format is easily written by first writing the matrix array of I's on the right side and then inserting the a's remembering to insert minus signs for off-diagonal terms.

Let us next compute the product of inertia for a pair of mutually perpendicular axes, Ok and Oq, as shown in Fig. 9.17. The direction cosines of Ok we shall take as l, m, and n, whereas the direction cosines of Oq we shall take as l', m', and n'. Since the axes are at right angles to each other, we know that

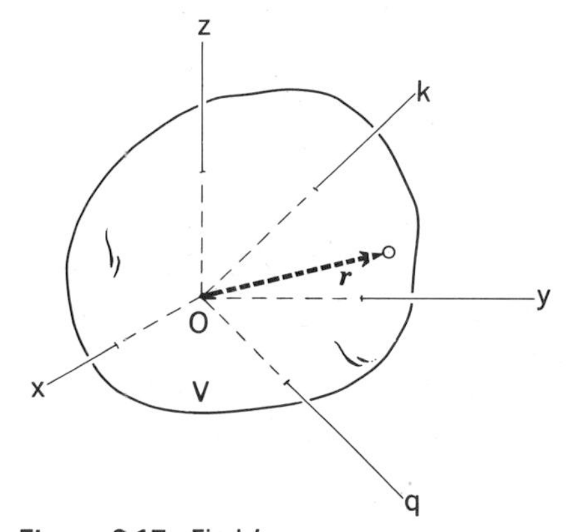

Figure 9.17. Find I_{kq}.

$$\boldsymbol{\epsilon}_k \cdot \boldsymbol{\epsilon}_q = 0$$

Therefore,

$$ll' + mm' + nn' = 0 \tag{9.15}$$

Noting that the coordinates of the mass element $\rho\, dv$ along the axes Ok and Oq are $\boldsymbol{r} \cdot \boldsymbol{\epsilon}_k$ and $\boldsymbol{r} \cdot \boldsymbol{\epsilon}_q$, respectively, we have, for I_{kq}:

$$I_{kq} = \iiint_V (\boldsymbol{r} \cdot \boldsymbol{\epsilon}_k)(\boldsymbol{r} \cdot \boldsymbol{\epsilon}_q)\rho\, dv$$

Using xyz components of $\boldsymbol{r}$ and the unit vectors, we have

$$I_{kq} = \iiint_V [(x\boldsymbol{i} + y\boldsymbol{j} + z\boldsymbol{k}) \cdot (l\boldsymbol{i} + m\boldsymbol{j} + n\boldsymbol{k})] \times [(x\boldsymbol{i} + y\boldsymbol{j} + z\boldsymbol{k}) \cdot (l'\boldsymbol{i} + m'\boldsymbol{j} + n'\boldsymbol{k})]\rho \, dv \quad (9.16)$$

Carrying out the dot products in the integrand above, we get the following result:

$$I_{kq} = \iiint_V (xl + ym + zn)(xl' + ym' + zn')\rho \, dv$$

Hence,

$$I_{kq} = \iiint_V (x^2ll' + y^2mm' + z^2nn' + xylm' + xzln' + yxml' + yzmn' + zxnl' + zynm')\rho \, dv \quad (9.17)$$

Noting from Eq. 9.15 that $(ll' + mm' + nn')$ is zero, we may for convenience add the term, $(-x^2 - y^2 - z^2)(ll' + mm' + nn')$, to the integrand in the equation above. After canceling some terms, we have

$$I_{kq} = \iiint_V (-x^2mm' - x^2nn' - y^2ll' - y^2nn' - z^2ll' - z^2mm' + xylm' + xzln' + yxml' + yzmn' + zxnl' + xynm')\rho \, dv$$

Collecting terms and bringing the direction cosines outside the integrations, we get

$$\begin{aligned} I_{kq} = {} & -ll' \iiint_V (y^2 + z^2)\rho \, dv - mm' \iiint_V (x^2 + z^2)\rho \, dv \\ & -nn' \iiint_V (y^2 + x^2)\rho \, dv + (lm' + ml') \iiint_V xy\rho \, dv \\ & + (ln' + nl') \iiint_V xz\rho \, dv + (mn' + nm') \iiint_V yz\rho \, dv \end{aligned} \quad (9.18)$$

Noting the definitions in Eq. 9.1, we can state the desired transformation:

$$\begin{aligned} I_{kq} = {} & -ll'I_{xx} - mm'I_{yy} - nn'I_{zz} + (lm' + ml')I_{xy} \\ & + (ln' + nl')I_{xz} + (mn' + nm')I_{yz} \end{aligned} \quad (9.19)$$

We can now rewrite the previous equation in a more useful and simple form using a's as direction cosines. Thus, noting that $l' = a_{qx}$, etc.,

$$\boxed{\begin{aligned} -I_{kq} = {} & I_{xx}a_{kx}a_{qx} - I_{xy}a_{kx}a_{qy} - I_{xz}a_{kx}a_{qz} \\ & -I_{yx}a_{ky}a_{qx} + I_{yy}a_{ky}a_{qy} - I_{yz}a_{ky}a_{qz} \\ & -I_{zx}a_{kz}a_{qx} - I_{zy}a_{kz}a_{qy} + I_{zz}a_{kz}a_{qz} \end{aligned}} \quad (9.20)$$

Again you will note that the right side can easily be set forth by first putting down the matrix array of I_{ij} and then inserting the a's with easily determined subscripts while remembering to insert minus signs for off-diagonal terms.

By making the axis Ok in Fig. 9.17 an x' axis at O and using the direction cosines for this axis ($a_{x'x}, a_{x'y}, a_{x'z}$), we can formulate $I_{x'x'}$ from Eq. 9.14. By a similar procedure, we can formulate $I_{y'y'}$ and $I_{z'z'}$ for reference $x'y'z'$ at 0 rotated arbitrarily relative to xyz. Also, by considering the Ok and Oq axes to be x' and y' axes, respectively, with $a_{x'x}$, $a_{x'y}$, and $a_{x'z}$ as direction cosines for the x' axis and $a_{y'x}$, $a_{y'y}$, and $a_{y'z}$ as direction cosines for the y' axis, we can evaluate $I_{x'y'}$ at O using Eq. 9.20. This approach can similarly be followed to find $I_{x'z'}$ and $I_{y'z'}$. Thus, employing Eqs. 9.14 and 9.20 as parent equations, we can develop equations for computing the nine inertia quantities for a reference $x'y'z'$ rotated arbitrarily relative to xyz at O in terms of the nine known inertia quantities for reference xyz at O. Thus, once the nine inertia quantities are known for one reference at some point, they can be determined for *any* reference at that point. We say that the inertia terms *transform* from one set of components for xyz at some point O to another set of components for $x'y'z'$ at O by means of certain transformations formed from Eqs. 9.14 and 9.20.

We now define a symmetric,[7] *second-order tensor as a set of nine components*

$$\begin{pmatrix} A_{xx} & A_{xy} & A_{xz} \\ A_{yx} & A_{yy} & A_{yz} \\ A_{zx} & A_{zy} & A_{zz} \end{pmatrix}$$

which transforms with a rotation of axes according to the following parent equations. For the diagonal terms,

$$\begin{aligned} A_{kk} = {} & A_{xx}a_{kx}^2 + A_{xy}a_{kx}a_{ky} + A_{xz}a_{kx}a_{kz} \\ & + A_{yx}a_{ky}a_{kx} + A_{yy}a_{ky}^2 + A_{yz}a_{ky}a_{kz} \\ & + A_{zx}a_{kz}a_{kx} + A_{zy}a_{kz}a_{ky} + A_{zz}a_{kz}^2 \end{aligned} \tag{9.21}$$

For the off-diagonal terms,

$$\begin{aligned} A_{kq} = {} & A_{xx}a_{kx}a_{qx} + A_{xy}a_{kx}a_{qy} + A_{xz}a_{kx}a_{kz} \\ & + A_{yx}a_{ky}a_{qx} + A_{yy}a_{ky}a_{qy} + A_{yz}a_{ky}a_{kz} \\ & + A_{zx}a_{kz}a_{qx} + A_{zy}a_{kz}a_{qy} + A_{zz}a_{kz}a_{qz} \end{aligned} \tag{9.22}$$

On comparing Eqs. 9.21 and 9.22, respectively, with Eqs. 9.14 and 9.20, we can conclude that the array of terms

$$I_{ij} = \begin{pmatrix} I_{xx} & -I_{xy} & -I_{xz} \\ -I_{yx} & I_{yy} & -I_{yz} \\ -I_{zx} & -I_{zy} & I_{zz} \end{pmatrix} \tag{9.23}$$

is a second-order tensor.

You will learn that because of the common transformation law identifying certain quantities as tensors, there will be extremely important common characteristics for these quantities which set them apart from other quantities. Thus, in order to learn these common characteristics in an efficient way and to understand them better, we do become involved with tensors as an entity in the engineering sciences, physics, and applied mathematics. You will soon be confronted with the stress and strain tensors in your courses in strength of materials

[7] The word "symmetric" refers to the condition $A_{12} = A_{21}$, etc., that is required if the transformation equation is to have the form given. We can have nonsymmetric second-order tensors, but since they are less common in engineering work, we shall not concern ourselves here with such possibilities.

and in solid mechanics. And in electromagnetic theory and nuclear physics, you will be introduced to the quadrupole tensor.[8]

EXAMPLE 9.5

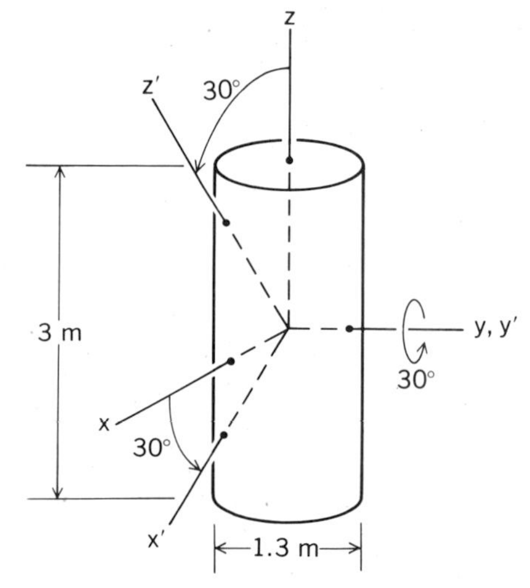

Figure 9.18. Find $I_{z'z'}$ and $I_{x'z'}$.

Find $I_{z'z'}$ and $I_{x'z'}$ for the cylinder shown in Fig. 9.18. The reference $x'y'z'$ is found by rotating about the y axis an amount 30°, as shown in the diagram. The mass of the cylinder is 100 kg.

It is simplest to first get the inertia tensor components for reference xyz. Thus, using formulas from the inside cover pages we have

$$I_{zz} = \frac{1}{2}Mr^2 = \frac{1}{2}(100)\left(\frac{1.3}{2}\right)^2 = 21.13 \text{ kg-m}^2$$

$$I_{xx} = I_{yy} = \frac{1}{12}M(3r^2 + h^2)$$

$$= \frac{1}{12}(100)\left[(3)\left(\frac{1.3}{2}\right)^2 + 3^2\right]$$

$$= 85.6 \text{ kg-m}^2$$

Noting that the xyz coordinate planes are planes of symmetry, we can conclude that

$$I_{xz} = I_{yx} = I_{yz} = 0$$

Next, evaluate the direction cosines of the z' and the x' axes relative to xyz. Thus,

For z' axis:

$$a_{z'x} = \cos 60° = .500$$

$$a_{z'y} = \cos 90° = 0$$

$$a_{z'z} = \cos 30° = .866$$

For x' axis:

$$a_{x'x} = \cos 30° = .866$$

$$a_{x'y} = \cos 90° = 0$$

---

[8]Vectors may be defined in terms of the way components of the vector for a new reference are related to the components of the old reference at a point A. Thus, for any direction $\boldsymbol{n}$, we have for component A_n:

$$A_n = A_x a_{nx} + A_y a_{ny} + A_z a_{nz} \tag{a}$$

Using Eq. (a), we can find components of vector $\boldsymbol{A}$ with respect to $x'y'z'$ rotated arbitrarily relative to xyz. Thus, all vectors must transform in accordance with Eq. (a) on rotation of the reference. Obviously, the vector, as seen from this point of view, is a special, simple case of the second-order tensor. We say, accordingly, that vectors are *first-order tensors*.

As for scalars, there is clearly no change in value when there is a rotation of axes at a point. Thus,

$$T(x', y', z') = T(x, y, z) \tag{b}$$

for $x'y'z'$ rotated relative to xyz. Scalars are a special form of tensor when considered from a transformation point of view. In fact, they are called *zero-order tensors*.

$$a_{x'z} = \cos 120^\circ = -.500$$

First, we employ Eq. 9.14 to get $I_{z'z'}$.

$$I_{z'z'} = (85.6)(.500)^2 + (21.23)(.865)^2$$
$$= 37.32 \text{ kg-m}^2$$

Finally, we employ Eq. 9.20 to get $I_{x'z'}$.

$$-I_{x'z'} = (85.6)(.500)(.866) + (21.13)(.866)(-.500)$$

Therefore,

$$I_{x'z'} = -27.92 \text{ kg-m}^2$$

Problems

In the following problems, use the formulas for moments and products of inertia at the mass center to be found in the inside cover pages.

9.20. What are the moments and products of inertia for the xyz and $x'y'z'$ axes for the cylinder?

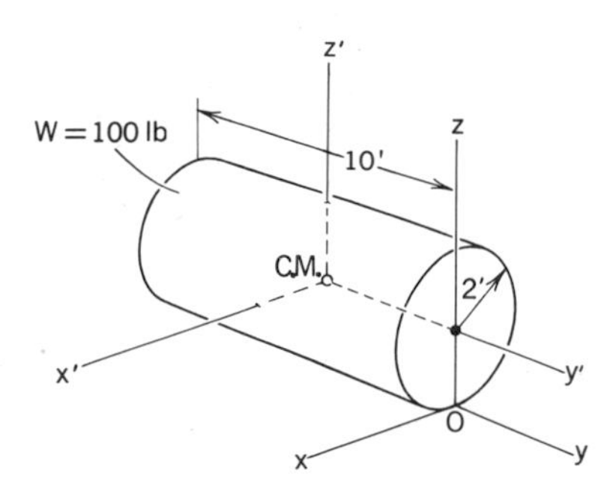

Figure P.9.20

9.21. For the uniform block, compute the inertia tensor at the center of mass, at point a, and at point b for axes parallel to the xyz reference. Take the mass of the body as M kg.

9.22. Determine $I_{xx} + I_{yy} + I_{zz}$ as a function of x, y, and z for all points in space for the uniform rectangular parallelepiped. Note that xyz has its origin at the center of mass and is parallel to the sides.

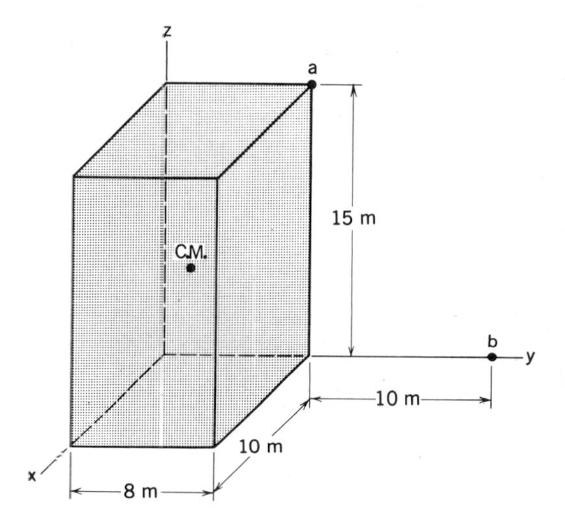

Figure P.9.21

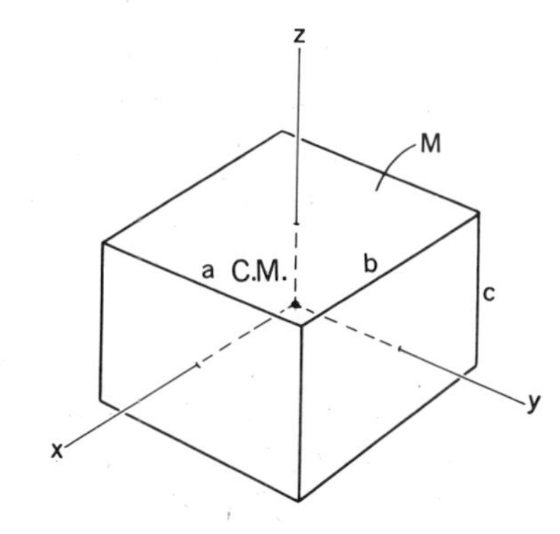

Figure P.9.22

9.23. A thin plate weighing 100 N has the following mass moments of inertia at mass center O:

$$I_{xx} = 15 \text{ kg-m}^2$$
$$I_{yy} = 13 \text{ kg-m}^2$$
$$I_{xy} = -10 \text{ kg-m}^2$$

What are the moments of inertia $I_{x'x'}$, $I_{y'y'}$, and $I_{z'z'}$ at point P having the position vector:

$$\boldsymbol{r} = .5\boldsymbol{i} + .2\boldsymbol{j} + .6\boldsymbol{k} \text{ m}$$

Also determine $I_{x'z'}$ at P.

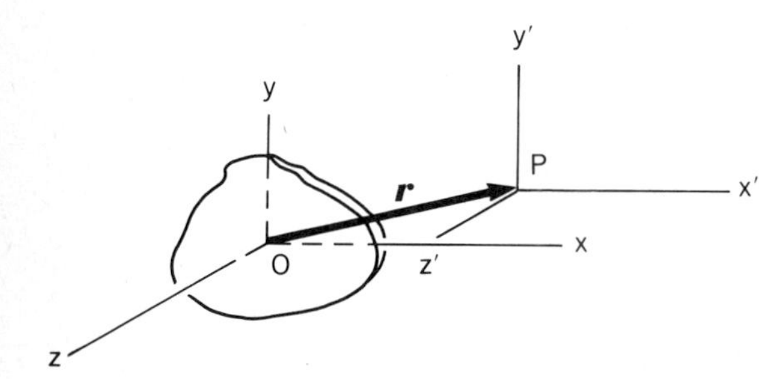

Figure P.9.23

9.24. A crate with its contents weighs 20 kN and has its center of mass at

$$\boldsymbol{r}_c = 1.3\boldsymbol{i} + 3\boldsymbol{j} + .8\boldsymbol{k} \text{ m}$$

It is known that at corner A,

$$I_{x'x'} = 5500 \text{ kg-m}^2$$
$$I_{x'y'} = -1500 \text{ kg-m}^2$$

for primed axes parallel to xyz. At point B, find $I_{x''x''}$ and $I_{x''y''}$ for double-primed axes parallel to xyz.

Figure P.9.24

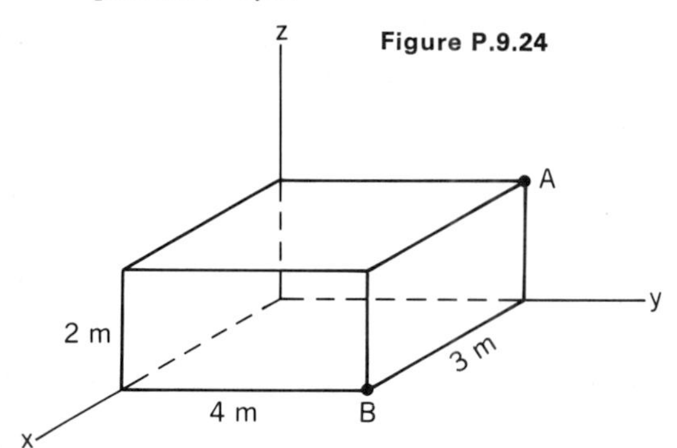

9.25. A cylindrical crate and its contents weigh 500 N. The center of mass is at

$$\boldsymbol{r}_c = .6\boldsymbol{i} + .7\boldsymbol{j} + 2\boldsymbol{k} \text{ m}$$

It is known that at A,

$$(I_{yy})_A = 85 \text{ kg-m}^2$$
$$(I_{yz})_A = -22 \text{ kg-m}^2$$

Find I_{yy} and I_{zy} at B.

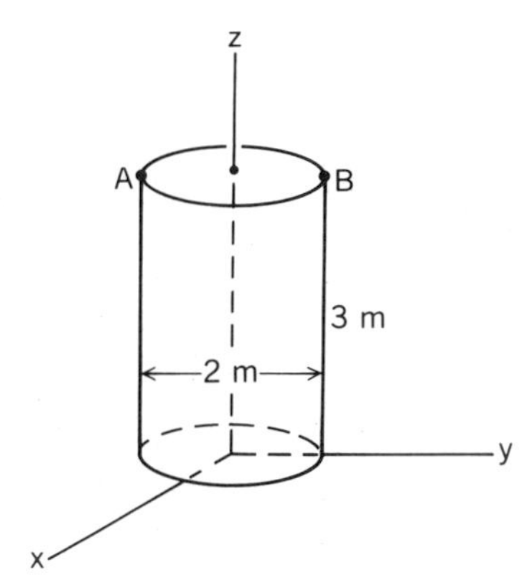

Figure P.9.25

9.26. A block having a uniform density of 5 grams/cm³ has a hole of diameter 40 mm cut out. What are the principal moments of inertia at point A at the centroid of the right face of the block?

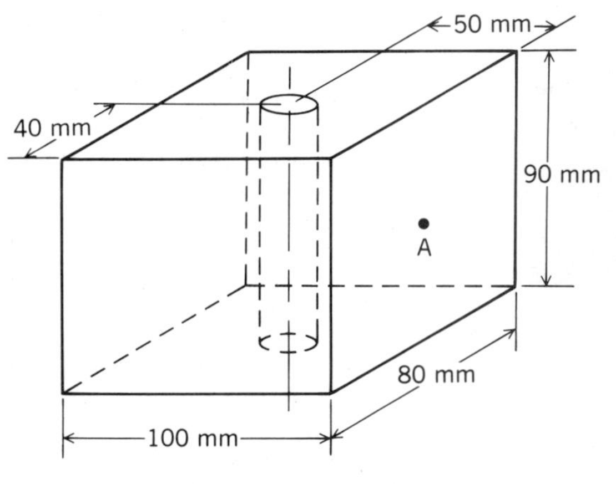

Figure P.9.26

9.27. Find maximum and minimum moment of inertia at point A. The block weighs 20 N and the cone weighs 14 N.

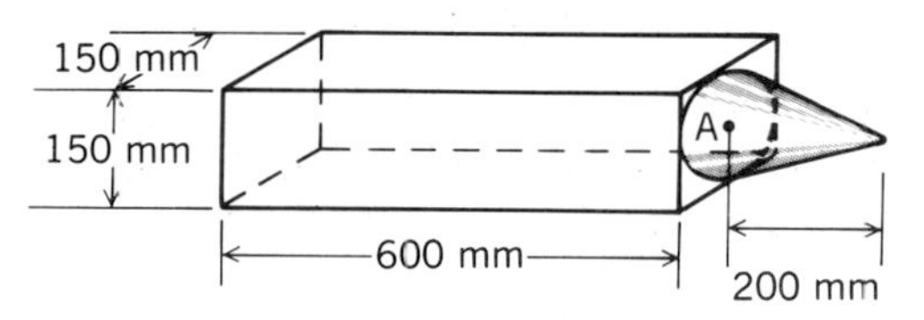

Figure P.9.27

9.28. Solid spheres C and D each weighing 25 N and having radius of 50 mm are attached to a thin solid rod weighing 30 N. Also, solid spheres E and G each weighing 20 N and having radii of 30 mm are attached to a thin rod weighing 20 N. The rods are attached to be orthogonal to each other. What are the principal moments of inertia at point A?

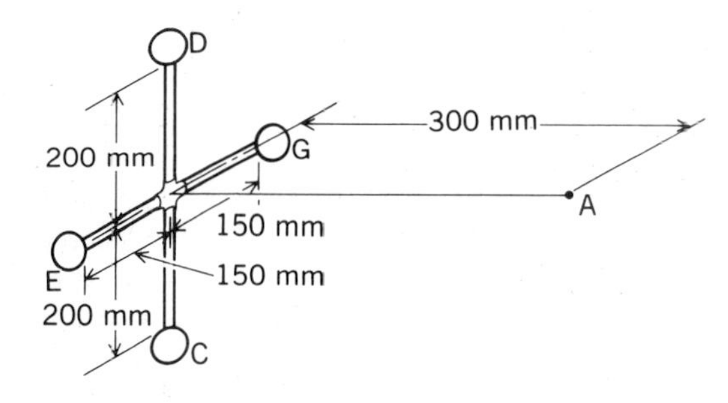

Figure P.9.28

9.29. A cylinder is shown having a conical cavity oriented parallel to the axis A–A and a cylindrical cavity oriented normal to A–A. If the density of the material is 7200 kg/m³, what is I_{AA}?

Figure P.9.29

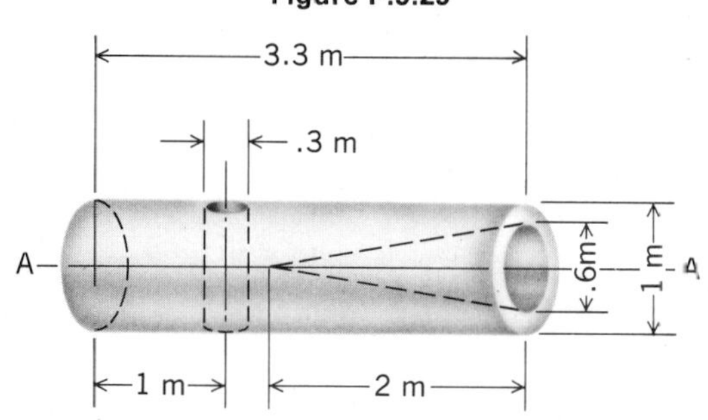

9.30. A flywheel is made of steel having a specific weight of 490 lb/ft³. What is the moment of inertia about its geometric axis? What is the radius of gyration?

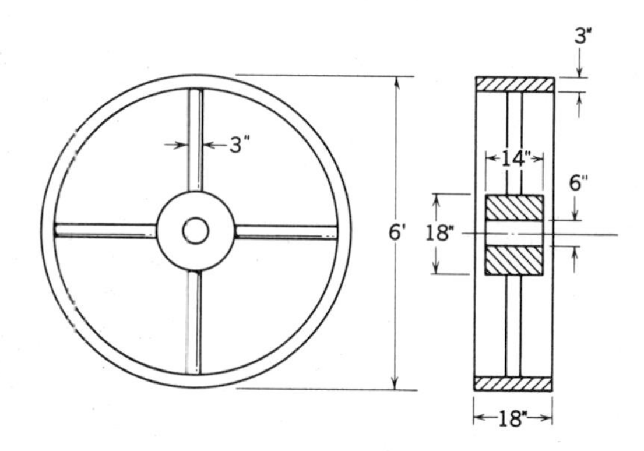

Figure P.9.30

9.31. Compute I_{yy} and I_{xy} for the right circular cylinder, which has a mass of 50 kg, and the square rod, which has a mass of 10 kg, when the two are joined together so that the rod is radial to the cylinder.

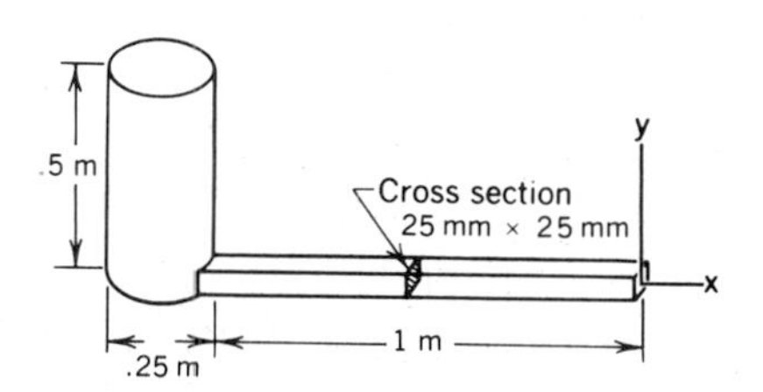

Figure P.9.31

9.32. Compute the moments and products of inertia for the xy axes. The specific weight is 490 lb/ft³ throughout.

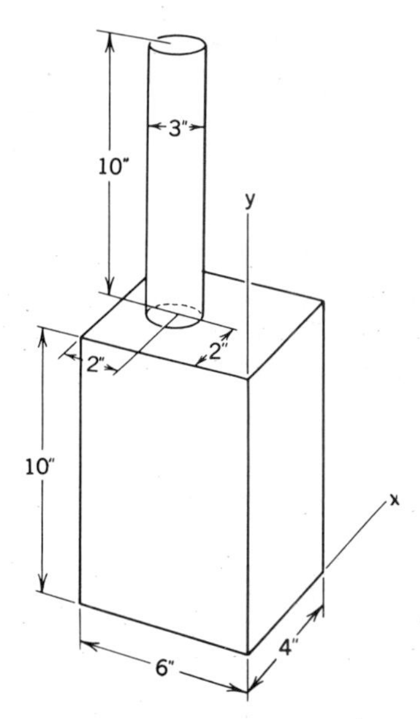

Figure P.9.32

9.33. A disc A is mounted on a shaft such that its normal is oriented 10° from the centerline of the shaft. The disc has a diameter of 2 ft, is 1 in. in thickness, and weighs 100 lb. Compute the moment of inertia of the disc about the centerline of the shaft.

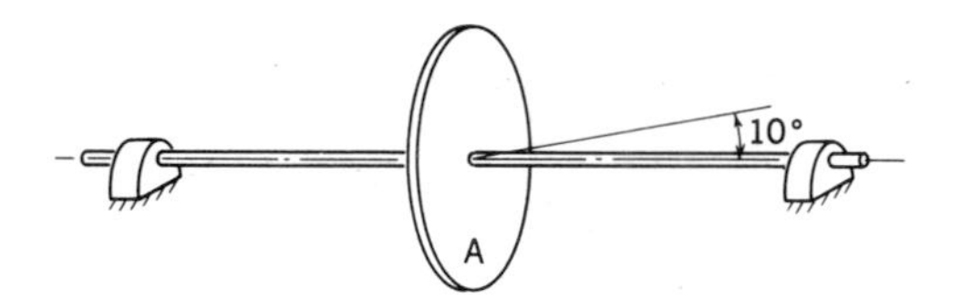

Figure P.9.33

9.34. A gear B having a mass of 25 kg rotates about axis C–C. If the rod A has a mass distribution of 7.5 kg/m, compute the moment of inertia of A and B about the axis C–C.

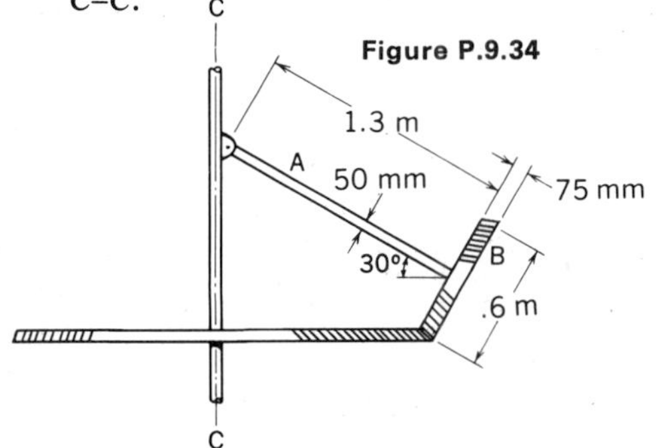

Figure P.9.34

9.35. A block weighing 100 N is shown. Compute the moment of inertia about the diagonal D–D.

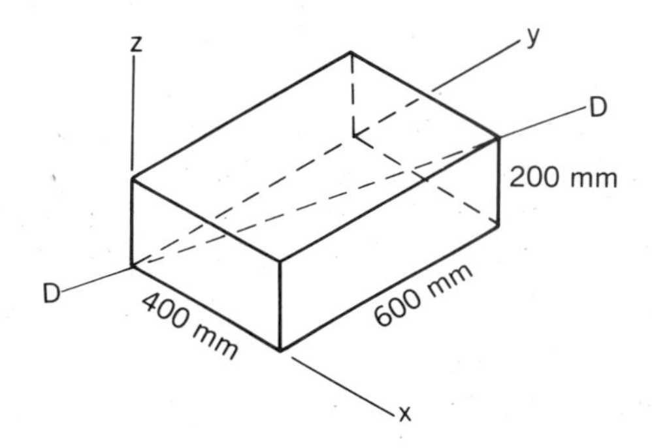

Figure P.9.35

9.36. A solid sphere A of diameter 1 ft and weight 100 lb is connected to the shaft B–B by a solid rod weighing 2 lb/ft and having a diameter of 1 in. Compute $I_{z'z'}$ for the rod and ball.

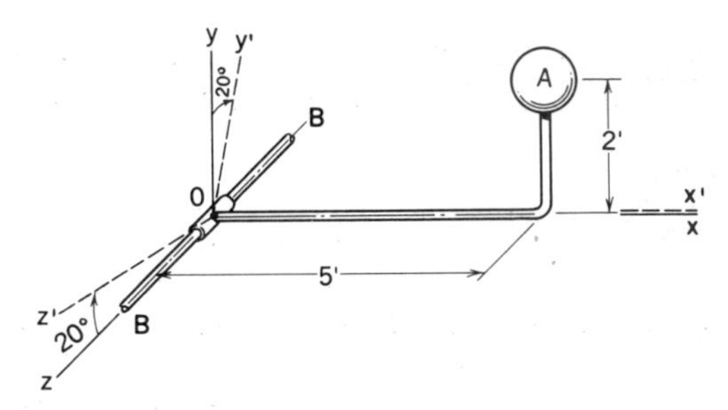

Figure P.9.36

9.37. In Problem 9.13, we found the following results for the thin plate:

$$I_{xx} = I_{yy} = .1019 \text{ grams-m}^2$$

$$I_{xy} = .0649 \text{ grams-m}^2$$

Find all components for the inertia tensor for reference $x'y'z'$. Axes $x'y'$ lie in mid-plane of the plate.

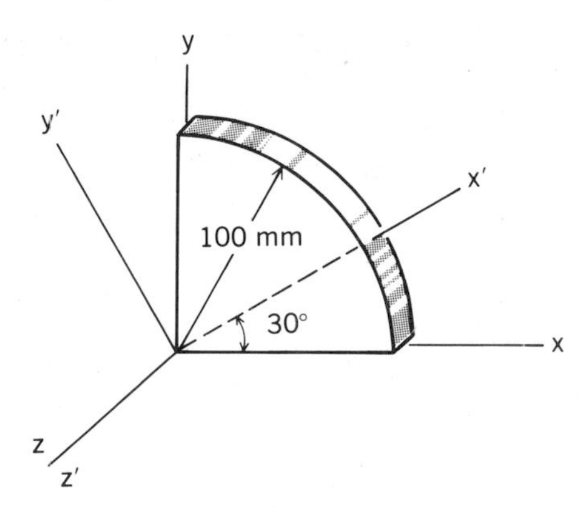

Figure P.9.37

9.38. A bent rod weighs .1 N/mm. What is I_{nn} for

$$\boldsymbol{\epsilon}_n = .30\boldsymbol{i} + .45\boldsymbol{j} + .841\boldsymbol{k}?$$

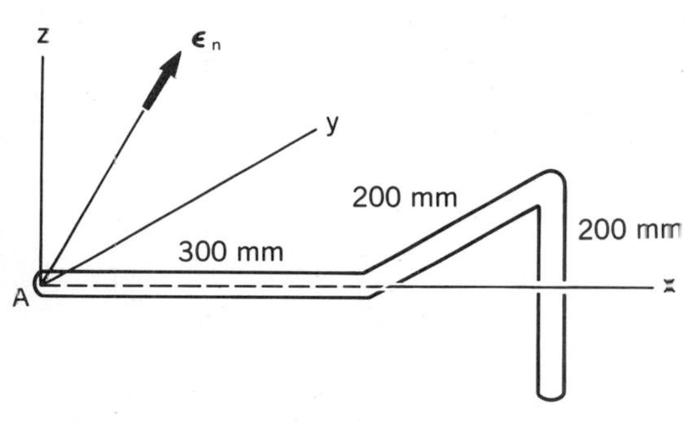

Figure P.9.38

9.39. Evaluate the matrix of direction cosines for the primed axes relative to the unprimed axes.

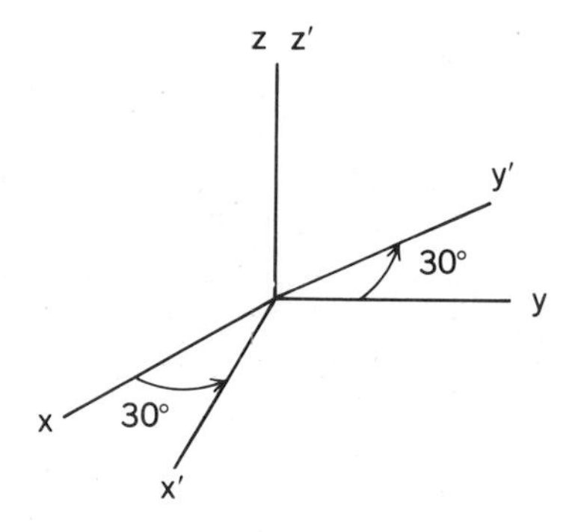

Figure P.9.39

$$a_{ij} = \begin{pmatrix} a_{x'x} & a_{x'y} & a_{x'z} \\ a_{y'x} & a_{y'y} & a_{y'z} \\ a_{z'x} & a_{z'y} & a_{z'z} \end{pmatrix}$$

9.40. The block is uniform in density and weighs 10 N. Find $I_{y'z'}$.

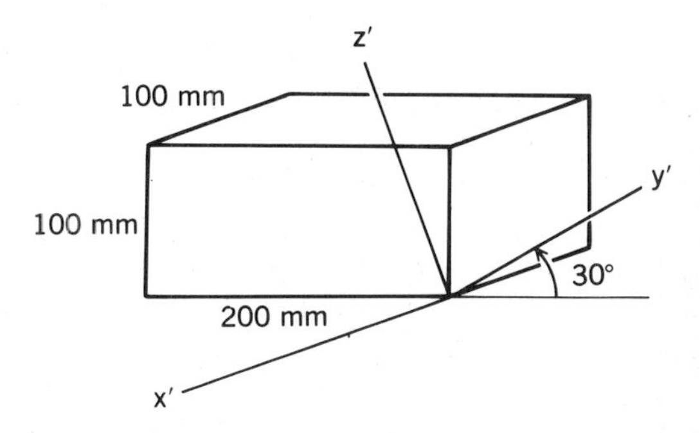

Figure P.9.40

***9.41.** A thin rod of length 300 mm and weight 12 N is oriented relative to $x'y'z'$ such that

$$\boldsymbol{\epsilon}_n = .4\boldsymbol{i}' + .3\boldsymbol{j}' + .866\boldsymbol{k}'$$

What is $I_{x'y'}$?

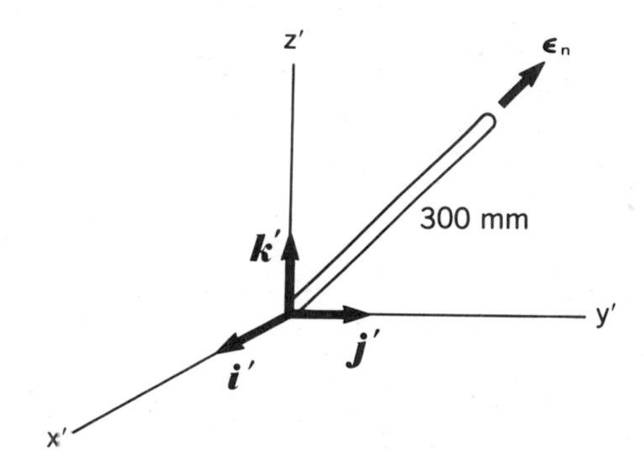

Figure P.9.41

***9.42.** Show that the transformation equation for the inertia tensor components at a point when there is a rotation of axes (i.e., Eqs. 9.14 and 9.20) can be given as follows:

$$I_{kq} = \sum_j \sum_i a_{ki} a_{qj} I_{ij}$$

where k can be x', y', or z' and q can be x', y', or z', and where i and j go from x to y to z. The equation above is a compact definition of *second-order tensors*. Remember that in the inertia tensor you must have a minus sign in front of each product of inertia term (i.e., $-I_{xy}$, $-I_{yz}$, etc.). (*Hint:* Let $i = x$; then sum over j; then let $i = y$ and sum again over j; etc.)

***9.43.** In Problem 9.42, express the transformation equation to get $I_{y'z'}$ in terms of the inertia tensor components for reference xyz having the same origin as $x'y'z'$.

*9.6 The Inertia Ellipsoid and Principal Moments of Inertia

Equation 9.14 gives the moment of inertia of a body about an axis k in terms of the direction cosines of that axis measured from an orthogonal reference with an origin O on the axis, and in terms of six independent inertia quantities for this reference. We wish to explore the nature of the variation of I_{kk} at a point O in space as the direction of k is changed. (The k axis and the body are shown in Fig. 9.19, which we shall call the physical diagram.) To do this, we will employ a geometric representation of moment of inertia at a point that is developed in the following manner. Along the axis k, we lay off as a distance the quantity OA given by the relation

$$OA = \frac{d}{\sqrt{I_{kk}/M}} \tag{9.24}$$

where d is any arbitrary constant that has a dimension of length that will render OA dimensionless, as the reader can verify. The term $\sqrt{I_{kk}/M}$ is the *radius of gyration* and was presented earlier. To avoid confusion, this operation is shown in another diagram, called the inertia diagram (Fig. 9.20), where the new ξ, η, and ζ axes are *parallel* to x, y, and z axes of the physical diagram. Considering all possible directions of k, we observe that some surface will be formed about the point O', and this surface is related to the shape of the body through Eq. 9.14. We can express the equation of this surface quite readily. Suppose that we call ξ, η, and ζ the coordinates of point A. Since $O'A$ is

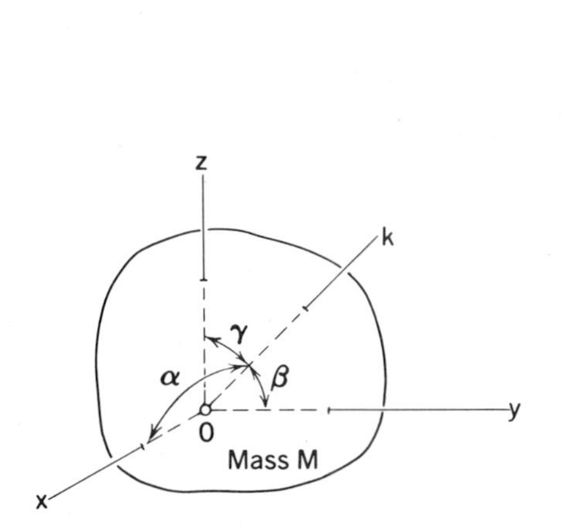

Figure 9.19. Physical diagram.

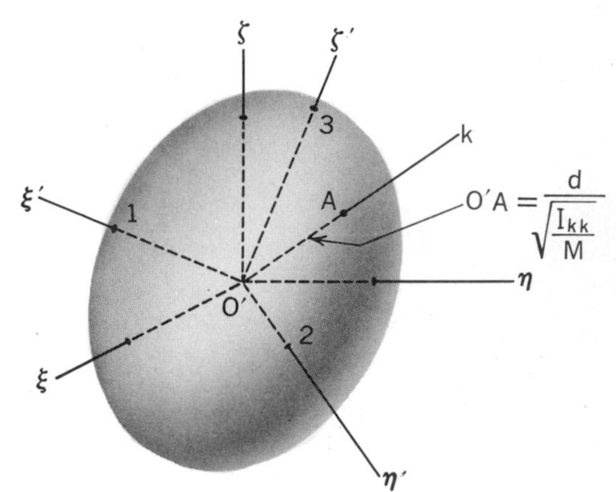

Figure 9.20. Inertia diagram.

parallel to the line k and thus has the direction cosines a_{kx}, a_{ky}, and a_{kz} that are associated with this line, we can say that

$$
\begin{aligned}
a_{kx} &= \frac{\xi}{O'A} = \frac{\xi}{d\sqrt{M/I_{kk}}} \\
a_{ky} &= \frac{\eta}{O'A} = \frac{\eta}{d\sqrt{M/I_{kk}}} \\
a_{kz} &= \frac{\zeta}{O'A} = \frac{\zeta}{d\sqrt{M/I_{kk}}}
\end{aligned}
\tag{9.25}
$$

Now replace the direction cosines in Eq. 9.13, using the relations above:

$$
\begin{aligned}
I_{kk} = &\frac{\xi^2}{Md^2/I_{kk}} I_{xx} + \frac{\eta^2}{Md^2/I_{kk}} I_{yy} + \frac{\zeta^2}{Md^2/I_{kk}} I_{zz} \\
&+ 2\frac{\xi\eta}{Md^2/I_{kk}}(-I_{xy}) + 2\frac{\xi\zeta}{Md^2/I_{kk}}(-I_{zz}) + 2\frac{\eta\zeta}{Md^2/I_{kk}}(-I_{yz})
\end{aligned}
\tag{9.26}
$$

We can see that I_{kk} cancels out of the preceding equation, leaving an equation involving the coordinates ξ, η, and ζ of the surface and the inertia terms of the body itself. Rearranging the terms, we then have

$$
\frac{\xi^2}{Md^2/I_{xx}} + \frac{\eta^2}{Md^2/I_{yy}} + \frac{\zeta^2}{Md^2/I_{zz}} + \frac{2\xi\eta}{Md^2}(-I_{xy}) + \frac{2\xi\zeta}{Md^2}(-I_{xz}) + \frac{2\eta\zeta}{Md^2}(-I_{yz}) = 1
\tag{9.27}
$$

Considering analytic geometry, we know that the surface is that of an ellipsoid (see Fig. 9.20), and is thus called the *ellipsoid of inertia*. The distance squared from O' to any point A on the ellipsoid is inversely proportional to the moment of inertia (see Eq. 9.24) about an axis in the body at O having the same direction as $O'A$. We can conclude that the inertia tensor for any point of a body can be represented geometrically by such a second-order surface, and this surface may be thought of as analogous to the arrow used to represent a vector graphically. The size, shape, and inclination of the ellipsoid will vary for each point in space for a given body. (Since all second-order tensors may be represented by second-order surfaces, you will, if you study elasticity, also encounter the ellipsoids of stress and strain.[9])

An ellipsoid has three orthogonal axes of symmetry, which have a common point at the center, O'. In the diagram, these axes are shown as $O'1$, $O'2$, and $O'3$. We pointed out that the shape and inclination of the ellipsoid of inertia depend on the mass distribution of the body about the *origin* of the xyz reference, and they have nothing to do with the choice of *the orientation of the xyz* (and hence the $\xi\eta\zeta$) reference at the point. We can therefore imagine that the xyz reference (and hence the $\xi\eta\zeta$ reference) can be chosen to have directions that coincide with the aforementioned symmetric axes, $O'1$,

[9]See I. H. Shames, *Mechanics of Deformable Solids*, Prentice-Hall, Inc., Englewood Cliffs, N.J., 1964, Chap. 2. Also, Krieger Publishing Co., N.Y., 1979.

$O'2$, and $O'3$. If we call such references $x'y'z'$ and $\xi'\eta'\zeta'$, respectively, we know from analytic geometry that Eq. 9.27 becomes

$$\frac{(\xi')^2}{Md^2/I_{x'x'}} + \frac{(\eta')^2}{Md^2/I_{y'y'}} + \frac{(\zeta')^2}{Md^2/I_{z'z'}} = 1 \tag{9.28}$$

where ξ', η', and ζ' are coordinates of the ellipsoidal surface relative to the new reference, and $I_{x'x'}$, $I_{y'y'}$, and $I_{z'z'}$ are mass moments of inertia of the body about the new axes. We can now draw several important conclusions from this geometrical construction and the accompanying equations. One of the symmetrical axes of the ellipsoid above is the longest distance from the origin to the surface of the ellipsoid, and another axis is the smallest distance from the origin to the ellipsoidal surface. Examining the definition in Eq. 9.24, we must conclude that the minimum moment of inertia for the point O must correspond to the axis having the maximum length, and the maximum moment of inertia must correspond to the axis having the minimum length. The third axis has an intermediate value that makes the sum of the moment of inertia terms equal to the sum of the moment of inertia terms for all orthogonal axes at point O, in accordance with Eq. 9.2. In addition, Eq. 9.28 leads us to conclude that $I_{x'y'} = I_{y'z'} = I_{x'z'} = 0$. That is, the products of inertia of the mass about these axes must be zero. Clearly, these axes are the *principal axes* of inertia at the point O.

Since the preceding operations could be carried out at any point in space for the body, we can conclude that:

> *At each point there is a set of principal axes having the extreme values of moments of inertia for that point and having zero products of inertia.*[10] *The orientation of these axes will vary continuously from point to point throughout space for the given body.*

All second-order tensor quantities have the properties discussed above for the inertia tensor. By transforming from the original reference to the principal reference, we change the inertia tensor representation from

$$\begin{pmatrix} I_{xx} & (-I_{xy}) & (-I_{xz}) \\ (-I_{yx}) & I_{yy} & (-I_{yz}) \\ (-I_{zx}) & (-I_{zy}) & I_{zz} \end{pmatrix} \quad \text{to} \quad \begin{pmatrix} I_{x'x'} & 0 & 0 \\ 0 & I_{y'y'} & 0 \\ 0 & 0 & I_{z'z'} \end{pmatrix} \tag{9.29}$$

In mathematical parlance, we have "diagonalized" the tensor by the preceding operations.

9.7 Closure

In this chapter, we first introduced the nine components comprising the inertia tensor. Next, we considered the case of the very thin flat plate in which the xy axes form the midplane of the plate. We found that the mass-inertia terms $(I_{xx})_M$, $(I_{yy})_M$, and $(I_{xy})_M$ for the plate are proportional respectively to $(I_{xx})_A$, $(I_{yy})_A$, and $(I_{xy})_A$, the second moments and product of area of the plate surface. As a result, we could set forth the

---

[10] A general procedure for computing principal moments of inertia is set forth in Appendix III.

concept of principal axes for the inertia tensor as an extension of the work in Chapter 8. Thus, we pointed out that for these axes the products of inertia will be zero. Furthermore, one principal axis corresponds to the maximum moment of inertia at the point while another of the principal axes corresponds to the minimum moment of inertia at the point. We pointed out that for bodies with two orthogonal planes of symmetry, the principal axes at any point on the line of intersection of the planes of symmetry must be along this line of intersection and normal to this line in the planes of symmetry.

Those readers who studied the starred sections from Section 9.5 onward will have found proofs of the extensions set forth earlier about principal axes from Chapter 8. Even more important is the disclosure that the inertia tensor components change their values when the axes are rotated at a point in exactly the same way as many other physical quantities having nine components. Such quantities are called second-order tensors. Because of the common transformation equation for such quantities, they have many important identical properties, such as principal axes. In your course in strength of materials you should learn that stress and strain are second-order tensors and hence have principal axes.[11] Additionally, you will find that a two-dimensional stress distribution called *plane stress* is related to the stress tensor exactly as the moments and products of area are related to the inertia tensor. The same situation exists with strain. Consequently, there are Mohr's-circle constructions for plane stress and the corresponding case for strain (plane strain). Thus, by taking the extra time to consider the mathematical considerations of Sections 9.5 and 9.6, you will find unity between Chapter 9 and some very important aspects of strength of materials to be studied later in your program.

In Chapter 10, we shall introduce another approach to studying equilibrium beyond what we have used thus far. This approach is valuable for certain important classes of statics problems and at the same time forms the groundwork for a number of advanced techniques that many students will study later in their programs.

Review Problems

9.44. Find I_{zz} for the body of revolution having uniform density of .2 N/mm³. The radial distances out from the z axis to the surface is given as

$$r^2 = -4\,z \text{ mm}^2$$

where z is in millimeters. (*Hint:* Make use of the formula for the moment of inertia about the axis of a disc, $\frac{1}{2}Mr^2$.)

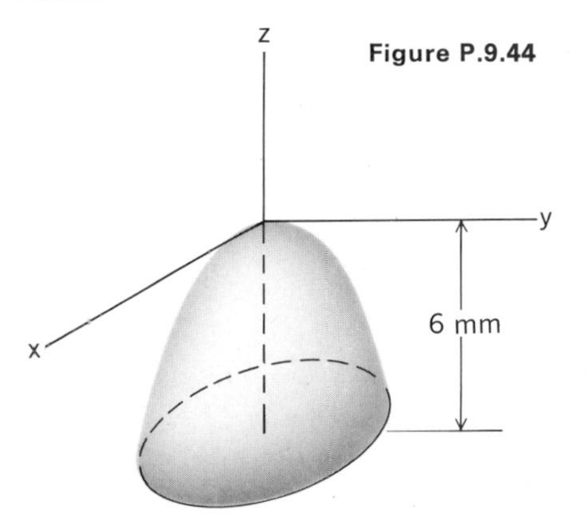

Figure P.9.44

[11] See I. H. Shames, *Introduction to Solid Mechanics*, Prentice-Hall, Inc., Englewood Cliffs, N.J., 1975.

9.45. In Problem 9.44, determine I_{zz} without using the disc formula but using multiple integration instead.

9.46. What are the inertia tensor components for the thin plate about axes xyz? The plate weighs 2 N.

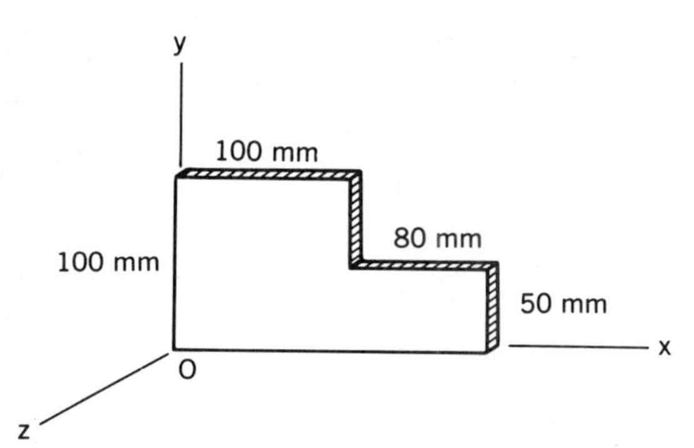

Figure P.9.46

9.47. In Problem 9.46, what are the principal axes and the principal moments of inertia for the intertia tensor at O?

9.48. What are the principal mass moments of inertia at point O? Block A weighs 15 N. Rod B weighs 6 N and solid sphere C weighs 10 N. The density in each body is uniform. The diameter of the sphere is 50 mm.

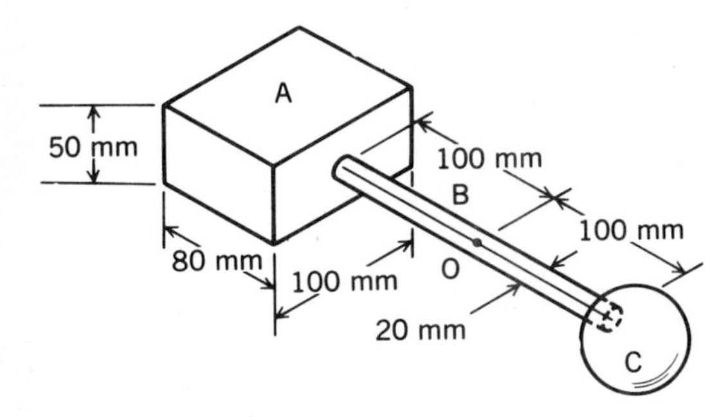

Figure P.9.48

9.49. The block has a density of 15 kg/m³. Find the moment of inertia about axis AB.

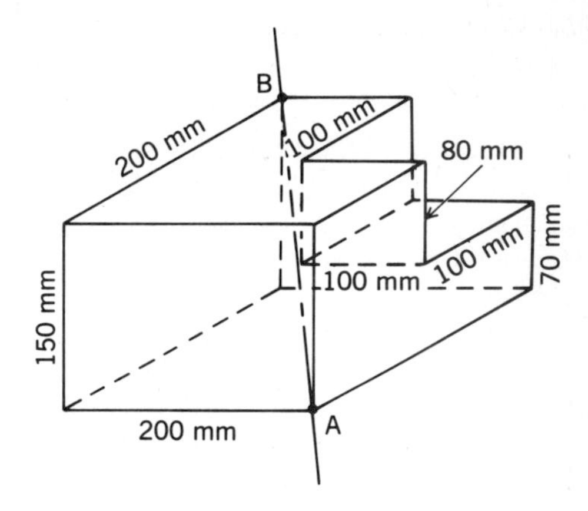

Figure P.9.49

9.50. A crate and its contents weighs 10 kN. The center of mass of the crate and its contents is at

$$\boldsymbol{r}_c = .40\boldsymbol{i} + .30\boldsymbol{j} + .60\boldsymbol{k} \text{ m}$$

If at A we know that

$$I_{yy} = 800 \text{ kg-m}^2$$
$$I_{yz} = 500 \text{ kg-m}^2$$

find I_{yy} and I_{yz} at B.

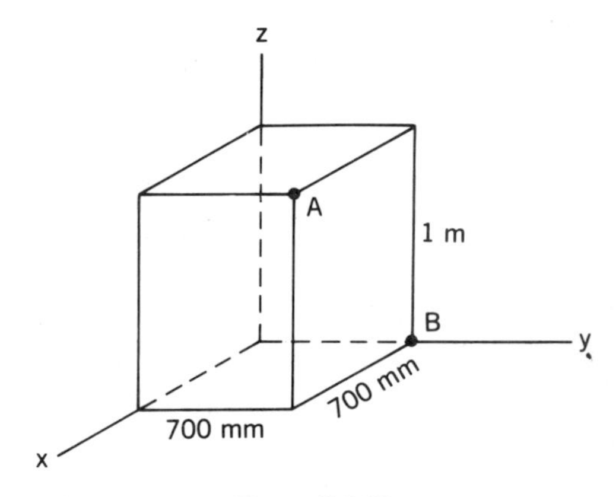

Figure P.9.50

9.51. A semicylinder weighs 50 N. What are the principal moments of inertia at O? What is the product of inertia $I_{y'x'}$? What conclusion can you draw about the direction of principal axes at O?

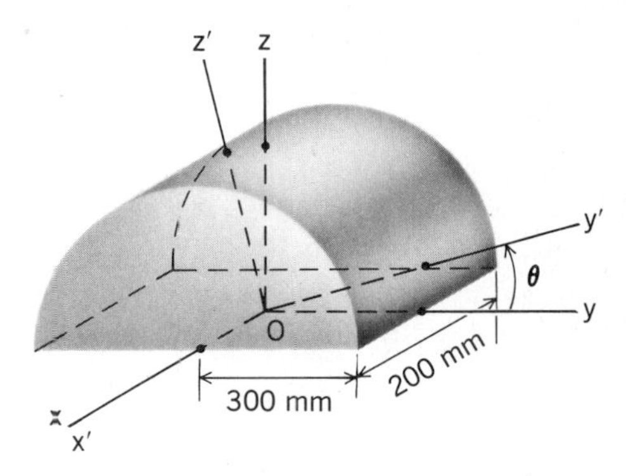

Figure P.9.51

10
*Methods of
Virtual Work
and Stationary
Potential Energy

10.1 Introduction

In the study of statics thus far, we have followed the procedure of isolating a body to expose certain unknown forces and then formulating either scalar or vector equations of equilibrium that include *all* the forces acting on the body. At this time, alternative methods of expressing conditions of equilibrium, called the *method of virtual work* and, allied to it, the *method of stationary potential energy*, will be presented. These methods will yield equilibrium equations equivalent to those of preceding sections. Furthermore, these new equations include only certain forces on a body, and accordingly in some problems will provide a more simple means of solving for desired unknowns.

Actually, we are making a very modest beginning into a vast field of endeavor called *variational mechanics* or *energy methods* with important applications to both rigid-body and deformable-body solid mechanics. Indeed, more advanced studies in these fields will surely center around these methods.[1]

A central concept for energy methods is the work of a force. A differential amount of work dW_k of a force $\boldsymbol{F}$ acting on a particle equals the component of this force in the direction of movement of the particle times the differential displacement of the particle:

$$dW_k = \boldsymbol{F} \cdot d\boldsymbol{r} \tag{10.1}$$

[1]For a treatment of energy methods for deformable solids, see C. Dym and I. H. Shames, *Solid Mechanics—A Variational Approach*, McGraw-Hill Book Company, New York, 1973.

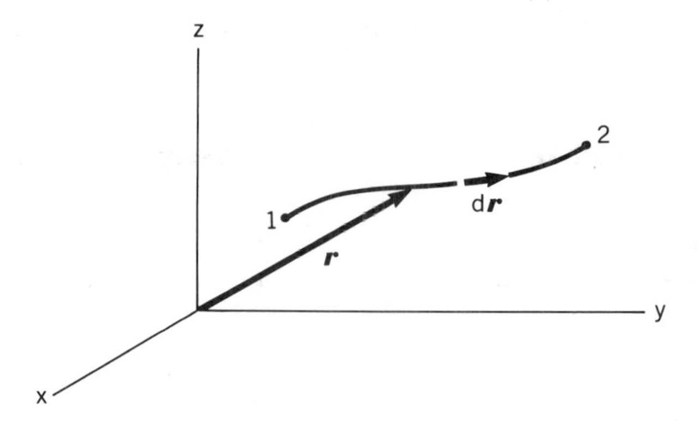

Figure 10.1. Path of particle on which F does work.

And the work W_k on a particle by force F when the particle moves along some path (see Fig. 10.1) from point 1 to point 2 is then

$$W_k = \int_{r_1}^{r_2} F \cdot dr \tag{10.2}$$

Note that the value and direction of F can vary along the path. This fact must be taken into account during the integration. We shall have more to say about the concept of work in later sections.[2]

Part A
Method of Virtual Work

10.2 Principle of Virtual Work for a Particle

For our introduction to the principle of virtual work, we will first consider a particle acted on by external loads $K_1, K_2, \ldots, K_n$, whose resultant force pushes the particle against a rigid constraining surface S in space (Fig. 10.2). This surface S is assumed to be frictionless and will thus exert a constraining force N on the particle which is normal to S. The forces K_i are called *active forces* in connection with the method of

---

[2]We could have defined work as

$$W_k = \int_{t_1}^{t_2} F \cdot V\, dt$$

where V is the velocity of the point of application of the force. When the force acts on a particular particle, the result above becomes $\int_{r_1}^{r_2} F \cdot dr$, where r is the position vector of the particle. There are times when the force acts on continually *changing* particles as time passes. The more general formulation above can then be used effectively.

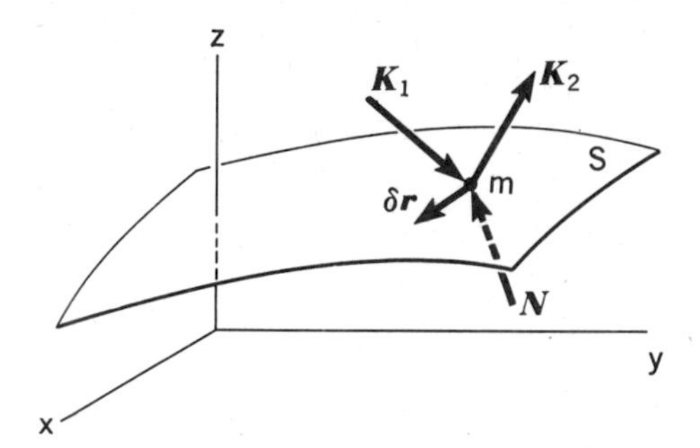

Figure 10.2. Particle on a frictionless surface.

virtual work, while $\boldsymbol{N}$ retains the identification of a *constraining force* as used previously. Employing the re ultant active force $\boldsymbol{K}_R$, we can give the necessary and sufficient[3] conditions for eq ıilibrium for the particle as

$$\boldsymbol{K}_R + \boldsymbol{N} = \boldsymbol{0} \tag{10.3}$$

We shall now prove that we can express the necessary and sufficient conditions of equilibrium in yet another way. Let us imagine that we give the particle an infinitesimal hypothetical displacement that is consistent with the constraints (i.e., along the surface), while keeping the forces $\boldsymbol{K}_R$ and $\boldsymbol{N}$ constant. Such a displacement is termed a *virtual displacement*, and will be denoted by $\delta\boldsymbol{r}$, in contrast to a real infinitesimal displacement, $d\boldsymbol{r}$, which might actually occur during a time interval dt. We can then take the dot product of the vector $\delta\boldsymbol{r}$ with the force vectors in the equation above:

$$\boldsymbol{K}_R \cdot \delta\boldsymbol{r} + \boldsymbol{N} \cdot \delta\boldsymbol{r} = 0 \tag{10.4}$$

Since $\boldsymbol{N}$ is normal to the surface and $\delta\boldsymbol{r}$ is tangential to the surface, the corresponding scalar product must be zero, leaving

$$\boldsymbol{K}_R \cdot \delta\boldsymbol{r} = 0 \tag{10.5}$$

The expression $\boldsymbol{K}_R \cdot \delta\boldsymbol{r}$ is called the *virtual work* of the system of forces and is denoted as δW. Thus, the virtual work by the active forces on a particle with frictionless constraints is *necessarily* zero for a particle in equilibrium for any virtual displacement consistent with the frictionless constraints.

We shall now show that this statement is also *sufficient* to ensure equilibrium for the case of a particle initially at rest (relative to an inertial reference) at the time of application of the active loads. To demonstrate this, *assume that Eq. 10.5 holds but that the particle is not in equilibrium.* If the particle is not in equilibrium, it must move in a direction that corresponds to the direction of the resultant of all forces acting on the particle. Consider that $d\boldsymbol{r}$ represents the initial displacement during the time interval dt. The work done by the forces must exceed zero for this movement. Since the normal force $\boldsymbol{N}$ cannot do work for this displacement,

$$\boldsymbol{K}_R \cdot d\boldsymbol{r} > 0 \tag{10.6}$$

[3]The sufficiency condition applies to an initially stationary particle.

However, we can choose a *virtual* displacement $\delta\mathbf{r}$ to be used in Eq. 10.5 that is *exactly* equal to the proposed $d\mathbf{r}$ stated above, and so we see that, by admitting nonequilibrium, we arrive at a result (10.6) that is in *contradiction to the starting known condition* (Eq. 10.5). We can then conclude that the conjecture that the particle is not in equilibrium is false. Thus, Eq. 10.3 is not only a necessary condition of equilibrium, but, for an initially stationary particle, is in itself sufficient for equilibrium. Thus, Eq. 10.5 is completely equivalent to the equation of equilibrium, 10.3.

We can now state the principle of virtual work for a particle.

> *The necessary and sufficient condition for equilibrium of an initially stationary particle with frictionless constraints requires that the virtual work for all virtual displacements consistent with the constraints be zero.*[4]

The case of a particle that is not constrained is a special case of the situation discussed above. Here $\mathbf{N} = \mathbf{0}$, so that Eq. 10.5 is applicable for *all* infinitesimal displacements as a criterion for equilibrium.

10.3 Principle of Virtual Work for Rigid Bodies

We now examine a rigid body in equilibrium acted on by active forces $\mathbf{K}_i$ and constrained without the aid of friction (Fig. 10.3). The constraining forces $\mathbf{N}_i$ arise from direct contact with other immovable bodies (in which case the constraining forces are oriented normal to the contact surface) or from contact with immovable bodies through pin and ball-joint connections. We shall consider the body to be made up of elementary particles for the purposes of discussion.

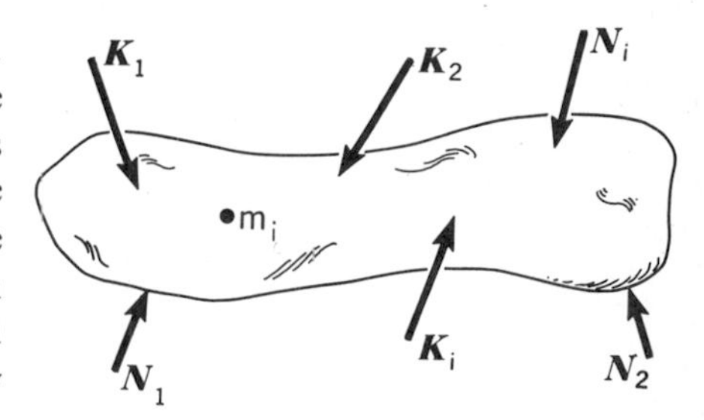

Figure 10.3. Rigid body with active forces and ideal constraining forces.

Now consider such a particle of mass m_i. Active loads, external constraining forces, and forces from other particles may possibly be acting on the particle. The forces from other particles are internal forces $\mathbf{S}_i$ which

---

[4]This test breaks down for a particle that is moving. Consider a particle constrained to move in a circular path in a horizontal plane, as shown in the diagram. The particle is moving with constant speed. There are no active forces, and we consider the constraints as frictionless. The work for a vir-

tual displacement consistent with the constraints at any time t gives us a zero result. Nevertheless, the particle is not in equilibrium, since clearly there is at time t an acceleration toward the center of curvature. Thus, we had to restrict the sufficiency condition to particles that are initially stationary.

maintain the rigidity of the body. Using the resultants of these various forces on the particle, we may state from Newton's law that the necessary and sufficient[5] condition for equilibrium of the ith particle is

$$(\boldsymbol{K}_R)_i + (\boldsymbol{N}_R)_i + (\boldsymbol{S}_R)_i = \boldsymbol{0} \tag{10.7}$$

Now, we give the particle a virtual displacement $\delta \boldsymbol{r}_i$ that is consistent with the exterior constraints and with the condition that the body is rigid. Taking the dot product of the vectors in the equation above with $\delta \boldsymbol{r}_i$, we get

$$(\boldsymbol{K}_R)_i \cdot \delta \boldsymbol{r}_i + (\boldsymbol{N}_R)_i \cdot \delta \boldsymbol{r}_i + (\boldsymbol{S}_R)_i \cdot \delta \boldsymbol{r}_i = 0 \tag{10.8}$$

Clearly, $(\boldsymbol{N}_R)_i \cdot \delta \boldsymbol{r}_i$ must be zero, because $\delta \boldsymbol{r}_i$ is normal to $\boldsymbol{N}_i$ for constraint stemming from direct contact with immovable bodies or because $\delta \boldsymbol{r}_i = \boldsymbol{0}$ for constraint stemming from pin and ball-joint connections with immovable bodies. Let us sum the equations of the form 10.8 for all the particles that are considered to make up the body. We have, for n particles,

$$\sum_{i=1}^{n} (\boldsymbol{K}_R)_i \cdot \delta \boldsymbol{r}_i + \sum_{i=1}^{n} (\boldsymbol{S}_R)_i \cdot \delta \boldsymbol{r}_i = 0 \tag{10.9}$$

Let us now consider in more detail the internal forces in order to show that the second quantity on the left-hand side of the equation above is zero. The force on m_i from particle m_j will be equal and opposite to the force on particle m_j from particle m_i, according to Newton's third law. The internal forces on these particles are shown as $\boldsymbol{S}_{ij}$ and $\boldsymbol{S}_{ji}$ in Fig. 10.4. The first subscript identifies the particle on which a force acts, while the second subscript identifies the particle exerting this force. We can then say that

$$\boldsymbol{S}_{ij} = -\boldsymbol{S}_{ji} \tag{10.10}$$

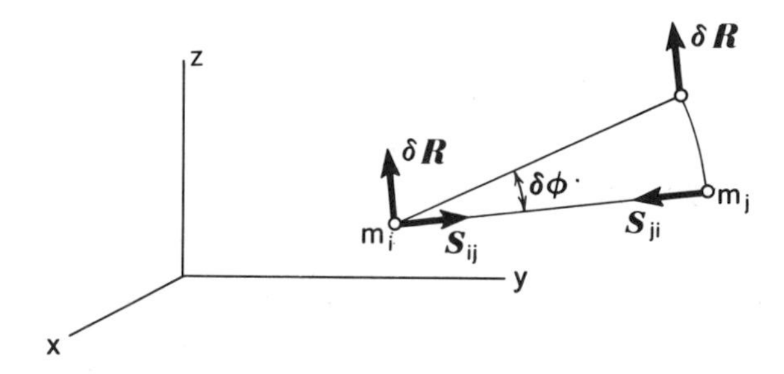

Figure 10.4. Two particles of a body undergoing displacement δR and rotation $\delta \phi$.

Any virtual motion we give to any pair of particles must maintain a constant distance between the particles. This requirement stems from the rigid body condition and will be true if:

1. Both particles are given the same displacement $\delta \boldsymbol{R}$.

---

[5]The sufficiency requirement again applies to an initially stationary particle.

2. The particles are rotated $\delta\phi$ relative to each other.[6]

We now consider the general case where both motions are present: that is, both m_i and m_j are given a virtual displacement $\delta\boldsymbol{R}$, and furthermore, m_j is rotated through some angle $\delta\phi$ about m_i (Fig. 10.4). The work done in the rotation must be zero, since $\boldsymbol{S}_{ji}$ is at right angles to the motion of the mass m_j. Also, the work done on each particle during the equal displacement of both masses must be equal and opposite since the forces move through equal displacements and are themselves equal and opposite. The mutual effect of all particles of the body is of the type described. Thus, we can conclude that the internal work done for a rigid body during a virtual displacement is zero. Hence, a *necessary* condition for equilibrium is

$$\sum_{i=1}^{n} (\boldsymbol{K}_R)_i \cdot \delta\boldsymbol{r}_i = \delta W = 0 \tag{10.11}$$

Thus, the virtual work done by active forces on a rigid body having frictionless constraints during virtual displacements consistent with the constraints is zero if the body is in equilibrium.

We can readily prove that Eq. 10.11 is a *sufficient* condition for equilibrium of an initially stationary body by reasoning in the same manner that we did in the case of the single particle. We shall *state first that Eq. 10.11 is valid for a body*. If the body is not in equilibrium, it must begin to move. Let us say that each particle m_i moves a distance $d\boldsymbol{r}_i$ consistent with the constraints under the action of the forces. The work done on particle m_i is

$$(\boldsymbol{K}_R)_i \cdot d\boldsymbol{r}_i + (\boldsymbol{N}_R)_i \cdot d\boldsymbol{r}_i + (\boldsymbol{S}_R)_i \cdot d\boldsymbol{r}_i > 0 \tag{10.12}$$

But $(\boldsymbol{N}_R)_i \cdot d\boldsymbol{r}_i$ is necessarily zero because of the nature of the constraints. When we sum the terms in the equation above for all particles, $\sum_i (\boldsymbol{S}_R)_i \cdot d\boldsymbol{r}$ must also be zero because of the condition of rigidity of the body. Therefore, we may state that the supposition of no equilibrium leads to the following inequality:

$$\sum_{i=1}^{n} (\boldsymbol{K}_R)_i \cdot d\boldsymbol{r}_i > 0 \tag{10.13}$$

But we can conceive a virtual displacement $\delta\boldsymbol{r}_i$ *equal* to $d\boldsymbol{r}_i$ for each particle to be used in Eq. 10.11, thus bringing us to a contradiction between this equation and Eq. 10.13. Since we have taken Eq. 10.11 to apply, we conclude that the supposition of non-equilibrium which led to Eq. 10.13 must be invalid, and so the body must be in equilibrium. This logic proves the sufficiency condition for the principle of virtual work in the case of a rigid body with ideal constraints that is initially stationary at the time of application of the active forces.

Consider now *several* movable rigid bodies that are interconnected by smooth pins and ball joints or that are in direct frictionless contact with each other (Fig. 10.5). Some of these bodies are also ideally constrained by immovable rigid bodies in the manner described above. Again, we may examine the system of particles m_i making

---

[6] The virtual displacements $\delta\boldsymbol{r}_i$ of each of the two particles must then be the result of the superposition of $\delta\boldsymbol{R}$ and $\delta\phi$.

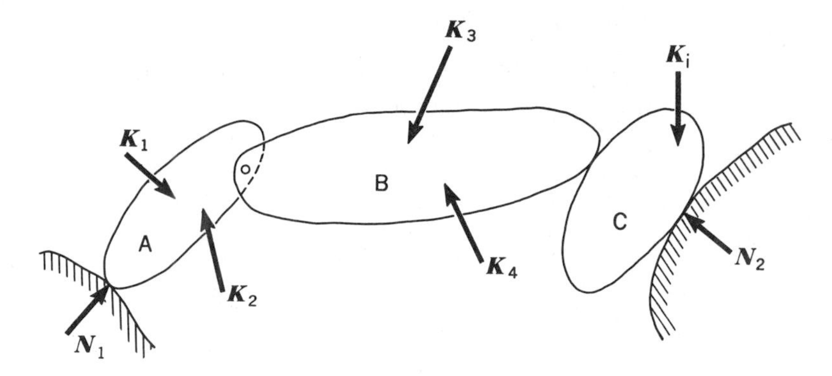

Figure 10.5. System of ideally constrained rigid bodies acted on by forces $\boldsymbol{K}_i$.

up the various rigid bodies. The only new kind of force to be considered is a force at the connecting point between bodies. The force on one such particle on body A will be equal and opposite to the force on the corresponding particle in body B at the contact point; and so on. Since such pairs of contiguous particles have the same virtual displacement, clearly the virtual work at all connecting points between bodies is zero for any virtual displacement of the system consistent with the constraints. Hence, we can say *for a system of initially stationary rigid bodies, the necessary and sufficient condition of equilibrium is that the virtual work of the active forces be zero for all possible virtual displacements consistent with the constraints.* We may then use the following equation instead of equilibrium:

$$\sum_i (\boldsymbol{K}_R)_i \cdot \delta \boldsymbol{r}_i = \delta W = 0 \tag{10.14}$$

where $(\boldsymbol{K}_R)_i$ are the active forces on the system of rigid bodies and $\delta \boldsymbol{r}_i$ are the movements of the application of these forces during a virtual displacement of the system consistent with the constraints.

10.4 Degrees of Freedom and the Solution of Problems

We have developed equations sufficient for equilibrium of initially stationary systems of bodies by using the concept of virtual work for virtual displacements consistent with the constraints. These equations do not involve reactions or connecting forces, and when these forces are not of interest, the method is quite useful. Thus, we may solve for as many unknown *active* forces as there are *independent* equations stemming from virtual displacements. Then our prime interest is to know how many independent equations can be written for a system stemming from virtual displacements.

For this purpose, we define *the number of degrees of freedom of a system as the number of generalized coordinates*[7] *which is required to fully specify the configuration of the system*. Thus, for the pendulum in Fig. 10.6, which is restricted to move in a plane, one *independent* coordinate θ locates the pendulum. Hence, this system has but one degree of freedom. We may ask: Can't we specify x and y of the bob, and thus aren't there two degrees of freedom? The answer is no, because when we specify x or y, the other coordinate is *determined* since the pendulum support, being inextensible, must sweep out a known circle as shown in the diagram. In Fig. 10.7, the piston and crank arrangement, the four-bar linkage,[8] and the balance require only one coordinate and thus have but one degree of freedom. On the other hand, the double pendulum has two degrees of freedom and a particle in space has three degrees of freedom. The number of degrees of freedom may usually be readily determined by inspection.

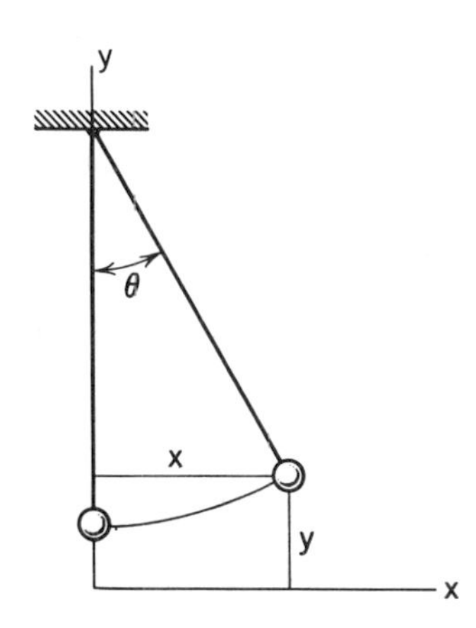

Figure 10.6. Plane pendulum.

Since each degree of freedom represents an independent coordinate, we can, for an n-degree-of-freedom system, institute n unique virtual displacements by varying each coordinate separately. This procedure will then give n independent equations of equilibrium from which n unknowns related to the active forces can be solved. We shall examine several problems to illustrate the method of virtual work and its advantages.

Before considering the examples, we wish to point out that a torque $\boldsymbol{M}$ undergoing a virtual displacement $\delta\boldsymbol{\phi}$ does an amount of virtual work δW equal

$$\delta W = \boldsymbol{M} \cdot \delta\boldsymbol{\phi} \tag{10.15}$$

The proof of this is asked for in Problem 10.30.

EXAMPLE 10.1

In Fig. 10.8 is a device for compressing metal scrap, namely a compactor. A horizontal force $\boldsymbol{P}$ is exerted on joint B. The piston at C then compresses the scrap material. For a given force $\boldsymbol{P}$ and a given angle θ, what is the force F developed on the scrap by the piston C? Neglect the friction between the piston and the cylinder wall, and consider the pin joints to be ideal.

We see by inspection that one coordinate θ describes the configuration of the system. The device therefore has one degree of freedom. We shall neglect the weight

[7] *Generalized coordinates* are any set of *independent* numbers that can fully specify the configuration of a system. Generalized coordinates can include any of usual coordinates, such as Cartesian coordinates or cylindrical coordinates, but need not. We shall only consider these cases where the usual coordinates serve as the generalized coordinates.

[8] The fourth bar is the base.

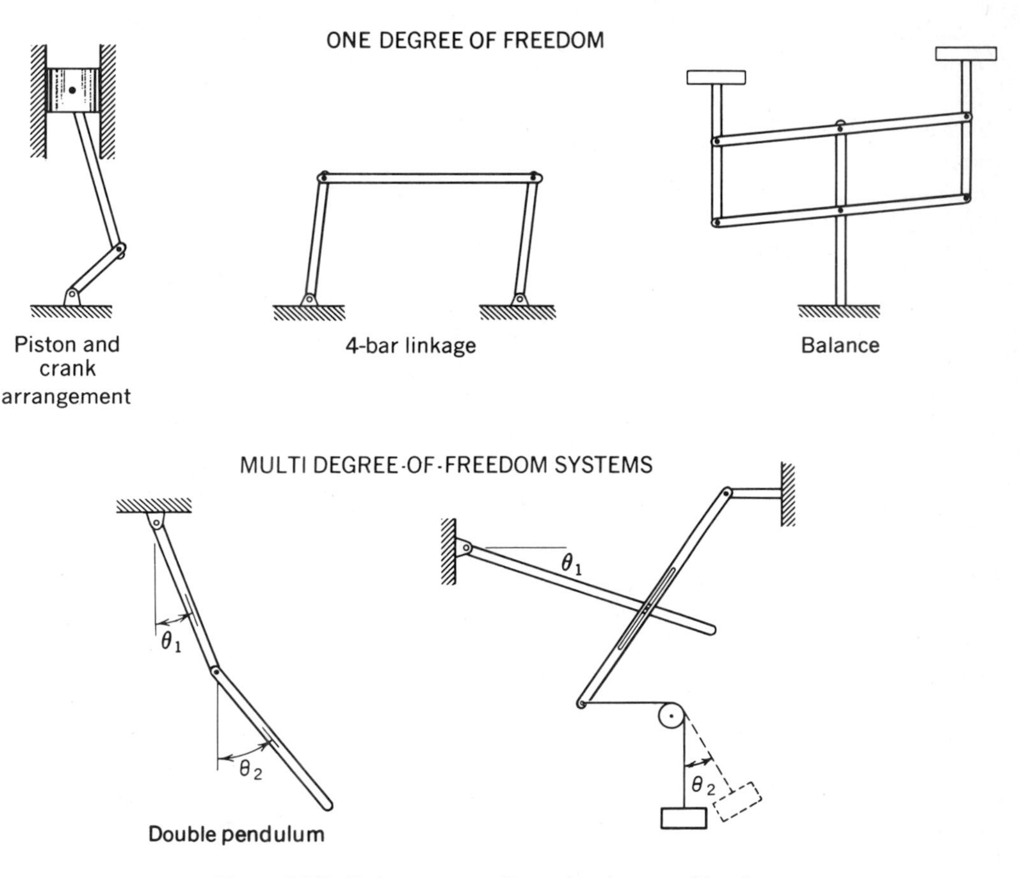

Figure 10.7. Various systems illustrating degrees of freedom.

of the members, and so only two active forces are present, P and F. By assuming a virtual displacement $\delta\theta$, we will involve in the principle of virtual work only those quantities that are of interest to us, P, F, and θ.[9] Let us then compute the virtual work of the active forces.

Force *P*. The virtual displacement $\delta\theta$ is such that force P has a motion in the horizontal direction of $(l\,\delta\theta\cos\theta)$ as can readily be deduced from Fig. 10.9 by elementary trigonometric considerations. There is yet another way of deducing this horizontal motion, which, sometimes, is more desirable. Using an xy coordinate system at A as shown in Figs. 10.8 and 10.9, we can say for joint B:

$$y_B = l\sin\theta \tag{a}$$

Now take the differential of both sides of the equation to get

$$dy_B = l\cos\theta\,d\theta \tag{b}$$

[9]If we had used a free-body approach, we would have had to bring in force components at A and at C, and we would have had to dismember the system. To appreciate the method of virtual work even for this simple problem, we urge you to at least set up the problem by the use of free-body diagrams.

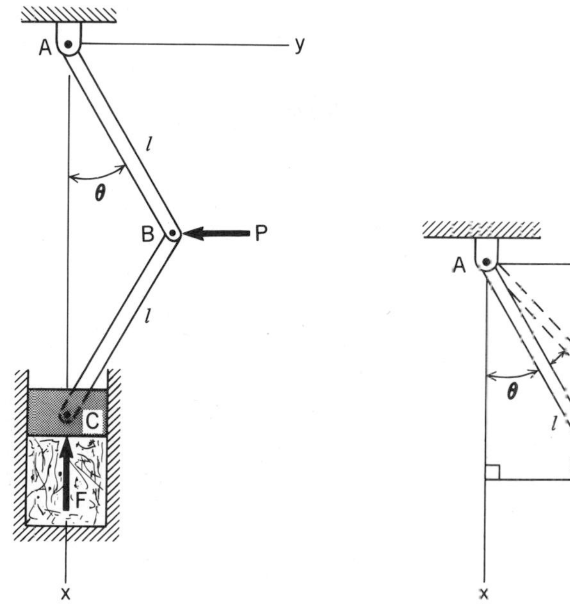

Figure 10.8. Compacting device.

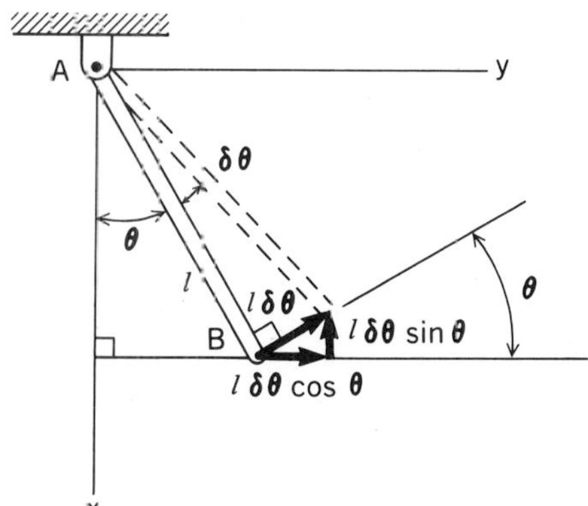

Figure 10.9. Virtual movement of leg *AB*.

A differential of a quantity A, namely dA, is very similar to a variation of the quantity, δA. The former might actually take place in a process; the latter takes place in the mind of the engineer. Nevertheless, the relation between differential quantities should be the same as the relation between varied quantities. Accordingly, from Eq. (b), we can say:

$$\delta y_B = l \cos\theta \, \delta\theta \tag{c}$$

Note that the same horizontal movement of B for $\delta\theta$ is thus computed as at the outset using trigonometry.

For the variation $\delta\theta$ chosen, the force P has opposite direction as δy_B, and so this virtual-work contribution is negative. Thus, we have

$$\delta W_P = -Pl \cos\theta \, \delta\theta \tag{d}$$

Force *F*. We can use the differential approach to get the virtual displacement of piston C. That is,

$$x_c = l \cos\theta + l \cos\theta = 2l \cos\theta$$

$$dx_c = -2l \sin\theta \, d\theta$$

Therefore,

$$\delta x_c = -2l \sin\theta \, \delta\theta \tag{e}$$

Since the force F is in the same direction as δx_c, we should have a positive result for work. Accordingly, we have

$$\delta W_c = F(2l \sin\theta) \, \delta\theta \tag{f}$$

We may now employ the principle of virtual work which recall is sufficient here for ensuring equilibrium. Thus, we can say that

$$-Pl\cos\theta\,\delta\theta + F(2l\sin\theta)\,\delta\theta = 0 \tag{g}$$

Canceling $l\,\delta\theta$ and solving for F, we get

$$F = \frac{P}{2\tan\theta} \tag{h}$$

For any given values of P and θ, we now know the amount of compressive force that the compactor can develop.

In the next example, we consider a case where we cannot conveniently use the differential approach in arriving at the virtual displacements.

EXAMPLE 10.2

A tractor with a bulldozer (Fig. 10.10) is used to push at constant speed an earthmover. If the tractor force on the earthmover is 150 kN, what is the hydraulic ram force F at joint B?

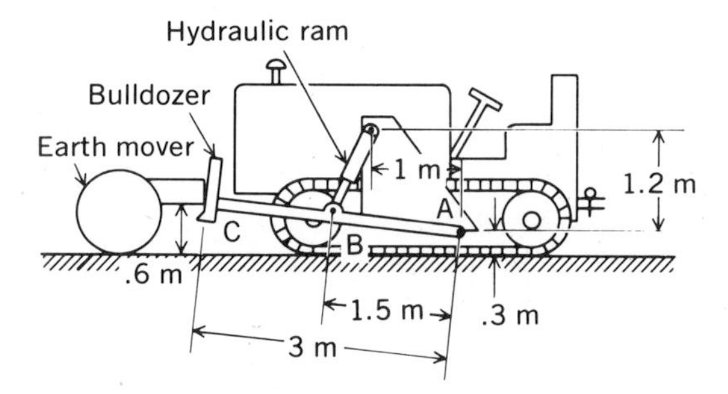

Figure 10.10. Tractor with bulldozer.

Member ABC (the bulldozer) has a single degree of freedom relative to the tractor frame. We show this member diagrammatically in Fig. 10.11, with independent coordinate θ and active forces F and 150 kN. The 1.492 m dimension in Fig.

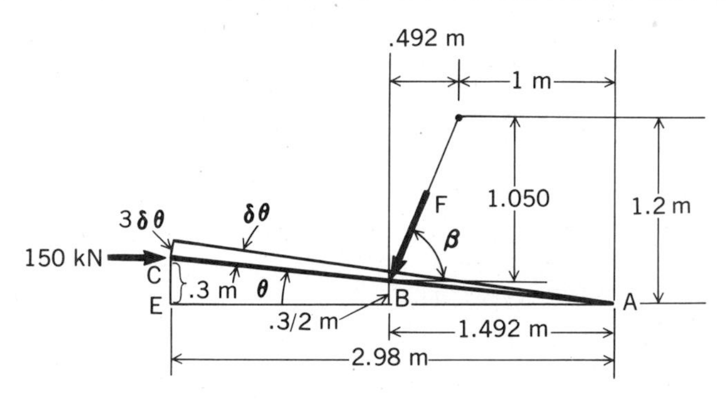

Figure 10.11. Bulldozer portion of device.

10.11 is determined using the Pythagorean theorem. That is, first noting in Fig. 10.10 that $\overline{AB} = 1.5$ m, we have on observing Fig. 10.11:

$$\left[\overline{AB}^2 - \left(\frac{.6 - .3}{2}\right)^2\right]^{\frac{1}{2}} = (1.5^2 - .15^2)^{\frac{1}{2}} = 1.492 \text{ m}$$

The .492 m dimension in the upper part of Fig. 10.11 follows directly. Furthermore, $\overline{AE} = 2(1.492) = 2.98$ m. The angle θ is next easily determined.

$$\theta = \tan^{-1}\frac{.3}{2.98} = 5.74°$$

The 1.050 m dimension can next be determined. Finally the angle β is computed as

$$\beta = \tan^{-1}\frac{1.050}{.492} = 64.9°$$

In computing the virtual work of the ram force F, it is simplest not to use the differential approach that worked so nicely in the preceding example, but instead to use simple geometry and trigonometry. The reason for the geometrical approach here is due to the fact that F (see Fig. 10.12) is not in a convenient coordinate direction in relation to θ as was the case for the active forces in the earlier cases. We can now use the method of virtual work as follows.

$$\delta W = (150)[(3)(\delta\theta) \sin \theta] - F[1.5\delta\theta \cos (90 - \theta - \beta)] = 0$$
$$(150)(3)(\delta\theta) \sin (5.74°) - F[1.5\delta\theta \cos (90 - 5.74 - 64.9)] = 0$$

Therefore,

$$F = 31.8 \text{ kN}$$

The decision as to using differentials or as to using simple geometry and trigonometry depends on the problem.

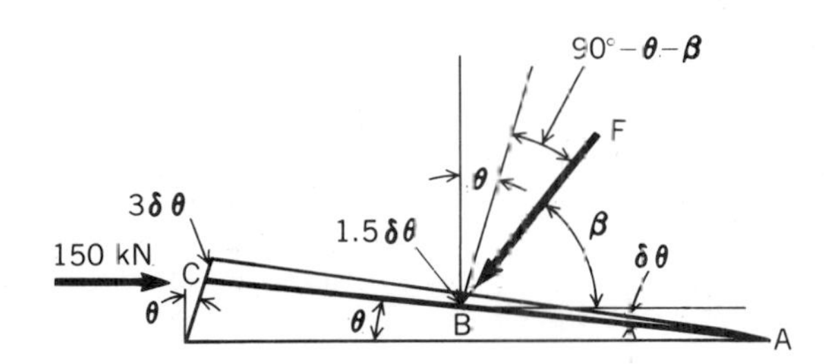

Figure 10.12. Give AC virtual displacement $\delta\theta$.

*EXAMPLE 10.3

A hydraulic-lift platform for loading trucks in shown in Fig. 10.13(a). Only one side of the system is shown; the other side is identical. If the diameter of the piston in the cylinder is 4 in., what pressure p is needed to support a load W of 5000 lb when $\theta = 60°$? The following additional data apply:

$$l = 24 \text{ in.}$$
$$d = 60 \text{ in.}$$
$$e = 10 \text{ in.}$$

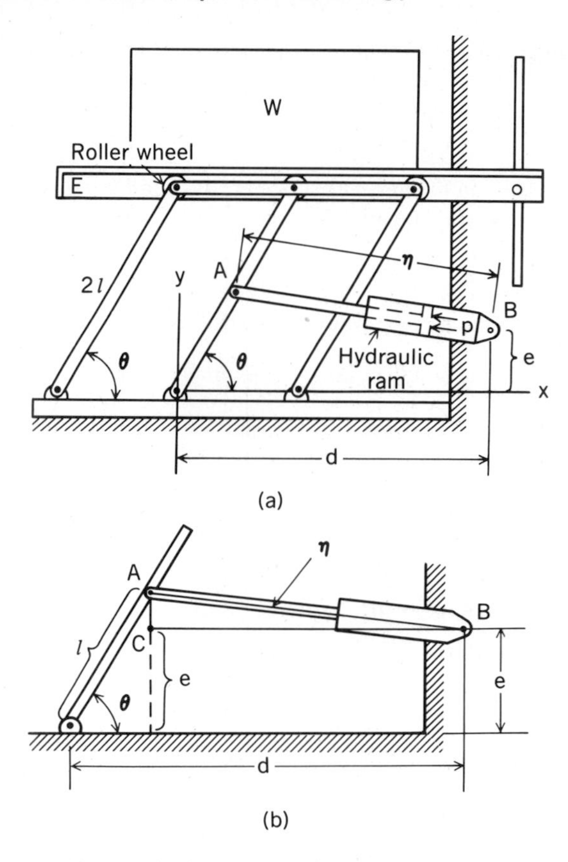

Figure 10.13. Pneumatic loading platform.

Pin A is at the center of the rod.

We have here a system with one degree of freedom characterized by the angle θ. The active forces that do work during a virtual displacement $\delta\theta$ are the weight W and the force from the hydraulic ram. Accordingly, the virtual movements of both the platform and joint A of the pump must be found. Note first using reference xy:

$$y_E = 2l \sin\theta$$

Therefore,

$$\delta y_E = 2l \cos\theta \, \delta\theta \tag{a}$$

For the ram force, we want the movement of pin A in the direction of the axis of the pump, namely $\delta\eta$, where η is shown in Fig. 10.13(a). Observing Fig. 10.13(b) we can say for η:

$$\begin{aligned} \eta^2 &= \overline{AC}^2 + \overline{CB}^2 \\ &= [(l \sin\theta) - e]^2 + (d - l\cos\theta)^2 \end{aligned} \tag{b}$$

Hence, we have

$$2\eta \, \delta\eta = 2(l\sin\theta - e)(l\cos\theta)\,\delta\theta + 2(d - l\cos\theta)(l\sin\theta)\,\delta\theta \tag{c}$$

Solving $\delta\eta$, we get

$$\delta\eta = \frac{l}{\eta}[(l\sin\theta - e)\cos\theta + (d - l\cos\theta)\sin\theta]\,\delta\theta$$

$$= \frac{l}{\eta}(l\sin\theta\cos\theta - e\cos\theta + d\sin\theta - l\sin\theta\cos\theta)\,\delta\theta \qquad \text{(d)}$$

$$= \frac{l}{\eta}(d\sin\theta - e\cos\theta)\,\delta\theta$$

The principle of virtual work is now applied to ensure equilibrium. Thus, considering one side of the system and using half the load, we have

$$-\frac{W}{2}(\delta y_B) + \left[p\frac{\pi(4^2)}{4}\right]\delta\eta = 0$$

Hence,

$$-(2500)(2l\cos\theta\,\delta\theta) + p(4\pi)\left[\frac{l}{\eta}(d\sin\theta - e\cos\theta)\right]\delta\theta = 0 \qquad \text{(e)}$$

The value of η at the configuration of interest may be determined from Eq. (b). Thus,

$$\eta^2 = [(24)(.866) - 10]^2 + [60 - (24)(.5)]^2$$

Therefore,

$$\eta = 49.2 \text{ in}$$

Now canceling $\delta\theta$ and substituting known data into Eq. (e), we may then determine p for equilibrium:

$$-(2500)(2)(24)(.5) + p(4\pi)\left\{\frac{24}{49.2}[(60)(.866) - (10)(.5)]\right\} = 0$$

Therefore,

$$p = 208 \text{ psi} \qquad \text{(f)}$$

In a few of the homework problems, you have to use simple kinematics of a cylinder rolling without slipping (see Fig. 10.14). You will recall from physics that the cylinder is actually rotating about the point of contact A. If the cylinder rotates an angle $\delta\theta$, then $\delta C = -r\,\delta\theta$. We shall consider kinematics of rigid bodies in detail later in the text.

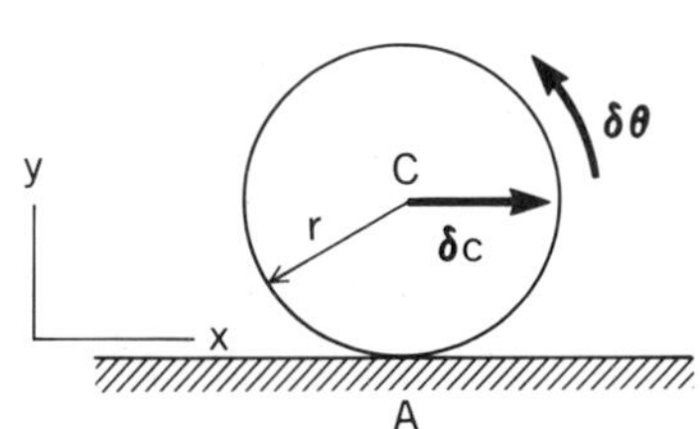

Figure 10.14. Cylinder rolling without slipping.

In concluding this section, we wish to point out that the method of virtual work is actually *not* restricted to ideal systems. Furthermore, it is permissible to give virtual displacements that *violate* one or more constraints. We then proceed by considering those friction forces and torques that perform virtual work as active forces. And, where a constraint is violated, we consider the corresponding constraining force or torque to be active. We point out that the method of virtual work offers generally no advantage in situations where there is friction and where constraints are violated. Furthermore, the extensions of virtual work to other useful theories are primarily restricted to ideal systems. Accordingly we shall consider only ideal systems and shall take virtual displacements that do not violate constraints.

Problems

10.1. How many degrees of freedom do the following systems possess? What coordinates can be used to locate the system?

(a) A rigid body not constrained in space.
(b) A rigid body constrained to move along a plane surface.
(c) The board *AB* in the diagram (a).
(d) The spherical bodies shown in diagram (b) may slide along shaft *C–C*, which in turn rotates about axis *E–E*. Shaft *C–C* may also slide along *E–E*. The spindle *E–E* is on a rotating platform. Give the number of degrees of freedom and coordinates for a sphere, shaft *C–C*, and spindle *E–E*.

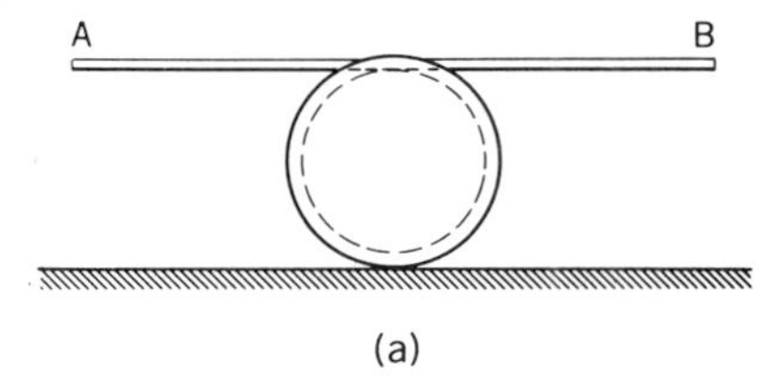

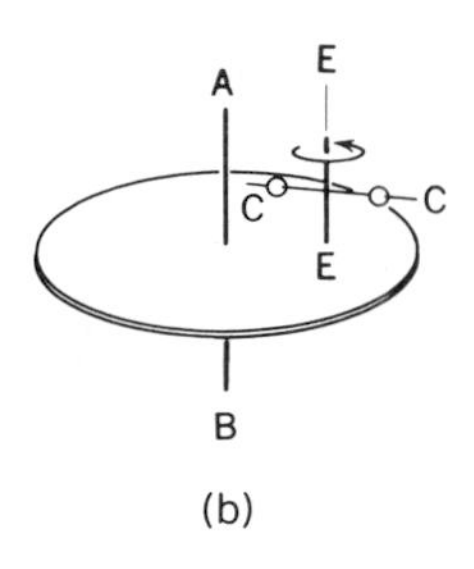

Figure P.10.1

10.2. A parking-lot gate arm weighs 150 N. Because of the taper, the weight can be regarded as concentrated at a point 1.25 m from the pivot point. What is the solenoid force to lift the gate? What is the solenoid force if a 300-N counterweight is placed .25 m to left of the pivot point?

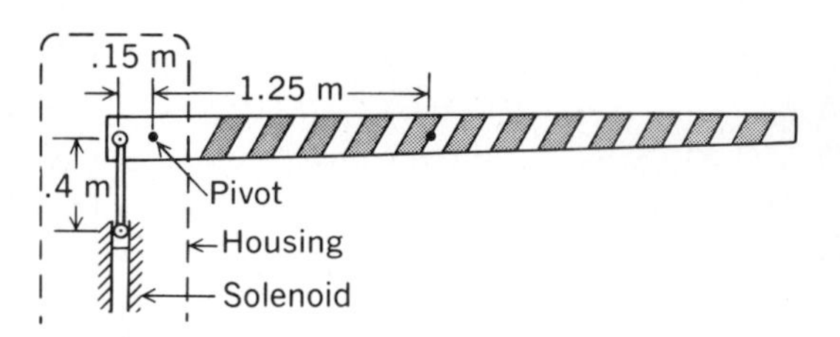

Figure P.10.2

10.3. What is the longest portion of pipe weighing 400 lb/ft that can be lifted without tipping the 12,000-lb tractor?

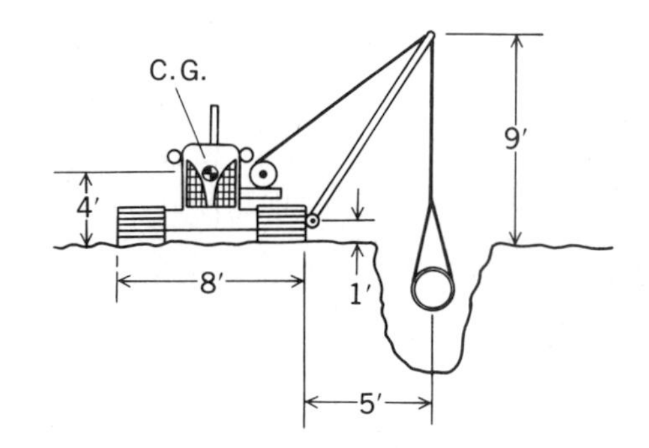

Figure P.10.3

10.4. If $W_1 = 100$ N and $W_2 = 150$ N, find the angle θ for equilibrium.

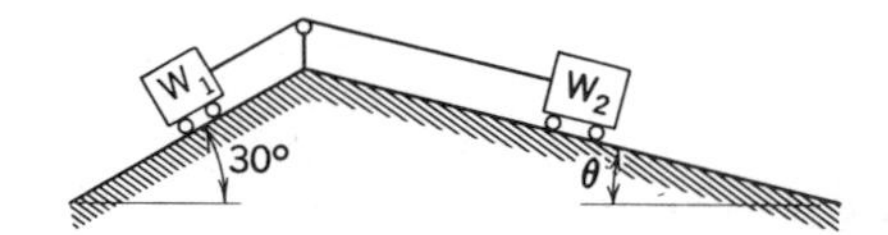

Figure P.10.4

10.5. The triple pulley sheave and the double pulley sheave weigh 150 N and 100 N, respectively. What rope force is necessary to lift a 3500-N engine?

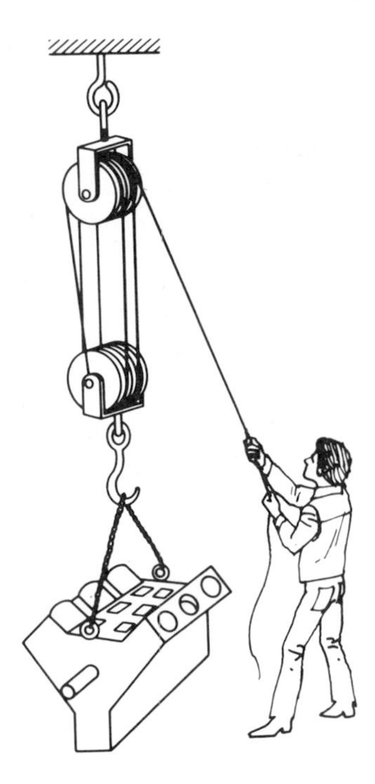

Figure P.10.5

10.6. What weight W can be lifted with the A-frame hoist in the position shown if the cable tension is T?

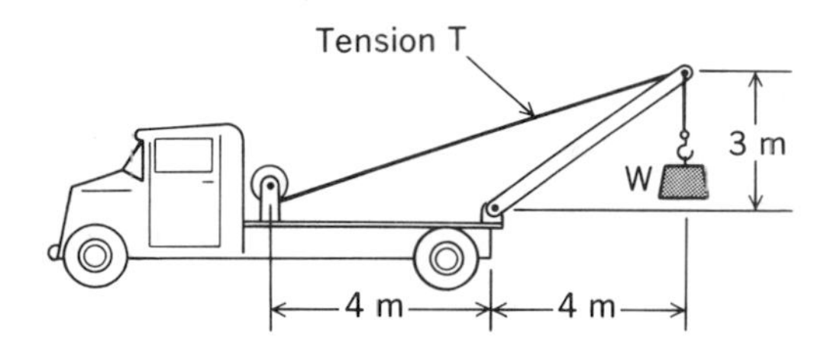

Figure P.10.6

10.7. A small hoist has a lifting capacity of 20 kN. What is the maximum cable tension?

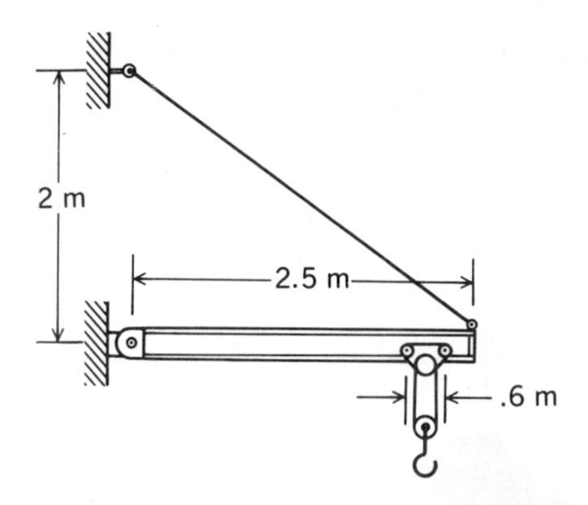

Figure P.10.7

10.8. If $W = 1000$ N and $P = 300$ N, find the angle θ for equilibrium.

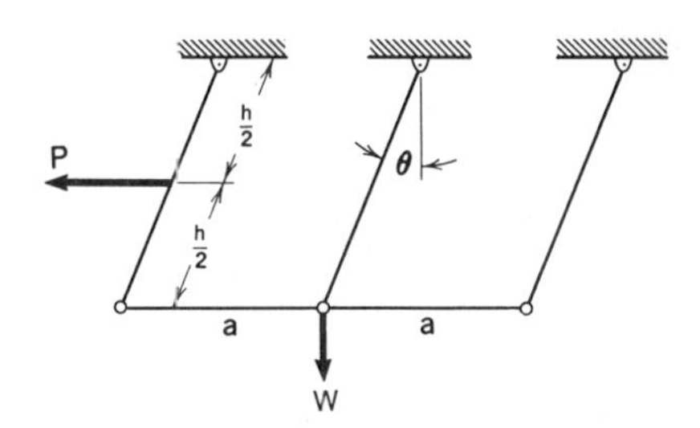

Figure P.10.8

10.9. What is the tension in the cables of a 10-ft-wide 12-ft-long 6000-lb drawbridge when the bridge is first raised? When the bridge is at 45°?

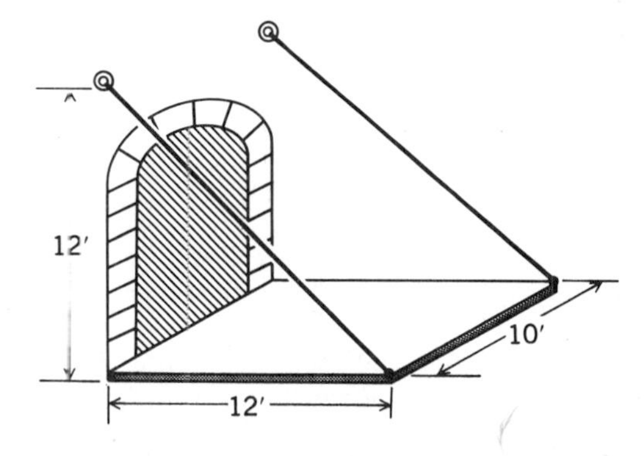

Figure P.10.9

10.10. Assuming frictionless contacts, determine the magnitude of P for equilibrium.

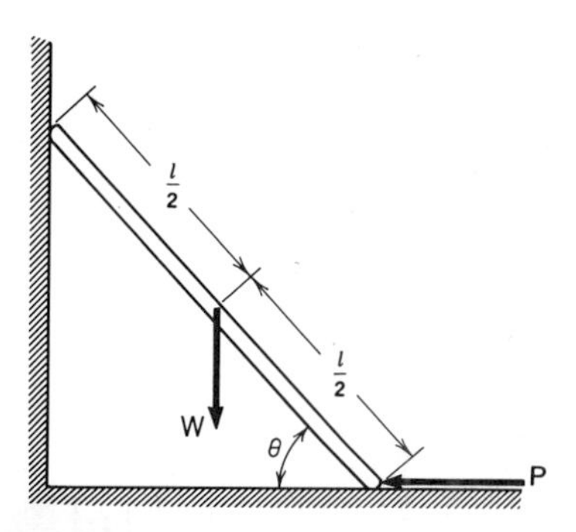

Figure P.10.10

10.11. A rock crusher is shown in action. If $p_1 =$ 50 psig and $p_2 = 100$ psig, what is the force on the rock at the configuration shown? The diameter of the pistons is 4 in.

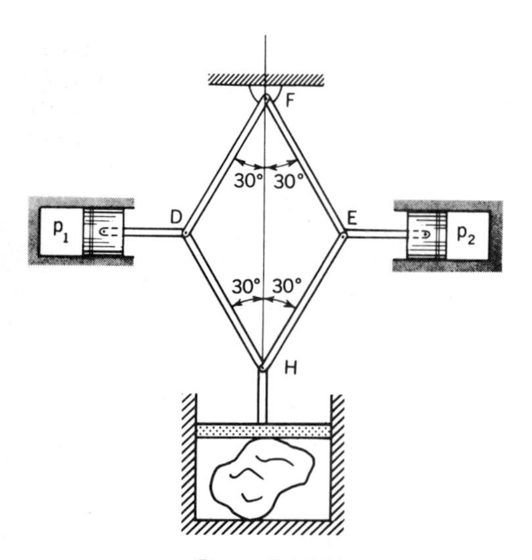

Figure P.10.11

10.12. A 20-lb-ft torque is applied to a scissor jack. If friction is disregarded throughout, what weight can be maintained in equilibrium? Take the pitch of the screw threads to be .3 in. in opposite senses. All links are of equal length, 1 ft.

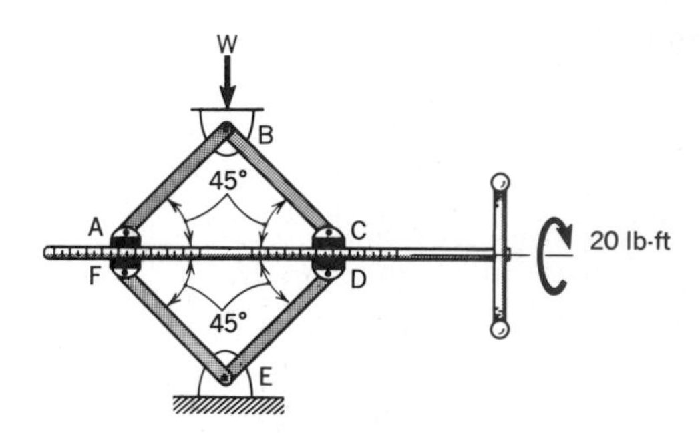

Figure P.10.12

10.13. The 5000-lb van of an airline food catering truck rises straight up until its floor is level with the airplane floor. What is the ram force in that position?

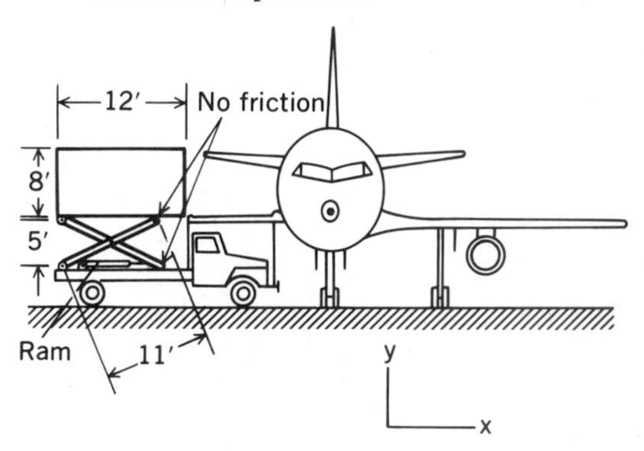

Figure P.10.13

10.14. What are the cable tensions when the arms of the power shovel are in the position shown? Arm AC weighs 13 kN, arm DF weighs 11 kN, and the shovel and payload weigh 9 kN.

Figure P.10.14

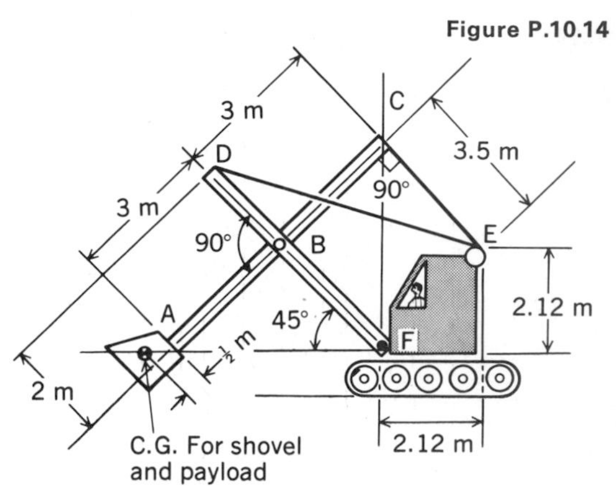

10.15. A hydraulically actuated gate in a 2-m-square water-carrying tunnel under a dam is held in place with a vertical beam *AC*. What is the force in the hydraulic ram if the specific weight of water is 9818 N/m³? (*Hint:* See the statement just before Problem 4.59 for assistance in computing the resultant force from the water on the gate.)

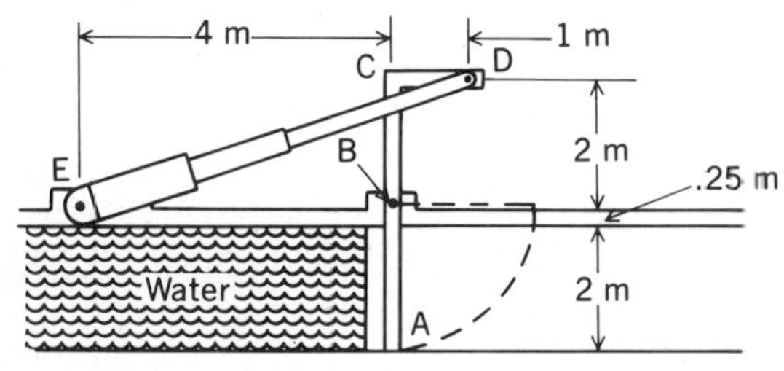

Figure P.10.15

10.16. Find the angle β for equilibrium in terms of the parameters given in the diagram. Neglect friction and the weight of the beam.

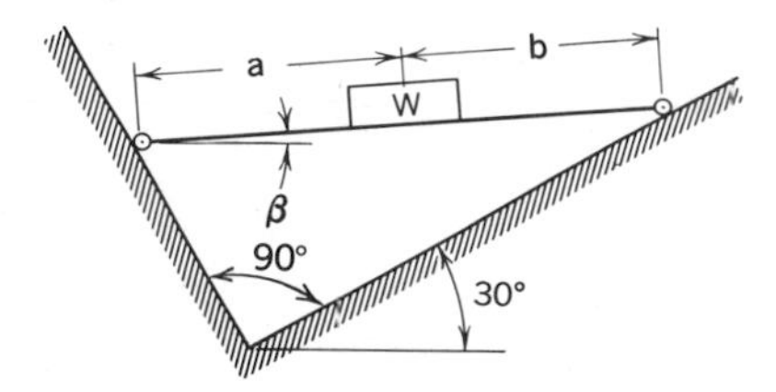

Figure P.10.16

10.17. Do Problem 5.55 by the method of virtual work.

10.18. Do Problem 5.56 by the method of virtual work.

10.19. What is the relation among *P*, *Q*, and θ for equilibrium?

Figure P.10.19

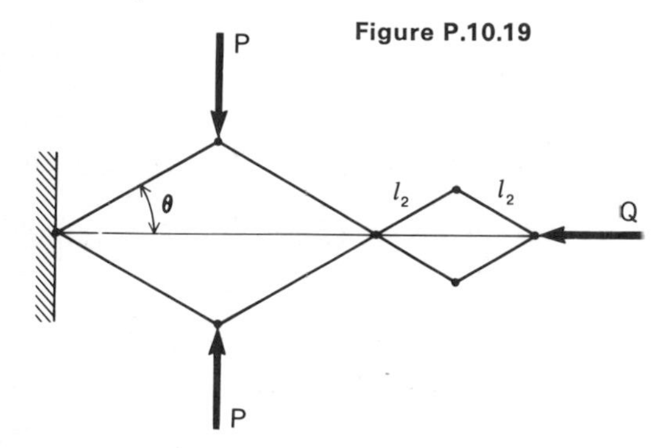

10.20. A paper collater is shown with the weight *Q* of the collated papers equal to .2 N. The collater rests on a smooth surface and, accordingly, can slide on this surface with no resistance. What force *P* is needed to keep the system in equilibrium for the position shown?

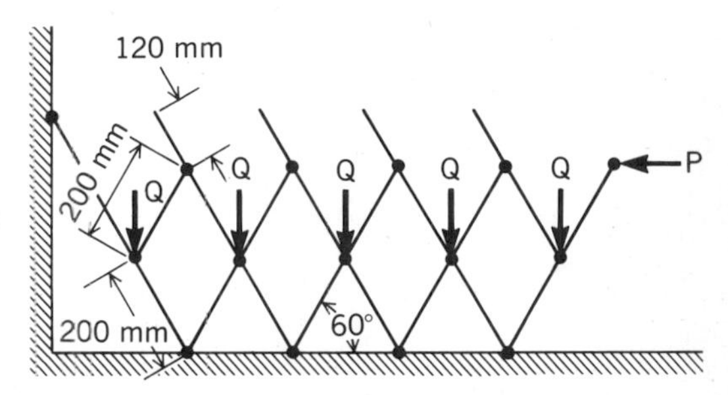

Figure P.10.20

10.21. A stepped cylinder of weight 500 lb is connected to vehicle *A* weighing 300 lb and to sheave *B* weighing 50 lb. Sheave *B* supports a weight *C*. What is the value of the weight of *C* for equilibrium? Neglect friction.

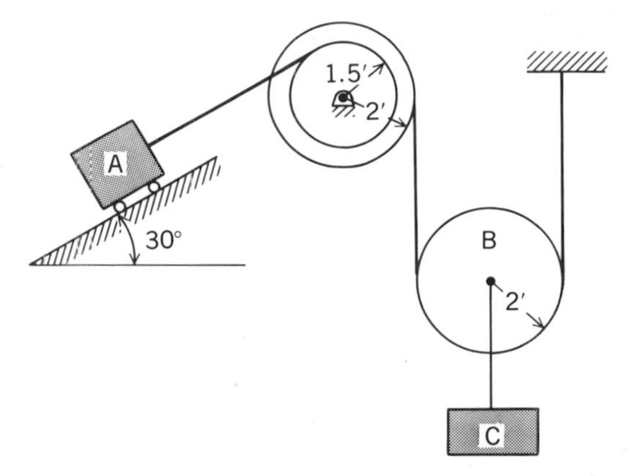

Figure P.10.21

10.22. Do the first part of Problem 5.71 by the method of virtual work.

10.23. Compute the weight *W* that can be lifted by the *differential pulley* system for an applied force *F*. Neglect the weight of the lower pulley.

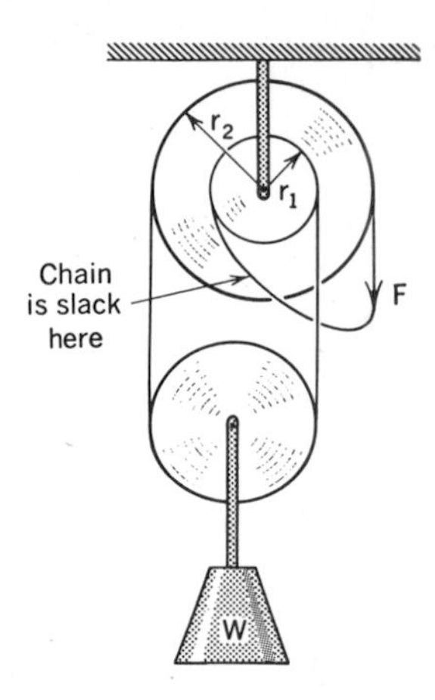

Figure P.10.23

10.24. The pressure p driving a piston of diameter 100 mm is 1 N/mm². At the configuration shown, what weight W will the system hold if we neglect friction?

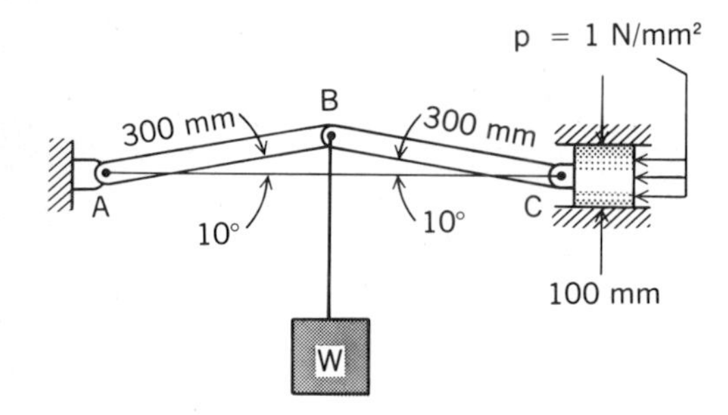

Figure P.10.24

10.25. Blocks A and B weigh 200 N and 150 N, respectively. They are connected at their base by a light cord. At what position θ is there equilibrium if we disregard friction?

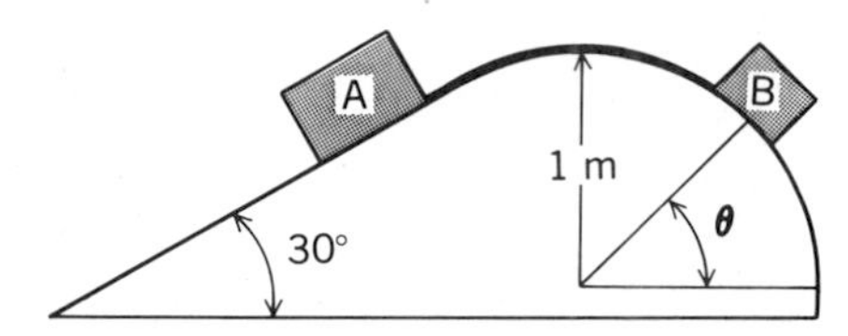

Figure P.10.25

10.26. If A weighs 500 N, and if B weighs 100 N, determine the proper weight of C for equilibrium.

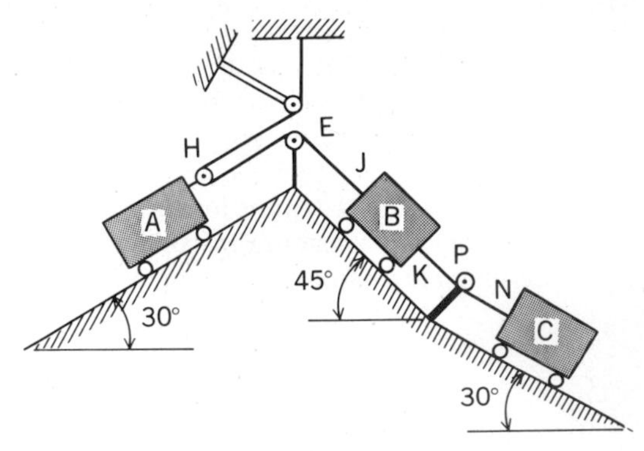

Figure P.10.26

10.27. An embossing device imprints an image at D on metal stock. If a force F of 200 N is exerted by the operator, what is the force at D on the stock? The lengths of AB and BC are each 150 mm.

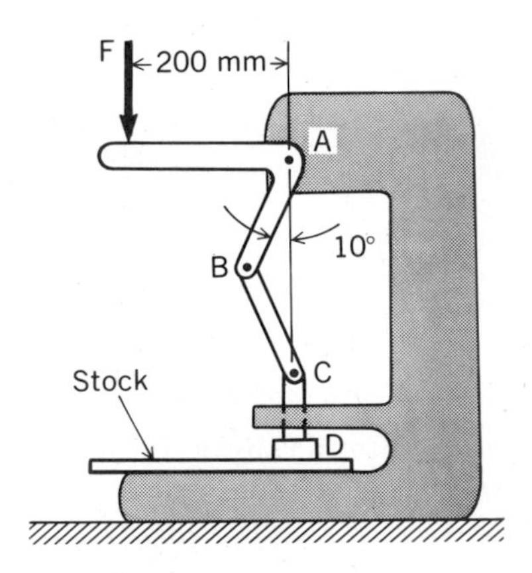

Figure P.10.27

10.28. A support system holds a 500-N load. Without the load, $\theta = 45°$ and the spring is not compressed. If K for the spring is 10,000 N/m, how far down d will the 500-N load depress the upper platform if the load is applied slowly and carefully? Neglect all other weights. $DB = BE = AB = CB = 400$ mm. (*Note:* The force from the spring is K times its contraction.)

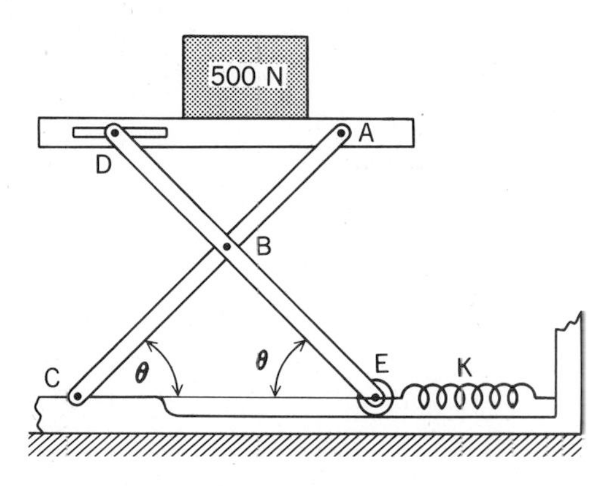

Figure P.10.28

10.29. Rod ABC is connected through a pin and slot to a sleeve which slides on a vertical rod. Before the weight W of 100 N is applied at C, the rod is inclined at an angle of 45°. If K of the spring is 8000 N/m, what is the angle θ for equilibrium? The length of AB is 300 mm and the length of BC is 200 mm when $\theta = 45°$. Neglect friction and all weights other than W. (*Note:* The force from the spring is K times its contraction.)

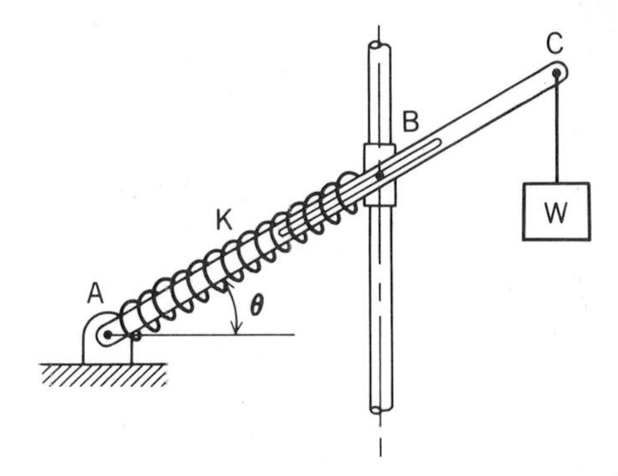

Figure P.10.29

10.30. Show that the virtual work of a couple moment M for a rotation $\delta\phi$ is given as:

$$\delta W = \boldsymbol{M} \cdot \delta\boldsymbol{\phi}$$

(*Hint:* Decompose $\boldsymbol{M}$ into components normal to and collinear with $\delta\boldsymbol{\phi}$.)

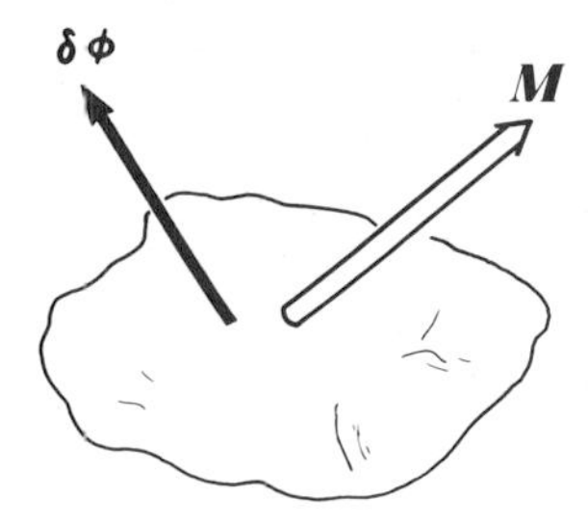

Figure P.10.30

Part B
Method of Total Potential Energy

10.5 Conservative Systems

We shall restrict ourselves in this section to certain types of active forces. This restriction will permit us to arrive at some additional very useful relations.

Consider first a body acted on only by gravity force W as an active force and moving along a frictionless path from position 1 to position 2, as shown in Fig. 10.15. The work done by gravity, W_{1-2}, is then

$$\begin{aligned} W_{1-2} = \int_1^2 \boldsymbol{F} \cdot d\boldsymbol{r} = \int_1^2 (-W\boldsymbol{j}) \cdot d\boldsymbol{r} = -W\int_1^2 dy &= -W(y_2 - y_1) \\ &= W(y_1 - y_2) \end{aligned} \qquad (10.16)$$

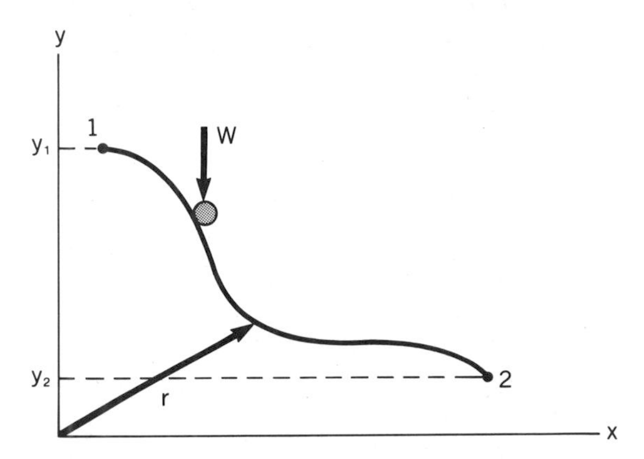

Figure 10.15. Particle moving along frictionless path.

Note that the work done *does not depend* on the path, but depends only on the positions of the *end points* of the path. *Force fields which are functions of position and whose work like gravity is independent of the path are called conservative force fields.* In general, we can say for a conservative force field $\boldsymbol{F}(x, y, z)$ that, along a path between positions 1 and 2, the work is analogous to that in Eq. 10.16

$$W_{1-2} = \int_1^2 \boldsymbol{F} \cdot d\boldsymbol{r} = V_1(x, y, z) - V_2(x, y, z) \tag{10.17}$$

where V, a scalar function evaluated at the end points, is called the *potential energy function.*[10] We may rewrite the equation above as follows:

$$-\int_1^2 \boldsymbol{F} \cdot d\boldsymbol{r} = V_2(x, y, z) - V_1(x, y, z) = \Delta V \tag{10.18}$$

From the result above we can say that *the change in potential energy*, ΔV, associated with a force field is *the negative of the work done by this force field in going from position 1 to position 2 along any path.* For any *closed* path, the work done by a conservative force field $\boldsymbol{F}$ is

$$\oint \boldsymbol{F} \cdot d\boldsymbol{r} = 0 \tag{10.19}$$

How is the potential energy function V related to $\boldsymbol{F}$? To answer this query, consider that an arbitrary infinitesimal path $d\boldsymbol{r}$ starts from point 1. We can then give Eq. 10.18 as

$$\boldsymbol{F} \cdot d\boldsymbol{r} = -dV \tag{10.20}$$

Expressing the dot product on the left side in terms of components, and expressing dV as a total differential, we get

$$F_x\,dx + F_y\,dy + F_z\,dz = -\left(\frac{\partial V}{\partial x}\,dx + \frac{\partial V}{\partial y}\,dy + \frac{\partial V}{\partial z}\,dz\right) \tag{10.21}$$

---

[10]The context of any discussion should make clear whether V refers to potential energy, or to speed, or to volume.

We can conclude from the equation above that since $d\boldsymbol{r}$ is arbitrary

$$F_x = -\frac{\partial V}{\partial x}$$
$$F_y = -\frac{\partial V}{\partial y} \qquad (10.22)$$
$$F_z = -\frac{\partial V}{\partial z}$$

Or, in other words,

$$\begin{aligned} \boldsymbol{F} &= -\left(\frac{\partial V}{\partial x}\boldsymbol{i} + \frac{\partial V}{\partial y}\boldsymbol{j} + \frac{\partial V}{\partial z}\boldsymbol{k}\right) \\ &= -\left(\frac{\partial}{\partial x}\boldsymbol{i} + \frac{\partial}{\partial y}\boldsymbol{j} + \frac{\partial}{\partial z}\boldsymbol{k}\right)V \\ &= -\textbf{grad}\, V = -\boldsymbol{\nabla} V \end{aligned} \qquad (10.23)$$

The operator we have introduced is called the *gradient* operator and is given as follows for rectangular coordinates:

$$\textbf{grad} \equiv \boldsymbol{\nabla} \equiv \frac{\partial}{\partial x}\boldsymbol{i} + \frac{\partial}{\partial y}\boldsymbol{j} + \frac{\partial}{\partial z}\boldsymbol{k} \qquad (10.24)$$

We can now say, as an alternative definition, that *a conservative force field must be a function of position and expressible as the gradient of a scalar field function.* The inverse to this statement is also valid. *That is, if a force field is a function of position and the gradient of a scalar field, it must then be a conservative force field.*

Two examples of conservative force fields will now be presented and discussed.

Constant Force Field. If the force field is constant at all positions, it can always be expressed as the gradient of a scalar function of the form $V = -(ax + by + cz)$, where a, b, and c are constants. The constant force field, then, is $\boldsymbol{F} = a\boldsymbol{i} + b\boldsymbol{j} + c\boldsymbol{k}$.

In limited changes of position near the earth's surface (a common situation), we can consider the gravitational force on a particle of mass, m, as a constant force field given by $-mg\boldsymbol{k}$. Thus, the constants for the general force field given above are $a = b = 0$ and $c = -mg$. Clearly, $V = mgz$ for this case.

Force Proportional to Linear Displacements. Consider a body limited by constraints to move along a straight line. Along this line a force is developed directly proportional to the displacement of the body from some point on the line. If this line is the x axis, we give this force as

$$\boldsymbol{F} = -Kx\boldsymbol{i} \qquad (10.25)$$

where x is the displacement from the point. The constant K is a positive number, so that, with the minus sign in this equation, a positive displacement x from the origin means that the force is negative and is then directed back to the origin. A displacement in the negative direction from the origin (negative x) means that the force is positive and is directed again toward the origin. Thus, the force given above is a

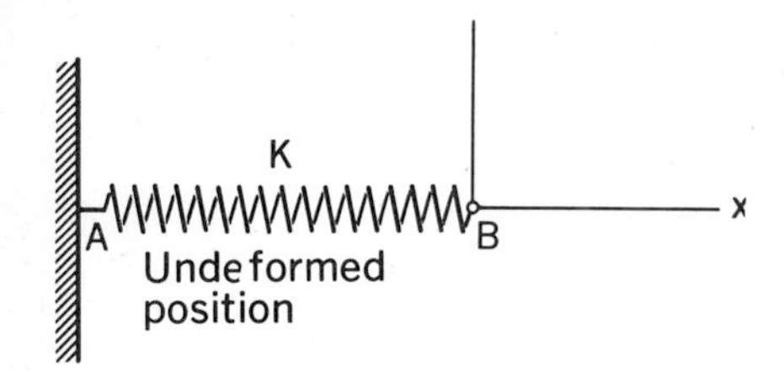

Figure 10.16. Linear spring.

restoring force about the origin. An example of this force is that of a linear spring (Fig. 10.16). The force that the spring exerts will be directly proportional to the amount of elongation or compression in the x direction beyond the unextended position which is taken at the origin of the x axis. Furthermore, the force is a restoring force. The constant K in this situation is called the *spring constant*.

The change in potential energy due to the displacements from the origin to some position x, therefore, is

$$V = \frac{Kx^2}{2} \tag{10.26}$$

The *change* in potential energy has been defined as the *negative* of the work done by a conservative force as we go from one position to another. Clearly, the potential energy change is then *directly equal* to the work done by the *reaction* to the conservative force during this displacement. In the case of the spring, the reaction force would be the force *from* the surroundings acting *on* the spring at point B (Fig. 10.16). During extension or compression of the spring from the undeformed position, this force (from the surroundings) clearly must do a positive amount of work. This work must as noted above equal the potential energy change. We now note that we can consider this work (or in other words the change in potential energy) to be a measure of the energy *stored* in the spring. That is, when allowed to return to its original position, the spring will do this amount of positive work *on* the surroundings at B, provided that the return motion is slow enough to prevent oscillations, etc. The reason for employing the name "potential energy" for V may now be more apparent.

10.6 Condition of Equilibrium for a Conservative System

Let us now consider a system of rigid bodies that is ideally constrained and acted on by conservative active forces. For a virtual displacement from a configuration of equilibrium, the virtual work done by the active forces, which are maintained constant during the virtual displacement, must be zero. We shall now show that the condition of equilibrium can be stated in yet another way for this system.

Specifically, suppose that we have n conservative forces acting on the system of bodies. The increment of work for a real infinitesimal movement of the system can be given as follows:

$$\begin{aligned} dW &= \sum_{i=1}^{n} \boldsymbol{F}_i \cdot d\boldsymbol{r}_i \\ &= \sum_{i=1}^{n} \left[-\left(\frac{\partial V_i}{\partial x_i}\boldsymbol{i} + \frac{\partial V_i}{\partial y_i}\boldsymbol{j} + \frac{\partial V_i}{\partial z_i}\boldsymbol{k} \right) \right] \cdot (dx_i\boldsymbol{i} + dy_i\boldsymbol{j} + dz_i\boldsymbol{k}) \\ &= \sum_{i=1}^{n} \left[-\left(\frac{\partial V_i}{\partial x_i} dx_i + \frac{\partial V_i}{\partial y_i} dy_i + \frac{\partial V_i}{\partial z_i} dz_i \right) \right] \end{aligned}$$

$$= -\sum_{i=1}^{n} dV_i = -d\left(\sum_{i=1}^{n} V_i\right) = -dV$$

where V without subscripts refers to *total* potential energy. By treating $\delta \boldsymbol{r}_i$ like $d\boldsymbol{r}_i$ in the equations above, we can express the virtual work δW as

$$\begin{aligned}
\delta W &= \sum_{i=1}^{n} \boldsymbol{F}_i \cdot \delta \boldsymbol{r}_i \\
&= \sum_{i=1}^{n} \left[-\left(\frac{\partial V_i}{\partial x}\boldsymbol{i} + \frac{\partial V_i}{\partial y}\boldsymbol{j} + \frac{\partial V_i}{\partial z}\boldsymbol{k}\right)\right] \cdot (\delta x_i \boldsymbol{i} + \delta y_i \boldsymbol{j} + \delta z_i \boldsymbol{k}) \\
&= \sum_{i=1}^{n} \left[-\left(\frac{\partial V_i}{\partial x_i}\delta x_i + \frac{\partial V_i}{\partial y_i}\delta y_i + \frac{\partial V_i}{\partial z_i}\delta z_i\right)\right] \\
&= -\sum_{i=1}^{n} \delta V_i = -\delta\left(\sum_{i=1}^{n} V_i\right) = -\delta V
\end{aligned}$$

But we know that for equilibrium $\delta W = 0$, and so we can similarly say for equilibrium:

$$\boxed{\delta V = 0} \tag{10.27}$$

Mathematically, this means that *the potential energy has a stationary or an extremum value at a configuration of equilibrium*, or, putting it another way, *the variation of V is zero at a configuration of equilibrium.*[11] Thus, we have another criterion which we may use to solve problems of equilibrium for conservative force systems with ideal constraints.

To use this formulation for solving problems, we proceed in the following manner. First, determine the potential energy of the system using a convenient set of

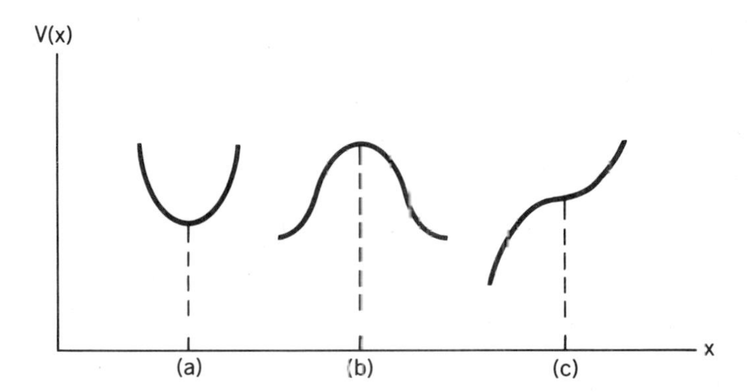

Figure 10.17. Stationary or extremum points.

[11]To further understand this, consider V as a function of only one variable, x. A *stationary* value (or, as we may say, an extremum) might be a local minimum (a in Fig. 10.17), a local maximum (b in the figure), or an inflection point (c in the figure). Note for these points that for a differential movement, δx, there is zero first-order change in V (i.e., $\delta V = 0$).

independent coordinates to locate the system. Then, take the variation, δ, of the potential energy. This operation is, for our purposes, the same as taking the differential. Thus, suppose that V is a function of independent variables $q_1, q_2, \ldots, q_n$, thereby having n degrees of freedom. The variation of V becomes

$$\delta V = \frac{\partial V}{\partial q_1}\delta q_1 + \frac{\partial V}{\partial q_2}\delta q_2 + \ldots + \frac{\partial V}{\partial q_n}\delta q_n \tag{10.28}$$

For equilibrium, we set this variation equal to zero according to Eq. 10.27. For the right side of the equation above to be zero, the coefficient of each δq_i must be zero, since the δq_i are independent of each other. Thus,

$$\boxed{\begin{aligned} \frac{\partial V}{\partial q_1} &= 0 \\ \frac{\partial V}{\partial q_2} &= 0 \\ &\vdots \\ \frac{\partial V}{\partial q_n} &= 0 \end{aligned}} \tag{10.29}$$

We now have n independent equations, which we can now solve for n unknowns. This method of approach is illustrated in the following examples.

EXAMPLE 10.4

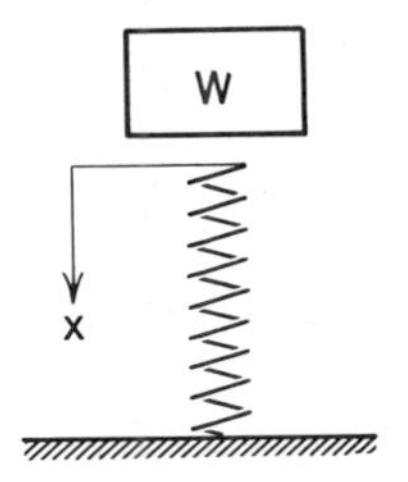

Figure 10.18. Mass placed on a linear spring.

A block weighing W lb is placed slowly on a spring having a spring contant of K lb/ft (see Fig. 10.18). Calculate how much the spring is compressed at the equilibrium configuration.

This is a simple problem and could be solved by using the definition of the spring constant, but we shall take advantage of the simplicity to illustrate the preceding comments. Notice only conservative forces act on W—gravity and the spring force. Using the unextended position of the spring as the datum for gravitational potential energy and measuring x from this position we have, for the potential energy of the system:

$$V = -Wx + \tfrac{1}{2}Kx^2$$

Consequently, for equilibrium, we have since there is only one degree of freedom

$$\frac{dV}{dx} = -W + Kx = 0$$

Solving for x, we have

$$x = \frac{W}{K}$$

EXAMPLE 10.5

A mechanism shown in Fig. 10.19 consists of two weights W, four pinned linkage rods of length a, and a spring K connecting the linkage rods. The spring is unextended when $\theta = 45°$. If friction and the weights of the linkage rods are negligible, what are the equilibrium configurations for the system of linkage rods and weights?

Only conservative forces can perform work on the system, and so we may use the stationary potential-energy criterion for equilibrium. We shall compute the potential energy as a function of θ (clearly, there is but one degree of freedom) using the configuration $\theta = 45°$ as the source of datum levels for the various energies. Observing Fig. 10.20, we can say that

$$V = -2Wd + \tfrac{1}{2}K(2d)^2 \tag{a}$$

As for the distance d, we can say (see Fig. 10.20)

$$d = a\cos 45° - a\cos\theta \tag{b}$$

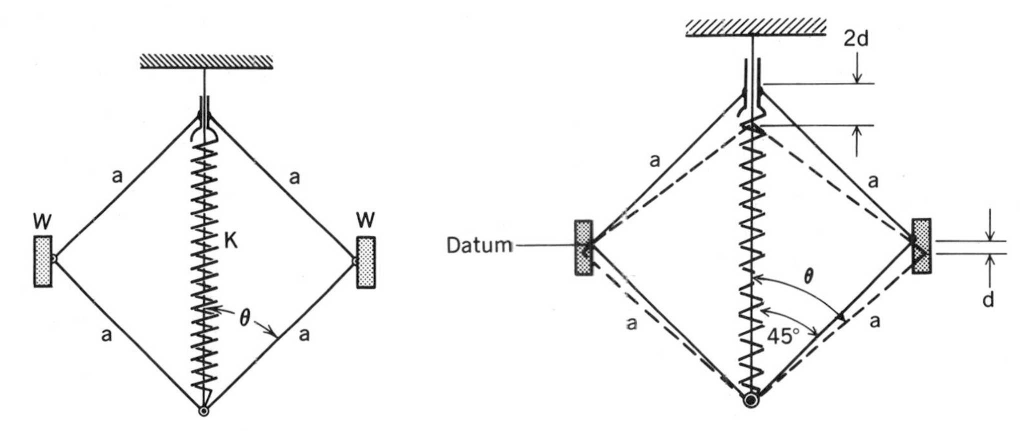

Figure 10.19. A mechanism.

Figure 10.20. Movement of mechanism as determined by θ.

Hence, we have, for Eq. (a),

$$V = -2Wa(\cos 45° - \cos\theta) + \tfrac{1}{2}K4a^2(\cos 45° - \cos\theta)^2$$

For equilibrium, we require that

$$\frac{dV}{d\theta} = 0 = -2Wa\sin\theta + 4Ka^2(\cos 45° - \cos\theta)(\sin\theta) \tag{c}$$

We can then say

$$\sin\theta\left[-W - 2Ka\left(\cos\theta - \frac{1}{\sqrt{2}}\right)\right] = 0 \tag{d}$$

We have here two possibilities for satisfying the equation. First, $\sin\theta = 0$ is a solution, so we may say that $\theta_1 = 0$ (this may not be mechanically possible) is a configuration of equilibrium. Clearly, another solution can be reached by setting the bracketed terms equal to zero:

$$-W - 2Ka\left(\cos\theta - \frac{1}{\sqrt{2}}\right) = 0$$

Therefore,

$$\cos\theta = \frac{1}{\sqrt{2}} - \frac{W}{2Ka} \qquad \text{(e)}$$

The solution for θ then is

$$\theta_2 = \cos^{-1}\left(\frac{1}{\sqrt{2}} - \frac{W}{2Ka}\right) \qquad \text{(f)}$$

We have here two possible equilibrium configurations.

Problems

10.31. A 50-kg block is placed carefully on a spring. The spring is nonlinear. The force to deflect the spring a distance x mm is proportional to the square of x. Also, we know that 5 N deflects the spring 1 mm. By method of minimum potential energy, what will be the compression of the spring? Check the result using simple calculation based on the behavior of the spring.

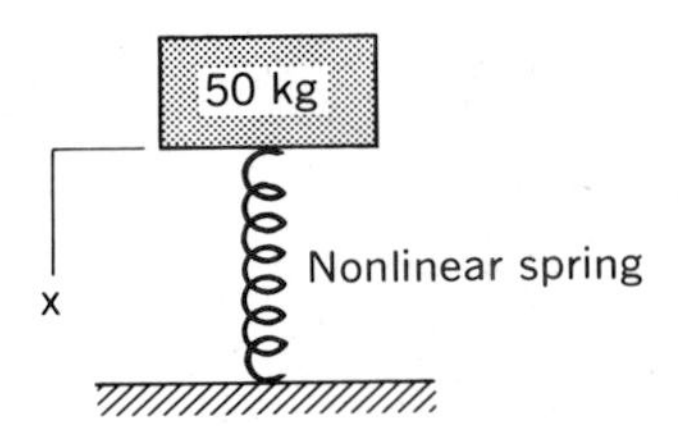

Figure P.10.31

10.32. A cylinder of radius 2 ft has wrapped around it a light, inextensible cord which is tied to a 100-lb block B on a 30° inclined surface. The cylinder A is connected to a *torsional spring*. This spring requires a torque of 1000 ft-lb/rad of rotation and it is linear and, of course, restoring. If B is connected to A when the torsional spring is unstrained, and if B is allowed to move slowly down the incline, what distance d do you allow it to move to reach an equilibrium configuration? Use the method of stationary potential energy and then check the result by more elementary reasoning.

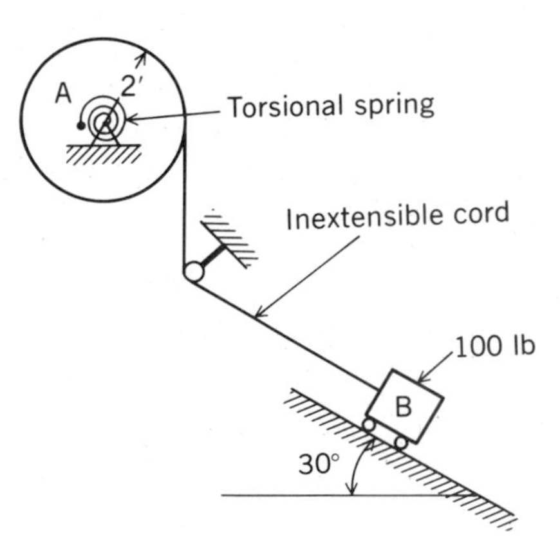

Figure P.10.32

10.33. Find the equilibrium configurations for the system of equal bars W of length 3 m and mass 25 kg. The spring is unstretched when the bars are horizontal and has a spring constant of 1500 N/m.

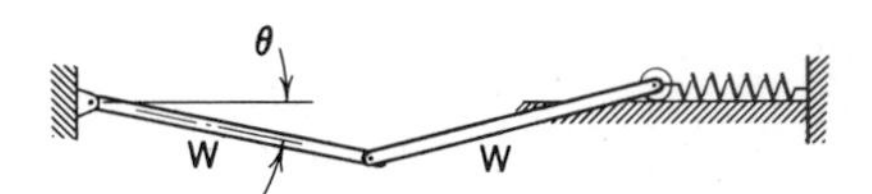

Figure P.10.33

10.34. The springs of the mechanism are unstretched when $\theta = \theta_0$. Show that $\theta =$ 17.10° when the weight W is added. Take $W = 500$ N, $a = .3$ m, $K_1 = 1$ N/mm, $K_2 = 2$ N/mm, and $\theta_0 = 45°$. Neglect the weight of the members.

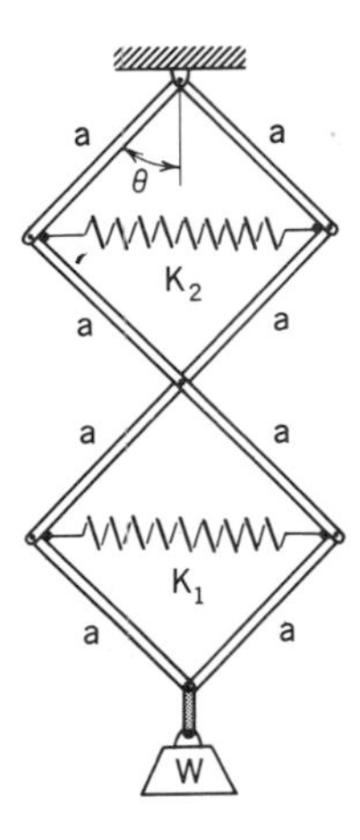

Figure P.10.34

10.35. At what elevation h must body A be for equilibrium? Neglect friction. (*Hint:* What is the differential relation between θ and l to the positions of the blocks along the surface? Integrate to get the relations themselves.)

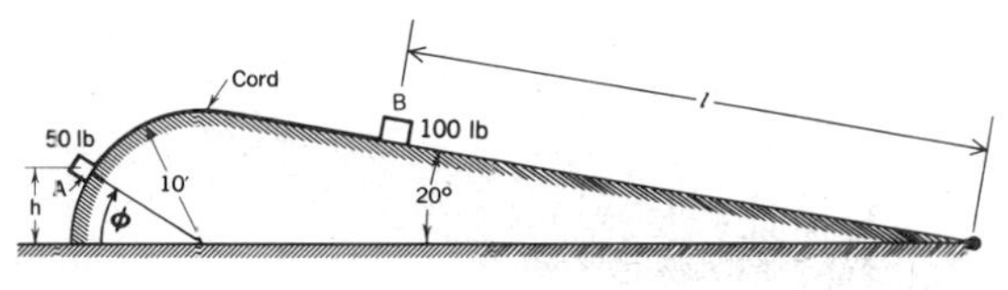

Figure P.10.35

10.36. Show that the position of equilibrium is $\theta = 77.3°$ for the 20-kg rod AB. Neglect friction.

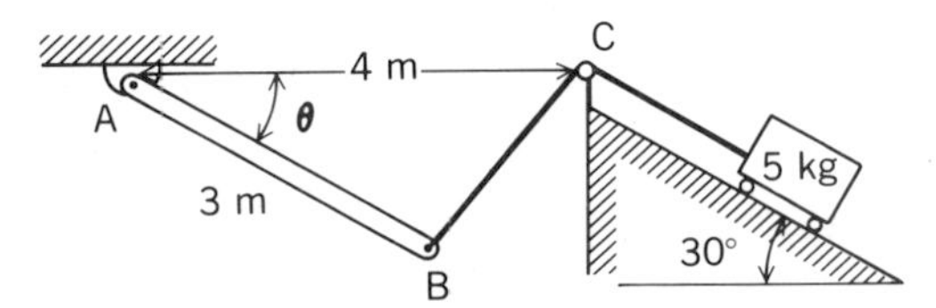

Figure P.10.36

10.37. A beam BC of length 15 ft and weight 500 lb is placed against a spring (which has a spring constant of 10 lb/in.) and smooth walls and allowed to come to rest. If the end of the spring is 5 ft away from the vertical wall when it is not compressed, show by energy methods that the amount that the spring will be compressed is .889 ft.

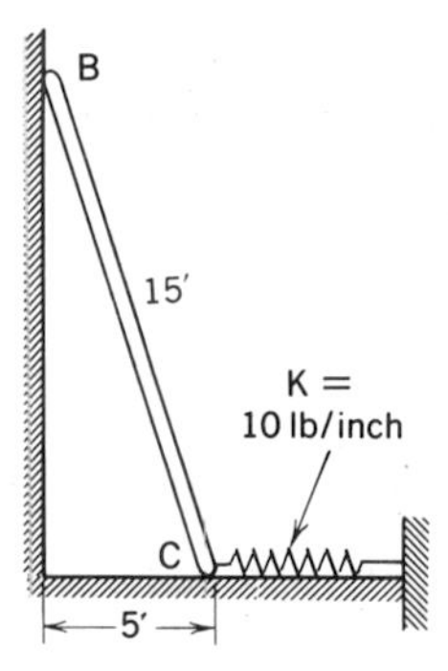

Figure P.10.37

10.38. Light rods AB and BC support a 500-N load. End A of rod AB is pinned, whereas end C is on a roller. A spring having a spring constant of 1000 N/m is connected to A and C. The spring is unstretched when $\theta = 45°$. Show that the force in the spring is 1066 N when the 500-N load is being supported.

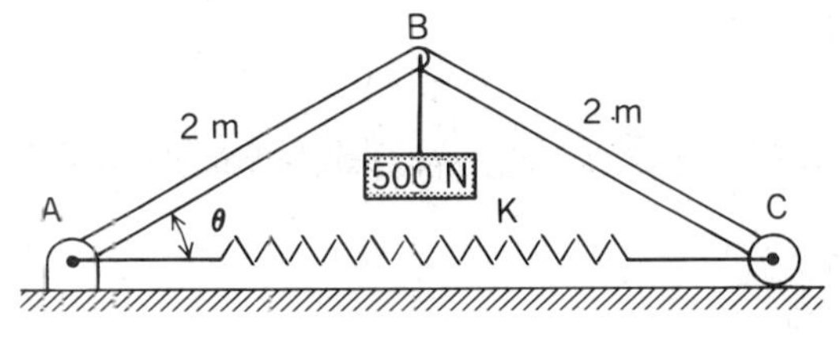

Figure P.10.38

10.39. Work Problem 10.28 using the method of total potential energy.

10.40. Work Problem 10.29 using the method of total potential energy.

10.41. Do Problem 10.25 by the method of total potential energy. (*Hint:* Consider a length of cord on a circular surface. Use the top part of the surface as a datum.)

10.42. If member AB is 10 ft and member BC is 13 ft, show that the angle θ corresponding to equilibrium is 34.5° if the spring constant K is 10 lb/in. Neglect the weight of the members and friction everywhere. Take $\theta = 30°$ for the configuration where the spring is unstretched.

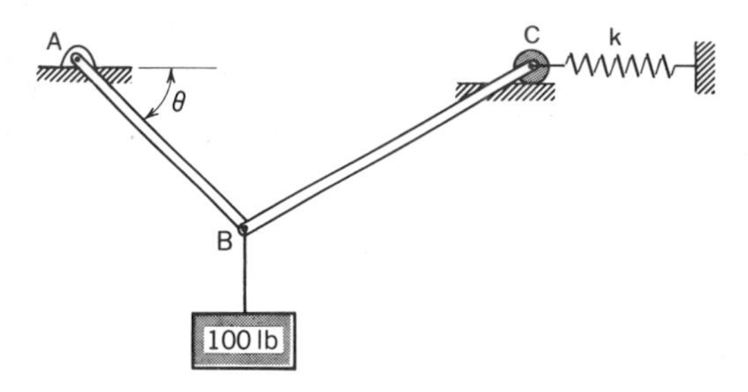

Figure P.10.42

10.43. A combination of spring and torsion-bar suspension is shown. The spring has a spring constant of 150 N/mm. The torsion bar is shown on end at A and has a torsional resistance to rotation of rod AB of 5000 N-m/rad. If the load is zero, the vertical spring is of length 450 mm, and rod AB is horizontal. What is the angle α when the suspension supports a weight of 5 kN? Rod AB is 400 mm in length.

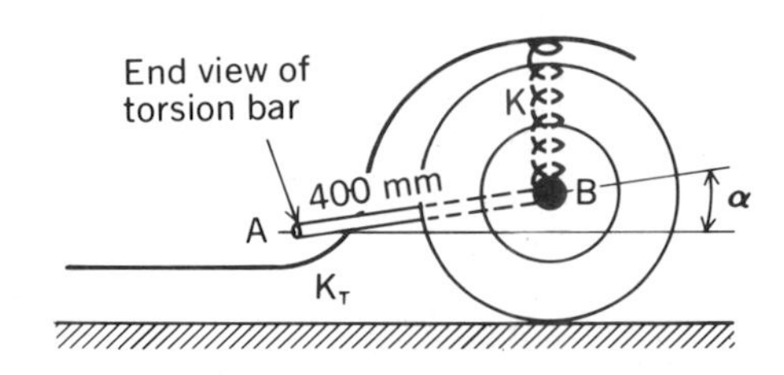

Figure P.10.43

10.44. Light rods AB and CB are pinned together at B and pass through frictionless bearings D and E. These bearings are connected to the ground by ball-and-socket connections and are free to rotate about these joints. Springs, each having a spring constant K = 800 N/m, restrain the rods as shown. The springs are unstretched when $\theta = 45°$. Show that the deflection of B is .440 m when a 500-N load is attached slowly to pin B. The rods are each 1 m in length, and each unstretched spring is .250 m in length. Neglect the weight of the rods.

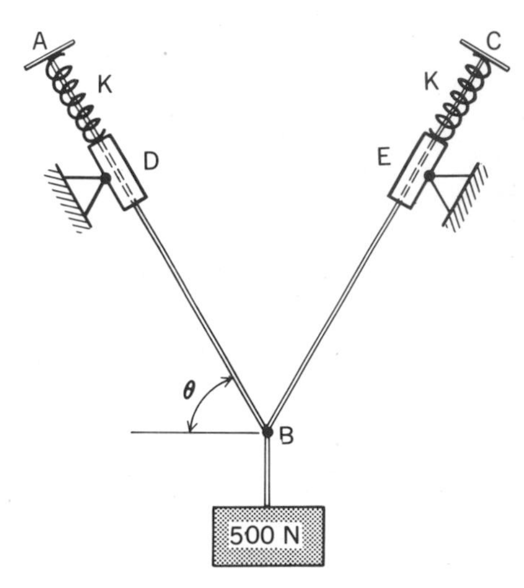

Figure P.10.44

10.45. Do Problem 10.26 by the method of total potential energy. (*Hint:* Use E as a datum and get lengths EJ, KP, and PN in terms of length HE, including unknown constants.)

10.46. An elastic band is originally 1 m long. Applying a tension force of 30 N, the band will stretch .8 m in length. What deflection a does a 10-N load induce on the band when the load is applied slowly at the center of the band? Consider the force vs. elongation of the band to be linear like a spring. (*Hint:* If you consider half of the band, you double the "spring constant.")

Figure P.10.46

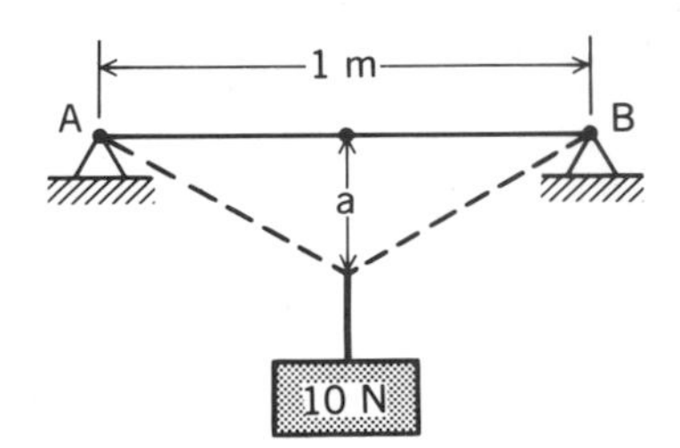

10.47. In Problem 10.46, the band is first stretched and then tied while stretched to supports A and B so that there is an initial tension in the band of 15 N. What is then the deflection a caused by the 10-N load?

10.48. A rubber band of length .7 m is stretched to connect to points A and B. A tension force of 40 N is thereby developed in the band. A 20-N weight is then attached to the band at C. Find the distance a that point C moves downward if the 20-N weight is constrained to move vertically downward along a frictionless rod. (*Hint:* If you consider part of the band, the "spring constant" for it will be greater than that of the whole band.)

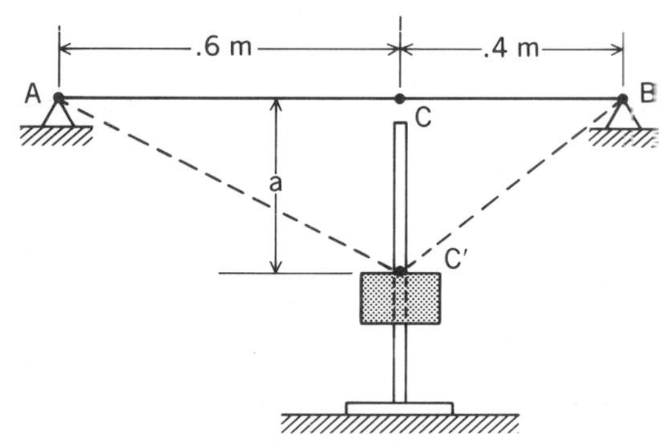

Figure P.10.48

10.49. The spring connecting bodies A and B has a spring constant K of 3 N/mm. The unstretched length of the spring is 450 mm. If body A weighs 60 N and body B weighs 90 N, what is the stretched length of the spring for equilibrium? (*Hint:* V will be a function of two variables.)

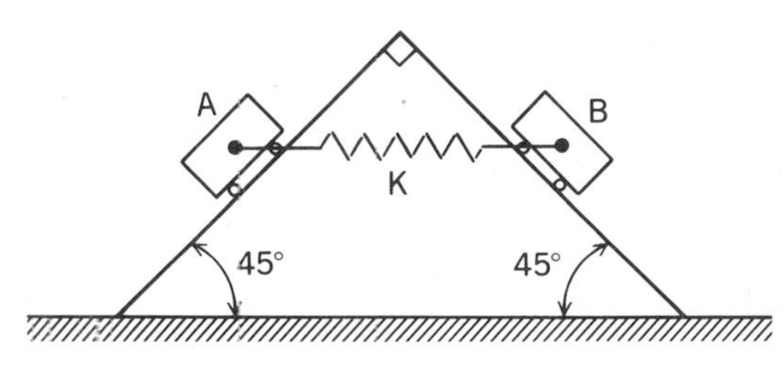

Figure P.10.49

10.7 Stability

Consider a cylinder resting on various surfaces (Fig. 10.21). If we neglect friction, the only active force is that of gravity. Thus, we have here conservative systems for which Eq. 10.27 is valid. The only virtual displacement for which contact with the surfaces is maintained is along the path. In each case, dy/dx is zero. Thus, for an infinitesimal virtual displacement, the first-order change in elevation is zero. Hence, the change in potential energy is zero for the first-order considerations. The bodies, therefore, are in *equilibrium*, according to the previous section. However, distinct physical differences exist between the states of equilibrium of the four cases.

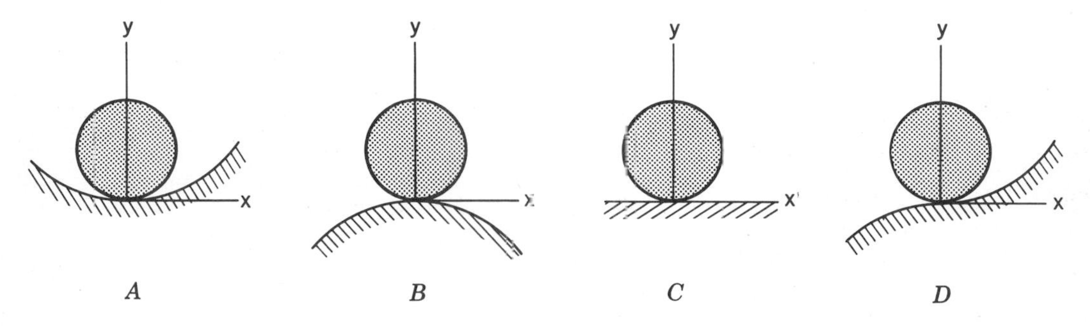

Figure 10.21. Different equilibrium configurations.

Case A. The equilibrium here is said to be *stable* in that an actual displacement from this configuration is such that the forces tend to return the body to its equilibrium configuration. Notice that the potential energy is at a *minimum* for this condition.

Case B. The equilibrium here is said to be *unstable* in that an actual displacement from the configuration is such that the forces aid in increasing the departure from the equilibrium configuration. The potential energy is at a *maximum* for this condition.

Case C. The equilibrium here is said to be *neutral.* Any displacement means that another equilibrium configuration is established. The potential energy is a constant for all possible positions of the body.

Case D. This equilibrium state is considered *unstable* since any displacement to the left of the equilibrium configuration will result in an increasing departure from this position.

How can we tell whether a system is stable or unstable at its equilibrium configuration other than by physical inspection, as was done above? Consider again a simple situation where the potential energy is a function of only one space coordinate x. That is, $V = V(x)$. We can expand the potential energy in the form of a Maclaurin series about the position of equilibrium.[12] Thus,

$$V = V_{\text{eq}} + \left(\frac{dV}{dx}\right)_{\text{eq}} x + \frac{1}{2!}\left(\frac{d^2V}{dx^2}\right)_{\text{eq}} x^2 + \dots \tag{10.30}$$

We know from Eq. 10.29 applied to one variable that at the equilibrium configuration $(dV/dx)_{\text{eq}} = 0$. Hence, we can restate the equation above:

$$V - V_{\text{eq}} = \Delta V = \frac{1}{2!}\left(\frac{d^2V}{dx^2}\right)_{\text{eq}} x^2 + \frac{1}{3!}\left(\frac{d^3V}{dx^3}\right)_{\text{eq}} x^3 + \dots \tag{10.31}$$

For small enough x, say x_0, the sign of ΔV will be determined by the sign of the first term in the series, $(1/2!)(d^2V/dx^2)_{\text{eq}}x^2$.[13] For this reason this term is called the *dominant* term in the series. Hence, the sign of $(d^2V/dx^2)_{\text{eq}}$ is vital in determining the sign of ΔV for small enough x. If $(d^2V/dx^2)_{\text{eq}}$ is positive, then ΔV is positive for any value of x smaller than x_0. This means that V is a *local minimum* at the equilibrium configuration, and we have *stable equilibrium.*[14] If $(d^2V/dx^2)_{\text{eq}}$ is negative, then V is a *local maximum* at the equilibrium configuration and we have *unstable equilibrium.* Finally, if $(d^2V/dx^2)_{\text{eq}}$ is zero, we must investigate the next higher-order derivative in the expansion, and so forth.

For cases where the potential energy is known in terms of several variables, the determination of the kind of equilibrium for the system is correspondingly more complex. For example, if the function V is known in terms of x and y, we have from the calculus of several variables:

For minimum potential energy and therefore for stability:

---

[12]Note that in a Maclaurin series the coefficients of the independent variable x are evaluated at $x = 0$, which for us is the equilibrium position. We denote this position with the subscript eq.

[13]As x gets smaller than unity, x^2 will become increasingly larger than x^3 and powers of x higher than 3. Hence, depending on the values of derivatives of V at equilibrium, there will be a value of x—say x_0—for which the first term in the series will be larger than the sum of all other terms for values of $x < x_0$.

[14]That is, if the body is displaced a distance $x < x_0$, the body will return to equilibrium on release.

$$\frac{\partial V}{\partial x} = \frac{\partial V}{\partial y} = 0 \tag{10.32a}$$

$$\left(\frac{\partial^2 V}{\partial x\, \partial y}\right)^2 - \frac{\partial^2 V}{\partial x^2}\frac{\partial^2 V}{\partial y^2} < 0 \tag{10.32b}$$

$$\frac{\partial^2 V}{\partial x^2} + \frac{\partial^2 V}{\partial y^2} > 0 \tag{10.32c}$$

For maximum potential energy and therefore for instability:

$$\frac{\partial V}{\partial x} = \frac{\partial V}{\partial y} = 0 \tag{10.33a}$$

$$\left(\frac{\partial^2 V}{\partial x\, \partial y}\right)^2 - \frac{\partial^2 V}{\partial x^2}\frac{\partial^2 V}{\partial y^2} < 0 \tag{10.33b}$$

$$\frac{\partial^2 V}{\partial x^2} + \frac{\partial^2 V}{\partial y^2} < 0 \tag{10.33c}$$

The criteria become increasingly more complex for three and more independent variables.

EXAMPLE 10.6

A thick plate whose bottom edge is that of a circular arc of radius R is shown in Fig. 10.22. The center of gravity of the plate is a distance h above the ground when the plate is in the vertical position as shown in the diagram. What relation must be satisfied by h and R for stable equilibrium?

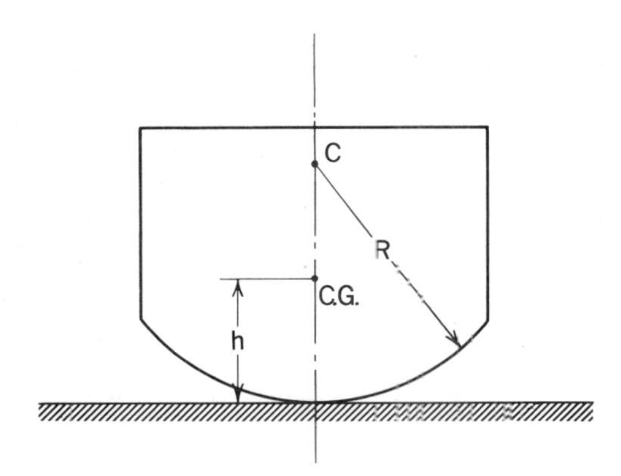

Figure 10.22. Plate with circular bottom edge.

The plate has one degree of freedom under the action of gravity and we can use the angle θ (Fig. 10.23) as the independent coordinate. We can express the potential energy V of the system relative to the ground as a function of θ in the following manner (see Fig. 10.24):

$$V = W[R - (R - h)\cos\theta] \tag{a}$$

where W is the weight of the plate. Clearly, $\theta = 0$ is a position of equilibrium since

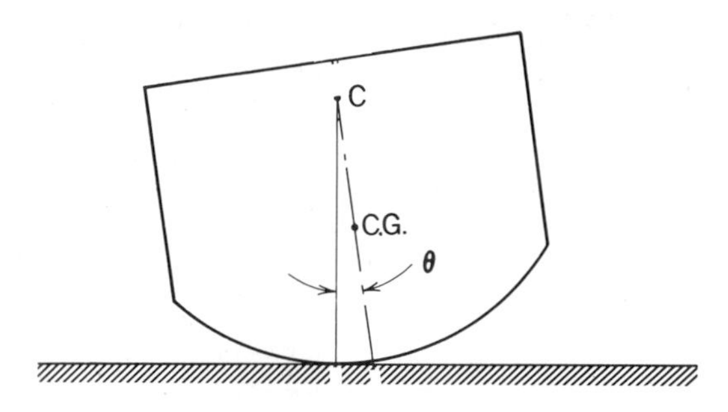

Figure 10.23. One degree of freedom.

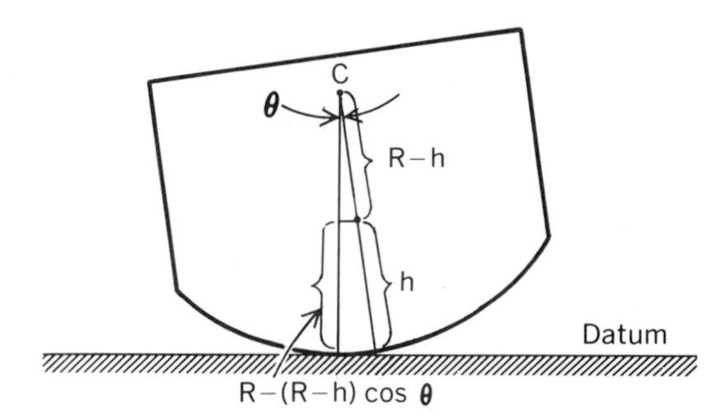

Figure 10.24. Position of C.G.

$$\left(\frac{dV}{d\theta}\right)_{\theta=0} = [W(R-h)\sin\theta]_{\theta=0} = 0 \tag{b}$$

Now consider $d^2V/d\theta^2$ at $\theta = 0$. We have

$$\left(\frac{d^2V}{d\theta^2}\right)_{\theta=0} = W(R-h) \tag{c}$$

Clearly, when $R > h$, $(d^2V/d\theta^2)_{\theta=0}$ is positive, and so this is the desired requirement for stable equilibrium.

Problems

10.50. A rod AB is connected to the ground by a frictionless ball-and-socket connection at A. The rod is free to rest on the inside edge of a plate as shown in the diagram. The square $abcd$ has its center directly over A. The curve efg is a semicircle. Without resorting to mathematical calculations, identify positions on this inside edge where equilibrium is possible for the rod AB. Describe the nature of the equilibrium and supply supporting arguments. Assume the edge of plate is frictionless.

Figure P.10.50

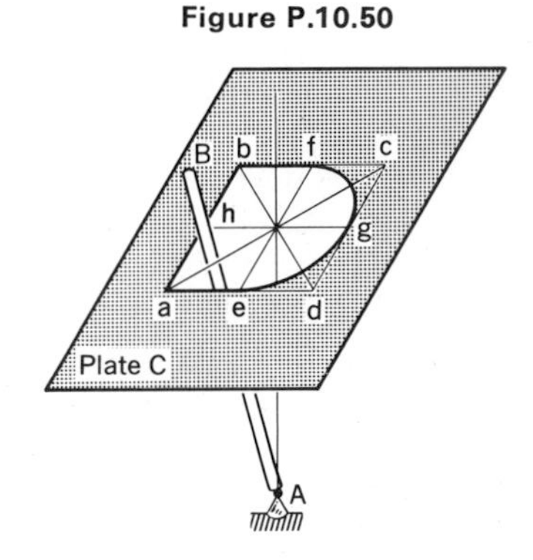

10.51. In Problem 10.50, show mathematically that position h is a position of unstable equilibrium for the rod.

10.52. Rod AB is supported by a frictionless ball-and-socket joint at A and leans against the inside edge of a plate. What is the nature of the equilibrium position a for the rod? Assume that the edge of the plate is frictionless.

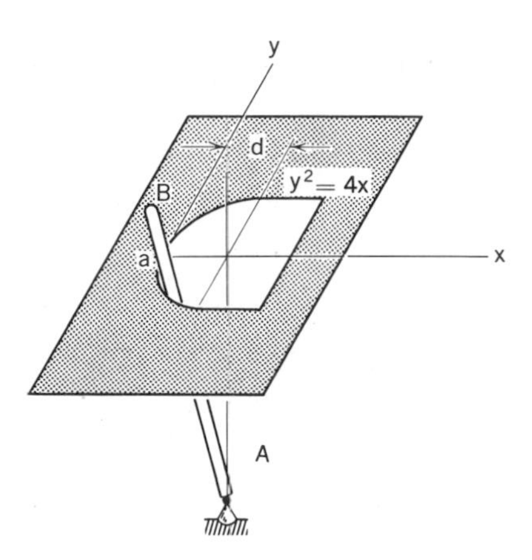

Figure P.10.52

10.53. Consider that the potential energy of a system is given by the formulation: $V = 8x^3 + 6x^2 - 7x$. What are the equilibrium positions? Indicate whether these positions are stable or not.

10.54. A section of a cylinder is free to roll on a horizontal surface. If γ of a triangular portion of the cylinder is 180 lb/ft³ and that of the semicircular portion of the cylinder is 100 lb/ft³, is the configuration shown in the diagram in stable equilibrium?

Figure P.10.54

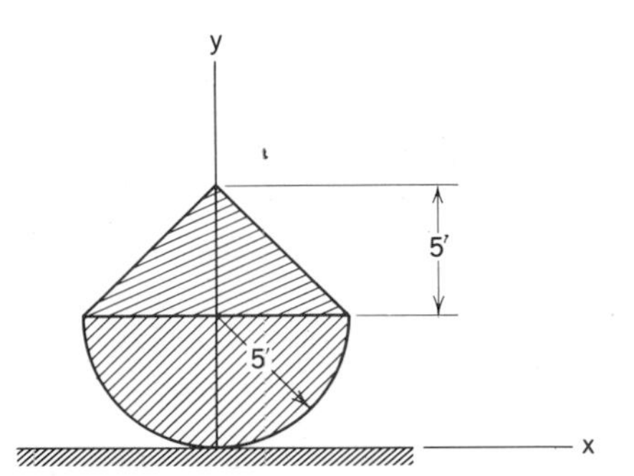

10.55. A system of springs and rigid bodies AB and BC is acted on by a weight W through a pin connection at A. If K is 50 N/mm, what is the range of the value of W so that the system has an unstable equilibrium configuration when the rods AB and BC are collinear? Neglect the weight of the rods.

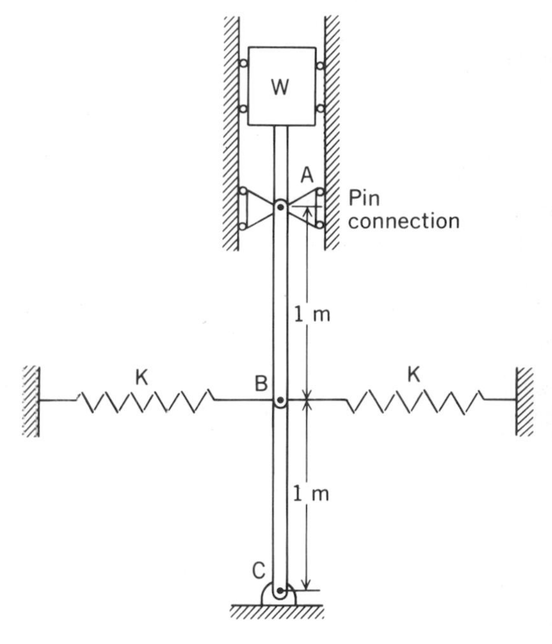

Figure P.10.55

10.56. A weight W is welded to a light rod AB. At B there is a torsional spring for which it takes 500 ft-lb to rotate 1 rad. The torsional spring is linear and restoring and is, for rotation, the analog of the ordinary linear spring for extension or contraction. If the torsional spring is unstrained when the rod is vertical, what is the largest value of W for which we have stable equilibrium in the vertical direction?

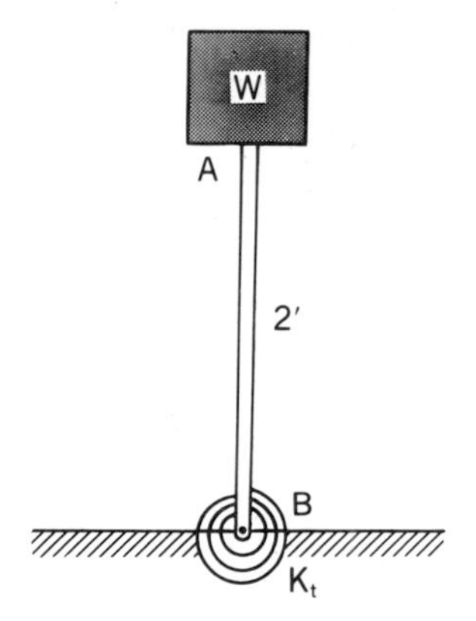

Figure P.10.56

10.57. A light rod AB is pinned to a block of weight W at A. Also at A are two identical springs K. Show that, for W less than $2Kl$, we have stable equilibrium in the vertical position and, for $W > 2Kl$, we have unstable equilibrium. The value $W = 2Kl$ is called a *critical load* for reasons that will be seen in Problem 10.58.

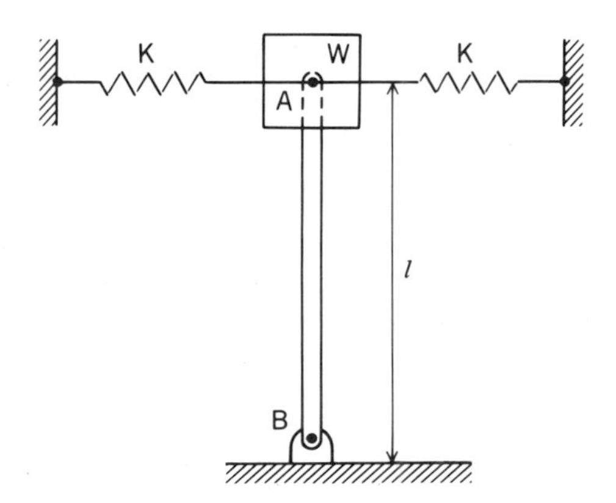

Figure P.10.57

10.58. In Problem 10.57, apply a small transverse force F to body A as shown. Compute the horizontal deflection δ of point A for a position of equilibrium by using ordinary statics as developed in earlier chapters. Now show that when $W = 2Kl$ (i.e., the critical weight), the deflection δ mathematically blows up to infinity. This shows that, even if $W < 2Kl$ and we have stable equilibrium with $F = 0$, we get increasingly very large deflections as the weight W approaches its critical value and a side load F, however small, is introduced. The study of stability of equilibrium configuration therefore is an important area of study in mechanics. Most of you will encounter this in the strength of materials course.

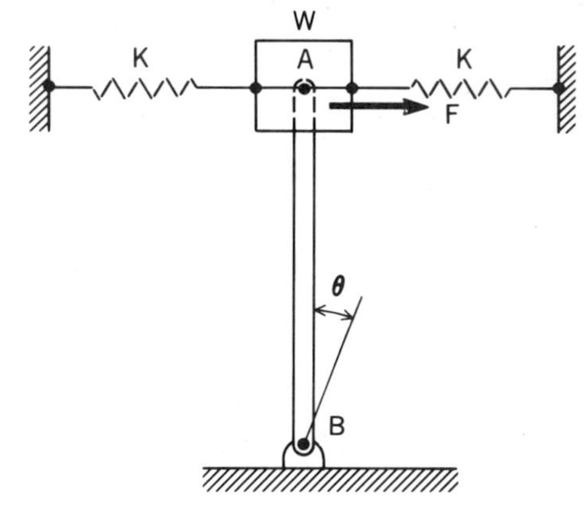

Figure P.10.58

10.59. Cylinders A and B have semicircular cross sections. Cylinder A supports a rectangular solid shown as C. If $\rho_A = 1600$ kg/m^3 and $\rho_C = 800$ kg/m^3, ascertain whether arrangement shown is in stable equilibrium. (*Hint:* Make use of point O in computing V.)

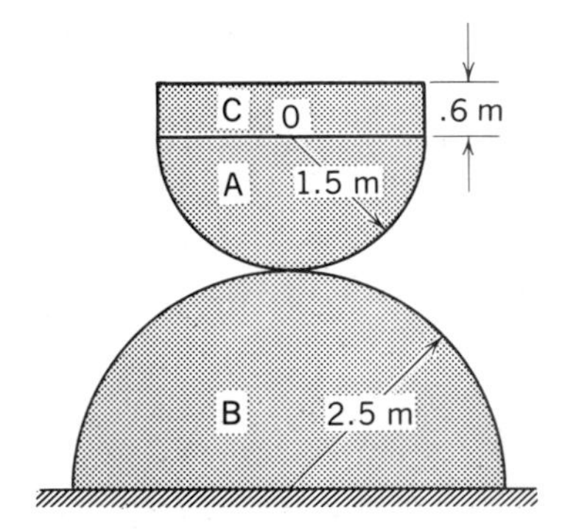

Figure P.10.59

10.8 Closure

In this chapter, we have taken an approach that differs radically from the approach used earlier in the text. In earlier chapters, we isolated a body for the purpose of writing equilibrium equations using all the forces acting on the body. This is the approach we often call *vectorial mechanics*. In this chapter, we have compared mathematically the equilibrium configuration with admissible neighboring configurations. We concluded that the equilibrium configuration was one from which there is zero virtual work under a virtual displacement. Or, equivalently for conservative active forces, the equilibrium configuration was the configuration having the minimum potential energy when compared to admissible configurations in the neighborhood. We call such an approach *variational mechanics*. The variational mechanics point of view is no doubt strange to you at this stage of study and far more subtle and mathematical than the vectorial mechanics approach.

Shifts like the one from the more physically acceptable vectorial mechanics to the more abstract variational mechanics take place in other engineering sciences. Variational methods and techniques are used in the study of plates and shells, elasticity, quantum mechanics, orbital mechanics, statistical thermodynamics, and electromagnetic theory. The variational methods and viewpoints thus are important and even vital in more advanced studies in the engineering sciences, physics, and applied mathematics.

Review Problems

10.60. At what position must the operator of the counterweight crane locate the 50-kN counterweight when he lifts the 10-kN load of steel?

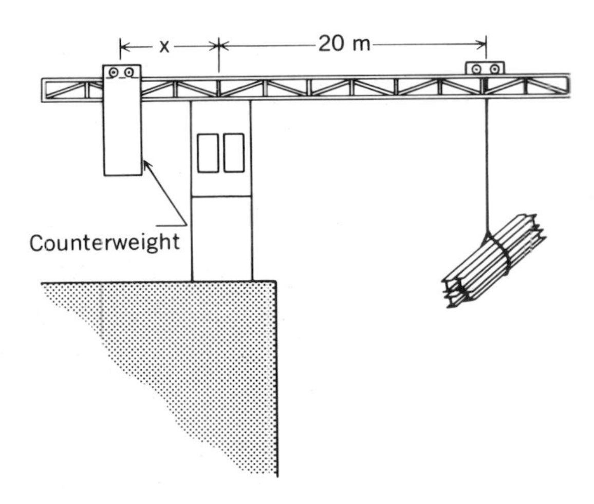

Figure P.10.60

10.61. What is the relation between P and Q for equilibrium?

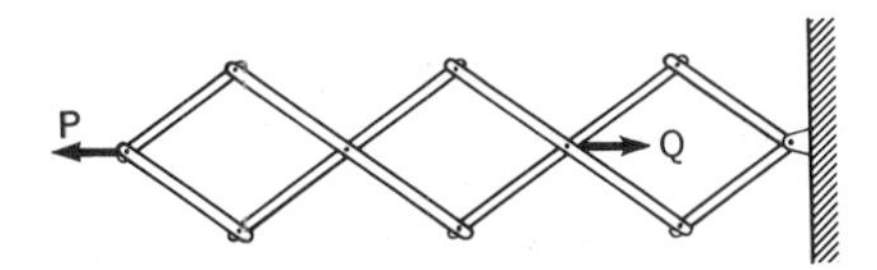

Figure P.10.61

10.62. A 50 lb-ft torque is applied to a press. The pitch of the screw is .5 in. If there is no friction on the screw, and if the base of the screw can rotate frictionlessly in a base plate A, what is the force P imposed by the base plate on body B?

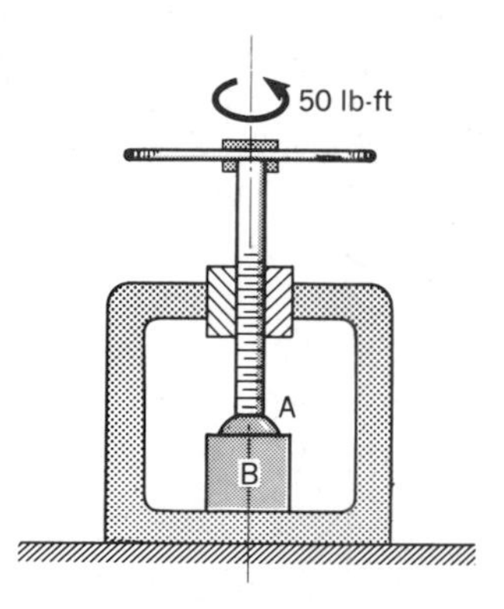

Figure P.10.62

10.63. The spring is unstretched when $\theta = 30°$. At any position of the pendulum, the spring remains horizontal. If the spring constant is 50 lb/in., at what position will the system be in equilibrium?

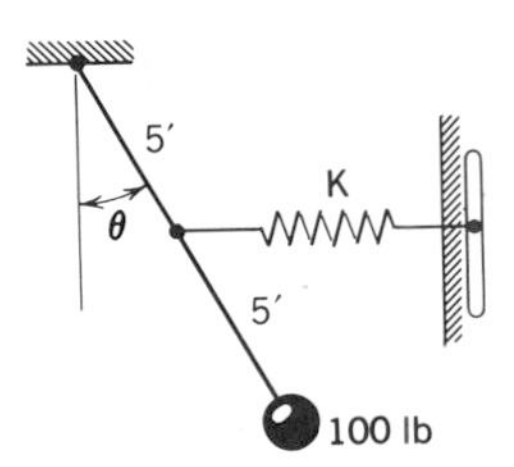

Figure P.10.63

10.64. If the springs are unstretched when $\theta = \theta_0$, find the angle θ when the weight W is placed on the system. Use the method of minimizing potential energy.

Figure P.10.64

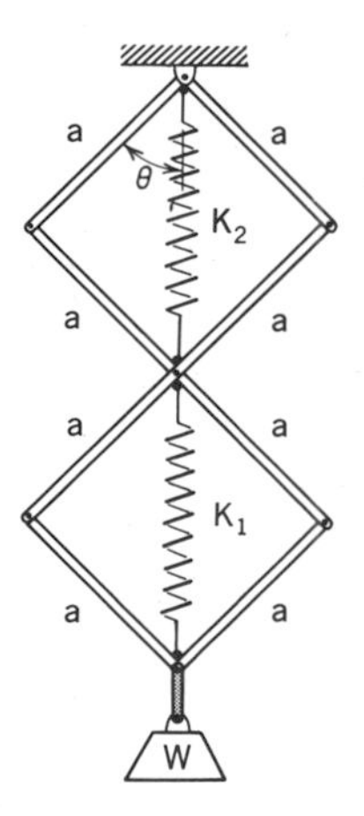

10.65. A mass M of 20 kg slides with no friction along a vertical rod. Two internal springs K_1 of spring constant 2 N/mm and an external spring K_2 of spring constant 3 N/mm restrain the weight W. If all springs are unstrained at $\theta = 30°$, show that the equilibrium configuration θ is 27.8°.

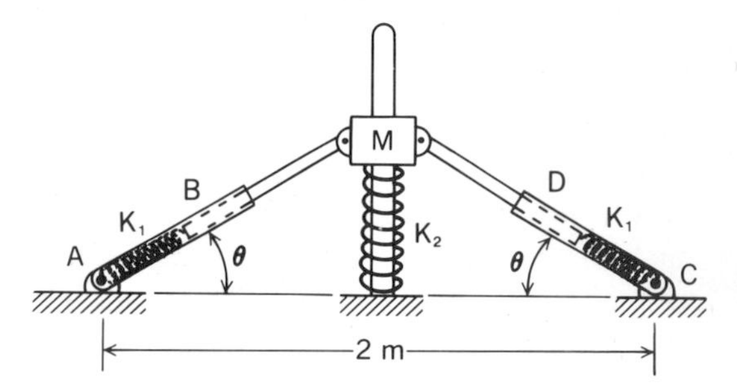

Figure P.10.65

10.66. When rod AB is in the vertical position, the spring attached to the wheel by a flexible cord is unstretched. Determine all the possible angles θ for equilibrium. Show which are stable and which are not stable. The spring has a spring constant of 8 lb/in.

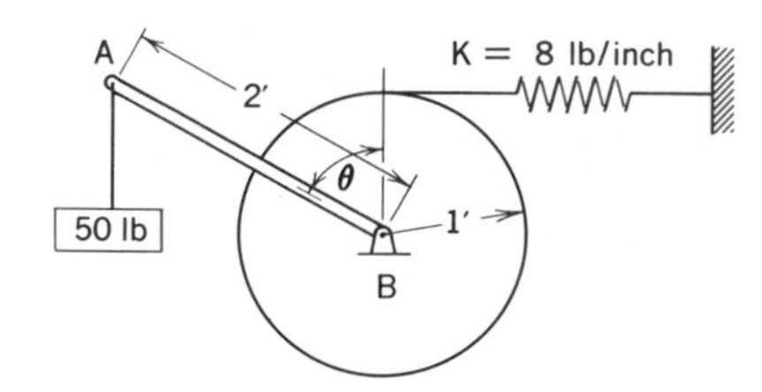

Figure P.10.66

10.67. Two identical rods are pinned together at B and are pinned at A and C. At B there is a torsional spring requiring 500 N-m/rad of rotation. What is the maximum weight W that each rod can have for a case of stable equilibrium when the rods are collinear?

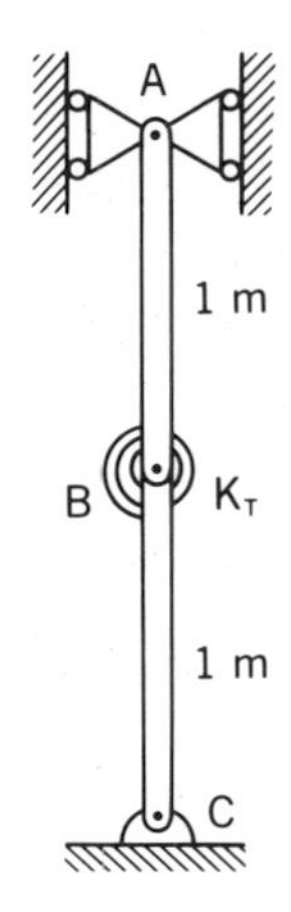

Figure P.10.67

10.68. A rectangular solid body of height h rests on a cylinder with a semicircular section. Set up criteria for stable and unstable equilibrium in terms of h and R for the position shown.

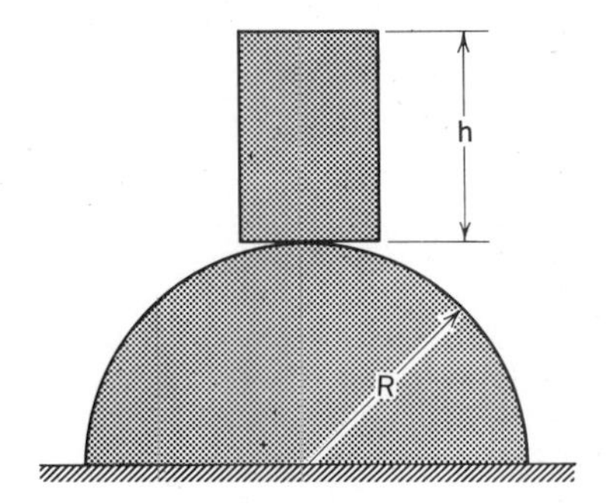

Figure P.10.68

VOLUME II

DYNAMICS

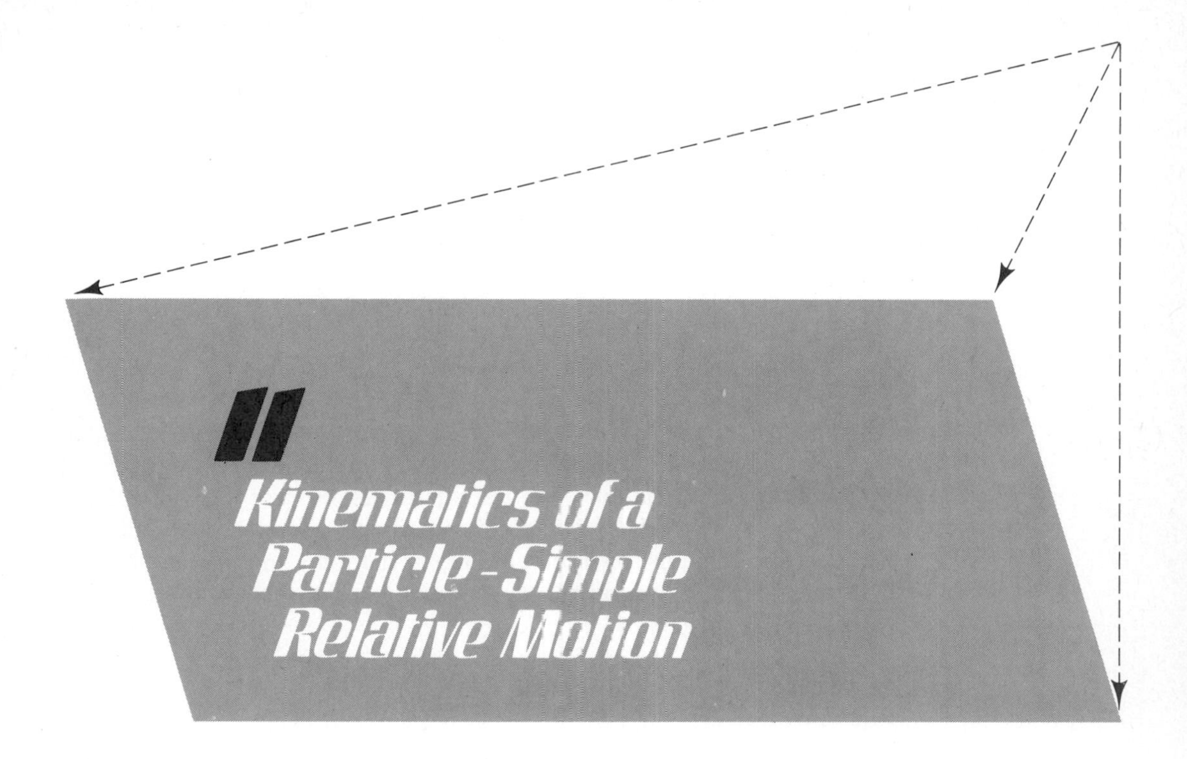

11.1 Introduction

Kinematics is that phase of mechanics concerned with the study of the motion of particles and rigid bodies without consideration of what has caused the motion. We can consider kinematics as the geometry of motion. Once kinematics is mastered, we can smoothly proceed to the relations between the factors causing the motion and the motion itself. The latter area of study is called *dynamics*. Dynamics can be conveniently separated into the following divisions, most of which we shall study in this text:

1. Dynamics of a single particle. (You will remember from our chapters on statics that a particle is an idealization having no volume but having mass.)
2. Dynamics of a system of particles. This follows division 1 logically and forms the basis for the motion of continuous media such as fluid flow and rigid-body motion.
3. Dynamics of a rigid body. A large portion of this text is concerned with this important part of mechanics.
4. Dynamics of a system of rigid bodies.
5. Dynamics of a continuous deformable medium.

Clearly, from our opening statements, the particle plays a vital role in the study of dynamics. What is the connection between the particle, which is a completely hypothetical concept, and the finite bodies encountered in physical problems? Briefly the relation is this: In many problems, the size and shape of a body are not relevant in the discussion of certain aspects of its motion; only the mass of the object is significant for

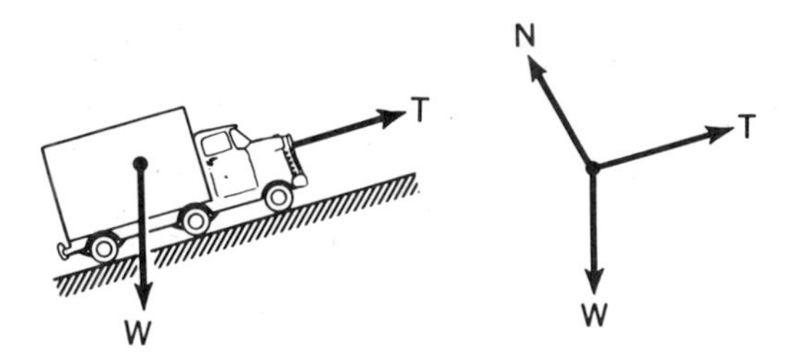

Figure 11.1. Truck considered as a particle.

such computations. For example, in towing a truck up a hill, as shown in Fig. 11.1, we would only be concerned with the mass of the truck and not with its shape or size (if we neglect forces from the wind, etc., and the rotational effects of the wheels). The truck can just as well be considered a particle in computing the necessary towing force.

We can present this relationship more precisely in the following manner. As will be learned in the next chapter (Section 12.10), the equation of motion of the center of mass of any body can be formed by:

1. Concentrating the entire mass at the mass center of the body.
2. Applying the total resultant force acting on the body to this hypothetical particle.

When the motion of the mass center characterizes all we need to know of the motion of the body, we employ the particle concept (i.e., we find the motion of the mass center). Thus, if all points of a body have the same velocity at any time t (this is called *translatory motion*), we need only know the motion of the mass center to fully characterize the motion. (This was the case for the truck, where the rotational inertia of the tires was neglected.) If, additionally, the size of a body is small compared to its trajectory (as in planetary motion, for example), the motion of the center of mass is all that might be needed, and so again we can use the particle concept for such bodies.

Part A
General Notions

11.2 Differentiation of a Vector with Respect to Time

In the study of statics, we dealt with vector quantities. We found it convenient to incorporate the directional nature of these quantities in a certain notation and set of operations. We called the totality of these very useful formulations "vector algebra." We shall again expand our thinking from scalars to vectors—this time for the operations of differentiation and integration with respect to any scalar variable t (such as time).

For scalars, we are concerned only with the variation in magnitude of some quantity that is changing with time. The scalar definition of the time derivative, then, is given as

$$\frac{df(t)}{dt} = \lim_{\Delta t \to 0} \left[\frac{f(t + \Delta t) - f(t)}{\Delta t}\right] \tag{11.1}$$

This operation leads to another function of time, which can once more be differentiated in this manner. The process can be repeated again and again, for suitable functions, to give higher derivatives.

In the case of a vector, the variation in time may be a change in magnitude, a change in direction, or both. The formal definition of the derivative of a vector $\boldsymbol{F}$ with respect to time has the same form as Eq. 11.1:

$$\frac{d\boldsymbol{F}}{dt} = \lim_{\Delta t \to 0} \left[\frac{\boldsymbol{F}(t + \Delta t) - \boldsymbol{F}(t)}{\Delta t}\right] \tag{11.2}$$

If $\boldsymbol{F}$ has no change in direction during the time interval, this operation differs little from the scalar case. However, when $\boldsymbol{F}$ changes in direction, we find for the derivative of $\boldsymbol{F}$ a new vector, having a magnitude as well as a direction, that is different from $\boldsymbol{F}$ itself. This directional consideration can be somewhat troublesome.

Let us consider the rate of change of the position vector for a reference xyz of a particle with respect to time; this rate is defined as the *velocity vector*, $\boldsymbol{V}$, of the particle relative to xyz. Following the definition given by Eq. 11.2, we have

$$\frac{d\boldsymbol{r}}{dt} = \lim_{\Delta t \to 0} \left[\frac{\boldsymbol{r}(t + \Delta t) - \boldsymbol{r}(t)}{\Delta t}\right]$$

The position vectors given in brackets are shown in Fig. 11.2. The subtraction between the two vectors gives rise to the displacement vector $\Delta\boldsymbol{r}$, which is shown as a chord connecting two points Δs apart along the trajectory of the particle. Hence, we can say that

$$\frac{d\boldsymbol{r}}{dt} = \lim_{\Delta t \to 0} \left(\frac{\Delta \boldsymbol{r}}{\Delta t}\right) = \lim_{\Delta t \to 0} \left(\frac{\Delta \boldsymbol{r}}{\Delta s}\frac{\Delta s}{\Delta t}\right)$$

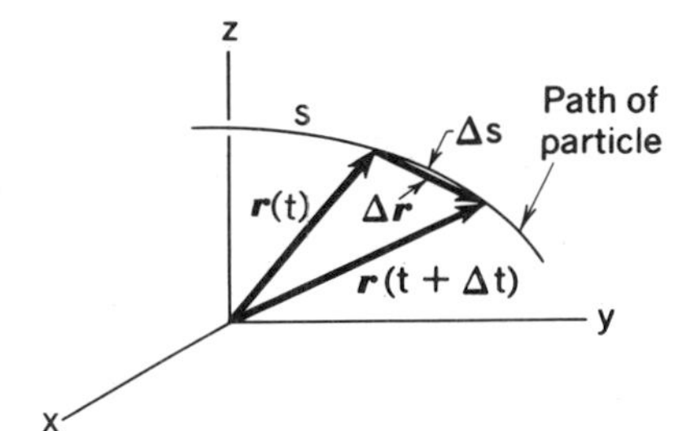

Figure 11.2. Particle at times t and $t + \Delta t$.

where we have multiplied and divided by Δs in the last expression. As Δt goes to zero, the direction of $\Delta\boldsymbol{r}$ approaches tangency to the trajectory at position $\boldsymbol{r}(t)$ and approaches Δs in magnitude. Consequently, in the limit, $\Delta\boldsymbol{r}/\Delta s$ becomes a unit vector $\boldsymbol{\epsilon}_t$ tangent to the trajectory. That is,

$$\frac{d\boldsymbol{r}}{ds} = \boldsymbol{\epsilon}_t \tag{11.3}$$

We can then say

$$\frac{d\boldsymbol{r}}{dt} = \boldsymbol{V} = \lim_{\Delta t \to 0} \left[\left(\frac{\Delta s}{\Delta t}\right)\left(\frac{\Delta \boldsymbol{r}}{\Delta s}\right)\right] = \frac{ds}{dt}\boldsymbol{\epsilon}_t \tag{11.4}$$

Therefore, $d\boldsymbol{r}/dt$ leads to a vector having a magnitude equal to the speed of the particle and a direction tangent to the trajectory. Keep in mind that there can be any angle between the position vector and the velocity vector. Students seem to want to limit this angle to 90°, which actually restricts you to a circular path. The acceleration vector of a particle can then be given as

$$\boldsymbol{a} = \frac{d\boldsymbol{V}}{dt} = \frac{d^2\boldsymbol{r}}{dt^2} \tag{11.5}$$

The differentiation and integration of vectors $\boldsymbol{r}$, $\boldsymbol{V}$, and $\boldsymbol{a}$ will concern us throughout the text.

Part B
Velocity and Acceleration Calculations

11.3 Introductory Remark

As you know from statics, we can express a vector in many ways. For instance, we can use rectangular components, or, as we will shortly explain, we can use cylindrical components. In evaluating derivatives of vectors with respect to time, we must proceed in accordance with the manner in which the vector has been expressed. In Part B of this chapter, we will therefore examine certain differentiation processes that are used extensively in mechanics. Other differentiation processes will be examined later at appropriate times.

We have already carried out a derivative operation in Section 11.2 directly on the vector $\boldsymbol{r}$. You will see in Section 11.5 that the approach used gives the derivative in terms of *path variables*. This approach will be one of several that we shall now examine with some care.

11.4 Rectangular Components

Consider first the case where the position vector $\boldsymbol{r}$ of a moving particle is expressed for a given reference in terms of rectangular components in the following manner:

$$\boldsymbol{r}(t) = x(t)\boldsymbol{i} + y(t)\boldsymbol{j} + z(t)\boldsymbol{k} \tag{11.6}$$

where $x(t)$, $y(t)$, and $z(t)$ are scalar functions of time. The unit vectors $\boldsymbol{i}, \boldsymbol{j}$, and $\boldsymbol{k}$ are fixed in magnitude and direction at all times, and so we can obtain $d\boldsymbol{r}/dt$ in the following straightforward manner:

$$\frac{d\boldsymbol{r}}{dt} = \boldsymbol{V}(t) = \frac{dx(t)}{dt}\boldsymbol{i} + \frac{dy(t)}{dt}\boldsymbol{j} + \frac{dz(t)}{dt}\boldsymbol{k} = \dot{x}(t)\boldsymbol{i} + \dot{y}(t)\boldsymbol{j} + \dot{z}(t)\boldsymbol{k} \tag{11.7}$$

A second differentiation with respect to time leads to the acceleration vector:

$$\frac{d^2\boldsymbol{r}}{dt^2} = \boldsymbol{a} = \ddot{x}(t)\boldsymbol{i} + \ddot{y}(t)\boldsymbol{j} + \ddot{z}(t)\boldsymbol{k} \tag{11.8}$$

By such a procedure, we have formulated velocity and accleration vectors in terms of components parallel to the coordinate axes.

Up to this point, we have formulated the rectangular velocity components and the rectangular acceleration components, respectively, by differentiating the position vector once and twice with respect to time. Quite often, we know the acceleration vector of a particle as a function of time in the form

$$\boldsymbol{a}(t) = \ddot{x}(t)\boldsymbol{i} + \ddot{y}(t)\boldsymbol{j} + \ddot{z}(t)\boldsymbol{k} \tag{11.9}$$

and wish to have for this particle the velocity vector or the position vector or any of their components at any time. We then integrate the time function $\ddot{x}(t)$, $\ddot{y}(t)$, and $\ddot{z}(t)$, remembering to include a constant of integration for each integration. For example, consider $\ddot{x}(t)$. Integrating once, we obtain the velocity component $V_x(t)$ as follows:

$$V_x(t) = \int \ddot{x}(t)\, dt + C_1 \tag{11.10}$$

where C_1 is the constant of integration. Knowing V_x at some time t_0, we can determine C_1 by substituting t_0 and $(V_x)_0$ into the equation above and determining C_1. Similarly, for $x(t)$ we obtain from the above:

$$x(t) = \int \left[\int \ddot{x}(t)\, dt \right] dt + C_1 t + C_2 \tag{11.11}$$

where C_2 is the second constant of integration. Knowing x at some time t, we can determine C_2 from Eq. 11.11. The same procedure involving additional constants applies to the other acceleration components.

We now illustrate the procedures described above in the following series of problems.

EXAMPLE 11.1

Pins A and B must always remain in the vertical slot of yoke C, which moves to the right at a constant speed of 6 ft/sec in Fig. 11.3. Furthermore, the pins cannot leave the elliptic slot. (a) What is the speed at which the particles approach each other when the yoke slot is at $x = 5$ ft? (b) What is the rate of change of speed toward each other when the yoke slot is at $x = 5$ ft?

The equation of the elliptic path in which the pins must move is seen by inspection to be

$$\frac{x^2}{10^2} + \frac{y^2}{6^2} = 1 \tag{a}$$

Clearly, if coordinates (x, y) are to represent the coordinates of pin B, they must be time functions such that for any time t the values $x(t)$ and $y(t)$ satisfy Eq. (a). Also, $\dot{x}(t)$ and $\dot{y}(t)$ must be such that pin B moves at all times in the elliptic path. We

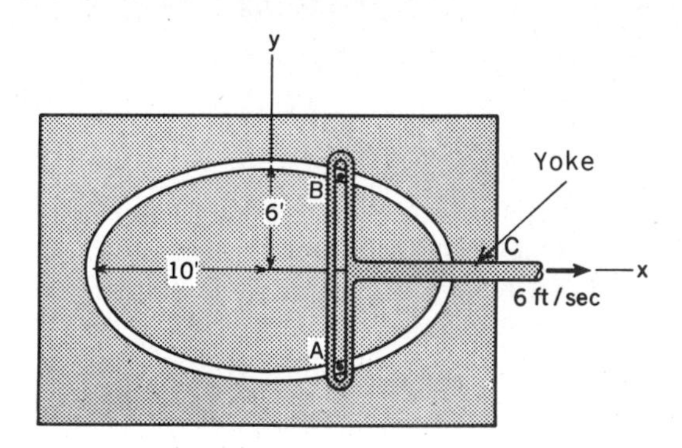

Figure 11.3. Pin slides in slot and yoke.

can satisfy this requirement by first differentiating Eq. (a) with respect to time. Canceling the factor 2, we obtain

$$\frac{x\dot{x}}{10^2} + \frac{y\dot{y}}{6^2} = 0 \tag{b}$$

Now $x(t)$, $y(t)$, $\dot{x}(t)$, and $\dot{y}(t)$ must satisfy Eq. (b) for all values of t to ensure that B remains in the elliptic path.

We can now proceed to solve part (a) of this problem. We know that pin B must have a velocity $\dot{x} = 6$ ft/sec because of the yoke. When $x = 5$ ft, we know from Eq. (a) that

$$\frac{5^2}{10^2} + \frac{y^2}{6^2} = 1 \tag{c}$$

$$y = 5.20 \text{ ft}$$

Now going to Eq. (b), we can solve for $\dot{y}$ at the instant of interest.

$$\frac{(5)(6)}{10^2} + \frac{(5.20)(\dot{y})}{6^2} = 0$$

Therefore,

$$\dot{y} = -2.08 \text{ ft/sec}$$

Thus, particle B moves downward with a speed of 2.08 ft/sec. Clearly, pin A must move upward with a speed also of 2.08 ft/sec. The pins approach each other at the instant of interest at a speed of 4.16 ft/sec.

To get the acceleration $\ddot{y}$ of pin B, we first differentiate Eq. (b) with respect to time.

$$\frac{x\ddot{x} + \dot{x}^2}{10^2} + \frac{y\ddot{y} + \dot{y}^2}{6^2} = 0 \tag{d}$$

The accelerations $\ddot{x}$ and $\ddot{y}$ must satisfy the equation above. Since the yoke moves at constant speed, we can say immediately that $\ddot{x} = 0$. And using for x, y, $\dot{x}$, and $\dot{y}$ known quantities for the configuration of interest, we can solve for $\ddot{y}$ from Eq. (d). Thus,

$$\frac{0 + 6^2}{10^2} + \frac{5.20\ddot{y} + 2.08^2}{6^2} = 0$$

Therefore,

$$\ddot{y} = -3.32 \text{ ft/sec}^2$$

Pin B must be accelerating downward at a rate of 3.32 ft/sec² while pin A accelerates upward at the same rate. The pins accelerate toward each other, then, at a rate of 6.64 ft/sec² at the configuration of interest.

In the motion of particles near the earth's surface, such as the motion of shells or ballistic missiles, we can often simplify the problem by neglecting air resistance and taking the acceleration of gravity g as constant (32.2 ft/sec² or 9.81 m/sec²). For such a case (see Fig. 11.4), we know immediately that $\ddot{y}(t) = -g$ and $\ddot{x}(t) = \ddot{z}(t) = 0$. On integrating these accelerations, we can often determine useful information as to velocities or positions at certain times of interest in the problem. We illustrate this procedure in the following examples.

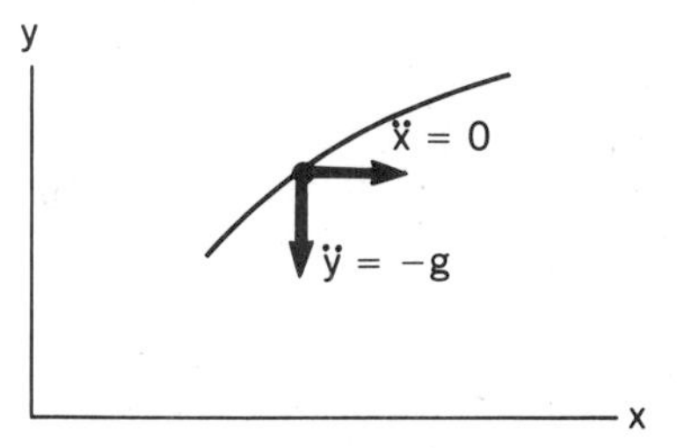

Figure 11.4. Simple ballistic motion of a shell.

EXAMPLE 11.2

A shell is fired from a hill 500 ft above a plain. The angle α of firing (see Fig. 11.5) is 15° above the horizontal, and the muzzle velocity V_0 is 3000 ft/sec. At what horizontal distance, d, will the shell hit the plain if we neglect friction of the air? What is the maximum height of the shell above the plain? Finally, determine the trajectory of the missile [i.e., find $x = f(y)$].

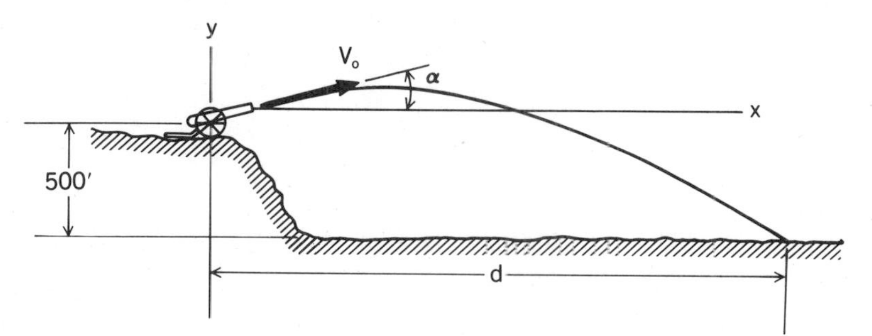

Figure 11.5. Ballistics problem: find d.

We know immediately that

$$\ddot{y}(t) = -32.2 \text{ ft/sec}^2 \tag{a}$$

$$\ddot{x}(t) = 0 \tag{b}$$

We need not bother with $\ddot{z}(t)$, since the motion is coplanar with $\dot{z}(t) = z = 0$ at all times. Integrating the equations above, we get

$$V_y(t) = -32.2t + C_1$$

$$V_x(t) = C_2$$

We shall take $t = 0$ at the instant the cannon is fired. At this instant, we know V_y and V_x and can determine C_1 and C_2. Thus,

$$V_y(0) = 3000 \sin 15° = (-32.2)(0) + C_1$$

Therefore,

$$C_1 = V_y(0) = 776 \text{ ft/sec}$$

$$V_x(0) = 3000 \cos 15° = C_2$$

Therefore,

$$C_2 = V_x(0) = 2900 \text{ ft/sec}$$

We can give the velocity components of the shell now as follows:

$$V_y(t) = -32.2t + 776 \text{ ft/sec} \tag{c}$$

$$V_x(t) = 2900 \text{ ft/sec} \tag{d}$$

Thus, the horizontal velocity is constant. Integrating once again, we get the coordinates x and y of the shell:

$$y(t) = -32.2\frac{t^2}{2} + 776t + C_3 \tag{e}$$

$$x(t) = 2900\,t + C_4 \tag{f}$$

When $t = 0$, $y = x = 0$. Thus, from Eqs. (e) and (f), we clearly see that $C_3 = C_4 = 0$. The coordinates of the shell are then

$$y(t) = -16.1t^2 + 776t \tag{g}$$

$$x(t) = 2900t \tag{h}$$

To determine distance d, first find the time t for the impact of the shell on the plain. That is, set $y = -500$ in Eq. (g) and solve for the time t. Thus,

$$-500 = -16.1t^2 + 776t$$

Therefore,

$$16.1t^2 - 776t - 500 = 0$$

Using the quadratic formula, we get for t:

$$t = 48.8 \text{ sec}$$

Substituting this value of t into Eq. (h), we get

$$d = (2900)(48.8) = 141{,}500 \text{ ft}$$

To get the maximum height y_{max} above the plane, first find the time t when $V_y = 0$. Thus, from Eq. (c) we get

$$0 = -32.2t + 776$$

Therefore,

$$t = 24.1 \text{ sec}$$

Now substitute $t = 24.1$ sec into Eq. (g). This gives us y_{max}.

$$y_{max} = -(16.1)(24.1)^2 + 776(24.1)$$

$$= 9350 \text{ ft}$$

Finally, to get the trajectory of the shell, solve for t in Eq. (h) and substitute this into Eq. (g). We then have

$$y = -16.1\left(\frac{x}{2900}\right)^2 + 776\left(\frac{x}{2900}\right)$$

Therefore,

$$y = -1.917 \times 10^{-6}x^2 + .268x \qquad \text{(i)}$$

Clearly, the trajectory is that of a *parabola*.

EXAMPLE 11.3

A gun emplacement is shown on a cliff in Fig. 11.6. The muzzle velocity of the gun is 1000 m/sec. At what angle α must the gun point in order to hit target A shown in the diagram? Neglect friction again.

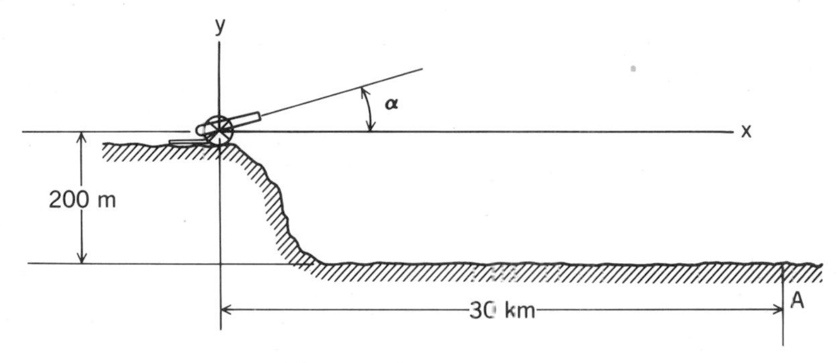

Figure 11.6. Find α to hit A.

Newton's law for the shell is given as follows for a reference xy having its origin at the gun.

$$\ddot{y}(t) = -9.81$$

$$\ddot{x}(t) = 0$$

Integrating, we get

$$\dot{y}(t) = V_y(t) = -9.81t + C_1 \qquad \text{(a)}$$

$$\dot{x}(t) = V_x(t) = C_2 \qquad \text{(b)}$$

When $t = 0$, we have $\dot{y} = 1000 \sin \alpha$ and $\dot{x} = 1000 \cos \alpha$. Applying these conditions to Eqs. (a) and (b), we solve for C_1 and C_2. Thus,

$$1000 \sin \alpha = 0 - C_1$$

Therefore,

$$C_1 = 1000 \sin \alpha$$

$$1000 \cos \alpha = C_2$$

Therefore,

$$C_2 = 1000 \cos \alpha$$

Hence, we have

$$\dot{y}(t) = -9.81t + 1000 \sin \alpha$$

$$\dot{x}(t) = 1000 \cos \alpha$$

Integrating again, we get

$$y(t) = -9.81\frac{t^2}{2} + 1000 \sin \alpha \, t + C_3$$

$$x(t) = 1000 \cos \alpha \, t + C_4$$

When $t = 0$, $x = y = 0$. Hence, it is clear that $C_3 = C_4 = 0$. Thus, we have

$$y = -4.905t^2 + 1000 \sin \alpha\, t \tag{c}$$

$$x = 1000 \cos \alpha\, t \tag{d}$$

To get the trajectory, we solve for t in Eq. (d) and substitute into Eq. (c).

$$\begin{aligned} y &= -4.905\frac{x^2}{(1000 \cos \alpha)^2} + 1000 \sin \alpha \frac{x}{1000 \cos \alpha} \\ &= -4.905 \times 10^{-6}\frac{x^2}{\cos^2 \alpha} + x \tan \alpha \end{aligned} \tag{e}$$

When $x = 30$ km (i.e., 30,000 m), $y = -200$ m. Hence, we have on substituting these data in Eq. (e):

$$-200 = -4.905 \times 10^{-6}\frac{(30{,}000)^2}{\cos^2 \alpha} + 30{,}000 \tan \alpha$$

Replace $1/\cos^2 \alpha$ by $\sec^2 \alpha = (1 + \tan^2 \alpha)$:

$$-200 = -4.905 \times 10^{-6}(30{,}000)^2(1 + \tan^2 \alpha) + 30{,}000 \tan \alpha$$

Therefore,

$$\tan^2 \alpha - 6.796 \tan \alpha + .955 = 0 \tag{f}$$

Using the quadratic formula, we find the following angles:

$$\alpha_1 = 8.17^\circ$$

$$\alpha_2 = 81.44^\circ$$

There are thus two possible firing angles that will permit the shell to hit the target, as shown in Fig. 11.7.

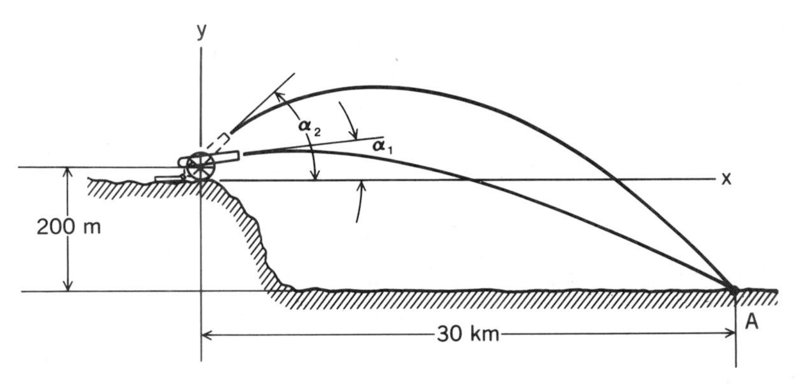

Figure 11.7. Two firing angles are possible.

We do not always know the variation of the position vector with time in the form of Eq. 11.6. Furthermore, it may be that the components of velocity and acceleration that we desire are not those parallel to a fixed Cartesian reference. The evaluation of $\boldsymbol{V}$ and $\boldsymbol{a}$ for certain other circumstances will be considered in the following sections.

11.5 Velocity and Acceleration in Terms of Path Variables

We have formulated velocity and acceleration for the case where the rectangular coordinates of a particle are known as functions of time. We now explore another approach in which the formulations are carried out in terms of the path variables of the particle, that is, in terms of geometrical parameters of the path and the speed and the rate of change of speed of the particle along the path. These results are particularly useful when a particle moves along a path that we know a priori (such as the case of a roller coaster).

As a matter of fact, in Section 11.2 (Eq. 11.4) we expressed the velocity vector in terms of path variables in the following form:

$$\boldsymbol{V} = \frac{ds}{dt}\boldsymbol{\epsilon}_t \tag{11.12}$$

where ds/dt represents the speed along the path and $\boldsymbol{\epsilon}_t = d\boldsymbol{r}/ds$ is the unit vector tangent to the path (and hence collinear with the velocity vector). The acceleration becomes

$$\frac{d\boldsymbol{V}}{dt} = \boldsymbol{a} = \frac{d^2s}{dt^2}\boldsymbol{\epsilon}_t + \frac{ds}{dt}\frac{d\boldsymbol{\epsilon}_t}{dt} \tag{11.13}$$

Replace $d\boldsymbol{\epsilon}_t/dt$ in this expression by $(d\boldsymbol{\epsilon}_t/ds)(ds/dt)$, the validity of which is assured by the chain rule of differentiation. We then have

$$\boldsymbol{a} = \frac{d^2s}{dt^2}\boldsymbol{\epsilon}_t + \left(\frac{ds}{dt}\right)^2 \frac{d\boldsymbol{\epsilon}_t}{ds} \tag{11.14}$$

Before proceeding further, let us consider the unit vector $\boldsymbol{\epsilon}_t$ at two positions that are Δs apart along the path of the particle as shown in Fig. 11.8. If Δs is small enough, the unit vectors $\boldsymbol{\epsilon}_t(s)$ and $\boldsymbol{\epsilon}_t(s + \Delta s)$ can be considered to intersect and thus to form a plane. If $\Delta s \longrightarrow 0$, these unit vectors then form a *limiting plane*, which we shall call the *osculating plane*.[1] The plane will have an orientation that depends on the position s on the path of the particle.

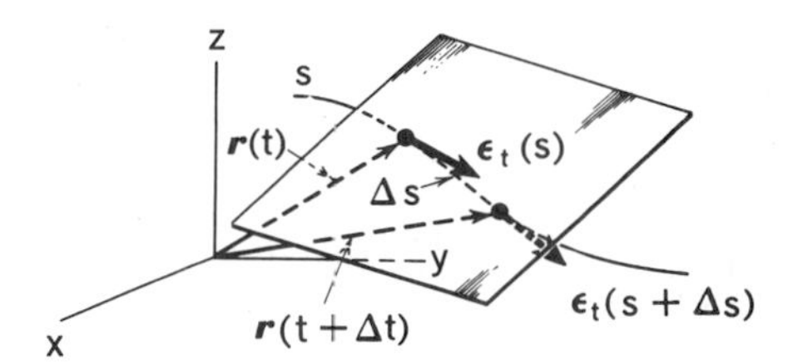

Figure 11.8. Osculating plane.

The osculating plane at $\boldsymbol{r}(t)$ is illustrated in Fig. 11.8. Having defined the osculating plane, let us continue discussion of Eq. 11.14.

Since we have not formally carried out the differentiation of a vector with respect to a spatial coordinate, we shall carry out the derivative $d\boldsymbol{\epsilon}_t/ds$ needed in Eq. 11.14 from the basic definition. Thus,

---

[1]From the definition, it should be apparent that the osculating plane at position s along a curve is actually *tangent* to the curve at position s. Since osculate means to kiss, the plane "kisses" the curve, as it were, at s.

$$\frac{d\boldsymbol{\epsilon}_t}{ds} = \lim_{\Delta s \to 0}\left[\frac{\boldsymbol{\epsilon}_t(s+\Delta s) - \boldsymbol{\epsilon}_t(s)}{\Delta s}\right] = \lim_{\Delta s \to 0}\left(\frac{\Delta \boldsymbol{\epsilon}_t}{\Delta s}\right) \tag{11.15}$$

The vectors $\boldsymbol{\epsilon}_t(s)$ and $\boldsymbol{\epsilon}_t(s + \Delta s)$ are shown in Fig. 11.9(a) along the path and are also shown with $\Delta\boldsymbol{\epsilon}_t$ as a vector triangle in Fig. 11.9(b). As pointed out earlier, for small enough Δs the lines of action of the unit vectors $\boldsymbol{\epsilon}_t(s)$ and $\boldsymbol{\epsilon}_t(s + \Delta s)$ will intersect to form a plane as shown in Fig. 11.9(a). Now in this plane, draw normal lines to the aforementioned vectors at the respective positions s and $s + \Delta s$. These lines will intersect

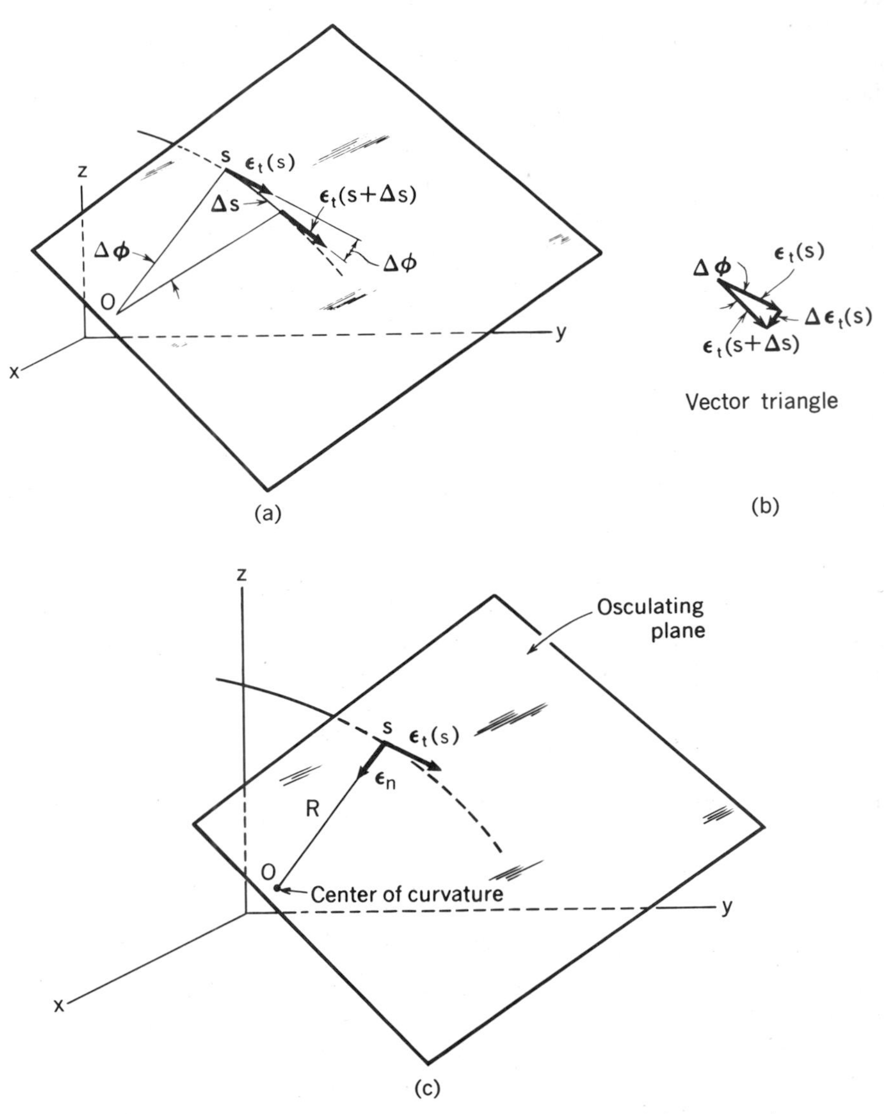

Figure 11.9. Development of the osculating plane and the center of curvature.

at some point O, as shown in the diagram. Next, consider what happens to the planes and to point O as $\Delta s \longrightarrow 0$. Clearly, the limiting plane is our osculating plane at s [see Fig. 11.9(c)]. Furthermore, the limiting position arrived at for point O is *in the osculating plane* and is called the *center of curvature* for the path at s. The distance between O and s is denoted as R and is called the *radius of curvature*. Finally, the vector $\Delta \boldsymbol{\epsilon}_t$, in the limit as $\Delta s \longrightarrow 0$, ends up in the osculating plane normal to the path at s and directed toward the center of curvature. The unit vector collinear with the limiting vector for $\Delta \boldsymbol{\epsilon}_t$ is denoted as $\boldsymbol{\epsilon}_n$ and is called the *principal normal vector*.

With the limiting *direction* of $\Delta \boldsymbol{\epsilon}_t$ established, we next evaluate the *magnitude* of $\Delta \boldsymbol{\epsilon}_t$ as an approximate value that becomes correct as $\Delta s \longrightarrow 0$. Observing the vector triangle in Fig. 11.9(b), we can accordingly say:

$$|\Delta \boldsymbol{\epsilon}_t| \approx |\boldsymbol{\epsilon}_t| \, \Delta\phi = \Delta\phi \tag{11.16}$$

Next, we note in Fig. 11.9(a) that the lines from point O to the points s and $s + \Delta s$ along the trajectory form the same angle $\Delta\phi$ as is between the vectors $\boldsymbol{\epsilon}_t(s)$ and $\boldsymbol{\epsilon}_t(s + \Delta s)$ in the vector triangle, and so we can say:

$$\Delta\phi = \frac{\Delta s}{Os} \approx \frac{\Delta s}{R}$$

Hence, we have for Eq. 11.16:

$$|\Delta \boldsymbol{\epsilon}_t| \approx \frac{\Delta s}{R}$$

We thus have the magnitude of $\Delta \boldsymbol{\epsilon}_t$ established in an approximate manner. Using $\boldsymbol{\epsilon}_n$, the principal normal at s, to approximate the direction of $\Delta \boldsymbol{\epsilon}_t$ we can write

$$\Delta \boldsymbol{\epsilon}_t \approx \frac{\Delta s}{R} \boldsymbol{\epsilon}_n$$

If we use this result in the limiting process of Eq. 11.15 (where it becomes exact), the evaluation of $d\boldsymbol{\epsilon}_t/ds$ becomes

$$\frac{d\boldsymbol{\epsilon}_t}{ds} = \lim_{\Delta s \to 0} \left(\frac{\Delta \boldsymbol{\epsilon}_t}{\Delta s} \right) = \lim_{\Delta s \to 0} \left[\frac{(\Delta s/R)\boldsymbol{\epsilon}_n}{\Delta s} \right] = \frac{\boldsymbol{\epsilon}_n}{R} \tag{11.17}$$

When we substitute Eq. 11.17 into Eq. 11.14, the acceleration vector becomes

$$\boxed{\boldsymbol{a} = \frac{d^2 s}{dt^2} \boldsymbol{\epsilon}_t - \frac{(ds/dt)^2}{R} \boldsymbol{\epsilon}_n} \tag{11.18}$$

We thus have two components of acceleration: *one component in a direction tangent to the path and one component in the osculating plane at right angles to the path and pointing toward the center of curvature*. These components are of great importance in some problems.

For the special case of a *plane curve*, we learned in analytic geometry that the radius of curvature R is given by the relation

$$R = \frac{\left[1 + \left(\frac{dy}{dx}\right)^2\right]^{3/2}}{\left|\frac{d^2y}{dx^2}\right|} \tag{11.19}$$

Furthermore, in the case of a plane curve, the osculating plane at every point clearly must correspond to the plane of the curve, and the computation of unit vectors $\boldsymbol{\epsilon}_n$ and $\boldsymbol{\epsilon}_t$ is quite simple, as will be illustrated in Example 11.5.

How do we get the principal normal vector $\boldsymbol{\epsilon}_n$, the radius of curvature, and the direction of the osculating plane for a three-dimensional curve? One procedure is to evaluate $\boldsymbol{\epsilon}_t$ as a function of s and then differentiate this vector with respect to s. Accordingly, from Eq. 11.17 we can then determine $\boldsymbol{\epsilon}_n$ as well as R. We establish the direction of the osculating plane by taking the cross product of $\boldsymbol{\epsilon}_n \times \boldsymbol{\epsilon}_t$, to get a unit vector normal to the osculating plane. This vector is called the *binormal vector*.

EXAMPLE 11.4

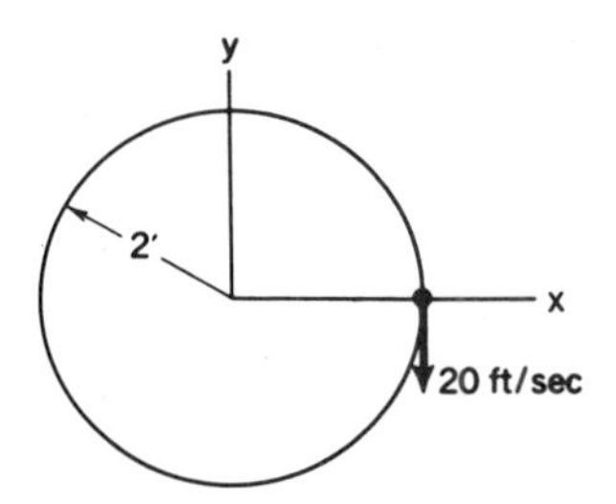

Figure 11.10. Particle on circular path.

A particle is moving along a circular path in the xy plane (Fig. 11.10). When the particle crosses the x axis, it has an acceleration along the path of 5 ft/sec² and is moving with the speed of 20 ft/sec in the negative y direction. What is the total acceleration of the particle?

Clearly, the osculating plane must be the plane of the path. Hence, R is 2 ft, as is shown in the diagram. We need simply to employ Eq. 11.18 for the desired result. Thus,

$$\boldsymbol{a} = 5\boldsymbol{\epsilon}_t + \frac{20^2}{2}\boldsymbol{\epsilon}_n \text{ ft/sec}^2$$

For the xy reference, the acceleration is

$$\boldsymbol{a} = -5\boldsymbol{j} - 200\boldsymbol{i} \text{ ft/sec}^2$$

EXAMPLE 11.5

A particle is moving along a parabolic path given as $y = 1.22\sqrt{x}$ (see Fig. 11.11) with x and y in meters. At position A, the particle has a speed of 3 m/sec and has a rate of change of speed of 3 m/sec² along the path. What is the acceleration vector of the particle at this position?

We first find $\boldsymbol{\epsilon}_t$ by noting from the diagram that

$$\boldsymbol{\epsilon}_t = \cos\alpha\, \boldsymbol{i} + \sin\alpha\, \boldsymbol{j} \tag{a}$$

where

$$\tan\alpha = \frac{dy}{dx} = \frac{d}{dx}(1.22\sqrt{x}) = \frac{.610}{\sqrt{x}} \tag{b}$$

At the position of interest, we have

$$\tan\alpha = \frac{.610}{\sqrt{1.5}} = \frac{1}{2}$$

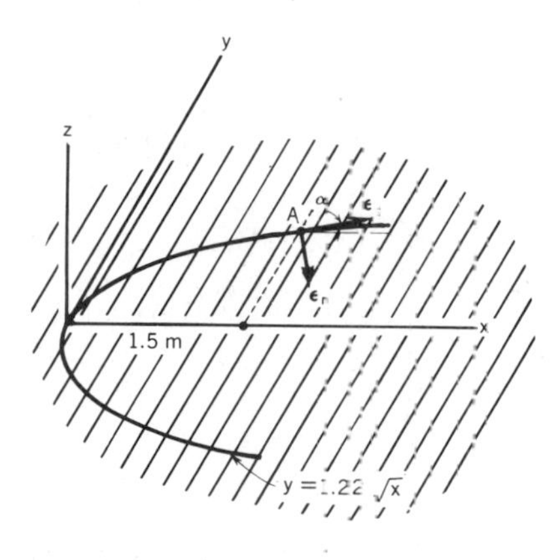

Figure 11.11. Particle on a parabolic path.

Therefore,

$$\alpha = 26.5°$$

Hence,

$$\boldsymbol{\epsilon}_t = .895\boldsymbol{i} + .446\boldsymbol{j} \tag{c}$$

As for $\boldsymbol{\epsilon}_n$, we see from the diagram that

$$\boldsymbol{\epsilon}_n = \sin\alpha\,\boldsymbol{i} - \cos\alpha\,\boldsymbol{j}$$

Therefore,

$$\boldsymbol{\epsilon}_n = .446\boldsymbol{i} - .895\boldsymbol{j} \tag{d}$$

Next, employing Eq. 11.19, we can find R. We shall need the following results for this step:

$$\frac{dy}{dx} = .610x^{-1/2} \tag{e}$$

$$\frac{d^2y}{dx^2} = -.305x^{-3/2} \tag{f}$$

Substituting Eqs. (e) and (f) into Eq. 11.19, we have for R:

$$R = \frac{[1 + (.610x^{-1/2})^2]^{3/2}}{.305x^{-3/2}} \tag{g}$$

At the position of interest, $x = 1.5$, we get

$$R = 8.40 \text{ m} \tag{h}$$

We can now give the desired acceleration vector. Thus, from Eq. 11.18, we have

$$\begin{aligned} \boldsymbol{a} &= 3(.895\boldsymbol{i} + .446\boldsymbol{j}) + \frac{9}{8.40}(.446\boldsymbol{i} - .895\boldsymbol{j}) \\ &= 3.16\boldsymbol{i} + .379\boldsymbol{j} \text{ m/sec}^2 \end{aligned} \tag{i}$$

Problems

11.1. A mass is supported by four springs. The mass is given a vibratory movement in the horizontal (x) direction and simultaneously a vibratory movement in the vertical (y) direction. These motions are given as follows:

$$x = 2 \sin 2t \text{ mm/sec}$$

$$y = 2 \cos (2t + .3) \text{ mm/sec}$$

What is the value of the acceleration vector at $t = 4$ sec? How many g's of acceleration does this correspond to?

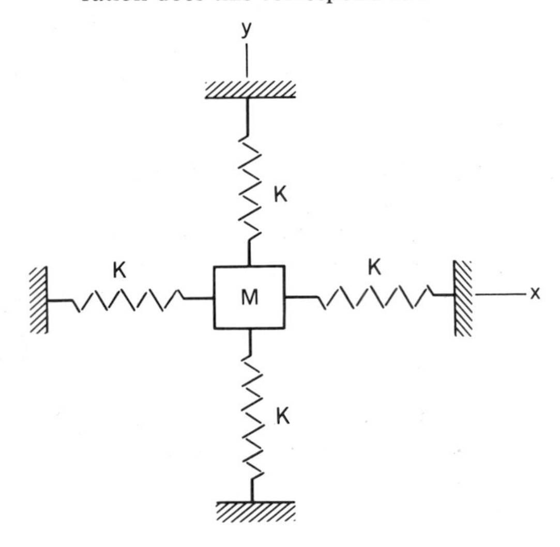

Figure P.11.1

11.2. A particle moves along a plane circular path of radius r equal to 1 ft. The position OA is given as a function of time as follows:

$$\theta = 6 \sin 5t \text{ rad}$$

where t is in seconds. What are the rectangular components of velocity for the particle at time $t = \frac{1}{5}$ sec?

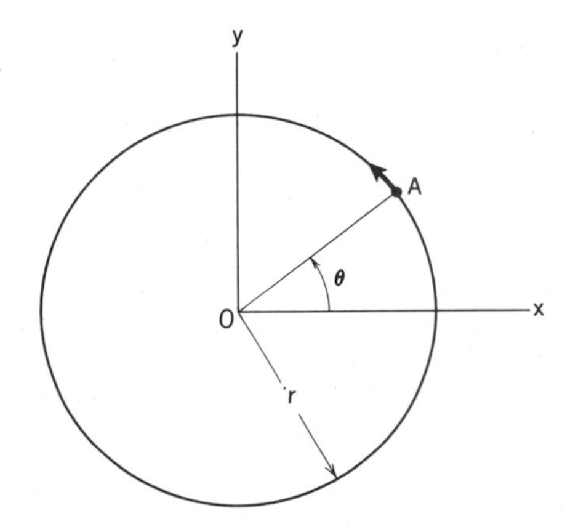

Figure P.11.2

11.3. A particle with an initial position vector $\boldsymbol{r} = 5\boldsymbol{i} + 6\boldsymbol{j} + \boldsymbol{k}$ m has an acceleration imposed on it, given as

$$\boldsymbol{a} = 6t\boldsymbol{i} + 5t^2\boldsymbol{j} + 10\boldsymbol{k} \text{ m/sec}^2$$

If the particle has zero velocity initially, what are the acceleration, velocity, and position of the particle when $t = 10$ sec?

11.4. The position of a particle at times $t = 10$ sec, $t = 5$ sec, and $t = 2$ sec is known to be, respectively:

$$\boldsymbol{r}(10) = 10\boldsymbol{i} + 5\boldsymbol{j} - 10\boldsymbol{k} \text{ ft}$$

$$\boldsymbol{r}(5) = 3\boldsymbol{i} + 2\boldsymbol{j} + 5\boldsymbol{k} \text{ ft}$$

$$\boldsymbol{r}(2) = 8\boldsymbol{i} - 20\boldsymbol{j} + 10\boldsymbol{k} \text{ ft}$$

What is the acceleration of the particle at time $t = 5$ sec if the acceleration vector has the form

$$\boldsymbol{a} = C_1 t\boldsymbol{i} + C_2 t^2\boldsymbol{j} + C_3 \ln t\boldsymbol{k} \text{ ft/sec}^2$$

where C_1, C_2, and C_3 are constants and t is in seconds?

11.5. A highly idealized diagram is shown of an *accelerometer*, a device for measuring the acceleration component of motion along a certain direction—in this case the indicated x direction. A mass B is constrained in the accelerometer case so that it can only move against linear springs in the x direction. When the accelerometer case accelerates in this direction, the mass assumes a displaced position, shown dashed, at a distance δ from its original position. This configuration is such that the force in the springs gives the mass B the acceleration corresponding to that of the accelerometer case. The shift δ of the mass in the case is picked up by an electrical sensor device and is plotted as a function of time. The damping fluid present eliminates extraneous oscillations of the mass. If a plot of a_x versus time has the form shown, what is the speed of the body after 10 sec, 30 sec, and 45 sec? The acceleration a_x is measured in g's—i.e., in units of 32.2 ft/sec² or 9.81 m/sec². Assume that the body starts from rest at $x = 0$.

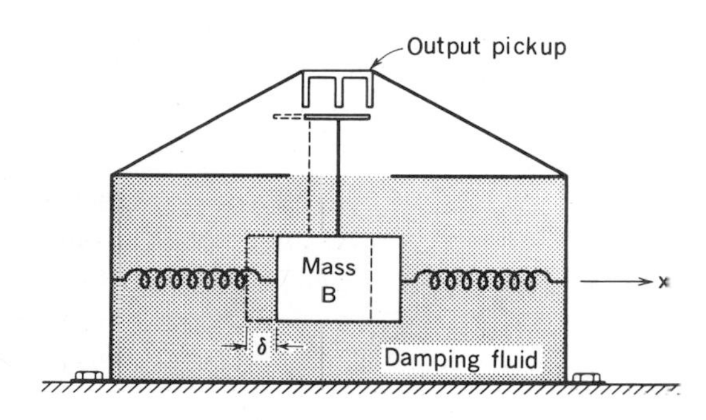

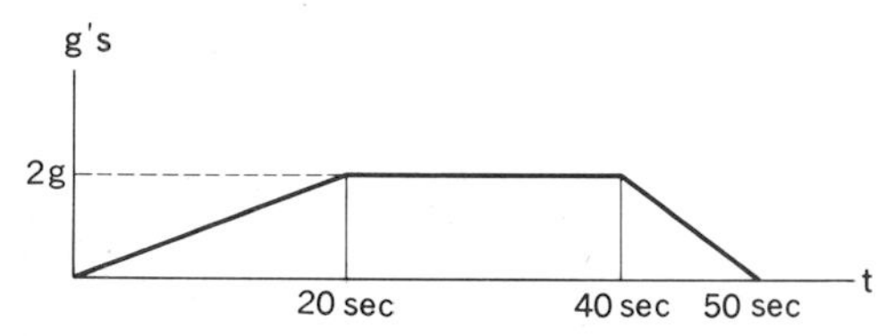

Figure P.11.5

11.6. The position vector of a particle is given as

$$r = 6t\boldsymbol{i} + (5t + 10)\boldsymbol{j} + 6t^2\boldsymbol{k} \text{ m}$$

What is the acceleration of the particle at $t = 3$ sec? What distance has been traveled by the particle during this time? (*Hint:* Let $dr = \sqrt{dx^2 + dy^2 + dz^2}$ and divide and multiply by dt in second half of problem. Look up integration form $\int \sqrt{a^2 + t^2}\, dt$ in Appendix I.

11.7. In Example 11.1, what is the acceleration vector for pin B if the yoke C is accelerating at the rate of 10 ft/sec² at the instant of interest?

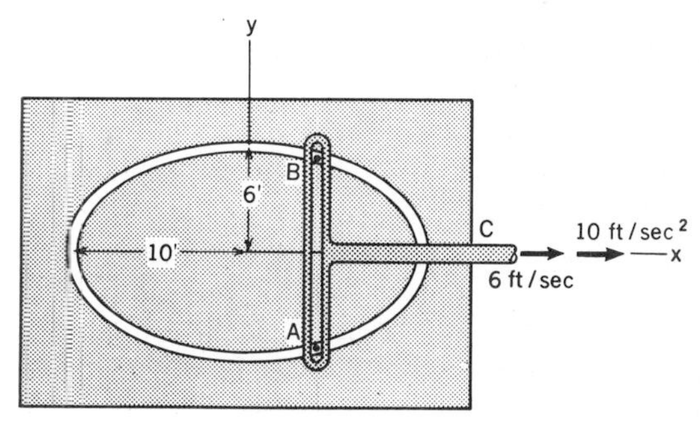

Figure P.11.7

11.8. Particles A and B are confined to always be in a circular groove of radius 5 ft. At the same time, these particles must also be in a slot which has the shape of a parabola. The slot is shown dashed at time $t = 0$. If the slot moves to the right at a constant speed of 3 ft/sec, what are the speed and rate of change of speed of particles toward each other at $t = 1$ sec?

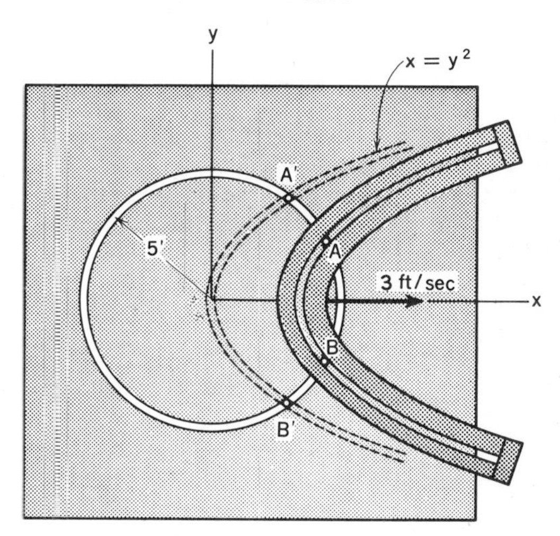

Figure P.11.8

11.9. The face of a cathode ray tube is shown. An electron is made to move in the horizontal (x) direction due to electric fields in the cathode tube with the following motion:

$$x = A \sin \omega t \text{ mm/sec}$$

Also, the electron is made to move in the vertical direction with the following motion:

$$y = A \sin (\omega t + \alpha) \text{ mm/sec}$$

Show that for $\alpha = \pi/2$, the trajectory on the screen is that of a *circle* of radius A mm. If $\alpha = \pi$, show that the trajectory is that of a *straight line* inclined at 45° to the xy axes. Finally, give the formulations for the directions of velocity and acceleration of the electron in the xy plane.

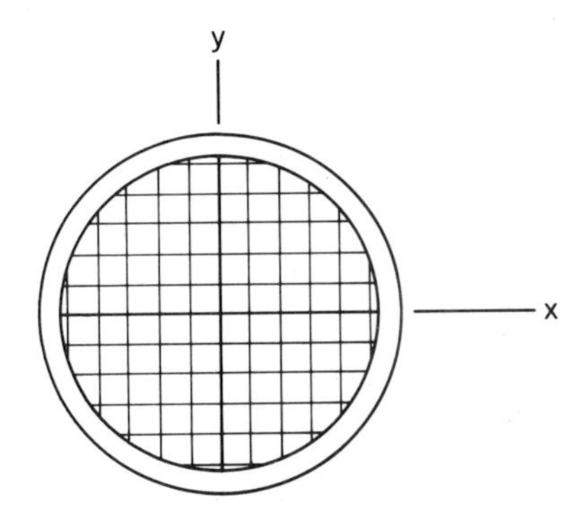

Figure P.11.9

11.10. A stunt motorcyclist is to attempt a "jump" over a deep chasm. The distance between jump-off point and landing is 100 m. As technical advisor to this stunt man, what speed do you tell him to exceed at the jump-off point A? The cycle is highly streamlined to minimize wind resistance. Give the result in km/hr.

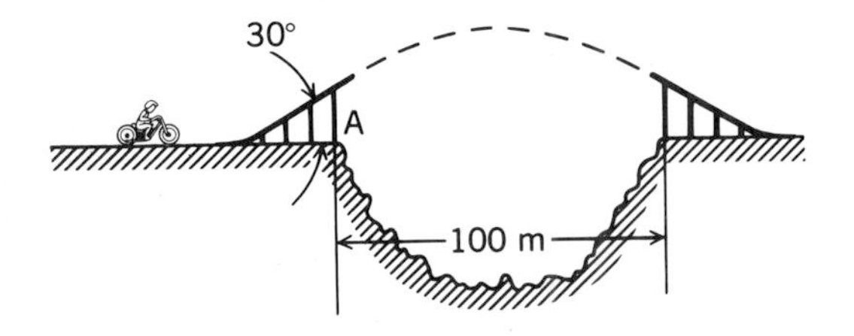

Figure P.11.10

11.11. A charged particle is shot at time $t = 0$ at an angle of 45° with a speed 10 ft/sec. If an electric field is such that the body has an acceleration $-200t^2\mathbf{j}$ ft/sec², what is the equation for the trajectory? What is the value of d for impact?

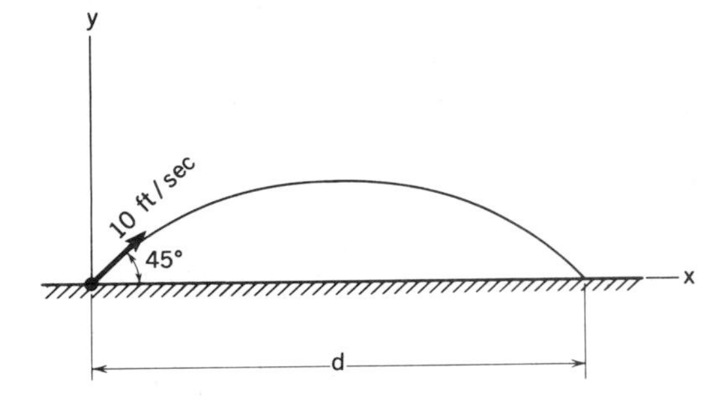

Figure P.11.11

11.12. A projectile is fired at a speed of 1000 m/sec at an angle ϵ of 40° measured from an inclined surface which is at an angle ϕ of 20° from the horizontal. If we neglect friction, at what distance along the incline does the projectile hit the incline?

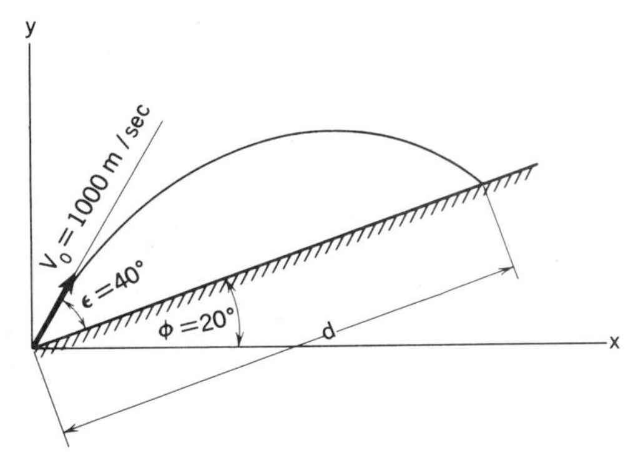

Figure P.11.12

11.13. Grain is being blown into an open train container at a speed V_0 of 20 ft/sec. What should the minimum and maximum elevations d be to ensure that all the grain gets into the train? Neglect friction and winds.

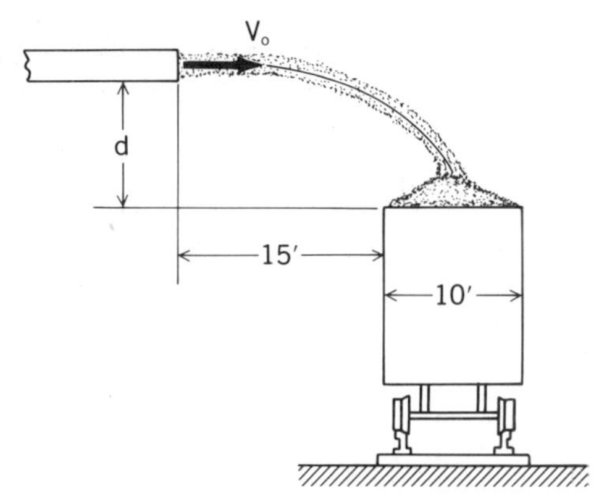

Figure P.11.13

11.14. The engine room of a freighter is on fire. A fire-fighting tugboat has drawn alongside and is directing a stream of water to go into the stack of the freighter. If the speed of the jet of water is 70 ft/sec, what angle α is needed to accomplish the task? (*Hint:* Only one α will result in water getting into the stack.)

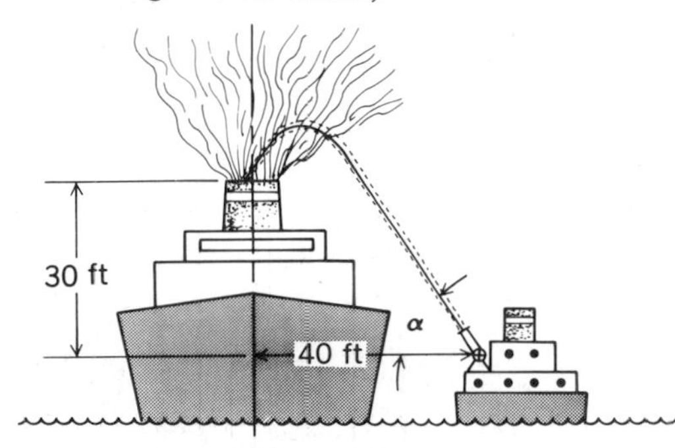

Figure P.11.14

11.15. A rocket-powered test sled slides over rails. This test sled is used for experimentation on the ability of man to undergo large persistent accelerations. To brake the sled from high speeds, small scoops are lowered to deflect water from a stationary tank of water placed near the end of the run. If the sled is moving at a speed of 100 km/hr at the instant of interest, compute h and d of the deflected stream of water as seen from the sled. Assume no loss in speed of the water relative to the scoop. Consider the sled as an inertial reference at the instant of interest and attach xy reference to the sled.

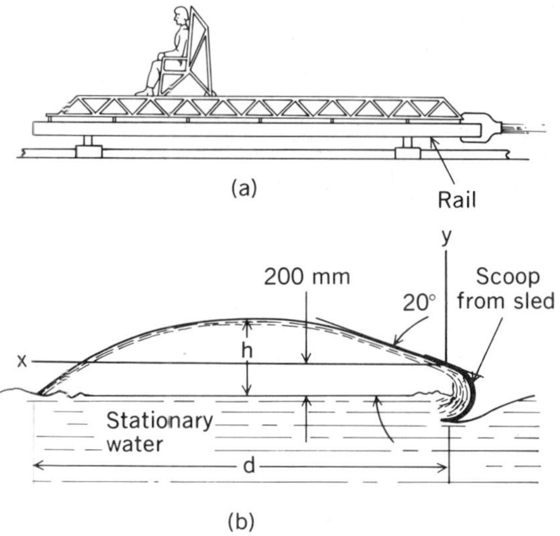

Figure P.11.15

11.16. A sportsman in a valley is trying to shoot a deer on a hill. He quickly estimates the distance of the deer along his line of sight as 500 yd and the height of the hill as 100 yd. His gun has a muzzle velocity of 3000 ft/sec. If he has no graduated sight, how many feet above the deer should he aim his rifle in order to hit it? (Neglect friction.)

11.17. A fireman is directing water from a hose into the broken window of a burning house. The velocity of the water is 15 m/sec as it leaves the hose. What are the angles α needed to do the job?

Figure P.11.17

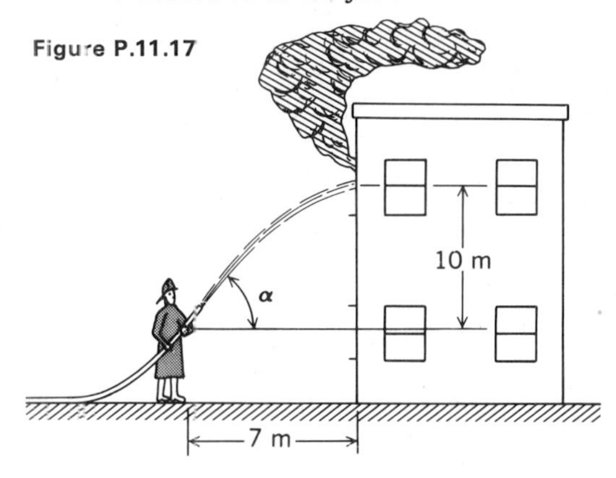

11.18. An archer in a Jeep is chasing a deer. The Jeep moves at 30 mi/hr and the deer moves at 15 mi/hr along the same direction. At what inclination must the arrow be shot if the deer is 100 yd ahead of the Jeep and if the initial speed of the arrow is 200 ft/sec relative to archer? (Neglect friction.)

11.19. A fighter plane is directly over an antiaircraft gun at time $t = 0$. The plane has a speed V_1 of 500 km/hour. A shell is fired at $t = 0$ in an attempt to hit the plane. If the muzzle velocity V_0 is 1000 m/sec, how many meters d should the gun be aimed ahead of the plane to hit it? What is the time of impact?

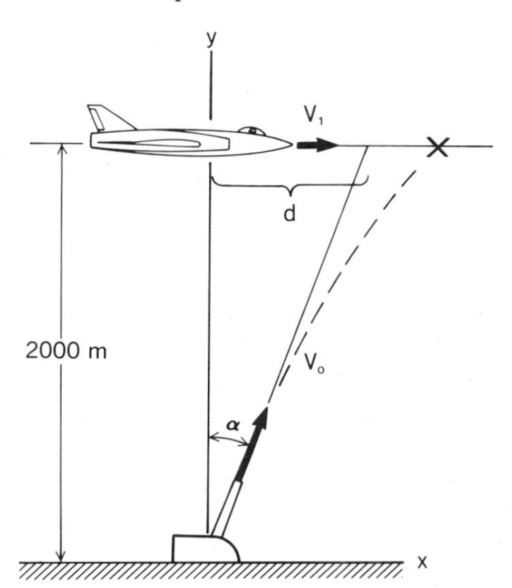

Figure P.11.19

11.20. A destroyer is making a run at full speed of 75 km/hr. When abreast of a missile site target, it fires two shells. The target is 12,000 m from the destroyer. If the muzzle velocity is 400 m/sec, what is the angle of firing α with the horizontal that the computer must set the guns? Also, what angle β must the turret be rotated relative to the line of sight at the instant of firing? (*Hint:* To hit target, what must V_y of the shell be? Result: $\alpha = 23.7°$ and $\beta = 3.26°$.)

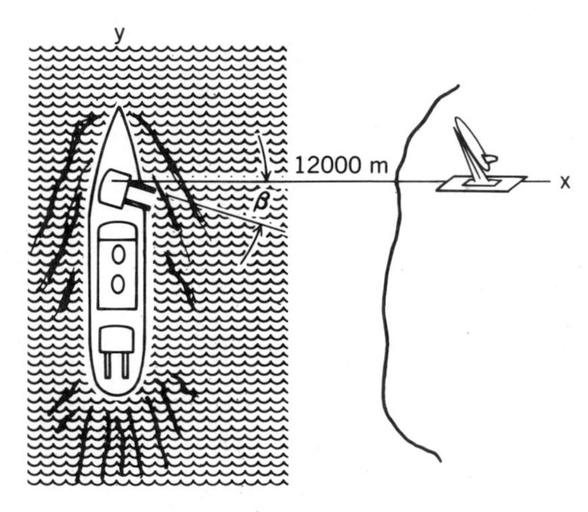

Figure P.11.20

***11.21.** A Jeep with an archer is moving at a speed of 30 mi/hr. At 100 yd distance and moving at right angles to the Jeep is a deer running at a speed of 15 mi/hr. If the initial speed of the arrow shot by the archer to bag the deer is 200 ft/sec, what inclination α must the shot have with the horizontal and what angle β must the shot have relative to the line AB?

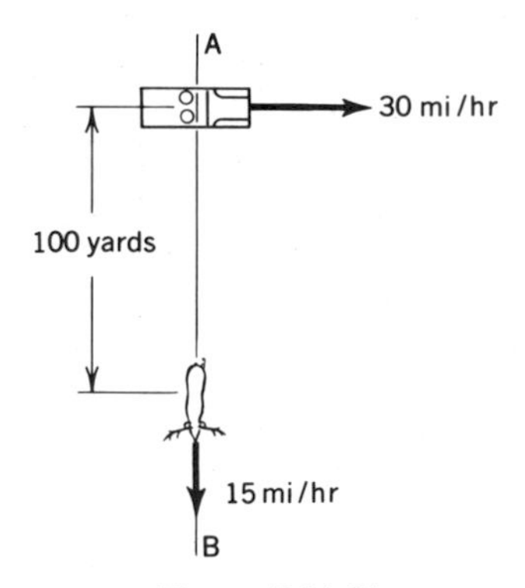

Figure P.11.21

11.22. A particle moves with a constant speed of 5 ft/sec along the path. Compute the acceleration at points 1, 2, and 3.

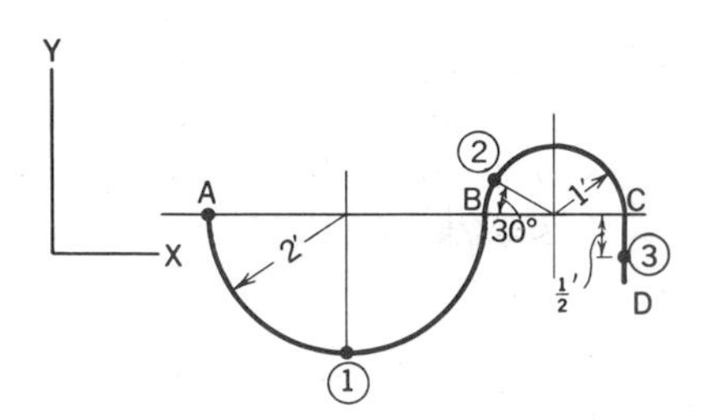

Figure P.11.22

11.23. If, in Problem 11.22 the speed is 5 ft/sec only at point A, and it increases 5 ft/sec for each foot traveled, compute the acceleration at points 1, 2, and 3.

11.24. A car is moving at a speed of 88 km/hr along a highway. At a curve in the highway, the radius of curvature is 1300 m. What is the acceleration of the car? To decrease this acceleration by 30%, what must its speed be?

11.25. A high-speed train is running at 100 km/hr. It goes into a curve having a minimum radius of curvature of 2000 m. What is the acceleration that sitting passengers are subjected to? If the radius of curvature were to be doubled, at what constant speed could the train then go with the same acceleration?

11.26. An amusement park ride consists of a cockpit in which a passenger is strapped in a seated position. The cockpit rotates about A with angular speed ω. The average person's head is 10 ft from the axis of rotation at A. We know that if a person's head is subjected to an acceleration of 3 g's or more in a direction from shoulders to head for any length of time, he/she will be uncomfortable and perhaps black out. What, then, is the maximum value of ω in rpm to prevent these effects, using a safety factor of 3? (*Hint:* You will soon learn that the speed in a circular path is $R\omega$.)

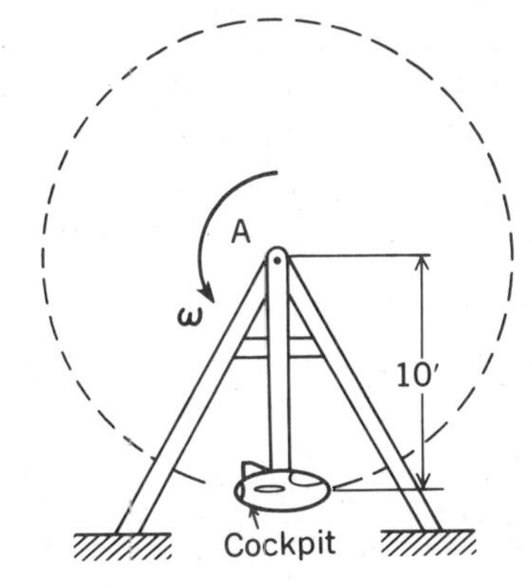

Figure P.11.26

11.27. A motorcyclist is moving along a circular path having a radius of curvature of 400 m. He is increasing his speed along the path at the rate of 5 km/hr/sec. If he enters the curve at a speed of 48 km/hr, what is his total acceleration after traveling 10 sec along this path?

11.28. What is the direction of the normal vector and the value of the radius of curvature at a position a of the curve?

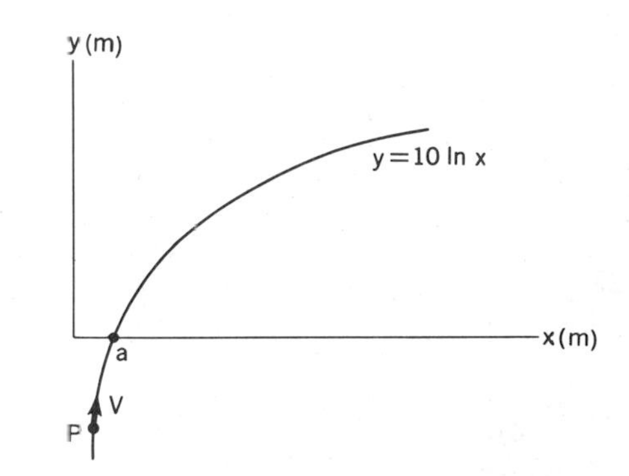

Figure P.11.28

11.29. A particle P moves with constant speed V along the curve $y = 10 \log x$ m. At what position x does the particle have the maximum acceleration? What is the value of this acceleration if $V = 1$ m/sec?

11.30. A particle moves with a constant speed of 3 m/sec along the path. What is the acceleration $\boldsymbol{a}$ at position $x = 1.5$ m? Give the rectangular components of $\boldsymbol{a}$.

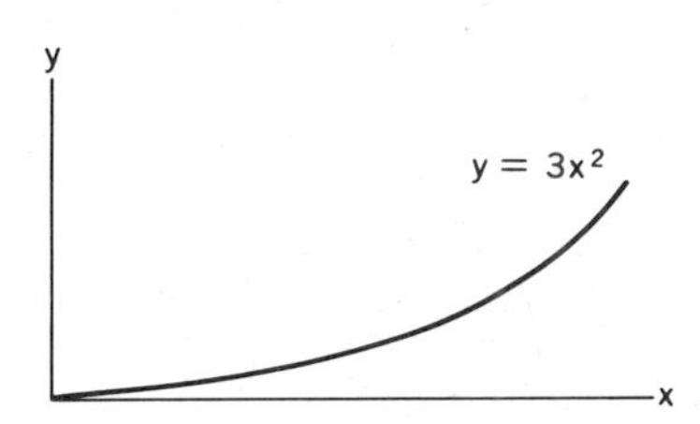

Figure P.11.30

11.31. A particle moves along a sinusoidal path. If the particle has a speed of 10 ft/sec and a rate of change of speed of 5 ft/sec² at A, what is the *magnitude* of the acceleration? What is the magnitude and direction of the acceleration of the particle at B, if it has a speed of 20 ft/sec and a rate of change of speed of 3 ft/sec² at this point?

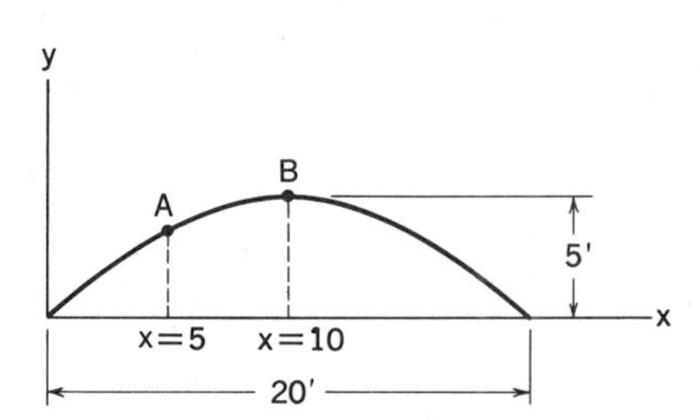

Figure P.11.31

11.32. A passenger plane is moving at constant speed of 200 km/hr in a holding pattern at a constant elevation. At the instant of interest, the angle β between the velocity vector and the x axis is 30°. The vector is known through on-board gyroscopic instrumentation to be changing at the rate $\dot{\beta}$ at $-5°$/sec. What is the radius of curvature of the path at this instant?

$$\left[Hint: \boldsymbol{a} = \frac{d(V\boldsymbol{\epsilon}_t)}{dt} = \frac{V^2}{R}\boldsymbol{\epsilon}_n.\right]$$

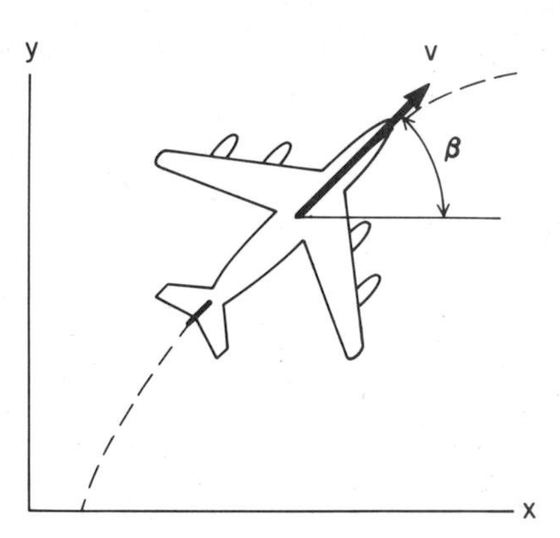

Figure P.11.32

11.33. At what position along the ellipse shown does the normal vector have a set of direction cosines (.707, .707, 0)? Recall that the equation for an ellipse in position shown is $x^2/a^2 + y^2/b^2 = 1$.

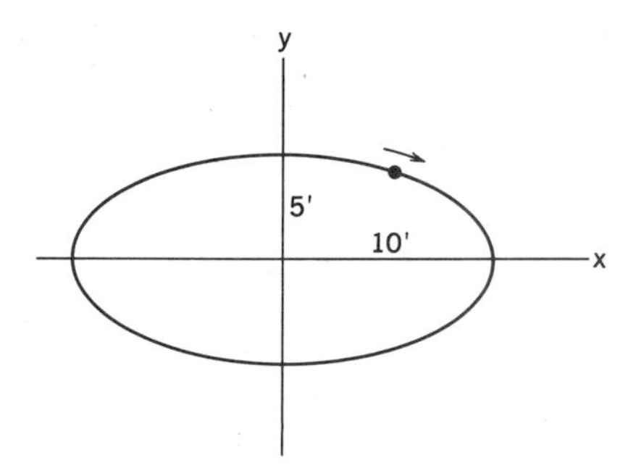

Figure P.11.33

11.34. A particle moves along a path given as

$$y = 3x^2 \text{ ft}$$

The projection of the particle along the x axis varies as $.2t^2$ ft (where t is in seconds) starting at the origin at $t = 0$. What are the acceleration components normal and tangential to the path at $t = 2$ sec? What is the radius of curvature at this point?

11.35. A particle moves along a path $y^2 = 10x$ with x and y in meters. The distance traversed along this path starting from the origin is given by S such that

$$S = \frac{t}{2} + \frac{t^2}{100} \text{ m}$$

where t is measured in units of seconds. What are the normal and tangential acceleration components of the particle when $y = 10$ m? $\left(\textit{Hint: } \int \sqrt{y^2 + a^2}\, dy\right)$ is presented in Appendix I. Also, note that $ds = \sqrt{dx^2 + dy^2} = \sqrt{\left(\frac{dx}{dy}\right)^2 + 1}\, dy$.

11.36. Show by arguments similar to those used in the text for deriving the relation $d\boldsymbol{\epsilon}_t/ds = (1/R)\boldsymbol{\epsilon}_n$ that $d\boldsymbol{\epsilon}_n/ds = -(1/R)\boldsymbol{\epsilon}_t$.

11.37. (a) For coplanar paths in the xy plane, find the formula for $\dot{\boldsymbol{a}}$, that is, the "jerk."

(b) If a particle moves on a plane circular path of radius 5 m at a speed of 5 m/sec, and if the rate of change of speed is 2 m/sec², what is $\dot{\boldsymbol{a}}$ for the particle if the second derivative of its speed along the path is 10 m/sec³? (*Hint:* Use the result of Problem 11.36.)

11.6 Cylindrical Coordinates

The final method we shall consider for evaluating the velocity and acceleration of a particle brings us back to considering coordinates of the particle as time functions, as we did at the outset of this study. Now we shall employ cylindrical coordinates, and we shall evaluate velocity and acceleration in components having certain directions that are associated with the cylindrical coordinates of the particle. Thus, particle P in

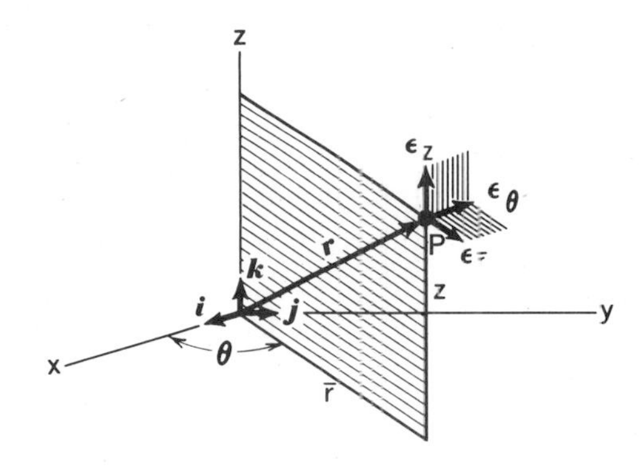

Figure 11.12. Cylindrical coordinates.

Fig. 11.12 is located by specifying cylindrical coordinates θ, $\bar{r}$, and z.[2] The transformation equations between Cartesian and cylindrical coordinates are

$$x = \bar{r}\cos\theta, \qquad \bar{r} = (x^2 + y^2)^{1/2}$$
$$y = \bar{r}\sin\theta, \qquad \theta = \tan^{-1}\frac{y}{x} \tag{11.22}$$

Unit vectors are associated with these coordinates and are given as:

[2]The notation $\bar{r}$ is used to distinguish it from r, which, according to previous definitions in statics, is the magnitude of $\boldsymbol{r}$, the position vector.

$\boldsymbol{\epsilon}_z$, which is parallel to the z axis and, for practical purposes, is the same as $\boldsymbol{k}$. This is considered to be the *axial direction.*

$\boldsymbol{\epsilon}_{\bar{r}}$, which is normal to the z axis, pointing out from the axis, and is identified as the *radial direction* from z.

$\boldsymbol{\epsilon}_\theta$, which is normal to the plane formed by $\boldsymbol{\epsilon}_{\bar{r}}$ and $\boldsymbol{\epsilon}_z$ and has a sense in accordance with the right-hand-screw rule for the permutation $z, \bar{r}, \theta$. We call this the *transverse direction.*

Note that $\boldsymbol{\epsilon}_{\bar{r}}$ *and* $\boldsymbol{\epsilon}_\theta$ *will change direction as the particle moves relative to the xyz reference.* Thus, these unit vectors are generally *functions of time*, whereas $\boldsymbol{\epsilon}_z$ is a constant vector.

Using previously developed concepts, we can express the velocity and acceleration of the particle relative to the xyz reference *in terms of components always in the transverse, radial, and axial directions and can use cylindrical coordinates exclusively in the process.* This information is most useful, for instance, in turbomachine studies (i.e., for centrifugal pumps, compressors, jet engines etc.), where, if we take the z axis as the axis of rotation, the axial components of fluid acceleration are used for thrust computation while the transverse components are important for torque considerations. It is these components that are meaningful for such computations and not components parallel to some xyz reference.

The position vector $\boldsymbol{r}$ of the particle determines the direction of the unit vectors $\boldsymbol{\epsilon}_{\bar{r}}$ and $\boldsymbol{\epsilon}_\theta$ at any time t and can be expressed as

$$\boldsymbol{r} = \bar{r}\boldsymbol{\epsilon}_{\bar{r}} + z\boldsymbol{\epsilon}_z \tag{11.23}$$

To get the desired velocity, we differentiate $\boldsymbol{r}$ with respect to time:

$$\frac{d\boldsymbol{r}}{dt} = \boldsymbol{V} = \bar{r}\dot{\boldsymbol{\epsilon}}_{\bar{r}} + \dot{\bar{r}}\boldsymbol{\epsilon}_{\bar{r}} + \dot{z}\boldsymbol{\epsilon}_z$$

Our task here is to evaluate $\dot{\boldsymbol{\epsilon}}_{\bar{r}}$. On consulting Fig. 11.12, we see clearly that changes in direction of $\boldsymbol{\epsilon}_{\bar{r}}$ occur only when the θ coordinate of the particle changes. Hence, remembering that the magnitude of $\boldsymbol{\epsilon}_{\bar{r}}$ is always constant, we have for $\dot{\boldsymbol{\epsilon}}_{\bar{r}}$ using the chain rule:

$$\frac{d\boldsymbol{\epsilon}_{\bar{r}}}{dt} = \frac{d\boldsymbol{\epsilon}_{\bar{r}}}{d\theta}\frac{d\theta}{dt} = \frac{d\boldsymbol{\epsilon}_{\bar{r}}}{d\theta}\dot{\theta} \tag{11.24}$$

To evaluate $d\boldsymbol{\epsilon}_{\bar{r}}/d\theta$, we have shown in Fig. 11.13(a) the vector $\boldsymbol{\epsilon}_{\bar{r}}$ for a given $\bar{r}$ and z at positions corresponding to θ and $(\theta + \Delta\theta)$. In Fig. 11.13(b), furthermore, we have formed a vector triangle from these vectors and, in this way, we have shown the vector $\Delta\boldsymbol{\epsilon}_{\bar{r}}$ (i.e., the change in $\boldsymbol{\epsilon}_{\bar{r}}$ during a change in the coordinate θ). From the vector triangle, we see that

$$|\Delta\boldsymbol{\epsilon}_{\bar{r}}| \approx |\boldsymbol{\epsilon}_{\bar{r}}|\,\Delta\theta = \Delta\theta \tag{11.25}$$

Furthermore, as $\Delta\theta \longrightarrow 0$ we see, on consulting Fig. 11.13, that the *direction* of $\Delta\boldsymbol{\epsilon}_{\bar{r}}$ approaches that of the unit vector $\boldsymbol{\epsilon}_\theta$, and so we can approximate $\Delta\boldsymbol{\epsilon}_{\bar{r}}$ as

$$\Delta\boldsymbol{\epsilon}_{\bar{r}} \approx |\Delta\boldsymbol{\epsilon}_{\bar{r}}|\,\boldsymbol{\epsilon}_\theta \approx \Delta\theta\,\boldsymbol{\epsilon}_\theta \tag{11.26}$$

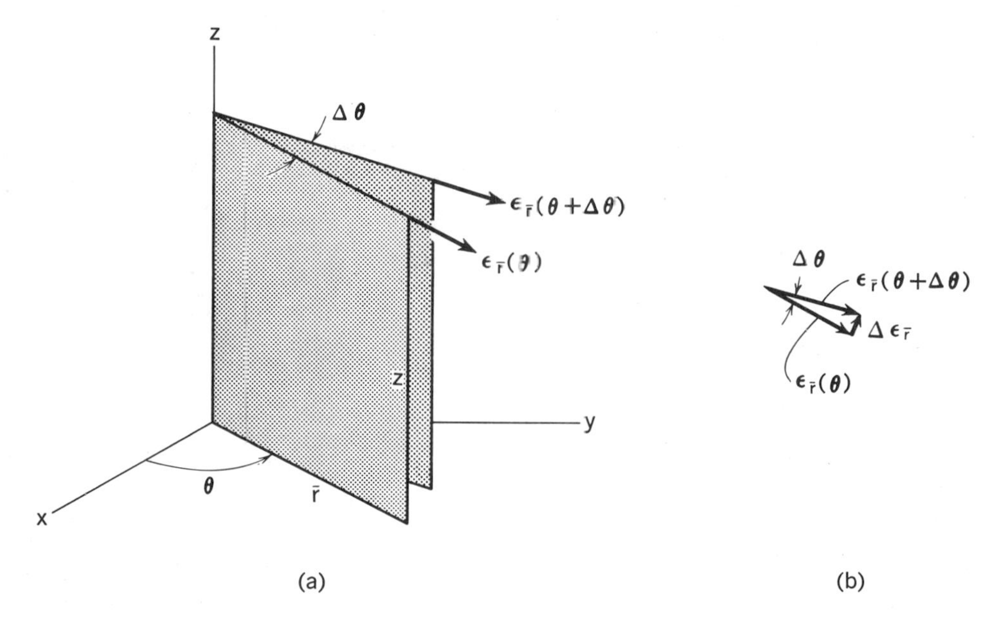

Figure 11.13. Change of unit vector $\epsilon_{\bar{r}}$.

where we have used Eq. 11.25 in the last step. Going back to Eq. 11.24, we utilize the preceding result to write

$$\frac{d\epsilon_{\bar{r}}}{dt} = \left(\frac{d\epsilon_{\bar{r}}}{d\theta}\right)\dot{\theta} \approx \left(\frac{\Delta\epsilon_{\bar{r}}}{\Delta\theta}\right)\dot{\theta} \approx \left[\frac{(\Delta\theta)\epsilon_\theta}{\Delta\theta}\right]\dot{\theta} = \dot{\theta}\epsilon_\theta \tag{11.27}$$

In the limit, as $\Delta\theta \longrightarrow 0$, all the previously made approximations become exact statements and we accordingly have

$$\frac{d\epsilon_{\bar{r}}}{dt} = \dot{\theta}\epsilon_\theta \tag{11.28}$$

The velocity of particle P is, then,

$$\boxed{V = \dot{\bar{r}}\epsilon_{\bar{r}} + \bar{r}\dot{\theta}\epsilon_\theta + \dot{z}\epsilon_z} \tag{11.29}$$

To get the acceleration relative to xyz in terms of cylindrical coordinates and radial, transverse, and axial components, we simply take the time derivative of the velocity vector above:

$$a = \frac{dV}{dt} = \ddot{\bar{r}}\epsilon_{\bar{r}} + \dot{\bar{r}}\dot{\epsilon}_{\bar{r}} + \dot{\bar{r}}\dot{\theta}\epsilon_\theta + \bar{r}\ddot{\theta}\epsilon_\theta + \bar{r}\dot{\theta}\dot{\epsilon}_\theta + \ddot{z}\epsilon_z \tag{11.30}$$

We must next evaluate $\dot{\epsilon}_\theta$. Like $\epsilon_{\bar{r}}$, the vector ϵ_θ can vary only when a change in the coordinate θ causes a change in direction of this vector, as has been shown in Fig. 11.14(a). The vectors $\epsilon_\theta(\theta)$ and $\epsilon_\theta(\theta + \Delta\theta)$ have been shown in a vector triangle in Fig. 11.14(b) and here we have shown $\Delta\epsilon_\theta$, the change of the vector ϵ_θ as a result of the change in coordinate θ. We can then say, using the chain rule,

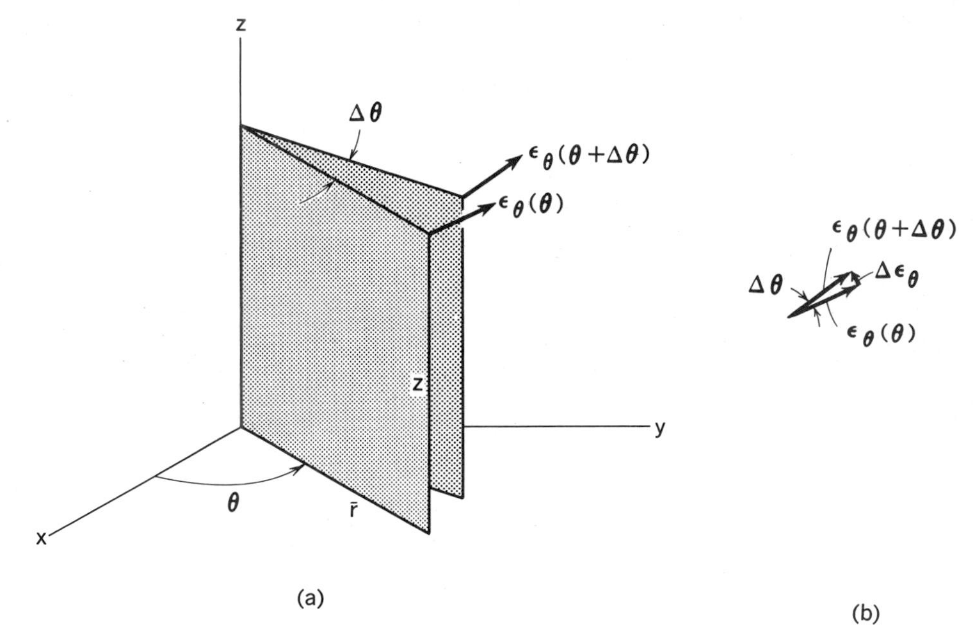

Figure 11.14. Change of unit vector ϵ_θ.

$$\frac{d\epsilon_\theta}{dt} = \frac{d\epsilon_\theta}{d\theta}\dot{\theta} \approx \frac{\Delta\epsilon_\theta}{\Delta\theta}\dot{\theta} \tag{11.31}$$

As $\Delta\theta \longrightarrow 0$, the direction of $\Delta\epsilon_\theta$ becomes that of $-\epsilon_{\bar{r}}$ and the magnitude of $\Delta\epsilon_\theta$, on consulting the vector triangle, clearly approaches $\Delta\theta$. Thus, $\Delta\epsilon_\theta$ becomes approximately $-\Delta\theta\, \epsilon_{\bar{r}}$. In the limit, we then get for Eq. 11.31:

$$\dot{\epsilon}_\theta = -\dot{\theta}\epsilon_{\bar{r}} \tag{11.32}$$

Using Eqs. 11.28 and 11.32, we find that Eq. 11.30 now becomes

$$\boldsymbol{a} = \ddot{\bar{r}}\epsilon_{\bar{r}} + \dot{\bar{r}}\dot{\theta}\epsilon_\theta + \dot{\bar{r}}\dot{\theta}\epsilon_\theta + \bar{r}\ddot{\theta}\epsilon_\theta - \bar{r}\dot{\theta}^2\epsilon_{\bar{r}} + \ddot{z}\epsilon_z$$

Collecting components, we write

$$\boxed{\boldsymbol{a} = (\ddot{\bar{r}} - \bar{r}\dot{\theta}^2)\epsilon_{\bar{r}} + (\bar{r}\ddot{\theta} + 2\dot{\bar{r}}\dot{\theta})\epsilon_\theta + \ddot{z}\epsilon_z} \tag{11.33}$$

Thus, we have accomplished the desired task. A similar procedure can be followed to reach corresponding formulations for spherical coordinates. By now you should be able to produce the preceding equations readily from the foregoing basic principles and should by no means attempt to memorize them.

For motion in a *circle* in the *xy* plane, note that $\dot{\bar{r}} = \dot{z} = 0$, and $\bar{r} = r$. We get the following simplifications:

$$\boldsymbol{V} = r\dot{\theta}\epsilon_\theta \tag{11.34a}$$

$$\boldsymbol{a} = r\ddot{\theta}\epsilon_\theta - r\dot{\theta}^2\epsilon_r \tag{11.34b}$$

Furthermore, the unit vector $\boldsymbol{\epsilon}_\theta$ is tangent to the path, and the unit vector $\boldsymbol{\epsilon}_r$ is normal to the path and points away from the center of curvature. Therefore, when we compare Eq. 11.34b with those stemming from considerations of path variables (Section 11.5), clearly for circular motion in the xy coordinate plane of a right-hand triad:

$$|r\ddot{\theta}| = \left|\frac{d^2 s}{dt^2}\right| \qquad |r\dot{\theta}^2| = \left|\frac{V^2}{r}\right| \qquad \begin{cases} \boldsymbol{\epsilon}_\theta = \boldsymbol{\epsilon}_t & \text{(for counterclockwise motion as seen from } +z)^3 \\ \boldsymbol{\epsilon}_\theta = -\boldsymbol{\epsilon}_t & \text{(for clockwise motion as seen from } +z) \\ \boldsymbol{\epsilon}_r = -\boldsymbol{\epsilon}_n \end{cases} \tag{11.35}$$

Thus, Eqs. 11.34b and 11.18 are equally useful for quickly expressing the acceleration of a particle moving in a circular path. You probably remember these formulas from earlier physics courses and may want to use them in the ensuing work of this chapter.

EXAMPLE 11.6

A *towing tank* is a device used for evaluating the drag and stability of ship hulls. Scaled models are moved by a rig along the water at carefully controlled speeds and attitudes while measurements are being made. Usually, the water is contained in a long narrow tank with the rig moving overhead along the length of the tank. However, another useful setup consists of a rotating radial arm (see Fig. 11.15) which gives the model a transverse motion. A radial motion along the arm is another degree of freedom possible for the model in this system.

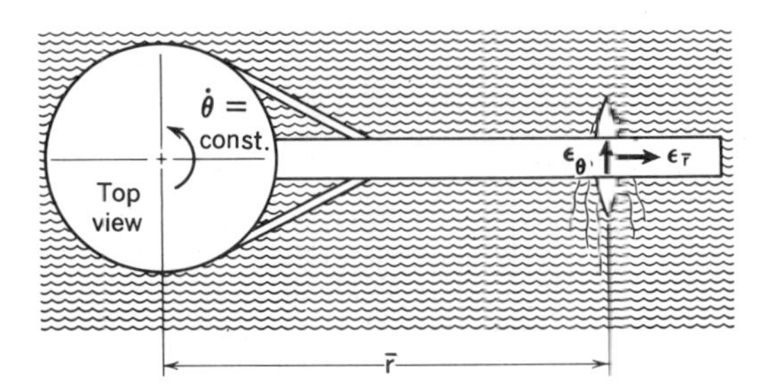

Figure 11.15. Circular towing tank.

Consider the case where a model is being moved out radially so that in one revolution of the main beam it has gone, at constant speed relative to the main beam, from position $\bar{r} = 3.3$ m to $\bar{r} = 4$ m. The angular speed of the beam is 3 rpm. What is the acceleration of the hull model relative to the water when $\bar{r} = 4$ m?

In order to find the radial speed of the model, note that one revolution of the arm corresponds to a time τ evaluated as:

$$\tau = \frac{1}{\frac{3}{60}} = 20 \text{ sec}$$

[3]The sense of $\boldsymbol{\epsilon}_t$ is that of the velocity of the particle, whereas the sense of $\boldsymbol{\epsilon}_\theta$ is determined by the reference xyz. For this reason a multiplicity of relations between these unit vectors exists.

Hence, we can say for $\dot{\bar{r}}$:

$$\dot{\bar{r}} = \frac{4 - 3.3}{\tau} = .035 \text{ m/sec}$$

We can now readily describe the motion of the system at the instant of interest with cylindrical coordinates as follows:

$$\bar{r} = 4 \text{ m}, \qquad \dot{\theta} = 3\left(\frac{2\pi}{60}\right) = .314 \text{ rad/sec}$$

$$\dot{\bar{r}} = .035 \text{ m/sec}, \qquad \ddot{\theta} = 0$$

$$\ddot{\bar{r}} = 0, \qquad \ddot{z} = 0$$

Using Eq. 11.33, we may now evaluate the acceleration vector.

$$\begin{aligned} \boldsymbol{a} &= [0 - (4)(.314)^2]\boldsymbol{\epsilon}_r + [0 + (2)(.035)(.314)]\boldsymbol{\epsilon}_\theta + [0]\boldsymbol{\epsilon}_z \\ &= -.394\boldsymbol{\epsilon}_r + .022\boldsymbol{\epsilon}_\theta \text{ m/sec}^2 \end{aligned}$$

Finally,

$$|\boldsymbol{a}| = .395 \text{ m/sec}^2$$

Problems

11.38. A car is moving along a circular track of radius 40 ft. The position S along the path is given as

$$S = 3t^2 + \frac{t^3}{6} \text{ ft}$$

The time t is given in seconds. What are the angular velocity and angular acceleration of the car at $t = 5$ sec?

11.39. A point P fixed on a rotating plate has an acceleration in the x direction of -10 m/sec². If r for the point is 1 m, what is the angular acceleration of the plate? The angular speed at the instant of interest is 2 rad/sec counterclockwise.

11.40. A flat disc A with a rubber surface is driven by bevel gears having diameters $D_1 = 8$ in. and $D_2 = 3$ in. A second rubber disc B of diameter $D_3 = 2$ in. is turned by the friction contact with A. We thus have a *friction drive* system. At the instant of interest, $\omega = 5$ rad/sec and $\dot{\omega} = 3$ rad/sec². If wheel B is moved downward at a speed $V_B = 3$ ft/sec at the instant of interest, what is the rotational speed Ω and the rate of change of rotational speed $\dot{\Omega}$ of the small disc B? Slipping between B and A occurs only in the radial direction of disc A. The distance r is 4 in. at the instant of interest.

Figure P.11.39

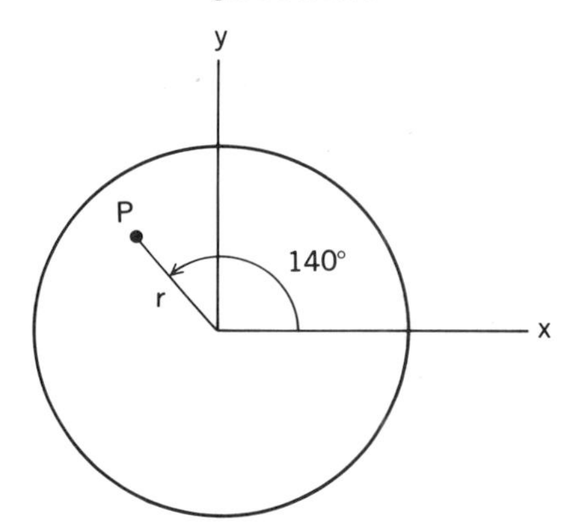

Figure P.11.40

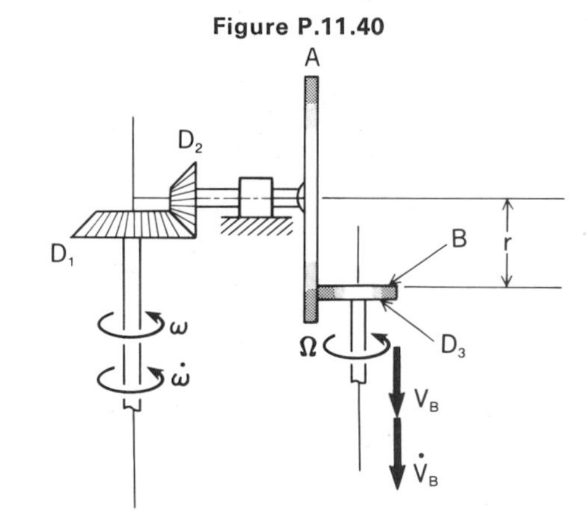

11.41. A vertical member rotates in accordance with:

$$\omega = 3 \sin (.1t) \text{ rad/sec}$$

with t in seconds. Attached to CD is a system of rods HI and FG pinned together at K and of length 200 mm and GA and IA of length 100 mm pinned together as shown. At the end of A is a stylus which scribes a curve on plate J. The angle β of the system is given as

$$\beta = 1.3 - \frac{t}{10} \text{ rad}$$

with t in seconds. What are the radial and transverse velocity and acceleration components of the stylus at time $t = 5$ sec about axis G–G? (*Note:* Pin F is fixed but pin H moves vertically in a slot as shown.)

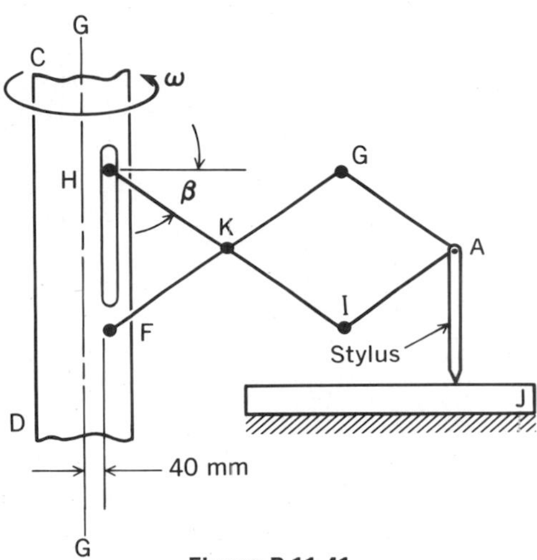

Figure P.11.41

11.42. A particle moves with a constant speed of 5 ft/sec along a straight line having direction cosines $l = .5$, $m = .3$. What are the cylindrical coordinates when $|\boldsymbol{r}| = 20$ ft? What are the axial and transverse velocities of the particle at this position?

Figure P.11.42

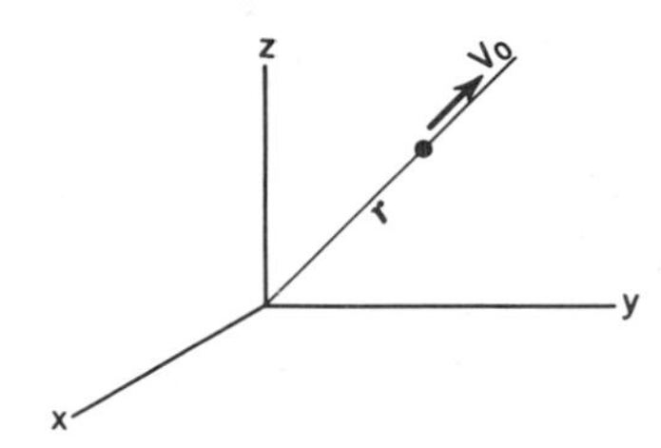

11.43. A wheel is rotating at time t with an angular speed ω of 5 rad/sec. At this instant, the wheel also has a rate of change of angular speed of 2 rad/sec^2. A body B is moving along a spoke at this instant with a speed of 3 m/sec relative to the spoke and is increasing in its speed at the rate of 1.6 m/sec^2. These data are given when the spoke, on which B is moving, is vertical and when B is .6 m from the center of the wheel, as shown in the diagram. What are the velocity and acceleration of B at this instant relative to the fixed reference xyz?

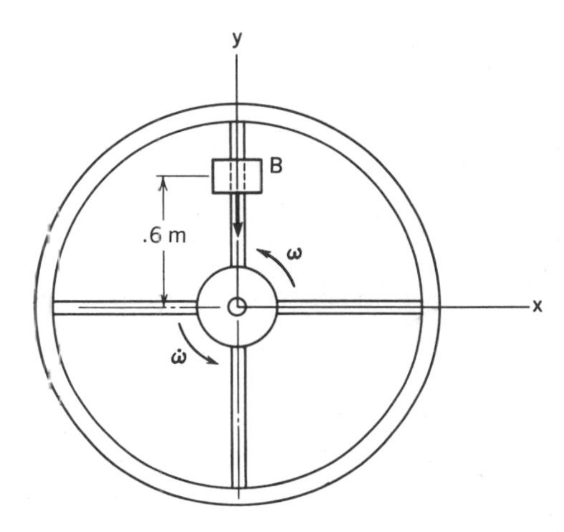

Figure P.11.43

11.44. A plane is shown in a dive-bombing mission. It has at the instant of interest a speed of 485 km/hr and is increasing its speed downward at a rate of 81 km/hr/sec. The propeller is rotating at 150 rpm and has a diameter of 4 m. What is the velocity of the tip of the propeller shown at A and its acceleration at the instant of interest?

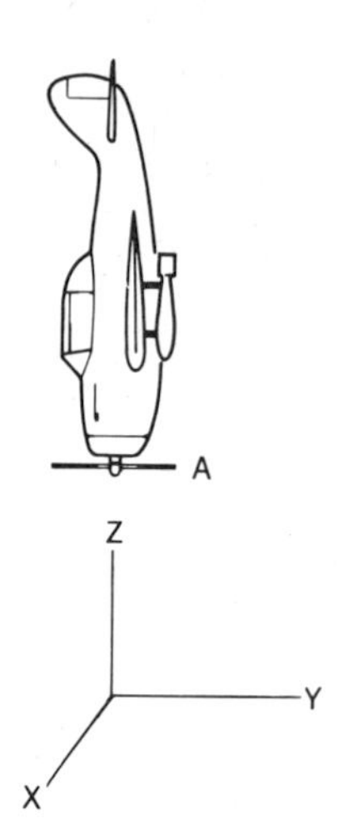

Figure P.11.44

11.45. A wheel of diameter 2 ft is rotated at a speed of 2 rad/sec and is increasing its rotational speed at the rate of 3 rad/sec². It advances along a screw having a pitch of .5 in. What is the acceleration of elements on the rim in terms of cylindrical components?

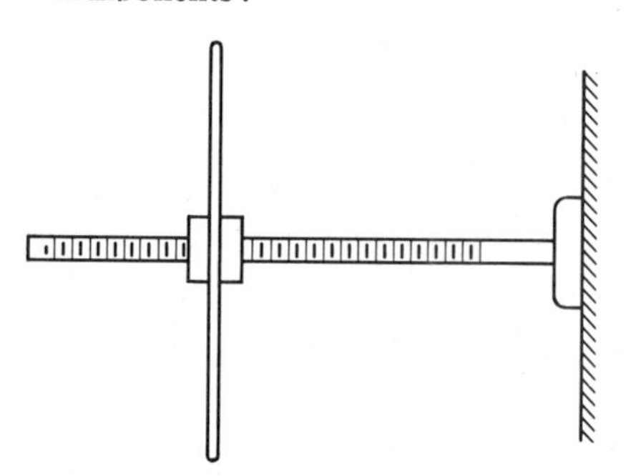

Figure P.11.45

11.46. A wheel with a threaded hub is rotating at an angular speed of 80 rpm on a right-hand screw having a pitch of .5 in. At the instant of interest, the rate of change of angular speed is 20 rpm/sec. A sleeve A is advancing along a spoke at this instant with a speed of 5 ft/sec and a rate of increase of speed of 5 ft/sec². The sleeve is 2 ft from the centerline at O at the instant of interest. What are the velocity vector and the acceleration vector of the sleeve at this instant? Use cylindrical coordinates.

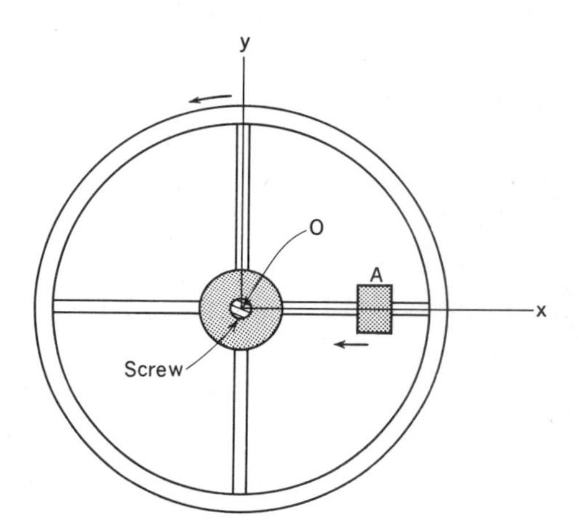

Figure P.11.46

11.47. A simple garden sprinkler is shown. Water enters at the base and leaves at the end at a speed of 3 m/sec as seen from the rotor of the sprinkler. Furthermore, it leaves upward relative to the rotor at an angle of 60° as shown in the diagram. The rotor has an angular speed ω of 2 rad/sec. As seen from the ground, what are the axial, transverse, and radial velocity and acceleration components of the water just as it leaves the rotor?

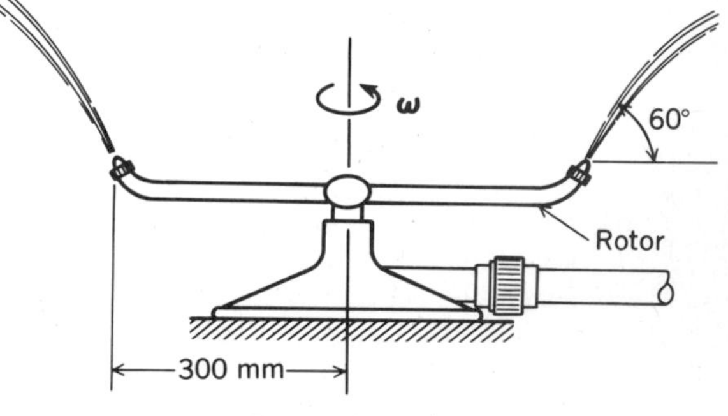

Figure P.11.47

11.48. In Chapter 1, you learned that Newton's gravitational law is

$$F = G\frac{m_1 m_2}{r^2} \qquad \text{(a)}$$

where G is the universal gravitational constant and r is the distance between the centers of mass of the bodies. Consider a particle on the earth's surface at the equator and show that

$$GM = gR^2 \tag{b}$$

where M is the mass of the earth and R is the radius of the earth (3960 mi). What is the speed of a space satellite relative to XYZ in a circular orbit about the earth? The radius of the orbit is 6000 mi. XYZ is an inertial reference about which the earth rotates once a day.

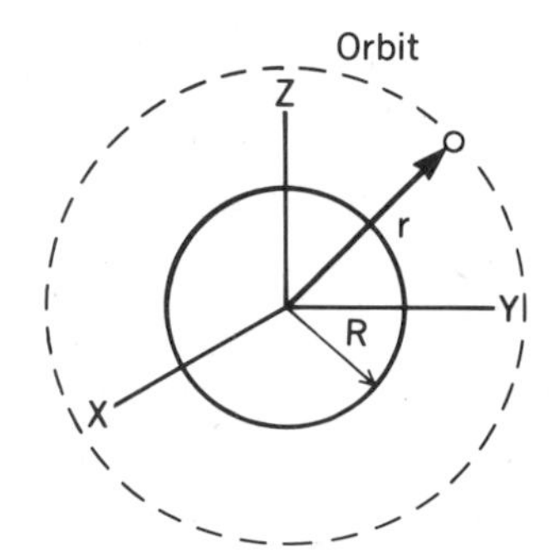

Figure P.11.48

11.49. The acceleration of gravity on the surface of Mars is .385 times the acceleration of gravity on earth. The radius R of Mars is about .532 times that of the earth. What is the time of flight of one cycle for a satellite in a circular parking orbit 803 miles from the surface of Mars? [Note Eq. (b) of Problem 11.48.]

11.50. A threaded rod rotates with angular speed $\dot{\theta} = .315t^2$ rad/sec. On the rod is a nut which rotates relative to the rod at the rate $\omega = .2t^2$ rad/sec. When $t = 0$, the nut is at a distance 2 ft from A. What is the velocity and acceleration of the nut at $t = 10$ sec? The thread has a pitch of .2 in. Give results in radial and transverse directions.

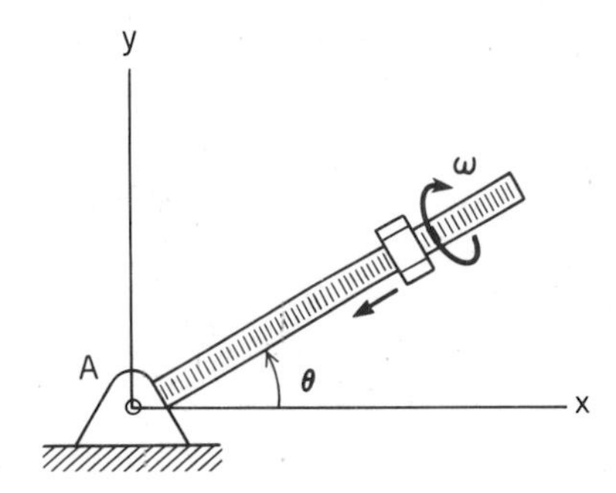

Figure P.11.50

11.51. A fire truck has a telescoping boom holding a fireman as shown in the diagram. At time t, the boom is extending at the rate of .6 m/sec and increasing its rate of extension at .3 m/sec². Also, at time t, $r = 10$ m and $\beta = 30°$. If a velocity component of the man of 3.3 m/sec vertically is desired, what should $\dot{\beta}$ be? Also, if a vertical acceleration component of the man of 1.7 m/sec² is desired, what should $\ddot{\beta}$ be?

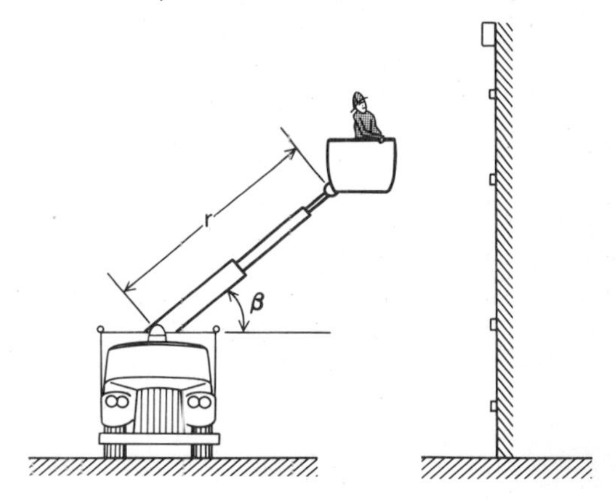

Figure P.11.51

11.52. Underwater cable is being laid from an ocean-going ship. The cable is unwound from a large spool A at the rear of the ship. The cable must be laid so that is *not dragged* on the ocean bottom. If the ship is moving at a speed of 3 knots, what is the necessary angular speed ω of the spool A when the cable is coming off at a radius of 3.2 m. What is the average rate of change of ω for the spool required for proper operation? The cable has a diameter of 150 mm.

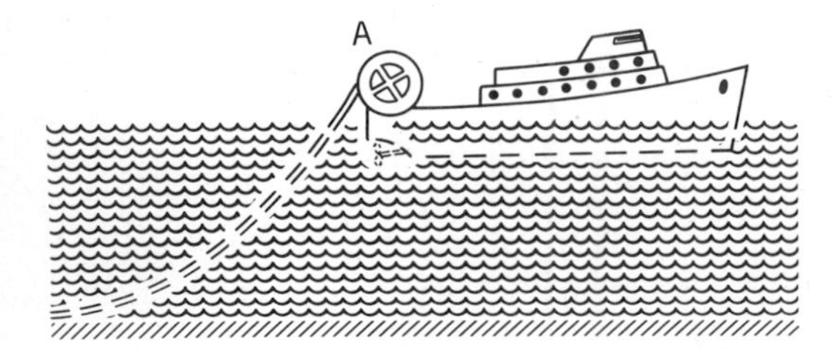

Figure P.11.52

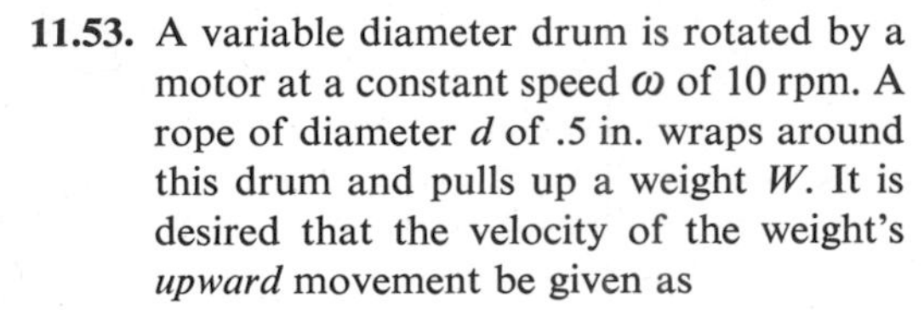

11.53. A variable diameter drum is rotated by a motor at a constant speed ω of 10 rpm. A rope of diameter d of .5 in. wraps around this drum and pulls up a weight W. It is desired that the velocity of the weight's *upward* movement be given as

$$\dot{X} = .4 + \frac{t^2}{8000} \text{ ft/sec}$$

where for $t = 0$ the rope is just about to start wrapping around the drum at $Z = 0$. What should the radius $\bar{r}$ of the drum be as a function of Z to accomplish this? What are the velocity components $\dot{Y}$ and $\dot{Z}$ of the weight W when $t = 100$ sec?

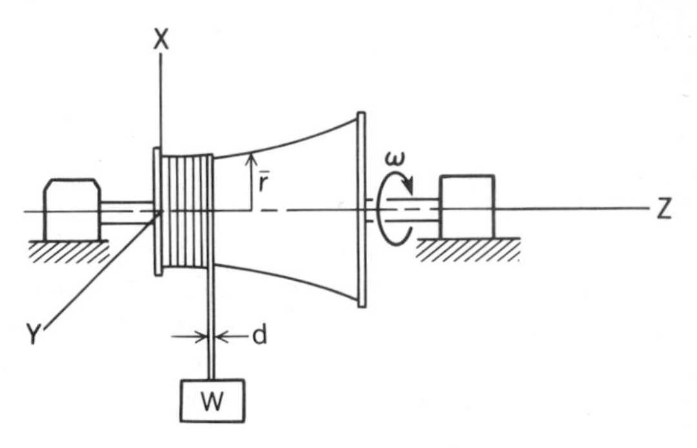

Figure P.11.53

11.54. Plastic sheet is being rolled from slotted wheel A onto roll B. The roll B rotates at a constant angular speed ω of 25 rpm. The thickness ϵ of the sheet of plastic is .25 in. At time t', $r_2 = 2$ ft and $r_1 = 1.7$ ft. What is the angular velocity Ω and angular acceleration $\dot{\Omega}$ of wheel A at time t'? Consider the plastic sheet to be inextensible. What are the transverse and radial acceleration of points on the plastic sheet on the periphery of roll A at time t'?

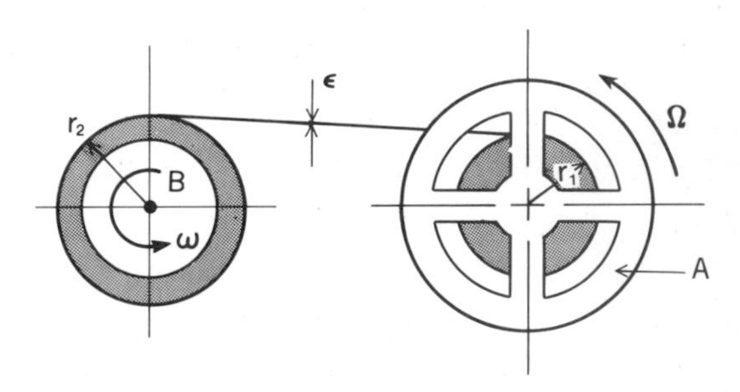

Figure P.11.54

Part C Simple Kinematical Relations and Applications

11.7 Simple Relative Motion

Up to now, we have considered only a single reference in our kinematical considerations. There are times when two or more references may be profitably employed in

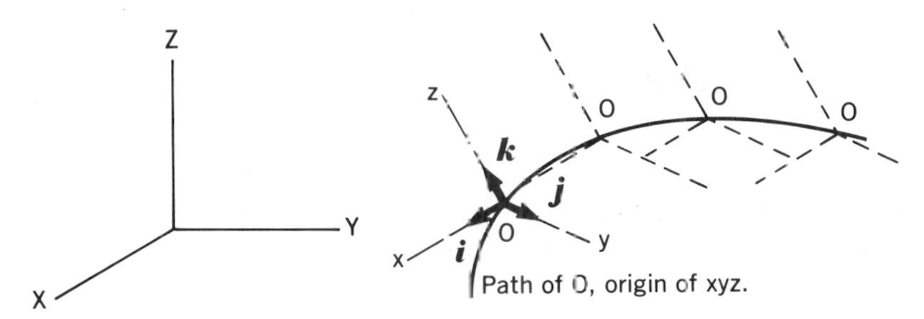

Figure 11.16. Axes *xyz* are translating relative to *XYZ*.

describing the motion of a particle. We shall consider in this section a very simple case that will fulfill our needs in the early portion of the text.

As a first step, consider two references *xyz* and *XYZ* (Fig. 11.16) moving in such a way that the direction of the axes of *xyz* always retain the same orientation relative to *XYZ* such as has been suggested by the dashed references giving successive positions of *xyz*. Such a motion of *xyz* relative to *XYZ* is called *translation.*

Suppose now that we have a vector $\boldsymbol{A}(t)$ which varies with time. Now in the general case, the time variation of $\boldsymbol{A}$ will depend on from which reference we are observing the time variation. For this reason, we often include subscripts to identify the reference relative to which the time variation is taken. Thus, we have $(d\boldsymbol{A}/dt)_{xyz}$ and $(d\boldsymbol{A}/dt)_{XYZ}$ as time derivatives of $\boldsymbol{A}$ as seen from the *xyz* and *XYZ* axes, respectively. How are these derivatives related for axes *xyz* and *XYZ* that are translating relative to each other? For this purpose, consider $(d\boldsymbol{A}/dt)_{XYZ}$. We will decompose $\boldsymbol{A}$ into components parallel to the *xyz* axes and so we have

$$\left(\frac{d\boldsymbol{A}}{dt}\right)_{XYZ} = \left[\frac{d}{dt}(A_x\boldsymbol{i} + A_y\boldsymbol{j} + A_z\boldsymbol{k})\right]_{XYZ} \tag{11.36}$$

where A_x, A_y, and A_z are the scalar components of $\boldsymbol{A}$ along the *xyz* axes. Because *xyz* translates relative to *XYZ* (see Fig. 11.16), the unit vectors of *xyz* which we have denoted as $\boldsymbol{i}, \boldsymbol{j}$, and $\boldsymbol{k}$ are *constant vectors* as seen from *XYZ*. That is, whereas these vectors may change their lines of action, they *do not* change *magnitude* and *direction* as seen from *XYZ* and are thus constant vectors as seen from *XYZ*. We then have, for the equation above:

$$\left(\frac{d\boldsymbol{A}}{dt}\right)_{XYZ} = \left(\frac{dA_x}{dt}\right)_{XYZ}\boldsymbol{i} + \left(\frac{dA_y}{dt}\right)_{XYZ}\boldsymbol{j} + \left(\frac{dA_z}{dt}\right)_{XYZ}\boldsymbol{k} \tag{11.37}$$

But A_x, A_y, and A_z are *scalars* and a time derivative of a scalar, as you may remember from the calculus, is not dependent on a reference of observation. We could readily replace $(dA_x/dt)_{XYZ}$ by $(dA_x/dt)_{xyz}$, etc., with no change in meaning—or we could leave off the subscripts entirely for these terms. Thus, we can say now:

$$\left(\frac{d\boldsymbol{A}}{dt}\right)_{XYZ} = \left(\frac{dA_x}{dt}\right)\boldsymbol{i} + \left(\frac{dA_y}{dt}\right)\boldsymbol{j} + \left(\frac{dA_z}{dt}\right)\boldsymbol{k} \tag{11.38}$$

Now consider $(d\boldsymbol{A}/dt)_{xyz}$. Again, decomposing $\boldsymbol{A}$ into components along the *xyz* axes and noting that $\boldsymbol{i}, \boldsymbol{j}$, and $\boldsymbol{k}$ are constant vectors as seen from *xyz*, we can conclude that

$$\begin{aligned}\left(\frac{dA}{dt}\right)_{xyz} &= \left(\frac{dA_x}{dt}\right)_{xyz} i + \left(\frac{dA_y}{dt}\right)_{xyz} j + \left(\frac{dA_z}{dt}\right)_{xyz} k \\ &= \left(\frac{dA_x}{dt}\right) i + \left(\frac{dA_y}{dt}\right) j + \left(\frac{dA_z}{dt}\right) k\end{aligned} \tag{11.39}$$

where as discussed earlier we have dropped the xyz subscripts. Observing Eqs. 11.38 and 11.39, we conclude that

$$\left(\frac{dA}{dt}\right)_{XYZ} = \left(\frac{dA}{dt}\right)_{xyz} \tag{11.40}$$

We can conclude that

$$\left(\frac{d}{dt}\right)_{XYZ} = \left(\frac{d}{dt}\right)_{xyz} \tag{11.41}$$

That is, the *derivative of a vector is the same for all reference axes which are translating relative to each other.*

Note in the discussion that the fact that the unit vectors of xyz were *constant* relative to XYZ resulted in the simple relation 11.41. If xyz were *rotating* relative to XYZ, the unit vectors of xyz would not be constant as seen from XYZ and a more complex relationship would exist between $(dA/dt)_{XYZ}$ and $(dA/dt)_{xyz}$. We shall develop this relationship later in the text.

A pair of references xyz and XYZ are shown now in Fig. 11.17 moving in translation relative to each other. The *velocity* of any particle P depends on the reference

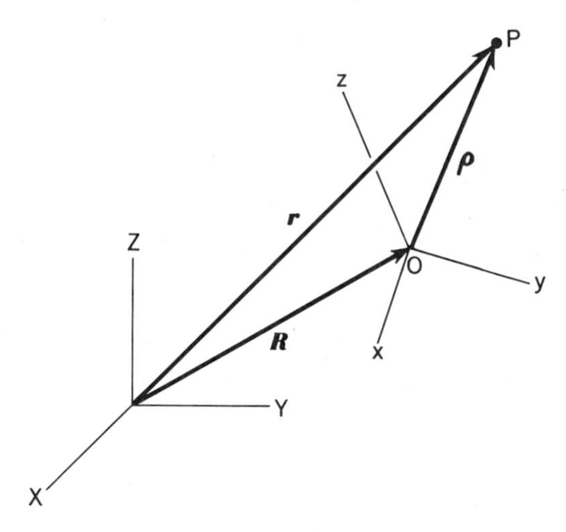

Figure 11.17. Axes xyz are translating relative to XYZ.

from which the motion is observed. More precisely, we say that the velocity of particle P relative to reference XYZ is the time rate of change of the position vector r for this reference, where this rate of change is viewed from the XYZ reference. This can be stated mathematically as

$$V_{XYZ} = \left(\frac{dr}{dt}\right)_{XYZ} \tag{11.42}$$

Similarly, for the velocity of particle P as seen from reference xyz, we have

$$\boldsymbol{V}_{xyz} = \left(\frac{d\boldsymbol{\rho}}{dt}\right)_{xyz} \tag{11.43}$$

where we now use position vector $\boldsymbol{\rho}$ for reference xyz and view the change from the xyz reference (see Fig. 11.17). By the same token, $(d\boldsymbol{R}/dt)_{XYZ}$ is the velocity of the origin of the xyz reference as seen from XYZ. Since all points of the xyz reference have the same velocity relative to XYZ at any time t for this case (translation of xyz), we can say that $(d\boldsymbol{R}/dt)_{XYZ}$ is the velocity of reference xyz as seen from XYZ.

From Fig. 11.17 we can relate position vectors $\boldsymbol{\rho}$ and $\boldsymbol{r}$ by the equation

$$\boldsymbol{r} = \boldsymbol{R} + \boldsymbol{\rho} \tag{11.44}$$

Now take the time rate of change of these vectors as seen from XYZ. We get

$$\left(\frac{d\boldsymbol{r}}{dt}\right)_{XYZ} = \left(\frac{d\boldsymbol{R}}{dt}\right)_{XYZ} + \left(\frac{d\boldsymbol{\rho}}{dt}\right)_{XYZ} \tag{11.45}$$

The term on the left side of this equation is $\boldsymbol{V}_{XYZ}$, as indicated earlier, and we shall use the notation $\dot{\boldsymbol{R}}$ for $(d\boldsymbol{R}/dt)_{XYZ}$. We can replace the last term by the derivative $(d\boldsymbol{\rho}/dt)_{xyz}$ in accordance with Eq. 11.41 since the axes are in translation relative to each other. But $(d\boldsymbol{\rho}/dt)_{xyz}$ is simply $\boldsymbol{V}_{xyz}$, the velocity of P relative to xyz. Thus, we have

$$\boxed{\boldsymbol{V}_{XYZ} = \boldsymbol{V}_{xyz} + \dot{\boldsymbol{R}}} \tag{11.46}$$

By the same reasoning, we can show that the acceleration of particle P is related to references XYZ and xyz as follows[4]:

$$\boxed{\boldsymbol{a}_{XYZ} = \boldsymbol{a}_{xyz} + \ddot{\boldsymbol{R}}} \tag{11.47}$$

Equations 11.46 and 11.47 convey the physically simple picture that the motion of a particle relative to XYZ is the sum of the motion of the particle relative to xyz plus the motion of xyz relative to XYZ.

It must be kept clearly in mind that the equations which we have developed apply only to references which have a *translatory* motion relative to each other. In Chapter 15 we shall consider references which have arbitrary motion relative to each other. (Since a reference is a rigid system, we shall need to examine at that time the kinematics of rigid bodies in order to develop these general considerations of relative motion.) The equations presented here will then be special cases.

---

[4]As you no doubt will anticipate, the acceleration of a particle as seen from reference XYZ is

$$\boldsymbol{a}_{XYZ} = \left(\frac{d\boldsymbol{V}_{XYZ}}{dt}\right)_{XYZ}$$

Similarly, we have for $\boldsymbol{a}_{xyz}$,

$$\boldsymbol{a}_{xyz} = \left(\frac{d\boldsymbol{V}_{xyz}}{dt}\right)_{xyz}$$

How can we make use of multiple references? In many problems the motion of a particle is known relative to a given rigid body, and the motion of this body is known relative to the ground or other convenient reference. We can fix a reference xyz to the body, and if the body is in translation relative to the ground, we can then employ the given relations presented in this section to express the motion of the particle relative to the ground.

If, in ensuing chapters, we talk about the "motion of particles relative to a point," such as, for example, the center of mass of the system, then it will be understood that this motion is relative to a *hypothetical reference* moving with the center of mass in a *translatory manner* or, in other words, relative to a nonrotating observer moving with the center of mass.

We illustrate these remarks in the following examples.

EXAMPLE 11.7

A truck is moving at a speed of 100 km/hr at time t (see Fig. 11.18) and is accelerating at the rate of 2 km/hr/sec. At this instant, a car is moving at 70 km/hr and is decelerating at the rate of $.2g$. What is the velocity and acceleration of the car relative to a passenger in the truck at the instant of interest?

Because we know information relative to the ground, fix reference XYZ *to the ground* and because we want to observe from the truck, fix xyz *to the truck* as shown in Fig. 11.19. Now consider the velocity of the *car*. As seen from the ground, it is denoted as $(V_{\text{car}})_{XYZ}$ and, as seen from the truck, it is denoted as $(V_{\text{car}})_{xyz}$. We can say for these velocities

$$(V_{\text{car}})_{XYZ} = (V_{\text{car}})_{xyz} + \dot{\boldsymbol{R}} \qquad \text{(a)}$$

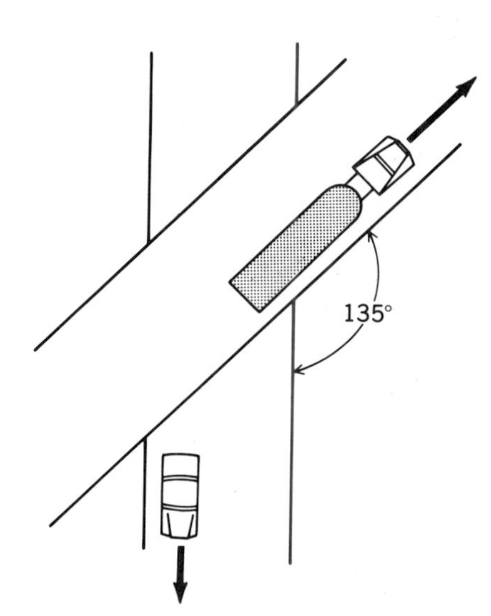

Figure 11.18. Find the motion of the car relative to the truck.

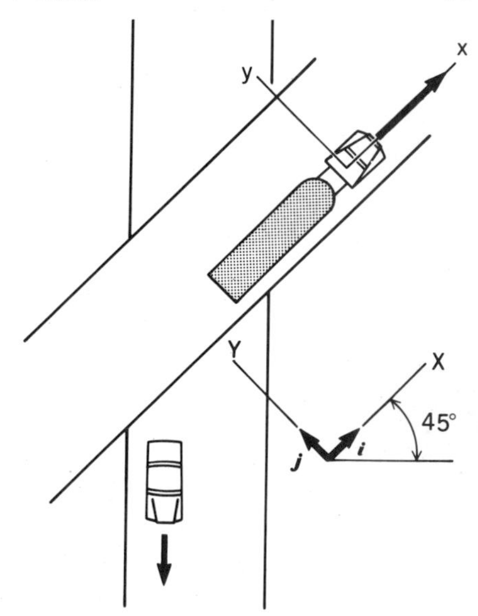

Figure 11.19. Fix xy to truck; XY to ground.

where clearly, $\dot{\boldsymbol{R}}$ is the velocity of the origin of xyz (the truck) relative to XYZ (the ground). Note that

$$\begin{aligned}(\boldsymbol{V}_{\text{car}})_{XYZ} &= 70(-.707\boldsymbol{j} - .707\boldsymbol{i}) \text{ km/hr} \\ \dot{\boldsymbol{R}} &= 100\boldsymbol{i} \text{ km/hr}\end{aligned} \tag{b}$$

Substituting into Eq. (a), we get on solving for $(\boldsymbol{V}_{\text{car}})_{xyz}$:

$$\begin{aligned}(\boldsymbol{V}_{\text{car}})_{xyz} &= 70(-.707\boldsymbol{j} - .707\boldsymbol{i}) - 100\boldsymbol{i} \\ &= -149.5\boldsymbol{i} - 49.49\boldsymbol{j} \text{ km/hr}\end{aligned} \tag{c}$$

The result above is then the velocity of the car relative to the truck.

Now consider the acceleration of the *car* relative to the ground, $(\boldsymbol{a}_{\text{car}})_{XYZ}$, and relative to the truck, $(\boldsymbol{a}_{\text{car}})_{xyz}$. We can relate these vectors as follows:

$$(\boldsymbol{a}_{\text{car}})_{XYZ} = (\boldsymbol{a}_{\text{car}})_{xyz} + \ddot{\boldsymbol{R}} \tag{d}$$

where $\ddot{\boldsymbol{R}}$ is the acceleration of the origin of xyz (the truck) relative to XYZ (the ground). Note that

$$\begin{aligned}(\boldsymbol{a}_{\text{car}})_{XYZ} &= (.2)(9.81)(.707\boldsymbol{i} + .707\boldsymbol{j}) \text{ m/sec}^2 \\ \ddot{\boldsymbol{R}} &= (2)\left(\frac{1000}{3600}\right)\boldsymbol{i} \text{ m/sec}^2\end{aligned}$$

Substituting the results above into Eq. (d), we get on solving for $(\boldsymbol{a}_{\text{car}})_{xyz}$:

$$\begin{aligned}(\boldsymbol{a}_{\text{car}})_{xyz} &= (.2)(9.81)(.707\boldsymbol{i} + .707\boldsymbol{j}) - (2)\left(\frac{1000}{3600}\right)\boldsymbol{i} \\ &= .8316\boldsymbol{i} + 1.387\boldsymbol{j} \text{ m/sec}^2\end{aligned}$$

This is then the acceleration of the car relative to the truck.

EXAMPLE 11.8

A jet airliner is shown in Fig. 11.20 flying at a speed of 600 mi/hr in a translatory manner relative to the ground reference XYZ. At the instant of interest, a downdraft causes the plane to accelerate downward at a rate of 50 mi/hr/sec. While this is happening, the pilot cuts back on the throttle so that the plane is decelerating in the Y direction at the rate of 30 mi/hr/sec. Thus, the plane has an acceleration given as

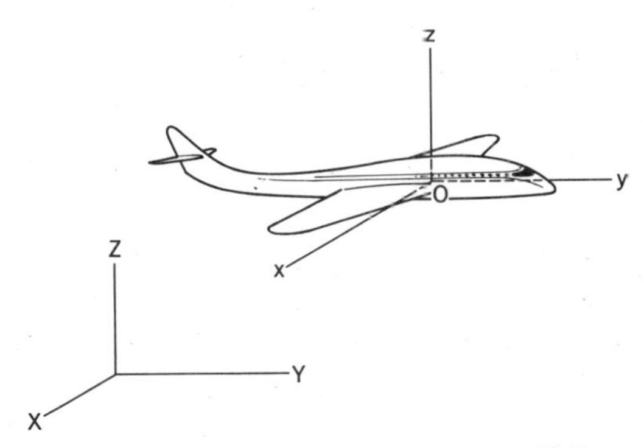

Figure 11.20. Plane translates relative to XYZ.

$$\boldsymbol{a} = -50\boldsymbol{k} - 30\boldsymbol{j} \text{ mi/hr/sec} \qquad \text{(a)}$$

while maintaining a translatory attitude. While this is happening, a solenoid is operated to close a valve gate which weighs $\frac{1}{2}$ lb. What is the force on the valve gate from the plane at the instant when the valve gate is moving downward relative to the airplane at a speed of 10 ft/sec and accelerating downward relative to the plane at a rate of 16.1 ft/sec²?

We must find the acceleration of the valve relative to the ground reference XYZ, which may be taken in the problem to be an inertial reference. This information will permit us to use the familiar form of Newton's law. It will be convenient in this undertaking to *fix* a reference xyz, having the same unit vectors as reference XYZ, to the airplane at any convenient location (see Fig. 11.20). We can then say for the motion of the valve gate relative to xyz:

$$\boldsymbol{a}_{xyz} = -16.1\boldsymbol{k} \text{ ft/sec}^2 \qquad \text{(b)}$$

The acceleration of O, the origin of xyz relative to XYZ, is

$$\ddot{\boldsymbol{R}} = -50\boldsymbol{k} - 30\boldsymbol{j} \text{ mi/hr/sec} \qquad \text{(c)}$$

Since the references are translating relative to each other, we can employ Eq. 11.47 to get $\boldsymbol{a}_{XYZ}$, the acceleration of the valve gate relative to inertial space. Thus,

$$\boldsymbol{a}_{XYZ} = (-50\boldsymbol{k} - 30\boldsymbol{j})\left(\frac{5280}{3600}\right) + (-16.1\boldsymbol{k})$$

$$= -44\boldsymbol{j} - 89.5\boldsymbol{k} \text{ ft/sec}^2$$

We can now employ Newton's law in the form

$$\boldsymbol{F} = m\boldsymbol{a}_{XYZ} \qquad \text{(d)}$$

Thus, denoting the total force from the airplane as $\boldsymbol{F}_{\text{plane}}$, and remembering that the gate valve weighs $\frac{1}{2}$ lb, we have

$$\boldsymbol{F}_{\text{plane}} - \tfrac{1}{2}\boldsymbol{k} = \frac{\frac{1}{2}}{g}(-44\boldsymbol{j} - 89.5\boldsymbol{k}) \qquad \text{(e)}$$

where $-\frac{1}{2}\boldsymbol{k}$ is the force of gravity. Solving for $\boldsymbol{F}_{\text{plane}}$, we get

$$\boldsymbol{F}_{\text{plane}} = -.684\boldsymbol{j} - .890\boldsymbol{k} \text{ lb} \qquad \text{(f)}$$

Problems

11.55. Two wheels rotate about stationary axes each at the same angular velocity, $\dot{\theta} = 5$ rad/sec. A particle A moves along the spoke of the larger wheel at the speed V_1 of 5 ft/sec relative to the spoke and at the instant shown is decelerating at the rate of 3 ft/sec² relative to the spoke. What are the velocity and acceleration of particle A as seen by an observer on the hub of the smaller wheel? What are the velocity and acceleration of particle A to an observer on the hub of the smaller wheel if the axis of the larger wheel moves at the instant of interest to the left with a speed of 10 ft/sec while decelerating at the rate of 2 ft/sec²? Both wheels maintain equal angular speeds.

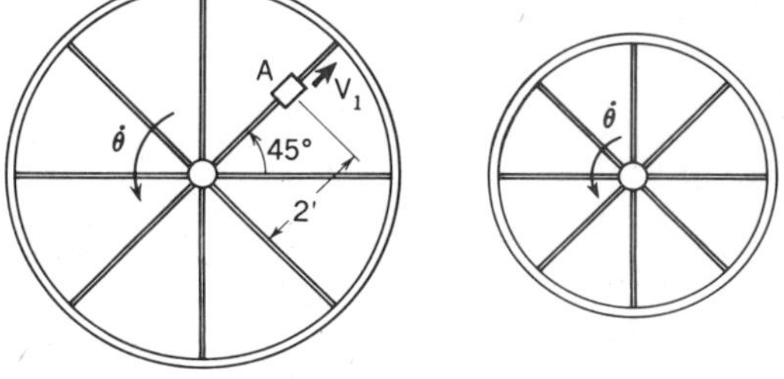

Figure P.11.55

11.56. Four particles of equal mass undergo coplanar motion in the xy plane with the following velocities:

$$V_1 = 2 \text{ m/sec}$$
$$V_2 = 3 \text{ m/sec}$$
$$V_3 = 2 \text{ m/sec}$$
$$V_4 = 5 \text{ m/sec}$$

We showed in Section 8.3 that the velocity of the center of mass can be found as follows:

$$(\sum_i m_i)V_c = \sum_i m_i V_i$$

where V_c is the velocity of the center of mass. What are the velocities of the particles relative to the center of mass?

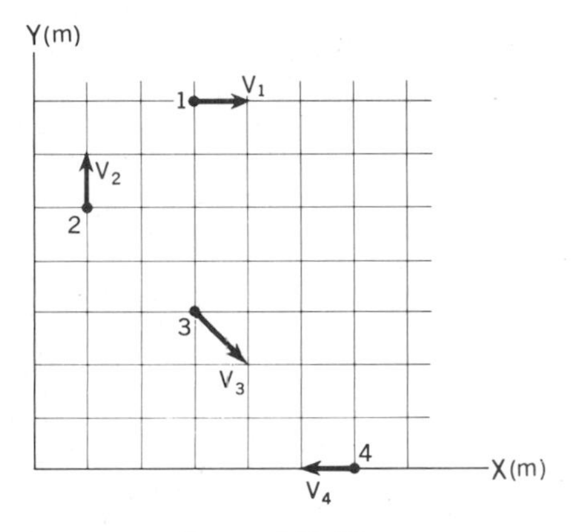

Figure P.11.56

11.57. A sled, used by researchers to test man's ability to perform during large accelerations over extended periods of time, is powered by a small rocket engine in the rear and slides on lubricated tracks. If the sled is accelerating at $6g$, what force does the man need to exert on a 3-ounce body to give it an acceleration relative to the sled of

$$30\boldsymbol{i} + 20\boldsymbol{j} \text{ ft/sec}^2$$

Figure P.11.57

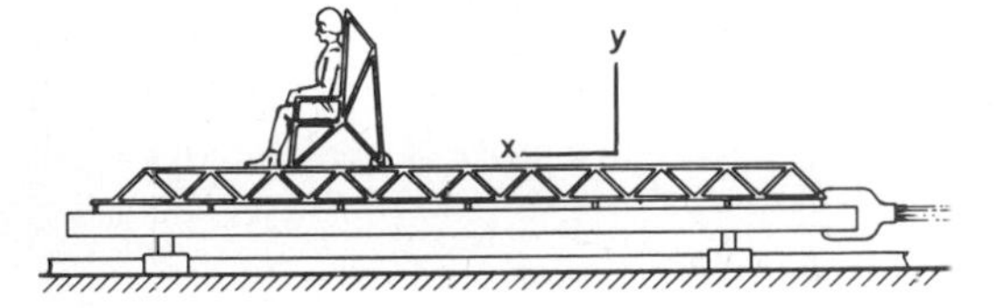

11.58. On the sled of Problem 11.57 is a device (see the diagram) on which mass M rotates about a horizontal axis at an angular speed ω of 5000 rpm. If the inclination θ of the arm BM is maintained at 30° with the vertical plane C–C, what is the total force on the mass M at the instant it is in its uppermost position? The sled is undergoing an acceleration of $5g$. Take M as having a mass of .15 kg.

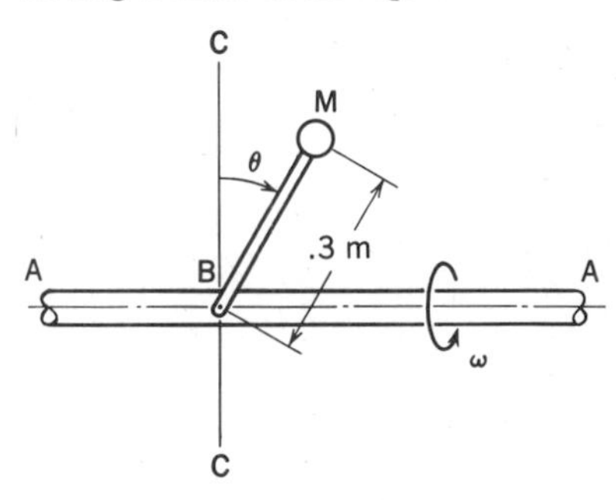

Figure P.11.58

11.59. A vehicle, wherein a mass M of 1 lbm rotates with an angular speed ω equal to 5 rad/sec, moves with a speed V given as $V = 5 \sin \Omega t$ ft/sec relative to the ground with t in seconds. When $t = 1$ sec, the rod AM is in the position shown. At this instant, what is the total force exerted by the mass M along the axis of rod AM if $\Omega = 3$ rad/sec?

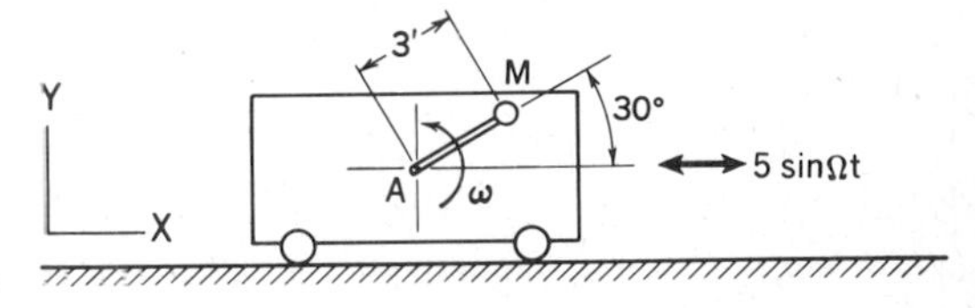

Figure P.11.59

11.60. In Problem 11.59, what is the frequency of oscillation, Ω, of the vehicle and the value of ω if, at the instant shown, there is a force on the mass M given as

$$\boldsymbol{F} = 25\boldsymbol{i} - 35\boldsymbol{j} \text{ lb}$$

11.61. A cockpit C is used to carry a worker for service work on road lighting systems. The cockpit is moved always in a translatory manner relative to the ground. If the angular speed ω of arm AB is 1 rad/min when $\theta = 30°$, what are the velocity and acceleration of any point in the cockpit body relative to the truck? At this instant, what are the velocity and acceleration, relative to the truck, of a particle moving with a horizontal speed V of .5 ft/sec and with a rate of increase of speed of .02 ft/sec² both relative to the cockpit?

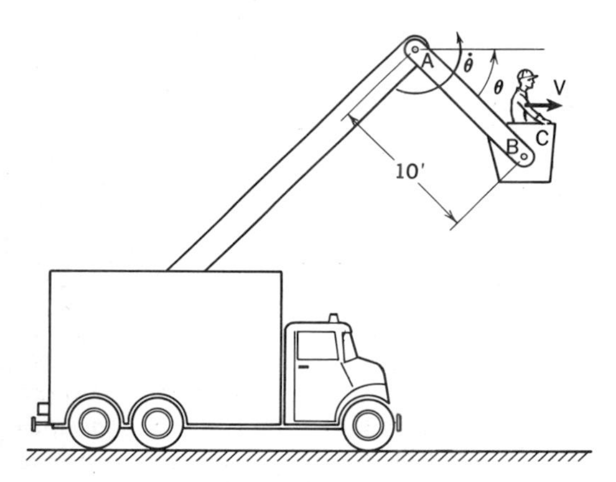

Figure P.11.61

11.62. A ferris wheel rotates at the instant of interest with an angular speed $\dot{\theta}$ of .5 rad/sec and is increasing its angular speed at the rate of .1 rad/sec². A ball is thrown from the ground to an occupant at A. The ball arrives at the instant of interest with a speed relative to the ground given as

$$V_{XYZ} = -10\boldsymbol{j} - 2\boldsymbol{k} \text{ ft/sec}$$

What are the velocity and the acceleration of the ball relative to the occupant at seat A provided that this seat is not "swinging"? The radius of the wheel is 20 ft.

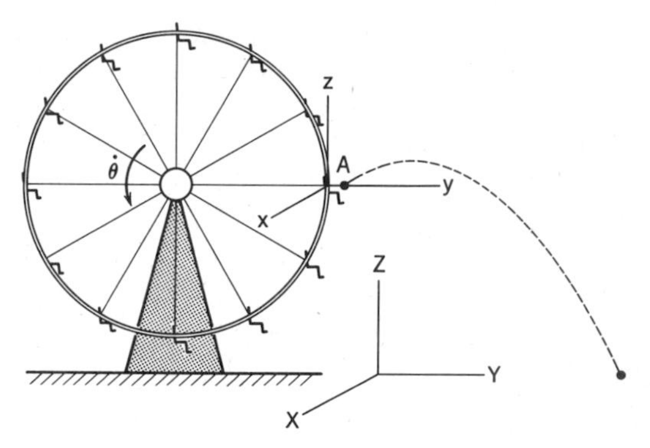

Figure P.11.62

11.63. A rocket moves at a speed of 700 m/sec and accelerates at a rate of 5 g relative to the ground reference XYZ. The products of combustion at A leave the rocket at a speed of 1700 m/sec relative to the rocket and are accelerating at the rate of 30 m/sec² relative to the rocket. What are the speed and acceleration of an element of the combustion products as seen from the ground? The rocket moves along a straight-line path whose direction cosines for the XYZ reference are $l = .6$ and $m = .6$.

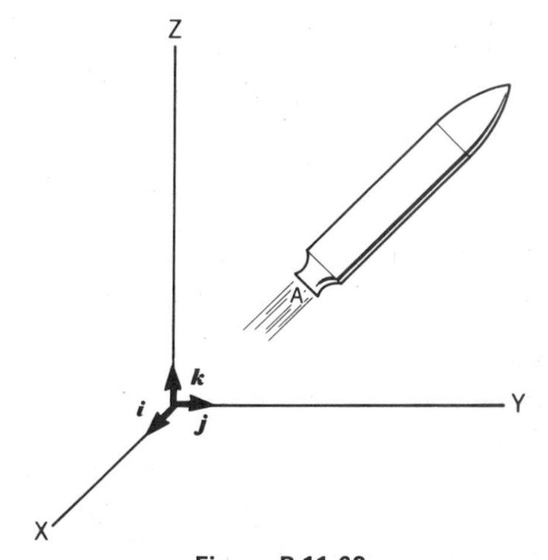

Figure P.11.63

11.64. In a steam turbine, steam is expanded through a stationary nozzle at a speed V_0 of 3000 m/sec at an angle of 30°. The steam impinges on a series of blades mounted all around the periphery of a cylinder which is rotating at a speed ω of 5000 rpm. The steam impinges on the blades at a radial distance of 1.20 m from the axis of rotation of the cylinder. What angle α should the left side of the blades have for the steam to enter the region between blades most smoothly? [*Hint:* Let xyz move with a blade in a translatory manner relative to the ground (xyz is thus not entirely fixed to the blade and hence does not rotate).]

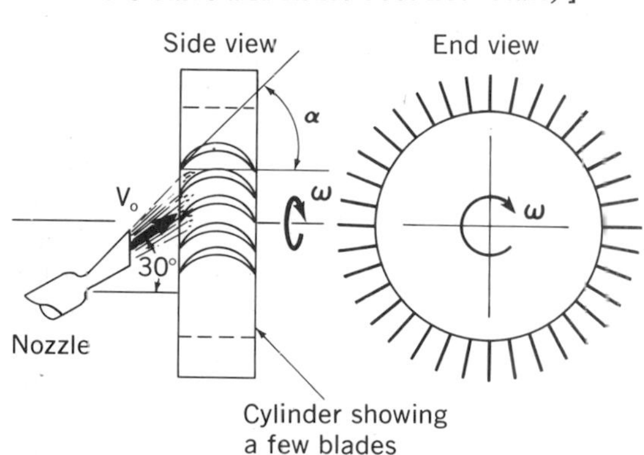

Figure P.11.64

11.65. A sailboat moves at a speed V_0 of 8 mi/hr relative to a stationary reference XY. The wind is moving uniformly at a speed V_1 relative to XY in the direction shown. On top of the mast, is a direction vane responding to the wind relative to the boat. If this vane points in a direction of 170° from the x axis, what is the velocity V_1 of the wind?

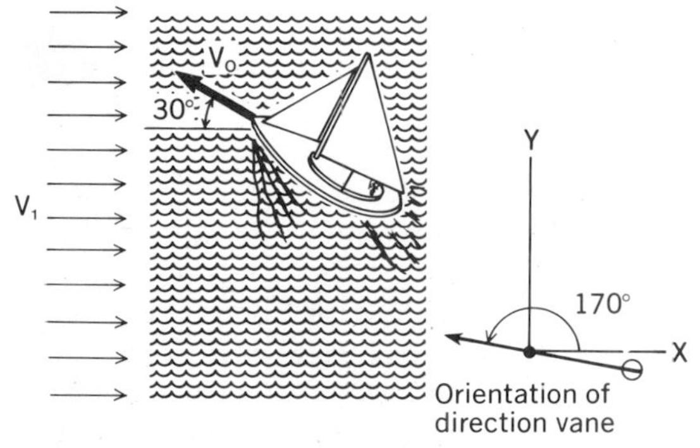

Figure P.11.65

11.66. A boat is about to depart from point A on the shore of a river which has a uniform velocity V_0 of 5 ft/sec. If the boat can move at the rate of 15 ft/sec relative to the water, and if we want to move along a straight path from A to B, how long will it take to go from A to B? At what angle β should the boat be aimed relative to the water?

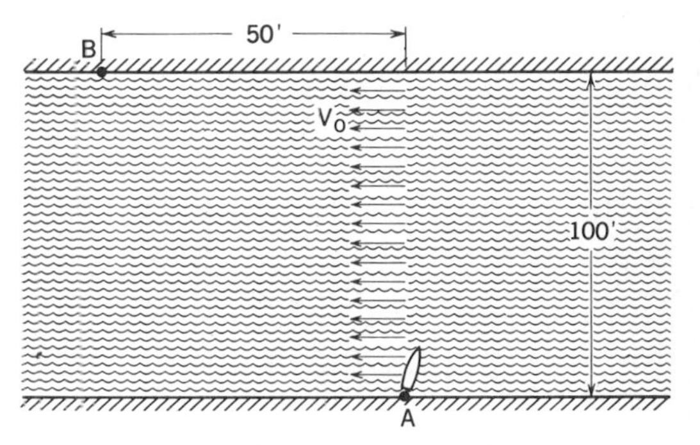

Figure P.11.66

11.67. A jet passenger plane is moving at a speed V_0 of 800 km/hr. A storm region extending 4 km in width is reported 15 km due east of its position. The region is moving NW at a speed V_1 of 100 km/hr. At what maximum angle α from due N can the plane fly to just miss the storm front?

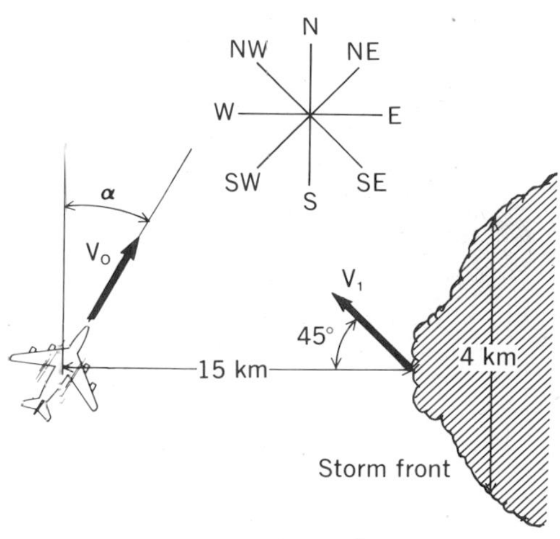

Figure P.11.67

11.68. A light plane is approaching a runway in a cross-wind. This cross-wind has a uniform speed V_0 of 33 mi/hr. The plane has a velocity component V_1 parallel to the ground of 70 mi/hr relative to the wind at an angle β of 30°. The rate of descent is such that the plane will touch down somewhere along A–A. Will this touchdown occur on the runway or off the runway for the data given?

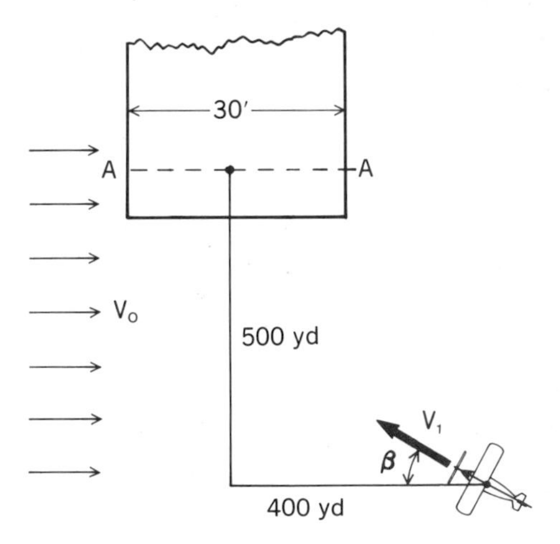

Figure P.11.68

11.8 Closure

In this chapter, we have presented, first, a few general comments on differentiation and integration of vectors. We then carried out differentiations in a variety of ways. In the first case, the vector $\boldsymbol{r}$ was expressed in terms of rectangular scalar components and the fixed unit vectors $\boldsymbol{i}, \boldsymbol{j}$, and $\boldsymbol{k}$. The procedure for finding $\dot{\boldsymbol{r}}$ and $\ddot{\boldsymbol{r}}$ in terms of rectangular scalar components is straightforward and involves only the familiar differentiation operations of scalar calculus. We next considered the kinematics of a particle moving along some given path. Here, we obtained $\dot{\boldsymbol{r}}$ and $\ddot{\boldsymbol{r}}$ in terms of speeds and rates of changes of speeds of the particle along the path with component directions no longer fixed in space but instead related at each point along the path to the geometry of the path. For this reason, we brought in certain concepts of differential geometry such as the osculating plane, the normal vector, etc. Finally, we computed $\dot{\boldsymbol{r}}$ and $\ddot{\boldsymbol{r}}$ in terms of cylindrical coordinates with component directions always in the radial, transverse, and axial directions. Clearly, the radial and transverse directions are not fixed in space and change as the particle moves about.

In carrying out various derivatives of unit vectors which are not fixed in space, such as $\boldsymbol{\epsilon}_r$ and $\boldsymbol{\epsilon}_\theta$, we went through a simple limiting process in arriving at the desired results. You should be able now to think through these limiting processes to arrive at the simple derivatives of unit vectors that you will need in the early part of the text. Later, in the study of kinematics of a rigid body, we present simple straightforward formal procedures for this purpose.

We next investigated the relations between velocities and accelerations of a particle, as seen from different references, which are translating relative to each other. We called such motions simple relative motion. Later, when we undertake rigid-body motion, we shall consider the case involving references moving arbitrarily relative to each other. It is vital to remember that we must measure $\boldsymbol{a}$ relative to an *inertial reference* when we employ Newton's law in the form $\boldsymbol{F} = m\boldsymbol{a}$. We may at times find it convenient to employ two references in this connection where one reference is the inertial reference needed for the desired acceleration vector. This situation is illustrated in Example 11.8.

In Chapter 12, we shall consider the *dynamics* of motion of a particle. We shall then have ample opportunity to employ the kinematics of Chapter 11.

Review Problems

11.69. A particle at position (3, 4, 6) ft at time $t_0 = 1$ sec is given a constant acceleration having the value $6\boldsymbol{i} + 3\boldsymbol{j}$ ft/sec². If the velocity at the time t_0 is $16\boldsymbol{i} + 20\boldsymbol{j} + 5\boldsymbol{k}$ ft/sec, what is the velocity of the particle 20 sec later? Also give the position of the particle.

11.70. A pin is confined to slide in a circular slot of radius 6 m. The pin must also slide in a straight slot which moves to the right at a constant speed, V, of 3 m/sec while maintaining a constant angle of 30° with the horizontal. What are the velocity and acceleration of the pin A at the instant shown?

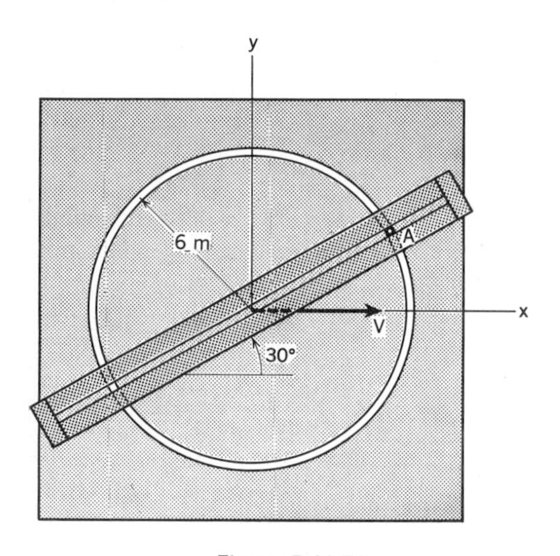

Figure P.11.70

11.71. A light line attached to a streamlined weight A is "shot" by a line rifle from a small boat C to a large boat D in heavy seas. The weight must travel a distance of 20 yd horizontally and reach the larger boat's deck, which is 20 ft higher than the deck of boat C. If the angle α of firing is 40°, what minimum velocity V_0 is needed? At the instant of firing, boat C is dipping down into the water at a speed of 5 ft/sec. Assume that the larger boat remains essentially fixed at constant level.

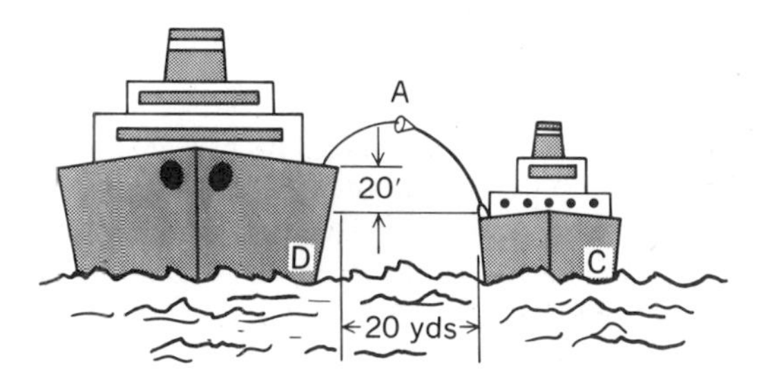

Figure P.11.71

11.72. A projectile is fired at an angle of 60° as shown. At what elevation y does it strike the hill whose equation has been estimated as $y = 10^{-5}x^2$ m? Neglect air friction and take the muzzle velocity as 1000 m/sec.

Figure P.11.72

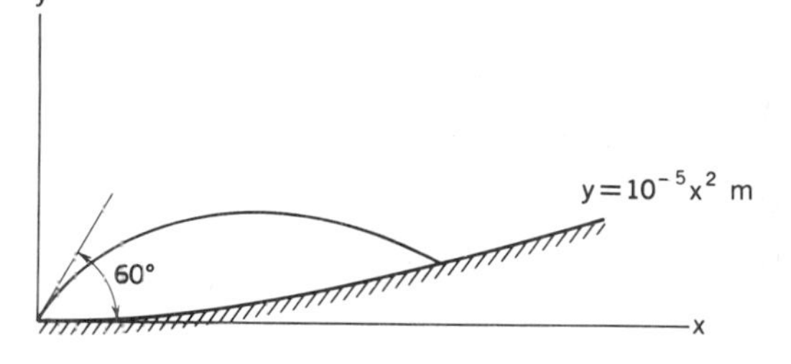

11.73. Pilots of fighter planes wear special suits designed to prevent blackouts during a severe maneuver. These suits tend to keep the blood from draining out of the head when the head is accelerated in a direction from shoulders to head. With this suit, a flier can take 5 g's of acceleration in the aforementioned direction. If a flier is diving at a speed of 1000 km/hr, what is the minimum radius of curvature that he can manage at pullout without suffering bad physiological effects?

11.74. A particle moves with constant speed of 1.5 m/sec along a path given as $x = y^2 - \ln y$ m. Give the acceleration vector of the particle in terms of rectangular component when the particle is at position $y = 3$ m. Do the problem by using path coordinate techniques and then by Cartesian-component techniques. How many g's of acceleration is the particle subject to?

11.75. A mechanical "arm" for handling radioactive materials is shown. The distance $\bar{r}$ can be varied by telescoping action of the arm. The arm can be rotated about the vertical axis A–A. Finally, the arm can be raised or lowered by a worm gear drive (not shown). What is the velocity and acceleration of the object C if the end of the arm moves out radially at a rate of 1 ft/sec while the arm turns at a speed ω of 2 rad/sec. Finally, the arm is raised at a rate of 2 ft/sec. The distance $\bar{r}$ at the instant of interest is 5 ft. What is the acceleration in the direction $\boldsymbol{\epsilon} = .8\boldsymbol{i} + .6\boldsymbol{j}$?

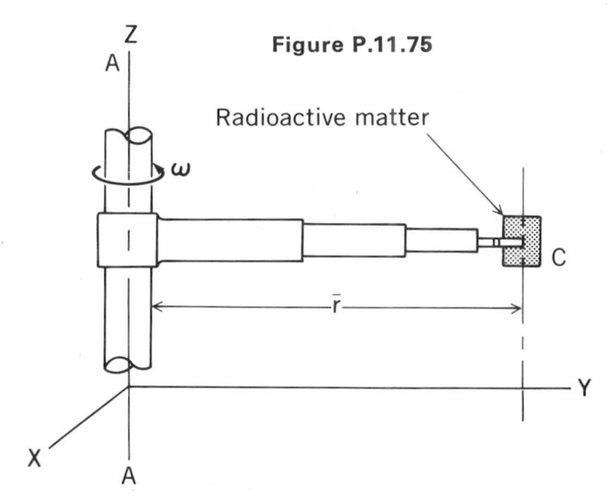

Figure P.11.75

11.76. A top-section view of a water sprinkler is shown. Water enters at the center from below and then goes through four passageways in an impeller. The impeller is rotating at constant speed ω of 8 rpm. As seen from the impeller, the water leaves at a speed of 10 ft/sec at an angle of 30° relative to r. What is the velocity and acceleration as seen from the ground of the water as it leaves the impeller and becomes free of the impeller? Give results in the radial, axial, and transverse directions. Use one reference only.

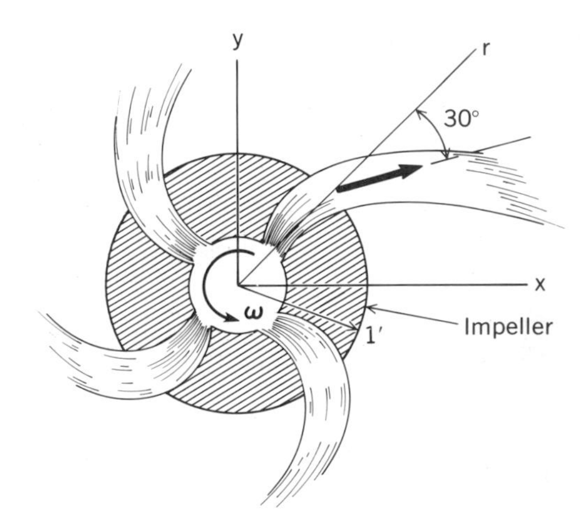

Figure P.11.76

11.77. A luggage dispenser at an airport resembles a pyramid with six flat segments as sides as shown in the diagram. The system rotates with an angular speed ω of 2 rpm. Luggage is dropped from above and slides down the faces to be picked up by travelers at the base.

A piece of luggage is shown on a face. It has just been dropped at the position indicated. It has at this instant zero velocity as seen from the rotating face but has at this instant and thereafter an acceleration of $.2g$ along the face. What is the total acceleration, as seen from the ground, of the luggage as it reaches the base at B? Use one reference only.

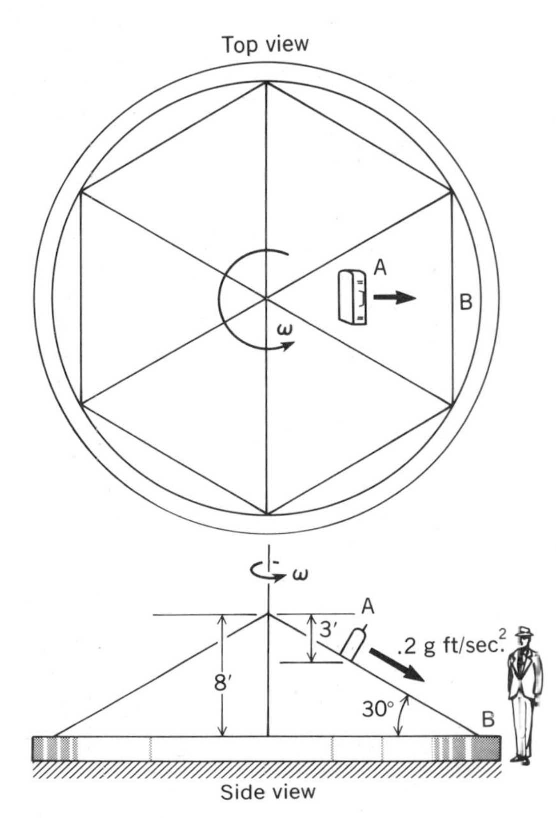

Figure P.11.77

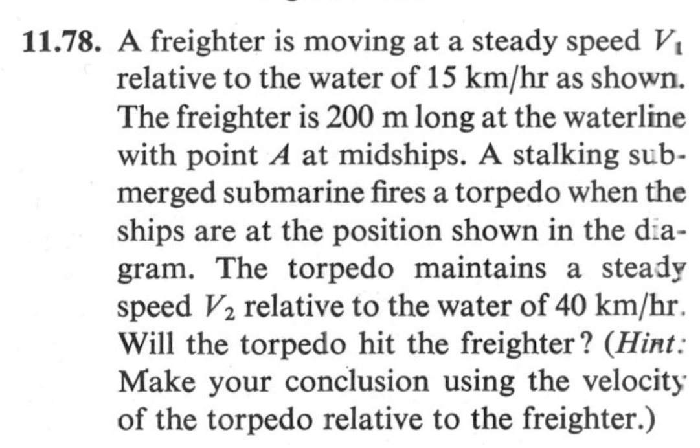

11.78. A freighter is moving at a steady speed V_1 relative to the water of 15 km/hr as shown. The freighter is 200 m long at the waterline with point A at midships. A stalking submerged submarine fires a torpedo when the ships are at the position shown in the diagram. The torpedo maintains a steady speed V_2 relative to the water of 40 km/hr. Will the torpedo hit the freighter? (*Hint:* Make your conclusion using the velocity of the torpedo relative to the freighter.)

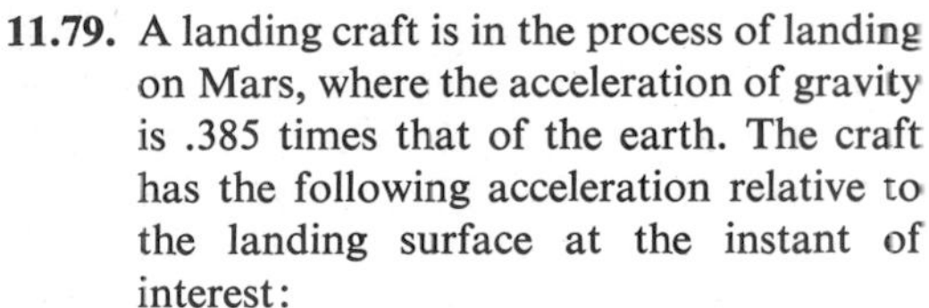

11.79. A landing craft is in the process of landing on Mars, where the acceleration of gravity is .385 times that of the earth. The craft has the following acceleration relative to the landing surface at the instant of interest:

$$\boldsymbol{a} = .2g\boldsymbol{i} + .4g\boldsymbol{j} - 2g\boldsymbol{k} \text{ m/sec}^2$$

where g is the acceleration of gravity on

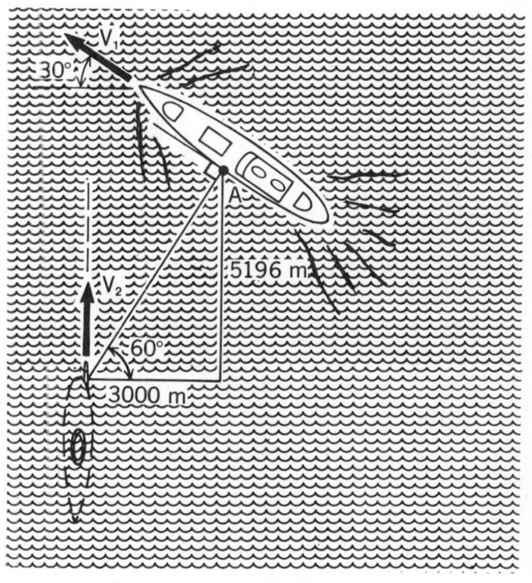

Figure P.11.78

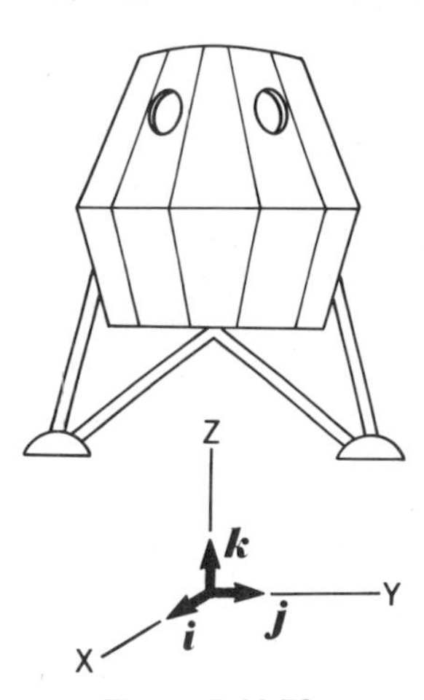

Figure P.11.79

the earth. At this instant, an astronaut[5] is raising a hand camera weighing 3 N on the earth. If he is giving the camera an upward acceleration of 3 m/sec² relative to the landing craft, what force must the astronaut exert on the camera at the instant of interest?

[5]It is estimated that it would cost $100 billion to land a man on Mars.

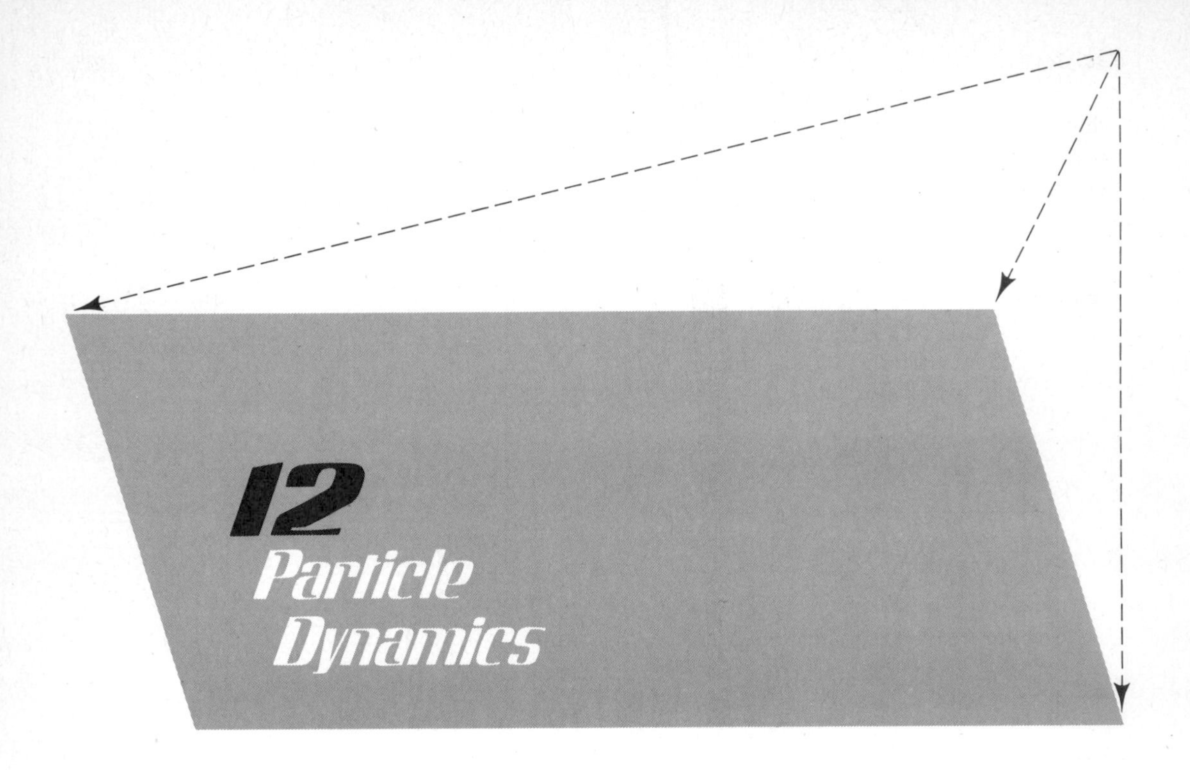

12.1 Introduction

In Chapter 11, we examined the geometry of motion—the kinematics of motion. In particular, we considered various kinds of coordinate systems: rectangular coordinates, cylindrical coordinates, and path coordinates. In this chapter, we shall consider Newton's law for the three coordinate systems mentioned above, as applied to the motion of a particle.

Before embarking on this study, we shall review notions concerning units of mass presented earlier in Chapter 1. Recall that a pound mass (lbm) is the amount of matter attracted by gravity at a specified location on the earth's surface by a force of 1 pound (lbf). A slug, on the other hand, is the amount of matter that will accelerate relative to an inertial reference at the rate of 1 ft/sec² when acted on by a force of 1 lbf. Note that the slug is defined via Newton's law, and therefore the slug is the proper unit to be used in Newton's law. The relation between the pound mass (lbm) and the slug is

$$M(\text{slugs}) = \frac{M(\text{lbm})}{32.2} \tag{12.1}$$

Note also that the weight of a body in pounds force near the earth's surface will numerically equal the mass of the body in pounds mass. It is important in using Newton's law that the mass of the body in pounds mass be properly converted into slugs via Eq. 12.1.

In SI units, recall that a kilogram is the mass that accelerates relative to an inertial reference at the rate of 1 meter/sec² when acted on by a force of 1 newton (which is about one fifth of a pound). If the weight W of a body is given in terms of newtons, we must divide by 9.81 to get the mass in kilograms needed for Newton's law. That is,

$$M(\text{kg}) = \frac{W(\text{N})}{9.81} \tag{12.2}$$

We are now ready to consider Newton's law in rectangular coordinates.

Part A
Rectangular Coordinates; Rectilinear Translation

12.2 Newton's Law for Rectangular Coordinates

In rectangular coordinates, we can express Newton's law as follows:

$$\begin{aligned} F_x &= ma_x = m\frac{dV_x}{dt} = m\frac{d^2x}{dt^2} \\ F_y &= ma_y = m\frac{dV_y}{dt} = m\frac{d^2y}{dt^2} \\ F_z &= ma_z = m\frac{dV_z}{dt} = m\frac{d^2z}{dt^2} \end{aligned} \tag{12.3}$$

If the motion is known relative to an inertial reference, we can easily solve for the rectangular components of the resultant force on the particle. The equations to be solved are just algebraic equations. The *inverse* of this problem, wherein the forces are known over a time interval and the motion is desired during this interval, is not so simple. For the inverse case, we must get involved generally with integration procedures.

In the next section, we shall consider situations in which the resultant force on a particle has the same direction and line of action at all times. The resulting motion is then confined to a straight line and is usually called *rectilinear translation*.

12.3 Rectilinear Translation

For rectilinear translation, we may consider the line of action of the motion to be collinear with one axis of a rectilinear coordinate system. Newton's law is then one of the equations of the set 12.3. We shall use the x axis to coincide with the line of action of the motion. The resultant force F (we shall not bother with the x subscript here) can be a constant, a function of time, a function of speed, a function of position, or any combination of these. At this time, we shall examine some of these cases, leaving others to Chapter 19, where, with the aid of the students' knowledge in differential equations,[1] we shall be more prepared to consider them.

Case 1. Force Is a Function of Time or a Constant. A particle of mass m acted on by a time-varying force $F(t)$ is shown in Fig. 12.1. The plane on which the body moves is frictionless. The force of gravity is equal and opposite to the normal force from the

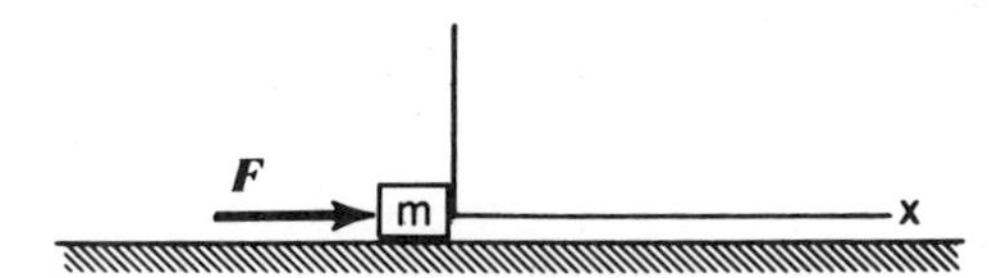

Figure 12.1. Rectilinear translation.

plane so that $F(t)$ is the resultant force acting on the mass. Newton's law can then be given as follows:

$$F(t) = m\frac{d^2x}{dt^2}$$

Therefore,

$$\frac{d^2x}{dt^2} = \frac{F(t)}{m} \tag{12.4}$$

Knowing the acceleration in the x direction, we can readily solve for $F(t)$.

The inverse problem, where we know $F(t)$ and wish to determine the motion, requires integration. For this operation, the function $F(t)$ must be piecewise continuous.[2] To integrate, we rewrite Eq. 12.4 as follows:

$$\frac{d}{dt}\left(\frac{dx}{dt}\right) = \frac{F(t)}{m}$$

$$d\left(\frac{dx}{dt}\right) = \frac{F(t)}{m}\,dt$$

Now integrating both sides we get

$$\frac{dx}{dt} = V = \int \frac{F(t)}{m}\,dt + C_1 \tag{12.5}$$

where C_1 is a constant of integration. Integrating once again after bringing dt from the left side of the equation to the right side, we get

$$x = \int\left[\int \frac{F(t)}{m}\,dt\right]dt + C_1 t + C_2 \tag{12.6}$$

We have thus found the velocity of the particle and its position as functions of time to within two arbitrary constants. These constants can be readily determined by having the solutions yield a certain velocity and position at given times. Usually,

[1] Most students studying dynamics will concurrently be taking a course in differential equations.
[2] That is, the function has only a finite number of finite discontinuities.

these conditions are specified at time $t = 0$ and are then termed *initial conditions.* That is, when $t = 0$,

$$V = V_0 \quad \text{and} \quad x = x_0 \tag{12.7}$$

These equations can be satisfied by substituting the initial conditions into Eqs. 12.5 and 12.6 and solving for the constants C_1 and C_2.

Although the preceding discussion centered about a force that is a function of time, the procedures apply directly to a force that is a constant. The following examples illustrate the procedures set forth.

EXAMPLE 12.1

A 100-lb body is initially stationary on a 45° incline as shown in Fig. 12.2(a). The coefficient of dynamic friction μ_d between the block and incline is .5. What distance along the incline must the weight slide before it reaches a speed of 40 ft/sec?

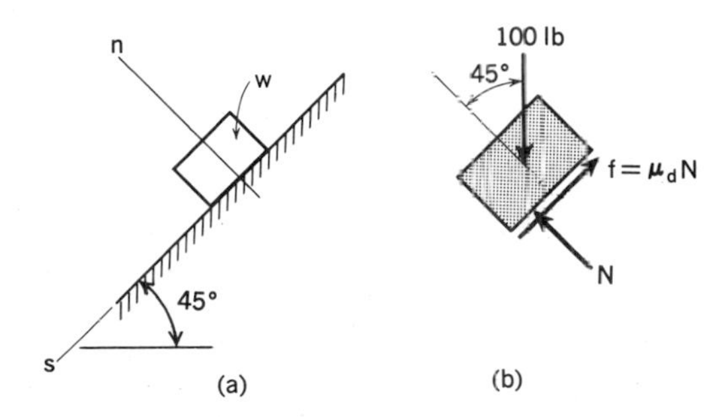

Figure 12.2. Body slides on an incline.

A free-body diagram is shown in Fig. 12.2(b). Since the acceleration is zero in the direction normal to the incline, we have from equilibrium that

$$100 \cos 45° = N = 70.7 \text{ lb} \tag{a}$$

Now applying Newton's law in a direction along the incline, we have

$$\frac{100}{g}\frac{d^2s}{dt^2} = 100 \sin 45° - \mu_d N$$

Therefore,

$$\frac{d^2s}{dt^2} = 11.38 \tag{b}$$

Rewriting Eq. (b) we have

$$d\left(\frac{ds}{dt}\right) = 11.38\, dt$$

Integrating, we get

$$\frac{ds}{dt} = 11.38t + C_1 \tag{c}$$

$$s = 11.38\frac{t^2}{2} + C_1 t + C_2 \tag{d}$$

When $t = 0$, $s = ds/dt = 0$, and thus $C_1 = C_2 = 0$. When $ds/dt = 40$ ft/sec, we have for t from Eq. (c) the result

$$40 = 11.38t$$

Therefore,

$$t = 3.51 \text{ sec}$$

Substituting this value of t in Eq. (d), we can get the distance traveled to reach the speed of 40 ft/sec as follows:

$$s = 11.38\frac{(3.51)^2}{2} = 70.4 \text{ ft} \tag{e}$$

EXAMPLE 12.2

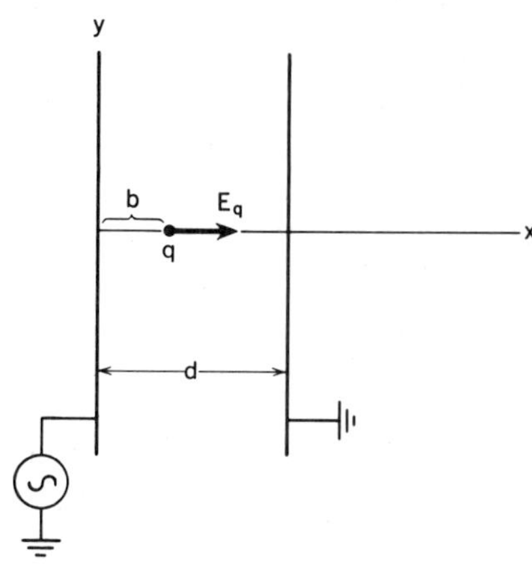

Figure 12.3. Charged particle between condenser plates.

A charged particle is shown in Fig. 12.3 at time $t = 0$ between large parallel condenser plates separated a distance d in a vacuum. A time-varying voltage given as

$$V = 6 \sin \omega t \tag{a}$$

is applied to the plates. What is the motion of the particle if it has a charge q and if we do not consider gravitational action?

As we learned in physics, the field E becomes for this case

$$E = \frac{V}{d} \tag{b}$$

The force on the particle is accordingly qE and the resulting motion is that of rectilinear translation. Using Newton's law, we have

$$\frac{d^2x}{dt^2} = q\frac{6 \sin \omega t}{md} \tag{c}$$

Rewriting Eq. (c), we have

$$d\left(\frac{dx}{dt}\right) = q\frac{6 \sin \omega t}{md}\,dt$$

Integrating, we get

$$\frac{dx}{dt} = -\frac{6q}{\omega md}\cos \omega t + C_1 \tag{d}$$

$$x = -\frac{6q}{\omega^2 md}\sin \omega t + C_1 t + C_2 \tag{e}$$

Applying the initial conditions $x = b$ and $dx/dt = 0$ when $t = 0$, we see that $C_1 = 6q/m\omega d$ and $C_2 = b$. Thus, we get

$$x = -\frac{6q}{\omega^2 md}\sin \omega t + \frac{6q}{m\omega d}t + b \tag{f}$$

Case 2. Force Is a Function of Speed. We next consider the case where the resultant force on the particle depends only on the value of the speed of the particle. An example of such a force is the aerodynamic drag force on an airplane or missile.

We can express Newton's law in the following form:

$$\frac{dV}{dt} = \frac{F(V)}{m} \tag{12.8}$$

where $F(V)$ is a piecewise continuous function representing the force in the positive x direction. If we rearrange the equation in the following manner (this is called *separation of variables*):

$$\frac{dV}{F(V)} = \frac{1}{m}\,dt$$

we can integrate to obtain

$$\int \frac{dV}{F(V)} = \frac{1}{m}t + C_1 \tag{12.9}$$

The result will give t as a function of V. However, we will generally prefer to solve for V in terms of t. The result will then have the form

$$V = H(t, C_1)$$

where H is a function of t and the constant of integration C_1. A second integration may now be performed by replacing V by dx/dt and bringing dt over to the right side of the equation. We get

$$x = \int H(t, C_1)\,dt + C_2 \tag{12.10}$$

The constants of integration are determined from the initial conditions of the problem.

EXAMPLE 12.3

A high-speed land racer (Fig. 12.4) is moving at a speed of 100 m/sec. The resistance to motion of the vehicle is primarily due to aerodynamic drag, which for this speed can be approximated as $.2V^2$ N with V in m/sec. If the vehicle has a mass of 4000 kg, what distance will it coast before its speed is reduced to 70 m/sec?

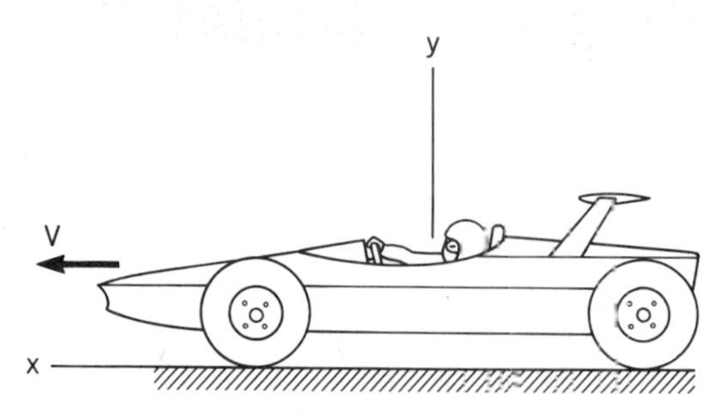

Figure 12.4. High-speed racer.

We have, using Newton's law for this case,

$$\frac{dV}{dt} = -\frac{.2V^2}{4000} = -5 \times 10^{-5}V^2 \tag{a}$$

Separating the variables, we get

$$\frac{dV}{V^2} = -5 \times 10^{-5}dt \tag{b}$$

Integrating, we have

$$-\frac{1}{V} = -5 \times 10^{-5}t + C_1 \tag{c}$$

Taking $t = 0$ when $V = 100$, we get $C_1 = -1/100$. Replacing V by dx/dt, we have next

$$\frac{1}{V} = \frac{dt}{dx} = 5 \times 10^{-5}t + \frac{1}{100} \tag{d}$$

Separating variables once again, we get

$$\frac{dt}{5 \times 10^{-5}t + (1/100)} = dx$$

Integrating, we get

$$\ln\left(5 \times 10^{-5}t + \frac{1}{100}\right) = 5 \times 10^{-5}x + C_2$$

When $t = 0$, we take $x = 0$ and so $C_2 = \ln(1/100)$. We then have on combining the log terms:

$$\ln(5 \times 10^{-3}t + 1) = 5 \times 10^{-5}x \tag{e}$$

Substitute $V = 70$ in Eq. (d); solve for t. We get $t = 85.7$ sec. Finally, find x for this time from Eq. (e). Thus,

$$\ln[(5 \times 10^{-3})(85.7) + 1] = 5 \times 10^{-5}x$$

Therefore,

$$x = 7.13 \text{ km}$$

The distance traveled is then 7.13 km.

Case 3. Force Is a Function of Position. As the final case of this series, we now consider the rectilinear motion of a body under the action of a force that is expressible as a function of position. Perhaps the simplest example of such a case is the frictionless mass–spring system shown in Fig. 12.5. The body is shown at a position where the spring is unstrained. The horizontal force from the spring at all positions of the body clearly will be a function of position x.

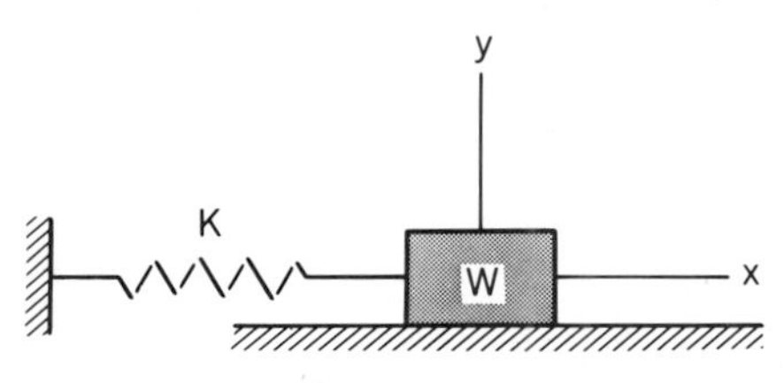

Figure 12.5. Mass–spring system.

Newton's law for position-dependent forces can be given as

$$m\frac{dV}{dt} = F(x) \tag{12.11}$$

We cannot separate the variables for this form of the equation as in previous cases since there are three variables. However, by using the chain rule of differentiation, we can change the left side of the equation to a more desirable form in the following manner:

$$m\frac{dV}{dt} = m\frac{dV}{dx}\frac{dx}{dt} = mV\frac{dV}{dx}$$

We can now separate the variables in Eq. 12.11 as follows:

$$mV\,dV = F(x)\,dx$$

Integrating, we get

$$\frac{mV^2}{2} = \int F(x)\,dx + C_1 \tag{12.12}$$

Solving for V and using dx/dt in its place, we get

$$\frac{dx}{dt} = \left[\frac{2}{m}\int F(x)\,dx + C_1\right]^{1/2}$$

Separating variables and integrating again, we get

$$t = \int \frac{dx}{\left[\frac{2}{m}\int F(x)\,dx + C_1\right]^{1/2}} + C_2 \tag{12.13}$$

For a given $F(x)$, V and x can accordingly be evaluated as functions of time from Eqs. 12.12 and 12.13. The constants of integration C_1 and C_2 are determined from the initial conditions.

A very common force that occurs in many problems is the *linear restoring force*. Such a force occurs when a body W is constrained by a linear spring (see Fig. 12.5). The force from such a spring will be proportional to x measured from a position of W corresponding to the undeformed configuration of the system. Consequently, the force will have a magnitude of $|Kx|$, where K, called the *spring constant*, is the force needed on the spring per unit elongation or compression of the spring. Furthermore, when x has a positive value, the spring force points in the negative direction, and when x is negative, the spring force points in the positive direction; that is, it always points toward the position $x = 0$ for which the spring is undeformed. The spring force is for this reason called a *restoring* force and must be expressed as $-Kx$ to give the proper direction for all values of x.

For a *nonlinear* spring, K will not be constant but will be a function of the elongation or shortening of the spring. The spring force is then given as

$$F_{\text{spring}} = -\int_0^x K(x)\,dx \tag{12.14}$$

In the following example and in the homework problems, we examine certain limited aspects of mass–spring systems to illustrate the formulations of case 3 and to familiarize us with springs in dynamic systems. A more complete study of spring–mass systems will be made in Chapter 19. The motion of such systems, we shall later learn, centers about some stationary point. That is, the motion is *vibratory* in nature.

We shall study vibrations in Chapter 19, wherein time-dependent and velocity-dependent forces are present in addition to the linear restoring force. We are deferring this topic so as to make maximal use of your course in differential equations that you are most likely studying concurrently with dynamics. It is important to understand, however, that even though we defer vibration studies until later, such studies are not something apart from the general particle dynamics undertaken in this chapter.

EXAMPLE 12.4

A cart A (see Fig. 12.6) having a mass of 200 kg is held on an incline so as to just touch an undeformed spring whose spring constant K is 50 N/mm. If body A is released very slowly, what distance down the incline must A move to reach an equilibrium configuration? If body A is released suddenly, what is its speed when it reaches the aforementioned equilibrium configuration for a slow release?

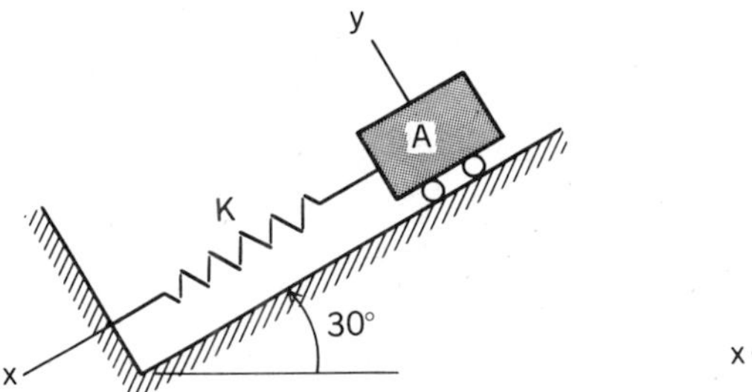

Figure 12.6. Cart–spring system.

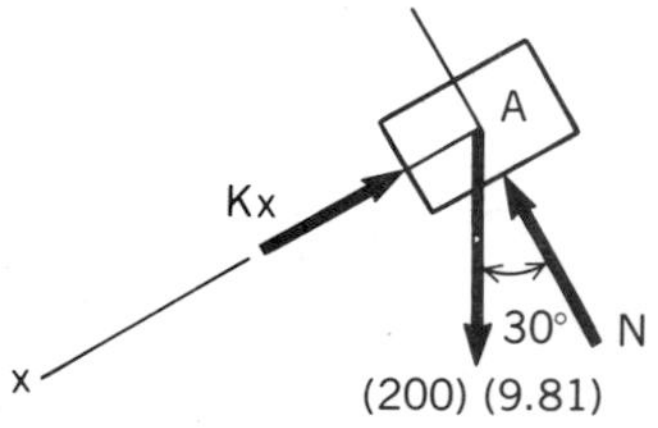

Figure 12.7. Free-body diagram of cart.

As a first step, we have shown a free body of the vehicle in Fig. 12.7. To do the first part of the problem, all we need do is utilize the definition of the spring constant. Thus, if δ represents the compression of the spring, we can say:

$$K = \frac{F}{\delta}$$

Therefore,

$$\delta = \frac{F}{K} = \frac{(200)(9.81)\sin 30°}{50}$$

$$= 19.62 \text{ mm}$$

Thus, the spring will be compressed .01962 m by the cart if it is allowed to move down the incline very slowly.

For the case of the quick release, we use *Newton's* law. Thus, using x in meters so that K is (50)(1000) N/m:

$$200\ddot{x} = (200)(9.81)\sin 30° - (50)(1000)(x)$$

Therefore,

$$\ddot{x} = 4.905 - 250x$$

Rewriting $\ddot{x}$, we have

$$V\frac{dV}{dx} = 4.905 - 250x$$

Separating variables and integrating,

$$\frac{V^2}{2} = 4.905x - 125x^2 + C_1$$

To determine the constant of integration C_1, we set $x = 0$ when $V = 0$. Clearly, $C_1 = 0$. As a final step, we set $x = .01962$ m and solve for V.

$$\begin{aligned} V &= \{2[(4.905)(.01962) - (125)(.01962)^2]\}^{1/2} \\ &= .310 \text{ m/sec} \end{aligned}$$

The following example illustrates an interesting device used by the U.S. Navy to test small devices for high, prolonged acceleration. Hopefully, the length of the problem will not intimidate you. Those hardy souls who go ahead to study it will see use made of the gas laws presented in your elementary chemistry courses.

*EXAMPLE 12.5

An *air gun* is used to test the ability of small devices to withstand high accelerations. A floating piston A (Fig. 12.8), on which the device to be tested is mounted, is held at position C while region D is filled with highly compressed air. Region E is initially at atmospheric pressure but is entirely sealed from the outside. When "fired," a quick-release mechanism releases the piston and it accelerates rapidly toward the other end of the gun, where the trapped air in E "cushions" the motion so that the piston will begin eventually to return. However, as it starts back, the high pressure developed in E is released through valve F and the piston only returns a short distance.

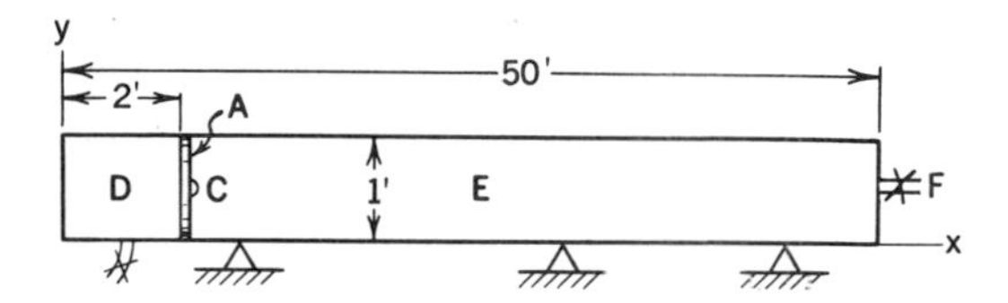

Figure 12.8. Air gun.

Suppose that the piston and its test specimen have a combined mass of 2 lbm and the pressure initially in the chamber D is 1000 psig (above atmosphere). Compute the speed of the piston at the halfway point of the air gun if we make the simple assumption that the air in D expands according to pv = constant and the air in E is compressed also according to pv = constant.[3] Note that v is the specific volume

[3]You should recall from your earlier work in physics and chemistry that we are using here the isothermal form of the equation of state for a perfect gas. Two factors of caution should be pointed out relative to the use of this expression. First, at the high pressures involved in part of the expansion, the perfect gas model is only an approximation for the gas, and so the equation of state of a perfect gas which gives us pv = constant is only approximate. Furthermore, the assumption of isothermal expansion gives only an approximation of the actual process. Perhaps a better approximation is to assume an adiabatic expansion (i.e., no heat transfer). This is done in Problem 12.36.

(i.e., the volume per unit mass). Take v of this fluid at D to be initially .207 ft^3/lbm and v in E to be initially 13.10 ft^3/lbm. Neglect the inertia of the air.

The force on the piston results from the pressures on each face, and we can show that this force is a function of x (see Fig. 12.8 for reference axes). Thus, examining the pressure p_D first for region D, we have, from initial conditions,

$$(p_D v_D)_0 = [(1000 + 14.7)(144)](.207) = 30{,}300 \tag{a}$$

Furthermore, the mass of air D given as M_D is determined from initial data as

$$M_D = \frac{(V_D)_0}{(v_D)_0} = \frac{(2)(\pi/4)}{.207} = 7.58 \text{ lbm} \tag{b}$$

where $(V_D)_0$ is the volume of the air in D initially. Using the results of Eqs. (a) and (b), we now have for p_D at any position x of the piston:

$$p_D = \frac{30{,}300}{v_D} = \frac{30{,}300}{V_D/M_D} = \frac{30{,}300}{(\pi/4)x/7.58}$$

Therefore,

$$p_D = \frac{293{,}000}{x} \tag{c}$$

We can similarly get p_E as a function of x for region E. Thus,

$$(p_E v_E)_0 = (14.7)(144)(13.10) = 27{,}700$$

and

$$M_E = \frac{(V_E)_0}{(v_E)_0} = \frac{(48)(\pi/4)}{13.10} = 2.88 \text{ lbm}$$

Hence,

$$p_E = \frac{27{,}700}{v_E} = \frac{27{,}700}{V_E/M_E} = \frac{27{,}700}{(\pi/4)(50 - x)/2.88}$$

Therefore,

$$p_E = \frac{101{,}600}{50 - x}$$

Now we can write *Newton's* law for this case. Noting that V without subscripts is velocity and not volume,

$$MV\frac{dV}{dx} = \frac{\pi 1^2}{4}(p_D - p_E) = \frac{\pi}{4}\left(\frac{293{,}000}{x} - \frac{101{,}600}{50 - x}\right) \tag{d}$$

where M is the mass of piston and load. Separating variables and integrating, we get

$$\frac{MV^2}{2} = \frac{\pi}{4}[293{,}000 \ln x + 101{,}600 \ln (50 - x)] + C_1 \tag{e}$$

To get the constant C_1, set $V = 0$ when $x = 2$ ft. Hence,

$$C_1 = -\frac{\pi}{4}(293{,}000 \ln 2 + 101{,}600 \ln 48)$$

Therefore,

$$C_1 = -468{,}000$$

Substituting C_1 in Eq. (e), we get

$$V = \left(\frac{2}{M}\right)^{1/2}\left\{\frac{\pi}{4}[293{,}000 \ln x + 101{,}600 \ln (50 - x)] - 468{,}000\right\}^{1/2}$$

We may rewrite this as follows noting that $M = 2$ lbm/g:

$$V = 566[23 \ln x + 7.98 \ln (50 - x) - 46.8]^{1/2}$$

At $x = 25$ ft, we then have for V the desired result:

$$V = 566(23 \ln 25 + 7.98 \ln 25 - 46.8)^{1/2}$$
$$= 4120 \text{ ft/sec}$$

12.4 A Comment

We have in Part A considered only rectilinear motions of particles. Actually in Chapter 11, we considered the coplanar motion of particles having a constant acceleration of gravity in the y direction and zero acceleration in the x direction. These were the *ballistic* problems. We treated them earlier in Chapter 11 because the considerations were primarily kinematic in nature. In this chapter, they correspond to the coplanar motion of a particle having a constant force in the y direction along with an initial velocity component in this direction, plus a zero force in the x direction, with a possible initial velocity component in this direction. Therefore, in the context of Chapter 12 we would have integrated two scalar equations of Newton's law in rectangular components (Eqs. 12.3) for a single particle.

Problems

12.1. A particle of mass 1 slug is moving in a constant force field given as

$$\boldsymbol{F} = 3\boldsymbol{i} + 10\boldsymbol{j} - 5\boldsymbol{k} \text{ lb}$$

The particle starts from rest at position (3, 5, −4). What is the position and velocity of the particle at time $t = 8$ sec? What is the position when the particle is moving at a speed of 20 ft/sec?

12.2. A particle of mass m is moving in a constant force field given as

$$\boldsymbol{F} = 2m\boldsymbol{i} - 12m\boldsymbol{j} \text{ N}$$

Give the vector equation for $\boldsymbol{r}(t)$ of the particle if, at time $t = 0$, it has a velocity V_0 given as

$$\boldsymbol{V}_0 = 6\boldsymbol{i} + 12\boldsymbol{j} + 3\boldsymbol{k} \text{ m/sec}$$

Also, at time $t = 0$, it has a position given as

$$\boldsymbol{r}_0 = 3\boldsymbol{i} + 2\boldsymbol{j} + 4\boldsymbol{k} \text{ m}$$

What are coordinates of the body at the instant that the body reaches its maximum height, y_{max}?

12.3. A block is permitted to slide down an inclined surface. The coefficient of friction is .05. If the velocity of the block is 30 ft/sec on reaching the bottom of the incline, how far up was it released and how many seconds has it traveled?

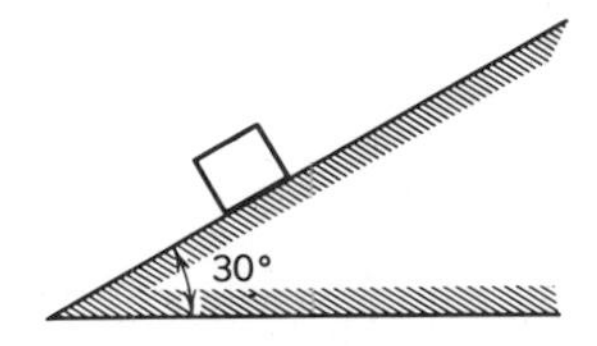

Figure P.12.3

12.4. An arrow is shot upward with an initial speed of 80 ft/sec. How high up does it go and how long does it take to reach the maximum elevation if we neglect friction?

12.5. A mass D at $t = 0$ is moving to the left at a speed of .6 m/sec relative to the ground on a belt that is moving at constant speed to the right at 1.6 m/sec. If there is coulombic friction present with $\mu_d = .3$, how long does it take before the speed of D relative to the belt is .3 m/sec to the left?

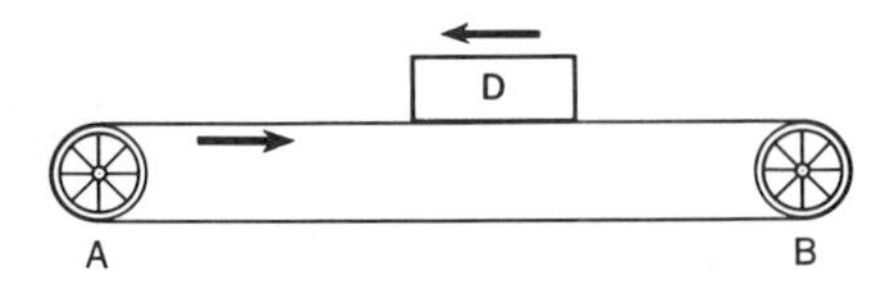

Figure P.12.5

12.6. Do Problem 12.5 with the belt system inclined 15° with the horizontal so that end B is above end A.

12.7. A drag racer can develop a torque of 200 ft-lb on each of the rear wheels. If we assume that this maximum torque is maintained and that there is no wind friction, what is the time to travel a quarter mile from a standing start? What is the speed of the vehicle at the quarter-mile mark? The weight of the racer and the driver altogether is 1600 lb. For simplicity, neglect the rotational effects of the wheels.

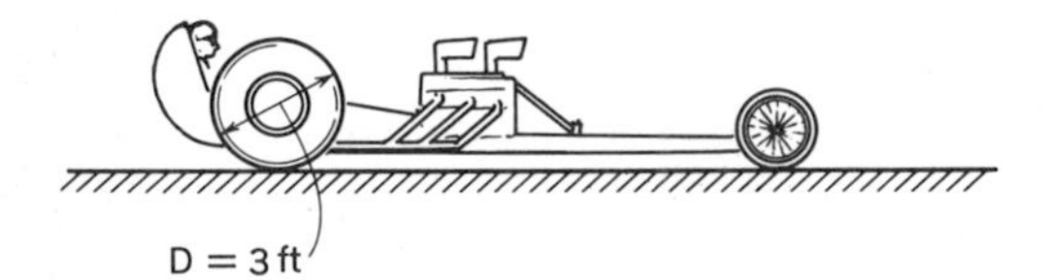

Figure P.12.7

12.8. A truck is moving down a 10° incline. The driver strongly applies his brakes to avoid a collision and the truck decelerates at the steady rate of 1 m/sec². If the static coefficient of friction μ between the load W and the truck trailer is .3, will the load slide or remain stationary relative to the truck trailer? The weight of W is 4500 N and it is not held to the truck by cables.

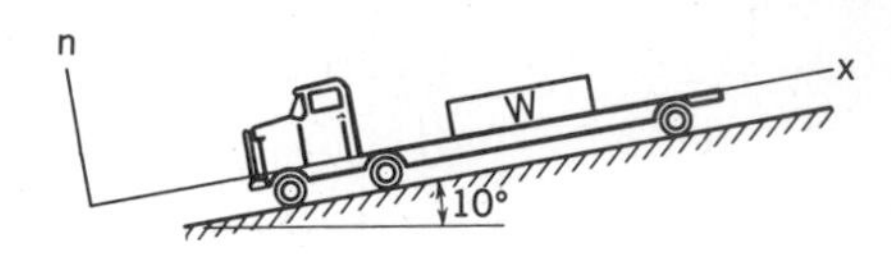

Figure P.12.8

12.9. A simple device for measuring reasonably uniform accelerations is the pendulum. Calibrate θ of the pendulum for vehicle accelerations of 5 ft/sec², 10 ft/sec², and 20 ft/sec². The bob weighs 1 lb. The bob is connected to a post with a flexible string.

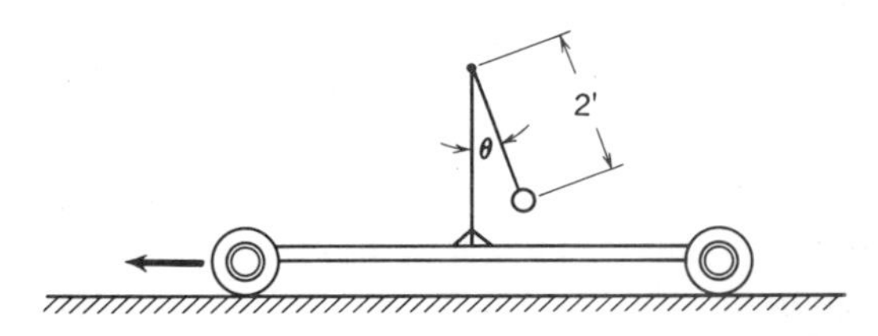

Figure P.12.9

12.10. A piston is being moved through a cylinder. The piston is moved at a constant speed V_p of .6 m/sec relative to the ground by a force F. The cylinder is free to move along the ground on small wheels. There is a coulombic friction force between the piston and the cylinder such that $\mu_d = .3$. What distance d must the piston move relative to the ground to advance .1 m along the cylinder if the cylinder is stationary at the outset? The piston has a mass of 2.5 kg and the cylinder has a mass 5 kg.

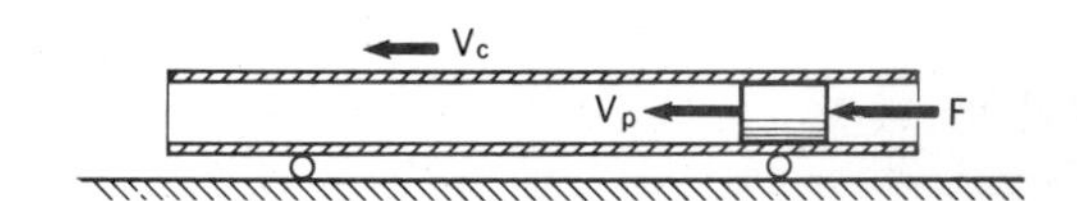

Figure P.12.10

12.11. A small body M of mass 1 kg slides along a wire from A to B. There is coulombic friction between the mass M and the wire. The coefficient of friction is .4. How long does it take to go from A to B?

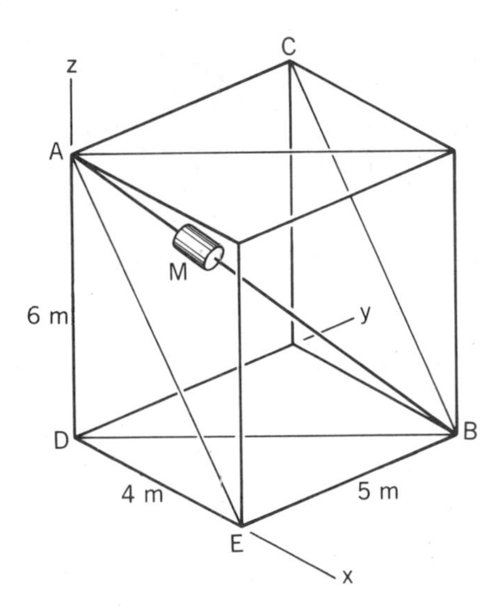

Figure P.12.11

12.12. A force F is applied to a system of light pulleys to pull body A. If F is 10 kN and A has a mass of 5000 kg, what is the speed of A after 1 sec starting from rest?

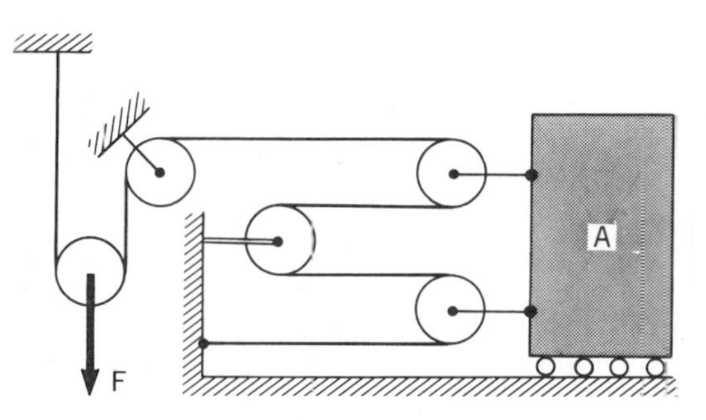

Figure P.12.12

12.13. A force represented as shown acts on a body having a mass of 1 slug. What is the position and velocity at $t = 30$ sec if the body starts from rest at $t = 0$?

Figure P.12.13

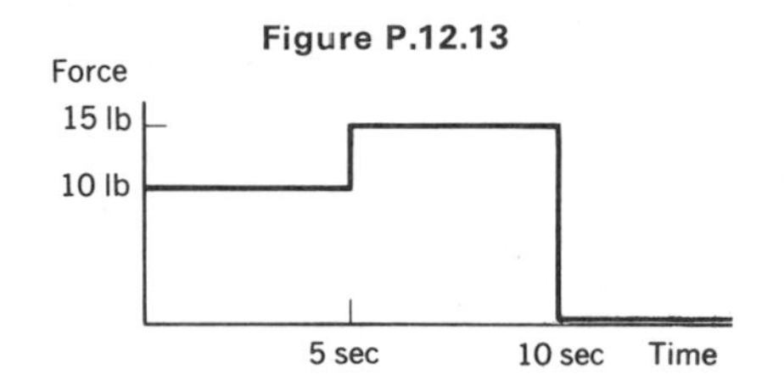

12.14. A body of mass 1 kg is acted on by a force as shown in the diagram. If the velocity of the body is zero at $t = 0$, what is the velocity and distance traversed when $t = 1$ min? The force acts for only 45 sec.

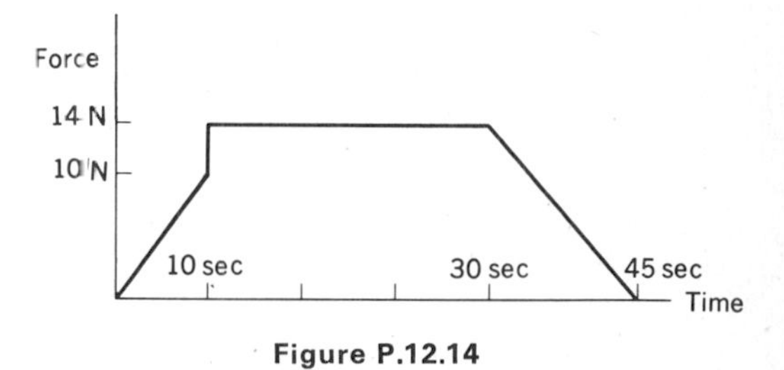

Figure P.12.14

12.15. Three coupled streetcars are moving down an incline at a speed of 20 km/hr when the brakes are applied for a panic stop. All the wheels lock except for car B, where due to a malfunction all the brakes on the front end of the car do not operate. How far does the system move and what are the forces in the couplings between the cars? Each streetcar weighs 220 kN and the coefficient of dynamic friction μ_d between wheel and rail is .30.

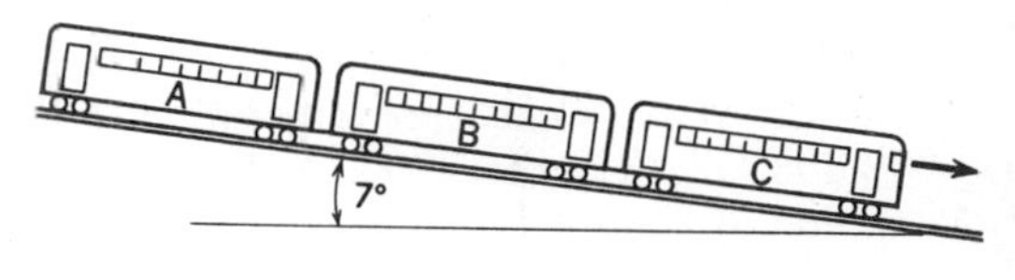

Figure P.12.15

12.16. A body having a mass of 30 lbm is acted on by a force given by

$$F = 30t^2 + e^{-t} \text{ lb}$$

If the velocity is 10 ft/sec at $t = 0$, what is the body's velocity and the distance traveled when $t = 2$ sec?

12.17. A body of mass 10 kg is acted on by a force in the x direction, given by the relation $F = 10 \sin 6t$ N. If the body has a velocity of 3 m/sec when $t = 0$ and is at position $x = 0$ at that instant, what is the position reached by the body from the origin at

$t = 4$ sec? Sketch the displacement-versus-time curve.

12.18. A force given as 5 sin $3t$ lb acts on a mass of 1 slug. What is the position of the mass at $t = 10$ sec? Determine the total distance traveled. Assume the motion started from rest.

12.19. A block A of mass 500 kg is pulled by a force of 10,000 N as shown. A second block B of mass 200 kg rests on small frictionless rollers on top of block A. A wall prevents block B from moving to the left. What is the speed of block A after 1 sec starting from a stationary position? The coefficient of friction μ_d is .4 between A and the horizontal surface.

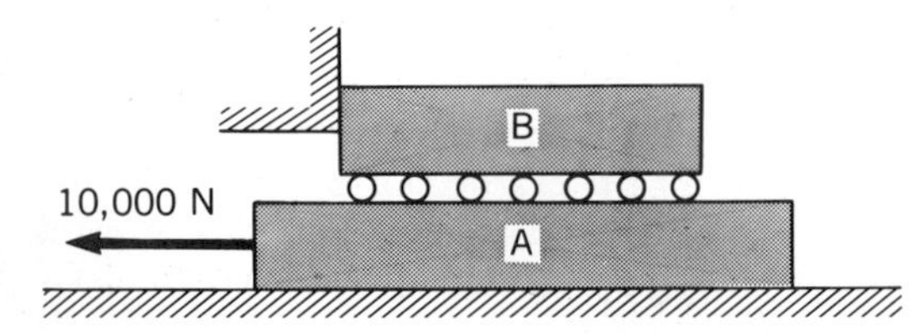

Figure P.12.19

12.20. Block B weighing 500 N rests on block A which weighs 300 N. The dynamic coefficient of friction between contact surfaces is .4. At wall C there are rollers whose friction we can neglect. What is the acceleration of body A when a force F of 5000 N is applied? Compute the acceleration after the bodies are moving relative to each other.

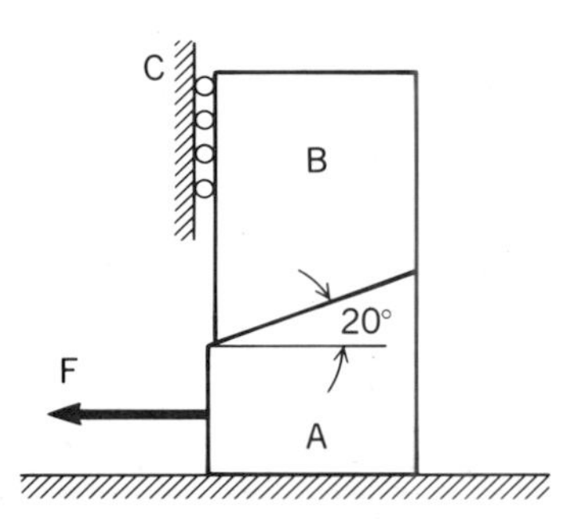

Figure P.12.20

12.21. A body A of mass 1 lbm is forced to move by the device shown. What total force is exerted on the body at time $t = 6$ sec? What is the maximum total force on the body, and when is the first time this force is developed after $t = 0$? What is $|\boldsymbol{F}|$ at this time?

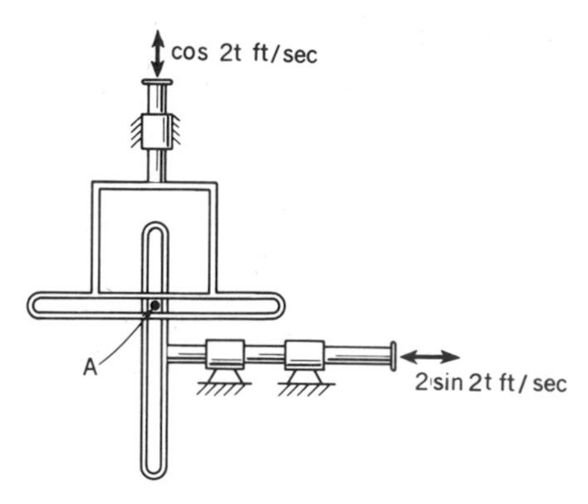

Figure P.12.21

12.22. Do Problem 12.10 for the case where there is viscous friction between piston and cylinder given as 150 N/m/sec of relative speed. Also, what is the maximum distance l the piston can advance relative to the cylinder?

12.23. The high-speed aerodynamic drag on a car is $.02V^2$ lb with V in ft/sec. If the initial speed is 100 mi/hr, how far will the car move before its speed is reduced to 60 mi/hr? The mass of the car is 2000 lbm.

12.24. A block slides on a film of oil. The resistance to motion of the block is proportional to the speed of the block relative to the incline at the rate of 7.5 N/m/sec. If the block is released from rest, what is the *terminal speed*? What is the distance moved after 10 sec?

Figure P.12.24

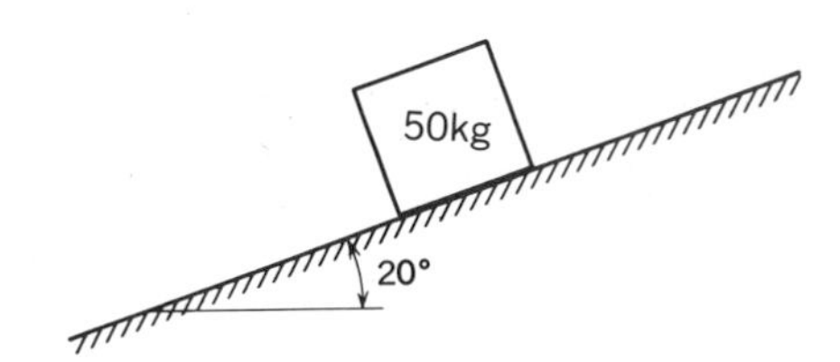

12.25. A poison dart gun is shown. The cross-sectional area inside the tube is 1 in.^{2}. The dart being blown weighs 3 oz. The dart gun bore has a viscous resistance given as .3 oz per unit velocity in ft/sec. The hunter applies a constant pressure p at the mouth of the gun. Express the relation between p, V (velocity), and t. What constant pressure p is needed to cause the dart to reach a speed of 60 ft/sec in 2 sec? Assume the dart gun is long enough.

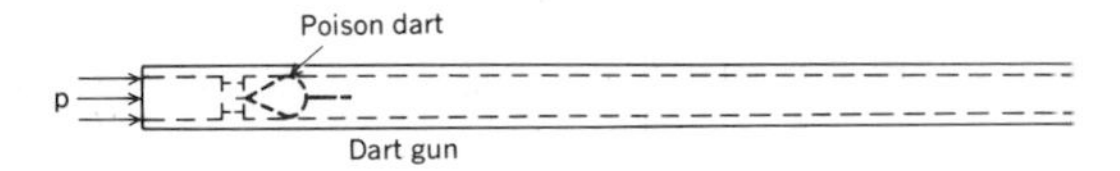

Figure P.12.25

12.26. Using the diagram for Problem 12.5, assume that there is a lubricant between the body D of mass 5 lbm and the belt such that there is a viscous friction force given as .1 lb per unit relative velocity between the body and the belt. The belt moves at a uniform speed of 5 ft/sec to the right and initially the body has a speed to the left of 2 ft/sec relative to ground. At what time later does the body have a zero instantaneous velocity relative to the ground?

12.27. In Problem 12.26, assume that the belt system is inclined 20° from the horizontal with end B above end A. What minimum belt speed is required so that a body of mass M moving downward will come to a permanent halt relative to the ground? For this belt speed, how long does it take for the body to slow down to half of its initial speed of 2 ft/sec relative to the ground?

12.28. The largest of the supertankers in the world today is the *S.S. Globtik London*, having a weight when fully loaded of 476,292 tons. The thrust needed to keep this ship moving at 10 knots is 50 kN. If the drag on the ship from the water is proportional to the speed, how long will it take for this ship to slow down from 10 knots to 5 knots after the engines are shut down? (The answer may make you wonder about the safety of such ships.)

12.29. A cantilever beam is shown. It is observed that the vertical deflection of the end A is directly proportional to a vertical tip load F provided that this load is not too excessive. A body B of mass 200 kg, when attached to the end of the beam with F removed, causes a deflection of 5 mm there after all motion has ceased. What is the speed of this body if it is attached suddenly to the beam and has descended 3 mm?

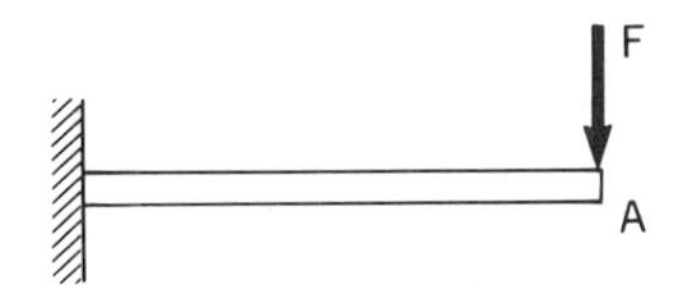

Figure P.12.29

12.30. The spring shown is nonlinear. That is, K is not a constant, but is a function of the extension of the spring. If $K = 2x + 3$ lb/in. with x measured in inches, what is the speed of the mass when $x = 0$ after it is released from a state of rest at a position 3 in. from the equilibrium position? The mass of the body is 1 slug.

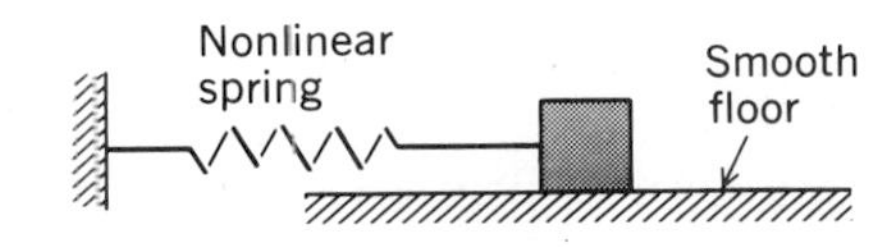

Figure P.12.30

12.31. A particle of mass m is subject to the following force field:

$$\boldsymbol{F} = m\boldsymbol{i} + 4m\boldsymbol{j} + 16m\boldsymbol{k} \text{ lb}$$

In addition, it undergoes a frictional force $\boldsymbol{f}$ given as

$$\boldsymbol{f} = -m\dot{x}\boldsymbol{i} - m\dot{y}\boldsymbol{j} + 2m\dot{z}\boldsymbol{k} \text{ lb}$$

The particle is stationary at the origin at time $t = 0$. What is the position of the particle at time $t = 1$ sec?

12.32. For $m = 1$ slug and $K = 10$ lb/in., what is the speed at $x = 1$ in. if a force of 5 lb in the x direction is applied suddenly to the mass–spring system and then maintained constant? Neglect the mass of the spring and friction.

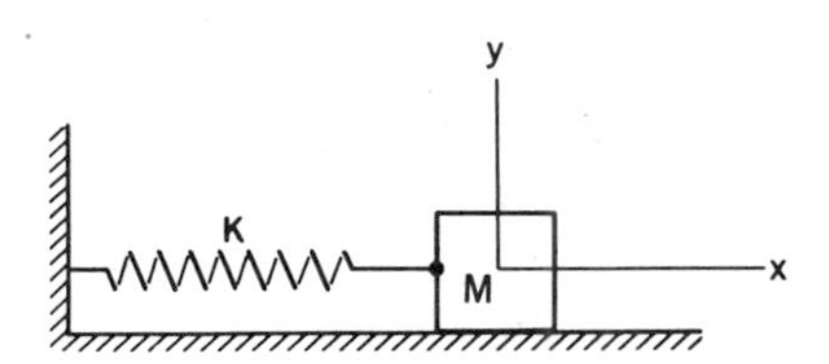

Figure P.12.32

12.33. A rod B of mass 500 kg rests on a block A of mass 50 kg. A force F of 10,000 N is applied suddenly to block A at the position shown. If the coefficient of friction μ_d is .4 for all contact surfaces, what is the speed of A when it has moved 3 m to the end of the rod?

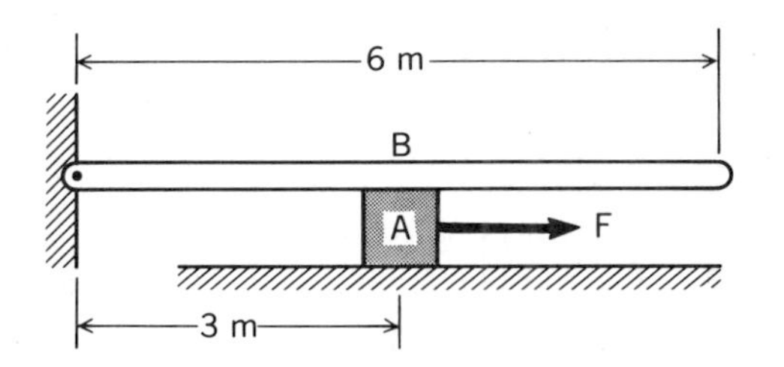

Figure P.12.33

12.34. Shown is a simply supported beam. You will learn in strength of materials that a vertical force F applied at the center causes a deflection δ at the center given as

$$\delta = \frac{1}{48}\frac{FL^3}{EI}$$

If a mass of 200 lbm, fastened to the beam at its midpoint, is suddenly released, what will its speed be when the deflection is $\frac{1}{8}$ in.? Neglect the mass of the beam. The length of beam L is 20 ft, Young's modulus E is 30×10^6 psi, and the moment of inertia of the cross section I is 20 in.^{4}.

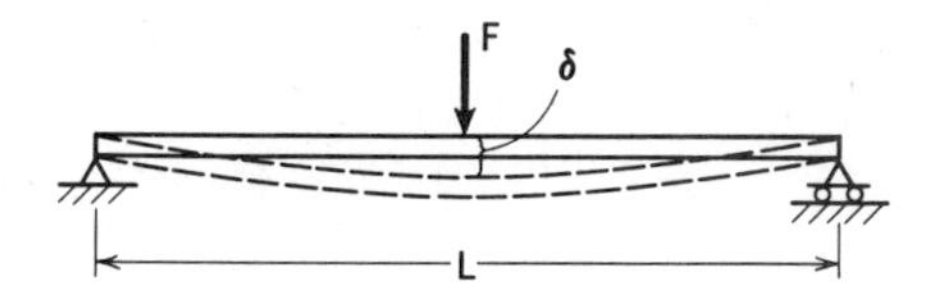

Figure P.12.34

12.35. A piston is shown maintaining air at a pressure of 8 psi above that of the atmosphere. If the piston is allowed to accelerate to the left, what is the speed of the piston after it moves 3 in.? The piston assembly has a mass of 3 lbm. Assume that the air expands *adiabatically* (i.e., with no heat transfer). This means that at all times $pV^k =$ constant, where V is the volume of the gas and k is a constant which for air equals 1.4. Neglect the inertial effects of the air.

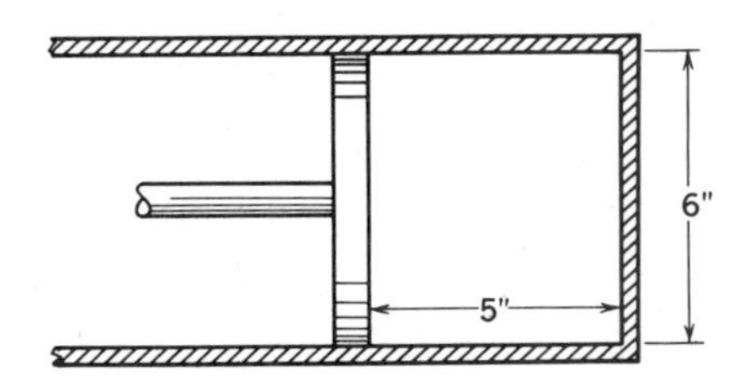

Figure P.12.35

***12.36.** In Example 12.5 assume that there are adiabatic expansions and compressions of the gases (i.e., that $pv^k =$ constant with $k = 1.4$). Compare the results for speed of the piston. Explain why your result should be higher or lower than for the isothermal case.

12.37. Body A and body B are connected by an inextensible cord as shown. If both bodies are released simultaneously, what distance do they move in $\frac{1}{2}$ sec? Take $M_A = 25$ kg and $M_B = 35$ kg. The coefficient of friction μ_d is .3.

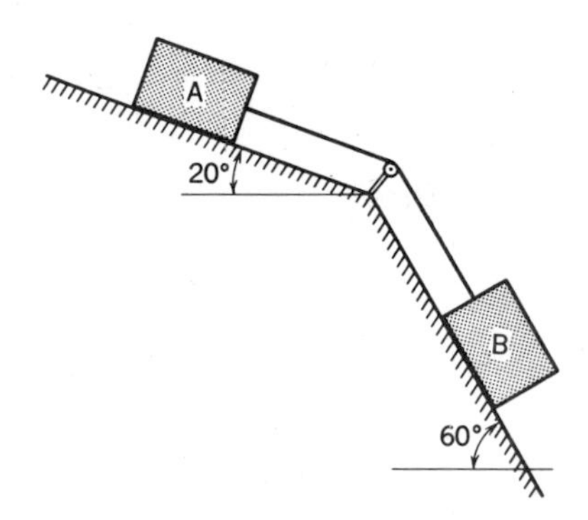

Figure P.12.37

12.38. The system shown is released from rest. What distance does the body C drop in 2 sec? The cable is inextensible. The coefficient of dynamic friction μ_d is .4 for contact surfaces of bodies A and B.

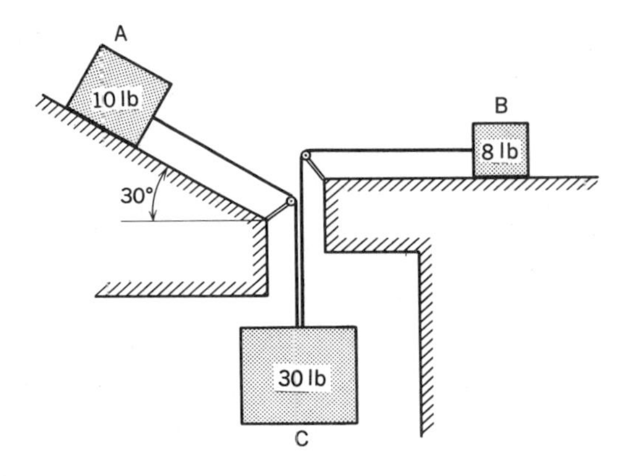

Figure P.12.38

12.39. Do Problem 12.38 for the case where there is viscous damping for the contact surfaces of bodies A and B given as $.5V$ lb, with V in ft/sec.

12.40. Two bodies A and B are shown having masses of 40 kg and 30 kg, respectively. The cables are inextensible. Neglecting the inertia of the cable and pulleys at C and D, what is the speed of the block B 1 sec after the system has been released from rest? The coefficient friction μ_d for the contact surface of body A is .3. (*Hint:* From your earlier work in physics, recall that pulley D is instantaneously rotating about point a and hence point c moves at a speed that is twice that of point b.)

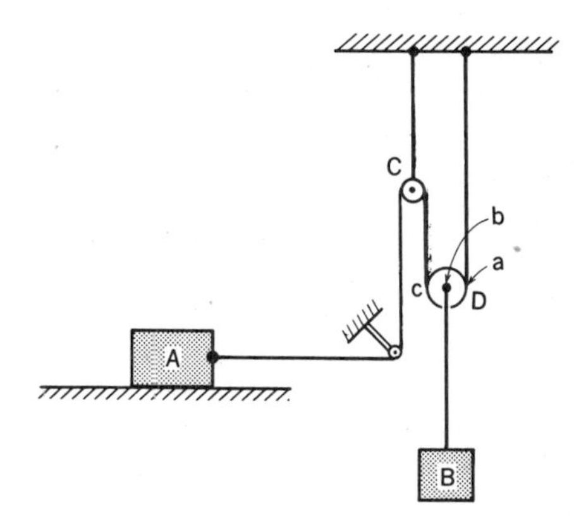

Figure P.12.40

12.41. Bodies A, B, and C have weights respectively 100 lb, 200 lb, and 150 lb. If released from rest, what are the respective speeds of the bodies after 1 sec? Neglect the weight of pulleys.

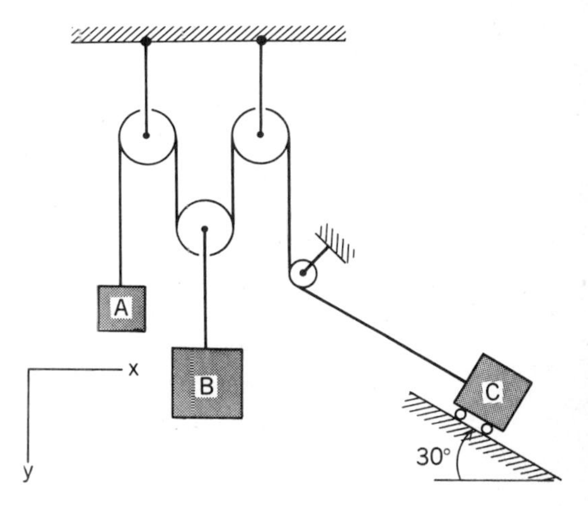

Figure P.12.41

12.42. A system of light pulleys and inextensible wire connects bodies A, B, and C as shown. If the coefficient of friction between C and the support is .4, what is the acceleration of each body? Take M_A as 100 kg, M_B as 300 kg, and M_C as 80 kg.

12.43. Do Problem 12.40 for the case where there is a viscous friction for the contact surface of body A given as $.4V$ N with V in m/sec.

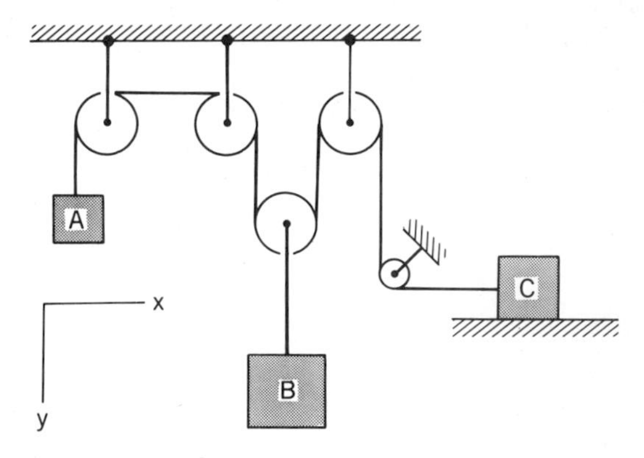

Figure P.12.42

Part B Cylindrical Coordinates; Central Force Motion

12.5 Newton's Law for Cylindrical Coordinates

In cylindrical coordinates we can express Newton's law as follows:

$$F_r = m(\ddot{\bar{r}} - \bar{r}\dot{\theta}^2) \tag{12.15a}$$

$$F_\theta = m(\bar{r}\ddot{\theta} + 2\dot{\bar{r}}\dot{\theta}) \tag{12.15b}$$

$$F_z = m\ddot{z} \tag{12.15c}$$

If the motion is known, it is a simple matter to ascertain the force components using Eqs. 12.15. The inverse problem of determining the motion given the forces is particularly difficult in this case. The reason for this difficulty, as you may have already learned in your differential equations course, is that Eqs. 12.15a and 12.15b are *nonlinear*[4] for all force functions. For this reason, we cannot present integration procedures as in Part A. The following example will serve to illustrate the kind of problem we are able to solve with the methods thus far presented in this chapter.

EXAMPLE 12.6

A platform shown in Fig. 12.9 has a constant angular velocity ω equal to 5 rad/sec. A mass B of 2 kg slides in a frictionless chute attached to the platform. The mass is connected via a light inextensible cable to a linear spring having a spring

[4]A differential equation is nonlinear if the dependent variable and its derivatives form powers greater than unity or form products anywhere in the equation.

constant K of 20 N/m. A swivel connector at A allows the cable to turn freely relative to the spring. The spring is unstretched when the mass B is at the center C of the platform. If the mass B is released at $r = 200$ mm from a stationary position relative to the platform, what is its speed relative to the platform when it has moved to position $r = 400$ mm? What is the transverse force on the body B at this position?

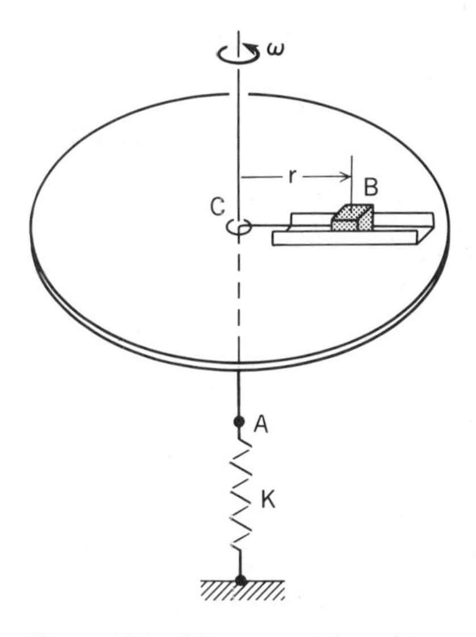

Figure 12.9. Slider on rotating platform.

We have here a coplanar motion for which cylindrical coordinates are most useful. Because the motion is coplanar, we can use r instead of $\bar{r}$ with no ambiguity. Applying Eq. 12.15a first, we have

$$-20r = 2(\ddot{r} - 25r)$$

Therefore,

$$\ddot{r} = 15r \qquad \text{(a)}$$

As in Example 12.4, we can replace $\ddot{r}$ so as to allow for a separation of variables.

$$\ddot{r} \equiv \frac{dV_r}{dt} \equiv \frac{dV_r}{dr}\frac{dr}{dt} \equiv V_r \frac{dV_r}{dr} = 15r$$

Therefore,

$$V_r \, dV_r = 15r \, dr$$

Integrating, we get

$$\frac{V_r^2}{2} = \frac{15r^2}{2} + C_1 \qquad \text{(b)}$$

To determine C_1, note that, when $r = .20m$, $V_r = 0$. Hence,

$$C_1 = -\frac{.600}{2}$$

Equation (b) then becomes

$$V_r^2 = 15r^2 - .600 \qquad \text{(c)}$$

When $r = .40$ m, we get for V_r from Eq. (c):

$$V_r = 1.342 \text{ m/sec} \qquad \text{(d)}$$

This is the desired velocity relative to the platform.

To get the transverse force F_θ, go to Eq. 12.15b. Substituting the known data into the equation, we have

$$F_\theta = 2[(.40)(0) + (2)(1.342)(5)]$$
$$= 26.84 \text{ N}$$

This is the force on the mass B. The reaction to this force is the desired force.

Although you will be asked to solve problems similar to the preceding example, the main use of cylindrical coordinates in Part B of this chapter will be for gravitational central force motion. We shall first present the basic physics underlying this motion expressing certain salient characteristics of the motion, and then we shall arrive at a point where we can effectively employ cylindrical coordinates to describe the motion.

12.6 Central Force Motion—An Introduction

At this time, we shall consider the motion of a particle on which the resultant force is always *directed toward some point fixed in inertial space*. Such forces are termed *central forces* and the resulting motion of this particle is called *central force motion*. A simple example of this is a small body such as a space vehicle moving in space with its own propulsion system off in the vicinity of a large planet (see Fig. 12.10). Away from the planet's atmosphere, this vehicle will experience no frictional forces, and, if no other astronautical bodies are reasonably close, the only force acting on the vehicle will be the gravitational attraction of the fixed planet.[5] This force is directed toward the center of the planet and, from the gravitational law, is given as

$$\boldsymbol{F} = -G\frac{M_{\text{planet}}m_{\text{body}}}{r^2}\hat{\boldsymbol{r}} \tag{12.16}$$

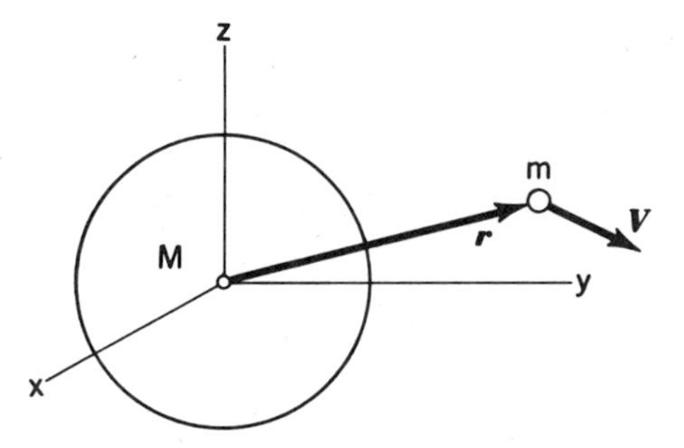

Figure 12.10. Body *m* moving about a planet.

In the ensuing problems for this chapter and also for Chapter 14, we shall need to compute the quantity GM in the equation above. For this purpose, note that, for any particle of mass m at the surface of any planet of mass M and radius R, by the law of gravitation:

$$W = mg = \frac{GMm}{R^2}$$

where g is the acceleration of gravity at the surface of the planet. Solving for GM, we get

$$GM = gR^2 \tag{12.17}$$

---

[5]We are neglecting drag developed from collisions of the space vehicle with solar dust particles.

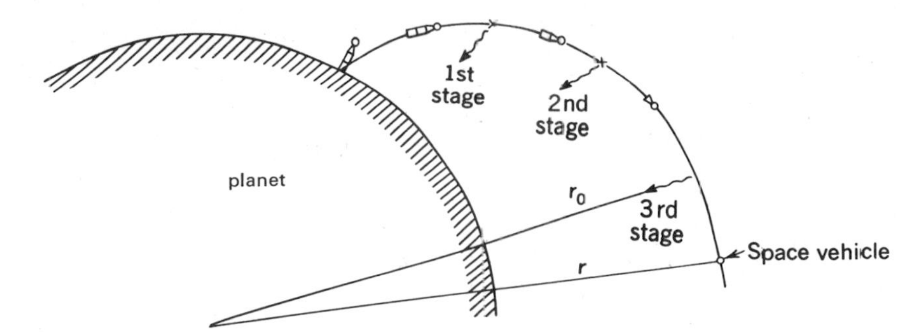

Figure 12.11. Launching a space vehicle.

Thus, knowing g and R for a planet, it is a simple matter to find GM needed for orbit calculations around this planet.

As pointed out earlier, the motion of a space vehicle with power off is an important example of a central force motion—more precisely a *gravitational* central force motion. The vehicle is usually launched from a planet and accelerated to a high speed outside the planet's atmosphere by multistage rockets (see Fig. 12.11). The velocity at the final instant of powered flight is called the *burnout* velocity. After burnout, the vehicle undergoes gravitational central force motion. Depending on the position and velocity at burnout, the vehicle can go into an orbit around the earth (elliptic and circular orbits are possible), or it can depart from the earth's influence on a parabolic or a hyperbolic trajectory. In all cases, the motion must be coplanar.

In the following sections, we shall make a careful detailed study of gravitational central force trajectories. Those who do not have the time for such a detailed study of the trajectories can still make many useful and interesting calculations in Chapter 14 using energy and momentum methods that we shall soon undertake.

*12.7 Gravitational Central Force Motion

For gravitational central force motion, we shall employ an inertial reference xy in the plane of the trajectory with the origin of the reference taken at the point P toward which the central force is directed (see Fig. 12.12). We shall use cylindrical coordinates r and θ for describing the motion. Because $z = 0$ at all times, these coordinates are also called polar coordinates. Since the motion is coplanar in plane xy, we can delete the overbar used previously for r with no danger of ambiguity.

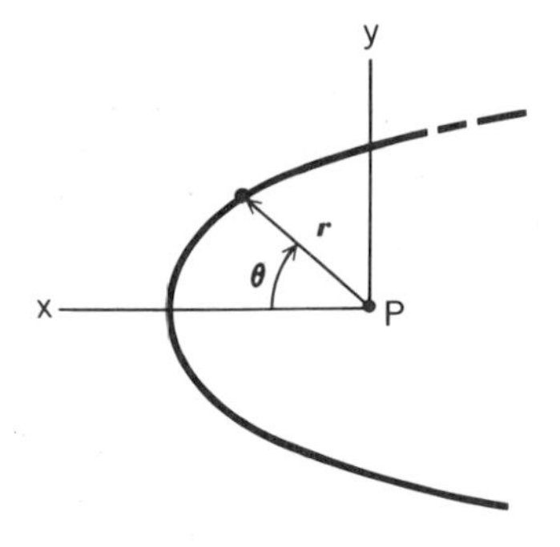

Figure 12.12. xy is inertial reference in plane of the trajectory.

Let us consider *Newton's* law for a body of mass m which is moving near a star of mass M:

$$m\frac{dV}{dt} = -G\frac{Mm}{r^2}\hat{\boldsymbol{r}} \tag{12.18}$$

Canceling m and using cylindrical coordinates and components, we can express the equation above in the following manner:

$$(\ddot{r} - r\dot{\theta}^2)\boldsymbol{\epsilon}_r + (r\ddot{\theta} + 2\dot{r}\dot{\theta})\boldsymbol{\epsilon}_\theta = -\frac{GM}{r^2}\hat{\boldsymbol{r}} \tag{12.19}$$

Since $\boldsymbol{\epsilon}_r$ and $\hat{\boldsymbol{r}}$ are identical vectors, the scalar equations of the preceding equation become

$$\ddot{r} - r\dot{\theta}^2 = -GM/r^2 \tag{12.20a}$$

$$r\ddot{\theta} + 2\dot{r}\dot{\theta} = 0 \tag{12.20b}$$

Equation 12.20b can be expressed in the form

$$\frac{1}{r}\frac{d}{dt}(r^2\dot{\theta}) = 0 \tag{12.21}$$

as you can readily verify. We can conclude from Eq. 12.21 that

$$r^2\dot{\theta} = \text{constant} = C \tag{12.22}$$

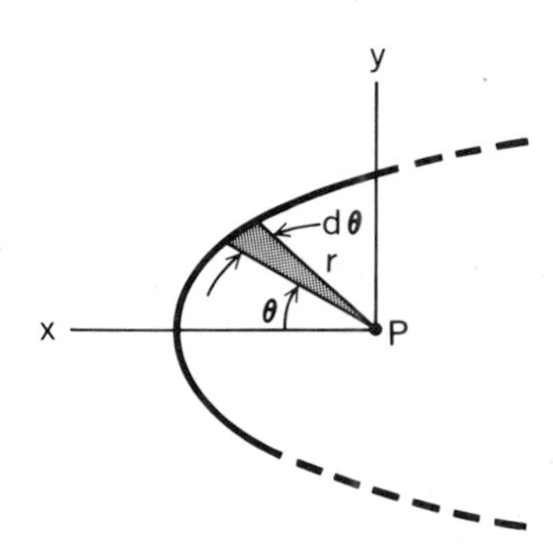

Figure 12.13. Particle sweeps out area.

Equation 12.22 leads to an important conclusion. To establish this, consider the area swept out by r during a time dt which in Fig. 12.13 is the shaded area. By considering this area to be that of a triangle, we can express it as

$$dA = \frac{r^2\,d\theta}{2}$$

Dividing through by dt, we have

$$\frac{dA}{dt} = \frac{r^2\dot{\theta}}{2}$$

Now dA/dt is the rate at which area is being swept out by r; it is called *areal velocity*. And, since $r^2\dot{\theta}$ is a constant for each gravitational central force motion, we can conclude that the areal velocity is a constant for each gravitational central force motion. (This is Kepler's second law.) This means that when r is decreased, $\dot{\theta}$ must increase, etc. The constant, understand, will be different for each different trajectory.

In order to determine the general trajectory, we replace the ind pendent variable t of Eq. 12.20a. Consider first the time derivatives of r:

$$\dot{r} = \frac{dr}{dt} = \frac{dr}{d\theta}\frac{d\theta}{dt} = \frac{d\theta}{dt}\frac{dr}{d\theta} = \frac{C}{r^2}\frac{dr}{d\theta} \tag{12.23}$$

where we have used Eq. 12.22 to replace $d\theta/dt$. Next, consider $\ddot{r}$ in a similar manner:

$$\ddot{r} = \frac{d\dot{r}}{dt} = \frac{d}{dt}\left(\frac{C}{r^2}\frac{dr}{d\theta}\right) = \frac{d}{d\theta}\left(\frac{C}{r^2}\frac{dr}{d\theta}\right)\frac{d\theta}{dt} \tag{12.24}$$

Again, using Eq. 12.22 to replace $d\theta/dt$, we get

$$\ddot{r} = \left[\frac{d}{d\theta}\left(\frac{C}{r^2}\frac{dr}{d\theta}\right)\right]\frac{C}{r^2} \tag{12.25}$$

For convenience, we now introduce a new dependent variable, $u = 1/r$, into the right side of this equation

$$\begin{aligned}\ddot{r} &= \left[\frac{d}{d\theta}\left(Cu^2\frac{d(1/u)}{d\theta}\right)\right]Cu^2 \\ &= \left\{\frac{d}{d\theta}\left[Cu^2\left(-\frac{1}{u^2}\right)\frac{du}{d\theta}\right]\right\}Cu^2 \\ &= -C^2u^2\frac{d^2u}{d\theta^2}\end{aligned}$$

By replacing $\ddot{r}$ in this form in Eq. 12.20a and $\dot{\theta}^2$ in the form C^2u^4 from Eq. 12.22, and finally, r by $1/u$, we get

$$-C^2u^2\frac{d^2u}{d\theta^2} - C^2u^3 = -GMu^2$$

Canceling terms and dividing through by C^2, we have

$$\frac{d^2u}{d\theta^2} + u = \frac{GM}{C^2} \tag{12.26}$$

This is a simple differential equation that you may have already studied in your differential equation course. Specifically, it is a second-order differential equation with constant coefficients and a constant driving function GM/C^2. We want to find the most general function $u(\theta)$ which when substituted into the differential equation satisfies the differential equation—i.e, renders it an identity. The theory of differential equations indicates that this general solution is composed of two parts. They are:

1. The general solution of the differential equation with the right side of the differential equation set equal to zero and hence given as

$$\frac{d^2u}{d\theta^2} + u = 0 \tag{12.27}$$

   This solution is called the *complementary* (or *homogeneous*) *solution*, u_c.
2. *Any* solution u_p that satisfies the full differential equation. This part is called the *particular solution.*

The desired general solution is then the sum of the complementary and particular solutions. It is a simple matter to show by substitution that the function $A \sin\theta$ satisfies Eq. 12.27 for any value of A. This is similarly true for $B \cos\theta$ for any value of B. The theory of differential equations tells us that there are two independent functions for the solution of Eq. 12.27. The general complementary solution is then

$$u_c = A \sin\theta + B \cos\theta \tag{12.28}$$

where A and B are arbitrary constants of integration. Considering the full differential equation (Eq. 12.26), we see by inspection furthermore that a particular solution is

$$u_p = \frac{GM}{C^2} \tag{12.29}$$

The general solution to the differential equation (Eq. 12.26) is then

$$u = \frac{GM}{C^2} + A \sin \theta + B \cos \theta \tag{12.30}$$

By simple trigonometric considerations, we can put the complementary solution in the equivalent form, $D \cos (\theta - \beta)$, where D and β are then the constants of integration.[6] We then have as an alternative formulation for u ($= 1/r$):

$$u = \frac{1}{r} = \frac{GM}{C^2} + D \cos (\theta - \beta) \tag{12.31}$$

You may possibly recognize this equation as the general *conic equation* in polar coordinates with the focus at the origin. In your analytic geometry class, you probably saw the following form for the general conic equation.[7]

$$\frac{1}{r} = \frac{1}{\epsilon p} + \frac{1}{p} \cos (\theta - \beta) \tag{12.37}$$

---

[6]By expanding $[D \cos (\theta - \beta)]$ as $[(D \cos \beta) \cos \theta + (D \sin \beta) \sin \theta]$, we see, since D and β are arbitrary, that $[D \cos (\theta - \beta)]$ is equivalent to $[A \cos \theta + B \sin \theta]$, where A and B are arbitrary.

[7]A *conic section* is the locus of all points whose distance from a *fixed point* has a *constant ratio* to the distance from a *fixed line*. The fixed point is called the *focus* (or focal point) and the line is termed the *directrix*. In Fig. 12.14 we have shown point P, a directrix DD, and a focus O. For a conic section to be traced by P, it must move in a manner that keeps the ratio $r/\overline{DP}$, called the *eccentricity*, a fixed number. Clearly, for every acceptable position P, there will be a mirror image position P' (see the diagram) about a line normal to the directrix and going through the focal point O. Thus, the conic section will be *symmetrical* about axis OC.

Using the letter ϵ to represent the eccentricity, we can say:

$$\frac{r}{\overline{DP}} \equiv \epsilon = \frac{r}{p + r \cos \eta} \tag{12.32}$$

where p is the distance from the focus to the directrix. Replacing $\cos \eta$ by $-\cos (\theta - \beta)$, where β (see Fig. 12.14) is the angle between the x axis and the axis of symmetry, we then get

$$\frac{r}{p - r \cos (\theta - \beta)} = \epsilon \tag{12.33}$$

Now, rearranging the terms in the equation, we arrive at a standard formulation for conic sections:

$$\boxed{\frac{1}{r} = \frac{1}{\epsilon p} + \frac{1}{p} \cos (\theta - \beta)} \tag{12.34}$$

To understand the significance of the eccentricity ϵ, let us consider conic sections in terms of a reference xy, where x is the axis of symmetry (i.e., consider $\beta = 0$ in preceding formulations and refer to Fig. 12.15). Equation 12.34 can be expressed for these rectangular coordinates in the following manner:

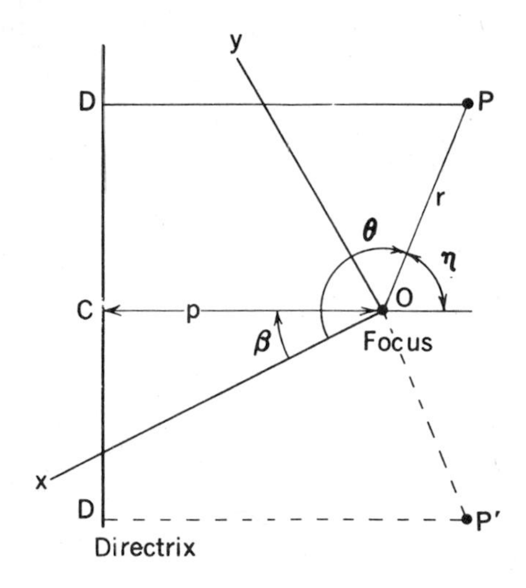

Figure 12.14. $r/\overline{DP} = \epsilon \equiv$ constant for conic section.

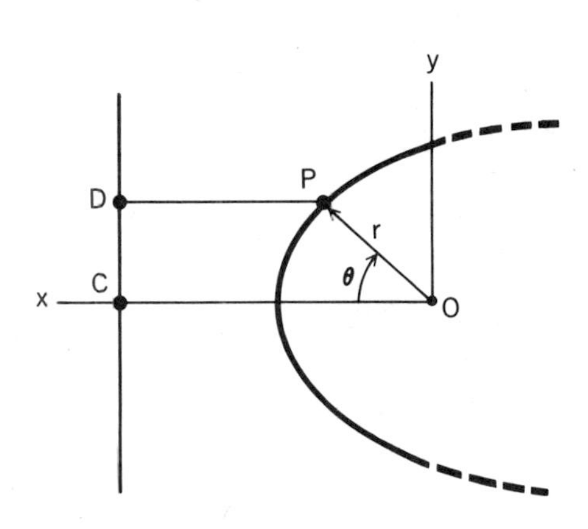

Figure 12.15. Case for $\beta = 0$; x axis is axis of symmetry.

where ϵ is the *eccentricity*, p is the distance from the *focus* to the *directrix*, and β is the angle between the x axis and the axis of symmetry of the conic section.

Comparing Eqs. 12.31 and 12.37, we see that

$$p = \frac{1}{D} \tag{12.38a}$$

$$\epsilon = \frac{DC^2}{GM} \tag{12.38b}$$

From our knowledge of conic sections, we can then say that if

$$\frac{DC^2}{GM} > 1, \text{ the trajectory is a hyperbola} \tag{12.39a}$$

$$\frac{DC^2}{GM} = 1, \text{ the trajectory is a parabola} \tag{12.39b}$$

$$\frac{1}{\sqrt{x^2 + y^2}} = \frac{1}{\epsilon p} + \frac{1}{p}\frac{x}{\sqrt{x^2 + y^2}} \tag{12.35}$$

Simple algebraic manipulation permits us to put the preceding equation into the following form:

$$(1 - \epsilon^2)x^2 + y^2 + 2p\epsilon^2 x - \epsilon^2 p^2 = 0 \tag{12.36}$$

If $\epsilon > 1$, the coefficients of x^2 and y^2 are different in sign and unequal in value. The equation then represents a *hyperbola*.

If $\epsilon = 1$, only one of the squared terms remains and we have a *parabola*.

If $\epsilon < 1$, the coefficients of the squared terms are unequal but have the same sign. The curve is that of an *ellipse*.

If $\epsilon = 0$, clearly we have a *circle* since the coefficients of the squared terms are equal in value and sign.

In Appendix III, we discuss in more detail the particular case of the ellipse.

$$\frac{DC^2}{GM} < 1, \text{ the trajectory is an ellipse} \tag{12.39c}$$

$$\frac{DC^2}{GM} = 0, \text{ the trajectory is a circle} \tag{12.39d}$$

Clearly, DC^2/GM, the eccentricity, is an extremely important quantity. We shall next look into the practical applications of the preceding general theory to problems in space mechanics.

*12.8 Applications to Space Mechanics

We shall now employ the theory set forth in the previous section to study the motion of space vehicles—a problem of great present-day interest. We shall assume that at the end of powered flight the position $\boldsymbol{r}_0$ and velocity $\boldsymbol{V}_0$ of the vehicle are known from rocket calculations. The reference employed will be an inertial reference at the center of the planet and so the reference will translate with the planet relative to the "fixed stars." Accordingly, the earth will rotate one cycle per day for such a reference. We know that the trajectory of the body will form a plane fixed in inertial space and so, for convenience, we take the xy plane of the reference to be the plane of the trajectory. It is the usual practice to choose the x axis to be the axis of symmetry for the trajectory. If there is a *zero radial velocity* component at "burnout," then the launching clearly occurs at a position along the axis of symmetry of the trajectory (i.e., along the x axis). This case has been shown in Fig. 12.16, wherein the subscript 0 denotes launch data. If, on the other hand, a radial component $(V_r)_0$ is present at burnout, then the launch condition occurs at some position θ_0 from the x axis, as shown in Fig. 12.17. We generally do not know θ_0 a priori, since its value depends on the equation of the trajectory. Finally, the angle α shown in the diagram will be called the *launching angle* in the ensuing discussion.

Since the x axis has been chosen to be the axis of symmetry, the equation of motion of the vehicle after powered flight is given in terms of arbitrary constants C and D by Eq. 12.31 with the angle β set equal to zero. Thus, we have

$$\frac{1}{r} = \frac{GM}{C^2} + D\cos\theta \tag{12.40}$$

The problem is to find the constants C and D from launching data. We shall illustrate this step in the examples following this section. Note that when these constants are evaluated, the value of the eccentricity $\epsilon = DC^2/GM$ is then available so that we can state immediately the general characteristics of the trajectory.

Furthermore, if the vehicle goes into orbit, we can readily compute the orbital time τ for one cycle around a planet. We know from the theory that the aerial velocity is constant and given as

$$\frac{dA}{dt} = \frac{r^2\dot{\theta}}{2} = \text{constant} \tag{12.41}$$

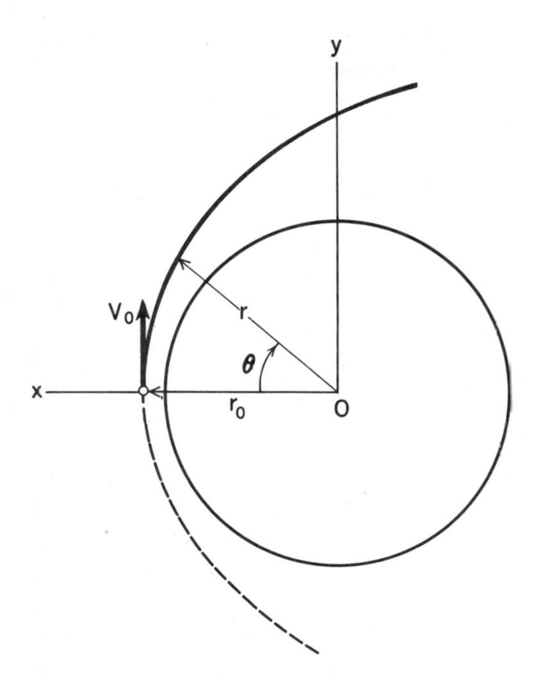

Figure 12.16. Launching at axis of symmetry.

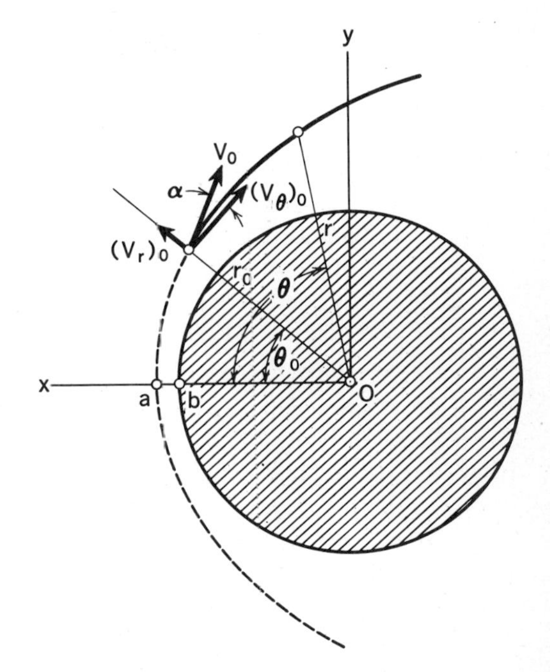

Figure 12.17. Burnout with radial velocity present.

But $r^2\dot{\theta}$ equals the constant C in accordance with Eq. 12.22. Hence,

$$dA = \frac{C}{2}\,dt \tag{12.42}$$

The area swept out for one cycle is the area of an ellipse given as πab, where a and b are the semimajor and semiminor diameters of the ellipse, respectively. Hence, we have on integrating Eq. 12.42:

$$A = \pi ab = \int_0^\tau \frac{C}{2}\,dt = \frac{C}{2}\tau$$

Therefore,

$$\tau = \frac{2\pi ab}{C} \tag{12.43}$$

We have shown in Appendix III that

$$a = \frac{\epsilon p}{1 - \epsilon^2} \tag{12.44a}$$

$$b = a(1 - \epsilon^2)^{1/2} \tag{12.44b}$$

Replacing p by $1/D$ in accordance with Eq. 12.38a, we then get

$$a = \frac{\epsilon}{D(1 - \epsilon^2)} \tag{12.45a}$$

$$b = a(1 - \epsilon^2)^{1/2} = \frac{\epsilon}{D(1 - \epsilon^2)^{1/2}} \tag{12.45b}$$

Thus, we can get the orbital time τ quite easily once the constants of the trajectory, D and C, are evaluated.

To illustrate many of the previous general remarks in a most simple manner, we now examine the special case where, as shown in Fig. 12.18, various launchings (i.e., burnout conditions) are made from a given point a such that the launching angle $\alpha = 0$. Clearly, $(V_r)_0 = 0$ for these cases and the launching axis corresponds to the axis of symmetry of the various trajectories. Only V_0 will be varied in this discussion.

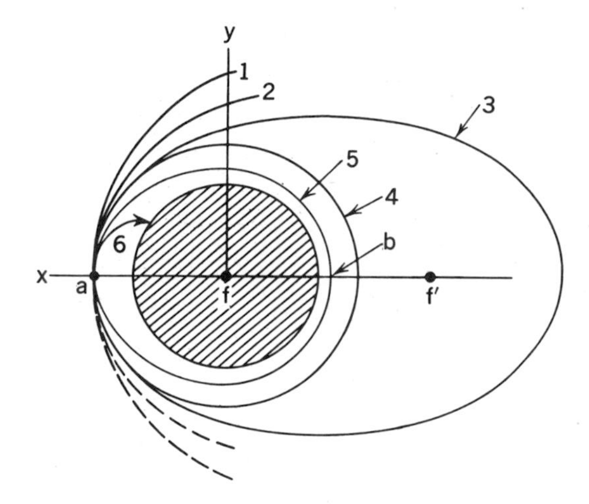

Figure 12.18. Various launchings from the earth or some other planet.

The constants C and D are readily available for these trajectories. Thus, we have from Eq. 12.22:

$$C = r^2\dot{\theta} = rV_\theta = r_0V_0 \tag{12.46}$$

And from Eq. 12.40, setting $r = r_0$ when $\theta = \theta_0 = 0$, we get, on solving for D:

$$D = \frac{1}{r_0} - \frac{GM}{C^2} = \frac{1}{r_0} - \frac{GM}{r_0^2V_0^2} \tag{12.47}$$

Since C and D above, for a given r_0, depend only on V_0, we conclude that the eccentricity here is dependent only on V_0 for a given r_0.

If V_0 is so large that DC^2/GM exceeds unity, the vehicle will have the trajectory of a hyperbola (curve 1) and will eventually leave the influence of the earth. If V_0 is decreased to a value such that the eccentricity is unity, the trajectory becomes a parabola (curve 2). Since a further decrease in the value of V_0 will cause the vehicle to orbit, curve 2 is the limiting trajectory with our launching conditions for outer-space flight. The launching velocity for this case is accordingly called the *escape velocity* and is denoted as $(V_0)_E$. We can solve for $(V_0)_E$ for this launching by substituting for C and D from Eqs. 12.46 and 12.47 into the equation $DC^2/GM = 1$. We get

$$(V_0)_E = \sqrt{\frac{2GM}{r_0}} \tag{12.48}$$

a result that is correct for more general launching conditions (i.e., for cases where launching angle $\alpha \neq 0$). Thus, launching a vehicle with a speed equaling or exceeding the value above for a given r_0 will cause the vehicle to leave the earth until such time as the vehicle is influenced by other astronomical bodies or by its own propulsion system. If V_0 is less than the escape velocity, the vehicle will move in the trajectory of an ellipse (curve 3). The closest point to the earth is called *perigee*; the farthest point is called *apogee*. Clearly, these points lie along the axis of symmetry. Such an orbiting vehicle is often called a space satellite. (Kepler, in his famous first law of planetary motion, explained the motion of planets about the sun in this same manner.) One focus for the aforementioned conic curves is at the center of the planet. Another focus f' now moves in from infinity for the satellite trajectories. As the launching speed is decreased, f' moves toward f. When the foci coincide, the trajectory is clearly a circle and, as pointed out earlier, the eccentricity ϵ is zero. Accordingly, the constant D must be zero (the constant C clearly will not be zero) and, from Eq. 12.47, the speed for a *circular* orbit $(V_0)_C$ is

$$(V_0)_C = \sqrt{\frac{GM}{r_0}} \tag{12.49}$$

For launching velocities less than the preceding value for a given r_0, the eccentricity becomes negative and the focus f' moves to the left of the earth's center. Again, the trajectory is that of an ellipse (curve 5). However, the satellite will now come closer to the earth at position b, which now becomes the perigee, than at the launching position, which up to now had been the minimum distance from the earth.[8] If friction is encountered, the satellite will slow up, spiral in toward the atmosphere, and either burn up or crash. If V_0 is small enough, the missile will not go into even a temporary orbit but will plummet to the earth (curve 6). However, for a reasonably accurate description of this trajectory, we must consider friction from the earth's atmosphere. Since this type of force is a function of the velocity of the missile and is not a central force, we cannot use the results here in such situations for other than approximate calculations.

---

[8]Note that with the positive x axis going through perigee, r is *minimum* when $\theta = 0$. From Eq. 12.40, we can conclude for this case (θ is measured here from perigee) that, to minimize r, the constant D must be positive. The eccentricity must then be positive for θ measured from perigee. If the positive x axis goes through apogee, then r is *maximum* when $\theta = 0$. From Eq. 12.40 we can conclude that D must be negative for this case (θ is here measured from apogee). Thus, the eccentricity is negative for θ measured from apogee. This is clearly the case for curve 5.

EXAMPLE 12.7

The first American satellite, the Vanguard, was launched at a velocity of 18,000 mi/hr at an altitude of 400 mi (see Fig. 12.19). If the "burnout" velocity of the last stage is parallel to the earth's surface, compute the maximum altitude from the earth's surface that the Vanguard satellite will reach. Consider the earth to be perfectly spherical with a radius of 3960 mi (r_0 is therefore 4360 mi).

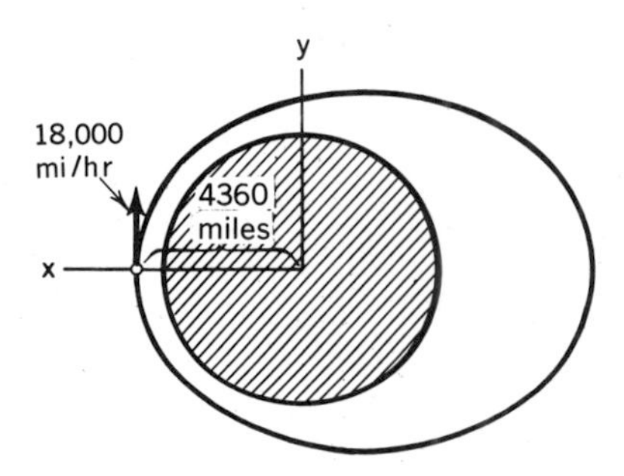

Figure 12.19. Launching of the Vanguard satellite.

We must now compute the quantities GM, C, and D from the initial data and other known data. To determine GM, we employ Eq. 12.17 and in terms of units of miles and hours we get

$$GM = (32.2)\left(\frac{3600^2}{5280}\right)(3960)^2 = 1.239 \times 10^{12} \text{ mi}^3/\text{hr}^2 \tag{a}$$

The constant C is readily determined directly from initial data as

$$C = r_0 V_0 = (4360)(18{,}000) = 7.85 \times 10^7 \text{ mi}^2/\text{hr} \tag{b}$$

Finally, the constant D is available from Eq. 12.47:

$$D = \frac{1}{r_0} - \frac{GM}{C^2} = \frac{1}{4360} - \frac{1.239 \times 10^{12}}{(7.85 \times 10^7)^2} = .283 \times 10^{-4} \text{ mi}^{-1} \tag{c}$$

The eccentricity DC^2/GM can now be computed as

$$\epsilon = \frac{DC^2}{GM} = \frac{(.283 \times 10^{-4})(7.85 \times 10^7)^2}{(1.239 \times 10^{12})} \doteq .1408 \tag{d}$$

The Vanguard will thus definitely not escape into outer space.

The trajectory of this motion is formed from Eq. 12.40:

$$\frac{1}{r} = \frac{1.239 \times 10^{12}}{(7.85 \times 10^7)^2} + .283 \times 10^{-4} \cos\theta$$

Therefore,

$$\frac{1}{r} = 2.01 \times 10^{-4} + .283 \times 10^{-4} \cos\theta \tag{e}$$

We can compute the maximum distance from the earth's surface by setting $\theta = \pi$ in the equation above:

$$\frac{1}{r_{\text{max}}} = (2.01 - .283) \times 10^{-4} = 1.727 \times 10^{-4} \text{ mi}^{-1}$$

Therefore,

$$r_{\text{max}} = 5790 \text{ mi} \tag{f}$$

By subtracting 3960 miles from this result, we find that the highest point in the trajectory is 1830 mi from the earth's surface.

EXAMPLE 12.8

In Example 12.7, first compute the escape velocity and then the velocity for a circular orbit at burnout.

Using Eq. 12.48, we have for the escape velocity:

$$(V_0)_E = \sqrt{\frac{2GM}{r_0}} = \left[\frac{2(1.239)(10^{12})}{4360}\right]^{1/2}$$
$$= 23{,}840 \text{ mi/hr}$$

For a circular orbit, we have from Eq. 12.49:

$$(V_0)_C = \sqrt{\frac{GM}{r_0}} = 16{,}860 \text{ mi/hr}$$

Thus, the Vanguard is almost in a circular orbit.

EXAMPLE 12.9

Determine the orbital time in Example 12.7 for the Vanguard satellite.

We employ Eqs. 12.44 for the semimajor and semiminor axes of the elliptic orbit. Thus, recalling that $p = 1/D$ we have

$$a = \frac{\epsilon}{D(1-\epsilon^2)} = \frac{.1408}{.283 \times 10^{-4}(1 - .1408^2)}$$
$$= 5080 \text{ mi}$$
$$b = a(1-\epsilon^2)^{1/2} = 5080(1 - .1408^2)^{1/2}$$
$$= 5030 \text{ mi}$$

Therefore, from Eq. 12.43 we have for the orbital time:

$$\tau = \frac{\pi ab}{C/2} = \frac{(\pi)(5080)(5030)}{7.85 \times 10^7/2}$$
$$= 2.05 \text{ hr} = 122.7 \text{ min}$$

EXAMPLE 12.10

A space vehicle is in a circular "parking" orbit around the planet Venus, 320 km above the surface of this planet. The radius of Venus is 6160 km, and the escape velocity at the surface is 1.026×10^4 m/sec. A retro-rocket is fired to slow the vehicle so that it will come within 32 km of the planet. If we consider that the rocket changes the speed of the vehicle over a comparatively short distance of its travel, what is this change of speed? What is the speed of the vehicle at its closest position to the surface of Venus?

We show the vehicle in a circular parking orbit in Fig. 12.20. We shall consider that the retro-rockets are fired at position A so as to establish a new elliptic orbit with apogee at A and perigee at B.

As a first step, we shall compute GM using the escape-velocity equation 12.48. Thus, we have

$$V_E = \sqrt{\frac{2GM}{R}}$$

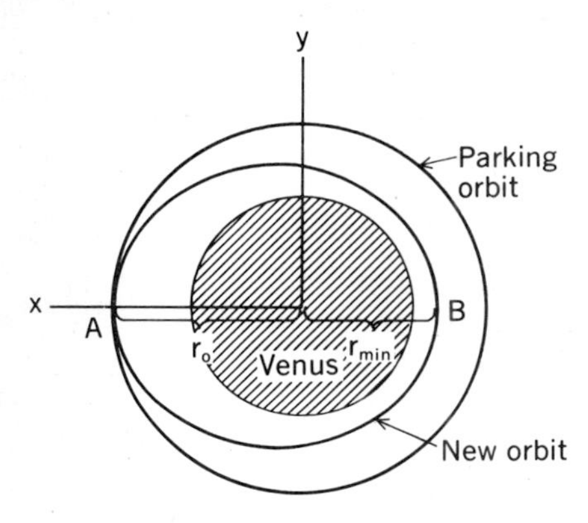

Figure 12.20. Change of orbit.

Therefore,

$$GM = \frac{V_E^2 R}{2} = \left[\left(\frac{1.026 \times 10^4}{1000}\right)(3600)\right]^2\left(\frac{.6160}{2}\right)$$
$$= 4.20 \times 10^{12} \text{ km}^3/\text{hr}^2$$

The equation for the *new elliptic orbit* is given as

$$\frac{1}{r} = \frac{GM}{C^2} + D\cos\theta \tag{a}$$

Note that when

$$\theta = 0, \qquad r = r_0 = 6480 \text{ km} \tag{b}$$
$$\theta = \pi, \qquad r = r_{\min} = 6192 \text{ km} \tag{c}$$

To determine the constant C, we subject Eq. (a) to the conditions (b) and (c). Thus,

$$\frac{1}{6480} = \frac{4.20 \times 10^{12}}{C^2} + D \tag{d}$$

$$\frac{1}{6192} = \frac{4.20 \times 10^{12}}{C^2} - D \tag{e}$$

Adding these equations, we eliminate D and can solve for C. Thus,

$$\frac{8.40 \times 10^{12}}{C^2} = \frac{1}{6480} + \frac{1}{6192}$$

Therefore,

$$C = 1.631 \times 10^8 \text{ km}^2/\text{hr}$$

Accordingly, for the new orbit,

$$r_0 V_0 = 1.631 \times 10^8$$

Therefore,

$$V_0 = 25{,}168 \text{ km/hr}$$

For the *circular* parking orbit the velocity V_c is

$$V_c = \sqrt{\frac{GM}{r_0}} = \sqrt{\frac{4.20 \times 10^{12}}{6480}}$$
$$= 25{,}458 \text{ km/hr}$$

The change in velocity that the retro-rocket must induce is then

$$\Delta V = 25{,}168 - 25{,}458 = -290 \text{ km/hr}$$

The velocity at the apogee at B is easily computed since

$$r_B V_B = C = 1.631 \times 10^8$$

Therefore,

$$V_B = 1.631 \times 10^8/6192 = 26{,}300 \text{ km/hr}$$

Now let us consider more general launching conditions where the launching angle α is not zero (see Fig. 12.21). The constant C is still easily evaluated in terms of launching data as $r_0(V_\theta)_0$. To get D, we write Eq. 12.40 for launching conditions. Thus,

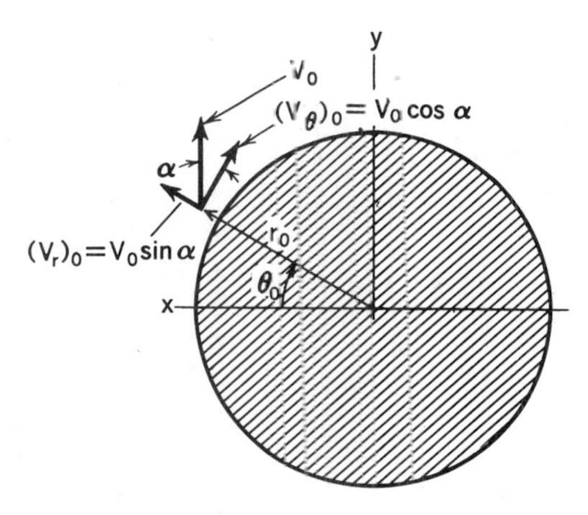

Figure 12.21. Launch with radial velocity.

$$\frac{1}{r_0} = \frac{GM}{C^2} + D\cos\theta_0 \tag{12.50}$$

The value of θ_0 is not yet known. Thus, we have two unknown quantities in this equation, namely D and θ_0. Differentiating Eq. 12.40 with respect to time and solving for $\dot{r}$, we get

$$\dot{r} = Dr^2\dot{\theta}\sin\theta = DC\sin\theta \tag{12.51}$$

Noting that $\dot{r}$ is equal to V_r and submitting the preceding equation to launching conditions, we then form a second equation for the evaluation of the unknown constants D and θ_0. Thus,

$$(V_r)_0 = DC\sin\theta_0 \tag{12.52}$$

Rearranging Eq. 12.50, we have

$$\frac{1}{r_0} - \frac{GM}{C^2} = D\cos\theta_0 \tag{12.53}$$

Divide both sides of Eq. 12.52 by C. Now, squaring Eqs. 12.52 and 12.53 and adding terms, we get for the constant D the result[9]

$$D = \left\{\left(\frac{1}{r_0} - \frac{GM}{C^2}\right)^2 + \left[\frac{(V_r)_0}{C}\right]^2\right\}^{1/2} \tag{12.54}$$

Having taken the positive root for D, we note (see footnote, p. 483) that θ is to be measured from perigee. The eccentricity is

$$\epsilon = \frac{C^2}{GM}\left\{\left(\frac{1}{r_0} - \frac{GM}{C^2}\right)^2 + \left[\frac{(V_r)_0}{C}\right]^2\right\}^{1/2} \tag{12.55}$$

First, bringing C^2 into the bracket and then replacing C by $r_0(V_\theta)_0$ in the entire equation, we get the eccentricity conveniently in terms of launching data:

---

[9]The student has the option of formulating Eqs. 12.52 and 12.53 for each problem and finding D from these equations, or he can use Eq. 12.54 directly. In some of the homework problems we shall ask you to do both.

$$\epsilon = \frac{r_0(V_\theta)_0}{GM}\left\{(V_r)_0^2 + \left[(V_\theta)_0 - \frac{GM}{r_0(V_\theta)_0}\right]^2\right\}^{1/2} \tag{12.56}$$

We can show, using the preceding formulations, that the equation for the escape velocity developed earlier, namely

$$V_E = \sqrt{\frac{2GM}{r}}$$

is valid for any launching angle α. Remember that V_E in this equation is measured from a reference xyz at the center of the planet translating in inertial space. The velocity attainable by a rocket system relative to the planet's surface does not depend on the position of firing on the earth, but depends primarily on the rocket system and trajectory of flight. However, the velocity attainable by a rocket system relative to the aforementioned reference xyz *does* depend on the position of firing on the planet's surface. This position, accordingly, is important in determining whether an escape velocity can be reached. The extreme situations of a launching at the equator and at the North Pole are shown in Fig. 12.22 and should clarify this point. Note that the motion of the planet's surface adds to the final vehicle velocity at the equator, but that no such gain is achieved at the North Pole.

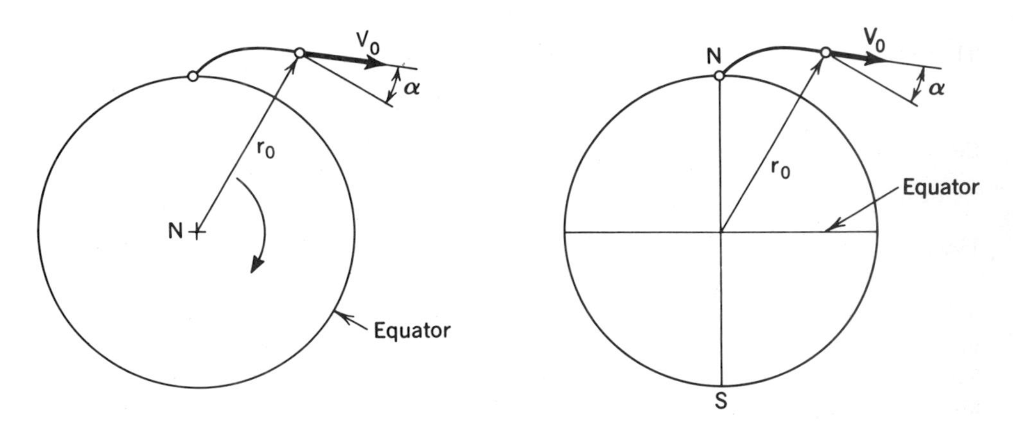

Figure 12.22. Launching at equator and North Pole.

EXAMPLE 12.11

Suppose that the Vanguard satellite in Example 12.7 is off course by an angle $\alpha = 5°$ at the time of launching but otherwise has the same initial data. Determine whether the satellite goes into orbit. If so, determine the maximum and minimum distances from the earth's surface.

The initial data for the launching are

$$r_0 = 4360 \text{ mi}, \qquad V_0 = 18{,}000 \text{ mi/hr}$$

Hence,

$$\begin{aligned}(V_r)_0 = (18{,}000) \sin \alpha &= (18{,}000)(0.0872)\\ &= 1569 \text{ mi/hr}\end{aligned}$$

$$(V_\theta)_0 = (18{,}000)\cos\alpha = (18{,}000)(0.996)$$
$$= 17{,}930 \text{ mi/hr}$$

To determine whether we have an orbit, we would have to show first that the eccentricity ϵ is less than unity. This condition would preclude the possibility of an escape from the earth. Furthermore, we must be sure that the perigee of the orbit is far enough from the earth's surface to ensure a reasonably permanent orbit. Actually, for both questions we need only calculate r for $\theta = 0$ and $\theta = \pi$. An infinite value of one of the r's will mean that we have an escape condition, and a value not sufficiently large will mean a crash or a decaying orbit due to atmospheric friction.

Using the value of GM as 1.239×10^{12} mi^3/hr^2 from Example 12.7, and using Eq. 12.54 for the constant D, we can express the trajectory of the satellite (Eq. 12.40) as

$$\frac{1}{r} = \frac{1.239 \times 10^{12}}{[(4360)(17{,}930)]^2} + \left\{\left[\frac{1}{4360} - \frac{1.239 \times 10^{12}}{[(4360)(17{,}930)]^2}\right]^2 + \left[\frac{1569}{(4360)(17{,}930)}\right]^2\right\}^{1/2} \cos\theta$$

Therefore,

$$\frac{1}{r} = 2.03 \times 10^{-4} + 3.33 \times 10^{-5} \cos\theta \text{ mi}^{-1} \qquad \text{(a)}$$

Set $\theta = 0$:

$$\frac{1}{r(0)} = 20.3 \times 10^{-5} + 3.33 \times 10^{-5} \text{ mi}^{-1}$$

Hence,

$$r(0) = 4230 \text{ mi}$$

Thus, after being launched at a position 400 mi above the earth's surface, the satellite comes within 270 mi of the earth as a result of a 5° change in the launching angle. This satellite, therefore, must be launched almost parallel to the earth if it is to attain a reasonably permanent orbit.

Now, setting $\theta = \pi$, we get

$$\frac{1}{r(\pi)} = 20.3 \times 10^{-5} - 3.33 \times 10^{-5}$$

Hence,

$$r_{max} = 5893 \text{ mi}$$

Obviously, the maximum distance from the earth's surface is 1933 mi.

Problems

12.44. A device used at amusement parks consists of a circular room that is made to revolve about its axis of symmetry. People stand up against the wall, as shown in the diagram. After the whole room has been brought up to speed, the floor is lowered. What minimum angular speed is required to ensure that a person will not slip down the wall when the floor is lowered? Take $\mu_s = .3$.

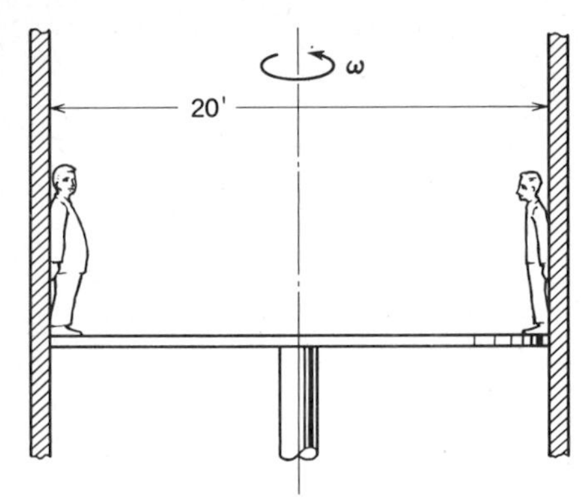

Figure P.12.44

12.45. A flywheel is rotating at a speed of $\omega =$ 10 rad/sec and has at this instant a rate of change of speed $\dot{\omega}$ of 5 rad/sec². A solenoid at this instant moves a valve toward the centerline of the flywheel at a speed of 1.5 m/sec and is decelerating at the rate of .6 m/sec². The valve has a mass of 1 kg and is .3 m from the axis of rotation at the time of interest. What is the total force on the valve?

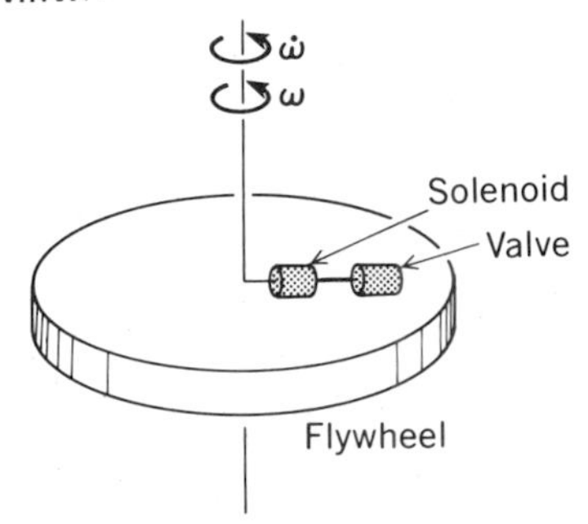

Figure P.12.45

12.46. A conical pendulum of length l is shown. The pendulum is made to rotate at a constant angular speed of ω about the vertical axis. Compute the tension in the cord if the pendulum bob has weight W. What is the distance of the plane of the trajectory of the bob from the support at O?

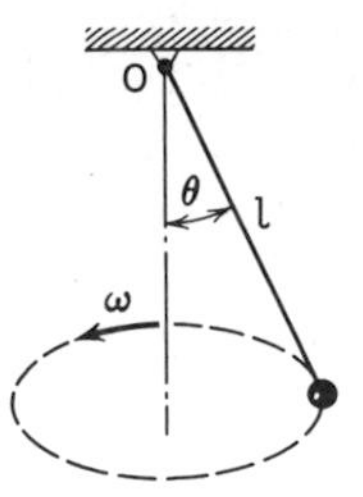

Figure P.12.46

12.47. A shaft AB rotates at an angular velocity of 100 rpm. A body E of mass 10 kg can move without friction along rod CD fixed to AB. If the body E is to remain stationary relative to CD at any position along CD, how must the spring constant K vary? The distance r_0 from the axis is the unstretched length of the spring.

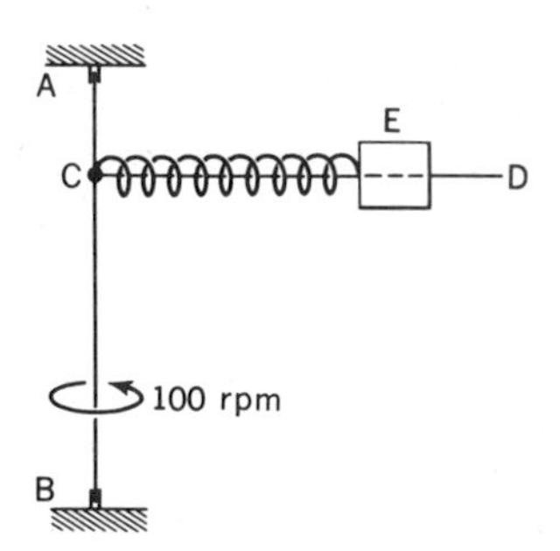

Figure P.12.47

12.48. A device called a *flyball governor* is used to regulate the speed of such devices as steam engines and turbines. As the governor is made to rotate through a system of gears by the device to be controlled, the balls will attain a configuration given by the angle θ, which is dependent on both the angular speed ω of the governor and the force P acting on the collar bearing at A. The up-and-down motion of the bearing at A in response to a change in ω is then used to open or close a valve to regulate the speed of the device. Find the angular velocity required to maintain the configuration of the flyball governor for $\theta = 30°$. Neglect friction.

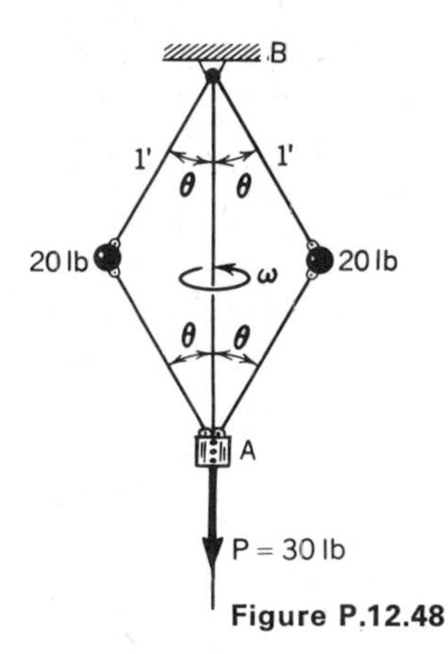

Figure P.12.48

12.49. A platform rotates at 2 rad/sec. A body C weighing 450 N rests on the platform and is connected by a flexible weightless cord to a mass weighing 225 N, which is prevented from swinging out by part of the platform. At what range of value of x will bodies C and B remain stationary relative to the platform? The coefficient of friction for all surfaces is .4.

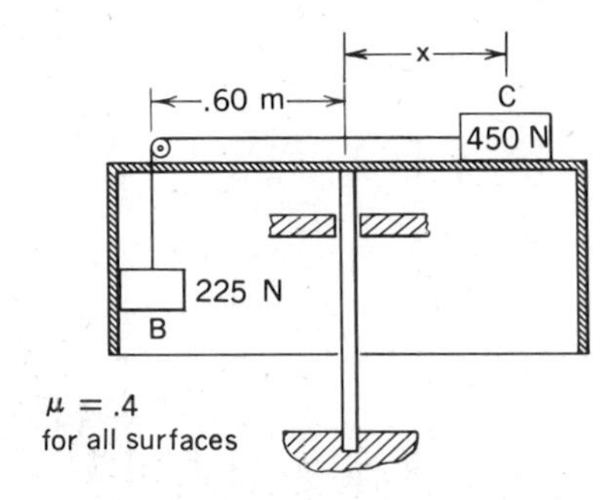

Figure P.12.49

12.50. A particle moves under gravitational influence about a body M, the center of which can be taken as the origin of an inertial reference. The mass of the particle is 50 slugs. At time t, the particle is at a position 4500 mi from the center of M with direction cosines $l = .5, m = -.5, n = .707$. The particle is moving at a speed of 17,000 mi/hr along the direction $\boldsymbol{\epsilon}_t = .8\boldsymbol{i} + .2\boldsymbol{j} + .566\boldsymbol{k}$. What is the direction of the normal to the plane of the trajectory?

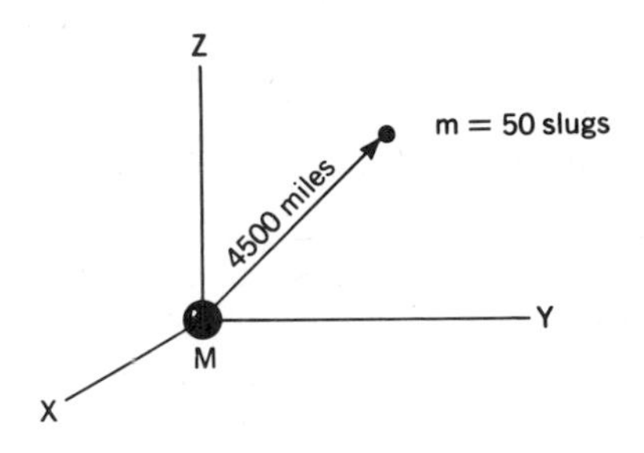

Figure P.12.50

12.51. If the position of the particle in Problem 12.50 were to reach a distance of 4300 mi from the center of body M, what would the transverse velocity V_θ of the particle be?

12.52. Use Eqs. 12.38b and 12.40 to show that if the eccentricity is zero, the trajectory must be that of a circle.

12.53. A satellite has at one time during its flight around the earth a radial component of velocity 3200 km/hr and a transverse component of 25,600 km/hr. If the satellite is at a distance of 7040 km from the center of the earth, what is its areal velocity?

12.54. Compute the escape velocity at a position 8000 km from the center of the earth. What speed is needed to maintain a circular orbit at that distance from the earth's center? Derive the equation for the speed needed for a circular orbit directly from Newton's law without using information about eccentricities, etc.

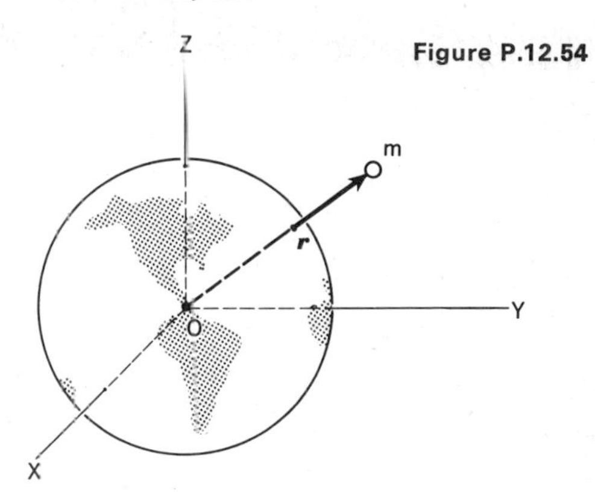

Figure P.12.54

12.55. What is the velocity and altitude of a *communications* satellite that remains in the same position above the equator relative to the earth's surface?

12.56. A satellite is launched and attains a velocity of 19,000 mi/hr relative to the center of the earth at a distance of 240 mi from the earth's surface. The satellite has been guided into a path that is parallel to the earth's surface at burnout.

(a) What kind of trajectory will it have?
(b) What is its farthest position from the earth's surface?
(c) If it is in orbit, compute the time it takes to go from the minimum point (perigee) to the maximum point (apogee) from the earth's surface.
(d) What is the minimum escape velocity for this position of launching?

12.57. A rocket system is capable of giving a satellite a velocity of 35,200 km/hr relative to the earth's surface at an elevation of 320 km above the earth's surface. What would be its maximum distance h from the surface of the earth if it were launched (1) from the North Pole region or (2) from the equator, utilizing the spin of the earth as an aid?

12.58. The acceleration of gravity on the planet Mars is about .385 times the acceleration of gravity on earth, and the radius of Mars is about .532 times that of the earth. What is the escape velocity from Mars at a position 100 mi from the surface of the planet?

12.59. In 1971 Mariner 9 was placed in orbit around Mars with an eccentricity of .5. At the lowest point in the orbit, Mariner 9 is 320 km from the surface of Mars.

(a) Compute the maximum velocity of the space vehicle relative to the center of Mars.
(b) Compute the time of one cycle.

Use the data in Problem 12.58 for Mars.

12.60. A man is in orbit around the earth in a space-shuttle vehicle. At his lowest position, he is moving with a speed of 18,500 mi/hr at an altitude of 200 mi. When he wants to come back to earth, he fires a retro-rocket straight ahead when he is at the aforementioned lowest position and slows himself down. If he wishes subsequently to get within 50 mi from the earth's surface during the first cycle after firing his retro-rocket, what must his decrease in velocity be? (Neglect air resistance.)

12.61. A space vehicle is to change from a circular parking orbit 320 km above the surface of Venus to one that is 1620 km above this surface. This motion will be accomplished by two firings of the rocket system of the vehicle. The first firing causes the vehicle to attain an apogee that is 1620 km above the surface of Venus. At this apogee, a second firing is accomplished so as to achieve the desired circular orbit. What is the change in speed demanded for each firing if the thrust is maintained in each instance over a small portion of the trajectory of the vehicle? Neglect friction. The radius of Venus is 6160 km, and the escape velocity at the surface is 1.026×10^4 m/sec^2.

12.62. The Pioneer 10 space vehicle approaches the planet Jupiter with a trajectory having an eccentricity of 3. The vehicle comes to within 1000 mi of the surface of Jupiter. What is the speed of the vehicle at this instant? The acceleration of gravity of Jupiter is 90.79 ft/sec^2 at the surface and the radius is 43,400 mi.

12.63. If the moon has a motion about the earth that has an eccentricity of .0549 and a period of 27.3 days, what is the closest distance of the moon to the earth in its trajectory?

12.64. The satellite Hyperion about the planet Saturn has a motion with an eccentricity known to be .1043. At its closest distance from Saturn, Hyperion is 1.485×10^6 km away (measured from center to center). What is the period of Hyperion about

Saturn? The acceleration of gravity of Saturn is 13.93 m/sec² at its surface. The radius of Saturn is 57,600 km.

12.65. Two satellite stations, each in a circular orbit around the earth, are shown. A small vehicle is shot out of the station at *A* tangential to the trajectory in order to "hit" station *B* when it is at a position *E* 120° from the *x* axis as shown in the diagram. What is the velocity of the vehicle relative to station *A* when it leaves? The circular orbits are 200 miles and 400 miles respectively from the earth's surface.

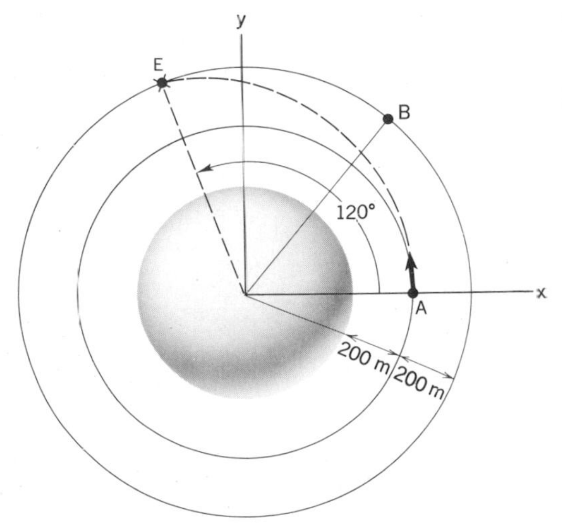

Figure P.12.65

12.66. In Problem 12.65, determine the total velocity of the vehicle as it arrives at *E* as seen by an observer in the satellite *B*. The values of *C* and *D* for the vehicle from Problem 12.65 are 7.292×10^7 mi²/hr and 7.373×10^{-6} mi^{-1}, respectively.

12.67. The Viking I space probe is approaching Mars. When it is 80,650 km from the center of Mars, it has a speed of 16,130 km/hr with a component (V_r) toward the center of Mars of 15,800 km/hr. Does Viking I crash into Mars, go into orbit, or have one pass in the vicinity of Mars? If there is no crash, how close to Mars does it come? The acceleration of gravity on the surface of Mars is 4.13 m/sec², and its radius is 3400 km. Do not use formula for *D* as given by Eq. 12.54, but work from the trajectory equations.

12.68. A meteor is moving at a speed of 20,000 mi/hr relative to the center of the earth when it is 350 mi from the surface of the earth. At that time, the meteor has a radial velocity component of 4000 mi/hr toward the center of the earth. How close does it come to the earth's surface? Do this problem without the aid of Eq. 12.54.

12.69. Do Problem 12.68 with the aid of Eq. 12.54.

12.70. The moon's radius is about .272 times that of the earth, and its acceleration of gravity at the surface is .165 times that of the earth at the earth's surface. A space vehicle approaches the moon with a velocity component toward the center of the moon of 3200 km/hr and a transverse component of 8000 km/hr relative to the center of the moon. The vehicle is 3200 km from the center of the moon when it has these velocity components. Will the vehicle go into orbit around the moon if we consider only the gravitational effect of the moon on the vehicle? If it goes into orbit, how close will it come to the surface of the moon? If not, does it collide with the moon? Do this problem without the aid of Eq. 12.54.

12.71. Do Problem 12.70 with the aid of Eq. 12.54.

12.72. Assume that a satellite is placed into orbit about a planet that has the same mass and diameter as the earth but no atmosphere. At the minimum height of its trajectory, the satellite has an elevation of 645 km from the planet's surface and a velocity of 29,800 km/hr. To observe the planet more closely, we send down a smaller satellite from the main body to within 16 km of this planet. The "subsatellite" is given a velocity component toward the center of the planet when the main satellite is at its lowest position. What is this radial velocity, and what is the eccentricity of the trajectory of the subsatellite? What is a better way to get closer to the planet?

12.73. Suppose that you are on a planet having no atmosphere. This planet rotates once every 6 hr about its axis relative to an inertial reference XYZ at its center. The planet has a radius of 1600 km, and the acceleration of gravity at the surface is 7 m/sec^2. A bullet is fired by a man at the equator in a direction normal to the surface of the planet as seen by this man. The muzzle velocity of the gun is 1500 m/sec. What is the eccentricity of the trajectory and the maximum height h of the bullet above the surface of the planet?

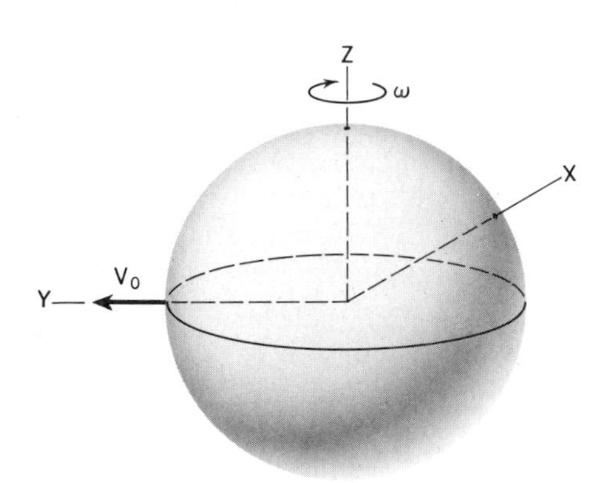

Figure P.12.73

***12.74.** A satellite is launched at A. We wish to determine the time required, Δt, to get to position B. Show that for this calculation we can employ the formulation

$$\Delta t = \frac{1}{C}\int_{\theta_0}^{\theta_B} r^2\, d\theta$$

For integration purposes, show that the formulation above becomes

$$\Delta t = \frac{1}{C}\int_{\theta_0}^{\theta_B} \frac{d\theta}{[(GM/C^2) + D\cos\theta]^2}$$

Carry out the integration.

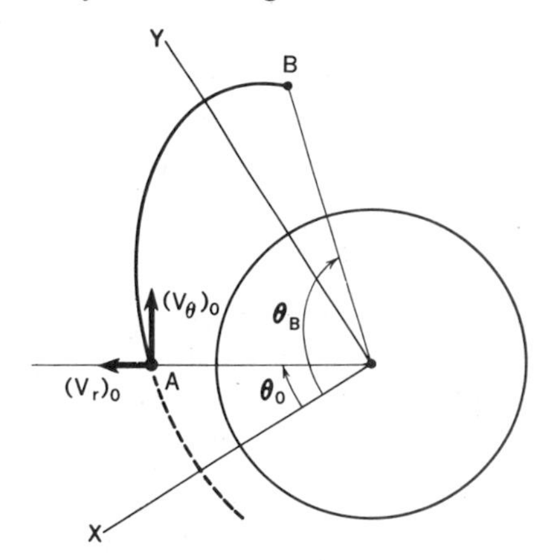

Figure P.12.74

***12.75.** A satellite is launched at a speed of 20,000 mi/hr relative to the earth's center at an altitude of 340 mi above the earth's surface. The guidance system has malfunctioned, and the satellite has a direction 20° up from the tangent plane to the earth's surface. Will the satellite go into orbit? Give the time required for one cycle if it goes into orbit or the time it takes before it strikes the earth after firing. Neglect friction in both cases. (See Problem 12.74 before doing this problem.)

Part C
Path Variables

12.9 Newton's Law for Path Variables

We can express Newton's law for path variables as follows:

$$F_t = m\frac{d^2s}{dt^2} \tag{12.57a}$$

$$F_n = m\frac{(ds/dt)^2}{R} \tag{12.57b}$$

Notice that the second of these equations is always nonlinear, as discussed in Section 12.5.[10] This condition results from both the squared term and the radius of curvature R. It is therefore difficult to integrate this differential equation. Accordingly, we shall be restricted to reasonably simple cases. We now illustrate the use of the preceding equations.

EXAMPLE 12.12

A portion of a roller coaster that one finds in an amusement park is shown in Fig. 12.23(a). The portion of the track shown is coplanar. The curve from A to the right is that of a parabola, given as

$$(y - 100)^2 = 100x \tag{a}$$

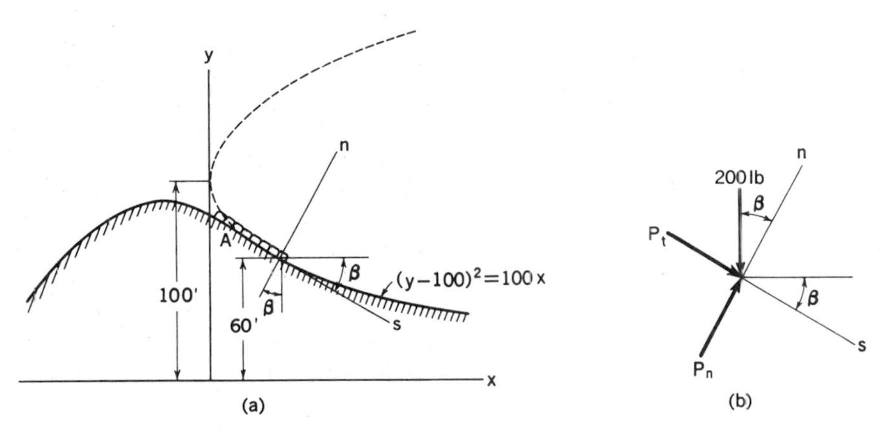

Figure 12.23. Roller coaster trajectory.

with x and y in feet. If the train of cars is moving at a speed of 40 ft/sec when the front car is 60 ft above the ground, what is the total normal force exerted by a 200-lb occupant of the front car on the seat and floor of the car?

[10]Equation 12.57a could also be nonlinear, depending on the nature of the function F_t.

Since we require only the force F normal to the path, we need only be concerned with a_n. Thus, we have

$$a_n = \frac{(ds/dt)^2}{R} = \frac{40^2}{R} \tag{b}$$

We can compute R from analytic geometry as follows:

$$R = \frac{[1 + (dy/dx)^2]^{3/2}}{|d^2y/dx^2|} \tag{c}$$

wherein from Eq. (a) we have

$$\frac{dy}{dx} = \frac{50}{y - 100} \tag{d}$$

$$\frac{d^2y}{dx^2} = -\frac{50}{(y - 100)^2}\frac{dy}{dx} = -\frac{2500}{(y - 100)^3} \tag{e}$$

Substituting into Eq. (c), we have

$$R = \frac{\left[1 + \left(\frac{50}{y - 100}\right)^2\right]^{3/2}}{|2500/(y - 100)^3|}$$

At the position of interest, we get

$$R = \frac{\left(1 + \frac{2500}{1600}\right)^{3/2}}{(2500/64{,}000)} = 105 \text{ ft} \tag{f}$$

Accordingly, we now have for F_n, as required by Newton's law:

$$F_n = \frac{W}{g}a_n = \frac{200}{g}\frac{1600}{105} = 94.6 \text{ lb} \tag{g}$$

Note that F_n is the total force component normal to the trajectory needed by the occupant for maintaining his motion on the given trajectory. This force component comes from the action of gravity and the forces from the seat and floor of the car. These forces have been shown in Fig. 12.23(b), where P_n and P_t are the normal and tangential force components from the car acting on the occupant. The resultant of this force system must, accordingly, have a component along n equal to 94.6 lb. Thus,

$$-200 \cos \beta + P_n = 94.6 \tag{h}$$

To get β, note with the help of Eq. (d) that

$$\tan^{-1}\left(\frac{dy}{dx}\right)_{y=60} = \tan^{-1}\frac{50}{-40} = -51.3° \tag{i}$$

Therefore,

$$\beta = 51.3°$$

Substituting into Eq. (h) and solving for P_n, we get

$$P_n = 200 \cos 51.3° + 94.6$$
$$= 220 \text{ lb}$$

This is the force component *from* the vehicle *onto* the passenger. The reaction to this force is the force component *from* the passenger *onto* the vehicle.

Part D
A System of Particles

12.10 The General Motion of a System of Particles

Let us examine a system of n particles (Fig. 12.24) that has interactions between the particles for which *Newton's third law* of motion (action equals reaction) applies. *Newton's second law* for any particle (let us say the ith particle) is then

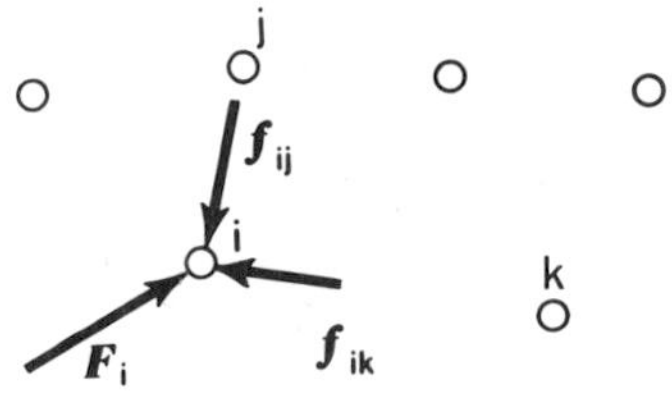

$$m_i \frac{d^2 \boldsymbol{r}_i}{dt^2} = \boldsymbol{F}_i + \sum_{\substack{j=1 \\ i \neq j}}^{n} \boldsymbol{f}_{ij} \tag{12.58}$$

Figure 12.24. Forces on ith particle of the system.

where $\boldsymbol{f}_{ij}$ is the force on particle i from particle j and is thus considered an *internal* force for the system of particles. Clearly, the $j = i$ term of the summation must be deleted since the ith particle cannot exert force on itself. The force $\boldsymbol{F}_i$ represents the resultant force on the ith particle from the forces *external* to the system of particles.

If these equations are added for all n particles, we have

$$\sum_{i=1}^{n} m_i \frac{d^2 \boldsymbol{r}_i}{dt^2} = \sum_{i=1}^{n} \boldsymbol{F}_i + \sum_{i=1}^{n} \sum_{j=1}^{n} \boldsymbol{f}_{ij} \tag{12.59}$$

Carrying out the double summation and excluding terms with repeated indexes, such as $\boldsymbol{f}_{11}$, $\boldsymbol{f}_{22}$, etc., we find that for each term with any one set of indexes there will be a term with the reverse of these indexes present. For example, for the force $\boldsymbol{f}_{12}$, a force $\boldsymbol{f}_{21}$ will exist. Considering the meaning of the indexes, we see that $\boldsymbol{f}_{ij}$ and $\boldsymbol{f}_{ji}$ represent action and reaction forces between a pair of particles. Thus, as a result of Newton's third law, the double summation in Eq. 12.59 should add up to zero. Newton's second law for a system of particles then becomes:

$$\boldsymbol{F} = \sum_{i=1}^{n} m_i \frac{d^2 \boldsymbol{r}_i}{dt^2} = \frac{d^2}{dt^2} \sum_{i=1}^{n} m_i \boldsymbol{r}_i \tag{12.60}$$

where $\boldsymbol{F}$ now represents the vector sum of all the *external* forces acting on *all* the particles of the system.

To make further useful simplifications, we use the first moment of mass of a system of n particles about a point A given as

$$\text{first moment vector} \equiv \sum_{i=1}^{n} m_i \boldsymbol{r}_i$$

where $\boldsymbol{r}_i$ represents the position vector from the point A to the ith particle (Fig. 12.25). As explained in Chapter 8, we can find a position, called the *center of mass* of the sys-

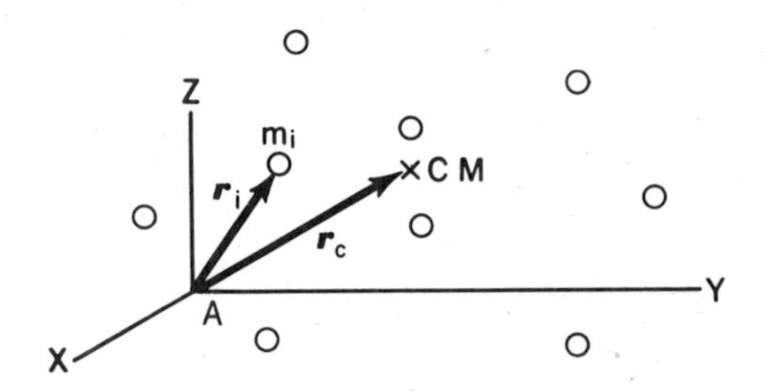

Figure 12.25. Center of mass of system.

tem, with position vector $\boldsymbol{r}_c$, where the entire mass of the system of particles can be concentrated to give the correct first moment. Thus,

$$\boldsymbol{r}_c \sum_{i=1}^{n} m_i = \sum_{i=1}^{n} m_i \boldsymbol{r}_i$$

Therefore,

$$\boldsymbol{r}_c = \frac{\sum m_i \boldsymbol{r}_i}{\sum m_i} = \frac{\sum m_i \boldsymbol{r}_i}{M} \tag{12.61}$$

Let us reconsider Newton's law using the center-of-mass concept. To do this, replace $\sum m_i \boldsymbol{r}_i$ by $M\boldsymbol{r}_c$ in Eq. 12.60. Thus,

$$\boldsymbol{F} = \frac{d^2}{dt^2}(M\boldsymbol{r}_c) = M\frac{d^2\boldsymbol{r}_c}{dt^2} \tag{12.62}$$

We see that *the center of mass of any aggregate of particles has a motion that can be computed by methods already set forth, since this is a problem involving a single hypothetical particle of mass M.* You will recall that we have alluded to this important relationship several times earlier to justify the use of the particle concept in the analysis of many dynamics problems. We must realize for such an undertaking that $\boldsymbol{F}$ is the total *external* force acting on *all* the particles.

EXAMPLE 12.13

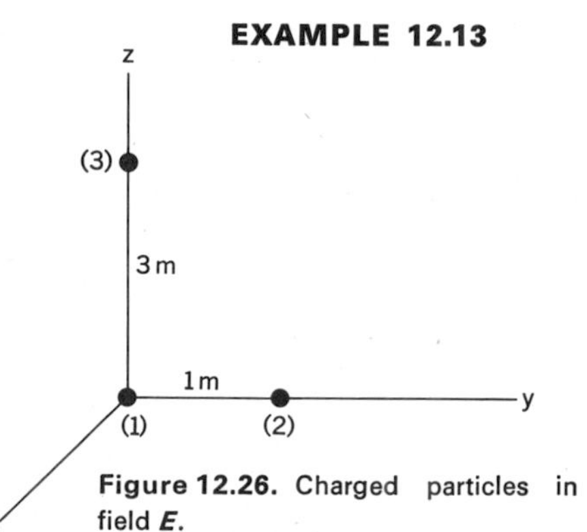

Figure 12.26. Charged particles in field $\boldsymbol{E}$.

Three charged particles in a vacuum are shown in Fig. 12.26. Particle 1 has a mass of 10^{-5} kg and a charge of 4×10^{-5} C (coulombs) and is at the origin at the instant of interest. Particles 2 and 3 each have a mass of 2×10^{-5} kg and a charge of 5×10^{-5} C and are located respectively at the instant of interest 1 m along the y axis and 3 m along the z axis. An electric field $\boldsymbol{E}$ given as

$$\boldsymbol{E} = 2x\boldsymbol{i} + 3z\boldsymbol{j} + 3(y + z^2)\boldsymbol{k} \text{ N/C} \tag{a}$$

is imposed from the outside. Compute: (a) the position of the center of mass for the system, (b) the acceleration of the center of mass, and (c) the acceleration of particle 1.

To get the position of the center of mass, we merely equate moments of the masses about the origin with that of a particle having a mass equal to the sum of masses of the system. Thus,

$$(1 + 2 + 2) \times 10^{-5}\boldsymbol{r}_c = (2 \times 10^{-5})\boldsymbol{j} + (2 \times 10^{-5})3\boldsymbol{k}$$

Therefore,

$$\boldsymbol{r}_c = .4\boldsymbol{j} + 1.2\boldsymbol{k} \text{ m} \tag{b}$$

To get the acceleration of the mass center, we must find the sum of the *external* forces acting on the particles. Two external forces act on each particle: the force of gravity and the electrostatic force from the external field. Recall from physics that this electrostatic force is given as $q\boldsymbol{E}$, where q is the charge on the particle. Hence, the total external force for each particle is given as follows:

$$\boldsymbol{F}_1 = -(9.81)(10^{-5})\boldsymbol{k} + \boldsymbol{0} \text{ N} \tag{c}$$

$$\boldsymbol{F}_2 = -(9.81)(2 \times 10^{-5})\boldsymbol{k} + (5 \times 10^{-5})(3\boldsymbol{k}) \text{ N} \tag{d}$$

$$\boldsymbol{F}_3 = -(9.81)(2 \times 10^{-5})\boldsymbol{k} + (5 \times 10^{-5})(9\boldsymbol{j} + 27\boldsymbol{k}) \text{ N} \tag{e}$$

The sum of these forces $\boldsymbol{F}_T$ is

$$\boldsymbol{F}_T = 45 \times 10^{-5}\boldsymbol{j} + 100.9 \times 10^{-5}\boldsymbol{k} \text{ N} \tag{f}$$

Accordingly, we have for $\ddot{\boldsymbol{r}}_c$:

$$\begin{aligned} \ddot{\boldsymbol{r}}_c &= \frac{45 \times 10^{-5}\boldsymbol{j} + 100.9 \times 10^{-5}\boldsymbol{k}}{5 \times 10^{-5}} \\ &= 9\boldsymbol{j} + 20.2\boldsymbol{k} \text{ m/sec}^2 \end{aligned} \tag{g}$$

Finally, to get the acceleration of particle 1, we must include the coulombic forces from particles 2 and 3. As you learned in physics, this force is given between two particles a and b with charges q_a and q_b as follows:

$$f_{\text{coul}} = -\frac{q_a q_b}{4\pi\epsilon_0 r^2}\hat{\boldsymbol{r}}$$

where $\hat{\boldsymbol{r}}$ is the unit vector between the particles, and ϵ_0 is the dielectric constant equal to 8.854×10^{-12} F/m (farads per meter) for a vacuum. Note that the coulombic force is repulsive between like charges. The total coulombic force $\boldsymbol{F}_c$ from particles 2 and 3 is

$$\begin{aligned} \boldsymbol{F}_C &= -\frac{(4 \times 10^{-5})(5 \times 10^{-5})}{(4\pi\epsilon_0)(1^2)}\boldsymbol{j} - \frac{(4 \times 10^{-5})(5 \times 10^{-5})}{(4\pi\epsilon_0)(3^2)}\boldsymbol{k} \\ &= -18\boldsymbol{j} - 2\boldsymbol{k} \text{ N} \end{aligned} \tag{h}$$

The total force acting on particle 1 is then

$$(\boldsymbol{F}_1)_T = \underbrace{-(9.81)(10^{-5})\boldsymbol{k}}_{\substack{\text{from}\\\text{weight}}} + \underbrace{\boldsymbol{0}}_{\substack{\text{from}\\\text{external}\\\text{field}}} + \underbrace{(-18\boldsymbol{j} - 2\boldsymbol{k})}_{\substack{\text{from}\\\text{internal}\\\text{field}}} \text{ N} \tag{i}$$

Clearly, the internal field dominates here.
Newton's law then gives us

$$\begin{aligned} \ddot{\boldsymbol{r}}_1 &= \frac{-18\boldsymbol{j} - 2\boldsymbol{k}}{10^{-5}} \\ &= -18 \times 10^5\boldsymbol{j} - 2 \times 10^5\boldsymbol{k} \text{ m/sec}^2 \end{aligned} \tag{j}$$

We see here from Eqs. (g) and (j) that although the particles tend to "scramble" away from each other due to very strong internal coulombic forces, the center of mass accelerates slowly by comparison.

Problems

12.76. A warrior of old is turning a sling in a vertical plane. A rock of mass .3 kg is held in the sling prior to releasing it against an enemy. What is the minimum speed ω to hold the rock in the sling?

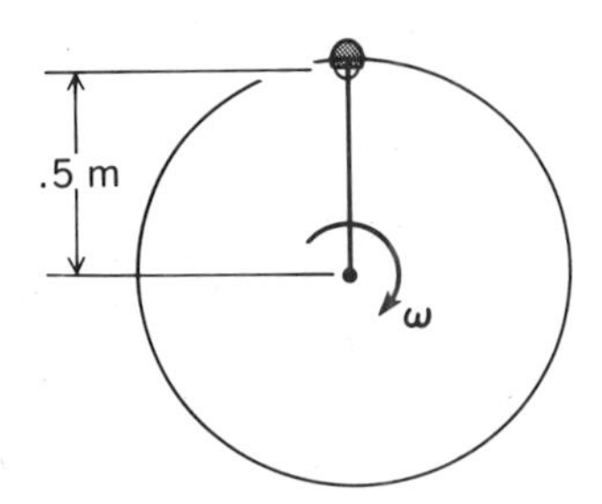

Figure P.12.76

12.77. A car is traveling at a speed of 55 mi/hr along a banked highway having a radius of curvature of 500 ft. At what angle should the road be banked in order that a zero friction force is needed for the car to go around this curve?

12.78. A car weighing 20 kN is moving at a speed V of 60 km/hr on a road having a vertical radius of curvature of 200 m as shown. At the instant shown, what is the maximum deceleration possible from the brakes along the road for the vehicle if the coefficient of dynamic friction between tires and the road is .55?

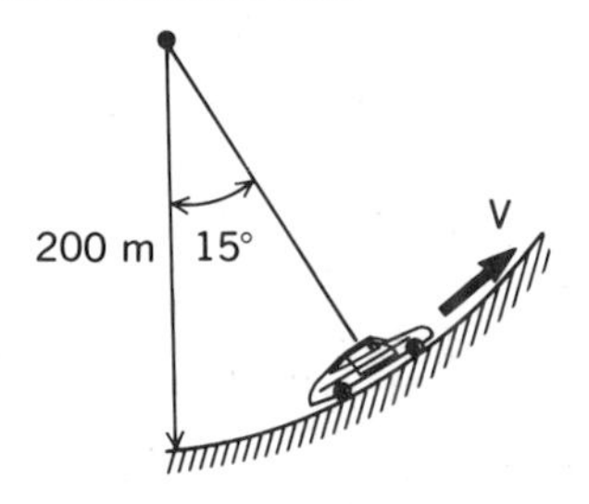

Figure P.12.78

12.79. A particle moves at uniform speed of 1 m/sec along a plane sinusoidal path given as

$$y = 5 \sin \pi x \text{ m}$$

What is the position between $x = 0$ and $x = 1$ m for the maximum force normal to the curve? What is this force if the mass of the particle is 1 kg?

12.80. A catenary curve is formed by the cable of a suspension bridge. The equation of this curve relative to the axes shown can be given as

$$y = \frac{a}{2}(e^{ax} + e^{-ax}) = a \cosh ax$$

with x and y in feet. A small one-passenger vehicle is designed to move along the catenary to facilitate repair and painting of the bridge. Consider that the vehicle moves at uniform speed of 10 ft/sec along the curve. If the vehicle and passenger have a combined mass of 250 lbm, what is the force normal to the curve as a function of position x?

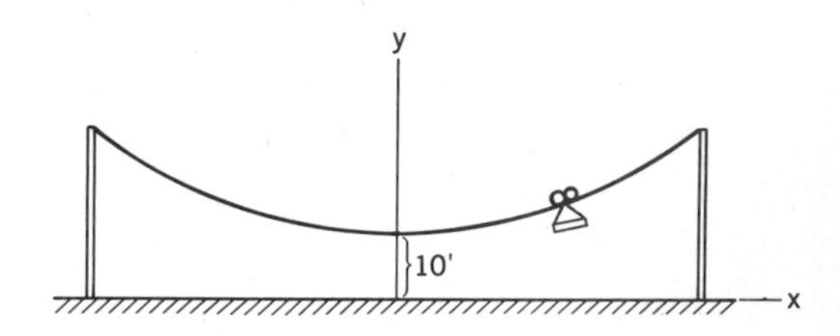

Figure P.12.80

12.81. A rod CD rotates with shaft G–G at an angular speed ω of 300 rpm. A sleeve A of mass 500 g slides on CD. If no friction is present between A and CD, what is the distance S for no relative motion between A and CD?

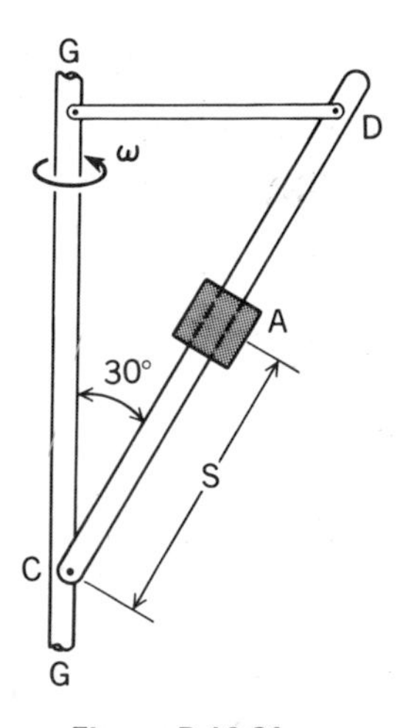

Figure P.12.81

12.82. In Problem 12.81, what is the range of values for S for which A will remain stationary relative to CD if there is coulombic friction between A and CD such that $\mu_s = .4$?

12.83. A circular rod EB rotates at constant angular speed ω of 50 rpm. A sleeve A of mass 2 lbm slides on the circular rod. At what position θ will sleeve A remain stationary relative to the rod EB if there is no friction?

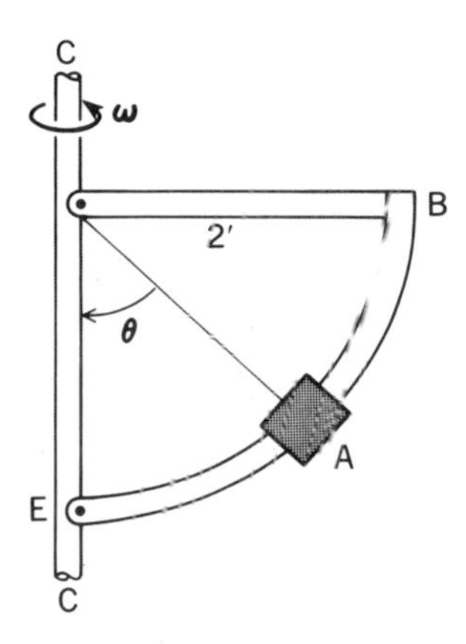

Figure P.12.83

12.84. In Problem 12.83 assume that there is coulombic friction between A and EB with $\mu_s = .3$. Show that the minimum value of θ for which the sleeve will remain stationary relative to the rod is 75.45°.

12.85. Three bodies have the following weights and positions at time t:

$$W_1 = 10 \text{ lb}, \quad x_1 = 6 \text{ ft}, \quad y_1 = 10 \text{ ft}, \quad z_1 = 10 \text{ ft}$$

$$W_2 = 5 \text{ lb}, \quad x_2 = 5 \text{ ft}, \quad y_2 = 6 \text{ ft}, \quad z_2 = 0$$

$$W_3 = 8 \text{ lb}, \quad x_3 = 0, \quad y_3 = -4 \text{ ft}, \quad z_3 = 0$$

Determine the position vector of the center of mass at time t. Determine the velocity of the center of mass if the bodies have the following velocities:

$$\boldsymbol{V}_1 = 6\boldsymbol{i} + 3\boldsymbol{j} \text{ ft/sec}$$
$$\boldsymbol{V}_2 = 10\boldsymbol{i} - 3\boldsymbol{k} \text{ ft/sec}$$
$$\boldsymbol{V}_3 = 6\boldsymbol{k} \text{ ft/sec}$$

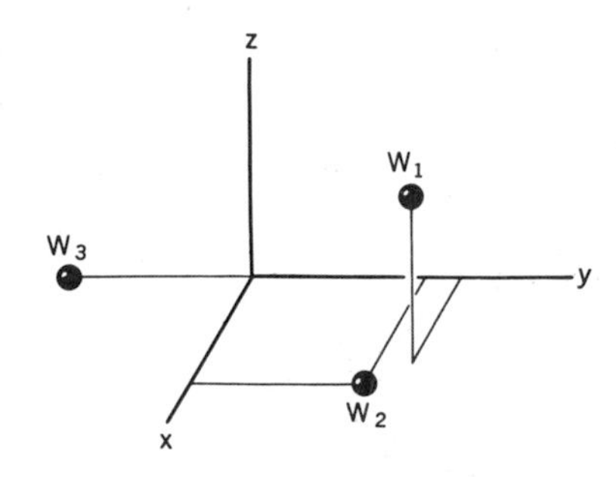

Figure P.12.85

12.86. In Problem 12.85, the following external forces act on the respective particles:

$$\boldsymbol{F}_1 = 6t\boldsymbol{i} + 3\boldsymbol{j} - 10\boldsymbol{k} \text{ lb} \quad \text{(particle 1)}$$
$$\boldsymbol{F}_2 = 15\boldsymbol{i} - 3\boldsymbol{j} \text{ lb} \quad \text{(particle 2)}$$
$$\boldsymbol{F}_3 = \boldsymbol{0} \text{ lb} \quad \text{(particle 3)}$$

What is the acceleration of the center of mass, and what is its position after 10 sec from that given initially? From Problem 12.85 at $t = 0$:

$$r_C = 3.70\boldsymbol{i} + 4.26\boldsymbol{j} + 4.35\boldsymbol{k} \text{ ft}$$

$$V_C = 4.78\boldsymbol{i} + 1.304\boldsymbol{j} + 1.435\boldsymbol{k} \text{ ft/sec}$$

12.87. The following data for a system of particles are given at time $t = 0$:

$M_1 = 50$ kg at position $(1, 1.3, -3)$ m

$M_2 = 25$ kg at position $(-.6, 1.3, -2.6)$ m

$M_3 = 5$ kg at position $(-2.6, 5.3, 1)$ m

The particles are acted on by the following respective external forces:

$\boldsymbol{F}_1 = 50\boldsymbol{j} + 10t\boldsymbol{k}$ N (particle 1)

$\boldsymbol{F}_2 = 50\boldsymbol{k}$ N (particle 2)

$\boldsymbol{F}_3 = 5t^2\boldsymbol{i}$ N (particle 3)

What is the velocity of M_1 relative to the mass center after 5 sec, assuming that at $t = 0$, the particles are at rest?

***12.88.** Given the following force field:

$$\boldsymbol{F} = -2x\boldsymbol{i} + 3\boldsymbol{j} - z\boldsymbol{k} \text{ lb/slug}$$

which is the force on any particle in the field per unit mass of the particle? If we have two particles initially stationary in the field with position vectors

$$\boldsymbol{r}_1 = 3\boldsymbol{i} + 2\boldsymbol{j} \text{ ft}$$

$$\boldsymbol{r}_2 = 4\boldsymbol{i} - 2\boldsymbol{j} + 4\boldsymbol{k} \text{ ft}$$

what is the velocity of each particle relative to the center of mass of the system after 2 sec have elapsed? Each particle has a weight of .1 oz.

12.89. A stationary uniform block of ice is acted on by forces that maintain constant magnitude and direction at all times. If

$F_1 = (25\,g)$ N

$F_2 = (10\,g)$ N

$F_3 = (15\,g)$ N

what is the velocity of the center of mass of the block after 10 sec? Neglect friction. The density of ice is 56 lbm/ft³.

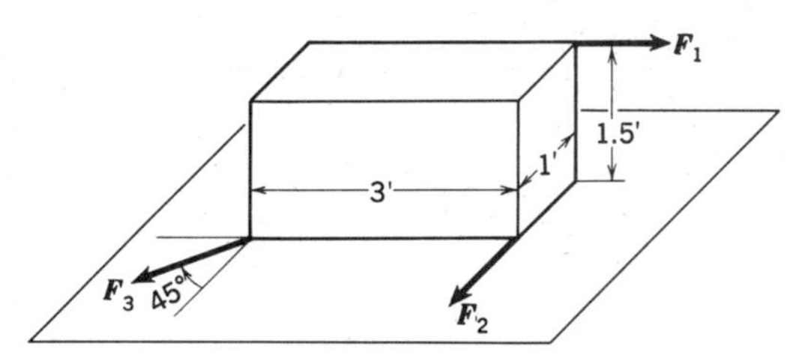

Figure P.12.89

12.90. A space vehicle decelerates downward (Z direction) at 1613 km/hr/sec while moving in a translatory manner relative to inertial space. Inside the vehicle is a rod BC rotating in the plane of the paper at a rate of 50 rad/sec relative to the vehicle. Two masses rotate at the rate of 20 rad/sec around BC on rod EF. The masses are each 300 mm from C. Determine the force transmitted at C between BC and EF if the mass of each of the rotating bodies is 5 kg and the mass of rod EF is 1 kg. BC is in the vertical position at the time of interest. Neglect gravity.

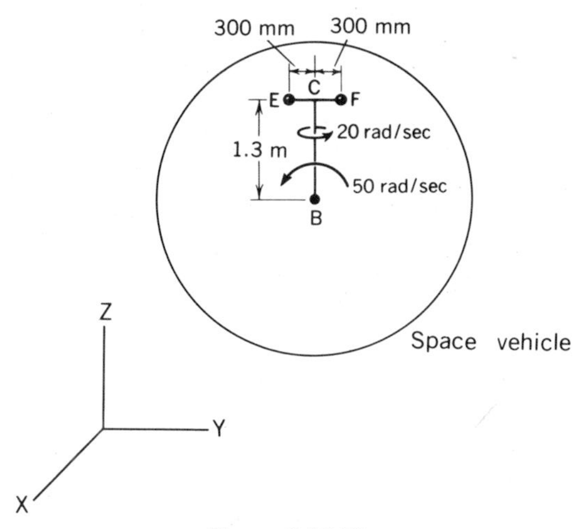

Figure P.12.90

12.91. A young man is standing in a canoe awaiting a young lady. The man weighs 200 lb and is at the far end of the canoe, which also weighs 200 lb. When the young lady appears, he scrambles forward to greet her, but, when he has moved the 20 ft to the forward end of the canoe, to his surprise he finds (not having studied mechanics) that he cannot reach her. How far is the tip of the canoe from the dock when our "gallant" has made the 20-ft dash? The canoe is in no way tied to the pier, and there are no currents in the water. (*Hint:* Does the moment of mass of the man–canoe system change about any fixed point?)

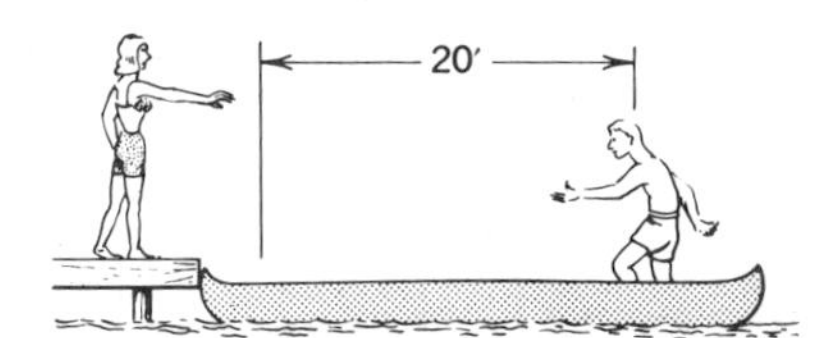

Figure P.12.91

12.92. Two men climb aboard a barge at A to shift a load with the aid of a fork lift. The barge has a mass of 20,000 kg and is 10 m long. The load consists of four containers each with a mass of 1300 kg and each having a length of 1 m. The men shift the containers to the opposite end of the barge, put the fork lift where they found it, and prepare to step off the barge at A, where they came on. If the barge has not been constrained and if we neglect water friction, currents, wind, and so on, how far has the barge shifted its position? The fork lift has a mass of 1000 kg. See the hint of Problem 12.91.

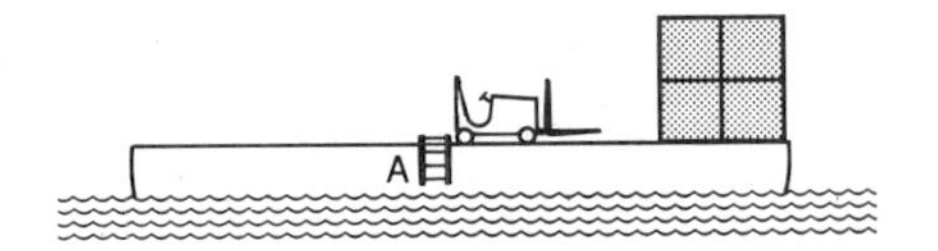

Figure P.12.92

12.93. An astronaut on a space walk pulls a mass A of 100 kg toward him and shortens the distance d by 5 m. If the astronaut weighs 660 N on earth, how far does the mass A move from its original position? Neglect the mass of the cord. See the hint of Problem 12.91.

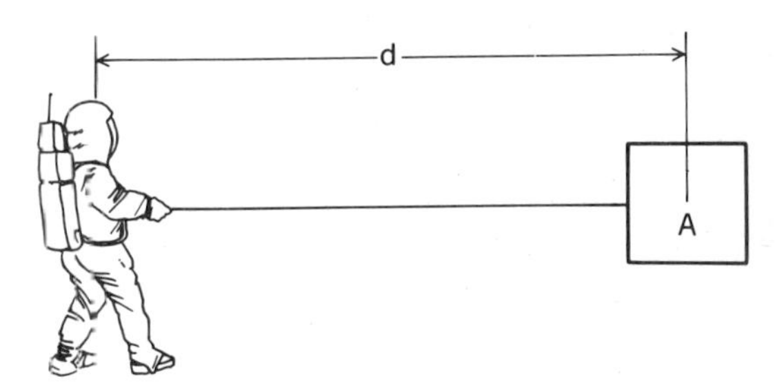

Figure P.12.93

12.94. Two identical adjacent tanks are each 10 ft long, 5 ft high, and 5 ft wide. Originally, the left tank is completely full of water while the right tank is empty. Water is pumped by an internal pump from the left tank to the right tank. At the instant of interest, the rate of flow Q is 20 ft³/sec, while $\dot{Q}$ is 5 ft³/sec². What horizontal force on the tanks is needed at this instant from the foundation? Assume that the water surface in the tanks remain horizontal. The specific weight of water is 62.4 lb/ft³.

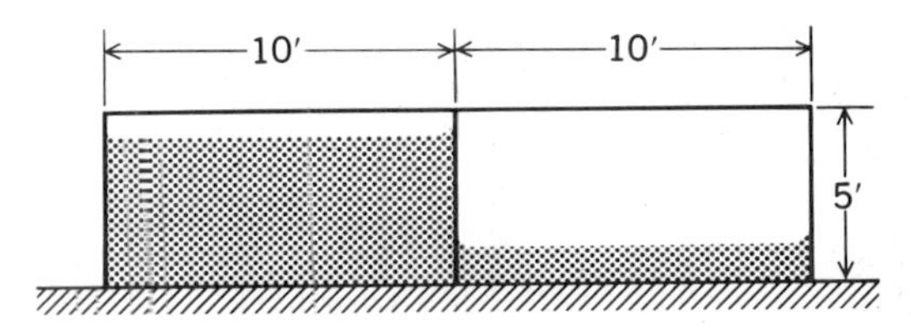

Figure P.12.94

12.11 Closure

In this chapter, we integrated Newton's law for various coordinate systems. Also, with the aid of the mass center concept, we formulated Newton's law for any aggregate of particles. In the next two chapters, we shall present alternative procedures for more efficient treatment of certain classes of dynamics problems for particles. You will note that, since the new concepts are all derived from Newton's law, whatever can be solved by these new methods could also be solved by the methods we have already presented. A separate and thorough study of these topics is warranted by the gain in insight into dynamics and the greater facility in solving problems that can be achieved by examining these alternative methods and their accompanying concepts. As in this chapter, we will make certain generalizations applicable to any aggregate of particles.

Review Problems

12.95. A block A of mass 10 kg rests on a second block B of mass 8 kg. A force F equal to 100 N pulls block A. The coefficient of friction between A and B is .5; between B and the ground, .1. What is the speed of block A relative to block B in $\frac{1}{10}$ sec if the system starts from rest?

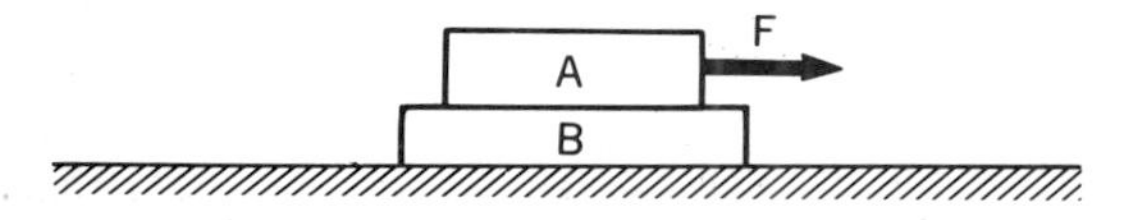

Figure P.12.95

***12.96.** A block B slides from A to F along a rectangular chute where there is coulombic friction on the faces of the chute. The coefficient of dynamic friction is .4. The bottom face of the chute is parallel to face $EACF$ and the other two faces are perpendicular to $EACF$. The body weighs 5 lb. How long does it take B to go from A to F starting from rest?

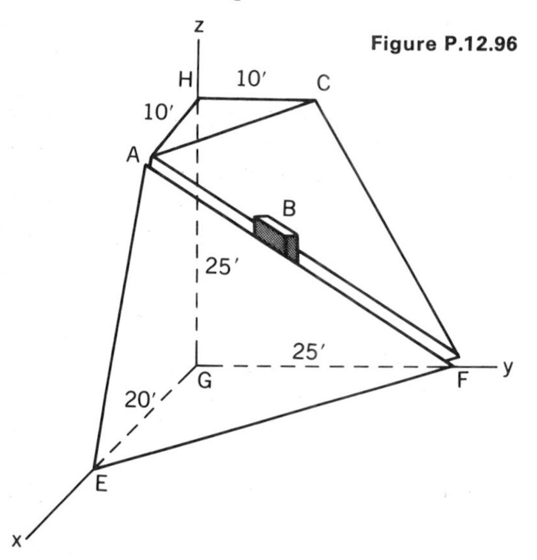

Figure P.12.96

12.97. A tugboat is pushing a barge at a steady speed of 8 knots. The thrust from the tugboat needed for this motion is 800 lb. The barge with load weighs 100 tons. If the water resistance to the barge is proportional to the speed of the barge, how long will it take the barge to slow to 5 knots after the tugboat ceases to push? (*Note*: 1 knot equals 1.152 mi/hr.)

Figure P.12.97

12.98. A spring requires a force x^2 N for a deflection of x mm, where x is the deflection of the spring from the undeformed geometry. Because the deflection is not proportional to x to the first power, the spring is called a *nonlinear* spring. If a 100-kg block is suddenly released on the undeformed spring, what is the speed of the block after it has descended 10 mm?

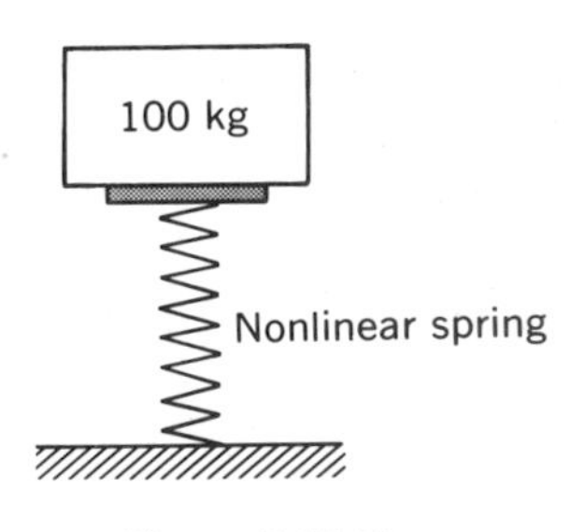

Figure P.12.98

12.99. Weights A and B are held by light pulleys. If released from rest, what is the speed of each weight after 1 sec? Weight A is 10 lb and weight B is 40 lb.

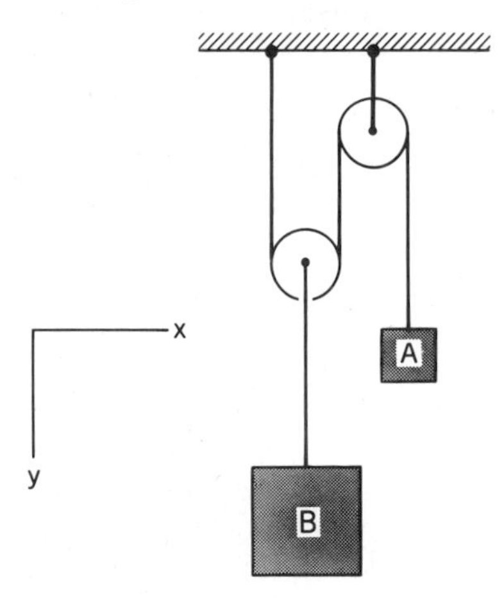

Figure P.12.99

12.100. The following data are given for the fly-ball governor (read Problem 12.48 for details on how the governor works):

$$l = .3 \text{ m}$$
$$D = 50 \text{ mm}$$
$$\omega = 300 \text{ rpm}$$
$$\theta = 45°$$

What is the force P acting on frictionless collar A if each ball has a mass of 1 kg and we neglect the weight of all other moving members of the system?

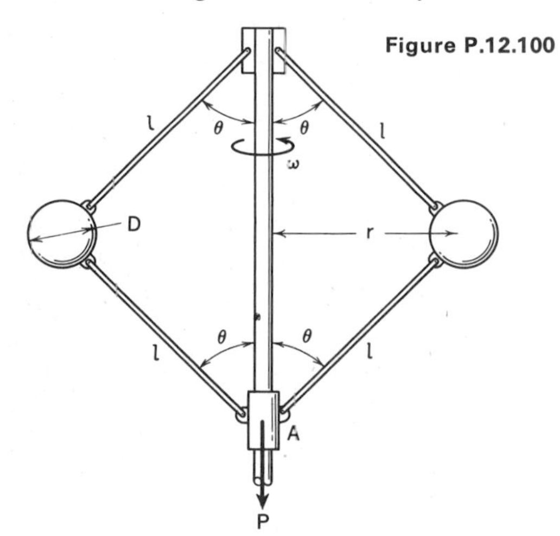

Figure P.12.100

12.101. A spy satellite to observe the United States is put into a circular orbit about the North and South Poles. The satellite is to make 10 cycles/day (24 hr). What must be the distance from the surface of the earth for this satellite?

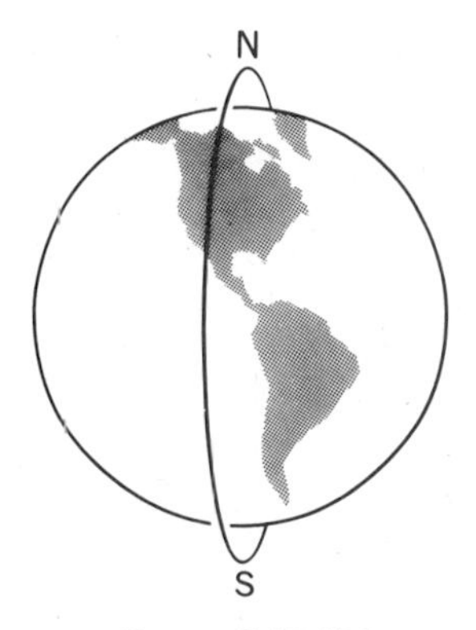

Figure P.12.101

12.102. A skylab is in a circular orbit about the earth at a distance of 500 km above the earth's surface. A space shuttle has rendezvoused with the skylab and now, wishing to depart, decouples and fires its rockets to move more slowly than the skylab. If the rockets are fired over a short time interval, what should the relative speed between the space shuttle and skylab be at the end of rocket fire if the space shuttle is to come as close as 100 km to the earth's surface in subsequent ballistic (rocket motors off) flight?

12.103. A space vehicle is launched at a speed of 19,000 mi/hr relative to the earth's center at a position 250 mi above the earth's surface. If the vehicle has a radial velocity component of 3000 mi/hr toward the earth's center, what is the eccentricity of the trajectory? What is the maximum elevation above the earth's surface reached by the vehicle? Do not use Eq. 12.54.

12.104. A skier is moving down a hill at a speed of 30 mi/hr when he is at the position shown. If the skier weighs 180 lb, what total force do his skis exert on the snow surface? Assume that the coefficient of friction is .1. The hill can be taken as a parabolic surface.

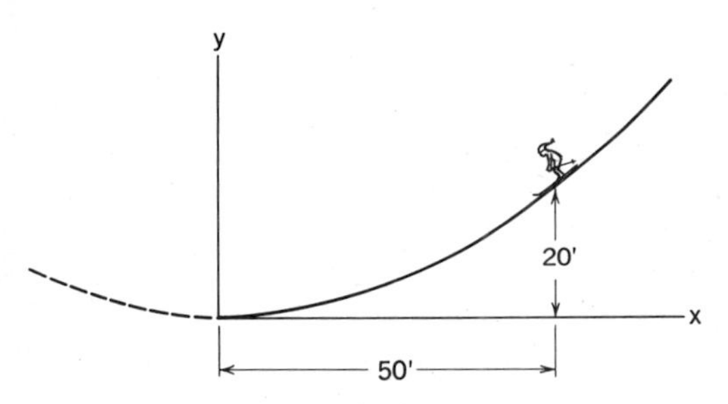

Figure P.12.104

12.105. A submarine is moving at constant speed of 15 knots below the surface of the ocean. The sub is at the same time descending downward while remaining horizontal with an acceleration of .023 g. In the submarine a flyball governor operates with weights having a mass each of 500 g. The governor is rotating with speed ω of 5 rad/sec. If at time t, $\theta = 30°$, $\dot{\theta} = .2$ rad/sec, and $\ddot{\theta} = 1$ rad/sec², what is the force developed on the support of governor system as a result solely of the motion of the weights at this instant?

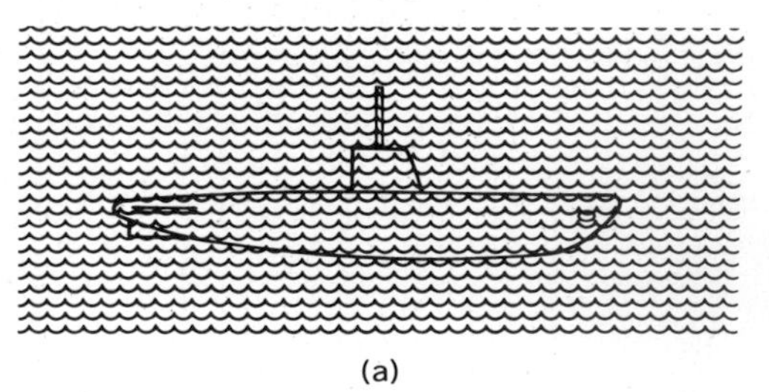

(a)

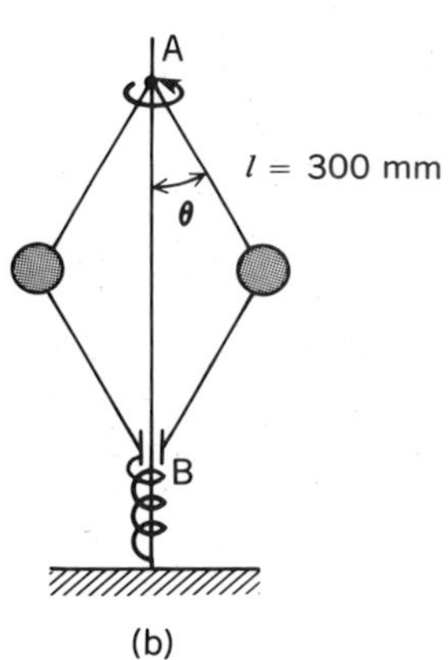

(b)

Figure P.12.105

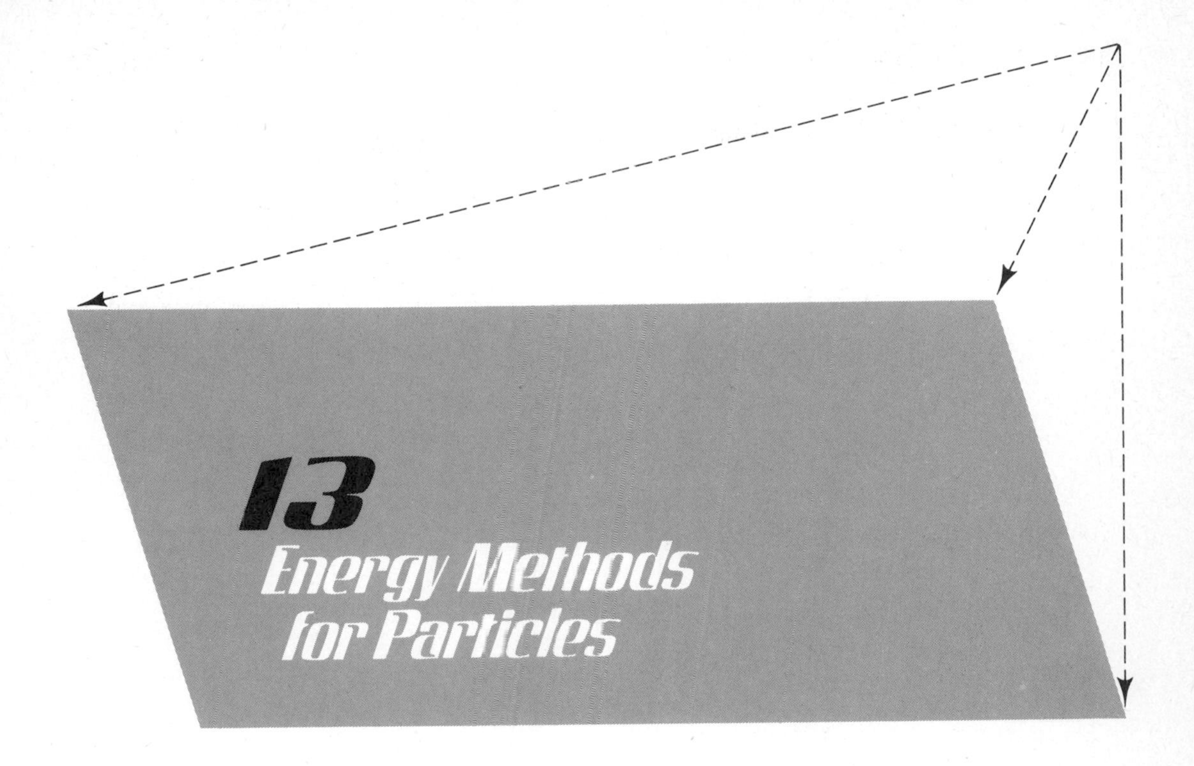

13 Energy Methods for Particles

Part A
Analysis for a Single Particle

13.1 Introduction

In Chapter 12, we integrated the differential equation derived from Newton's law to yield velocity and position as functions of time. At this time, we shall present an alternative procedure, that of the method of energy, and we shall see that certain classes of problems can be more easily handled by this method in that we shall not need to integrate a differential equation.

To set forth the basic equation underlying this approach, we start with Newton's law for a particle moving relative to an inertial reference, as shown in Fig. 13.1. Thus,

$$\boldsymbol{F} = m\frac{d^2\boldsymbol{r}}{dt^2} = m\frac{d\boldsymbol{V}}{dt} \tag{13.1}$$

Multiply each side of this equation by $d\boldsymbol{r}$ as a dot product and integrate from r_1 to r_2 along the path of motion:

$$\int_{r_1}^{r_2} \boldsymbol{F} \cdot d\boldsymbol{r} = m\int_{r_1}^{r_2} \frac{d\boldsymbol{V}}{dt} \cdot d\boldsymbol{r} = m\int_{t_1}^{t_2} \frac{d\boldsymbol{V}}{dt} \cdot \frac{d\boldsymbol{r}}{dt}\,dt$$

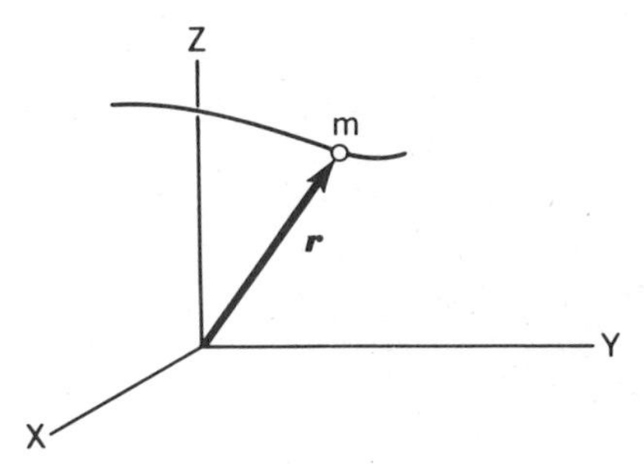

Figure 13.1. Particle moving relative to an inertial reference.

In the last integral, we multiplied and divided by dt, thus changing the variable of integration to t. Since $d\boldsymbol{r}/dt = \boldsymbol{V}$, we then have

$$\int_{r_1}^{r_2} \boldsymbol{F} \cdot d\boldsymbol{r} = m \int_{t_1}^{t_2} \left(\frac{d\boldsymbol{V}}{dt} \cdot \boldsymbol{V}\right) dt = \frac{1}{2} m \int_{t_1}^{t_2} \frac{d}{dt}(\boldsymbol{V} \cdot \boldsymbol{V})\, dt$$

$$= \frac{1}{2} m \int_{t_1}^{t_2} \frac{d}{dt} V^2\, dt = \frac{1}{2} m \int_{V_1}^{V_2} d(V^2)$$

On carrying out the integration, we arrive at the familiar equation

$$\int_{r_1}^{r_2} \boldsymbol{F} \cdot d\boldsymbol{r} = \tfrac{1}{2} m(V_2^2 - V_1^2) \tag{13.2}$$

where the left side is the well-known expression for *work* (to be denoted at times as W_{1-2})[1] and the right side is clearly the change in *kinetic energy* as the mass moves from position $\boldsymbol{r}_1$ to $\boldsymbol{r}_2$.

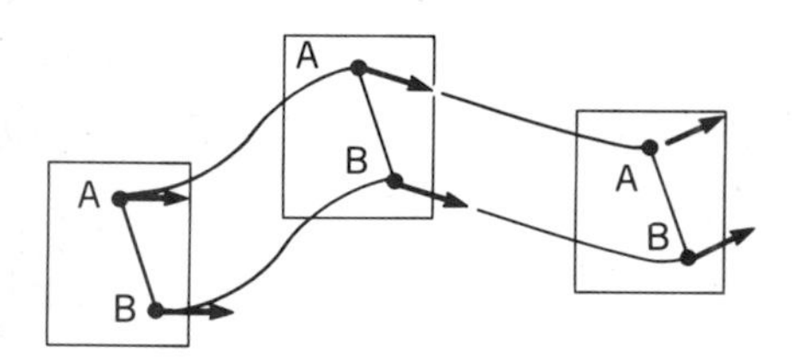

Figure 13.2. Translating body.

We shall see in Section 13.7 that for any system of particles, including, of course, rigid bodies, we get a work–energy equation of the form 13.2, where the velocity is that of the mass center, the force is the resultant external force on the system, and the path of integration is that of the mass center. Clearly, then, we can use a single particle model (and consequently Eq. 13.2) for:

1. *A rigid body moving without rotation.* Such a motion was discussed in Chapter 11 and is called *translation.* Note that lines in a translating body remain parallel to their original directions, and points in the body move over a path which has identically the same form for all points. This condition is illustrated in Fig. 13.2 for two points A and B. Furthermore, each point in the body has at any instant of time t the same velocity as any other point. Clearly the motion of the center of mass fully characterizes the motion of the body and Eq. 13.2 will be used often for this situation.
2. Sometimes for *a body whose size is small compared to its trajectory.* Here the paths of points in the body differ very little from that of the mass center and knowing where the center of mass is tells us with sufficient accuracy all we need to know about the position of the body. However, keep in mind that the *velocity* and *acceleration* relative to the center of mass of a part of the body may be *very large,*

[1]It is important to note that the work done by a force system depends on the path over which the forces move, except in the case of conservative forces to be considered in Section 13.3. Thus, W_{1-2} is called a *path function* in thermodynamics. However, kinetic energy depends only on the instantaneous state of motion of the particle and is independent of the path. Kinetic energy is called accordingly a *point function* in thermodynamics.

In the next section, we shall present a more general definition of work.

irrespective of how small the body may be compared to the trajectory of its center of mass. Then, information about the velocity and acceleration of this part of the body would require a more detailed consideration beyond a simple one-particle model centered around the center of mass.

Thus, as in our considerations of Newton's law in Chapter 12, when the motion of the mass center characterizes with sufficient accuracy what we want to know about the motion of a body, we use a particle at the mass center for energy considerations.

Next, suppose that we have a component of Newton's law in one direction, say the x direction:

$$F_x \boldsymbol{i} = m \frac{dV_x}{dt} \boldsymbol{i}$$

Taking the dot product of each side of this equation with $dx\boldsymbol{i} + dy\boldsymbol{j} + dz\boldsymbol{k}$ $(= d\boldsymbol{r})$, we get, after integrating in the manner set forth at the outset:

$$\int_{x_1}^{x_2} F_z \, dx = \frac{m}{2}[(V_x)_2^2 - (V_x)_1^2] \tag{13.3a}$$

Similarly,

$$\int_{y_1}^{y_2} F_y \, dy = \frac{m}{2}[(V_y)_2^2 - (V_y)_1^2] \tag{13.3b}$$

$$\int_{z_1}^{z_2} F_z \, dz = \frac{m}{2}[(V_z)_2^2 - (V_z)_1^2] \tag{13.3c}$$

Thus, the foregoing equations demonstrate that the work done on a particle in any direction equals the change in kinetic energy associated with the component of velocity in that direction.

Instead of employing Newton's law, we can now use the energy equations developed in this section for solving certain classes of problems. This energy approach is particularly handy when velocities are desired and forces are functions of position. However, please understand that any problem solvable with the energy equation can be solved from Newton's law; the choice between the two is mainly a question of convenience and the manner in which the information is given.

EXAMPLE 13.1

An automobile is moving at 60 mi/hr (see Fig. 13.3) when the driver jams on his brakes and goes into a skid in the direction of motion. The car weighs 4000 lb, and the dynamic coefficient of friction between the rubber tires and the concrete road is .60. How far, l, will the car move before stopping?

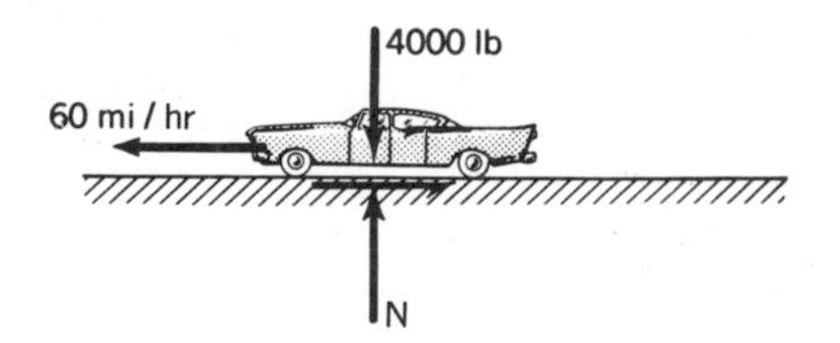

Figure 13.3. Car moving with brakes locked.

A constant friction force acts, which from Coulomb's law is $\mu N = (.60)(4000) = 2400$ lb. This force is the only force performing work, and clearly it is changing the kinetic energy of the vehicle from that corresponding to the speed of 60 mi/hr (or 88 ft/sec) to zero. (You will learn in thermodynamics that this work facilitates a transfer of kinetic energy of the vehicle to an increase of internal energy of the vehicle, the road, and the air.) From the work–energy equation 13.2, we get[2]

$$-2400l = \frac{1}{2}\frac{4000}{g}(0 - 88^2)$$

Hence,

$$l = 200 \text{ ft}$$

(Perhaps every driver should solve this problem periodically.)

EXAMPLE 13.2

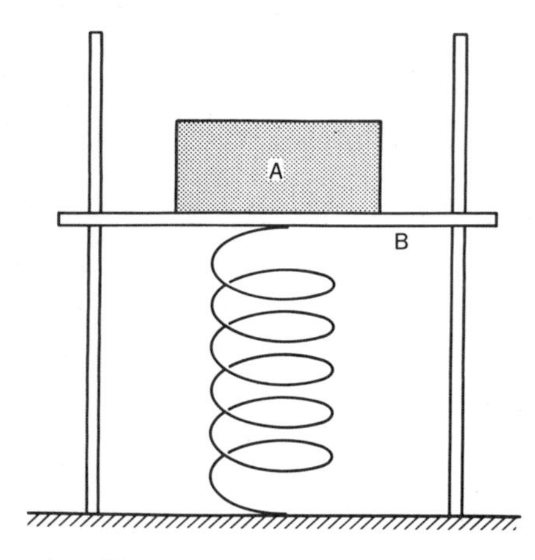

Figure 13.4. Preloaded nonlinear spring.

Shown in Fig. 13.4 is a light platform B guided by vertical rods. The platform is positioned so that the spring has been compressed 10 mm. In this configuration a body A weighing 100 N is placed on the platform and released suddenly. If the guide rods give a total resistance force f to downward movement of the platform of 5 N, what is the largest distance that the weight falls? The spring used here is a *nonlinear* spring requiring $.5x^2$ N of force for a deflection of x mm.

We take as the position of interest for the body the location δ below the initial configuration where the body A reaches zero velocity for the first time after having been released. The change in kinetic energy over the interval of interest is accordingly zero. Thus, zero net work is done by the forces acting on the body A during displacement δ. These forces comprise the force of gravity, the friction force from the guides, and finally the force from the spring. Using as the origin for our measurements the *undeformed* end position of the spring, we can say:

$$\int_{10}^{(10+\delta)} \boldsymbol{F} \cdot d\boldsymbol{r} = \int_{10}^{(10+\delta)} (W_A - f - .5x^2)\, dx$$
$$= \int_{10}^{(10+\delta)} (100 - 5 - .5x^2) dx = 0 \tag{a}$$

Integrating, we get

$$95\delta - \frac{.5}{3}[(10 + \delta)^3 - 10^3] = 0 \tag{b}$$

Therefore,

$$\delta^3 + 30\delta^2 - 270\delta = 0 \tag{c}$$

One solution to Eq. (c) is $\delta = 0$. Clearly, no work is done if there is no deflection. But this solution has no meaning for this problem since the force in the spring is only $.5x^2 = .5(10)^2 = 50$ N, when the weight of 100 N is released. Therefore, there must be a nonzero positive value of δ that satisfies the equation and has physical

[2]Note that the sign of the work done is negative since the friction force is opposite in sense to the motion.

meaning. Factoring out one δ from the equation, we then set the resulting quadratic expression equal to zero. Two roots result and the positive root $\delta = 7.25$ mm is the one with physical meaning.

In the following example we deal with two bodies which can be considered as particles, rather than with one body as has been the case in the previous examples. We shall deal with these bodies separately in this example. Later in the chapter, we shall consider *systems* of particles, and in that context we will be able to consider this problem as a system of particles with less work needed to reach a solution.

EXAMPLE 13.3

In Fig. 13.5, we have shown bodies A and B interconnected through a block and pulley system. Body B has a mass of 100 kg, whereas body A has a mass of 900 kg. Initially the system is stationary with B held at rest. What speed will B have when it reaches the ground at a distance $h = 3$ m below after being released? Neglect the masses of the pulleys and the rope. Consider the rope to be inextensible.

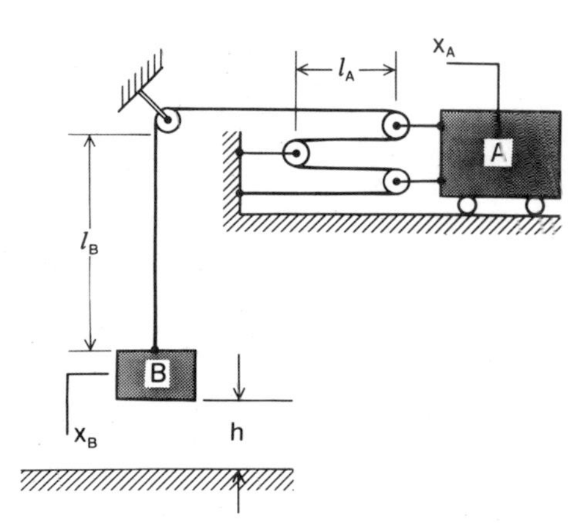

Figure 13.5. System of blocks and pulleys.

You will note from Fig. 13.5 that, as the bodies move, only the distances l_B and l_A change; the other distances involving the ropes do not change. And because the rope is taken as inextensible, we conclude that at all times

$$l_B + 4l_A = \text{constant} \tag{a}$$

Differentiating with respect to time, we can find that

$$\dot{l}_B + 4\dot{l}_A = 0$$

Therefore,

$$\dot{l}_B = -4\dot{l}_A \tag{b}$$

On inspecting Fig. 13.5, you should have no trouble in concluding that $\dot{l}_A = -V_A$ and that $\dot{l}_B = V_B$. Hence, from Eq. (b), we can conclude that

$$V_B = 4V_A \tag{c}$$

Next take the differential of Eq. (a):

$$dl_B + 4dl_A = 0$$

Therefore,

$$dl_B = -4dl_A \tag{d}$$

Note that $dx_B = dl_B$ and that $dx_A = -dl_A$. Hence, we see from Eq. (d) that a movement Δ_A of body A results in a movement, $4\Delta_A$, of body B:

$$\Delta_B = 4\Delta_A \tag{e}$$

With these kinematical conclusions as a background, we are now ready to proceed with the work–energy considerations.

For this purpose, we have shown a free-body diagram of body B in Fig. 13.6. The work–energy equation for body B can then be given as follows:

$$(100\,g - T)h = \tfrac{1}{2}\,100V_B^2$$

Therefore,

$$(981 - T)(3) = \tfrac{1}{2}(100)V_B^2 \tag{f}$$

Now consider the free-body diagram of body A in Fig. 13.7. The work–energy equation for body A is then

$$(4T)(\Delta_A) = \tfrac{1}{2}\,900V_A^2 \tag{g}$$

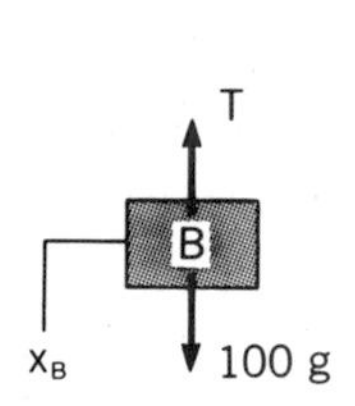

Figure 13.6. Free-body diagram of B.

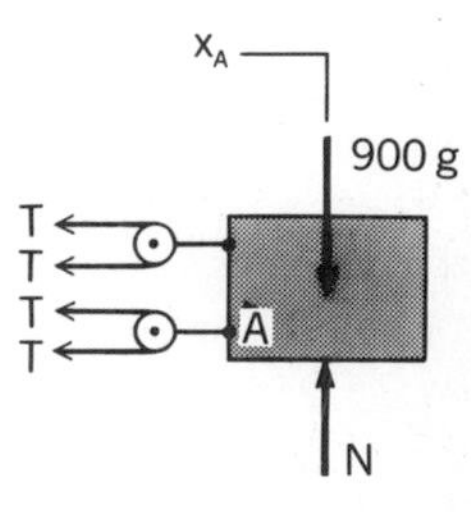

Figure 13.7. Free-body diagram of A.

But according to Eq. (e),

$$\Delta_A = \frac{\Delta_B}{4} = \frac{h}{4} = \frac{3}{4}\text{ m} \tag{h}$$

And according to Eq. (c),

$$V_A = \tfrac{1}{4}\,V_B \tag{i}$$

Substituting the results from Eqs. (h) and (i) into (g), we get

$$(4T)\left(\frac{3}{4}\right) = \frac{1}{2}900\left(\frac{V_B^2}{16}\right) \tag{j}$$

Adding Eqs. (f) and (j), we can eliminate T to form the following equation with V_B as the only unknown:

$$(981)(3) = \tfrac{1}{2}(V_B^2)(100 + \tfrac{900}{16})$$

Therefore,

$$V_B = 6.14\text{ m/sec downward}$$

Hence,

$$V_A = 1.534\text{ m/sec to the left}$$

13.2 Power Considerations

The rate at which work is performed is called *power* and is a very useful concept for engineering purposes. Employing the notation W_k to represent work, we have

$$\text{power} = \frac{dW_k}{dt} \tag{13.4}$$

Since dW_k for any given force $\boldsymbol{F}_i$ is $\boldsymbol{F}_i \cdot d\boldsymbol{r}_i$, we can say that the power being developed by a system of n forces at time t is, for a reference xyz,

$$\text{power} = \frac{\sum_{i=1}^{n} \boldsymbol{F}_i \cdot d\boldsymbol{r}_i}{dt} = \sum_{i=1}^{n} \boldsymbol{F}_i \cdot \boldsymbol{V}_i \tag{13.5}$$

where $\boldsymbol{V}_i$ is the velocity of the point of application of the ith force at time t as seen from reference xyz.[3]

In the following example we shall illustrate the use of the power concept. Note, however, that we shall find use of Newton's law advantageous in certain phases of the computation.

EXAMPLE 13.4

In hilly terrain, motors of an electric train are sometimes advantageously employed as brakes, particularly on downhill runs. The motors are reversed to act as generators so that, during braking, energy is returned to the central power source from the train. In this way, we save the energy lost when employing conventional brakes—a considerable saving in every round trip. Such a train consisting of a single car is shown in Fig. 13.8 moving down a 15° incline at an initial speed of 3 m/sec.

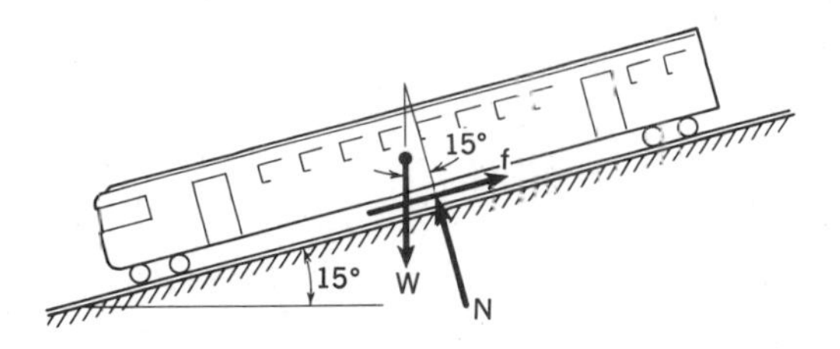

Figure 13.8. Train moving downhill with generators acting as brakes.

---

[3]We could have defined work W_K in terms of power as follows:

$$W_K = \int_{t_1}^{t_2} (\boldsymbol{F} \cdot \boldsymbol{V})\, dt$$

When the force acts on a particular particle, the result above becomes the familiar $\int_{r_1}^{r_2} \boldsymbol{F} \cdot d\boldsymbol{r}$, where $\boldsymbol{r}$ is the position vector of the particle. There are times when the force acts on *continuously changing* particles as time passes (see Section 13.8). The more general formulation above can then be used effectively.

This car has a mass of 20,000 kg and has a cogwheel drive. If the conductor maintains an adjustment of the fields in his generators so as to develop a constant power output of 50 kW, how long does it take before the car moves at the rate of 5 m/sec? Neglect the wind resistance and rotational effects of the wheels. The mechanical efficiency of the generators is 90%.

We have shown all the forces acting on the car in the diagram. *Newton's* law along the direction of the incline can be given as

$$W \sin 15° - f = M\frac{dV}{dt} \tag{a}$$

where f is the traction force developed by the generator action. Multiplying by V to get power, we get

$$W \sin 15° V - fV = MV\frac{dV}{dt} \tag{b}$$

If the mechanical efficiency of the generators (i.e., the power output divided by the power input) is .90, we can compute fV, which is the power delivered to the generators, in the following manner:

$$\frac{\text{generator output}}{.90} = fV \tag{c}$$

Hence,

$$fV = \frac{(50)(1000)}{.90} = 55{,}560 \text{ W} \tag{d}$$

Equation (b) can now be given as[4]

$$(20{,}000)(9.81)(.259)V - 55{,}560 = 20{,}000\ V\frac{dV}{dt}$$

Therefore,

$$2.54V - 2.78 = V\frac{dV}{dt} \tag{e}$$

We can separate the variables as follows:

$$dt = \frac{V\,dV}{2.54\ V - 2.78} \tag{f}$$

Integrating, using formulas in Appendix I, we get

$$t = \frac{1}{2.54^2}[2.54V - 2.78 + 2.78 \ln (2.54V - 2.78)] + C \tag{g}$$

To get the constant of integration C, note that when $t = 0$, $V = 3$ m/sec. Hence,

$$0 = \frac{1}{2.54^2}\{(2.54)(3) - 2.78 + 2.78 \ln [(2.54)(3) - 2.78]\} + C$$

Therefore,

$$C = -1.430$$

We thus have for Eq. (g):

$$t = \frac{1}{2.54^2}[(2.54V - 2.78) + 2.78 \ln (2.54V - 2.78)] - 1.430$$

[4]One watt is 1 J/sec, which in turn is 1 N-m/sec.

When $V = 5$ m/sec, we get for the desired value of t:

$$t = \frac{1}{2.54^2}\{(2.54)(5) - 2.78 + 2.78 \ln [(2.54)(5) - 2.78]\} - 1.430$$
$$= 1.0963 \text{ sec}$$

Problems

13.1. What value of constant force P is required to bring the 100-lb body, which starts from rest, to a velocity of 30 ft/sec in 20 ft? Neglect friction.

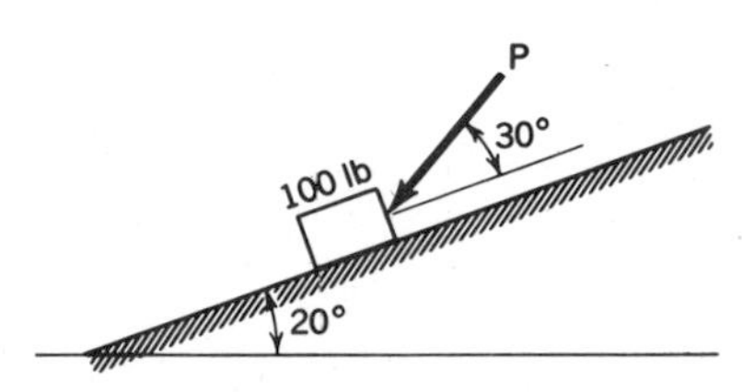

Figure P.13.1

13.2. A light cable passes over a frictionless pulley. Determine the velocity of the 100-lb block after it has moved 30 ft from rest. Neglect the inertia of the pulley.

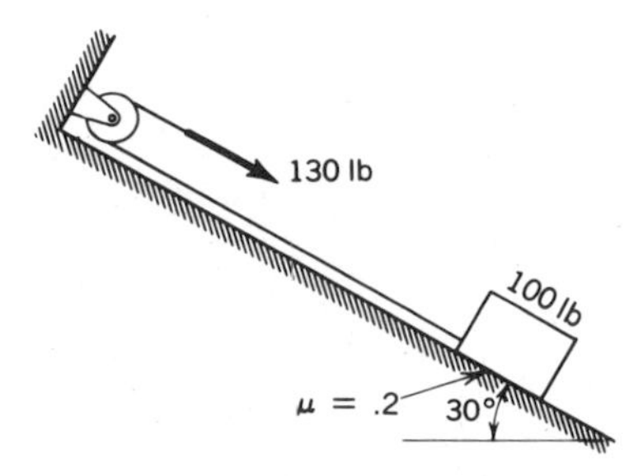

Figure P.13.2

13.3. In Problem 13.2, the pulley has a radius of 1 ft and has a resisting torque at the bearing of 10 lb-ft. Neglect the inertia of the pulley and the mass of the cable. Compute the kinetic energy of the 100-lb block after it has moved 30 ft from rest.

13.4. A light cable is wrapped around two drums fixed between a pair of blocks. The system has a mass of 50 kg. If a 250-N tension is exerted on the free end of the cable, what is the velocity change of the system after 3 m of travel down the incline? The body starts from rest. Take μ_d for all surfaces as .05.

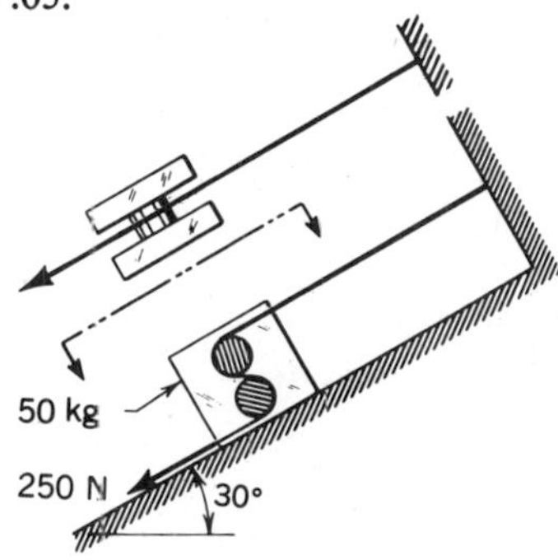

Figure P.13.4

13.5. A 50-kg mass on a spring is moved so that it extends the spring 50 mm from its unextended position. If the coefficient of friction between the mass and the supporting surface is .3,

(a) What is the velocity of the mass as it returns to the undeformed configuration of the spring?

(b) How far will the spring be compressed when the mass stops instantaneously before starting to the left?

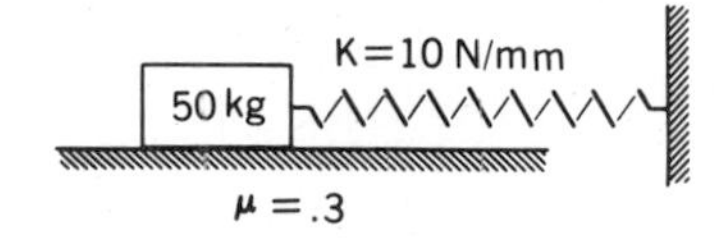

Figure P.13.5

13.6. A truck–trailer is shown carrying three crushed junk automobile cubes each weighing 2500 lb. An electromagnet is used to pick up the cubes as the truck moves by. Suppose the truck starts at position 1 by applying a constant 600 in.-lb total torque on the drive wheels. The magnet picks up only one cube C during the process. What will the velocity of the truck be when it has moved a total of 100 ft? The truck unloaded weighs 5000 lb and has a tire diameter of 18 in. Neglect the rotational effects of the tires and wind friction.

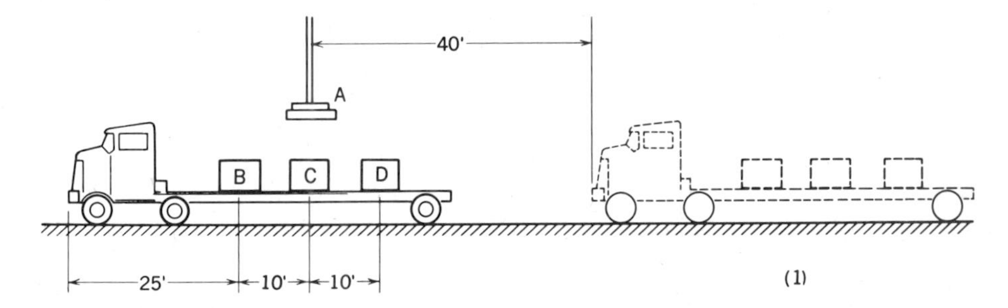

Figure P.13.6

13.7. Do Problem 13.6 if the first cube B and the last cube D are removed as they go by the magnet.

13.8. A passenger ferry is shown moving into its dock to unload passengers. As it approaches the dock, it has a speed of 3 knots (1 knot = .563 m/sec). If the pilot reverses his engines just as the front of the ferry comes abreast of the first pilings at A, what constant reverse thrust will stop the ferry just as it reaches the ramp B? The ferry weighs 4450 kN. Assume that the ferry does not hit the side pilings and undergoes no resistance from them. Neglect the drag of the water.

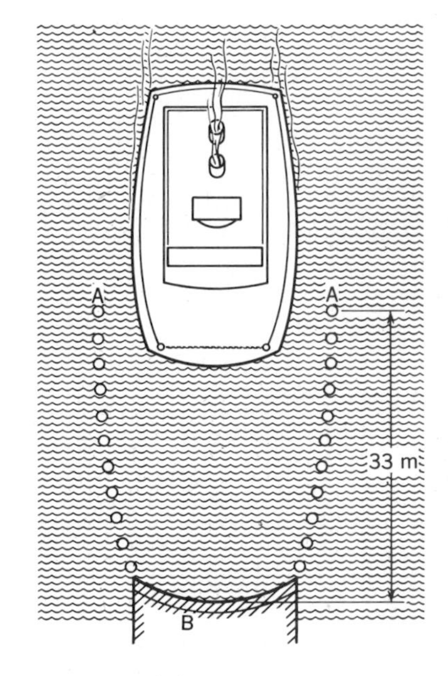

Figure P.13.8

13.9. Do Problem 13.8 assuming that the ferry rubs against the pilings as a result of a poor entrance and undergoes a resistance against its forward motion given as

$$f = 9(x + 50) \text{ N}$$

where x is measured in meters from the first pilings at A to the front of the ferry.

13.10. A freight car weighing 90 kN is rolling at a speed of 1.7 m/sec toward a spring-stop system. If the spring is nonlinear such that it develops a $.0450x^2$-kN force for a deflection of x mm, what is the maximum deceleration that the car A undergoes?

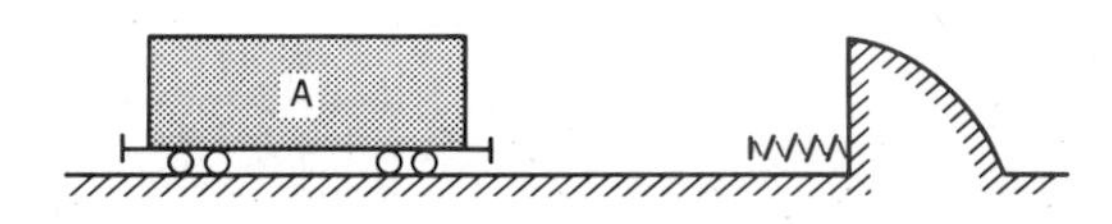

Figure P.13.10

13.11. A particle of mass 10 lbm is acted on by the following force field:

$$\boldsymbol{F} = 5x\boldsymbol{i} + (16 + 2y)\boldsymbol{j} + 20\boldsymbol{k} \text{ lb}$$

When it is at the origin, the particle has a velocity $\boldsymbol{V}_0$ given as

$$\boldsymbol{V}_0 = 5\boldsymbol{i} + 10\boldsymbol{j} + 8\boldsymbol{k} \text{ ft/sec}$$

What is its kinetic energy when it reaches position (20, 5, 10) while moving along a frictionless path? Does the shape of the path between the origin and (20, 5, 10) affect the result?

13.12. A plate AA is held down by screws C and D so that a force of 245 N is developed in each spring. Mass M of 100 kg is placed on plate AA and released suddenly. What is the maximum distance that plate AA descends if the plate can slide freely down the vertical guide rods? Take $K = 3600$ N/m.

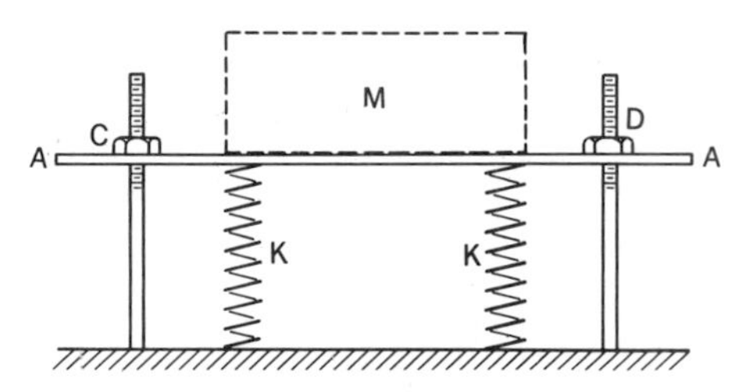

Figure P.13.12

13.13. A 200-lb block is dropped on the system of springs. If $K_1 = 600$ lb/ft and $K_2 = 200$ lb/ft, what is the maximum force developed on the body?

Figure P.13.13

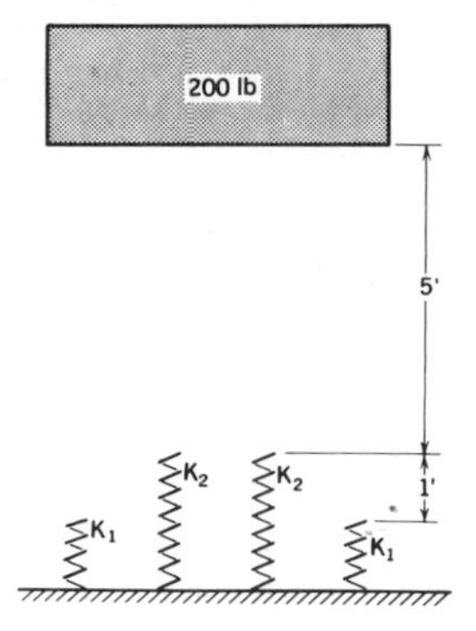

13.14. A block weighing 50 lb is shown on an inclined surface. The block is released at the position shown at a rest condition. What is the maximum compression of the spring? The spring has a spring constant K of 10 lb/in., and the coefficient of friction between the block and the incline is .3.

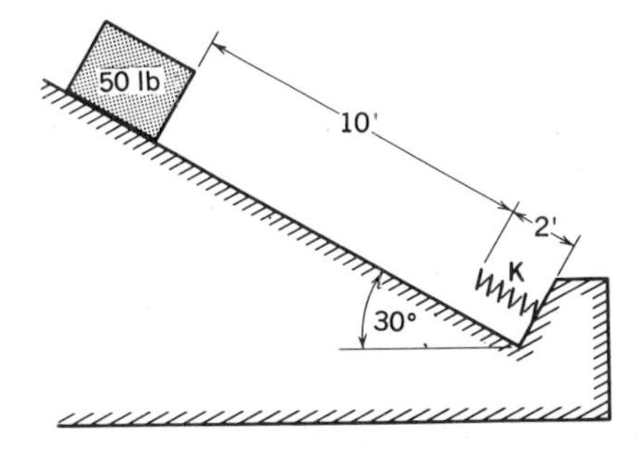

Figure P.13.14

13.15. A classroom demonstration unit is used to illustrate vibrations and interactions of bodies. Body A has a mass of .5 kg and is moving to the left at a speed of 1.6 m/sec at the position indicated. The body rides on a cushion of air supplied from the tube B through small openings in the tube. If there is a constant friction force of .1 N, what speed will A have when it returns to the position shown in the diagram? There are two springs at C, each having a spring constant of 15 N/m.

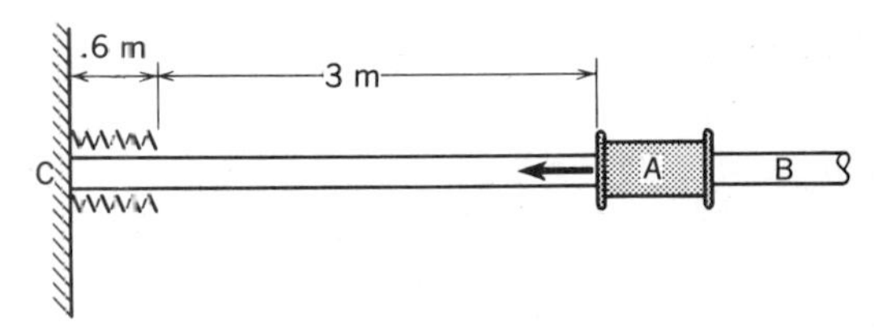

Figure P.13.15

13.16. An electron moves in a circular orbit in a plane at right angles to the direction of a uniform magnetic field B. If the strength of B is slowly changed so that the radius of the orbit is halved, what is the ratio of the final to the initial angular speed of the electron? Explain the steps you take. The force $\boldsymbol{F}$ on a charged particle is $q\boldsymbol{V} \times \boldsymbol{B}$, where q is the charge and $\boldsymbol{V}$ is the velocity of the particle.

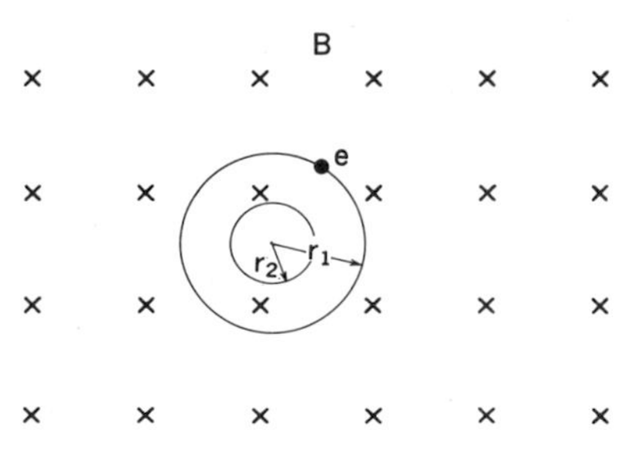

Figure P.13.16

13.17. A light rod CD rotates about pin C under the action of constant torque T of 1000 N-m. Body A having a mass of 100 kg slides on the horizontal surface for which the coefficient of friction is .4. If rod CD starts from rest, what angular speed is attained in one complete revolution? The entire weight of A is born by the horizontal surface.

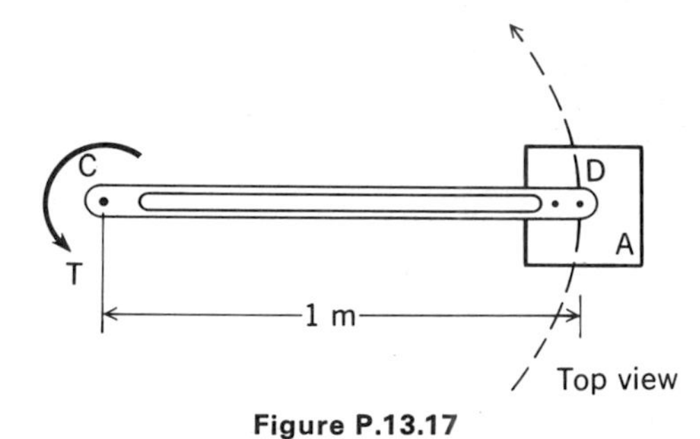

Figure P.13.17

13.18. An astronaut is attached to his orbiting space laboratory by a light wire. The astronaut is propelled by a small attached compressed air device. The propulsive force is in the direction of the man's height from foot to head. When the wire is extended its full length of 20 ft, the propulsion system is started, giving the astronaut a steady push of 5 lb. If this push is at right angles to the wire at all times, what speed will the astronaut have in one revolution about A? The weight on earth of the astronaut plus equipment is 250 lb. The mass of the laboratory is large compared to that of the man and his equipment.

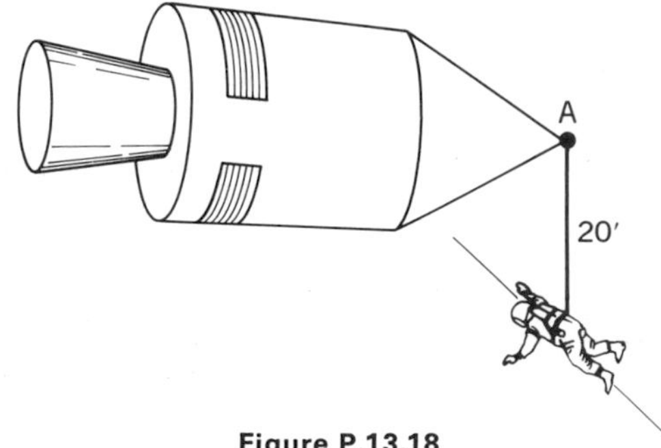

Figure P.13.18

13.19. Body A, having a mass of 100 kg, is connected to body B by an inextensible light cable. Body B has a mass of 80 kg and is on small wheels. The coefficient of friction between A and the horizontal surface is .2. If the system is released from rest, how far d must B move along the incline before reaching a speed of 2 m/sec?

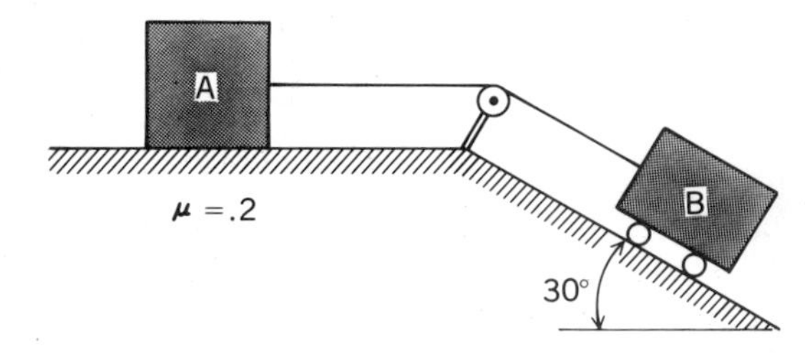

Figure P.13.19

13.20. A conveyor has drum D driven by a torque of 50 ft-lb. Bodies A and B on the conveyor each weigh 30 lb. The coefficient of friction between the conveyor belt and the conveyor bed is .2. If the conveyor starts from rest, how fast along the conveyor do A and B move after traveling 2 ft? Drum C rotates freely, and the tension in the belt on the underside of the conveyor is 20 lb. The diameter of both drums is 1 ft. Neglect the mass of drums and belt.

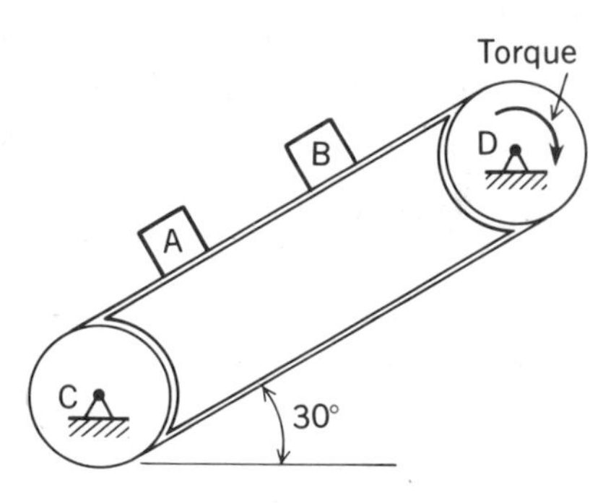

Figure P.13.20

13.21. Bodies A and B are connected to each other through two light pulleys. Body A has a mass of 500 kg, whereas body B has a mass of 200 kg. A constant force F of value 10,000 N is applied to body A whose surface of contact has a dynamic coefficient of friction equal to .4. If the system starts from rest, what distance d does B ascend before it has a speed of 2 m/sec? (*Hint:* Considering pulley E, we have instantaneous rotation about point e. Hence, $V_b = \frac{1}{2}V_c$.)

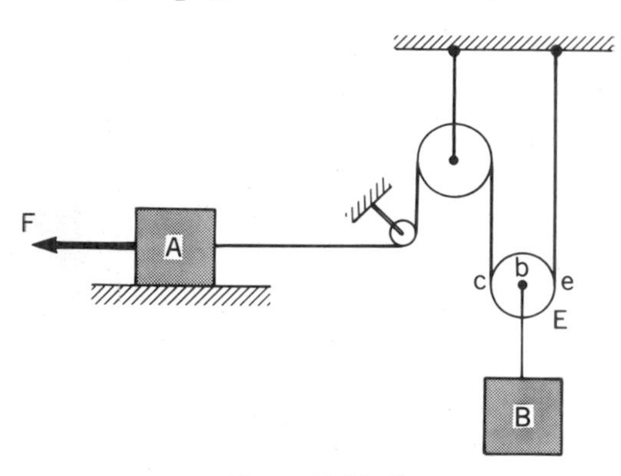

Figure P.13.21

13.22. A rope tow for skiers is shown pulling 20 skiers up a 20° incline. The driving pulley A has a diameter of 5 ft. The idler pulley B rotates freely. The system has been stopped to allow a fallen skier to untangle himself. The driving pulley starts from rest and is given a torque of 2000 ft-lb. With this torque, what distance d do skiers move before their speed is 15 ft/sec? The tension on the slack side of the tow can be taken as zero. The coefficient of friction between skis and slope is .15 and the average weight of the skiers is 150 lb. Neglect the mass of the rope and the pulleys.

Figure P.13.22

13.23. A uniform block A has a mass of 25 kg. The block is hinged at C and is supported by a small block B as shown in the diagram. A constant force F of 400 N is applied to block B. What is the speed of B after it moves 1.6 m? The mass of block B is 2.5 kg and the coefficient of friction for all contact surfaces is .3.

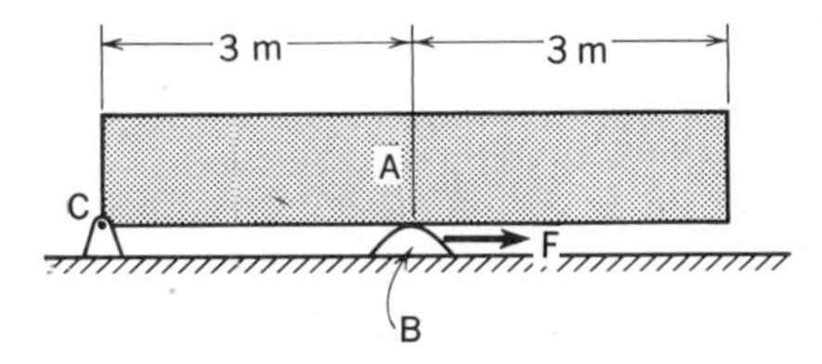

Figure P.13.23

13.24. A triangular block of uniform density and total weight 100 lb rests on a hinge and on a movable block B. If a constant force F of 150 lb is exerted on the block B, what will be its speed after it moves 10 ft? The mass of block B is 10 lbm, and the coefficient of friction for all contact surfaces is .3.

Figure P.13.24

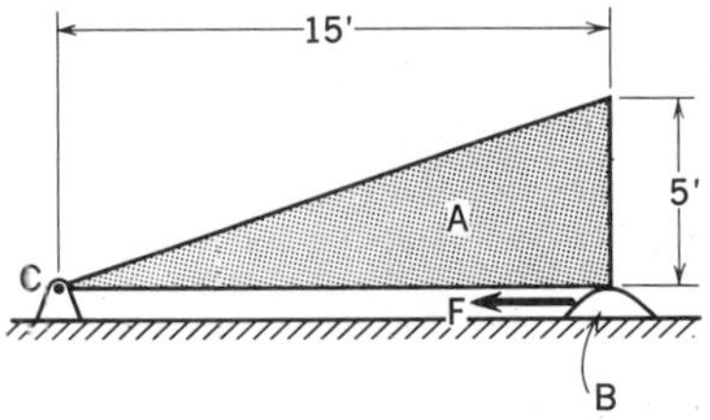

13.25. Block A weighs 200 lb and block B weighs 150 lb. If the system starts from rest, what is the speed of block B after it moves 1 ft? Neglect the weight of the pulleys.

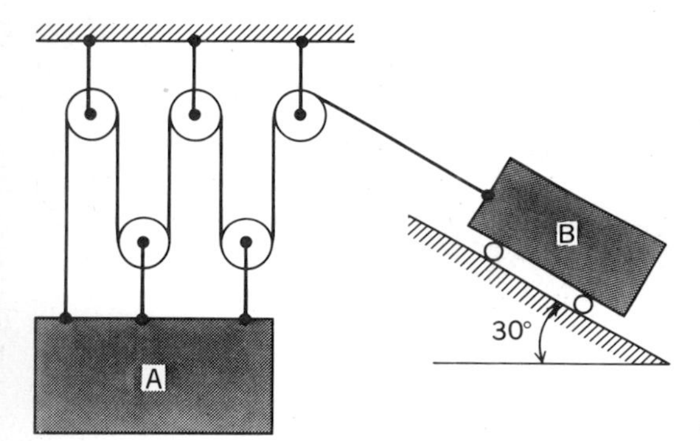

Figure P.13.25

13.26. A weight W is to be lowered by a man. He lets the rope slip through his hands while maintaining a tension of 130 N on the rope. What is the maximum weight W that he can handle if the weight is not to exceed a speed of 5 m/sec starting from rest and dropping 3 m? Use the coefficients of friction shown in the diagram. Neglect the mass of the rope. There are three wraps of rope around the post.

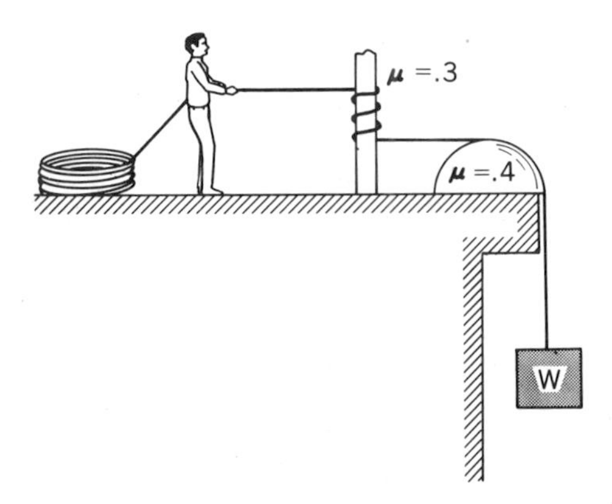

Figure P.13.26

13.27. A vehicle B is being let down a 30° incline. The vehicle is attached to a weight A that restrains the motion. Vehicle B weighs 2000 lb. What should the minimum restraining weight A be if, after starting from rest, the system does not exceed 8 ft/sec after moving 10 ft? There are two wraps around the post.

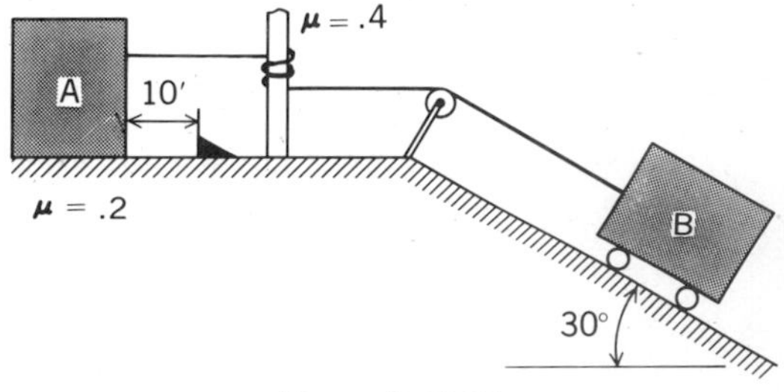

Figure P.13.27

***13.28.** A spiral path is given parametrically in terms of the parameter τ as follows:

$$x_p = A \sin \eta\tau \text{ ft}$$
$$y_p = A \cos \eta\tau \text{ ft}$$
$$z_p = C\tau \text{ ft}$$

where A, η, and C are known constants. A particle P of mass 1 lbm is released from a position of rest 1 ft above the xy plane. The particle is constrained by a spring ($K = 2$ lb/ft) coiled around the path. The spring is unstretched when P is released. Neglect friction and find how far P drops. Take $\eta = \pi/2$, $A = C = 1$.

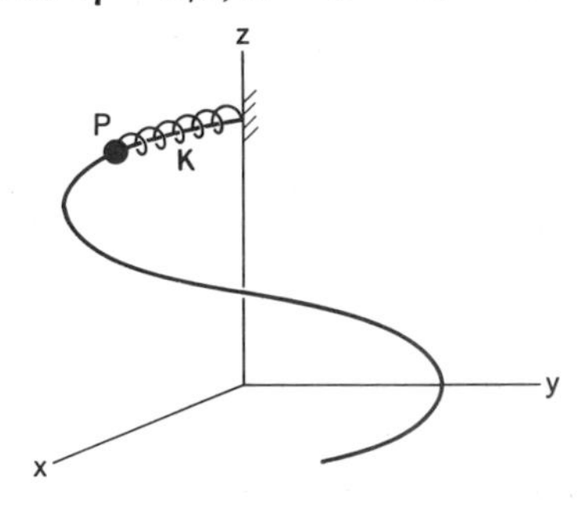

Figure P.13.28

13.29. A body A of mass 1 lbm is moving at time $t = 0$ with a speed V of 1 ft/sec on a smooth cylinder as shown. What is the speed of the body when it arrives at B? Take $r = 2$ ft.

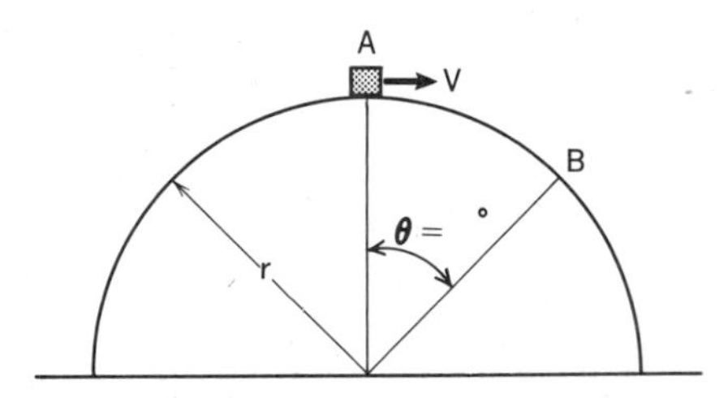

Figure P.13.29

***13.30.** Set up an integro-differential equation (involving derivatives and integrals) for θ in Problem 13.29 if there is Coulombic friction with $\mu = .2$.

13.31. At what angle θ does body A of Problem 13.29 leave the circular surface?

***13.32.** Show that the work–energy equation for a particle can be expressed in the following way:

$$\int_0^x F\,dx = \int_0^V V\,d(mV)$$

Integrating the right side by parts,[5] and using relativistic mass $m_0/\sqrt{1 - V^2/c^2}$, where m_0 is the *rest mass* and c is the speed of light, show that a relativistic form of this equation can be given as

$$\int_0^x F\,dx = \frac{m_0c^2}{\sqrt{1 - V^2/c^2}} - m_0c^2$$
$$= mc^2 - m_0c^2$$

so that the *relativistic kinetic energy* is

$$\text{KE} = mc^2 - m_0c^2$$

***13.33.** By combining the kinetic energy as given in Problem 13.32 and m_0c^2 to form E, the *total energy*, we get the famous formula of Einstein:

$$E = mc^2$$

in which energy is equated with mass. How much energy is equivalent to 6×10^{-8} lbm of matter? How high could a weight of 100 lb be lifted with such energy?

13.34. A 100-lb boy climbs up a rope in gym in 10 sec and slides down in 4 sec after he reaches uniform speed downward. What is the horsepower developed by the boy going up? What is the average horsepower dissipated on the rope by the boy going down after reaching uniform speed? The distance moved before reaching uniform speed downward is 2 ft.

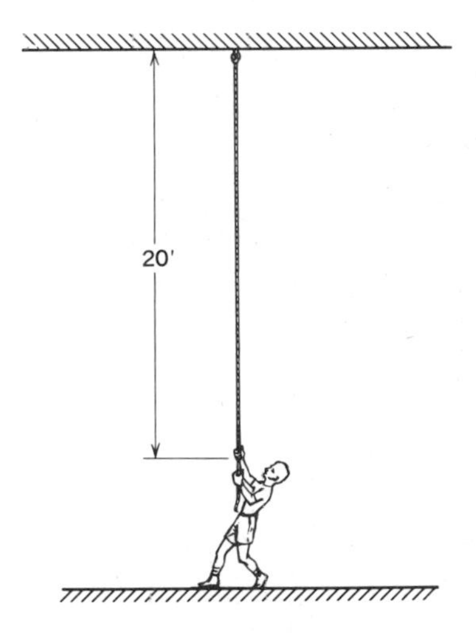

Figure P.13.34

13.35. An aircraft carrier is shown in the process of launching an airplane via a catapult mechanism. Before leaving the catapult, the plane has a speed of 192 km/hr relative to the ship. If the plane is accelerating at the rate of $1g$ and if it has a mass of 18,000 kg, what horsepower is being developed by the catapult system at the end of launch on the plane if we neglect drag? The thrust from the jet engines of the plane is 100,000 N.

---

[5] To *integrate by parts*, note that

$$d(uv) = u\,dv + v\,du$$

Now integrate these terms:

$$\int_1^2 d(uv) = \int_1^2 u\,dv + \int_1^2 v\,du$$

Therefore,

$$\int_1^2 u\,dv = (uv)\Big|_1^2 - \int_1^2 v\,du$$

The last formulation is called integration by parts.

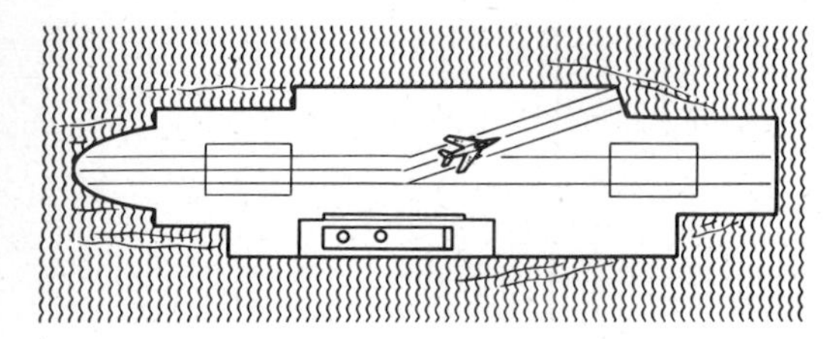

Figure P.13.35

13.36. An automobile engine under test is rotating at 4400 rpm and develops a torque on a dynamometer of 40 N-m. What is the horsepower developed by the engine? If the dynamometer has a mechanical efficiency of .90, what is the kilowatt output of the generator? (*Hint:* The work of a torque equals the torque times the angle of rotation in radians.)

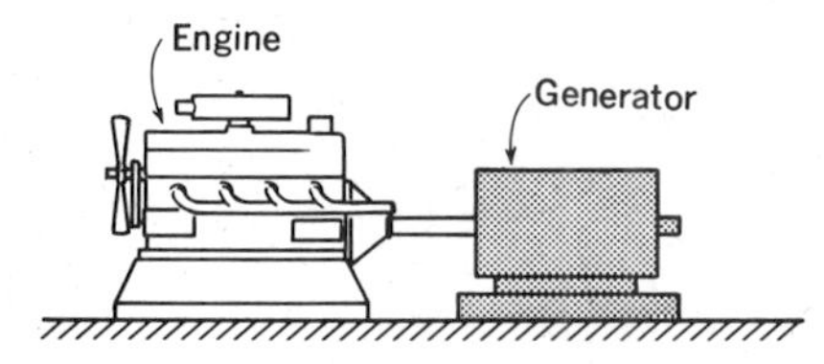

Figure P.13.36

13.37. A rocket is undergoing static thrust tests in a test stand. A thrust of 300,000 lb is developed while 300 gal of fuel (specific gravity .8) is burned per second. The exhaust products of combustion have a speed of 5000 ft/sec relative to the rocket. What power is being developed on the rocket? What is the power developed on the exhaust gases? (1 gal = .1337 ft³.)

13.38. A 15-ton streetcar accelerates from rest at a constant rate a_0 until it reaches a speed V_1, at which time there is zero acceleration. The wind resistance is given as κV^2. Formulate expressions for power developed for the stated ranges of operation.

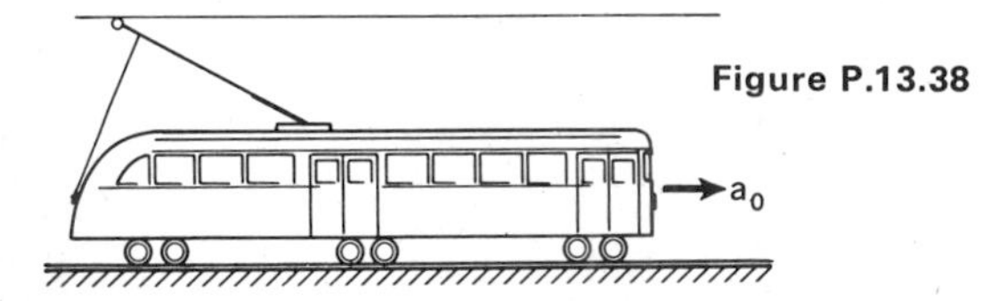

Figure P.13.38

13.39. What is the maximum horsepower that can be developed on a streetcar weighing 133.5 kN? The car has a coefficient of static friction of .20 between wheels and rail and a drag given as $32V^2$ N, where V is in m/sec. All wheels are drive wheels.

13.40. A 7500-kg streetcar starts from rest when the conductor draws 5 kW of power from the line. If this input is maintained constant and if the mechanical efficiency of the motors is 90%, how long does the streetcar take to reach a speed of 10 km/hr? Neglect wind resistance. (1 kW = 1.341 hp.)

13.41. A children's boat ride can be found in many amusement parks. Small boats each weighing 100 lb are rotated in a tank of water. If the system is rotating with a speed $\dot{\theta}$ of 10 rpm, what is the kinetic energy of the system? Assume that each boat has two 60-lb children on board and that the kinetic energy of the supporting structure can be accounted for by "lumping" an additional 30 lbm into each boat. If a wattmeter indicates that 4 kW of power is being absorbed by the motor turning the system, what is the drag for each boat? Take the mechanical efficiency of the motor to be 80%. (1 kW = 1.341 hp.)

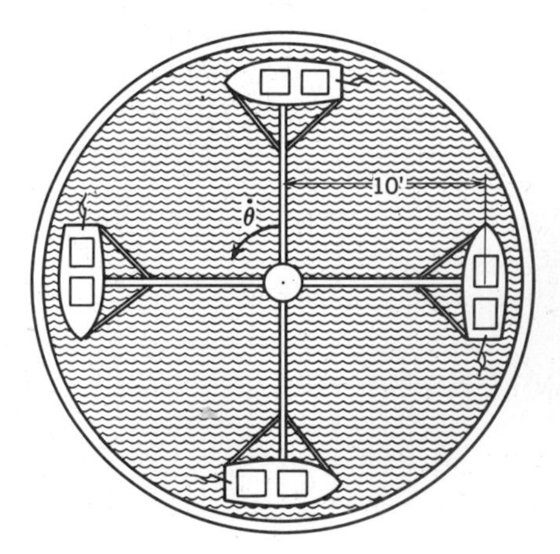

Figure P.13.41

13.42. A 180-lb man runs up an escalator while it is not in operation in 10 sec. What is the horsepower developed by the man? If the escalator is moving at a speed of 2 ft/sec and carrying, on the average, 2000 people per hour, what is the power requirement on the driving motor assuming that the average weight of a passenger is 150 lb? Take the mechanical efficiency of the drive system to be 80%. Assume that passengers enter and leave at the same speed of 2 ft/sec and that there are equal numbers of passengers on the escalator at any one time.

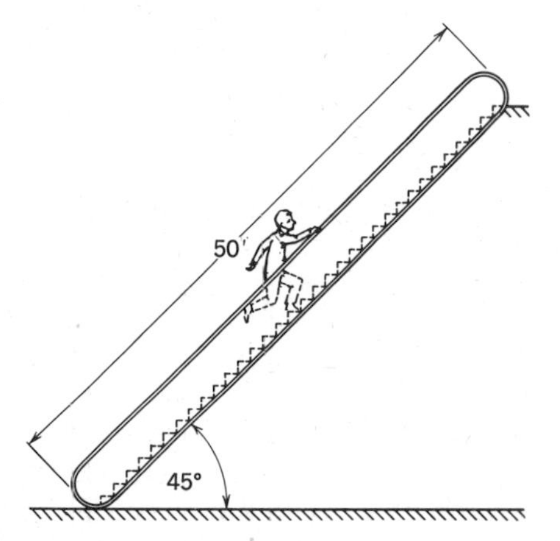

Figure P.13.42

13.43. Grain is coming out of a hopper at the rate of 7200 kg/hr and falls onto a conveyor system that takes the grain into a bin. The conveyor belt moves at a steady speed of 2 m/sec. What power in watts is needed to operate the system for a efficiency of .6? What power is needed if we double the belt speed?

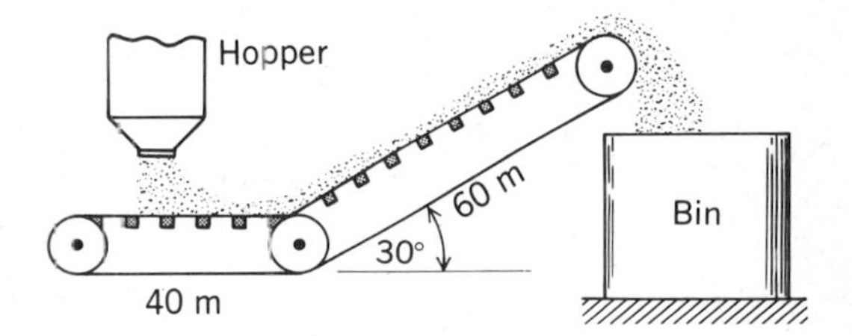

Figure P.13.43

13.44. A self-propelled vehicle A has a weight of $\frac{1}{4}$ ton. A gasoline engine develops torque on the drive wheels to help move A up the incline. A counterweight B of 300 lb is also shown in the diagram. What horsepower is needed when A is moving up at a speed of 2 ft/sec and has an acceleration of 3 ft/sec²? Neglect the weight of the pulley. (*Hint:* The pulley rolls along cord dg without slipping. It therefore has an instantaneous center of rotation at d. What does this mean about the relative value of velocity of point b on the pulley and point a?)

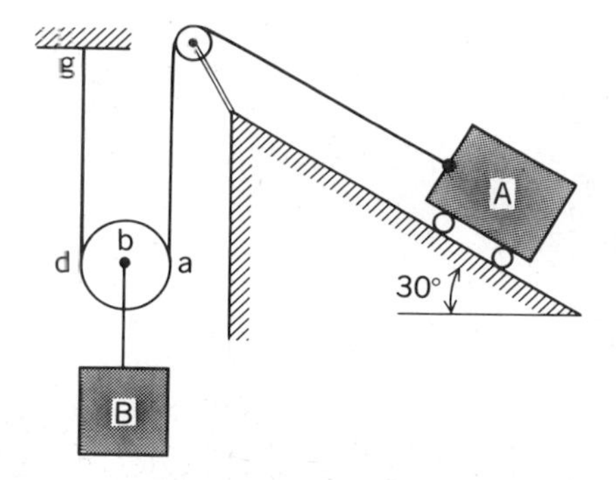

Figure P.13.44

13.3 Conservative Force Field

In Section 10.9, we discussed an important class of forces called conservative forces. For convenience, we shall now repeat this discussion.

Consider first a body acted on only by gravity W as an active force (i.e., a force that can do work) and moving along a frictionless path from position 1 to position 2, as shown in Fig. 13.9. The work done by gravity W_{1-2} is then

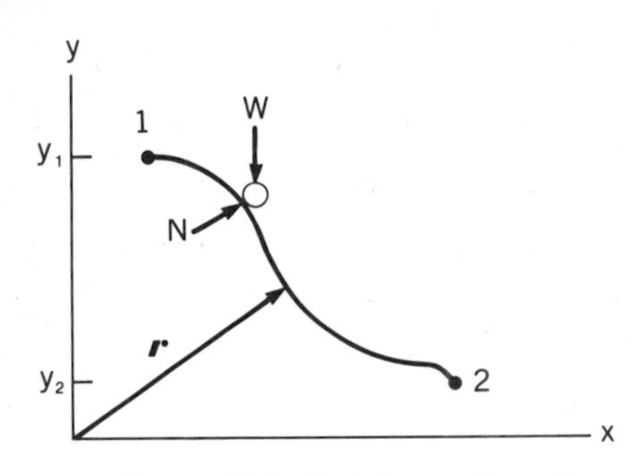

Figure 13.9. Particle moving along frictionless path.

$$W_{1-2} = \int_1^2 \boldsymbol{F} \cdot d\boldsymbol{r} = \int_1^2 (-W\boldsymbol{j}) \cdot d\boldsymbol{r}$$
$$= -W \int_1^2 dy = -W(y_2 - y_1) \tag{13.6}$$
$$= W(y_1 - y_2)$$

Note that the work done *does not depend* on the path, but depends only on the positions of the end points of the path. *Force fields whose work like gravity is independent of the path are called conservative force fields.* In general, we can say for conservative force field $\boldsymbol{F}(x, y, z)$ that, along a path between positions 1 and 2, the work is

$$W_{1-2} = \int_1^2 \boldsymbol{F} \cdot d\boldsymbol{r} = V_1(x, y, z) - V_2(x, y, z) \tag{13.7}$$

where V is a function of position of the end points and is called the *potential energy function.*[6] We may rewrite Eq. 13.7 as follows:

$$-\int_1^2 \boldsymbol{F} \cdot d\boldsymbol{r} = V_2(x, y, z) - V_1(x, y, z) = \Delta V \tag{13.8}$$

Note that the potential energy, $V(x, y, z)$, depends on the reference xyz used or, as we shall often say, the *datum* used. However, the *change* in potential energy, ΔV, is *independent* of the datum used.[7] Since we shall be using the change in potential energy, the datum is arbitrary and is chosen for convenience. From Eq. 13.8, we can say that *the change in potential energy*, ΔV, of a force field is *the negative of the work done by this force field on a particle in going from position 1 to position 2 along any path.* For any *closed* path, clearly the work done by a conservative force field $\boldsymbol{F}$ is

$$\oint \boldsymbol{F} \cdot d\boldsymbol{r} = 0 \tag{13.9}$$

How is the potential energy function V related to $\boldsymbol{F}$? To answer this query, consider that an infinitesimal path $d\boldsymbol{r}$ starts from point 1. We can than give Eq. 13.8 as

$$\boldsymbol{F} \cdot d\boldsymbol{r} = -dV \tag{13.10}$$

Expressing the dot product on the left side in terms of components, and expressing dV as a total differential, we get

$$F_x\, dx + F_y\, dy + F_z\, dz = -\left(\frac{\partial V}{\partial x} dx + \frac{\partial V}{\partial y} dy + \frac{\partial V}{\partial z} dz\right) \tag{13.11}$$

---

[6]We shall also use the notation P.E. or simply PE for V. Note that the letter V has also been used for volume and velocity. The context of discussion should make clear what V represents. We use this notation since it is very common.

[7]Thus, considering Eq. 13.6, the value of y itself for a particle at any time depends on the position of the origin O of the xyz reference. However, changing the position of O but keeping the same direction of the xyz axes (i.e., *changing the datum*) does not affect the value of $y_2 - y_1$.

We can conclude from this equation that

$$F_x = -\frac{\partial V}{\partial x}$$
$$F_y = -\frac{\partial V}{\partial y} \qquad (13.12)$$
$$F_z = -\frac{\partial V}{\partial z}$$

In other words,

$$\begin{aligned} \boldsymbol{F} &= -\left(\frac{\partial V}{\partial x}\boldsymbol{i} + \frac{\partial V}{\partial y}\boldsymbol{j} + \frac{\partial V}{\partial z}\boldsymbol{k}\right) \\ &= -\left(\frac{\partial}{\partial x}\boldsymbol{i} + \frac{\partial}{\partial y}\boldsymbol{j} + \frac{\partial}{\partial z}\boldsymbol{k}\right)V \qquad (13.13) \\ &= -\textbf{grad}\ V = -\boldsymbol{\nabla} V \end{aligned}$$

The operator **grad** or **∇** that we have introduced is called the *gradient* operator[8] and is given as follows for rectangular coordinates:

$$\textbf{grad} \equiv \boldsymbol{\nabla} \equiv \left(\frac{\partial}{\partial x}\boldsymbol{i} + \frac{\partial}{\partial y}\boldsymbol{j} + \frac{\partial}{\partial z}\boldsymbol{k}\right) \qquad (13.14)$$

We can now say as an alternative definition that a *conservative force field must be a function of position and expressible as the gradient of a scalar function.* The *inverse* to this statement is also valid. That is, *if a force field is a function of position and the gradient of a scalar field, it must then be a conservative force field.*

Two examples of conservative force fields will now be presented and discussed.

Constant Force Field. If the force field is constant at all positions, it can always be expressed as the gradient of a scalar function of the form $V = -(ax + by + cz)$, where a, b, and c are constants. The constant force field, then, is $\boldsymbol{F} = a\boldsymbol{i} + b\boldsymbol{j} + c\boldsymbol{k}$.

In limited changes of position near the earth's surface (a common situation), we can consider the gravitational force on a particle of mass, m, as a constant force field given by $-mg\boldsymbol{k}$ (or $-W\boldsymbol{k}$). Thus, the constants for the general force field given above are $a = b = 0$ and $c = -mg$. Clearly, $V = mgz$ for this case.

Force Proportional to Linear Displacements. Consider a body limited by constraints to move along a straight line. Along this line is developed a force directly proportional to the displacement of the body from some position O at $x = 0$ along the line. Furthermore, this force is always directed toward point O; it is then termed a *restoring* force. We can give this force as

$$\boldsymbol{F} = -Kx\boldsymbol{i} \qquad (13.15)$$

---

[8]The gradient operator comes up in many situations in engineering and physics. In short, the gradient represents a *driving action*. Thus, in the present case, the gradient is a driving action to cause mass to move. And, the gradient of temperature causes heat to flow. Finally, the gradient of electric potential causes electric charge to flow.

where x is the displacement from point O. An example of this force is that of the linear spring (Fig. 13.10) discussed in Section 12.3. The potential energy of this force field is

$$V = \frac{Kx^2}{2} \tag{13.16}$$

What is the physical meaning of the term V? Note that the change in potential energy has been defined (see Eq. 13.8) as the *negative* of the work done by a conservative force as the particle on which it acts goes from one position to another. Clearly, the change in the potential energy is then *directly equal* to the work done by the *reaction* to the conservative force during this displacement. In the case of the *spring*, the reaction force would be the force *from* the surroundings acting *on* the spring at point B (Fig. 13.10). During extension or compression of the spring from the undeformed position, this force (from the surroundings) does a *positive* amount of work. This work can be considered as a measure of the energy *stored* in the spring. Why? Because when allowed to return to its original position, the spring will do this amount of positive work *on* the surroundings at B, provided that the return motion is slow enough to prevent oscillations; and so on. Clearly then, since V equals work of the surroundings on the spring, then V is in effect the stored energy in the spring. In a general case, V is the energy stored in the force field as measured from a given datum.

In previous chapters, several additional force fields were introduced: the gravitational central force field, the electrostatic field, and the magnetic field. Let us see which we can add to our list of conservative force fields.

Consider first the central gravitational force field where particle m, shown in Fig. 13.11, experiences a force given by the equation

$$\mathbf{F} = -G\frac{Mm}{r^2}\hat{\mathbf{r}} \tag{13.17}$$

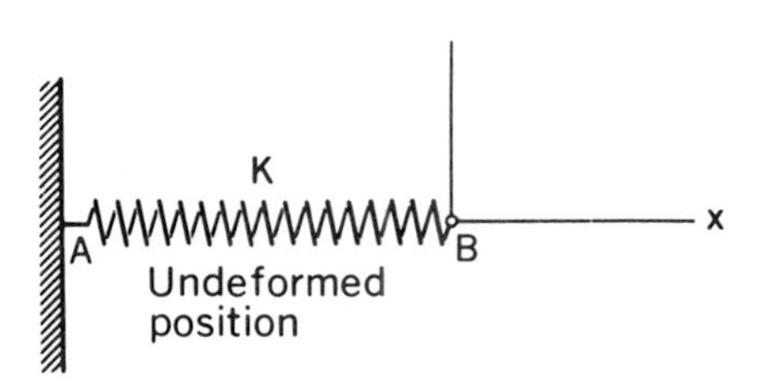

Figure 13.10. Linear spring.

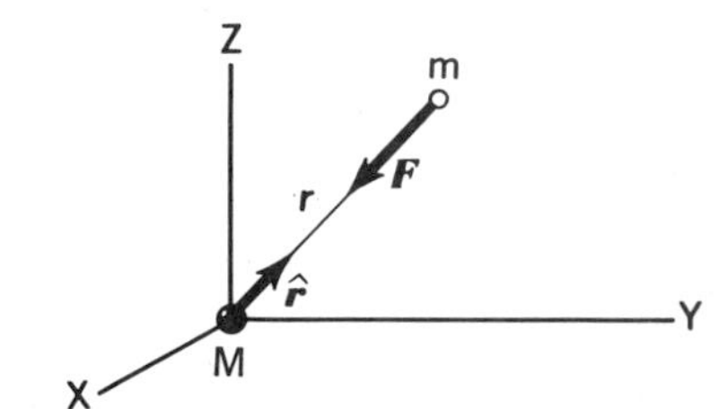

Figure 13.11. Central force on *m*.

Clearly, this force field is a function of spatial coordinates and can easily be expressed as the gradient of a scalar function in the following manner:

$$\mathbf{F} = -\mathbf{grad}\left(-\frac{GMm}{r}\right) \tag{13.18}$$

Hence, this is a conservative force field. The potential energy is then

$$PE = -\frac{GMm}{r} \tag{13.19}$$

Next, the force on a particle of unit positive charge from a particle of charge q_1 is given by Coulomb's law as

$$E = \frac{q_1}{4\pi\epsilon_c r^2}\hat{r} \tag{13.20}$$

Since this equation has the same form as Eq. 13.17 (i.e., is also a function of $1/r^2$) we see immediately that the force field from q_1 is conservative. The potential energy per unit charge is then

$$PE = \frac{q_1}{4\pi\epsilon_0 r} \tag{13.21}$$

The remaining field introduced was the magnetic field where $\boldsymbol{F} = q\boldsymbol{V} \times \boldsymbol{B}$. For this field, the force on a charged particle depends on the velocity of the particle. The condition that the force be a function of position is not satisfied, therefore, and the magnetic field does *not* form a conservative force field.

13.4 Conservation of Mechanical Energy

Let us now consider the motion of a particle upon which only a conservative force field does work. We start with Eq. 13.2:

$$\int_{r_1}^{r_2} \boldsymbol{F} \cdot d\boldsymbol{r} = \tfrac{1}{2}mV_2^2 - \tfrac{1}{2}mV_1^2 \tag{13.22}$$

Using the definition of potential energy, we replace the left side of the equation in the following manner:

$$(\text{PE})_1 - (\text{PE})_2 = \tfrac{1}{2}mV_2^2 - \tfrac{1}{2}mV_1^2 \tag{13.23}$$

Rearranging terms, we reach the following useful relation:

$$\boxed{(\text{PE})_1 + \tfrac{1}{2}mV_1^2 = (\text{PE})_2 + \tfrac{1}{2}mV_2^2} \tag{13.24}$$

Since positions 1 and 2 are arbitrary, obviously *the sum of the potential energy and the kinetic energy for a particle remains constant at all times during the motion of the particle.* This statement is sometimes called the *law of conservation of mechanical energy for conservative systems*. The usefulness of this relation can be demonstrated by the following examples.

EXAMPLE 13.5

A particle is dropped with zero initial velocity down a frictionless chute (Fig. 13.12). What is the magnitude of its velocity if the vertical drop during the motion is h ft?

For small trajectories, we can assume a uniform force field $-mg\boldsymbol{j}$. Since this is the only force that can perform work on the particle (the normal force from the chute does no work), we can employ the conservation-of-mechanical-energy equation. If we take position 2 as a datum, we then have from Eq. 13.24:

$$mgh + 0 = 0 + m\frac{V_2^2}{2}$$

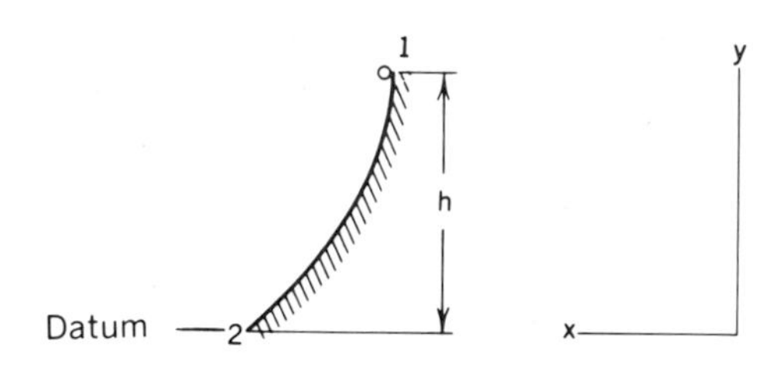

Figure 13.12. Particle on frictionless shute.

Solving for V_2, we get

$$V_2 = \sqrt{2gh}$$

The advantages of the energy approach for conservative fields become apparent from this problem. That is, not all the forces need be considered in computing velocities, and the path, however complicated, is of no concern. If friction were present, a nonconservative force would perform work, and we would have to go back to the general relation given by Eq. 13.2 for the analysis.

EXAMPLE 13.6

A mass is dropped onto a spring that has a spring constant K and a negligible mass (see Fig. 13.13). What is the maximum deflection δ?

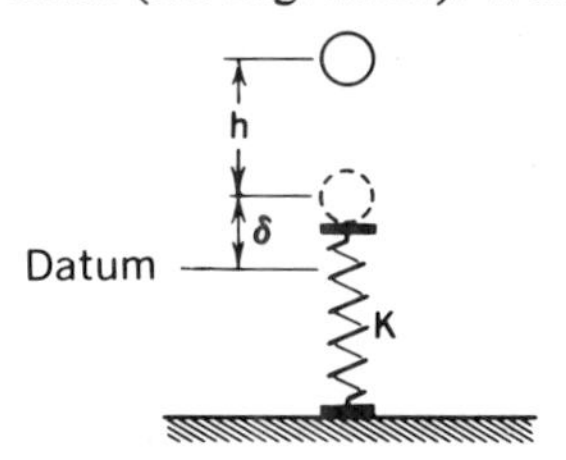

Figure 13.13. Mass dropped on spring.

In this problem, only conservative forces act on the body as it falls. Using the lowest position of the body as a datum, we see that the body falls a distance $h + \delta$. We shall equate the mechanical energies at the uppermost and lowest positions of the body. Thus,

$$\underbrace{mg(h+\delta)}_{\text{PE gravity}} + \underbrace{0}_{\text{PE spring}} + \underbrace{0}_{\text{KE}} = \underbrace{0}_{\text{PE gravity}} + \underbrace{\tfrac{1}{2}K\delta^2}_{\text{PE spring}} + \underbrace{0}_{\text{KE}} \tag{a}$$

Rearranging the terms,

$$\delta^2 - \frac{2mg}{K}\delta - \frac{2mgh}{K} = 0 \tag{b}$$

We may solve for a physically meaningful δ from this equation by using the quadratic formula.

EXAMPLE 13.7

A block A of mass 200 g slides on a frictionless surface as shown in Fig. 13.14. The spring constant K_1 is 25 N/m and initially, at the position shown, it is stretched .40 m. An elastic cord connects the top support to point C on A. It has a spring constant K_2 of 10.26 N/m. Furthermore, the cord disconnects from C at the instant that C reaches point G. If A is released from rest at the indicated position, what value of θ corresponds to the position B where A just loses contact with the surface? The elastic cord is initially unstretched.

We have conservative forces peforming work on A so we have *conservation of mechanical energy*. Using the datum at B and using l_0 as the unstretched length of the spring, we then say that

$$mgz_1 + \frac{mV_1^2}{2} + \frac{1}{2}K_1\,\delta_1^2 = mgz_2 + \frac{mV_2^2}{2} + \frac{1}{2}K_1\,\delta_2^2 + \frac{1}{2}K_2(\overline{CG})^2$$

where the last term is the energy in the elastic cord when it disconnects at G. Therefore,

$$\begin{aligned}(.200)(9.81)[(.92)(.707) + (.92)(.707) + (.92)\sin\theta] + 0 + \tfrac{1}{2}(25)(.40)^2 \\ = 0 + \tfrac{1}{2}(.20)V_2^2 + \tfrac{1}{2}(25)(.92 - l_0)^2 + \tfrac{1}{2}(10.26)(.94)^2\end{aligned} \tag{a}$$

To get l_0, examine the initial configuration of the system. With an initial stretch of .40 m for the spring, we can say:

$$\begin{aligned}l_0 &= [(.92)(.707) + (.92)(.707)] - .40 \\ &= .901 \text{ m}\end{aligned}$$

Equation (a) can then be written as

$$V_2^2 = .1490 + 18.05\sin\theta \tag{b}$$

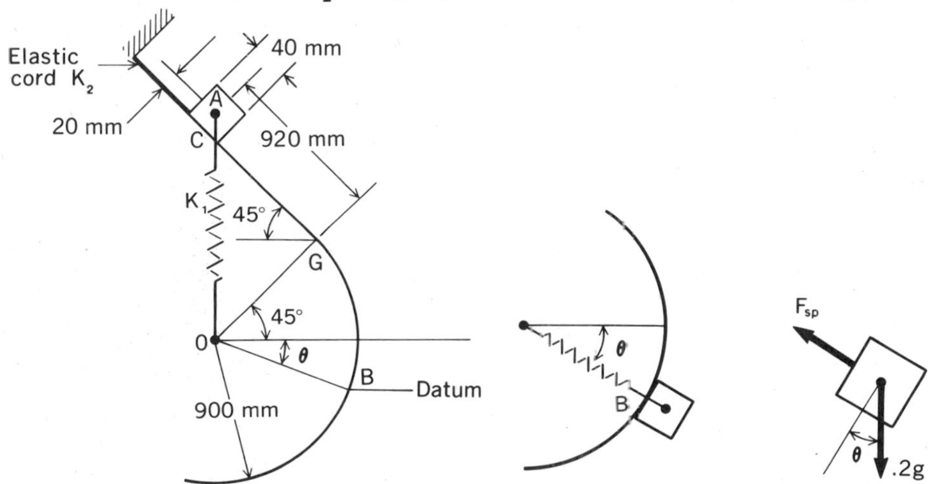

Figure 13.14. Mass *A* slides along frictionless surface.

Figure 13.15. Contact is first lost at θ.

We now use *Newton's law* at the point of interest B where A just loses contact. This condition is shown in detail in Fig. 13.15, where you will notice that the contact force N has been taken as zero and thus deleted from the free body diagram. In the radial direction, we have using path coordinates

$$-F_{sp} + (.200)(g)\sin\theta = -.200\left(\frac{V_2^2}{.92}\right)$$

Therefore,

$$-(25)(.92 - .901) + (.200)(9.81)\sin\theta = -\frac{V_2^2}{4.60}$$

This equation can be written as

$$V_2^2 = 2.20 - 9.03 \sin\theta \qquad \text{(c)}$$

Solving Eqs. (b) and (c) simultaneously for θ, we get

$$\theta = 4.34°$$

13.5 Alternative Form of Work–Energy Equation

With the aid of the material in Section 13.4, we shall now set forth an alternative energy equation which has much physical appeal and which resembles the *first law of thermodynamics* as used in other courses. Let us take the case where certain of the forces acting on a particle are conservative while others are not. Remember that for conservative forces the negative of the change in potential energy between positions 1 and 2 equals the work done by these forces as the particle goes from position 1 to position 2 along any path. Thus, we can restate Eq. 13.2 in the following way:

$$\int_1^2 \boldsymbol{F} \cdot d\boldsymbol{r} - \Delta(\text{PE})_{1,2} = \Delta(\text{KE})_{1,2} \qquad (13.25)$$

where the integral represents the work of the *nonconservative* forces. Calling this integral W_{1-2}, we then have, on rearranging the equation:

$$\boxed{\Delta(\text{KE} + \text{PE}) = W_{1-2}} \qquad (13.26)$$

In this form, we say that the work of *nonconservative* forces goes into changing the kinetic energy plus the potential energy for the particle. Since potential energies of such common forces as linear restoring forces, coulombic forces, and gravitational forces are so well known, the formulation above is useful in solving problems if it is understood thoroughly and applied properly.[9]

EXAMPLE 13.8

Three coupled streetcars (Fig. 13.16) are moving at the speed of 32 km/hr down a 7° incline. Each car has a weight of 198 kN. The cars must stop within 50 m beyond the position where the brakes are fully applied so as to cause the wheels to lock.

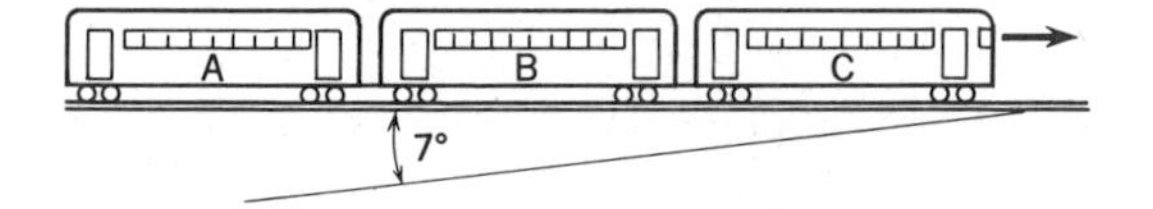

Figure 13.16. Coupled streetcars.

What is the maximum number of brake failures that can be tolerated and still satisfy this specification? Assume that the weight is loaded equally among the wheels and that we have 24 brake systems, one for each wheel. Take $\mu_d = .45$.

---

[9]Equation 13.26, you may notice, is actually a form of the first law of thermodynamics for the case of no heat transfer.

The friction force f on any one wheel where the brake has operated is ascertained from Coulomb's law as

$$f = \frac{198{,}000 \cos 7^\circ}{8}(.45) = 11{,}050 \text{ N}$$

We now consider the work–energy relation 13.26 for the case where a minimum number of good brakes, n, just causes the trains to stop in 50 m. We shall neglect the kinetic energy due to rotation of the rather small wheels. This assumption permits us to use a single particle to represent the three cars, wherein this particle moves a distance of 50 m. Using the end configuration of the train as the datum for potential energy of gravity, we have for Eq. 13.26:

$$\Delta\text{KE} + \Delta\text{PE} = W_{1-2}$$

$$\left(0 - 3\left\{\frac{1}{2}\frac{198{,}000}{g}\left[\frac{(32)(1000)}{3600}\right]^2\right\}\right) + [0 - (3)(198{,}000)(50)\sin 7^\circ]$$

$$= -(n)(11{,}050)(50)$$

$$n = 10.89$$

The number of brake failures that can accordingly be tolerated is $24 - 11 = 13$.

Problems

13.45. A railroad car traveling 5 km/hr runs into a stop at a railroad terminal. A vehicle having a mass of 1800 kg is held by a linear restoring force system that has an equivalent spring constant of 20,000 N/m. If the railroad car is assumed to stop suddenly and if the wheels in the vehicle are free to turn, what is the maximum force developed by the spring system? Neglect rotational inertia of the wheels.

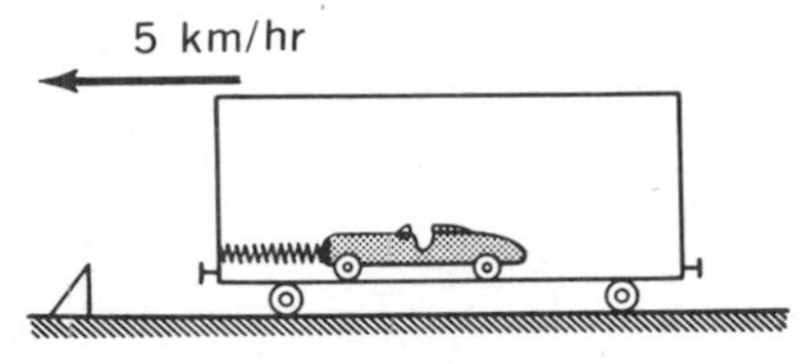

Figure P.13.45

13.46. A mass of one slug is moving at a speed of 50 ft/sec along a horizontal frictionless surface which later inclines upward at an angle 45°. A spring of constant $K = 5$ lb/in. is present along the incline. How high does the mass move?

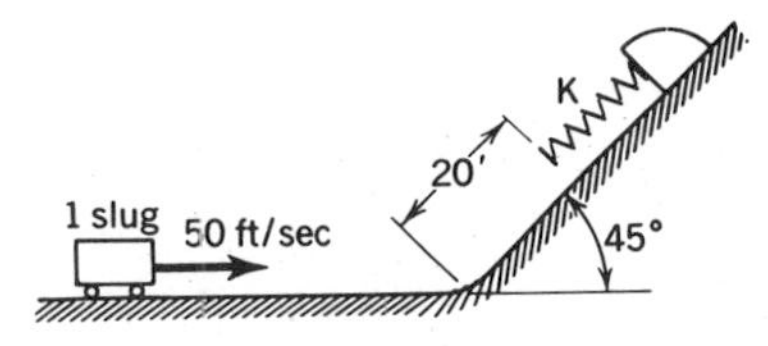

Figure P.13.46

13.47. A block weighing 10 lb is released from rest where the springs acting on the body are horizontal and have a tension of 10 lb each. What is the velocity of the block after it has descended 4 in. if each spring has a spring constant $K = 5$ lb/in.?

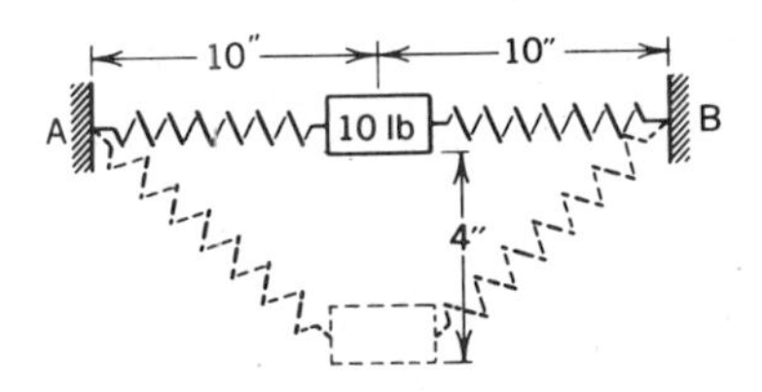

Figure P.13.47

13.48. A nonlinear spring develops a force given as $.06x^2$ N, where x is the amount of compression of the spring in millimeters. Does such a spring develop a conservative force? If so, what is the potential energy stored in the spring for a deflection of 60 mm?

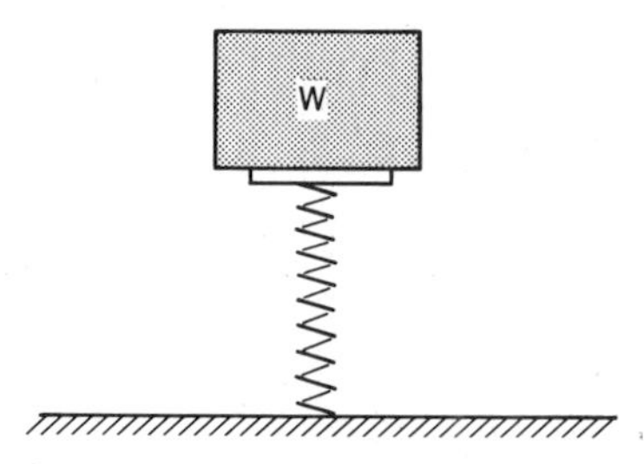

Figure P.13.48

13.49. In Problem 13.48, a weight W of 225 N is released suddenly from rest on the nonlinear spring. What is the maximum deflection of the spring?

13.50. A ski jumper moves down the ramp aided only by gravity. If the skier moves 33 m in the horizontal direction and is to land very smoothly at B, what must be the angle θ for the landing incline? Neglect friction.

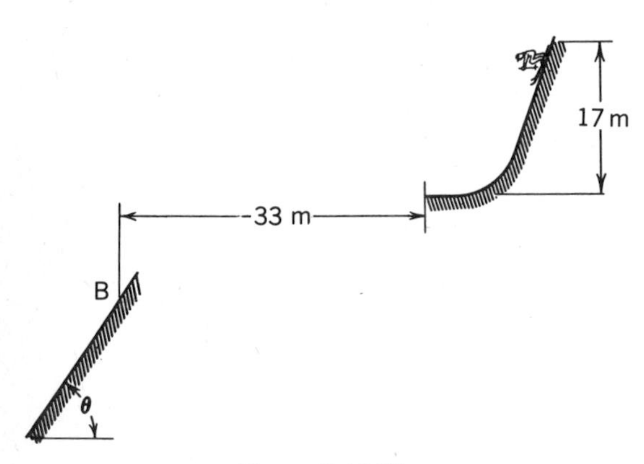

Figure P.13.50

13.51. A vector that you will learn more about in fluid mechanics and electromagnetic theory is the *curl* vector, which is defined for rectangular coordinates as

$$\textbf{curl}\ V(x, y, z) = \left(\frac{\partial V_z}{\partial y} - \frac{\partial V_y}{\partial z}\right)\boldsymbol{i} + \left(\frac{\partial V_x}{\partial z} - \frac{\partial V_z}{\partial x}\right)\boldsymbol{j} + \left(\frac{\partial V_y}{\partial x} - \frac{\partial V_x}{\partial y}\right)\boldsymbol{k}$$

(When the curl is applied to a fluid velocity field V as above, the resulting vector field is twice the angular velocity field of infinitesimal elements in the flow.) Show that if $\boldsymbol{F}$ is expressible as $\nabla\phi(x, y, z)$, then it must follow that **curl** $\boldsymbol{F} = \boldsymbol{0}$. The converse is also true, namely that *if* **curl** $\boldsymbol{F} = \boldsymbol{0}$, *then* $\boldsymbol{F} = \nabla\phi(x, y, z)$ *and is thus a conservative force field.*

13.52. Determine whether the following force fields are conservative or not.

(a) $\boldsymbol{F} = (10z + y)\boldsymbol{i} + (15yz + x)\boldsymbol{j} + \left(10x + \frac{15y^2}{2}\right)\boldsymbol{k}$

(b) $\boldsymbol{F} = (z \sin x + y)\boldsymbol{i} + (4yz + x)\boldsymbol{j} + (2y^2 - 5 \cos x)\boldsymbol{k}$

See Problem 13.51 before doing this problem.

13.53. Given the following conservative force field:

$$\boldsymbol{F} = (10z + y)\boldsymbol{i} + (15yz + x)\boldsymbol{j} + \left(10x + \frac{15y^2}{2}\right)\boldsymbol{k}\ \text{N}$$

find the force potential to within an arbitrary constant. What work is done by the force field on a particle going from $\boldsymbol{r}_1 = 10\boldsymbol{i} + 2\boldsymbol{j} + 3\boldsymbol{k}$ m to $\boldsymbol{r}_2 = -2\boldsymbol{i} + 4\boldsymbol{j} - 3\boldsymbol{k}$ m? [*Hint:* Note that if $\partial\phi/\partial x$ equals some function $(xy^2 + z)$, then we can say on integrating that

$$\phi = \frac{x^2y^2}{2} + zx + g(y, z)$$

where $g(y, z)$ is an arbitrary function of y and z. Note we have held y and z constant during the integration.]

13.54. If the following force field is conservative,

$$\boldsymbol{F} = (5z \sin x + y)\boldsymbol{i} + (4yz + x)\boldsymbol{j} + (2y^2 - 5 \cos x)\boldsymbol{k}\ \text{lb}$$

(where x, y, and z are in ft), find the force

potential up to an arbitrary constant. What is the work done on a particle starting at the origin and moving in a circular path of radius 2 ft to form a semicircle along the positive x axis? (See the hint in Problem 13.53.)

13.55. Masses A and B (both are 75 kg) are constrained to move in frictionless slots. They are connected by a light bar of length .3 m, and weight B is connected by two springs of equal spring constant $K = 900$ N/m. The springs are unstretched when the connecting bar is vertical. What is the velocity of B when A descends a distance of 25 mm?

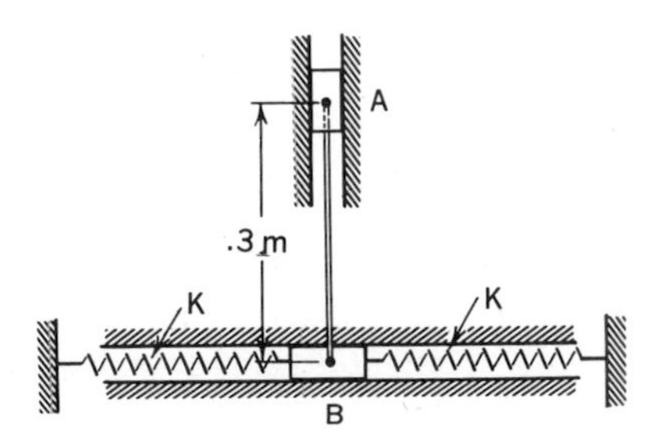

Figure P.13.55

13.56. A body A can slide in a frictionless manner along rod CD. At the position shown, the spring along CD has been compressed 6 in. and A is at a distance of 4 ft from D. The spring connecting A to E has been elongated 1 in. What is the speed of A after it moves 1 ft? The spring constants are $K_1 = 1.0$ lb/in. and $K_2 = .5$ lb/in. The mass of A is 30 lbm.

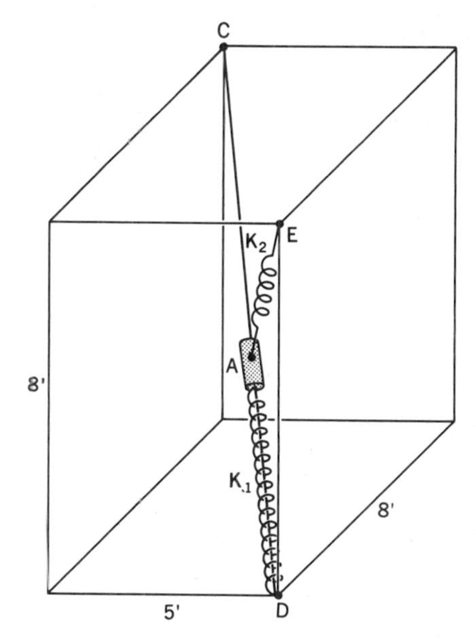

Figure P.13.56

13.57. A collar A of mass 10 lbm slides on a frictionless tube. The collar is connected to a linear spring whose spring constant K is 5.0 lb/in. If the collar is released from rest at the position shown, what is its speed when the spring is at elevation EF? The spring is stretched 3 in. at the initial position of the collar.

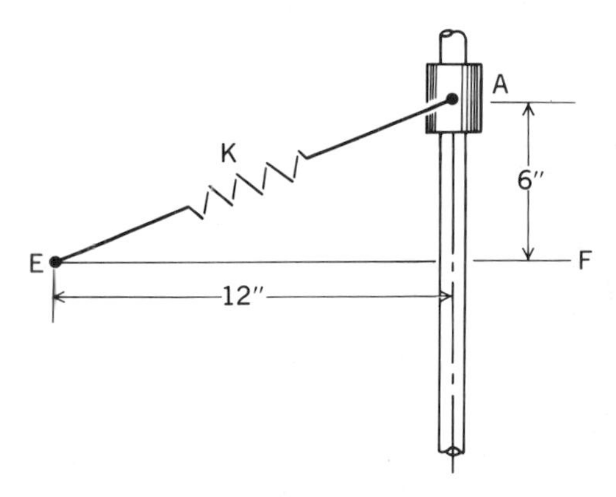

Figure P.13.57

13.58. A collar A having a mass of 5 kg can slide without friction on a pipe. If released from rest at the position shown, where the spring is unstretched, what speed will the collar have after moving 50 mm? The spring constant is 2000 N/m.

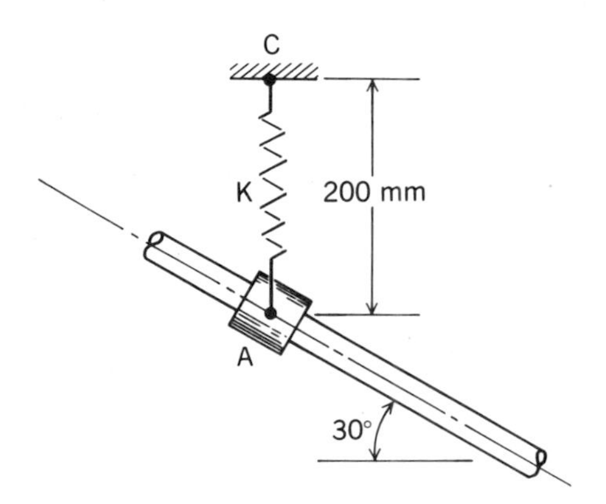

Figure P.13.58

13.59. A slotted rod A is moving to the left at a speed of 2 m/sec. Pins are moved to the left by this rod. These pins must slide in a slot under the rod as shown in the diagram. The pins are connected by a spring having a spring constant K of 1500 N/m. The spring is unstretched in the configuration shown. What distance d do the pins reach before stopping instantaneously? The mass of the slotted rod is 10 kg. The spring is held in the slotted rod so as not to buckle outward.

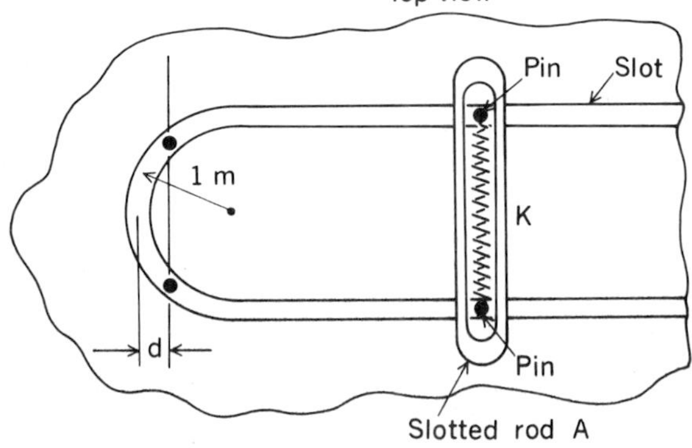

Figure P.13.59

13.60. The top view of a slotted bar of mass 30 lbm is shown. Two pins guided by the slotted bar ride in slots which have the equation of a hyperbola $xy = 5$, where x and y are in feet. The pins are connected by a linear spring having a spring constant K of 5 lb/in. When the pins are 2 ft from the y axis, the spring is stretched 8 in. and the slotted bar is moving to the right at a speed of 2 ft/sec. What is $\dot{V}$ of the bar? *Hint:* Differentiate energy equation.

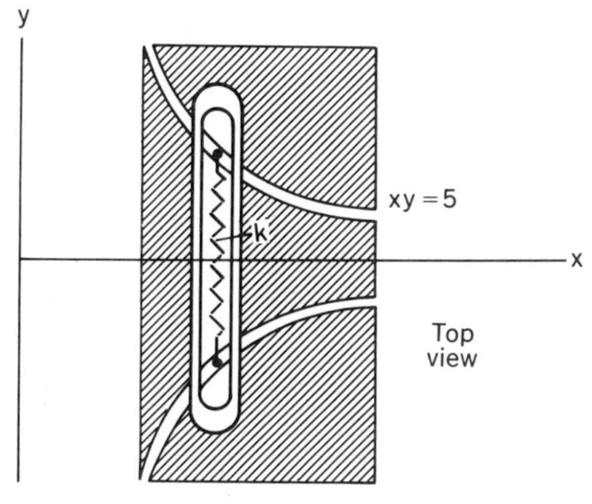

Figure P.13.60

13.61. In Problem 13.60, what is the speed of the slotted bar when $x = 2.20$ ft?

13.62. Perhaps many of you as children constructed toy guns from half a clothespin, a wooden block, and bands of rubber cut from the inner tube of an automobile tire [see diagram (a)]. Rubber band A holds the half-clothespin to the wooden "gun stock." The "ammunition" is a rubber band B held by the clothespin at C by friction and stretched to go around the block at the other end. The rubber band B when laid flat as in (b) has a length of 7 in. To "load the ammunition" takes a force of 20 lb at C. If the gun is pointed upward, estimate how high the fired rubber band will go when "fired" if it weighs 4 oz. To "fire" the gun you push lowest part of clothespin toward the nail (see diagram) to release at C.

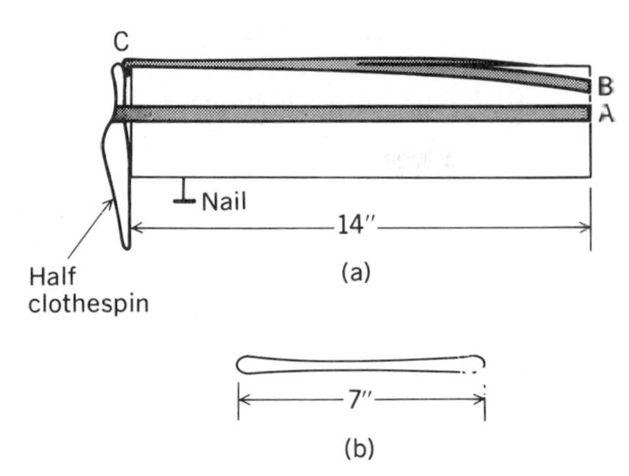

Figure P.13.62

13.63. A collar B having a mass of 100 g moves along a frictionless curved rod in a vertical plane. A light rubber band connects B to a fixed point A. The rubber band is 250 mm in length when unstretched. A force of 30 N is required to extend the band 50 mm. If the collar is released from rest, what maximum distance can d be so that the upward normal force on the rod at C is no greater than 20 N?

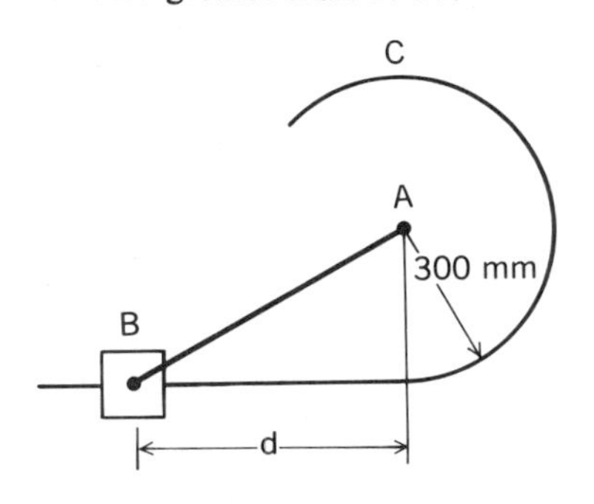

Figure P.13.63

13.64. When your author was a graduate student he built a system for examining the effects of high-speed moving loads over elastically supported beams (see the diagram). A "vehicle" slides along a slightly lubricated square tube guide. At the base of the vehicle is a spring-loaded light wheel which will run over the beam (not shown). The vehicle is catapulted to a high speed by a stretched elastic cord (shock cord) which is pulled back from position A–A to the position B shown prior to "firing." At A–A the shock cord is elongated 10 in., while at the firing position it is elongated 30 in. A force of 10 lb is required for each inch of elongation of the cord. If the cord weighs a total of 1.5 lb and the vehicle weighs 10 oz, what is the speed of the vehicle when the cord reaches A–A after firing? Take into account in some reasonable way the kinetic energy of the cord, but neglect friction.

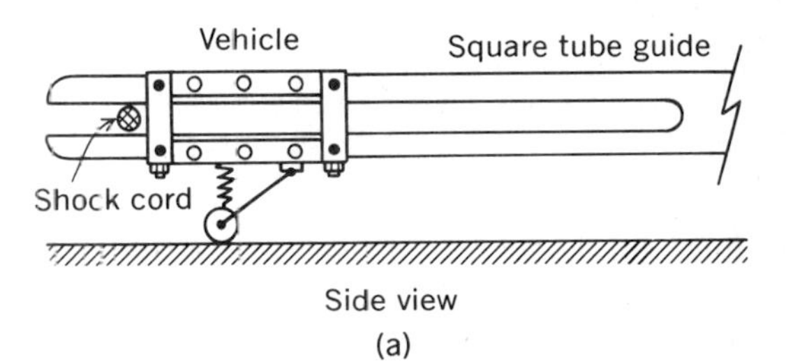

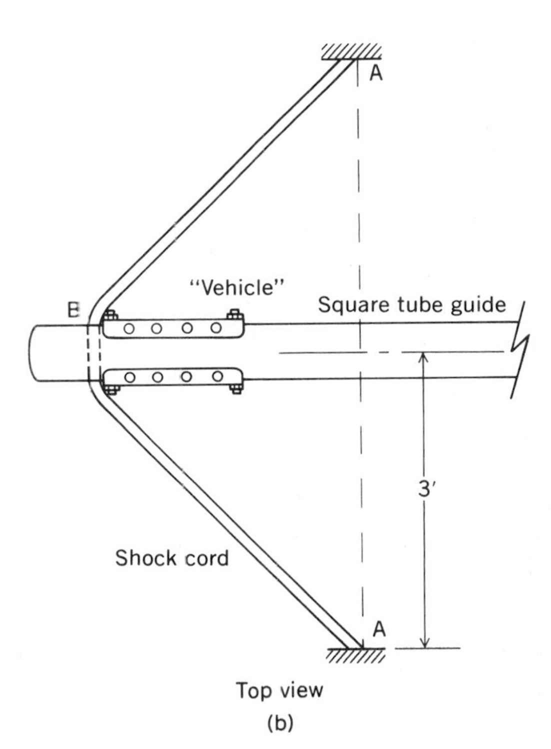

Figure P.13.64

13.65. A meteor has a speed of 56,000 km/hr when it is 320,000 km from the center of the earth. What will be its speed when it is 160 km from the earth's surface?

13.66. Do Problem 13.2 using energy equation in the usual form of the first law of thermodynamics.

13.67. Do Problem 13.5 using energy equation in the usual form of the first law of thermodynamics.

13.68. Do Problem 13.14 using energy equation in the usual form of the first law of thermodynamics.

13.69. Do Problem 13.15 using energy equation in the usual form of the first law of thermodynamics.

13.70. A constant-torque electric motor A is hoisting a weight W of 30 lb. An inextensible cable connects the weight W to the motor over a stationary drum of diameter $D = 1$ ft. The diameter d of the motor drive is 6 in., and the delivered torque is 150 lb-ft. The coefficient of friction between the drum and cable is .2. If the system is started from rest, what is the speed of the weight W after it has been raised 5 ft?

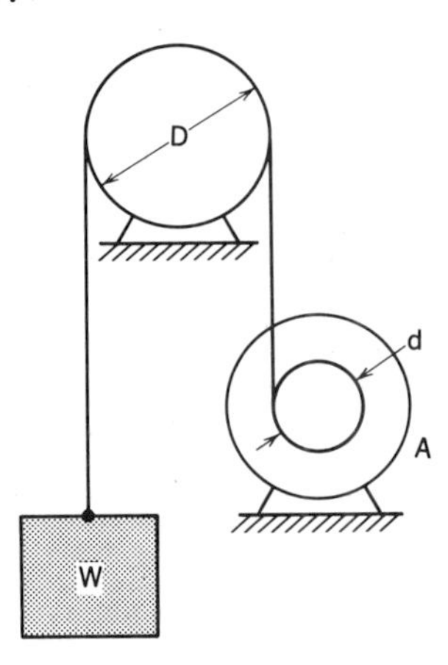

Figure P.13.70

13.71. A body A, weighing 10 lb, can slide along a fixed rod B–B. A spring is connected between fixed point C and the mass. AC is 2 ft in length when the spring is unextended. If the body is released from rest at the configuration shown, what is its speed when it reaches the y axis? Assume that a constant friction force of 6 oz acts on the body A. The spring constant K is 1 lb/in.

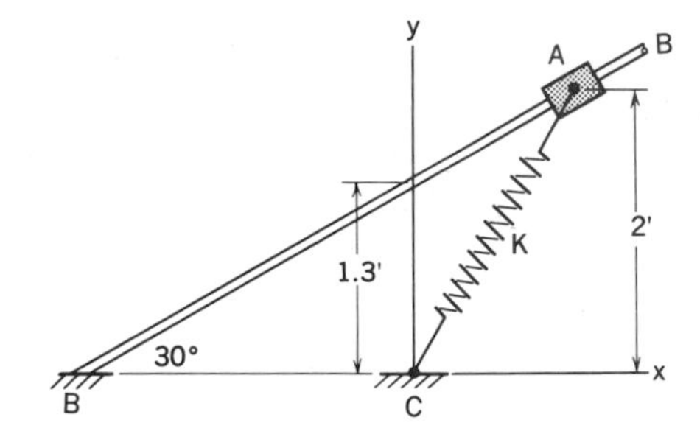

Figure P.13.71

13.72. A body A is released from rest on a vertical circular path as shown. If a constant resistance force of 1 N acts along the path, what is the speed of the body when it reaches B? The mass of the body is .5 kg and the radius r of the path is 1.6 m.

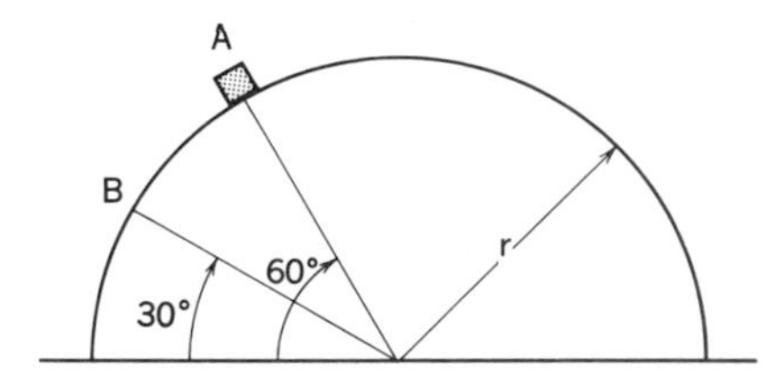

Figure P.13.72

13.73. In ordnance work a very vital test for equipment is the *shock test*, in which a piece of equipment is subjected to a certain level of acceleration of short duration. A common technique for this test is the *drop test*. The specimen is mounted on a rigid carriage, which upon release is dropped along guide rods onto a set of lead pads resting on a heavy rigid anvil.

The pads deform and absorb the energy of the carriage and specimen. We estimate through other tests that the energy E absorbed by a pad versus compression distance δ is given as shown, where the curve can be taken as a parabola. For four such pads, each placed directly on the anvil, and a height h of 3 m, what is the compression of the pads? The carriage and specimen together weigh $50g$ N. Neglect the friction of the guides. (*Note:* 1 J = 1 N-m.)

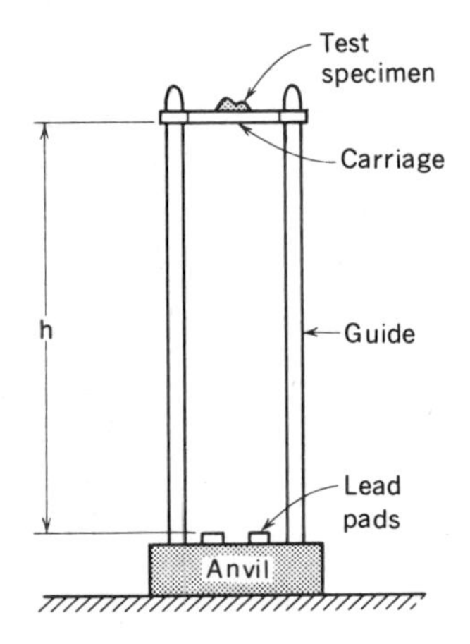

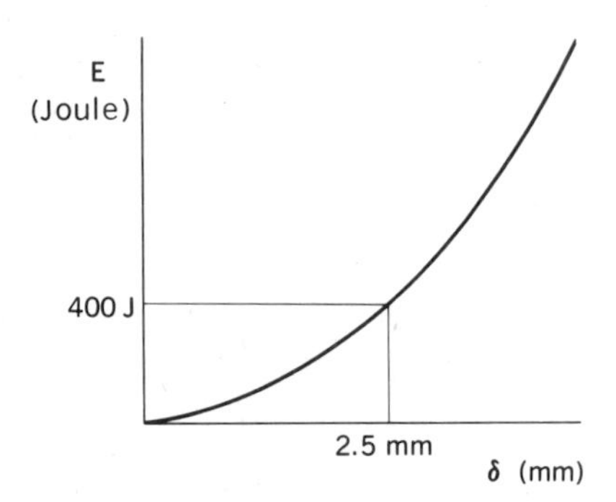

Figure P.13.73

13.74. Two bodies are connected by an inextensible cord over a frictionless pulley. If released from rest, what velocity will they reach when the 500-lb body has dropped 5 ft?

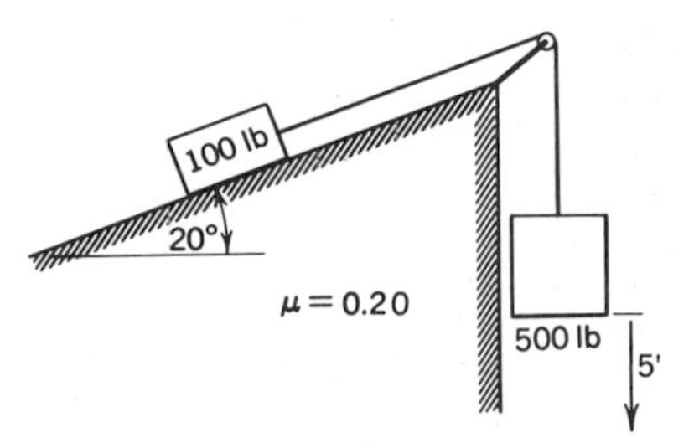

Figure P.13.74

13.75. Suppose in Example 13.8 that only the brakes on train A operate and lock. What is the distance d before stopping? Also, determine the force in each coupling of the system.

13.76. A large constant force F is applied to a body of weight W resting on an inclined surface for which the coefficient of dynamic friction is μ_d. The body is acted on by a spring having a spring constant K. If initially the spring is compressed a distance δ, compute the velocity of the body in terms of F and the other parameters that are given, when the body has moved from rest a distance up the incline of $\frac{3}{2}\delta$.

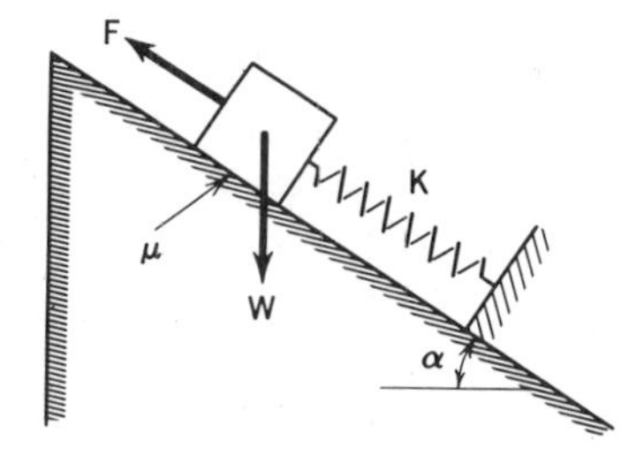

Figure P.13.76

Part B
Systems of Particles

13.6 Work–Energy Equations

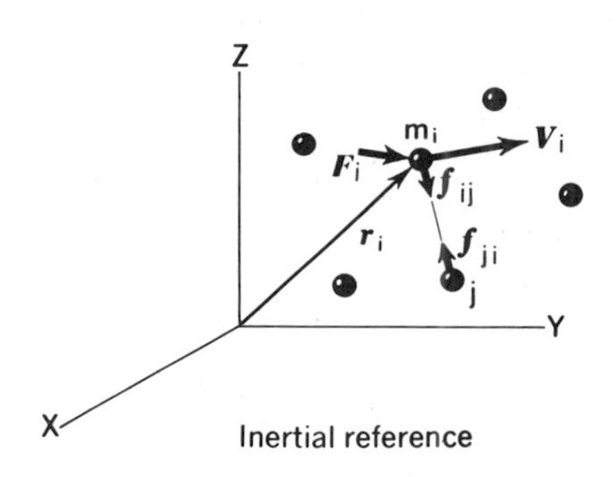

Figure 13.17. System of particles.

We shall now examine a system of particles from an energy viewpoint. A general aggregate of n particles is shown in Fig. 13.17. Considering the ith particle, we can say, by employing Eq. 13.2:

$$\int_1^2 \boldsymbol{F}_i \cdot d\boldsymbol{r}_i + \int_1^2 \left(\sum_{\substack{j=1\\ j\neq i}}^{n} \boldsymbol{f}_{ij}\right) \cdot d\boldsymbol{r}_i = (\tfrac{1}{2}m_i V_i^2)_2 - (\tfrac{1}{2}m_i V_i^2)_1 \tag{13.27}$$

where, as in Chapter 12, $\boldsymbol{f}_{ij}$ is the force from the jth particle onto the ith particle, as illustrated in the diagram, and is thus an internal force. In contrast, $\boldsymbol{F}_i$ represents the total external force on the ith particle. In words, Eq. 13.27 says that for a displacement between $\boldsymbol{r}_1$ and $\boldsymbol{r}_2$ along some path, the energy relations for the ith particle are:

$$\text{external work} + \text{internal work} = (\text{change in kinetic energy relative to } XYZ) \tag{13.28}$$

Furthermore, we can adopt the point of view set forth in Section 13.5 and identify conservative forces, both external and internal, so as to utilize potential energies for these forces in the energy equation. To qualify as a conservative force, an internal force would have to be a function of only the spatial configuration of the system and expressible as the gradient of a scalar function. Clearly, forces arising from (1) the gravitational attraction between the particles, (2) the electrostatic forces from electric charges on the particles, and (3) elastic connectors between the particles (such as springs) are all conservative internal forces. We now sum Eqs. 13.27 for all the particles in the system. We do *not* get a cancellation of contributions of the internal forces as we did for Newton's law because we are now adding the *work* done by each internal force on each particle. And even though we have pairs of internal forces that are equal and opposite, the *movements* of the corresponding particles in general are *not* equal. The result is that the work done by a pair of equal and opposite internal forces is not generally zero. However, in the case of a *rigid body*, the contact forces between pairs of particles making up the body have the same motion, and so in this case the internal work is *zero* from such forces.[10] We can then say that

$$\Delta(\text{KE} + \text{PE}) = W_{1-2} \tag{13.29}$$

[10] We shall show this more directly in Chapter 17.

where W_{1-2} represents the net work done by *internal and external* nonconservative forces, and PE represents the total potential energy of the conservative *internal and external* forces. As pointed out earlier, since we are employing the *change* in potential energy, the datums chosen for measuring PE are of little significance here.[11] For instance, any convenient datum for measuring the potential energy due to gravity of the earth yields the same result for the term ΔPE.

Let us now consider the action of gravity on a system of particles. The potential energy relative to a datum plane, xy, for such a system (see Fig. 13.18) is simply

$$\text{PE} = \sum_i m_i g z_i$$

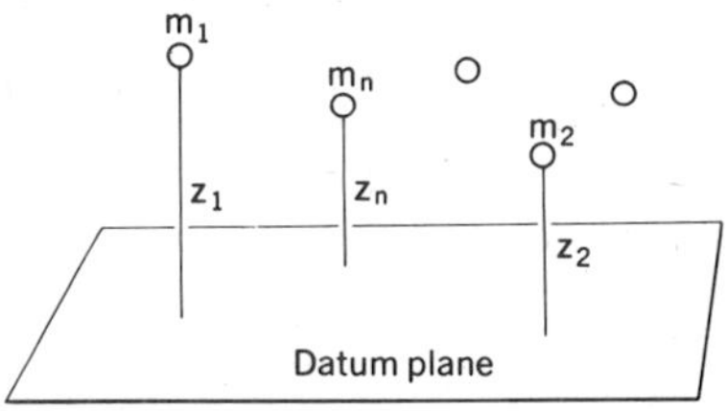

Note that the right side of this equation represents the first moment of the weight of the system about the xy plane. This quantity can be given in terms of the center of gravity and the entire weight of W as follows:

Figure 13.18. Particles above reference plane.

$$\text{PE} = W z_c$$

where z_c is the vertical distance from the datum plane to the center of gravity. Note that if g is constant, the center of gravity corresponds to the center of mass. And so for any system of particles, the change in potential energy is readily found by concentrating the entire weight at the center of gravity or, as is almost always the case, at the center of mass.

EXAMPLE 13.9

In Fig. 13.19, two blocks have weights W_1 and W_2, respectively. They are connected by a flexible, *elastic* cable of negligible mass which has an equivalent spring constant of K_1. Body 1 is connected to the wall by a spring having a spring constant K_2 and slides along a horizontal surface for which the dynamic coefficient of friction with the body is μ. Body 2 is supported initially by some external agent so that, at the outset of the problem, the spring and cable are unstretched. What is the total kinetic energy of the system when, after release, body 2 has moved a distance d_2 and body 1 has moved a smaller distance d_1?

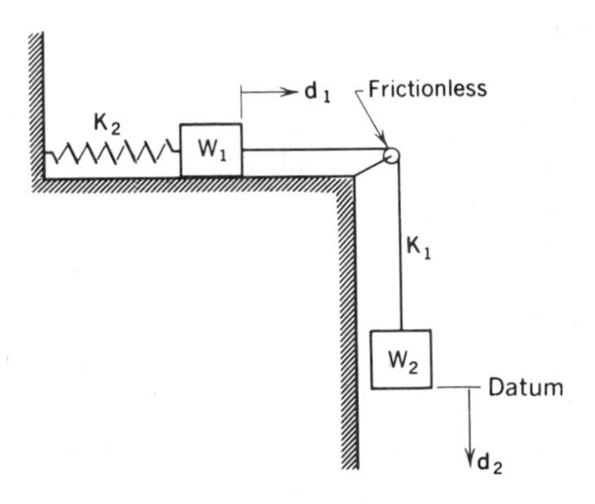

Figure 13.19. Elastically connected bodies.

Use Eq. 13.29. Only one nonconservative force exists in the system, the external friction force on body 1. Therefore, the work term of the equation becomes

[11]One precaution in this regard must again be brought to your attention. You will remember that in the spring-force formula, $-Kx$, the term x represents the elongation or contraction of the spring from the *undeformed* condition. This condition must not be violated in the potential-energy expression $\frac{1}{2}Kx^2$.

$$W_{1-2} = -W_1 \mu d_1 \tag{a}$$

Three conservative forces are present; the spring force and the gravitational force are *external* and the force from the elastic cable is *internal*. (We neglect mutual gravitational forces between the bodies.) Using the initial position of W_2 as the datum for gravitational potential energy, we have, for the total change in potential energy:

$$\Delta\text{PE} = [\tfrac{1}{2}K_2 d_1^2 + \tfrac{1}{2}K_1(d_2 - d_1)^2 - W_2 d_2]$$

We can compute the desired change in kinetic energy from Eq. 13.29 as

$$\Delta\text{KE} = -W_1 \mu d_1 - \tfrac{1}{2}K_2 d_1^2 - \tfrac{1}{2}K_1(d_2 - d_1)^2 + W_2 d_2 \tag{b}$$

As an additional exercise, you should arrive at this result by using the basic Eq. 13.28, where you cannot rely on familiar formulas for potential energies.

13.7 Kinetic Energy Expression Based on Center of Mass

In this and the next section, we shall introduce the center of mass into our discussion in order to develop useful expressions for the kinetic energy of an aggregate. Also, we shall develop the work–energy equation for the center of mass set forth at the outset of this chapter.

Consider a system of n particles, shown in Fig. 13.20. The total kinetic energy relative to xyz of a system of particles can be given as

$$\text{KE} = \sum_{i=1}^{n} \tfrac{1}{2} m_i V_i^2 \tag{13.30}$$

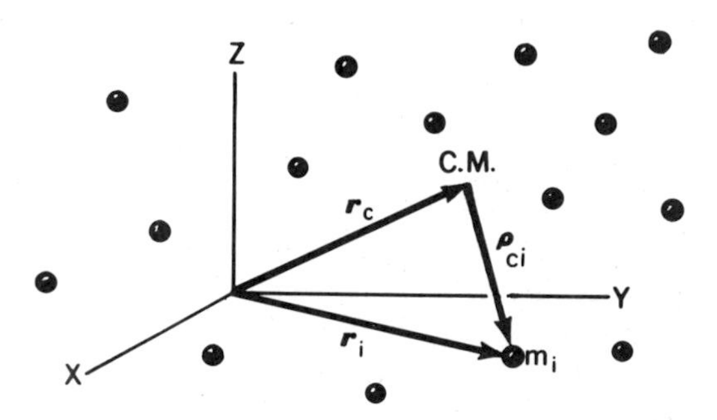

Figure 13.20. System of particles with center of mass.

We shall now express Eq. 13.30 in another way by introducing the mass center. Note in the diagram we have employed the vector $\boldsymbol{\rho}_{ci}$ as the displacement vector from the center of mass to the ith particle. We can accordingly say:

$$\boldsymbol{r}_i = \boldsymbol{r}_c + (\boldsymbol{r}_i - \boldsymbol{r}_c) = \boldsymbol{r}_c + \boldsymbol{\rho}_{ci} \tag{13.31}$$

Differentiating with respect to time, we get

$$\dot{r}_i = \dot{r}_c + (\dot{r}_i - \dot{r}_c) = \dot{r}_c + \dot{\boldsymbol{\rho}}_{ci}$$

Therefore,

$$V_i = V_c + \dot{\boldsymbol{\rho}}_{ci} \tag{13.32}$$

From our earlier discussions on simple relative motion we can say that $\dot{\boldsymbol{\rho}}_{ci}$ the is motion of the ith particle *relative to the mass center.*[12] Substituting the relation above into the expression for kinetic energy, Eq. 13.30, we get

$$\text{KE} = \sum_{i=1}^{n} \tfrac{1}{2} m_i (V_c + \dot{\boldsymbol{\rho}}_{ci})^2 = \sum_{i=1}^{n} \tfrac{1}{2} m_i (V_c + \dot{\boldsymbol{\rho}}_{ci}) \cdot (V_c + \dot{\boldsymbol{\rho}}_{ci})$$

Carrying out the dot product, we have

$$\text{KE} = \tfrac{1}{2} \sum_{i=1}^{n} m_i V_c^2 + \sum_{i=1}^{n} m_i V_c \cdot \dot{\boldsymbol{\rho}}_{ci} + \tfrac{1}{2} \sum_{i=1}^{n} m_i \dot{\boldsymbol{\rho}}_{ci}^2 \tag{13.33}$$

Since V_c is common for all values of the summation index, we can extract it from the summation operation, and this leaves

$$\text{KE} = \tfrac{1}{2}\Big(\sum_{i=1}^{n} m_i\Big) V_c^2 + V_c \cdot \Big(\sum_{i=1}^{n} m_i \dot{\boldsymbol{\rho}}_{ci}\Big) + \tfrac{1}{2} \sum_{i=1}^{n} m_i \dot{\boldsymbol{\rho}}_{ci}^2 \tag{13.34}$$

Perform the following replacements:

$$\sum_{i=1}^{n} m_i \text{ by } M, \quad \text{and} \quad \sum_{i=1}^{n} m_i \dot{\boldsymbol{\rho}}_{ci} \quad \text{by } \frac{d}{dt} \sum_{i=1}^{n} m_i \boldsymbol{\rho}_{ci}$$

We then have

$$\text{KE} = \tfrac{1}{2} M V_c^2 + V_c \cdot \frac{d}{dt} \sum_{i=1}^{n} m_i \boldsymbol{\rho}_{ci} + \tfrac{1}{2} \sum_{i=1}^{n} m_i \dot{\boldsymbol{\rho}}_{ci}^2 \tag{13.34}$$

But the expression

$$\sum_{i=1}^{n} m_i \boldsymbol{\rho}_{ci}$$

represents the first moment of mass of the system about the center of mass for the system. Clearly by definition, this quantity must always be zero. The expression for kinetic energy becomes

$$\boxed{\text{KE} = \tfrac{1}{2} M V_c^2 + \tfrac{1}{2} \sum_{i=1}^{n} m_i \dot{\boldsymbol{\rho}}_{ci}^2} \tag{13.35}$$

Thus, we see that the *kinetic energy* for some reference *can be considered to be composed of two parts: (1) the kinetic energy of the total mass moving relative to the reference*

---

[12]Note that

$$\dot{\boldsymbol{\rho}}_{ci} = \dot{r}_i - \dot{r}_c$$

That is, $\dot{\boldsymbol{\rho}}_{ci}$ is the *difference* between the velocity of the ith particle and that of the mass center. This is then the velocity of the particle *relative* to the center of the mass (i.e., relative to a reference translating with c or to a nonrotating observer moving with c.)

with the velocity of the mass center, plus (2) the kinetic energy of the motion of the particles relative to the mass center.

EXAMPLE 13.10

A hypothetical vehicle is moving at speed V_0 in Fig. 13.21. On this vehicle are two bodies each of mass m sliding along a horizontal rod at a speed v relative to the rod. This rod is rotating at an angular speed ω rad/sec relative to the vehicle. What is the kinetic energy of the two bodies relative to the ground (XYZ) when they are at a distance r from point A?

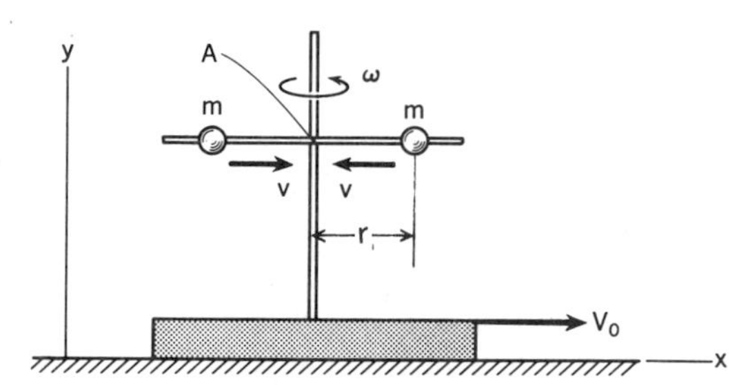

Figure 13.21. Moving device.

Clearly, the center of mass corresponds to point A and is thus moving at a speed V_0 relative to the ground. Hence, we have as part of the kinetic energy the term

$$\tfrac{1}{2}MV_c^2 = mV_0^2 \tag{a}$$

The velocity of each ball relative to the center of mass is easily formed using cylindrical components. Thus, imagining a reference xyz at A translating with the vehicle relative to XYZ, we have for the velocity of each ball relative to xyz:

$$\dot{\rho}^2 = \dot{r}^2 + (\omega r)^2 = v^2 + (\omega r)^2 \tag{b}$$

The total kinetic energy of the two masses relative to the ground is then

$$\text{KE} = mV_0^2 + m[v^2 + (\omega r)^2] \tag{c}$$

In Example 13.10, we considered a case where the bodies involved constituted a finite number of *discrete* particles. In the next example, we consider a case where we have a *continuum* of particles forming a rigid body. The formulation given by Eq. 13.35 can still be used but now, instead of summing for a finite number of discrete particles, we must *integrate* to account for the infinite number of infinitesimal particles comprising the system. We are thus taking a glimpse, for a simple case, of rigid-body dynamics to be studied later in the text. Those that do not have time for studying these energy problems in detail will be able to solve simple but useful rigid-body dynamics problems on the basis of this example as well as later examples in this chapter.

EXAMPLE 13.11

A uniform hoop of radius R is rolling without slipping such that O, the mass center, moves at a speed V (Fig. 13.22). If the hoop weighs W lb, what is the kinetic energy of the hoop relative to the ground?

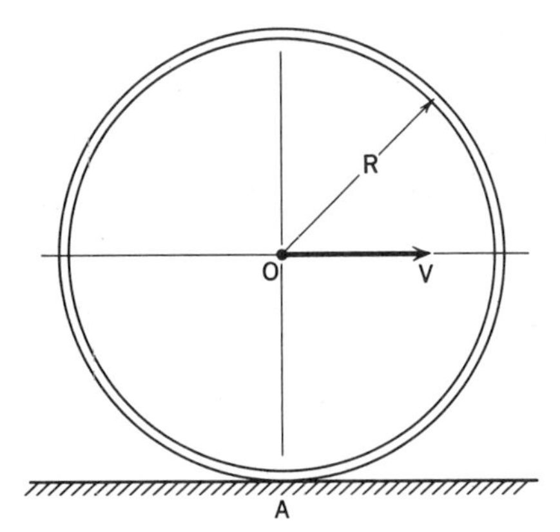

Figure 13.22. Rolling hoop.

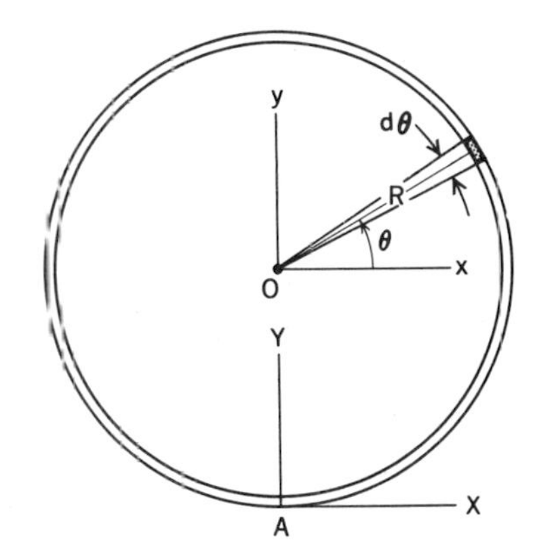

Figure 13.23. xy translates with O relative to XY.

Clearly, the hoop cannot be considered as a finite number of discrete finite particles as in previous example, and so we must consider an infinity of infinitesimal contiguous particles. It is simplest to employ here the center-of-mass approach. The main problem then is to find the kinetic energy of the hoop relative to the mass center O, that is, relative to a reference xy translating with the mass center as seen from the ground reference XY (see Fig. 13.23). The motion relative to xy is clearly simple rotation; accordingly, we must find the angular velocity of the hoop for this reference. The no slipping condition means that the point of contact of the hoop with the ground has instantaneously a zero velocity. Observe the motion from a stationary reference XY. As you may have learned in physics, and as will later be shown, the body has a *pure instantaneous rotational* motion about the point of contact. The angular velocity ω for this motion is then easily evaluated by considering point O rotating about the instantaneous center of rotation A. Thus,

$$\omega = \frac{V}{R} \tag{a}$$

Since reference xy *translates* relative to reference XY, an observer on xy sees the *same angular velocity* ω for the hoop as the observer on XY. Accordingly, we can now readily evaluate the second term on the right side of Eq. 13.35. As particles, use elements of the hoop which are $R\,d\theta$ in length, as shown in Fig. 13.23, and which have a mass per unit length of $W/(2\pi Rg)$. We then have, on replacing summation by integration, the result

$$\frac{1}{2}\sum_{i=1}^{n} m_i \dot{\rho}_{ci}^2 = \frac{1}{2}\int_0^{2\pi}\left[\left(\frac{W}{g2\pi R}\right)(R\,d\theta)\right](\omega R)^2 = \frac{1}{2}\int_0^{2\pi}\left[\frac{W}{(g)(2\pi R)}(R\,d\theta)\right]\left(\frac{V}{R}R\right)^2$$
$$= \frac{1}{2}\frac{W}{g}V^2 \tag{b}$$

The kinetic energy of the hoop is then in accordance with Eq. 13.35:

$$\text{KE} = \frac{1}{2}\frac{W}{g}V^2 + \frac{1}{2}\frac{W}{g}V^2 = \frac{W}{g}V^2 \tag{c}$$

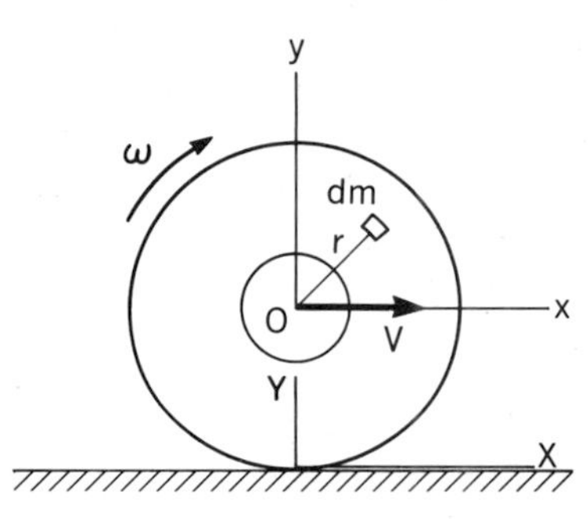

Figure 13.24. Rolling cylinder.

Suppose that the body were a generalized cylinder (see Fig. 13.24) such as a tire of radius R having O as the center of mass with axisymmetrical distribution of mass about the axis at O. Then, we would express Eq. (b) as follows:

$$\frac{1}{2}\sum_{i}^{n} m_i \dot{\rho}_{ci}^2 = \frac{1}{2}\iiint_M (dm)(r\omega)^2 \tag{d}$$

You will recall from Chapter 9 that

$$\iiint r^2\, dm$$

is the *second moment* of inertia of the body taken about the z axis at O. That is,

$$I_{zz} = \iiint r^2\, dm$$

Thus, we have for the kinetic energy of such a body:

$$\text{KE} = \frac{1}{2}\frac{W}{g}V^2 + \frac{1}{2}I_{zz}\omega^2 \tag{e}$$

You may also recall from Chapter 9 that we could employ the *radius of gyration* k to express I_{zz} as follows:

$$I_{zz} = k^2 M \tag{f}$$

Hence, Eq. (e) can be given as

$$\text{KE} = \frac{1}{2}\frac{W}{g}V^2 + \frac{1}{2}k^2 M\omega^2$$

We shall examine the kinetic energy formulations of rigid bodies carefully in a later chapter. Here, we have used certain familiar results from physics pertaining to kinematics of plane motion of a nonslipping rolling rigid body. For a more general undertaking, we shall have to carefully consider more general aspects of kinematics of rigid-body motion. This will be done in Chapter 15.

13.8 Work–Kinetic Energy Expressions Based on Center of Mass

The work–kinetic energy expressions of Section 13.6 were developed for a system of particles without regard to the mass center. We shall now introduce this point into the work–kinetic energy formulations. You will recall that Newton's law for the mass center of any system of particles is

$$\boldsymbol{F} = M\ddot{\boldsymbol{r}}_c \tag{13.36}$$

where $\boldsymbol{F}$ is the total *external* force on the system of particles. By the same development as presented in Section 13.1, we can readily arrive at the following equation:

$$\boxed{\int_1^2 \boldsymbol{F} \cdot d\boldsymbol{r}_c = (\tfrac{1}{2}MV_c^2)_2 - (\tfrac{1}{2}MV_c^2)_1} \tag{13.37}$$

It is *vital* to understand from the left side of Eq. 13.37 that the external *forces must all move with the center of mass* for the computation of the proper work term in this equation.[13] We wish next to point out that the single particle model represents a special case of the use of Eq. 13.37. Specifically, the single-particle model represents the case where the motion of the center of mass of a body sufficiently describes the motion of the body and where the external forces on the body essentially move with the center of mass of the body. Such cases were set forth in Section 13.2.

Before proceeding to the examples, let us consider for a moment the case of the cylinder rolling without slipping down an incline (see Fig. 13.25). We shall consider the cylinder as an *aggregate of particles* which form a rigid body—namely a cylinder. When using such an approach, we require that *all the forces both external and internal must move with their respective points of application.* Let us then consider the external work done on the particles making up the cylinder other than the work done by gravity. Clearly, only particles on the *rim* of the cylinder are acted on by external forces other than gravity. Consider one such particle during one rotation of the cylinder. This particle will have acting on it a friction force f and a normal force N at the *instant* when the particle is in *contact* with the inclined surface. The particle will have *zero* external force (except for gravity) at all other positions during the cycle. Now, at the instant of this contact, the normal force N has zero velocity because of the rigidity of the bodies. Therefore, N transmits no power and does no work on the particle during the cycle under consideration. Also, the friction force f has zero velocity at the instant of contact with the particle because of the *no slipping* condition. Accordingly, f transmits no power and does no work on the particle during this cycle.[14] This result must be true for each and every particle on the rim of the cylinder. Thus, clearly, f and N do no work when the cylinder rolls down the incline. Also, because of the rigidity of the body the internal forces do no work as pointed out earlier. Thus, only gravity does work.

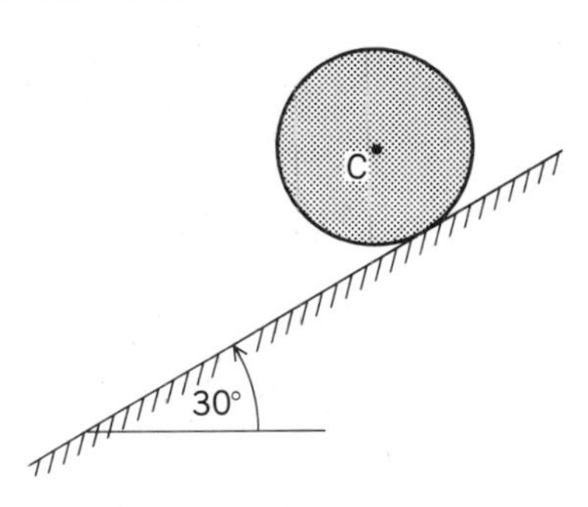

Figure 13.25. Cylinder on incline.

---

[13]This is in direct contrast to the work–energy equation for a system of particles wherein each external force moves with its *actual* point of application. Also, only external forces are involved for the center-of-mass approach, in contrast to the system of particles where internal forces are also involved. Note that in Examples 13.1, 13.2, and 13.3 we were using a particle approach and thus were really considering the motion of the center of mass. The friction forces then moved with the center of mass.

[14]Observe the figures for Problems 13.86 and 13.96. Consider the two cylinders and the block as simply an *aggregate of particles*. If there is no slipping between the block and the cylinders, the velocities of the particles on the block and the cylinders at the points of contact between these bodies have the *same* velocity at any time t. Furthermore, the friction force on the cylinder from the block is *equal* and *opposite* to the friction force on the block from the cylinder at the point of contact. We can then conclude that there is a net of zero work done by the friction forces between block and cylinders when considering them as an aggregate of particles.

However, in considering the motion of the *center of mass* C of the cylinder in Fig. 13.26, we note that force f now *moves* with C and hence *does* work.

EXAMPLE 13.12

A cylinder with a mass of 25 kg is released from rest on an incline, as shown in Fig. 13.25. The diameter of the cylinder is .60 m. If the cylinder rolls without slipping, compute the speed of the centerline O after it has moved 1.6 m along the incline. Also, ascertain the friction force acting on the cylinder. Use the result from Problem 13.85 that the kinetic energy of a cylinder rotating about its own stationary axis is $\frac{1}{4}MR^2\omega^2$, where ω is the angular speed in rad/sec.

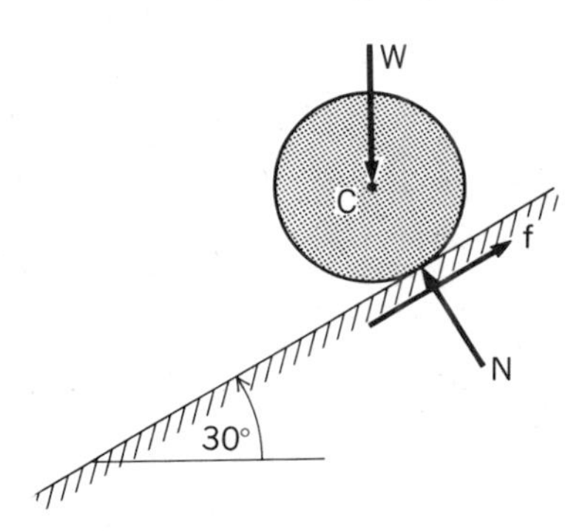

Figure 13.26. Free-body diagram of cylinder.

In Fig. 13.26 we have shown the free body of the cylinder. We proceed to use the work–energy equation for an aggregate of particles. Recall that we can concentrate the weight at the center of gravity (Section 13.6). Accordingly, using the lowest position as a datum and noting from our earlier discussion that the friction force f does no work we have

$$\Delta(\text{PE} + \text{KE}) = W_{1-2}$$

$$[0 - (25)(9.81)(1.6) \sin 30°] + \{[\tfrac{1}{2}(25)\, V_c^2 + \tfrac{1}{4}(25)(.30)^2(\omega^2)] - 0\} = 0 \quad \text{(a)}$$

where the kinetic energy of the cylinder is given as the kinetic energy of the mass taken at the mass center (straight line of motion of C) plus kinetic energy of the cylinder relative to the center of mass (pure rotation about centerline). Noting from Example 13.11 that

$$\omega = \frac{V}{R} = \frac{V}{.30}$$

we substitute into Eq. (a) and solve for V_c. We get

$$V_c = 3.23 \text{ m/sec} \quad \text{(b)}$$

Now to find f, we *consider the motion of the mass center of the cylinder*. This means that we use Eq. 13.37 for the center of mass. Now *all external forces must move with the center of mass*; thus, f does work. Since the center of mass moves along a path always at right angles to N, this force still does no work. Accordingly, we can say:

$$-f(1.6) + W(1.6 \sin 30°) = \tfrac{1}{2}MV_c^2$$

$$-f(1.6) + (25)(9.81)(1.6) \sin 30° = \tfrac{1}{2}(25)(3.23^2)$$

$$f = 41.1 \text{ N}$$

Problems

13.77. A chain of total length L is released from rest on a smooth support as shown. Determine the velocity of the chain when the last link moves off the horizontal surface.

In this problem, neglect friction. Also, do not attempt to account for centrifugal effects stemming from the chain links rounding the corner.

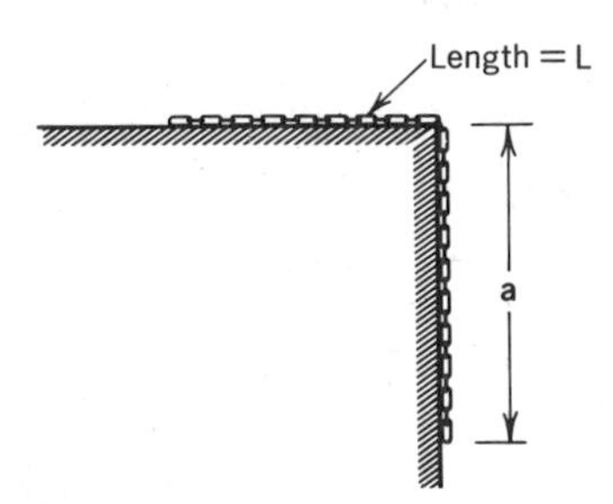

Figure P.13.77

13.78. A chain is 50 ft long and weighs 100 lb. A force P of 80 lb has been applied at the configuration shown. What is the speed of the chain after force P has moved 10 ft? The coefficient of friction between the chain and the supporting surface is .3. Give an approximate analysis.

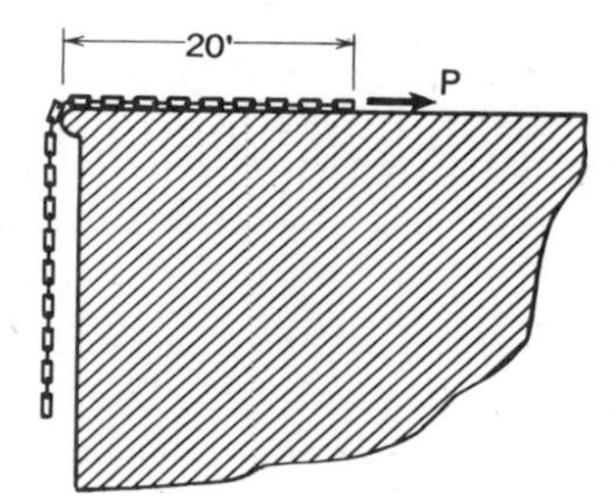

Figure P.13.78

13.79. A bullet of weight W_1 is fired into a block of wood weighing W_2 lb. The bullet lodges in the wood, and both bodies then move to the dashed position indicated in the diagram before falling back. Compute the amount of internal work done during the action. Discuss the effects of this work. The bullet has a speed V_0 before hitting the block. Neglect the mass of the supporting rod and friction at A.

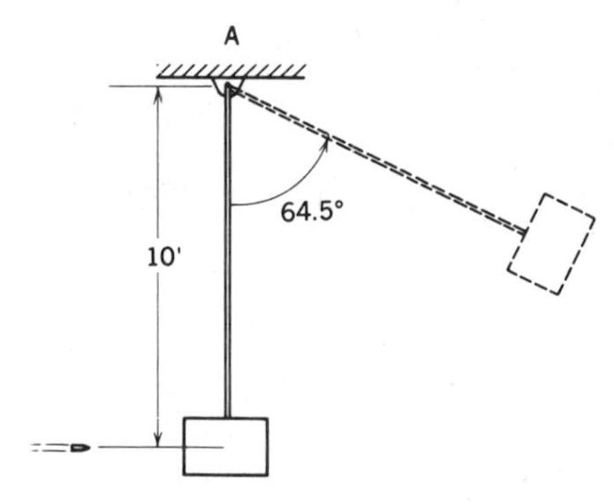

Figure P.13.79

13.80. Three weights A, B, and C slide frictionlessly along the system of connected rods. The bodies are connected by a light, flexible, inextensible wire that is directed by frictionless small pulleys at E and F. If the system is released from rest, what is its speed after it has moved 300 mm? Employ the following data for the body masses:

Body A: 5 kg

Body B: 4 kg

Body C: 7.5 kg

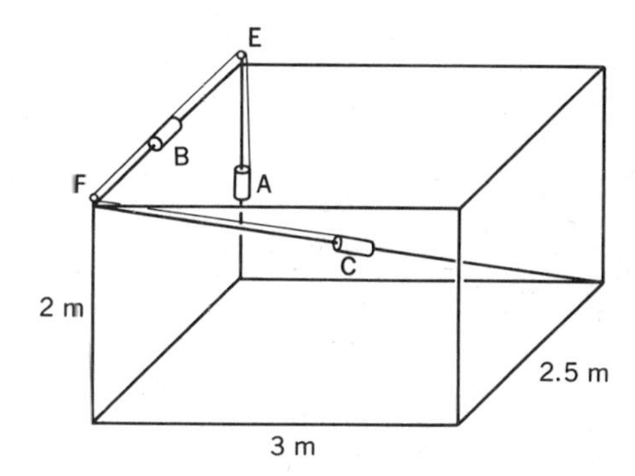

Figure P.13.80

13.81. Bodies E and F slide in frictionless grooves. They are interconnected by a light, flexible, inextensible cable (not shown). What is the speed of the system after it has moved 2 ft? The weights of bodies E and F are 10 lb and 20 lb, respectively. B is equidistant from A and C. E remains in top groove.

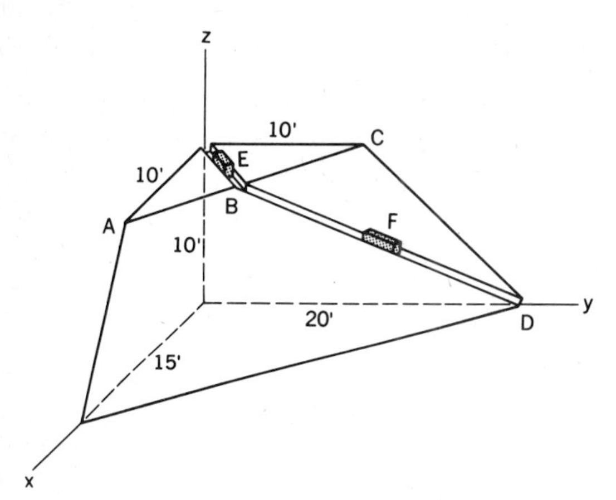

Figure P.13.81

13.82. A device is mounted on a platform that is rotating with an angular speed of 10 rad/sec. The device consists of two masses (each is .1 slug) rotating on a spindle with an angular speed of 5 rad-sec relative to the platform. The masses are moving radially outward with a speed of 10 ft/sec, and the entire platform is being raised at a speed of 5 ft/sec. Compute the kinetic energy of the system of two particles when they are 1 ft from the spindle.

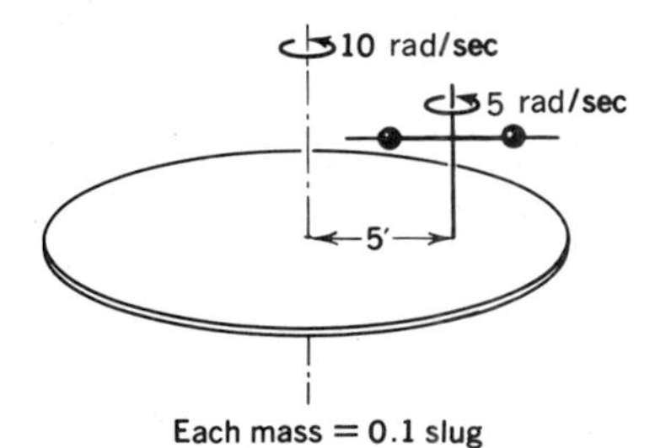

Figure P.13.82

13.83. A hoop, with four spokes, rolls without slipping such that the center C moves at a speed V of 1.7 m/sec. The diameter of the hoop is 3.3 m and the weight per unit length of the rim is 14 N/m. The spokes are uniform rods also having a weight per unit length of 14 N/m. Assume that rim and spokes are thin. What is the kinetic energy of the body?

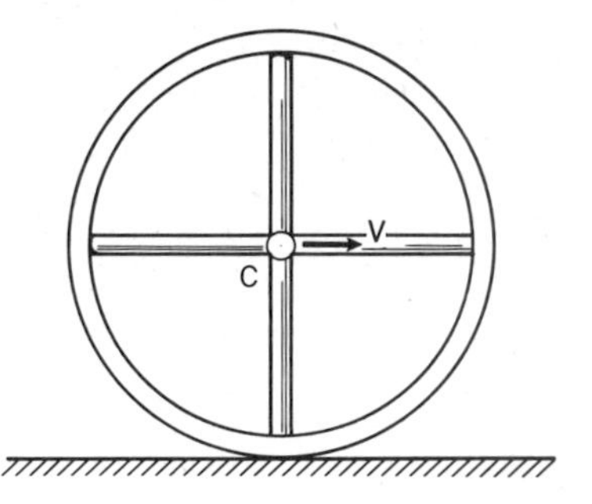

Figure P.13.83

13.84. A tank is moving at the speed V of 16 km/hr. What is the kinetic energy of each of the treads for this tank if they each have a mass per unit length of 300 kg/m?

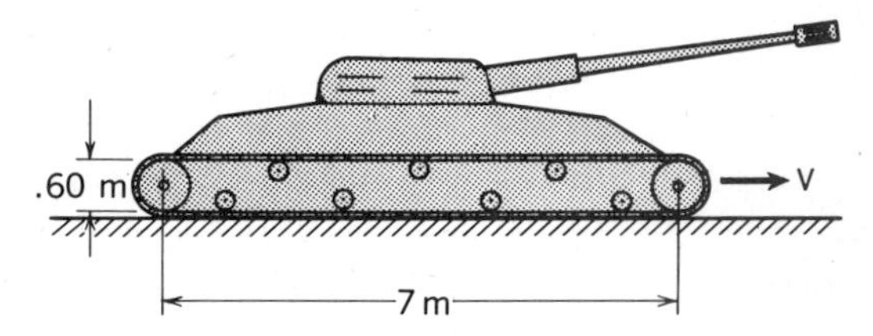

Figure P.13.84

13.85. A cylinder of radius R rotates about its own axis with an angular speed of ω. If the total mass is M, show that the kinetic energy is $\frac{1}{4}MR^2\omega^2$.

13.86. Cylinders B and C each weigh 100 lb and have a diameter of 2 ft. Body A, weighing 300 lb, rides on these cylinders. If there is no slipping anywhere, what is the kinetic energy of the system when the body A is moving at a speed V of 10 ft/sec? Use result of Problem 13.85 and see footnote 14 on page 545.

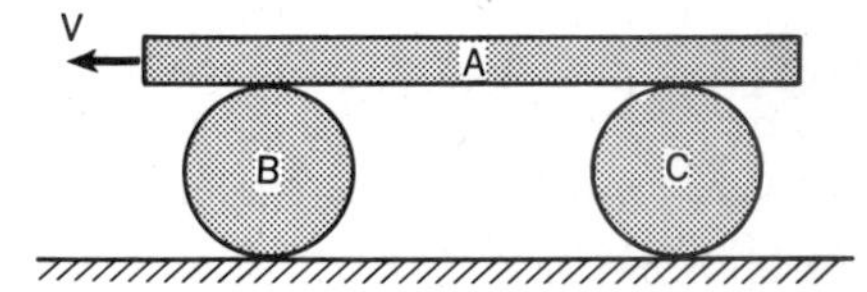

Figure P.13.86

13.87. A pendulum has a bob with a comparatively large uniform disc of diameter 2 ft and mass M of 3 lbm. At the instant shown, the system has an angular speed $\dot{\theta}$ of .3 rad/sec. If we neglect the mass of the rod, what is the kinetic energy of the pendulum at this instant? What error is incurred if one considers the bob to be a particle as we have done earlier for smaller bobs? Use the result of Problem 13.85.

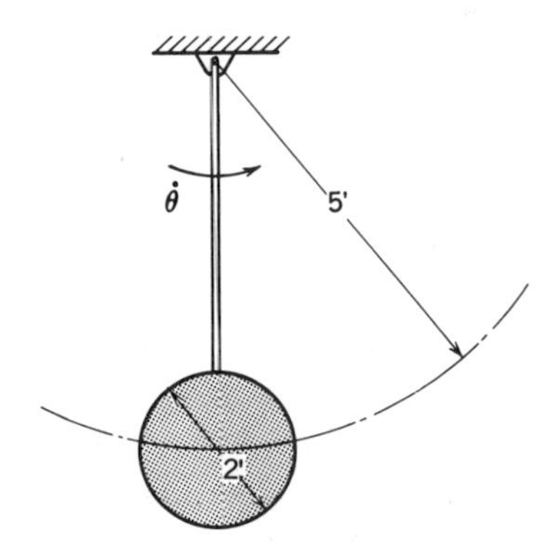

Figure P.13.87

13.88. In Problem 13.87, compute the maximum angle that the pendulum rises.

13.89. Do Example 13.3 by treating as an aggregate of particles.

13.90. Do Problem 13.25 by treating as an aggregate of particles.

13.91. Do Problem 13.19 by treating as an aggregate of particles.

13.92. Do Problem 13.21 by treating as an aggregate of particles.

13.93. A constant force F is applied to the axis of a cylinder, as shown, causing the axis to increase its speed from 1 ft/sec to 3 ft/sec in 10 ft without slipping. What is the friction force acting on the cylinder? The cylinder weighs 100 lb.

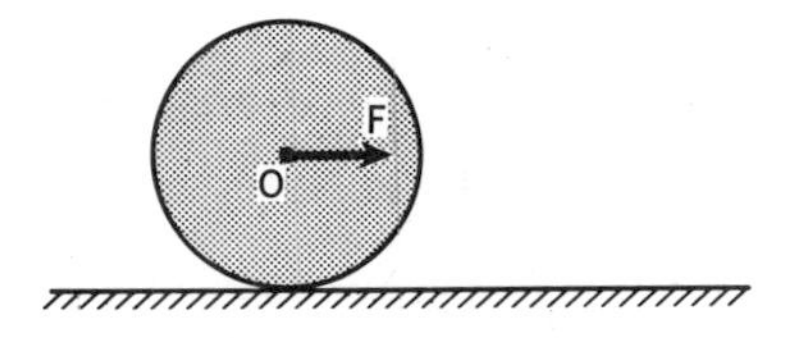

Figure P.13.93

13.94. A cylinder with a mass of 25 kg is released from rest on an incline, as shown. The inner diameter D of the cylinder is 300 mm. If the cylinder rolls without slipping, compute the speed of the centerline O after the cylinder has moved 1.6 m along the incline. Ascertain the friction force acting on the cylinder. The radius of gyration k at O is $.30/\sqrt{2}$ m.

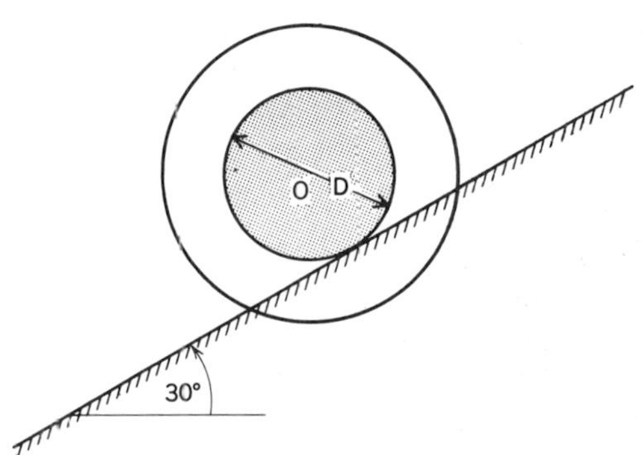

Figure P.13.94

13.95. A uniform cylinder having a diameter of 2 ft and a weight of 100 lb rolls down a 30° incline without slipping, as shown. What is the speed of the center after it has moved 20 ft? Compare this result with that for the case when there is no friction present. (*Hint:* Use the result of Problem 13.85.)

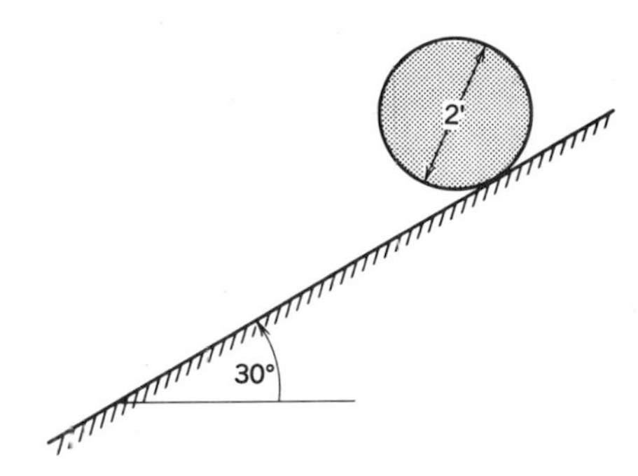

Figure P.13.95

13.96. Cylinders A and B each have a mass of 25 kg and a diameter of 300 mm. Block C, riding on A and B, has a mass of 100 kg.

If the system is released from rest at the configuration shown, what is the speed of C after the cylinders have made half a revolution? Use the result of Problem 13.85 and see footnote 14 on page 545.

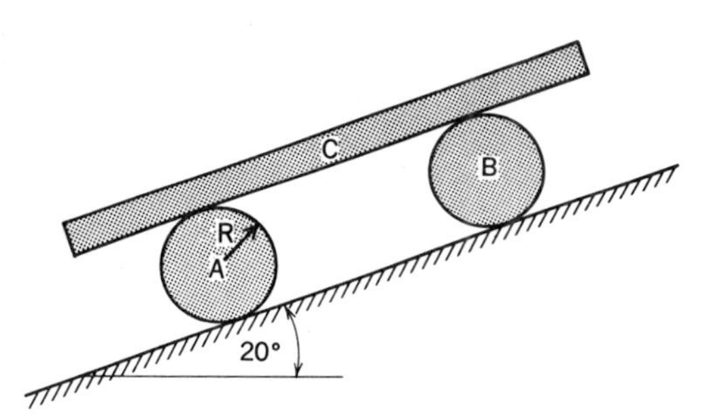

Figure P.13.96

13.97. Shown are two identical blocks A and B, each weighing 50 lb. A force F of 100 lb is applied to the lower block, causing it to move to the right. Block A, however, is restrained by the wall C. If block B reaches a speed of 10 ft/sec in 2 ft starting from rest at the position shown in the diagram, what is the restraining force from the wall? The coefficient of friction between B and the ground surface is .3. Do this problem first by using Eq. 13.28. Then check the result by using separate free-body diagrams, and so on.

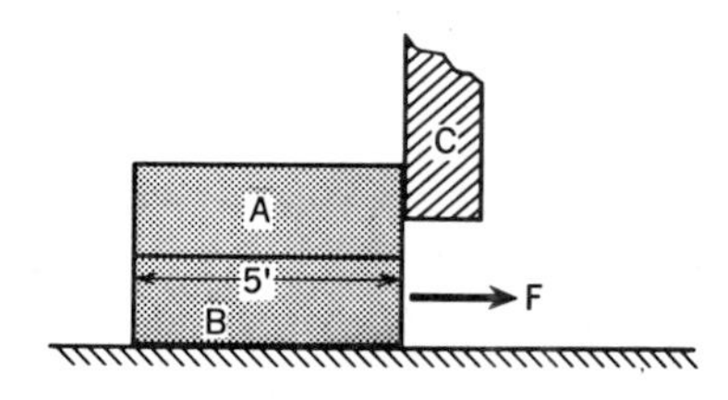

Figure P.13.97

13.98. What is the tension T to accelerate the end of the cable downward at the rate of 1.5 m/sec^2? From body C, weighing $50g$ N, is lowered a body D weighing $12.5g$ N at the rate of 1.5 m/sec^2 relative to body C. Neglect the inertia of pulleys A and B and the cable. (*Hint:* From earlier courses in physics, recall that pulley B is rotating instantaneously about point e, and hence point b has an acceleration half that of point f. We will consider such relations carefully at a later time.)

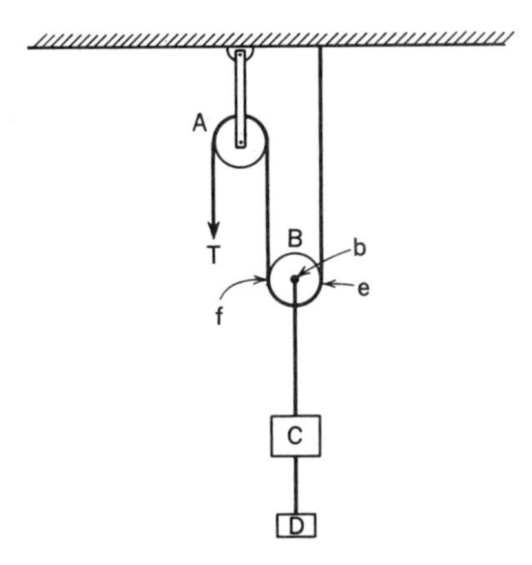

Figure P.13.98

13.9 Closure

In this chapter, we presented the energy method as applied to particles. In Part A, we presented three forms of the energy equation applied to a *single* particle. The basic equation was

$$\int_1^2 \boldsymbol{F} \cdot d\boldsymbol{r} = \tfrac{1}{2}(MV^2)_2 - \tfrac{1}{2}(MV^2)_1 \tag{13.38}$$

For the case of only conservative forces acting, we presented the equation for the *conservation of mechanical energy*:

$$(\text{PE})_1 + (\text{KE})_1 = (\text{PE})_2 + (\text{KE})_2 \tag{13.39}$$

Finally, for both conservative and nonconservative forces, we presented an equation resembling the *first law of thermodynamics* as it is usually employed:

$$\Delta(\text{PE} + \text{KE}) = W_{1-2} \tag{13.40}$$

In Part B, we considered a *system of particles* and presented the above equation again, but this time the work and potential-energy terms are from both *internal* and *external* force systems.[15] Furthermore, all work and potential-energy terms are evaluated by using the *actual movement* of the points of application of internal and external forces.

Next, we presented the work–energy equation for the *center of mass* of any system of particles:

$$\int_1^2 \boldsymbol{F} \cdot d\boldsymbol{r}_c = \tfrac{1}{2}(MV_c^2)_2 - \tfrac{1}{2}(MV_c^2)_1 \tag{13.41}$$

where $\boldsymbol{F}$, the resultant *external force, moves with the center of mass* in the computation of the work expression. We pointed out that the *single-particle* model is a *special case* of the use of Eq. 13.41 applicable when the motion of the center of mass of a body sufficiently describes the motion of a body and where the external forces on the body essentially move with the center of mass of the body.

To illustrate the use of the work–energy equation for a system of particles, we considered various elementary plane motions of simple rigid bodies. A more extensive treatment of the energy method applied to rigid bodies is found in Chapter 18.

We now turn to yet another useful set of relations derived from Newton's law, namely the methods of linear impulse-momentum and angular impulse-momentum for a particle and systems of particles.

Review Problems

13.99. A tractor exerts a force of 800 lb on a block A which has a coefficient of friction with block B of .7. Block B has a coefficient of friction of .2 with the ground. If block A weighs 400 lb and block B weighs 600 lb, what is the speed of the block A when, after starting from rest, the tractor has moved 2 ft? What is the acceleration of block B?

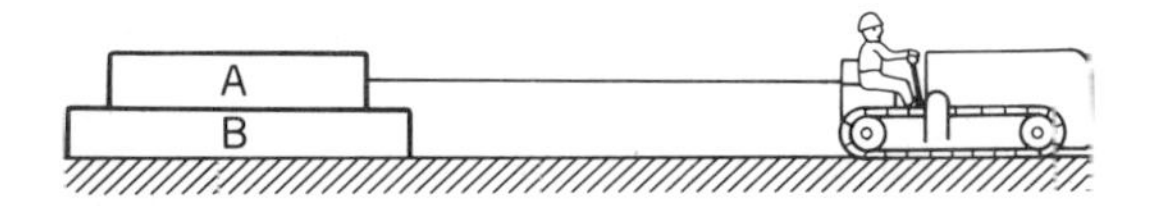

Figure P.13.99

13.100. A body A is released from a condition of rest on a frictionless circular surface. The body then moves on a horizontal surface CD whose coefficient of friction with the body is .2. A spring having a spring constant $K = 900$ N/m is positioned at C as shown in the diagram. How much will the spring be compressed? The body has a mass of 5 kg.

---

[15]As will be seen in Chapter 14, this equation for a system of particles is the *only one* that involves internal forces. Note, however, that for a *rigid body* the internal forces *do no work*.

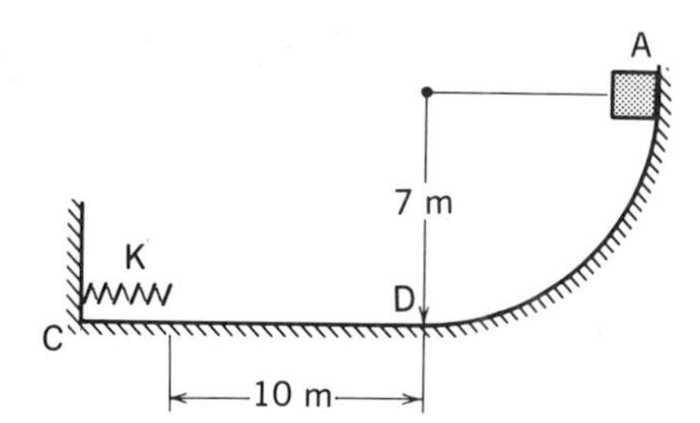

Figure P.13.100

13.101. Vehicle *B*, weighing 25 kN, is to go down a 30° incline. The vehicle is connected to body *A* through light pulleys and a capstan. What should body *A* weigh if starting from rest it restricts body *B* to a speed of 5 m/sec when *B* moves 3 m? There are two wraps of rope around the capstan.

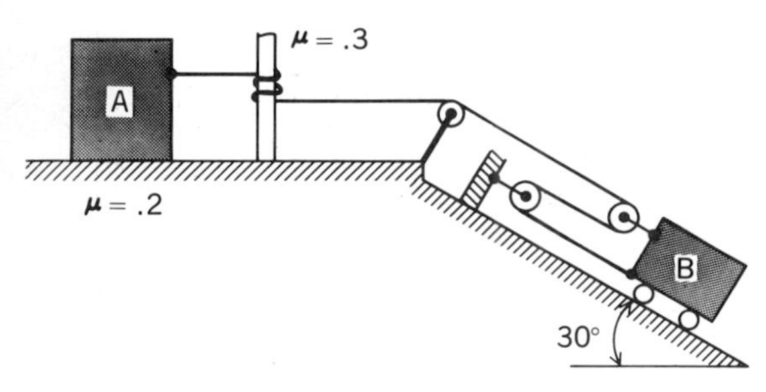

Figure P.13.101

13.102. A jet passenger plane is moving along the runway for a takeoff. If each of its four engines is developing 44.5 kN of thrust, what is the horsepower developed when the plane is moving at a speed of 240 km/hr?

13.103. Block *B*, with a mass of 200 kg, is being pulled up an incline. A motor *C* pulls on one cable, developing 4 hp. The other cable is connected to a counterweight *A* having a mass of 150 kg. If *B* is moving at a speed of 2 m/sec, what is its acceleration? *Hint:* Start with Newton's Law for *A* and *B*.

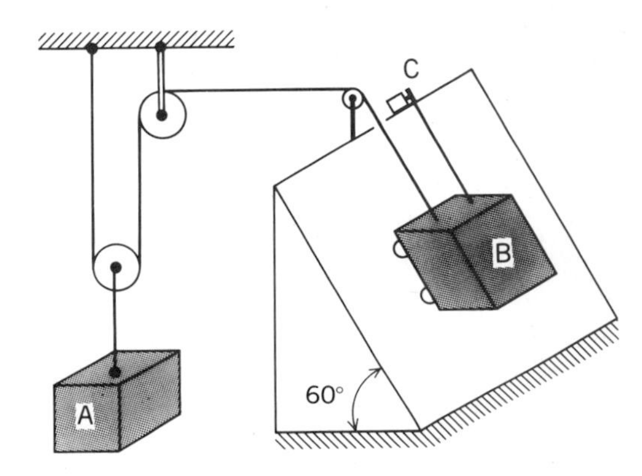

Figure P.13.103

13.104. A block *G* slides along a frictionless path as shown. What is the minimum initial speed that *G* should have along the path if it is to remain in contact when it gets to *A*, the uppermost position of the path? The block weighs 9 N. What is the normal force on the path when for the condition described the block is at position *B*?

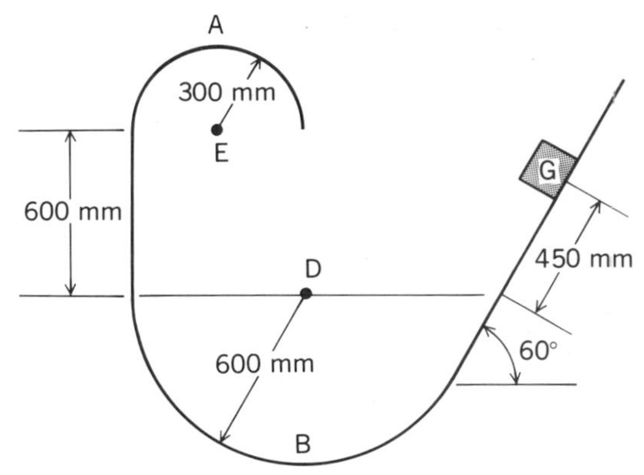

Figure P.13.104

13.105. A body *B* of mass 60 kg slides in a frictionless slot on an inclined surface as shown. An elastic cord connects *B* to *A*. The cord has a "spring constant" of 360 N/m. If the body *B* is released from rest from a position where the elastic cord is unstretched, what is body *B*'s speed after it moves .3 m?

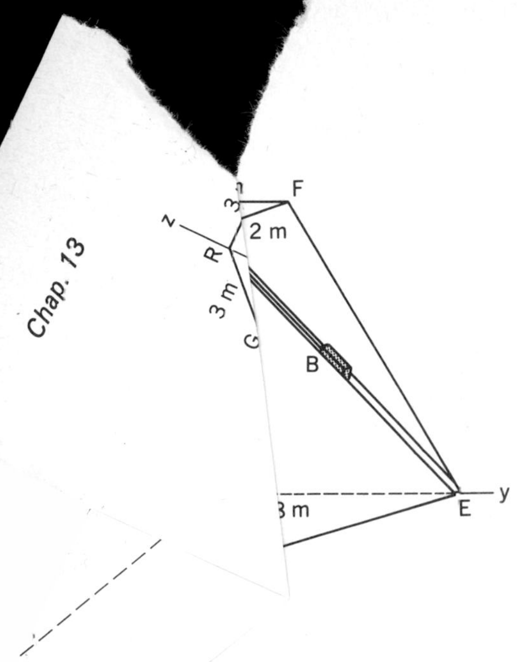

ıgure P.13.105

…ollar slides on a frictionless tube as shown. The spring is unstretched when in the horizontal position and has a spring constant of 1.0 lb/in. What is the minimum weight of A to just reach A' when released from rest from the position shown in the diagram? What is the force on the tube when A has traveled half the distance to A'?

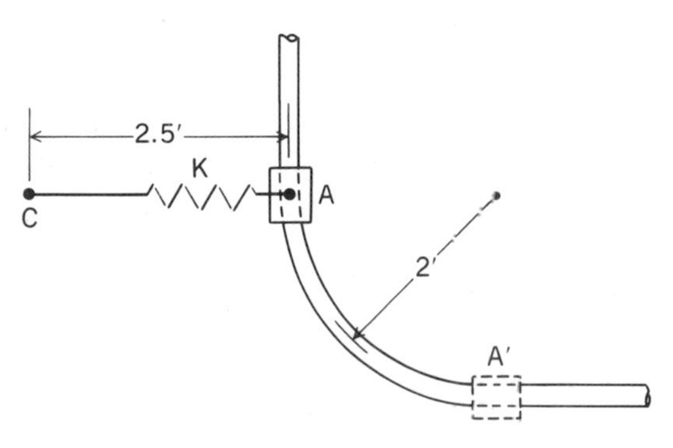

Figure P.13.106

13.107. Three blocks are connected by an inextensible flexible cable. The blocks are released from a rest configuration with the cable taut. If A can only fall a distance h equal to 2 ft, what is the velocity of bodies C and B after each has moved a distance of 3 ft? Each body weighs 100 lb. The coefficient of dynamic friction for body C is .3 and for body B is .2.

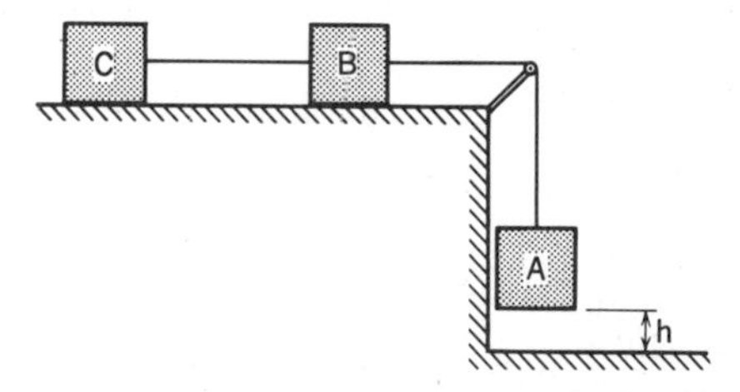

Figure P.13.107

13.108. A 15-kg vehicle has two bodies (each with mass 1 kg) mounted on it, and these bodies rotate at an angular speed of 50 rad/sec relative to the vehicle. If a 500-N force acts on the vehicle for a distance of 17 m, what is the kinetic energy of the system, assuming that the vehicle starts from rest and the bodies in the vehicle have constant rotational speed? Neglect friction and the inertia of the wheels.

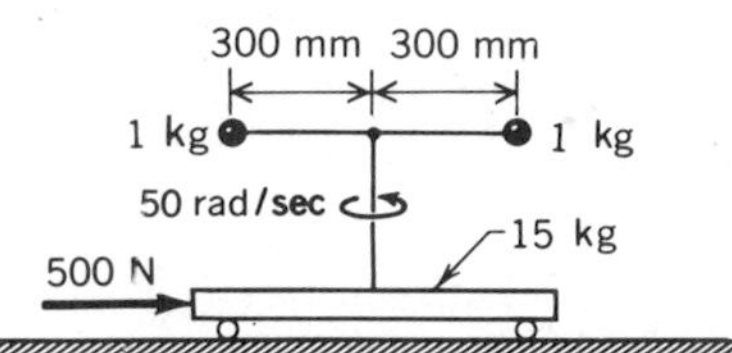

Figure P.13.108

13.109. Two identical solid cylinders each weighing 100 N support a load A weighing 50 N. If a force F of 300 N acts as shown, what is the speed of the vehicle after moving 5 m? Also, what is the total friction force on each wheel? Neglect the mass of the supporting system connecting the cylinders. Note that the kinetic energy of the angular motion of a cylinder about its own axis is $\frac{1}{4}MR^2\omega^2$. The system starts from rest.

Figure P.13.109

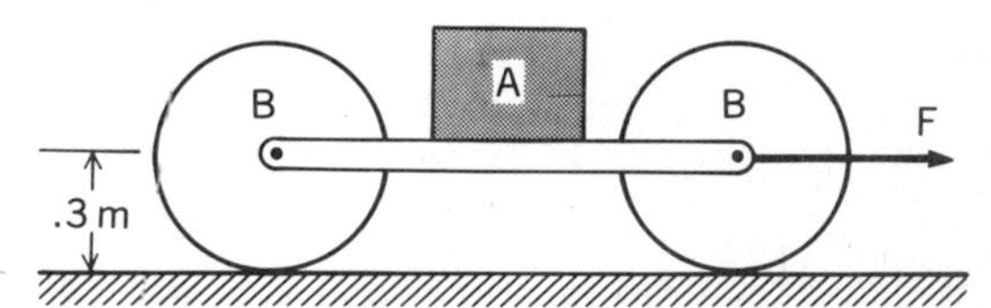

14 Methods of Momentum for Particles

Part A
Linear Momentum

14.1 Impulse and Momentum Relations for a Particle

In Section 12.3, we integrated differential equations of motion for particles that are acted upon by forces which are functions of time. In this chapter, we shall again consider such problems and shall present alternative formulations, called *methods of momentum*, for handling certain of these problems in a convenient and straightforward manner. We start by considering Newton's law for a particle:

$$\boldsymbol{F} = m\frac{d\boldsymbol{V}}{dt} \tag{14.1}$$

Multiply both sides by dt and integrate from some initial time t_i to some final time t_f:

$$\boxed{\int_{t_i}^{t_f} \boldsymbol{F}\,dt = \int_{t_i}^{t_f} m\frac{d\boldsymbol{V}}{dt}\,dt = m\boldsymbol{V}_f - m\boldsymbol{V}_i} \tag{14.2}$$

Note first that this is a vector equation, in contrast to the work–kinetic energy equation 13.2. The integral

$$\int_{t_i}^{t_f} \boldsymbol{F}\, dt$$

which we shall denote as $\boldsymbol{I}$, is called the *impulse* of the force $\boldsymbol{F}$ during the time interval $t_f - t_i$, whereas $m\boldsymbol{V}$ is the *linear-momentum vector* of the particle. Equaton 14.2, then, states that *the impulse $\boldsymbol{I}$ over a time interval equals the change in linear momentum of a particle during that time interval.* As we shall demonstrate later, the impulse of a force may be known even though the force itself is not known.

Finally, you must remember that to produce an impulse, a force need only exist for a time interval. Sometimes we use the work integral so much that we tend to think—erroneously—that a stationary force does not produce an impulse.

We now illustrate the use of the impulse-momentum equation.

EXAMPLE 14.1

A particle initially at rest is acted on by a force whose variation with time is shown graphically in Fig. 14.1. If the particle has a mass of 1 slug and is constrained to move rectilinearly in the direction of the force, what is the speed after 15 sec?

From the definition of the impulse, the area under the force–time curve will, in the one-dimensional example, equal the impulse magnitude. Thus, we simply compute this area between the times $t = 0$ and $t = 15$ sec:

$$\text{impulse} = \underset{\text{area 1}}{\tfrac{1}{2}(10)(10)} + \underset{\text{area 2}}{(5)(15)} = 125 \text{ lb-sec}$$

The final velocity, then, is given as

$$125 = (1)(V_f) - 0$$

Therefore,

$$V_f = 125 \text{ ft/sec}$$

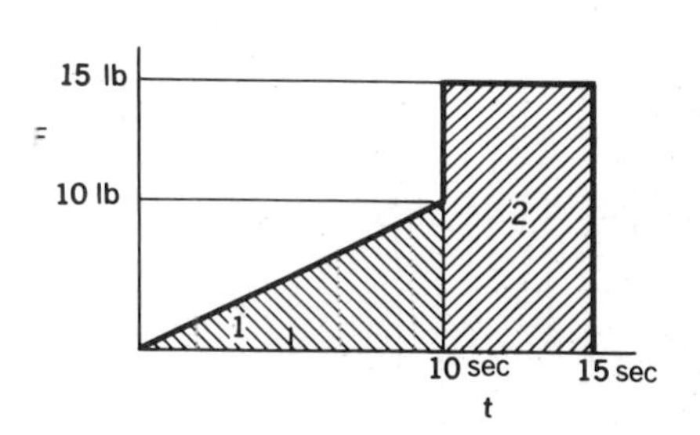

Figure 14.1. Force-versus-time plot.

Note that the impulse-momentum equation is useful when the force variation during a time interval is not a curve that can be conveniently expressed mathematically. The impulse which is the area under an F versus t curve can then be found with the help of a *planimeter*, thus permitting a quick solution of the velocity change during the time interval.[1]

EXAMPLE 14.2

A particle A with a mass of 1 kg has an initial velocity $\boldsymbol{V}_0 = 10\boldsymbol{i} + 6\boldsymbol{j}$ m/sec. After particle A strikes particle B, the velocity becomes $\boldsymbol{V} = 16\boldsymbol{i} - 3\boldsymbol{j} + 4\boldsymbol{k}$ m/sec.

[1]A planimeter is a mechanical device for measuring the area of a plane region bounded by an arbitrary curve.

If the time of encounter is 10 msec, what average force was exerted on the particle A? What is the change of linear momentum of particle B?

The impulse $\boldsymbol{I}$ acting on A is immediately determined by computing the change in linear momentum during the encounter:

$$\boldsymbol{I}_A = (1)(16\boldsymbol{i} - 3\boldsymbol{j} + 4\boldsymbol{k}) - (1)(10\boldsymbol{i} + 6\boldsymbol{j})$$
$$= 6\boldsymbol{i} - 9\boldsymbol{j} + 4\boldsymbol{k} \text{ N-sec}$$

Since

$$\int_{t_i}^{t_f} \boldsymbol{F}_A \, dt = (\boldsymbol{F}_{av})_A \, \Delta t$$

the average force $(\boldsymbol{F}_{av})_A$ becomes

$$(\boldsymbol{F}_{av})_A(0.010) = 6\boldsymbol{i} - 9\boldsymbol{j} + 4\boldsymbol{k}$$

Therefore,

$$(\boldsymbol{F}_{av})_A = 600\boldsymbol{i} - 900\boldsymbol{j} + 400\boldsymbol{k} \text{ N}$$

On the basis of the principle that action equals reaction, an equal but opposite average force must act on the object B during the 10-msec time interval. Thus, the impulse on particle B is $-\boldsymbol{I}_A$. Equating this impulse to the change in linear momentum, we get

$$\Delta(m\boldsymbol{V})_B = -\boldsymbol{I}_A = -6\boldsymbol{i} + 9\boldsymbol{j} - 4\boldsymbol{k} \text{ N-sec}$$

During impacts where the exact force variation is unknown, the impulse momentum principle is very useful. We shall examine impacts in more detail in a later section.

EXAMPLE 14.3

Two bodies, 1 and 2, are connected by an inextensible and weightless cord (Fig. 14.2). Initially, the bodies are at rest. If the dynamic coefficient of friction is μ for body 1 on the surface inclined at angle α, compute the velocity of the bodies at any time t before body 1 has reached the end of the incline.

Since only constant forces exist and since a time interval has been specified, we can use momentum considerations advantageously. The free-body diagrams of bodies 1 and 2 are shown in Fig. 14.3. Equilibrium considerations lead to the conclusion that $N_1 = W_1 \cos\alpha$, so the friction force f_1 is

$$f_1 = \mu N_1 = \mu W_1 \cos\alpha$$

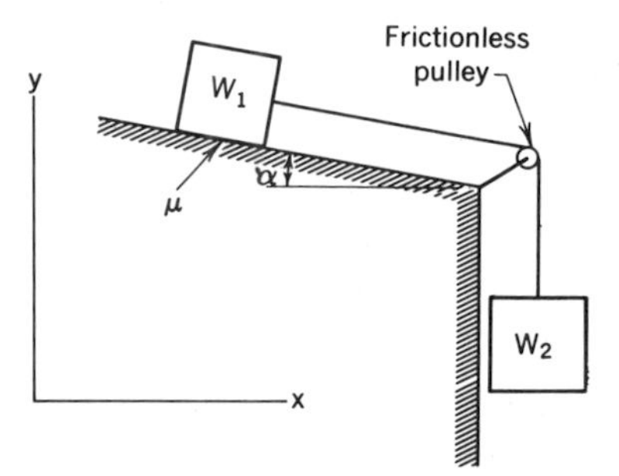

Figure 14.2. Two bodies connected by a cord.

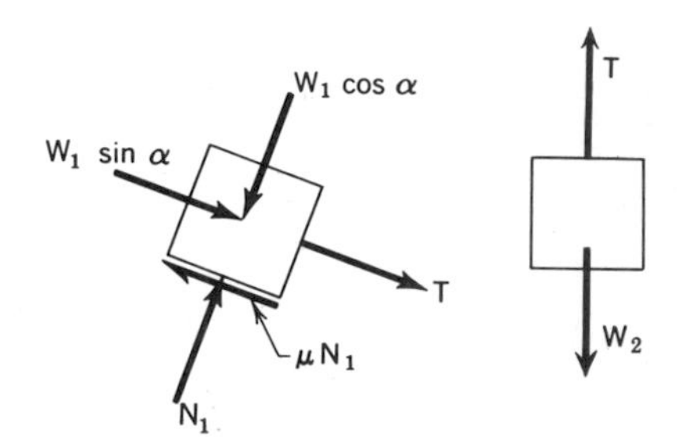

Figure 14.3. Free-body diagrams of W_1 and W_2.

For body 1, take the component of the linear impulse–momentum equation along the incline:

$$\int_0^t (-\mu W_1 \cos \alpha + W_1 \sin \alpha + T)\, dt = \frac{W_1}{g}(V - 0)$$

Carrying out the integration, we have

$$(-\mu W_1 \cos \alpha + W_1 \sin \alpha + T)\, t = \frac{W_1}{g} V \tag{a}$$

For body 2, we have for the momentum equation in the vertical direction:

$$\int_0^t (W_2 - T)\, dt = \frac{W_2}{g}(V - 0)$$

where, because of the inextensible property of the cable and the frictionless condition of the pulley, the magnitudes of the velocity V and the force T are the same for bodies 1 and 2. Integrating the equation above, we write

$$(W_2 - T)t = \frac{W_2}{g} V \tag{b}$$

By adding Eqs. (a) and (b), we can eliminate T and solve for the desired unknown V. Thus,

$$(-\mu W_1 \cos \alpha + W_1 \sin \alpha + W_2)t = \frac{V}{g}(W_1 + W_2)$$

Therefore,

$$V = \frac{gt}{W_1 + W_2}(W_2 + W_1 \sin \alpha - \mu W_1 \cos \alpha) \tag{c}$$

Note that we have used considerations of linear momentum for a *single* particle each time in solving this problem.

EXAMPLE 14.4

A conveyor belt is moving from left to right at a constant speed V of 1 ft/sec in Fig. 14.4. Two hoppers drop objects onto the belt at the total rate n of 4 per second. The objects each have a weight W of 2 lb and fall a height h of 1 ft before landing on the conveyor belt. Farther along the belt (not shown) the objects are removed by personnel so that, for steady-state operation, the number N of objects on the belt at any time is 10. If the dynamic coefficient of friction between belt and conveyor bed is .2, estimate the average difference in tension $T_2 - T_1$ of the belt to maintain this operation. The weight of the belt on the conveyor bed is 10 lb.

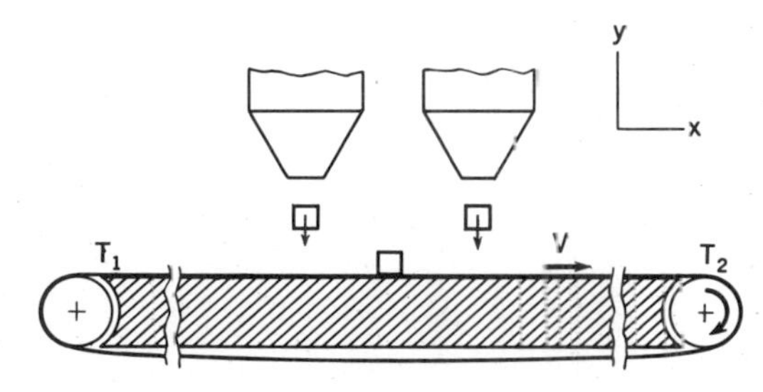

Figure 14.4. Objects falling on moving conveyor.

We shall superimpose the following effects to get the desired result.

1. A friction force from the bed onto the belt results from the static weight of the ten objects riding on the belt and the weight of the portion of belt on the bed.
2. A friction force from the bed onto the belt results from the force in the y direction needed to change the *vertical* linear momentum of the falling objects from a value corresponding to the free-fall velocity just before impact ($\sqrt{2gh}$) to a value of zero after impact.
3. Finally, the belt must supply a force in the x direction to change the *horizontal* linear momentum of the falling objects from a value of zero to a value corresponding to the speed of the belt.

Thus, we have for the first contribution, which we denote at ΔT_1, the following result:

$$\Delta T_1 = (NW + 10)\mu = [(10)(2) + 10](.2) = 6 \text{ lb} \tag{a}$$

As for the second contribution, we can only compute an average value $(\Delta T_2)_{av}$ by noting that each impacting object is given a vertical change in linear momentum equal to

$$\begin{aligned}\text{vertical change in linear momentum per object} &= \frac{W}{g}(\sqrt{2gh}) \\ &= \frac{2}{g}\sqrt{(2g)(1)} \\ &= .498 \text{ lb-sec}\end{aligned}$$

where we have assumed a free fall starting with zero velocity at the hopper. For four impacts per second, we have as the total vertical change in linear momentum per second the value 1.994 lb-sec. The average vertical force during the 1-sec interval to give the impulse needed for this change in linear momentum is clearly 1.994 lb. Since this result is correct for every second, 1.994 lb is the average normal force that the bed of the conveyor must transmit to the belt for arresting the vertical motion of the falling objects. The desired $[(\Delta T)_2]_{av}$ for the belt arising from friction is accordingly given as

$$[(\Delta T)_2]_{av} = (\mu)(1.994) = .399 \text{ lb} \tag{b}$$

Finally, for the last contribution $[(\Delta T)_3]_{av}$, we note that the belt must give in the horizontal direction for each impacting object a change in linear momentum having the value

$$\begin{aligned}\text{horizontal change in linear momentum per object} &= \left(\frac{W}{g}\right)(1) \\ &= .0621 \text{ lb-sec}\end{aligned}$$

For four impacts per second we have as the total horizontal change in linear momentum developed by the belt during 1 sec the value .248 lb-sec. The average horizontal force during 1 sec needed for this change in linear momentum is clearly .248 lb. Thus, we have

$$[(\Delta T)_3]_{av} = .248 \text{ lb} \tag{c}$$

The total average difference in tension is then

$$(\Delta T)_{av} = 6 + .399 + .248 = 6.65 \text{ lb} \tag{d}$$

14.2 Linear-Momentum Considerations for a System of Particles

In Section 14.1, we considered impulse-momentum relations for a single particle. Although Examples 14.3 and 14.4 involved more than one particle, nevertheless the impulse-momentum considerations were made on one particle at a time. We now wish to set forth impulse-momentum relations for a *system* of particles.

Let us accordingly consider a system of n particles. We may start with Newton's law as developed previously for a system of particles:

$$\boldsymbol{F} = \sum_{j=1}^{n} m_j \frac{d\boldsymbol{V}_j}{dt} \tag{14.3}$$

Since we know that the internal forces cancel, $\boldsymbol{F}$ must be the *total external* force on the system of n particles. Multiplying by dt, as before, and integrating between t_i and t_f, we write

$$\boxed{\int_{t_i}^{t_f} \boldsymbol{F}\, dt = \boldsymbol{I}_{\text{ext}} = \left(\sum_{j=1}^{n} m_j \boldsymbol{V}_j\right)_f - \left(\sum_{j=1}^{n} m_j \boldsymbol{V}_j\right)_i} \tag{14.4}$$

Thus, we see that *the impulse of the total external force on the system of particles during a time interval equals the sum of the changes of the linear momentum vectors of the particles during the time interval.*

We now consider an example.

EXAMPLE 14.5

A 3-ton truck is moving at a speed of 60 mi/hr. [See Fig. 14.5(a).] The driver suddenly applies his brakes at time $t = 0$ so as to lock his wheels in a panic stop. Load A weighing 1 ton breaks loose from its ropes and at time $t = 4$ sec is sliding *relative to the truck* at a speed of 3 ft/sec. What is the speed of the truck at that time? Take μ_d between the tires and pavement to be .4.

Since we do not know the nature of the forces between the truck and load A while the latter is breaking loose, it is easiest to consider the *system* of two particles comprising the truck and the load simultaneously whereby the aforementioned forces become internal and are not considered. Accordingly, we have shown the system with all the external loads in Fig. 14.5(b). Clearly, $N = (4)(2000) = 8000$ lb and the friction force is $(.4)(8000)$ 3200 lb. We now employ Eq. 14.4 in the x direction as follows:

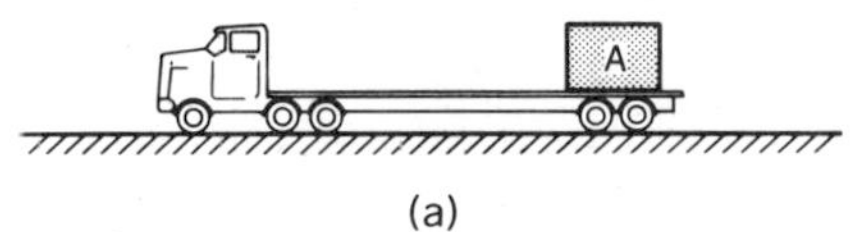

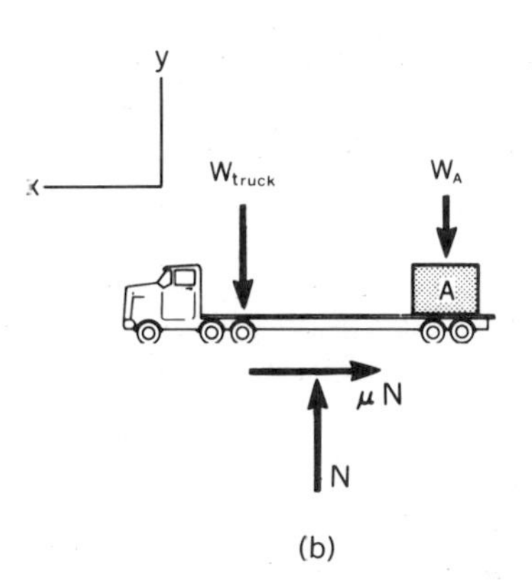

Figure 14.5. Truck undergoing panic stop.

$$\int_{t_1}^{t_2} F_x\,dt = (\sum_j m_j V_j)_2 - (\sum_j m_j V_j)_1$$

$$\int_0^4 (-3200)\,dt = \left[\frac{(3)(2000)}{g} V_2 + \frac{(1)(2000)}{g}(V_2 + 3)\right] \tag{a}$$

$$- \left[\frac{4(2000)}{g} \frac{(60)(5280)}{3600}\right]$$

Note that the first quantity inside the first brackets on the right side of Eq. (a) is the momentum of the truck at $t = 4$ sec, and the second quantity inside the same brackets is the momentum of the load at this instant. We may readily solve for V_2:

$$V_2 = 35.7 \text{ ft/sec}$$

Introducing *mass-center* quantities into Eq. 14.4 is easy and sometimes advantageous. You will remember that

$$M\boldsymbol{r}_c = \sum_{j=1}^{n} m_j \boldsymbol{r}_j \tag{14.5}$$

Differentiating with respect to time, we get

$$M\boldsymbol{V}_c = \sum_{j=1}^{n} m_j \boldsymbol{V}_j \tag{14.6}$$

Thus, we see from this equation that *the total linear momentum of a system of particles equals the linear momentum of a particle that has the total mass of the system and that moves with the velocity of the mass center.* Using Eq. 14.6 to replace the right side of Eq. 14.4, we can say:

$$\boxed{\int_{t_i}^{t_f} \boldsymbol{F}\,dt = \boldsymbol{I}_{\text{ext}} = M(\boldsymbol{V}_c)_f - M(\boldsymbol{V}_c)_i} \tag{14.7}$$

Thus, *the total external impulse on a system of particles equals the change in linear momentum of a hypothetical particle having the mass of the entire aggregate and moving with the mass center.*

When the separate motions of the individual particles are reasonably simple, as a result of constraints, and the motion of the mass center is not easily available, then Eq. 14.4 can be employed for linear-momentum considerations as was the case for Example 14.5. On the other hand, when the motions of the particles individually are very complex and the motion of the mass center of the system is reasonably simple, then clearly Eq. 14.7 can be of great value for linear-momentum considerations. Also, as in the case of energy considerations, we note that the single-particle model is really a *special case* of the center-of-mass formulation above, wherein the motion of the center of mass of a body describes sufficiently the motion of the body in question.

EXAMPLE 14.6

A truck in Fig. 14.6 has two rectangular compartments of identical size for the purpose of transporting water. Each compartment has the dimensions 20 ft $\times$ 10 ft

× 8 ft. Initially, tank A is full and tank B is empty. A pump in tank A begins to pump water from A to at the rate Q_1 of 10 cfs (cubic feet per second) and 10 sec later is delivering wa at the rate Q_2 of 30 cfs. If the level of the water in the tanks remains horizontal, at is the average horizontal force needed to restrain the truck from moving dur this interval?

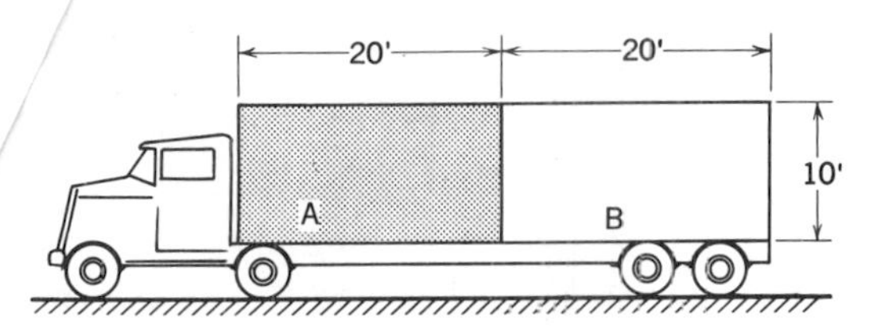

Figure 14.6. Truck with tank compartments.

n this setup, the mass center of the water in the tanks is moving from left to t and moving nonuniformly during the time interval of interest. We show the ater in Fig. 14.7 at some time t where the level in tank A has dropped an amount

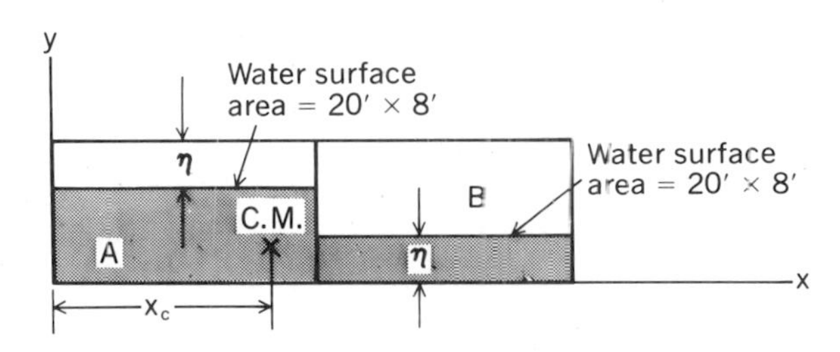

Figure 14.7. Compartments showing flow of water.

η while, by conservation of mass, the level in tank B has risen exactly the same amount η. The position x_c of the center of mass at this instant can be readily calculated in terms of η. Thus, using the basic definition of the center of mass, we can say:

$$Mx_c = (M_A)(x_A) + (M_B)(x_B)$$

$$[(20)(8)(10)](\rho)(x_c) = [(20)(8)(10 - \eta)](\rho)(10) + [(20)(8)(\eta)](\rho)(30) \qquad \text{(a)}$$

Since we are interested in the time rate of change of x_c so that we can profitably employ Eq. 14.7, we next differentiate with respect to time as follows:

$$[(20)(8)(10)](\rho)(\dot{x}_c) = -[(20)(8)\dot{\eta}](\rho)(10) + [(20)(8)\dot{\eta}](\rho)(30) \qquad \text{(b)}$$

But $(20)(8)\dot{\eta}$ is the volume of flow[2] from tank A to tank B at time t. Using Q to represent this volume flow, we get for the equation above:

$$[(20)(8)(10)](\rho)\dot{x}_c = -(\rho)(10)Q + (\rho)(30)Q = 20(\rho)(Q)$$

[2]Remember that 20 ft × 8 ft is the area of the top water surface in each tank, as shown in Fig. 14.7.

Solving for $\dot{x}_c$, we have

$$\dot{x}_c = \frac{1}{80} Q \tag{c}$$

Now consider the momentum equation in the x direction the water using the center of mass. We can say from Eq. 14.7:

$$\int_0^{10} F\,dt = [(M\dot{x}_c)_2 - (M\dot{x}_c)_1]$$

Therefore,

$$\begin{aligned}(F_{av})(10) &= [(20)(8)(10)(\rho)][(\dot{x}_c)_2 - (\dot{x}_c)_1] \\ &= [(20)(8)(10)(\rho)][\tfrac{1}{80}(Q_2 - Q_1)]\end{aligned}$$

where we have used Eq. (c) in the last step. Putting in $Q_2 = 30$ cfs and $Q_1 =$ we then get for the average force during the 10-sec interval of interest on using 62.4/g slugs/ft³:

$$F_{av} = 77.5 \text{ lb}$$

This is the average horizontal force that the truck exerts on the water. Clearly, this force is also what the ground must exert on the truck in the horizontal direction to prevent motion of the truck during the water transfer operation.

If the total external force on a system of particles is zero, it is clear from the previous discussion that there can be no change in the linear momentum of the system. This is the principle of *conservation of linear momentum*, which means, furthermore, that *with a zero total impulse on an aggregate of particles, there can be no change in the velocity of the mass center*. If at some time t_0 the velocity of the mass center of such a system of particles is zero, then this velocity must remain zero if the impulse on the system of particles is zero. That is, no matter what movements and gyrations the elements of the system may have, they must be such that the center of mass must remain stationary. We reached the same conclusion in Chapter 12, where we found from Newton's law that if the total external force on a system of particles is zero, then the acceleration of the center of mass is zero.[3]

14.3 Impulsive Forces

Let us now examine the action involved in the explosion of a bomb that is initially suspended from a wire, as shown in Fig. 14.8. First, consider the situation *directly after* the explosion has been set off. Since very large forces are present from expanding gases, a *fragment* of the bomb receives an appreciable impulse during this short time interval. Also, directly after the explosion, the gravitational forces are no longer counteracted by the supporting wire, so there is an additional impulse acting on the fragment. But since the gravitational force is small compared to forces from the explosion, the gravitational impulse on a fragment can be considered negligibly small for the short period of time under discussion compared to that of the expanding gases on the fragment. A plot of explosive force and the force of gravity on a fragment is

[3]Problems 12.91, 12.92, and 12.93 are examples of this condition.

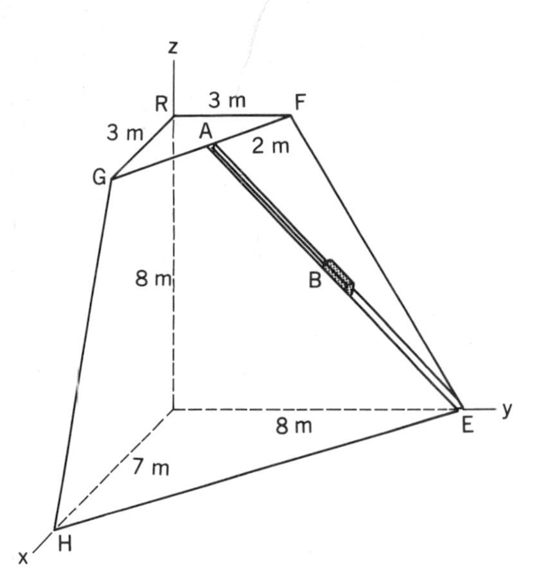

Figure P.13.105

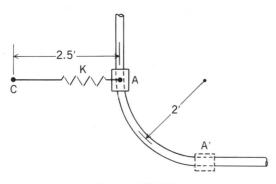

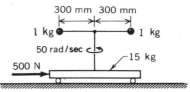

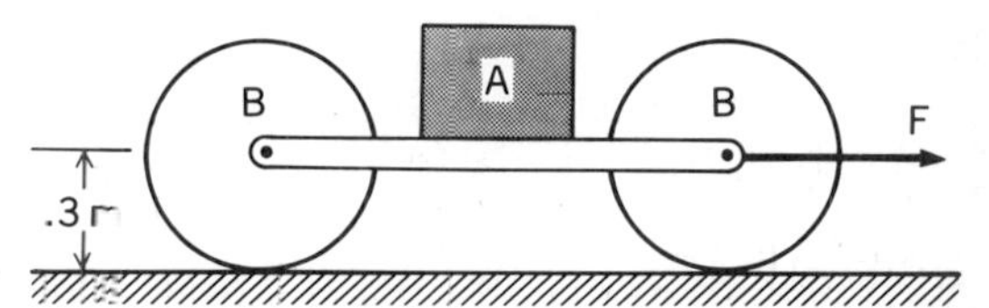

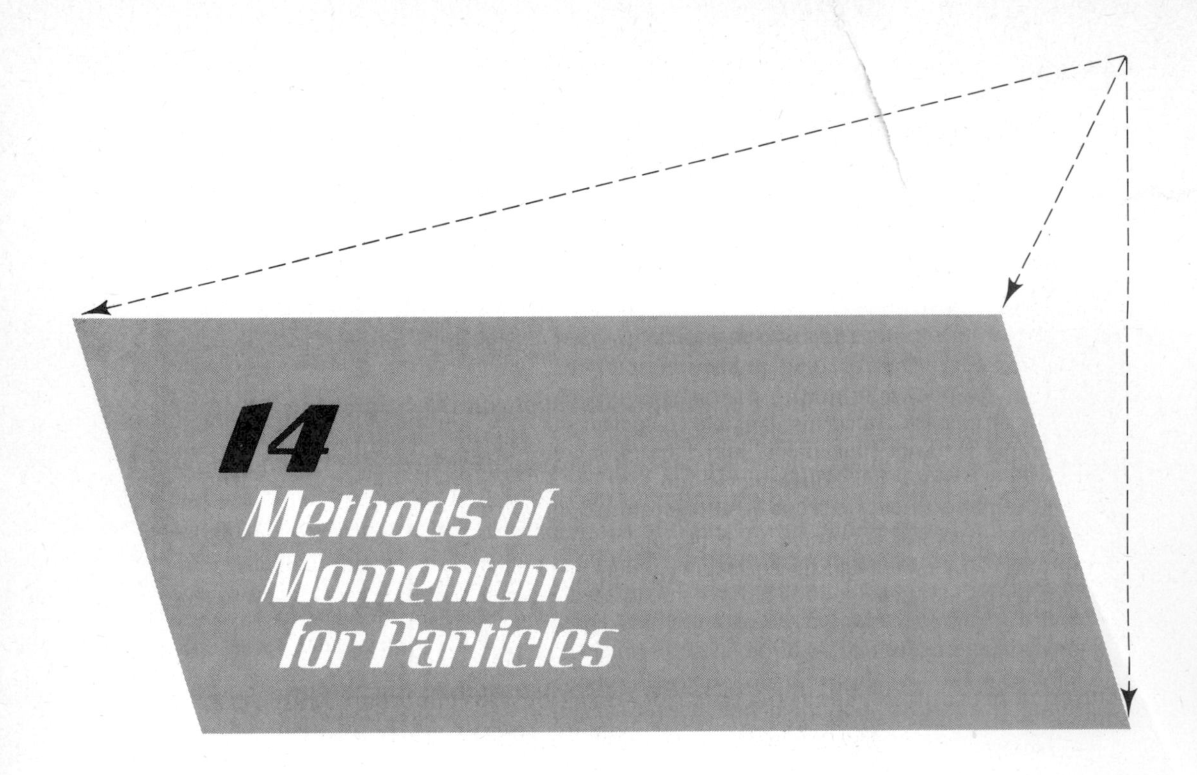

$$M\mathbf{r}_c = \sum_{j=1}^{n} m_j \mathbf{r}_j \tag{14.5}$$

$$M\mathbf{V}_c = \sum_{j=1}^{n} m_j \mathbf{V}_j \tag{14.6}$$

$$\boxed{\int_{t_i}^{t_f} \mathbf{F}\,dt = \mathbf{I}_{\text{ext}} = M(\mathbf{V}_c)_f - M(\mathbf{V}_c)_i} \tag{14.7}$$

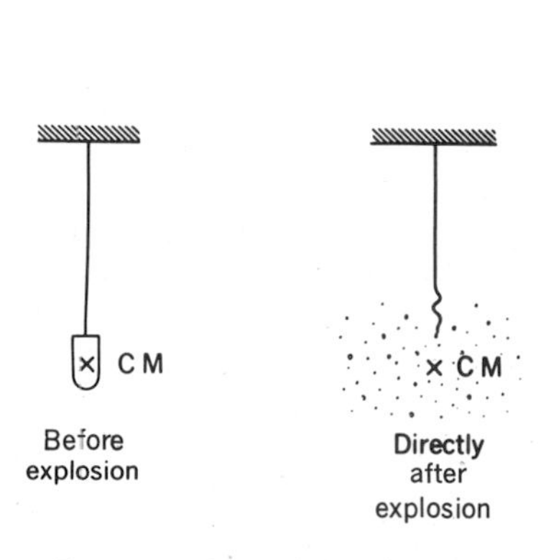

Figure 14.8. Exploding bomb.

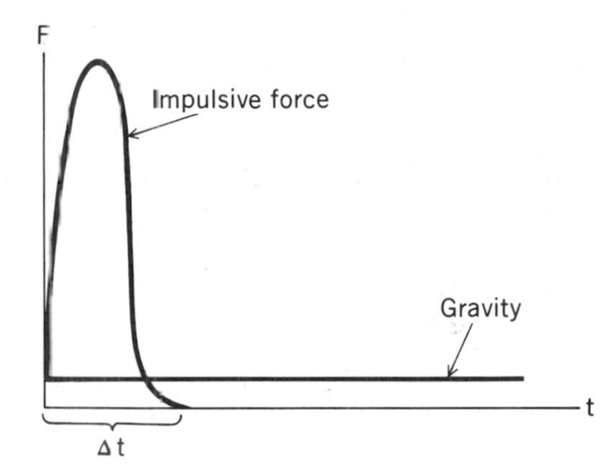

Figure 14.9. Plot of explosive force and gravity force.

shown in Fig. 14.9. It is clear from this diagram that an impulse for the explosive force for a very short time Δt can be significant, whereas the impulse from gravity during the same short time is by comparison negligible. Forces that act over a very short time but have nevertheless appreciable impulse are called *impulsive forces*. In actions involving very small time intervals, we need only consider impulsive forces. Furthermore, during a very short time Δt an impulsive force acting on a particle can change the velocity of the particle in accordance with the impulse-momentum equation an appreciable amount while the particle undergoes very little change in position during the time Δt. It is simplest in many cases to consider the *change in velocity of a particle from an impulsive force to occur over zero distance.*

Up to now, we have only considered a fragment of the bomb. Now let us consider all the fragments of the bomb taken as a system of particles. Since the explosive action is *internal* to the bomb, the action causes impulses that for any direction have equal and opposite counterparts, and thus *the total impulse on the bomb due to the explosion is zero*. We can thus conclude that *directly after* the explosion *the center of mass of the bomb has not moved appreciably* despite the high velocity of the fragments in all directions, as illustrated in Fig. 14.8. As time progresses beyond the short time interval described above, the gravitational impulse increases and has significant effect. If there were no friction, the center of mass would descend from the position of support as a freely falling body under this action of gravity.

The following problem will illustrate these ideas.

EXAMPLE 14.7

A 9000-N idealized cannon with a recoil spring ($K = 4000$ N/m) fires a 45-N projectile with a muzzle velocity of 625 m/sec at an angle of 50° (Fig. 14.10). Determine the maximum compression of the spring.

The firing of the cannon takes place in a very short time interval. The force on the projectile and the force on the cannon from the explosion are impulsive forces. As a result, the cannon can be considered to achieve a recoil velocity instantaneously without having moved appreciably. However, like the exploding bomb, the impulse

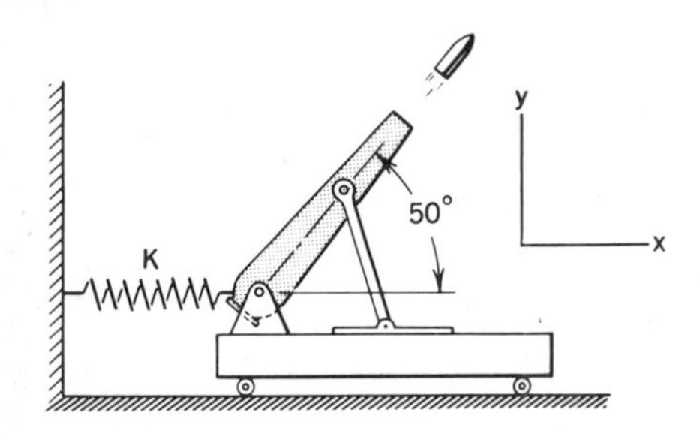

Figure 14.10. Idealized cannon.

on the cannon *plus* projectile is zero, as a result of the firing process. Since the linear momentum of the cannon plus projectile is zero just before firing, this linear momentum must be zero directly after firing. Thus, just after firing, we can say for the x direction:

$$(MV_x)_{\text{cannon}} + (MV_x)_{\text{projectile}} = 0 \qquad \text{(a)}$$

Using V_c for the cannon velocity,

$$\frac{9000}{g} V_c + \frac{45}{g}[(625)(\cos 50°) + V_c] = 0$$

Solving for V_c, we get

$$V_c = -2.00 \text{ m/sec} \qquad \text{(b)}$$

After this initial impulsive action, which results in an instantaneous velocity being imparted to the cannon, the motion of the cannon is then impeded by the spring. We may now use *conservation of mechanical energy* for a particle in this phase of motion of the cannon. Denoting δ as the maximum deflection of the spring, we can say:

$$\frac{1}{2}\frac{9000}{g}(2.00)^2 = \frac{1}{2}(4000)(\delta^2)$$

Therefore,

$$\delta = .958 \text{ m}$$

EXAMPLE 14.8

For target practice, a 9-N rock is thrown into the air and fired on by a pistol. The pistol bullet, of mass 57 g and moving with a speed of 312 m/sec, strikes the rock as it is descending vertically at a speed of 6.25 m/sec. [See Fig. 14.11(a).] Both the velocity of the bullet and the rock are parallel to the xy plane. After the bullet hits the rock, the rock breaks up into two pieces, A weighing 5.78 N and B weighing 3.22 N. What is the velocity of B after collision for the given postcollision velocities of the bullet and piece A shown in Fig. 14.11(b)? The indicated 219-m/sec and 25-

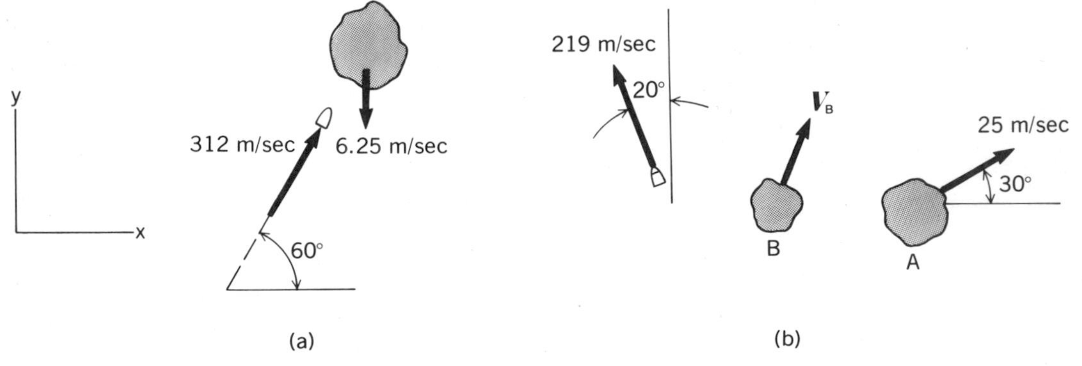

Figure 14.11. Bullet striking a rock.

m/sec velocities are in the xy plane. If we neglect wind resistance, how high up does the center of mass of the rock and bullet system rise after collision?

Linear momentum is conserved during the collision, so we can equate linear momenta directly before and directly after collision. Thus,

$$(.057)(312)(.5\boldsymbol{i} + .866\boldsymbol{j}) + \frac{9}{g}(-6.25\boldsymbol{j})$$
$$= (.057)(219)(-\sin 20°\boldsymbol{i} + \cos 20°\boldsymbol{j})$$
$$+ \frac{5.78}{g}25(.866\boldsymbol{i} + .5\boldsymbol{j}) + \frac{3.22}{g}[(V_B)_x\boldsymbol{i} + (V_B)_y\boldsymbol{j}]$$

We may solve for the desired quantities $(V_B)_x$ and $(V_B)_y$ to get

$$(V_B)_x = 1.235 \text{ m/sec}$$
$$(V_B)_y = -28.7 \text{ m/sec}$$

We now compute the velocity of the center of mass just before collision. Thus,

$$MV_c = \left(\frac{9}{g} + .057\right)V_c = \frac{9}{g}(-6.25)\boldsymbol{j} + .057(312)(.5\boldsymbol{i} + .866\boldsymbol{j})$$

Therefore,

$$V_c = 9.125\boldsymbol{i} + 9.92\boldsymbol{j} \text{ m/sec}$$

Hence, for the center of mass there is an initial velocity upward of 9.92 m/sec just before collision. Directly after collision, the center of mass still has this upward speed. But now considering larger time intervals, we must take into account the action of gravity which gives the center of mass a downward acceleration of 9.81 m/sec^2. Thus,

$$\ddot{y}_c = -9.81$$
$$\dot{y}_c = -9.81t + C_1$$
$$y_c = -9.81\frac{t^2}{2} + C_1 t + C_2$$

When $t = 0$, $\dot{y}_c = 9.92$ and we take $y_c = 0$ for convenience. Hence, we have

$$\dot{y}_c = -9.81t + 9.92 \quad \text{(a)}$$
$$y_c = -9.81\frac{t^2}{2} + 9.92t \quad \text{(b)}$$

Set $\dot{y}_c$ in (a) equal to zero and solve for t. We get

$$t = 1.011 \text{ sec}$$

Substitute this value of t in Eq.(b) and solve for y_c, which now gives the desired maximum elevation of the center of mass after collision. Thus,

$$(y_c)_{\max} = 5.01 \text{ m} \quad \text{(c)}$$

Problems

14.1. A body weighing 100 lb reaches an incline of 30° while it is moving at 50 ft/sec. If the coefficient of friction is .3, how long before the body stops?

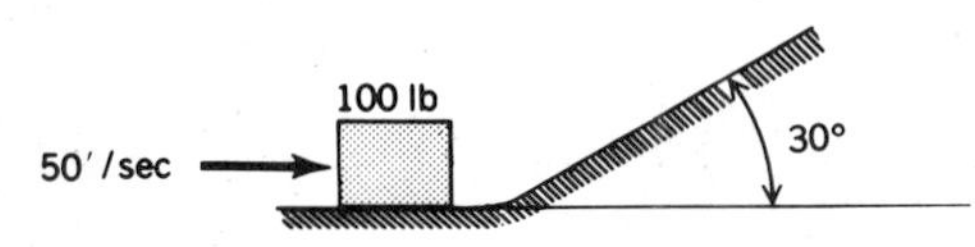

Figure P.14.1

14.2. A particle of mass 1 kg is initially stationary at the origin of a reference. A force having a known variation with time acts on the particle. That is,

$$\boldsymbol{F}(t) = t^2\boldsymbol{i} + (6t + 10)\boldsymbol{j} + 1.6t^3\boldsymbol{k}\ \text{N}$$

where t is in seconds. After 10 sec, what is the velocity of the body?

14.3. A unidirectional force acting on a particle of mass 16 kg is plotted. What is the velocity of the particle at 40 sec? Initially, the particle is at rest.

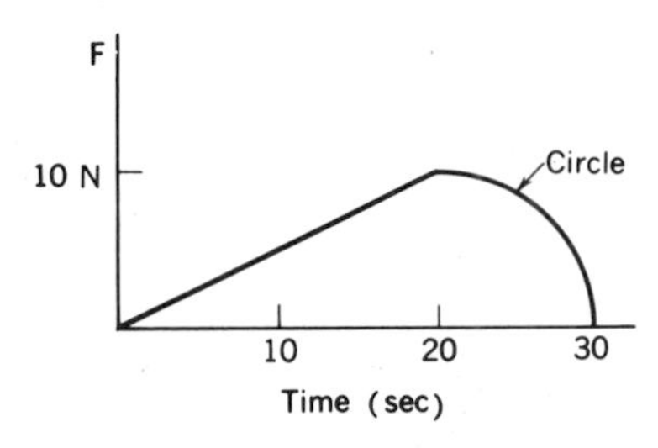

Figure P.14.3

14.4. A 100-lb block is acted on by a force P which varies with time as shown. What is the speed of the block after 80 sec? Assume that the block starts from rest and neglect friction.

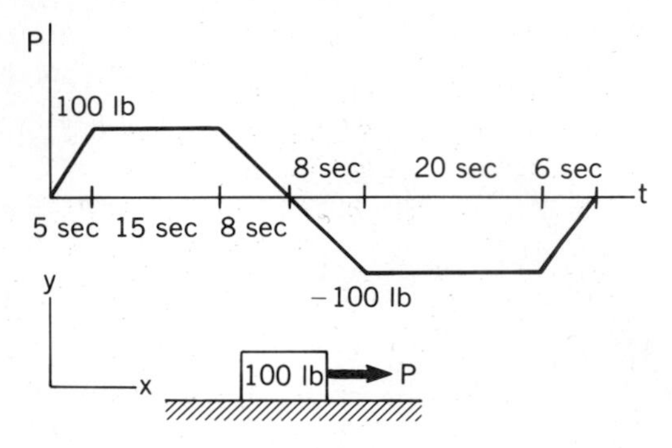

Figure P.14.4

14.5. If the coefficient of static friction is .5 in Problem 14.4 and the coefficient of dynamic friction is .3, what is the speed of the block after 28 sec?

14.6. A body is dropped from rest. (a) Determine the time required for it to acquire a velocity of 16 m/sec. (b) Determine the time needed to increase its velocity from 16 m/sec to 23 m/sec.

14.7. A body having a mass of 5 lbm is acted on by the following force:

$$\boldsymbol{F} = 8t\boldsymbol{i} + (6 + 3\sqrt{t})\boldsymbol{j} - (16 + 3t^2)\boldsymbol{k}\ \text{lb}$$

where t is in seconds. What is the velocity of the body after 5 sec if the initial velocity is

$$\boldsymbol{V}_1 = 6\boldsymbol{i} + 3\boldsymbol{j} - 10\boldsymbol{k}\ \text{ft/sec?}$$

14.8. A body with a mass of 16 kg is required to change its velocity from $\boldsymbol{V}_1 = 2\boldsymbol{i} + 4\boldsymbol{j} - 10\boldsymbol{k}$ m/sec to a velocity $\boldsymbol{V}_2 = 10\boldsymbol{i} - 5\boldsymbol{j} + 20\boldsymbol{k}$ m/sec in 10 sec. What average force $\boldsymbol{F}$ over this time interval will do the job?

14.9. In Problem 14.8, determine the force as a function of time for the case where force varies linearly with time starting with a zero value.

14.10. A hockey puck moves at 30 ft/sec from left to right. The puck is intercepted by a player who whisks it at 80 ft/sec toward goal A, as shown. The puck is also rising from the ice at a rate of 10 ft/sec. What is the impulse on the puck, whose weight is 5 oz?

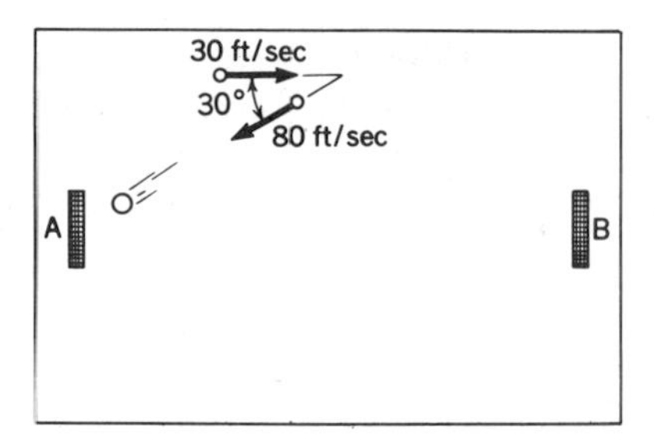

Figure P.14.10

14.11. Do Problem 12.5 by methods of momentum.

14.12. Do Problem 12.6 by methods of momentum.

14.13. A commuter train made up of two cars is moving at a speed of 80 km/hr. The first car has a mass of 20,000 kg and the second 15,000 kg.

(a) If the brakes are applied simultaneously to both cars, determine the minimum time the cars travel before stopping. The coefficient of static friction between the wheels and rail is .3.

(b) If the brakes on the first car only are applied, determine the time the cars travel before stopping and the force F transmitted between the cars.

14.14. Compute the velocity of the bodies after 10 sec if they start from rest. The cable is inextensible, and the pulleys are frictionless. For the contact surfaces, $\mu_d = .2$.

Figure P.14.14

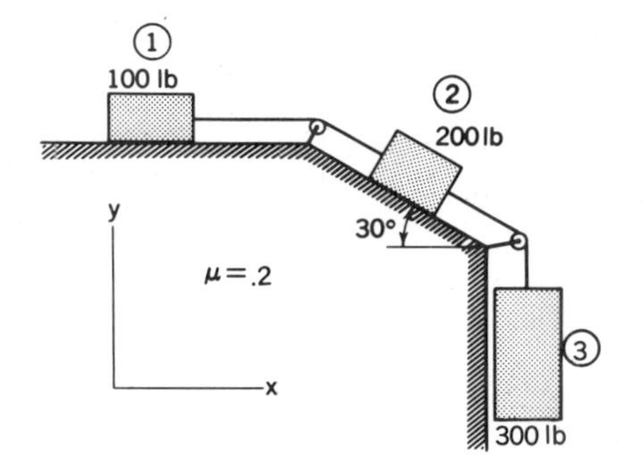

14.15. Determine the velocity of body A and body B after 3 sec if the system is released from rest. Neglect friction and the inertia of the pulleys.

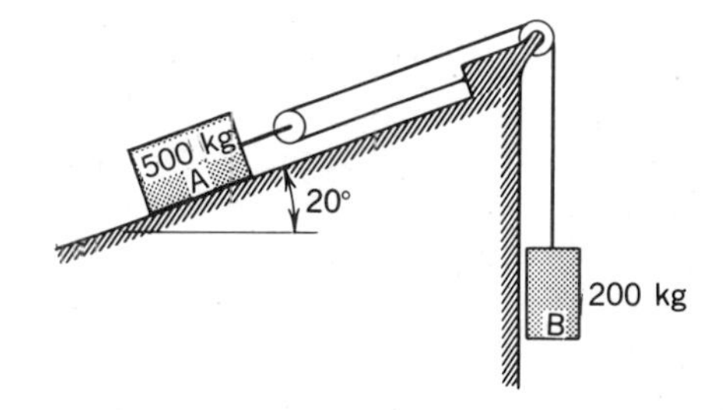

Figure P.14.15

14.16. Some top-flight tennis players hit the ball on service at the instant the ball is at the top of its trajectory when thrown by the free hand. The ball is often given a speed V of 100 mi/hr by the racket. If the time of contact with the racket is .03 sec, what is the average force magnitude on the ball from the racket during this time interval? The ball weighs 1.5 oz.

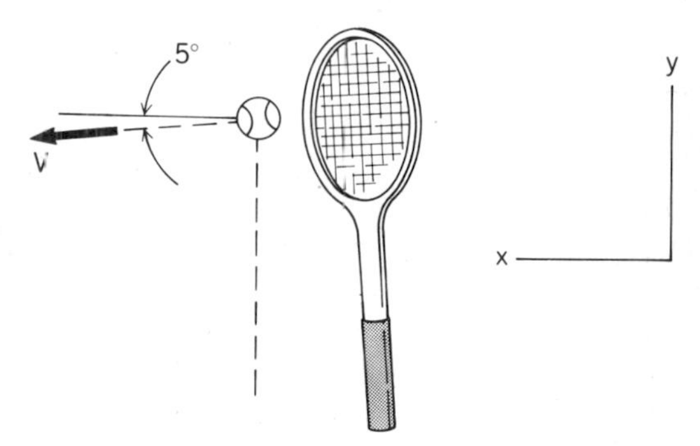

Figure P.14.16

14.17. A torpedo boat weighing 100,000 lb moves at 40 knots (1 knot = 6080 ft/hr) away from an engagement. To go even faster, all four 50-caliber machine guns are ordered to fire simultaneously toward the rear. Each weapon fires at a muzzle velocity of 3000 ft/sec and fires 3000 rounds per minute. Each slug weighs 2 oz. How much is the average force on the boat increased by this action? Neglect the rate of change of the total mass of boat.

14.18. A vertical conveyor has sprocket A as the driver, and sprocket B turns freely. The bodies to be lifted are pushed onto the conveyor by a plunger C and are taken off from the conveyor at D as shown in the diagram. If the belt runs at 2000 mm/sec and the bodies being transported each has a mass of 250 g, what average torque is required by the driving sprocket A? On the average, 40 bodies are on the conveyor at any time.

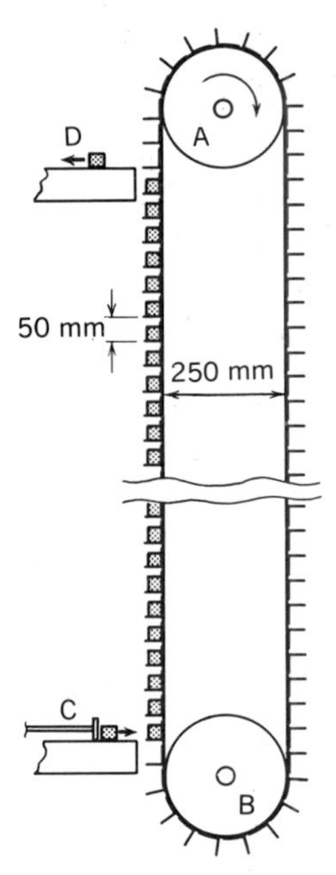

Figure P.14.18

14.19. A conveyor A is feeding boxes onto a conveyor B. Each box weighs 2 lb and lands on conveyor B with a downward-speed component of 3 ft/sec. Conveyor belt A has a speed of .2 ft/sec. If conveyor B runs at a speed of 5 ft/sec and if five boxes land per second on the average, what net average force T_2 must be exerted on the conveyor belt B to slide it over its bed? At any time, 50 boxes are on belt B. Take $\mu_d = .2$ for all surfaces. Neglect the weight of conveyor belt B.

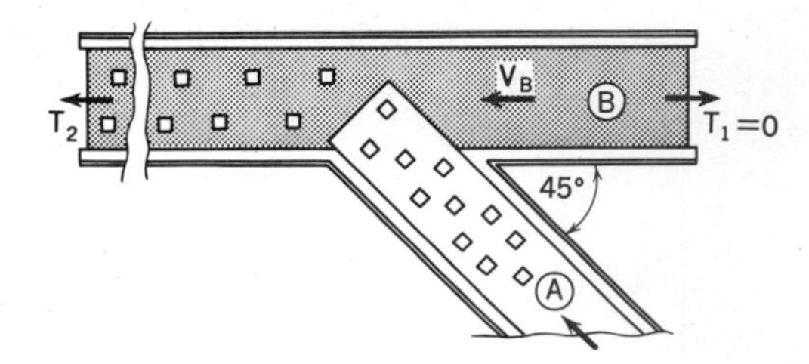

Figure P.14.19

14.20. An idealized one-dimensional pressure wave (i.e., pressure is a function of one coordinate and time) generated by an explosion travels at a speed V of 1200 ft/sec, as shown at time $t = 0$. The peak pressure of this wave is 5 psia. What impulse per square foot is delivered to a wall oriented at right angles to the x axis? The wave is reflected from the wall, and the pressure at the wall is double the incoming pressure at all times. Do the problem for two time intervals corresponding to the interval (a) from when the wave front first touches the wall to when the peak reaches the wall and (b) from when the peak hits the wall and to when the end of the wave reaches the wall.

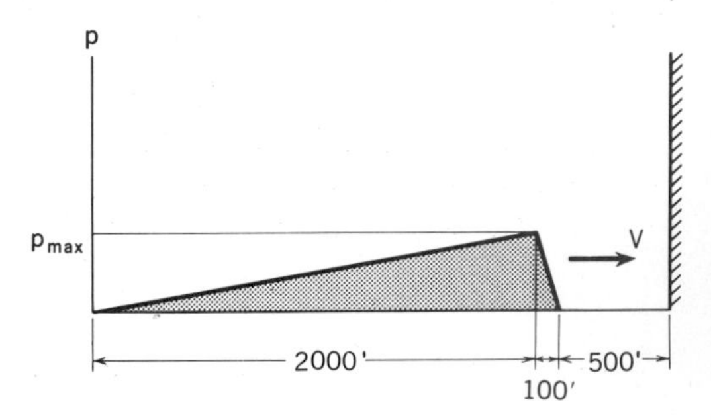

Figure P.14.20

14.21. Blocks A and B move on frictionless surfaces. The blocks are interconnected with a light bar. Body A weighs 30 lb; the weight of body B is not known. A constant force F of 100 lb is applied at the configuration shown. If a speed of 25 ft/sec is reached by A after 1 sec, what impulse is developed on the vertical wall?

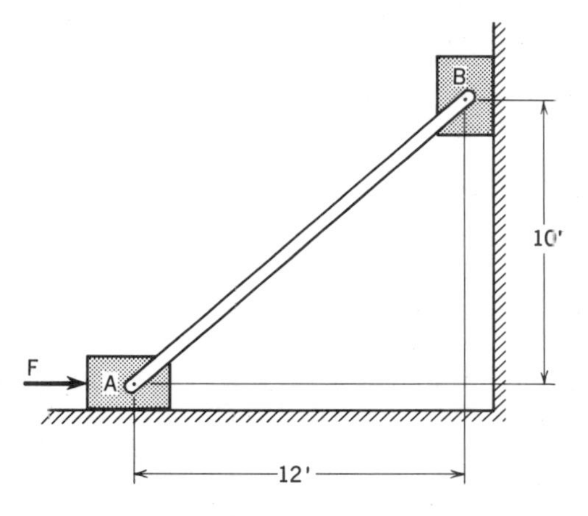

Figure P.14.21

14.22. In Problem 14.21, compute the impulse on the horizontal surface. A moves 4 ft in 1 sec and $W_B = 20$ lb.

***14.23.** A chain of wrought iron, with length of 7 m and a mass of 100 kg, is held so that it just touches the support AB. If the chain is released, determine the total impulse during 2 sec in the vertical direction experienced by the support if the impact is plastic (i.e., the chain does not bounce up) and if we move the support so that the links land on the platform and not on each other? (*Hint:* Note that any chain *resting* on AB delivers a vertical impulse. Also check to see if the entire chain lands on AB before 2 sec.)

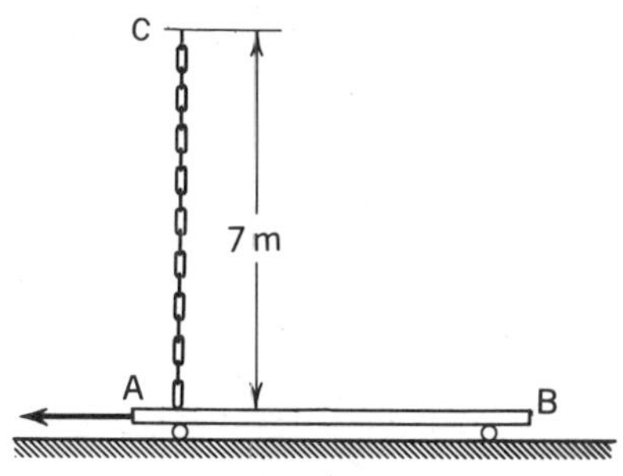

Figure P.14.23

14.24. A three-seater racing scull is poised for a start. The scull weighs 300 lb, and each occupant weighs about 150 lb. We want to know the speed of the scull after 2 sec. At the sound of the starting gun, each man exerts a 30-lb constant push on the water from each oar in the direction of the axis of the boat. At the 2-sec mark, each man is moving to the right relative to the hull with a speed of 1 ft/sec. Neglect the inertia of the oars as well as water and air friction.

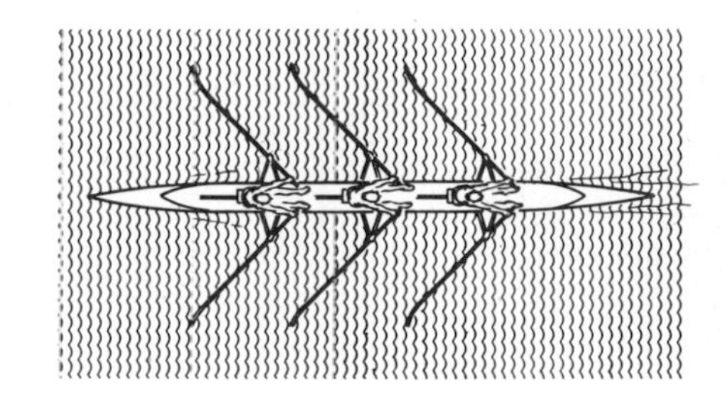

Figure P.14.24

14.25. An antitank airplane fires two 90-N projectiles at a tank at the same time. The muzzle velocity of the guns is 1000 m/sec relative to the plane. If the plane before firing weighs 65 kN and is moving with a velocity of 320 km/hr, compute the change in its speed when it fires the two projectiles.

14.26. A toboggan has just entered the horizontal part of its run. It carries three people weighing 120 lb, 180 lb, and 150 lb, respectively. Suddenly, a pedestrian weighing 200 lb strays onto the course and is turned end for end by the toboggan, landing safely among the riders. Since the toboggan path is icy, we can neglect friction with the toboggan path for all actions described here. If the toboggan is traveling at a speed of 35 mph just before collision occurs, what is the speed after the collision when the pedestrian has become a rider? The toboggan weighs 30 lb.

Figure P.14.26

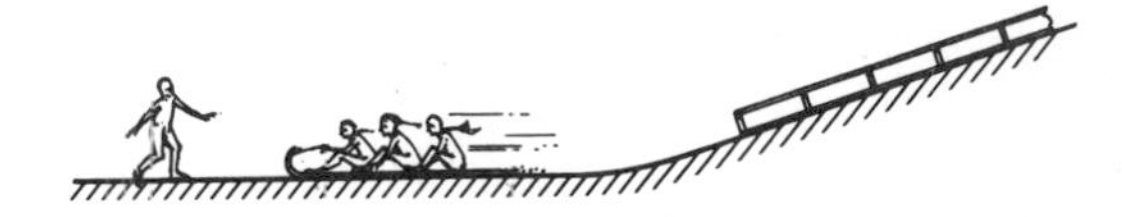

14.27. An 890-N rowboat containing a 668-N man is pushed off the dock by an 800-N man. The speed that is imparted to the boat is .30 m/sec by this push. The man then leaps into the boat from the dock with a speed of .60 m/sec relative to the dock in the direction of motion of the boat. When the two men have settled down in the boat and before rowing commences, what is the speed of the boat? Neglect water resistance.

Figure P.14.27

14.28. Two vehicles connected with an inextensible cable are rolling along a road. Vehicle B, using a winch, draws A toward it so that the relative speed is 5 ft/sec at $t = 0$ and 10 ft/sec at $t = 20$ sec. Vehicle A weighs 2000 lb and vehicle B weighs 3000 lb. Each vehicle has a rolling resistance that is .01 times the vehicle's weight. What is the speed of A relative to the ground at $t =$ 20 sec if A is initially moving to the right at a speed of 10 ft/sec?

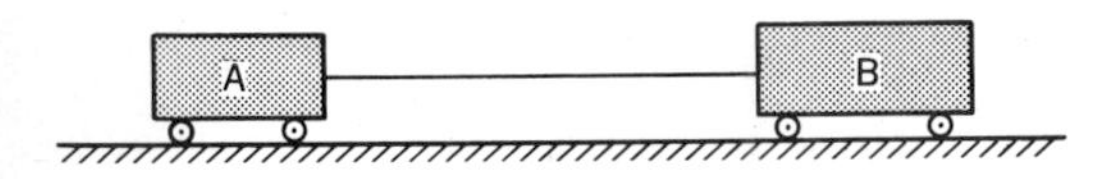

Figure P.14.28

14.29. Two trucks are shown moving up at 10° incline. Truck A weighs 26.7 kN and is developing a 13.30-kN driving force on the road. Truck B weighs 17.8 kN and is connected with an inextensible cable to truck A. By operating a winch b, truck B approaches truck A with a constant acceleration of .3 m/sec². If at time $t = 0$ both trucks have a speed of 10 m/sec, what are their speeds at time $t = 15$ sec?

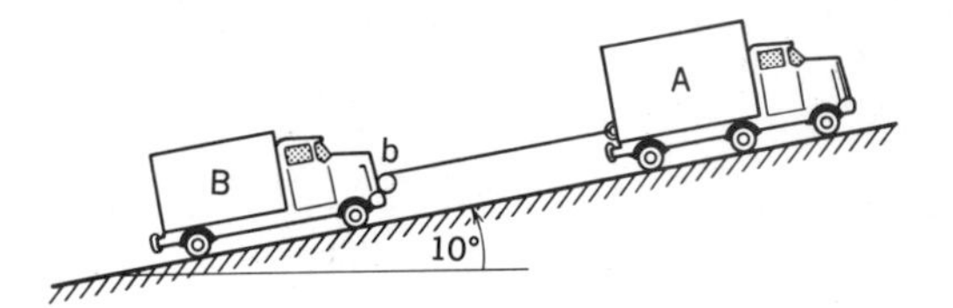

Figure P.14.29

14.30. Treat Example 14.3 as a two-particle system in the impulse-momentum considerations. Verify the results of Example 14.3 for V. (Be sure to include *all* external forces for the system.)

14.31. Do Problem 14.15 by considering a system of particles. (Be sure to include *all* external forces for the system of bodies A and B.)

14.32. A 40-kN truck is moving at the speed of 40 km/hr carrying a 15-kN load A. The load is restrained only by friction with the floor of the truck where there is a dynamic coefficient of friction .2. The driver suddenly jams his brakes on so as to lock all wheels for 1.5 sec. At the end of this interval, the brakes are released. What is the final speed V of the truck neglecting wind resistance and rotational inertia of the wheels after load A stops slipping? The dynamic coefficient of friction between the tires and the road is .4.

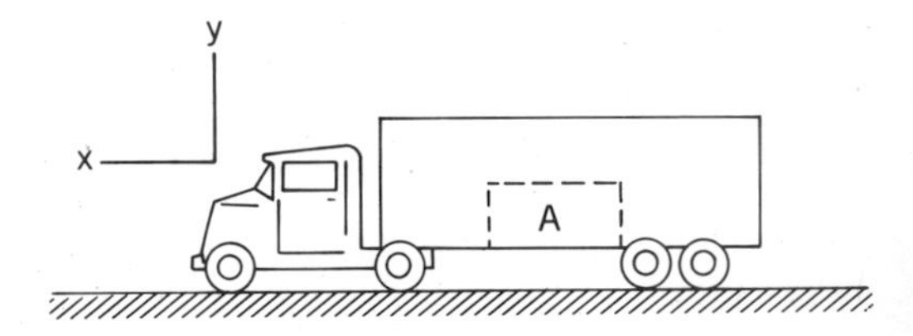

Figure P.14.32

14.33. A 1300-kg Jeep is carrying three 100-kg passengers. The Jeep is in four-wheel drive and is under test to see what maximum speed is possible in 5 sec from a start on an icy road surface for which $\mu_s = .1$. Compute $V_{\max}$ at $t = 5$ sec.

Figure P.14.33

14.34. Two adjacent tanks A and B are shown. Both tanks are rectangular with a width of 4 m. Gasoline from tank A is being pumped into tank B. When the level of tank A is .7 m from the top, the rate of flow Q from A to B is 300 liters/sec, and 10 sec later it is 500 liters/sec. What is the average horizontal force from the fluids onto the tank during this 10-sec time interval? The density of the gasoline is $.8 \times 10^3$ kg/m^3. Tank A is originally full and tank B is originally empty.

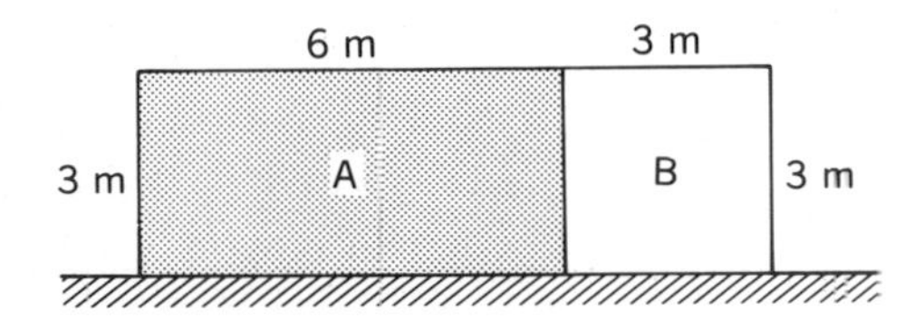

Figure P.14.34

***14.35.** Two tanks A and B are shown. Tank A is originally full of water ($\rho = 62.4$ lbm/ft^3), while tank B is empty. Water is pumped from A to B. If initially 100 cfs of water is being pumped and if this flow increases at the rate of 10 cfs/sec^2 for 30 sec thereafter, what is the average vertical force onto the tanks from the water during this time period, aside from the static dead weight of the water?

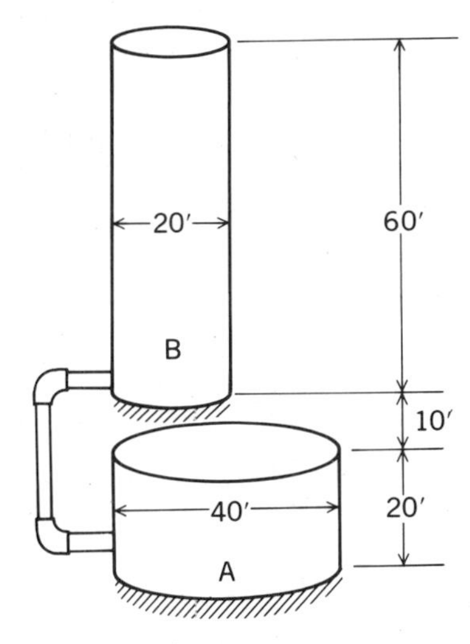

Figure P.14.35

14.36. A device to be detonated is shown in (a) suspended above the ground. Ten seconds after detonation, there are four fragments having the following masses and position vectors relative to reference XYZ:

$$m_1 = 5 \text{ kg},$$
$$\boldsymbol{r}_1 = 1000\boldsymbol{i} + 2000\boldsymbol{j} + 900\boldsymbol{k} \text{ m}$$
$$m_2 = 3 \text{ kg},$$
$$\boldsymbol{r}_2 = 800\boldsymbol{i} + 1800\boldsymbol{j} + 2500\boldsymbol{k} \text{ m}$$
$$m_3 = 4 \text{ kg},$$
$$\boldsymbol{r}_3 = 400\boldsymbol{i} + 1000\boldsymbol{j} + 2000\boldsymbol{k} \text{ m}$$
$$m_4 = 6 \text{ kg},$$
$$\boldsymbol{r}_4 = X_4\boldsymbol{i} + Y_4\boldsymbol{j} + Z_4\boldsymbol{k}$$

Find the position $\boldsymbol{r}_4$ if the center of mass of the device is at position $\boldsymbol{r}_0$, wherein

$$\boldsymbol{r}_0 = 600\boldsymbol{i} + 1200\boldsymbol{j} + 2300\boldsymbol{k} \text{ m}$$

Neglect wind resistance.

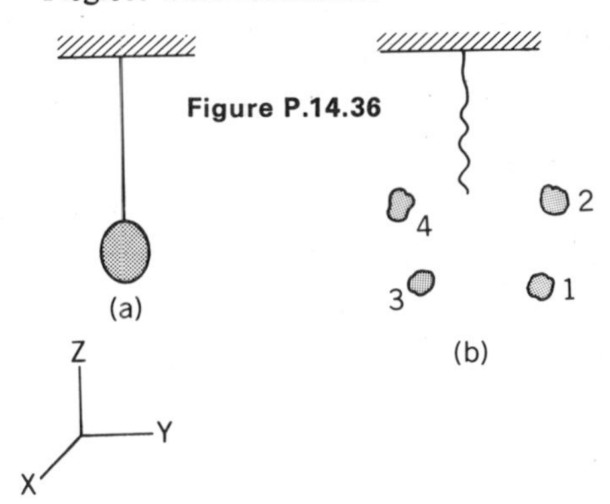

Figure P.14.36

14.37. A device to be detonated with a small charge is suspended in space [see Fig. (a)]. Directly after detonation, four fragments are formed moving away from the point of suspension. The following information is known about these fragments.

$$m_1 = 1 \text{ lbm},$$
$$V_1 = 200\boldsymbol{i} - 100\boldsymbol{j} \text{ ft/sec}$$
$$m_2 = 2 \text{ lbm},$$
$$V_2 = 125\boldsymbol{i} + 180\boldsymbol{j} - 100\boldsymbol{k} \text{ ft/sec}$$
$$m_3 = 1.6 \text{ lbm},$$
$$V_3 = -200\boldsymbol{i} + 150\boldsymbol{j} + 180\boldsymbol{k} \text{ ft/sec}$$
$$m_4 = 3.2 \text{ lbm}$$

What is the velocity V_4?

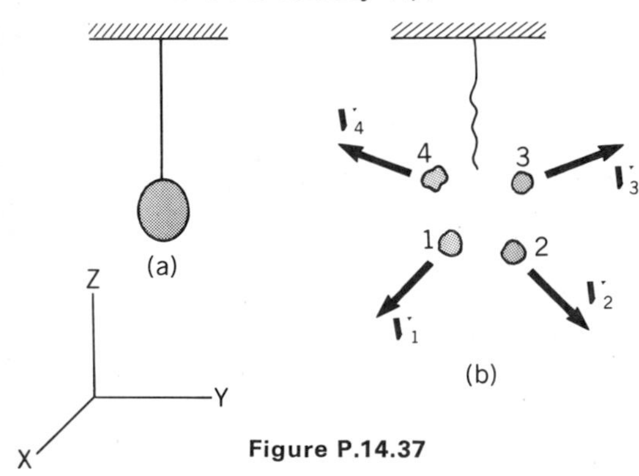

Figure P.14.37

14.38. A hawk is a predatory bird which often attacks smaller birds in flight. A hawk having a mass of 1.3 kg is swooping down on a sparrow having a mass of 150 g. Just before seizing the sparrow with its claws, the hawk is moving downward with a speed V_H of 20 km/hr. The sparrow is moving horizontally at a speed V_S of 15 km/hr. Directly after seizure, what is the speed of the hawk and its prey? What is the loss in kinetic energy in joules?

14.39. The principal mode of propulsion of an octopus is to take in water through the mouth and then after closing the inlet to eject the water to the rear. If a 5-lb octopus after taking in 1 lb of water is moving at a speed of 3 ft/sec, what is its speed directly after ejecting the water? The water is ejected at an average speed to the rear of 10 ft/sec relative to the initial speed of the octopus. What horsepower is being developed on the octopus from the above action if it occurs in 1 sec?

***14.40.** In the *fission* process in a nuclear reactor, a ^{235}U nucleus first absorbs or captures a neutron [see Fig. (a)]. A short time later, the ^{235}U nucleus breaks up into fission products plus neutrons, which may subsequently be captured by other ^{235}U nuclei and maintain a *chain reaction.* Energy is released in each fission. In Fig. (b) we have shown the results of a possible fission. The following information is known for this fission:

	Mass No.	Kinetic Energy (MeV)	Direction of V
Product A	138	E	$\boldsymbol{\epsilon}_A = .3\boldsymbol{i} - .2\boldsymbol{j} + .98\boldsymbol{k}$
Product B	96	90	$\boldsymbol{\epsilon}_B = l_B\boldsymbol{i} + m_B\boldsymbol{j} + n_B\boldsymbol{k}$
Neutron 1	1	10	$\boldsymbol{\epsilon}_1 = .6\boldsymbol{i} + .8\boldsymbol{j}$
Neutron 2	1	10	$\boldsymbol{\epsilon}_2 = .4\boldsymbol{i} - .6\boldsymbol{j} - .693\boldsymbol{k}$

What is the energy E of product A in MeV and what is the vector $\boldsymbol{\epsilon}_B$ for the velocity of product B? Assume that before fission the nucleus of ^{235}U plus captured neutrons is stationary. (*Hint:* You do not have to actually convert MeV to joules or atomic number to kilograms to carry out the problem.)

Figure P.14.40

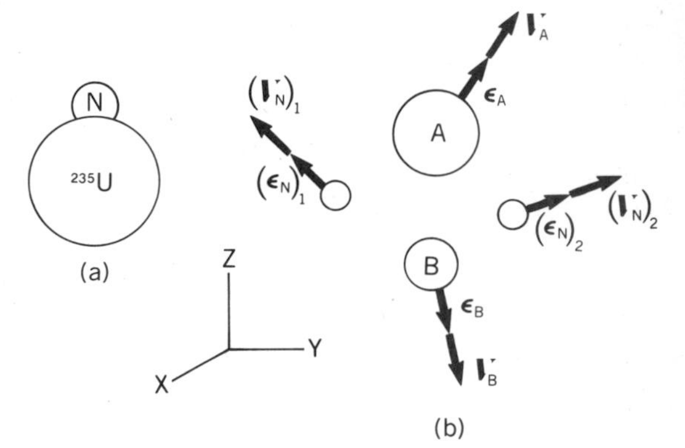

14.4 Impact

In Section 14.3, we discussed impulsive forces. We shall in this section discuss in detail an action in which impulsive forces are present. This situation occurs when two bodies collide and remain intact. The time interval during collision is very small, and comparatively large forces are developed on the bodies during the small time interval. This action is called *impact*. For such actions with such short time intervals, the force of gravity generally causes a negligible impulse. The impact forces on the colliding bodies are always equal and opposite to each other, so the net impulse on the *pair* of bodies during collision is *zero*. This means that the total linear momentum directly after impact (postimpact) equals the total linear momentum directly before impact (preimpact).

We shall consider at this time two types of impact for which certain definitions are needed. We shall call the normal to the *plane of contact* during the collision of two bodies the *line of impact*. If the centers of mass of the two colliding bodies lie along the line of impact, the action is called *central impact* and is shown for the case of two spheres in Fig. 14.12.[4] If, in addition, the velocity vectors of the mass centers approaching the collision are collinear with the line of impact, the action is called *direct central impact*. This action is illustrated by V_1 and V_2 in the diagram. Should one (or both) of the velocities have a line of action not collinear with the line impact—for example, V_1' and/or V_2'—the action is termed *oblique central impact*.

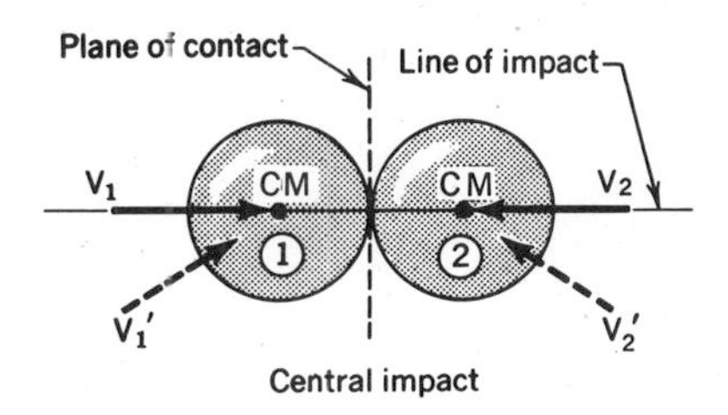

Figure 14.12. Cenrtal impact of two spheres.

In either case, linear momentum is conserved during the short time interval from directly before the collision (indicated with the subscript i) to directly after the collision (indicated with subscript f). That is,

$$(m_1 V_1)_i + (m_2 V_2)_i = (m_1 V_1)_f + (m_2 V_2)_f \tag{14.8}$$

In the *direct-central-impact* case for *smooth* bodies, this equation becomes a single scalar equation since $(V_1)_f$ and $(V_2)_f$ are collinear with the line of impact. Usually, the initial velocities are known and the final values are desired, which means that we have for this case one scalar equation involving two unknowns. Clearly, we must know more about the manner of interaction of the bodies, since Eq. 14.8 as it stands is valid for materials of any deformability (e.g., putty or hardened steel) and takes no account of such important considerations. Thus, we cannot consider the bodies undergoing impact only as particles as has been the case thus far, but must, in addition, consider them as deformable bodies of finite size in order to generate enough information to solve the problem at hand.

For the *oblique-impact case*, we can write components of the linear-momentum equation along the line of impact and along two other directions at right angles to the line of impact. If we know the initial velocities, then we have six unknown final

[4]Noncentral or *eccentric impact* is examined in Chapter 17 for the case of plane motion.

velocity components and only three equations. Thus, we need even more information to establish fully the final velocities after this more general type of impact. We now consider each of these cases in more detail in order to establish these additional relations.

Case 1. Direct Central Impact. Let us first examine the direct-central-impact case. We shall consider the period of collision to be made up of two subintervals of time. The *period of deformation* refers to the duration of the collision, starting from initial contact of the bodies and ending at the instant of maximum deformation. During this period, we shall consider that impulse $\int D\,dt$ acts oppositely on each of the bodies. The second period, covering the time from the maximum deformation condition to the instant at which the bodies just separate, we shall term the *period of restitution*. The impulse acting oppositely on each body during this period we shall indicate as $\int R\,dt$. If the bodies are *perfectly elastic*, they will reestablish their initial shapes during the period of restitution (if we neglect the internal vibrations of the bodies), as shown in Fig. 14.13(a). When the bodies do not reestablish their initial shapes [Fig. 14.13(b)], we say that *plastic deformation* has taken place.

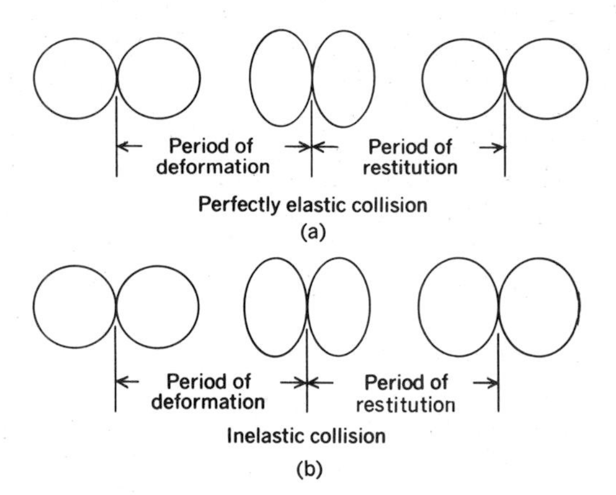

Figure 14.13. Collision process.

The ratio of the impulse during the restitution period $\int R\,dt$ to the impulse during the deformation period $\int D\,dt$ is a number ϵ, which depends mainly on the physical properties of the bodies in collision. We call this number the *coefficient of restitution*. Thus,

$$\epsilon = \frac{\text{impulse during restitution}}{\text{impulse during deformation}} = \frac{\int R\,dt}{\int D\,dt} \tag{14.9}$$

We must strongly point out that the coefficient of restitution depends also on the size, shape, and approach velocities of the bodies before impact. These dependencies result from the fact that plastic deformation is related to the magnitude and nature of the force distributions in the bodies and also to the rate of loading. However, values of ϵ

have been established for different materials and can be used for approximate results in the kind of computations to follow. We shall now formulate the relation between the coefficient of restitution and the initial and final velocities of the bodies undergoing impact.

Let us consider *one* of the bodies during the two phases of the collision. If we call the velocity at the maximum deformation condition $(V)_D$, we can say for mass 1:

$$\int D\,dt = [(m_1V_1)_D - (m_1V_1)_i] = -m_1[(V_1)_i - (V_1)_D] \tag{14.10}$$

During the period of restitution, we find that

$$\int R\,dt = -m_1[(V_1)_D - (V_1)_f] \tag{14.11}$$

Dividing Eq. 14.11 by Eq. 14.10, canceling out m_1, and noting the definition in Eq. 14.9, we can say:

$$\epsilon = \frac{(V_1)_D - (V_1)_f}{(V_1)_i - (V_1)_D} \tag{14.12}$$

A similar analysis for the other mass (2) gives

$$\epsilon = \frac{(V_2)_D - (V_2)_f}{(V_2)_i - (V_2)_D} = \frac{(V_2)_f - (V_2)_D}{(V_2)_D - (V_2)_i} \tag{14.13}$$

In this expression, we have changed the sign of numerator and denominator. At the intermediate position at the end of deformation and the begining of restitution the masses have essentially the same velocity. Thus, $(V_1)_D = (V_2)_D$. Since the quotients in Eqs. 14.12 and 14.13 are equal to each other, we can add numerators and denominators to form another equal quotient, as you can demonstrate yourself. Noting the abovementioned equality of the V_D terms, we have the desired result:

$$\boxed{\epsilon = -\frac{(V_2)_f - (V_1)_f}{(V_2)_i - (V_1)_i} = -\frac{\text{relative velocity of separation}}{\text{relative velocity of approach}}} \tag{14.14}$$

This equation involves the coefficient ϵ, which is presumably known or estimated, and the initial and final velocities of the bodies undergoing impact. Thus, with this equation we can solve for the final velocities of the bodies after collision when we use the linear-momentum equation 14.8, for the case of direct central impact.

During a *perfectly elastic* collision, the impulse for the period of restitution equals the impulse for the period of deformation,[5] so the coefficient of restitution is *unity* for this case. For nonelastic collisions, the coefficient of restitution is less than unity since the impulse is diminished on restitution as a result of the failure of the bodies to resume their original geometries. For a *perfectly plastic* impact, $\epsilon = 0$ [i.e., $(V_2)_f = (V_1)_f$] and the bodies remain in contact.

[5]The impulses are equal because during the period of restitution the body can be considered to undergo identically the reverse of the process corresponding to the deformation period. Thus, from a thermodynamics point of view, we are considering the elastic impact to be a *reversible* process.

Case 2. Oblique Central Impact. Let us now consider the case of oblique central impact. The velocity components along the line of impact can be related by the scalar component of the linear-momentum equation 14.8 in this direction and also by Eq. 14.14, where velocity components along the line of impact are used and where the coefficient of restitution may be considered (for smooth bodies) to be the same as for the direct-central-impact case. If we know the initial conditions, we can accordingly solve for those velocity components after impact in the direction of the line of impact. As for the other components of velocity, we can say that for smooth bodies, these velocity components are unaffected by the collision, since no impulses act in these directions on either body. That is, the velocity components normal to the line of impact for each body are the same immediately after impact as before. Thus, the final velocity components of both bodies can be established, and the motions of the bodies can be determined within the limits of the discussion. The following examples are used to illustrate the use of the preceding formulations.

EXAMPLE 14.9

Two billiard balls (of the same size and mass) collide with the velocities of approach shown in Fig. 14.14. For a coefficient of restitution of .90, what are the final velocities of the balls directly after they part? What is the loss in kinetic energy?

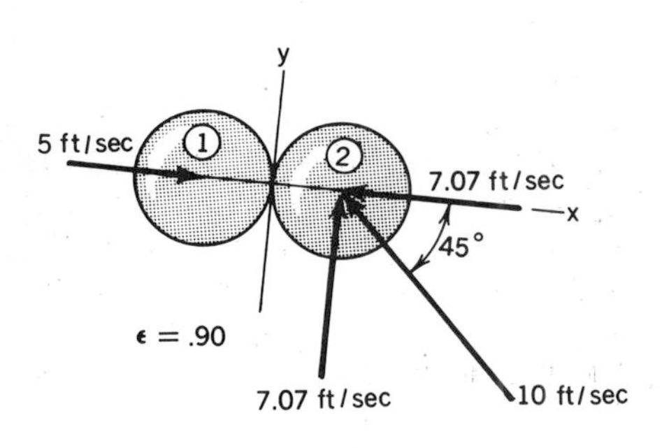

Figure 14.14. Oblique central impact.

A reference is established so that the x axis is along line of impact and the y axis is in the plane of contact such that the reference plane is parallel to the billiard table. The approach velocities have been decomposed into components along these axes. The velocity components $(V_1)_y$ and $(V_2)_y$ are unchanged during the action. Along the line of impact, linear-momentum considerations lead to

$$5m - 7.07m = m[(V_1)_x]_f + m[(V_2)_x]_f \qquad \text{(a)}$$

Using the coefficient-of-restitution relation (Eq. 14.14), we have

$$\epsilon = .90 = -\frac{[(V_2)_x]_f - [(V_1)_x]_f}{-7.07 - 5} \qquad \text{(b)}$$

We thus have two equations, (a) and (b), for the unknown components in the x direction. Simplifying these equations, we have

$$[(V_1)_x]_f + [(V_2)_x]_f = -2.07 \qquad \text{(c)}$$

$$[(V_1)_x]_f - [(V_2)_x]_f = -10.86 \qquad \text{(d)}$$

Adding, we get

$$[(V_1)_x]_f = -6.47 \text{ ft/sec}$$

Solving for $[(V_2)_x]_f$ in Eq. (c), we write

$$[(V_2)_x]_f - 6.47 = -2.07$$

Therefore,

$$[(V_2)_x]_f = 4.40 \text{ ft/sec}$$

The final velocities after collision are then

$$(V_1)_f = -6.47\boldsymbol{i} \text{ ft/sec}$$
$$(V_2)_f = 4.40\boldsymbol{i} + 7.07\boldsymbol{j} \text{ ft/sec}$$

The loss in kinetic energy is given as

$$(\text{KE})_i - (\text{KE})_f = (\tfrac{1}{2}m5^2 + \tfrac{1}{2}m10^2) - [\tfrac{1}{2}m6.47^2 + \tfrac{1}{2}m(7.07^2 + 4.40^2)]$$
$$\Delta\text{KE} = \tfrac{1}{2}m[25 + 100 - (41.9 + 50.0 + 19.33)] = 6.89m \text{ ft-lb}$$

Please note that mechanical energy is conserved *only* if ϵ is unity (i.e., a perfectly elastic impact). For all other cases, there is always dissipation of mechanical energy into heat and permanent deformation. However, *all* impacts involve conservation of linear momentum for the system.

*14.5 Collision of a Particle with a Massive Rigid Body

In Section 14.4, we employed conservation-of-momentum considerations and the concept of the coefficient of restitution to examine the impact of two smooth bodies of comparable size. Now we shall extend this approach to include the impact of a spherical body with a much larger and more massive *rigid* body, as shown in Fig. 14.15.

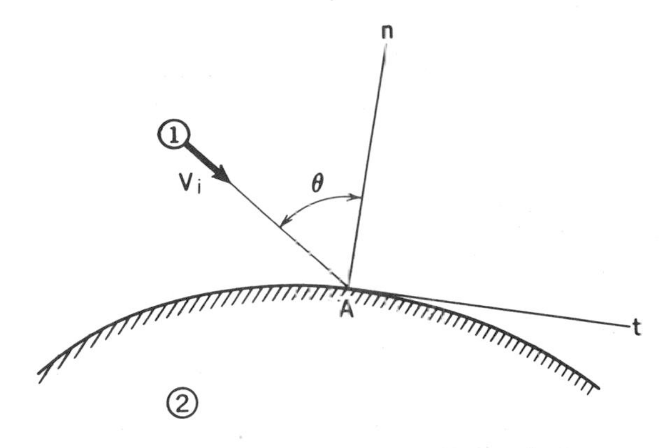

Figure 14.15. Small body collides with large body.

The procedure we shall follow is to consider the massive body to be a spherical body of *infinite* mass with a radius equal to the local radius of curvature of the surface of the massive body at the point of contact A. This condition is shown in Fig. 14.16. The line of impact then becomes identical with the normal n to the surface of the massive body at the point of impact. Note that the case we show in the diagram corresponds to oblique central impact. With no friction, clearly only the components along the line of impact n can change as a result of impact. But in this case, the velocity of the sphere representing the massive body must undergo no change in value after impact because of its infinite mass.[6] We cannot make good use here of the conserva-

[6]Otherwise, there would be an infinite change in momentum for this sphere.

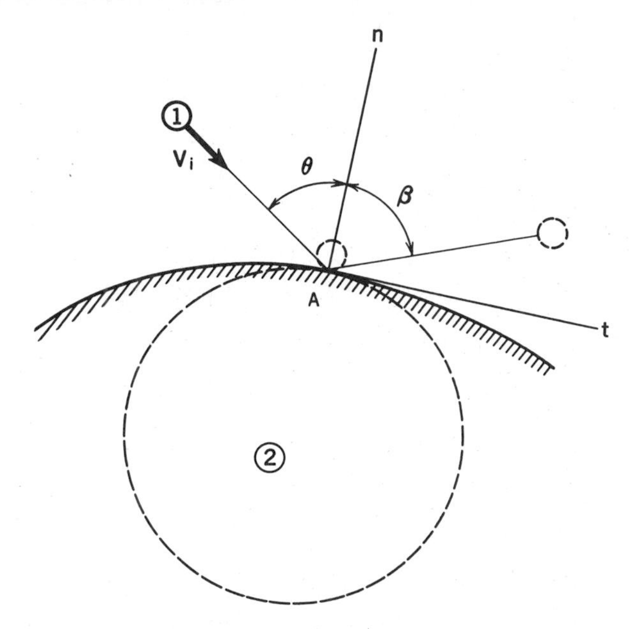

Figure 14.16. Angle of incidence and angle of reflection.

tion of the linear-momentum equation in the n direction because the infinite mass of the hypothetical body (2) will render the equation indeterminate. However, we can use Eq. 14.14, assuming we have a coefficient of restitution ϵ for the action. Noting that the velocity of the massive body *does not change*, we accordingly get

$$\epsilon = -\frac{[(V_1)_n]_f - [(V_2)_n]}{[(V_1)_n]_i - [(V_2)_n]} \tag{14.15}$$

Thus, knowing the velocities of the bodies before impact, as well as the quantity ϵ, we are able to compute the velocity of the particle after impact. If the collision is perfectly elastic, $\epsilon = 1$, and we see from Eq. 14.15 that for a stationary massive body

$$[(V_1)_n]_i = -[(V_1)_n]_f$$

This means that the angle of incidence θ (see Fig. 14.16) equals the angle of reflection β. For $\epsilon < 1$ (i.e., for inelastic collision), the angle of reflection β will clearly exceed θ.

We now illustrate the use of these formulations.

EXAMPLE 14.10

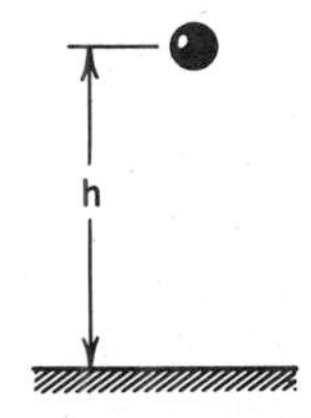

Figure 14.17. Ball dropped on concrete floor.

A ball is dropped onto a concrete floor from height h (Fig. 14.17). If the coefficient of restitution is .90 for the action, to what height h' will the ball rise on the rebound?

Here the massive body has an infinite radius at the surface. Furthermore, we have a direct central impact. Accordingly, from Eq. 14.15 we have

$$\epsilon = -\frac{(V)_f - 0}{(V)_i - 0} = -\frac{\sqrt{2gh'}}{\sqrt{2gh}}$$

Solving for h', we get

$$h' = \epsilon^2 h = .81h$$

EXAMPLE 14.11

A satellite in the form of a sphere with radius R [Fig. 14.18(a)] is moving above the earth's surface in a region of highly rarefied atmosphere. We wish to estimate the drag on the satellite. Neglect the contribution from the antennas.

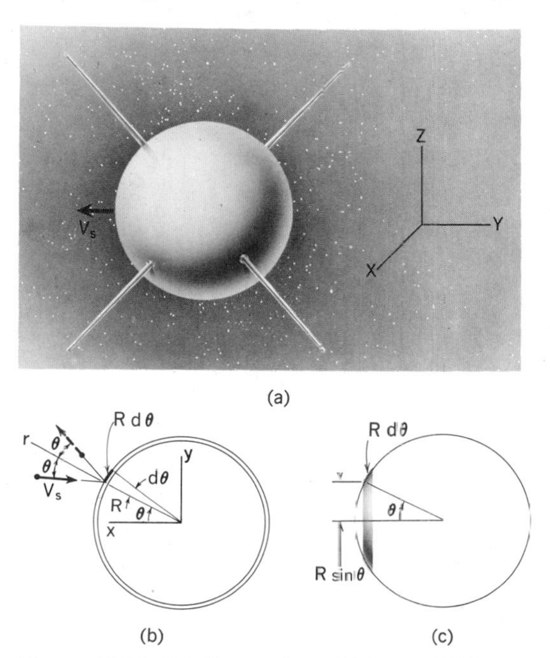

Figure 14.18. Satellite moving at high speed V_s in space.

In this highly rarefied atmosphere, we shall assume that the average spacing of the molecules is large enough relative to the satellite that we cannot use the continuum approach of fluid dynamics, wherein matter is assumed to be continuously distributed. Instead, we must consider collisions of the individual molecules with the satellite, which is a noncontinuum approach, as discussed in Section 1.7. The mass per molecule is m slugs and the number density of the molecules is n molecules/ft³. Since the satellite is moving with a speed V_s much greater than the speed of the molecules (the molecules move at about the speed of sound), we can assume that the molecules are stationary relative to inertial space reference XYZ and that only the satellite is moving. Furthermore, we assume that when the satellite hits a molecule there is an elastic, frictionless collision.

To study this problem, we have shown a section of the satellite in Fig. 14.18(b). A reference xyz is fixed to the satellite at its center. We shall consider this reference also to be an inertial reference—a step which for small drag will introduce little error for the ensuing calculations. Relative to this reference, the molecules approach the satellite with a horizontal velocity V_s, as shown for one molecule. They then collide with the surface with an angle of incidence measured by the polar coordinate θ. Finally, they deflect with an equal angle of reflection of θ. The component of the impulse given to the molecule in the x direction $(I_{mol})_x$ is

$$(I_{mol})_x = (mV_s \cos 2\theta) - (-mV_s) = mV_s(1 + \cos 2\theta) \tag{a}$$

This is the impulse component that would be given to any molecule hitting a strip which is $R\,d\theta$ in width and which is revolved around the x axis as shown in Fig. 14.18(c). The number of such collisions per second for this strip can readily be calculated as follows:

collisions for strip per second

$$= \begin{bmatrix}\text{projected area}\\ \text{of the strip}\\ \text{in } x \text{ direction}\end{bmatrix}\begin{bmatrix}\text{distance the}\\ \text{strip moves}\\ \text{in 1 sec}\end{bmatrix}\begin{bmatrix}\text{number of}\\ \text{molecules per}\\ \text{unit volume}\end{bmatrix}$$

$$= [(R\,d\theta \cos\theta)(2\pi R \sin\theta)][V_s][n]$$

$$= 2\pi R^2 n V_s \sin\theta \cos\theta\, d\theta \tag{b}$$

The impulse component dI_x provided by the strip in 1 sec is the product of the right sides of Eqs. (a) and (b). Thus,

$$dI_x = 2\pi mnR^2V_s^2(\sin\theta\cos\theta)(1+\cos 2\theta)\,d\theta \tag{c}$$

Noting that $2\sin\theta\cos\theta = \sin 2\theta$, we have

$$dI_x = \pi mnR^2V_s^2(\sin 2\theta + \sin 2\theta \cos 2\theta)\,d\theta$$

$$= \pi mnR^2V_s^2\left(\sin 2\theta + \frac{\sin 4\theta}{2}\right)d\theta$$

Integrating from $\theta = 0$ to $\theta = \pi/2$,[7] we get the total impulse for 1 sec by the sphere:

$$I_x = \pi mnR^2V_s^2\left(\int_0^{\pi/2}\sin 2\theta\,d\theta + \tfrac{1}{2}\int_0^{\pi/2}\sin 4\theta\,d\theta\right)$$

$$= \pi mnR^2V_s^2\left(-\tfrac{1}{2}\cos 2\theta\Big|_0^{\pi/2} - \tfrac{1}{8}\cos 4\theta\Big|_0^{\pi/2}\right) \tag{d}$$

$$= \pi mnR^2V_s^2(1+0) = \pi mnR^2V_s^2$$

The average force needed to give this impulse by the satellite is clearly $\pi mnR^2V_s^2$, and so the reaction to this force is the desired drag.

Problems

14.41. Two cylinders move along a rod in a frictionless manner. Cylinder A has a mass of 10 kg and moves to the right at a speed of 3 m/sec, while cylinder B has a mass of 5 kg and moves to the left at a speed of 2.5 m/sec. What is the speed of cylinder B after impact for a coefficient of restitution ϵ of .8? What is the loss in kinetic energy?

[7]We integrate only up to $\pi/2$ because collisions take place only on the *front* part of the sphere. This is so since, in our model, the molecules are moving only from left to right toward the sphere.

Figure P.14.41

14.42. In Problem 14.41, what coefficient of restitution is needed for body A to be stationary after impact?

14.43. Two smooth cylinders of identical radius roll toward each other such that their centerlines are perfectly parallel. Cylinder A has a mass of 10 kg, and cylinder B has a mass of 7.5 kg. What is the speed at which cylinder A moves directly after collision for a coefficient of restitution $\epsilon = .75$?

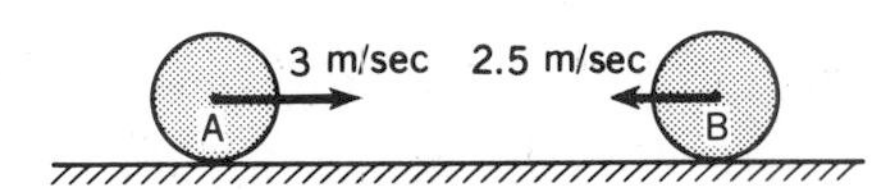

Figure P.14.43

14.44. Cylinder A, weighing 10 lb, moves toward cylinder B, weighing 40 lb, at the speed of 20 ft/sec. Mass B is attached to a spring having a spring constant K equal to 10 lb/in. If the collision has a coefficient of restitution $\epsilon = .9$, what is the maximum deflection δ of the spring? Assume that there is no friction along the rod and that the spring has negligible mass.

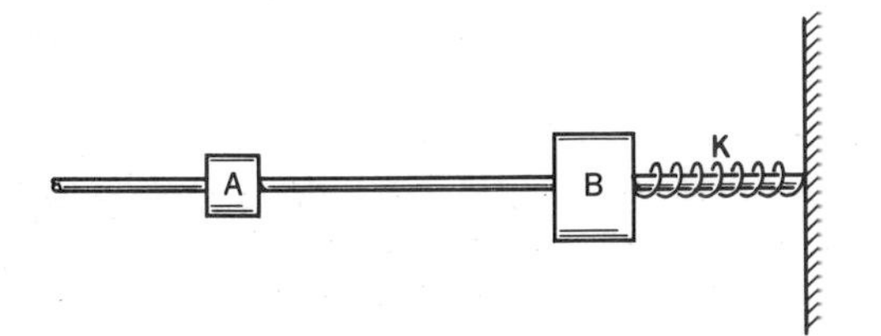

Figure P.14.44

14.45. Do Problem 14.44 for the case where there is a perfectly plastic impact and the spring is nonlinear such that $5x^{3/2}$ lb of force is required for a deflection of x inches.

14.46. Assume a perfectly plastic impact as the 5-kg body falls from a height of 2.6 m onto a plate of mass 2.5 kg. This plate is mounted on a spring having a spring constant of 1772 N/m. Neglect the mass of the spring as well as friction, and compute the maximum deflection of the spring after impact.

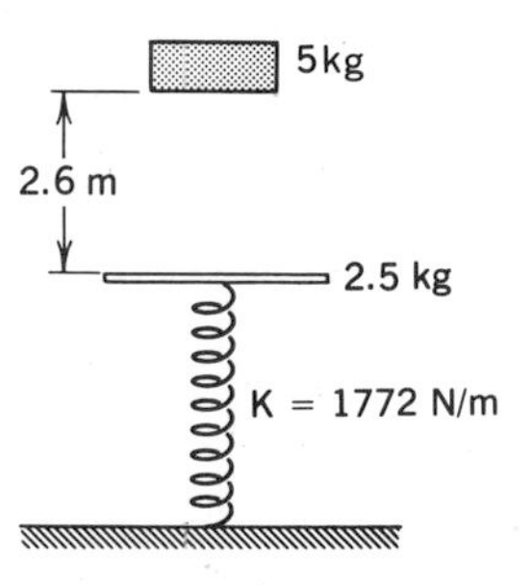

Figure P.14.46

14.47. Identical spheres B, C, and D lie along a straight line on a frictionless surface. Sphere A, which is identical to the others, moves toward the other spheres at a speed V_A in a direction collinear with the centers of the spheres. For perfectly elastic collisions, what are the final velocities of the bodies?

14.48. In Problem 14.47: (a) What is the final velocity of sphere D if $\epsilon = .80$ for all spheres and $V_A = 50$ ft/sec? (b) Set up a relation for the speed of the $(n + 1)$th sphere in terms of the speed of the nth sphere, again for $\epsilon = .80$ and $V_A = 50$ ft/sec.

14.49. A spherical mass M_1 of 20 lbm is held at an angle θ_1 of 60° before being released. It strikes mass M_2 of 10 lbm with an impact having a coefficient of restitution equal to .75. Mass M_2 is held by a light rod of length 2 ft at the end of which is a torsional spring requiring 500 ft-lb per radian of rotation. The spring has no torque when l_2 is vertical. What is the maximum rotation of l_2 after impact? The length of $l_1 =$ 18 in. (*Hint:* The work of a couple C rotating on angle $d\theta$ is $C\,d\theta$. A trial-and-error solution for θ_2 will be necessary.)

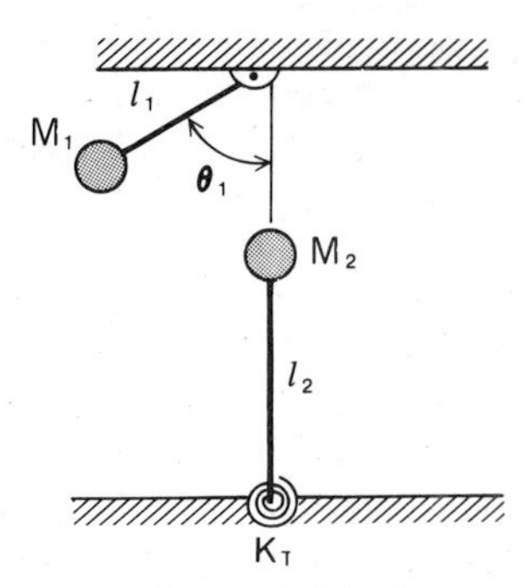

Figure P.14.49

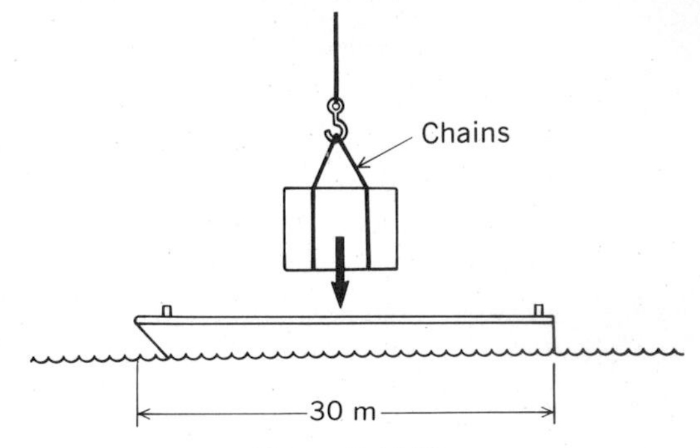

Figure P.14.51

14.50. Cylinder A, weighing 20 lb, is moving at a speed of 20 ft/sec when it is at a distance 10 ft from cylinder B, which is stationary. Cylinder B weighs 15 lb and has a coefficient of friction with the rod on which it rides of .3. Cylinder A has a coefficient of friction of .1 with the rod. What is the coefficient of restitution if cylinder B comes to rest after collision at a distance 12 ft to the right of the initial position?

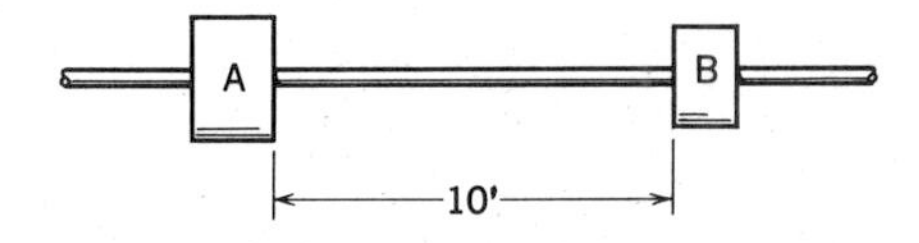

Figure P.14.50

14.51. A load is being lowered at a speed of 2 m/sec into a barge. The barge weighs 1000 kN, and the load weighs 100 kN. If the load hits the barge at 2 m/sec and the collision is plastic, what is the maximum depth that the barge is lowered into the water, assuming that the position of loading is such as to maintain the barge in a horizontal position? The width of the barge is 10 m. What are the weaknesses (if any) of your analysis? The density of water is 1000 kg/m^3.

14.52. A tractor–trailer weighing 50 kN without a load carries a 10-kN load A as shown. The driver jams on his brakes until they lock for a panic stop. The load A breaks loose from its ropes. When the truck has stopped the load is 3 m from the left end of the trailer wall (see diagram) and is moving at a speed of 4 m/sec relative to the truck. The coefficient of dynamic friction between the load A and the trailer is .2 and between the tires and road is .5. If there is a plastic impact between A and the trailer and the driver keeps his brakes locked, how far d does the truck then move?

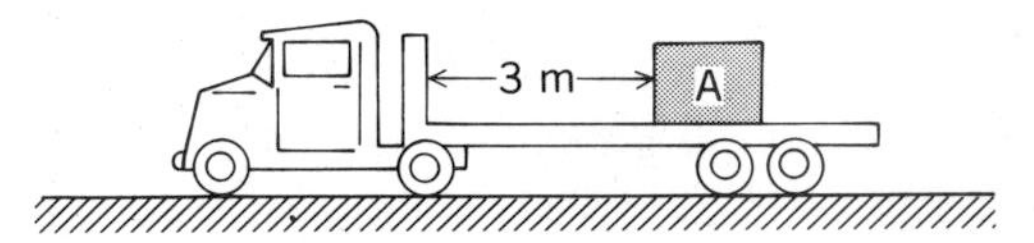

Figure P.14.52

14.53. A pile driver is used to drive a pile A into the ground. The device consists of a piston C on which there is a pressure p from steam. The piston is connected to 1000-lb hammer B. This system is driven so as to drop a distance h of 2 ft to impact on the pile A, weighing 400 lb. If an outside force of 25,000 lb is required to move the pile downward, what distance d will the pile move for a drop involving no contribution from p (which is then 0 psig). The impact is plastic. The weight of the piston plus the connecting rod is 100 lb.

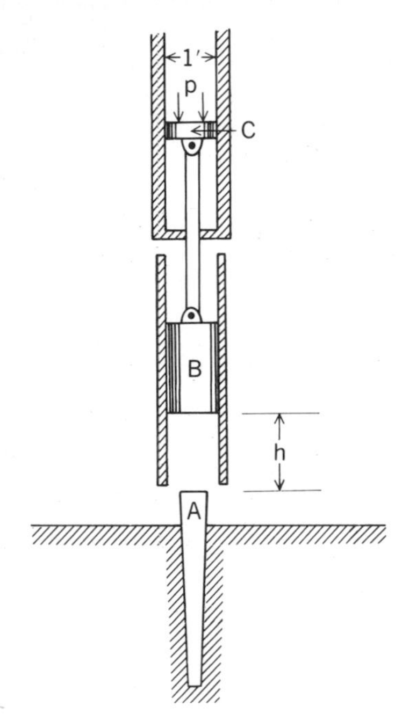

Figure P.14.53

14.54. Do Problem 14.53 for a constant pressure $p = 50$ psig.

14.55. A thin disc A weighing 5 lb translates along a frictionless surface at a speed of 20 ft/sec. The disc strikes a square stationary plate B weighing 10 lb at the center of a side. What are the velocity and direction of motion of the plate and the disc after collision? Assume that the surfaces of the plate and disc are smooth. Take $\epsilon = .7$.

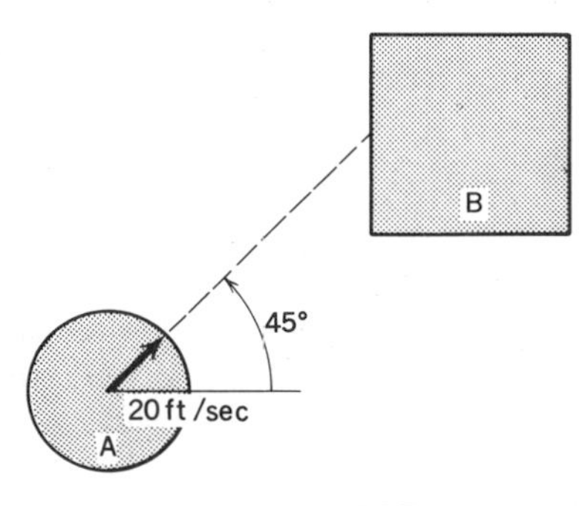

Figure P.14.55

14.56. In Problem 14.55, at the instant of contact between the bodies a clamping device firmly connects the bodies together so as to form one rigid unit. Find the velocity of the center of mass of the system after impact.

14.57. The theory of collisions of *subatomic* particles is called the theory of *scattering*. The coefficient-of-restitution concept presented for macroscopic bodies in this chapter cannot be used. However, conservation of momentum can be used.

A neutron N shown moving with a speed V_0 strikes a stationary proton P. After collision the velocity of the neutron is V_N and that of the proton is V_P, as shown in the diagram. For an *elastic* collision, prove that $\phi + \theta = \pi/2$. (*Hint:* Use the vector polygon concept and the Pythagorean theorem. Also, take the masses of proton and neutron to be equal.)

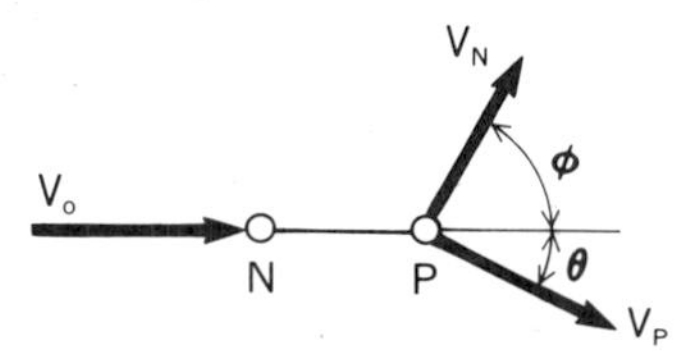

Figure P.14.57

14.58. A neutron N is moving toward a stationary helium nucleus H_e (atomic number 2) with kinetic energy 10 MeV. If the collision is inelastic, causing a loss of 20% of the kinetic energy, what is the angle θ after collision? See the first paragraph (only) of Problem 14.57. [*Hint:* There is no need (if one is clever) to have to convert the atomic number to kilograms.]

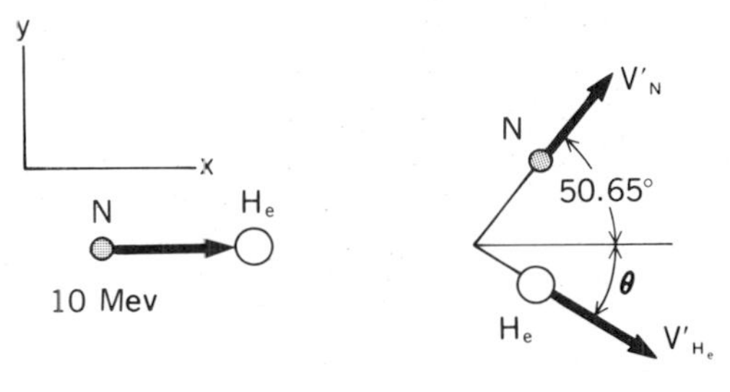

Figure P.14.58

14.59. A ball is thrown against a floor at an angle of 60° with a speed at impact of 16 m/sec. What is the angle of rebound α if $\epsilon = .7$? Neglect friction.

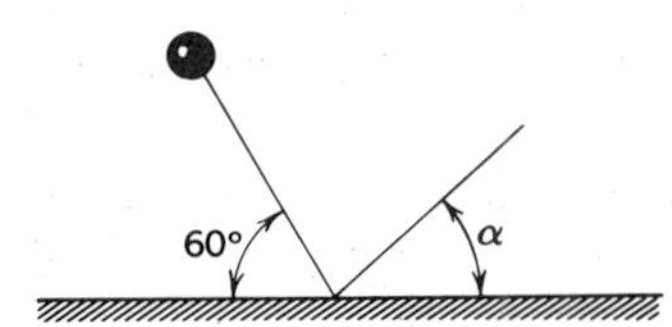

Figure P.14.59

14.60. A ball strikes the xy plane of a handball court at $\boldsymbol{r} = 3\boldsymbol{i} + 7\boldsymbol{j}$ ft. The ball has initially a velocity $\boldsymbol{V}_1 = -10\boldsymbol{i} - 10\boldsymbol{j} - 15\boldsymbol{k}$ ft/sec. The coefficient of restitution is .8. Determine the final velocity $\boldsymbol{V}_2$ after it bounds off the xy, yz, and xz planes once. Neglect gravity and friction.

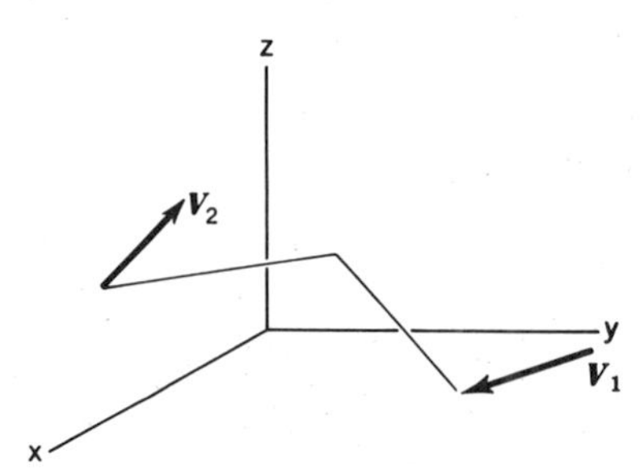

Figure P.14.60

14.61. A space vehicle in the shape of a cone–cylinder is moving at a speed V km/sec, many times the speed of sound through highly rarefied atmosphere. If each molecule of the gas has a mass m kg and if there are, on the average, n molecules per cubic meter, compute the drag on the cone–cylinder. The cone half-angle is 30°. Take the collision to be perfectly elastic. (*Hint:* The volume that the cone sweeps out in a unit time equals the volume that the projected area of the cone sweeps out. Can you readily show this?)

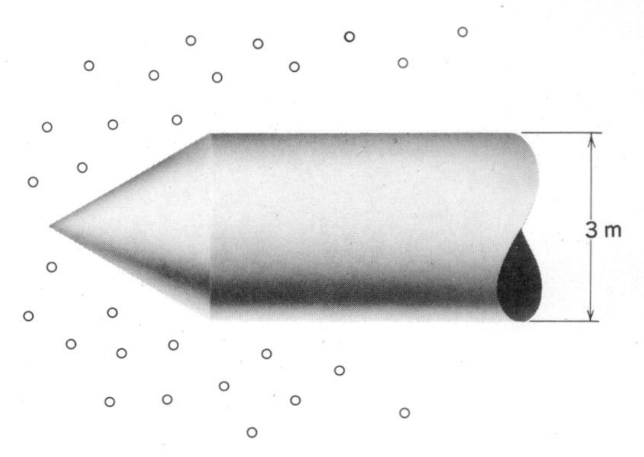

Figure P.14.61

14.62. Do Problem 14.61 for a case where the collisions are assumed to be inelastic. Assume the coefficient of restitution to be .8.

14.63. A double-wedge airfoil section for a space glider is shown. If the glider moves in highly rarefield atmosphere at a speed V many times greater than the speed of sound, what is the drag per unit length of this airfoil? Assume the collision to be perfectly elastic. There are n molecules per ft^3, each having a mass m in slugs. See the hint in Problem 14.61.

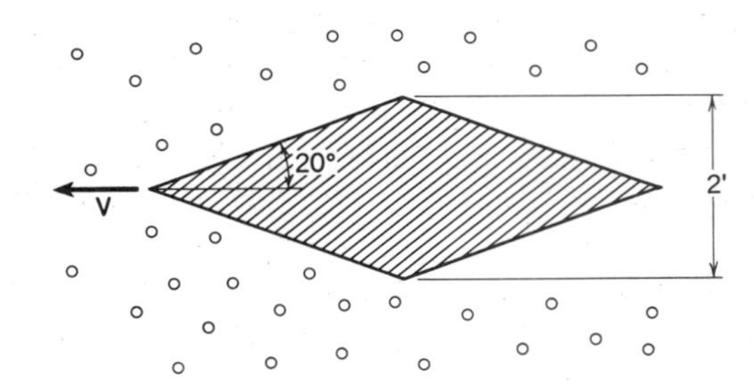

Figure P.14.63

14.64. A small elastic ball is dropped from a height of 5 m onto a rigid cylindrical body having a radius of 1.5 m. At what position on the x axis does the ball land?

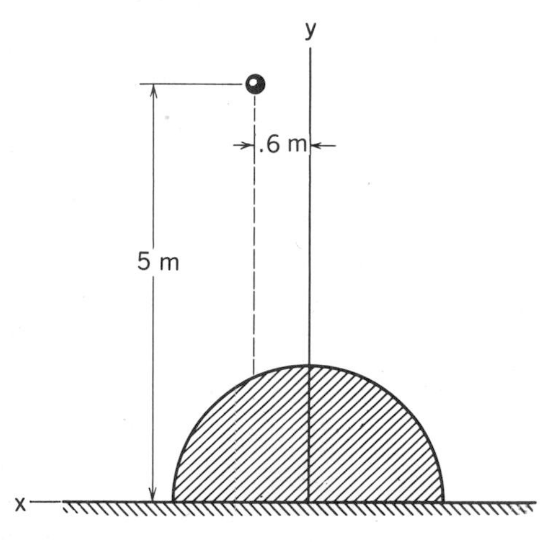

Figure P.14.64

14.65. Do Problem 14.64 for an inelastic impact with $\epsilon = .6$.

14.66. A small elastic sphere is dropped from position (2, 3, 30) ft onto a hard spherical body having a radius of 5 ft positioned so that the z axis of the reference shown is along a diameter. For a perfectly elastic collision, give the speed of the small sphere directly after impact.

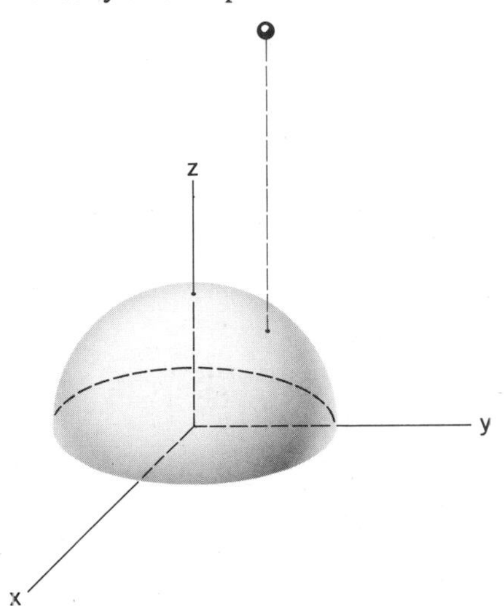

Figure P.14.66

14.67. Do Problem 14.66 for an inelastic impact with $\epsilon = .6$.

14.68. A bullet hits a smooth, hard, massive two-dimensional body whose boundary has been shown as a parabola. If the bullet strikes 1.5 m above the x axis and if the collision is perfectly elastic, what is the maximum height reached by the bullet as it ricochets? Neglect air resistance and take the velocity of the bullet on impact as 700 m/sec with a direction that is parallel to the x axis.

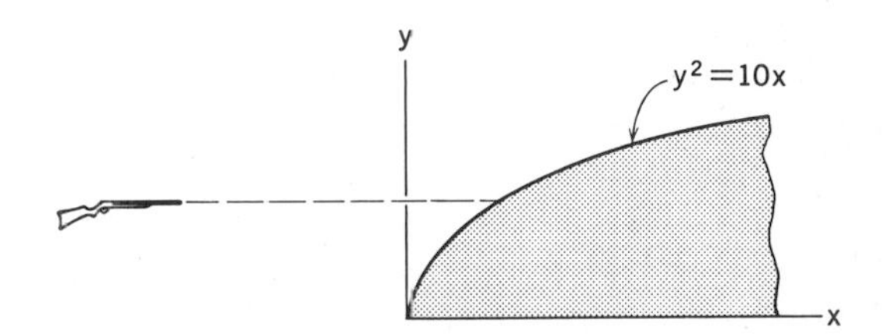

Figure P.14.68

14.69. In Problem 14.68, assume an inelastic impact with $\epsilon = .6$. At what position along x does the bullet strike after the impact?

14.70. Consider a parallel beam of light having an energy flux of S watts/m^2, shining normal to a flat surface that completely absorbs the energy. You learned in physics that an impulse dI is developed on the surface during time dt given by the formula

$$dI = \frac{S}{c}\,dt\,dA$$

where c is the speed of light in vacuo in m/sec. If the surface reflects the light, then we have an impulse dI developed on the surface given as

$$dI = 2\frac{S}{c}\,dt\,dA$$

Compute the force stemming from the reflection of light shining normal to a perfectly reflecting mirror having an area of 1 m^2. The light has an energy flux S of 20 W/m^2. Take the speed $c = 3 \times 10^8$ m/sec. What is the radiation pressure p_{rad} on the mirror?

*14.71. The Echo satellite when put into orbit is inflated to a 45-m-diameter sphere having a skin made up of a laminate of aluminum over mylar over aluminum. This skin is highly reflectant of light. Because of the small mass of this satellite, it may be affected by small forces such as that stemming from the reflection of light. If a parallel beam of light having an energy density S of .50 W/mm² impinges on the Echo satellite, what total force is developed on the satellite from this source? From physics (see Problem 14.70), the radiation pressure, p_{rad}, on a reflecting surface from a beam of light inclined by $\theta°$ from the normal to the surface is

$$p_{rad} = 2\frac{S}{c}\cos^2\theta$$

The pressure is in the direction of the incident radiation.

Part B
Moment of Momentum

14.6 Moment-of-Momentum Equation for a Single Particle

At this time, we shall introduce another auxiliary statement that follows from Newton's law and that will have great value when extended to the case of a rigid body. We start with Newton's law for a particle in the following form:

$$\boldsymbol{F} = \frac{d}{dt}(m\boldsymbol{V}) = \dot{\boldsymbol{P}}$$

where the symbol $\boldsymbol{P}$ represents the linear momentum of the particle. We next take the moment of each side of the equation about a point a in space (see Fig. 14.19):

$$\boldsymbol{\rho}_a \times \boldsymbol{F} = \boldsymbol{\rho}_a \times \dot{\boldsymbol{P}} \tag{14.16}$$

If this point a is positioned at a fixed location in XYZ, we can simplify the right side of Eq. 14.16. Accordingly, examine the expression $(d/dt)(\boldsymbol{\rho}_a \times \boldsymbol{P})$:

$$\frac{d}{dt}(\boldsymbol{\rho}_a \times \boldsymbol{P}) = \boldsymbol{\rho}_a \times \dot{\boldsymbol{P}} + \dot{\boldsymbol{\rho}}_a \times \boldsymbol{P} \tag{14.17}$$

But the expression $\dot{\boldsymbol{\rho}}_a \times \boldsymbol{P}$ can be written as $\dot{\boldsymbol{\rho}}_a \times m\dot{\boldsymbol{r}}$. The vectors $\boldsymbol{\rho}_a$ and $\boldsymbol{r}$ are measured in the same reference from a fixed point a to the particle and from the origin to the particle, respectively (see Fig. 14.20). They are thus different at all times to the extent of a constant vector $\overrightarrow{Oa}$. Note that

$$\boldsymbol{r} = \overrightarrow{Oa} + \boldsymbol{\rho}_a$$

Therefore,

$$\dot{\boldsymbol{r}} = \dot{\boldsymbol{\rho}}_a$$

Accordingly, the expression $\dot{\boldsymbol{\rho}}_a \times m\dot{\boldsymbol{r}}$ is zero. Thus, Eq. 14.17 becomes

$$\frac{d}{dt}(\boldsymbol{\rho}_a \times \boldsymbol{P}) = \boldsymbol{\rho}_a \times \dot{\boldsymbol{P}} \tag{14.18}$$

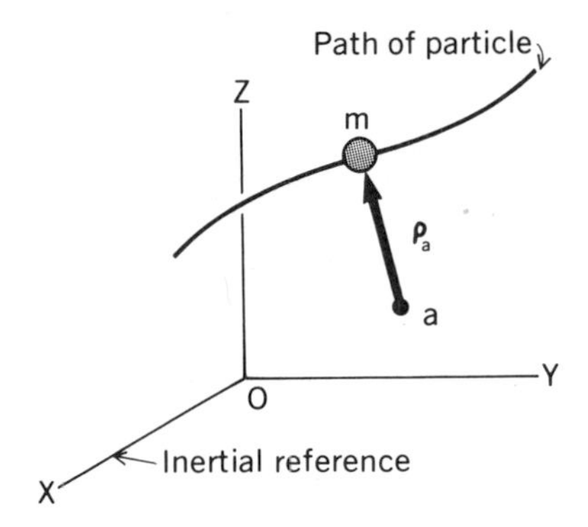

Figure 14.19. Point a fixed in inertial space.

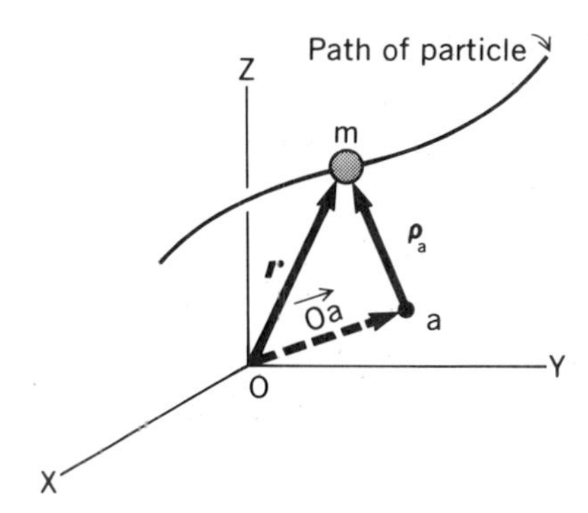

Figure 14.20. Position vectors to m and a.

and Eq. 14.16 can be written in the form

$$\boldsymbol{\rho}_a \times \boldsymbol{F} = \boldsymbol{M}_a = \frac{d}{dt}(\boldsymbol{\rho}_a \times \boldsymbol{P}) = \dot{\boldsymbol{H}}_a$$

Therefore,

$$\boxed{\boldsymbol{M}_a = \dot{\boldsymbol{H}}_a} \tag{14.19}$$

where $\boldsymbol{H}_a$ *is the moment about point a of the linear momentum vector*. Also, $\boldsymbol{H}$ is termed the *angular momentum vector*. Equation 14.19, then, states that *the moment* $\boldsymbol{M}_a$ *of the resultant force on a particle about a point a, fixed in an inertial reference, equals the time rate of change of the moment about point a of the linear momentum of the particle relative to the inertial reference*. This is the desired alternative form of Newton's law.

The scalar component of Eq. 14.19 along some axis, say the z axis, can be useful. Thus,

$$M_z = \dot{H}_z$$

where M_z is the torque of the total external force about the z axis and H_z is the moment of the momentum (or angular momentum) about the z axis.

EXAMPLE 14.12

A boat containing a man is moving near a dock (see Fig. 14.21). He throws out a light line and lassos a piling on the dock at A. He starts drawing in on the line so that when he is in the position shown in the diagram, the line is taut and has a length of 25 ft. His speed V_1 is 5 ft/sec in a direction normal to the line. If the net horizontal force F on the boat from tension in the line and from water resistance is maintained at 50 lb essentially in the direction of the line, what is the component of his velocity toward piling A (i.e., V_A) after the man has pulled in 3 ft of line? The boat and the man have a combined weight of 350 lb.

We may consider the boat and man as a particle for which we can apply Eq. 14.19. Thus,

$$\boldsymbol{M}_A = \dot{\boldsymbol{H}}_A \tag{a}$$

Clearly, here $\boldsymbol{M}_A = \boldsymbol{0}$ since $\boldsymbol{F}$ goes through A at all times. Thus, $\boldsymbol{H}_A$ is a constant—that is, the angular momentum about A must be constant. Observing Fig. 14.22, we can say accordingly

$$\boldsymbol{r}_1 \times m\boldsymbol{V}_1 = \boldsymbol{r}_2 \times m\boldsymbol{V}_2$$

$$(25)(m)(5) = (22)(m)(V_2)_t$$

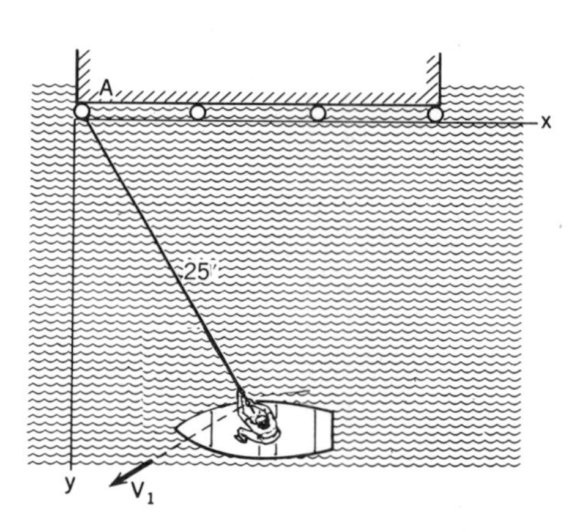

Figure 14.21. Man pulls toward piling.

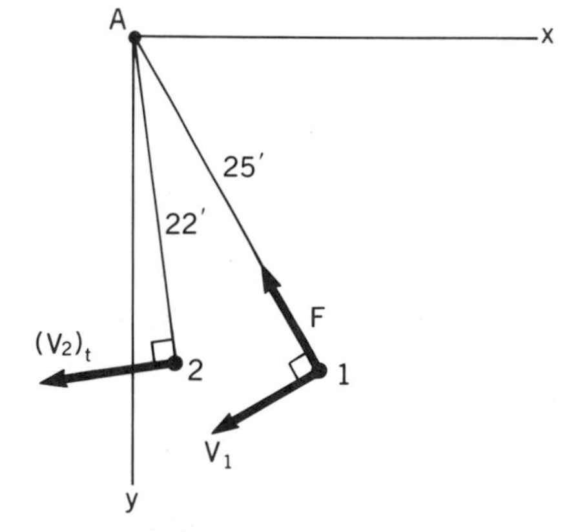

Figure 14.22. Boat at positions 1 and 2.

Therefore,

$$(V_2)_t = 5.68 \text{ ft/sec} \tag{b}$$

We need more information to get the desired result V_A toward the piling. We have not yet used the fact that $F = 50$ lb. Accordingly, we now employ the *work–kinetic energy equation* from Chapter 13. Thus,

$$\int_1^2 \boldsymbol{F} \cdot d\boldsymbol{s} = \left(\frac{1}{2}MV^2\right)_2 - \left(\frac{1}{2}MV^2\right)_1$$

$$(50)(3) = \frac{1}{2}\frac{350}{g}(V_2)^2 - \frac{1}{2}\frac{350}{g}(25)$$

Therefore,

$$V_2 = 7.25 \text{ ft/sec} \tag{c}$$

Now V_2 is the total velocity of the boat at position 2. To get the desired component V_A toward the piling, we can say, using Eqs. (b) and (c):

$$V_2^2 = (V_2)_t^2 + V_A^2$$

$$(7.25)^2 = (5.68)^2 + V_A^2$$

Therefore,

$$V_A = 4.51 \text{ ft/sec}$$

Many problems of space mechanics can be solved by using energy and angular-momentum methods of this and the preceding chapter without considering the detailed trajectory equations of Chapter 12. Let us therefore set forth some salient factors concerning the motion of a space vehicle moving in the vicinity of a planet or star with the engine shut off and with negligible friction from the outside.[8]

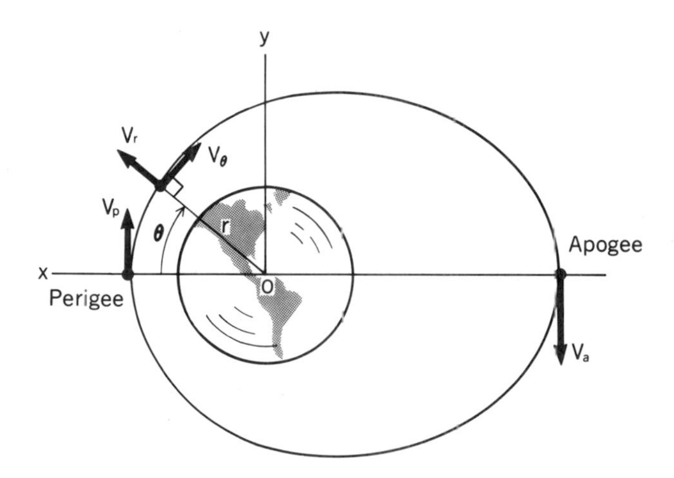

Figure 14.23. Elliptic orbit with perigee and apogee.

After the space vehicle has been propelled at great speed by its rocket engines to a position outside the planet's atmosphere (the final powered velocity is called the *burnout* velocity), the vehicle then undergoes plane, gravitational, central-force motion. If it continues to go around the planet, the vehicle is said to go into *orbit* and the trajectory is that of a circle or that of an ellipse. If, on the other hand, the vehicle escapes from the influence of the planet, then the trajectory will either be a parabola or a hyperbola. In the case of an elliptic orbit, the position closest to the surface of the planet is called *perigee* (see Fig. 14.23) and the position farthest from the surface of the planet is called *apogee*. Notice that at apogee and perigee the velocity vectors $\boldsymbol{V}_a$ and $\boldsymbol{V}_p$ of the vehicle are parallel to the surface of the planet and so at these points (and only at these points)

$$V = V_\theta; \qquad V_r = 0$$

In the case of a *circular* orbit of radius r and velocity V_c, we can use Newton's law and the gravitational law to state

$$\frac{GMm}{r^2} = m(r\omega^2)$$

[8]Those readers who have studied Sections 12.7 and 12.8 have already gone into these factors in considerable depth.

where M is the mass of the planet and ω is the angular speed of the radius vector to the vehicle. Replacing the acceleration term $r\omega^2$ by V_c^2/r and solving for V_c, we get

$$V_c = \sqrt{\frac{GM}{r}} \tag{14.20}$$

Knowing GM and R, we can readily compute the speed V_c for a particular circular orbit. In Section 12.6 we showed that GM can be easily computed using the relation

$$GM = gR^2 \tag{14.21}$$

where g is the acceleration of gravity at the surface of the planet and R is the radius of the planet.

In gravitational central-force motion, only the conservative force of gravity is involved, and so we must have *conservation of mechanical energy*. Furthermore, since this force is directed to O, the center of the planet, at all times (see Fig. 14.23), then the moment about O of the gravitation force must be zero. As a consequence, we must have *conservation of angular momentum* about O.[9]

We shall illustrate in the next example the dual use of the conservation-of-angular-momentum principle and the conservation-of-mechanical-energy principle for space mechanics problems. In the homework problems you will be asked to solve again some of the space problems of Chapter 12 using the principles above without getting involved with the trajectory equations. Such problems, you will then realize, are sometimes more easily solved by using the two principles discussed above than by using the trajectory equations.

EXAMPLE 14.13

A space-shuttle vehicle on a rescue mission (see Fig. 14.24) is sent into a circular orbit at a distance 1200 km from the earth's surface. A spacecraft meanwhile is in an elliptic orbit in the same plane with an apogee of 10,000 km from the center of the earth. The spacecraft cannot start its engines to effect a splashdown, so the space shuttle has the mission of docking with the spacecraft and rescuing the occupants. The timing of insertion of the circular orbit of the space shuttle has so been chosen that the space shuttle by firing its rockets at the position shown can, by the proper change of speed, reach apogee at the same time and same location as does the crippled space vehicle. At this time, docking procedures can be carried out. Considering that the rocket engines of the space-shuttle vehicle operate during a *very short distance* of travel[10] to achieve the proper velocity V_0 for the mission, determine the change in speed that the space shuttle must achieve. The radius of the earth is 6373 km.

---

[9]Those who have studied the trajectory equations of Chapter 12 might realize that

$$C = rV_\theta = \text{constant}$$

is actually a statement of the conservation of angular momentum since mrV_θ is the moment about O of the linear momentum relative to O.

[10]During this part of the flight we do *not* have central-force motion.

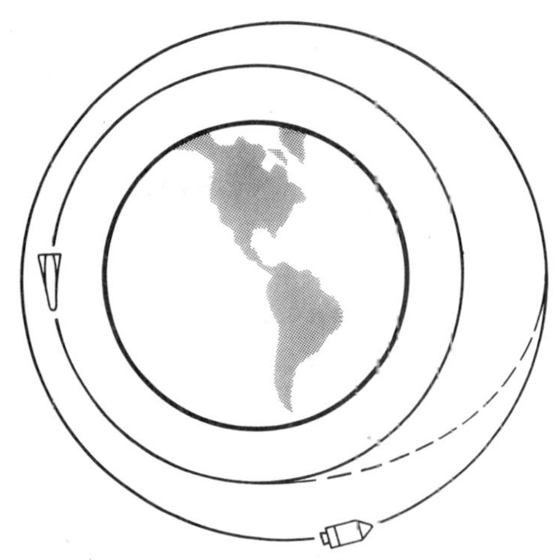

Figure 14.24. Rescue mission for space-shuttle vehicle.

We shall first compute GM. Thus, working with kilometers and hours,

$$GM = gR^2 = \left[(9.81)\left(\frac{3600^2}{1000}\right)\right](6373)^2 = 5.16 \times 10^{12} \text{ km}^3/\text{hr}^2$$

The velocity for the circular orbit for the space shuttle is then

$$V_c = \sqrt{\frac{GM}{r}} = \sqrt{\frac{5.16 \times 10^{12}}{7573}} = 26{,}105 \text{ km/hr} \tag{1}$$

From *conservation of angular momentum* for the space-shuttle rescue orbit we can say:

$$mr_0 V_0 = m(rV)_{\text{apogee}}$$

$$(7573)(V_0) = (10{,}000)(V)_{\text{apogee}}$$

Therefore,

$$V_0 = 1.320 V_{\text{apogee}} \tag{2}$$

where V_0 is the speed of the space shuttle just *after* firing rockets. Next, we use the principle of *conservation of mechanical energy* for the rescue orbit, Thus,

$$-\frac{GMm}{r_0} + \frac{mV_0^2}{2} = -\frac{GMm}{r} + \frac{mV_{\text{apogee}}^2}{2}$$

$$-\frac{5.16 \times 10^{12}}{7573} + \frac{V_0^2}{2} = -\frac{5.16 \times 10^{12}}{10{,}000} + \frac{V_{\text{apogee}}^2}{2} \tag{3}$$

Substitute for V_{apogee} using Eq. (2) and solve for V_0. We get

$$V_0 = 27{,}861 \text{ km/hr} \tag{4}$$

Hence, using Eq. (1), we can say:

$$\Delta V = 27{,}861 - 26{,}105 = 1756 \text{ km/hr}$$

Problems

14.72. A particle rotates at 30 rad/sec along a frictionless surface at a distance 2 ft from the center. A flexible cord restrains the particle. If this cord is pulled so that the particle moves inward at a velocity of 5 ft/sec, what is the magnitude of the total velocity when the particle is 1 ft from the center?

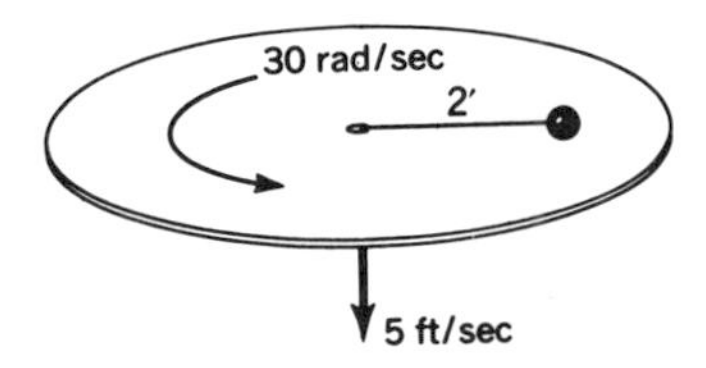

Figure P.14.72

14.73. A satellite has an apogee of 7128 km. It is moving at a speed of 36,480 km/hr. What is the transverse velocity of the satellite when $r = 6970$ km?

14.74. A system is shown rotating freely with an angular speed ω of 2 rad/sec. A mass A of 1.5 kg is held at a against a spring such that the spring is compressed 100 mm. If the device a holding the mass in position is suddenly removed, determine how far toward the vertical axis of the system the mass will move. The spring constant K is .531 N/mm. Neglect all friction and inertia of the bars. The spring is not connected to the mass.

Figure P.14.74

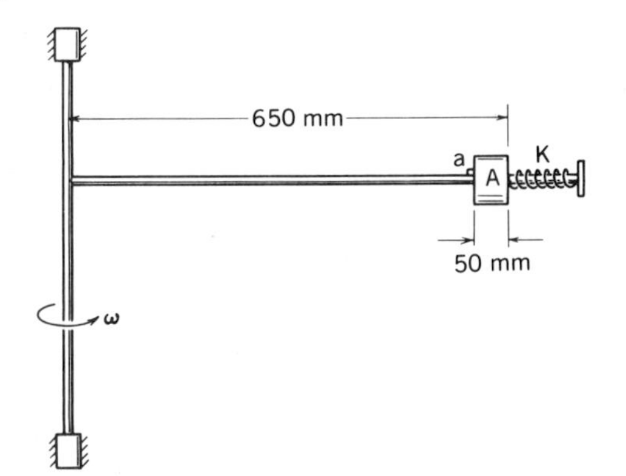

14.75. Do Problem 14.74 for the case where there is coulombic friction between the mass A and the horizontal rod with a constant μ_d equal to .4.

14.76. A body A weighing 10 lb is moving at a speed of V_1 of 20 ft/sec on a frictionless surface. An elastic cord AO, which has a length l of 20 ft, becomes taut but not stretched at the position shown in the diagram. What is the radial speed toward O of the body when the cord is stretched 2 ft? The cord has an equivalent spring constant of .3 lb/in.

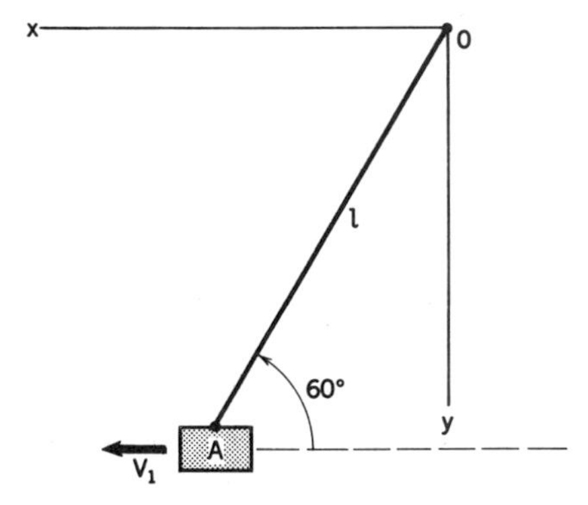

Figure P.14.76

14.77. A small ball B weighing 2 lb is rotating about a vertical axis at a speed ω_1 of 15 rad/sec. The ball is connected to bearings on the shaft by light inextensible strings having a length l of 2 ft. The angle θ_1 is 30°. What is the angular speed ω_2 of the ball if bearing A is moved up 6 in.?

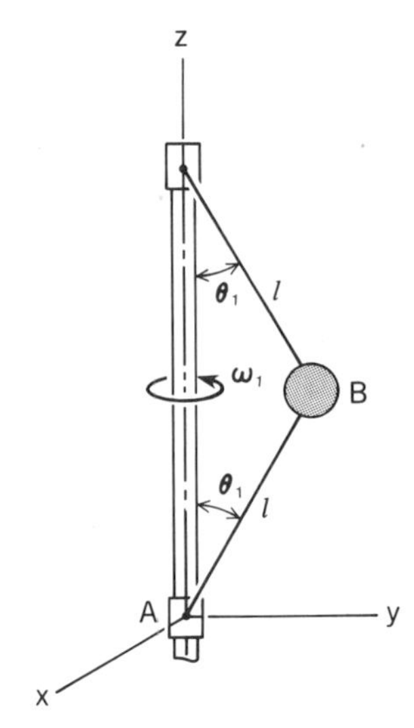

Figure P.14.77

14.78. A mass m of 1 kg is swinging freely about the z axis at a speed ω_1 of 10 rad/sec. The length l_1 of the string is 250 mm. If the tube A through which the connecting string passes is moved down a distance d of 90 mm, what is ω_2 of the mass? You should get a fourth-order equation for ω_2 which has as the desired root $\omega_2 = 21.05$ rad/sec.

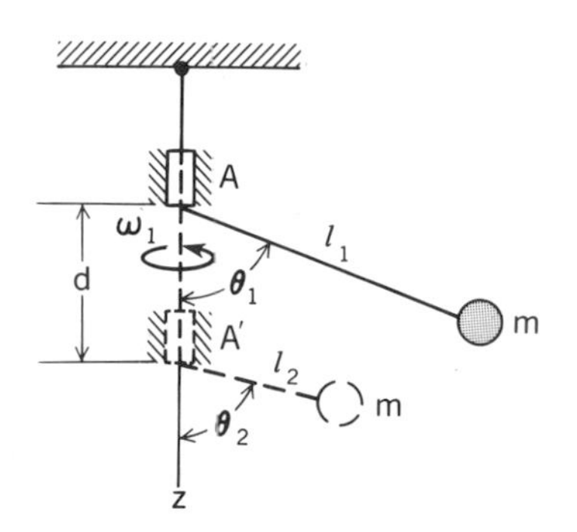

Figure P.14.78

14.79. A small 2-lb ball B is rotating at angular speed ω_1 of 10 rad/sec about a horizontal shaft. The ball is connected to the bearings with light elastic cords which when unstretched are each 12 in. in length. A force of 15 lb is required to stretch the cord 1 in. The distance d_1 between the bearings originally is 20 in. If bearing A is moved to shorten d by 6 in., what is the angular velocity ω_2 of the ball? Neglect the effects of gravity and the mass of the elastic cords. (*Hint:* You should arrive at a transcendental equation for θ_2 whose solution is 54.49°.)

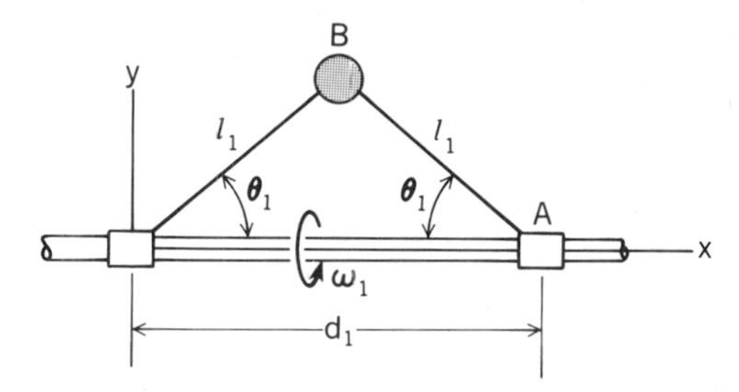

Figure P.14.79

14.80. Do Problem 12.20 using the principles of conservation momentum and conservation of mechanical energy.

14.81. In Problem 12.72, find the radial velocity by using the method of conservation of angular momentum and mechanical energy.

14.82. Do Problem 12.68 by the method of conservation of angular momentum and mechanical energy.

14.83. In Problem 12.73, find the height of the bullet above the surface of the planet by the methods of conservation of angular momentum and mechanical energy.

14.84. In Problem 12.103, find the maximum elevation above the earth's surface by the methods of conservation of angular momentum and mechanical energy.

14.85. Do Problem 12.61 by methods of conservation of angular momentum and mechanical energy. (*Hint:* The escape velocity $= \sqrt{2GM/r} = \sqrt{2}\, V_c$.)

14.86. Do Problem 12.57 using the principles of conservation of angular momentum and mechanical energy.

14.87. A space station is in a circular parking orbit around the earth at a distance of 5000 mi from the center. A projectile is fired in a direction tangential to the trajectory of the space station with a speed of 5000 mi/hr relative to the space station. What is the maximum distance from earth reached by the projectile?

14.88. A skylab is in a circular orbit about the earth 500 km above the earth's surface. A space-shuttle vehicle has rendezvoused with the skylab and now, after disengaging from the skylab, its rocket engines are fired so as to move the vehicle with a speed of 800 m/sec relative to the skylab in the opposite direction to that of the skylab. Assume that the firing of the rocket takes place over a short distance and does not affect the skylab. What speed would the space-shuttle vehicle have when it encounters appreciable atmosphere at about 50 km above the earth's surface? What is the radial velocity at this position?

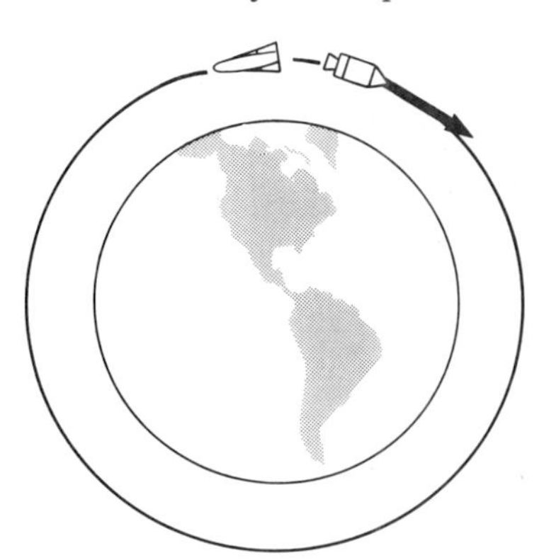

Figure P.14.88

14.89. A space probe is approaching Mars. When the probe is 50,000 mi from the center of Mars it has a speed V_0 of 10,000 mi/hr with a component $(V_r)_0$ toward the center of Mars of 9800 mi/hr. How close does the probe come to the surface of Mars? If retro-rockets are fired at this lowest position A, what change in speed is needed to alter the trajectory into a circular orbit as shown? The acceleration of gravity at the surface of Mars is 12.40 ft/sec², and the radius R of the planet is 2107 mi.

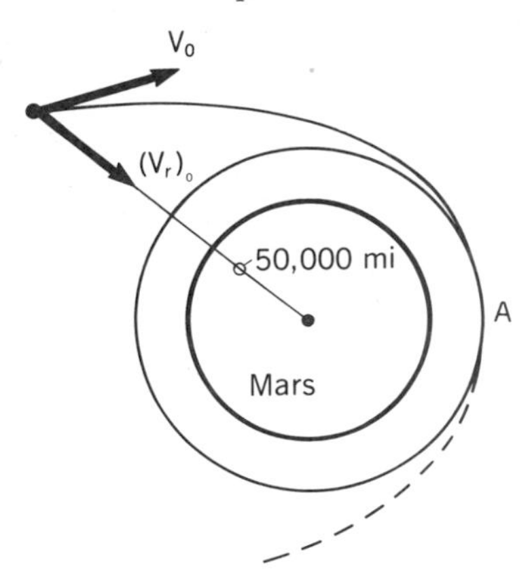

Figure P.14.89

14.90. In Problem 14.89, a midcourse correction is to be made to get the probe within 1000 mi from the surface of Mars. If V_0 at $r_0 = 50{,}000$ is still to be 10,000 mi/hr, what should be the radial velocity component $(V_r)_0$?

14.91. The Apollo command module is in a circular parking orbit about the moon at a distance of 161.0 km above the surface of the moon. The lunar exploratory module is to detach from the command module. The lunar-module rockets are fired briefly to give a velocity V_0 relative to the command module in the opposite direction. If the lunar module is to have a transverse velocity of 1500 m/sec when it is 80 km from the surface of the moon before rockets are fired again, what must V_0 be? What is the radial velocity at this position? The radius of the moon is 1733 km, and the acceleration of gravity is 1.700 m/sec² at the surface.

Figure P.14.91

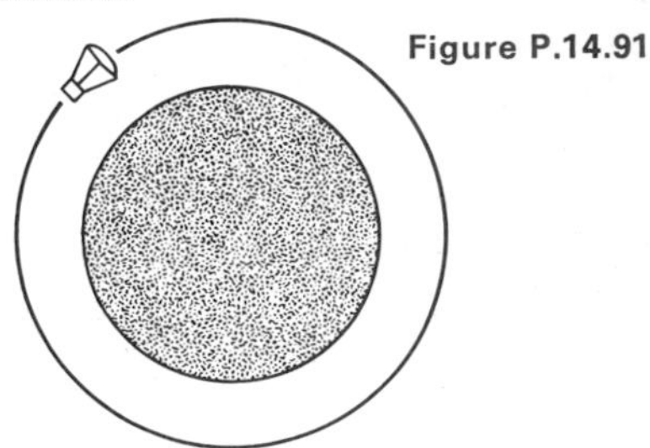

14.7 Moment-of-Momentum Equations for a System of Particles

We shall now develop the moment-of-momentum equations for an aggregate of particles. The resulting equations will be of vital importance when we apply them to rigid bodies in later chapters. We shall consider a number of cases.

Case 1. Fixed Reference Point in Inertial Space. An aggregate of n particles and an inertial reference are shown in Fig. 14.25. The moment of momentum equation for the ith particle is now written about the origin of this reference:

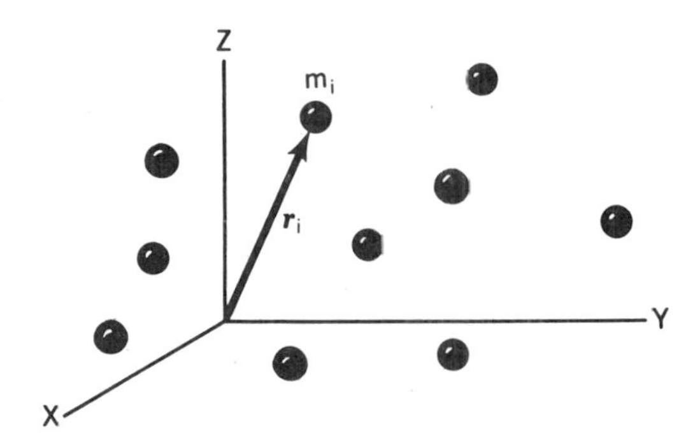

Figure 14.25. System of n particles.

$$\boldsymbol{r}_i \times \boldsymbol{F}_i + \boldsymbol{r}_i \times \left(\sum_{\substack{j=1 \\ i \neq j}}^{n} \boldsymbol{f}_{ij} \right) = \frac{d}{dt}(\boldsymbol{r}_i \times \boldsymbol{P}_i) \tag{14.22}$$

where, as usual, $\boldsymbol{f}_{ij}$ is the internal force from the jth particle on the ith particle. We now sum this equation for all n particles:

$$\sum_{i=1}^{n} \boldsymbol{r}_i \times \boldsymbol{F}_i + \sum_{i=1}^{n} \sum_{j=1}^{n} (\boldsymbol{r}_i \times \boldsymbol{f}_{ij}) = \frac{d}{dt}\left[\sum_{i=1}^{n} (\boldsymbol{r}_i \times \boldsymbol{P}_i) \right] = \dot{\boldsymbol{H}}_{\text{total}} \tag{14.23}$$

where the summation operation has been put before the differentiation on the right side (permissible because of the distributive property of differentiation with respect to addition). For any pair of particles, the internal forces will be equal and opposite and collinear (see Fig. 14.26). Hence, the forces will have a zero moment about the origin. (This result is most easily understood by remembering that, for purposes of taking moments about a point, forces are transmissible.) We can then conclude that the expression

$$\sum_{i=1}^{n} \sum_{j=1}^{n} (\boldsymbol{r}_i \times \boldsymbol{f}_{ij})$$

in this equation is zero. Realizing that $\sum_i \boldsymbol{r}_i \times \boldsymbol{F}_i$ is the total moment of the external forces about the origin, we have as a result for Eq. 14.23:

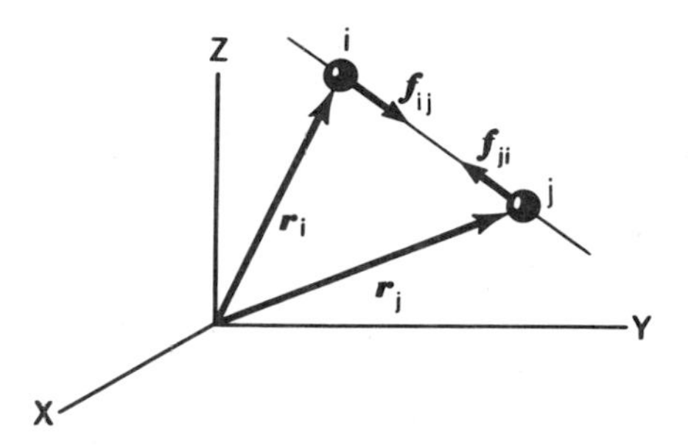

Figure 14.26. Internal equal and opposite forces.

$$\boxed{\boldsymbol{M}_a = \dot{\boldsymbol{H}}_a} \tag{14.24}$$

Thus, *the total moment* $\boldsymbol{M}$ *of external forces acting on an aggregate of particles about a point a fixed in an inertial reference* (the point in the development was picked as the origin merely for convenience) *equals the time rate of change of the total moment of the linear momentum relative to the inertial reference, where this moment is taken about the aforementioned point a.*[11]

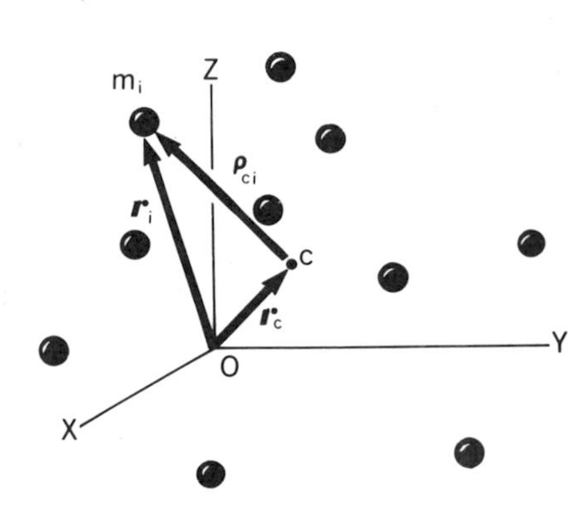

Figure 14.27. *c* is center of mass of aggregate.

We may express Eq. 14.24 in a different form by considering the center of mass. In a manner analogous to kinetic energy of an aggregate of particles, we can show that the angular momentum of an aggregate of particles about a fixed point can be given in terms of the angular momentum relative to the center of mass plus the angular momentum of the center of mass relative to the fixed point.

Accordingly, consider the center of mass c of an aggregate of particles as shown in Fig. 14.27. For the ith particle, we can say:

$$\boldsymbol{r}_i = \boldsymbol{r}_c + \boldsymbol{\rho}_{ci} \tag{14.25}$$

The angular momentum for the aggregate of particles about O is then

$$\begin{aligned}\boldsymbol{H}_0 &= \sum_i (\boldsymbol{r}_c + \boldsymbol{\rho}_{ci}) \times \boldsymbol{P}_i \\ &= \sum_i (\boldsymbol{r}_c + \boldsymbol{\rho}_{ci}) \times [(m_i)(\dot{\boldsymbol{r}}_c + \dot{\boldsymbol{\rho}}_{ci})]\end{aligned} \tag{14.26}$$

Carry out the cross product and extract $\boldsymbol{r}_c$ from the summations:

$$\boldsymbol{H}_0 = \boldsymbol{r}_c \times M\dot{\boldsymbol{r}}_c + \boldsymbol{r}_c \times \sum_i m_i\dot{\boldsymbol{\rho}}_{ci} + (\sum_i m_i\boldsymbol{\rho}_{ci}) \times \dot{\boldsymbol{r}}_c + \sum_i \boldsymbol{\rho}_{ci} \times m_i\dot{\boldsymbol{\rho}}_{ci} \tag{14.27}$$

But since c is the center of mass, it follows that

$$\sum_i m_i\boldsymbol{\rho}_{ci} = \boldsymbol{0}$$

$$\sum_i m_i\dot{\boldsymbol{\rho}}_{ci} = \boldsymbol{0}$$

Going back to Eq. 14.27, we see that the second and third terms are to be deleted and we get the desired result for $\boldsymbol{H}_0$:

$$\boldsymbol{H}_0 = \boldsymbol{r}_c \times M\dot{\boldsymbol{r}}_c + \sum_i \boldsymbol{\rho}_{ci} \times m_i\dot{\boldsymbol{\rho}}_{ci} = \boldsymbol{r}_c \times M\dot{\boldsymbol{r}}_c + \boldsymbol{H}_c$$

[11]Point a could also be moving with a constant velocity V_0 relative to inertial reference XYZ. However, a would then be fixed in another inertial reference $X'Y'Z'$ which is translating with respect to XYZ at a speed V_0.

where $\boldsymbol{H}_c$ is the moment about the center of mass of the linear momentum seen from the center of mass for the aggregate.[12] This may be rewritten and expressed for *any* fixed point a where, using $\boldsymbol{r}_{ac}$ as the position vector from fixed point a to the center of mass c, we have

$$\boldsymbol{H}_a = \boldsymbol{H}_c + \boldsymbol{r}_{ac} \times M\dot{\boldsymbol{r}}_{ac} \tag{14.28}$$

Note that $\dot{\boldsymbol{r}}_{ac}$ is the velocity of c relative to fixed point a and is thus equal to the velocity $\boldsymbol{V}_c$ of the mass center relative to XYZ. Thus, we can express Eq. (14.28) as

$$\boxed{\boldsymbol{H}_a = \boldsymbol{H}_c + \boldsymbol{r}_{ac} \times M\boldsymbol{V}_c} \tag{14.29}$$

Furthermore, we have for $\dot{\boldsymbol{H}}_a$:

$$\dot{\boldsymbol{H}}_a = \dot{\boldsymbol{H}}_c + \boldsymbol{r}_{ac} \times M\dot{\boldsymbol{V}}_c$$

where we have used the fact that $\dot{\boldsymbol{r}}_{ac} = \boldsymbol{V}_c$ to delete one expression. We may now restate Eq. 14.24 for a fixed point a as follows on replacing $\dot{\boldsymbol{H}}_a$ using the above equation. Using $\boldsymbol{a}_c$ for $\dot{\boldsymbol{V}}_c$ we have the desired result:

$$\boxed{\boldsymbol{M}_a = \dot{\boldsymbol{H}}_c + \boldsymbol{r}_{ac} \times M\boldsymbol{a}_c} \tag{14.30}$$

Case 2. Reference Point at Center of Mass. We need only use Eq. 14.30 for this purpose. We express the left side of this equation using the left side of Eq. 14.23 and replacing $\boldsymbol{r}_i$ in the first expression using Eq. 14.25. Thus, for point a at the origin O of XYZ, whereupon $\boldsymbol{r}_{ac}$ becomes simply $\boldsymbol{r}_c$, we have on replacing $\boldsymbol{a}_c$ by $\ddot{\boldsymbol{r}}_c$

$$\sum_i (\boldsymbol{r}_c + \boldsymbol{\rho}_{ci}) \times \boldsymbol{F}_i + \sum_j \sum_i \boldsymbol{r}_i \times \boldsymbol{f}_{ij} = \dot{\boldsymbol{H}}_c + \boldsymbol{r}_c \times M\ddot{\boldsymbol{r}}_c$$

The internal forces $\boldsymbol{f}_{ij}$ give zero contribution in this equation as explained earlier and we have

$$\boldsymbol{r}_c \times \sum_i \boldsymbol{F}_i + \sum_i \boldsymbol{\rho}_{ci} \times \boldsymbol{F}_i = \boldsymbol{r}_c \times M\ddot{\boldsymbol{r}}_c + \dot{\boldsymbol{H}}_c$$

From Newton's law for the center of mass, we know that $\sum \boldsymbol{F}_i = M\ddot{\boldsymbol{r}}_c$ and so the first terms on the left and right sides of the equation above cancel. The remaining expression on the left side of the equation is the moment about the center of mass of the external forces. We then get

$$\boxed{\boldsymbol{M}_c = \dot{\boldsymbol{H}}_c} \tag{14.31}$$

We thus get the same formulation for the center of mass as for a fixed point in inertial space. Please note that $\boldsymbol{H}_c$ is taken relative to the center of mass c.

[12]That is, as seen from a reference translating with c relative to XYZ—in other words, as seen by a nonrotating observer moving with c.

Case 3. Point Accelerating Toward the Mass Center. There is yet a third point of interest to be considered and that is a point a accelerating toward or away from the mass center of the aggregate (Fig. 14.28). For such a point, we can again give the same simple equation presented for cases 1 and 2. Thus,

$$\boxed{\boldsymbol{M}_a = \dot{\boldsymbol{H}}_a} \tag{14.32}$$

where $\boldsymbol{H}$ is taken relative to point a. We have asked for the derivation of this equation in Problem 14.112.

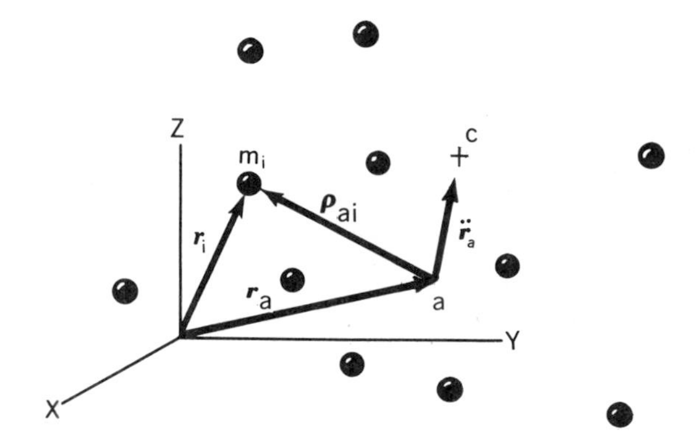

Figure 14.28. Point a accelerates toward or away from c.

The component of the equation $\boldsymbol{M}_a = \dot{\boldsymbol{H}}_a$ for any one of the three cases in say the x direction,

$$M_x = \dot{H}_x$$

can be very useful. Here, M_x is the torque about the x axis, and H_x is the moment of momentum (angular momentum) about the x axis. We now examine such a problem in the following example.

EXAMPLE 14.14

A heavy chain of length 20 ft lies on a light plate A which is freely rotating at an angular speed of 1 rad/sec (see Fig. 14.29). A channel C acts as a guide for the chain on the plate, and a stationary pipe acts as a guide for the chain below the plate. What is the speed of the chain after it moves 5 ft starting from rest relative to the platform? Neglect friction, the angular momentum of the plate, and the angular momentum of the vertical section of the chain about its own axis. The chain weight per unit length, w, is 10 lb/ft.

We shall first apply the *moment of momentum* equation about point D for the chain and plate. Taking the component of this equation along the z axis, we can say:

$$M_z = (\dot{H})_z \tag{a}$$

Clearly $M_z = 0$, and so we have conservation of angular momentum. That is,

$$H_z = \text{constant}$$
$$(H_z)_1 = (H_z)_2 \tag{b}$$

where 1 and 2 refer to the initial condition and the condition after the chain moves 5 ft. We can then say:

$$\int_0^{10} r(V_\theta)_1\left(\frac{w}{g}\,dr\right) = \int_0^5 r(V_\theta)_2\left(\frac{w}{g}\,dr\right)$$

$$\int_0^{10} (r)(\omega_1 r)\left(\frac{w}{g}\,dr\right) = \int_0^5 r(\omega_2 r)\left(\frac{w}{g}\,dr\right)$$

$$(1)\left(\frac{w}{g}\right)\int_0^{10} r^2 dr = (\omega_2)\left(\frac{w}{g}\right)\int_0^5 r^2\,dr$$

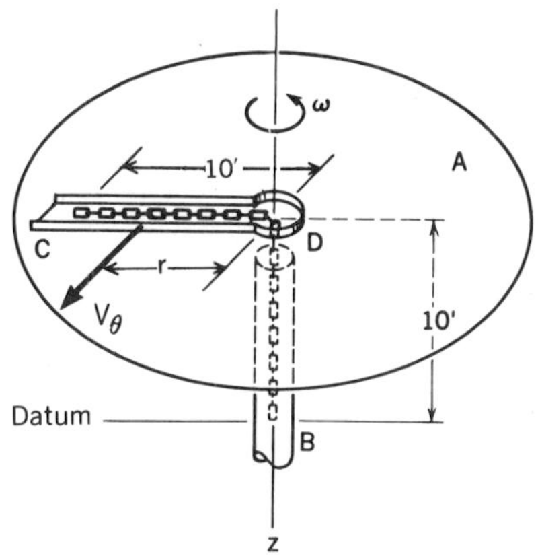

Figure 14.29. Sliding chain.

Therefore,

$$\omega_2 = 8 \text{ rad/sec} \tag{c}$$

To find the speed of movement of the chain, we must next go to *energy* considerations. Because only conservative forces are acting here, we may employ the *conservation-of-mechanical-energy* principle. In so doing, we shall use as a datum the end of the chain B at the initial condition (see Fig. 14.30). We can then say:

$$(\text{PE})_1 = (10)(w)(10) + (10)(w)(5) = 1500 \text{ ft-lb}$$

Observing Fig. 14.30, we can say for condition 2:

$$(\text{PE})_2 = (5)(w)(10) + (10)(w)(5) - (5)(w)(2.5)$$
$$= 875 \text{ ft-lb}$$

As for kinetic energy, we have

$$(\text{KE})_1 = \frac{1}{2}\left(10\frac{w}{g}\right)(V_{\text{channel}}^2)$$
$$+\frac{1}{2}\left(10\frac{w}{g}\right)V_{\text{pipe}}^2 + \frac{1}{2}\int_0^{10}(r\omega_1)^2\frac{w}{g}\,dr$$

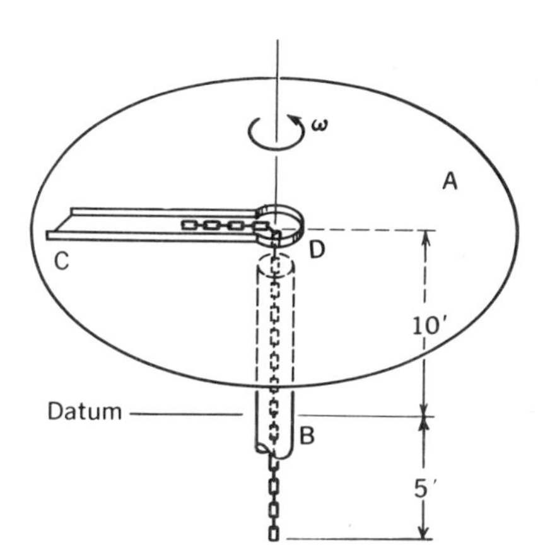

Figure 14.30. Chain after motion of 5 ft.

where the first two expressions give the kinetic energy from the motion relative to the channel and pipe, respectively. The last expression is the kinetic energy due to rotation of that part of the chain that is in the channel. Clearly, $V_{\text{channel}} = V_{\text{pipe}} = 0$ initially, and so we have

$$(\text{KE})_1 = \frac{1}{2}(1)^2\left(\frac{10}{g}\right)\int_0^{10} r^2\,dr = 51.8 \text{ ft-lb}$$

Furthermore, at condition 2, we have (see Fig. 14.30)

$$(\text{KE})_2 = \frac{1}{2}\left(5\frac{w}{g}\right)(V_{\text{channel}}^2)_2 + \frac{1}{2}\left(15\frac{w}{g}\right)(V_{\text{pipe}}^2)_2 + \frac{1}{2}\int_0^5 (r\omega_2)^2\frac{w}{g}\,dr$$

Note that $(V_{\text{channel}})_2 = (V_{\text{pipe}})_2$. Simply calling this quantity V_2, we have

$$(\text{KE})_2 = \frac{1}{2}(20)\left(\frac{w}{g}\right)(V_2^2) + \frac{1}{2}(8^2)\left(\frac{w}{g}\right)\int_0^5 r^2\,dr$$
$$= 3.11V_2^2 + 414$$

We can now state

$$(\text{PE})_1 + (\text{KE})_1 = (\text{PE})_2 + (\text{KE})_2$$
$$1500 + 51.8 = 875 + (3.11V_2^2 + 414)$$

Therefore,

$$V_2 = 9.19 \text{ ft/sec}$$

We can conclude that the chain is moving at a speed of 9.19 ft/sec along the channel and down the stationary pipe and that the plate A is rotating at an angular speed of 8 rad/sec.

Much time will be spent later in the text in applying $\boldsymbol{M}_a = \dot{\boldsymbol{H}}_a$ to a rigid body. There, the rigid body is considered to be made up of an infinite number of contiguous elements. Summations then give way to integration, and so on. The final equations of this section accordingly are among the most important in mechanics.

In the homework assignments we have included, as in Chapter 13, several very simple rigid-body problems to illustrate the use of the equation $\boldsymbol{M}_a = \dot{\boldsymbol{H}}_a$ and to give an early introduction to rigid-body mechanics.[13] We now illustrate such a problem.

EXAMPLE 14.15

A uniform cylinder of radius 400 mm and mass 100 kg is acted on at its center by a force of 500 N (see Fig. 14.31). What is the friction force f? Take $\mu_s = .2$.

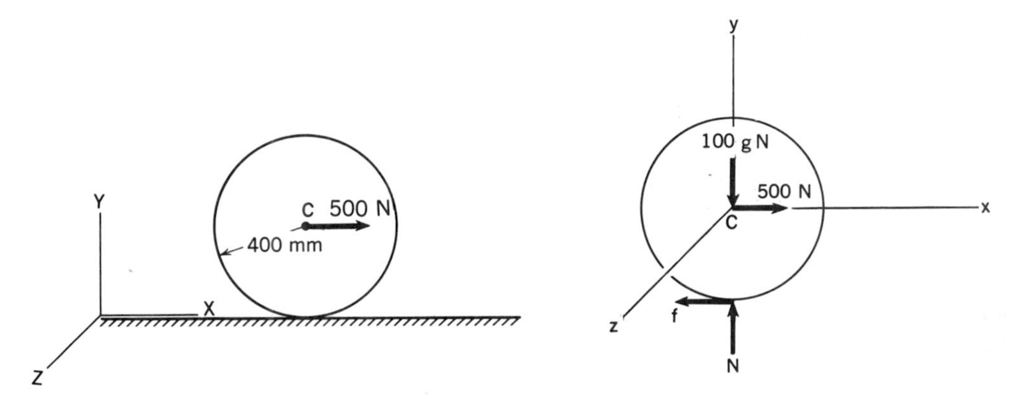

Figure 14.31. Rolling cylinder.

Figure 14.32. Free body.

We have shown a free-body diagram of the cylinder in Fig. 14.32. A reference xyz with origin at C translates with the cylinder. We first apply *Newton's law* relative

[13]The instructor may wish not to get into rigid-body dynamics at this time. This can be done with no loss in continuity.

to inertial reference XYZ. Thus, for the X direction we have for the center of mass C:

$$500 - f = 100\ddot{X}_C \tag{a}$$

Next, we write the *moment-of-momentum* equation about the z axis which goes through the center of mass. Thus, noting that we have simple circular motion about the z axis for all particles of the cylinder and observing Fig. 14.33:

$$M_z = \frac{d}{dt}(H_z)$$

$$-(f)(.40) = \frac{d}{dt}\left[\int_0^{2\pi}\int_0^{.40} \underbrace{(\rho r\, dr\, d\theta\, l)}_{dm}(r)\underbrace{(r\omega)}_{V_\theta}\right]$$

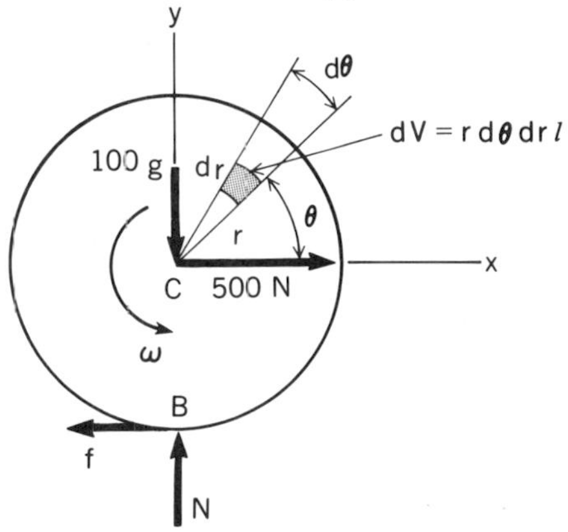

where ρ is the mass density of the cylinder and l is the thickness of the cylinder. Evaluating the integral and differentiating with respect to time, we get

$$.40f = -(\rho l)(2\pi)(\dot{\omega})\left(\frac{.40^4}{4}\right)$$

$$f = -.1005(\rho l)\dot{\omega} \tag{b}$$

Figure 14.33. Element dV in cylinder. Unknown ω shown as positive.

We can determine ρl as follows from geometry:

$$M = 100 = (\rho l)[\pi(.40)^2]$$

$$\rho l = 198.9 \text{ kg/m}^2 \tag{c}$$

We now need another independent equation. This equation can be found from *kinematics*. Thus, assuming a *no-slipping* condition, we have pure instantaneous rotation about point B. From your work in physics (we shall later prove this) we can say for point C of the cylinder:

$$(.40)\omega = -\dot{X}$$

Therefore,

$$(.40)\dot{\omega} = -\ddot{X} \tag{d}$$

Substituting for ρl and $\dot{\omega}$ in Eq. (b) using Eqs. (c) and (d), we get

$$f = (-.1005)(198.9)\left(-\frac{\ddot{X}}{.40}\right) \tag{e}$$

Now solve for $\ddot{X}$ from Eq. (a) and substitute into Eq. (e):

$$f = (.1005)(198.9)\left(\frac{5 - .01f}{.40}\right)$$

Solving for f, we get

$$f = 166.6 \text{ N}$$

We must now check to see whether our no-slipping assumption is valid. The maximum possible friction force clearly is

$$f_{max} = (100)(9.81)(.2) = 196.2 \text{ N}$$

which is greater than the actual friction force, so that the no-slip assumption is consistent with our results.

Problems

14.92. A system of particles is shown at time t moving in the xy plane. The following data apply:

$m_1 = 1$ kg, $V_1 = 5\boldsymbol{i} + 5\boldsymbol{j}$ m/sec

$m_2 = 0.7$ kg, $V_2 = -4\boldsymbol{i} + 3\boldsymbol{j}$ m/sec

$m_3 = 2$ kg, $V_3 = -4\boldsymbol{j}$ m/sec

$m_4 = 1.5$ kg, $V_4 = 3\boldsymbol{i} - 4\boldsymbol{j}$ m/sec

(a) What is the total linear momentum of the system?
(b) What is the linear momentum of the center of mass?
(c) What is the total moment of momentum of the system about the origin and about point (2, 6)?

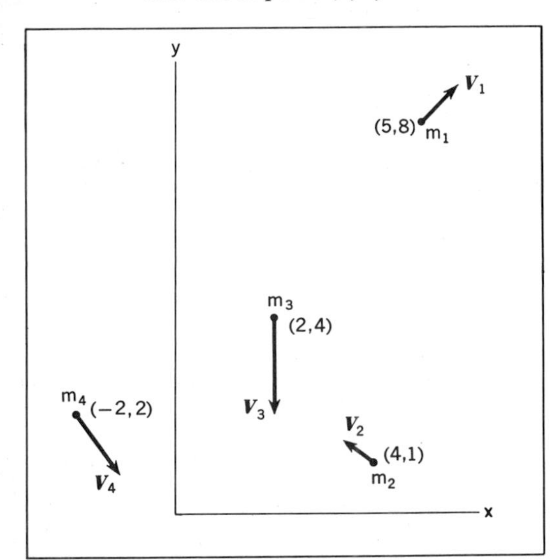

Figure P.14.92

14.93. A system of particles at time t has the following velocities and masses:

$V_1 = 20$ ft/sec, $m_1 = 1$ lbm

$V_2 = 18$ ft/sec, $m_2 = 3$ lbm

$V_3 = 15$ ft/sec, $m_3 = 2$ lbm

$V_4 = 5$ ft/sec, $m_4 = 1$ lbm

Determine (a) the total linear momentum of the system, (b) the angular momentum of the system about the origin, and (c) the angular momentum of the system about point a.

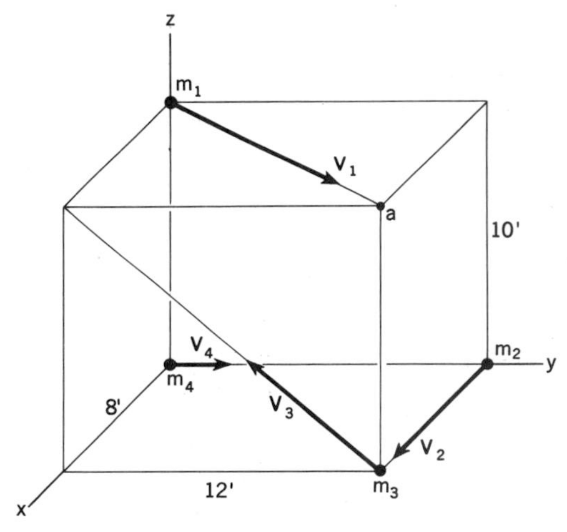

Figure P.14.93

14.94. A system of particles at time t_1 has masses $m_1 = 2$ lbm, $m_2 = 1$ lbm, $m_3 = 3$ lbm and locations and velocities as shown in Fig. (a). The same system of masses is shown in (b) at time t_2. What is the total linear impulse on the system during this time interval? What is the total angular impulse $\int \boldsymbol{M}\, dt$ during this time interval about the origin?

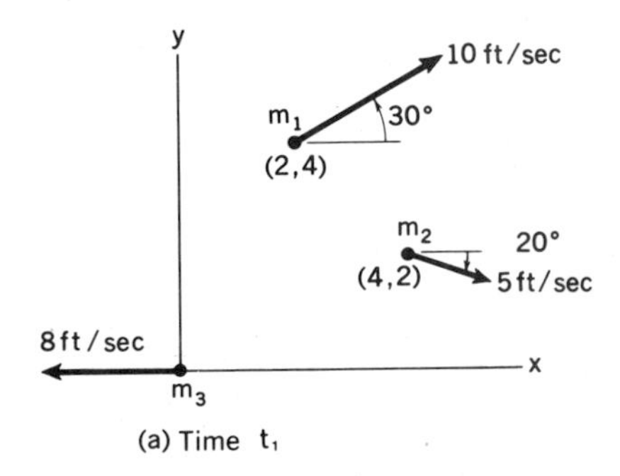

Figure P.14.94(a)

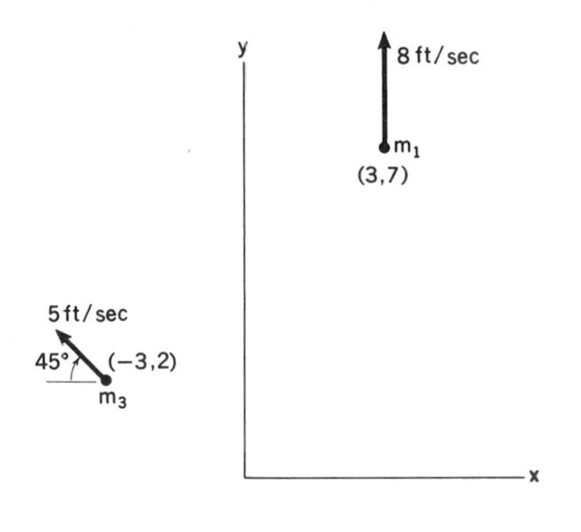

Figure P.14.94(b)

14.95. Two masses slide along bar AB at a constant speed of 1.5 m/sec. Bar AB rotates freely about axis CD. Consider only the mass of the sliding bodies to determine the angular acceleration of AB when the bodies are 1.5 m from CD if the angular velocity at that instant is 10 rad/sec.

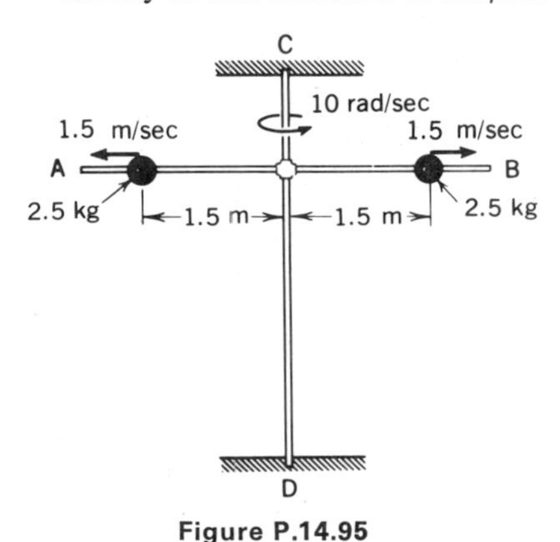

Figure P.14.95

14.96. A mechanical system is composed of three identical bodies A, B, and C each of mass 3 lbm moving along frictionless bars 120° apart on a wheel. Each of these bodies is connected with an inextensible cord to the freely hanging weight D. The connection of the cords to D is such that no torque can be transmitted to D. Initially, the three masses A, B, and C are held at a distance of 2 ft from the centerline while the wheel rotates at 3 rad/sec. What is the angular speed of the wheel and the velocity of descent of D if, after release of the radial bodies, body D moves 1 ft? Assume that body D is initially stationary (i.e., is not rotating). Body D weighs 100 lb.

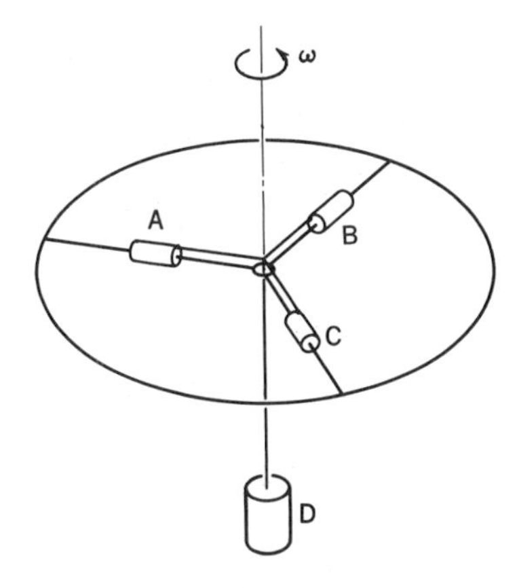

Figure P.14.96

14.97. Two sets of particles a, b, and c, d (each particle of mass m) are moving along two shafts AB and CD, which are, in turn, rigidly attached to a crossbar EF. All particles are moving at a constant speed V_1 away from EF, and their positions at the moment of interest are as shown. The system is rotating about G, and a constant torque of magnitude T is acting in the plane of the system. Assume that all masses other than the concentrated masses are negligible and that the angular velocity of the system at the instant of discussion is ω. Determine the instantaneous angular acceleration in terms of m, T, ω, s_1, and s_2.

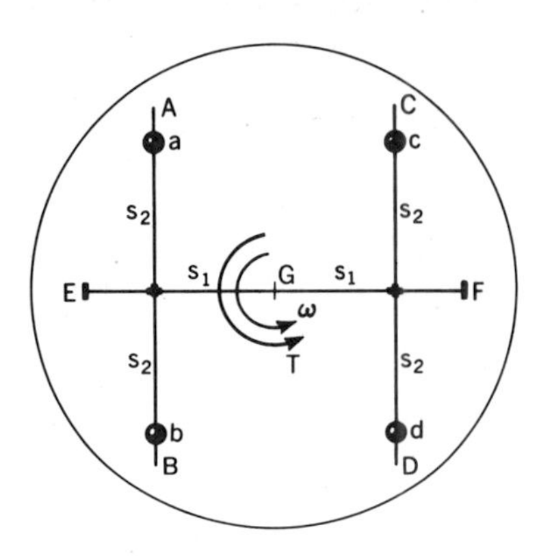

Figure P.14.97

14.98. Compute the angular momentum about O of a uniform rod, of length $L = 3$ m and mass per unit length m of 7.5 kg/m, at the instant when it is vertical and has an angular speed ω of 3 rad/sec.

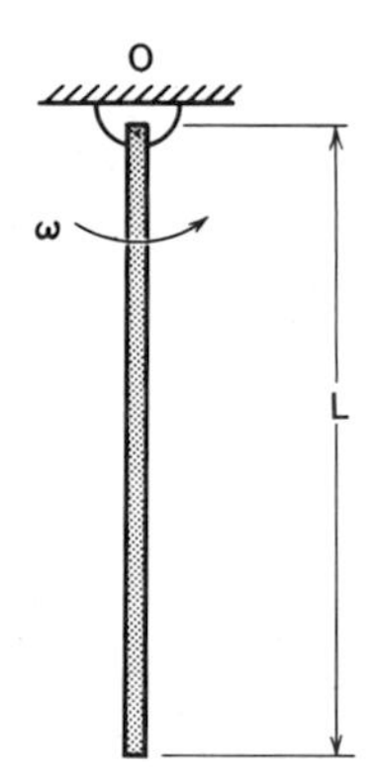

Figure P.14.98

14.99. A wheel consisting of a thin rim and four thin spokes is shown rotating about its axis at a speed ω of 2 rad/sec. The radius of the wheel R is 2 ft and the weight per unit length of rim and spoke is 2 lb/ft. What is the moment of momentum of the wheel about O? What is the total linear momentum?

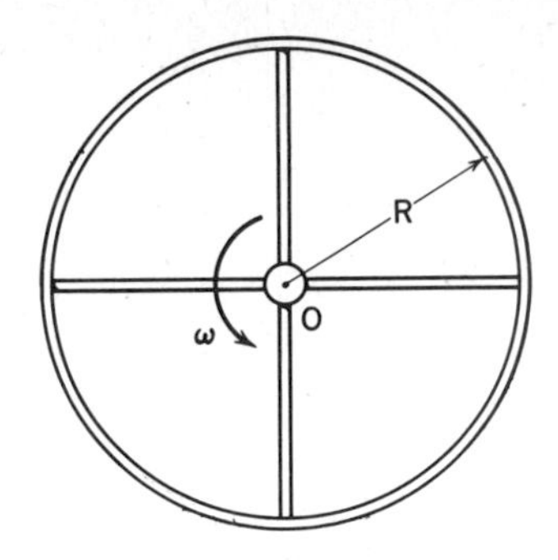

Figure P.14.99

14.100. Two uniform cylinders A and B are connected as shown. The density of the cylinders is 10,000 kg/m^3, and the system is rotating at a speed ω of 10 rad/sec about its geometric axis. What is the angular momentum of the body?

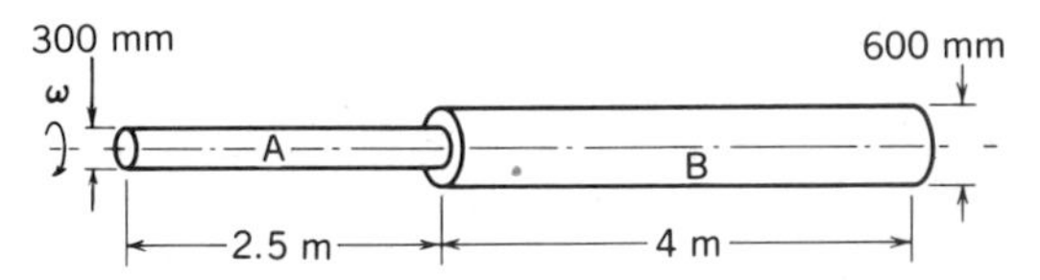

Figure P.14.100

14.101. A closed container is full of water. By rotating the container for some time and then suddenly holding the container stationary, we develop a rotational motion of the water, which, you will learn in fluid mechanics, resembles a vortex. If the velocity of the fluid elements is zero in the radial direction and is given as $10/r$ ft/sec in the transverse direction, what is the angular momentum of the water?

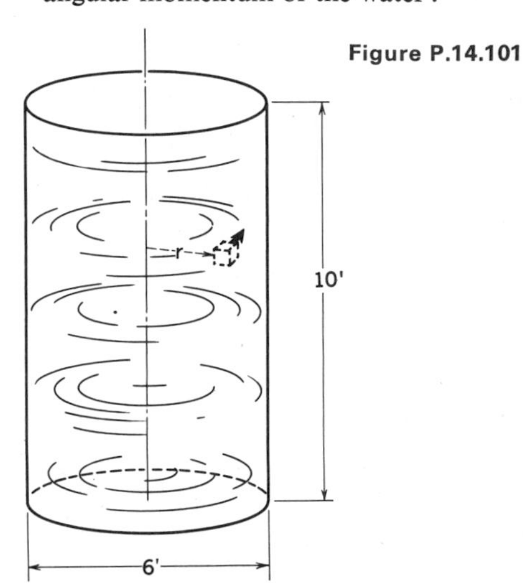

Figure P.14.101

14.102. A canal with a rectangular cross section is shown having a width of 30 ft and a depth of 5 ft. The velocity of the water is assumed to be zero at the banks and to vary parabolically over the section as shown in the diagram. If δ is the radial distance from the centerline of the channel, the transverse velocity V_θ is given as

$$V_\theta = \frac{1}{20}(225 - \delta^2) \text{ ft/sec}$$

What is the angular momentum H_0 about O at any time t of the water in the circular portion of the canal (i.e., between the x and y axes)? The radial component V_r is zero.

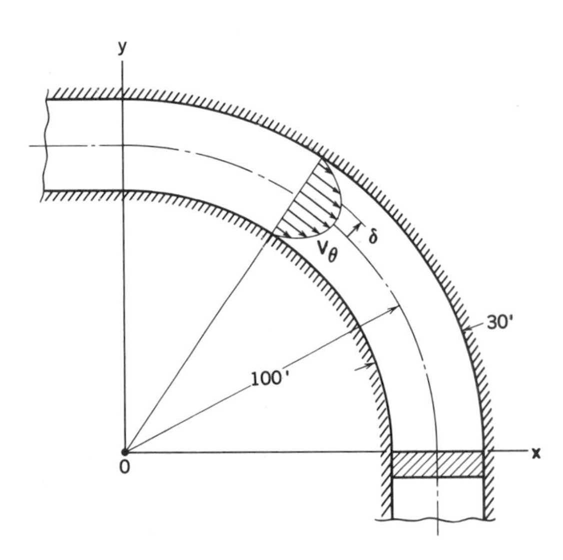

Figure P.14.102

14.103. A hoop with mass per unit length 6.5 kg/m rests flat on a frictionless surface. A 500-N force is suddenly applied. What is the angular acceleration of the hoop? What is the acceleration of the mass center?

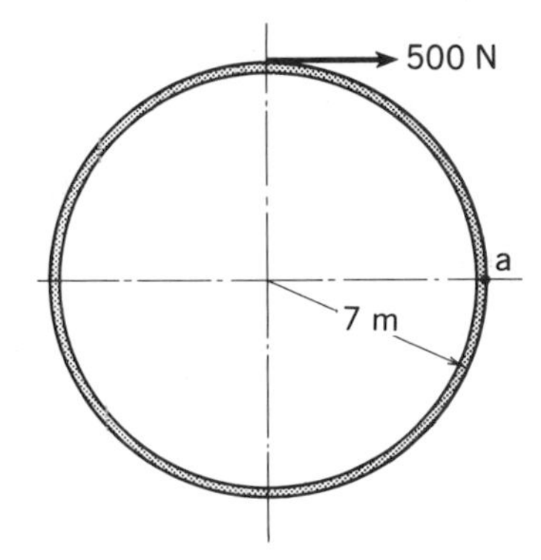

Figure P.14.103

14.104. Do Problem 14.103 for the case where a force given as:

$$F = 50i + 75j \text{ N}$$

is applied at point a instead of the 500-N force.

14.105. A cylinder weighing 50 lb rests on a frictionless surface. Two forces are applied simultaneously as shown in the diagram. What is the angular acceleration of the cylinder? What is the acceleration of the mass center?

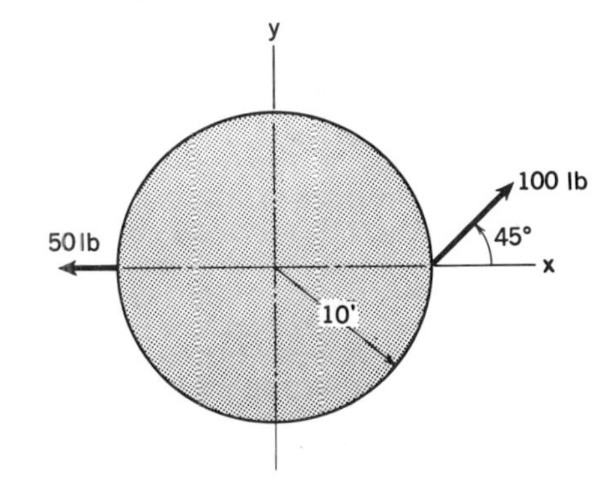

Figure P.14.105

14.106. A uniform hoop rolls without slipping down a 30° incline. The hoop material weighs 5 lb/ft and has a radius R of 4 ft. What is the angular acceleration of the hoop?

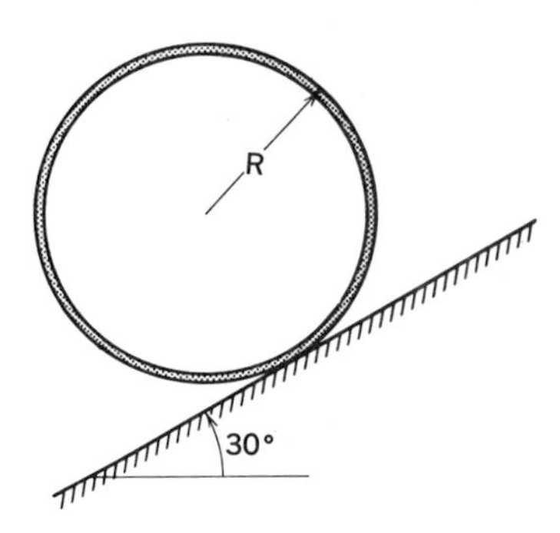

Figure P.14.106

14.107. A uniform cylinder of radius 1 m rolls without slipping down a 30° incline. What is the angular acceleration of the cylinder if it has a mass of 50 kg?

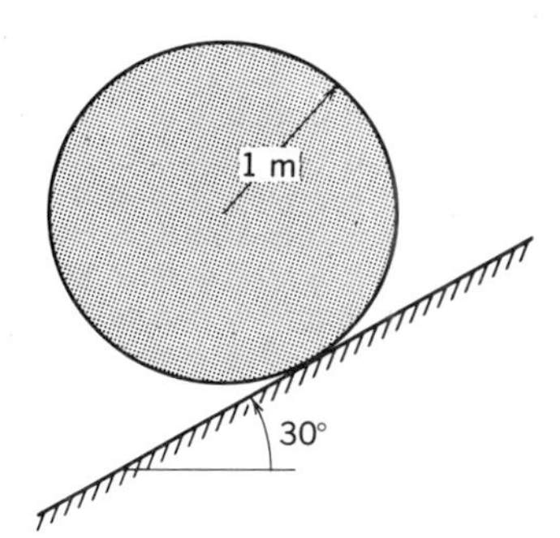

Figure P.14.107

14.108. A cylinder of length 3 m and a mass of 45 kg is acted on by a torque $T = (11.25t + 21t^2)$ N-m where t is in seconds about its geometric axis. What is the angular speed after 10 sec? The cylinder is at rest when its torque is applied.

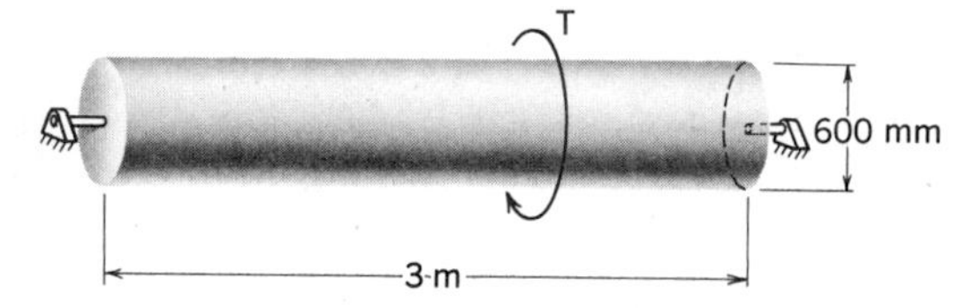

Figure P.14.108

14.109. A constant torque of 800 N-m is applied to a uniform cylinder of radius 400 mm and mass 50 kg. A 1.500-kN weight is attached to the cylinder with a light cable. What is the acceleration of W?

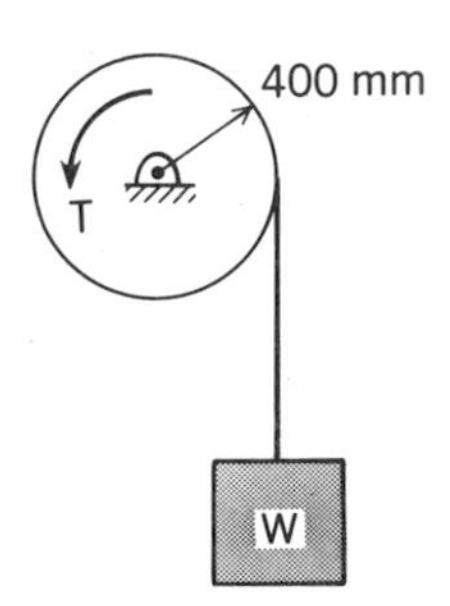

Figure P.14.109

14.110. In Problem 14.109, the torque T is $T = (300 + .2t^2)$ N-m, where t is in seconds. When $t = 0$, the system is at rest. Determine the acceleration of W at the instants when it has zero velocity for $t > 0$.

14.111. A constant torque T of 500 in.-lb is applied to a uniform cylinder of radius 1 ft. A light inextensible cable is wrapped partly around a similar cylinder and is then connected to a block W weighing 100 lb. What is the acceleration of W if the cable does not slip on the cylinders? Take $\mu_s = .3$ for block. $W_A = W_B = 100$ lb.

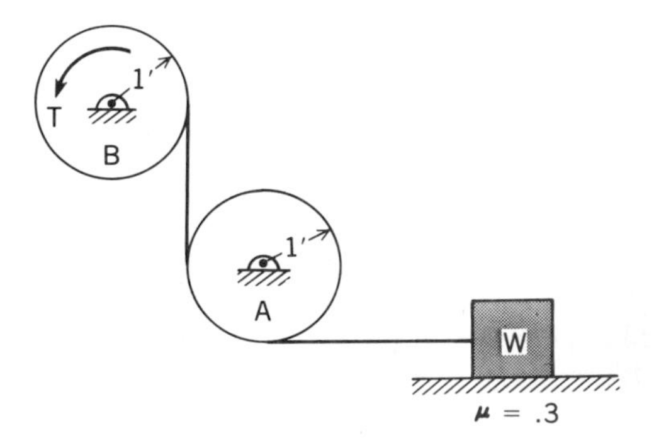

Figure P.14.111

14.112. Consider an aggregate of particles with c as the mass center and point A accelerating toward or away from c. Start with the expression for $\dot{H}$ about O given as

$$M_0 = \dot{H}_0 = \frac{d}{dt}[\sum_i r_i \times m_i\dot{r}_i] = \sum_i r_i \times \dot{P}_i$$
$$= \sum_i (r_A + \rho_{Ai}) \times \dot{P}$$

Formulate M_0 in terms of F_i and use Newton's law to eliminate terms. Next show from the resulting equation that

$$M_A = (\sum_i m_i \rho_{Ai}) \times \ddot{r}_A + \sum_i \rho_{Ai} \times m_i \ddot{\rho}_{Ai}$$

Replace $\sum_i m_i \rho_{Ai}$ by $M\rho_{Ac}$. Explain why it follows then that

$$M_A = \dot{H}_A$$

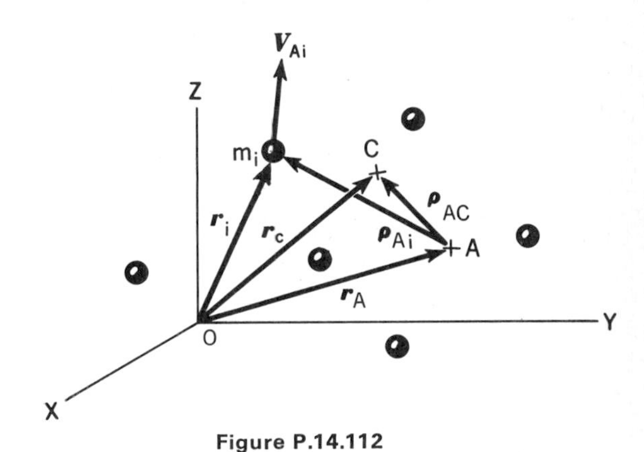

Figure P.14.112

14.8 Closure

One of the topics studied in this chapter is the impact of bodies under certain restricted conditions. For such problems, we can consider the bodies as particles before and after impact, but during impact the bodies act as deformable media for which a particle model is not meaningful or sufficient. By making an elementary picture of the action, we introduce the coefficient of restitution to yield additional information we need to determine velocities after impact. This is an empirical approach, so our analyses are limited to simple problems. To handle more complex problems or to do the simple ones more precisely, we would have to make a more rational investigation of the deformation actions taking place during impact—that is, a continuum approach to part of the problem would be required. However, we cannot make a careful study of the deformation aspects in this text since the subject of high-speed deformation of solids is a difficult one that is still under careful study by engineers and physicists.

In our study of moment of momentum for a system of particles, we set forth one of the key equations of mechanics, $M_A = \dot{H}_A$, and we introduced in the examples several considerations whose more careful and complete study will occupy a good portion of the remainder of the text. Thus, in Example 14.15 we have "in miniature," as it were, the major elements involved in the study of much rigid-body dynamics. Recall that we employed Newton's law for the mass center and the moment-of-momentum equation about the mass center to reach the desired results. In so doing, however, we had to make use of certain elementary kinematical ideas from our earlier work in physics. Accordingly, to prepare ourselves for rigid-body dynamics in Chapters 16 and 17, we shall devote ourselves in Chapter 15 to a rather careful examination of the general kinematics of a rigid body.

Although we shall be much concerned in Chapter 15 with the kinematics of rigid bodies, we shall not cease to consider particles. You will see that an understanding of rigid-body kinematics will permit us to formulate very powerful relations for the general relative motions of a particle involving references that move in any arbitrary manner with respect to each other.

Review Problems

14.113. A disc is rotated in the horizontal plane with a constant angular speed ω of 30 rad/sec. A body A with a mass of .4 lbm is moved in a frictionless slot at a uniform speed of 1 ft/sec relative to the platform by a force F as shown. What is the linear momentum of the body relative to the ground reference XY when $r = 2$ ft and $\theta = 45°$? What is the impulse developed on the body as it goes from $r = 2$ ft to $r = 1$ ft? Neglect the mass of the disc.

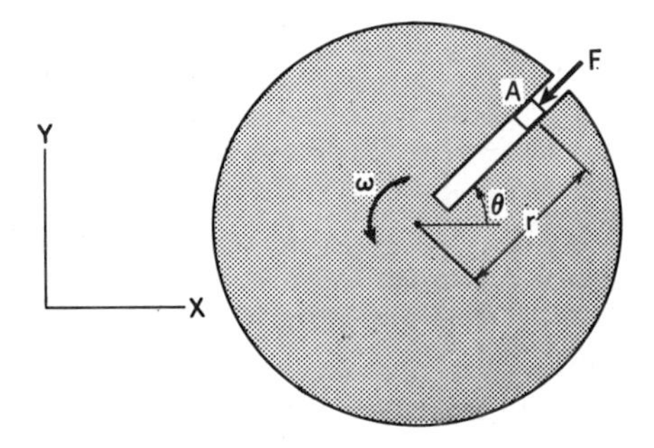

Figure P.14.113

14.114. Three bodies are towed by a force $F = (100 + 50e^{-t})$ lb as shown. If $W_1 = 30$ lb, $W_2 = 60$ lb, and $W_3 = 50$ lb, what is the speed 5 sec after the application of the given force? The dynamic coefficient of friction is .3 for all surfaces.

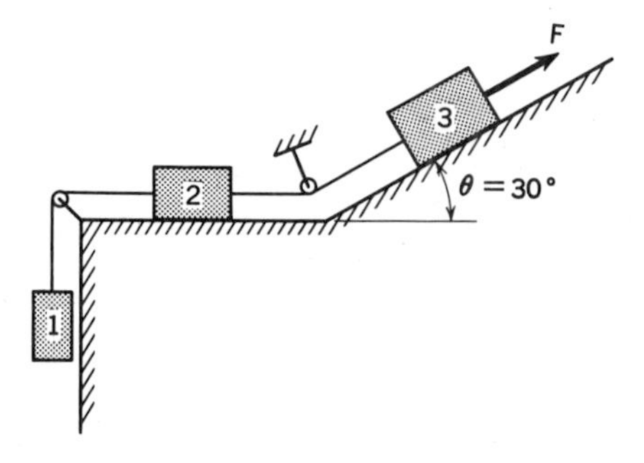

Figure P.14.114

14.115. A tugboat weighing 100 tons is moving toward a stationary barge weighing 200 tons and carrying a load C weighing 50 tons. The tug is moving at 5 knots and its propellors are developing a thrust of 5000 lb when it contacts the barge. As a result of the soft padding at the nose of the tug, consider that there is plastic impact. If the load C is not tied in any way to the barge and has a dynamic coefficient of friction of .1 with the slippery deck of the barge, what is the speed V of the barge 2 sec after the tug first contacts the barge? The load C slips during a 1-sec interval at the beginning of the contact.

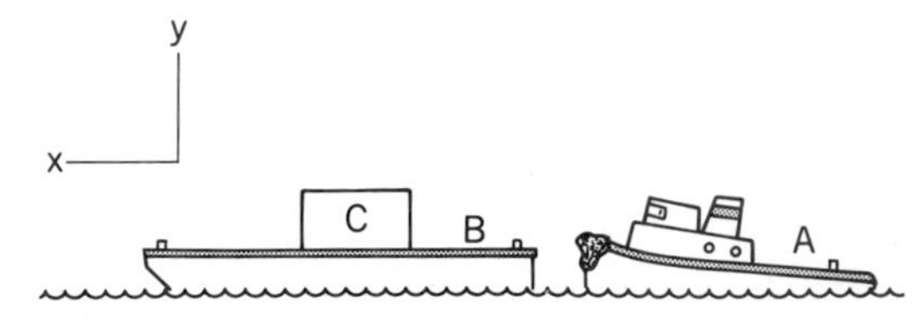

Figure P.14.115

14.116. A hopper drops small cylinders each weighing 10 N onto a conveyor belt which is moving at a speed of 3 m/sec. At the top, the cylinders are dropped off as shown. If at any time t there are 14 cylinders on the belt and if 10 cylinders are dropped per second from a hopper from a height of 300 mm above the belt, what average torque is needed to operate the conveyor? The weight of the belt that is on the conveyor bed is 100 N. The coefficient of friction μ_d between the belt and the bed is .3. The radius of the driving cylinder is 300 mm. Neglect bearing friction.

Figure P.14.116

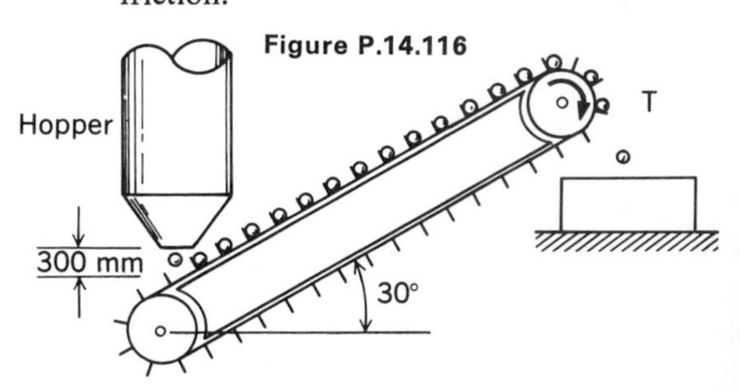

14.117. A body A weighing 2 tons is allowed to slide down an incline on a barge as shown. Body A moves a distance of 25 ft along the incline before it is stopped at B. If we neglect water resistance, how far does the barge shift in the horizontal direction? If the maximum speed of body A relative to the incline of the barge is 2 ft/sec, what is the maximum speed of the barge relative to the water? The weight of the barge is 20 tons.

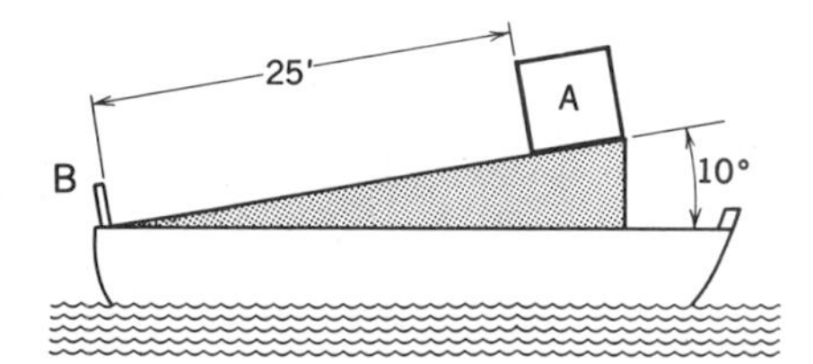

Figure P.14.117

14.118. A water droplet of diameter 2 mm is falling in the atmosphere at the rate of 2 m/sec. As a result of an updraft, a second water droplet of diameter 1 mm impinges on the aforementioned droplet. The velocity of the second droplet just prior to impingement is $3\boldsymbol{i} + 1\boldsymbol{j}$ m/sec. After impingement three droplets are formed moving parallel to xy plane. We have the following information

$$D_1 = .6 \text{ mm}, \quad V_1 = 2 \text{ m/sec}, \quad \theta_1 = 45°$$

$$D_2 = 1.2 \text{ mm}, \quad V_2 = 1 \text{ m/sec}, \quad \theta_2 = 30°$$

Find D_3, V_3, and θ_3.

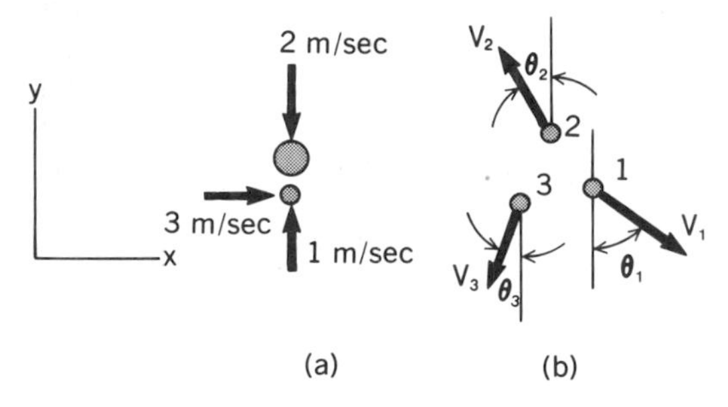

Figure P.14.118

14.119. If the coefficient of restitution is .8 for the two spheres, what are the maximum angles from the vertical that spheres will reach after the first impact? Neglect the mass of the cables.

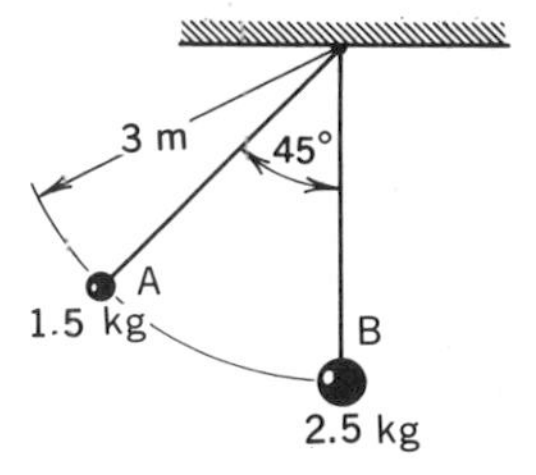

Figure P.14.119

14.120. Thin discs A and B slide along a frictionless surface. Each disc has a radius of 25 mm. Disc A has a mass of 85 g, whereas disc B has a mass of 227 g. What are the speeds of the discs after collision for $\epsilon = .7$? Assume that the discs slide on a frictionless surface.

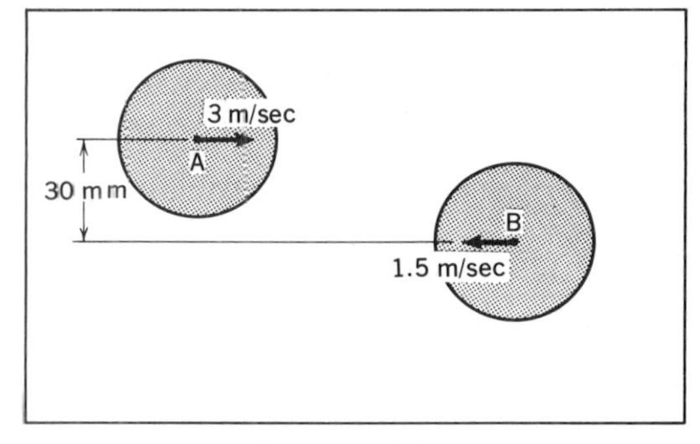

Figure P.14.120

14.121. A BB is shot at the hard, rigid surface. The speed of the pellet is 300 ft/sec as it strikes the surface. If the direction of the velocity for the pellet is given by the following unit vector:

$$\boldsymbol{\epsilon} = -.6\boldsymbol{i} - .8\boldsymbol{k}$$

what is the final velocity vector of the pellet for a collision having $\epsilon = .7$?

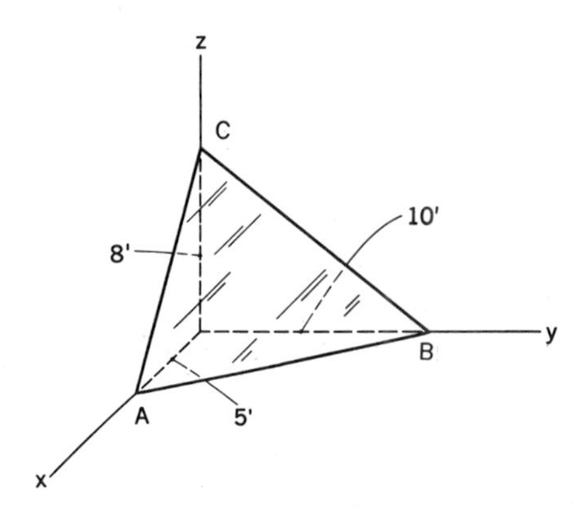

Figure P.14.121

14.122. Identical thin masses A and B slide on a light horizontal rod that is attached to a freely turning light vertical shaft. When the masses are in the position shown in the diagram, the system rotates at a speed ω of 5 rad/sec. The masses are released suddenly from this position and move out toward the identical springs, which have a spring constant $K = 800$ lb/in. Set up the equation for the compression δ of the spring once all motions of the bodies relative to the rod have damped out. The mass of each body is 10 lbm. Neglect the mass of the rods and coulombic friction. Show that $\delta = 1.004$ in. satisfies your equation.

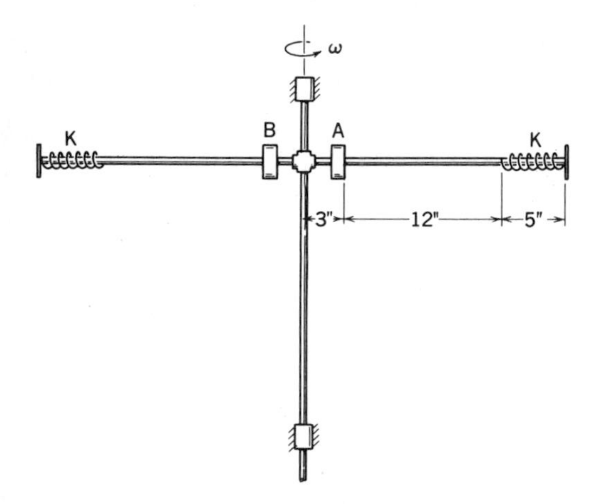

Figure P.14.122

14.123. A spacecraft has a burnout velocity V_0 of 8,300 m/sec at an elevation of 80 km above the earth's surface. The launch angle α is 15°. What is the maximum elevation h from the earth's surface for the spacecraft?

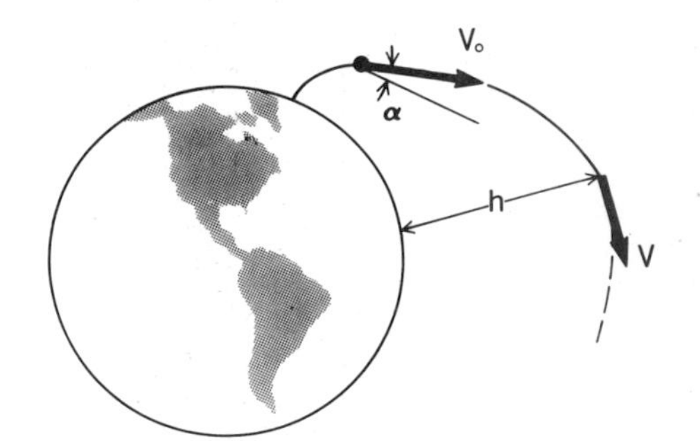

Figure P.14.123

14.124. A set of particles, each having a mass of $\frac{1}{2}$ slug, rotates about axis A–A. The masses are moving out radially at a constant speed of 5 ft/sec at the same time that they are rotating about the A–A axis. When they are 1 ft from A–A, the angular velocity is 5 rad/sec and at that instant a torque is applied in the direction of motion which varies with time t in seconds as

$$\text{torque} = (6t^2 + 10t) \text{ lb-ft}$$

What is the angular velocity when the masses have moved out radially at constant speed to 2 ft?

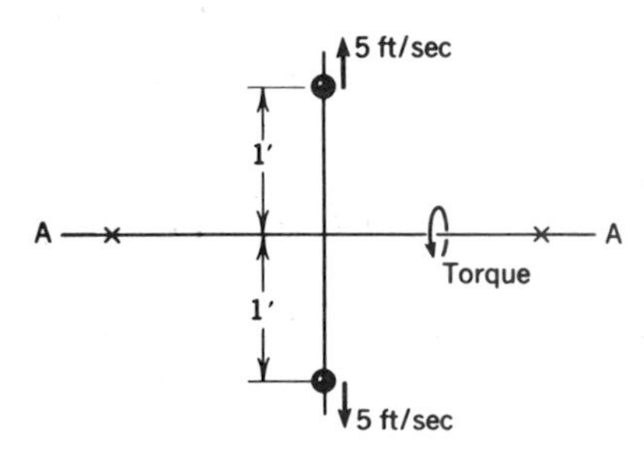

Figure P.14.124

14.125. A uniform rod with a mass of 7 kg/m rests flat on a frictionless surface. A force of 250 N acts on the rod as shown in the diagram. What is the angular acceleration of the rod? What is the acceleration of the mass center?

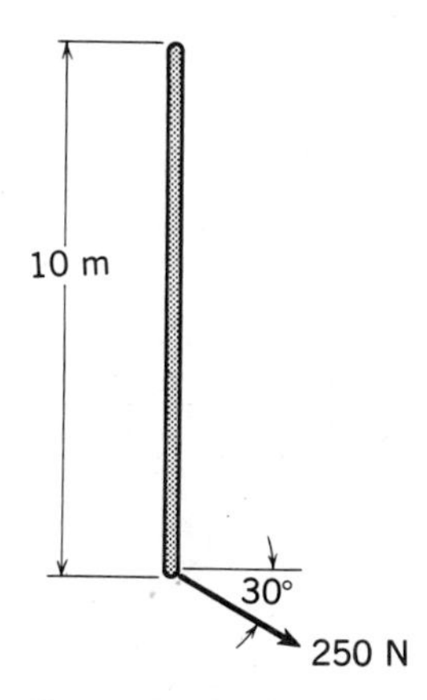

Figure P.14.125

15 Kinematics of Rigid Bodies: Relative Motion

15.1 Introduction

In Chapter 11, we studied the kinematics of a particle. During virtually all of this study only a single reference was used. However, at the end of that chapter, we briefly introduced the use of two references—the case of *simple relative motion* involving two references *translating* with respect to each other.

One of the things we shall do in this chapter is to generalize the formulations for multireference analysis. There are two reasons for doing this. First, we shall be able to analyze complicated motions in a more simple systematic way by using several references. Second, the motion of a particle is often known relative to a moving body (such as an airplane), to which we can fix a reference xyz, while the motion of the plane (and hence xyz) is known relative to an inertial reference XYZ (such as the ground). Now Newton's law, $\boldsymbol{F} = m\boldsymbol{a}$, is valid only for an inertial reference. Hence, to use Newton's law for the particle we must express the acceleration of the particle relative to the inertial reference directly. Accordingly, for practical reasons we must become involved in multireference systems.

A reference is a rigid body, and, before we can set forth multireference considerations, we must first study the kinematics of a rigid body. In so doing, we will also set the stage for our main effort in the remaining portion of the text involving the dynamics of rigid bodies.

15.2 Translation and Rotation of Rigid Bodies

For purposes of dynamics, a rigid body is considered to be composed of a continuous distribution of particles having fixed distances between each other. We shall profitably define once again two simple types of motion of a rigid body:

> **Translation.** As pointed out in Chapter 11, if a body moves so that all the particles have at time t the same velocity relative to some reference, the body is said to be in *translation* relative to this reference at this time. The velocity of a translating body can vary with time and so can be represented as $V(t)$. Accordingly, translational motion does not necessarily mean motion along a straight line. For example, the body shown in Fig. 15.1 is in translation over the interval indicated because at each instant, each particle in the body has a common velocity. A characteristic of translational motion is that any straight line in the body always retains an orientation parallel to its *original* direction during such a motion.

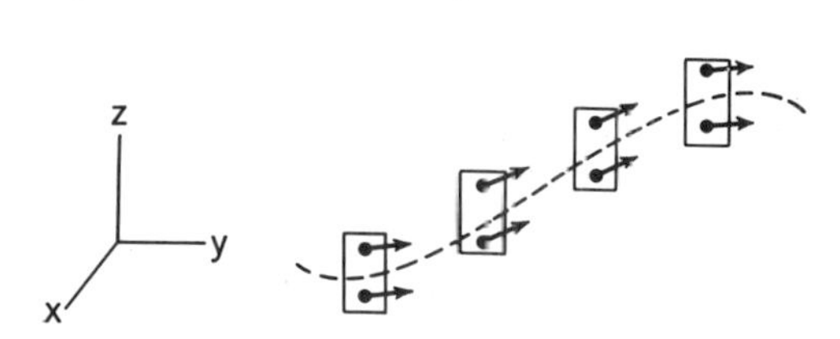

Figure 15.1. Translation of a body.

> **Rotation.** If a rigid body moves so that along some straight line all the particles of the body, or a hypothetical extension of the body, have zero velocity relative to some reference, the body is said to be in *rotation* relative to this reference. The line of stationary particles is called the *axis of rotation.*

We shall now consider how we measure the rotation of a body. A single revolution is defined as the amount of rotation in either a clockwise or a counterclockwise direction about the axis of rotation that brings the body back to its original position. Partial revolutions can conveniently be measured by observing *any* line segment such as AB in the body (Fig. 15.2) from a viewpoint M-M directed along the axis of rotation. In Fig. 15.3, we have shown this view of AB at the beginning of the partial rotation as seen along the axis of rotation, as well as the view $A'B'$ at the end of the partial rotation. The angle β that these lines form will be the same for the initial and final projections viewed along the axis of rotation of *any* line segment so examined in the partial rotation of the rigid body. Accordingly, the angle β so formed during a partial rotation is the measure of rotation.

In Chapter 1, we pointed out that finite rotations, although they have a magnitude and a direction along the axis of rotation, are not vectors. The superposition of rota-

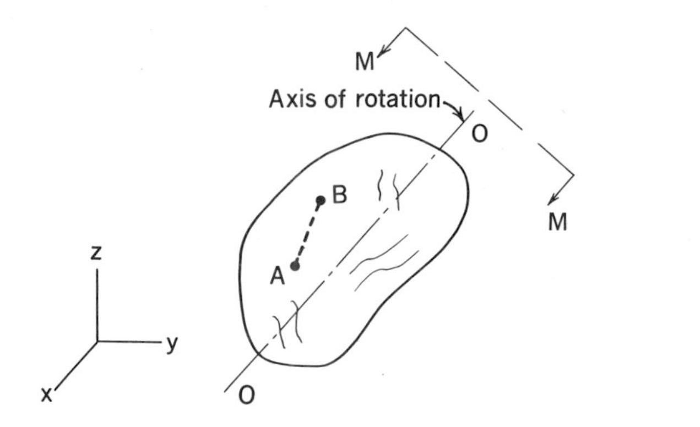

Figure 15.2. Rotation of a body.

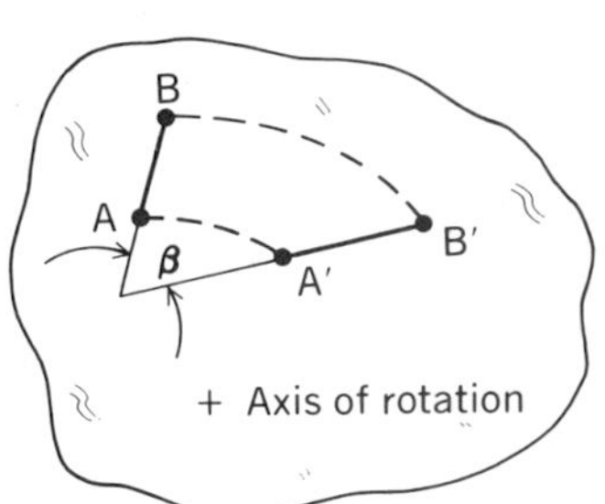

Figure 15.3. Measure of a partial rotation.

tions is not commutative, and therefore rotations do not add according to the parallelogram law, which, you will recall, is a requirement of all vector quantities. However, we can show (see Appendix IV) that as rotations become infinitesimal, they satisfy in the limit the commutative law of addition, so that infinitesimal rotations $d\beta$ are vector quantities. Therefore, the *angular velocity* is a vector quantity having a magnitude $d\beta/dt$ with an orientation parallel to the axis of rotation and a sense in accordance with the right-hand-screw rule. We shall employ $\boldsymbol{\omega}$ to represent the angular velocity vector. Note that this definition does not prescribe the line of action of this vector, for the line of action may be considered at positions other than the axis of rotation. The line of action depends on the situation (as will be discussed in later sections).

15.3 Chasle's Theorem

We have just considered two simple motions of a body, translation and rotation. We shall now demonstrate that at each instant, the motion of any rigid body can be thought of as the superposition of both a translational and a rotational motion.

Consider for simplicity a body moving in a plane. Positions of the body are shown tinted at times t and $(t + \Delta t)$ in Fig. 15.4. Let us select any point B of the body. Imagine that the body is displaced without rotation from its position at time t to the position at time $(t + \Delta t)$ so that point B reaches its correct final position B'. The displacement vector for this translation is shown at $\Delta \boldsymbol{R}_B$. To reach the correct orientation for $(t + \Delta t)$, we must now rotate the body an angle $\Delta\phi$ about an axis of rotation which is normal to the plane and which passes through point B'.

What changes would occur had we chosen some other point C for such a procedure? Consider Fig. 15.5, where we have included an alternative procedure by translating the body so that point C reaches the correct final position C'. Next, we must rotate the body an amount $\Delta\phi$ about an axis of rotation which is normal to the plane and which passes through C' in order to get to the final orientation of the body. Thus, we have indicated two routes. We conclude from the diagram that the displacement $\Delta \boldsymbol{R}_C$

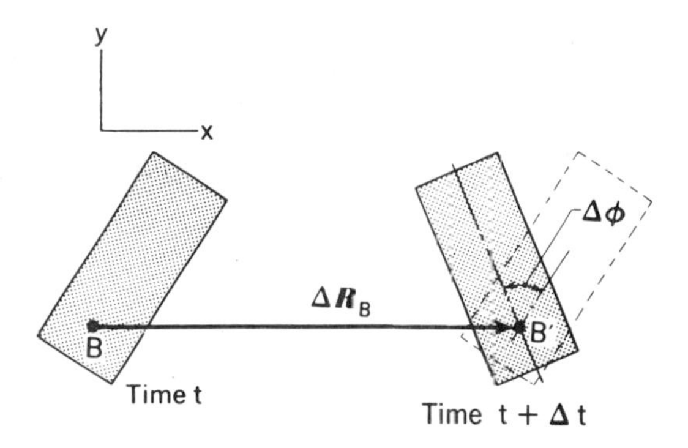

Figure 15.4. Translation and rotation of a rigid body.

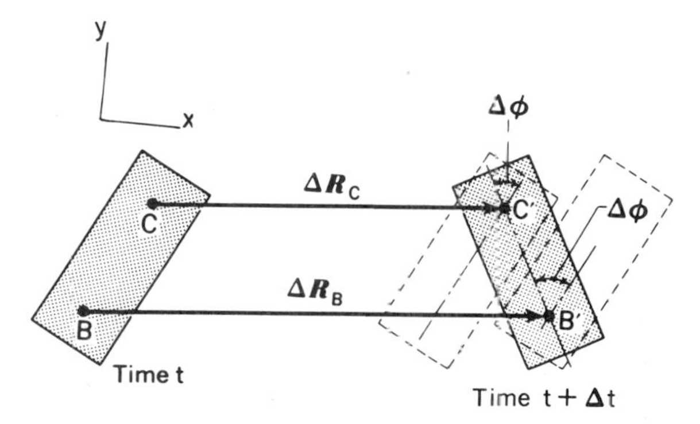

Figure 15.5. Translation and rotation of a rigid body using points *B* and *C*.

differs from $\Delta \boldsymbol{R}_B$, but there is no difference in the amount of rotation $\Delta\phi$. Thus, in general, $\Delta \boldsymbol{R}$ *and the axis of rotation will depend on the point chosen, while the amount of rotation* $\Delta\phi$ *will be the same for all such points.*

Consider now the ratios $\Delta \boldsymbol{R}/\Delta t$ and $\Delta\phi/\Delta t$. These quantities can be regarded as an average translational speed and an average rotational speed, respectively, of the body, which we could superpose to get from the initial position to the final position in the time Δt. Thus, $\Delta \boldsymbol{R}/\Delta t$ and $\Delta\phi/\Delta t$ represent an average measure of the motion during the time interval Δt. *If we go to the limit by letting* $\Delta t \longrightarrow 0$*, we have instantaneous translational and angular velocities which, when superposed, give the instantaneous motion of the body.* The displacement vector of the chosen point B in the previous discussion represents the translation of the body during the time Δt. Furthermore, the chosen point B undergoes no other motion during Δt other than that occurring during translation. Thus, we can conclude that, in the limit, the *translational velocity* used for the body corresponds to the *actual instantaneous* velocity of the chosen point B at time t. The angular velocity $\boldsymbol{\omega}$ to be used in the movement of the body, as described above, is the same vector for *all* points B chosen. Accordingly, $\boldsymbol{\omega}$ is the *instantaneous angular* velocity of the body. Since $\boldsymbol{\omega}$ can have any line of action depending on the position of

B, we can consider $\boldsymbol{\omega}$ as a *free* vector. We have thus far considered the movement of the body along a plane surface. The same conclusions can be reached for the general motion of an arbitrary rigid body in space. We can then make the following statements for the description of the general motion of a rigid body relative to some reference at time t. These statements comprise *Chasle's theorem.*

1. Select any point B in the body. Assume that all particles of the body have at the time t a velocity equal to $\boldsymbol{V}_B$, the actual velocity of the point B.
2. Superpose a pure rotational velocity $\boldsymbol{\omega}$ about an axis of rotation going through point B.

With the proper $\boldsymbol{\omega}$, the actual instantaneous motion of the body is determined, and $\boldsymbol{\omega}$ will be the same for all points B which might be chosen. Thus, only the translational velocity and the axis of rotation change when different points B are chosen. However, clearly understand that the *actual instantaneous axis of rotation* at time t is the one going through those points of the body having zero velocity at time t.

15.4 Derivative of a Vector Fixed in a Moving Reference

Two references XYZ and xyz move arbitrarily relative to each other in Fig. 15.6. Assume we are observing xyz from XYZ. Since a reference is a rigid system, we can apply Chasle's theorem to reference xyz. Thus, to fully describe the motion of xyz relative to XYZ, we choose the origin O, and we superpose a translation velocity $\dot{\boldsymbol{R}}$, equal to the velocity of O, on a rotational velocity $\boldsymbol{\omega}$ with an axis of rotation through O.

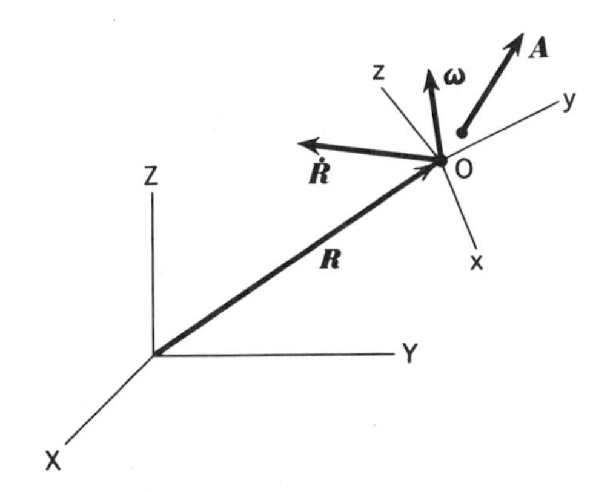

Figure 15.6. Vector $\boldsymbol{A}$ fixed in xyz moving relative to XYZ.

Now suppose that we have a vector $\boldsymbol{A}$ of fixed length and of fixed orientation as seen from reference xyz. We say that such a vector is "fixed" in reference xyz. Clearly,

The foregoing result gives the time rate of change of a vector $\boldsymbol{A}$ fixed in reference xyz moving arbitrarily relative to reference XYZ. From this result, we see that $(d\boldsymbol{A}/dt)_{XYZ}$ depends only on the vectors $\boldsymbol{\omega}$ and $\boldsymbol{A}$ and not on their lines of action. Thus, we can conclude that the time rate of change of $\boldsymbol{A}$ fixed in xyz is not altered when:

1. The vector $\boldsymbol{A}$ is fixed at some other location in xyz provided the vector itself is not changed.
2. The actual axis of rotation of the xyz system is shifted to a new parallel position.

We can differentiate the terms in Eq. 15.1 a second time. We thus get

$$\left(\frac{d^2\boldsymbol{A}}{dt^2}\right)_{XYZ} = \left(\frac{d\boldsymbol{\omega}}{dt}\right)_{XYZ} \times \boldsymbol{A} + \boldsymbol{\omega} \times \left(\frac{d\boldsymbol{A}}{dt}\right)_{XYZ} \tag{15.2}$$

Using Eq. 15.1 to replace $(d\boldsymbol{A}/dt)_{XYZ}$ and using $\dot{\boldsymbol{\omega}}$ to replace $(d\boldsymbol{\omega}/dt)_{XYZ}$, since the reference being used for this derivative is clear,[2] we get

$$\boxed{\left(\frac{d^2\boldsymbol{A}}{dt^2}\right)_{XYZ} = \dot{\boldsymbol{\omega}} \times \boldsymbol{A} - \boldsymbol{\omega} \times (\boldsymbol{\omega} \times \boldsymbol{A})} \tag{15.3}$$

You can compute higher-order derivatives by continuing the process. We suggest that only Eq. 15.1 be remembered and that all subsequent higher-order derivatives be evaluated when needed.

In this discussion thus far, we have considered a vector $\boldsymbol{A}$ fixed in a reference xyz. But a reference xyz is a rigid system and can be considered a *rigid body*. Thus, the words "*fixed in a reference xyz*" in the previous discussion can be replaced by the words "*fixed in a rigid body*." The angular velocity $\boldsymbol{\omega}$ used in Eq. 15.1 is then the angular velocity of the rigid body in which $\boldsymbol{A}$ is fixed. We shall illustrate this condition in the following examples, which you are urged to study very carefully. An understanding of these examples is vital for attaining a good working grasp of rigid-body dynamics.

As an aid in carrying out computations involving the triple cross product, we wish to point out that the product

$$\omega_1\boldsymbol{k} \times (\omega_1\boldsymbol{k} \times C\boldsymbol{j}) = -\omega_1^2 C\boldsymbol{j}$$

That is, the product is minus the product of the scalars and has a direction corresponding to the last unit vector, $\boldsymbol{j}$. Remembering this will greatly facilitate your computations.

Additionally, consider a situation where the angular velocity of body A relative to body B is given as $\boldsymbol{\omega}_1$, while the angular velocity of body B relative to the ground is $\boldsymbol{\omega}_2$. What is the *total* angular velocity $\boldsymbol{\omega}_T$ of body A relative to the ground? In such a case, we must remember that the angular velocity $\boldsymbol{\omega}_1$ of body A *relative* to body B is

[2]When it is clear from the discussion what reference is involved for a time derivative, we shall use the dot to indicate a time derivative.

actually the *difference* between the angular velocity $\boldsymbol{\omega}_T$ of body A as seen from the ground and the angular velocity $\boldsymbol{\omega}_2$ of body B as seen from the ground. Thus,

$$\boldsymbol{\omega}_1 = \boldsymbol{\omega}_T - \boldsymbol{\omega}_2$$

Solving for $\boldsymbol{\omega}_T$ we get

$$\boldsymbol{\omega}_T = \boldsymbol{\omega}_1 + \boldsymbol{\omega}_2$$

We see from above that to get the total angular velocity $\boldsymbol{\omega}_T$ we add the various relative angular velocities.

EXAMPLE 15.1

A disc C is mounted on a shaft AB in Fig. 15.9. The shaft and disc rotate with a constant angular speed ω_2 of 10 rad/sec relative to the platform to which bearings A and B are attached. Meanwhile, the platform rotates at a constant angular speed ω_1 of 5 rad/sec relative to the ground in a direction parallel to the Z axis of the ground reference XYZ. What is the angular velocity vector $\boldsymbol{\omega}$ for the disc C relative to XYZ? What are $(d\boldsymbol{\omega}/dt)_{XYZ}$ and $(d^2\boldsymbol{\omega}/dt^2)_{XYZ}$?

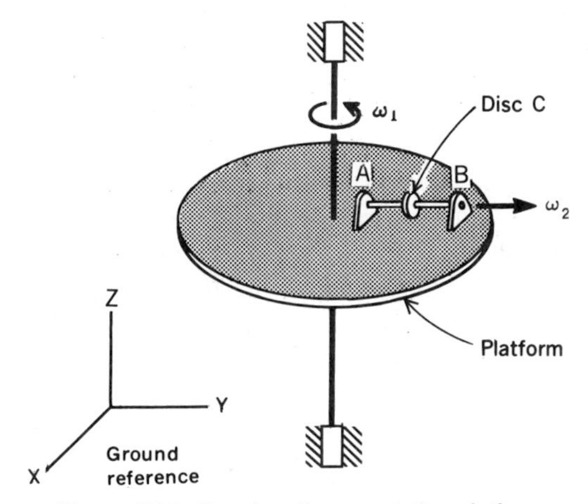

Figure 15.9. Rotating disc on rotating platform.

The total angular velocity $\boldsymbol{\omega}$ of the disc relative to the ground is easily given at all times as follows:

$$\boldsymbol{\omega} = \boldsymbol{\omega}_1 + \boldsymbol{\omega}_2 \text{ rad/sec} \tag{a}$$

At the instant of interest as depicted by Fig. 15.9, we have for $\boldsymbol{\omega}$:

$$\boldsymbol{\omega} = 5\boldsymbol{k} + 10\boldsymbol{j} \text{ rad/sec}$$

To get the first time derivative of $\boldsymbol{\omega}$, we go back to Eq. (a), which is always valid and hence can be differentiated with respect to time. Using a dot to represent the time derivative as seen from XYZ, we have

$$\dot{\boldsymbol{\omega}} = \dot{\boldsymbol{\omega}}_1 + \dot{\boldsymbol{\omega}}_2 \tag{b}$$

Consider now the vector $\boldsymbol{\omega}_2$. Note that this vector is constrained in direction to be always collinear with the axis AB of the bearings. This clearly is a physical requirement. Also, since ω_2 is of constant value, we may think of the vector $\boldsymbol{\omega}_2$ as *fixed* to

the time rate of change of A as seen from reference xyz must be zero. We can express this statement mathematically as

$$\left(\frac{dA}{dt}\right)_{xyz} = 0$$

However, as seen from XYZ, the time rate of change A will *not* necessarily be zero. To evaluate $(dA/dt)_{XYZ}$, we make use of Chasle's theorem in the following manner:

1. Consider the *translational* motion $\dot{R}$. This motion does not alter the direction of A as seen from XYZ. Also, the magnitude of A is fixed; thus, vector A cannot change as a result of this motion.[1]
2. We next consider *solely* a pure rotation about a stationary axis collinear with ω and passing through point O.

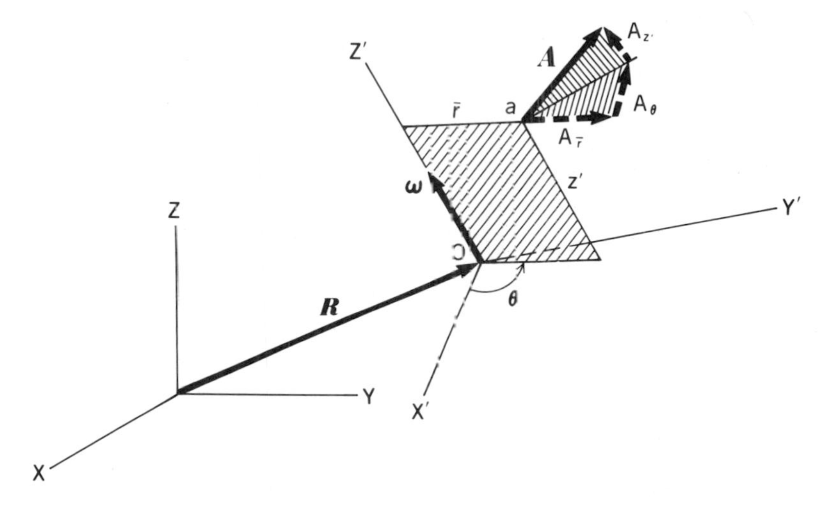

Figure 15.7. Cylindrical components for vector ***A***.

To best observe this rotation, we shall employ at O a *stationary* reference $X'Y'Z'$ positioned so that Z' coincides with the axis of rotation. This reference is shown in Fig. 15.7. Now the vector A is rotating at this instant about the Z' axis. We have shown cylindrical coordinates to the end of A (i.e., at point a), and have shown cylindrical components $A_{\bar{r}}$, A_θ, and $A_{Z'}$. In Fig. 15.8, we have shown point a with unit vectors $\epsilon_{\bar{r}}$, ϵ_θ, and $\epsilon_{Z'}$ for cylindrical coordinates at this point. We can accordingly express A as

$$A = A_{\bar{r}}\epsilon_{\bar{r}} + A_\theta\epsilon_\theta + A_{Z'}\epsilon_{Z'}$$

[1]The *line of action* of A, however, will change as seen from XYZ. But a change of line of action does not signify a change in the vector, as pointed out in Chapter 1 on the discussion of equality of vectors.

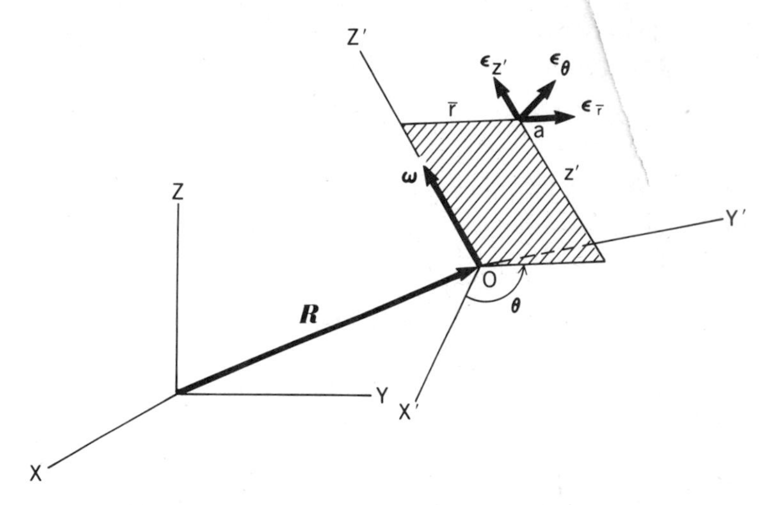

Figure 15.8. Unit vectors for cylindrical coordinates.

Clearly, as A rotates about Z', the values of the cylindrical scalar components of A for $X'Y'Z'$, namely A_r, A_θ, and $A_{Z'}$, do not change. Hence, as seen from $X'Y'Z'$, $\dot{A}_r = \dot{A}_\theta = \dot{A}_Z = 0$. Also, noting that $\dot{\boldsymbol{\epsilon}}_{Z'} = \mathbf{0}$, we can say that

$$\left(\frac{dA}{dt}\right)_{X'Y'Z'} = A_r\left(\frac{d\boldsymbol{\epsilon}_r}{dt}\right)_{X'Y'Z'} + A_\theta\left(\frac{d\boldsymbol{\epsilon}_\theta}{dt}\right)_{X'Y'Z'}$$

We have already evaluated the time derivatives of the unit vectors for cylindrical coordinates. Hence, using Eqs. 11.28 and 11.32 and noting that $\dot{\theta}$ corresponds to ω, we have

$$\left(\frac{dA}{dt}\right)_{X'Y'Z'} = A_r\omega\boldsymbol{\epsilon}_\theta - A_\theta\omega\boldsymbol{\epsilon}_r$$

But the right hand side is simply the cross product of $\boldsymbol{\omega}$ and A as you can see by carrying out the cross product with cylindrical components. Thus,

$$\begin{aligned}\boldsymbol{\omega} \times A &= \omega\boldsymbol{\epsilon}_{Z'} \times (A_r\boldsymbol{\epsilon}_r + A_\theta\boldsymbol{\epsilon}_\theta + A_{Z'}\boldsymbol{\epsilon}_{Z'}) \\ &= \omega A_r\boldsymbol{\epsilon}_\theta - \omega A_\theta\boldsymbol{\epsilon}_r\end{aligned}$$

We conclude that

$$\left(\frac{dA}{dt}\right)_{X'Y'Z'} = \boldsymbol{\omega} \times A$$

Since $X'Y'Z'$ is stationary relative to XYZ, we would observe the same time derivative from the former reference as from the latter reference. That is, $(d/dt)_{XYZ} = (d/dt)_{X'Y'Z'}$, and we can conclude that

$$\boxed{\left(\frac{dA}{dt}\right)_{XYZ} = \boldsymbol{\omega} \times A} \tag{15.1}$$

The foregoing result gives the time rate of change of a vector A fixed in reference xyz moving arbitrarily relative to reference XYZ. From this result, we see that $(dA/dt)_{XYZ}$ depends only on the vectors ω and A and not on their lines of action. Thus, we can conclude that the time rate of change of A fixed in xyz is not altered when:

1. The vector A is fixed at some other location in xyz provided the vector itself is not changed.
2. The actual axis of rotation of the xyz system is shifted to a new parallel position.

We can differentiate the terms in Eq. 15.1 a second time. We thus get

$$\left(\frac{d^2A}{dt^2}\right)_{XYZ} = \left(\frac{d\omega}{dt}\right)_{XYZ} \times A + \omega \times \left(\frac{dA}{dt}\right)_{XYZ} \tag{15.2}$$

Using Eq. 15.1 to replace $(dA/dt)_{XYZ}$ and using $\dot{\omega}$ to replace $(d\omega/dt)_{XYZ}$, since the reference being used for this derivative is clear,[2] we get

$$\boxed{\left(\frac{d^2A}{dt^2}\right)_{XYZ} = \dot{\omega} \times A + \omega \times (\omega \times A)} \tag{15.3}$$

You can compute higher-order derivatives by continuing the process. We suggest that only Eq. 15.1 be remembered and that all subsequent higher-order derivatives be evaluated when needed.

In this discussion thus far, we have considered a vector A fixed in a reference xyz. But a reference xyz is a rigid system and can be considered a *rigid body*. Thus, the words "*fixed in a reference xyz*" in the previous discussion can be replaced by the words "*fixed in a rigid body*." The angular velocity ω used in Eq. 15.1 is then the angular velocity of the rigid body in which A is fixed. We shall illustrate this condition in the following examples, which you are urged to study very carefully. An understanding of these examples is vital for attaining a good working grasp of rigid-body dynamics.

As an aid in carrying out computations involving the triple cross product, we wish to point out that the product

$$\omega_1 k \times (\omega_1 k \times Cj) = -\omega_1^2 Cj$$

That is, the product is minus the product of the scalars and has a direction corresponding to the last unit vector, j. Remembering this will greatly facilitate your computations.

Additionally, consider a situation where the angular velocity of body A relative to body B is given as ω_1, while the angular velocity of body B relative to the ground is ω_2. What is the *total* angular velocity ω_T of body A relative to the ground? In such a case, we must remember that the angular velocity ω_1 of body A *relative* to body B is

---

[2]When it is clear from the discussion what reference is involved for a time derivative, we shall use the dot to indicate a time derivative.

actually the *difference* between the angular velocity $\boldsymbol{\omega}_T$ of body A as seen from the ground and the angular velocity $\boldsymbol{\omega}_2$ of body B as seen from the ground. Thus,

$$\boldsymbol{\omega}_1 = \boldsymbol{\omega}_T - \boldsymbol{\omega}_2$$

Solving for $\boldsymbol{\omega}_T$ we get

$$\boldsymbol{\omega}_T = \boldsymbol{\omega}_1 + \boldsymbol{\omega}_2$$

We see from above that to get the total angular velocity $\boldsymbol{\omega}_T$ we add the various relative angular velocities.

EXAMPLE 15.1

A disc C is mounted on a shaft AB in Fig. 15.9. The shaft and disc rotate with a constant angular speed ω_2 of 10 rad/sec relative to the platform to which bearings A and B are attached. Meanwhile, the platform rotates at a constant angular speed ω_1 of 5 rad/sec relative to the ground in a direction parallel to the Z axis of the ground reference XYZ. What is the angular velocity vector $\boldsymbol{\omega}$ for the disc C relative to XYZ? What are $(d\boldsymbol{\omega}/dt)_{XYZ}$ and $(d^2\boldsymbol{\omega}/dt^2)_{XYZ}$?

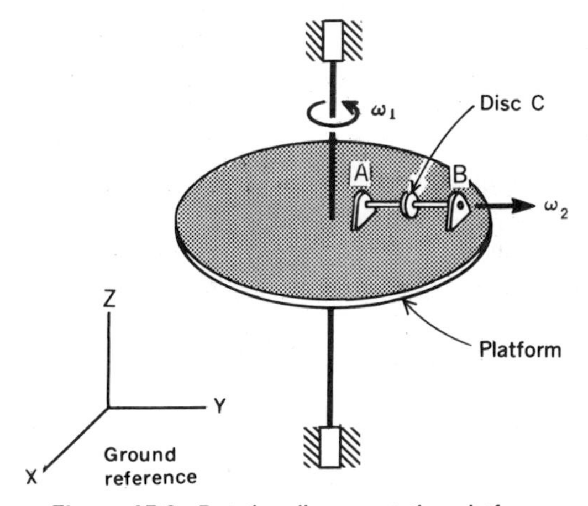

Figure 15.9. Rotating disc on rotating platform.

The total angular velocity $\boldsymbol{\omega}$ of the disc relative to the ground is easily given at all times as follows:

$$\boldsymbol{\omega} = \boldsymbol{\omega}_1 + \boldsymbol{\omega}_2 \text{ rad/sec} \tag{a}$$

At the instant of interest as depicted by Fig. 15.9, we have for $\boldsymbol{\omega}$:

$$\boldsymbol{\omega} = 5\boldsymbol{k} + 10\boldsymbol{j} \text{ rad/sec}$$

To get the first time derivative of $\boldsymbol{\omega}$, we go back to Eq. (a), which is always valid and hence can be differentiated with respect to time. Using a dot to represent the time derivative as seen from XYZ, we have

$$\dot{\boldsymbol{\omega}} = \dot{\boldsymbol{\omega}}_1 + \dot{\boldsymbol{\omega}}_2 \tag{b}$$

Consider now the vector $\boldsymbol{\omega}_2$. Note that this vector is constrained in direction to be always collinear with the axis AB of the bearings. This clearly is a physical requirement. Also, since ω_2 is of constant value, we may think of the vector $\boldsymbol{\omega}_2$ as *fixed* to the platform along AB. Therefore, since the platform has an angular velocity of $\boldsymbol{\omega}_1$ relative to XYZ, we can say:

$$\dot{\boldsymbol{\omega}}_2 = \boldsymbol{\omega}_1 \times \boldsymbol{\omega}_2 \tag{c}$$

As for $\boldsymbol{\omega}_1$, note that as seen from XYZ this vector is a constant vector, and so at all times $\dot{\boldsymbol{\omega}}_1 = \mathbf{0}$. Hence, Eq. (b) can be written as follows:

$$\dot{\boldsymbol{\omega}} = \boldsymbol{\omega}_1 \times \boldsymbol{\omega}_2 \tag{d}$$

This equation is valid at all times and so can be differentiated again. At the instant of interest as depicted by Fig. 15.9, we have for $\dot{\boldsymbol{\omega}}$:

$$\dot{\boldsymbol{\omega}} = 5\boldsymbol{k} \times 10\boldsymbol{j} = -50\boldsymbol{i} \text{ rad/sec}^2 \tag{e}$$

To get $\ddot{\boldsymbol{\omega}}$, we now differentiate (d) with respect to time. Noting Eq. (c), we have

$$\begin{aligned} \ddot{\boldsymbol{\omega}} &= \dot{\boldsymbol{\omega}}_1 \times \boldsymbol{\omega}_2 + \boldsymbol{\omega}_1 \times \dot{\boldsymbol{\omega}}_2 \\ &= \mathbf{0} + \boldsymbol{\omega}_1 \times (\boldsymbol{\omega}_1 \times \boldsymbol{\omega}_2) \end{aligned} \tag{f}$$

where we have used the fact that $\dot{\boldsymbol{\omega}}_1 = \mathbf{0}$ at all times. At the instant of interest, we have

$$\ddot{\boldsymbol{\omega}} = 5\boldsymbol{k} \times (5\boldsymbol{k} \times 10\boldsymbol{j}) = -250\boldsymbol{j} \text{ rad/sec}^3$$

EXAMPLE 15.2

In Example 15.1, consider a position vector $\boldsymbol{\rho}$ between two points on the rotating disc (see Fig. 15.10). The length of $\boldsymbol{\rho}$ is 100 mm and, at the instant of interest, is in the vertical direction. What are the first and second time derivatives of $\boldsymbol{\rho}$ at this instant as seen from the ground reference?

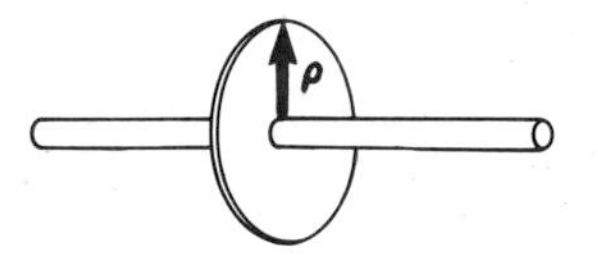

Figure 15.10. Displacement vector $\boldsymbol{\rho}$ in disc.

It should be obvious that the vector $\boldsymbol{\rho}$ is fixed to the disc which has at all times an angular velocity relative to XYZ equal to $\boldsymbol{\omega}_1 + \boldsymbol{\omega}_2$. Hence, at all times we can say:

$$\dot{\boldsymbol{\rho}} = (\boldsymbol{\omega}_1 + \boldsymbol{\omega}_2) \times \boldsymbol{\rho} \tag{a}$$

At the instant of interest, we have

$$\dot{\boldsymbol{\rho}} = (5\boldsymbol{k} + 10\boldsymbol{j}) \times 100\boldsymbol{k} = 1000\boldsymbol{i} \text{ mm/sec} \tag{b}$$

To get the second derivative of $\boldsymbol{\rho}$, go back to Eq. (a) and differentiate:

$$\ddot{\boldsymbol{\rho}} = (\dot{\boldsymbol{\omega}}_1 + \dot{\boldsymbol{\omega}}_2) \times \boldsymbol{\rho} + (\boldsymbol{\omega}_1 + \boldsymbol{\omega}_2) \times \dot{\boldsymbol{\rho}}$$

Noting that $\dot{\boldsymbol{\omega}}_1 = 0$ at all times and, as discussed in Example 15.1, that $\boldsymbol{\omega}_2$ is fixed in the platform, we can say:

$$\ddot{\boldsymbol{\rho}} = (0 + \boldsymbol{\omega}_1 \times \boldsymbol{\omega}_2) \times \boldsymbol{\rho} + (\boldsymbol{\omega}_1 + \boldsymbol{\omega}_2) \times \dot{\boldsymbol{\rho}} \tag{c}$$

At the instant of interest we have, on noting Eq. (b):

$$\begin{aligned} \ddot{\boldsymbol{\rho}} &= (5\boldsymbol{k} \times 10\boldsymbol{j}) \times 100\boldsymbol{k} + (5\boldsymbol{k} + 10\boldsymbol{j}) \times 1000\boldsymbol{i} \text{ mm/sec}^2 \\ &= 10\boldsymbol{j} - 10\boldsymbol{k} \text{ m/sec}^2 \end{aligned}$$

Although we shall later formally examine the case of the time derivative of vector $\boldsymbol{A}$ as seen from XYZ when $\boldsymbol{A}$ is *not fixed* in a body or a reference xyz, we can handle such cases less formally with what we already know. We illustrate this in the following example.

EXAMPLE 15.3

For the disc in Fig. 15.9, $\omega_2 = 6$ rad/sec and $\dot{\omega}_2 = 2$ rad/sec², both relative to the platform at the instant of interest. At this instant, $\omega_1 = 2$ rad/sec and $\dot{\omega}_1 = -3$ rad/sec² for the platform relative to the ground. Find the angular acceleration vector $\dot{\boldsymbol{\omega}}$ for the disc relative to the ground at the instant of interest.

The angular velocity of the disc relative to the ground at all times is

$$\boldsymbol{\omega} = \boldsymbol{\omega}_1 + \boldsymbol{\omega}_2 \tag{a}$$

For $\dot{\boldsymbol{\omega}}$, we can then say

$$\dot{\boldsymbol{\omega}} = \dot{\boldsymbol{\omega}}_1 + \dot{\boldsymbol{\omega}}_2 \tag{b}$$

It is apparent on inspecting Fig. 15.11 that at all times $\boldsymbol{\omega}_1$ is vertical, and so we can say:

$$\dot{\boldsymbol{\omega}}_1 = \frac{d}{dt_{XYZ}}(\omega_1 \boldsymbol{k}) = \dot{\omega}_1 \boldsymbol{k} \tag{c}$$

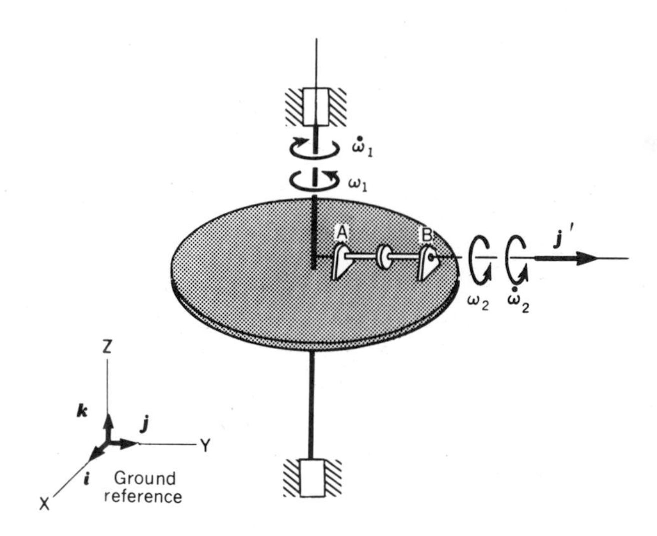

Figure 15.11. Unit vector $\boldsymbol{j}'$ fixed to platform.

However, $\boldsymbol{\omega}_2$ is changing direction and, most importantly, is changing magnitude. Because of the latter, $\boldsymbol{\omega}_2$ cannot be considered fixed in a reference or a rigid body for purposes of computing $\dot{\boldsymbol{\omega}}_2$. To get around this difficulty, we fix a unit vector $\boldsymbol{j}'$ onto the *platform* as shown in Fig. 15.11. We know the angular velocity of this unit vector; it is $\boldsymbol{\omega}_1$ at all times. We can then express $\boldsymbol{\omega}_2$ in the following manner, which is valid at all times:

$$\boldsymbol{\omega}_2 = \omega_2 \boldsymbol{j}' \tag{d}$$

We can differentiate the above with respect to time as follows:

$$\dot{\boldsymbol{\omega}}_2 = \dot{\omega}_2 \boldsymbol{j}' + \omega_2 \dot{\boldsymbol{j}}'$$

But $\boldsymbol{j}'$ is *fixed* to the platform which has angular velocity $\boldsymbol{\omega}_1$ relative to XYZ at all times. Hence, we have for the above,

$$\dot{\boldsymbol{\omega}}_2 = \dot{\omega}_2 \boldsymbol{j}' + \omega_2(\boldsymbol{\omega}_1 \times \boldsymbol{j}') \quad \text{(e)}$$

Thus, Eq. (b) then can be given as

$$\dot{\boldsymbol{\omega}} = \dot{\omega}_1 \boldsymbol{k} + \dot{\omega}_2 \boldsymbol{j}' + \omega_2(\boldsymbol{\omega}_1 \times \boldsymbol{j}')$$

This expression is valid at all times and could be differentiated again. At the instant of interest, we can say, noting that $\boldsymbol{j}' = \boldsymbol{j}$ at this instant,

$$\dot{\boldsymbol{\omega}} = -3\boldsymbol{k} + 2\boldsymbol{j} + 6(2\boldsymbol{k} \times \boldsymbol{j})$$
$$= -12\boldsymbol{i} + 2\boldsymbol{j} - 3\boldsymbol{k} \text{ rad/sec}^2$$

Problems

15.1. Is the motion of the cabin of a ferris wheel rotational or translational if the wheel moves at uniform speed and the occupants cause no disturbances? Why?

15.2. A cylinder rolls without slipping down an inclined surface. What is the actual axis of rotation at any instant? Why? How is this axis moving?

15.3. A reference xyz is moving such that the origin O has at time t a velocity relative to reference XYZ given as

$$\boldsymbol{V}_0 = 6\boldsymbol{i} + 12\boldsymbol{j} + 13\boldsymbol{k} \text{ ft/sec}$$

The xyz reference has an angular velocity $\boldsymbol{\omega}$ relative to XYZ at time t given as

$$\boldsymbol{\omega} = 10\boldsymbol{i} + 12\boldsymbol{j} + 2\boldsymbol{k} \text{ rad/sec}$$

What is the time rate of change relative to XYZ of a directed line segment $\boldsymbol{\rho}$ going from position (3, 2, −5) to (−2, 4, 6) in xyz? What is the time rate of change relative to XYZ of position vectors $\boldsymbol{i}'$ and $\boldsymbol{k}'$?

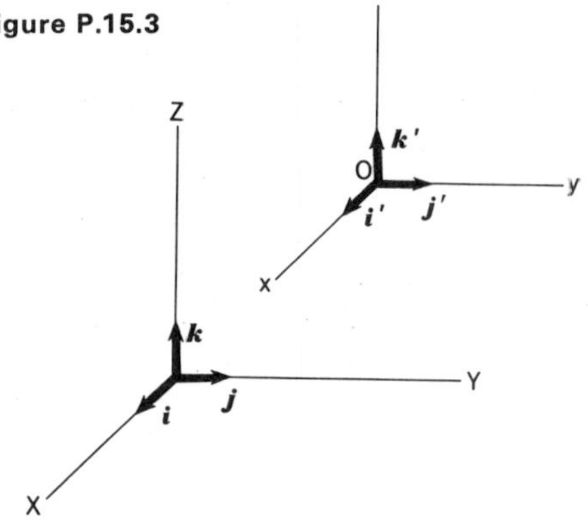

Figure P.15.3

15.4. A reference xyz is moving relative to XYZ with a velocity of the origin given at time t as

$$\boldsymbol{V}_0 = 6\boldsymbol{i} + 4\boldsymbol{j} + 6\boldsymbol{k} \text{ m/sec}$$

The angular velocity of reference xyz relative to XYZ is

$$\boldsymbol{\omega} = 3\boldsymbol{i} + 14\boldsymbol{j} + 2\boldsymbol{k} \text{ rad/sec}$$

What is the time rate of change as seen from XYZ of a directed line segment $\boldsymbol{\rho}_{1,2}$ in xyz going from position 1 to position 2 where the position vectors in xyz for these points are, respectively,

$$\boldsymbol{\rho}_1 = 2\boldsymbol{i}' + 3\boldsymbol{j}' \text{ m}$$

$$\boldsymbol{\rho}_2 = 3\boldsymbol{i}' - 4\boldsymbol{j}' + 2\boldsymbol{k}' \text{ m}$$

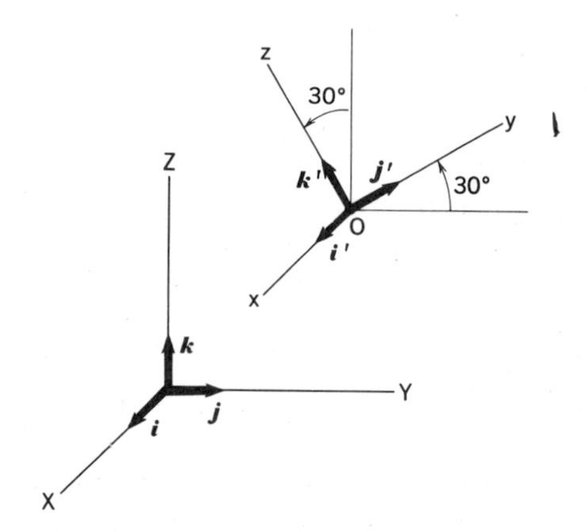

Figure P.15.4

15.5. Find the second derivatives as seen from XYZ of the vector $\boldsymbol{\rho}$ and the unit vector $\boldsymbol{i}'$ specified in Problem 15.3. The angular acceleration of xyz relative to XYZ at the instant of interest is

$$\dot{\boldsymbol{\omega}} = 5\boldsymbol{i} + 2\boldsymbol{j} + 3\boldsymbol{k} \text{ rad/sec}^2$$

15.6. Find the second derivative as seen from XYZ of the vector $\boldsymbol{\rho}_{1,2}$ specified in Problem 15.4. Take the angular acceleration of xyz relative to XYZ at the instant of interest as

$$\dot{\boldsymbol{\omega}} = 15\boldsymbol{i} - 2\boldsymbol{k} \text{ rad/sec}^2$$

15.7. A platform is rotating with a constant speed ω_1 of 10 rad/sec relative to the ground. A shaft is mounted on the platform and rotates relative to the platform at a speed ω_2 of 5 rad/sec. What is the angular velocity of the shaft relative to the ground? What are the first and second time derivatives of the angular velocity of the shaft relative to the ground?

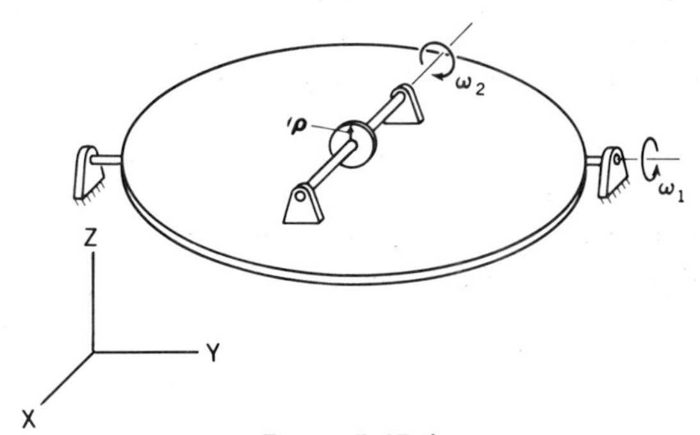

Figure P.15.7

15.8. In Problem 15.7, what are the first and second time derivatives of a directed line segment $\boldsymbol{\rho}$ in the disc at the instant that the system has the geometry shown? The vector $\boldsymbol{\rho}$ is of length 10 mm.

15.9. A tank is maneuvering its gun into position. At the instant of interest, the turret A is rotating at an angular speed $\dot{\theta}$ of 2 rad/sec relative to the tank and is in position $\theta = 20°$. Also, at this instant, the gun is rotating at an angular speed $\dot{\phi}$ of 1 rad/sec relative to the turret and forms an angle $\phi = 30°$ with the horizontal plane. What are $\boldsymbol{\omega}$, $\dot{\boldsymbol{\omega}}$, and $\ddot{\boldsymbol{\omega}}$ of the gun relative to the ground?

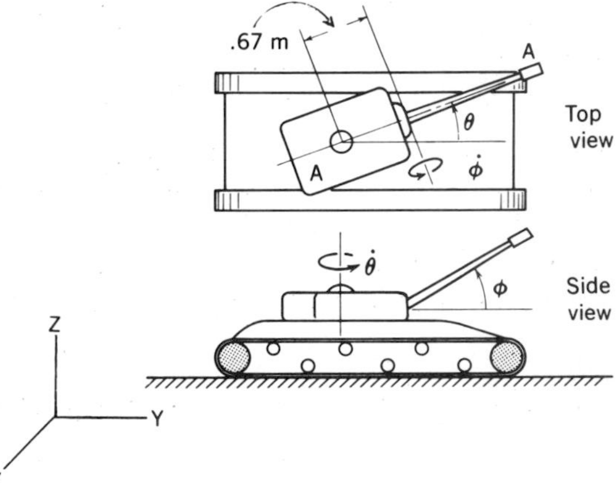

Figure P.15.9

15.10. In Problem 15.9, determine $\boldsymbol{\omega}$ and $\dot{\boldsymbol{\omega}}$ assuming that the tank is also rotating about the vertical axis at a rate of .2 rad/sec relative to the ground in a clockwise direction as viewed from above.

15.11. A particle is made to move at constant speed V equal to 10 m/sec along a straight groove on a plate B. The plate rotates at a constant angular speed ω_2 equal to 3 rad/sec relative to a platform C while the platform rotates with a constant angular speed ω_1 of 5 rad/sec relative to the ground reference XYZ. Find the first and second derivatives of V as seen from the ground reference.

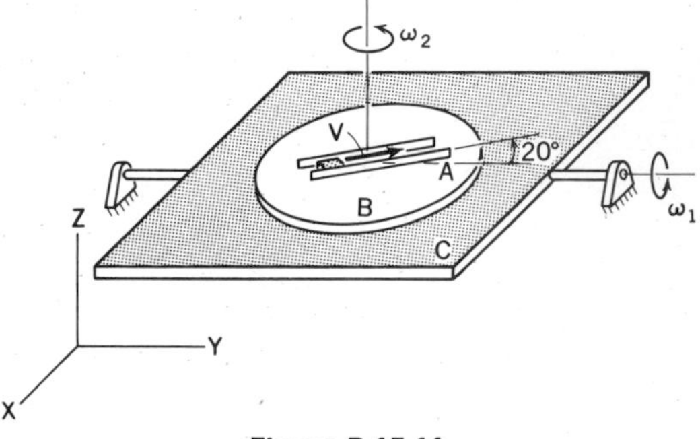

Figure P.15.11

15.12. A jet fighter plane has just taken off and is retracting its landing gear. At the end of its run on the ground, the plane is moving at a speed of 200 km/hr. If the diameter of the tires is 460 mm and if we neglect the loss of angular speed of the wheels due to wind friction after the plane is in the air, what is the angular speed $\boldsymbol{\omega}$ and the angular acceleration $\dot{\boldsymbol{\omega}}$ of the left wheel (under the wing) at the instant shown in the diagram? Take $\omega_2 = .4$ rad/sec and $\dot{\omega}_2$ is .2 rad/sec² at the instant of interest.

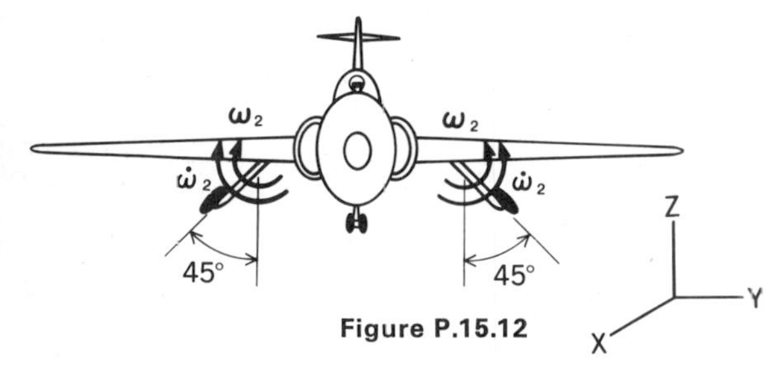

Figure P.15.12

15.13. A truck is carrying a cockpit for a worker who repairs overhead road fixtures. At the instant shown in the diagram, the base D is rotating with constant speed ω_2 of 1 rad/sec relative to the truck. Arm AB is rotating at constant angular speed ω_1 of 2 rad/sec relative to DA. Cockpit C is rotating relative to AB so as to always keep the man upright. What are $\boldsymbol{\omega}$, $\dot{\boldsymbol{\omega}}$, and $\ddot{\boldsymbol{\omega}}$ of arm AB relative to the ground at the instant of interest? The truck is stationary.

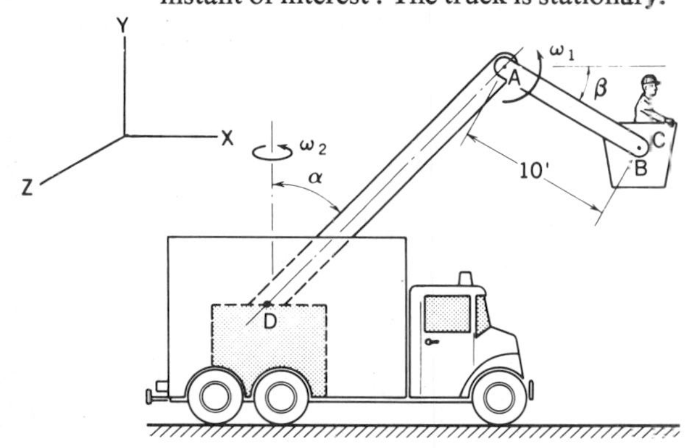

Figure P.15.13

15.14. An electric motor M is mounted on a plate A which is welded to a shaft D. The motor has a constant angular speed ω_2 relative to plate A of 1750 rpm. Plate A at the instant of interest is in a vertical position as shown and is rotating with an angular speed ω_1 equal to 100 rpm and a rate of change of angular speed $\dot{\omega}_1$ equal to 30 rpm/sec—all relative to the ground. The normal projection of the centerline of the motor shaft onto the plate A is at an angle of 45° with the edge of the plate FE. Compute the first and second time derivatives of $\boldsymbol{\omega}$, the angular velocity of the motor, as seen from the ground.

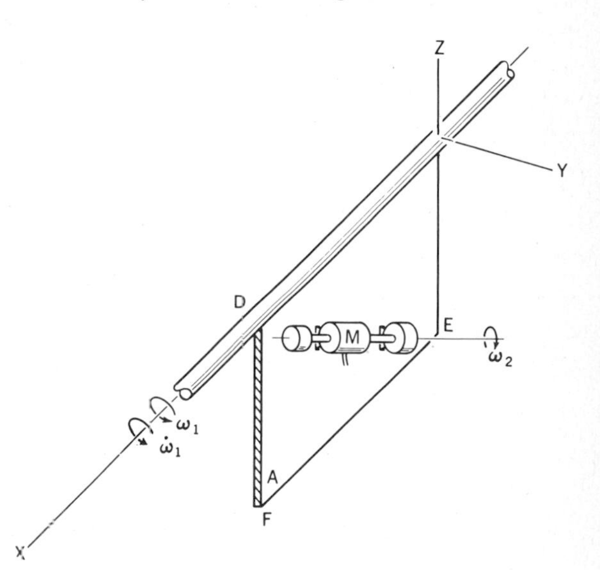

Figure P.15.14

15.15. A racing car is moving at a constant speed of 200 mi/hr when the driver turns his front wheels at an increasing rate, $\dot{\omega}_1$, of .02 rad/sec². If $\omega_1 = .0168$ rad/sec at the instant of interest, what are $\boldsymbol{\omega}$ and $\dot{\boldsymbol{\omega}}$ of the front wheels at this instant? The diameter of the tires is 30 in.

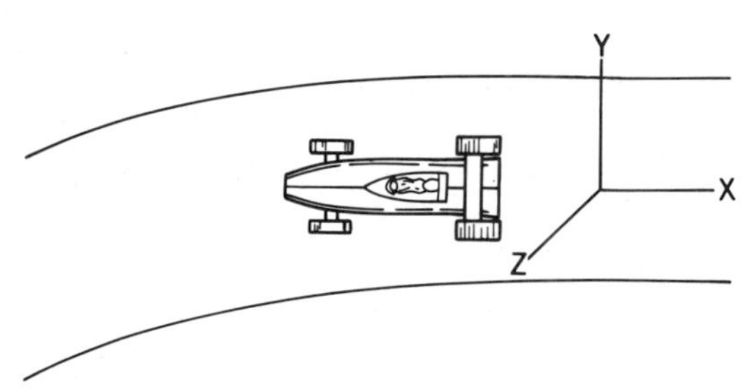

Figure P.15.15

15.16. A cone is rolling without slipping such that its centerline rotates at the rate ω_1 of 5 revolutions per second about the Z axis. What is the angular velocity $\boldsymbol{\omega}$ of the body relative to the ground? What is the angular acceleration vector for the body?

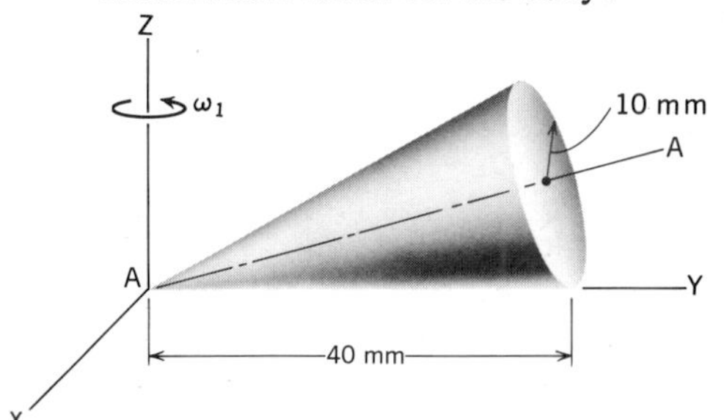

Figure P.15.16

15.17. A small cone A is rolling without slipping inside a large conical cavity B. What is the angular velocity $\boldsymbol{\omega}$ of cone A relative to the large cone cavity B if the centerline of A undergoes an angular speed ω_1 of 5 rotations per second about the Z axis?

Figure P.15.17

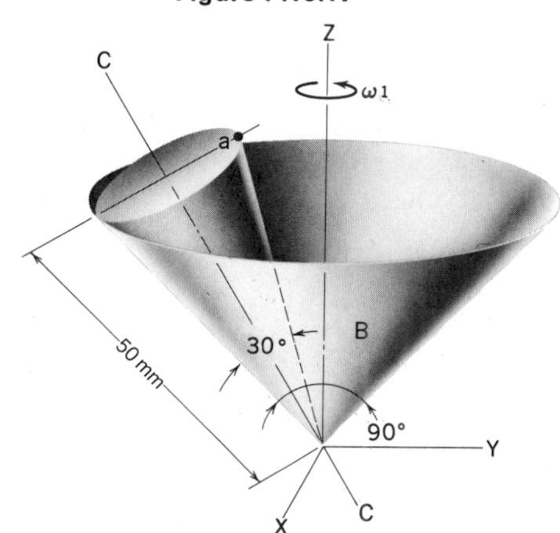

15.18. An amusement park ride consists of a stationary vertical tower with arms that can swing outward from the tower and at the same time can rotate about the tower. At the ends of the arms, cockpits containing passengers can rotate relative to the arms. Consider the case where cockpit A rotates at angular speed ω_2 relative to arm BC, which rotates at angular speed ω_1 relative to the tower. If θ is fixed at 90°, what are the total angular velocity and the angular acceleration of the cockpit relative to the ground? Use $\omega_1 = .2$ rad/sec and $\omega_2 = .6$ rad/sec.

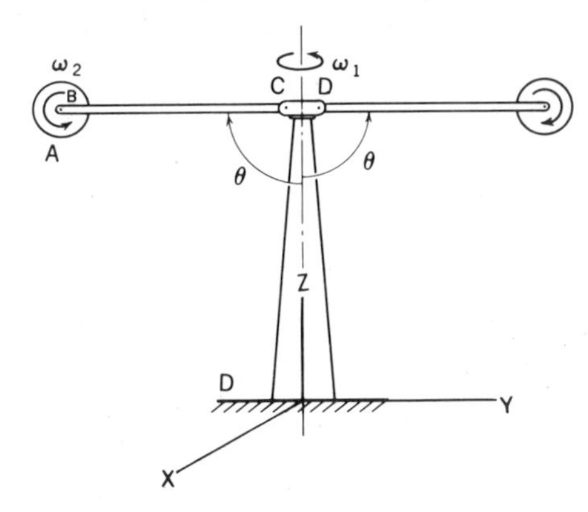

Figure P.15.18

15.19. In Problem 15.18, find $\dot{\boldsymbol{\omega}}$ of the cockpit for the case where $\dot{\theta} = \omega_3 = .8$ rad/sec at the instant that $\theta = 90°$.

The last four problems of this set are designed for those students who have studied Example 15.3.

15.20. In Problem 15.18, find $\dot{\boldsymbol{\omega}}$ of the cockpit A for the case where $\dot{\omega}_1 = .2$ rad/sec² and $\dot{\omega}_2 = .3$ rad/sec².

15.21. In Problem 15.13, find $\dot{\boldsymbol{\omega}}$ of beam AB relative to the ground if at the instant shown the following data apply:

$$\omega_1 = .3 \text{ rad/sec}$$
$$\dot{\omega}_1 = .2 \text{ rad/sec}^2$$
$$\omega_2 = .6 \text{ rad/sec}$$
$$\dot{\omega}_2 = -.1 \text{ rad/sec}^2$$

15.22. In Problem 15.9, find the angular acceleration vector $\dot{\boldsymbol{\omega}}$ for the gun barrel, if, for the instant shown in the diagram, the following data apply:

$$\dot{\phi} = .30 \text{ rad/sec}, \qquad \theta = 20°$$
$$\ddot{\phi} = .26 \text{ rad/sec}^2, \qquad \phi = 30°$$
$$\dot{\theta} = .17 \text{ rad/sec}$$
$$\ddot{\theta} = -.34 \text{ rad/sec}^2$$

15.23. In Problem 15.11, find $\dot{V}$ if at the instant shown in the diagram:

$$\omega_1 = 5 \text{ rad/sec}$$
$$\dot{\omega}_1 = 10 \text{ rad/sec}^2$$
$$\omega_2 = 2 \text{ rad/sec}$$
$$\dot{\omega}_2 = 3 \text{ rad/sec}^2$$
$$V = 10 \text{ m/sec}$$
$$\dot{V} = 5 \text{ m/sec}^2$$

15.5 Applications of the Fixed-Vector Concept

In Section 15.4, we considered the time derivative, as seen from a reference XYZ, of a vector $\boldsymbol{A}$ fixed in a rigid body or fixed in reference xyz. The result was a simple formula:

$$\dot{\boldsymbol{A}} = \boldsymbol{\omega} \times \boldsymbol{A}$$

where $\boldsymbol{\omega}$ is the angular velocity relative to XYZ of the body or the reference in which $\boldsymbol{A}$ is fixed. In this section, we shall use the preceding formula for a vector connecting two points a and b in a rigid body (see Fig. 15.12). This vector, which we denote as $\boldsymbol{\rho}_{ab}$, clearly is fixed in the rigid body. The body in accordance with Chasle's theorem has a velocity $\dot{\boldsymbol{R}}$ relative to XYZ corresponding to some point O in the body plus an angular velocity $\boldsymbol{\omega}$ relative to XYZ with the axis of rotation going through O. We can then say on observing from XYZ:

$$\dot{\boldsymbol{\rho}}_{ab} = \boldsymbol{\omega} \times \boldsymbol{\rho}_{ab} \tag{15.4}$$

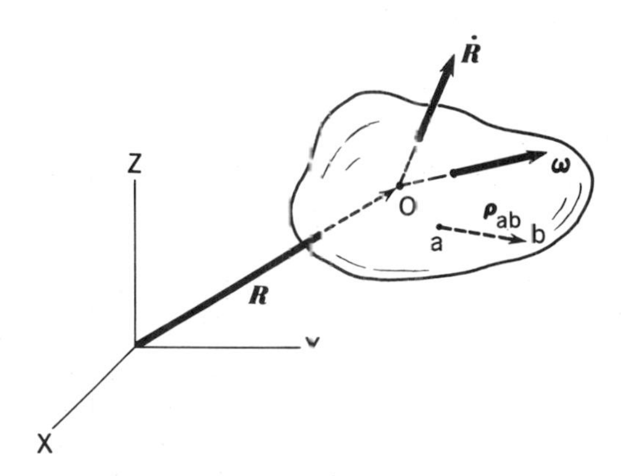

Figure 15.12. $\boldsymbol{\rho}_{ab}$ fixed in rigid body.

Now consider position vectors at a and b as shown in Fig. 15.13. We can say:

$$\boldsymbol{r}_a + \boldsymbol{\rho}_{ab} = \boldsymbol{r}_b$$

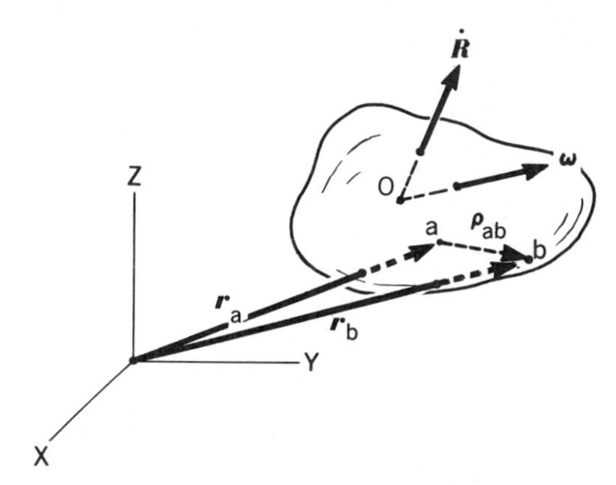

Figure 15.13. Insert position vectors.

Taking the time derivative as seen from XYZ, we have

$$\left(\frac{d\boldsymbol{r}_a}{dt}\right)_{XYZ} + \left(\frac{d\boldsymbol{\rho}_{ab}}{dt}\right)_{XYZ} = \left(\frac{d\boldsymbol{r}_b}{dt}\right)_{XYZ}$$

This equation can be written as

$$\left(\frac{d\boldsymbol{\rho}_{ab}}{dt}\right)_{XYZ} = \boldsymbol{V}_b - \boldsymbol{V}_a \tag{15.5}$$

Since $(d\boldsymbol{\rho}_{ab}/dt)_{XYZ}$ is the *difference* between the velocity of point b and that at point a as noted above, we can say that $(d\boldsymbol{\rho}_{ab}/dt)_{XYZ}$ is the velocity of point b *relative* to point a.[3] Next, using Eq. 15.4 to replace $(d\boldsymbol{\rho}_{ab}/dt)_{XYZ}$, we have, on rearranging terms, a very useful equation:

$$\boxed{\boldsymbol{V}_b = \boldsymbol{V}_a + \boldsymbol{\omega} \times \boldsymbol{\rho}_{ab}} \tag{15.6}$$

In using the foregoing equation, we must be sure that we get the sequence of subscripts correct on $\boldsymbol{\rho}$ since a change in ordering brings about a change in sign (i.e., $\boldsymbol{\rho}_{ab} = -\boldsymbol{\rho}_{ba}$). This equation is a statement of the physically obvious result that *the velocity of particle b of a rigid body as seen from XYZ equals the velocity of any other particle a of this body as seen from XYZ plus the velocity of particle b relative to particle a.*

Differentiating Eq. 15.6 again, we can get a relation involving the acceleration vectors of two points on a rigid body:

$$\boldsymbol{a}_b = \boldsymbol{a}_a + \left(\frac{d\boldsymbol{\omega}}{dt}\right)_{XYZ} \times \boldsymbol{\rho}_{ab} + \boldsymbol{\omega} \times \left(\frac{d\boldsymbol{\rho}_{ab}}{dt}\right)_{XYZ}$$

Hence, we have on using Eq. 15.4 in the last expression

$$\boxed{\boldsymbol{a}_b = \boldsymbol{a}_a + \dot{\boldsymbol{\omega}} \times \boldsymbol{\rho}_{ab} + \boldsymbol{\omega} \times (\boldsymbol{\omega} \times \boldsymbol{\rho}_{ab})} \tag{15.7}$$

---

[3]That is, $(d\boldsymbol{\rho}_{ab}/dt)_{XYZ}$ is the velocity of b as seen by an observer translating relative to XYZ with point a, i.e., as seen by a nonrotating observer moving with a.

We have thus formulated relations between *the motions of two points of a rigid body as seen from a single reference*. Such relations can be very useful in the study of machine elements.

Before going to the examples, let us now consider the case of a circular cylinder rolling *without slipping* (see Fig. 15.14). The point of contact A of the cylinder with the ground has instantaneously *zero velocity* and hence we have pure instantaneous rotation at any time t about an instantaneous axis of rotation at the line of contact. The velocity of any point B of the cylinder can then easily be found by using Eq. 15.6 for points B and A. Thus:

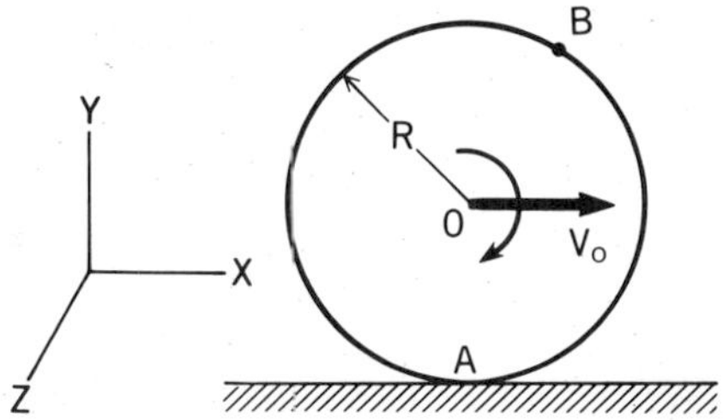

Figure 15.14. Cylinder rolling without slipping on a flat surface.

$$\boldsymbol{V}_B = \boldsymbol{V}_A + \boldsymbol{\omega} \times \boldsymbol{\rho}_{AB}$$

Therefore,

$$\boldsymbol{V}_B = \boldsymbol{0} + \boldsymbol{\omega} \times \boldsymbol{\rho}_{AB}$$

From the above equation it is clear that for computing the velocity of any point on the cylinder we can think of the cylinder as *hinged* at the point of contact. In particular for point O, the center of the cylinder, we get from above:

$$\boldsymbol{V}_0 = -\omega R\boldsymbol{i}$$

If the velocity V_0 is known, clearly the angular velocity has a magnitude of V_0/R.

Another way of relating V and ω is to realize that the distance s that O moves must equal the length of circumference coming into contact with the ground. That is, measuring θ from the X axis to the Y axis:

$$s = -R\theta$$

Differentiating we get:

$$V_0 = -R\dot{\theta} = -R\omega$$

thus reproducing the previous result. Differentiating again, we get

$$a_0 = -R\ddot{\theta} = -R\alpha \tag{15.8}$$

relating now the acceleration of O and the angular acceleration α. Clearly, the acceleration vector for O must be parallel to the ground. Again, for computing $\boldsymbol{a}_0$, we have a simple situation.

Next, let us determine the acceleration vector for the *point of contact* A of the cylinder. Thus, we can say for points A and O:

$$\boldsymbol{a}_O = \boldsymbol{a}_A + \dot{\boldsymbol{\omega}} \times \boldsymbol{\rho}_{AO} + \boldsymbol{\omega} \times (\boldsymbol{\omega} \times \boldsymbol{\rho}_{AO})$$

Therefore,

$$-R\ddot{\theta}\boldsymbol{i} = \boldsymbol{a}_A + \ddot{\theta}\boldsymbol{k} \times R\boldsymbol{j} + \dot{\theta}\boldsymbol{k} \times (\dot{\theta}\boldsymbol{k} \times R\boldsymbol{j}) \tag{15.9}$$

Carrying out the products:

$$-R\ddot{\theta}\boldsymbol{i} = \boldsymbol{a}_A - R\ddot{\theta}\boldsymbol{i} - R\dot{\theta}^2\boldsymbol{j}$$

Therefore,

$$a_A = R\dot{\theta}^2 j \tag{15.10}$$

We see that *point A is accelerating upward toward the center of the cylinder.*[4] This information will be valuable for us in Chapter 16 when we study rigid-body dynamics.

EXAMPLE 15.4

Wheel D rotates at an angular speed ω_1 of 2 rad/sec counterclockwise in Fig. 15.15. Find the angular speed ω_A of gear A relative to the ground at the instant shown in the diagram.

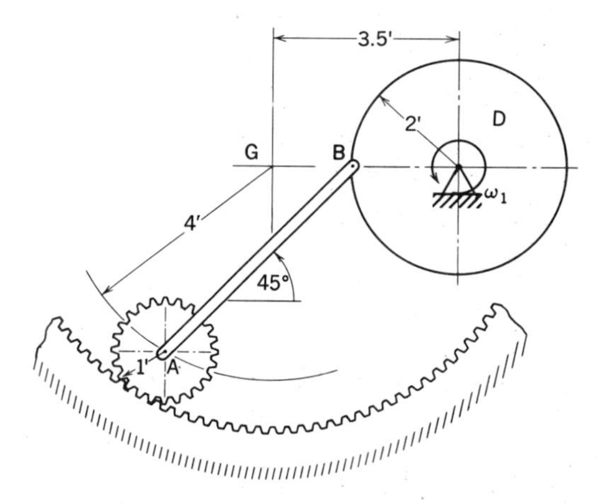

Figure 15.15. Two-dimensional device.

We have information about two points of one of the rigid bodies, namely AB, of the device. At B, the velocity must be downward with the value of $(\omega_1)(r_D) =$ 4 ft/sec as shown in Fig. 15.16. Furthermore, since point A must travel a circular path of radius GA we know that A has velocity V_A with a direction at right angles to GA. Accordingly, the angle between V_A and the horizontal must be $(90° - \alpha - 45°)$ $= (45° - \alpha)$. If we can determine velocity V_A, we can get the desired angular speed of gear A immediately.

Before examining rigid body AB, we have some geometrical steps to take. Considering triangle GAB in Fig. 15.16, we can first solve for α using the law of sines as follows:

$$\frac{GA}{\sin(\measuredangle GBA)} = \frac{GB}{\sin\alpha}$$

Therefore,

$$\frac{4}{\sin 45°} = \frac{1.5}{\sin\alpha} \tag{a}$$

[4]This conclusion must apply also to a sphere rolling without slipping on a flat surface.

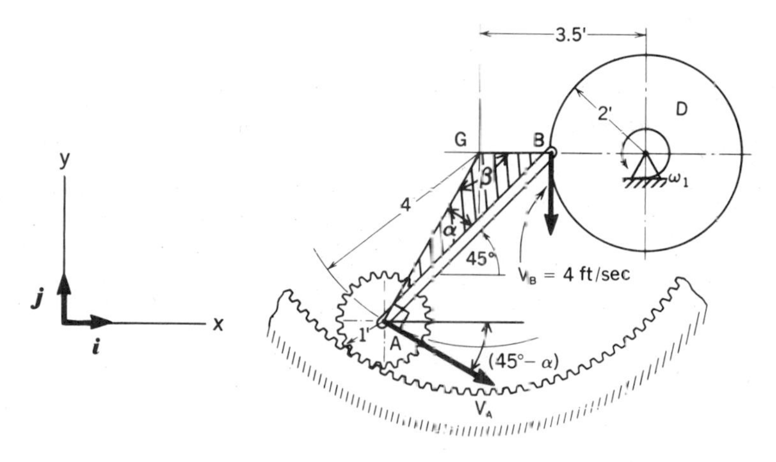

Figure 15.16. Velocity vectors for two points of a rigid body shown.

Solving for α, we get

$$\alpha = 15.37° \tag{b}$$

The angle β is then easily evaluated considering the angles in the triangle GBA as

$$\begin{aligned} \beta &= 180° - \alpha - \measuredangle GBA \\ &= 180° - 15.37° - 45° = 119.6° \end{aligned} \tag{c}$$

Finally, we can determine AB of the triangle, again using the law of sines. Thus,

$$\frac{AB}{\sin \beta} = \frac{GA}{\sin 45°}$$

$$\frac{AB}{\sin 119.6°} = \frac{4}{.707}$$

Solving for AB, we get

$$AB = 4.92 \text{ ft} \tag{d}$$

We now can consider bar AB. For the points A and B on this body, we can say[5]:

$$\boldsymbol{V}_A = \boldsymbol{V}_B + \boldsymbol{\omega}_{AB} \times \boldsymbol{\rho}_{BA}$$

Noting that the motion is coplanar and that $\boldsymbol{\omega}_{AB}$ must then be normal to the plane of motion, we have

$$\begin{aligned} &V_A[\cos (45° - \alpha)\boldsymbol{i} - \sin (45° - \alpha)\boldsymbol{j}] \\ &\qquad = -4\boldsymbol{j} + \omega_{AB}\boldsymbol{k} \times 4.92(-\cos 45°\boldsymbol{i} - \sin 45°\boldsymbol{j}) \end{aligned}$$

Inserting the value $\alpha = 15.37°$, we then get the following vector equation:

$$V_A(.869)\boldsymbol{i} - V_A(.494)\boldsymbol{j} = -4\boldsymbol{j} - 3.48\omega_{AB}\boldsymbol{j} + 3.48\omega_{AB}\boldsymbol{i} \tag{e}$$

[5]Our practice will be to consider unknown angular velocities as *positive*. The sign for the unknown angular velocity coming out of the computations will then correspond to the *actual convention* sign for the angular velocity.

The scalar equations are

$$\begin{aligned} .869V_A &= 3.48\omega_{AB} \\ -.494V_A &= -4 - 3.48\omega_{AB} \end{aligned} \tag{f}$$

Solving, we get[6]

$$\begin{aligned} V_A &= -10.66 \text{ ft/sec} \\ \omega_{AB} &= -2.66 \text{ rad/sec} \end{aligned} \tag{g}$$

Thus, point A moves in a direction *opposite* to that shown in Fig. 15.16. We now can readily evaluate ω_A, which clearly must have a value of

$$\omega_A = \frac{V_A}{r_A} = \frac{10.66}{1} = 10.66 \text{ rad/sec}$$

in the counterclockwise direction.

EXAMPLE 15.5

In the device in Fig. 15.17, find the angular velocities and angular accelerations of both bars.

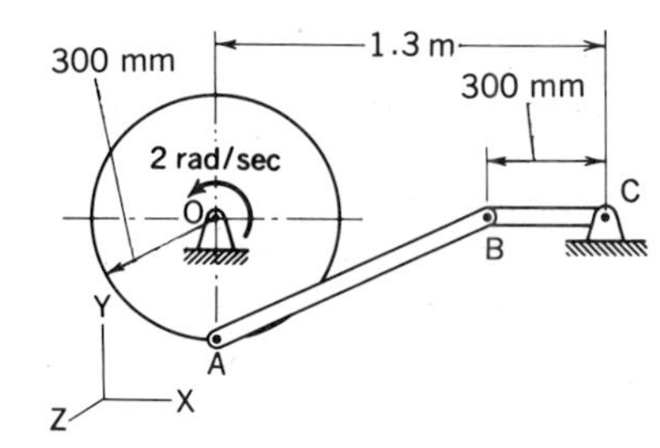

Figure 15.17. Two-dimensional device.

We shall consider points A and B of bar AB. Note first that at the instant shown:

$$\mathbf{V}_B = -(.300)(\omega_{BC})\mathbf{j} \text{ m/sec} \tag{a}$$

$$\begin{aligned} \mathbf{V}_A &= (2)(.300)\mathbf{i} \\ &= .600\mathbf{i} \text{ m/sec} \end{aligned} \tag{b}$$

Noting that $\boldsymbol{\omega}_{AB}$ must be oriented in the Z direction because we have plane motion in the XY plane, we have for Eq. 15.6:

$$\mathbf{V}_B = \mathbf{V}_A + \boldsymbol{\omega}_{AB} \times \boldsymbol{\rho}_{AB}$$

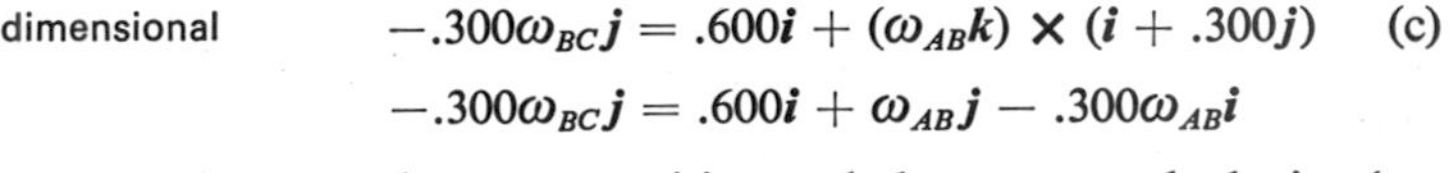

$$-.300\omega_{BC}\mathbf{j} = .600\mathbf{i} + (\omega_{AB}\mathbf{k}) \times (\mathbf{i} + .300\mathbf{j}) \tag{c}$$

$$-.300\omega_{BC}\mathbf{j} = .600\mathbf{i} + \omega_{AB}\mathbf{j} - .300\omega_{AB}\mathbf{i}$$

Note we have assumed ω_{BC} and ω_{AB} as positive and thus counterclockwise (see footnote below). The scalar equations are:

$$\begin{aligned} .600 &= .300\omega_{AB} \\ -.300\omega_{BC} &= \omega_{AB} \end{aligned} \tag{d}$$

We then get

$$\begin{aligned} \omega_{AB} &= 2 \text{ rad/sec} \\ \omega_{BC} &= -6.67 \text{ rad/sec} \end{aligned} \tag{e}$$

Therefore, ω_{AB} is counterclockwise while ω_{BC} must be clockwise.

---

[6]By having assumed ω_{AB} as positive and thus *counterclockwise* for the reference xy employed, we conclude from the presence of the minus sign that the assumption is wrong and that ω_{AB} must be *clockwise* for the reference used. It is significant to note that as a result of the initial positive assumption, the result $\omega_{AB} = -2.66$ rad/sec gives at the same time the *correct convention sign* for the actual angular velocity for the reference used.

Let us now turn to the angular acceleration considerations for the bars. We consider separately now points A and B of bar AB. Thus,

$$\boldsymbol{a}_A = (r\omega^2)\boldsymbol{j} = (.300)(2^2)\boldsymbol{j} = 1.200\boldsymbol{j}\ \text{m/sec}^2$$

$$\begin{aligned}\boldsymbol{a}_B &= \rho_{BC}\omega_{BC}^2\boldsymbol{i} + \rho_{BC}\dot{\omega}_{BC}(-\boldsymbol{j})\\ &= (.300)(-6.67^2)\boldsymbol{i} - .300\dot{\omega}_{BC}\boldsymbol{j}\\ &= 13.33\boldsymbol{i} - .300\dot{\omega}_{BC}\boldsymbol{j}\end{aligned}$$

Again, we have assumed $\dot{\omega}_{BC}$ positive and thus counterclockwise. Considering bar AB, we can say for Eq. 15.7:

$$\boldsymbol{a}_B = \boldsymbol{a}_A + \dot{\boldsymbol{\omega}}_{AB} \times \boldsymbol{\rho}_{AB} + \boldsymbol{\omega}_{AB} \times (\boldsymbol{\omega}_{AB} \times \boldsymbol{\rho}_{AB}) \qquad \text{(f)}$$

Noting that $\dot{\boldsymbol{\omega}}_{AB}$ must be in the Z direction, we have for the foregoing equation:

$$13.33\boldsymbol{i} - .300\dot{\omega}_{BC}\boldsymbol{j} = 1.200\boldsymbol{j} + \dot{\omega}_{AB}\boldsymbol{k} \times (\boldsymbol{i} + .300\boldsymbol{j}) + (2\boldsymbol{k}) \times [2\boldsymbol{k} \times (\boldsymbol{i} + .300\boldsymbol{j})] \qquad \text{(g)}$$

The scalar equations are

$$17.33 = -.300\dot{\omega}_{AB}$$

$$-.300\dot{\omega}_{BC} = \dot{\omega}_{AB}$$

We get

$$\dot{\omega}_{AB} = -57.8\ \text{rad/sec}^2$$

$$\dot{\omega}_{BC} = 192.6\ \text{rad/sec}^2$$

Clearly, for the reference used, $\dot{\omega}_{AB}$ must be clockwise and $\dot{\omega}_{BC}$ must be counterclockwise.

EXAMPLE 15.6

In Example 15.5, find the *instantaneous axis of rotation* for the rod AB.

The intersection of the instantaneous axis of rotation with the xy plane will be a point E in a hypothetical rigid-body extension of bar AB having zero velocity at the instant of interest. We can accordingly say:

$$\boldsymbol{V}_E = \boldsymbol{V}_A + \boldsymbol{\omega}_{AB} \times \boldsymbol{\rho}_{AE}$$

Therefore,

$$\boldsymbol{0} = .60\boldsymbol{i} + (2\boldsymbol{k}) \times (\Delta x\boldsymbol{i} + \Delta y\boldsymbol{j}) \qquad \text{(a)}$$

where Δx and Δy are the components of the directed line segment from point A to the center of rotation E. The scalar equations are:

$$0 = .60 - 2\,\Delta y$$

$$0 = 2\,\Delta x$$

Clearly, $\Delta y = .3$ and $\Delta x = 0$. Thus, the center of rotation is point O.

We could have easily deduced this result by inspection in this case. The velocity of each point of bar AB must be at *right angles* to a line from the center of rotation to the point. The velocity of point A is in the horizontal direction and the velocity of point B is in the vertical direction. Clearly, point O is the only point from which lines to points A and B are normal to the velocities at these points.

*EXAMPLE 15.7

A wheel E is rotating about a fixed axis at a constant angular speed ω_1 of 5 rad/sec in Fig. 15.18. A bar CD is held by the wheel at D by a ball-joint connection and is guided along AB by a collar at C having a second ball-joint connection with CD, as shown in the diagram. Compute the velocity of C.

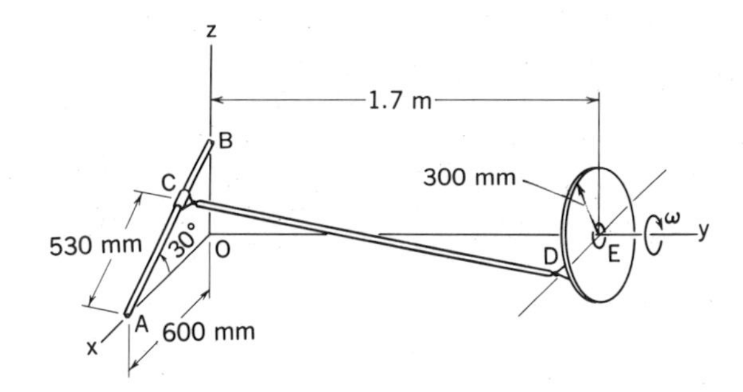

Figure 15.18. Three-dimensional device.

We shall need the vector $\boldsymbol{\rho}_{DC}$. Thus,

$$\begin{aligned}\boldsymbol{\rho}_{DC} &= \boldsymbol{r}_C - \boldsymbol{r}_D \\ &= [(.600 - .530\cos 30°)\boldsymbol{i} + .530\sin 30°\boldsymbol{k}] - (1.7\boldsymbol{j} + .300\boldsymbol{i}) \\ &= -.1590\boldsymbol{i} - 1.7\boldsymbol{j} + .265\boldsymbol{k} \text{ m}\end{aligned}$$

Now employ Eq. 15.6 for rod CD. Thus,

$$\boldsymbol{V}_C = \boldsymbol{V}_D + \boldsymbol{\omega}_{CD} \times \boldsymbol{\rho}_{DC}$$

Therefore,

$$\begin{aligned}V_C(\cos 30°\boldsymbol{i} - \sin 30°\boldsymbol{k}) &= (5)(.30)\boldsymbol{k} + (\omega_x\boldsymbol{i} + \omega_y\boldsymbol{j} + \omega_z\boldsymbol{k}) \\ &\quad \times (-.1590\boldsymbol{i} - 1.7\boldsymbol{j} + .265\boldsymbol{k}) \\ V_C(.866\boldsymbol{i} - .500\boldsymbol{k}) &= 1.50\boldsymbol{k} - 1.7\omega_x\boldsymbol{k} - .265\omega_x\boldsymbol{j} \\ &\quad + .1590\omega_y\boldsymbol{k} + .265\omega_y\boldsymbol{i} - .1590\omega_z\boldsymbol{j} \\ &\quad + 1.7\omega_z\boldsymbol{i}\end{aligned}$$

The scalar equations are:

$$.866V_C = .265\omega_y + 1.7\omega_z \tag{a}$$

$$0 = -.265\omega_x - .1590\omega_z \tag{b}$$

$$-.500V_C = 1.50 - 1.7\omega_x + .1590\omega_y \tag{c}$$

From these equations, we cannot solve for ω_x, ω_y, and ω_z because the spin of CD about its own axis (allowed by the ball joints) can have *any value* without affecting the velocity of slider C. However, we can determine V_C, as we shall now demonstrate.

In Eq. (b), solve for ω_x in terms of ω_z.

$$\omega_x = -.600\omega_z \tag{d}$$

In Eq. (a), solve for ω_y in terms of ω_z:

$$\omega_y = 3.27V_C - 6.415\omega_z \tag{e}$$

Substitute for ω_x and ω_y in Eq. (c) using the foregoing results:

$$-.500V_C = 1.50 - 1.7(-.600\omega_z) + .1590(3.27V_C - 6.415\omega_z)$$

Therefore,

$$-1.020V_C = 1.5 + 1.020\omega_z - 1.020\omega_z$$
$$V_C = -1.471 \text{ m/sec}$$

Hence,

$$\boldsymbol{V}_C = -1.471(\cos 30°\boldsymbol{i} - \sin 30°\boldsymbol{k})$$
$$= -1.274\boldsymbol{i} + .7355\boldsymbol{k} \text{ m/sec}$$

Before going on to the next section, we wish to point out a simple relation that will be of use in the remainder of the chapter. Suppose that you have a moving particle whose position vector $\boldsymbol{r}$ has a magnitude that is constant (see Fig. 15.19). This position vector, however, has an angular velocity $\boldsymbol{\omega}$ relative to xyz. We wish to know the velocity

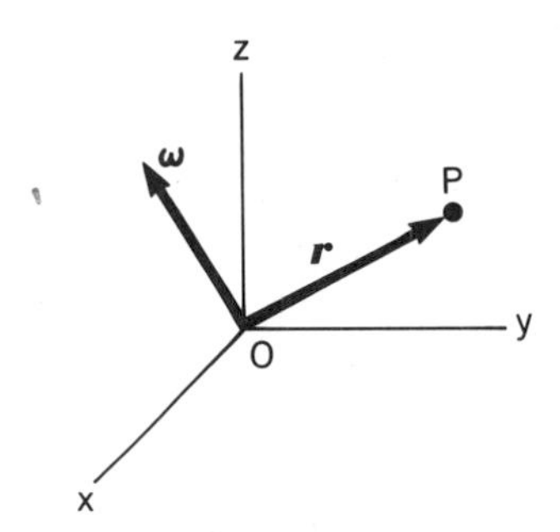

Figure 15.19. Position vector $\boldsymbol{r}$ has constant magnitude but rotates relative to xyz with angular velocity $\boldsymbol{\omega}$.

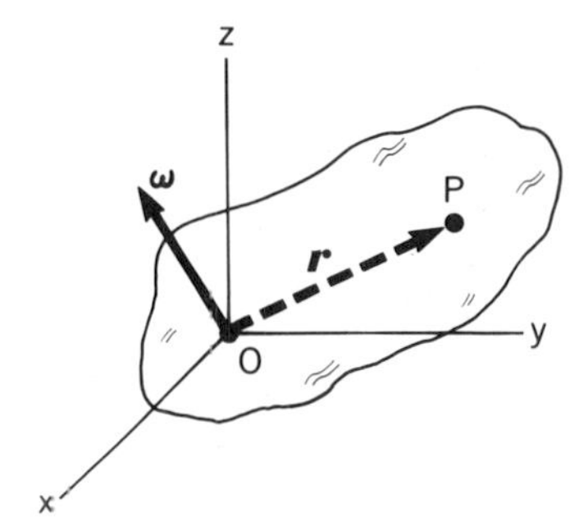

Figure 15.20. P now considered as a point in a rigid body attached at O and having angular velocity $\boldsymbol{\omega}$.

of the particle P relative to xyz. We could imagine for this purpose that particle P is part of a rigid body attached to xyz at O and rotating with angular velocity $\boldsymbol{\omega}$. This situation is shown in Fig. 15.20. Using Eq. 15.6, we can then say:

$$\boldsymbol{V}_P = \boldsymbol{V}_O + \boldsymbol{\omega} \times \boldsymbol{\rho}_{CP}$$

But $\boldsymbol{V}_O = \boldsymbol{0}$ and $\boldsymbol{\rho}_{OP}$ is simply $\boldsymbol{r}$. Hence, we have

$$\boldsymbol{V}_p = \boldsymbol{\omega} \times \boldsymbol{r}$$

We thus have a simple formula for the velocity of a particle moving at a fixed distance from the origin of xyz. This velocity is simply the cross product of the angular velocity $\boldsymbol{\omega}$ of the position vector about xyz times the position vector.

Problems

15.24. A body is spinning about an axis having direction cosines $l = .5$, $m = .5$, and $n = .707$. The angular speed is 50 rad/sec. What is the velocity of a point in the body having a position vector $\boldsymbol{r} = 6\boldsymbol{i} + 4\boldsymbol{j}$ ft?

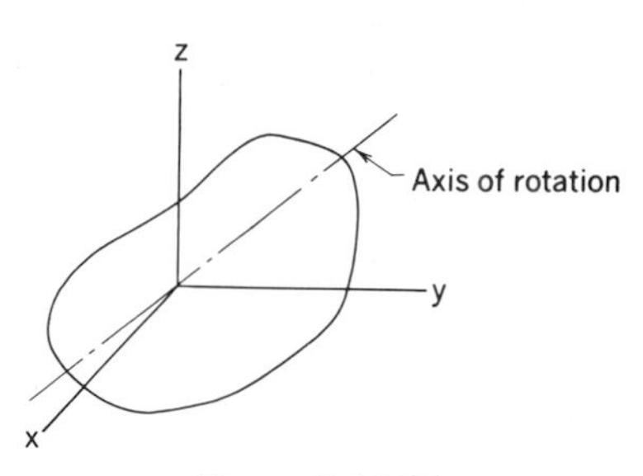

Figure P.15.24

15.25. In Problem 15.24, what is the relative velocity between a point in the body at position $x = 10$ m, $y = 6$ m, $z = 3$ m and a point in the body at position $x = 2$ m, $y = -3$ m, $z = 0$ m?

15.26. If the body in Problem 15.24 is given an additional angular velocity $\boldsymbol{\omega}_2 = 6\boldsymbol{j} + 10\boldsymbol{k}$ rad/sec, what is the direction of the axis of rotation? Compute the velocity at $\boldsymbol{r} = 10\boldsymbol{j} + 3\boldsymbol{k}$ ft if the actual axis of rotation goes through the origin.

15.27. A wheel is rolling along at 17 m/sec without slipping. What is the angular speed? What is the velocity of point B on the rim of the wheel at the instant shown?

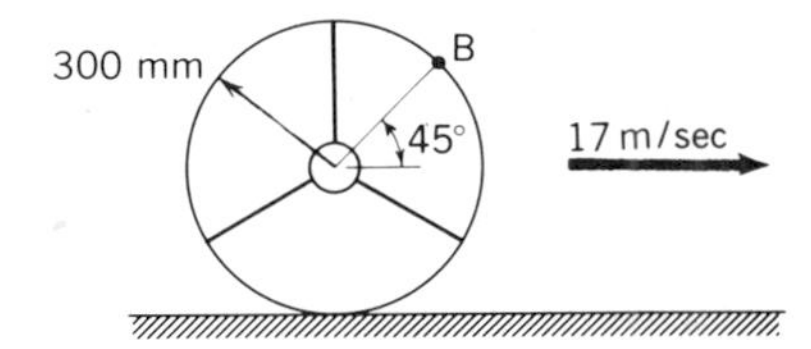

Figure P.15.27

15.28. A flexible cord is wrapped around a spool and is pulled at a velocity of 10 ft/sec relative to the ground. If there is no slipping at C, what is the velocity of points O and D at the instant shown?

Figure P.15.28

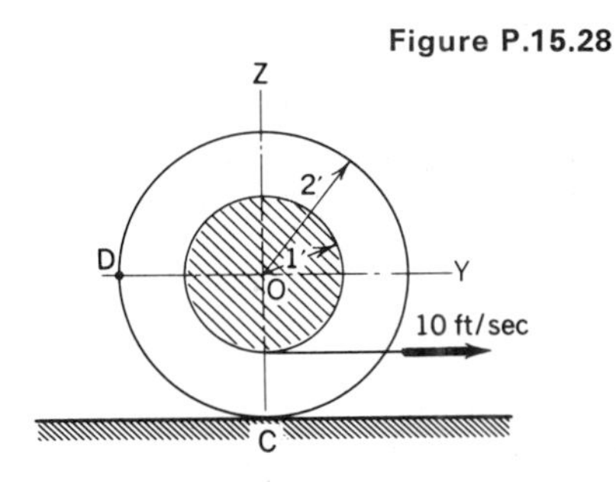

15.29. A piston P is shown moving downward at the constant speed of 1 ft/sec. What is the speed of slider A at the instant of interest?

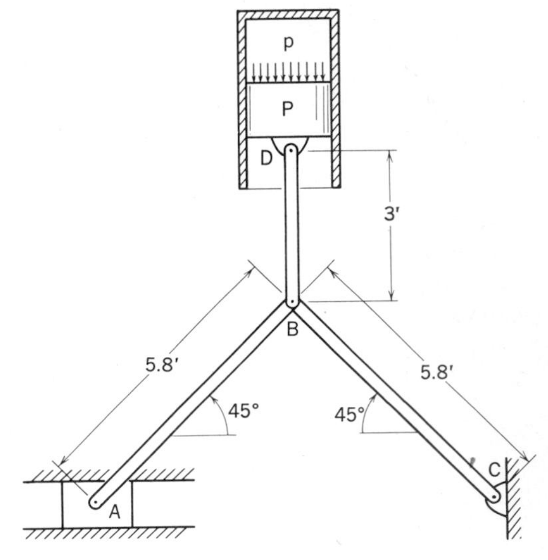

Figure P.15.29

15.30. A rod AB is 1 m in length. If the end A slides down the surface at a speed V_A of 3 m/sec, what is the angular speed of AB at the instant shown?

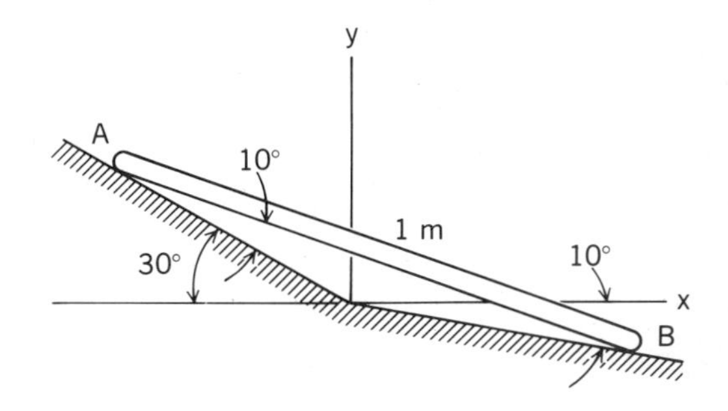

Figure P.15.30

15.31. A plate moves along a horizontal surface. Components of the velocity for three corners are:

$$(V_A)_x = 2 \text{ m/sec}$$
$$(V_B)_y = -3 \text{ m/sec}$$
$$(V_C)_y = 5 \text{ m/sec}$$

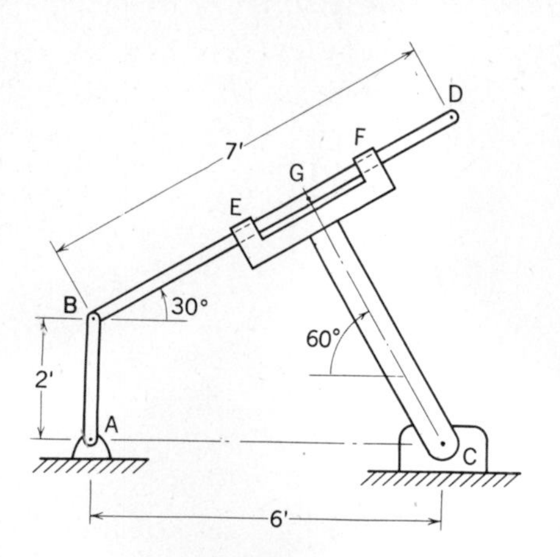

Figure P.15.36

15.37. A wheel rotates with an angular speed of 20 rad/sec. A connecting rod connects points A on the wheel with a slider at B. Compute the angular velocity of the connecting rod and the velocity of the slider when the apparatus is in the position shown in the diagram.

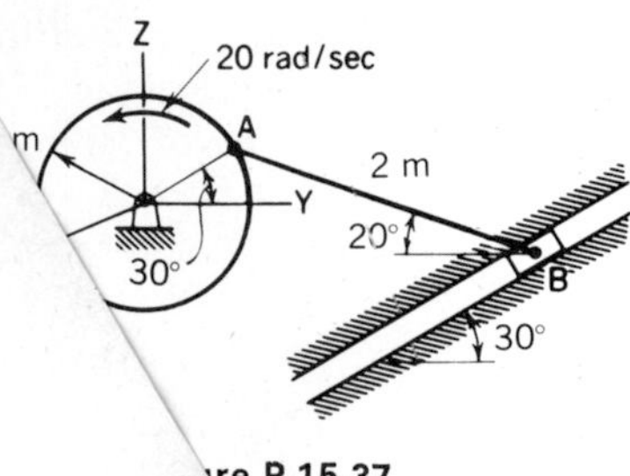

re P.15.37

7, if $V_B = 14.30$ m/sec
rad/sec, where is the
rotation of connect-

, and crankshaft
schematically.
0 rpm. At the
velocity of
and what
nnecting

speed of
What is
and body
f point D.
f the veloc-

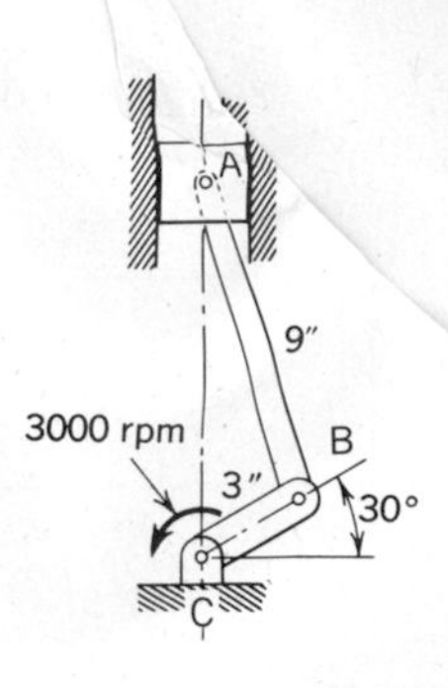

Figure P.15.39

15.40. Member AB is rotating at a constant speed of 4 rad/sec in a counterclockwise direction. What is the angular velocity of bar BC for the position shown in the diagram? What is the velocity of point D at the center of bar BC? Bar BC is 3 ft in length.

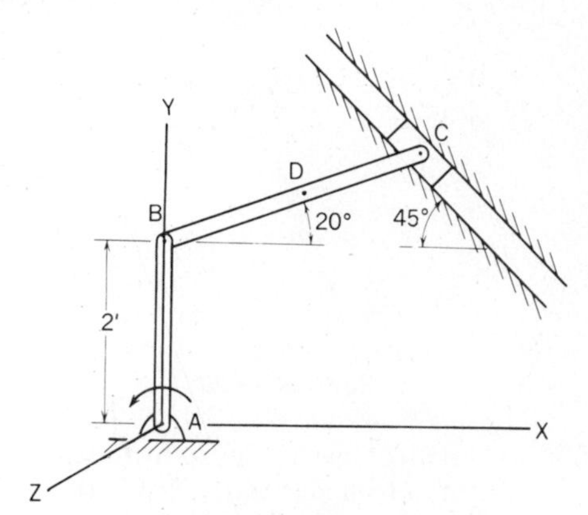

Figure P.15.40

15.41. In Problem 15.40, determine in the simplest manner the instantaneous axis of rotation for bar BC.

15.42. Suppose that bar AB of Problem 15.40 has an angular velocity of 3 rad/sec counterclockwise and a counterclockwise angular acceleration of 5 rad/sec^2. What is the angular acceleration of bar BC, which is 3 ft in length?

15.43. A rod moves in the plane of the paper in such a way that end A has a speed of 3 m/sec. What is the velocity of point B of the rod when the rod is inclined at 45°

to the horizontal? B is at the upper sup-

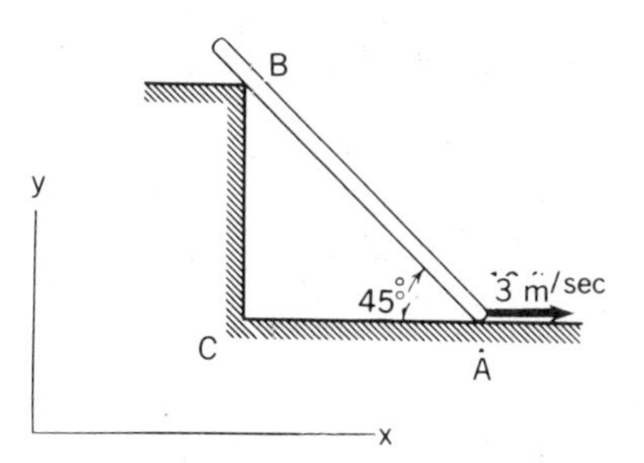

Figure P.15.43

15.44. In Problem 15.43, locate by inspection the instantaneous axis of rotation. Take $CA = 2$ m.

15.45. A rod is moving on a horizontal surface and is shown at time t. What is V_y of end A and ω of the rod at the instant shown? (*Hint:* Use the fact that the rod is inextensible.)

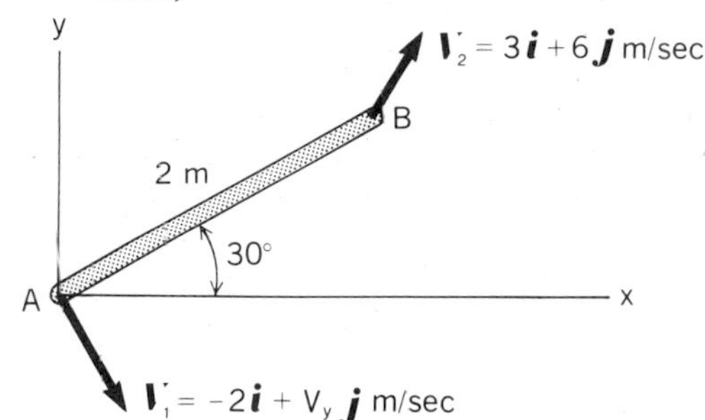

Figure P.15.45

15.46. A plate $ABCD$ moves on a horizontal surface. At time t corners A and B have the following velocities:

$$V_A = 3i + 2j \text{ m/sec}$$

$$V_B = (V_B)_x i + 5j \text{ m/sec}$$

Find the location of the instantaneous axis of rotation.

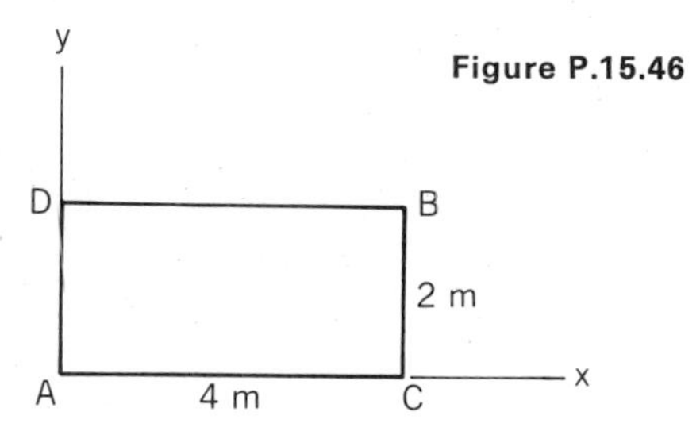

Figure P.15.46

15.47. Find the velocity and acceleration relative to the ground of pin B on the wheel. The wheel rolls without slipping. Also, find the angular velocity and angular acceleration of the slotted bar in which the pin B of the wheel slides when θ of the bar is 30°.

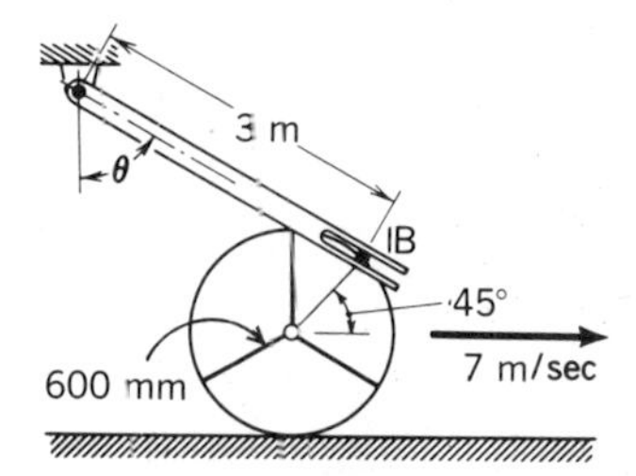

Figure P.15.47

15.48. If $\omega_1 = 5$ rad/sec and $\dot{\omega}_1 = 3$ rad/sec² for bar CD, compute the angular velocity and angular acceleration of the gear D relative to the ground. Solve the problem using Eqs. 15.6 and 15.7, and then check the result by considering simple circular motion of point D.

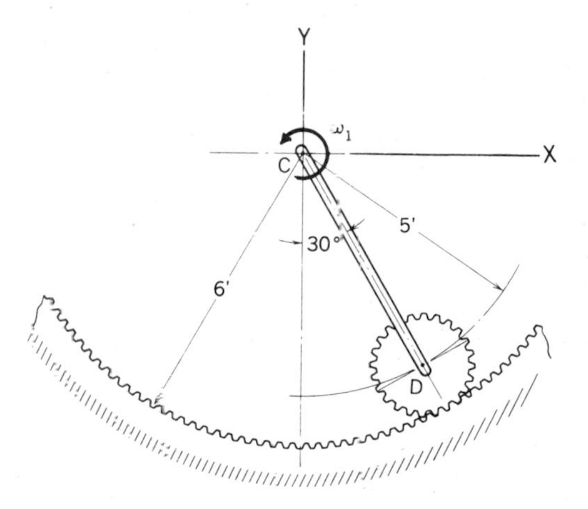

Figure P.15.48

15.49. A mechanism with two sliders is shown. Slider A at the instant of interest has a speed of 3 m/sec and is accelerating at the rate of 1.7 m/sec². If member AB is 2.5 m in length, what are the angular velocity and angular acceleration for this member?

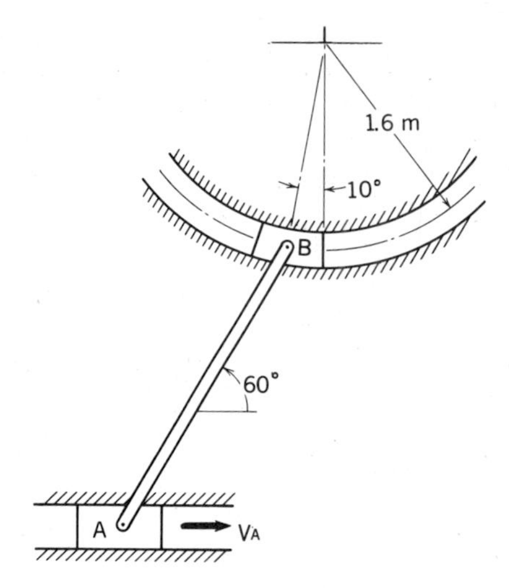

Figure P.15.49

15.50. In Problem 15.49 find the instantaneous center of rotation of bar AB if V_A is 2.7 m/sec.

15.51. The velocity of corner A of the block is known to be at time t:

$$V_A = 10i + 4j - 3k \text{ m/sec}$$

The angular speed about edge $\overline{AD}$ is 2 rad/sec, and the angular speeds about the diagonals $\overline{AF}$ and $\overline{HE}$ are known to be 3 rad/sec and 6 rad/sec, respectively. What is the velocity of corner B at this instant?

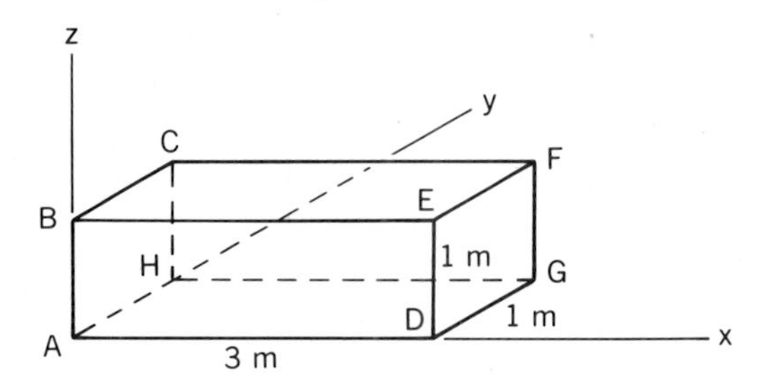

Figure P.15.51

15.52. A rigid sphere is moving in space. The velocities for two points A and B on the surface have the values at time t:

$$V_A = 6i + 3j + 2k \text{ ft/sec}$$

$$V_B = (V_B)_x i + 6j - 4k \text{ ft/sec}$$

The position vectors for points A and B are at time t:

$$r_A = 10i + 15j + 12k \text{ ft}$$

$$r_B = 7i + 20j + 18k \text{ ft}$$

What is the angular velocity of the sphere? At the instant of interest the sphere has zero spin about axis $\overline{AB}$.

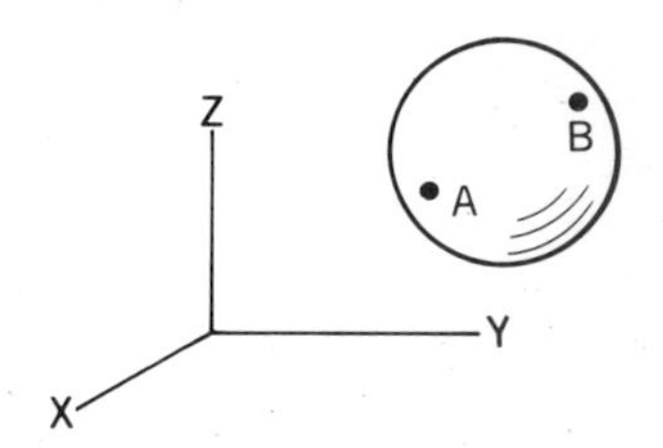

Figure P.15.52

15.53. A conveyor element moves down the incline at a speed of 5 ft/sec. A shaft and platform move with the conveyor element but have a spin of .5 rad/sec about the centerline AB. Also, the shaft swings in the XZ plane at a speed ω_1 of 1 rad/sec. What is the velocity and acceleration of point D on the platform at the instant it is in the YZ plane, as shown in the diagram? Note that at the instant of interest AB is vertical.

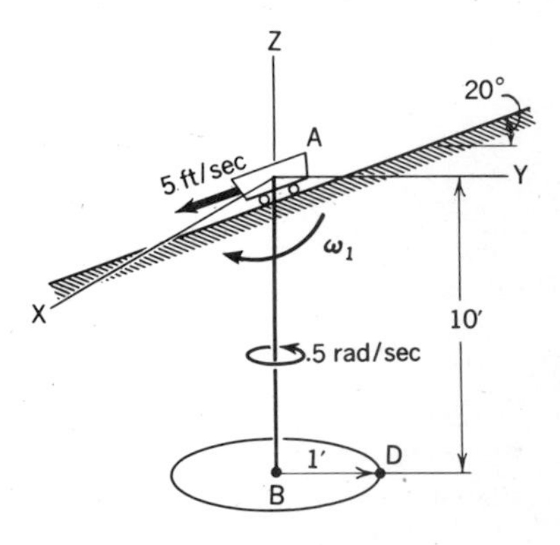

Figure P.15.53

15.54. A conveyor element moves down an incline at a speed of 15 m/sec. A plate hangs down from the conveyor element and, at the instant of interest shown in the dia-

What is the angular speed of the block, and what is the velocity of corner D?

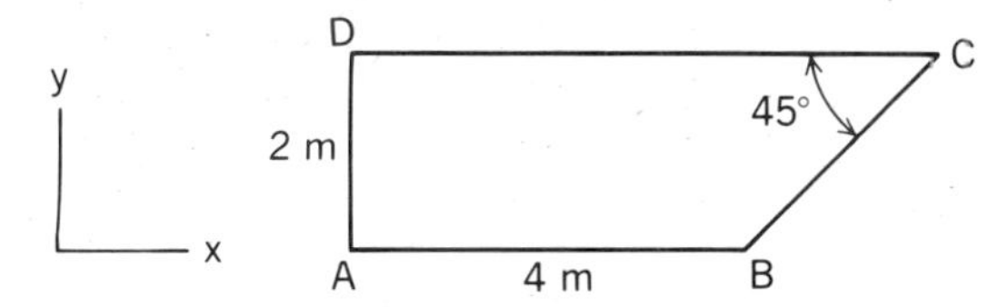

Figure P.15.31

15.32. Rod DC has an angular speed ω_1 of 5 rad/sec at the configuration shown. What is the angular speed of bar AB?

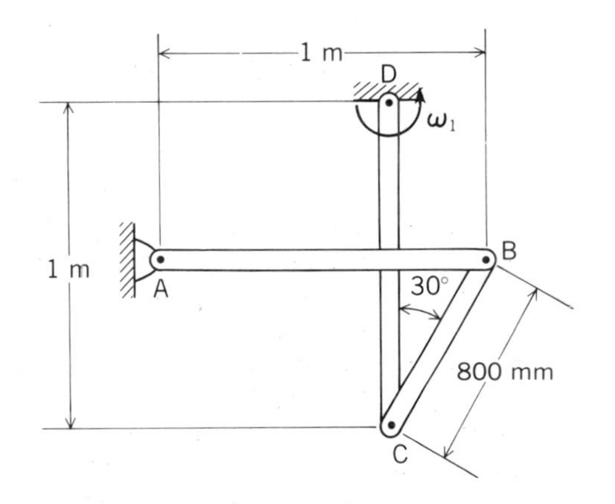

Figure P.15.32

15.33. A system of meshing gears includes gear A, which is held stationary. Rod AC rotates with a speed ω_1 of 5 rad/sec. What is the angular speed of gear C? The gears have the following diameters:

$$D_A = 600 \text{ mm}$$

$$D_B = 350 \text{ mm}$$

$$D_C = 200 \text{ mm}$$

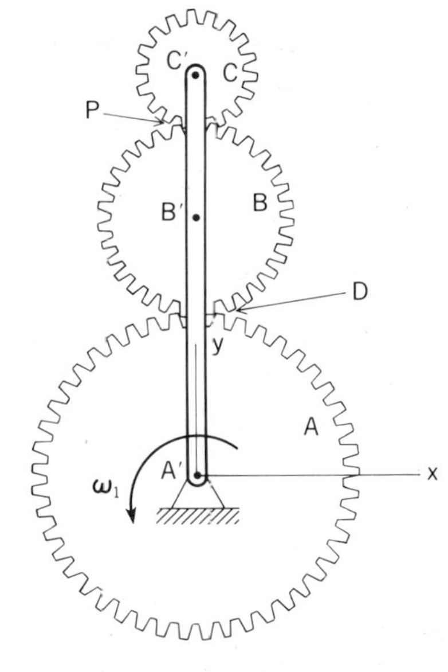

Figure P.15.33

15.34. In Problem 15.33, take $\omega_1 = 10$ rad/sec. If gear C is to translate, what angular speed should gear A have?

15.35. A bar moves in the plane of the page so that end A has a velocity of 7 m/sec and decelerates at a rate of 3.3 m/sec². What are the velocity and acceleration of point C when BA is at 30° to the horizontal?

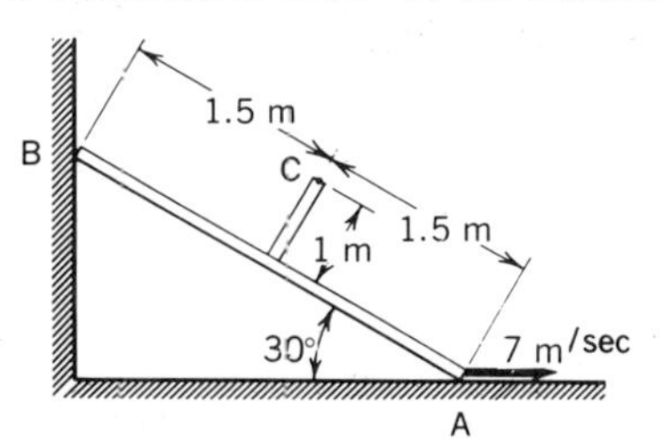

Figure P.15.35

15.36. Bar AB is rotating at a constant speed of 5 rad/sec clockwise in a device. What is the angular velocity of bar BD and body EFC? Determine the velocity of point D. (*Hint:* What is the direction of the velocity of point G?)

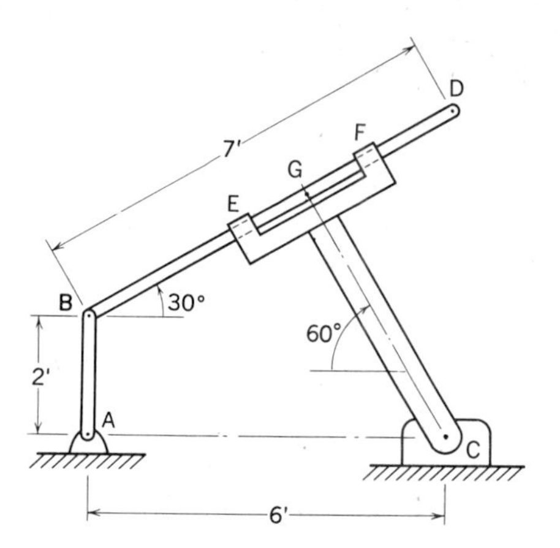

Figure P.15.36

15.37. A wheel rotates with an angular speed of 20 rad/sec. A connecting rod connects points A on the wheel with a slider at B. Compute the angular velocity of the connecting rod and the velocity of the slider when the apparatus is in the position shown in the diagram.

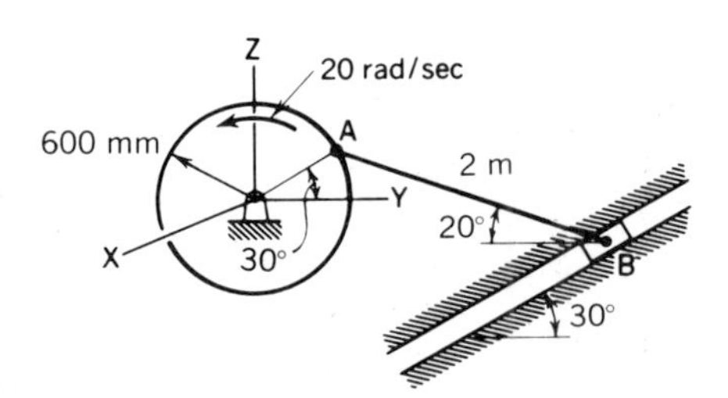

Figure P.15.37

15.38. In Problem 15.37, if $V_B = 14.30$ m/sec and $\omega_{AB} = -9.33$ rad/sec, where is the instantaneous axis of rotation of connecting rod AB?

15.39. A piston, connecting rod, and crankshaft of an engine are represented schematically. The engine is rotating at 3000 rpm. At the position shown, what is the velocity of pin A relative to the engine block and what is the angular velocity of the connecting rod AB?

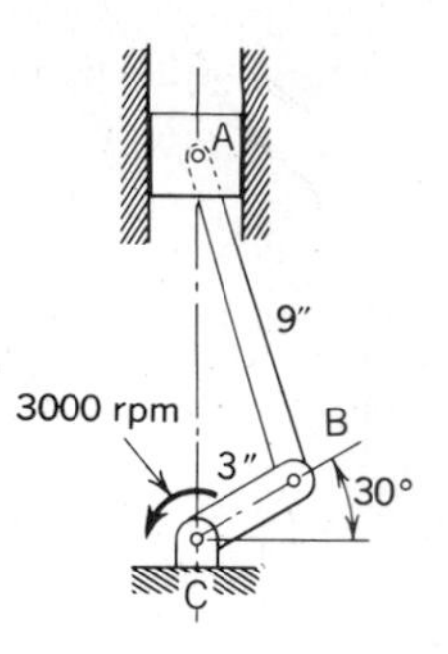

Figure P.15.39

15.40. Member AB is rotating at a constant speed of 4 rad/sec in a counterclockwise direction. What is the angular velocity of bar BC for the position shown in the diagram? What is the velocity of point D at the center of bar BC? Bar BC is 3 ft in length.

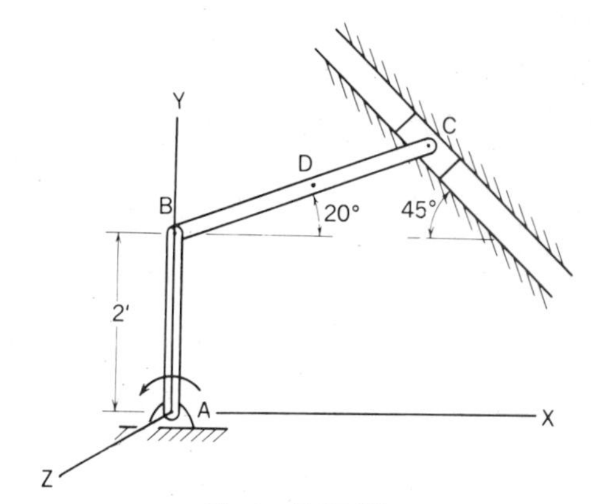

Figure P.15.40

15.41. In Problem 15.40, determine in the simplest manner the instantaneous axis of rotation for bar BC.

15.42. Suppose that bar AB of Problem 15.40 has an angular velocity of 3 rad/sec counterclockwise and a counterclockwise angular acceleration of 5 rad/sec². What is the angular acceleration of bar BC, which is 3 ft in length?

15.43. A rod moves in the plane of the paper in such a way that end A has a speed of 3 m/sec. What is the velocity of point B of the rod when the rod is inclined at 45°

to the horizontal? B is at the upper sup-

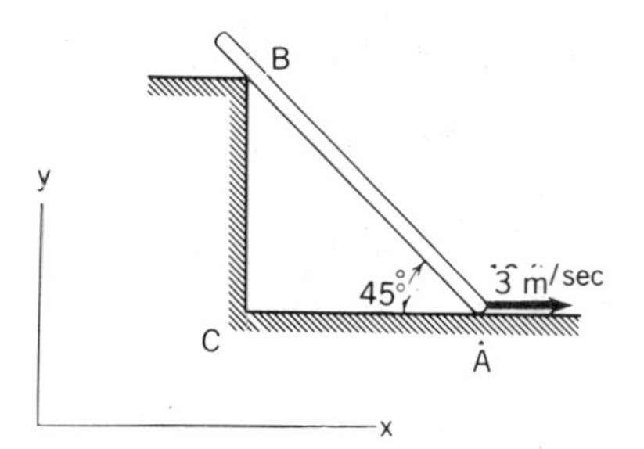

Figure P.15.43

15.44. In Problem 15.43, locate by inspection the instantaneous axis of rotation. Take $CA = 2$ m.

15.45. A rod is moving on a horizontal surface and is shown at time t. What is V_y of end A and ω of the rod at the instant shown? (*Hint:* Use the fact that the rod is inextensible.)

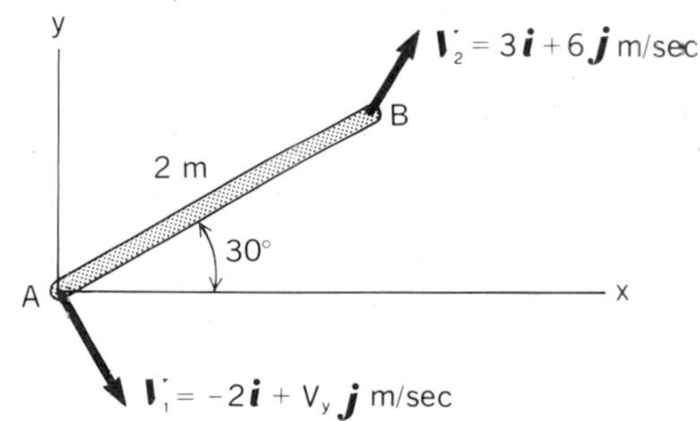

Figure P.15.45

15.46. A plate $ABCD$ moves on a horizontal surface. At time t corners A and B have the following velocities:

$$V_A = 3i + 2j \text{ m/sec}$$

$$V_B = (V_B)_x i + 5j \text{ m/sec}$$

Find the location of the instantaneous axis of rotation.

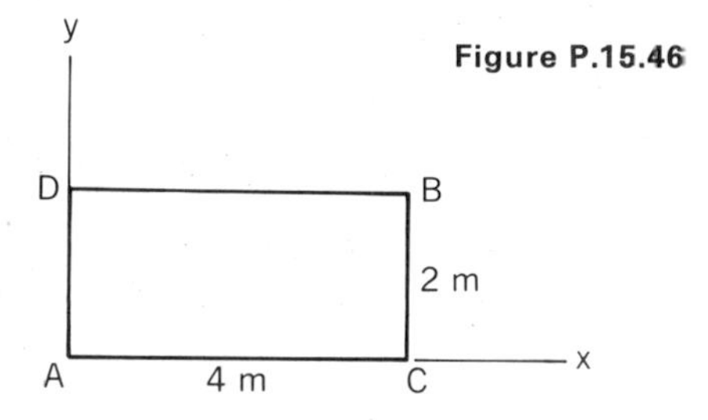

Figure P.15.46

15.47. Find the velocity and acceleration relative to the ground of pin B on the wheel. The wheel rolls without slipping. Also, find the angular velocity and angular acceleration of the slotted bar in which the pin B of the wheel slides when θ of the bar is 30°.

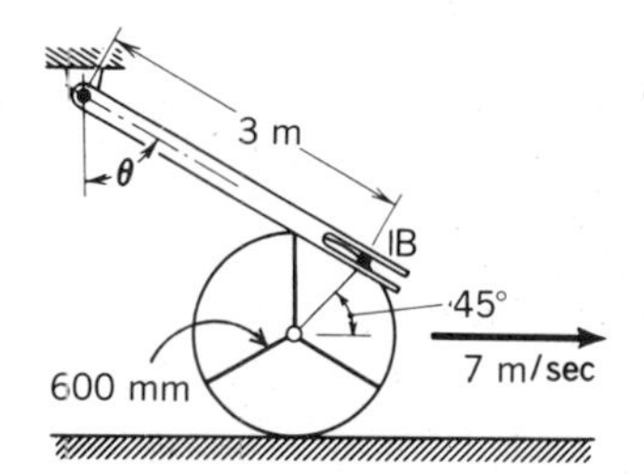

Figure P.15.47

15.48. If $\omega_1 = 5$ rad/sec and $\dot{\omega}_1 = 3$ rad/sec^2 for bar CD, compute the angular velocity and angular acceleration of the gear D relative to the ground. Solve the problem using Eqs. 15.6 and 15.7, and then check the result by considering simple circular motion of point D.

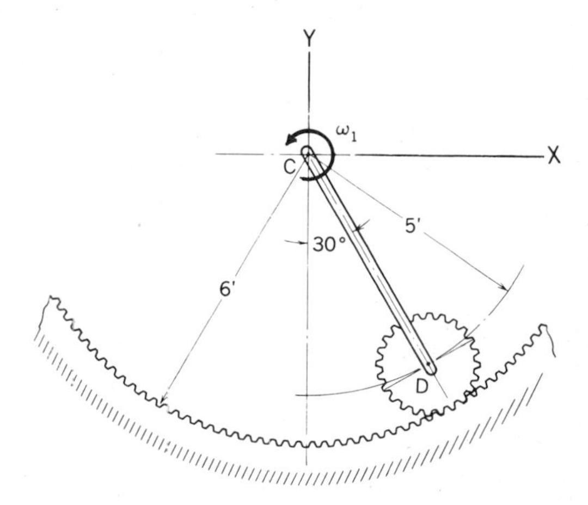

Figure P.15.48

15.49. A mechanism with two sliders is shown. Slider A at the instant of interest has a speed of 3 m/sec and is accelerating at the rate of 1.7 m/sec^2. If member AB is 2.5 m in length, what are the angular velocity and angular acceleration for this member?

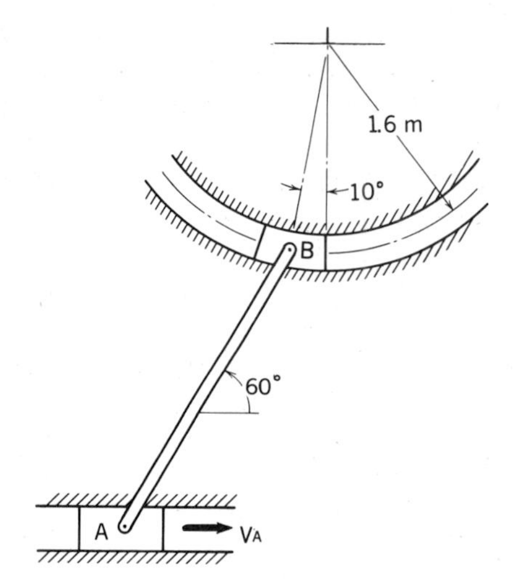

Figure P.15.49

15.50. In Problem 15.49 find the instantaneous center of rotation of bar AB if V_A is 2.7 m/sec.

15.51. The velocity of corner A of the block is known to be at time t:

$$V_A = 10\boldsymbol{i} + 4\boldsymbol{j} - 3\boldsymbol{k} \text{ m/sec}$$

The angular speed about edge $\overline{AD}$ is 2 rad/sec, and the angular speeds about the diagonals $\overline{AF}$ and $\overline{HE}$ are known to be 3 rad/sec and 6 rad/sec, respectively. What is the velocity of corner B at this instant?

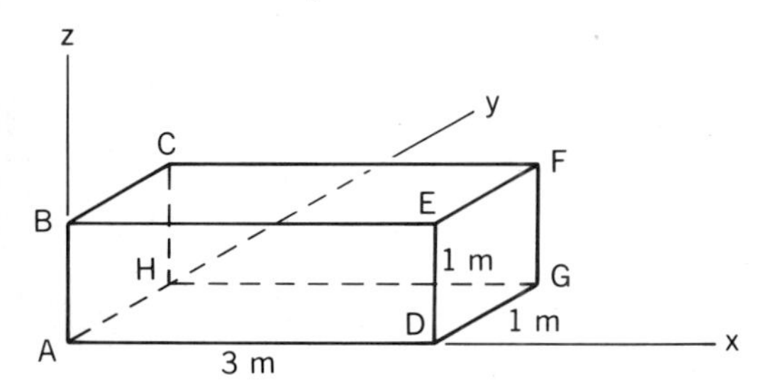

Figure P.15.51

15.52. A rigid sphere is moving in space. The velocities for two points A and B on the surface have the values at time t:

$$V_A = 6\boldsymbol{i} + 3\boldsymbol{j} + 2\boldsymbol{k} \text{ ft/sec}$$

$$V_B = (V_B)_x\boldsymbol{i} + 6\boldsymbol{j} - 4\boldsymbol{k} \text{ ft/sec}$$

The position vectors for points A and B are at time t:

$$\boldsymbol{r}_A = 10\boldsymbol{i} + 15\boldsymbol{j} + 12\boldsymbol{k} \text{ ft}$$

$$\boldsymbol{r}_B = 7\boldsymbol{i} + 20\boldsymbol{j} + 18\boldsymbol{k} \text{ ft}$$

What is the angular velocity of the sphere? At the instant of interest the sphere has zero spin about axis $\overline{AB}$.

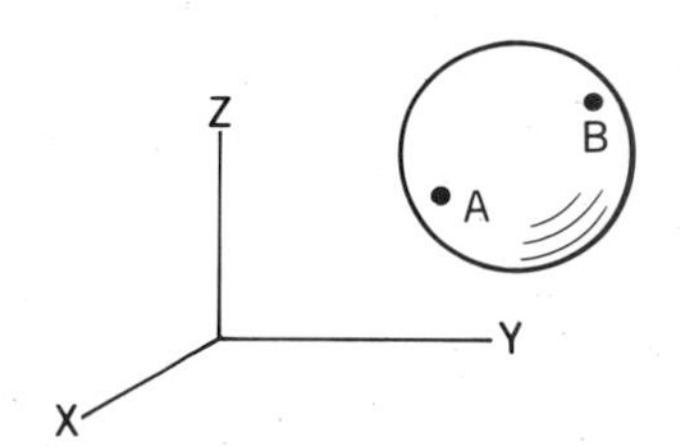

Figure P.15.52

15.53. A conveyor element moves down the incline at a speed of 5 ft/sec. A shaft and platform move with the conveyor element but have a spin of .5 rad/sec about the centerline AB. Also, the shaft swings in the XZ plane at a speed ω_1 of 1 rad/sec. What is the velocity and acceleration of point D on the platform at the instant it is in the YZ plane, as shown in the diagram? Note that at the instant of interest AB is vertical.

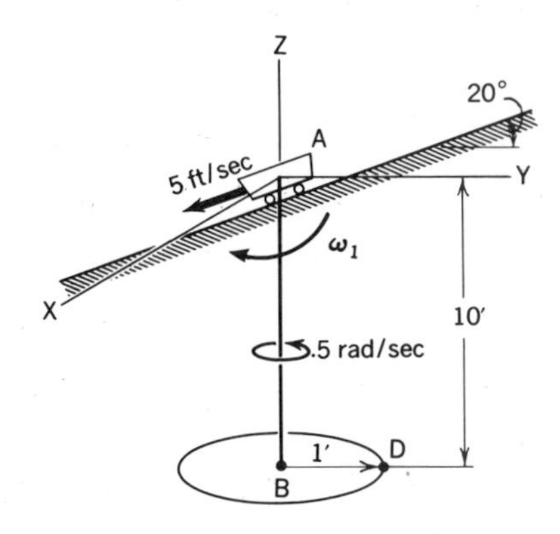

Figure P.15.53

15.54. A conveyor element moves down an incline at a speed of 15 m/sec. A plate hangs down from the conveyor element and, at the instant of interest shown in the diagram, is spinning about AB at the rate of 5 rad/sec. Also, the axis AB swings in the YZ plane at the rate ω_1 of 10 rad/sec and $\dot{\omega}_1 = 3$ rad/sec² at the instant of interest. Find the velocity and acceleration of point D at the instant shown.

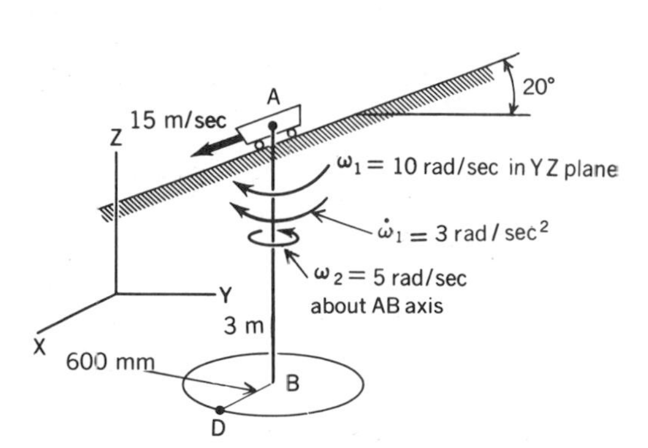

Figure P.15.54

15.55. Two stationary half-cylinders F and I are shown, on which roll cylinders G and H. If the motion is such that line BA has an angular speed of 2 rad/sec clockwise, what is the angular speed and the angular acceleration of cylinder H relative to the ground? The cylinders roll without slipping.

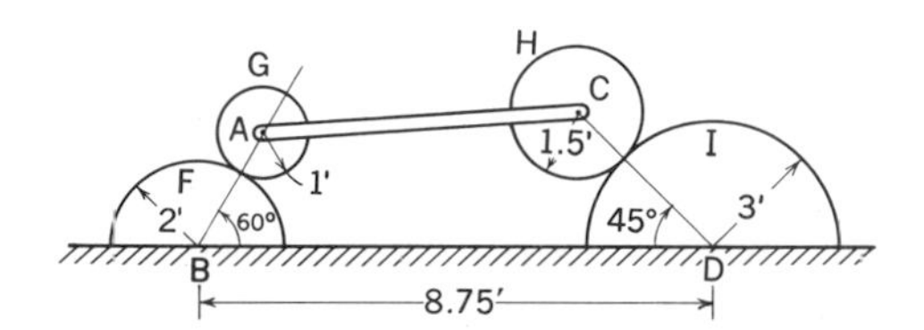

Figure P.15.55

15.56. In Problem 15.55, assume that cylinder G is rotating at a speed of 5 rad/sec clockwise as seen from the ground. What is the speed and rate of change of speed of point C relative to the ground? Assume that no slipping occurs.

15.57. A wheel D of radius $R_1 = 6$ in. rotates at a speed $\omega_1 = 5$ rad/sec as shown. A second wheel C is connected to wheel D by connecting rod AB. What is the angular speed of wheel C at the instant shown? The radius $R_2 = 12$ in. The wheels are separated by a distance $d = 2$ ft. At A and at B there are ball-and-socket connections.

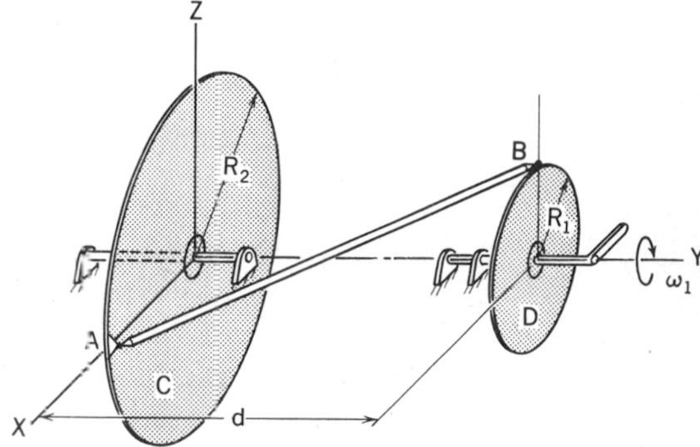

Figure P.15.57

***15.58.** A bar AB can slide along members CD and FG of a rigid structure. If A is moving at a speed of 300 mm/sec along CD toward D and is at this instant a distance of 300 mm from C, what is the speed of B along FG? At A and B there are ball-and-socket-joint connections.

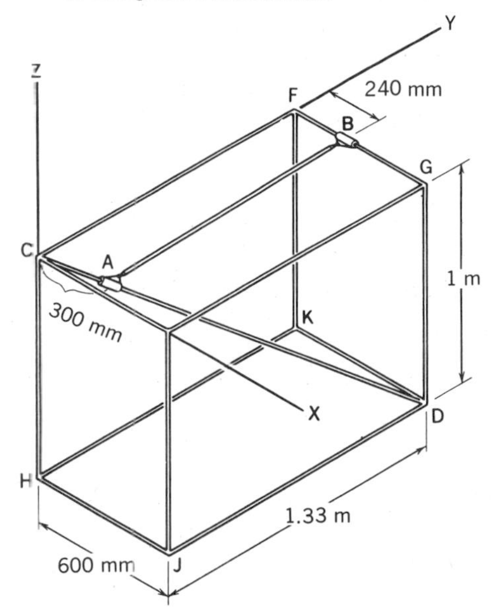

Figure P.15.58

*15.6 General Relationship Between Time Derivatives of a Vector for Different References

In Section 15.4, we considered the time derivatives of a vector A "fixed" in a reference xyz moving arbitrarily relative to XYZ. Our conclusions were:

$$\left(\frac{dA}{dt}\right)_{xyz} = 0$$

$$\left(\frac{dA}{dt}\right)_{XYZ} = \boldsymbol{\omega} \times A$$

We now wish to extend these considerations to include time derivatives of a vector A which is not necessarily fixed in reference xyz. Primarily, our intention in this section is to relate time derivatives of such vectors A as seen both from reference xyz and from XYZ, two references moving arbitrarily relative to each other.

For this purpose, consider Fig. 15.21, where we show a moving particle P with a position vector $\boldsymbol{\rho}$ in reference xyz. Reference xyz moves arbitrarily relative to reference XYZ with translational velocity $\dot{R}$ and angular velocity $\boldsymbol{\omega}$ in accordance with Chasle's theorem. We shall now form a relation between $(d\boldsymbol{\rho}/dt)_{xyz}$ and $(d\boldsymbol{\rho}/dt)_{XYZ}$. We shall then extend this result so as to relate the time derivative of any vector A as seen from any two references.

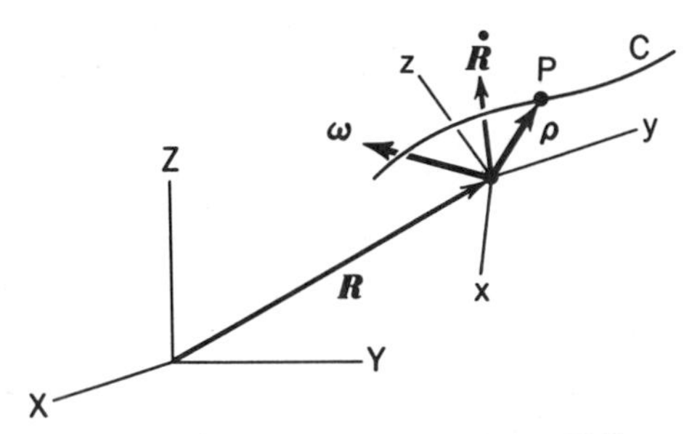

Figure 15.21. xyz moves relative to XYZ.

To reach the desired results effectively, we shall express the vector $\boldsymbol{\rho}$ in terms of components parallel to the xyz reference:

$$\boldsymbol{\rho} = x\boldsymbol{i} + y\boldsymbol{j} + z\boldsymbol{k} \tag{15.11}$$

where $\boldsymbol{i}, \boldsymbol{j}$, and $\boldsymbol{k}$ are unit vectors for reference xyz. Differentiating this equation with respect to time for the xyz reference, we have[7]

$$\left(\frac{d\boldsymbol{\rho}}{dt}\right)_{xyz} = \dot{x}\boldsymbol{i} + \dot{y}\boldsymbol{j} + \dot{z}\boldsymbol{k} \tag{15.12}$$

If we next take the derivative of $\boldsymbol{\rho}$ with respect to time for the XYZ reference, we must remember that $\boldsymbol{i}, \boldsymbol{j}$, and $\boldsymbol{k}$ of Eq. 15.11 generally will each be a function of time,

[7]Note that $\dot{x}$, $\dot{y}$, and $\dot{z}$ are time derivatives of scalars and accordingly there is no identification with any reference as far as the time derivative operation is concerned.

since these vectors will generally have some rotational motion relative to the XYZ reference. Thus, if dots are used for the time derivatives:

$$\left(\frac{d\boldsymbol{\rho}}{dt}\right)_{XYZ} = (\dot{x}\boldsymbol{i} + \dot{y}\boldsymbol{j} + \dot{z}\boldsymbol{k}) + (x\dot{\boldsymbol{i}} + y\dot{\boldsymbol{j}} + z\dot{\boldsymbol{k}}) \tag{15.13}$$

The unit vector $\boldsymbol{i}$ is a vector *fixed* in reference xyz, and accordingly $\dot{\boldsymbol{i}}$ equals $\boldsymbol{\omega} \times \boldsymbol{i}$. The same conclusions apply to $\boldsymbol{j}$ and $\boldsymbol{k}$. The last expression in parentheses can then be stated as

$$\begin{aligned} x\dot{\boldsymbol{i}} + y\dot{\boldsymbol{j}} + z\dot{\boldsymbol{k}} &= x(\boldsymbol{\omega} \times \boldsymbol{i}) - y(\boldsymbol{\omega} \times \boldsymbol{j}) + z(\boldsymbol{\omega} \times \boldsymbol{k}) \\ &= \boldsymbol{\omega} \times (x\boldsymbol{i}) + \boldsymbol{\omega} \times (y\boldsymbol{j}) + \boldsymbol{\omega} \times (z\boldsymbol{k}) \\ &= \boldsymbol{\omega} \times (x\boldsymbol{i} + y\boldsymbol{j} + z\boldsymbol{k}) = \boldsymbol{\omega} \times \boldsymbol{\rho} \end{aligned} \tag{15.14}$$

In Eq. 15.13 we can replace $(\dot{x}\boldsymbol{i} + \dot{y}\boldsymbol{j} + \dot{z}\boldsymbol{k})$ by $(d\boldsymbol{\rho}/dt)_{xyz}$, in accordance with Eq. 15.12, and $(x\dot{\boldsymbol{i}} + y\dot{\boldsymbol{j}} + z\dot{\boldsymbol{k}})$ by $\boldsymbol{\omega} \times \boldsymbol{\rho}$, in accordance with Eq. 15.14. Hence,

$$\left(\frac{d\boldsymbol{\rho}}{dt}\right)_{XYZ} = \left(\frac{d\boldsymbol{\rho}}{dt}\right)_{xyz} + \boldsymbol{\omega} \times \boldsymbol{\rho} \tag{15.5}$$

We can generalize the preceding result for any vector $\boldsymbol{A}$:

$$\boxed{\left(\frac{d\boldsymbol{A}}{dt}\right)_{XYZ} = \left(\frac{d\boldsymbol{A}}{dt}\right)_{xyz} + \boldsymbol{\omega} \times \boldsymbol{A}} \tag{15.16}$$

where, you must remember, $\boldsymbol{\omega}$ is the angular velocity of the xyz reference relative to the XYZ reference. Note that Eq. 15.1 is a special case of Eq. 15.16 since for $\boldsymbol{A}$ fixed in xyz, $(d\boldsymbol{A}/dt)_{xyz} = \boldsymbol{0}$. We shall have much use for this relationship in succeeding sections.

*15.7 Relationship Between Velocities of a Particle for Different References

We shall now define the velocity of a particle again in the presence of several references:

> The velocity of a particle relative to a reference is the derivative as seen from this reference of the position vector of the particle in the reference.

In Fig. 15.21, the velocities of the particle P relative to the XYZ and the xyz references are, respectively,[8]

$$\boldsymbol{V}_{XYZ} = \left(\frac{d\boldsymbol{r}}{dt}\right)_{XYZ}, \qquad \boldsymbol{V}_{xyz} = \left(\frac{d\boldsymbol{\rho}}{dt}\right)_{xyz} \tag{15.17}$$

We should point out that $\boldsymbol{V}_{XYZ}$ can be expressed in components parallel to the xyz

[8]Generally, we have employed $\boldsymbol{r}$ as a position vector and $\boldsymbol{\rho}$ as a displacement vector. With two references, we shall often used $\boldsymbol{\rho}$ to denote a position vector for one of the references.

reference at any time t, while $\boldsymbol{V}_{xyz}$ may be expressed in components parallel to the XYZ reference at any time t.

Now, we shall relate these velocities by first noting that

$$\boldsymbol{r} = \boldsymbol{R} + \boldsymbol{\rho} \tag{15.18}$$

Differentiating with respect to time for the XYZ reference, we have

$$\left(\frac{d\boldsymbol{r}}{dt}\right)_{XYZ} \equiv \boldsymbol{V}_{XYZ} = \left(\frac{d\boldsymbol{R}}{dt}\right)_{XYZ} + \left(\frac{d\boldsymbol{\rho}}{dt}\right)_{XYZ} \tag{15.19}$$

The term $(d\boldsymbol{R}/dt)_{XYZ}$ is clearly the velocity of the origin of the xyz reference relative to the XYZ reference, according to our definitions, and we denote this velocity as $\dot{\boldsymbol{R}}$. The term $(d\boldsymbol{\rho}/dt)_{XYZ}$ can be replaced, by use of Eq. 15.15, in which $(d\boldsymbol{\rho}/dt)_{xyz}$ is the velocity of the particle relative to the xyz reference. Denoting $(d\boldsymbol{\rho}/dt)_{xyz}$ simply as $\boldsymbol{V}_{xyz}$, we find that the foregoing equation then becomes the desired relation:

$$\boxed{\boldsymbol{V}_{XYZ} = \boldsymbol{V}_{xyz} + \dot{\boldsymbol{R}} + \boldsymbol{\omega} \times \boldsymbol{\rho}} \tag{15.20}$$

We shall adopt the understanding that $\boldsymbol{\omega}$ *without* subscripts represents the angular velocity of xyz relative to XYZ. This $\boldsymbol{\omega}$ always goes into the last expression of Eq. 15.20.

Note that in Sections 15.4 and 15.5 we considered the motion of *two* particles in a rigid body as seen from a single reference. Now we are considering the motion of a *single* particle as seen from *two* references.

The multireference approach can be very useful. For instance, we could know the motion of a particle relative to some device, such as a rocket, to which we attach a reference xyz. Furthermore, from telemetering devices, we know the translational and rotational motion (Chasle's theorem) of the rocket (and hence xyz) relative to an inertial reference XYZ. It is often important to know the motion of the aforementioned particle relative directly to the inertial reference. The multireference approach clearly is invaluable for such problems.

We now illustrate the use of Eq. 15.20. We shall proceed in a particular methodical way which we encourage you to follow in your homework problems. In these examples, we remind you that $\boldsymbol{\omega}$ without subscripts refers to the angular velocity of xyz relative to XYZ. And, we shall use the dot to represent differentiation with respect only to XYZ.

EXAMPLE 15.8

An airplane moving at 200 ft/sec is undergoing a roll of 2 rad/min (Fig. 15.22). When the plane is horizontal, an antenna is moving out at a speed of 8 ft/sec relative to the plane and is at a position of 10 ft from the centerline of the plane. If we assume that the axis of roll corresponds to the centerline, what is the velocity of the antenna end relative to the ground when the plane is horizontal?

A *stationary reference XYZ on the ground* is shown in the diagram. A moving reference *xyz is fixed to the plane* with the x axis along the axis of roll and the y axis collinear with the antenna. We can say for this system:

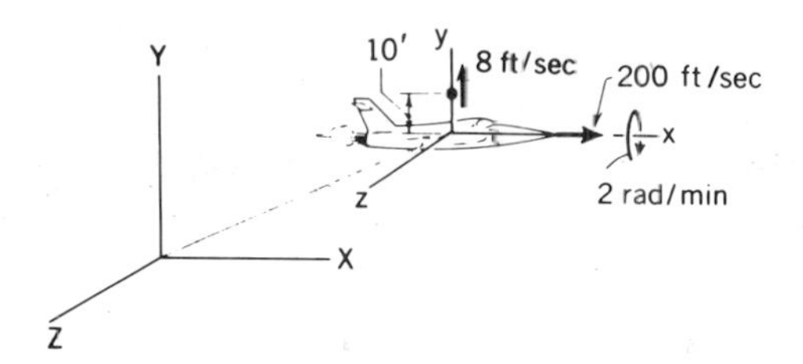

Figure 15.22. *xyz* fixed to plane; *XYZ* fixed to ground.

A. Motion of particle (antenna end) relative to xyz[9]

$$\boldsymbol{\rho} = 10\boldsymbol{j} \text{ ft}$$
$$V_{xyz} = 8\boldsymbol{j} \text{ ft/sec}$$

B. Motion of xyz **(moving reference) relative to** XYZ **(fixed reference)**

$$\dot{\boldsymbol{R}} = 200\boldsymbol{i} \text{ ft/sec}$$
$$\boldsymbol{\omega} = -\tfrac{2}{60}\boldsymbol{i} = -\tfrac{1}{30}\boldsymbol{i} \text{ rad/sec}$$

We now employ Eq. 15.20 to get

$$\begin{aligned} V_{XYZ} = V_{xyz} &= \dot{\boldsymbol{R}} + \boldsymbol{\omega} \times \boldsymbol{\rho} \\ &= 8\boldsymbol{j} + 200\boldsymbol{i} + \left(-\frac{\boldsymbol{i}}{30}\right) \times (10\boldsymbol{j}) \\ &= 200\boldsymbol{i} + 8\boldsymbol{j} - \tfrac{1}{3}\boldsymbol{k} \text{ ft/sec} \end{aligned}$$

EXAMPLE 15.9

A tank is moving up an incline with a speed of 10 km/hr in Fig. 15.23. The turret is rotating at a speed ω_1 of 2 rad/sec relative to the tank, and the gun barrel is being raised (rotating) at a speed ω_2 of .3 rad/sec relative to the turret. What is the velocity of point A of the gun barrel relative to the tank and relative to the ground? The gun barrel is 3 m in length. We proceed as follows (see Fig. 15.24).

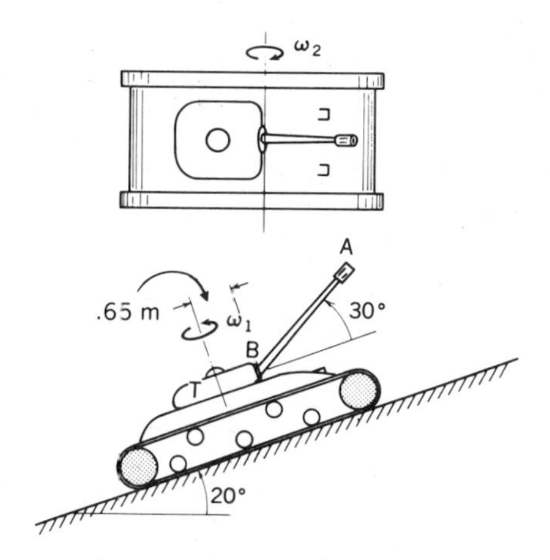

Figure 15.23. Tank with turret and gun barrel in motion.

Fix xyz to turret.
Fix XYZ to tank.

A. Motion of particle relative to xyz

$$\boldsymbol{\rho} = 3(\cos 30°\boldsymbol{j} + \sin 30°\boldsymbol{k}) = 2.60\boldsymbol{j} + 1.50\boldsymbol{k} \text{ m}$$

[9]Note that since the corresponding axes of the references are parallel to each other at the instant of interest, the unit vectors $\boldsymbol{i}, \boldsymbol{j}$, and $\boldsymbol{k}$ apply to either reference at the instant of interest.

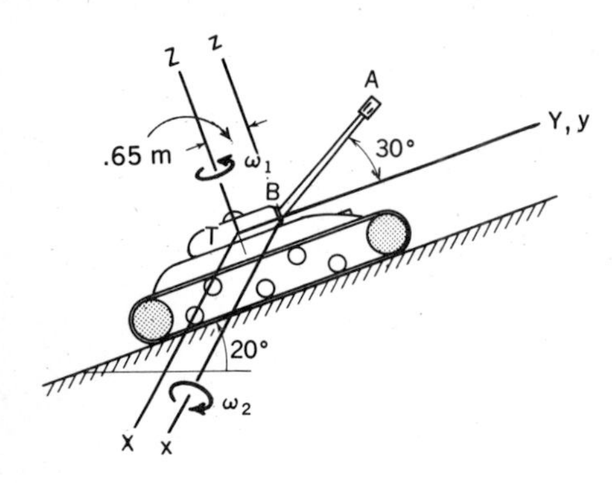

Figure 15.24. *xyz* fixed to turret; *XYZ* fixed to tank.

Since $\boldsymbol{\rho}$ is fixed in the gun barrel, which has an angular velocity $\boldsymbol{\omega}_2$ relative to xyz, we have

$$\boldsymbol{V}_{xyz} = \left(\frac{d\boldsymbol{\rho}}{dt}\right)_{xyz} = \boldsymbol{\omega}_2 \times \boldsymbol{\rho} = (-3\boldsymbol{i}) \times (2.60\boldsymbol{j} + 1.5\boldsymbol{k})$$

$$= -.780\boldsymbol{k} + .45\boldsymbol{j} \text{ m/sec}$$

B. Motion of *xyz* relative to *XYZ*

$$\boldsymbol{R} = .65\boldsymbol{j}$$

Since $\boldsymbol{R}$ is fixed in the turret, which is rotating with angular speed $\boldsymbol{\omega}_1$ relative to XYZ, we have

$$\dot{\boldsymbol{R}} = \boldsymbol{\omega}_1 \times \boldsymbol{R} = 2\boldsymbol{k} \times .65\boldsymbol{j} = -1.3\boldsymbol{i} \text{ m/sec}$$

$$\boldsymbol{\omega} = \boldsymbol{\omega}_1 = 2\boldsymbol{k} \text{ rad/sec}$$

We can now substitute into the basic equation relating $\boldsymbol{V}_{xyz}$ to $\boldsymbol{V}_{XYZ}$. That is,

$$\begin{aligned}\boldsymbol{V}_{XYZ} &= \boldsymbol{V}_{xyz} + \dot{\boldsymbol{R}} + \boldsymbol{\omega} \times \boldsymbol{\rho} \\ &= (-.780\boldsymbol{k} + .45\boldsymbol{j}) - 1.3\boldsymbol{i} + (2\boldsymbol{k}) \times (2.60\boldsymbol{j} + 1.50\boldsymbol{k}) \\ &= -6.5\boldsymbol{i} + .45\boldsymbol{j} - .780\boldsymbol{k} \text{ m/sec}\end{aligned}$$

This result is the desired velocity of A relative to the tank. Since the tank is moving with a speed of (10)(1000)/(3600) = 2.78 m/sec relative to the ground, we can say that A has a velocity relative to the ground given as

$$\begin{aligned}\boldsymbol{V}_{\text{ground}} &= \boldsymbol{V}_{XYZ} + 2.78\boldsymbol{j} \\ &= -6.5\boldsymbol{i} + 3.23\boldsymbol{j} - .780\boldsymbol{k} \text{ m/sec}\end{aligned}$$

EXAMPLE 15.10

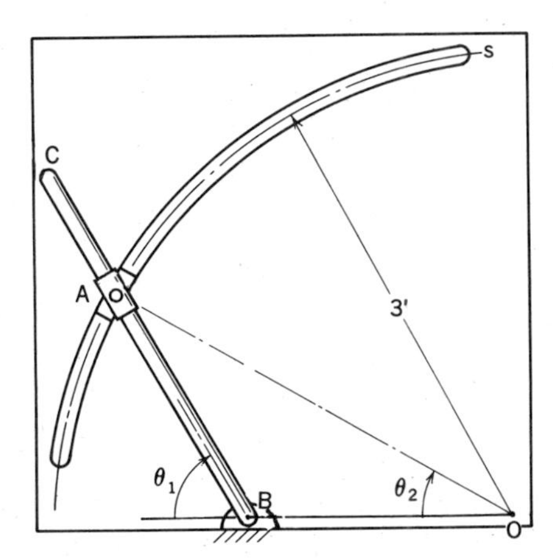

Figure 15.25. Slider mechanism.

A rod CB rotates in a clockwise direction at the rate $\dot{\theta}_1$ of 2 rad/sec in Fig. 15.25. At A, a collar slides on this rod but is also connected by a pin to a slider moving in a circular slot having a radius of 3 ft. When $\theta_1 = 60°$, determine the speed of A relative to the ground and relative to the rod. The angle θ_2 at the instant of interest is given as 30°.

We need to know the distance $\overline{AB}$. From Fig. 15.25, we consider triangle AOB. Using the law of sines, we can say:

$$\frac{\overline{AO}}{\sin(\pi - \theta_1)} = \frac{\overline{AB}}{\sin\theta_2}$$

$$\overline{AB} = \frac{3 \sin 30°}{\sin 120°} = 1.732 \text{ ft}$$

We now are ready to proceed with the analysis.

Fix xyz to rod BC at A. Fix XYZ to ground.

These references are shown in Fig. 15.26. Note, $\boldsymbol{i}, \boldsymbol{j}$ and $\boldsymbol{k}$ correspond to axes XYZ.

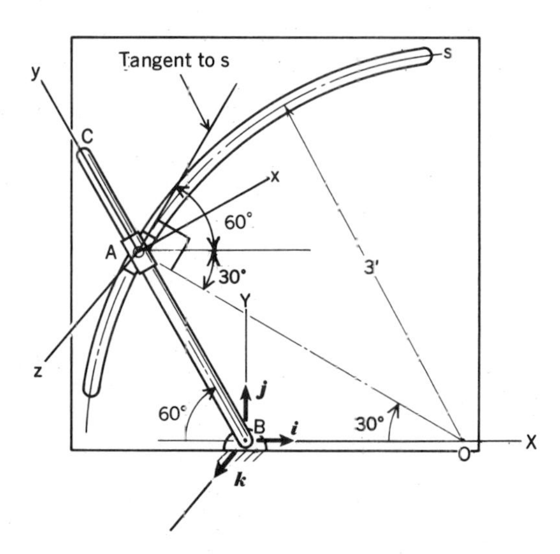

Figure 15.26. xyz fixed to rod BC at A; XYZ fixed to ground.

A. Motion of slider relative to xyz

$$\boldsymbol{\rho} = \boldsymbol{0}$$

$$\boldsymbol{V}_{xyz} = V_{xyz}(-\cos 60°\boldsymbol{i} + \sin 60°\boldsymbol{j})$$

$$= V_{xyz}(-.500\boldsymbol{i} + .866\boldsymbol{j}) \text{ ft/sec}$$

This result stems from the fact that, as seen from xyz, the slider at all times must be moving in the direction of the rod BC.

B. Motion of xyz relative to XYZ

$$\boldsymbol{R} = 1.732(-\cos 60°\boldsymbol{i} + \sin 60°\boldsymbol{j})$$

$$= -.866\boldsymbol{i} + 1.500\boldsymbol{j} \text{ ft}$$

$$\dot{\boldsymbol{R}} = (\dot{\theta}_1\boldsymbol{k}) \times \boldsymbol{R} = (-2\boldsymbol{k}) \times (-.866\boldsymbol{i} + 1.500\boldsymbol{j})$$

$$= 1.732\boldsymbol{j} + 3\boldsymbol{i} \text{ ft/sec}$$

$$\boldsymbol{\omega} = -2\boldsymbol{k} \text{ rad/sec}$$

Hence,

$$\boldsymbol{V}_{XYZ} = \boldsymbol{V}_{xyz} + \dot{\boldsymbol{R}} + \boldsymbol{\omega} \times \boldsymbol{\rho}$$

Noting that V_{XYZ} must be tangent to circular slot (see Fig. 15.26), we have

$$V_{XYZ}(\cos 60°\boldsymbol{i} + \sin 60°\boldsymbol{j})$$
$$= V_{xyz}(-.500\boldsymbol{i} + .866\boldsymbol{j}) + 1.732\boldsymbol{j} + 3\boldsymbol{i} + (-2\boldsymbol{k}) \times (\boldsymbol{0})$$

Therefore,

$$.500V_{XYZ}\boldsymbol{i} + .866V_{XYZ}\boldsymbol{j} = -.500V_{xyz}\boldsymbol{i} + .866V_{xyz}\boldsymbol{j} + 1.732\boldsymbol{j} + 3\boldsymbol{i}$$

The scalar equations are

$$.500V_{XYZ} = -.500V_{xyz} + 3$$
$$.866V_{XYZ} = .866V_{xyz} + 1.732$$

Solving simultaneously, we get

$$V_{XYZ} \equiv \text{speed relative to ground} = 4 \text{ ft/sec}$$
$$V_{xyz} \equiv \text{speed relative to rod} = 2 \text{ ft/sec}$$

Problems

15.59. A space laboratory, in order to simulate gravity, rotates relative to inertial reference XYZ at a rate ω_1. For occupant A to feel comfortable, what should ω_1 be? Clearly, at the center room B, there is close to zero gravity for zero-g experiments. A conveyor along one of the spokes transports items from the living quarters at the periphery to the zero-gravity laboratory at the center. In particular, a particle D has a velocity toward B of 5 m/sec relative to the space station. What is its velocity relative to the inertial reference XYZ?

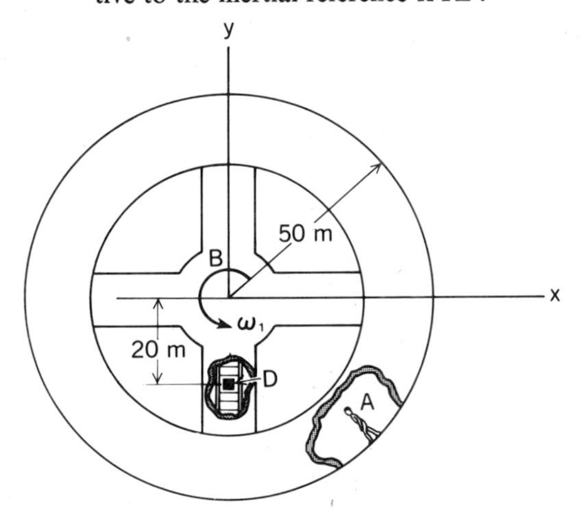

Figure P.15.59

15.60. Bodies a and b slide away from each other each with a constant velocity of 5 ft/sec along the axis C–C mounted on a platform. The platform rotates relative tô the ground reference XYZ at an angular velocity of 10 rad/sec about axis E–E and has an angular acceleration of 5 rad/sec² relative to the ground reference XYZ at the time when the bodies are at a distance $r = 3$ ft from E–E. Determine the velocity of particle b relative to the ground reference.

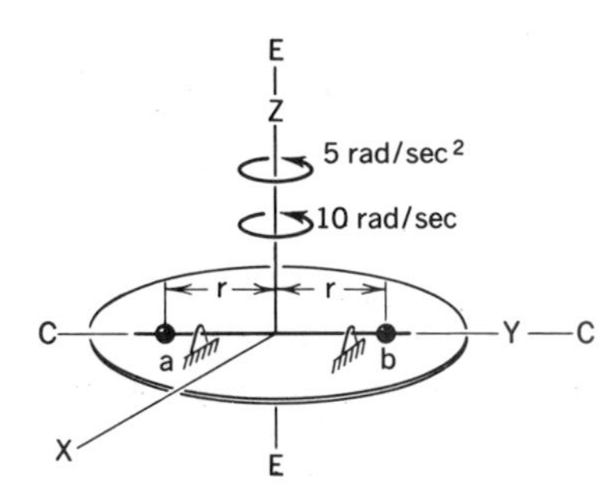

Figure P.15.60

15.61. A particle rotates at a constant angular speed of 10 rad/sec on a platform, while the platform rotates with a constant angular speed of 50 rad/sec about axis A–A. What is the velocity of the particle P at the instant the platform is in the XY plane and the radius vector to the particle forms an angle of 30° with the Y axis as shown?

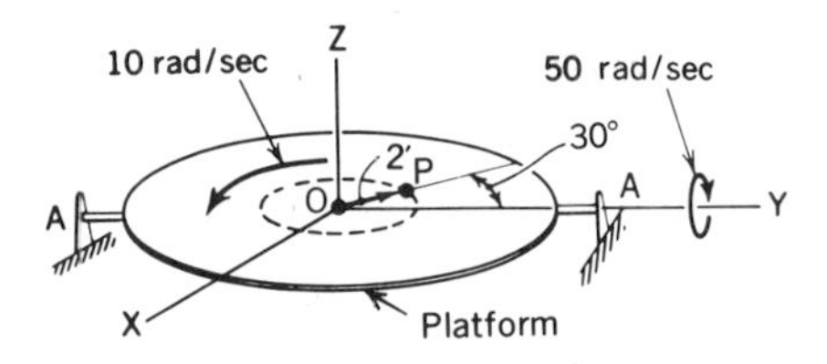

Figure P.15.61

15.62. A platform A is rotating with constant angular speed ω_1 of 1 rad/sec. A second platform B rides on A, contains a row of test tubes, and has a constant angular speed ω_2 of .2 rad/sec relative to the platform A. A third platform C is in no way connected with platforms A and B. E on platform C is positioned above A and B and carries dispensers of chemicals which are electrically operated at proper times to dispense drops into the test tubes held by B below. What should the angular speed ω_3 be for platform E if it is to dispense a drop of chemical having a zero tangential velocity relative to the test tube below?

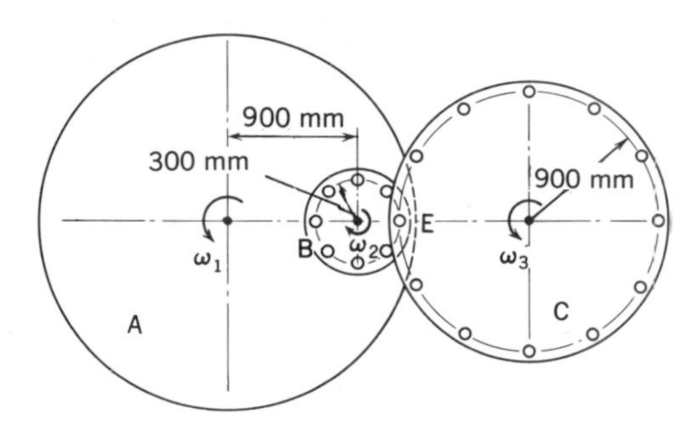

Figure P.15.62

15.63. In an amusement park ride, the cockpit containing two occupants can rotate at an angular speed ω_1 relative to the main arm. The arm can rotate with angular speed ω_2 relative to the ground. For the position shown in the diagram and for $\omega_1 = 2$ rad/sec and $\omega_2 = .2$ rad/sec, find the velocity of point A (corresponding to the position of the eyes of an occupant) relative to the ground.

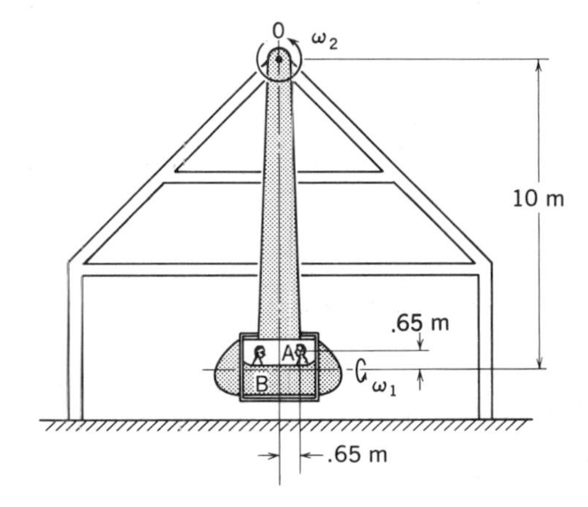

Figure P.15.63

15.64. A water sprinkler has .4 cfs (cubic ft/sec) of water fed into the base. The sprinkler turns at the rate ω_1 of 1 rad/sec. What is the speed of the jet of water relative to the ground at the exits? The outlet area of the nozzle cross section is .75 in^2. (*Hint:* The volume of flow through a cross section is VA, where V is the velocity and A is the area of the cross section.)

Figure P.15.64

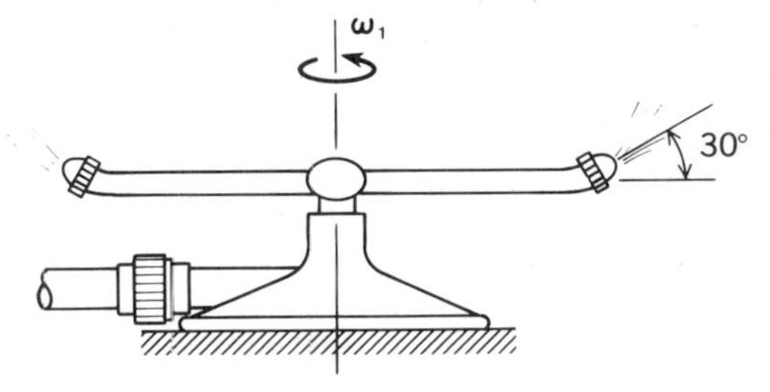

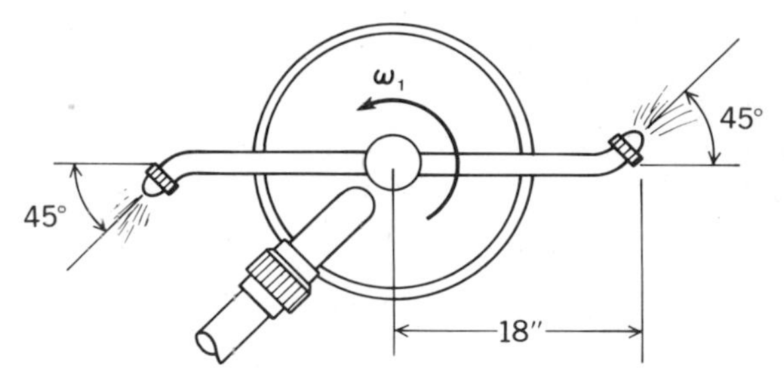

15.65. We can show that Eq. 15.6 is actually a special case of Eq. 15.20. For this purpose, consider a rigid body moving relative to *XYZ*. Choose two points *a* and *b* in the body. The body has a translational velocity corresponding to the velocity of point *a* and a rotational velocity $\boldsymbol{\omega}$ as shown in the diagram. Now embed a reference *xyz* into the body with origin at point *a*. Next, use this diagram and consider point *b* to show that Eq. 15.20 can be reformulated to be identical to Eq. 15.6.

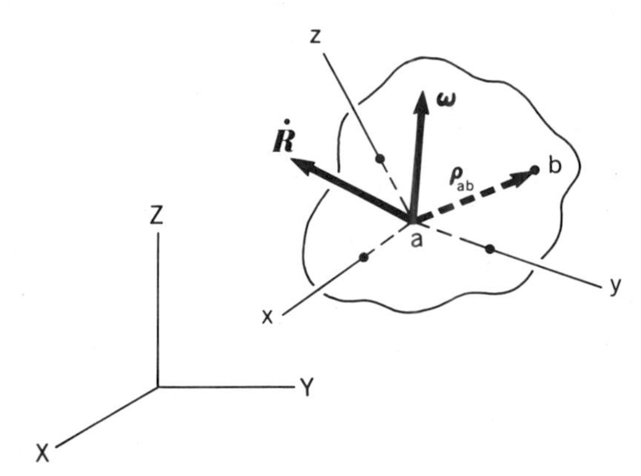

Figure P.15.65

15.66. A simple-impulse type of turbomachine called a *Pelton* water wheel has a single jet of water issuing out of a nozzle and impinging on the system of buckets attached to a wheel. The runner, which is the assembly of buckets and wheel, has a radius of *r* to the center of the buckets. The shape of the bucket is also shown where a horizontal midsection of the bucket has been taken. Note that the jet is split in two parts by the bucket and is rotated relative to the bucket in the horizontal plane as measured by β. If we neglect gravity and friction, the speed of the water relative to the bucket is unchanged during the action. Suppose that 8 liters of water per second flow through the nozzle, whose cross-sectional area at the exit is 2000 mm². If $r = 1$ m, what should ω_1 be (in rpm) for the water to have zero velocity relative to the ground in the *Y* direction when it comes off the bucket? Take $\beta = 10°$. (Why is it desirable to have the exit velocity equal to zero in the *Y* direction?) See the hint of Problem 15.64.

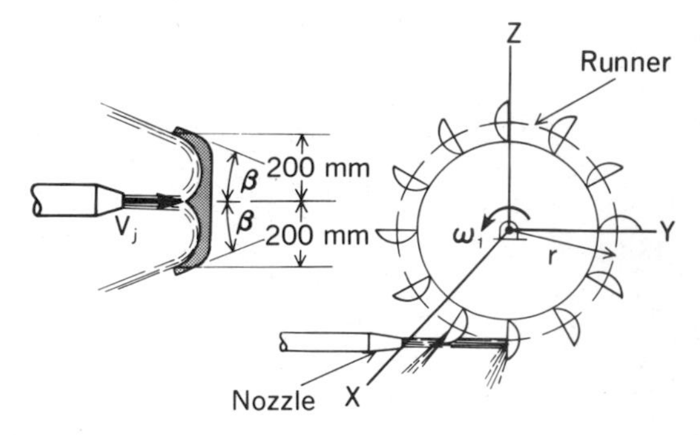

Figure P.15.66

15.67. A propeller-driven airplane is moving at a speed of 130 km/hr. Also, it is undergoing a yaw rotation of $\frac{1}{4}$ rad/sec and is simultaneously undergoing a loop rotation of $\frac{1}{4}$ rad/sec. The propeller is rotating at the rate of 100 rpm with a sense in the positive *Y* direction. What is the velocity of the tip of the propeller *a* relative to the ground at the instant that the plane is horizontal as shown? The propeller is 3 m in total length and at the instant of interest the blade is in a vertical position.

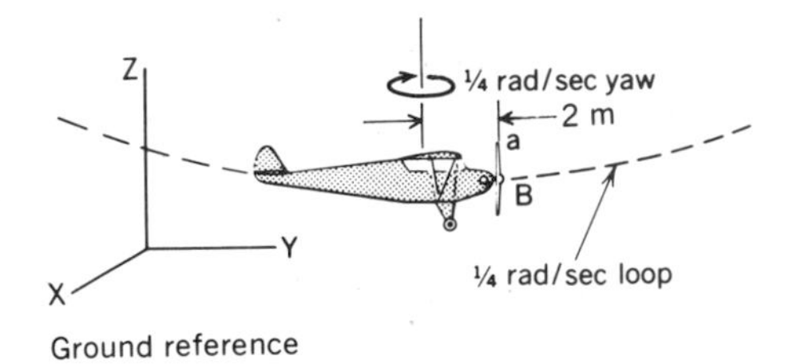

Figure P.15.67

15.68. A crane moves to the right at a speed of 5 km/hr. The boom *OB*, which is 15 m long, is being raised at an angular speed ω_2 relative to the cab of .4 rad/sec, while the cab is rotating at an angular speed ω_1 of .2 rad/sec relative to the base. What is the velocity of pin B relative to the ground at the instant when OB is at an angle of 35° with the ground? The axis of rotation O of the boom is 1 m from the axis of rotation A–A of the cab, as shown in the diagram.

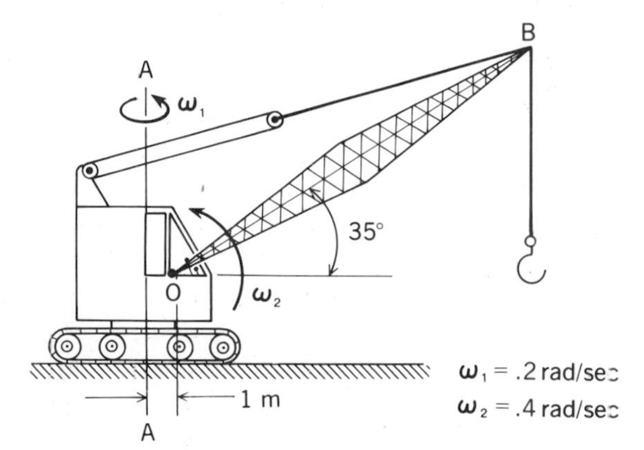

Figure P.15.68

15.69. A power shovel main arm AC rotates with angular speed ω_1 of .3 rad/sec relative to the cab. Arm ED rotates at a speed ω_2 of .4 rad/sec relative to the main arm AC. The cab rotates about axis A–A at a speed ω_3 of .15 rad/sec relative to the tracks which are stationary. What is the velocity of point D, the center of the shovel, at the instant of interest shown in the diagram? AB has a length of 5 m and BD has a length of 4 m.

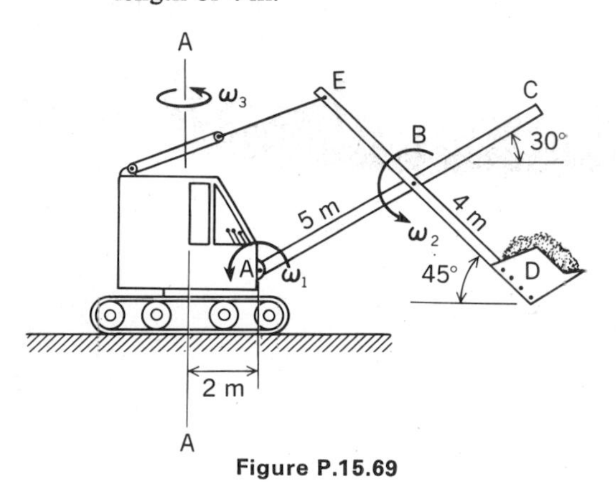

Figure P.15.69

15.70. An antiaircraft gun is shown in action. The values of ω_1 and ω_2 are .3 rad/sec and .6 rad/sec, respectively. At the instant shown, what is the velocity of a projectile *normal* to the direction of the gun barrel when it just leaves the gun barrel as seen from the ground?

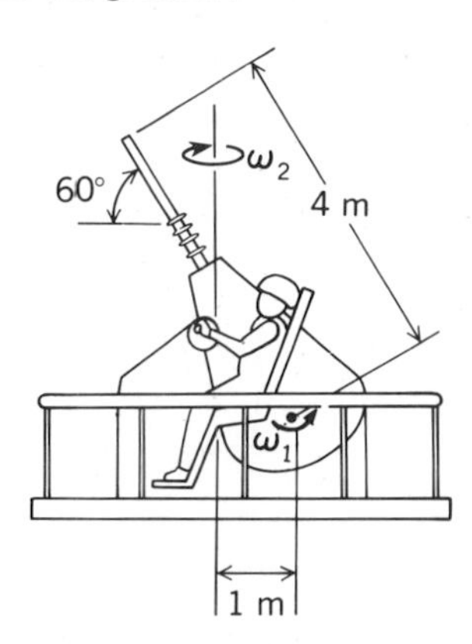

Figure P.15.70

1571. A cone is rolling without slipping about the Z axis such that its centerline rotates at the rate ω_1 of 5 rad/sec. Use a multireference approach to determine the total angular velocity of the body relative to the ground.

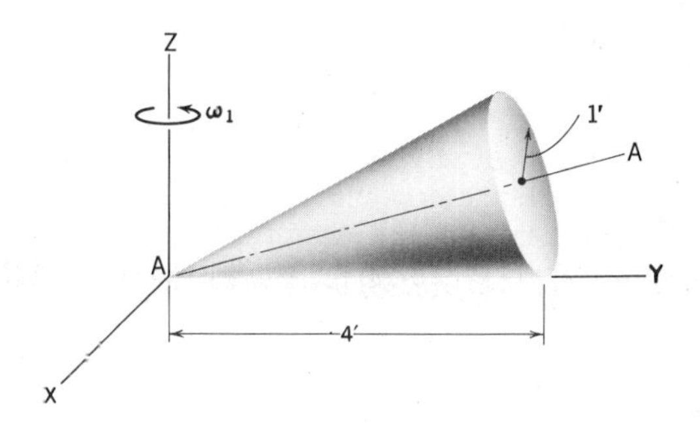

Figure P.15.71

15.72. Find the velocity of gear tooth A relative to the ground reference XYZ. Note that ω_1 and ω_2 are both relative to the ground. Bevel gear A is free to rotate in the collar at C. Take $\omega_1 = 2$ rad/sec and $\omega_2 = 4$ rad/sec.

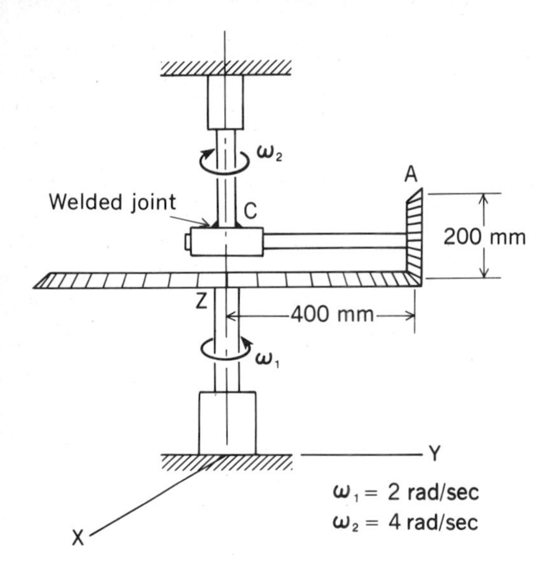

Figure P.15.72

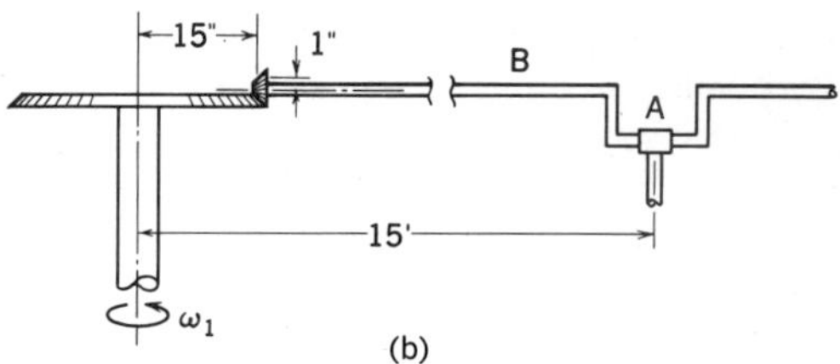

Figure P.15.73

15.73. In a merry-go-round, the main platform rotates at the rate ω_1 of 10 revolutions per minute. A set of 45° bevel gears causes B to rotate at an angular speed $\dot{\theta}$ relative to the platform. The horse is mounted on AB, which slides in a slot at C and is moved at A by shaft B, as indicated in the diagram, where part of the merry-go-round is shown. If $AB = 1$ ft and $AC = 15$ ft, compute the velocity of point C relative to the platform. Then, compute the velocity of point C relative to the ground. Take $\theta = 45°$ at the instant of interest. What is the angular velocity of the horse relative to the platform and relative to the ground at the instant of interest?

15.74. Rod BO rotates at a constant angular speed $\dot{\theta}$ of 5 rad/sec clockwise. A collar A on the rod is pinned to a slider C, which moves in the groove shown in the diagram. When $\theta = 60°$, compute the speed of the collar A relative to the ground. What is the speed of collar A relative to the rod?

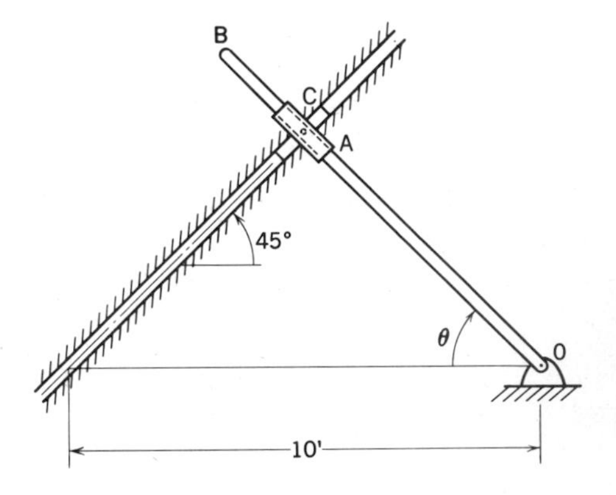

Figure P.15.74

15.75. Work Problem 15.74 assuming that pin O is on rollers moving to the right at a speed of 3 ft/sec relative to the ground. In addition, OB rotates at a constant angular speed $\dot{\theta}$ of 5 rad/sec clockwise.

15.76. Rod AD rotates at a constant speed $\dot{\theta}$ of 2 rad/sec. Collar C on the rod DA is constrained to move in the circular groove shown in the diagram. When the rod is at the position shown, compute the speed of collar C relative to the ground. What is the speed of collar C relative to the rod AD? Point A is stationary.

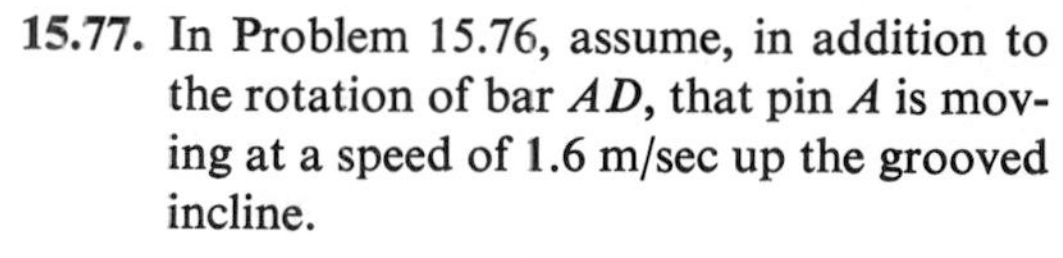

15.77. In Problem 15.76, assume, in addition to the rotation of bar AD, that pin A is moving at a speed of 1.6 m/sec up the grooved incline.

15.78. Rod AC is connected to a gear D and is guided by a bearing B. Bearing B is stationary but can rotate in the plane of the gears. If the angular speed of AC is 5 rad/sec clockwise, what is the angular speed of gear D relative to the ground? The diameter of gear D is 2 ft.

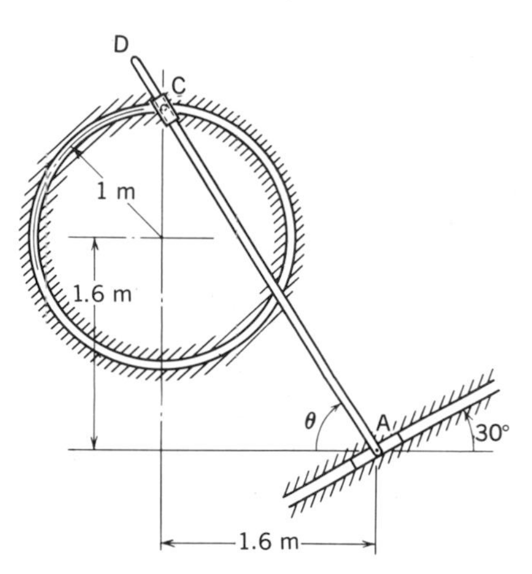

Figure P.15.76

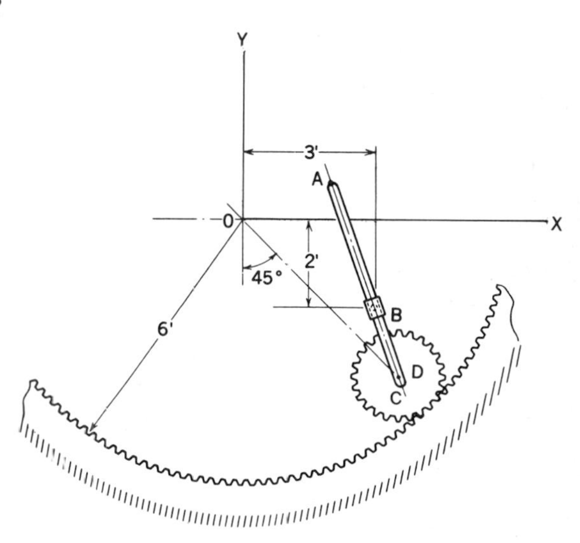

Figure P.15.78

*15.8 Acceleration of a Particle for Different References

The acceleration of a particle relative to a coordinate system is simply the time derivative, as seen from the coordinate system, of the velocity relative to the coordinate system. Thus, observing Fig. 15.27, we can say:

$$\begin{aligned} \boldsymbol{a}_{XYZ} &= \left(\frac{d}{dt}\boldsymbol{V}_{XYZ}\right)_{XYZ} = \left(\frac{d^2\boldsymbol{r}}{dt^2}\right)_{XYZ} \\ \boldsymbol{a}_{xyz} &= \left(\frac{d}{dt}\boldsymbol{V}_{xyz}\right)_{xyz} = \left(\frac{d^2\boldsymbol{\rho}}{dt^2}\right)_{xyz} \end{aligned} \tag{15.21}$$

This notation may at first seem cumbersome to you, but it will soon be simplified.

Let us now relate the acceleration vectors of a particle for two references moving

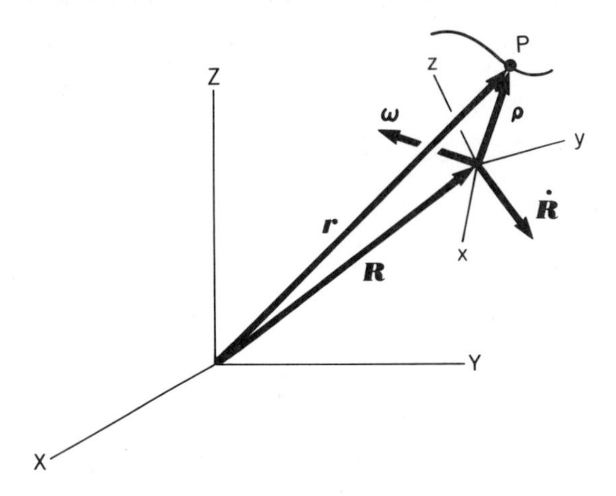

Figure 15.27. *xyz* moves arbitrarily relative to *XYZ*.

arbitrarily relative to each other. We do this by differentiating with respect to time the terms in Eq. 15.20 for the XYZ reference. Thus,

$$\left(\frac{d\boldsymbol{V}_{XYZ}}{dt}\right)_{XYZ} \equiv \boldsymbol{a}_{XYZ} = \left(\frac{d\boldsymbol{V}_{xyz}}{dt}\right)_{XYZ} + \ddot{\boldsymbol{R}} + \left[\frac{d}{dt}(\boldsymbol{\omega} \times \boldsymbol{\rho})\right]_{XYZ} \tag{15.22}$$

Now carry out the derivative of the cross product using the product rule.

$$\boldsymbol{a}_{XYZ} = \left(\frac{d\boldsymbol{V}_{xyz}}{dt}\right)_{XYZ} + \ddot{\boldsymbol{R}} + \boldsymbol{\omega} \times \left(\frac{d\boldsymbol{\rho}}{dt}\right)_{XYZ} + \left(\frac{d\boldsymbol{\omega}}{dt}\right)_{XYZ} \times \boldsymbol{\rho} \tag{15.23}$$

To introduce more physically meaningful terms, we can replace

$$\left(\frac{d\boldsymbol{V}_{xyz}}{dt}\right)_{XYZ} \quad \text{and} \quad \left(\frac{d\boldsymbol{\rho}}{dt}\right)_{XYZ}$$

using Eq. 15.16 in the following way:

$$\left(\frac{d\boldsymbol{V}_{xyz}}{dt}\right)_{XYZ} = \left(\frac{d\boldsymbol{V}_{xyz}}{dt}\right)_{xyz} + \boldsymbol{\omega} \times \boldsymbol{V}_{xyz}$$

$$\left(\frac{d\boldsymbol{\rho}}{dt}\right)_{XYZ} = \left(\frac{d\boldsymbol{\rho}}{dt}\right)_{xyz} + \boldsymbol{\omega} \times \boldsymbol{\rho}$$

Substituting into Eq. 15.23, we get

$$\boldsymbol{a}_{XYZ} = \left(\frac{d\boldsymbol{V}_{xyz}}{dt}\right)_{xyz} + \boldsymbol{\omega} \times \boldsymbol{V}_{xyz} + \ddot{\boldsymbol{R}} + \boldsymbol{\omega} \times \left(\frac{d\boldsymbol{\rho}}{dt}\right)_{xyz} + \boldsymbol{\omega} \times (\boldsymbol{\omega} \times \boldsymbol{\rho}) + \left(\frac{d\boldsymbol{\omega}}{dt}\right)_{XYZ} \times \boldsymbol{\rho}$$

You will note that $(d\boldsymbol{V}_{xyz}/dt)_{xyz}$ is $\boldsymbol{a}_{xyz}$; that $(d\boldsymbol{\rho}/dt)_{xyz}$ is $\boldsymbol{V}_{xyz}$; and that $(d\boldsymbol{\omega}/dt)_{XYZ}$ is $\dot{\boldsymbol{\omega}}$. Hence, rearranging terms, we have

$$\boxed{\boldsymbol{a}_{XYZ} = \boldsymbol{a}_{xyz} + \ddot{\boldsymbol{R}} + 2\boldsymbol{\omega} \times \boldsymbol{V}_{xyz} + \dot{\boldsymbol{\omega}} \times \boldsymbol{\rho} + \boldsymbol{\omega} \times (\boldsymbol{\omega} \times \boldsymbol{\rho})} \tag{15.24}$$

where $\boldsymbol{\omega}$ and $\dot{\boldsymbol{\omega}}$ are the angular velocity and acceleration, respectively, of the xyz reference relative to the XYZ reference. The vector $2(\boldsymbol{\omega} \times \boldsymbol{V}_{xyz})$ is called the *Coriolis acceleration vector*; we shall examine its interesting effects in Section 15.10.

Although Eq. 15.24 may seem somewhat terrifying at first, you will find that, by using it, problems that would otherwise be tremendously difficult can readily be carried out in a systematic manner. *You should keep in mind when solving problems that any of the methods developed in Chapter 11 can be used for determining the motion of the particle relative to the xyz reference or for determining the motion of the origin of xyz relative to the XYZ reference.* We shall now examine several problems, in which we shall use the notation, $\boldsymbol{\omega}_1, \boldsymbol{\omega}_2$, etc., to denote the various angular velocities involved. The notation, $\boldsymbol{\omega}$ (i.e., without subscripts), however, will be reserved to represent the angular velocity of the xyz reference relative to the XYZ reference.

EXAMPLE 15.11

A stationary truck is carrying a cockpit for a worker who repairs overhead fixtures. At the instant shown in Fig. 15.28, the base D is rotating at angular speed ω_2 of .1 rad/sec with $\dot{\omega}_2 = .2$ rad/sec^2 relative to the truck. Arm AB is rotating at angular speed ω_1 of .2 rad/sec with $\dot{\omega}_1 = .8$ rad/sec^2 relative to DA. Cockpit C is rotating relative to AB so as to always keep the man upright. What are the velocity and acceleration vectors of the man relative to the ground if $\alpha = 45°$ and $\beta = 30°$ at the instant of interest? Take $DA = 13$ m.

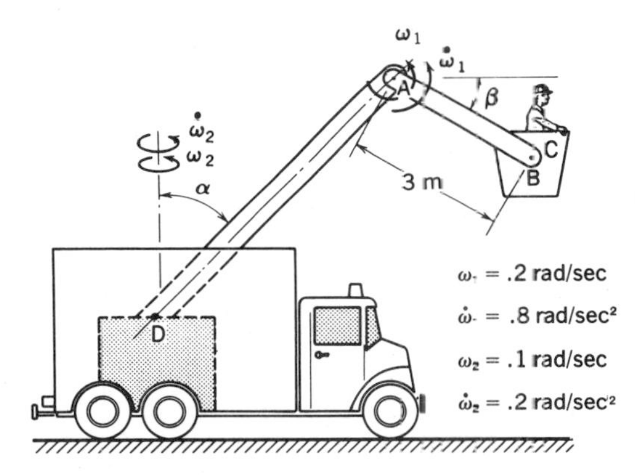

Figure 15.28. Truck with moving cockpit.

Because of the rotation of the cockpit C relative to arm AB to keep the man vertical, clearly, each particle in that body including the man has the same motion as point B of arm AB. Therefore, we shall concentrate our attention on this point.

Fix xyz to arm DA.
Fix XYZ to truck.

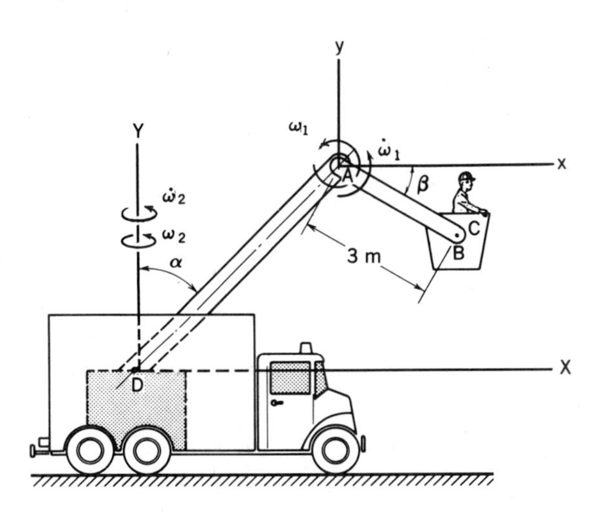

Figure 15.29. *xyz* fixed to *DA*; *XYZ* fixed to truck.

This situation is shown in Fig. 15.29.

A. Motion of *B* relative to *xyz*

$$\boldsymbol{\rho} = 3(\cos\beta\boldsymbol{i} - \sin\beta\boldsymbol{j}) = 2.60\boldsymbol{i} - 1.5\boldsymbol{j}\text{ m}$$

Since $\boldsymbol{\rho}$ is fixed in AB, which has angular velocity $\boldsymbol{\omega}_1$ relative to xyz, we have

$$\begin{aligned} \boldsymbol{V}_{xyz} &= \boldsymbol{\omega}_1 \times \boldsymbol{\rho} = (.2\boldsymbol{k}) \times (2.60\boldsymbol{i} - 1.5\boldsymbol{j}) \\ &= .520\boldsymbol{j} + .3\boldsymbol{i}\text{ m/sec} \end{aligned}$$

$$\boldsymbol{a}_{xyz} = \left(\frac{d\boldsymbol{\omega}_1}{dt}\right)_{xyz} \times \boldsymbol{\rho} + \boldsymbol{\omega}_1 \times \left(\frac{d\boldsymbol{\rho}}{dt}\right)_{xyz}$$

As seen from xyz, only the value of $\boldsymbol{\omega}_1$ and not its direction is changing. Also note that $(d\boldsymbol{\rho}/dt)_{xyz} = \boldsymbol{V}_{xyz}$. Hence,

$$\begin{aligned} \boldsymbol{a}_{xyz} &= (.8\boldsymbol{k}) \times (2.60\boldsymbol{i} - 1.5\boldsymbol{j}) + (.2\boldsymbol{k}) \times (.520\boldsymbol{j} + .3\boldsymbol{i}) \\ &= 1.09\boldsymbol{i} + 2.14\boldsymbol{j}\text{ m/sec}^2 \end{aligned}$$

B. Motion of *xyz* relative to *XYZ*

$$\boldsymbol{R} = 13(.707\boldsymbol{i} + .707\boldsymbol{j}) = 9.19\boldsymbol{i} + 9.19\boldsymbol{j}\text{ m}$$

Since $\boldsymbol{R}$ is fixed in DA, and since DA rotates with angular velocity $\boldsymbol{\omega}_2$ relative to XYZ, we have

$$\begin{aligned} \dot{\boldsymbol{R}} &= \boldsymbol{\omega}_2 \times \boldsymbol{R} = (.1\boldsymbol{j}) \times (9.19\boldsymbol{i} + 9.19\boldsymbol{j}) \\ &= -.919\boldsymbol{k}\text{ m/sec} \end{aligned}$$

$$\begin{aligned} \ddot{\boldsymbol{R}} &= \dot{\boldsymbol{\omega}}_2 \times \boldsymbol{R} + \boldsymbol{\omega}_2 \times \dot{\boldsymbol{R}} \\ &= (.2\boldsymbol{j}) \times (9.19\boldsymbol{i} + 9.19\boldsymbol{j}) + (.1\boldsymbol{j}) \times (-.919\boldsymbol{k}) \\ &= -1.838\boldsymbol{k} - .0919\boldsymbol{i}\text{ m/sec}^2 \end{aligned}$$

$$\boldsymbol{\omega} = \boldsymbol{\omega}_2 = .1\boldsymbol{j}\text{ rad/sec}$$

$$\dot{\boldsymbol{\omega}} = \dot{\boldsymbol{\omega}}_2 = .2\boldsymbol{j}\text{ rad/sec}^2$$

Hence,

$$
\begin{aligned}
\boldsymbol{V}_{XYZ} &= \boldsymbol{V}_{xyz} + \dot{\boldsymbol{R}} + \boldsymbol{\omega} \times \boldsymbol{\rho} \\
&= .520\boldsymbol{j} + .3\boldsymbol{i} - .919\boldsymbol{k} + (.1\boldsymbol{j}) \times (2.60\boldsymbol{i} - 1.5\boldsymbol{j}) \\
&= .3\boldsymbol{i} + .520\boldsymbol{j} - 1.179\boldsymbol{k} \text{ m/sec}
\end{aligned}
$$

$$
\begin{aligned}
\boldsymbol{a}_{XYZ} &= \boldsymbol{a}_{xyz} + \ddot{\boldsymbol{R}} + 2\boldsymbol{\omega} \times \boldsymbol{V}_{xyz} + \dot{\boldsymbol{\omega}} \times \boldsymbol{\rho} + \boldsymbol{\omega} \times (\boldsymbol{\omega} \times \boldsymbol{\rho}) \\
&= 1.096\boldsymbol{i} + 2.14\boldsymbol{j} - 1.838\boldsymbol{k} - .0919\boldsymbol{i} \\
&\quad + 2(.1\boldsymbol{j}) \times (.520\boldsymbol{j} + .3\boldsymbol{i}) + (.2\boldsymbol{j}) \times (2.60\boldsymbol{i} - 1.5\boldsymbol{j}) \\
&\quad + (.1\boldsymbol{j}) \times [(.1\boldsymbol{j}) \times (2.60\boldsymbol{i} - 1.5\boldsymbol{j})] \\
&= .978\boldsymbol{i} + 2.14\boldsymbol{j} - 2.42\boldsymbol{k} \text{ m/sec}
\end{aligned}
$$

Notice that the essential aspects of the analysis come in the consideration of parts A and B of the problem, while the remaining portion involves direct substitution and vector algebraic operations.

EXAMPLE 15.12

A wheel rotates with an angular speed ω_2 of 5 rad/sec on a platform which rotates with a speed ω_1 of 10 rad/sec relative to the ground as shown in Fig. 15.30. A valve gate A moves down the spoke of the wheel, and when the spoke is vertical the valve gate has a speed of 20 ft/sec, an acceleration of 10 ft/sec² along the spoke, and is 1 ft from the shaft centerline of the wheel. Compute the velocity and acceleration of the valve gate relative to the ground at this instant.

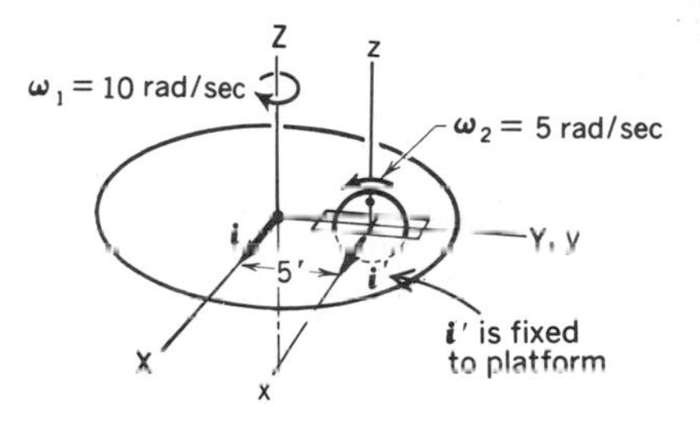

Figure 15.30. *xyz* fixed to wheel; *XYZ* fixed to ground.

Fix *xyz* to wheel.
Fix *XYZ* to ground.

A. Motion of particle relative to *xyz*

$$
\begin{aligned}
\boldsymbol{\rho} &= \boldsymbol{k} \text{ ft} \\
\boldsymbol{V}_{xyz} &= -20\boldsymbol{k} \text{ ft/sec} \\
\boldsymbol{a}_{xyz} &= -10\boldsymbol{k} \text{ ft/sec}^2
\end{aligned}
$$

B. Motion of *xyz* relative to *XYZ*

$$\boldsymbol{R} = 5\boldsymbol{j} \text{ ft}$$

Since $\boldsymbol{R}$ is fixed to the platform:

$$
\begin{aligned}
\dot{\boldsymbol{R}} &= \boldsymbol{\omega}_1 \times \boldsymbol{R} = (-10\boldsymbol{k}) \times (5\boldsymbol{j}) = 50\boldsymbol{i} \text{ ft/sec} \\
\ddot{\boldsymbol{R}} &= \dot{\boldsymbol{\omega}}_1 \times \boldsymbol{R} + \boldsymbol{\omega}_1 \times \dot{\boldsymbol{R}} \\
&= 0 + (-10\boldsymbol{k}) \times (50\boldsymbol{i}) = -500\boldsymbol{j} \text{ ft/sec}^2 \\
\boldsymbol{\omega} &= \boldsymbol{\omega}_2 + \boldsymbol{\omega}_1 = 5\boldsymbol{i} - 10\boldsymbol{k} \text{ rad/sec} \\
\dot{\boldsymbol{\omega}} &= \dot{\boldsymbol{\omega}}_2 + \dot{\boldsymbol{\omega}}_1
\end{aligned}
$$

Note that $\boldsymbol{\omega}_2$ is of constant magnitude but, because of the bearings of the wheel, $\boldsymbol{\omega}_2$ must rotate with the platform. In short, we can say that $\boldsymbol{\omega}_2$ is *fixed* to the platform and so $\dot{\boldsymbol{\omega}}_2 = \boldsymbol{\omega}_1 \times \boldsymbol{\omega}_2$. Hence,

$$\begin{aligned}\dot{\boldsymbol{\omega}} &= \boldsymbol{\omega}_1 \times \boldsymbol{\omega}_2 + \mathbf{0} \\ &= (-10\boldsymbol{k}) \times (5\boldsymbol{i}) = -50\boldsymbol{j} \text{ rad/sec}^2\end{aligned}$$

We then have

$$\begin{aligned}\boldsymbol{V}_{XYZ} &= \boldsymbol{V}_{xyz} + \dot{\boldsymbol{R}} + \boldsymbol{\omega} \times \boldsymbol{\rho} \\ &= -20\boldsymbol{k} + 50\boldsymbol{i} + (5\boldsymbol{i} - 10\boldsymbol{k}) \times \boldsymbol{k} \\ &= 50\boldsymbol{i} - 5\boldsymbol{j} - 20\boldsymbol{k} \text{ ft/sec}\end{aligned}$$

Also,

$$\begin{aligned}\boldsymbol{a}_{XYZ} &= \boldsymbol{a}_{xyz} + \ddot{\boldsymbol{R}} + 2\boldsymbol{\omega} \times \boldsymbol{V}_{xyz} + \dot{\boldsymbol{\omega}} \times \boldsymbol{\rho} + \boldsymbol{\omega} \times (\boldsymbol{\omega} \times \boldsymbol{\rho}) \\ &= -10\boldsymbol{k} - 500\boldsymbol{j} + 2(5\boldsymbol{i} - 10\boldsymbol{k}) \times (-20\boldsymbol{k}) \\ &\quad + (-50\boldsymbol{j}) \times \boldsymbol{k} + (5\boldsymbol{i} - 10\boldsymbol{k}) \times [(5\boldsymbol{i} - 10\boldsymbol{k}) \times \boldsymbol{k}] \\ &= -100\boldsymbol{i} - 300\boldsymbol{j} - 35\boldsymbol{k} \text{ ft/sec}^2\end{aligned}$$

EXAMPLE 15.13

In Example 15.12, the wheel accelerates at the instant under discussion with $\dot{\omega}_2 = 5$ rad/sec², and the platform accelerates with $\dot{\omega}_1 = 10$ rad/sec². Find the velocity and acceleration of the valve gate A.

If we review the contents of parts A and B of Example 15.12, it will be clear that only $\dot{\boldsymbol{\omega}}$ is affected by the fact that $\dot{\omega}_1 = 10$ rad/sec² and $\dot{\omega}_2 = 5$ rad/sec². In this regard, consider $\boldsymbol{\omega}_2$. It is no longer of constant value and cannot be considered as *fixed* in the platform. However we can express $\boldsymbol{\omega}_2$ as $\omega_2\boldsymbol{i}'$ *at all times*, wherein $\boldsymbol{i}'$ is *fixed* in the platform as shown in Fig. 15.30. Thus, we can say for $\boldsymbol{\omega}$:

$$\boldsymbol{\omega} = \omega_2\boldsymbol{i}' + \boldsymbol{\omega}_1$$

Therefore,

$$\begin{aligned}\dot{\boldsymbol{\omega}} &= \dot{\omega}_2\boldsymbol{i}' + \omega_2\dot{\boldsymbol{i}}' + \dot{\boldsymbol{\omega}}_1 \\ &= 5\boldsymbol{i}' + 5(\boldsymbol{\omega}_1 \times \boldsymbol{i}') - 10\boldsymbol{k} \\ &= 5\boldsymbol{i}' + 5(-10\boldsymbol{k}) \times \boldsymbol{i}' - 10\boldsymbol{k}\end{aligned}$$

At the instant of interest, $\boldsymbol{i}' = \boldsymbol{i}$. Hence,

$$\dot{\boldsymbol{\omega}} = 5\boldsymbol{i} - 50\boldsymbol{j} - 10\boldsymbol{k} \text{ rad/sec}^2$$

Hence, we use the above $\dot{\boldsymbol{\omega}}$ in part B of Example 15.12 and compute $\boldsymbol{a}_{XYZ}$ accordingly. We leave the details to the reader.

An understanding of Examples 15.11, 15.12, and 15.13 involving two angular velocities of component parts is sufficient for most of the homework problems of this section covering a wide range of applications. In the next example, we have three angular velocities to deal with. Although we have starred this example, we urge you to examine it carefully if time allows. It is an interesting problem, and comprehension of the three different analyses given will ensure a strong grasp of multireference kinematics.[10]

*EXAMPLE 15.14

To simulate the flight conditions of a space vehicle, engineers have developed the *centrifuge*, shown diagrammatically in Fig. 15.31. A main *arm*, 40 ft long, rotates about the *A–A* axis. The pilot sits in a *cockpit*, which can rotate about axis *C–C*. The *seat* for the pilot can rotate inside the cockpit about an axis shown at the point *B*. These rotations are controlled by a computer that is set to simulate certain maneuvers corresponding to the entry and exit from the earth's atmosphere, malfunctions of the control system, and so on. When a pilot sits in the cockpit, his head has the position shown in Fig. 15.31, 3 ft from *B*. At the instant of interest, the main arm is rotating at 10 rpm and accelerating at 5 rpm². The cockpit is rotating at a constant speed about *C–C* relative to the main arm at 10 rpm. Finally, the seat is rotating at a constant speed of 5 rpm relative to the cockpit. How many *g*'s acceleration relative to the ground is the pilot's head subjected to?[11]

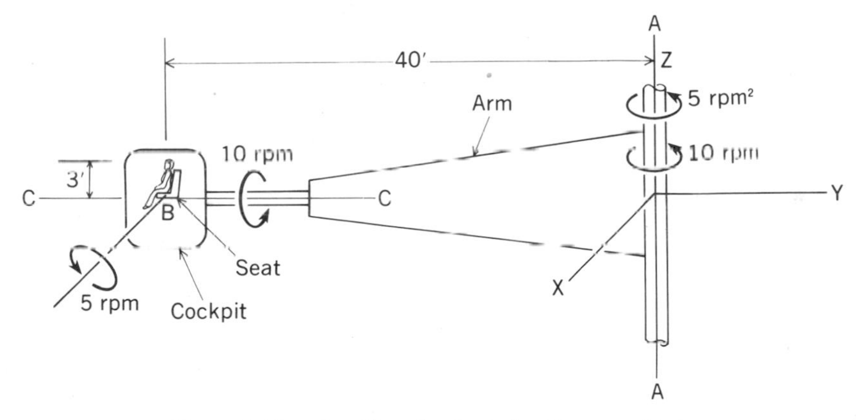

Figure 15.31. Centrifuge for simulating flight conditions.

In Fig. 15.32 the arm of the centrifuge rotates relative to the ground at an angular velocity of ω_1. The cockpit meanwhile rotates relative to the arm with angular speed ω_2. Finally, the seat rotates relative to the cockpit at an angular speed ω_3. For constant ω_2, we see that, because of bearings in the arm, the vector $\boldsymbol{\omega}_2$ is "fixed" in the arm. Also, for constant ω_3, because of bearings in the cockpit, the vector $\boldsymbol{\omega}_3$ is "fixed" in the cockpit. Before we examine the acceleration of pilot's head, note that at the instant of interest:

---

[10]Example 15.14 was given as two homework problems in both the first and second editions of this text. They were so instructive that for this edition the author decided to move the problems into the main text.

[11]A *g* of acceleration is an amount of acceleration equal to that of gravity (32.2 ft/sec² or 9.81 m/sec²). Thus, a 4*g* acceleration is equivalent to an acceleration of 128.8 ft/sec².

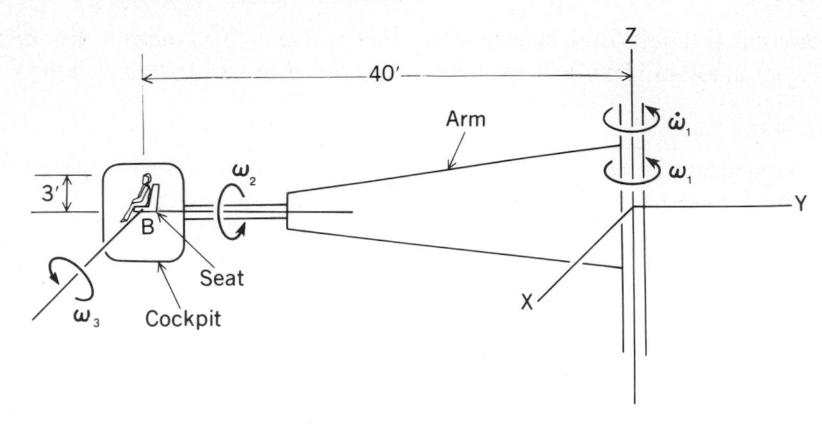

Figure 15.32. Centrifuge listing ω's.

$$\omega_1 = \omega_2 = 10 \text{ rpm} = 1.048 \text{ rad/sec}$$
$$\dot{\omega}_1 = 5 \text{ rpm}^2 = .00873 \text{ rad/sec}^2$$
$$\omega_3 = 5 \text{ rpm} = .524 \text{ rad/sec}$$

We shall do this problem using three different kinds of moving references xyz.

ANALYSIS I

Fix xyz to arm.
Fix XYZ to ground.

Note in Fig. 15.33 that xyz and the arm to which it is fixed are shown dark. Note also that the axes xyz and XYZ are parallel to each other at the instant of interest.

A. Motion of particle relative to xyz

$$\boldsymbol{\rho} = 3\boldsymbol{k} \text{ ft}$$

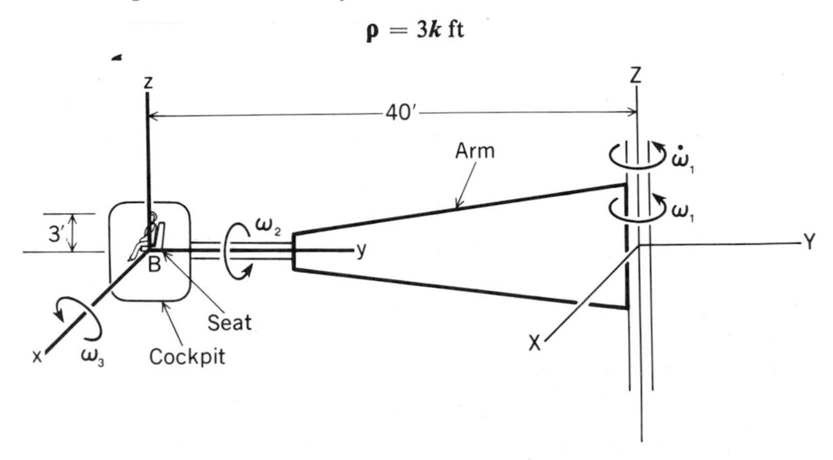

Figure 15.33. Centrifuge with xyz fixed to arm.

Note that $\boldsymbol{\rho}$ is "fixed" to the seat and that the seat has an angular velocity of $(\boldsymbol{\omega}_2 + \boldsymbol{\omega}_3)$ relative to the arm and thus to xyz. Hence,

$$\begin{aligned} \boldsymbol{V}_{xyz} &= (\boldsymbol{\omega}_2 + \boldsymbol{\omega}_3) \times \boldsymbol{\rho} \\ &= (1.048\boldsymbol{j} + .524\boldsymbol{i}) \times 3\boldsymbol{k} = 3.14\boldsymbol{i} - 1.572\boldsymbol{j} \text{ ft/sec} \end{aligned}$$

$$\boldsymbol{a}_{xyz} = \left(\frac{d\boldsymbol{V}_{xyz}}{dt}\right)_{xyz} = \left[\frac{d}{dt}_{xyz}(\boldsymbol{\omega}_2 + \boldsymbol{\omega}_3)\right] \times \boldsymbol{\rho} + (\boldsymbol{\omega}_2 + \boldsymbol{\omega}_3) \times \left(\frac{d\boldsymbol{\rho}}{dt}\right)_{xyz}$$

Clearly, relative to the arm, and thus to xyz, $\boldsymbol{\omega}_2$ is constant. And $\boldsymbol{\omega}_3$ is fixed in the cockpit that has an angular velocity of $\boldsymbol{\omega}_2$ relative to xyz. Thus, we have

$$\begin{aligned} \boldsymbol{a}_{xyz} &= (0 + \boldsymbol{\omega}_2 \times \boldsymbol{\omega}_3) \times \boldsymbol{\rho} + (\boldsymbol{\omega}_2 + \boldsymbol{\omega}_3) \times \boldsymbol{V}_{xyz} \\ &= (1.048\boldsymbol{j} \times .524\boldsymbol{i}) \times 3\boldsymbol{k} + (1.048\boldsymbol{j} + .524\boldsymbol{i}) \times (3.14\boldsymbol{i} - 1.572\boldsymbol{j}) \\ &= -4.12\boldsymbol{k} \text{ ft/sec}^2 \end{aligned}$$

B. Motion of *xyz* relative to *XYZ*

$$\boldsymbol{R} = -40\boldsymbol{j} \text{ ft}$$

Note that $\boldsymbol{R}$ is fixed in the arm, which has an angular velocity $\boldsymbol{\omega}_1$ relative to XYZ. Hence,

$$\begin{aligned} \dot{\boldsymbol{R}} &= \boldsymbol{\omega}_1 \times \boldsymbol{R} = 1.048\boldsymbol{k} \times (-40\boldsymbol{j}) = 41.9\boldsymbol{i} \text{ ft/sec} \\ \ddot{\boldsymbol{R}} &= \boldsymbol{\omega}_1 \times \dot{\boldsymbol{R}} + \dot{\boldsymbol{\omega}}_1 \times \boldsymbol{R} \\ &= 1.048\boldsymbol{k} \times 41.9\boldsymbol{i} + .00873\boldsymbol{k} \times (-40\boldsymbol{j}) \\ &= 43.9\boldsymbol{j} + .349\boldsymbol{i} \text{ ft/sec}^2 \\ \boldsymbol{\omega} &= \boldsymbol{\omega}_1 = 1.048 \text{ rad/sec} \\ \dot{\boldsymbol{\omega}} &= \dot{\omega}_1\boldsymbol{k} = .00873\boldsymbol{k} \text{ rad/sec}^2 \end{aligned}$$

We can now substitute into the following equation:

$$\boldsymbol{a}_{XYZ} = \boldsymbol{a}_{xyz} + \ddot{\boldsymbol{R}} + 2\boldsymbol{\omega} \times \boldsymbol{V}_{xyz} + \dot{\boldsymbol{\omega}} \times \boldsymbol{\rho} + \boldsymbol{\omega} \times (\boldsymbol{\omega} \times \boldsymbol{\rho})$$

Therefore,

$$\boldsymbol{a}_{XYZ} = 3.64\boldsymbol{i} + 50.5\boldsymbol{j} - 4.12\boldsymbol{k} \text{ ft/sec}^2$$

$$|\boldsymbol{a}_{XYZ}| = \frac{\sqrt{3.64^2 + 50.5^2 + 4.12^2}}{32.2} = 1.578g$$

ANALYSIS II

Fix xyz to cockpit. Fix XYZ to ground.

This situation is shown in Fig. 15.34.

A. Motion of particle relative to *xyz*

$$\boldsymbol{\rho} = 3\boldsymbol{k} \text{ ft}$$

Note that $\boldsymbol{\rho}$ is fixed to the seat, which has an angular velocity of $\boldsymbol{\omega}_3$ relative to the cockpit and thus relative to xyz. Hence,

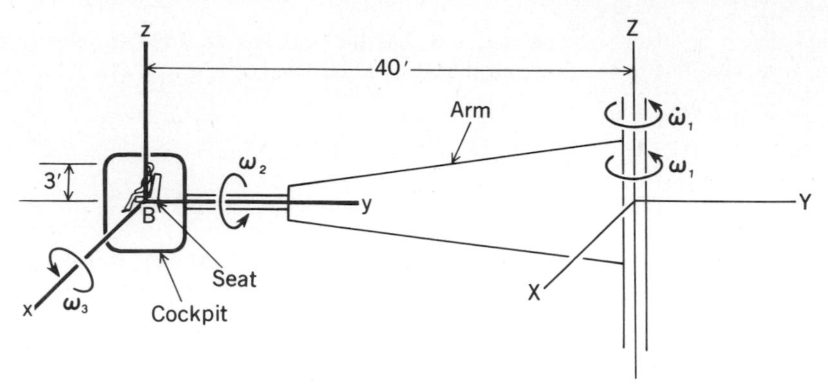

Figure 15.34. Centrifuge with *xyz* fixed to cockpit.

$$V_{xyz} = \boldsymbol{\omega}_3 \times \boldsymbol{\rho} = .524\boldsymbol{i} \times 3\boldsymbol{k} = -1.572\boldsymbol{j}\text{ ft/sec}$$

$$\boldsymbol{a}_{xyz} = \left(\frac{d\boldsymbol{\omega}_3}{dt}\right)_{xyz} \times \boldsymbol{\rho} + \boldsymbol{\omega}_3 \times \left(\frac{d\boldsymbol{\rho}}{dt}\right)_{xyz}$$

But $\boldsymbol{\omega}_3$ is constant as seen from the cockpit and thus from *xyz*. Hence,

$$\boldsymbol{a}_{xyz} = \boldsymbol{0} \times \boldsymbol{\rho} + \boldsymbol{\omega}_3 \times \boldsymbol{V}_{xyz} = .524\boldsymbol{i} \times (-1.572\boldsymbol{j})$$
$$= -.824\boldsymbol{k}\text{ ft/sec}^2$$

B. Motion of *xyz* relative to *XYZ*. The origin of *xyz* in this analysis has the same motion as the origin of *xyz* in the previous analysis. Thus, we use the results of analysis I for $\boldsymbol{R}$ and its time derivatives.

$$\boldsymbol{R} = -40\boldsymbol{j}\text{ ft}$$
$$\dot{\boldsymbol{R}} = 41.9\boldsymbol{i}\text{ ft/sec}$$
$$\ddot{\boldsymbol{R}} = 43.9\boldsymbol{j} + .349\boldsymbol{i}\text{ ft/sec}^2$$
$$\boldsymbol{\omega} = \boldsymbol{\omega}_1 + \boldsymbol{\omega}_2 = 1.048\boldsymbol{j} + 1.048\boldsymbol{k}\text{ rad/sec}$$
$$\dot{\boldsymbol{\omega}} = \dot{\boldsymbol{\omega}}_1 + \dot{\boldsymbol{\omega}}_2$$

We are given $\dot{\omega}_1$ about the Z axis and $\boldsymbol{\omega}_2$ is fixed in the arm, which is rotating with angular velocity $\boldsymbol{\omega}_1$ relative to the XYZ reference. Hence,

$$\dot{\boldsymbol{\omega}} = \dot{\omega}_1\boldsymbol{k} + \boldsymbol{\omega}_1 \times \boldsymbol{\omega}_2 = .00873\boldsymbol{k} + (1.048\boldsymbol{k} \times 1.048\boldsymbol{j})$$
$$= -1.098\boldsymbol{i} + .00873\boldsymbol{k}\text{ rad/sec}^2$$

We can now substitute into the key equation, 15.24:

$$\boldsymbol{a}_{XYZ} = \boldsymbol{a}_{xyz} + \ddot{\boldsymbol{R}} + 2\boldsymbol{\omega} \times \boldsymbol{V}_{xyz} + \dot{\boldsymbol{\omega}} \times \boldsymbol{\rho} + \boldsymbol{\omega} \times (\boldsymbol{\omega} \times \boldsymbol{\rho})$$
$$= 3.64\boldsymbol{i} + 50.5\boldsymbol{j} - 4.12\boldsymbol{k}\text{ ft/sec}^2$$
$$|\boldsymbol{a}_{XYZ}| = 1.578g$$

ANALYSIS III

Fix *xyz* to seat.
Fix *XYZ* to ground.

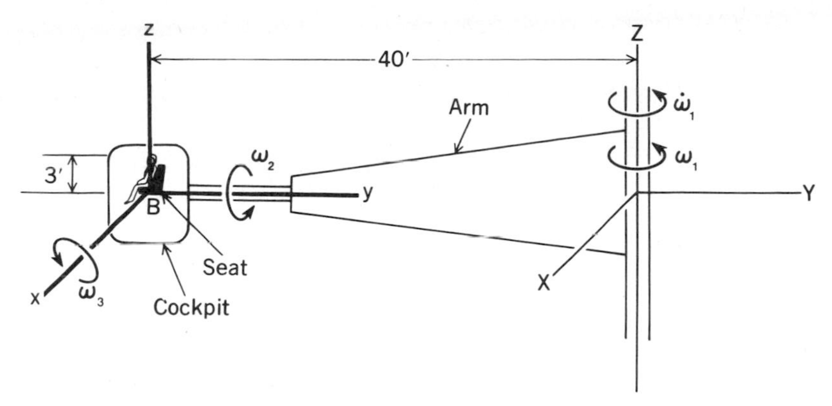

Figure 15.35. Centrifuge with *xyz* fixed to seat.

This situation is shown in Fig. 15.35.

A. Motion of particle relative to *xyz*

$$\boldsymbol{\rho} = 3\boldsymbol{k} \text{ ft}$$

Since the particle is fixed to the seat and is thus fixed in xyz, we can say:

$$\boldsymbol{V}_{xyz} = \boldsymbol{0}$$

$$\boldsymbol{a}_{xyz} = \boldsymbol{0}$$

B. Motion of *xyz* relative to *XYZ*. Again, the origin of xyz has identically the same motion as in the previous analyses. Thus, we have the same results as before for $\boldsymbol{R}$ and its derivatives.

$$\boldsymbol{R} = -40\boldsymbol{j} \text{ ft}$$

$$\dot{\boldsymbol{R}} = 41.9\boldsymbol{i} \text{ ft/sec}$$

$$\ddot{\boldsymbol{R}} = 43.9\boldsymbol{j} + .349\boldsymbol{i} \text{ ft/sec}^2$$

$$\boldsymbol{\omega} = \boldsymbol{\omega}_1 + \boldsymbol{\omega}_2 + \boldsymbol{\omega}_3 = 1.048\boldsymbol{k} + 1.048\boldsymbol{j} + .524\boldsymbol{i}$$

$$\dot{\boldsymbol{\omega}} = \dot{\boldsymbol{\omega}}_1 + \dot{\boldsymbol{\omega}}_2 + \dot{\boldsymbol{\omega}}_3$$

Note that $\dot{\omega}_1$ is given. Also, $\boldsymbol{\omega}_2$ is fixed in the arm, which rotates with angular speed $\boldsymbol{\omega}_1$ relative to XYZ. Finally, $\boldsymbol{\omega}_3$ is fixed in the cockpit, which has an angular velocity $\boldsymbol{\omega}_2 + \boldsymbol{\omega}_1$ relative to XYZ. Thus,

$$\begin{aligned}\dot{\boldsymbol{\omega}} &= \dot{\omega}_2\boldsymbol{k} + \boldsymbol{\omega}_1 \times \boldsymbol{\omega}_2 + (\boldsymbol{\omega}_1 + \boldsymbol{\omega}_2) \times \boldsymbol{\omega}_3 \\ &= .00873\boldsymbol{k} + (1.048\boldsymbol{k} \times 1.048\boldsymbol{j}) + (1.048\boldsymbol{k} + 1.048\boldsymbol{j}) \times (.524\boldsymbol{i}) \\ &= -1.098\boldsymbol{i} + .549\boldsymbol{j} - .540\boldsymbol{k}\end{aligned}$$

We now go to the basic equation, 15.24.

$$\boldsymbol{a}_{XYZ} = \boldsymbol{a}_{xyz} + \ddot{\boldsymbol{R}} + 2\boldsymbol{\omega} \times \boldsymbol{V}_{xyz} + \dot{\boldsymbol{\omega}} \times \boldsymbol{\rho} + \boldsymbol{\omega} \times (\boldsymbol{\omega} \times \boldsymbol{\rho})$$

Substituting, we get

$$\boldsymbol{a}_{XYZ} = 3.64\boldsymbol{i} + 50.5\boldsymbol{j} - 4.12\boldsymbol{k} \text{ ft/sec}^2$$

$$|\boldsymbol{a}_{XYZ}| = 1.578g$$

Problems

15.79. A truck has a speed V of 20 mi/hr and an acceleration $\dot{V}$ of 3 mi/hr/sec at time t. A cylinder of radius equal to 2 ft is rolling without slipping at time t such that relative to the truck it has an angular speed ω_1 and angular acceleration $\dot{\omega}_1$ of 2 rad/sec and 1 rad/sec², respectively. Determine the velocity and acceleration of the center of the cylinder relative to the ground.

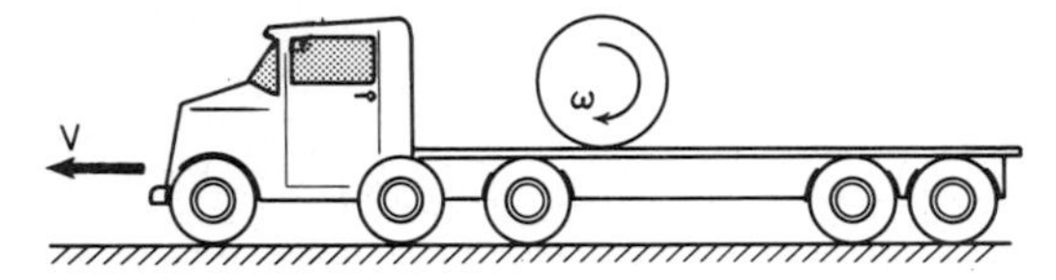

Figure P.15.79

15.80. A wheel rotates with an angular speed ω_2 of 5 rad/sec relative to a platform, which rotates with a speed ω_1 of 10 rad/sec relative to the ground as shown. A collar moves down the spoke of the wheel, and, when the spoke is vertical, the collar has a speed of 20 ft/sec, an acceleration of 10 ft/sec² along the spoke, and is positioned 1 ft from the shaft centerline of the wheel. Compute the velocity and acceleration of the collar relative to the ground at this instant.

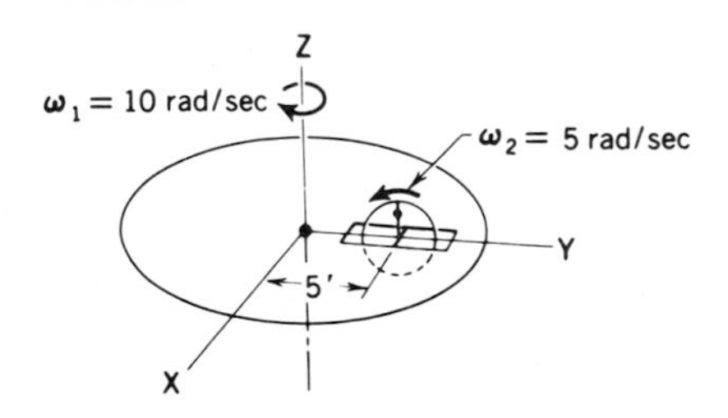

Figure P.15.80

15.81. In Problem 15.60, determine the acceleration of the particle at the instant of interest.

15.82. In Problem 15.61, find the acceleration of the particle P relative to the ground reference.

15.83. In Problem 15.63, find the acceleration of point A relative to the ground.

15.84. In Problem 15.67, find the acceleration of the tip of the propeller relative to the ground reference. Take the yaw rotation to be zero and the loop rotation radius r to be 500 m.

15.85. In Problem 15.68, find the acceleration of point B relative to the ground.

15.86. In Problem 15.68, find the acceleration of point B for the following data at the instant of interest shown in the diagram.

$$\omega_1 = .2 \text{ rad/sec}$$
$$\dot{\omega}_1 = -.1 \text{ rad/sec}^2$$
$$\omega_2 = .4 \text{ rad/sec}$$
$$\dot{\omega}_2 = .3 \text{ rad/sec}^2$$

15.87. In Problem 15.70, determine the acceleration of the top tip of the gun.

15.88. In Problem 15.63, find the acceleration of point A for the configuration shown. Take $\omega_1 = 2$ rad/sec, $\dot{\omega}_1 = 3$ rad/sec², $\omega_2 = .1$ rad/sec, and $\dot{\omega}_2 = 2$ rad/sec². How many g's of acceleration is this point subject to?

15.89. In Problem 15.69, find the acceleration of D relative to the ground. (*Hint:* Use two position vectors to get $\boldsymbol{\rho}$.)

15.90. Find the acceleration of gear tooth A relative to the ground in Problem 15.72.

15.91. In Problem 15.80, the wheel accelerates at the instant under discussion with 5 rad/sec² relative to the platform, and the platform increases its angular speed at 10 rad/sec² relative to the ground. Find the velocity and acceleration of the collar.

15.92. As with the velocity equation 15.20, we can easily show that Eq. 15.7, relating accelerations between two points on a rigid body, is actually a special case of Eq. 15.24. Thus, consider the diagram showing a rigid body moving arbitrarily relative to XYZ. Choose two points a and b in the body and embed a reference xyz in the body with the origin at a. Now express the acceleration of point b as seen from the two references. Show how this equation can be reformulated as Eq. 15.7.

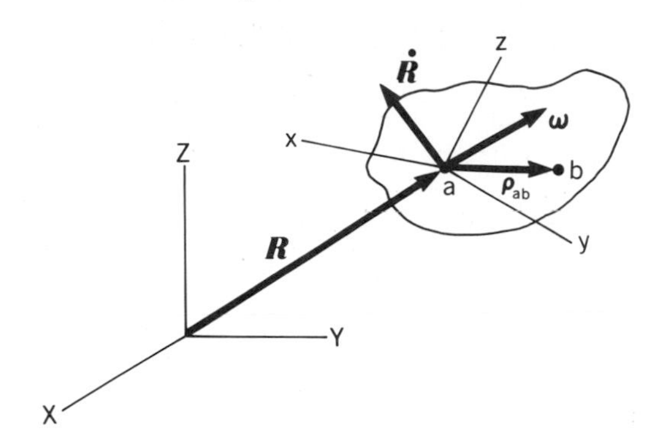

Figure P.15.92

15.93. Solve Problem 15.69 for the following data:

$$\omega_1 = .3 \text{ rad/sec}$$
$$\dot{\omega}_1 = .2 \text{ rad/sec}^2$$
$$\omega_2 = .40 \text{ rad/sec}$$
$$\dot{\omega}_2 = .10 \text{ rad/sec}^2$$
$$\omega_3 = .15 \text{ rad/sec}$$
$$\dot{\omega}_3 = -.2 \text{ rad/sec}^2$$

15.94. In Problem 15.70, find the acceleration relative to the ground of the projectile normal to the gun barrel at the instant that the projectile just leaves the barrel. Use the following data:

$$\omega_1 = .3 \text{ rad/sec}$$
$$\dot{\omega}_1 = .2 \text{ rad/sec}^2$$
$$\omega_2 = -.6 \text{ rad/sec}$$
$$\dot{\omega}_2 = -.4 \text{ rad/sec}^2$$

15.95. In Problem 15.76, find the magnitude of the acceleration of collar C relative to the ground for the following data at the instant shown:

$$\dot{\theta} = 2 \text{ rad/sec}$$
$$\ddot{\theta} = 4 \text{ rad/sec}^2$$

15.96. A truck is moving at a constant speed $V = 1.7$ m/sec at time t. The truck loading compartment has at this instant a constant angular speed $\dot{\theta}$ of .1 rad/sec at an angle $\theta = 45°$. A cylinder of radius of 300 mm rolls relative to the compartment at a speed ω_1 of 1 rad/sec, accelerating at a rate $\dot{\omega}_1$ of .5 rad/sec² at time t. What are the velocity and acceleration of the center of the cylinder relative to the ground at time t? The distance d at time t is 5 m.

Figure P.15.96

***15.97.** In Example 15.14, suppose at the instant of interest that there is an angular acceleration $\dot{\omega}_2 = .3$ rad/sec² of the cockpit relative to the arm and that there is an acceleration $\dot{\omega}_3 = .2$ rad/sec² of the seat relative to the cockpit. Find the number of g's to which the pilot's head is subjected. Follow analysis I in the example. (*Hint:* The angular velocity $\boldsymbol{\omega}_3$ can always be expressed as $\omega_3\hat{\mathbf{c}}$, where $\hat{\mathbf{c}}$ is a unit vector *fixed to the cockpit* having a direction along x axis at the instant of interest.)

*15.9 A New Look at Newton's Law

The proper form of Newton's law has been presented as

$$\boldsymbol{F} = m\boldsymbol{a}_{XYZ} \tag{15.25}$$

where the acceleration is measured relative to an inertial reference. There are times when the motion of a particle is known and makes sense only relative to a noninertial reference. Such a case would arise, for example, in an airplane or rocket, where machine elements must move in a certain way relative to the vehicle in order to function properly. Therefore, the motion of the machine element relative to the vehicle is known. If, however, the vehicle is undergoing a severe maneuver relative to inertial space, we cannot use Eq. 15.25 with the acceleration of the machine element measured relative to the vehicle. This is so since the vehicle is not at that instant an inertial reference, and to disregard this fact will lead to erroneous results. In such problems, the motion of the vehicle may be known relative to inertial space, and we can employ to good advantage the multireference analysis of the previous section. Attaching the reference xyz to the vehicle and XYZ to inertial space, we can then use Newton's law in the following way:

$$\boldsymbol{F} = m[\boldsymbol{a}_{xyz} + \ddot{\boldsymbol{R}} + 2\boldsymbol{\omega} \times \boldsymbol{V}_{xyz} + \dot{\boldsymbol{\omega}} \times \boldsymbol{\rho} + \boldsymbol{\omega} \times (\boldsymbol{\omega} \times \boldsymbol{\rho})] \tag{15.26}$$

Clearly, the bracketed expression is the required quantity $\boldsymbol{a}_{XYZ}$ needed for Newton's law. It is the usual practice to write Eq. 15.26 in the following form:

$$\boxed{\boldsymbol{F} - m[\ddot{\boldsymbol{R}} + 2\boldsymbol{\omega} \times \boldsymbol{V}_{xyz} + \dot{\boldsymbol{\omega}} \times \boldsymbol{\rho} + \boldsymbol{\omega} \times (\boldsymbol{\omega} \times \boldsymbol{\rho})] = m\boldsymbol{a}_{xyz}} \tag{15.27}$$

This equation may now be considered as Newton's law written for a *noninertial* reference xyz. The terms $-m\ddot{\boldsymbol{R}}$, $-m(2\boldsymbol{\omega} \times \boldsymbol{V}_{xyz})$, and so on, are then considered as forces and are termed *inertial forces*. Thus, we can take the viewpoint that for a noninertial reference, xyz, we can still say force $\boldsymbol{F}$ equals mass times acceleration, $\boldsymbol{a}_{xyz}$, provided that we include with the applied force $\boldsymbol{F}$ all the inertial forces. Indeed, we shall adopt this viewpoint in this text. The inertial force $-2m\boldsymbol{\omega} \times \boldsymbol{V}_{xyz}$ is the very interesting *Coriolis force*, which we shall later discuss in some detail.

The inertial forces result in baffling actions that are sometimes contrary to our intuition. Most of us during our lives have been involved in actions where the reference used (knowingly or not) has been with sufficient accuracy an inertial reference, usually the earth's surface. We have, accordingly, become conditioned to associating an acceleration proportional to, and in the same direction as, the applied force. Occasions do arise when we finds ourselves relating our motions to a reference that is highly noninertial. For example, fighter pilots and stunt pilots carry out actions in a cockpit of a plane while the plane is undergoing severe maneuvers. Unexpected results frequently occur for flyers if they use the cockpit interior as a reference for their actions. Thus, to move their hands from one position to another relative to the cockpit sometimes requires an exertion that is not the one anticipated, causing considerable confusion. The next example will illustrate this, and the sections that follow will explore further some of these interesting effects.[12]

EXAMPLE 15.15

The plan view of a rotating platform is shown in Fig. 15.36. A man is seated at the position labeled A and is facing point O of the platform. He is carrying a mass of $\frac{1}{50}$ slug at the rate of 10 ft/sec in a direction straight ahead of him (i.e., toward the center of the platform). If this platform has an angular speed of 10 rad/sec and an angular acceleration of 5 rad/sec² relative to the ground at this instant, what force $\boldsymbol{F}$ must he exert to cause the mass to accelerate 5 ft/sec² toward the center?

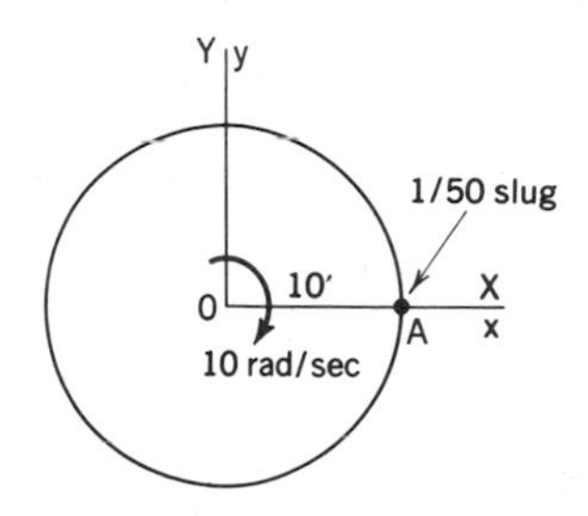

Figure 15.36. Rotating platform.

For purposes of determining inertial forces, we proceed as follows:

Fix xyz to platform.
Fix XYZ to ground.

A. Motion of mass relative to xyz reference

$$\boldsymbol{\rho} = 10\boldsymbol{i}\text{ ft},\qquad \boldsymbol{V}_{xyz} = -10\boldsymbol{i}\text{ ft/sec},\qquad \boldsymbol{a}_{xyz} = -5\boldsymbol{i}\text{ ft/sec}^2$$

B. Motion of xyz relative to XYZ

$$\dot{\boldsymbol{R}} = 0,\qquad \ddot{\boldsymbol{R}} = 0,\qquad \boldsymbol{\omega} = -10\boldsymbol{k}\text{ rad/sec},\qquad \dot{\boldsymbol{\omega}} = -5\boldsymbol{k}\text{ rad/sec}^2$$

Hence,

$$\boldsymbol{a}_{XYZ} = -5\boldsymbol{i} + 2(-10\boldsymbol{k}) \times (-10\boldsymbol{i}) + (-5\boldsymbol{k}) \times 10\boldsymbol{i} + (-10\boldsymbol{k}) \times (-10\boldsymbol{k} \times 10\boldsymbol{i})$$

Therefore,

$$\boldsymbol{a}_{XYZ} = -5\boldsymbol{i} + 200\boldsymbol{j} - 50\boldsymbol{j} - 1000\boldsymbol{i}$$

Employing Newton's law (Eq. 15.27) for the mass, we get

$$\boldsymbol{F} - \tfrac{1}{50}(200\boldsymbol{j} - 50\boldsymbol{j} - 1000\boldsymbol{i}) = \tfrac{1}{50}(-5\boldsymbol{i})$$

Solving for $\boldsymbol{F}$, we get

$$\boldsymbol{F} = 3\boldsymbol{j} - 20.1\boldsymbol{i}\text{ lb}$$

---

[12]At this juncture we should remind ourselves that "physical feel" or "intuition" is really a direct consequence of past experiences. For this reason, many of the things you will later formally learn will initially be at variance with your physical feel and intuition. Thus, certain phenomena occurring in supersonic fluid flow will seem very strange, since your direct experience with fluid flow (faucets, swimming, and so on) has been entirely subsonic. Because you have not moved with speeds approaching the speed of light and because you have not been prowling around the nucleus of an atom, you will find the tenets of relativity theory and quantum mechanics absolutely bizarre. Should you have little feel for the Coriolis force at this time, do not be unduly concerned (unless you have spent a lot of time moving about high-speed merry-go-rounds). We must in such instances rely on the theory. Working with the theory, we can often build up a strong "physical feel" in the new areas.

This force $\boldsymbol{F}$ is the *total* external force on the mass. Since the man must exert this force and also withstand the pull of gravity (the weight) in the $-\boldsymbol{k}$ direction, the force exerted by the man on the mass is

$$\boldsymbol{F}_{\text{man}} = 3\boldsymbol{j} - 20.1\boldsymbol{i} + \frac{g}{50}\boldsymbol{k} \text{ lb} \tag{a}$$

If the platform were *not* rotating at all, it could serve as an inertial reference. Then, we would have for the total external force $\boldsymbol{F}'$:

$$\boldsymbol{F}' = \tfrac{1}{50}(-5i) = -\tfrac{1}{10}\boldsymbol{i} \text{ lb}$$

The force exerted by the man, $\boldsymbol{F}'_{\text{man}}$, is then

$$\boldsymbol{F}'_{\text{man}} = -\frac{1}{10}\boldsymbol{i} + \frac{g}{50}\boldsymbol{k} \text{ lb} \tag{b}$$

This force is considerably different from that given in Eq. (a).

As a matter of interest, we note that aviators of World War I were required to carry out such maneuvers on a rapidly rotating and accelerating platform so as to introduce them safely to these "peculiar" effects.

*15.10 The Coriolis Force

Of great interest is the Coriolis force, defined in Section 15.9, particularly as it relates to certain terrestrial actions. For many of our problems, the earth's surface serves with sufficient accuracy as an inertial reference. However, where the time interval of interest is large (such as in the flight of rockets, or the flow of rivers, or the movement of winds and ocean currents), we must consider such a reference as noninertial in certain instances and accordingly, when using Newton's law, we must include some or all of the inertial forces given in Eq. 15.27. For such problems (as you will recall from Chapter 12), we often use an inertial reference that has an origin at the center of the earth (see Fig. 15.37) with the Z axis collinear with the N-S axis of the earth and moving such that the earth rotates one revolution per 24 hr relative to the reference. Thus, the reference approaches a translatory motion about the sun. To a high degree of accuracy, it is an inertial reference.

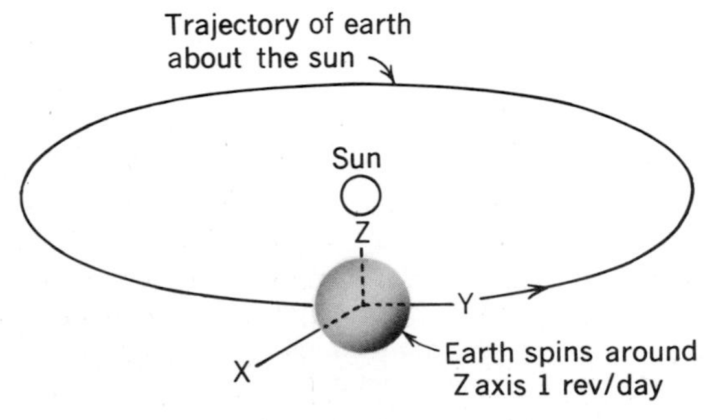

Figure 15.37. Proposed inertial reference.

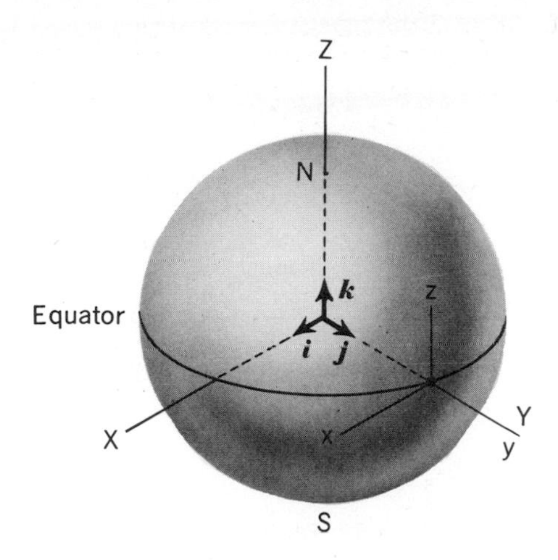

Figure 15.38. *xyz* fixed to earth.

We start by considering particles that are stationary relative to the earth. We choose a reference xyz fixed to the earth at the equator as shown in Fig. 15.38. The angular velocity of xyz fixed anywhere on the earth's surface can readily be evaluated as follows:

$$\boldsymbol{\omega} = \frac{2\pi}{(24)(3600)}\boldsymbol{k} = 7.27 \times 10^{-5}\boldsymbol{k} \text{ rad/sec}$$

Newton's law, in the form of Eq. 15.27, for a "stationary" particle positioned at the origin of xyz simplifies to

$$\boldsymbol{F} - m\ddot{\boldsymbol{R}} = \boldsymbol{0} \tag{15.28}$$

since $\boldsymbol{\rho}$, $\boldsymbol{V}_{xyz}$, and $\boldsymbol{a}_{xyz}$ are zero vectors. Let us next evaluate the inertial force, $-m\ddot{\boldsymbol{R}}$, for the particle, using $R = 3960$ mi:

$$\begin{aligned} -m\ddot{\boldsymbol{R}} &= -m(-|\boldsymbol{R}|\omega^2\boldsymbol{j}) = m(3960)(5280) \times (7.27 \times 10^{-5})^2\boldsymbol{j} \\ &= m(.1105)\boldsymbol{j} \text{ lb} \end{aligned}$$

Clearly, this is a "centrifugal force," as we learned in physics. Note in Fig. 15.39 that the direction of this force is collinear with the gravitational force on a particle, but with opposite sense. Note further that the centrifugal force has a magnitude that is $(.1105\text{ m}/32.2\text{ m}) \times 100 = .34$ of 1% of the gravitational force at the indicated location. Thus, clearly, in the usual engineering problems, such effects are neglected.

Assume that the particle is restrained by a flexible cord. In accordance with to Eq. 15.28, the external force $\boldsymbol{F}$ (which includes gravitational attraction and the force from the cord) and the centrifugal force add up to zero, and hence these forces are in equilibrium. They are shown in Fig. 15.39, in which $\boldsymbol{T}$ represents the contribution of the cord. Clearly, a force $\boldsymbol{T}$ radially out from the center of the earth will restrain the particle, and so the direction of the flexible cord will point toward the center of the earth. On the other hand, at a nonequatorial location this will not be true. The gravity

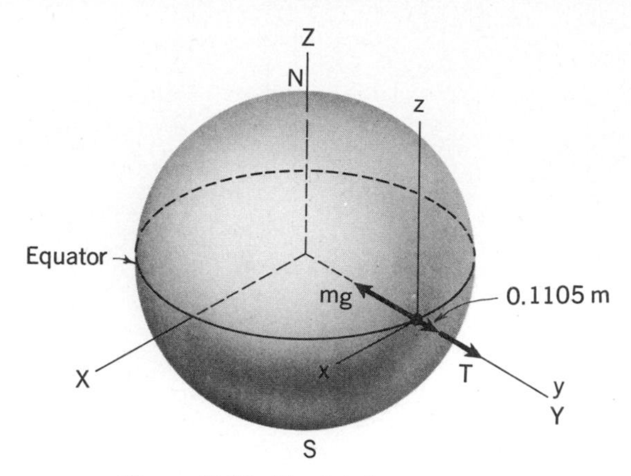

Figure 15.39. Plumb bob at equator.

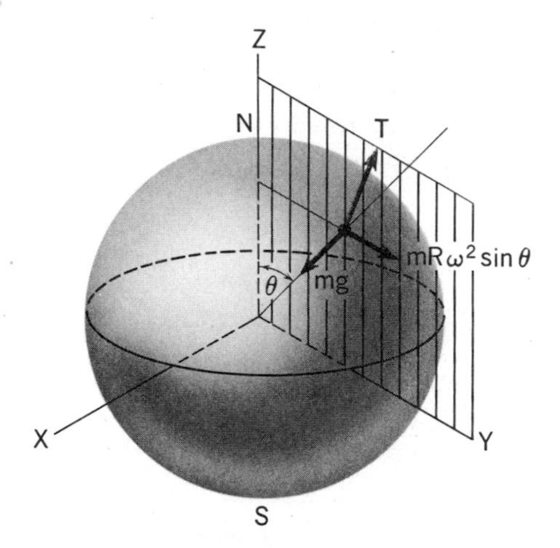

Figure 15.40. Plumb bob does not point to center of earth.

force points toward the center of the earth (see Fig. 15.40), but the centrifugal force—now having the value $m[R(\sin\theta)\omega^2]$—points radially out from the Z axis, and thus T, the restraining force, must be inclined somewhat from a direction toward the center of the earth. Therefore, except at the equator or at the poles (where the centrifugal force is zero), a *plumb bob* does not point directly toward the center of the earth. This deviation is very small and is negligible for most but not all engineering work.

Consider now a body that is held above the earth's surface so as to always be above the same point on the earth's surface. (The body thus moves with the earth with the same angular motion). If the body is released we have what is called a *free fall*. The body will attain initially a downward velocity $\boldsymbol{V}_{xyz}$ relative to the earth's surface (see Fig. 15.41). Now in addition to a centrifugal force described earlier, we have a *Coriolis force* given as

$$\boldsymbol{F}_{\text{Coriolis}} = -2m\boldsymbol{\omega} \times \boldsymbol{V}_{xyz}$$

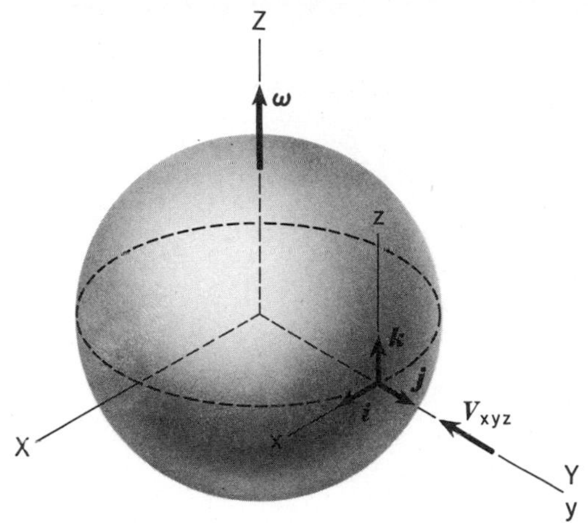

Figure 15.41. Free fall at the equator.

In Fig. 15.42 we have shown the $\boldsymbol{\omega}$ and V_{xyz} vectors. The Coriolis force must point to the right as you should verify (do not forget the minus sign). If we dropped a mass from a position in xyz above a target, therefore, the mass as a result of the Coriolis force would curve slightly away from the target (see Fig. 15.43), even if there were no friction, wind, etc., to complicate matters. Furthermore, the induced motion in the x direction itself induces Coriolis-force components of a smaller order in the y direction, and so forth. You will surely begin to appreciate how difficult a "free fall" can really become when great precision is attempted.

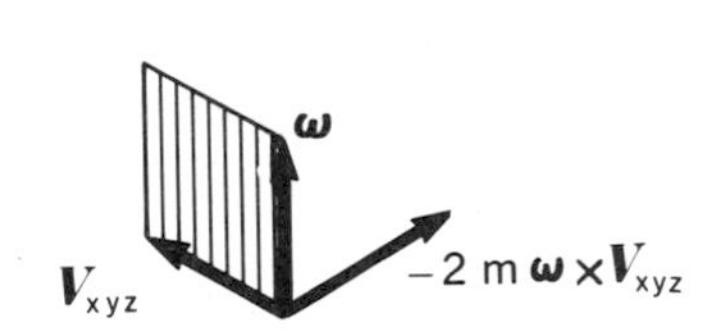

Figure 15.42. Direction of Coriolis force.

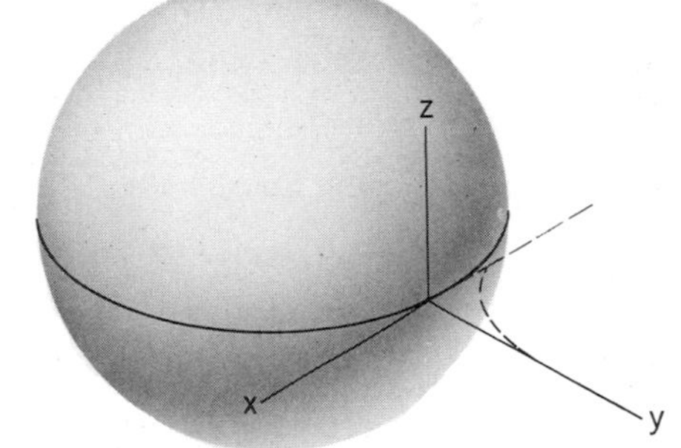

Figure 15.43. Primary Coriolis effect on free fall.

Finally, consider a current of air or a current of water moving in the Northern Hemisphere. In the absence of a Coriolis force, the fluid would move in the direction of the pressure drop. In Fig. 15.44 the pressure drop has been shown for simplicity along a meridian line pointing toward the equator. For fluid motion in this direction, a Coriolis force will be present in the negative y direction and so the fluid will follow the dashed-line path BA. The prime induced motion is to the right of the direction of flow developed by the pressure alone. By similar argument, you can demonstrate that, in the Southern Hemisphere, the Coriolis force induces a motion to the left of the flow

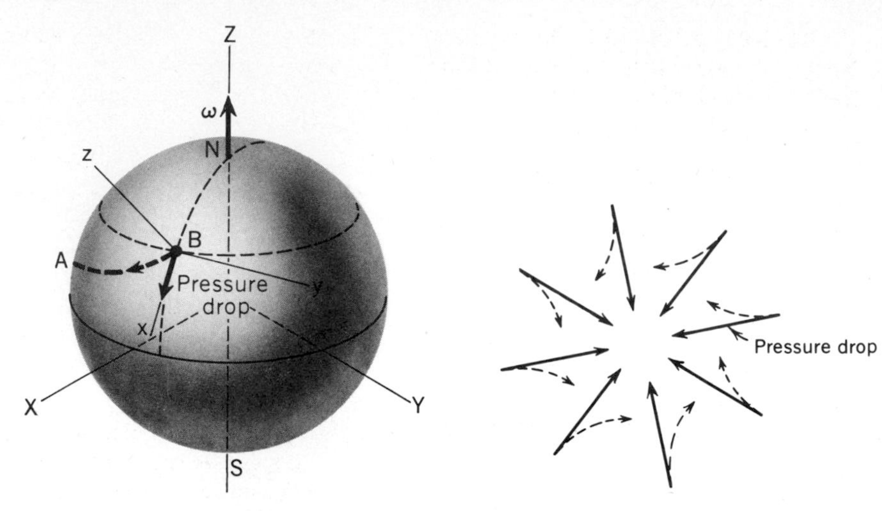

Figure 15.44. Coriolis effect on wind.

Figure 15.45. Beginning of whirlpool.

that would be present under the action of the pressure drop alone. Such effects are of significance in meteorology and oceanography.[13]

The conclusions in the preceding paragraph explain why cyclones and whirlpools rotate in a counterclockwise direction in the Northern Hemisphere and a clockwise direction in the Southern Hemisphere. In order to start, a whirlpool or cyclone needs a low-pressure region with pressure increasing radially outward. The pressure drops are shown as full lines in Fig. 15.45. For such a pressure distribution, the air will begin to move radially inward. As this happens, the Coriolis force causes the fluid in the Northern Hemisphere to swerve to the right of its motion, as indicated by the dashed lines. This result is the beginning of a counterclockwise motion. You can readily demonstrate that in the Southern Hemisphere a clockwise rotation will be induced.

Problems

15.98. A reference xyz is attached to a space probe, which has the following motion relative to an inertial reference XYZ at a time t when the corresponding axes of the references are parallel:

$$\ddot{\boldsymbol{R}} = 100\boldsymbol{j} \text{ m/sec}^2$$

$$\boldsymbol{\omega} = 10\boldsymbol{i} \text{ rad/sec}$$

$$\dot{\boldsymbol{\omega}} = -8\boldsymbol{k} \text{ rad/sec}^2$$

If a force $\boldsymbol{F}$ given as

$$\boldsymbol{F} = 500\boldsymbol{i} + 200\boldsymbol{j} - 300\boldsymbol{k} \text{ N}$$

acts on a particle of mass 1 kg at position

$$\boldsymbol{\rho} = .5\boldsymbol{i} + .6\boldsymbol{j} \text{ m}$$

what is the acceleration vector relative to the probe? The particle has a velocity $\boldsymbol{V}$ relative to xyz of

$$\boldsymbol{V} = 10\boldsymbol{i} + 20\boldsymbol{j} \text{ m/sec}$$

---

[13]Keep in mind that the Coriolis force in these situations is small but, because it persists during long time intervals and because the resultant of other forces is often also small, this force usually must carefully be taken into account in studies of meteorology and oceanography.

15.99. In the space probe of Problem 15.98, what must the velocity vector $\boldsymbol{V}_{xyz}$ of the particle be to have the acceleration

$$\boldsymbol{a}_{xyz} = 495.2\boldsymbol{i} + 100\boldsymbol{j} \text{ m/sec}^2$$

if all other conditions are the same? Is there a component of $\boldsymbol{V}_{xyz}$ that can have any value for this problem?

15.100. A mass A weighing 4 oz is made to rotate at a constant angular speed of $\omega_2 = 15$ rad/sec relative to a platform. This motion is in the plane of the platform, which, at the instant of interest, is rotating at an angular speed $\omega_1 = 10$ rad/sec and decelerating at a rate of 5 rad/sec² relative to the ground. If we neglect the mass of the rod supporting the mass A, what are the axial force and shear force at the base of the rod (i.e., at 0)? The rod at the instant of interest is shown in the diagram. The shear force is the total force acting on a section of the member in a direction *tangent* to the section.

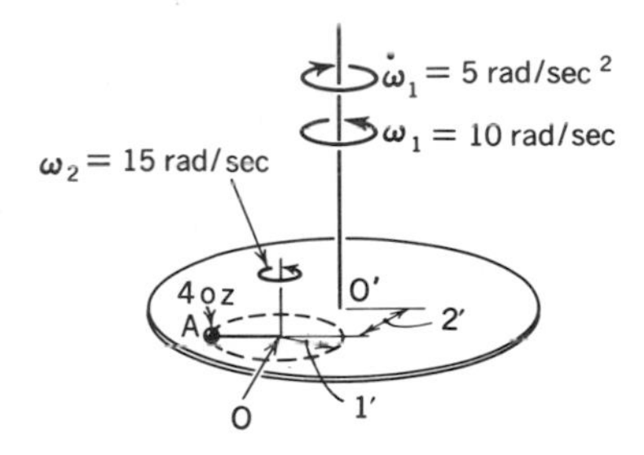

Figure P.15.100

15.101. In Problem 15.79, what is the total external force acting on the cylinder for the case when

$$V = 5 \text{ ft/sec}$$
$$\dot{V} = -2 \text{ ft/sec}^2$$
$$\omega_1 = 2 \text{ rad/sec}$$
$$\dot{\omega}_1 = 1 \text{ rad/sec}^2\,?$$

The mass of the cylinder is 100 lbm.

15.102. In Problem 15.96, what is the sum of the forces on the cylinder for the following data:

$$V = 15 \text{ ft/sec}$$
$$\dot{V} = -3 \text{ ft/sec}^2$$
$$\theta = 60°$$
$$\dot{\theta} = .2 \text{ rad/sec}$$
$$\ddot{\theta} = .1 \text{ rad/sec}^2$$
$$d = 15 \text{ ft}$$

The cylinder weighs 50 lb and has a clockwise angular speed and a clockwise angular acceleration relative to the compartment of 2 rad/sec and 3 rad/sec², respectively.

15.103. A truck is moving at constant speed V of 10 mi/hr. A crane AB is at time t at $\theta = 45°$ with $\dot{\theta} = 1$ rad/sec and $\ddot{\theta} = .2$ rad/sec². Also at time t, the base of AB rotates with speed $\omega_1 = 1$ rad/sec relative to the truck. If AB is 30 ft in length, what is the axial force along AB as a result of mass M of 100 lbm at B?

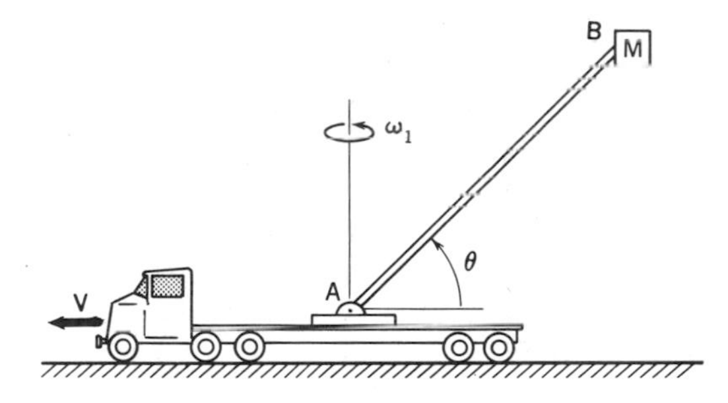

Figure P.15.103

15.104. An exploratory probe shot from the earth is returning to the earth. On entering the earth's atmosphere, it has a constant angular velocity component ω_1 of 10 rad/sec about an axis normal to the page and a constant component ω_2 of 50 rad/sec about the vertical axis. The velocity of the probe at the time of interest is 1300 m/sec vertically with a deceleration of 160 m/sec². A small sphere is rotating at $\omega_3 = 5$ rad/sec inside the probe, as shown. At the time of interest, the probe is oriented so that the trajectory of the sphere in the probe is in the plane of the page and the arm is vertical. What are the axial force in the arm and the bending moment at its base (neglect the mass of the arm) at this instant of time, if the sphere has a mass of 300 g? (The bending moment is the couple moment acting on the cross section of the beam.)

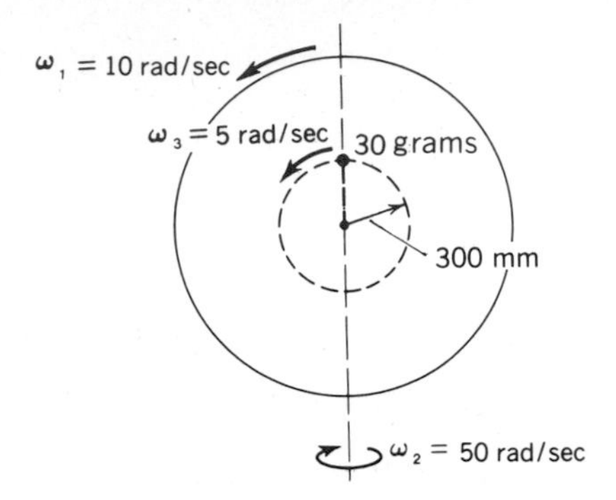

Figure P.15.104

15.105. A clutch assembly is shown. Rods AB are pinned to a disc at B, which rotates at an angular speed $\omega_1 = 1$ rad/sec and $\dot{\omega}_1 = 2$ rad/sec² at time t. These rods extend through a rod EF, which rotates with the rods and at the same time is moving to the left with a speed V of 1 m/sec. At the instant shown, corresponding to time t, what is the axial force on the member AB as a result of the motion of particle A having a mass of .6 kg?

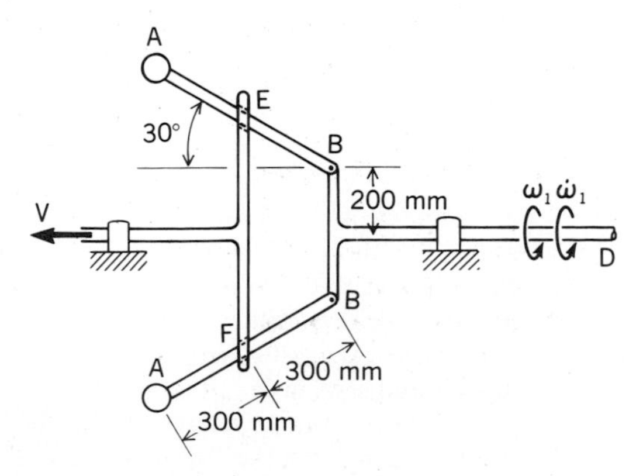

Figure P.15.105

15.106. A flyball governor is shown. The weights C and D each have a mass of 200 g. At the instant of interest, $\theta = 45°$ and the system is rotating about axis AB at a speed ω_1 of 2 rad/sec. At this instant, collar B is moving upward at a speed of .5 m/sec. If we neglect the mass of the members, find the axial forces in the members at the instant of interest. What is the total shear force F_s on the members? (The shear force is the force component tangent to the cross section of the member.)

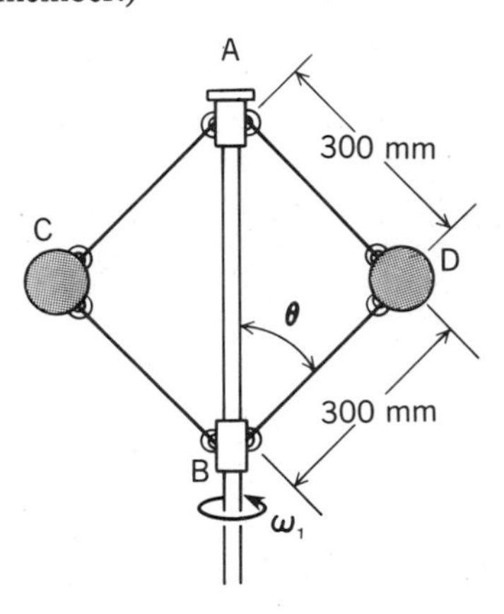

Figure P.15.106

15.107. A man throws a ball weighing 3 oz from one side of a rotating platform to a man diametrically opposite, as shown. What is the Coriolis acceleration and force on the ball? Relative to the platform, in what direction does the ball tend to go as a result of the Coriolis force?

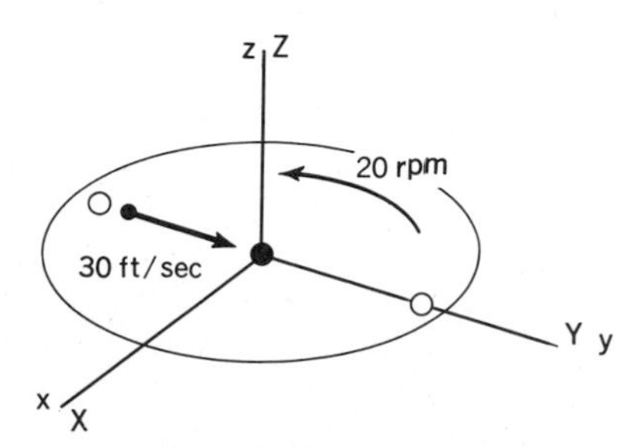

Figure P.15.107

15.108. A river flows at 2 ft/sec average velocity in the Northern Hemisphere at a latitude of 40° in the north–south direction. What is the Coriolis acceleration of the water relative to the center of the earth? What is the Coriolis force on 1 lbm of water?

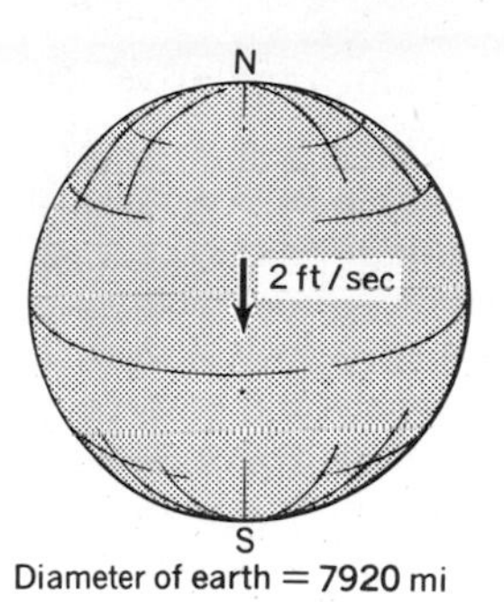

Figure P.15.108

15.11 Closure

In this chapter, we first presented Chasle's theorem for describing the motion of a rigid body. Making use of Chasle's theorem for describing the motion of a reference xyz moving relative to a second reference XYZ, we presented next a simple differentiation formula for vectors $\boldsymbol{A}$ *fixed* in the reference xyz or a rigid body. Thus,

$$\left(\frac{d\boldsymbol{A}}{dt}\right)_{XYZ} = \boldsymbol{\omega} \times \boldsymbol{A}$$

where $\boldsymbol{\omega}$ is the angular velocity of xyz or the rigid body relative to XYZ.

We next considered *two points fixed in a rigid body in the presence of a single reference.* We can relate velocities and accelerations of the points relative to the aforementioned reference as follows:

$$\boldsymbol{V}_b = \boldsymbol{V}_a + \boldsymbol{\omega} \times \boldsymbol{\rho}_{ab}$$

$$\boldsymbol{a}_b = \boldsymbol{a}_a + \dot{\boldsymbol{\omega}} \times \boldsymbol{\rho}_{ab} + \boldsymbol{\omega} \times (\boldsymbol{\omega} \times \boldsymbol{\rho}_{ab})$$

where $\boldsymbol{\omega}$ is the angular velocity of the body relative to the single reference. These relations can be valuable in studies of kinematics of machine elements.

We then considered *one particle in the presence of two references xyz and XYZ.* We expressed the velocity and acceleration as seen from the two references as follows:

$$\boldsymbol{V}_{XYZ} = \boldsymbol{V}_{xyz} + \dot{\boldsymbol{R}} + \boldsymbol{\omega} \times \boldsymbol{\rho}$$

$$\boldsymbol{a}_{XYZ} = \boldsymbol{a}_{xyz} + \ddot{\boldsymbol{R}} + 2\boldsymbol{\omega} \times \boldsymbol{V}_{xyz} + \dot{\boldsymbol{\omega}} \times \boldsymbol{\rho} + \boldsymbol{\omega} \times (\boldsymbol{\omega} \times \boldsymbol{\rho})$$

In computing $\boldsymbol{V}_{xyz}$, $\boldsymbol{a}_{xyz}$, $\dot{\boldsymbol{R}}$, and $\ddot{\boldsymbol{R}}$, we use the various techniques presented in Chapter 11 for computing the velocity and acceleration of a particle relative to a given reference. Thus, use can be made of Cartesian components, path components, and cylindrical components as presented in that chapter. We then explored some interesting and often unexpected effects that occur when we use a noninertial reference. You will have occasion to use these important formulations in your basic studies of solid and fluid mechanics as well as in your courses in kinematics of machines and machine design.

Now that we can express the motion of a rigid body in terms of a velocity vector

$\dot{\boldsymbol{R}}$ and an angular velocity vector $\boldsymbol{\omega}$, our next job will be to relate these quantities with the forces acting on the body. You may recall from your physics course and from the end of Chapter 14 that for a body rotating about a fixed axis in an inertial reference, we could relate the torque T and the angular acceleration α as

$$T = I\alpha$$

where I is the mass moment of inertia of the body about the axis of rotation. In Chapter 16, we shall see that this motion is a special case of plane motion, which itself is a special case of general motion.

Review Problems

15.109. A light plane is circling an airport at constant elevation. The radius R of the path $= 3$ km and the speed of the plane is 120 km/hr. The propeller of the plane is rotating at 100 rpm relative to the plane in a clockwise sense as seen by the pilot. What are $\boldsymbol{\omega}$, $\dot{\boldsymbol{\omega}}$, and $\ddot{\boldsymbol{\omega}}$ of the propeller as seen from the ground at the instant shown in the diagram?

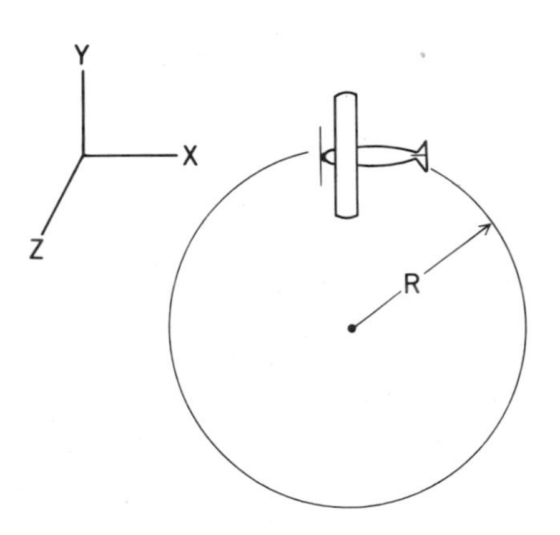

Figure P.15.109

***15.110.** In Problem 15.7, find the angular acceleration $\dot{\boldsymbol{\omega}}$ of the disc for the configuration shown in the diagram, if, at the instant shown, the following data apply:

$$\omega_1 = 3 \text{ rad/sec}$$
$$\dot{\omega}_1 = 2 \text{ rad/sec}^2$$
$$\omega_2 = -10 \text{ rad/sec}$$
$$\dot{\omega}_2 = -4 \text{ rad/sec}^2$$

15.111. A slider A has at the instant of interest a speed V_A of 3 m/sec with a deceleration of 2 m/sec². Compute the angular velocity and angular acceleration of bar AB at the instant of interest. What is position of the instantaneous axis of rotation of bar AB?

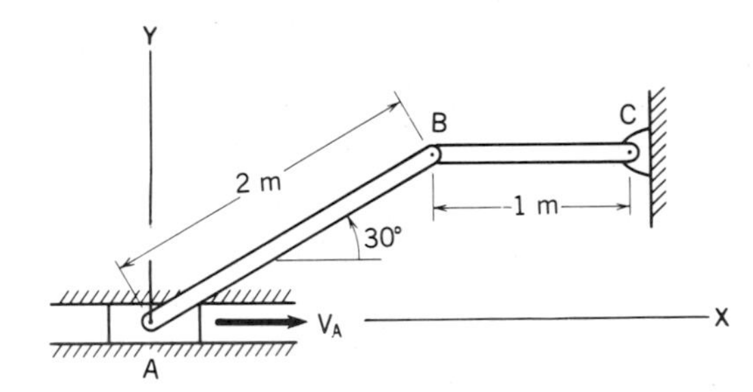

Figure P.15.111

15.112. A cylinder C rolls without slipping on a half-cylinder D. Rod BA is 7 m long and is connected at A to a slider which at the instant of interest is moving in a groove at the speed V of 3 m/sec and increasing its speed at the rate of 2 m/sec². What is the angular speed and the angular acceleration of cylinder C relative to the ground?

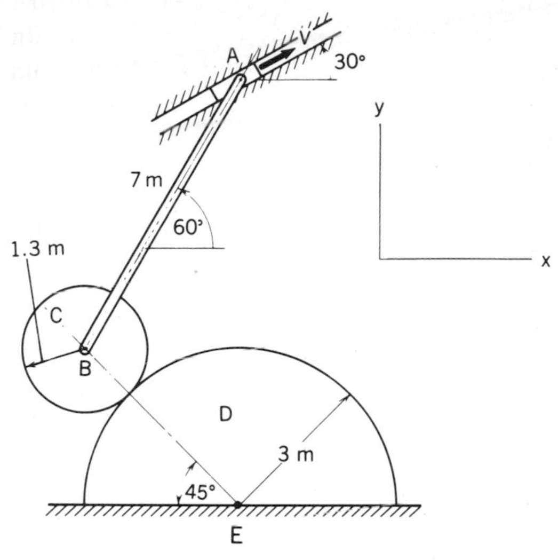

Figure P.15.112

***15.113.** A wheel is rotating with a constant angular speed ω_1 of 10 rad/sec relative to a platform, which in turn is rotating with a constant angular speed ω_2 of 5 rad/sec relative to the ground. Find the velocity and acceleration relative to the ground at a point b on the wheel at the instant when it is directly vertically above point a.

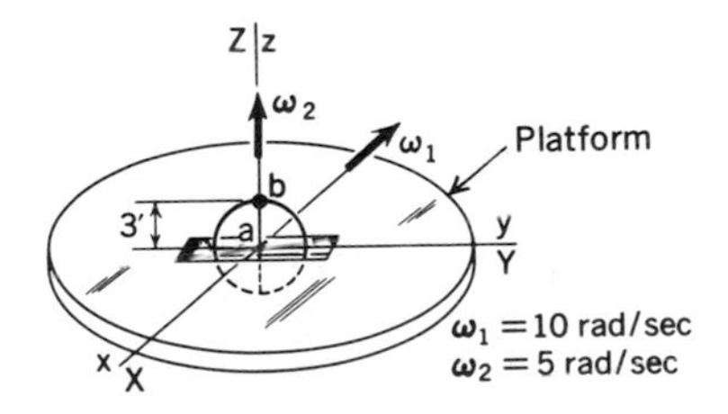

Figure P.15.113

***15.114.** Solve Problem 15.113 for the case where ω_1 is increasing in value at the rate of 5 rad/sec² and where ω_2 is increasing in value at the rate of 10 rad/sec².

***15.115.** A barge is shown with a derrick arrangement. The main beam AB is 40 ft in length. The whole system at the instant of interest is rotating with a speed ω_1 of 1 rad/sec and an acceleration $\dot{\omega}_1$ of 2 rad/sec² relative to the barge. Also, at this instant $\theta = 45°$, $\dot{\theta} = 2$ rad/sec, and $\ddot{\theta} = 1$ rad/sec². What are the velocity and acceleration of point B relative to the barge?

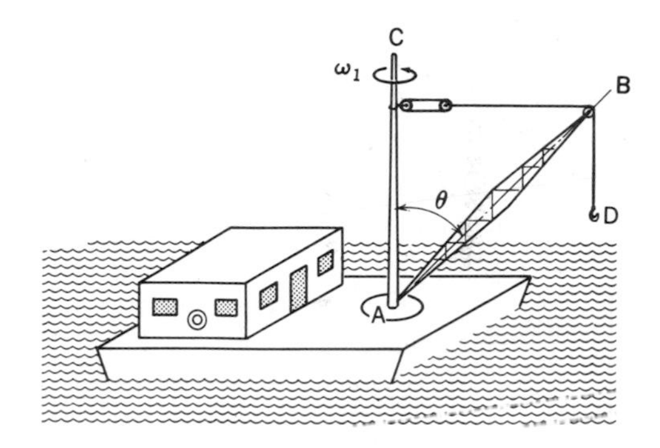

Figure P.15.115

***15.116.** In Problem 15.74, find the acceleration of the collar C relative to the ground if for the configuration shown:

$$\dot{\theta} = 5 \text{ rad/sec}$$

$$\ddot{\theta} = 8 \text{ rad/sec}^2$$

***15.117.** In Example 15.11, find axial force for the beam AB at A resulting from cockpit C, which weighs (with occupant) $136g$ N. The following data apply:

$$\beta = 20°$$

$$\alpha = 60°$$

$$\omega_1 = .2 \text{ rad/sec}$$

$$\omega_2 = .1 \text{ rad/sec}$$

16 Kinetics of Plane Motion of Rigid Bodies

16.1 Introduction

In kinematics we learned that the motion of a rigid body at any time t can be considered to be a superposition of a translational motion and a rotational motion. The translational motion may have the actual instantaneous velocity of any point of the body, and the angular velocity of the rotation, $\boldsymbol{\omega}$, then has its axis of rotation through the chosen point. A convenient point is, of course, the center of mass of the rigid body. The translatory motion can then be found from particle dynamics. You will recall that the motion of the center of mass of any aggregate of particles (this includes a rigid body) is related to the total external force by the equation

$$\boldsymbol{F} = M\dot{\boldsymbol{V}}_c \tag{16.1}$$

where M is the total mass of the aggregate. Integrating this equation, we get the motion of the center of mass. To ascertain fully the motion of the body, we must next find $\boldsymbol{\omega}$. As we saw in Chapter 14,

$$\boldsymbol{M}_A = \dot{\boldsymbol{H}}_A \tag{16.2}$$

for any system of particles where the point A about which moments are to be taken can be (1) the mass center, (2) a point fixed in an inertial reference, or (3) a point accelerating toward or away from the mass center. For these points, we shall later show that the angular velocity vector $\boldsymbol{\omega}$ is involved in the equation above when it is applied to rigid bodies. Also, the inertia tensor will be involved. After we find the motion of the mass center from Eq. 16.1 and the angular velocity $\boldsymbol{\omega}$ from Eq. 16.2, we get the instantaneous motion by letting the entire body have the velocity $\boldsymbol{V}_c$ plus the angular velocity $\boldsymbol{\omega}$, with the axis of rotation going through the center of mass.

16.2 Moment-of-Momentum Equations

Consider now a rigid body wherein each particle of the body moves parallel to a plane. Such a body is said to be in *plane motion* relative to this plane. We shall consider that axes XY are in the aforementioned plane in the ensuing discussion. The Z axis is then normal to the velocity vector of each point in the body. Furthermore, we consider only the situation where XYZ is an *inertial reference*. A body undergoing plane motion relative to XYZ as described above is shown in Fig. 16.1.

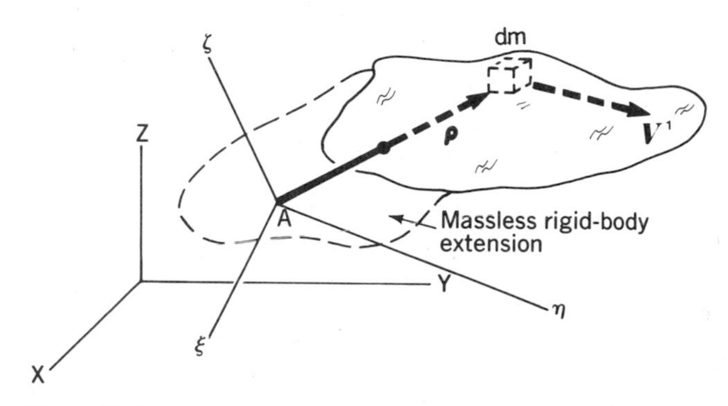

Figure 16.1. Body undergoing plane motion parallel to XY plane. $\xi\eta\zeta$ translates with A.

Choose some point A in this body or in a hypothetical massless extension of this body. An element dm of the body is shown at a position $\boldsymbol{\rho}$ from A. The velocity $\boldsymbol{V}'$ of dm relative to A is simply the velocity of dm relative to a reference $\xi\eta\zeta$, which *translates* with A relative to XYZ. Similarly, the linear momentum of dm relative to A (i.e., $\boldsymbol{V}'\,dm$) is the linear momentum of dm relative to $\xi\eta\zeta$ translating with A. We can now give the moment of this momentum (i.e., the angular momentum) $d\boldsymbol{H}_A$ about A as

$$d\boldsymbol{H}_A = \boldsymbol{\rho} \times \boldsymbol{V}'\,dm = \boldsymbol{\rho} \times \left(\frac{d\boldsymbol{\rho}}{dt}\right)_{\xi\eta\zeta} dm$$

But since A is fixed in the body (or in a hypothetical massless extension of the body) the vector $\boldsymbol{\rho}$ must be fixed in the body and, accordingly,

$$\left(\frac{d\boldsymbol{\rho}}{dt}\right)_{\xi\eta\zeta} = \boldsymbol{\omega} \times \boldsymbol{\rho}$$

where $\boldsymbol{\omega}$ is the angular velocity of the body relative to $\xi\eta\zeta$. However, since $\xi\eta\zeta$ translates relative to XYZ, $\boldsymbol{\omega}$ *is the angular velocity of the body relative to* XYZ as well. Hence, we can say:

$$d\boldsymbol{H}_A = \boldsymbol{\rho} \times (\boldsymbol{\omega} \times \boldsymbol{\rho})\,dm \tag{16.3}$$

Note that the angular velocity $\boldsymbol{\omega}$ for plane motion relative to the XY plane must have a direction *normal* to the XY plane.

We now *fix* a reference xyz to the body at point A such that the z axis is normal to the plane of motion while the other axes have arbitrary orientation (see Fig. 16.2).

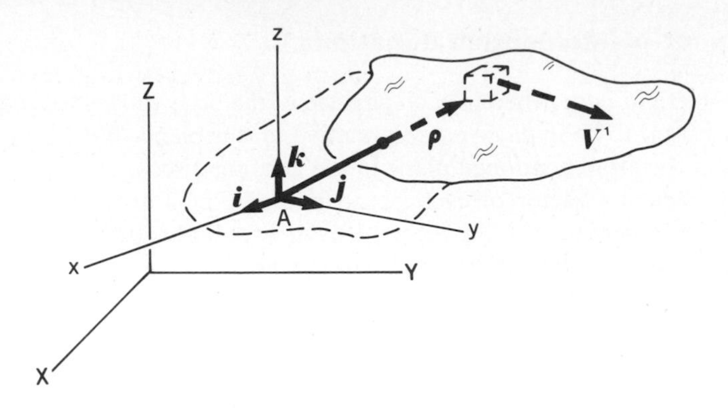

Figure 16.2. *xyz* fixed to body at *A*.

Note that the z axis will remain normal to XY as the body moves because of the plane motion restriction. Next, we evaluate Eq. 16.3 in terms of components relative to xyz as follows:

$$(dH_A)_x\boldsymbol{i} + (dH_A)_y\boldsymbol{j} + (dH_A)_z\boldsymbol{k} = (x\boldsymbol{i} + y\boldsymbol{j} + z\boldsymbol{k}) \times [(\omega\boldsymbol{k}) \times (x\boldsymbol{i} + y\boldsymbol{j} + z\boldsymbol{k})]\, dm$$

The scalar equations resulting from the foregoing vector equations are

$$(dH_A)_x = -\omega xz\, dm$$
$$(dH_A)_y = -\omega yz\, dm$$
$$(dH_A)_z = \omega(x^2 + y^2)\, dm$$

Integrating over the entire body, we get[1]

$$(H_A)_x = -\iiint_M \omega xz\, dm = -\omega \iiint_M xz\, dm = -\omega I_{xz}$$
$$(H_A)_y = -\iiint_M \omega yz\, dm = -\omega \iiint_M yz\, dm = -\omega I_{yz} \qquad (16.4)$$
$$(H_A)_z = \iiint_M \omega(x^2 + y^2)\, dm = \omega \iiint_M (x^2 + y^2)\, dm = \omega I_{zz}$$

We now have the angular momentum components for reference xyz at A. Note that, because xyz is *fixed* to the body, the inertia terms I_{xz}, I_{yz}, and I_{zz} must be *constants*.

In order to employ the moment-of-momentum equation, $\boldsymbol{M}_A = \dot{\boldsymbol{H}}_A$, at time t, we next restrict point A of the body to be any one of the following three cases[2]:

[1] Note that any massless extension of the rigid body has zero density and hence does not contribute to the integration.

[2] The "body" in this discussion includes the hypothetical, massless, rigid-body extension as well as the actual body.

1. Point A of the body is the *center of mass* of the body.
2. Point A of the body is *fixed* or moving with *constant velocity* at time t in inertial reference XYZ (i.e., point A has zero acceleration at time t relative to XYZ).
3. Point A of the body is *accelerating* toward or away from the mass center at time t.

Using the results of Section 15.6 involving the relation between derivatives of vectors as seen from different references, we next can say, using references XYZ and xyz:

$$\left(\frac{d\boldsymbol{H}_A}{dt}\right)_{XYZ} = \left(\frac{d\boldsymbol{H}_A}{dt}\right)_{xyz} + \boldsymbol{\omega} \times \boldsymbol{H}_A$$

where $\boldsymbol{\omega}$ is the angular velocity of xyz and thus the body relative to XYZ. Hence, the moment-of-momentum equation can be stated as follows:

$$\boldsymbol{M}_A = \left(\frac{d\boldsymbol{H}_A}{dt}\right)_{xyz} + \boldsymbol{\omega} \times \boldsymbol{H}_A \tag{16.5}$$

Using Eq. 16.4 for the components of $\boldsymbol{H}_A$, we get for this equation:

$$\boldsymbol{M}_A = \frac{d}{dt_{xyz}}(-\omega I_{xz}\boldsymbol{i} - \omega I_{yz}\boldsymbol{j} + \omega I_{zz}\boldsymbol{k}) + \omega \boldsymbol{k} x(-\omega I_{xz}\boldsymbol{i} - \omega I_{yz}\boldsymbol{j} + \omega I_{zz}\boldsymbol{k})$$

Noting that $\boldsymbol{i}, \boldsymbol{j}$, and $\boldsymbol{k}$ are constant vectors as seen from xyz, as are the inertia terms, we get

$$\boldsymbol{M}_A = -\dot{\omega} I_{xz}\boldsymbol{i} - \dot{\omega} I_{yz}\boldsymbol{j} + \dot{\omega} I_{zz}\boldsymbol{k} - \omega^2 I_{xz}\boldsymbol{j} + \omega^2 I_{yz}\boldsymbol{i}$$

The scalar forms of the equation above are then

$$(M_A)_x = -I_{xz}\dot{\omega} + I_{yz}\omega^2 \tag{16.6a}$$

$$(M_A)_y = -I_{yz}\dot{\omega} - I_{xz}\omega^2 \tag{16.6b}$$

$$(M_A)_z = I_{zz}\dot{\omega} \tag{16.6c}$$

These are the *general angular momentum equations for plane motion*. The last equation is probably familiar to you from your work in physics. There you expressed it as

$$T = I\alpha \tag{16.7}$$

or as

$$T = I\ddot{\theta} \tag{16.8}$$

We shall now consider special cases of plane motion, starting with the most simple case and going toward the most general case. However, please remember that for *all* plane motions relative to an inertial reference, the moment of the forces about the z axis at A *always* equals $I_{zz}\dot{\omega}$. The other two equations of 16.6 may get simplified for various special plane motions.

16.3 Pure Rotation of a Body of Revolution About Its Axis of Revolution

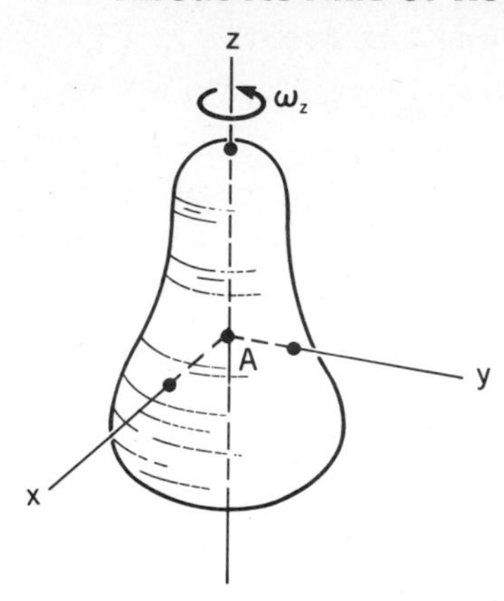

Figure 16.3. Rigid uniform body of revolution.

A uniform body of revolution is shown in Fig. 16.3. If the body undergoes pure rotation about the axis of revolution in inertial space, we then have plane motion parallel to any plane for which the axis of revolution is a normal. A reference xyz is fixed to the body such that the z axis is collinear with the axis of revolution. Since all points along the axis of revolution are fixed in inertial space, we can choose for the origin of reference xyz any point A along this axis. The x and y axes forming a righthanded triad then have arbitrary orientation. Clearly, the plane zy is a plane of symmetry for this body, and the x axis is normal to this plane of symmetry. From our work in Chapter 9, recall[3] that, as a consequence, $I_{xy} = I_{xz} = 0$. Similarly, with y normal to a plane of symmetry, xz, we conclude that $I_{yx} = I_{yz} = 0$. Hence, xyz are principal axes. Returning to Eq. 16.6, we find that only one equation of the set has nonzero moment, and that is the familiar equation

$$M_z = I_{zz}\dot{\omega}_z \tag{16.9}$$

The other pair of equations from 16.6 yield

$$\begin{aligned} M_x &= 0 \\ M_y &= 0 \end{aligned} \tag{16.10}$$

Since the center of mass of the body is stationary at all times (it is on the axis of rotation), we can say from *Newton's law*:

$$\begin{aligned} \sum F_x &= 0 \\ \sum F_y &= 0 \\ \sum F_z &= 0 \end{aligned} \tag{16.11}$$

Thus, the applied forces at any time t, the supporting forces, and the weight of the body of revolution satisfy *all* the equations of equilibrium *except* for motion about the axis of revolution where Eq. 16.9 applies.

Notice that the key equation (16.9) has the *same form* as Newton's law for *rectilinear translation* of a particle along an axis, say the x axis. We write both equations together as follows:

$$M_z = I_{zz}\ddot{\theta} \tag{16.12a}$$

$$F_x = M\ddot{x} \tag{16.12b}$$

[3]We pointed out in Chapter 9 that if an axis, such as the x axis, is normal to a plane of symmetry, then the products of inertia with x as a subscript must be zero. This is similarly true for other axes normal to a plane of symmetry.

In Chapter 12 we integrated Eq. 16.12b for various kinds of force functions: time functions, velocity functions, and position functions. The same techniques used then to integrate Eq. 16.12b can now be used to integrate Eq. 16.12a, where the moment functions can also be time functions, angular velocity functions, and angular position functions.

We illustrate these possibilities in the following three examples. The first example involves a torque which in part is a function of angular position θ.

EXAMPLE 16.1

A stepped cylinder having a radius of gyration $k = .40$ m and a mass of 200 kg is shown in Fig. 16.4. The cylinder supports a weight W of mass 100 kg with an inextensible cord and is restrained by a linear spring whose constant K is 2 N/mm. What is the angular acceleration of the stepped cylinder when it has rotated 10° after it is released from a state of rest? The spring is initially unstretched. What are the supporting forces at this time?

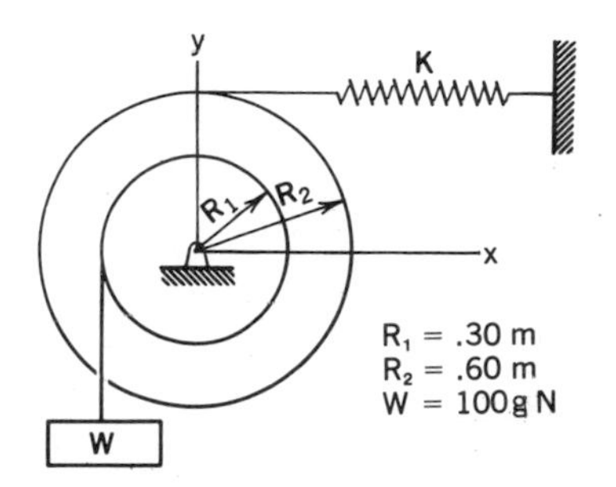

Figure 16.4. Stepped cylinder. xy is stationary.

We have shown free body diagrams of the stepped cylinder and the weight W in Fig. 16.5. A tension T from the cord is shown acting both on the weight W and the stepped cylinder. We can apply the *moment-of-momentum* equation about the center-line of the cylinder:

$$TR_1 - KR_2^2\theta = I\ddot{\theta} = (Mk^2)\ddot{\theta}$$

Therefore,

$$T(.30) - [(2)(1000)](.60)^2\theta = (200)(.40)^2\ddot{\theta} \quad \text{(a)}$$

where θ is the rotation of the cylinder from a position corresponding to the unstretched condition of the spring. Now considering the weight W, which is in a translatory motion, we can say, from *Newton's law*, taking reference xy as stationary

$$T - W = M\ddot{y}$$

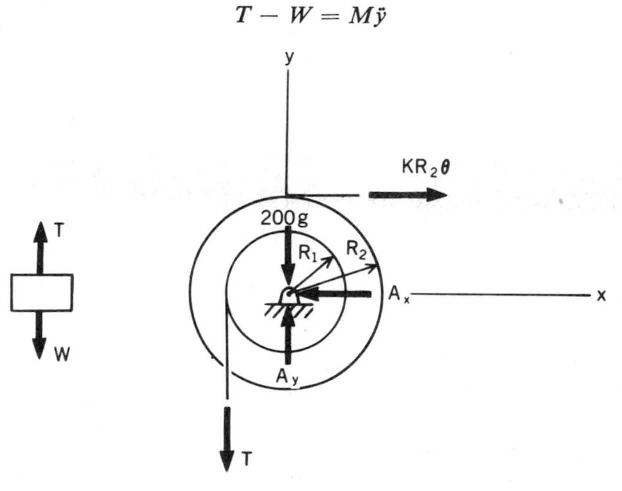

Figure 16.5. Free-body diagrams of components.

Therefore,

$$T - (100)(9.81) = 100\ddot{y} \tag{b}$$

From *kinematics* we note that

$$R_1\ddot{\theta} = -\ddot{y}$$

Therefore,

$$.30\ddot{\theta} = -\ddot{y} \tag{c}$$

Substituting for T in Eq. (a) using Eq. (b) and for $\ddot{y}$ using Eq. (c), we then have

$$[(100)(9.81) + (100)(-.30\ddot{\theta})](.30) - [(2)(1000)](.60)^2\theta = (200)(.40)^2\ddot{\theta} \tag{d}$$

When $\theta = (10°)(2\pi/360°) = .1745$ rad, we get for $\ddot{\theta}$ from Eq. (d) the desired result:

$$\ddot{\theta} = 4.11 \text{ rad/sec}^2$$

From Eqs. (b) and (c), we then have for T:

$$T = 981 - 123.3 = 858 \text{ N}$$

We next use *Newton's law* for the center of mass A of the stepped cylinder. Thus, considering Fig. 16.5 we can say, realizing that the center of mass is in equilibrium:

$$-858 - 200g + A_y = 0$$

$$A_y = 2820 \text{ N}$$

$$-A_x + [(2)(1000)](.60)(.1745) = 0$$

$$A_x = 209 \text{ N}$$

It should be clear on examining Fig. 16.4 that the motion of the cylinder, after W is released from rest, will be rotational oscillation. This motion ensues because the spring develops a restoring torque much as the spring in the classic spring–mass system (Fig. 16.6) supplies a restoring force. We shall study torsional oscillation or

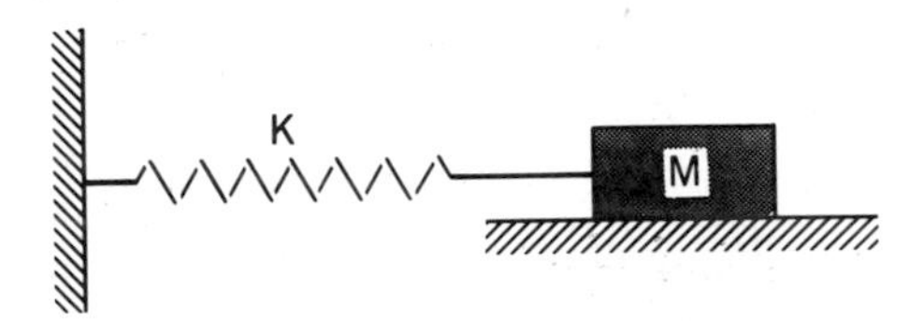

Figure 16.6. Classic spring–mass system.

vibration in Chapter 19 when we consider the vibration of particles in rectilinear translation. The key concepts and mathematical techniques for both motions you will find to be identical.

The torque in the next example is, in part, a function of time.

EXAMPLE 16.2

A mechanically powered windlass is shown in Fig. 16.7. A torque T drives gear C, which in turn drives gear B and drum A. A mass D of 800 kg is being raised by the windlass. The torque T is given as

$$T = 300 + 15t \text{ N-m}$$

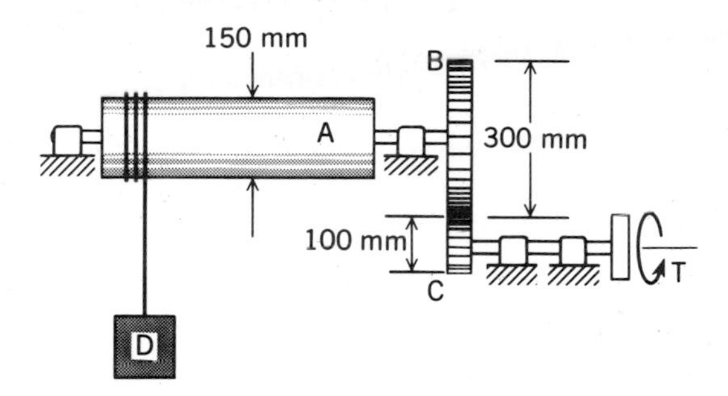

Figure 16.7. Powered windlass.

where t is given in seconds. If the system starts from rest, what distance d does body D move in 5 sec? The combined radius of gyration of drum A, gear B, and the connecting shaft is 200 mm, and the total mass of these bodies is 100 kg. The radius of gyration of gear C and associated shaft is 80 mm, with a total mass of 10 kg.

We have what can be considered as two interconnected bodies of revolution in pure rotation plus a third body in translation. We show the free-body diagram of each of these three bodies in Fig. 16.8. The *moment-of-momentum* equation for the driving system [Fig. 16.8(b)] is then

$$T - (F)(.050) = (10)(.08)^2\ddot{\theta}_1 \tag{a}$$

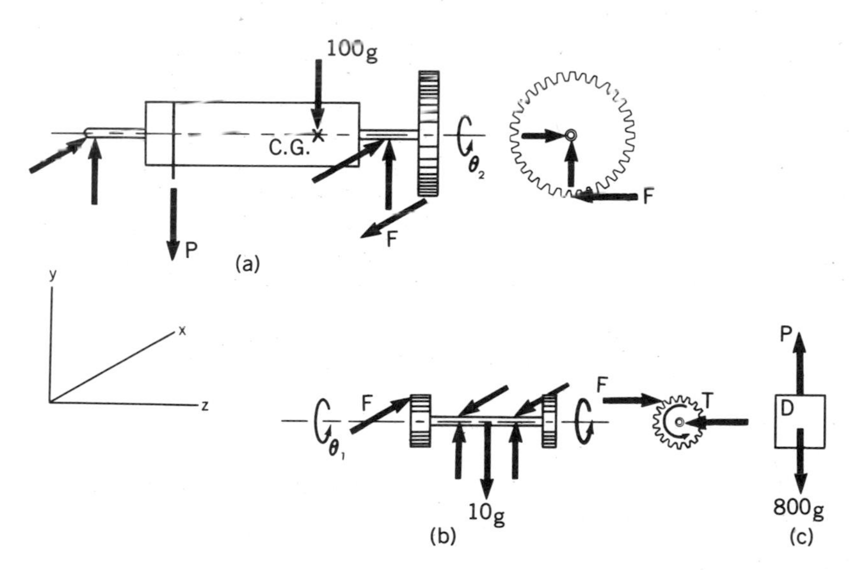

Figure 16.8. Free-body diagrams.

Similarly, for the windlass system [Fig. 16.8(a)] we have

$$P\left(\frac{.15}{2}\right) - F(.15) = (100)(.20)^2\ddot{\theta}_2 \tag{b}$$

Finally, *Newton's law* for body D is

$$P - 800g = 800\ddot{y} \tag{c}$$

From *kinematics*, we can say:

$$\ddot{\theta}_2 = -\frac{1}{3}\ddot{\theta}_1 \tag{d}$$

$$\ddot{y} = -\left(\frac{.15}{2}\right)\ddot{\theta}_2 \tag{e}$$

Replacing $\ddot{\theta}_1$ and $\ddot{\theta}_2$ in terms of $\ddot{y}$ in Eqs. (a) and (b), and replacing P in Eq. (b), using Eq. (c), we have for Eqs. (a) and (b)

$$T - (.05)F = (10)(.08)^2[(40)\ddot{y}] \tag{f}$$

$$(800g + 800\ddot{y})\left(\frac{.15}{2}\right) - F(.15) = (100)(.20)^2\left(-\frac{2}{.15}\ddot{y}\right) \tag{g}$$

Now solving for F in Eq. (f) and substituting into Eq. (g), we get

$$(800g + 800\ddot{y})\left(\frac{.15}{2}\right) - 3[T - (10)(.08)^2 40\ddot{y}] = (100)(.20)^2\left(-\frac{2}{.15}\ddot{y}\right) \tag{h}$$

This equation can be simplified to

$$T - 196.2 = 40.3\ddot{y}$$

Finally, putting in the proper time function for T and rearranging, we get

$$40.3\ddot{y} = 15t + 103.8$$

We can integrate this equation twice as follows:

$$40.3\dot{y} = \frac{15t^2}{2} + 103.8t + C_1$$

$$40.3y = \frac{15t^3}{6} + 103.8\frac{t^2}{2} + C_1 t + C_2$$

Taking $y = 0$ at $t = 0$ and noting that $\dot{y} = 0$ at $t = 0$, we see that $C_1 = C_2 = 0$. The distance d then is found by substituting $t = 5$ sec in the last equation.

$$d = \frac{1}{40.3}\left[\frac{15}{6}(5^3) + \frac{103.8}{2}(5^2)\right]$$

$$= 39.9 \text{ m}$$

16.4 Pure Rotation of a Body with Two Orthogonal Planes of Symmetry

Consider next a uniform body having *two* orthogonal planes of symmetry. Such a body is shown in Fig. 16.9, where in (a) we have shown the aforementioned planes of symmetry and in (b) we have shown a view along the intersection of the planes of symmetry. We shall consider pure rotation of such a body about an axis collinear with the intersection of the planes of symmetry, which we take as the z axis. The origin A can be taken anywhere along the axis of rotation, and the x and y axes are taken in the planes of symmetry, as shown in the diagram. We leave it to the reader to show

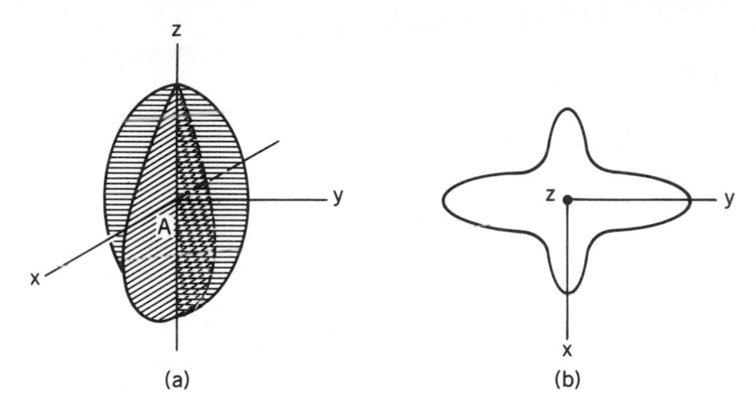

Figure 16.9. Body with two orthogonal planes of symmetry.

that the identical equations apply to this case as to the previous case of a body of revolution.

The torque in the next example is a function of angular speed.

EXAMPLE 16.3

A thin-walled shaft is shown in Fig. 16.10. On it are welded identical plates A and B, each having a mass of 10 kg. Also welded onto the shaft at right angles to A and B are two identical plates C and D, each having a mass of 6 kg. The thin-walled shaft is of diameter 100 mm and has a mass of 15 kg. The wind resistance to rotation of this system is given for small angular velocities as $.2\dot{\theta}$ N-m, with $\dot{\theta}$ in rad/sec. What is the time required for the system to reach 100 rpm if a torque T of 5 N-m is applied? What are the forces on the bearings G and E when this speed is reached?

The *moment-of-momentum* equation about the axis of rotation is, if we use formulas for I_{zz} of a rectangular body plus the parallel-axis theorem:

$$5 - .2\dot{\theta}$$
$$= \{(15)(.05)^2 + 2[\tfrac{1}{12}(10)(.30^2 + .01^2) + 10(.20)^2] + 2[\tfrac{1}{12}(6)(.10^2 + .01^2) + (6)(.10)^2]\}\ddot{\theta}$$

This becomes

$$5 - .2\dot{\theta} = 1.118\,\frac{d\dot{\theta}}{dt} \qquad \text{(a)}$$

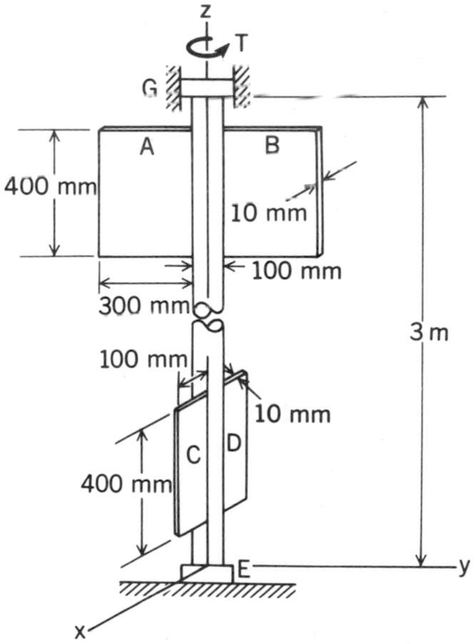

Figure 16.10. Device with rotational frictions.

We can separate the variables as follows:

$$\frac{1.118\,d\dot{\theta}}{5 - .2\dot{\theta}} = dt \qquad \text{(b)}$$

Now make the following substitution of variable:

$$5 - .2\dot{\theta} = \eta$$

Therefore, taking the differential

$$-.2d\dot{\theta} = d\eta$$

Hence, we have for Eq. (b):

$$-5.59\frac{d\eta}{\eta} = dt$$

Integrate to get

$$-5.59 \ln \eta = t + C_1$$

Hence, replacing η

$$-5.59 \ln (5 - .2\dot{\theta}) = t + C_1 \quad \text{(c)}$$

When $t = 0$, $\dot{\theta} = 0$, and we have for C_1:

$$-5.59 \ln (5) = C_1$$

Therefore,

$$C_1 = -8.99$$

Hence, Eq. (c) becomes

$$-5.59 \ln (5 - .2\dot{\theta}) = t - 8.99$$

Let $\dot{\theta} = (100/60)(2\pi) = 10.47$ rad/sec in the above equation. The desired time t for this speed to be reached then is

$$t = 3.03 \text{ sec}$$

We now consider the supporting forces for the system. For reasons set forth in Section 16.3 we know that

$$M_x = 0$$
$$M_y = 0$$

for the other moment of momentum equations. Also, from *Newton's law*,

$$\sum F_x = 0$$
$$\sum F_y = 0$$
$$\sum F_z = 0$$

for the center of mass. Clearly, the dead weights of bodies (in the z direction) give rise to a constant supporting force of $[2(10 + 6) + 15]g = 461$ N at bearing E. All other forces are zero.

16.5 Pure Rotation of Slablike Bodies

We now consider bodies that have a *single* plane of symmetry, such as is shown in Fig. 16.11. Such bodies we shall call *slablike* bodies. We have oriented the body in Fig. 16.8 so that the plane of symmetry is parallel to the XY plane. We shall now consider the pure rotation of such a body about a fixed axis normal to the XY plane and going through any point A in the plane of symmetry of the body. We fix a reference xyz at point A with xy in the plane of symmetry and z along the axis of rotation. The angular velocity $\boldsymbol{\omega}$ is then along the z axis. Since z is normal to the plane of symmetry, it is clear immediately that $I_{zx} = I_{zy} = 0$. And so Eqs. 16.6 become for this case:

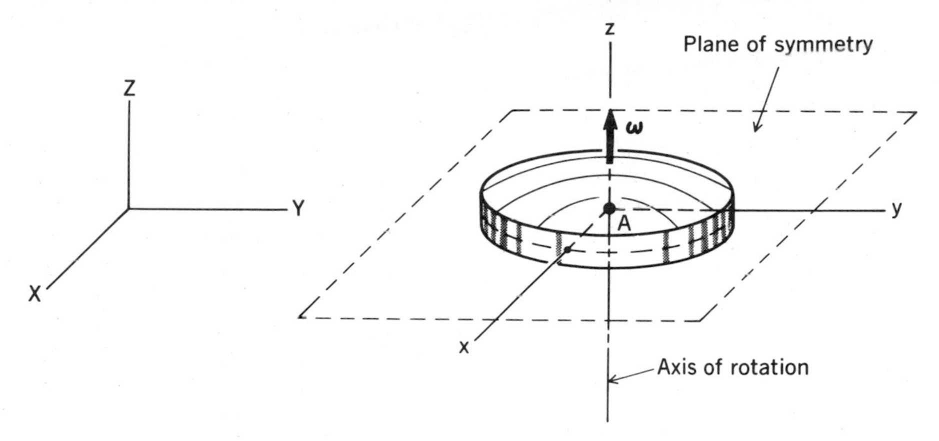

Figure 16.11. Slablike body undergoing pure rotation.

$$
\begin{aligned}
M_x &= 0 \\
M_y &= 0 \\
M_z &= I_{zz}\dot{\omega}
\end{aligned}
\qquad (16.13)
$$

If the center of mass is not at a position along the axis of rotation, then we no longer have equilibrium conditions for the center of mass. However, through *Newton's law* we can relate the external forces on the body to the acceleration of the mass center. We may then have to use the *kinematics* of rigid-body motion to yield enough equations to solve the problem. We now illustrate this case.

In the examples to follow, we shall often use a reference xyz as a stationary inertial reference rather than using XYZ as was the case in Chapter 15.

EXAMPLE 16.4

A uniform rod of weight W and length L supported by a pin connection at A and a wire at B is shown in Fig. 16.12. What is the force on pin A at the instant that the wire is released? What is the force at A when the rod has rotated 45°?

Part A. A free-body diagram of the rod is shown in Fig. 16.13 at the instant that the wire is released at B. The *moment-of-momentum* equation about the axis of rotation at A, on using the formula for I of a rod about a transverse axis at the end, yields

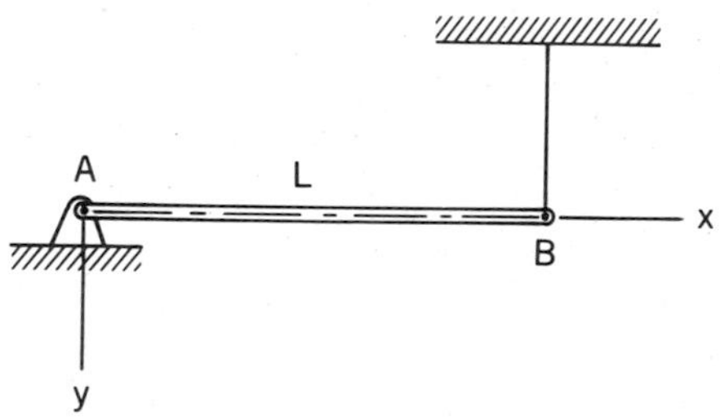

Figure 16.12. Rod supported by wire.

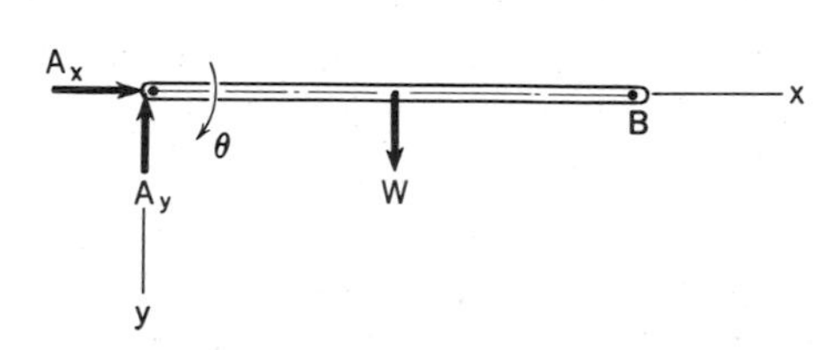

Figure 16.13. Wire suddenly cut.

$$\frac{WL}{2} = I\ddot{\theta} = \frac{1}{3}\left(\frac{W}{g}\right)L^2\ddot{\theta}$$

Therefore,

$$\ddot{\theta} = \frac{3}{2}\frac{g}{L} \quad \text{at time } t = 0 \tag{a}$$

Using simple *kinematics* of plane circular motion, we can now determine the acceleration of the mass center at $t = 0$:

$$\ddot{x} = 0, \qquad \ddot{y} = \frac{L}{2}\ddot{\theta} = \frac{3}{4}g \tag{b}$$

where we have used Eq. (a) in the last step. Now express *Newton's law* for the mass center:

$$\frac{W}{g}\ddot{x} = A_x, \qquad \frac{W}{g}\ddot{y} = W - A_y$$

Accordingly, at time $t = 0$ we have, on noting Eq. (b):

$$A_x = 0, \qquad A_y = \tfrac{1}{4}W \tag{c}$$

Thus, we see that at the instant of releasing the wire there is a upward force of $\frac{1}{4}W$ on the left support.

Part B. We next express the *moment-of-momentum* equation for the rod at any arbitrary position θ. Observing Fig. 16.14(a), we get

$$\frac{WL}{2}\cos\theta = \frac{1}{3}\frac{W}{g}L^2\ddot{\theta}$$

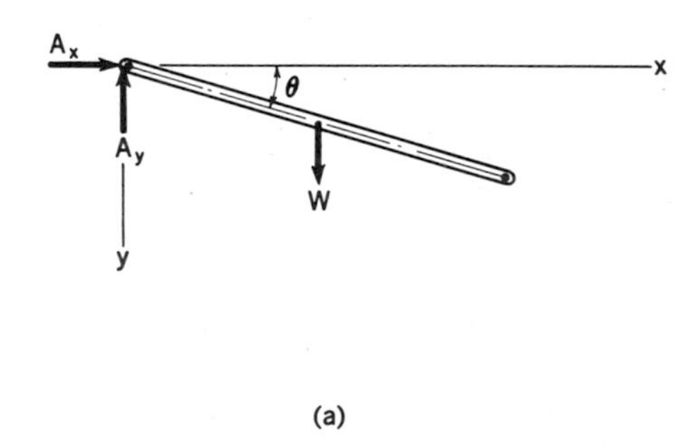

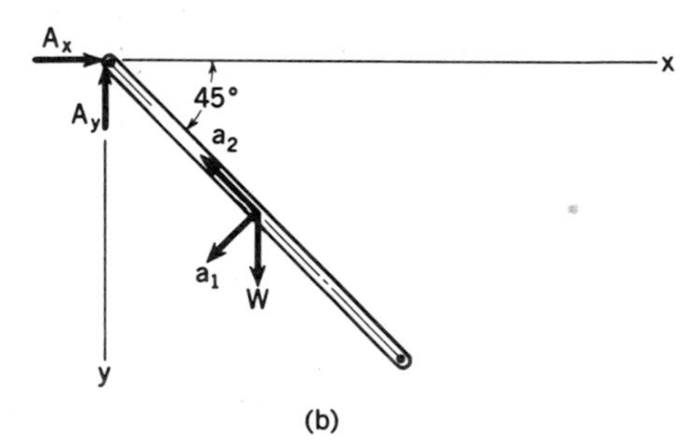

Figure 16.14. (a) Rod at position θ; (b) rod at $\theta = 45°$.

Therefore,

$$\ddot{\theta} = \frac{3}{2}\frac{g}{L}\cos\theta \tag{d}$$

Consequently, at $\theta = 45°$ we have

$$\ddot{\theta} = (1.5)(.707)\frac{g}{L} = 1.060\frac{g}{L} \tag{e}$$

We shall also need $\dot{\theta}$, and accordingly we now rewrite Eq. (d) as follows:

$$\ddot{\theta} \equiv \left(\frac{d\dot{\theta}}{d\theta}\right)\left(\frac{d\theta}{dt}\right) \equiv \left(\frac{d\dot{\theta}}{d\theta}\right)(\dot{\theta}) = \frac{3}{2}\frac{g}{L}\cos\theta \tag{f}$$

Separating variables, we get

$$\dot{\theta}\,d\dot{\theta} = \frac{3}{2}\frac{g}{L}\cos\theta\,d\theta$$

which we integrate to get

$$\frac{\dot{\theta}^2}{2} = \frac{3}{2}\frac{g}{L}\sin\theta + C$$

When $\theta = 0$, $\dot{\theta} = 0$; accordingly, $C = 0$. We then have

$$\dot{\theta}^2 = 3\frac{g}{L}\sin\theta \qquad \text{(g)}$$

At the instant of interest, we get for $\dot{\theta}^2$:

$$\dot{\theta}^2 = 3\frac{g}{L}(.707) = 2.12\frac{g}{L} \qquad \text{(h)}$$

For $\theta = 45°$, we can now give the acceleration component a_1 of the center of mass directed normal to the rod and component a_2 directed along the rod [see Fig. 16.14(b)]. From *kinematics* we can say, using Eqs. (e) and (h):

$$\begin{aligned} a_1 &= \frac{L}{2}\ddot{\theta} = \frac{L}{2}\left(1.060\frac{g}{L}\right) = .530g \\ a_2 &= \frac{L}{2}(\dot{\theta})^2 = \frac{L}{2}\left(2.12\frac{g}{L}\right) = 1.060g \end{aligned} \qquad \text{(i)}$$

Now, employing *Newton's law* for the mass center, we have

$$A_x = \frac{W}{g}(-a_1 \sin 45° - a_2 \cos 45°)$$

Therefore, using Eqs. (i)

$$A_x = -1.124W$$

Also, from *Newton's law*

$$-A_y + W = \frac{W}{g}(-a_2 \sin 45° + a_1 \cos 45°)$$

Therefore,

$$A_y = 1.375W \qquad \text{(j)}$$

Consider next the case of a body undergoing pure rotation about an axis which, for some point A in the body (or massless hypothetical extension of the body), is a *principal* axis (see Fig. 16.15). For a reference xyz at A with z collinear with the axis

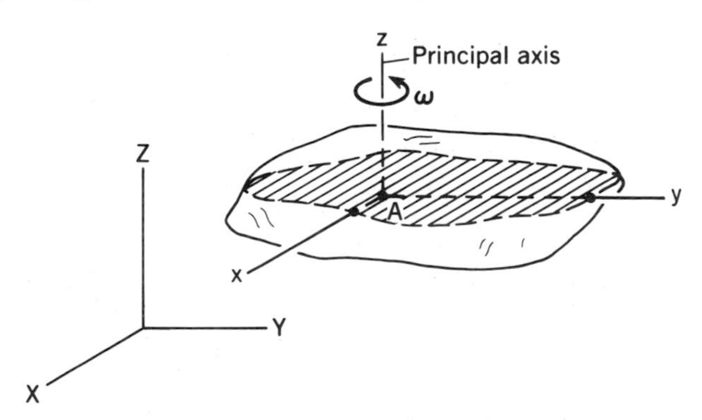

Figure 16.15. Axis z is a principal axis for point A.

of rotation, it is clear that $I_{zx} = I_{zy} = 0$, and hence the moment of momentum equations simplify to the exact same forms as presented here for the rotation of slablike bodies.

Problems

16.1. A shaft and disc of steel having a density of 7626 kg/m³ are subjected to a constant torque T of 67.5 N-m, as shown. After 1 min, what is the angular velocity of the system? How many revolutions have occurred during this interval? Neglect friction of the bearings. Use $\frac{1}{2}Mr^2 = I$ for disc.

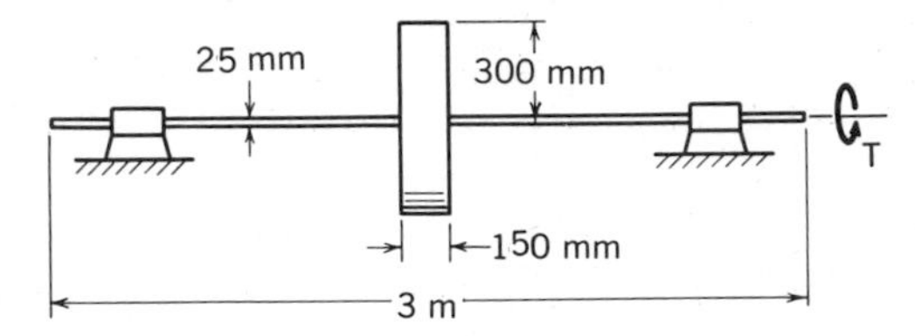

Figure P.16.1

16.2. In Problem 16.1 include wind and bearing friction losses by assuming that they are proportional to angular speed $\dot{\theta}$. If the disc will halve its speed after 5 min from a speed of 300 rpm when there is no external applied torque, what is the resisting torque at 600 rpm?

16.3. A 400-lb flywheel is shown. A 100-lb block is held by a light cable wrapped around the hub of the flywheel the diameter of which is 1 ft. If the initially stationary weight descends 3 ft in 5 sec, what is the radius of gyration of the flywheel?

Figure P.16.3

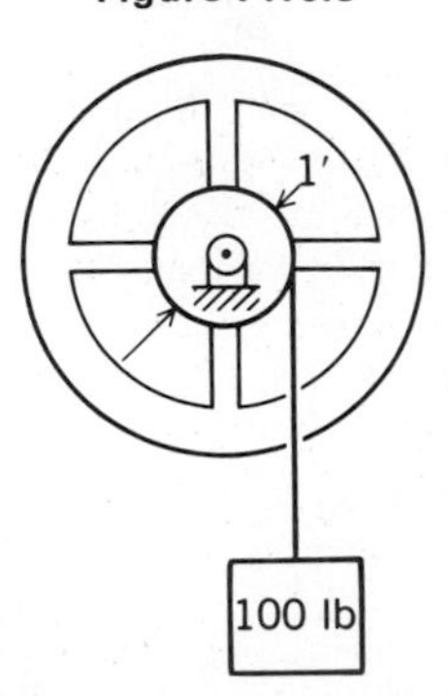

16.4. A stepped cylinder has the dimensions $R_1 = .30$ m, $R_2 = .65$ m, and the radius of gyration, k, is .35 m. The mass of the stepped cylinder is 100 kg. Weights A and B are connected to the cylinder. If weight B is of mass 80 kg and weight A is of mass 50 kg, how far does A move in 5 sec? In which direction does it move?

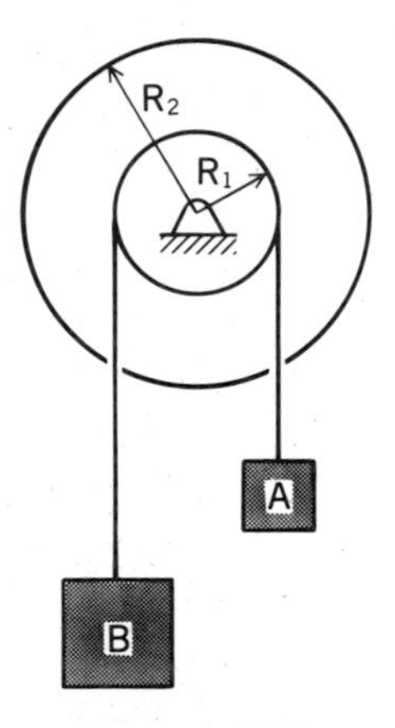

Figure P.16.4

16.5. Two discs E and F of diameter 1 ft rotate in frictionless bearings. Disc F weighs 100 lb and rotates with angular speed ω_1 of 10 rad/sec, whereas disc E weighs 30 lb and rotates with angular speed ω_2 of 5 rad/sec. Neglecting the angular momentum of the shafts, what is the total angular momentum of the system relative to the ground? Use Eq. 16.4 to compute H from first principles. Consider that the discs are forced together along the axis of rotation. What is the common angular velocity when friction has reduced relative motion between the discs to zero?

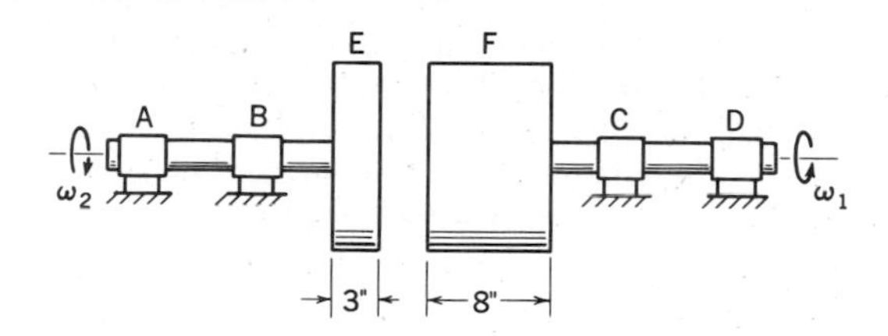

Figure P.16.5

16.6. In Problem 16.5, if $\mu_s = .2$ between the discs, and it takes 30 sec for them to reach the same angular speed of 6.54 rad/sec, what is the constant normal force required to bring the discs together in such a manner?

16.7. A shaft A is rotating at an angular speed of 1750 rpm when the hand brake is applied. If the shaft and its parts have a radius of gyration of 200 mm and a mass of 500 kg, how long does it take to halve the speed of the shaft? The dynamic coefficient of friction between belt and shaft is .3.

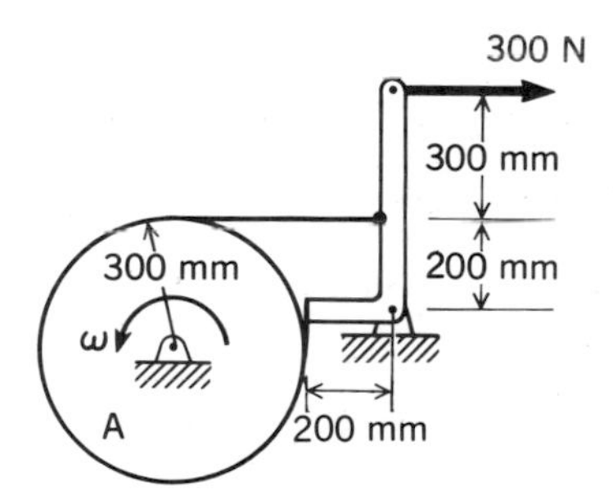

Figure P.16.7

16.8. A plunger A is connected to two identical gears B and C, each weighing 10 lb. The plunger weighs 40 lb. How far does the plunger drop in 1 sec if released from rest?

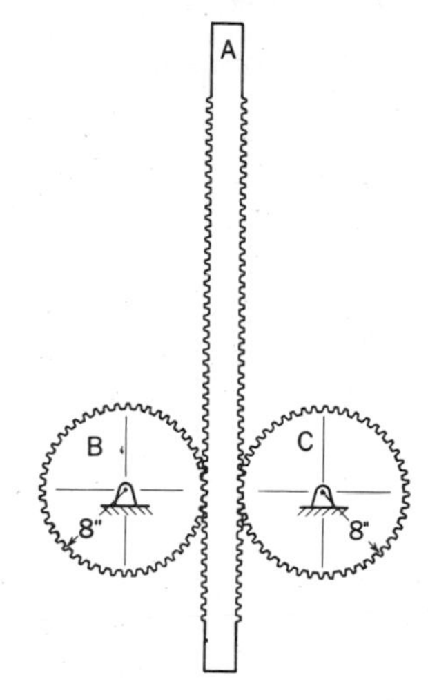

Figure P.16.8

16.9. A pulley A and its rotating accessories have a mass of 1000 kg and a radius of gyration of .25 m. A simple hand brake is applied as shown using a force P. If the dynamic coefficient of friction between belt and pulley is .2, what must force P be to change ω from 1750 rpm to 300 rpm in 60 sec?

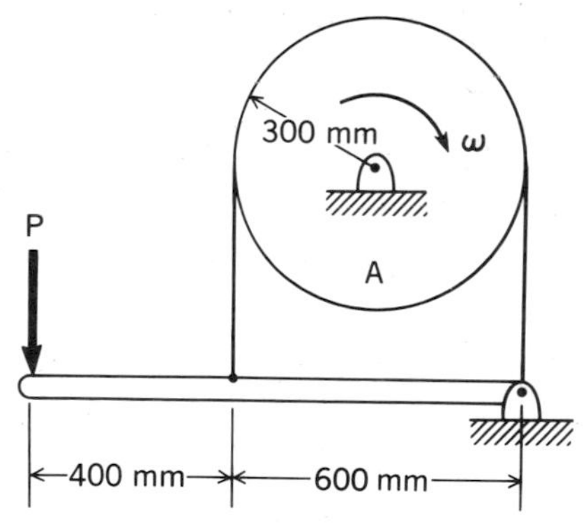

Figure P.16.9

16.10. A flywheel is shown. There is a viscous damping torque due to wind and bearing friction which is known to be $-.04\omega$ N-m, where ω is in rad/sec. If a torque $T =$ 100 N-m is applied, what is the speed in 5 min after starting from rest? The mass of the wheel is 500 kg, and the radius of gyration is .50 m.

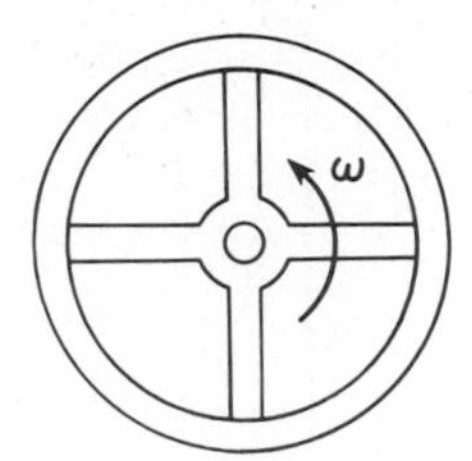

Figure P.16.10

16.11. The dynamic coefficient of friction for contact surfaces E and G is .2 and for A is .3. If a force P of 250 lb is applied, what will be the tension in the cord HB? Start by assuming no slipping at A. Check your assumption at the end of the calculation.

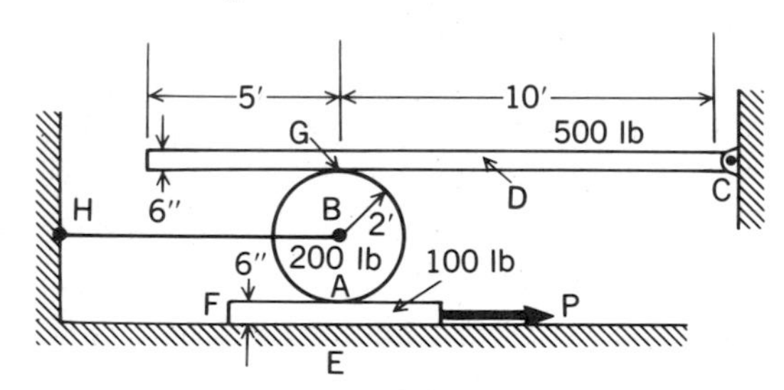

Figure P.16.11

16.12. A torque T of 15 N-m is applied to a rod AB as shown. At B there is a pin which slides in a frictionless slot in a disc E whose mass is 10 kg and whose radius of gyration is 300 mm. If the system is at rest at the instant shown, what is the angular acceleration of the rod and the disc? The rod has a mass of 18 kg.

Figure P.16.12

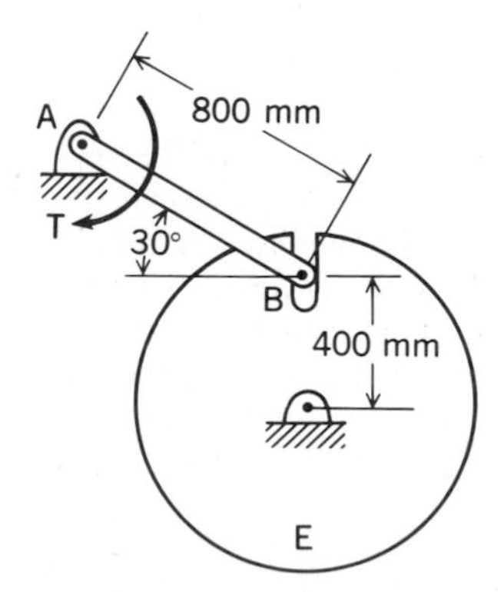

16.13. Cylinder A has an angular speed ω_1 of 3 rad/sec when it is lowered onto cylinder B, which has an angular speed ω_2 of 5 rad/sec before contact is made. What are the final angular velocities of the cylinders resulting from friction at the surfaces of contact? The mass of A is 500 lbm and of B is 400 lbm. If $\mu_d = .3$ for the contact surface of the cylinders and if the normal force transmitted from A to B is 10 lb, how long does it take for the cylinders to reach a constant speed?

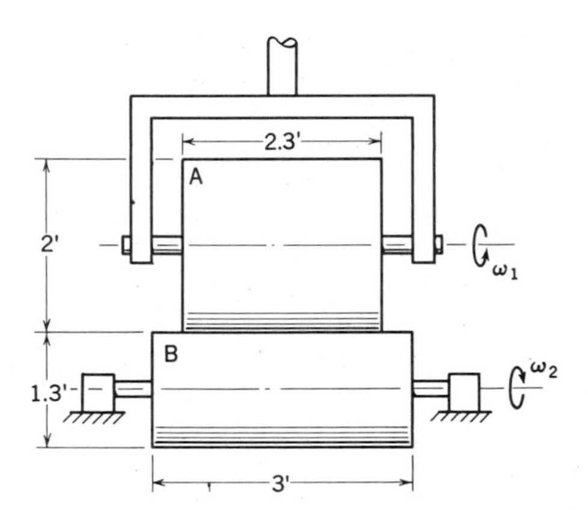

Figure P.16.13

16.14. A driving cylinder A has a torque T_A of 30 N-m applied to it while the driven cylinder B has a resisting torque T_B of 10 N-m. Cylinder B has a mass of 15 kg, a radius of gyration of 100 mm, and a diameter of 400 mm. Cylinder A has a mass of 50 kg, a radius of gyration of 200 mm, and a diameter of 800 mm. Rod AB is a light rod connecting the cylinders. What is the angular acceleration of cylinder A at the instant shown if the system is stationary at this instant? Member AB is 1 m long.

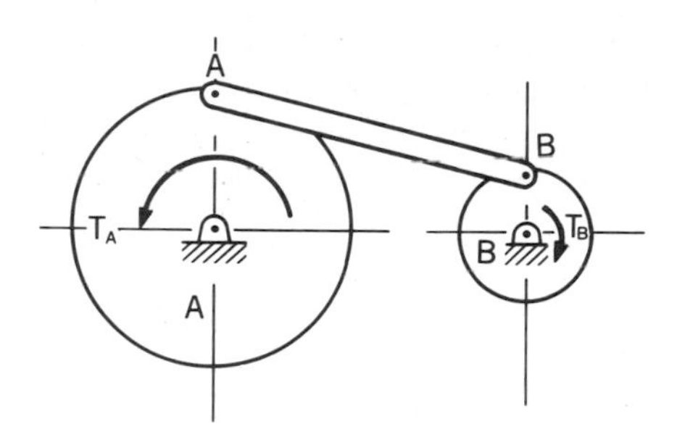

Figure P.16.14

16.15. Do Problem 16.14 for the case where the angular velocity of B is 20 rad/sec clockwise at the instant of interest.

16.16. A torque T of 100 N-m is applied to a wheel D having a mass of 50 kg, a diameter of 600 mm, and a radius of gyration of 280 mm. The wheel D is attached by a light member AB to a slider C having a mass of 30 kg. If the system is at rest at the instant shown, what is the acceleration of slider C? What is the axial force in member AB? Neglect friction everywhere, and neglect the inertia of the member AB.

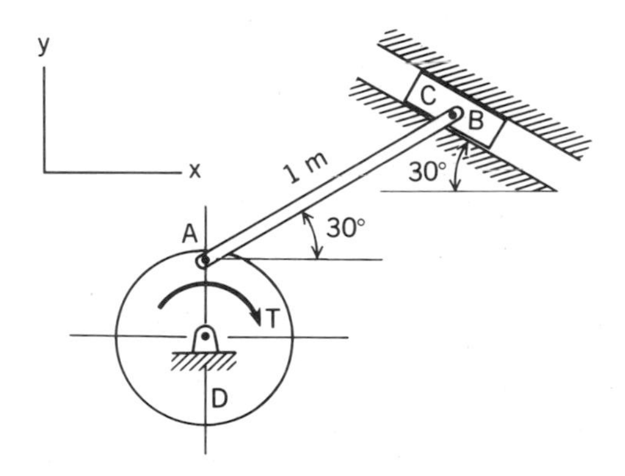

Figure P.16.16

16.17. Do Problem 16.16 for the case where at the instant of interest $\omega_D = 2$ rad/sec counterclockwise.

16.18. A torque T of 50 N-m is applied to the device shown. The bent rods are of mass per unit length 5 kg/m. Neglecting the inertia of the shaft, how many rotations does the system make in 10 sec? Are there forces coming onto the bearings other than from the dead weights of the system?

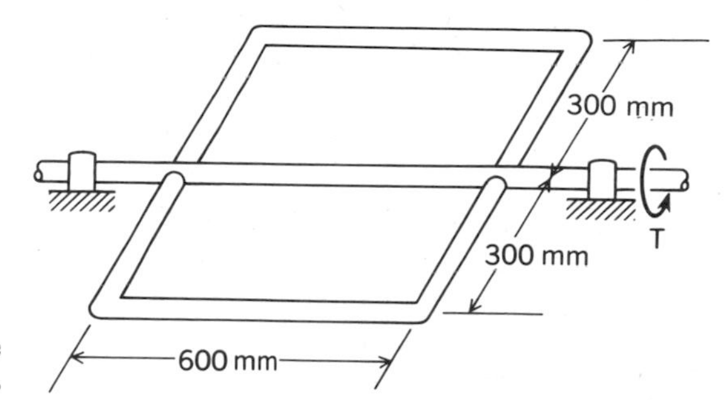

Figure P.16.18

16.19. An idealized torque-versus-angular-speed curve for a shunt, direct-current motor is shown as curve A. The motor drives a pump which has a resisting torque-versus-speed curve shown in the diagram as curve B. Find the angular speed of the system as a function of time, after starting, over the range of speeds given in the diagram. Take the moment of inertia of motor, connecting shaft, and pump to be I.

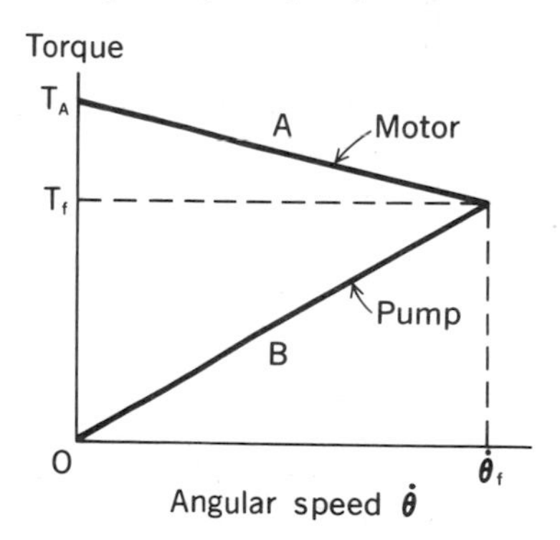

Figure P.16.19

16.20. Rods of length L have been welded onto a rigid drum A. The system is rotating at a speed ω of 5000 rpm. By this time, you may have studied stress in a rod in your strength of materials class. In any case, the stress is the normal force per unit area of cross section of the rod. If the cross-sectional area of the rod is 2 in.^{2} and the mass per unit length is 5 lbm/ft, what is the normal stress τ_{rr} on a section at any position r? The length L of the rods is 2 ft. What is τ_{rr} at $r = 1.5$ ft? Consider the upper rod when it is vertical.

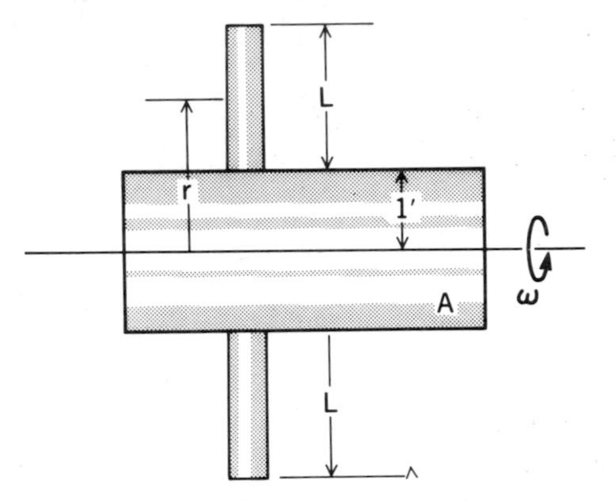

Figure P.16.20

***16.21.** In Problem 16.20, consider that the mass per unit length varies linearly from 5 lbm/ft at $r = 1$ ft (at the bottom of the rod) to 6 lbm/ft at $r = 3$ ft (at the top of the rod). Find τ_{rr} at any position r and then compute τ_{rr} for $r = 1.5$.

16.22. A plate weighing 3 lb/ft^2 is supported at A and B. What are the force components at B at the instant support A is removed?

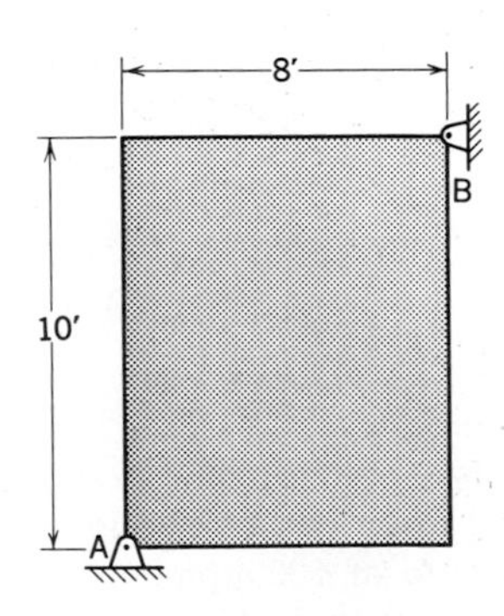

Figure P.16.22

16.23. When the uniform rigid bar is horizontal, the spring at C is compressed 3 in. If the bar weighs 50 lb, what is the force at B when support A is removed suddenly? The spring constant is 50 lb/in.

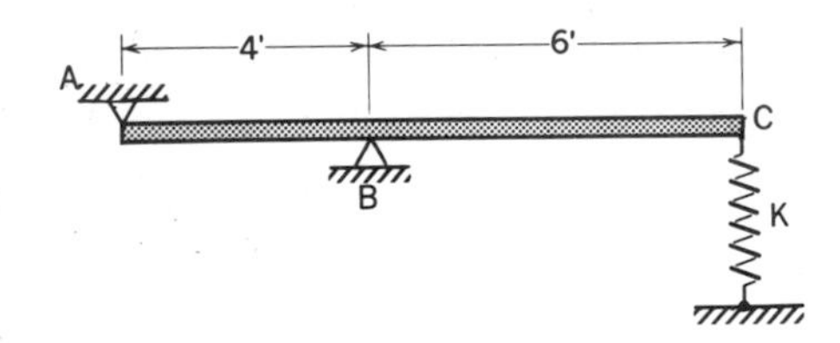

Figure P.16.23

16.24. An electric motor E drives a light shaft through a coupling D which transmits only torque. A disc A is on the shaft and has its center of gravity 200 mm from the geometric center of the disc as shown in the diagram. A torque T given as

$$T = .005t^2 + .03t \text{ N-m}$$

is applied to the shaft from the motor (t is in seconds measured from when the system is at rest). What are the force components on the bearings when $t = 30$ sec? The disc has a mass of 15 kg and a radius of gyration of 250 mm about an axis going through the center of gravity. Take the disc at the position shown at the instant

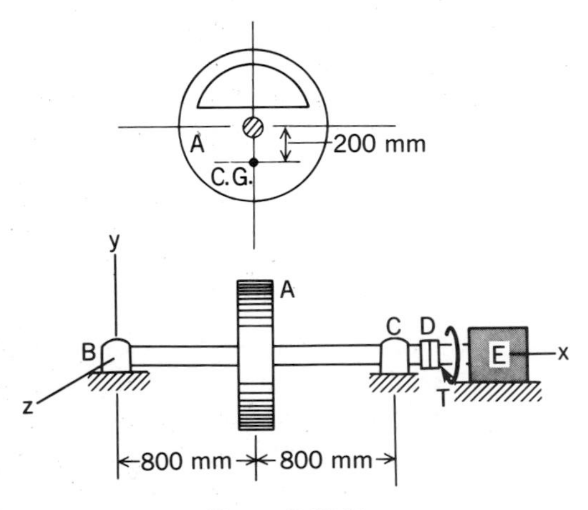

Figure P.16.24

16.25. A single cam A is mounted on a shaft as shown. The cam has a mass of 10 kg and has a center of mass 300 mm from the centerline of the shaft. Also, the cam has a radius of gyration of 180 mm about an axis through the center of mass. The shaft has a mass per unit length of 10 kg/m and has a diameter of 30 mm. A torque T given as

$$T = .001t^2 + 10 \text{ N-m}$$

is applied at coupling D (t is in seconds). What are the force components in the bearings after 25 sec if the cam has the position shown in the diagram at this instant?

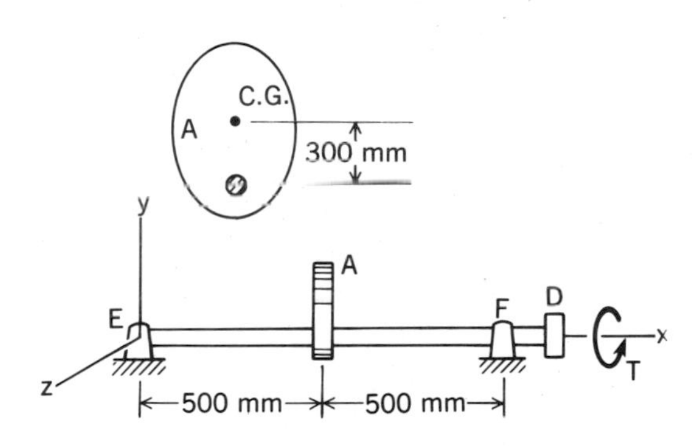

Figure P.16.25

16.26. A circular plate is rotating at a speed of 10,000 rpm. A hole has been cut out of the plate so that the beam of light is allowed through for very short intervals of time. Such a device is called a *chopper*. If the plate weighed 20.00 lb originally and the material removed for the hole weighed 3 oz, what are the forces in the bearings A and B from the circular plate for the instant shown?

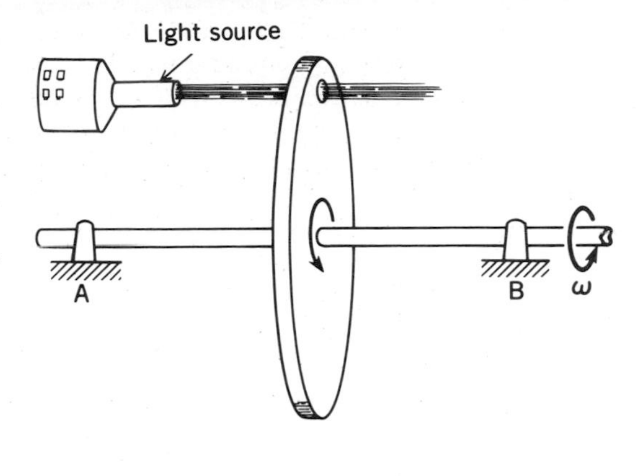

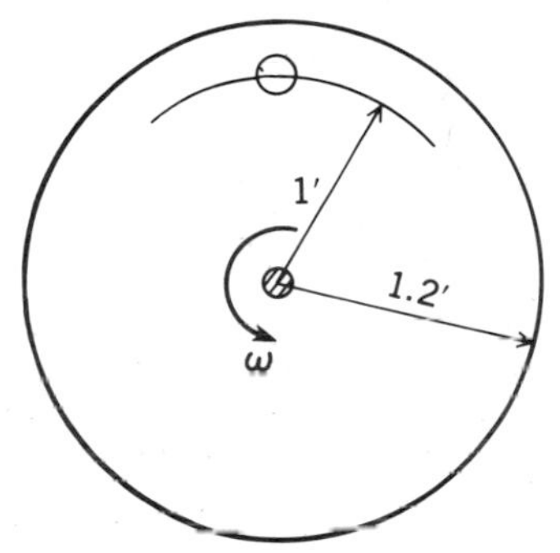

Figure P.16.26

16.27. Part of a conveyor system is shown. A link belt is meshed around portions of gears A and B. The belt has a mass of 5 kg/m. Furthermore, each gear has a mass of 3 kg and a radius of 300 mm. If a force of 100 N is applied at one end as shown, what is the maximum possible force T that can be transmitted at the other end if $\ddot{\theta}_A = 5$ rad/sec²? The gears turn freely.

Figure P.16.27

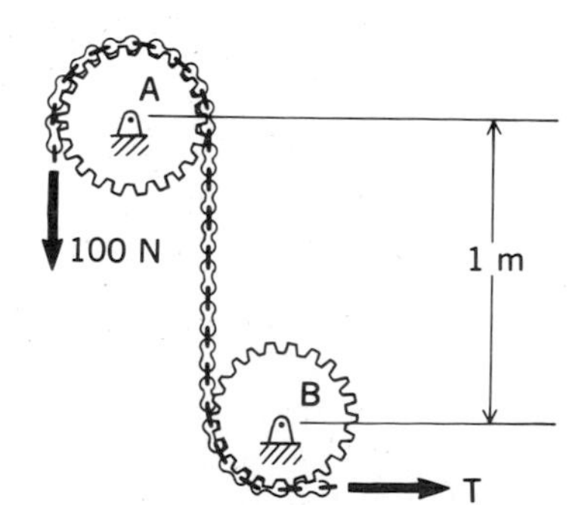

16.28. A box C weighing 150 lb rests on a conveyor belt. The driving drum B has a mass of 100 lbm and a radius of gyration of 4 in. The driven drum A has a mass of 70 lbm and a radius of gyration of 3 in. The belt weighs 3 lb/ft. Supporting the belt on the top side is a set of 20 rollers each with a mass of 3 lbm, a diameter of 2 in., and a radius of gyration of .8 in. If a torque T of 50 ft/lb is developed on the driving drum, what distance does C travel in 1 sec starting from rest? Assume that no slipping occurs.

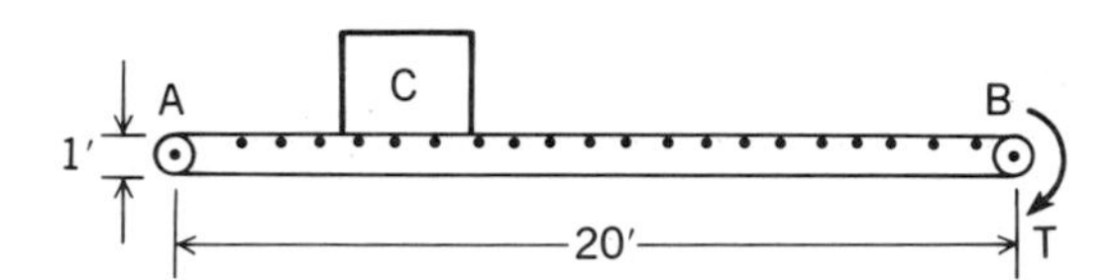

Figure P.16.28

16.29. A uniform slender member is supported by a hinge at A. A force P is applied at an angle α with the horizontal. What value should P have and at what distance d should it be applied to result in zero reactive forces at A at the configuration shown if $\alpha = 45°$? The weight of the member AB is W. What is the angular acceleration of the bar for these conditions at the instant of interest?

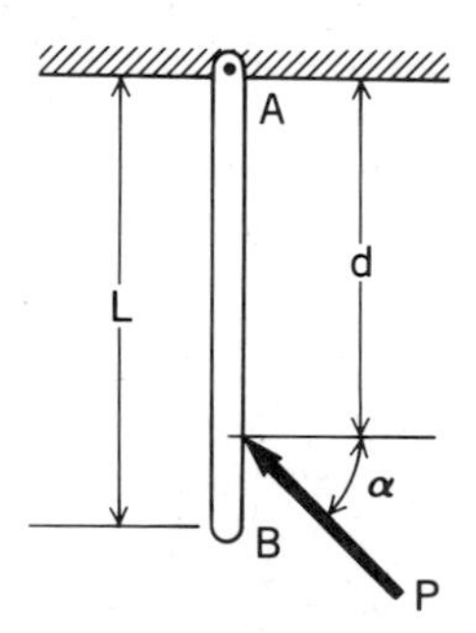

Figure P.16.29

***16.30.** A rod AB is welded to a rod CD, which in turn is welded to a shaft as shown. The shaft has the following angular motion at time t:

$$\omega = 10 \text{ rad/sec}$$
$$\dot{\omega} = 40 \text{ rad/sec}^2$$

What are the shear force, axial force, and bending moment along CD at time t as a function of r? The rods have a mass per unit length of 5 kg/m. Neglect gravity.

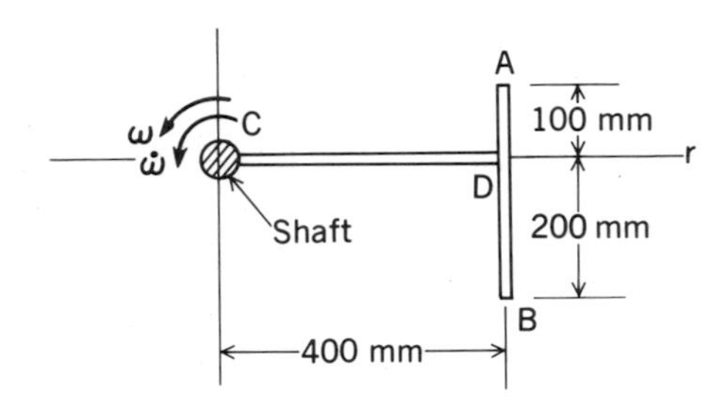

Figure P.16.30

16.31. A four-bar linkage is shown (the ground is the fourth linkage). Each member is 300 mm long and has a mass per unit length of 10 kg/m. A torque T of 5 N-m is applied each to bars AB and DC. What is the angular acceleration of bars AB and CD?

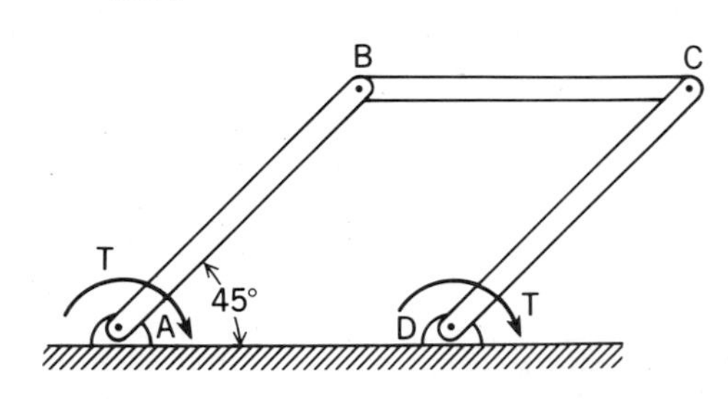

Figure P.16.31

16.32. A platform A has a torque T applied about its axis of rotation. The platform has a mass of 1000 kg and a radius of gyration of 2 m. A block B rests on the platform but is prevented from sliding off by small stops C and D. The block has a mass of 1 kg and has dimensions 200 mm $\times$ 200 mm $\times$ 200 mm. The center of mass of the block is at its geometric center. If a torque $T = 20t^2 + 50t$ N-m is applied with t in seconds, when and how does the block first tip? (Because of the small size and mass of B, consider the system to be a slablike body.)

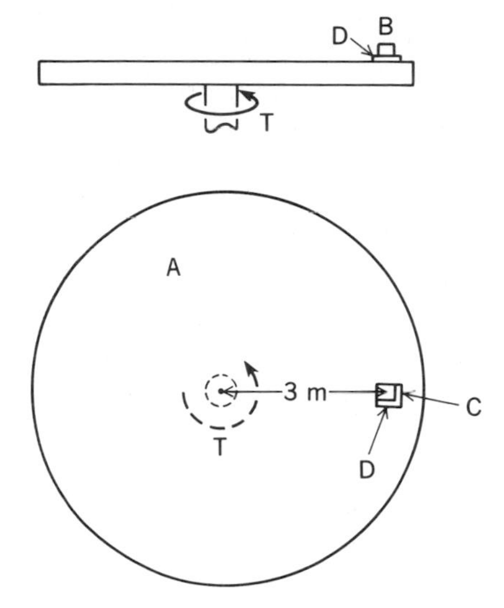

Figure P.16.32

16.6 Rolling Slablike Bodies

We now consider the rolling without slipping of slablike bodies such as cylinders, spheres, or plane gears. As we have indicated in Chapter 15, the point of contact of the body has instantaneously *zero velocity*, and we have *pure instantaneous rotation about this contact point*. We pointed out that for getting velocities of points on a rolling body, we could imagine that there is a *hinge* at the point of contact. Also, the acceleration of the *center* of a rolling sphere or cylinder can be computed using the simple formula $-R\ddot{\theta}$. Finally, you can readily show that if the angular speed is zero, we can compute the acceleration of any point in the cylinder or sphere by again imagining a *hinge* at the point of contact. For other cases, we must use more detailed kinematics, as discussed in Chapter 15.

A very important conclusion we reached in Chapter 15 for cylinders and spheres was that the acceleration of the contact point on the cylinder or sphere is *toward the geometric center of the cylinder or sphere*. If the center of mass of the body lies anywhere along the line AO from the contact point A to the geometric center O, then clearly we can use Eq. 16.6 for the point A. This action is justified since point A is then an example of case 3 in Section 16.2 (A accelerates toward the mass center). Thus, for the body in Fig. 16.16 we can use $T = I\alpha$ about the point of contact A of the cylinder at the instant shown. However, in Fig. 16.17 we cannot do this because the point of contact A of the cylinder is *not* accelerating toward the center of mass as in the previous case. We can use $T = I\alpha$ about the *center of mass* in the latter case.

We shall now examine a problem involving rolling without slipping. The equations of motion, you can readily deduce, are the same as in the previous section.

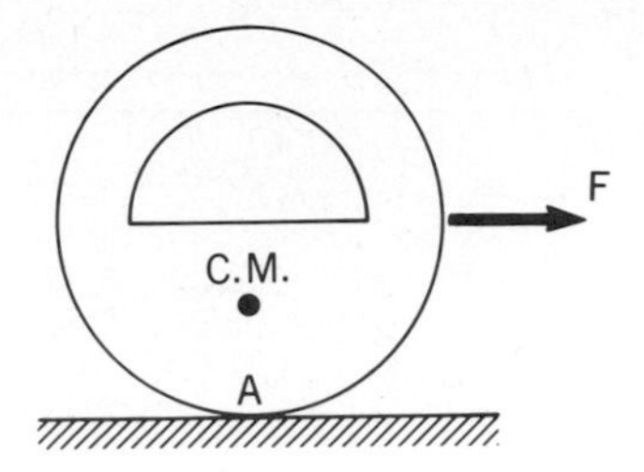

Figure 16.16. Point A accelerates toward center of mass.

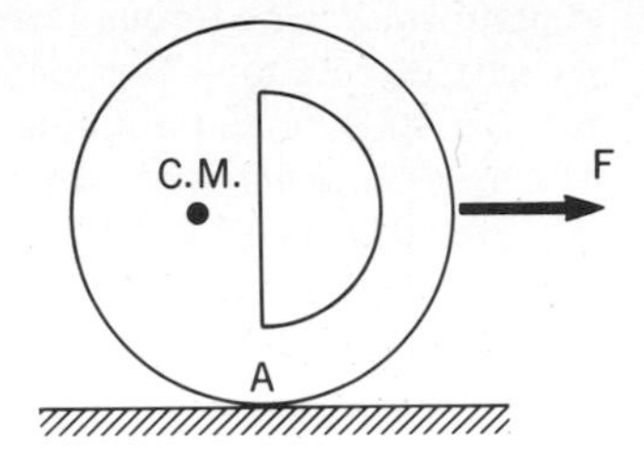

Figure 16.17. Point A does not accelerate toward center of mass.

EXAMPLE 16.5

A steam roller is shown going up a 5° incline in Fig. 16.18. Wheels A have a radius of gyration of 1.5 ft and a weight each of 500 lb, whereas roller B has a radius of gyration of 1 ft and a weight of 5000 lb. The vehicle, minus the wheels and roller but including the operator, has a weight of 7000 lb with a center of mass positioned as shown in the diagram. The steam roller is to accelerate at the rate of 1 ft/sec². In part A of the problem, we are to determine the torque T_{eng} from the engine onto the drive wheels.

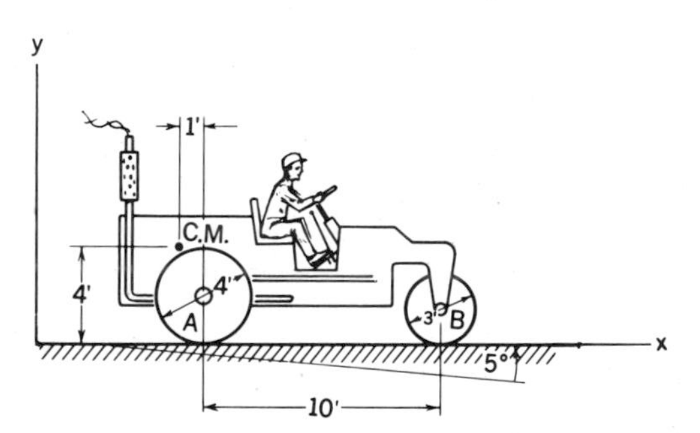

Figure 16.18. Steam roller moving up incline.

Part A. In Fig. 16.19, we have shown free-body diagrams of the drive wheels and the roller. Note we have combined the two drive wheels into a single 1000-lb wheel. In each case, the point of contact on the wheel accelerates toward the mass center of the wheel and we can put to good use the *moment-of-momentum* equation (16.9) for the points of contact on the cylinders. Accordingly, for body A, we have

$$(A_x + 1000 \sin 5°)(2) - T_{\text{eng}} = \frac{1000}{g}(1.5^2 + 2^2)\ddot{\theta}_A \qquad \text{(a)}$$

where we have employed the parallel-axis theorem in computing the moment of inertia about the line of contact at C. Similarly, for the roller, we have

$$(5000 \sin 5° - B_x)(1.5) = \frac{5000}{g}(1^2 + 1.5^2)\ddot{\theta}_B \qquad \text{(b)}$$

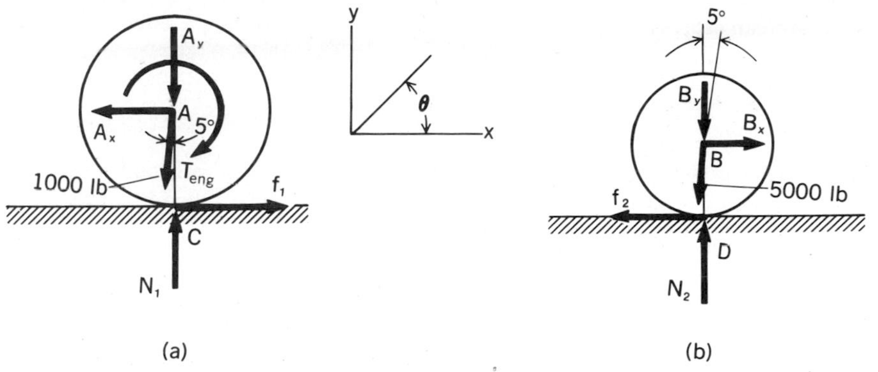

Figure 16.19. Free-body diagrams of driving wheels and roller.

We have here two equations and no fewer than five unknowns. By considering the free body of the vehicle minus wheels shown diagrammatically in Fig. 16.20, we can say, from *Newton's law:*

$$A_x - B_x - 7000 \sin 5° = \frac{7000}{g}(1) \tag{c}$$

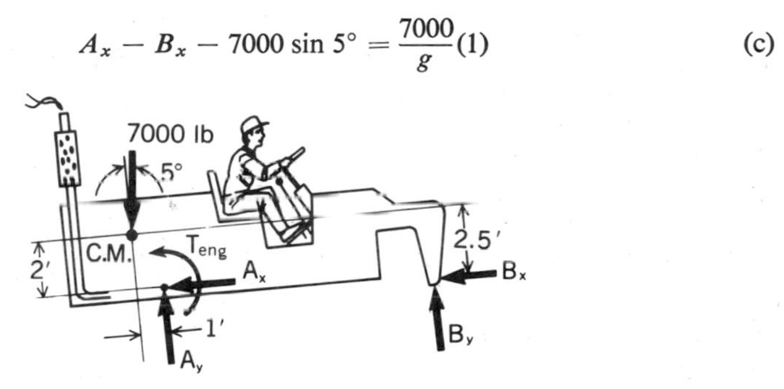

Figure 16.20. Free-body diagram of vehicle without wheels and roller.

Finally, from *kinematics* we can say:

$$\begin{aligned} \ddot{\theta}_A &= -\frac{\ddot{x}}{r_A} = -\frac{1}{2} = -.5 \\ \ddot{\theta}_B &= -\frac{\ddot{x}}{r_B} = -\frac{1}{1.5} = -.667 \end{aligned} \tag{d}$$

where $\ddot{x} = 1$ ft/sec² is the acceleration of the vehicle up the incline. We can now readily solve the equations. We get B_x directly from Eq. (b) on replacing $\ddot{\theta}_B$ by $-.667$. Next, we get A_x from Eq. (c). Finally, going back to Eq. (a), we can solve for T_{eng}. The results are:

$$\begin{aligned} T_{eng} &= 3250 \text{ ft-lb} \\ A_x &= 1487 \text{ lb} \\ B_x &= 660 \text{ lb} \end{aligned}$$

Part B. Determine next the normal forces N_1 and N_2 at the wheels and roller, respectively.

We can express *Newton's law* for the wheels and roller in the direction normal to the incline by using the free-body diagrams of Fig. 16.19. Thus,

$$N_1 - A_y - 1000 \cos 5° = 0 \tag{e}$$

$$N_2 - B_y - 5000 \cos 5° = 0 \tag{f}$$

Next, we consider the free body of the vehicle without the wheels and roller (Fig. 16.20). *Newton's law* in the y direction for the center of mass then becomes

$$A_y + B_y - 7000 \cos 5° = 0 \tag{g}$$

The *moment-of-momentum* equation about the center of mass of the vehicle without wheels and roller is

$$A_x(2) - B_x(2.5) + A_y(1) + B_y(11) + T_{\text{eng}} = 0 \tag{h}$$

We have four equations in four unknowns. Solve for A_y in Eq. (g), and substitute into Eq. (h). Inserting known values for A_x, B_x, and T_{eng}, we have

$$(1487)(2) - (660)(2.5) + 7000 \cos 5° - B_y + 11B_y + 3250 = 0$$

Therefore,

$$B_y = -1155 \text{ lb}$$

Now from Eq. (g) we get A_y:

$$A_y = 7000 \cos 5° + 1154.5 = 8128 \text{ lb}$$

Finally, from Eqs. (e) and (f) we get N_1 and N_2.

$$N_1 = 8128 + 1000 \cos 5° = 9120 \text{ lb}$$

$$N_2 = -1155 + 5000 \cos 5° = 3830 \text{ lb}$$

Hence, on each wheel we have a normal force of 4560 lb, and for the roller we have a normal force of 3830 lb.

*EXAMPLE 16.6

A gear A weighing 100 N is connected to a stepped cylinder B (see Fig. 16.21) by a light rod DC. The stepped cylinder weighs 1 kN and has a radius of gyration of 250 mm along its centerline. The gear A has a radius of gyration of 120 mm along its centerline. A force $F = 1500$ N is applied to the gear at D. What is the compressive force in member DC if, at the instant that F is applied, the system is stationary?

Noting that DC is a two-force compressive member, we draw the free-body diagrams for the gear and the stepped cylinder in Fig. 16.22. The *moment-of-momentum* equations about the contact points for both bodies (points a and b, respectively) are

$$-(100 + 1500)(.15) + DC(\sin 20°)(.15) = \left(\frac{100}{9.81}\right)[(.12)^2 + (.15)^2]\ddot{\theta}_A$$

$$(DC \sin 20° + 1000)(\sin 30°)(.10) - (DC \cos 20°)(.30 + .10 \cos 30°)$$
$$= \frac{1000}{9.81}[(.25)^2 + (.10)^2]\ddot{\theta}_B$$

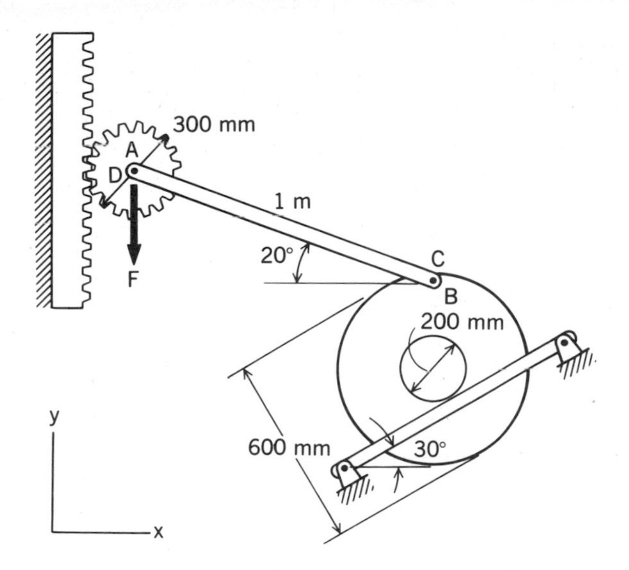

Figure 16.21. Stepped cylinder connected to a gear by a light rod.

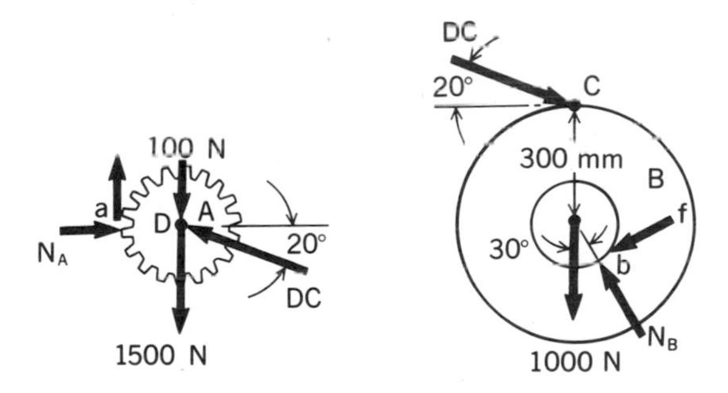

Figure 16.22. Free-body diagrams of gear and cylinder.

These equations simplify to the following pair:

$$.0513DC - 240 = .376\ddot{\theta}_A \tag{a}$$

$$-.3462DC + 50 = 7.39\ddot{\theta}_B \tag{b}$$

Clearly, we need an equation from *kinematics* at this time. Considering rod DC, we can say:[4]

$$\boldsymbol{a}_C = \boldsymbol{a}_D + \dot{\boldsymbol{\omega}}_{DC} \times \boldsymbol{\rho}_{DC} + \boldsymbol{\omega}_{DC} \times (\boldsymbol{\omega}_{DC} \times \boldsymbol{\rho}_{DC})$$

$$\ddot{\theta}_B\boldsymbol{k} \times [(.30 + .10 \cos 30°)\boldsymbol{j} - (.10 \sin 30°)\boldsymbol{i}]$$
$$= \ddot{y}_D\boldsymbol{j} + \dot{\omega}_{DC}\boldsymbol{k} \times (\cos 20°\boldsymbol{i} - \sin 20°\boldsymbol{j}) + \boldsymbol{0}$$

---

[4]Note that because cylinder B has zero angular velocity, we can imagine it to be hinged at b for computing $\boldsymbol{a}_C$.

The scalar equations are

$$-.3866\ddot{\theta}_B = .342\dot{\omega}_{DC} \tag{c}$$

$$-.05\ddot{\theta}_B = \ddot{y}_D + .940\dot{\omega}_{DC} \tag{d}$$

Also, from *kinematics* we can say, considering gear A:

$$\ddot{y}_D = .15\ddot{\theta}_A \tag{e}$$

Multiply Eq. (c) by .940/.342 and rewrite Eq. (d) with $\ddot{y}_D$ replaced by using Eq. (e):

$$-1.063\ddot{\theta}_B = .940\dot{\omega}_{DC}$$

$$-.05\ddot{\theta}_B - .15\ddot{\theta}_A = .940\dot{\omega}_{DC}$$

Subtracting, we get

$$-1.013\ddot{\theta}_B + .15\ddot{\theta}_A = 0$$

Therefore,

$$\ddot{\theta}_A = 6.75\ddot{\theta}_B \tag{f}$$

Solving Eq. (a), (b), and (f) simultaneously gives us for DC the result

$$DC = 1511 \text{ N (compression)}$$

16.7 General Plane Motion of a Slablike Body

We now consider *general plane motion* of slablike bodies. The motion to be studied will be parallel to the plane of symmetry. Accordingly, using the center of mass the angular velocity vector $\boldsymbol{\omega}$ will be normal to the plane of symmetry, and, in accordance with Chasle's theorem, will be taken to pass through the center of mass. The translational velocity vector $\boldsymbol{V}_c$ will be parallel to the plane of symmetry. We fix a reference at the center of mass of the body such that the xy plane coincides with the plane of symmetry as shown in Fig. 16.23. Note that the actual instantaneous axis of rotation is shown. For the same reasons used in Section 16.5 for slablike bodies, the *moment-of-momentum* equations become

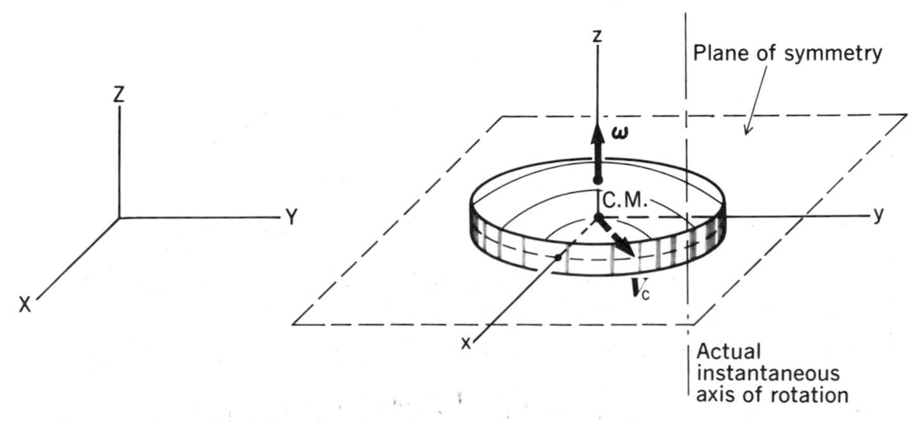

Figure 16.23. Slablike body undergoing general plane motion parallel to XY.

$$M_x = 0 \tag{16.14a}$$

$$M_y = 0 \tag{16.14b}$$

$$M_z = I_{zz}\dot{\omega} \tag{16.14c}$$

Furthermore, considering the center of mass, we must have *equilibrium* in the z direction.

$$\sum F_z = 0 \tag{16.15}$$

while the full form of *Newton's law* holds in the x and y directions. We now illustrate the use of these equations in the following examples.

EXAMPLE 16.7

A stepped cylinder having a weight of 450 N and a radius of gyration k of 300 mm is shown in Fig. 16.24(a). The radii R_1 and R_2 are, respectively, 300 mm and 600 mm. A total pull T equal to 180 N is exerted on the ropes attached to the inner cylinder. What is the ensuing motion? The coefficients of static and dynamic friction between the cylinder and the ground are, respectively, .1 and .08.

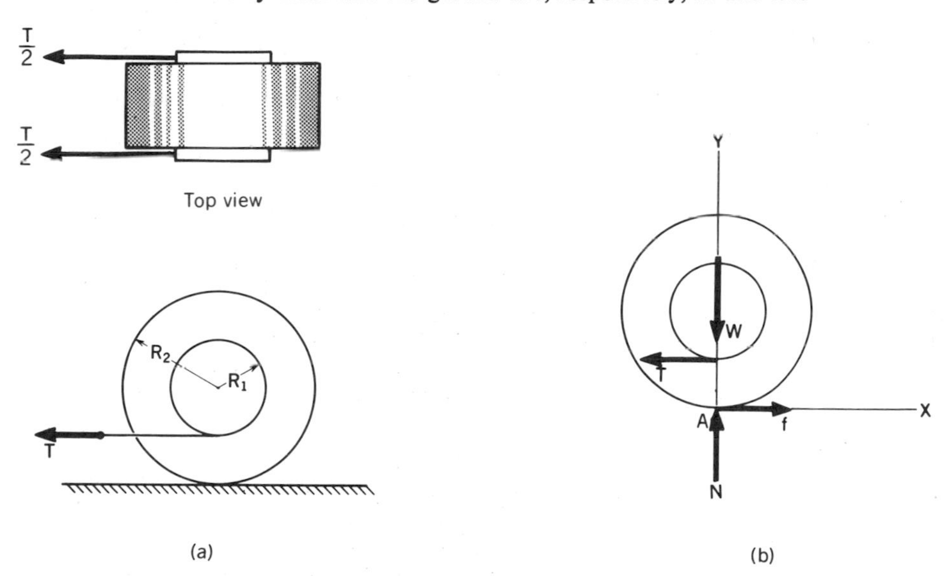

Figure 16.24. (a) Stepped cylinder; (b) free-body diagram of cylinder. xy is stationary.

A free-body diagram of the cylinder is shown in Fig. 16.24(b). Let us assume first that there is no slipping at the contact surface. We have then pure rotation about the contact point A, and we can say for the *moment of momentum* about the axis of contact:

$$T(R_2 - R_1) = \left(\frac{W}{g}k^2 + \frac{W}{g}R_2^2\right)\ddot{\theta} \tag{a}$$

wherein we have used the parallel-axis theorem for moment of inertia. Inserting numerical values, we can solve directly for $\ddot{\theta}$ at the instant that the force T is applied.

Thus,

$$(180)(.30) = \left[\frac{450}{g}(.30)^2 + \frac{450}{g}(.60)^2\right]\ddot{\theta}$$

Therefore,

$$\ddot{\theta} = 2.62 \text{ rad/sec}^2 \tag{b}$$

Now employ *Newton's law* for the mass center. In the x direction we get

$$-180 + f = \frac{W}{g}\ddot{x} \tag{c}$$

Using *kinematics*, we note that

$$\ddot{x} = -R_2\ddot{\theta} = -.60\ddot{\theta} \tag{d}$$

Substituting into Eq. (c) for $\ddot{x}$ using Eq. (d) and putting in known numerical values, we can solve for f:

$$f = 180 - \frac{450}{9.81}[(.60)(2.62)] = 107.9 \text{ N}$$

Thus, for no slipping, we must be able to develop a friction force of 107.9 N. The maximum friction force that we can have, however, is, according to Coulomb's law,

$$f_{\text{max}} = W\mu_s = (450)(.1) = 45 \text{ N} \tag{e}$$

Accordingly, we must conclude that the cylinder *does* slip, and we must reexamine the problem as a general plane-motion problem.

Using $\mu_d = .08$, we now take f to be 36 N and employ the *moment-of-momentum* equation for the mass center. We then have (see Fig. 16.24 (b))

$$fR_2 - TR_1 = \frac{W}{g}k^2\ddot{\theta} \tag{f}$$

Inserting numerical values, we get for $\ddot{\theta}$:

$$\ddot{\theta} = -7.85 \text{ rad/sec}^2 \tag{g}$$

Now, using *Newton's law* in the x direction for the mass center, we get

$$-T + f = \frac{W}{g}\ddot{x} \tag{h}$$

Inserting numerical values, we get for $\ddot{x}$:

$$\ddot{x} = -3.14 \text{ m/sec}^2 \tag{i}$$

Thus, the cylinder has a linear acceleration of 3.14 m/sec^2 to the left and an angular acceleration of 7.85 rad/sec^2 in the clockwise direction. Equations (g) and (i) are valid at all times, so we can integrate them if we like to get θ and x at any time t.

EXAMPLE 16.8

A 4.905-kN flywheel rotating at a speed ω of 200 rpm (see Fig. 16.25) breaks away from the steam engine that drives it and falls on the floor. If the coefficient of dynamic friction between the floor and the flywheel surface is .4, at what speed will the flywheel axis move after 2 sec? At what speed will it hit the wall A? The radius of gyration of the flywheel is 1 m and its diameter is 2.30 m. Do not consider effects

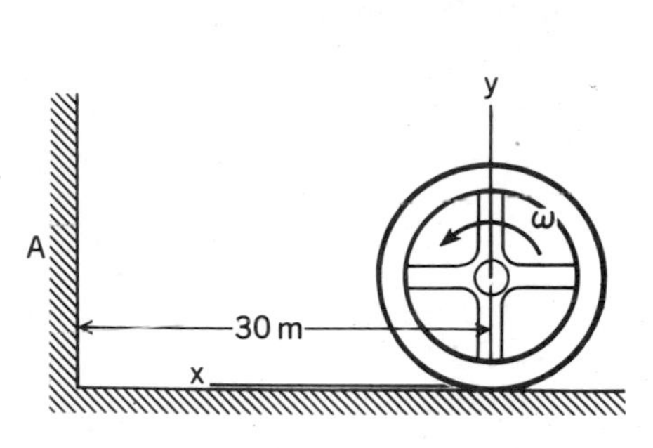

Figure 16.25. Runaway flywheel at initial position.

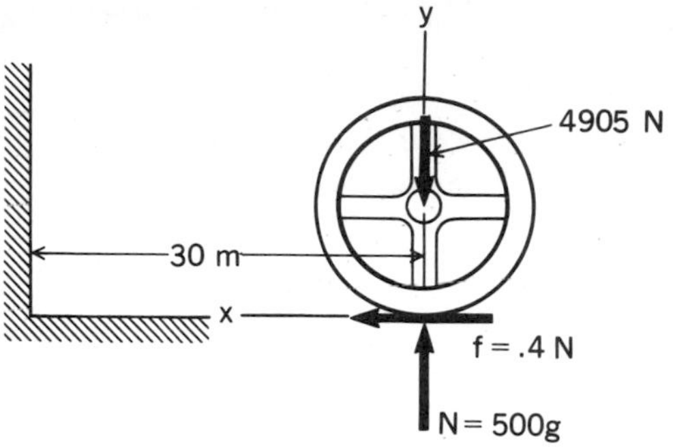

Figure 16.26. *xy* fixed at initial position.

of bouncing in your analysis. Neglect rolling resistance (Section 7.7) and wind friction losses.

We assume slipping occurs when the flywheel first touches the floor (see Fig. 16.26). *Newton's law* for the center of mass of the flywheel is

$$(.4)N = \left(\frac{4905}{9.81}\right)\ddot{x}$$

Therefore,

$$\ddot{x} = 3.92 \text{ m/sec}^2$$

Integrate twice:

$$\dot{x} = 3.92t + C_1 \qquad \text{(a)}$$

$$x = 1.962t^2 + C_1 t + C_2 \qquad \text{(b)}$$

At $t = 0$, $\dot{x} = 0$ and $x = 0$. Hence, $C_1 = 0$ and $C_2 = 0$. The *moment-of-momentum* equation for the center of mass is next given.

$$(.4)(N)\left(\frac{2.30}{2}\right) = \left(\frac{4905}{9.81}\right)(1)^2\ddot{\theta}$$

Therefore,

$$\ddot{\theta} = 4.51 \text{ rad/sec}^2$$

Integrate twice:

$$\dot{\theta} = 4.51t + C_3 \qquad \text{(c)}$$

$$\theta = 2.26t^2 + C_3 t + C_4 \qquad \text{(d)}$$

When $t = 0$, $\theta = 0$, and $\dot{\theta} = -200(2\pi/60) = -20.94$ rad/sec. Hence, $C_3 = -20.94$ and $C_4 = 0$.

We now ask when does the slipping stop? Clearly, it stops when there is zero velocity of the point of contact of the cylinder. From *kinematics* we have for this condition:

$$\dot{x} + \left(\frac{2.30}{2}\right)\dot{\theta} = 0 \qquad \text{(e)}$$

Substituting from Eq. (a) and (c) for $\dot{x}$ and $\dot{\theta}$, respectively, we have for Eq. (e):

$$3.92t + \left(\frac{2.30}{2}\right)(4.51t - 20.94) = 0$$

Therefore,

$$t = 2.64 \text{ sec}$$

Since we get a time here greater than zero, we can be assured that the initial slipping assumption is valid. The position $x_{\text{N.S.}}$ at the time of initial no-slipping is deduced from Eq. (b). Thus,

$$x_{\text{N.S.}} = 1.962(2.64)^2 = 13.67 \text{ m}$$

Accordingly, the flywheel hits the wall *after* it starts rolling without slipping. At $t = 2$ sec, there is still slipping, and we can use Eq. (a) to find $\dot{x}$ at this instant. Thus,

$$(\dot{x})_2 = (3.92)(2) = 7.85 \text{ m/sec}$$

The speed, once there is no further slipping, is constant, and so the speed at the wall is found by using $t = 2.64$ sec in Eq. (a). Thus,

$$(\dot{x})_{\text{wall}} = (3.92)(2.64) = 10.35 \text{ m/sec}$$

EXAMPLE 16.9

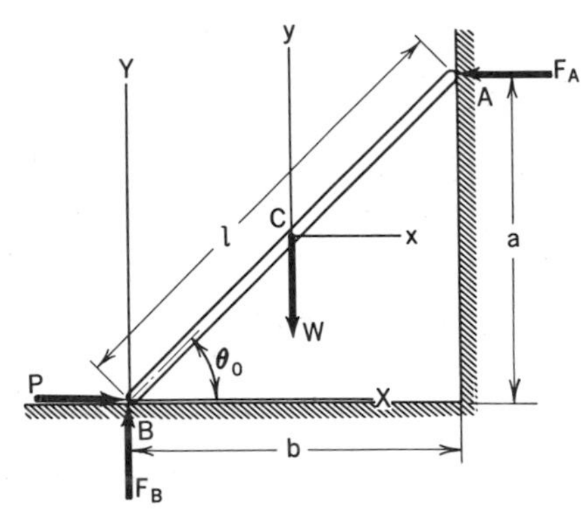

Figure 16.27. Rod on frictionless wall and floor.

A rigid rod, AB, slides against a frictionless wall and floor (Fig. 16.27). The rod has a weight W and is made to move to the right in the plane of the page by a force P as shown at B. What is the differential equation for the angular acceleration of the rod in terms of θ?

This is a case of plane motion of a slablike body. We can use the *moment-of-momentum* equation about an axis normal to the plane of motion at the center of mass:

$$M_z = I_{zz}\dot{\omega}_z = I_{zz}\ddot{\theta} \tag{a}$$

Putting in the proper moments and using $(M/12)l^2$ as I_{zz}, we have for this equation:

$$P\frac{l}{2}\sin\theta - F_B\frac{l}{2}\cos\theta + F_A\frac{l}{2}\sin\theta = \frac{M}{12}l^2\ddot{\theta} \tag{b}$$

We have here three unknowns, F_B, F_A, and $\ddot{\theta}$, so we must consider other possible equations. The motion of the center of mass can be used for this purpose. From *Newton's law*, we have

$$P - F_A = \frac{W}{g}(a_c)_x \tag{c}$$

$$F_B - W = \frac{W}{g}(a_c)_y \tag{d}$$

Two more equations have been set forth, but we do not know $(a_c)_x$ and $(a_c)_y$ (i.e., the acceleration components of the center of mass). Let us turn next to *kinematical* considerations. The acceleration of the mass center can be given in terms of the acceleration of point B by employing Eq. 15.7:

$$\boldsymbol{a}_C = \boldsymbol{a}_B + \dot{\boldsymbol{\omega}} \times \boldsymbol{\rho}_{BC} + \boldsymbol{\omega} \times (\boldsymbol{\omega} \times \boldsymbol{\rho}_{BC}) \tag{e}$$

In rectangular components, we get

$$(a_C)_x\boldsymbol{i} + (a_C)_y\boldsymbol{j} = a_B\boldsymbol{i} + \ddot{\theta}\boldsymbol{k} \times \left(\frac{l}{2}\sin\theta\boldsymbol{j} + \frac{l}{2}\cos\theta\boldsymbol{i}\right) + \dot{\theta}\boldsymbol{k} \times \left[\dot{\theta}\boldsymbol{k} \times \left(\frac{l}{2}\sin\theta\boldsymbol{j} + \frac{l}{2}\cos\theta\boldsymbol{i}\right)\right] \quad \text{(f)}$$

Considering the y scalar equation (and thus omitting the unknown a_B), we have for $(a_C)_y$:

$$(a_C)_y = \ddot{\theta}\frac{l}{2}\cos\theta - \dot{\theta}^2\frac{l}{2}\sin\theta \quad \text{(g)}$$

We can get $(a_C)_x$ by similarly considering points C and A of the rod. Thus,

$$\boldsymbol{a}_C = \boldsymbol{a}_A + \dot{\boldsymbol{\omega}} \times \boldsymbol{\rho}_{AC} + \boldsymbol{\omega} \times (\boldsymbol{\omega} \times \boldsymbol{\rho}_{AC}) \quad \text{(h)}$$

Therefore,

$$(a_C)_x\boldsymbol{i} + (a_C)_y\boldsymbol{j} = a_A\boldsymbol{j} + \ddot{\theta}\boldsymbol{k} \times \left(-\frac{l}{2}\sin\theta\boldsymbol{j} - \frac{l}{2}\cos\theta\boldsymbol{i}\right) + \dot{\theta}\boldsymbol{k} \times \left[\dot{\theta}\boldsymbol{k} \times \left(-\frac{l}{2}\sin\theta\boldsymbol{j} - \frac{l}{2}\cos\theta\boldsymbol{i}\right)\right] \quad \text{(i)}$$

Considering now the x scalar equation (and thus omitting the unknown a_A), we then have

$$(a_C)_x = \ddot{\theta}\frac{l}{2}\sin\theta + \dot{\theta}^2\frac{l}{2}\cos\theta \quad \text{(j)}$$

Substituting the acceleration components from Eqs. (g) and (j) in Eqs. (c) and (d), respectively, we get, with Eq. (b), a system of three simultaneous equations in three unknowns. They are

$$P - F_A = \frac{Wl}{2g}(\ddot{\theta}\sin\theta + \dot{\theta}^2\cos\theta) \quad \text{(k)}$$

$$F_B - W = \frac{Wl}{2g}(\ddot{\theta}\cos\theta - \dot{\theta}^2\sin\theta) \quad \text{(l)}$$

$$P\sin\theta - F_B\cos\theta + F_A\sin\theta = \frac{Ml}{6}\ddot{\theta} \quad \text{(m)}$$

In Eq. (m), note that we have canceled out $l/2$. Solving for F_A and F_B in Eqs. (k) and (l), respectively, and substituting into Eq. (m), we then get the equation

$$P\sin\theta - \left[W + \frac{Wl}{2g}(\ddot{\theta}\cos\theta - \dot{\theta}^2\sin\theta)\right]\cos\theta + \left[P - \frac{Wl}{2g}(\ddot{\theta}\sin\theta + \dot{\theta}^2\cos\theta)\right]\sin\theta = \frac{Ml}{6}\ddot{\theta}$$

Collecting and rearranging the terms, we have

$$\left[\frac{Wl}{2g}(\cos^2\theta + \sin^2\theta) + \frac{Ml}{6}\right]\ddot{\theta} = 2P\sin\theta - W\cos\theta$$

Noting that $(\cos^2\theta + \sin^2\theta)$ is unity and replacing M by W/g, we have finally:

$$\ddot{\theta} = \frac{3}{2}\frac{g}{Wl}(2P\sin\theta - W\cos\theta) \quad \text{(n)}$$

Note from Eq. (n) that the angular acceleration does not depend on the angular velocity (the $\dot{\theta}$ terms canceled). [The integration of θ as a function of time from Eq. (n) involves elliptic integrals.]

As in Section 16.5, we can generalize the results of this section to include the plane motion of a body having at the center of mass a *principal* axis z normal to the plane of motion XY. Clearly, for axes xyz at A, we get the same equations of motion as for the slablike body.

Problems

16.33. A stepped cylinder is released from a rest configuration where the spring is stretched 200 mm. A constant force F of 360 N acts on the cylinder, as shown. The cylinder has a mass of 146 kg and has a radius of gyration of 1 m. What is the friction force at the instant the stepped cylinder is released? Take $\mu_s = .3$ for the coefficient of friction. The spring constant K is 270 N/m.

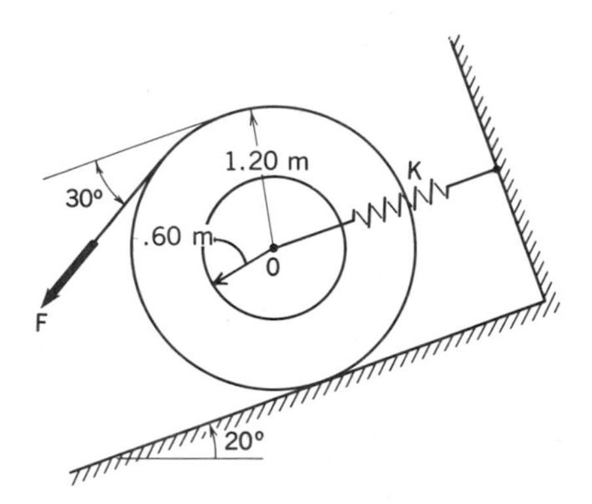

Figure P.16.33

16.34. A stepped cylinder is held on an incline with an inextensible cord wrapped around the inner cylinder and an outside agent (not shown). If the tension T on the cord at the instant that the cylinder is released by the outside agent from the position shown is 100 lb, what is the initial angular acceleration? What is the acceleration of the mass center? Use the following data:

$$W = 300 \text{ lb}$$

$$\text{radius of gyration} = 3 \text{ ft}$$

$$R_1 = 2 \text{ ft}$$

$$R_2 = 4 \text{ ft}$$

$$\mu_s = .1$$

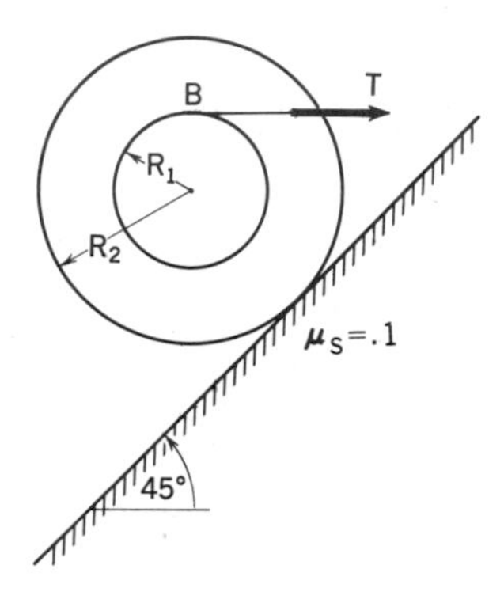

Figure P.16.34

16.35. The cylinder shown is acted on by a 100-lb force. At the contact point A, there is viscous friction such that the friction force is given as

$$f = .05V_A$$

where V_A is the velocity of the cylinder at the contact point in ft/sec. The weight of the cylinder is 30 lb, and the radius of gyration is 1 ft. Set up a third-order differential equation for finding the position of O as a function of time.

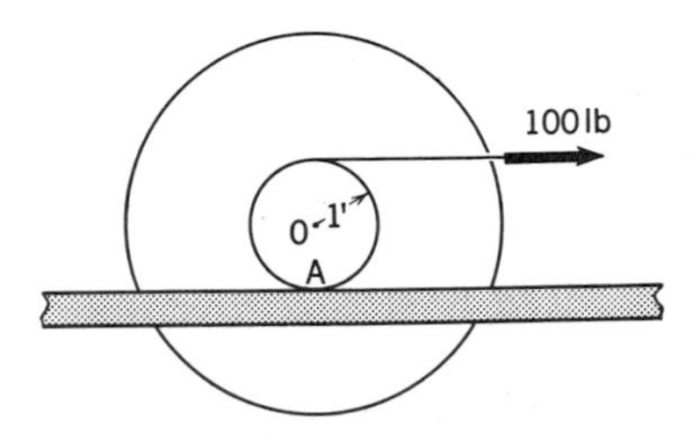

Figure P.16.35

16.36. The cylinder shown weighs 445 N and has a radius of gyration of .27 m. What is the minimum coefficient of friction at A that will prevent the body from moving? Using half of this coefficient of friction, how far d does point O move in 1.2 sec if the cylinder is released from rest?

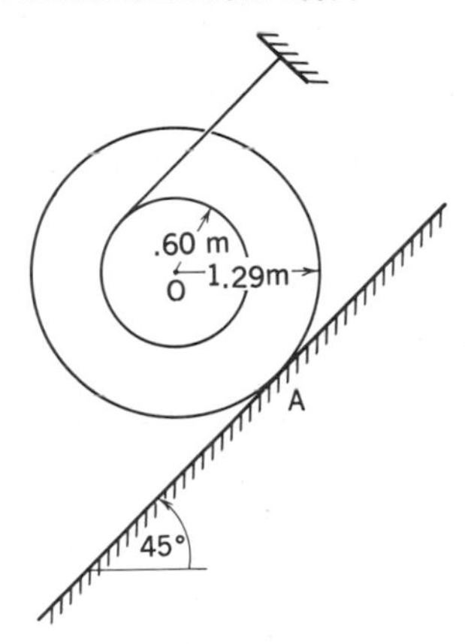

Figure P.16.36

16.37. A semicircular cylinder A is shown. The diameter of A is 1 ft, and the weight is 100 lb. What is the angular acceleration of A at the position shown if at this instant there is no slipping and the semicylinder is stationary?

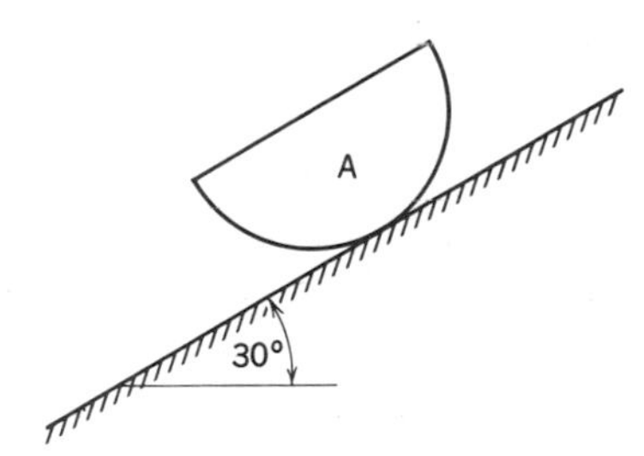

Figure P.16.37

16.38. A 20 kg bar AB connects two gears G and H. These gears each have a mass of 5 kg and a diameter of 300 mm. A torque T of 5 N-m is applied to gear H. What is the angular speed of H after 20 sec if the system starts from rest? The system is in a horizontal plane. Bar AB is 2 m long.

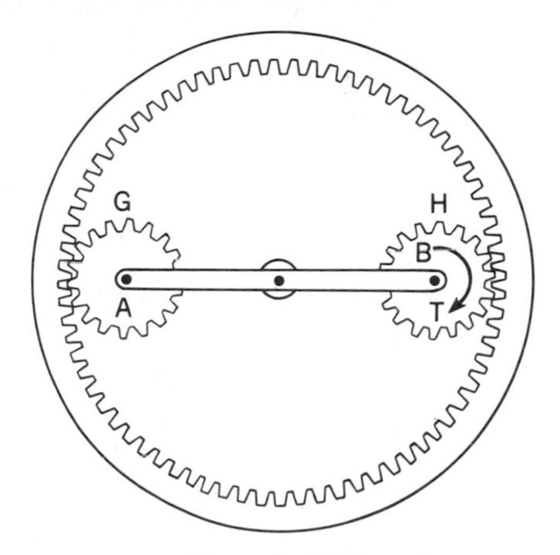

Figure P.16.38

16.39. A bar C weighing 445 N rolls on cylinders A and B, each weighing 223 N. What is the acceleration of bar C when the 90-N force is applied as shown? There is no slipping.

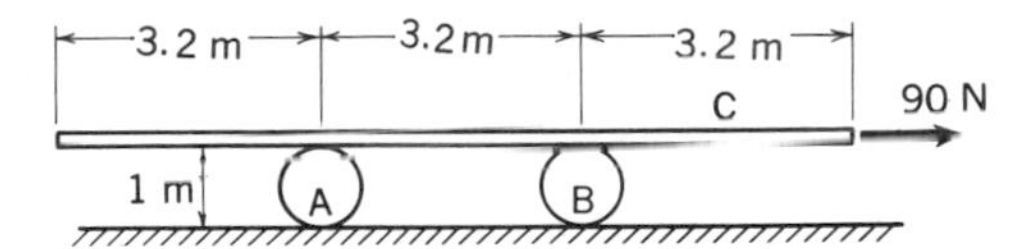

Figure P.16.39

16.40. In Problem 16.39, at what position of the bar relative to the wheels does slipping first occur after the force is applied? Take $\mu_s = .2$ for the bottom contact surface and $\mu_s = .1$ for the contact surface between bar and cylinders. From Problem 16.39, $\ddot{x}_c = 1.442\ \text{m/sec}^2$.

16.41. A platform B, of weight 30 lb and carrying block A of weight 100 lb, rides on gears D and E as shown. If each gear weighs 30 lb, what distance will platform B move in 1 sec after the application of a 100-lb force as shown?

Figure P.16.41

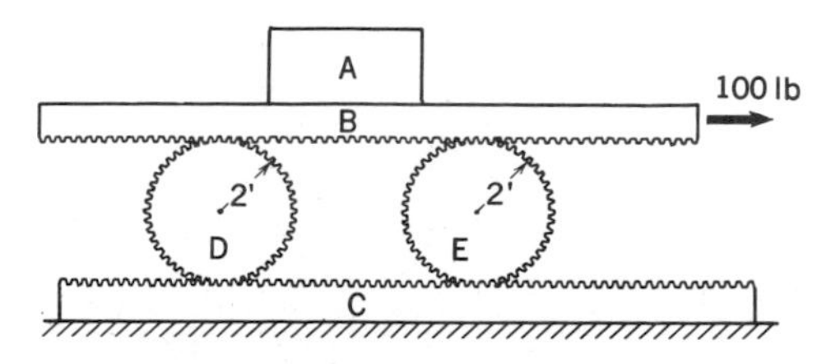

16.42. A crude cart is shown. A horizontal force P of 100 lb is applied to the cart. The coefficient of static friction between wheels and ground is .6. If $D = 3$ ft, what is the acceleration of the cart to the right? The wheels weigh 50 lb each. Neglect friction in the axle bearings. The total weight of cart with load is 322 lb.

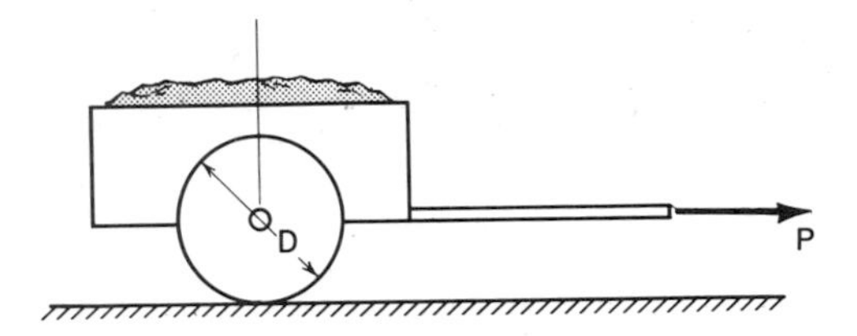

Figure P.16.42

16.43. What minimum force component P is required to cause the cart in Problem 16.42 to move so that the wheels slip rather than roll without slipping?

16.44. A pulley system is shown. Sheave A has a mass of 25 kg and has a radius of gyration of 250 mm. Sheave B has a mass of 15 kg and has a radius of gyration of 150 mm. If released from rest, what is the acceleration of the 50-kg block? There is no slipping.

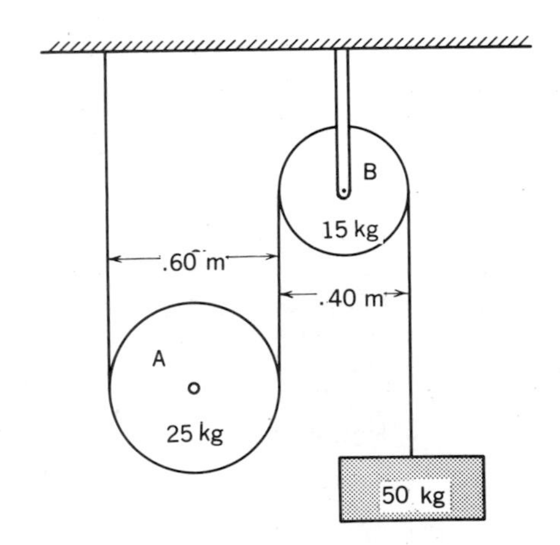

Figure P.16.44

16.45. A steam locomotive drive system is shown. Each drive wheel weighs 5 kN and has a radius of gyration of 400 mm. At the instant shown, a pressure $p = .50$ N/mm^2 above atmospheric acts on the piston to drive the train backward. If the train is moving at 1 m/sec backward at the instant shown, what is its acceleration? Members AB and BC are to be considered stiff but light in comparison to other parts of the engine. Also, the piston assembly can be considered light. Only the driving car is in action in this problem. It has one driving system, such as is shown on each side. It has two additional wheels of the size and mass shown above on each side plus additional small wheels whose rotational inertia we shall neglect. The drive train minus its eight large wheels has a weight of 150 kN. Assume no slipping, and neglect friction in the piston assembly.

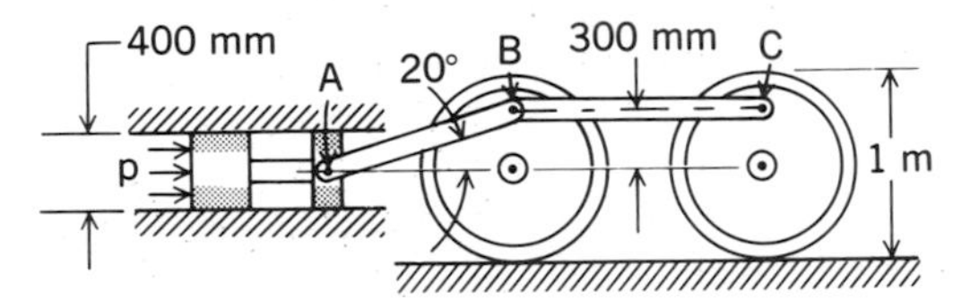

Figure P.16.45

16.46. A system of interconnected gears is shown. Gear B rotates about a fixed axis, and gear D is stationary. If a torque T of 2.5 N-m is applied to gear B at the configuration shown, what is the angular acceleration of gear A? Gear A has a mass of 1.36 kg while gear B has a mass of 4.55 kg. The system is in a vertical orientation relative to the ground. What vertical force is transmitted to stationary gear D?

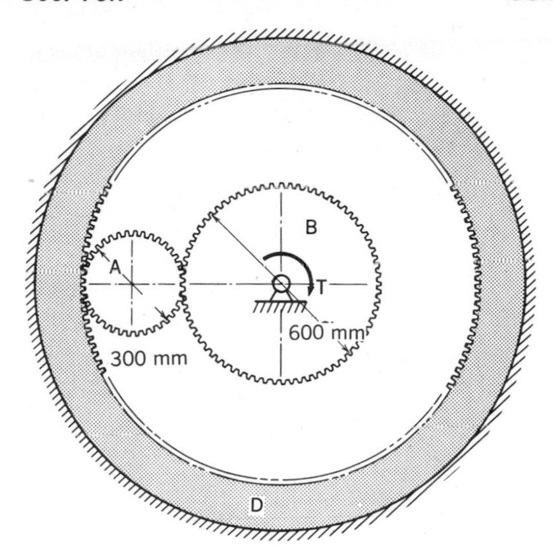

Figure P.16.46

16.47. A solid semicircular cylinder of weight W and radius R is released from rest from the position shown. What is the friction force at that instant?

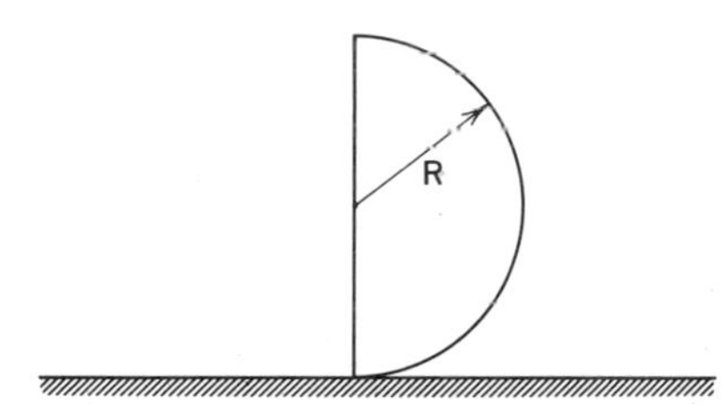

Figure P.16.47

16.48. A cylinder is shown made up of two semi-cylinders A and B weighing 15 lb and 30 lb, respectively. If the cylinder has a diameter of 3 in., what is the angular acceleration when it is released from a stationary configuration at the position shown? Assume no slipping.

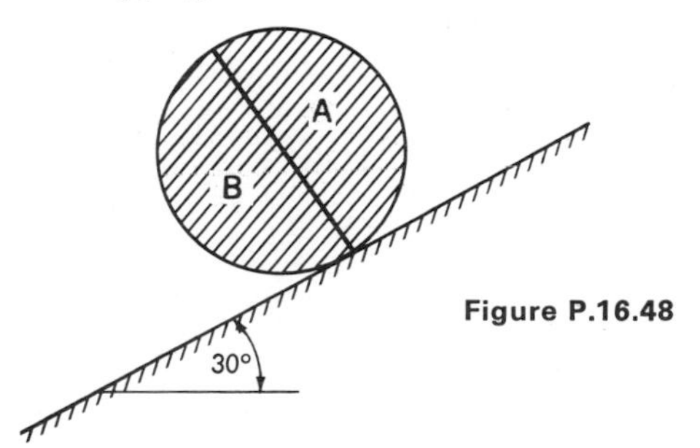

Figure P.16.48

16.49. A thin-walled cylinder is shown held in position by a cord AB. The cylinder has a mass of 10 kg and has an outside diameter of 600 mm. What are the normal and friction forces at the contact point A at the instant that cord AB is cut? Assume that no slipping occurs.

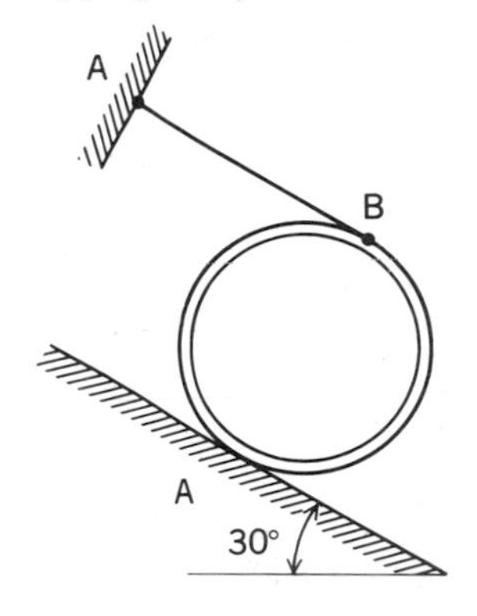

Figure P.16.49

16.50. A bent rod $CBEF$ is welded to a shaft. At the ends C and F are identical gears G and H, each of mass 3 kg and radius of gyration 70 mm. The gears mesh with a large stationary gear D. A torque T of 50 N-m is applied to the shaft. What is the angular speed of the shaft in 10 sec? The bent rod has a mass per unit length of 5 kg/m. Will there be forces on the bearings of the shaft other than those from gravity? Why?

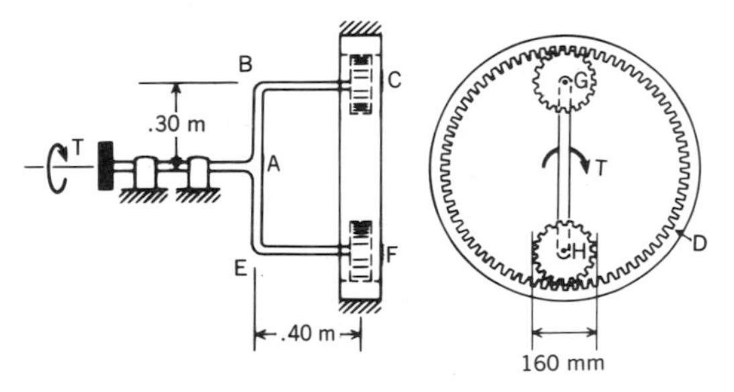

Figure P.16.50

16.51. A tractor and driver have a mass of 1350 kg. If a total torque T of 300 N-m is developed on the two drive wheels by the motor, what is the acceleration of the tractor? The large drive wheels each have a mass of 90 kg, a diameter of 1 m, and a radius of gyration of 400 mm. The small wheels each have a mass 20 kg and have a diameter of 300 mm with a radius of gyration of 100 mm.

Figure P.16.51

16.52. A block B weighing 100 lb rides on two identical cylinders C and D weighing 50 lb each as shown. On top of block B is a block A weighing 100 lb. Block A is prevented from moving to the left by a wall. If we neglect friction between A and B and between A and the wall and we consider no slipping at the contact surfaces of the wheels, what is the angular speed of the wheels in 2 sec for $P = 80$ lb?

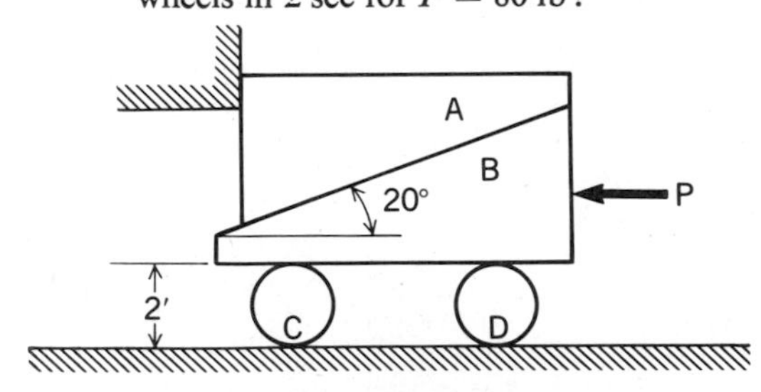

Figure P.16.52

16.53. A cable is wrapped around two pulleys A and B. A tension T is applied to the end of the cable at G. Each pulley weighs 5 lb and has a radius of gyration of 4 in. The diameter of the pulleys is 12 in. A body C weighing 100 lb is supported by pulley B. Suspended from body C is a body D weighing 25 lb. Body D is lowered from body C so as to accelerate at the rate of 5 ft/sec^2 relative to body C. What tension T is then needed to pull the cable downward at G at the increasing rate of 5 ft/sec^2?

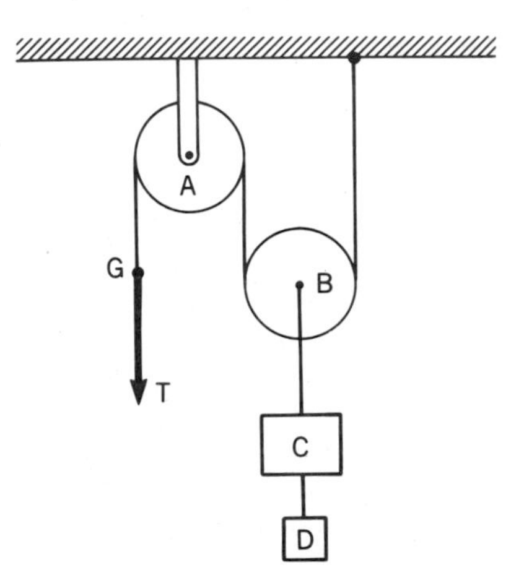

Figure P.16.53

16.54. A cylinder A is acted on by a torque T of 1000 N-m. The cylinder has a mass of 75 kg and a radius of gyration of 400 mm. A light rod CD connects cylinder A with a second cylinder B having a mass of 50 kg and a radius of gyration of 200 mm. What is the force in member CD when torque T is applied? The system is stationary at the instant the torque is applied. Assume no slipping of cylinder C along the incline.

Figure P.16.54

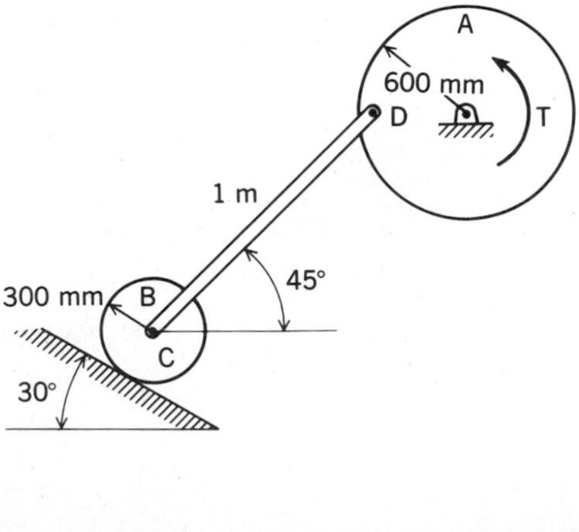

16.55. A ring rests on a smooth surface shown as seen from above. The ring has a mean radius of 2 m and a mass of 30 kg. A force of 400 N is applied to the ring at B. What is the acceleration of point A on the ring?

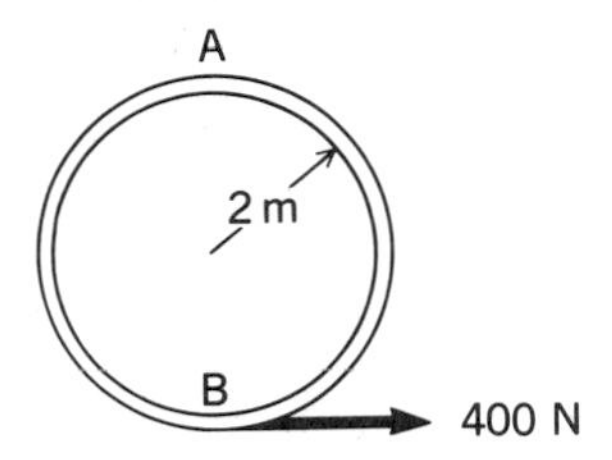

Figure P.16.55

16.56. A block having a mass of 300 kg rests on a smooth surface as shown from above. A 500-N force is applied at an angle of 45° to the side. What is the acceleration of point B of the block when the force is applied? The center of mass of the block coincides with the geometric center.

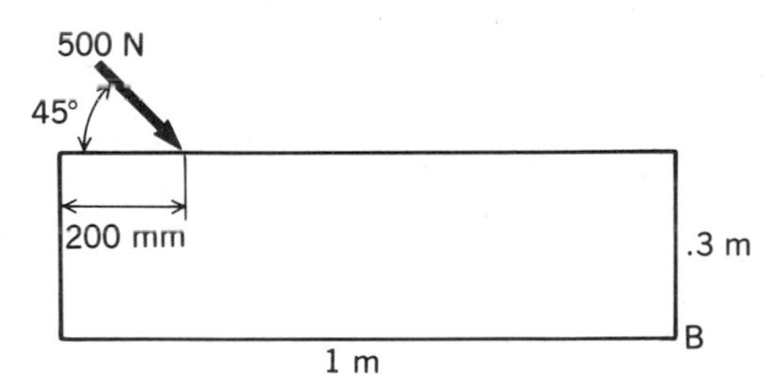

Figure P.16.56

16.57. A bent rod rests on a smooth surface. The rod has a mass of 20 kg. What is the acceleration of point A when a force $P = 100$ N is applied?

Figure P.16.57

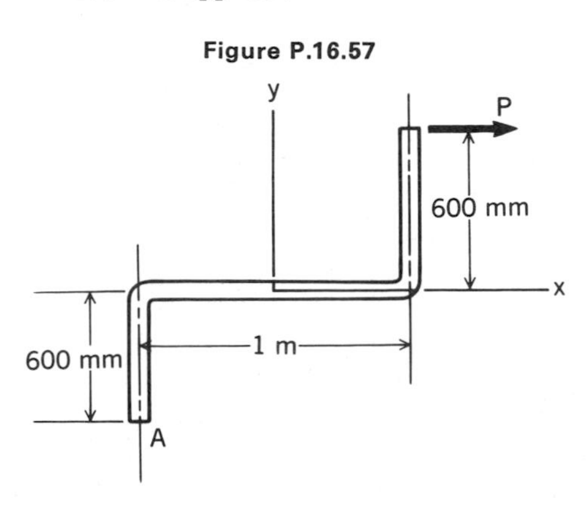

16.58. A constant force of 100 lb is exerted on a rope wrapped around a 50-lb cylinder. How high does the cylinder rise in 2 sec? How many rotations has the cylinder had at that time? Neglect initial frictional effects by the ground support.

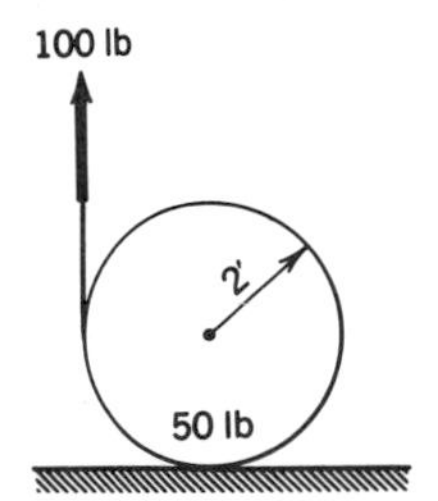

Figure P.16.58

16.59. A bar of weight W equal to 100 lb is at rest on a horizontal surface S at the instant that a force P equal to 60 lb is applied. Show that the center of rotation at the instant that the force P is applied is 3.27 ft from left end. The coefficient of friction μ_s equals .2. The length of the bar is 10 ft.

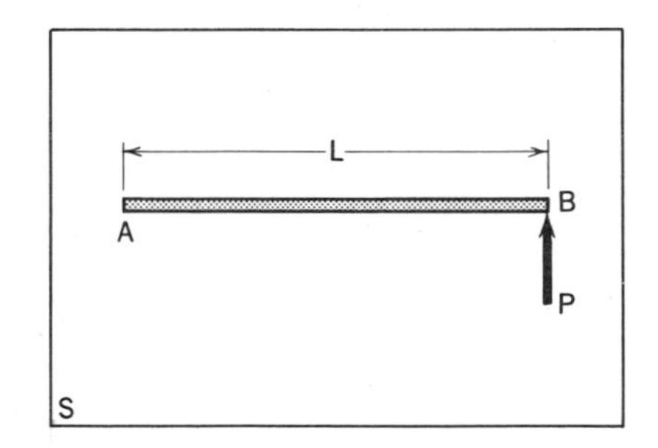

Figure P.16.59

16.60. A cart B is given a constant acceleration of 5 m/sec². On the cart is a cylinder A having a mass of 5 kg and a diameter of 600 mm. If initially everything is stationary, how far does the cylinder move relative to the cart in 1.5 sec? Assume no slipping.

Figure P.16.60

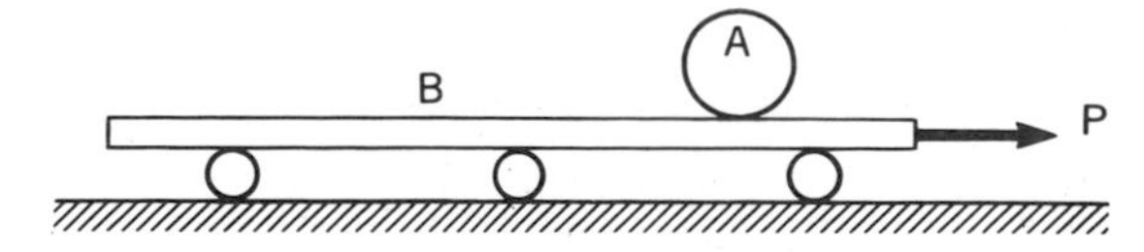

16.61. In Problem 16.60, what force P is needed to uniformly accelerate the cart so that the cylinder A moves 1 m in 2 sec relative to the cart. Cart B has a mass of 10 kg. Neglect the inertia of the small rollers supporting the cart, and assume there is no slipping.

16.62. A wedge B is shown with a cylinder A of mass 20 kg and diameter 500 mm on the incline. The wedge is given a constant acceleration of 20 m/sec² to the right. How far d does the cylinder move in $\frac{1}{2}$ sec relative to the incline if there is no slipping? The system starts from rest.

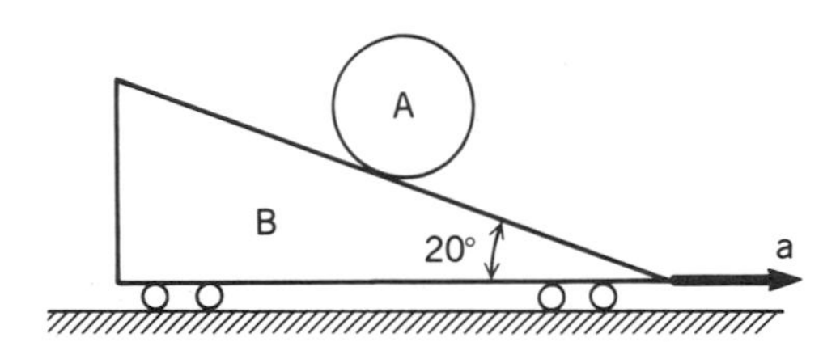

Figure P.16.62

16.63. A block weighing 100 N is held by three inextensible guy wires. What are the forces in wires AC and BD at the instant that wire EC is cut?

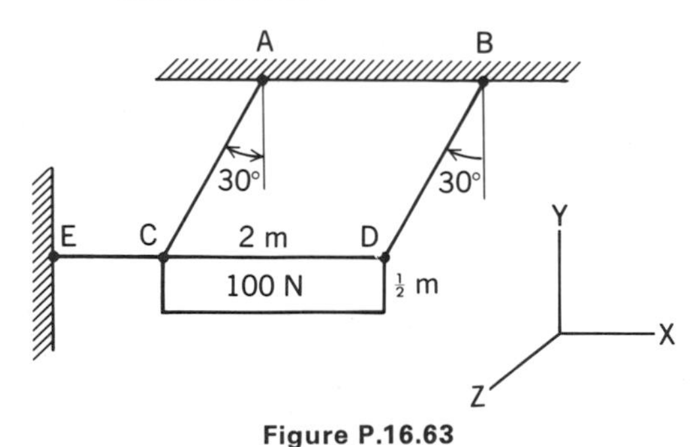

Figure P.16.63

16.64. A bowler releases his ball at a speed of 3 m/sec. If the ball has a diameter of 250 mm, what spin ω should be put on the ball so there is no slipping? If he puts on only half of this spin ω, keeping the same speed of 3 m/sec, what is the final speed of the ball? What is the speed after .3 sec? The ball has a mass of 1.8 kg and a coefficient of friction with the floor of .1. Neglect wind resistance and rolling resistance (as discussed in Section 7.7).

Figure P.16.64

16.65. In Problem 16.64 suppose ω is 1.4 times the value $\omega = 24$ rad/sec needed for no slipping. What is the speed of the ball when it strikes the pins? What is the speed of the ball after .3 sec?

16.66. A tug is pushing on the side of a barge which is loaded with sand and which has a total weight of 50 tons. The tug generates a 3-kN force which is always normal to the barge. If the barge rotates 5° in 20 sec, what is the moment of inertia of the barge at the center of mass which we take at the geometric center of the barge? The distance d for this maneuver is 5 m. What is the acceleration of the center of mass of the barge at $t = 35$ sec? Neglect the resistance of the water.

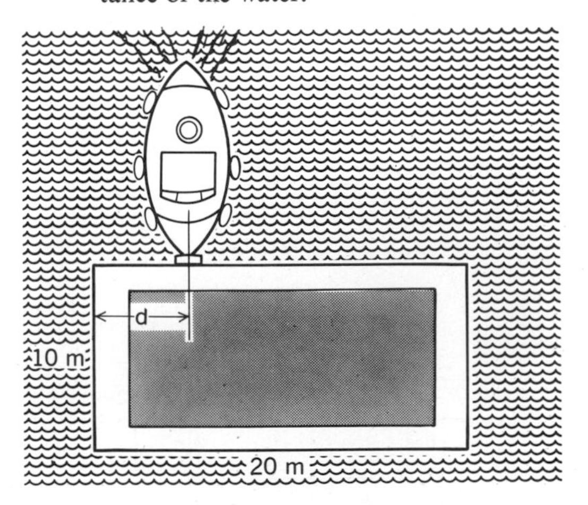

Figure P.16.66

16.67. A 10-m I-beam having a mass of 400 kg is being pulled by an astronaut with his space propulsion rig as shown. The force is 40 N and is always in the same direction. Initially, the beam is stationary relative to the astronaut, and the connecting cord is at right angles to the beam. Consider the beam to be a long slender rod. What is the angular acceleration when beam has rotated 15°? What is the position of the center of mass after 10 sec? (Can you integrate the differential equation for θ to get familiar functions? Explain.)

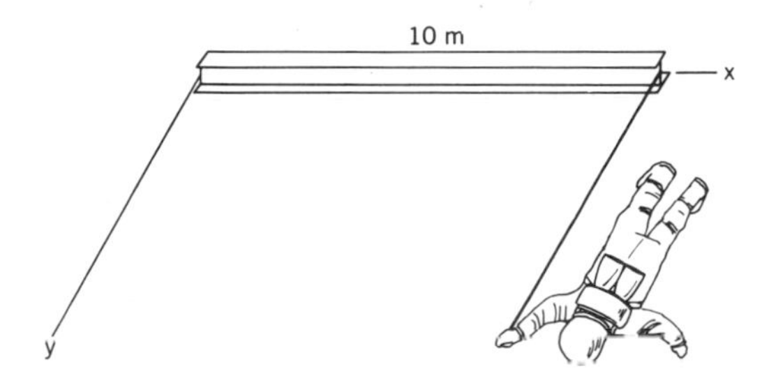

Figure P.16.67

16.68. A rectangular box having a mass of 20 kg is being transported on a conveyor belt. The center of gravity of the box is 150 mm above the conveyor belt, as shown. What is the maximum starting torque T for which the box will not tip? The belt has a mass per unit length of 2 kg/m, and the driving and driven drums have a mass of 5 kg each and a radius of gyration of 130 mm. The coefficient of friction between the belt and conveyor bed is .2.

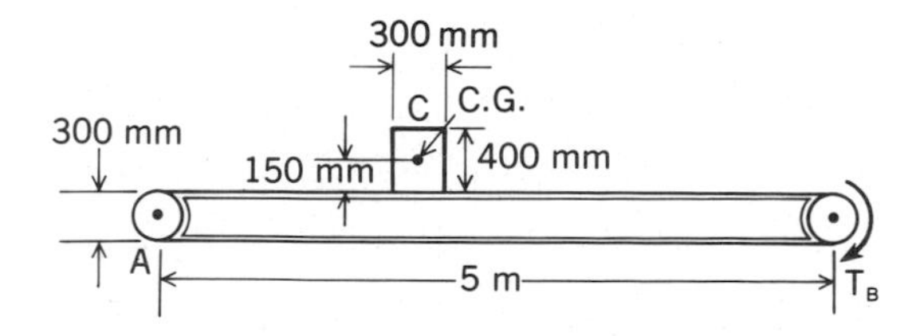

Figure P.16.68

16.69. A ring is shown supported by wire AB and a smooth surface. The ring has a mass of 10 kg and a mean radius of 2 m. A body D having a mass of 3 kg is fixed to the ring as shown. If the wire is severed, what is the acceleration of body D?

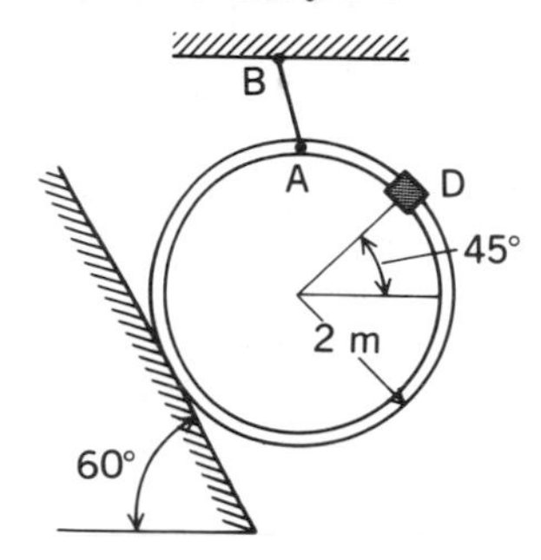

Figure P.16.69

16.70. If the rod shown is released from rest at the configuration shown, what are the supporting forces at A and B at that instant? The rod weighs 100 lb and is 10 ft long. The static coefficient of friction is .2 for all surface contacts.

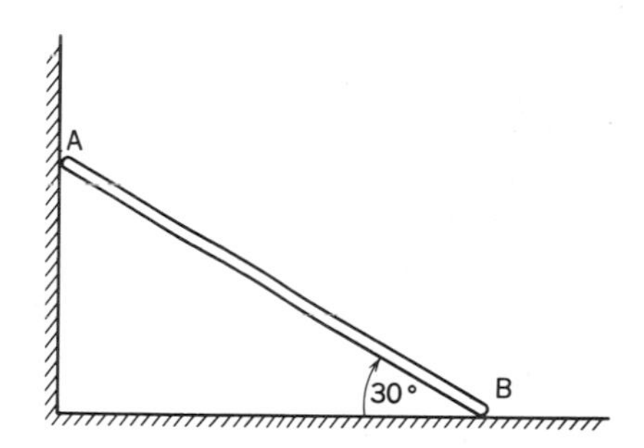

Figure P.16.70

16.71. Do Problem 16.70 for the case where end A is moving downward at a speed of 10 ft/sec at the instant shown and where $\mu_d = .2$.

16.72. In Problem 16.70, find by inspection the instantaneous axis of rotation for the rod. What are the magnitude and direction of the acceleration vector for the axis of rotation at the instant the rod is released? We know from Problem 16.70 that $\dot{\omega} = 3.107$

rad/sec² and $\boldsymbol{a}_c = 7.77\boldsymbol{i} - 13.45\boldsymbol{j}$ ft/sec² for the center of mass.

16.73. A rod AB of length 3 m and weight 445 N is shown immediately after it has been released from rest. Compute the tension of wires EA and DB at this instant.

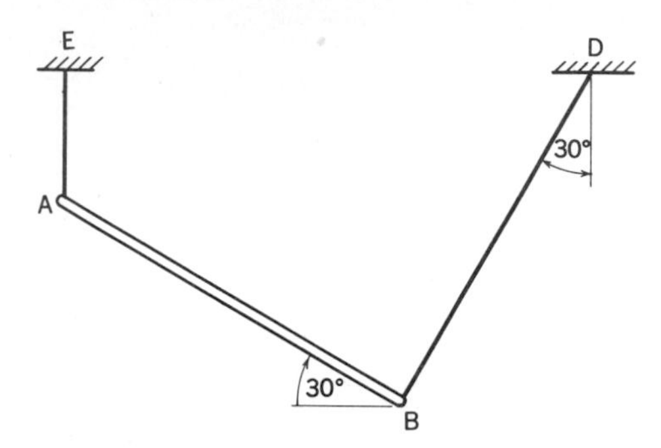

Figure P.16.73

16.74. Rod AB is released from the configuration shown. What are the supporting forces at this instant if we neglect friction? The rod weighs 200 lb and is 20 ft in length.

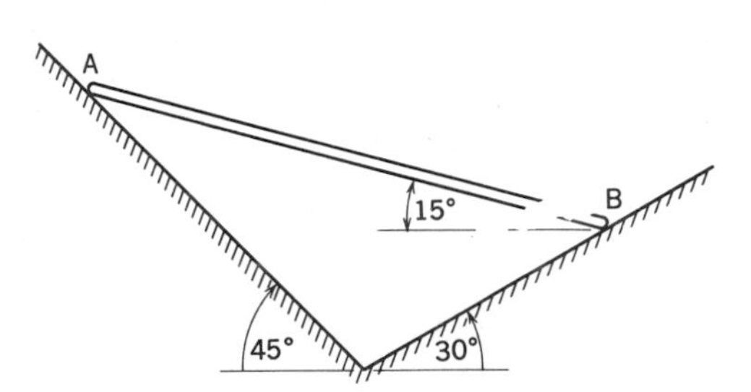

Figure P.16.74

16.75. A cylinder weighing 100 lb with a radius of 1 ft is held fixed on an incline that is rotating at $\frac{1}{2}$ rad/sec. The cylinder is released when the incline is at position θ equal to 30°. If the cylinder is 20 ft from the bottom A at the instant of release, what is the initial acceleration of the center of the cylinder relative to the incline? There is no slipping.

Figure P.16.75

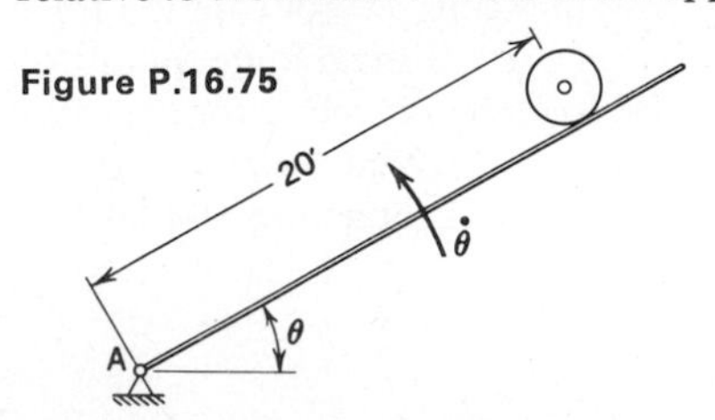

16.76. In Problem 16.75, $\ddot{\theta} = 2$ rad/sec² at the instant of interest, and the cylinder is rolling downward at a speed ω_1 of 3 rad/sec relative to the incline. What is the acceleration of the center of the cylinder relative to the incline at the instant of interest? There is no slipping.

16.77. Two identical bars, each having a mass of 9 kg, hang freely from the vertical. A force of 45 N is applied at the center of the upper bar AB. What are the angular accelerations of the bars?

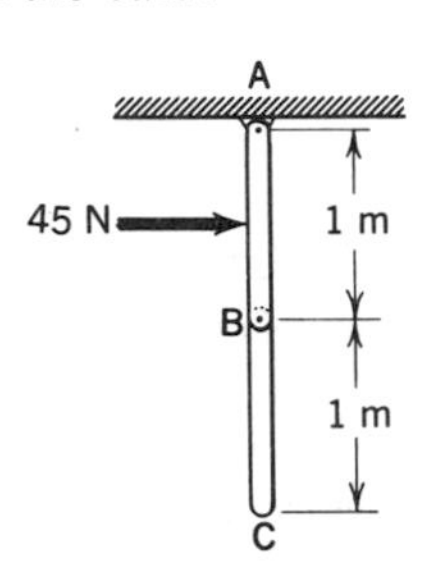

Figure P.16.77

16.78. A rod BA rotates at a constant speed ω of 100 rpm. It drives a rod BC, having a mass of 5 kg, which in turn moves a gear D having a mass of 3 kg and a radius of gyrations of 200 mm. The diameter of the gear is 450 mm. At the instant shown, what are the forces transmitted by the pins at C and at B?

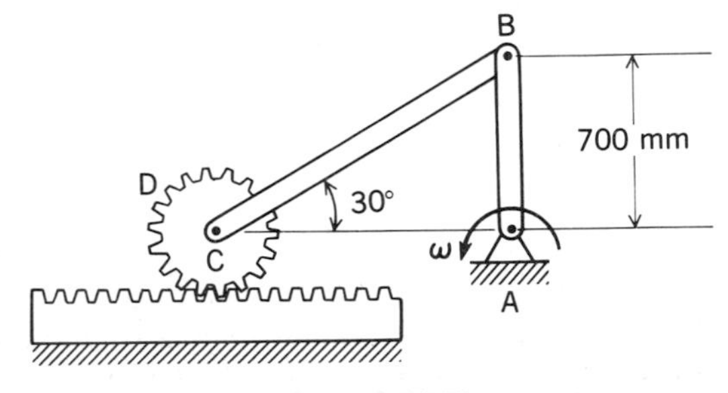

Figure P.16.78

16.79. A 1-m rod AB weighing 10 kg is suspended at one end by a cord BC and at the other end rides on an inclined surface on small wheels. Initially the wheels are being held at the position shown. At the instant the wheels are released, what is the angular acceleration of rod AB? Neglect friction on the inclined surface.

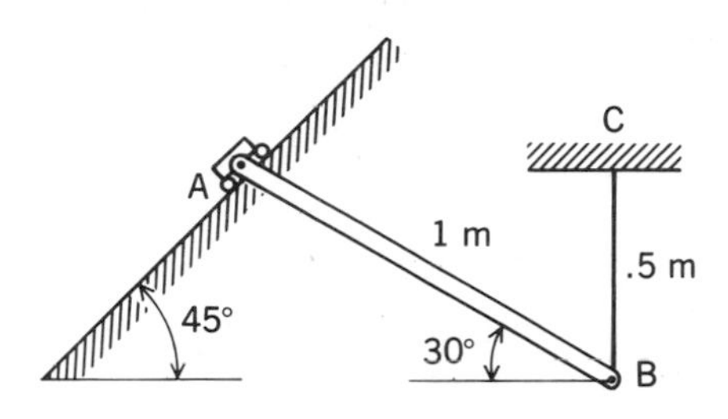

Figure P.16.79

16.80. Identical bars AB and BC are pinned as shown with frictionless pins. Each bar is 2.3 m in length and has a mass of 9 kg. A force of 450 N is exerted at C when the bars are inclined at 60°. What is the angular acceleration of the bars?

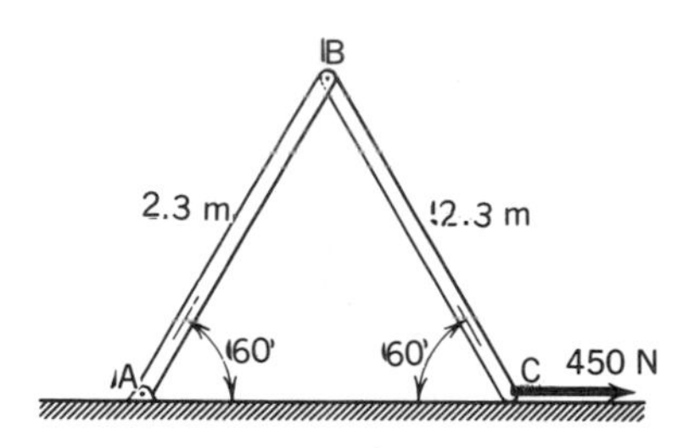

Figure P.16.80

16.81. A compressor is shown. Member AB is rotating at a constant speed ω_1 of 100 rpm. Member BC has a mass of 2 kg and piston C has a mass of 1 kg. The pressure p on the piston is 10,000 Pa. At the instant shown, what are the forces transmitted by pins B and C?

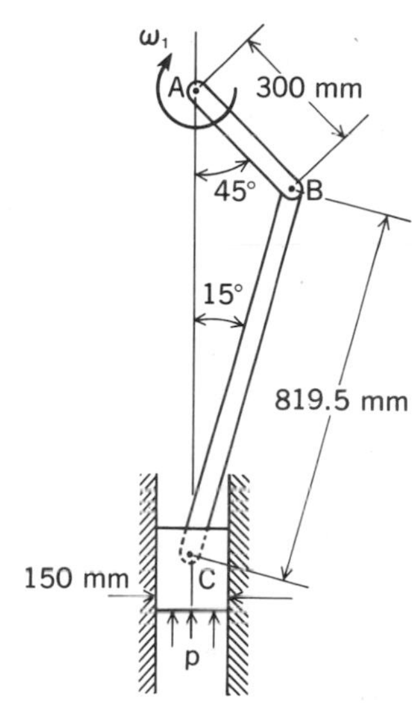

Figure P.16.81

16.8 Pure Rotation of an Arbitrary Rigid Body

We now consider a body having an arbitrary distribution of mass rotating about an axis of rotation fixed in inertial space. We consider this axis to be the z axis. We can take the origin of xyz anywhere along the z axis since all such points are fixed in inertial space. The *moment-of-momentum equations to be used will now be the general equations 16.6* since I_{zx} and I_{zy} will generally not equal zero. If the center of mass is along the z axis, then it obviously has no acceleration, and so we can apply the rules of statics to the center of mass. For other cases we shall often need to use Newton's law for the center of mass. In this regard it will be helpful to note from the definition of the center of mass that for a system of rigid bodies such as is shown in Fig. 16.28

$$M\mathbf{r}_c = \sum_{i=1}^{n} m_i \mathbf{r}_i \tag{16.16}$$

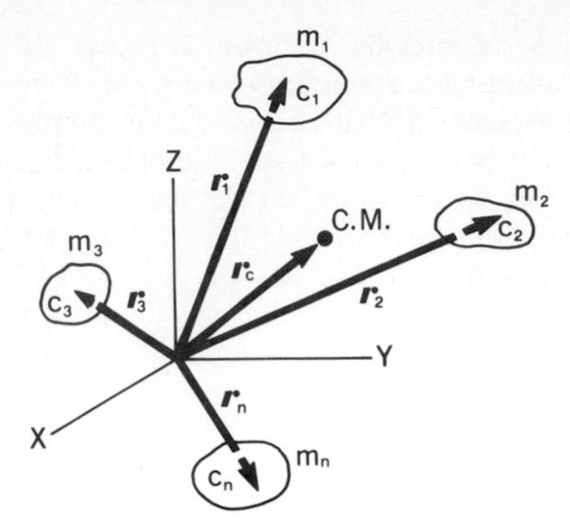

Figure 16.28. m rigid bodies having a total mass M.

where m_i is the mass of the ith rigid body, $\boldsymbol{r}_i$ is the position vector to the center of mass of the ith rigid body, M is the total mass, and $\boldsymbol{r}_c$ is the position vector to the center of mass of the system. We can then say on differentiating:

$$M\dot{\boldsymbol{r}}_c = \sum_{i=1}^{n} m_i \dot{\boldsymbol{r}}_i \tag{16.17}$$

$$M\ddot{\boldsymbol{r}}_c = \sum_{i=1}^{n} m_i \ddot{\boldsymbol{r}}_i \tag{16.18}$$

In *Newton's law* for the mass center of a system of rigid bodies, we conclude that we can use the centers of mass of the component parts of the system as given on the right side of Eq. 16.18 rather than the center of mass of the total mass.

EXAMPLE 16.10

A shaft has protruding arms each of which weighs 40 N/m (see Fig. 16.29). A torque T gives the shaft an angular acceleration $\dot{\omega}$ of 2 rad/sec². At the instant shown in the diagram, ω is 5 rad/sec. If the shaft without arms weighs 180 N, compute the vertical and horizontal forces at bearings A and B (see Fig. 16.30). Note that we have numbered the various arms for convenient identification.

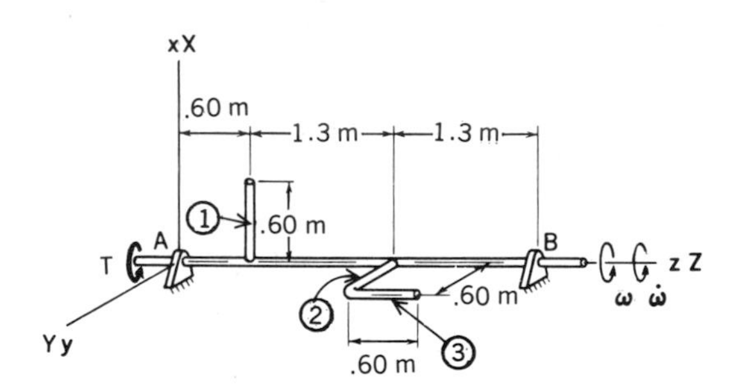

Figure 16.29. Rotating shaft with arms.

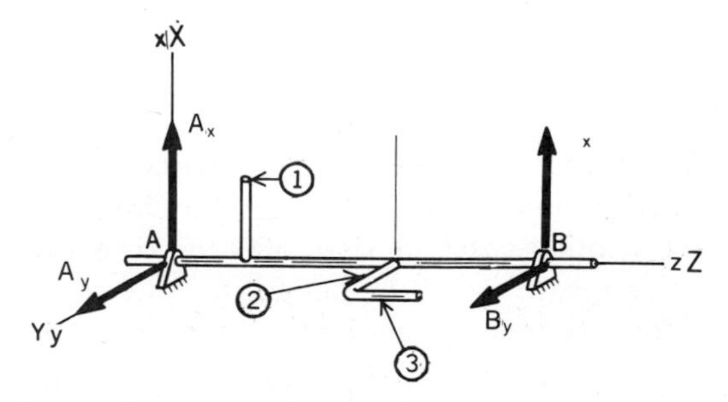

Figure 16.30. Supporting forces.

We first fix a reference xyz to the shaft at A. Also at A we fix an inertial reference XYZ to the ground. We can directly use Eqs. 16.6a and 16.6b about point A. For this reason, we shall compute the required products of inertia of the shaft system for reference xyz. Accordingly, using the parallel-axis theorem, we have:

$$(I_{xz})_{\text{arm}(1)} = 0 + \frac{(40)(.60)}{g}[(.6)(.3)] = .440 \text{ kg-m}^2$$

$$(I_{xz})_{\text{arm}(2)} = 0 + \frac{(40)(.60)}{g}[(1.9)(0)] = 0$$

$$(I_{xz})_{\text{arm}(3)} = 0 + \frac{(40)(.60)}{g}[(2.20)(0)] = 0$$

$$(I_{xz})_{\text{shaft}} = 0 + 0 = 0$$

Hence, for the system, $I_{xz} = .440$ kg-m².

We next consider I_{zy}. Accordingly, we have

$$(I_{zy})_{\text{arm}(1)} = 0 + \frac{(40)(.60)}{g}[(.60)(0)] = 0$$

$$(I_{zy})_{\text{arm}(2)} = 0 + \frac{(40)(.60)}{g}[(1.9)(0.30)] = 1.394 \text{ kg-m}^2$$

$$(I_{zy})_{\text{arm}(3)} = 0 + \frac{(40)(.60)}{g}[(2.20)(.60)] = 3.23 \text{ kg-m}^2$$

$$(I_{zy})_{\text{shaft}} = 0 + 0 = 0$$

Hence, for the system $I_{zy} = 4.62$ kg-m².

We can now employ Eq. 16.6 to get M_x and M_y about point A. Thus, we have

$$M_x = -(2)(.440) + (5^2)(4.62) = 114.7 \text{ N-m} \qquad \text{(a)}$$

$$M_y = -(2)(4.62) - (5^2)(.440) = -20.2 \text{ N-m} \qquad \text{(b)}$$

Summing moments of all the forces acting on the system about the y axis at A, we can say (see Fig. 16.30):

$$M_y = -20.2 = -(40)(.60)(.60) - (40)(.60)(1.9)$$
$$- (40)(.60)(2.2) - (180)(1.6) + (B_x)(3.2)$$

Therefore,

$$B_x = 118.9 \text{ N} \qquad \text{(c)}$$

Summing moments about the x axis at A, we can say:

$$M_x = 114.7 = -B_y(3.2)$$

Therefore,

$$B_y = -35.8 \text{ N} \qquad \text{(d)}$$

We next use *Newton's law* considering the three arms to be three particles at their mass centers as has been shown in Fig. 16.31. In the x direction, we have, using Eq. 16.18 and noting that each of the aforementioned particles has circular motion:

$$118.9 + A_x - 180 - (3)[(40)(.60)] = -\frac{(40)(.60)}{g}(.30)(5^2)$$
$$- \frac{(40)(.60)}{g}(.30)(2) - \frac{(40)(.60)}{g}(.60)(2)$$

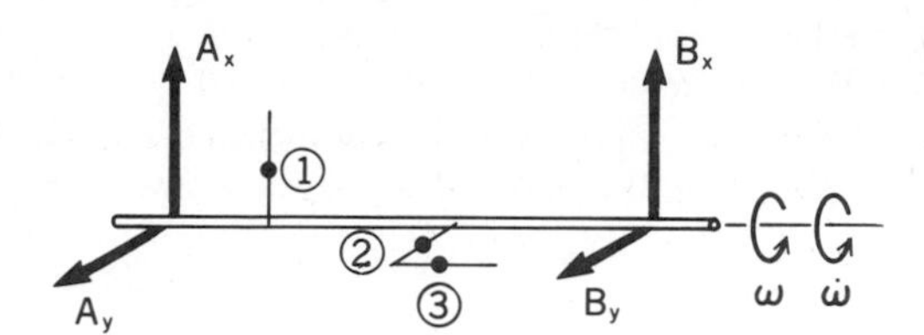

Figure 16.31. Arms replaced by mass centers.

Therefore,

$$A_x = 110.3 \text{ N} \tag{e}$$

In the y direction, we can say similarly:

$$A_y - 35.8 = \frac{(40)(.60)}{g}(.30)(\dot{\omega}) - \frac{(40)(.60)}{g}(.30)\omega^2 - \frac{(40)(.60)}{g}(.60)(\omega^2)$$

Therefore,

$$A_y = -17.78 \text{ N} \tag{f}$$

The forces acting on the shaft are shown in Fig. 16.32. The reactions to these forces are then the desired forces on the bearings. In the z direction it should be clear there is no force on the bearings.

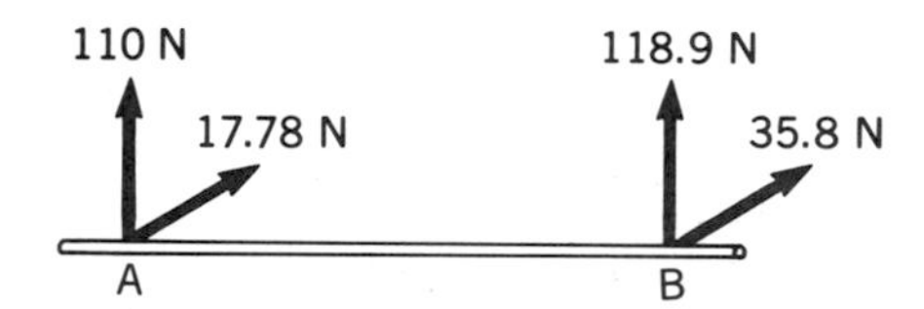

Figure 16.32. Forces on shaft.

If, in the last example, we had ignored the constant forces of gravity, we would have determined forces at bearings A and B that are due entirely to the motion of the body. Forces computed in this way are called *dynamic forces*. If the body were rotating with constant speed ω, these forces would clearly have constant values in the x and y directions. Since the xy axes are rotating with the body relative to the ground reference XYZ, such dynamic forces must also rotate about the axis of rotation with the speed ω of the body. This means that, in any *fixed* direction normal to the shaft at a bearing, there will be a *sinusoidal force variation* with a frequency corresponding to the angular rotation of the shaft. Such forces can induce vibrations of large amplitude in the structure or support if a natural frequency or multiple of a natural frequency is reached in these bodies.[5] When a shaft creates rotating forces on the bearings by virtue of its own rotation, the shaft is said to be *unbalanced*. We shall set up criteria for balancing a rotating body in the next section.

[5]Natural frequencies will be discussed in Chapter 19.

*16.9 Balancing

We shall now set forth the criteria for the condition of dynamic balance in a rotating body. Then, we shall set forth the requirements needed to achieve balance in a rotating body. Consider some arbitrary rigid body rotating with angular speed ω and a rate of change of angular speed $\dot{\omega}$ about axis AB (Fig. 16.33). We shall set up equations for

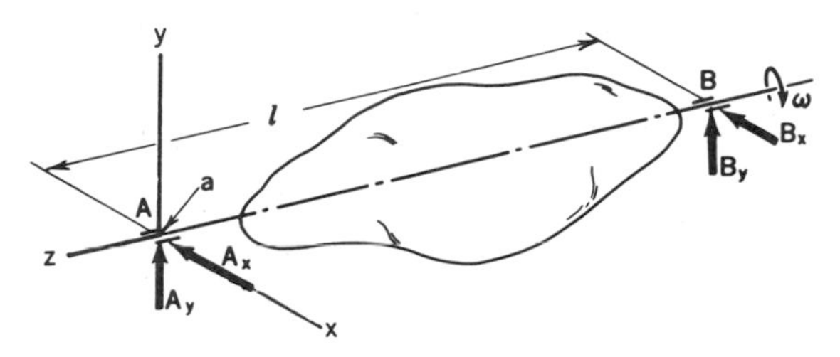

Figure 16.33. Rotating body.

determining the supporting forces at the bearings. Consider point a on the axis of rotation at the bearing A and establish a set of axes xyz fixed to the rotating body with the z axis corresponding to the axis of rotation. The x and y axes are chosen for convenience. Using the *moment of momentum* equations (a) and (b) in Eq. 16.6 and including only dynamic forces, we get for point a:

$$B_y l = -I_{xz}\dot{\omega} + I_{yz}\omega^2 \tag{16.19a}$$

$$B_x l = -I_{yz}\dot{\omega} - I_{xz}\omega^2 \tag{16.19b}$$

If the axis of rotation is a *principal axis* at bearing A, then $I_{yz} = I_{xz} = 0$ in accordance with the results of Chapter 9. The dynamic forces at bearing B are then zero.

Next, we shall show that if in addition the *center of mass lies along the axis of rotation*, this axis is a principal axis for *all points* along its line of action. In Fig. 16.34 a

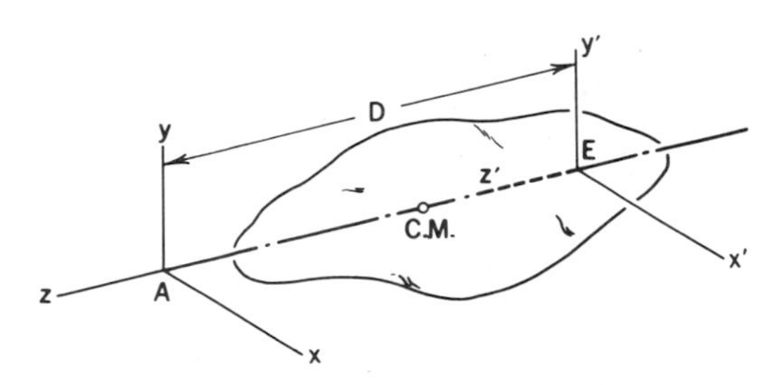

Figure 16.34. Reference $x'y'z'$ fixed at E.

set of axes $x'y'z'$ parallel to the xyz axes has been set up at an arbitrary point E along the axis of rotation. We can see from the arrangement of the axes that for any element of the body dm:

$$y' = y, \qquad x' = x, \qquad z' = D + z \tag{16.20}$$

Also, we know for the xyz reference that

$$I_{xz} = \int_M xz\, dm = 0, \qquad I_{yz} = \int_M yz\, dm = 0 \tag{16.21}$$

And if the center of mass is along the centerline, we can say:

$$\int_M y\, dm = \int_M y'\, dm = My_c = 0 \tag{16.22}$$

$$\int_M x\, dm = \int_M x'\, dm = Mx_c = 0$$

We shall now show that all products of inertia involving the z' axis at E are zero under these conditions and, consequently, that the z' axis is a principal axis at E. Substituting from Eqs. 16.20 into 16.21, we get

$$\int_M x'(z' - D)\, dm = 0 \tag{a}$$

$$\int_M y'(z' - D)\, dm = 0 \tag{b}$$

If we carry out the multiplication in the integrand of the above Eqs. (a) and (b), we get

$$\int_M x'z'\, dm - D \int_M x'\, dm = 0 \tag{c}$$

$$\int_M y'z'\, dm - D \int_M y'\, dm = 0 \tag{d}$$

As a result of Eq. 16.22, the second integrals of Eqs. (c) and (d) are zero, and we conclude that the products of inertia $I_{x'z'}$ and $I_{y'z'}$ are zero. Now the xy axes and hence the $x'y'$ axes can have any orientation as long as they are normal to the axis of rotation. This means that at E the z' axis yields a zero product of inertia for all axes normal to it. As a result of our deliberations of Chapter 9, we can conclude that z' is a principal axis for point E. And since E is any point on the axis of rotation, we can say the following:

> *If the axis of rotation is a principal axis at any point along its line of action and if the center of mass is on the axis of rotation, then the axis of rotation is a principal axis at all points along its line of action.*

We now consider Fig. 16.35, where reference $x'y'z'$ is set up at bearing B. We can next employ the moment-of-momentum equation (16.6) for these axes at B. We get

$$-A_y l = -I_{x'z'}\dot{\omega} + I_{y'z'}\omega^2$$

$$-A_x l = -I_{y'z'}\dot{\omega} - I_{x'z'}\omega^2$$

With the z axis a principal axis at A and the center of mass along the axis of rotation, the z' axis at B must be a principal axis, and hence $I_{y'z'} = I_{x'z'} = 0$. The dynamic forces at bearing A, therefore, are zero. The rotating system is thus balanced.

We can now conclude that *for a rotating system to be dynamically balanced, it is necessary and sufficient (1) that at any point along the axis of rotation this axis is a*

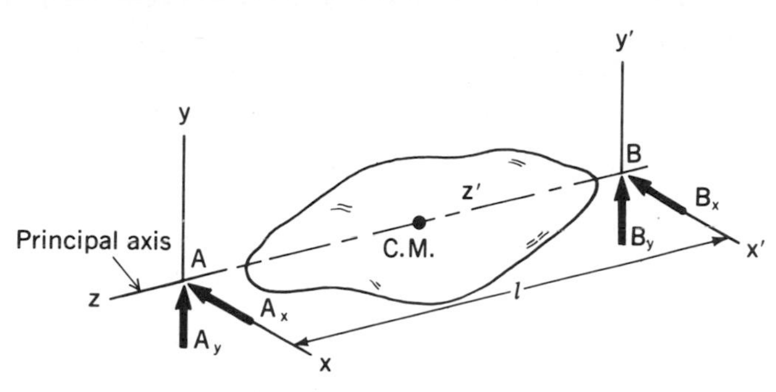

Figure 16.35. Reference $x'y'z'$ fixed at B.

principal axis and (2) *that the center of mass is along the the axis of rotation.* We next illustrate how we can make use of these results to balance a rotating body.

EXAMPLE 16.11

A rotating member carries two weights, $W_1 = 5$ lb and $W_2 = 8$ lb, at radial distances $r_1 = 1$ ft and $r_2 = 1\frac{1}{2}$ ft, respectively. The weights and a reference xyz fixed to the shaft are shown in Fig. 16.36. They are to be balanced by two other weights W_3 and W_4 (shown dashed), which are to be placed in the balancing planes A and B, respectively. If the weights are placed in these planes at a distance of 1 ft from the axis of rotation, determine the value of these weights and their position relative to the xyz reference.

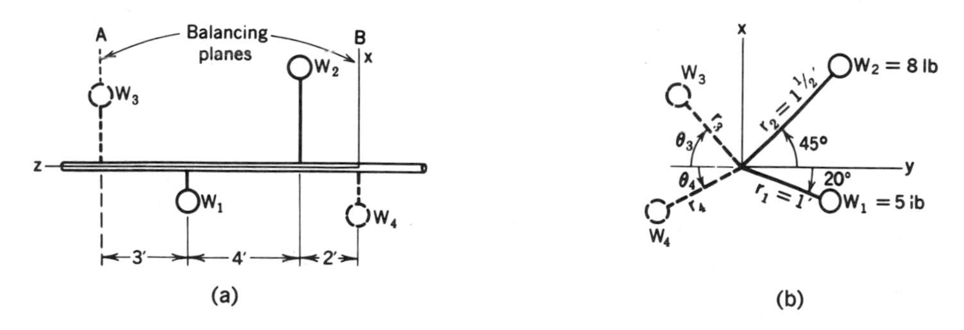

Figure 16.36. Rotating system to be balanced.

We have two unknown weights and two unknown angles, that is, four unknowns [see Fig. 16.36(b)], to evaluate in this problem. The condition that the mass center be on the centerline yields the following relations[6]:

---

[6]We are considering the weights to be particles in this discussion. In some homework problems you will be asked to balance rotating systems for which the particle model will not be proper. You will then have to carry out integrations and/or employ the formulas and transfer theorems for first moments of mass and products of inertia.

$\int_M y\,dm = 0$:

$$\frac{W_1}{g}r_1 \cos 20° + \frac{W_2}{g}r_2 \cos 45° - \frac{W_3}{g}r_3 \cos \theta_3 - \frac{W_4}{g}r_4 \cos \theta_4 = 0$$

$\int_M x\,dm = 0$:

$$-\frac{W_1}{g}r_1 \sin 20° + \frac{W_2}{g}r_2 \sin 45° + \frac{W_3}{g}r_3 \sin \theta_3 - \frac{W_4}{g}r_4 \sin \theta_4 = 0$$

When the numerical values of r_1, r_2, etc., are inserted, these equations become

$$W_3 \cos \theta_3 + W_4 \cos \theta_4 = 13.18 \quad \text{(a)}$$

$$W_3 \sin \theta_3 - W_4 \sin \theta_4 = -6.77 \quad \text{(b)}$$

Now require that the products of inertia I_{yz} and I_{xz} be zero for the xyz reference positioned so that xy is in the balancing plane B:

$I_{xz} = 0$:

$$\frac{W_1}{g}(6)(-r_1 \sin 20°) + \frac{W_2}{g}(2)(r_2 \sin 45°) + \frac{W_3}{g}(9)(r_3 \sin \theta_3) = 0 \quad \text{(c)}$$

$I_{yz} = 0$:

$$\frac{W_1}{g}(6)(r_1 \cos 20°) + \frac{W_2}{g}(2)(r_2 \cos 45°) + \frac{W_3}{g}(9)(-r_3 \cos \theta_3) = 0 \quad \text{(d)}$$

Equations (c) and (d) can be put in the form

$$9W_3 \sin \theta_3 = -6.71 \quad \text{(e)}$$

$$9W_3 \cos \theta_3 = 45.2 \quad \text{(f)}$$

Dividing Eq. (f) into Eq. (e), we get

$$\tan \theta_3 = -.1486$$

$$\theta_3 = 171.5° \quad \text{or} \quad 351.6°$$

and so, from Eq. (e),

$$W_3 = -\frac{6.71}{9}\frac{1}{\sin \theta_3} = 5.08 \text{ lb}$$

In order to have a positive weight W_3, we chose θ_3 to be 351.6° rather than 171.5°. Now we return to Eqs. (a) and (b). We can then say, on substituting known values of W_3 and θ_3:

$$W_4 \cos \theta_4 = 13.18 - 5.01 = 8.16 \quad \text{(g)}$$

$$W_4 \sin \theta_4 = 6.77 - .75 = 6.02 \quad \text{(h)}$$

Dividing Eq. (g) into Eq. (h), we get

$$\tan \theta_4 = .738$$

$$\theta_4 = 36.4° \quad \text{or} \quad 216.4°$$

Hence, from Eq. (g), we have

$$W_4 = \frac{8.16}{\cos \theta_4} = 10.14 \text{ lb}$$

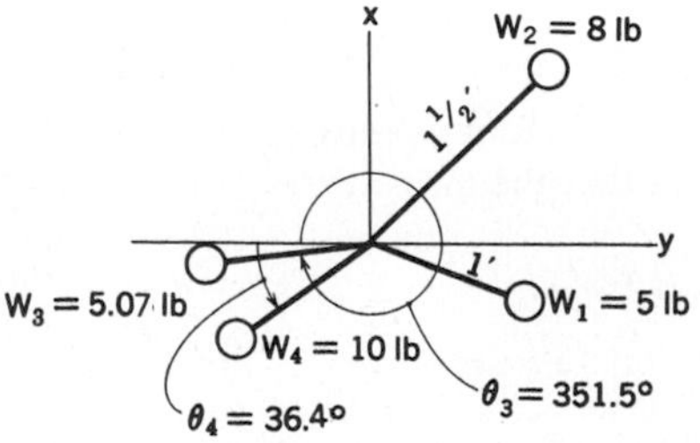

Figure 16.37. Balanced system.

if we use $\theta_4 = 36.4°$ rather than $216.4°$ to prevent a negative W_4. The final orientation of the balanced system is shown in Fig. 16.37.

Problems

16.82. A shaft shown supported by bearings A and B is rotating at a speed ω of 3 rad/sec. Identical blocks C and D weighing 30 lb each are attached to the shaft by light structural members. What are the bearing reactions in the x and y directions if we neglect the weight of the shaft?

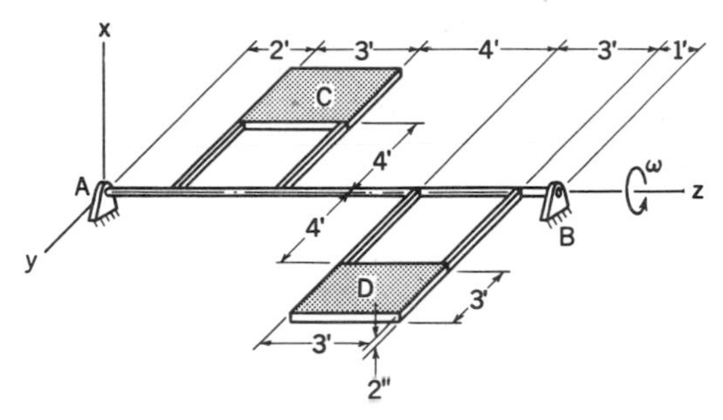

Figure P.16.82

16.83. Shaft AB is rotating at a constant speed ω of 20 rad/sec. Two rods having a weight of 10 N each are welded to the shaft and support a disc D weighing 30 N. What are the supporting forces at the instant shown?

Figure P.16.83

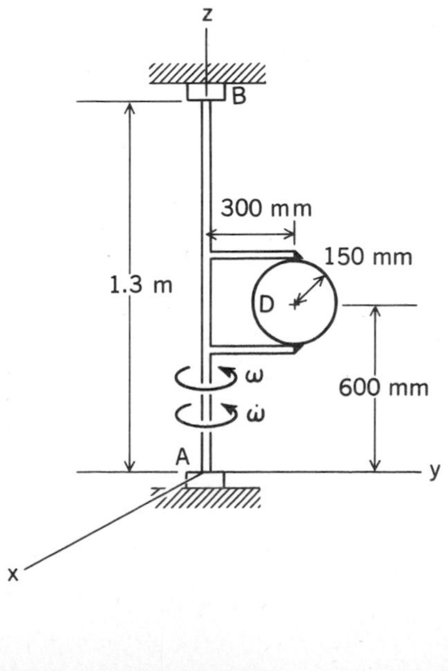

16.84. Do Problem 16.83 for the case in which $\omega = 20$ rad/sec and $\dot{\omega} = 38$ rad/sec^2 at the instant of interest as shown.

16.85. A uniform wooden panel is shown supported by bearings A and B. A 100-lb weight is connected with an inextensible cable to the panel at point G over a light pulley D. If the system is released from rest at the configuration shown, what is the angular acceleration of the panel, and what are the forces at the bearings? The panel weighs 60 lb.

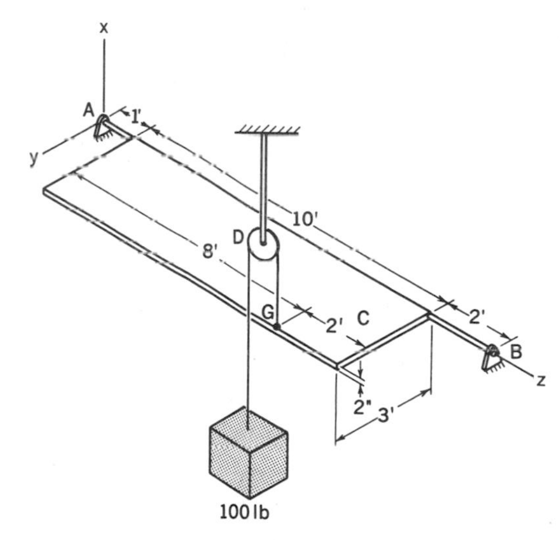

Figure P.16.85

16.86. Do Problem 16.85 when there is a frictional torque at the bearings of 10 ft-lb and the moment of inertia of the pulley is 10 lbm-ft^2 with radius 1 ft.

16.87. A thin rectangular plate weighing 50 N is rotating about its diameter at a speed ω of 25 rad/sec. What are the supporting forces in the x and y directions at the instant shown when the plate is parallel to the yz plane?

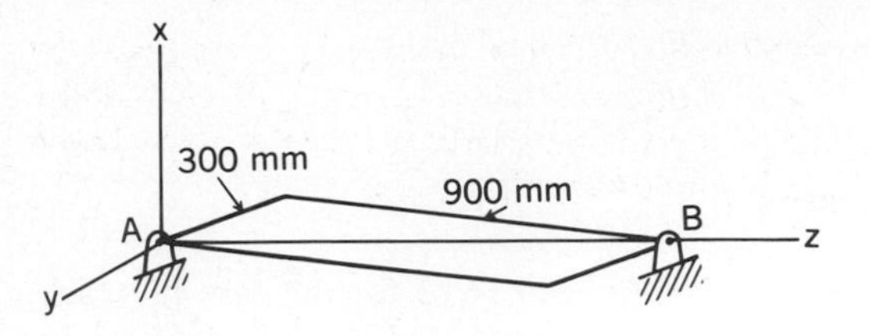

Figure P.16.87

16.88. A shaft is shown rotating at a speed of 20 rad/sec. What are the supporting forces at the bearings? The rods welded to the shaft weigh 40 N/m. The shaft weighs 80 N.

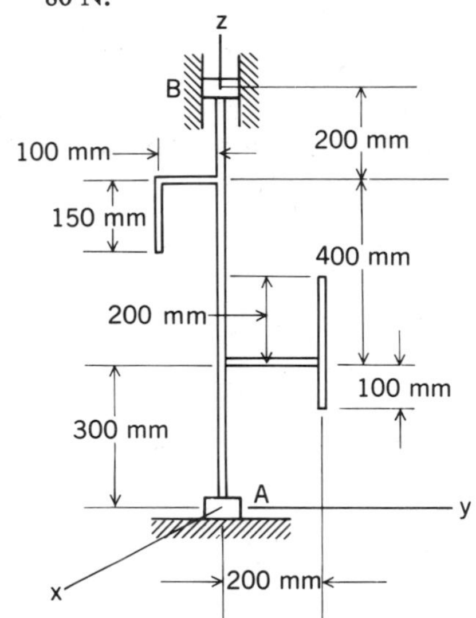

Figure P.16.88

16.89. A cylinder is shown mounted at an angle of 30° to a shaft. The cylinder weighs 400 N. If a torque T of 20 N-m is applied, what is the angular acceleration of the system? What are the supporting forces in the x and y directions at the configuration shown wherein the system is stationary? Neglect the mass of the shaft. The centerline of the cylinder is in the xz plane at the instant shown.

16.90. A bent shaft has applied to it a torque T given as

$$T = 10 + 5t \text{ N-m}$$

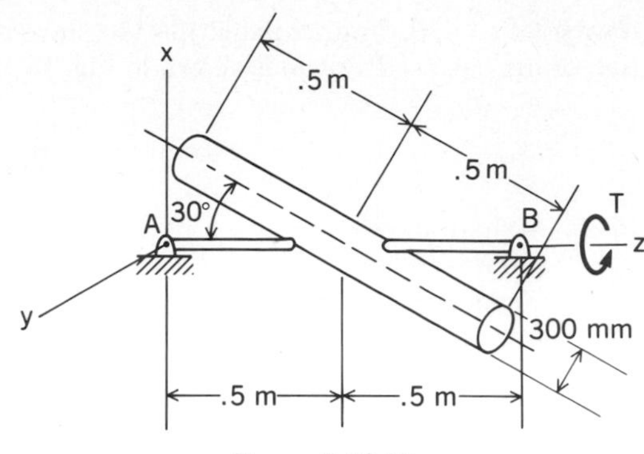

Figure P.16.89

where t is in seconds. What are the supporting forces at the bearings in the x and y directions when $t = 3$ sec? The shaft is made from a rod 20 mm in diameter and weighing 70 N/m.

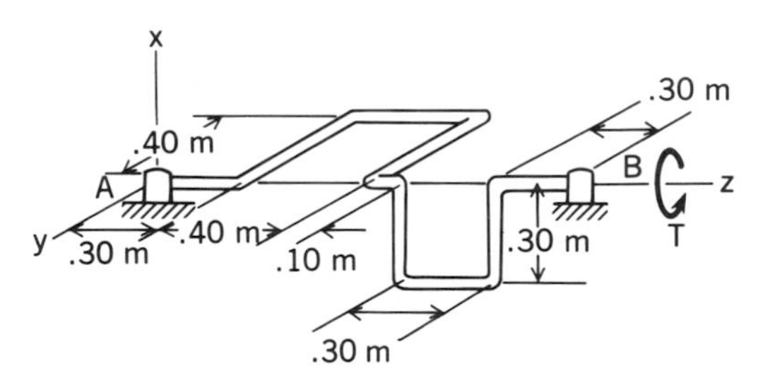

Figure P.16.90

***16.91.** A shaft has an angular velocity ω of 10 rad/sec and an angular acceleration of $\dot{\omega}$ of 5 rad/sec^2 at the instant of interest. What are the bending moments at this instant about the x and y axes just to the right of the bearing at A. Also, what are the twisting moment and shear forces there? The shaft and attached rods have a diameter of 20 mm and a weight per unit length of 50 N/m.

Figure P.16.91

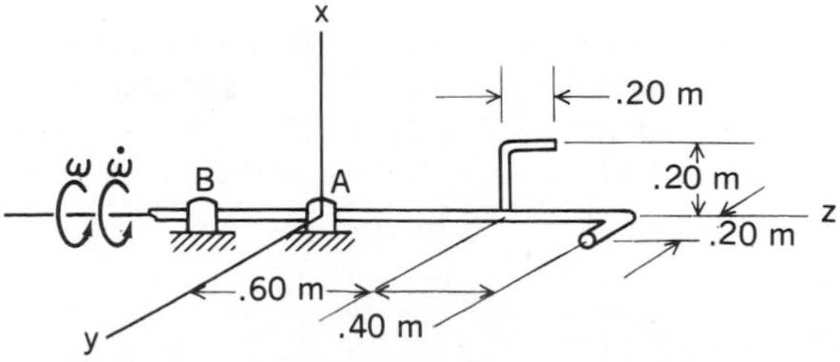

x

.20 m

.20 m

ω $\dot{\omega}$ B A

z

.20 m

.60 m

.40 m

y

16.92. Balance the system in planes A and B at a distance 1 ft from centerline. Use two weights.

16.93. Balance the system in Problem 16.92 by using a weight in plane A of $1\frac{1}{2}$ lb and a weight in plane B of 1 lb. You may choose suitable radii in these planes.

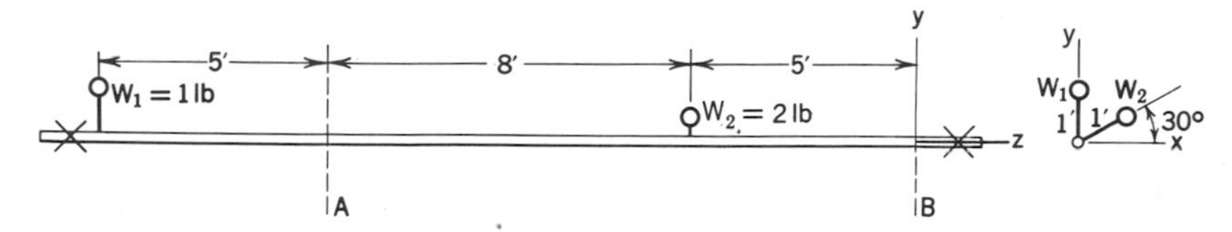

Figure P.16.92

16.94. Balance the shaft of Problem 16.83 using rods weighing 50 N/m and welded to the shaft normal to the centerline at the bearings A and B. Determine the lengths of these rods and their orientations relative to the xy axes.

16.95. A disc and a cylinder are mounted on a shaft. The disc has been mounted eccentrically so that the center of mass is $\frac{1}{2}$ in. from the centerline of the shaft. Balance the shaft using balancing planes 5 ft in from bearing A and 3 ft in from bearing B, respectively. The balancing masses each weigh 3 lb and can be regarded as particles. Give the proper position of the balancing masses in these planes.

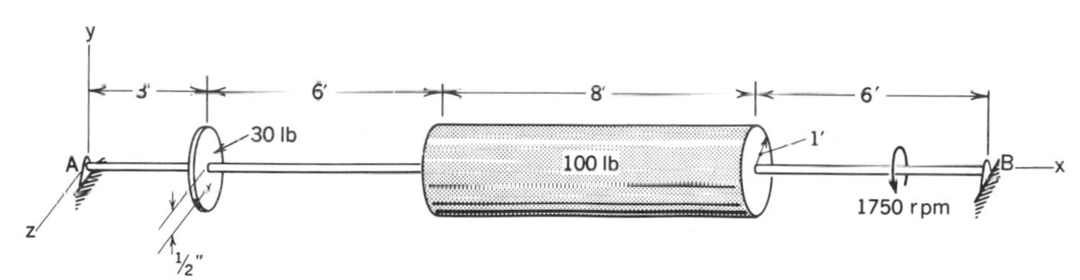

Figure P.16.95

16.96. Balance the shaft described in Problem 16.95 by removing a small chunk of metal from each of the end faces of the 100-lb cylinder at a position 10 in. from the shaft centerline. What are the weights of these chunks and what are their orientations?

16.97. Balance the shaft in Problem 16.88 using balancing planes at the bearings A and B. At bearing B use a small balancing sphere of weight 30 N and at bearing B use a rod having a weight per unit length of 35 N/m.

16.98. A disc is shown mounted off-center at B on a shaft CD that rotates with angular speed ω. The disc weighs 50 lb and has a diameter of 6 ft. Balance the system using two rods, each weighing 10 lb/ft having a diameter of 2 in., attached normal to the shaft at position 1 ft in from bearing C and 2 ft in from bearing D. The diameter of the shaft is 2 in. Determine the lengths of these rods and their inclination.

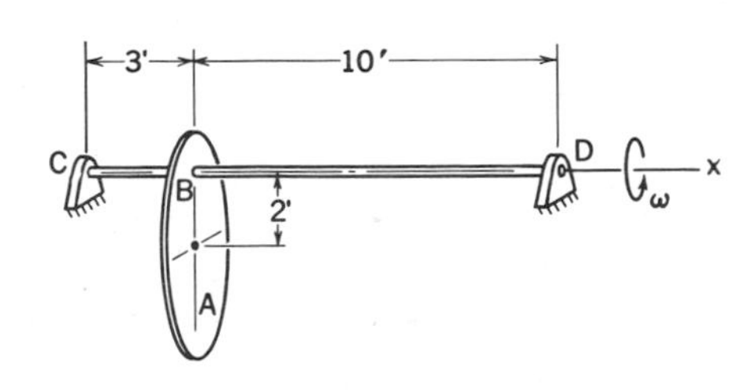

Figure P.16.98

16.99. Balance the shaft shown for Problem 16.90. Use a balancing plane at bearing A and one at bearing B. Attach a circular plate normal to the shaft at each bearing, and cut a hole with diameter 60 mm at the proper position in the plate to balance the system. The plates are 30 mm thick and have a specific weight of 8×10^{-5} N/mm³.

16.10 Closure

In this chapter, we have developed the moment-of-momentum equations for plane motion of a rigid body. We applied this equation to various cases of plane motion starting from the simplest case and going to the most difficult case. Many problems of engineering interest can be taken as plane-motion problems; the results of this chapter are hence quite important. The use of $\boldsymbol{M}_A = \dot{\boldsymbol{H}}_A$ applied to three-dimensional motion of a rigid body is considered in Chapter 18 (starred chapter). Students who cover that chapter will find the development of the key equations (the Euler equations) very similar to the development of Eqs. 16.6, the key equations for plane motion.

Recall next that in Chapter 13 we considered the work–energy equations for the plane motion of simple bodies in the process of rolling without slipping. We did this to help illustrate the use of the work–energy equations for an aggregate of particles. Also, this undertaking served to motivate a more detailed study of kinematics of rigid bodies and to set forth in miniature the more general procedures to come later. We are therefore now ready to examine energy methods as applied to rigid bodies in a more general way. This will be done in Part A of Chapter 17. We shall develop the work–energy equations for three-dimensional motion and apply them to all kinds of motions, including plane motions. The student should not have difficulty in going directly to the general case; indeed, a better understanding of the subject should result.

In Part B of Chapter 17, we shall consider the impulse-momentum equations for rigid bodies. This will be an extension of the useful impulse-momentum methods discussed in Chapter 14 for particles and aggregates of particles. Again, we shall be able to go to the general case and then apply the results to three- and two-dimensional motions (plane motions).

Review Problems

16.100. A circular plate A is used in electric meters to damp out rotations of a shaft by rotating in a bath of oil. The plate and its shaft have a mass of 300 g and a radius of gyration of 100 mm. If the shaft and plate slow down from 30 rpm to 20 rpm in 5 sec, what angular acceleration can be developed by a .005 N-m torque when ω of the shaft is 10 rpm? Assume that the damping torque is proportional to the angular speed.

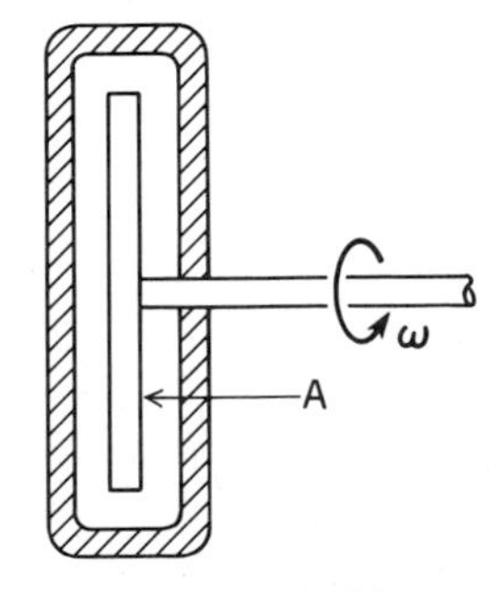

Figure P.16.100

16.101. A motor B drives a gear C which connects with gear D to driven device A. The top system of shaft and gear has a moment of inertia I_1 about the axis of rotation of 3 lbm-ft², whereas the bottom system of device A and gear D has a moment of inertia about its axis of rotation of I_2 equal to 1 lbm-ft². If motor B develops a torque given as

$$T = 30 - .02t^2 \text{ in.-lb}$$

with t in seconds, what is the angular speed of gear D 6 sec after starting from a stationary configuration? How many revolutions has it undergone during this time interval?

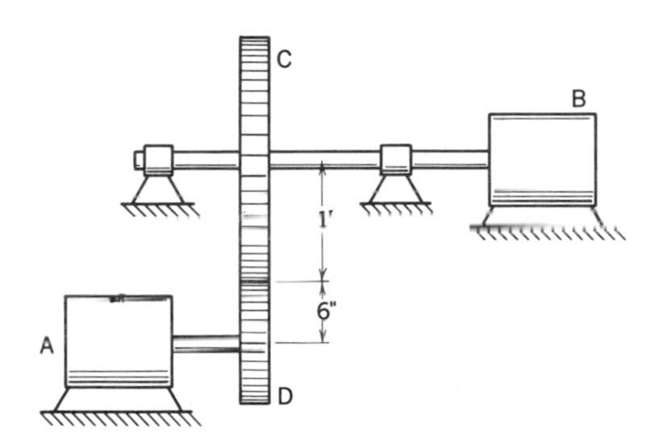

Figure P.16.101

16.102. A 50-kg container A is being transported by a conveyor as shown. A torque T of 200 N-m is applied to the driving drum. Both driving and driven drums have each a mass of 10 kg and a radius of gyration of 130 mm. The belt has a mass per unit length of 3 kg/m and a coefficient of friction of .3 with the conveyor bed. What is the acceleration of container A?

Figure P.16.102

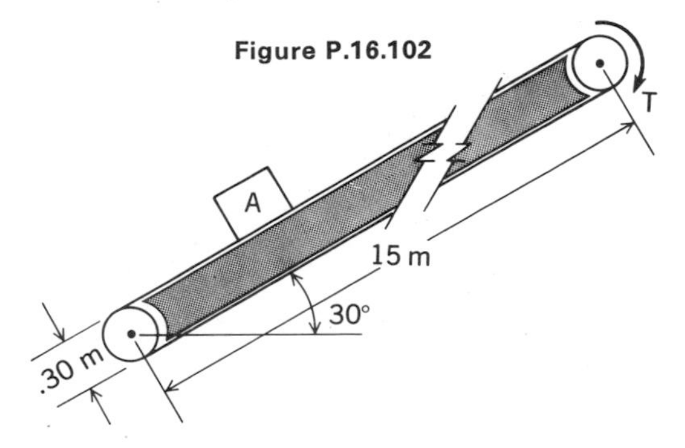

16.103. A four-bar linkage is shown. A torque T of 10 N-m is applied to member AB, which at the instant shown is rotating clockwise at a rate of speed of 3 rad/sec. Bars AB and CD are 300 mm long. Bar BC is a circular arc of radius 400 mm and length 450 mm. All bars have a mass of 10 kg per meter. What is the angular acceleration of bar AB at the instant shown?

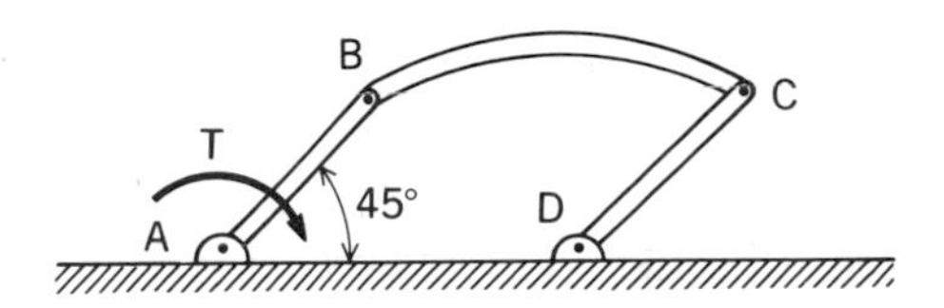

Figure P.16.103

16.104. Bar AB having a mass of 20 kg is connected to two gears at its ends. Each such gear has a diameter of 300 mm and a mass of 5 kg. The two aforementioned gears mesh with a stationary gear E. If a torque of 100 N-m is applied to the bar AB, what is the angular acceleration of the small gears?

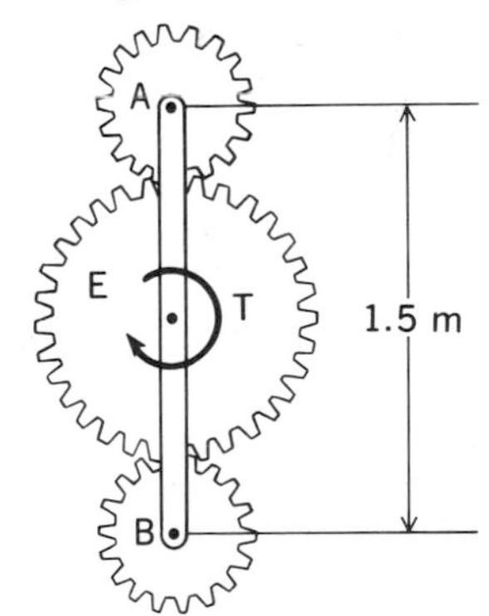

Figure P.16.104

16.105. An astronaut is assembling a space laboratory in outer space. At the instant shown, he has an H-beam made of composite materials[7] coplanar with two other similar beams. The members are sta-

[7]Made up of plastics and various kinds of fibers.

tionary relative to each other. Using a small rocket system that he is wearing, he develops a force F of 200 N always normal to the beam A and in the plane of the beams. At what position d should he exert this force to have the beam A parallel to the tops of beams C and D in 2 sec? What is the acceleration of the center of mass of A at time $t = 1$ sec? Beam A has a mass 200 kg and is 10 m long. Consider beam A to be a long slender rod.

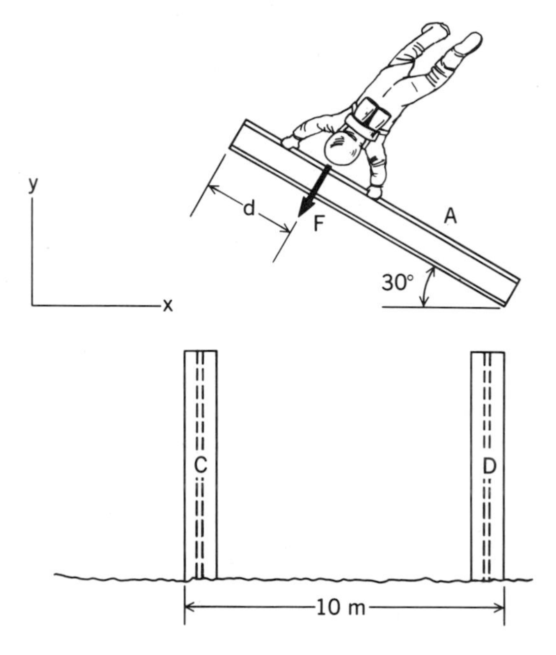

Figure P.16.105

16.106. Rod AB, of length 20 ft and weight 200 lb, is released from rest at the configuration shown. AC is a weightless wire, and the incline at B is frictionless. What is the tension in the wire AC at this instant?

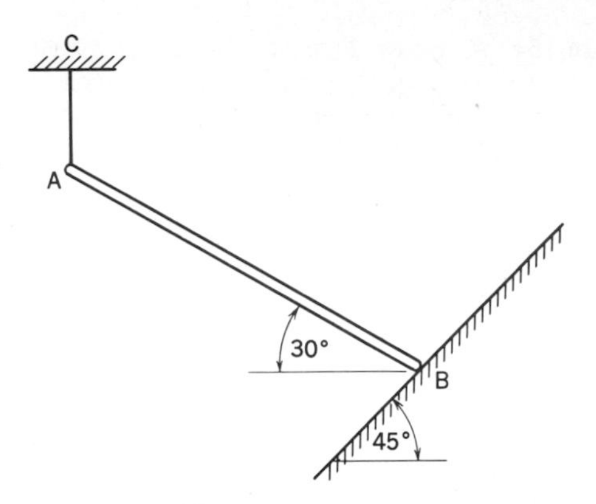

Figure P.16.106

16.107. Rod AB, of length 1 m and mass 10 kg, is pinned to a disc D having a mass of 20 kg and a diameter of .5 m. A torque $T = 15$ N-m is applied to the disc. What are the angular accelerations of the rod AB and disc at this instant?

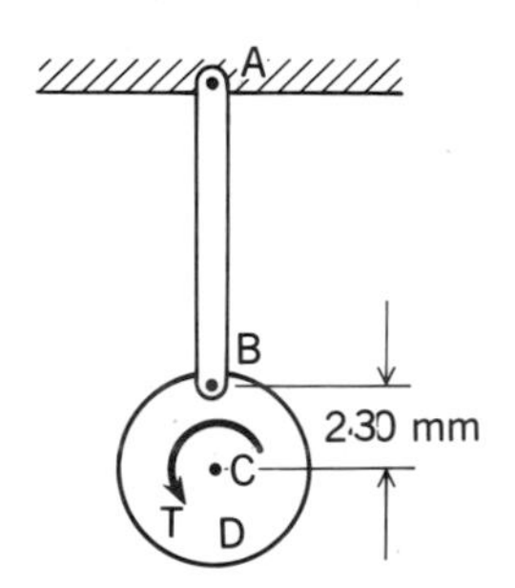

Figure P.16.107

16.108. A four-bar linkage is shown. Bar AB has the following angular motion at the instant shown.

$$\omega_1 = 2 \text{ rad/sec}$$
$$\dot{\omega}_1 = 3 \text{ rad/sec}^2$$

What are the supporting forces at D at this instant for the following data?

$$\text{mass of } AB = 4 \text{ kg}$$
$$\text{mass of } BC = 3 \text{ kg}$$
$$\text{mass of } CD = 6 \text{ kg}$$

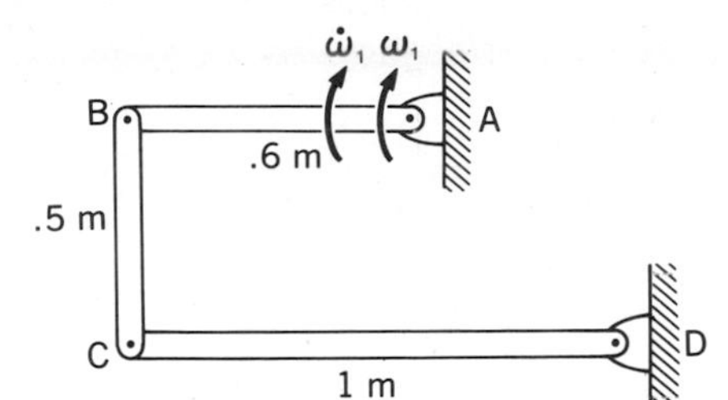

Figure P.16.108

***16.109.** A vertical shaft rotates with angular speed ω of 5 rad/sec in bearings A and D. A uniform plate B weighing 50 lb is attached to the shaft as is a disc C weighing 30 lb as shown in the diagram. What are the bearing reactions at the configuration shown? The shaft weighs 20 lb and the thickness of disc and plate is 2 in.

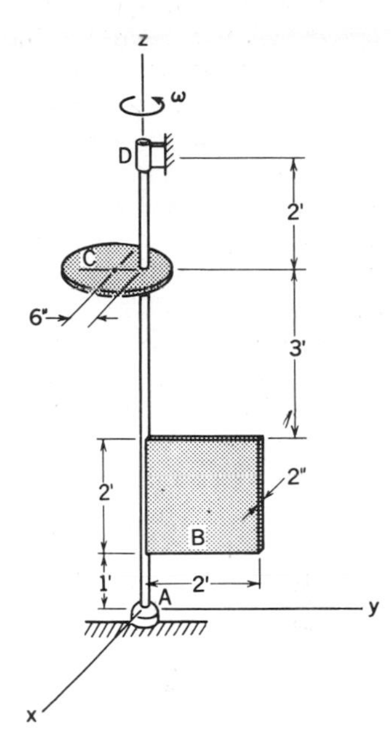

Figure P.16.109

***16.110.** Do Problem 16.89 for the case where $\omega = 5$ rad/sec at the instant of interest.

***16.111.** In Example 16.10 balance the rotating system by properly placing 36-N spherical masses in balancing planes 300 mm inside from the bearings A and B. Assume that the balancing masses are particles.

17 Energy and Impulse-Momentum Methods for Rigid Bodies

17.1 Introduction

Let us pause to reflect on where we have been thus far in dynamics and where we are about to go. In Chapter 11, you will recall, we worked directly with *Newton's law* and integrated it several times to consider the motion of a *particle*. Then, in Chapters 13 and 14, we formulated certain useful integrated forms from Newton's law and thereby presented the *energy methods* and the *linear impulse-momentum* methods also for a *particle*. At the end of Chapter 14, we derived the important *angular momentum* equation, $\boldsymbol{M}_A = \dot{\boldsymbol{H}}_A$. In Chapter 16, we returned to Newton's law and along with the angular momentum equation, $\boldsymbol{M}_A = \dot{\boldsymbol{H}}_A$, carried out integrations to solve *plane motion* problems of rigid bodies. In the present chapter, we shall again consider *energy methods* and *linear impulse-momentum methods*—this time for the *general motion* of rigid bodies. In addition, we shall use a certain integrated form of the angular momentum equation $\boldsymbol{M}_A = \dot{\boldsymbol{H}}_A$, namely the *angular impulse-momentum* equation. These equations at times will be applied to a single rigid body. At other times, we shall apply them to several interconnected rigid bodies considered as a whole. When we do the latter, we say we are dealing with a *system* of rigid bodies. We shall consider energy methods first.

Part A
Energy Methods

17.2 Kinetic Energy of a Rigid Body

First, we shall derive a convenient expression for the kinetic energy of a rigid body. We have already found (Section 13.7) that the kinetic energy of an aggregate of particles relative to any reference is the sum of two parts, which we list again as:

1. The kinetic energy of a hypothetical particle that has a mass equal to the total mass of the system and a motion corresponding to that of the mass center of the system, plus
2. The kinetic energy of the particles relative to the mass center.

Mathematically,

$$\text{KE} = \tfrac{1}{2}M|\dot{\mathbf{r}}_c|^2 + \tfrac{1}{2}\sum_{i=1}^{n} m_i|\dot{\boldsymbol{\rho}}_i|^2 \tag{17.1}$$

where $\boldsymbol{\rho}_i$ is the position vector from the mass center to the ith particle.

Let us now consider the foregoing equation as applied to a rigid body which is a special "aggregate of particles" (Fig. 17.1). In such a case, the velocity of any particle relative to the mass center becomes

$$\dot{\boldsymbol{\rho}}_i = \boldsymbol{\omega} \times \boldsymbol{\rho}_i \tag{17.2}$$

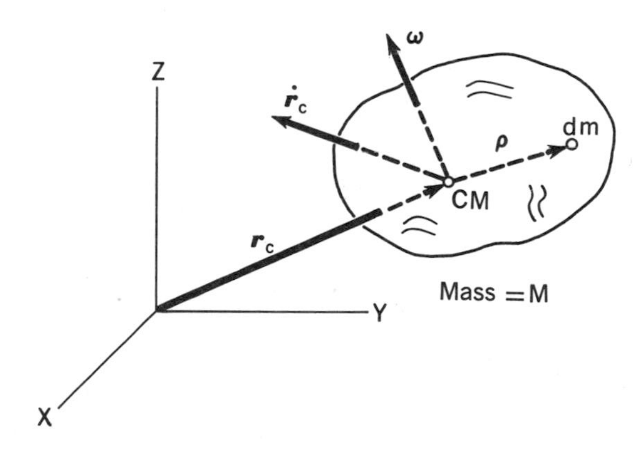

Figure 17.1. Rigid body.

where $\boldsymbol{\omega}$ is the angular velocity of the body relative to reference XYZ in which we are computing the kinetic energy. For the rigid body, the discrete particles of mass m_i become a continuum of infinitesimal particles each of mass dm, and the summation in Eq. 17.1 then becomes an integration. Thus, we can say for the rigid body, replacing $|\dot{\boldsymbol{r}}_c|^2$ by V_c^2.

$$\text{KE} = \tfrac{1}{2}MV_c^2 + \tfrac{1}{2}\iiint_M |\boldsymbol{\omega} \times \boldsymbol{\rho}|^2 \, dm \tag{17.3}$$

where $\boldsymbol{\rho}$ represents the position vector from the center of mass to any element of mass dm. Let us now choose a set of orthogonal directions xyz at the center of mass, so we can carry out the preceding integration in terms of the scalar components of $\boldsymbol{\omega}$ and $\boldsymbol{\rho}$. This step is illustrated in Fig. 17.2. We first express the integral in Eq. 17.3 in the following manner:

$$\iiint_M |\boldsymbol{\omega} \times \boldsymbol{\rho}|^2 \, dm = \iiint_M (\boldsymbol{\omega} \times \boldsymbol{\rho}) \cdot (\boldsymbol{\omega} \times \boldsymbol{\rho}) \, dm \tag{17.4}$$

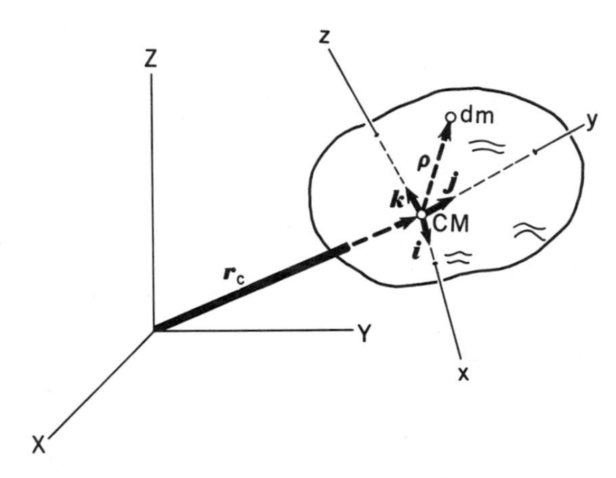

Figure 17.2. Put *xyz* at center of mass.

Inserting the scalar components, we get:

$$\iiint_M |\boldsymbol{\omega} \times \boldsymbol{\rho}|^2 \, dm = \iiint \{[(\omega_x \boldsymbol{i} + \omega_y \boldsymbol{j} + \omega_z \boldsymbol{k}) \times (x\boldsymbol{i} + y\boldsymbol{j} + z\boldsymbol{k})] \cdot [(\omega_x \boldsymbol{i} + \omega_y \boldsymbol{j} + \omega_z \boldsymbol{k}) \times (x\boldsymbol{i} + y\boldsymbol{j} + z\boldsymbol{k})]\} \, dm$$

Carrying out first the cross products and then the dot product in the integrand and collecting terms, we form the following relation:

$$\begin{aligned}\iiint_M |\boldsymbol{\omega} \times \boldsymbol{\rho}|^2 \, dm = {} & \left[\iiint_M (z^2 + y^2) \, dm\right]\omega_x^2 + \left[\iiint_M xy \, dm\right]\omega_x\omega_y + \left[\iiint_M xz \, dm\right]\omega_x\omega_z \\ & + \left[\iiint_M yx \, dm\right]\omega_y\omega_x + \left[\iiint_M (x^2 + z^2) \, dm\right]\omega_y^2 + \left[\iiint_M yz \, dm\right]\omega_y\omega_z \\ & + \left[\iiint_M zx \, dm\right]\omega_z\omega_x + \left[\iiint_M zy \, dm\right]\omega_z\omega_y + \left[\iiint_M (x^2 + y^2) \, dm\right]\omega_z^2\end{aligned}$$

You will recognize that the integrals are the components of the inertia tensor for the *xyz* reference. Thus,[1]

$$\iiint_M |\boldsymbol{\omega} \times \boldsymbol{\rho}|^2\, dm = I_{xx}\omega_x^2 + I_{xy}\omega_x\omega_y + I_{xz}\omega_x\omega_z$$
$$+ I_{yx}\omega_y\omega_x + I_{yy}\omega_y^2 + I_{yz}\omega_y\omega_z$$
$$+ I_{zx}\omega_z\omega_x + I_{zy}\omega_z\omega_y + I_{zz}\omega_z^2$$

We can now give the kinetic energy of a rigid body in the following form:

$$\begin{aligned} \text{KE} = \tfrac{1}{2}MV_c^2 + \tfrac{1}{2}(I_{xx}\omega_x^2 &+ I_{xy}\omega_x\omega_y + I_{xz}\omega_x\omega_z \\ + I_{yx}\omega_y\omega_x &+ I_{yy}\omega_y^2 + I_{yz}\omega_y\omega_z \\ + I_{zx}\omega_z\omega_x &+ I_{zy}\omega_z\omega_y + I_{zz}\omega_z^2) \end{aligned} \tag{17.5}$$

Note that the first expression on the right side of the preceding equation is the kinetic energy of translation of the rigid body, while the second expression is the kinetic energy of rotation of the rigid body about its center of mass. If principal axes are chosen, Eq. 17.5 becomes

$$\text{KE} = \tfrac{1}{2}MV_c^2 + \tfrac{1}{2}(I_{xx}\omega_x^2 + I_{yy}\omega_y^2 + I_{zz}\omega_z^2) \tag{17.6}$$

Note that for this condition the kinetic energy terms for rotation have the same form as the kinetic energy term that is due to translation, with the moment of inertia corresponding to mass and angular velocity corresponding to linear velocity.

As a special case, we shall consider the calculation of kinetic energy of any rigid body undergoing *pure rotation* relative to *XYZ* with angular velocity $\boldsymbol{\omega}$ about an actual axis of rotation[2] going through part of the body or through a rigid hypothetical massless extension of the body. We have shown this situation in Fig. 17.3, where the *Z*

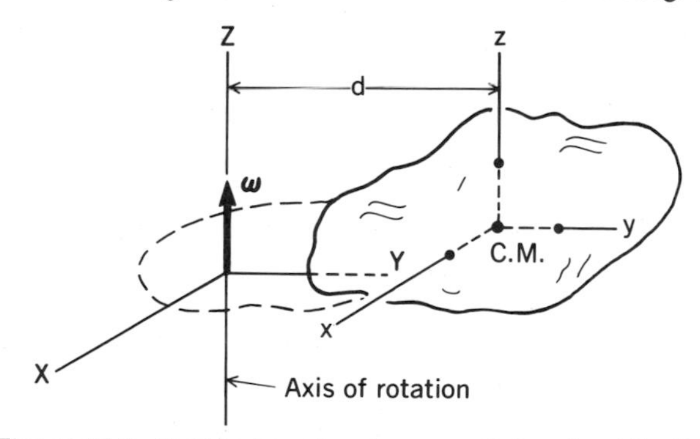

Figure 17.3. Rigid body undergoing pure rotation about *Z* axis.

[1]Note that we have deliberately used a matrixlike array for ease in remembering the formulation.

[2]An actual axis of rotation in *XYZ* is a line along which the velocity relative to *XYZ* is zero.

axis is chosen to be collinear with the $\boldsymbol{\omega}$ vector and the axis of rotation. The reference xyz at the center of mass is chosen parallel to XYZ. Clearly, $\omega_x = \omega_y = 0$ and $\omega_z = \omega$ so Eq. 17.5 becomes

$$\text{KE} = \tfrac{1}{2}MV_c^2 + \tfrac{1}{2}I_{zz}\omega^2$$

where I_{zz} is about an axis which goes through the mass center parallel to Z. Note that $V_c = \omega d$, where d is the distance between the axis of rotation Z and the z axis at the center of mass. We then have

$$\begin{aligned}\text{KE} &= \tfrac{1}{2}(Md^2)\omega^2 + \tfrac{1}{2}I_{zz}\omega^2\\ &= \tfrac{1}{2}(I_{zz} + Md^2)\omega^2\end{aligned}$$

But the bracketed expression is the moment of inertia of the body about the axis of rotation Z. Denoting this moment of inertia simply as I, we get for the kinetic energy:

$$\boxed{\text{KE} = \tfrac{1}{2}I\omega^2} \qquad (17.7)$$

This simple expression for pure rotation is completely analogous to the kinetic energy of a body in pure translation.

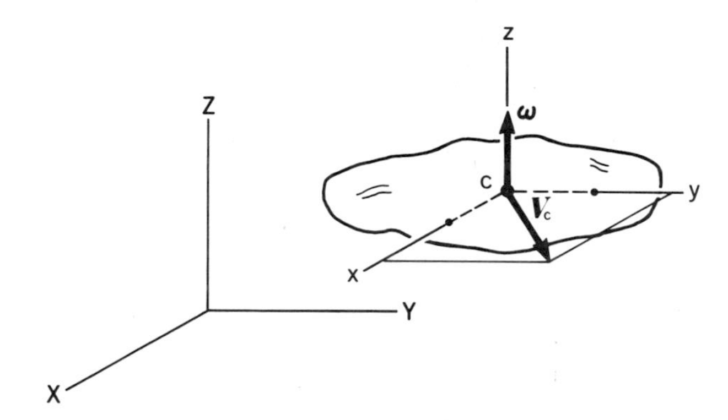

Figure 17.4. Plane motion relative to the *XY* plane.

For a body undergoing *general plane motion* (see Fig. 17.4) parallel to the XY plane, where xyz are taken at the center of mass and oriented parallel to XYZ, we get from Eq. 17.5:

$$\boxed{\text{KE} = \tfrac{1}{2}MV_c^2 + \tfrac{1}{2}I_{zz}\omega_z^2} \qquad (17.8)$$

We now illustrate the calculation of the kinetic energy in the following example.

EXAMPLE 17.1

Compute the kinetic energy of the crank system in the configuration shown in Fig. 17.5. Piston A weighs 2 lb, rod AB is 2 ft long and weighs 5 lb, and flywheel D

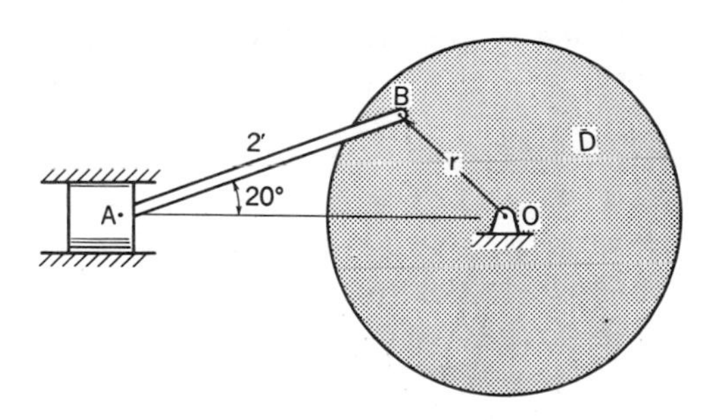

Figure 17.5. Crank system.

weighs 100 lb with a radius of gyration of 1.2 ft. The radius r is 1 ft. At the instant of interest, piston A is moving to the right at a speed V of 10 ft/sec.

We have here a translatory motion (piston A), a plane motion (rod AB), and a pure rotation (flywheel D). Thus, for piston A we have for the kinetic energy:

$$(\text{KE})_A = \frac{1}{2}MV^2 = \frac{1}{2}\left(\frac{2}{32.2}\right)(10^2) = 3.11 \text{ ft-lb} \tag{a}$$

For the rod AB, we must first consider *kinematical* aspects of the motion. For this purpose we have shown rod AB again in Fig. 17.6, where V_A is the known velocity

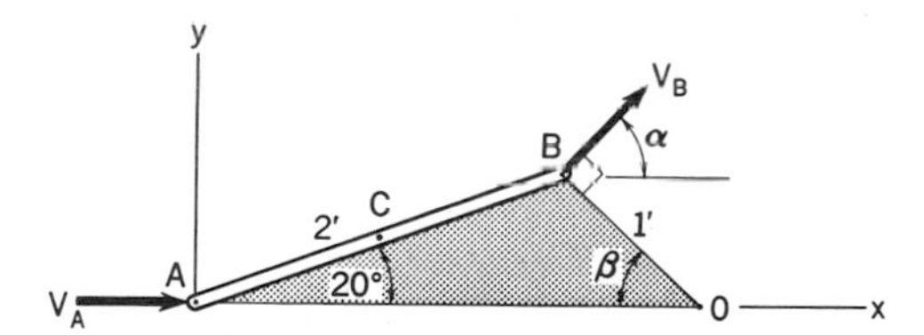

Figure 17.6. Kinematics of rod *AB*.

of point A and V_B is the velocity vector for point B oriented at an angle α such that V_B is perpendicular to OB. We can readily find α for the configuration of interest by trigonometric considerations of triangle ABO. To do this, we use the law of sines to first compute the angle β:

$$\frac{2}{\sin\beta} = \frac{1}{\sin 20^\circ}$$

Therefore,

$$\beta = 43.2^\circ \tag{b}$$

Because V_B is at right angles to OB, we have for the angle α:

$$\alpha = 90^\circ - \beta = 46.8^\circ \tag{c}$$

From kinematics of a rigid body we can now say:

$$\boldsymbol{V_B} = \boldsymbol{V_A} + (\omega_{AB}\boldsymbol{k}) \times \boldsymbol{\rho_{AB}}$$

Hence,

$$\begin{aligned} V_B(\cos\alpha\boldsymbol{i} + \sin\alpha\boldsymbol{j}) &= 10\boldsymbol{i} + \omega_{AB}\boldsymbol{k} \times (2\cos 20^\circ\boldsymbol{i} + 2\sin 20^\circ\boldsymbol{j}) \\ V_B(.684\boldsymbol{i} + .729\boldsymbol{j}) &= 10\boldsymbol{i} + 1.879\omega_{AB}\boldsymbol{j} - .684\omega_{AB}\boldsymbol{i} \end{aligned} \tag{d}$$

From this we solve for ω_{AB} and V_B. Thus,

$$V_B = 10.53 \text{ ft/sec}$$
$$\omega_{AB} = 4.09 \text{ rad/sec} \tag{e}$$

To get the velocity of the mass center C of AB, we proceed as follows:

$$\begin{aligned} V_C &= V_A + (\omega_{AB}\boldsymbol{k}) \times \boldsymbol{\rho}_{AB} \\ &= 10\boldsymbol{i} + 4.09\boldsymbol{k} \times (.940\boldsymbol{i} + .342\boldsymbol{j}) \\ &= 10\boldsymbol{i} + 3.84\boldsymbol{j} - 1.399\boldsymbol{i} = 8.60\boldsymbol{i} + 3.84\boldsymbol{j} \text{ ft/sec} \end{aligned} \tag{f}$$

We can now calculate $(\text{KE})_{AB}$, the kinetic energy of the rod:

$$\begin{aligned} (\text{KE})_{AB} &= \frac{1}{2}M_{AB}V_c^2 + \frac{1}{2}I_{zz}\omega_{AB}^2 \\ &= \frac{1}{2}\left(\frac{5}{32.2}\right)(8.60^2 + 3.84^2) + \frac{1}{2}\left(\frac{1}{12}\frac{5}{32.2}2^2\right)(4.09^2) \\ &= 7.32 \text{ ft-lb} \end{aligned} \tag{g}$$

Finally, we consider the flywheel D. The angular speed ω_D can easily be computed using V_B of Eq. (e). Thus,

$$\omega_D = \frac{V_B}{r} = \frac{10.53}{1} = 10.53 \text{ rad/sec} \tag{h}$$

Accordingly, we get for $(\text{KE})_C$:

$$(\text{KE})_C = \left[\frac{1}{2}\left(\frac{100}{32.2}\right)(1.2^2)\right](10.53^2) = 248 \text{ ft-lb} \tag{i}$$

The total kinetic energy of the system can now be given as

$$\begin{aligned} \text{KE} &= (\text{KE})_A + (\text{KE})_{AB} + (\text{KE})_D \\ &= 3.11 + 7.32 + 248 = 258 \text{ ft-lb} \end{aligned} \tag{j}$$

17.3 Work–Energy Relations

We presented in Chapter 13 the work–energy relation for a *single* particle m_i in a *system* of n particles (see Fig. 17.7) to reach the following equation:

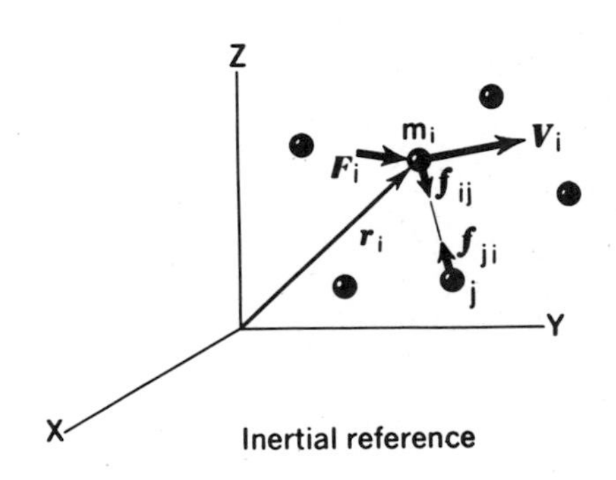

Figure 17.7. System of particles.

$$\int_1^2 \boldsymbol{F}_i \cdot d\boldsymbol{r}_i + \int_1^2 \sum_{\substack{j=1 \\ j\neq i}}^{n} \boldsymbol{f}_{ij} \cdot d\boldsymbol{r}_i = \tfrac{1}{2}(m_iV_i^2)_2 - \tfrac{1}{2}(m_iV_i^2)_1 = (\Delta\text{KE})_i \tag{17.9}$$

where $\boldsymbol{f}_{ij}$ is the force from particle j onto particle i and is an internal force. (Note that since a particle cannot exert a force on itself, $\boldsymbol{f}_{ii} = \boldsymbol{0}$.) Now consider that the particle m_i is part of a rigid body, as shown in Fig. 17.8. From Newton's third law we can say that

$$\boldsymbol{f}_{ij} = -\boldsymbol{f}_{ji} \tag{17.10}$$

It might be intuitively obvious to the reader that for any motion of a rigid body the totality of internal forces $\boldsymbol{f}_{ij}$ can do no work. If not, read the following proof to verify this claim.

Suppose the rigid body moves an infinitesimal amount. We employ Chasle's theorem, whereby we give the entire body a displacement $d\boldsymbol{r}$ corresponding to the actual displacement of particle m_i (see Fig. 17.8). The total work done by $\boldsymbol{f}_{ij}$ and $\boldsymbol{f}_{ji}$ is clearly zero for this displacement as a result of Eq. 17.10. In addition, we will have a rotation $d\boldsymbol{\phi}$ about an axis of rotation going through m_i. We can decompose $d\boldsymbol{\phi}$ into orthogonal components, such that one

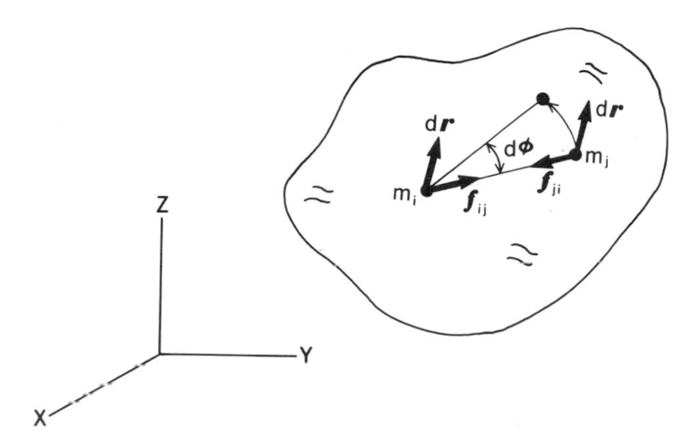

Figure 17.8. Particles of a rigid body.

component $d\phi_1$ is along the line between m_i and m_j (and thus collinear with $\boldsymbol{f}_{ij}$) and two components are at right angles to this line (see Fig. 17.9). Clearly, the work done by the forces $\boldsymbol{f}_{ij}$ and $\boldsymbol{f}_{ji}$ for $d\phi_1$ is zero. Also, the movement of m_j for the other components of $d\boldsymbol{\phi}$ is at right angles to $\boldsymbol{f}_{ji}$, and again there is no work done. Consequently, the work done by $\boldsymbol{f}_{ij}$ and $\boldsymbol{f}_{ji}$ is

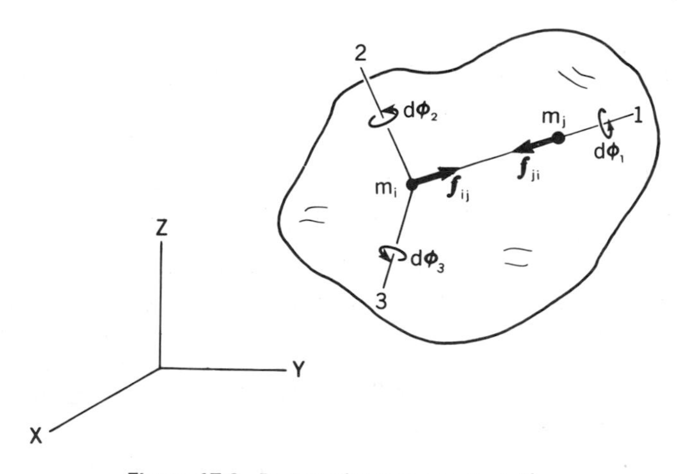

Figure 17.9. Rectangular components of $d\boldsymbol{\phi}$.

zero during the total infinitesimal movement. And since a finite movement is a sum of such infinitesimal movements, the work done for a finite movement of m_i and m_j is zero. But a rigid body consists of *pairs* of interacting particles such as m_i and m_j. Hence, on summing Eq. 17.9 for all particles of a rigid body, we can conclude that the *work done by forces internal to a rigid body for any rigid-body movement is always zero.*

We must clearly point out here that although the internal forces *in* a rigid body can do no work, forces *between rigid bodies* of a *system* of rigid bodies *can* do a net amount of work even though Newton's third law applies and even though these forces are *internal to the system*. We shall say more about this later when we discuss systems of rigid bodies.

We accordingly compute the work done on a rigid body in moving from configuration I to configuration II by summing the work terms for all the *external* forces. Thus, for the body shown in Fig. 17.10, we can express the work between I and II in the following manner:

$$(\text{work})_{\text{I, II}} = \int_{\substack{\text{I}\\ \text{path 1}}}^{\text{II}} \boldsymbol{F}_1 \cdot d\boldsymbol{s}_1 + \int_{\substack{\text{I}\\ \text{path 2}}}^{\text{II}} \boldsymbol{F}_2 \cdot d\boldsymbol{s}_2 + \dots + \int_{\substack{\text{I}\\ \text{path } n}}^{\text{II}} \boldsymbol{F}_n \cdot d\boldsymbol{s}_n \qquad (17.11)$$

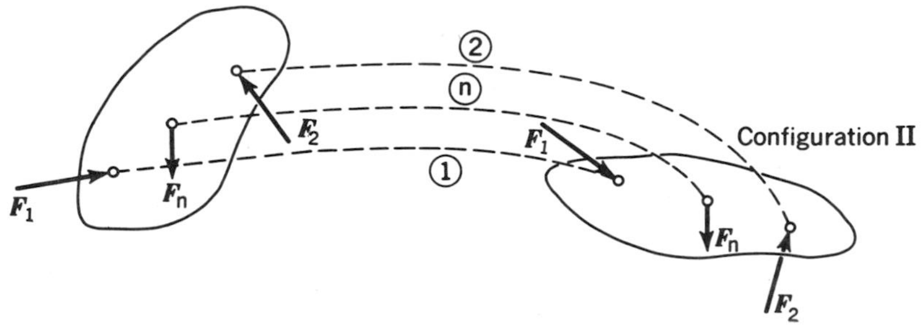

Figure 17.10. Rigid body moves from configuration I to configuration II.

In this equation, we must remember, the dot products of *nonconservative* forces are to be integrated over the *actual paths* along which the *points of application of the forces on the rigid body move*. We must take into account the variations of direction and magnitude of these nonconservative forces along their paths. For conservative forces we can use the concept of potential energy.

Although we can treat *couples* as sets of discrete forces in the foregoing manner, it is often useful to take advantage of the special properties of couples and to treat them separately. It should be a simple matter for you to show (see Fig. 17.11) that a torque T about an axis upon rotating an angle $d\theta$ about the axis does an amount of work dW_K given as

$$dW_K = T\,d\theta$$

Dividing and multiplying by dt, we can say further that

$$dW_K = T\frac{d\theta}{dt}\,dt = T\dot{\theta}\,dt$$

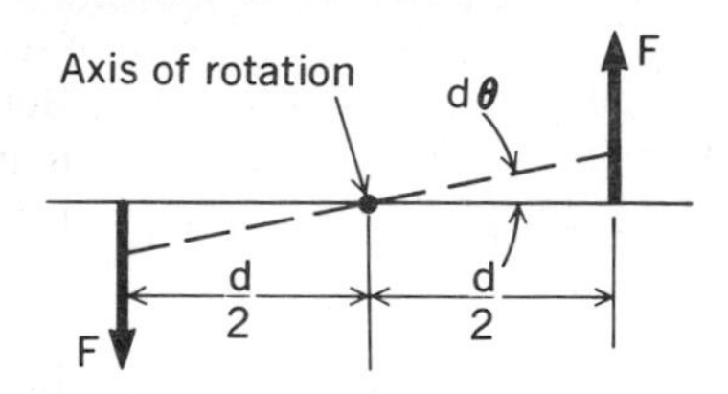

$$dW_k = F\frac{d}{2}\,d\theta + F\frac{d}{2}\,d\theta = (Fd)\,d\theta = T\,d\theta$$

Figure 17.11. Work of torque T about an axis rotating an angle $d\theta$ about the axis.

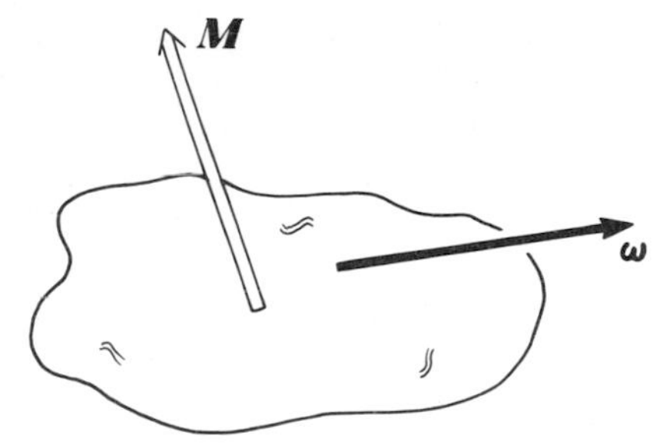

Figure 17.12. $dW_K = \boldsymbol{M} \cdot \boldsymbol{\omega}\, dt$.

Integrating, we get

$$W_K = \int_{t_1}^{t_2} T\dot{\theta}\, dt \tag{17.12}$$

In this case the torque T and angular speed $\dot{\theta}$ are about the same axis. The generalization of Eq. 17.12 for any moment $\boldsymbol{M}$ and any angular velocity $\boldsymbol{\omega}$ (see Fig. 17.12) then is

$$W_K = \int_{t_1}^{t_2} \boldsymbol{M} \cdot \boldsymbol{\omega}\, dt \tag{17.13}$$

We thus have formulations for finding the work done by external forces and couples on a rigid body. For the conservative forces, we know from Chapter 13 that we can use for work a quantity that is the change in potential energy from I to II without having to specify the path taken.

Using this information for computing work, we can then say for any rigid body:

$$\boxed{W_K \text{ from I to II} = \Delta\text{KE}} \tag{17.14}$$

where W_K is the work by *external* forces.

If there are only *conservative* external forces present, we can also say for the rigid body:

$$\boxed{(\text{PE})_{\text{I}} + (\text{KE})_{\text{I}} = (\text{PE})_{\text{II}} + (\text{KE})_{\text{II}}} \tag{17.15}$$

If both conservative *and* nonconservative external forces are present, we can say:

$$\boxed{\text{nonconservative } W_K \text{ from I to II} = \Delta\text{PE} + \Delta\text{KE}} \tag{17.16}$$

These three equations parallel the three we developed for a particle in Chapter 13.

The foregoing equations are expressed for a *single* rigid body. For a *system* of *interconnected rigid* bodies, we distinguish between two types of internal forces. They are

1. Forces internal to any rigid body of the system.
2. Forces *between* rigid bodies of the system.

For a system of bodies, as in the case of a single rigid body, forces of category 1 can do no work. However, if the forces *between two bodies* of a system do not move the same distance over the same path, then there may be a net amount of work done on the system by these internal forces. We must include such work contributions when employing Eqs. 17.14–17.16 *for a system of interconnected rigid bodies.*[3] Example 17.4 is an example of this situation.

EXAMPLE 17.2

Neglect the weight of the cable in Fig. 17.13, and find the speed of the 450-N weight after it has moved 1.7 m along the incline from a position of rest. The static coefficient of friction along the incline is .32, and the dynamic coefficient of friction is .30. Consider the pulley to be a uniform cylinder.

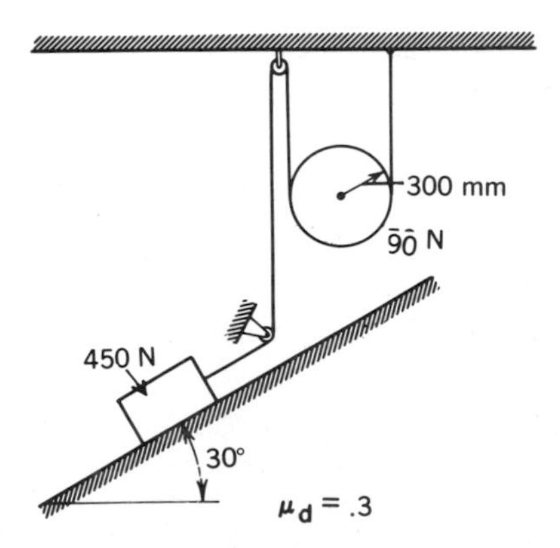

Figure 17.13. Pulley system.

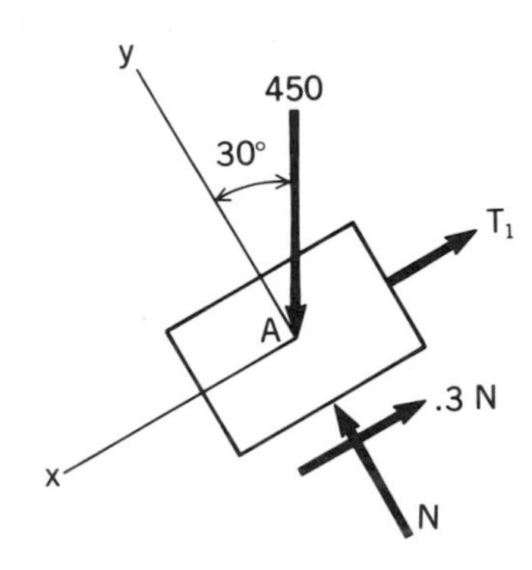

Figure 17.14. Free body of block.

We must first decide which way the block moves along the incline. To overcome friction and move down the incline, the block must create a force in the cable exceeding $\frac{1}{2}(90) = 45$ N. Considering the block alone (see Fig. 17.14), we can readily conclude that the minimum force T_1 from the block for this action is

$$(T_1)_{\min} = -(.32)\,N + 450 \sin 30°$$
$$= -(.32)(450) \cos 30° + 450 \sin 30° = 100.3 \text{ N}$$

Hence, the block goes down the incline.

We now use the *work–kinetic energy* equation separately for each body. Thus, for the block we have

$$(450 \sin 30°)(1.7) - (450)(\cos 30°)(.3)(1.7) - T_1(1.7) = \frac{1}{2}\frac{450}{g}V_A^2$$

[3]Recall from Chapter 13, that Eq. 17.16 and hence Eqs. 17.14 and 17.15 are valid for *any* aggregate of particles provided we include the work of internal forces both conservative and nonconservative.

Therefore,

$$T_1 = 108.1 - 13.49V_A^2 \qquad \text{(a)}$$

We now consider the cylinder for which the free-body diagram is displayed in Fig. 17.15. Note that the cylinder is in effect rolling without slipping along the right supporting cable. Hence, T_2 does no work as explained in Chapter 13.[4] Thus, the *work–kinetic energy* equation is as follows using the formula $\frac{1}{2}Mr^2$ for I of the cylinder:

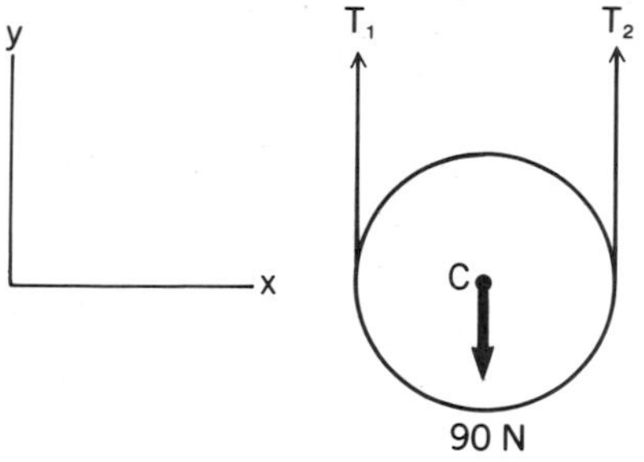

$$T_1(1.7) - 90\left(\frac{1.7}{2}\right) = \frac{1}{2}\frac{90}{g}V_C^2 + \frac{1}{2}\left[\frac{1}{2}\left(\frac{90}{g}\right)(.30)^2\right]\omega^2$$

y, x, T_1, T_2, C, 90 N

Figure 17.15. Free body of cylinder.

Therefore,

$$T_1 = 45 + 2.70V_C^2 + .1214\omega^2 \qquad \text{(b)}$$

From *kinematics* we can conclude on inspection that

$$V_C = \tfrac{1}{2}V_A \qquad \text{(c)}$$

$$\omega = -\frac{V_A}{.60} \qquad \text{(d)}$$

Subtracting Eq. (b) from Eq. (a) to eliminate T_1, and then substituting from Eq. (c) and Eq. (d) for V_C and ω, we then get for V_A:

$$V_A = 2.09 \text{ m/sec} \qquad \text{(e)}$$

This problem can readily be solved as a system, i.e., without disconnecting the bodies. You are asked to do this in Problem 17.26.

In the previous example, we considered the problem to be composed of two *discrete* bodies. We proceeded by expressing equations for each body separately. In the following example, we will consider a *system* of bodies expressing equations for the whole system directly. In this problem, the forces between any two bodies of the system have the *same velocity*; consequently, from Newton's third law, we can conclude that these forces contribute zero net work. We shall illustrate a case where this condition is not so in Example 17.5.

EXAMPLE 17.3

A conveyor is moving a weight W of 64.4 lb in Fig. 17.16. Cylinders A and B have a diameter of 1 ft and weigh 32.2 lb each. Also, they each have a radius of gyration of .4 ft. Rollers C, D, E, F, and G each have a diameter of 3 in., weigh 10 lb each, and have a radius of gyration of 1 in. What constant torque will increase the speed of W from 1 ft/sec of 3 ft/sec in 5 ft of travel? There is no slipping at any of the rollers and drums. The belt weighs 25 lb.

We shall use the *work–kinetic energy* relation 17.14 for this problem. Only external forces and torques do work for the system; the interactive forces between bodies

[4]Recall that the point of contact of the cylinder has zero velocity, and hence the friction force (in this case, T_2) transmits no power to the cylinder.

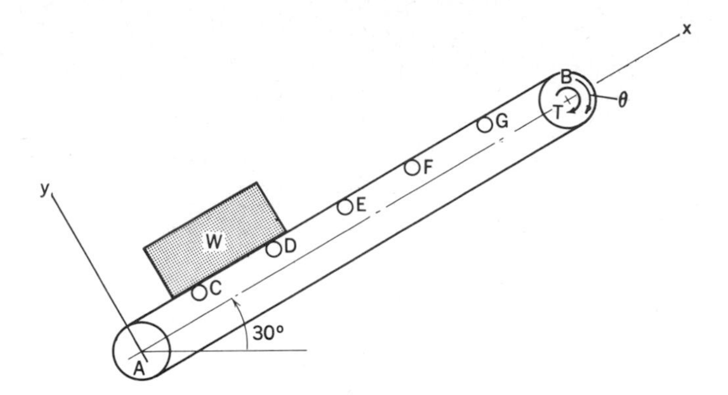

Figure 17.16. Conveyor moving weight W.

do work in amounts that clearly cancel each other because of the condition of no slipping. Hence, measuring θ of the cylinder clockwise:

$$T\theta - W(5)\sin 30° = (\text{KE})_2 - (\text{KE})_1 \tag{a}$$

where T is the applied torque. The general expression for the kinetic energy is

$$\text{KE} = 5\left(\frac{1}{2}I_{\text{roll}}\omega_{\text{roll}}^2\right) + 2\left(\frac{1}{2}I_{\text{cyl}}\omega_{\text{cyl}}^2\right) + \frac{1}{2}M_{\text{belt}}V_{\text{belt}}^2 + \frac{1}{2}\frac{W}{g}V_{\text{weight}}^2 \tag{b}$$

From *kinematics* we can say:

$$\omega_{\text{roll}} = -\frac{V_{\text{belt}}}{(\frac{1}{2})(\frac{3}{12})} = -8V_{\text{belt}}$$

$$\omega_{\text{cyl}} = -\frac{V_{\text{belt}}}{(\frac{1}{2})(1)} = -2V_{\text{belt}}$$

$$V_{\text{weight}} = V_{\text{belt}}$$

Using these results, we can give the kinetic energy at the end and at the beginning of the interval of interest as

$$(\text{KE})_2 = 5\left\{\left(\frac{1}{2}\right)\left(\frac{10}{g}\right)\left(\frac{1}{12}\right)^2[(8)(3)]^2\right\} + 2\left\{\left(\frac{1}{2}\right)\left(\frac{32.2}{g}\right)(.4)^2[(2)(3)]^2\right\}$$
$$+ \frac{1}{2}\left(\frac{25}{g}\right)(3^2) + \left(\frac{1}{2}\right)\left(\frac{64.4}{g}\right)(3^2)$$
$$= 21.4 \text{ ft-lb}$$

$$(\text{KE})_1 = 5\left\{\left(\frac{1}{2}\right)\left(\frac{10}{g}\right)\left(\frac{1}{12}\right)^2[(8)(1)]^2\right\} + 2\left\{\left(\frac{1}{2}\right)\left(\frac{32.2}{g}\right)(.4)^2[(2)(1)]^2\right\}$$
$$+ \frac{1}{2}\left(\frac{25}{g}\right)(1^2) + \frac{1}{2}\left(\frac{64.4}{g}\right)(1^2)$$
$$= 2.37 \text{ ft-lb}$$

Substituting these results into Eq. (a), we get

$$T\theta - (64.4)(5)(\sin 30°) = 21.4 - 2.37 \tag{c}$$

From *kinematics* again we can say for θ, on considering the rotation of a cylinder and the distance traveled by the belt:

$$(r_{cyl})(\theta) = 5$$

Therefore,

$$\theta = \frac{5}{\frac{1}{2}} = 10 \text{ rad}$$

Substituting back into Eq. (c), we can then solve for the desired torque T:

$$T = \tfrac{1}{10}[21.4 - 2.37 + (64.4)(5)(\sin 30°)]$$
$$= 18.00 \text{ ft-lb}$$

In the next example, we have a case of internal forces between bodies that satisfy Newton's third law but do not have identical velocities.

EXAMPLE 17.4

A diesel-powered electric train moves up a 7° grade in Fig. 17.17. If a torque of 750 N-m is developed at each of its six pairs of drive wheels, what is the increase of speed of the train after it moves 100 m? Initially, the train has a speed of 16 km/hr. The train weighs 90 kN. The drive wheels have a diameter of 600 mm. Neglect the rotational energy of the drive wheels.

We shall consider the train as a *system of bodies* including the 8 pairs of wheels and the body. We have shown the train in Fig. 17.18 with the external forces, W, N, and f. In addition, we have shown certain internal torques M.[5] The torques shown

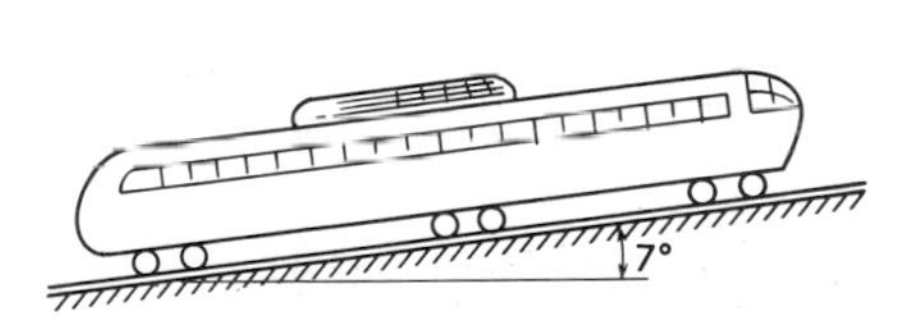

Figure 17.17. Diesel–electric train.

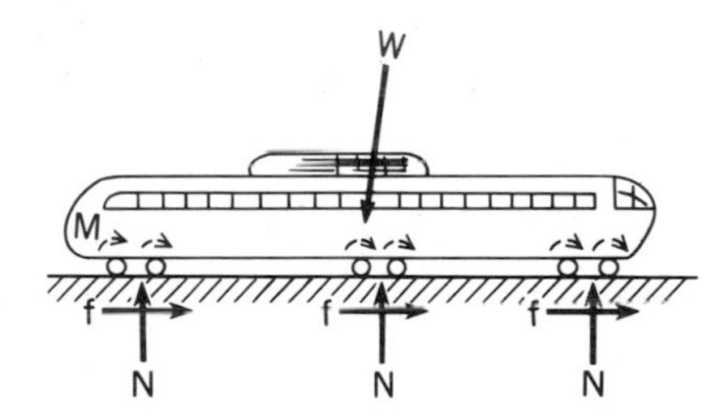

Figure 17.18. External and internal forces and torques.

act on the *rotors* of the motors, and, as the train moves, these torques rotate and accordingly do work. The *reactions* to these torques are equal and opposite to M according to Newton's third law and act on the *stators* of the motors (i.e., the field coils). The stators are stationary, and so the reactions to M do *no* work as the train moves. Thus, we have an example of equal and opposite internal forces between bodies of a system performing a nonzero net amount of work. We now employ Eq. 17.16. Thus,

$$\Delta\text{PE} + \Delta\text{KE} = W_K \quad \text{(a)}$$

Using the initial configuration as the datum, we have[6]

---

[5]Figure 17.18, accordingly, is *not* a free-body diagram.

[6]Recall that for a *rolling* body with *no slipping*, the friction force does *no* work. If we were considering the *center of mass* of the train, then f *would move* with the center of mass and then do work as we shall see in Example 17.5.

$$[(90{,}000)(100 \sin 7^\circ - 0)] + \left\{\frac{1}{2}\frac{90{,}000}{g}V^2 - \frac{1}{2}\frac{90{,}000}{g}\left[\frac{(16)(1000)}{3600}\right]^2\right\} = (6)(750)(\theta) \qquad \text{(b)}$$

where θ is the clockwise rotation of the rotor in radians. Assuming direct drive from rotor to wheel, we can compute θ as follows for the 100-m distance over which the train moves:

$$\theta = \underbrace{\left(\frac{100}{2\pi r}\right)}_{\text{rev}} \underbrace{(2\pi)}_{\text{rad/rev}} = \frac{100}{.30} \qquad \text{(c)}$$

Substituting into Eq. (b) and solving for V, we get

$$V = 10.38 \text{ m/sec}$$

Hence,

$$\Delta V = \frac{(10.38)(3600)}{1000} - 16 = 21.4 \text{ km/hr} \qquad \text{(d)}$$

In Chapter 13, we also developed a work–energy equation involving the *mass center* of any system of particles. You will recall that

$$\boxed{\int_1^2 \boldsymbol{F} \cdot d\boldsymbol{r}_c = \tfrac{1}{2}(MV_c^2)_2 - \tfrac{1}{2}(MV_c^2)_1} \qquad \text{(17.17a)}$$

where $\boldsymbol{F}$ is the *total external force* (only!) which hypothetically *moves with the center of mass*. This equation applies to a rigid body. Note that an external torque makes no work contribution here since equal and opposite forces each having identical motion (that of the mass center) can do no net amount of work. For a system of *interconnected rigid bodies*, we can say:

$$\boxed{\int_1^2 \boldsymbol{F} \cdot d\boldsymbol{r}_c = [\sum_i \tfrac{1}{2}M_i(V_c)_i^2]_2 - [\sum_i \tfrac{1}{2}M_i(V_c)_i^2]} \qquad \text{(17.17b)}$$

The force $\boldsymbol{F}$ includes *only external forces* (internal forces between interconnecting bodies are equal and opposite and must move with the mass center of the system; hence they contribute no work to the left side of the foregoing equation). On the other hand, external friction forces on wheels rolling without slipping must move with the mass center of the system in this formulation and thus can *do work*, in contrast to the previous approach, in which the mass center is not used. On the right side, we have summed the kinetic energies of each of the mass centers of the constituent bodies of the system.[7]

We now illustrate the use of Eq. 17.17b.

[7]We discussed this topic in Section 16.8.

EXAMPLE 17.5

We shall now return to Example 17.4 and work it by using the *center of mass* of the train. A free-body diagram of the train is shown in Fig. 17.19. Thus, only external forces are shown. Using Eq. 17.17b, we can say:

$$[6f - (90{,}000) \sin 7°](100) = \frac{1}{2}\frac{90{,}000}{g}(V^2) - \frac{1}{2}\frac{90{,}000}{g}\left[\frac{(16)(1000)}{3600}\right]^2 \qquad \text{(a)}$$

Note that *all* the external forces including f are assumed to *move with the center of mass of the train.*

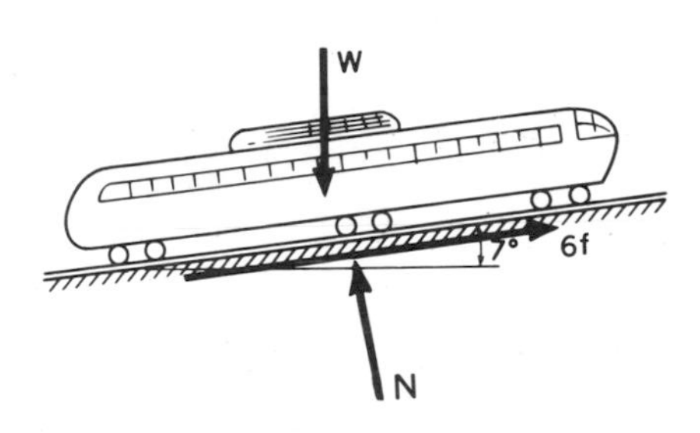

Figure 17.19. Free-body diagram of train.

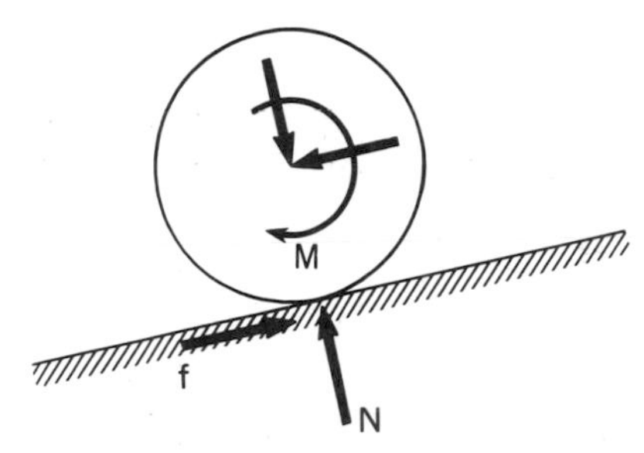

Figure 17.20. Free body of drive wheel.

To determine the friction force f, we examine the free body of a wheel as shown in Fig. 17.20. Since we are neglecting the energy of rotation of the wheel, the total torque about the axle must be zero. Hence, we can say:

$$(f)(r) - M = 0$$

Therefore,

$$f = \frac{750}{.30} = 2500 \text{ N} \qquad \text{(b)}$$

Substituting into Eq. (a), we get

$$V = 10.38 \text{ m/sec}$$
$$\Delta V = 21.4 \text{ km/hr}$$

Problems

In the following problems, neglect friction unless otherwise instructed.

17.1. A uniform solid cylinder of radius 2 ft and weight 200 lb rolls without slipping down a 45° incline and drags the 100-lb block B with it. What is the kinetic energy of the system if block B is moving at a speed of 10 ft/sec? Neglect the mass of connecting agents between the bodies.

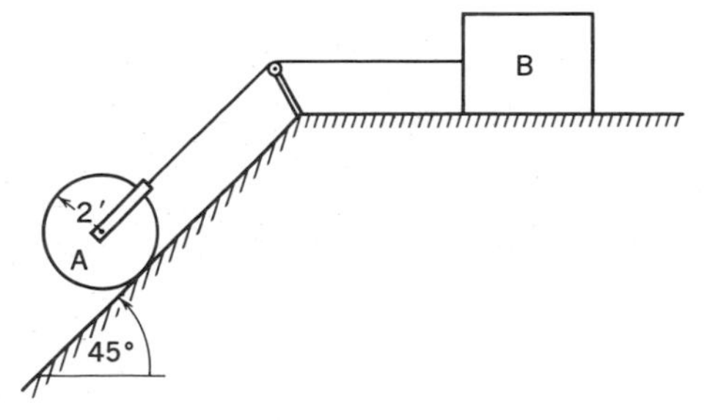

Figure P.17.1

17.2. A steam roller with driver weighs 5 tons. Wheel A weighs 1 ton and has a radius of gyration of .8 ft. Drive wheels B have a total mass of 1000 lb and a radius of gyration of 1.6 ft. If the steam roller is coasting at a speed of 5 ft/sec with motor disconnected, what is the total kinetic energy of the system?

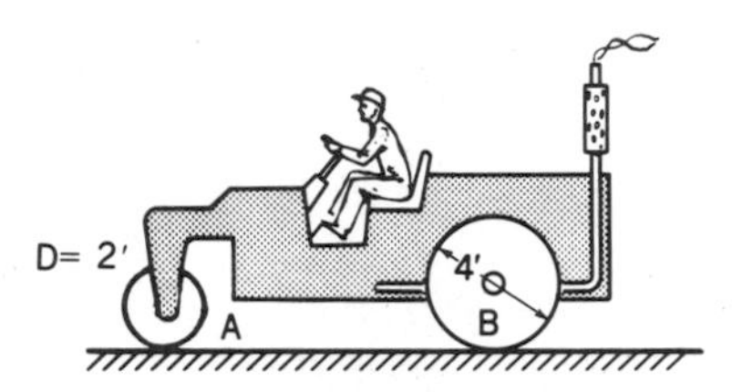

Figure P.17.2

17.3. A thin disc weighing 450 N is suspended from an overhead conveyor moving at a speed of 10 m/sec. If the disc rotates at a speed of 5 rad/sec in the plane of the page (i.e., ZY plane), compute the kinetic energy of the disc relative to the ground.

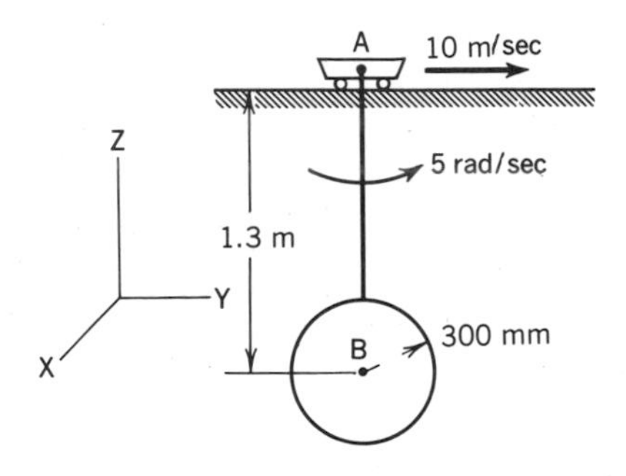

Figure P.17.3

17.4. Two slender rods CD and EA are pinned together at B. Rod EA is rotating at a speed ω equal to 2 rad/sec. Rod CD rides in a vertical slot at D. For the configuration shown in the diagram, compute the kinetic energy of the rods. Rod CD weighs 50 N and rod EA weighs 80 N.

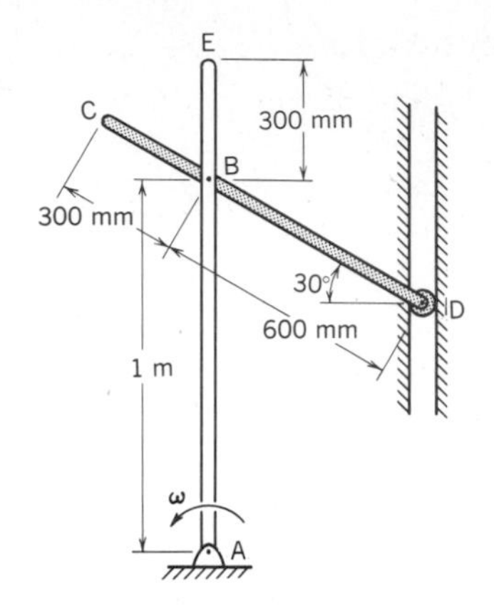

Figure P.17.4

17.5. Consider the connecting rod AB to be a slender rod weighing 2 lb, and compute its kinetic energy for the data given.

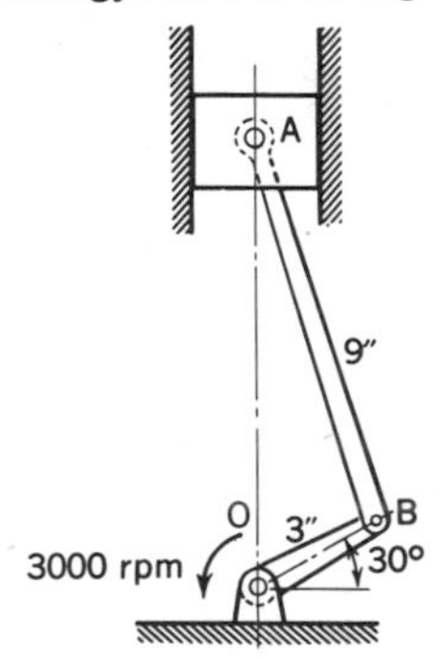

Figure P.17.5

17.6. Identical rods CB and AB are pinned together at B. Rod BC is pinned to a block D weighing 225 N. Each rod is 600 mm in length and weighs 45 N. Rod BA rotates counterclockwise at a constant speed ω of 3 rad/sec. Compute the kinetic energy of the system when BA is oriented (a) at an angle of 60° with the vertical and (b) at an angle of 90° with the vertical (the latter position is shown dashed in the diagram).

Figure P.17.6

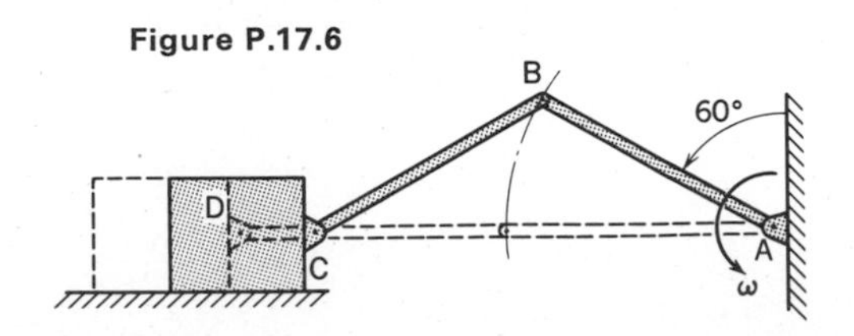

17.7. Find the kinetic energy of the rotating system described in Example 16.10. The diameter of the shaft is 50 mm.

17.8. The centerline of gear A rotates about axis M–M at an angular speed ω_1 of 3 rad/sec. The mean diameter of gear A is 1 ft. If gear A has a mass of 10 lbm, what is the kinetic energy of the gear? Consider the gear to be a disc.

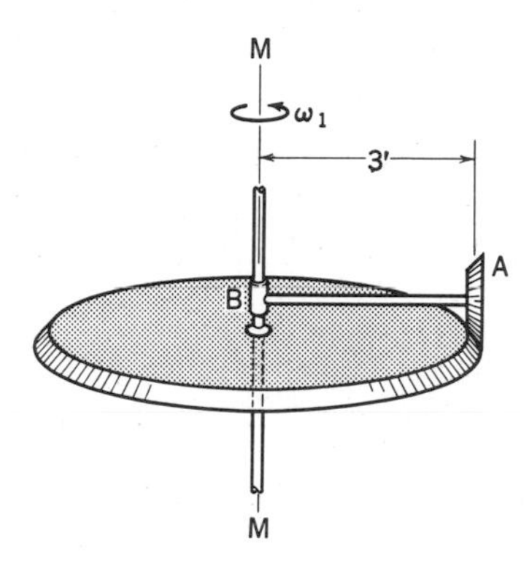

Figure P.17.8

17.9. A cone B weighing 20 lb rolls without slipping inside a conical cavity C. The cone has a length of 10 ft. The centerline of the cone rotates with an angular speed ω_1 of 5 rad/sec about the Y axis. Compute the kinetic energy of the cone.

Figure P.17.9

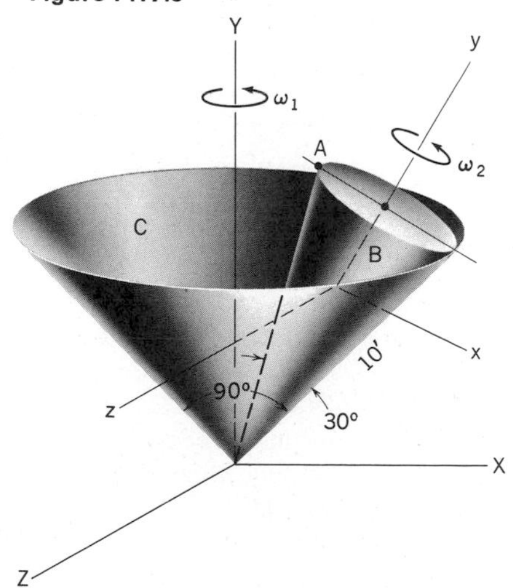

17.10. A uniform cylinder has a radius r and a weight W_1. A weight W_2, which we shall consider a particle because of its small physical dimensions, is placed at G a distance a from O, the center of the disc, such that OG is vertical. What is the angular velocity of the cylinder when, after it is released from rest, the point G reaches its lowest elevation, as shown at the right? The cylinder rolls without slipping.

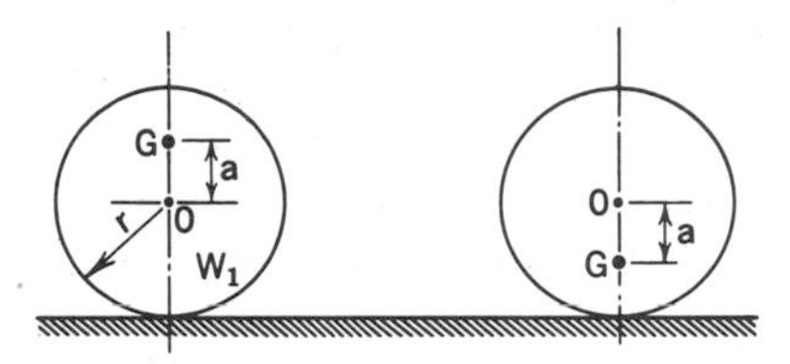

Figure P.17.10

17.11. A homogeneous solid cylinder of radius 300 mm is shown with a fine wire held fixed at A and wrapped around the cylinder. If the cylinder is released from rest, what will its velocity be when it has dropped 3 m?

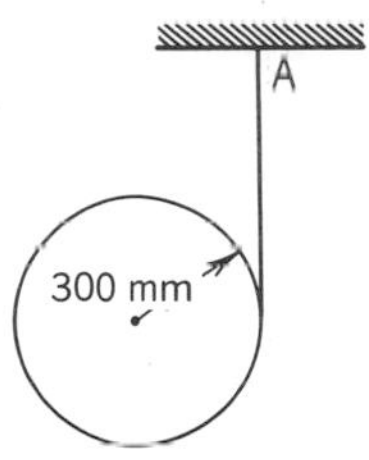

Figure P.17.11

17.12. Three identical bars, each of length l and weight W, are connected to each other and to a wall with smooth pins at A, B, C, and D. A spring having spring constant K is connected to the center of bar BC at E and to a pin at F, which is free to slide in the slot. Compute the angular speed $\dot{\theta}$ as a function of time if the system is released from rest when AB and DC are at right angles to the wall. The spring is unstretched at the outset of the motion. Neglect friction.

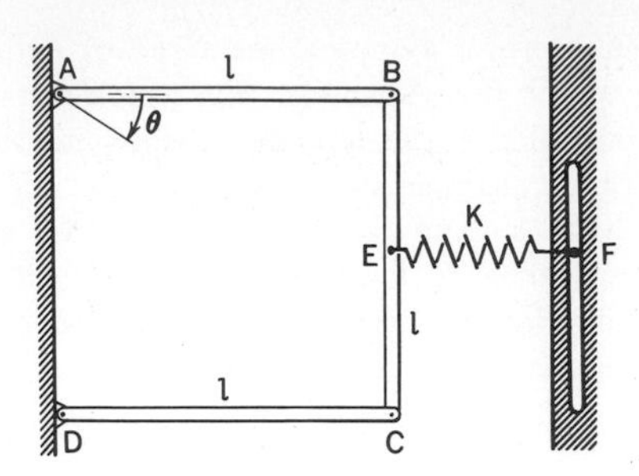

Figure P.17.12

17.13. A 3-m rod AB weighing 225 N is guided at A by a slot and at B by a smooth horizontal surface. Neglect the mass of the slider at A, and find the speed of B when A has moved 1 m along the slot after starting from a rest configuration shown in the diagram.

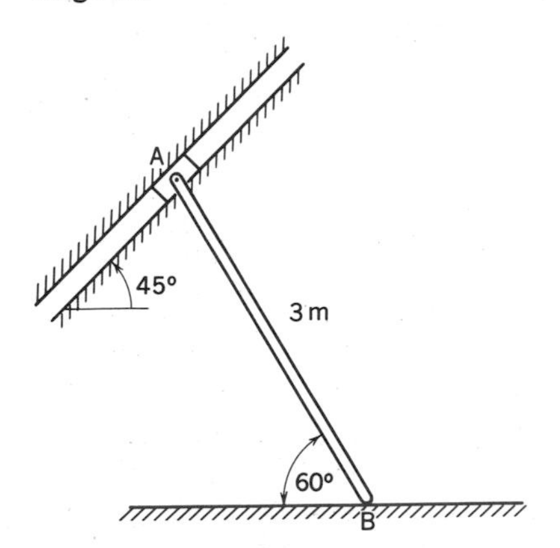

Figure P.17.13

17.14. A stepped cylinder has radii of 600 mm for the smaller radius and 1.3 m for the larger radius. A rectangular block A weighing 225 N is welded to the cylinder at B. The spring constant K is .18 N/mm. If the system is released from a configuration of rest, what is the angular speed of the cylinder after it has rotated 90°? The radius of gyration for the stepped cylinder is 1 m and its mass is 36 kg. The spring is unstretched in the position shown.

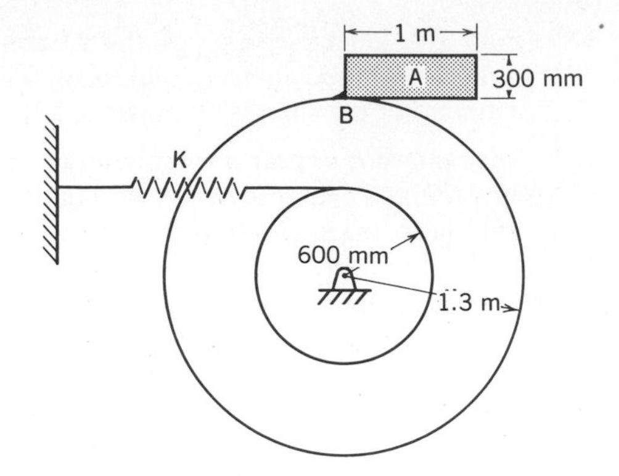

Figure P.17.14

17.15. A cylinder of diameter 2 ft is composed of two semicylinders C and D weighing 50 lb and 80 lb, respectively. Bodies A and B, weighing 20 lb and 50 lb, respectively, are connected by a light, flexible cable that runs over the cylinder. If the system is released from rest for the configuration shown, what is the speed of B when the cylinder has rotated 90°? Assume no slipping.

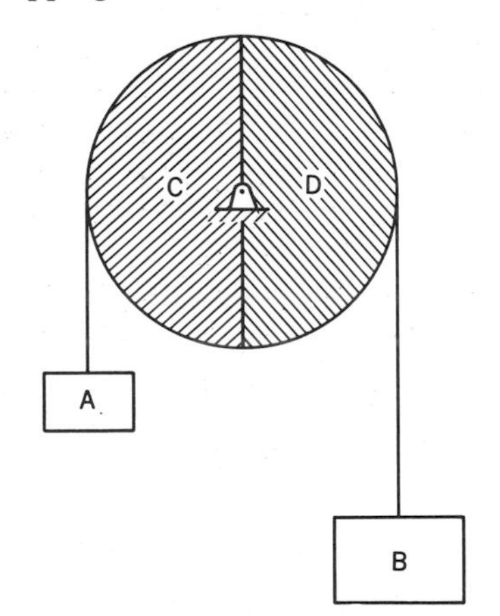

Figure P.17.15

17.16. A force F of 450 N acts on block A weighing 435 N. Block A rides on identical uniform cylinders B and C, each weighing 290 N and having a radius of 300 mm. If there is no slipping, what is the speed of A after it moves 1 m?

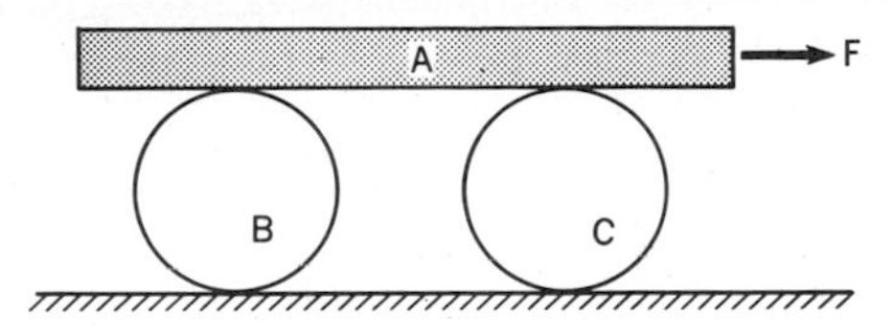

Figure P.17.16

17.17. Four identical rods, each of length $l =$ 1.3 m and weight 90 N, are connected at the frictionless pins A, B, C, and D. A compression spring of spring constant $K = 5.3$ N/mm connects pins B and C, and a weight W_2 of 450 N is supported at pin D. The system is released from a configuration where $\theta = 45°$. If the spring is not compressed at that configuration, show that the maximum deflection of the weight W_2 is .1966 m.

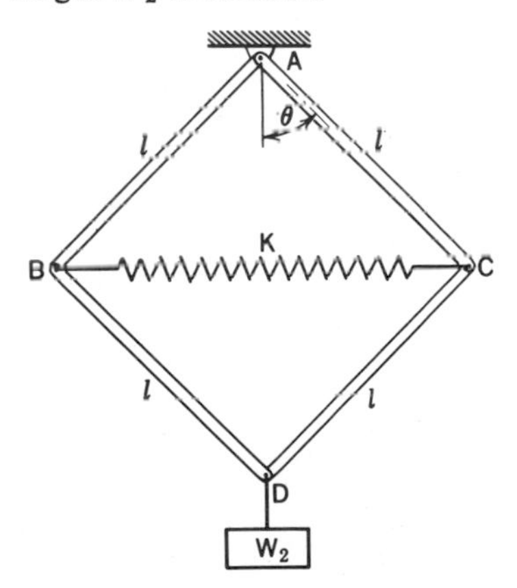

Figure P.17.17

17.18. A stepped cylinder weighing 30 lb with a radius of gyration of 1 ft is connected to a 50-ft chain weighing 100 lb. The chain hangs down from the horizontal surface a distance of 10 ft when the system is released. Determine the speed of the chain when 30 additional feet of chain have come off the horizontal surface. The coefficient of friction between the chain and the horizontal surface is .2, and the smaller diameter of the stepped cylinder is 4 in.

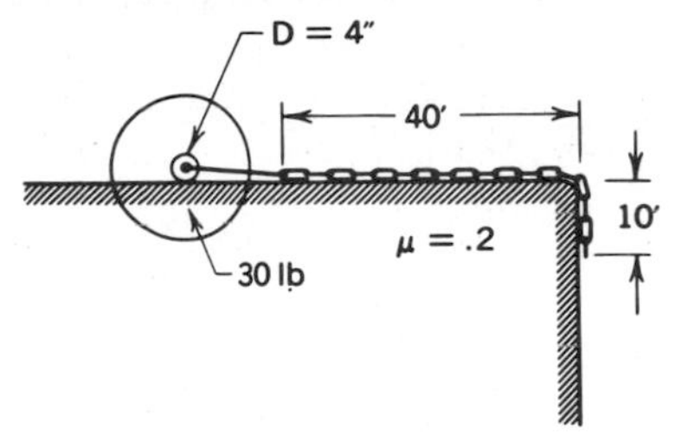

Figure P.17.18

17.19. The linkage system rests on a frictionless plane. The lengths AB, BF, etc., are each 300 mm, and the bars, all of the same stock, weigh 67.5 N/m. A force F of 450 N is applied at D. What is the speed of D after it moves 300 mm? The system is stationary in the configuration shown. The view of the linkage system is from above.

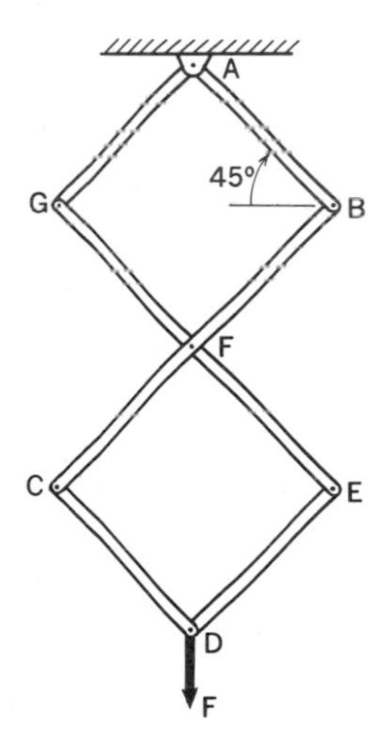

Figure P.17.19

17.20. Work Problem 17.19 for the case where the system is in a vertical plane.

17.21. A belt weighing 10 lb is mounted over two pulleys of diameter 1 ft and 2 ft, respectively. The radius of gyration and weight for pulley A are 6 in. and 50 lb, respectively, and for pulley B are 9 in. and 200 lb, respectively. A constant torque of 20 lb-in. is applied to pulley A. After 30 revolutions of pulley A, what will its angular speed be if the system starts from rest? There is no slipping between belts and pulleys, and pulley B turns freely.

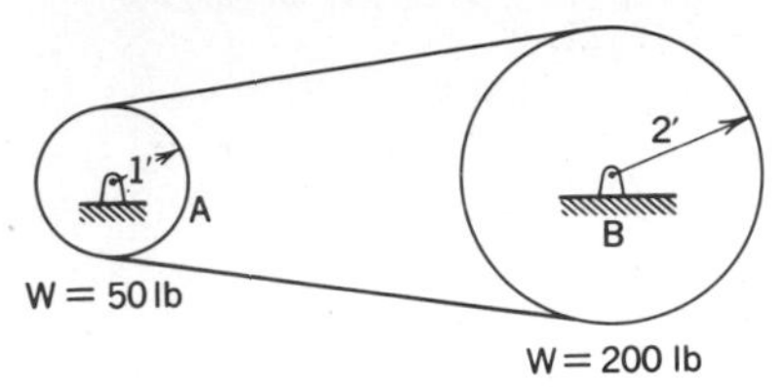

Figure P.17.21

17.22. Two identical members, AB and BC, are pinned together at B. Also member BC is pinned to the wall at C. Each member weighs 32.2 lb and is 20 ft long. A spring having a spring constant $K = 20$ lb/ft is connected to the centers of the members. A force $P = 100$ lb is applied to member AB at A. If initially the members are inclined 45° to the ground and the spring is unstretched, what is $\dot{\beta}$ after A has moved 2 ft?

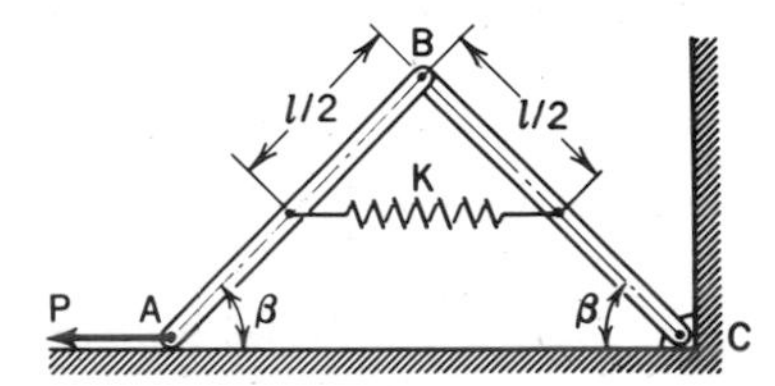

Figure P.17.22

17.23. A flexible cord of total length 50 ft and weighing 50 lb is pinned to a wall at A and is wrapped around a cylinder having a radius of 4 ft and weighing 30 lb. A 50-lb force is applied to the end of the cord. What is the speed of the cylinder after the end of the cord has moved 10 ft? The system starts from rest in the configuration shown in the diagram. Neglect potential energy considerations arising from the sag of the upper cord.

Figure P.17.23

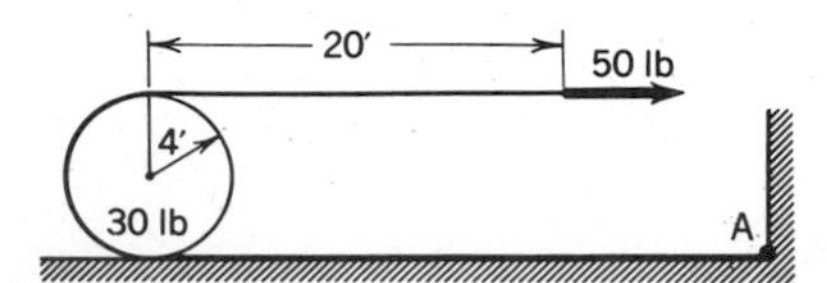

17.24. In Problem 17.5, suppose that an average pressure of 20 psig exists in the cylinder. What is the rpm after the crankshaft has rotated 60° from position shown? The crank rod OB weighs 1 lb and has a radius of gyration about O of 2 in. The diameter of the piston is 4 in., and its weight is 8 oz. The crankshaft is rotating at 3000 rpm at the position shown. Take the center of mass of the crank rod at the midpoint of OB.

17.25. A right circular cone of weight 32.2 lb, height 4 ft, and cone angle 20° is allowed to roll without slipping on a plane surface inclined at an angle of 30° to the horizontal. The cone is started from rest when the line of contact is parallel to the X axis. What is the angular speed of the centerline of the cone when it has its maximum kinetic energy?

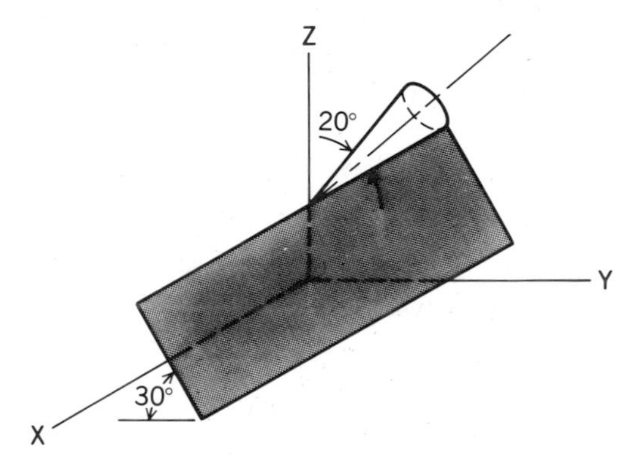

Figure P.17.25

17.26. Work Example 17.2 by considering the system to be the block, pulley, and cable.

17.27. A weight W_1 is held with a light flexible wire. The wire runs over a stationary semicylinder of radius R equal to 1 ft. A pulley weighing 32.2 lb and having a radius of gyration of unity rides on the wire and supports a weight W_2 of 16.1 lb. If W_1 weighs 128.8 lb and the coefficient of friction for the semicylinder and wire is .2, what is the drop in the weight W_1 for an increase in speed of 5 ft/sec of weight W_1 starting from rest? The diameter d of the small pulley is 1 ft.

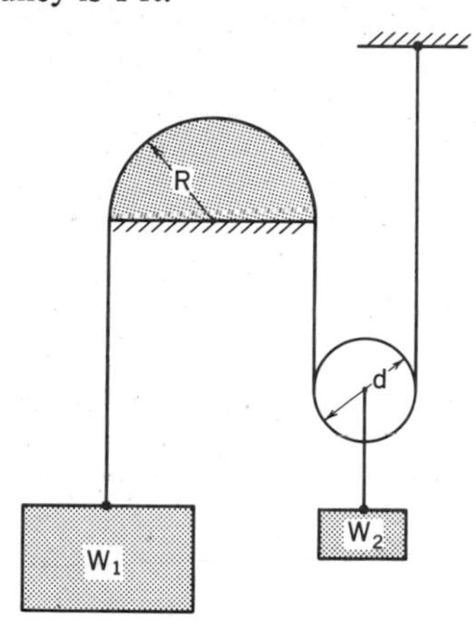

Figure P.17.27

17.28. A solid uniform block A moves along two frictionless angle-iron supports at a speed of 6 m/sec. One of the supports is inclined at an angle of 20° from the horizontal at B and causes the block to rotate about its front lower edge as it moves to the right of B. What is the speed of the block after it moves 300 mm to the right of B (measured horizontally)? The block weighs 450 N. Consider that no binding occurs between the block and the angle-iron supports.

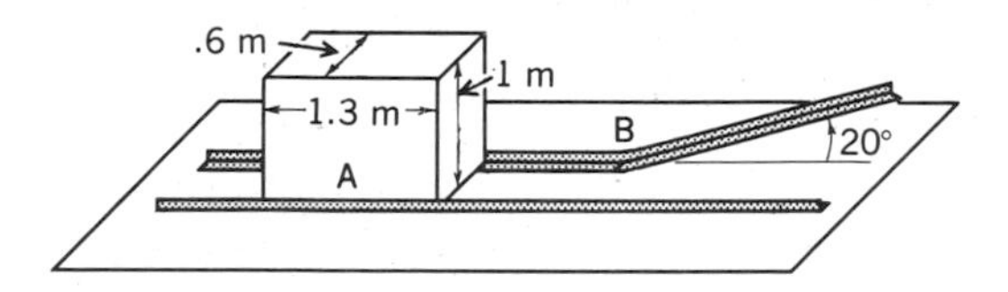

Figure P.17.28

17.29. In Problem 17.28, will the block reach an instantaneous zero velocity and then slide back or will it tip over onto face A?

17.30. A torque $T = .30$ N-m is applied to a bevel gear B. Bevel gear D meshes with gear B and drives a pump A. Gear B has a radius of gyration of 150 mm and a weight of 50 N, whereas gear D and the impeller of pump A have a combined radius of gyration of 50 mm and a weight of 100 N. Show that the *work of the contact forces between gears B and D is zero.* Next, find how many revolutions of gear B are needed to get the pump up to 200 rpm from rest. Treat the problem as a system of bodies.

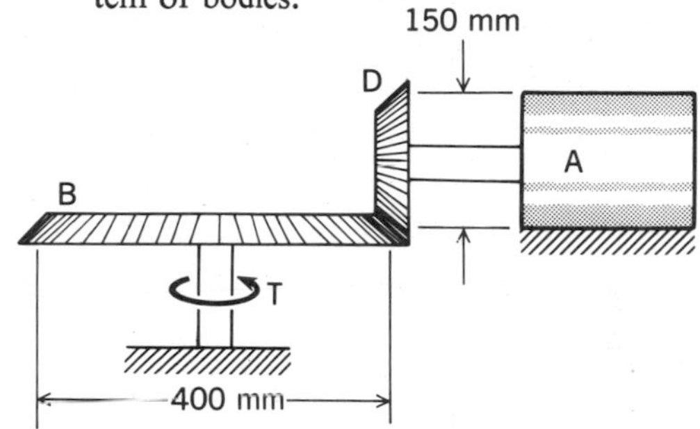

Figure P.17.30

17.31. A torque T of .5 N-m acts on worm gear E, which meshes with gear A, which drives a gear train. After five revolutions of gear H, what is its angular speed? One revolution of worm gear E corresponds to .2 revolution of gear A. Use the following data:

	k (mm)	W(N)	D (mm)
A	30	10	100
B	30	10	100
C	100	40	300
H	200	100	500

Neglect inertia of the worm gear, and consider an energy loss of 10% of the input due to friction. Note the italicized statement of Problem 17.30, and treat the problem as a system of bodies.

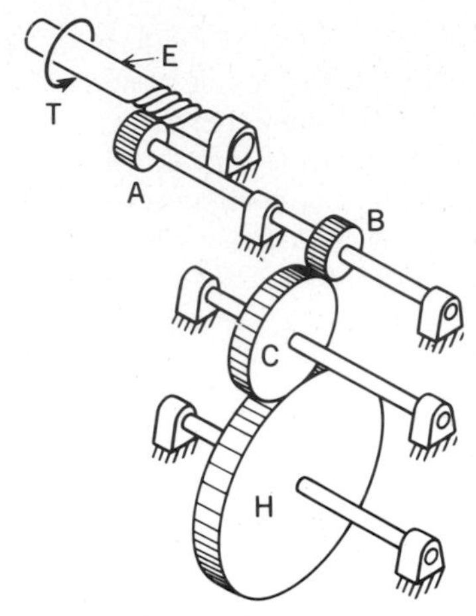

Figure P.17.31

17.32. A vehicle for traversing swampland has four-wheel drive and (with passenger) weighs 22.5 kN. Each wheel weighs 2 kN and has a radius of gyration of 800 mm. If each wheel gets a torque of 100 N-m, what is the speed after 20 m of travel starting from rest? Do not use the center-of-mass approach for the whole system.

Figure P.17.32

17.33. Work Problem 17.32 using the center-of-mass approach for the whole system. What are the friction forces on the wheels from the ground?

17.34. An electric train (one car) uses its motors as electric generators for braking action. Suppose that this train is moving down a 15° incline at a speed initially of 10 m/sec and, during the next 100 m, the generators develop 1.5 kW-hr of energy. What is the speed of the train at the end of this interval? The train with passengers weighs 200 kN. Each of the eight wheels weighs 900 N and has a radius of gyration of 250 mm and a diameter of 600 mm. Neglect wind resistance, and consider that there is no slipping. The efficiency of the generators for developing power is 90%. (*Hint:* One watt is 1 N-m/sec.) Do not use center of mass approach.

17.35. Work Problem 17.34 using the center of mass approach for the whole train. Also, find the average friction force from the rail onto the wheels. Consider that each wheel is attached to a generator.

17.36. A windlass has a rotating part which weighs 75 lb and has a radius of gyration of 1 ft. When the suspended weight of 20 lb is dropping at a speed of 20 ft/sec, a 100-lb force is applied to the lever at *A*. This action applies the brake shoe at *B*, where there is a coefficient of friction of .5. How far will the 20-lb weight drop before stopping?

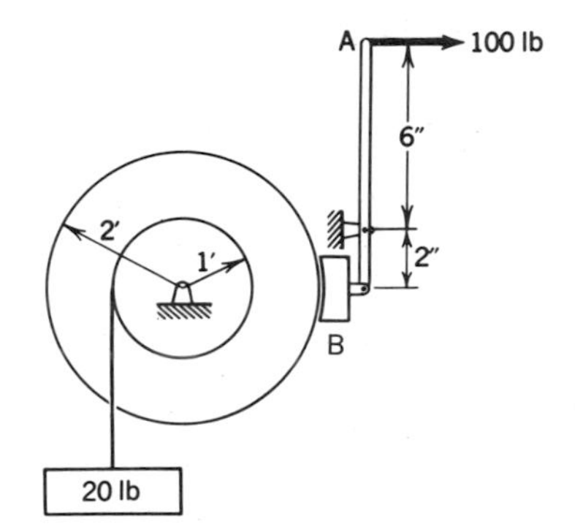

Figure P.17.36

17.37. A square-threaded screw has a diameter of 50 mm and is inclined 45° to the horizontal. The pitch of the thread is 5 mm, and it is single-threaded. A body *A* weighing 290 N and having a radius of gyration of 300 mm screws onto the shaft. A torque *T* of 45 N-m is applied to *A* as shown. What is the angular speed of *A* after three revolutions starting from a rest configuration? Neglect friction.

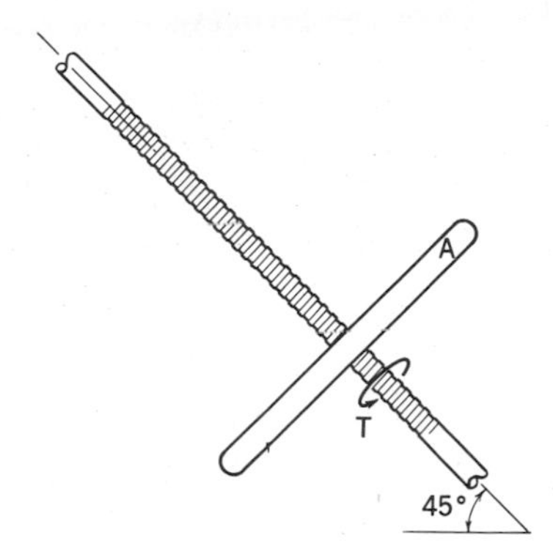

Figure P.17.37

17.38. A uniform block A weighing 64.4 lb is pulled by a force P of 50 lb as shown. The block moves along the rails on small, light wheels. One rail descends at an angle of 15° at point B. If the force P always remains horizontal, what is the speed of the block after it has moved 5 ft in the horizontal direction? The block is stationary at the position shown. Assume that the block does not tilt forward.

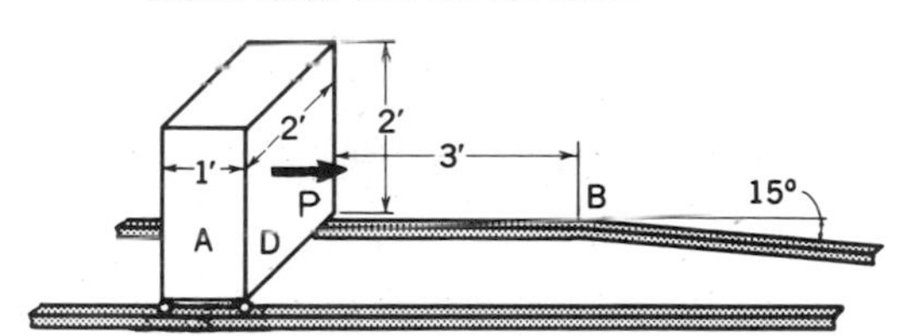

Figure P.17.38

***17.39.** A solid uniform rod AB connects two light slider bearings A and B, which move in a frictionless manner along the indicated guide rods. The rod AB has a mass of 150 lbm and a diameter of 2 in. Smooth ball-joint connections exist between the rod and the bearings. If the rod is released from rest at the configuration shown, what is the speed of the bearing A when it has dropped 2 ft?

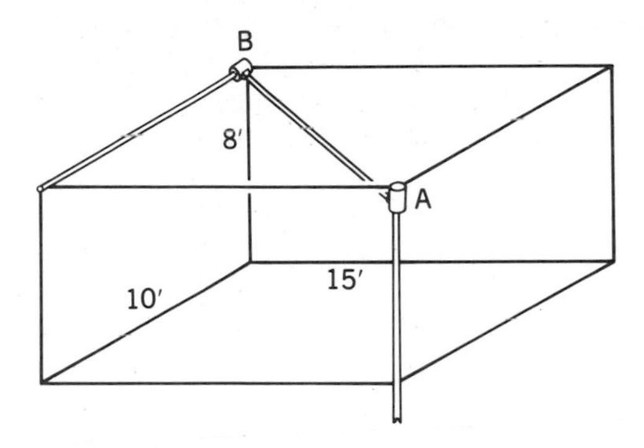

Figure P.17.39

***17.40.** A rod weighing 90 N is guided by two slider bearings A and B. Smooth ball joints connect the rod to the bearings. A force F of 45 N acts on bearing A. What is the speed of A after it has moved 200 mm? The system is stationary for the configuration shown. Neglect friction.

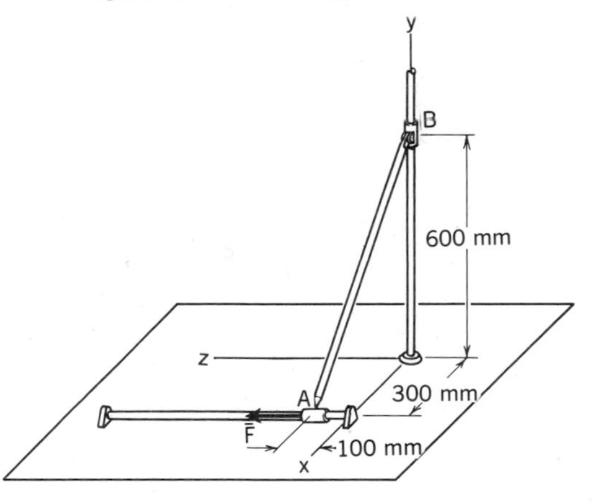

Figure P.17.40

Part B Impulse-Momentum Methods

17.4 Angular Momentum of a Rigid Body About Any Point in the Body

We shall now formulate an expression for the angular momentum $\boldsymbol{H}$ of a rigid body about a point in the rigid body. For this purpose, we choose a point A in a rigid body or hypothetical massless extension of the rigid body as shown in Fig. 17.21. An element of mass dm at a position $\boldsymbol{\rho}$ from A is shown. The velocity $\boldsymbol{V}'$ of dm relative to A is

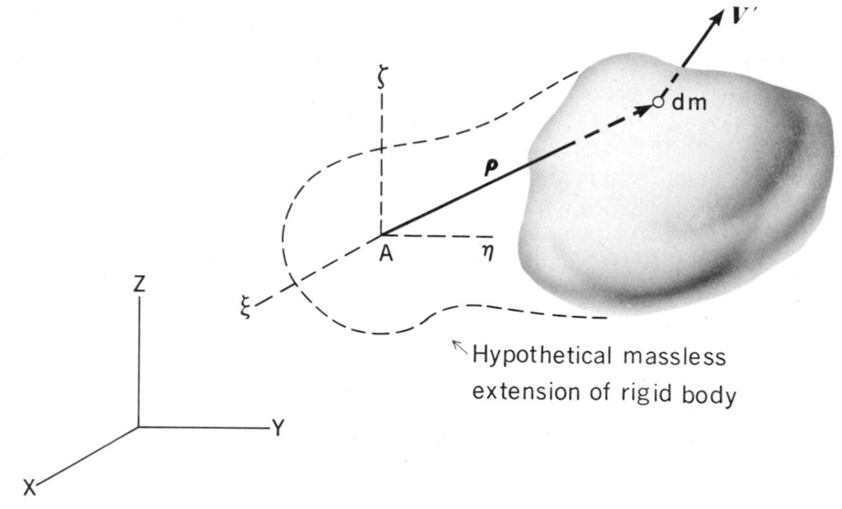

Figure 17.21. Velocity of *dm* relative to point *A*.

simply the velocity of dm relative to reference $\xi\eta\zeta$ which *translates* with A relative to XYZ. Similarly, the linear momentum of dm relative to A is the linear momentum of dm relative to a reference $\xi\eta\zeta$ translating with A. We can now give the angular momentum $d\boldsymbol{H}_A$ for element dm about A as

$$d\boldsymbol{H}_A = \boldsymbol{\rho} \times \boldsymbol{V}'\,dm = \boldsymbol{\rho} \times \left(\frac{d\boldsymbol{\rho}}{dt}\right)_{\xi\eta\zeta} dm \tag{17.18a}$$

But since A is fixed in the body (or in the hypothetical massless extension of the body), the vector $\boldsymbol{\rho}$ must be fixed in the body. Accordingly, $(d\boldsymbol{\rho}/dt)_{\xi\eta\zeta} = \boldsymbol{\omega} \times \boldsymbol{\rho}$, where $\boldsymbol{\omega}$ is the angular velocity of the body relative to $\xi\eta\zeta$. However, since $\xi\eta\zeta$ translates with respect to XYZ, *$\boldsymbol{\omega}$ is also the angular velocity of the body relative to XYZ as well*. Hence, we can say:

$$d\boldsymbol{H}_A = \boldsymbol{\rho} \times (\boldsymbol{\omega} \times \boldsymbol{\rho})\,dm \tag{17.18b}$$

We shall find it convenient to express Eq. 17.18 in terms of orthogonal components. For this purpose, imagine a reference xyz having the origin at A and any arbitrary orientation relative to XYZ,[8] as shown in Fig. 17.22. We next decompose

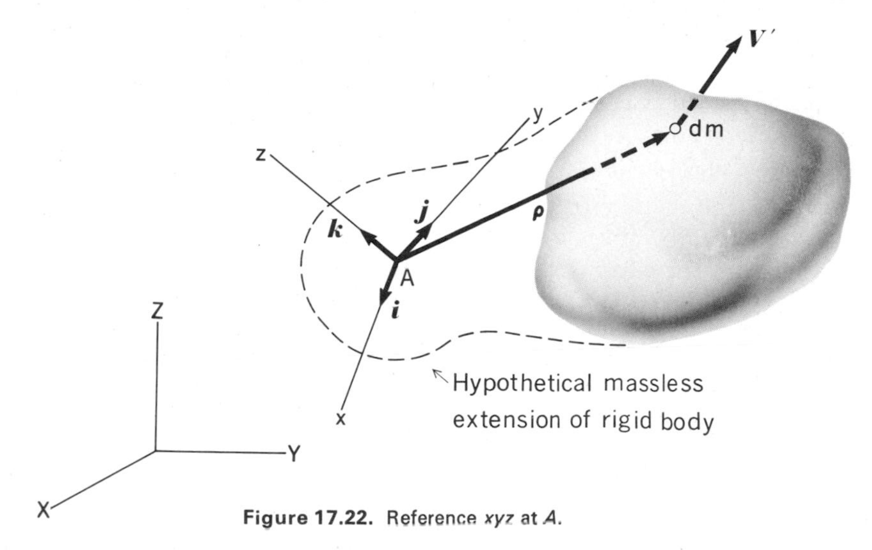

Figure 17.22. Reference xyz at A.

each of the vectors in Eq. 17.18b into rectangular components in the $\mathbf{i}$, $\mathbf{j}$, and $\mathbf{k}$ directions associated with the x, y, and z axes, respectively. Thus,

$$d\mathbf{H}_A = (dH_A)_x\mathbf{i} + (dH_A)_y\mathbf{j} + (dH_A)_z\mathbf{k} \tag{17.19a}$$

$$\boldsymbol{\rho} = x\mathbf{i} + y\mathbf{j} + z\mathbf{k} \tag{17.19b}$$

$$\boldsymbol{\omega} = \omega_x\mathbf{i} + \omega_y\mathbf{j} + \omega_z\mathbf{k} \tag{17.19c}$$

We then have for Eq. 17.18b:

$$(dH_A)_x\mathbf{i} + (dH_A)_y\mathbf{j} + (dH_A)_z\mathbf{k} = (x\mathbf{i} + y\mathbf{j} + z\mathbf{k}) \times [(\omega_x\mathbf{i} + \omega_y\mathbf{j} + \omega_z\mathbf{k}) \times (x\mathbf{i} + y\mathbf{j} + z\mathbf{k})]\, dm \tag{17.20}$$

Carrying out the cross products and collecting terms, we have

$$(dH_A)_x = \omega_x(y^2 + z^2)\, dm - \omega_y xy\, dm - \omega_z xz\, dm \tag{17.21a}$$

$$(dH_A)_y = -\omega_x yx\, dm + \omega_y(x^2 + z^2)\, dm - \omega_z yz\, dm \tag{17.21b}$$

$$(dH_A)_z = -\omega_x zx\, dm - \omega_y zy\, dm + \omega_z(x^2 + y^2)\, dm \tag{17.21c}$$

If we integrate these relations for all the mass elements dm of the rigid body, we see that the components of the inertia tensor for point A appear:

---

[8] At this time we can forget about the axes ξ, η, and ζ. They only become necessary when we ask the question: What is the velocity or linear momentum of a particle relative to point A. To repeat, the velocity or linear momentum of a particle relative to point A is the velocity or momentum relative to a reference $\xi\eta\zeta$ translating with point A as seen from XYZ or, in other words, relative to a nonrotating observer moving with A.

$$(H_A)_x = I_{xx}\omega_x - I_{xy}\omega_y - I_{xz}\omega_z \tag{17.22a}$$
$$(H_A)_y = -I_{yx}\omega_x + I_{yy}\omega_y - I_{yz}\omega_z \tag{17.22b}$$
$$(H_A)_z = -I_{zx}\omega_x - I_{zy}\omega_y + I_{zz}\omega_z \tag{17.22c}$$

We thus have components of the angular momentum vector $\boldsymbol{H}_A$ for a rigid body about point A in terms of an arbitrary set of directions x, y, and z at point A.

We now illustrate the calculation of $\boldsymbol{H}_A$ in the following example.

EXAMPLE 17.6

A disc B has a mass M and is rotating around centerline E–E in Fig. 17.23 at a speed ω_1 relative to E–E. Centerline E–E, meanwhile, has an angular speed ω_2 about the vertical axis. Compute the angular momentum of the disc about point A.

The angular velocity of the disc relative to point A and thus relative to the ground is

$$\boldsymbol{\omega} = \omega_1 \boldsymbol{i} + \omega_2 \boldsymbol{j} \tag{a}$$

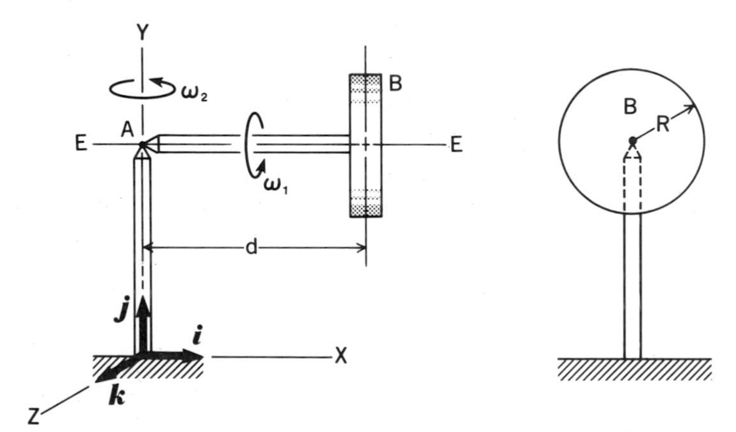

Figure 17.23. Rotating disc.

For a set of axes xyz with the origin at A and parallel to the inertial reference XYZ, we can say (see Fig. 17.24):

$$\omega_x = \omega_1, \qquad \omega_y = \omega_2, \qquad \omega_z = 0 \tag{b}$$

The inertia tensor for the disc taken at A is next presented.

$$\begin{array}{lll} I_{xx} = \frac{1}{2}MR^2 & I_{xy} = 0 & I_{xz} = 0 \\ I_{yx} = 0 & I_{yy} = \frac{1}{4}MR^2 + Md^2 & I_{yz} = 0 \\ I_{zx} = 0 & I_{zy} = 0 & I_{zz} = \frac{1}{4}MR^2 + Md^2 \end{array} \tag{c}$$

Note that the product-of-inertia terms are zero because the xy and the xz planes are planes of symmetry. Clearly, moments of inertia for the xyz axes are principal

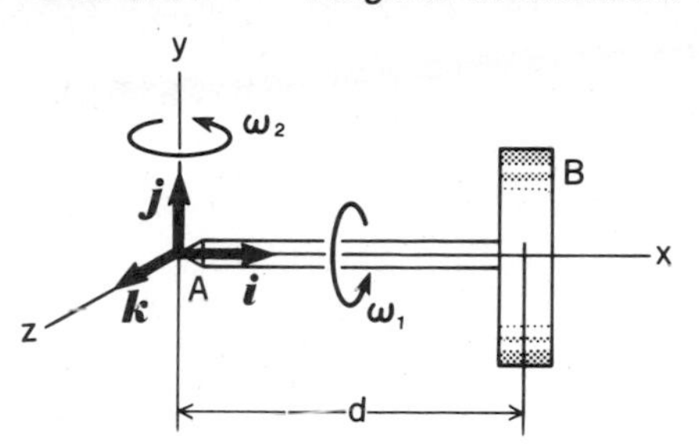

Figure 17.24. *xyz* axes at *A*.

moments of inertia. Now going to Eq. 17.22, we have

$$(H_A)_x = \frac{MR^2}{2}\omega_1$$
$$(H_A)_y = (\tfrac{1}{4}MR^2 + Md^2)\omega_2 \qquad \text{(d)}$$
$$(H_A)_z = 0$$

As seen in Example 17.6, when *xyz* are *principal* axes, $\boldsymbol{H}_A$ simplifies to

$$\boldsymbol{H}_A = I_{xx}\omega_x\boldsymbol{i} + I_{yy}\omega_y\boldsymbol{j} + I_{zz}\omega_z\boldsymbol{k} \qquad (17.23)$$

which is analogous to the linear momentum vector $\boldsymbol{P}$. Thus,

$$\boldsymbol{P} = MV_x\boldsymbol{i} + MV_y\boldsymbol{j} + MV_z\boldsymbol{k}$$

Note that mass plays the same role as does I, and $\boldsymbol{V}$ plays the same pole as does $\boldsymbol{\omega}$.

In Chapter 16, we considered with some care the plane motion of a slablike body, the motion being parallel to the plane of symmetry of the body. We have shown such a case in Fig. 17.25. A reference xyz is shown at A with xy at the midplane of the body and with the z axis oriented normal to the plane of motion. Recall now that for a

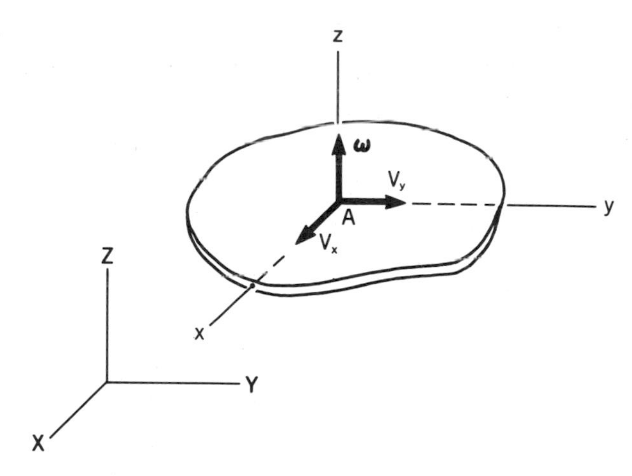

Figure 17.25. Slablike body in plane motion.

slablike body the plane xy must be a plane of symmetry, and consequently that $I_{zx} = I_{zy} = 0$. Also, the only nonzero component of $\boldsymbol{\omega}$ is ω_z. Going back to Eq. 17.22, we see that only $(H_A)_z$ is nonzero, with the result

$$(H_A)_z = (I_{zz})_A\omega_z \qquad (17.24)$$

Because of the importance of plane motion of slablike bodies, we shall often use the foregoing simple formula.

We leave it for you to show that the Eq. 17.24 also applies to a body of revolution rotating about its axis of symmetry in inertial space, where z is taken along this axis. Also, Eq. 17.24 is valid for a body having two orthogonal planes of symmetry rotating about an axis corresponding to the intersection of these planes of symmetry in inertial space, where z is taken along this axis.

17.5 Impulse-Momentum Equations

You will recall that Newton's law for the center of mass of any body is

$$\boldsymbol{F} = M\frac{d\boldsymbol{V}_c}{dt} = \frac{d}{dt}(M\boldsymbol{V}_c)$$

where $\boldsymbol{F}$ is the total *external* force on the body. The corresponding impulse-momentum equation can then be given as

$$\boxed{\int_{t_1}^{t_2} \boldsymbol{F}\,dt = \boldsymbol{I}_{\text{lin}} = (M\boldsymbol{V}_C)_2 - (M\boldsymbol{V}_C)_1} \qquad (17.25)$$

where $\boldsymbol{I}_{\text{lin}}$ is the *linear impulse*. For a system of n rigid bodies we have, for the foregoing equation:

$$\boxed{\int_{t_1}^{t_2} \boldsymbol{F}\,dt = \boldsymbol{I}_{\text{lin}} = \left[\sum_{i=1}^{n} M_i(\boldsymbol{V}_C)_i\right]_2 - \left[\sum_{i=1}^{n} M_i(\boldsymbol{V}_C)_i\right]_1} \qquad (17.26)$$

where $\boldsymbol{F}$ is the total *external* force on the system, M_i is the mass of the ith body, and $(\boldsymbol{V}_C)_i$ is the velocity of the center of mass of the ith body. We are justified in forming the preceding equation as a result of Eq. 16.17.

For the *angular impulse-momentum equation*, we consider points A which are part of the rigid body or massless extension of the rigid body and which, in addition, are either:

1. The center of mass.
2. A point fixed or moving at constant $\boldsymbol{V}$ in inertial space.
3. A point accelerating toward or away from the center of mass.

In such cases we can say:

$$\boldsymbol{M}_A = \dot{\boldsymbol{H}}_A$$

where $\boldsymbol{H}_A$ is given by Eq. 17.22. Integrating with respect to time, we then get the desired *angular impulse-momentum equation*:

$$\boxed{\int_{t_1}^{t_2} \boldsymbol{M}_A\,dt = \boldsymbol{I}_{\text{ang}} = (\boldsymbol{H}_2)_A - (\boldsymbol{H}_1)_A} \qquad (17.27)$$

where $\boldsymbol{I}_{\text{ang}}$ is the *angular impulse*. We now illustrate the use of the linear impulse- and angular impulse-momentum equations.

EXAMPLE 17.7

A thin bent rod is sliding along a smooth surface (Fig. 17.26). The center of mass has the velocity

$$V_C = 10\boldsymbol{i} + 15\boldsymbol{j} \text{ m/sec}$$

and the angular speed ω is 5 rad/sec counterclockwise. At the configuration shown, the rod is given two simultaneous impacts as a result of a collision. These impacts have the following impulse values:

$$\int_{t_1}^{t_2} F_1\, dt = 5 \text{ N-sec}$$

$$\int_{t_1}^{t_2} F_2\, dt = 3 \text{ N-sec}$$

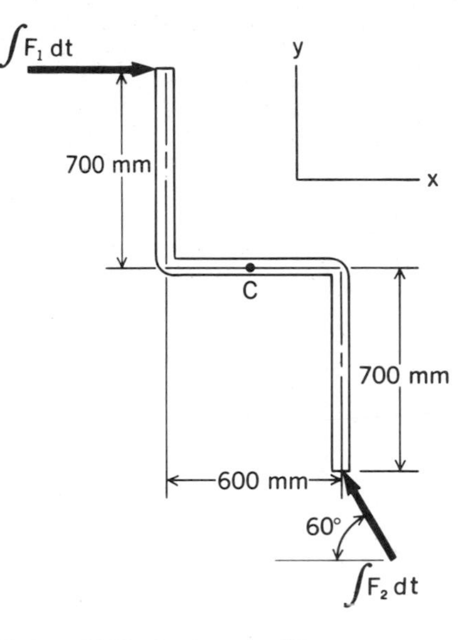

Figure 17.26. Bent rod slides on smooth horizontal surface.

What is the angular speed of the rod and the linear velocity of the mass center, directly after the impact? The rod weighs 35 N/m.

The velocity of the mass center after the impact can easily be determined using Eq. 17.25. Thus, we have

$$5\boldsymbol{i} + 3 \sin 60°\boldsymbol{j} - 3 \cos 60°\boldsymbol{i}$$

$$= (.7 + .7 + .6)\left(\frac{35}{g}\right)(V_2 - 10\boldsymbol{i} - 15\boldsymbol{j})$$

Solving for V_2:

$$V_2 = 10.49\boldsymbol{i} + 15.36\boldsymbol{j} \text{ m/sec} \tag{a}$$

For the angular velocity, we use Eq. 17.27 simplified for the case of plane motion of a slablike body. Again using the center of mass, we have for Eq. (17.27):

$$\int_1^2 \boldsymbol{M}\, dt = (I_{zz}\omega_2 - I_{zz}\omega_1)\boldsymbol{k} \tag{b}$$

Putting in numerical data and canceling $\boldsymbol{k}$, we get

$$-(5)(.70) + (3) \sin 60°(.30) - (3) \cos 60°(.70) = I_{zz}(\omega_2 - 5) \tag{c}$$

We next compute I_{zz} at C:

$$I_{zz} = \frac{1}{12}\left[\frac{35}{g}(.60)\right](.60)^2 + 2\left[\frac{1}{12}\left(\frac{35}{g}\right)(.70)(.70)^2 + \frac{35}{g}(.70)(.30^2 + .35^2)\right] \tag{d}$$

$$= 1.330 \text{ kg-m}^2$$

Going back to Eq. (c), we can now give ω_2:

$$\omega_2 = 2.16 \text{ rad/sec}$$

EXAMPLE 17.8

A solid block weighing 300 N is suspended from a wire (see Fig. 17.27) and is stationary when a horizontal impulse $\int F\,dt$ equal to 100 N-sec is applied to the body as a result of an impact. What is the velocity of corner A of the block just after impact? Does the wire remain taut?

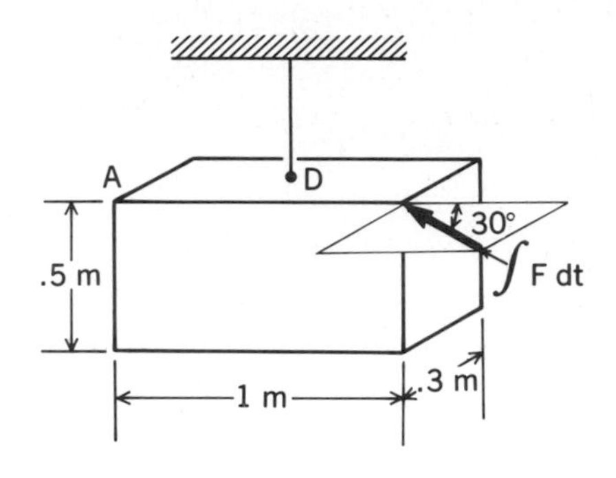

Figure 17.27. Stationary block under impact.

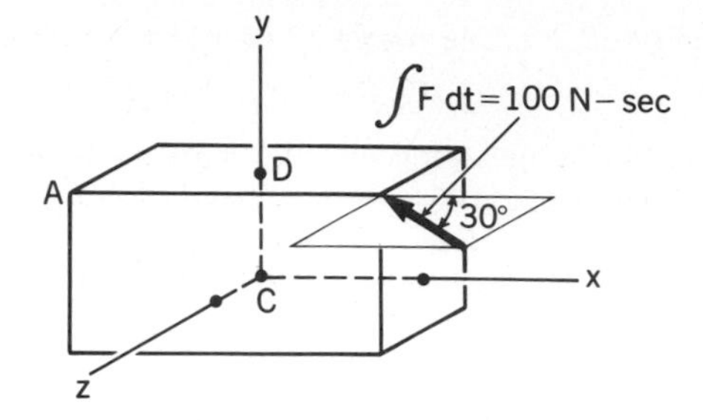

Figure 17.28. *xyz* at center of mass.

For the linear momentum, we can say for the center of mass (see Fig. 17.28) in the z, x, and y directions:

$$-100 \sin 30° = \frac{300}{g}[(V_c)_z - 0]$$

$$-100 \cos 30° = \frac{300}{g}[(V_c)_x - 0] \tag{a}$$

$$0 = \frac{300}{g}[(V_c)_y - 0]$$

From these equations, we get

$$\boldsymbol{V_C} = -2.83\boldsymbol{i} - 1.635\boldsymbol{k} \text{ m/sec} \tag{b}$$

For the angular momentum about C, we can say, on noting that xyz are principal axes:

$$-100 (\sin 30°)(.25) = I_{xx}\omega_x - 0$$

$$100 \sin 30°(.5) - 100 \cos 30°\left(\frac{.3}{2}\right) = I_{yy}\omega_y - 0 \tag{c}$$

$$100 (\cos 30°)(.25) = I_{zz}\omega_z - 0$$

Note that

$$I_{xx} = \frac{1}{12}\left(\frac{300}{g}\right)(.3^2 + .5^2) = .866 \text{ kg-m}^2$$

$$I_{yy} = \frac{1}{12}\left(\frac{300}{g}\right)(1^2 + .3^2) = 2.78 \text{ kg-m}^2 \tag{d}$$

$$I_{zz} = \frac{1}{12}\left(\frac{300}{g}\right)(1^2 + .5^2) = 3.19 \text{ kg-m}^2$$

We then get for $\boldsymbol{\omega}$ from Eq. (c):

$$\boldsymbol{\omega} = -14.43\boldsymbol{i} + 4.32\boldsymbol{j} + 6.79\boldsymbol{k} \text{ rad/sec}$$

Hence, for the velocity of point A, we have

$$\begin{aligned}\boldsymbol{V_A} &= \boldsymbol{V_C} + \boldsymbol{\omega} \times \boldsymbol{\rho}_{CA} \\ &= -2.83\boldsymbol{i} - 1.635\boldsymbol{k} + (-14.43\boldsymbol{i} + 4.32\boldsymbol{j} + 6.79\boldsymbol{k}) \times (-.5\boldsymbol{i} + .25\boldsymbol{j} + .15\boldsymbol{k}) \\ &= -3.88\boldsymbol{i} - 1.230\boldsymbol{j} - 3.08\boldsymbol{k} \text{ m/sec}\end{aligned}$$

Finally, to decide if wire remains taut, find the velocity of point D after impact.

$$\begin{aligned} V_D &= V_C + \boldsymbol{\omega} \times \boldsymbol{\rho}_{CD} \\ &= -2.83\boldsymbol{i} - 1.635\boldsymbol{k} + (-14.43\boldsymbol{i} + 4.32\boldsymbol{j} + 6.79\boldsymbol{k}) \times (.25\boldsymbol{j}) \\ &= -4.53\boldsymbol{i} - 5.24\boldsymbol{k} \text{ m/sec} \end{aligned}$$

Since there is zero velocity component in the y direction stemming from the given impulse, we can conclude that the wire remains taut.

In the following example we consider a problem involving a system of interconnected bodies.

EXAMPLE 17.9

A tractor weighs 2000 lb, including the driver (Fig. 17.29). The large drive wheels each weigh 200 lb with a radius of 2 ft and a radius of gyration of 1.8 ft. The small wheels weigh 40 lb each, with a radius of 1 ft and a radius of gyration of 10 in. The tractor is pulling a bale of cotton weighing 300 lb. The coefficient of friction between the bale and the ground is .2. What torque is needed on the drive wheels from the motor to go from 5 ft/sec to 10 ft/sec in 25 sec? Assume that the tires do not slip.

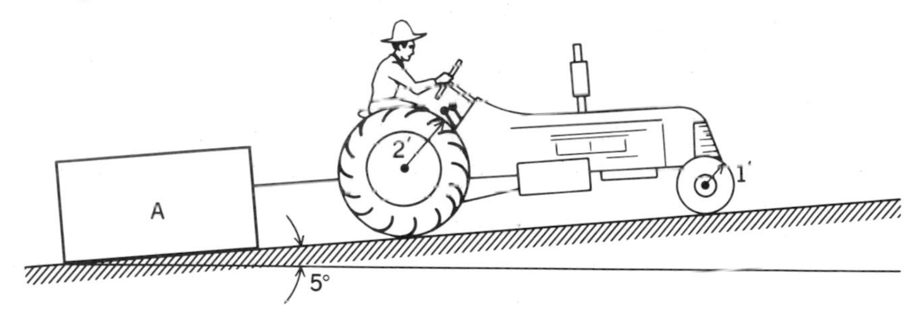

Figure 17.29. Tractor pulling bale of cotton.

We have shown a free-body diagram of the system in Fig. 17.30. Noting on inspection that $N_1 = 300 \cos 5°$, we can give the *linear momentum equation* for the system in the x direction as

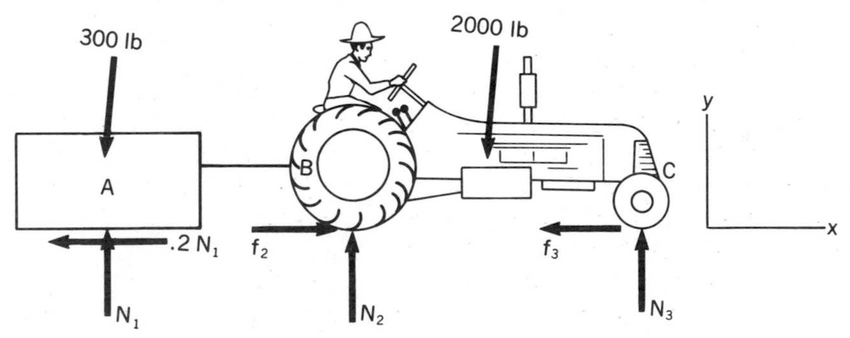

Figure 17.30. Free-body diagram of system.

$$[f_2 - f_3 - (.2)(300)\cos 5° - 300\sin 5° - 2000\sin 5°](25) = \frac{2300}{g}(10-5)$$

Therefore,

$$f_2 - f_3 = 275 \qquad \text{(a)}$$

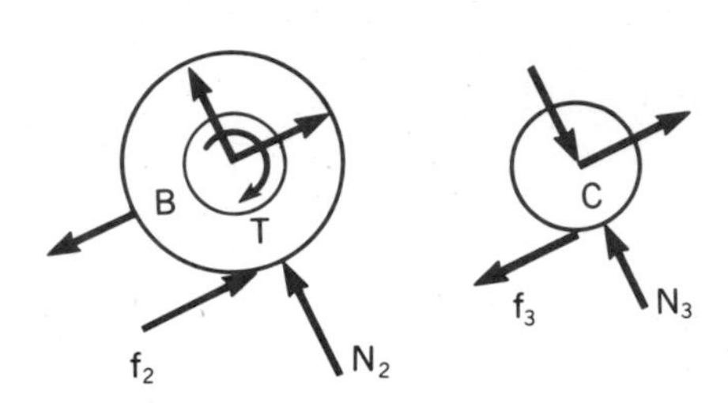

Figure 17.31. Free-body diagrams of wheels.

We next consider free-body diagrams of the wheels in Fig. 17.31. The *impulse-angular momentum* equation for the drive wheels then can be given for the center of mass as

$$[-T + f_2(2)](25) = \frac{400}{g}(1.8)^2[(\omega_B)_2 - (\omega_B)_1]$$

Noting from *kinematics* that $\omega_B = -V/2$, we have

$$-T + 2f_2 = -4.025 \qquad \text{(b)}$$

The *impulse-angular momentum* equation about the center of mass for the front wheels is then

$$-(f_3)(1)(25) = \frac{80}{g}\left(\frac{10}{12}\right)^2[(\omega_C)_2 - (\omega_C)_1]$$

Noting from *kinematics* again that $\omega_C = -V$, we get from the equation above:

$$f_3 = .345 \text{ lb} \qquad \text{(c)}$$

From Eq. (a) we may now solve for f_2. Thus,

$$f_2 = .345 + 275 = 275.3 \text{ lb}$$

Finally, from Eq. (b) we get the desired torque T:

$$T = 4.025 + (2)(275.3)$$
$$= 555 \text{ ft-lb}$$

In Example 17.9, there was no obvious convenient stationary point or stationary axis which could be considered as part of a rigid-body extension of *all* the bodies at any time. Therefore, in order to use the formulas for $\boldsymbol{H}$ given by Eq. 17.22, we considered rigid bodies *separately*. In the following example, we have a case where there is a stationary axis present which can be considered as part of (or a hypothetical rigid-body extension of) all bodies in the system at the instants of interest. And for this reason, we shall consider the angular momentum equation for the entire system using this stationary axis. Also, if the torque about such a common axis for a system of bodies is zero, then the angular momentum of the system about the aforestated axis must be *conserved*. In the example to follow, we shall also illustrate conservation of angular momentum about an axis for such a case.

EXAMPLE 17.10

A flyball-governor apparatus (Fig. 17.32) consists of four identical arms (solid rods) each of weight 10 N and two spheres of weight 18 N and radius of gyration 30 mm about a diameter. At the base and rotating with the system is a cylinder B of weight 20 N and radius of gyration along its axis of 50 mm. Initially, the system is rotating at a speed ω_1 of 500 rpm for $\theta = 45°$. A force F at the base B maintains

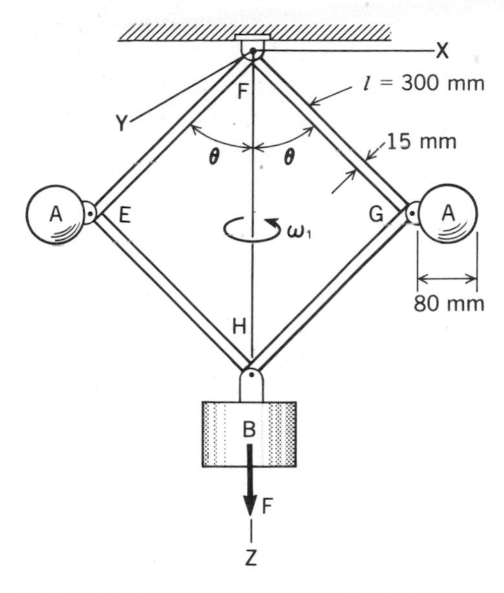

Figure 17.32. Flyball governor apparatus; Y axis is normal to page.

the configuration shown. If the force is changed so as to decrease θ from 45° to 30°, what is the angular velocity of the system?

Clearly, there is zero torque from external forces about the stationary axis FH which we take as a Z axis at all times. Hence, we have conservation of angular momentum about this axis at all times. And, since the axis is an axis of rotation for all bodies of the system,[9] we can use Eq. 17.22 for computing H about FH for all bodies in the system. As a first step we shall need I_{ZZ} for the members of the system.

Consider first member FG, which is shown in Fig. 17.33. The axes $\xi\eta\zeta$ are principal axes of inertia for the rod at F. The η axis is collinear with the Y axis, and these are normal to the page. The axes XYZ are reached by $\xi\eta\zeta$ by rotating $\xi\eta\zeta$ about the η axis an angle θ. Using the transformation equations for I_{ZZ} (see Eq. 9.13), we can say:

$$\begin{aligned}(I_{ZZ})_{FG} &= I_{\xi\xi}\left[\cos\left(\frac{\pi}{2}+\theta\right)\right]^2 + I_{\eta\eta}\left(\cos\frac{\pi}{2}\right)^2 \\ &\quad + I_{\zeta\zeta}(\cos\theta)^2 \\ &= \left[\frac{1}{3}\frac{10}{g}(.30)^2\right]\sin^2\theta + \\ &\quad \left[\frac{1}{2}\frac{10}{g}(.0075)^2\right]\cos^2\theta \end{aligned} \qquad \text{(a)}$$

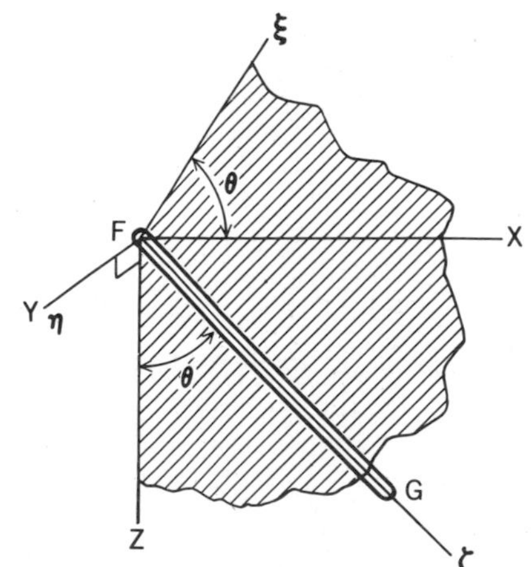

Figure 17.33. $\xi\eta\zeta$ are principal axes of FG at F.

[9]That is, at any time t the stationary axis FH either is part of a rigid body directly or is part of a hypothetical extension of a rigid body for all bodies of the system.

For the sphere we have

$$(I_{ZZ})_{\text{sph}} = \frac{18}{g}(.03)^2 + \frac{18}{g}[(.30)\sin\theta + .04]^2 \tag{b}$$

Finally, for cylinder B we have

$$(I_{ZZ})_{\text{cyl}} = \frac{20}{g}(.05)^2 \tag{c}$$

Conservation of angular momentum about the Z axis then prescribes the following:

$$[4(I_{ZZ})_{FG} + 2(I_{ZZ})_{\text{sph}} + (I_{ZZ})_{\text{cyl}}]_{\theta=45^\circ}\frac{(500)(2\pi)}{60}$$

$$= [4(I_{ZZ})_{FG} + 2(I_{ZZ})_{\text{sph}} + (I_{ZZ})_{\text{cyl}}]_{\theta=30^\circ}\,\omega_2$$

Substituting from Eqs. (a), (b), and (c), we have

$$\left(4\left[\frac{1}{3}\frac{10}{g}(.30)^2(.707)^2 + \frac{1}{2}\frac{10}{g}(.0075)^2(.707)^2\right] + 2\left\{\frac{18}{g}(.03)^2 + \frac{18}{g}\left[(.30)(.707)\right.\right.\right.$$

$$\left.\left.\left. + .04\right]^2\right\} + \frac{20}{g}(.05)^2\right)\frac{(500)(2\pi)}{60} = \left(4\left[\frac{1}{3}\frac{10}{g}(.30)^2(.5)^2 + \frac{1}{2}\frac{10}{g}(.0075)^2(.866)^2\right]\right.$$

$$\left. + 2\left\{\frac{18}{g}(.03)^2 + \frac{18}{g}[(.30)(.5) + .04]^2\right\} + \frac{20}{g}(.05)^2\right)\omega_2$$

Therefore,

$$\omega_2 = 92.4 \text{ rad/sec} = 883 \text{ rpm}$$

Before closing the section, we note that we have worked with *fixed points* or axes in inertial space and with the *mass center*. What about a point *accelerating toward the mass center*? A common example of such a point is the point of contact A of a cylinder rolling without slipping on a circular arc with the center of mass of the cylinder coinciding with the geometric center of the cylinder. We can then say:

$$\boldsymbol{M}_A = \dot{\boldsymbol{H}}_A$$

The question then arises: Can we form the familiar angular momentum equation from above about point A? In other words, is the following equation valid for plane motion?

$$\int_{t_1}^{t_2} M_A\,dt = I_A\omega_2 - I_A\omega_1 \tag{17.28}$$

The reason we might hesitate to do this is that the point of contact *continually changes* during a time interval when the cylinder is rolling. However, we have asked you to prove in Problem 17.62 that for rolling without slipping along a circular or straight path, the equation above is still valid.[10] At times it can be very useful.

Problems

17.41. A uniform cylinder C of radius 1 ft and thickness 3 in. rolls without slipping at its center plane on the stationary platform B such that the centerline of CD makes 2 revolutions per second relative to the platform. What is the angular momentum

[10]The statement is actually valid for point A when there is rolling of the cylinder on a *general* path.

vector for the cylinder about the center of mass of the cylinder? The cylinder weighs 64.4 lb.

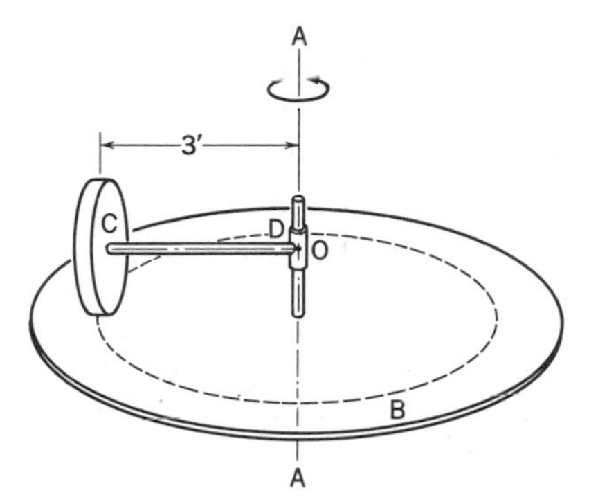

Figure P.17.41

17.42. In Problem 17.41, find the angular momentum of the disc about the stationary point O along the vertical axis A–A.

17.43. A platform rotates at an angular speed of ω_1, while a cylinder of radius r and length a mounted on the platform rotates relative to the platform at an angular speed of ω_2. When the axis of the cylinder is collinear with the stationary Y axis, what is the angular momentum vector of the cylinder about the center of mass of the cylinder? The mass of the cylinder is M.

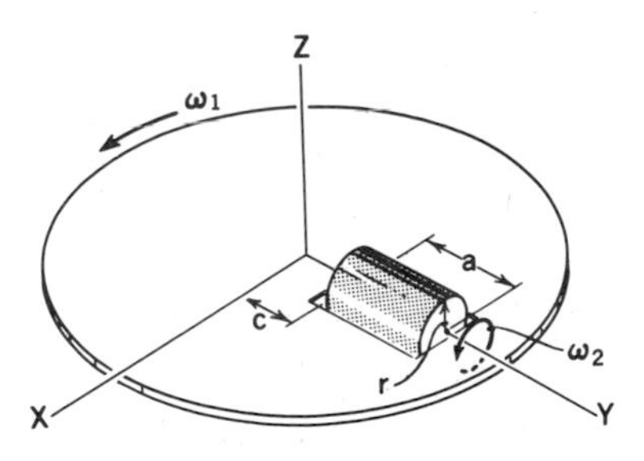

Figure P.17.43

17.44. A disc A rotates relative to an inclined shaft CD at the rate ω_2 of 3 rad/sec while shaft CD rotates about vertical axis FE at the rate ω_1 of 4 rad/sec relative to the ground. What is the angular momentum of the disc about its mass center as seen from the shaft CD? What is the angular momentum of the disc about its mass center as seen from the ground? The disc weighs 290 N.

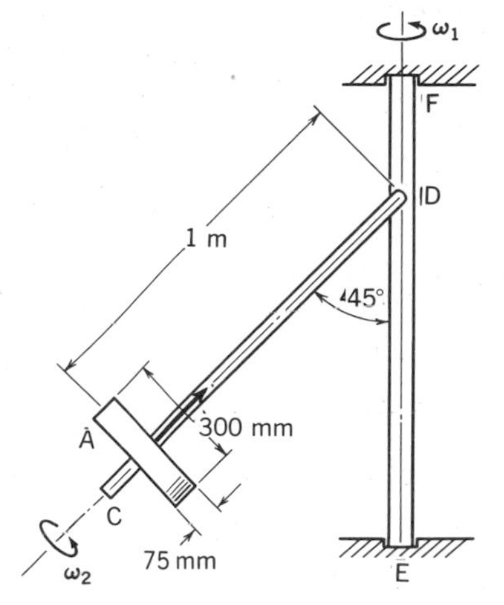

Figure P.17.44

17.45. Work Problem 16.5 by methods of momentum.

17.46. A flywheel having a mass of 1000 kg is brought to speed of 200 rpm in 360 sec by an electric motor developing a torque of 60 N-m. What is the radius of gyration of the wheel?

Figure P.17.46

17.47. Work Problem 16.7 by method of angular momentum.

17.48. Work Problem 16.9 by method of angular momentum.

17.49. A light plane is coming in for a landing at a speed of 100 km/hr. The wheels have zero rotation just before touching the runway. If $\frac{1}{10}$ the weight of the plane is maintained by the upward force of the runway for the first second, what is the approximate length of the skid mark left by the wheel on the runway? The wheels weigh 100 N and have a radius of gyration of 180 mm and a diameter of 450 mm. The plane weighs with load 8000 N. The coefficient of friction between the tire and runway is .3.

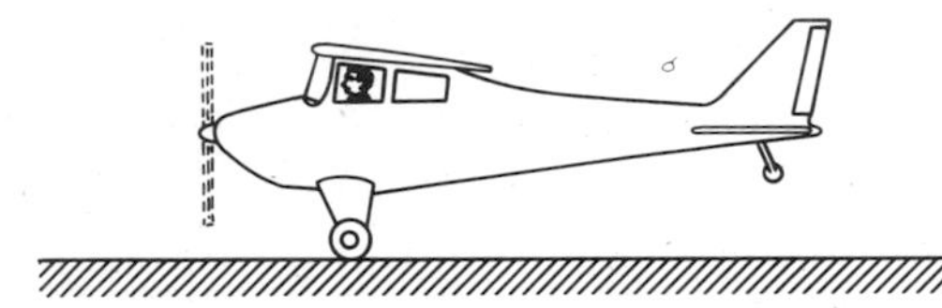

Figure P.17.49

17.50. Work Problem 17.49 for the case where the upward force from the ground on the plane during the first second after touchdown is

$$N = 8000t^2 \text{ N}$$

where t is in seconds after touchdown.

17.51. A circular conveyor carries cylinders a from position A through a heat treatment furnace. The cylinders are dropped onto the conveyor at A from a stationary position above and picked up at B. The conveyor is to turn at an average speed ω of 2 rpm. The cylinders are dropped onto the conveyor at the rate of 9 per minute. If the resisting torque due to friction is 1 N-m, what average torque T is needed to maintain the prescribed angular motion? Each cylinder weighs 300 N and has a radius of gyration of 150 mm about its axis.

Figure P.17.51

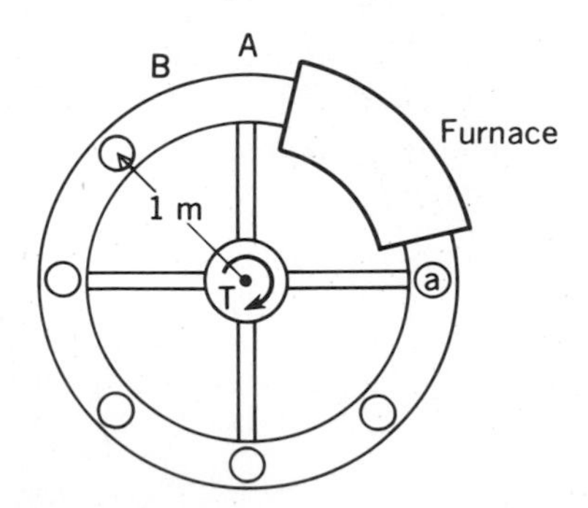

17.52. A circular *towing tank* has a main arm A which has a mass of 1000 kg and a radius of gyration of 1 m. On the arm rides the model support B, having a mass of 200 kg and having a radius of gyration about the vertical axis at its mass center of 600 mm. If a torque T of 50 N-m is developed on A when B is at position $r = 1.8$ m, what will be the angular speed 5 sec later if B moves out at a constant radial speed of .1 m/sec. The initial angular speed of the arm A is 2 rpm. Neglect the drag of the model.

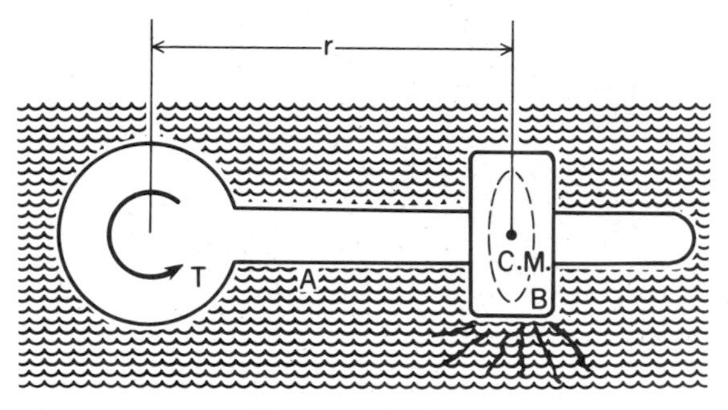

Figure P.17.52

17.53. A steam roller with driver weighs 5 tons. Wheel A weighs 2 tons and has a radius of gyration of .8 ft. Drive wheels B have a total weight of 1 ton and a radius of gyration of 1.5 ft. If a total torque of 400 ft-lb is developed by the engine on the drive wheels, what is the speed of the steam roller after 10 sec starting from rest? There is no slipping.

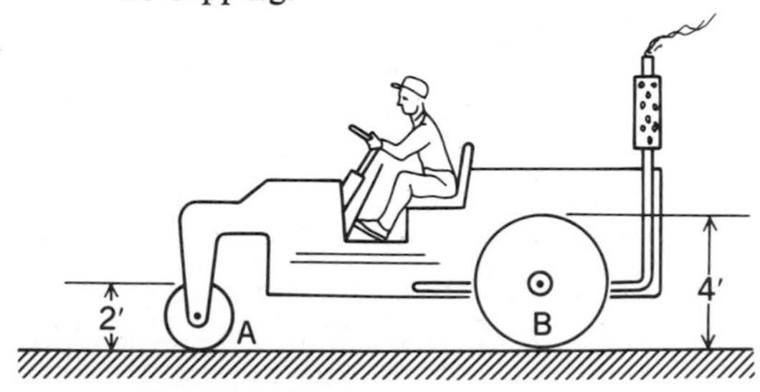

Figure P.17.53

17.54. An electric motor D drives gears C, B, and device A. The diameters of gears C and B are 6 in. and 16 in., respectively. The mass of A is 200 lbm. The combined mass of the motor armature and gear C is 50 lbm, while the radius of gyration of this combination is 8 in. Also, the mass of B is 20 lbm. If a constant counterclockwise torque of 60 lb-ft is developed on the armature of the motor, what is the speed of A in 2 sec after starting from rest? Neglect the inertia of the small wheels under A.

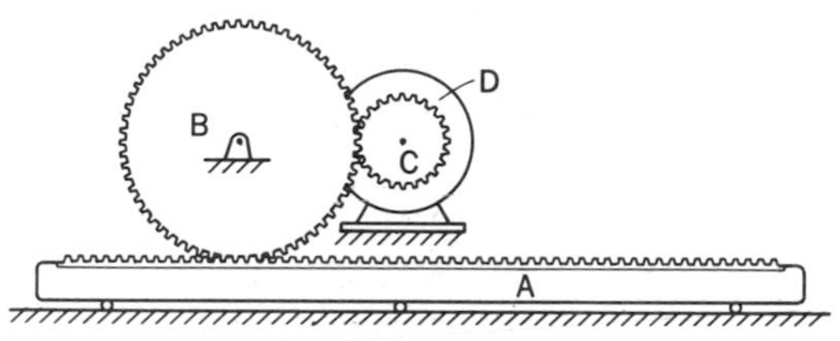

Figure P.17.54

17.55. A conveyor is moving a weight W of 64.4 lb. Cylinders A and B have a diameter of 1 ft and weigh 32.2 lb each. Also, they each have a radius of gyration of .8 ft. Rollers C, D, E, F and G each have a diameter of 3 in., weigh 10 lb each, and have a radius of gyration of 2 in. What constant torque T will increase the speed of W from 3 ft/sec to 5 ft/sec in 3 seconds? The belt weighs 50 lb.

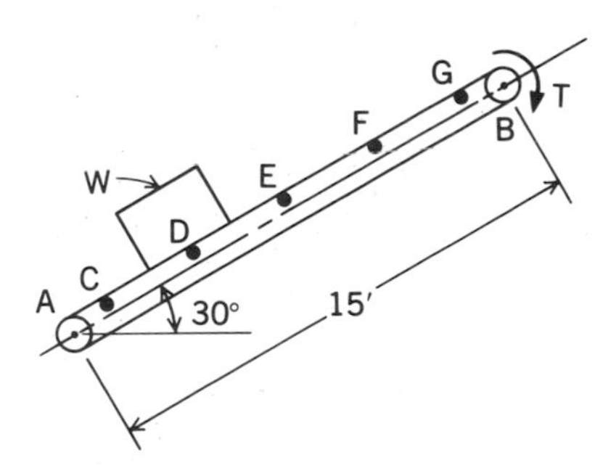

Figure P.17.55

17.56. A rectangular block A is rotating freely at a speed ω of 200 rpm about a light hollow shaft. Attached to A is a circular rod B which can rotate out from the block A about a hinge at C. This rod weighs 20 N. When the system is rotating at the speed ω of 200 rpm, the rod B is vertical as shown. If the catch at the upper end of B releases so that B falls to a horizontal orientation (shown as dashed in the diagram), what is the new angular velocity? The block A weighs 60 N. Neglect the inertia of the shaft.

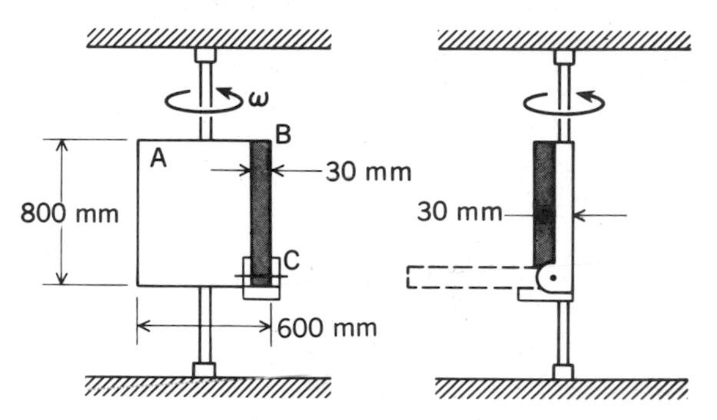

Figure P.17.56

17.57. A space laboratory is in orbit and has an angular velocity ω of .5 rad/sec relative to inertial space so as to have a "gravitational" force for the living quarters on the outside ring. The space lab has a mass of 4500 kg and a radius of gyration about its axis of 20 m. Two space ships C and D, each of mass 1000 kg, are shown docked so as also to get the benefit of "gravity." The centers of mass of each vehicle is .7 m from the wall of the space lab. The radius of gyration of each vehicle about an axis at the mass center parallel to the axis of the space lab is .5 m. Small rocket engines A and B are turned on to develop a thrust each of 250 N. How long should they be on to increase the angular speed so as to have 1g of gravity at the outer radius of the space lab?

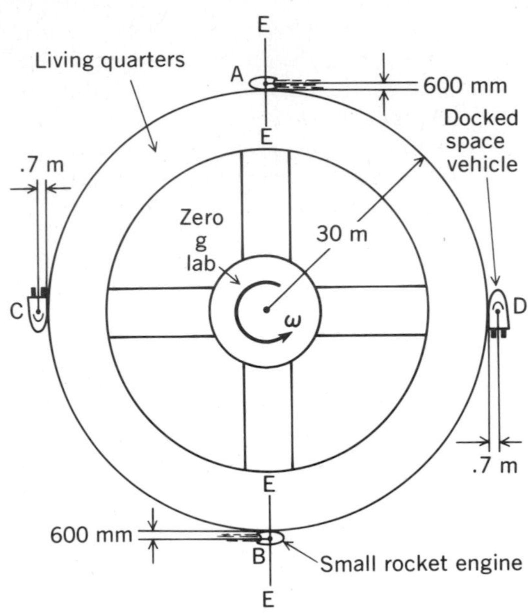

Figure P.17.57

17.58. In Problem 17.57 the space lab and docked vehicles C and D are rotating at a speed ω for $1g$ gravity at the outer periphery of the space lab. The space vehicles are detached simultaneously. In this detachment process the space lab produces an impulsive torque on each vehicle so that each vehicle ceases to rotate relative to inertial space. For how long and in what directions should the small rockets A and B of the space lab be fired to bring the angular speed ω of the space lab back to normal? The rockets can be rotated about axes E–E and, for this maneuver, they are developing a thrust of 100 N each.

17.59. A turbine is rotating freely with a speed ω of 6000 rpm. A blade breaks off at its base at the position shown. What is the velocity of the center of mass of the blade just after the fracture? Does the blade have an angular velocity just after fracture assuming that no impulsive torques or forces occur at fracture? Explain.

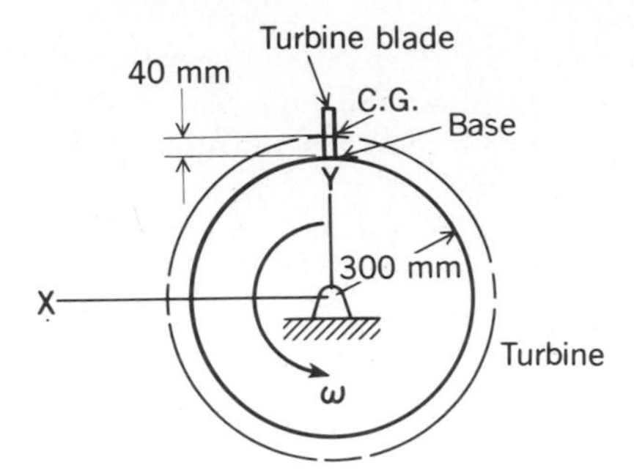

Figure P.17.59

17.60. Two rods are welded to a drum which has an angular velocity of 2000 rpm. The rod A breaks at the base at the position shown. If we neglect wind friction, how high up does the center of mass of the rod go? What is the angular orientation of the rod at the instant that the center of mass reaches its apex? Assume that there are no impulsive torques or forces at fracture. The Y axis is vertical. Neglect friction.

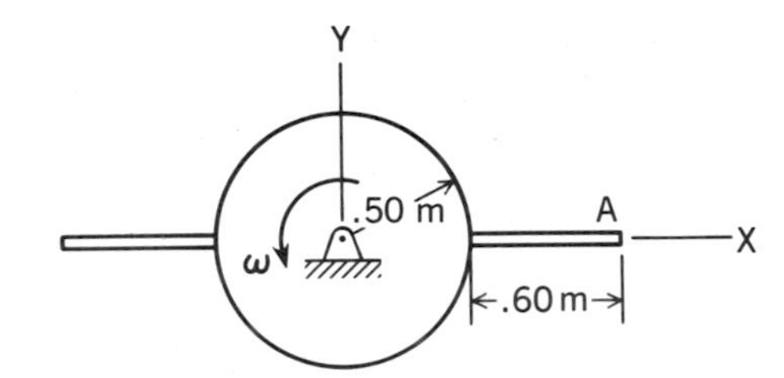

Figure P.17.60

17.61. A stepped cylinder is released on a 45° incline where the dynamic coefficient of friction at the contact is .2 and the static coefficient of friction is .22. What is the angular speed of the stepped cylinder after 4 sec? The stepped cylinder has a weight of 500 N and a radius of gyration about its axis of 250 mm. Be sure to check to see if the cylinder moves at all!

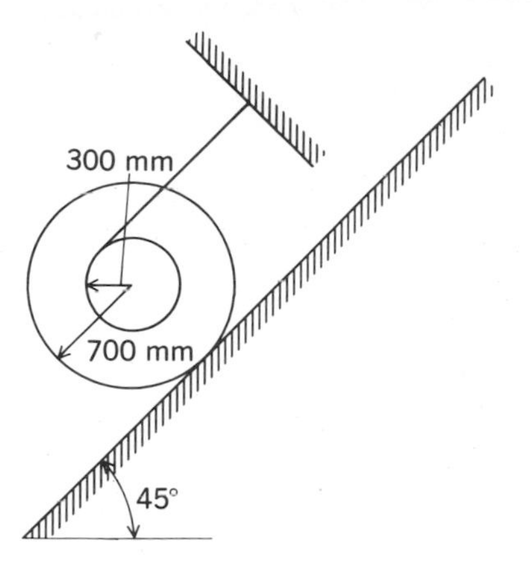

Figure P.17.61

17.62. Prove that you can apply the angular momentum equation about the contact point A on a cylinder rolling without slipping on a circular (and hence including a straight) path. A force P always normal to OA acts on the cylinder as does a couple moment T and weight W. Specifically, prove that

$$\int_{t_1}^{t_2} (2Pr + W \sin\theta\, r + T)\, dt = M(k^2 + r^2)(\omega_2 - \omega_1) \quad \text{(a)}$$

where ω is the total angular speed of the cylinder. [*Hint:* Express *angular momentum* equation about C and then, from *Newton's law* using the cylindrical component in the transverse direction, show on integrating that

$$\int_{t_1}^{t_2} (f + P + W \sin\theta)\, dt = -M(R - r)(\dot{\theta}_2 - \dot{\theta}_1)$$

where f is the friction force at the point of contact. Now let xy rotate with line OC. From *kinematics* first show that $R\theta = -r\phi$, where ϕ is the rotation of the cylinder relative to xy. Then, show that $\omega = -[(R - r)/r]\dot{\theta}$. From these three considerations, you should readily be able to derive Eq. (a).]

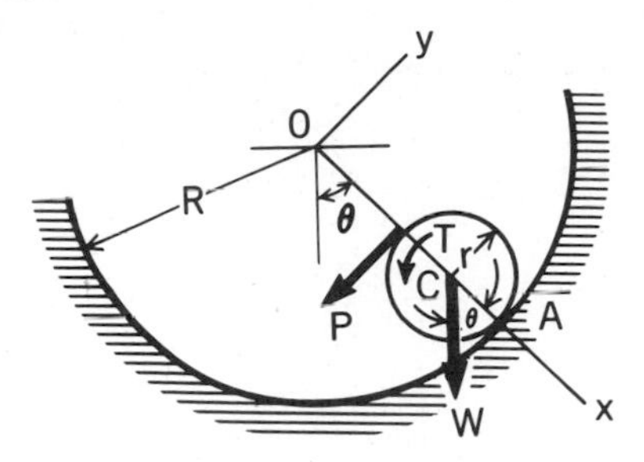

Figure P.17.62

17.63. A slab A weighing 2 kN rides on two rollers each having a mass of 50 kg and a radius of gyration of 200 mm. If the system starts from rest, what minimum constant force T is needed to prevent the slab from exceeding the speed of 3 m/sec down the incline in 4 sec? There is no slipping.

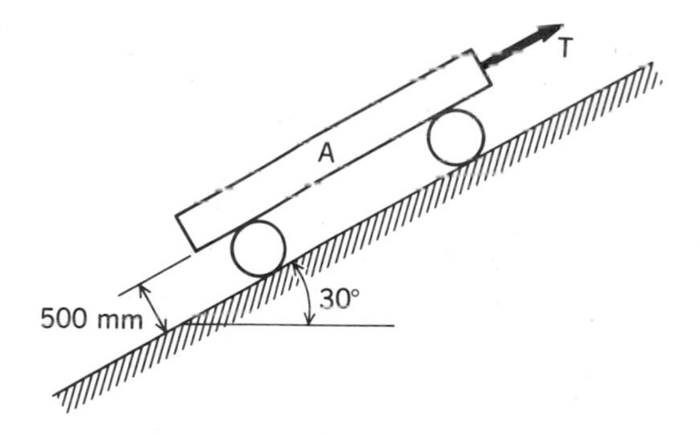

Figure P.17.63

17.64. A main gear A rotates about a fixed axis and meshes with four identical floating gears B. The floating gears, in turn, mesh with the stationary gear F. If a torque T of 200 N-m is applied to the main gear A, what is its angular speed in 5 sec? The following data apply:

$$M_A = 100 \text{ kg}, \qquad k_A = 250 \text{ mm}$$
$$M_B = 20 \text{ kg}, \qquad k_B = 40 \text{ mm}$$

The system is horizontal.

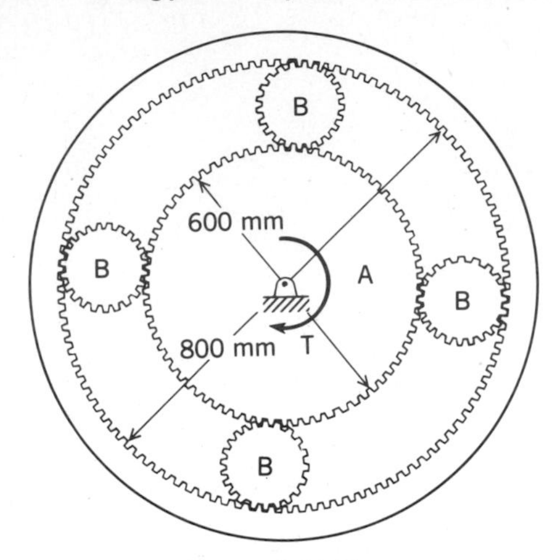

Figure P.17.64

17.65. Two spheres, each of weight 2 N and diameter 30 mm, slide in smooth troughs inclined at an angle of 30°. A torque T of 2 N-m is applied as shown for 3 sec and is then zero. How far d up the inclines do the spheres move? The support system exclusive of the spheres has a weight of 10 N and a radius of gyration for the axis of rotation of 100 mm. Neglect friction and wind-resistance losses. Treat the spheres as particles.

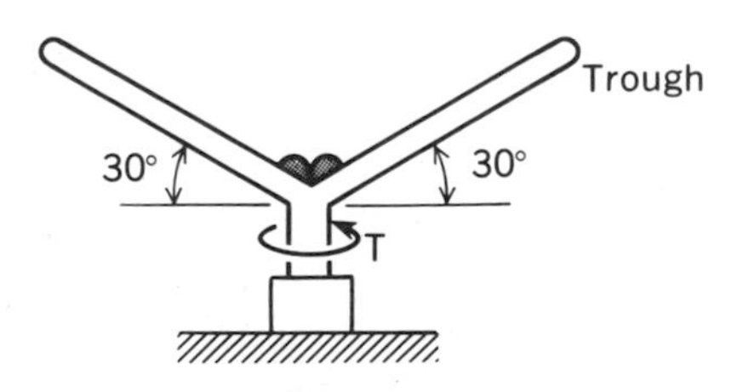

Figure P.17.65

17.66. Two small spheres, each of weight 3 N, slide on a smooth rod curved in the shape of a parabola as shown. The curved rod is mounted on a platform on which a torque T of 1.5 N-m is applied for 2 sec. What is the angular speed of the system after 2 sec? Neglect friction. The curved rod and the platform have a mass of 5 kg and a radius of gyration of 200 mm.

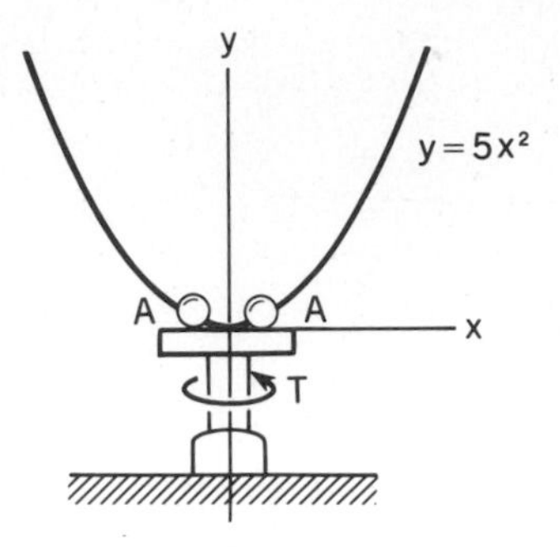

Figure P.17.66

***17.67.** A disc of mass 3 kg is suspended between two wires. A bullet is fired at the disc and lodges at point A, for which $r = 100$ mm and $\theta = 45°$. If the bullet is traveling at a speed of 600 m/sec before striking the disc, what is the angular velocity of the system directly after the bullet gets lodged? The bullet weighs 1 N. (*Caution:* What point in the disc should you work with? It is not 0!)

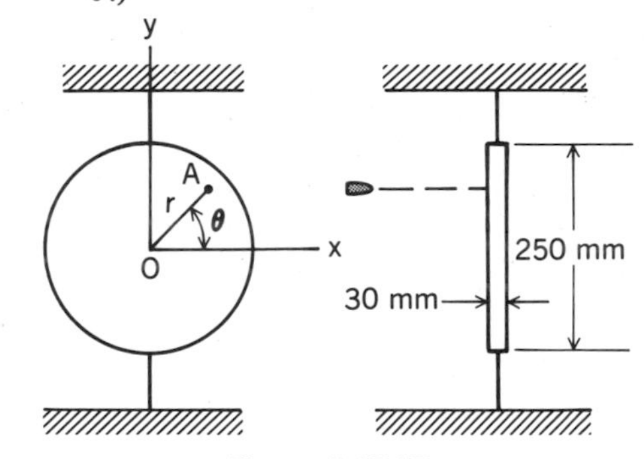

Figure P.17.67

***17.68.** A device with thin walls contains water. The device is supported on a platform on which a torque T of .5 N-m is applied for 2 sec and is zero thereafter. What is the angular velocity ω of the system at the end of 2 sec assuming that, as a result of low viscosity, the water has no rotation relative to the ground in the vertical tubes. The system of tubes and platform have a weight of 50 N and a radius of gyration of 200 mm. (*Hint:* The pressure in the liquid is equal at all times to the specific weight

γ times the distance d below the free surface of the liquid.) Note that for water $\gamma = 9810\ \text{N/m}^3$.

***17.69.** In Problem 17.68 what is the final angular speed of the system when viscosity has had its full effect and there is no movement of the water relative to the container walls? Disregard other frictional effects.

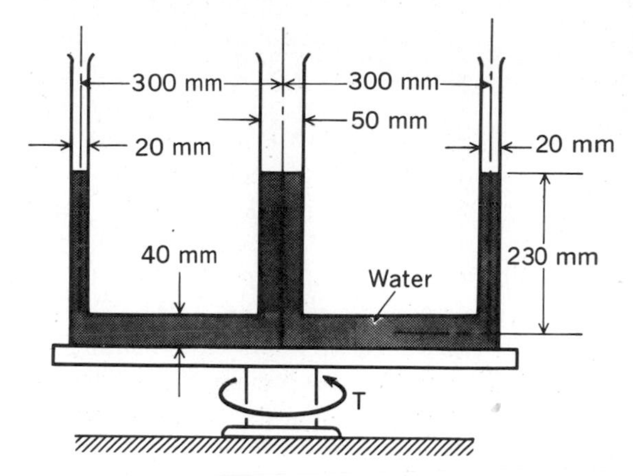

Figure P.17.68

*17.6 Impulsive Forces and Torques: Eccentric Impact

In Chapter 14, we introduced the concept of an *impulsive force*. Recall that an impulsive force F acts over a very short time interval Δt but has a very high value during this interval such that the impulse $\int_0^{\Delta t} F\,dt$ is significant. The impulse of other ordinary forces (not having very high peaks during Δt) is usually neglected for the short interval Δt. The same concept applies to torques, so that we have *impulsive torques*. The impulsive force and impulsive torque concepts are most valuable for the consideration of impact of bodies. Here, the collision forces and torques are impulsive while the other forces, such as gravity forces, have negligible impulse during collision.

In Chapter 14, we considered the case of *central impact* between bodies. Recall that for such problems the mass centers of the colliding bodies lie along the line of impact.[11] At this time, we shall consider the *eccentric impact* of *slab-like* bodies undergoing plane motion such as shown in Fig. 17.34. For eccentric impact, at least one of the mass centers does not lie along line of impact. The bodies in Fig. 17.34 have just begun contact whereby point A of one body has just touched point B of the other body. The velocity of point A just before contact (preimpact) is given as $(\boldsymbol{V}_A)_i$, while the velocity just before contact for point B is $(\boldsymbol{V}_B)_i$. (The i stands for "initial," as in earlier work.) We shall consider only *smooth bodies*, so that the impul-

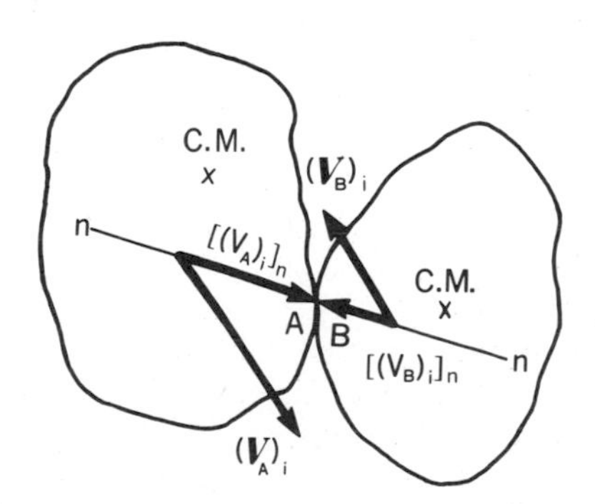

Figure 17.34. Eccentric impact between two bodies.

[11]The line of impact is normal to the plane of contact between the bodies.

sive forces acting on the body at the point of contact are *collinear* with the line of impact. As a result of the impulsive forces, there is a *period of deformation*, as in our earlier studies, and a *period of restitution*. In the period of deformation, the bodies are deforming, while in the period of restitution there is a complete or partial recovery of the original geometries. At the end of the period of deformation, the points A and B have the *same velocity* and we denote this velocity as $\mathbf{V}_D$. Directly after impact (post-impact), the velocities of points A and B are denoted as $(\mathbf{V}_A)_f$ and $(\mathbf{V}_B)_f$, respectively, where the subscript f is used to connote the final velocity resulting solely from the impact process. We shall be able to use the linear impulse-momentum equation and the angular impulse-momentum equation to relate the velocities, both linear and angular, for preimpact and postimpact states. These equations do not take into account the nature of the material of the colliding bodies, and so additional information is needed for solving these problems. For this reason, we use the ratio between the impulse on each body during the period of restitution, $\int R\,dt$, and the impulse on each body during the period of deformation, $\int D\,dt$. As in central impact, the ratio is a number ϵ, called the *coefficient of restitution*, which depends primarily on the materials of the bodies in collision. Thus,

$$\epsilon = \frac{\int R\,dt}{\int D\,dt} \tag{17.29}$$

We shall now show that the components along the line of impact n–n of $\mathbf{V}_A$ and $\mathbf{V}_B$, taken at pre- and postimpact, are related to ϵ by the very same relation that we had for central impact. That is,

$$\boxed{\epsilon = -\frac{[(V_B)_f]_n - [(V_A)_f]_n}{[(V_B)_i]_n - [(V_A)_i]_n}} \tag{17.30}$$

Recall that the numerator represents the relative velocity of separation along n of the points of contact, whereas the denominator represents the relative velocity of approach along n of these points.

We shall first consider the case where the bodies are in *no way constrained* in their plane of motion; we can then neglect all impulses except that coming from the impact. We now consider the body having contact point A. In Fig. 17.35 we have shown this body with impulse $\int D\,dt$ acting. Using the component of the linear momentum equation along line of impact n–n, we can say for the center of mass:

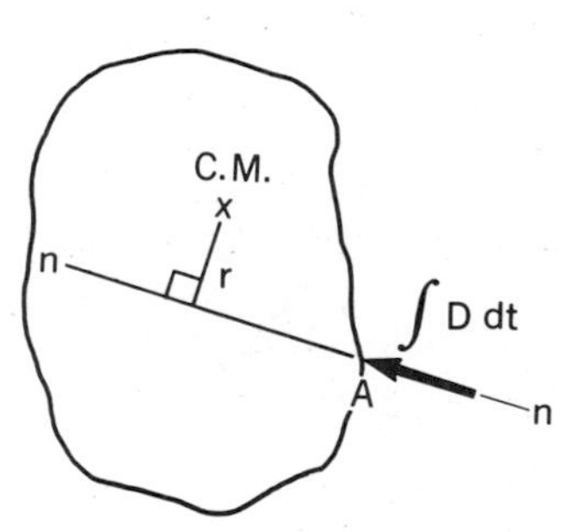

Figure 17.35. Impulse acting on one of the bodies.

$$\int D\,dt = M[(V_C)_D]_n - M[(V_C)_i]_n \tag{17.31}$$

where $(V_C)_i$ refers to the preimpact velocity of the

center of mass and where $(V_C)_D$ is the velocity of the center of mass at the end of the deformation period. Similarly, for the period of restitution, we have

$$\int R\,dt = M[(V_C)_f]_n - M[(V_C)_D]_n \tag{17.32}$$

For angular momentum, we can say for the deformation period using r as the distance from the center of mass to n–n:

$$r\int D\,dt = I\omega_D - I\omega_i \tag{17.33}$$

Similarly, for the period of restitution:

$$r\int R\,dt = I\omega_f - I\omega_D \tag{17.34}$$

Now, substitute the right sides of Eqs. 17.31 and 17.32 into Eq. 17.29. We get on cancellation of M:

$$\epsilon = \frac{[(V_C)_f]_n - [(V_C)_D]_n}{[(V_C)_D]_n - [(V_C)_i]_n} = \frac{[(V_C)_D]_n - [(V_C)_f]_n}{[(V_C)_i]_n - [(V_C)_D]_n} \tag{17.35}$$

Next, substitute for the impulses in Eq. 17.29 using Eqs. 17.33 and 17.34. We get on canceling only I:

$$\epsilon = \frac{r\omega_f - r\omega_D}{r\omega_D - r\omega_i} = \frac{r\omega_D - r\omega_f}{r\omega_i - r\omega_D} \tag{17.36}$$

Adding the numerators and denominators of Eqs. 17.35 and 17.36, we can then say on rearranging the terms:

$$\epsilon = \frac{\{[(V_C)_D]_n + r\omega_D\} - \{[(V_C)_f]_n + r\omega_f\}}{\{[(V_C)_i]_n + r\omega_i\} - \{[(V_C)_D]_n + r\omega_D\}} \tag{17.37}$$

We pause now to consider the *kinematics* of the motion. We can relate the velocities of points A and C on the body (see Fig. 17.36) as follows:

$$\mathbf{V}_A = \mathbf{V}_C + \boldsymbol{\omega} \times \boldsymbol{\rho}_{CA} \tag{17.38}$$

Note that the magnitude of $\boldsymbol{\rho}_{CA}$ is R as shown in the diagram. Since $\boldsymbol{\omega}$ is normal to the plane of symmetry of the body and thus to $\boldsymbol{\rho}_{CA}$, the value of the last term in Eq. 17.38 is $R\omega$ with a direction normal to R as has been shown in Fig. 17.36. The components of the vectors in Eq. 17.38 in direction n can then be given as follows:

$$(V_A)_n = (V_C)_n + \omega R\cos\theta \tag{17.39}$$

Since $R\cos\theta = r$ (see Fig. 17.36), we conclude that

$$(V_A)_n = (V_C)_n + r\omega \tag{17.40}$$

With the preceding result applied to the initial condition (i), the final condition (f), and the intermediate condition (D), we can now go back to Eq. 17.37 and replace the expressions inside the braces ({ }) by the left side of Eq. 17.40 as follows:

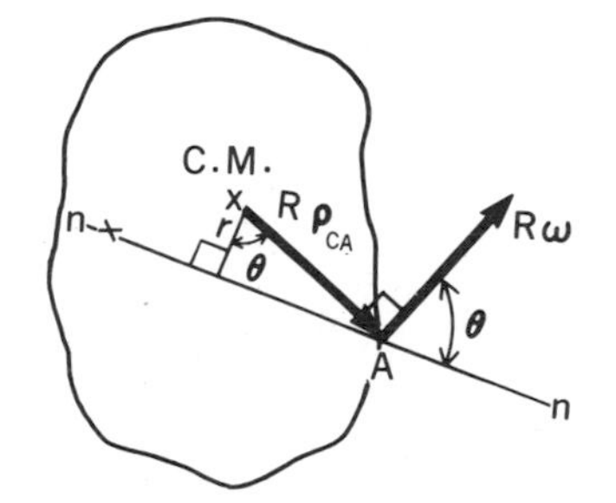

Figure 17.36. Slab with $|\boldsymbol{\rho}_{CA}|$ as R.

$$\epsilon = \frac{[(V_A)_D]_n - [(V_A)_f]_n}{[(V_A)_i]_n - [(V_A)_D]_n} \tag{17.41}$$

A similar process for the body having contact point B will yield the preceding equation with subscript B replacing subscript A:

$$\epsilon = \frac{[(V_B)_D]_n - [(V_B)_f]_n}{[(V_B)_i]_n - [(V_B)_D]_n} = \frac{[(V_B)_f]_n - [(V_B)_D]_n}{[(V_B)_D]_n - [(V_B)_i]_n} \tag{17.42}$$

Now add the numerators and denominators of the right side of Eq. 17.41 and the extreme right side of Eq. 17.42. Noting that

$$[(V_A)_D]_n = [(V_B)_D]_n \tag{17.43}$$

we get Eq. 17.30, thus demonstrating the validity of that equation.

Let us next consider the case where one or both bodies undergoing impact is constrained to *rotate about a fixed axis*. We have shown such a body in Fig. 17.37 where O is the axis of rotation and point A is the contact point. If an impulse is developed at the point of contact A (we have shown the impulse during the period of deformation), then clearly there will be an impulsive force at O, as shown in the diagram. We shall employ the angular impulse-momentum equation about the fixed point O. Thus, we have for the period of deformation and the period of restitution:

$$r\int D\,dt = I_0\omega_D - I_0\omega_i$$

$$r\int R\,dt = I_0\omega_f - I_0\omega_D$$

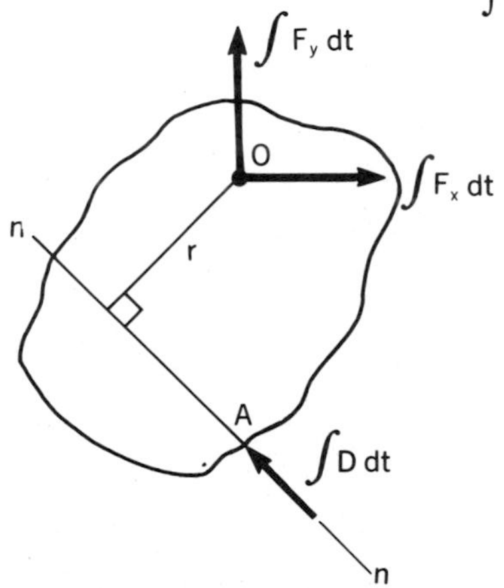

Figure 17.37. Impact for a body under constraint at O.

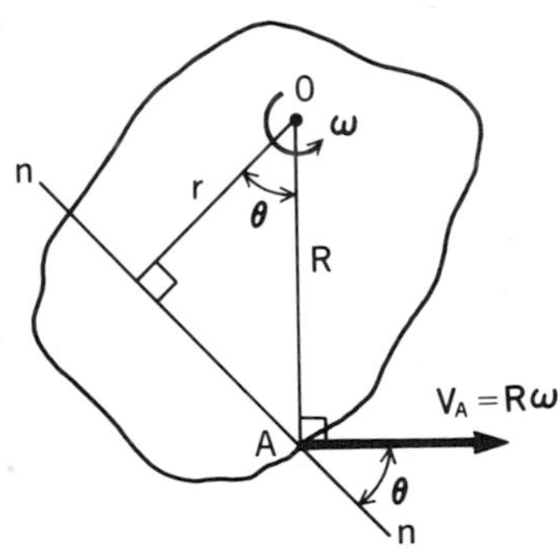

Figure 17.38. Velocity of point A is $R\omega$.

Now solve for the impulses in the equation above, and substitute into Eq. 17.29. Canceling I_0, the moment of inertia about the axis of rotation at O, we get

$$\epsilon = \frac{r\omega_f - r\omega_D}{r\omega_D - r\omega_i} = \frac{r\omega_D - r\omega_f}{r\omega_i - r\omega_D} \tag{17.44}$$

In Fig. 17.38 we see that

$$V_A = R\omega$$

Therefore,

$$(V_A)_n = R\omega \cos\theta = r\omega$$

Using the above result in Eq. 17.44, we get

$$\epsilon = \frac{[(V_A)_D]_n - [(V_A)_f]_n}{[(V_A)_i]_n - [(V_A)_D]_n} \tag{17.45}$$

But this expression is identical to Eq. 17.41. And by considering the second body, which is either free or constrained to rotate, we get an equation corresponding to Eq. 17.42. We can then conclude that Eq. 17.30 is valid for impact where one or both bodies are constrained to rotate about a fixed axis.

In a typical impact problem, the motion of the bodies preimpact is given and the motion of the bodies postimpact is desired. Thus, there could be four unknowns—two velocities of the mass centers of the bodies plus two angular velocities. The required equations for solving the problem are formed from linear and angular momentum considerations of the bodies taken separately or taken as a system. Only impulsive forces are taken into consideration during the time interval spanning the impact. If the bodies are considered separately, we simply use the formulations of Section 17.5, remembering to observe Newton's third law at the point of impact between the two bodies. Furthermore, we must use the coefficient of restitution equation 17.30. Generally, kinematic considerations are also needed to solve the problem. When there are no other impulsive forces other than those occurring at the point of impact, it might be profitable to consider the bodies as one system. Then, clearly, as a result of Newton's third law, we must have *conservation of linear momentum* relative to an inertial reference, and also we must have *conservation of angular momentum* about *any one axis* fixed in inertial space.

We now illustrate these remarks in the following series of examples.

EXAMPLE 17.11

A rectangular plate A weighing 20 N has two identical rods weighing 10 N each attached to it (see Fig. 17.39). The plate moves on a plane smooth surface at a

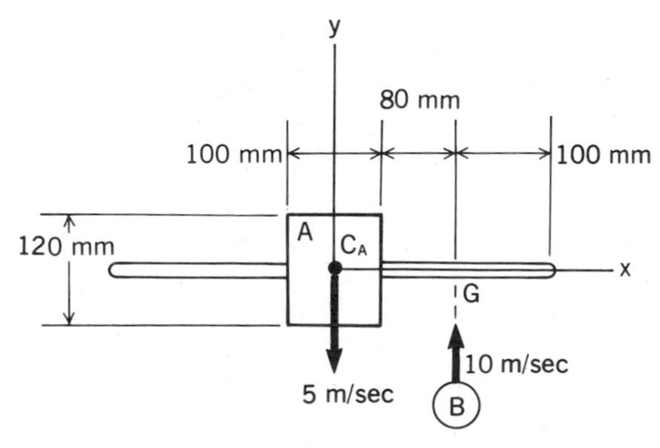

Figure 17.39. Colliding bodies.

speed of 5 m/sec. Moving oppositely at 10 m/sec is disc B, weighing 10 N. A perfectly elastic collision ($\epsilon = 1$) takes place at G. What is the speed of the center of mass of the plate just after collision (postimpact)? Solve the problem two ways: consider the bodies separately and consider the bodies as a system.

Solution 1. We can consider the disc to be a particle since this body will only translate. Let point G be the point of contact on the rod. We then have, for Eq. 17.30:

$$\epsilon = 1 = -\frac{(V_G)_f - (V_B)_f}{-5 - 10}$$

Therefore,

$$(V_G)_f - (V_B)_f = 15 \tag{a}$$

Now, consider linear and angular momentum for each of the bodies. For this purpose we have shown the bodies in Fig. 17.40 with only impulsive forces acting. We might call such a diagram an "impulsive free-body diagram." We then see that C_A must move in the plus or minus y direction after impact. We can then say for body A, using *linear impulse-momentum* and *angular impulse-momentum* equations (the latter about the center of mass):

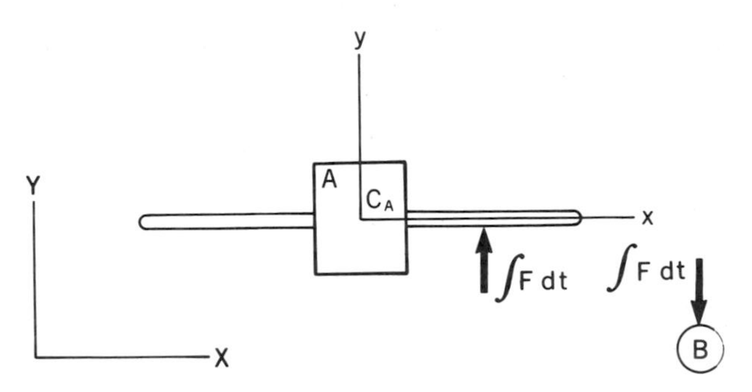

Figure 17.40. Impulsive free-body diagrams.

$$\int F\,dt = \frac{40}{g}[(V_{C_A})_f - (-5)] \tag{b}$$

$$(.13)\int F\,dt = \left[\frac{1}{12}\left(\frac{20}{g}\right)(.10^2 + .12)^2 + 2\left(\frac{1}{12}\right)\left(\frac{10}{g}\right)(.18)^2 + 2\left(\frac{10}{g}(.14)^2\right](\omega_A - 0) = .0496\omega_A \tag{c}$$

For body B, we have for the *linear impulse-momentum* equation:

$$-\int F\,dt = \frac{10}{g}[(V_B)_f - 10] \tag{d}$$

From *kinematics* we have for the plate with arms:

$$(V_G)_f = (V_{C_A})_f + (\omega_A)(.13) \tag{e}$$

We have now a complete set of equations which we can proceed to solve. We get:

$$(V_{C_A})_f = -.305 \text{ m/sec}$$

Solution 2. Equation (a) above from the coefficient of restitution and Eq. (e) above from kinematics also apply to solution 2. Substituting for $(V_G)_f$ in Eq. (a) of solution 1 using Eq. (e) of solution 1, we get for solution 2:

$$(V_{C_A})_f + .13\omega_A - (V_B)_f = 15 \qquad \text{(a)}$$

Conservation of linear momentum for the system leads to the requirement in the y direction that

$$\frac{40}{g}(-5) + \frac{10}{g}(10) = \frac{40}{g}(V_{C_A})_f + \frac{10}{g}(V_B)_f$$

Therefore,

$$4(V_{C_A})_f + (V_B)_f = -10 \qquad \text{(b)}$$

Also, *angular momentum is conserved* about any fixed z axis. We choose the axis at the position corresponding to C_A at the time of impact. Noting that I for the plate and arms is .0496 kg-m² from solution 1 and noting that B can be considered as a particle, we have

$$\underbrace{\frac{10}{g}(10)(.13)}_{\substack{\text{angular}\\ \text{momentum}\\ \text{preimpact}}} = \underbrace{(.0496)\omega_A + \frac{10}{g}(V_B)_f(.13)}_{\substack{\text{angular}\\ \text{momentum}\\ \text{postimpact}}}$$

Therefore,

$$\omega_A = -2.67(V_B)_f + 26.7 \qquad \text{(c)}$$

Solving Eqs. (a), (b), and (c) simultaneously we get:

$$(V_{C_A})_f = -.305 \text{ m/sec}$$

We leave it to you to demonstrate that there has been *conservation of mechanical energy* during this perfectly elastic impact. We could have used this fact in place of Eq. (a) in solution 1.

Note that there was some saving of time and labor in using the system approach throughout for the preceding problem wherein a rigid body, namely the plate and its arms, collided with a body, the disc, which could be considered as a particle. In problems involving two colliding rigid bodies neither of which can be considered a particle, we must consider the bodies separately since the system approach does not yield a sufficient number of independent equations as you can yourself demonstrate.

In the preceding example, the bodies were not constrained except to move in a plane. If one or both colliding bodies is pinned, the procedure for solving the problem may be a little different than what was shown in Example 17.10. Note that there will be unknown supporting impulsive forces at the pin of any pinned body. If we are not interested in the supporting impulsive force for a pinned body, we only consider *angular momentum* about the pin for that pinned body; in this way the undesired unknown supporting impulsive forces at the pin do not enter the calculations. Other than this one factor, the calculations are the same as in the previous example.

You will recall from momentum considerations of particles that we considered the collision of a comparatively small body with a very massive one. We could not use the conservation of momentum equation for the collision between such bodies since the velocity change of the massive body went to zero as the mass (mathematically speaking) went to infinity, thus producing an indeterminacy in our idealized formulations. We shall next illustrate the procedure for the collision of a very massive body with a much smaller one. You will note that we cannot consider a system approach for linear or angular momentum conservation for the same reasons set forth in Chapter 14.

EXAMPLE 17.12

A 20-lb rod AB is dropped onto a massive body (Fig. 17.41). What is the angular velocity of the rod postimpact for the following conditions:

A. Smooth floor; elastic impact.
B. Rough floor (no slipping); plastic impact.

In either case, the velocity of end B preimpact is

$$V_B = \sqrt{2gh} = \sqrt{(2)(32.2)(2)} = 11.35 \text{ ft/sec}$$

We now consider each case separately.

Case A. Equation 17.30 can be used here. Thus,

$$\epsilon = 1 = -\frac{[(V_B)_f]_n - 0}{-11.35 - 0}$$

Therefore,

$$[(V_B)_f]_n = 11.35 \text{ ft/sec} \qquad \text{(a)}$$

Next, considering rod AB in Fig. 17.42 we have for *linear impulse-* and *angular impulse-momentum* considerations (the latter about an axis at the center of mass):

$$\int F\,dt = \frac{20}{g}[(V_C)_f - (-11.35)] \qquad \text{(b)}$$

$$-(2)(\cos 30°)\int F\,dt = \frac{1}{12}\left(\frac{20}{g}\right)(4^2)\omega_f \qquad \text{(c)}$$

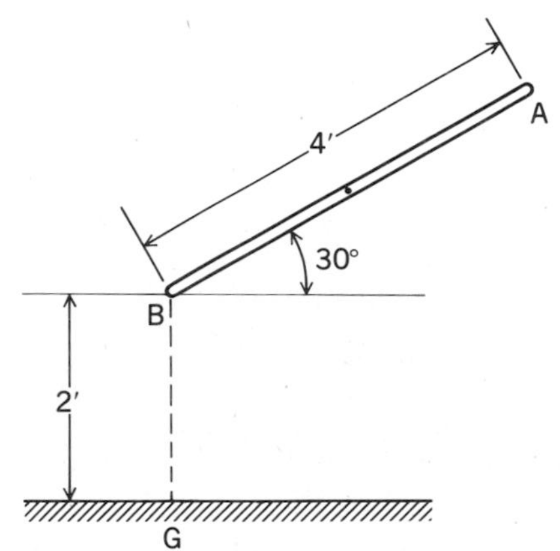

Figure 17.41. Falling rod on a massive body.

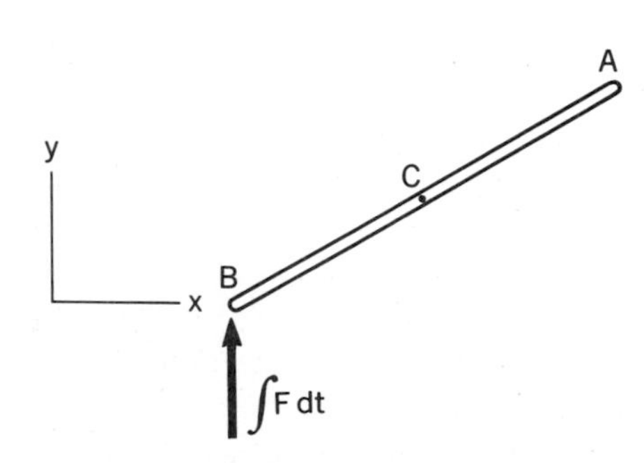

Figure 17.42. Impulsive free-body diagram.

From *kinematics*, we have[12]

$$(V_B)_f = (V_C)_f + \boldsymbol{\omega}_f \times \boldsymbol{\rho}_{CB}$$

$$[(V_B)_x]_f\boldsymbol{i} + 11.35\boldsymbol{j} = (V_C)_f\boldsymbol{j} + \omega_f\boldsymbol{k} \times (2)(-.866\boldsymbol{i} - .5\boldsymbol{j})$$

$$[(V_B)_x]_f\boldsymbol{i} + 11.35\boldsymbol{j} = (V_C)_f\boldsymbol{j} - 1.732\omega_f\boldsymbol{j} + \omega_f\boldsymbol{i}$$

Hence, the scalar equations are

$$11.35 = (V_C)_f - 1.732\omega_f \qquad \text{(d)}$$

$$[(V_B)_x]_f = \omega_f \qquad \text{(e)}$$

We now have a complete set of equations considering $\int F\,dt$ as an unknown.

Solving, we have for the desired unknowns:

$$\omega_f = -9.07 \text{ rad/sec}$$

$$[(V_B)_x]_f = -9.07 \text{ ft/sec}$$

You can demonstrate that energy has been conserved in this action. We could have used this fact in lieu of Eq. (a) for this problem.

Case B. Here we have no slipping on the rough surface and zero vertical movement of point B. Accordingly, we have shown rod BA with vertical and horizontal impulses in Fig. 17.43. The *linear momentum equations* for the center of mass then are

$$\int F_1\,dt = \frac{20}{g}\{[(V_C)_y]_f - (-11.35)\} \qquad \text{(f)}$$

$$\int F_2\,dt = \frac{20}{g}\{[(V_C)_x]_f - 0\} \qquad \text{(g)}$$

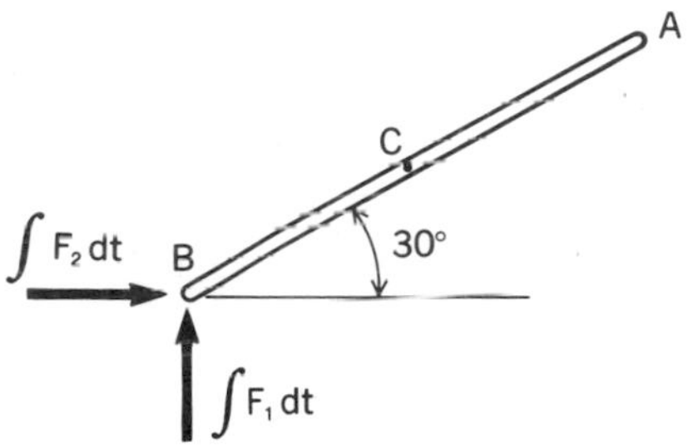

Figure 17.43. Impulsive free-body diagram.

The *angular impulse-momentum equation* about the center of mass is

$$-2\,(\cos 30^\circ)\int F_1\,dt + 2\,(\sin 30^\circ)\int F_2\,dt = \frac{1}{12}\left(\frac{20}{g}\right)(4^2)\omega_f \qquad \text{(h)}$$

From *kinematics*, noting that at postimpact there is pure rotation about point B, we have

$$[(V_C)_y]_f = 2(\cos 30^\circ)(\omega_f) = 1.732\omega_f \qquad \text{(i)}$$

$$[(V_C)_x]_f = -2(\sin 30^\circ)(\omega_f) = -\omega_f \qquad \text{(j)}$$

Substitute for $\int F_1\,dt$ and $\int F_2\,dt$ from Eqs. (f) and (g) into Eq. (h). We get on employing Eqs. (i) and (j):

$$-(2)(.866)\left(\frac{20}{g}\right)(1.732\omega_f + 11.35) + (2)(.5)\left(\frac{20}{g}\right)(-\omega_f) = \frac{1}{12}\left(\frac{20}{g}\right)(4^2)\omega_f$$

Therefore,

$$\omega_f = -3.69 \text{ rad/sec}$$

---

[12] Note that since the initial velocity of C is vertical and since the impulsive force is vertical, the final velocity of C must be vertical. This fact is used in the kinematics.

Problems

17.70. A rod AB slides on a smooth surface at a speed of 10 m/sec and hits a disc D moving at a speed of 5 m/sec. What is the post-impact velocity of point A for a coefficient of restitution $\epsilon = .8$? The rod weighs 30 N and the disc weighs 8 N. Solve by considering AB and D separately.

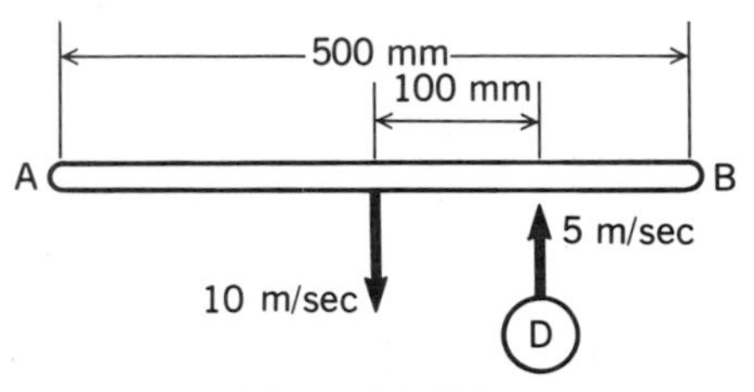

Figure P.17.70

17.71. Solve Problem 17.70 by considering the two bodies as a system.

17.72. Rods AB and HD translate on a horizontal frictionless surface. When they collide we have a coefficient of restitution of .7. Rod HD weighs 70 N and rod AB weighs 40 N. What is the postimpact angular speed of AB?

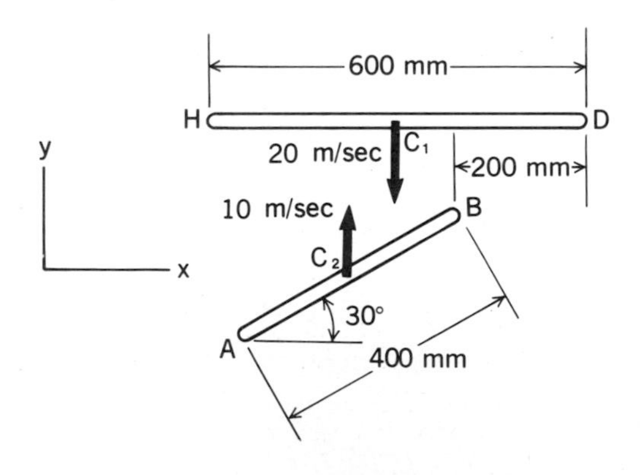

Figure P.17.72

17.73. Two slender rods 20 lb each with small 45° protuberances are shown on a smooth horizontal surface. The rods are identical except for the position of the protuberances. Rod A moves to the left at a speed of 10 ft/sec while rod B is stationary. If the protuberance surfaces are smooth, and an impact having $\epsilon = .8$ occurs, what should d be in order for the postimpact angular speed of A to be 5 rad/sec? Neglect the protuberance for determination of moments of inertia and centers of mass.

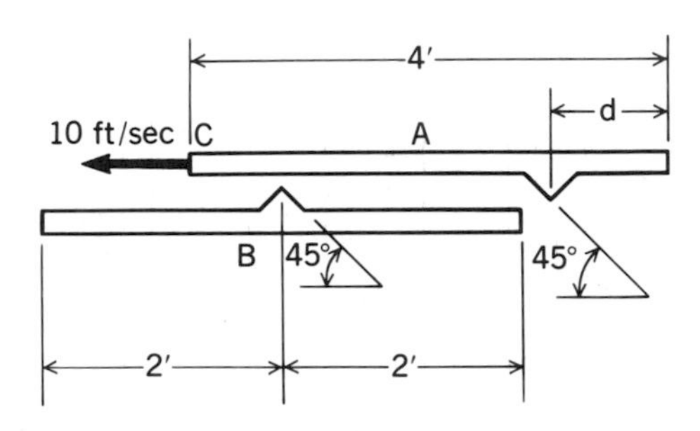

Figure P.17.73

17.74. A 22-N sphere moving at a speed of 10 m/sec hits the end of a 1-m rod having a mass of 10 kg. The coefficient of restitution for the impact is .9. What is the post-impact angular velocity of the rod if it is stationary just before impact? The rod is pinned at O.

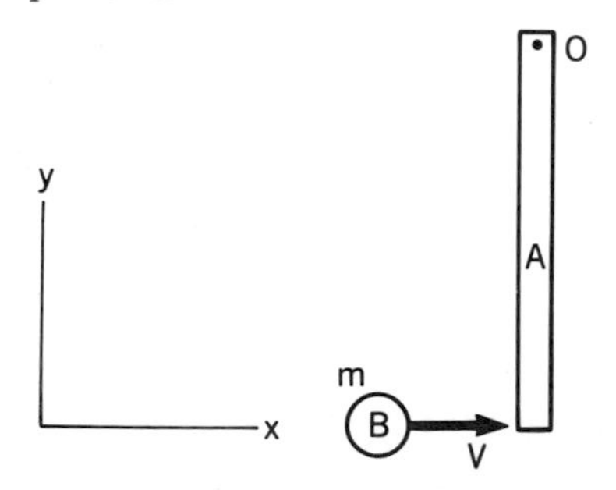

Figure P.17.74

17.75. A rod A weighing 50 N rotates freely about a hinge at a speed of 10 rad/sec on a frictionless surface just prior to hitting a disc B weighing 20 N and moving at a speed of 5 m/sec. If the coefficient of restitution is .8, what is the postimpact angular speed of

A? The contact surface between the rod and the disc is smooth.

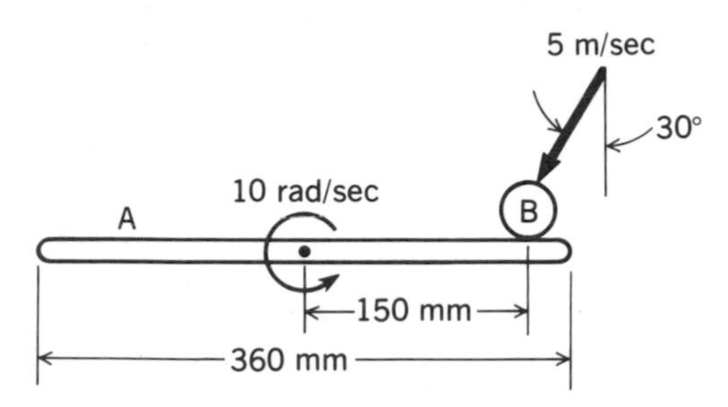

Figure P.17.75

17.76. A stiff bent rod is dropped so that end A strikes a heavy table D. If the impact is plastic and there is no sliding at A, what is the postimpact speed of end B? The rod weighs per unit length 30 N/m.

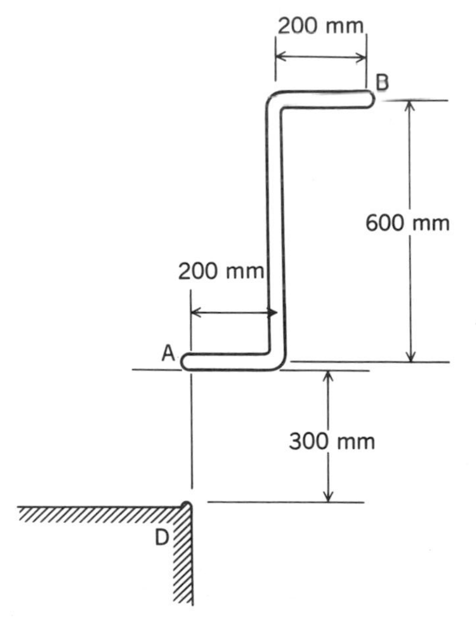

Figure P.17.76

17.77. Solve Problem 17.76 for an elastic impact at A. Take the surface of contact to be smooth. Demonstrate that energy has been conserved.

17.78. A drumstick and a wooden learner's drum are shown in (a) of the diagram. At the beginning of a drum roll, the drumstick is horizontal. The action of the drummer's hand is simulated by force components F_x and F_y, which make A a stationary axis of rotation. Also, there is a constant couple M of .5 N-m. If the action starts from a stationary position shown, what is the frequency of the drum roll for perfectly elastic collision between the drumstick and the learner's drum? The mass of the drumstick is 200 g. Idealize the drumstick as a uniform slender rod.

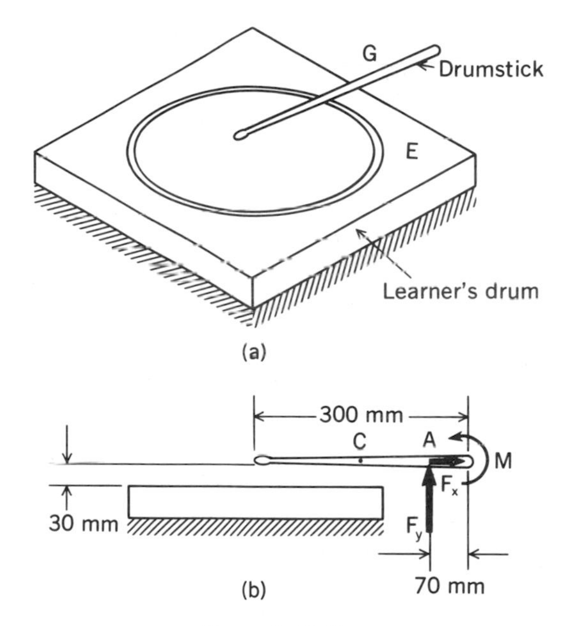

Figure P.17.78

17.79. A block of ice 1 ft × 1 ft × $\frac{1}{2}$ ft slides along a surface at a speed of 10 ft/sec. The block strikes stop D in a plastic impact. Does the block turn over after the impact? What is the highest angular speed reached? Take $\gamma = 62.4$ lb/ft².

Figure P.17.79

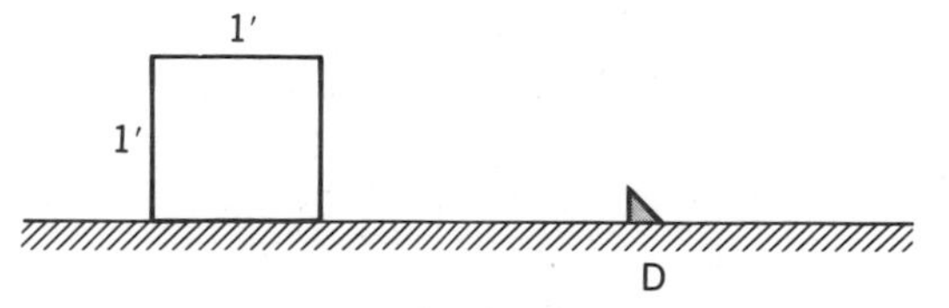

17.80. A horizontal rigid rod is dropped from a height 10 ft above a heavy table. The end of the rod collides with the table. If the coefficient of impact ϵ between the end of the rod and the corner of the table is .6, what is the postimpact angular velocity of the rod? Also, what is the velocity of the center of mass postimpact? The rod is 1 ft in length and weighs 1.5 lb.

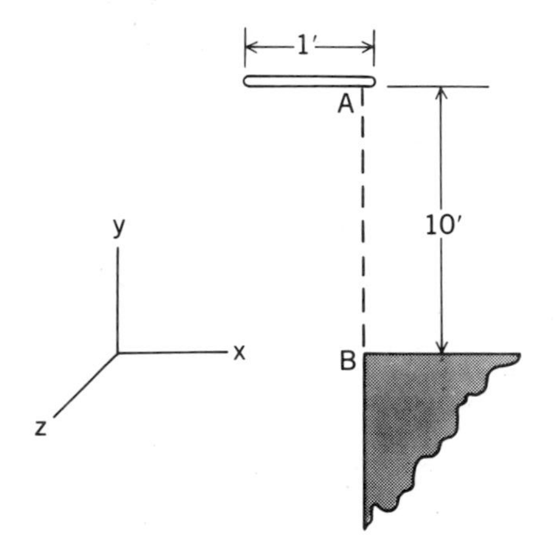

Figure P.17.80

17.81. A rod A is dropped from a height of 120 mm above B. If an elastic collision takes place at B, at what time Δt later does a second collision with support D take place? The rod weighs 40 N.

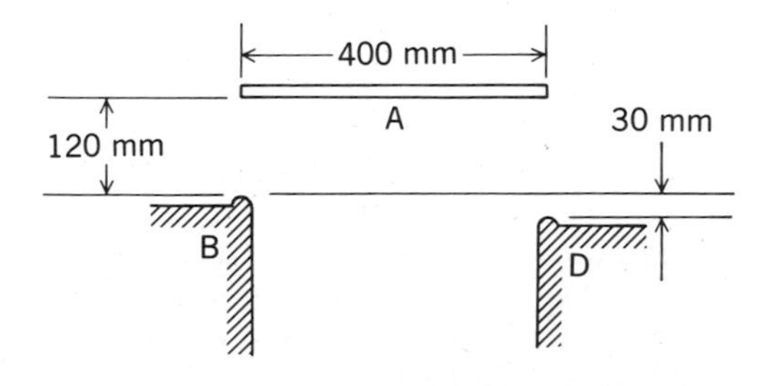

Figure P.17.81

17.82. A packing crate falls on the side of a hill at a place where there is smooth rock. If $\epsilon = .8$, determine the speed of the center of the crate just after impact. The crate weighs 400 N and has a radius of gyration about an axis through the center and normal to the face in the diagram of $\sqrt{2}$ m.

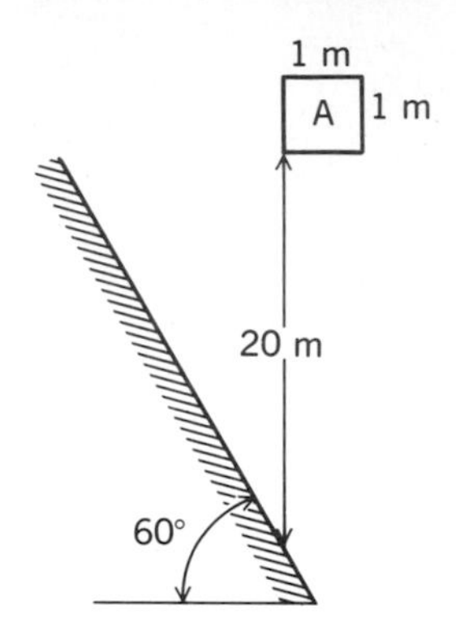

Figure P.17.82

17.83. An arrow moving essentially horizontally at a speed of 20 m/sec impinges on a stationary wooden block which can rotate freely about light rods DE and GH. The arrowhead weighs 1.5 N, and the shaft, which is 250 mm long, weighs .7 N. If the block weighs 10 N, what is the angular velocity of the block just after the arrowhead becomes stuck in the wood? The arrowhead sinks into the wood at point A such that the shaft sticks out 250 mm from the surface of the wood.

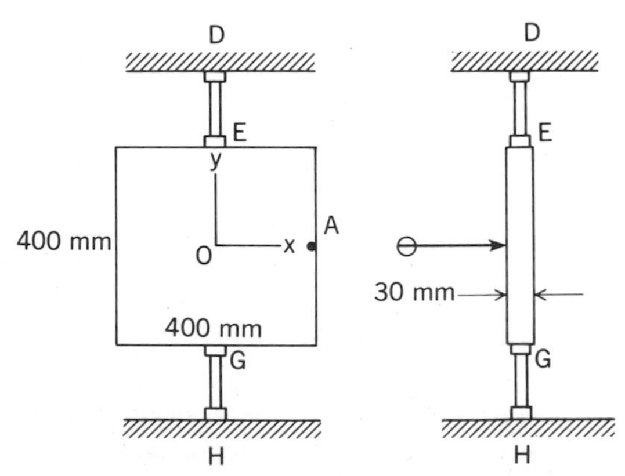

Figure P.17.83

17.84. A bullet weighing $\frac{1}{2}$ N is fired at a wooden block weighing 60 N. What is the angular speed of the block after the bullet has lodged in the block at the right end? Two light hollow rods support the block at the longitudinal midplane. The bullet has a velocity of 400 m/sec in a direction along the longitudinal midplane of the block. (*Hint:* Where is the instantaneous axis of rotation postimpact?)

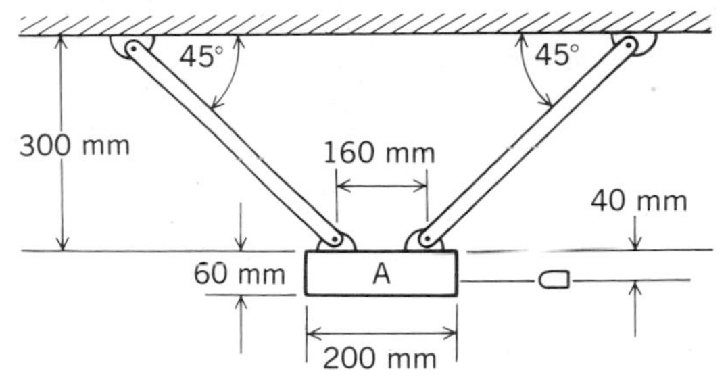

Figure P.17.84

17.7 Closure

Having integrated Newton's law and the special form of the moment of momentum equation $\boldsymbol{M}_A = \dot{\boldsymbol{H}}_A$ for plane motion of slablike bodies in Chapter 16, we then went on in Chapter 17 to study energy methods in Part A and linear impulse-momentum and angular impulse-momentum methods in Part B. In these undertakings, we went directly to the general case and solved three-dimensional problems as well as plane-motion problems. As a useful example of impulse-momentum methods, we then considered eccentric impact of slablike bodies in plane motion.

To complete our study of rigid-body motion, we next return to Chapter 16 to consider the moment-of-momentum equation, $\boldsymbol{M}_A = \dot{\boldsymbol{H}}_A$—this time, for general three-dimensional motion. Accordingly, in Chapter 18 we shall develop the famous Euler equation of motion for a rigid body and we shall then be able to examine interesting problems associated with the gyroscope.

Review Problems

17.85. Gear E rotates at an angular speed ω of 5 rad/sec and drives four smaller "floating" gears A, B, C, and D, which roll within stationary gear F. What is the kinetic energy of the system if gear E weighs 50 lb and each of the small gear weighs 10 lb?

17.86. A homogeneous rectangular parallelepiped weighing 200 N rotates at 20 rad/sec about a main diagonal held by bearings A and B, which are mounted on a vehicle moving at a speed 20 m/sec. What is the kinetic energy of the rectangular parallelepiped relative to the ground?

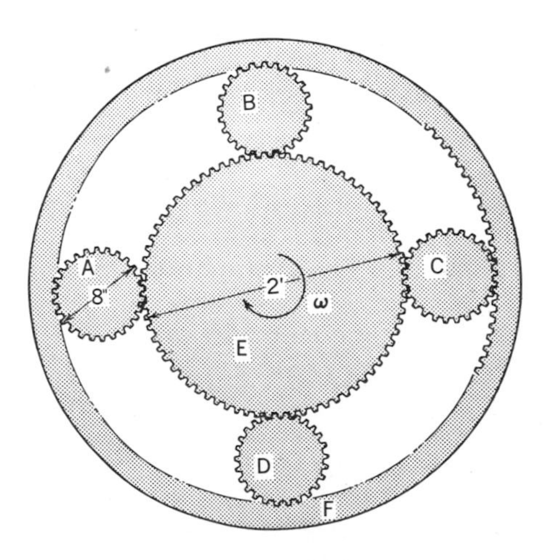

Figure P.17.85

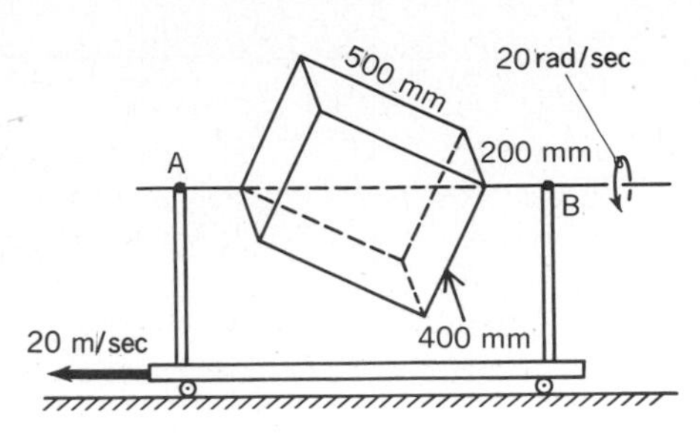

Figure P.17.86

Figure P.17.88

17.87. A stepped cylinder weighs 100 lb and has a radius of gyration of 4 ft. A 50-lb block A is welded to the cylinder. If the spring is unstretched in the configuration shown and has a spring constant K of .5 lb/in., what is the angular speed of the cylinder after it rotates 90°? Assume that the cylinder rolls without slipping.

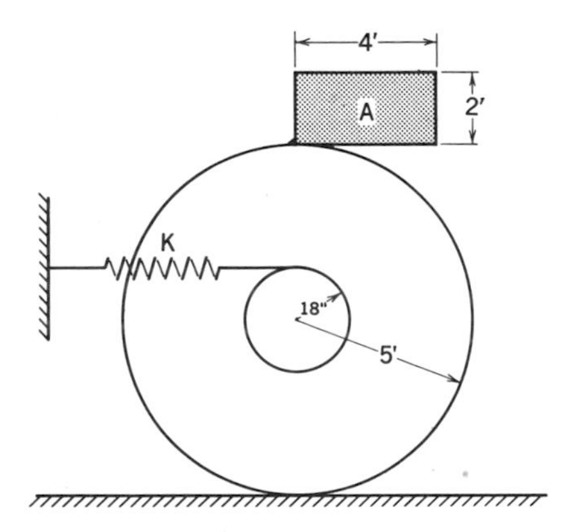

Figure P.17.87

17.88. A tractor with driver weighs 3000 lb. If a torque of 200 ft-lb is developed on each of the drive wheels by the motor, what is the speed of the tractor after it moves 10 ft? The large drive wheels each weigh 200 lb and have a diameter of 3 ft and a radius of gyration of 1 ft. The small wheels each weigh 40 lb and have a diameter of 1 ft with a radius of gyration of .4 ft. Do not use the center-of-mass approach for the whole system.

17.89. Work Problem 17.88 using the center of mass of the whole system. Find the friction forces on each wheel.

17.90. What is the angular momentum vector about the center of mass of a homogeneous rectangular parallelepiped rotating with an angular velocity of 10 rad/sec about a main diagonal? The sides of the rectangular parallelepiped are 1 ft, 2 ft, and 4 ft, as shown, and the weight is 4000 lb.

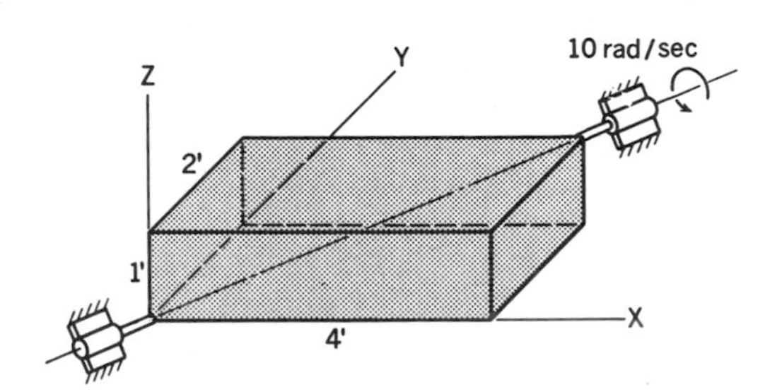

Figure P.17.90

17.91. A force P of 200 N is applied to a cart. The cart minus the four wheels weighs 150 N. Each wheel has a weight of 50 N, a diameter of 400 mm, and a radius of gyration of 150 mm. The load A weighs 500 N. If the cart starts from rest, what is its speed in 20 sec? The wheels roll without slipping.

Figure P.17.91

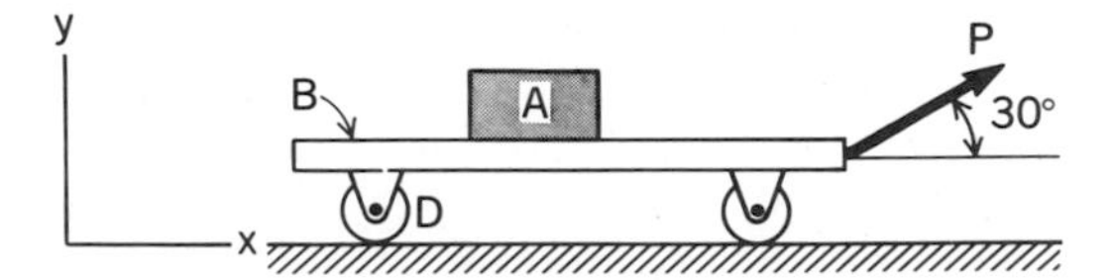

17.92. A device consists of two identical rectangular plates each weighing 100 N welded to a rod weighing 50 N. The device rests horizontally on a smooth surface. As a result of a collision, a horizontal impulse of 30 N-sec is delivered to point D. What is the velocity of point E on plate A postimpact?

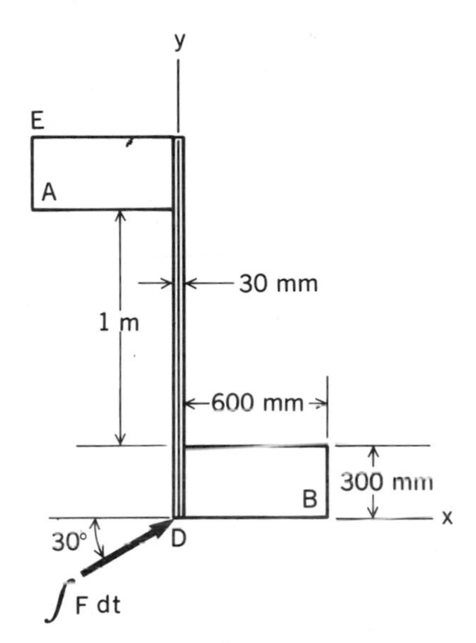

Figure P.17.92

17.93. A cylinder A of weight 150 N and radius of gyration 100 mm is placed onto a conveyor belt which is moving at the constant speed of $V_B = 10$ m/sec. Give the speed of the axis of the cylinder at time $t = 5$ sec.

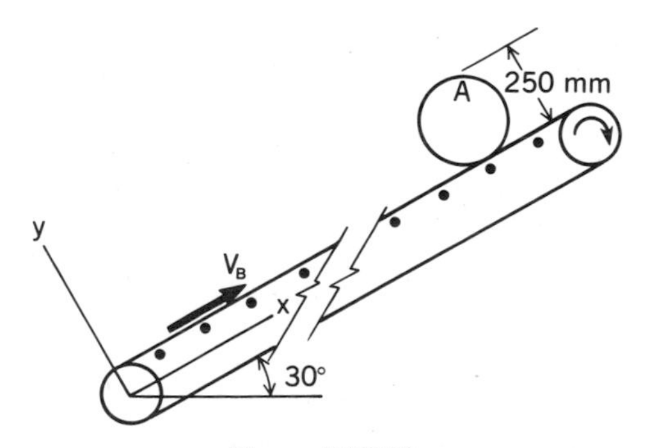

Figure P.17.93

The coefficient of friction between the cylinder and the belt is .5.

17.94. A rectangular rod A rests on a smooth horizontal surface. A disc B moves toward the rod at a speed of 10 m/sec. What is the postimpact angular velocity of the rod for a coefficient of restitution of .6? The rod weighs 50 N, and the disc B weighs 8 N.

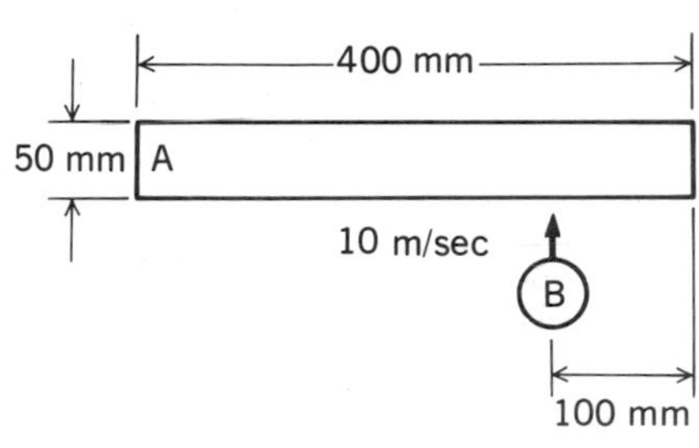

Figure P.17.94

17.95. A block B weighs 10 lb and is dropped onto the end of a rectangular rod A weighing 20 lb. What is the angular speed of rod A postimpact for $\epsilon = .7$? The rod is pinned at its center as shown.

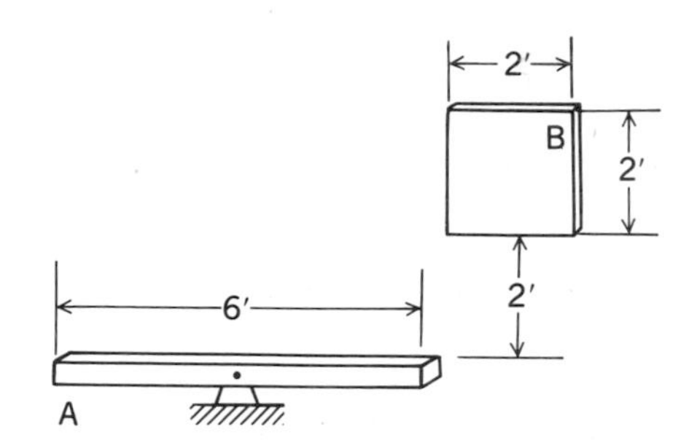

Figure P.17.95

17.96. A rod AB is held at a position $\theta = 45°$ and released. HA is a light wire to which the rod is attached. When AB is vertical, it strikes a stop E and undergoes an elastic collision. What position d of the stop will result in a zero postimpact velocity of point A? The rod has a mass $M = 10$ kg. Support E is immovable.

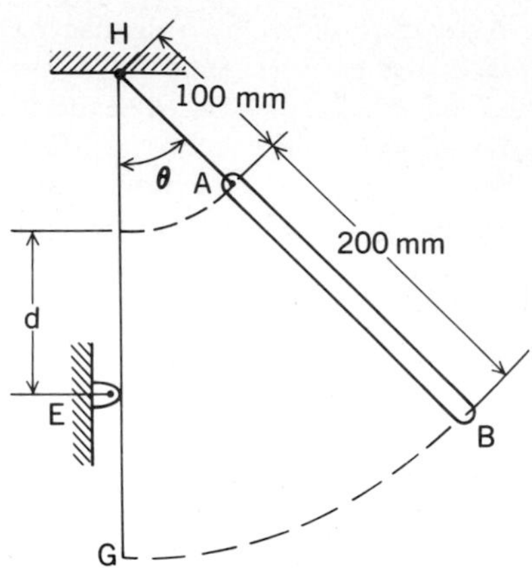

Figure P.17.96

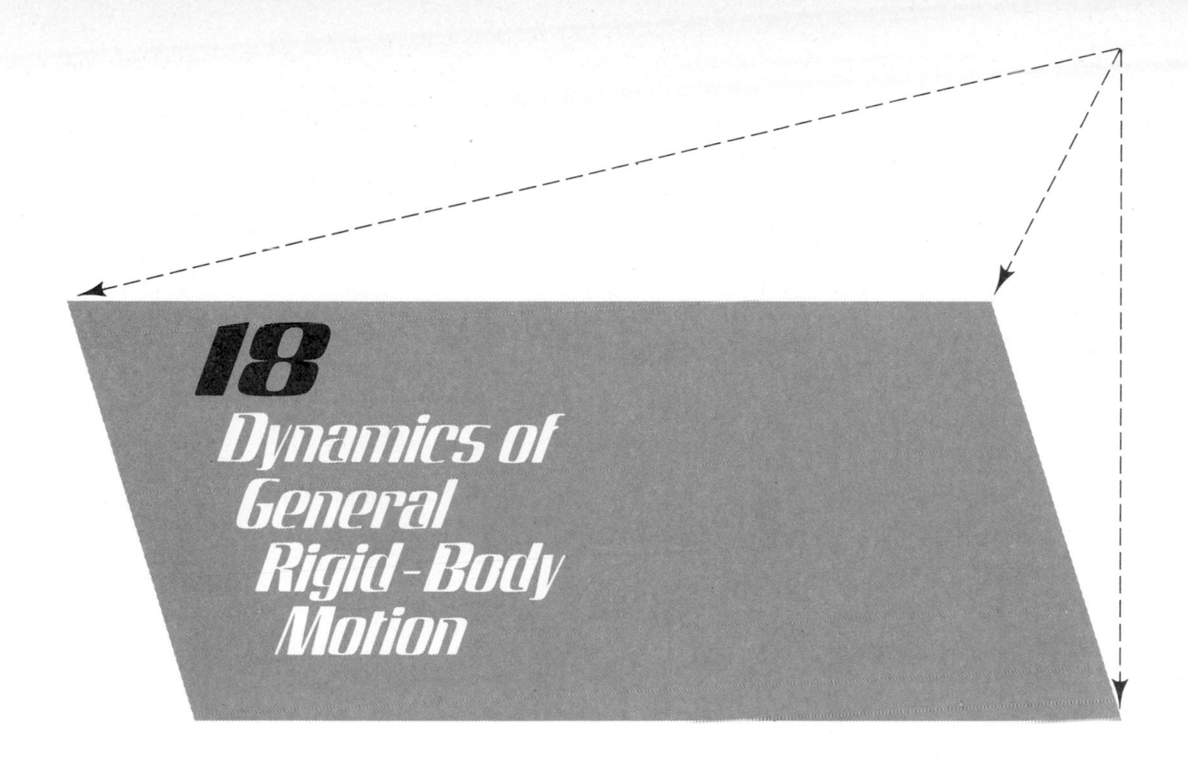

18.1 Introduction

In Section 16.2, recall that we set forth the angular momentum vector $\boldsymbol{H}_A$ about point A for a rigid body having arbitrary motion relative to reference XYZ. Recall that point A was fixed in the rigid body or in a hypothetical extension of the rigid body. We determined components of $\boldsymbol{H}_A$ along an *arbitrary* set of axes xyz at A. The following result was found at time t:

$$\begin{aligned} (H_A)_x &= \quad I_{xx}\omega_x - I_{xy}\omega_y - I_{xz}\omega_z \\ (H_A)_y &= -I_{yx}\omega_x + I_{yy}\omega_y - I_{yz}\omega_z \\ (H_A)_z &= -I_{zx}\omega_x - I_{yz}\omega_y + I_{zz}\omega_z \end{aligned} \tag{18.1}$$

where $\boldsymbol{\omega}$ is the angular velocity of the body relative to XYZ at time t. Now the axes xyz served *only* to give a set of directions for $\boldsymbol{H}_A$ at time t. The reference xyz could have any angular velocity $\boldsymbol{\Omega}$ relative to XYZ.

Note that the angular velocity $\boldsymbol{\Omega}$ did not enter the formulations for $\boldsymbol{H}_A$ at time t. The reason for this result is that the values of the components of $\boldsymbol{H}_A$ along xyz at time t do not in any way depend on angular velocity $\boldsymbol{\Omega}$—they depend only on the instantaneous orientation of xyz at time t. To illustrate this point, suppose that we have two sets of axes xyz and $x'y'z'$ at point A in Fig. 18.1. At time t they coincide as has been shown in the diagram, but xyz has zero angular velocity relative to XYZ, whereas $x'y'z'$ has an angular velocity $\boldsymbol{\Omega}$ relative to XYZ. Clearly, one can say at time t:

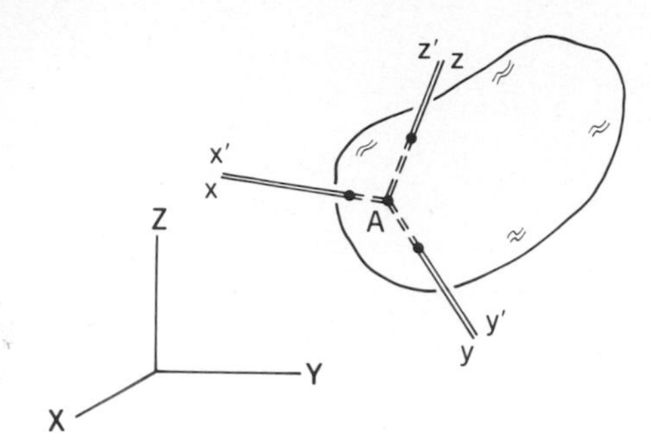

Figure 18.1. xyz and $x'y'z'$ coincide at time t.

$$(H_A)_x = (H_A)_{x'}$$
$$(H_A)_y = (H_A)_{y'}$$
$$(H_A)_z = (H_A)_{z'}$$

However, *the time derivatives* of the corresponding components will *not* be equal to each other at time t. Accordingly, in the next section, where $\dot{\boldsymbol{H}}_A$ is treated, we must properly account for $\boldsymbol{\Omega}$.

18.2 Euler's Equations of Motion

We shall now restrict point A of Section 18.1 further, as we did in Chapter 17, by considering only those points for which the equation $\boldsymbol{M}_A = \dot{\boldsymbol{H}}_A$ is valid:

1. The mass center.
2. Points fixed or moving with constant $\boldsymbol{V}$ at time t in inertial space (i.e., points having zero acceleration at time t relative to inertial reference XYZ).[1]
3. A point accelerating toward or away from the mass center.

We learned in Chapter 15 that derivatives of a vector as seen from different references could be related as follows:

$$\left(\frac{d\boldsymbol{A}}{dt}\right)_{XYZ} = \left(\frac{d\boldsymbol{A}}{dt}\right)_{xyz} + \boldsymbol{\omega} \times \boldsymbol{A}$$

where $\boldsymbol{\omega}$ is the angular velocity of xyz relative to XYZ. We shall employ this equation in the basic moment-of-momentum equation to shift the observation reference from XYZ to xyz in the following way:

$$\boxed{\boldsymbol{M}_A = \left(\frac{d\boldsymbol{H}_A}{dt}\right)_{XYZ} = \left(\frac{d\boldsymbol{H}_A}{dt}\right)_{xyz} + \boldsymbol{\Omega} \times \boldsymbol{H}_A} \tag{18.2}$$

The idea now is to choose the angular velocity $\boldsymbol{\Omega}$ of reference xyz at A in such a way that $(d\boldsymbol{H}_A/dt)_{xyz}$ is most easily evaluated. With this accomplished, the next step is to attempt the integration of the resulting differential equation.

With regard to attempts at integration, we point out at this early stage that Eq. (18.2) is valid only as long as point A is one of the three qualified points just discussed. Clearly, if A is the mass center, then Eq. 18.2 is valid at all times and can be integrated with respect to time provided that the mathematics are not too difficult. However, if for cases (2) and (3) point A qualifies only at time t, then Eq. 18.2 is valid only at time t and accordingly cannot be integrated. If, on the other hand, for case (2), there is an

[1] Clearly, points along an axis rotation fixed in inertial space would qualify.

axis of rotation fixed in inertial space, then Eq. 18.2 is valid at all times for any point A along the axis of rotation and accordingly can be integrated. We have already done this in Chapter 16. If, furthermore, the axis of rotation always goes through the chosen point A but does not have a fixed orientation in inertial space (see Fig. 18.2), we can again use Eq. 18.2 at all times and attempt to integrate it with respect to time. The carrying out of such integrations may be quite difficult, however.[2]

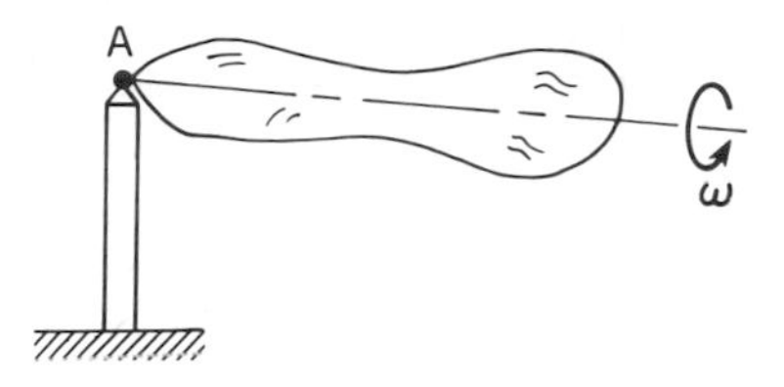

Figure 18.2. Axis of rotation goes through fixed point A.

Returning to Eq. 18.2, we can work directly with this equation selecting a reference xyz for each problem to yield the simplest working equation. On the other hand, we can develop Eq. 18.2 further for certain classes of references xyz. For example, we could have xyz translate relative to XYZ. This would mean that $\mathbf{\Omega} = \mathbf{0}$ so that Eq. 18.2 would seem to be more simple for such cases.[3] However, the body will be rotating relative to xyz and the moments and products of inertia measured about xyz will then be time functions. Since the computation of these terms as time functions is generally difficult, such an approach has limited value. On the other hand, the procedure of *fixing xyz* in the body (as we did for the case of plane motion in Chapter 16) does lead to very useful forms of Eq. 18.2, and we shall accordingly examine these equations with great care. Note first that the moments and products of inertia will be constants for this case and that $\mathbf{\Omega} = \boldsymbol{\omega}$. Hence, we have

$$\boxed{\mathbf{M}_A = \left(\frac{d\mathbf{H}_A}{dt}\right)_{xyz} + \boldsymbol{\omega} \times \mathbf{H}_A} \tag{18.3}$$

Employing components for all vectors along axes xyz and utilizing Eq. 18.1, we get on dropping the subscript A:

$$\begin{aligned} M_x\mathbf{i} + M_y\mathbf{j} + M_z\mathbf{k} = \frac{d}{dt_{xyz}}[(\omega_x I_{xx} - \omega_y I_{xy} - \omega_z I_{xz})\mathbf{i} + (-\omega_x I_{yx} + \omega_y I_{yy} - \omega_z I_{yz})\mathbf{j} \\ + (-\omega_x I_{zx} - \omega_y I_{zy} + \omega_z I_{zz})\mathbf{k}] + \boldsymbol{\omega} \times [(\omega_x I_{xx} - \omega_y I_{xy} - \omega_z I_{xz})\mathbf{i} \\ + (-\omega_x I_{yx} + \omega_y I_{yy} - \omega_z I_{yz})\mathbf{j} + (-\omega_x I_{zx} - \omega_y I_{zy} + \omega_z I_{zz})\mathbf{k}] \end{aligned}$$

In the first expression on the right side of the foregoing equation, the vectors $\mathbf{i}, \mathbf{j}$, and $\mathbf{k}$ are constant vectors as seen from xyz. Also, because xyz is fixed in the body, the moments and products of inertia are constant. Only ω_x, ω_y, and ω_z, the angular velocity components of the rigid body, are time functions. We then can say:

$$\begin{aligned} M_x\mathbf{i} + M_y\mathbf{j} + M_z\mathbf{k} = (\dot{\omega}_x I_{xx} - \dot{\omega}_y I_{xy} - \dot{\omega}_z I_{xz})\mathbf{i} + (-\dot{\omega}_x I_{yx} + \dot{\omega}_y I_{yy} - \dot{\omega}_z I_{yz})\mathbf{j} \\ + (-\dot{\omega}_x I_{zx} - \dot{\omega}_y I_{zy} + \dot{\omega}_z I_{zz})\mathbf{k} + (\omega_x I_{xx} - \omega_y I_{xy} - \omega_z I_{xz})(\boldsymbol{\omega} \times \mathbf{i}) \\ + (-\omega_x I_{yx} + \omega_y I_{yy} - \omega_z I_{yz})(\boldsymbol{\omega} \times \mathbf{j}) + (-\omega_x I_{zx} - \omega_y I_{zy} + \omega_z I_{zz})(\boldsymbol{\omega} \times \mathbf{k}) \end{aligned}$$

---

[2] In a later section we shall examine this case in some detail.

[3] We shall later find it advantageous not to fix xyz to the body for certain problems.

Carrying out the cross products, collecting terms, and expressing as scalar equations, we get

$$M_x = \dot{\omega}_x I_{xx} + \omega_y \omega_z (I_{zz} - I_{yy}) + I_{xy}(\omega_z \omega_x - \dot{\omega}_y) - I_{xz}(\dot{\omega}_z + \omega_y \omega_x) - I_{yz}(\omega_y^2 - \omega_z^2) \quad (18.4a)$$

$$M_y = \dot{\omega}_y I_{yy} + \omega_z \omega_x (I_{xx} - I_{zz}) + I_{yz}(\omega_x \omega_y - \dot{\omega}_z) - I_{yx}(\dot{\omega}_x + \omega_z \omega_y) - I_{zx}(\omega_z^2 - \omega_x^2) \quad (18.4b)$$

$$M_z = \dot{\omega}_z I_{zz} + \omega_x \omega_y (I_{yy} - I_{xx}) + I_{zx}(\omega_y \omega_z - \dot{\omega}_x) - I_{zy}(\dot{\omega}_y + \omega_x \omega_z) - I_{xy}(\omega_x^2 - \omega_y^2) \quad (18.4c)$$

This is indeed a formidable set of equations. (Equation 16.6 is a special case of this equation found by setting $\omega_y = \omega_x = \dot{\omega}_y = \dot{\omega}_x = 0$ for plane motion.) However, if we choose reference xyz so that it coincides with the *principal axes* of the body at point A, then it is clear that the products of inertia are all zero in the system of equations above, and this fact enables us to simplify the equations considerably. The resulting equations given below are the famous *Euler equations* of motion. Note that these equations relate the angular velocity and the angular acceleration to the moment of the external forces about the point A.

$$M_x = I_{xx}\dot{\omega}_x + \omega_y \omega_z (I_{zz} - I_{yy}) \quad (18.5a)$$

$$M_y = I_{yy}\dot{\omega}_y + \omega_z \omega_x (I_{xx} - I_{zz}) \quad (18.5b)$$

$$M_z = I_{zz}\dot{\omega}_z + \omega_x \omega_y (I_{yy} - I_{xx}) \quad (18.5c)$$

In both sets of Eqs. 18.4 and 18.5, we have three simultaneous first-order differential equations. If the motion of the body about point A is known, we can easily compute the required moments about point A. On the other hand, if the moments are known functions of time and the angular velocity is desired, we have the difficult problem of solving simultaneous nonlinear differential equations for the unknowns ω_x, ω_y, and ω_z. However, in practical problems, we often know some of the angular velocity and acceleration components from constraints or given data, so, with the restrictions mentioned earlier, we can sometimes integrate the equations readily as we did in Chapter 16 for plane motion. At other times we use them to solve for certain desired *instantaneous values* of the unknowns.

We shall now discuss the use of Euler's equations.

18.3 Application of Euler's Equations

In this section, we shall apply the Euler equations to a number of problems. Before taking up these problems, let us first carefully consider how to express the components $\dot{\omega}_x$, $\dot{\omega}_y$, and $\dot{\omega}_z$ for use in Euler's equations. First note that $\dot{\omega}_x$, $\dot{\omega}_y$, and $\dot{\omega}_z$ are time derivatives as seen from XYZ of the components of $\boldsymbol{\omega}$ along reference xyz. A possible procedure then is to express ω_x, ω_y, and ω_z first in a way that ensures that these quantities are correctly stated over a time interval rather than at some instantaneous configuration. Once this is done, we can simply differentiate these scalar quantities with respect to time in this interval to get $\dot{\omega}_x$, $\dot{\omega}_y$, and $\dot{\omega}_z$.

To illustrate this procedure, consider in Fig. 18.3 the case of a block E rotating about rod AB, which in turn rotates about vertical axis CD. A reference xyz is fixed to the block at the center of mass so as to coincide with the principal axes of the block

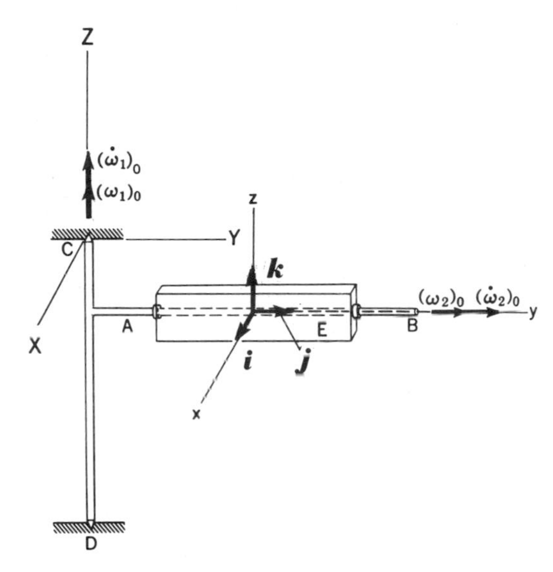

Figure 18.3. Block at time t_0.

at the center of mass. When the block is vertical as shown, the angular speed and rate of change of angular speed relative to AB have the known values $(\omega_2)_0$ and $(\dot{\omega}_2)_0$, respectively. At that instant, AB has an angular speed and rate of change of angular speed about axis CD of known values $(\omega_1)_0$ and $(\dot{\omega}_1)_0$, respectively. We can immediately give the angular velocity components at the instant shown as follows:

$$\omega_x = 0$$
$$\omega_y = (\omega_2)_0$$
$$\omega_z = (\omega_1)_0$$

But to get the quantities $(\dot{\omega}_x)_0$, $(\dot{\omega}_y)_0$, and $(\dot{\omega}_z)_0$ for the instant of interest[4] we must first express ω_x, ω_y, and ω_z as *general functions of time* in order to permit differentiation with respect to time.

To do this differentiating, we have shown the system at some arbitrary time in

[4]You may be tempted to say $(\dot{\omega}_x)_0 = 0$, $(\dot{\omega}_y)_0 = (\dot{\omega}_2)_0$, and $(\dot{\omega}_z)_0 = (\dot{\omega}_1)_0$ by just inspecting the diagram. You will now learn that this will not be correct.

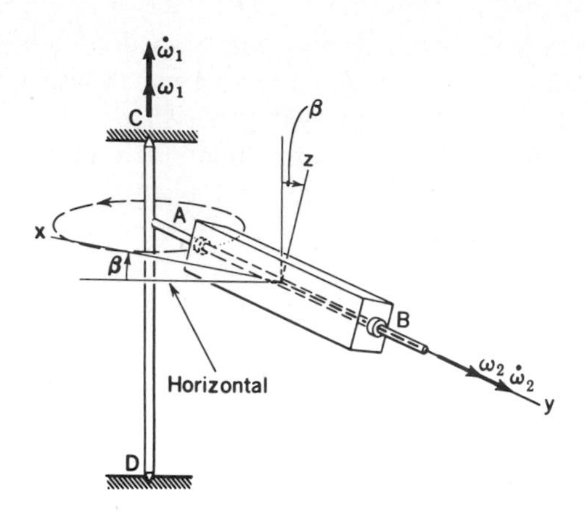

Figure 18.4. Block at time *t*.

Fig. 18.4. Note that the x axis is at some angle β from the horizontal. When β becomes zero, we arrive back at the configuration of interest, and ω_1, $\dot{\omega}_1$, ω_2, and $\dot{\omega}_2$ become known values $(\omega_1)_0$, $(\dot{\omega}_1)_0$, $(\omega_2)_0$, and $(\dot{\omega}_2)_0$, respectively. The angular velocity components for this arbitrary situation are:

$$\omega_x = \omega_1 \sin \beta, \qquad \omega_y = \omega_2, \qquad \omega_z = \omega_1 \cos \beta \tag{18.6}$$

Since these relations are valid over the time interval of interest, we can differentiate them with respect to time and get

$$\begin{aligned} \dot{\omega}_x &= \dot{\omega}_1 \sin \beta + \omega_1 \cos \beta \dot{\beta} \\ \dot{\omega}_y &= \dot{\omega}_2 \\ \dot{\omega}_z &= \dot{\omega}_1 \cos \beta - \omega_1 \sin \beta \dot{\beta} \end{aligned} \tag{18.7}$$

It should be clear upon inspecting the diagram that $\dot{\beta} = -\omega_2$, and so the preceding terms become

$$\begin{aligned} \dot{\omega}_x &= \dot{\omega}_1 \sin \beta - \omega_1 \omega_2 \cos \beta \\ \dot{\omega}_y &= \dot{\omega}_2 \\ \dot{\omega}_z &= \dot{\omega}_1 \cos \beta + \omega_1 \omega_2 \sin \beta \end{aligned} \tag{18.8}$$

If we now let β become zero, we reach the configuration of interest and we get from Eqs. 18.6 and 18.7 the proper values of the angular velocity components and their time derivatives at this configuration:

$$\begin{aligned} \omega_x &= 0, & \dot{\omega}_x &= -(\omega_1)_0(\omega_2)_0 \\ \omega_y &= (\omega_2)_0, & \dot{\omega}_y &= (\dot{\omega}_2)_0 \\ \omega_z &= (\omega_1)_0, & \dot{\omega}_z &= (\dot{\omega}_1)_0 \end{aligned} \tag{18.9}$$

Actually, we do not have to employ such a procedure for the evaluation of these quantities. There is a simple direct approach that can be used, but we must preface

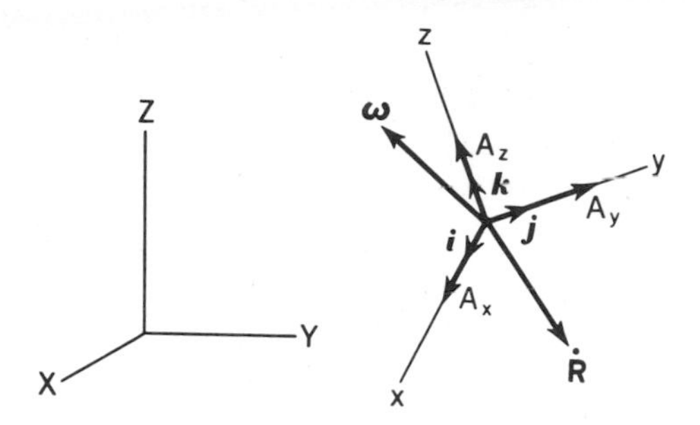

Figure 18.5. Components of ***A*** along *xyz*.

the discussion of this method by some general remarks about the time derivative, as seen from the *XYZ* axes, of a vector ***A*** expressed in terms of components always parallel to the *xyz* reference, which moves relative to *XYZ* (Fig. 18.5). We can then say, considering ***i***, ***j***, and ***k*** as unit vectors for reference *xyz*:

$$\begin{aligned}\left(\frac{d\boldsymbol{A}}{dt}\right)_{XYZ} &= \frac{d}{dt_{XYZ}}(A_x\boldsymbol{i} + A_y\boldsymbol{j} + A_z\boldsymbol{k}) \\ &= \dot{A}_x\boldsymbol{i} + \dot{A}_y\boldsymbol{j} + \dot{A}_z\boldsymbol{k} + A_x(\boldsymbol{\omega} \times \boldsymbol{i}) + A_y(\boldsymbol{\omega} \times \boldsymbol{j}) + A_z(\boldsymbol{\omega} \times \boldsymbol{k})\end{aligned} \tag{18.10}$$

If we decompose the vector $(d\boldsymbol{A}/dt)_{XYZ}$ into components parallel to the *xyz* axes at time *t* and carry out the cross products on the right side in terms of *xyz* components, then we get, after collecting terms and expressing the results as scalar equations:

$$\left[\left(\frac{d\boldsymbol{A}}{dt}\right)_{XYZ}\right]_x = \dot{A}_x + A_z\omega_y - A_y\omega_z \tag{18.11a}$$

$$\left[\left(\frac{d\boldsymbol{A}}{dt}\right)_{XYZ}\right]_y = \dot{A}_y + A_x\omega_z - A_z\omega_x \tag{18.11b}$$

$$\left[\left(\frac{d\boldsymbol{A}}{dt}\right)_{XYZ}\right]_z = \dot{A}_z + A_y\omega_x - A_x\omega_y \tag{18.11c}$$

We can learn an important lesson from these equations. If you take the time derivative of a vector ***A*** with respect to a reference *XYZ* and express the *components* of this vector parallel to the axes of a reference *xyz* rotating relative to *XYZ* (these are the terms on the left side of the above equations), then the results are in general *not the same* as *first* taking the components of the vector ***A*** along the directions *xyz* and *then* taking time derivatives of these scalars. Thus,

$$\left[\left(\frac{d\boldsymbol{A}}{dt}\right)_{XYZ}\right]_x \neq \frac{d(A_x)}{dt} = \frac{d}{dt}A_x = \dot{A}_x \quad \text{etc.} \tag{18.12}$$

How does this result relate to our problem where we are considering $\dot{\omega}_x$, $\dot{\omega}_y$, and $\dot{\omega}_z$? Clearly, these expressions are time derivatives of the components of the vector $\boldsymbol{\omega}$ along the moving *xyz* axes, and so in this respect they correspond to the terms on the right side of inequality 18.12. Let us then consider vector ***A*** to be $\boldsymbol{\omega}$ and examine Eq. 18.11:

$$\left[\left(\frac{d\boldsymbol{\omega}}{dt}\right)_{XYZ}\right]_x = \dot{\omega}_x + \omega_z\omega_y - \omega_y\omega_z$$

$$\left[\left(\frac{d\boldsymbol{\omega}}{dt}\right)_{XYZ}\right]_y = \dot{\omega}_y + \omega_x\omega_z - \omega_z\omega_x \tag{18.13}$$

$$\left[\left(\frac{d\boldsymbol{\omega}}{dt}\right)_{XYZ}\right]_z = \dot{\omega}_z + \omega_y\omega_x - \omega_x\omega_y$$

We see that terms on the right side cancel for this case, leaving us

$$\left[\left(\frac{d\boldsymbol{\omega}}{dt}\right)_{XYZ}\right]_x = \dot{\omega}_x$$

$$\left[\left(\frac{d\boldsymbol{\omega}}{dt}\right)_{XYZ}\right]_y = \dot{\omega}_y \tag{18.14}$$

$$\left[\left(\frac{d\boldsymbol{\omega}}{dt}\right)_{XYZ}\right]_z = \dot{\omega}_z$$

We see that for the vector $\boldsymbol{\omega}$ (i.e., the angular velocity of the xyz reference relative to the XYZ reference), we have an exception to the rule stated earlier (Eq. 18.12). Here is the one case where the derivative of a vector as seen from one set of axes XYZ has components along the directions of another set of axes xyz rotating relative to XYZ, wherein these components are respectively equal to the simple time derivatives of the scalar components of the vector along the xyz directions. In other words, you can take the derivative of $\boldsymbol{\omega}$ first from the XYZ axes and then take scalar components along xyz, or you can take scalar components along xyz first and then take simple time derivatives of the scalar components, and the results are the same.

If we fully understand the exceptional nature of Eq. 18.14, we can compute $\dot{\omega}_x$, $\dot{\omega}_y$, and $\dot{\omega}_z$ in a straightforward manner by simply first determining $(d\boldsymbol{\omega}/dt)_{XYZ}$ and then taking the components. This is a step that we have practiced a great deal in kinematics. For instance, for the problem introduced at the outset of this discussion, we see by inspecting Fig. 18.3 that at all times

$$\boldsymbol{\omega} = \omega_2\boldsymbol{j} + \omega_1\boldsymbol{k}_1 \tag{18.15}$$

where $\boldsymbol{k}_1$ is the unit vector in the fixed Z direction. Now differentiate with respect to time for the XYZ reference:

$$\dot{\boldsymbol{\omega}} = \dot{\omega}_2\boldsymbol{j} + \omega_2\dot{\boldsymbol{j}} + \dot{\omega}_1\boldsymbol{k}_1 \tag{18.16}$$

But $\boldsymbol{j}$ is fixed in rod AB that is rotating with angular velocity $\omega_1\boldsymbol{k}_1$ relative to XYZ. We then get

$$\dot{\boldsymbol{\omega}} = \dot{\omega}_2\boldsymbol{j} + \omega_2(\omega_1\boldsymbol{k}_1) \times \boldsymbol{j} + \dot{\omega}_1\boldsymbol{k}_1 \tag{18.17}$$

When the xyz axes are parallel to the XYZ axes, the unit vector $\boldsymbol{k}$ becomes the same as the unit vector $\boldsymbol{k}_1$, and ω_1, ω_2, etc., become known values $(\omega_1)_0$, $(\omega_2)_0$, etc. We then get for that configuration:

$$\dot{\boldsymbol{\omega}}_0 = (\dot{\omega}_2)_0\boldsymbol{j} - (\omega_1)_0(\omega_2)_0\boldsymbol{i} + (\dot{\omega}_1)_0\boldsymbol{k} \tag{18.18}$$

The components of this equation give the desired values of $\dot{\omega}_x$, $\dot{\omega}_y$, and $\dot{\omega}_z$ at the instant of interest. Thus, we have

$$\dot{\omega}_x = -(\omega_1)_0(\omega_2)_0$$
$$\dot{\omega}_y = (\dot{\omega}_2)_0 \tag{18.19}$$
$$\dot{\omega}_z = (\dot{\omega}_1)_0$$

In most of the following examples we shall proceed by the second method discussed here. That is, we shall get $\dot{\omega}_x$, $\dot{\omega}_y$, and $\dot{\omega}_z$ using the following formulations:

$$\dot{\omega}_x = \left[\left(\frac{d\boldsymbol{\omega}}{dt}\right)_{XYZ}\right]_x \tag{18.20a}$$

$$\dot{\omega}_y = \left[\left(\frac{d\boldsymbol{\omega}}{dt}\right)_{XYZ}\right]_y \tag{18.20b}$$

$$\dot{\omega}_z = \left[\left(\frac{d\boldsymbol{\omega}}{dt}\right)_{XYZ}\right]_z \tag{18.20c}$$

EXAMPLE 18.1

In Fig. 18.6 a thin disc of radius $R = 4$ ft and weight 322 lb rotates at an angular speed ω_2 of 100 rad/sec relative to a platform. The platform rotates with an angular speed ω_1 of 20 rad/sec relative to the ground. Compute the bearing reactions at A and B. Neglect the mass of the shaft and assume that bearing A restrains the system in the radial direction.

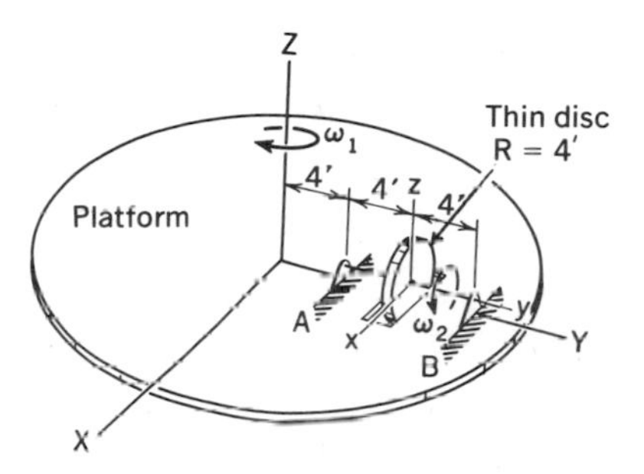

Figure 18.6. Rotating disc on platform.

Clearly, we shall need to use *Euler's equations* as part of the solution to this problem, and so we *fix a reference xyz to the center of mass* of the disc as shown in Fig. 18.6. *XYZ is fixed to the ground.* In using Euler's equations, the key step is to get the angular velocity components and their time derivatives for the body as seen from XYZ. Accordingly, we have

$$\boldsymbol{\omega} = \boldsymbol{\omega}_1 + \boldsymbol{\omega}_2 = -20\boldsymbol{k} + 100\boldsymbol{j} \text{ rad/sec}$$

Hence, we have for the xyz components:

$$\omega_x = 0$$
$$\omega_y = 100 \text{ rad/sec}$$
$$\omega_z = -20 \text{ rad/sec}$$

Next, we have

$$\dot{\boldsymbol{\omega}} = \dot{\boldsymbol{\omega}}_1 + \dot{\boldsymbol{\omega}}_2 = \boldsymbol{0} + \boldsymbol{\omega}_1 \times \boldsymbol{\omega}_2$$
$$= (-20\boldsymbol{k}) \times (100\boldsymbol{j}) = 2000\boldsymbol{i} \text{ rad/sec}^2$$

We thus have for the xyz components:

$$\dot{\omega}_x = 2000 \text{ rad/sec}^2$$
$$\dot{\omega}_y = 0$$
$$\dot{\omega}_z = 0$$

Before going to the Euler equations, we shall need the principal moments of inertia of the disc. Using formulas from the inside covers, we get

$$I_{yy} = \frac{1}{2}\frac{W}{g}R^2 = \frac{1}{2}(10)(16) = 80 \text{ slug-ft}^2$$

$$I_{xx} = I_{zz} = \frac{1}{4}\frac{W}{g}R^2 = 40 \text{ slug-ft}^2$$

We can now substitute into the *Euler* equations.

$$M_x = (40)(2000) + (100)(-20)(40 - 80) = 160{,}000 \text{ ft-lb}$$

$$M_y = (80)(0) + (0)(-20)(40 - 40) = 0$$

$$M_z = (40)(0) + (0)(100)(80 - 40) = 0$$

Now the moment components above are generated by the bearing-force components (see Fig. 18.7). Hence, we can say:

$$M_x = 160{,}000 = 4B_z - 4A_z \quad \text{(a)}$$

$$M_y = 0 \qquad = 0 \quad \text{(b)}$$

$$M_z = 0 \qquad = 4A_x - 4B_x \quad \text{(c)}$$

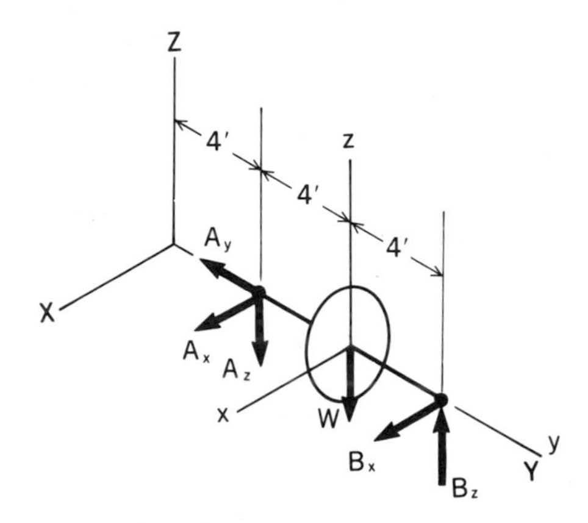

Figure 18.7. Bearing forces.

We have effectively two equations for four unknowns. We next use *Newton's* law for the mass center:

$$A_z + B_z - 322 = 0$$

$$A_x + B_x = 0 \quad \text{(d)}$$

$$-A_y = -\left(\frac{322}{g}\right)(8)(20)^2$$

The first two equations are equilibrium equations. The third equation relates the radial force A_y, from the bearing A, and the radial acceleration of the center of mass of the disc which you will notice is in simple circular motion about the Z axis. We now have enough equations for all the unknowns. It is then a simple matter to evaluate the forces from the bearings. They are:

$$A_x = B_x = 0$$
$$A_y = 32{,}000 \text{ lb}$$
$$B_z = 20{,}161 \text{ lb}$$
$$A_z = -19{,}839 \text{ lb}$$

The *reactions* to these forces are the desired forces *onto* the bearings.

In Example 18.1, you may have been surprised at the large value of the bearing forces in the z direction. Actually, if we did not include the weight of the disc, then A_z and B_z would have formed a sizable couple. This couple stems from the fact that a body having a high angular momentum about one axis is made to move such that the aforementioned axis rotates about yet a second axis. Such a couple is called a *gyroscopic couple*. It occurs in no small measure for the front wheels of a car that is steered while moving at high speeds. It occurs in the jet engine of a plane that is changing its direction of flight. You will have opportunity to investigate these effects in the homework problems.

EXAMPLE 18.2

A cylinder AB is rotating in bearings mounted on a platform (Fig. 18.8). The cylinder has an angular speed ω_2 and a rate of change of speed $\dot{\omega}_2$, both quantities being relative to the platform. The platform rotates with an angular speed ω_1 and has a rate of change of speed $\dot{\omega}_1$, both quantities being relative to the ground. Compute the moment of the supporting forces of the cylinder AB about the center of mass of the cylinder in terms of the aforementioned quantities and the moments of inertia of the cylinder.

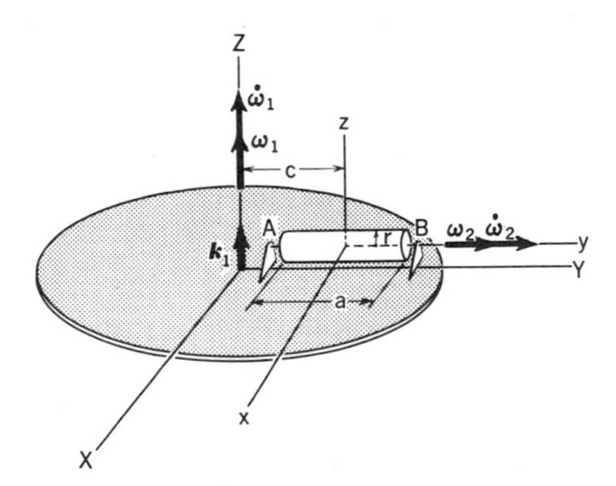

Figure 18.8. Rotating cylinder on platform.

We shall do this problem by two methods, one using axes fixed to the body and using Euler's equations, and the other using axes fixed to the platform and using Eq. 18.2.

Method I: Reference fixed to cylinder. In Fig. 18.8 we have fixed axes xyz to the cylinder at the mass center. To get components of $\boldsymbol{M}$ parallel to the inertial reference, we consider the problem when the xyz reference is parallel to the XYZ reference. The angular velocity vector $\boldsymbol{\omega}$ for the body is then

$$\boldsymbol{\omega} = \boldsymbol{\omega}_1 + \boldsymbol{\omega}_2 = \omega_1 \boldsymbol{k} + \omega_2 \boldsymbol{j}$$

Hence,

$$\begin{aligned} \omega_x &= 0 \text{ rad/sec} \\ \omega_y &= \omega_2 \text{ rad/sec} \\ \omega_z &= \omega_1 \text{ rad/sec} \end{aligned} \tag{a}$$

Also, we can say:

$$\begin{aligned} \dot{\boldsymbol{\omega}} = \dot{\boldsymbol{\omega}}_1 + \dot{\boldsymbol{\omega}}_2 &= \dot{\omega}_1 \boldsymbol{k} + \left(\frac{d}{dt}\right)_{XYZ} (\omega_2 \boldsymbol{j}) \\ &= \dot{\omega}_1 \boldsymbol{k} + \dot{\omega}_2 \boldsymbol{j} + \omega_2(\boldsymbol{\omega}_1 \times \boldsymbol{j}) \\ &= \dot{\omega}_1 \boldsymbol{k} + \dot{\omega}_2 \boldsymbol{j} + \omega_2(\omega_1 \boldsymbol{k} \times \boldsymbol{j}) \\ &= \dot{\omega}_1 \boldsymbol{k} + \dot{\omega}_2 \boldsymbol{j} - \omega_1 \omega_2 \boldsymbol{i} \end{aligned}$$

Accordingly, we have at the instant of interest

$$\begin{aligned} \dot{\omega}_x &= -\omega_1 \omega_2 \text{ rad/sec}^2 \\ \dot{\omega}_y &= \dot{\omega}_2 \text{ rad/sec}^2 \\ \dot{\omega}_z &= \dot{\omega}_1 \text{ rad/sec}^2 \end{aligned} \tag{b}$$

The *Euler equations* then become

$$M_x = I_{xx}(-\omega_1 \omega_2) + \omega_1 \omega_2 (I_{zz} - I_{yy}) \tag{c}$$

$$M_y = I_{yy} \dot{\omega}_2 + 0 \tag{d}$$

$$M_z = I_{zz} \dot{\omega}_1 + 0 \tag{e}$$

Since $I_{zz} = I_{xx}$, we see that $I_{xx}(-\omega_1 \omega_2)$ cancels $\omega_1 \omega_2 I_{zz}$ in Eq. (c), and we then have the desired result:

$$\boldsymbol{M} = - I_{yy} \omega_1 \omega_2 \boldsymbol{i} + I_{yy} \dot{\omega}_2 \boldsymbol{j} + I_{zz} \dot{\omega}_1 \boldsymbol{k} \tag{f}$$

Method II: Reference fixed to platform. We shall now do this problem by having xyz at the mass center C of the cylinder again, but now fixed to the platform. In other words, the cylinder rotates relative to the xyz reference with angular speed ω_2. Keeping this in mind, we can still refer to Fig. 18.8.

Obviously, we cannot use Euler's equations here and must return to Eq. 18.2.

$$\boldsymbol{M}_c = \left(\frac{d\boldsymbol{H}_C}{dt}\right)_{xyz} + \boldsymbol{\Omega} \times \boldsymbol{H}_C \tag{a}$$

Because the cylinder is a body of revolution about the y axis, the products of inertia I_{xy}, I_{xz}, and I_{yz} are always zero, and I_{xx}, I_{yy}, and I_{zz} are *constants* at all times. Were these conditions not present, this method of approach would be very difficult, since we would have to ascertain the time derivatives of these inertia terms. Thus, using Eq. 18.1, we see that

$$\begin{aligned} \left(\frac{d\boldsymbol{H}_C}{dt}\right)_{xyz} &= \left(\frac{d}{dt}\right)_{xyz} (H_x \boldsymbol{i} + H_y \boldsymbol{j} + H_z \boldsymbol{k})_C \\ &= \left(\frac{d}{dt}\right)_{xyz} (I_{xx} \omega_x \boldsymbol{i} + I_{yy} \omega_y \boldsymbol{j} + I_{zz} \omega_z \boldsymbol{k}) \end{aligned} \tag{b}$$

$$= \left(\frac{d}{dt}\right)_{xyz} (0\boldsymbol{i} + I_{yy}\omega_2\boldsymbol{j} + I_{zz}\omega_1\boldsymbol{k})$$

Note that $\boldsymbol{i}, \boldsymbol{j}$, and $\boldsymbol{k}$ are constants as seen from xyz. Only ω_1 and ω_2 are time functions and undergo simple time differentiation of scalars. Thus,

$$\left(\frac{d\boldsymbol{H}_C}{dt}\right)_{xyz} = I_{yy}\dot{\omega}_2\boldsymbol{j} + I_{zz}\dot{\omega}_1\boldsymbol{k} \tag{c}$$

Noting further that $\boldsymbol{\Omega} = \omega_1\boldsymbol{k}$, we have on substituting Eq. (c) into Eq. (a):

$$\begin{aligned} \boldsymbol{M} &= I_{yy}\dot{\omega}_2\boldsymbol{j} + I_{zz}\dot{\omega}_1\boldsymbol{k} + \omega_1\boldsymbol{k} \times (H_x\boldsymbol{i} + H_y\boldsymbol{j} + H_z\boldsymbol{k}) \\ &= I_{yy}\dot{\omega}_2\boldsymbol{j} + I_{zz}\dot{\omega}_1\boldsymbol{k} + \omega_1\boldsymbol{k} \times (0\boldsymbol{i} + I_{yy}\omega_2\boldsymbol{j} + I_{zz}\omega_1\boldsymbol{k}) \qquad \text{(d)} \\ &= -I_{yy}\omega_1\omega_2\boldsymbol{i} + I_{yy}\dot{\omega}_2\boldsymbol{j} + I_{zz}\dot{\omega}_1\boldsymbol{k} \end{aligned}$$

This equation is identical to the one obtained using method I.

Problems

18.1. The moving parts of a jet engine consist of a *compressor* and a *turbine* connected to a common shaft. Suppose that this system is rotating at a speed ω_1 of 10,000 rpm and the plane it is in moves at a speed of 600 mi/hr in a circular loop of radius 2 mi. What is the direction and magnitude of the gyroscopic torque transmitted to the plane from the engine through bearings A and B? The engine has a weight of 200 lb and a radius of gyration about its axis of rotation of 1 ft. The radius of gyration at the center of mass for an axis normal to the centerline is 1.5 ft. What advantage is achieved by using two oppositely turning jet engines instead of one large one?

Figure P.18.1

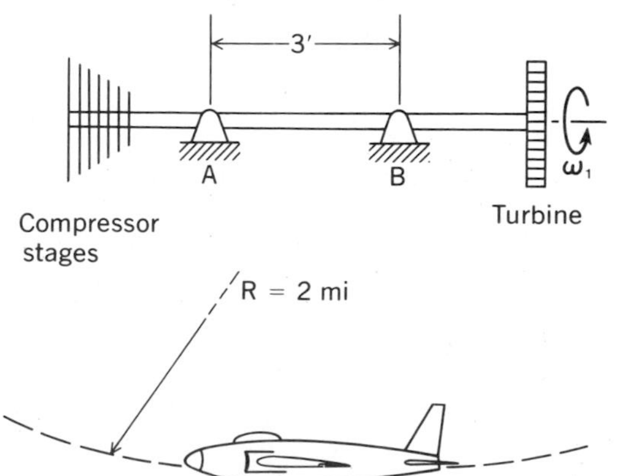

18.2. A space capsule (unmanned) is tumbling in space due to malfunctioning of its control system such that at time t, $\omega_1 = 3$ rad/sec, $\omega_2 = 5$ rad/sec, and $\omega_3 = 4$ rad/sec. At this instant, small jets are creating a torque T of 30 N-m. What are the angular acceleration components at this instant? The vehicle is a body of revolution with $k_z = 1$ m, $k_y = k_x = 1.6$ m, and has a mass of 1000 kg.

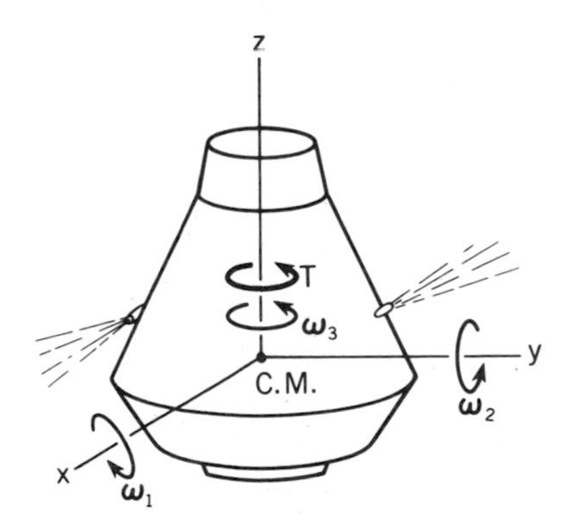

Figure P.18.2

18.3. The left front tire of a car is moving at 55 mi/hr along an unbanked road along a circular path having a mean radius of 150 yd. The tire is 26 in. in diameter. The rim plus tire weighs 30 lb and has a combined radius of gyration of 9 in. about its axis. Normal to the axis, the radius of gyration is 7.5 in. What is the gyroscopic torque on the bearings of this front wheel coming solely from the motion of the front wheel?

18.4. In Problem 18.3, suppose that the driver is turning the front wheel at a rate ω_3 of .2 rad/sec at the instant where the radius of curvature of the path is 150 yd. What is then the torque needed on the wheel solely from the motion of the wheel?

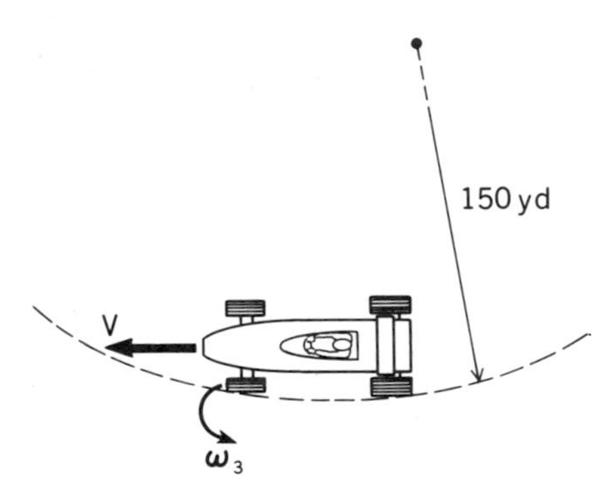

Figure P.18.4

18.5. A student is holding a rapidly rotating wheel in front of him. He is standing on a platform that can turn freely. If the student exerts a torque M_1 as shown, what begins to happen initially? What happens a little later?

Figure P.18.5

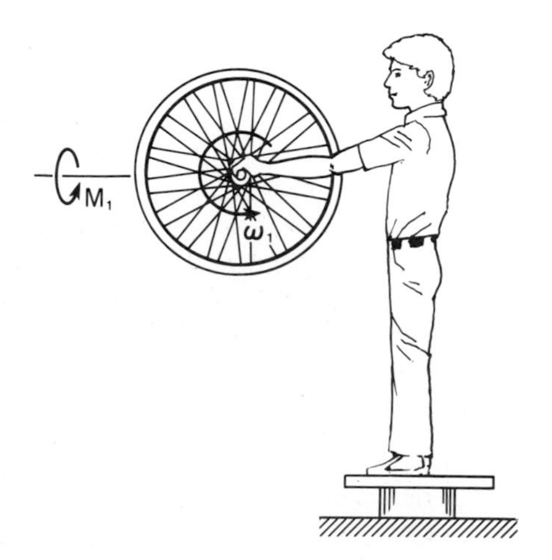

18.6. A motor is mounted on a rotating platform having an angular velocity ω_1 of 2 rad/sec. The motor drives two fans at the rate ω_2 of 1750 rpm relative to the platform. The fans plus armature of the motor have a total weight of 100 N and a radius of gyration along the axis of 200 mm. About the Z axis, the radius of gyration is 300 mm. What is the torque coming onto the bearings of the motor as a result of the motion?

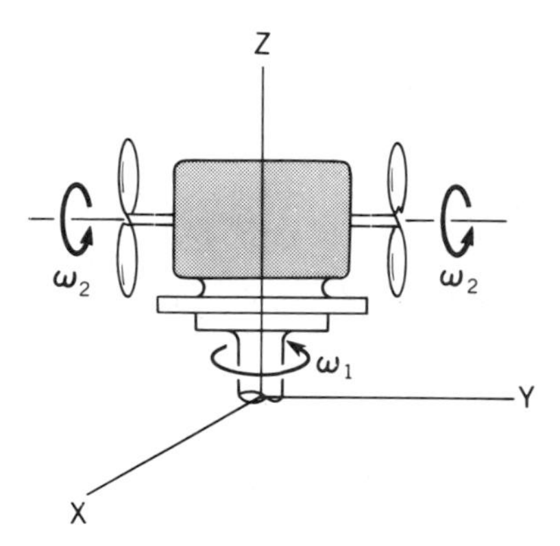

Figure P.18.6

18.7. A lug wrench is translating in inertial space inside an orbiting space vehicle. A torque $\boldsymbol{T} = 2\boldsymbol{i} + 6\boldsymbol{j} + 5\boldsymbol{k}$ N-m is exerted at the center of the wrench. What is the acceleration of end A at this instant? Approximate the wrench as two slender rods. The wrench weighs 12 N.

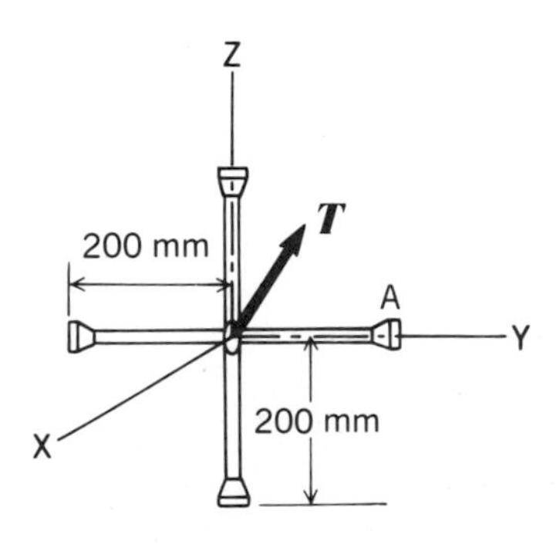

Figure P.18.7

18.8. Work Problem 18.7 for the case where

$$\boldsymbol{\omega} = 3\boldsymbol{i} - 2\boldsymbol{j} + 4\boldsymbol{k} \text{ rad/sec}$$

about an axis of rotation through the center of mass at the instant when $\boldsymbol{T}$ is applied.

18.9. A propeller-driven airplane is at the bottom of a loop of radius 2000 ft and traveling at 350 mi/hr. The propeller consists of two identical blades at right angles, weighs 322 lb, has a radius of gyration of 2 ft about its axis of rotation, and is rotating at 1200 rpm. If the propeller rotates counterclockwise as viewed from the rear of the plane, compute the torques coming onto the propeller at the bearings from the motion if one blade is vertical and the other is horizontal at the time of interest.

***18.10.** A thin disc weighing 32.2 lb rotates on rod AB at a speed ω_2 of 100 rad/sec in a clockwise direction looking from B to A. The radius of the disc is 1 ft, and the disc is located 10 ft from the centerline of the shaft CD, to which rod AB is fixed. Shaft CD rotates at $\omega_1 = 50$ rad/sec in a counterclockwise direction as we look from C to D. Find the tensile force, bending moment, and shear force on rod AB at the end A due to the disc.

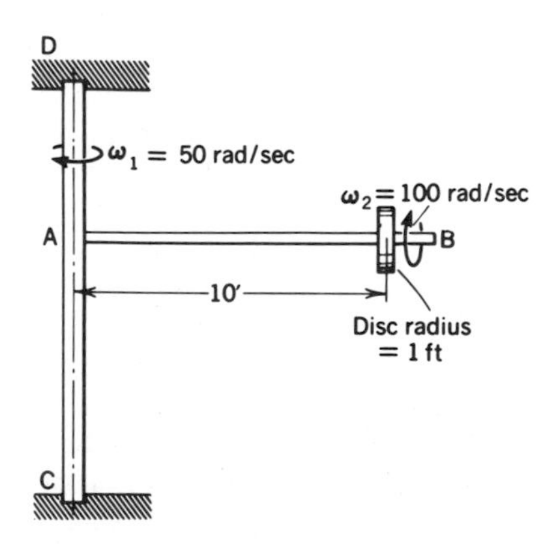

Figure P.18.10

18.11. The turbine in a ship is parallel to the longitudinal axis of the ship and is rotating at a rate ω_1 of 800 rpm counterclockwise as viewed from stern to bow. The turbine weighs 445 kN and has radii of gyration at the mass center of 1 m about axes normal to the centerline and of 300 mm about the centerline. The ship has a pitching motion, which is approximately sinusoidal, given as

$$\theta = .3 \sin \frac{2\pi}{15} t \text{ rad}$$

The amplitude of the pitching is thus .3 rad and the period is 15 sec. Determine the moment as a function of time about the mass center needed for the motion of the turbine.

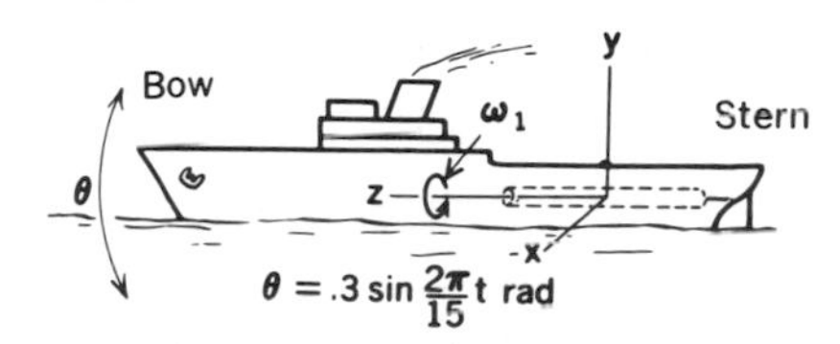

Figure P.18.11

18.12. Explain how the roll of a ship can be stabilized by the action of a heavy rapidly spinning disc (gyroscope) rotating in a set of bearings in the ship as shown.

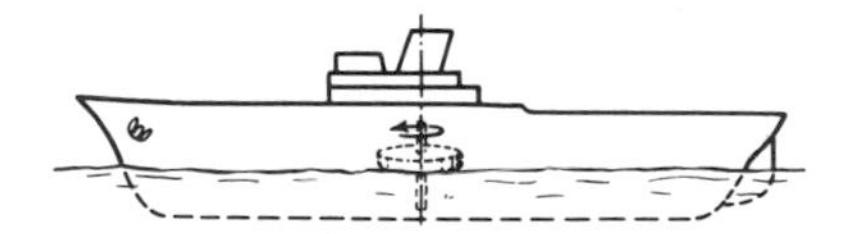

Figure P.18.12

18.13. A 10-kg disc rotates with speed $\omega_1 = 10$ rad/sec relative to rod AB. Rod AB rotates with speed $\omega_2 = 4$ rad/sec relative to the vertical shaft, which rotates with speed $\omega_3 = 2$ rad/sec relative to the ground. What is the torque coming onto the bearings at B due to the motion at a time when $\theta = 60°$? Take $\dot{\omega}_1 = \dot{\omega}_2 = \dot{\omega}_3 = 0$.

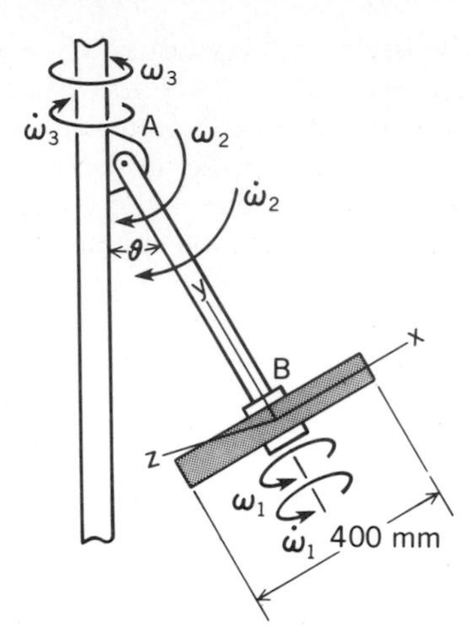

Figure P.18.13

18.14. Solve Problem 18.13 for the case where $\dot{\omega}_1 = 2$ rad/sec², $\dot{\omega}_2 = 3$ rad/sec², and $\dot{\omega}_3 = 4$ rad/sec² in the directions shown.

18.15. A man is seated in a centrifuge of the type described in Example 15.14. If $\omega_1 = 2$ rad/sec, $\omega_2 = 3$ rad/sec, and $\omega_3 = 4$ rad/sec, what torque must the seat develop about the center of mass of the man as a result of the motion? The man weighs 700 N and has the following radii of gyration as determined by experiment while sitting in the seat:

$$k_x = 600 \text{ mm}$$
$$k_y = 500 \text{ mm}$$
$$k_z = 150 \text{ mm}$$

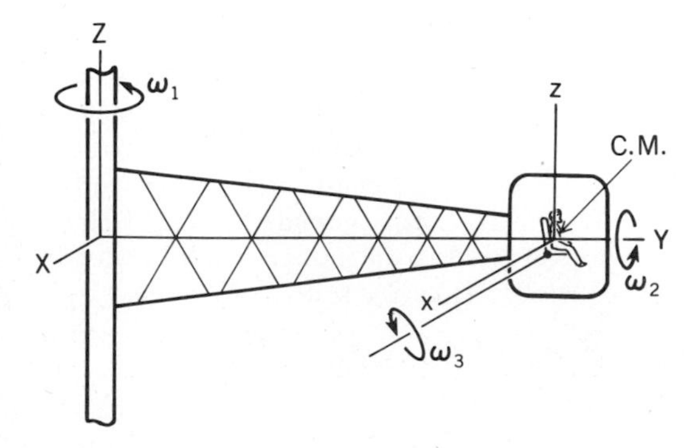

Figure P.18.15

***18.16.** Work Problem 18.15 for the following data:

$$\dot{\omega}_1 = 3 \text{ rad/sec}^2$$
$$\dot{\omega}_2 = -2 \text{ rad/sec}^2$$
$$\dot{\omega}_3 = 4 \text{ rad/sec}^2$$

The other data are the same.

18.17. A thin disc has its axis inclined to the vertical by an angle θ and rolls without slipping with an angular speed ω_1 about the supporting rod held at B with a ball-and-socket joint. If $l = 10$ ft, $r = 2$ ft, $\theta = 45°$, and $\omega_1 = 10$ rad/sec, compute the angular velocity of the rod BC about axis O–O. If the disc weighs 40 lb, what is the total moment about point B from all forces acting on the system? Neglect the mass of the rod OC. (*Hint:* Use a reference xyz at B when two of the axes are in the plane of O–O and OC.)

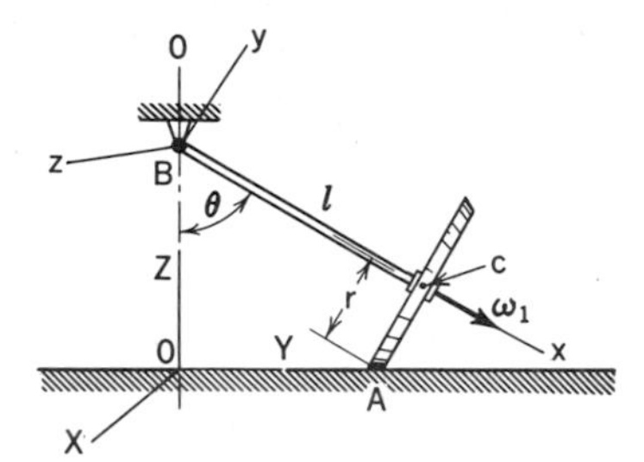

Figure P.18.17

18.18. A gage indicator CD in an instrument is 20 mm long and is rotating relative to platform A at the rate $\omega_1 = .3$ rad/sec. The platform A is fixed to a space vehicle which is rotating at speed $\omega_2 = 1$ rad/sec about vertical axis LM. At the instant shown, what are the moment components from the bearings E and G about the center of mass of CD needed for the motion of CD and EG? The indicator weighs .25 N, and the shaft EG weighs .30 N, has a diameter of 1 mm, and a length of 10 mm.

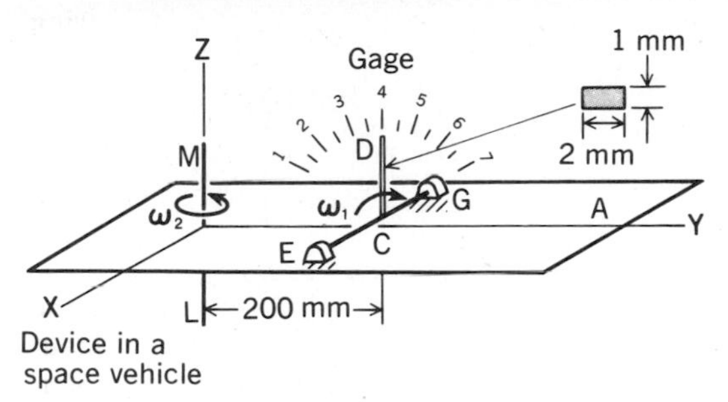

Figure P.18.18

18.19. An eight-bladed fan is used in a wind tunnel to drive the air. The angular velocity ω_1 of the fan is 120 rpm. At the instant of interest, each blade is rotating about its own axis z (in order to change the angle of attack of the blade) such that $\omega_2 = 1$ rad/sec and $\dot{\omega}_2$ is .5 rad/sec². Each blade weighs 200 N and has the following radii of gyration at the mass center (C.M.):

$$k_x = 600 \text{ mm}$$

$$k_y = 580 \text{ mm}$$

$$k_z = 100 \text{ mm}$$

Consider only the dynamics of the system (and not the aerodynamics) to find the torque required about the z axis of the blade. If a couple moment of 4 N-m is developed in the x direction at the base of the blade, what force component F_y is needed at the base of the blade for the motion described?

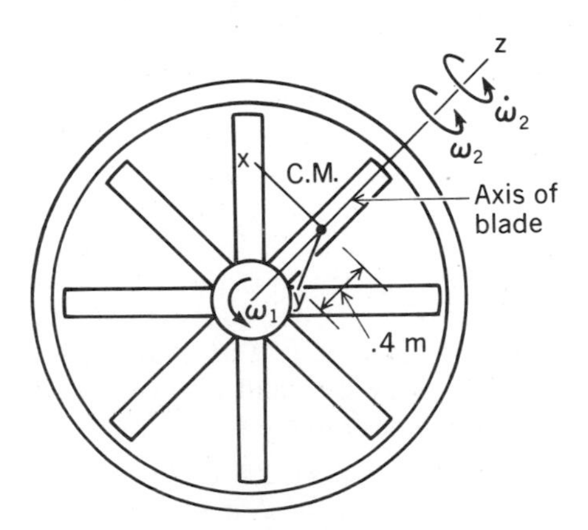

Figure P.18.19

18.20. A swing-wing fighter plane is moving with a speed of Mach 1.3. (Note that Mach 1 corresponds to a speed of about 1200 km/hr.) The pilot at the instant shown is swinging his wings back at the rate ω_1 of .3 rad/sec. At the same time, he is rolling at a rate ω_2 of .6 rad/sec and is performing a loop as shown. The wing weighs 6.5 kN and has the following radii of gyration at the center of mass:

$$k_x = .8 \text{ m}$$

$$k_y = 4 \text{ m}$$

$$k_z = 6 \text{ m}$$

What is the moment that the fuselage must develop about the center of mass of the wing to accomplish the dynamics of the described motions? Do not consider aerodynamics.

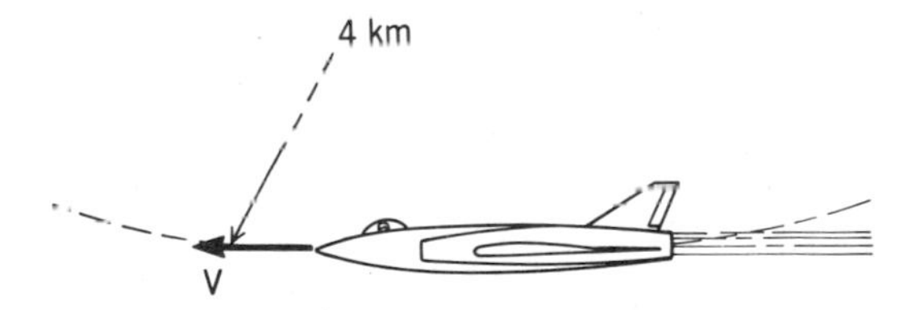

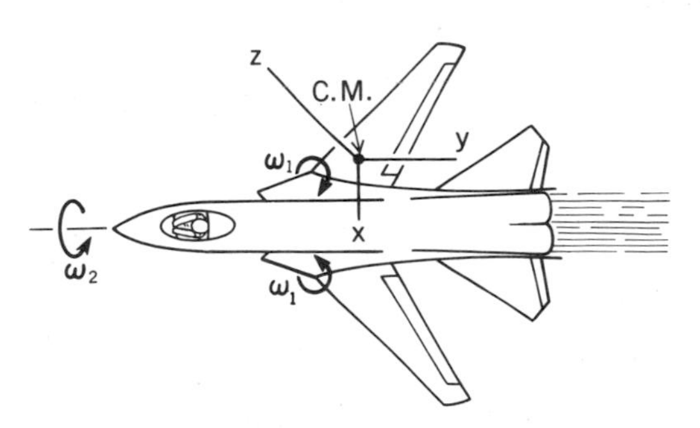

Figure P.18.20

18.21. An orbiting skylab is rotating at an angular speed ω_1 of .4 rad/sec to give a "gravitational" effect for the living quarters in the outer annulus. A many-bladed fan is mounted as shown on the outer "floor." The fan blades rotate at a speed ω_2 of

200 rpm and the base rotates at a speed ω_3 of 1 rad/sec with $\dot{\omega}_3$ equal to .6 rad/sec^2 all relative to the skylab. At the instant of interest $\boldsymbol{\omega}_1$, $\boldsymbol{\omega}_2$, and $\boldsymbol{\omega}_3$ are orthogonal to each other. If the blade has a mass of 100 grams and if the radius of gyration about its axis is 200 mm, what is the torque coming onto the fan due solely to the motion? Take the transverse radii of gyration of the fan to be 120 mm.

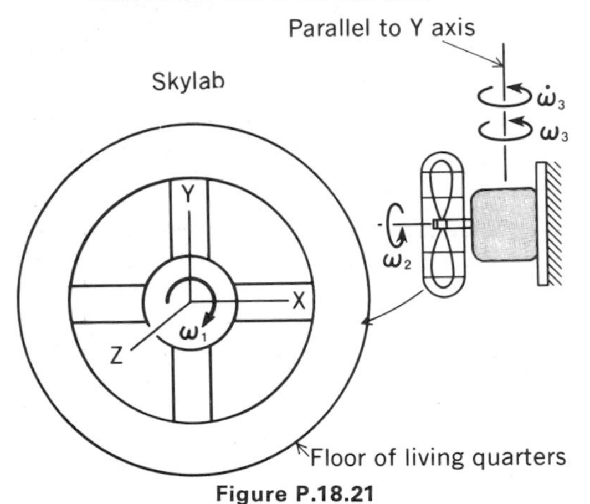

Figure P.18.21

18.22. You will learn in fluid mechanics that when air moves across a rotating cylinder a force is developed normal to the axis of the cylinder (the *Magnus effect*). In 1926 Flettner used this principle to "sail" a vessel across the Atlantic. Two cylinders were kept at a constant rotational speed by a motor of $\omega_1 = 200$ rpm relative to the ship. Suppose that in rough seas the ship is rolling about the axis of the ship with a speed ω_2 of .8 rad/sec. What couple moment components must the ship transmit to the *base* of the cylinder as a result only of the motion of the ship? Each cylinder weighs 700 lb and has a radius of gyration along the axis of 2 ft and normal to the axis at the center of mass of 10 ft.

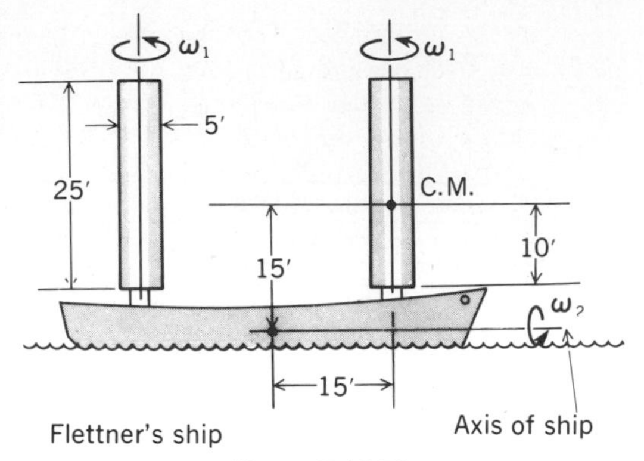

Figure P.18.22

***18.23.** Part of a clutch system consists of identical rods AB and AC rotating relative to a shaft at the speed $\omega_2 = 3$ rad/sec while the shaft rotates relative to the ground at a speed $\omega_1 = 40$ rad/sec. As a result of the motion what are the bending moment components at the base of each rod if each rod weighs 10 N?

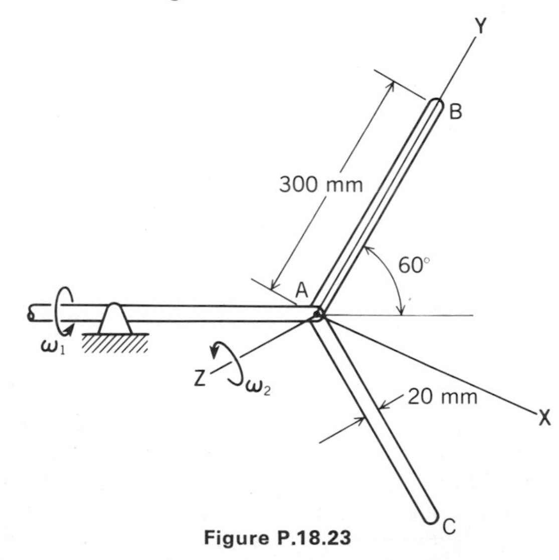

Figure P.18.23

***18.24.** Solve Problem 18.23 for the case when $\dot{\omega}_1 = 2$ rad/sec^2 and $\dot{\omega}_2 = 6$ rad/sec^2 at the instant of interest.

18.25. A rod D weighing 2 N and having a length of 200 mm and a diameter of 10 mm rotates relative to platform K at a speed

ω_1 of 120 rad/sec. The platform is in a space vehicle and turns at a speed ω_2 of 50 rad/sec relative to the vehicle. The vehicle rotates at a speed ω_3 of 30 rad/sec relative to inertial space about the axis shown. What are the bearing forces at A and B normal to the axis AB resulting from the motion of rod D? What torque T is needed about AB to maintain a constant value of ω_1 at the instant of interest?

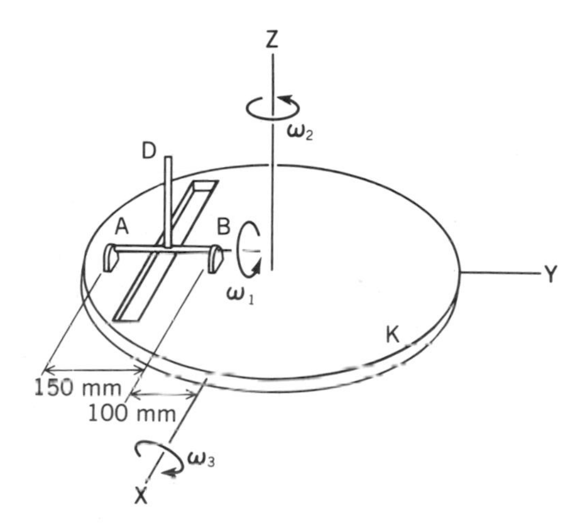

Figure P.18.25

18.26. Solve Problem 18.6 using a reference at the center of mass of the armature but not fixed to the armature.

18.27. In Problem 18.10, find the moment about the C.M. of the disc using a reference xyz fixed to the rod AB and not to the disc. Take $\dot{\omega}_1 = 10$ rad/sec^2 and $\dot{\omega}_2 = 30$ rad/sec^2.

18.28. Work Problem 18.11 using reference xyz fixed to the ship with the origin at the center of mass of the turbine. Do not use Euler's equations.

18.29. Solve Problem 18.13 using a set of axes xyz at the center of mass of the disc but fixed to arm AB.

18.30. Work Problem 18.13 by using a reference xyz fixed to the arm at the center of mass of the disc for the case where $\dot{\omega}_1 = 2$ rad/sec^2, $\dot{\omega}_2 = 3$ rad/sec^2, and $\dot{\omega}_3 = 4$ rad/sec^2 in directions shown.

18.4 Necessary and Sufficient Conditions for Equilibrium of a Rigid Body

In this chapter, we have employed Newton's law at the mass center as well as the equation $\boldsymbol{M}_A = \dot{\boldsymbol{H}}_A$ and from it we derived Euler's equations for rigid bodies. We can now go back to our work in Chapter 5 and put on firm ground the fact that $\boldsymbol{M} = \boldsymbol{0}$ and $\boldsymbol{F} = \boldsymbol{0}$ are necessary conditions for equilibrium of a rigid body. (You will recall we accepted these equations for statics at that time by intuition, pending a proof to come later.)

A particle is in equilibrium, you will recall, if it is stationary or moving with constant speed along a straight line in inertial space. To be in a state of equilibrium, every point in a rigid body must accordingly be stationary or be moving at uniform speed along straight lines in inertial space. The rigidity requirement thus limits a rigid body in equilibrium to translational motion along a straight line at constant speed in inertial space. This means that $\dot{\boldsymbol{V}}_C = \boldsymbol{0}$ and $\boldsymbol{\omega} = \boldsymbol{0}$ for equilibrium and so, from Newton's law and Euler's equations, we see that $\boldsymbol{F} = \boldsymbol{M}_A = \boldsymbol{0}$ are *necessary* conditions for equilibrium.

For a sufficiency proof, we go the other way. For a body initially in equilibrium, the conditions $\boldsymbol{F} = \boldsymbol{M}_A = \mathbf{0}$[5] ensures that equilibrium will be maintained. More specifically, we shall start with a body in equilibrium at time t and apply a force system satisfying the preceding conditions. We address ourselves to the question: Does the body stay in a state of equilibrium? According to Newton's law there will be no change in the velocity of the mass center since $\boldsymbol{F} = \mathbf{0}$. And, with $\boldsymbol{\omega} = \mathbf{0}$ at time t, Euler's equations lead to the result for $\boldsymbol{M}_C = \mathbf{0}$, that $\dot{\omega}_x = \dot{\omega}_y = \dot{\omega}_z = 0$. Thus, the angular velocity must remain zero. With the velocity of the center of mass constant, and with $\dot{\boldsymbol{\omega}} = \mathbf{0}$ in inertial space we know that the body remains in equilibrium. Thus, if a body is *initially in equilibrium*, the condition $\boldsymbol{F} = \mathbf{0}$ and $\boldsymbol{M}_A = \mathbf{0}$ is *sufficient* for maintaining equilibrium.

18.5 Three-Dimensional Motion About a Fixed Point; Euler Angles

We shall now examine the motion of selected rigid bodies constrained to have a point fixed in an inertial reference (Fig. 18.9). This topic will lead to study of an important device—the gyroscope.

We now show that angles of rotation θ_x, θ_y, and θ_z, along orthogonal axes x, y, and z are not highly suitable for measuring the orientation of a body with a fixed point. Thus, in Fig. 18.10(a) consider a conical surface and observe a straight line OA

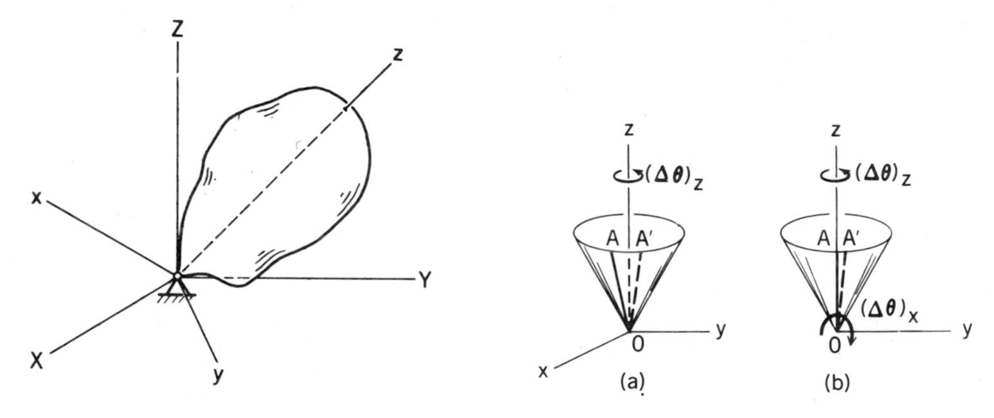

Figure 18.9. Body with a fixed point.

Figure 18.10. $\Delta\theta_z$ causes $\Delta\theta_x$.

which is on this surface and in the xz plane. Rotate the cone an angle $\Delta\theta_z$ thus causing OA to move to OA' shown as dashed. In Fig. 18.10(b), view line OA along a line of sight corresponding to the x axis. As a result of the rotation $\Delta\theta_z$ about the z axis, there will clearly be a rotation $\Delta\theta_x$ of this line about the x axis. Thus, we see that θ_x and θ_z are *mutually dependent* and thus not suitable for our use. This result stems from the fact that we are using directions that have a fixed mutual relative orientation. We now

[5]We have shown in statics that if $\boldsymbol{F} = \mathbf{0}$ and $\boldsymbol{M}_A = \mathbf{0}$ about some point A in inertial space, then $\boldsymbol{M} = \mathbf{0}$ about any point in inertial space.

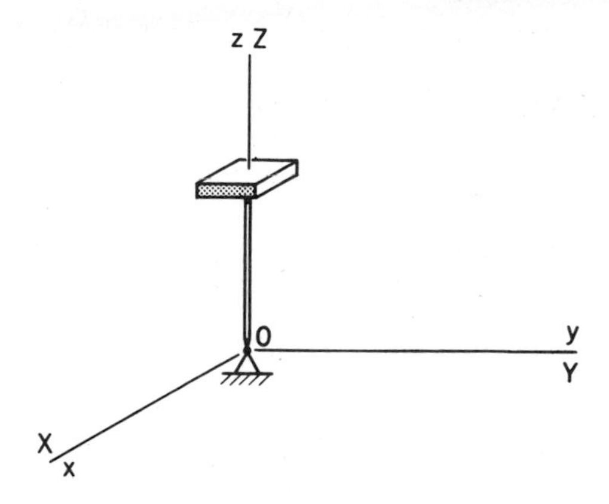

Figure 18.11. Rigid body.

introduce a set of rotations that are *independent*. And, not unexpectedly, the axes for these rotations will *not* have a fixed relative orientation.

Accordingly, consider the rigid body shown in Fig. 18.11. We shall specify a sequence of three independent rotations in the following manner:

1. Keep a reference xyz fixed in the body, and rotate the body about the Z axis through an angle ψ shown in Fig. 18.12.

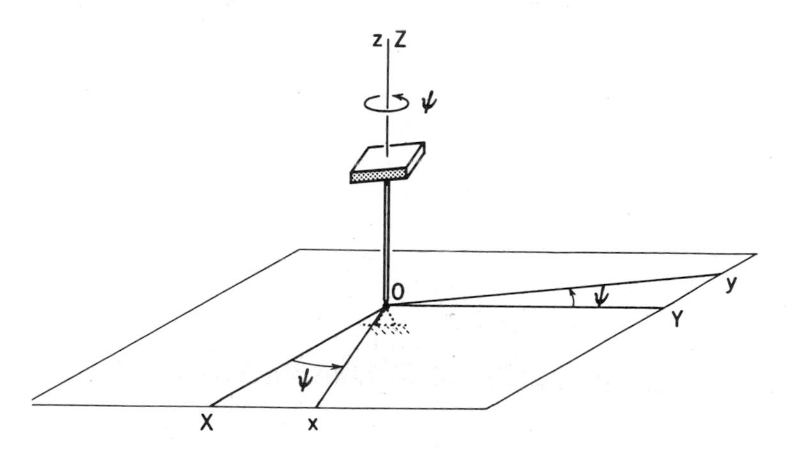

Figure 18.12. First rotation about Z.

2. Now rotate the body about the x axis through an angle θ to reach the configuration in Fig. 18.13. Note that the z, Z, and y axes form a plane I normal to the XY plane and normal also to the x axis. The axis of rotation for this rotation (x axis) is called the *line of nodes.*
3. Finally, rotate the body an angle ϕ about the z axis. We provide the option here of detaching the xyz reference from the body for this movement (see Fig. 18.14), in

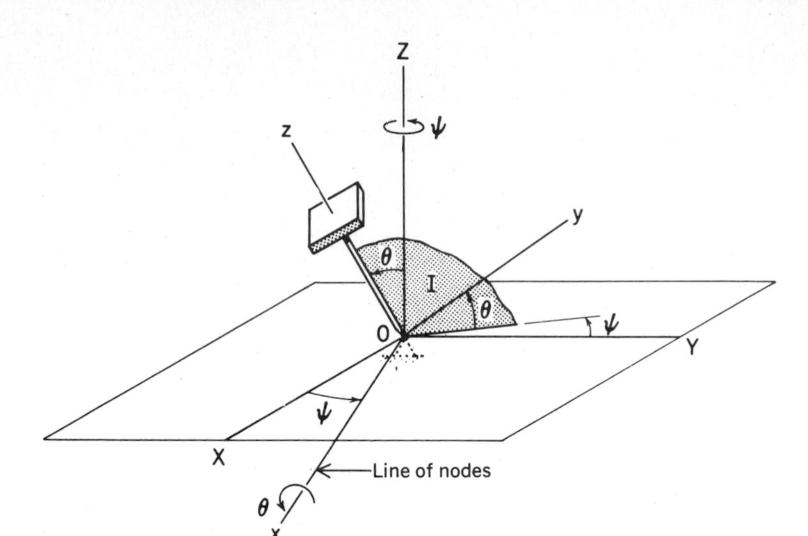

Figure 18.13. Second rotation about x.

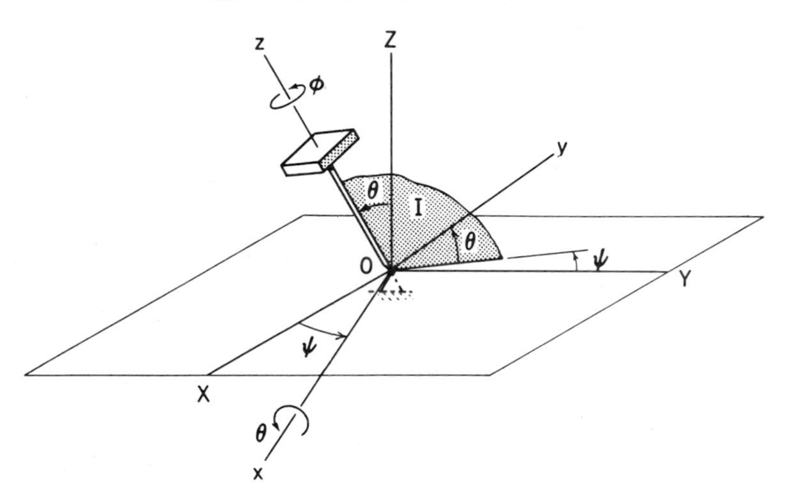

Figure 18.14. Third rotation about z.

which case the body rotates an angle ϕ relative to xyz, and the x axis remains collinear with the line of nodes. Or, we can permit the reference xyz to remain fixed in the body (see Fig. 18.15), in which case we can use components of $\boldsymbol{M}$ and $\boldsymbol{\omega}$ along these axes in employing Euler's equations. The line of nodes is now identified simply as the normal to plane I containing z and Z.

We thus arrive at the desired orientation. Positive rotations in each case are those taken as counterclockwise as one looks to the origin O along the axis of rotation. (Thus, we have performed three positive rotations here.)

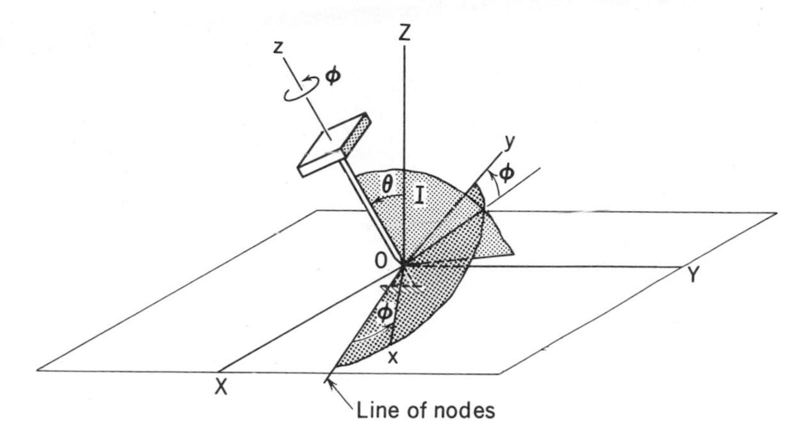

Figure 18.15. Axes *xyz* fixed to body.

We call these angles the *Euler angles*, and we assign the following names.

$$\psi = \text{angle of precession}$$

$$\theta = \text{angle of nutation}$$

$$\phi = \text{angle of spin}$$

Furthermore, the z axis is usually called the *body axis*, and the Z axis is often called the *axis of precession*. The line of nodes then is normal to the body axis and the precession axis.

We have shown that the position of a body moving with one point fixed can be established by three independent rotations given in a certain sequence. For an infinitesimal change in position, this situation would be a rotation $d\psi$ about the Z axis, $d\theta$ about the line of nodes, and $d\phi$ about the body axis z. Because these rotations are infinitesimal, they can be construed as vectors, and the order mentioned above is no longer required. The limiting ratios of these changes in angles with respect to time give rise to three angular velocity vectors (Fig. 18.16), which we express in the following manner:

$\dot{\boldsymbol{\psi}}$, directed along the Z axis
$\dot{\boldsymbol{\theta}}$, directed along the line of nodes
$\dot{\boldsymbol{\phi}}$, directed along the z body axis

Note that the nutation velocity vector $\dot{\boldsymbol{\theta}}$ is always normal to plane I, and consequently the nutation velocity vector is always normal to the spin velocity vector $\dot{\boldsymbol{\phi}}$ and the precession velocity vector $\dot{\boldsymbol{\psi}}$. However, the spin velocity vector $\dot{\boldsymbol{\phi}}$ will generally *not* be at right angles with the precession velocity $\dot{\boldsymbol{\psi}}$, and so this system of angular velocity vectors generally is *not* an orthogonal system.

Finally, it should be clear that the reference *xyz* moves with the body during precession and nutation motion of the body, but we can choose that it not move

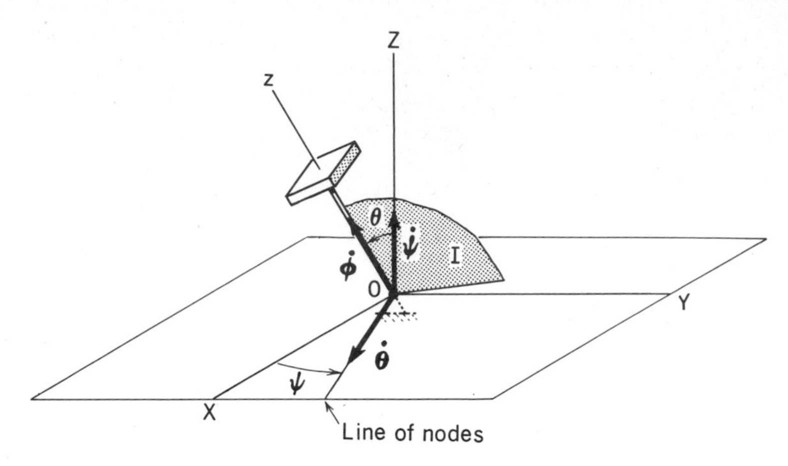

Figure 18.16. Precession, nutation, and spin velocity vectors.

(relative to XYZ) during a spin rotation. Hence, while the body has the angular velocity $\dot{\boldsymbol{\phi}} + \dot{\boldsymbol{\psi}} + \dot{\boldsymbol{\theta}}$ at any time t, the reference xyz would have, for the aforestated condition, an angular velocity, denoted as $\boldsymbol{\Omega}$, equal to $\dot{\boldsymbol{\psi}} + \dot{\boldsymbol{\theta}}$. The velocity of the disc relative to xyz is, accordingly, $\dot{\phi}\boldsymbol{k}$ for this case.

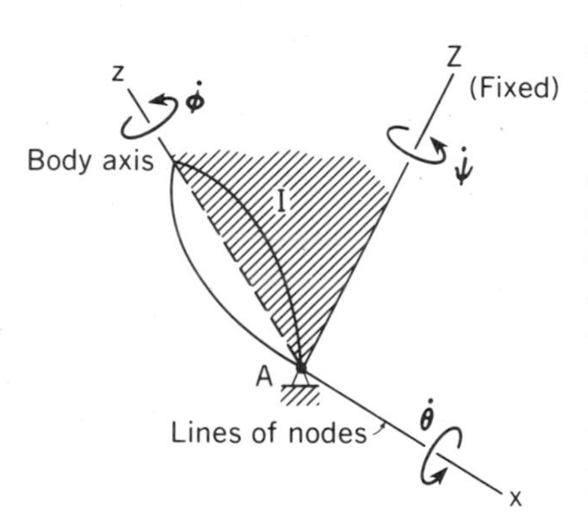

Figure 18.17. Line of nodes always normal to plane I.

Consider a body with two orthogonal planes of symmetry forming an axis in the body. The body is moving about a fixed point A on the axis (Fig. 18.17). How do we decide what axes to use to describe the motion in terms of spin, precession, and nutation? First, we take the axis of the body to be the z axis; the angular speed about this axis is then the spin $\dot{\phi}$. This step is straightforward for the bodies that we shall consider. The next step is not. By inspection find a Z axis in a *fixed* direction so as to form with the aforementioned body axis z a plane I whose angular speed about the Z axis is either known or is sought. Such an axis Z is then the *precession* axis about which we have for the body axis z a precession speed $\dot{\psi}$. The *line of nodes* is then the axis which at all times remains *normal* to plane I containing the body axis z and the precession axis Z. The nutation speed $\dot{\theta}$ finally is the angular speed component of the body axis z about the line of nodes.

18.6 Equations of Motion Using Euler Angles

Consider next a body having a shape such that, at any point along the body axis z, the moments of inertia for all axes *normal* to the body axis have the same value I'. Such would, of course, be true for the special case of a body of revolution having the z axis as the axis of symmetry.[6] We shall consider the motion of such a body about a fixed point O, which is somewhere along the axis z (see Fig. 18.18). Reference xyz at O has the same nutation and precession motion as the body but will be chosen such that the body rotates with an angular speed $\dot{\phi}$ relative to it. Since the reference is not fixed to the body, we cannot use Euler's equations but must go back to the equation $\boldsymbol{M}_0 = \dot{\boldsymbol{H}}_0$, which when carried out in terms of components parallel to the xyz reference becomes

$$\boldsymbol{M}_0 = \left(\frac{d\boldsymbol{H}_0}{dt}\right)_{xyz} + \boldsymbol{\Omega} \times (H_x\boldsymbol{i} + H_y\boldsymbol{j} + H_z\boldsymbol{k}) \tag{18.21}$$

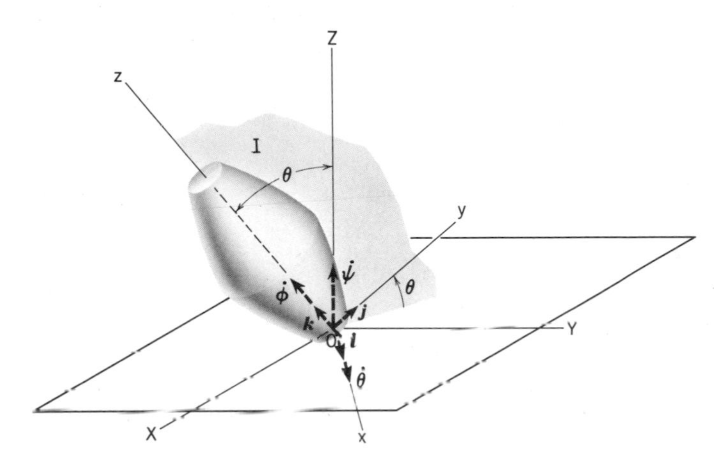

Figure 18.18. Body moving about fixed point O.

Since the xyz axes remain at *all times* principal axes, we have

$$H_x = I'\omega_x, \qquad H_y = I'\omega_y, \qquad H_z = I\omega_z \tag{18.22}$$

where I is the moment of inertia about the axis of symmetry and I' is the moment of inertia about an axis normal to the z axis at O. Considering Fig. 18.18, we can see by inspection that the angular velocity of the body relative to XYZ is at *all times* given by components parallel to xyz as follows:

$$\omega_x = \dot{\theta} \tag{18.23a}$$

$$\omega_y = \dot{\psi} \sin\theta \tag{18.23b}$$

$$\omega_z = \dot{\phi} + \dot{\psi}\cos\theta \tag{18.23c}$$

Hence, the components of the angular momentum at all times are:

$$\begin{aligned}(H_x) &= I'\dot{\theta}\\ (H_y) &= I'\dot{\psi}\sin\theta\\ (H_z) &= I(\dot{\phi} + \dot{\psi}\cos\theta)\end{aligned} \tag{18.24}$$

---

[6]You will be asked to show in Problem 18.41 that, if I_{xx}, I_{yy}, and I_{zz} are principal axes and $I_{xx} = I_{yy} = I'$, then $I'_{xx} = I'_{yy} = I'$ for *any* axes x', y' formed by rotating xyz about the z axis. Thus, homogeneous cylinders having regular cross sections such as squares or octagons would meet the requirements of this section.

We then have for $\boldsymbol{H}_0$:

$$\boldsymbol{H}_0 = I'\dot{\theta}\boldsymbol{i} + I'\dot{\psi}\sin\theta\boldsymbol{j} + I(\dot{\phi} + \dot{\psi}\cos\theta)\boldsymbol{k}$$

Remembering that $\boldsymbol{i}$, $\boldsymbol{j}$, and $\boldsymbol{k}$ are constants as seen from xyz, we can say:

$$\begin{aligned}\left(\frac{d\boldsymbol{H}_0}{dt}\right)_{xyz} &= \left(\frac{d}{dt}\right)_{xyz}[I'\dot{\theta}\boldsymbol{i} + I'\dot{\psi}\sin\theta\boldsymbol{j} + I(\dot{\phi} + \dot{\psi}\cos\theta)\boldsymbol{k}] \\ &= I'\ddot{\theta}\boldsymbol{i} + I'(\ddot{\psi}\sin\theta + \dot{\psi}\dot{\theta}\cos\theta)\boldsymbol{j} + I(\ddot{\phi} + \ddot{\psi}\cos\theta - \dot{\psi}\dot{\theta}\sin\theta)\boldsymbol{k}\end{aligned} \tag{18.25}$$

As for the angular velocity of reference xyz, we have on considering Eq. 18.24 with $\dot{\phi}$ deleted because xyz is not fixed to the body as far as spin is concerned:

$$\boldsymbol{\Omega} = \dot{\theta}\boldsymbol{i} + \dot{\psi}\sin\theta\boldsymbol{j} + \dot{\psi}\cos\theta\boldsymbol{k} \tag{18.26}$$

Consequently, we have

$$\begin{aligned}\boldsymbol{\Omega}\times\boldsymbol{i} &= -\dot{\psi}\sin\theta\boldsymbol{k} + \dot{\psi}\cos\theta\boldsymbol{j} \\ \boldsymbol{\Omega}\times\boldsymbol{j} &= \dot{\theta}\boldsymbol{k} - \dot{\psi}\cos\theta\boldsymbol{i} \\ \boldsymbol{\Omega}\times\boldsymbol{k} &= -\dot{\theta}\boldsymbol{j} + \dot{\psi}\sin\theta\boldsymbol{i}\end{aligned} \tag{18.27}$$

Substituting the results from Eqs. 18.25, 18.24, and 18.27 into Eq. 18.21, we get

$$\begin{aligned}M_x\boldsymbol{i} + M_y\boldsymbol{j} + M_z\boldsymbol{k} = {} & I'\ddot{\theta}\boldsymbol{i} + I'(\ddot{\psi}\sin\theta + \dot{\psi}\dot{\theta}\cos\theta)\boldsymbol{j} \\ & + I(\ddot{\phi} + \ddot{\psi}\cos\theta - \dot{\psi}\dot{\theta}\sin\theta)\boldsymbol{k} \\ & + I'\dot{\theta}(-\dot{\psi}\sin\theta\boldsymbol{k} + \dot{\psi}\cos\theta\boldsymbol{j}) \\ & + I'\dot{\psi}\sin\theta(\dot{\theta}\boldsymbol{k} - \dot{\psi}\cos\theta\boldsymbol{i}) \\ & + I(\dot{\phi} + \dot{\psi}\cos\theta)(-\dot{\theta}\boldsymbol{j} + \dot{\psi}\sin\theta\boldsymbol{i})\end{aligned} \tag{18.28}$$

The corresponding scalar equations are:

$$M_x = I'\ddot{\theta} + (I - I')(\dot{\psi}^2\sin\theta\cos\theta) + I\dot{\phi}\dot{\psi}\sin\theta \tag{18.29a}$$

$$M_y = I'\ddot{\psi}\sin\theta + 2I'\dot{\theta}\dot{\psi}\cos\theta - I(\dot{\phi} + \dot{\psi}\cos\theta)\dot{\theta} \tag{18.29b}$$

$$M_z = I(\ddot{\phi} + \ddot{\psi}\cos\theta - \dot{\psi}\dot{\theta}\sin\theta) \tag{18.29c}$$

The foregoing equations are valid at all times for the motion of a homogeneous body having $I_{xx} = I_{yy} = I'$ moving about a fixed point on the axis of the body. Clearly, these equations are also applicable for motion about the center of mass for such bodies. Note that the equations are nonlinear and, except for certain special cases, are very difficult to integrate. They are, of course, very useful as they stand when computer methods are to be employed.

As a special case, we shall now consider a motion involving a constant nutation angle θ, a constant spin speed $\dot{\phi}$, and a constant precession speed $\dot{\psi}$. Such a motion is termed *steady precession.*

To determine the torque $\boldsymbol{M}$ for a given steady precession, we set $\dot{\theta}$, $\ddot{\theta}$, $\ddot{\phi}$, and $\ddot{\psi}$ equal to zero in Eq. 18.29. Accordingly, we get the following result:

$$M_x = [I(\dot{\phi} + \dot{\psi} \cos \theta) - I'\dot{\psi} \cos \theta]\, \dot{\psi} \sin \theta \tag{18.30a}$$

$$M_y = 0 \tag{18.30b}$$

$$M_z = 0 \tag{18.30c}$$

We see that for such a motion, we require a *constant torque about the line of nodes* as given by Eq. 18.30a. Noting that $\dot{\phi} + \dot{\psi} \cos \theta = \omega_z$ from Eq. 18.23c, this torque may also be given as

$$M_x = (I\omega_z - I'\dot{\psi} \cos \theta)\dot{\psi} \sin \theta \tag{18.31}$$

Examining Fig. 18.18, we can conclude that for the body to maintain a constant spin speed $\dot{\phi}$ about its body axis (i.e., relative to xyz) while the body axis (and also xyz) is rotating at constant speed $\dot{\psi}$ about the Z axis at a fixed angle θ, we require a constant torque having a value dependent on the motion of the body as well as the values of the moments of inertia of the body, and having a direction *always normal to the body and precession axes* (i.e., normal to plane I). Intuitively you may feel that such a torque should cause a rotation about its own axis (the *torque axis*) and should thereby change θ. Instead, the torque causes a rotation $\dot{\psi}$ of the body axis about an axis *normal* to the torque axis. As an example, consider the special case where θ has been chosen as 90° for motion of a disc about its center of mass (see Fig. 18.19). In accordance with Eq. 18.31, we have as a required torque for a steady precession the result

$$M_x = I\omega_z\dot{\psi} = I\dot{\phi}\dot{\psi} \tag{18.32}$$

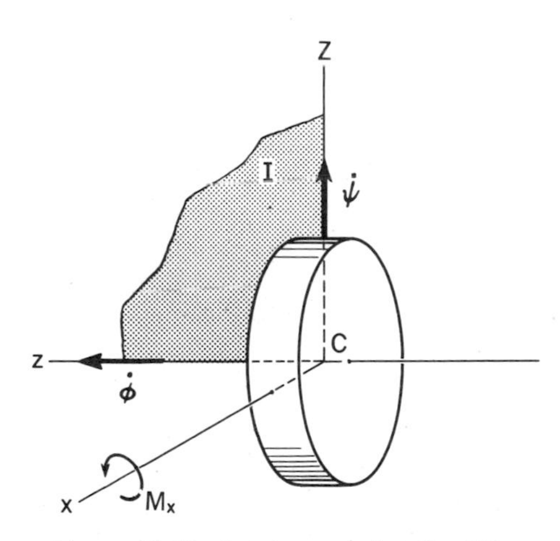

Figure 18.19. Steady precession; θ = 90°.

Here the proper torque about the line of nodes maintains a steady rotation of the spin axis z about an axis (the Z axis), which is at *right angles* both to the torque and the spin axes. Because of this unexpected phenomenon, toy manufacturers have developed various gyroscopic devices to surprise and delight children (as well as their parents). Here is yet another case where relying solely on intuition may lead to highly erroneous conclusions.

We should strongly point out that steady precessions are not easily initiated. We must have, at the start, simultaneously the proper precession and spin speeds as well as the proper θ for the given applied torque. If these conditions are not properly met initially, a complicated motion ensues.

EXAMPLE 18.3

A *single-degree-of-freedom gyro* is shown in Fig. 18.20. The spin axis of disc E is held by a gimbal A which can rotate about bearings C and D. These bearings are supported by the gyro case, which in turn is generally clamped to the vehicle to be

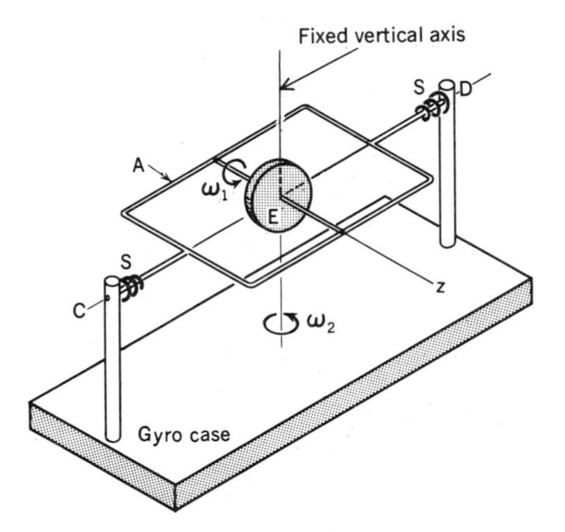

Figure 18.20. Single-degree-of-freedom gyro.

guided. If the gyro case rotates about a vertical axis (i.e., normal to its base) while the rotor is spinning, the gimbal A will tend to rotate about CD in an attempt to align with the vertical. When gimbal A is resisted from rotation about CD by a set of torsional springs S with a combined torsional spring constant given as K_t, the gyro is called a *rate gyro*. If the rotation of the gyro case is constant (at speed ω_2), the gimbal A assumes a fixed orientation relative to the vertical as a result of the restraining springs and a damper (not shown). About the body axis z there is a constant angular rotation of the rotor of ω_1 rad/sec maintained by a motor (not shown). This angular rotation is clearly is the spin speed $\dot{\phi}$. Next, note in Fig. 18.21 that the z axis and the fixed vertical axis form a plane (plane I) which has a known angular speed ω_2 about this fixed vertical axis. Clearly, this fixed vertical axis will be our precession axis, and the precession speed $\dot{\psi}$ equals ω_2. The line of nodes has also been shown; it must at all times normal to plane I and is thus collinear with axis C–D of gimbal A. With θ fixed, we have a case of steady precession.

Given the following data:

$$I = 3 \times 10^{-4} \text{ slug-ft}^2$$

$$I' = 1.5 \times 10^{-4} \text{ slug-ft}^2$$

$$\dot{\phi} = 20{,}000 \text{ rad/sec}$$

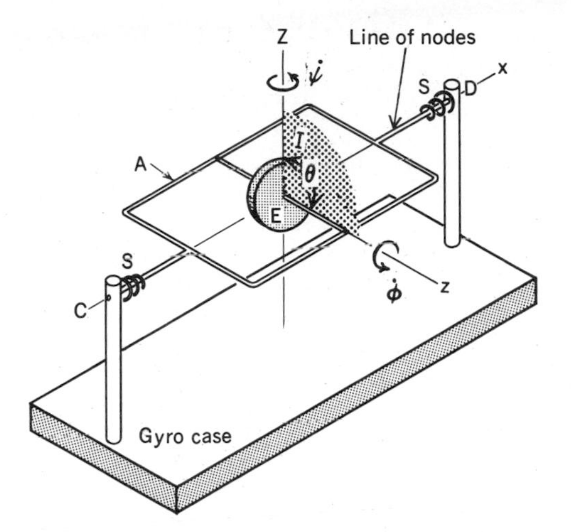

Figure 18.21. Gyro showing line of nodes and precession axis.

$$K_t = 4.95 \text{ ft-lb/rad}$$
$$\dot{\psi} = 1 \text{ rad/sec}$$

what is θ for the condition of steady precession? The torsional springs are unstretched when $\theta = 90°$. We have for Eq. 18.30a:

$$M_x = K_t\left(\frac{\pi}{2} - \theta\right) = [I(\dot{\phi} + \dot{\psi}\cos\theta) - I'\dot{\psi}\cos\theta]\dot{\psi}\sin\theta \qquad \text{(a)}$$

Putting in numerical values, we have

$$(4.95)\left(\frac{\pi}{2} - \theta\right) = [3 \times 10^{-4}(2 \times 10^4 + \cos\theta) - 1.5 \times 10^{-4}\cos\theta]\sin\theta$$

Therefore,

$$4.95 \times 10^4\left(\frac{\pi}{2} - \theta\right) = (6 \times 10^4 + 1.5\cos\theta)\sin\theta$$

We can neglect the term $(1.5\cos\theta)$, and so we have

$$4.95\left(\frac{\pi}{2} - \theta\right) = 6\sin\theta$$

Therefore,

$$\frac{\pi}{2} - \theta = 1.212\sin\theta \qquad \text{(b)}$$

Solving by trial and error, we get

$$\theta = 43° \qquad \text{(c)}$$

The way the *rate gyro* is used in practice is to maintain θ close to 90° by a small motor. The torque M_x developed to maintain this angle is measured, and from Eq. 18.30a we have available the proper $\dot{\psi}$, which tells us of the rate of rotation of the gyro case and hence the rate of rotation of the vehicle about an axis normal to the gyro case. Now $\dot{\psi}$ need not be constant as was the case in this problem. If it does not change very rapidly, the results from Eq. 18.30a can be taken as instantaneously valid even though the equation, strictly speaking, stems from steady precession where $\dot{\psi}$ should be constant.

EXAMPLE 18.4

A *two-degree-of-freedom gyroscope* is shown in Fig. 18.22. The rotor E rotates in gimbal A, which in turn rotates in gimbal C. Note that the axes b–b of the rotor and a–a of the gimbal C are always at right angles to each other. Gimbal C is held by bearings c supported by the gyro case. Axes c–c and a–a must always be at right angles to each other, as can easily be seen from the diagram. This kind of suspension of the rotor is called a *Cardan suspension.* If the bearings at a, b, and c are frictionless, a torque cannot be transmitted from the gyro case to the rotor.[7] The rotor is said to be *torque-free* for this case.

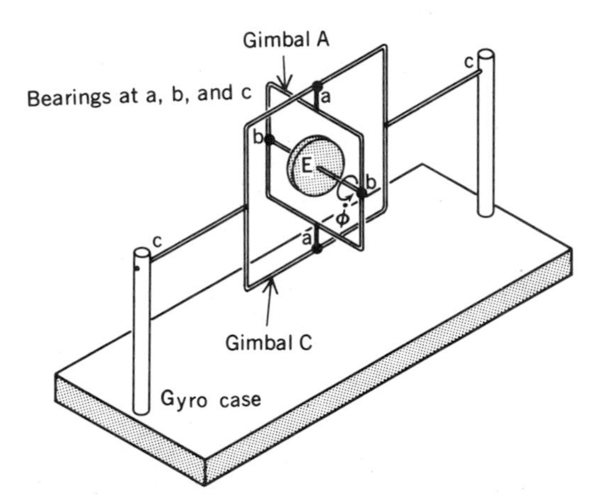

Figure 18.22. Two-degree-of-freedom gyro.

If the rotor is given a rapid spin velocity in a given direction in inertial space (such as toward the North Star), then for the ideal case of frictionless bearings the rotor will maintain this direction even though the gyro case is given rapid and complicated motions in inertial space. This constancy of direction results since no torque can be transmitted to the rotor to alter the direction of its angular momentum. Thus, the two-dimensional gyro gives a fixed direction in inertial space for purposes of guidance of a vehicle such as a missile. In use, the gyro case is rigidly fixed to the frame of the missile, and measurements of the orientation of the missile are accomplished by having pickoffs mounted between the gyro case and the outer gimbal and between the outer and inner gimbals.

The presence of some friction in the gyro bearings is, of course, inevitable. The counteraction of this friction when possible and, when not, the accounting for its action is of much concern to the gyro engineer. Suppose that the gyro has been given a motion such that the spin axis b–b (see Fig. 18.23) has an angular speed ω_1 about axis c–c of .1 rev/sec while maintaining a fixed orientation of 85° with axis c–c. The gyro case is stationary, and the spin speed $\dot{\phi}$ of the disc relative to gimbal A is

[7]That is, except for the singular situation where the gimbal axes are coplanar.

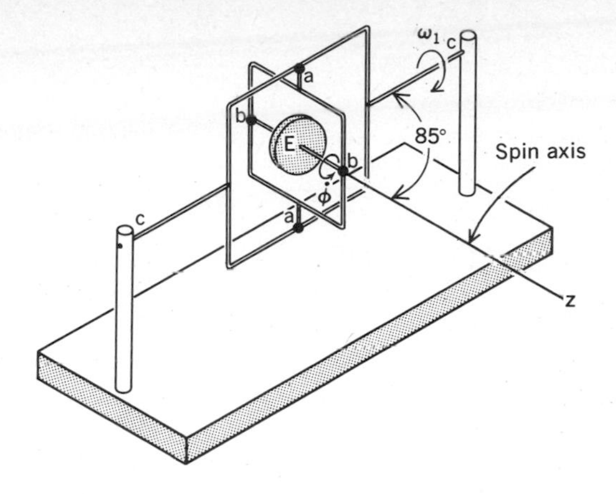

Figure 18.23. Two-degree-of-freedom gyro with body axis *z*.

10,000 rpm. What torque must be developed on the rotor for this motion? From what bearings must such a torque arise? The radius of gyration for the disc is 50 mm for the axis of symmetry and 38 mm for the transverse axes at the center of mass. The weight of the disc is 4.5 N.

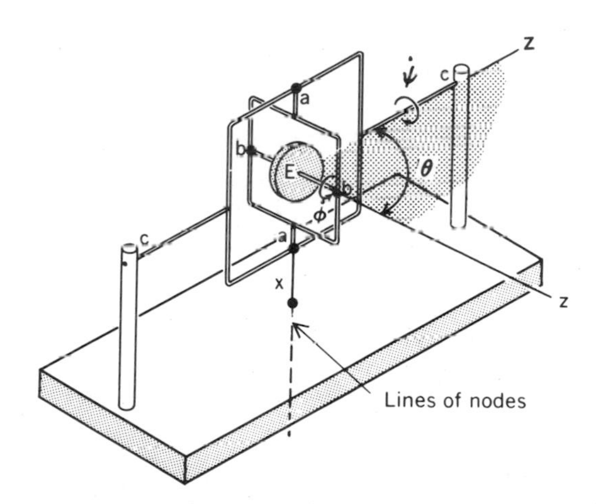

Figure 18.24. Axis *c–c* is precession axis.

Note in Fig. 18.24 that the spin axis z and fixed axis c–c form a plane that has a known angular speed ω_1 about axis c–c. Clearly, c–c then can be taken as the precession axis Z; the precession speed $\dot{\psi} = \omega_1$ is then .10 rev/sec. The line of nodes x is along axis a–a at all times. With $\theta = 85°$ at all times, we have steady precession. A constant torque M_x is required to maintain this motion. We can solve for M_x as follows:

$$\begin{aligned} M_x &= [I(\dot{\phi} + \dot{\psi}\cos\theta) - I'\dot{\psi}\cos\theta]\dot{\psi}\sin\theta \\ &= \left\{\frac{4.5}{9.81}(.05)^2\left[10{,}000\frac{2\pi}{60} + (.1)(2\pi)\cos 85°\right]\right. \\ &\quad \left. - \frac{4.5}{9.81}(.038)^2(.1)(2\pi)\cos 85°\right\}(.1)(2\pi)\sin 85° \\ &= .752 \text{ N-m} \end{aligned}$$

Thus, bearings along the *a–a* axis interconnecting the two gimbals are developing the frictional torque.

Problems

18.31. The z axis coincides initially with the centerline of the block. The block is given the following rotations in the sequence listed: (a) $\psi = 30°$, (b) $\theta = 45°$, (c) $\phi = 20°$. What are the projections of the centerline OA along the XYZ axes in the final position?

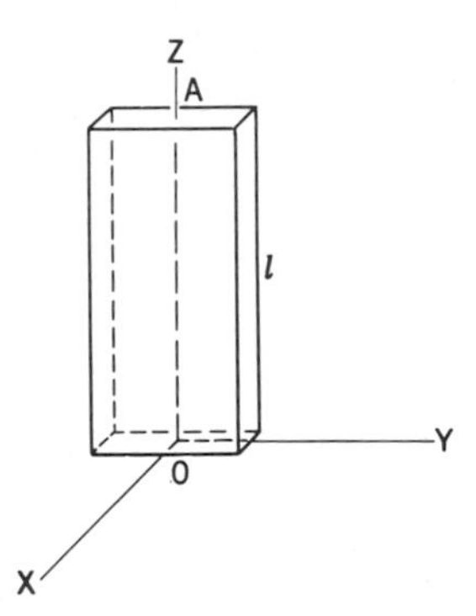

Figure P.18.31

18.32. A disc A of mean diameter 300 mm rolls without slipping so that its centerline rotates an angular speed ω_1 of 2 rad/sec about the Z axis. What are the precession, nutation, and spin angular velocity components?

Figure P.18.32

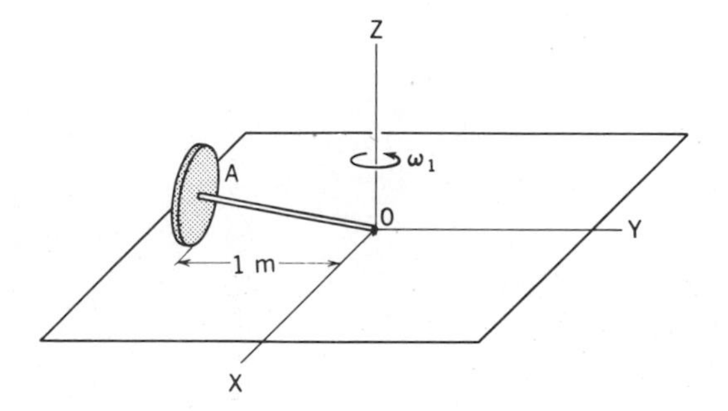

18.33. A body has the following components of angular velocity:

$$\dot{\phi} = 10 \text{ rad/sec}$$
$$\dot{\theta} = 5 \text{ rad/sec}$$
$$\dot{\psi} = 2 \text{ rad/sec}$$

when the following Euler angles are known to be

$$\psi = 45°, \qquad \theta = 30°$$

What is the magnitude of the total angular velocity?

18.34. If the disc shown were to be undergoing regular precession as shown at the rate of .3 rad/sec, what would have to be the spin velocity $\dot{\phi}$? The disc weighs 90 N. Neglect the mass of the rod.

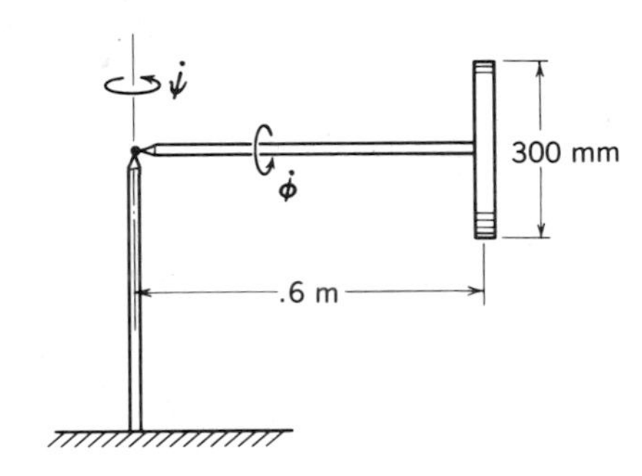

Figure P.18.34

18.35. In Problem 18.34, explain how you institute such a motion. Would you get the steady precession if for the computed $\dot{\phi}$ you merely released the disc from a horizontal configuration of the disc centerline?

18.36. A 20-lb cylinder having a radius of 1 ft and a length of $\frac{1}{4}$ ft is connected by a 2 ft rod to a fixed point O where there is a ball-joint connector. The cylinder spins about its own centerline at a speed of 50 rad/sec. What external torque about O is required for the cylinder to precess uniformly at a rate of $\frac{1}{2}$ rad/sec about the Z axis at an inclination of 45° to the Z axis? (Compute the torque when the centerline of the cylinder is in the XZ plane.)

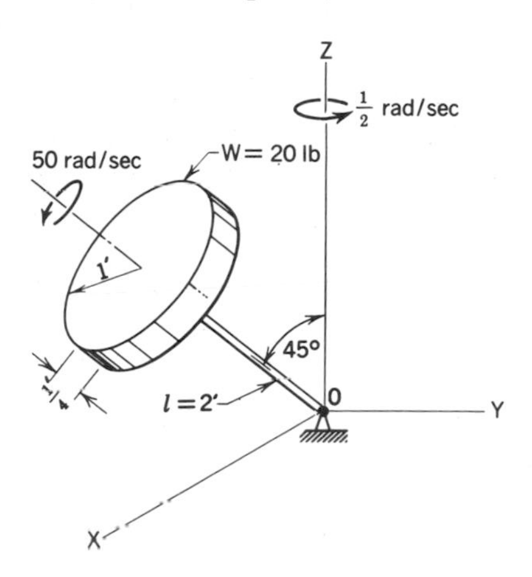

Figure P.18.36

18.37. The centerline of the rod rotates uniformly in a horizontal plane with a constant torque of 2.29 N-m applied about O. Each cylinder weighs 225 N and has a radius of 300 mm. The discs rotate on a bar AB with a speed ω_1 of 5000 rpm. Bar AB is held at O by a ball-joint connection. The applied torque is always perpendicular to AB and can only rotate about the vertical axis. What is the precession speed of the system?

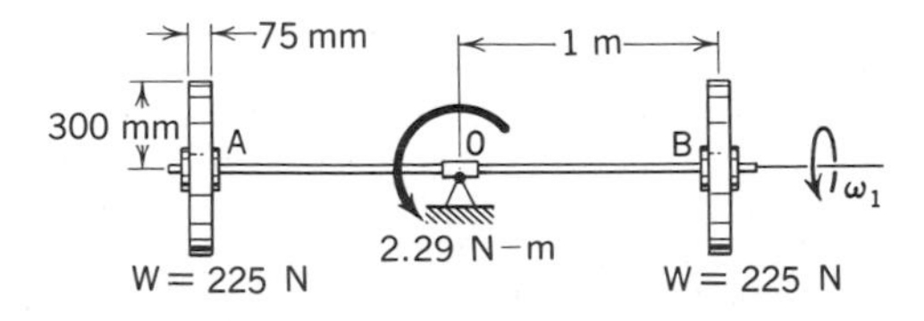

Figure P.18.37

18.38. (a) In Problem 18.37, consider the disc at B to have an angular speed of 5000 rpm and the disc at A to have a speed of 2500 rpm. What is the precession speed for the condition of steady precession?

(b) If the disc at A and the disc at B have angular speeds of 5000 rpm in opposite directions, what is the initial motion of the system when a torque perpendicular to AB is suddenly applied?

18.39. Two discs A and B roll without slipping at their midplanes. Light shafts cd and ef connect the discs to a centerpost which rotates at an angular speed ω_1 of 2 rad/sec. If each disc weighs 20 lb, what total force downward is developed by the discs on the ground support?

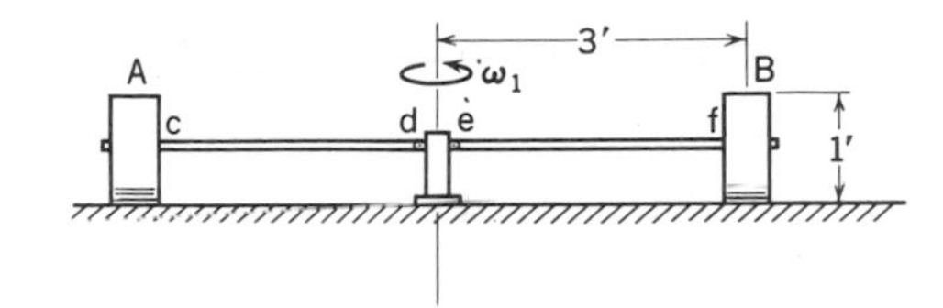

Figure P.18.39

18.40. A disc is spinning about its centerline with speed $\dot{\phi}$ while the centerline is precessing uniformly at fixed angle θ about the vertical axis. The mass of the disc is M. Consider the evaluation of $\dot{\psi}$, and show that such a state of regular precession is possible if

$$\omega_z^2 > \frac{4I'Mgl}{I^2}$$

Also show that there are, for every θ, two possible precession speeds. In particular, show that as ω_z gets very large, the following precessional speeds are possible:

$$\dot{\psi}_1 = \frac{I\omega_z}{I'\cos\theta}, \qquad \dot{\psi}_2 = \frac{Mgl}{I\omega_z}$$

(*Hint:* Consider Eq. 18.31 for first part of proof. Then use a power expansion of the root when evaluating $\dot{\psi}$.)

Figure P.18.40

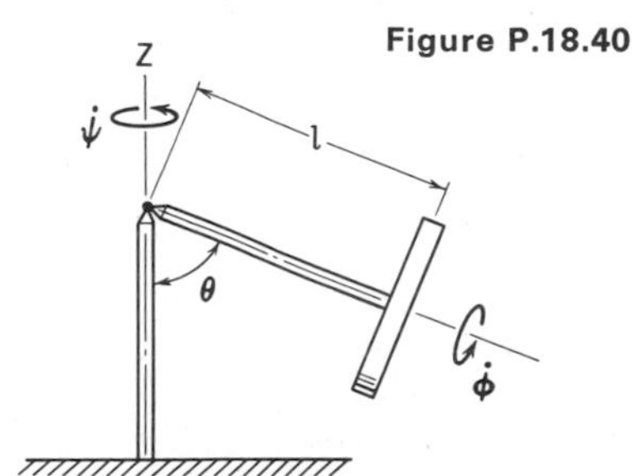

18.41. A uniform prism has a square cross section. Prove that $I_{\eta\eta}$ for any angle θ equals $I_{xx} = I_{yy}$. Thus, in order to use $I' = I_{xx} = I_{yy}$, as was done in the development in Section 18.6, the body need not be a body of revolution. Show in general that if $I_{xx} = I_{yy}$ and if xyz are principal axes, then $I_{\eta\eta} = I_{xx} = I_{yy}$.

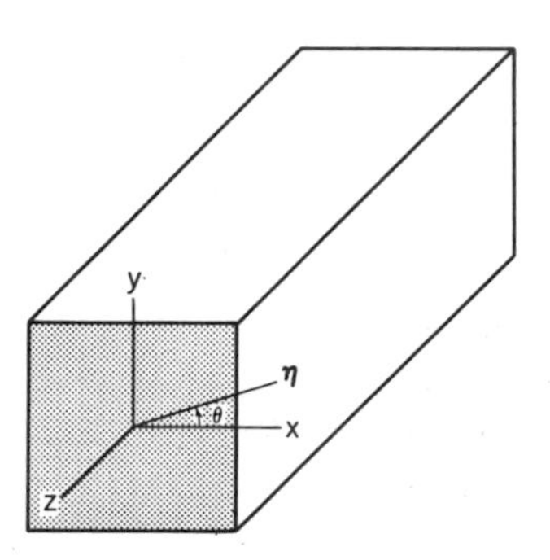

Figure P.18.41

18.42. A block weighing 50 N and having a square cross section is spinning about its body axis at a rate ω_1 of 100 rad/sec. For an angle $\alpha = 30°$, what is the precession speed $\dot{\psi}$ of the axis of the block? Neglect the weight of rod AB. See Problem 18.41 before proceeding.

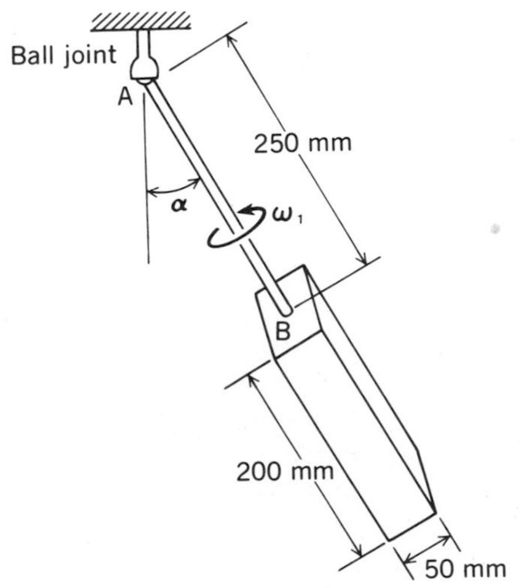

Figure P.18.42

18.43. In Problem 18.42, at what angle α does a steady rotation of the axis of the block occur about the indicated vertical axis at the rate of 6 rad/sec counterclockwise looking down from above? The angular speed ω_1 is 100 rad/sec.

18.44. A single-degree-of-freedom gyro is mounted on a vehicle moving at constant speed V of 100 ft/sec on a track which is coplanar and is circular, having a mean radius of 200 ft. The disc weighs 1 lb and has a radius of 2 in. It is turning at a speed of 20,000 rpm relative to the gimbal. If the gimbal maintains a rotated position of 15° with the horizontal, what is the equivalent torsional spring constant about axis A–A for the gimbal suspension?

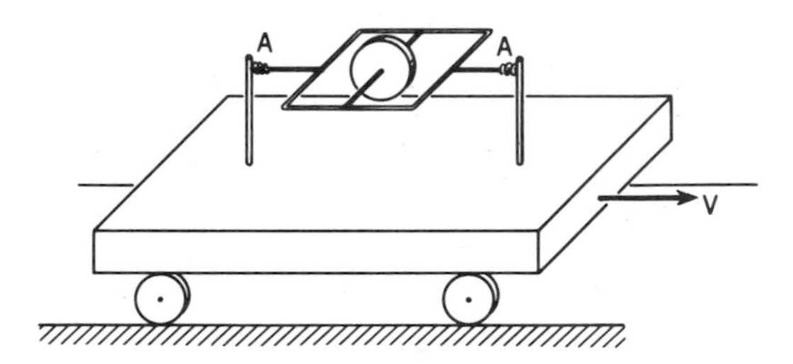

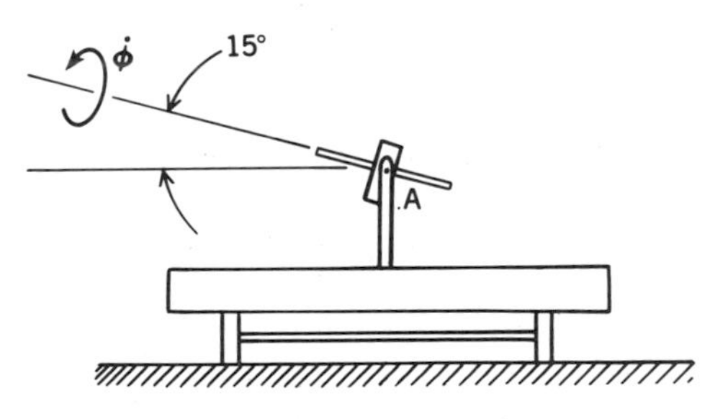

Figure P.18.44

18.45. In Problem 18.44, where we found that $K_t = 1.663$ ft-lb/rad, suppose that the speed of the vehicle were adjusted to 50 ft/sec (i.e., half its given speed). What would then be the position of the gimbal for steady-state precession?

18.46. A plate is rotating about shaft CD at a constant speed ω_1 of 2 rad/sec. A single-degree-of-freedom gyro is mounted on the plate. The gyro has mounted on it through a gimbal a disc of weight 3.4 N and radius 50 mm. The disc has a rotational speed ω_2 of 10,000 rpm relative to the gimbal. If the gimbal is to be maintained parallel to the plate, what torque is required? If the bearings are frictionless, explain how this torque is developed. What is the direction of the line of nodes for steady precession?

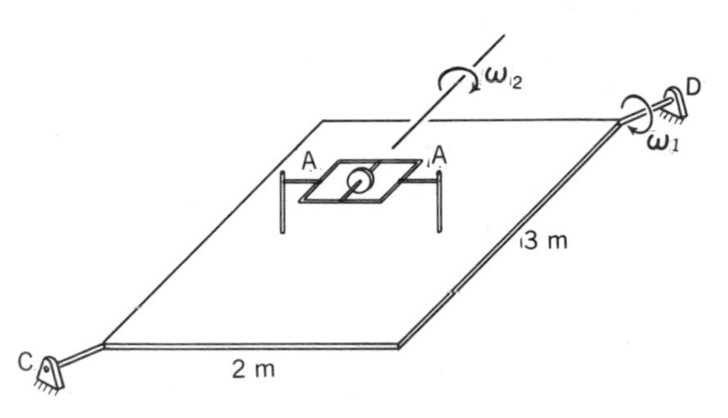

Figure P.18.46

***18.47.** A solid sphere of diameter 200 mm and weight 50 N is spinning about its body axis at a speed $\dot{\phi}_0$ of 100 rad/sec while its axis is held stationary at an angle θ_0 of 120°. The sphere is released suddenly. What is the precession speed $\dot{\psi}$ when θ has increased by 10°? Use the fact that

$$H_Z = \text{const.} = I'\dot{\psi}\sin^2\theta + I\,\omega_z\cos\theta$$

(This equation can be reached by projecting H_x, H_y, and H_z in the Z direction.)

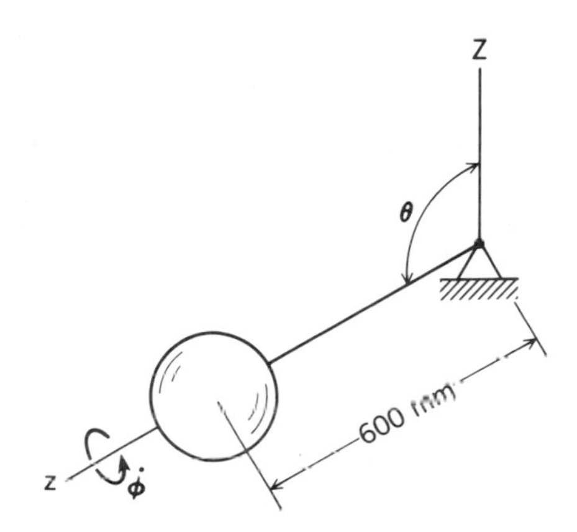

Figure P.18.47

*18.7 Torque-Free Motion

We shall now consider possible *torque-free motion* for a body having $I_{xx} = I_{yy} = I'$ for axes normal to the body axis. An example of a device that can approach such motion is the two-degree-of-freedom gyroscope described in Example 18.4.

Let us now examine the equations of torque-free motion. First, consider the general relation

$$\boldsymbol{M}_C = \dot{\boldsymbol{H}}_C \tag{18.33}$$

Since $\boldsymbol{M}_C$ is zero, $\boldsymbol{H}_C$ must be constant. Thus,

$$\boldsymbol{H}_C = \boldsymbol{H}_0 \tag{18.34}$$

where $\boldsymbol{H}_0$ is the *initial* angular momentum about the mass center. We shall first assume and later justify in this section that all torque-free motions will be *steady precessions* about an axis going through the center of mass and directed *parallel to the vector* $\boldsymbol{H}_0$. Accordingly, we choose Z to pass through the center of mass and to have a direction corresponding to that of $\boldsymbol{H}_0$, as shown in Fig. 18.25. The axis x' normal to axes z and Z forming plane I is then the line of nodes and y' is in plane I. The reference xyz is *fixed* to the body and hence spins about the z axis. Axes xy then rotate in the $x'y'$

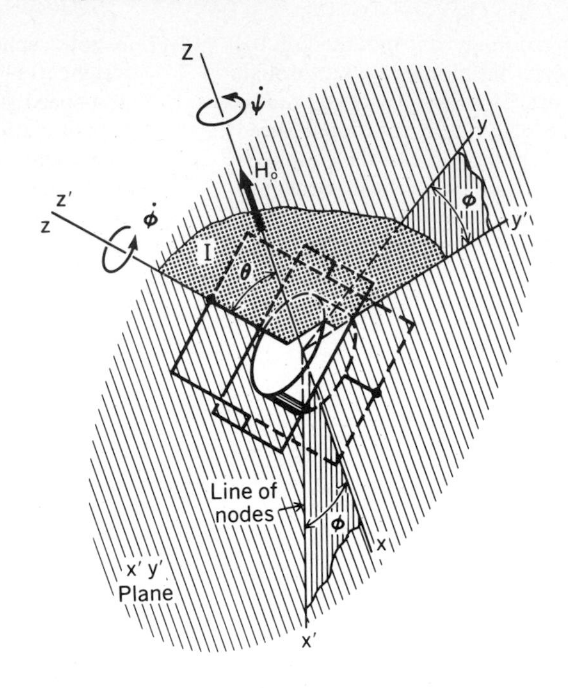

Figure 18.25. Two-degree-of-freedom gyroscope illustrating torque-free motion. Outer gimbal support is not shown.

plane as shown in the diagram. Using Fig. 18.25, we can then express $\boldsymbol{H}_0$ in terms of its x, y, and z components in the following way at all times:

$$\boldsymbol{H}_0 = H_0 \sin\theta \sin\phi \boldsymbol{i} + H_0 \sin\theta \cos\phi \boldsymbol{j} + H_0 \cos\theta \boldsymbol{k} \tag{18.35}$$

Since xyz are principal axes, we can also state:

$$\boldsymbol{H}_0 = I'\omega_x \boldsymbol{i} + I'\omega_y \boldsymbol{j} + I\omega_z \boldsymbol{k} \tag{18.36}$$

Comparing Eqs. 18.35 and 18.36, we then have

$$\omega_x = \frac{H_0 \sin\theta \sin\phi}{I'} \tag{18.37a}$$

$$\omega_y = \frac{H_0 \sin\theta \cos\phi}{I'} \tag{18.37b}$$

$$\omega_z = \frac{H_0 \cos\theta}{I} \tag{18.37c}$$

By using the preceding formulations for ω_x, ω_y, and ω_z, we can write *Euler's equations* using fixed axes xyz in a form that includes the constant H_0:

$$I' \frac{d}{dt}\left(\frac{H_0 \sin\theta \sin\phi}{I'}\right) + \frac{I - I'}{I'I} H_0^2 \sin\theta \cos\theta \cos\phi = 0 \tag{18.38a}$$

$$I' \frac{d}{dt}\left(\frac{H_0 \sin\theta \cos\phi}{I'}\right) + \frac{I' - I}{I'I} H_0^2 \sin\theta \cos\theta \sin\phi = 0 \tag{18.38b}$$

$$I\frac{d}{dt}\left(\frac{H_0 \cos\theta}{I}\right) = 0 \tag{18.38c}$$

From Eq. 18.38c, it is then clear that

$$\frac{H_0 \cos\theta}{I} = \text{constant} \tag{18.39}$$

Thus, since H_0 and I are constant, we can conclude from this equation that *the nutation angle* is a fixed angle θ_0.[8] Now consider Eq. 18.38b, using the fact that $\theta = \theta_0$. Canceling H_0 and carrying out the differentiation, we get

$$-(\sin\theta_0)(\sin\phi)(\dot\phi) + \frac{I' - I}{I'I} H_0 \sin\theta_0 \cos\theta_0 \sin\phi = 0$$

Therefore,

$$\dot\phi = \frac{I' - I}{I'I} H_0 \cos\theta_0 \tag{18.40}$$

Thus, the *spin speed*, $\dot\phi$, is constant.

To get the precession speed $\dot\psi$, note from Fig. 18.25 that

$$\omega_z = \dot\phi + \dot\psi \cos\theta_0$$

Now equate the right side of this equation with the right side of Eq. 18.37c:

$$\dot\phi + \dot\psi \cos\theta_0 = \frac{H_0 \cos\theta_0}{I} \tag{18.41}$$

Substituting for $\dot\phi$ from Eq. 18.40 and solving for $\dot\psi$, we get

$$\dot\psi = \frac{H_0}{I} - H_0 \frac{I' - I}{I'I}$$

Collecting terms, we have

$$\dot\psi = \frac{H_0}{I}\left(1 - \frac{I' - I}{I'}\right) = \frac{H_0}{I'} \tag{18.42}$$

The results of the discussion for torque-free motion of the body of revolution can then be given as

$$\theta = \theta_0 \tag{18.43a}$$

$$\dot\psi = \frac{H_0}{I'} \tag{18.43b}$$

$$\dot\phi = \frac{I' - I}{I'I} H_0 \cos\theta_0 \tag{18.43c}$$

Since $\dot\theta$, $\ddot\theta$, $\ddot\psi$, and $\ddot\phi$ are all zero, Eqs. 18.43 depict a case of steady precession, and so the assumption made earlier to this effect is completely consistent with the results

[8] We now see that taking $\boldsymbol{Z}$ to be collinear with $\boldsymbol{H}_0$ at the outset is justified.

emerging from Euler's equation. Thus, we can consider the assumption and the ensuing conclusions as correct.

Hence, if a body of revolution is torque-free—as, for example, in the case illustrated in Fig. 18.26, where the center of mass is fixed and where the body has initially any angular momentum vector $\boldsymbol{H}_0$, then at all times the angular momentum $\boldsymbol{H}$ is constant and equals $\boldsymbol{H}_0$. Furthermore, the body will have a regular precession that

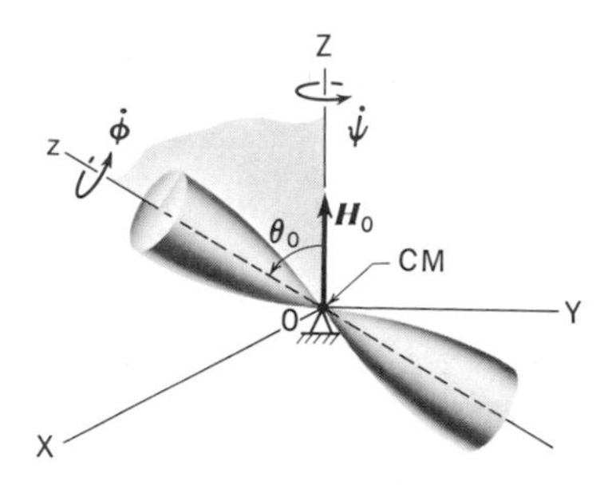

Figure 18.26. Body of revolution is also symmetric about C.M. where it is supported.

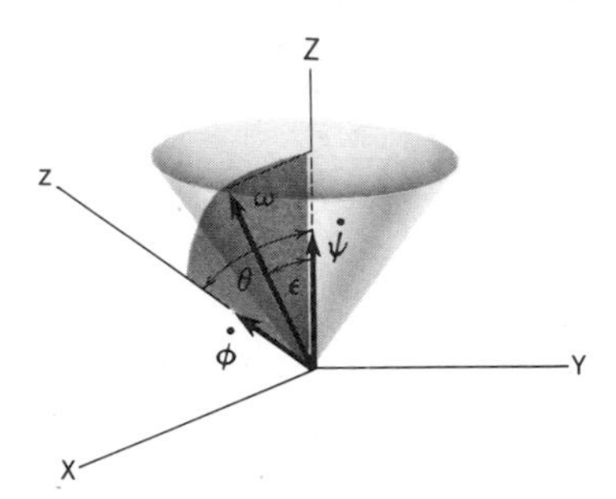

Figure 18.27. Vector $\boldsymbol{\omega}$ sweeps out a conical surface.

consists of a constant angular velocity $\dot{\psi}$ of the centerline about a Z axis collinear with $\boldsymbol{H}_0$ at a fixed inclination θ_0 from Z. Finally, there is a constant spin speed $\dot{\phi}$ about the centerline. Thus, two angular velocity vectors $\dot{\boldsymbol{\psi}}$ and $\dot{\boldsymbol{\phi}}$, are present, and the *total* angular velocity $\boldsymbol{\omega}$ is at an inclination of ϵ from the Z axis (see Fig. 18.27) and precesses with angular speed $\dot{\psi}$ about the Z axis. This must be true, since the direction of one component of $\boldsymbol{\omega}$, namely $\dot{\boldsymbol{\phi}}$, precesses in this manner while the other component, $\dot{\boldsymbol{\psi}}$, is fixed in the Z direction. The vector $\boldsymbol{\omega}$ then can be considered to continuously sweep out a cone, as illustrated in Fig. 18.27.

We now illustrate the use of the basic formulations for torque-free motion.

EXAMPLE 18.5

A cylindrical space capsule in orbit is shown in Fig. 18.28. A quarter section of the cylinder can be opened about AB to a test configuration as shown in Fig. 18.29. At the earth's surface, the end plates of the capsule each have a weight of 300 lb and the cylindrical portion weighs 1800 lb. In the closed configuration, the center of mass of the capsule is at the geometric center.

If at time t the capsule is placed in a *test configuration* (Fig. 18.29) with a total instantaneous angular speed ω of 2 rad/sec in the z direction in inertial space,[9] what will be the precession axis for the capsule and the rate of precession of the capsule when door C subsequently is closed by an internal mechanism?

---

[9]The initial motion with just rotation about the z axis can be only an *instantaneous* motion at the instant when the device is placed into the shown configuration. At this instant, $\boldsymbol{H}_C$ is at some angle relative to the z axis [as you will soon see in Eq. (d)] and is rotating about z at an angular speed $\boldsymbol{\omega}$. Subsequent motion, being torque-free, requires $\boldsymbol{H}_C$ to be *constant* and this, in turn, means that the z axis will thereafter have to be rotating about $\boldsymbol{H}_C$.

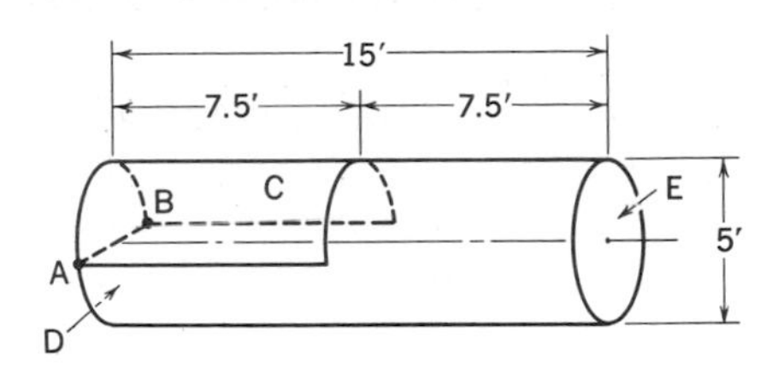

Figure 18.28. Space capsule.

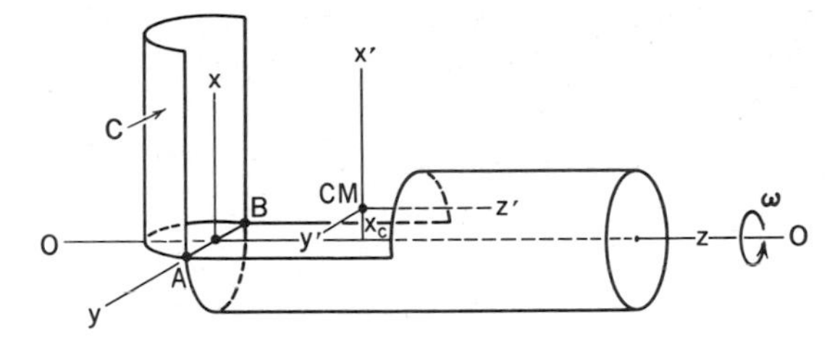

Figure 18.29. Space capsule in initial test configuration.

During a change of configuration, there is a zero net torque on the system so that $\boldsymbol{H}_C$, the angular momentum about the center of mass, is not changed. As a first step we shall compute $\boldsymbol{H}_C$ using data for the instantaneous test configuration. For this calculation we shall need the position of the center of mass.

A reference xyz has been fixed to the system as shown in Fig. 18.29, and the system has been decomposed into simple portions in Fig. 18.30, for convenience in carrying out ensuing calculations. Employing formulas for positions of centers of mass as given in the inside covers, we have for moments of mass about the origin of xyz:

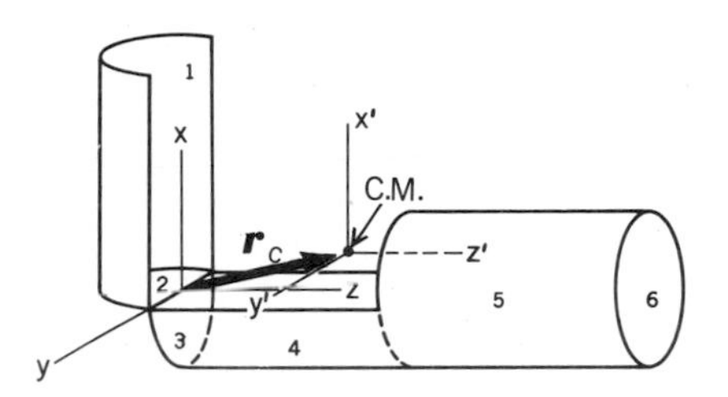

Figure 18.30. Space capsule subdivided into simple shapes.

$$\frac{2400}{g}\boldsymbol{r}_C = \sum_{i=1}^{6} M_i(\boldsymbol{r}_C)_i$$

$$= \frac{450}{g}\left(\frac{15}{4}\boldsymbol{i} - \frac{5}{\pi}\boldsymbol{k}\right) + \frac{150}{g}\left[-(.424)\left(\frac{5}{2}\right)(\boldsymbol{k})\right] + \frac{150}{g}\left[-(.424)\left(\frac{5}{2}\right)\boldsymbol{i}\right]$$

$$+ \frac{450}{g}\left(\frac{15}{4}\boldsymbol{k} - \frac{5}{\pi}\boldsymbol{i}\right) + \frac{900}{g}\left(\frac{3}{4}\right)(15)\boldsymbol{k} + \frac{300}{g}(15)\boldsymbol{k}$$

Therefore,

$$\boldsymbol{r}_C = 6.44\boldsymbol{k} + .338\boldsymbol{i} \text{ ft} \tag{a}$$

To get $\boldsymbol{H}_C$, we next set up a second reference $x'y'z'$ at the center of mass as shown in Fig. 18.29. In accordance with Eq. 18.1, we have for $\boldsymbol{H}_C$, on noting that the only nonzero component of $\boldsymbol{\omega}$ is in the z' direction:

$$\boldsymbol{H}_C = -I_{x'z'}\omega\boldsymbol{i} - I_{y'z'}\omega\boldsymbol{j} + I_{z'z'}\omega\boldsymbol{k} \tag{b}$$

To compute $I_{z'z'}$ we proceed as follows, using the decomposed capsule parts of Fig. 18.30 and employing transfer theorems for moments of inertia:

$$(I_{z'z'})_1 = \frac{1}{2}\left(\frac{450}{g}\right)\left(2.5^2 + \frac{7.5^2}{6}\right) + \frac{450}{g}(3.75 - .338)^2 = 272 \text{ slugs-ft}^2$$

$$(I_{z'z'})_2 = \frac{1}{4}\left(\frac{150}{g}\right)(2.5^2) + \frac{150}{g}(.338)^2 = 7.81$$

$$(I_{z'z'})_3 = \left\{\frac{1}{2}\left(\frac{150}{g}\right)(2.5)^2 - \frac{150}{g}[(.424)(2.5)]^2\right\}$$

$$+\frac{150}{g}[.338 + (.424)(2.5)]^2 = 18.43$$

$$(I_{z'z'})_4 = \left[\frac{450}{g}(2.5)^2 - \frac{450}{g}\left(\frac{5}{\pi}\right)^2\right] + \frac{450}{g}\left(.338 + \frac{5}{\pi}\right)^2 = 104.0$$

$$(I_{z'z'})_5 = \frac{900}{g}(2.5)^2 + \frac{900}{g}(.338)^2 = 177.9$$

$$(I_{z'z'})_6 = \frac{1}{2}\left(\frac{300}{g}\right)(2.5)^2 + \frac{300}{g}(.338)^2 = 30.2$$

Accordingly, we have for $I_{z'z'}$:

$$I_{z'z'} = \sum_{i=1}^{6} (I_{zz})_i = 610 \text{ slugs-ft}^2 \tag{c}$$

We proceed in a similar manner to compute $I_{z'x'}$. We will illustrate this computation for portion 1 of the system and then give the total result. Thus, employing the parallel axis theorem, we have

$$(I_{z'x'})_1 = 0 + \frac{450}{g}\left(-6.44 - \frac{5}{\pi}\right)\left(\frac{7.5}{2} - .338\right) = -383 \text{ slug-ft}^2$$

It is important to remember that the transfer distances (with proper signs) used for computing $(I_{z'x'})_1$ using the parallel-axis theorem are measured *from $x'y'z'$, about which you are making the calculation, to the center of mass of body 1*. For the entire system we have, by similar calculation:

$$I_{z'x'} = -329 \text{ slug-ft}^2$$

Because of symmetry, furthermore:

$$I_{z'y'} = 0$$

From Eq. (b) we then have for $\boldsymbol{H_C}$:

$$\begin{aligned} \boldsymbol{H_C} &= 329\omega \boldsymbol{i} + 610\omega \boldsymbol{k} \\ &= 658\boldsymbol{i} + 1220\boldsymbol{k} \end{aligned} \tag{d}$$

Since the precession axis is collinear with $\boldsymbol{H_C}$, we have thus established this axis (Z axis), as is shown in Fig. 18.31. A reference $\xi\eta\zeta$ has been set up at the center of mass for the closed configuration. The body axis ζ remains at a fixed angle θ with the Z axis and precesses about it with an angular speed $\dot{\psi}$ given in accordance with Eq. 18.43b as

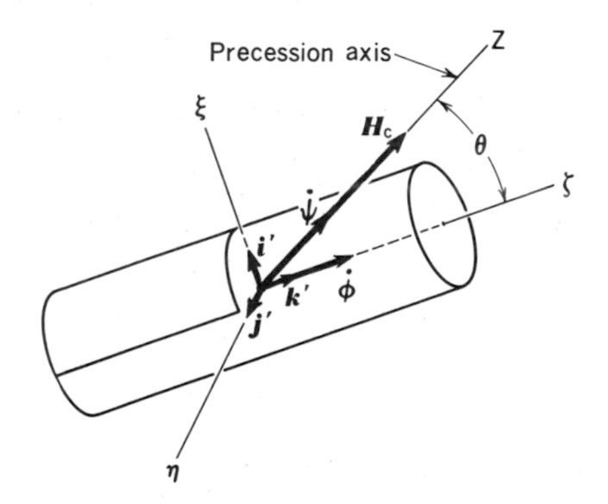

Figure 18.31. Space capsule in final configuration.

$$\begin{aligned} \dot{\psi} &= \frac{H_C}{I_{\xi\xi}} \\ &= \frac{\sqrt{658^2 + 1220^2}}{\frac{1}{2}\left(\frac{1800}{g}\right)\left(2.5^2 + \frac{15^2}{6}\right) + 2\left(\frac{1}{4}\right)\left(\frac{300}{g}\right)(2.5^2) + 2\left(\frac{300}{g}\right)(7.5)^2} \\ &= \frac{1386}{2300} = .603 \text{ rad/sec} \end{aligned} \tag{e}$$

We can now state that the body axis will precess around an axis collinear with $\boldsymbol{H}_C$ as given by Eq. (d) with a speed of .603 rad/sec. However, we are unable to determine θ and $\dot{\phi}$ with what information we now have available. We need more information as to the way the door was closed. Thus, knowing how much net work was done in the configuration change from the work energy relations, we can write another equation. And, with Eqs. 18.37 and Eq. 18.40, we can compute θ and $\dot{\phi}$. We shall present several problems with that information available in the homework exercises.

Let us now examine Eqs. 18.43 for the special case where $I = I'$. Here the spin velocity $\dot{\phi}$ must be zero, leaving only *one* angular motion, $\dot{\psi}$, the precession. This rotation is about the Z axis so the direction of angular velocity $\boldsymbol{\omega}$ of the body corresponds to $\boldsymbol{H}_0$. Since this is a body of revolution, the moment-of-inertia condition for this case ($I = I'$) means that the moments of inertia for principal axes x, y, and z are mutually equal, and we can verify from Eq. 9.13 that *all* axes inclined to the xyz reference have the same moment of inertia I (and all therefore are principal axes at the point). Thus, the body, if homogeneous, would be a sphere, a cube, any regular polyhedron, or any body that possesses point symmetry. *No matter how we launch this body, the angular momentum* $\boldsymbol{H}$ *will be equal to* $I\boldsymbol{\omega}$ *and will thus always coincide with the direction of angular velocity* $\boldsymbol{\omega}$. This situation can also be shown analytically as follows:

$$\boldsymbol{H} = H_x\boldsymbol{i} + H_y\boldsymbol{j} + H_z\boldsymbol{k} \tag{18.44}$$

For principal axes, we have

$$\boldsymbol{H} = \omega_x I_{xx}\boldsymbol{i} + \omega_y I_{yy}\boldsymbol{j} + \omega_z I_{zz}\boldsymbol{k} \tag{18.45}$$

If $I_{xx} = I_{yy} = I_{zz} = I$, we have for the foregoing

$$\boldsymbol{H} = I(\omega_x\boldsymbol{i} + \omega_y\boldsymbol{j} + \omega_z\boldsymbol{k}) = I\boldsymbol{\omega} \tag{18.46}$$

indicating that $\boldsymbol{H}$ and $\boldsymbol{\omega}$ must be collinear. The situation just described represents the case of a thrown baseball or basketball.

There are *two* situations for the case $I_{xx} = I_{yy}$ in which $\boldsymbol{H}$ and $\boldsymbol{\omega}$ are collinear. Examining Eq. 18.43, we thus see that if θ_0 is 90°, then $\dot{\phi} = \dot{\theta} = 0$, leaving only precession $\dot{\psi}$ along the Z axis. The Z axis for the analysis corresponds to the direction of $\boldsymbol{H}$. We thus see that since $\boldsymbol{\omega} = \dot{\boldsymbol{\psi}}$ then $\boldsymbol{\omega}$ is collinear with $\boldsymbol{H}$. This case corresponds to a proper "drop kick" or "place kick" of a football [Fig. 18.32(a)] wherein the body axis z is at right angles to the Z axis.

The other case consists of $\theta_0 = 0$. This means that $\dot{\boldsymbol{\phi}}$ and $\dot{\boldsymbol{\psi}}$ have the same direction—that is, along the Z direction, which then means that $\boldsymbol{\omega}$ and $\boldsymbol{H}$ again are collinear. This case corresponds to a good football pass [Fig. 18.32(b)]. *For all other motions of bodies where* $I_{xx} = I_{yy}$, *the angular velocity vector* $\boldsymbol{\omega}$ *will not have the direction of angular momentum* $\boldsymbol{H}_0$.

Upon further consideration, we can make a simple model of torque-free motion. Start with the fixed cone described earlier (see Fig. 18.27) about the Z axis, where the cone surface

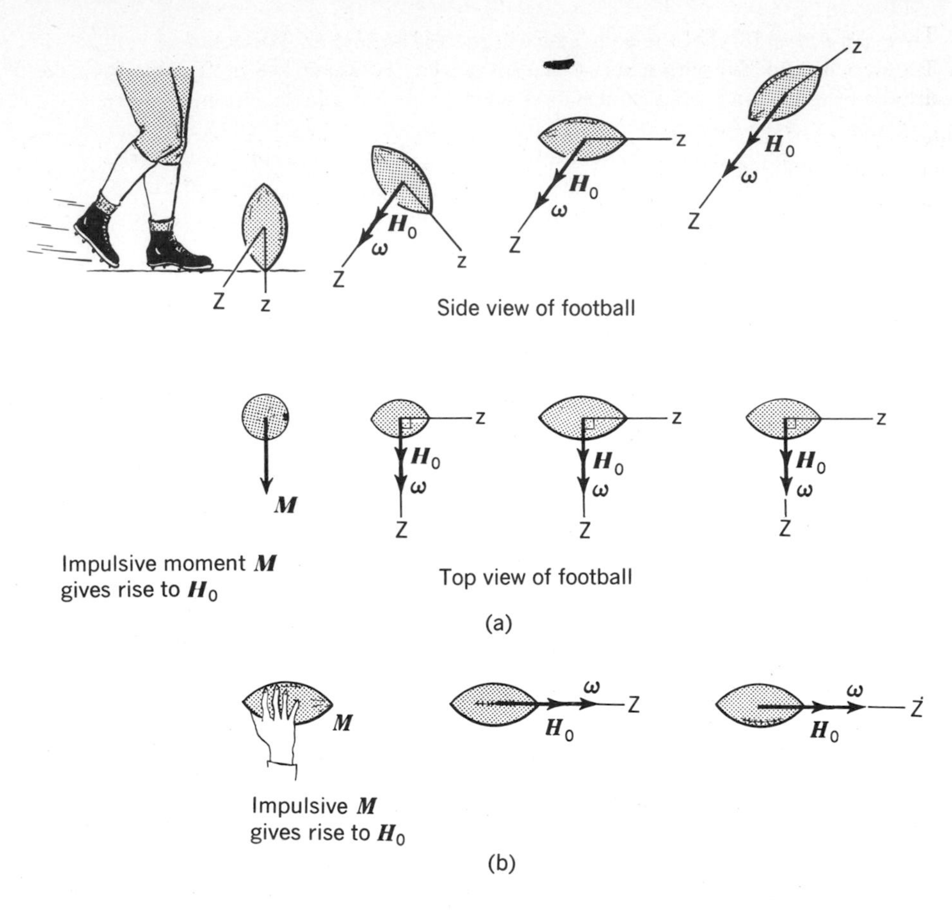

Figure 18.32. Two cases where $\boldsymbol{H}$ and $\boldsymbol{\omega}$ are collinear.

is that swept out by the total angular velocity vector $\boldsymbol{\omega}$ of the torque-free body. Now consider a second cone about the spin axis z of the torque-free body (see Fig. 18.33) in direct contact with the initial stationary cone. Rotate the second cone about its axis with a speed and sense corresponding to $\dot{\phi}$ of the torque-free body and impose a no-slipping condition between the cones. Clearly, the second cone will precess about the Z axis at some speed $\dot{\psi}'$. Also, the total angular velocity $\boldsymbol{\omega}'$ of the moving cone will lie along the line of contact between cones and is thus collinear with $\boldsymbol{\omega}$. We shall now show that the speed $\dot{\psi}'$ of the cone model equals $\dot{\psi}$ of the torque-free body and, as a consequence, that $\boldsymbol{\omega}'$ for the moving cone equals $\boldsymbol{\omega}$ of the torque-free body. We know now that:

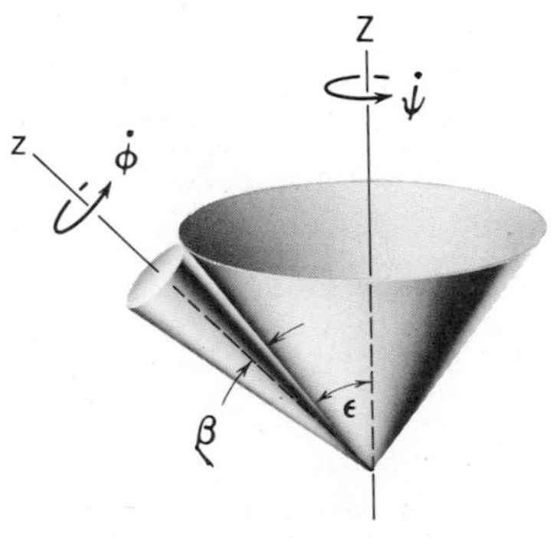

Figure 18.33. Rolling-cone model.

1. $\dot{\phi}$ is the same for both the device and the physical case.

2. The direction of resultant angular velocity is the same for both cases.
3. The direction of the precession velocity must be the same for both cases (i.e., the Z direction).

This situation is shown in Fig. 18.34. Note that $\dot{\boldsymbol{\phi}}$ is the same in both the physical case and the mechanical model and that the directions of $\boldsymbol{\omega}$ and $\boldsymbol{\omega}'$ as well as $\dot{\boldsymbol{\psi}}$ and $\dot{\boldsymbol{\psi}}'$ are, respectively, the same for both diagrams. Accordingly, when we consider the construction of the parallelogram of vectors, we see that the vectors $\boldsymbol{\omega}$ and $\boldsymbol{\omega}'$ as well as $\dot{\boldsymbol{\psi}}$ and $\dot{\boldsymbol{\psi}}'$ must necessarily be equal for both the physical case and the model, respectively.

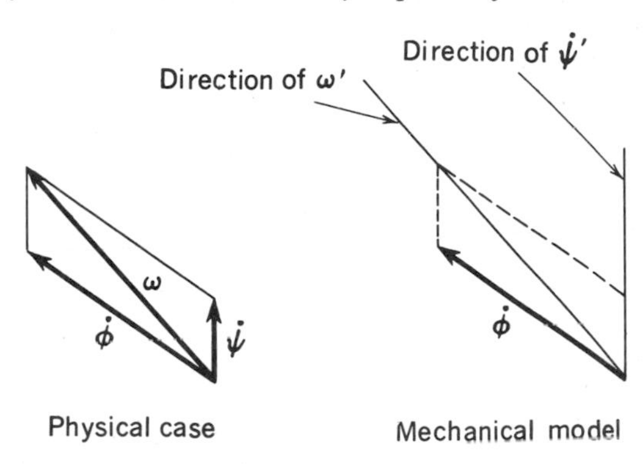

Figure 18.34. Angular vector diagrams.

We shall now investigate more carefully the relation between the senses of rotation for corresponding angular velocities between the model and the physical case for certain classes of geometries of the physical body.

1. $I' > I$. From Eq. 18.43c, we see that when θ_0 is less than $\pi/2$ rad, $\dot{\phi}$ is positive for this case.[10] Thus, the spin must be counterclockwise as we look along the z axis toward the origin. From Eq. 18.43b, we see that $\dot{\psi}$ is positive and thus counterclockwise as we look toward the origin along Z. Clearly, from these stipulations, the rolling-cone model shown in Fig. 18.36 has the proper motion for this case. The motion is termed *regular precession*.
2. $I' < I$. Here, the spin $\dot{\phi}$ will be negative for a nutation angle less than 90° as stipulated by Eq. 18.43c. However, the precession $\dot{\phi}$ must still be positive in accordance with Eq. 18.43b. The rolling-cone model thus far presented clearly cannot give these proper senses, but if the moving cone is *inside* the stationary cone (Fig. 18.35) we have motion that is consistent with the relations in Eq. 18.43. Such motion is called *retrograde precession*.

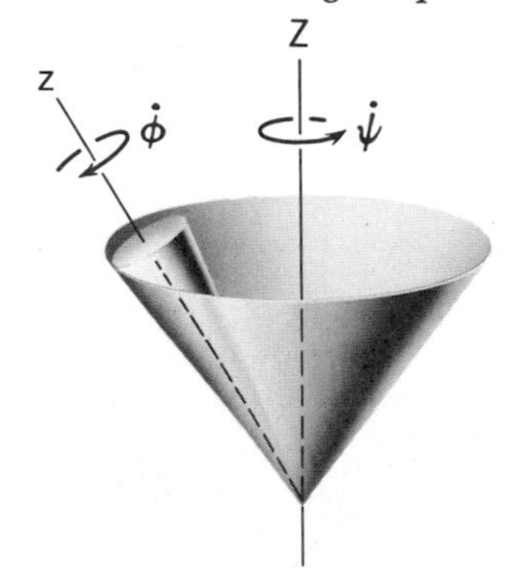

Figure 18.35. Retrograde precession.

EXAMPLE 18.6

The space capsule of Example 18.5 is shown again in Fig. 18.36 rotating about its axis of symmetry z in

[10] Recall from Eq. 18.35 that H_0 is just a magnitude. Also, note that this case corresponds to what has been shown in Fig. 18.27—namely, the case we have just discussed.

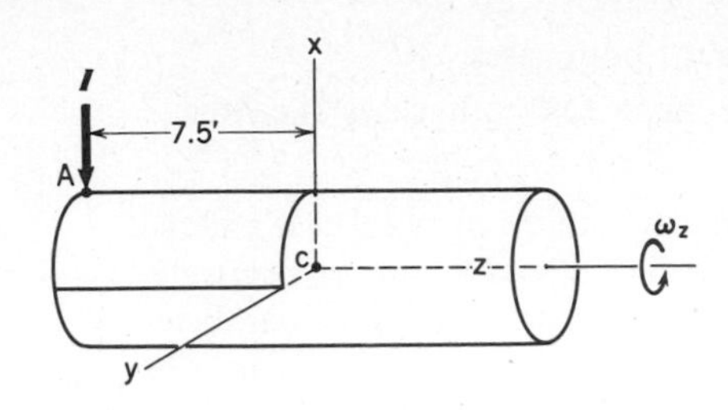

Figure 18.36. Space capsule with impulse.

inertial space with an angular speed ω_z of 2 rad/sec. As a result of an impact with a meteorite, the capsule is given an impulse $\boldsymbol{I}$ of 20 lb-sec at position A as shown in the diagram. Ascertain the postimpact motion.

The impact will give the cylinder an angular impulse:

$$\boldsymbol{I}_{\text{ang}} = (20)(7.5)\boldsymbol{j} = 150\boldsymbol{j} \text{ slug-ft}^2/\text{sec} \tag{a}$$

From the *angular impulse-momentum* equation we can say for the impact:

$$\boldsymbol{I}_{\text{ang}} = \boldsymbol{H}_2 - \boldsymbol{H}_1$$

Therefore,

$$150\boldsymbol{j} = \boldsymbol{H}_2 - I_{zz}\omega_z\boldsymbol{k}$$

The postimpact angular momentum is then

$$\begin{aligned} \boldsymbol{H}_2 &= 150\boldsymbol{j} + I_{zz}\omega_z\boldsymbol{k} \\ &= 150\boldsymbol{j} + \left\{2\left[\frac{1}{2}\left(\frac{300}{g}\right)(2.5)^2\right] + \frac{1800}{g}(2.5)^2\right\}2\boldsymbol{k} \\ &= 150\boldsymbol{j} + 815\boldsymbol{k} \text{ slug-ft}^2/\text{sec} \end{aligned} \tag{b}$$

Since the ensuing motion is *torque-free*, we have thus established the direction of the precession axis (Z). This axis has been shown in Fig. 18.37 coinciding with $\boldsymbol{H}_2$ in the yz plane. Furthermore, we can give $\boldsymbol{H}_2$ postimpact as follows remembering that xyz are principal axes:[11]

$$\boldsymbol{H}_2 = 150\boldsymbol{j} + 815\boldsymbol{k} = \omega_x I_{xx}\boldsymbol{i} + \omega_y I_{yy}\boldsymbol{j} + \omega_z I_{zz}\boldsymbol{k} \tag{c}$$

Hence, from Eq. (c) we have at postimpact:

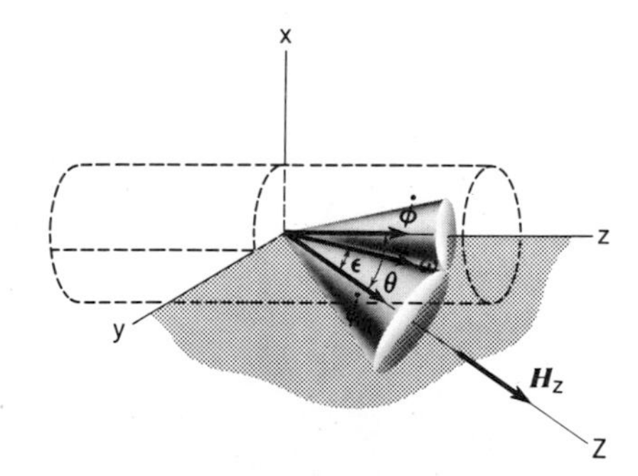

Figure 18.37. Cone model; Z is in yz plane.

$$\omega_x = 0$$

$$\omega_y = \frac{150}{I_{yy}} = \frac{150}{2300} = .0652 \text{ rad/sec}$$

$$\omega_z = \frac{815}{I_{zz}} = \frac{815}{407.5} = 2 \text{ rad/sec}$$

where for I_{yy} we used the value of $I_{\xi\xi}$ as computed in Eq. (e) of Example 18.5. Hence,

$$\boldsymbol{\omega} = .0652\boldsymbol{j} + 2\boldsymbol{k} \text{ rad/sec} \tag{d}$$

We can now make good use of the cone model representing the motion. Accordingly, in Fig. 18.37 corresponding to postimpact we have shown two cones, one about the body axis z (this is the moving cone) and one about the Z axis (this is the stationary cone). The line of contact between the cones coincides with the total angular velocity vector $\boldsymbol{\omega}$. The capsule must subsequently have the same motion as

[11] Remember that the *position* of a body is considered *not* to change during the action of an impulsive force whereas its linear and angular *momenta do* change. Therefore, the xyz axes postimpact have not moved from the position corresponding to preimpact.

the moving cone as it rolls with an angular speed $\dot{\phi} = 2$ rad/sec about its own axis without slipping on the stationary cone having Z as an axis. We can easily compute the angles θ and ϵ using Eqs. (b) and (d). Thus:

$$\tan\theta = \frac{H_y}{H_z} = \frac{150}{815} = 0.1840$$

Therefore,

$$\theta = 10.43^\circ$$

$$\epsilon = \theta - \tan^{-1}\frac{\omega_y}{\omega_z} = 10.43^\circ - \tan^{-1}\frac{0.0652}{2}$$

Therefore,

$$\epsilon = 8.56^\circ$$

In Fig. 18.38 we have shown ω, $\dot{\phi}$, and $\dot{\psi}$ in the yz plane corresponding to postimpact. Knowing $\dot{\phi}$, ω, ϵ, and θ, we can easily compute $\dot{\psi}$. Thus, using the law of sines, we get

$$\frac{\dot{\psi}}{\sin(\theta - \epsilon)} = \frac{\omega}{\sin(\pi - \theta)}$$

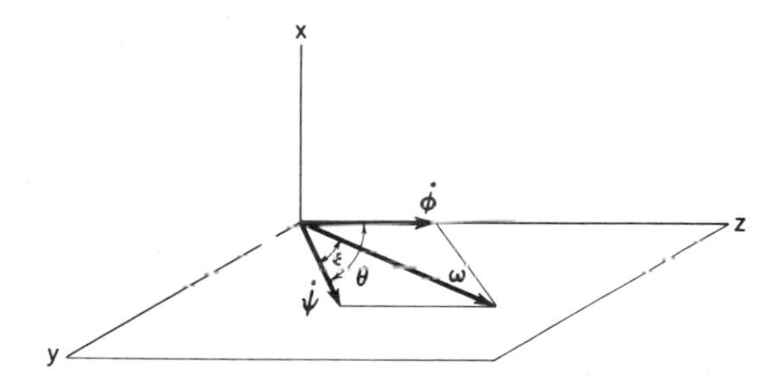

Figure 18.38. Angular-velocity diagram.

Therefore,

$$\dot{\psi} = \frac{\sin(\theta - \epsilon)\,\omega}{\sin\theta} = \frac{(.0326)(2^2 + .0652^2)^{1/2}}{.1810} = .360 \text{ rad/sec}$$

Thus, we can say that the body continues to spin at 2 rad/sec about its axis, but now the axis precesses about the indicated Z direction at the rate of .360 rad/sec.

Problems

18.48. A dynamical model of a device in orbit consists of a 2000-lbm cylindrical shell A of uniform thickness and a disc B rotating relative to the cylinder at a speed ω_2 of 5000 rad/min. The disc B is 1 ft in diameter and has a mass of 100 lbm. The cylinder is rotating at a speed ω_1 of 10 rad/min about axis D–D in inertial space. If the shaft FF about which B rotates is made to line up with D–D by an internal mechanism, what is the final angular momentum vector for the system? Neglect the mass of all bodies except the disc B and the shell.

Figure P.18.48

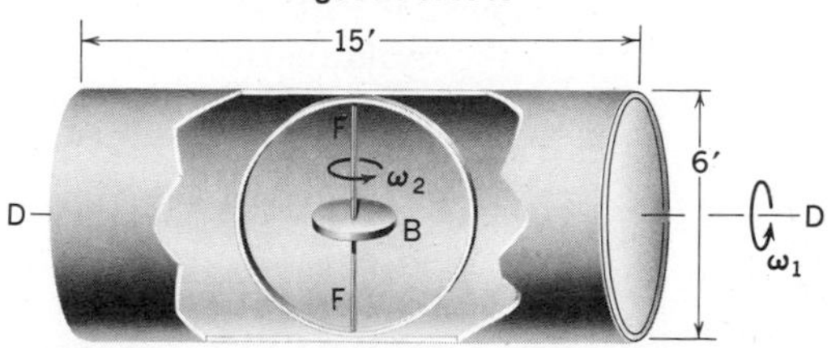

18.49. A solar energy power unit is in orbit having been given an angular speed ω_1 equal to .2 rad/sec about the z axis at time t. Vane B is identical to vane A but is rotated 90° from vane A. By an internal mechanism, vane B is rotated about its axis to be parallel to vane A. What is the new angular speed of the system after this adjustment has taken place? What is the final direction of $\boldsymbol{\omega}$? The vanes on earth each weigh 200 lb and can be considered as uniform blocks. The radii of gyration for the configuration corresponding to vane A are as follows at the mass center C.M.:

$$k_x = .5 \text{ ft}$$
$$k_y = 5 \text{ ft}$$
$$k_z = 3.5 \text{ ft}$$

The unit D can be considered as equivalent to a uniform sphere of weight on earth of 300 lb and radius 1 ft. (*Advice:* This is a simple problem despite seeming complexity.)

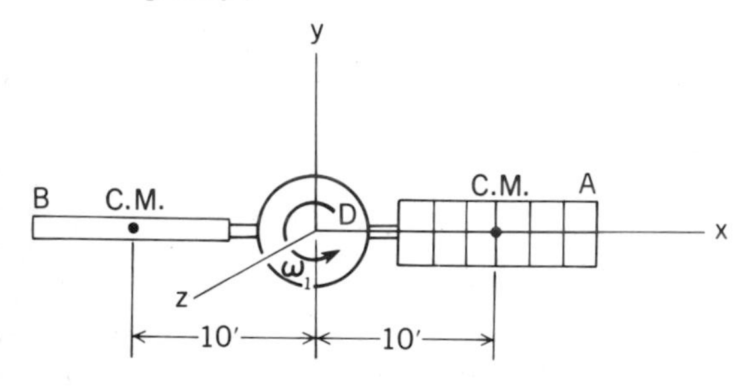

Figure P.18.49

18.50. A projectile is shot out of a weapon in such a manner that it has an angular velocity $\boldsymbol{\omega}$ at a known angle α from the centerline as it leaves the weapon. Using the cone model, draw a picture depicting the ensuing motion. Denote θ on this diagram, and indicate the direction of $\boldsymbol{H}$. Assume that the spin $\dot{\phi}$ about the axis of symmetry is known, as are the moments of inertia at the C.M. Set up formulations leading to the evaluation of the rate of precession of z about $\boldsymbol{H}$ and the angle θ between z and $\boldsymbol{H}$. *Hint:* Use trigonometry as well as two of Eqs. 18.43.

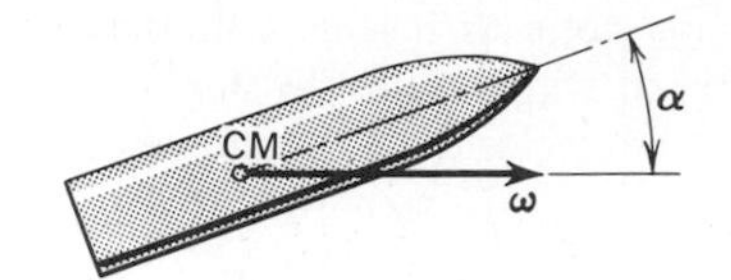

Figure P.18.50

18.51. A space capsule is rotating about its axis of symmetry in inertial space with an angular speed ω_1 of 2 rad/sec. As a result of an impact with a meteorite at point A, an impulse $\boldsymbol{I}$ of 40 lb-sec is developed. Find the axis of precession and the precession velocity for postimpact motion. What are the spin velocity and nutation angle? The mass of the capsule is 3000 lbm. The radius of gyration for the axis of symmetry is 2 ft, whereas the radius of gyration for transverse axes at the center of mass is 2.5 ft.

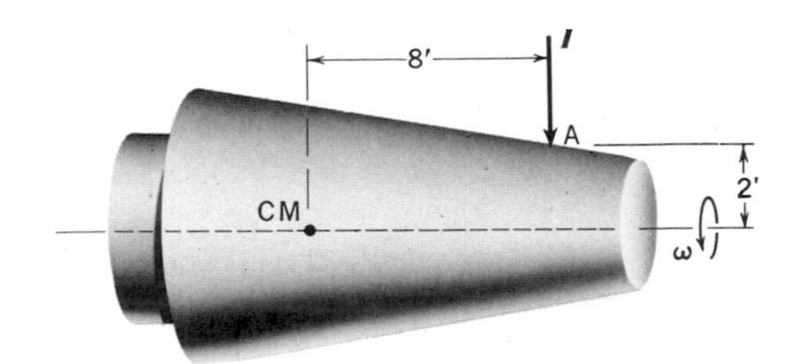

Figure P.18.51

18.52. Work Problem 18.51 for the case where $\boldsymbol{I}$ is inclined to the right an angle of 45°.

18.53. A rocket casing is in orbit. The casing has a spin of 5 rad/sec about its axis of symmetry. The axis of symmetry is oriented 30° from the precession axis as shown in the diagram. What is the precession speed and the angular momentum of the casing? The casing has a mass of 909 kg, a radius of gyration of .60 m about the axis of symmetry, and a radius of gyration about transverse axes at its center of mass of 1 m.

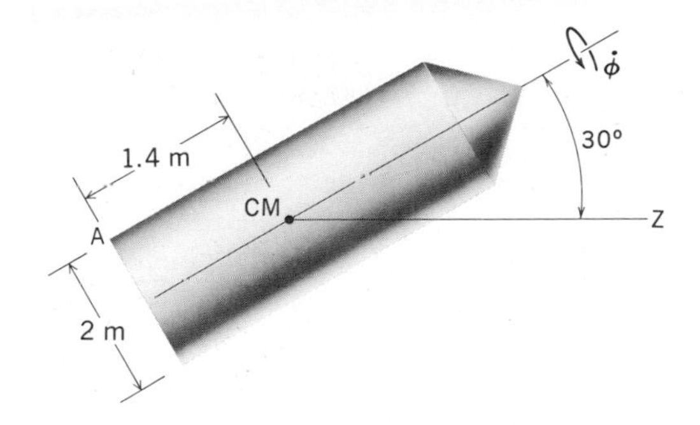

Figure P.18.53

18.54. In Problem 18.53 assume that an impulse in the vertical direction is developed at point A as a result of an impact with a meteorite. If the impulse from the impact is 133 N-sec, what are the new precession axis and the rate of precession after impact? H_0 from Problem 18.53 is 2952 kg-m²/sec.

18.55. An object representing dynamically a space device is made of three homogeneous blocks A, B, and C each of specific weight on earth of 6075 N/m³. Blocks A and C are identical and are hinged along aa and bb. At the configuration shown, the system is in orbit and is made to rotate instantaneously about an axis parallel to RR at a speed ω_1 of 3 rad/sec. The block C is then closed by an internal mechanism. What are the subsequent precession axis and rate of precession?

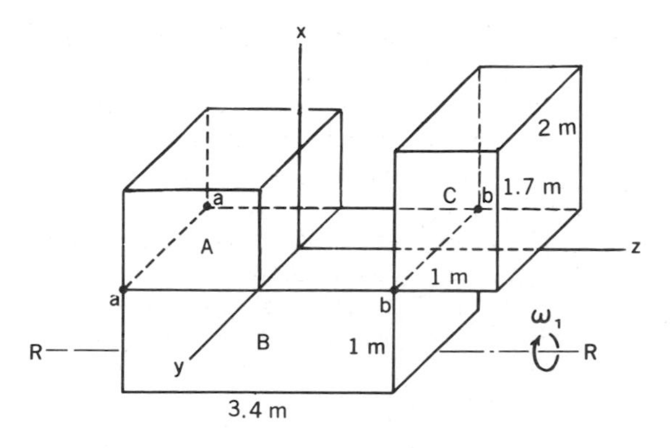

Figure P.18.55

18.56. A space vehicle has zero rotation relative to inertial space. A jet at A is turned on to give a thrust of 50N for .8 sec. Identify the body axis and the subsequent line of nodes. Then, give the nutation angle, the spin speed $\dot{\phi}$, and the precession speed $\dot{\psi}$. The vehicle weighs 10 kN and has a radius of gyration $k_z = 1$ m and, transverse to the z axis at the center of mass, $k' = .8$ m. Consider the thrust to be impulsive.

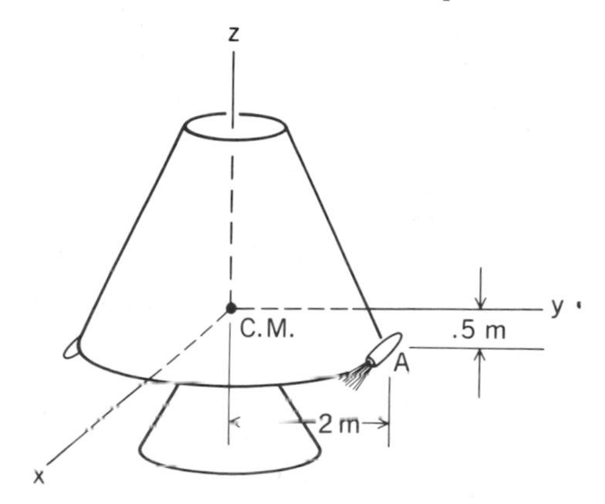

Figure P.18.56

18.57. An intermediate-stage rocket engine is separated from the first stage by activating exploding bolts. The angular velocity of the spent engine is given as

$$\boldsymbol{\omega} = 2\boldsymbol{i} + 3\boldsymbol{j} + .2\boldsymbol{k} \text{ rad/sec}$$

What is the spin speed $\dot{\phi}$, the precession speed $\dot{\psi}$, and the nutation speed $\dot{\theta}$? Give direction of Z axis. Identify the line of nodes. Note that $I_{zz} = 1500$ kg-m² and $I_{xx} = I_{yy} = 2000$ kg-m².

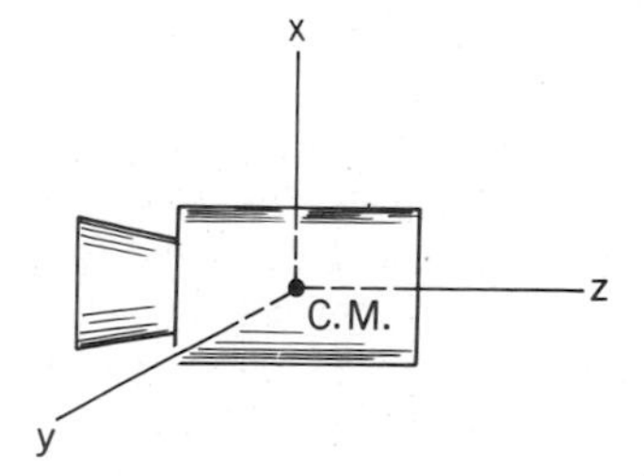

Figure P.18.57

18.58. In Problem 18.55, if 15 N-m of mechanical energy is added to the bodies from a battery in the closure process, determine the nutation angle θ and spin speed $\dot{\phi}$. The following results are available from Problem 18.55:

$$\left.\begin{aligned} I &= 5615 \text{ kg-m}^2 \\ I' &= 10{,}921 \text{ kg-m}^2 \end{aligned}\right\} \text{closed configuration}$$

$$H_0 = 22{,}270 \text{ kg-m}^2/\text{sec}$$

$$I_{zz} = 6877 \text{ kg-m}^2\} \text{ open configuration}$$

The initial kinetic energy is 31,000 N-m. (*Hint:* Make use of work-energy equation and Eq. 18.37).

18.59. In Example 18.5, if 20 ft-lb of work is done on the system from an internal power source when going from a test configuration to a closed configuration, determine the nutation angle θ and the spin speed $\dot{\phi}$ for closed configuration. (*Hint:* Make use of work-energy equation and Eq. 18.37.)

18.8 Closure

This chapter brings to a close our study of the motion of rigid bodies. In the final chapter of this text we shall for the most part go back to particle mechanics to consider the dynamics of particles constrained to move about a fixed point in a small domain. This is the study of small vibrations (alluded to in Chapter 12) which we have held in abeyance so as to take full advantage of your course work in differential equations.

Review Problems

18.60. A plane just after takeoff is flying at a speed V of 200 km/hr and is in the process of retracting its wheels. The back wheels (under wings) are being rotated at a speed ω_1 of 3 rad/sec and at the instant of interest have rotated 30° as shown in the diagram. The plane is rising by following a circular trajectory of radius 1000 m. If at the instant shown, $\dot{V}$ is 50 km/hr/sec, what is the total moment coming onto the bearings of the wheel from the motion of the wheel? The diameter of the wheel is 600 mm and its weight is 900 N. The radius of gyration along its axis is 250 mm and transverse to its axis is 180 mm. Neglect wind and bearing friction.

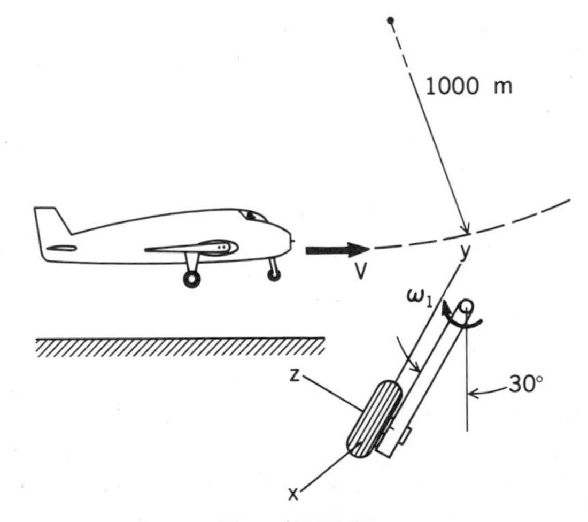

Figure P.18.60

18.61. A disc A weighing 10 N and of diameter 100 mm rotates with constant speed $\omega_1 =$ 15 rad/sec relative to G. (A motor on G, not shown, ensures this constant speed.) The shaft of motor B on C rotates with constant speed $\omega_2 = 8$ rad/sec relative to C and causes platform G to rotate relative to C. Finally, platform C rotates with angular speed $\omega_3 = 3$ rad/sec relative to the ground. What are the supporting forces on bearings H and K of the disc A on platform G? Bearing K alone supports disc A in the axial direction (i.e., it acts as a thrust bearing in addition to being a regular bearing).

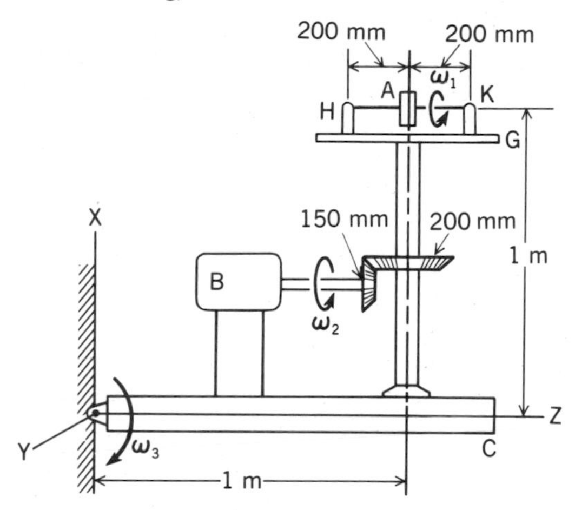

Figure P.18.61

18.62. Discs A and B are rolling without slipping at their centerlines against an upper surface D. Each disc weighs 40 lb, and each spins about a shaft which connects to a centerpost E rotating at an angular speed ω_1. If a total of 20 lb is developed upward on D, what is ω_1?

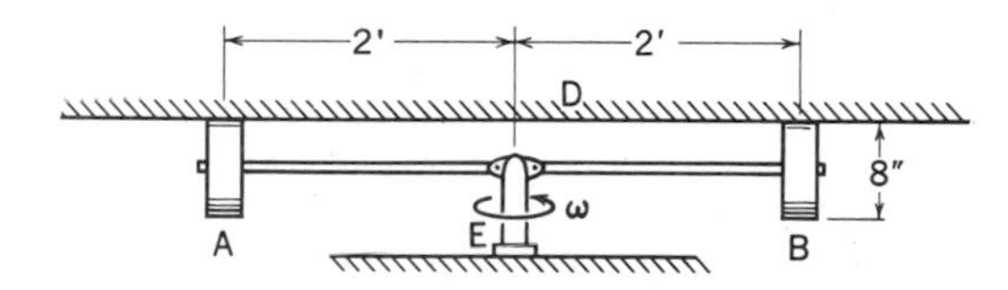

Figure P.18.62

18.63. A right circular cone weighing 20 N is spinning like a top about a fixed point O at a speed $\dot{\phi}$ of 15,000 rad/sec. What are two possible precession speeds for $\theta = 30°$?

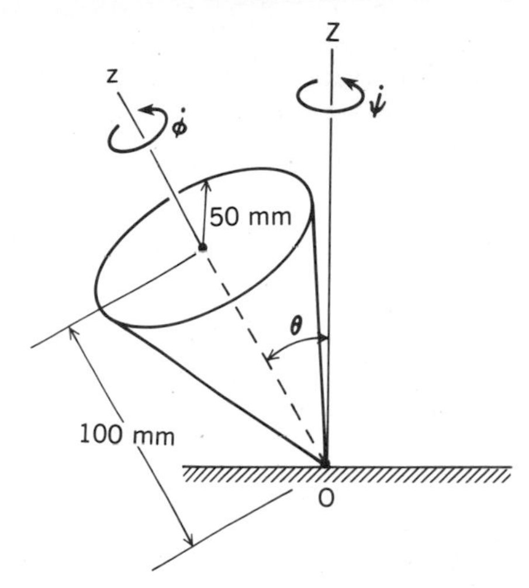

Figure P.18.63

18.64. A submerged submarine is traveling at 20 knots in a circular path. A single-degree-of-freedom gyroscope (rate gyro) turns 5° against two torsional springs each having a torsional spring constant of 1 N-m/rad. The gyro shown in (b) is viewed from the rear (from stern to bow on the submarine). What is the radius of curvature R of the path of the sub from this reading? This disc weighs 4 N, has a radius of 50 mm, and rotates at 20,000 rpm.

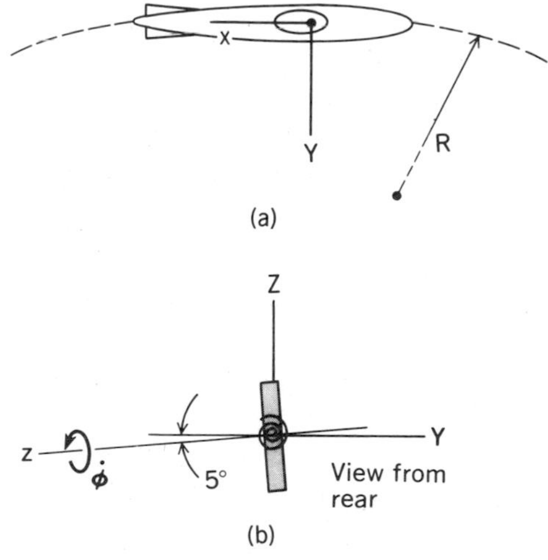

Figure P.18.64

18.65. Work Problem 18.56 for the case where, before activation of the jet, the vehicle has the following angular velocity:

$$\boldsymbol{\omega} = .20\boldsymbol{i} + .3\boldsymbol{j} + .15\boldsymbol{k} \text{ rad/sec}$$

18.66. In Problem 18.49, the initial angular velocity is:

$$\boldsymbol{\omega} = .3\boldsymbol{i} + .2\boldsymbol{j} + .4\boldsymbol{k} \text{ rad/sec}$$

What are ω_x, ω_y, and ω_z after vane B has been rotated parallel to vane A by an internal mechanism?

18.67. A research space capsule is in orbit having imposed on it an instantaneous angular speed ω of .4 rad/sec. The capsule consists of a spherical unit D which can be considered a uniform sphere of radius .60 m and weight 800 N. Arms A and B extend from the sphere and consist of cylinders which pick up and record dust particle collisions. Each such unit weighs 500 N and has the following radii of gyration at the mass center, C.M.

$$k \text{ (lateral axes)} = .80 \text{ m}$$

$$k \text{ (longitudinal axis)} = .20 \text{ m}$$

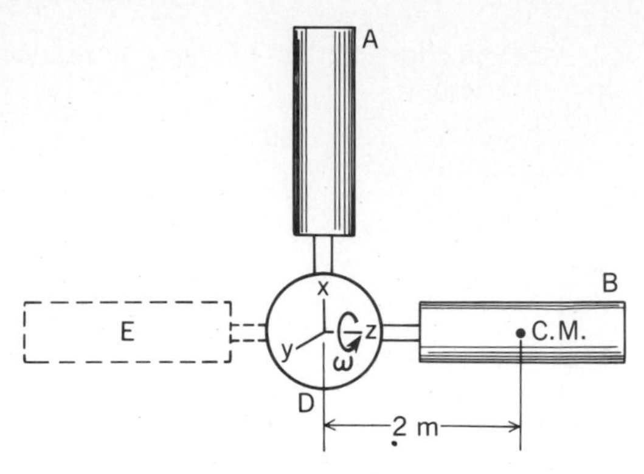

Figure P.18.67

If arm A is rotated 90° to position E by an internal mechanism, what is the angular precession speed of the system? Give the direction of Z about which there is precession. All weights are as on earth.

18.68. In Problem 18.67, 20 N-m of energy is used to cause the change in configuration. What is the spin speed and nutation angle? From a previous solution, $\dot{\psi} = .2064$ rad/sec. (*Hint:* Make use of work-energy equation and Eq. 18.37.)

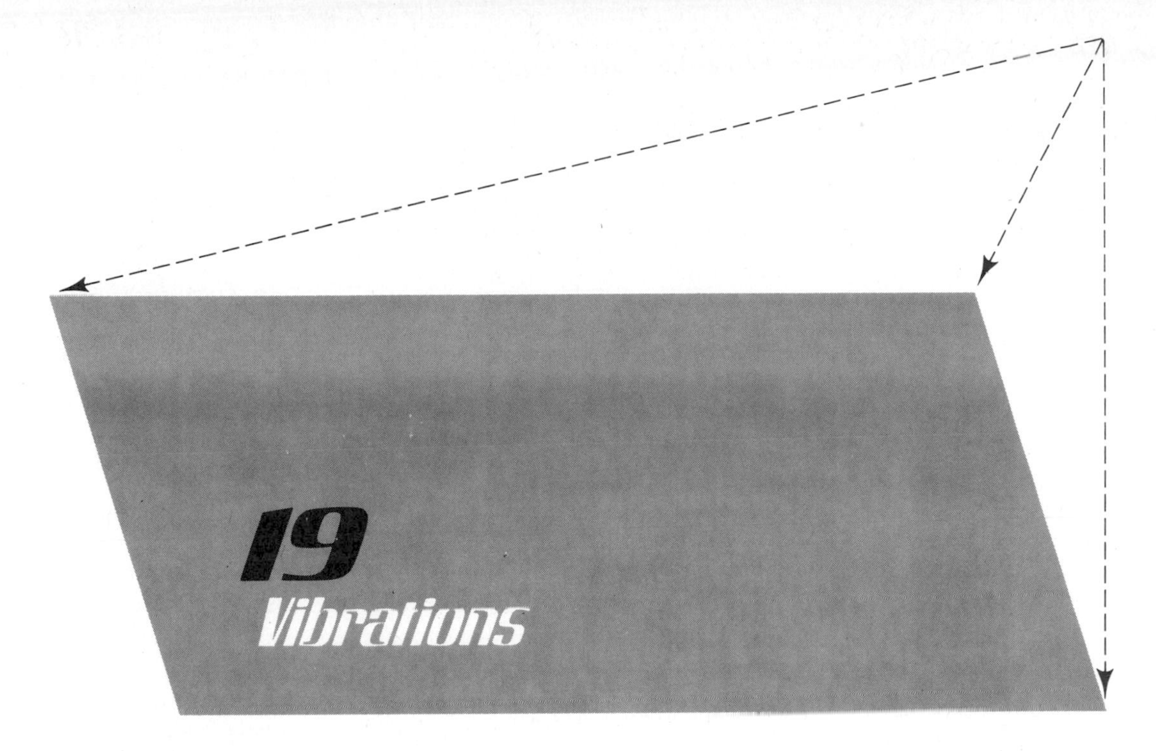

19.1 Introduction

You will recall that in Chapter 12 we said we would defer a more general examination of particle motion about a fixed point until the very end of the text. We do this to take full advantage of any course in differential equations that you might be taking simultaneously with this course. Accordingly, we shall now continue the work begun in Chapter 12.

19.2 Free Vibration

Let us begin by reiterating what we have done earlier leading to the study of vibrations. Recall that we examined the case of a particle in rectilinear translation acted on either by a constant force, a force given as a function of time, a force that is a function of speed, or, finally, a force that is a function of position. In each case we could separate the variables and effect a quadrature to arrive at the desired algebraic equations, including constants of integration. In particular we considered, as a special case of a force given as a function of position, the linear restoring force resulting from the action (or equivalent action) of a linear spring. Thus, for the spring–mass system shown in Fig. 19.1, the differential equation of motion was shown to be

$$\frac{d^2x}{dt^2} + \frac{K}{m}x = 0 \tag{19.1}$$

where K is the spring constant and where x is measured from the static equilibrium

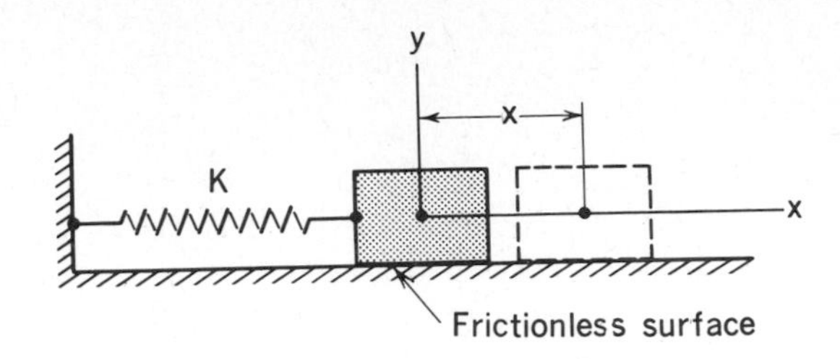

Figure 19.1. Spring–mass system.

position of the mass. You will now recognize this equation from your studies in mathematics as a second-order, linear differential equation with constant coefficients.

Instead of rearranging the equation to effect a quadrature, as we did in the previous case,[1] we shall take a more general viewpoint toward the solving of differential equations.

To solve a differential equation, we must find a function of time, $x(t)$, which when substituted into the equation satisfies the equation (i.e., reduces it to an identity $0 = 0$). We can either guess at $x(t)$ or use a formal procedure. You have learned in your differential equations course that the most general solution of the above equation will consist of a linear combination of two functions that cannot be written as multiples of each other (i.e., the functions are linearly independent). There will also be two arbitrary constants of integration. Thus, $C_1 \cos \sqrt{K/m}\, t$ and $C_2 \sin \sqrt{K/m}\, t$ will satisfy the equation, as we can readily demonstrate by substitution, and are independent in the manner described. We can therefore say

$$x = C_1 \cos \sqrt{\frac{K}{m}} t + C_2 \sin \sqrt{\frac{K}{m}} t \tag{19.2}$$

where C_1 and C_2 are the aforementioned constants of integration to be determined from the initial conditions.

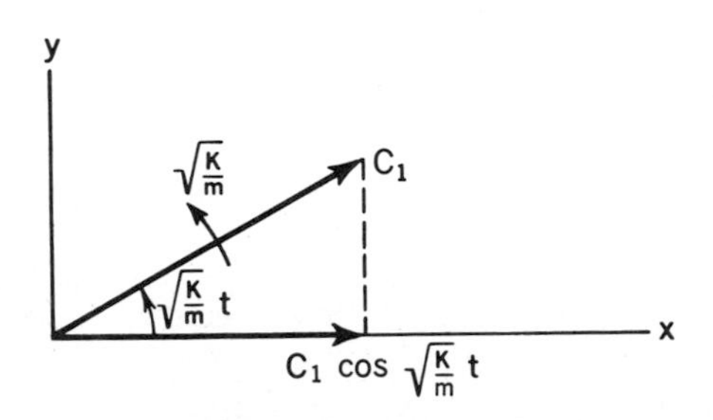

Figure 19.2. Phasor representation.

We can conveniently represent each of the above functions by employing rotating vectors of magnitudes that correspond to the coefficients of the functions. This representation is shown in Fig. 19.2, where, if the vector C_1 rotates counterclockwise with an angular velocity of $\sqrt{K/m}$ radians per unit time and if C_1 lies along the x axis at time $t = 0$, then the projection of this vector along the x axis represents one of the functions of Eq. 19.2, namely $C_1 \cos \sqrt{K/m}\, t$. Vectors used in this manner are called *phasors*.

Consider now the function $C_2 \sin \sqrt{K/m}\, t$, which we can replace by $C_2 \cos (\sqrt{K/m}\, t - \pi/2)$, as we learned in elementary trigonometry. The phasor representation for this function, therefore, would be a vector of magnitude C_2 that rotates with

[1]Recall that this can be done by replacing d^2x/dt^2 by $(dV/dx)(dx/dt)$, which is simply $V(dV/dx)$.

Amplitude. The largest displacement attained by the body during a cycle is the amplitude. In this case, the amplitude corresponds to the coefficient C_3.

Phase angle. The phase angle is the angle between the phasor and the x axis when $t = 0$ (i.e., the angle β).

A plot of the motion as a function of time is presented in Fig. 19.5, where certain of these various quantities are shown graphically.

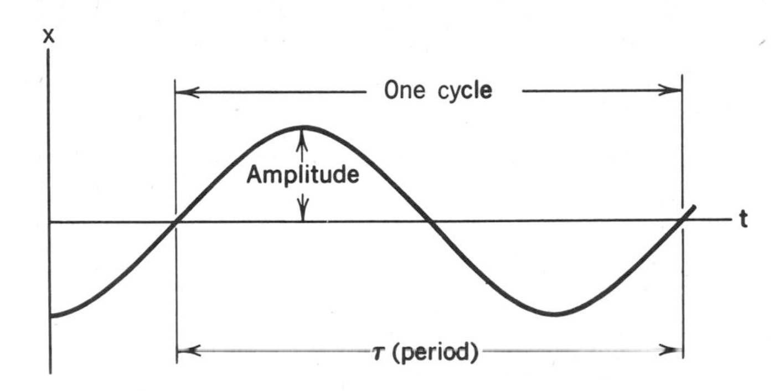

Figure 19.5. Plot of harmonic motion.

It is usually easier to use the earlier form of solution, Eq. 19.2, rather than Eq. 19.4 in satisfying initial conditions. Accordingly, the position and velocity can be given as

$$x = C_1 \cos\sqrt{\frac{K}{m}}t + C_2 \sin\sqrt{\frac{K}{m}}t$$

$$V = -C_1\sqrt{\frac{K}{m}}\sin\sqrt{\frac{K}{m}}t + C_2\sqrt{\frac{K}{m}}\cos\sqrt{\frac{K}{m}}t$$

The initial conditions to be applied to these equations are:

$$\text{when } t = 0, \qquad x = x_0, \quad V = V_0$$

Substituting, we get

$$x_0 = C_1, \qquad V_0 = C_2\sqrt{\frac{K}{m}}$$

Therefore, the motion is given as

$$x = x_0 \cos\sqrt{\frac{K}{m}}t + \frac{V_0}{\sqrt{K/m}}\sin\sqrt{\frac{K}{m}}t \tag{19.6a}$$

$$V = -x_0\sqrt{\frac{K}{m}}\sin\sqrt{\frac{K}{m}}t + V_0\cos\sqrt{\frac{K}{m}}t \tag{19.6b}$$

We can generalize from these results by noting that any agent supplying a linear restoring force for all rectilinear motions of a mass can take the place of the spring in

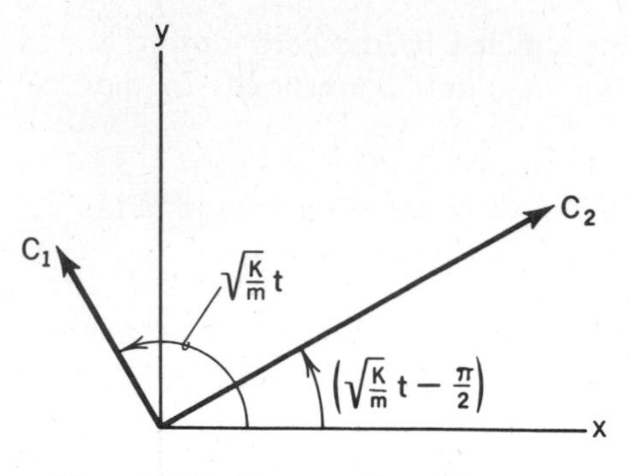

Figure 19.3. Phasors $\pi/2$ out of phase.

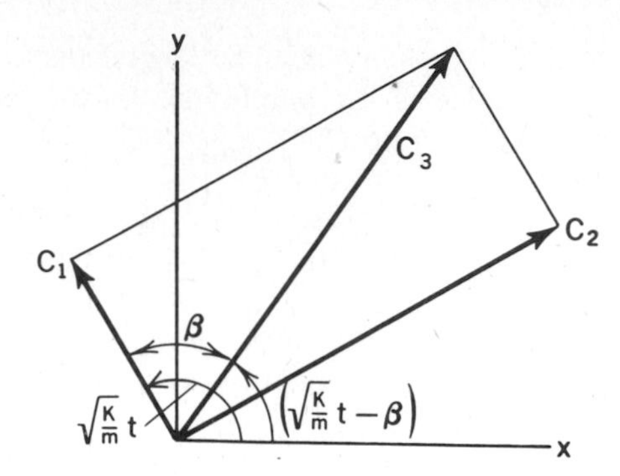

Figure 19.4. Vector sum of phasors.

angular velocity $\sqrt{K/m}$ and that is out of phase by $\pi/2$ with the phasor C_1 (Fig. 19.3). Thus, the projection of C_2 on the x axis is the other function of Eq. 19.2. Clearly, because vectors C_1 and C_2 rotate at the same angular speed, we can represent the combined contribution by simply summing the vectors and considering the projection of the resulting single vector along the x axis. This summation is shown in Fig. 19.4 where vector C_3 replaces the vectors C_1 and C_2. Now we can say:

$$C_3 = \sqrt{C_1^2 + C_2^2}, \qquad \beta = \tan^{-1}\frac{C_2}{C_1} \tag{19.3}$$

Because C_1 and C_2 are arbitrary constants, C_3 and β are also arbitrary constants. Consequently, we can replace the solution given by Eq. 19.2 by another equivalent form:

$$x = C_3 \cos\left(\sqrt{\frac{K}{m}}t - \beta\right) \tag{19.4}$$

From this form, you probably recognize that the motion of the body is *harmonic motion*. In studying this type of motion, we shall use the following definitions:

Cycle. The cycle is that portion of a motion (or series of events in the more general usage) which, when repeated, forms the motion. On the phasor diagrams, a cycle would be the motion associated with one revolution of the rotating vector.

Frequency. The number of cycles per unit time is the frequency. The frequency is equal to $\sqrt{K/m}/2\pi$ for the above motion, because $\sqrt{K/m}$ has units of radians per unit time. Often $\sqrt{K/m}$ is termed the *natural frequency* of the system in radians per unit time or, when divided by 2π, in cycles per unit time.

Period. The period, τ, is the time of one cycle, and is therefore the reciprocal of frequency. That is,

$$\tau = \frac{2\pi}{\sqrt{K/m}} \tag{19.5}$$

the preceding computations. We must remember, however, that to behave this way the agent must have negligible mass. Thus, we can associate with such agents an *equivalent spring constant* K_e, which we can ascertain if we know the static deflection δ permitted by the agent on application of some known force F. We can then say:

$$K_e = \frac{F}{\delta}$$

Once we determine the equivalent spring constant, we immediately know that the natural frequency of the system is $(1/2\pi)\sqrt{K_e/m}$ cycles per unit time. This natural frequency is the number of cycles the system will repeat in a unit time if some initial disturbance is imposed on the mass. Note that this natural frequency depends only on the "stiffness" of the system and on the mass of the system and is not dependent on the amplitude of the motion.[2]

We shall now consider several problems in which we can apply what we have just learned about harmonic motion.

EXAMPLE 19.1

A mass weighing 45 N is placed on the spring shown in Fig. 19.6 and is released very slowly, extending the spring a distance of 50 mm. What is the natural frequency of the system? If the mass is given a velocity instantaneously of 1.60 m/sec down from the equilibrium position, what is the equation for displacement as a function of time?

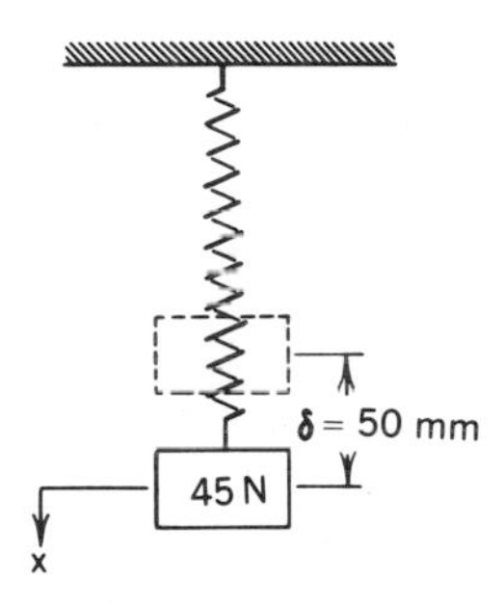

Figure 19.6. x measured from static deflection position.

The spring constant is immediately available by the equation

$$K = \frac{F}{\delta} = \frac{45 \text{ N}}{50 \text{ mm}} = .9 \text{ N/mm}$$

The equation of motion for the mass can be written for a reference whose origin is at the static equilibrium position shown in the diagram. Thus,

$$m\frac{d^2x}{dt^2} = W - K(x + \delta)$$

where δ is the distance from the unextended position of the spring to the origin of the reference. However, from our initial equation, $\delta = F/K = W/K$. Therefore, we have

$$m\frac{d^2x}{dt^2} = W - K\left(x + \frac{W}{K}\right) = -Kx$$

and the equation becomes identical to Eq. 19.1:

$$\frac{d^2x}{dt^2} + \frac{K}{m}x = 0 \qquad \text{(a)}$$

[2]Actually, when the amplitude gets comparatively large, the spring ceases to be linear, and the motion does depend on the amplitude. Our results do not apply for such a condition.

Thus, the motion will be an oscillation about the position of static equilibrium, which is an extended position of the spring. Measuring x from the *static equilibrium position* and considering the spring force as $-Kx$, we can thus disregard the weight on the body in writing Newton's law for the body to reach Eq. (a) most directly.

Accordingly, we can use the results stemming from our main discussion. Employing the notation ω_n as the natural frequency in units of radians per unit time, we have

$$\omega_n = \sqrt{\frac{K}{m}} = \sqrt{\frac{\left(\frac{.9\text{ N}}{\text{mm}}\right)\left(\frac{1000\text{ mm}}{\text{m}}\right)}{\frac{45}{g}\text{ kg}}} = \sqrt{\frac{\left(\frac{.9\text{ N}}{\text{mm}}\right)\left(\frac{1000\text{ mm}}{\text{m}}\right)}{\frac{45}{g}\left(\frac{\text{N}}{\text{m/sec}^2}\right)}} = 14.01\text{ rad/sec}$$

The motion is now given by the equations

$$x = C_1 \sin 14.01t + C_2 \cos 14.01t$$

$$\dot{x} = 14.01C_1 \cos 14.01t - 14.01C_2 \sin 14.01t$$

From the specified initial conditions, we know that when $t = 0$, $x = 0$, and $\dot{x} =$ 1.6 m/sec. Therefore, the constants of integration are

$$C_2 = 0, \qquad C_1 = \frac{1.60}{14.01} = .1142$$

The desired equation, then, is

$$x = .1142 \sin 14.01t \text{ m}$$

EXAMPLE 19.2

A body weighing 22 N is positioned in Fig. 19.7 on the end of a slender cantilever beam whose mass we can neglect in considering the motion of the body at its end.

If we know the geometry and the composition of the cantilever beam, and if the deflection involved is small, we can compute from strength of materials the deflection of the end of the beam that results from a vertical load there. This deflection is directly proportional to the load. In this case, suppose that we have computed a deflection of 12.5 mm for a force of 4 N (see Fig. 19.8). What would be the natural frequency of the body weighing 22 N for small oscillations in the vertical direction?

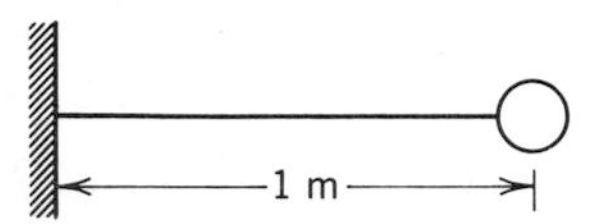

Figure 19.7. Slender cantilever beam with weight at end.

Figure 19.8. Beam acts as linear spring.

Because the motion is restricted to small amplitudes, we can consider the mass to be in rectilinear motion in the vertical direction in the same manner as the mass on the spring in the previous case. The beam now supplies the linear restoring force. The formulations of this section are once again applicable. The equivalent spring constant is found to be

$$K_e = \frac{F}{\delta} = \frac{4}{12.5} = .32 \text{ N/mm}$$

The natural frequency for vibration of the 22-N weight at the end of the cantilever is then

$$\omega_n = \sqrt{\frac{(.32)(1000)}{22/9.81}} = 11.94 \text{ rad/sec}$$

Before starting out on the exercises, we wish to point out results of Problem 19.2 that will be of use to you. In that problem, you are asked to show that the equivalent spring constant for springs in *parallel* and subject to the same deflection [see Fig. P.19.2(a)] is simply the sum of the spring constants of the springs. That is,

$$K_e = K_1 + K_2 \tag{19.7}$$

For springs in *series* [see Fig. P.19.2(b)] we have

$$\frac{1}{K_e} = \frac{1}{K_1} + \frac{1}{K_2} \tag{19.8}$$

Also, we wish to point out that, for small values of θ, we can approximate $\sin\theta$ by θ and $\cos\theta$ by unity. To justify this, expand $\sin\theta$ and $\cos\theta$ as power series and retain first terms.

Problems

19.1. If a 5-kg mass causes an elongation of 50 mm when suspended from the end of a spring, determine the natural frequency of the spring–mass system.

19.2. (a) Show that the spring constant is doubled if the length of the spring is halved.

(b) Show that two springs having spring constants K_1 and K_2 have a combined spring constant of $K_1 + K_2$ when connected in parallel, and have a combined spring constant whose reciprocal is $1/K_1 + 1/K_2$ when combined in series.

19.3. A mass M of 2 kg rides on a vertical frictionless guide rod. With only one spring K_1, the natural frequency of the system is 2 rad/sec. If we want to increase the natural frequency threefold, what must the spring constant K_2 of a second spring be?

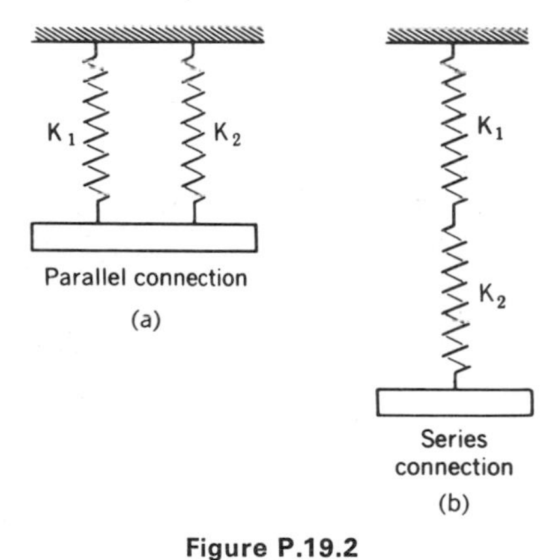

Figure P.19.2

Figure P.19.3

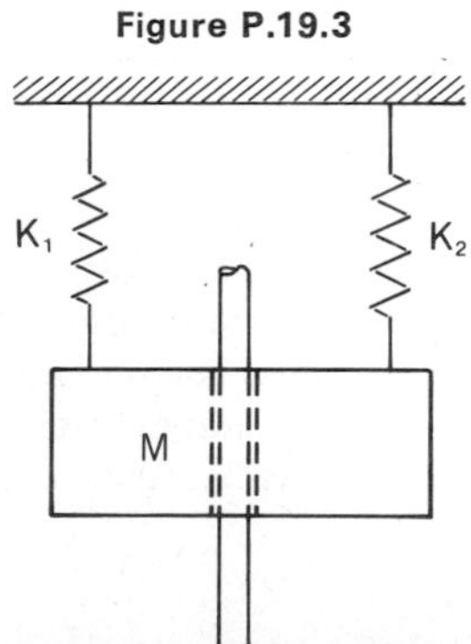

19.4. A mass M of 100 g rides on a frictionless guide rod. If the natural frequency with spring K_1 attached is 5 rad/sec, what must K_2 be to increase the natural frequency to 8 rad/sec?

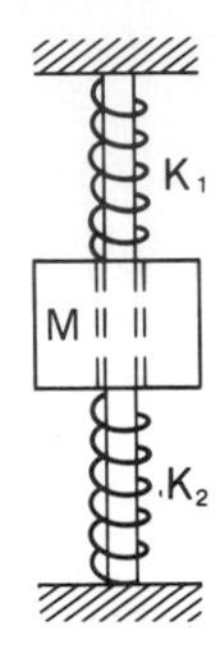

Figure P.19.4

19.5. For small oscillations, what is the natural frequency of the system in terms of a, b, K, and W? (Neglect the mass of the rod.)

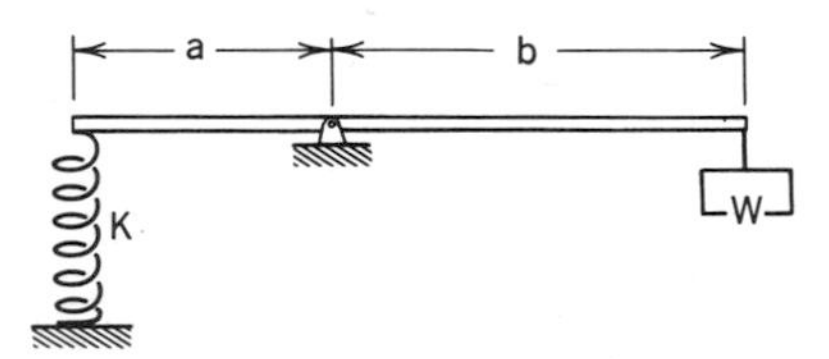

Figure P.19.5

19.6. A rod is supported on two rotating grooved wheels. The contact surfaces have a coefficient of friction of μ_d. Explain how the rod will oscillate in the horizontal direction if it is disturbed in that direction. Compute the natural frequency of the system.

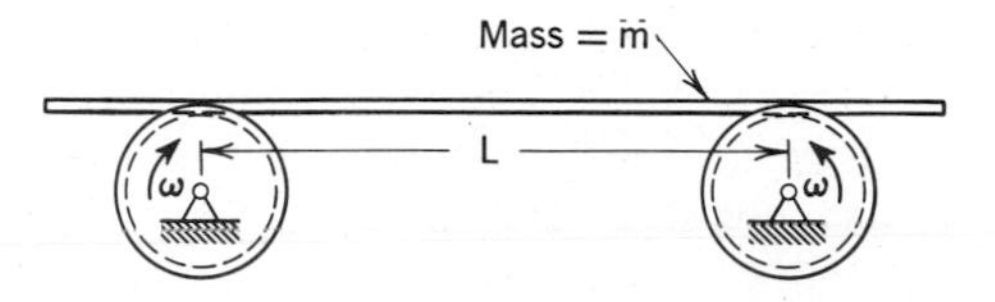

Figure P.19.6

19.7. A mass is held so it just makes contact with a spring. If the mass is released suddenly from this position, give the amplitude, frequency, and the center position of the motion. First use the *undeformed position* to measure x. Then, do the problem using x' from the *static equilibrium* position.

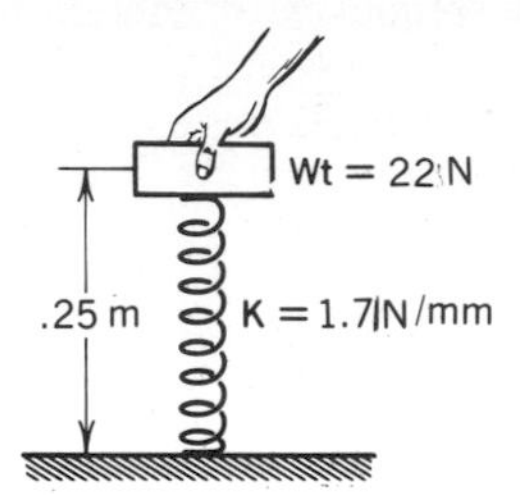

Figure P.19.7

19.8. A *hydrometer* is a device to measure the *specific gravity* of liquids. The hydrometer weighs .36 N, and the diameter of the cylindrical portion above the base is 6 mm. If the hydrometer is disturbed in the vertical direction, what is the frequency of vibration in cycles/sec as it bobs up and down? Recall from *Archimedes' principle* that the buoyant force equals the weight of the water displaced. Water has a density of 1000 kg/m³.

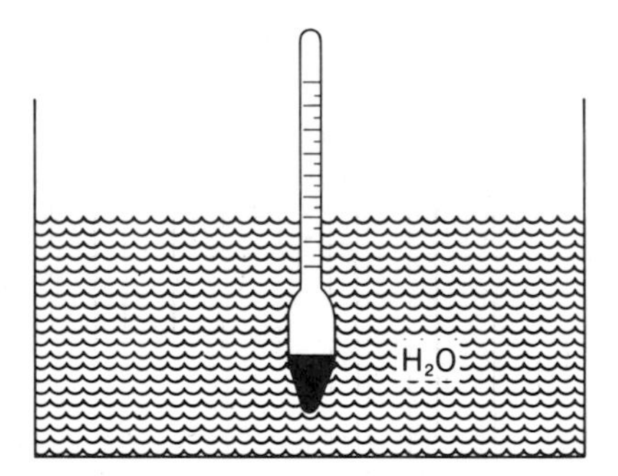

Figure P.19.8

***19.9.** The hydrometer of Problem 19.8 is used to test the specific gravity of battery acid in a car battery. What is the period of oscillation in this case? (*Hint:* Note if hydrometer goes down, the water surface will have to rise a certain amount simultaneously. The battery acid has a density of 1100 kg/m³.)

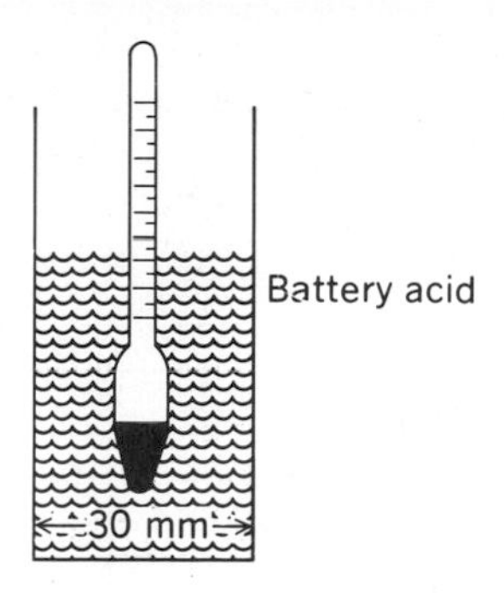

Figure P.19.9

19.10. A 30-kg block is suspended from two light wires. What is the frequency in cycles/sec at which the block will swing back and forth in the x direction if it is *slightly* disturbed in this direction? (*Hint:* For small θ, $\sin\theta \approx \theta$ and $\cos\theta \approx 1$.)

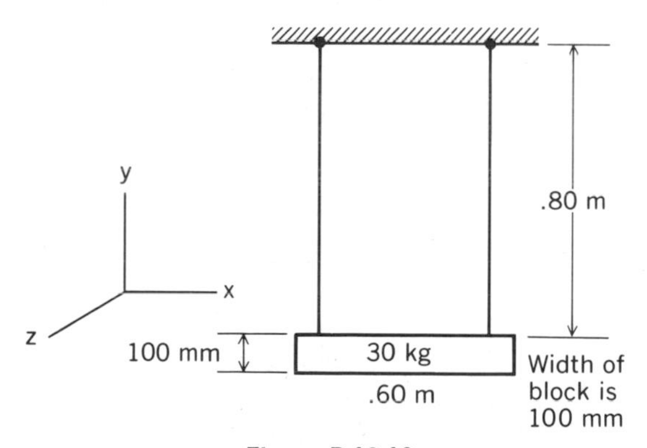

Figure P.19.10

19.11. In Problem 19.10, what is the period of small oscillations for a small disturbance that causes the block to move in the z direction?

19.12. What is the natural frequency of motion for block A for small oscillation? Consider BC to have negligible mass and body A to be a particle. When body A is attached to the rod, the static deflection is 25 mm. The spring constant K_1 is 1.75 N/mm. Body A weighs 110 N. What is K_2?

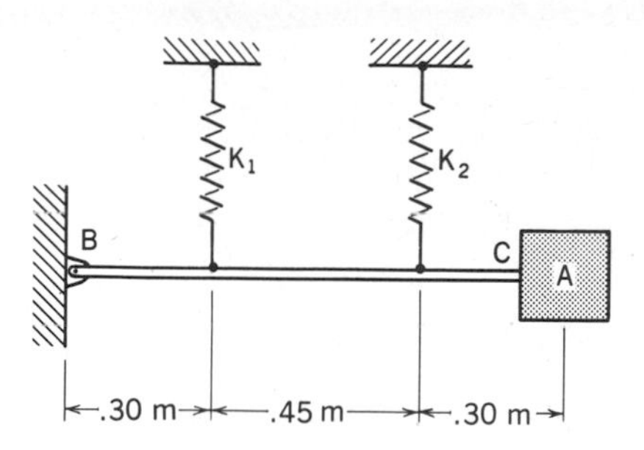

Figure P.19.12

19.13. If bar ABC is of negligible mass, what is the natural frequency of free oscillation of the block for small amplitude of motion? The springs are identical, each having a spring constant K of 25 lb/in. The weight of the block is 10 lb. The springs are unstretched when AB is oriented vertically as shown in the diagram.

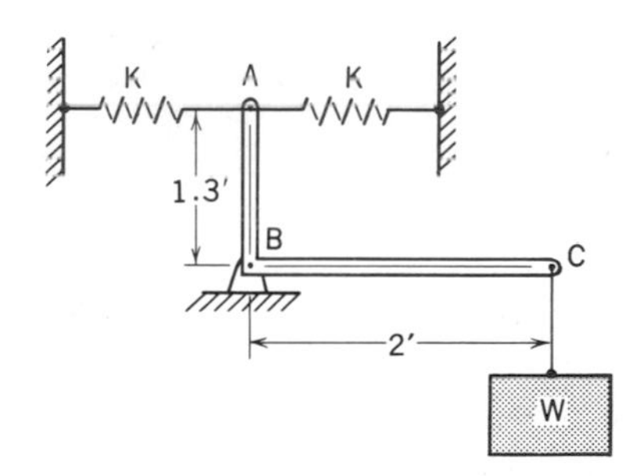

Figure P.19.13

19.14. Work Problem 19.13 for the case where the springs are both stretched 1 in. when AB is vertical.

19.15. What are the differential equation of motion about the static-equilibrium configuration and the natural frequency of motion of body A for small motion of BC? Neglect inertial effects from BC. The following data apply:

$$K_1 = 15 \text{ lb/in.}$$
$$K_2 = 20 \text{ lb/in.}$$
$$K_3 = 30 \text{ lb/in.}$$
$$W_A = 30 \text{ lb}$$

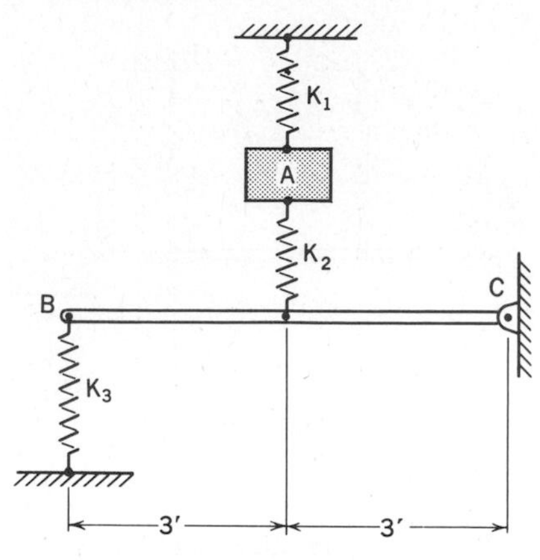

Figure P.19.15

19.16. A horizontal platform is rotating with a uniform angular speed of ω rad/sec. On the platform is a rod CD on which slides a cylinder A having weight W. The cylinder is connected to C through a linear spring having a spring constant K. What is the equation of motion for A relative to the platform after it has been disturbed? What is the natural frequency of oscillation? Take r_0 as the unstretched length of spring.

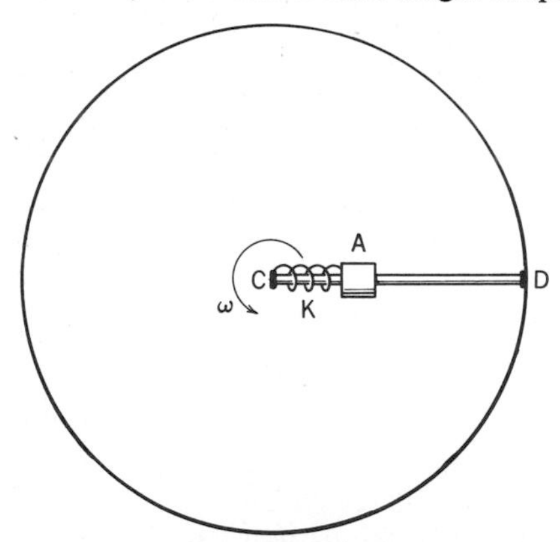

Figure P.19.16

19.17. A rigid body A rests on a spring with stiffness K equal to 8.80 N/mm. A lead pad B falls onto the block A with a speed on impact of 7 m/sec. If the impact is perfectly plastic, what are the frequency and amplitude of the motion of the system, provided that the lead pad sticks to A at all times? Take $W_A = 134$ N and $W_B = 22$ N. What is the distance moved by A in .02 sec? (*Caution:* Be careful about the initial conditions.)

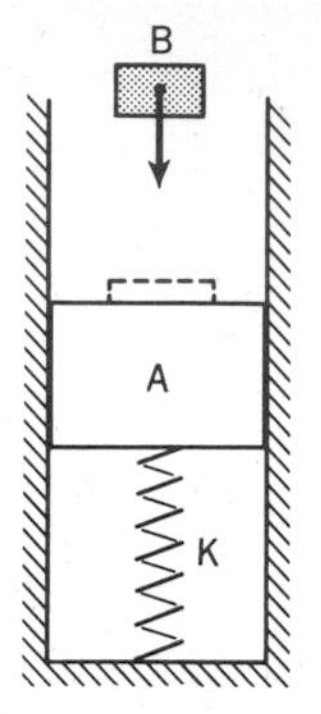

Figure P.19.17

19.18. A small sphere of weight 5 lb is held by taut elastic cords on a frictionless plane. If 50 lb of force is needed to cause an elongation of 1 in. for each cord, what is the natural frequency of small oscillation of the weight in a transverse direction? Also, determine the natural frequency of the weight in a direction along the cord for small oscillations. Neglect the mass of the cord. The tension in the cord in the configuration shown in 100 lb.

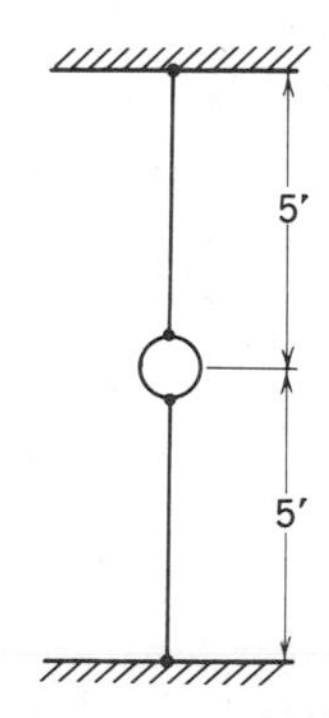

Figure P.19.18

19.19. Body A weighs 445 N and is connected to a spring having a spring constant K_1 of 3.50 N/mm. At the right of A is a second spring having a spring constant K_2 of 8.80 N/mm. Body A is moved 150 mm to the left from the configuration of static equilibrium shown in the diagram, and it is released from rest. What is the period of oscillation for the body? (*Hint:* Work with half a cycle.)

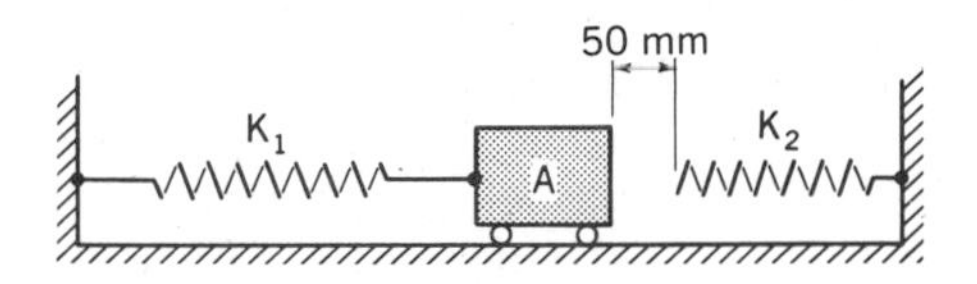

Figure P.19.19

19.20. Body A, weighing 50 lb, has a speed of 20 ft/sec to the left. If there is no friction, what is the period of oscillation of the body for the following data:

$$K_1 = 20 \text{ lb/in.}$$
$$K_2 = 10 \text{ lb/in.}$$

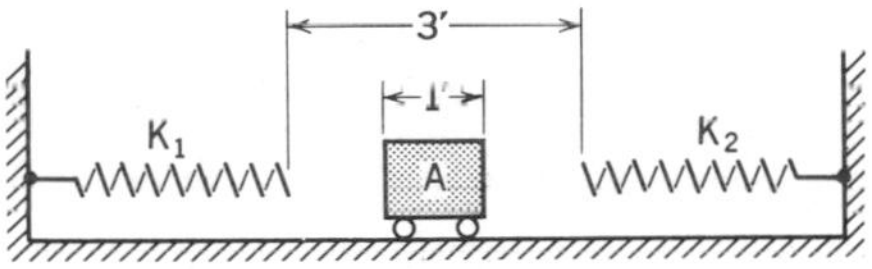

Figure P.19.20

19.21. A spherical body A of mass 2 kg is attached by a light rod to a shaft BC which is inclined by an angle of 30°. For small oscillations, what is the natural frequency of the system? [*Hint:* Recall that the moment about an axis n is $(\boldsymbol{r} \times \boldsymbol{F}) \cdot \hat{\boldsymbol{n}}$.]

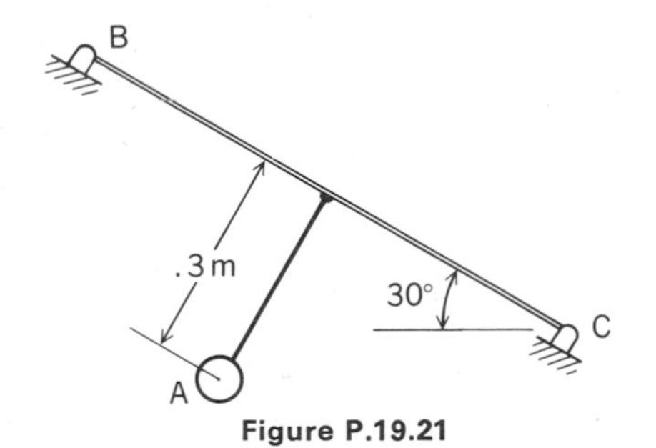

Figure P.19.21

***19.22.** A cube A, .25 m on a side, has a specific gravity of 1.10 and is attached to a cone having a specific gravity of .8. What is the equation for up-and-down motion of the system? Neglect the mass and buoyant force for rod CD. For very small oscillations, what is the approximate natural frequency?

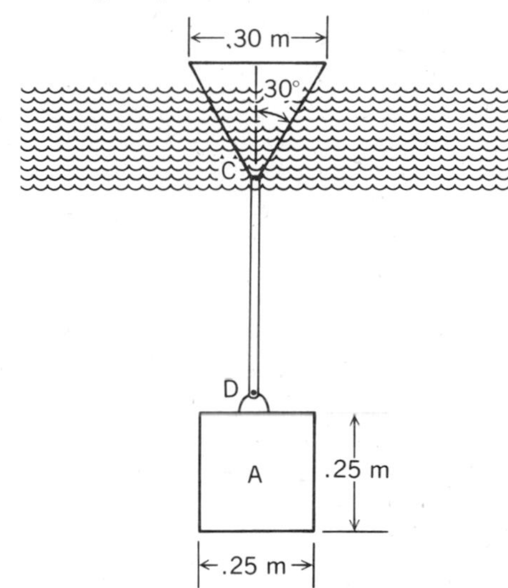

Figure P.19.22

19.3 Torsional Vibration

We showed in Section 16.2 that, for a body constrained to rotate about an axis fixed in inertial space, the angular momentum equation about the fixed axis is

$$I_{zz}\dot{\omega}_z = I_{zz}\ddot{\theta} = M_z \tag{19.9}$$

Numerous homework problems involved the determination of $\dot{\theta}$ and θ for applied torques which either were constant, varied with time, varied with speed $\dot{\theta}$, or, finally varied with position θ. The analyses paralleled very closely the corresponding cases for rectilinear translation of Chapter 12. Primarily, the approach was that of separation of variables and then that of carrying out one or more quadratures.

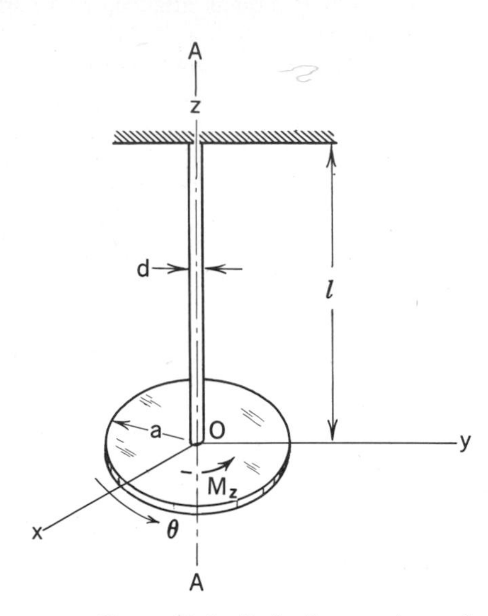

Figure 19.9. Shaft–disc analog of spring–mass system.

Paralleling the case of the linear restoring force in rectilinear translation is the important case where M_z is a *linear restoring torque*. For example, consider a circular disc attached to the end of a light shaft as shown in Fig. 19.9. Note that the upper end of the shaft is fixed. If the disc is twisted by an external agent about the centerline A–A of the shaft, then the disc will rotate essentially as a rigid body, whereas the shaft, since it is so much thinner and longer, will twist and supply a restoring torque on the disc that tries to bring the disc back to its initial position. In considering the possible motions of such a system disturbed in the aforementioned manner, we idealize the problem by *lumping* all elastic action into the shaft and all inertial effects into the disc. We know from strength of materials that for a circular shaft of constant cross section the amount of twist θ induced by torque M_z is, in the elastic range of deformation,

$$\theta = \frac{M_z L}{GJ} \tag{19.10}$$

where G is the shear modulus of the shaft material, J is the polar moment of area of the shaft cross section, and L is the length of the shaft. We can set forth the concept of a torsional spring constant K_t given as

$$K_t = \frac{M_z}{\theta} \tag{19.11}$$

For the case at hand, we have

$$K_t = \frac{GJ}{L} \tag{19.12}$$

Thus, the thin shaft has the same role in this discussion as the light linear spring of Section 19.2. Employing Eq. 19.11 for M_z and using the proper sign to ensure that we have a restoring action, we can express Eq. 19.9 as follows:

$$\ddot{\theta} + \frac{K_t}{I_{zz}}\theta = 0 \tag{19.13}$$

Notice that this equation is identical in form to Eq. 19.1. Accordingly, all the conclusions developed in that discussion apply with the appropriate changes in notation. Thus, the disc, once disturbed by being given an angular motion, will have a *torsional* natural oscillation frequency of $\sqrt{K_t/I_{zz}}$ radians per unit time. The equation of motion for the disc is

$$\theta = C_1 \cos \sqrt{\frac{K_t}{I_{zz}}}t + C_2 \sin \sqrt{\frac{K_t}{I_{zz}}}t \tag{19.14}$$

where C_1 and C_2 are constants of integration to be determined from initial conditions. Thus, for $\theta = \theta_0$ and $\dot{\theta} = \dot{\theta}_0$ at $t = 0$ we have

$$\theta = \theta_0 \cos \sqrt{\frac{K_t}{I_{zz}}}t + \frac{\dot{\theta}_0}{\sqrt{K_t/I_{zz}}} \sin \sqrt{\frac{K_t}{I_{zz}}}t \tag{19.15}$$

In the example just presented, the linear restoring torque stemmed from a long thin shaft. There could be other agents that can develop a linear restoring torque on a system otherwise free to rotate about an axis fixed in inertial space. We then talk about an *equivalent* torsional spring constant. We shall illustrate such cases in the following examples.

EXAMPLE 19.3

What are the equation of motion and the natural frequency of oscillation for small amplitude of a simple plane pendulum shown in Fig. 19.10? The pendulum rod may be considered massless.

Because the pendulum bob is small compared to the radius of curvature of its possible trajectory of motion, we may consider it as a particle. The pendulum has one degree of freedom, and we can use θ as the independent coordinate.[3] Notice from the diagram that there is a restoring torque about point A developed by gravity given as

$$M_z = -WL \sin \theta \tag{a}$$

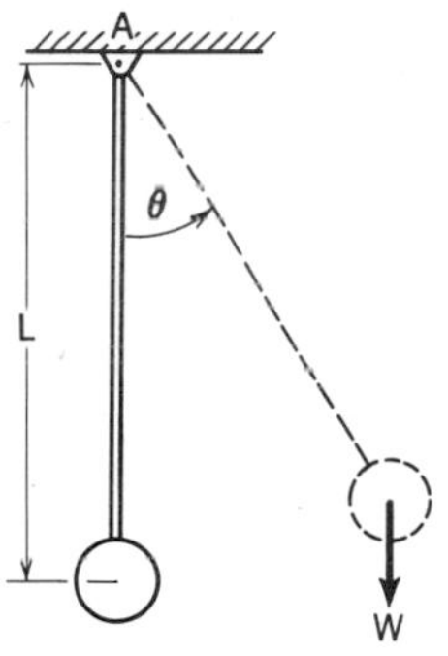

Figure 19.10. Pendulum.

where W is the weight of the bob. If the amplitude of the motion θ is very small, we can replace $\sin \theta$ by θ and so for this case we have a linear restoring torque given as

$$M_z = -WL\theta \tag{b}$$

We then have an equivalent torsional spring constant for the system

$$K_t = WL \tag{c}$$

The equation of possible *small-amplitude* motions for the pendulum is given as

$$-WL\theta = (ML^2)\ddot{\theta} \tag{d}$$

where we have used the *moment-of-momentum* equation about the fixed point A. Rearranging terms, we get

$$\ddot{\theta} + \frac{WL}{ML^2}\theta = 0 \tag{e}$$

---

[3]One degree of freedom means that one independent coordinate locates the system.

Noting that $W = Mg$, we have

$$\ddot{\theta} + \frac{g}{L}\theta = 0 \tag{f}$$

Accordingly, the natural frequency of oscillation is

$$\omega_n = \sqrt{\frac{g}{L}} \text{ rad/sec} \tag{g}$$

The equation of motion for this system is

$$\theta = C_1 \cos \sqrt{\frac{g}{L}}t + C_2 \sin \sqrt{\frac{g}{L}}t \tag{h}$$

where C_1 and C_2 are computed from known conditions at some time t_0.

EXAMPLE 19.4

A stepped disc is shown in Fig. 19.11 supporting a weight W_1 while being constrained by a linear spring having a spring constant K. The mass of the stepped disk is M and the radius of gyration about its geometric axis is k. What is the equation of motion for the system if the disc is rotated a small angle θ_1 counterclockwise from its static-equilibrium configuration and then suddenly released from rest? Assume the cord holding W_1 is weightless and perfectly flexible.

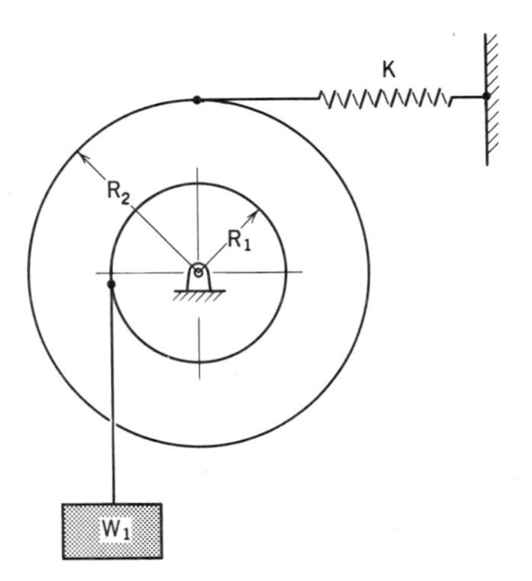

Figure 19.11. Stepped disc.

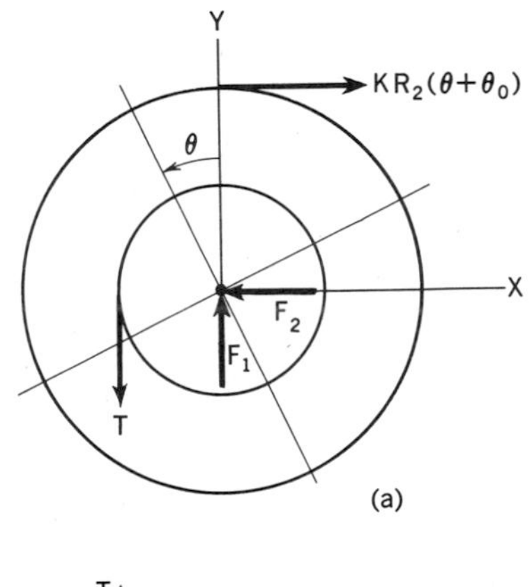

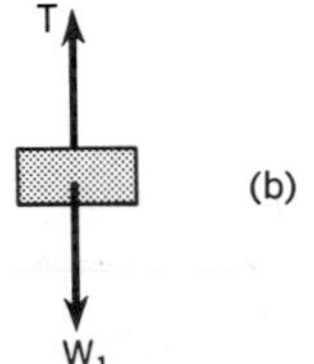

Figure 19.12. Free-body diagrams.

If we measure θ from the static-equilibrium position as shown in Fig. 19.12(a) the spring is stretched an amount $R_2(\theta + \theta_0)$ wherein θ_0 is the amount of rotation induced by the weight W_1 to reach the static-equilibrium configuration. Consequently, applying the angular momentum equation to the stepped disc about the axis of rotation, we get

$$R_1 T - KR_2^2(\theta + \theta_0) = Mk^2\ddot{\theta} \tag{a}$$

Next consider the suspended weight W_1. Clearly we have only translation for this body, for which *Newton's law* gives us

$$-T + W_1 = \frac{W_1}{g}R_1\ddot{\theta} \tag{b}$$

where we have made the assumption that the cord is always taut and is inextensible. We may replace T in Eq. (a) using Eq. (b) as follows:

$$R_1 W_1 - \frac{W_1}{g}R_1^2\ddot{\theta} - KR_2^2(\theta + \theta_0) = Mk^2\ddot{\theta} \tag{c}$$

Rearranging terms, we get

$$\left(Mk^2 + \frac{W_1}{g}R_1^2\right)\ddot{\theta} + KR_2^2\theta = R_1 W_1 - KR_2^2\theta_0 \tag{d}$$

Considering the *static-equilibrium* configuration of the system, we see on summing moments about the axis of rotation that the right side of the equation above is zero. Accordingly, we have for Eq. (d):

$$\ddot{\theta} + \frac{KR_2^2}{Mk^2 + (W_1/g)R_1^2}\theta = 0 \tag{e}$$

We can say immediately that the natural torsional frequency of the system is

$$\omega_n = \sqrt{\frac{KR_2^2}{Mk^2 + (W_1/g)R_1^2}} \text{ rad/sec} \tag{f}$$

The equation of motion is then

$$\theta = C_1 \cos\sqrt{\frac{KR_2^2}{Mk^2 + (W_1/g)R_1^2}}\,t + C_2 \sin\sqrt{\frac{KR_2^2}{Mk^2 + (W_1/g)R_1^2}}\,t \tag{g}$$

Submitting Eq. (g) to the initial conditions to determine C_1 and C_2, we get

$$\theta = \theta_1 \cos\sqrt{\frac{KR_2^2}{Mk^2 + (W_1/g)R_1^2}}\,t \tag{h}$$

It is important to note that we could have reached Eq. (e) more directly by realizing that, when θ is measured from the *equilibrium configuration*, the moment from the weight W_1 is already counteracted by the moment from the stretch $R_2\theta_0$ of the spring and accordingly only the moment from the force $-R_2K\theta$ from further stretch of the spring and the moment from the inertial force $-(W_1/g)R_1\ddot{\theta}$ of the hanging weight from Eq. (b) need be considered in the angular-momentum equation (a). Thus, we have from this viewpoint:

$$-\frac{W_1}{g}R_1^2\ddot{\theta} - KR_2^2\theta = Mk^2\ddot{\theta} \tag{i}$$

Rearranging, we have

$$\ddot{\theta} + \frac{KR_2^2}{Mk^2 + (W_1/g)R_1^2}\theta = 0 \tag{j}$$

Accordingly, we arrive at very same differential equations in a more direct manner. We can again conclude as in Example 19.1 that *when the coordinate is measured from an equilibrium configuration we can forget about contributions of torques that are present for the equilibrium configuration and include only new torques developed when there is a departure from the equilibrium configuration.*

Before you start on the problems, we wish to point out that shafts directly connected to each other (see the shafts on the right side of the disc in Fig. P.19.23) are analogous to springs in *series* as far as the equivalent torsional spring constant is concerned. On the other hand, shafts on opposite sides of the disc are analogous to springs in *parallel* as far as the equivalent torsional spring constant is concerned. You should have no trouble justifying these observations.

Problems

19.23. Compute the equivalent torsional spring constant of the shaft on the disc.

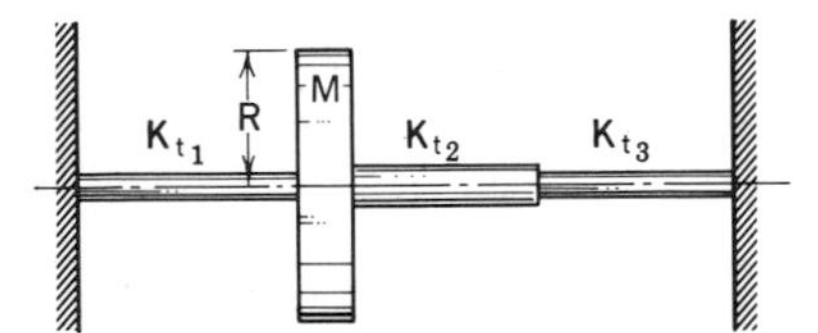

Figure P.19.23

19.24. What is the equivalent torsional spring constant on the disc from the shafts? The modulus of elasticity G for the shafts is 10×10^{10} N/m². What is the natural frequency of the system? If the disc is twisted 10° and then released, what will its angular position be in 1 sec? Neglect the mass of the shafts. The disc weighs 143 N.

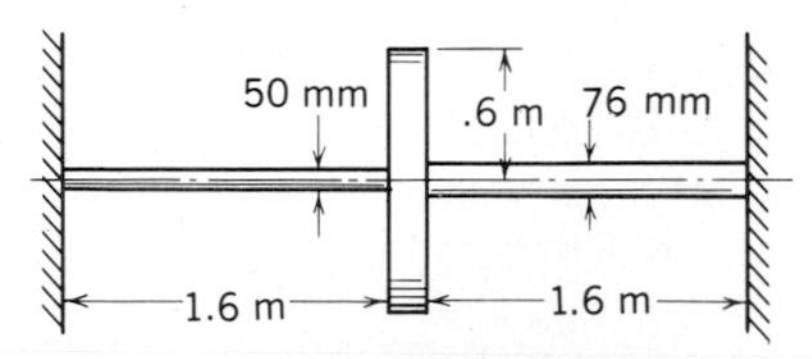

Figure P.19.24

19.25. A small pendulum is mounted in a rocket that is accelerating upward at the rate of $3g$. What is the natural frequency of rotation of the pendulum if the bob has a mass of 50 g? Neglect the weight of the rod. Consider small oscillations.

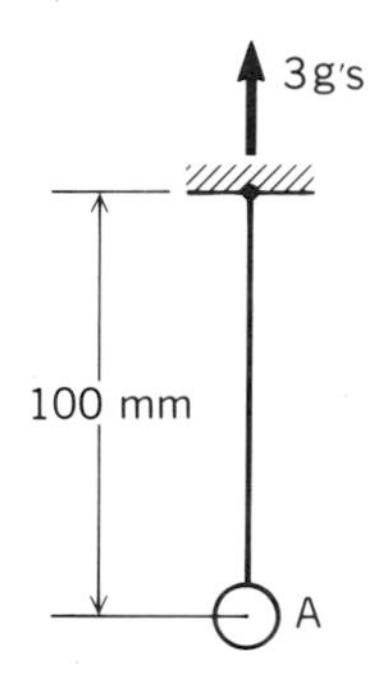

Figure P.19.25

19.26. Work Problem 19.25 for the case where the rocket is decelerating at $.6g$ in the vertical direction.

19.27. What is the natural frequency for small oscillations of the compound pendulum supported at A?

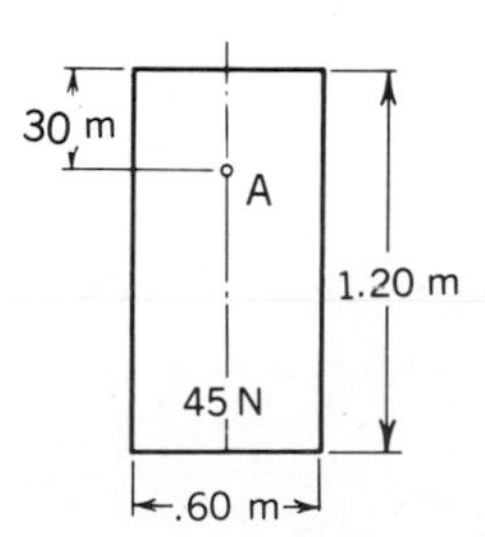

Figure P.19.27

19.28. A slender rod weighing 140 N is held by a frictionless pin at A and by a spring having a spring constant of 8.80 N/mm at B.

(a) What is the natural frequency of oscillation for small vibrations?

(b) If point B of the rod is depressed 25 mm at $t = 0$ from the static-equilibrium position, what will its position be when $t = .02$ sec?

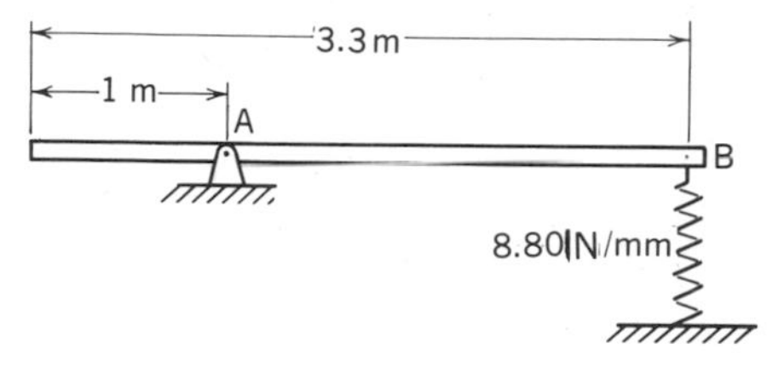

Figure P.19.28

19.29. What is the natural frequency of the pendulum shown for small oscillations? Take into account the inertia of the rod whose mass is m. Also, consider the bob to be a sphere of diameter D and mass M, rather than a particle.

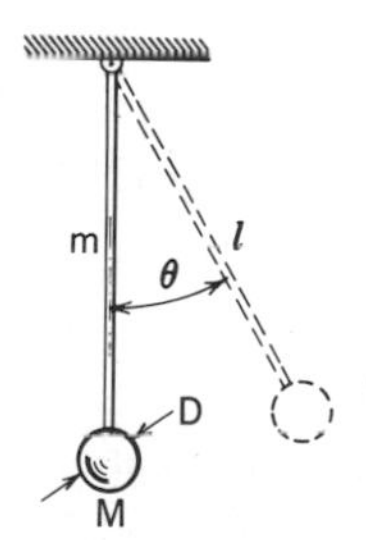

Figure P.19.29

19.30. A cylinder of mass M and radius R is connected to identical springs and rotates without friction about O. For small oscillations, what is the natural frequency? The cord supporting W_1 is wrapped around the cylinder.

Figure P.19.30

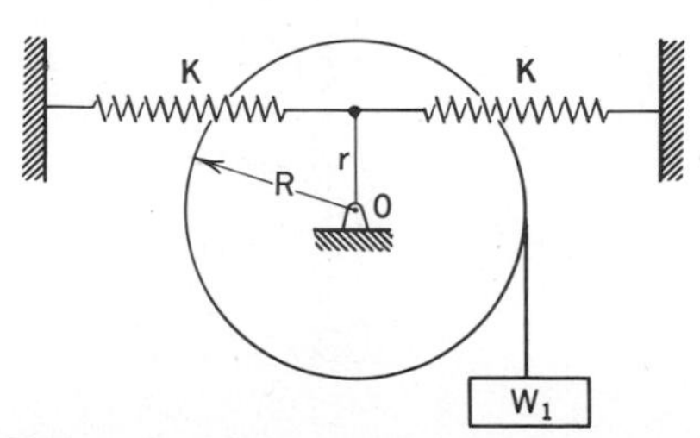

19.31. The author's 22-ft Columbia sailboat is suspended by straps from a crane. The boat is made to swing freely about support A in the xy plane (the plane of the page). What is the radius of gyration about the z axis at the center of gravity if the period of oscillation is 5 sec? The boat has a mass of 1000 kg. Neglect the weight of supporting wires and belts.

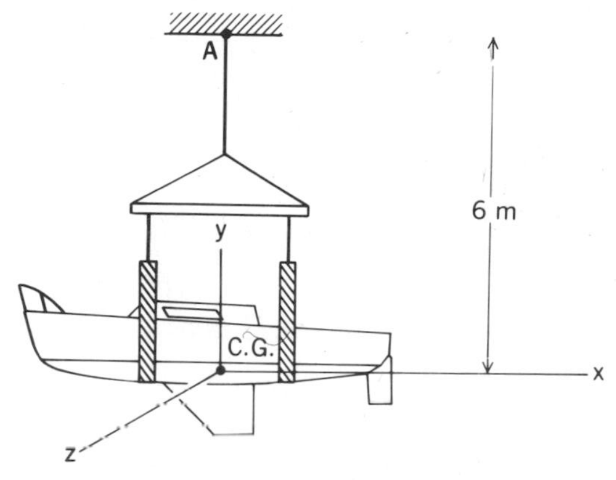

Figure P.19.31

19.32. A uniform bar of length L and weight W is suspended by strings. What is the differential equation of motion for small torsional oscillation about a vertical axis at the center of mass at C? What is the natural frequency?

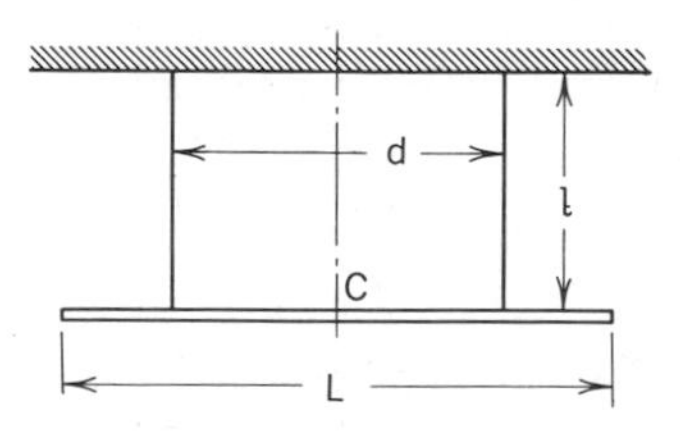

Figure P.19.32

19.33. In Problem 19.12, do not consider body A to be a particle, and compute the natural frequency of the system for small vibrations. Take the dimension of A to be that of a 150-mm cube. If ω_n for the particle approach is 19.81 rad/sec, what percentage error is incurred using the particle approach?

19.34. Gears A and B weighing 50 lb and 80 lb, respectively, are fixed to supports C and D as shown. If the shear modulus G for the shafts is 15×10^6 psi, what is the natural frequency of oscillation for the system?

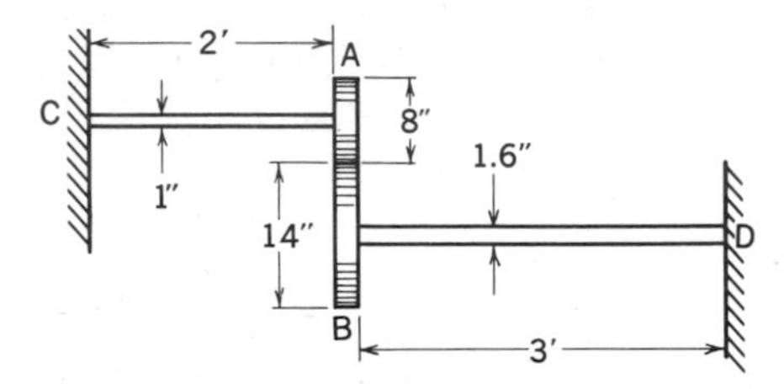

Figure P.19.34

19.35. A four-bar linkage, $ABCD$, is disturbed slightly so as to oscillate in the xy plane. What is the frequency of oscillation if each bar has a mass of 5.0 g/mm?

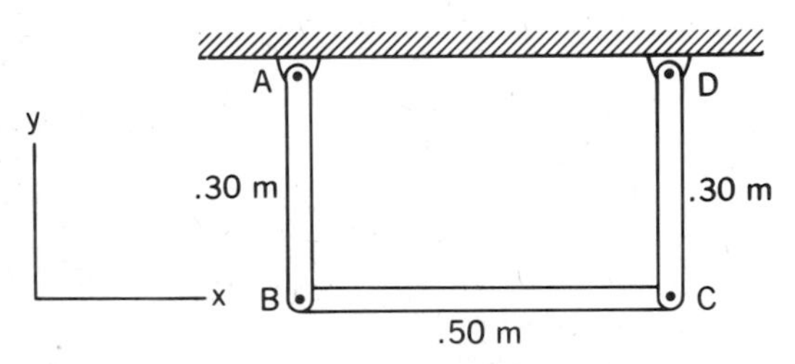

Figure P.19.35

19.36. A plate A weighing 1 kN is attached to a rod CD. If at the instant that the rod CD is torsionally unstrained, the plate has an angular speed of 2 rad/sec about the centerline of CD, what is the amplitude of twist developed by the rod? Take $G = 6.90 \times 10^{10}$ N/m² for the rod.

Figure P.19.36

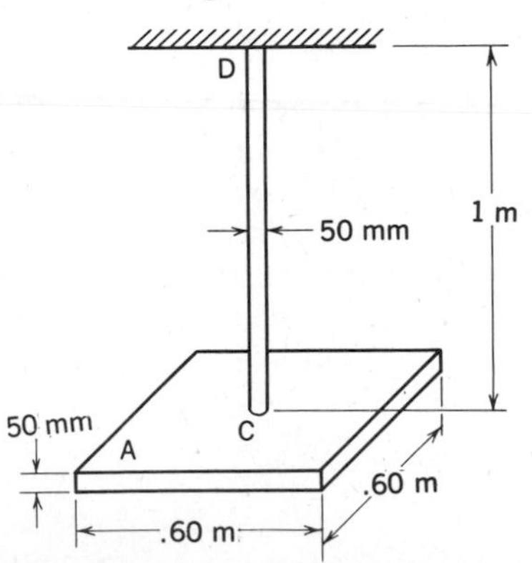

19.37. A block having a uniform density of 300 lb/ft³ is suspended by a fixed shaft of length 3 ft as shown. If the area of the top surface of the block is 5 ft², what are the values of a and b for extreme values of natural torsional frequency of the system? The shear modulus G for the shaft is 15×10^6 psi. Compute the natural frequency for the extreme cases.

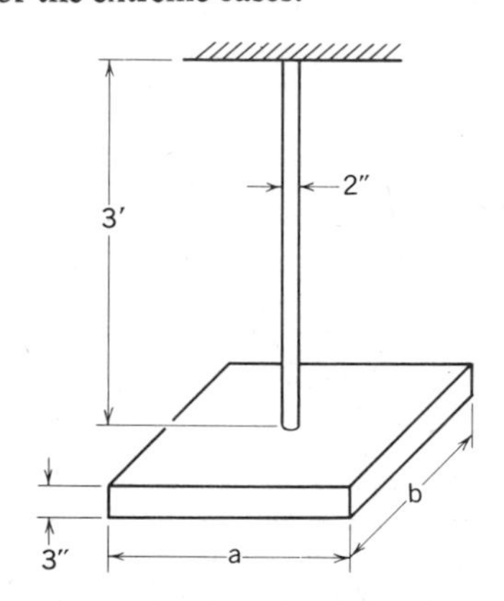

Figure P.19.37

19.38. A rod of weight W and length L is restrained in the vertical position by two identical springs having spring constant K. A vertical load P acts on top of the rod. What value of P, in terms of W, L, and K, will cause the rod to have a natural frequency of oscillation about A approaching zero for small oscillations? What does this signify physically? (*Hint:* For small θ we may take $\cos \theta = 1$.)

Figure P.19.38

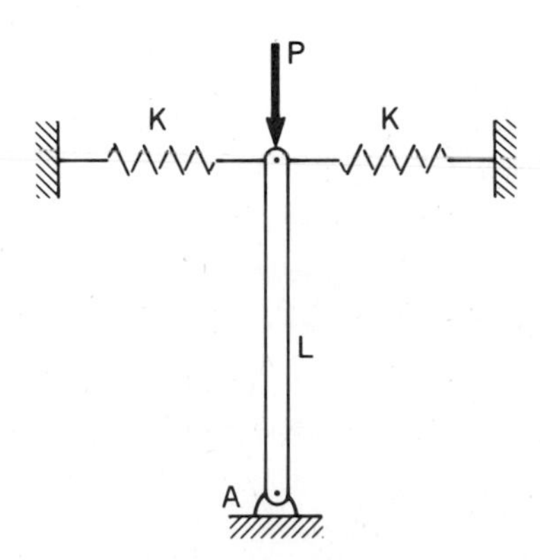

19.39. A 1-m rod weighing 60 N is maintained in a vertical position by two identical springs having each a spring constant of 50 N/mm. What vertical force P will cause the natural frequency of the rod about A to approach zero value for small oscillations? (*Hint:* For small θ we can say that $\cos\theta = 1$.)

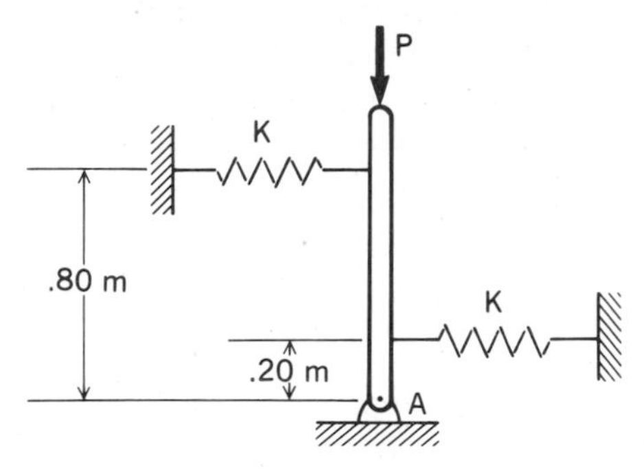

Figure P.19.39

19.40. What is the natural frequency of torsional vibration for the stepped cylinder? The mass of the cylinder is 45 kg, and the radius of gyration is .46 m. The following data also apply:

$$D_1 = .30 \text{ m}$$
$$D_2 = .60 \text{ m}$$
$$K_1 = .875 \text{ N/mm}$$
$$K_2 = 1.8 \text{ N/mm}$$
$$W_A = 178 \text{ N}$$

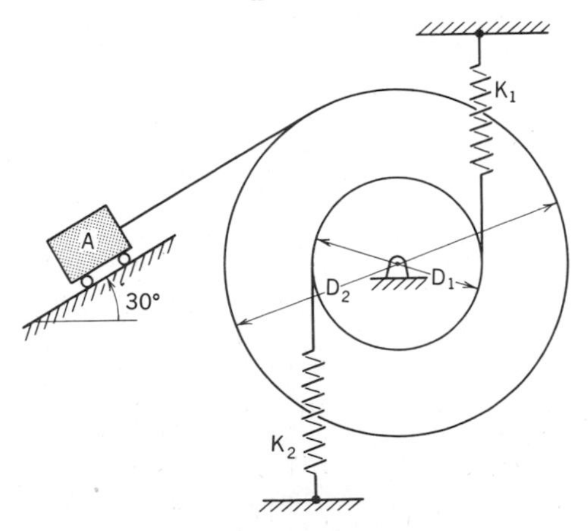

Figure P.19.40

19.41. A disc A weighs 445 N and has a radius of gyration of .45 m about its axis of symmetry. Note that the center of gravity does not coincide with the geometric center. What are the amplitude of oscillation and frequency of oscillation if, at the instant that the center of gravity is directly below B, the disc is rotating at a speed of .01 rad/sec counterclockwise?

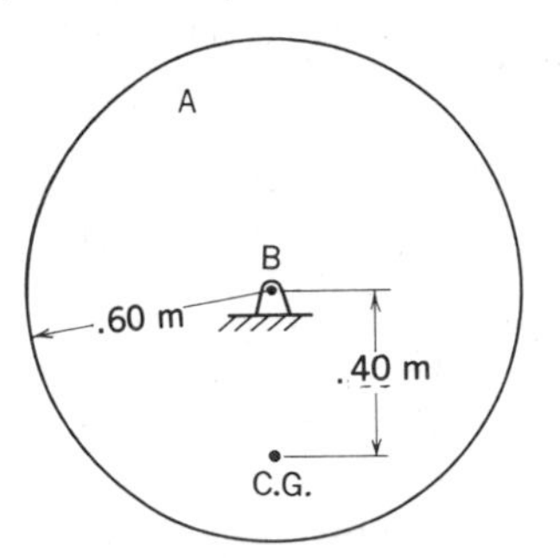

Figure P.19.41

19.42. A disc B is suspended by a flexible wire. The tension in the top wire is 4450 N. If the disc is observed to have a period of lateral oscillation of .2 sec for very small amplitude and a period of torsional oscillation of 5 sec, what is the radius of gyration of the disc about its geometric axis? The torsional spring constant for each of the wires is 1470 N-mm/rad.

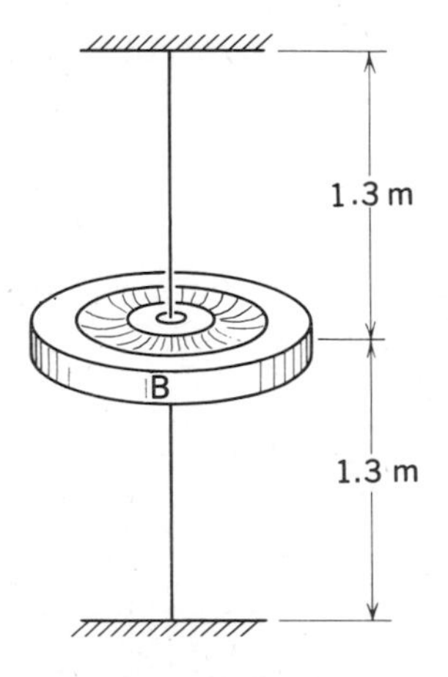

Figure P.19.42

19.43. Two discs are forced together such that, at the point of contact, a normal force of 50 lb is transmitted from one disc to the other. Disc A weighs 200 lb and has a radius of gyration of 1.4 ft about C, whereas disc B weighs 50 lb and has a radius of gyration about D of 1 ft. What is the natural frequency of oscillation for the system, if disc A is rotated 10° counterclockwise and then released? The center of gravity of B coincides with the geometric center.

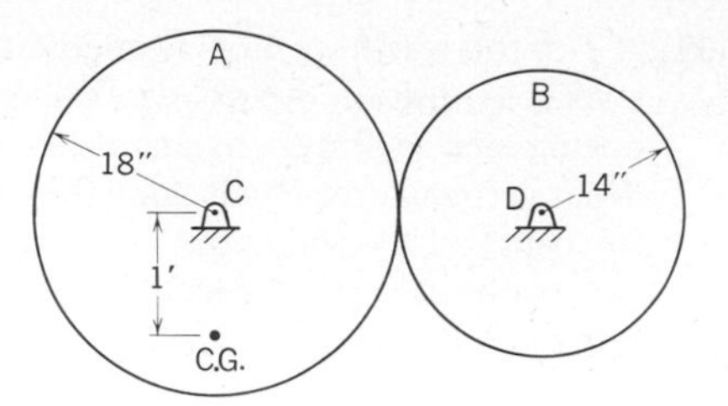

Figure P.19.43

19.44. In Problem 19.43, find the minimum coefficient of friction for no slipping between the discs. From Problem 19.43, $\omega_n = 3.68$ rad/sec.

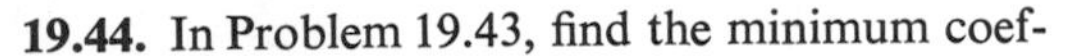

*19.4 Examples of Other Free-Oscillating Motions

In the previous sections, we examined the rectilinear translation of a rigid body under the action of a linear restoring force as well as the pure rotation of a rigid body under the action of a linear restoring torque. In this section, we shall first examine a body with one degree of freedom undergoing *plane motion* governed by a differential equation of motion of the form given in the previous section. The dependent variable for such a case varies harmonically with time, and we have a *vibratory plane motion.* Consider the following example.

EXAMPLE 19.5

Shown in Fig. 19.13(a) on an inclined plane is a uniform cylinder maintained in a position of equilibrium by a linear spring having a spring constant K. If the cylinder rolls without slipping, what is the equation of motion when it is disturbed from its equilibrium position?

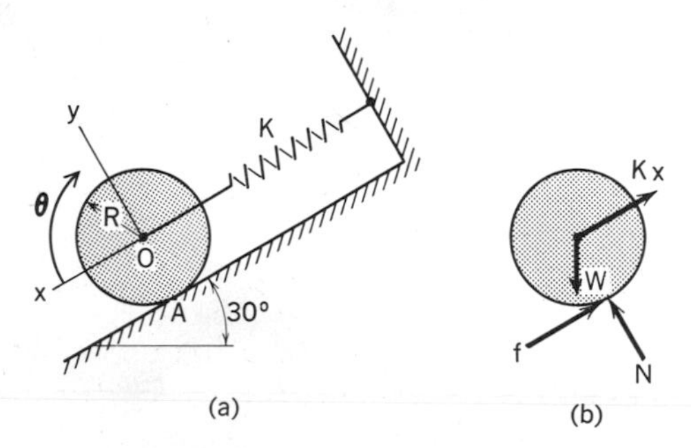

Figure 19.13. Plane-motion vibration.

We have here a case of plane motion about a configuration of equilibrium. We shall measure the displacement x of the center of mass from the equilibrium position and accordingly shall need to consider only those forces and torques developed as the cylinder departs from this position. Accordingly, we have for *Newton's law* for the mass center [see Fig. 19.13(b)]:

$$-f - Kx = M\ddot{x} \tag{a}$$

Now employ the *angular-momentum* equation about the geometric axis of the cylinder at O. Using θ to measure the rotation of the cylinder about this axis from the equilibrium configuration, we get

$$-fR = \tfrac{1}{2}MR^2\ddot{\theta} \tag{b}$$

Noting from *kinematics* that $\ddot{x} = -R\ddot{\theta}$ as a result of the no-slipping condition, we have for Eq. (b):

$$fR = \frac{1}{2}MR^2\left(\frac{\ddot{x}}{R}\right)$$

Therefore,

$$f = \tfrac{1}{2}M\ddot{x} \tag{c}$$

Substituting for f in Eq. (a) using this result, we have

$$M\ddot{x} = -\tfrac{1}{2}M\ddot{x} - Kx$$

Therefore,

$$\ddot{x} + \frac{2}{3}\frac{K}{M}x = 0 \tag{d}$$

We could also have arrived at the differential equation above by noting that we have instantaneous pure rotation about the line of contact A as a result of the no-slipping condition. Thus, the *angular momentum* equation can be used as follows about the point of contact on the cylinder:

$$(\tfrac{1}{2}MR^2 + MR^2)\ddot{\theta} = KxR \tag{e}$$

Noting as before that $\ddot{\theta} = -\ddot{x}/R$, we get

$$-\frac{3}{2}MR^2\frac{\ddot{x}}{R} = KxR$$

Therefore,

$$\ddot{x} + \frac{2}{3}\frac{K}{M}x = 0 \tag{f}$$

We may solve the differential equation to give us

$$x = x_0 \cos\sqrt{\frac{2}{3}\frac{K}{M}}t + \frac{\dot{x}_0}{\sqrt{\frac{2}{3}K/M}}\sin\sqrt{\frac{2}{3}\frac{K}{M}}t \tag{g}$$

where x_0 and $\dot{x}_0$ are the initial position and speed of the center of mass, respectively. Since $\theta = -x/R$ (we have here only one degree of freedom[4] as a result of the no-slipping condition), we have for θ from Eq. (g):

$$\theta = \frac{-x_0}{R}\cos\sqrt{\frac{2}{3}\frac{K}{M}}t - \frac{\dot{x}_0}{R\sqrt{\frac{2}{3}K/M}}\sin\sqrt{\frac{2}{3}\frac{K}{M}}t \tag{h}$$

[4]See footnote 3 on page 853.

*19.5 Energy Methods

Up to now, the procedure has been primarily to work with Newton's law or the angular-momentum equation in reaching the differential equation of interest. There is an alternative approach to the handling of free vibration problems that may be very useful in dealing with simple systems and in setting up approximate calculations for more complex systems. Suppose we know for a one-degree-of-freedom system that only linear restoring forces and torques do work during possible motions of the system. Then, the agents developing such forces are *conservative* force agents and may be considered to store potential energy. You will recall from Section 13.3 that the *total mechanical energy* for such systems is *conserved*. Thus, we have

$$\text{PE} + \text{KE} = \text{constant} \tag{19.16}$$

Also, we know from our present study that the system must oscillate harmonically when disturbed and then allowed to move freely with only the linear restoring agents doing work. Thus, if κ is the independent coordinate measured from the static-equilibrium configuration, we have

$$\kappa = A \sin(\omega_n t + \beta) \tag{19.17}$$

Hence,

$$\dot{\kappa} = A\omega_n \cos(\omega_n t + \beta) \tag{19.18}$$

Now at the instant when $\kappa = 0$, we are at the static-equilibrium position and the potential energy of the system is a minimum. Since the total mechanical energy must be conserved at all times once such a motion is under way, it is also clear that the kinetic energy must be at a maximum at that instant. If we take the lowest potential energy as zero, then we have for the total mechanical energy simply the maximum kinetic energy. Also, when the body is undergoing a change in direction of its motion at the outer extreme position, the kinetic energy is zero instantaneously, and accordingly the potential energy must be a maximum and equal to the total mechanical energy of the system. Thus, we can equate the maximum potential energy with the maximum kinetic energy.

$$(\text{KE})_{\text{max}} = (\text{PE})_{\text{max}} \tag{19.19}$$

In computing the $(\text{KE})_{\text{max}}$, we will involve $(\dot{\kappa})_{\text{max}}$ and hence $A\omega_n$, whereas for the $(\text{PE})_{\text{max}}$ we will involve $(\kappa)_{\text{max}}$ and hence A. In this way we can set up quickly an equation for ω_n, the natural frequency of the system. For example, if we have the simple linear spring–mass system of Fig. 19.1, we can say:

$$(\text{PE}) = \tfrac{1}{2}Kx^2$$

Therefore,

$$(\text{PE})_{\text{max}} = \tfrac{1}{2}K(x_{\text{max}})^2 = \tfrac{1}{2}KA^2$$

where we have made use of our knowledge that $x = A \sin(\omega_n t + \beta)$. And, noting that $\dot{x} = A\omega_n \cos(\omega_n t + \beta)$, we have

$$(\text{KE})_{\text{max}} = \tfrac{1}{2}M(\dot{x}_{\text{max}})^2 = \tfrac{1}{2}M(A\omega_n)^2$$

Now, equating these expressions, we get

$$\tfrac{1}{2}KA^2 = \tfrac{1}{2}M(A\omega_n)^2$$

Therefore,

$$\omega_n = \sqrt{\frac{K}{M}}$$

which is the expected result. We next illustrate this approach in a more complex problem.

EXAMPLE 19.6

A cylinder of radius r and weight W rolls without slipping along a circular path of radius R as shown in Fig. 19.14. Compute the natural frequency of oscillation for small oscillation.

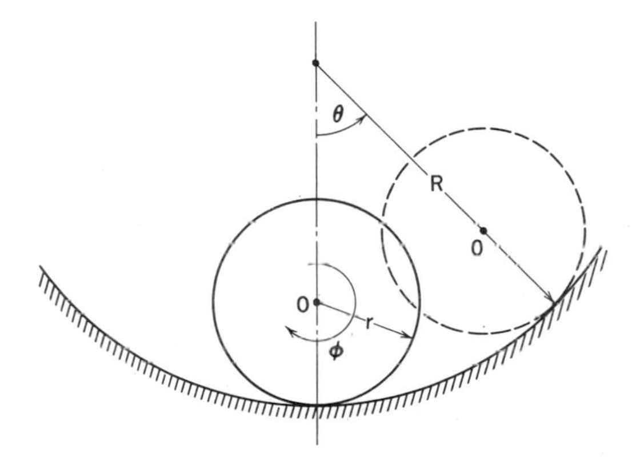

Figure 19.14. Cylinder rolls without slipping.

This system has one degree of freedom. We can use ϕ, the angle of rotation of the cylinder about its axis of symmetry, as the independent coordinate, or we may use θ as shown in the diagram. To relate these variables for no slipping we may conclude, on observing the motion of point O, that for small rotation:

$$(R - r)\theta = r\phi$$

Therefore,

$$\theta = \frac{r}{R - r}\phi \tag{a}$$

The only force that does work during the possible motions of the system is the force of gravity W. The torque developed by W about the point of contact for a given θ is easily determined after examining Fig. 19.15 to be

$$\text{torque} = Wr \sin\theta = Wr \sin\left(\frac{r}{R - r}\phi\right) \tag{b}$$

This is a restoring torque, and if we limit ourselves to *small oscillations* it becomes $W[r^2\phi/(R - r)]$, which is clearly a linear restoring torque. Because the force

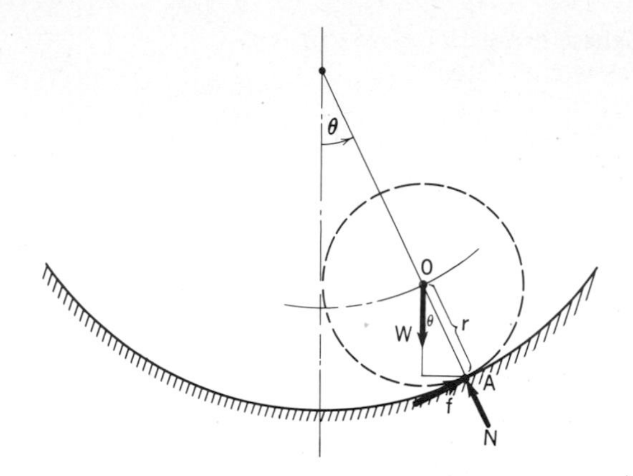

Figure 19.15. Free-body diagram of cylinder.

doing work on the cylinder is *conservative*, and because it results in a *linear restoring torque*, we can employ the energy formulation of this section.

The motion may be considered to be given as follows:

$$\theta = C \sin (\omega_n t + \beta)$$

or

$$\phi = \frac{R - r}{r} C \sin (\omega_n t + \beta) \tag{c}$$

Expressing the maximum potential and kinetic energies and using C for θ_{max} and the lowest position of O as the datum, we have:

$$\begin{aligned} (PE)_{max} &= W(R - r)(1 - \cos \theta_{max}) \\ &= W(R - r)(1 - \cos C) \end{aligned} \tag{d}$$

$$\begin{aligned} (KE)_{max} &= \frac{1}{2}\frac{W}{g}(R - r)^2\dot{\theta}^2_{max} + \frac{1}{4}\frac{W}{g}r^2\dot{\phi}^2_{max} \\ &= \frac{1}{2}\frac{W}{g}(R - r)^2(C\omega_n)^2 + \frac{1}{4}\frac{W}{g}r^2\left(\frac{R - r}{r}C\omega_n\right)^2 \end{aligned} \tag{e}$$

We have used Eq. (c) in the last equation. Expanding cos C in a power series and retaining the first two terms $(1 - C^2/2)$, we then get, on equating the right sides of the equations above:

$$W(R - r)\frac{C^2}{2} = \frac{W}{2g}\left[(R - r)^2 + \frac{(R - r)^2}{2}\right]\omega_n^2 C^2$$

Therefore,

$$\omega_n = \sqrt{\frac{g}{\frac{3}{2}(R - r)}} \tag{f}$$

Problems

19.45. A cylinder of diameter 1 m is shown. The center of gravity of the cylinder is .30 m from the geometric center, and the radius of gyration is .60 m at the center of mass. What is the natural frequency of oscillation for small vibrations without slipping?

The cylinder weighs 220 N. Work the problem by two methods.

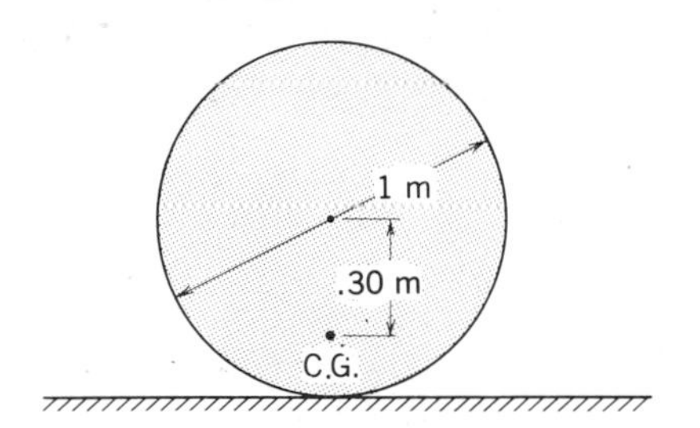

Figure P.19.45

19.46. A stepped cylinder is maintained along the incline by a spring having a spring constant K. What is the formulation for the natural frequency of oscillation for the system? What is the maximum friction force? Take the weight of the cylinder as W and the radius of gyration about the geometric centerline O as k. The initial conditions are $\theta = \theta_0$ at $t = 0$, and $\dot{\theta} = \dot{\theta}_0$ at $t = 0$. There is no slipping.

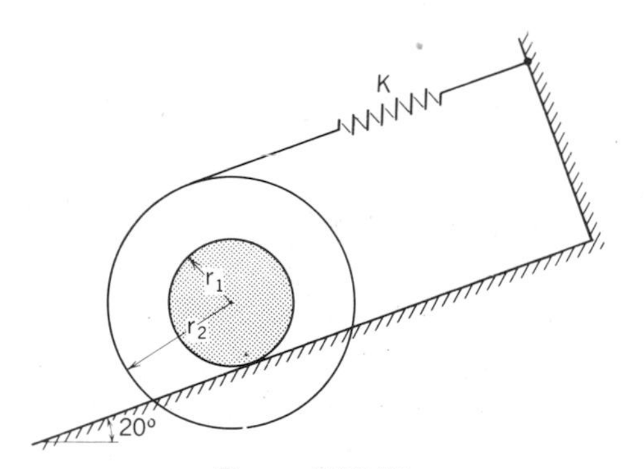

Figure P.19.46

19.47. Two masses are attached to a light rod. The rod rides on a frictionless horizontal rail. $M_1 = 45$ kg and $M_2 = 14$ kg. What is the natural frequency of oscillation of the system if a small impulsive torque is applied to the system when it is in a rest configuration? Consider the masses as particles. (*Hint:* Consider motion about the center of mass of the system.)

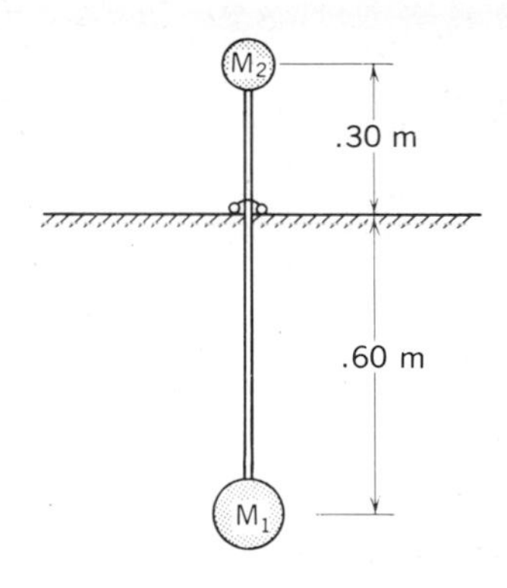

Figure P.19.47

19.48. Work Problem 19.30 by energy methods.

19.49. Work Problem 19.13 by energy methods.

19.50. Work Problem 19.40 by energy methods.

19.51. A *manometer* used for measuring pressures is shown. If the mercury has a length L in the tube, what is the formulation for the natural frequency of movement of the mercury?

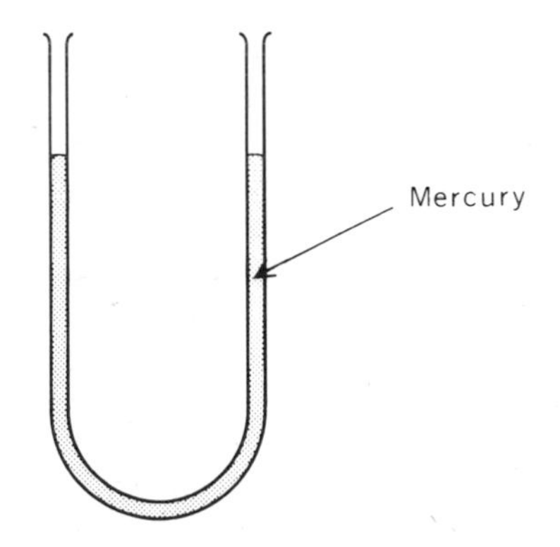

Figure P.19.51

19.52. An *inclined manometer* is often used for more accurate pressure measurements. If the mercury in the tube has a length L, what is the natural frequency of oscillation of the mercury in the tube?

Figure P.19.52

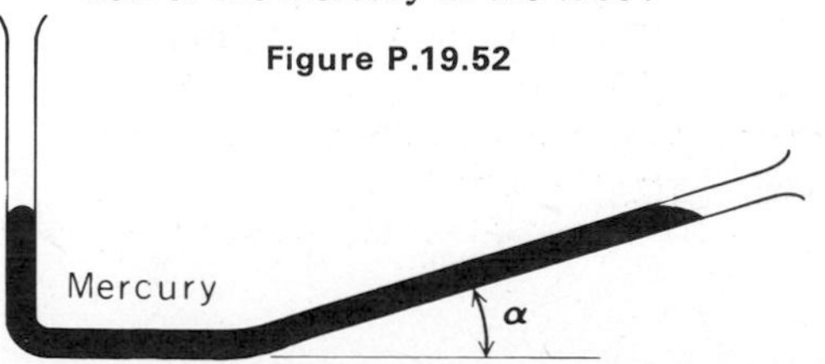

19.53. A *differential* manometer is used for measuring high pressures. What is the natural frequency of oscillation of the mercury?

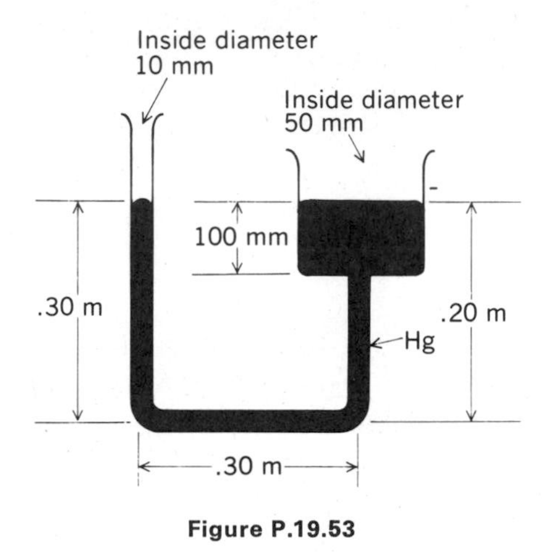

Figure P.19.53

19.54. A stepped cylinder rides on a circular path. For small oscillations, what is the natural frequency? Take the radius of gyration about the geometric axis O as k and the weight of the cylinder as W.

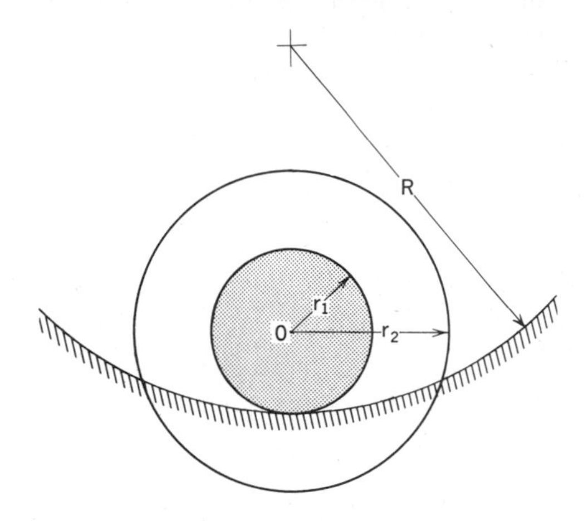

Figure P.19.54

19.6 Linear Restoring Force and a Force Varying Sinusoidally with Time

We shall now consider the case of a sinusoidal force acting on a spring–mass system (Fig. 19.16). The sinusoidal force has a frequency of ω (not to be confused with ω_n, the natural frequency) and an amplitude of F_0. At time $t = 0$, the mass will be assumed to have some known velocity and position, and we shall investigate the ensuing motion.

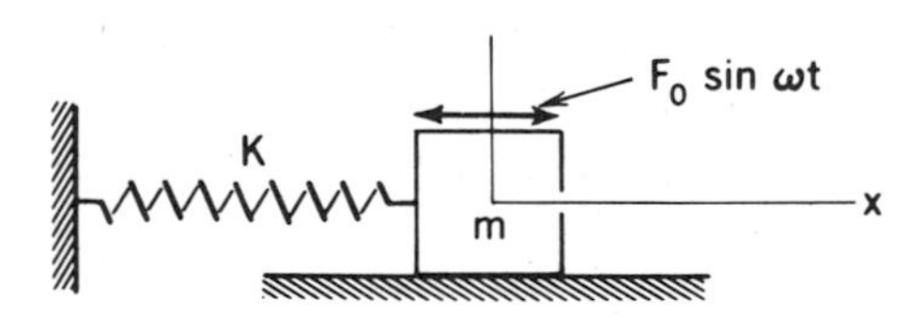

Figure 19.16. Spring–mass system with harmonic disturbance.

Measuring the position x from the unextended position of the spring, we have for *Newton's law*:

$$m\frac{d^2x}{dt^2} = -Kx + F_0 \sin \omega t \tag{19.20}$$

Rearranging the equation so that the dependent variable and its derivatives are on the left-hand side and dividing through by m, we get the standard form:

$$\frac{d^2x}{dt^2} + \frac{K}{m}x = \frac{F_0}{m} \sin \omega t \tag{19.21}$$

single phasor, and hence the motion is not harmonic. The two parts of the motion are termed the *transient* part, corresponding to the complementary solution, and the *steady-state* part, corresponding to the particular solution, having frequencies ω_n and ω, respectively. With the introduction of friction (next section), we shall see that the transient part of the motion dies out while the steady state persists as long as there is a disturbance present.

Let us now consider the steady-state part of the motion in Eq. 19.24. Dividing numerator and denominator by K/m, we have for this motion, which we denote as x_p:

$$x_p = \frac{F_0/K}{1 - (\omega^2 m/K)} \sin \omega t = \frac{F_0/K}{1 - (\omega/\omega_n)^2} \sin \omega t \tag{19.27}$$

It will be useful to study with respect to ω/ω_n the variation of the magnitude of the steady-state amplitude x_p for $F_0/K = 1$, namely

$$\left| \frac{1}{1 - (\omega/\omega_n)^2} \right|$$

shown plotted in Fig. 19.17. As the forcing frequency approaches the natural frequency, the term

$$\left| \frac{1}{1 - (\omega/\omega_n)^2} \right|$$

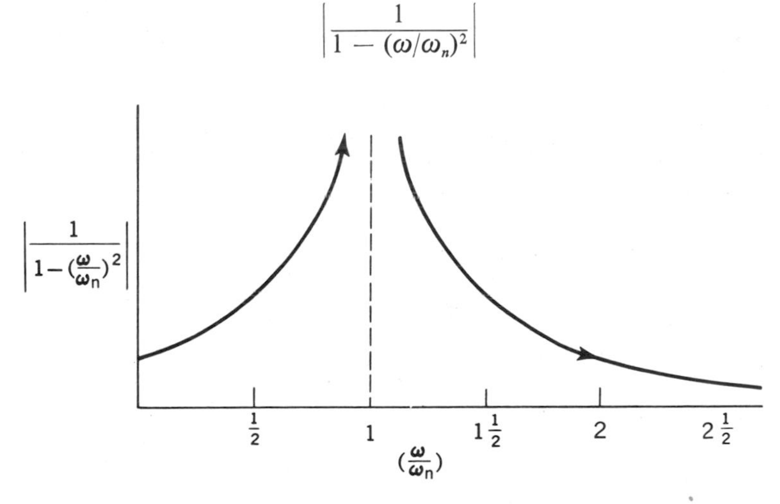

Figure 19.17. Plot shows amplitude variation of steady-state motion versus (ω/ω_n) for $F_0/K = 1$.

goes to infinity, and thus the amplitude of the forced vibration approaches infinity. This is the condition of *resonance*. Under such circumstances, friction, which we neglect here but which is always present, may limit the amplitude. If not, when very large amplitudes are developed, the properties of the restoring element do not remain linear, so that the theory which predicts infinite amplitudes is inapplicable. Thus, the linear, frictionless formulations cannot yield correct amplitudes at resonance in real problems. The condition of resonance, however, does indicate that large amplitudes

If the right-hand side is zero the equation is termed *homogeneous*. This was the equation studied in Section 19.5. If any function of t or constant appears on the right side, as in the case above, the equation is *nonhomogeneous*.

The general solution of a nonhomogeneous differential equation of this type is found by getting the general solution of the corresponding homogeneous equation and then finding a *particular solution* that satisfies the full equation. The sum of these solutions, then, is the general solution of the equation. Often, the solution for the homogeneous equation is termed the *complementary solution*.

In this case, we have already ascertained the complementary solution:

$$x_c = C_1 \sin \sqrt{\frac{K}{m}}t + C_2 \cos \sqrt{\frac{K}{m}}t \tag{19.22}$$

To get a particular solution x_p, we can see by inspection that a function of the form $x_p = C_3 \sin \omega t$ will give a solution if the constant C_3 is chosen properly. Substituting this function into Eq. 19.21, we thus have

$$-C_3\omega^2 \sin \omega t + \frac{K}{m}C_3 \sin \omega t = \frac{F_0}{m} \sin \omega t$$

Clearly, the value of C_3 must be

$$C_3 = \frac{F_0/m}{K/m - \omega^2} \tag{19.23}$$

We can now express the general solution of the differential equation at hand:

$$x = C_1 \sin \sqrt{\frac{K}{m}}t + C_2 \cos \sqrt{\frac{K}{m}}t + \frac{F_0/m}{K/m - \omega^2} \sin \omega t \tag{19.24}$$

Note that there are two arbitrary constants which are determined from the initial conditions of the problem. Do not use the results of Eq. 19.6 for these constants, because *we must now include the particular solution in ascertaining the constants*. When $t = 0$, $x = x_0$ and $\dot{x} = \dot{x}_0$. We apply these conditions to Eq. 19.24:

$$\begin{aligned} x_0 &= C_2 \\ \dot{x}_0 &= C_1\sqrt{\frac{K}{m}} + \frac{F_0/m}{K/m - \omega^2}\omega \end{aligned} \tag{19.25}$$

Solving for the constants, we get

$$\begin{aligned} C_2 &= x_0 \\ C_1 &= \frac{\dot{x}_0}{\sqrt{K/m}} - \frac{\omega F_0/m}{(K/m - \omega^2)\sqrt{K/m}} \end{aligned} \tag{19.26}$$

Returning to Eq. 19.24, notice that we have the superposition of two harmonic motions—one with a frequency equal to $\sqrt{K/m}$, the natural frequency ω_n of the system, and the other with a frequency ω of the "driving function" (i.e., the nonhomogeneous part of the equation). The frequencies ω and ω_n are not the same in the general case. The phasor representation then leads us to the fact that since the rotating vectors have different angular speeds, the resulting motion cannot be represented by a

are to be expected. Furthermore, these amplitudes can be dangerous, because large force concentrations will be present in parts of the restoring system as well as in the moving body and may result in disastrous failures. It is therefore important in most situations to avoid resonance. If a disturbance corresponding to the natural frequency is present and cannot be eliminated, we may find it necessary to change either the stiffness or the mass of a system in order to avoid resonance.

From Fig. 19.17 we can conclude that the amplitude will become small as the frequency of the disturbance becomes very high. Also, considering the amplitude C_3 for steady-state motion (Eq. 19.23), we see that below resonance the sign of this expression is positive, and above resonance it is negative, indicating that below resonance the motion is *in phase* with the *disturbance* and above resonance the motion is directly 180° *out of phase* with the *disturbance*.

EXAMPLE 19.7

A motor mounted on springs is constrained by the rollers to move only in the vertical direction (Fig. 19.18). The assembly weighs 2.6 kN and when placed carefully on the springs causes a deflection of 2.5 mm. Because of an unbalance in the rotor, a disturbance results that is approximately sinusoidal in the vertical direction with a frequency equal to the angular speed of the rotor. The amplitude of this disturbance is 130 N when the motor is rotating at 1720 rpm. What is the steady-state motion of this system under these circumstances if we neglect the mass of the springs, the friction, and the inertia of the rollers?

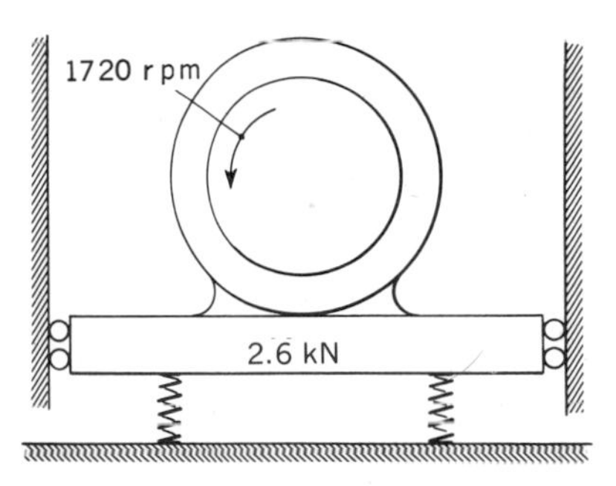

Figure 19.18. Motor with unbalanced rotor.

The spring constant for the system is

$$K = \frac{2600}{2.5} = 1040 \text{ N/mm} = 1.040 \times 10^6 \text{ N/m}$$

and the natural frequency becomes

$$\omega_n = \sqrt{\frac{1.040 \times 10^6}{2600/9.81}}$$

$$= 62.6 \text{ rad/sec} = 9.97 \text{ cycles/sec}$$

The steady-state motion is

$$x_p = \frac{F_0/K}{1 - (\omega/\omega_n)^2} \sin \omega t$$

$$= \frac{130/1.040 \times 10^6}{1 - [1720/(60)(9.97)]^2} \sin \frac{1720}{60}(2\pi)t$$

$$= -1.720 \times 10^{-5} \sin 180.1t \text{ m}$$

$$= -.01720 \sin 180.1t \text{ mm}$$

Note that the driving frequency is above the natural frequency. In starting up motors and turbines, we must sometimes go through a natural frequency of the system, and it is wise to get through this zone as quickly as possible to prevent large amplitudes from building up.

EXAMPLE 19.8

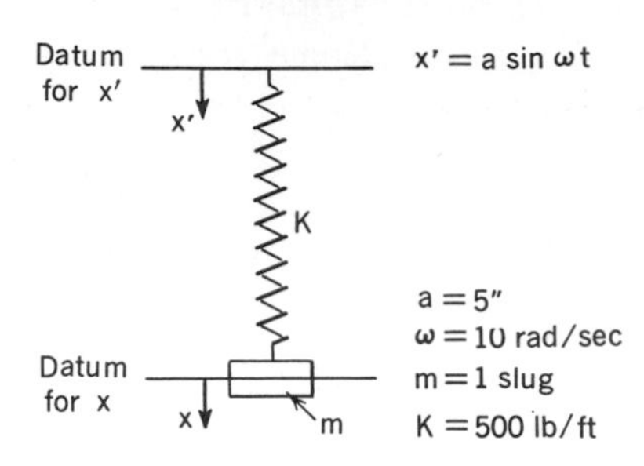

Figure 19.19. Spring–mass system with disturbance.

A mass on a spring is shown in Fig. 19.19. The support of the spring at x' is made to move with harmonic motion in the vertical direction by some external agent. This motion is expressed as $a \sin \omega t$. If at $t = 0$ the mass is displaced in a downward position a distance of 1 in. and if it has at this instant a speed downward of 3 in./sec, what is the position of the mass at $t = 5$ sec? Take $a = 5$ in., $\omega = 10$ rad/sec, $K = 500$ lb/ft, and $m = 1$ slug.

Let us express Newton's law for the mass. Note that the extension of the spring is $x - x'$. Hence,

$$m\frac{d^2x}{dt^2} = -K(x - x')$$

Replacing x' by the known function of time, we get, upon rearranging the terms:

$$\frac{d^2x}{dt^2} + \frac{K}{m}x = \frac{Ka}{m}\sin \omega t$$

This is the same form as Eq. 19.21 for the case where the disturbance is exerted on the mass directly. The solution, then, is

$$x = C_1 \sin \sqrt{\frac{K}{m}}t + C_2 \cos \sqrt{\frac{K}{m}}t + \frac{a}{1 - (\omega/\sqrt{K/m})^2}\sin \omega t$$

Putting in the numerical values of $\sqrt{K/m}$, etc., we have

$$x = C_1 \sin 22.4t + C_2 \cos 22.4t + 6.24 \sin 10t \text{ in.}$$

Now impose the initial conditions to get

$$1 = C_2$$

$$3 = 22.4C_1 + (6.24)(10)$$

Therefore,

$$C_1 = -2.65$$

The motion, then, is given as

$$x = -2.65 \sin 22.4t + \cos 22.4t + 6.24 \sin 10t \text{ in.}$$

When $t = 5$ sec, the position of the mass relative to the lower datum is given as

$$(x)_5 = -2.65 \sin (22.4)(5) + \cos (22.4)(5) + 6.24 \sin 50 = 1.177 \text{ in.}$$

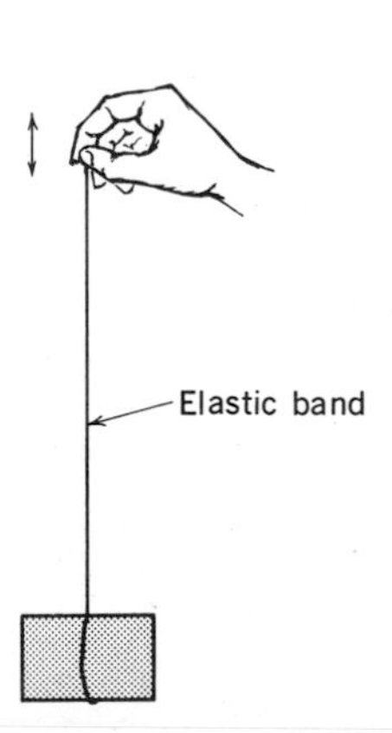

Figure 19.20. Simple resonance and change of phase demonstration.

You may approximate the setup of this problem profitably with an elastic band supporting a small body as shown in Fig. 19.20. By oscillating the free end of the band with varying frequency from low frequency to high frequency, you can demonstrate the rapid change of phase between the disturbance and the excited motion as you pass through resonance. Thus, at low frequencies both motions will be in phase and at frequencies well above resonance the motion will be close to being 180° out of phase. Without friction this change, according to the mathematics, is discontinuous, but with the presence of friction (i.e., in a real case) there is actually a smooth, although sometimes rapid, transition between both extremes.

Problems

19.55. A mass is held by three springs. Assume that the rolling friction on the floor is negligible, as are the inertial effects of the rollers. The spring constants are:

$$K_1 = 30 \text{ lb/in.}$$

$$K_2 = 20 \text{ lb/in.}$$

$$K_3 = 10 \text{ lb/in.}$$

A sinusoidal force having an amplitude of 5 lb and a frequency of $10/\pi$ cycles/sec acts on the body in the direction of the springs. What is the steady-state amplitude of the motion of the body?

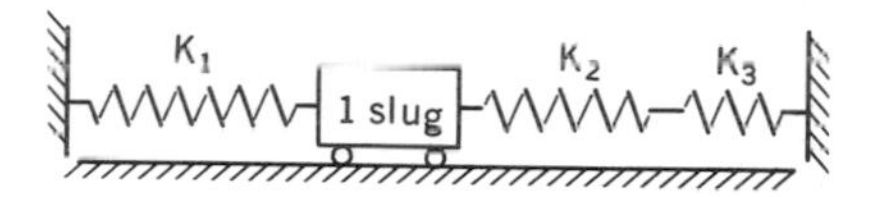

Figure P.19.55

19.56. In Problem 19.55, the initial conditions are:

(a) The initial position of the body is 3 in. to the right of the static-equilibrium position.
(b) The initial velocity is zero.
(c) At $t = 0$, the sinusoidal disturbing force has a value of 5 lb in the positive direction.

Find the position of the body after 3 sec.

19.57. A sinusoidal force, with amplitude F of 22 N and frequency $1/2\pi$ cycles/sec, acts on a body having a mass of 15 kg. Meanwhile, the wall moves with a motion given as 8 cos t mm. For a spring constant $K =$ 8.8 N/mm, what is the amplitude of the steady-state motion?

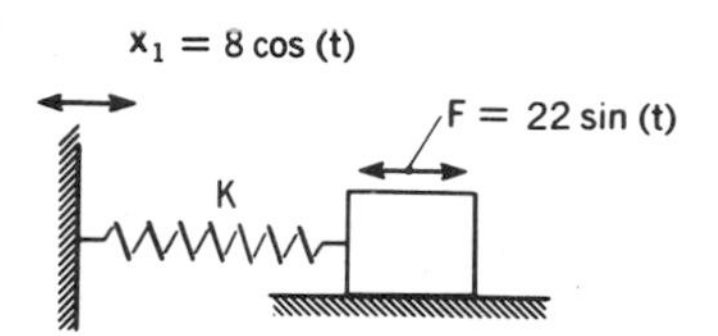

Figure P.19.57

19.58. In Problem 19.57, suppose that the disturbing force F is 22 sin $(t + \pi/4)$. What is the amplitude of the steady-state motion?

19.59. A torque $T = A \sin \omega t$ is applied to the disc. Express the solution for the transient torsional motion and the steady-state torsional motion in terms of arbitrary constants of integration. Take the shear modulus of elasticity of the shaft as G. (*Hint:* Recall that K_t for a shaft is GJ/L.)

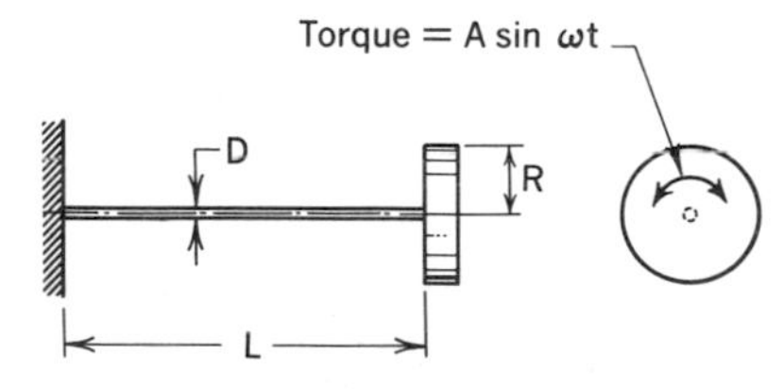

Figure P.19.59

19.60. A *vibrograph* is a device for measuring the amplitude of vibration in a given direction. The apparatus is bolted to the machine to be tested. A seismic mass M in the vibrograph rides along a rod CD under constraint of a linear spring of spring constant K. If the machine being tested has a harmonic motion $\bar{x}$ of frequency ω in the direction of C-D, then M will have a steady-state oscillatory frequency also of frequency ω. The motion of M relative to the vibrograph is given as x' and is recorded on the rotating drum. Show that the amplitude of motion of the machine is

$$\left|\frac{(\omega/\omega_n)^2 - 1}{(\omega/\omega_n)^2}\right|$$

times the amplitude of the recorded motion x', where $\omega_n = \sqrt{K/M}$.

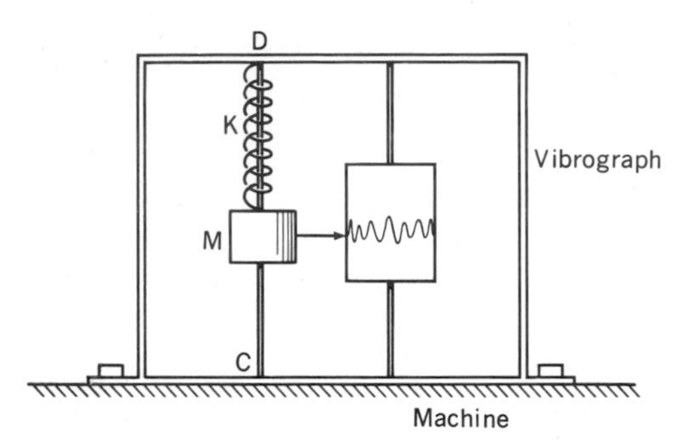

Figure P.19.60

19.61. A vibrograph is attached rigidly to a diesel engine for which we want to know the vibration amplitude. If the seismic spring–mass system has a natural frequency of 10 cycles/sec, and if the seismic mass vibrates relative to the vibrograph with an amplitude of 1.27 mm when the diesel is turning over at 1000 rpm, what is the amplitude of vibration of the diesel in the direction of the vibrograph? The seismic mass weighs 4.5 N. See Problem 19.60 before doing this problem.

19.62. Explain how you could devise an instrument to measure torsional vibrations of a shaft in a manner analogous to the way the vibrograph measures linear vibrations of a machine. Such instruments are in wide use and are called *torsiographs*. What would be the relation of the amplitude of oscillations as picked up by your apparatus to that of the shaft being measured? See Problem 19.60 before doing this problem.

19.63. A trailer of weight W moves over a washboard road at a constant speed V to the right. The road is approximated by a sinusoid of amplitude A and wavelength L. If the wheel B is small, the center of the wheel will have a motion x closely resembling the aforementioned sinusoid. If the trailer is connected to the wheel through a linear spring of stiffness K, formulate the steady-state equation of motion x' for the trailer. List all assumptions. What speed causes resonance?

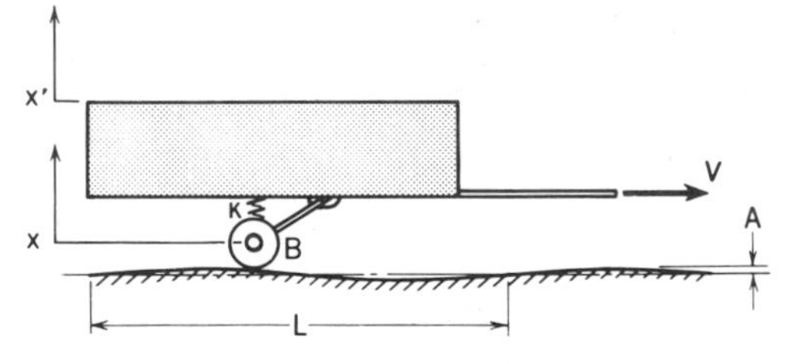

Figure P.19.63

19.64. In Problem 19.63, compute the amplitude of motion of the trailer for the following data:

$$W = 5.34 \text{ kN}$$
$$V = 16 \text{ km/hr}$$
$$K = 43.8 \text{ N/mm}$$
$$L = 10 \text{ m}$$
$$A = 100 \text{ mm}$$

What is the resonance speed V_{res} for the case? From Problem 19.63, we have

$$x'_p = \frac{A}{|1 - (2\pi V/L)^2(W/gK)|} \sin\frac{2\pi Vt}{L}$$

19.65. A cantilever beam of length L has an electric motor A weighing 100 N fastened to the end. The tip of the cantilever beam descends 1.2 mm when the motor is attached. If the *center of mass* of the armature of the motor is a distance 2 mm from the axis of rotation of the motor, what is the amplitude of vibration of the motor when it is rotating at 1750 rpm? The armature weighs 40 N. Neglect the mass of the beam.

Figure P.19.65

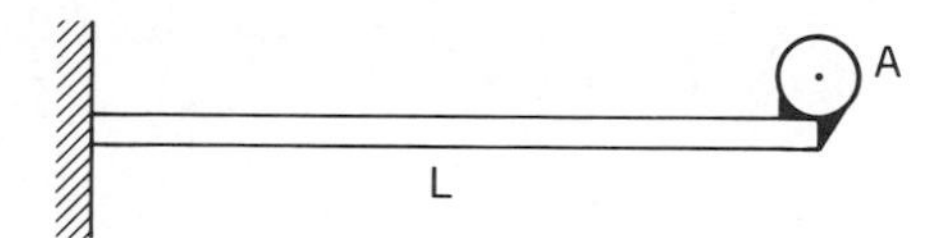

19.66. Suppose that a 2-N block is glued to the top of the motor in Problem 19.65, where the maximum strength of the bond is $\frac{1}{2}$ N. At what minimum angular speed ω of the motor will the block fly off?

19.67. An important reason for mounting rotating and reciprocating machinery on springs is to decrease the transmission of vibration to the foundation supporting the machine. Show that the amplitude of force transmitted to the ground, F_{TR}, for such cases is

$$F_0\left|\frac{1}{1-(\omega/\omega_n)^2}\right|$$

where F_0 is the disturbing force from the machine. The factor $|1/[1-(\omega/\omega_n)^2]|$ is called the *relative transmission factor*. Show that, unless the springs are soft, ($\omega_n < \omega/\sqrt{2}$), the use of springs actually increases the transmission of vibratory forces to the foundation.

19.68. In Example 19.7, what is the amplitude of the force transmitted to the foundation? What must K of the spring system be to decrease the amplitude by one-half? See Problem 19.67.

19.69. A machine weighing W N contains a reciprocating mass of weight w N having a vertical motion relative to the machine given approximately as $x' = A \sin \omega t$. The machine is mounted on springs having a total spring constant K. This machine is guided so that it can move only in the vertical direction. What is the differential equation of motion for this machine? What is the formulation for the amplitude of the machine for steady-state operation?

19.70. A mass M of .5 kg is suspended from a stiff rod AB via a spring whose spring constant K is 100 N/m. The end of rod AB is given a vertical sinusoidal motion $\delta_A = 2 \sin 14t$ mm, with t in seconds. What is the maximum force on the rod at C long after the motion has started?

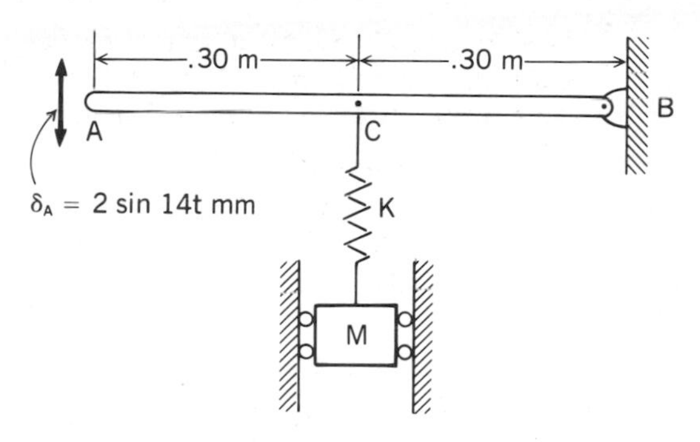

Figure P.19.70

19.71. In Problem 19.70, what range of frequencies of the motion δ_A must be excluded to keep the maximum force at C less than 7 N? Consider only steady-state motion. (*Hint:* Note that below resonance disturbance and motion are in phase whereas above resonance they are 180° out of phase. Therefore, x_p in Eq. 19.27 will be positive below resonance and negative above resonance.)

19.72. A bob B of weight W is suspended from a vehicle A which is made to have a motion $x_A = \delta \sin \omega t$. If δ is very small, what should ω be so that bob B has an amplitude of motion equal to 1.5δ?

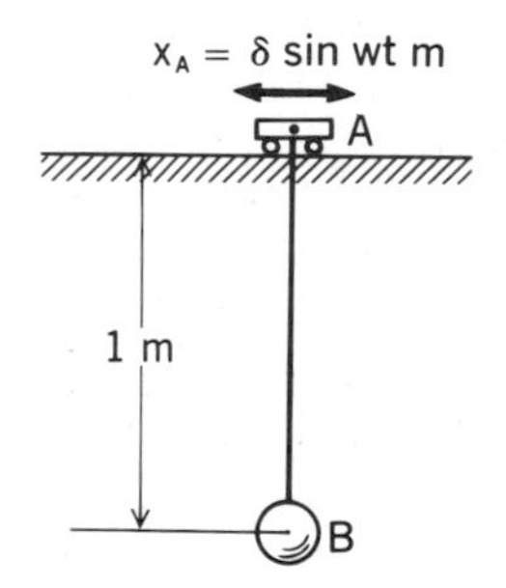

Figure P.19.72

19.73. Two spheres each of mass $M = 2$ kg are welded to a light rod that is pinned at B. A second light rod AC is welded to the first rod. At A we apply a disturbance $F_0 \sin \omega t$. At the other end C, there is a restraining spring which is unstretched when AC is horizontal. If the amplitude of steady-state rotation of the system is to be kept below .02 rad, what ranges of frequencies ω are permitted? The following data apply:

$$l = 300 \text{ mm}$$
$$K = 7.0 \text{ N/mm}$$
$$F_0 = 10 \text{ N}$$
$$a = 100 \text{ mm}$$

See the hint in Problem 19.71.

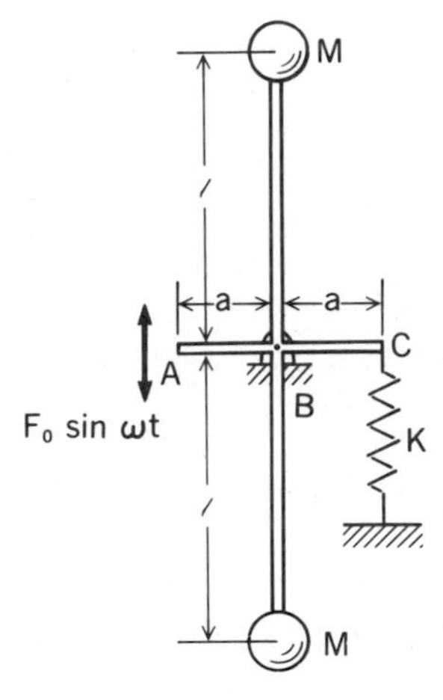

Figure P.19.73

19.74. In Problem 19.73, what is the angle of rotation of the system 10 sec after the application of the sinusoidal load? The system is stationary at time $t = 0$. Take $\omega = 13$ rad/sec.

19.75. A rod of length L and weight W is suspended from a light support at A. This support is given a movement $x_A = \delta \sin \omega t$, where δ is very small compared to L. At what frequency, ω, should A be moved if the amplitude of motion of tip, B, is to be 1.5δ?

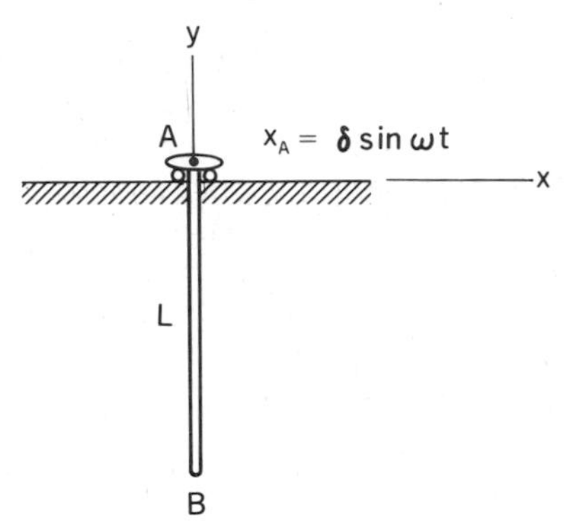

Figure P.19.75

19.7 Linear Restoring Force with Viscous Damping

We shall now consider the case in which a special type of friction is present. In the chapters on statics, you will recall, we considered coulombic or dry friction for the cases of sliding and impending motion. This force was proportional to the normal force at the interface of contact and dependent on the material of the bodies. At this time, we shall consider the case of bodies separated from each other by a thin film of fluid. The frictional force (called a *damping* force) is independent of the material of the bodies but depends on the nature of the fluid and is proportional for a given fluid to the relative velocity of the two bodies separated by the film. Thus,

$$f = -c\left(\frac{dx}{dt}\right)_{\text{rel}} \tag{19.28}$$

where c is called the *coefficient of damping*. The minus sign indicates that the frictional force opposes the motion (i.e., the friction force must always have a sign opposite to that of the relative velocity).

In Fig. 19.21 is shown the spring–mass model with damping present. We shall

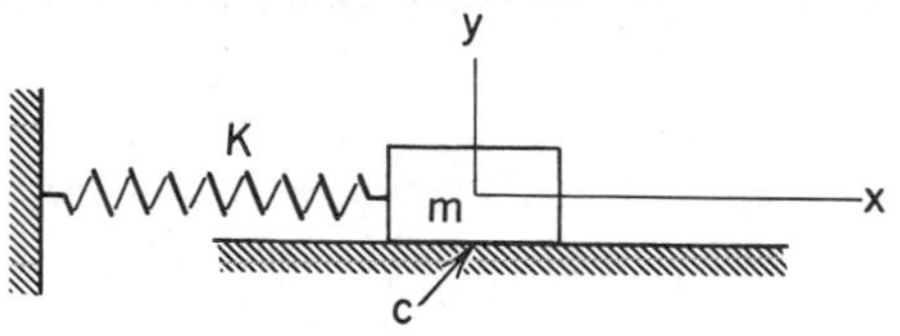

Figure 19.21. Spring–mass system with damping.

investigate possible motions consistent with a set of given initial conditions. The differential equation of motion is

$$m\frac{d^2x}{dt^2} = -Kx - c\frac{dx}{dt}$$

In standard form, we get

$$\frac{d^2x}{dt^2} + \frac{c}{m}\frac{dx}{dt} + \frac{K}{m}x = 0 \qquad (19.29)$$

This is a homogeneous, second-order, differential equation with constant coefficients. We shall expect two independent functions with two arbitrary constants to form the general solution to this equation. Because of the presence of the first derivative in the equation, we cannot use sines or cosines for trial solutions, since the first derivative changes their form and prevents a cancellation of the time function. Instead, we use e^{pt}, where p is determined so as to satisfy the equation. Thus, let

$$x = C_1 e^{pt}$$

Substituting, we get

$$C_1 p^2 e^{pt} + \frac{c}{m}C_1 p e^{pt} + \frac{K}{m}C_1 e^{pt} = 0$$

Canceling out $C_1 e^{pt}$, we get

$$p^2 + \frac{c}{m}p + \frac{K}{m} = 0$$

Solving for p, we write

$$p = \frac{-c/m \pm \sqrt{(c/m)^2 - 4K/m}}{2} = -\frac{c}{2m} \pm \sqrt{\left(\frac{c}{2m}\right)^2 - \frac{K}{m}} \qquad (19.30)$$

It will be helpful to consider three cases here.

Case A

$$\frac{c}{2m} > \sqrt{\frac{K}{m}}$$

Here the value p is *real*. Using both possible values of p and employing C_1 and C_2 as arbitrary constants, we get

$$x = C_1 \exp\{[-(c/2m) + \sqrt{(c/2m)^2 - K/m}]t\}$$
$$+ C_2 \exp\{[-(c/2m) - \sqrt{(c/2m)^2 - K/m}]t\} \qquad (19.31)$$

Rearranging terms in Eq. 19.31 we get the following standard form of solution:

$$x = \exp\,[-(c/2m)t]\{C_1 \exp\,[\sqrt{(c/2m)^2 - K/m}\, t] + C_2 \exp\,[-\sqrt{(c/2m)^2 - K/m}\, t]\} \tag{19.32}$$

Since $c/2m > \sqrt{(c/2m)^2 - K/m}$, we see from Eq. 19.32 that the first exponential dominates and so as the time t increases, the motion can only be that of an exponential of *decreasing* amplitude. Thus, there can be no oscillation. The motion is illustrated in Fig. 19.22 and is called *overdamped* motion.

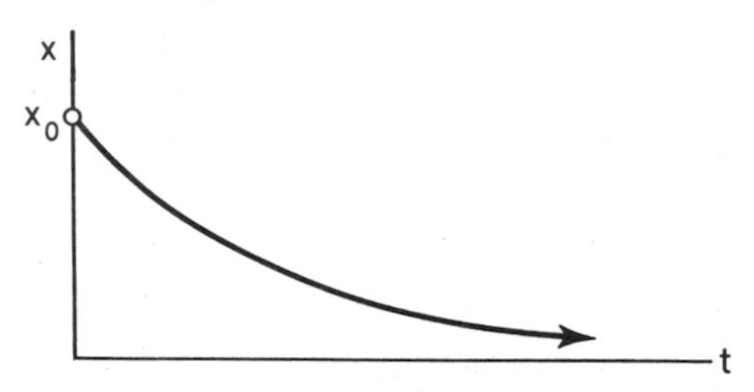

Figure 19.22. Overdamped motion.

Case B

$$\frac{c}{2m} < \sqrt{\frac{K}{m}}$$

This means that we have a negative quantity under the root in Eq. 19.30. Extracting $\sqrt{-1} = i$, we can then write p as follows:

$$p = -\frac{c}{2m} \pm i\sqrt{\frac{K}{m} - \left(\frac{c}{2m}\right)^2}$$

The solution then becomes

$$x = \exp\,[-(c/2m)t]\{C_1 \exp\,[i\sqrt{K/m - (c/2m)^2}\, t] + C_2 \exp\,[-i\sqrt{K/m - (c/2m)^2}\, t]\} \tag{19.33}$$

From complex-number theory, we know that $e^{i\theta}$ may be replaced by $\cos\theta + i\sin\theta$ and thus the equation above can be put in the form

$$\begin{aligned} x = \exp\,[-(c/2m)t]\Bigg\{C_1\Bigg[\cos\sqrt{\frac{K}{m} - \left(\frac{c}{2m}\right)^2}\, t + i \sin\sqrt{\frac{K}{m} - \left(\frac{c}{2m}\right)^2}\, t\Bigg] \\ + C_2\Bigg[\cos\sqrt{\frac{K}{m} - \left(\frac{c}{2m}\right)^2}\, t - i \sin\sqrt{\frac{K}{m} - \left(\frac{c}{2m}\right)^2}\, t\Bigg]\Bigg\} \end{aligned} \tag{19.34}$$

Collecting terms and replacing sums and differences of arbitrary constants by other arbitrary constants, we get the result:

$$x = \exp\,[-(c/2m)t]\Bigg[C_3 \cos\sqrt{\frac{K}{m} - \left(\frac{c}{2m}\right)^2}\, t + C_4 \sin\sqrt{\frac{K}{m} - \left(\frac{c}{2m}\right)^2}\, t\Bigg] \tag{19.35}$$

The quantity in brackets represents a harmonic motion which has a frequency less than the free undamped natural frequency of the system. The exponential term to the left of the brackets, then, serves to decrease continually the amplitude of this motion. A plot of the displacement against time for this case is illustrated in Fig. 19.23, where

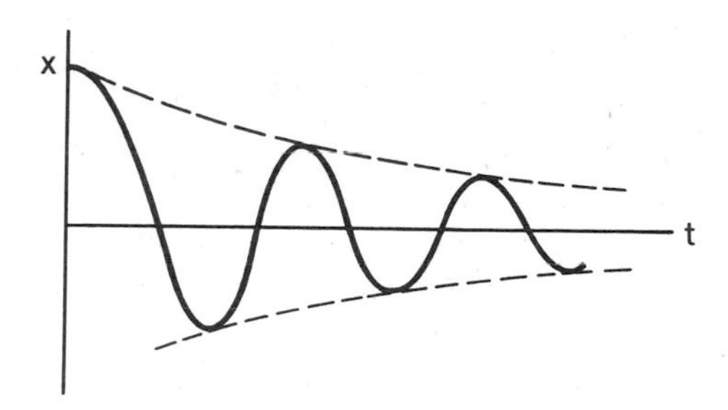

Figure 19.23. Underdamped motion.

the upper dashed envelope corresponds in form to the exponential function $e^{-(c/2m)t}$. We call this motion *underdamped* motion.

Case C

$$\frac{c}{2m} = \sqrt{\frac{K}{m}}$$

Since this is the dividing line between the overdamped case and one in which oscillation is possible, the motion is termed a *critically damped motion*. We have here *identical* roots for p given as

$$p = -\frac{c}{2m} \tag{19.36}$$

and accordingly for such a case the general solution to Eq. 19.29 according to the theory of differential equation is then

$$x = (C_1 + C_2 t)\, e^{-(c/2m)t} \tag{19.37}$$

First we see from this equation that we do *not* have an oscillatory motion. Also, you will recall from the calculus that as t goes to infinity an exponential of the form e^{-At}, with A a positive constant, goes to zero faster than Ct goes to infinity. Accordingly, Fig. 19.22 can be used to picture the plot of x versus t for this case.

The damping constant for this case is called the *critical* damping constant and is denoted as c_{cr}. The value of c_{cr} clearly is

$$c_{cr} = 2\sqrt{Km} \tag{19.38}$$

It should be clear that, for a damping constant less than c_{cr}, we will have underdamped motion while for a damping constant greater than c_{cr} we will have overdamped motion.

In all the preceding cases for damped free vibration, the remaining step for a complete evaluation of the solution is to compute the arbitrary constants from the initial conditions of the particular problem. Note that in discussing damped motion we shall consider the "natural frequency" of the system to be that of the corresponding *undamped* case and shall refer to the actual frequency of the motion as the frequency of free, damped motion.

EXAMPLE 19.9

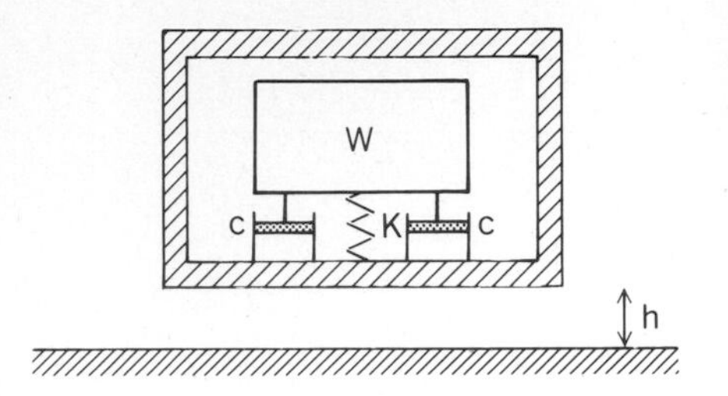

Figure 19.24. Packaging to reduce breakage.

Springs and dashpots are used in packaging delicate equipment in crating so that during transit the equipment will be protected from shocks. In Fig. 19.24, we have shown a piece of equipment whose weight W is 500 N. It is supported in a crate by one spring and two dashpots (or shock absorbers). The value of K for the spring is 30 N/mm and the coefficient of damping, c, is 1 N/mm/sec for each dashpot.

The crate is held above a rigid floor at a height h of 150 mm. It is then released and allowed to hit the floor in a plastic impact. What is the maximum deflection of W relative to the crate?

As a first step, we compute the *critical damping* to find what regime we are in.

$$c_{cr} = 2\sqrt{Km} = 2\sqrt{(30)(1000)(500)/g} = 2473 \text{ N/m/sec} \tag{a}$$

The total damping coefficient for our case is

$$c_{\text{total}} = (2)(1)(1000) = 2000 \text{ N/m/sec}$$

We are therefore *underdamped*. The motion is then given as follows:

$$x = e^{-(c/2m)t}\left[C_3 \cos\sqrt{\frac{K}{m} - \left(\frac{c}{2m}\right)^2}\,t + C_4 \sin\sqrt{\frac{K}{m} - \left(\frac{c}{2m}\right)^2}\,t\right] \tag{b}$$

Note that

$$\frac{c}{2m} = \frac{2000}{(2)(500)/9.81} = 19.62 \text{ sec}$$

$$\frac{K}{m} = \frac{(30)(1000)}{500/9.81} = 589 \text{ sec}^{-2}$$

Hence,

$$x = e^{-19.62t}(C_3 \cos 14.27t + C_4 \sin 14.27t) \tag{c}$$

When $t = 0$, at the instant of impact, take

$$x = 0 \quad \text{and} \quad \dot{x} = \sqrt{2gh} = \sqrt{(2)(9.81)(.15)} = 1.716 \text{ m/sec}$$

The first condition renders $C_3 = 0$. For the second condition, note first that

$$\dot{x} = e^{-19.62t}[C_4(14.27) \cos 14.27t] - 19.62e^{-19.62t}(C_4 \sin 14.27t)$$

For the second condition ($\dot{x} = 1.716$ at $t = 0$) we get

$$1.716 = C_4(14.27)$$

Therefore,

$$C_4 = .1202$$

Thus, we have for x:

$$x = .1202e^{-19.62t} \sin 14.27t \tag{d}$$

$$\dot{x} = e^{-19.62t}(1.716 \cos 14.27t - 2.358 \sin 14.27t) \tag{e}$$

Set $\dot{x} = 0$ and solve for t in order to get the maximum deflection of W.

$$1.716 \cos 14.27t - 2.358 \sin 14.27t = 0$$

Therefore,

$$\tan 14.27t = .7274$$

The smallest t satisfying the equation above is

$$t = .0441 \text{ sec}$$

The value of x for this time is from Eq. (d):

$$x = .1202e^{-(19.62)(.0441)} \sin [14.27(.0441)]$$
$$= .0298 \text{ m} = 29.8 \text{ mm}$$

Hence, W moves a maximum distance of 29.8 mm downward after impact.

EXAMPLE 19.10

A block W of 200 N (see Fig. 19.25) moves on a film of oil which is .1 mm in thickness under the block. The area of the bottom surface of the block is 2×10^4 mm². The spring constant K is 2 N/m. If the weight is pulled in the x direction and released, what is the nature of the motion?

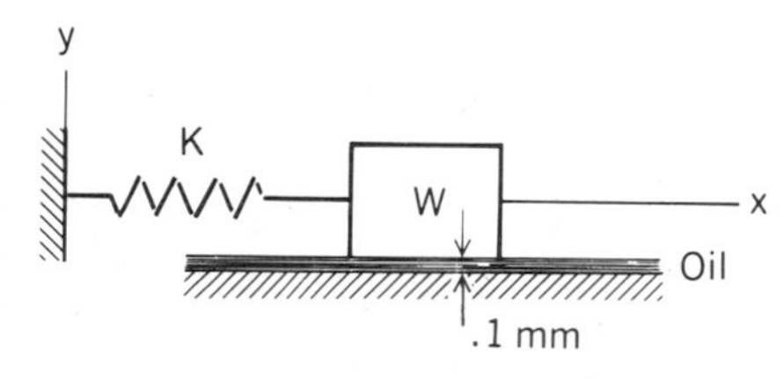

Figure 19.25. Spring–mass on film of oil.

You may have learned in physics that friction force per unit area (i.e., shear stress) on the block W from the oil is given by *Newton's viscosity law* as:

$$\tau = \mu\left(\frac{\partial V}{\partial y}\right)_{\text{block}} \tag{a}$$

where τ is the shear stress (force per unit area), μ is the *coefficient of viscosity*, and $\partial V/\partial y$ is the slope of the velocity profile at the block surface (see Fig. 19.26). Now the oil will stick to the surfaces of the block W and the ground surface. And so we can approximate the velocity profile as shown in Fig. 19.27, where we have used a straight-line profile connecting zero velocity at the bottom and velocity $\dot{x}$ of the block W at the top. Such a procedure gives good results when the film of oil is thin as in the present case. The desired slope $(\partial V/\partial y)_{\text{block}}$ is then approximated as

$$\left(\frac{\partial V}{\partial y}\right)_{\text{block}} = \frac{\dot{x}}{.0001} \tag{b}$$

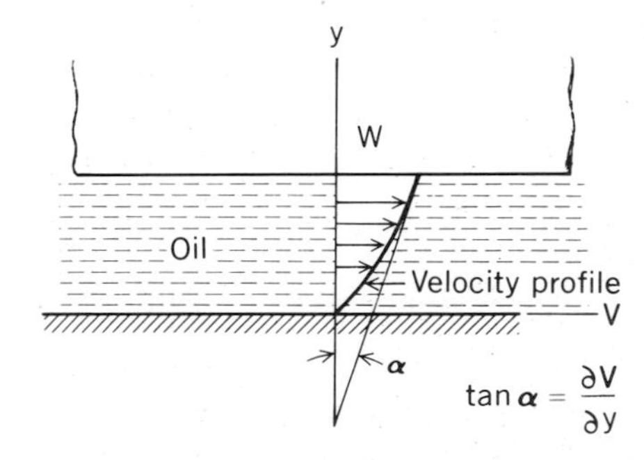

Figure 19.26. Slope of velocity profile at bottom of W.

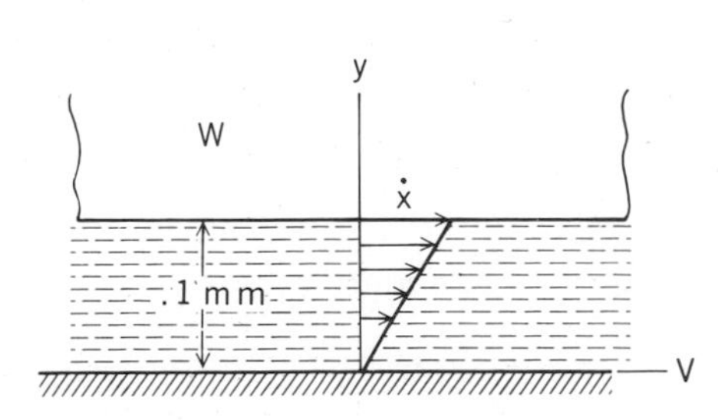

Figure 19.27. Approximate velocity profile.

The coefficient of viscosity can be found in handbooks. For our case, let us say that $\mu = .0080$ N-sec/m².

It is now an easy matter to compute the coefficient of damping c. Thus, the friction force is

$$f = \tau A = -\left[(.0080)\left(\frac{\dot{x}}{.0001}\right)\right](2 \times 10^4/10^6) \qquad \text{(c)}$$
$$= -1.600\dot{x} \text{ N}$$

Thus, $c = 1.600$. The critical damping for the problem is

$$c_{cr} = 2\sqrt{Km} = 2\sqrt{(2)\left(\frac{200}{g}\right)} = 12.77$$

Thus, motion clearly will be *underdamped*. The frequency of oscillation is then

$$\omega = \sqrt{\frac{K}{m} - \left(\frac{c}{2m}\right)^2}$$
$$= \sqrt{\frac{2}{200/g} - \left[\frac{1.600}{(2)(200)/g}\right]^2}$$
$$= .311 \text{ rad/sec} = .0495 \text{ cycle/sec}$$

Similar problems involving viscous friction of lubricants and oils are given in the homework section.

*19.8 Linear Restoring Force, Viscous Damping, and a Harmonic Disturbance

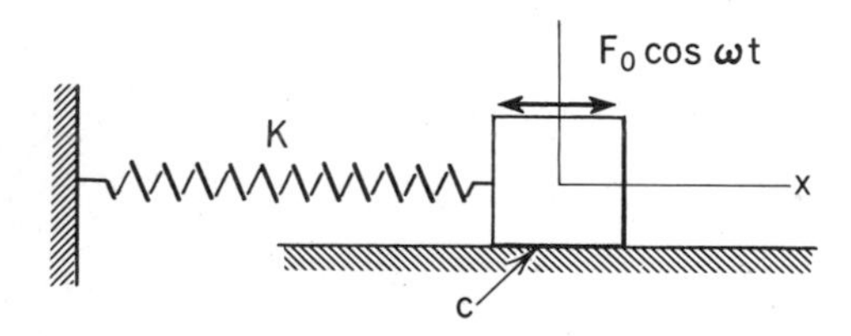

Figure 19.28. Spring–mass system with damping.

In the spring–mass problem shown in Fig. 19.28 we include driving function $F_0 \cos \omega t$ along with viscous damping. The differential equation in the standard form then becomes

$$\frac{d^2x}{dt^2} + \frac{c}{m}\frac{dx}{dt} + \frac{K}{m}x = \frac{F_0}{m}\cos \omega t \qquad (19.39)$$

Equation 19.39 is a nonhomogeneous equation. The general solution will be the homogeneous solution worked out in Section 19.7, plus any particular solution of Eq. 19.39.

Because there is a first derivative on the left side of the equation, we cannot expect a particular solution of the form $D \cos \omega t$ to satisfy the differential equation. Instead, from the method of *undetermined coefficients* we shall try the following:

$$x_p = D \sin \omega t + E \cos \omega t \qquad (19.40)$$

The constants D and E are to be adjusted to facilitate a solution. Substituting into the differential equation, we write

$$-D\omega^2 \sin \omega t - E\omega^2 \cos \omega t + \frac{c}{m}\omega D \cos \omega t - \frac{c}{m}\omega E \sin \omega t$$
$$+ \frac{K}{m}D \sin \omega t + \frac{K}{m}E \cos \omega t = \frac{F_0}{m}\cos \omega t$$

Collecting the terms, we have

$$\left(-D\omega^2 - \frac{c}{m}\omega E + \frac{K}{m}D\right)\sin\omega t + \left(-\frac{F_0}{m} - E\omega^2 + \frac{c}{m}\omega D + \frac{K}{m}E\right)\cos\omega t = 0$$

We set each coefficient of the time functions equal to zero and thus get two simultaneous equations in the unknowns E and D:

$$-D\omega^2 - \frac{\omega c}{m}E + \frac{K}{m}D = 0$$

$$-\frac{F_0}{m} - E\omega^2 + \frac{\omega c}{m}D + \frac{K}{m}E = 0$$

Rearranging and replacing K/m by ω_n^2, we get

$$D(\omega^2 - \omega_n^2) + E\left(\frac{\omega c}{m}\right) = 0$$

$$D\left(-\frac{\omega c}{m}\right) + E(\omega^2 - \omega_n^2) = -\frac{F_0}{m}$$

Using Cramer's rule, we see that the constants D and E become

$$D = \frac{\begin{vmatrix} 0 & \omega c/m \\ -F_0/m & \omega^2 - \omega_n^2 \end{vmatrix}}{\begin{vmatrix} \omega^2 - \omega_n^2 & \omega c/m \\ -\omega c/m & \omega^2 - \omega_n^2 \end{vmatrix}} = \frac{(F_0/m)(\omega c/m)}{(\omega^2 - \omega_n^2)^2 + (\omega c/m)^2}$$

$$E = \frac{\begin{vmatrix} \omega^2 - \omega_n^2 & 0 \\ -\omega c/m & -F_0/m \end{vmatrix}}{\begin{vmatrix} \omega^2 - \omega_n^2 & \omega c/m \\ -\omega c/m & \omega^2 - \omega_n^2 \end{vmatrix}} = \frac{(F_0/m)(\omega_n^2 - \omega^2)}{(\omega^2 - \omega_n^2)^2 + (\omega c/m)^2}$$

The entire solution can then be given as

$$x = x_c + \frac{(F_0/m)(\omega_n^2 - \omega^2)}{(\omega^2 - \omega_n^2)^2 + (\omega c/m)^2}\cos\omega t + \frac{F_0\omega c/m^2}{(\omega^2 - \omega_n^2)^2 + (\omega c/m)^2}\sin\omega t \qquad (19.41)$$

The constants of integration are present in the complementary solution x_c and are determined by the initial condition to which the *entire* solution given above is subject.

The complementary solution here is a *transient* in the true sense of the word, because it dies out in the manner explained in Section 19.7. The particular solution is a harmonic motion with the same frequency as the disturbance. Only the amplitude of this motion is affected by the damping present. Note that, mathematically, the amplitude of the steady-state motion cannot become infinite with damping present unless F_0 becomes infinite.

We now write the steady-state solution in the following way:

$$x_p = \frac{(F_0/m)(\omega_n^2 - \omega^2)}{(\omega^2 - \omega_n^2)^2 + (\omega c/m)^2}\cos\omega t + \frac{(F_0/m)(\omega c/m)}{(\omega^2 - \omega_n^2)^2 + (\omega c/m)^2}\sin\omega t \qquad (19.42)$$

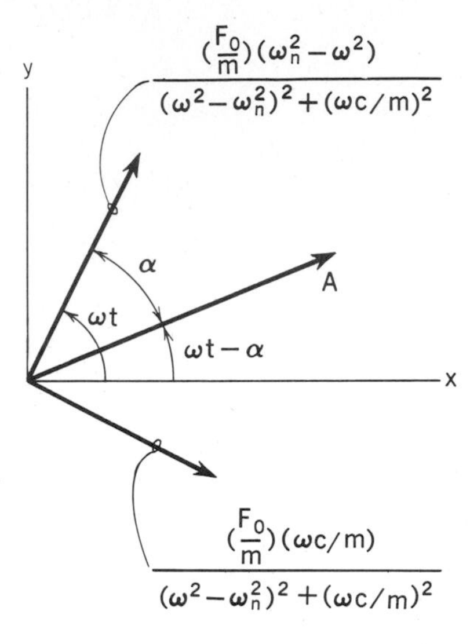

Figure 19.29. Phasor diagram.

We can represent this formulation in a phasor diagram as shown in Fig. 19.29. It should be clear that we can give x_p in the following form:

$$x_p = A \cos(\omega t - \alpha) \tag{19.43}$$

where the amplitude A is given as

$$\begin{aligned} A &= \left\{\left[\frac{(F_0/m)(\omega^2 - \omega_n^2)}{(\omega^2 - \omega_n^2)^2 + (\omega c/m)^2}\right]^2 + \left[\frac{(F_0/m)(\omega c/m)}{(\omega^2 - \omega_n^2)^2 + (\omega c/m)^2}\right]^2\right\}^{1/2} \\ &= \frac{F_0}{m}\frac{\sqrt{(\omega^2 - \omega_n^2)^2 + (\omega c/m)^2}}{(\omega^2 - \omega_n^2)^2 + (\omega c/m)^2} \\ &= \frac{F_0/m}{\sqrt{(\omega^2 - \omega_n^2)^2 + (\omega c/m)^2}} \\ &= \frac{F_0}{\sqrt{(m\omega^2 - K)^2 + (\omega c)^2}} \end{aligned} \tag{19.44}$$

and where α, the phase angle, is given as

$$\begin{aligned} \alpha &= \tan^{-1}\left[\frac{(F_0/m)(\omega c/m)}{(\omega^2 - \omega_n^2)^2 + (\omega c/m)^2} \cdot \frac{(\omega^2 - \omega_n^2)^2 + (\omega c/m)^2}{(F_0/m)(\omega_n^2 - \omega^2)}\right] \\ &= \tan^{-1}\frac{\omega c}{K - m\omega^2} \end{aligned} \tag{19.45}$$

We may express the amplitude A in yet another form by dividing numerator and denominator in Eq. 19.44 by K and by recalling from Eq. 19.38 that the ratio $2\sqrt{Km}/c_{cr}$ is unity. Thus, we get

$$\begin{aligned} A &= \frac{F_0/K}{\sqrt{\left[\left(\frac{\omega}{\omega_n}\right)^2 - 1\right]^2 + \frac{1}{K^2}\left(\frac{2\sqrt{Km}}{c_{cr}}\right)^2(\omega c)^2}} \\ &= \frac{\delta_{st}}{\sqrt{\left[\left(\frac{\omega}{\omega_n}\right)^2 - 1\right]^2 + \left[2\left(\frac{c}{c_{cr}}\right)\left(\frac{\omega}{\omega_n}\right)\right]^2}} \end{aligned} \tag{19.46}$$

where $F_0/K = \delta_{st}$ is the static deflection. The term

$$\frac{1}{\sqrt{\left[\left(\frac{\omega}{\omega_n}\right)^2 - 1\right]^2 + \left[2\left(\frac{c}{c_{cr}}\right)\left(\frac{\omega}{\omega_n}\right)\right]^2}}$$

is called the *magnification factor* which is a dimensionless factor giving the amplitude of steady-state motion per unit static deflection. Accordingly, this factor for a given system is useful for examining the effects of frequency changes or damping changes on the steady-state vibration amplitude. A plot of the magnification factor versus ω/ω_n for various values of c/c_{cr} is shown in Fig. 19.30. We see from this plot that small

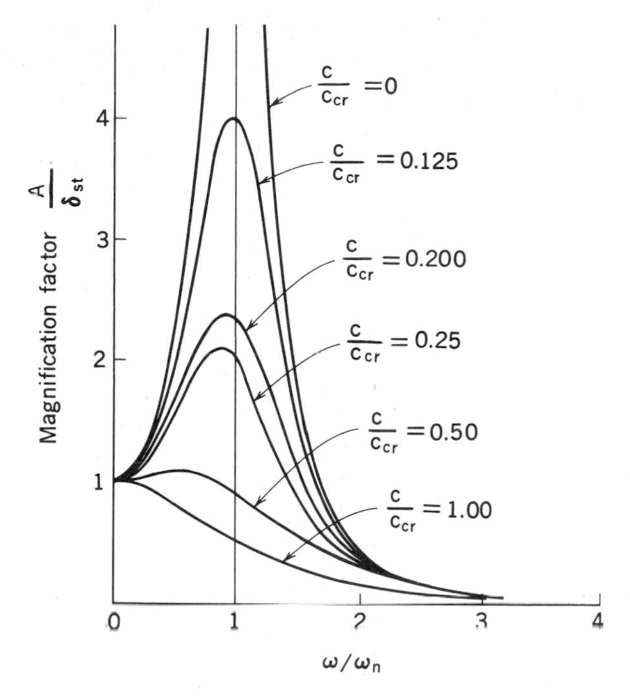

Figure 19.30. Magnification factor plot.

vibrations result when ω is kept far from ω_n. Additionally, note that maximum amplitude does not occur at resonance but actually at frequencies somewhat below resonance. Only when the damping goes to zero does the maximum amplitude occur at resonance. However, for light damping we can usually consider that when $\omega/\omega_n = 1$, we have an amplitude very close to the maximum amplitude possible for the system.

EXAMPLE 19.11

A *vibrating table* is a machine that can be given harmonic oscillatory motion over a range of amplitudes and frequencies. It is used as a test apparatus for imposing a desired sinusoidal motion on a device.

In Fig. 19.31 is shown a vibrating table with a device bolted to it. The device has in it a body B of mass 16.1 lbm supported by two springs each of stiffness equal

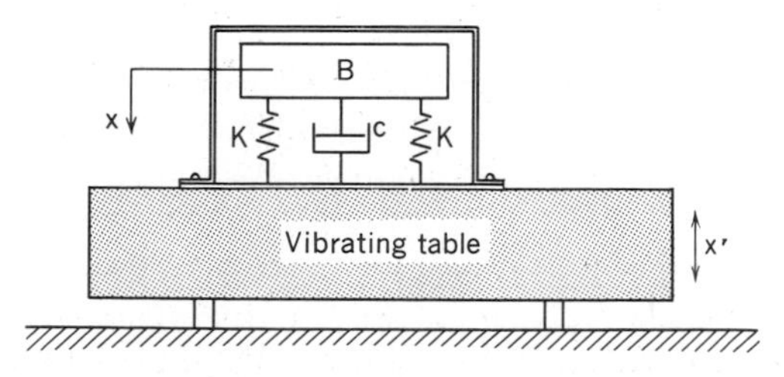

Figure 19.31. A device on a vibrating table.

to 30 lb/in. and a dashpot having a damping constant c equal to 6 lb/ft/sec. If the table has been adjusted for a vertical motion x' given as $\sin 40t$ in. with t in seconds, compute:

1. The steady-state amplitude of motion for body B.
2. The maximum number of g's acceleration that body B is subjected to for steady-state motion.
3. The maximum force that body B exerts on the vibrating table during steady-state motion.

Measuring the vertical position of body B from the static-equilibrium position with coordinate x, we have from Newton's law:

$$M\ddot{x} + c(\dot{x} - \dot{x}') + K(x - x') = 0 \tag{a}$$

Using $P \sin \omega t$ to represent x' for now, we have

$$\ddot{x} + \frac{c}{M}\dot{x} + \frac{K}{M}x = \frac{cP\omega}{M}\cos \omega t + \frac{KP}{M}\sin \omega t \tag{b}$$

Letting $cP\omega = F_1$ and $KP = F_2$, we have

$$\ddot{x} + \frac{c}{M}\dot{x} + \frac{K}{M}x = \frac{F_1}{M}\cos \omega t + \frac{F_2}{M}\sin \omega t \tag{c}$$

Using a phasor-diagram representation, we can combine the forcing functions into one expression as follows:

$$\frac{F_1}{M}\cos \omega t + \frac{F_2}{M}\sin \omega t = \frac{\sqrt{F_1^2 + F_2^2}}{M}\cos(\omega t - \alpha)$$

$$= \frac{R}{M}\cos(\omega t - \alpha)$$

where $\alpha = \tan^{-1}(F_2/F_1)$ and where $R = \sqrt{F_1^2 + F_2^2}$. Thus, we have

$$\ddot{x} + \frac{c}{M}\dot{x} + \frac{K}{M} = \frac{R}{M}\cos(\omega t - \alpha) \tag{d}$$

Except for the phase angle α, Eq. (d) is identical in form to Eq. 19.39. Clearly, if we are interested only in the steady-state amplitude, the phase angle α is of no consequence. Hence, we can use the result given by Eq. 19.44 with R taking the place of F_0. Thus, for the amplitude A of mass B, we have

$$A = \frac{R}{\sqrt{(M\omega^2 - K)^2 + (\omega c)^2}} \tag{e}$$

The following numerical values apply:

$$M = \tfrac{1}{2} \text{ slug}, \quad K = (12)(60) \text{ lb/ft}, \quad \omega = 40 \text{ rad/sec},$$

$$c = 6 \text{ lb/ft/sec}, \quad P = \tfrac{1}{12} \text{ ft}$$

$$R = \sqrt{F_1^2 + F_2^2} = [(\omega c P)^2 + (KP)^2]^{1/2}$$

$$= \{[(40)(6)(\tfrac{1}{12})]^2 + [(12)(60)(\tfrac{1}{12})]^2\}^{1/2} = 63.2$$

Hence, we have for A:

$$A = \frac{63.2}{\sqrt{(800 - 720)^2 + [(40)(6)]^2}} \tag{f}$$

$$= .25 \text{ ft} = 3 \text{ in.}$$

To get the maximum steady-state acceleration[5] for body B we compute $|(\ddot{x}_p)|_{\max}$. Thus, we have from Eq. 19.43:

$$\begin{aligned} \ddot{x}_p &= -A\omega^2 \cos(\omega t - \alpha) \\ |(\ddot{x}_p)|_{\max} &= \omega^2 A \\ &= (1600)(.25) = 400 \text{ ft/sec}^2 \qquad \text{(g)} \\ &= \frac{400}{32.2}g = 12.41g \end{aligned}$$

The maximum force transmitted to the body by the springs and dashpot during steady-state motion is established clearly when the body B has its greatest acceleration in the upward direction. We have for the maximum force F_B on noting that $\dot{x}_p = 0$ when $\ddot{x}_p$ is maximum:

$$\begin{aligned} F_B &= W_B + \frac{W_B}{g}(\ddot{x}_p)_{\max} \qquad \text{(h)} \\ &= 16.1 + (\tfrac{1}{2})(400) = 216 \text{ lb} \end{aligned}$$

If there were no spring–dashpot system between B and the vibratory table, the maximum force transmitted to B would be

$$\begin{aligned} F_B - W_B &+ \frac{W_B}{g}(\ddot{x}')_{\max} \qquad \text{(i)} \\ &= 16.1 + (\tfrac{1}{2})[(\tfrac{1}{12})(40)^2] = 82.8 \text{ lb} \end{aligned}$$

We see from Eq. (f) that the amplitude of the induced motion on B is three times what it would be if there were no spring–damping system present to separate B from the table. And from Eqs. (h) and (i) we see that the presence of the spring–damping system has resulted in a considerable *increase* in force acting on body B. Now the use of springs and dashpots for suspending or packaging equipment is generally for the purpose of reducing—not increasing—the amplitude of forces acting on the suspended body. The reason for the increase in these quantities for the disturbing frequency of 40 rad/sec is the fact that the natural frequency of the system is 37.8 rad/sec, thus putting us just above resonance. To protect the body B for disturbances of 40 rad/sec, we must use considerably softer springs.

As an exercise at the end of the section you will be asked in Problem 19.90 to compute K for permitting only a maximum of $\frac{1}{2}$ in. amplitude of vibration for this problem.

*19.9 Oscillatory Systems with Multi-Degrees of Freedom

We shall concern ourselves here with a very simple system that has two degrees of freedom, and we shall be able to generalize from this simple case. In the system of masses shown in Fig. 19.32, the masses are equal, as are the spring constants of the outer springs. We neglect friction, windage, etc. How can we describe the motion of the masses subsequent to any imposed set of initial conditions?

---

[5]We wish to remind you that the *steady-state* solution corresponds to the *particular* solution $\ddot{x}_p$.

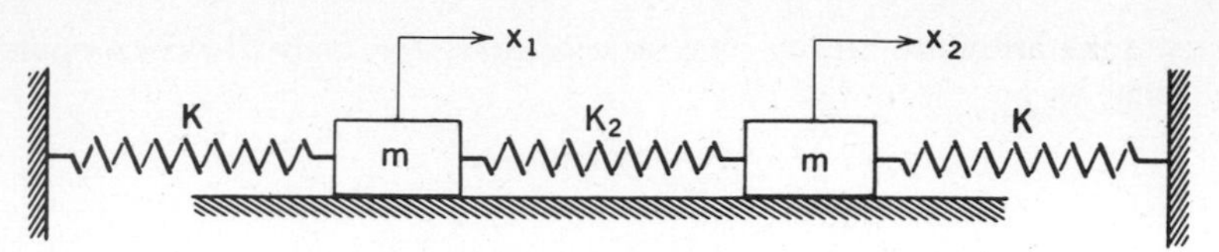

Figure 19.32. Two-degree-of-freedom system.

We first express Newton's law for each mass. To do this, imagine the masses at any position x_1, x_2 measured from the equilibrium configuration, and then compute the forces. Assume for convenience that $x_1 > x_2$. The spring K_2 is in compression for this supposition, and hence it produces a negative force on the mass at x_1 and a positive force on the mass at x_2. The equations of motion then are:

$$m\frac{d^2x_1}{dt^2} = -Kx_1 - K_2(x_1 - x_2) \tag{19.47a}$$

$$m\frac{d^2x_2}{dt^2} = -Kx_2 + K_2(x_1 - x_2) \tag{19.47b}$$

If you imagine that the masses are at any other nontrivial position, you will still arrive at the above equations.

Because both dependent variables appear in both differential equations, they are termed *simultaneous* differential equations. We rearrange the equations to the following standard form:

$$\frac{d^2x_1}{dt^2} + \frac{K}{m}x_1 + \frac{K_2}{m}(x_1 - x_2) = 0 \tag{19.48a}$$

$$\frac{d^2x_2}{dt^2} + \frac{K}{m}x_2 - \frac{K_2}{m}(x_1 - x_2) = 0 \tag{19.48b}$$

Finding a solution is equivalent to finding two functions of time $x_1(t)$ and $x_2(t)$, which when substituted into Eqs. 19.48 (a) and (b) reduce each equation to an identity. Only second derivatives and zeroth derivatives appear in these equations, and we would thus expect that sine or cosine functions of time would yield a possible solution. And since both x_1 and x_2 appear in the same equation, these time functions must be of the same form. A trial solution, therefore, might be:

$$x_1 = C_1 \sin(pt + \alpha) \tag{19.49a}$$

$$x_2 = C_2 \sin(pt + \alpha) \tag{19.49b}$$

where C_1, C_2, α, and p are as yet undetermined. Substituting into Eq. 19.48 and canceling out the time function, we get:

$$-C_1p^2 + \frac{K}{m}C_1 + \frac{K_2}{m}(C_1 - C_2) = 0 \tag{19.50a}$$

$$-C_2p^2 + \frac{K}{m}C_2 - \frac{K_2}{m}(C_1 - C_2) = 0 \tag{19.50b}$$

Rearranging the above equations, we write:

$$\left(-p^2 + \frac{K}{m} + \frac{K_2}{m}\right)C_1 - \frac{K_2}{m}C_2 = 0 \tag{19.51a}$$

$$-\frac{K_2}{m}C_1 + \left(-p^2 + \frac{K}{m} + \frac{K_2}{m}\right)C_2 = 0 \tag{19.51b}$$

One way of ensuring the satisfaction of this equation is to have $C_1 = 0$ and $C_2 = 0$. This means, from Eqs. 19.49 (a) and (b), that x_1 and x_2 are always zero, which corresponds to the static equilibrium position. While this is a valid solution, since this static equilibrium is a possible motion, the result is trivial. We now ask: Is there a means of satisfying these equations without setting C_1 and C_2 equal to zero?

To answer this, solve for C_1 and C_2 in terms of the coefficients, as if they were unknowns in the above equations. Using Cramer's rule, we then have:

$$\begin{aligned} C_1 &= \frac{\begin{vmatrix} 0 & -K_2/m \\ 0 & -p^2 + K/m + K_2/m \end{vmatrix}}{\begin{vmatrix} -p^2 + K/m + K_2/m & -K_2/m \\ -K_2/m & -p^2 + K/m + K_2/m \end{vmatrix}} \\ C_2 &= \frac{\begin{vmatrix} -p^2 + K/m + K_2/m & 0 \\ -K_2/m & 0 \end{vmatrix}}{\begin{vmatrix} -p^2 + K/m + K_2/m & -K_2/m \\ -K_2/m & -p^2 + K/m + K_2/m \end{vmatrix}} \end{aligned} \tag{19.52}$$

Notice that the determinant in the numerator is in each case zero. If the denominator is other than zero, we must have the trivial solution $C_1 = C_2 = 0$, the significance of which we have just discussed. A *necessary* condition for a *nontrivial* solution is that the denominator also be zero, for then we get the indeterminate form 0/0 for C_1 and C_2. Clearly, C_1 and C_2 can then have possible values other than zero, and so the required condition for a nontrivial solution is:

$$\begin{vmatrix} -p^2 + K/m + K_2/m & -K_2/m \\ -K_2/m & -p^2 + K/m + K_2/m \end{vmatrix} = 0 \tag{19.53}$$

Carrying out this determinant multiplication, we get:

$$\left(-p^2 + \frac{K}{m} + \frac{K_2}{m}\right)^2 = \left(\frac{K_2}{m}\right)^2 \tag{19.54}$$

Taking the roots of both sides, we have:

$$-p^2 + \frac{K}{m} + \frac{K_2}{m} = \pm\frac{K_2}{m} \tag{19.55}$$

Two values of p^2 satisfy the necessary condition we have imposed. If we use the positive roots, the values of p are:

$$p_1 = \sqrt{\frac{K}{m}}$$
$$p_2 = \sqrt{\frac{K}{m} + \frac{2K_2}{m}} \qquad (19.56)$$

where p_1 and p_2 are found for the plus and minus cases, respectively, of the right side of Eq. 19.55.

Let us now return to Eqs. 19.51 (a) and (b) to ascertain what further restrictions we may have to impose to ensure a solution, because these equations form the criterion for acceptance of a set of functions as solutions. Employing $\sqrt{K/m}$ for p in Eq. 19.51 (a), we have:

$$\left(-\frac{K}{m} + \frac{K}{m} + \frac{K_2}{m}\right)C_1 - \left(\frac{K_2}{m}\right)C_2 = 0 \qquad (19.57)$$

From this equation we see that when we use this value of p it is necessary that $C_1 = C_2$ to satisfy the equation. The same conclusions can be reached by employing Eq. 19.51(b). We can now state a permissible solution to the differential equation. Using A as the amplitude in place of $C_1 = C_2$, we have:

$$x_1 = A \sin\left(\sqrt{\frac{K}{m}}\,t + \alpha\right) \qquad (19.58a)$$

$$x_2 = A \sin\left(\sqrt{\frac{K}{m}}\,t + \alpha\right) \qquad (19.58b)$$

If we examine the second value of p, we find that for this value it is required that $C_1 = -C_2$. Thus if we use B for C_1, and use β as the arbitrary value in the sine function, another possible solution is:

$$x_1 = B \sin\left(\sqrt{\frac{K}{m} + \frac{2K_2}{m}}\,t + \beta\right)$$
$$x_2 = -B \sin\left(\sqrt{\frac{K}{m} + \frac{2K_2}{m}}\,t + \beta\right) \qquad (19.59)$$

Let us consider each of these solutions. In the first case, the motions of both masses are in phase with each other, have the same amplitude, and thus move together with simple harmonic motion with a natural frequency $\sqrt{K/m}$. For this motion, the center spring is not extended or compressed, and, since the mass of the spring has been neglected, it has no effect on this motion. This explains why the natural frequency has such a simple formulation.

The second possible independent solution is one in which the amplitudes are equal for both masses but the masses are 180° out of phase. Each mass oscillates harmonically with a natural frequency greater than the preceding motion. Since the masses move in opposite directions in the manner described, the center of the middle spring must be stationary for this motion. It is as if each mass were vibrating under the action of a spring of constant K and the action of half the length of a spring with a spring constant K_2 (Fig. 19.33), which explains why the natural frequency for this motion is $\sqrt{(K + 2K_2)/m}$. (It will be left for you to demonstrate in an exercise that halving the length of the spring doubles the spring constant.)

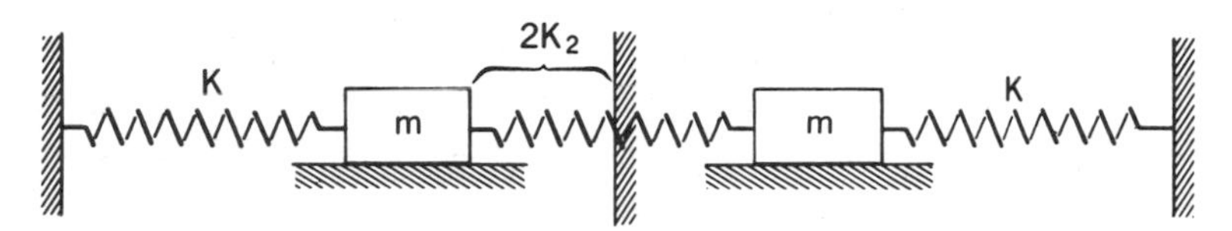

Figure 19.33. Bodies 180° out of phase.

Each of these motions as given by Eqs. 19.58 and 19.59 is called a natural *mode*. The first mode refers to the motion of lower natural frequency, and the second mode identifies the one with the higher natural frequency. It is known from differential equations that the general solution is the sum of the two solutions presented:

$$\begin{aligned} x_1 &= \left[A \sin\left(\sqrt{\frac{K}{m}}\,t + \alpha\right)\right] + \left[B \sin\left(\sqrt{\frac{K}{m} + \frac{2K_2}{m}}\,t + \beta\right)\right] \\ x_2 &= \underbrace{\left[A \sin\left(\sqrt{\frac{K}{m}}\,t + \alpha\right)\right]}_{\text{first mode of motion}} + \underbrace{\left[-B \sin\left(\sqrt{\frac{K}{m} + \frac{2K_2}{m}}\,t + \beta\right)\right]}_{\text{second mode of motion}} \end{aligned} \tag{19.60}$$

Four constants are yet to be determined: A, B, α, and β. These are the constants of integration and are determined by the initial conditions of the motion—that is, the velocity and position of each mass at time $t = 0$.

From this discussion we can make the following conclusions. The general motion of the system under study is the superposition of two modes of motion of harmonic nature that have distinct natural frequencies with amplitudes and phase angles that are evaluated to fit the initial conditions. Thus the basic modes are the "building blocks" of the general free motion.

If the masses, as well as the springs, were unequal, the analysis would still produce two natural frequencies and mode shapes, but these would neither be as simple as the special case we have worked out nor, perhaps, as intuitively obvious.

As we discussed in the first paragraph of this section, two natural frequencies correspond to the two degrees of freedom. In the general case of n degrees of freedom, there will be n natural frequencies, and the general free vibrations will be the superposition of n modes of motion that have proper amplitudes and are phased together in such a way that they satisfy $2n$ initial conditions.

Problems

19.76. A body of weight W N is suspended between two springs. Two identical dashpots are shown. Each dashpot resists motion of the block at the rate of c N/m/sec. What is the equation of motion for the block? What is c for critical damping?

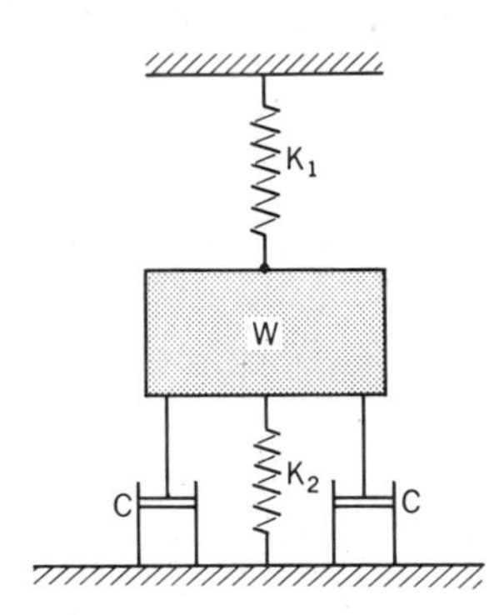

Figure P.19.76

19.77. In Problem 19.76, the following data apply:

$$W = 445 \text{ N}$$
$$K_1 = 8.8 \text{ N/mm}$$
$$K_2 = 14.0 \text{ N/mm}$$
$$c = 825 \text{ N/m/sec}$$

Is the system underdamped, overdamped, or critically damped? If the weight W is released 150 mm above its static-equilibrium configuration, what are the speed and position of the block after .1 sec? What force is transmitted to the foundation at that instant?

19.78. The damping constant c for the body is $\frac{1}{2}$ lb/ft/sec. If, at its equilibrium position, the body is suddenly given a velocity of 10 ft/sec to the right, what will the frequency of its motion be? What is the position of the mass at $t = 5$ sec?

Figure P.19.78

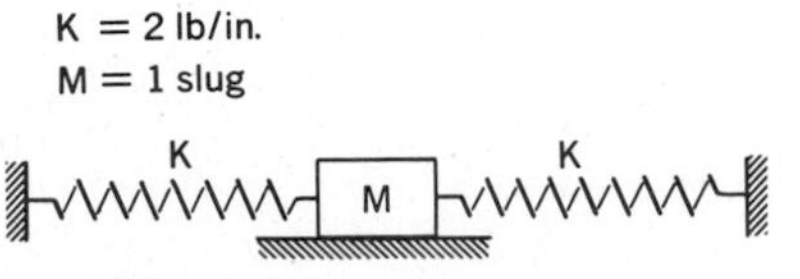

K = 2 lb/in.
M = 1 slug

K M K

19.79. The damping in Problem 19.78 is increased so that it is twice the critical damping. If the mass is released from a position 3 in. to the right of equilibrium, how far from the equilibrium position is it in 5 sec? Theoretically, does it ever reach the equilibrium position?

19.80. A plot of a free damped vibration is shown. What should the constant C_3 be in Eq. 19.35 for this motion? Show that $\ln(x_1/x_2)$, where x_1 and x_2 are two succeeding peaks, can be given as $(c/4m)\tau$. The expression $\ln(x_1/x_2)$ is called the *logarithmic decrement* and is used in vibration work.

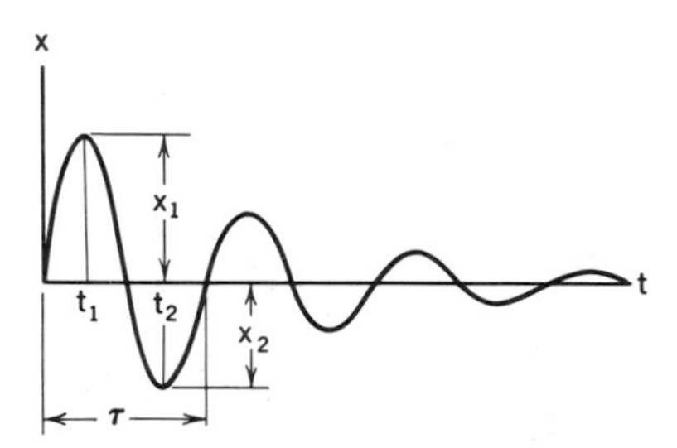

Figure P.19.80

19.81. A rod of length $2\frac{1}{4}$ m and weight 220 N is shown in the static-equilibrium position supported by a spring of stiffness $K = 14$ N/mm. The rod is connected to a dashpot having a damping force c of 69 N/m/sec. If an impulsive torque gives the rod an angular speed clockwise of $\frac{1}{2}$ rad/sec at the position shown, what is the position of point A at $t = .2$ sec?

Figure P.19.81

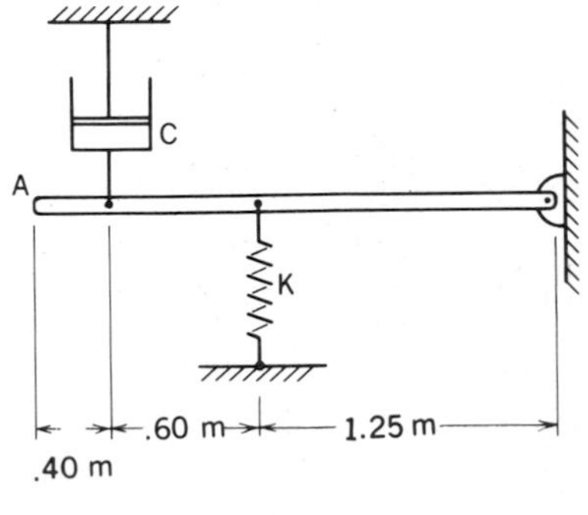

C
A
K
.60 m 1.25 m
.40 m

19.82. A spherical ball of weight 134 N is welded to a vertical light rod which in turn is welded at B to a horizontal rod. A spring of stiffness $K = 8.8$ N/mm and a damper c having a value 179 N/m/sec are connected to the horizontal rod. If A is displaced 75 mm to the right, how long does it take for it to return to its vertical configuration?

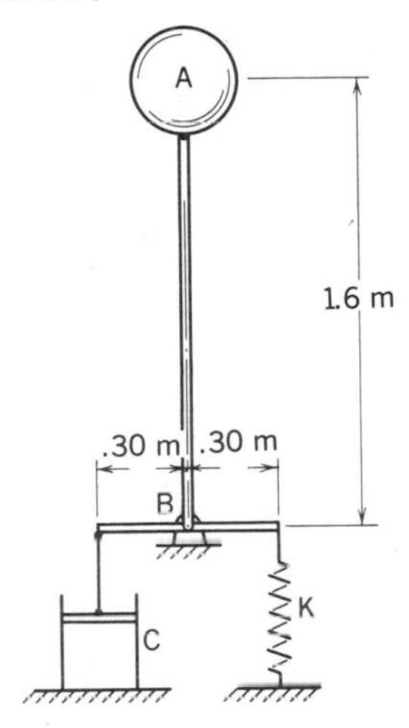

Figure P.19.82

19.83. Cylinder A of weight 200 N slides down a vertical cylindrical chute. A film of oil of thickness .1 mm separates the cylinder from the chute. If the air pressure before and behind the cylinder is maintained at the same value of 15 psig, what is the maximum velocity that the cylinder can attain by gravity? The coefficient of viscosity of the oil is .00800 N-sec/m².

Figure P.19.83

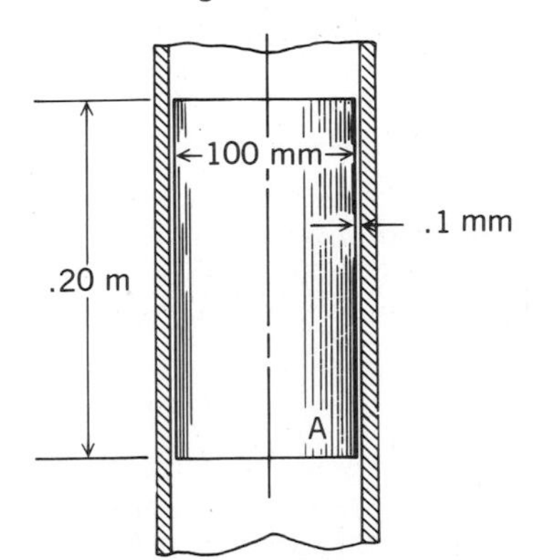

19.84. A disc A with a mass of 5 kg is constrained during rotation about its axis by a torsional spring having a constant K_T equal to 2×10^{-5} N-m/rad. The disc is in a journal having a diameter 2 mm larger than the disc. Oil having a viscosity .0085 N-sec/m² fills the outer space between disc and journal. If we assume a linear profile for the oil film, what is the frequency of oscillation of the disc if it is rotated from its equilibrium position and then released? The diameter of the disc is 40 mm and its length is 30 mm. The oil acts only on the disc's outer periphery.

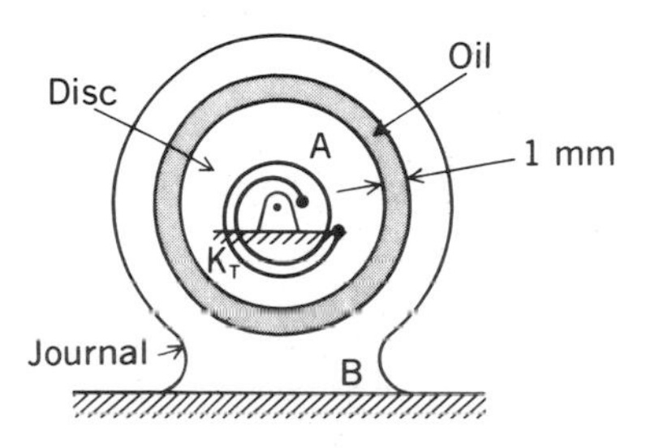

Figure P.19.84

19.85. A block W weighing 60 N is released from rest at a configuration 100 mm above its equilibrium position. It rides on a film of oil whose thickness is .1 mm and whose coefficient of viscosity is .00950 N-sec/m². If K is 50 N/m, how far down the incline will the block move? The block is .20 m on each edge.

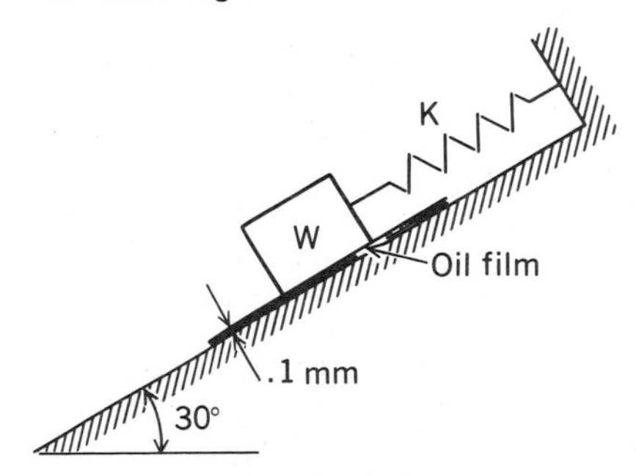

Figure P.19.85

***19.86.** In Problem 19.85, set up two simultaneous equations to determine the spring constant K so that, after W is released, W comes back to its equilibrium position with no oscillation. As a short project, solve for K using a computer.

19.87. A disc B of diameter 100 mm rotates in a stationary housing filled with oil of viscosity .00600 N-sec/m². The disc and its shaft have a mass of 30 g and a radius of gyration of 20 mm. The shaft and disc connect to a device that supplies a linear restoring torque of .05 N-mm/rad. Use a linear velocity profile for the oil and find how long each oscillation of the disc about its axis takes.

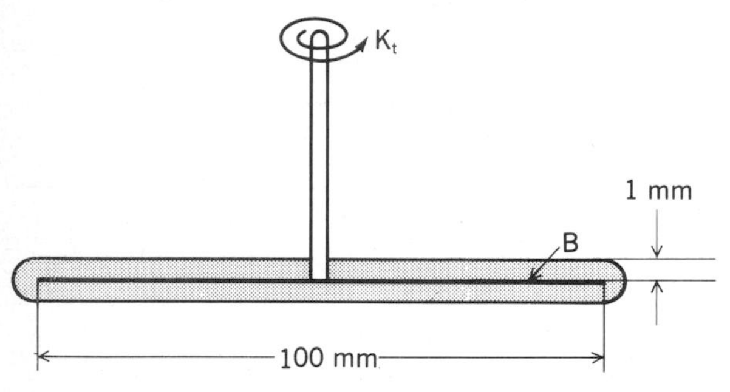

Figure P.19.87

19.88. Examine the case of the spring–mass system with viscous damping for a sinusoidal forcing function given as $F_0 \sin \omega t$. Go through the steps in the text leading up to Eq. 19.41 for this case.

19.89. A force $F = 35 \sin 2t$ N acts on a block having a weight of 285 N. A spring having stiffness K of 550 N/m and a dashpot having a damping factor c of 68 N-sec/m are connected to the body. What is the amplitude of steady-state motion for the body, and what is the maximum force transmitted to the wall?

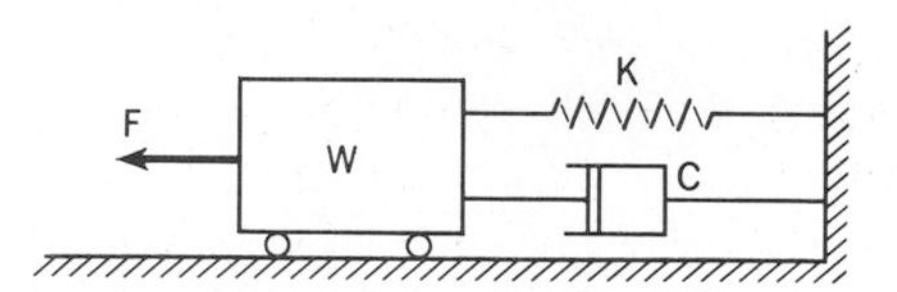

Figure P.19.89

19.90. In Example 19.11, compute K for an amplitude of steady-state vibration of $\frac{1}{2}$ in.

19.91. A block of mass M rests on two springs, the total spring constant of which is K. Also, there is a dashpot of constant c. A small sphere of mass m is attached to M and is made to rotate at a speed of ω. The distance from the center of rotation to the sphere is r. Derive the equation of motion for the block first by considering the motions of M and m separately as single masses. Show that you could reach the same equation of motion by lumping the masses M and m into one body of mass $(M + m)$ on which a sinusoidal disturbance equal to $mr\omega^2 \sin \omega t$ (from the rotating sphere) is applied.

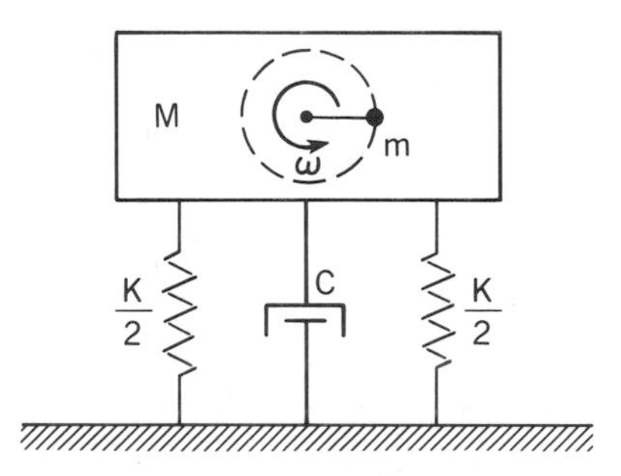

Figure P.19.91

19.92. A platform weighing 222 N deflects the spring 50 mm when placed carefully on the spring. A motor weighing 22 N is then clamped on top of the platform and rotates an eccentric mass m which weighs 1 N. The mass m is displaced 150 mm from the axis of rotation and rotates at an angular speed of 28 rad/sec. The viscous damping present causes a resistance to the motion of the platform of 275 N-m/sec. What is the steady-state amplitude of the motion of the platform? See Problem 19.91 before doing this problem.

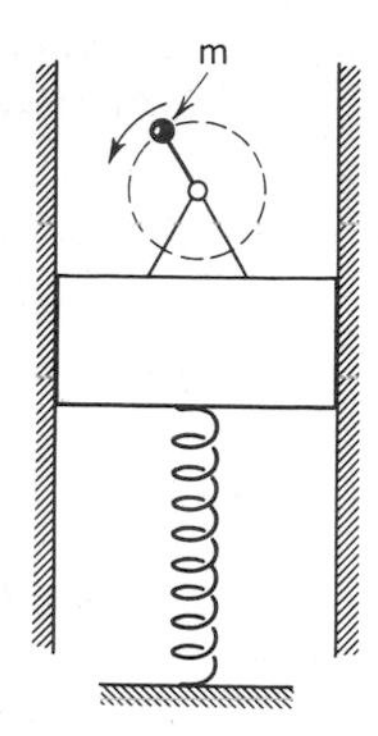

Figure P.19.92

19.93. A body weighing 143 N is connected by a light rod to a spring of stiffness K equal to 2.6 N/mm and to a dashpot having a damping factor c. Point B has a given motion x' of 30.5 sin t mm with t in seconds. If the center of A is to have an amplitude of steady-state motion of 20 mm, what must c be?

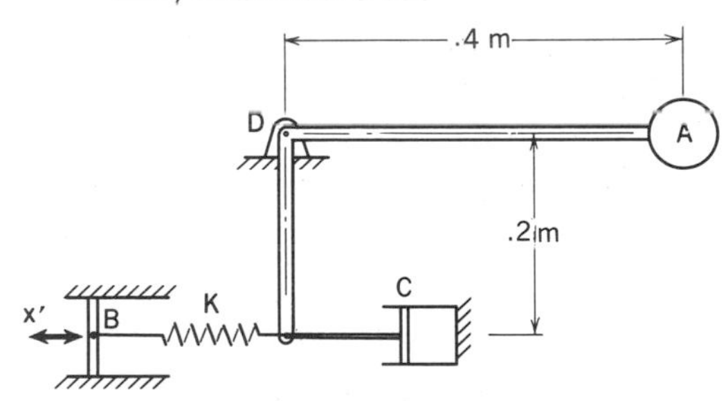

Figure P.19.93

***19.94.** In Problem 19.15, if we include the inertial effects of rod BC, how many degrees of freedom are there? If rod BC weighs 5 lb, set up the differential equations of motion for the system.

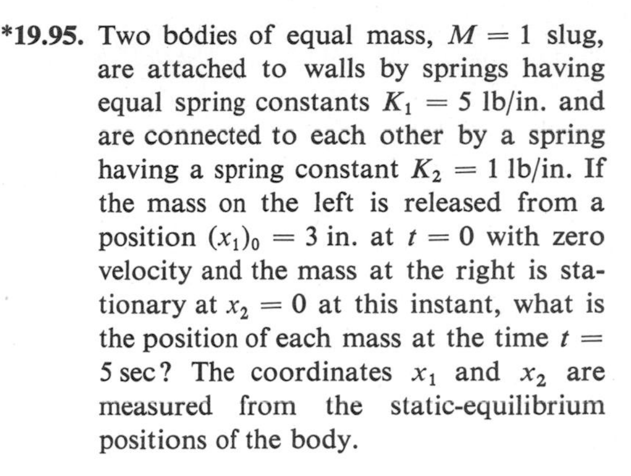

***19.95.** Two bodies of equal mass, $M = 1$ slug, are attached to walls by springs having equal spring constants $K_1 = 5$ lb/in. and are connected to each other by a spring having a spring constant $K_2 = 1$ lb/in. If the mass on the left is released from a position $(x_1)_0 = 3$ in. at $t = 0$ with zero velocity and the mass at the right is stationary at $x_2 = 0$ at this instant, what is the position of each mass at the time $t = 5$ sec? The coordinates x_1 and x_2 are measured from the static-equilibrium positions of the body.

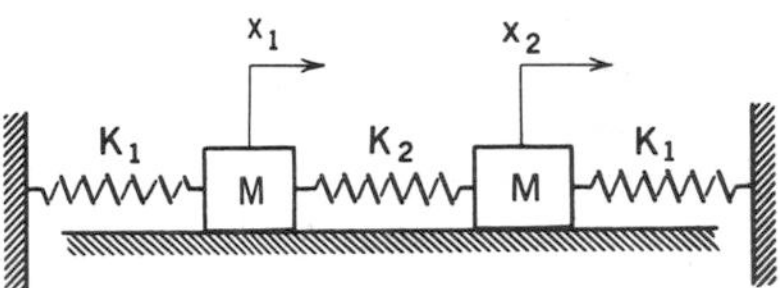

Figure P.19.95

***19.96.** Let K_2 be very small compared to K_1. Assume one mass has been released at $t = 0$ from a position displaced from equilibrium with zero velocity, while the other mass is released from the equilibrium position at that instant with zero velocity. Show that one mass will have a maximum velocity while the other will have a minimum velocity and that there will be a continual transfer of kinetic energy from one mass to the other at a frequency equal to the beat frequency of the natural frequencies of the system. $\left(\textit{Hint:}\text{ Study the phasors } A\cos\sqrt{\frac{K_1}{M}}\,t \text{ and } A\cos\sqrt{\frac{K_1}{M}+\frac{2K_2}{M}}\,t.\right)$

19.10 Closure

This introductory study of vibrations brings to a close the present study of particle and rigid-body mechanics. As you progress to the study of deformable media in your courses in solid and fluid mechanics you will find that particle mechanics and, to a lesser extent rigid-body mechanics, will form cornerstones for these disciplines. And in your studies involving the design of machines and the performance of vehicles you will find rigid-body mechanics indispensible.

It should be realized, however, that we have by no means said the last word on particle and rigid-body mechanics. More advanced studies will emphasize the variational approach introduced in statics. With the use of the calculus of variations, such topics as Hamilton's principle, Lagrange's equation,[6] and Hamilton–Jacobi theory will be presented and you will then see a greater unity between mechanics and other areas of physics such as electromagnetic theory and wave mechanics. Also the special theory of relativity will most surely be considered.

Finally, in your studies of modern physics you will come to more fully understand the limitations of classical mechanics when you are introduced to quantum mechanics.

Review Problems

19.97. If $K_1 = 2K_2 = 1.8K_3$, what should K_3 be for a period of free vibration of .2 sec? The mass M is 3 kg.

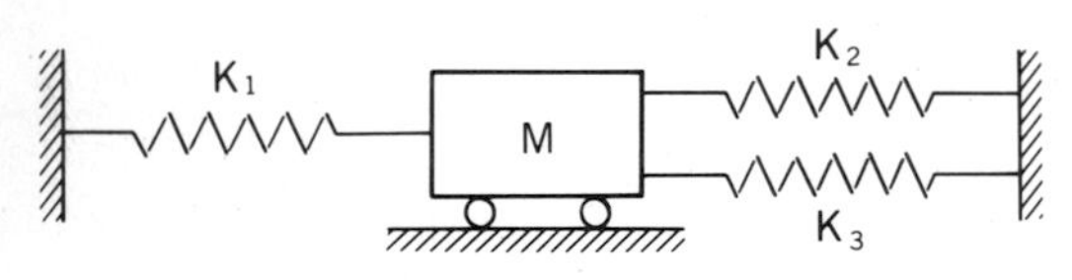

Figure P.19.97

19.98. In Problem 19.55, determine the natural frequency of the system. If the mass is deflected 2 in. and then released, determine the displacement from equilibrium after 3 sec. Finally, determine the *total* distance traveled during this time.

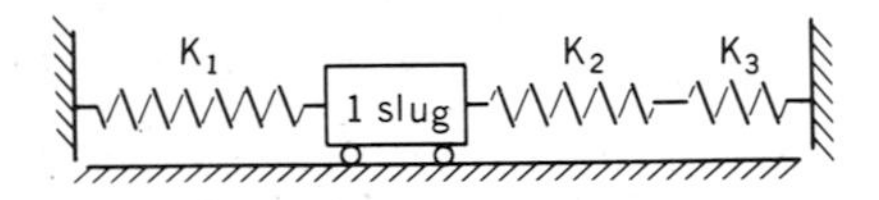

Figure P.19.98

19.99. Find the natural frequency of motion of body A for small motion of rod BD when we neglect the inertial effects of rod BD. The spring constant K_2 is .9 N/mm and the spring constant K_1 is 1.8 N/mm. The weight of block A is 178 N. Neglect friction everywhere. Rod BD weighs 44 N.

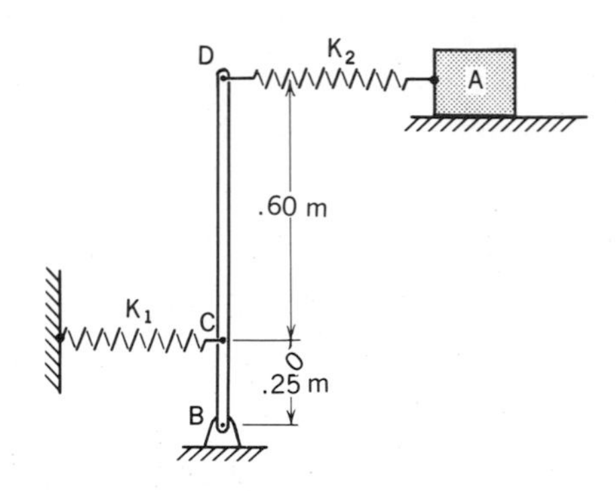

Figure P.19.99

19.100. What is the equivalent spring constant for small oscillations about the shaft AB? Neglect all mass except the block at B, which weighs 100 lb. The shear modulus of elasticity G for the shaft is 15×10^6 psi. What is the natural frequency of the system for torsional oscillation of small amplitude?

19.101. What is the radius of gyration of the speedboat about a vertical axis going through the center of gravity, if it is noted that the boat will swing about this vertical axis one time per second? The mass of the boat is 500 kg.

[6]Some of these topics and others are treated in the author's text, written with C. L. Dym, *Solid Mechanics—A Variational Approach* (New York: McGraw-Hill Book Company, 1973).

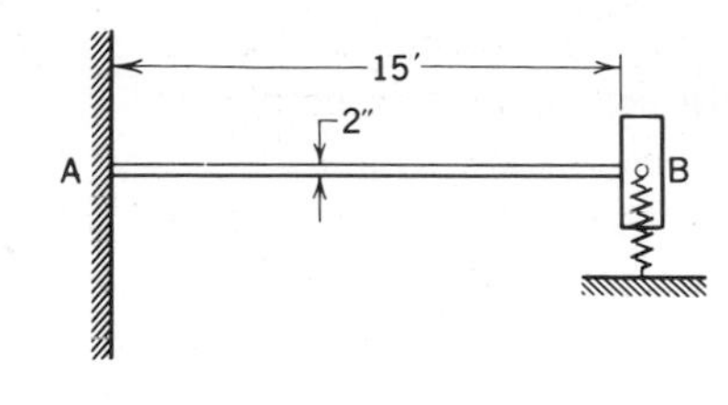

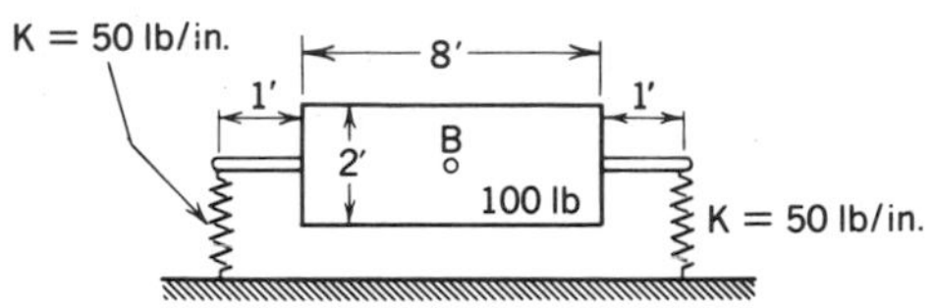

Figure P.19.100

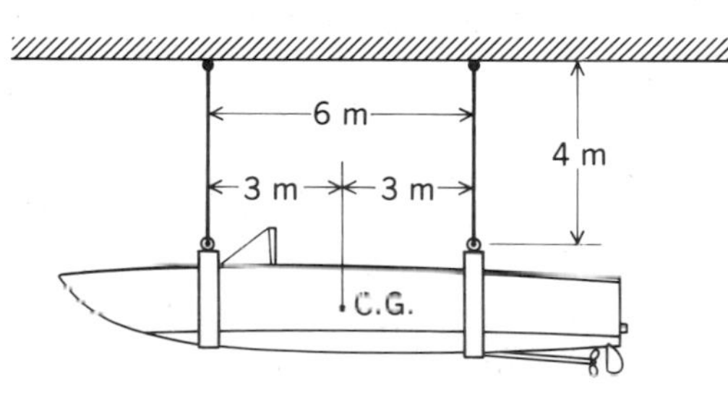

Figure P.19.101

19.102. A rod of length L and mass M is suspended from a frictionless roller. If a small impulsive torque is applied to the rod when it is in a state of rest, what is the natural frequency of oscillation about this state of rest?

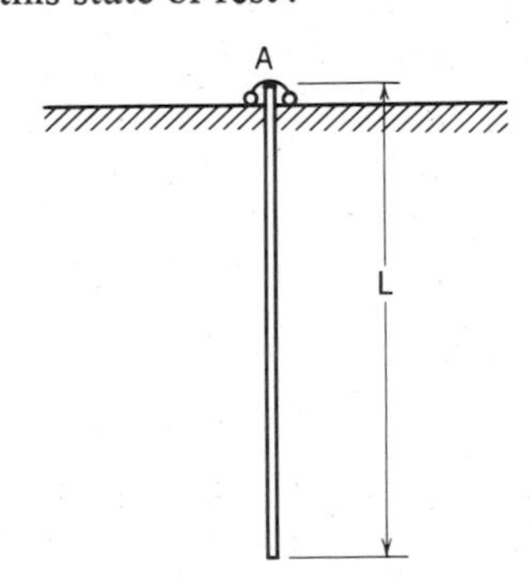

Figure P.19.102

***19.103.** Solve Problem 19.12 by energy methods.

19.104. A block is acted on by a force F given as

$$F = 90 + 22 \sin 80t \text{ N}$$

and is found to oscillate, after transients have died out, with an amplitude of .5 mm about a position 50 mm to the left of the static-equilibrium position corresponding to the condition when no force is present. What is the weight of the body?

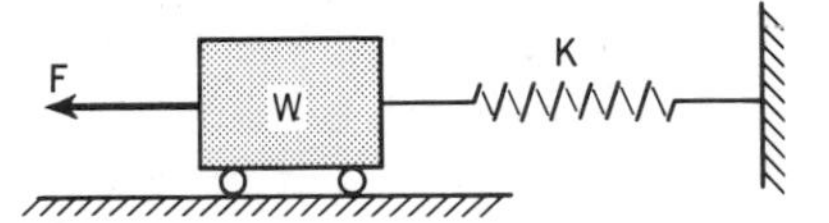

Figure P.19.104

19.105. A light stiff rod ACB is made to rotate about C such that θ varies sinusoidally with $\omega = 20$ rad/sec. What should the amplitude of rotation θ_0 be to cause a steady-state amplitude of vibration of M to be 20 mm? The following data apply:

$$M = 3 \text{ kg}$$
$$K = 2.5 \text{ N/mm}$$
$$l = 1 \text{ m}$$

What is the one essential difference between the motions of the two masses?

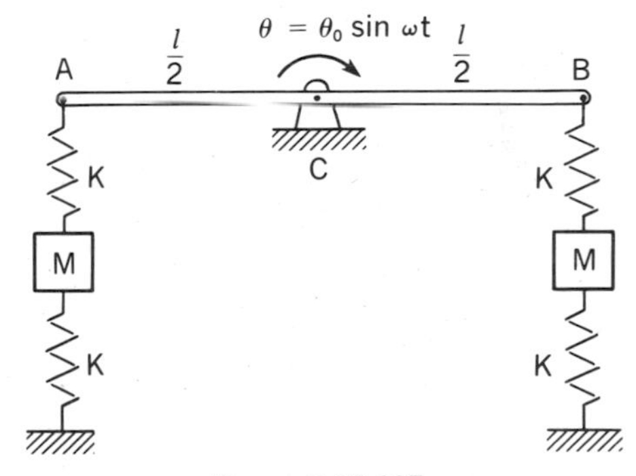

Figure P.19.105

19.106. In Problem 19.105, what is the angle θ after 5 sec? System is stationary at the outset and AB is horizontal. Take $\omega =$ 39 rad/sec and θ_0 as .02 rad.

19.107. A body rests on a conveyor moving with a speed of 5 ft/sec. If the damping constant is 2 lb/ft/sec, determine the equilibrium force in the spring. If the body is displaced 3 in. to the left from the equilibrium position, what is the time for the mass to pass through the equilibrium position again?

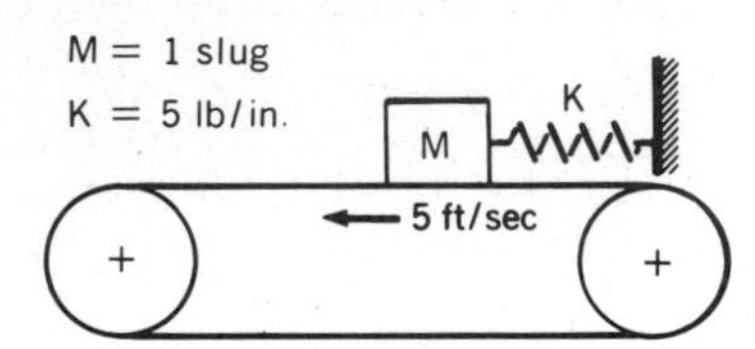

Figure P.19.107

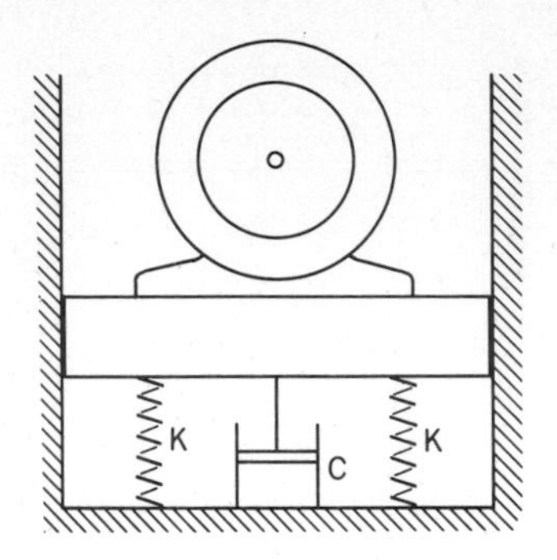

Figure P.19.108

***19.108.** A motor is mounted on two springs of stiffness 8.8 N/mm each and a dashpot having a coefficient c of 96 N-sec/m. The motor weighs 222 N. The armature of the motor weighs 89 N with a center of mass 5 mm from the geometric centerline. If the machine rotates at 1750 rpm, what is the amplitude of motion in the vertical direction of the motor? Determine the maximum force transmitted to the ground.

***19.109.** In Problem 19.108, what is the resonant condition for the system? What is the amplitude of motion for this case? To what value must c be changed if the amplitude at this motor speed is to be halved?

Appendices

Appendix I
Integration Formulas

1. $\int \frac{x\,dx}{a+bx} = \frac{1}{b^2}[a + bx - a\ln(a+bx)]$

2. $\int \frac{dx}{a^2 - x^2} = \frac{1}{2a}\ln\left(\frac{a+x}{a-x}\right)$

3. $\int \sqrt{x^2 \pm a^2}\,dx = \frac{1}{2}[x\sqrt{x^2 \pm a^2} \pm a^2 \ln(x + \sqrt{x^2 \pm a^2})]$

4. $\int \sqrt{a^2 - x^2}\,dx = \frac{1}{2}\left(x\sqrt{a^2 - x^2} + a^2 \sin^{-1}\frac{x}{a}\right)$

5. $\int x\sqrt{a^2 - x^2}\,dx = -\frac{1}{3}\sqrt{(a^2 - x^2)^3}$

6. $\int x\sqrt{a + bx}\,dx = -\frac{2(2a - 3bx)\sqrt{(a+bx)^3}}{15b^2}$

7. $\int x^2\sqrt{a^2 - x^2}\,dx = -\frac{x}{4}\sqrt{(a^2 - x^2)^3} + \frac{a^2}{8}\left(x\sqrt{a^2 - x^2} + a^2\sin^{-1}\frac{x}{a}\right)$

8. $\int x^2\sqrt{a^2 \pm x^2}\,dx = \frac{x}{4}\sqrt{(x^2 \pm a^2)^3} \mp \frac{a^2}{8}x\sqrt{x^2 \pm a^2} - \frac{a^4}{8}\ln(x + \sqrt{x^2 \pm a^2})$

9. $\int \frac{dx}{\sqrt{a^2 - x^2}} = \sin^{-1}\frac{x}{a}$

10. $\int \frac{dx}{\sqrt{x^2 + a^2}} = \ln(x + \sqrt{x^2 + a^2}) = \sinh^{-1}\frac{x}{a}$

11. $\int x^m e^{ax}\,dx = \frac{x^m e^{ax}}{a} - \frac{m}{a}\int x^{m-1}e^{ax}\,dx$

12. $\int x^m \ln x\,dx = x^{m+1}\left[\frac{\ln x}{m+1} - \frac{1}{(m+1)^2}\right]$

13. $\int \sin^2\theta\,d\theta = \frac{1}{2}\theta - \frac{1}{4}\sin 2\theta$

14. $\int \cos^2\theta\,d\theta = \frac{1}{2}\theta + \frac{1}{4}\sin 2\theta$

15. $\int \sin^3 \theta \, d\theta = -\frac{1}{3} \cos \theta (\sin^2 \theta + 2)$

16. $\int \cos^m \theta \sin \theta \, d\theta = -\frac{\cos^{m+1} \theta}{m+1}$

17. $\int \sin^m \theta \cos \theta \, d\theta = \frac{\sin^{m+1} \theta}{m+1}$

18. $\int \sin^m \theta \, d\theta = -\frac{\sin^{m-1} \theta \cos \theta}{m} + \frac{m-1}{m} \int \sin^{m-2} \theta \, d\theta$

19. $\int \theta^2 \sin \theta \, d\theta = 2\theta \sin \theta - (\theta^2 - 2) \cos \theta$

20. $\int \theta^2 \cos \theta \, d\theta = 2\theta \cos \theta + (\theta^2 - 2) \sin \theta$

21. $\int \theta \sin^2 \theta \, d\theta = \frac{1}{4}[\sin \theta (\sin \theta - 2\theta \cos \theta) + \theta^2]$

22. $\int \sin m\theta \cos m\theta \, d\theta = -\frac{1}{4m} \cos 2m\theta$

23. $\int \frac{d\theta}{(a + b \cos \theta)^2} = \frac{1}{(a^2 - b^2)} \left(\frac{-b \sin \theta}{a + b \cos \theta} + \frac{2a}{\sqrt{a^2 - b^2}} \tan^{-1} \frac{\sqrt{a^2 - b^2} \tan \frac{\theta}{2}}{a + b} \right)$

Appendix II Computation of Principal Moments of Inertia

We now turn to the problem of computing the principal moments of inertia and the directions of the principal axes for the case where we do not have planes of symmetry. It is unfortunate that a careful study of this important calculation is beyond the level of this text. However, we shall present enough material to permit the computation of the principal moments of inertia and the directions of their respective axes.

The procedure that we shall outline is that of extremizing the mass moment of inertia at a point where the inertia-tensor components are known for a reference xyz. This will be done by varying the direction cosines l, m, and n of an axis k so as to extremize I_{kk} as given by Eq. 9.13. We accordinaly set the differential of I_{kk} equal to zero as follows:

$$\begin{aligned} dI_{kk} = \; & 2lI_{xx}\,dl + 2mI_{yy}\,dm + 2nI_{zz}\,dn \\ & - 2lI_{xy}\,dm - 2mI_{xy}\,dl - 2lI_{xz}\,dn \\ & - 2nI_{xz}\,dl - 2mI_{yz}\,dn - 2nI_{yz}\,dm = 0 \end{aligned} \tag{II.1}$$

Collecting terms and canceling the factor 2, we get

$$(lI_{xx} - mI_{xy} - nI_{zz})\,dl + (-lI_{xy} + mI_{yy} - nI_{yz})\,dm + (-lI_{xz} - mI_{yz} + nI_{zz})\,dn = 0 \tag{II.2}$$

If the differentials dl, dm, and dn were *independent* we could set their respective coefficients equal to zero to satisfy the equation. However, they are not independent because the equation

$$l^2 + m^2 + n^2 = 1 \tag{II.3}$$

must at all times be satisfied. Accordingly, the differentials of the direction cosines must be related as follows[1]:

$$l\,dl + m\,dm + n\,dn = 0 \tag{II.4}$$

We can of course consider any two differentials as independent. The third is then established in accordance with the equation above.

We shall now introduce the *Lagrange multiplier* λ to facilitate the extremizing process. This constant is an arbitrary constant at this stage of the calculation. Multiplying Eq. II.4 by λ and subtracting Eq. II.4 from Eq. II.2 we get when collecting terms:

$$\begin{aligned} [(I_{xx} - \lambda)l - I_{xy}m - I_{xz}n]\,dl + [-I_{xy}l + (I_{yy} - \lambda)m - I_{yz}n]\,dm & \\ + [-I_{xz}l - I_{yz}m + (I_{zz} - \lambda)n]\,dn = 0 & \end{aligned} \tag{II.5}$$

Let us next consider that m and n are independent variables and consider the value of λ so chosen that the coefficient of dl is zero. That is,

$$(I_{xx} - \lambda)l - I_{xy}m - I_{xz}n = 0 \tag{II.6}$$

---

[1]We are thus extremizing I_{kk} in the presence of a constraining equation.

With the first term Eq. II.5 disposed of in this way, we are left with differentials dm and dn, which are independent. Accordingly, we can set their respective coefficients equal to zero in order to satisfy the equation. Hence, we have in addition to Eq. II.6 the following equations:

$$\begin{aligned} -I_{xy}l + (I_{yy} - \lambda)m - I_{yz}n &= 0 \\ -I_{xz}l - I_{yz}m + (I_{zz} - \lambda)n &= 0 \end{aligned} \tag{II.7}$$

A necessary condition for the solution of a set of direction cosines l, m, and n, from Eqs. II.6 and II.7, which does not violate Eq. II.3[2] is that the determinant of these variables be zero. Thus:

$$\begin{vmatrix} I_{xx} - \lambda & -I_{xy} & -I_{xz} \\ -I_{xy} & I_{yy} - \lambda & -I_{yz} \\ -I_{xz} & -I_{yz} & I_{zz} - \lambda \end{vmatrix} = 0 \tag{II.8}$$

This results in a cubic equation for which we can show there are three real roots for λ. Substituting these roots into any two of Eqs. II.6 and II.7 plus Eq. II.3, we can determine three direction cosines for each root. These are the direction cosines for the principal axes measured relative to xyz. We could get the principal moments of inertia next by substituting a set of these direction cosines into Eq. 9.13 and solving for I_{kk}. However, that is not necessary, since it can be shown that the three Lagrange multipliers *are* the principal moments of inertia.

[2]This precludes the possibility of a trivial solution $l = m = n = 0$.

Appendix III Additional Data for the Ellipse

If we restrict our attention to the case of an ellipse, as shown in Fig. III.1, we can compute the length of the major diameter (usually called the major axis) by solving for r from Eq. 12.34 with β set equal to zero, separately for $\theta = 0$ and for $\theta = \pi$, and then adding the results.

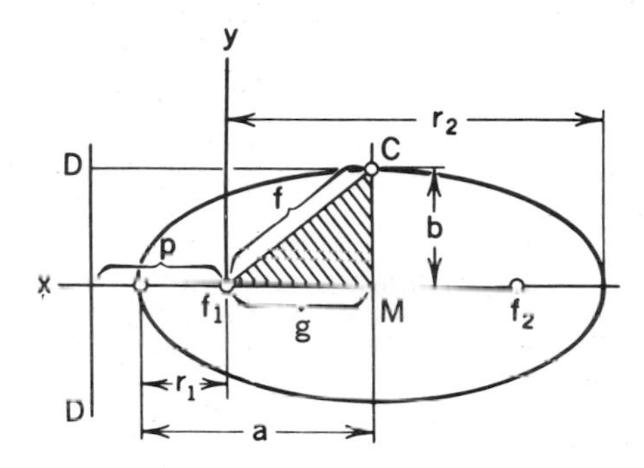

Figure III.1. Ellipse.

Thus,

$$r_1 = \frac{\epsilon p}{1+\epsilon}, \qquad r_2 = \frac{\epsilon p}{1-\epsilon} \tag{III.1}$$

Therefore,

$$r_1 + r_2 = 2a = \epsilon p\left(\frac{1}{1+\epsilon} + \frac{1}{1-\epsilon}\right)$$

Solving for a, we get

$$\boxed{a = \frac{\epsilon p}{1-\epsilon^2}} \tag{III.2}$$

The term a is the semimajor diameter. To determine the semiminor diameter, b, we consider point C on the trajectory in Fig. III.1. Distance r_c is indicated as f in the diagram, and the distance from the focus f_1 to the center at M is g. Using the basic definition of a conic, we can say:

$$\frac{f}{DC} = \frac{f}{p+g} = \epsilon \tag{III.3}$$

Noting the shaded right triangle in the diagram, we can write

$$f = \sqrt{b^2 + g^2} \tag{III.4}$$

By substituting Eq. III.4 into Eq. III.3, squaring both sides, and rearranging, we get

$$b^2 + g^2 = \epsilon^2(p + g)^2 \tag{III.5}$$

Observing Fig. III.1 and noting Eq. III.1, we can express the distance g as follows:

$$g = a - r_1 = a - \frac{\epsilon p}{1 + \epsilon} \tag{III.6}$$

Substituting Eq. III.6 into Eq. III.5, we get

$$b^2 + \left(a - \frac{\epsilon p}{1 + \epsilon}\right)^2 = \epsilon^2\left(p + a - \frac{\epsilon p}{1 + \epsilon}\right)^2 \tag{III.7}$$

From Eq. III.2 we see that

$$p = \frac{a}{\epsilon}(1 - \epsilon^2) \tag{III.8}$$

Substituting Eq. III.8 into Eq. III.7, we have

$$b^2 + \left[a - \frac{a(1 - \epsilon^2)}{1 + \epsilon}\right]^2 = \epsilon^2\left[\frac{a}{\epsilon}(1 - \epsilon^2) + a - \frac{a(1 - \epsilon^2)}{1 + \epsilon}\right]^2$$

Canceling terms wherever possible, we get the desired result on noting Eq. III.8:

$$\boxed{b = a\sqrt{1 - \epsilon^2}} \tag{III.9}$$

Finally, we can show by straightforward integration that the area of the ellipse is given as

$$\boxed{A = \pi ab} \tag{III.10}$$

Appendix IV Proof that Infinitesimal Rotations Are Vectors

You will recall that finite rotations did not qualify as vectors, even through they had magnitude and direction, because they did not combine according to the parallelogram law. Specifically the fact that the combination of finite rotations was not commutative disqualified them as vectors. We shall here show that, in the limit, as rotations become vanishingly small they *do* combine in a commutative manner and accordingly can then be considered as vectors.

Accordingly, consider Fig. IV.1 showing a rigid body with point P at position $\boldsymbol{r}$ measured from stationary reference XYZ. If the body undergoes a small but finite rotation $\Delta\phi$ about axis A–A, point P goes to P', as has been in the diagram. We can express the magnitude of $\Delta\boldsymbol{r}$ between P and P' as follows:

$$|\Delta\boldsymbol{r}| \approx |\boldsymbol{r}| \sin\theta \, \Delta\phi \tag{IV.1}$$

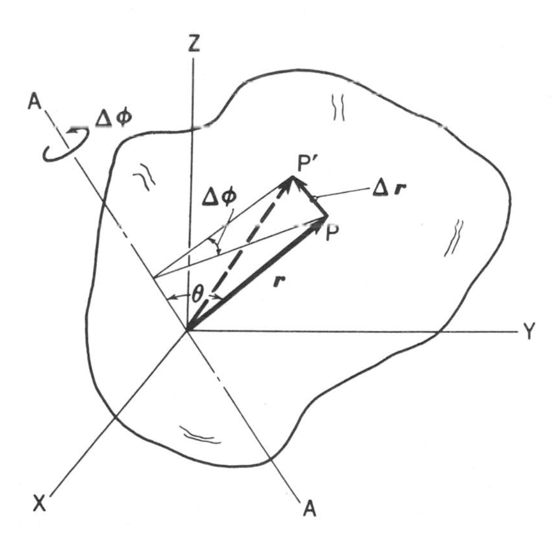

Figure IV.1. Rigid body undergoes rotation $\Delta\phi$.

If we assume, for the moment, that $\Delta\phi$ is a vector having a direction along the axis of rotation consistent with the right-hand rule, we may express the equation above as a vector equation:

$$\Delta\boldsymbol{r} \approx \Delta\boldsymbol{\phi} \times \boldsymbol{r} \tag{IV.2}$$

In the limit as $\Delta\boldsymbol{\phi} \longrightarrow 0$ the relation above becomes exact.

Now consider two arbitrary, small, but finite rotations represented by proposed vectors $\Delta\boldsymbol{\phi}_1$ and $\Delta\boldsymbol{\phi}_2$. For the first rotation we get a displacement for point P given as

$$\Delta\boldsymbol{r}_1 \approx \Delta\boldsymbol{\phi}_1 \times \boldsymbol{r} \tag{IV.3}$$

And for a second successive rotation we get for point P:

$$\begin{aligned} \Delta\boldsymbol{r}_2 &\approx \Delta\boldsymbol{\phi}_2 \times (\boldsymbol{r} + \Delta\boldsymbol{r}_1) \\ &\approx \Delta\boldsymbol{\phi}_2 \times (\boldsymbol{r} + \Delta\boldsymbol{\phi}_1 \times \boldsymbol{r}) \end{aligned} \tag{IV.4}$$

The total displacement for point P is then

$$\begin{aligned} \Delta\boldsymbol{r}_1 + \Delta\boldsymbol{r}_2 &\approx \Delta\boldsymbol{\phi}_1 \times \boldsymbol{r} + \Delta\boldsymbol{\phi}_2 \times (\boldsymbol{r} + \Delta\boldsymbol{\phi}_1 \times \boldsymbol{r}) \\ &\approx \Delta\boldsymbol{\phi}_1 \times \boldsymbol{r} + \Delta\boldsymbol{\phi}_2 \times \boldsymbol{r} + \Delta\boldsymbol{\phi}_2 \times (\Delta\boldsymbol{\phi}_1 \times \boldsymbol{r}) \end{aligned} \tag{IV.5}$$

As the rotations become vanishingly small, we can replace the approximate equality sign by an exact equality sign and we can drop the last expression in the equation above as second order. We then have on collecting terms:

$$d\boldsymbol{r}_1 + d\boldsymbol{r}_2 = (d\boldsymbol{\phi}_1 + d\boldsymbol{\phi}_2) \times \boldsymbol{r} \tag{IV.6}$$

We see from the above equation that the total displacement of any point P for successive infinitesimal rotations is *independent* of the order of these rotations. Thus, superposition of vanishingly small rotations is commutative and we can now fully accept $d\boldsymbol{\phi}$ as a vector.

Answers to Problems

Answers to Problems

The answers to even numbered problems are presented here as well as the answers to *all* review problems. The author has used the same coordinate axes that were used in the text in various sections. If perchance you use a different set of axes from the author's (for example, if your x axis corresponds to the author's y axis, etc.), you should still have no difficulty in deciding, after comparison of the numbers and a moment's thought, whether there is agreement or not.

CHAPTER 2

2.2 38.5 N at 66.6°

2.6 $B = 22.9$ N; D at 66.4°

2.8 221 N; 685 N

2.10 17.66 kg

2.12 66.5 lb

2.14 36.4 lb; 81.51 lb

2.16 .590 F; .630 F

2.18 134.5 lb; 109.8 lb

2.20 37.4 lb; $l = .267$, $m = .535$, $n = -.802$

2.22 $\gamma = 68.6°$
$OA = 17.10$ m
$OB = 43.3$ m
$EC = 18.24$ m

2.24 $F_A = 56.6$ N comp.
$F_B = 70.7$ N comp.
$F_C = 42.4$ N tens.

2.26 $25.7\boldsymbol{i}+24.7\boldsymbol{j}+16\boldsymbol{k}$ lb

2.28 $A = \pm 5\sqrt{2}\,\boldsymbol{i} \mp 5\sqrt{2}\,\boldsymbol{j}$

2.30 $\hat{f} = .465\boldsymbol{i}+.814\boldsymbol{j}+.349\boldsymbol{k}$
$F = 46.5\boldsymbol{i}+81.4\boldsymbol{j}+34.9\boldsymbol{k}$ N

2.32 $-.467$; -10.5

2.34 $D = 10\boldsymbol{i}-.769\boldsymbol{j}-3.77\boldsymbol{k}$

2.38 $16q$ joules

2.40 $A = 2.5$ N; $\alpha = 45.7°$

2.42 -75 ft²; 95.9°

2.44 52.2°

2.46 $18\boldsymbol{i}+20\boldsymbol{j}-42\boldsymbol{k}$; 47

2.50 $n = .804\boldsymbol{i}-.465\boldsymbol{j}+.372\boldsymbol{k}$

2.52 $A_1 - 500\boldsymbol{k}-800\boldsymbol{j}$ m²
$A_\epsilon = 640$ m²

2.54 Results should be the same.

2.56 2600 ft³

2.58 $\begin{cases} .303n - .256m = 0 \\ .256l - 1.610n = 0 \\ 1.610m - .303l = 0 \end{cases}$

$l = .971$; $m = .1828$; $n = .1545$

2.59 35.7 N

2.60 55.95°; 45.6°

2.61 130.7 km

2.62 57.1 N; 342 N; 971 N

2.63 209 lb; $l = .479$, $m = .866$, $n = -.1437$

2.64 32.4 ft-lb

2.65 $F_z = -80.1 \times 10^{-13}$ N
$F_x = F_y = 0$

2.66 $-156\boldsymbol{i}+94\boldsymbol{j}+28\boldsymbol{k}$ m²/sec

2.67 251 ft²

CHAPTER 3

3.2 $\varrho = 4i - 16j - 3k$ ft

3.4 $r = 6i + 10.16j - 2.4k$ m

3.6 $r = \pm 2\sqrt{2z}\,j + zk$

3.8 $-35{,}000i + 10{,}000j - 1333k$ yd

3.10 149,000 ft-lb

3.12 $M_A = 2820$ lb-ft
$M_B = 1879$ lb-ft

3.14 $M = 300$ kN–m

3.16 $M = -84i + 94j - 46k$ N–m

3.18 $M_A = M_B = \frac{10}{\sqrt{3}} ak - \frac{10}{\sqrt{3}} aj$ lb-ft
$M_E = \frac{-10}{\sqrt{3}} ai + \frac{10}{\sqrt{3}} ak$ lb-ft

3.20 $M = -6971j + 4024i$ kN-m

3.22 $M_A = .60i$ kN-m
$T_A = .60$ kN-m

3.24 -18.19 N-m

3.26 $F = 146.2$ lb
$M_x = 77.0$ lb-ft

3.28 5769 lb-ft

3.30 -56.3 kN-m; 26.25 kN-m

3.32 43.5 lb; 128.3 ft-lb

3.34 $d_{max} = 2$ ft

3.36 60 lb; 39 lb

3.38 $-50i$; 0

3.40 60 lb; 127.3 lb-in.

3.42 $M_A = M_P = -261i - 261j$ N-m

3.44 $M_P = 48i - 36j - 226k$ N-m
$M_\epsilon = -146.9$ N-m

3.46 -27.2 N-m

3.48 3635 lb-in. at 68.2° to horizontal

3.50 $35.35i + 22.4j + 80.1k$ N-m

3.52 $10{,}600i + 4500j - 5000k$ m
$|r| = 12{,}550$ m

3.53 $M_A = -2165i + 850j - 6500k$ lb-ft
$M_B = -11{,}230i - 4330j - 4330k$ lb-ft

3.54 3725 N-m

3.55 120 ft-lb; 160 ft-lb; 60 lb

3.56 1750 N-m

3.57 83.7 lb-ft

3.58 $C = 600i - 600j - 800k$

3.59 1079.5 N-m

CHAPTER 4

4.2 $F = 15$ kN
$C = 30$ kN-m

4.4 $F = 150$ N
$C = 187.5$ N-m

4.6 $F = -10j$ kN
$M_A = 8k$ kN-m
$M_B = 22.7k$ kN-m

4.8 Move 200-lb force 5 ft to left.

4.10 $F = -150k - 200j$ N
$C = 34i + 26k$ N-m

4.12 $F = 20i - 60j + 30k$ N
$C_A = 900i + 680j + 760k$ N-m

4.14 $F_A = -440k$ lb
$C_A = -1950i - 2080j$ ft-lb

4.16 $F = -16.77i + 9.68j + 134.2k$ kN
$C = 1936i - 3354j$ N-m

4.18 $F = -25i - 43.3k$ lb
$C_A = -1210i - 1106j + 700k$ ft-lb
$M_y = -1106$ ft-lb

4.20 $F = -100k$ kN
$(C_R)_1 = -120i - 600j$ kN-m
$(C_R)_2 = -120i + 1400j$ kN-m

4.22 $F_R = -31.8i - 238j$ lb
$C_R = -690i + 775j - 1297k$ lb-ft

4.24 $F_R = -250i - 2300j$ lb
$\bar{x} = 15.66$ ft

4.26 $F_R = 37.5i - 495j$ lb
$\bar{x} = 18.66$ ft

4.28 $F_R = 39.4i + 74.7j$ N
$\bar{x} = .669$ m

4.30 (a) $F_R = -100k$ m
$\bar{x} = 2.5$ m; $\bar{y} = 2.2$ m
(b) $F_R = \mathbf{0}$
$C_R = 280i + 450j$ N-m

4.32 $F_R = \mathbf{0}$
$C_R = 860i + 900j$ N-mm

4.34 $F = -43000j$ lb
$\bar{x} = 23.2$ ft

4.38 $F = 50{,}625k$ oz.

4.40 $\bar{x} = 6.67$ m

4.42 $\bar{z} = h/4$

4.46 2.37×10^8 kN

4.48 $\bar{x} = 2.615$ m; $\bar{y} = 1.598$ m

4.50 $F = -37.5k$ kN
$\bar{x} = .844$ m, $\bar{y} = 1.067$ m } from left front corner
5.25 kN or 3.5 m^3

4.52 61.1 lb-ft
$\bar{x} = 12.33$ ft
$F = 4950$ lb

4.54 $F_R = -P_0 \dfrac{ab}{2} k$

$\bar{y} = \dfrac{2}{3} a$

4.56 315 K
$\bar{x} = 48.0$ ft; $\bar{y} = 21.7$ ft

4.58 $F_R = 45{,}000$ N
$\bar{x} = 13.33$ m; $\bar{y} = 5.185$ m

4.60 13,230 lb

4.62 $F_R = 1.699 \times 10^{10}$ N
24.04 m from base

4.64 $\bar{x} = 19.4$ ft
$\bar{y} = 8.12$ ft

4.66 $F = 2200$ lb
$\bar{x} = 4.136$ ft from origin

4.68 $F = 720$ lb
$\bar{x} = 14$ ft

4.70 $F = 86.6j + 50k$ lb
9.42 ft above B

4.71 $F = -400j$ kN
$M_1 = -900k$ kN-m
$M_2 = -300k$ kN-m

4.72 $F = 10i + j - 2k$ lb
$C = 13i - 22j + 32k$ ft-lb

4.73 $F_R = 4.70i - 10.09j$ kN
$M_R = -.3i + 3.23k$ kN-m

4.74 $F_R = -90j$ lb
$\bar{x} = 21.2$ ft

4.75 $F_R = 1033$ kN
$\bar{x} = 2.62$ m

4.76 $F_R = 58{,}900$ lb
$\bar{x} = 3.92$ ft from left of post

4.77 $F_R = -120j$ N
$\bar{x} = 6.97$ m; $\bar{z} = .471$ m

4.78 $\bar{x} = 1.20$ ft; $\bar{y} = 1.125$ ft;

$\bar{z} = \dfrac{t}{2}$ ft

4.79 $F = 91.5$ kN
Distance above A is .49 m.

4.80 $\bar{y} = 1.299$ ft

4.81 $\bar{x} = 5.32$ m

4.82 $F = -1550$ lb
$\bar{x} = 20.04$ ft

CHAPTER 5

5.18 $T_{BA} = 37.9$ lb
$T_{BC} = 26.8$ lb

5.20 9000 N
$T_1 = 13{,}135$ N
$T_2 = 13{,}126$ N

5.24 $BN = 4905$ N
$F = 1380$ N

5.26 $2T\left[\left(\dfrac{T}{5000} + \dfrac{1}{2}\right) - \left(\dfrac{1}{2}\right)^2\right]^{1/2} =$

$\left(\dfrac{T}{5000} + \dfrac{1}{2}\right)(1000)$

5.28 7.46 mm

5.30 (a) $T = 2121$ lb
$H = 3000$ lb; $V = 3000$ lb
(b) $T = 1148$ lb
$H = 2121$ lb; $V = 5121$ lb

5.32 $DB = 353$ lb
$AC = 289$ lb
$EC - 394$ lb

5.34 63.75 kN-m

5.36 260 N

5.38 $A_y = B_y = 68{,}170$ N
$A_x = 6366$ N

5.40 $A = 260$ tons
$B = 276$ tons
$C = 259$ tons
$D = 0$

5.42 $B_x = 288$ N
$B_y = 433$ N
$M_B = 252$ N-m

5.44 $A_x = 650$ N
$A_y = -138.9$ N
$M_A = -1500$ N-m

5.46 $A_x = 388$ N
$A_y = -31$ N
$B_x = 388$ N
$B_y = 431$ N

5.48 $A_x = -377$ lb
$A_y = 244$ lb
$C_y = 206$ lb

5.50 $T = 55.3$ ft-lb

5.52 $T = 11.36$ ft-lb

5.54 $T = 443$ N-m

5.56 $T_E = -40$ N-m

5.58 $B_x = 5620$ N
$B_y = -27{,}620$ N
$T_D = 25{,}900$ N

5.62 (a) $R = 4.69$ kN
$F = 5.11$ kN
(b) $R = 3.75$ kN
$F = 8.05$ kN

5.64 24 ft

5.66 $D_x = 0$; $D_y = 4.1\ K$;
$C_y = 2.9\ K$

5.68 320 N; 530 N; 2025 N-m

5.70 73.5 kN; 106.5 kN

5.72 1.521 m

5.74 $F_1 = 7.33$ kN
$F_2 = .667$ kN
$F_3 = 12.00$ kN

5.76 $F_A = -625$ lb
$F_B = 6250$ lb
$F = 5625$ lb

5.78 $F_1 = W$
$C_2 = .6W$
$F_3 = .1316C_1$

5.80 $A_x = 0$ $B_y = -100$ kN
$A_y = -600$ kN $B_z = 0$
$A_z = 200$ kN $C_y = 600$ kN

5.82 $A = -10\boldsymbol{i} + 3\boldsymbol{j} + 500\boldsymbol{k}$ N
$M_A = -560\boldsymbol{i} - 4300\boldsymbol{j} + 10\boldsymbol{k}$ N-m

5.84 $F = 16.67$ lb $B_x = 0$
$A_x = 0$ $B_y = 63.3$ lb
$A_y = 53.3$ lb

5.86 $T_G = 7490$ N $A_y = 21{,}020$ N
$T_K = 15{,}060$ N $A_z = -1521$ N
$A_x = -587$ N

5.88 $P = 20{,}800$ N $B_z = 150$ N
$B_x = 25{,}300$ N $A_y = -8913$ N
$B_y = 1670$ N $A_x = 8705$ N

5.90 $T_E = 1688$ N
$T_D = 2840$ N

5.92 $T_R = 1928$ N $l = .0828$
$T_P = 709$ N $m = .782$
$T_A = 2129$ N $n = .618$

5.94 $A = 36.7\boldsymbol{i} + 8.73\boldsymbol{j} + 15.91\boldsymbol{k}$ N

5.96 750 N

5.98 69.3 kN

5.100 $T_1 = 649$ N
$T_2 = 0$
$A = 591$ N

5.102 $CE = 4.36$ kN
$A_x = .800$ kN
$A_y = 196.7$ kN

5.104 2440 N

5.106 $F_A = 2617$ N
$F_B = -617$ N
$C = 2801$ N

5.108 $CD = 600$ lb
$F = 1620$ lb

5.110 $T = .975$ compression
$F = 1.378$ N upward

5.112 $C_y = 0$ $A = 312.5$ lb
$C_x = 600$ lb $D = 1033$ lb
$C_z = -320$ lb

5.114 208 psi

5.116 $\boldsymbol{A} = 433\mathrm{i} - 533\boldsymbol{j} - 19.00\boldsymbol{k}$ lb
$\boldsymbol{E} = 240\boldsymbol{k}$ lb
$\boldsymbol{H} = -433\boldsymbol{i} + 433\boldsymbol{j} - 520\boldsymbol{k}$
$\boldsymbol{K} - \boldsymbol{0}$

5.118 $F_{BC} = 4963$ N
$F_{EB} = 6541$ N
$F_{GE} = 4920$ N

5.119 $G_x = 750$ lb
$G_y = 650$ lb

5.120 $T = 30$ ft-tons
$B_y = 6.8$ tons

5.121 $F_x = A_x = 806$ lb
$F_y = A_y = 500$ lb

5.122 335 N-m

5.123 $R_1 = 41$ kN
$R_2 = 88$ kN
$R_3 = 141$ kN

5.124 $\boldsymbol{B} = 12.50\boldsymbol{i} + 100\boldsymbol{j} - 18.75\boldsymbol{k}$ lb
$\boldsymbol{A} = 12.50\boldsymbol{i} - 18.75\boldsymbol{k}$ lb

5.125 $F_{CD} = 1694$ lb
$F_{EF} = 4931$ lb

5.126 $R = 7930$ lb
$C = 9640$ ft-lb

5.127 $\theta = 43.73°$

5.128 $AB = 800$ lb
$E_x = 5000$ lb
$E_y = 6928$ lb
$M_E = 5000$ ft-lb

5.129 $E_x = -23.2$ kN
$E_y = -30.18$ kN
$D = 46.4$ kN

CHAPTER 6

6.2 $CD = 3606$ C. $ED = 2000$ T. $EF = 2000$ T. $CE = 2000$ T. $AF = 4808$ T. $AB = 2670$ C. $BF = 0$ $BC = 2670$ C. $FC = 1202$ T.

6.4 $CH = 12{,}480$ lb C.
$AB = DE = 18{,}720$ lb C.
$EF = AJ = 24{,}960$ lb C.
$AI = EG = 26{,}474$ lb T.
$DG = BI = 18{,}720$ lb C.
$DH = HB = 8{,}825$ lb T.
$HI = HG = 18{,}720$ lb T.

6.6 $AB = DE = 26.8$ kN C.
$AG = FE = 19$ kN T.
$BG = DF = 3.14$ kN T.
$GC = CF = 3.39$ kN C.
$BC = CD = 20$ kN C.
$GF = 21.36$ kN C.

6.8 $AB = 353$ kN C.
$AC = 250$ kN T.
$LK = 250$ kN T.
$LB - 100$ kN T.
$BC = 316$ kN C.
$BK = 70.8$ kN T.
$KJ = 300$ kN T.
$CK = 49.9$ kN T.
$CD = 304$ kN C.
$CJ = .251$ kN C.
$DE = 304$ kN C.
$DJ = 100.4$ kN T.

6.10 $JI = 25$ kN T.
$AI = 44.2$ kN T.
$IH = 56.25$ kN T.
$BH = 8.84$ kN C.
$CD = 50$ kN C.
$HD = 8.84$ kN T.
$GD = 6.25$ kN C.
$AJ = 31.25$ kN C.
$AB = 56.25$ kN C.
$BI = 6.25$ kN T.
$BC = 50$ kN C.
$HC = 0$
$HG = 43.75$ kN T.
$DE = 43.75$ kN C.
$GE = 61.9$ kN T.
$EF = 43.75$ kN C.
$GF = 0$

6.12 $CD = 583$ lb C.
$AC = BC = AD = 0$
$BD = 1158$ lb T.

6.14 $BA = BC = BE = AF = EF =$
$CE = AE = 0$; $AC = 14.14$ kN T.
$CD = 14$ kN T.
$DE = 2$ kN
$AD = 24.5$ kN C.

6.16 $CB = 250$ N C.
$BE = 500$ N C.

6.18 $KU = 15$ kN T.

6.20 $CD = 7.07$ kN C.
$DG = 0$
$HG = 25$ kN T.

6.22 $LK = 0$
$KJ = 2000$ lb T.

6.24 $HE = 4.72$ K T.
$FH = 4.72$ K T.
$FE = 2.5$ K T.
$FC = 0$

6.26 $FI = .558$ K C.
$EH = 1.118$ K C.
$DH = 1.118$ K T.

6.28 $EG = 916{,}000$ lb C.
$FH = 836{,}000$ lb T.
$IJ = 144{,}400$ lb T.

6.30 $\underline{0 < x < 5}$
$V = M = 0$
$\underline{5 < x \leqslant 10}$
$V = 500$ lb
$M = -500x + 2500$ ft-lb
$\underline{10 \leqslant x < 25}$
$V = 200 + 30x$ lb
$M = -15x^2 - 200x + 1000$ ft-lb

6.32 (a) $V = -737.5$ lb
$M = 3690$ ft-lb
(b) $V = 262.5$ lb
$M = 6850$ ft-lb
(c) $V = 262.5$ lb
$M = 1312$ ft-lb

6.34 $\underline{0 < s < 14.14}$
$V = -37.1$ lb
$H = -37.1$ lb
$M = 37.1s$ ft-lb
$\underline{14.14 < s \leqslant 29.14}$
$V = -52.5$ lb
$H = 0$
$M = 52.5s - 217$ ft-lb

$\underline{29.14 \leqslant s < 44.14}$
$V = 10s - 343.9$ lb
$H = 0$
$M = -5s^2 + 344s - 4462$ ft-lb
$\underline{44.14 < s < 58.28}$
$V = 68.9$ lb
$H = -68.9$ lb
$M = -68.9s + 4018$ ft-lb

6.36 Section CD
$\underline{0 \leqslant s < 10}$
$V_x = 0$
$H = 0$
$V_y = -20s$ N
$M_x = -10s^2$ N-m
$M_y = M_z = 0$
Section BC
$\underline{10 < s < 20}$
$V_y = 200$ N
$H = V_z = 0$
$M_x = 1000$ N-m
$M_y = 0$
$M_z = 200s - 2000$ N-m
Section AB
$\underline{20 < s < 30}$
$V_x = 0$
$V_z = -1000$ N
$H = 200$ N
$M_x = -1000s + 21{,}000$ N-m
$M_y = 0$
$M_z = 2000$ N-m

6.38 $\underline{0 < \theta < \pi / 4}$
$V = -70.7 \sin\theta$ N
$H = -70.7 \cos\theta$ N
$M = 424 - 424 \cos \theta$ N-m
$\underline{(\pi / 4) < \theta < (\pi / 2)}$
$V = 29.3 \sin \theta$ N
$H = 29.3 \cos \theta$ N
$M = 176 \cos \theta$ N-m

6.40 Section AB
$H = 0$
$V_x = 0$
$V_z = 11.36s - 262$ lb
$M_x = -5.68s^2 + 262s - 2194$ ft-lb
$M_y = 410$ ft-lb
$M_z = 0$
Section BC
$V = -130.0 + 5.64s$ lb
$H = 227.5 - 9.87s$ lb
$M_{z'} = -123.3$ ft-lb
$M_{x'} = -70.47$ ft-lb
Section CD
$V_z = -262 + 11.36s$
$M_y = 0$
$M_x = -3021 + 262s - 5.68s^2$ ft-lb

6.42 $\underline{0 < x < 5}$
$V = M = 0$
$\underline{5 < x \leqslant 10}$
$V = 500$ lb
$M = -500x + 2500$ ft-lb
$\underline{10 \leqslant x < 25}$
$V = 30x + 200$ lb
$M = -15x^2 - 200x + 1000$ ft-lb

6.44 $\underline{0 < x < 10}$
$V = -737.5$ lb
$M = 737.5x$ ft-lb
$\underline{10 < x < 25}$
$V = 262.5$ lb
$M = -262.5x + 10{,}000$ ft-lb
$\underline{25 < x < 40}$
$V = 262.5$ lb
$M = -262.5x + 10{,}500$ ft-lb

6.46 $\underline{0 < x \leqslant 5}$
$V = -1333$ N
$M = 1333x$ N
$\underline{5 \leqslant x \leqslant 15}$
$V = 20x^2 + 200x - 2833$ N
$M = 1333x - 200\,(x-5)^2 + \dfrac{20}{3}(x-5)^3$ N-m
$\underline{15 \leqslant x < 25}$
$V = 667$ N
$M = 667x + 23{,}340$ N-m

6.54 $M_{\max} = 400 / \pi^2$ ft-lb

6.56 (a) 2 ft from left end
(b) Just to left of 8000-lb force

6.58 $\underline{0 < x < 30}$
$V = 1000$ lb
$M = -1000x$ ft-lb
$\underline{30 < x \leqslant 60}$
$V = -500 + \dfrac{5}{3}(x-30)^2$ lb
$M = 500x - 45{,}000 - \dfrac{5}{9}(x-30)^3$ ft-lb
$\underline{60 \leqslant x < 90}$
$V = 1000$ lb
$M = -1000x + 30{,}000$ ft-lb

$\underline{90 < x < 120}$
$V = 0$
$M = -60{,}000$ ft-lb

6.60 $y = .000702x^2 + .08727x$
$T_{max} = 2.02 \times 10^6$ N

6.62 $$T_{max} = \frac{5l^2\sqrt{l^1 + (3h)^2}}{6h} \text{ N}$$

6.64 300 ft; 89.7 ft

6.66 11.34 m; $T_{max} = 840$ N

6.68 53 ft; 432.6 ft

6.70 AB: $\alpha_1 = 56.3°$; 18 ft
BC: $\alpha_2 = 45°$; 28.3 ft
DE: $\alpha_3 = 45°$; 28.3 ft

6.71 $DB = BE = AE = 0$
$BC = 45.3$ kN C.
$DC = 32$ kN C.
$DE = 32$ kN C.
$BA = 45.3$ kN T.

6.72 $AB = 85$ kN C.
$AH = 73.0$ kN T.
$HG = 78.6$ kN T.
$HB = 29.2$ kN C.
$BC = 85.0$ kN C.
$BG = 0$
$CD = BC = 85.0$ kN C.
$CG = 58.3$ kN T.
Others via symmetry.

6.73 $CD = 1500$ lb T.
$CA = CE = 1060$ lb C.
$AE = 1250$ lb T.
$ED = 1030$ lb C.
$AD = AE = 1250$ lb T.

6.74 $BG = 5.59$ K C.
$BF = 7.50$ K T.
$CE = 15$ K T.

6.75 $\underline{0 \leqslant x < 3}$
$V = 10x$ N
$M = -5x^2$ N-m
$\underline{3 < x < 9}$
$V = 10x - 19.44$ N
$M = -5x^2 + 19.44\,(x-3)$ N-m
$\underline{9 < x < 12}$
$V = 70.6$ N
$M = 19.44\,(x-3) - 90(x-4.5) + 500$ N-m

6.76 $\underline{0 \leqslant x < 2}$
$V = 0$
$M = 0$
$\underline{2 < x \leqslant 5}$
$V = 0$
$M = 300$ N-m
$\underline{5 \leqslant x < 12}$
$$V = \frac{500}{14}(x-5)^2 \text{ N}$$
$$M = 300 - \frac{500}{42}(x-5)^3 \text{ N-m}$$

6.77 $\underline{0 \leqslant x < 10}$
$V = M = 0$
$\underline{10 < x < 20}$
$V = -650$ lb
$M = 650\,(x-10)$ ft-lb
$\underline{20 < x < 40}$
$V = 350$ lb
$M = 650(x-10) - 1000(x-20)$ ft-lb
$\underline{40 < x < 65}$
$V = 0$
$M = -500$ ft-lb
$M = 0$ occurs at 28.6 ft to right of left support.

6.78 $R_1 = 518$ N
$R_2 = 432$ N
$V_{max} = -518$ N
$M_{max} = 1354$ N-m

6.79 $$y = h\left(1 - \cos\frac{\pi x}{l}\right)$$

6.80 $\alpha = 30.7°$
$W = 122.6$ lb

CHAPTER 7

7.4 600 lb; yes

7.6 151.0 lb; 1390 lb

7.8 $F = 1176$ lb; $T = 15.59$ ton-ft

7.10 1447 N

7.12 $\beta = 58.8°$

7.14 8.33 lb

7.16 $T = 229.8$ lb; no slipping

7.18 $f = 115.5$ N; $\mu = .578$

7.22 C moves 1.5 m to left

7.24 3520 lb

7.28 $$c = \frac{.483a + .1294d}{.978\mu_s + .259} - d$$

7.30 $T = 38.5$ kN-m
$P = 44.5$ kN

7.32 15.8 in.

7.34 $\mu_s = .620$

7.36 1.753 m

7.38 6.81 ft

7.40 .991 in.

7.42 50 mm

7.44 $\alpha = 7.26°$

7.48 43.3 N-mm

7.50 2.73P N-mm

7.52 2144 N

7.54 3 turns

7.56 $\mu = .292$

7.58 $F_1 = .652W$
$F_2 = .348W$

7.60 197.1 ft-lb

7.62 Large wheel: 313 ft-lb
Small wheel: 156.6 ft-lb

7.64 20 lb
T at right wheel: 53.2 lb
T at left wheel: 33.2 lb

7.68 217 N

7.70 3687 N

7.72 $T_1 = 513$ N
$T_1{}^1 = 324$ N
37% error

7.74 315 N-mm

7.76 1906 in.-lb

7.78 (a) $M_z = 6790$ N-mm
(b) $M_z = 7599$ N-mm

7.82 381 N-mm

7.83 128.2 N

7.84 50 lb

7.85 $\theta = 30.96°$
$T = 292$ lb

7.86 $M_2 = 840$ N-m
$M_1 = 16.80$ N-m
105 N

7.87 3333 N

7.88 267 N

7.89 94.7 mm

7.90 1065 in.-lb

7.91 (a) $T_1 = 272$ lb
$P = 13.22$ lb
(b) $T_1 = 272$ lb
$P = 163.2$ lb

7.92 20 in.-lb
$T_1 = 18.65$ lb
$T_2 = 8.65$ lb
$T_3 = 32.8$ lb
$T_4 = 12.77$ lb

7.93 40.7 N-m

7.94 74.7 N

CHAPTER 8

8.4 $M_x = 125$ ft^3
$M_y = 200$ ft^3

8.10 $x_c = 1.712$ mm
$y_c = 3.75$ mm

8.12 $x_c = 3.79$ mm
$y_c = 7.58$ mm
$M_{AA} = 234$ mm^3

8.14 $x_c = y_c = 424$ mm

8.16 $x_c = 1.717$ m
$y_c = .1722$ m

8.18 $x_c = 2.89$ m
$y_c = 2.67$ m

8.20 $y_c = 315.7$ mm

8.22 $x_c = 40.1$ mm
$y_c = 19.51$ mm

8.24 $\bar{x}_c = 12.92$ ft

8.26 $x_c = a/2$
$y_c = \frac{2}{3}b$
$z_c = c/3$

8.28 $z_c = \frac{3}{8}a$

8.30 $\mathbf{r}_c = 1.179\mathbf{i} + .955\mathbf{j} + .284\mathbf{k}$ m

8.32 $x_c = 2.57$ m

8.34 $x_c = 8.22$ m
$y_c = z_c = 0$

8.36 <u>Center of volume</u>
$x_c = -9.75$ mm
$y_c = 122.6$ mm
$z_c = 0$
<u>Center of mass and gravity</u>
$\bar{x} = -32.8$ mm
$\bar{y} = 447$ mm
$\bar{z} = 0$

8.38 $x_c = 67.65$ mm
$y_c = 608$ mm
$z_c = 493$ mm

8.40 $\mathbf{r}_c = .742\mathbf{j} - .1720\mathbf{k}$ m

8.42 $y_c = \frac{2}{3}\frac{b}{\pi}$

8.44 $A = 1751 \text{ in.}^2$
$V = 5242 \text{ in.}^3$

8.46 $A = .8624 \text{ m}^2$
$V = .0633 \text{ m}^3$

8.48 $I_{xx} = 11{,}426 \text{ ft}^4$
$I_{yy} = 137{,}100 \text{ ft}^4$
$I_{xy} = 34{,}280 \text{ ft}^4$

8.50 $I_{xx} = 122.7 \text{ m}^4$
$I_{yy} = 122.7 \text{ m}^4$

8.52 27.3 ft^4

8.54 .763

8.56 $I_{xx} = 1.758$
$I_{yy} = 4.45$
$I_{xy} = 0$

8.58 $I_{xx} = .357 \text{ ft}^4$
$I_{yy} = .232 \text{ ft}^4$
$I_{xy} = -.0625 \text{ ft}^4$

8.60 $I_{xy} = 5906 \text{ ft}^4$

8.62 270 ft^4

8.66 $I_{xx} = I_{yy} = \dfrac{b-a}{12}(a^3 + ab^2 + a^2b + b^3)$
$I_{xy} = 0$

8.68 $I_{xx} = 11.28 \times 10^4 \text{ mm}^4$
$I_{yy} = 8.36 \times 10^4 \text{ mm}^4$
$I_{x_cx_c} = 4.298 \text{ mm}^4$
$I_{y_cy_c} = 8.36 \times 10^4 \text{ mm}^4$

8.70 $I_{x_cx_c} = 1.652 \times 10^7 \text{ mm}^4$
$I_{y_cy_c} = 6.58 \times 10^7 \text{ mm}^4$
$I_{x_cy_c} = 0$

8.72 $I_{xx} = 3.29 \times 10^7 \text{ mm}^4$
$I_{yy} = 2.89 \times 10^6 \text{ mm}^4$
$I_{xy} = 6.90 \times 10^6 \text{ mm}^4$
$I_{x_cx_c} = 9.85 \times 10^6 \text{ mm}^4$
$I_{y_cy_c} = 1.607 \times 10^6 \text{ mm}^4$
$I_{x_cy_c} = 1.457 \times 10^6 \text{ mm}^4$

8.74 $I_{x'x'} = 833 \text{ mm}^4$
$I_{y'y'} = 833 \text{ mm}^4$
$I_{x'y'} = 0$

8.76 $I_{xx} = 91.9 \text{ ft}^4$
$I_{yy} = 1455 \text{ ft}^4$
$I_{xy} = -96.5 \text{ ft}^4$
$(I_p)_a = 1547 \text{ ft}^4$
$(I_p)_b = 1547 \text{ ft}^4$

8.80 $I_1 = 109.2 \text{ mm}^4$
$I_2 = 2367 \text{ mm}^4$

8.82 $I_1 = 1299 \text{ mm}^4$
$I_2 = 79.1 \text{ mm}^4$

8.85 $y_c = \dfrac{3}{8} \text{ m}$
$x_c = \pi / 2 \text{ m}$
$M_{x'} = -2.55 \text{ m}^3$
$M_{y'} = 4.04 \text{ m}^3$

8.86 $y_c = 1.671 \text{ m}$

8.87 (a) Center of volume:
$r_C = 15\boldsymbol{i} + 17.72\boldsymbol{j} + 13.72\boldsymbol{k}$ mm
(b) Center of mass and center of gravity:
$\boldsymbol{r} = 15\boldsymbol{i} + 16.71\boldsymbol{j} + 15.39\boldsymbol{k}$ mm

8.88 $x_c = y_c = \dfrac{4}{3}\dfrac{r}{\pi}$

8.89 $A = 155.4 \text{ m}^2$
$V = 117.8 \text{ m}^3$

8.90 $I_{x'x'} = 6044 \text{ m}^4$
$I_{y'y'} = 27{,}590 \text{ m}^4$
$I_{x'y'} = 14{,}820 \text{ m}^4$

8.91 $I_{x_cx_c} = 83.25 \text{ in.}^4$
$I_{y_cy_c} = 19.25 \text{ in.}^4$
$I_{x_cy_c} = 0$

8.92 $I_1 = 16.53 \text{ ft}^4$
$I_2 = 1130 \text{ ft}^4$

8.93 6294 m^4

CHAPTER 9

9.2 $I_{xx} = 4333 \text{ lbm-ft}^2$
$I_{x'x'} = 15{,}333 \text{ lbm-ft}^2$

9.4 $I_{xx} = \dfrac{1}{12}M(a^2 + l^2)$
$I_{yy} = \dfrac{1}{12}M(a^2 + b^2)$
$I_{zz} = \dfrac{1}{12}M(b^2 + l^2)$
$I_{xy} = 0$

9.6 $I_{BB} = \dfrac{1}{2}Mr^2$

9.8 $I_{xx} = 3.95 \times 10^8 \text{ gram-mm}^2$
$I_{zz} = 2.67 \times 10^7 \text{ gram-mm}^2$

9.10 $I_{zz} = 1.723 \times 10^6 \text{ gramm-mm}^2$

9.12 $I_{yy} = 3959 \text{ kg-mm}^2$

9.14 $1.004 \times 10^9 \text{ mm}^4$
$20{,}500 \text{ kg-mm}^2$
3820 kg-mm^2
2390 kg-mm^2

9.16 $I_1 = 37.0 \text{ kg-mm}^2$
$I_2 = 166.8 \text{ kg-mm}^2$
$I_3 = 204 \text{ kg-mm}^2$

9.20 $I_{x'x'} = I_{z'z'} = 29.0$ slug-ft²
$I_{y'y'} = 6.21$ slug-ft²
$I_{x'y'} = I_{x'z'} = I_{y'z'} = 0$
$I_{xx} = 119.1$ slug-ft²
$I_{yy} = 18.63$ slug-ft²
$I_{zz} = 106.6$ slug-ft²
$I_{xy} = I_{zx} = 0$
$I_{yz} = -31.1$ slug-ft²

9.22 $(M/6)(a^2+b^2+c^2)+2M(x^2+y^2+c^2)$

9.24 $I_{x''x''} = 3870$ kg-m²
$I_{x''y''} = 4615$ kg-m²

9.26 1.258×10^{-6} kg-m²
1.239×10^{-6} kg-m²
3.91×10^{-5} kg-m²

9.28 1.650 kg-m²
1.417 kg-m²
.432 kg-m²

9.30 28,400 lbm-ft²
2.54 ft

9.32 $I_{xx} = 8093$ lbm-in.²
$I_{yy} = 1603$ lbm-in.²
$I_{xy} = 1282$ lbm-in.²

9.34 50.8 kg-m²

9.36 $I_{z'z'} = 3026$ lbm-ft²

9.38 .499 kg-m²

9.40 −.01206 kg-m²

9.44 362 kg-mm²

9.45 362 kg-mm²

9.46 $I_{xx} = 534$ kg-mm²
$I_{yy} = 1659$ kg-mm²
$I_{zz} = 2190$ kg-mm²
$I_{xy} = 568$ kg-mm²
$I_{xy} = I_{yz} = \approx 0$

9.47 2190 kg-mm²
1896 kg-mm²
297 kg-mm²
$\alpha_1 = 22.65°$
$\alpha_2 = 112.65°$
z axis

9.48 .0493 kg-m²
.00187 kg-m²
.0503 kg-m²

9.49 .00295 kg-m²

9.50 $I_{yy} = 1075$ kg-m²
$I_{yz} = 92.35$ kg-m²

9.51 $I_{y'x'} = 0$ for all θ

CHAPTER 10

10.2 $S = 1250$ N
$S = 750$ N

10.4 $\theta = 19.48°$

10.6 $W = .351\ T$

10.8 $\theta = 8.53°$

10.10 $P = (W/2)\cot\beta$

10.12 2513 lb

10.14 $T_{CE} = 7.42$ kN
$T_{DE} = 29.3$ kN

10.16 $\tan\beta = \dfrac{1}{\sqrt{3}}[(a-3b)/(b+a)]$

10.18 $T' = 40$ N-m

10.20 $P = .0545$ N

10.22 $W = 270$ lb

10.24 $W = 2770$ N

10.26 $C = 108.6$ N

10.28 $d = 72.0$ mm

10.32 $d = .2$ ft

10.40 $\theta = 19.22°$

10.46 $a = .358$ m

10.48 $a = .1126$ m

10.52 When $d > 2$, stable equilibrium.
When $d < 2$, unstable equilibrium.

10.56 $W_{max} = 250$ lb

10.60 $x = 4$ m

10.61 $Q = 3P$

10.62 $P = 7540$ lb

10.63 $\theta = 27.7°$

10.64 $\cos\theta = \dfrac{\cos\theta_0(aK_1 + aK_2) + W}{aK_1 + aK_2}$

10.66 $\theta = 0$ (unstable)
$\theta = 28.1°$ (stable)

10.67 $W_{max} = 1000$ N

10.68 $R > (h/2)$ for stable equilibrium.
$R < (h/2)$ for unstable equilibrium.

CHAPTER 11

11.2 $\dot{x} = 15.30$ ft / sec
$\dot{y} = 5.35$ ft / sec

11.4 $\boldsymbol{a} = .6765\boldsymbol{i} - 1.2075\boldsymbol{j} - .3177\boldsymbol{k}$
ft / sec

11.6 12 m / sec²; 60.9 m

11.8 2.02 ft / sec; 1.522 ft / sec²

11.10 121.2 km / hr

11.12 7.42×10^4 m

11.14 $\alpha = 81.4°$

11.16 4.22 ft

11.18 6.28°

11.20 $\alpha = 23.75°$
$\beta = 3.26°$

11.22 $\mathbf{a}_1 = 12.5\mathbf{j}$ ft / sec²
$\mathbf{a}_2 = 21.65\mathbf{i} - 12.5\mathbf{j}$ ft / sec²
$\mathbf{a}_3 = \mathbf{0}$

11.24 (a) .460 m / sec²
(b) 73.6 km / hr

11.26 $\omega = 1028$ rpm

11.28 $\hat{\boldsymbol{\varepsilon}}_n = .995\mathbf{i} - .0995\mathbf{j}$
$R = 101.5$ m

11.30 $\mathbf{a} = .00803\mathbf{i} - .0723\mathbf{j}$ m / sec²

11.32 637 m

11.34 $a_n = 5.48$ ft / sec²
$a_t = 28.3$ ft / sec²
$R = 67.4$ ft

11.38 1.063 rad / sec ; .275 rad / sec²

11.40 $\Omega = 53.3$ rad / sec
$\dot{\Omega} = 72$ rad / sec²

11.42 $\mathbf{r} = 11.66\boldsymbol{\varepsilon}_{\bar{r}} + 16.25\boldsymbol{\varepsilon}_z$ ft
$\mathbf{V} = 2.92\boldsymbol{\varepsilon}_{\bar{r}} + 4.06\boldsymbol{\varepsilon}_z$ ft / sec

11.44 $\mathbf{V} = 31.4\boldsymbol{\varepsilon}_\theta + 134.7\boldsymbol{\varepsilon}_z$ m / sec
$\mathbf{a} = -493\boldsymbol{\varepsilon}_{\bar{r}} + 22.5\boldsymbol{\varepsilon}_z$ m / sec²

11.46 $\mathbf{V} = -5\boldsymbol{\varepsilon}_{\bar{r}} + 16.76\boldsymbol{\varepsilon}_\theta + .0556\boldsymbol{\varepsilon}_z$
ft / sec
$\mathbf{a} = -145.4\boldsymbol{\varepsilon}_{\bar{r}} - 79.6\boldsymbol{\varepsilon}_\theta + .01389\boldsymbol{\varepsilon}_z$
ft / sec²

11.48 14,370 mi / hr

11.50 $\mathbf{V} = -.01062\boldsymbol{\varepsilon}_{\bar{r}} + 12.27\boldsymbol{\varepsilon}_\theta$ ft / sec
$\mathbf{a} = -77.3\boldsymbol{\varepsilon}_{\bar{r}} + 1.093\boldsymbol{\varepsilon}_\theta$ ft / sec²

11.52 $\omega = .482$ rad / sec
$\dot{\omega} = 17.35 \times 10^{-4}$ rad / sec²

11.54 $\Omega = 3.08$ rad / sec
$\dot{\Omega} = .03187$ rad / sec²
$\mathbf{a} = -16.127\boldsymbol{\varepsilon}_{\bar{r}} - .008715\boldsymbol{\varepsilon}_\theta$
ft / sec²

11.56 $(\mathbf{V}_1)_{\text{rel}} = 2.397\mathbf{i} - .397\mathbf{j}$ m / sec
$(\mathbf{V}_2)_{\text{rel}} = .397\mathbf{i} + 2.60\mathbf{j}$ m / sec
$(\mathbf{V}_3)_{\text{rel}} = 1.811\mathbf{i} - 1.811\mathbf{j}$ m / sec
$(\mathbf{V}_4)_{\text{rel}} = -4.60\mathbf{i} - .397\mathbf{j}$ m / sec

11.58 $\mathbf{F} = -10{,}680\mathbf{i} + 7.36\mathbf{j}$ N

11.60 $\omega = 27.4$ rad / sec
$\Omega = 553$ rad / sec

11.62 $\mathbf{V} = -10\mathbf{j} - 12\mathbf{k}$ ft / sec
$\mathbf{a} = 5\mathbf{j} - 34.2\mathbf{k}$ ft / sec²

11.64 $\alpha = 18.55°$

11.66 $t = 6.76$ sec
$\beta = 80.5°$

11.69 $\mathbf{V} = 136\mathbf{i} + 80\mathbf{j} + 5\mathbf{k}$ ft / sec
$\mathbf{r} = 1523\mathbf{i} + 1004\mathbf{j} + 106\mathbf{k}$ ft

11.70 $\dot{y} = -1.298$ m / sec
$\ddot{y} = -.1869$ m / sec²

11.71 $V_0 = 63.9$ ft / sec

11.72 $y = .0432$ km

11.73 1.573 km

11.74 $\mathbf{a} = .00433\mathbf{i} - .02452\mathbf{j}$ m / sec²
$.00255g$

11.75 $\mathbf{V} = 1\boldsymbol{\varepsilon}_{\bar{r}} + 10\boldsymbol{\varepsilon}_\theta + 2\boldsymbol{\varepsilon}_z$ ft / sec
$\mathbf{a} = -10\boldsymbol{\varepsilon}_{\bar{r}} + 4\boldsymbol{\varepsilon}_\theta$ ft / sec²
$a_\varepsilon = -15.2$ ft / sec²

11.76 $\mathbf{V} = 8.66\boldsymbol{\varepsilon}_{\bar{r}} - 4.16\boldsymbol{\varepsilon}_\theta$ ft / sec
$\mathbf{a} = -17.32\boldsymbol{\varepsilon}_{\bar{r}} - 72.05\boldsymbol{\varepsilon}_\theta - 32.2\boldsymbol{\varepsilon}_z$
ft / sec²

11.77 $a = 7.21$ ft / sec²

11.78 No; the torpedo misses the freighter.

11.79 $\mathbf{F} = .600\mathbf{i} + 1.199\mathbf{j} - 3.93\mathbf{k}$ N

CHAPTER 12

12.2 $\mathbf{r} = 10\mathbf{i} + 8\mathbf{j} + 7\mathbf{k}$ m

12.4 99.4 ft; 2.485 sec

12.6 6.26 sec

12.10 $d = .01021$ m

12.12 4 m / sec

12.14 435 m / sec; 16,490 m

12.16 96.8 ft / sec; 64.1 ft

12.18 17.22 ft; 17.22 ft

12.20 77.53 m / sec²

12.22 $d = 13.86$ mm
$l = 20$ mm

12.24 22.4 m / sec; 108.1 m

12.26 .522 sec

12.28 7.48 hr

12.30 1.9365 ft / sec

12.32 $V = 0$

12.34 .342 ft / sec

12.36 $V = 3130$ ft / sec

12.38 38.0 ft

12.40 .309 m / sec

12.42 $\ddot{y}_A = .504$ m / sec²
$\ddot{y}_B = 3.61$ m / sec²
$\ddot{x}_C = -7.72$ m / sec²

12.44 31.3 rpm

12.46 $T = \dfrac{W}{g} l\omega^2$
$d = \dfrac{g}{\omega^2}$

12.48 9.64 rad / sec

12.50 $\hat{\varepsilon}_n = -.594\boldsymbol{i} + .396\boldsymbol{j} + .701\boldsymbol{k}$

12.54 $V_E = 35{,}921$ km / hr
$V_C = 25{,}397$ km / hr

12.56 (a) $\varepsilon = .224$; elliptical trajectory
(b) 2661 mi
(c) 1.123 hr
(d) 24,290 mi / hr

12.58 11,322 mi / hr

12.60 1401 mi / hr

12.62 194,500 mi / hr

12.64 548 hr

12.66 484 mi / hr

12.68 56.1 mi

12.70 Rocket comes within 1156 km of moon's surface and then escapes.

12.72 6210 km / hr; $\varepsilon = .326$

12.76 $\omega = 4.43$ rad / sec

12.78 $a_t = -8.51$ m / sec²

12.80 $(F_n)_T =$

$$\frac{250}{g}\left[\frac{(1 + a^4 \sinh^2 ax)^{3/2}}{a^3 \cosh ax}\right]^{-1}(100)$$

$+ 250 \cos [\tan^{-1} (a^2 \sinh ax)]$

12.82 From 15.66 mm to 138.0 mm

12.86 $\ddot{r}_C = 1.4[(6t + 15)\boldsymbol{i} - 10\boldsymbol{k}]$ ft / sec²
$\Delta r = 2498\boldsymbol{i} + 13.05\boldsymbol{j} - 686\boldsymbol{k}$ ft

12.88 $(V_1)_C = -487\boldsymbol{i} + 15{,}450\boldsymbol{j} + 84.2\boldsymbol{k}$ ft / sec
$(V_2)_C = -581\boldsymbol{i} + 15{,}450\boldsymbol{j} + 319\boldsymbol{k}$ ft / sec

12.90 −30.8 kN

12.92 1.566 m

12.94 96.9 lb

12.95 .1171 m / sec

12.96 2.375 sec

12.97 48.9 sec

12.98 .435 m / sec

12.99 $\dot{y}_A = -16.1$ ft / sec
$\dot{y}_B = 8.05$ ft / sec

12.100 199.5 N

12.101 2727 km

12.102 408 km / hr

12.103 2980 mi

12.104 224 lb

12.105 .0652 N downward

CHAPTER 13

13.2 34.8 ft / sec

13.4 5.91 m / sec

13.6 6.04 ft / sec

13.8 19.6 kN

13.10 −45.2 m / sec²

13.12 .1363 m

13.14 1.518 ft

13.16 2

13.18 12.72 ft / sec

13.20 11.31 ft / sec

13.22 19.02 ft

13.24 72.3 ft / sec

13.26 121.0 kN

13.28 .288 ft

13.30 $2(1 - \cos\theta) - .4 \sin\theta -$

$$\frac{1}{16.1}\int_0^t \dot{\theta}^3 dt - \frac{1}{64.4}(4\theta^2 - 1) = 0$$

13.34 .364 hp; .818 hp

13.36 24.7 hp; 16.57 kW

13.38 $932a_0 V + \varkappa V_1^2$
$\varkappa V_1^3$

13.40 6.44 sec

13.42 1.157 hp; 6.70 hp

13.44 .558 hp

13.46 24.8 ft up the incline

13.48 4.32 N-m

13.50 44.15°

13.52 (a) Conservative.
(b) Nonconservative.

13.54 $\Phi = -5z \cos x + xy + 2y^2 z +$ constant
$W_K = 0$

13.56 3.89 ft / sec

13.58 .389 m / sec

13.60 10.36 ft / sec²

13.62 23.3 ft

13.64 114.9 ft / sec

13.66 34.8 ft / sec

13.68 $\delta = 1.518$ ft
13.70 55.8 ft / sec
13.72 2.85 m / sec
13.74 15.50 ft / sec
13.76 $V = \left\{3\delta g\left[\frac{F}{W} - \mu\cos\alpha - \sin\alpha + \frac{K\delta}{4W}\right]\right\}^{1/2}$
13.78 9.83 ft / sec
13.80 .753 m / sec
13.82 265 ft-lb
13.84 9.41×10^4 N-m
13.86 582 ft-lb
13.88 6.84°
13.90 3.78 ft / sec
13.92 $d = .312$ m
13.94 2.29 m / sec; $f = 81.75$ N
13.96 2.58 m / sec
13.98 321 N
13.99 $V_A = 12.94$ ft / sec
$a_x = 4.29$ ft / sec²
13.100 $\delta = .727$ m
13.101 9.81 N
13.102 14,500 hp
13.103 2.3 m / sec²
13.104 3.60 m / sec; 58.5 N
13.105 2.00 m / sec
13.106 17.63 lb; 36.3 N
13.107 2.32 ft / sec
13.108 8713 N-m
13.109 9.17 m / sec; $f = 85.7$ N

CHAPTER 14

14.2 $V = 333i + 400j + 4000k$ m / sec
14.4 177.1 ft / sec to left
14.6 1.631 sec; .714 sec
14.8 $F_{avg} = 12.8i + 14.40j + 48.0k$ N
14.10 $I = -.964i - .388j + .097k$ lb-sec
14.12 6.26 sec
14.14 185.4 ft / sec
14.16 14.23 lb
14.18 14.76 N-m
14.20 1200 lb-sec
14.22 59.3 lb-sec
14.24 14.86 ft / sec
14.26 24.7 mi / hr
14.28 6.56 ft / sec
14.32 18.81 km / hr
14.34 72.0 N
14.36 $r_4 = 300i + 367j + 2094k$ m
14.38 18 km / hr; 3.24J
14.40 $E = 58.96$ MeV
$\hat{\varepsilon}_B = -.325i + .1875j - .974k$
14.42 Coefficient of restitution is .636
14.44 .773 ft
14.46 .354 m
14.48 (a) 36.45 ft / sec
(b) $(V_{n+1})_f = .9(V_n)_i$
14.50 $\varepsilon = .455$
14.52 $d = 11.97$ mm
14.54 6.65 in.
14.56 $V_C = 4.71i + 4.71j$ ft / sec
14.58 44.95°
14.60 $V = 8i + 8j + 12k$ ft / sec
14.62 Drag $= 2.99\, mnV^2$
14.64 $x = 8.49$ m
14.66 41.3 ft / sec
14.68 7564 m
14.70 13.33×10^{-8} N
$p_{rad} = 13.33 \times 10^{-8}$ Pa
14.72 120.1 ft / sec
14.74 .279 m
14.76 10.27 ft / sec
14.80 1401 mi / hr
14.82 $h = 60$ mi
14.84 29714 mi
14.86 (a) $h_{max} = 20{,}984$ km
(b) $h_{max} = 43{,}391$ km
14.88 26,579 km / hr; 4243 km / hr
14.90 9965 mi / hr
14.92 (a) $6.70i - 6.90j\ \frac{\text{km-m}}{\text{sec}}$
(b) $6.70i - 6.90j\ \frac{\text{km-m}}{\text{sec}}$
(c) $-16.80\ \frac{\text{km-m}^2}{\text{sec}}$
(d) $37.2\ \frac{\text{km-m}^2}{\text{sec}}$

14.94 $I = -.0480i + .350j$ lb-sec
$I_{ang} = 2.53k$ slug-ft² / sec

14.96 $\omega = 12$ rad / sec
$V = 7.09$ ft / sec

14.98 $202.5 \dfrac{\text{kg-m}^2}{\text{sec}}$

14.100 $5288 \dfrac{\text{kg-m}^2}{\text{sec}}$

14.102 34.4×10^6 slug-ft² / sec

14.104 $\dot{\omega} = .0375$ rad / sec²
$\dot{V}_C = .1749i + .262j$ m / sec²

14.106 2.01 rad / sec²

14.108 593 cycles / sec

14.110 $\dot{y} = 21.0$ m / sec

14.113 $P = -.536i + .518j$ slugs-ft / sec
$I = .745i - .210j$ lb-sec

14.114 27.7 ft / sec

14.115 1.645 knots

14.116 69.6 N-m

14.117 $\delta = 2.24$ ft
$V_B = .1791$ ft / sec

14.118 $V_3 = 2.35$ m / sec
$d_3 = 1.918$ mm
$\theta_3 = -12.39°$

14.119 $\theta_A = 5.47°$
$\theta_B = 29.9°$

14.120 $V_A = 2.73$ m / sec
$V_B = 1.014$ m / sec

14.121 $V = 161i + 170.8j - 26.1k$ ft / sec

14.122 $\delta = 1.004$ in.

14.123 3000 km

14.124 $\omega = 1.304$ rad / sec

14.125 $\dot{\omega} = 1.856$ rad / sec²
$\dot{V}_C = 3.09i - 1.786j$ m / sec

CHAPTER 15

15.4 $-10.66i + 7.31j - 35.2k$ m / sec

15.6 $-521i + 108.7j + 65.2k$ m / sec²

15.8 $\dot{\varrho} = 100i + 50j$ mm / sec
$\ddot{\varrho} = -1250k$ mm / sec²

15.10 $\omega = .940i + .342j + 1.8k$ rad / sec
$\dot{\omega} = 1.692j - .616i$ rad / sec²

15.12 $\omega = -.4i + 170.8j - 170.8k$ rad / sec
$\dot{\omega} = -.2i - 68.3j - 68.3k$ rad / sec²

15.14 $\dot{\omega} = 3.14i + 1356j$ rad / sec²
$\ddot{\omega} = 407j + 14{,}237k$ rad / sec³

15.16 $\omega = -121.6j$ rad / sec
$\dot{\omega} = 3823j$ rad / sec²

15.18 $\omega = .2k + .6i$ rad / sec
$\dot{\omega} = -.12j$ rad / scc²
$\ddot{\omega} = -.024i$ rad / sec³

15.20 $\dot{\omega} = .3i + .12j + .2k$ rad / sec²

15.22 $\dot{\omega} = .227i + .1369j - .34k$ rad/sec²

15.24 $-141i + 212j - 50k$ ft / sec

15.26 $\hat{\varepsilon} = .414i + .513j + .752k$
$V = -361i - 75j + 250k$ ft / sec

15.28 $V_0 = 20j$ ft/sec
$V_D = 20j + 20k$ ft / sec

15.30 1.041 rad / sec

15.32 2.89 rad / sec

15.34 6.67 rad / sec

15.36 $\omega_{BD} = 1.190$ rad / sec
$\omega_{EFC} = 1.190$ rad / sec
$V_D = 5.84i + 7.21j$ ft / sec

15.38 $y_c = 1.634$ m
$z_c = .944$ m

15.40 $\omega_{BC} = 4.46$ rad / sec
$V_D = 10.29i + 6.29j$ ft / sec

15.42 $\dot{\omega}_{AB} = 39.7$ rad / sec²

15.44 Directly above A at a distance of 4 m.

15.46 $x_c = -2.67$ m
$y_c = 4$m

15.48 $\omega = -25$ rad / sec
$\dot{\omega} = -15$ rad / sec²

15.50 4.92 m directly below A

15.52 $\omega = -.686i - .857j + .372k$ rad / sec

15.54 $V = -41.1j - 5.13k$ m / sec
$a = -15i - 9j + 240k$ m / sec²

15.56 $V_C = 4.92$ ft / sec
$a_C = 34.8$ ft / sec²

15.58 1.695 m / sec

15.60 $V_{XYZ} = 5j - 30i$ ft / sec

15.62 $\omega_3 = -1.267$ rad / sec

15.64 39.32 ft / sec

15.66 $\omega_1 = 18.95$ rpm

15.68 $V = -2.05i + 4.92j - 2.66k$ m / sec

15.70 $V_N = -.908i - .525j - .6k$ m / sec

15.72 $V = 4.00i$ m / sec

15.74 37.9 ft / sec relative to ground
9.81 ft / sec relative to rod

15.76 $V_1 = 3.76$ m / sec
$V_2 = 7.17$ m / sec

15.78 9.03 rad / sec

15.80 $V = 50i - 5j - 20k$ ft / sec
$a = -100i - 300j - 35k$ ft / sec^2

15.82 $a = 2600i - 173.2j - 1732k$
ft / sec^2

15.84 $a = -161.9k$ m / sec^2

15.86 $a = -5.08i + 2.31j + 2.70k$
m / sec^2

15.88 $a = 1.950i - 18.69j - 1.207k$
m / sec^2

15.90 $a = -25.6j - 57.6k$ m / sec^2

15.94 $a_\perp = -5.07i - 2.93j - 1.647k$
m / sec^2

15.96 $V = -1.113i + .1202j$ m / sec
$a = .0969i - .1859j$ m / sec^2

15.98 $a = 495i + 164j - 700k$ m / sec^2

15.100 Axial force = 4.77 lb
Shear force = $1.592i - .25k$ lb

15.102 $|F| = 7.95$ lb

15.104 Axial force = 30.7 N
Bending moment = 27 N-m

15.106 $F_{BD} = -1.184$ N C.
$F_{AD} = 1.591$ N T.
$F_{shear} = -.20$ N

15.108 $F_{Cor} = -5.81 \times 10^{-6} j$ lb

15.109 $\omega = -10.47i + .01111k$ rad / sec
$\dot{\omega} = -.1163j$ rad / sec^2
$\ddot{\omega} = .001292i$ rad / sec^3

15.110 $\dot{\omega} = 2j - 4i + 30k$ rad / sec^2

15.111 $\omega_{AB} = 3$ rad / sec
$\dot{\omega}_{AB} = -44.6$ rad / sec^2
$x_c = 0$
$y_c = 1$

15.112 $\omega_c = 2.07$ rad / sec
$\dot{\omega}_c = -1.799$ rad / sec^2

15.113 $V = 30j$ ft / sec
$a = -300k - 300i$ ft / sec^2

15.114 $V = 30j$ ft / sec
$a = 300i + 15j - 300k$ ft / sec^2

15.115 $V = 56.6i - 56.6j - 28.3k$ ft / sec
$a = -113.2i - 141.5j - 169.8k$
ft / sec^2

15.116 $a = 162.2$ ft / sec^2

15.117 491 N C.

CHAPTER 16

16.2 2.12 N-m

16.4 A drops 16.70 m

16.6 .672 lb

16.8 12.88 ft

16.10 2189 rpm

16.12 $\ddot{\theta}_{AB} = -16.07$ rad / sec^2
$\ddot{\theta}_E = 16.07$ rad / sec^2

16.14 $\ddot{\theta}_A = 3.84$ rad / sec^2

16.16 7.63 m / sec^2
$F_{AB} = 163.4$ N C.

16.18 552 rotations

16.20 $\tau_{rr} = 2.5\,(3\text{-}r)(12{,}770 + 4257r)$
71,800 psi

16.22 $F_x = 87.7$ lb
$F_y = 169.7$ lb

16.24 $B_y = C_y = 2244$ N
$B_z = C_z = 5.27$ N

16.26 $A_y = B_y = 3200$ lb

16.28 4.24 ft

16.30 Axial force = $100 - 250r^2$ N
Shear force = $-40 + 100r^2$ N
Bending moment =
$2.55 + (.44 - r)[-32 + 50r^2]$
$-16.67(.4 - r)^3$ N-m

16.32 Block tips in radial direction at
$t = 9.17$ sec.

16.34 $a = -12.13$ ft / sec^2
$\ddot{\theta} = -1.035$ rad / sec^2

16.36 $\mu = .317$
$d = -2.08$ m

16.38 -205 rad / sec

16.40 After C moves .7165 m to
to the right.

16.42 8.66 ft / sec^2

16.44 -5.32 m / sec^2

16.46 $\ddot{\theta}_A = 1.986$ rad / sec^2
6.77 N

16.48 $\ddot{\theta} = 99.6$ rad / sec^2

16.50 356 rad / sec
No dynamic forces.

16.52 9.32 rad / sec

16.54 2275 N C.

16.56 $a_B = 1.470\boldsymbol{i} - .205\boldsymbol{j}$ m / sec²

16.58 64.4 ft; 20.5 rotations

16.60 3.75 m

16.62 1.288 m up the incline

16.64 (a) $\omega = 24$ rad / sec
(b) 2.57 m / sec
(c) 2.71 m / sec

16.66 3.44×10^7 kg-m²
$a = .00575\boldsymbol{i} - .0657\boldsymbol{j}$ m / sec²

16.68 81.45 N-m

16.70 $N_A = 34.4$ lb
$F_A = 6.87$ lb
$N_B = 51.2$ lb
$F_B = 10.24$ lb

16.72 $a = 15.54\boldsymbol{i} - 26.9\boldsymbol{j}$ ft / sec²

16.74 $N_A = 111.2$ lb
$N_B = 139.5$ lb

16.76 $\ddot{x} = -6.73$ ft / sec²

16.78 $C_x = 238$ N
$B_x = 349$ N
$B_y = -241$ N
$C_y = 98.2$ N

16.80 $\ddot{\theta}_{AB} = -18.34$ rad / sec²
$\ddot{\theta}_{BC} = 18.34$ rad / sec²

16.82 $A_x = 24.85$ lb
$A_y = 27.8$ lb
$B_x = 32.3$ lb
$B_y = -24.85$ lb

16.84 $A_x = -25.0$ N
$A_y = 254$ N
$B_x = -21.5$ N
$B_y = -235$ N

16.86 $A_x = 20.05$ lb
$A_y = 0$
$B_x = 4.60$ lb
$B_y = 0$

16.88 $A_x = 0$
$A_y = -69.6$ N
$A_z = 110$ N
$B_x = 0$
$B_y = 28.2$ N

16.90 $A_x = 1103$ N
$A_y = 7234$ N
$B_x = 2164$ N
$B_y = 1897$ N

16.92 $W_3 = 1.089$ lb
$\theta_3 = 192.2°$
$W_4 = 1.892$ lb
$\theta_4 = 249.3°$

16.94 $L_A = .508$ m
$\theta_A = \pi$
$L_B = .471$ m
$\theta_B = \pi$

16.96 $W_C = 2.625$ lb
$\theta_C = 180°$
$W_D = 1.125$ lb
$\theta_D = 0°$

16.98 $L_E = 4$ ft
$\theta_E = 0°$
$L_F = 2$ ft
$\theta_F = 0°$

16.100 1.580 rad / sec²

16.101 $\dot{\theta}_D = -136.9$ rad / sec
$\theta_D = 65.6$ revolutions

16.102 4.47 m / sec²

16.103 −43.8 rad / sec²

16.104 −41.0 rad / sec²

16.105 $d = 2.82$ m
$a_C = .383\boldsymbol{i} - .924\boldsymbol{j}$ m / sec²

16.106 60.4 lb

16.107 $\ddot{\theta}_{AB} = -3.81$ rad / sec²
$\ddot{\theta}_D = 19.31$ rad / sec²

16.108 $D_x = 6.96$ N
$D_y = 31.2$ N

16.109 $A_x = 0$
$A_y = -21.8$
$A_z = 100$ lb
$D_x = 0$
$D_y = -5.345$ lb

16.110 $A_x = 234$ N
$A_y = 21.9$ N
$B_x = 165.7$ N
$B_y = 21.9$ N

16.111 $\theta_C = 226.2°$
$r_C = .256$ m
$\theta_D = 267°$
$r_D = .416$ m

CHAPTER 17

17.2 4630 ft-lb

17.4 24.5 N-m

17.6 4.95 N-m

17.8 18.95 ft-lb

17.10 $\omega = \left[\dfrac{8W_2 ag}{3W_1 r^2 + 2W_2 (r-a)^2} \right]^{1/2}$

17.12 $\dot{\theta} =$

$$\left[\frac{4W \sin\theta - Kl(1-\cos\theta)^2}{(5/3)\dfrac{W}{g} l} \right]^{1/2} \text{rad / sec}$$

17.14 2.79 rad / sec

17.16 3.68 m / sec

17.18 7.99 ft / sec

17.20 5.87 m / sec

17.22 .861 rad / sec

17.24 4490 rpm

17.26 2.09 m / sec

17.28 5.47 m / sec

17.30 4.84 revolutions

17.32 2.38 m / sec

17.34 4.60 m / sec

17.36 2.11 ft

17.38 15.72 ft / sec

17.40 12.09 m / sec

17.42 $H = 12\pi j + 74\pi k \dfrac{\text{slug-ft}^2}{\text{sec}}$

17.44 H normal to axis =
1.941 kg-m² / sec
H along axis = .510 kg-m² / sec

17.46 $k = 1.016$ m

17.48 $P = 678$ N

17.50 .990 m

17.52 2.73 rpm

17.54 −27.3 ft / sec

17.56 109.9 rpm

17.58 $t = 2.00$ sec

17.60 1431 m; 82.8°

17.64 −108.7 rad / sec

17.66 9.90 rad / sec

17.68 4.12 rad / sec

17.70 −11.04 m / sec

17.72 $\omega_{AB} = -165.2$ rad / sec
$\omega_{HD} = 24.2$ rad / sec

17.74 $\omega_f = 4.23$ rad / sec
$(V_B)_f = 3.71$ ft / sec

17.76 $V_B = -1.028j + 1.542i$ m / sec

17.78 15.05 per second

17.80 $\omega_f = 60.9$ rad / sec

17.82 $V_C = 12.85i - 12.39j$ m / sec

17.84 21.4 rad / sec

17.85 15.52 ft-lb

17.86 4160 N-m

17.87 .935 rad / sec

17.88 7.29 ft / sec

17.89 $f_1 = 126.0$ lb
$f_2 = 2.12$ lb

17.90 $H = -14{,}550i - 24{,}700j - 14{,}530k \dfrac{\text{lbm-ft}^2}{\text{sec}}$

17.91 35.3 m / sec

17.92 $V_E = -.312i - .434j$ m / sec

17.93 $V = -11.05$ m / sec
$\omega = 168.4$ rad / sec

17.94 $\omega_A = 5.55$ rad / sec

17.96 $d = .1246$ m

CHAPTER 18

18.2 $\dot{\omega}_x = 12.19$ rad / sec²
$\dot{\omega}_y = 7.31$ rad / sec²
$\dot{\omega}_z = .030$ rad / sec²

18.4 $-1.684j$ ft-lb

18.6 $149.8i$ N-m

18.8 $a = -122.6i - 5j + 22.9k$ m / sec²

18.10 Shear force = 32.2 lb
Tensile force = 25,000 lb
Moment = 2178 ft-lb

18.14 $M = 8.35i - .586j + 3.59k$ N-m

18.16 $M = -148.8i + 300j - 108.6k$ N-m

18.18 Moment about the center of mass of CD due to motion of CD:
$M_{CD} = -3.18 \times 10^{-9}$ N-m
Moment about the center of mass EG due to motin of EG:
$M_{EG} = -1.147 \times 10^{-9}$ N-m

18.20 $M = -2462i - 1106j - 2212k$ N-m

18.22 $M = 1474i$

18.24 $T_x = 3.72$ N-m
$T_y = -.00524$ N-m
$T_z = -20.95$ N-m

18.26 $M = 149.4i$ N-m

18.28 $M = -2390 \sin .419t\, i + 4.29 \times 10^4 \cos .419t\, j$

18.30 $M = 8.35i - .586j + 3.59k$ N-m

18.32 $\dot{\psi} = 2$ rad / sec
$\dot{\theta} = 0$
$\dot{\phi} = -13.33$ rad / sec

18.34 1744 rad / sec

18.36 $T_x = -24.9$ ft-lb

18.38 $\dot{\phi} = .00282$ rad / sec
Result is an angular acceleration in the direction of the torque equal to .0488 rad / sec.^2

18.42 $\dot{\phi} = -5.43$ rad / sec;
5.82 rad / sec

18.44 1.663 ft-lb / rad

18.46 .755 N-m
From the posts in a direction normal to the plate.

18.48 $H = 32.4i + 93.2k \dfrac{\text{lbm-ft}^2}{\text{sec}}$

18.50
$$\frac{\dot{\psi}}{\sin\alpha} = \frac{\dot{\phi}}{\sin(\theta-\alpha)} \qquad (1)$$
$$\omega^2 = \dot{\psi}^2 + \dot{\phi}^2 + 2\dot{\phi}\dot{\psi}\cos\theta \qquad (2)$$
$$\dot{\phi} = \frac{I_{xx} - I_{zz}}{I_{zz}}\dot{\psi}\cos\theta \qquad (3)$$

18.52 $\dot{\psi} = 1.314$ rad / sec
$\dot{\phi} = 2$ rad / sec
$\theta = 12.9°$

18.54 $\dot{\psi} = 3.26$ rad / sec

18.56 $\theta = 13.98°$
$\dot{\phi} = -.0442$ rad / sec
$\dot{\psi} = .1264$ rad / sec

18.58 $\theta = 128.3°$
$\dot{\phi} = -1.194$ rad / sec

18.60 $M = -29.5i + 3186j + .409k$ N-m

18.61 $K = -.1262i + .1433j + 9.17k$ N
$H = -.700i + .1433j$ N

18.62 15.54 rad / sec

18.63 2310 rad / sec or .0654 rad / sec

18.64 $R = 62.7$ m

18.65 $\theta = 47.3°$
$\dot{\phi} = -.1283$ rad / sec
$\dot{\psi} = .525$ rad / sec

18.66 $\omega_x = .3$ rad / sec
$\omega_y = .1895$ rad / sec
$\omega_z = .423$ rad / sec

18.67 $\dot{\psi} = .2064$ rad / sec
Z in direction of z axis

18.68 $\dot{\phi} = 2.05$ rad / sec
$\theta = 73.15°$

CHAPTER 19

19.4 $K_2 = 3.90$ N-m

19.6
$$\omega_n = \sqrt{\frac{2\mu g}{L}} \text{ rad/sec}$$

19.8 .438 cycles / sec

19.10 .557 cycles / sec

19.12 19.81 rad / sec; 8.34 N / mm

19.14 28.6 rad / sec

19.16
$$\ddot{r} + \left(\frac{K}{m} - \omega^2\right) r = \frac{Kr_0}{m}$$
$$\omega_n = \sqrt{\frac{K}{m} - \omega^2}$$

19.18 1605 rad / sec; 87.9 rad / sec

19.20 1.420 sec

19.22 $\ddot{x} + .284x + .01292x^2 + .0001956x^3 = 0$
$\omega = .0848$ cycles / sec

19.24 $(K_t)_{eq} = 2.43 \times 10^5$ N-m / rad
$\omega_n = 304$ rad / sec
$\theta = -7.42°$

19.26 $\omega_n = 1$ cycle / sec

19.28 7.88 cycles / sec
$\theta = -.342°$

19.30
$$\omega_n = \sqrt{\frac{2Kr^2}{R^2\,[(M/2) - (W_1/g)]}}$$

19.32
$$\frac{1}{12}\frac{W}{g}L^2\ddot{\theta} + \frac{Wd^2}{4l}\theta = 0$$
$$\omega_n = \sqrt{\frac{3d^2g}{lL^2}}$$

19.34 235 rad / sec

19.36 1.377°

19.38 $P = [(2KL - W/2)]$

19.40 2.32 rad / sec

19.42 $k = .518$ m

19.44 $\mu = 0.809$

19.46 $\omega_n = \sqrt{\dfrac{K(r_1+r_2)^2}{(W/g)(k^2+r_1^2)}}$

$f_{max} = \dfrac{[I_{00}\,\omega^2 - K(r_1+r_2)(r_2)]\,A}{r_1}$

19.48 $\omega_n = \sqrt{\dfrac{2Kr^2}{\frac{1}{2}MR^2 + (W_1/g)R^2}}$

19.50 2.32 rad / sec

19.52 $\omega_n = \sqrt{\dfrac{(3+\sin\alpha)\,g}{L}}$

19.54 $\omega_n = \sqrt{\dfrac{g}{(R-r_1)[1+(k^2/r^2)]}}$

19.56 $x = .004185$ ft

19.58 .00992 m

19.64 110.7 mm
$V = 51.4$ km / hr

19.66 $\omega = 15.95$ rad / sec

19.68 17.89 N
$K = 553 \times 10^3$ N / m

19.70 9.805 N

19.72 1.808 rad / sec

19.74 $-6.37°$

19.76 $m\ddot{x} + 2c\dot{x} + (K_1+K_2)\,x = 0$

$c_{cr} = \sqrt{\dfrac{W}{g}\,(K_1+K_2)}$

19.78 $x(5) = -.211$ in.
$f \cong 6.92$ rad / sec

19.82 $t = .321$ sec

19.84 $\omega = .0225$ cycle / sec

19.86 $e^{-.311\tau}\left[\left(-.05 - \dfrac{.0155}{\alpha}\right)e^{-\alpha\tau} + \left(-.05 + \dfrac{.0155}{\alpha}\right)e^{-\alpha\tau}\right] = 0 \quad \text{(I)}$

$e^{-.311\tau}\,[(-.05\,\alpha - .01555)\,e^{\alpha\tau} + (-.05 + .01555)e^{-\alpha\tau}]$

$-.311e^{-.311\tau}\left[\left(-.05 - \dfrac{.01555}{\alpha}\right)e^{\alpha\tau} + \left(-.05 + \dfrac{.01555}{\alpha}\right)e^{\alpha\tau}\right] = 0 \quad \text{(II)}$

19.90 8.73 lb / in.

19.92 .000705 m

19.94 2 degrees of freedom
$\ddot{x} + 49.6x - 39.7\theta = 0$
$\ddot{\theta} - 752x + 677\theta = 0$

19.97 $K_3 = 799$ N / m

19.98 1.972 in.; 80.03 in.

19.99 2.35 rad / sec

19.100 48.2 rad / sec

19.101 $k = .748$ m

19.102 $\omega_n = \sqrt{\dfrac{6g}{L}}$

19.103 $\omega = 19.80$ rad / sec

19.104 $W = 64.6$ N

19.105 $\theta_0 = .0608$ rad

19.106 .000638 rad

19.107 .221 sec

19.108 $A = .001454$ m
$(F_{tr})_{max} = 337$ N

19.109 33.3 rad / sec
.01573 m
$c = 192.0$ N-sec / m

Index

Index

SI UNIT PREFIXES

Multiplication Factor	Prefix	Symbol	Pronunciation	Term
1 000 000 000 000 = 10^{12}	tera	T	as in *terra*ce	one trillion
1 000 000 000 = 10^{9}	giga	G	jig ′a (*a* as in *a*bout)	one billion
1 000 000 = 10^{6}	mega	M	as in *mega*phone	one million
1 000 = 10^{3}	kilo	k	as in *kilo*watt	one thousand
100 = 10^{2}	hecto	h	heck ′toe	one hundred
10 = 10	deka	da	deck ′a (*a* as in *a*bout)	ten
0.1 = 10^{-1}	deci	d	as in *deci*mal	one tenth
0.01 = 10^{-2}	centi	c	as in *senti*ment	one hundredth
0.001 = 10^{-3}	milli	m	as in *mili*tary	one thousandth
0.000 001 = 10^{-6}	micro	μ	as in *micro*phone	one millionth
0.000 000 001 = 10^{-9}	nano	n	nan ′oh (*an* as in *an*t)	one billionth
0.000 000 000 001 = 10^{-12}	pico	p	peek ′oh	one trillionth

PROPERTIES OF VARIOUS AREAS

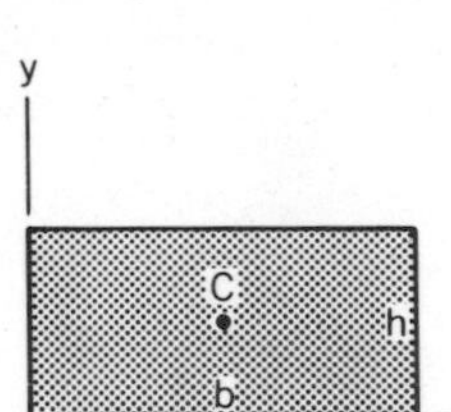

Rectangle

$$A = bh \qquad (I_{xx})_c = \frac{1}{12} bh^3$$

$$x_c = \frac{b}{2} \qquad (I_{yy})_c = \frac{1}{12} hb^3$$

$$y_c = \frac{h}{2} \qquad (I_{xy})_c = 0$$

$$J_c = \frac{1}{12} bh\,(b^2 + h^2)$$

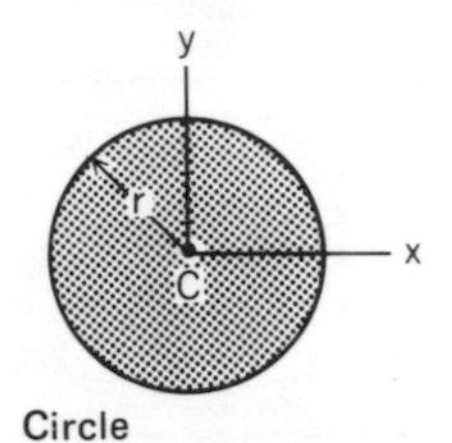

Circle

$$A = \pi r^2 \qquad (I_{xx})_c = \frac{1}{4} \pi r^4$$

$$x_c = 0 \qquad (I_{yy})_c = \frac{1}{4} \pi r^4$$

$$y_c = 0 \qquad J_c = \frac{1}{2} \pi r^4$$

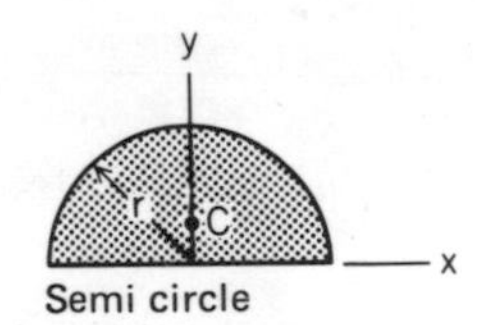

Semi circle

$$A = \frac{\pi r^2}{2} \qquad (I_{xx})_c = .00686\, d^4$$

$$x_c = 0 \qquad (I_{yy})_c = \frac{1}{8} \pi r^4$$

$$y_c = .424r$$

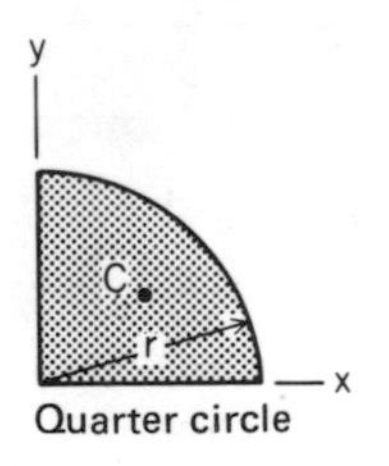

Quarter circle

$$A = \frac{\pi r^2}{4} \qquad (I_{xx})_c = .0549\, r^4$$

$$x_c = \frac{4r}{3\pi} \qquad (I_{yy})_c = .0549\, r^4$$

$$y_c = \frac{4r}{3\pi}$$

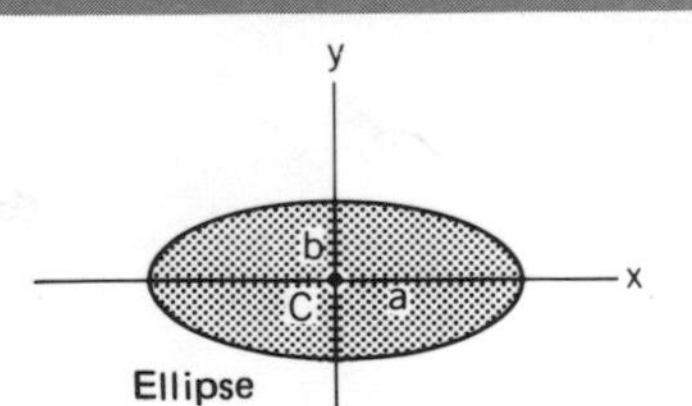

Ellipse

$$A = \pi ab \qquad (I_{xx})_c = \frac{\pi ab^3}{4}$$

$$(I_{yy})_c = \frac{\pi ba^3}{4}$$